Vitamin D

Vitamin D

Editor-in-Chief

DAVID FELDMAN

Division of Endocrinology, Gerontology, and Metabolism
Stanford University School of Medicine
Stanford, California 94305

Associate Editors

FRANCIS H. GLORIEUX

Genetics Unit
Shriners Hospital for Children and
Departments of Surgery and Pediatrics
Montréal, Québec H3G 1A6, Canada

J. WESLEY PIKE

Department of Molecular and Cellular Physiology
University of Cincinnati
Cincinnati, Ohio 45267

ACADEMIC PRESS

San Diego London Boston New York Sydney Tokyo Toronto

This book is printed on acid-free paper. 

Academic Press
a division of Harcourt Brace & Company
525 B Street, Suite 1900, San Diego, California 92101-4495, USA
http://www.apnet.com

Academic Press Limited
24-28 Oval Road, London NW1 7DX, UK
http://www.hbuk.co.uk/ap/

Library of Congress Cataloging-in-Publication Data

Vitamin D / edited by David Feldman, Francis H. Glorieux, J. Wesley Pike.
p. cm.
Includes index.
ISBN 0-12-252685-6 (alk. paper)
1. Vitamin D. I. Feldman, David, date. II. Glorieux, Francis H. III. Pike, J. Wesley.
QP772.V53V572 1997
612.3'99--dc21 97-23438
CIP

PRINTED IN THE UNITED STATES OF AMERICA
97 98 99 00 01 02 MM 9 8 7 6 5 4 3 2 1

Contents

Contributors

Numbers in parentheses indicate the pages on which the authors' contributions begin.

John S. Adams (903)
University of California, Los Angeles, School of Medicine, Burns and Allen Research Institute, Cedars-Sinai Medical Center, Los Angeles, California 90048

Judith E. Adams (619)
Department of Diagnostic Radiology, The University of Manchester, Manchester M13 9PT, England, United Kingdom

Jane E. Aubin (313)
Department of Anatomy and Cell Biology, University of Toronto, Toronto, Ontario M5S 1A8, Canada

Joshua D. Beck (209)
Department of Biochemistry and Molecular Biology, University of Medicine and Dentistry of New Jersey, New Jersey Medical School, Newark, New Jersey 07103

Norman H. Bell (521)
Department of Medicine, Medical University of South Carolina, Department of Veterans Affairs Medical Center, Charleston, South Carolina 29401

Ariane Berdal (423)
Laboratoire de Biologie-Odontologie, Faculté de Chirurgie Dentaire, Université Paris VII, Institut Biomédical des Cordeliers, 75270 Paris Cedex 06, France

Daniel D. Bikle (379)
Endocrine Research Unit, VA Medical Center, University of California, San Francisco Medical Services, San Francisco, California 94121

John P. Bilezikian (883)
Department of Medicine and Pharmacology, College of Physicians and Surgeons, Columbia University, New York, New York 10032

Ernst Binderup (1027)
Departments of Biochemistry and Chemical Research, Leo Pharmaceutical Products, DK-2750 Ballerup, Denmark

Lise Binderup (1027)
Departments of Biochemistry and Chemical Research, Leo Pharmaceutical Products, DK-2750 Ballerup, Denmark

Nicholas J. Bishop (533)
McGill University and Genetics Unit, Shriners Hospitals for Children Canadian Unit, Montréal, Québec H3G 1A6, Canada

Jean-Philippe Bonjour (499)
Division of Clinical Pathophysiology, WHO Collaborating Center for Osteoporosis and Bone Disease, Department of Internal Medicine, University Hospital, Geneva, Switzerland

Roger Bouillon (1183)
Laboratory for Experimental Medicine and Endocrinology (LEGENDO), Catholic University of Leuven, Belgium

Andrew Rael Bowman (797)
Department of Medicine, Albert Einstein Medical Center, Philadelphia, Pennsylvania 19141

Barbara D. Boyan (395)
Departments of Orthopedics, Periodontics, and Biochemistry, The University of Texas Health Science Center at San Antonio, San Antonio, Texas 78284

Thomas A. Brasitus (1141)
Department of Medicine, The University of Chicago Hospitals & Clinics, Chicago, Illinois 60637

Neil A. Breslau (607)
University of Texas Southwestern Medical Center at Dallas, Center for Mineral Metabolism and Clinical Research, and Baylor University, Dallas, Texas 75235

Alex J. Brown (849, 995)
Renal Division, Department of Medicine, Washington University School of Medicine, St. Louis, Missouri 63110

Anthony D. Care (437)
Institute of Biological Sciences, University of Wales, Aberystwyth SY23 3DD, United Kingdom

Thomas O. Carpenter (923)
Department of Pediatrics, Yale University School of Medicine, New Haven, Connecticut 06520

Susan Carswell (1197)
Cephalon, Inc., West Chester, Pennsylvania 19380

Kristina Casteels (1183)
Laboratory for Experimental Medicine and Endocrinology (LEGENDO), Catholic University of Leuven, Belgium

Joseph Caversazio (499)
Division of Clinical Pathophysiology, WHO Collaborating Center for Osteoporosis and Bone Disease, Department of Internal Medicine, University Hospital, Geneva, Switzerland

Fredriech K. W. Chan (883)
Department of Medicine, Queen Elizabeth Hospital, Hong Kong

Marie-Claire Chapuy (679)
INSERM Unit 403 and Department of Rheumatology and Bone Disease, Edouard Herriot Hospital, 69437 Lyon Cedex 03, France

Sylvia Christakos (209)
Department of Biochemistry and Molecular Biology, University of Medicine and Dentistry of New Jersey, New Jersey Medical School, Newark, New Jersey 07103

Fredric L. Coe (867)
Nephrology Section, The University of Chicago Pritzker School of Medicine, Chicago, Illinois 60637

Kay Colston (1107)
Steroid Biochemistry Group, Clinical Biochemistry, St. George's Hospital Medical School, London SW17 ORE, England

Juliet Compston (573)
University of Cambridge School of Clinical Medicine, Cambridge CB2 2QQ, United Kingdom

Nancy E. Cooke (87)
Departments of Medicine and Genetics, University of Pennsylvania, Philadelphia, Pennsylvania 19104

Elvis D. Daniels (663)
MRC Mineral Metabolism Research Unit, Department of Paediatrics, University of the Witwatersrand and Baragwanath Hospital, Soweto, Gauteng, South Africa

Michael Davies (831)
University of Manchester Bone Disease Research Center, Department of Medicine, Manchester Royal Infirmary, Manchester M13 9WL, United Kingdom

David D. Dean (395)
Department of Orthopedics, The University of Texas Health Science Center at San Antonio, San Antonio, Texas 78284

Hector F. DeLuca (3)
Department of Biochemistry, University of Wisconsin–Madison, Madison, Wisconsin 53706

Marc K. Drezner (733)
Departments of Medicine and Cell Biology and Sarah W. Stedman Nutrition Center, Duke University Medical Center, Durham, North Carolina 27710

Richard Eastell (695)
Department of Human Metabolism and Clinical Biochemistry, University of Sheffield, Sheffield S5 7AU, England, United Kingdom

Sol Epstein (797)
Division of Endocrinology and Metabolism, Albert Einstein Medical Center, and Temple University School of Medicine, Philadelphia, Pennsylvania 19141

Murray J. Favus (867)
Nephrology Section and Section of Endocrinology, The University of Chicago Pritzker School of Medicine, Chicago, Illinois 60637

David Feldman (179, 765, 1125)
Division of Endocrinology, Gerontology and Metabolism, Stanford University School of Medicine, Stanford, California 94305

Leonard P. Freedman (127)
Cell Biology and Molecular Biology Programs, Memorial Sloan-Kettering Cancer Center, New York, New York 10021

Masafumi Fukagawa (1227)
First Department of Internal Medicine, University of Tokyo School of Medicine, Bunkyo-ku, Tokyo 113, Japan

Robert F. Gagel (369)
Section of Endocrinology, University of Texas M. D. Anderson Cancer Center, Houston, Texas 77030

Marielle Gascon-Barré (41)
Département de Pharmacologie, Faculté de Médecine, Université de Montréal, and Centre de Recherche Clinique André-Viallet, Hôpital Saint-Luc, Montréal, Québec H2X 3J4, Canada

Francis H. Glorieux (293, 755)
Genetics Unit, Shriners Hospital for Children, and Departments of Surgery and Pediatrics, McGill University, Montréal, Québec H3G 1A6, Canada

Wagn Ole Godtfredsen (1027)
Departments of Biochemistry and Chemical Research, Leo Pharmaceutical Products, DK-2750 Ballerup, Denmark

Coleman Gross (1125)
Division of Endocrinology, Gerontology, and Metabolism, Stanford University School of Medicine, Stanford, California 94305

Anandarup Gupta (499)
Renal Division, Barnes-Jewish Hospital, St. Louis, Missouri 63110

John G. Haddad* (87)
Department of Medicine, University of Pennsylvania, Philadelphia, Pennsylvania 19104

Bernard P. Halloran (541)
Department of Medicine, University of California, San Francisco, and Divisions of Endocrinology and Geriatrics, and Veterans Affairs Medical Center, San Francisco, California 94121

Carol A. Haussler (149)
Department of Biochemistry, College of Medicine, The University of Arizona, Tucson, Arizona 85724

Mark R. Haussler (149)
Department of Biochemistry, College of Medicine, The University of Arizona, Tucson, Arizona 85724

Robert P. Heaney (485)
Creighton University, Omaha, Nebraska 68131

Johan N. M. Heersche (313)
Faculty of Dentistry, University of Toronto, Toronto, Ontario M5S 1A8, Canada

Helen L. Henry (57)
Department of Biochemistry, University of California at Riverside, Riverside, California 92521

Martin Hewison (447)
Department of Medicine, Queen Elizabeth Medical Centre, The University of Birmingham, Birmingham B15 2TH, United Kingdom

Michael F. Holick (33)
Vitamin D, Skin, and Bone Research Laboratory; Department of Medicine; Endocrinology, Nutrition and Diabetes Section; Boston Medical Center and Boston University School of Medicine, Boston, Massachusetts 02118

Bruce W. Hollis (587)
Departments of Pediatrics, Biochemistry, and Molecular Biology, Medical University of South Carolina, Charleston, South Carolina 29425

* We are all deeply saddened by John Haddad's untimely death, May 1997. He will be missed by all of us.

Ronald L. Horst (13)
U.S. Department of Agriculture, Agricultural Research Service, National Animal Disease Center, Metabolic Diseases and Immunology Research Unit, Ames, Iowa 50010

Keith Hruska (499)
Renal Division, Washington University School of Medicine, Barnes-Jewish Hospital, St. Louis, Missouri 63110

Jui-Cheng Hsieh (149)
Department of Biochemistry, College of Medicine, The University of Arizona, Tucson, Arizona 85724

Sven Johan Hyllner (209)
Department of Biochemistry and Molecular Biology, University of Medicine and Dentistry of New Jersey, New Jersey Medical School, Newark, New Jersey 07103

Karl L. Insogna (923)
Department of Medicine, Yale University School of Medicine, New Haven, Connecticut 06520

Glenville Jones (973)
Departments of Biochemistry and Medicine, Queen's University, Kingston, Ontario K7L 3N6, Canada

Peter W. Jurutka (149)
Department of Biochemistry, College of Medicine, The University of Arizona, Tucson, Arizona 85724

Masafumi Kitaoka (1227)
Division of Endocrinology and Metabolism, Showa General Hospital, Kodaira-shi, Tokyo 187, Japan

Lilia M. C. Koberle (883)
Health Sciences Department, Federal University, Sao Carlos, SP (UFSCAR), Brazil

H. Phillip Koeffler (1155)
Division of Hematology/Oncology, Department of Medicine, Cedars-Sinai Medical Center, University of California, Los Angeles School of Medicine, Los Angeles, California 90048

Knud Kragballe (1213)
Department of Dermatology, Marselisborg Hospital, University of Aarhus, 8000 Aarhus C, Denmark

Barbara E. Kream (201)
Department of Medicine, The University of Connecticut Health Center, Farmington, Connecticut 06030

Aruna V. Krishnan (179)
Division of Endocrinology, Gerontology and Metabolism, Stanford University School of Medicine, Stanford, California 94305

Noboru Kubodera (1071)
Chugai Pharmaceutical Co., Ltd., Tokyo 104, Japan

Rajiv Kumar (275)
Division of Nephrology, Mayo Clinic and Foundation, Rochester, Minnesota 55905

Kiyoshi Kurokawa (1227)
Tokai University Faculty of Medicine, Boseidai, Isehara-shi, Kanagawa 259-11, Japan
Jacques Lemire (1167)
Division of Pediatric Nephrology, Department of Pediatrics, University of California, San Diego, La Jolla, California 92093
Bryan D. Lemon (127)
Cell Biology and Molecular Biology Programs, Memorial Sloan-Kettering Cancer Center, New York, New York 10021
Alexander C. Lichtler (201)
Department of Pediatrics, The University of Connecticut Health Center, Farmington, Connecticut 06030
Barbara P. Lukert (789)
University of Kansas Medical Center, Kansas City, Kansas 66103
Peter J. Malloy (765)
Division of Endocrinology, Gerontology, and Metabolism, Stanford University School of Medicine, Stanford, California 94305
Chantal Mathieu (1183)
Laboratory for Experimental Medicine and Endocrinology (LEGENDO), Catholic University of Leuven, Belgium
E. Barbara Mawer (831)
University of Manchester Bone Disease Research Center, Department of Medicine, Manchester Royal Infirmary, Manchester M13 9WL, United Kingdom
B. May (69)
Department of Biochemistry, University of Adelaide, Adelaide 5001, South Australia
Pierre J. Meunier (679)
INSERM Unit 403 and Department of Rheumatology and Bone Disease, Edouard Herriot Hospital, 69437 Lyon Cedex 03, France
Devashis Mitra (521)
Department of Medicine, Medical University of South Carolina, Department of Veterans Affairs Medical Center, Charleston, South Carolina 29401
Roberta Morosetti (1155)
Instituto di Semeiotica Medica, Universita Cattolica del S. Cuore, Policlinico Agostino Gemelli, 00168 Rome, Italy
Nigel Morrison (713)
Genomics Research Centre, Griffith University, Gold Coast, Southport 4215, Queensland, Australia
Tally Naveh-Many (353)
Minerva Center for Calcium and Bone Metabolism, Nephrology Services, Hadassah University Hospital, Hebrew University Hadassah Medical School, Jerusalem, Israel 91120
Yasuho Nishii (1071)
Chugai Pharmaceutical Co., Ltd., Tokyo 104, Japan
Anthony W. Norman (233)
Department of Biochemistry and Division of Biomedical Sciences, University of California at Riverside, Riverside, California 92521
William H. Okamura (939)
Department of Chemistry, University of California, Riverside, Riverside, California 92521
J. Omdahl (69)
Department of Biochemistry and Molecular Biology, University of New Mexico, School of Medicine, Albuquerque, New Mexico 87131
Jeffrey L. H. O'Riordan (447)
Bone and Mineral Centre, University College London Medical School, London W1N 8AA, United Kingdom
A. Michael Parfitt (645)
Division of Endocrinology and Center for Osteoporosis and Metabolic Bone Disease, University of Arkansas for Medical Sciences, Little Rock, Arkansas 72205
Donna M. Peehl (1125)
Department of Urology, Stanford University School of Medicine, Stanford, California 94305
Sara Peleg (369, 1011)
Department of Medical Specialties, University of Texas M. D. Anderson Cancer Center, Houston, Texas 77030
John M. Pettifor (663)
MRC Mineral Metabolism Research Unit, Department of Paediatrics, University of the Witwatersrand and Baragwanath Hospital, Soweto, Gauteng, South Africa
J. Wesley Pike (105, 765)
Department of Molecular and Cellular Physiology, University of Cincinnati, Cincinnati, Ohio 45267
Huibert A. P. Pols (1089)
Department of Internal Medicine III, Erasmus University Medical School, 3015 GD Rotterdam, The Netherlands
Anthony A. Portale (541)
Departments of Medicine and Pediatrics, University of California, San Francisco, San Francisco, California 94143
Lawrence G. Raisz (789)
Division of Endocrinology and Metabolism, University of Connecticut Health Center, Farmington, Connecticut 06030
Satyanarayana G. Reddy (1045)
Brown University School of Medicine, Providence, Rhode Island
Timothy A. Reinhardt (13)
U.S. Department of Agriculture, Agricultural Research Service, National Animal Disease Center, Metabolic Diseases and Immunology Research Unit, Ames, Iowa 50010

Lenore S. Remus (149)
Department of Biochemistry, College of Medicine, The University of Arizona, Tucson, Arizona 85724

B. Lawrence Riggs (695)
Department of Endocrinology and Metabolism, Mayo Clinic and Foundation, Rochester, Minnesota 55905

Bernard L. Salle (533)
Department of Neonatology, Hôpital Édouard Herriot, 69437 Lyon, France

Katsuhiko Sato (1071)
Chugai Pharmaceutical Co., Ltd., Tokyo 104, Japan

Zvi Schwartz (395)
Departments of Orthopedics and Periodontics, The University of Texas Health Science Center at San Antonio, San Antonio, Texas 78284; and Hebrew University Hadassah Faculty of Dental Medicine, Jerusalem, Israel

Yoshiki Seino (305)
Department of Pediatrics, Okayama University Medical School, Okayama 700, Japan

Sanford H. Selznick (149)
Department of Biochemistry, College of Medicine, The University of Arizona, Tucson, Arizona 85724

Justin Silver (353)
Minerva Center for Calcium and Bone Metabolism, Nephrology Services, Hadassah University Hospital, Hebrew University Hadassah Medical School, Jerusalem, Israel 91120

Michael D. Sitrin (1141)
Department of Medicine, The University of Chicago Hospitals & Clinics, Chicago, Illinois 60637

Eduardo S. Slatopolsky (849, 995)
Renal Division, Department of Medicine, Washington University School of Medicine; Chromalloy American Kidney Center; and Barnes-Jewish Hospital, St. Louis, Missouri 63110

René St-Arnaud (293, 755)
Genetics Unit, Shriners Hospital for Children, and Departments of Surgery and Human Genetics, McGill University, Montréal, Québec H3G 1A6, Canada

Paula H. Stern (341)
Departments of Molecular Pharmacology and Biological Chemistry, Northwestern University Medical School, Chicago, Illinois 60611

George P. Studzinski (1045)
University of Medicine and Dentistry of New Jersey—New Jersey Medical School, Newark, New Jersey

Tatsuo Suda (329)
Department of Biochemistry, School of Dentistry, Showa University, Shinagawa-ku, Tokyo 142, Japan

Victor L. Sylvia (395)
Department of Orthopedics, The University of Texas Health Science Center at San Antonio, San Antonio, Texas 78284

Naoyuki Takahashi (329)
Department of Biochemistry, School of Dentistry, Showa University, Shinagawa-ku, Tokyo 142, Japan

Hiroyuki Tanaka (305)
Department of Pediatrics, Okayama University Medical School, Okayama 700, Japan

Monique Thomasset (223)
INSERM U.458, Aliée CNRS, Hôpital Robert Debré, 75019 Paris, France

Paul D. Thompson (149)
Department of Biochemistry, College of Medicine, The University of Arizona, Tucson, Arizona 85724

Susan Thys-Jacobs (883)
Department of Medicine, College of Physicians and Surgeons, Columbia University, New York, New York 10032

Milan R. Uskoković (1045)
Hoffmann-La Roche, Inc., Nutley, New Jersey 07110

Johannes P. T. M. van Leeuwen (1089)
Department of Internal Medicine III, Erasmus University Medical School, 3015 GD Rotterdam, The Netherlands

Marian R. Walters (463)
Department of Physiology, Tulane University School of Medicine, New Orleans, Louisiana 70112

Robert H. Wasserman (259)
Department of Physiology, College of Veterinary Medicine, Cornell University, Ithaca, New York 14853

G. Kerr Whitfield (149)
Department of Biochemistry, College of Medicine, The University of Arizona, Tucson, Arizona 85724

Michael P. Whyte (557)
Division of Bone and Mineral Diseases, Washington University School of Medicine at Barnes-Jewish Hospital, St. Louis, Missouri, and Metabolic Research Unit, Shriners Hospital for Children, St. Louis, Missouri 63110

Joseph E. Zerwekh (607)
University of Texas Southwestern Medical Center at Dallas, Center for Mineral Metabolism and Clinical Research, Dallas, Texas 75235

Gui-Dong Zhu (939)
Department of Chemistry, University of California, Riverside, Riverside, California 92521

Preface

Our reasons for deciding to publish an entire book devoted to vitamin D can be found in the rapid and extensive advances currently being made in this important field of research. Enormous progress in investigating many aspects of vitamin D, from basic science to clinical medicine, has been made in recent years. The ever-widening scope of vitamin D research has created new areas of inquiry so that even workers immersed in the field are not fully aware of the entire spectrum of current investigation. Our goal in planning this book was to bring the diverse scientific and clinical fields together in one definitive and up-to-date volume. It is our hope that this compendium on vitamin D will serve as both a resource for current researchers and a guide to stimulate and assist those in related disciplines to enter this field of research. In addition, we hope to provide clinicians and students with a comprehensive source of information for the varied and extensive material related to vitamin D.

The explosion of information in the vitamin D sphere has led to new insights into many different areas, and in our treatment of each subject in this book we have tried to emphasize the recent advances as well as the established concepts. The classic view of vitamin D action, as a hormone limited to calcium metabolism and bone homeostasis, has undergone extensive revision and amplification in the past few years. We now know that the vitamin D receptor (VDR) is present in most tissues of the body and that vitamin D actions, in addition to the classic ones, include important effects on an extensive array of other target organs. To cover this large number of subjects, we have organized the book in the following manner: Section I, the enzymes involved in vitamin D metabolism and the activities of the various metabolites; Section II, the mechanism of action of vitamin D, including rapid, nongenomic actions and the role of the VDR in health and disease; Section III, the effects of vitamin D and its metabolites on the various elements that constitute bone and the expanded understanding of vitamin D actions in multiple target organs, both classic and nonclassic; Section IV, the role of vitamin D in the physiology and regulation of calcium and phosphate metabolism and the multiplicity of hormonal, environmental, and other factors influencing vitamin D metabolism and action; and Sections V and VI, the role of vitamin D in the etiology and treatment of rickets, osteomalacia, and osteoporosis and the pathophysiological basis, diagnosis, and management of numerous clinical disorders involving vitamin D.

The recent recognition of an expanded scope of vitamin D action and the new investigational approaches it has generated were part of the impetus for developing this volume on vitamin D. It has become clear that in addition to the classic vitamin D actions, a new spectrum of vitamin D activities that include important effects on cellular proliferation, differentiation, and the immune system has been identified. This new information has greatly expanded our understanding of the breadth of vitamin D action and has opened for investigation a large number of new avenues of research that are covered in Sections VII and VIII of this volume. Furthermore, these recently recognized nonclassic actions have led to a consideration of the potential application of vitamin D therapy to a range of diseases not previously envisioned. This therapeutic potential has spawned the search for vitamin D analogs that might have a more favorable therapeutic profile, one that is less active in causing hypercalcemia and hypercalciuria while more active in a desired application such as inducing antiproliferation, prodifferentiation, or immunosuppression. Since 1α,25-dihydroxyvitamin D [1,25$(OH)_2$D] and its analogs are all presumed to act via a single VDR, a few years ago most of us probably would have thought that it was impossible to achieve a separation of these activities. Yet today, many analogs that exhibit different profiles of activity relative to $1,25(OH)_2D$ have already been produced and extensively studied. The development of analogs with an improved therapeutic index has opened another large and complex area of vitamin D research. This work currently encompasses three domains: (1) the design and synthesis of vitamin D analogs exhibiting a separation of actions with less hypercalcemic and more antiproliferative or immunosuppressive activity, (2) the interesting biological question of the mechanism(s) by which these analogs achieve their differential activity, and (3) the investigation of the potential therapeutic applications of these analogs to treat various disease states. These new therapeutic applications, from psoriasis to cancer, from immunosuppression to neurodegenerative diseases, have drawn into the field an expanded population of scientists and physicians interested in vitamin D.

Our goal in editing this book was to create a comprehensive resource on vitamin D that would be of use to a mix of researchers in different disciplines. To achieve this goal, we sought authors who had contributed greatly to their respective fields of vitamin D research. The book has a large number of chapters to accommodate many contributors and provide expertise in multiple areas. Introductory chapters in each section of the book are designed to furnish an overview of that area of vitamin D research, with other chapters devoted to a narrowly focused subject. Adjacent and closely related subjects are often covered in separate chapters by other authors. While this intensive style may occasionally create some redundancy, it has the advantage of allowing the reader to view the diverse perspectives of the different authors working in overlapping fields. In this regard, we have endeavored to provide many cross-references to guide the reader to related information in different chapters.

We express our thanks to Jasna Markovac (Editor-in-Chief), for encouraging us to develop this book and guiding us through the process; to Tari Paschall (Acquisitions Editor) and the Academic Press staff for their diligence, expertise, and patience in helping us complete this work. Most of all, we thank the authors for their contributions that have made this book possible.

DAVID FELDMAN

FRANCIS H. GLORIEUX

J. WESLEY PIKE

Abbreviations

ACTH	adrenocorticotropin
ADHR	autosomal dominant hypophosphatemic rickets
ADP	adenosine diphosphate
AHO	Albright's hereditary osteodystrophy
AIDS	acquired immunodeficiency syndrome
ALP	alkaline phosphatase
APC	antigen presenting cell
APD	aminohydroxypropylidene bisphosphonate
AR	androgen receptor
ATP	adenosine triphosphate
ATRA	all-*trans*-retinoic acid
B_{max}	maximum number of binding sites
bFGF	basic fibroblast growth factor
BFU	burst-forming unit
BGP	bone Gla protein (osteocalcin)
BLM	basal lateral membrane
BMAR	bone mineral apposition rate
BMD	bone mineral density
BMI	body mass index
BMP	bone morphogenetic protein
BMU	basic multicellular unit
bp	base pairs
BPH	benign prostatic hyperplasia
BSA	bovine serum albumin
$[Ca^{2+}]_i$	internal calcium ion molar concentration
CaBP	calcium binding protein
cAMP	cyclic AMP
CaR	calcium receptor
CAT	chloramphenicol acetyltransferase
CDK or Cdk	cyclin-dependent kinase
cDNA	complementary DNA
CFU	colony-forming unit
cGMP	cyclic GMP
CK-II	casein kinase-II
cM	centimorgans
CNS	central nervous system
CPBA	competitive protein binding assays
cpm	counts per minute
CRE	cAMP response element
CREB	cAMP response element binding protein
CRF	chronic renal failure
CsA	cyclosporin A
CSF	colony-stimulating factor
CT	calcitonin
CTR	calcitonin receptor
CTX	cerebrotendinous xanthomatosis
CYP	cytochrome P450
DAG	1,2-diacylglycerol
DBD	DNA binding domain
DBP	vitamin D binding protein
DCT	distal convoluted tubule
DEXA or DXA	dual energy X-ray absorptiometry
7-DHC	7-dehydrocholesterol

DHEA	dehydroepiandrosterone
DHT	dihydrotachysterol
DMSO	dimethyl sulfoxide
DR	direct repeat
E_1	estrone
E_2	estradiol
EAE	experimental autoimmune encephalitis
EBV	Epstein-Barr virus
EC_{50} or ED_{50}	effective concentration (dose) to cause 50% effect
ECF	extracellular fluid
EDTA	ethylenediaminetetraacetic acid
EGF	epidermal growth factor
ELISA	enzyme-linked immunosorbent assay
EMSA	electrophoretic mobility shift assay
ER	estrogen receptor
ERE	estrogen response element
FACS	fluorescence-activated cell sorting or sorter
FAD	flavin adenine dinucleotide
FCS	fetal calf serum
FDA	U.S. Food and Drug Administration
FMTC	familial medullary thyroid carcinoma
FS	Fanconi syndrome
FSK	forskolin
g	gram
g	acceleration due to gravity
G_0, G_1, G_2	gap phases of the cell cycle
GAG	glycosaminoglycan
GC-MS	gas chromatography-mass spectrometry
G-CSF	granulocyte colony-stimulating factor
GFR	glomerular filtration rate
GH	growth hormone
GHRH	growth hormone-releasing hormone
GM-CSF	granulocyte-macrophage colony-stimulating factor
GnRH	gonadotropin-releasing hormone
GR	glucocorticoid receptor
GRE	glucocorticoid response element
GRTH	generalized resistance to thyroid hormone
HHRH	hereditary hypophosphatemic rickets with hypercalciuria
HIV	human immunodeficiency virus
HPLC	high-performance liquid chromatography
HPV	human papilloma virus
hr	hour
HRE	hormone response element
Hsp	heat-shock protein
HSV	herpes simplex virus
HVDRR	hereditary vitamin D-resistant rickets
HVO	hypovitaminosis D osteopathy
IBMX	isobutylmethylxanthine
IC_{50}	concentration to inhibit 50% effect
ICA	intestinal calcium absorption
ICMA	immunochemiluminometric assay
IDDM	insulin-dependent diabetes mellitus
IDM	infants of diabetic mothers
IFN	interferon
Ig	immunoglobulin
IGF-I, -II	insulin-like growth factor type I, II
IGF-IR	IGF-I receptor
IGFBP	IGF binding protein
IL	interleukin (e.g., IL-1, IL-1β, etc.)
i.m.	intramuscular
i.p.	intraperitoneal
IP_3	inositol 1,4,5-trisphosphate
IRMA	immunoradiometric assay
IU	international units
i.v.	intravenous
K_d	dissociation constant
K_m	Michaelis constant
kb	kilobases
kbp	kilobase pairs
kDa	kilodaltons
LBD	ligand binding domain
LIF	leukemia inhibitory factor
LNH	late neonatal hypocalcemia
LOD	logarithm of the odds
LPS	lipopolysaccharide
LT	leukotriene
M	mitosis phase of cell cycle
M	molar
Mab	monoclonal antibody
MCR	metabolic clearance rate
M-CSF	macrophage colony-stimulating factor

MEN2 multiple endocrine neoplasia type 2
MGP matrix Gla protein
MHC major histocompatibility complex
min minute
MLR mixed lymphocyte reaction
MR mineralcorticoid receptor
MRI magnetic resonance imaging
mRNA messenger ribonucleic acid
MTC medullary thyroid carcinoma
NADH nicotinamide adenine dinucleotide
NADPH nicotinamide adenine dinucleotide phosphate
NBT nitroblue tetrazolium
NcAMP nephrogenous cAMP
NGF nerve growth factor
NHANES III National Health and Nutrition Examination Survey III
NHL Non-Hodgkin's lymphoma
NIDDM non-insulin-dependent diabetes mellitus
NIH National Institutes of Health
NK cell natural killer cell
NLS nuclear localization signal
NMR nuclear magnetic resonance
OCT 22-oxacalcitriol
ODF osteoclast differentiation factor
$1\alpha OHD_3$ 1α-hydroxyvitamin D_3
$25OHD_3$ 25-hydroxyvitamin D_3
$1,25(OH)_2D_3$ 1α,25-dihydroxyvitamin D_3
$24,25(OH)_2D_3$ 24,25-dihydroxyvitamin D_3
OSM oncostatin M
OVX ovariectomy
P_i inorganic phosphate
PA_2 phospholipase A_2
PAM pulmonary alveolar macrophage
PBL peripheral blood lymphocyte
PBS phosphate-buffered saline
PCNA proliferating cell nuclear antigen
PCR polymerase chain reaction
PDDR pseudovitamin D deficiency rickets
PDGF platelet-derived growth factor
PEIT percutaneous ethanol injection therapy
PEX *p*hosphate regulating gene with homologies to *e*ndopeptidases on the *X* chromosome
PG prostaglandin
PHA phytohemagglutinin
PHP pseudohypoparathyroidism
PKA protein kinase A
PKC protein kinase C
PKI protein kinase inhibitor
PLA_2 phospholipase A_2
PLC phospholipase C
PMA phorbol 12-myristate 13-acetate
PMCA plasma membrane calcium pump
p.o. oral
PPAR peroxisome proliferator-activated receptor
PRL prolactin
PSA prostate-specific antigen
PTH parathyroid hormone
PTHrP parathyroid hormone-related peptide
PTX parathyroidectomy
PUVA psoralen-ultraviolet A
QCT quantitative computerized tomography
QSAR quantitative structure-activity relationship
9-*cis*-RA 9-*cis*-retinoic acid
RA rheumatoid arthritis
RAR retinoic acid receptor
RARE retinoic acid response element
RCI relative competitive index
RDA recommended dietary allowance
RFLP restriction fragment length polymorphism
RIA radioimmunoassay
RNase ribonuclease
RPA ribonuclease protection assay
RRA radioreceptor assay
RT-PCR reverse transcriptase-polymerase chain reaction
RXR retinoid X receptor
RXRE retinoid X receptor response element
SD standard deviation
SDS sodium dodecyl sulfate

SE	standard error
SEM	standard error of the mean
SPF	sun protection factor
SSCP	single strand conformational polymorphism
SV40	simian virus 40
$t_{1/2}$	half-time
T_3	triiodothyronine
T_4	thyroxine
TBP	TATA binding protein
TC	tumoral calcinosis
TF	tubular fluid
TFIIB	general transcription factor IIB
TGF	transforming growth factor
TIO	tumor-induced osteomalacia
TK	thymidine kinase
TmP or TmP_i	tubular absorptive maximum for phosphorus
TNF	tumor necrosis factor
TPA	12-*O*-tetradecanoylphorbol-13-acetate
TPN	total parenteral nutrition
TPTX	thyroparathyroidectomized
TRAP	tartrate-resistant acid phosphatase
TRE	thyroid hormone response element
TRE	TPA response element
TRH	thyrotropin-releasing hormone
Trk	tyrosine kinase
TSH	thyrotropin
UF	ultrafilterable fluid
US	ultrasonography
USDA	U.S. Department of Agriculture
UTR	untranslated region
UV	ultraviolet
VDDR-I	vitamin D-dependent rickets type I (*see* PDDR)
VDR	vitamin D receptor
VDRE	vitamin D response element
VEGF	vascular endothelial growth factor
XLH	X-linked hypophosphatemia

Approximate Normal Range for Serum Values[a]

Measure	SI units	Conventional units	Conversion factor[b]
Ionized calcium	1–1.5 mmol/L	4–4.6 mg/dL	0.2495
Total calcium	2.2–2.6 mol/L	9–10.5 mg/dL	0.2495
Phosphorus, inorganic	1–1.5 mmol/L	3.0–4.5 mg/dL	0.3229
25OHD	25–130 nmol/L	10–52 ng/mL	2.496
$1,25(OH)_2D$	36–144 pmol/L	15–60 pg/mL	2.40

[a] Normal ranges differ in various laboratories and are provided only as a general guide.
[b] Conversion factor × conventional units = SI units.

Useful Equivalencies of Different Units

Vitamin D	1 μg = 40 IU
Calcium	1 mmol = 40 mg
Phosphorus	1 mmol = 30 mg

SECTION I

Chemistry and Metabolism

CHAPTER 1

Historical Overview

HECTOR F. DELUCA Department of Biochemistry, University of Wisconsin–Madison, Madison, Wisconsin

I. DISCOVERY OF THE VITAMINS

A. Early Nutritional Views

The field of nutrition was largely dominated in the nineteenth century by German chemists, led by Justus von Liebig [1]. They taught that adequacy of the diet could be described by an analysis of protein, carbohydrate, fat, and mineral. Thus, a diet containing 12% protein, 5% mineral, 10–30% fat, and the remainder as carbohydrate would be expected to support normal growth and reproduction. This view remained largely unchallenged until the very end of the nineteenth century and the beginning of the twentieth century [2–5]. However, evidence opposing this view began to appear. One of the first was the famous study of Eijkman who studied prisoners in the Dutch East Indies maintained on a diet of polished rice [6]. A high incidence of the neurological disorder beri-beri was recorded in these inmates. Eijkman found that either feeding whole rice or returning the hulls of the polished rice could eliminate beri-beri. Eijkman reasoned that polished rice contained a toxin that was somehow neutralized by the rice hulls. Later, a colleague, Grijns [7], revisited the question and correctly demonstrated that hulls contained an important and required nutrient that prevented beri-beri.

Other reports revealed that microorganic nutrients might be present. The development of scurvy in mariners was a common problem. This disease was prevented by the consumption of limes on British ships (hence, the term "Limey" to describe British sailors) and sauerkraut and fruits on other ships. This led Holst and Fröhlich to conclude that scurvy could be prevented by a nutrient present in these foods [8]. Experiments by Lunin, Magendie, Hopkins, and Funk showed that a diet of purified carbohydrate, protein, fat, and salt is unable to support growth and life of experimental animals [2–5]. This suggested that some unknown or vital factor present in natural foods was missing from the purified diets. Hopkins developed a growth test in which natural foods were found to support rapid growth of experimental animals whereas purified materials could not [3]. Funk had found similar results for the prevention of neuritis and reasoned that there were "vital amines" present in foods from natural sources and actually provided the basis for the term "vitamins" used later to describe essential micronutrients [5].

B. McCollum and Osborne and Mendel's Discovery of Vitamin A and B Complex

A key experiment demonstrating essential micronutrients was one carried out at the Wisconsin Agricultural

Experiment Station, engineered by Stephen Moulton Babcock and carried out by E. B. Hart supported by McCollum and Steenbock [9]. Herds of dairy cows were maintained on a diet composed individually only of corn, oats, or wheat or were fed a mixture of all of these grains, all receiving the same amount of carbohydrate, protein, fat, and salts and all providing equal analysis according to the German chemists [1]. The animals on the corn diet did very well, produced milk in large amounts, and reproduced normally. Those on the wheat diet failed to thrive and soon were unable to reproduce or lactate. The oat group was found to be intermediate between the corn and wheat groups, and the mixture approximated the growth and reproduction found with corn. Yet all these diets had the same proximate analysis.

The conclusion of the Wisconsin Experiment Station study was that there are unknown nutrients present in corn and not found in wheat that are essential for life and reproduction. This led E. B. Hart, Chairman of Biochemistry at Wisconsin, to conceive that a search for these nutrients must begin. Professor McCollum was asked to search for these nutrients using small experimental animals. McCollum and Davis demonstrated there was present in butter fat a substance that prevented xerophthalmia and was also required for growth. They termed this "a lipin-soluble growth factor" [10]. McCollum later named this factor "vitamin A" [11]. This substance was absent from lard and other fats but was found in large amounts in cod liver oil. In constructing the diets, McCollum obtained the carbohydrates and salts from milk whey which, unknown to him, supplied the vitamin B complex group of micronutrients that permitted him to observe a vitamin A deficiency. McCollum at Wisconsin [11] and Osborne and Mendel [12] at the Connecticut Experiment Station carried out experiments in which cod liver oil was used as a source of fat in the diet but the minerals were supplied from pure chemicals mixed to approximate the mineral composition of milk. Starch or sugar was used as the carbohydrate. These animals developed a different group of symptoms, namely, neuritis, which could be cured by the provision of the milk components. McCollum and Osborne and Mendel correctly concluded that this activity was due to a different micronutrient called "vitamin B." This ushered in the concept of the organic micronutrients known as vitamins.

C. History of Rickets

The disease rickets was very likely known in antiquity but was described in the fifteenth century as revealed by later writings. Whistler first provided a clear description of rickets in which the skeleton was poorly mineralized and deformed [13]. Rickets undoubtedly in ancient times appeared only on rare occasions and hence was not considered a problem. However, at the end of the nineteenth century, the industrial revolution had taken place: a highly agrarian population had become urbanized, and smoke from the industrial plants polluted the atmosphere. Thus, in low sunlight countries such as England, rickets appeared in epidemic proportions. In fact, it was known as the English Disease [14]. Some reports of the beneficial action of cod liver oil had appeared. However, they were not given scientific credence.

With the discovery of the vitamins, Sir Edward Mellanby in Great Britain began to reason that rickets might also be a disease caused by a dietary deficiency [15]. Mellanby fed dogs a diet composed primarily of oatmeal, which was the diet consumed where the incidence of rickets was the highest (i.e., Scotland). McCollum inadvertently maintained the dogs on oatmeal indoors and away from ultraviolet light. The dogs developed severe rickets. Learning from the experiments of McCollum, Mellanby provided cod liver oil to cure or prevent the disease. Mellanby could not decide whether the healing of rickets was due to vitamin A known to be present in the cod liver oil or whether it was a new and unknown substance. Therefore, the activity of healing rickets was first attributed to vitamin A.

D. Discovery of Vitamin D

McCollum, who had moved to Johns Hopkins from Wisconsin, continued his experiments on the fat-soluble materials. McCollum used aeration and heating of cod liver oil to destroy the vitamin A activity or the ability to support growth and to prevent xerophthalmia [16]. However, cod liver oil treated in this manner still retained the ability to cure rickets. McCollum correctly reasoned that the activity in healing rickets was due to a new and heretofore unknown vitamin which he termed "vitamin D." On the basis of the experiments of McCollum and of Mellanby, vitamin D became known as an essential nutrient.

II. DISCOVERY THAT VITAMIN D IS NOT A VITAMIN

At the same time that Sir Edward Mellanby was carrying out the experiments in dogs, Huldshinsky [17] and Chick *et al.* [18] independently found that rickets in children could be prevented or cured by exposing them to sunlight or to artificially induced ultraviolet light. Thus, the curious findings were that sunlight and ultraviolet light somehow equaled cod liver oil. These strange and divergent results required resolution.

Steenbock and Hart had noted the importance of sunlight in restoring positive calcium balance in goats [19]. At Wisconsin, with McCollum carrying out experiments in small experimental animals (i.e., rats), Steenbock was required to work with larger animals. Steenbock then began to study goats because they would consume less materials and could serve as better experimental animals than cows. Steenbock began to study the calcium balance of lactating goats and found that those goats maintained outdoors in the sunlight were found to be in positive calcium balance, whereas those maintained indoors lost a great deal of their skeletal calcium to lactation [19]. Steenbock and Hart, therefore, noted the importance of sunlight on calcium balance. This work then undoubtedly led Steenbock to realize that the ultraviolet healing properties described by Huldschinky might be related to the calcium balance experiments in goats. By irradiating the animals and diets, Steenbock and Black found that vitamin D activity could be induced and rickets could be cured [20]. A similar finding was reported soon thereafter by Hess and Weinstock [21]. Steenbock then traced this to the nonsaponifiable fraction of the lipids in foods [22]. He found that ultraviolet light activated an inactive substance to become a vitamin D active material. Thus, ultraviolet light could be used to irradiate foods, induce vitamin D activity, and fortify foods to eliminate rickets as a major medical problem. This discovery also made available a source of vitamin D for isolation and identification.

III. ISOLATION AND IDENTIFICATION OF NUTRITIONAL FORMS OF VITAMIN D

From irradiation of mixtures of plant sterols, Windaus and colleagues isolated a material that was active in healing rickets [23]. This substance was called "vitamin D_1," but its structure was not determined. Vitamin D_1 proved to be an adduct of tachysterol and vitamin D_2, and thus vitamin D_1 was actually an error in identification. The British group led by Askew was successful in isolating and determining the structure of the first vitamin D, vitamin D_2 or ergocalciferol, from irradiation of plant sterols [24]. A similar identification by the Windaus group confirmed the structure of vitamin D_2 [25]. Windaus and Bock also isolated the precursor of vitamin D_3 from skin, namely, 7-dehydrocholesterol [26]. Furthermore, 7-dehydrocholesterol was synthesized [27] and converted to vitamin D_3 (cholecalciferol) as identified by the Windaus group [28]. Thus, the structures of nutritional forms of vitamin D became known (Fig. 1). Windaus and Bock, having isolated 7-dehydrocholesterol from skin, provided the presumptive evidence that vitamin D_3 is the form of vitamin D produced in skin, a discovery that was later confirmed by the chemical identification of vitamin D_3 in skin by Esvelt *et al.* [29] and of a previtamin D_3 in skin by Holick *et al.* [30]. Synthetic vitamin D as produced by the irradiation process replaced the irradiation of foods as a means of fortifying foods with vitamin D and was also rapidly applied to a variety of diseases including rickets and tetany and in the provision to domestic animals such as chickens, cows, and pigs.

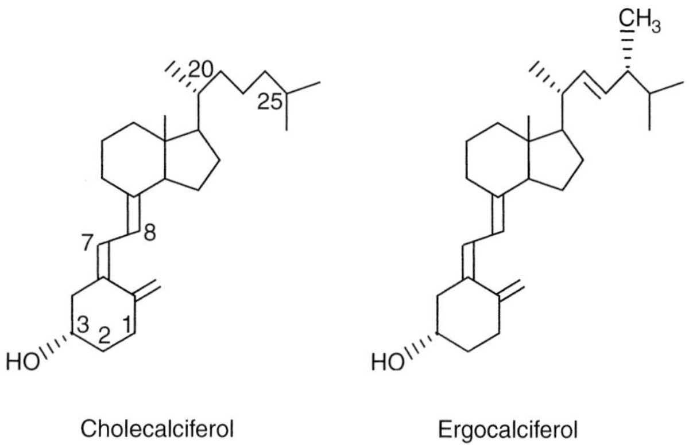

FIGURE 1 Nutritional forms of vitamin D.

Windaus' group provided chemical syntheses of the vitamin D compounds, confirming their structures and thus ending the era of the isolation and identification of nutritional forms of vitamin D and making them available for the treatment of disease. For his contributions, Windaus received the 1928 Nobel Prize in chemistry.

IV. DISCOVERY OF THE PHYSIOLOGICAL FUNCTIONS OF VITAMIN D

A. Intestinal Calcium and Phosphorus Absorption

Besides bone mineralization, the earliest discovered function of vitamin D is its important role in the absorption and utilization of calcium. The first report of this finding was in the early 1920s by Orr and colleagues [31]. Kletzien *et al.* [32] demonstrated that vitamin D plays an important role in the utilization of calcium from the diet, and a number of experiments were carried out on the utilization of calcium and phosphorus from cereal diets. Nicolaysen was responsible, however, for demonstrating unequivocally the role of vitamin D in the absorption of calcium and independently of phosphorus from the diet [33]. Nicolaysen also followed the early work of Kletzien *et al.* [32] in which animals adapted to a low calcium diet were better able to utilize calcium than were animals on an adequate calcium diet. This work was confirmed by Nicolaysen, who postulated the existence of an "endogenous factor" that would inform the intestine of the skeletal needs for calcium [34]. This endogenous factor later proved to be largely the active form of vitamin D, 1,25-dihydroxyvitamin D_3 [$1,25(OH)_2D_3$] [35].

B. Mobilization of Calcium from Bone

For many years, investigators have attempted to show that vitamin D plays a role directly on the mineralization process of the skeleton. However, early work by Howland and Kramer [36], later work by Lamm and Neuman [37], and more recent work by Underwood and DeLuca [38] demonstrated very clearly that vitamin D does not play a significant role in the actual mineralization process of the skeleton but that the failure to mineralize the skeleton in vitamin D deficiency is due to inadequate levels of calcium and phosphorus in the plasma. Thus, the action of vitamin D in mineralizing the skeleton and in preventing hypocalcemic tetany is the elevation of plasma calcium and phosphorus [39]. These discoveries laid to rest the concept of a role of vitamin D in mineralization. However, Carlsson [40] and Bauer *et al.* [41] were the first to realize that a major function of vitamin D is to induce the mobilization of calcium from bone when required. Thus, in animals on a low calcium diet, the rise in serum calcium induced by vitamin D is the result of actual mobilization of calcium from bone [42]. This important function is known to be essential for the provision of calcium to meet soft tissue needs, especially those of nerves and muscle, on a minute-to-minute basis when it is in insufficient supply from the diet. It is unknown whether the ability of vitamin D to mobilize calcium from bone is osteoclastic-mediated or whether it is due to a transport of calcium across osteoblastic and bone lining cell membranes into the plasma compartment [43]. It is clear, however, that both vitamin D and parathyroid hormone are required for this function [44]. Furthermore, it is clear that vitamin D plays an important role in osteoclastic-mediated bone resorption [45], which is certainly the first event in bone remodeling and an essential event in bone modeling [46]. The actions of vitamin D and parathyroid hormone on osteoclastic-mediated bone resorption are independent of one another [47].

C. Renal Reabsorption of Calcium and Phosphorus

A final site of vitamin D action to elevate plasma calcium is in the distal renal tubule. Although experiments were suggestive of a role for vitamin D in increasing renal tubule reabsorption of calcium, a clear demonstration of this did not occur until the late 1980s at the hands of Yamamoto *et al.* [48]. The renal tubule reabsorbs 99% of the filtered calcium even in the absence of vitamin D. However, reabsorption of the last 1% of the filtered load requires both vitamin D and parathyroid hormone. Thus, these agents work in concert in the renal reabsorption of calcium as well as in the mobilization of calcium from bone. Both agents are required to carry out this function.

D. Discovery of New Functions of Vitamin D

With discovery of the receptor for the vitamin D hormone (described in Section V,G below) came the surprising result that this receptor could be found in a variety of tissues not previously appreciated as targets of vitamin D action. It localizes in the distal renal tubule cells, enterocytes of the small intestine, bone lining cells, and osteoblasts in keeping with its known role in calcium metabolism [49,50]. However, its appearance in tissues such as parathyroid gland, islet cells of the pancreas, cells in bone marrow (i.e., promyelocytes), lymphocytes, and certain neural cells raised the question of whether the functions of vitamin D might be broader than previously anticipated [49,50]. As a result of those findings, new functions of vitamin D have been found. For example, vitamin D plays a role in causing differentiation of promyelocytes to monocytes and the subsequent coalescing of the monocytes into multinuclear osteoclast precursors and ultimately into active osteoclasts [51,52]. Suppression of parathyroid cell growth and suppression of parathyroid hormone gene expression represent other new vitamin D actions [53,54]. In keratinocytes of skin, vitamin D appears to play a role in suppression of growth and in cellular differentiation [55]. Likely, discoveries of many new functions of $1,25(OH)_2D_3$ will be made and are well on their way, as described in later chapters of this volume.

V. DISCOVERY OF THE HORMONAL FORM OF VITAMIN D

A. Early Work of Kodicek

The true pioneer of vitamin D metabolism was Egan Kodicek working at the Dunn Nutritional Laboratory in Cambridge. Kodicek used a bioassay at first to study the fate of the vitamin D molecule and found that much vitamin D was converted to biologically inactive products [56]. Clearly, however, this approach of assaying vitamin D activity following administration of known doses of vitamin D was of limited value in determining metabolism.

B. Radiolabeled Vitamin D Experiments

Professor Kodicek then began to synthesize radiolabeled vitamin D_2. Unfortunately, the degree of labeling was not sufficient to permit the administration of truly physiological doses of vitamin D. Nevertheless, Professor Kodicek continued investigations into this important area. At the conclusion of 10 years of work, he concluded that vitamin D was active without metabolic modification and that the metabolites that were found were biologically inactive [57]. This conclusion was reached even as late as 1967, when it was concluded that vitamin D_3 itself was the active form of vitamin D in the intestine [58]. However, chemical synthesis of vitamin D_3 of high specific activity in the laboratory of the author proved to be of key importance in the demonstration of biologically active metabolites [59]. By providing a truly physiological dose of vitamin D, it could be learned that the vitamin D itself disappeared and instead polar metabolites could be found in the target tissues [60]. The polar metabolites proved to be more biologically active and acted more rapidly than vitamin D itself [61]. Thus, presumptive evidence of conversion of vitamin D to active forms had been obtained as early as 1967.

C. Isolation and Identification of the Active Form of Vitamin D

By 1968, the first active metabolite of vitamin D was isolated in pure form and chemically identified as 25-hydroxyvitamin D_3 ($25OHD_3$) [62]. Its structure was confirmed by chemical synthesis [63] that provided it for study to the medical and scientific world. For a couple of years, 25OHD was visualized as the active form of vitamin D. However, when it was synthesized in radiolabeled form, it was found to be rapidly metabolized to yet more polar metabolites [64]. By this time, the Kodicek laboratory reawakened their interest in metabolism of vitamin D and began to study the metabolism of 1α-tritium-labeled vitamin D [65]. Furthermore, polar metabolites of vitamin D were found by Haussler, Myrtle, and Norman [66]. The Wisconsin group labeled these metabolites as peak 5 [64], the Norman group called it peak 4B [66], and Lawson, Wilson, and Kodicek described it as peak P [65]. Kodicek *et al.* claimed that the metabolite of vitamin D found in intestine was deficient in tritium at the 1 position [65]. However, Myrtle *et al.* reported that peak 4B did not lose its tritium [67]. Thus, the suggestion of a modification at the 1 position could not be confirmed. The DeLuca group, however, isolated the active metabolite from intestines of 1600 chickens given radiolabeled vitamin D, and, by means of mass spectrometric techniques and specific chemical reactions, the structure of the active form of vitamin D in the intestine was unequivocally demonstrated as $1,25(OH)_2D_3$ [68]. Of great importance was the finding of Fraser and Kodicek that the peak P metabolite could be produced by homogenates of chicken kidney and that anephric animals are unable to produce the peak P metabolite [69]. They correctly concluded that the site of synthesis of the active form of vitamin D is the kidney. The Wisconsin group then chemically synthesized both $1\alpha,25(OH)_2D_3$ [70] and $1\beta,25(OH)_2D_3$ [71] and provided unequivocal proof that the active form is $1\alpha,25(OH)_2D_3$. Furthermore, this group was able to synthesize $1\alpha OHD_3$, an important analog that assumed great importance as a therapeutic agent throughout the world [72].

D. Proof That $1,25(OH)_2D_3$ Is the Active Form of Vitamin D

Proof that $1,25(OH)_2D_3$ and not $25OHD_3$ is the active form was provided by experiments in which anephric animals respond to $1,25(OH)_2D_3$ by increasing intestinal absorption of calcium and bone calcium mobilization, whereas animals receiving $25OHD_3$ at physiological doses did not [73–75]. Furthermore, the experiment of nature, namely, vitamin D-dependency rickets type I, an autosomal recessive disorder, provided final proof [76]. This disease could be treated by physiological doses of synthetic $1,25(OH)_2D_3$, whereas large amounts of vitamin D_3 or $25OHD_3$ were needed to heal the rickets. Although the exact defect in this disease is yet unknown, it is believed to be a defect in the 1α-hydroxylase enzyme that converts 25OHD to the final active form (see Chapter 47). $25OHD_3$ at pharmacological doses likely acts as an analog of the final vitamin D hormone, $1,25(OH)_2D_3$ (Fig. 2).

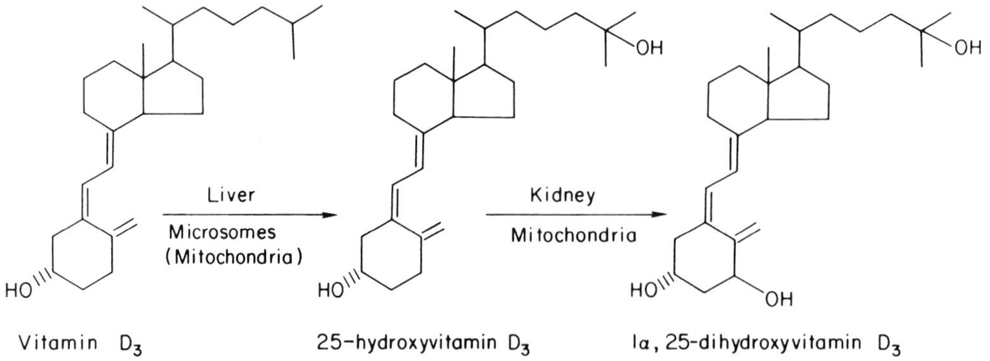

FIGURE 2 Activation of the vitamin D_3 molecule.

E. Discovery of the Vitamin D Endocrine System

Immediately after the identification of $1{,}25(OH)_2D_3$ as the active form of vitamin D came studies in which it could be shown that animals on a low calcium diet produce large quantities of $1{,}25(OH)_2D_3$, whereas those on a high calcium diet produce little or no $1{,}25(OH)_2D_3$ [77]. A reciprocal arrangement was found for the metabolite $24R{,}25(OH)_2D_3$. Thus, when calcium is needed, production of $1{,}25(OH)_2D_3$ is markedly stimulated and the 24-hydroxylation degradation reaction is suppressed. When adequate calcium is present, production of $1{,}25(OH)_2D_3$ is shut off and the 24-hydroxylation reaction is turned on. This discovery also satisfactorily provided evidence that $1{,}25(OH)_2D_3$ is the likely endogenous factor originally described by Nicolaysen *et al.* [34].

The next important step was the demonstration that it is parathyroid hormone that activates 1α-hydroxylation of $25OHD_3$ in the kidney [78]. Thus, parathyroidectomy eliminates the hypocalcemic stimulation of 1α-hydroxylation and suppression of 24-hydroxylation, whereas administration of parathyroid hormone restores that capability. Fraser and Kodicek also provided evidence that, in intact chickens, injection of parathyroid hormone stimulated the 1α-hydroxylation reaction [79]. Thus, the basic vitamin D endocrine system was largely discovered and reported in the early 1970s, being completed by 1974.

F. Other Metabolites of Vitamin D

During the course of identification of $1\alpha{,}25(OH)_2D_3$, $21{,}25(OH)_2D_3$ was reported as a metabolite, as was $25{,}26(OH)_2D_3$ [80,81]. However, the identification of $21{,}25(OH)_2D_3$ was in error and was corrected to $24{,}25(OH)_2D_3$, with the correct stereochemistry as $24R{,}25(OH)_2D_3$ [82]. Over the late 1970s and early 1980s, as many as 30 metabolites of vitamin D were identified [83]. These are covered in other chapters in this volume. Of great importance was the use of the fluoro derivatives of vitamin D to illustrate that the only activation pathway of vitamin D is 25-hydroxylation followed by 1-hydroxylation [84]. Thus, 24-difluoro-$25OHD_3$ supported all known functions of vitamin D for at least two generations of animals [85]. 24-Difluoro-$25OHD_3$ cannot be 24-hydroxylated. Furthermore, other fluoro derivatives such as 26,27-hexafluoro-$25OHD_3$ [86] and 23-difluoro-$25OHD_3$ [87] are all fully biologically active, illustrating that 26-hydroxylation, 24-hydroxylation, and 23-hydroxylation are not essential to the function of vitamin D.

G. Discovery of the Vitamin D Receptor

Zull and colleagues provided evidence that the function of vitamin D is blocked by transcription and protein inhibitors [88]. Thus, it became clear very early that a nuclear activity is required for vitamin D to carry out its functions. This work confirmed and extended the earlier work of Eisenstein and Passavoy [89]. With the discovery of the active forms of vitamin D came new attention to the idea that vitamin D may work through a nuclear mechanism. Thus, Haussler *et al.* reported vitamin D compounds to be associated with chromatin [66]. However, these experiments did not exclude the possibility that the vitamin D compounds might be bound to the nuclear membrane. The first clear demonstration of the existence of a vitamin D receptor was at the hands of Brumbaugh and Haussler [90]. Furthermore, the experiments of Kream *et al.* [91] provided strong and unequivocal evidence of the existence of a nuclear receptor for $1{,}25(OH)_2D_3$. Intense efforts toward purification of the receptor and its study appeared with the knowledge that it is an approximately 55,000 molecular weight receptor protein. In 1987, a partial cDNA sequence for the chicken vitamin D receptor was determined [92]. This was followed by isolation of the full coding sequence for the human [93] and rat [94,95] receptors.

From a historical point of view, one of the most important discoveries was vitamin D-dependency rickets type II [96], which is now known to be due to a defect in the receptor gene [97,98] (discussed in Chapter 48). This discovery essentially provided receptor knockout experiments in humans, allowing unequivocal proof of the essentially of the vitamin D receptor for the function of vitamin D. The nature of the receptor and how it functions are described in subsequent chapters along with current thinking on the molecular mechanism of action of $1,25(OH)_2D_3$.

Acknowledgments

This work was supported in part by a program project grant (No. DK14881) from the National Institutes of Health, a fund from the National Foundation for Cancer Research, and a fund from the Wisconsin Alumni Research Foundation.

References

1. Von Liebig J 1957 Animal Chemistry or Organic Chemistry in Its Application to Physiology and Pathology. In: Glass, HB (ed) A History of Nutrition. The Riverside Press, Cambridge, Massachusetts.
2. Lunin N 1881 Über di bedeuting der anorganischen salze fü die ernahrung des tieres. Z Physiol Chem **5**:31–39.
3. Hopkins G 1912 Feeding experiments illustrating the importance of accessory food factors in normal dietaries. J Physiol **44**:425–460.
4. Magendie F 1816 Ann Chim Phys. **3**:86–87. In: Glass HB (ed) A History of Nutrition by McCollum EV. The Riverside Press, Cambridge Massachusetts (1957).
5. Funk C 1911 The chemical nature of the substance that cures polyneurities in birds produced by a diet of polished rice. J Physiol (London) **43**:395–402.
6. Eijkman C 1897 Ein beri-beri änliche der hühner. Virchow's Arch **148**:523–527.
7. Grijns G 1957 Concerning polyneuritis geneeskundig gallinarum tijschrift voor ned indie 1901. In: McCollum EV, Glass HB (eds) A History of Nutrition. Houghton Mifflin, Boston, Massachusetts, p. 216.
8. Holst A. Frölich T 1907 Experimental studies relating ship-beriberi to scurvy. II. On the etiology of scurvy. J Hyg **7**:634–671.
9. Hart EB, McCollum EV, Steenbock H, Humphrey GC 1911 Physiological effect on growth and reproduction of rations balanced from restricted sources. Wis Agric Exp Sta Res Bull **17**:131–205.
10. McCollum EV, Davis M 1913 The necessity of certain lipins in the diet during growth. J Biol Chem **25**:167–131.
11. McCollum EV, Simmons N, Pitz W 1916 The relation of the unidentified dietary factors, the fat-soluble A and water-soluble B of the diet to the growth promoting properties of milk. J Biol Chem **27**:33–38.
12. Osborne TB, Mendel LB 1917 The role of vitamines in the diet. J Biol Chem **31**:149–163.
13. Whistler D 1645 De morbo puerli anglorum, quem patrio ideiomate indigenae vocant "the rickets" (lugduni, batavorum 1645). As cited by Smerdon GT, Daniel Whistler and the English Disease. A translation and biographical note. J. Hist Med **5**:397–415 (1950).
14. Hess A 1929 The history of rickets. In: Rickets, Including Osteomalacia and Tetany. Lea & Febiger, Philadelphia, Pennsylvania, pp. 22–37.
15. Mellanby E 1919 An experimental investigation on rickets. Lancet **1**:407–412.
16. McCollum EV, Simmonds N, Becker JE, Shipley PG 1922 An experimental demonstration of the existence of a vitamin which promotes calcium deposition. J Biol Chem **53**:293–298.
17. Huldshinsky K 1919 Heilung von rachitis durch künstalich höhensonne. Deut Med Wochenschr **45**:712–713.
18. Chick H, Palzell EJ, Hume EM 1923 Studies of rickets in Vienna 1919–1922. Medical Research Council, Special Report No. 77.
19. Steenbock H, Hart EB 1913 The influence of function on the lime requirements of animals. J Biol Chem **14**:59–73.
20. Steenbock H, Black A 1924 Fat-soluble vitamins. XVII. The induction of growth-promoting and calcifying properties in a ration by exposure to ultraviolet light. J Biol Chem **61**:405–422.
21. Hess AF, Weinstock M 1924. Antirachitic properties imparted to lettuce and to growing wheat by ultraviolet irradiation. Proc Soc Exp Biol Med **22**:5–6.
22. Steenbock H, Black A 1925 Fat-soluble vitamins. XXIII. The induction of growth-promoting and calcifying properties in fats and their unsaponifiable consituents by exposure to light. J Biol Chem **64**:263–298.
23. Windaus A, Linsert O 1928 Vitamin D_1. Ann Chem **465**:148.
24. Askew FA, Bourdillon RB, Bruce HM, Jenkins RGC, Webster TA 1931 The distillation of vitamin D. Proc R Soc **B107**:76–90.
25. Windaus A, Linsert O, Lüttringhaus A, Weidlich G 1932 Crystalline-vitamin D_2. Ann Chem **492**:226–241.
26. Windaus A, Bock F 1937 Über das provitamin aus dem sterin der schweineschwarte. Z Physiol Chem **245**:168–170.
27. Windaus A, Lettre H, Schenck F 1935 7-Dehydrocholesterol. Ann Chem **520**:98–107.
28. Windaus A, Schenck F, von Werder F 1936 Über das antirachitisch wirksame bestrahlungs-produkt aus 7-dehydro-cholesterin. Hoppe-Seylers Z Physiol Chem **241**:100–103.
29. Esvelt RP, Schnoes HK, DeLuca HF 1978 Vitamin D_3 from rat skins irradiated *in vitro* with ultraviolet light. Arch Biochem Biophys **188**:282–286.
30. Holick MF, MacLaughlin JA, Clark MB, Holick SA, Potts JT Jr, Anderson RR, Blank IH, Parrish JA, Elias P 1980 Photosynthesis of previtamin D_3 in human skin and the physiologic consequences. Science **210**:203–205.
31. Orr WJ, Holt LE Jr, Wilkins L, Boone FH 1923 The calcium and phosphorus metabolism in rickets, with special reference to ultraviolet rat therapy. Am J Dis Children **26**:362–372.
32. Kletzien SWF, Templin VM, Steenbock H, Thomas BH 1932 Vitamin D and the conservation of calcium in the adult. J Biol Chem **97**:265–280.
33. Nicolaysen R 1937 Studies upon the mode of action of vitamin D. III. The influence of vitamin D on the absorption of calcium and phosphorus in the rat. Biochem J **31**:122–129.
34. Nicolaysen R, Eeg-Larsen N, Malm OJ 1953 Physiology of calcium metabolism. Physiol Rev **33**:424–444.
35. Boyle IT, Gray RW, Omdahl JL, DeLuca HF 1972 Calcium control of the *in vivo* biosynthesis of 1,25-dihydroxyvitamin D_3: Nico-

laysen's endogenous factor. In: Taylor S (ed.) Endocrinology 1971. Heinemann Medical Books, London, pp. 468–476.
36. Howland J, Kramer B 1921 Calcium and phosphorus in the serum in relation to rickets. Am J Dis Children **22**:105–119.
37. Lamm M, Neuman WF 1958 On the role of vitamin D in calcification. Arch Pathol **66**:204–209.
38. Underwood JL, DeLuca HF 1984 Vitamin D is not directly necessary for bone growth and mineralization. Am J Physiol **246**:E493–E498.
39. DeLuca HF 1967 Mechanism of action and metabolic fate of vitamin D. Vit Horm **25**:315–367.
40. Carlsson A 1952 Tracer experiments on the effect of vitamin D on the skeletal metabolism of calcium and phosphorus. Acta Physiol Scand **26**:212–220.
41. Bauer GCH, Carlsson A, Lindquist B 1955 Evaluation of accretion, resorption, and exchange reactions in the skeleton. Kungl Fysiograf Sallskapets I. Lund Forhandlingar **25**:3–18.
42. Blunt JW, Tanaka Y, DeLuca HF 1968 The biological activity of 25-hydroxycholecalciferol, a metabolite of vitamin D_3. Proc Natl Acad Sci USA **61**:1503–1506.
43. Talmage RV, Grubb SA, Vander Wiel CJ, Doppelt SH, Norimatsu H 1978 The demand for bone calcium in maintenance of plasma calcium concentrations. Calcif Tissue Res **22**:73–91.
44. Garabedian M, Tanaka Y, Holick MF, DeLuca HF 1974 Response of intestinal calcium transport and bone calcium mobilization to 1,25-dihydroxyvitamin D_3 in thyroparathyroidectomized rats. Endocrinology **94**:1022–1027.
45. Raisz LG, Trummel CL, Holick MF, DeLuca HF 1972 1,25-Dihydroxycholecalciferol: A potent stimulator of bone resorption in tissue culture. Science **175**:768–769.
46. Frost HM 1966 Bone dynamics in osteoporosis and osteomalacia. Henry Ford Hospital Surgical Monograph Series, Thomas, Springfield, Illinois.
47. Stern PH, Halloran BP, DeLuca HF, Hefley TJ 1983 Responsiveness of vitamin D-deficient fetal rat limb bones to parathyroid hormone in culture. Am J Physiol **244**:E421–E424.
48. Yamamoto M, Kawanobe Y, Takahashi H, Shimazawa E, Kimura S, Ogata E 1984 Vitamin D deficiency and renal calcium transport in the rat. J Clin Invest **74**:507–513.
49. Stumpf WE, Sar M, Reid FA, Tanaka Y, DeLuca HF 1979 Target cells for 1,25-dihydroxyvitamin D_3 in intestinal tract, stomach, kidney, skin, pituitary and parathyroid. Science **206**:1188–1190.
50. Stumpf WE, Sar M, DeLuca HF 1981 Sites of action of 1,25 $(OH)_2$ vitamin D_3 identified by thaw-mount autoradiography. In: Cohn DV, Talmage RV, Matthews JL (eds), Hormonal Control of Calcium Metabolism. Excerpta Medica, Amsterdam, pp. 222–229.
51. Suda T, Takahashi N, Martin TJ 1992 Modulation of osteoclast differentiation. Endocr Rev **13**:66–80.
52. Suda T 1989 The role of 1α,25-dihydroxyvitamin D_3 in the myeloid cell differentiation. Proc Soc Exp Biol Med **191**:214–220.
53. Demay MB, Kiernan, MS, DeLuca HF, Kronenberg HM 1992 Sequences in the human parathyroid hormone gene that bind the 1,25-dihydroxyvitamin D_3 receptor and mediate transcriptional repression in response to 1,25-dihydroxyvitamin D_3. Proc Natl Acad Sci USA **89**:8097–8101.
54. Silver J 1994 Regulation of parathyroid hormone production by $1\alpha,25\text{-}(OH)_2D_3$ and its analogues. Their therapeutic usefulness in secondary hyperparathyroidism. In: Vitamin D and its Analogues. The Second International Forum on Calcified Tissue and Bone Metabolism. Chugai Pharmaceutical, Tokyo, pp. 60–63.
55. Smith EL, Walworth NC, Holick MF 1986 Effect of 1α,25-dihydroxyvitamin D_3 on the morphologic and biochemical differentiation of cultured human epidermal keratinocytes grown in serum-free conditions. J Invest Dermatol **86**:709–714.
56. Kodicek E 1956 Metabolic studies on vitamin D. In: Wolstenholme GWE, O'Connor CM (eds) Ciba Foundation Symposium on Bone Structure and metabolism. Little, Brown, and Co., Boston, Massachusetts, pp. 161–174.
57. Kodicek E 1960 The metabolism of vitamin D. In: Umbreit W. Molitor H (eds) Proceedings of the Fourth International Congress of Biochemistry. Pergamon, London, Vol. 11, pp. 198–208.
58. Haussler MR, Norman AW 1967 The subcellular distribution of physiological doses of vitamin D_3. Arch Biochem Biophys **118**:145–153.
59. Neville PF, DeLuca HF 1966 The synthesis of [1,2-^{3}H]vitamin D_3 and the tissue localization of a 0.25 μg (10 IU) dose per rat. Biochemistry **5**:2201–2207.
60. Lund J, DeLuca HF 1966 Biologically active metabolite of vitamin D_3 from bone, liver, and blood serum. J Lipid Res **7**:739–744.
61. Morii H, Lund J, Neville PF, DeLuca HF 1967 Biological activity of a vitamin D metabolite. Arch Biochem Biophys **120**:508–512.
62. Blunt JW, DeLuca HF, Schnoes HK 1968 25-Hydroxycholecalciferol. A biologically active metabolite of vitamin D_3. Biochemistry **7**:3317–3322.
63. Blunt JW, DeLuca HF 1969 The synthesis of 25-hydroxycholecalciferol. A biologically active metabolite of vitamin D_3. Biochemistry **8**:671–675.
64. DeLuca HF 1970 Metabolism and function of vitamin D. In: DeLuca HF, Suttie JW (eds) The Fat-Soluble Vitamins. Univ. of Wisconsin Press, Madison, pp. 3–20.
65. Lawson DEM, Wilson PW, Kodicek E 1969 Metabolism of vitamin D. A new cholecalciferol metabolite, involving loss of hydrogen at C-1, in chick intestinal nuclei. Biochem J **115**:269–277.
66. Haussler MR, Myrtle JF, Norman AW 1968 The association of a metabolite of vitamin D_3 with intestinal mucosa chromatin *in vivo*. J Biol Chem **243**:4055–4064.
67. Myrtle JF, Haussler MR, Norman AW 1970 Evidence for the biologically active form of cholecalciferol in the intestine. J Biol Chem **245**:1190–1196.
68. Holick MF, Schnoes HK, DeLuca HF, Suda T, Cousins RJ 1971 Isolation and identification of 1,25-dihydroxycholecalciferol. A metabolite of vitamin D active in intestine. Biochemistry **10**:2799–2804.
69. Fraser DR, Kodicek E 1970 Unique biosynthesis by kidney of a biologically active vitamin D metabolite. Nature **228**:764–766.
70. Semmler EJ, Holick MF, Schnoes HK, DeLuca HF 1972 The synthesis of 1α,25-dihydroxycholecalciferol—A metabolically active form of vitamin D_3. Tetrahedron Lett **40**:4147–4150.
71. Paaren HE, Schnoes HK, DeLuca HF 1977 Synthesis of 1β-hydroxyvitamin D_3 and 1β,25-dihydroxyvitamin D_3. J Chem Soc Chem Commun 890–892.
72. Holick MF, Semmler EJ, Schnoes HK, DeLuca HF 1973 1α-Hydroxy derivative of vitamin D_3: A highly potent analog of 1α,25-dihydroxyvitamin D_3. Science **180**:190–191.
73. Boyle IT, Miravet L, Gray RW, Holick MF, DeLuca HF 1972 The response of intestinal calcium transport to 25-hydroxy and 1,25-dihydroxy vitamin D in nephrectomized rats. Endocrinology **90**:605–608.
74. Holick MF, Garabedian M, DeLuca HF 1972 1,25-Dihydroxycholecalciferol: Metabolite of vitamin D_3 active on bone in anephric rats. Science **176**:1146–1147.
75. Wong RG, Norman AW, Reddy CR, Coburn JW 1972 Biologic effects of 1,25-dihydroxycholecalciferol (a highly active vitamin D metabolite) in acutely uremic rats. J Clin Invest **51**:1287–1291.
76. Fraser D, Kooh SW, Kind HP, Holick MF, Tanaka Y, DeLuca HF 1973 Pathogenesis of hereditary vitamin D dependent rickets: An inborn error of vitamin D metabolism involving defective

conversion of 25-hydroxyvitamin D to 1α,25-dihydroxyvitamin D. N Engl J Med **289**:817–822.
77. Boyle IT, Gray RW, DeLuca HF 1971 Regulation by calcium of in vivo synthesis of 1,25-dihydroxycholecalciferol and 21,25-dihydroxycholecalciferol. Proc Natl Acad Sci USA **68**:2131–2134.
78. Garabedian M, Holick MF, DeLuca HF, Boyle IT 1972 Control of 25-hydroxycholecalciferol metabolism by the parathyroid glands. Proc Natl Acad Sci USA **69**:1673–1676.
79. Fraser DR, Kodicek E 1973 Regulation of 25-hydroxycholecalciferol-1-hydroxylase activity in kidney by parathyroid hormone. Nature (New Biol) **241**:163–166.
80. Suda T, DeLuca HF, Schnoes HK, Ponchon G, Tanaka Y, Holick MF 1970 21,25-Dihydroxycholecalciferol. A metabolite of vitamin D_3 preferentially active on bone. Biochemistry **9**:2917–2922.
81. Suda T, DeLuca HF, Schnoes HK, Tanaka Y, Holick MF 1970 25,26-Dihydroxycholecalciferol, a metabolite of vitamin D_3 with intestinal calcium transport activity. Biochemistry **9**:4776–4780.
82. Tanaka Y, Frank H, DeLuca HF, Koizumi N, Ikekawa N 1975 Importance of the stereochemical position of the 24-hydroxyl to biological activity of 24-hydroxyvitamin D_3. Biochemistry **14**:3293–3296.
83. DeLuca HF, Schnoes HK 1983 Vitamin D: Recent advances. Annu Rev Biochem **52**:411–439.
84. Brommage R, DeLuca HF 1985 Evidence that 1,25-dihydroxyvitamin D_3 is the physiologically active metabolite of vitamin D_3. Endocr Rev **6**:491–511.
85. Brommage R, Jarnagin K, DeLuca HF, Yamada S, Takayama H 1983 1- but not 24-hydroxylation of vitamin D is required for skeletal mineralization in rats. Am J Physiol **244**:E298–E304.
86. Tanaka Y, Pahuja DN, Wichmann JK, DeLuca HF, Kobayashi Y, Taguchi T, Ikekawa N 1982 25-Hydroxy-26,26,26,27,27,27-hexafluorovitamin D_3: Biological activity in the rat. Arch Biochem Biophys **218**:134–141.
87. Nakada M, Tanaka Y, DeLuca HF, Kobayashi Y, Ikekawa N 1985 Biological activities and binding properties of 23,23-difluoro-25-hydroxyvitamin D_3 and its 1α-hydroxy derivative. Arch Biochem Biophys **241**:173–178.
88. Zull JE, Czarnowska-Misztal E, DeLuca HF 1965 Actinomycin D inhibition of vitamin D action. Science **149**:182–184.
89. Eisenstein R, Passavoy M 1964 Actinomycin D inhibits parathyroid hormone and vitamin D activity. Proc Soc Exp Biol Med **117**:77–79.
90. Brumbaugh PF, Haussler MR 1973 1α,25-Dihydroxyvitamin D_3 receptor: Competitive binding of vitamin D analogs. Life Sci **13**:1737–1746.
91. Kream BE, Reynolds RD, Knutson JC, Eisman JA, DeLuca HF 1976 Intestinal cytosol binders of 1,25-dihydroxyvitamin D_3 and 25-hydroxyvitamin D_3. Arch Biochem Biophys **176**:779–787.
92. McDonnell DP, Mangelsdorf DJ, Pike JW, Haussler MR, O'Malley BW 1987 Molecular cloning of complementary DNA encoding the avian receptor for vitamin D. Science **235**:1214–1217.
93. Baker AR, McDonnell DP, Hughes M, Crisp TM, Mangelsdorf DJ, Haussler MR, Pike JW, Shine J, O'Malley BW 1988 Cloning and expression of full-length cDNA encoding human vitamin D receptor. Proc Natl Acad Sci USA **85**:3294–3298.
94. Burmester JK, Maeda N, DeLuca HF 1988 Isolation and expression of rat 1,25-dihydroxyvitamin D_3 receptor cDNA. Proc Natl Acad Sci USA **85**:1005–1009.
95. Burmester JK, Wiese RJ, Maeda N, DeLuca HF 1988 Structure and regulation of the rat 1,25-dihydroxyvitamin D_3 receptor. Proc Natl Acad Sci USA **85**:9499–9502.
96. Brooks MH, Bell NH, Love L, Stern PH, Orfei E, Queener SF, Hamstra AJ, DeLuca HF 1978 Vitamin D-dependent rickets Type II. Resistance of target organs to 1,25-dihydroxyvitamin D. N Engl J Med **298**:996–999.
97. Hughes MR, Malloy PJ, Kieback DG, Kesterson RA, Pike JW, Feldman D, O'Malley BW 1988 Point mutations in the human vitamin D receptor gene associated with hypocalcemic rickets. Science **242**:1702–1706.
98. Wiese RJ, Goto H, Prahl JM, Marx SJ, Thomas M, Al-Aqeel A, DeLuca HF 1993 Vitamin D-dependency rickets Type II: Truncated vitamin D receptor in three kindreds. Mol Cell Endocrinol **90**:197–201.

Chapter 2

Vitamin D Metabolism

Ronald L. Horst and Timothy A. Reinhardt
U.S. Department of Agriculture, Agricultural Research Service, National Animal Disease Center, Metabolic Diseases and Immunology Research Unit, Ames, Iowa

I. INTRODUCTION

In 1919, when the field of experimental nutrition was still in its infancy, Sir Edward Mellanby conducted a classic experiment which for the first time associated the supplementation of various growth-promoting fats with the prevention of rickets [1]. He credited the cure to the presence of a fat soluble substance called vitamin A. McCollum *et al.* [2], however, later discovered that the factor responsible for healing rickets was distinct from vitamin A. McCollum named this new substance vitamin D. It was also during this period that the scientists realized that there were two antirachitic factors with distinct structures [3]. As discussed by Norman [3], the first factor to be identified was designated vitamin D_2 (also known as ergocalciferol), whereas the structure of vitamin D_3 (cholecalciferol) became evident some 4 to 5 years later. Vitamins D_3 and D_2 are used for supplementation of animal and human diets in the United States. Vitamin D_3 is the form of vitamin D that is synthesized by vertebrates, whereas vitamin D_2 is the major naturally occurring form of the vitamin in plants. Animals that bask in the sun such as amphibia, reptiles, and birds therefore synthesize sufficient endogenous vitamin D_3 to meet their daily needs. However, herbivores may have evolved utilizing vitamin D_2 as their predominant source.

This review focuses on the general control and function of key enzymes involved in the regulation of vitamin D_2 and vitamin D_3 metabolism. Species differences in vitamin D metabolism, as well as vitamin D toxicity, are also discussed. The reader is also directed toward a number of additional reviews regarding vitamin D metabolism and action [3–8]. This chapter gives an overview of vitamin D metabolism; critical steps are discussed in further detail in the subsequent chapters of this section. Metabolism of vitamin D analogues is covered in Chapter 58.

II. VITAMIN D METABOLISM

A. Overview

Contemporary views categorize vitamin D_3 not as a vitamin but, rather, as a pro-steroid hormone. This concept is supported by the fact that in mammals vitamin D_3 is derived from a cholesterol-like precursor found in the skin. The direct action of sunlight on this precursor, 7-dehydrocholesterol, results in cleavage of the B ring of the steroid structure that on thermoisomerization yields the characteristic secosteroid (see Chapter 3). The significance of vitamin D as a pro-steroid hormone became clearer in 1967 when Morii *et al.* [9] isolated a new metabolite of vitamin D_3 from rats that was as effective as vitamin D_3 in healing rickets, raising blood calcium, and increasing intestinal calcium transport. This compound acted more rapidly than vitamin D_3, requiring only 8 to 10 hr after oral administration to initiate its response. This metabolite was identified as 25-hydroxyvitamin D_3 ($25OHD_3$) [10]. The liver was demonstrated to be important in the production of this most abundant circulating form of vitamin D_3 which, under normal conditions, is present at 20 to 50 ng/ml [8]. Shortly following the discovery of $25OHD_3$, a number of laboratories showed that this metabolite is specifically hydroxylated at the 1α-position in the kidney to yield 1,25-dihydroxyvitamin D_3 [$1,25(OH)_2D_3$] [11–13]. The latter metabolite is now generally accepted as the hormonally active form of vitamin D_3. $1,25(OH)_2D_3$ circulates at approximately 1000-fold lower concentrations than $25OHD_3$ and is generally present at 20 to 65 pg/ml in normal human plasma [14].

This simplistic picture outlined for vitamin D_3 activation is complicated by the fact that vitamin D_3 can be oxidatively metabolized to a variety of products. Most of these numerous metabolites have no identifiable biological function, and indeed many have been isolated from animals fed abnormally high amounts of vitamin D_3. Nevertheless, the evidence collected to date indicates that 25-hydroxylated vitamin D_3 metabolites are preferentially metabolized at the side chain. In particular, carbon centers C-23, C-24, and C-26 are readily susceptible to further oxidation. Figure 1 illustrates products of these oxidative pathways. As indicated, these pathways are shared by both $25OHD_3$ and $1,25(OH)_2D_3$, and their importance is still a matter of controversy. For example, there is evidence that 24,25-

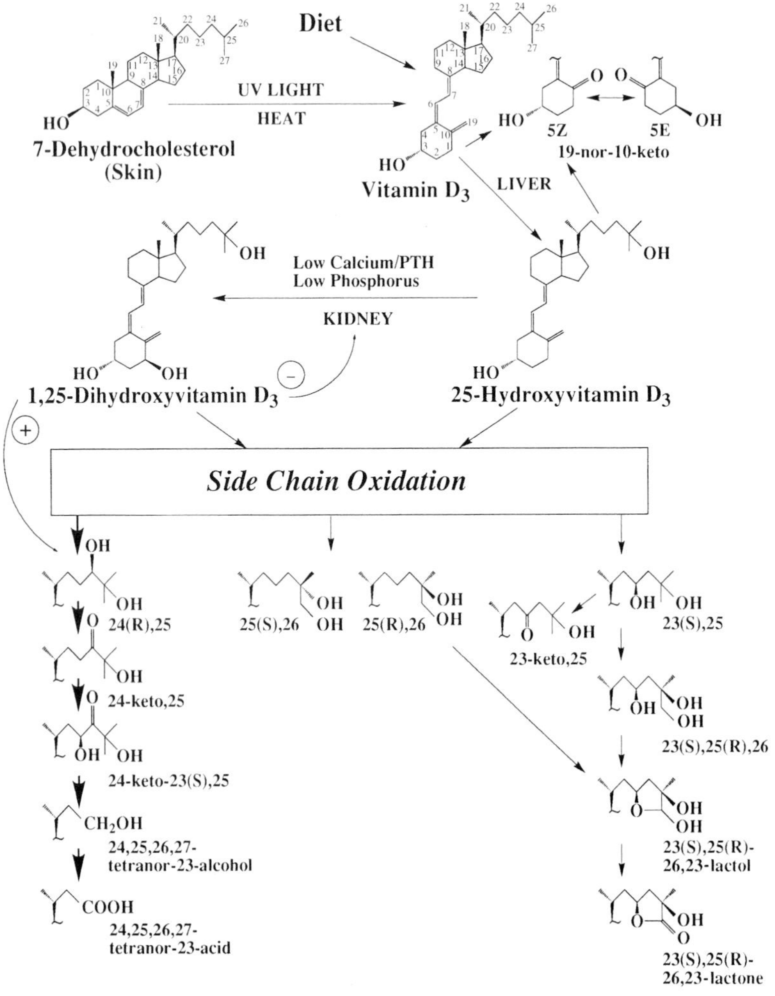

FIGURE 1 Pathways of vitamin D_3 metabolism.

dihydroxyvitamin D_3 [24,25$(OH)_2D_3$] may function to stimulate bone mineralization [15,16], suppress parathyroid hormone (PTH) secretion [17], and maintain embryonic development [18]. For the most part, however, these side chain modifications are generally considered to be catabolic in nature.

Although these side chain oxidative pathways yield metabolites that are considered "nonfunctional," the presence of these compounds in circulation could pose serious problems in the analysis for 25OHD_3 and 1,25$(OH)_2D_3$ [19]. Further complicating the issue of understanding vitamin D activation, catabolism, and metabolite analysis is the presence of vitamin D_2. Vitamin D_2 has been shown to contribute significantly to the overall vitamin D status in humans and other mammals consuming supplemental vitamin D_2 [20–22]. Vitamin D_2 can also be metabolized in a similar fashion to produce several metabolites analogous to the vitamin D_3 endocrine system, including the hormonally active form of vitamin D_2, 1,25-dihydroxyvitamin D_2 [1,25$(OH)_2D_2$] [23]. Simple inspection of the side chain, however, would imply that differences between metabolism of vitamin D_2 and vitamin D_3 may exist. The presence of unsaturation at carbon centers C-22/C-23, along with the additional methyl group at C-24, would seem to preclude the existence of the same metabolic pathways for the two vitamins. Figure 2 outlines the known pathways of vitamin D_2 metabolism that have been shown to date. Deviations in the vitamin D_2 and vitamin D_3 pathways are discussed in detail in the following sections.

B. 25-Hydroxylase

The 25-hydroxylation of vitamin D is the initial step in vitamin D activation. The enzyme responsible for production of this metabolite is located in the liver (see Chapter 4). Extrahepatic sources of 25-hydroxylation

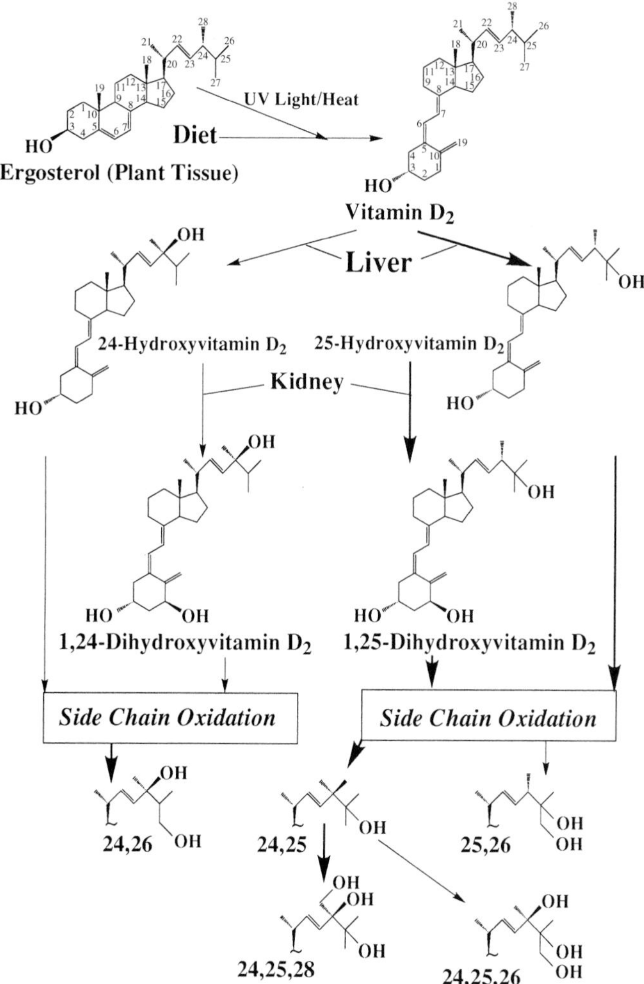

FIGURE 2 Pathways of vitamin D_2 metabolism.

have been described [24]; however, experiments with hepatectomized rats provided evidence that the liver is the major, if not the sole, physiologically relevant site of 25-hydroxylation of vitamin D [25]. Subsequent studies also described the existence of the 25-hydroxylase in both liver mitochondria and microsomes [26–30]. In early work, the microsomal enzyme was described as an enzyme of low capacity and high affinity and, therefore, the enzyme of greatest physiological importance [30]. In contrast, the mitochondrial enzyme was described as a high capacity, low affinity enzyme thought to be relevant only under conditions of high vitamin D concentration such as vitamin D toxicity [31]. Early evidence that the microsomal enzyme was the physiologically relevant enzyme came from experiments which suggested this enzyme could be regulated by vitamin D status [30]. It is now clear that liver production of 25-hydroxyvitamin D (25OHD) is not significantly regulated. 25OHD production is primarily dependent on substrate concentration. An important consequence of this lack of physiological regulation of 25OHD is that measurement of blood 25OHD is an excellent measure of vitamin D status.

The purification and cloning of putative liver 25-hydroxylases have been reviewed several times in the early 1990s [32–34] and are examined further in Chapter 4 in this book. Examination of the literature shows that most of the current focus is on the mitochondrial 25-hydroxylase designated CYP27. CYP27 is a cytochrome P450 capable of C-26(27) hydroxylation of sterols involved in bile acid synthesis and the 25-hydroxylation of vitamin D_3. The rat, rabbit, and human enzymes have been cloned [35–38]. The CYP27 clone has been expressed in COS cells [36,39], and its activity has been isolated from the mitochondria of these cells. The expressed enzyme was found to 27-hydroxylate cholestanetriol and 25-hydroxylate vitamin D_3. However, CYP27 does not 25-hydroxylate vitamin D_2 [36]. Rather, CYP27 was found to 24-hydroxylate and 26(27)-hydroxylate vitamin D_2. These activities could explain the presence of 24-hydroxyvitamin D_2 ($24OHD_2$), 1,24-dihydroxyvitamin D_2 [$1,24(OH)_2D_2$], and 24,26-dihydroxyvitamin D_2 [$24,26(OH)_2D_2$] [40–42] in the plasma of rats and cows. Since rats fed vitamin D-deficient diets and supplemented with physiological amounts of vitamin D_2 have $25OHD_2$ as their predominant monohydroxylated vitamin D_2 metabolite in the plasma [41], these data would suggest that CYP27 is not the physiologic enzyme responsible for the 25-hydroxylation of vitamin D.

The rat liver microsomal 25-hydroxylase (CYP2C11) has also been studied, but it has been shown to be male specific [43]. Also, data have been presented indicating that human microsomes do not possess 25-hydroxylase activity [44]. Therefore, conclusions regarding the importance of the CYP27 and other ostensible microsomal 25-hydroxylases require additional research.

Data by Axen *et al.* [45] suggest that a third liver 25-hydroxylase exists that is microsomal in origin. In the pig, this enzyme is present equally in males and females and is markedly different from CYP27 and CYP2C11 based on a terminal amino acid sequence. Most important is the finding that this pig microsomal enzyme 25-hydroxylates vitamins D_2 and D_3 equally. The 25-hydroxylation of vitamin D is not yet completely understood. Several enzymes may play a role in the 25-hydroxylation of vitamin D. Whether one enzyme is more physiologically relevant than others remains to be determined. Nevertheless, it is clear that mammals can use vitamin D_2 as a sole source of vitamin D. Therefore, any 25-hydroxylase proclaimed as the key enzyme(s) in the 25-hydroxylation of vitamin D must be capable of 25-hydroxylating vitamin D_2 as well as vitamin D_3.

C. 1α-Hydroxylase

In the late 1960s, $25OHD_3$ was believed to be the metabolically active form of vitamin D. However, the presence of a more polar metabolite, which accumulated in the intestinal mucosa chromatin of chicks administered ^3H-labeled vitamin D_3, suggested a new candidate for the active form of vitamin D [46]. Subsequent work by Lawson *et al.* [47] showed that during the formation of this metabolite the 1α-^3H label was lost. This led them to suggest that the new metabolite had an oxygen function inserted at C-1 in addition to the hydroxyl group at C-25. The enhanced biological activity of this new metabolite was evident before its structure could be determined [48–50]. Fraser and Kodicek [49] demonstrated that nephrectomy abolished production of the new metabolite, and this active vitamin D compound was synthesized by kidney mitochondria. In 1971, three laboratories identified the active form of vitamin D as $1,25(OH)_2D_3$ [12,13,51]. Subsequently, the vitamin D_2 form was also isolated and identified [23].

The 1α-hydroxylase is located in the inner mitochondrial membrane of the proximal convoluted tubule cells of the kidney [52] and is discussed in detail in Chapter 5. Extrarenal sites of 1α-hydroxylation have been reported in bone, liver, placenta, macrophages, and skin [53]. The physiological significance of these sites on systemic calcium metabolism is in doubt, as nephrectomy and/or severe renal failure results in very low to undetectable circulating $1,25(OH)_2D_3$ levels [54].

The regulation of $1,25(OH)_2D_3$ production is reciprocally regulated with respect to $24,25(OH)_2D_3$ [55]. Hypocalcemia caused by calcium-deficient diets, vitamin D deficiency, or pathological factors results in increased production of $1,25(OH)_2D_3$ [55–61]. This hypocalcemic-mediated induction of 1,25-dihydroxyvitamin D [$1,25(OH)_2D$] production is secondary to increased PTH. Administration of PTH to animals results in increased $1,25(OH)_2D_3$ production [58,62,63]. PTH treatment *in vitro* induces the 1α-hydroxylase in renal slices [57] and cultured kidney cells [62,64], and is cAMP dependent [57,62,64,65]. Thyroparathyroidectomy (TPTX) or parathyroidectomy (PTX) results in the loss of the ability to synthesize $1,25(OH)_2D_3$. In humans, acute administration of PTH or primary hyperparathyroidism results in increased production of $1,25(OH)_2D$ [14,66], which is evidenced by elevations in plasma $1,25(OH)_2D$. However, in animal studies where PTH was administered chronically to goats and calves, a transient rise in plasma $1,25(OH)_2D$ was observed followed by a rapid decline to nearly undetectable levels [67,68]. These results could be attributed to hypercalcemic feedback on the renal 1α-hydroxylase. When plasma calcium in these animals reached 13 mg/dl, $1,25(OH)_2D$ production appeared to cease. The same group conducted similar experiments in rats and showed that chronic PTH infusion did not result in a reduction of plasma $1,25(OH)_2D_3$, but rather a modest rise [69]. Clearly, there are varying degrees of direct calcium-mediated feedback on the renal 1α-hydroxylase. It is possible that species and age affect the set points at which plasma calcium becomes a direct negative regulator of $1,25(OH)_2D_3$ production.

In contrast to the indirect role of plasma calcium in inducing $1,25(OH)_2D_3$ production, the role of plasma phosphate appears more direct. As plasma phosphate declines, animals shift from $24,25(OH)_2D$ production to increased $1,25(OH)_2D_3$ production [61,70]. Since phosphate-deficient animals are hypercalcemic, serum PTH is down and therefore cannot be providing the signal to increase $1,25(OH)_2D_3$ production. Furthermore, TPTX phosphate-deficient animals produce $1,25(OH)_2D_3$ similarly to intact phosphate-deficient animals [60]. Gray *et al.* [71] demonstrated that hypophysectomy abolished the increase in plasma $1,25(OH)_2D$ concentrations that normally accompany dietary phosphate deprivation. They demonstrated that growth hormone or triiodothyronine replacement to hypophysectomized rats restored elevations in plasma $1,25(OH)_2D$ associated with low dietary phosphorus, therefore suggesting a permissive role of these hormones in regulation of the renal 1α-hydroxylase during phosphorus deficiency.

A direct negative effect of $1,25(OH)_2D$ on its own production has been reported. The inhibitory effect of $1,25(OH)_2D_3$ on renal 1α-hydroxylase activity occurs both *in vivo* and *in vitro* [58,72]. *In vivo,* this effect may be partially mediated through the ability of $1,25(OH)_2D_3$ to inhibit PTH secretion [73]. Similarly, vitamin D toxicity mildly inhibits renal 1α-hydroxylase activity [74]. Beckman *et al.* [74] observed (as have many others), that low calcium diets significantly induced the 1α-hydroxylase. They further demonstrated that administration of toxic doses of vitamin D to animals fed a calcium-deficient diet reduced 1α-hydroxylase activity by 90%. This result occurred in spite of the fact that these animals were hypocalcemic and had serum PTH levels equal to those of control animals receiving calcium-deficient normal vitamin D diets. These data suggest that high plasma concentrations of vitamin D metabolites may act directly to reduce 1α-hydroxylase activity.

There are many additional factors such as calcitonin (CT), acidosis, sex steroids, prolactin, growth hormone, glucocorticoids, thyroid hormone, and pregnancy that are potential regulators of $1,25(OH)_2D$ production. Discussion of these is beyond the scope of this general review of vitamin D metabolites. They have been reviewed by Kumar [75] and Henry [76] and are further discussed in Chapter 5.

The purification and/or cloning of the renal 1α-hydroxylase represents the Holy Grail of vitamin D metabolism [34]. The difficulties and challenges that have faced investigators in this field are discussed in Chapter 5.

D. 24-Hydroxylase

The 24-hydroxylation of $25OHD_3$ and $1,25(OH)_2D_3$ is the primary mechanism and the first step in a metabolic pathway to inactivate and degrade these vitamin D metabolites. Suda *et al.* [77] originally described a vitamin D_3 metabolite (Va), isolated from vitamin D-treated pigs, that was more polar than $25OHD_3$. They proposed that this metabolite was made exclusively in the kidney mitochondria and its formation could be enhanced by feeding diets high in strontium or calcium [78,79]. Holick *et al.* [80] later identified peak Va as $24,25(OH)_2D_3$.

We now know that the 24-hydroxylase is ubiquitous and may be present in every cell and tissue that contains the vitamin D receptor (VDR). In the kidney, the 24-hydroxylase is found on the inner mitochondrial membrane of the renal tubules [81]. The primary regulators of renal 24-hydroxylase activity are PTH and $1,25(OH)_2D_3$. Normal and TPTX animals receiving injections or infusions of $1,25(OH)_2D_3$ show marked increases in both renal 24-hydroxylase mRNA levels and activity [57,69,82,83]. Administration of PTH partially or completely blocks expression of 24-hydroxylase mRNA and activity in these animals [57,63,69,82,83]. PTH acts on the kidney via adenylate cyclase and cAMP, and it has been shown that infusions of cAMP *in vivo* block $1,25(OH)_2D_3$-mediated inductions of the renal 24-hydroxylase [82,84]. Animals on calcium-deficient diets have elevated plasma 1,25$(OH)_2$D concentrations, which are accompanied by suppressed or undetectable renal 24-hydroxylase activity [72,82], as well as reduced VDR concentrations [85].

The reasons for the inability of 1,25$(OH)_2D_3$ to up-regulate renal 24-hydroxylase *in vivo* are not clear. Iida *et al.* [86] have proposed that the down-regulation of renal VDR during calcium deficiency may be responsible for preventing the 1,25$(OH)_2D_3$-mediated induction of renal 24-hydroxylase. *In vivo* studies by Reinhardt and Horst [69], however, suggest that under these conditions PTH is probably the more important mediator of renal 24-hydroxylase regulation, rather than down-regulation of VDR. In their experiments, Reinhardt and Horst [69] showed that 1,25$(OH)_2D_3$ treatment of animals on normal calcium diets resulted in significant up-regulation of renal 24-hydroxylase as well as VDR. However, when PTH was infused simultaneously with 1,25$(OH)_2D_3$, VDR up-regulation was still observed (albeit to a lesser degree), whereas the 24-hydroxylase up-regulation was completely blocked. The importance of PTH in preventing the 1,25$(OH)_2D_3$-mediated up-regulation of the kidney 24-hydroxylase is also apparent by observation in aged rats. With advancing age, renal PTH receptors are down-regulated [87,88], while VDR remains unchanged [89]. The reduction in renal PTH receptors makes the kidney less responsive to PTH [58], which is associated with significant elevations in 24-hydroxylase mRNA [89,90]. These data suggest that renal responsiveness to PTH, not a decline in VDR, is the major physiological regulator of the kidney 24-hydroxylase.

In the intestine, 1,25$(OH)_2D_3$ is the primary regulator of 24-hydroxylase. *In vivo* administration of 1,25$(OH)_2D_3$ rapidly induces intestinal 24-hydroxylase activity [91]. This activity peaks by 6 hr post-injection, and rapidly declines thereafter to control levels 24 hr postinjection. Time course experiments show that 24-hydroxylase mRNA peaks 4 to 6 hr postinjection and then rapidly disappears [90]. This is in contrast to the renal 24-hydroxylase mRNA, which peaks 12 to 24 hr post-1,25$(OH)_2D_3$ treatment and declines much more slowly. Shinki *et al.* [82] proposed that the intestinal 24-hydroxylase was 100 times more sensitive to 1,25$(OH)_2D_3$ stimuli than the renal 24-hydroxylase. However, they examined 24-hydroxylase mRNA only 3 hr after a 1,25$(OH)_2D_3$ dose. Because renal 24-hydroxylase requires an additional 6 to 12 hr to reach peak expression, they likely underestimated the true sensitivity of the kidney to a 1,25$(OH)_2D_3$ dose. In contrast to their effect in the kidney, TPTX, PTH administration, or cAMP infusion does not affect intestinal expression of 24-hydroxylase induced by 1,25$(OH)_2D_3$ [82]. Animals fed low calcium diets, with the associated secondary hyperparathyroidism and high plasma 1,25$(OH)_2D_3$ concentrations, have marked inductions of both intestinal 24-hydroxylase mRNA and activity [82,91]. Another contrast between intestinal and renal 24-hydroxylase expression is seen in the aging rat model. Intestinal 24-hydroxylase mRNA and activity decline or change very little in the aged animal. This contrasts to the largely increased expression of renal 24-hydroxylase observed in the aged animal [89].

Calcitonin (CT) has been shown to be a potent suppressor of intestinal 24-hydroxylase expression [92]. In these experiments, Beckman *et al.* [74,92] showed that vitamin D toxicity was a potent inducer of 24-hydroxylase mRNA and enzyme expression in both intestine and kidney. They also showed that if the hypervitaminosis D_3-induced hypercalcemia was prevented by feeding low calcium diets, the intestinal 24-hydroxylase expression was enhanced 4-fold over hypercalcemic animals receiving the same toxic doses of vitamin D_3 but consuming a normal calcium diet. These observations prompted the examination of the possibility that CT released in response to the hypercalcemia may have suppressed the induced expression of the intestinal 24-hydroxylase. In their series of experiments, Beckman *et al.* [92] clearly demonstrated that CT was a potent suppressor of intestinal 24-hydroxylase activity. Conceivably, the CT-mediated suppression of 24-hydroxylase activity could enhance 1,25$(OH)_2$D-mediated activities by prolonging its half-life. This could exacerbate conditions that manifest hypercalcemia such as hypervitaminosis D by preventing 24-hydroxylation and catabolism of active vitamin D metabolites.

The 24-hydroxylase has been purified [93,94] and cloned [95,96], and the clone has been expressed [95,96]. Analysis of the amino acid sequence of the rat and human 24-hydroxylases showed that the sequences were 90% similar. The 21-amino acid heme binding region was found to be 100% identical [95]. Ohyama *et al.* [97] isolated the gene encoding the rat 24-hydroxylase. This single copy gene was approximately 15 kb and was composed of 12 exons. Several putative vitamin D response elements have been identified and are currently under study. Details of the purification and cloning of the 24-hydroxylase have been reviewed previously [34], and additional review of the molecular analysis and regulation of the 24-hydroxylase can be found in Chapter 6 in this book.

E. Physiological Role of 24-Hydroxylase

The major site for 24-hydroxylation appears to be the kidney. This is based on the observation that nephrectomy reduced or eliminated plasma 24,25$(OH)_2D_3$

[98]. However, nephrectomy also eliminates the production of $1,25(OH)_2D_3$, the primary stimulator of the 24-hydroxylase. Therefore, the possibility remains that $24,25(OH)_2D_3$ may reappear in plasma of nephrectomized subjects treated with therapeutic doses of $1,25(OH)_2D_3$. Early studies also suggested that $25OHD_3$ is the primary substrate for the 24-hydroxylase. Subsequent research, however, reported that the enzyme is present throughout the body and that its K_m favors $1,25(OH)_2D_3$ over $25OHD_3$ as a substrate for its action [82].

Napoli and Horst [99] suggested that the 24-hydroxylase, along with other metabolic steps, most likely represented mechanisms for terminating the cellular action of $1,25(OH)_2D_3$. In a review, Haussler [100] proposed a model for the cellular action of $1,25(OH)_2D_3$ in which he suggested that receptor-mediated, self-induced catabolism of $1,25(OH)_2D_3$ modulates the action of $1,25(OH)_2D_3$. The work of Lohnes and Jones [101] and Reddy and Tserng [102] provided further support for this proposal by showing the ubiquitous presence of catabolic pathways initiated by the 24-hydroxylase in $1,25(OH)_2D_3$ target tissues and the complete destruction of $1,25(OH)_2D_3$ by these pathways. In fact, the 24-hydroxylase has been shown to do more than just initiate this catabolic cascade. Akiyoshi-Shibata *et al.* [103] expressed the rat 24-hydroxylase cDNA in *Escherichia coli*. They found that this enzyme not only 24-hydroxylates but catalyzes the dehydrogenation of the 24-OH group and performs 23-hydroxylation resulting in 24-oxo-1,23,25-trihydroxyvitamin D_3 [24-oxo-$1,23,25(OH)_3D_3$]. Only the cleavage at C-23/C-24 resulting in the 24,25,26,27-tetranor-1OH,23COOHD$_3$ was not demonstrated. In similar experiments Beckman *et al.* [71] expressed the human C-24 hydroxylase in *Spodoptera frugiperda* insect cells. They found that the 24-hydroxylase catalyzed all of the metabolic steps reported by Akiyoshi-Shibata *et al.* [103]. Beckman *et al.* [104], however, demonstrated the production of 24,25,26,27 tetranor-23OHD$_3$ and, therefore, provided the first evidence that the 24-hydroxylase can indeed perform cleavage at C-23/C-24. They were unable to demonstrate the presence of 24,25,26,27-tetranor-23COOHD$_3$.

Direct evidence that self-induced metabolism (24-hydroxylase) of $1,25(OH)_2D_3$ suppresses the action of $1,25(OH)_2D_3$ on target cells was reported by Pols *et al.* [105,106] and Reinhardt and Horst [107,108]. Both laboratories showed that ketoconazole inhibited $1,25(OH)_2D_3$-induced metabolism. This inhibition resulted in increased specific accumulation of $1,25(OH)_2D_3$ in target cells and a significant increase in the cellular half-life of $1,25(OH)_2D_3$-occupied VDR [108]. A result of blocking the self-induced metabolism of $1,25(OH)_2D_3$ was up-regulation of the VDR. Reinhardt and Horst [107] extended these studies by demonstrating that self-induced metabolism of $1,25(OH)_2D_3$ in target cells limits the response of target cells to a primary $1,25(OH)_2D_3$ stimulus by reducing occupancy of VDR by $1,25(OH)_2D_3$ and by preventing VDR up-regulation. Additionally, their data showed that entry of $1,25(OH)_2D_3$ into the cell is restricted, due to extensive metabolism of the $1,25(OH)_2D_3$. In whole cell VDR assays, hormone is degraded so rapidly that VDR binding was prevented. Reinhardt *et al.* [109] confirmed the inhibitory effects of self-induced induction of the 24-hydroxylase on the cellular action of $1,25(OH)_2D_3$ *in vivo* by demonstrating that ketoconazole potentiates the $1,25(OH)_2D_3$ up-regulation of VDR in rat intestine and bone. Clearly, one of the primary roles of the 24-hydroxylase catabolic pathway is terminating the actions of $1,25(OH)_2D_3$. The role of 24-hydroxylase as an enzyme responsible for the production of a biologically active compound [$24,25(OH)_2D_3$] remains controversial.

F. Other Vitamin D_3 Derivatives Functionalized at C-24

In a series of experiments conducted by Wichmann *et al.* [110,111], a number of 24-hydroxylated derivatives were isolated from plasma of chicks made toxic with vitamin D_3. These metabolites (Fig. 3) included $24OHD_3$, 23,24,25-trihydroxyvitamin D_3 [$23,24,25(OH)_3D_3$], and 24,25,26-trihydroxyvitamin D_3 [$24,25,26(OH)_3D_3$] [110, 111]. The biological significance of these metabolites is unknown; however, it is likely that $23,24,25(OH)_3D_3$

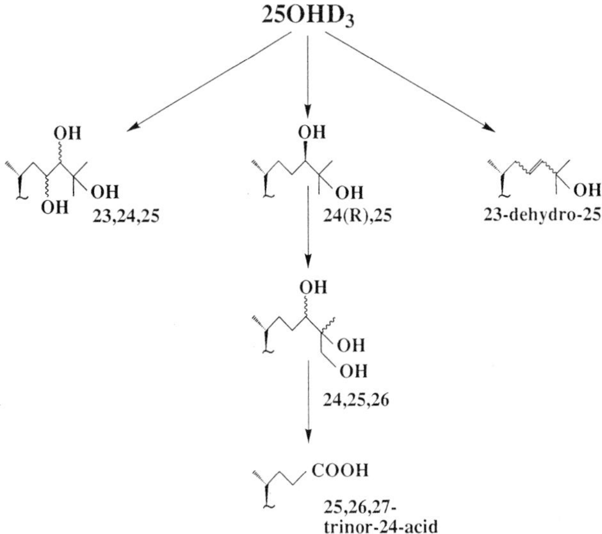

FIGURE 3 Metabolites of $25OHD_3$ functionalized at C-24.

and $24,25,26(OH)_3D_3$ are metabolites of $24,25(OH)_2D_3$. $24,25(OH)_2D_3$ is also the probable precursor to the formation of the side chain cleavage product 25,26,27-trinorvitamin D_3-24-carboxylic acid [112]. This metabolite has been shown to be a product of *in vitro* kidney perfusion using $25OHD_3$ as substrate. The analogous pathway, however, could not be demonstrated using $1,25(OH)_2D_3$ as substrate (S. Reddy, personal communications 1996). Another metabolite isolated in the experiments of Wichmann *et al.* [110] was 23-dehydro-$25OHD_3$. The immediate precursor, site(s), and biological activity of this compound are unknown. Plausible sources of the 23-dehydro compound are dehydration of $24,25(OH)_2D_3$ or 23,25-dihydroxyvitamin D_3 [$23,25(OH)_2D_3$]. It is not certain if any of the metabolites are important under physiological conditions.

G. C-24 Oxidized Vitamin D_2 Metabolites

The 24 position of vitamin D_2, in contrast to the similar position in vitamin D_3, can be considered to be highly reactive. It is a tertiary carbon as well as an allylic position, and the formation of a reactive intermediate (radical, cation) at this position would be highly stabilized. The proximity of this reactive center to the 25 position would afford the possibility of C-24-hydroxylation of vitamin D_2, but the presence of the C-24 methyl would preclude further oxidation to C-24-keto compounds as is known to occur in vitamin D_3 metabolism. Jones *et al.* [40] were the first to demonstrate C-24-oxidation when they isolated $24OHD_2$ from the plasma of male rats treated with 100 IU of radiolabeled vitamin D_2. Engstrom and Koszewski [113] have determined that production of $24OHD_2$ can occur in liver homogenates from a variety of species, and actually exceeds the formation of $25OHD_2$. Horst *et al.* [41] have shown that the concentration of $24OHD_2$ in plasma was about 20% that of $25OHD_2$ in rats receiving physiological doses of vitamin D_2, and was equivalent to $25OHD_2$ in rats receiving pharmacological doses of vitamin D_2. They also demonstrated that 1-hydroxylation of $24OHD_2$ to form $1,24(OH)_2D_2$ represented a minor but significant pathway for vitamin D_2 activation. In their experiments, they determined that $1,24(OH)_2D_2$ rivaled both $1,25(OH)_2D_2$ and $1,25(OH)_2D_3$ in biopotency.

Both $24OHD_2$ and $25OHD_2$ and their 1-hydroxylated metabolites can undergo subsequent hydroxylation to form 24,25-dihydroxyvitamin D_2 [$24,25(OH)_2D_2$] and 1,24,25-trihydroxyvitamin D_2 [$1,24,25(OH)_3D_2$]. The formation of $1,24,25(OH)_3D_2$ represented an unequivocal deactivation of the vitamin D_2 molecule [114]. Conversely, the comparable vitamin D_3 analogue, $1,24,25(OH)_3D_3$, maintains significant biological activity and must undergo further side chain oxidation to be rendered totally inactive [114].

H. 23-Hydroxylase

The discovery of a C-23 oxidative pathway emerged much later than the other pathways and was ushered in by the identification of 23(*S*),25(*R*)$25OHD_3$-26,23-lactone [115,116], 23(*S*),25(*R*)$1,25(OH)_2D_3$-26,23-lactone [117], and their respective precursors 23(*S*),$25(OH)_2D_3$ and 1,23(*S*),$25(OH)_3D_3$ [118,119]. The compound $25OHD_3$-26,23-lactone can be detected in plasma from normal rats, pigs, and chicks [119,120]. However, in several species this metabolite is not expressed unless animals are consuming excessive amounts of vitamin D_3 [121]. This metabolite has unique activity in that it is 3- to 5-fold more competitive than $25OHD_3$ for binding to the plasma vitamin D binding protein (DBP) [116]. It was, therefore, the first modification of $25OHD_3$ that led to enhanced binding to the plasma DBP. The metabolite $1,25(OH)_2D_3$-26,23-lactone has also been demonstrated under normal conditions [122], with elevated plasma concentrations occurring during exogenous administration of pharmacological amounts of $1,25(OH)_2D_3$ [123]. The major locus for formation of the C-23 hydroxylated derivatives appears to be the kidney. Horst *et al.* [121] and Napoli *et al.* [119] demonstrated that nephrectomy eliminated or greatly impaired the biosynthesis of $25OHD_3$-26,23-lactone when animals were treated with excess vitamin D_3 or $25OHD_3$. They showed that this response was due to the inability of the animals to synthesize 23(*S*),$25(OH)_2D_3$. However, when 23(*S*),$25(OH)_2D_3$ was given to nephrectomized animals, the synthesis of $25OHD_3$-26,23-lactone was restored. These data suggested that C-23-hydroxylation occurred predominantly, but not exclusively, in the kidney, whereas extrarenal tissues are quantitatively important in the pathway leading to $25OHD_3$-26,23-lactone synthesis, which includes formation of the lactone intermediates $23,25,26(OH)_3D_3$ [124] and $25OHD_3$-26,23-lactol [125].

Although ambiguities remain regarding the biological effects of C-24 oxidation, 23-hydroxylation appears to be clearly a deactivation event. 23-Hydroxylation is the first side chain modification of $25OHD_3$ noted to substantially reduce its affinity for the plasma DBP [126]. 23-Hydroxylation of $1,25(OH)_2D_3$ also leads to its increased plasma clearance and reduced VDR binding and biological activity [127].

The role of 23-hydroxylation as a primary oxidation event for the further metabolism of $25OHD_3$ and $1,25(OH)_2D_3$ is relatively minor to its role in the further metabolism of vitamin D_3 metabolites that have been previously oxidized at C-24. In other words, very little production of $23,25(OH)_2D_3$ or $1,23,25(OH)_3D_3$ would be expected under normal conditions. Rather, the convergences of the C-24 and C-23 oxidative pathways would lead predominantly to the formation of 24-keto-$23,25(OH)_2D_3$ and 24-keto-$1,23,25(OH)_3D_3$ [99], which subsequently cleave to form C-23 acids [102,128]. Therefore, as indicated in Fig. 1, the C-23 oxidative pathway can lead to two different patterns of side chain modifications for both $25OHD_3$ and $1,25(OH)_2D_3$. One pathway, which is relatively minor under physiological conditions and more predominant during hypervitaminosis D_3, leads through 23-hydroxylation to formation of the lactones whereas the other more physiologically significant pathway leads through 24-hydroxylation to 23-hydroxylation of 24-keto metabolites.

Other oxidized C-23 metabolites that have been identified include 23-keto derivatives of $25OHD_3$ and $1,25(OH)_2D_3$. 23-Keto-25-hydroxyvitamin D_3 was synthesized *in vitro* from $23(S)25(OH)_2D_3$ and $23(R),25(OH)_2D_3$, and it has unique properties in that it binds with 2-fold higher affinity than $25OHD_3$ for binding sites on the plasma DBP [129]. This affinity should be compared to that for $23(S),25(OH)_2D_3$, which binds with 6- to 10-fold less affinity. 23-Keto-$25OHD_3$ is also about 4-fold more competitive than $25OHD_3$ for binding to the VDR. 23-Ketonization is, therefore, the first example of a side chain modification enhancing the affinity of $25OHD_3$ for the VDR. This high affinity of 23-keto-$25OHD_3$ for VDR prompted biosynthesis of 23-keto-$1,25(OH)_2D_3$ to determine if this modification might enhance binding of $1,25(OH)_2D_3$ to VDR. Horst *et al.* [130] prepared this metabolite by incubating 23-keto-$25OHD_3$ in kidney homogenates prepared from vitamin D-deficient chicks. The major metabolite was isolated and identified as 23-keto-$1,25(OH)_2D_3$ and was shown to possess about 40% the activity of $1,25(OH)_2D_3$ for VDR binding. 23-Ketonization of $1,25(OH)_2D_3$, therefore, reduced the affinity of $1,25(OH)_2D_3$ to VDR rather than increased binding. Ohnuma *et al.* [131] and Mayer *et al.* [132] also produced 23-keto-$1,25(OH)_2D_3$, as well as 23-keto-$1,25,26(OH)_3D_3$, from $1,25(OH)_2D_3$. These products were prepared in intestinal homogenates from vitamin D_3- replete chicks given exogenous $1,25(OH)_2D_3$. Napoli *et al.* [133], however, could not demonstrate the presence of these metabolites from rat intestine *in vivo* or *in vitro*. Mayer *et al.* [132] suggested that formation of 23-keto-$1,25(OH)_2D_3$ was the first step in formation of the 1-hydroxy-23-carboxytetranorvitamin D_3 excretory product. However, more recent data [102] suggested that formation of the 1-hydroxy-23-carboxytetranorvitamin D_3 excretory product actually proceeds through 24-keto formation followed by 23-hydroxylation.

I. C-23 Oxidized Vitamin D_2 Metabolites

Because of the presence of the C-22 alkene, neither $25OHD_2$ nor $1,25(OH)_2D_2$ give rise to C-23 oxidized metabolites. Rather, oxidation at C-23 may take a different course. Perhaps the 22,23-epoxide would be formed, which could give rise to the 22,23-diol. The latter compound may form one or more ketones. In the absence of data, these points are speculative.

J. C-26 Hydroxylation

26-Hydroxylation of $25OHD_3$ and $1,25(OH)_2D_3$ produces $25,26(OH)_2D_3$ [134] and $1,25,26(OH)_3D_3$ [135, 136], respectively. The natural product was originally assigned the 25(*R*) configuration. However, Partridge *et al.* [137] gave the assignment as 25(*S*). Ikekawa *et al.* [138] later discovered that the naturally occurring $25,26(OH)_2D_3$ actually existed as a mixture of 25(*S*) and 25(*R*) isomers. Although this assignment seems somewhat trivial, it was important in unraveling a controversy that existed regarding the physiological precursor to the *in vivo* synthesis of $25OHD_3$-26,23-lactone. Hollis *et al.* [139] demonstrated that $25,26(OH)_2D_3$ isolated from *in vivo* sources could act as a precursor to the formation of the $25OHD_3$-26,23-lactone. Subsequent research, however, suggested that synthetic $25(S),26(OH)_2D_3$ (which at the time was thought to be the natural configuration) did not act as precursor to the formation of $25OHD_3$-26,23-lactone [140] but synthetic $25(R),26(OH)_2D_3$ could act a precursor [119,141]. As naturally occurring $25,26(OH)_2D_3$ is a mixture of the *R* and *S* isomers, this research validated the conclusion of Hollis *et al.* [139] suggesting that formation of $25OHD_3$-26,23-lactone could indeed proceed through $25,26(OH)_2D_3$. This pathway has been shown to be relatively minor, with the major pathway to $25OHD_3$-26,23-lactone synthesis proceeding through $23(S),25(OH)_2D_3$ [119, 124,141].

The major locus for the 26-hydroxylase is unknown. Blood concentrations of $25,26(OH)_2D_3$ are not depressed in nephrectomized humans or pigs [121, 142,143]. Therefore, production of these metabolites must take place at extrarenal sources. 26-Hydroxylase activity has, however, been demonstrated in microsomes isolated from rat and pig kidneys [144]. The only extrarenal source was reported in liver mitochondria [145]. The physiological role of the C-26 oxidative pathway remains elusive. However, $25,26(OH)_2D_3$ and $1,25,26(OH)_3D_3$ have been shown to possess biological activity with regard to stimulating bone calcium resorption and intestinal calcium absorption, albeit to a lesser degree than either $25OHD_3$ or $1,25(OH)_2D_3$ [134,146]. Therefore, it seems unlikely that 26-hydroxylation is essential for calcium uptake from the gut or release of calcium from bone.

K. C-26 and C-28 Oxidized Vitamin D_2 Metabolites

The compounds 25,26-dihydroxyvitamin D_2 [$25,26(OH)_2D_2$] and 1,25,26-trihydroxyvitamin D_2 [$1,25,26(OH)_3D_2$] have been chemically synthesized [147,148] (J. Barrish and M. R. Usković, unpublished data, 1988). The metabolite $25,26(OH)_2D_2$ has also been tentatively, but not exhaustively, identified from *in vivo* sources [120,149], but it could not be demonstrated in kidneys perfused with $25OHD_2$ [150]. Similarly, $1,25,26(OH)_3D_2$ could not be demonstrated either *in vivo* (R. L. Horst, personal observations, 1991) or *in vitro* [151]. 26-Hydroxylation has, however, been shown to occur when vitamin D_2 metabolites have been previously 24-hydroxylated. For example, when $24,25(OH)_2D_2$ and $1,24,25(OH)_3D_2$ were used as precursors, the metabolites $24,25,26(OH)_3D_2$ and $1,24,25,26(OH)_4D_2$, respectively, were produced in significant amounts [150,151]. Similarly, Koszewski *et al.* [42] and Jones *et al.* [152] demonstrated that C-26 hydroxylation was the major metabolic pathway for the further metabolism of $24OHD_2$ and $1,24(OH)_2D_2$. Oxidation at C-24, therefore, appears to be a prerequisite for C-26 oxidation of vitamin D_2 compounds. A similar situation also appears to exist for C-28 oxidation, as demonstrated by Reddy and co-workers [150,151], who isolated $24,25,28(OH)_3D_2$ and $1,24,25,28(OH)_4D_2$ from rat kidney perfusions. Through the use of various substrates, they were able to show that these new metabolites proceed through the initial formation of $24,25(OH)_2D_2$ and $1,24,25(OH)_3D_2$, respectively.

III. VITAMIN D TOXICITY

A. Overview

The first documented reports of vitamin D intoxication were made in the late 1920s by Kreitmeir and Moll [153] and Putscher [154]. These cases resulted from the ingestion of large quantities of vitamin D in the diet. Vitamin D intoxication, however, has never been reported following prolonged sunlight exposure. Holick *et al.* [155] suggested that nature has provided various control points which prevent the overproduction of vitamin D_3 by the skin. The most important point of control is the diversion of vitamin D_3 production from 7-dehydrocholesterol to non-biologically active overirradiation products such as lumisterol and tachysterol [155]. In addition, these authors suggested that skin pigmentation and latitude were also significant determinants (albeit to a lesser degree) which limit the cutaneous production of vitamin D_3.

As discussed earlier, once vitamin D is in circulation the conversion to 25OHD is relatively uncontrolled. Normally, 25OHD circulates at 30 to 50 ng/ml in most species [120]. However, when vitamin D is given in excess, plasma 25OHD can be elevated to concentrations of 1000 ng/ml or greater [156,157], while plasma $1,25(OH)_2D$ remains at or below normal concentrations [158]. When circulating at very high concentrations, 25OHD can compete effectively with $1,25(OH)_2D$ for binding to the VDR. Therefore, during vitamin D toxicosis, 25OHD can induce actions usually attributed to $1,25(OH)_2D$ [158]. High circulating 25OHD can, therefore, explain how humans with low circulating concentrations of $1,25(OH)_2D$ can show signs of vitamin D toxicity [158] and why anephric humans [who are incapable of producing $1,25(OH)_2D$] can become vitamin D toxic [159]. Clinical aspects of vitamin D toxicity are discussed in Chapter 54.

Although it is generally accepted that $1,25(OH)_2D$ is reduced during hypervitaminosis, a notable exception to this generalization is the ruminant. Horst and coworkers [160,161] have shown that vitamin D_3 intoxication initiated by giving 15 million IU of vitamin D_3 intramuscularly (i.m.) results in significant elevations in plasma $1,25(OH)_2D_3$. In contrast, pigs given the same i.m. dose showed a reduction in plasma $1,25(OH)_2D_3$, as was observed in other species [162]. Therefore, elevations in plasma $1,25(OH)_2D$ may play a significant role in the pathogenesis of vitamin D toxicity in ruminants.

B. Differences in Toxicity between Vitamins D_2 and D_3

Most research dealing with utilization of vitamins D_2 and D_3 assumes that these two forms are equally potent in most mammals. However, when large and potentially toxic doses were administered orally to rhesus monkeys [163] and horses [164], or were used in treating childhood osteodystrophy [165], vitamin D_2 presented fewer hypercalcemic side effects than vitamin D_3. In addition, $1OHD_2$, which is as effective as the $1OHD_3$ in standard bioassays, was shown to be 5- to 15-fold less toxic than $1OHD_3$ [166].

Studies by Horst *et al.* [41] provide some insight for the difference between vitamin D_2 and vitamin D_3 toxicity. They demonstrated that under physiological conditions the predominant monohydroxylated form of both vitamins D_2 and D_3 is 25OHD. In the vitamin D_2-dosed rats, $24OHD_2$ accounted for approximately 20% of the monohydroxylated metabolites, whereas $24OHD_3$ could not be detected in the vitamin D_3-dosed rats. When a modest superphysiological dose (800 IU/day) of vitamin D_3 was given to rats, $25OHD_3$ remained the predominant metabolite in vitamin D_3-dosed rats and was present at 26.3 ng/ml. Under these conditions, there was still no evidence for the presence of $24OHD_3$. However, when the same amount of vitamin D_2 was given, the concentrations of $24OHD_2$ (14.1 ng/ml) nearly matched those of $25OHD_2$ (15.9 ng/ml). Interestingly, the combined concentrations of $24OHD_2$ and $25OHD_2$ in the vitamin D_2-dosed animals (~30 ng/ml) was similar to the $25OHD_3$ concentration (26.3 ng/ml) in the vitamin D_3-dosed rat. In standard assays, $25OHD_2$ and $25OHD_3$ are equipotent at displacing ^3H-$1,25(OH)_2D_3$ from the calf thymus VDR. However, $24OHD_2$ has been shown to have at least a 2-fold lower affinity for binding to the calf thymus VDR (R. L. Horst unpublished data, 1996). Therefore, the reduced toxicity of vitamin D_2 is probably a result of diverting metabolism away from the production of $25OHD_2$ in favor of $24OHD_2$, which has a relatively limited affinity for VDR (a step necessary for the initiation of a biological response). Further differences between vitamins D_2 and D_3 were noted in their ability to up-regulate the VDR. Beckman *et al.* [167] noticed that VDR was significantly more enhanced in animals fed excess vitamin D_3 relative to those animals receiving an equivalent amount of vitamin D_2. Increased VDR would potentially accentuate toxic side effects by enhancing the responsiveness of intestinal tissue to the elevated 25OHD (see Chapter 11).

C. Factors Affecting Toxicity

The severity of the effects and pathogenic lesions in vitamin D intoxication depend on such factors as the type of vitamin D (vitamin D_2 versus vitamin D_3), the dose, the functional state of the kidneys, and the composition of the diet. Vitamin D toxicity is enhanced by a rich dietary supply of calcium and phosphorus, and it is reduced when the diet is low in calcium and phosphorus [168,169]. Toxicity is also reduced when the vitamin is accompanied by high intakes of vitamin A or by thyroxine injections [170]. The route of administration also influences toxicity. Parenteral administration of 15 million IU of vitamin D_3 in a single dose caused toxicity and death in many pregnant dairy cows [161]. On the other hand, oral administration of 20 to 30 million IU of vitamin D_2 daily for 7 days resulted in little or no toxicity in pregnant dairy cows [171]. Napoli *et al.* [172] have shown that rumen microbes are capable of metabolizing vitamin D to the inactive 10-keto-19-norvitamin D_3. Parenteral administration would circumvent the deactivation of vitamin D by rumen microbes and may partially explain the difference in toxicity between oral and parenteral vitamin D.

Various measures have been used in human medicine for treatment of vitamin D toxicity. These measures are mainly concerned with management of hypercalcemia. Vitamin D withdrawal is obviously indicated. It is usually not immediately successful, however, owing to the long plasma half-life of vitamin D (5 to 7 days) and 25OHD (20 to 30 days). This is in contrast to the short plasma half-life of $1OHD_3$ (1 to 2 days) and $1,25(OH)_2D_3$ (4 to 8 hr). Because intestinal absorption of calcium contributes to hypercalcemia, a prompt reduction in dietary calcium is indicated. Sodium phytase, an agent that reduces intestinal calcium absorption, has also been used successfully in vitamin D toxicity management in monogastrics [173]. This treatment would be of little benefit to ruminants because of the presence of rumen microbial phytases. There have also been reports that CT [174], glucagon [175], etidronate [176], and glucocorticoid therapy [177] reduce serum calcium levels or prevent the calcinosis resulting from vitamin D intoxication (see Chapter 54).

IV. SPECIES VARIATION IN VITAMIN D METABOLISM AND ACTION

Most concepts of vitamin D metabolism and function have been developed with the rat and/or chick as experimental models. Studying vitamin D metabolism is hampered by the paucity of data on the normal circulating levels of vitamin D metabolites in birds, mammals, and reptiles under normal conditions. Most recent research has focused on the analysis of 25OHD and $1,25(OH)_2D$ as indicators of vitamin D status or aberrant physiological states. Table I summarizes the concentrations of the two metabolites that have been reported for several species by various laboratories. Close inspection of the information suggests that some mammals (mole rat, wild wood vole, horse, and wild wood mouse) and aquatic species (lamprey, carp, halibut, and bullfrog) appear to have very low or undetectable concentrations of 25OHD, and yet these animals appeared to be normal with no evidence of vitamin D deficiency. It is questionable whether some of these species have a requirement for vitamin D. The damara mole rat, for example, is a subterranean herbivore that in its natural habitat has no access to any obvious source of vitamin D and consumes a diet of roots and tubers [178]. These animals exhibit a high apparent calcium absorption efficiency

TABLE I Plasma 25-Hydroxyvitamin D and 1,25-Dihydroxyvitamin D Concentrations in Several Species of Animals

	Concentration		
Species	25OHD ng/ml	1,25(OH)$_2$D pg/ml	Ref.
Human	32	31	143
Rhesus monkey	188	207	196
Rhesus monkey	50	95	197
Marmoset	90	400	197
Marmoset	64	640	198
Wild wood mouse	<5	<10	199
Wild bank vole	<5	<5	199
Mole rat	<2	17	178
Lamprey	ND[a]	274	200
Shark	ND	87	200
Leopard shark	56	3	201
Horned shark	33	6	201
Carp	ND	174	200
Bastard halibut	ND	192	200
Atlantic cod	<2	59	202
Bullfrog (mature)	2	21	200
Soft shelled turtle	16	12	200
Turkey	26	52	120
Chicken	27	21	120
Cow	43	38	120
Sheep	27	36	120
Pig	76	60	120
Horse	7	ND	203

[a] ND, Not done

(91%) and, like the horse and rabbit, actually use renal calcium excretion as the major regulator of calcium homeostasis [179,180]. In studies with rabbits consuming adequate amounts of calcium, it is very difficult to develop any overt or histological signs of vitamin D deficiency, and vitamin D may play a minor, if any, role in normal day-to-day functions in these animals [181].

In the wild, most animals do not have a dietary need for vitamin D, as sufficient vitamin D_3 can be synthesized in the skin on irradiation by sunlight. However, indoor confinement of humans and other animals has resulted in the diet becoming the main source of vitamin D, leading to considerable research to determine the amount of dietary vitamin D required to substitute for lack of exposure to sunlight. Photochemically produced vitamin D_3 enters the circulation and becomes immediately available, whereas dietary vitamin D_3 may undergo modifications prior to becoming available for use by the body. One species where significant modification of vitamin D occurs before absorption is the ruminant. Within 24 hr, as much as 80% of vitamin D can undergo metabolism *in vitro* in rumen incubation media [182]. At least four metabolites are produced by the rumen microbes [182,183]. Two of these metabolites have been identified as the cis (5*Z*) and trans (5*E*)-isomers of 10-keto-19-norvitamin D_3 (Fig. 1) [172]. The trans isomer has also been identified in cow plasma (R. L. Horst, unpublished data, 1983). Neither compound has agonistic activity with regard to promoting bone calcium resorption [184] or intestinal calcium absorption [172]. Rather, this novel metabolism is likely a detoxification process, as evidenced by the ability of ruminants to tolerate large oral doses of vitamin D_3 that would be toxic if given parenterally. The presence of the rumen, therefore, represents a major control point in vitamin D metabolism that may differ from monogastrics. Such a control point may have survival value, because the ruminant evolved as a grazing animal, with the opportunity for long periods of sunlight exposure, as well as consumption of large quantities of irradiated plants. If left uncontrolled, such a combination might result in vitamin D toxicity.

Shortly after the discovery of vitamin D, it seemed apparent that vitamins D_2 and D_3 had similar biological activities in most mammals and that birds and New World monkeys discriminated against vitamin D_2 in favor of vitamin D_3 [185,186]. More recent research, fostered by the discovery of sensitive analytical techniques and the availability of high specific activity ^{3}H-labeled vitamin D species, indicated that differences in the metabolism of vitamins D_2 and D_3 in mammals are perhaps widespread. Most notable were the apparent discrimination against vitamin D_2 by pigs [187], cows [183], and humans [188] and the apparent preference for vitamin D_2 by rats [187,189].

Vitamin D and its metabolites are transported in the blood of vertebrates attached to a specific protein commonly known as the vitamin D binding protein or DBP [190]. Baird *et al.* [191] have shown that protein binding increases the solubility of steroids and that the metabolic clearance rate of steroids is in part dependent on their binding to specific plasma proteins. Affinity of metabolites to the plasma transport proteins may, therefore, provide a means for determining which species would utilize vitamin D_2 poorly. For example, if the binding protein showed lower affinity toward $25OHD_2$ relative to $25OHD_3$, then one would predict that $25OHD_2$ would be removed from the circulation faster than $25OHD_3$. This is indeed the case for the chick. Hoy *et al.* [192] showed that chick discrimination against vitamin D_2 was probably a result of enhanced clearance of the vitamin D_2 metabolites $25OHD_2$ and $1,25\text{-}(OH)_2D_2$, and that the enhanced clearance was associated with weaker binding of these vitamin D_2 metabolites (relative to the vitamin D_3 forms) to DBP.

In one of the most comprehensive studies reported to date, Hay and Watson [193] studied the affinities of DBP for $25OHD_2$ and $25OHD_3$ in 63 vertebrate species. They found that the DBP in fish, reptiles, and birds discriminated against $25OHD_2$ in favor of $25OHD_3$, which is consistent (at least in birds) with the discrimination against vitamin D_2. One notable exception to this hypothesis, however, is the New World monkey. Hay and Watson [193] found that in New World monkeys, the plasma transport protein has equal affinity for $25OHD_2$ and $25OHD_3$, which is inconsistent with the well documented discrimination against vitamin D_2. Factors other than affinity of the binding protein for 25OHD are, therefore, important in determining how efficiently the different forms of vitamin D can be utilized by animals.

Another example of species discrimination against the different vitamin D forms is in the rat. However, in this species discrimination is against vitamin D_3 in favor of vitamin D_2 [187]. The rat DBP is known to have equal affinity for $25OHD_2$ and $25OHD_3$, but a lower affinity for vitamin D_2 relative to vitamin D_3 [194]. Reddy *et al.* [195] suggested that the lower affinity for vitamin D_2 resulted in its enhanced availability for liver 25-hydroxylation. Hence, in the presence of DBP, more $25OHD_2$ was made relative to $25OHD_3$ when equal amounts of vitamin D_2 or vitamin D_3 substrate were perfused into rat livers. This observation is consistent with the higher circulating concentrations of $25OHD_2$ observed in acute experiments with vitamin D-deficient rats dosed with equal amounts of vitamins D_2 and D_3 [187]. In the experiments conducted by Reddy *et al.* [194], if binding protein was eliminated from the perfusion media, equal amounts of $25OHD_2$ and $25OHD_3$ were synthesized. Collectively, these data suggest that discrimination against the different forms of vitamin D could likely result from variations in the affinity of DBP for the parent compound and/or one or more of their metabolites. Regardless of the mechanism for discrimination, it appears that these differences are present to afford animals the most efficient utilization of the most abundant antirachitic agents available in their environment.

V. CONCLUSIONS

Vitamin D metabolism still remains an exciting area of research with much more to be learned. Critical questions remain unanswered regarding complete elucidation of the vitamin D_2 metabolic pathway, and species differences between vitamins D_2 and D_3 metabolism are still virtually unexplored. The introduction of vitamin D analogs has also resulted in a totally different set of issues regarding their metabolism, tissue kinetics, mechanism of action, and potential therapeutic uses.

References

1. Mellanby E 1919 An experimental investigation on rickets. Lancet I **4985**:407–412.
2. McCollum EV, Simmonds N, Becker JE, Shipley PG 1922 Studies on experimental rickets. XXI. An experimental demonstration of the existence of a vitamin which promotes calcium deposition. J Biol Chem **53**:293–312.
3. Norman AW 1979 Vitamin D: The Calcium Homeostatic Hormone. In: Darby WJ (ed) Nutrition: Basic and Applied Science. Academic Press, New York.
4. DeLuca HF, Schnoes HK 1976 Metabolism and action of vitamin D. Annu Rev Biochem **45**:631–666.
5. Kodicek E 1974 The story of vitamin D from vitamin to hormone. Lancet I **7853**:325–329.
6. Haussler MR, McCain TA 1977 Basic and clinical concepts related to vitamin D metabolism and action. N Engl J Med **297**:974–983 and 1041–1050.
7. Lawson DEM 1978 Vitamin D. Academic Press, London.
8. Napoli JL, Horst RL 1984 In: Kumar R (ed) Vitamin D Metabolism. Martinus Nijhoff, Boston, pp. 91–123.
9. Morii H, Lund J, Neville PF, DeLuca HF 1967 Biological activity of a vitamin D metabolite. Arch Biochem Biophys **120**:508–512.
10. Blunt JW, DeLuca HF, Schnoes HK 1968 25-Hydroxycholecalciferol: A biologically active metabolite of vitamin D_3. Biochemistry **7**:3317–3322.
11. Holick MF, Schnoes HK, DeLuca HF, Suda T, Cousins RJ 1971 Isolation and identification of 1,25-dihydroxycholecalciferol. A metabolite of vitamin D active in intestine. Biochemistry **10**:2799–2804.
12. Lawson DEM, Fraser DR, Kodicek E, Morris HR, Williams DH 1971 Identification of 1,25-dihydroxycholecalciferol, a new kidney hormone controlling calcium metabolism. Nature **230**:228–230.
13. Norman AW, Myrtle JF, Midgett RJ, Nowicki HG 1971 1,25-Dihydroxycholecalciferol: Identification of the proposed active form of vitamin D_3 in the intestine. Science **173**:51–54.
14. Broadus AE, Horst RL, Lang R, Littledike ET, Rasmussen H 1980 The importance of circulating 1,25-dihydroxyvitamin D in the pathogenesis of hypercalciuria and renal-stone formation in primary hyperparathyroidism. N Engl J Med **302**:421–426.
15. Ornoy A, Goodwin D, Noff D, Edelstein S 1978 24,25-Dihydroxyvitamin D is a metabolite of vitamin D essential for bone formation. Nature **276**:517–519.
16. Bordier P, Rasmussen H, Marie P, Miravet L, Gueris J, Ryckwaert A 1978 Vitamin D metabolism and bone mineralization in man. J Clin Endocrinol Metab **46**:284–294.
17. Canterbury JM, Lerman S, Claflin AJ, Henry H, Norman A, Reiss E 1977 Inhibition of parathyroid hormone secretion by 25-hydroxycholecalciferol and 24,25-dihydroxycholecalciferol in the dog. J Clin Invest **78**:1375–1383.
18. Henry HL, Norman AW 1978 Vitamin D: Two dihydroxylated

metabolites are required for normal chicken egg hatchability. Science **201**:835–837.
19. Horst RL, Littledike ET, Riley JL, Napoli JL 1981 Quantitation of vitamin D and its metabolites and their plasma concentrations in five species of animals. Anal Biochem **116**:189–203.
20. Hartwell D, Hassager C, Christiansen C 1987 Effect of vitamin D_2 and vitamin D_3 on the serum concentrations of $1,25(OH)_2D_2$ and $1,25(OH)_2D_3$ in normal subjects. Acta Endocrinol **115**:378–384.
21. Hollis BW, Pittard WB 1984 Relative concentrations of 25-hydroxyvitamin D_2/D_3 and 1,25-dihydroxyvitamin D_2/D_3 in maternal plasma at delivery. Nutr Res **4**:27–32.
22. Reinhardt TA, Horst RL, Orf JW, Hollis BW 1984 A microassay for 1,25- dihydroxyvitamin D not requiring high performance liquid chromatography: Application to clinical studies. J Clin Endocrinol Metab **58**:91–98.
23. Jones G, Schnoes HK, DeLuca HF 1975 Isolation and identification of 1,25-dihydroxyvitamin D_2. Biochemistry **14**:1250–1256.
24. Tucker G, Gaganon R, Haussler M 1973 Vitamin D_3-25-hydroxylase: Tissue occurrence and apparent lack of regulation. Arch Biochem Biophys **155**:47–57.
25. Ponchon G, DeLuca HF 1969 Metabolites of vitamin D_3 and their biological activity. J Nutr **99**:157–167.
26. Bhattacharyya MH, DeLuca HF 1973 The regulation of rat liver calciferol-25-hydroxylase. J Biol Chem **248**:2969–2973.
27. Bhattacharyya MH, DeLuca HF 1973 Comparative studies on the 25-hydroxylation of vitamin D_3 and dihydrotachysterol. J Biol Chem **248**:2974–2977.
28. Bjorkhem I, Holmberg I 1979 On the 25-hydroxylation of vitamin D_3 in vitro studied with a mass fragmentographic technique. J Biol Chem **254**:9518–9524.
29. Bjorkhem I, Holmberg I 1978 Assay and properties of a mitochondrial 25-hydroxylase active on vitamin D_3. J Biol Chem **253**:842–849.
30. Madhok TC, DeLuca HF 1979 Characteristics of the rat liver microsomal enzyme system converting cholecalciferol into 25-hydroxycholecalciferol. Evidence for the participation of cytochrome P-450. Biochem J **184**:491–499.
31. Bjorkhem I, Holmberg I, Oftebro H, Pedersen JI 1980 Properties of a reconstituted vitamin D_3 25-hydroxylase from rat liver mitochondria. J Biol Chem **255**:5244–5249.
32. Okuda K 1992 Vitamin D_3 hydroxylases—Its occurrence, biological importance and biochemistry. In: Ruckpaul K, Rein H (eds) Cytochrome P-450 Dependent Biotransformation of Endogenous Substrates: Frontiers in Biotransformation, Vol. 6. Akademie-Verlag, Berlin, pp. 114–147.
33. Okuda K 1994 Liver mitochondrial p450 involved in cholesterol catabolism and vitamin activation. J Lipid Res **35**:361–372.
34. Okuda K-I, Usui E, Ohyama Y 1995 Recent progress in enzymology and molecular biology of enzymes involved in vitamin D metabolism. J Lipid Res **36**:1641–1652.
35. Andersson S, Davis DL, Dahlbach H, Jornvall H 1989 Cloning, structure, and expression of the mitochondrial cytochrome P-450 sterol 26-hydroxylase, a bile acid biosynthetic enzyme. J Biol Chem **264**:8222–8229.
36. Guo Y-D, Strugnell S, Back DW, Jones G 1993 Transfected human liver cytochrome P-450 hydroxylates vitamin D analogs at different side-chain positions. Proc Natl Acad Sci USA **90**:8668–8672.
37. Usui E, Noshiro M, Okuda K 1990 Molecular cloning of cDNA for vitamin D_3 25-hydroxylase from rat liver mitochondria. Fed Eur Biol Sci Lett **262**:135–138.
38. Cali J, Russell D 1991 Characterization of human sterol 27-hydroxylase. J Biol Chem **266**:7774–7778.
39. Usui E, Noshiro M, Ohyama Y, Okuda K 1990 Unique property of liver mitochondrial P450 to catalyze the two physiologically important reactions involved in both cholesterol catabolism and vitamin D activation. FEBS Lett **274**:175–177.
40. Jones G, Schnoes HK, Levan L, DeLuca HF 1980 Isolation and identification of 24-hydroxyvitamin D_2 and 24,25-dihydroxyvitamin D_2. Arch Biochem Biophys **202**:450–457.
41. Horst RL, Koszewski NJ, Reinhardt TA 1990 1α-Hydroxylation of 24-hydroxyvitamin D_2 represents a minor physiological pathway for the activation of vitamin D_2 in mammals. Biochemistry **29**:578–582.
42. Koszewski NJ, Reinhardt TA, Napoli JL, Beitz DC, Horst RL 1988 24,26-Dihydroxyvitamin D_2: A unique physiological metabolite of vitamin D_2. Biochemistry **27**:5785–5790.
43. Hayashi S, Usui E, Okuda K 1988 Sex-related difference in vitamin D_3 25-hydroxylase of rat liver mitochondria. J Biochem (Tokyo) **103**:863–866.
44. Saarem K, Bergseth S, Oftebro H, Petersen JI 1984 Subcellular localization of vitamin D_3 25-hydroxylation in human liver. J Biochem **259**:10936–10946.
45. Axen E, Bergman T, Wikvall K 1994 Microsomal 25-hydroxylation of vitamin D_2 and vitamin D_3 in pig liver. J Steroid Biochem Mol Biol **51**:97–106.
46. Haussler MR, Myrtle JF, Norman AW 1968 The association of a metabolite of vitamin D_3 with intestinal mucosa chromatin *in vivo*. J Biol Chem **243**:4055–4064.
47. Lawson DEM, Wilson PW, Kodicek E 1969 Metabolism of vitamin D. A new cholecalciferol metabolite, involving loss of hydrogen at C-1, in chick intestinal nuclei. Biochem J **115**:269–277.
48. Kodicek E, Lawson DEM, Wilson PW 1970 Biological activity of a polar metabolite of vitamin D_3. Nature **228**:763–764.
49. Fraser DR, Kodicek E 1970 Unique biosynthesis by kidney of a biologically active vitamin D metabolite. Nature **228**:764–766.
50. Haussler MR, Boyce DW, Littledike ET, Rasmussen H 1971 A rapidly acting metabolite of vitamin D_3. Proc Natl Acad Sci USA **68**:177–181.
51. Holick MF, Schnoes HK, DeLuca HF 1971 Identification of 1,25-dihydroxycholecalciferol, a form of vitamin D_3 metabolically active in the intestine. Proc Natl Acad Sci USA **68**:803–804.
52. Paulson SK, DeLuca HF 1985 Subcellular location and properties of rat renal 25-hydroxyvitamin D_3-1α-hydroxylase. J Biol Chem **260**:11488–11492.
53. Armbrecht HJ, Okuda K, Wongsurawat N, Nemani RK, Chen ML, Boltz MA 1992 Characterization and regulation of the vitamin D hydroxylases. J Steroid Biochem **43**:1073–1081.
54. Gray RW, Weber HP, Dominguez JH, Lemann J Jr 1974 The metabolism of vitamin D_3 and 25-hydroxyvitamin D_3 in normal and anephric humans. J Clin Endocrinol Metabol **39**:1045–1056.
55. DeLuca HF 1974 Vitamin D: The vitamin and the hormone. Fed Proc **33**:2211–2219.
56. Boyle I, Gray R, DeLuca HF 1971 Regulation by calcium of in vivo synthesis of 1,25-dihydroxycholecalciferol and 21,25-dihydroxycholecalciferol. Proc Natl Acad Sci USA **68**:2131–2137.
57. Armbrecht HJ, Wongsurawat N, Zenser TV, Davis BB 1984 Effect of PTH and $1,25\text{-}(OH)_2D_3$ on renal $25(OH)D_3$ metabolism, adenylate cyclase, and protein kinase. Am J Physiol **246**:E102-E107.
58. Armbrecht HJ, Wongsurawat N, Zenser TV, Davis BB 1982 Differential effects of parathyroid hormone on the renal 1,25-dihydroxyvitamin D_3 and 24,25-dihydroxyvitamin D_3 production of young and adult rats. Endocrinology **111**:1339–1344.
59. Horst RL, Eisman JA, Jorgensen NA, DeLuca HF 1977 Ade-

quate response of plasma 1,25-dihydroxyvitamin D to parturition in paretic (milk fever) dairy cows. Science **196**:662–663.

60. Hughes MR, Brumbaugh PF, Haussler MR, Wergedal JE, Baylink DJ 1975 Regulation of serum 1α,25-dihydroxyvitamin D_3 by calcium and phosphate in the rat. Science **190**:578–580.
61. Henry HL, Midgett RJ, Norman AW 1974 Regulation of 25-hydroxyvitamin D_3-1-hydroxylase in vivo. J Biol Chem **249**:7584–7592.
62. Henry HL 1979 Regulation of the hydroxylation of 25-hydroxyvitamin D_3 in vivo and in primary cultures of chick kidney cells. J Biol Chem **254**:2722–2729.
63. Omdahl JL 1978 Interaction of the parathyroid and 1,25-dihydroxyvitamin D_3 in the control of renal 25-hydroxyvitamin D_3 metabolism. J Biol Chem **253**:8474–8487.
64. Henry HL 1981 Insulin permits parathyroid hormone stimulation of 1,25-dihydroxyvitamin D_3 production in cultured kidney cells. Endocrinology **108**:733–735.
65. Armbrecht HJ, Forte LR, Wongsurawat N, Zenser TV, Davis BB 1984 Forskolin increases 1,25-dihydroxyvitamin D_3 production by rat renal slices in vitro. Endocrinology **114**:644–649.
66. Aarskog D, Aksnes L 1980 Acute response of plasma 1,25-dihydroxyvitamin D to parathyroid hormone. Lancet I **8164**:362–363.
67. Hove K, Horst RL, Littledike ET, Beitz DC 1984 Infusions of parathyroid hormone in ruminants: Hypercalcemia and reduced plasma 1,25-dihydroxyvitamin D concentrations. Endocrinology **114**:897–903.
68. Hustmyer FG, Beitz DC, Goff JP, Nonnecke BJ, Horst RL, Reinhardt TA 1994 Effects of *in vivo* administration of 1,25-dihydroxyvitamin D_3 on *in vitro* proliferation of bovine lymphocytes. J Dairy Sci **77**:3324–3330.
69. Reinhardt TA, Horst RL 1990 Parathyroid hormone down-regulates 1,25-dihydroxyvitamin D receptors (VDR) and VDR messenger ribonucleic acid *in vitro* and blocks homologous up-regulation of VDR *in vivo*. Endocrinology **127**:942–948.
70. Tanaka Y, DeLuca HF 1973 The control of 25-hydroxyvitamin D metabolism by inorganic phosphorus. Arch Biochem Biophys **154**:566–574.
71. Gray RW 1987 Evidence that somatomedins mediate the effect of hypophosphatemia to increase serum 1,25-dihydroxyvitamin D_3 levels in rats. Endocrinology 121:504–512.
72. Armbrecht HJ, Zenser TV, Davis BB 1982 Modulation of renal production of 24,25- and 1,25-dihydroxyvitamin D_3 in young and adult rats by dietary calcium, phosphorus, and 1,25-dihydroxyvitamin D_3. Endocrinology **110**:1983–1988.
73. Slatopolsky E, Lopez-Hilker S, Delmez J, Dusso A, Brown A, Martin KJ 1990 The parathyroid-calcitriol axis in health and chronic renal failure. Kidney Int **38**:S41–S47.
74. Beckman MJ, Johnson JA, Goff JP, Reinhardt TA, Beitz DC, Horst RL 1995 The role of dietary calcium in the physiology of vitamin D toxicity: Excess dietary vitamin D_3 blunts parathyroid hormone induction of kidney 1-hydroxylase. Arch Biochem Biophys **319**:535–539.
75. Kumar R 1984 Metabolism of 1,25-dihydroxyvitamin D_3. Physiol Rev **64**:478–504.
76. Henry HL 1992 Vitamin D hydroxylases. J Cell Biochem **49**:4–9.
77. Suda T, DeLuca HF, Schnoes HK, Ponchon G, Tanaka Y, Holick MF 1970 21,25-Dihydroxycholecalciferol: A metabolite of vitamin D_3 preferentially active on bone. Biochemistry **9**:2917–2922.
78. Boyle IT, Mivavet L, Gray RW, Holick MF, DeLuca HF 1972 The response of intestinal calcium transport to 25 hydroxy- and 1,25-dihydroxyvitamin D in nephrectomized rats. Endocrinology **90**:605–608.
79. Omdahl JL, Gray RW, Boyle IT, Knutson JT, DeLuca HF 1972 Regulation of metabolism of 25-hydroxycholecalciferol by kidney tissue in vitro by dietary calcium. Nature New Biol **237**:63–66.
80. Holick MF, Schnoes HK, DeLuca HF, Gray RW, Boyle IT, Suda T 1972 Isolation and identification of 24,25-dihydroxycholecalciferol, a metabolite of vitamin D_3 made in the kidney. Biochemistry **11**:4251–4255.
81. Iwata K, Yamamoto A, Satoh S, Ohyama Y, Tashiro Y, Setoguchi T 1995 Quantitative immunoelectron microscopic analysis of the localization and induction of 25-hydroxyvitamin D_3 24-hydroxylase in rat kidney. J Histochem Cytochem **43**:255–262.
82. Shinki T, Jin CH, Nishimura A, Nagai Y, Ohyama Y, Noshiro M, Okuda K, Suda T 1992 Parathyroid hormone inhibits 25-hydroxyvitamin D_3-24-hydroxylase mRNA expression stimulated by 1α,25-dihydroxyvitamin D_3 in rat kidney but not in intestine. J Biol Chem **267**:13757–13762.
83. Nishimura A, Shinki T, Jin CH, Ohyama Y, Noshiro M, Okuda K, Suda T 1994 Regulation of messenger ribonucleic acid expression of 1α,25-dihydroxyvitamin D_3-24-hydroxylase in rat osteoblasts. Endocrinology **134**:1794–1799.
84. Shigematsu T, Horiuchi N, Ogura Y, Miyahara T, Suda T 1986 Human parathyroid hormone inhibits renal 24-hydroxylase activity of 25-hydroxyvitamin D_3 by a mechanism involving adenosine 3′,5′-monophosphate in rats. Endocrinology **118**:1583–1589.
85. Goff JP, Reinhardt TA, Beckman MJ, Horst RL 1990 Contrasting effects of exogenous 1,25-dihydroxyvitamin D [1,25$(OH)_2$D] versus endogenous 1,25$(OH)_2$D, induced by dietary calcium restriction, on vitamin D receptors. Endocrinology **126**:1031–1035.
86. Iida K, Shinki T, Yamaguchi A, DeLuca HF, Kurokawa K, Suda T 1995 A possible role of vitamin D receptors in regulating vitamin D activation in the kidney. Proc Natl Acad Sci USA **92**:6112–6116.
87. Hanai H, Liang CT, Cheng L, Sacktor B 1989 Desensitization to parathyroid hormone in renal cells from aged rats is associated with alterations in G-protein activity. J Clin Invest **83**:268–277.
88. Hanai H, Brennan DP, Cheng L, Goldman ME, Chorev M, Levine MA, Sacktor B, Liang CT 1990 Down-regulation of parathyroid hormone receptors in renal membranes from aged rats. Am J Physiol **259**:F444–F450.
89. Johnson JA, Beckman MJ, Pansini-Porta A, Christakos S, Bruns ME, Beitz DC, Horst RL, Reinhardt TA 1995 Age and gender effects on 1,25-dihydroxyvitamin D_3-regulated gene expression. Exptl Gerontol **30**:631–643.
90. Matkovits T, Christakos S 1995 Variable *in vivo* regulation of rat vitamin D-dependent genes (osteopontin, Ca,Mg-adenosine triphosphatase, and 25-hydroxyvitamin D_3 24-hydroxylase): Implications for differing mechanisms of regulation and involvement of multiple factors. Endocrinology **136**:3971–3982.
91. Goff JP, Reinhardt TA, Engstrom GW, Horst RL 1992 Effect of dietary calcium or phosphorus restriction and 1,25-dihydroxyvitamin D administration on rat intestinal 24-hydroxylase. Endocrinology **131**:101–104.
92. Beckman MJ, Goff JP, Reinhardt TA, Beitz DC, Horst RL 1994 *In vivo* regulation of rat intestinal 24-hydroxylase: Potential new role of calcitonin. Endocrinology **135**:1951–1955.
93. Ohyama Y, Okuda K 1991 Isolation and characterization of a cytochrome P-450 from rat kidney mitochondria that catalyzes the 24-hydroxylation of 25-hydroxyvitamin D_3. J Biol Chem **266**:8690–8695.
94. Ohyama Y, Hayashi S, Okuda K 1989 Purification of 25-hydroxyvitamin D_3 24- hydroxylase from rat kidney mitochondria. Fed Eur Biol Sci Lett **255**:405–408.
95. Chen K-S, DeLuca HF 1995 Cloning of the human 1α,25-dihy-

droxyvitamin D-3 24-hydroxylase gene promoter and identification of two vitamin D-responsive elements. Biochim Biophys Acta **1263**:1–9.

96. Ohyama Y, Noshiro M, Okuda K 1991 Cloning and expression of cDNA encoding 25-hydroxyvitamin D_3 24-hydroxylase. Fed Eur Biol Sci Lett **278**:195–198.
97. Ohyama Y, Noshiro M, Eggertsen G, Gotoh O, Kato Y, Bjorkhem I, Okuda K 1993 Structural characterization of the gene encoding rat 25-hydroxyvitamin D_3 24-hydroxylase. Biochemistry **32**:76–82.
98. Horst RL, Littledike ET, Gray RW, Napoli JL 1981 Impaired 24,25-dihydroxyvitamin D production in anephric human and pig. J Clin Invest **67**:274–280.
99. Napoli JL, Horst RL 1983 C(24)- and C(23)-oxidation, converging pathways of intestinal 1,25-dihydroxyvitamin D_3 metabolism: Identification of 24-keto-1,23,25-trihydroxyvitamin D_3. Biochemistry **22**:5848–5853.
100. Haussler MR 1986 Vitamin D receptors: Nature and function. Annu Rev Nutr **6**:527–562.
101. Lohnes D, Jones G 1987 Side chain metabolism of vitamin D_3 in osteosarcoma cell line UMR-106. Characterization of products. J Biol Chem **262**:14394–14401.
102. Reddy GS, Tserng KY 1989 Calcitroic acid, end product of renal metabolism of 1,25-dihydroxyvitamin D_3 through C-24 oxidation pathway. Biochemistry **28**:1763–1769.
103. Akiyoshi-Shibata M, Sakaki Y, Ohyama Y, Noshiro M, Okuda K, Yabusaki Y 1994 Further oxidation of hydroxycalcidiol by calcidiol 24-hydroxylase. Eur J Biochem **224**:335–343.
104. Beckman MJ, Tadikonda P, Werner E, Prahl J, Yamada S, DeLuca HF 1996 Human 25-hydroxyvitamin D_3-24-hydroxylase, a multicatalytic enzyme. Biochemistry **35**:8465–8472.
105. Pols HAP, van Leeuwen JPTM, Schilte JP, Visser TJ, Birkenhäger JC 1988 Heterologous up-regulation of the 1,25-dihydroxyvitamin D_3 receptor by parathyroid hormone (PTH) and PTH-like peptide in osteoblast-like cells. Biochem Biophys Res Commun **156**:588–594.
106. Pols HAP, Birkenhager JC, Schilte JP, Visser TJ 1988 Evidence that the self-induced metabolism of 1,25-dihydroxyvitamin D-3 limits the homologous up-regulation of its receptor in rat osteosarcoma cells. Biochim Biophys Acta **970**:122–129.
107. Reinhardt TA, Horst RL 1989 Self-induction of 1,25-dihydroxyvitamin D_3 metabolism limits receptor occupancy and target tissue responsiveness. J Biol Chem **264**:15917–15921.
108. Reinhardt TA, Horst RL 1989 Ketoconazole inhibits self-induced metabolism of 1,25-dihydroxyvitamin D_3 and amplifies 1,25-dihydroxyvitamin D_3 receptor up-regulation in rat osteosarcoma cells. Arch Biochem Biophys **272**:459–465.
109. Reinhardt TA, Horst RL, Engstrom GW, Atkins KB 1988 Ketoconazole potentiates $1,25(OH)_2D_3$-directed up-regulation of $1,25(OH)_2D_3$-receptors in rat intestine and bone. In: Norman AW, Schaefer K, Grigoleit H-G, Herrath Dv (eds) Vitamin D: Molecular, Cellular and Clinical Endocrinology. de Gruyter, Berlin and New York, pp. 234–234.
110. Wichmann JK, Schnoes HK, DeLuca HF 1981 23,24,25-Trihydroxyvitamin D_3, 24,25,26-trihydroxyvitamin D_3, 24-keto-25-hydroxyvitamin D_3, and 23-dehydro-25-hydroxyvitamin D_3: New in vivo metabolites of vitamin D_3. Biochemistry **20**:7385–7391.
111. Wichmann J, Schnoes HK, DeLuca HF 1981 Isolation and identification of 24(*R*)-hydroxyvitamin D_3 from chicks given large doses of vitamin D_3. Biochemistry **20**:2350–2353.
112. DeLuca HF, Schnoes HK 1979 Recent developments in the metabolism of vitamin D_3. In: Norman AW, Schaefer K, Herrath Dv, Grigoleit HG, Coburn JW, DeLuca HF, Mawer EB, Suda T (eds) Vitamin D: Basic Research and Its Clinical Application. de Gruyter, Berlin and New York, pp. 445–458.
113. Engstrom GW, Koszewski NJ 1989 Metabolism of vitamin D_2 in pig liver homogenates: Evidence for a free radical reaction. Arch Biochem Biophys **270**:432–440.
114. Horst RL, Reinhardt TA, Ramberg CF, Koszewski NJ, Napoli JL 1986 24-Hydroxylation of 1,25-dihydroxyergocalciferol: An unambiguous deactivation process. J Biol Chem **261**:9250–9256.
115. Wichmann JK, DeLuca HF, Schnoes HK, Horst RL, Shepard RM, Jorgensen NA 1979 25-Hydroxyvitamin D_3 26,23-lactone: A new in vivo metabolite of vitamin D. Biochemistry **18**:4775–4780.
116. Horst RL 1979 25-OHD_3-26,23-lactone: A metabolite of vitamin D_3 that is 5 times more potent than 25-OHD_3 in the rat plasma competitive protein binding radioassay. Biochem Biophys Res Commun **89**:286–293.
117. Ohnuma N, Norman AW 1982 Production in vitro of 1α,25-dihydroxyvitamin D_3-26,23-lactone from 1α,25-dihydroxyvitamin D_3 by rat small intestinal mucosa homogenates. Arch Biochem Biophys **213**:139–147.
118. Napoli JL, Horst RL 1983 (23*S*)-1,23,25-Trihydroxycholecalciferol, an intestinal metabolite of 1,25-dihydroxycholecalciferol. Biochem J **214**:261–264.
119. Napoli JL, Pramanik BC, Partridge JJ, Uskoković MR, Horst RL 1982 23*S*,25-Dihydroxyvitamin D_3 as a circulating metabolite of vitamin D_3. Its role in 25-hydroxyvitamin D_3-26,23-lactone biosynthesis. J Biol Chem **257**:9634–9639.
120. Horst RL, Littledike ET 1982 Comparison of plasma concentrations of vitamin D and its metabolites in young and aged domestic animals. Comp Biochem Physiol **73B**:485–489.
121. Horst RL, Littledike ET 1980 25-OHD_3-26,23 lactone: Demonstration of kidney-dependent synthesis in the pig and rat. Biochem Biophys Res Commun **93**:149–154.
122. Ishizuka S, Ohba T, Norman AW 1988 1,25-$(OH)_2D_3$-26,23-lactone is a major metabolite of 1,25-$(OH)_2D_3$ under physiological conditions. In: Norman AW, Schaefer K, Grigoleit H-G, Herrath Dv (eds) Vitamin D: Molecular, Cellular and Clinical Endocrinology. de Gruyter, Berlin and New York, pp. 143–144.
123. Ishizuka S, Ishimoto S, Norman AW 1984 Isolation and identification of 1α,25-dihydroxy-24-oxovitamin D_3, 1α,25-dihydroxyvitamin D_3 26,23-lactone, and 1α,24(*S*),25-trihydroxyvitamin D_3: In vivo metabolites of 1α,25-dihydroxyvitamin D_3. Biochemistry **23**:1473–1478.
124. Napoli JL, Horst RL 1982 23,25,26-Trihydroxycholecalciferol. A 25-hydroxycholecalciferol-26,23-lactone precursor. Biochem J **206**:173–176.
125. Yamada S, Nakayama K, Takayama H, Shinki T, Takasaki Y, Suda T 1984 Isolation, identification, and metabolism of (23*S*,25*R*)-25-hydroxyvitamin D_3 26,23-lactol. A biosynthetic precursor of (23*S*,25*R*)-25-hydroxyvitamin D_3 26,23-lactone. J Biol Chem **259**:884–889.
126. Horst RL, Pramanik BC, Reinhardt TA, Shiuey SJ, Partridge JJ, Uskoković MR, Napoli JL 1982 Binding properties of 23*S*,25-dihydroxyvitamin D_3: An in vivo metabolite of vitamin D_3. Biochem Biophys Res Commun **106**:1006–1011.
127. Horst RL, Wovkulich PM, Baggiolini EG, Uskoković MR, Engstrom GW, Napoli JL 1984 (23*S*)-1,23,25-Trihydroxyvitamin D_3: Its biologic activity and role in 1,25-dihydroxyvitamin D_3 26,23-lactone biosynthesis. Biochemistry **23**:3973–3979.
128. Esvelt RP, Schnoes HK, DeLuca HF 1979 Isolation and characterization of 1α-hydroxy-23-carboxytetranorvitamin D: A major metabolite of 1,25-dihydroxyvitamin D_3. Biochemistry **18**:3977–3983.
129. Horst RL, Reinhardt TA, Pramanik BC, Napoli JL 1983 23-Keto-25-hydroxyvitamin D_3: A vitamin D_3 metabolite with high

affinity for the 1,25-dihydroxyvitamin D specific cytosol receptor. Biochemistry **22**:245–250.

130. Horst RL, Reinhardt TA, Napoli JL 1982 23-Keto-25-hydroxyvitamin D_3 and 23-keto-1,25-dihydroxyvitamin D_3: Two new vitamin D_3 metabolites with high affinity for the 1,25-dihydroxyvitamin D_3 receptor. Biochem Biophys Res Commun **107**:1319–1325.
131. Ohnuma N, Kruse JR, Popjak G, Norman AW 1982 Isolation and chemical characterization of two new vitamin D metabolites produced by the intestine: 1,25-Dihydroxy-23-oxo-vitamin D_3 and 1,25,26-trihydroxy-23-oxo-vitamin D_3. J Biol Chem **257**: 5097–5102.
132. Mayer E, Ohnuma N, Norman AW 1982 Isolation, chemical characterization and biological activity of 1,25-dihydroxy-23-oxo-vitamin D_3 and 1,25,26-trihydroxy-23-oxo-vitamin D_3, two metabolites of vitamin D. Curr Adv Skeletogen **589**:173–178.
133. Napoli JL, Pramanik BC, Royal PM, Reinhardt TA, Horst RL 1983 Intestinal synthesis of 24-keto-1,25-dihydroxyvitamin D_3. A metabolite formed *in vivo* with high affinity for the vitamin D cytosolic receptor. J Biol Chem **258**:9100–9107.
134. Suda T, DeLuca HF, Schnoes HK, Tanaka Y, Holick MF 1970 25,26-Dihydroxycholecalciferol, a metabolite of vitamin D_3 with intestinal calcium transport activity. Biochemistry **9**:4776–4780.
135. Reinhardt TA, Napoli JL, Beitz DC, Littledike ET, Horst RL 1981 A new in vivo metabolite of vitamin D_3: 1,25,26-Trihydroxyvitamin D_3. Biochem Biophys Res Commun **99**:302–307.
136. Reinhardt TA, Napoli JL, Pramanik B, Littledike ET, Beitz DC, Partridge JJ, Usković MR, Horst RL 1981 1α,25,26-Trihydroxyvitamin D_3: An in vivo and in vitro metabolite of vitamin D_3. Biochemistry **20**:6230–6235.
137. Partridge JJ, Shiuey SJ, Chadha HK, Baggiolini ET, Bolount JE, Usković MR 1981 Synthesis and structure proof of a vitamin D_3 metabolite, 25(*S*),26-dihydroxycholecalciferol. J Am Chem Soc **103**:1253–1255.
138. Ikekawa N, Koizumi N, Ohshima E, Ishizuka S, Takeshita T, Tanaka Y, DeLuca HF 1983 Natural 25,26-dihydroxyvitamin D_3 is an epimeric mixture. Proc Natl Acad Sci USA **80**:5286–5288.
139. Hollis BW, Roos BA, Lambert PW 1980 25,26-Dihydroxycholecalciferol: A precursor in the renal synthesis of 25-hydroxycholecalciferol-26,23-lactone. Biochem Biophys Res Commun **95**: 520–528.
140. Napoli JL, Horst RL 1981 25,26-Dihydroxyvitamin D_3 is not a major intermediate in 25-hydroxyvitamin D_3-26,23-lactone formation. Arch Biochem Biophys **212**:754–758.
141. Tanaka Y, DeLuca HF, Schnoes HK, Ikekawa N, Eguchi T 1981 23,25-Dihydroxyvitamin D_3: A natural precursor in the biosynthesis of 25-hydroxyvitamin D_3-26,23-lactone. Proc Natl Acad Sci USA **78**:4805–4808.
142. Horst RL, Shepard RM, Jorgensen NA, DeLuca HF 1979 The determination of 24,25-dihydroxyvitamin D and 25,26-dihydroxyvitamin D in plasma from normal and nephrectomized man. J Lab Clin Med **93**:277–285.
143. Shepard RM, Horst RL, Hamstra AJ, DeLuca HF 1979 Determination of vitamin D and its metabolites in plasma from normal and anephric man. Biochem J **182**:55–69.
144. Napoli JL, Okita RT, Masters BS, Horst RL 1981 Identification of 25,26-dihydroxyvitamin D_3 as a rat renal 25-hydroxyvitamin D_3 metabolite. Biochemistry **20**:5865–5871.
145. Bergman T, Postlind H 1991 Characterization of mitochondrial cytochromes P-450 from pig kidney and liver catalyzing 26-hydroxylation of 25-hydroxyvitamin D_3 C_{27} steroids. Biochem J **276**:427–432.
146. Tanaka Y, Schnoes HK, Smith CM, DeLuca HF 1981 1,25,26-Trihydroxyvitamin D_3 isolation, identification and biological activity. Arch Biochem Biophys **210**:104–109.
147. Williams DH, Morris DS, Gilhooly MA, Norris AF 1982 Synthesis of 25(R)-hydroxycholecalciferol-26,23(*S*)-lactone and of 25ξ,26-dihydroxyergocalciferol. In: Norman AW, Schaefer K, Herrath Dv, Grigoleit H-G (eds) Vitamin D: Chemical, Biochemical and Clinical Endocrinology of Calcium Metabolism. de Gruyter, Berlin and New York, pp. 1067–1072.
148. Mazur Y, Segev D, Jones G 1982 Synthesis and determination of absolute configuration of kidney metabolites of vitamin D_2. In: Norman AW, Schaefer K, Herrath Dv, Grigoleit H-G (eds) Vitamin D: Chemical, Biochemical and Clinical Endocrinology of Calcium Metabolism. de Gruyter, Berlin and New York, pp. 1101–1106.
149. Hay AWM, Jones G 1979 The elution profile of vitamin D_2 metabolites from Sephadex LH-20 columns. Clin Chem **25**:473–475.
150. Reddy GS, Tserng KY 1990 24,25,28-Trihydroxyvitamin D_2 and 24,25,26-trihydroxyvitamin D_2: Novel metabolites of vitamin D_2. Biochemistry **29**:943–949.
151. Reddy GS, Tserng KY 1986 Isolation and identification of 1,24,25-trihydroxyvitamin D_2, 1,24,25,28-tetrahydroxyvitamin D_2, and 1,24,25,26-tetrahydroxyvitamin D_2: New metabolites of 1,25-dihydroxyvitamin D_2 produced in rat kidney. Biochemistry **25**:5328–5336.
152. Jones G, Byford V, Makin H, Kremer R, Rice R, deGraffenried LA, Knutson JC, Bishop CW 1996 Anti-proliferative activity and target cell catabolism of the vitamin D analog 1α,24(*S*)-$(OH)_2D_2$ in normal and immortalized human epidermal cells. Biochem Pharmacol **52**:133–140.
153. Kreitmeir H, Moll T 1928 Hypervitaminose durch grosse dosen vitamin D. Munch Med Wochenschr **75**:637–639.
154. Putscher W 1929 Über Vigantolschädigung der Niere einum Kinde. Z Kinderheilkd **48**:269–273.
155. Holick MF, MacLaughlin JA, Doppelt SH 1981 Regulation of cutaneous provitamin D_3 photosynthesis in man: Skin pigment is not an essential regulator. Science **211**:590–593.
156. Clark MB, Potts JT 1977 25-Hydroxyvitamin D_3 regulation. Calcif Tissue Res **22**(Suppl. 1):29–34.
157. Shepard RM, DeLuca HF 1980 Plasma concentrations of vitamin D_3 and its metabolites in the rat as influenced by vitamin D_3 or 25-hydroxyvitamin D_3 intakes. Arch Biochem Biophys **202**:43–53.
158. Hughes MR, Baylink DJ, Jones PG, Haussler MR 1976 Radioligand receptor assay for 25-hydroxyvitamin D_2/D_3 and 1α,25-dihydroxyvitamin D_2/D_3: Application to hypervitaminosis D. J Clin Invest **58**:61–70.
159. Counts SJ, Baylink DJ, Shen FH, Sherrard DJ, Hickman RO 1975 Vitamin D intoxication in an anephric child. Ann Intern Med **82**:196–200.
160. Reinhardt TA, Conrad HR 1980 Mode of action of pharmacological doses of cholecalciferol during parturient hypocalcemia in dairy cows. J Nutr **110**:1589–1594.
161. Littledike ET, Horst RL 1982 Vitamin D_3 toxicity in dairy cows. J Dairy Sci **65**:749–759.
162. Horst RL, Reinhardt TA 1983 Vitamin D metabolism in ruminants and its relevance to the periparturient cow. J Dairy Sci **66**:661–678.
163. Hunt RD, Garcia FG, Walsh RJ 1972 A comparison of the toxicity of ergocalciferol and cholecalciferol in Rhesus monkeys (*Macaca mulatta*). J Nutr **102**:975–986.
164. Harrington DD, Page EH 1983 Acute vitamin D_3 toxicosis in horses: Case reports and experimental studies of the comparative toxicity of vitamins D_2 and D_3. J Am Vet Med Assoc **182**:1358–1369.
165. Hodson EM, Evans RA, Dunstan CR, Hills E, Wong SYP, Rosenberg AR, Roy LP 1985 Treatment of childhood renal osteo-

dystrophy with calcitriol or ergocalciferol. Clin Nephrol **24**:192–200.
166. Sjoden G, Smith C, Lingren U, DeLuca HF 1985 1-α-Hydroxyvitamin D_2 is less toxic than 1-α-hydroxyvitamin D_3 in the rat. Proc Soc Exp Biol Med **178**:432–436.
167. Beckman MJ, Horst RL, Reinhardt TA, Beitz DC 1990 Up-regulation of the intestinal 1,25-dihydroxyvitamin D receptor during hypervitaminosis D: A comparison between vitamin D_2 and vitamin D_3. Biochem Biophys Res Commun **169**:910–915.
168. Mortensen JT, Brinck P, Binderup L 1993 Toxicity of vitamin D analogues in rats fed diets with standard or low calcium contents. Pharmacol Toxicol **72**:124–127.
169. Hines TG, Jacobson NL, Beitz DC, Littledike ET 1985 Dietary calcium and vitamin D: Risk factors in the development of atherosclerosis in young goats. J Nutr **115**:167–178.
170. Payne JM, Manston R 1967 The safety of massive doses of vitamin D_3 in the prevention of milk fever. Vet Rec **81**:214–216.
171. Hibbs JW, Pounden WD 1955 Studies on milk fever in dairy cows. IV. Prevention by short-time, prepartum feeding of massive doses of vitamin D. J Dairy Sci **38**:65–72.
172. Napoli JL, Sommerfeldt JL, Pramanik BC, Gardner R, Sherry AD, Partridge JJ, Uskoković MR, Horst RL 1983 19-Nor-10-keto-vitamin D derivatives: Unique metabolites of vitamin D_3, vitamin D_2, and 25-hydroxyvitamin D_3. Biochemistry **22**:3636–3640.
173. Recker RR, Schoenfeld JL, Slatopolsky R, Goldsmith R, Brickman A 1979 25-Hydroxyvitamin D (calcidiol) on renal osteodystophy: Long term results of a six center trial. In: Norman AW, Schaefer K, Herrath Dv, Grigolet HG, Coburn JW, DeLuca HF, Mawer EB, Suda T (eds) Vitamin D: Basic Research and Its Clinical Application. de Gruyter, Berlin and New York, pp. 869–875.
174. West TET, Joffe M, Sinclair L, O'Riordan JLH 1971 Treatment of hypercalcemia with calcitonin. Lancet I **701**:675–678.
175. Ulbrych-Jabloonska A 1972 The hypocalcemic effect of glucagon in cases of hypercalcemia. Helv Paediatr Acta **27**:613–619.
176. Kingma JG, Roy PE 1990 Effect of ethane-1-hydroxy-1,1-diphosphonate on arterial calcinosis induced by hypervitaminosis D: A morphologic investigation. J Exp Pathol (Oxford) **71**:145–153.
177. Streck WF, Waterhouse C, Haddad JG 1979 Glucocorticoid effects in vitamin D intoxication. Arch Intern Med **139**:974–977.
178. Buffenstein R, Skinner DC, Yahav S, Moodley GP, Cavaleros M, Zachen D, Ross FP, Pettifor JM 1991 Effect of oral cholecalciferol supplementation at physiological and supraphysiological doses in naturally vitamin D_3-deficient subterranean damara mole rats (*Cryptomys damarensis*). J Endocrinol **131**:197–202.
179. Cheeke PR, Amberg JW 1973 Comparative calcium excretion by rats and rabbits. J Anim Sci **37**:450–454.
180. Schryver HF, Hintz HF, Lowe JE 1974 Calcium and phosphorus in the nutrition of the horse. Cornell Vet **64**:493–515.
181. Bourdeau JE, Schwer-Dymerski DA, Stern PH, Langman CB 1986 Calcium and phosphorus metabolism in chronically vitamin D-deficient laboratory rabbits. Miner Electrolyte Metab **12**:176–185.
182. Sommerfeldt JL, Horst RL, Littledike ET, Beitz DC 1979 In vitro degradation of cholecalciferol in rumen fluid. J Dairy Sci **62**:192–193.
183. Sommerfeldt JL, Napoli JL, Littledike ET, Beitz DC, Horst RL 1983 Metabolism of orally administered [^{3}H]ergocalciferol and [^{3}H]cholecalciferol by dairy calves. J Nutr **113**:2595–2600.
184. Stern PH, Horst RL, Gardner R, Napoli JL 1985 10-Keto or 25-hydroxy substitution confer equivalent in vitro bone-resorbing activity to vitamin D_3. Arch Biochem Biophys **236**:555–558.
185. Hunt RD, Garcia FG, Hegsted DM, Kaplinsky N 1967 Vitamin D_2 and vitamin D_3 in New World primates: Influence on calcium absorptions. Science **158**:943–947.
186. Steenbock H, Kletzien SWF, Haplin JG 1932 Reaction of chicken to irradiated ergosterol and irradiated yeast as contrasted with natural vitamin D of fish liver oils. J Biol Chem **97**:249–264.
187. Horst RL, Napoli JL, Littledike ET 1982 Discrimination in the metabolism of orally dosed ergocalciferol and cholecalciferol by the pig, rat and chick. Biochem J **204**:185–189.
188. Sebert JL, Garabedian M, deChasteigner C, Defrance D 1991 Comparative effects of equal doses of vitamin D_2 and vitamin D_3 for the correction of vitamin D deficiency in the elderly. In: Norman AW, Bouillon R, Thomasset M (eds) Vitamin D: Gene Regulation, Structure-Function Analysis and Clinical Application. de Gruyter, Berlin and New York, pp. 765–766.
189. Reddy GS, Ray R, Holick MF 1990 Serum 1,25-dihydroxyvitamin D_2 levels are two times higher than 1,25-dihydroxyvitamin D_3 levels in vitamin D-deficient rats, dosed acutely with equal amounts of vitamin D_2 and vitamin D_3. J Bone Miner Res **5**(Suppl 2):S265.
190. Haddad JG, Walgate J 1976 25-Hydroxyvitamin D transport in human plasma. J Biol Chem **251**:4803–4809.
191. Baird DT, Horton R, Longcope C, Tait JF 1969 Steroid hormone dynamics under steady state conditions. Recent Prog Horm Res **25**:611–664.
192. Hoy DA, Ramberg CF, Jr, Horst RL 1988 Evidence that discrimination against ergocalciferol by the chick is the result of enhanced metabolic clearance rates for its mono- and di-hydroxylated metabolites. J Nutr **118**:633–638.
193. Hay AWM, Watson G 1977 Vitamin D_2 in vertebrate evolution. Comp Biochem Physiol **56B**:375–380.
194. Horst RL, Reinhardt TA, Beitz DC, Littledike ET 1981 A sensitive competitive protein binding assay for vitamin D in plasma. Steroids **37**:581–592.
195. Reddy GS, Napoli JL, Hollis BW 1984 Preferential 25-hydroxylation of vitamin D_2 vs vitamin D_3 by rat liver is dependent upon vitamin D-binding protein. Calcif Tissue Intl **36**(4):524–529.
196. Arnaud SB, Young DR, Cann C, Reinhardt TA, Henrickson R 1985 Is hypervitaminosis D normal in the rhesus monkey? In: Norman AW, Schaefer K, Grigoleit H-G, Herrath Dv (eds) Vitamin D. Chemical, Biochemical and Clinical Update. de Gruyter, Berlin and New York, pp. 585–586.
197. Shinki T, Shiina Y, Takahashi N, Tanioka Y, Koizumi H, Suda T 1983 Extremely high circulating levels of 1α,25-dihydroxyvitamin D_3 in the marmoset, a New World monkey. Biochem Biophys Res Commun **114**:452–457.
198. O'Loughlin PD, Morris HA 1991 Duodenal transport of calcium in the marmoset. In: Norman AW, Bouillon R, Thomasset M (eds) Vitamin D: Gene Regulation, Structure–Function Analysis and Clinical Application. de Gruyter, Berlin and New York, pp. 345–346.
199. Shore RF, Hayes ME, Balment RJ, Mawer EB 1988 Serum $25(OH)D_3$ and $1,25(OH)_2D_3$ levels in wild and laboratory-bred wood mice and bank voles. In: Norman AW, Schaefer K, Grigoleit H-G, Herrath Dv (eds) Vitamin D. Molecular, Cellular and Clinical Endocrinology. de Gruyter, Berlin and New York, pp. 633–634.
200. Kobayashi T, Takeuchi A, Okano T 1991 An evolutionary aspect in vertebrates from the viewpoint of vitamin D_3 metabolism. In: Norman AW, Bouillon R, Thomasset M (eds) Vitamin D: Gene Regulation, Structure–Function Analysis and Clinical Application. de Gruyter, Berlin and New York, pp. 679–680.
201. Glowacki J, Deftos LJ, Mayer E, Norman AW, Henry H 1982 Chondrichthyes cannot resorb implanted bone and have calcium-regulating hormones. In: Norman AW, Schaefer K, Herrath Dv, Grigoleit H-G (eds) Vitamin D: Chemical, Biochemical and Clini-

cal Endocrinology of Calcium Metabolism. de Gruyter, Berlin and New York, pp. 613–615.

202. Sundell K, Bishop JE, Björnsson BT, Horst RL, Hollis BW, Norman AW 1991 Organ distribution of a high affinity receptor for $1,25(OH)_2$ vitamin D_3 in the Atlantic cod (*Gadus morhua*), a marine teleost. In: Norman AW, Bouillon R, Thomasset M (eds) Vitamin D. Gene Regulation, Structure–Function Analysis and Clinical Application. de Gruyter, Berlin and New York, pp. 114–115.

203. Horst RL, Koszewski NJ, Reinhardt TA 1988 Species variation of vitamin D metabolism and action: Lesson to be learned from farm animals. In: Norman AW, Schaefer K, Grigoleit H-G, Herrath Dv (eds) Vitamin D: Molecular, Cellular and Clinical Endocrinology. de Gruyter, Berlin and New York, pp. 93–101.

CHAPTER 3

Photobiology of Vitamin D

MICHAEL F. HOLICK Vitamin D, Skin, and Bone Research Laboratory; Department of Medicine; Endocrinology, Nutrition and Diabetes Section; Boston Medical Center and Boston University School of Medicine, Boston, Massachusetts

I. INTRODUCTION

Vitamin D is neither a vitamin nor a hormone when adequate exposure to sunlight is available to promote the synthesis of vitamin D in the skin [1]. First produced in ocean-dwelling phytoplankton and zooplankton, vitamin D has been made by life forms on earth for almost 1 billion years [2]. Although the physiological function of vitamin D in these lower life forms is uncertain, it is well recognized that the photosynthesis of vitamin D became critically important for land-dwelling vertebrates that required a mechanism to increase the efficiency of absorption of dietary calcium. The major physiological function of vitamin D in vertebrates is to maintain extracellular fluid concentrations of calcium and phosphorus within a normal range. Vitamin D accomplishes this by increasing the efficiency of the small intestine to absorb dietary calcium and phosphorus, and by stimulating the mobilization of calcium and phosphorus stores from the bone [3].

II. HISTORICAL PERSPECTIVE

It was the lack of appreciation of the beneficial effect of sunlight in preventing rickets that caused the widespread endemic outbreak of this bone deforming disease in the seventeenth through nineteenth centuries. As early as 1822, Sniadecki [4] had suggested that the high incidence of rickets that occurred in Warsaw, Poland, in the early 1800s was likely caused by the lack of adequate exposure to sunlight. This was based on his observations that whereas children who lived in Warsaw had a very high incidence of rickets, children who lived in the rural areas outside of the city did not suffer the same malady. Seventy years later Palm [5] also recognized that lack of sunlight was a common denominator associated with the high incidence of rickets in children living in the inner cities when compared to children living in underdeveloped countries. Both Sniadecki [4] and Palm [5] encouraged systematic sunbathing as a means of preventing and curing rickets, but their recommendations went unheeded.

Huldschinsky [6] was the first to clearly demonstrate that it was exposure of the skin to ultraviolet radiation that was responsible for the antirachitic activity of sunlight that had been postulated by Sniadecki [4] and Palm [5]. He exposed four rachitic children to radiation from a mercury arc lamp and reported dramatic reversal of rickets within 4 months. He further reasoned that the radiation responsible for producing the antirachitic factor was the same radiation that caused melanization of the skin. The intimate relationship between sunlight and its antirachitic properties in humans was finally demonstrated in 1921 when Hess and Unger [7] took rachitic children and exposed them to sunlight on the roof of a New York City hospital and demonstrated that within 4 months the rachitic lesions had healed.

The concept that rickets was caused by a nutritional deficiency was supported by the observation in 1827 by Bretonneau who treated a 15-month-old child with acute rickets with cod liver oil and noted the incredible speed at which the patient was cured [8]. His student Trousseau used liver oils from a variety of fish and aquatic mammals including herring, whales, and seals for the treatment of rickets and osteomalacia. In the early twentieth century, Mellanby [9] conducted classic studies where he demonstrated that rachitic beagles which were fed cod liver oil were cured of their bone-deforming disease. Originally, it was thought that the active antirachitic factor in cod liver oil was vitamin A. However, McCollum *et al.* [10] heated cod liver oil to destroy the vitamin A activity and demonstrated that the preparation maintained its antirachitic activity. He, therefore, coined the term "vitamin D" for the antirachitic factor.

The appreciation that exposure to sunlight could cure rickets in children prompted Steenbock and Black [11] and Hess and Weinstock [12] in 1924 to expose a variety of foods and other substances including vegetable oils, animal feed, lettuce, wheat, and grasses to ultraviolet radiation. They found that this process imparted antirachitic activity to all of the substances. These observations led Steenbock [13] to conclude that there would be great utility in irradiation of food substances for the prevention and cure of rickets in children. This concept initially led to the addition of provitamin D (ergosterol or 7-dehydrocholesterol) to milk and its subsequent irradiation to impart antirachitic activity. Once vitamin D (D without a subscript refers to either D_2 or D_3) was inexpensive to produce, vitamin D was directly added to milk. This simple concept led to the eradication of rickets as a significant health problem in the United States and other countries that followed this practice.

III. PHOTOBIOLOGY OF VITAMIN D

A. Photosynthesis of Vitamin D_3

When human skin is exposed to sunlight, it is the solar ultraviolet B photons with energies between 290 and 315 nm that are responsible for causing the photolysis of 7-dehydrocholesterol (provitamin D_3; the immediate precursor in the cholesterol biosynthetic pathway) to previtamin D_3 [1,14] (Fig. 1). This photochemical process occurs in the plasma membrane of skin cells; as a result, the thermodynamically unstable cis,cis isomer of previtamin D_3 is rapidly transformed by a rearrangement of double bonds to form vitamin D_3 [15] (Fig. 2). Approximately 50% of previtamin D_3 is converted to vitamin D_3 within 2 hr. As vitamin D_3 is formed in the membrane, its open flexible structure is likely jettisoned from the plasma membrane into the extracellular space. Once vitamin D_3 enters the extracellular fluid space, it is attracted to the vitamin D binding pro-

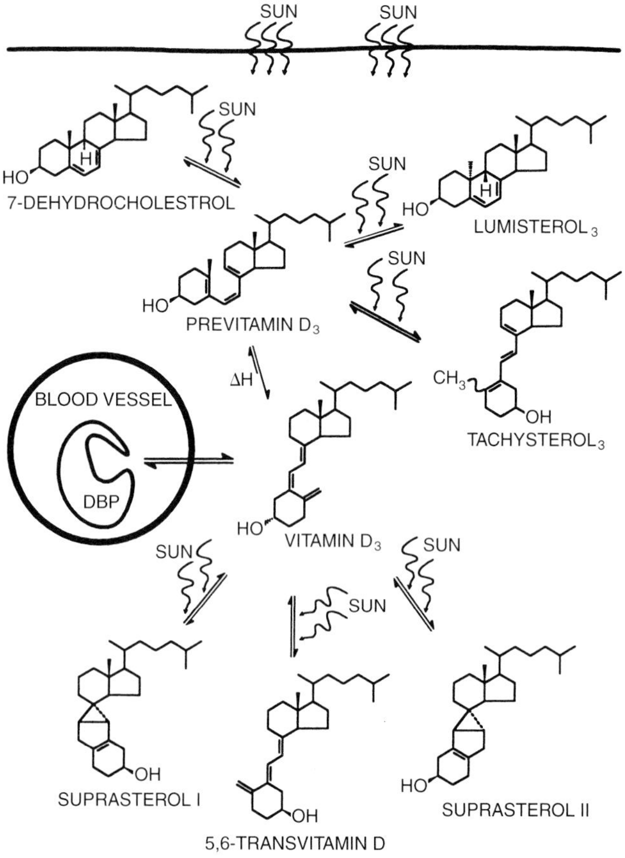

FIGURE 1 Photochemical events that lead to the production and regulation of vitamin D_3 in the skin. Reprinted with permission from Holick [1]. DBP is the plasma vitamin D binding protein.

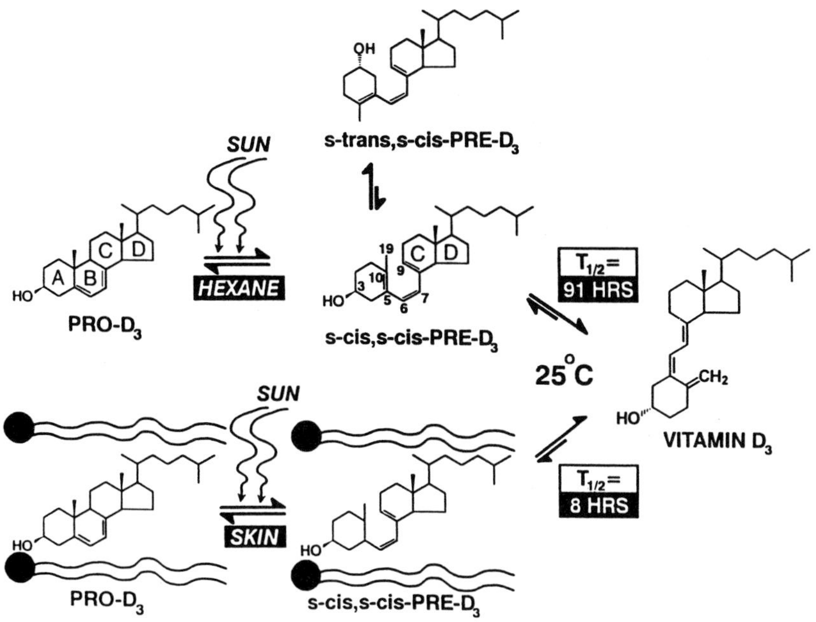

FIGURE 2 Photolysis of provitamin D_3 (pro-D_3) into previtamin D_3 (pre-D_3) and its thermal isomerization to vitamin D_3 in hexane and in lizard skin. In hexane, pro-D_3 is photolyzed to s-*cis*,s-*cis*-pre-D_3. Once formed, this energetically unstable conformation undergoes a conformational change to the s-*trans*,s-*cis*-pre-D_3. Only the s-*cis*,s-*cis*-pre-D_3 can undergo thermal isomerization to vitamin D_3. The s-cis,s-cis conformer of pre-D_3 is stabilized in the phospholipid bilayer by hydrophilic interactions between the 3β-hydroxyl group and the polar head of the lipids, as well as by van der Waals interactions between the steroid ring and side chain structure and the hydrophobic tail of the lipids. These interactions significantly decrease the conversion of the s-cis,s-cis conformer to the s-trans,s-cis conformer, thereby facilitating the thermal isomerization of s-cis,s-cis-pre-D_3 to vitamin D_3. Reprinted with permission from Holick *et al.* [15].

tein (DBP) in the circulation, and thus enters the dermal capillary bed.

B. Regulation of Vitamin D Synthesis by Sunlight

In the late 1960s, a theory was popularized that human skin pigmentation evolved to play a critically important role in regulating the cutaneous photosynthesis of vitamin D_3. Loomis [16] speculated that melanin pigmentation in humans evolved for the purpose of preventing excessive production of vitamin D_3 in the skin due to chronic excessive exposure to sunlight especially in peoples who live near the equator. Although it is likely that increased pigmentation was important for the prevention of skin cancer, there is little evidence that skin pigmentation evolved to prevent vitamin D intoxication. The reason for this is that once previtamin D_3 is formed in the skin, it can either isomerize to vitamin D_3 or can absorb solar ultraviolet B radiation and isomerize into biologically inactive photoisomers lumisterol and tachysterol [17] (Fig. 1). Furthermore, when vitamin D_3 is made, it can either enter into the circulation or absorb solar ultraviolet photons and isomerize to at least three photoproducts including suprasterol I, suprasterol II, and 5,6-trans-vitamin D_3 [18] (Fig. 1). Thus, sunlight itself can regulate the total output of vitamin D_3 in the skin by causing the photodegradation of previtamin D_3 and vitamin D_3. This is the likely explanation for why there are no reported cases of vitamin D intoxication-induced hypercalcemia from chronic excessive exposure to sunlight.

Melanin, on the other hand, provides the body with an effective natural sunscreen. Thus, increased melanin pigmentation will reduce the efficiency of the sun-mediated photosynthesis of previtamin D_3 [19]. This is particularly important in Blacks who live in northern latitudes and who ingest very little dietary vitamin D. It is the likely explanation for why Blacks have lower circulating concentrations of 25-hydroxyvitamin D_3 (25OHD) and are more prone to developing vitamin D deficiency [20] (see Chapter 33).

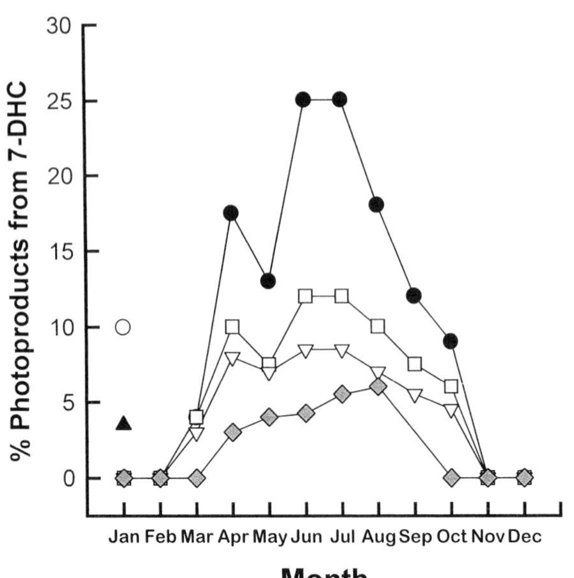

FIGURE 3 Photosynthesis of previtamin D_3 after exposure of 7-dehydrocholesterol (7-DHC) to sunlight. Measurements were as follows: in Boston (42°N) after 1 hr (■) and 3 hr (●); in Edmonton, Canada (52°N), after 1 hr (▽); in Los Angeles (34°N) (▲) and Puerto Rico (18°N) in January (○). Reprinted with permission from Webb *et al.* [22].

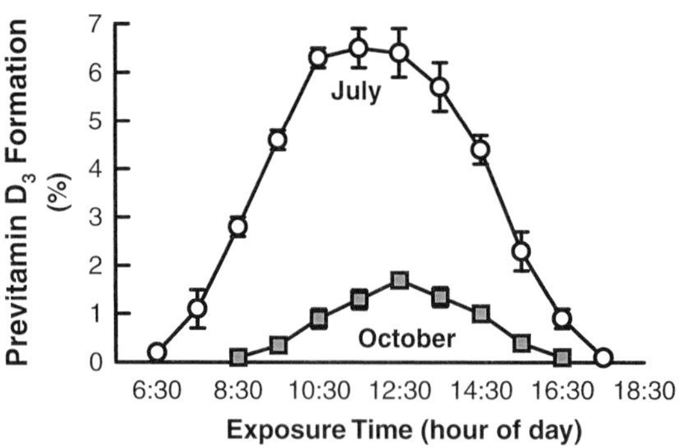

FIGURE 4 Effect of season and time of day on previtamin D_3 formation. Photosynthesis of previtamin D_3 was measured at various times on cloudless days in Boston in October (■) and July (○). Reprinted with permission [Lu Z, Chen TC, Holick MF 1992 Influence of season and time of day on the synthesis of vitamin D_3. In: Holick MF, Kligman A (eds) Proceedings of the Biologic Effects of Light Symposium. de Gruyter, Berlin, pp. 53–56].

C. Influence of Latitude, Season, and Time of Day on Vitamin D Synthesis

It was recognized at the beginning of the twentieth century that the incidence of rickets was much higher during the winter and early spring months and was less prevalent during the summer and autumn [21]. An evaluation of the effect of season on the cutaneous production of vitamin D_3 in Boston revealed that during the summer months of June and July, 7-dehydrocholesterol was most efficiently converted to previtamin D_3 [22] (Fig. 3). There was a gradual decline in the production of previtamin D_3 after August, and there was essentially no previtamin D_3 formed in human skin after November (Fig. 3). Previtamin D_3 photosynthesis commenced in the middle of March. To evaluate the influence of latitude on the cutaneous production of vitamin D_3, similar studies were conducted in Edmonton, Canada (52°N), Los Angeles (34°N), and San Juan, Puerto Rico (18°N). In Edmonton, the photosynthesis of previtamin D_3 essentially ceased by mid-October and did not resume until middle April. However, in Los Angeles and San Juan, the production of previtamin D_3 in the skin occurred throughout the year (Fig. 3).

The cutaneous production of vitamin D_3 was also evaluated in Boston every hour from the time the sun rose to the time the sun set in the middle of the month on cloudless days for a full year. It was found that during the summer, sunlight was capable of producing vitamin D_3 in the skin from 0700 to as late as 1700 hr Eastern Standard Time (EST) (Fig. 4). However, as the zenith angle of the sun increased in the spring and autumn, previtamin D_3 photosynthesis in the skin began at approximately 0900 and ceased at approximately 1600 EST [1].

D. Influence of Sunscreen Use, Melanin, Clothing, Glass, and Plastics on Vitamin D Synthesis

Any substance such as melanin, clothing, or a sunscreen that absorbs ultraviolet B radiation will reduce the cutaneous production of vitamin D_3 [1,23–25]. Sunscreen use is now widely accepted as an effective method for diminishing the damaging effects due to chronic excessive exposure to sunlight. Sunscreens are of great benefit because they can prevent sunburn, skin cancer, and skin damage associated with exposure to sunlight. The solar radiation that is most responsible for causing damage to the skin is the ultraviolet B (UVB) radiation. Because the major function of sunscreens is to absorb solar UVB radiation on the surface of the skin before it penetrates into the deeper viable layers, sunscreen use diminishes the total number of UVB photons that can reach the 7-dehydrocholesterol stores in the viable epidermis to form previtamin D_3. Topical application of a sunscreen will substantially diminish or completely prevent the cutaneous production of previtamin D_3 [25]. The topical application of a sunscreen with a sun protection factor (SPF) of 8, followed by whole-body exposure

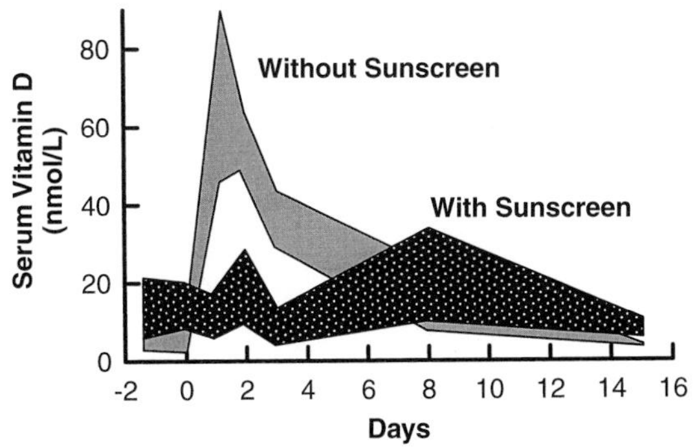

FIGURE 5 Effect of sunscreen on vitamin D status. Circulating concentrations of vitamin D were measured in young adults after application of a cream either with a sun protection factor of 8 or without sunscreen (topical placebo cream) following a single exposure to one minimal erythemal dose of simulated sunlight. Reprinted with permission from Matsuoka *et al.* [23].

to one minimal erythemal dose of simulated sunlight, prevented any significant increase in the blood levels of vitamin D_3 in healthy young adult volunteers. On the other hand, without the sunscreen, there was a marked 10- to 20-fold increase in circulating concentrations of vitamin D_3 [23] (Fig. 5). Similarly, since clothing absorbs ultraviolet B radiation, the wearing of clothing on body surfaces prevents the cutaneous production of vitamin D_3 in those covered surfaces [24] (Fig. 6). Thus, in cultures that require the covering of almost the entire body with clothing, such as Bedouins living in the Negev Desert, despite the intense sunlight environment, the Bedouin women have an increased risk of vitamin D deficiency and osteomalacia and their children are more prone to developing vitamin D deficiency rickets [26].

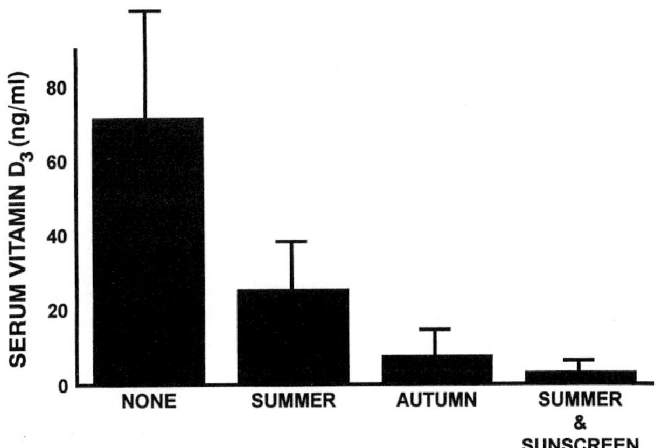

FIGURE 6 Effect of clothing and sunscreen on vitamin D status. Circulating concentrations of vitamin D were measured in human subjects who wore either no clothing, summer-type clothing, autumn-type clothing, or summer-type clothing and sunscreen 24 hr after a whole-body exposure to one minimal erythemal dose of ultraviolet B radiation. Reprinted with permission from Matsuoka *et al.* [24].

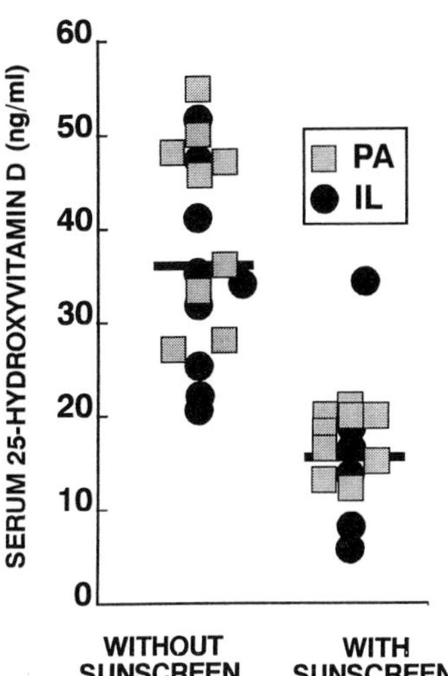

FIGURE 7 Effect of chronic sunscreen use on vitamin D status. Circulating concentrations of 25-hydroxyvitamin D were measured in adults from Pennsylvania (PA) and Illinois (IL) who always wore a sunscreen or never wore a sunscreen. Reprinted with permission from Matsuoka *et al.* [25].

Although sunscreens and clothing significantly diminish the cutaneous production of vitamin D, for children and young adults, there is little concern about their developing vitamin D deficiency by practicing sun protection. The main reason is that the casual every day limited exposure of sunlight to the face and hands is sufficient to provide the vitamin D requirement. However, elderly people, who are often very concerned about their health and appearance, will consistently apply a topical sunscreen on all sun-exposed areas and/or wear clothing over most sun-exposed areas before going outdoors. Since aging significantly decreases the capacity of the skin to produce vitamin D_3 because of the marked decline in 7-dehydrocholesterol in the epidermis, chronic sunscreen use by elderly people can increase the risk of vitamin D deficiency [1,23] (Fig. 7).

Because glass and Plexiglas efficiently absorb most if not all ultraviolet B photons, exposure of the skin to sunlight through glass or Plexiglas will not produce any vitamin D_3 (Fig. 8).

IV. ROLE OF SUNLIGHT AND DIETARY VITAMIN D FOR BONE HEALTH

There is ample evidence that, at latitudes where vitamin D_3 synthesis is diminished or absent during the winter months, there is a seasonal variation in the concentrations of the major circulating form of vitamin D, $25OHD_3$ [27–30]. However, children and young adults

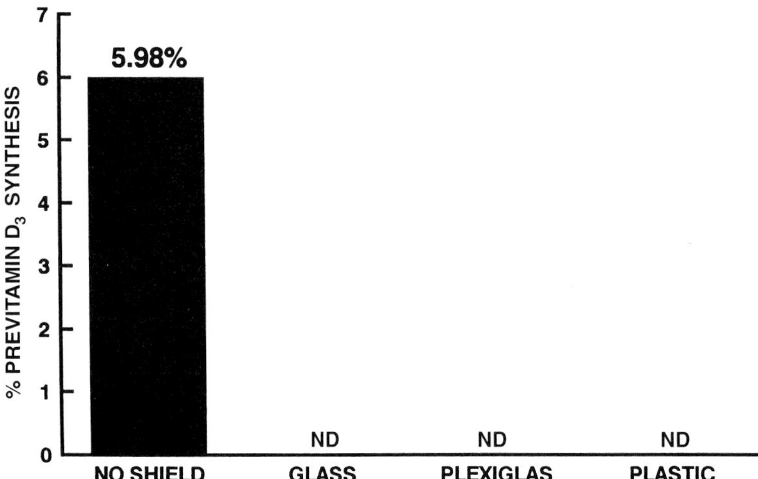

FIGURE 8 Prevention of previtamin D_3 formation by UV radiation by common light-shielding materials. The effects of glass, plastic, or Plexiglas (DuPont Chemical Company, Memphis, TN) placed between the simulated sunlight source and the provitamin D_3 were measured. ND, Not detectable. Reprinted with permission from Holick [1].

who received an ample supply of sunlight during the spring, summer, and fall are not at significant risk of developing vitamin D deficiency because they store excess cutaneously produced vitamin D in fat stores that is released at times of need [31]. However, the same cannot be said for elderly people. There is strong evidence that elderly who are either free living or residing in a nursing home are at high risk for developing vitamin D deficiency [32–40] (see Chapter 43). As discussed in Chapter 41, vitamin D deficiency causes a defect in the mineralization of the skeleton leading to osteomalacia. Vitamin D deficiency also causes secondary hyperparathyroidism that results in increased mobilization of calcium from the skeleton thereby exacerbating osteoporosis. There is a strong association between the risk of hip fracture and vitamin D deficiency [41,42] (Chapter 43).

V. RECOMMENDATIONS

Although there are various estimated nutrient requirements for vitamin D at various ages in different countries, in general, it is recommended that the intake of vitamin D be higher in neonatal and very young children (~400 IU /day) and 200 IU in adults. For children and young and middle-aged adults who have adequate exposure to sunlight, an RDA of 200 IU of vitamin D is reasonable. However, in the absence of any exposure to sunlight, the estimated nutrient requirement could be at least 3-fold higher. Young male submariners, who were not exposed to sunlight for 3 months, could not maintain their circulating 25OHD levels on a daily intake of 600 IU of vitamin D_3 [1]. Once the sailors left the submarine, they all markedly increased their 25OHD levels presumably due to sunlight exposure.

For the elderly, it is advisable that they receive suberythemal exposure to sunlight two to three times a week on their face, hands, and, when possible, arms and legs. If they cannot receive adequate exposure to sunlight to satisfy the requirement for vitamin D, they should take a daily multivitamin that contains 400 IU of vitamin D.

Acknowledgments

This work was supported in part by the following grants: General Clinical Research Center M01RR 00533, Photobiology R01 AR36963, NIH P01 AG04390, and NASA NAGW 4936.

References

1. Holick MF 1994 McCollum Award Lecture, Vitamin D—New horizons for the 21st century. Am J Clin Nutr **60**:619–630.
2. Holick MF 1989 Phylogenetic and evolutionary aspects of vitamin D from phytoplankton to humans. In: Pang PKT, Schreibman MP (eds) Vertebrate Endocrinology: Fundamentals and Biomedical Implications, Vol. 3. Academic Press, Orlando, Florida, pp. 7–43.
3. Holick MF 1995 Vitamin D: Photobiology, metabolism, and clinical applications. In: DeGroot LJ, Besser M, Burger HG, Jameon JL, Loriaux DL, Marshall JC, Odell WD, Potts JT, Rubenstein AH (eds) Endocrinology, 3rd Ed., Chap. 59. Saunders, Philadelphia, Pennsylvania, pp. 990–1013.
4. Sniadecki J 1939 Cited by W Mozolowski, Jerdrzej Sniadecki (1768–1883) on the cure of rickets. Nature **143**:121.

5. Palm TA 1980 The geographic distribution and etiology of rickets. Practitioner **45**:270–279 and 421–442.
6. Huldschinsky K 1919 Curing rickets by artificial UV-radiation (Heilung von Rachitis durch Kunstliche Hohensonne). Deut Med Wochenschr **45**:712–713 (in German).
7. Hess AF, Unger LF 1921 Cure of infantile rickets by sunlight. J Am Med Assoc **77**:39.
8. Mayer J 1957 Armand Trousseau and the arrow of time. Nutr Rev **15**:321–323.
9. Mellanby T 1918 The part played by an accessory factor in the production of experimental rickets. J Physiol **52**:11–14.
10. McCollum EF, Simmonds N, Becker JE, Shipley PG 1922 Studies on experimental rickets; and experimental demonstration of the existence of a vitamin which promotes calcium deposition. J Biol Chem **53**:293–312.
11. Steenbock H, Black A 1924 The reduction of growth-promoting and calcifying properties in a ration by exposure to ultraviolet light. J Biol Chem **61**:408–411.
12. Hess AF, Weinstock M 1924 Antirachitic properties imparted to inert fluids and green vegetables by ultraviolet irradiation. J Biol Chem **62**:301–313.
13. Steenbock H 1924 The induction of growth-prompting and calcifying properties in a ration exposed to light. Science **60**:224–225.
14. MacLaughlin JA, Anderson RR, Holick MF 1982 Spectral character of sunlight modulates photosynthesis of previtamin D_3 and its photoisomers in human skin. Science **216**:1001–1003.
15. Holick MF, Tian XQ, Allen M 1995 Evolutionary importance for the membrane enhancement of the production of vitamin D_3 in the skin of poikilothermic animals. Proc Natl Acad Sci USA **92**:3124–3126.
16. Loomis F 1967 Skin-pigment regulation of vitamin D biosynthesis in man. Science **157**:501–506.
17. Holick MF, MacLaughlin JA, Doppelt SH 1981 Factors that influence the cutaneous photosynthesis of previtamin D_3. Science **211**:590–593.
18. Webb AR, deCosta BR, Holick MF 1989 Sunlight regulates the cutaneous production of vitamin D_3 by causing its photodegradation. J Clin Endocrinol Metab **68**:882–887.
19. Clemens TL, Adams JS, Henderson SL, Holick MF 1982 Increased skin pigment reduces the capacity of the skin to synthesize vitamin D. Lancet **1**:74–76.
20. Bell NH, Greene A, Epstein S, Oexmann MJ, Shaw W, Shary J 1985 Evidence for alteration of the vitamin D endocrine system in Blacks. J Pediatr **76**:470–473.
21. Kassowitz M 1987 Tetany and autointoxication in infants (Tetani and autointoxication in kindersalter). Wien Med Presse **97**:139 (in Dutch).
22. Webb AR, Kline L, Holick MF 1988 Influence of season and latitude on the cutaneous synthesis of vitamin D_3: Exposure to winter sunlight in Boston and Edmonton will not promote vitamin D_3 synthesis in human skin. J Clin Endocrinol Metab **67**:373–378.
23. Matsuoka LY, Ide L, Wortsman J, MacLaughlin JA, Holick MF 1987 Sunscreens suppress cutaneous vitamin D_3 synthesis. J Clin Endocrinol Metab **64**:1165–1168.
24. Matsuoka LY, Wortsman J, Dannenberg MJ, Hollis BW, Lu Z, Holick MF 1992 Clothing prevents ultraviolet-B radiation-dependent photosynthesis of vitamin D_3. J Clin Endocrinol Metab **75**:1099–1103.
25. Matsuoka LY, Wortsman J, Hanifan N, Holick MF 1988 Chronic sunscreen use decreases circulating concentrations of 25-hydroxyvitamin D: A preliminary study. Arch Dermatol **124**: 1802–1804.
26. Sedrani SH, Al-Arabi KM, Abanny A, Elidrissy A 1990 Frequency of vitamin D deficiency rickets in Riyadh. In: Study of Vitamin D Status and Factors Leading to Its Deficiency in Saudi Arabia. King Saudi Univ. Press, Riyadh, pp. 281–285.
27. Feliciano ES, Ho ML, Specker BL, Falciglia G, *et al.* 1994 Seasonal and geographical variations in the growth rate of infants in China receiving increasing dosages of vitamin D supplements. J Tropic Pediatr **40**:162–165.
28. Oliveri MB, Ladizesky M, Mautalen CA, Alonso A, Martinez L 1993 Seasonal variations of 25-hydroxyvitamin D parathyroid hormone in Ushuaia (Argentina), the southernmost city of the world. Bone Miner **20**:99–108.
29. Raincho JA, del Arco C, Arteaga R, Herranz JL, 1989 Influence of solar irradiation on vitamin D levels in children on anticonvulsant drugs. Acta Neurol Scand **79**:296–299.
30. Specker BL, Valanis B, Hertzberg V, Edwards N, Tsang R 1985 Sunshine exposure and serum 25-hydroxyvitamin D concentrations in exclusively breast-fed infants. J Pediatr **107**:372–376.
31. Mawer EB, Backhouse J, Holman CA, Lumb GA, Stanbury SW 1972 The distribution and storage of vitamin D and its metabolites in human tissues. Clin Sci **43**:413–431.
32. Dawson-Hughes B, Dallal GE, Krall EA, Harris S, Sokoll LJ, Falconer G 1991 Effect of vitamin D supplementation on wintertime and overall bone loss in healthy postmenopausal women. Ann Intern Med **115**:505–512.
33. Hordon LD, Peacock M 1987 Vitamin D metabolism in women with femoral neck fracture. Bone Miner **2**:413–426.
34. Krall E, Sahyoun N, Tannenbaum S, Dallal G, Dawson-Hughes B 1989 Effect of vitamin D intake on seasonal variations in parathyroid hormone secretion in postmenopausal women. N Engl J Med **321**:1777–1783.
35. Lamberg-Allardt C, von Knorring J, Slatis P, Holmstrom T 1989 Vitamin D status and concentrations of serum vitamin D metabolites and osteocalcin in elderly patients with femoral neck fracture: A follow-up study. Eur J Clin Nutr **43**:355–361.
36. Lips P, Wiersinga A, van Ginkel FC, Jongen MJM, Netelenbos C, Hackeng WHL, Delmas PD, Van der Vijgh JF 1988 The effect of vitamin D supplementation on vitamin D status and parathyroid function in elderly subjects. J Clin Endocrinol Metab **67**:644–650.
37. McGrath N, Singh V, Cundy T 1993 Severe vitamin D deficiency in Auckland. New Zealand Med J **106**:524–526.
38. Ng K, St. John A, Bruce DG 1994 Secondary hyperparathyroidism, vitamin D deficiency and hip fracture: Importance of sampling times after fracture. Bone Miner **25**:103–109.
39. Ooms RE, Roos JC, Bezemer P, Van Der Vijgh WJF 1995 Prevention of bone loss by vitamin D supplementation in elderly women: A randomized double-blind trial. J Clin Endocrinol Metab **80**:1052–1058.
40. Rosen CJ, Morrison A, Zhou H, Storm D, Hunter SJ, Musgrave K, Chen T, Liu WW, Holick MF 1994 Elderly women in northern New England exhibit seasonal changes in bone mineral density and calciotropic hormones. Bone Miner **25**:83–92.
41. Aaron JE, Gallagher JC, Anderson J, Stasiak L, Longton E, Nordin B Nicholson M 1974 Frequency of osteomalacia and osteoporosis in fractures of the proximal femur. Lancet 230–233.
42. Chapuy MC, Arlot M, Duboeuf F, Brun J, Crouzet B, Arnaud S, Delmas P, Meunier P 1992 Vitamin D_3 and calcium to prevent hip fractures in elderly women. N Engl J Med **327**:1637–1642.

CHAPTER 4

The Vitamin D 25-Hydroxylase

MARIELLE GASCON-BARRÉ Département de Pharmacologie, Faculté de Médecine, Université de Montréal, and Centre de Recherche Clinique André-Viallet, Hôpital Saint-Luc, Montréal, Canada

I. INTRODUCTION

25-Hydroxyvitamin D (25OHD) is the first hydroxylated metabolite of vitamin D and the immediate precursor of the fully active and hormonal form of the vitamin, 1α,25-dihydroxyvitamin D [$1,25(OH)_2D$]. It was discovered by DeLuca and his group, who identified the liver as the first site of activation of vitamin D [1–3]. Over the past 25–30 years, the enzyme systems involved in the C-25 hydroxylation of vitamins D_3 and D_2 and several of their analogs have been the object of intense investigation by groups in North America, Europe, and Japan. The research has allowed the identification of two intra-hepatic organelles, the smooth endoplasmic reticulum (microsomes) and the mitochondrion, as sites possessing fully active but distinct vitamin D 25-hydroxylases.

The mitochondrial enzyme has been cloned [4–6] and its identity as a vitamin D 25-hydroxylase established with certainty. Moreover, its presence and activity have been positively identified in all species studied including the human [7]. In contrast, the microsomal enzyme, which received the attention of early workers in the field, has not yet been identified with certainty. It has proved to be an enzyme active in the oxidation of several endogenous and exogenous substances, but it is still believed by many, on the basis of the enzyme kinetics of the respective microsomal and mitochondrial enzymes, to be more physiologically relevant than the mitochondrial entity. Although present in laboratory and farm animals, its presence in the human liver remains, however, unresolved [8].

In this chapter, we review the most relevant research on the vitamin D 25-hydroxylases and address the specificity and regulation of each enzyme in the context of their dynamic functioning, including uptake of substrate and the intrahepatic regionalization of the enzyme systems.

II. HEPATIC UPTAKE OF VITAMIN D

The hepatic sequestration of vitamin D is the first prerequisite step toward its activation and subsequent C-25 hydroxylation by the liver. Under normal conditions, the fractional hepatic uptake of vitamin D_3 during a single pass across the rat and dog liver lies between approximately 40 and 60% [9–11], an uptake higher than that observed for all hydroxylated metabolites of vitamin D [9,12,13]. Its hepatic clearance is estimated at 357 ml min^{-1} in dogs [10]. Studies in which total uptake has been investigated for periods varying from 18 sec to 70 min have revealed that the liver does not accumulate significantly more vitamin D_3 than that observed during the first pass across the organ [12,14–18], suggesting a steady-state liver/serum concentration ratio on the order of 0.4 to 0.6 at physiological concentrations of the vitamin. In addition, data available indicate that there is no stringent regulation of uptake by the vitamin D status [12]. Both hepatocytes and nonparenchymal liver cells are able to sequester vitamin D_3, but Dueland *et al.* [15] found that only hepatocytes were able to transform vitamin D_3 into $25OHD_3$.

The uptake of vitamin D_3 is not regioselective within the hepatic acinus and not perturbed by the destruction of either the periportal (proximal) or perivenous (distal or pericentral) region [19], indicating that extraction of vitamin D_3 takes place according to the concentration gradient along the acinus. This observation suggests that the intrahepatic concentration of the vitamin should be higher in periportal than perivenous hepatocytes. This notion may be important in light of the reported intraacinar localization of the mitochondrial vitamin D 25-hydroxylase, as discussed in Section VI of this chapter, and the influence of regioselective liver diseases on the biosynthesis of $25OHD_3$ (see Chapter 51 for review). In addition, the hepatic extraction of vitamin D_3 has been shown to be independent of its hepatic venous or arterial route of delivery [20], indicating that circulating vitamin D of endogenous or exogenous origin should have similar hepatic availability. Although chylomicron remnants or low density lipoproteins (LDL) have been shown to optimize uptake of vitamin D *in vitro* [14], manipulation of the diet in order to obtain different proportions of putative vitamin D_3 carrier proteins *in vivo* were found not to significantly affect the hepatic extraction of vitamin D_3 [20], an observation also made *in vitro* by Ziv *et al.* [18].

III. THE MONOOXYGENASES ACTIVE IN THE HYDROXYLATION OF VITAMIN D AT C-25

A. Early Studies

Early studies established that the liver was the site of hydroxylation of vitamin D_3 at C-25 [3,21,22]. The first experimental evidence obtained indicated that the 25-hydroxylase was localized in both the microsomal and mitochondrial fractions of liver homogenates [23]. However, it was soon reported that the enzyme was almost entirely found in liver microsomes and that mitochondria exhibited no or very low activity toward vitamin D_3 [24]. It was also rapidly established that ^{18}O from molecular oxygen was incorporated at C-25, strongly suggesting that the microsomal enzyme was a mixed-function oxidase [25]. Firm demonstration of the presence of a mitochondrial vitamin D 25-hydroxylase only came several years later [7]. In 1978, studies by Suda's group [132] using isolated perfused rat liver preparations demonstrated that the activation of $1\alpha OHD_3$ was not regulated (K_m 2.0 μM) compared to that of the natural substrate, and that two distinct K_m values could be defined for vitamin D_3 (5.6 nM and 1 μM), suggesting the participation of more than one enzyme system in the 25-hydroxylation of vitamin D.

Evaluation of enzyme activity revealed that $25OHD_3$ production was also present in extrahepatic tissue [26,27]. At that time, support for the extrahepatic presence of a vitamin D_3 25-hydroxylase came from the *in vivo* studies of Ponchon *et al.* [2] and Olson *et al.* [3], who reported that hepatectomy markedly reduced but did not eliminate the plasma concentration of $25OHD_3$ after ^3H-vitamin D_3 injection. Indeed, they observed that about 10% of the radioactivity appearing in the plasma of intact rats was found in hepatectomized animals 4 hr after injection of the parent compound. These observations have now been confirmed at the molecular level with the demonstration of the presence of the mitochondrial vitamin D 25-hydroxylase in numerous tissues and organs. The participation of the extrahepatic enzyme in the production of 25OHD and/or its contribution to the circulating concentration of 25OHD under normal physiological conditions are not, however, presently known.

B. The Microsomal Enzyme

DeLuca's group was the first to report on the subcellular localization of the 25-hydroxylase. Indeed, in the mid-1970s, Battacharyya and DeLuca [24,28,29] observed that the 25-hydroxylase was present in the microsomal fraction of rat liver and that it was influenced by the vitamin D status; moreover, it required the presence of a cytosolic factor for optimum activity. They also noticed that 25-hydroxylation of the synthetic compound dihydrotachysterol$_3$ (DHT_3) was not regulated compared to that of the natural substrate vitamin D_3 [29]. In 1979, Madhok and DeLuca [30] confirmed the presence of the enzyme in rat liver microsomes, and purification and reconstitution of the enzyme was achieved in the Björkhem [31] and DeLuca [32] laboratories soon thereafter. The enzyme was shown to be a cytochrome P450 (P450, CYP) mixed-function oxidase requiring NADPH and NADPH cytochrome *c*/P450 reductase. Yoon and DeLuca [32] also reported that a soluble cytosolic factor was necessary for full reconstitution of enzyme activity (see Table I).

In 1983, Andersson *et al.* [33] purified to homogeneity a rat microsomal cytochrome P-450 active in C-25 hydroxylation of the bile acid intermediates 5β-cholestane-3α,7α,12α-triol (cholestane-triol, C-triol), 5β-cholestane-3α,7α-diol (C-diol), as well as vitamin D_3 and 1αOHD$_3$ but not vitamin D_2. A year later, Hayashi *et al.* [34] purified and partially sequenced a P450 active in 25-hydroxylation of vitamin D_3 and suggested that an inhibitor was removed from microsomes during the purification steps, as previously suggested by others [31,35]. Moreover, contrary to earlier observations, the reconstituted enzyme did not require the presence of a cytoplasmic factor. The partial amino acid sequence obtained indicated that the enzyme was different from P450 (PB-1) and P450 (MC-1) (now referred to as P4502B1 and 1A1, respectively, according to P450 nomenclature [36]). The same group later obtained an N-terminal amino acid sequence and reported that their purified P450 was also active on several endo- and xenobiotics (as reported by others [33,37]) as well as vitamin D_3 and 1αOHD$_3$ but not vitamin D_2 [38], further demonstrating the broad spectrum of activity of the enzyme.

Could the microsomal D 25-hydroxylase be CYP2C11? It is now recognized that the enzyme obtained by Hayashi *et al.* [38] had the same N-terminal amino acid sequence as P4502C11 [39] known to exist in male rat liver microsomes but not in those of females. This observation may well explain why microsomes (or P450 purified from the microsomal fraction) obtained from female livers have consistently been shown, as indicated in Table I, to exhibit significantly far less activity in C-25 hydroxylation of vitamin D_3 than preparations obtained from male counterparts [40–42]. An exception to this finding was reported by Thierry-Palmer *et al.* [43] at low concentration. Whether Yoon and DeLuca [32] and Andersson *et al.* [33] reported on the same isoenzyme as Hayashi *et al.* [38] is not known, as no amino acid sequence was proposed in the former studies. Andersson and Jörnvall [40] later proposed a partial sequence suggestive of the male-specific steroid 16α-hydroxylase, which is now known to be a P450 compatible with CYP2C11. These observations do not preclude the existence of P450s other than CYP2C11 in the activation of vitamin D_3 at C-25. Definitive identification of the microsomal vitamin D_3 25-hydroxylase will, however, require further kinetic, immunological, as well as molecular characterization.

C. The Mitochondrial Enzyme

1. THE C_{27} STEROL 27-HYDROXYLASE

Bile acid biosynthesis represents the most important pathway in the metabolism and excretion of cholesterol. Although several enzymes participate in oxidation of the cholesterol molecule, the C_{27} sterol hydroxylase catalyzes the first and probably the rate-limiting step in the oxidation of the side chain in the acidic bile acid biosynthesis pathway [44]. In accordance with IUPA sterol nomenclature [45] and the stereochemistry of the initial reaction catalyzed by the enzyme [46–48], Cali and Russell [6] termed the enzyme sterol 27-hydroxylase (EC 1.14.13.15) although the enzyme exhibits low substrate specificity and was referred to as the 26-hydroxylase in most earlier publications [49]. *CYP27* has been adopted as the gene symbol [36].

After considerable debate on the subcellular localization of the 27-hydroxylase [39,50,51], the enzyme was established in 1973 to be located in the inner membrane matrix of liver mitochondria. It was shown to display properties similar to that of P450 enzymes catalyzing hydroxylation in the mitochondrial membrane of the adrenal gland [52,53]. These observations were later confirmed by others [54–56]. Thus, sterol 27-hydroxylase is a mixed-function oxidase requiring the two mitochondrial-specific protein cofactors, ferrodoxin (adrenodoxin), an iron–sulfur protein [57], and ferrodoxin (adrenodoxin) reductase, an FAD-containing enzyme (Fig. 1) [58]. The enzyme catalyzes the C-27 hydroxylation of c-triol [55], c-diol [50], and cholesterol [59–61] as well as several other substrates involved in bile acid biosynthesis as discussed in Section IV of this chapter. Okuda *et al.* [61] using female rat liver mitochondria achieved final purification of the 27-hydroxylase and noticed that the enzyme exhibited no activity toward xenobiotics using benzphetamine, 7-ethoxycoumarin, and benzo[*a*]pyrene as substrates.

In 1975, Björkhem *et al.* [62] observed the presence of the 27-hydroxylase in human liver mitochondria, and

TABLE I The Microsomal Vitamin D 25-Hydroxylase

Species	Sex	Substrate	Hydroxylation site	Characteristics	Reference
Rat	Male	Vitamin D_3	C-25	K_m 180 nM^a (D-depleted), 444 nM^a (D-repleted)	Delvin *et al.* [127]
Rat	Male	Vitamin D_3	C-25	K_m 44 nM^a,	Madhok and DeLuca [30]
		DHT_3	C-25	360 nM^a	
Rat (phenobarbital pretreated)	Male	Vitamin D_3	C-25	108 to 205 pmol 30 min^{-1} 0.75 nmol $P450^{-1b}$	Björkhem *et al.* [31]
Rat	Male	Vitamin D_3	C-25	7.68 pmol hr^{-1} mg $protein^{-1b,c}$	Yoon and DeLuca [32]
Bovine		Vitamin D_3	C-25	4.3 pmol min^{-1} nmol $P450^{-1b}$	Hiwatashi *et al.* [37]
		$1\alpha OHD_3$	C-25	3 pmol min^{-1} nmol $P450^{-1b}$	
		Vitamin D_2	C-25	2 pmol min^{-1} nmol $P450^{-1b}$	
		Benzphetamine	Demethylation	921 pmol min^{-1} nmol $P450^{-1b}$	
Rat	Male	Vitamin D_3	C-25	3.5–4 pmol hr^{-1} 10^6 $hepatocytes^{-1}$ K_m 4 μM (D-depleted), 6 μM (D-repleted)	Dueland *et al.* [15]
Rat	Male	Vitamin D_3	C-25	335 pmol min^{-1} nmol $P450^{-1b}$	Andersson *et al.* [33]
		$1\alpha OHD_3$	C-25	1000 pmol min^{-1} nmol $P450^{-1b}$	
		Vitamin D_2	C-25	<10 pmol min^{-1} nmol $P450^{-1b}$	
		C-triol[e]	C-25	2600 pmol min^{-1} nmol $P450^{-1b}$	
		C-diol[f]	C-25	360 pmol min^{-1} nmol $P450^{-1b}$	
		Ethylmorphine	Demethylation	21,000 pmol min^{-1} nmol $P450^{-1b}$	
Rat	Male	Vitamin D_3	C-25	2.3 nmol min^{-1} mg $protein^{-1b}$ (152 pmol min^{-1} nmol P450)	Hayashi *et al.* [34]
Rat	Male	Vitamin D_3	C-25	0.21 nmol min^{-1} nmol $P450^{-1b}$	Hayashi *et al.* [38]
		$1\alpha OHD_3$	C-25	1.73 nmol min^{-1} nmol $P450^{-1b}$	
		Testosterone	C2α	9.34 nmol min^{-1} nmol $P450^{-1b}$	
			C-16α	8.36 nmol min^{-1} nmol $P450^{-1b}$	
		Dehydroepiandrosterone	C-16α	14.6 nmol min^{-1} nmol $P450^{-1b}$	
		Benzphetamine	Demethylation	15.6 nmol min^{-1} nmol $P450^{-1b}$	
Rat	Male	Vitamin D_3	C-25	195 pmol min^{-1} nmol $P450^{-1b}$	Andersson and Jörnvall [40]
		C-triol	C-25	580 pmol min^{-1} nmol $P450^{-1b}$	
		Testosterone	C-16α	1283 pmol min^{-1} nmol $P450^{-1b}$	
	Female	Vitamin D_3	C-25	9 pmol min^{-1} nmol $P450^{-1b}$	
		C-triol[e]	C-25	90 pmol min^{-1} nmol $P450^{-1b}$	
		C-diol[f]	C-25	188 pmol min^{-1} nmol $P450^{-1b}$	
Rat	Male	Vitamin D_3	C-25	335 pmol min^{-1} mg $protein^{-1b}$ (167 pmol min^{-1} nmol $P450^{-1}$)	Dählback and Wikvall [41]
	Female	Vitamin D_3	C-25	6 pmol min^{-1} mg $protein^{-1b}$ (4 pmol min^{-1} nmol $P450^{-1}$)	
Rat	Male	$1\alpha OHD_3$	C-25	0.152 nmol min^{-1} mg $protein^{-1a}$	Saarem and Pedersen [42]
	Female	$1\alpha OHD_3$	C-25	0.053 nmol min^{-1} mg $protein^{-1a}$	
Rat	Male	Vitamin D_3	C-25	37 pmol min^{-1} 10^6 $hepatocytes^{-1d}$ (D depleted) 29 pmol min^{-1} 10^6 $hepatocytes^{-1d}$ [$1,25(OH)_2D_3$ treated]	Benbrahim *et al.* [128]
Rat	Male	Vitamin D_3	C-25	0.46 pmol hr^{-1} mg $protein^{-1a,c}$	Thierry-Palmer *et al.* [43]
	Female		C-25	0.40 pmol hr^{-1} mg $protein^{-1a}$	

[a] Microsomes.
[b] Reconstituted system.
[c] In the presence of a cytosolic factor.
[d] Freshly isolated hepatocytes.
[e] 5β-Cholestane-3α, 7α, 12α-triol.
[f] 5β-Cholestane-3α, 7α, diol.

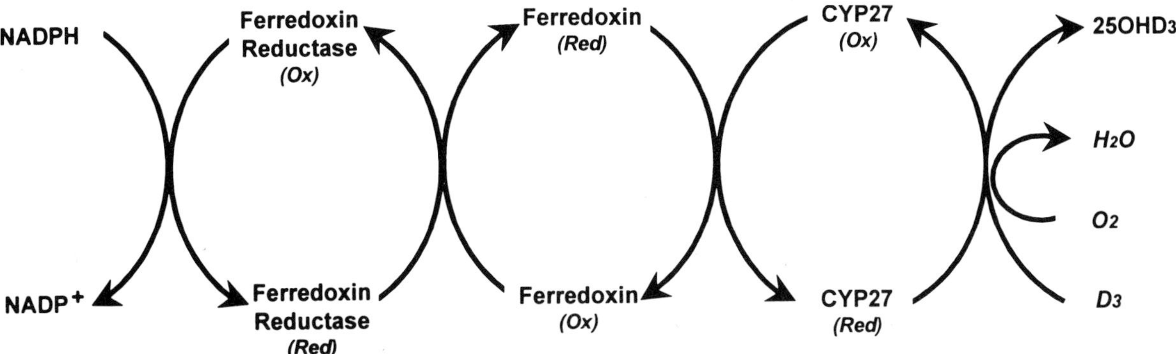

FIGURE 1 Mechanism of hydroxylation of vitamin D_3 at C-25 by the liver sterol 27-hydroxylase (CYP27), the mitochondrial vitamin D_3 25-hydroxylase. Ox, oxidated; red, reduced.

a few years later, the same group [7] was the first to report that rat and human mitochondria catalyzed the C-25 hydroxylation of vitamin D_3. At that time, however, they attributed the vitamin D_3 25-hydroxylation activity to an enzyme different from the cholestanetriol 27-hydroxylase.

2. Linkage between the C_{27} Sterol 27-Hydroxylase and the Hydroxylation of Vitamin D_3 at C-25

The initial observation of Björkhem and co-workers identifying the presence of a vitamin D_3 25-hydroxylase in rat liver mitochondria [7] awakened the interest of several groups active in the bile acid biosynthesis field [31,63,64]. In 1980, Björkhem and Holmberg [65] studied the activities of a P450 preparation solubilized from rat liver mitochondria obtained from rats which had been treated with phenobarbital or subjected to a rachitogenic diet. They observed that both phenobarbital and the rachitogenic diet increased C-25 hydroxylation of vitamin D_3 but had little effect on the C-27 hydroxylation of C-triol. In addition, inhibition experiments revealed only noncompetitive inhibition of vitamin D_3 on C-triol hydroxylation. These observations led the investigators to conclude that two distinct P450 species existed, each responsible for the hydroxylation of cholestanetriol and vitamin D_3. Dählback and Wikvall [66] purified rabbit liver mitochondria according to the procedure used to purify the sterol 27-hydroxylase and noticed that it exhibited vitamin D_3 25-hydroxylase activity with a turnover number of 395 pmol min^{-1} nmol $P450^{-1}$. Using an antibody raised against the rabbit liver mitochondrial 27-hydroxylase, Dählback [67] found that the antibody inhibited C-triol C-27 hydroxylation but not vitamin D_3 C-25 hydroxylation; Dählback concluded, as Björkhem and co-workers [65] had previously in 1980, that two P450 species were involved in these hydroxylations.

At the same time, Masumoto *et al.* [64] purified to homogeneity the vitamin D_3 25-hydroxylase from rat liver mitochondria and characterized it as a P450 catalyzing the C-25 hydroxylation of both vitamin D_3 and $1\alpha OHD_3$. These researchers proposed that a single enzyme was involved in the C-27 hydroxylation of 5β-cholestane-$3\alpha,7\alpha,12\alpha$-triol and in the C-25 hydroxylation of vitamin D_3, although the preparation exhibited much lower activity toward vitamin D_3 compounds than toward the C-27 hydroxylation of bile acid intermediates, as illustrated in Table II [61]. Ohyama *et al.* [68] pursued the hypothesis of Masumoto and co-workers [64] and found that vitamin D_3 competitively inhibited the 27-hydroxylation of C-triol, whereas C-triol inhibited vitamin D_3 25-hydroxylation; they concluded, on the basis of criteria proposed by Dixon and Webb [69], that both substrates were catalyzed at a common active site on a single protein. Meanwhile Oftebro *et al.* [70] and Saarem *et al.* [8,71] confirmed the original observation of Björkhem *et al.* [7] that a vitamin D_3 25-hydroxylase was present in human liver mitochondria and that C_{27} sterol 27-hydroxylation as well as vitamin D_3 25-hydroxylation activities were located, probably exclusively, in the inner mitochondrial membrane matrix [8,71].

3. The Mitochondrial Vitamin D_3 25-Hydroxylase is the C_{27} Sterol 27-Hydroxylase

Cloning of the rabbit [4], rat [5,72], and human [6,73] *CYP27* (chromosomal location 2q33-qter [74]) was achieved in 1989, 1990, and 1991, respectively. Cloning of the rabbit 27-hydroxylase by Andersson and co-workers [4] illustrated the hydrophobic nature of the protein (with 36% of the amino acids having either aromatic or hydrophobic side chains) and the presence of a conserved cysteine residue at position 444, which is believed to be the ligand for the heme moiety in P450 en-

TABLE II The Mitochondrial Vitamin D 25-Hydroxylase

Species	Sex	Substrate	Hydroxylation site	Characteristics	Reference
Rat untreated	—	Vitamin D_3	C-25	0.02 nmol min^{-1} nmol $P450^{-1a}$	Björkhem *et al.* [65]
		1αOHD_3	C-25	0.116 nmol min^{-1} nmol $P450^{-1a}$	
		DHT_3	C-25	0.01 nmol min^{-1} nmol $P450^{-1a}$	
Phenobarbital treated		C-triol[b]	C-26(27)	8.7 nmol min^{-1} nmol $P450^{-1a}$	
		Vitamin D_3	C-25	0.03 nmol min^{-1} nmol $P450^{-1a}$	
Rachitogenic diet		Vitamin D_3	C-25	0.064 nmol min^{-1} nmol $P450^{-1a}$	
Human	Male	Vitamin D_3		0.16 nmol min^{-1} nmol $P450^{-1a}$	Oftebro *et al.* [70]
		C-triol[b]		40 nmol min^{-1} nmol $P450^{-1a}$	
Human	Male and Female	Vitamin D_3	C-25	K_m $10^{-5}M^c$	Saarem *et al.* [8]
Rat	Male		C-25	K_m $2 \times 10^{-5}M^{c,d}$	Saarem *et al.* [71]
Human	—	Vitamin D_3	C-25	10 pmol min^{-1} mg $protein^{-1c}$	Holmberg *et al.* [129]
		Vitamin D_2	C-25	2 pmol min^{-1} mg $protein^{-1c}$	
Rat	Male	Vitamin D_3	C-25	0.093 nmol hr^{-1} mg $protein^{-1c}$	Saarem and Pedersen [42]
	Female			0.435 nmol hr^{-1} mg $protein^{-1c}$	
Rat	Male	Vitamin D_3	C-25	550 pmol min^{-1} mg $protein^{-1c}$	Dählback and Wikvall [41]
				(305 pmol min^{-1} nmol $P450^{-1c}$	
	Female		C-25	525 pmol min^{-1} nmol $P450^{-1c}$	
				350 pmol min^{-1} nmol $P450^{-1c}$	
Rabbit		Vitamin D_3	C-25	395 pmol min^{-1} nmol $P450^{-1a}$	Dählback and Wikvall [66]
		1αOHD_3	C-25	1200 pmol min^{-1} nmol $P450^{-1a}$	
		25OHD_3	C-1α	<10 pmol min^{-1} nmol $P450^{-1a}$	
		C-triol[b]	C-27	2500 pmol min^{-1} nmol $P450^{-1a}$	
Rat	Female	1αOHD_3	C-25	K_m 54 μM^a	Masumoto *et al.* [64]
				3.8 nmol min^{-1} nmol $P450^{-1a}$	
Rat	Female	C-triol[b]	C-27	K_m 6.3 μM^a	Okuda *et al.* [61]
Rat	Female	Vitamin D_3	C-25	0.36 nmol min^{-1} nmol $P450^{-1a}$	Ohyama *et al.* [68]
		1αOHD_3	C-25	1.4 nmol min^{-1} nmol $P450^{-1a}$	
		C-triol[b]	C-27	36 nmol min^{-1} nmol $P450^{-1a}$	
COS cells[d]		Vitamin D_2	C-25	N/D[a,e]	Guo *et al.* [73]
		Vitamin D_3	C-25	342 pmol hr^{-1} 10^6 $cells^{-1a}$	
		DHD_3	C-25	344 pmol hr^{-1} 10^6 $cells^{-1a}$	
		1αOHD_2	C-25	16 pmol hr^{-1} 10^6 $cells^{-1a}$	
		1αOHD_3	C-25	1328 pmol hr^{-1} 10^6 $cells^{-1a}$	
		1αOHDHT_3	C-25	1383 pmol hr^{-1} 10^6 $cells^{-1a}$	
E. coli[f]		Vitamin D_3	C-25	72 pmol min^{-1} nmol $P450^{-1a}$	Axén *et al.* [91]
		1αOHD_3	C-25	189 pmol min^{-1} nmol $P450^{-1a}$	
		25OHD_3	C-1α	24 pmol min^{-1} nmol $P450^{-1a}$	
		25OHD_3	C-27	19 pmol min^{-1} nmol $P450^{-1a}$	
		C-triol[b]	C-27	4337 pmol min^{-1} nmol $P450^{-1a}$	

[a] Reconstituted system.
[b] 5β-Cholestane-3α,7α,12α-triol.
[c] Mitochondria.
[d] In the presence of a cytosolic factor.
[e] N/D, Not detectable.
[f] Transfected with hCYP27.

zymes. Andersson's work also confirmed earlier biochemical observations indicating extrahepatic enzyme activity with the demonstration that *CYP27* mRNAs were present in multiple tissues (Table III).

The predicted amino acid sequence of the rabbit [4], rat [5], and human [6] mitochondrial vitamin D_3 25-hydroxylase contains a mitochondrial specific presequence of 32–36 amino acids on which the subcellular destination of the hemoprotein is dependent. This dependency was unequivocally illustrated by Sakaki *et al.*

TABLE III Expression Sites of *CYP27*[a]

Species	Location	Reference
Human	Liver, fibroblasts[b]	Cali and Russel [6]
	HepG2 hepatoma cells	Guo *et al.* [73]
	Kidney	Demers *et al.* [124]
Mouse	Liver, duodenum, calvaria, lung, skin, long bone, spleen, osteoblasts (primary culture)	Ichikawa *et al.* [125]
Rabbit	Liver, duodenum, adrenal, lung, kidney, spleen	Andersson *et al.* [4]
Rat	Liver[c]	Usui *et al.* [5], Twisk *et al.* [104][b]
	Ovaries	Su *et al.* [72]
	Kidney	Mullick *et al.* [77], Axén *et al.* [96]

[a] In addition to these sites, enzyme activity has been shown in human macrophages [130] and bovine endothelial cells [131].

[b] SV40 transformed.

[c] *In situ* hybridization illustrates a preferential perivenous localization within the acinus [104].

[75] using a modified *CYP27* whose mitochondrial targeting signal had been replaced by a microsomal N-terminal targeting signal. The construct resulted in the microsomal localization of a fully active enzyme, provided that adequate electron transfer was made available to the enzyme. The mature protein contains 444 to 501 amino acids depending on the species involved; the rat enzyme molecular weight has been estimated to be 51,182 [5]. Northern analyses demonstrate two mRNA sizes of approximately 1.9 and 2.3-kb for *CYP27* in liver and fibroblast cells [6,72]. Characterization of the 2.3-kb mRNA indicated a nucleotide sequence identical to the 1.9-kb mRNA in the protein coding region, except for an approximately 400-nucleotide extended sequence at the 5′ end [76]. The rat *CYP27* gene contains 11 exons of 80–415 nucleotides that are separated by 10 introns of 83 bases to about 10 kb [77]. Studies of the gene have revealed regulation by several hormones, as discussed below [76–80]. The protein sequence of the human enzyme has been reported to be 72% identical to rat CYP27 and 81% identical to rabbit CYP27 [5,6].

Heterologous cell systems confirmed the identity of sterol 27-hydroxylase and vitamin D_3 25-hydroxylase as well as the requirement for adrenodoxin and adrenodoxin reductase but not NADPH P450 reductase [4]. Indeed, Usui *et al.* [81] prepared an expression plasmid encoding the rat liver vitamin D_3 25-hydroxylase and transfected it in monkey COS cells. They observed both sterol C-27 and vitamin D_3 C-25 hydroxylation activities in a solubilized mitochondrial fraction supplemented with adrenodoxin and NADPH-adrenodoxin reductase. Akiyoshi-Shibata *et al.* [82] and Sakaki *et al.* [75] expressed the cDNA encoding the precursor protein of the rat liver vitamin D_3 25-hydroxylase in *Saccharomyces cerevisiae* and found that the mitochondrial fraction exhibited both vitamin D_3 25-hydroxylase and C-triol 27-hydroxylase activities. The data clearly confirmed the identity of the two enzymes and ruled out the possibility that the simian cell line contained endogenous 27-hydroxylase.

Mutation that affects either the expression of *CYP27* or its primary sequence leads to cerebrotendinous xanthomatosis (CTX) [83,84], an inherited disorder of sterol metabolism and storage characterized by atherosclerosis and progressive neurological dysfunction. In these patients, however, both normal [85] and abnormal [84,86,87] vitamin D and/or calcium metabolism have been reported, which gives credence to the concept that more than one enzyme is active in the 25-hydroxylation of vitamin D_3. Importantly, some CTX patients have been shown to retain residual 27-hydroxylase activity [39], whereas the human cell line Hep3B has been shown to efficiently activate $1\alpha OHD_3$ at C-25 despite the absence of any detectable *CYP27* transcript when evaluated by Northern analysis [88,89].

IV. SPECIFICITY OF THE STEROL 27-HYDROXYLASE: THE MITOCHONDRIAL VITAMIN D_3 25-HYDROXYLASE

The sterol 27-hydroxylase is a high-capacity enzyme involved in the hydroxylation of bile acid intermediates, but its substrate specificity is broad and extends beyond the field of bile acids. As discussed earlier, the purified enzyme has been shown to be active in both C-25 and C-27 hydroxylation of cholesterol [39,49,53], and to hydroxylate 5β-cholestane-$3\alpha,7\alpha$-diol and 5β-cholestane-$3\alpha,7\alpha,12\alpha$-triol at C-27. The enzyme has also been shown to carry out the oxidation of the product of the reaction, 5β-cholestane-$3\alpha,7\alpha,12\alpha,27$-tetrol, to $3\alpha,7\alpha,12\alpha$-trihydroxy-5β-cholestanoic acid [90] as well as to substitute for the classic pathway of bile acid biosynthesis involving an NAD-coupled alcohol dehydrogenase. The rabbit, rat, human, pig, and bovine enzyme hydroxylates vitamin D_3 compounds at C-25 as well as at other positions on the side chain of the molecule [5,72,73,91].

It has now been demonstrated that the 27-hydroxylase catalyzes the C-24, C-25, and C-27 hydroxylation of vitamin D_3, vitamin D_2, and related compounds [73,92]. Although vitamin D_3 substrates were found to be preferably hydroxylated at C-25, Guo *et al.* [73] observed that when the substrate exhibited the ergocalciferol side chain such as that found in vitamin D_2 or $1\alpha OHD_2$, a predominance of 24-hydroxyl metabolites occurred (metabolites which have also been found *in vivo* in rat, cow, and chicken [93–95]) with some production of 27-hydroxylated products. Surprisingly, $1\alpha OHD_2$ was only poorly hydroxylated at position 25 whereas vitamin D_2 was not hydroxylated as indicated in Table II. In addition, several laboratories have now shown that the enzyme prefers 1α-hydroxylated analogs of vitamin D_3 or D_2 over nonhydroxylated counterparts, and this includes the natural substrate vitamin D_3, which is hydroxylated at C-25 several times less efficiently than $1\alpha OHD_3$ (see Table II) [65,66,68,73]. The observation that the 27-hydroxylase is far more efficient in hydroxylating compounds exhibiting a 1α-hydroxyl group than toward the natural substrate is in full accordance with results of Ohyama *et al.* [68], who observed that the activity of the enzyme toward bile acid intermediates was proportional to the number of hydroxyl groups on the nucleus. Björkhem [49] also raised the hypothesis that the binding of nonpolar substrates such as vitamin D or cholesterol to the active site of the enzyme may differ from the corresponding binding of more polar substrates such as 5β-cholestane-$3\alpha,7\alpha,12\alpha$-triol.

Axén *et al.* [91] have reported that the 27-hydroxylase purified from pig and rabbit livers as well as recombinant human *CYP27* expressed in *Escherichia coli* or monkey COS cells were also able to catalyze the 1α-hydroxylation of $25OHD_3$, albeit at a much lower rate than that observed for the conversion of vitamin D_3 to $25OHD_3$. The data were confirmed in a subsequent paper a year later, leading the authors to put forward the provocative hypothesis that the mitochondrial vitamin D_3 25-hydroxylase could also be the $25OHD_3$ 1α-hydroxylase [96]. Interestingly, 1α-hydroxylase activity has been reported previously in pig [97], fish [98], and fetal rat livers [99], but whether the reported activity was due to the 27-hydroxylase is unknown at this time.

Jones and co-workers [100] have also shown that lengthening the side chain of the vitamin D molecule is very well tolerated in that the enzyme is able to continue to efficiently hydroxylate vitamin D_3 compounds including analogs of $1,25(OH)_2D_3$. Indeed, these studies reported by Dilworth *et al.* [100] led to the proposal of a model of interaction between enzyme and substrate where the 25-hydroxylase appears to be directed to its terminal hydroxylation site by the distance from the end of the side chain, as illustrated in Fig. 2. However, the ratio of C-25 to C-27 hydroxylation did change as the side chain was extended. Thus, the sterol 27-hydroxylase seems to tolerate not only extension and epimerization of the side chain [100,101] but also modification of the

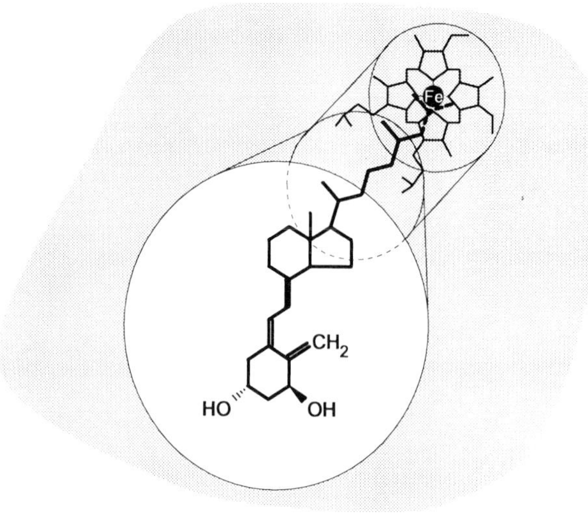

FIGURE 2 Model for the interaction of the vitamin D_3 25-hydroxylase (sterol 27-hydroxylase/CYP27) with its substrate. A two-chamber model is proposed that allows for accommodation of the vitamin D nucleus as well as its side chain. The model for CYP27 comprises a large nonspecific pocket for different sterol/secosterol ring structures and a more specific adjoining pocket for cholesterol-type side chains of varying length. The relative position of the heme group is illustrated by the porphyrin ring structure coordinating an iron atom. Reprinted from Dilworth *et al.* [100] with permission.

steroidal/vitamin D nucleus [4,73,91,102]. As proposed by Jones and co-workers [100], these data also suggest that the substrate binding pocket of the 27-hydroxylase can tolerate a variety of steroidal shapes either by accommodating only the terminal carbons of the side chain or by having a specific cleft for the side chain within a broader pocket for the vitamin D/steroidal ring structure.

V. REGULATION OF *CYP27* AND ENZYME ACTIVITY

Although the half-life of *CYP27* mRNA has been reported to be between 18 and 24 hr suggesting a slow response to regulators [103], several hormones and endogenous products have been shown to regulate the gene. They include bile acids, growth hormone, glucocorticoids, insulin, and the physiological state of the animal [72,77,79,80,104]. To date, however, most regulation studies on *CYP27* have been linked to the metabolism of cholesterol and bile acids, and few studies have addressed the effect of these hormones on the handling of vitamin D.

In the rat [78,79] but not the rabbit [105], *CYP27* is highly sensitive to the prevailing concentrations of bile acids, with increases in enzyme activity, steady-state mRNA level, and rate of gene transcription occurring following interruption of the enterohepatic circulation of bile acids [79,104]. A regioselective dynamic response of the liver on depletion of the bile acid pool is also observed in the rat [104] as reviewed in the next section.

The role of insulin on *CYP27* regulation was illustrated when bile acid production via the sterol 27-hydroxylase was shown to be inhibited up to 58% after 24 hr of incubation of hepatocytes with 140 nmol/liter insulin [80]. The decrease in enzyme activity could be explained by a concomitant reduction in *CYP27* mRNA level (−62%) as well as in transcriptional activity (−75%) [80]. Thus, physiological concentrations of insulin seem to down-regulate *CYP27* gene transcription through a direct effect of hormone on the hepatocytes [80], although the exact locus of interaction of insulin within the gene has not yet been studied.

Avadhani's group first reported regulation of *CYP27* by pituitary-regulated steroids, growth hormone, and the diurnal rhythm [79]. A 2-fold increase in enzyme activity was observed in the mid-dark compared to the mid-light period [79]. They subsequently showed that *CYP27* is regulated by glucocorticoids, as illustrated by a 7-fold induction in mRNA level in isolated hepatocytes exposed to dexamethasone [103]. Shayiq and Avadhani [76] also report that the relative levels of the 2.3- and 1.9-kb mRNA species in the rat are both tissue-specific and appear to be modulated by the physiological state of the animal. But transcripts were regulated by growth hormone in parallel with serine protease inhibitor mRNAs. For example, as shown earlier by Su *et al.* [72], there is a predominance of the 2.3-kb mRNA with negligible 1.9-kb mRNA detected by Northern blot analysis in the liver of immature rats. However, in hepatic tissue from mature females, nearly 20–25% of the hybridization is in the form of the 1.9-kb species. Avadhani's group later reported that the 2-kb mRNA (formerly referred to as 1.9 kb) was transcribed as an independent transcript driven by an immediate upstream promoter located within exon 2 of *CYP27* [77]. These observations may be important for the regulation of *CYP27* in mature and/or immature females in view of the fact that the mitochondrial 25-hydroxylase is the predominant enzyme able to transform vitamin D_3 into $25OHD_3$ in the female rat and seemingly the sole enzyme in humans [8].

Regulation of *CYP27* by vitamin D status has not yet been investigated. However, Axén *et al.* [96] reported that *CYP27* located in kidney and liver was affected by $1,25(OH)_2D_3$ administration but that kidney *CYP27* mRNA was decreased to a greater extent than that of the liver. The significance of these observations on the 25-hydroxylation of vitamin D_3 has not been studied, although earlier studies had raised the hypothesis that $1,25(OH)_2D_3$ might inhibit the production of 25OHD in human subjects [106]. Later studies indicated, however, that the decrease in 25OHD concentrations could be explained by an acceleration in metabolic clearance rate [107–109].

The effect of drug administration on vitamin D metabolism is addressed in Chapter 50. At the enzyme level, however, cyclosporin A has been shown to inhibit the C-25 hydroxylation of both vitamin D_3 and cholesterol through direct inhibition of the sterol 27-hydroxylase [110,111].

VI. INTRAACINAR LOCALIZATION OF *CYP27*

In the rat, *CYP27* is heterogeneously distributed within the liver acinus, with higher perivenous than periportal mRNA levels. *CYP27* gene transcription also favored this distribution when examined by Northern analysis of selectively isolated regional hepatocytes or *in situ* hybridization, as illustrated by the work of Twisk and co-workers presented in Fig. 3 [104]. Regulation studies have shown increases in production of 27-hydroxycholesterol, *CYP27* mRNA level, and gene transcription [79,104] as well as a recruitment of hepatocytes toward the mid to periportal region of the liver acinus [104] when the enterohepatic circulation was interrupted by a bile salt sequestrant (Fig. 3C). These data clearly indicate that *CYP27* is subject to feedback inhibition at the transcriptional level by bile acids returning via the portal blood. Both the intensity and the intrahepatic regionalization of the expression are altered albeit to a lower degree than *CYP7*, which encodes the enzyme catalyzing the 7α-hydroxylation of cholesterol in the neutral bile acid biosynthesis pathway [79,104].

The effect of vitamin D deficiency on *CYP27* gene expression and transcriptional regulation has not been investigated, although it is known that interruption of the enterohepatic circulation rapidly leads to vitamin D depletion secondary to malabsorption of the vitamin in the absence of an adequate amount of bile salts in the intestinal lumen [112]. Whether malabsorption-induced vitamin D depletion leads to a similar regulation of the 25-hydroxylation of vitamin D_3 as that observed on the 27-hydroxylation of cholesterol following bile acid depletion [79] remains to be investigated. Most studies on the regulation of vitamin D_3 metabolism by the mitochondrial P450 have reported little feedback inhibition by the product of the reaction. Contrary to their effect on the hydroxylation of cholesterol, however, bile acids do not seem to directly affect the total hepatic output of $25OHD_3$ *in vivo*. Indeed, by directly sampling the

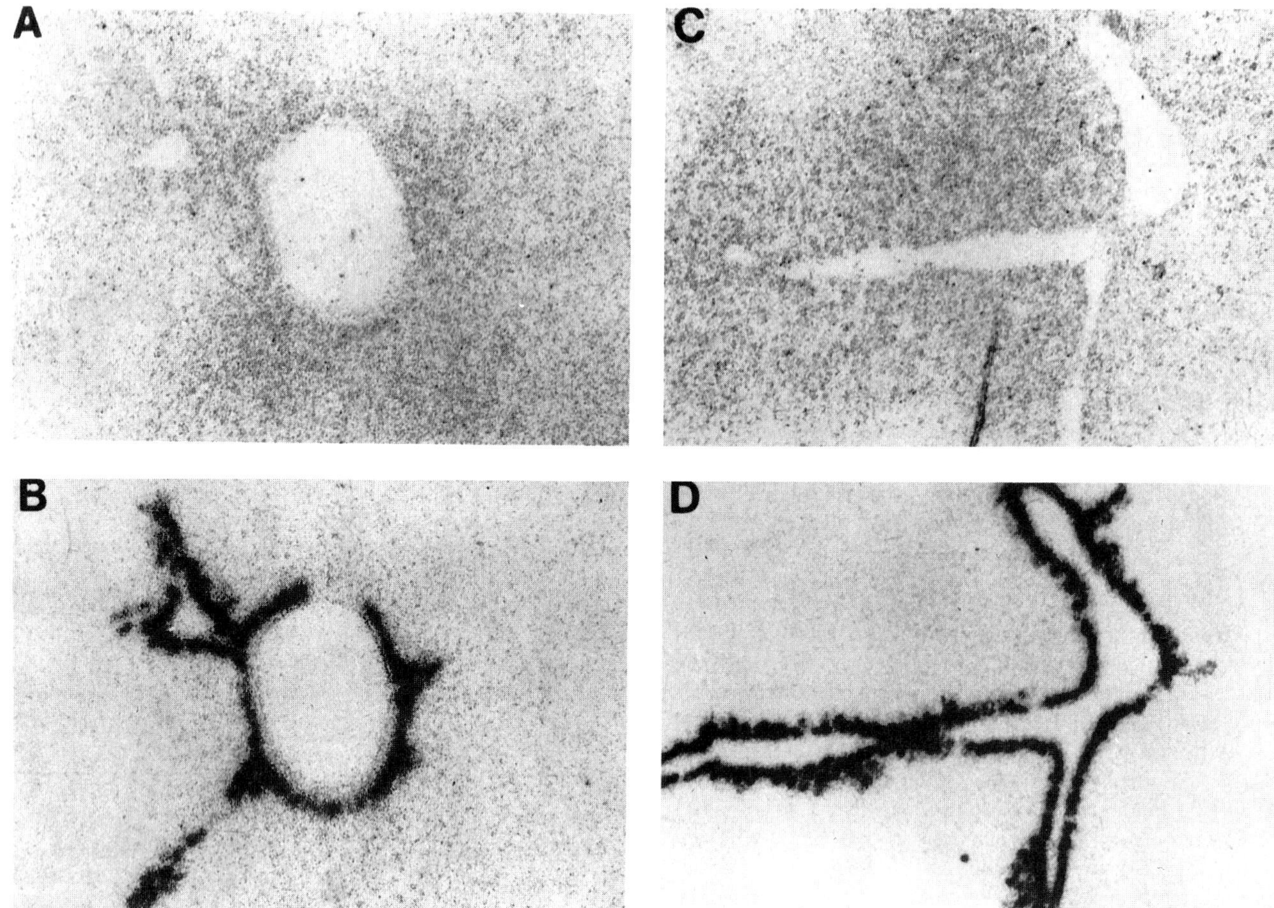

FIGURE 3 Hepatic intraacinar localization of *CYP27*. The localization of *CYP27* mRNA on serial liver sections were done by *in situ* hybridization. Liver sections were made from control (A) and colestid-treated rats (C) and were hybridized *in situ* with ^{35}S-labeled probes for *CYP27* (A,C), and glutamine synthetase (B,D). The latter enzyme was assessed as a positive identification of the perivenous zone. Sections depicted are of a central vein with surrounding hepatocytes. Reproduced from *The Journal of Clinical Investigation*, 1995, Volume 95, p. 1240 [104], by copyright permission of The American Society for Clinical Investigation.

hepatic effluent after intraportal vitamin D_3 injection, Plourde *et al.* [112] observed that the 25-hydroxylation of D_3 was similar in dogs with interruption of the enterohepatic circulation by choledococystostomy anastomosis and with biliary cirrhosis secondary to bile duct ligation. In contrast, Bolt *et al.* [113] have shown that cholestasis induced by bile duct ligation in rats significantly inhibited the 25-hydroxylation of D_3 and that addition of bile salts to liver homogenates also had an inhibitory effect on the enzyme, leaving open any definitive conclusion as to the effect of bile salts on the C-25 hydroxylation of vitamin D_3 by the mitochondrial and/or microsomal enzyme.

In a study where examination of the intrahepatic regionalization of the C-25 hydroxylation activity was investigated in vitamin D-depleted male rats, Gascon-Barré *et al.* [19] observed a slight but significant predominance of enzyme activity in unstimulated hepatocytes isolated from the periportal area compared to those obtained from the perivenous area. However, perivenous but not periportal hepatocytes could be stimulated by an increase in intracellular calcium concentration mediated by the calcium ionophore A23187. These observations suggest modulation of the 25-hydroxylase(s) within the liver but still leave open the question of the role of the mitochondrial and microsomal enzymes in relation to the intrahepatic regionalization of vitamin D_3 uptake, enzyme distribution, and regulation by hormones, endo- or xenobiotics, regioselective liver diseases, and vitamin D status.

VII. ONTOGENY OF THE VITAMIN D 25-HYDROXYLASE

To date, no attempt has been made to evaluate the presence of the mitochondrial vitamin D_3 25-hydroxylase in fetal or neonatal liver. It has been shown, however, that fetal rat liver can adequately sequester vitamin D_3 [114]. Moreover, when compared to values obtained in the late fetal period, vitamin D_3 uptake decreased significantly during the first 2 weeks post partum. Between days 14 and 22 post partum uptake increased dramatically (6-fold) and reached a capacity similar to that previously observed in the adult rat. The livers of 18- (3 days before birth), 19-, and 22-day-old rat fetuses hydroxylates vitamin D_3 at C-25, and the values were shown to remain unchanged at day 1 after birth [115,116]. In one study [115], the microsomal 25-hydroxylase activity was shown to sharply increase from days 2 to 11 and to remain unchanged until day 60 of chronological age, whereas in another study [116] maximal enzyme activity was reached only near weaning. It was also noticed that a cytosolic factor was necessary for optimal 25-hydroxylase activity in both fetuses, neonates, and adult rats. In that context, it has been suggested that cytosol contains a cellular vitamin D binding protein [117] that could facilitate access of substrate to the enzyme as proposed by Tsankova *et al.* [118].

In humans, 25OHD has been shown to be lower in cord than in maternal serum, although a significant relationship was observed between the two sites [119]. Premature infants have low circulating 25OHD concentrations compared to infants born at term [120,121] however, levels increase in the subsequent 4 days after birth [121], indicating fully operational vitamin D 25-hydroxylase activity in preterm infants.

VIII. SEX DIFFERENCES IN THE HYDROXYLATION OF VITAMIN D AT C-25

Despite the fact that the circulating $25OHD_3$ concentrations are similar in males and females, several studies indicate that both the microsomal and the mitochondrial enzymes may be gender specific and/or regulated by sex hormones. Indeed, the presently identified (but not exclusive) microsomal vitamin D_3 25-hydroxylase CYP2C11 is known to be male specific in the rat and rabbit but not in human [74]. Not surprisingly, the enzyme activity reported in the rat has been largely shown to be higher in males than females [40,41]. In addition, Hayashi *et al.* [122] concluded, on the basis of immunoprecipitation techniques, that the vitamin D_3 25-hydroxylase found in the male rat liver was a different entity from that found in the female, suggesting that a P450 distinct from CYP2C11 is also able to catalyze the 25-hydroxylation of vitamin D in the rodent.

The mitochondrial enzyme has been shown in the rat to be several times more active in females than in males [42,123]. In addition, Saarem and Pedersen [42] observed that both enzyme activities increased when estradiol was injected into male rats, whereas testosterone decreased the activities of both enzymes in female rats. Addya *et al.* [123] reported that the relative 25-hydroxylase level was reduced in castrated females but increased in castrated males. Regulation of the human enzyme by sex hormones is not presently known. However, the level of *CYP27* mRNA was found to be similar in normal adult liver samples obtained from men and women [124].

IX. CONCLUSIONS

In liver, both microsomes and mitochondria harbor a monooxygenase system active in the 25-hydroxylation of, endogenous vitamin D_3, as well as of several analogs of vitamin D, allowing the design of molecules with therapeutic potential that all can be activated by a single hydroxylation step in the liver. The sterol 27-hydroxylase is a multifunctional enzyme exhibiting high capacity, with a reported K_m for vitamin D_3 many times higher (micromolar range) than the circulating steady-state concentration of substrate *in vivo*. The dynamic liver regulation of enzyme distribution associated with uptake of substrate favoring selective acinar accumulation seems to create the necessary conditions within the hepatocyte microenvironment for efficient transformation of vitamin D into 25OHD *in vivo*. This is well exemplified in the human where the mitochondrial enzyme is claimed to be the unique vitamin D 25-hydroxylase. In addition, the presence of CYP27 in multiple extrahepatic tissues and organs and the metabolic defects associated with CTX are suggestive of an enzyme displaying pleiotropic function in the metabolism of cholesterol and perhaps of vitamin D, when the liver is unable to support the hydroxylation of the vitamin. Whether these sites participate in the toxicity of vitamin D or participate, in a tissue-specific manner, in the *in situ* delivery of 25OHD to the extrarenal 1α-hydroxylase, as observed *in vitro* [125] remains to be investigated.

The microsomal 25-hydroxylase was originally thought to be a high-specificity, low-capacity enzyme, but it has now been shown that the enzymes described to date also have low substrate specificity and exhibit, as the mitochondrial counterpart does, regulation by sex hormones and preferential intraacinar localization to the perivenous area (CYP2C11) [126]. Is CYP2C11 a microsomal vitamin D 25-hydroxylase? Groups in Japan [38] and Sweden [33] have independently purified an enzyme active in C-25 hydroxylation of vitamin D_3 that is compatible with the male-specific CYP2C11. Ambiguity still remains as to its role and relative importance in the hepatic output of 25OHD under normal conditions and during liver diseases or chronic drug administration. Confirmation will await further studies as all the molecular tools are now available to unequivocally define the role of CYP2C11 in the hepatic activation of vitamin D.

In conclusion, the liver is essential to the normal homeostasis of vitamin D. Indeed, it activates vitamin D of endogenous and exogenous origin, and it efficiently exports 25OHD to the systemic circulation where its concentration represents the best marker of vitamin D nutritional status, as discussed in later chapters.

References

1. Blunt JW, DeLuca HF, Schnoes HK 1968 25-Hydroxycholecalciferol. A biologically active metabolite of vitamin D_3. Biochemistry **7**:3317–3322.
2. Ponchon G, Kennan AL, DeLuca HF 1969 "Activation" of vitamin D by the liver. J Clin Invest **48**:2032–2037.
3. Olson EBJ, Knutson JC, Bhattacharyya MH, DeLuca HF 1976 The effect of hepatectomy on the synthesis of 25-hydroxyvitamin D_3. J Clin Invest **57**:1213–1220.
4. Andersson S, Davis DL, Dählback H, Jörnvall H, Russell DW 1989 Cloning, structure, and expression of the mitochondrial cytochrome P-450 sterol 26-hydroxylase, a bile acid biosynthetic enzyme. J Biol Chem **264**:8222–8229.
5. Usui E, Noshiro M, Okuda K 1990 Molecular cloning of cDNA for vitamin D_3 25-hydroxylase from rat liver mitochondria. FEBS Lett **262**:135–138.
6. Cali JJ, Russell DW 1991 Characterization of human sterol 27-hydroxylase. A mitochondrial cytochrome P-450 that catalyzes multiple oxidation reactions in bile acid biosynthesis. J Biol Chem **266**:7774–7778.
7. Björkhem I, Holmberg I 1978 Assay and properties of a mitochondrial 25-hydroxylase active on vitamin D_3. J Biol Chem **253**:842–849.
8. Saarem K, Bergseth S, Oftebro H, Pedersen JI 1984 Subcellular localization of vitamin D_3 25-hydroxylase in human liver. J Biol Chem **259**:10936–10960.
9. Gascon-Barré M, Huet PM 1982 Role of the liver in the homeostasis of calciferol metabolism in the dog. Endocrinology **110**:563–570.
10. Gascon-Barré M, Vallières S, Huet PM 1986 Influence of phenobarbital on the hepatic handling of [^{3}H]vitamin D_3 in the dog. Am J Physiol (Gastrointest Liver Physiol) **251**:G627–G635.
11. Gascon-Barré M, Gamache M 1991 Contribution of the biliary pathway to the homeostasis of vitamin D_3 and of 1,25-dihydroxyvitamin D_3. Endocrinology **129**:2335–2344.
12. Rojanasathit S, Haddad JG 1976 Hepatic accumulation of vitamin D_3 and 25-hydroxyvitamin D_3. Biochim Biophys Acta **421**:12–21.
13. Gascon-Barré M, Vallières S, Huet PM 1986 Uptake of the hormone 1,25-dihydroxyvitamin D_3 by the dog liver. Can J Physiol Pharmacol **64**:699–702.
14. Haddad JG, Jennings AS, Choon AWT 1988 Vitamin D uptake and metabolism by perfused rat liver: Influences of carrier proteins. Endocrinology **123**:498–504.
15. Dueland S, Holmberg I, Berg T, Pedersen JI 1981 Uptake and 25-hydroxylation of vitamin D_3 by isolated rat liver cells. J Biol Chem **256**:10430–10434.
16. Dueland S, Helgerud P, Pedersen JI, Berg T, Drevon CA 1983 Plasma clearance, transfer, and distribution of vitamin D_3 from intestinal lymph. Am J Physiol **245**:E326–E331.
17. Gascon-Barré M, Elbaz H 1985 Sequestration and microsomal C-25 hydroxylation of [^{3}H]vitamin D_3. Metabolism **34**:244–250.
18. Ziv E, Bar-On H, Silver J 1985 Vitamin D_3 uptake by the isolated perfused rat liver from lipoprotein fractions is separate from cholesterol and triglyceride uptake. Eur J Clin Invest **15**:95–99.
19. Gascon-Barré M, Vallières S, Benbrahim N 1992 C-25 hydroxylation of vitamin D_3 in the periportal and perivenous region of the hepatic acinus. Am J Physiol (Endocrinol Metab) **262**:E810–E817.
20. Gascon-Barré M, Huet PM, St-Onge Brault G, Brault A, Kassissia I 1988 Liver extraction of vitamin D_3 is independent of its venous or arterial route of delivery. Studies in isolated-perfused rat liver preparations. J Pharmacol Exp Ther **245**:975–981.
21. Horsting M, DeLuca HF 1969 *In vitro* production of 25-hydroxycholecalciferol. Biochem Biophys Res Commun **36**:251–256.
22. Ponchon G, DeLuca HF 1969 The role of the liver in the metabolism of vitamin D. J Clin Invest **48**:1273–1279.
23. DeLuca HF 1971 Metabolism and mechanism of action of 25-hydroxy-cholecalciferol. In: Nicholes G Jr, Wasserman RH (eds) Cellular Mechanisms for Calcium and Homeostasis. Academic Press, New York, pp. 426–430.
24. Bhattacharyya MH, DeLuca HF 1974 Subcellular location of rat liver calciferol-25-hydroxylase. Arch Biochem Biophys **160**:58–62.
25. Madhok TC, Schnoes HK, DeLuca HF 1978 Incorporation of oxygen-18 into the 25-position of cholecalciferol by hepatic cholecalciferol 25-hydroxylase. Biochem J **175**:479–482.
26. Tucker III G, Gagnon RE, Haussler MR 1973 Vitamin D_3-25-hydroxylase: Tissue occurrence and apparent lack of regulation. Arch Biochem Biophys **155**:47–57.
27. Bhattacharyya MH, DeLuca HF 1974 The regulation of calciferol-25-hydroxylase in the chick. Biochem Biophys Res Commun **59**:734–741.
28. Bhattacharyya MH, DeLuca HF 1973 The regulation of rat liver calciferol-25-hydroxylase. J Biol Chem **248**:2969–2973.
29. Bhattacharyya MH, DeLuca HF 1973 Comparative studies on the 25-hydroxylation of vitamin D_3 and dihydrotachysterol$_3$. J Biol Chem **248**:2974–2977.
30. Madhok TC, DeLuca HF 1979 Characteristics of the rat liver microsomal enzyme system converting cholecalciferol into 25-hydroxycholecalciferol. Biochem J **184**:491–499.
31. Björkhem I, Hansson R, Holmberg I, Wikvall K 1979 25-Hydroxylation of vitamin D_3 by a reconstituted system from rat liver microsomes. Biochem Biophys Res Commun **90**:615–622.
32. Yoon PS, DeLuca HF 1980 Resolution and reconstitution of soluble components of rat liver microsomal vitamin D_3-25-hydroxylase. Arch Biochem Biophys **203**:529–541.
33. Andersson S, Holmberg I, Wikvall K 1983 25-hydroxylation of C_{27}-steroids and vitamin D_3 by a constitutive cytochrome P-450 from rat liver microsomes. J Biol Chem **258**:6777–6781.
34. Hayashi SI, Noshiro M, Okuda K 1984 Purification of cytochrome P-450 catalyzing 25-hydroxylation of vitamin D_3 from rat liver microsomes. Biochem Biophys Res Commun **121**:994–1000.
35. Holmberg I 1984 Inhibition of reconstituted vitamin D_3 25-hydroxylase by a protein fraction from rat liver microsomes. Biochem Biophys Res Commun **123**:1209–1214.
36. Nelson DR, Kamataki T, Waxman DJ, Guengerich FP, Estabrook RW, Feyereisen R, Gonzalez FJ, Coon MJ, Gunsalus IC, Gotoh O, Okuda K, Nebert DW 1993 The P450 superfamily: Update on new sequences, gene mapping, accession numbers,

early trivial names of enzymes, and nomenclature. DNA Cell Biol **12**:1–51.
37. Hiwatashi A, Ichikawa Y 1980 Purification and organ-specific properties of cholecalciferol 25-hydroxylase system: Cytochrome P-450_{D25}-linked mixed function oxidase system. Biochem Biophys Res Commun **97**:1443–1449.
38. Hayashi SI, Noshiro M, Okuda K 1986 Isolation of a cytochrome P-450 that catalyzes the 25-hydroxylation of vitamin D_3 from rat liver microsomes. J Biochem (Tokyo) **99**:1753–1763.
39. Okuda KI 1994 Liver mitochondrial P450 involved in cholesterol catabolism and vitamin D activation. J Lipid Res **35**:361–372.
40. Andersson S, Jörnvall H 1986 Sex differences in cytochrome P-450-dependent 25-hydroxylation of C_{27}-steroids and vitamin D_3 in rat liver microsomes. J Biol Chem **261**:16932–16936.
41. Dählback H, Wikvall K 1987 25-Hydroxylation of vitamin D_3 in rat liver: Roles of mitochondrial and microsomal cytochrome P-450. Biochem Biophys Res Commun **142**:999–1005.
42. Saarem K, Pedersen JI 1987 Sex differences in the hydroxylation of cholecalciferol and of 5β-cholestane-3α,7α,12α-triol in rat liver. Biochem J **247**:73–78.
43. Thierry-Palmer M, Free AL, Nagappan PR, Butler AR 1995 Cholecalciferol 25-hydroxylation is similar in liver microsomes from male and female rats when cholecalciferol concentration is low. J Nutr **125**:104–111.
44. Björkhem I 1985 Mechanism of bile acid biosynthesis in mammalian liver. In: Danielsson H, Sjovall J (eds) Sterols and Bile Acids. Elsevier, Amsterdam pp. 231–278.
45. Popják G, Edmond J, Anet FAL, Easton NR Jr. 1977 Carbon-13 NMR studies on cholesterol biosynthesized from [^{13}C]mevalonates. J Am Chem Soc **99**:931–935.
46. Danielsson H 1960 On the oxidation of 3α,7α,12α-trihydroxycoprostane by mouse and rat liver homogenates. Acta Chem Scand **14**:348–352.
47. Shefer S, Cheng FW, Batta AK, Dayal B, Tint GS, Salen G, Mosbach EH 1978 Stereospecific side chain hydroxylations in the biosynthesis of chenodeoxycholic acid. J Biol Chem **253**:6386–6392.
48. Atsuta Y, Okuda K 1981 On the stereospecificity of cholestanetriol 26-monooxygenase. J Biol Chem **256**:9144–9146.
49. Björkhem I 1992 Mechanism of degradation of the steroid side chain in the formation of bile acids. J Lipid Res **33**:455–471.
50. Björkhem I, Gustafsson J 1973 Omega-hydroxylation of steroid side-chain in biosynthesis of bile acids. Eur J Biochem **36**:201–212.
51. Cronholm T, Johansson G 1970 Oxidation of 5β-cholestane-3α,7α,12α-triol by rat liver microsomes. Eur J Biochem **16**:373–381.
52. Taniguchi S, Hoshita N, Okuda K 1973 Enzymatic characteristics of CO-sensitive 26-hydroxylase system for 5β-cholestane-3α,7α,12α-triol in rat-liver mitochondria and its intramitochondrial localization. Eur J Biochem **40**:607–617.
53. Björkhem I, Gustafsson J 1974 Mitochondrial ω-hydroxylation of the cholesterol side chain. J Biol Chem **249**:2528–2535.
54. Okuda KI, Weber P, Ullrich V 1977 Photochemical action spectrum of the CO-inhibited 5β-cholestane-3α,7α,12α-triol 26-hydroxylase system. Biochem Biophys Res Commun **74**:1071–1076.
55. Sato R, Atsuta Y, Imai Y, Taniguchi S, Okuda KI 1977 Hepatic mitochondrial cytochrome P-450: Isolation and functional characterization. Proc Natl Acad Sci USA **74**:5477–5481.
56. Pedersen JI, Oftebro H, Vanngard T 1977 Isolation from bovine liver mitochondria of a soluble ferredoxin active in a reconstituted steroid hydroxylation reaction. Biochem Biophys Res Commun **76**:666–673.
57. Okamura T, John ME, Zuber MX, Simpson ER, Waterman MR 1985 Molecular cloning and amino acid sequence of the precursor form of bovine adrenodoxin: Evidence for a previously unidentified COOH-terminal peptide. Proc Natl Acad Sci USA **82**:5705–5709.
58. Hanukoglu I, Gutfinger T 1989 cDNA sequence of adrenodoxin reductase. Identification of NADP-binding sites in oxidoreductases. Eur J Biochem **180**:479–484.
59. Ugele B, Locher M, Burger HJ, Gebhardt R 1986 Is there a heterogeneity of liver parenchyma in taurocholate uptake? In: Paumgartner G, Stiehl A, Gerok W (eds) Bile Acids and the Liver. MTP Press, Lancaster, Boston, The Hague, and Dordrecht, pp. 153–160.
60. Pedersen JI, Godager HK 1978 Purification of NADPH-ferredoxin reductase from rat liver mitochondria. Biochim Biophys Acta **525**:28–36.
61. Okuda K, Masumoto O, Ohyama Y 1988 Purification and characterization of 5β-cholestane-3α,7α,12α-triol 27-hydroxylase from female rat liver mitochondria. J Biol Chem **263**:18138–18142.
62. Björkhem I, Gustafsson J, Johansson G, Persson B 1975 Biosynthesis of bile acids in man. Hydroxylation of the C_{27}-steroid side chain. J Clin Invest **55**:478–486.
63. Pedersen JI, Holmberg I, Björkhem I 1979 Reconstitution of vitamin D_3 25-hydroxylase activity with a cytochrome P-450 preparation from rat liver mitochondria. FEBS Lett **98**:394–398.
64. Masumoto O, Ohyama Y, Okuda K 1988 Purification and characterization of vitamin D 25-hydroxylase from rat liver mitochondria. J Biol Chem **263**:14256–14260.
65. Björkhem I, Holmberg I 1980 Properties of a reconstituted vitamin D_3 25-hydroxylase from rat liver mitochondria. J Biol Chem **255**:5244–5249.
66. Dählback H, Wikvall K 1988 25-Hydroxylation of vitamin D_3 by a cytochrome P-450 from rabbit liver mitochondria. Biochem J **252**:207–213.
67. Dählback H 1988 Characterization of the liver mitochondrial cytochrome P450 catalyzing the 26-hydroxylation of 5β-cholestane-3α,7α,12α-triol. Biochem Biophys Res Commun **157**:30–36.
68. Ohyama Y, Masumoto O, Usui E, Okuda K 1991 Multi-functional property of rat liver mitochondrial cytochrome P-450. J Biochem (Tokyo) **109**:389–393.
69. Dixon M, Webb EC 1979 The investigation of enzyme specificity. In: Dixon M, Webb EC (eds) Enzymes. Longmans London, pp. 232–234.
70. Oftebro H, Saarem K, Björkhem I, Pedersen JI 1981 Side chain hydroxylation of C_{27}-steroids and vitamin D_3 by a cytochrome P-450 enzyme system isolated from human liver mitochondria. J Lipid Res **22**:1254–1264.
71. Saarem K, Pedersen JI 1985 25-hydroxylation of 1α-hydroxyvitamin D-3 in rat and human liver. Biochim Biophys Acta **840**:117–126.
72. Su P, Rennert H, Shayiq RM, Yamamoto R, Zheng Y, Addya S, Strauss JFI, Avadhani NG 1990 A cDNA encoding a rat mitochondrial cytochrome P450 catalyzing both the 26-hydroxylation of vitamin D_3: Gonadotropic regulation of the cognate mRNA in ovaries. DNA Cell Biol **9**:657–665.
73. Guo YD, Strugnell S, Back DW, Jones G 1993 Transfected human liver cytochrome P-450 hydroxylates vitamin D analogs at different side-chain positions. Proc Natl Acad Sci USA **90**:8668–8672.
74. Nebert DW, McKinnon RA 1994 Cytochrome P450: Evolution and functional diversity. In: Boyer JL, Ockner RK (eds) Progress in Liver Diseases. Saunders, Philadelphia, Pennsylvania, pp. 63–97.
75. Sakaki T, Akiyoshi-Shibata M, Yabusaki Y, Ohkawa H 1992 Organelle-targeted expression of rat liver cytochrome P450c27

in yeast. Genetically engineered alteration of mitochondrial P450 into a microsomal form creates a novel functional electron transport chain. J Biol Chem **267**:16497–16502.

76. Shayiq RM, Avadhani NG 1992 Sequence complementarity between the 5′-terminal regions of mRNAs for rat mitochondrial cytochrome P-450c27/25 and a growth hormone-inducible serine protease inhibitor. A possible gene overlap. J Biol Chem **267**:2421–2428.
77. Mullick J, Addya S, Sucharov C, Avadhani NG 1995 Localization of a transcription promoter within the second exon of the cytochrome P-450c27/25 gene for the expression of the major species of two-kilobase mRNA. Biochemistry **34**:13729–13742.
78. Wiese RJ, Uhland-Smith A, Ross TK, Prahl JM, DeLuca HF 1992 Up-regulation of the vitamin D receptor in response to 1,25-dihydroxyvitamin D_3 results from ligand-induced stabilization. J Biol Chem **267**:20082–20086.
79. Vlahcevic ZR, Jairath SK, Heuman DM, Stravitz RT, Hylemon PB, Avadhani NG, Pandak WM 1996 Transcriptional regulation of hepatic sterol 27-hydroxylase by bile acids. Am J Physiol **270**:G646–G652.
80. Twisk J, Hoekman FM, Lehmann EM, Meijer P, Mager WH, Princen HMG 1995 Insulin suppresses bile acid synthesis in cultured rat hepatocytes by down-regulation of cholesterol 7α-hydroxylase and sterol 27-hydroxylase gene transcription. Hepatology **21**:501–510.
81. Usui E, Noshiro M, Ohyama Y, Okuka K 1990 Unique property of liver mitochondrial P450 to catalyze the two physiologically important reactions involved in both cholesterol catabolism and vitamin D activation. FEBS Lett **274**:175–177.
82. Akiyoshi-Shibata M, Usui E, Sakaki T, Yabusaki Y, Noshiro M, Okuda K, Ohkawa H 1991 Expression of rat liver vitamin D_3 25-hydroxylase cDNA in *Saccharomyces cerevisiae*. FEBS Lett **280**(2):367–370.
83. Cali JJ, Hsieh CL, Francke U, Russell DW 1991 Mutations in the bile acid biosynthetic enzyme sterol 27-hydroxylase underlie cerebrotendinous xanthomatosis. J Biol Chem **266**:7779–7783.
84. Leitersdorf E, Reshef A, Meiner V, Levitzki R, Schwartz SP, Dann EJ, Berkman N, Cali JJ, Klapholz L, Berginer VM 1993 Frameshift and splice-junction mutations in the sterol 27-hydroxylase gene cause cerebrotendinous xanthomatosis in Jews of Moroccan origin. J Clin Invest **91**:2488–2496.
85. Kuriyama M, Fujiyama J, Kubota R, Nakagawa M, Osame M 1993 [Letter to the editor]. Metab Clin Exp **42**:1497.
86. Berginer VM, Shany S, Alkalay D, Berginer J, Dekel S, Salen G, Tint GS, Gazit D 1993 Osteoporosis and increased bone fractures in cerebrotendinous xanthomatosis. Metab Clin Exp **42**:69–74.
87. Leitersdorf E, Safadi R, Meiner V, Reshef A, Björkhem I, Friedlander Y, Morkos S, Berginer VM 1994 Cerebrotendinous xanthomatosis in the Israeli Druze: Molecular genetics and phenotypic characteristics. Am J Hum Genet **55**:907–915.
88. Guo YD, Strugnell S, Jones G 1991 Identification of a human liver mitochondrial cytochrome P-450 cDNA corresponding to the vitamin D_3-25-hydroxylase. J Bone Miner Res **6**:S120 (abstract).
89. Strugnell S, Byford V, Makin HLJ, Moriarty RM, Gilardi R, LeVan LW, Knutson JC, Bishop CW, Jones G 1995 1α,24(*S*)-Dihydroxyvitamin D_2: A biologically active product of 1α-hydroxyvitamin D_2 made in the human hepatoma, Hep3B. Biochem J **310**:233–241.
90. Holmberg-Betsholtz I, Lund E, Björkhem I, Wikvall K 1993 Sterol 27-hydroxylase in bile acid biosynthesis. Mechanism of oxidation of 5β-cholestane-3α, 7α,12α,27-tetrol into 3α,7α,12α-trihydroxy-5β-cholestanoic acid. J Biol Chem **268**:11079–11085.
91. Axén E, Postlind H, Sjöberg H, Wikvall K 1994 Liver mitochondrial cytochrome P450 CYP27 and recombinant-expressed human CYP27 catalyze 1α-hydroxylation of 25-hydroxyvitamin D_3. Proc Natl Acad Sci USA **91**:10014–10018.
92. Arlazoroff A, Roitberg B, Werber E, Shidlo R, Berginer VM 1991 Epileptic seizure as a presenting symptom of cerebrotendinous xanthomatosis. Epilepsia **32**:657–661.
93. Jones G, Schnoes HK, Levan L, DeLuca HF 1980 Isolation and identification of 24-hydroxyvitamin D_2 and 24,25-dihydroxyvitamin D_2. Arch Biochem Biophys **202**:450–457.
94. Wichmann JK, Schnoes HK, DeLuca HF 1981 Isolation and identification of 24(*R*)-hydroxyvitamin D_3 from chicks given large doses of vitamin D_3. Biochemistry **20**:2350–2353.
95. Horst RL, Koszewski NJ, Reinhardt TA 1990 1α-Hydroxylation of 24-hydroxyvitamin D_2 represents a minor physiological pathway for the activation of vitamin D_2 in mammals. Biochemistry **29**:578–582.
96. Axén E, Postlind H, Wikvall K 1995 Effects on CYP27 mRNA expression in rat kidney and liver by 1α,25-dihydroxyvitamin D_3, 1a-hydroxylase activity. Biochem Biophys Res Commun **215**:136–141.
97. Hollis BW 1990 25-Hydroxyvitamin D_3-1α-hydroxylase in porcine hepatic tissue: Subcellular localization to both mitochondria and microsomes. Proc Natl Acad Sci USA **87**:6009–6013.
98. Takeuchi A, Okano T, Kobayashi T 1991 The existence of 25-hydroxyvitamin D_3-1α-hydroxylase in the liver of carp and bastard halibut. Life Sci **48**:275–282.
99. Takeuchi A, Okano T, Sekimoto H, Kobayashi T 1994 The enzymatic formation of 1α,25-dihydroxyvitamin D_3 from 25-hydroxyvitamin D_3 in the liver of fetal rats. Comp Biochem Physiol (C) **109C**:1–7.
100. Dilworth FJ, Scott I, Green A, Strugnell S, Guo YD, Roberts EA, Kremer R, Calverley MJ, Makin HLJ, Jones G 1995 Different mechanisms of hydroxylation site selection by liver and kidney cytochrome P450 species (CYP27 and CYP24) involved in vitamin D metabolism. J Biol Chem **270**:16766–16774.
101. Dilworth FJ, Strugnell S, Guo YD, Makin HLJ, Calverley MJ, Jones G 1994 Site and rate of hydroxylation of 1α-OH-D_3 analogs by CYP27 not altered by increasing length or changing orientation of vitamin D_3 side chain. In: Norman AW, Bouillon R, Thomasset M (eds) Vitamin D: A Pluripotent Steroid Hormone: Structural Studies, Molecular Endocrinology and Clinical Applications. de Gruyter, New York, pp. 131–132.
102. Usui E, Noshiro M, Ohyama Y, Okuda K 1990 Unique property of liver mitochondrial P450 to catalyze the two physiologically important reactions involved in both cholesterol catabolism and vitamin D activation. FEBS Lett **274**:175–177.
103. Stravitz RT, Vlahcevic ZR, Russell TL, Heizer ML, Avadhani NG, Hylemon PB 1996 Regulation of sterol 27-hydroxylase and an alternative pathway of bile acid biosynthesis in primary cultures of rat hepatocytes. J Steroid Biochem Mol Biol **57**:337–347.
104. Twisk J, Hoekman MFM, Mager WH, Moorman AFM, De Boer PAJ, Scheja L, Princen HMG, Gebhardt R 1995 Heterogenous expression of cholesterol 7α-hydroxylase and sterol 27-hydroxylase genes in the rat liver lobulus. J Clin Invest **95**:1235–1243.
105. Araya Z, Sjöberg H, Wikvall K 1995 Different effects on the expression of CYP7 and CYP27 in rabbit liver by cholic acid and cholestyramine. Biochem Biophys Res Commun **216**:868–873.
106. Bell NH, Shaw S, Turner RT 1984 Evidence that 1,25-dihydroxyvitamin D_3 inhibits the hepatic production of 25-hydroxyvitamin D in man. J Clin Invest **74**:1540–1544.
107. Halloran BP, Bikle DD, Levens MJ, Castro ME, Globus RK, Holton E 1986 Chronic 1,25-dihydroxyvitamin D_3 administration in the rat reduces the serum concentration of 25-hydroxyvitamin

D by increasing metabolic clearance rate. J Clin Invest **78**:622–628.

108. Haddad P, Gascon-Barré M, Brault G, Plourde V 1986 Influence of calcium or 1,25-dihydroxyvitamin D_3 supplementation on the hepatic microsomal and *in vivo* metabolism of vitamin D_3 in vitamin D-depleted rats. J Clin Invest **78**:1529–1537.
109. Clements MR, Johnson L, Fraser DR 1987 A new mechanisms for induced vitamin D deficiency in calcium deprivation. Nature **325**:62–65.
110. Princen HMG, Meijer P, Wolthers BG, Vonk RJ, Kuipers F 1991 Cyclosporine A blocks bile acid synthesis in cultured hepatocytes by specific inhibition of chenodeoxycholic acid synthesis. Biochem J **275**:501–505.
111. Dählback-Sjöberg H, Björkhem I, Princen HMG 1993 Selective inhibition of mitochondrial 27-hydroxylation of bile acid intermediates and 25-hydroxylation of vitamin D_3 by cyclosporin A. Biochem J **2933**:203–206.
112. Plourde V, Gascon-Barré M, Willems B, Huet PM 1988 Severe cholestasis leads to vitamin D depletion without perturbing its C-25 hydroxylation in the dog. Hepatology **8**:1577–1585.
113. Bolt MJG, Sitrin MD, Favus MJ, Rosenberg IH 1981 Hepatic vitamin D 25-hydroxylase: Inhibition by bile duct ligation or bile salts. Hepatology **1**:436–440.
114. Martial J, Plourde V, Gascon-Barré M 1985 Sequestration of vitamin D_3 by the fetal and neonatal rat liver. Biol Neonate **48**:21–28.
115. Plourde V, Haddad P, Gascon-Barré M 1985 Microsomal C-25 hydroxylation of [^{3}H]vitamin D_3 by the fetal and neonatal rat liver. Pediatr Res **19**:1206–1209.
116. Thierry-Palmer M, Cullins S, Rashada S, Gray TK, Free A 1986 Development of vitamin D_3 25-hydroxylase activity in rat liver microsomes. Arch Biochem Biophys **250**:120–127.
117. Miller ML, Ghazarian JG 1980 Studies on vitamin D_3 metabolism. Discrete liver cytosolic binding proteins for vitamin D_3 and 25-hydroxyvitamin D_3. Biochem Biophys Res Commun **96**:1619–1625.
118. Tsankova V, Visentin M, Cantoni L, Carelli M, Tacconi MT 1996 Peripheral benzodiazepine receptor ligands in rat liver mitochondria: Effect on 27-hydroxylation of cholesterol. Eur J Pharmacol **299**:197–203.
119. Delvin EE, Glorieux FH, Salle BL, David L, Varenne JP 1982 Control of vitamin D metabolism in preterm infants: Feto-maternal relationships. Arch Dis Child **57**:754–757.
120. Hillman LS, Hoff N, Salmons S, Martin L, McAlister WH, Haddad J 1985 Mineral homeostasis in very premature infants: Serial evaluation of serum 25-hydroxyvitamin D, serum minerals, and bone mineralization. J Pediatr **106**:970–980.
121. Salle BL, Glorieux FH, Delvin EE, David LS, Meunier G 1983 Vitamin D metabolism in preterm infants. Serial serum calcitriol values during the first four days of life. Acta Paediatr Scand **72**:203–206.
122. Hayashi SI, Usui E, Okuda K 1988 Sex-related difference in vitamin D_3 25-hydroxylase of rat liver microsomes. J Biochem (Tokyo) **103**:863–866.
123. Addya A, Zheng YM, Shayiq RM, Fan J, Avadhani NG 1991 Characterization of a female-specific hepatic mitochondrial cytochrome P-450 whose steady-state level is modulated by testosterone. Biochemistry **30**:8323–8330.
124. Demers C, Lapointe R, Valiquette L, Guo DY, Jones G, Gascon-Barré M 1995 Human liver expression of CYP-27, a gene encoding a cytochrome P-450 active in the C-25 hydroxylation of vitamin D_3. J Bone Miner Res **10**:S493(abstract)
125. Ichikawa F, Sato K, Nanjo M, Nishii Y, Shinki T, Takahashi N, Suda T 1995 Mouse primary osteoblasts express vitamin D_3 25-hydroxylase mRNA and convert 1α-hydroxyvitamin D_3 into 1a,25-dihydroxyvitamin D_3. Bone **16**:129–135.
126. Bühler R, Lindros KO, Nordling A, Johansson I, Ingelman-Sundberg M 1992 Zonation of cytochrome *P450* isozyme expression and induction in rat liver. Eur J Biochem **204**:407–412.
127. Delvin EE, Arabian A, Glorieux FH 1978 Kinetics of liver microsomal cholecalciferol 25-hydroxylase in vitamin D-depleted and -repleted rats. Biochem J **172**:417–422.
128. Benbrahim N, Dubé C, Vallières S, Gascon-Barré M 1988 The calcium ionophore A23187 is a potent stimulator of the vitamin D_3-25 hydroxylase in hepatocytes isolated from normocalcemic vitamin D-depleted rats. Biochem J **255**:91–97.
129. Holmberg I, Berlin T, Ewerth S, Björkhem I 1986 25-Hydroxylase activity in subcellular fractions from human liver. Evidence for different rates of mitochondrial hydroxylation of vitamin D_2 and D_3. Scand J Clin Lab Invest **46**:785–790.
130. Björkhem I, Andersson O, Diczfalusy U, Sevastik B, Xiu RJ, Duan C, Lund E 1994 Atherosclerosis and sterol 27-hydroxylase: Evidence for a role of this enzyme in elimination of cholesterol from human macrophages. Proc Natl Acad Sci USA **91**:8592–8596.
131. Reiss AB, Martin KO, Javitt NB, Martin DW, Grossi EA, Galloway AC 1994 Sterol 27-hydroxylase: High levels of activity in vascular endothelium. J Lipid Res **35**:1026–1030.
132. Fukushima M, Nishii Y, Suzuki M, Suda T 1978 Comparative studies on the 25-hydroxylations of cholecalciferol and 1α-hydroxycholecalciferol in perfused rat liver. Biochem J **170**:495–502.

CHAPTER 5

The 25-Hydroxyvitamin D 1α-Hydroxylase

HELEN L. HENRY Department of Biochemistry, University of California at Riverside, Riverside, California

I. OCCURRENCE AND CHARACTERISTICS OF 25OHD$_3$ 1α-HYDROXYLASE

A. The Kidney as the Site of Production of 1,25(OH)$_2$D$_3$

It is now well accepted that vitamin D is a precursor of the sterol hormone 1α,25-dihydroxyvitamin D$_3$ [1,25(OH)$_2$D$_3$]. The general pathway of production of 1,25(OH)$_2$D$_3$ is shown in Fig. 1. It has been appreciated for some time [1,2] that the kidney is the major site of production of circulating 1,25(OH)$_2$D$_3$, although as described below, other tissues and cell types have been shown to produce 1,25(OH)$_2$D$_3$ from 25-hydroxyvitamin D$_3$ (25OHD$_3$) under experimental conditions.

B. Characteristics of the Proteins Involved in the 1α-Hydroxylation of 25OHD$_3$

Enzymes that hydroxylate endogenous steroids in a stereospecific manner are called mixed-function oxidases because they reduce one atom of molecular oxygen to water and one atom to the hydroxyl group to be stereospecifically incorporated into the steroid (Fig. 2). Such reactions are important in the pathways of the production of androgens, estrogens, progestins, mineralcorticoids, and glucocorticoids. Indeed, the first step in the production of these steroid hormones, the cleavage of six carbons from the side chain of cholesterol, results from two successive hydroxylations at C-20 and C-22. Some mixed-function oxidases are microsomal (e.g., the 17α-hydroxylase/C-17–C-20 lyase and the 21-hydroxylase), and some (e.g., cholesterol side chain cleavage and the 11β-hydroxylase) are located in the inner mitochondrial membrane. The 25OHD$_3$ 1α-hydroxylase is believed to belong to the latter group since its activity is located in kidney mitochondria.

For mitochondrial mixed-function oxidases, the electrons for the reduction of molecular oxygen are ultimately derived from NADPH, most likely generated intramitochondrially from tricarboxylic acid cycle intermediates. *In vitro* assays of 1α-hydroxylase activity in isolated mitochondria have used either malate [3] or other citric acid cycle intermediates [4] as a source of reducing equivalents. The current model for mitochondrial steroid hydroxylases involves the sequential transfer of electrons from NADPH to NADPH-ferredoxin

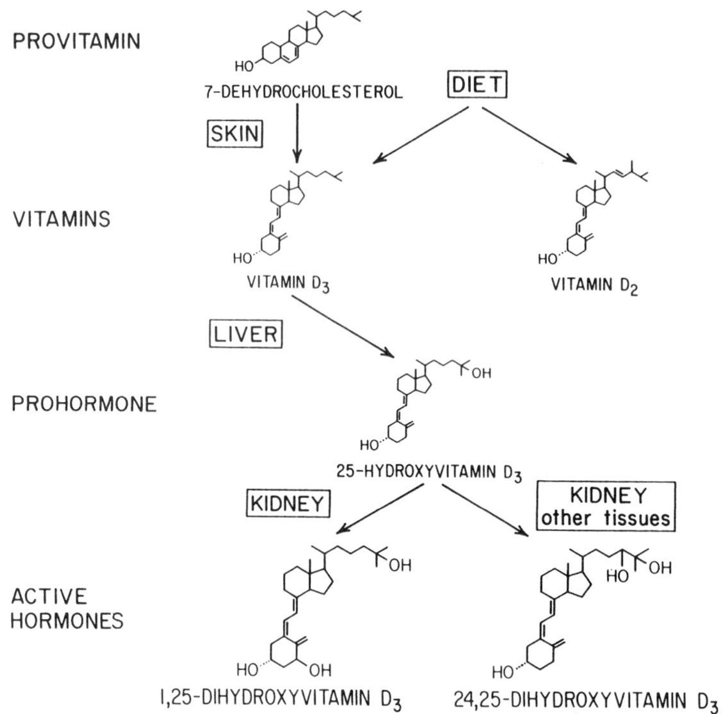

FIGURE 1 Basic metabolic pathway for the major circulating vitamin D metabolites. The two renal hydroxylases that metabolize $25OHD_3$ are, in general, regulated in a reciprocal fashion to one another.

reductase, a 50-kDa protein that, in the case of adrenal and gonadal steroid hydroxylases, is loosely associated with the inner mitochondrial membrane [5]. The electrons are then passed to ferredoxin, a 11-kDa mitochondrial matrix protein that shuttles between the NADPH-ferredoxin reductase and the terminal component of the

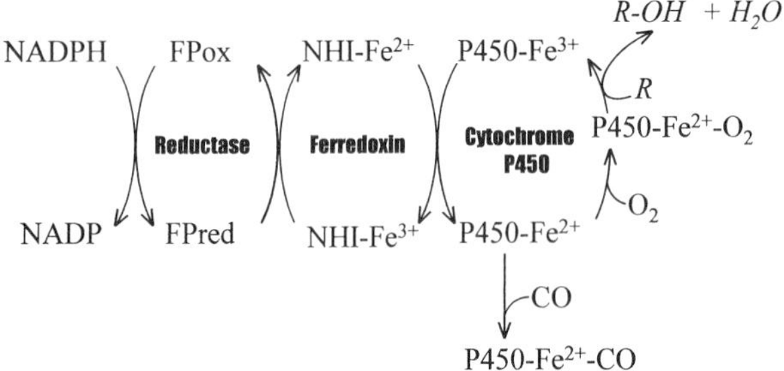

FIGURE 2 Electron transport chain for mitochondrial steroid hydroxylases. The general class of mitochondrial mixed-function oxidases, of which the $25OHD_3$ 1α-hydroxylase is a member, consists of three components in or associated with the inner mitochondrial membrane. Cytochrome P450 reduces molecular oxygen to water and to the hydroxyl group to be incorporated into the steroid; it confers the stereospecificity of the hydroxylation reaction. In its reduced form, cytochrome P450 binds carbon monoxide and in this form absorbs light of 450 nm. The flavoprotein (FP) is NADPH-ferredoxin reductase, and the nonheme iron (NHI) protein is ferredoxin. Ox, oxidized; red, reduced.

hydroxylase machinery, cytochrome P450 (so named because of its distinct spectral characteristics when carbon monoxide is bound to it). As discussed below, each of these three proteins has been the subject of study in the context of the 1α-hydroxylation of $25OHD_3$.

1. NADPH-FERREDOXIN REDUCTASE

The flavoprotein NADPH-ferredoxin reductase, which accepts electrons from intramitochondrially generated NADPH, has been purified from mammalian adrenal glands [6,7] and from pig kidney mitochondria [8] and does not appear to differ significantly in size (~52 kDa) or enzymatic characteristics from those reductases from other mammalian tissues that carry out hydroxylations in the production of endogenous steroids such as the adrenal gland and the corpus luteum. The porcine kidney NADPH-ferredoxin reductase is also immunologically similar to the bovine adrenal protein. cDNAs for bovine and human NADPH-ferredoxin reductase have been cloned and shown to code for a mitochondrial leader sequence in addition to the mature protein [9].

2. FERREDOXIN

a. Characteristics of Mitochondrial Ferredoxin Vertebrate mitochondrial ferredoxins are nonheme iron–sulfur proteins of 114–128 amino acids, which on electrophoresis generally migrate at 11–12 kDa. cDNA analysis indicates a mitochondrial leader sequence of 58–62 amino acids, depending on the species. Some typical sequences of vertebrate ferredoxins are shown in Fig. 3. The ferredoxins involved in the mitochondrial hydroxylation of endogenous steroids are matrix proteins that shuttle between NADPH-ferredoxin reductase and cytochrome P450 to deliver, one at a time, the two electrons required for the reduction of molecular oxygen. The ferredoxins from mammalian adrenal glands have been studied most extensively [10,11], although the protein has been isolated from several other steroidogenic tissues [12].

Chick kidney ferredoxin was first partially purified and shown to be required for reconstitution of 1α-hydroxylase activity in solubilized mitochondrial preparations in the mid-1970s [13]. This protein cross-reacted with antibodies against bovine adrenal ferredoxin, but no molecular weight information was given. A molecular weight of 11,900 was subsequently reported for ferreodoxin isolated from chick kidney mitochondria, and no effect of vitamin D status was observed on the final amount of purified protein obtained [14]. In preparations from porcine kidney and bovine adrenal mitochondria, two forms of ferredoxin with distinct N-terminal sequences were reported, one evidently a mature protein and one suggested to be a precursor, containing a mitochondrial signaling sequence [15]. The physiological significance of the latter iron–sulfur protein and its involvement in renal $25OHD_3$ metabolism have not yet been elucidated.

cDNAs encoding mitochondrial ferredoxin from bovine [16] and human [17] adrenal glands have been cloned. Porcine kidney ferredoxin cDNA was found to be very similar to mammalian adrenal ferredoxins and identical to porcine adrenal ferredoxin. Similarly, the cDNA for chick kidney ferredoxin has been cloned [18] and shown to be virtually identical to the chick testis protein [19]. These results from both pigs and chicks suggest that a single gene encodes ferredoxin for the

```
        1                                                  50
Chkfdx                                                   PAVR
Pigfdx      MAVRLL RVASAALGDT AVRWQPLVGP RAGNRGPGGS IWLGLGGRAA
Bovadx      MAARLL RVASAALGDT AGRWRLLVRP RAGAGGLRGS RGPGLGGGAV
Humadx  MAAAGGARLL RAASAVLGGP AGRWLHHAGS RAGSSGLLRN RGPG..GSAE

        51         |                                      100
Chkfdx  TLRPLSLSSR AACSSEDKIT VHFINRDGDK LTAKGKPGDS LLDVVVENNL
Pigfdx  AARTLSLSAR AWSSSEDKIT VHFINRDGKT LTTQGKVGDS LLDVVIENNL
Shpfdx               SSSEDKVT VNFINRDGET LTTKGKVGDS LLDVVVENNL
Bovadx  ATRTLSVSGR AQSSSEDKIT VHFINRDGET LTTKGKIGDS LLDVVVQNNL
Humadx  ASRSLSVSAR ARSSSEDKIT VHFINRDGET LTTKGKVGDS LLDVVVENNL

        101    *      *  *•                   • •    •     •150
Chkfdx  DIDGFGACEG TLACSTCHLI FEDHIFEKLD AITDEEMDML DLAYGLTETS
Pigfdx  DIDGFGACEG TLACSTCHLI FEDHIFEKLE AITDEENDML DLAYGLTDRS
Shpfdx  DIDGFGACEG TLACSTCHLI FEQHIYEKLE AITDEENDML DLAYGLTDRS
Bovadx  DIDGFGACEG TLACSTCHLI FEQHIFEKLE AITDEENDML DLAYGLTDRS
Humadx  DIDGFGACEG TLACSTCHLI FEDHIYEKLD AITDEENDML DLAYGLTDRS

        151   *                     Δ           190
Chkfdx  RLGCQICLKK SMDNMTVRVP EAVADARQSV DLSKNS
Pigfdx  RLGCQICLTK AMDNMTVRVP EAVADARESI DLGKNSSKLE
Shpfdx  RLGCQICLTK AMDNMTVRVP DAVSDARESI DMGMNSSKIE
Bovadx  RLGCQICLTK AMDNMTVRVP DAVSDARESI DMGMNSSKIE
Humadx  RLGCQICLTK SMDNMTVRVP ETVADARQSI DVGKTS
```

FIGURE 3 Comparison of amino acid sequences of ferredoxins from chicken, pig, sheep, bovine, and human. The beginning of each mature peptide is marked by a vertical line above the first residue. Stars mark positions of the four cysteines that coordinate the iron–sulfur center. Dots mark highly conserved residues that have been identified as critical for interactions between ferredoxin and NADPH-ferredoxin reductase and cytochrome P450. The bold glutamate in the chick sequence is the only deviation from these conserved amino acids.

various tissues in which it is expressed. The existence of a single gene for ferredoxin is further supported by Southern blot analysis of bovine [20] and chick genomic DNA [18]. In the human, two ferredoxin genes have been identified, but their coding regions for the mature protein are identical [21]. Taken together the available evidence indicates that a single ferredoxin protein serves as the electron shuttle for all mitochondiral steroid hydroxylases in the species which have been examined thus far.

As is the case for ferredoxins in other steroidogenic tissues, chick kidney ferredoxin is a mitochondrial matrix protein [22]. This submitochondrial localization is consistent with its function as an electron carrying shuttle between the NADPH-ferredoxin reductase and cytochrome $P450_{1\alpha}$. Reconstitution assays in a number of laboratories have demonstrated the absolute requirement of 1α-hydroxylase activity for ferredoxin [14,15,22]. Chick kidney ferredoxin has been overexpressed in *Escherichia coli* and shown to be approximately twice as active as bovine adrenal ferredoxin in supporting 1α-hydroxylation in a reconstituted assay system [23], further supporting the similarities that the 1α-hydroxylase shares with other mitochondrial mixed-function oxidases which hydroxylate endogenous steroids.

b. Regulation of Mitochondrial Ferredoxin Levels Mammalian ferredoxins and their mRNAs have been shown to be developmentally and hormonally regulated [24–27]. For example, hypophysectomy decreases and replacement treatment with adrenocorticotropin (ACTH) increases adrenal ferredoxin mRNA levels in the rat [25,28]. In a study of several cell types, ferredoxin mRNA was stimulated by appropriate pituitary tropic hormones, possibly through a cAMP-dependent mechanism, and ferredoxin mRNA levels decreased in the testis and adrenal during fetal development [24].

Chick kidney mitochondrial ferredoxin mRNA is upregulated approximately 40% in the vitamin D-deficient state [18]. Although kidney mitochondrial ferredoxin is also presumably required for the 24-hydroxylation of vitamin D, which is, in general, regulated reciprocally to the 1α-hydroxylase, 1α-hydroxylase activity in vitamin D deficiency is much higher than is 24-hydroxylase activity in vitamin D repletion in the chick. Therefore, increased mRNA levels of ferredoxin mRNA in vitamin D deficiency is not likely to be involved in the reciprocal regulation of these two enzymes but rather is due to the increased demand of maximal 1α-hydroxylase activity brought about by vitamin D deficiency.

Ferredoxin phosphorylation has been suggested as a posttranslational mechanism for the regulation of its steroid hydroxylation activity in the adrenal gland and kidney. Adrenal ferredoxin can be phosphorylated *in vitro* by cAMP-dependent protein kinase, with the serine at position 88 reported to be the most likely site of phosphorylation [29]. However, no convincing evidence of phosphorylated adrenal ferredoxin *in vivo* has appeared, nor has a role been established for such a phosphoferredoxin in the physiological regulation of adrenal or gonadal steroidogenesis.

Phosphorylation of chick kidney ferredoxin by a mitochondrial cAMP-stimulated protein kinase was reported to be associated with changes in 1α- and $24R$-hydroxylase activities; however, the differences were small, and no statistical analysis of the data were given [30]. Parathyroid hormone (PTH) decreased phosphorylation of ferredoxin in rat kidney slices, which was accompanied by increased 1α-hydroxylase activity [31], but in this study the calcium ionophore A23187 stimulated phosphorylation of ferredoxin but did not affect 1α-hydroxylase activity.

More recently, ferredoxin from primary cultures of chick kidney cells has been shown to exist as a phosphoprotein and to be rapidly dephosphorylated by treatment of the cells with the phorbol ester 12-*O*-tetradecanoylphorbol-13-acetate (TPA) [22]. This dephosphorylation is accompanied by decreased 1α-hydroxylase activity, suggesting a role of ferredoxin phosphorylation and dephosphorylation in the regulation of 1α-hydroxylation of $25OHD_3$. However, the molecular mechanisms involved remain to be determined.

3. Cytochrome P450

The evidence that the 1α-hydroxylase is a member of the class of cytochrome P450-dependent steroid hydroxylases was obtained in the mid-1970s and consists of the following: (1) solubilized preparations of chick kidney mitochondria were shown to contain cytochrome P450 and to carry out 1α-hydroxylation of $25OHD_3$ [32]; (2) more directly, it was shown that inhibition of 1α-hydroxylase activity in isolated chick kidney mitochondria by carbon monoxide was reversed specifically by light of 450 nm [3].

There have been sporadic reports (Table I) of the partial purification of 1α-hydroxylase cytochrome P450 or preparation of antibodies against this protein. An apparently electrophoretically pure bovine kidney cytochrome P450, with a molecular weight of 57,000, retained putative 1α-hydroxylase activity [33]; however, Hiwatashi *et al.* [33] did not provide confirmation of the identity of the product of $25OHD_3$ hydroxylation, nor did they indicate the staining procedure used in the electrophoretic analysis. A single report of monoclonal antibodies recognizing 1α- and 24-hydroxylase activities appeared [34], but no subsequent work has appeared from this group. In the early 1990s, there were two

TABLE I Reports of Biochemical Identification of 1α-Hydroxylase Cytochrome P450

Species	Tissue	Subcellular localization	MW[a]	Activity (pmol/min/mg protein)	Ref.
Chick	Kidney	Mitochondria	NR[b]	100	33
Chick	Kidney	Mitochondria	57,000	NR	30
Chick	Kidney	Mitochondria	57,000	NR	36
Chick	Kidney	Mitochondria	59,000	245	35
Chick	Kidney	Mitochondria	58,000	278	38
Bovine	Kidney	Mitochondria	52,000	<1	34
Porcine	Kidney	Mitochondria	NR	153	37
Porcine	Liver	Mitochondria	54,000	30	42
Porcine	Liver	Mitochondria/ microsomes	NR	<1	40
Rat	Liver	Microsomes	NR	<1	41

[a] Reported molecular weight.
[b] NR, Not reported.

TABLE II Enzymatic Characteristics of the 1α-Hydroxylase

Fraction	V_{max} (pmol/min/mg)	K_m ($\times 10^{-8}$ M)
Isolated mitochondria	16	0.1
Solubilized ammonium sulfate precipitate	12	2.8

reports purporting to have purified the 1α-hydroxylase cytochrome P450 from either vitamin D-repleted or vitamin D-deficient chick kidney mitochondria [35,36]. The purified proteins had molecular weights of 59,000 and 57,000 and were similar, but not identical, in amino acid composition. Amino-terminal sequence analysis also revealed differences between the proteins isolated by the two laboratories, suggesting that one or both was not, indeed, the 1α-hydroxylase. In 1990, 1α-hydroxylase activity was reported to have been separated from steroid side chain hydroxylating enzymes in pig kidney mitochondria [37]. More recently, a 54-kDa protein with 1α-hydroxylase activity and appropriate spectral characteristics was isolated from chick kidney [38]. No sequence information was given to aid in the resolution of previous conflicting reports. Finally, in an abstract [39], the cloning of the 1α-hydroxylase from a rat kidney cDNA library was reported; this study was based on low-stringency hybridization screening with a 24-hydroxylase cDNA. The latter report, if confirmed by appropriate control experiments, will usher in a new era with respect to the study of the 1α-hydroxylase.

In addition to the lack of subsequent definitive structural and functional information regarding kidney mitochondrial preparations containing 1α-hydroxylase activity described above, endeavors to purify a convincing 1α-hydroxylase have been further confounded by reports of 1α-hydroxylase activity measured *in vitro* in preparations from mammalian liver microsomes [40,41] and mitochondria [42]. The latter enzyme has been identified as CYP27, a liver cytochrome P450 involved in bile acid synthesis. mRNA levels for CYP27 are down-regulated by $1,25(OH)_2D_3$ in both the liver and the kidney [43]. The relationship, if any, of these cytochrome P450s to the authentic 1α-hydroxylase remains to be determined.

To summarize the state of 1α-hydroxylase cytochrome P450 purification and cloning at the present time, it is clear that this has been an area of research in the vitamin D field which, while intensively pursued, has as yet met with little tangible success. Whether this is due to the very low abundance of the protein, its low activity on solubilization (see Table II), or the possibility that the 1α-hydroxylation of $25OHD_3$ is promiscuously catalyzed by several proteins, remains to be determined. What is required is the identification of a previously unknown cytochrome P450 that occurs primarily if not exclusively in the kidney, as opposed to other major organs which have been shown to be devoid of 1α-hydroxylase activity such as the intestine and the liver, and whose catalytic activity can be demonstrated to be the 1α-hydroxylation of $25OHD_3$. It is certainly possible that more than one gene for 1α-hydroxylase activity exists.

C. Distribution of the 1α-Hydroxylase

1. Species Distribution

The 1α-hydroxylase, not surprisingly, occurs widely among vertebrate species [44]. None of the species examined in this comparison has since been shown to lack the renal 1α-hydroxylase. Thus, it can be concluded that this hydroxylation reaction is widely distributed among vertebrates and that the ability to produce $1,25(OH)_2D_3$ arose early in vertebrate evolution.

There has been a special interest in the occurrence of vitamin D and its metabolites in fish, due in part to the historical importance of cod liver oil as a source of vitamin D and in part to the question of how fish produce vitamin D under conditions of limited ultraviolet light. Plasma levels of vitamin D metabolites have been measured in several species of fish [45–48]. Fish liver and kidney homogenates have been reported to convert $25OHD_3$ to $1,25(OH)_2D_3$ [44,47,49,50]. More recently [51] both phytoplankton and zooplankton were reported to have abundant quantities of provitamin D and vitamin D and were suggested to be the dietary source of vitamin D for fish.

Another interesting comparative aspect of vitamin D metabolism is posed by the naked mole rat. This nocturnal, cave-dwelling animal with no access to ultraviolet light nevertheless has both the ability to synthesize $1,25(OH)_2D_3$ and receptors for the hormone [52,53]. Thus, although these animals are usually vitamin D deficient, they maintain the enzymatic machinery necessary to produce the active hormonal form and the receptor necessary to respond to it.

The two most common experimental models for the study of the 1α-hydroxylase are the chick, due to the ease of controlling vitamin D status in this species, and the rat, which has been and continues to be the subject of studies of the physiological regulation of calcium homeostasis. Because 1α-hydroxylase activity is highest in the vitamin D-deficient state and 24-hydroxylase activity is present when animals are vitamin D replete (see below), vitamin D status is an important consideration when choosing a model in which to investigate the 1α-hydroxylase. Bovine and porcine kidney have been used as a source of tissue for purification of one or more of the protein components of the 1α-hydroxylase owing to the abundant availability of these tissues. However, the vitamin D status of these animals is not controlled, and studies carried out with renal tissue from them are complicated by the presence of $25OHD_3$ 24-hydroxylase and possibly other renal mitochondrial cytochrome P450 enzymes.

2. EXTRARENAL 1α-HYDROXYLATION OF $25OHD_3$

The kidney was identified as the site of production of circulating $1,25(OH)_2D_3$ in the early 1970s [1,2]. Physiological experiments, for example, studies in nephrectomized animals [54] and clinical experience with patients suffering from chronic renal failure [55,56], strongly supported this concept. Thus, the kidney was the focus of investigation of the 1α-hydroxylase for the decade following its identification as the site of $1,25(OH)_2D_3$ synthesis. In the early 1980s reports began appearing that various bone cell preparations [57,58] and placental decidual cells [59,60] could produce $1,25(OH)_2D_3$ from $25OHD_3$, although the enzymatic production of $1,25(OH)_2D_3$ by trophoblastic tissues has been questioned [61]. In addition, many studies [62–66] have led to the understanding that cells of the lymphohematopoeitic system, which had also been found to be target cells of $1,25(OH)_2D_3$ [67], could produce $1,25(OH)_2D_3$. It is now generally accepted that the major contribution to circulating levels of $1,25(OH)_2D_3$ is made by the proximal tubular cells of the renal cortex [68,69] and that the $1,25(OH)_2D_3$ produced in cells of the lymphohematopoietic system most likely serves autocrine or paracrine functions (see Chapters 29 and 55). It should be noted that 1α-hydroxylation of $25OHD_3$ in the latter cell types does not respond to the regulatory influences involved in calcium homeostasis, that control the renal enzyme activity. It is possible that the gene for this 1α-hydroxylase differs from that found in kidney. Thus, both the production and biological functions of $1,25(OH)_2D_3$ should be considered to be either endocrine or autocrine/paracrine in nature, depending on the site of synthesis under investigation.

II. REGULATION OF 1α-HYDROXYLASE ACTIVITY

A. Regulation of 1α-Hydroxylase Activity in Isolated Kidney Mitochondria

Many early studies of the 1α-hydroxylase examined its regulation in mitochondria isolated from chick or mammalian kidneys (reviewed by Ross and Henry [70]). The regulating agents investigated included calcium and phosphate ions, both of which are required for some activity but, at high concentrations, inhibit 1α-hydroxylase activity. In mammals, a serum protein, reported to be the vitamin D binding protein (DBP) that binds the substrate $25OHD_3$ and limits its availability to the mitochondrial enzyme, was identified [71]. Studies in primary cultures of chick kidney cells confirmed that serum, presumably due to the presence of DBP, reduces $25OHD_3$ entry into cells for further metabolism [72]. Exogenous phospholipids have also been shown to dramatically affect 1α-hydroxylase activity in isolated mitochondria; interestingly, the fatty acid content of the phospholipids appeared to be the dominant factor in the effectiveness of the inhibitory ability of the phospholipid [73]. In retrospect, although investigation of the effects of various agents on 1α-hydroxylase activity in isolated mitochondria have made valuable contributions to our understanding of how the enzyme behaves in these experimental conditions, they have not shed a great deal of light on the basic mechanisms underlying physiological regulation of the 1α-hydroxylase.

An exception to the above assessment has been the synthesis of analogs designed to specifically inhibit the 1α-hydroxylase (Fig. 4). Substitution at C-2 in the A ring with oxygen resulted in the effective inhibitors 2-oxa-3-deoxy-$25OHD_3$ and 2-oxa-9(11)-dehydro-3-deoxy-$25OHD_3$ [74,75]. More recently, A-ring sulfur an-

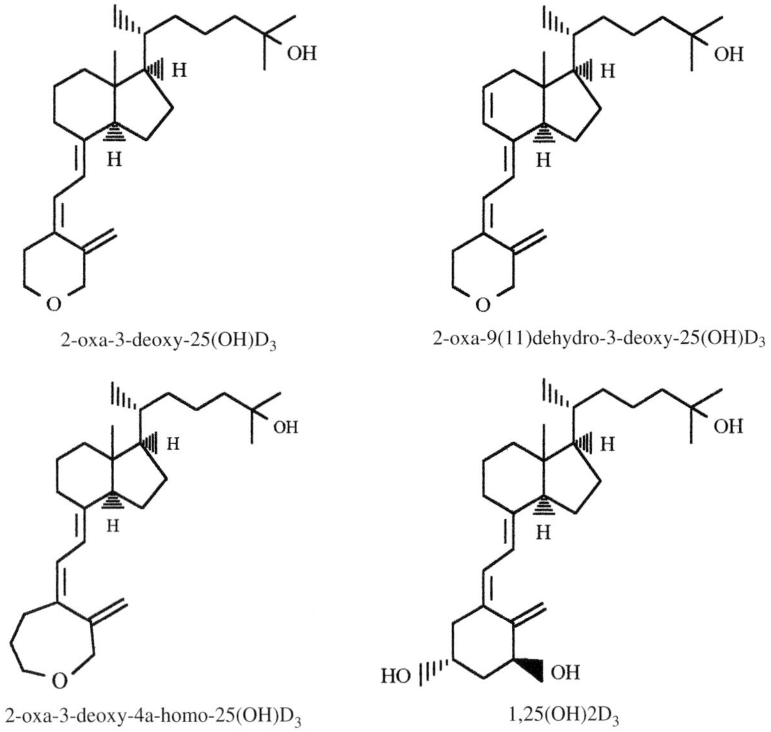

FIGURE 4 Structures of effective 1α-hydroxylase inhibitors.

alogs have been shown to be effective inhibitors of the 1α-hydroxylase, particularly if the oxygen functional groups are in the 3β and 1α positions as in 3-deoxy-3-thia-1α,25$(OH)_2D_3$ [76]. In the latter group of compounds, there was a good correlation between the ability of a compound to inhibit the 1α-hydroxylase and its relative affinity for the chick intestinal vitamin D receptor (VDR), suggesting that the two proteins show similar requirements for the configuration of the A ring. Interestingly, however, modification of the side chain, as in the arocalciferols [77], resulted in analogs that had relatively high affinities for the intestinal VDR but were completely ineffective as inhibitors of the 1α-hydroxylase. These results suggest that the VDR is more tolerant of changes in the side chain than is the 1α-hydroxylase, which, it should be remembered, absolutely requires the 25-hydroxyl group since the parent vitamin D is not 1α-hydroxylated.

B. Regulation of 1α-Hydroxylase Activity in Whole Animals and in Cell Culture

1. VITAMIN D STATUS

It was recognized at the time of the localization of the production of 1,25$(OH)_2D_3$ in the kidney that, in both avian and mammalian species, vitamin D-deficient animals have higher 1α-hydroxylase activity than do vitamin D-replete animals [1,78,79]. The ability of 1,25$(OH)_2D_3$ to replicate this effect in chick kidney cell culture was demonstrated [72,80]. These *in vivo* and *in vitro* effects are depicted quantitatively in Fig. 5 along with the well-established induction of 24-hydroxylase discussed in Chapter 6.

When 1,25$(OD)_2D_3$ is removed from cell culture medium [72] or when its administration *in vivo* is terminated [78], 1α-hydroxylase activity returns to the high levels characteristic of the vitamin D-deficient state, indicating the ready reversibility of the inhibitory effect of 1,25$(OH)_2D_3$ on kidney 1α-hydroxylase activity. Furthermore, the same experiments showed that 24-hydroxylase activity induction was blocked by inhibitors of protein and RNA synthesis [72]. Numerous reports of the up-regulation of 24-hydroxylase mRNA in many cell types by 1,25$(OH)_2D_3$ have appeared [82–86] (see Chapter 6).

As shown in Table III, the effects of 1,25$(OH)_2D_3$ and cycloheximide are not additive, suggesting that they may both be acting through the same mechanism and that 1,25$(OH)_2D_3$ may down-regulate a specific protein required for 1α-hydroxylase activity. The more pronounced response to 1,25$(OH)_2D_3$ at 2 hr suggests an additional posttranslational mechanism. However, it

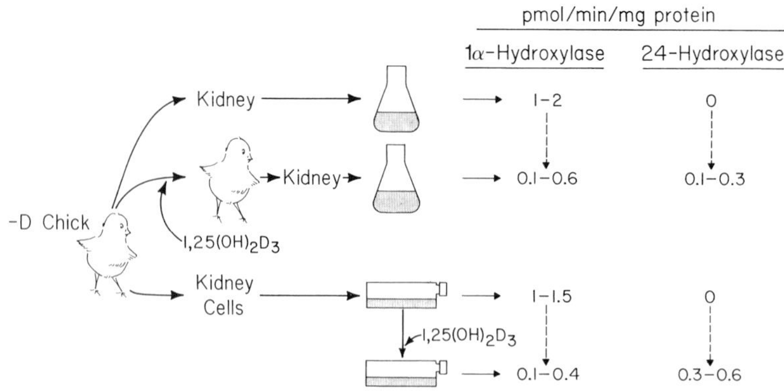

FIGURE 5 Effect of $1,25(OH)_2D_3$ on $25OHD_3$ metabolism *in vivo* or in cell culture. Treatment of chicks with $1,25(OH)_2D_3$ results in decreased 1α-hydroxylase activity and increased 24-hydroxylase activity measured subsequently in kidney homogenates. The same results are obtained when kidney cells in culture are treated with $1,25(OH)_2D_3$.

should be kept in mind that these experiments are indirect. Although it is tempting to speculate that the effect of $1,25(OH)_2D_3$ on 1α-hydroxylase activity is mediated by a transcriptional down-regulation of the message for its specific cytochrome P450 in a manner analogous to up-regulation of the 24-hydroxylase, the reagents to test this hypothesis directly are not yet available. Nevertheless, it is clear that vitamin D [and thus $1,25(OH)_2D_3$] status is one of the most powerful influences on the renal levels of the 1α-hydroxylase activity.

2. PARATHYROID HORMONE

Parathyroidectomy diminishes and administration of PTH increases 1α-hydroxylase activity in kidney tissue, in both avian and mammalian species [87–92]. PTH has also been shown to stimulate 1α-hydroxylase activity in cultured avian [93–95] and mammalian [96] kidney cells. This effect of PTH is mediated at least in part by the cAMP signaling pathway, as its effects can be mimicked by forskolin [97]. Although specific proteins with altered phosphorylation states in response to PTH have been identified [98], it has not yet been determined whether they play a physiological role in the regulation of the 1α-hydroxylation of $25OHD_3$.

In addition to the cAMP signaling pathway, protein kinase C is also involved in the regulation of the renal metabolism of $25OHD_3$ by both the 1α-hydroxylase and the 24-hydroxylase. In chick kidney cells, the phorbol ester protein kinase C (PKC) activator TPA decreases 1α-hydroxylase activity transiently [22] and over a 4-hr incubation period [99]. The latter decrease in 1α-hydroxylase activity is associated with increased 24-hydroxylase activity [99]. The abilities of TPA and $1,25(OH)_2D_3$ to decrease 1α-hydroxylase activity in cultured chick kidney cells are additive, indicating that they operate through distinct mechanisms. In perfused rat kidney proximal tubules, on the other hand, PKC activation by TPA resulted in a transient increase in 1α-hydroxylase activity and a sustained increase when the calcium ionophore A23187 was added with TPA [100]. PTH has been reported to stimulate the PKC pathway in rat renal cells [101]. The relative physiological importance of the cAMP and PKC signaling pathways in the control of 1α-hydroxylase activity has not yet been fully resolved. Furthermore, the fact that TPA activation of PKC decreases 1α-hydroxylase activity in chick kidney cells, both in the short term [22] and over longer times [99], while increasing its activity in rat preparations [100], suggests that the involvement of PKC in the regulation of 1α-hydroxylase may be more complex than initially appreciated.

TABLE III Effect of $1,25(OH)_2D_3$ and Cycloheximide on 1α-Hydroxylase in Cell Culture[a]

	1α-Hydroxylase activity remaining (%)		
Time (hr)	$1,25(OH)_2D_3$	Cycloheximide	$1,25(OH)_2D_3$ + cycloheximide
2	58 ± 2	80 ± 1	60 ± 3
8	48 ± 2	51 ± 3	44 ± 4

[a] Chick kidney cell cultures were incubated for the indicated time with $1,25(OH)_2D_3$ (10^{-7} *M*) or cycloheximide (5 μg/ml) prior to the addition of radiolabeled substrate for a 30-min assay. Results are given as the percentage of control activity (no additions) with the mean ± SD of four determinations.

3. OTHER AGENTS INFLUENCING 1α-HYDROXYLASE ACTIVITY

The antimycotics ketoconazole and miconazole were first reported to inhibit 24-hydroxylase [102], and 1α-hydroxylase [103] and were subsequently shown to be effective in decreasing serum levels of $1,25(OH)_2D_3$ and active calcium transport in rats [104]. Clinical studies have suggested lowered serum $1,25(OH)_2D_3$ levels in individuals being treated with these antifungal agents [105], a situation reminiscent of their effects on other steroid hydroxylases [102,106]. In one report, ketoconazole was used to ameliorate hypercalcemia resulting from acute active tuberculosis [107].

Many hormones of other endocrine systems have been tested for their effects, or lack thereof, on $1,25(OH)_2D_3$ production *in vivo* or in cell culture. These include estrogens, prolactin, dexamethasone, and calcitonin. The effects of estrogen in chicks and Japanese quail [108–111], however, could not be confirmed as direct ones in cell culture [112]. The reported effects of prolactin in cell culture [113] likewise have not been confirmed. Dexamethasone exhibited a slight inhibitory effect on 1α-hydroxylase activity in cultured chick kidney cells [114]. Finally, calcitonin, which in early reports appeared to stimulate 1α-hydroxylase activity *in vivo* (reviewed by Henry and Norman [115]), has not been confirmed to directly affect enzyme activity in chick [115] or rat [101] cells *in vitro*. The *in vitro* effects of these hormones are summarized in Table IV.

III. SUMMARY

In summary, studies *in vivo* and in cell culture support the view that the two most important physiological regulators of 1α-hydroxylase activity are $1,25(OD)_2D_3$ itself, which suppresses the activity of the enzyme that produces it, and parathyroid hormone, which up-regulates this activity. The former probably operates through gene transcriptional mechanisms, although this remains to be demonstrated; the latter mediates at least part of its effects through cAMP. The agenda for the study of the 1α-hydroxylase over the next few years includes the elucidation of the molecular mechanisms underlying the substantial observations in the existing literature regarding its physiological regulation.

TABLE IV Summary of Several Regulatory Influences on $25OHD_3$ Metabolism *in Vitro*

Hormone	Effect on $25OHD_3$ 1α-hydroxylase[a]
$1\alpha,25(OH)_2D_3$	↓
Parathyroid hormone	↑
Calcitonin	↑ ↓ →
Estrogens	→
Prolactin	↑
Glucocorticoids	↓

[a] Arrows indicate the direction of the effect of the indicated treatment of fresh renal tubule preparations or cultured cells. Single arrows indicate strong consensus among work from several investigators, whereas multiple arrows indicate mixed results.

References

1. Fraser DR, Kodicek E 1970 Unique biosynthesis by kidney of a biologically active vitamin D metabolite. Nature **288**:764–766.
2. Norman AW 1971 Evidence for a new kidney produced hormone, 1,25-dihydroxycholecalciferol, the proposed biologically active form of vitamin D. Am J Clin Nutr **24**:1346–1351.
3. Henry HL, Norma AW 1974 Studies on calciferol metabolism. IX. Renal 25-hydroxy-vitamin D_3-1-hydroxylase. Involvement of cytochrome P-450 and other properties. J Biol Chem **249**:7529–7535.
4. Hagenfeldt Y, Pedersen JI, Bjorkhem I 1988 Properties of guinea-pig kidney 25-hydroxyvitamin D_3-1α-hydroxylase assayed by isotope dilution–mass spectrometry. Biochem J **250**:521–526.
5. Churchill PF, deAlvare LR, Kimura T 1978 Topological studies of the steroid hydroxylase complexes in bovine adrenocortical mitochondria. J Biol Chem **253**:4924–4929.
6. Chu JW, Kimura T 1973 Studies on the adrenal steroid hydroxylases. Molecular and catalytic properties of adrenodoxin reductase (a flavoprotein). J Biol Chem **248**:2089–2094.
7. Brandt ME, Vickery LE 1992 Expression and characterization of human mitochondrial ferredoxin reductase in *Escherichia coli*. Arch Biochem Biophys **294**:735–740.
8. Gnanaiah W, Omdahl JL 1986 Isolation and characterization of pig kidney mitochondrial ferredoxin: $NADP^+$ oxidoreductase. J Biol Chem **261**:12649–12654.
9. Solish SB, Picado-Leonard J, Morel Y, Kuhn RW, Mohandas TK, Hanukoglu I, Miller WL 1988 Human adrenodoxin reductase: Two mRNAs encoded by a single gene on chromosome 17cen-> q25 are expressed in steroidogenic tissues. Proc Natl Acad Sci USA **85**:7104–7108.
10. Sakihama N, Hiwatashi A, Miyatake A, Shin M, Ichikawa Y 1988 Isolation and purification of mature bovine adrenocortical ferredoxin with an elongated carboxyl end. Arch Biochem Biophys **264**:23–29.
11. Cupp JR, Vickery LE 1989 Adrenodoxin with a COOH-terminal deletion (des 116–128) exhibits enhanced activity. J Biol Chem **264**:1602–1607.
12. Coughlan VM, Cupp JR, Vickery LE 1988 Purification and characterization of human placental ferredoxin. Arch Biochem Biophys **264**:376–382.
13. Pedersen JI, Ghazarian JG, Orme-Johnson NR, DeLuca HF 1976 Isolation of chick renal mitochondria ferredoxin active in the 25-hydroxyvitamin D_3-1 alpha-hydroxylase system. J Biol Chem **251**:3933–3941.
14. Yoon PS, DeLuca HF 1980 Purification and properties of chick renal mitochondrial ferredoxin. Biochemistry **19**:2165–2171.

15. Driscoll WJ, Omdahl JL 1989 Characterization and N-terminal amino acid sequence of multiple ferredoxins in kidney and adrenal ferredoxins. Eur J Biochem **185**:181–187.
16. Okamura M, Maliyakal EJ, Zuber MX, Simpson ER, Waterman MR 1985 Molecular cloning and amino acid sequence of the precursor form of bovine adrenodoxin: Evidence for a previously unidentified N-terminal peptide. Proc Natl Acad Sci USA **82**:5705–5709.
17. Picado-Leonard J, Voutilainen R, Kao L, Chung B, Strauss JF, Miller WL 1988 Human adrenodoxin: Cloning of three cDNAs and cycloheximide enhancement in JEG-3 cells. J Biol Chem **263**:3240–3244.
18. Blanchard RK, Henry HL 1996 Chick kidney ferredoxin: Complementary DNA cloning and vitamin D effects on mRNA levels. Comp Biochem Physiol (B) **114B**:337–344.
19. Kagimoto K, McCarthy JL, Waterman MR, Kagimoto M 1988 Deduced amino acid sequence of mature chick testis ferredoxin. Biochem Biophys Res Commun **155**:379–383.
20. Kagimoto M, Kagimoto K, Simpson ER, Waterman MR 1988 Transcription of the bovine adrenodoxin gene produces two species of mRNA of which only one is translated into adrenodoxin. J Biol Chem **263**:8925–8928.
21. Chang C, Wu D-A, Mohandas TK, Chung B-C 1990 Structure, sequence, chromosomal location and evolution of the human ferredoxin gene. DNA Cell Biol **9**:205–212.
22. Tang C, Kain SR, Henry HL 1993 The phorbol ester 12-*O*-tetradecanoyl-phorbol-13-acetate stimulates the dephosphorylation of mitochondrial ferredoxin in cultured chick kidney cells. Endocrinology **133**:1823–1829.
23. Tang C, Henry HL 1993 Overexpression in *Escherichia coli* and affinity purification of chick kidney ferredoxin. J Biol Chem **268**:5069–5075.
24. Voutilaninen R, Picado-Leonard J, DiBlasio AM, Miller WL 1988 Hormonal and developmental regulation of adrenodoxin messenger ribonucleic acid in steroidogenic tissues. J Clin Endocrinol Metab **66**:383–388.
25. John ME, John TS, Boggaram V, Simpson ER, Waterman MR 1986 Transcriptional regulation of steroid hydroxylase genes by corticotropin. Proc Natl Acad Sci USA **83**:4715–4719.
26. Kramer RE, Anderson CM, Peterson JA, Simpson ER, Waterman MR 1982 Adrenodoxin biosynthesis by bovine adrenal cells in monolayer culture. J Biol Chem **257**:14921–14925.
27. Miller WL 1988 Molecular biology of steroid hormone synthesis. Endocr Rev **9**:295–318.
28. Simpson ER, Waterman MR 1983 Regulation by ACTH of steroid hormone biosynthesis in the adrenal cortex. Can J Biochem Cell Biol **61**:692–707.
29. Monnier N, Defaye G, Chambaz M 1987 Phosphorylation of bovine adrenodoxin: Structural study and enzymatic activity. Eur J Biochem **169**:147–153.
30. Mandel ML, Moorthy B, Ghazarian JG 1990 Reciprocal post-translational regulation of renal 1 alpha- and 24-hydroxylases of 25-hydroxyvitamin D_3 by phosphorylation of ferredoxin. mRNA-directed cell-free synthesis and immunoisolation of ferredoxin. Biochem J **266**:385–392.
31. Siegel N, Wongsurawat N, Armbrecht HJ 1986 Parathyroid hormone stimulates dephosphorylation of the renodoxin component of the 25-hydroxyvitamin D_3-1α-hydroxylase from rat renal cortex. J Biol Chem **261**:16998–17003.
32. Ghazarian JG, Jefcoate CR, Knutson JC, Orme-Johnson W, DeLuca HF 1974 Mitochondrial cytochrome P450. A component of chick kidney 25-hydroxycholecalciferol-1 alpha-hydroxylase. J Biol Chem **249**:3026–3033.
33. Hiwatashi A, Nishii Y, Ichikawa Y 1982 Purification of cytochrome P-450-1α (25-hydroxyvitamin D_3-1α hydroxylase) of bovine kidney mitochondria. Biochem Biophys Res Commun **105**:320–327.
34. Bort R, Crivello JF 1988 Characterization of monoclonal antibodies specific to bovine renal vitamin D hydroxylases. Endocrinology **123**:2491–2498.
35. Burgos-Trinidad M, Ismail R, Ettinger RA, Prahl JM, DeLuca HF 1992 Immunopurified 25-hydroxyvitamin D 1α-hydroxylase and 1,25-dihydroxyvitamin D 24-hydroxylase are closely related but distinct enzymes. J Biol Chem **267**:3498–3505.
36. Moorthy B, Mandel ML, Ghazarian JG 1991 Amino-terminal sequence homology of two chick kidney mitochondrial proteins immunoisolated with monoclonal antibodies to the cytochrome P450 of the 25-hydroxyvitamin D_3-1α hydroxylase. J Bone Miner Res **6**:199–204.
37. Postlind H 1990 Separation of the cytochromes P-450 in pig kidney mitochondria catalyzing 1α, 24- and 26-hydroxylations of 25-hydroxyvitamin D_3. Biochem Biophys Res Commun **168**:261–266.
38. Wakino S, Meguro M, Suzuki H, Saruta T, Ogishima T, Ishimura Y, Shinki T, Suda T 1996 Evidence for 54-kD protein in chicken kidney as a cytochrome P450 with a high molecular activity of 25-hydroxyvitamin D_3 1α-hydroxylase. Gerentology **42**:67–77.
39. St. Arnaud R, Glorieux F 1996 Molecular cloning and characterization of a cDNA for vitamin D 1α-hydroxylase. J Bone and Miner Res **11**:S124.
40. Hollis BW 1990 25-Hydroxyvitamin D_3-1α-hydroxylase in porcine hepatic tissue: Subcellular localization to both mitochondria and microsomes. Proc Natl Acad Sci USA **87**:6009–6013.
41. Negrea LA, Slatopolsky E, Dusso AS 1995 1,25-Dihydroxyvitamin D synthesis in rat liver microsomes. Horm Metab Res **27**:461–464.
42. Axén E, Postlind H, Sjöberg H, Wikvall K 1994 Liver mitochondrial cytochrome P450 CYP27 and recombinant-expressed human CYP27 catalyze 1α-hydroxylation of 25-hydroxyvitamin D_3. Proc Natl Acad Sci USA **91**:10014–10018.
43. Axén E, Postlind H, Wikvall K 1995 Effects on CYP27 mRNA expression in rat kidney and liver by 1α,25-dihydroxyvitamin D_3, a suppressor of renal 25-hydroxyvitamin D_3 1α-hydroxylase activity. Biochem Biophys Res Commun **215**:136–141.
44. Henry HL, Norman AW 1975 Presence of renal 25-hydroxyvitamin-D-1-hydroxylase in species of all vertebrate classes. Comp Biochem Physiol **50B**:431–434.
45. Nahm TH, Lee SW, Fausto A, Sonn Y, Avioli LV 1979 25OHD, a circulating vitamin D metabolite in fish. Biochem Biophys Res Commun **89**:396–402.
46. Egaas E, Lambertsen G 1979 Naturally occurring vitamin D_3 in fish products analyzed by high performance liquid chromatography using vitamin D_2 as an internal standard. J Am Oil Chem Soc **56**:188.
47. Takeuchi A, Okano T, Kobayashi T 1991 The existence of 25-hydroxyvitamin D_3-1α-hydroxylase in the liver of carp and bastard halibut. Life Sci **48**:275–282.
48. Sunita Rao D, Raghuramulu N 1995 Vitamin D and its related parameters in freshwater wild fishes. Comp Biochem Physiol **111A**:191–198.
49. Oizumi K, Monder C 1972 Localization and metabolism of 1,2-^{3}H-vitamin D_3 and 26,27-^{3}H-25-hydroxycholecalciferol in goldfish (*Carassius auratus* L.). Comp Biochem Physiol **42**:523–532.
50. Brown PB, Robinson EH 1992 Vitamin D studies with channel catfish (*Ictalurus punctatus*) reared in calcium-free water. Comp Biochem Physiol **103A**:213–219.
51. Rao DS, Raghuramulu N 1996 Food chain as origin of vitamin D in fish. Comp Biochem Physiol **114A**:15–19.

52. Buffenstein R, Sergeev IN, Pettifor JM 1993 Vitamin D hydroxylases and their regulation in a naturally vitamin D-deficient subterranean mammal, the naked mole rat (*Heterocephalus glaber*). J Endocrinol **138**:59–64.
53. Sergeev IN, Buffenstein R, Pettifor JM 1993 Vitamin D receptors in a naturally vitamin D-deficient subterranean mammal, the naked mole rat (*Heterocephalus glaber*): Biochemical characterization. Gen Comp Endocrinol **90**:338–345.
54. Hartenbower DL, Stella FJ, Norman AW, Friedler RM, Coburn JW 1977 Impaired vitamin D metabolism in acute uremia. J Lab Clin Med **90**:760–766.
55. Mawer EB, Taylor CM, Backhouse J, Lumb GA, Stanbury SW 1973 Failure of formation of 1,25-dihydroxycholecalciferol in chronic renal insufficiency. Lancet **1**:626–628.
56. Eisman JA, Hamstra AJ, Kream BE, DeLuca HF 1976 1,25-Dihydroxyvitamin D in biological fluids: A simplified and sensitive assay. Science **193**:1021–1023.
57. Turner RT, Puzas JE, Forte MD, Lester GE, Gray TK, Howard GA, Baylink DJ 1980 *In vitro* synthesis of 1α,25-dihydroxycholecalciferol and 24,25-dihydroxycholecalciferol by isolated calvarial cells. Proc Natl Acad Sci USA **77**:5720–5724.
58. Puzas JE, Turner RT, Howard GA, Brand JS, Baylink DJ 1987 Synthesis of 1,25-dihydroxycholecalciferol and 24,25-dihydroxycholecalciferol by calvarial cells: Characterization of the enzyme systems. Biochem J **245**:333–338.
59. Delvin EE, Arabian A 1987 Kinetics and regulation of 25-hydroxycholecalciferol 1-alpha-hydroxylase from cells isolated from human term decidua. Eur J Biochem **164**:659–662.
60. Glorieux FH, Arabian A, Delvin EE 1995 Pseudo-vitamin D deficiency. Absence of 25-hydroxyvitamin D 1α-hydroxylase activity in human placenta decidual cells. J Clin Endocrinol Metab **80**:2255–2258.
61. Hollis BW, Iskersky VN, Chang MK 1989 *In vitro* metabolism of 25-hydroxyvitamin D_3 by human trophoblastic homogenates, mitochondria, and microsomes: Lack of evidence for the presence of 25-hydroxyvitamin D_3-1α-hydroxylase and 24*R*-hydroxylase. Endocrinology **125**:1224–1230.
62. Reichel H, Bishop JE, Koeffler HP, Norman AW 1991 Evidence for 1,25-dihydroxyvitamin D_3 production by cultured porcine alveolar macrophages. Mol Cell Endocrinol **75**:163–167.
63. Koeffler HP, Reichel H, Munker R, Barbers R, Norman AW 1985 Interaction of 1,25-dihydroxyvitamin D_3 and the hematopoietic system. In: Gale RP, Golde DW (eds) Leukemia: Recent Advances in Biology and Treatment. Alan R. Liss, New York, pp. 399–413.
64. Mawer EB, Hayes ME 1992 Renal and extra-renal synthesis of 1,25-dihydroxy vitamin D. Prog Endocrinol **87**:382–386.
65. Cadranel J, Garabedian M, Milleron B, Guillozo H, Akoun G, Hance AJ 1990 1,25$(OH)_2D_3$ production by T lymphocytes and alveolar macrophages recovered by lavage from normocalcemic patients with tuberculosis. J Clin Invest **85**:1588–1593.
66. Pryke AM, Duggan C, White CP, Posen S, Mason RS 1990 Tumor necrosis factor-alpha induces vitamin D-1-hydroxylase activity in normal human alveolar macrophages. J Cell Physiol **142**(3):652–656.
67. Abe E, Miyaura C, Sakagami H, Takeda M, Konno K, Yamazaki T, Yoshiki S, Suda T 1981 Differentiation of mouse myeloid leukemia cells induced by 1α,25-dihydroxyvitamin D_3. Proc Natl Acad Sci USA **78**:4990–4994.
68. Brunette MG, Chan M, Ferriere C, Roberts KD 1978 Site of 1,25$(OH)_2$ vitamin D_3 synthesis in the kidney. Nature **276**:287–289.
69. Kawashima H, Torikai S, Kurokawa K 1981 Calcitonin selectively stimulates 25-hydroxyvitamin D_3-1 alpha-hydroxylase in proximal straight tubules of rat kidney. Nature **291**:327–329.
70. Ross FP, Henry HL 1980 Recent advances in the understanding of the metabolism and functions of vitamin D. Clin Orthop Relat Res **149**:249–267.
71. Ghazarian JG, Kream B, Botham KM, Nickells MW, DeLuca HF 1978 Rat plasma 25-hydroxyvitamin D_3 binding protein: An inhibitor of the 25-hydroxyvitamin D_3-1-alpha-hydroxylase. Arch Biochem Biophys **189**:212–220.
72. Henry HL 1979 Regulation of the hydroxylation of 25-hydroxyvitamin D_3 *in vivo* and in primary cultures of chick kidney cells. J Biol Chem **254**:2722–2729.
73. Cunningham NS, Lee BS, Henry HL 1986 The renal metabolism of 25-hydroxyvitamin D_3: A possible role for phospholipids. Biochim Biophys Acta **881**:480–488.
74. Henry HL, Fried S, Shen G-Y, Barrack SA, Okamura WH 1991 Effect of three A-ring analogs of 1α-dihydroxyvitamin D_3 on 25-OH-D_3-1α-hydroxylase in isolated mitochondria and on 25-hydroxyvitamin D_3 metabolism in cultured kidney cells. J Steroid Biochem Mol Biol **38**:775–779.
75. Daniel D, Middleton R, Henry HL, Okamura WH 1996 Inhibitors of 25-hydroxyvitamin D_3-1α-hydroxylase: A-ring oxa analogs of 25-hydroxyvitamin D_3. J Org Chem **61**:5617–5625.
76. Muralidharan KR, Rowland-Goldsmith M, Lee AS, Park G, Norman AW, Henry HL, Okamura WH 1997 Inhibitors of 25-hydroxyvitamin-D_3-1-hydroxylase: Thiavitamin D analogs and biological evaluation. J Steroid Biochem Mol Biol In Press.
77. Figadère B, Norman AW, Henry HL, Koeffler HP, Zhou JY, Okamura WH 1991 Arocalciferols: Synthesis and biological evaluation of aromatic side-chain analogues of 1α,25-dihydroxyvitamin D_3. J Med Chem **34**:2452–2463.
78. Henry HL, Midgett RJ, Norman AW 1974 Studies on calciferol metabolism X: Regulation of 25-hydroxyvitamin D_3-1-hydroxylase, *in vivo*. J Biol Chem **249**:7584–7592.
79. Warner M, Tenenhouse A 1985 Regulation of renal vitamin D hydroxylase activity in vitamin D deficient rats. Can J Physiol **63**(8):978–982.
80. Spanos E 1983 Regulation of the renal 25OHD_3 hydroxylases by vitamin-D_3 metabolites and analogs in primary chick kidney cell cultures. Acta Endocrinol **104**:25–26.
81. Johnson EW, Eller PM, Jafek BW, Norman AW 1992 Calbindin-like immunoreactivity in two peripheral chemosensory tissues of the rat: Taste buds and the vomeronasal organ. Brain Res **572**:319–324.
82. Ichikawa F, Sato K, Nanjo M, Nishii Y, Shinki T, Takahashi N, Suda T 1995 Mouse primary osteoblasts express vitamin D_3 25-hydroxylase mRNA and convert 1α,25-dihydroxyvitamin D_3. Bone **16**:129–135.
83. Chen K-S, Prahl JM, DeLuca HF 1993 Isolation and expression of human 1,25-dihydroxyvitamin D_3 24-hydroxylase cDNA. Proc Natl Acad Sci USA **90**:4543–4547.
84. Armbrecht HJ, Hodam TL 1994 Parathyroid hormone and 1,25-dihydroxyvitamin D synergistically induce the 1,25-dihydroxyvitamin D-24-hydroxylase in rat UMR-106 osteoblast-like cells. Biochem Biophys Res Commun **205**:674–679.
85. Koyama H, Inaba M, Nishizawa Y, Ishimura E, Imanishi Y, Hino M, Furuyama T, Takagi H, Morii H 1994 Potentiated 1,25$(OH)_2D_3$-induced 24-hydroxylase gene expression in uremic rat intestine. Am J Physiol **267**:F926–F930.
86. Nishimura A, Shinki A, Cheng HJ 1994 Regulation of messenger ribonucleic acid expression of 1α,25-dihydroxyvitamin D_3-24-hydroxylase in rat osteoblast. Endocrinology **134**:1794–1799.
87. Fraser DR, Kodicek E 1973 Regulation of 25-hydroxycholecalcif-

erol-1-hydroxylase activity in kidney by parathyroid hormone. Nature New Biol **241**:163–166.
88. Baksi SN, Kenny AD 1979 Acute effects of parathyroid extract on renal vitamin D hydroxylases in Japanese quail. Pharmacology **18**:169–174.
89. Booth BE, Tsai HC, Morris RC 1985 Vitamin D status regulates 25-hydroxyvitamin-D_3-1-alpha-hydroxylase and its responsiveness to parathyroid hormone in the chick. J Clin Invest **75**:155–161.
90. Tanaka Y, Lorenc RS, DeLuca HF 1975 The role of 1,25-dihydroxyvitamin D_3 and parathyroid hormone in the regulation of chick renal 25-hydroxyvitamin D_3-24-hydroxylase. Arch Biochem Biophys **171**:521–526.
91. Walker AT, Stewart AF, Korn EA, Shiratori T, Mitnick MA, Carpenter TO 1990 Effect of parathyroid hormone-like peptides on 25-hydroxyvitamin D-1alpha-hydroxylase activity in rodents. Am J Physiol **258**(2):E297–E303.
92. Liang CT, Balakir RA, Bames J, Sacktor B 1984 Responses of chick cells to parathyroid hormone: Effect of vitamin D. Am J Physiol **246**:C401–C406.
93. Spanos E, Brown DJ, Macintyre I 1979 Regulation of 25-OHD_3 metabolism by parathyroid hormone in primary chick kidney cell cultures. FEBS Lett **105**:31–34.
94. Rost CR, Bikle DD, Kaplan RA 1981 In vitro stimulation of 25-hydroxycholecalciferol 1alpha-hydroxylation by parathyroid hormone in chick kidney slices: Evidence for a role for adenosine 3′5′-monophosphate. Endocrinology **108**:1002–1006.
95. Henry HL 1981 Insulin permits parathyroid hormone stimulation of 1,25-dihydroxyvitamin D_3 production in cultured kidney cells. Endocrinology **108**:733–735.
96. Tanaka Y, DeLuca HF 1984 Rat renal 25-hydroxyvitamin-D_3 1- and 24-hydroxylases: Their *in vivo* regulation. Am J Physiol **246**:E168–E173.
97. Henry HL 1985 Parathyroid hormone modulation of 25-hydroxyvitamin D_3 metabolism by cultured chick kidney cells is mimicked and enhanced by forskolin. Endocrinology **116**: 503–507.
98. Noland TA, Henry HL 1983 Protein phosphorylation in chick kidney: Response to parathyroid hormone, cyclic AMP, calcium and phosphatidyl serine. J Biol Chem **258**:538–546.
99. Henry HL, Luntao EM 1989 Interactions between intracellular signals involved in the regulation of 25-hydroxyvitamin D_3 metabolism. Endocrinology **124**:2228–2234.
100. Ro HK, Tembe V, Favus MJ 1992 Evidence that activation of protein kinase-C can stimulate 1,25-dihydroxyvitamin D_3 secretion by rat proximal tubules. Endocrinology **131**:1424–1428.
101. Dunlay R, Hruska KA 1990 PTH receptor coupling to phospholipase C is an alternative pathway of signal transduction in bone and kidney. Am J Physiol **258**:F223–F231.
102. Loose DS, Kan PB, Hirst MA, Marcas RA, Feldman D 1983 Ketoconazole blocks adrenal steroidogenesis by inhibiting cytochrome P450-dependent enzymes. J Clin Invest **71**:1495–1499.
103. Henry HL 1985 Effect of ketoconazole and micronazole on 25-hydroxyvitamin D_3 metabolism by cultured chick kidney cells. J Steroid Biochem Mol Biol **23**:991–994.
104. Boass T, Toverud SU 1995 Suppression of circulating calcitriol and duodenal active Ca transport by ketoconazole in pregnant rats. Am J Physiol Endocrinol Metab **269**:E934–E939.
105. Glass AR, Eil C 1986 Ketoconazole-induced reduction in serum 1,25-dihydroxyvitamin D. J Clin Endocrinol Metab **63**(3):766–769.
106. Kowal J 1983 The effect of ketoconazole on steroidogenesis in cultured mouse adrenal cortex tumor cells. Endocrinology **112**:1541–1543.
107. Saggese G, Bertelloni S, Baroncelli GI, Di Nero G 1993 Ketoconazole decreases the serum ionized calcium and 1,25-dihydroxyvitamin D levels in tuberculosis-associated hypercalcemia. Am J Dis Child **147**:270–273.
108. Castillo L, Tanaka Y, DeLuca HF, Sunde ML 1977 The stimulation of 25-hydroxyvitamin D_3-1alpha-hydroxylase by estrogen. Arch Biochem Biophys **179**:211–217.
109. Tanaka Y, Castillo L, DeLuca HF 1976 Control of renal vitamin D hydroxylases in birds by sex hormones. Proc Natl Acad Sci USA **73**:2701–2705.
110. Baksi SN, Kenny AD 1980 Estradiol-induced stimulation of 25-hydroxyvitamin D_3-1-hydroxylase in vitamin D-deficient Japanese quail. Pharmacology **20**:298–303.
111. Pike JW, Spanos E, Coloston KW, Macintyre I, Haussler MR 1978 Influence of estrogen on renal vitamin D hydroxylases and serum 1-alpha,25-$(OH)_2D_3$ in chicks. Am J Physiol:E338–E343.
112. Henry HL 1981 25$(OH)D_3$ metabolism in kidney cell cultures: Lack of a direct effect of estradiol. Am J Physiol **240**:E119–E124.
113. Spanos E, Brown DJ, Stevenson JC, Macintyre I 1981 Stimulation of 1,25-dihydroxycholecalciferol production by prolactin and related peptides in intact renal cell preparations in vitro. Biochim Biophys Acta **672**:15.
114. Henry HL 1986 Effect of dexamethasone on 25-hydroxyvitamin D_3 metabolism in chick kidney cell cultures. Endocrinology **118**:940–944.
115. Henry HL, Norman AW 1984 Vitamin D: Metabolism and biological action. Annu Rev Nutr **4**:493–520.

CHAPTER 6

The 25-Hydroxyvitamin D 24-Hydroxylase

J. OMDAHL Department of Biochemistry and Molecular Biology, University of New Mexico, School of Medicine, Albuquerque, New Mexico

B. MAY Department of Biochemistry, University of Adelaide, Adelaide, South Australia

I. BACKGROUND

Vitamin D is a secosteroid whose molecular action is dependent on the selective addition of specific hydroxyl groupings. Most metabolites of 25-hydroxyvitamin D_3 (25OHD_3) consist of a combinatorial arrangement of hydroxyl groups at C-1, C-24, or C-26/C-27. Enzymes responsible for the hydroxyl modifications are members of the cytochrome P450 (CYP) superfamily of mixed-function oxidases. Hydroxylation at C-25 and C-26/C-27 is mediated by a liver 27-hydroxylase enzyme (i.e., cytochrome $P450_{27}$ or CYP27) [1–3] that is discussed in detail in Chapter 4. The 1- and 24-hydroxylated metabolites occupy central positions in the bioactivation and metabolic clearance pathway for 25OHD_3. Hydroxylation at C-1 is directed by a 1-hydroxylase enzyme (i.e., cytochrome $P450_1$ or CYP1) that is responsible for synthesis of the hormonally active 1,25-dihydroxyvitamin D_3 [1,25$(OH)_2D_3$] [4,5], which is discussed in detail in Chapter 5. Most cellular actions of 1,25$(OH)_2D_3$ are mediated by alterations in the transcription of vitamin D-dependent genes. The secosteroid binds to the ligand-dependent vitamin D receptor (VDR). In turn, the receptor–hormone complex binds to promoter-specific vitamin D response elements (VDREs), usually as a heterodimer with the retinoid X receptor (RXR) [6–8]. Promoter-specific binding of the VDR–RXR complex results in altered expression of the attendant gene [9] (Fig. 1). Details on the mechanism of action for VDR are extensively covered in Section II of this book.

Although 1,25$(OH)_2D_3$ is the active hormone, the major product of 25OHD_3 is 24,25$(OH)_2D_3$, a dihydroxyvitamin D metabolite whose biosynthesis is catalyzed by a 24-hydroxylase enzyme (i.e., cytochrome $P450_{24}$ or CYP24) [10,11]. CYP1 (1-hydroxylase) and CYP24 (24-hydroxylase) are regulated enzymes whose major expression occurs in the kidney. These enzyme activities are frequently regulated in a diametric manner. For example, CYP1 activity is up-regulated during high parathyroid hormone (PTH)/low calcium states, whereas CYP24 is down-regulated. In contrast, high vitamin D/high calcium states result in low CYP1 and high CYP24 expression. The regulatory and expression linkages between these enzymes are fundamental to understanding the vitamin D regulatory pathway and will serve as a point of reference in this chapter as well as other sections in this book in which various aspects of vitamin D metabolism are discussed extensively.

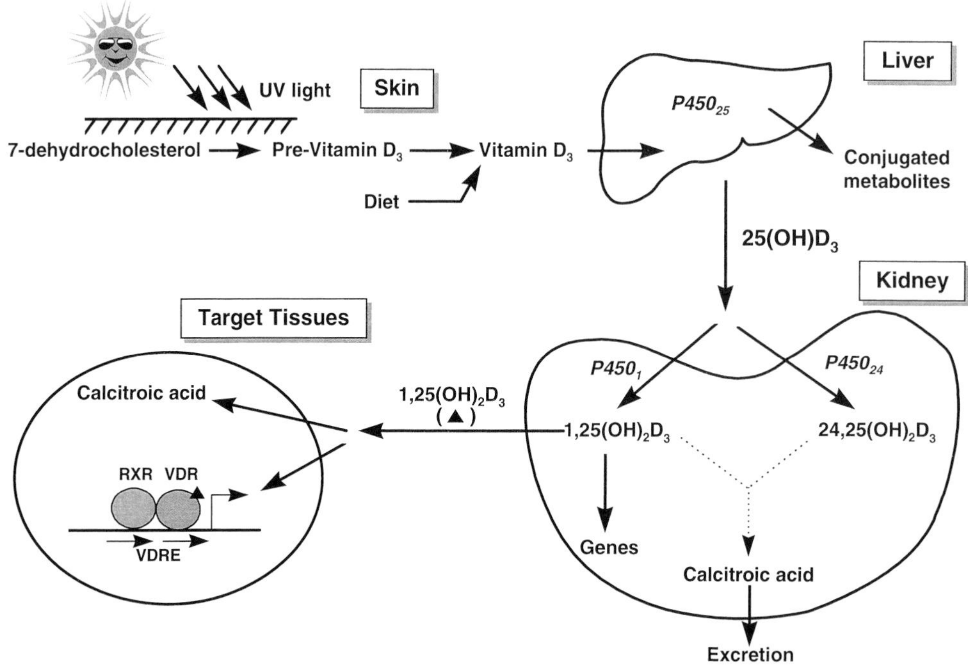

FIGURE 1 General diagram of vitamin D_3 metabolism describing the biosynthesis and metabolism of vitamin D_3 and its major metabolites, and the cellular action of $1,25(OH)_2D_3$.

A. Enzyme Characterization

CYP24 (i.e., 25-hydroxyvitamin D 24-hydroxylase) is a 53-kDa enzyme of the vitamin D pathway that catalyzes 24(*R*)-hydroxylations [12]. It requires NADPH and oxygen, and functions as a mixed-function oxidase [13]. The enzyme is an integral mitochondrial membrane hemoprotein that is part of a mini-electron transport chain consisting of NADPH-ferredoxin reductase, ferredoxin (an iron–sulfur protein), and the terminal cytochrome $P450_{24}$ [14] (Fig. 2). Speed is not an attribute of the 24-hydroxylase enzyme, for which turnover numbers of 6–20 (mol/min/mol P450) have been recorded for the rat enzyme [15,16]. The K_m for CYP24 is about 2.5–5 μM when $25OHD_3$ is substrate and over 10-fold lower when the enzyme acts on $1,25(OH)_2D_3$ [14–16]. On the basis of substrate affinity, the preferred substrate is $1,25(OH)_2D_3$. However, the biosynthesis of $24,25(OH)_2D_3$ is expected to be higher than $1,24,25(OH)_3D_3$ owing to the lower CYP24 turnover number for $1,25(OH)_2D_3$ versus $25OHD_3$ (i.e., 6 versus 20 mol/min/mol P450) [15] and a concentration of $25OHD_3$ that exceeds $1,25(OH)_2D_3$ by nearly 1000-fold. Such concentration and kinetic parameters are consistent with $24,25(OH)_2D_3$ being the major circulating dihydroxyvitamin D metabolite. Expression of cytochrome CYP24 activity requires substrate metabolites to be 25-hydroxylated [17]; and in this regard, the enzyme is not active toward xenobiotics [15] or reactants from steroid hormone pathways. CYP24 does not react

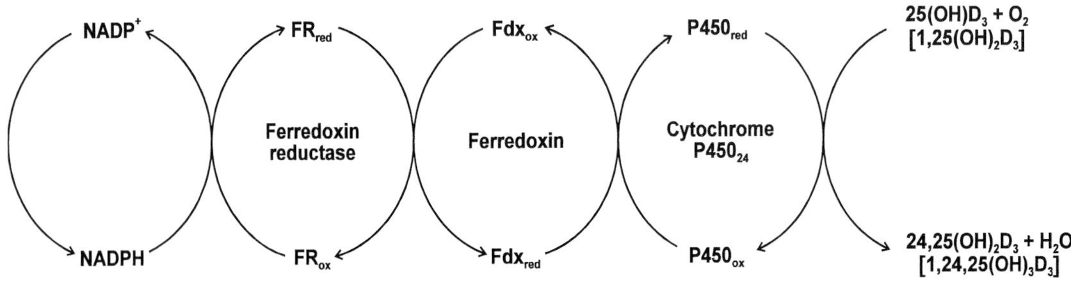

FIGURE 2 Cytochrome P450 system for the 24-hydroxylase enzyme. Illustrated is the sequential electron transfer from the NADPH source, through ferredoxin reductase (FR) and ferredoxin (Fdx) to the terminal $P450_{24}$, which directs the 24-hydroxylation of substrate molecules. ox, oxidized; red, reduced.

immunologically with antibodies to several liver P450s including CYP27, the 25-hydroxylase of the vitamin D pathway, even though the two vitamin D enzymes share about 30% sequence similarity [18].

B. Enzyme Pathways

CYP24 can direct the synthesis of 24,25$(OH)_2D_3$ [17] and 1,24,25$(OH)_3D_3$. These two vitamin D metabolites express less biological activity than 1,25$(OH)_2D_3$ although 24,25$(OH)_2D_3$ is suggested to have focused actions in bone and cartilage (see relevant chapters in Section III of this book). Both metabolites also represent initial reactants in the 24-oxidation pathway that lead to metabolite inactivation via generation of 23- or 24-COOH end products [19–22]. Acquisition of a 24-hydroxyl function is required for entry into the oxidation pathway, which is followed by oxidation of the 24-hydroxyl group to a keto (oxo) function (Fig. 3). Subsequent steps involve the 23-hydroxylation of the 24-oxo metabolites and oxidative cleavage between C-23 and C-24 followed by oxidation to 23-COOH-24,25,26,27-tetranor-D_3 or calcitroic acid from 24,25$(OH)_2D_3$ or 1,25$(OH)_2D_3$, respectively. The 23- and 24-hydroxylase activities are usually coexpressed in the same target cell, with 24-hydroxylase constituting the major activity. However, guinea pigs demonstrate an opposite activity profile in which 23-hydroxylase is the major expressed activity [23]. An investigation with partially purified 24-hydroxylase enzyme from pig kidney documented copurification of the 23- and 24-hydroxylase activities [16], which was consistent with one enzyme expressing both activities. Recombinant cytochrome $P450_{24}$ was subsequently shown to express both 23- and 24-hydroxylase activities [16]. It is suggested from previous studies that the preferential expression of 23- or 24-hydroxylase activity in different isoforms could be due to point residue variations that alter the substrate binding/catalytic sites and, thereby, enzyme specificity. In that respect, the enzyme with lowest activity would seemingly represent the rate-limiting step in the side-chain oxidation pathway. It is evident, therefore, that cytochrome $P450_{24}$ (CYP24) is capable of directing initial reactions in the side chain oxidation of 25OHD_3 and 1,25$(OH)_2D_3$ metabolites to acidic end products (Fig. 3). Furthermore, from studies with the recombinant human CYP24, it has been suggested that the enzyme may also be capable of directing the final steps involved with side-chain cleavage and metabolite oxidation [24].

II. GENE STRUCTURE AND LOCATION

A. Isolation of cDNA Clones

Purification of the rat kidney mitochondrial 24-hydroxylase enzyme and raising of a rabbit polyclonal

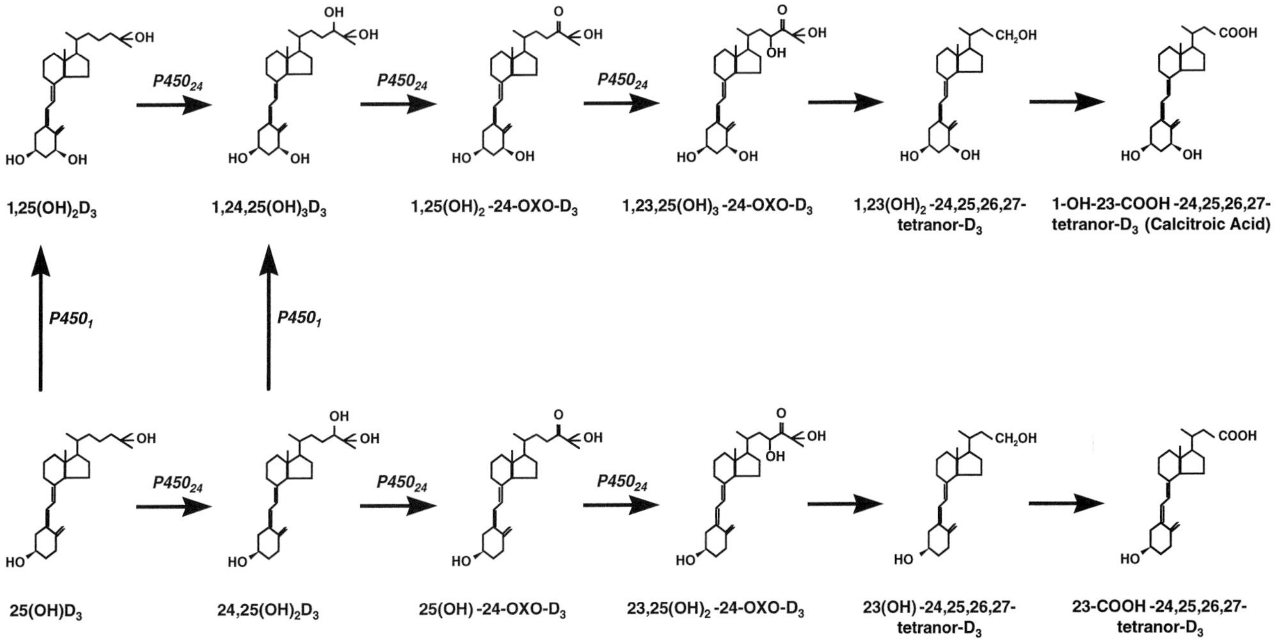

1,25$(OH)_2D_3$ → ($P450_{24}$) 1,24,25$(OH)_3D_3$ → ($P450_{24}$) 1,25$(OH)_2$-24-OXO-D_3 → ($P450_{24}$) 1,23,25$(OH)_3$-24-OXO-D_3 → 1,23$(OH)_2$-24,25,26,27-tetranor-D_3 → 1-OH-23-COOH-24,25,26,27-tetranor-D_3 (Calcitroic Acid)

$P450_1$ (from 25(OH)D_3 and 24,25$(OH)_2D_3$)

25(OH)D_3 → ($P450_{24}$) 24,25$(OH)_2D_3$ → ($P450_{24}$) 25(OH)-24-OXO-D_3 → ($P450_{24}$) 23,25$(OH)_2$-24-OXO-D_3 → 23(OH)-24,25,26,27-tetranor-D_3 → 23-COOH-24,25,26,27-tetranor-D_3

FIGURE 3 The 24-oxidation pathway. Two pathways are illustrated for 1,25$(OH)_2D_3$ and 24,25$(OH)_2D_3$ metabolism to side chain cleaved 23-COOH end products.

antibody [15,25] were a milestone achievements that led to the isolation of a cDNA clone for rat 24-hydroxylase by immunological screening [26]. The mRNA contains an open reading frame of 514 amino acid residues that includes the mitochondrial signal sequence. A comparison of the presequence with the N-terminal amino acid sequence of the mature enzyme pinpointed the mitochondrial proteolytic cleavage site between amino acid residues 35 and 36. The signal peptide is long by comparison and contains a typical amphipathic arrangement of hydrophobic and charged amino acids. The cDNA clone was confirmed as coding for the rat 24-hydroxylase by expression studies in COS-7 monkey kidney cells and measurement of catalytic activity and antibody cross-reactivity [26]. Full-length cDNA clones were subsequently isolated for the human [27] and mouse [28] 24-hydroxylase proteins. The human 24-hydroxylase showed about 80% amino acid identity with both the rat and mouse homologs, whereas the latter two were 95% identical. The deduced amino acid sequence of the 24-hydroxylase was not more than 30% identical with any P450 sequence reported to date, and it was concluded that the enzyme belongs to a novel P450 family. The highest homology (30% identity) was with the liver mitochondrial enzyme P450LMT25 of the bile acid pathway, which can catalyze the 27-hydroxylation of sterols and the 25- and 27-hydroxylation of vitamin D_3 [18].

B. Isolation of Genomic Clones

Genomic clones for rat [29,30] and human [31] 24-hydroxylase have been isolated. The rat gene is a single copy gene that spans about 15 kb and contains 12 exons [29]. The intron–exon arrangement of the gene most closely resembles that of the CYP11 family [29]. The transcription start site for the rat 24-hydroxylase gene and a likely TATA box have been located, and several possible control elements have been identified in the promoter including VDREs and CCAAT, GC, and TATA binding sites (Fig. 4) [29,30]. Similar binding sites for transcription factors are also present in the promoter for human 24-hydroxylase [31]. Such sites are typical of a ubiquitously expressed gene that can respond to $1,25(OH)_2D_3$ through action of the VDR.

C. Gene Localization

The gene for human 24-hydroxylase has been localized by fluorescence *in situ* hybridization to the long arm of chromosome 20 at q13.2-q13.3 [32] and in another study at q13.1 [33]. Other genes that map in this region include endothelin, cystatin, and guanine nucleotide binding protein subunit. There are no known diseases that can be attributed to a defect in the 24-hydroxylase gene. The genetic alterations associated with vitamin D-dependency type I rickets (VDDR-I or PDDR) and X-linked hypophosphatemia, the most prevalent form of inherited rickets in humans, are linked to chromosome 12q14 [33,34] and the X chromosome, respectively, and hence cannot be ascribed to a defective 24-hydroxylase gene. In mice, the 24-hydroxylase gene has been localized to chromosome 2 and forms part of the conserved syntenic group between mouse chromosome 2 and human chromosome 20 [35].

D. Gene Knock-out

A knock-out model has been developed for the mouse *CYP24* gene (R. St-Arnaud and F. Glorieux, personal communication, 1996). The *CYP24* gene was targeted for null mutation by using a construct in which exons 9 and 10 (exon 10 codes for the heme-binding region) were deleted and replaced by the phosphoglycerate kinase neomycin (PGK-neo) selection cassette. Following germ line transmission of the mutant gene, the heterozygous animals appeared normal and were fertile. Mice homozygous for the *CYP24* gene mutation were also fertile; however, they displayed a high frequency of perinatal death (~50%). Homozygous animals that survived displayed serum calcium, phosphate, and $1,25(OH)_2D_3$ levels that were normal, as was the microscopic anatomy for the kidney, liver, spleen, bone, and intestine. Homozygotes born of homozygous fe-

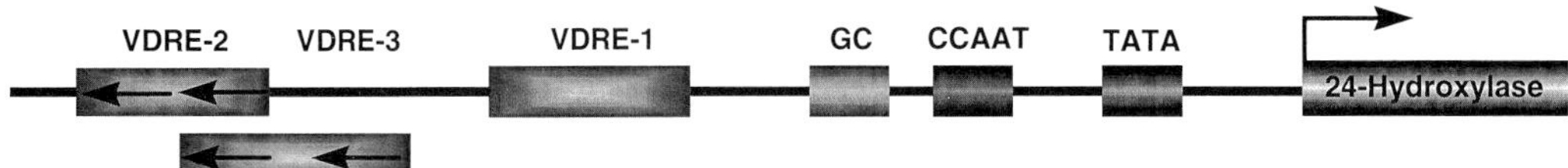

FIGURE 4 Promoter elements in the rat 24-hydroxylase gene. The immediate promoter (about 300 bp) of the rat 24-hydroxylase gene and sequential locations of TATA, CCAAT, and GC boxes together with three putative vitamin D response elements are shown.

males, however, displayed abnormal intramembranous bone formation with no apparent alteration of the growth plate. Such results suggest a seminal role for the *CYP24* gene in the developmental regulation of intramembranous bone formation, which is discussed in detail in Chapter 18. These results could arise as a result of either the requirement for $24{,}25(OH)_2D_3$ or alternatively the lack of $1{,}25(OH)_2D_3$ catabolism, or both.

III. CELLULAR EXPRESSION AND REGULATION

A. Cellular Expression

The 24-hydroxylase enzyme displays a broad tissue distribution. A major route of enzyme induction involves the combined action of $1{,}25(OH)_2D_3$ and VDR to increase transcription of the *CYP24* gene. Consequently, cellular expression of CYP24 is linked tightly to the coexpression of VDR. Most cells that contain VDR, therefore, express a basal level of CYP24 or respond to increased $1{,}25(OH)_2D_3$ levels by inducing biosynthesis of the enzyme, particularly in kidney and small intestine [36–38]. Basal expression under normal calcium–homeostatic conditions has been documented for kidney, intestine, bone, placenta, skin (keratinocytes), and macrophages.

Particular attention has been given to expression of the enzyme in the kidney, which is the major site of CYP24 activity. In adult control rats, basal *CYP24* mRNA is detected predominantly in the kidney. As this tissue is the major site of $1{,}25(OH)_2D_3$ synthesis, it seems likely that the high basal expression in kidney represents induction by endogenous $1{,}25(OH)_2D_3$. In support of this concept, there is no renal *CYP24* mRNA in vitamin D-deficient rats. In the kidney, enzyme expression has been localized to the proximal tubule and does not display the more general distribution as noted for VDR [39]. CYP24 expression in bone occurs in osteoblasts and may play an important role in bone mineral dynamics. Placenta is active in calcium transport, although a role for CYP24 has yet to be defined. However, placenta expresses 24-hydroxylase activity [40], and *CYP24* mRNA has been detected by reverse transcriptase–polymerase chain reaction (RT-PCR; J. Omdahl, unpublished data, 1996). Using the rat developmental model, Matkovits and Christakos [38] observed low levels of renal *CYP24* mRNA at birth that remained unchanged for 2–3 weeks, after which a substantial increase in mRNA occurred. The enzyme mRNA content continued to increase when measured over a 20-month period. The increase in enzyme message levels corresponded with high levels of VDR mRNA and elevated levels of serum $1{,}25(OH)_2D_3$, which indicated that developmental expression of CYP24 is controlled by $1{,}25(OH)_2D_3$-dependent transcriptional induction.

Mechanistic studies of $1{,}25(OH)_2D_3$ action and CYP24 expression have relied on a host of cell culture models. Investigations with cultured cells have revealed induction of the *CYP24* gene in many different cell types. Enzyme activity and/or mRNA analyzed by Northern blots has been detected in primary and transformed cultured cell lines including chick [41] and rat kidney primary cultures [42], monkey kidney JTC-12 cells [43], human colon cancer cells [44,45], human promyelocytic leukemia HL-60 cells [46], human prostate carcinoma cells [47], osteoblastic cells [48], skin fibroblasts [49], lymphocytes [50], human trophoblasts [40], rat glial cells [51], and human keratinocytes [52]. In most of these studies, cells incubated in a serum free-medium possessed little if any 24-hydroxylase activity or mRNA, but the levels were greatly increased by $1{,}25(OH)_2D_3$ addition. The induction of *CYP24* mRNA in primary cultures of rat renal tubular cells was prevented by actinomycin D, indicating that the increase in *CYP24* mRNA was due to gene transcription [42]. Cycloheximide also inhibited the induction of mRNA in these cells, demonstrating a continued need for protein synthesis [42], but the identity or role of such proteins is not known.

The induction of CYP24 in many different cell types is in keeping with the almost ubiquitous distribution of VDR [53,54]. It seems that a wide range of VDR-containing cells are targets for $1{,}25(OH)_2D_3$ action in which the *CYP24* gene is simultaneously induced in response to hormone. The widespread distribution of 24-hydroxylase supports the contention that a major role for the enzyme involves regulating the local concentration and action of $1{,}25(OH)_2D_3$.

B. Regulation of Expression

Consistent with the general tissue distribution of CYP24, a broad spectrum of regulatory agents act to control cellular expression of the enzyme. Steroid and peptide hormones function through the transcription and signal transduction pathways to alter CYP24 activity. The enzyme level is also changed in response to age and several genetic disorders of mineral metabolism.

1. Steroid Hormones

a. Vitamin D Metabolites The strongest up-regulator of the 24-hydroxylase enzyme is the secosteroid $1{,}25(OH)_2D_3$ although high levels of $24{,}25(OH)_2D_3$ and other metabolites, particularly $1{,}24{,}25(OH)_3D_3$, can induce expression of CYP24 [42]. All of these metabolites are thought to function through binding to VDR [55], but the current discussion is limited to the more active $1,25(OH)_2D_3$, which displays the highest affinity for VDR.

In conjunction with the transcription factor VDR, the $1,25(OH)_2D_3$ hormone functions in target cells to induce transcription of the *CYP24* gene. In this regard, the hormone can also augment its inductive action by up-regulating VDR [56]. Expression of functional 24-hydroxylase enzyme occurs several hours following gene induction [57]. Increased 24-hydroxylase activity results in a greater turnover of $1,25(OH)_2D_3$ via the C-24 oxidation pathway. This hormone feedback action appears to function by a paracrine/autocrine mechanism to regulate cellular $1,25(OH)_2D_3$ in concert with systemic calcium requirements. Mechanistic studies on the action of $1,25(OH)_2D_3$ to up-regulate CYP24 expression have implicated a role for the protein kinase C (PKC) pathway. Treatment of cells with 12-*O*-tetradecanoylphorbol-13-acetate (TPA or PMA), a PKC stimulator, enhances the action of $1,25(OH)_2D_3$ to induce CYP24 expression [42]. PKC inhibitors were active in inhibiting the inductive action of $1,25(OH)_2D_3$, which is suggestive of a molecular role for PKC in mediating up-regulation of the *CYP24* gene. The molecular aspects of the action of TPA are discussed in Section IV of this chapter. The action of $1,25(OH)_2D_3$ to induce *CYP24* gene expression may also be augmented by the hormone's function to lower protein kinase inhibitor (PKI) activity [58]. PKI is an inhibitor of cAMP protein kinase A (PKA), and, therefore, a decrease in its activity could function to alter PKA pathway activity, as discussed below for peptide hormones.

b. Sex Steroids The previously mentioned increase in rat renal *CYP24* mRNA with age [38] was observed to be significantly higher in females. In addition, removal of the ovaries resulted in increased renal expression of *CYP24* mRNA [38]. This raises the possibility that lowered ovarian hormone levels may function to up-regulate renal *CYP24* gene expression and thus act to reduce ambient levels of $1,25(OH)_2D_3$ with age. In this regard, the sex steroids estrogen and testosterone have been reported in avian models to stimulate CYP1 activity while inhibiting CYP24 expression [59,60]. More recent work in rats has demonstrated a similar action of estrogen to suppress renal 24-hydroxylase activity [39]. The lowered CYP24 activity correlated with a decrease in renal VDR mRNA that was enhanced by combined treatment with estrogen and testosterone. The decrease in CYP24 activity and mRNA expression was attributed to lowered VDR levels. However, it is unlikely that CYP24 expression can be explained by changes in VDR content alone [39]. For example, the promoter for the *CYP24* gene contains putative estrogen response elements [29] that could function to directly regulate gene expression.

2. Peptide Hormones

a. Parathyroid Hormone It is evident from a number of *in vivo* studies that parathyroid hormone (PTH) functions to increase $1,25(OH)_2D_3$ production and suppress CYP24 activity in the kidney. Production of $24,25(OH)_2D_3$ was shown to be increased in thyroparathyroidectomized rats and could be prevented by administration of parathyroid extract [61]. Subsequently, it was demonstrated that cAMP administration to thyroparathyroidectomized rats mimicked the effect of PTH [62]. The inhibitory action of PTH on basal and $1,25(OH)_2D_3$-induced expression of CYP24 has also been demonstrated with kidney cultures. For example, PTH addition to rat renal slices [63] or cultured chick kidney cells [64,65] decreased basal $24,25(OH)_2D_3$ production whereas in monkey kidney cells PTH lowered the induction of 24-hydroxylase activity following $1,25(OH)_2D_3$ treatment [43]. PTH and $1,25(OH)_2D_3$ function in a reciprocal manner to regulate renal CYP24 expression, with the observed enzyme expression representing a net regulatory action of the two hormones [57].

The proximal convoluted tubule is the major site for both $1,25(OH)_2D_3$ synthesis and expression of CYP24 [67,68]. Because VDR is also present in these cells, one might expect the synthesis and release of $1,25(OH)_2D_3$ to be suppressed by the induction of CYP24 by the secosteroid. However, $1,25(OH)_2D_3$ continues to be synthesized in the proximal convoluted tubule cells. An insight to this dilemma has been obtained through tubule microdissection techniques in rats with different circulating PTH levels [39]. Elevated PTH, which is known to stimulate $1,25(OH)_2D_3$ synthesis, functions to inhibit proximal tubule expression of VDR and *CPY24* mRNA. Limiting VDR expression would lead to lowered transcription of the *CYP24* gene and 24-hydroxylase enzyme. Consequently, $1,25(OH)_2D_3$ production and release would be the preferred activity in proximal convoluted tubules during high PTH states. However, VDR and *CYP24* mRNA were not down-regulated in the adjoining cortical collecting ducts [39]. How PTH functions to inhibit specifically VDR expression in convoluted tubules remains to be determined. The mechanism is not dependent on the $1,25(OH)_2D_3$ status, but rather it would appear to involve a direct action of PTH. The hormone interaction with G-protein-coupled receptor can lead to stimulation of the PKA and PKC pathways [69]. As cAMP and forskolin mimic the inhibitory effect of PTH on renal 24-hydroxylase expression [36,37,63,64], the action of PTH may occur via the PKA pathway. Whether PTH also has a direct effect on the expression of the CYP24 gene is not known. Further work is required to clarify the molecular mechanism by which PTH regulates expression of VDR and CYP24 in the proximal convoluted tubules.

The action of PTH to impede *CYP24* mRNA induction by $1,25(OH)_2D_3$ is not observed in the intestine due to the lack of PTH receptors [36–38]. The major regulators of intestinal CYP24 expression are cellular VDR and changes in ambient $1,25(OH)_2D_3$ [37]. Cellular calcium does not appear to impact expression of the enzyme [70]. Intestinal CYP24 does not contribute significantly to ambient $24,25(OH)_2D_3$ levels [70] that predominantly follow renal 24-hydroxylase activity. Rather, induction of intestinal CYP24 may function to down-regulate transient elevations in cellular $1,25(OH)_2D_3$ through the 24-oxidation pathway. Such an action is consistent with the transient expression of intestinal CYP24 in response to acute changes in $1,25(OH)_2D_3$ levels [36,37,70].

Bone osteoblasts contain PTH receptors and VDR; however, PTH does not function to down-regulate CYP24 expression as observed in the kidney. For example, *CYP24* mRNA in calvaria is not repressed in rats with elevated PTH levels [71]. Rather, several studies have shown PTH to act synergistically with $1,25(OH)_2D_3$ in the up-regulation of CYP24 in bone cells [72,73]. Armbrecht and Hodam [72] found that treatment of rat UMR-106 osteoblastic cells with PTH and $1,25(OH)_2D_3$ resulted in a marked synergistic induction of *CYP24* mRNA. This action of PTH could be mimicked by forskolin [72]. Related studies from Krishnan *et al.* [73] showed that PTH pretreatment enhanced the $1,25(OH)_2D_3$-dependent induction of *CYP24* mRNA in MC3T3-E1 mouse calvaria-derived cells and rat UMR-106–01 cells. It was also reported that PTH up-regulated VDR [73], by an apparent posttranscriptional mechanism, and the elevated VDR was suggested as a means for increased induction of the 24-hydroxylase. Whether up-regulation of VDR in osteoblasts by PTH underlies the observed synergism remains to be confirmed. An alternative explanation for the synergy could involve PTH acting via cAMP to increase transcription of the *CYP24* gene through a cAMP response element. Two sites resembling cAMP response elements have been identified in the rat promoter, but their functionality has not been examined [74]. However, this mechanism is not easily reconciled with the observation that PTH alone has no effect on *CYP24* mRNA expression. In other studies with rat osteoblastic cell cultures [71], *CYP24* mRNA was induced by $1,25(OH)_2D_3$, but no synergy was observed with PTH. This may be due to time-dependent aspects of the measurements or a cell line dependency [72].

The physiological rationale for the synergistic action of PTH and $1,25(OH)_2D_3$ on induction of bone *CYP24* expression seems perplexing when considered in the context of the renal actions for PTH. However, increased bone CYP24 activity due to synergism between PTH and $1,25(OH)_2D_3$ could result in elevated bone $24,25(OH)_2D_3$ and the associated increase in bone mineralization [75], an action that would function to counterregulate the osteoclastic bone resorptive function of PTH.

b. Calcitonin Calcitonin (CT) acts on the kidney proximal straight tubule to up-regulate $1,25(OH)_2D_3$ production, a parallel action to that previously documented for PTH in the proximal convoluted tubule [76–79]. Recognizing the similar vitamin D regulatory actions for PTH and CT, it seems reasonable that CT could act to down-regulate CYP24 expression, as discussed above for PTH. Consistent with this regulatory pattern, CT has been shown to function in the intestine to suppress CYP24 mRNA expression and 24-hydroxylase activity [80]. Although CT functions to up-regulate renal $1,25(OH)_2D_3$ production, it remains to be determined if CT also functions in kidney to regulate *CYP24* gene expression.

c. Insulin and IGF-1 Serum $1,25(OH)_2D_3$ is lowered and $24,25(OH)_2D_3$ elevated in young patients [81] and rats [82,83] that are insulin deficient. The decrease in $1,25(OH)_2D_3$ is associated with an inhibition of intestinal calcium absorption that can be corrected by treatment with $1,25(OH)_2D_3$ [84] or insulin [85]. Changes in metabolite levels are due mainly to alterations in enzymes of the biosynthetic pathway as illustrated by depressed renal $1,25(OH)_2D_3$ synthesis and increased $24,25(OH)_2D_3$ production and *CYP24* mRNA in diabetic rats [82,83]. Serum PTH levels are normal in the diabetic rat, but the capacity of PTH to stimulate renal $1,25(OH)_2D_3$ production and inhibit $24,25(OH)_2D_3$ synthesis is reduced in diabetes [82]. In this regard, the action of PTH to increase 24-hydroxylase enzymatic activity in UMR-106 osteoblast-like cells is potentiated by insulin [86]. Why PTH expresses opposite CYP24 regulatory actions in bone cells compared to kidney is not understood currently. Nevertheless, it is evident that the regulatory action of PTH in bone and renal tissue is refractory in diabetes and full action of the hormone requires insulin. Serum calcitonin is also reduced in diabetic rats [82], and the lowered hormone level could contribute to up-regulation of CYP24 expression [87]. Whether insulin also functions through the signal transduction pathway to regulate transcription of the *CYP1* and *CYP24* genes remains to be evaluated.

It appears that the catabolic state of diabetes favors an increased level of circulating $24,25(OH)_2D_3$, whereas insulin-mediated anabolism is associated with a lowered metabolite level. However, the mechanism for altered vitamin D metabolism in diabetes is not known with certainty. Part of the mechanism for increased $24,25(OH)_2D_3$ productivity in diabetes would appear to involve age along with the degree and duration of insulin deficiency. This is most evident in adult and elderly type I and type II diabetic patients who display a decrease in serum $24,25(OH)_2D_3$ rather than the higher level observed in younger patients and animals [81–83,88,89]. Why serum $24,25(OH)_2D_3$ decreases with age and duration of insulin deficiency is not understood, although adaptive responses resulting in increased cellular actions for insulin, CT, or PTH could result in lowered $24,25(OH)_2D_3$ levels.

In a preliminary study [90], growth hormone (GH) has been observed to decrease serum $24,25(OH)_2D_3$ levels in GH-deficient children supplemented with hormone. Because GH functions through insulin-like growth factor-1 (IGF-1), it was suggested that this pathway could link the similar actions of insulin and GH to suppress circulating $24,25(OH)_2D_3$ levels.

3. Genetic Disorders

a. Hereditary Vitamin D-Dependent Rickets Types I and II The hereditary disorder vitamin D-dependent rickets type I (VDDR-I or PDRR) is characterized by a defect in the renal 1-hydroxylase (CYP1) and low circulating $1,25(OH)_2D_3$ levels [91]. However, the circulating $24,25(OH)_2D_3$ level appears to be unaffected in patients with VDDR-I. In contrast, patients with hereditary vitamin D-dependent rickets type II (VDDR-II or HVDRR) have a mutated VDR and display impaired $1,25(OH)_2D_3$ induction of CYP24 activity in cultured VDRR-II skin fibroblasts [92]. This abnormality can be linked to a decrease in circulating $24,25(OH)_2D_3$ levels [93,94]. Consequently, the lesion in synthesis of functional VDR in VDDR-II can result in lowered expression of vitamin D-dependent genes including *CYP24.* These disorders are discussed in detail in Chapters 47 and 48 of this volume.

b. X-Linked Hypophosphatemia The X-linked hypophosphatemia (*hyp*) mouse is a model for the most prevalent form of inherited rickets in humans, X-linked hypophosphatemia [95]. The regulation of vitamin D metabolism is perturbed by the X-linked mutation, and there is evidence for a role of renal 24-hydroxylase in this abnormal regulation. In normal mice on a low phosphate diet, there is a rise in serum $1,25(OH)_2D_3$ but no change in renal C-24 oxidation products [96]. In contrast, *hyp* mice on a low phosphate diet exhibit reduced serum $1,25(OH)_2D_3$ and a dramatic increase in renal C-24 oxidation products [96]. It has been shown that an increase of renal 24-hydroxylase protein and mRNA underlies the increased C-24 oxidation in *hyp* mice, which is specific for the kidney [97]. Following $1,25(OH)_2D_3$ treatment, induction of renal 24-hydroxylase protein and mRNA were seen with both the normal and *hyp* mice on a standard diet. However, the induction in phosphate-deprived animals was lowered in normal mice, whereas *hyp* mice maintained the elevated level of CYP24 expression. Although the *hyp* gene has not been identified, these results raise the possibility that the gene product at the X locus is a repressor for 24-hydroxylase gene transcription that is active under phosphate deprivation in normal mice but inactive as a mutant form in *hyp* mice. The inverse relationship between the level of $1,25(OH)_2D_3$ in the serum of *hyp* mice deprived of phosphate and renal 24-hydroxylase activity supports the proposal that this enzyme is important for maintaining physiological concentrations of $1,25(OH)_2D_3$ by regulating the metabolic turnover rate of the hormone. This is discussed further in Chapter 46.

IV. REGULATION OF GENE EXPRESSION

A. Rat Promoter

Considerable progress has been made in understanding the molecular basis of $1,25(OH)_2D_3$-dependent induction of the 24-hydroxylase gene. Using nested sets of upstream promoter sequences linked to reporter gene constructs, it was possible to identify specific promoter regions that are required for the inductive action of $1,25(OH)_2D_3$ [30,74,98,99]. Studies have focused on the proximal promoter region between −298 and +74, although regulatory regions are also present upstream of −298 [30,98,99]. Within the proximal promoter region, two functional vitamin D response elements (VDREs) have been identified on the antisense strand [30,74,98,99]. Half-site mutational analysis was used to verify the sequence and function of the two VDREs [30,100]. A proximal VDRE (designated VDRE-1) was identified at −136/−150 with the sequence 5′-AGGTGAgtgAGGGCG-3′ whereas a more distal VDRE (VDRE-2) was identified at −244/−258 and had the sequence 5′-GGTTCAgcgGGTGCG-3′. VDREs consisting of two 6-bp half-sites and separated by 3 bp belong to the DR3 family of VDREs. Although a common sequence motif is evident between different VDREs, sequence similarity for VDRE-1 and VDRE-2 is greatest with osteocalcin and osteopontin, respectively. On the basis of current sequence information

TABLE I VDRE Sequences

Gene	DR3 type	Position
Rat osteocalcin	GGGTGAatgAGGACA	−455 to −441
Human osteocalcin	GGGTGAacgGGGGCA	−500 to −486
Mouse osteopontin	GGTTCAcgaGGTTCA	−757 to −743
Rat calbindin-D_{9K}	GGGTGTcggAAGCCC	−488 to −472
Rat $P450_{24}$	AGGTGAgtgAGGGCG	−136 to −150
Rat $P450_{24}$	GGTTCAgcgGGTGCG	−244 to −258
Human $P450_{24}$	AGGTGAgcgAGGGCG	−155 to −169
Human $P450_{24}$	GAGTTCaccGGGTGT	−276 to −290

(Table I), the presence of two VDREs has been documented only for the *CYP24* gene promoters, and the dual VDRE arrangement was suggestive of a coordinated regulation between VDRE-1 and VDRE-2. Although this dual arrangement of VDREs may be unique to CYP24, it is not uncommon for multiple response elements to exist in genes activated by other nuclear receptors or transcription factors.

1. BINDING OF VDR–RXR COMPLEX

Binding of nuclear proteins to VDRE-1 and VDRE-2 has been investigated by gel mobility shift analysis using radiolabeled VDRE probes for each site and intestinal or kidney cell nuclear extracts [74,98,99] or *in vitro* transcribed VDR [95]. Nuclear extracts devoid of VDR did not show specific complex formation with the DNA oligomers containing VDRE sequences. However, nuclear extracts containing VDR formed a specific protein complex that comigrated with a standard VDR–VDRE complex. Participation of VDR in the VDRE-1 and -2 complexes was established by use of a competitive VDR antibody and *in vitro* synthesized VDR [98,100]. Initial studies established the binding of VDR as an RXR complex at VDRE-1, which was verified and extended to include VDRE-2 through supershift analysis with a specific monoclonal RXR antibody [100]. When analyzed separately, VDRE-2 showed about 4- to 5-fold higher affinity for the VDR–RXR complex than VDRE-1. This is perhaps not surprising, as one hexameric half-site of VDRE-2 is identical to that in the osteopontin VDRE, which is known to bind the VDR–RXR protein complex strongly [101].

2. COOPERATIVITY BETWEEN VDREs

Protein complex formation with the two VDREs showed different kinetics of binding, and yet it was not known how these differences were expressed in the context of native promoter. Therefore, promoter constructs with mutated response elements were expressed in various cell lines in order to determine the contribution of each VDRE in the context of the natural rat promoter and to investigate whether there is transcriptional cooperation between the two sites [100]. The two VDREs were specifically altered by site-directed mutagenesis either individually or in combination within the −298 to +74 bp region of native promoter. Promoter–luciferase reporter gene constructs were introduced into COS-1 (monkey kidney), ROS 17/2.8 (rat osteosarcoma), and JTC-12 (monkey kidney proximal tubular) cells. In response to $1,25(OH)_2D_3$, the wild-type construct gave about an 18-fold level of induction in COS-1 cells, whereas mutation of VDRE-1 or VDRE-2 sites reduced expression to 3-fold and 6-fold respectively (Fig. 5). There was no induction when both VDREs were mutated. These data verified that both VDREs are functional [99], with a greater contribution provided from VDRE-1. There is transcriptional synergism between the VDREs, as the wild-type induction (18-fold) was greater than the sum of the individual contributions of VDRE-1 (6-fold) and VDRE-2 (3-fold). Similar results were seen with the three different cell lines (Fig. 5). In COS-1 cells, the maximum wild type level of induction obtained with 10^{-7} *M* $1,25(OH)_2D_3$ was retained with concentrations as low as 10^{-10} *M*, establishing the functionality of the VDRE at physiological $1,25(OH)_2D_3$

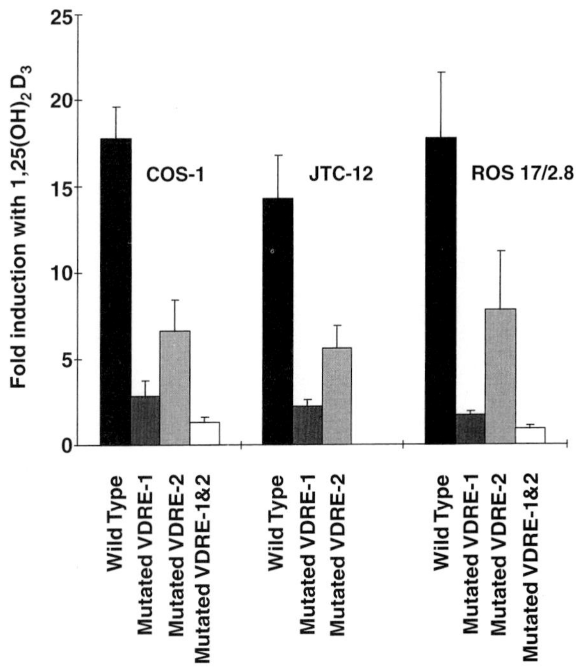

FIGURE 5 Wild-type and mutated promoter expressions for DR3 VDREs. Verification of response element function for VDRE-1 and VDRE-2 was achieved using mutational analysis in rat −298/+74 promoter–luciferase constructs expressed in COS-1, JTC-12, and ROS 17/2.8 cells [100].

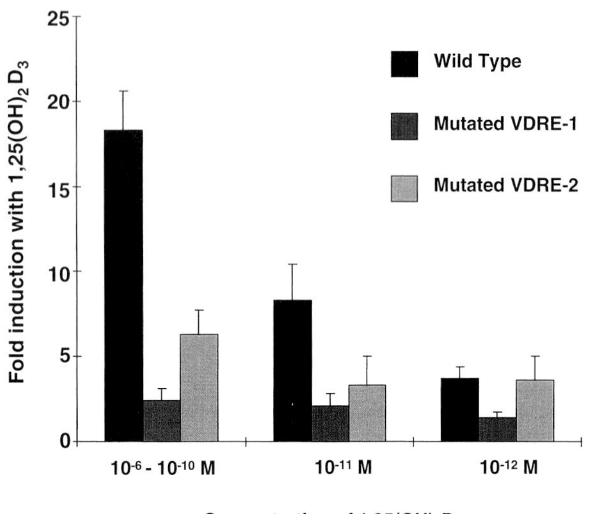

FIGURE 6 Dose–response of promoter construct to $1{,}25(OH)_2D_3$ treatment. Wild-type and mutated VDRE-1 and -2 were evaluated in COS-1 cells for their inductive response to a 10^6 range of $1{,}25(OH)_2D_3$ concentration [100].

concentrations (Fig. 6). It is evident from data for the rat *CYP24* gene that transcriptional activation synergism occurs in different cell types, over a wide range of $1{,}25(OH)_2D_3$ concentrations, in which the proximal VDRE is more active and primarily responsible for induction at limiting levels of $1{,}25(OH)_2D_3$. Studies on the *CYP24* gene also illustrated the importance of interpreting gel mobility shift assays in the context of native promoter regulatory activity.

The finding that VDRE-1 rather than VDRE-2 contributed more substantially to $1{,}25(OH)_2D_3$-dependent induction was somewhat unexpected, as gel shift data indicated that VDRE-2 had a higher affinity for the VDR–RXR complex. The greater distance of VDRE-2 from the transcription machinery may lower its transactivation capacity, however, this seems unlikely, because removal of the sequence between the VDREs does not affect $1{,}25(OH)_2D_3$ induction in transient assays [99]. Liu and Freedman [102] have reported that $1{,}25(OH)_2D_3$ induction directed by a VDRE-containing artificial promoter is increased considerably by the presence of a transcription factor bound nearby. On this basis, it could be proposed that one or more transcription factors associate with the 24-hydroxylase promoter and cooperate with the VDR–RXR complex at VDRE-1, thus imparting increased transcriptional activity to this VDRE. Again, this concept has been firmly established with other promoters, therefore, its applicability to a vitamin D responsive gene is to be expected.

As stated earlier, multiple copies of binding sites for steroid hormone receptors frequently occur in promoters of hormone responsive genes and can act synergistically [103], however, the 24-hydroxylase promoter may be at this time the only naturally occurring vitamin D promoter with two functional VDREs. Notable features of the VDREs in the rat promoter are their positions and location on the antisense strand. The closeness of the VDREs to the transcription initiation site is in contrast to other DR3-type VDREs such as rat osteocalcin, mouse osteopontin, and rat calbindin-D9K, which are 450 bp or more upstream. Only two other VDREs have been reported on the antisense strand. One VDRE is located in the PTH gene promoter, which acts in a negative fashion [104], and the other is a putative VDRE located at about −1200 in the chicken carbonic anhydrase-II gene [105].

The observed 2-fold synergism between the VDREs in the rat promoter is only moderate, but it results in an overall induction level of 14- to 18-fold in the different cell types and constitutes the most active $1{,}25(OH)_2D_3$ responsive promoter identified to date. Transcriptional synergism between the VDREs of the 24-hydroxylase promoter could have important biological consequences by ensuring the rapid removal of hormone when levels are sufficiently high to cause hypercalcemia and accelerated bone resorption [106]. It is possible that while VDRE-1 would respond preferentially at a slight excess of hormone, VDRE-2 would be activated at high levels, with the resulting synergism ensuring rapid detoxification of hormone. The molecular mechanism underlying this synergism remains to be elucidated. Cooperative binding of the VDR–RXR protein complexes on each VDRE is a possibility, although Zierold *et al.* [99] could not detect such an interaction using gel mobility shift analysis. An alternative mechanism could involve the independent binding of the VDR–RXR complexes to the VDREs, with synergism resulting from subsequent cooperative interactions with the basal transcription machinery [107] (Fig. 7). In this context, there is evidence that VDR can interact with the general transcription factor TFIIB [108] and that both VDR and RXR can interact with a transcriptional intermediary factor (TIF1) [109].

3. DR6-Type DNA Sequences

A DR6-type sequence has been identified in the rat promoter on the antisense strand at position −249/−323 [110]. This sequence contains two hexameric half-sites separated by 6 bp and shares a common half-site with VDRE-2. Studies by Kahlen and Carlberg [110] have shown that the DR6-type sequence directs $1{,}25(OH)_2D_3$ induction, but only when fused to the thymidine kinase (TK) promoter and when its transient regulation is examined in ROS 17/2.8 cells. To determine if this type of element is functional in the native promoter, analysis

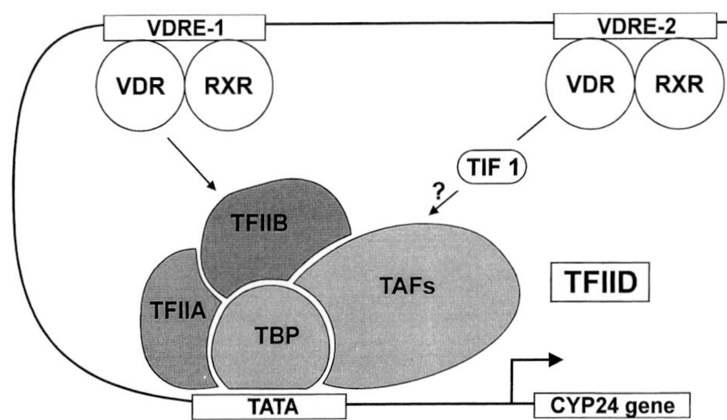

FIGURE 7 Model for $1,25(OH)_2D_3$-induced transcriptional synergism. Shown diagrammatically is an arrangement whereby VDRE-1 and VDRE-2 are brought into the vicinity of the transcription complex and function cooperatively to up-regulate expression of the rat 24-hydroxylase gene. Illustrated are the known interactions of VDR with the general transcription factor TFIIB and the transcriptional intermediary factor TIF1. Other adapter molecules that have not been characterized are most likely involved in the synergistic response. TBP, TATA-binding protein; TAFs, transcription activation factors.

of the mutated −298 construct was conducted in several cell lines. When the DR6 response element was specifically mutated, leaving VDRE-1 and VDRE-2 intact, the level of expression of the mutant construct was the same as the wild type in all cell lines (Fig. 8) [100]. Also, there

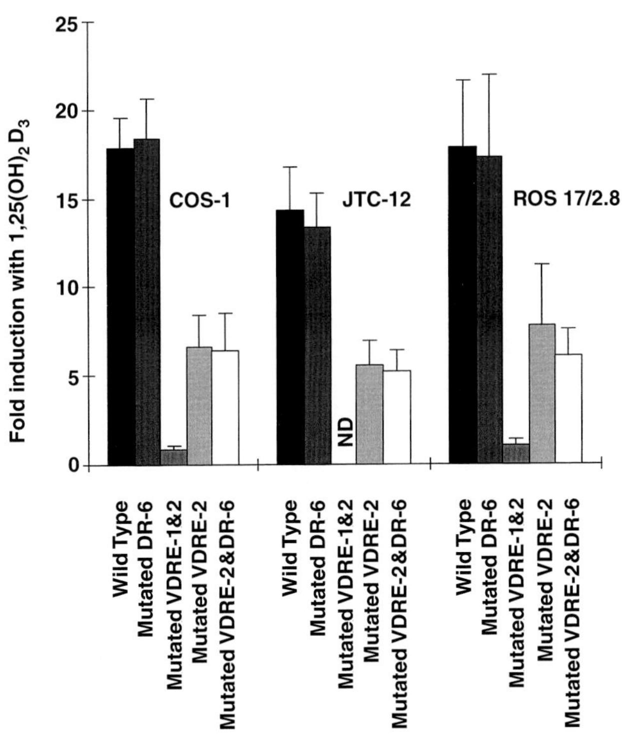

FIGURE 8 Mutational analysis of the putative DR6 VDRE. The functionality of a putative DR6-like sequence was evaluated by mutational analysis of −298/+74 promoter–luciferase constructs in COS-1, JTC-12, and ROS 17/2.8 cells [100].

was no induction when both VDRE-1 and VDRE-2 were mutated leaving the DR6 response element intact. In addition, the induction of a construct with both VDRE-2 and the DR6 response element mutated was the same as when only VDRE-2 was mutated. Therefore, the DR6 sequence was inactive in its native promoter context when tested in three different cell lines (Fig. 8). Because the DR6 sequence is responsive to hormone when fused to the TK promoter [110], it is apparent that promoter environment is important, which emphasizes the need to examine the functionality of a putative VDRE in the context of its native promoter. Other DR6-type elements have been reported in the promoters of the osteocalcin and fibronectin genes [8], but their activity in the native promoter has not yet been substantiated. One could predict, on the basis of the previous study, that they are nonfunctional.

4. Protein Kinase C and Regulation

Members of the protein kinase C (PKC) family of serine/threonine kinases [111] are activated by diacylglycerol in response to specific external signals, and, through altering the phosphorylation status of nuclear proteins, these kinases can influence gene expression. As mentioned in Section III, there is evidence of a role for PKC in regulating 24-hydroxylase expression. The involvement of PKC has been inferred from the use of PKC inhibitors and the phorbol ester 12-*O*-tetradecanoylphorbol-13-acetate (TPA), also referred to as phorbol 12-myristate 13-acetate (PMA), which mimics diacylglycerol and is a potent activator of PKC.

In initial studies, Henry [112] and Henry and Luntao [113] reported that TPA treatment of chick kidney primary cultures resulted in an increase in basal $24,25(OH)_2D_3$ production and a decrease in $1,25(OH)_2D_3$ synthesis. Studies with mouse renal tubule cells also established that TPA treatment increased basal $24,25(OH)_2D_3$ production, which was prevented by PKC inhibitors [114]. The induction of renal 24-hydroxylase activity by TPA was very rapid, and it was suggested that TPA may have a nongenomic effect, with its action being mediated through phosphorylation and activation of certain mitochondrial proteins. Mandla *et al.* [114] found that TPA treatment of kidney tubules isolated from *hyp* mice did not stimulate $24,25(OH)_2D_3$ production; because renal 24-hydroxylase and PKC were elevated in these mice relevant to normal mice, it was argued that PKC may be involved in the abnormal expression of 24-hydroxylase in these animals [114].

In more recent studies, Chen *et al.* [42] used a specific rat cDNA clone to investigate the effect of TPA on the synthesis of *CYP24* mRNA by rat renal primary cultures. When cells were treated with TPA alone, there was no effect on *CYP24* mRNA levels. However, pretreatment of cells with $1,25(OH)_2D_3$ for several hours, followed by TPA addition, resulted in a substantial synergistic response compared with $1,25(OH)_2D_3$ alone. A PKC inhibitor, when added at the time of TPA addition, blocked the TPA-induced increase, which implicated PKC in the regulatory mechanism. Similar results were seen with rat intestinal epithelial cells, and a marked synergistic increase in the mRNA level for 24-hydroxylase was observed following the addition of TPA to $1,25(OH)_2D_3$-pretreated cells [115,116]. In these studies with rat renal and intestinal cells, the synergy between TPA and $1,25(OH)_2D_3$ resulted in an induction of *CYP24* mRNA beyond that attainable with $1,25(OH)_2D_3$ alone [42,115]. Other studies carried out with rat intestinal epithelial cells established that the synergistic action of TPA was not due to an alteration in the content or binding affinity of VDR [116]. Although a TPA synergistic response is observed in kidney cells, Armbrecht and Hodam [72] have shown that in the osteoblastic cell line UMR-106 *CYP24* mRNA can be induced by $1,25(OH)_2D_3$, but there was no effect of TPA in either the presence or absence of $1,25(OH)_2D_3$. Whether there is tissue specificity involved in the synergistic response remains to be determined.

It is well established that TPA treatment of cells can alter gene expression through increased levels of active AP-1, a specific transcription factor complex composed of Jun and Fos family members [117,118]. Activated PKC, in response to TPA, can stimulate the ERK/MAP kinase signaling pathway leading to c-*fos* mRNA induction and elevated AP-1 [118]. If the *CYP24* gene promoter contains a binding site(s) for AP-1, then the observed transcriptional synergy could be explained by a cooperative interaction between AP-1 and VDR–RXR complex bound to the promoter (Fig. 9). The rat 24-hydroxylase promoter–luciferase constructs have been used in transient expression experiments to investigate the role of TPA in the $1,25(OH)_2D_3$-dependent induction of gene expression [119] (J. Omdahl, unpublished data, 1996). Short-term TPA treatment of various cell lines leads to minimal induction of luciferase activity; however, TPA pretreatment results in a marked synergistic induction by $1,25(OH)_2D_3$ compared with hormone alone. The synergistic effect of TPA as well as the inductive action of $1,25(OH)_2D_3$ alone can be prevented by treatment of the cells with PKC inhibitors. Experiments are in progress to delineate control regions of the promoter that may regulate the TPA response (e.g., sequences that approximate AP-1 elements). In similar studies, Pike *et al.* [119] have expressed the human 24-hydroxylase promoter construct in mammalian cells; they also observed a modest increase in basal ex-

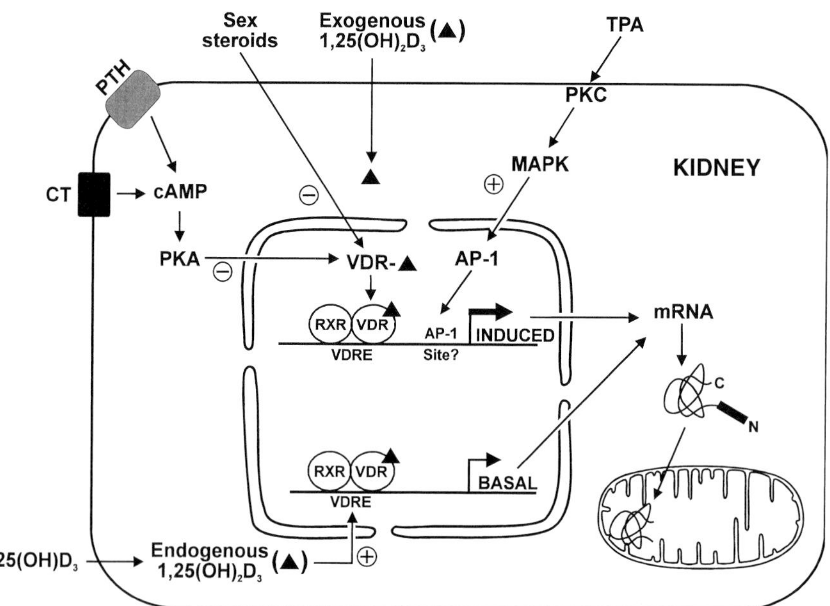

FIGURE 9 Summary diagram of factors affecting expression of the 24-hydroxylase gene in the kidney. The top part illustrates the negative (down) regulation of *CYP24* (24-hydroxylase) gene expression by calcitonin (CT) and parathyroid hormone (PTH) via the cAMP-dependent protein kinase (PKA) pathway and the negative regulation by sex steroids. The positive (up) regulation is shown for TPA via the protein kinase C (PKC) and mitogen-activated protein kinase (MAPK) pathway. The basal level of enzyme expression is shown at bottom, in which regulatory agents have been omitted. The 24-hydroxylase gene product is localized to the inner mitochondrial membrane.

pression with TPA in the absence of $1,25(OH)_2D_3$ but strong synergy in the presence of hormone. Deletion experiments established the location of the responsive region to the first 250 bp of the promoter. However, mutations in a putative AP-1 site adjacent to the proximal VDRE (−169/−155) did not alter the TPA response, and these authors concluded that the responsive element lies elsewhere.

Although the effect of TPA is typically mediated by AP-1, it is possible that phosphorylation of other proteins by PKC may play a role. For example, different transcription factors involved in regulation of *CYP24* or VDR gene expression could be involved in mediating the action of the phorbol ester. There is considerable evidence that the phosphorylation status of the VDR is important for its activity [120,121], and work with PKC inhibitors suggests that a PKC-mediated phosphorylation of VDR may be important for the transactivation function of VDR [122]. Of particular interest to regulation of *CYP24* gene expression is a study on an androgen-induced promoter. Transcriptional synergism between androgen and TPA was observed, but it was not due to either increased synthesis of androgen receptor or its phosphorylation. Rather, it appeared to be due to a direct interaction between AP-1 and androgen receptor [123]. Overall, current data strongly suggest that TPA through PKC plays a role in regulating the 24-hydroxylase enzyme in kidney and intestinal cells; however, the identity and site of PKC action are not known, and the mechanism of its function at the transcriptional level remains to be elucidated. Whether PKC also regulates basal levels of 24-hydroxylase activity through phosphorylation of transcription factors or through proteins associated with the enzyme catalysis mechanism, as suggested by the studies of Mandla *et al.* [114], remains an interesting possibility.

B. Human Promoter

Two VDREs for the human promoter have been located by Chen and DeLuca [31] on the antisense strand at −169/−155 and at −290/−276 that are similar to the VDREs in the rat promoter (Fig. 10). The proximal VDRE was identical in sequence to the corresponding region in the rat, but the distal VDRE deviated considerably in sequence from its rat counterpart (Table I). To determine functionality, deletion promoter fragments containing one or both VDREs were fused to a TK/chloramphenicol acetyltransferase (CAT) reporter and transient expression performed in ROS 17/2.8 cells [31]. These experiments established the importance of each VDRE, with a greater contribution from the proximal VDRE as found in the rat promoter. Although there was no evidence for synergism between the VDREs from these deletion experiments, this question should be explored further by specific mutations in the putative VDRE sites. A potential DR6-type sequence also exists in the human promoter. However, this putative response element is unlikely to be a functional element because of the presence of T residues at the second and fifth positions of the two half-sites (a base pattern that is absent from known VDREs; see Table I) as well as the aberrant 6 bp spacing.

SUMMARY

The 24-hydroxylase enzyme (CYP24) is a widely distributed component of the vitamin D pathway. The enzyme is regulated by a spectrum of hormones and functions to synthesize 24-hydroxylated metabolites, with preferential action in promoting bone mineralization and possibly other undisclosed cellular functions that

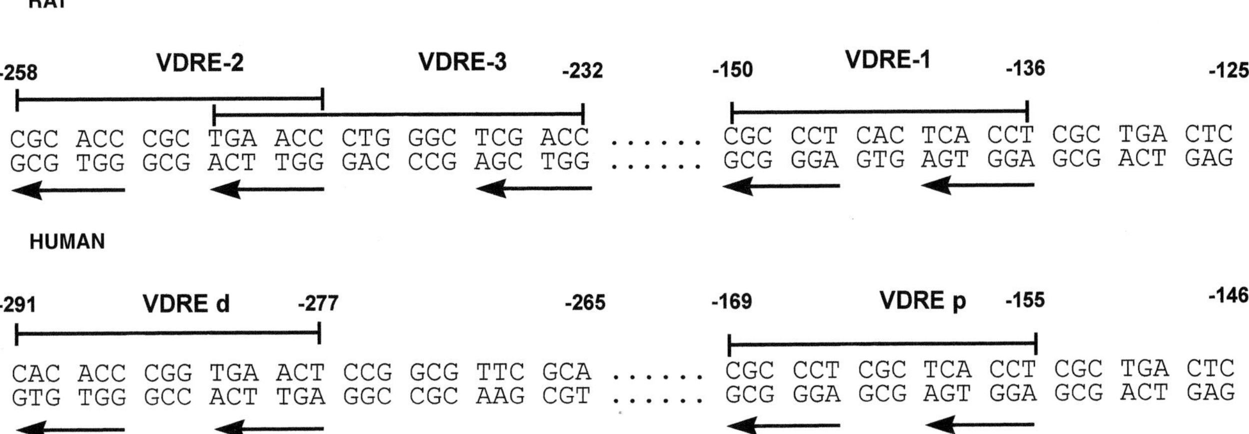

FIGURE 10 Rat and human promoter comparison. The VDREs for the rat and human promoters are compared by sequence and location within the early promoter region of the 24-hydroxylase gene. p, proximal; d, distal.

may become evident through the use of gene knockout models. CYP24 also plays a central role in directing the metabolic turnover of several 25-hydroxylated vitamin D metabolites. In this regard, the enzyme expresses both 23- and 24-hydroxylase activities, which catalyze critical steps in the 24-oxidation pathway for the metabolic degradation of $1,25(OH)_2D_3$ and $24,25(OH)_2D_3$. The importance of the 24-hydroxylase enzyme in regulating circulating $1,25(OH)_2D_3$ is most evident in the mouse model for X-linked hypophosphatemia in which the $1,25(OH)_2D_3$ level is lowered due to a high set point for CYP24.

The *CYP24*-gene contains a strong promoter for upregulation by $1,25(OH)_2D_3$. Two VDREs are present in the immediate upstream region of the gene and function in a coordinated manner to regulate 24-hydroxylase enzyme expression in direct response to the ambient $1,25(OH)_2D_3$ level. Mechanistic studies have shown CYP24 expression to be controlled at the transcription level. This inductive action can be enhanced through activation of the PKC pathway, which may involve the phosphorylation of nuclear factors associated with transcriptional activation of the *CYP24* gene.

Acknowledgements

The authors recognize the expert assistance of Chris Matthew, Eva Quintana, and Deeantha Gutierrez who greatly facilitated the preparation of this chapter. Also invaluable were correspondences and discussions with David Kerry, Howard Morris, Prem Dwivedi, Satya Reddy, Francis Glorieux, and Ronald Horst on subject matter contained in this chapter.

References

1. Masumoto O, Ohyama Y, Okuda K 1988 Purification and characterization of vitamin D 25-hydroxylase from rat liver mitochondria. J Biol Chem **263**:14256–14260.
2. Axen E, Bergman T, Wikvall K 1992 Purification and characterization of a vitamin D_3 25-hydroxylase from pig liver microsomes. Biochem J **287**:725–731.
3. DeLuca HF, Suda T, Schnoes HK, Tanaka Y, Holick MF 1970 25,26-Dihydroxycholecalciferol, a metabolite of vitamin D_3 with intestinal calcium transport activity. Biochemistry **9**:4776–4780.
4. Ghazarian JG, Schnoes HK, DeLuca HF 1973 Mechanism of 25-hydroxycholecalciferol 1α-hydroxylation: Incorporation of oxygen-18 into the 1α-position of 25-hydroxycholecalciferol. Biochemistry **12**:2555–2558.
5. Ghazarian JG, Jefcoate CR, Knutson JC, Orme-Johnson WH, DeLuca HF 1974 Mitochondrial cytochrome P450: A component of chick kidney 25-hydrocholecalciferol-1alpha-hydroxylase. J Biol Chem **249**:3026–3033.
6. Umesono K, Muakami KK, Thompson CC, Evans RM 1991 Direct repeats as selective response elements for the thyroid hormone, retinoic acid and vitamin D_3 receptors. Cell **65**:1255–1266.
7. Ozono K, Liao J, Kerner SA, Scott RA, Pike JW 1990 The vitamin D-responsive element in the human osteocalcin gene. Association with a nuclear proto-oncogene enhancer. J Biol Chem **265**:21881–21888.
8. Sone T, Kerner S, Pike JW 1991 Vitamin D receptor interaction with specific DNA association as a 1,25-dihydroxyvitamin D_3-modulated heterodimer. J Biol Chem **266**:23296–23305.
9. Whitfield GK, Hsieh JC, Jurutka PW, Selznick SH, Haussler CA, MacDonald PN, Haussler MR 1995 Genomic actions of 1,25-dihydroxyvitamin D_3. J Nutr **125**:1690S–1694S.
10. Holick MF, Schnoes HK, DeLuca HF, Gray RW, Boyle IT, Suda T 1972 Isolation and identification of 24,25-dihydroxycholecalciferol, a metabolite of vitamin D_3 made in the kidney. Biochemistry **11**:4251–4255.
11. Madhok TC, Schnoes HK, DeLuca HF 1977 Mechanism of 25-hydroxyvitamin D_3 24-hydroxylation: Incorporation of oxygen-18 into the 24 position of 25-hydroxyvitamin D_3. Biochemistry **16**:2142–2145.
12. Tanaka Y, DeLuca HF, Ikekawa N, Morisaki M, Koizumi N 1975 Determination of stereochemical configuration of the 24-hydroxyl group of 24,25-dihydroxyvitamin D_3 and its biological importance. Arch Biochem Biophys **170**:620–626.
13. Knutson JC, DeLuca HF 1974 25-Hydroxyvitamin D_3-24-hydroxylase: Subcellular location and properties. Biochemistry **13**:1543–1548.
14. Gray RW, Omdahl JL, Ghazarian JG, Horst RL 1990 Induction of 25-OH-vitamin D_3 24- and 23-hydroxylase activities in partially purified renal extracts from pigs given exogenous $1,25\text{-}(OH)_2D_3$. Steroids **55**:395–398.
15. Ohyama Y, Okuda K 1991 Isolation and characterization of a cytochrome P-450 from rat kidney mitochondria that catalyzes the 24-hydroxylation of 25-hydroxyvitamin D_3. J Biol Chem **266**:8690–8695.
16. Akiyoshi-Shibata M, Sakaki T, Ohyama Y, Noshiro M, Okuda K, Yabusaki Y 1994 Further oxidation of hydroxycalcidiol by calcidiol 24-hydroxylase: A study with the mature enzyme expressed in *Escherichia coli*. Eur J Biochem **224**:335–343.
17. Tanaka Y, Castillo L, DeLuca HF 1977 The 24-hydroxylation of 1,25-dihydroxyvitamin D_3. J Biol Chem **252**:1421–1424.
18. Usui E, Noshiro M, Okuda K 1990 Molecular cloning of cDNA for vitamin D_3 25-hydroxylase from rat liver mitochondria. FEBS Lett **262**:135–138.
19. Horst RL, Reinhardt TA, Ramberg CF, Koszewski NJ, Napoli JL 1985 24-Hydroxylation of 1,25-dihydroxyergocalciferol: An unambiguous deactivation process. J Biol Chem **261**:9250–9256.
20. Esvelt RP, Schnoes HK, DeLuca HF 1979 Isolation and characterization of 1 alpha-hydroxy-23-carboxytetranor vitamin D: A major metabolite of 1,25-dihydroxyvitamin D_3. Biochemistry **18**:3977.
21. DeLuca HF, Schnoes HK 1979 Recent development in the metabolism of vitamin D. In: Norman AW, Schaefer K, Herrath DV, Grigoleit H-G, Coburn JW, DeLuca HF, Mawer EB, Suda T (eds) Vitamin D: Basic Research and Its Clinical Application. de Gruyter, Berlin and New York, pp. 445–458.
22. Reddy GS, Tserng KY 1988 Calcitroic acid, end product of renal metabolism of 1,25-dihydroxyvitamin D_3 through C-24 oxidation pathway. Biochemistry **28**:1763–1769.
23. Pedersen JI, Hagenfeldt Y, Bjorkhem I 1988 Assay and properties of 25-hydroxyvitamin D_3 23-hydroxylase: Evidence that 23,25-dihydroxyvitamin D_3 is a major metabolite in 1,25-dihydroxyvitamin D_3-treated or fasted guinea pigs. Biochem J **250**:527–532.
24. Beckman MJ, Tadikonda P, Werner E, Prahl J, Yamada S, De-

Luca HF 1996 Human 25-hydroxyvitamin D_3-24-hydroxylase, a multicatalytic enzyme. Biochemistry **35**:8465–8472.

25. Ohyama Y, Hayashi S, Okuda K 1989 Purification of 25-hydroxyvitamin D_3 24 hydroxylase from rat kidney mitochondria. FEBS Lett **255**:405–408.
26. Ohyama Y, Noshiro M, Okuda K 1991 Cloning and expression of cDNA encoding 25-hydroxyvitamin D_3 24-hydroxylase. FEBS Lett **278**:195–198.
27. Chen K-S, Prahl JM, DeLuca HF 1993 Isolation and expression of human 1,25-dihydroxyvitamin D_3 24-hydroxylase cDNA. Proc Natl Acad Sci USA **90**:4543–4547.
28. Itoh S, Yoshimura T, Iemura O, Yamada E, Tsujikawa K, Kohama Y, Mimura T 1995 Molecular cloning of 25-hydroxyvitamin D-3 24-hydroxylase (Cyp-24) from mouse kidney: Its inducibility by vitamin D-3. Biochim Biophys Acta **1264**:26–28.
29. Ohyama Y, Noshiro M, Eggertsen G, Gotoh O, Bjorkhem I, Okuda K 1992 Structural characterization of the gene encoding rat 25-hydroxyvitamin D_3 24-hydroxylase. Biochemistry **32**:76–82.
30. Hahn CN, Kerry DM, Omdahl JL, May BK 1994 Identification of a vitamin D responsive element in the promoter of the rat cytochrome P45024 gene. Nucleic Acids Res **22**:2410–2416.
31. Chen KS, DeLuca HF 1995 Cloning of the human 1 alpha,25-dihydroxyvitamin D-3 24-hydroxylase gene promoter and identification of two vitamin D-responsive elements. Biochim Biophys Acta **1263**:1–9.
32. Hahn CN, Baker E, Laslo P, May BK, Omdahl JL, Sutherland GR 1993 Localization of the human vitamin D 24-hydroxylase gene (CYP24) to chromosome 20q13.2-q13.3. Cytogenet Cell Genet **62**:192–193.
33. Labuda M, Lemieux N, Tihy F, Prinster C, Glorieux FH 1993 Human 25-hydroxyvitamin D 24-hydroxylase cytochrome P450 subunit maps to a different chromosomal location than that of pseudovitamin D-deficient rickets. J Bone Miner Res **8**:1397–1406.
34. Labuda M, Morgan K, Glorieux FH 1990 Mapping autosomal recessive vitamin D dependency type I to chromosome 12q14 by linkage analysis. Am J Hum Genet **47**:28–36.
35. Malas S, Peters J, Abbott C 1994 The genes for endothelin 3, vitamin D 24-hydroxylase and melanocortin 3 receptor map to distal mouse chromosome 2 in the region of conserved synteny with chromosome 20. Mammalian Genome **5**:577–579.
36. Armbrecht HJ , Boltz MA 1991 Expression of 25-hydroxyvitamin D 24-hydroxylase cytochrome P450 in kidney and intestine: Effect of 1,25-dihydroxyvitamin D and age. FEBS Lett **292**:17–20.
37. Shinki T, Jin Ch, Nishimura A, Nagai Y, Ohyama Y, Noshiro M, Okuda K, Suda T 1992 Parathyroid hormone inhibits 25-hydroxyvitamin D_3-24-hydroxylase mRNA expression stimulated by 1,25-dihydroxyvitamin D_3 in rat kidney but not in intestine. J Biol Chem **267**:13757–13762.
38. Matkovits T, Christakos S 1995 Variable in vivo regulation of rat vitamin D-dependent genes (osteopontin, Ca,Mg-adenosine triphosphatase and 25-hydroxyvitamin D_3 24-hydroxylase): Implications for differing mechanisms of regulation and involvement of multiple factors. Endocrinology **136**:3971–3982.
39. Iida K, Yamaguchi A, DeLuca HF, Kurokawa K, Suda T 1995 A possible role of vitamin D receptors in regulating vitamin D activation in the kidney. Proc Natl Acad Sci USA **92**:6112–6116.
40. Rubin LP, Yeung B, Vouros P, Vilner LM, Reddy GS 1993 Evidence for human placental synthesis of 24,25-dihydroxyvitamin D_3 and 23,25-dihydroxyvitamin D3. Pediatr Res 34:98–104.
41. Henry HL, Dutta C, Cunningham N, Blanchard R, Penny R, Tang C, Marchetto G, Chou S-Y 1992 The cellular and molecular regulation of $1,25(OH)_2D_3$ production. J Steroid Biochem Mol Biol 41:401–407.
42. Chen ML, Boltz MA, Armbrecht HJ 1993 Effects of 1,25-dihydroxyvitamin D_3 and phorbol ester on 25-hydroxyvitamin D_3 24-hydroxylase cytochrome P450 messenger ribonucleic acid levels in primary cultures of rat renal cells. Endocrinology **132**:1782–1788.
43. Matsumoto T, Kawanobe Y, Ogata E 1985 Regulation of 24,25-dihydroxyvitamin D-3 production by 1,25-dihydroxyvitamin D-3 and synthetic human parathyroid hormone fragment 1–34 in a cloned monkey kidney cell line (JTC-12). Biochim Biophys Acta **845**:358–365.
44. Tomon M, Tenenhouse HS, Jones G 1990 Expression of 25-hydroxyvitamin D_3-24-hydroxylase activity in CaCo-2 cells: An in vitro model of intestinal vitamin D catabolism. Endocrinology 126:2868–2875.
45. Kane KF, Langman MJ, Williams GR 1996 Antiproliferative responses to two human colon cancer cell lines to vitamin D_3 are differently modified by 9-*cis*-retinoic acid. Cancer Res **56**:623–632.
46. Inaba M, Burgos-Trinidad M, DeLuca HF 1991 Characteristics of the 25-hydroxyvitamin D_3-and 1,25-dihydroxyvitamin D_3-24-hydroxylase(s) from HL-60 cells. Arch Biochem Biophys **284**:257–263.
47. Skowronski RJ, Peehl DM, Feldman D 1995 Actions of vitamin D3 analogs on human prostate cancer cell lines: Comparison with 1,25-dihydroxyvitamin D_3. Endocrinology **136**:20–26.
48. Makin G, Lohnes D, Byford V, Ray R, Jones G 1989 Target cell metabolism of 1,25-dihydroxyvitamin D_3 to calcitroic acid: Evidence for a pathway in kidney and bone involving 24-oxidation. Biochem J **262**:173–180.
49. Eil C, Liberman UA, Marx SJ 1986 The molecular basis for resistance to 1,25-dihydroxyvitamin D: Studies in cells cultured from patients with hereditary hypocalcemic $1,25(OH)_2D_3$-resistant rickets. Adv Exp Med Biol **196**:407–22.
50. Morgan JW, Reddy GS, Uskokovic MR, May BK, Omdahl JL, Maizel AL, Sharma S 1994 Functional block for 1,25-dihydroxyvitamin D_3-mediated gene regulation in human B lymphocytes. J Biol Chem 269:13437–13443.
51. Naveilhan P, Neveu I, Baudet C, Ohyama KY, Brachet P, Wion D 1993 Expression of 25(OH) vitamin D_3 24-hydroxylase gene in glial cells. Neureport **5**:255–257.
52. Chen ML, Heinrich G, Ohyama YI, Okuda K, Omdalh JL, Chen TC, Holick MF 1994 Expression of 25-hydroxyvitamin D_3-24-hydroxylase mRNA in cultured human keratinocytes. Proc Soc Exp Biol Med **207**:57–61.
53. Pike JW 1991 Vitamin D_3 receptors: Structure and function in transcription. Annu Rev Nutr **11**:189–216.
54. Hannah SS, Norman AW 1994 1 alpha,$25(OH)_2$ vitamin D_3-regulated expression of the eukaryotic genome. Nutr Rev **52**:376–382.
55. Uchida M, Ozono K, Pike JW 1994 Activation of the human osteocalcin gene by 24*R*,25-dihydroxyvitamin D_3 occurs through the vitamin D receptor and the vitamin D-responsive element. J Bone Miner Res **9**:1981–1987.
56. Uhland-Smith A, DeLuca HF 1993 The necessity for calcium for increased renal vitamin D receptor in response to 1,25-dihydroxyvitamin D. Biochim Biophys Acta **1176**:321–326.
57. Omdahl JL 1980 Direct modulation of 25-hydroxyvitamin D_3 hydroxylation in kidney tubules by 1,25-dihydroxyvitamin D_3. Biochem Biophys Res Commun **253**:8474–8478.
58. Henry HL, Al-Abdaly FA, Noland TAJ 1983 Cyclic AMP-dependent protein kinase and its endogenous inhibitor protein

in tissue distribution and effect of vitamin D status in the chick. Comp Biochem Physiol **74**:715–718.

59. Baksi SN, Kenny AD 1977 Vitamin D_3 metabolism in immature Japanese quail: Effects of ovarian hormones. Endocrinology **101**:1216–1220.
60. Tanaka Y, Castillo L, DeLuca HF 1976 Control of renal vitamin D hydroxylases in birds by sex hormones. Proc Natl Acad Sci USA **73**:2701–2705.
61. Garabedian M, Holick MF, DeLuca HF, Boyle IT 1972 Control of 25-hydroxycholecalciferol metabolism by parathyroid glands. Proc Natl Acad Sci USA **69**:1673–1676.
62. Shigematsu T, Horiuchi N, Ogura Y, Miyahara T, Suda T 1986 Human parathyroid hormone inhibits renal 24-hydroxylase activity of 25-hydroxyvitamin D_3 by a mechanism involving adenosine 3′,5′-monophosphate in rats. Endocrinology **118**:1583–1589.
63. Armbrecht HJ, Wongsurawat N, Zenser TV, Davis BB 1984 Effect of PTH and $1{,}25(OH)_2D_3$ on renal $25(OH)D_3$ metabolism, adenylate cyclase and protein kinase. Am J Physiol **246**:E102–E107.
64. Henry HL 1985 Parathyroid hormone modulation of 25-hydroxyvitamin D_3 metabolism by cultured chick kidney cells is mimicked and enhanced by forskolin. Endocrinology **116**:503–507.
65. Henry HL, Dutta C, Cunningham N, Blanchard R, Penny R, Tang C, Marchetto G, Chou S-Y 1992 The cellular and molecular regulation of $1{,}25(OH)_2D_3$ production. J Steroid Biochem Mol Biol **41**:401–407.
66. Iida K, Taniguchi S, Kurokawa K 1993 Distribution of 1,25-dihydroxyvitamin D_3 receptor and 25-hydroxyvitamin D_3-24-hydroxylase mRNA expression along rat nephron segments. Biochem Biophys Res Commun **194**:659–664.
67. Kumar R, Schaefer J, Roche PC 1994 Immunolocalization of calcitriol receptor, 24-hydroxylase cytochrome P-450 and calbindin D28K in human kidney. Am J Physiol **266**:F477–485.
68. Iwata K, Yamamoto A, Satoh S, Ohyama Y, Tashiro Y, Setoguchi T 1995 Quantitative immunoelectron microscopic analysis of the localization and induction of 25-hydroxyvitamin D_3 24-hydroxylase in rat kidney. J Histol Cytol **43**:255–262.
69. Azarani A, Goltzman D, Orlowski J 1995 Parathyroid hormone and parathyroid hormone-related peptide inhibit the apical Na/K exchanger NHE-3 isoform in renal cells (OK) via a dual signaling cascade involving protein kinase A and C. J Biol Chem **270**: 20004–20010.
70. Lemay J, Demers C, Hendy GN, Delvin EE, Gascon-Barre M 1995 Expression of the 1,25-dihydroxyvitamin D_3-24-hydroxylase in rat intestine: Response to calcium, vitamin D_3 and calcitriol administration in vivo. J Bone Miner Res **10**:1148–1157.
71. Nishimura A, Shinki T, Jin CH, Ohyama Y, Noshiro M, Okuda K, Suda T 1994 Regulation of messenger ribonucleic acid expression of 1 alpha,25-dihydroxyvitamin D_3 24-hydroxylase in rat osteoblasts. Endocrinology **134**:1794–1799.
72. Armbrecht HJ, Hodam TL 1994 Parathyroid hormone and 1,25-dihydroxvitamin D synergistically induce the 1,25-dihydroxyvitamin D-24-hydroxylase in rat UMR-106 osteoblast-like cells. Biochem Biophys Res Commun **205**:674–679.
73. Krishnan AV, Cramer SD, Bringhurst FR, Feldman D 1995 Regulation of 1,25-dihydroxyvitamin D_3 receptors by parathyroid hormone in osteoblastic cells: Role of second messenger pathways. Endocrinology **136**:705–712.
74. Zierold C, Darwish HM, DeLuca HF 1994 Identification of a vitamin D-response element in the rat calcidiol (25-hydroxyvitamin D_3) 24-hydroxylase gene. Proc Natl Acad Sci USA **91**: 900–902.
75. Ono T, Tanaka H, Yamate T, Nagai Y, Nakamura T, Seino Y 1996 24*R*,25-Dihydroxyvitamin D_3 promotes bone formation without causing excessive resorption in hypophosphatemic mice. Endocrinology **137**:2633–2637.
76. Kawashima H, Kurokawa K 1983 Unique hormonal regulation vitamin D metabolism in the mammalian kidney. Miner Electrolyte Metab **9**:227–235.
77. Horiuchi N, Takahashi H, Matsumoto T, Takahashi N, Shimazawa E, Suda T, Ogata E 1979 Salmon calcitonin-induced stimulation of 1-alpha,25-dihydroxycholecalciferol synthesis in rats involving a mechanism independent of adenosine 3′-5′-cyclic monophosphate. Biochem J **184**:269–275.
78. Kawashima H, Torikai S, Kurokawa K 1981 Calcitonin selectively stimulates 25-hydroxyvitamin D_3-1-alpha-hydroxylase in proximal straight tubule of rat kidney. Nature **291**:327–329.
79. Armbrecht HJ, Wongsurawat N, Paschal RE 1986 Effect of age on renal responsiveness to parathyroid hormone and calcitonin in rats. J Endocrinol **114**:173–178.
80. Beckman MJ, Goff JP, Reinhardt TA, Beitz DC, Horst RL 1994 *In vivo* regulation of rat intestinal 24-hydroxylase: Potential new role of calcitonin. Endocrinology **135**:1951–1955.
81. Frazer TE, White NH, Hough S, Santiago JV, McGee BR, Bryce G, Mallon J, Avioli LV 1981 Alterations in circulating vitamin D metabolites in the young insulin-dependent diabetic. J Clin Endocrinol Metab 53:1154–1159.
82. Wongsurawat N, Armbrecht HJ, Zenser TV, Davis BB, Thomas ML, Forte LR 1983 1,25-Dihydroxyvitamin D_3 and 24,25-dihydroxyvitamin D_3 production by isolated renal slices is modulated by diabetes and insulin in the rat. Diabetes **32**:302–306.
83. Ishimura E, Nishizawa Y, Koyama H, Shoji S, Inaba M, Morii H 1995 Impaired vitamin D metabolism and response in spontaneously diabetic GK rats. Miner Electrolyte Metab **21**:205–210.
84. Schneider LE, Omdahl JL, Schedl HP 1976 Effects of vitamin D and its metabolites on calcium transport in the diabetic rat. Endocrinology **99**:793–799.
85. Schneider LE, Nowosielski LM, Schedl HP 1977 Insulin treatment of diabetic rats: Effects on duodenal calcium absorption. Endocrinology **100**:67–73.
86. Armbrecht HJ, Wongsurawat VJ, Hodam TT, Wongsurawat N 1996 Insulin markedly potentiates the capacity of parathyroid hormone to increase expression of 25-hydroxyvitamin D_3-24-hydroxylase in rat osteoblast cells in the presence of 1,25-dihydroxyvitamin D_3. FEBS Lett **393**:77–80.
87. Wongsurawat N, Armbrecht HJ, Siegel NA 1991 Effects of diabetes mellitus on parathyroid hormone-stimulated protein kinase activity, ferredoxin phosphorylation, and renal 1,25-dihydroxyvitamin D production. J Lab Clin Med **117**:319–324.
88. Ishida H, Seino Y, Matsukura S, Ikeda M, Yawata M, Yamashita G, Ishizuka S, Imura H 1985 Diabetic osteopenia and circulating levels of vitamin D metabolites in type 2 (noninsulin-dependent) diabetes. Metabolism **34**:797–801.
89. Christiansen C, Christensen MS, McNair P, Nielsen B, Madsbad S 1982 Vitamin D metabolites in diabetic patients: Decreased serum concentration of 24,25-dihydroxyvitamin D. Scand J Clin Lab Invest **42**:487–491.
90. Wei, S, Kubo T, Tanaka H, Ono T, Moriwake T, Kanzaki S, Seino Y 1996 Growth hormone increases serum 1,25-dihydroxyvitamin D levels and decreases 24,25-dihydroxyvitamin D levels in children with growth hormone deficiency. J Bone Miner Res **11**:M541.
91. Mandla S, Jones G, Tenenhouse HS 1992 Normal 24-hydroxylation of vitamin D metabolites in patients with vitamin D-dependency rickets type I: Structural implications for the vitamin D hydroxylases. J Clin Endocrinol Metab **74**:814–820.
92. Gamblin G, Liberman U, Eil C, Downs RW Jr., DeGrange D,

Marx SJ 1985 Defective induction of 25-hydroxyvitamin D_3-24-hydroxylase by 1,25-dihydroxyvitamin D_3 in cultured skin fibroblasts. J Clin Invest **75**:954–960.

93. Takeda E, Yokota I, Kawakami I, Hashimoto T, Kuroda Y, Arase S 1989 Two siblings with vitamin-D-dependent rickets type II: No recurrence of rickets for 14 years after cessation of therapy. Eur J Pediatr **149**:54–57.
94. Takeda E, Yokota I, Saijo T, Kawakami I, Ito M, Kuroda Y 1990 Effect of long-term treatment with massive doses of 1 alpha-hydroxyvitamin D_3 on calcium–phosphate balance in patients with vitamin D-dependent rickets type II. Acta Paediatr Jpn **32**:39–43.
95. Rasmussen H, Tenenhouse HS 1989 Hypophosphatemias. In: Scriver CR, Beaudet AL, Sly WS, Valle D (eds) The Metabolic Basis of Inherited Disease, 6th Ed. McGraw-Hill, New York, pp. 2581–2601.
96. Tenenhouse HS, Glenville J 1990 Abnormal regulation of renal vitamin D catabolism by dietary phosphate in murine X-linked hypophosphatemic rickets. J Clin Invest **85**:1450–1455.
97. Roy S, Martel J, Ma S, Tenenhouse HS 1994 Increased renal 25-hydroxyvitamin D_3-24-hydroxylase messenger ribonucleic acid and immunoreactive protein in phosphate-deprived Hyp mice: A mechanism for accelerated 1,25-dihydroxyvitamin D_3 catabolism in X-linked hypophosphatemic rickets. Endocrinology **134**:1761–1767.
98. Ohyama Y, Ozono K, Uchida M, Shinki T, Kato S, Suda T, Yamamoto O, Noshiro M, Kato Y 1994 Identification of a vitamin D-responsive element in the 5′-flanking region of the rat 25-hydroxyvitamin D_3 24-hydroxylase gene. J Biol Chem **269**:10545–10550.
99. Zierold C, Hisham MD, DeLuca HF 1995 Two vitamin D response elements function in the rat 1,25-dihydroxyvitamin D 24-hydroxylase promoter. J Biol Chem **270**:1675–1678.
100. Kerry DM, Dwivedi PP, Hahn CN, Morris HA, Omdahl JL, May BK 1996 Transcriptional synergism between vitamin D-response elements in the rat 25-hydroxyvitamin D_3 24-hydroxylase (CYP24) promoter. J Biol Chem **271**:29715–29721.
101. Noda M, Vogel RL, Craig AM, Prahl J, DeLuca HF, Denhardt DT 1990 Identification of a DNA sequence responsible for binding of the 1,25-dihydroxyvitamin D_3 receptor and 1,25 dihydroxyvitamin D_3 enhancement of mouse secreted phosphoprotein 1 (Spp-1 of osteopontin) gene expression. Proc Natl Acad Sci USA **87**:9995–9999.
102. Liu M, Freedman LP 1994 Transcriptional synergism between the vitamin D_3 receptor and other nonreceptor transcription factors. Mol Endocrinol **8**:1593–1604.
103. Tsai SY, Tsai M-J, O'Malley BW 1989 Cooperative binding of steroid hormone receptors contributes to transcriptional synergism at target enhancer elements. Cell **57**:443–448.
104. Demay MB, Kiernan MS, DeLuca HF, Kronenberg HM 1992 Sequences in the human parathyroid hormone gene that bind the 1,25-dihydroxyvitamin D_3 receptor and mediate transcriptional repression in response to 1,25-dihydroxyvitamin D_3. Proc Natl Acad Sci USA **89**:8097–8101.
105. Quelo I, Kahlen J-P, Rascle A, Jurdic P, Carlberg C 1994 Identification and characterization of a vitamin D_3 response element of chicken carbonic anhydrase-II. DNA **13**:1181–1187.
106. Smith R 1993 Heritable metabolic bone diseases, chondrodysplasias and skeletal poisons. In: Nordin BEC, Need AG, Morris HA (eds) Metabolic Bone and Stone Disease, 3rd Ed. Churchill Livingston, Edinburgh, Scotland, pp. 213–248.
107. Chi T, Lieberman P, Ellwood K, Carey M 1995 A general mechanism for transcriptional synergy by eukaryotic activators. Nature **377**:254–257.
108. MacDonald PN, Sherman DR, Dowd DR, Jefcoat SC Jr, DeLisle RK 1995 The vitamin D receptor interacts with general transcription factor IIB. J Biol Chem **270**:4748–4752.
109. Le Douarin B, Zechel C, Garnier J-M, Lutz Y, Tora L, Pierrat B, Heery D, Gronemeyer H, Chambon P, Losson R 1995 The N-terminal part of TIF1, a putative mediator of the ligand-dependent and activation function (AF-2) of nuclear receptors, is fused to B-raf in the oncogenic protein T18. EMBO J **14**:2020–2033.
110. Kahlen J-P, Carlberg C 1994 Identification of a vitamin D receptor homodimer-type response element in the rat calcitriol 24-hydroxylase gene promoter. Biochem Biophys Res Commun **202**:1366–1372.
111. Jaken S 1996 Protein kinase C isozymes and substrates. Curr Opin Cell Biol **8**:168–173.
112. Henry HL 1986 Influence of a tumor promoting phorbol ester on the metabolism of 25-hydroxyvitamin D_3. Biochem Biophys Res Commun **139**:495–500.
113. Henry HL, Luntao EM 1989 Interactions between intracellular signals involved in the regulation of 25-hydroxyvitamin D_3 metabolism. Endocrinology **124**:2228–2234.
114. Mandla S, Boneh A, Tenenhouse HS 1990 Evidence for protein kinase C involvement in the regulation of renal 25-hydroxyvitamin D_3-24-hydroxylase. Endocrinology **127**:2639–2647.
115. Armbrecht HJ, Hodam TL, Boltz MA, Chen ML 1993 Phorbol ester markedly increases the sensitivity of intestinal epithelial cells to 1,25-dihydroxyvitamin D_3. FEBS Lett **327**:13–16.
116. Koyama H, Inaba M, Nishizawa Y, Ohno S, Morii H 1994 Protein kinase C is involved in 24-hydroxylase gene expression induced by 1,25-$(OH)_2D_3$ in intestinal epithelial cells. J Cell Biochem **55**:230–240.
117. Deng T, Karin M 1994 c-Fos transcriptional activity stimulated by H-Ras-activated protein kinase distinct from JNK and ERK. Nature **371**:171–175.
118. Karin M 1995 The regulation of AP-1 activity by mitogen-activated protein kinases. J Biol Chem **270**:16483–16486.
119. Pike JW, Kerner SA, Jin CH, Allegretto EA, Elgort M 1995 Direct activation of the human 25-hydroxyvitamin D_3-24-hydroxylase promoter by 1,25-$(OH)_2D_3$ and PMA: Identification of cis elements and transactivators. J Bone Miner Res **10**:S144.
120. Darwish HM, Burmester JK, Moss VE, DeLuca HF 1993 Phosphorylation is involved in transcriptional activation by the 1,25-dihydroxyvitamin D_3 receptor. Biochim Biophys Acta **1167**:29–36.
121. Matkovits T, Christakos S 1995 Ligand occupancy is not required for vitamin D receptor and retinoid receptor-mediated transcriptional activation. Mol Endocrinol **9**:232–242.
122. Desai RK, van Wijnen AJ, Stein JL, Stein GS, Lian JB 1995 Control of 1,25-dihydroxyvitamin D_3 receptor-mediated enhancement of osteocalcin gene transcription: Effects of perturbing phosphorylation pathways by okadaic acid and staurosporine. Endocrinology **136**:5685–5693.
123. de Ruiter PE, Teuwen R, Dijkema R, Brinkmann AO 1995 Synergism between androgens and protein kinase-C on androgen-regulated gene expression. Mol Cell Endocrinol **110**:R1–R6.

CHAPTER 7

Vitamin D Binding Protein

NANCY E. COOKE Departments of Medicine and Genetics, University of Pennsylvania, Philadelphia, Pennsylvania

JOHN G. HADDAD* Department of Medicine, University of Pennsylvania, Philadelphia, Pennsylvania

I. INTRODUCTION

Vitamin D binding protein (DBP) [1] (reviewed by Cooke and Haddad [2,3]) binds and transports vitamin D and its metabolites through serum. The function of DBP, as well as that of three other major plasma steroid binding proteins, namely, thyroid binding globulin, cortisol binding globulin, and sex hormone binding globulin, have been assumed to reflect the "free hormone hypothesis" [4]. This hypothesis states that only free (uncomplexed) hormones are biologically active, whereas the remainder are trapped in the vascular/extracellular compartments by their corresponding steroid-binding proteins. These sterol/protein complexes provide a readily available reservoir of hormones that can be made available to the free pool of biologically active hormone by dissociation [5].

Originally identified in 1959 as a polymorphic protein by serum electrophoresis, DBP was referred to as the group-specific component of serum (Gc-globulin) [6]. At that time its function was not known. This polymorphic protein became useful in population genetics [7] and in forensic medicine [8], and it is still referred to as Gc-globulin in the literature of those two fields. DBP is structurally related to albumin and α-fetoprotein, and the DBP gene is a member of the linked albumin and α-fetoprotein gene family. Until recently, binding, solubilization, and serum transport of the vitamin D sterols were considered to be the predominant roles of DBP. Now it is recognized that DBP has a spectrum of biological activities. The multifunctional properties of DBP make it unique among the other members of its multigene family as well as among the other serum steroid binding proteins. The structure, evolution, and genetics of DBP, along with these less recognized as well as the classic activities of DBP, are reviewed in the following sections.

II. THE DBP GENE FAMILY

A. Structural Features

Serum DBP is a highly polymorphic, monomeric, serum glycoprotein of approximately 58 kDa [9]. Its exact size is dependent on its glycosylation state. Since the

* John G. Haddad died unexpectedly on May 22, 1997, at the age of 59. This chapter is dedicated to his memory. His contributions to the vitamin D field were many; his colleagues, students, and family will miss him greatly.

initial reports of the human DBP cDNA structure [10,11], DBP cDNAs from the rat [12], mouse [13], and rabbit [14] have been cloned and sequenced. The amino acid structure of DBP was first predicted from these cDNA sequences. Human DBP mRNA contains 1690 nucleotides and encodes a 458-amino acid secreted protein after cleavage of its 16-amino acid signal peptide. The position of signal peptidase cleavage was assigned based on alignment of the amino terminus of the mature protein, determined by amino acid sequencing [15,16], with the sequence of the primary translation product as determined from the cDNA. The predicted mature DBP proteins in rat, mouse, and rabbit are each 460 amino acids long, two residues longer than the human sequence. Rat DBP has 77% amino acid identity to human DBP. Mouse DBP has 78% identity to human and 91% to rat, whereas rabbit DBP has 84, 79, and 78% sequence identity to the human, rat, and mouse proteins, respectively. There is a single N-linked glycosylation consensus sequence in human, rat, and mouse, which is not present in rabbit DBP. The sequences have been obtained for the human [17] and rat [18] DBP structural genes as well. Rat and human DBP genes span 35 and 42 kb, respectively. Both contain 13 exons. The signal sequence of the protein is encoded on the first exon, whereas the last exon contains the entire noncoding 3′-untranslated region. Overall the structures of the DBP gene, mRNA, and protein have been well-conserved among mammals.

A striking pattern of similarity exists among the genes and proteins of the DBP, albumin, and α-fetoprotein family [10–18]. The three RNAs and encoded proteins have sequence similarities of approximately 40% at the nucleotide level and 23% at the amino acid level. Moreover, there is a nearly identical periodic positioning of the 28 cysteine residues in DBP (Fig. 1), albumin, and α-fetoprotein. Two extra cysteines in DBP, located at amino acid positions 13 and 59, that are not present in albumin or α-fetoprotein may be essential for some of the unique biological functions of DBP. The shared cysteine residues predict a characteristic set of intramolecular disulfide bonds that define the secondary structure in albumin, α-fetoprotein, and DBP. This structure has been confirmed by crystallographic analysis of albumin [19,20]. The internal disulfide bonds define three internally homologous domains. The carboxy-terminal domain of DBP is 124 amino acids shorter than those of albumin and α-fetoprotein. This truncation is due to the loss of the exons homologous to 12 and 13 in the 15-exon albumin and α-fetoprotein genes, resulting in the 13-exon DBP gene [17,18]. With the exception of these two missing exons, the positions at which the introns interrupt the coding regions of the three genes are highly similar. These striking structural parallels firmly establish the inclusion of DBP in the albumin and α-fetoprotein multigene family.

B. Evolutionary Features

1. Chromosomal Localization and Linkages

The human DBP gene was assigned to chromosome 4 by analysis of somatic cell hybrids and sublocalized to bands 4q11-q13 by *in situ* hybridization to metaphase chromosome spreads [21]. This sublocalization overlapped with the position of albumin and α-fetoprotein, which had been mapped to the 4q11-q22 region [22]. On the basis of recombination frequencies in families with informative albumin mutations, the DBP and albumin genes were predicted to be located within 1.5 centimorgans (cM) of one another [23,24]. These three genes have been assigned to chromosome 13 in the rat [25] and to chromosome 5 in the mouse [13]. Mouse α-fetoprotein and DBP were shown to be less than 3.3 cM apart on the basis of an interspecific backcross between *Mus domesticus* and *M. spretus* [26]. The rodent chromosomes encoding DBP are syntenic with human chromosome 4, thus demonstrating conservation of the DBP, α-fetoprotein, albumin linkage in three species. A linkage between DBP and albumin has also been noted in horses and in chickens [27,28]. In humans, physical mapping has placed DBP two radiation hybrid bins proximal to albumin [29], and the two genes also appear to be linked by a single bridging yeast artificial chromosome (YAC) clone [30]. Confirmative physical mapping between DBP and albumin has yet to be provided, and pulsed-field gel restriction enzyme mapping has been unable to place these genes on unique restriction fragments [26]. These data predict that the DBP gene lies upstream of the albumin and α-fetoprotein genes (Fig. 2).

In contrast to the substantial distance between DBP and albumin, the albumin and α-fetoprotein genes are tightly linked in the mammalian species studied so far. Albumin is located only 14 to 20 kb upstream from the 5′ end of the α-fetoprotein gene [31–33]. Both genes are in the same transcriptional orientation. A fourth gene in the DBP/albumin/α-fetoprotein family has been discovered by two groups and named afamin by one group [34] and α-albumin by the other [35,36]. The function of α-albumin has not been determined. The rat afamin/α-albumin gene is located 10 kb downstream from the α-fetoprotein locus, so that the order of the four genes from 5′ to 3′ with respect to the direction of albumin transcription is DBP, albumin, α-fetoprotein, α-albumin (Fig. 2). α-Albumin is selectively expressed in the liver during late stages of development, suggesting that it may be a phylogenetic intermediate between α-

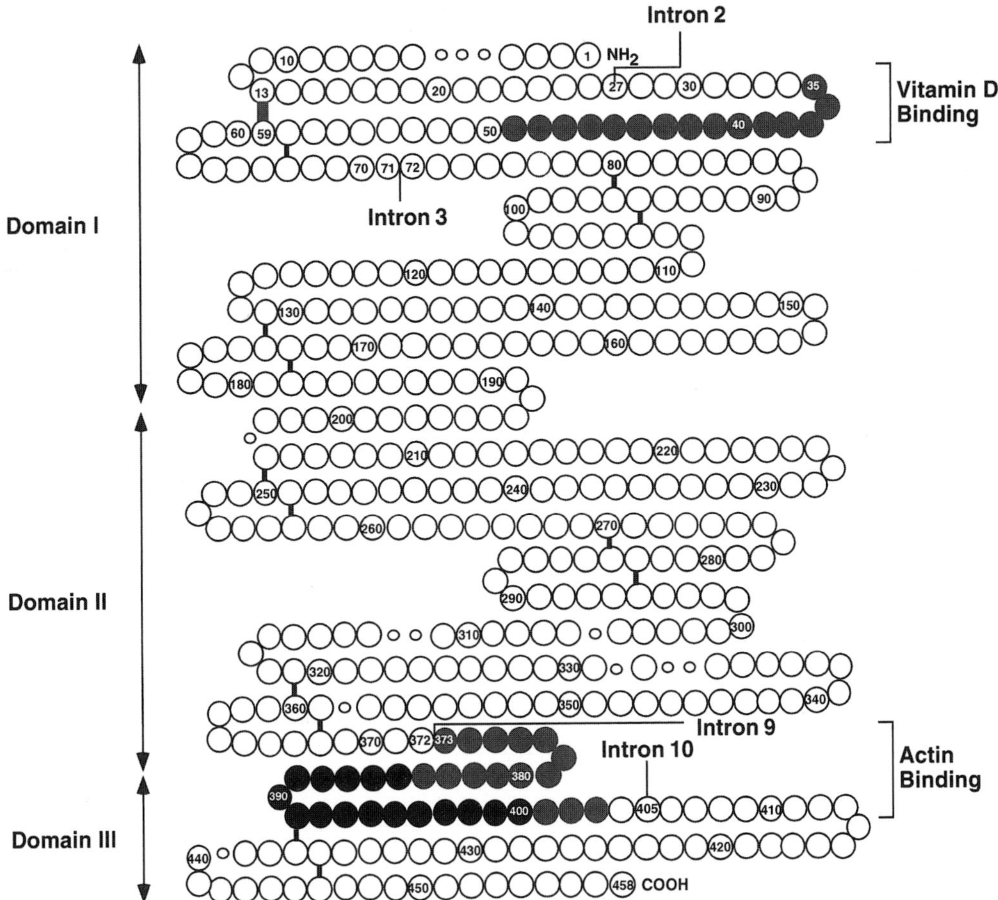

FIGURE 1 Human DBP. The three internal domains defined by amino acid sequence similarities are indicated at left. The third domain is truncated in DBP but is full length in albumin and α-fetoprotein. Amino acid positions are numbered. The smaller circles mark the positions of amino acid residues present in albumin and α-fetoprotein but absent in DBP [10]. Disulfide bond positions are indicated, and a disulfide bond that forms a loop containing the sterol binding domain, not present in either albumin or α-fetoprotein, is indicated by a distinctive line between residues 13 and 59. The experimentally determined 25OHD and actin binding domains are shaded [126]. The core of the actin binding domain, indicated in black, contains an actin consensus binding sequence [161]. The positions where flanking introns interrupt the coding regions around each binding domain indicate that each binding domain is encoded by a single exon [17].

fetoprotein and albumin. The α-albumin protein is 33% identical to α-fetoprotein, 29% to albumin, and 19% to DBP. The cysteine residues are highly conserved in position and in number. Overall its predicted structure would be most similar to α-fetoprotein. It has been suggested that this additional gene serves as a backup for albumin, possibly accounting for the lack of a phenotype in hereditary analbuminemia [35]. The human homolog of α-albumin was isolated as a serum protein migrating on sodium dodecyl sulfate (SDS)–polyacrylamide gels with an apparent molecular mass of 87 kD, and its gene was assigned to human chromosome 4. Its concentration in adult serum is 30 μg/ml compared to 40 mg/ml for albumin, 50 ng/ml for α-fetoprotein, and 350 μg/ml for DBP [34]. Thus, DBP is a member of a linked multigene family encoding at least four serum proteins, each of which is predominantly expressed in the fetal and/or adult liver.

2. Evolution Within the Gene Family

A number of models have been proposed to account for the evolution of the DBP multigene family and for the shared, triplicated internal domain structure. In one model, the ancestral internal domain is encoded by four exons that subsequently triplicated, creating the present gene encoding a three-domain protein (Fig. 3). This domain triplication likely predated vertebrate evolution [37,38]. The separation of a unique DBP gene from this precursor occurred about 560–600 million years ago [39]. As predicted by this date, DBP-like proteins have

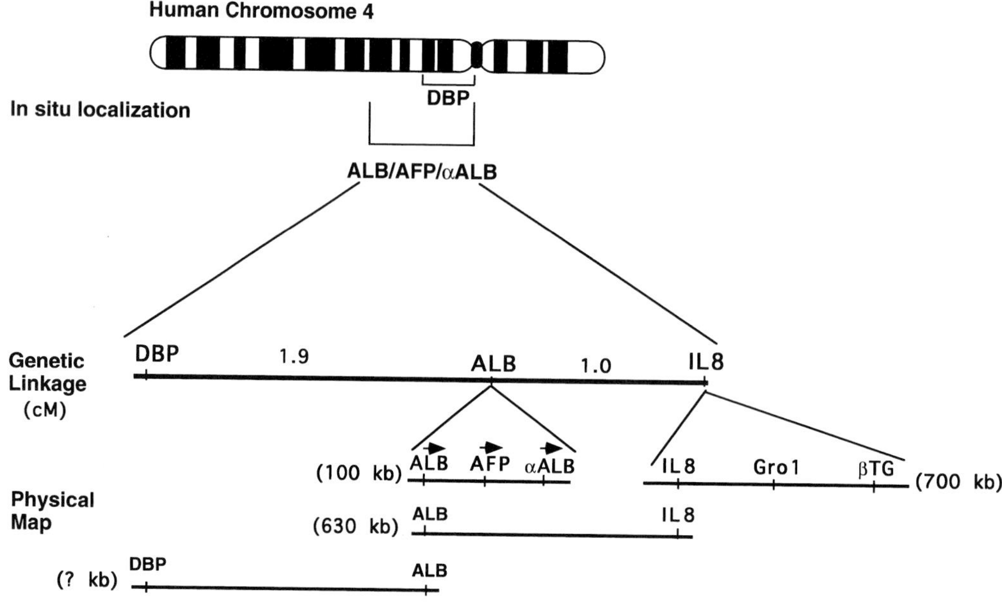

FIGURE 2 The linked DBP multigene family. The chromosomal localizations of DBP, albumin (ALB), α-fetoprotein (AFP), and α-albumin (αALB) as determined by *in situ* localizations are indicated [21,22]. Recombination analysis suggested a genetic linkage between DBP and albumin [23,107]. More recent recombination maps predicted that the gene for interleukin-8 (IL-8) is located between DBP and albumin [162] (not as shown). Physical mapping has linked albumin and α-fetoprotein, and α-albumin in the rat, all in the same transcriptional orientation as indicated by the arrowheads [35]. Physical mapping of the locus is incomplete, but the existing YAC contig [30] suggests that the DBP, albumin, and IL-8 genes are arranged as drawn. Details of the IL-8 locus are shown for comparison (Gro1, melanoma growth-stimulatory activity; βTG, β-thromboglobulin) [163]. The physical maps are not drawn to scale.

been detected in the plasma of fish, amphibians, reptiles, and birds [40]. It has been postulated that albumin and α-fetoprotein genes duplicated from the precursor gene and began to diverge about 280 million years ago, just after the time of the amphibian/reptile divergence,

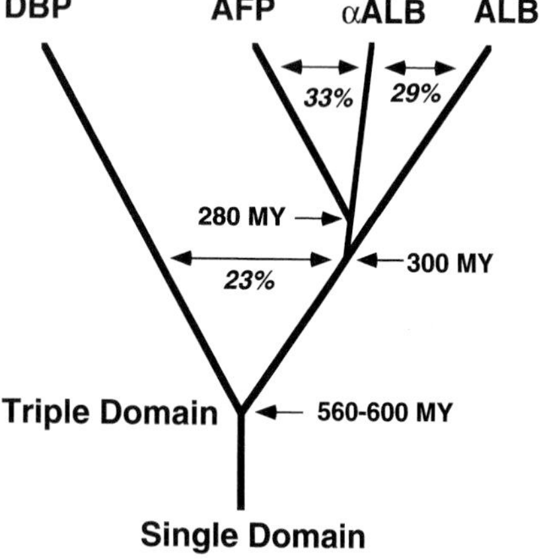

FIGURE 3 Evolution of the DBP gene family. Amino acid similarities among the rat genes are indicated in percentages [12,35]. Data such as these are the basis for predicting the evolutionary times of divergence, indicated in millions of years (MY) [39].

about 350 million years ago. The absence of a larval α-fetoprotein gene in amphibians and the existence of α-fetoprotein in chickens is consistent with this conclusion [39]. The discovery of α-albumin and its closest sequence identity to α-fetoprotein has led to the following evolutionary proposal. The emergence of α-albumin may have predated α-fetoprotein, occurring about 300 million years ago, as evidenced by the existence of a duplicated albumin gene in the adult amphibian. This was followed by duplication of α-albumin. One of the two gene copies evolved to become the fetal α-fetoprotein, first present in reptiles. This new α-fetoprotein gene may then have fulfilled a nutritive purpose in the emerging extraembryonic yolk sac [35]. Regardless of the precise details of this series of evolutionary events, DBP appears to be the oldest member of this ancient and well-conserved gene family.

3. DBP IN VERTEBRATE EVOLUTION

Serum proteins binding vitamin D are present in a wide variety of reptiles and amphibians, as well as fish [40,41]. In the emydid turtle family a surprising sequence identity was noted between its high affinity thyroxine binding protein, (TBP) and the amino terminus of mammalian DBP (68% identity, 87% with conservative substitutions) [42]. This turtle "DBP" binds 25-hydroxyvitamin D (25OHD) and is considered the major high affinity transport protein for vitamin D in the turtle *Trachemys scripta*. Thyroxine and vitamin D bind to separate sites on this single binding protein, and the levels of TBP/DBP are regulated by thyroxine [43]. Comparative studies in many species have suggested that the high affinity thyroxine binding site is a derived characteristic of DBP that is possibly unique to the Emydidae [41]. A study in two cartilaginous fish, *Tilapia nilotica* and *Cyprinus carpio*, indicated that the former lacked a high affinity serum binding protein for 25OHD, whereas the latter had a high affinity 25OHD binding protein that also interacted with actin, as expected for an authentic DBP precursor [44]. It has been suggested that lipoproteins may be utilized for vitamin D transport in the absence of a specific transport protein [38], and this may account for vitamin D transport in *Tilapia*. Thus, not all vertebrates with calcified skeletons have a specific DBP, and a discrete DBP appears to have emerged during the evolution of cartilaginous fish.

C. Gene Regulation

1. Sites and Timing of Expression

The vast majority of serum DBP is synthesized by the liver [10–12,45]. The DBP and albumin genes are also expressed in a wide variety of other tissues, but at much lower levels. Although each of the four genes in this family are predominantly expressed in the hepatocyte, each is activated in a distinctive developmental profile. The α-fetoprotein gene is initially transcribed by the yolk sac early in embryogenesis, and subsequently by the fetal liver [35,46–48]. DBP and albumin appear in fetal liver in parallel. In humans, DBP is detectable by the tenth to thirteenth week of gestation, and cord blood samples from preterm and term infants show parallel increases in DBP and albumin levels [49,50]. DBP and albumin mRNA are first detectable at 13–14 days of gestation in the rat and rise in parallel until adulthood. Rat α-albumin is not detectable in the fetal liver but is present in the postnatal liver [35]. α-Fetoprotein is uniquely turned off postnatally, while expression of the other three genes climbs to their peak levels in the adult [51]. Reactivation of α-fetoprotein expression in the adult liver or other gastrointestinal tissues is often an indication of neoplastic dedifferentiation.

DBP is expressed in a number of nonhepatic tissues. Rat DBP transcripts are readily detectable by Northern analysis in kidney, testis, and fat of adults, as well as in the 18-day fetal yolk sac. DBP cDNAs cloned from rat kidney have been sequenced and are identical to rat liver DBP cDNA. This indicates that kidney DBP mRNA is a common transcript from a single DBP gene and is not alternatively processed [45]. Although it is not known whether the kidney DBP mRNA is translated, there is preliminary evidence for nonhepatic synthesis of DBP protein in a human osteogenic sarcoma cell line [52]. At the higher sensitivity of reverse transcriptase–polymerase chain reaction (RT–PCR), DBP, albumin, and α-fetoprotein transcripts are all detectable in a wide variety of adult rat tissues, whereas in certain human cell lines such as unstimulated U937 monocytes, B lymphocytes, and T lymphocytes there are no detectable DBP transcripts [53,54]. In the rat, the DBP gene may be expressed in all tissues via leaky transcription [55] perhaps dependent on ubiquitous transcription factors. Alternatively, the ubiquitous expression of rat DBP may reflect its expression in a stromal element common to all positive tissues. The function of DBP gene expression in nonhepatic tissues is not known at present.

2. Transcriptional Regulation within the DBP Multigene Family

The albumin gene has served as a standard model for studying the establishment and maintenance of tissue-specific gene expression in liver [56]. Similarly, the postnatal extinction of α-fetoprotein gene expresssion has served as a model for the study of developmentally regulated gene silencing [51]. In contrast, very little is known about the regulation of the DBP gene and less about the α-albumin gene. The common tissue-specific hepatic expression patterns suggest that mechanisms coordinating the expression of the entire family are likely to exist. However, the overall observed physical organization of the four genes is not fully defined, and there is a general absence of studies successfully addressing the possibility of coordinate transcriptional controls.

Much of the work on the expression of the albumin and α-fetoprotein genes has focused on the study of their proximal promoter elements in transfected cells in culture. Unfortunately, in the case of hepatocytes, both transformed liver cell lines and primary cultured hepatocytes have proved to be particularly inadequate model systems for the study of liver gene expression. The rates of transgene transcription in such cultured cells are quite low, varying from 1 to 30% of endogenous liver transcription rates [57], and in some cases gene regulation is reversed in the cell lines compared to regulation *in vivo*. For example, insulin negatively regulates albumin mRNA in H4-IIE-C3 hepatoma cells, whereas it has the opposite effect *in vivo* [58]. Various hepatoma lines have also proved to be deficient in key transcription factors [59]. Despite these problems, cell culture models have been useful in defining certain and presumably important proximal promoter elements.

From such studies in cell models, the following conclusions can be made about albumin proximal promoter regulation. A relatively short 5′-flanking region of about 150 bases including the TATA box in the promoter, are sufficient to direct hepatocyte-specific transcription following transient transfections [60]. This region binds six transcription factors in close proximity: HNF-1, C/EBP (two sites), the transcription factor albumin D binding protein (also known as DBP) that is expressed in a circadian rhythm [61], CTF/NF-1-related factors, as well as an ubiquitous factor, NF-Y [62]. HNF-1 is the dominant positive transcription factor in this complex, compact, proximal promoter region [63]. Furthermore, the HNF-1 binding site closest to the promoter appears to be of particular importance because it is conserved and similar in position in all albumin promoters that have been sequenced, from *Xenopus* to human [64].

When studied *in vivo* in transgenic animals, the tissue specificity of albumin gene expression requires, in addition to its proximal promoter, an enhancer element lying 10 kb upstream from its transcriptional start site [65]. Three transcription factor binding sites in the −10 kb enhancer [66] appear to define liver specificity: the "eE" site binds members of the C/EBP family of liver-enriched transcription factors [67], the "eG" site binds members of the hepatocyte nuclear factor 3 (HNF-3) family of transcription factors [68], and the "eH" site binds NF-1/CTF and HNF-3 [69]. Thus, the hepatocyte-specific expression of albumin requires the binding and interplay of at least three transcriptional proteins on the enhancer with the six proximal promoter factors and the basic transcription machinery.

α-Fetoprotein gene transcription has also been studied in detail [70,71 and references within]. The α-fetoprotein gene shares certain features with albumin, such as proximal promoter regulation by HNF-1 and C/EBP, but has unique developmentally and hormonally regulated features. α-Fetoprotein is expressed abundantly in the visceral endoderm of the yolk sac and in the fetal liver, and less well in the fetal gut. The proximal promoter of α-fetoprotein (250 bp) is dependent on transcription factors HNF-1 and C/EBP in liver and gut. There are three enhancers positioned between 2.5, 5.0, and 6.5 kb upstream of the promoter in the mouse α-fetoprotein gene that are required for high level *in vivo* expression of the gene [72]. After birth, α-fetoprotein gene transcription rapidly declines [71]. A dominant-negative element located between −838 and −250 bp functions to repress α-fetoprotein gene expression in adult pericentric hepatocytes and is capable of turning off a linked, heterologous albumin gene in transgenic mice during the same developmental window [51,73]. This repressor is also active in the intestinal goblet cells and enteroendocrine cells [74]. The α-fetoprotein gene, but not albumin, is regulated by glucocorticoids [75,76]. In the mouse, this regulation is negative and dependent on antagonism between AP-1 and the glucocorticoid receptor, and this effect occurs during embryonic development.

The control of the DBP gene has not been explored in as great detail as albumin and α-fetoprotein. The proximal promoter of DBP contains an unusual TGTAAA motif in the position of the typical TATAAA sequence in the evolutionarily conserved albumin and α-fetoprotein promoters. Consensus binding sites in DBP, also common to albumin or α-fetoprotein, suggest that DBP should be regulated by HNF-1 [18,64]. It has been confirmed that liver-specific DBP gene expression is dependent on three functional HNF-1 binding sites as shown by gel-shift analyses, DNase I footprinting, and mutational analyses [77]. In addition, a novel negative regulatory element has been found in the proximal promoter region [77]. Ultimately, it is expected that at least some of the regulatory features that DBP shares with albumin and α-fetoprotein will be correlated with shared regulatory elements of their genes.

Transcription of DBP appears to respond to a number of hormonal stimuli. Human DBP gene expression may be regulated by estrogens. DBP levels rise during pregnancy then fall during the first 2 weeks of lactation as well as postmenopausally, although no variability is observed during the menstrual cycle [78–81]. DBP may also be positively regulated by growth hormone, but this has only been observed using pharmacological doses of recombinant human growth hormone [82]. When studied in the Hep3B human hepatoma cell line, glucocorticoids and epidermal growth factor positively regulate DBP gene expression. In addition, interleukin-6 (IL-6) increases DBP mRNA 2-fold and transforming growth factor-β (TGFβ) decreases DBP mRNA in a dose-dependent fashion up to 5-fold [83,84]. Consistent with the observed glucocorticoid regulation, the DBP promoter has been noted to contain putative glucocorticoid response element consensus binding sites [64]. From these observations it appears that DBP is not a simple constitutive gene, but undergoes hormonal and growth factor regulated alterations in expression.

D. Polymorphisms of DBP/Gc-Globulin Protein and Gene

There are three common polymorphisms in the structure of DBP in human populations. Each of these protein isoforms is encoded by one of three codominant alleles of the DBP gene known as Gc1F, Gc1S, or Gc2, with F and S referring to the relatively fast or slow migration observed after gel electrophoresis [6,85]. Two protein isoforms are produced by the Gc1F and Gc1S alleles, both of which are found as a mixed population of proteins containing or lacking a single *N*-acetylneuraminic acid on a threonine residue at position 420 [86]. In contrast, the Gc2 allele which contains a lysine at 420 produces a single protein, as lysine cannot be glycosylated. There are a number of electrophoretic techniques in use for differentiating these protein isoforms, including agar electrophoresis [6], microimmunoelectrophoresis [87], prolonged immunoelectrophoresis in agar gels [88], starch gel electrophoresis [89], isoelectric focusing [90,91], and perhaps even native polyacrylamide gel electrophoresis [92]. The key amino acid differences among these three alleles, both occurring in exon 11 of the human DBP gene, are summarized in Table I [10–11,93,94].

The DBP gene contains an *Alu* middle repetitive element located at the end of DBP intron 8 that results in four gene polymorphisms. These polymorphisms are created by differences in the size of the *Alu* poly(A) tract. Each of these polymorphisms is equally distributed among the three common alleles [95]. Although the alleles can be distinguished at the DNA level by restriction digestions (for examples, see Table 1), isoelectric focusing remains the best initial approach to assigning protein isoform phenotypes, and it is commonly used for forensic purposes. The expression of various combinations of these three alleles in a given individual results in the six common DBP phenotypes.

In addition to the three common alleles, there are also over 124 rare variant alleles described worldwide, making the DBP locus among the most polymorphic known [7,96]. The geographic occurrence of these variants often corresponds to patterns of human population migrations and thus are of anthropological interest [97].

TABLE I Molecular Basis for the Three Common DBP/Gc Alleles

Phenotype	Amino acid position	Amino acid	Nucleotides	New restriction site created
GC*2	416	Aspartic acid	GAT	
GC*1F	416	Aspartic Acid	GAT	
GC*1S	416	Glutamic acid	GAG	*Hae*III
GC*2	420	Lysine	AAG	*Sty*I
GC*1F	420	Threonine	ACG	
CG*1S	420	Threonine	ACG	

The molecular basis for some of these rare variants is now being determined by sequencing of exons amplified by the polymerase chain reaction [98,99]. DBP is also polymorphic among other mammals including primates, cattle, horses, rats, and chickens [27,28,100–102]. A number of genetic studies have suggested the presence of a DBP null allele [103,104], but these alleles have not been clearly distinguished from low expressing "pseudo-silent" alleles [105]. A homozygous, DBP-deficient mouse line has been generated by targeting a mutation to the DBP gene by homologous recombination. This DBP/Gc null animal is viable and fertile [106], suggesting that a null allele should be detectable in human populations.

III. DBP PROTEIN AND ITS ACTIONS

A. Synthesis and Concentration

Originally recognized as the electrophoretic group of Gc isotypes described above, DBP was thought to be an acute phase reactant [1,2,107,108]. Subsequently it was found to bind vitamin D sterols, and the identity of Gc-globulin and DBP was confirmed [1,2,108]. Initial studies revealed similar vitamin D sterol binding among the common genetic isoforms [93,108], although more recent studies have suggested that isoforms with more basic isoelectric points may display lower affinity for vitamin D sterols [109] and that 25OHD and 1,25-dihydroxyvitamin D [$1,25(OH)_2D$] binding may be diminished by mono- and polyunsaturated fatty acids, which also bind DBP [110]. This binding has suggested that DBP may also be a plasma carrier of fatty acids for ultimate transfer to peroxisome proliferator-activated receptors (PPAR), newly discovered members of the steroid receptor family [111]. The synthesis of a wide variety of vitamin D analogs has led to studies of their binding to DBP and the vitamin D receptor [112–114].

DBP circulates at 4–8 μM concentration in normal human plasma (Table II) [115]. The plasma half-life of DBP is 2.5 to 3 days in humans [116], and it appears to be completely degraded in a variety of tissues because no intermediate-sized forms are detectable in plasma [117]. The daily production rate of DBP, estimated from kinetic studies in humans, is approximately 10 mg/kg body weight [116]. Increased DBP titers are seen in pregnant subjects and those receiving oral estrogen [115,118], although normal titers are frequently seen during transcutaneous (patch) administration of estradiol. Decreased DBP levels can be seen in plasma from patients with advanced liver disease or those with massive proteinuria [115].

TABLE II Features and Binding Characteristics of DBP

Features		
Isoelectric point	4.5–4.8	
Electrophoretic migration	Post-albumin, inter-α-globulin	
Size	58 kDa, single chain glycoprotein	
Plasma concentration	4–8 μM (232–464 μg/ml)	
Plasma half-life	2.5–3.0 days	
Daily production rate	~10 mg/kg	
Altered plasma levels	~10 kg	
Increase	Oral estrogen, pregnancy	
Decrease	Severe nephrotic syndrome, liver disease	
Vitamin D Sterol binding		
Plasma capacity	mol/mol (2.4 mg D sterols/ liter)	
Normal sterol occupancy	2%	
Affinity (K_a)	25OHD, 24,25(OH)$_2$D, 25,26(OH)$_2$D	5×10^8 M^{-1}
	1,25(OH)$_2$D, Vitamin D	4×10^7 M^{-1}
Sterol not bound to DBP	"Free"	Albumin bound
25OHD	0.04%	12%
1,25(OH)$_2$D	0.40%	15%
Binding competition		
Polyunsaturated fatty acids	Yes	
Cholesterol	No	
Polysaturated fatty acids	No	
Steroid hormones	No	
Actin Binding		
Plasma capacity	mol/mol (270 mg/liter)	
Normal actin occupancy	0–? mg/liter	
Affinity (K_a)	2×10^9 M^{-1}	
Removal sites	G-actin/DBP complexes, hepatic Kupffer cells; F-actin–DBP complexes, hepatic sinusoidal endothelium	
DBP–actin complexes	Seen with tissue injuries, inflammation	
C5A cochemotaxis and macrophage activation		
DBP cell binding	Lymphocytes, neutrophils, monocytes, macrophages	
DBP cell affinity	K_a 10^7 M^{-1} and 10^6 M^{-1} (lymphocytes)	
DBP "activation"	By sialidase and β-galactosidase actions	
C5a enhancement	DBP binding to macrophages, neutrophils	
Cell processing	Some evidence of internalization, proteolysis	
Macrophage effects	Increase in Fc receptor expression, increase in superoxide anion production	

B. Vitamin D Sterol Binding

High affinity binding of 25OHD (5×10^8 M^{-1}) and 1,25(OH)$_2$D (4×10^7 M^{-1}) is recognized for DBP [2], with estimates of binding affinity, capacity, and "free" sterol forms calculated by the law of mass action [119], using adsorptive separations and resin-uptake assays [120], and a solid-phase technique [121]. Direct measurements of bound and free sterol moieties have been performed with the centrifugo-filtration methodology [118,122–124]. One stereospecific binding site for vitamin D sterols is present on one molecule of DBP. Sterol affinity ligands and chemical or proteolytic digestions have led to the recognition that the sterol binding domain of DBP is located in its amino-terminal region (Fig. 1) [125,126]. This location was confirmed by *in vitro* transcription and translation of a mutated rat kidney DBP cDNA encoding a peptide truncated in its carboxy terminus. This truncated DBP retained full sterol binding features [126]. The region containing the vitamin D binding domain is formed into a disulfide-bonded loop that is not found in albumin, α-fetoprotein, or α-albumin but is unique to DBP (Fig. 1), suggesting that this loop may be important for sterol binding.

There is high interest in vitamin D analogs because of their differential activities in regulating cytokine and parathyroid hormone (PTH) production, cellular proliferation, and differentiation [112–114]. These analogs display variable binding affinities toward DBP and the vitamin D receptor that alter the kinetics of the egress of the sterol analogs from plasma and ingress to target and metabolizing tissues. Such altered metabolism is of major importance in the applicability of analogs for *in vivo* clinical uses [113,114]. Also see Chapter 59 in this book.

In addition to the classic concept of vitamin D sterol entry into cells containing the vitamin D receptor, apparent surface vitamin D receptor-like binders of 1,25(OH)$_2$D are under study [127,128], and their features relative to the vitamin D receptor and DBP will be of interest. As a variety of studies have indicated cell associations by albumin [129], α-fetoprotein [130,131], and DBP [53,132], the possibilities of a DBP-mediated presentation, targeted delivery, or cell removal [133] of vitamin D sterols by DBP are worthy of consideration.

The remarkably high sterol binding capacity of plasma DBP (5 μM; 2 mg 25OHD/liter) has led to the conclusion that DBP constitutes a circulating reservoir for vitamin D sterols and thus regulates their tissue ingress. Certainly, the presence of DBP in high titer is consistent with intense serum retention of sterols derived from cutaneous cholecalciferol production, which occurs in an uncertain and intermittent fashion. Less than 5% of DBP binding sites are occupied in sera from normal persons [108,115]. Because vitamin D intoxication occurs at vitamin D sterol levels far below the binding capacity of DBP, it is likely that potent ligands, such as $1,25(OH)_2D$, are displaced from DBP by the more avidly bound 25OHD [134].

Although few natural foods contain much vitamin D, dietary sources can complement endogenous synthesis of vitamin D_3. Studies of plasma 25OHD, following various routes of administered vitamin D, indicated that oral ingestion caused quicker increases but less-sustained plasma levels of 25OHD compared to parenteral delivery in oily vehicle [134]. It had been surmised that DBP was likely to be the carrier of vitamin D_3 from skin [135], whereas chylomicrons in the intestinal lymph carried ingested vitamin D [108]. Direct analyses of endogenously synthesized cholecalciferol in plasma have shown DBP to be its major carrier, whereas other low density plasma proteins play a larger role in the transport of ingested vitamin D [136] (Fig. 4). It appears, therefore, that DBP plays an important role in the sustained, effective metabolism of vitamin D to 25OHD by its binding and slow release of bound vitamin D to the liver for 25-hydroxylation.

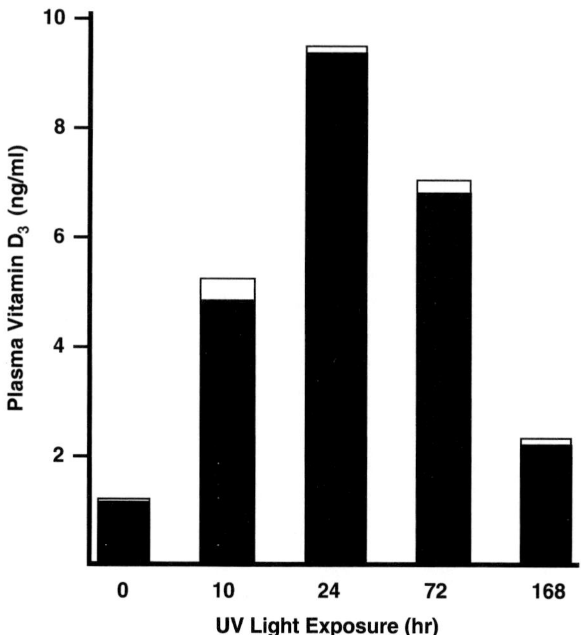

FIGURE 4 Vitamin D endogenously synthesized in the skin is primarily transported in the blood bound to serum DBP. The mean plasma vitamin D_3 concentration in seven subjects exposed to whole body UVB light is shown (white bars). The distribution of vitamin D_3 was determined on a density (d) gradient where $d > 1.30$ suggests binding to DBP and $d < 1.30$ suggests lipoprotein binding. Note that the vast majority of vitamin D_3 is associated with the $d > 1.30$ fraction (black bars). Actin affinity chromatography removed D_3 from the $d > 1.30$ fraction, confirming that the binder was DBP [136].

C. Actin Binding

A "big" DBP isoform, initially thought to be an artifact of plasma-contaminated tissues [2], was subsequently shown to be a 1:1 complex of DBP with monomeric or globular actin (G-actin) [2,137,138]. Later studies revealed that these proteins bound with high affinity (K_a 2×10^9 M^{-1}) and that the interaction was unaffected by sterol binding to DBP [139]. The binding was not covalent, being subject to disruption by ionic detergents, high salt concentrations, or heat destruction of actin [138]. Since the two proteins are identified with intracellular or extracellular compartments, the functional significance of their high affinity association remains puzzling.

Biophysical analyses revealed that DBP sequestered G-actin, thereby preventing its polymerization to fibrous or F-actin [140]. The monomeric actin sequestration was an activity also recognized for cellular proteins such as profilin and DNase I [139,141]. Injections of DBP into PtK2 cells led to depolymerization of the cytoskeleton and endosomal localization of the DBP [142]. Because the injected cells survived and regained their microfilament systems over the following 12–24 hr, the intracellular DBP was likely degraded by lysosomal proteases. A complementary effect has been seen when DBP and gelsolin are added to actin filaments *in vitro* [140,143]. Gelsolin severs actin filaments into oligomers, thereby increasing the number of growing and depolymerizing filament sites. DBP, working as a G-actin sump, prevents filament elongation and enhances filament depolymerization. The teamwork by these two proteins has been shown to inhibit actin-stimulated platelet aggregation [143].

Complexes of gelsolin–actin and DBP–actin have been found in plasma of patients undergoing tissue injuries and inflammation, such as hepatic necrosis, trophoblastic emboli, and lung injury [144]. It has therefore been proposed that DBP and gelsolin constitute an extracellular "actin-scavenger system" that protects the organism against actin toxicity [144]. At normal human plasma concentrations, DBP could sequester 270 mg of G-actin per liter of plasma. Gelsolin's G-actin binding activity would add to that capacity. Natural examples of saturation of this system have yet to be identified, perhaps because of the extensive and rapid tissue injuries required to overload this high capacity and rapid turnover system. Serum DBP levels in patients with fulminant hepatic failure may have some value in predicting survival from the disease [145]. The *in vivo* injection of increasing amounts of G-actin into the inferior venae cavae of rats, however, led to saturation and the consequent appearance of actin-filament thromboemboli in the microcirculation of the lungs and hearts of these animals [146]. In addition, endothelial cell disruption and edema were observed, resembling the lung pathology seen after tissue injuries such as crush/ischemia and sepsis [147].

Kinetic analyses of the plasma disposition of injected G-actin have led to conflicting results concerning plasma half-lives in rats and rabbits [2]. In careful studies in rats, Hermannsdoerfer and co-workers, as well as others [148,149], have identified separate plasma removal mechanisms for G-actin and F-actin. F-actin was removed more rapidly by hepatic sinusoidal endothelial cells, whereas G-actin–DBP complexes were bound to and removed by Kupffer cells [148]. Prior activation of Kupffer cells by endotoxin enhanced the plasma clearance of G-actin.

The actin binding domain of DBP has been studied by limited proteolysis and G-actin affinity chromatography [126]. Carboxy-terminal peptides with overlapping regions were analyzed for sequence and indicated that actin binding occurs within residues 373–403, a region containing a loose actin-binding consensus sequence at 385–400 (Fig. 1). The 373–404 region also contained the lowest sequence similarity to the other members of the DBP gene family, albumin and α-fetoprotein, neither of which exhibit high affinity actin binding.

D. Cell Associations

Many observations of cell–DBP incubations were directed to the disposition of radioactive vitamin D sterols between media and cells [108]. Without exception, extracellular DBP limits cellular access by 25OHD or $1,25(OH)_2D$. The presence of DBP in the extracellular medium prevents entry of 25OHD into kidney cells containing the 25OHD 1α-hydroxylase. This "inhibition" of hydroxylation can be overcome by saturating DBP with its preferred ligand, 25OHD, also the substrate of the 25OHD 1α-hydroxylase, thereby permitting competitive protein binding assays of the enzyme product, $1,25(OH)_2D$ [150]. Similarly, cell entry of $1,25(OH)_2D$ was limited by extracellular DBP, thwarting analyses of vitamin D receptor occupancy by the hormone as well as studies of the effects of the hormone on target tissues [108].

Direct analyses and localization of DBP itself, however, has led to the identification of an interesting array of cell associations, involving lymphocytes, monocytes, macrophages, neutrophils, and trophoblasts. When careful analyses were made, it became apparent that cell surface DBP was acquired from external sources [53]. Specific, saturable binding of DBP has been demonstrated on normal and malignant B lymphocytes, revealing sites with dissociation constants (K_d) of 10^{-7} and 10^{-6} *M* [132]. Evidence for cellular uptake of DBP was presented. Only 10% of the internalized DBP was released intact, in contrast to an apparent recycling of α-fetoprotein in similar experiments [131]. Others propose that DBP enters neutrophils and that a 12 to 14-kDa DBP subunit can be found on the surfaces of these cells [151]. Colocalization of DBP and immunoglobulin Fc receptors on lymphocytes has been found during immunocytochemical studies [152]. Collectively, the findings argue that selective and possibly functional cell associations do occur, and that cell surface DBP might not simply be the result of nonspecific binding.

1. Macrophage Activation

Provocative findings have been presented that link DBP with macrophage activation, leading to increased Fc receptor activity [153,154], superoxide production, and enhanced osteoclast differentiation and action [155,156]. Mixed cell cultures of B and T lymphocytes, along with macrophages, require a serum factor to enhance Fc receptor activity on the macrophage surface [153]. A modified form of DBP resulting from glycosidase action on the protein is thought to be the macrophage-activating factor [154]. Exoenzymes on B lymphocytes (galactosidase) and T lymphocytes (sialidase) are presumed to sequentially modify DBP, generating this potent macrophage activator [154,155]. Treatment of DBP *in vitro* with immobilized sialidase and galactosidase has also been reported to produce a potent stimulator of macrophage Fc receptor expression [153] and superoxide production [155].

2. Osteoclastogenesis and Development

Osteoclast progenitor cells are blood-borne cells of the monocyte–macrophage lineage. Evidence has been presented that glycosidase-treated DBP may have a role in osteoclast development and function [156]. Osteopetrosis is a heterogeneous disorder of bone characterized by excessive bone accumulation and a variety of immune defects. In the rat, two mutations known as osteopetrosis (*op*) and incisors absent (*ia*) display reduced bone resorptive activity. The *op* rat exhibits severe skeletal sclerosis with reduced numbers of osteoclasts that are large and structurally abnormal. The *ia* rat has milder disease with increased numbers of osteoclasts that appear to lack a ruffled border [156].

Yamamoto and co-workers have reported deficient β-galactosidase activity by *op* rat B lymphocytes [155]. Neither lysophospholipids nor alkylglycerols were able to stimulate macrophage activation in *op* rats, and basal peritoneal macrophage populations were much greater in wild-type versus *op* rats. Further, inflammatory lipid products produced an 8-fold increase of wild-type peritoneal macrophages but no increase in the *op* animals.

When β-galactosidase was injected into *op* rats, macrophage numbers and superoxide activity increased markedly [155]. Schneider and colleagues treated both *op* and *ia* rats, from birth to 2 weeks of age, with glycosidase-treated DBP in 200 pg subcutaneous doses every 4 days [156]. The skeletons of these animals were examined and revealed improved bone resorption in both mutant groups. Osteoclast numbers increased in *op* rats, and, strikingly, the majority of osteoclasts in *ia* rat bones appeared to be normal in appearance and to contain ruffled borders [156]. These findings suggest that DBP may have a focal role in macrophage activation, osteoclastogenesis, and osteoclast activity. However, because mice devoid of DBP do not appear to display osteopetrosis [106], further studies are needed to understand how DBP might function in these cell systems.

3. C5A COCHEMOTAXIS

A serum protein that enhances C5a-stimulated chemotaxis of leukocytes was identified as DBP [2]. Initially, DBP was thought to bind to C5a or C5a-desArg [157], possibly protecting the peptide from enzymatic degradation. Some work points to a possible interaction of DBP with neutrophils and dimers of the complement peptide [157]. It was shown that pre-incubation of DBP with neutrophils is sufficient to enhance C5a-stimulated chemotaxis, and that DBP–complement peptide binding is not required [157]. The mechanism(s) of enhancement of neutrophil response to C5a by DBP is unknown, but several studies convincingly document that such an effect exists [151,158,159]. DBP may be internalized and processed by responsive cells [151], findings reminiscent of those reported by Esteban and co-workers [132]. Could the macrophage activation by DBP be related to the cochemotaxin and/or costimulatory (with C5a) effects of DBP? Further studies, certainly including those on the nature of DBP binding to cells and the possibility of DBP-stimulated cell signaling, are needed.

E. DBP Functions—Overview

The evidence to date supports the role of DBP in vitamin D sterol transport by providing a high affinity, high capacity, high turnover carrier that can regulate the cellular ingress of vitamin D and its metabolites (Table II).The high affinity of DBP for G-actin, the presence of DBP–actin complexes in plasma during injury/inflammation, the coordinate action of DBP with gelsolin on actin filament disassembly, and escorted delivery of G-actin to removal by Kupffer cells all point to its role in an actin scavenger system useful in preventing actin toxicity in extracellular fluid. Convincing evidence for actions of cell-associated DBP in macrophage activation, C5a cochemotaxis, and, possibly, osteoclast development, warrants our attention and continued scrutiny. Other plasma steroid binding proteins function in several ways beyond steroid transport [160], and it now seems clear that DBP also displays multifunctional capabilities.

References

1. Daiger SP, Schanfield MS, Cavalli-Sforza LL 1975 Human group-specific component (Gc) proteins bind vitamin D and 25-hydroxyvitamin D. Proc Natl Acad Sci USA **72**:2076–2080.
2. Cooke NE, Haddad JG 1989 Vitamin D binding protein (Gc-globulin). Endocr Rev **10**:294–307.
3. Cooke NE, Haddad JG 1995 Vitamin D binding protein (Gc-globulin): update 1995. In Negro-Vilar A (ed) Endocrine Reviews Monographs, Vol. 4: Hormonal Regulation of Bone and Mineral Metabolism, The Endocrine Society, Bethesda, MD. pp. 125–128.
4. Mendel CM 1989 The free hormone hypothesis: A physiologically based mathematical model. Endocr Rev **10**:232–274.
5. Rosner W 1990 The functions of corticosteroid-binding globulin and sex hormone-binding globulin: Recent advances. Endocr Rev **11**:80–91.
6. Hirschfeld J, Jonsson B, Rasmusson M 1959 Immune-electrophoretic demonstration of qualitative differences in human sera and their relation to the haptoglobins. Acta Pathol Microbiol **47**:160–168.
7. Cleve H, Constans J 1988 The mutants of the vitamin-D binding protein: More than 120 variants of the Gc/DBP system. Vox Sanguinis **54**:215–225.
8. Westwood WA, Werret DJ 1986 Group-specific component: A review of the isoelectric focusing methods and auxiliary methods available for the separation of its phenotypes. Forensic Sci Int **32**:135–150.
9. Haddad JG, Walgate J 1976 25-Hydroxyvitamin D transport in human plasma: Isolation and partial characterization of calcifidiol-binding protein. J Biol Chem **251**:4803–4809.
10. Cooke NE, David EV 1985 Serum vitamin D binding protein is a third member of the albumin and α-fetoprotein gene family. J Clin Invest **76**:2420–2424.
11. Yang F, Brune JL, Naylor SL, Cupples RL, Naberhaus KH, Bowman BH 1985 Human group-specific component (Gc) is a member of the albumin family. Proc Natl Acad Sci USA **82**:7994–7998.
12. Cooke NE 1986 Rat vitamin D binding protein: Determination of the full-length primary structure from cloned cDNA. J Biol Chem **261**:3441–3450.
13. Yang F, Bergeron JM, Linehan LA, Lalley PA, Sakaguchi AY, Bowman BH 1990 Mapping and conservation of the group-specific component gene in mouse. Genomics **7**:509–516.
14. Osawa M, Tsuji T, Yukawa N, Saito T, Takeichi S 1994 Cloning and sequence analysis of cDNA encoding rabbit vitamin D-binding protein (Gc globulin). Biochem Mol Biol Int **34**:1003–1009.
15. Schoentgen F, Metz-Boutigue MH, Jolles J, Constans J, Jolles P 1986 Complete amino acid sequence of human vitamin D-binding protein (group-specific component): Evidence of a three-fold internal homology as in serum albumin and alpha-fetoprotein. Biochim Biophys Acta **871**:189–198.
16. Litwiller R, Fass R, Kumar R 1987 The amino acid sequence of

the NH_2-terminal portion of the rat and human vitamin D binding protein: Evidence for a high degree of homology between rat and human vitamin D binding protein. Life Sci **38**:2179–2184.
17. Witke FW, Gibbs PEM, Zielinski R, Yang F, Bowman BH, Dugaiczyk A 1993 Complete structure of the human Gc gene: Differences and similarities between members of the albumin gene family. Genomics **16**:751–754.
18. Ray K, Wang XK, Zhao M, Cooke NE 1991 The rat vitamin D binding protein (Gc-globulin) gene: Structural analysis, functional and evolutionary correlations. J Biol Chem **266**:6221–6229.
19. Brown JR 1976 Structural origins of mammalian albumin. Fed Proc **35**:2141–2144.
20. Min X, Carter DC 1992 Atomic structure and chemistry of human serum albumin. Nature **358**:209–215.
21. Cooke NE, Willard HF, David EV, George DL 1986 Direct regional assignment of the gene for vitamin D binding protein (Gc-globulin) to human chromosome 4q11–q13 and identification of an associated DNA polymorphism. Hum Genet **73**:225–229.
22. Harper ME, Dugaiczyk A 1983 Linkage of the evolutionarily-related serum albumin and α-fetoprotein genes with q11–q22 of human chromosome 4. Am J Hum Genet **35**:565–572.
23. Weitkamp LR, Remwick JR, Berger J, Shreffler DC, Drachman O, Wuhrmann F, Braend M, Franglen G 1970 Additional data and summary for albumin-Gc linkage in man. Hum Hered **20**:1–7.
24. Kaarsalo E, Melartin L, Blumberg BS 1967 Autosomal linkage between the albumin and Gc loci in humans. Science **158**:123–125.
25. Cooke NE, Levan G, Szpirer J 1987 The rat vitamin D binding protein (Gc-globulin) gene is syntenic with the rat albumin and α-fetoprotein genes on chromosome 14. Cytogenet Cell Genet **44**:98–100.
26. Guan N-J, Arhin G, Leung J, Tilghman SM 1996 Linkage between vitamin D-binding protein and α-fetoprotein in the mouse. Mammalian Genome **7**:103–106.
27. Weitkamp LR, Allen PZ 1979 Evolutionary conservation of equine Gc alleles and of mammalian Gc/albumin linkage. Genetics **92**:1347–2272.
28. Juneja RK, Sandberg K, Andersson L, Gahne B 1982 Close physical linkage between albumin and vitamin D binding protein (Gc) loci in chicken: A 300 million year old linkage group. Genet Res Cambridge **40**:95–98.
29. Stanford Human Genome Center 1996 Chromosome 4 Radiation Hybrid Map Internet (http://shge.stanford.edu/mapping).
30. Stanford Human Genome Center 1996 Chromosome 4 YAC STS-Content Map Internet (http://shge.stanford.edu/mapping).
31. Ingram RS, Scott RW, Tilghman SM 1981 α-Fetoprotein and albumin genes are in tandem in the mouse genome. Proc Natl Acad Sci USA **78**:4694–4698.
32. Chevrette M, Guertin M, Turcotte B, Belanger L 1987 The rat α1-fetoprotein gene: Characterization of the 5′-flanking region and tandem organization with the albumin gene. Nucleic Acids Res **15**:1338–1339.
33. Urano Y, Sakai M, Watanabe K, Tamaoki T 1984 Tandem arrangement of the albumin and alpha-fetoprotein genes in the human genome. Gene **32**:255–261.
34. Lichenstein HS, Lyons DE, Wurfel MM, Johnson DA, McGinley MD, Leidli JC, Trollinger DB, Mayer JP, Wright SD, Zukowski MM 1994 Afamin is a new member of the albumin, α-fetoprotein, and vitamin D-binding protein gene family. J Biol Chem **269**:18149–18154.
35. Belanger L, Roy S, Allard D 1994 New albumin gene 3′ adjacent to the α1-fetoprotein locus. J Biol Chem **269**:5481–5484.
36. Allard D, Gilbert S, Lamontagne A, Hamel D, Belanger L 1995 Identification of rat α-albumin and cDNA cloning of its human ortholog. Gene **153**:287–288.
37. Eiferman FA, Young PR, Scott RW, Tilghman SM 1981 Intragenic amplification and divergence in the mouse a-fetoprotein gene. Nature **294**:713–718.
38. Gibbs EM, Dugaiczyk A 1987 Origin of structural domains of the serum albumin gene family and a predicted structure of the gene for vitamin D-binding protein. Mol Biol Evol **4**:364–379.
39. Nardelli-Haeflinger DN, Moskaitis JE, Schoenberg DR, Wahli W 1989 Amphibian albumins as members of the albumin, alpha-fetoprotein, vitamin D-binding protein multigene family. J Mol Evol **29**:344–354.
40. Hay AWM, Watson G 1976 The plasma transport of 25-hydroxycholecalciferol in fish, amphibians, reptiles and birds. Comp Biochem Physiol **53**:167–172.
41. Licht P 1994 The relation of the dual thyroxine/vitamin D-binding protein (TBP/DBP) of emydid turtles to vitamin D-binding proteins of other vertebrates. Gen Comp Endocrinol **94**:215–224.
42. Licht P, Moore MF 1994 Structure of a reptilian plasma thyroxine binding protein indicates homology to vitamin D binding protein (Gc-globulin). Arch Biochem Biophys **309**:47–51.
43. Pavgi S, Licht P 1992 Measurement of plasma thyroxine binding protein in relation to thyroidal condition in the turtle *Trachemys scripta* by radioimmunoassay. Gen Comp Endocrinol **85**:147–155.
44. Van Baelen H, Allewaert K, Bouillon R 1988 New aspects of the plasma carrier protein for 25-hydroxycholecalciferol in vertebrates. Ann NY Acad Sci **538**:60–68.
45. McLeod JF, Cooke NE 1989 The vitamin D-binding protein, α-fetoprotein, albumin multigene family: Detection of transcripts in multiple tissues. J Biol Chem **264**:21760–21769.
46. Gitlin D, Perricelli A 1970 Synthesis of serum albumin, prealbumin, α-fetoprotein, and transferrin by the human yolk sac. Nature **228**:995–997.
47. Tilghman SM, Belayew A 1982 Transcriptional control of the murine albumin/α-fetoprotein locus during development. Proc Natl Acad Sci USA **79**:5254–5257.
48. Muglia L, Locker J 1984 Developmental regulation of albumin and α-fetoprotein gene expression in the rat. Nucl Acids Res **12**:6751–6762.
49. Melartin L, Hirvonen T, Kaarsalo E, Toivanen P 1966 Group-specific component and transferrins in human fetal sera. Scand J Haematol **3**:117–122.
50. Hillman L, Haddad JG 1983 Serial analyses of serum vitamin D-binding protein in preterm infants from birth to postconceptual maturity. J Clin Endocrinol Metab **56**:189–191.
51. Vacher J, Tilghman SM 1990 Dominant negative regulation of the mouse α-fetoprotein gene in adult liver. Science **250**:1732–1735.
52. Adrian GS, Yang F, Graves DT, Buchanan JM, Bowman BH 1990 Expression of transferrin and vitamin D-binding protein genes in an osteogenic sarcoma cell line. Exp Cell Res **186**:385–389.
53. Guoth M, Murgia A, Siegel D, Smith R, Cooke N, Haddad J 1990 Cell-surface vitamin D-binding protein (Gc-globulin) is acquired from plasma. Endocrinology **127**:2313–2321.
54. Sabbatini A, Petrini M, Mattii L, Arnaud P, Galbraith RM 1993 Vitamin D binding protein is produced by human monocytes. FEBS Lett **323**:89–92.
55. Chelly J, Concordet JP, Kaplan JC, Kahn A 1989 Illegitimate transcription: transcription of any gene in any cell type. Proc Natl Acad Sci USA **86**:2617–2621.
56. McPherson CE, Shim E-Y, Friedman DS, Zaret KS 1993 An

active tissue-specific enhancer and bound transcription factors existing in a precisely positioned nucleosomal array. Cell **75**:387–398.

57. Clayton DF, Weiss M, Darnell JE 1985 Liver-specific RNA metabolism in hepatoma cells: Variations in transcription rates and mRNA levels. Mol Cell Biol **5**:2633–2641.
58. Straus DS, Takemoto CD 1987 Insulin negatively regulates albumin mRNA at the transcriptional and posttranscriptional level in rat hepatoma cells. J Biol Chem **262**:1955–1960.
59. Friedman AD, Landschulz WH, McKnight SL 1989 CCAAT/enhancer binding protein activates the promoter of the serum albumin gene in cultured hepatoma cells. Genes Dev **3**:1314–1322.
60. Heard JM, Herbomel P, Ott MO, Mottura-Rollier A, Weiss M, Yaniv M 1987 Determinants of rat albumin promoter tissue specificity analyzed by an improved transient expression system. Mol Cell Biol **7**:2425–2434.
61. Wuarin J, Schibler U 1990 Expression of the liver-enriched transcriptional activator protein DBP follows a stringent circadian rhythm. Cell **63**:1257–1266.
62. Lichsteiner S, Wuarin J, Schibler U 1987 The interplay of DNA-binding proteins on the promoter of the mouse albumin gene. Cell **51**:963–973.
63. Maire P, Wuarin J, Schibler U 1989 The role of *cis*-acting promoter elements in tissue-specific albumin gene expression. Science **244**:343–346.
64. Cooke NE, McLeod JF, Wang X, Ray K 1991 Vitamin D binding protein: Genomic structure, functional domains and mRNA expression in tissues. J Steroid Mol Biol **40**:787–793.
65. Pinkert CA, Ornitz DM, Brinster RL, Palmiter RD 1987 An albumin enhancer located 10 kb upstream functions along with its promoter to direct efficient liver-specific expression in transgenic mice. Genes Dev **1**:268–276.
66. Liu JK, Bergman Y, Zaret KS 1988 The mouse albumin promoter and a distal upstream site are simultaneously DNase I hypersensitive in liver chromatin and bind similar liver-abundant factors *in vitro*. Genes Dev **2**:528–541.
67. Descombes P, Chojkier M, Lichsteiner M, Falvey E, Schibler U 1990 LAP, a novel member of the C/EBP gene family, encodes a liver-enriched transcriptional activator protein. Genes Dev **4**:1541–1551.
68. Lai E, Prezioso VR, Smith E, Litvin O, Costa RH, Darnell JE 1990 HNF3A, a hepatocyte-enriched transcription factor of novel structure is regulated transcriptionally. Genes Dev **4**:1427–1436.
69. Jackson DA, Rowader KE, Stevens K, Jiang C, Milos P, Zaret K 1993 Modulation of liver-specific transcription by interactions between hepatocyte nuclear factor 3 and nuclear factor 1 binding DNA in close apposition. Mol Cell Biol **13**:2401–2410.
70. Millonig JH, Emerson JA, Levorse JM, Tilghman SM 1995 Molecular analysis of the distal enhancer of the mouse α-fetoprotein gene. Mol Cell Biol **15**:3848–3856.
71. Camper SA, Godbout R, Tilghman SM 1989 The developmental regulation of albumin and α-fetoprotein gene expression. Prog Nucleic Acid Res Mol Biol **36**:131–143.
72. Hammer RE, Krumlauf R, Camper SA, Brinster RL, Tilghman SM 1987 Diversity of α-fetoprotein gene expression in mice is generated by a combination of separate enhancer elements. Science **235**:53–58.
73. Emerson JA, Vacher J, Carillo LA, Tilghman SM, Tyner AL 1992 The zonal expression of α-fetoprotein transgenes in the liver of adult mice. Dev Dynamics **195**:55–56.
74. Cirillo LA, Emerson JA, Vacher J, Tyner AL 1995 Developmental regulation of α-fetoprotein expression in intestinal epithelial cells of transgenic mice. Dev Biol **168**:395–405.
75. Poliard A, Labib B, Poiret M, Foiret D, Danan J-L 1990 Regulation of the rat α-fetoprotein gene expression in liver: Both the promoter region and an enhancer element are liver-specific and negatively modulated by dexamethasone. J Biol Chem **265**:2137–2141.
76. Zhang X-K, Dong J-M, Chiu J-F 1991 Regulation of α-fetoprotein gene expression by antagonism between AP-1 and the glucocorticoid receptor at their overlapping binding site. J Biol Chem **266**:8248–8254.
77. Song YH, Cooke NE 1996 Vitamin D-binding protein gene transcription is regulated by hepatic-nuclear factor 1. 10th International Congress of Endocrinology, June 12–15, San Francisco, California, p. 935.
78. Wilson SG, Retallack RW, Kent JC, Worth GK, Gutteridge DH 1990 Serum free 1,25-dihydroxyvitamin D and the free 1,25-dihydroxyvitamin D index during a longitudinal study of human pregnancy and lactation. Clin Endocrinol **32**:613–622.
79. Hartwell D, Riis BJ, Christiansen C 1990 Changes in vitamin D metabolism during natural and medical menopause. J Clin Endocrinol Metab **71**:127–132.
80. Prince RL, Dick I, Garcia-Webb P, Retallack RW 1990 The effects of menopause on calcitriol and parathyroid hormone: Responses to a low dietary calcium stress test. J Clin Endocrinol Metab **70**:1119–1123.
81. Nielsen HK, Brixen K, Bouillon R, Mosekilde L 1990 Changes in biochemical markers of osteoblastic activity during the menstrual cycle. J Clin Endocrinol Metab **70**:1431–1437.
82. Brixen K, Nielsen HK, Bouillon R, Flyvbjerg A, Mosekilde L 1992 Effects of short-term growth hormone treatment on PTH, calcitriol, thyroid hormones, insulin and glucagon. Acta Endocrinol **127**:331–336.
83. Egawa T, Ito H, Nakamura H, Yamamoto H, Kishimoto S 1992 Hormonal regulation of vitamin D-binding protein production by a human hepatoma cell line. Biochem Int **28**:551–557.
84. Guha C, Osawa M, Werner PA, Galbraith RM, Paddock GV 1995 Regulation of human Gc (vitamin D-binding) protein levels: Hormonal and cytokine control of gene expression *in vitro*. Hepatology **21**:1675–1681.
85. Constans J, Viau M 1977 Group-specific component: Evidence for two subtypes of the Gc1 gene. Science **198**:1070–1071.
86. Cleve H, Patutschnick W 1979 Neuraminidase treatment reveals sialic acid differences in certain genetic variants of the Gc system (vitamin D-binding protein). Hum Genet **47**:193–198.
87. Scheidegger JG 1955 Une micro-methode de l'immunoelectrophorese. Int Arch Allergy Appl Immunol **7**:103–110.
88. Reinskou T 1963 A heterogeneity of the fast moving component of the Gc-system. Acta Pathol Microbiol **59**:526–532.
89. Parker WC, Cleve H, Bearn AG 1963 Determination of phenotypes in the human group-specific component (Gc) system by starch gel electrophoresis. Am J Hum Genet **15**:353–367.
90. Constans J, Viau M, Cleve H, Jaeger G, Quilici JC, Palisson MJ 1978 Analysis of the Gc polymorphism in human populations by isoelectrofocusing on polyacrylamide gels. Demonstration of subtypes of the Gc1 allele and of additional Gc variants. Hum Genet **41**:53–60.
91. Emerson DL, Galbraith RM, Arnaud P 1984 Electrophoretic demonstration of interactions between group-specific component (vitamin D binding protein), actin, and 25-cholecalciferol. Electrophoresis **5**:22–26.
92. Tang WX, Bazaraa HM, Magiera H, Cooke NE, Haddad JG 1996 Electrophoretic mobility shift assay identifies vitamin D binding protein (Gc-globulin) in human, rat, and mouse sera. Anal Biochem **237**:245–251.
93. Braun A, Bichlmaier R, Cleve H 1992 Molecular analysis of the

gene for the human vitamin D-binding protein (group-specific component): Allelic differences of the common genetic GC types. Hum Genet **89**:401–406.
94. Cooke NE, Murgia A, McLeod JF 1988 Vitamin D-binding protein: Structure and pattern of expression. Ann NY Acad Sci **538**:49–59.
95. Braun A, Bichlmaier R, Muller B, Cleve H 1993 Molecular evaluation of an *Alu* repeat including a polymorphic variable poly(dA) (*AluVpA*) in the vitamin D binding protein (DBP) gene. Hum Genet **90**:526–532.
96. Lai LYC, Constans J, Archer GT 1990 A new cathodal Gc variant in Australia. Ann Hum Biol **17**:245–248.
97. Chen LZ, Easteal S, Board PG, Summers KM, Bhatia KK, Kirk RL 1990 Albumin-vitamin D-binding protein haplotypes in Asian-Pacific populations. Hum Genet **85**:89–97.
98. Kofler A, Braun A, Jenkins T, Serjeantson SW, Cleve H 1995 Characterization of mutants of the vitamin-D-binding protein/group specific component: GC aborigine (1A1) from Australian aborigines and South African blacks, and 2A9 from south Germany. Vox Sanguinis **68**:50–54.
99. Yuasa I, Kofler A, Braun A, Umetsu K, Bichlmaier R, Kammerer S, Cleve H 1995 Characterization of mutants of the vitamin D-binding protein/group-specific component: Molecular evolution of GC*1A3 and GC*1A3, common in some Asian populations. Hum Genet **95**:507–512.
100. Constans J, Gouaillare C, Bouissou C, Dugoujon JM 1987 Polymorphism of the vitamin D binding protein (DBP) among primates: An evolutionary analysis. Am J Phys Anthropol **73**:365–377.
101. Bouquet Y, Van de Weghe A, Van Zeveren A, Varewyck H 1986 Evolutionary conservation of the linkage between the structural loci for serum albumin and vitamin D binding protein (Gc) in catttle. Anim Genet **17**:175–182.
102. Bender K, Cleve H, Gunther E 1981 A previously described serum protein polymorphism in the rat identified as Gc ("vitamin D-binding protein"). Anim Blood Groups Biochem Genet **12**:31–36.
103. Dykes D, Polesky H, Cox E 1981 Isoelectric focusing of Gc (vitamin D binding globulin) in parentage testing. Hum Genet **58**:174–175.
104. Yoshifumi Y, Nishimoto H, Ikemoto S 1989 Interstitial deletion of the proximal long arm of chromosome 4 associated with father–child incompatibility within the Gc-system: Probable reduced gene dosage effect and partial piebald trait. Am J Med Genet **32**:520–523.
105. Vavrusa B, Cleve H, Constans J 1983 A deficiency mutant of the Gc system. Hum Genet **65**:102–107.
106. Thornton PS, Monks BR, Hu Y, Haddad JG, Liebhaber SA, Cooke NE 1995 Generation of a mouse line null for vitamin D binding protein by targeted homologous recombination. J Bone Miner Res **10**:S494 (abstract).
107. Weitkamp LR, Rucknagel DL, Gershowitz H 1966 Genetic linkage between structural loci for albumin and group specific component (Gc). Am J Hum Genet **18**:559–571.
108. Haddad JG 1987 Traffic, binding, and cellular access of vitamin D sterols. In Peck WA, (ed) Bone and Mineral Research, Vol. 5. Elsevier, New York, pp. 281–308.
109. Arnaud J, Constans J 1993 Affinity differences for vitamin D metabolites associated with the genetic isoforms of the human serum carrier protein (DBP). Hum Genet **92**:183–188.
110. Bouillon R, Xiang DZ, Convents R, Van Baelen H 1992 Polyunsaturated fatty acids decrease the apparent affinity of vitamin D metabolites for human vitamin D-binding protein. J Steroid Biochem Mol Biol **42**:855–861.
111. Green S, Wahli W 1994 Peroxisome proliferator-activated receptors: Finding the orphan a home. Mol Cell Endocrinol **100**:149–153.
112. Pols HAP, Birkenhager JC, van Leeuven JPTM 1994 Vitamin D analogues: From molecule to clinical application. Clin Endocrinol **40**:285–291.
113. Bishop JE, Collins ED, Okamura WH, Norman AW 1994 Profile of ligand specificity of the vitamin D binding protein for 1α,25-dihydroxyvitamin D_3 and its analogs. J Bone Miner Res **9**:1277–1288.
114. Bikle DD 1992 Vitamin D: New actions, new analogs, new therapeutic potential. Endocr Rev **13**:765–784.
115. Haddad JG 1992 Clinical aspects of measurements of plasma vitamin D sterols and the vitamin D binding protein. In Coe FL, Favus MJ (eds) Disorders of Bone and Mineral Metabolism. Raven, New York, pp. 195–216.
116. Kawakami M, Blum CB, Ranakrishman R, Dell RB, Goodman DS 1981 Turnover of the plasma binding protein for vitamin D and its metabolites in normal human subjects. J Clin Endocrinol Metab **53**:1110–1116.
117. Haddad JG, Fraser DR, Lawson DEM 1981 Vitamin D binding protein: Turnover and fate in the rabbit. J Clin Invest **67**:1550–1560.
118. Bikle DD, Gee E, Halloran B, Haddad JG 1984 Free 1,25$(OH)_2$D levels in serum from normal subjects, pregnant subjects and subjects with liver disease. J Clin Invest **74**:1966–1971.
119. Barsano CP, Baumann G 1989 Simple algebraic and graphic methods for the apportionment of hormone (and receptor) into bound and free fractions in binding equilibria; or how to calculate bound and free hormone. Endocrinology **124**:1101–1106.
120. Vieth R 1994 Simple method for determing specific binding capacity of vitamin D-binding protein and its use to calculate the concentration of "free" 1,25-dihydroxyvitamin D. Clin Chem **40**:435–441.
121. Teegarden D, Meredith SC, Sitrin MD 1991 Determination of the affinity of vitamin D metabolites to serum vitamin D binding protein using an assay employing lipid-coated polystyrene beads. Anal Biochem **199**:293–299.
122. Bikle DD, Siiteri BK, Ryzen E, Haddad JG 1985 Serum protein binding of 1,25$(OH)_2$D: A re-evaluation by direct measurement of free metabolite levels. J Clin Endocrinol Metab **61**:969–975.
123. Bikle DD, Gee E, Halloran BP, Kowalski MA, Ryzen E, Haddad JG 1986 Assessment of the free fraction of 25(OH)D in serum and its regulation by albumin and the vitamin D binding protein. J Clin Endocrinol Metab **63**:954–959.
124. Bikle DD, Halloran BP, Ryzen E, Kowalski MA, Haddad JG 1986 Free 25(OH)D levels are normal in subjects with liver disease and reduced total 25(OH)D levels. J Clin Invest **78**:748–752.
125. Ray R, Bouillon R, Van Baelen H, Holick MF 1991 Photoaffinity labeling of human serum vitamin D binding protein and chemical cleavages of the labeled protein: Identification of an 11.5 kDa peptide containing the putative 25-hydroxyvitamin D_3 binding site. Biochemistry **30**:7638–7642.
126. Haddad JG, Hu YZ, Kowalski MA, Laramore C, Ray K, Robzyk P, Cooke NE 1992 Identification of the sterol- and actin-binding domains of plasma vitamin D binding protein (Gc globulin). Biochemistry **31**:7174–7181.
127. Baran DT, Ray R, Sorensen AM, Honeyman T, Holick MF 1994 Binding characteristics of a membrane receptor that recognizes 1α,25-dihydroxyvitamin D_3 and its epimer, 1β,25-dihydroxyvitamin D_3. J Cell Biochem **56**:510–517.
128. Baran DT 1994 Nongenomic actions of the steroid hormone 1α,25-dihydroxyvitamin D_3. J Cell Biochem **56**:303–306.

129. Ghinea N, Eskenasy M, Simionescu M, Simionescu N 1989 Endothelial albumin binding proteins are membrane-associated components exposed on the cell surface. J Biol Chem **264**:4755–4758.
130. Suzuki Y, Zeng CQY, Alpert E 1992 Isolation and partial characterization of a specific alpha-fetoprotein receptor in human monocytes. J Clin Invest **90**:1530–1536.
131. Esteban C, Trojan J, Macho A, Mishal A, Lagarge-Frayssine CH, Uriel J 1993 Activation of an α-fetoprotein/receptor pathway in human normal and malignant peripheral blood mononuclear cells. Leukemia **7**:1807–1816.
132. Esteban C, Geuskens M, Ena JM, Mishal Z, Macho A, Torres JM, Uriel J 1992 Receptor-mediated uptake and processing of vitamin D-binding protein in human B-lymphoid cells. J Biol Chem **14**:10177–10183.
133. Thompson EB 1995 Membrane transporters of steroid hormones. Curr Biol **5**:730–732.
134. Pettifor JM, Bikle DD, Cavaleros M, Zachen D, Kamdar MC, Ross FP 1995 Serum levels of free 1,25-dihydroxyvitamin D in vitamin D toxicity. Ann Intern Med **122**:511–513.
135. Holick MF, MacLaughlin JA, Doppelt SH 1981 Regulation of cutaneous previtamin D_3 photosynthesis in man: Skin pigment is not an essential regulator. Science **211**:590–593.
136. Haddad JG, Matsuoka LY, Hollis BW, Hu YZ, Wortsman J 1993 Human plasma transport of vitamin D after its endogenous synthesis. J Clin Invest **91**:2552–2555.
137. Van Baelen H, Bouillon R, DeMoor P 1980 Vitamin D binding protein (Gc-globulin) binds actin. J Biol Chem **225**:2270–2272.
138. Cooke NE, Walgate J, Haddad JG 1979 Human serum binding protein for vitamin D and its metabolites: II. Specific, high affinity association with a protein in nucleated tissue. J Biol Chem **254**:5965–5971.
139. McLeod J, Kowalski MA, Haddad JG 1989 Interactions among serum vitamin D binding protein, monomeric actin, profilin and profilactin. J Biol Chem **264**:1260–1267.
140. Lees A, Haddad JG, Lin S 1984 Brevin and DBP comparison of the effects of two serum proteins on actin assembly and disassembly. Biochemistry **23**:3038–3047.
141. Korn ED 1982 Actin polymerization and its regulation by proteins from nonmuscle cells. Physiol Rev **62**:672–737.
142. Sanger JM, Dabiri G, Mittal B, Kowalski MA, Haddad JG, Sanger JW 1990 Disruption of microfilament organization in living non-muscle cells by microinjection of plasma vitamin D binding protein or DNAse I. Proc Natl Acad Sci USA **87**:5474–5478.
143. Vasconcellos CA, Lind SE 1993 Coordinated inhibition of actin-induced platelet aggregation by plasma gelsolin and vitamin D-binding protein. Blood **12**:3648–3657.
144. Lee WM, Galbraith RM 1992 The extracellular actin-scavenger system and actin toxicity. N Engl J Med **326**:1335–1341.
145. Lee WM, Galbraith RM, Watt GH, Hughes RD, McIntire DD, Hoffman BJ, Williams R 1995 Predicting survival in fulminant hepatic failure using serum Gc protein concentrations. Hepatology **21**:101–105.
146. Haddad JG, Harper KD, Guoth M, Pietra GG, Sanger JW 1990 Angiopathic consequences of saturating the plasma scavenger system for actin. Proc Natl Acad Sci USA **87**:1381–1385.
147. Rinaldo JE, Rogers RM 1986 Adult respiratory distress syndrome. N Engl J Med **315**:578–579.
148. Hermannsdoerfer AJ, Heeb GT, Fenstel PJ, Estes JE, Keenan CJ, Minnear FL, Selden L, Giunta C, Flor JR, Blumenstock, FA 1993 Vascular clearance and organ uptake of G- and F-actin in the rat. Am J Physiol **265**:G1071–G1081.
149. Flor J, Blumenstock FA, Estes J, Gershman LC, Selden L 1995 Binding of vitamin D-binding protein and G-actin to isolated endotoxin stimulated rat Kupffer cells. FASEB J **9**:A366.
150. Tanaka Y, DeLuca HF 1980 Measurement of mammalian 25-hydroxyvitamin D3, 24*R*- and 1α-hydroxylase. Proc Natl Acad Sci USA **78**:196–199.
151. Kew RR, Fisher JA, Webster RO 1995 Co-chemotactic effect of Gc-globulin (DBP) for C5a: Transient conversion into an active co-chemotaxin by neutrophils. J Immunol **155**:5369–5374.
152. Petrini M, Emerson DL, Galbraith RM 1983 Linkage between surface immunoglobulin and cytoskeleton of B-lymphocytes may involve Gc protein. Nature **306**:73–74.
153. Yamamoto N, Homma S, Haddad JG, Kowalski MA 1991 Vitamin D_3 binding protein required for *in vitro* activation of macrophages after dodecylglycerol treatment of mouse peritoneal cells. Immunology **74**:420–424.
154. Naraparaju VR, Yamamoto N 1994 Roles of β-galactosidase of B lymphocytes and sialidase of T lymphocytes in inflammation-primed activation of macrophages. Immunol Lett **43**:143–148.
155. Yamamoto N, Lindsay DD, Naraparajn VR, Ireland RA, Popoff SN 1994 A defect in the inflammation-primed macrophage-activation cascade in osteopetrotic rats. J Immunol **152**:5100–5107.
156. Schneider GB, Benis KA, Flay NW, Ireland RA, Popoff SN 1995 Effects of vitamin D binding protein-macrophage activating factor (DBP-MAF) infusion on bone resorption in two osteopetrotic mutations. Bone **6**:657–662.
157. Perez DH 1994 Gc globulin (vitamin D-binding protein) increases binding of low concentrations of C51 des Arg to human polymorphonuclear leukocytes: An explanation for its cochemotaxin activity. Inflammation **18**:215–220.
158. Kew RR, Mollison KW, Webster RO 1992 Binding of Gc globulin (DBP) to C5a or C5a des Arg is not necessary for co-chemotactic activity. J Leukocyte Biol **58**:55–58.
159. Kew RR, Webster RO 1988 Gc-globulin (DBP) enhances the neutrophil chemotactic activity of C5a and C5a des Arg. J Clin Invest **82**:364–369.
160. Hammond GL 1995 Potential functions of plasma steroid-binding proteins. Trends Endocrinol Metab **6**:298–304.
161. Tellam RL, Morton DJ, Clarke FM 1989 A common theme in the amino acid sequences of actin and many actin-binding proteins. Trends Biochem Sci **14**:130–133.
162. Fan J-B, DeYoung J, Lagace R, Lina RA, Xu Z, Murray JC, Buetow KH, Weissenbach J, Goold RD, Cox DR, Myers RM 1994 Isolation of yeast artificial chromosome clones from 54 polymorphic loci mapped with high odds on human chromosome 4. Hum Mol Genet **3**:243–246.
163. Tunnacliffe A, Majumdar S, Yan B, Poncz M 1992 Genes for β-thromboglobulin and platelet factor 4 are closely linked and form part of a cluster of related genes on chromosome 4. Blood **79**:2896–2900.

SECTION II

Mechanisms of Action

CHAPTER 8

The Vitamin D Receptor and Its Gene

J. WESLEY PIKE Department of Molecular and Cellular Physiology, University of Cincinnati, Cincinnati, Ohio

I. INTRODUCTION

Studies in the early 1960s by Jensen and co-workers on the actions of the steroid hormone estrogen suggested that this important small molecular weight lipophilic cholesterol derivative localized in the tissue of target organs such as the uterus [1]. These studies supported the idea that estrogen and perhaps hormones of similar chemical composition [2] might localize within the nuclei of appropriate tissues via a binding protein or receptor that functioned to modify gene expression. These early pioneering investigations into the mechanism of action of estrogenic hormones, followed closely by related observations in the progesterone, androgen, and glucocorticoid fields, formed the basis for what is now believed to be the steroid hormone genomic mechanism of action. The intervening years have borne witness to incredible advances in our understanding of certain aspects of this mechanism and have clearly validated the hypotheses put forth by early workers in the field [3–8]. In 1997, we know that sex and adrenal hormones bind to specific intracellular receptor proteins which in turn function to transduce the environmental information carried by these small hormonal signals to the cell nucleus where they modify gene expression. Most of the genes that encode these hormone receptors have been molecularly cloned with the ensuing insight that each is part of a common gene family of proteins with related structural similarity. Interestingly, members of this family of nuclear receptor genes also transduce vertebrate signals generated by hormonal forms of vitamin D (1,25-dihydroxyvitamin D) and vitamin A (retinoic acid and 9-*cis*-retinoic acid) as well as thyroid hormone, fatty acids, nutritional metabolites, and invertebrate signals such as the insect molting hormone ecdysone [8]. As a direct result of the cloning of this receptor family, substantial progress has been made in understanding how these hormonal signals activate their transcription factor receptors, the mechanisms by which they bind to the regulatory regions of genes, the location and nature of the DNA binding sites, and the mechanisms by which they modulate transcriptional initiation.

An hypothesis that vitamin D functioned to modulate gene expression in a manner analogous to that of the steroid hormones emerged in 1968 [9]. This hypothesis predated the actual discovery in 1971 of the active principle of vitamin D_3, the dihydroxylated metabolite 1,25-dihydroxyvitamin D_3 [1,25$(OH)_2D_3$], which represented the culmination of work by numerous investigators in the nutrition field [10–12]. A model describing the mechanism by which vitamin D is proposed to function is illustrated in Fig. 1. The lipophilic nature of the small vitamin D molecule and its capacity to localize in target tissues as well as the fact that vitamin D responses were sensitive to transcriptional inhibitors provided an initial basis for the hypothesis. The hypothesis was strongly supported in subsequent studies during the early 1970s which defined the exquisite and highly complex nature of the regulation of the synthesis and production of $1,25(OH)_2D_3$ by the renal enzyme 25-hydroxyvitamin D_3 1α-hydroxylase (see Chapter 5) [13]. Perhaps the most significant observation was the discovery of a binding protein in target tissues such as the intestine that appeared to be responsible for nuclear localization of what eventually was determined to be the active vitamin D hormone, $1,25(OH)_2D_3$ [14]. This initial observation led to further characterization of the "binding protein" and to the discovery that it bound $1,25(OH)_2D_3$ with extremely high affinity and specificity, was localized in expected target tissues, and was a nuclear macromolecule capable of binding to DNA. It was eventually purified from chicken and porcine intestines in the early 1980s [15,16]. The antibodies generated from this purified material provided the means whereby the chicken vitamin D receptor (VDR) was cloned in 1987, revealing it and subsequent mammalian homologs of the gene to be members of the steroid, thyroid, and vitamin A receptor family of transcription factors [17]. The cloning of the VDR established a clear mechanistic relationship between the actions of the steroid and thyroid hormones and vitamin D, and ushered in a new era of research on the molecular mechanisms by which $1,25(OH)_2D_3$ and the VDR function to modulate gene expression.

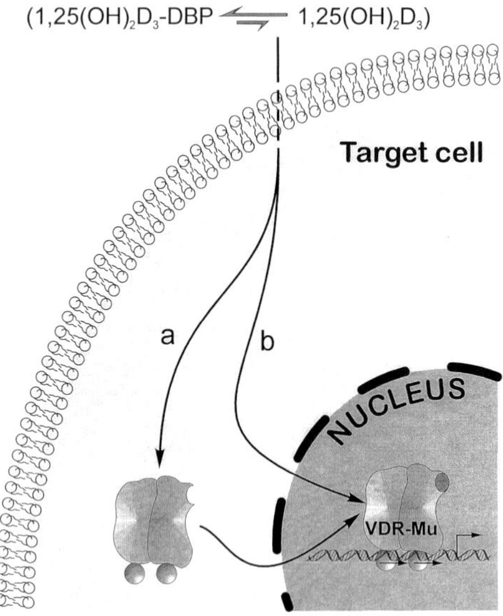

FIGURE 1 Model for the molecular mechanism of action of the vitamin D hormone. $1,25(OH)_2D_3$ dissociates from serum vitamin D binding protein (DBP), enters the cell by diffusion, and interacts with the vitamin D receptor. Activation by the ligand leads to interaction of the VDR with responsive genes and modulation of gene expression. VDR-Mu or VDR modulatory unit comprises one VDR molecule and an associated protein which is exemplified by but not restricted to a retinoid X receptor isoform. (a) An early model wherein the VDR, shown located in the cytoplasm, undergoes cytoplasmic to nuclear translocation on ligand activation and eventually binds to the regulatory region of a modulated gene. (b) Current view of the location of the VDR wherein the receptor is located in the nucleus and following ligand activation binds to the regulatory region of a vitamin D modulated gene.

In this chapter, we provide both an historical overview as well as contemporary consideration of the central role played by the VDR in mediating the actions of the vitamin D hormone. We describe the discovery of the receptor, certain features of its distribution, its subsequent biochemical characterization, and the cloning of both the structural gene and its chromosomal counterpart. In addition, we provide an overview of the structure of the VDR as well as its function in the regulation of gene expression. The reader is referred to Chapter 9 by Freedman and Lemon and to Chapter 10 by Haussler *et al.*, for additional important aspects of the structure of the VDR and molecular mechanisms by which it mediates vitamin D actions. Additional chapters that consider features of VDR activity include Chapters 48, 59, and 60.

II. DISCOVERY OF THE VITAMIN D RECEPTOR

Although several lines of evidence had suggested that vitamin D or an active metabolite of the vitamin functioned to regulate the expression of genes, the first successful studies that hinted at the existence of a binding protein or receptor (i.e., the VDR) were carried out in 1969 by Haussler and Norman [14]. In these studies, chickens were dosed with increasing concentrations of tritiated forms of vitamin D. Fifteen hours following dosing, the intestines were removed and utilized to prepare a crude nuclear chromatin fraction that had been demonstrated to preferentially accumulate labeled vitamin (or a metabolite of the vitamin). Tritium label was extracted from the nuclear chromatin preparation with 0.3 *M* KCl and quantitated. This experiment demonstrated preferential uptake of active vitamin into the intestinal chromatin fraction and provided further evidence that the process was a saturable one at low concentrations of administered active vitamin and likely mediated by a specific protein component.

Following these pioneering studies of the uptake of

vitamin D into intestinal cell target tissues, more definitive evidence for the VDR began to emerge. In studies by Brumbaugh and Haussler [18] and Lawson and Wilson [19], the protein nature of the receptor was established through proteolytic digestion studies and equilibrium sedimentation analysis. Further evaluation of the binding properties of the receptor by Brumbaugh and Haussler [20] suggested the affinity of the protein for labeled $1,25(OH)_2D_3$ to be in the low nanomolar range. Finally, *in vitro* experiments carried out in 1975 enabled the conclusion that the cytosol-derived VDR, which displayed a sedimentation coefficient (S) of about 3.5, could bind to chromatin fractions in the presence of the hormonal ligand [20]. These studies collectively provided definitive support for the existence of the VDR and prompted the suggestion that a cytoplasmic VDR translocated to the nucleus on ligand activation (Fig. 1a).

III. CHARACTERIZATION OF THE VDR

A. Tissue Distribution of the VDR

Over the years, the chicken has been utilized to investigate many of the novel calcium and phosphorus regulating mechanisms of vitamin D and in particular those involving the actions of the vitamin on intestinal calcium transport. Thus, it is not surprising that the initial discovery of the VDR was made from extracts prepared from chicken intestine. Nevertheless, the known actions of vitamin D during that time were not limited to avian species, nor were they limited to the intestine. Other known target organs for vitamin D included the kidney, bone, and the parathyroid glands. The discovery of the VDR in mammalian intestine and in each of the above tissues from both avian as well as mammalian species unfolded over the next several years. Kream *et al.* [21] overcame several technical problems inherent to cellular extracts of rat intestine and first provided evidence for the existence of VDR in a mammalian tissue. Its properties were similar to those of the chicken receptor, although equilibrium sedimentation experiments supported the notion that mammalian receptors might exhibit reduced molecular mass relative to the chicken protein. Evidence also accumulated for the presence of receptors in both avian and mammalian parathyroid glands [22,23], bone [24], and kidneys [25,26]. The discovery of receptors in these tissues was clearly facilitated technically by the availability of tritium-labeled $1,25(OH)_2D_3$ of increasing specific activity which provided enhanced sensitivity over earlier preparations of tracer compound. As initial studies on the localization and characterization of the VDR were entirely dependent on the availability of high specific activity radioisotopic $1,25(OH)_2D_3$, this availability represented a major determining factor in the discovery and subsequent characterization of the VDR.

The advent of high specific activity $1,25(OH)_2D_3$ prompted increased efforts toward discovering VDRs and thus identifying new cellular targets in tissues beyond those described above. This effort focused not only on animal tissues but on primary cells and cultured cell lines. This effort was also prompted by emerging evidence that the biological effects of $1,25(OH)_2D_3$ in tissues and cells extended beyond that of calcium and phosphorus homeostasis to include, among others, a role for $1,25(OH)_2D_3$ in cellular proliferation and differentiation [27,28]. As a consequence, VDRs were discovered in tissues such as pancreas [23,29], placenta [23], pituitary [30], ovary [31], testis [32], mammary gland [33], and heart [34]. The range of cultured cell lines defined as targets for $1,25(OH)_2D_3$ were similarly extensive, and now include cells of fibroblastic, osteoblastic, myoblastic, hematopoietic, and lymphopoietic origin as well as cells derived from normal kidney, intestine, and skin. A partial listing of vitamin D target tissues and cells is documented in Table I. Of a more fundamental nature, they also include cells of both tumorogenic and nontumorogenic origin [35–38].

The biological role of $1,25(OH)_2D_3$ in many of these tissues and cells is considered more extensively in several additional chapters in this book (including Chapter

TABLE I Cellular and Tissue Distribution of the VDR

System	Tissue
Gastrointestinal	Esophagus, stomach, small intestine, large intestine, colon
Hepatic	Liver parenchyma cells
Renal	Kidney, urethra
Cardiovascular	Cardiac muscle
Endocrine	Parathyroid gland, thyroid, adrenal, pituitary
Exocrine	Parotid gland, sebaceous gland
Reproductive	Testis, ovary, placenta, uterus, endometrium, yolk sac, avian chorioallantoic membrane, avian shell gland
Immune	Thymus, bone marrow, B cells, T cells
Respiratory	Lung alveolar cells
Musculoskeletal	Bone osteoblasts and osteocytes, cartilage chondrocytes, striated muscle
Epidermis/appendage	Skin, breast, hair follicles
Central nervous system	Brain neurons
Connective tissue	Fibroblasts, stroma

30). It is important to note, however, that perhaps the most interesting biological effects of vitamin D metabolites are their antiproliferative and prodifferentiating capacities [39]. These effects highlight a potential therapeutic role for $1,25(OH)_2D_3$ and its analogs as anticancer agents, an area considered in more depth in Chapters 64–68, and in the regulation of the immune system [40,41], considered in Chapters 69 and 70. Antiproliferative effects of vitamin D are currently useful in treating human psoriasis, a hyperproliferative disorder of skin [42]. The fundamental basis for these effects relate to the fact that normal skin is also an important target for vitamin D action [39,43], as outlined in more detail in Chapter 72. In any event, it is now clear that the vitamin D receptor is widely and perhaps ubiquitously distributed in vertebrate tissues, although both receptor-positive and receptor-negative cell types can be identified.

In addition to the utilization of binding and sedimentation assays to identify VDR in tissue extracts, three other techniques have been employed. Scintillant-enhanced autoradiography with tritiated $1,25(OH)_2D_3$ was employed early on by Zile *et al.* [44] and Jones and Haussler [45] to detect receptors in the intestine. Subsequent studies revealed cellular targets to include duodenal, ilial, and jejunal segments of the rat intestine, stomach, kidney, skin, and pituitary and several neurons in the brain [46,47]. The presence of $1,25(OH)_2D_3$ (or a metabolite of the administered compound) in these tissues through autoradiographic means supported the notion that the tissue was a target for vitamin D action but did not provide direct evidence for the presence of VDR. Following the production of monoclonal antibodies to purified chicken VDR that cross-reacted with VDR from other species, the technique of immunocytochemistry was applied to the detection of receptors in mammalian tissues and cells. As a consequence, the existence of the VDR protein was identified in a broad range of tissues and cells, largely confirming and extending the identity of direct cellular targets for vitamin D action [48–51]. The third and most recent general approach used to evaluate tissues and cells for the presence of receptor has been to utilize a VDR cDNA probe or known nucleotide sequence of the VDR cDNA to detect mRNA transcripts. The availability of such probes closely followed the cloning of the VDR in 1987 and 1988 (to be discussed in Section IV). However, in this general approach an often forgotten assumption is that VDR mRNA transcripts give rise to the production of VDR protein. This assumption must be rigorously proven in each case. Despite this caveat, Northern blot [17], *in situ* hybridization [52], and polymerase chain reaction [53] analyses have been applied successfully to the identification of tissue and cellular VDRs. The use of the latter techniques have been particularly helpful in the study of the transcriptional regulation of VDR production, discussed in detail in Chapter 11 and elsewhere in this text.

B. Subcellular Distribution of the VDR

Nuclear localization of the $1,25(OH)_2D_3$-liganded VDR has never been fundamentally questioned. However, the subcellular location of the ligand-free receptor capable of binding the incoming $1,25(OH)_2D_3$ signal has been controversial. Although early studies suggested that the unliganded VDR was cytosolic (see Fig. 1a and Brumbaugh and Haussler [20]), studies in the early 1980s suggested that the VDR might be loosely associated with the nuclear fraction [54,55]. Largely through the application of immunocytochemistry, the latter suggestion is now favored. Thus, as depicted in Fig. 1b, the VDR is believed to reside in the nucleus prior to activation by $1,25(OH)_2D_3$, similar to most other members of the nuclear receptor family. Although ligand-activated VDRs are believed to be bound to genomic DNA and perhaps more specifically to unique sequences of DNA lying adjacent to the promoters for $1,25(OH)_2D_3$-modulated genes, binding sites for unliganded receptors are currently unknown.

C. Biochemical Properties and Organization of the VDR

The VDR exists in relatively low abundance in target tissues and cultured cells at a concentration consistent with the fact that it is a potent transcription regulatory molecule [56,57]. Estimates of receptor abundance range from under 500 to over 25,000 copies of VDR/cell (10 to 100 fmol/mg protein) depending on the cell type or cell line examined, and up to 1 pmol/mg protein in certain tissue extracts. These estimates of abundance are based on the capacity of extracts to bind $1,25(OH)_2D_3$ and thus presumably reflect active functional receptor. The wide range in VDR abundance suggests that those cells with higher VDR content may be more highly responsive to the $1,25(OH)_2D_3$ hormone than those with lower levels of expression. Although the latter concept is intuitive and is supported by some evidence, it is important to note that numerous other factors also play an important role in individual cellular responsivity to the hormone. These factors include cellular capacity to internalize and subsequently metabolize $1,25(OH)_2D_3$, differential activation events that may modulate VDR activity in a cell- or tissue-specific manner (perhaps by phosphorylation of VDR), the nature, availability, and concentration of numerous partner proteins that are required for gene activation, and finally the accessibility and inducibility of specific genes. These as well as additional events contribute significantly to the sensitivity and biological responsivity of a particular cell to $1,25(OH)_2D_3$.

Immediately following the discovery of the VDR protein, both physical and functional properties of the VDR emerged, several of which are listed in Table II. With the exception of molecular mass (to be discussed below), no evidence developed to suggest that the VDR differed significantly in biochemical properties from cell to cell or from species to species. Sedimentation analysis revealed a protein of 3 to 3.7 S that exhibited an elongated shape. Gel filtration estimates of protein size ranged from 50,000 to 70,000 daltons depending on species. Perhaps the most important biochemical and functional property of the VDR was its capacity to bind $1,25(OH)_2D_3$ with both high affinity and selectivity [58–62]. In this regard, numerous experiments were performed that led to the determination of an equilibrium dissociation constant of 10^{-10} M for the natural ligand $1,25(OH)_2D_3$. VDR also binds $1,25(OH)_2D_3$ precursors as well as other metabolites of vitamin D with a range of substantially lower affinities [60,61]. The contribution of both the 25-hydroxyl and the 1α-hydroxyl groups on the $1,25(OH)_2D_3$ molecule in specific high affinity binding to VDR has been studied extensively [61].

Although numerous additional properties of the receptor emerged in the late 1970s, the discovery that the VDR exhibited DNA binding capabilities consistent with its role as a nuclear transcription factor represented a considerable advance [56]. A much more precise understanding of the properties of VDR DNA binding emerged following the identification of specific DNA binding sites (vitamin D response elements, VDREs) located adjacent to the promoter for vitamin D-inducible genes (see below). Nevertheless, the finding that the VDR bound to nonspecific DNA not only set the stage for ensuing studies aimed at a preliminary understanding of the structural organization and function of the VDR, but provided the initial means whereby the VDR could be isolated in quantities of sufficient purity to generate valuable immunological reagents. These reagents were ultimately useful in further characterization of the receptor and in the molecular cloning of its structural gene.

TABLE II Biochemical Properties of the VDR and Its RNA Transcript

VDR Protein
- Molecular weight: 50,000 (human) to 60,000 (chicken)
- Sedimentation coefficient (S): 3.2 (human, rat) to 3.7 (chicken)
- Equilibrium dissociation constant: $1\text{–}2 \times 10^{-10}$ M
- Isoelectric point: 6.2
- Phosphoprotein: Serine phosphorylated at residue 208 (human) in response to $1,25(OH)_2D_3$
- Lability: proteolytically sensitive
- Tissue abundance: 10 to 100 fmol of VDR/mg protein

Human VDR mRNA
- 4.8 kb (human), 3.0 kb (chicken)

Two important observations on the DNA binding properties the VDR were made that reflected the role of $1,25(OH)_2D_3$ in the receptor activation process [62,63]. First, VDR was capable of binding DNA in the absence of ligand, an observation that suggested that, unlike the latent DNA binding properties of the sex steroid receptors, the VDR was fully capable of binding to DNA in the absence of $1,25(OH)_2D_3$. Second, the "affinity" of the receptor for DNA was quantitatively increased following complex formation with $1,25(OH)_2D_3$. The latter property implied that the structure of the VDR or perhaps the composition of the active receptor was transformed in the presence of the hormonal ligand. More recent studies described later in this chapter as well as in Chapters 9 and 10 more precisely define the nature of the effects of ligand on the VDR and its DNA binding capabilities.

Purification of the VDR initially from chicken intestine [15,56] and later from porcine intestine (16) led to the development of monoclonal antibodies to the VDR. These reagents proved essential in further immunocharacterization of the VDR, provided alternative receptor isolation methods, and ultimately offered the means whereby the VDR gene was cloned. Western blotting analysis revealed for the first time that the precise molecular mass of VDR varied from species to species, ranging from 60,000 Da in the chicken [64] to approximately 50,000 Da in humans [65]. Immunological analyses of this type also revealed that the VDR comprised A an B isoforms in the chicken but an apparently single form in mammalian derived tissues [64]. The two forms of the chicken VDR are now known to arise from the existence of alternative start sites in the mRNA, although the functional consequence of two such proteins with differing amino-terminal extensions remains unknown [66,67]. Finally, immunoprecipitation of the VDR from *in vitro* translated tissue mRNA revealed that whereas the chicken receptor protein was larger than that from mammalian cells, the mRNA transcripts were considerably smaller [65]. The differential size of these transcripts has been confirmed through the use of hybridization techniques incorporating the cDNA for the VDR as a probe.

Initial insights into the structural organization of the VDR emerged using the techniques of limited proteolytic digestion and relied on both the above immunologic probes and tritiated $1,25(OH)_2D_3$. Initial studies by Pike [68] suggested that the VDR epitope for the anti-VDR monoclonal antibody 9A7 lay in or adjacent to a region or domain responsible for DNA binding. Several of the antibodies prepared by Dame *et al.* (16) likewise exhibited similar characteristics. Confronted with the possibility of identifying regions of the receptor responsible for DNA binding and/or $1,25(OH)_2D_3$ binding. Allegretto and co-workers [69,70] employed limited digestion with trypsin to cleave the $1,25(OH)_2D_3$-bound VDR into two domains, one a large fragment of 30 to 40 kDa that retained prebound $1,25(OH)_2D_3$ and a second of 16 to 20 kDa that retained the capacity to bind DNA. The ability to detect the latter fragment through immunologic reactivity with the 9A7 antibody together with the loss of immunologic reactivity in the hormone-binding fragment confirmed a relationship between the epitope for 9A7 and VDR DNA binding. More importantly, it suggested that the two functional domains were separable through a proteolytically sensitive "hinge." The inability to identify the hormone binding domain following trypsin digestion of unliganded receptor also argued strongly for a substantial ligand-induced change in receptor conformation. This finding substantiated earlier observations in tissue extracts of increased stability and decreased lability of VDR following introduction of $1,25(OH)_2D_3$. The ability of carboxypeptidase to reduce modestly the overall mass of the receptor and at the same time release receptor-bound $1,25(OH)_2D_3$ suggested that the DNA binding domain was located amino terminal to the hormone binding domain [71]. As described in the next section, these preliminary insights into the structural organization of the VDR proved correct following its molecular cloning.

IV. STRUCTURAL GENE FOR THE VDR

A. Cloning of the VDR

Molecular cloning of receptors for virtually all the known steroid, thyroid, and vitamin hormones occurred during the latter half of the 1980s [4,5]. Successful cloning of the glucocorticoid receptor gene followed by the cloning of receptor genes for estrogen, progesterone, vitamin D, and androgen was due primarily to the development of antibodies directed against individual receptors. These immunological probes, coupled with the use of a newly introduced λ phage protein expression vector in 1983 [72], provided the technical methodology by which rare mRNAs could be recovered from large cDNA libraries. In addition, sequence analysis of the first cloned receptor cDNAs revealed regions that exhibited significant sequence similarity, particularly in a domain that proved to be associated with DNA binding. Sequence similarities within this region suggested the possibility that related transcripts might be recovered through low stringency hybridization screening techniques. Indeed, the receptor genes for several known ligands such as retinoic acid and aldosterone as well as genes encoding a large number of unknown receptors (orphan receptors) were rapidly recovered by these means [5]. Interestingly, examination of several of these unknown receptors led to the later discovery of novel ligands and the identification of new hormonal systems. Examples are the discovery of 9-*cis*-retinoic acid as the activating ligand for the retinoid X receptors [73], prostaglandin J2 as an activating ligand for the peroxisome proliferator-activated receptor γ [74], and farnesol as a potential stimulator of an orphan protein now termed farnesol-activated receptor [75]. Interestingly, the sequence within the DNA binding domain of the glucocorticoid receptor was first noted to exhibit a high degree of similarity with the viral oncogene v-*erbA* [76]. The cloning of c-*erbA* led to the discovery that this gene encoded the first of two thyroid hormone receptor genes [77].

The anti-VDR monoclonal antibody 9A7 was utilized by McDonnell *et al.* [17] to screen randomly primed chicken intestinal cDNA expression libraries prepared in the viral expression system. A single cDNA clone was selected that produced a protein that exhibited immunological cross-reactivity not only with the screening probe but with an additional anti-VDR antibody as well. The DNA sequence of this cDNA clone and several additional clones recovered through cross-hybridization revealed them to contain a sequence that exhibited a high degree of similarity to a domain located in the glucocorticoid, estrogen, and progesterone receptor genes. This domain was initially believed to be related to that found in the transcription factor TFIIIA and was hypothesized to be responsible for receptor DNA binding. An important repeating module within this region that occurred twice in the domain for the receptors but multiple times in TFIIIA was a zinc-coordinated DNA binding finger structure. It is now known that there exists minimal structural relatedness between the zinc fingers of TFIIIA and the DNA binding domain of the receptors [78], although it is clear that both regions are responsible for their respective DNA binding functions (see below). The presence of this domain and its reactivity to the 9A7 antibody (known to interact in or near the DNA binding domain of the VDR) [62] led to initial confidence that these cDNAs represented a portion of the transcript encoding the VDR. Subsequent hybridization-selected *in vitro* translation techniques using these clones substantiated the authenticity of the cDNA clones [17].

The recovery of these cDNAs constituted the molecular cloning of the VDR. More importantly, they provided the first direct evidence of a structural relationship between the VDR and other bona fide members of the steroid receptor family of genes [79]. The recovery of the first cDNA for the VDR from the chicken enabled subsequent recovery of full-length VDR cDNA transcripts from human [80] tissue sources. The rat intestinal VDR was also cloned independently using monoclonal antibody selection by Burmester *et al.* [81]. Subsequently, sequences of the VDR have been reported from mouse [82], japanese quail [83], and *Xenopus* [84]. Recovery of a cDNA transcript from the human leukemia HL-60 cell line [85] revealed virtual identity to that of the original human VDR cDNA cloned by Baker *et al.* [80]. This provided important evidence that the VDR involved in cellular differentiation was not different from that involved in calcium metabolism.

A comparison of the sequences of the VDR from the above reported cDNAs has revealed that, in addition to having several domains of homology with other members of the nuclear receptor family, the VDR is also highly conserved across tissue sources and species. The overall sequence similarity of rat, mouse, and avian receptors to that of human VDR is 79, 86, and 66%, respectively. However, within specific domains such as the DNA binding domain this rises to above 95%. One substantial difference, noted early based on receptor protein size, is the variability in the extent of the residues lying amino terminal to the DNA binding domain. This region varies from 21 amino acids in the human VDR (the smallest of the VDRs) to approximately 57 amino acids in the chicken protein [71]. Additional inserts within the hinge region are also evident in the rat. The two proteins observed in tissues derived from the chicken can now be accounted for by the presence of two alternative start sites in the chicken mRNA [66] (the first at nucleotide 47 and the second at nucleotide 67, which corresponds to the most 5′ ATG in the human VDR transcript identified by Baker *et al.* [80]).

Two initiation sites also appear to be present in the human cDNA sequence; the second start site, however, lies only three codons downstream of the first [80]. Whether both are used to produce two proteins of almost equivalent mass (424 and 427 amino acids) is unknown. Interestingly, a polymorphism exists in the VDR gene in the human population [86,87]. This polymorphism is a single base pair transition that occurs in the first initiation site reported by Baker *et al.* [80] (ATG to ACG) The result of this polymorphism in humans is the apparent production of a smaller VDR of 424 amino acids. This polymorphism (resulting in a smaller protein) occurs in homozygous form in approximately 37% of the human population and in heterozygous form (one allele containing both initiation sites and the second containing only the downstream site) in approximimtately 48% of the population [88] (Chapter 45). Whether the two proteins exhibit different properties and/or different functions remains an open question. This study also suggests that the polymorphism may be associated with bone mineral density [88]. Both initiation sites are present in the mouse VDR cDNA, but only the upstream site occurs in the rat sequence.

Finally, the size of the cloned cDNAs supports the earlier contention that the avian and human VDR mRNAs differed significantly in size. Thus, whereas the chicken VDR mRNA is approximately 3 kb in length, the human transcript is approximately 4.8 kb in length [17,80]. The fundamental difference in size between chicken and human transcripts lies in the presence of a large 3′ noncoding region of unknown function in the human transcript. Molecular cloning of VDR transcripts thus confirmed a number of initial observations made at the protein level. As discussed in the next section, the cloning also confirmed and extended the original hypothesis that the VDR was a member of the steroid receptor family and enabled significant structure–function analyses to be conducted.

B. VDR Is a Member of the Intracellular Receptor Gene Family

The cloning of glucocorticoid [76] and estrogen [89] receptors in 1985 and 1986 represented the first of a long series of successful efforts to clone each of the known intracellular receptor genes. There now exist over 150 cloned members of this intracellular receptor gene family [7]. The size of this particular gene family eclipses that of any other currently known transcription factor group. It suggests that the common structural motifs within this family that include DNA binding domains paired with activity-regulating domains under the control of chemically diverse small signaling molecules have been highly successful in evolution. These hormonal ligand-activated transcription factors control an incredibly wide range of biological processes that include both growth and differentiation functions in the developing animal as well as a wide array of physiological and homeostatic functions in the adult.

Figure 2 delineates the currently known vertebrate and invertebrate (*Caenorhabditis elegans*, drosophila) members of the nuclear receptor family. Although little is known of the functions of receptors in *C. elegans*, receptors in drosophila play a significant role in morphogenesis [90,91]. At least one of these receptors is acti-

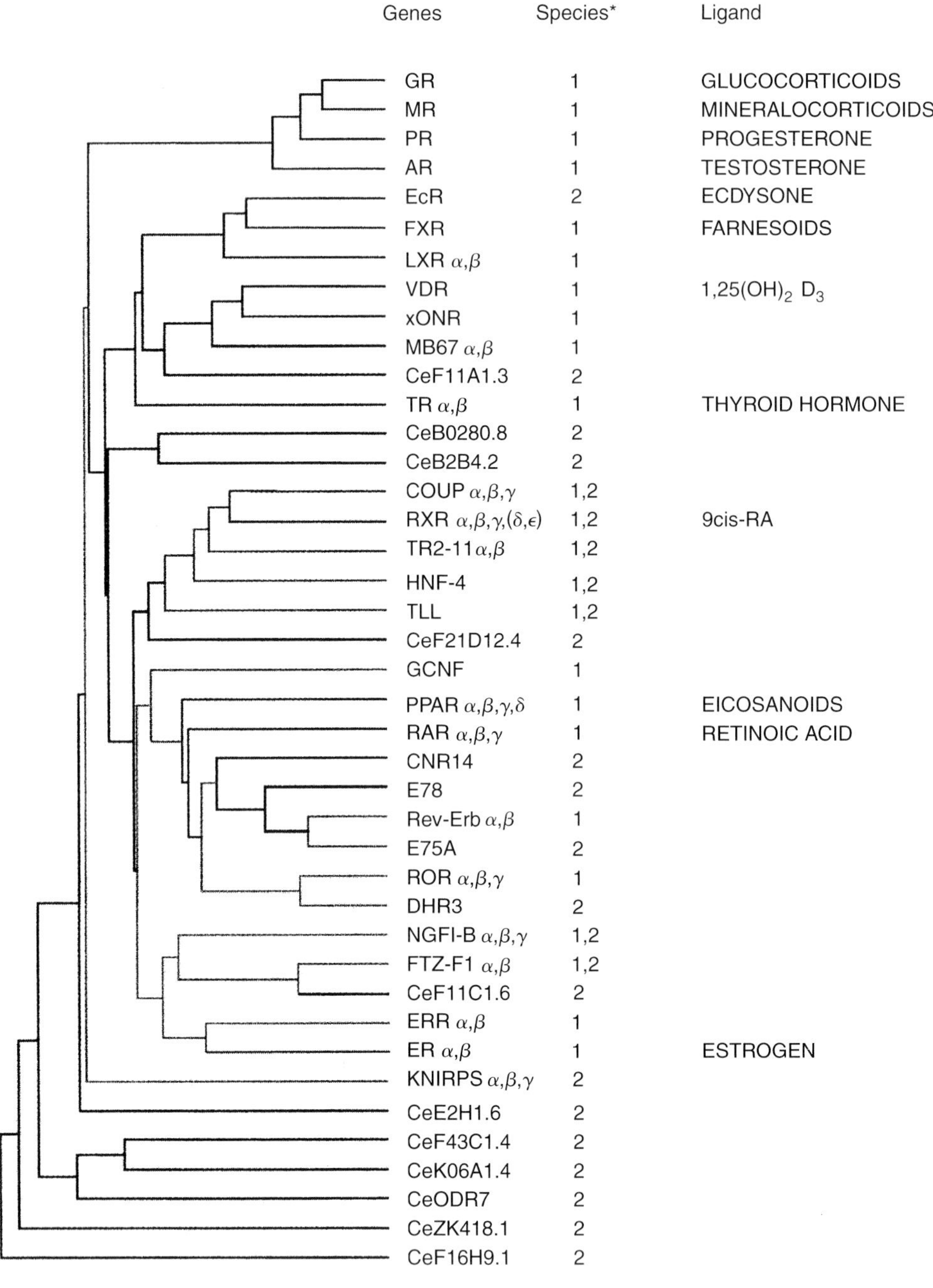

FIGURE 2 Members of the nuclear receptor superfamily of genes, illustrating the relationship between cloned family members based on multiple sequence alignments. This figure represents a modification of that in Mangelsdorf *et al.* [8] wherein further details can be obtained.

vated by ecdysone, a hormone that has long been known to function in metamorphosis [92]. The nuclear receptor family also can be seen in Fig. 2 to include genes for which small signaling molecules are known to exist, such as estrogen, progesterone, thyroid hormone, and vitamin D, as well as orphan receptor genes for which ligands do not exist or have not been discovered. Within the former category, not all the small activators represent endocrine hormones. Certain of these small molecules are specifically synthesized within the cell, such as retinoic acid and prostaglandin J2; others represent intermediates derived from metabolic pathways within cells, such as farnesol [75] or arachadonic acid [74], which may act in either autocrine or paracrine fashion. Although many of the receptors are encoded by a single gene (glucocorticoid receptor, VDR), others are represented by several subtypes, each of which is produced by a separate unique gene. For example, three genes encode the retinoic acid receptor (RARα, RARβ, and RARγ). In addition, isoforms of certain gene products also exist, often the result of alternative splicing and/or promoter usage. The nuclear receptors can be classified further based on the nature of the DNA binding sites with which they interact and on their unique mode of DNA binding. The reader is referred here to other reviews on the steroid receptor family for details [3–8] as well as summaries in Chapters 9, 10, 48, 59, and 60. The interaction of the VDR with DNA is discussed in particular detail in Chapter 9 and is discussed only briefly in the following section.

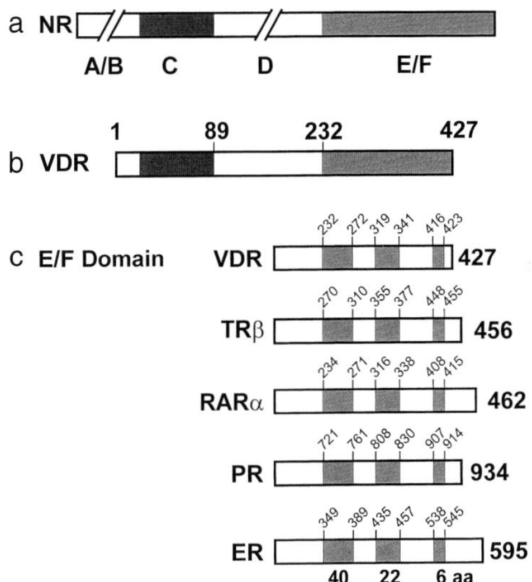

FIGURE 3 Functional domain structure of the nuclear receptor superfamily. (a) The nuclear receptors (NR) are separated into five regions designated A/B, C, D, and E/F. (b) Residue boundaries of corresponding regions within the VDR. (c) Three regions of sequence similarity and residue boundaries within the E/F domain of several nuclear receptors. Shown are thyroid receptor β, (TRβ), retinoid receptor α (RARα), progesterone receptor (PR), and estrogen receptor (ER). Functions associated with the domains include transactivation (A/B), DNA binding (C), flexible hinge (D), and dimerization, ligand binding, transactivation, and repression (E/F).

C. Domains of the VDR

1. Overview

The cloning of the estrogen receptor in 1986 led to a general designation of receptor segments to include A, B, C, D, E, and F domains [89]. As illustrated in Fig. 3a, segment A/B includes residues amino terminal to the DNA binding domain. The C region comprises the highly conserved DNA binding domain. The hinge region, which lies between the C domain and the ligand binding domain, is designated the D domain. Finally, the carboxy-terminal region that contains the ligand binding domain in the activated receptors is termed the E or E/F domain. Three regions of sequence similarity among members of the nuclear receptor family exist within the E region. The F domain is not conserved and exhibits extensive variability.

Figure 3b depicts the domain structure of mammalian VDRs. As can be seen, the A/B domain is highly abbreviated relative to other members of the nuclear receptor family, particularly those for the sex and adrenal steroids. The C region that comprises the DNA binding domain of the VDR represents the most highly conserved domain across all the nuclear receptors [93]. This domain is the hallmark of the nuclear receptor family. The D domain within the VDR appears to link, in a highly flexible fashion, the DNA binding and hormone binding domains [93]. Conservation here is low among the VDRs from different species and is not conserved in either length or sequence with that of other members of the nuclear receptor family [80,81]. Finally, the E/F domain contains the $1,25(OH)_2D_3$ binding function of the VDR [93]. In addition to binding ligand, this domain serves as a highly complex protein–protein interface for a series of additional proteins of varied function [94–96]. These features highlight the role of the VDR as a recruitment center for other transcription factors that contribute to the DNA binding function of the VDR as well as its transcription-regulating functions. As these biological activities are consistent with that of all the members of the receptor gene family, it is not surprising that this extended region contains several subdomains, depicted in Fig. 3c, that exhibit moderate conservation across the entire transcription factor gene family. Like the A/B domain, the F domain is ill-defined within the VDR.

2. DNA Binding Domain

Domain C of the VDR encodes the DNA binding domain. This domain consists of two similar modules, each comprising a zinc-coordinated finger structure. Each zinc atom is tetrahedrally coordinated through four highly conserved cysteine residues and serves to stabilize the finger structure itself. As stated earlier, these finger modules are structurally unrelated to the zinc fingers found in TFIIIA in which the zinc atom is coordinated through two cysteines and two histidines [78,97]. Although the two zinc modules of the VDR appear to be highly related structurally, they are not equivalent topologically because of differing chirality of the residues in each module that coordinate the zinc atom [97]. More importantly, the function of each of these modules in DNA binding is known to be substantially different. Thus, although it is possible that the two exons that encode these modules (see Section VI) evolved from a common ancestral gene through duplication and then diverged as a result of differing selective pressures, the more likely possibility is that the two modules evolved independently. Whereas the amino-terminal module functions to direct specific DNA binding in the major grove of the DNA binding site, the carboxy-terminal module serves as a dimerization interface for interaction with a partner protein [98,99]. In the case of the VDR, at least one of these protein partners is retinoid X receptor (see Section V). As the three-dimensional structure of the DNA-binding domain of several of the receptors has been determined through both nuclear magnetic resonance (NMR) spectroscopy and X-ray crystallography, our understanding of the structural organization of these modules as well as the mechanisms by which they function to interact with DNA is now well advanced [100–105].

3. Ligand Binding Domain

The E/F region of the VDR represents a multifunctional domain that exerts absolute regulatory control on the DNA binding as well as transcription-modifying properties of the VDR. The switch that converts this latent transcription factor into an active gene regulator is $1,25(OH)_2D_3$. Indeed, $1,25(OH)_2D_3$ binding is hypothesized to induce conformational changes in the ligand binding domain of the VDR, much like that of all other small molecule hormones in this class. It is these conformational changes that presumably are responsible for the reduction in proteolytic sensitivity observed by Allegretto and co-workers [69,70]. A more sensitive version of the proteolytic digestion assay has been developed, and application of this assay to an analysis of VDR structural domains has confirmed that VDR binding to $1,25(OH)_2D_3$ results in the appearance of a proteolytically resistant 34-kDa polypeptide largely comprising the E/F domain of the VDR [106].

Whether ligand-induced conformational changes are restricted to the E/F region of the VDR is unknown. Conversion to the active form following hormone binding includes increased formation of dimers that comprise the fundamental DNA binding subunit structure of the VDR as well as exposure of additional regions of the molecule which ultimately allow contact with the core transcriptional machinery. Much is known regarding the former; little is currently known regarding the latter. In addition, it is likely that other protein surfaces are affected that play a direct role in modifying both negatively and positively the activity of the receptor in perhaps cell-specific and gene promoter-specific ways.

The complexity of the ligand-regulated domain coupled to the mechanistic similarities by which the family of nuclear receptors modifies gene expression lead to a prediction that several regions of homology should exist. As observed in Fig. 3c, at least three regions of the VDR E/F domain exhibit significant sequence similarity to E/F domains of other nuclear receptors; these regions are in fact conserved among all family members [107]. Functional mapping studies have suggested that amino acids in the two most amino-terminal regions of homology are essential for dimer formation by the VDR [95,108].

The E/F region has been structurally elucidated through determination of the three-dimensional structure of the ligand binding domains of the retinoid X receptor (RXRα) [109], RARγ [110], and thyroid receptor (TRα1) [111]. The latter two receptors were crystallized in the presence of ligand (holodomains), whereas the RXRα structure was determined in the absence of ligand (apodomain). Twelve α helices (H1–H12) arranged as an antiparallel α-helical sandwich comprise the bulk of the structure of each of the receptors (see Chapter 9 for details). It is likely that the VDR will be arranged in a structurally similar although not identical manner. These three-dimensional structures support the idea that H9 and H10 are essential for the formation of dimers of RAR, RXR, and TR. Although functional studies support the essentiality of these helical sequences in RAR, RXR, and TR interactions, H9 and H10 may not be sufficient for VDR dimerization. This conclusion is supported by a more complete evaluation of the dimerization properties of the carboxy-terminal E/F region of the VDR [95]. These results suggest that, although significant insights will be gained through structural modeling of the VDR after other members of the nuclear receptor family, true insights will require direct structural determination of the VDR.

4. Activation Functions

An additional function inherent to the E/F region of the VDR is an activation function termed AF-2. This function lies within the smallest and most carboxy terminal third homology domain of the receptors and virtually at the carboxy terminus (amino acids 416 to 423) of the VDR (see Fig. 3c) [112,113]. The core of this domain function appears to be associated with H12, although it is clear from activity studies involving mutagenesis that additional components including those in the amino-terminal homology domain of the E/F regions also play a role [95]. An important observation regarding H12 is the clear repositioning of this α-helix back on the hydrophobic core of the E domain after ligand binding.

The interaction between H1 and other α-helices may account for loss of transcriptional capacity following mutagenesis of residues well upstream of H12. Determination of the three-dimensional structure of the VDR will no doubt answer these important questions.

5. LIGAND BINDING AND ACTIVATION

Binding of $1,25(OH)_2D_3$ alters substantially the conformation of the E/F region of the VDR. This hypothesis was suggested by very early studies demonstrating a decrease in the lability of the VDR [114] following ligand binding. More recent studies suggest that the presence of $1,25(OH)_2D_3$ stabilizes the VDR [115,116] through the demonstration that hormone binding increases the resistance of the receptor to proteolytic degradation [106]. Despite these indirect observations, the actual structural rearrangement that occurs on ligand occupancy can only be inferred on the basis of rearrangements that occur in crystallized holoreceptors. Likewise, the nature of the ligand-binding pocket of the VDR remains undefined. Loss of function studies demonstrate that mutagenesis of a number of amino acids throughout the entire E/F region can produce an alteration in $1,25(OH)_2D_3$ binding. This suggests that the three-dimensional binding pocket comprises many segments spanning the entire carboxy-terminal domain.

Because the function of $1,25(OH)_2D_3$ is to act as a small molecular switch capable of receptor activation, it should not be surprising that the binding of ligand would induce sweeping conformational changes in the E/F domain. The exact positioning of the $1,25(OH)_2D_3$ molecule within the ligand pocket awaits the solution of the VDR structure. It should be anticipated, however, that at least some of the residue contact sites that serve to stabilize the natural hormone within the pocket will not be identical to those that stabilize the binding of lower affinity vitamin D metabolites such as $24R,25(OH)_2D_3$ or synthetic analogs such as 20-epi-$1,25(OH)_2D_3$ and 1,25-dihydroxy-16-ene-23-yne-vitamin D_3 (see Chapter 60). The theoretical result of occupancy of the VDR by ligands other than $1,25(OH)_2D_3$ is a spectrum of receptor conformations potentially capable of unique and perhaps selective biological actions. The potential for this to occur is described in more detail in Chapters 59 and 60 and may account, at least in part, for the interesting and selective biological actions of numerous new metabolites and analogs. Although still theoretical for the VDR, the concept of ligand-induced conformational specificity that results in unique biological actions is now well established for several of the sex steroids [117,118]. The discovery of ligands with unique properties has resulted in broad-based therapeutic opportunities.

V. FUNCTIONAL ANALYSIS OF THE VDR

A. Osteocalcin Regulation as a Model

Osteocalcin is a small abundant noncollagenous bone protein whose extract function remains unclear. Genetic ablation of the osteocalcin gene suggests that this osteoblast-specific protein contributes to the density and structural integrity of bone [119]. Despite the uncertainty surrounding the function of osteocalcin, a broad number of cytokines, growth factors, and systemic hormones control its expression. One of the most potent regulators of osteocalcin production is $1,25(OH)_2D_3$ [120,121]. This fact together with the cloning of the human osteocalcin gene and its promoter in 1986 [122] provided a unique blend of opportunities for researchers to study the molecular determinants through which $1,25(OH)_2D_3$ modulated the expression of this gene.

Initial investigations demonstrated that the activity of $1,25(OH)_2D_3$ on the osteocalcin promoter was direct [93,123,124]. Thus, introduction of a plasmid containing a large upstream fragment of the human osteocalcin promoter (fused to the reporter gene chloramphenicol acetyltransferase) into osteoblast-like osteosarcoma cells revealed that the activity of the chimeric gene was sensitive to $1,25(OH)_2D_3$. An abbreviated version of this upstream sequence containing less than several hundred base pairs exhibited only basal activity [125]. Kerner *et al.* [123] initially localized the cis-acting element to a region approximately 500 bp upstream of the transcriptional start site. This study and an additional one by Ozono *et al.* [126] helped define the first vitamin D response element (VDRE) as a directly repeated hexanucleotide sequence separated by three base pairs. Parallel studies using the rat osteocalcin gene promoter led to a similar conclusion regarding the organizational motif of the VDRE [127–129]. These studies collectively provided the first insight regarding a specific DNA sequence that mediated vitamin D-inducible action.

Since these experiments were carried out, several additional genes have been explored for their sensitivity to $1,25(OH)_2D_3$, including genes for mouse osteopontin [130], mouse calbindin-D_{28K} [131], rat calbindin D_{9K} [132], rat [133,134] and human [135] 25-hydroxyvitamin D_3 24-hydroxylase (two apparent VDREs), and human p21 [136]. A list of the sequences that have been shown to mediate vitamin D action as well as their locations within the promoters are documented in Table III. It is clear from inspection of these sequences that a "typical" VDRE is comprised of two hexad repeats separated by

Table III Location and Sequence of Positive Natural Vitamin D Response Elements

Gene	Location	Nucleotide sequence
Rat osteocalcin	−460/−446	GGGTGA atg AGGACA
Human osteocalcin	−499/−485	GGGTGA acg GGGGCA
Mouse osteopontin	−757/−743	GGTTCA cga GGTTCA
Rat calbindin-D_{9K}	−489/−475	GGGTGT cgg AAGCCC
Mouse calbindin-D_{28K}	−198/−183	GGGGGA tgtg AGGAGA
Rat 24-hydroxylase	−150/−136 (proximal)	AGGTGA gtg AGGGCG
	−258/−244 (distal)	GGTTCA gcg GGTGCG
Human 24-hydroxylase	−169/−155 (proximal)	AGGTGA gcg AGGGCG
	−291/−277 (distal)	AGTTCA ccg GGTGTG
Rat Pit 1	−67/−52	AGTTCA gcga AGTTCA
Human p21	−779/−765	AGGGAG att GGTTCA

a spacer of 3 bp. Whereas the sequence of the "spacer" appears not be conserved, the general consensus hexad is AGGTCA or preferentially GGTTCA. Considerable variability in these hexad sequences is apparent, however.

The above efforts to define *cis*-acting elements that mediate vitamin D action contributed in part to the current view of hormone response elements [137]. The overall nature of these response elements permits classification of the DNA sites and thus the nuclear receptor family into three categories: palindromic half-sites that interact with sex steroid receptors, directly repeated half-sites that interact with the small receptors represented by the VDR, retinoic acid, and thyroid receptors, and single half-sites that mediate the actions of monomeric receptors such as NGFI-B [8]. As discussed below, whether a receptor functions on a repeated half-site as a homodimer or heterodimer further categorizes DNA binding. The reader is referred elsewhere [3–8] for complete review of the nature of these DNA binding sites.

B. VDR Binding to Specific DNA *in Vitro*

Two domains within the VDR are required for high affinity DNA binding *in vitro*, the C domain (DNA binding domain per se) and the E region (carboxy-terminal ligand-binding domain). This conclusion is based on an extensive battery of mutations that have been introduced into the VDR by numerous investigators. Thus, point mutations that lead to amino acid changes in the zinc finger modules [138–145] as well as mutations that alter or delete residues across a majority of the carboxy terminus [93,95,108,112] can block DNA binding. Interestingly, the molecular basis for abrogation of DNA binding by mutations in the two regions is different. In the first case, alterations in the DNA binding domain directly prevent interaction with DNA (see Chapter 9 for details of specific DNA binding). In the second instance, carboxy-terminal mutations block the ability of the receptor to form dimers that are in turn capable of DNA binding of high affinity, selectivity, and cooperativity [95,108].

The requirement that the VDR must form dimers in order to interact with DNA was suggested by the repeated half-site nature of VDREs. The surprising finding, however, was that the VDR bound to DNA not as a dimer but rather as a heterodimer (see Fig. 4). Liao *et al.* [146] and Sone *et al.* [115,138,147] observed that, although the VDR derived from mammalian cell extracts was fully capable of binding to DNA *in vitro*, the production of the VDR either through *in vitro* transcription/translation reactions or through recombinant means from nonmammalian cell sources such as yeast failed to produce a DNA binding-competent VDR. The addition of mammalian cell extract to yeast extract-derived VDRs, however, led to recovery of VDR DNA binding, suggesting the necessity for a DNA binding facilitator. The requirement for this factor, which could be found in a variety of tissues and cell sources [147], was confirmed by others [148,149]. This factor(s) was termed nuclear accessory factor (NAF). Nuclear accessory factors were simultaneously discovered for other nuclear receptors including the thyroid receptor and the retinoic acid receptor. In 1991 and 1992, Yu *et al.* [150],

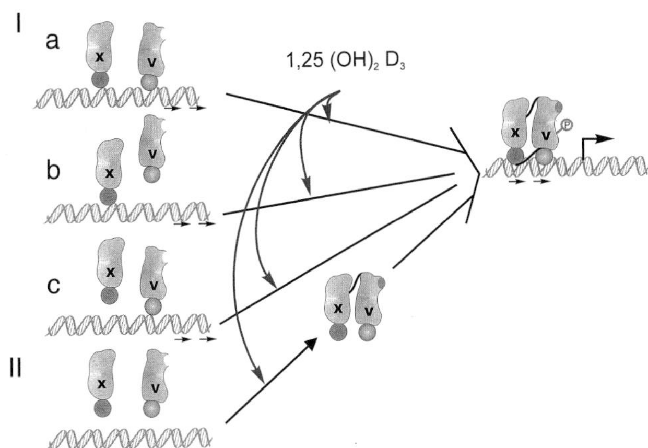

FIGURE 4 Role for $1,25(OH)_2D_3$ in the formation of the functional VDR heterodimer. Monomers of VDR (v) and a partner protein such as RXR (x) are either associated with DNA (I) or free in the nucleus (II). Binding of $1,25(OH)_2D_3$ to the VDR increases the affinity of the receptor for its partner, leading to the formation of dimers that associate with specific binding sites within vitamin D responsive gene promoters. Short arrows indicate DNA sequence half-sites for VDR binding in a responsive gene promoter, and P represents a phosphorylation site on serine-208 of human VDR.

Leid *et al.* [151], Zhang *et al.* [152], and Kliewer *et al.* [153] discovered that a previously cloned member of the nuclear receptor family, namely, retinoid X receptor (RXR), was the likely protein partner for the above nuclear receptors as well as for VDR. Thus, NAF was defined as the cumulative partnering activity of the RXRα, RXRβ, and/or RXRγ subtypes expressed in a given cell type.

Considerable efforts have gone into defining the nature, function, and complexity of receptor heterodimer DNA binding activity, and the reader is referred to reviews that deal with this complex issue [4,7]. It is now widely believed, however, that the two protein partners represent subunits that contribute to the activation process despite the fact that only the signaling partner (in this case the VDR) is activated by ligand. True dimerization between the VDR and RXR has been confirmed through definition of the VDR dimerization domain [95,108]. Jin *et al.* [95] utilized an extensive series of internal deletions of the VDR to define two regions essential for interaction with RXR. These regions coincide with the two moderately conserved subregions of homology located in the E/F domain of the VDR. Interestingly, although study by Perlmann *et al.* [154] suggests that a small region of 40 amino acids lying within the second E/F homology domain (corresponding to H9 and H10 of the crystal structure of the RXR ligand binding domain) is sufficient in RXR to form dimers with the retinoic acid and the thyroid hormone receptors, this same region is not sufficient to permit formation of RXR–VDR heterodimers. This finding suggests that the domains responsible for interaction between RXR and certain of the other signaling partners may be different. The deletion analysis described by Jin *et al.* [95], however, was unable to distinguish between structural abnormalities induced by the deletion and the actual dimerization domain itself. Thus, elucidation of the three-dimensional structure of VDR–RXR ligand binding domain dimers will still be required to define precisely the interaction between the two proteins. The existence of a single permissive partner protein that functions as a master or at least a central regulator for several endocrine systems suggests that substantial cross talk may exist between these systems.

C. Polarity of DNA Binding

The asymmetric nature of natural VDREs (direct repeats) coupled with the heteromeric nature of the receptor activation unit (VDR–RXR) indicates that the two receptor subunits must bind to the VDRE with a defined polarity. Studies by Jin and Pike [94] and Freedman and co-workers [140] addressed this question in detail. Through the use of chimeric receptors and chimeric response elements, it is now clear that RXR binds to the upstream 5′ half-element and the VDR binds to the downstream 3′ half-element of VDREs oriented on the DNA sense strand as illustrated in Fig. 4. This organization is consistent with the relative polarity noted for both RXR–thyroid receptor [155] and RXR–retinoic acid receptor heterodimers [156] bound to their respective response elements. For further details regarding the DNA binding polarity of VDR the reader is referred to Chapter 9.

D. Transactivation by the VDR

Early studies that introduced the human osteocalcin promoter into cell lines containing the VDR enabled evaluation of the elements which act *in cis* to mediate vitamin D action. Nevertheless, it was the capacity to introduce both the osteocalcin promoter and the VDR into a cell line initially devoid of endogenous VDR that established an absolute requirement for that protein in vitamin D action. Studies by McDonnell *et al.* [93] first demonstrated that whereas the human osteocalcin gene promoter was unresponsive to $1,25(OH)_2D_3$ when introduced into a VDR negative cell line (CV-1), introduction of an expression vector for the VDR permitted the recovery of vitamin D response. This biological assay enabled further examination of the transcriptional activity of the VDR, particularly as it related to DNA binding and dimerization functions. Indeed, each of the mutations within the VDR that were examined for DNA binding and dimerization *in vitro* led to transcriptionally inactive receptors in intact cells. These studies therefore confirm the crucial nature of both functional activities of the VDR in transactivation. This assay was also essential in establishing the inactive nature of many of the mutant VDRs that were identified in the human syndrome of hereditary $1,25(OH)_2D_3$-resistant rickets (HVDRR) considered in more detail in Chapter 48 [143–145].

The transactivation assay described above was essential in characterizing many features of the VDR. However, the ubiquitous expression of endogenous NAF (RXR) in mammalian cell lines precluded an experiment designed to unequivocally demonstrate a requirement for NAF in VDR function in intact cells. As a result of this, Jin and Pike [94] utilized yeast to recreate the VDR transcriptional response unit and to test for the requirement of RXRs in VDR-induced transcription. Although eukaryotic in nature, yeast do not express either VDR or the RXR genes. Yeast therefore represent a potential opportunity to evaluate a requirement for RXR by introducing each of the receptors into this cell background via recombinant means. Whereas VDR exhibited little capacity to activate a chimeric gene promoter that contained a fused osteocalcin VDRE sequence upstream of the yeast Cyc-1 promoter, the introduction of an RXR expression vector dramatically stimulated the ability of VDR to activate transcription. This study, as well as additional efforts that examined the role of the RXR ligand 9-*cis*-retinoic acid [141,157] on VDR function, lends additional support to the notion that RXR represents an essential partner for the VDR in intact cells.

The participation of the VDR in transcriptional activation likely requires additional regulatory proteins. Unique transactivation domains have been identified within many eukaryotic transcription factors, and the nuclear receptor superfamily is no exception. These domains represent interactive surfaces on the receptor or protein that facilitate the protein–protein interactions necessary for contact with the core promoter machinery of a gene. Indeed, mutations that compromise the transcriptional regulating capacity of the VDR without abrogating either DNA binding or dimerization have been defined [95,112]. These mutations lie at the extreme carboxy terminus of the VDR [112] and may be analogous to the activation function 2 (AF-2) domain of other members of the receptor family [113]. Activation is not restricted to this region, however. Jin *et al.* [95] utilized the power of a yeast selection system to define single mutations within the VDR that selectively compromise transactivation. These mutations lie within the two most amino-terminal homology domains found in the E region of the VDR. How these mutations relate to the proposed AF-2 region of the VDR remain unknown. A protein(s) that interacts with the AF-2 domain of the VDR and presumably mediates contact with the core promoter has yet to be discovered. The additional observation that the core promoter transcription factor TFIIB can associate with the VDR [96,158] suggests that the biochemical mechanisms by which the VDR contributes regulatory inputs into the basal transcriptional apparatus will be multifaceted and complex. The reader is referred to additional details regarding the interaction of the VDR with potential comodulators and the core transcription factor TFIIB in Chapters 9 and 10.

E. Role of $1,25(OH)_2D_3$ in VDR Activation

Most of the nuclear receptors that have well-defined biological roles are activated by ligands, the vast majority of which are true endocrine hormones. Despite the knowledge that this activation event occurs, an understanding of the mechanisms by which an otherwise latent transcription factor is converted to an active form is only beginning to emerge. Nevertheless, ligand-induced receptor conformational changes provide the theoretical basis for activation and likely drive many if not all the downstream events inherent to the acquisition of transcriptional activity. A key event precipitated by activating ligands for the sex steroid receptor subfamily is the dissociation of an inhibitory complex composed of various heat-shock proteins [159]. This permits homodimer formation by the receptor and subsequent DNA binding [3]. Other events such as induction of phosphorylation, interaction with other transcription factors, and contact with the core transcriptional machinery also probably occur as a result of structural changes in the protein. The VDR, in contrast, is not found associated with heat-shock proteins.

The discovery that the VDR requires a dimerization partner (NAF/RXR) for VDRE binding and that this interaction is a ligand-modulated phenomenon led to the hypothesis that at least one role for $1,25(OH)_2D_3$ might be to promote an increase in affinity of the VDR for its partner. Sone *et al.* [138] provided initial support for this hypothesis by demonstrating that in the absence of DNA the affinity of the VDR for NAF/RXR increased 9-fold in response to $1,25(OH)_2D_3$. This increase in affinity of the VDR for NAF almost certainly reflects a conformational change in the VDR induced by $1,25(OH)_2D_3$ (discussed earlier in Section III). The effect of ligand on dimerization has been confirmed between VDR and RXR partners through surface plasmon resonance techniques [160]. These experiments collectively support a fundamental role for $1,25(OH)_2D_3$ in promoting formation of an active VDR–RXR heterodimer as outlined in the model in Fig. 4. Whether the receptors are bound to DNA or free in the nucleoplasm prior to heterodimer formation is unknown, as are other downstream events that are required for eventual gene activation. Interestingly, the VDR appears not to require ligand for activation in yeast [94,95]. This observation hints at the existence in mammalian cells of an inhibitor, analogous to those which regulate the steroid receptors, or a transcriptional repressor that might be released on ligand binding. It remains for future studies to define additional important events that are initiated through binding of $1,25(OH)_2D_3$ to VDR.

VI. THE HUMAN VDR CHROMOSOMAL GENE

A. Organization of the Gene

The cloning of the VDR structural gene in 1987 precipitated over a decade of highly productive research on the mechanism of action of vitamin D. Interestingly, unlike many of the other members of the nuclear receptor family which are products of multiple genes, the VDR itself remains the apparent product of a single gene. The human VDR gene is reported to lie on chromosome 12 [161]. The initial organization of the intron–exon structure of human VDR chromosomal gene corresponding to the sequence of the VDR reported by Baker *et al.* [80] was determined in 1988 [143,162]. Additional efforts have defined the complete structure of the gene [163].

As depicted in Fig. 5, restriction mapping of several λ clones and a series of four recovered human cosmid clones coupled to nucleotide sequence analysis of relevant portions of these clones revealed a gene spanning over 75 kb of DNA. Eight exons comprise the coding sequence of the VDR protein. The first of these is exon 2, which contains the most proximal 3 bp of the 5′ noncoding sequence, the translation start site, and nucleotide sequence that encodes the first zinc finger module. Exon 3 which lies approximately 15 kb downstream encodes the second zinc finger module. Exons 4, 5, and 6 encode the D region or hinge. Exons 6, 7, 8, and 9 enclode a portion of the hinge and the carboxy-terminal E/F region as well as approximately 3200 nucleotides of 3′ noncoding sequence.

A view of the 5′ end of this large gene and its promoter is beginning to emerge. Two short exons that lie upstream of exon 2 account for the known 5′ noncoding sequence reported by Baker *et al.* [80]. These exons, termed 1a and 1c, are 77 and 81 bp in length. The promoter lies immediately upstream of exon 1a and is clearly characterized by its GC-rich nature and the absence of a TATA box. Interestingly, an exon of 121 bp not found in the originally reported sequence of Baker *et al.* [80] and termed exon 1b is located 4.5 kb downstream of exon 1a. Variable use of exons 1b and 1c leads to the production of alternatively spliced mRNAs whose nature and function remain unknown. The relationship between these exons, the full-length mRNA, and the VDR protein itself is illustrated in Fig. 5. Although examination of the promoter for this gene is only beginning, the apparent complexity within the 5′ end of the gene suggests the possibility that important determinants of expression of this gene will be located not only upstream but also within intron sequences. The latter will be technically difficult to identify. Despite this, it is clear that the human chromosomal gene for the VDR is not unlike other steroid receptor genes in size, exon organization, and promoter complexity.

B. Polymorphisms within the VDR Gene

Genetic polymorphisms have been identified within the human VDR gene. The first represents a C to T transition in the translation initiation site located in exon 2 [87]. The presence of a C in this position results in the initiation of translation at a site three amino acids downstream and in the production of a gene product three amino acids shorter. The distribution of this polymorphism in the human population and a potential relationship between the frequency of this polymorphism and bone mineral density was discussed in Section III of this chapter [88]. Genetic polymorphisms have also been defined within introns between exons 7 and 8 as well as within the 3′ noncoding region of exon 9 [164]. These polymorphisms appear also to correlate with bone mineral density in several human populations and are hypothesized to be predictive for osteoporosis. They are not, however, located in regions of the gene that might affect the structure of the protein. Although these observations are highly controversial at present, it is anticipated that the validity of this hypothesis as well as a mechanistic basis for the correlation will be forthcoming in the near future. This topic and associated references are considered in depth in Chapter 45.

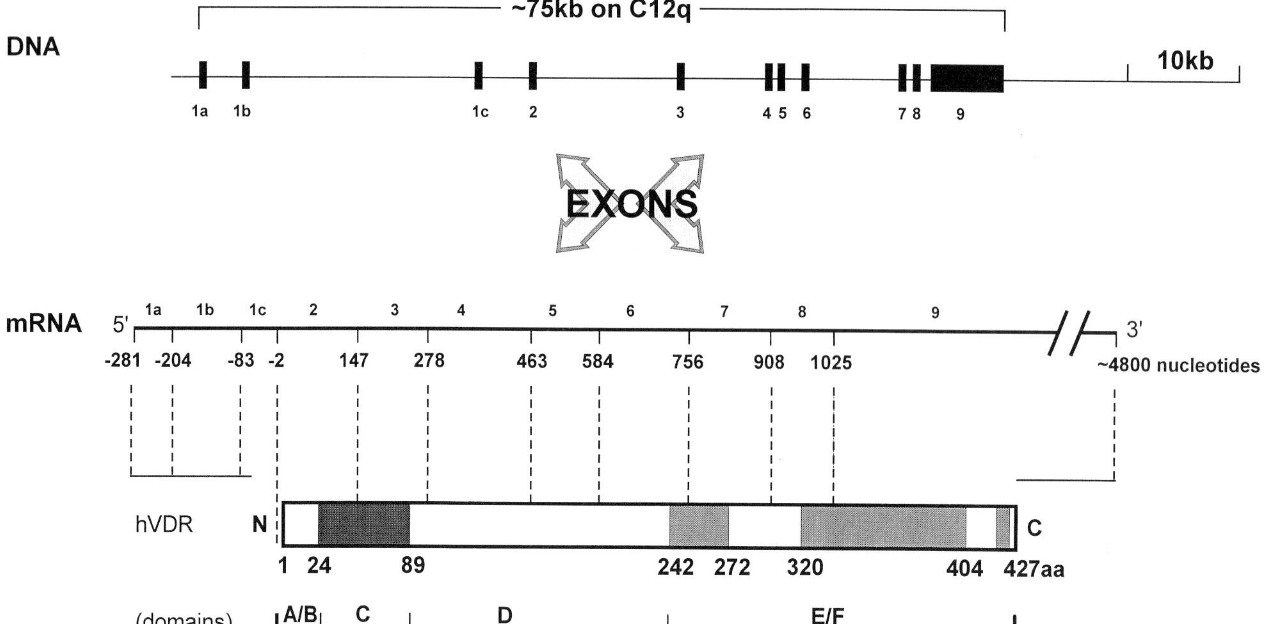

FIGURE 5 Structural organization of the human chromosomal VDR gene. The human VDR gene locus (DNA) is composed of 11 exons (1a, 1b, 1c, 2 through 9) spanning approximately 75 kb of DNA. A 10-kb scale is indicated at right. The location of exons relative to the mRNA transcript of ~4800 nucleotides (mRNA) and the encoded VDR protein of 427 amino acids (hVDR) is illustrated. With regard to the hVDR mRNA, negative numbers indicate 5′ noncoding nucleotides and positive numbers indicate protein encoding nucleotides beginning with +1 indicated by Baker *et al.* [80]. Numbers below the hVDR protein indicate the amino acid residue boundaries of shaded homology domains. Regions of functionality are designated A/B, C, D, and E/F as in Fig. 3.

VII. CONCLUDING COMMENTS

The basic elements of the mechanism of action of vitamin D have been defined. The pace of exploration into the actions of vitamin D has accelerated enormously since the mid 1980s largely as a result of the molecular cloning of the VDR in 1987 but also as a result of the cloning and availability of vitamin D target genes. As described in this chapter, we have gained considerable insight into the structure of the VDR and its compartmentalization into definable functional domains. The availability of recombinant clones has allowed investigation of the interaction of the VDR with vitamin D-inducible gene promoters and definition of vitamin D responsive elements or VDREs. Further studies have revealed that the VDR requires a protein partner for DNA binding in the form of RXR, a central regulator of several additional nuclear receptors. Although additional research will be necessary, a preliminary understanding of the role of receptor and its ligand in the regulation of transcription has begun to emerge. These insights are currently being utilized to gain an understanding of the tissue-selective mechanisms of action of a new generation of vitamin D analogs under consideration as therapeutic agents for a broad range of indications that include skin diseases, immunologic disorders, and cancer. The cloning of the VDR enabled the recovery of its chromosomal gene. Subsequent investigations into the nature of the human syndrome of hereditary $1,25(OH)_2D_3$-resistant rickets (HVDRR), made possible through a complete characterization of the VDR gene itself [143], revealed the underlying cause to be mutations in the gene so that dysfunctional proteins are produced. This discovery, together with the reported genetic ablation of the VDR gene in mice [165] that clearly mimics physiologically the syndrome of HVDRR, confirms the central role of the VDR in the regulation of mineral metabolism.

The future holds much promise. It is likely that the three-dimensional structure of the VDR will be determined, revealing its organization, the nature of the ligand binding site, and the changes that are induced in the protein on ligand binding. It is likely that new proteins which play a role in the vitamin D activation pathway by facilitating interaction of the receptor with the core gene promoter elements will be identified and cloned. It is at this level that a better understanding of the selective actions of vitamin D analogs will emerge. Also to emerge will be a better understanding of the regulation of the VDR at both the transcriptional and posttranslational levels and its contribution to the mechanism of vitamin D action. Perhaps more important than these molecular details is the likelihood that we will achieve a better understanding of how vitamin D controls directly as well as indirectly the expression of broad networks of genes which in turn are responsible for cellular and tissue activities. An area of particular focus will almost certainly be how the vitamin D hormone controls cell proliferation and differentiation. Thus, while much progress has been made, there is much yet to understand.

References

1. Jensen EV, Jacobson HI 1962 Basic guides to the mechanism of estrogen action. Recent Prog Horm Res **18**:387–401.
2. O'Malley BW, McGuire WL, Kohler PO, Korenman SG 1969 Studies on the mechanism of steroid hormone regulation of specific proteins. Recent Prog Horm Res **25**:105–160.
3. Beato M 1989 Gene regulation by steroid hormones. Cell **56**:335–344.
4. Evans RM 1988 The steroid and thyroid hormone receptor superfamily. Science **240**:889–895.
5. O'Malley BW 1990 The steroid receptor superfamily: More excitement predicted for the future. Mol Endocrinol **4**:363–369.
6. Beato M, Herrliche P, Schutz G 1995 Steroid hormone receptors: Many actors in search of a plot. Cell **83**:851–857.
7. Mangelsdorf DJ, Evans RM 1995 The RXR heterodimer and orphan receptors. Cell **83**:841–850.
8. Mangelsdorf DJ, Thummel C, Beato M, Herrliche P, Schultz G, Umesono K, Blumberg B, Kastner P, Mark M, Chambon P, Evans RM 1995 The nuclear receptor superfamily: The second decade. Cell **83**:835–839.
9. Norman AW 1968 The mode of action of vitamin D. Biol Rev **243**:4055–4064.
10. Lawson DEM, Wilson PW, Kodicek E, Morrison HR, Williams DH 1971 Identification of 1,25-dihydroxycholecalciferol, a new kidney hormone controlling calcium metabolism. Nature **230**:228–230.
11. Norman AW, Myrtle JF, Midgett RJ, Nowicki HG, Williams V, Popjak G 1971 1,25-Dihydroxycholecalciferol: Identification of the proposed active form of vitamin D in the intestine. Science **173**:51–54.
12. Holick MF, Schnoes HK, DeLuca HF, Suda T, Cousins RJ 1971 Isolation and identification of 1,25-dihydroxycholecalciferol. A metabolite of vitamin D active in intestine. Biochemistry **10**:2799–2804.
13. Fraser DR, Kodicek E 1970 Unique biosynthesis by kidney of a biologically active vitamin D metabolite. Nature **228**:764–766.
14. Haussler MR, Norman AW 1969 Chromosomal receptor for a vitamin D metabolite. Proc Natl Acad Sci USA **62**:155–162.
15. Pike JW, Marion SL, Donaldson CA, Haussler MR 1983 Serum and monoclonal antibodies against the chick intestinal receptor for 1,25-dihydroxyvitamin D_3: Generation by a preparation enriched in a 64,000 dalton protein. J Biol Chem **258**:1289–1296.
16. Dame MC, Pierce EA, Prahl JM, Hayes CE, DeLuca HF 1986 Monoclonal antibodies to the porcine intestinal receptor for 1,25-dihydroxyvitamin D_3: Interaction with distinct receptor domains. Biochemistry **25**:4523–4534.
17. McDonnell DP, Mangelsdorf DJ, Pike JW, Haussler MR, O'Malley BW 1987 Molecular cloning of complementary DNA encoding the avian receptor for vitamin D. Science **235**:1214–1217.
18. Brumbaugh PF, Haussler MR 1974 1,25-Dihydroxycholecalciferol receptors in intestine. Temperature-dependent transfer of the hormone to chromatin via a specific receptor. J Biol Chem **249**:1258–1262.
19. Lawson DEM, Wilson PW 1974 Intranuclear localization and receptor proteins for 1,25-dihydroxycholecalciferol in chick intestine. Biochem J **144**:573–583.
20. Brumbaugh PF, Haussler MR 1975 Specific binding of 1,25-dihydroxycholecalciferol to nuclear components of chick intestine. J Biol Chem **250**:1588–1594.
21. Kream BE, Yamada Y, Schnoes HK, DeLuca HF 1977 Specific cytosol-binding protein for 1,25-dihydroxyvitamin D_3 in rat intestine. J Biol Chem **254**:9488–9491.
22. Brumbaugh PF, Hughes MR, Haussler MR 1975 Cytoplasmic and nuclear components for 1,25-dihydroxyvitamin D_3 in chick parathyroid glands. Proc Natl Acad Sci USA **72**:4871–4875.
23. Pike JW, Gooze LL, Haussler MR 1980 Biochemical evidence for 1,25-dihydroxyvitamin D receptor macromolecules in parathyroid, pancreas, pituitary, and placental tissues. Life Sci **26**:407–414.
24. Kream BE, Jose M, Yamada S, DeLuca HF 1977 A specific high affinity binding macromolecule for 1,25-dihydroxyvitamin D_3 in fetal bone. Science **197**:1086–1088.
25. Chandler JS, Pike JW, Haussler MR 1979 1,25-Dihydroxyvitamin D_3 receptors in rat kidney cytosol. Biochem Biophys Res Commun **90**:1057–1063.
26. Colston K, Feldman D 1980 Nuclear translocation of the 1,25-dihydroxycholecalciferol receptor in mouse kidney. J Biol Chem **255**:7510–7513.
27. Abe E, Miyaura C, Sakagami H, Takeda M, Konno K, Yamazaki T, Yashiki S, Suda T 1981 Differentiation of mouse myeloid leukemia cells induced by 1,25-dihydroxyvitamin D_3. Proc Natl Acad Sci USA **78**:4990–4994.
28. Shavit ZB, Teitlebaum SL, Reitsma P, Hall A, Pegg LE, Trial J, Kahn AJ 1983 Induction of monocytic differentiation and bone resorption by 1,25-dihydroxyvitamin D_3. Proc Natl Acad Sci USA **80**:5907–5911.
29. Christakos S, Norman AW 1981 Biochemical characterization of 1,25-dihydroxyvitamin D_3 receptors in chick pancreas and kidney cytosols. Endocrinology **108**:140–149.
30. Haussler MR, Manolagas SC, Deftos L 1980 Evidence for a 1,25-dihydroxyvitamin D_3 receptor-like macromolecule in rat pituitary. J Biol Chem **255**:5007–5010.
31. Dokoh S, Donaldson CA, Marion SL, Pike JW, Haussler MR 1983 The ovary: A target organ for 1,25-dihydroxyvitamin D_3. Endocrinology **112**:200–206.
32. Merke J, Kreusser W, Bier B, Ritz E 1983 Demonstration and characterization of a testicular receptor for 1,25-dihydroxyvitamin D_3 in the rat. Eur J Biochem **130**:303–308.
33. Colston K, Hirst M, Feldman D 1980 Organ distribution of the cytoplasmic 1,25-dihydroxycholecalciferol receptor in various mouse tissues. Endocrinology **107**:1916–1922.
34. Walters MR, Wicker DC, Riggle PC 1986 1,25-Dihydroxyvitamin D_3 receptor identified in the rat heart. J Mol Cell Cardiol **18**:67–72.
35. Colston K, Colston MJ, Fieldsteel AH, Feldman D 1982 1,25-Dihydroxyvitamin D receptors in human epithelial cancer cell lines. Cancer Res **42**:856–859.
36. Colston K, Colston MJ, Feldman D 1981 1,25-Dihydroxyvitamin D and malignant melanoma: The presence of receptors and inhibition of cell growth in culture. Endocrinology **108**:1083–1086.
37. Eisman JA, Martin TJ, MacIntyre I, Frampton RJ, Moseley JM, Whitehead R 1980 1,25-Dihydroxyvitamin D_3 receptors in a cultured human breast cancer cell line (MCF-7). Biochem Biophys Res Commun **93**:9–15.

38. Haussler MR 1986 Vitamin D receptor: Nature and function. Annu Rev Nutr **6**:527–562.
39. Feldman D, Chen T, Hirst M, Colston K, Karasek M, Cone C 1980 Demonstration of 1,25-dihydroxyvitamin D receptors in human skin biopsies. J Clin Endocrinol Metab **51**:1463–1465.
40. Provvedini DM, Tsoukas CD, Deftos LJ, Manolagas SC 1983 1,25-Dihydroxyvitamin D_3 receptors in human leucocytes. Science **221**:1181–1183.
41. Bhalla AK, Amento EP, Clemens TL, Holick MF, Krane SM 1983 Specific high affinity receptors for 1,25-dihydroxyvitamin D_3 in human peripheral blood mononuclear cells: Presence in monocytes and induction in T lymphocytes following activation. J Clin Endocrinol Metab **57**:1308–1310.
42. Krageballe K 1992 Treatment of psoriasis with calcipotriol and other vitamin D analogues. J Am Acad Dermatol **27**:1001–1008.
43. Simpson RU, DeLuca HF 1980 Characterization of a receptor-like protein for 1,25-dihydroxyvitamin D_3 in rat skin. Proc Natl Acad Sci USA **77**:5822–5827.
44. Zile M, Barsness EC, Yamada S, Schnoes HK, DeLuca HF 1976 Localization of 1,25-dihydroxyvitamin D_3 in intestinal nuclei *in vivo*. Arch Biochem Biophys **186**:15–24.
45. Jones PG, Haussler MR 1979 Scintillation autoradiography localization of 1,25-dihydroxyvitamin D_3 in chick intestine. Endocrinology **104**:313–321.
46. Stumpf WE, Sar M, Reid FA, Tanaka Y, DeLuca HF 1979 Target cells for 1,25-dihydroxyvitamin D_3 in intestinal tract, stomach, kidney, skin, pituitary, and parathyroids. Science **206**:1188–1190.
47. Stumpf WE, Sar M, Clark SA, DeLuca HF 1982 Brain target sites for 1,25-dihydroxyvitamin D_3. Science **215**:1403–1405.
48. Clemens TL, Garrett KP, Zhou XY, Pike JW, Haussler MR, Dempster DW 1988 Immunocytochemical localization of the 1,25-dihydroxyvitamin D_3 receptor in target cells. Endocrinology **122**:1224–1230.
49. Berger U, Wilson P, McClelland RA, Colston K, Haussler MR, Pike JW, Coombes RC. 1988 Immunocytochemical detection of 1,25-dihydroxyvitamin D_3 receptors in normal human tissues. J Clin Endocrinol Metab **67**:607–613.
50. Milde P, Merke J, Ritz E, Haussler MR, Rauterberg EW 1989 Immunocytochemical detection of 1,25-dihydroxyvitamin D_3 receptors by monoclonal antibodies: A comparison of four immunoperoxidase methods. J Histochem Cytochem **37**:1609–1617.
51. Berger U, Wilson P, McClelland RA, Colston K, Haussler MR, Pike JW, Coombes RC 1987 Immunocytochemical detection of 1,25-dihydroxyvitamin D_3 receptors in breast cancer. Cancer Res **47**:6793–6799.
52. Nevah-Many T, Marx R, Keshet E, Pike JW, Silver J 1990 Regulation of 1,25-dihydroxyvitamin D_3 receptor gene expression by 1,25-dihydroxyvitamin D_3 in the parathyroid in vivo. J Clin Invest **86**:1968–1975.
53. Mocharla H, DeTogni P, Yu XP, White P, Jilka RJ, Manolagas SC 1994 In search for an association between vitamin D receptor polymorphism and VDR mRNA expression in peripheral blood mononuclear cells of normal volunteers. J Bone Miner Res **9**(Suppl. 1), S416.
54. Walters MR, Hunziker W, Norman AW 1980 Unoccupied 1,25-dihydroxyvitamin D_3 receptors. Nuclear/cytosol ratios depend on ionic strength. J Biol Chem **255**:6799–6805.
55. Walters MR, Hunziker W, Norman AW 1981 1,25-Dihydroxyvitamin D_3 receptors: Intermediates between triiodothyronine and steroid hormone receptors. Trends Biochem Sci **6**:268–271.
56. Pike JW, Haussler MR 1979 Purification of chicken intestinal receptor for 1,25-dihydroxyvitamin D_3. Proc Natl Acad Sci USA **76**:5488–5494.
57. Haussler MR, Pike JW, Chandler JS, Manolagas SC, Deftos LJ 1981 Molecular actions of 1,25-dihydroxyvitamin D_3: New cultured cell models. Ann NY Acad Sci **372**:502–517.
58. Mellon WS, DeLuca HF 1979 An equilibrium and kinetic study of 1,25-dihydroxyvitamin D_3 binding to chicken intestinal cytosol employing high specific activity 1,25-dihydroxy[^{3}H-26,27] vitamin D_3. Arch Biochem Biophys **197**:90–95.
59. Wecksler WR, Norman AW 1980 A kinetic and equilibrium binding study of 1,25-dihydroxyvitamin D_3 with its cytosol receptor from chick intestinal mucosa. J Biol Chem **255**:3571–3574.
60. Kream BE, Jose JL, DeLuca HF 1977 The chick intestinal cytosol binding protein for 1,25-dihydroxyvitamin D_3: A study of analog binding. Arch Biochem Biophys **179**:462–468.
61. Wecksler WR, Okamura WH, Norman AW 1978 Quantitative assessment of the structural requirements for the interaction of 1,25-dihydroxyvitamin D_3 with its chick intestinal mucosa receptor system. J Steroid Biochem **9**:929–937.
62. Pike JW, Haussler MR 1983 Association of 1,25-dihydroxyvitamin D_3 with cultured 3T6 mouse fibroblast. J Biol Chem **258**:8554–8560.
63. Hunziker W, Walters MR, Bishop JE, Norman AW 1983 Unoccupied and *in vitro* and *in vivo* occupied 1,25-dihydroxyvitamin D_3 intestinal receptors. Multiple biochemical forms and evidence for transformations. J Biol Chem **258**:8642–8648.
64. Pike JW, Sleator N, Haussler MR 1987 Chicken intestinal receptor for 1,25-dihydroxyvitamin D_3. Immunologic characterization and homogeneous isolation of a 60,000-dalton protein. J Biol Chem **262**:1305–1311.
65. Mangelsdorf DJ, Pike JW, Haussler MR 1987 Avian and mammalian receptors for 1,25-dihydroxyvitamin D_3: *In vitro* translation to characterize size and hormone-dependent regulation. Proc Natl Acad Sci USA **84**:354–358.
66. Pike JW, Kesterson RA, Scott RA, Kerner SA, McDonnell DP, O'Malley BW 1988 Vitamin D receptors: Molecular structure of the protein and its chromosomal gene. In: Normal AW, Schaefer K, Grigoleit H-G, Herrath Dv (eds) Vitamin D: Molecular, Cellular and Clinical Endocrinology. de Gruyter, Berlin, pp. 215–224.
67. Elaroussi MA, Prahl JM, DeLuca HF 1992 The avian vitamin D receptors: Primary structures and their origins. Proc Natl Acad Sci USA **91**:11596–11600.
68. Pike JW, 1984 Monoclonal antibodies to chick intestinal receptors for 1,25-dihydroxyvitamin D_3: Interaction and effects of binding on receptor function. J Biol Chem **259**:1167–1173.
69. Allegretto EA, Pike JW 1985 Trypsin cleavage of chick 1,25-dihydroxyvitamin D_3 receptors. Generation of discrete polypeptides which retain hormone but are unreactive to DNA and monoclonal antibody. J Biol Chem **260**:10139–10145.
70. Allegretto EA, Pike JW, Haussler MR 1987 Immunological detection of unique proteolytic fragments of the chick 1,25-dihydroxyvitamin D_3 receptor. Distinct 20-kDa DNA binding and 45-kDa hormone-binding species. J Biol Chem **262**:1312–1319.
71. Allegretto EA, Pike JW, Haussler MR 1987 C-Terminal proteolysis of the avian 1,25-dihydroxyvitamin D_3 receptor. Biochem Biophys Res Commun **147**:479–485.
72. Young RA, Davis RW 1983 Efficient isolation of genes using antibody probes. Proc Natl Acad Sci USA **80**:1194–1198.
73. Heyman RA, Mangelsdorf DJ, Dyck JA, Stein RB, Eichele G, Evans RM, Thaller C 1992 9-*cis*-Retinoic acid is a high affinity ligand for the retinoid X receptor. Cell **68**:397–406.
74. Forman BM, Tontonoz P, Chen C, Brun RP, Speigelman BM, Evans RM 1995 15-Deoxy-$\Delta^{12,14}$-prostaglandin J2 is a ligand for the adipocyte determination factor PPARγ. Cell **83**:803–812.
75. Forman BM, Goode E, Chen J, Oro AE, Bradley DJ, Perlman T, Noonan DJ, Burka LT, McMorris T, Lamph WW, Evans RM,

Weinberger C 1995 Identification of a nuclear receptor that is activated by farnesol metabolites. Cell **81**:687–693.
76. Hollenberg SM, Weinberger C, Ong ES, Cerelli G, Ono AE, Lebo R, Thompson EB, Rosenfeld MG, Evans RM 1985 Primary structure and expression of a functional human glucocorticoid receptor cDNA. Nature **318**:635–641.
77. Weinberger C, Thonpson CC, Ong ES, Lebo R, Gruol DJ, Evans RM 1986 The c-*erbA* gene encodes a thyroid hormone receptor. Nature **324**:641–646.
78. Berg JM 1989 DNA binding specificity and steroid receptors. Cell **57**:1065–1068.
79. McDonnell DP, Pike JW, O'Malley BW 1988 Vitamin D receptor: A primitive steroid receptor related to the thyroid hormone receptor. J Steroid Biochem **30**:41–47.
80. Baker AR, McDonnell DP, Hughes MR, Crisp TM, Mangelsdorf DJ, Haussler MR, Shine J, Pike JW, O'Malley BW 1988 Molecular cloning and expression of human vitamin D receptor complementary DNA: Structural homology with thyroid hormone receptor. Proc Natl Acad Sci USA **85**:3294–3298.
81. Burmester JK, Maeda N, DeLuca HF 1988 Isolation and expression of rat 1,25-dihydroxyvitamin D receptor cDNA. Proc Natl Acad Sci USA **85**:1005–1009.
82. Kamei Y, Kawada T, Fukuwatari T, Ono T, Kato S, Sugimoto E 1995 Cloning and sequence of the gene encoding the mouse vitamin D receptor. Gene **152**:281–282.
83. Elaroussi MA, Prahl JM, DeLuca HF 1992 The avian vitamin D receptors: Primary structures and their origins. Proc Natl Acad Sci USA **91**:11596–11600.
84. Li YC, Bergwitz C, Juppner H, Demay MB 1996 Cloning and characterization of the vitamin D receptor from *Xenopus laevis*. J Bone Miner Res **11**(Suppl. 1): S161.
85. Goto H, Chen KS, Prahl JM, DeLuca HF 1992 A single receptor identical to that for intestinal. T47D cells mediates the action of 1,25-dihydroxyvitamin D_3 in HL-60 cells. Biochim Biophys Acta **1132**:103–108.
86. Kesterson RA, Pike JW 1991 Unpublished.
87. Saijo T, Ito M, Takeda E, Huq AHM, Naito E, Yokoto I, Sone T, Pike JW, Kuroda Y 1991 A unique mutation in vitamin D receptor gene in three Japanese patients with vitamin D-dependent rickets type II: Utility of single-strand conformation polymorphism analysis for heterozygous carrier detection. Am J Hum Genet **49**:668–673.
88. Minamitani K, Takahashi Y, Minagawa T, Soeya T, Watanabe T, Yasuda T, Nimi H 1996 Exon 2 polymorphism in the human vitamin D receptor gene is a predictor of peak bone mineral density. J Bone Miner Res **11**(Suppl. 1):S207.
89. Green S, Walter P, Kumar V, Krust A, Bornert JM, Argos P, Chambon P 1986 Human oestrogen receptor cDNA: Sequence expression and homology to v-*erbA*. Nature **320**:134–139.
90. Nauber U, Pankratz MJ, Kienlin A, Seifert E, Klemm U, Jacle H 1988 Abdominal segmentation of the drosophila embryo requires a hormone receptor-like protein encoded by the gap gene *knirps*. Nature **336**:489–492.
91. Oro AE, Ong ES, Margolis JS, Posakony JW, McKeown M, Evans RM 1988 The drosophila gene *knirps*-related is a member of the steroid receptor gene superfamily. Nature **336**:493–496.
92. Koelle MR, Talbot WS, Segraves WA, Bender MT, Cherbas P, Hogness DS 1991 The drosophila EcR gene encodes an ecdysone receptor, a new member of the steoid receptor superfamily. Cell **67**:59–77.
93. McDonnell DP, Scott RA, Kerner RA, O'Malley BW, Pike JW 1989 Functional domains of the human vitamin D receptor regulate osteocalcin gene expression. Mol Endocrinol **3**:635–644.
94. Jin CH, Pike JW 1996 Human vitamin D receptor dependent transactivation in *Saccharomyces cerevisiae* requires retinoid X receptor. Mol Endocrinol **10**:196–205.
95. Jin CH, Kerner SA, Hong MH, Pike JW 1996 Transcriptional activation and dimerization functions of the human vitamin D receptor. Mol Endocrinol **10**:945–957.
96. MacDonald PN, Sherman DR, Dowd DR, Jefcoat SC, Jr, DeLisle RK 1995 The vitamin D receptor interacts with general transcription factor IIB. J Biol Chem **270**:4748–4752.
97. Berg JM 1988 Proposed structure for the zinc-binding domains from transcription factor IIIA and related proteins. Proc Natl Acad Sci USA **85**:99–102.
98. Mader S, Kumar V, deVereneuil H, Chambon P 1989 Three amino acids of the oestrogen receptor are essential to its ability to distinguish an oestrogen from a glucocorticoid responsive receptor. Nature **338**:271–274.
99. Umesono K, Evans RM 1989 Determinants of target gene specificity for steroid/thyroid hormone receptors. Cell **57**:1139–1146.
100. Schwabe JWR, Neuhaus D, Rhodes D 1990 Solution structure of the DNA binding domain of the oestrogen receptor. Nature **348**:458–461.
101. Hard T, Kellenbach E, Boelens R, Maler BA, Dahlman K, Freedman LP, Carlstedt-Duke J, Yamamoto KR, Gustafsson JA, Kaptein R 1990 Solution structure of the glucocorticoid receptor DNA binding domain. Science **249**:157–160.
102. Luisi BF, Xu W, Otwinowski Z, Freedman LP, Yamamoto KR, Sigler PB 1991 Crystallographic analysis of the interaction of the glucocorticoid receptor with DNA. Nature **352**:497–505.
103. Schwabe JWR, Chapman L, Finch JT, Rhodes D 1993 The crystal structure of the oestrogen receptor DNA binding domain bound to DNA: How receptors discriminate between their response elements. Cell **75**:567–578.
104. Lee MS, Kliewer SA, Provencal J, Wright PE, Evans RM 1993 Structure of the retinoid X receptor α DNA binding domain: A helix required for homodimeric DNA binding. Science **260**:1117–1121.
105. Rastinejad F, Perlmann T, Evans RM, Sigler PB 1995 Structural determinants of nuclear receptor assembly on DNA direct repeats. Nature **375**:203–211.
106. Peleg S, Sastry M, Collins ED, Bishop JE, Norman AW 1995 Distinct conformational changes induced by 20-epi analogues of 1,25-dihydroxyvitamin D_3 are associated with enhanced activation of the vitamin D receptor. J Biol Chem **270**:10551–10558.
107. Wang LH, Tsai SY, Cook RG, Beattie WG, Tsai M-J, O'Malley BW 1989 COUP transcription factor is a member of the steroid receptor superfamily. Nature **340**:163–166.
108. Nakajima S, Hsieh JC, MacDonald PN, Galligan MA, Haussler CA, Whitfield GK, Haussler MR 1994 The C-terminal region of the vitamin D receptor is essential to form a complex with a receptor auxillary factor required for high affinity binding to the vitamin D responsive element. Mol Endocrinol **8**:159–172.
109. Bourguet W, Ruff D, Chambon P, Gronemeyer H, Moras D 1995 Crystal structure of the ligand binding domain of the human nuclear receptor RXRα. Nature **375**:377–382.
110. Renaud JP, Natacha R, Ruff M, Vivat V, Chambon P, Gronemyer H, Moras D 1995 Crystal structure of the RARγ ligand binding domain bound to all-*trans*-retinoic acid. Nature **378**:681–689.
111. Wagner RL, Apriletti JW, McGrath ME, West BL, Baxter JD, Fletterick RJ 1995 A structural role for hormone in the thyroid hormone receptor. Nature **378**:690–697.
112. Whitfield GK, Hsieh JC, Nakajima S, MacDonald PN, Thompson PD, Jurutka PW, Haussler CA, Haussler MR 1995 A highly conserved region in the hormone-binding domain of the human vitamin D receptor contains residues vital for heterodimerization

with retinoid X receptor and for transcriptional activation. Mol Endocrinol **9**:1166–1179.

113. Danielian PS, White R, Lees JA, Parker MG 1992 Identification of a conserved region required for hormone dependent transcriptional activation by steroid hormone receptors. EMBO J **11**:1025–1033.
114. McCain TA, Haussler MR, Hughes MR, Okrent D 1978 Partial purification of the chick 1,25-dihydroxyvitamin D receptor. FEBS Lett **86**:65–70.
115. Sone T, McDonnell DP, O'Malley BW, Pike JW 1990 Expression of the human vitamin D receptor in *Saccharomyces cerevisiae*: Purification properties and generation of polyclonal antibodies. J Biol Chem **265**:21997–22003.
116. Santiso-Mere D, Sone T, Hilliard GM, Pike JW, McDonnell DP 1993 Positive regulation of the vitamin D receptor by its cognate ligand in heterologous expression systems. Mol Endocrinol **7**:833–839.
117. Tzukerman MT, Esty A, Santiso-Mere D, Danielian P, Parker MG, Stein RB, Pike JW, McDonnell DP 1994 Human estrogen receptor transcriptional capacity is determined by both cellular and promoter context and mediated by two functionally distinct intramolecular regions. Mol Endocrinol **8**:21–30.
118. McDonnell DP, Clemm DL, Hermann T, Goldman ME, Pike JW 1995 Analysis of estrogen receptor function *in vitro* reveals three distinct classes of anti-estrogens. Mol Endocrinol **9**:659–669.
119. Ducy P, Desbois C, Boyce B, Pinero G, Story B, Dunstan C, Smith E, Bonadio J, Goldstein S, Gundberg C, Bradley A, Karsenty G 1996 Increased bone formation in osteocalcin-deficient mice. Nature **382**:448–452.
120. Price PA, Baukol SA 1980 1,25-Dihydroxyvitamin D_3 increases synthesis of the vitamin D-dependent bone proteins by osteosarcoma cells. J Biol Chem **225**:11660–11663.
121. Pan LC, Price PA 1986 1,25-Dihydroxyvitamin D_3 stimulates transcription of bone gla protein. J Bone Miner Res **1**(Suppl. 1):A20.
122. Celeste AJ, Rosen V, Buecker JL, Kriz R, Wang EA, Wozney JM 1986 Isolation of the human gene for bone gla protein utilizing mouse and rat cDNA clones. EMBO J **5**:1885–1890.
123. Kerner SA, Scott RA, Pike JW 1989 Sequence elements in the human osteocalcin gene confer basal activation and inducible response to hormonal vitamin D_3. Proc Natl Acad Sci USA **86**:4455–4459.
124. Morrison NA, Shine J, Fragonas J-C, Verkest V, McMenemy ML, Eisman JA 1989 1,25-Dihydroxyvitamin D_3 responsive element and glucocorticoid repression in the osteocalcin gene. Science **246**:1158–1161.
125. Yoon K, Rutledge SJC, Buenaga RF, Rodan GA 1988 Characterization of the rat osteocalcin gene: Stimulation of promoter activity by 1,25-dihydroxyvitamin D3. Biochemistry **27**:8521–8526.
126. Ozono K, Liao J, Scott RA, Kerner SA, Pike JW 1990 The vitamin D responsive element in the human osteocalcin gene: Association with a nuclear proto-oncogene enhancer. J Biol Chem **265**:21881–21888.
127. Demay MB, Gerardi JM, DeLuca HF, Kronenberg HM 1990 DNA sequences in the rat osteocalcin gene that bind the 1,25-dihydroxyvitamin D receptor and confer responsiveness to 1,25-dihydroxyvitamin D_3. Proc Natl Acad Sci USA **87**:369–373.
128. Lian JB, Stewart C, Puchacz E, Mackowiak S, Shalhoub V, Colart D, Sambetti G, Stein G 1989 Structure of the rat osteocalcin gene and regulation of vitamin D-dependent expression. Proc Natl Acad Sci USA **86**:1143–1147.
129. Terpening CM, Haussler MR, Jurutka PW, Galligan MA, Komm BS, Haussler MR 1991 The vitamin D responsive element in the rat bone gla protein gene is an imperfect direct repeat that cooperates with other *cis* elements in 1,25-dihydroxyvitamin D_3 mediated transcriptional activation. Mol Endocrinol **5**:373–385.
130. Noda M, Vogel RL, Craig AM, Prahl J, DeLuca HF, Denhardt D 1990 Identification of a DNA sequence responsible for binding of the 1,25-dihydroxyvitamin D_3 receptor and 1,25-dihydroxyvitamin D_3 enhancement of mouse secreted phosphoprotein 1 (Supp-1 or osteopontin) gene expression. Proc Natl Acad Sci USA **87**:9995–9999.
131. Gill RK, Christakos S 1993 Identification of sequence elements in the mouse calbindin-D28K gene that confer 1,25-dihydroxyvitamin D- and butyrate inducible responses. Proc Natl Acad Sci USA **90**:2984–2988.
132. Darwish HM, DeLuca HF 1992 Identification of a 1,25-dihydroxyvitamin D_3 response element in the 5′ flanking region of the rat calbindin D-9K gene. Proc Natl Acad Sci USA **89**:603–607.
133. Ohyama Y, Ozono K, Uchida M, Shinki T, Kato S, Suda T, Yamamoto O, Noshiro M, Kato Y 1994 Identification of a vitamin D responsive element in the 5′-flanking region of the rat 25-hydroxyvitamin D_3 24-hydroxylase gene. J Biol Chem **269**:10545–10550.
134. Zierold C, Darwish HM, DeLuca HF 1995 Two vitamin D response elements function in the rat 1,25-dihydroxyvitamin D_3-24 hydroxylase promoter. J Biol Chem **269**:1675–1678.
135. Chen KS, DeLuca HF 1995 Cloning of the human 1,25-dihydroxyvitamin D_3-24-hydroxylase gene promote and identification of two vitamin D responsive elements. Biochim Biophys Acta **1263**:1–9.
136. Liu M, Lee M-H, Cohen M, Freedman LP 1996 Transcriptional activation of the p21 gene by vitamin D_3 leads to the differentiation of the myelomonocytic cell line U937. Genes Dev **10**:142–153.
137. Umesono K, Murikami KK, Thompson CC, Evans RM 1991 Direct repeats as selective response elements for the thyroid hormone, retinoic acid, and vitamin D_3 receptors. Cell **65**:1225–1266.
138. Sone T, Kerner SA, Pike JW 1991 Vitamin D receptor interaction with specific DNA. Association as a 1,25-dihydroxyvitamin D_3-modulated heterodimer. J Biol Chem **266**:23296–23305.
139. Freedman LP, Towers T 1991 DNA binding properties of the vitamin D receptor zinc fingers region. Mol Endocrinol **5**:1815–1826.
140. Towers T, Luisi BL, Asianov A, Freedman LP 1993 DNA target selectivity by the vitamin D receptor: Mechanism for dimer binding to an asymmetric repeat element. Proc Natl Acad Sci USA **90**:6310–6314.
141. Lemon B, Freedman LP 1996 Selective effects of ligands on vitamin D_3 receptor and retinoid X receptor mediated gene activation *in vivo*. Mol Cell Biol **16**:1006–1016.
142. Nishikawa J, Kitaura M, Matsumoto M, Imagawa M, Nishihara T 1994 Difference and similarity of DNA sequence recognized by VDR homodimer and VDR/RXR heterodimer. Nucleic Acids Res **22**:2902–2907.
143. Hughes MR, Malloy PJ, Kieback DG, Kesterson RA, Pike JW, Feldman D, O'Malley BW 1988 Point mutations in the human vitamin D receptor gene associated with hypocalcemic rickets. Science **242**:1702–1705.
144. Sone T, Scott R, Hughes M, Malloy P, Feldman D, O'Malley BW, Pike JW 1989 Mutant vitamin D receptors which confer hereditary resistance to 1,25-dihydroxyvitamin D_3 in humans are transcriptionally inactive *in vitro*. J Biol Chem **264**:20230–20234.
145. Hughes MR, Malloy PJ, O'Malley BW, Pike JW, Feldman D 1991 Genetic defects of the 1,25-dihydroxyvitamin D receptor. J Receptor Res **11**:699–716.

146. Liao J, Ozono K, Sone T, McDonnell DP, Pike JW 1990 Vitamin D receptor interaction with specific DNA requires a nuclear protein and 1,25-dihydroxyvitamin D_3. Proc Natl Acad Sci USA **87**:9751–9755.
147. Sone T, Ozono K, Pike JW 1991 A 55-kilodalton accessory factor facilitates vitamin D receptor DNA binding. Mol Endocrinol **5**:1578–1586.
148. MacDonald PN, Haussler CA, Terpening CM, Galligan MA, Reeder MC, Whitfield GK, Haussler MR 1991 Baculovirus-mediated expression of the human vitamin D receptor: Functional characterization, vitamin D response element interactions and evidence for a receptor auxiliary factor. J Biol Chem **266**:18808–18813.
149. Ross TK, Moss VE, Prahl JM, DeLuca HF 1992 A nuclear protein essential for binding of rat 1,25-dihydroxyvitamin D_3 receptor to its response elements. Proc Natl Acad Sci USA **89**:256–260.
150. Yu V, Delsert C, Andersen B, Holloway JM, Devary OV, Naar AM, Kim SY, Boutin JM, Glass CK, Rosenfeld MG 1991 RXRβ: A coregulator that enhances binding of retinoic acid, thyroid hormone, and vitamin D receptors to their cognate response elements. Cell **67**:1251–1266.
151. Leid M, Kastner P, Lyons R, Nakshatro H, Saunders M, Zacharewski T, Chen J-Y, Staub A, Garnier J-M, Mader S, Chambon P 1992 Purification, cloning and RXR identity of the HeLa cell factor with which RAR or TR heterodimerizes to bind target sequences efficiently. Cell **68**:377–395.
152. Zhang XK, Hoffman B, Tran PB, Graupner G, Pfahl M 1992 Retinoid X receptor is an auxiliary protein for thyroid hormone and retinoic acid receptors. Nature **355**:441–446.
153. Kliewer SA, Umesono K, Mangelsdorf DJ, Evans RM 1992 Retinoid X receptor interacts with nuclear receptors in retinoic acid, thyroid, and vitamin D signaling. Nature **355**:446–449.
154. Perlmann T, Umesono K, Rangarajan PN, Forman BM, Evans RM 1996 Two distinct dimerization interfaces differentially modulate target gene specificity of nuclear hormone receptors. Mol Endocrinol **10**:958–966.
155. Perlmann T, Rangarajan PN, Umesono K, Evans RM 1993 Determinants for selective RAR and TR recognition of direct repeat HREs. Genes Dev **7**:1411–1422.
156. Kurokawa B, Yu VC, Naar A, Kyakumoto S, Han Z, Silverman S, Rosenfeld MG Glass CK 1993 Differential orientation of the DNA-binding domain and carboxy terminal dimerization interface regulate binding site selection by nuclear receptor heterodimers. Genes Dev **7**:1423–1435.
157. MacDonald PN, Dowd DR, Nakajima S, Galligan MA, Reeder MC, Haussler CA, Ozato K, Haussler MR 1993 Retinoid X receptors stimulate and 9-*cis*-retinoic acid inhibits 1,25-dihydroxyvitamin D_3 activated expression of the rat osteocalcin gene. Mol Cell Biol **13**:5907–5917.
158. Blanco JCG, Wang I-M, Tsai SY, Tsai M-J, O'Malley BW, Jurutka PW, Haussler MR, Ozato K 1995 Transcription factor TFIIB and the vitamin D receptor cooperatively active ligand-dependent transcription. Proc Natl Acad Sci USA **92**:1535–1539.
159. Pratt W, Jolly DJ, Pratt DV, Hollenberg SM, Giguere V, Cadepon FM, Schweizer-Groyer G, Cartelli MG, Evans RM, Baulieu EE 1988 A region in the steroid-binding domain determines formation of the non-DNA binding glucocorticoid receptor complex. J Biol Chem **263**:267–273.
160. Cheskis B, Freedman LP 1996 Modulation of nuclear receptor interactions by ligands; Kinetic analysis using surface plasmon resonance. Biochemistry **35**:3309–3318.
161. Szpirer S, Szpirer C, Riviere M, Levan G, Maryen P, Cassiman JJ, Wiese R, DeLuca HF 1991 The SP1 transcription factor gene (SP1) and the 1,25-dihydroxyvitamin D receptor gene (VDR) are colocalized on human chromosomal arm 12q and rat chromosome 7. Genomics **11**:168–173.
162. Kesterson RA 1993 Regulatory elements in the human osteocalcin gene promoter direct bone-specific expression in transgenic mice. Ph.D. Dissertation. Baylor College of Medicine, Houston, Texas.
163. Miyamoto K, Kesterson RA, Yamamoto H, Nishiwaki E, Tatsumi S, Taketani Y, Morita K, Pike JW, Takeda E 1997 Organization of the human vitamin D receptor chromosomal gene and its promoter. Mol Endocrinol **11,** in press.
164. Morrison NA, Qi JC, Tokita A, Kelly PJ, Crofts L, Nguyen TV, Sambrook PN, Eisman JA 1994 Prediction of bone density from vitamin D receptor alleles. Nature **367**:284–287.
165. Yoshizawa T, Handa Y, Uematsu Y, Sekine K, Takeda S, Yoshihara Y, Kawakami T, Sato H, Alioka K, Tanimoto K, Fukamizu A, Masushige S, Matsumoto T, Kato S 1996 Disruption of the vitamin D receptor in the mouse. J Bone Miner Res **11**(Suppl. 1):S62.

CHAPTER 9

Structural and Functional Determinants of DNA Binding and Dimerization by the Vitamin D Receptor

LEONARD P. FREEDMAN AND BRYAN D. LEMON
Cell Biology and Molecular Biology Programs, Memorial Sloan-Kettering Cancer Center, New York, New York

I. INTRODUCTION

Protein dimerization facilitates cooperative, high affinity interactions with DNA. Many nuclear hormone receptors, including the human 1,25-dihydroxyvitamin D receptor (VDR), bind *in vitro* either as homodimers or as heterodimers with retinoid X receptors (RXR) to half-site repeats that are stabilized by protein–protein interactions mediated by residues within both the DNA and ligand binding domains. In this chapter, we summarize functional and structural results that demonstrate how these two separate dimerization regions of VDR serve to influence both DNA binding site selectivity as well as target gene activation, and the central role 1,25-dihydroxyvitamin D_3 [$1,25(OH)_2D_3$] plays in modulating these functions.

The regulation of gene transcription involves both the binding of regulatory proteins to specific DNA sequences as well as the formation of critical protein–protein associations. Dimerization can greatly influence DNA binding, since dimerization facilitates cooperative, high affinity interactions between the proteins and DNA. Dimerization can also yield considerable regulatory complexity because members of a given superfamily often bind to DNA as homo- and/or heterodimers. Conceivably, different combinations of family members should yield distinct binding and regulatory properties. This behavior is illustrated by members of the Jun–Fos and ATF–CREB families [containing the basic/leucine zipper (b-zip) domain] and a large family of proteins that dimerize and bind DNA through a basic, helix–loop–helix (bHLH) motif.

The vast majority of the members of the nuclear receptor superfamily, including the VDR, bind DNA and regulate transcription as homo- or heterodimers. Reflecting this, nuclear hormone receptors recognize and bind to target sites that are organized as imperfect repeats; the so-called half-sites of the repeat serve as targets for receptor dimers. Steroid receptors bind cooperatively and with high affinity as homodimers to palindromic half-sites (reviewed by Freedman and Luisi [1]). In contrast, a subgroup of nuclear receptors, to which VDR belongs, appear to recognize half-site targets organized as direct-repeat elements as homodimers or preferentially as heterodimers with a common partner, the RXR [2]. Here, spacing between the two half-sites appears to be a more important criterion for DNA binding selectivity than sequence differences in the half-sites: the vitamin D, thyroid hormone, and retinoic acid receptors all recognize a hexameric core half-site sequence (consensus: 5′-PuGG/TTCA-3′) (Pu, purine) but preferentially bind this element when direct repeats (DRs) of it are spaced by three, four, and five nucleotides, respectively [3,4]. Analogous to the mechanism of steroid receptor–DNA interactions, binding site discrimination of differentially spaced direct repeat elements is mediated principally by an asymmetric dimer interface formed between DNA binding domain subunits [5–8] and are enhanced by a dimerization region colocalized to the C-terminal ligand binding domain [9,10], as discussed in detail in this chapter.

This chapter focuses on the interrelated functions of DNA binding and dimerization by VDR, and how $1,25(OH)_2D_3$, as the active ligand for this receptor, can modulate these activities. Although no three-dimensional structures are as yet available for VDR, we refer extensively to solved crystallographic structures of other nuclear hormone receptors, given that these proteins exhibit high sequence and presumably structural conservation. However, the fact that this is an extrapolation should be kept in mind regarding the structure of the VDR, until the actual crystallographic structure of the protein has been determined directly.

II. STRUCTURE OF THE DNA BINDING DOMAIN

A. Tertiary Structure

The DNA binding domain (hereafter referred to as DBD) is the most highly conserved region among the nuclear receptors. Several amino acids are invariant throughout the family, including eight cysteines that have been shown to tetrahedrally coordinate two Zn ions [11] (Fig. 1), as well as particular basic residues that make direct contacts with bases in receptor DBD–DNA cocrystals (see below). Zinc coordination is essential to hold the DBD in a structurally and functionally active conformation. Mutations in coordinating cysteines, or removal of Zn from the DBD, renders the domain functionally inactive for DNA binding and transactivation; removal of Zn also renders the DBD highly protease sensitive, consistent with an unfolded state [11–13].

The three-dimensional structures of the DNA binding domains of the glucocorticoid, estrogen, and retinoid

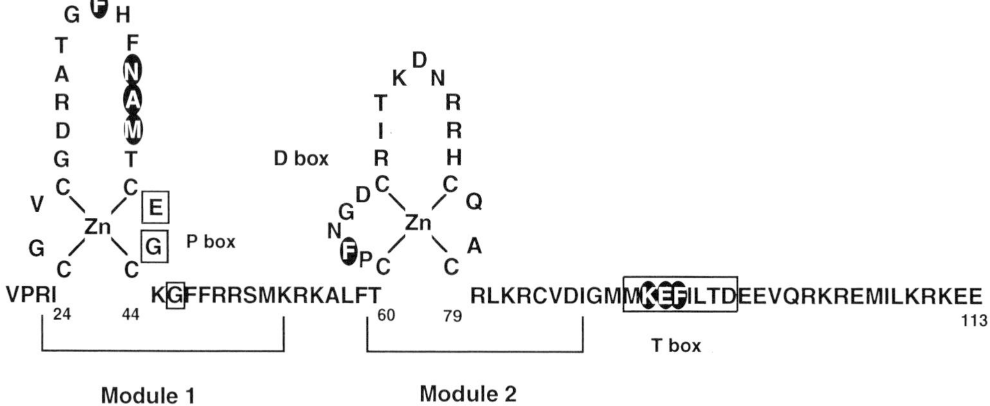

FIGURE 1 The primary sequence of the DNA-binding domain of the human vitamin D_3 receptor. Modules 1 and 2 are based on structural and functional data that define these two subdomains (see text). Residues shown to be important in target site discrimination (i.e., the P-box) are boxed. VDR residues that are proposed here to form an asymmetric dimerization interface (see text) are encircled and darkened. The "T box" refers to a short region first identified to direct dimerization of RXRβ to its DNA target [26], and contains three residues (Lys, Glu, Phe) that have been proposed to participate together with Asn37 in a RXR–VDR dimer interface on DNA [7,19].

X receptors (GR, ER, and RXR, respectively) have been established by nuclear resonance spectroscopy (NMR) [12–16] and X-ray crystallography [17,18]. In addition, the crystallographic structure of thyroid hormone receptor (TR)–RXR DBD heterodimer bound to DNA has been solved [19]. The salient feature of the nuclear receptor DNA binding domain is its Zn coordination sites. In agreement with stoichiometry determinations and spectroscopic evaluation [11], two Zn ions are each tetrahedrally coordinated by four conserved cysteines to stablize two peptide loops and cap amino termini of two amphipathic α helices (Fig. 2). The metal binding sites differ structurally and functionally from those found in the other eukaryotic transcription factors bearing "Zn fingers," such as TFIIIA, ADR1, and Xfin, where the metal is coordinated by two histidines and two cysteines [20]. It also differs from the Zn coordination site of GAL4 [21], where two metal ions share cysteine ligands. The nuclear receptor Zn finger subdomains are therefore referred to as "Zn modules" to distinguish them from the other Zn-bearing structures. As mentioned, cysteines coordinating the Zn and other residues supporting the fold of the domain are conserved throughout the family, which strongly suggests that the Zn module structures are also conserved among the receptors.

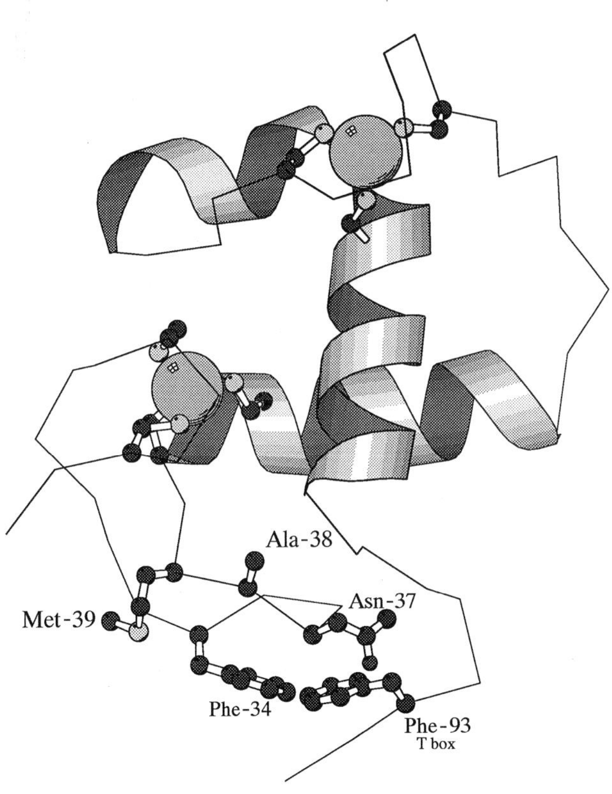

FIGURE 2 Three-dimensional model of the VDR DBD. Helices are depicted by ribbons, and the Cα backbone trace by lines. The model is based on the existing crystal structures for the GR, ER, and RXR–DBDs (see text). The two zinc ions are represented by spheres. The recognition α-helix of module 1, which lies in the major groove of the DNA, is oriented such that its axis is on the horizon. The principal helix of module 2 lies vertically. In an asymmetric dimerization alignment on a direct repeat response element (i.e., DR3), Phe-93 of the T box (see Fig. 1) may pack against Asn-37 and Phe-34, which is supported by Ala-38 and, possibly, Met-39. The hydrophobic interactions may orient the T box to make intermolecular contacts. See Section II,B of the text for details.

Although they appear to be structurally similar, the two Zn modules of the nuclear receptor DBD appear to serve different functions. The amino-terminal module exposes an α-helix to the major groove of DNA and directly contacts the bases of the target site. The carboxy-terminal module forms a dimerization interface, which for the symmetrically oriented steroid receptors is mediated principally through contacts made by residues in a short region referred to as the D box (Fig. 1) for its postulated role in dimerization [22]. The dimerization interface for VDR and other receptors binding direct, asymmetric DNA response elements is more complex, as discussed in detail below in Section II,B. The loops of both Zn modules make phosphate backbone contacts. Although structurally distinct, the Zn fingers, GAL4 Zn center, and the nuclear receptor Zn modules do share the general feature of serving to stabilize and orient an α helix for interaction with the DNA target through major groove contacts. In the case of the Zn fingers and the GAL4 Zn center, the recognition helix lies within the loop region, whereas it begins at the carboxy-terminal end of the loop in the nuclear receptor modules.

The apparent structural similarity of the two Zn modules gives the DBD the appearance of having an approximate structural repeat. Moreover, the modules are encoded by separate exons [23], which suggests that they may have arisen by gene duplication and then evolved with different functions. However, the modules are actually topologically nonequivalent, and thus their relationship may not be so simple. The topology is defined here as the chirality of Zn-coordinating residues, which can be *R* or *S*. The amino-terminal Zn module, like the Zn fingers [20] has the *S* configuration. The carboxy-terminal module, however, has the mirror image *R* configuration about the metal, and it cannot be changed to that found for the amino-terminal module without breaking a Zn–S bond. Despite their structural similarities, the two Zn modules may not have arisen by gene duplication unless one refolded catastrophically about the Zn ion in the very early states of its evolution.

The two amphipathic helices of the nuclear receptor DBD pack together to form a hydrophobic core that is important for maintaining the globular fold of the domain (Fig. 2) [17]. The residues in and supporting this core are very strongly conserved in the superfamily.

Functionally nonconservative changes here result in loss of function of GR *in vivo* and *in vitro* [24], and in VDR a substitution here is associated with hereditary vitamin D-resistant rickets (see Section II,D).

The NMR structures of the DBD in the absence of DNA and the crystal structures of the protein–DNA complex agree quite well. There is a difference, however, in the secondary structure of the carboxyl-terminal module. In the published crystal structures of GR and ER, there is a short α helix in the loop region of this module that makes phosphate backbone and dimer interface contacts. This helix is not present in the NMR structures, and it may be nucleated by DNA binding. There is precedent for this effect: it appears that interaction with nucleic acid may nucleate secondary structure in the DNA binding domains of GAL4 and leucine zipper transcription factors [20,25]. It is not clear if folding of the short α helix occurs in the context of full-length nuclear receptors, which may already exist as dimers before binding to DNA.

A major difference in the structures of the GR and ER DBDs in comparison to the RXR–TR DBD complex, and by extension VDR, is the presence of an extended third α helix immediately C-terminal to module 2 of TR (VDR residues 91–114; Fig. 1). This region, first described by Wilson *et al.* [26] as the "A-box" in NGFI-B, appears to make no tertiary contacts with the rest of the DBD, but rather lies across the minor groove corresponding to the spacer nucleotides between the two hexameric half-sites, where it makes extensive contacts with DNA (see Fig. 4A, color insert). As discussed in the next section, *in vitro* mutagenesis experiments predicted that this third helix plays a key role in setting up the dimer interface of the VDR DBD as it binds to a directly repeating vitamin D response element (VDRE) [7].

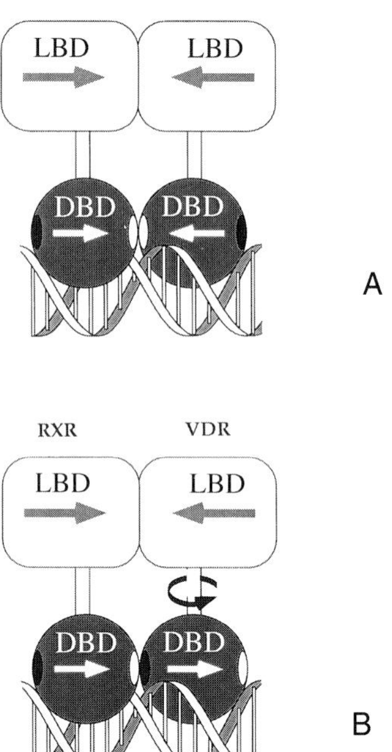

FIGURE 3 Mode of binding of steroid receptors to inverted repeats (A) versus binding of the RXR–VDR heterodimer to direct repeats (B). The orientation of the recognition element in the major groove is implied by the arrows of the DBD. The symmetrical dimerization region of the steroid receptor DBD is indicated by white ovals, and the heterologous dimerization region of RXR–VDR is indicated by dark and white ovals. The ligand binding domain (LBD), which has a stronger dimerization function, forms a symmetrical dimerization interface (see Fig. 7). The two domains may be linked by a flexible tether.

B. Asymmetry of the DNA Binding Domain Dimer Interface

The VDR belongs to the subgroup of nonsteroid receptors that all appear to recognize the estrogen response element core half-site AGGTCA, or slight modifications thereof, and therefore may distinguish targets by recognizing the relative orientation and spacing of two such sites rather than particular sequences within the half-site. This group also includes the thyroid hormone receptor, retinoic acid receptors (RAR), and many orphan receptors. In many cases, the response elements are arranged as direct repeats, which initially suggested that the proteins may bind as asymmetrical dimers (i.e., in a head-to-tail orientation). This is in contrast to the steroid receptors, which form symmetrical homodimers on their DNA targets (Fig. 3). Consistent with this fundamental difference in binding strategies, when the D box of the GR DBD is changed to that of VDR, cooperative binding to a palindromic glucocorticoid response element (GRE) is abolished [27]. This was the first evidence in addition to the orientation of the direct repeat response element suggesting that VDR (and the related receptors) does not form symmetrical dimers on the DNA target. Instead, VDR and other receptors appear to form asymmetrical dimers stabilized by protein–protein interactions.

The consensus high affinity VDR recognition element comprises two hexameric half-sites arranged as a direct repeat with spacing of 3 bp (DR3) [21] (see Table I). Using the GR DBD to model VDR interactions with such a target indicates that two VDR DBD monomers bound at this site lie on the same surface of the DNA and can make favorable protein–protein interactions; consistent with this, the VDR binds the DR3 target cooperatively [7]. Loss of this cooperative binding by the introduction of site-specific mutations of residues in the VDR DBD thought to be involved in this interface or any deviation in the 3-bp spacing suggested a model

TABLE I. Positive VDREs

Gene	Position	Sequence	Refs.
Human osteocalcin	−499 to −485	GGGTGAacgGGGGCA	47–49
Rat osteocalcin	−460 to −446	GGGTGAatgAGGACA	50,51
Mouse osteopontin	−757 to −743	GGTTCAcgaGGTTCA	38
Mouse calbindin-D_{28K}	−198 to −183	GGGGGAtgtg AGGAGA	52
Rat calbindin-D_{9K}	−489 to −475	GGGTGTcggAAGCCC	53
Rat 24-hydroxylase	−151 to −137 −239 to −245	AGGTGAgtgAGGGCG CGCACCcgcTGAACC	45,54–56
Avian integrin β3	−770 to −756	GAGGCAgaaGGGAGA	57
Human p21	−779 to −765 −570 to −556	AGGGAGattGGTTCA AGGTGCtccAGGTGC	46 and L. P. Freedman, unpublished
Synthetic DR3		PuG$\frac{G}{T}$TCAnnnPuG$\frac{G}{T}$TCA	3,6,7,39,40

in which residues in the carboxy terminus of the DNA binding domain make specific asymmetric contacts with residues in a segment of the amino-terminal module of a neighboring receptor [7] (Fig. 4; see color insert). Similar models of such an asymmetric interface were proposed for RXR–TR and RXR–RAR heterodimers bound to DR4 and DR5 binding elements, respectively [5,6,8]. The validation of these predictions came with the cocrystal structure of the RXR–TR DBD complexed to a DR4 DNA binding site [19]. This structure shows an extensive head-to-tail subunit arrangement involving nonreciprocal regions of each protein forming the interface (Fig. 4A; see color insert). Residues of the second zinc module of RXR (bound to the upstream half-site; see Section II,C) interact with residues of TR from both the first zinc module and the region just C-terminal to the second zinc module (the T box [28]). Although this scheme can be generalized to VDR it is interesting to note that residues contributing to the RXR–TR interface as seen in the crystal structure and those predicted for VDR in the mutagenesis experiments involve many nonconserved positions. This suggests that each DBD requires a unique set of specific interactions which are likely imposed by the spacer length between the two half-sites. As mentioned, it is the spacer length which appears to largely confer DNA binding specificity to this group of nuclear receptors, and it is the structural details of the asymmetric dimer interface that serve to provide a explanation for this specificity.

In lieu of a VDR DBD crystal structure, Rastinejad and colleagues have modeled VDR and RXR bound to a DR3 VDRE [19] based on the RXR–TR DBD crystal structure. As seen in Fig. 4B (see color insert), when half-sites are spaced by three nucleotides, VDR and RXR can make a series of favorable contacts at the dimer interface. A key residue is Asn-37, in the loop of the first zinc module (Figs. 1 and 2), which makes side chain contacts with two Arg residues and a Gln that are clustered in the tip of the second zinc module loop of RXR [Fig. 4B (color insert), N14 for VDR, R48 and R52 for RXR]. Interestingly, Asn-37 is unique to VDR and thus may be a key discriminator in setting up an asymmetric dimer interface with RXR on a DR3 binding site. Mutagenesis data are consistent with this, since a change of Asn-37 to Gly (the residue most often seen at this position in the nuclear receptor superfamily) alters the half-site spacer discrimination of VDR so that the receptor binds weakly to a DR3 and instead now prefers a DR5 element (a retinoic acid response element) [7]. In addition, Lys-91 and Glu-92 in the T-box region of VDR make salt bridges with RXR residues, thereby further supporting the dimer interface. As depicted in Fig. 4B (see color insert), the T box modeled for VDR has been adjusted relative to that of the TR crystal structure; this adjustment in fact permits the Lys-91 and Glu-92 (K68 and E69 in Fig. 4B; see color insert) contacts with RXR, and also results in a readjustment of the third α helix so that it does not sterically hinder RXR–VDR dimerization at the interface. This last point reinforces the notion that each receptor forms a unique asymmetric dimer that confers optimal registry with the half-site spacing of the DNA element, such that the correct and optimal contacts are made between amino acids within the DBD and key bases of each half-site. The end result is a receptor–DNA interaction that is both of high affinity and strongly cooperative [29]. In the case of VDR, this spacing is 3 bp. Several putative VDREs have been described in the literature that do not conform to this spacing (i.e., see Schrader *et al.* [30]). These elements must somehow be able to induce VDR to bind in an alternative conformation; otherwise, it is unclear from a structural perspective how the receptor could bind to such a DNA element with sufficiently high affinity so as to make a functionally relevant interaction.

C. Polarity of RXR–VDR DNA Binding

The RXR–TR DBD crystal structure also provides a three-dimensional explanation of a prediction several groups had made concerning the polarity of the receptor heterodimer bound to DNA. Given the fact that the DNA half-site binding specificity for RXR and its partner receptors, including VDR, are similar if not identical (see Section II,E,2) and that these receptors typically bind response elements as heterodimeric species, the question was raised as to whether RXR needs to bind to the upstream or downstream half-site, that is, whether polarity is imposed on the heterodimeric complex. By changing the DNA binding half-site specificity to that of GR together with introducing GRE half-sites in the upstream or downstream repeat element, several groups found that, in fact, there is strict polarity imposed on RXR–TR and RXR–RAR DNA binding, such that RXR always binds upstream [5,6,8]. This same polarity restriction was described for a RXR–VDR heterodimer on a DR3 [31]. From the RXR–TR DBD structure, it is clear that as TR lines up on a half-site, the extended third α helix is oriented upstream and is thereby able to make extensive minor groove contacts with the half-site spacer bases, stabilizing the TR DBD to make extensive major groove contacts on the downstream half-site (Fig. 4A; see color insert). The RXR DBD lacks this third α helix and therefore could not make these interactions if it were positioned downstream. As bound upstream, RXR accommodates TR on a DR4 and, as modeled in Fig. 4B (see color insert), VDR on a DR3. If the polarity were reversed, the critical contacts required of the third α helix of TR or VDR would not occur.

D. Stereochemistry of VDR Dysfunction in Hereditary Vitamin D-Resistant Rickets

Many clinically diagnosed diseases arise from mutations in the genes encoding the steroid or nuclear hormone receptors. These have been characterized primarily in VDR, TR, and some sex hormone receptors, where it appears they are generally not lethal but are readily identified from characteristic clinical abberrations. Where clinical aberrations have been studied, it has emerged that the mutations responsible for the aberrant phenotype occur principally in the DNA binding or ligand binding domains of the receptors.

The effects of mutations in the VDR DBD in hereditary 1,25-dihydroxyvitamin D_3-resistant rickets (HVDRR), the first syndrome for which mutations in a nuclear receptor were identified [32] (see Chapter 48 for a full discussion of HVDRR), can be understood with reference to the crystal structures of the receptor DBD–DNA complexes. As seen thus far in this chapter, given the structural similarity of the DBDs of the nuclear receptors, this modeling is likely to be sufficiently accurate to explain the mutational effects. Many of the substitutions in the VDR DBD involve residues that are conserved throughout the superfamily and, based on the structures of the GR–, ER–, and RXR–TR/DNA complexes, are contacting bases or the phosphate backbone of the DNA.

Residues Arg-50, Arg-73, and Arg-80 of VDR (see Fig. 1) are conserved at the corresponding positions throughout the nuclear receptor superfamily, and all three play important roles in the GR DBD–GRE, ER DBD–ERE, and RXR–TR DBD–TRE complexes (ERE and TRE are estrogen and thyroid hormone response elements). Each of these residues independently occurred as Gln in VDR of three patients with HVDRR [32–34]. In all cases, the substitution cannot support the same interactions made by the arginines. Arg-50 donates hydrogen bonds to the bases of the conserved sequence core of the VDRE, whereas Gln at this position could not reach the base. Arg-73 interacts with a DNA phosphate and is buttressed in this interaction by a hydrogen bond with Asp-29, a residue that is absolutely conserved in the corresponding position throughout the superfamily. Substitution of the Arg with Gln would allow one but not both of the above interactions to be made. Arg-80 forms a phosphate backbone contact, and a Gln side chain here would not be sufficiently long to duplicate this interaction.

The Gly-33 residue is strongly conserved in the nuclear receptor superfamily, and it is substituted by Asp in VDR from a patient suffering from HVDRR [35]. Gly-33 is probably required for two purposes. First, because it has no side chain, Gly permits the peptide backbone to assume a special conformation that orients His-35 such that it can donate a hydrogen bond to the DNA phosphate and Phe-36 to pack against the hydrophobic interior of the domain. Asp at the same position of VDR might not seriously affect the peptide fold, but it would electrostatically repel the phosphate backbone and prevent a close approach to the DNA. It is interesting to note in this regard that in the sequences for the mouse RXRβ receptor and the *Caenorhabditis elegans* nuclear receptor-like gene, the Gly occurs as cysteine, yet these receptors bind DNA [26]. Probably in these cases, the loop region of the module has a different backbone conformation from that found in GR

DBD and ER DBD, and it is possible that compensating substitutions may have occurred elsewhere in the protein to permit the formation of a favorable complex.

E. Specificity of VDR–DNA Interactions

1. Structural Considerations and Predictions

Nuclear receptors tend to fall into subgroups that, within a group, recognize the same response element "core" sequence or modifications thereof. The receptor DNA binding domain contains a region called the P box, located at the amino terminus of the recognition α helix of the first zinc module (Fig. 1); this box can be used to predict core preference (reviewed by Luisi *et al.* [36]). For example, receptors carrying the GS--V motif (i.e., glucocorticoid, progesterone, mineralocorticoid, and androgen receptors) all recognize a glucocorticoid response element (GRE; 5′-AG**A**ACA-3′ with high affinity. Receptors carrying the EG--A/G/S motif at the corresponding positions (i.e., estrogen, vitamin D_3, thyroid hormone, retinoic acid, ecdysone receptors, and many orphan receptors) all appear to bind to the AG**GT**CA core sequence (see Table I). As mentioned above, within this latter group, target-site discrimination may be achieved by recognition of the relative orientation and spacing of two such sites, rather than half-site sequence differences.

On the basis of mutagenesis and molecular modeling, it was predicted that the Glu within the recognition helix of the ER DBD accepts a hydrogen bond from the amino group of the C complementary to the third G in the ag**G**tca ERE core complement half-site (tga**C**ct), and it appears to be a critical component of the ability of a given receptor to discriminate a GRE half-site from an ERE [37]. The existence of this contact has been confirmed with the solving of ER DBD–ERE crystal structure, in addition to making water-mediated contacts to the A of tg**A**cct [18]. The identical contacts are conserved in the TR–RXR DBD crystal structure complex [19]. Presumably, the P-box Glu (at residue 42; see Fig. 1) in VDR is making the same key discriminating side chain contact. In addition, the conserved Lys in the first zinc module (Lys-45 in VDR) has been found to make side chain contacts in all three cocrystal structures, contacting both the G and T of ag**GT**ca. This is likely to be a conserved contact, given that Lys is found at this position in all members of the superfamily.

The P-box classification clearly serves only as a coarse approximation, with residues outside the conventional P box also playing a role in sequence discrimination. This is illustrated by the DNA binding properties of purified VDR, which binds well to a sequence derived from the mouse secreted phosphoprotein 1 gene (Spp-1), also known as osteopontin [38] (Table I). This VDRE is composed of a direct repeat with a three base separation: 5′-GGTTCAcgaGGTTCA-3′. Although VDR and ER share a similar P box, purified VDR binds the above target better than a direct repeat of the estrogen response element half-site: 5′-AGGTCAnnnAGGTCA-3′ [7,27,39,40]. As mentioned, the P box of VDR contains the Glu (Glu-42) that usually indicates a preference for a C in the ERE type half-site (tga**C**ct). In contrast, the above VDRE has an A at the corresponding position (tga**A**cc). Glu can make a hydrogen bonding interaction with the N-6 of the A, just as it does with the N-4 of the C in the ER/ERE complex. Freedman *et al.* [39] tested this hypothesis by changing the A–T base pair to an inosine–5-methylcytosine base pair. The effect of the substitution is to exchange the amino group of the pyrimidine for a carbonyl, which cannot interact favorably with the carboxylate of the Glu at physiological pH. As anticipated, the substitution weakened VDR binding. A second-site revertant was designed by changing the Glu to Gln. It was reasoned that the side chain of the Gln, which can either donate or accept a hydrogen bond, could make a compensatory interaction with the carbonyl group in the inosine–5-methylcytosine base pair. The activity was indeed restored by this "second-site" substitution, supporting the role of the Glu/A interaction in recognizing the target DNA. Additional interactions with the bases probably are directed at the side chain of Arg-49. It is interesting to note that an Arg occurs at the same position in TR and RAR. This effect may hint at subtle but important differences in the interaction of VDR with DNA relative to other nuclear receptors, which will hopefully be revealed by structural analyses.

2. Functional Vitamin D Response Elements

a. Binding Site Selection Although a variety of spacings and orientations have been reported [30], the high affinity, functional binding site for VDR is the DR3. A structural explanation for this selectivity has been outlined in the preceding sections. Functionally, the primacy of the DR3 VDRE has been best demonstrated by binding site selection experiments as well as simple comparisons of the sequences of existing, functional VDREs. In binding site selection, an oligonucleotide duplex containing a stretch of randomized DNA sequence is subjected to multiple rounds of protein binding followed by selection (i.e., by immunocoprecipitation, pull-down of a VDR fusion protein, or extraction from a gel mobility shift) and polymerase chain reaction (PCR) amplification. Using this method, three groups independently found that the most frequently recovered sequence for both VDR alone and RXR–RXR comprised a DR3 [6,39,40]. Selection and enrichment of DR1, DR2, DR4, and DR5 elements by VDR were extremely rare. Moreover, the highest affinity half-site sequence was AGTTCA, rather than the core AGGTCA element, and presumably because of this very high affinity, VDR homodimers could bind with affinity constants in the nanomolar range to this particular DR3 element. This is consistent with reports demonstrating VDR homodimer binding to the osteopontin VDRE (half-site sequence GGTTCA) but exclusively RXR–VDR heterodimeric binding to the related osteocalcin VDRE [6,39–42], suggesting that the latter VDRE is an intrinsically weaker binding site for VDR, requiring the more stable, higher affinity RXR–VDR heterodimer. It should be pointed out that Carlberg and colleagues erroneously reported the opposite observation for VDR homo- and heterodimeric binding to these two VDREs [43].

b. Natural VDREs A large number of genes and proteins have been reported to be up- or down-regulated by $1,25(OH)_2D_3$ in a variety of cellular contexts [44]. However, a rather limited number of functional VDREs are known. Functionality is defined here as a response element that has been demonstrated to (1) mediate $1,25(OH)_2D_3$ induction of transcription in the context of a natural promoter *in vivo* such that a mutation within the element will abolish this induction; (2) recognize and bind RXR–VDR specifically and with high affinity *in vitro*; and (3) confer $1,25(OH)_2D_3$ responsiveness to a heterologous promoter. This last criterion may not always be observed, since VDREs, as is true for many enhancer elements, may be more complex than short simple sequences; in these cases, the DR3 VDRE may represent the minimal functional element for $1,25(OH)_2D_3$ responsiveness.

A list of positive VDREs is shown in Table I [45–57]. Several of these VDREs are discussed individually in detail in other chapters in this volume. With the exception of the mouse osteopontin VDRE, all the VDREs are rather degenerate, imperfect repeats and, with the exception of the mouse calbindin-D_{28K}, are all spaced by 3 bp. In addition, some $1,25(OH)_2D_3$-regulated promoters may have more than one functional VDRE. This appears to be the case for the rat 24-hydroxylase promoter [45] and the human $p21^{WAF1,CIP1}$ promoter [46]. For $p21^{WAF1,CIP1}$, a mutation of either one of the two functional VDREs appears to cripple or abolish $1,25(OH)_2D_3$ induction ([46] and N. Skakcel and L. P. Freedman, 1996, unpublished experiments). In the rat 24-hydroxylase promoter, both VDREs are required for full responsiveness when fused to a heterologous promoter [45]. Thus each VDRE is necessary but not sufficient to mediate the full $1,25(OH)_2D_3$ induction of the linked target gene.

A number of negative VDREs have been described. Negative regulation by nuclear receptors may occur via a number of molecular mechanisms, including direct DNA binding and negative interference, indirect binding by tethering to a DNA-bound positive transcription factor, or protein–protein interactions entirely off the DNA (for review, see Miner and Yamamoto [58]). An example of DNA binding-mediated repression by VDR is the down-regulation of the parathyroid hormone (PTH) gene. The human PTH negative VDRE appears to be a single heptamer, AGGTTCA, but it is not clear whether additional flanking sequences are required for full $1,25(OH)_2D_3$-mediated repression [59]. In contrast to this unusual element, the avian PTH gene promoter, judging from *in vitro* footprinting with RXR–VDR, contains a strong DR3 VDRE, 5′-GGGTCAggaGGGTGT-3′, localized to positions −74 to −60 upstream of the transcription start site [60]. A very similar element (5′-GGGTGGAgagGGGTGA-3′) localized within a region of the parathyroid hormone-related peptide (PTHP) gene promoter (−1121 to −1075) appears to also act as a negative VDRE [61]. Interestingly when mutations were introduced into the 3′ half-site of the avian PTH VDRE that created a perfect DR3 (GGGTCA), ethylation interference footprints were nearly identical to those generated with the human osteocalcin VDRE, and in transient transfections this mutated negative VDRE conferred strongly positive responsiveness to $1,25(OH)_2D_3$ (N. J. Koszewski and J. Russell, 1996, personal communication). These results indicate that the particular sequence comprising the VDRE somehow influences VDR function, perhaps by acting as an allosteric effector of receptor-mediated activation or repression of transcription.

A seemingly more complex mode of negative regulation by VDR is exemplified by the interleukin-2 (IL-2) gene promoter. An underlying mechanism mediating $1,25(OH)_2D_3$ suppression of activated T cells may be the down-regulation of the expression of factors required for T-cell activation, such as IL-2. A key region for the activation of the IL-2 promoter appears to be the target of VDR down-regulation [62]. This portion of the promoter, at −270, contains a composite binding site for three positive trans-acting factors that mediate inducible IL-2 activation, NFATp, c-Jun, and c-Fos. This suggests that the negative effect by VDR involves a mechanism that interferes with the binding of one or all of these positive factors. By combining partially purified proteins *in vitro*, Alroy *et al.* [62] demonstrated the loss of the bound NFATp–AP-1–DNA complex on inclusion of VDR or RXR–VDR, but not with VDR mutants that cannot repress IL-2 transcription *in vivo*. Order of addition and off-rate experiments indicated that RXR–VDR blocks NFATp–AP-1 complex formation and then stably associates with the NF-AT-1 DNA element [62], indicating that direct DNA binding by VDR (with RXR) is important for the regulation. However, mutations in a putative VDR binding site in this region of the IL-2 promoter do not have dramatic effects on the down-regulation (I. Alroy, C. Bromleigh, and L. P. Freedman, 1995, unpublished data), although VDR–RXR will bind this site specifically *in vitro* and mutations in the VDR DBD abolish both *in vitro* DNA binding and *in vivo* $1,25(OH)_2D_3$ repression. These results suggest that the mechanism for repression is more complex than simply direct DNA binding by the receptor. A similar situation has been reported for the $1,25(OH)_2D_3$-mediated repression of the bone sialoprotein gene (BSP). A putative DR3 VDRE has been localized to overlap the BSP TATA box, but mutations introduced to this element do not significantly alter the transcriptional repression [63]. In both the IL-2 and BSP examples, other distal sites may be involved, or DNA binding by VDR may be secondary to receptor interactions with key activators.

III. THE LIGAND BINDING DOMAIN AND VDR DIMERIZATION

A. Homo- versus Heterodimerization

Steroid receptors (other than VDR) typically reside in the cytoplasm or nucleus of cells in an inactive state complexed to heat-shock proteins, primarily hsp70 and hsp90. On ligand binding, these receptors undergo conformational changes which release the receptors from heat-shock protein constraints and enhance homodimerization, nuclear localization (for GR and MR), and DNA binding (for review, Truss and Beato [64] and Tsai and O'Malley [65]). We have already discussed in detail the contributions of the DBD to dimerization. Interestingly, with one apparent exception, namely, progesterone receptor (PR) [66], ligand is not required for *in vitro* DNA binding by steroid receptors, which further implicates a role of heat-shock proteins or other inhibitors to maintain steroid receptors in an inactive state in the cell.

In contrast, VDR, TR, RARs, RXRs, peroxisome proliferator-activated receptors (PPARs) and many orphan receptors with as yet unidentified ligands reside constitutively within the cell nucleus and are apparently not associated with heat-shock proteins [67]. As discussed in Section II, this subfamily of nuclear receptors can form both homodimeric and heterodimeric complexes with RXRs. There are three isotypes of RXR (α, β, and γ), and each appears to have restricted expression patterns, with RXR-β being the most ubiquitous [68,69]. Early reports of factors which potentiated binding of VDR to the osteocalcin VDRE from extracts of nuclei isolated from kidney, liver, and intestine were subsequently identified as RXRs [34,70–73]. VDR apparently associates with all three RXR isotypes on the 24-hydroxylase VDRE; however, its transcriptional regulation in response to $1,25(OH)_2D_3$ may be restricted to complexes with RXRα and RXRγ, whereas VDR primarily associates with the RXRα and RXRβ isotypes in transcriptional regulation of the osteocalcin gene VDRE [74,75]. Similar RXR isotype preferences have been found for other nuclear receptors and their particular target genes [76].

The dimerization state on unliganded nuclear receptors in a cell remains unclear. However, using surface plasmon resonance (SPR), Cheskis and co-workers demonstrated that VDR and RXR can form homodimers and heterodimers in the absence of DNA *in vitro* [77]. Another group has shown that overexpressed TR and VDR (but, interestingly, not RAR or PPAR) can inhibit the constitutive response of a GAL4 DBD–RXR LBD–VP16 chimera from a GAL4 UAS-linked reporter plasmid in transfected cells. Because overexpression of VDR or TR can block VP16 transactivation, the authors interpret this to mean that VDR and TR can associated with RXR *in vivo* in the absence of ligand as well as stabilize DBD–response element intereactions [78]. Dimerization by nuclear receptors influences the recognition of DNA targets and confers specificity for a defined spacing between two directly repeating hexameric sequences, as discussed at length in Section II,B. The ability of nuclear receptors to form heterodimers portends cross talk between different ligands to potentially effect a range of physiological processes.

B. Influences of Ligand Binding on Dimerization

1. LIGANDS INDUCE TRANSCRIPTIONALLY ACTIVE DIMERIC COMPLEXES

As mentioned above, VDR and RXR can associate as homodimers and heterodimers *in vitro* in the absence of ligands and DNA. DNA binding stabilizes these "solution" (i.e., off DNA) interactions and induces a conformational change in VDR [41]. Highly purified recombinant VDR, expressed in *Escherichia coli*, can bind strongly to the VDRE from the osteopontin gene both as a homodimer and as a heterodimer complexed with RXR as measured by gel mobility shift [7,39–42]. It was therefore unclear to these authors which complex was responsible for regulating vitamin D_3-dependent gene expression from this VDRE *in vivo*. A high affinity ligand for RXR, 9-*cis*-retinoic acid (9-*cis*-RA), was demonstrated to enhance RXR homodimer DNA binding to DR1 elements (retinoid X response elements, RXREs) *in vitro* and to enhance transactivation from RXRE-linked reporters *in vivo* [79–81]. Later, it was additionally shown that 9-*cis*-RA also increased RXR–homodimer affinity in the absence of DNA [77]. Given that each receptor in an RXR–VDR heterodimer has an identified natural ligand, it was conceivable that each ligand might have a distinct effect on complex stability and resulting functions of the receptors.

By gel mobility shift and gel filtration analysis, VDR homodimer binding to the osteopontin VDRE is actually decreased by $1,25(OH)_2D_3$, and the primary VDR DNA-bound species is a liganded monomer [41]. Interestingly, $1,25(OH)_2D_3$ destabilized preformed VDR homodimer complexes in gel filtration experiments. It was later determined by SPR that these observations were primarily an effect of a decreased affinity of VDR for itself, as its K_D increased nearly 4-fold with approximately equal contributions to association and dissociation rates [77]. Importantly, addition of RXR in the absence of $1,25(OH)_2D_3$ to a preformed VDR homodimer could not significantly induce heterodimer formation, implying that one of the main roles of $1,25(OH)_2D_3$ could be to destabilize DNA-bound VDR homodimers. Whether VDR–VDR homodimers might form in the presence of almost ubiquitous RXR is unknown. In addition, $1,25(OH)_2D_3$ enhanced the formation of RXR–VDR DNA-bound heterodimers, which have additionally been observed to undergo a conformational change by an increase in gel shift mobility and by a decreased sensitivity to limited protease digestion [41,82,83; B. D. Lemon and L. P. Freedman, 1996, unpublished results]. These observations were later attributed to a 7-fold decrease in the K_D for RXR–VDR–DNA complexes resulting primarily from an increased association rate of solution interactions and a decreased dissociation rate of DNA-bound complexes [77]. This observation confirms the earlier discovery that $1,25(OH)_2D_3$ was capable of increasing the affinity of the VDR for a nuclear accessory factor by 8- to 9-fold in the absence of DNA [83a]. This nuclear accessory factor was revealed to be various mixtures of RXR.

Ligand effects on VDR dimerization were also examined by SPR with a series of vitamin D_3 analogs that more potently induce differentiation of myeloid-leukemic cell lines, such as HL-60, with less hypercalcemia than $1,25(OH)_2D_3$ in animal models. Of three analogs examined, two exhibited similar effects in enhancing RXR–VDR dimerization and decreased associations in combination with 9-*cis*-RA [84]. However, one such compound, $1,25(OH)_2$-16-ene-23-yne-26,27-hexafluoro-D_3, appeared to enhance RXR–VDR associations in the absence of DNA to a lesser extent than $1,25(OH)_2D_3$, but enhanced more tightly associated RXR–VDR complexes with DNA [84]. Additional studies have supported the notion that analogs are acting directly at the level of the receptor by demonstrating enhanced RXR–VDR DNA binding and resistance to protease digestion at lower effective concentrations of analogs relative to $1,25(OH)_2D_3$ [85–87]. These results imply that specific vitamin D_3 analogs might promote receptor conformations which preferentially stabilize RXR–VDR complexes on DNA, resulting in enhanced target gene activation (see Chapters 59 and 60 for a full discussion of the molecular basis of analog effects on VDR).

As mentioned above, 9-*cis*-RA was found to enhance RXR homodimer DNA binding to RXREs. 9-*cis*-RA also destabilized RXR–VDR heterodimer formation and DNA binding to the osteopontin VDRE by gel mobility shift [41]. Subsequently, it was determined that this was primarily an effect of decreased RXR–VDR association rates by 9-*cis*-RA. These effects could be attributed to a decreased available pool of RXR, given that 9-*cis*-RA stabilized RXR homodimers. As summarized in Fig. 5, these *in vitro* results suggested that $1,25(OH)_2D_3$ destabilizes inactive and potentially transcriptionally repressive DNA-bound VDR homodimers (formed in the absence of RXR) and enhances the formation of transcriptionally active RXR–VDR heterodimers, whereas 9-*cis*-RA has the opposing effect of shifting the equilibrium away from RXR–VDR heterodimers to RXR homodimers.

A similar model was proposed for thyroid hormone signaling, as it decreased TR association and 9-*cis*-RA inhibited RXR–TR binding to thyroid hormone responsive elements (TREs) [88–91]. Here, however, in contrast to VDR, strong evidence exists for the repressive nature of the TR homodimer. Retinoid signaling, on the other hand, may be distinct, as 9-*cis*-RA does not appear to significantly affect RXR–RAR heterodimer binding to a retinoic acid responsive element (RARE) [88]. This is possibly because 9-*cis*RA is a panagonist for RXRs and RARs; it is apparently equipotent to all-*trans*-RA for RARE transactivation and therefore would stabilize high affinity RAR–RXR heterodimers [79].

2. In Vivo Studies on $1,25(OH)_2D_3$–Retinoic Acid Cross Talk

The model derived from the *in vitro* studies described above (Fig. 5) was testable in cell culture by transient transfection experiments. Lemon and Freedman devised a system to investigate ligand-mediated effects

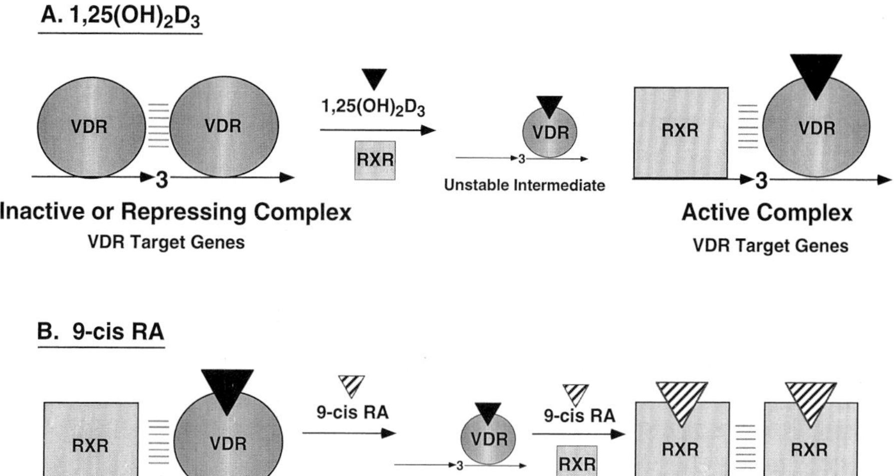

FIGURE 5 Model for ligand-modulated effects on dimerization by VDR and RXR. (A) $1{,}25(OH)_2D_3$ reduces the affinity of VDR homodimers and enhances RXR–VDR heterodimer affinity for VDRE (DR3) target DNA to transactivate VDRE-linked genes. Importantly, it is the RXR–VDR heterodimer that is the active complex from DR3 VDREs (see text, Section III,B,2) (B) 9-*cis*-RA reduces the affinity of RXR–VDR heterodimers by enhancing RXR homodimer affinity for RXRE (DR1) target DNA, thereby attenuating VDRE-linked gene expression and enhancing RXRE-linked gene expression. Dimeric complexes bound to VDRE and RXRE targets are shown and represent the net effects on equilibrium associations of VDR and RXR on and off DNA by addition of $1{,}25(OH)_2D_3$ and 9-*cis*-RA as determined by biochemical and cell-culture assays [31,41,77,96] (see also other references described in Section III,B).

on transactivation from VDRE-linked chloramphenicol acetyltransferase and RXRE-linked luciferase reporter plasmids singly or simultaneously introduced into cells in response to various combinations of overexpressed and endogenous receptors and specific ligands [31]. RXR enhanced $1{,}25(OH)_2D_3$-dependent stimulation from a VDRE-regulated reporter by both endogenous and overexpressed VDR, and 9-*cis*-RA both attenuated a single VDRE-containing reporter by approximately 30% and strongly attenuated a reporter containing two tandem VDREs by approximately 70%. When both a VDRE-containing reporter and an RXRE-containing reporter were simultaneously transfected, 9-*cis*-RA again attenuated transactivation from the VDRE-regulated reporter, and $1{,}25(OH)_2D_3$ attenuated the response to 9-*cis*-RA from the RXRE-regulated reporter. The attenuation of $1{,}25(OH)_2D_3$ responsiveness by 9-*cis*-RA from the VDRE-regulated reporter and of 9-*cis*-RA responsiveness by $1{,}25(OH)_2D_3$ from the RXRE-regulated reporter was apparent only when VDR was overexpressed relative to RXR. Linearly increased expression of RXR abolished the attenuation, and in some cells incurred a costimulation that was interpreted to be from RXR–RXR homodimers binding to and transactivating from the VDRE [31]. In addition, overexpression of RXR abolished the $1{,}25(OH)_2D_3$-dependent attenuation of 9-*cis*-RA responsiveness from the RXRE-regulated reporter.

Taken together, these results were consistent with the previously described *in vitro* results and suggested that the RXR–VDR heterodimer was the transcriptionally active receptor complex from this VDRE *in vivo*, and that the two reporters, and by inference VDRE- and RXRE-regulated genes, exhibit a competition for limiting concentrations of RXR in a cell. A similar RXR enhancement and 9-*cis*-RA-dependent attenuation of $1{,}25(OH)_2D_3$-dependent transactivation was reported earlier for the osteocalcin VDRE (which appears only to bind an RXR–VDR heterodimer), and from thyroid hormone-dependent transactivation from TRE-containing reporter [34,70,74,92,93]. Other groups have also observed costimulation by 9-*cis*-RA on overexpression of RXR relative to VDR from both the osteocalcin VDRE and a nonconventional palindromic VDRE in insect cells, and similarly through the rat growth hormone TRE composed of both an inverted palindrome and a direct repeat [30,94].

Additional support for the RXR–VDR heterodimer as the active receptor complex from a VDRE-linked reporter was derived from similar cotransfection experiments using Ligand Pharmaceuticals' RXR- and RAR-selective retinoids. Only the RXR-selective retinoid LG153, and not the RAR-selective retinoid LG272 (or TTNPB), could attenuate the $1,25(OH)_2D_3$-dependent transactivation from the VDRE. LG153 was able to induce transcription of an RXRE-regulated reporter, and this activation was attenuated by $1,25(OH)_2D_3$ [31]. An RAR-selective retinoid was shown to be the only requirement for RXR–RAR signaling from an RARE [95].

A cell-free transcription system using a G-free cassette-based assay with crude nuclear extracts, which is both completely dependent on exogenously provided highly purified recombinant VDR and RXR from *E. coli* or baculovirus expression systems and is responsive to ligands, has been described [96]. With this system, greater than 20-fold stimulation of transcription by receptors and physiologically responsive concentrations of $1,25(OH)_2D_3$ is observed. This system confirms that it is the RXR–VDR heterodimer which is the transcriptionally active receptor complex from the osteopointin VDRE, as ligand-dependent enhancement of transcription is observed only when both receptors are added. Similarly, the RXR–RAR and RXR–TR heterodimer complexes are required for transcriptional enhancement by RAR-selective retinoids and thyroid hormone in cell-free systems [97,98].

C. Corepressors and Promoter Occupancy

A few of the nuclear receptors, notably TR, RAR, and COUP-TF, have been shown to actively repress basal transcription either through direct contact with members of the general transcription machinery or by association with "corepressor" proteins, such as N-CoR or SMRT (also called thyroid and retinoid receptor associated repressors or TRAPs), based on *in vitro* and cell culture studies [78,99–107]. 5MRT-N-CoR appear to interact with a conserved domain in the "hinge region" of RAR and TR called a corepressor (CoR) box [103]. The consensus sequence of the CoR box is (S/T)x(E/Q)xx(D/E)LIxxxxxAHxxT, and it is located just C-terminal to the T/A box at the end of the DBD [103]. Interestingly, RXR does not contain a conserved CoR box (RXRα contains only the S and A residues, whereas RXRβ and RXRγ have only the A residue) and does not appear to interact with corepressors. VDR does contain a CoR box motif with two notable exceptions, SxExx**RI**IxxxxxAHxxT, of which the R was found to function in place of D/E in TR/RAR mutants, and a conserved I for L would not be expected to disrupt the interaction [103]. However, even possessing a near consensus CoR box, VDR does not appear to associate with the known corepressors. Nevertheless, there has been one report of basal repression mediated by VDR *in vivo* [108]. In contrast, some degree of ligand-independent transcriptional enhancement by RXR–VDR can be detected in the cell-free transcription system [96].

Whereas transactivation *in vivo* is completely dependent on ligand, a plausible explanation for the observed constitutive enhancement of transcription by RXR–VDR, as well as other steroid and nuclear receptors *in vitro*, including GR, PR, RAR, TR, and RXR [97,109–112], is the lack of chromatin assembly in the crude nuclear extracts. By footprinting analysis, DNase I hypersensitive sites are generated upstream of the calbindin-D_{28K} gene in isolated nuclei from $1,25(OH)_2D_3$-treated chicken intestinal cells [113], and *in vivo* occupancy of the osteocalcin promoter element by an RXR–VDR heterodimer also requires $1,25(OH)_2D_3$ [114]. Similarly, occupancy of the RARβ promoter RARE by an RXR–RAR heterodimer appears to require ligand [115]. However, *in vivo* occupancy may be more complicated, as another report describes occupancy of the TRβA promoter by both unliganded and liganded RXR–TR in *Xenopus* oocytes with an apparent functional requirement for the presence of the heterodimer during replication-coupled chromatin assembly [116]. In addition, members of the SWI/SNF family of chromatin remodelers have been shown to be essential for transactivation by steroid receptors in yeast [117]. Taken together, these results would suggest that chromatin antagonizes transactivation by nuclear receptors, and their potential association with nucleosome assembly machinery or SWI/SNF mediators (for review, see Kingston *et al.* [118]) might relieve this repression *in vivo*.

D. Dimerization Effects on the Ligand Binding Domain: Predictions Derived from Three-Dimensional Structures

1. Ligand Binding-Induced Changes in Tertiary Structure

The ligand binding domain (LBD) of the steroid/nuclear receptor superfamily contains a series of heptad repeats that form α helices (H1–H12) [9,10,119]. At the extreme C terminus of the LBD, a highly conserved ligand-dependent activation domain, called AF-2, was identified biochemically for many of the receptors, including VDR [120–122]. It is schematically presented in Fig. 6 as a helical wheel and will be discussed in more detail below. The amphipathic α helix (AH) formed by

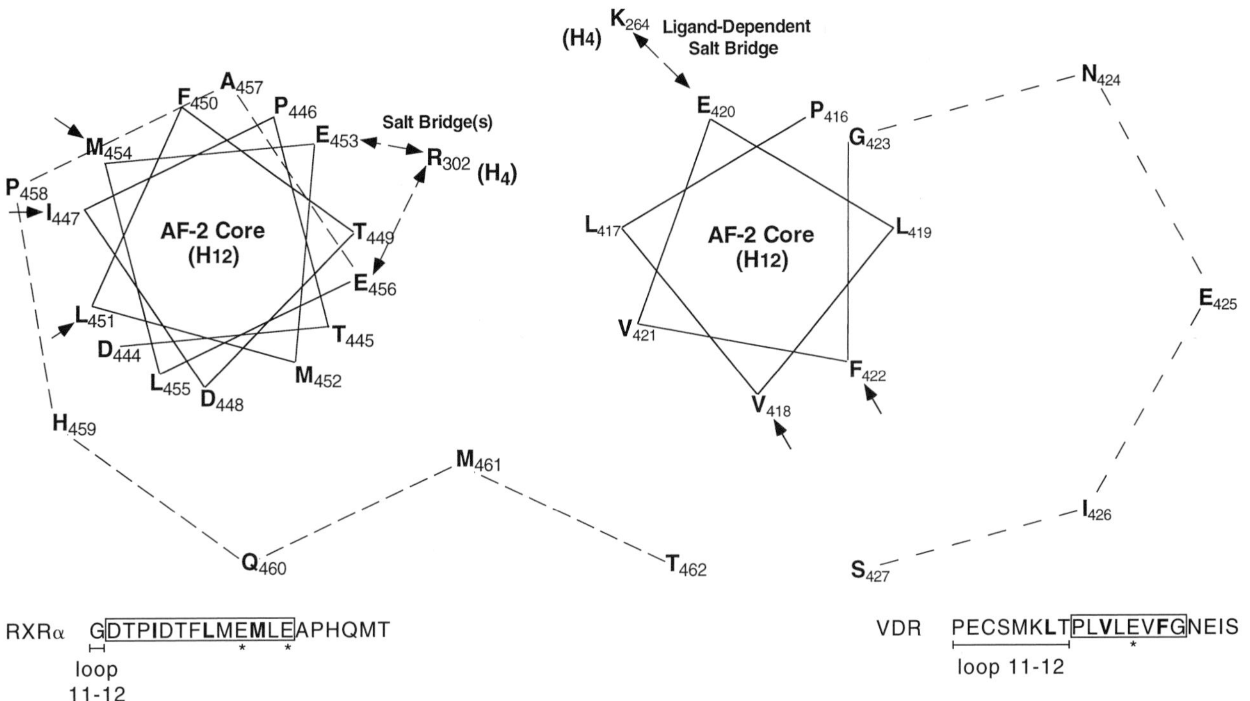

FIGURE 6 Schematic representation of RXR and VDR AF-2 domains. Helical wheels and sequences for AF-2 core residues of RXR (left) and VDR (right) were based on sequence alignment [131]. Solid lines (top) and boxed sequences (bottom) represent residues predicted to form the AF-2 core (H12), dashed lines represent contiguous C-terminal residues predicted to be disordered, broken arrows (top) and asterisks (bottom) depict predicted salt bridges, solid arrows (top) and boldface type (bottom) represent residues predicted to be in close proximity to ligands [i.e., 9-*cis*-RA for RXR and $1,25(OH)_2D_3$ for VDR], and extra sequence to the left of boxed sequence is that predicted to form loop 11–12. See Section III,D of the text for details.

H12, the core of AF-2 in some of the nuclear receptors, was able to constitutively activate when fused to a heterologous DBD (GAL4 or ER) much like previously described acidic, glutamine-rich, or proline-rich activation domains [81,121–123]. It was once thought that in the absence of ligand, AF-2 might be imbedded in an inactive state in the hydrophobic LBD core and, on ligand binding, was unmasked and exposed to directly contact components of the transcriptional machinery or the many putative cofactors/potential transcriptional mediators which have been found to associate with the receptors in a ligand-dependent manner [124–130]. Many of these interactions appeared to be mediated through AF-2, because disruption or deletion of this subdomain abrogated many of these interactions. However, from structural models, the effect on AF-2 seems to be more complicated than once thought.

Recently, X-ray crystallography has led to an elucidation of three LBD structures: unliganded (apo) RXRα LBD, liganded (holo) TRα1 LBD, and liganded (holo) RARγ LBD. The holo-TRα1 LBD structure has many similarities with the holo-RARγ LBD structure, and inferences to general structural changes resulting from ligand binding relative to the apo-RXRα LBD have been drawn by comparison [9,10,119]. Of important note, the apo-RXRα LBD is a dimer with 2-fold symmetry [primarily mediated by hydrophobic contacts between H10 and to a lesser extent H9 and loop 7–8 (between H7 and H8)], but the (holo) TRα1 and RARγ LBDs are monomers, consistent with observations from the biochemical methods described above. Overall, the apo-RXRα LBD (hereafter, apo-LBD) and holo-RARγ LBD (hereafter, holo-LBD) structures are very similar, both adopting an "antiparallel α-helical sandwich," and they have been extensively compared to give a picture of how the ligand effects distinct changes in the tertiary structure of the LBD [131].

The apo-LBD and holo-LBD secondary structural elements can be nearly superimposed, with the exceptions of the Ω loop (between helices H1/H2 and H3) and helices H11 and H12 [10,131]. A model of the potential structure of an apo-RXR–VDR LBD heterodimer based on the apo-RXRα LBD homodimer crystal structure is presented in Fig. 7A. Several major structural changes occur on ligand binding that result in a holo-LBD which is more compact than the apo-LBD, consistent with resistance to protease digestion and changes in mobility or elution profiles by biochemical assays

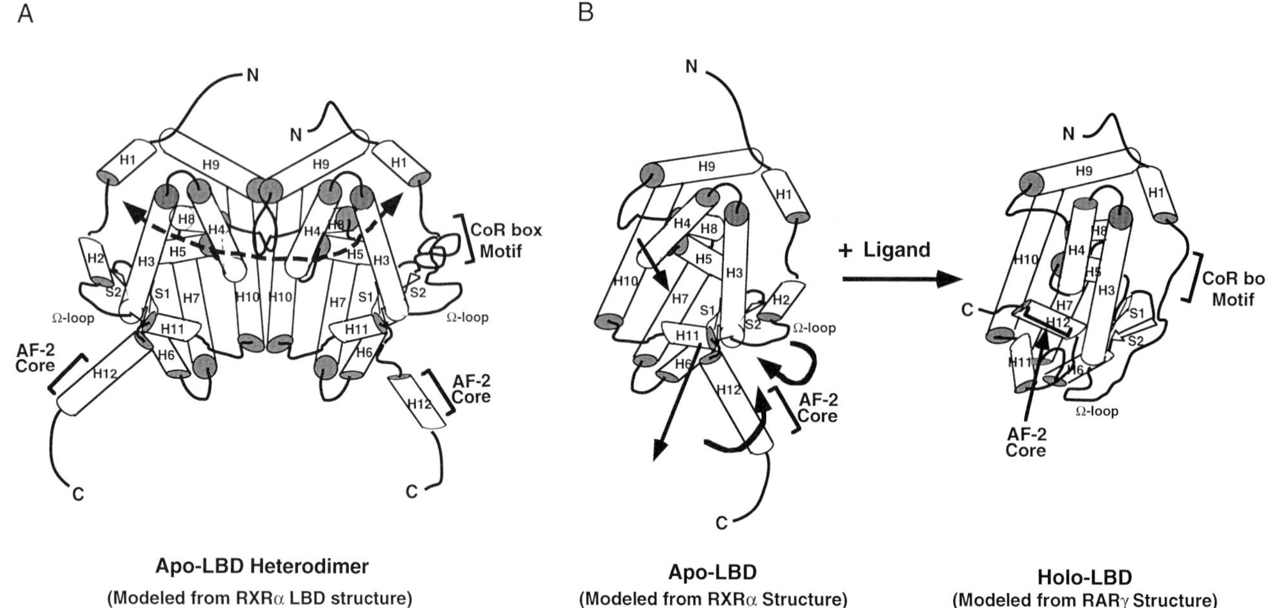

FIGURE 7 Structural representations of RXR and VDR LBDs and the conformational changes induced by ligand binding. (A) Three-dimensional model for an unliganded (apo) RXR (left) and VDR (right) LBD heterodimer based on structural determinations of an RXRα homodimer (9). Note primary contacts mediated by heptad 9 (α helix H10) and minor contacts by H9. In addition, the heterodimerization interface (so-called signature domain) determined biochemically and in cell culture (described in Section III,D,2,a) is found in H3, loop 3–4, and H4. A broken arrow is depicted to represent potential folding of the LBDs to accommodate closer contact by these regions. Also note the existence of an extended stem in VDR between H1 and H3 (H2 is only present in RXR) before the Ω loop which contains the putative corepressor binding site, or CoR box motif (described in Section III,C), represented as a disordered array. (B) Three-dimensional model for potential structural transitions in the VDR LBD mediated by $1,25(OH)_2D_3$ binding based on comparisons of the superimposable structural determinations of the apo-RXRα homodimer and the holo-RARγ monomer complexed to all-*trans*-RA [9,10,131]. Note arrows depicting changes (described in Section III.D) in specific regions, most importantly, the folding down of H11 and folding back of H12 (the AF-2 core) toward the hydrophobic center of the LBD to come into closer proximity to H4, and the flip of the Ω loop (N-terminal to H3) to reposition under H6 and stabilize H11/H12 transitions.

described above. The most prominent structural transitions include a 70° tilt of H11 away from the hydrophobic core of the LBD, a folding back of H12 toward the core, a 30° tilt of H6, and a "flip" of the Ω loop to become repositioned under H6 by a "hinged-lid motion," thereby bending the N-terminal part of H3 toward the hydrophobic core, which also stabilizes the shift in H11 and H12 (Fig. 7B). Collectively, these structural rearrangements form a "ligand binding pocket" or "cavity" created by clustal residues in H1, H3, H5, a β turn between H5 and H6, loop 6–7 (between H6 and H7), H11, loop 11–12, and H12 which envelopes the ligand by a so-called mouse trap mechanism [10,119,131]. Whether an analogous rearrangement in response to $1,25(OH)_2D_3$ occurs must await elucidation of the structure of the VDR LBD.

Renaud and co-workers describe a requirement for a salt bridge between highly conserved residues in H4 (K264 in RARγ corresponding to K264 in human VDR) and H12 (E414 in RARγ corresponding to E420 in VDR) (shown in Fig. 6) on ligand binding to prevent H12 protrusion from the LBD hydrophobic core into a solvent pocket (depicted in Fig. 6). Interestingly, a K264A RARγ mutant binds ligand and DNA and heterodimerizes with RXR with wild-type affinity, but it is dominant negative for transactivation in cell culture [10]. The converse mutations E412P in RARα and E401Q in TRα give identical results with the additional abrogation of the ability of TRα to repress transcription in the absence of ligand [121,122]. In addition, a mutation in H3 of VDR, M226R, immediately adjacent to a cluster of residues found to be in close proximity to all-*trans*-RA in the RARγ structure, displays wild-type DNA binding and heterodimerization affinities, but it is dominant negative for transactivation in cell culture and exhibits a K_D for $1,25(OH)_2D_3$ greater than 3 orders of magnitude higher than wild type (B. D. Lemon, B. Cheskis, and L. P. Freedman, 1996, unpublished results). Taken together, it is likely although not proven that the crystal structure of human VDR will have a similar fold to those described for RARγ and TRα when complexed with ligand.

2. Quaternary Implications for Ligand Binding

a. Dimerization as Determined by Specific Domains A core group of highly conserved hydrophobic residues peppered with nonconserved residues in H3 and that of the ligand binding domain (corresponding to residues F244 to L263 in VDR), called the "signature" domain, exists between clusters of residues in close proximity to the ligand in the RARγ structure and is putatively responsible for ligand specification [131] (Fig. 7). This domain was found to be essential for heterodimerization of TR, RAR, and VDR with RXR [132]. Deletion or disruption of this domain by specific point mutations in RAR, TR, and VDR abrogated heterodimerization and transactivation but did not appreciably affect ligand affinity or DNA binding [132,133]. Deletion of this region in RXR produced the same effect, but point mutations had minimal effects, suggesting that the RXR dimer interface may be flexible with many potential contact points [132]. Additionally, deletion of the ninth heptad (corresponding to H10) in RXR abolishes both dimerization and transactivation [134].

As described above, H10 provides the major dimerization contacts for the RXR LBD homodimer interface; however, taking into account the observations from mutations within the signature domain, the heterodimer interface is likely to involve other regions of the LBD(s) (Fig. 7A). The AF-2 regions of VDR and RXR are not required for DNA binding or homo- or heterodimerization, as a mutant truncated after M412, called VDRΔAF-2, and a mutant with randomly inserted residues after L414, called VDR-AF-2RND exhibited wild-type affinities when complexed with one another, wild-type RXR, or RXRΔC2 (mouseRXRβ truncated after L427 described by Leng *et al.* [81]) (B. D. Lemon and L. P. Freedman, 1996, unpublished results). However, these VDR mutants acted as dominant negatives in transactivation assays in cell culture and are inactive in cell-free transcription, probably as a consequence of their severely compromised ligand affinities [96] (B. D. Lemon and L. P. Freedman, 1996, unpublished results). A similar result was described for a VDR truncation after R402 [135]. In contrast, a VDR truncation mutant after A381 and several point mutations within K382–L390 in the N-terminal part of H10 (i.e., ninth heptad) abolished heterodimerization and ligand binding, consistent with predictions based on the crystal structures previously described [135].

b. Role of Dimerization in Transactivation: Allosteric Influence on the LBD As mentioned before, RXR is a common "silent" (i.e., nonsignaling) partner with VDR, RAR, TR, and PPAR, and it occupies the upstream half-site with the signaling partner occupying the downstream half-site [5,31,136]. Interestingly, RXR can become a signaling partner with silent orphan receptors LXR, NGFI-B (Nurr77), and NURR1 either when tethered to the DNA-bound orphan receptor or, uncharacteristically for signaling partners, when it retains its upstream polarity when bound to DNA in certain cases [137–139]. Heterodimerization and/or ligand binding may induce an allosteric change in the LBD of the dimeric partner(s). RXR appears to confer ligand dependence on NGFI-B/NURR1, which usually transactivates constitutively as a monomer from an extended half-site, by repressing transcription from a reporter in the absence of RXR-specific ligands and transactivating in the presence of RXR-specific ligands [26,137,138]. Another RXR partner, FXR (farnesoid X-activated receptor), is dependent on RXR for transactivation in response to farnesol metabolites which act synergistically on the FXR–RXR heterodimer in the presence of RXR-specific ligands [140].

DNA might also act as an allosteric effector to enhance dimerization and induce conformational changes of nuclear receptors. As we have reiterated, the primary determinant of steroid/nuclear receptor dimerization is the symmetric interface of the LBD, which is modulated by ligand binding. Cooperative enhancement of dimerization is achieved through the second dimerization interface in the DBD: symmetric for steroid receptors binding to inverted repeats, and asymmetric for nuclear receptors binding to direct repeats, as described in Sections II,A and II,B. DNA therefore stabilizes high affinity interactions leading to activation or repression of linked genes. Although the most important complex for $1,25(OH)_2D_3$-dependent transcription from characterized genes is an RXR : VDR heterodimer, this does not preclude the possibility of transactivation by other complexes. If the DBD of VDR is replaced by a heterologous DBD, such as the GR DBD (called VGV), $1,25(OH)_2D_3$-dependent transactivation can occur by a VGV homodimer from a GRE-linked reporter in mammalian cell culture, thereby demonstrating that VDR has intrinsic activation properties [31]. Unlike the case for VDR homodimers, VGV–DNA interactions are apparently not reduced by $1,25(OH)_2D_3$, perhaps as a result of the intrinsically strong symmetrical interface attributed to the palindromic organization of the GRE target. Therefore, it seems that if DNA can stabilize lower affinity complexes, perhaps acting as an allosteric effector, it is conceivable that receptor dimers might transactivate from nonconventional response element in response to ligands.

Certain nonconventional VDREs, including some resembling steroid responsive elements as inverted palindromes with an extended spacer, have been reported that appear to be transactivated by VDR dimers other than RXR–VDR, such as VDR homodimers and RAR–VDR and TR–VDR heterodimers [43,141–143]. In one of these reports, a TR–VDR heterodimer complex appeared to bind to an atypical DR4 element with VDR–TR polarity and to an atypical DR3 element with TR–VDR polarity. As with RXR–VDR heterodimers, response to either triiodothyronine (T_3) or 1,25$(OH)_2D_3$ from these elements occurred when the cognate receptor was oriented over the downstream half-site [142]. DNA binding and transactivation by nonconventional dimer pairs may be stabilized by interactions with promoter proximal nonreceptor factors or by specific interactions with non-half-site DNA bases within the spacer or surrounding DNA, as has been described for some orphan receptors [26]. DNA may also act as an allosteric effector in the conversion of a ligand-dependent transactivating RXR–VDR heterodimer to a ligand-dependent repressor by inducing a conformational change on receptor binding to specific DNA sequences, such as the negative VDREs discussed in Section II,E.

Nuclear receptor dimerization can also affect receptor/ligand affinity. In addition to restricted DNA binding polarity by RXR–NR heterodimers (where NR represents VDR, RAR, TR, orphans, etc.) directed by determinants within the DNA binding domain as described in Section II,C, many RXR partners restrict ligand binding by RXR. In both the absence and presence of RAR-specific ligands, RAR inhibits RXR-specific ligand binding by RXR when bound to a DR1 (RAR–RXR polarity—inactive) and DR5 (RXR–RAR polarity–RAR-specific signaling) response elements [136,137]. The RAR LBD is solely responsible for this inhibition as expressed, and purified RAR and RXR LBDs exhibit the same results. TR also exhibits the same suppression of ligand binding by RXR, but in this case ligand binding by TR is required [137]. In contrast, RXR appears to enhance 1,25$(OH)_2D_3$ binding by VDR, and VDR does not inhibit 9-*cis*-RA binding by RXR, since 9-*cis*-RA can destabilize preformed RXR–VDR heterodimers [41]. Therefore, it appears that the LBDs of specific partners interact with RXR in slightly different ways. The various associations and structural transitions modulated by ligand binding and dimerization establish a complex network that allows for diverse responsiveness to signals with a limited number of involved players. Further crystallographic comparisons of apo- and holoreceptors in the context of heterodimeric pairs will be necessary to determine exactly how these changes are manifested, and how they might affect interactions with various transcriptional components, including the general transcriptional machinery, possible cofactors or transcriptional mediators, and repressors.

c. Dimerization in Context: Stabilization of Auxiliary Components Even without a heterodimer structure complexed to ligand, it is possible to make some predictions from biochemical studies. Although there is no general requirement for the AF-2 domain(s) in dimerization or DNA binding, there is an apparent need for not only the AF-2 domain of the signaling dimeric partner, but also for the AF-2 region of a "silent" partner for transactivation. Thus, there is likely an activation surface composed of the two activation domains of a dimer pair, proposed here to be $(AF\text{-}2)_2$. Evidence for this includes a functional requirement for the AF-2 core or "C2" region of RXRα and RXRβ for transactivation from RARE- and VDRE-linked reporters in response to all-*trans*-RA and 1,25$(OH)_2D_3$, respectively [144]. In addition, there is also a requirement for the RXR AF-2 region for *in vivo* occupancy of the βRARE by an RXR–RAR heterodimer in response to all-*trans*-RA [144]. Another study corroborated these results by overexpressing the VDR, RAR, or TR LBDs with either a GAL4 DBD–RXR LBD or GAL4 DBD–RXRΔC2 equivalent in transfects cells. The RXR AF-2 domain was required for transactivation from a GAL4 UAS-containing reporter plasmid with these constructs, and from DR3, -4, and -5 series reporters with overexpressed full-length VDR, TR, and RAR receptors in response to ligands [145]. In addition, deletion of the AF-2 domain in RXR increased the association of RXR–RAR heterodimers with the corepressor SMRT. A cell-free transcription system for RXR–RAR signaling described a requirement for the AF-2 domain of RAR, but not for that of RXR [97]. If VDR were to interact with as yet unidentified corepressor proteins, the requirement for the AF-2 region in RXR for 1,25$(OH)_2D_3$-dependent signaling would be more clear. Another attractive and not necessarily inconsistent model that could account for these apparent discrepancies would be if the $(AF\text{-}2)_2$ activation domain were responsible for direct interaction with chromatin or SWI/SNF remodelers *in vivo*. As mentioned above, these complexes are known to be essential for steroid receptor signaling in yeast, for which there would be no requirement in cell-free systems in the absence of chromatin-assembled templates [117].

It is likely that additional regions of the LBD other than AF-2 comprise the activation surface(s). As described earlier, candidates for these other areas would include those immediately surrounding the AF-2 core, such as exposed residues on the surfaces of H3, H4, loop 11–12, and the insertion between H1 and H3 comprising the stem and Ω loop (see Fig. 7). This is particularly attractive for VDR as this region contains nearly 40 or 50 more residues [30% of which are glutamine (Q), proline (P), or acidic aspartate/glutamate (D/E), and another 30% of which are serine/threonine (S/T) that, if phosphorylated, could contribute to the overall acidity of the region] than the corresponding regions of TR and RAR, respectively, by sequence alignment. These residues could additionally account for the lack of association of VDR with the known corepressors N-CoR and SMRT, as this region is immediately adjacent to the proposed CoR box in H1 and may block or alter the conformation of H1. Support for the notion that other residues and multiple surfaces are involved in formation of an activation or repressive domain includes the requirement for residues in H1 and AF-2 for association with corepressors by TR and RAR [102,103,146]. Similarly, the interaction of both TR and VDR with TFIIB was dependent on residues in H1 and the stem/Ω-loop region [147–149]. The VDR–TFIIB interaction was enhanced by $1,25(OH)_2D_3$ when complexed with RXR on VDRE-linked promoter DNA [but apparently not with glutathione-*S*-transferase (GST)–VDR alone or with VDR in a two-hybrid assay in yeast] and cooperatively enhanced transcription in a cell-free system and in transactivation assays in cell culture, supporting the requirement for a common activation domain between the dimer pair [148,149].

As described above, several members of the steroid/nuclear receptor superfamily have been shown to have ligand-dependent associations with putative cofactors that are not members of the general transcriptional machinery. Few of these interacting proteins, with the notable exceptions of SRC-1 and CBP, have been demonstrated to have bona fide transcriptional coactivator activity with nuclear/steroid receptor family members [127,129]. Two of these putative coactivators, mSUG-1 and TIF-1, have been demonstrated to interact with VDR in a yeast two-hybrid assay [130]; however, no functional activity has yet been assigned to them in VDR-mediated transcription, and the role of SUG-1/TRIP-1 as a potential transcription mediator may be more complex than originally thought, as it was found to be a component of the proteosome (J. D. Fondell and R. G. Roeder, 1996, personal communication). One possibility for more restricted regulatory effects by nuclear receptors is the relative availability of cofactors required for transcriptional activation, and how these cofactors affect the structure and function of AF-2 and/or the entire ligand binding domain [150]. It is likely that the RXR–VDR dimer pair comprises a common activation domain which is rearranged upon dimerization and/or ligand binding, and may be altered by associations with regulatory proteins or adaptors to affect the stability of complexes and DNA binding. It will be necessary to examine in fine detail the intermolecular contacts mediated by putative cofactors/repressors with the RXR–VDR dimer, both biochemically and structurally, to gain a further understanding of the mechanisms by which this class of activators stimulates transcription of linked genes in response to $1,25(OH)_2D_3$.

Acknowledgments

The authors are indebted to current and previous laboratory colleagues who have contributed to many of the conceptual points discussed in this chapter, in particular, Iris Alroy, Boris Cheskis, and Terri Towers. We are also grateful to the structural biologists whose insights have increased our understanding of how VDR binds DNA: Ben Luisi, Fraydoon Rastinjead, John Schwabe, and Paul Sigler. We thank Fraydoon Rastinjead for contributing Fig. 4. Work in our laboratory described in this review is supported by National Institutes of Health Grant DK45460 and the Human Frontiers Science Program. L.P.F. is a Scholar of the Leukemia Society of America.

References

1. Freedman LP, Luisi BF 1993 On the mechanism of DNA binding by nuclear hormone receptors: A structural and functional perspective. J Cell Biochem **51**:140–150.
2. Mangelsdorf DJ, Evans RM 1995 The RXR heterodimers and orphan receptors. Cell **83**: 841–850.
3. Umesono K, Murakami KK, Thompson CC, Evans RM 1991 Direct repeats as selective response elements for the thyroid hormone, retinoic acid, and vitamin D_3 receptors. Cell **65**:1255–1266.
4. Naar AM, Boutin JM, Lipkin SM, Yu VC, Holloway JM, Glass CK 1991 The orientation and spacing of core DNA-binding motifs dictate selective transcriptional responses to three nuclear receptors. Cell **65**:1267–1279.
5. Perlmann T, Rangarajan PN, Umesono K, Evans RM 1993 Determinants for selective RAR and TR recognition of direct repeat HREs. Genes Dev **7**:1411–1422.
6. Kurokawa R, Yu V, Naar A, Kyakumoto S, Han Z, Silverman S, Rosenfeld MG, Glass CK 1993 Differential orientations of the DNA binding domain and C-terminal interface regulate binding site selection by nuclear receptor heterodimers. Genes Dev **7**:1423–1435.
7. Towers TL, Luisi BL, Asianov A, Freedman LP 1993 DNA target selectivity by the vitamin D_3 receptor: Mechanism for dimer binding to an asymmetric repeat element. Proc Natl Acad Sci USA **90**:6310–6314.
8. Zechel C, Shen X-Q, Chen J-Y, Chen Z-P, Chambon P, Groyenmeyer H 1994 The dimerization interfaces formed between the DNA binding domains of RXR, RAR, and TR determine binding specificity and polarity to direct repeats. EMBO J **13**:1425–1433.
9. Bourguet W, Ruff M, Chambon P, Groyenmeyer H, Moras D 1995 Crystal structure of the ligand-binding domain of the human nuclear receptor RXR-α. Nature **375**:377–388.
10. Renaud J-P, Rochel N, Ruff M, Vivat V, Chambon P, Groyenmeyer H, Moras D 1995 Crystal structure of the RAR-γ ligand-binding domain bound to all-*trans*-retinoic acid. Nature **378**:681–689.
11. Freedman LP, Luisi BF, Korszun ZR, Basavappa R, Sigler PJ,

Yamamoto KR 1988 The function and structure of the metal coordination sites within the glucocorticoid receptor DNA binding domain. Nature **334**:543–546.
12. Mader S, Kumar V, deVereneuil H, Chambon P 1989 Three amino acids of the oestrogen receptor are essential to its ability to distinguish an oestrogen from a glucocorticoid-responsive receptor. Nature **338**:271–274.
13. Schena M, Freedman LP, Yamamoto KR 1989 Mutations in the glucocorticoid zinc finger region that distinguish interdigitated DNA binding and transcriptional enhancement activities. Genes Dev **3**:1590–1601.
14. Hard T, Kellenbach E, Boelens R, Maler BA, Dahlman K, Freedman LP, Carlstedt-Duke J, Yamamoto KR, Gustafsson JA, Kaptein R 1990 Solution structure of the glucocorticoid receptor DNA binding domain. Science **249**:157–160.
15. Schwabe JWR, Neuhaus D, Rhodes D 1990 Solution structure of the DNA binding domain of the oestrogen receptor. Nature **348**:458–461.
16. Lee MS, Kliewer SA, Provencal J, Wright PE, Evans RM 1993 Structure of the retinoid X receptor alpha DNA binding domain: A helix required for homodimeric DNA binding. Science **260**:1117–21.
17. Luisi BF, Xu W, Otwinowski Z, Freedman LP, Yamamoto KR, Sigler PB 1991 Crystallographic analysis of the interaction of the glucocorticoid receptor with DNA Nature **352**:497–505.
18. Schwabe JWR, Chapman L, Finch JT, Rhodes D 1993 The crystal structure of the oestrogen receptor DNA-binding domain bound to DNA: How receptors discriminate between their response elements. Cell **75**:567–578.
19. Rastinejad F, Perlmann T, Evans RM, Sigler PB 1995 Structural determinants of nuclear receptor assembly on DNA direct repeats. Nature **375**:203–211.
20. Berg JM 1988 Proposed structure for the zinc-binding domains from transcription factor IIIA and related proteins. Proc Natl Acad Sci USA **85**:99–102.
21. Marmonstein R, Carey M, Ptashne M, Harrison SC 1992 DNA recognition by GAL4: Structure of a protein : DNA complex. Nature **356**:408–414.
22. Umesono K, Evans RM 1989 Determinants of target gene specificity for steroid/thyroid hormone receptors. Cell **57**:1139–1146.
23. Ponglikimongkol M, Green S, Chambon P 1988 Genomic organization of the human oestrogen receptor gene. EMBO J **7**:3385–3388.
24. Schena M, Freedman LP, Yamamoto KR 1989 Mutations in the glucocorticoid receptor zinc finger region that distinguish interdigitated DNA binding and transcriptional enhancement activities. Genes Dev **3**:1590–1601.
25. Patel L, Abate C, Curran T 1990 Altered protein conformation on DNA binding by Fos and Jun. Nature **347**:572–575.
26. Wilson TE, Paulsen RE, Padgett KA, Milbrandt J 1992 Participation of non-zinc finger residues in DNA binding by two orphan receptors. Science **256**:107–110.
27. Freedman LP, Towers T 1991 DNA binding properties of the vitamin D_3 receptor zinc finger region. Mol Endocrinol **5**:1815–1826.
28. Wilson TE, Fahrner TJ, Johnston M, Milbrandt J 1991 Identification of the DNA binding site of NFGI-B by genetic selection in yeast. Science **252**:1297–1300.
29. Xu W, Alroy I, Freedman LP, Sigler PB 1994 The stereochemistry of specific steroid receptor–DNA interactions. Cold Spring harbor Symp Quant Biol **58**:133–139.
30. Schrader M, Nayeri S, Kahlen J-P, Muller KM, Carlberg C 1995 Natural vitamin D_3 response elements formed by inverted palindromes: Polarity-directed ligand sensitivity of VDR–RXR heterodimer-mediated transactivation. Mol Cell Biol **15**:1154–1161.
31. Lemon BD, Freedman LP 1996 Selective effects of ligands on vitamin D_3 receptor- and retinoid X receptor-mediated gene activation in vivo. Mol Cell Biol **16**:1006–1016.
32. Hughes MR, Malloy PJ, Kieback DG, Kesterson RA, Pike JW, Feldman D, O'Malley BW 1988 Point mutations in the human vitamin D receptor gene associated with hypocalcemic rickets. Science **242**:1702–1705.
33. Sone T, Marx SJ, Liberman UA, Pike JW 1990 A unique point mutation in the human vitamin D receptor chromosomal gene confers hereditary resistance to 1,25-dihydroxyvitamin D_3. Mol Endocrinol **4**:623–631.
34. Sone T, Kerner S, Pike JW 1991 Vitamin D receptor interaction with specific DNA. J Biol Chem **266**:23296–23305.
35. Hughes MR, Malloy PJ, O'Malley BW, Pike JW, Feldman D 1991 Genetic defects of the 1,25-dihydroxyvitamin D_3 receptor. J Receptor Res **11**:699–716.
36. Luisi BF, Schwabe JWR, Freedman LP 1994 The steroid/nuclear receptors: From three-dimensional structure to complex function. In: Litwak G (ed) Vitamins and Hormones, Vol. 49. Academic Press, San Diego, pp. 1–47.
37. Alroy I, Freedman LP 1992 DNA binding analysis of glucocorticoid receptor specificity mutants. Nucleic Acids Res **20**:1045–1052.
38. Noda M, Vogel RL, Craig AM, Prahl J, DeLuca HF, Denhardt DT 1990 Identification of a DNA sequence responsible for binding of the 1,25-dihydroxyvitamin D_3 receptor and enhancement of mouse secreted phosphoprotein 1 (Spp-1 or osteopontin) gene expression. Proc Natl Acad Sci USA **87**:9995–9999.
39. Freedman LP, Arce V, Perez Fernandez R 1994 DNA sequences that act as high affinity targets for the vitamin D_3 receptor in the absence of the retinoid X receptor. Mol Endocrinol **8**:265–273.
40. Nishikawa J, Kitaura M, Matsumoto M, Imagawa M, Nishihara T 1994 Difference and similarity of DNA sequence recognized by VDR homodimer and VDR/RXR heterodimer. Nucleic Acids Res **22**:2902–2907.
41. Cheskis B, Freedman LP 1994 Ligand modulates the conversion of DNA-bound vitamin D_3 receptor (VDR) homodimers into VDR–retinoid X receptor heterodimers. Mol Cell Biol. **14**:3329–3338.
42. Koszewski NJ, Reinhardt TA, Horst RL 1996 Vitamin D receptor homodimer interactions with the murine osteopontin response element. J Steroid Biochem Mol Biol **59**:377–388.
43. Carlberg C, Bendik I, Wyss A, Meier E, Sturzenbecker LJ, Grippo JF, Hunziker W 1993 Two nuclear signalling pathways for vitamin D. Nature **361**:657–660.
44. Minghetti PP, Norman AW 1988 $1{,}25(OH)_2$-vitamin D_3 receptors: Gene regulation and genetic circuitry. FASEB J **2**:3043–3053.
45. Zierold C, Darwish HM, DeLuca HF 1995 Two vitamin D response elements function in the rat 1,25-dihydroxyvitamin D 24-hydroxylase promoter. J Biol Chem **270**:1675–1678.
46. Liu M, Lee M-H, Cohen M, Freedman LP 1996 Transcriptional activation of the p21 gene by vitamin D_3 leads to the differentiation of the myelomonocytic cell line U937. Genes Dev **10**:142–153.
47. Kerner SA, Scoot RA, Pike JW 1989 Sequence elements in the human osteocalcin gene confer basal activation and inducible response to hormonal vitamin D_3. Proc Natl Acad Sci USA **86**:4455–4459.
48. Morrison NA, Shine J, Fragonas JC, Verkest V, McMenemy ML, Eiseman JA 1989 1,25-dihydroxyvitamin D_3-responsive element

and glucocorticoid repression in the osteocalcin gene. Science **246**:1158–1161.
49. Ozono K, Liao J, Kerner SA, Scott RA, Pike JW 1990 The vitamin D responsive element in the human osteocalcin gene. J Biol Chem **265**:21881–21888.
50. Owen TA, Bortell R, Yocum SA, Smock SL, Zhang M, Abate C., Shalhoub V, Aronin N, Wijnen KLV, Stein JL, Curran T, Lian JB, Stein GS 1990 Coordinate occupancy of AP-1 sites in the vitamin D-responsive and CCAAT box elements by Fos-Jun in the osteocalcin gene: Model for phenotype suppression of transcription. Proc Natl Acad Sci USA **87**:9990–9994.
51. Demay MB, Gerardi JM, DeLuca HF, Kronenberg HM 1990 DNA sequences in the rat osteocalcin gene that bind the 1,25-dihydroxyvitamin D_3 receptor and confer responsiveness to 1,25-dihydroxyvitamin D_3. Proc Natl Acad Sci USA **87**:369–373.
52. Gill RK, Christakos S 1993 Identification of sequence elements in mouse calbindin-D28k gene that confer 1,25-dihydroxyvitamin D_3- and butyrate-inducible responses. Proc Natl Acad Sci USA **90**(7):2984–2988.
53. Darwish HM, DeLuca HF 1992 Identification of a 1,25-dihydroxyvitamin D_3-response element in the 5′ flanking region of the rat calbindin D-9k gene. Proc Natl Acad Sci USA **89**:603–607.
54. Zierold C, Darwish HM, DeLuca HF 1994 Identification of a vitamin D-response element in the rat calcidiol (25-hydroxy-vitamin D_3) 24-hydroxylase gene. Proc Natl Acad Sci USA **91**: 900–904.
55. Ohyama Y, Ozono K, Uchida M, Skinki T, Kato S, Suda T, Yamamoto O, Noshiro M, Kato Y 1994 Identification of a vitamin D responsive-element in the 5′ flanking region of the rat 25-hydroxyvitamin D_3 24-hydroxylase gene. J Biol Chem **269**:10545–10550.
56. Hahn C, Kerry D, Omdahl J, May B 1994 Identification of a vitamin D responsive element in the promoter of the rat P450 gene. Nucleic Acids Res **22**:2410–2416.
57. Cao X, Ross FP, Zhang L, MacDonald PN, Chappel J, Teitelbaum SF 1993 Cloning of the promoter for the avian integrin $\beta 3$ subunit gene and its regulation by 1,25-dihydroxyvitamin D_3. J Biol Chem **268**:27371–27380.
58. Miner JN, Yamamoto KR 1991 Regulatory crosstalk at composite response elements. Trends Biochem Sci **16**:423–426.
59. Demay MB, Kiernan MS, DeLuca HF, Kronenberg HM 1992 Characterization of the 1,25-dihydroxyvitamin D_3 receptor interactions with target sequences in the rat osteocalcin gene. Mol Endocrinol **5**:557–562.
60. Liu SM, Koszewski N, Lopez M, Malluche HH, Olivera A, Russell J 1996 Characterization of a response element in the 5′-flanking region of the avian (chicken) PTH gene that mediates negative regulation of gene transcription by 1,25-dihydroxyvitamin D_3 and binds the vitamin D_3 receptor. Mol Endocrinol **10**:206–215.
61. Kremer R, Sebag M, Champigny C, Meerovitch K, Hendy GN, White J, Goltzman D 1996 Identification and characterization of 1,25-dihydroxyvitamin D_3-responsive repressor sequences in the rat parathyroid hormone-related peptide gene. J Biol Chem **271**:16310–16316.
62. Alroy I, Towers TL, Freedman LP 1995 Transcriptional repression of the interleukin-2 gene by vitamin D_3: Direct inhibition of NFAT-AP1 complex formation by a nuclear hormone receptor. Mol Cell Biol **15**:5789–5799.
63. Sodek J, Kim RH, Ogata Y, Li J, Yamauchi M, Zhang Q, Freedman LP 1995 Regulation of bone sialoprotein gene transcription by steroid hormones. Connect Tissue Res **32**:209–217.
64. Truss M, Beato M 1993 Steroid hormone receptors: Interaction with deoxyribonucleic acid and transcription factors. Endocr Rev **14**:459–479.
65. Tsai M-J, O'Malley BW 1994 Molecular mechanisms of action of steroid/thyroid receptor superfamily members. Annu Rev Biochem **63**:451–486.
66. Bagchi MK, Tsai SY, Tsai M-J, O'Malley BW 1990 Identification of a functional intermediate in receptor activation in progesterone-dependent cell-free transcription. Nature **345**:547–550.
67. Dalman FC, Sturzenberger LJ, Levin A, Lucas DA, Perdew GH, Petkovich M, Chambon P, Grippo JF, Pratt WB 1991 Retinoic acid receptor belongs to a subclass of nuclear receptors that do not form "docking" complexes with hsp90. Biochemistry **30**:5605–5608.
68. Mangelsdorf DJ, Borgmeyer U, Heyman RA, Zhou JY, Ong ES, Oro AE, Kakizuka A, Evans RM 1992 Characterization of three RXR genes that mediate the action of 9-*cis*-retinoic acid. Genes Dev **6**:329–344.
69. Mangelsdorf DJ, Umesono K, Evans RM 1994 The retinoid receptors In: Spron MB, Roberts AB, Goodman DS (eds) The Retinoids: Biology, Chemistry, and Medicine, 2nd Ed. Raven, New York, pp. 319–349.
70. Liao J, Ozono K, Sone T, McDonnell DP, Pike JW 1990 Vitamin D receptor interaction with specific DNA requires a nuclear protein and 1,25-dihydroxyvitamin D_3. Proc Natl Acad Sci USA **87**:9751–9755.
71. Sone T, Ozono K, Pike JW 1991 A 55-kilodalton accessory factor facilitates vitamin D receptor DNA binding. Mol Endocrinol **5**:1578–1586.
72. Ross TK, Moss, VE, Prahl JM, DeLuca HF 1992 A nuclear protein essential for binding of rat 1,25-dihydroxyvitamin D_3 receptor to its response elements. Proc Natl Acad Sci USA **89**:256–260.
73. Munder M, Herzberg IM, Zierold C, Moss VE, Hanson K, Clagett-Dame M, DeLuca HF 1995 Identification of the porcine intestinal accessory factor that enables DNA sequence recognition by vitamin D receptor. Proc Natl Acad Sci USA **92**:2795–2799.
74. MacDonald PN, Dowd DR, Nakajama S, Galligan MA, Reeder MC, Haussler CA, Ozato K, Haussler MR 1993 Retinoid X receptors stimulate and 9-*cis*-retinoic acid inhibits 1,25-dihydroxyvitamin D_3-activated expression of the rat osteocalcin gene. Mol Cell Biol **13**:5907–5917.
75. Kephart DD, Walfish PG, DeLuca H, Butt TR 1996 Retinoid X receptor isotype identity directs human vitamin D receptor heterodimer transactivation from the 24-hydroxylase vitamin D response elements in yeast. Mol Endocrinol **10**:408–419.
76. Husmann M, Hoffmann B, Stump DG, Chytil F, Pfahl M 1992 A retinoic acid response element from the rat CRBPI promoter is activated by an RAR/RXR heterodimer. Biochem Biophys Res Commun **187**:1558–1564.
77. Cheskis B, Freedman LP 1996 Modulation of nuclear receptor interactions by ligands: Kinetic analysis using surface plasmon resonance. Biochemistry **35**:3309–3318.
78. Qi JS, Desai-Yajnik V, Greene ME, Raaka BM, Samuels HH 1995 The ligand-binding domains of the thyroid hormone/retinoid receptor gene subfamily function in vivo to mediate heterodimerization, gene silencing, and transactivation. Mol Cell Biol **15**:1817–1825.
79. Heyman RA, Mangelsdorf DJ, Dyck JA, Stein RB, Eichele G, Evans RM, Thaller C 1992 9-*cis*-Retinoic acid is a high affinity ligand for the retinoid X receptor. Cell **68**:397–406.
80. Zhang XK, Lehmann J, Hoffmann B, Dawson MI, Cameron J, Graupner G, Hermann T, Tran P, Pfahl M 1992 Homodimer

formation of retinoid X receptor induced by 9-*cis*-retinoic acid. Nature **358**:587–591.
81. Leng X, Blanco J, Tsai SY, Ozato K, O'Malley BW, Tsai M-J 1995 Mouse retinoid X receptor contains a separable ligand-binding and transactivation domain in its E region. Mol Cell Biol **15**:255–263.
82. Keidel S, LeMotte P, Apfel C 1994 Different agonist- and antagonist-induced conformational changes in retinoic acid receptors analyzed by protease mapping. Mol Cell Biol **14**:287–298.
83. Leid M 1994 Ligand-induced alteration of the protease sensitivity of retinoid X receptor α. J Biol Chem **269**:14175–14181.
83a. Sone T, Pike JW 1991 Vitamin D receptor interaction with specific DNA: Association as a 1,25-dihydroxyvitamin D_3 heterodimer. J Biol Chem **266**:23296–23305.
84. Cheskis B, Lemon BD, Uskokovic M, Lomedico PT, Freedman LP 1995 Vitamin D_3-retinoid X receptor dimerization, DNA binding, and transactivation are differentially affected by analogues of 1,25 dihydroxyvitamin D_3. Mol Endocrinol **9**:1814–1824.
85. Imai Y, Pike JW, Koeffler HP 1995 Potent vitamin D_3 analogs: Their abilities to enhance transactivation and to bind to the vitamin D_3 response element. Leuk Res **19**:147–158.
86. Peleg S, Sastry M, Collins ED, Bishop JE, Norman AW 1995 Distinct conformational changes are induced by 20-epi analogues of 1,25-dihydroxyvitamin D_3 and associated with enhanced activation of the vitamin D receptor. J Biol Chem **270**:10551–10558.
87. Sasaki H, Harada H, Handa Y, Morino H, Suzawa M, Shimpo E, Katsumata T, Masuhiro Y, Matsuda K, Ebihara K, Ono T, Masushige S, Kato S 1995 Transcriptional activity of a fluorinated vitamin D analog on VDR–RXR-mediated gene expression. Biochemistry **34**:370–377.
88. Lehmann JM, Jong L, Fanjul A, Cameron JF, Lu XP, Haefner P, Dawson MI, Pfahl M 1992 Retinoids selective for retinoid X receptor response pathways. Science **258**:1944–1946.
89. Andersson ML, Nordstrom K, Demczuk S, Harbers M, Vennstrom B 1992 Thyroid hormone alters the DNA binding properties of chicken thyroid hormone receptors alpha and beta. Nucleic Acids Res **20**:4803–4810.
90. Yen PM, Sugawara A, Chin WW 1992 Triiodothyronine (T3) differentially affects T3-receptor/retinoic acid receptor/retinoid X receptor heterodimer binding to DNA. J Biol Chem **267**:23248–23252.
91. Miyamoto T, Suzuki S, DeGroot LJ 1993 High affinity and specificity of dimeric binding of thyroid hormone receptors to DNA and their ligand-dependent dissociation. Mol Endocrinol **7**:224–231.
92. Lehmann JM, Zhang X-K, Graupner G, Lee M-O, Hermann T, Hoffmann B, Pfahl M 1993 Formation of retinoid X receptor homodimers leads to repression of T3 response: Hormonal cross talk by ligand-induced squelching. Mol Cell Biol **13**:7698–7707.
93. Leng X, Blanco J, Tsai SY, Ozato K, O'Malley BW, Tsai M-J 1994 Mechanisms for synergistic activation of thyroid hormone receptor and retinoid X receptor on different response elements. J Biol Chem **269**:31436–31442.
94. Rosen ED, O'Donnell AL, Koenig RJ 1992 Ligand-dependent synergy of thyroid hormone and retinoid X receptors. J Biol Chem **267**:22010–22013.
95. Xiao J-H, Durand B, Chambon P, Voorhees JJ 1995 Endogenous retinoic acid receptor (RAR)–retinoid X receptor (RXR) heterodimers are the major functional forms regulating retinoid responsive elements in adult human keratinocytes. J Biol Chem **270**:3001–3011.
96. Lemon BD, Fondell JD, Freedman LP 1997 Retinoid X receptor: Vitamin D_3 receptor heterodimers promote stable preinitiation complex formation and direct 1,25-dihydroxyvitamin D_3-dependent cell-free transcription. Mol Cell Biol **17**:1923–1937.
97. Valcarcel R, Holz H, Jimenez CG, Barettino D, Stunnenberg HG 1994 Retinoid-dependent in vitro transcription mediated by the RXR/RAR heterodimer. Genes Dev **8**:3068–3079.
98. Fondell JD, Ge H, Roeder RG 1996 Ligand induction of a transcriptionally active thyroid hormone receptor co-activator complex. Proc Natl Acad Sci USA **93**:8329–8333.
99. Cooney AJ, Leng X, Tsai SY, O'Malley BW, Tsai MJ 1993 Multiple mechanisms of chicken ovalbumin upstream promoter transcription factor-dependent repression of transactivation by the vitamin D, thyroid hormone, and retinoic acid receptors. J Biol Chem **268**:4152–4160.
100. Fondell JD, Roy AL, Roeder RG 1993 Unliganded thyroid hormone receptor inhibits formation of a functional preinitiation complex: Implications for active repression. Genes Dev **7**:1400–1410.
101. Burris T, Nawaz Z, Tsai M, O'Malley B 1995 A nuclear hormone receptor-associated protein that inhibits transactivation by the thyroid hormone and retinoic acid receptors. Proc Natl Acad Sci USA **92**:9525–9529.
102. Chen JD, Evans RM 1995 A transcriptional co-repressor that interacts with nuclear hormone receptors. Nature **377**:454–457.
103. Horlein AJ, Naar AM, Heinzel T, Torchia J, Gloss B, Kurokawa R, Ryan A, Kamei Y, Soderstrom M, Glass CK, Rosenfeld MG 1995 Ligand-independent repression by the thyroid hormone receptor mediated by a nuclear receptor corepressor. Nature **377**:397–403.
104. Piedrafita FJ, Bendik I, Ortiz MA, Pfahl M 1995 Thyroid hormone receptor homodimers can function as ligand-sensitive repressors. Mol Endocrinol **9**:563–578.
105. Fondell J, Brunel F, Hisatake K, Roeder R 1996 Unliganded thyroid hormone receptor alpha can target TATA-binding protein for transcriptional repression. Mol Cell Biol **16**:281–287.
106. Leng X, Cooney AJ, Tsai SY, Tsai M-J 1996 Molecular mechanisms of COUP-TF mediated transcriptional repression: Evidence for transrepression and active repression. Mol Cell Biol **16**:2332–2340.
107. Tong G-X, Jeyakumar M, Tanen MR, Bagchi MK 1996 Transcriptional silencing by unliganded thyroid hormone receptor-β requires a soluble corepressor that interacts with the ligand binding domain of the receptor. Mol Cell Biol **16**:1909–1920.
108. Yen PM, Liu Y, Sugawara A, Chin WW 1996 Vitamin D receptors repress basal transcription and exert dominant negative activity on triiodothyronine-mediated transcriptional activity. J Biol Chem **271**:10910–10916.
109. Freedman LP, Yoshinaga SK, Vanderbilt JN, Yamamoto KR 1989 In vitro transcription enhancement by purified derivatives of the glucocorticoid receptor. Science **245**:298–301.
110. Klein-Hitpass L, Tsai SY, Weigel NL, Allan GF, Riley D, Rodriguez R, Schrader WT, Tsai M-J, O'Malley BW 1990 The progesterone receptor stimulates cell-free transcription by enhancing the formation of a stable preinitiation complex Cell **60**:247–257.
111. Schmitt J, Stunnenberg HG 1993 The glucocorticoid receptor hormone binding domain mediates transcriptional activation in the absence of ligand. Nucleic Acids Res **21**:2673–2681.
112. Lee IJ, Driggers PH, Medin JA, Nikoderm VM, Ozato K 1994 Recombinant thyroid hormone receptor and retinoid X-receptor stimulate ligand-dependent transcription in vitro. Proc Natl Acad Sci USA **91**:1647–1651.
113. Cancela L, Ishida H, Bishop JE, Norman AW 1992 Local chromatin changes accompany the expression of the calbindin-D28k gene: Tissue specificity and effect of vitamin D activation. Mol Endocrinol **6**:468–475.

114. Breen EC, van Wijnen AJ, Lian JB, Stein GS, Stein JL 1994 In vivo occupancy of the vitamin D responsive element in the osteocalcin gene supports vitamin D-dependent transcriptional upregulation in intact cells. Proc Natl Acad Sci USA **91**:12902–12906.
115. Dey A, Minucci S, Ozato K 1994 Ligand-dependent occupancy of the retinoic acid receptor $\beta 2$ promoter in vivo. Mol Cell Biol **14**:8191–8202.
116. Wong J, Shi Y-B, Wolffe AP 1995 A role for nucleosome assembly in both silencing and activation of the *Xenopus* TRβA gene by the thyroid hormone receptor. Genes Dev **9**:2696–2711.
117. Yoshinaga SK, Peterson CL, Herskowitz I, Yamamoto KR 1992 Roles of SW11, SW12 and SW13 proteins for transcriptional enhancement by steroid receptors. Science **258**:1598–1604.
118. Kingston RE, Bunker CA, Imbalzano A 1996 Repression and activation by multiprotein complexes that alter chromatin structure. Genes Dev **10**:905–920.
119. Wagner RL, Apriletti JW, McGrath ME, West BL, Baxter JD, Fletterick RJ 1995 A structural role for hormone in the thyroid hormone receptor. Nature **378**:690–697.
120. Danielian PS, White R, Lees JA, Parker MG 1992 Identification of a conserved region required for hormone dependent transcriptional activation by steroid hormone receptors. EMBO J **11**:1025–1033.
121. Barettino D, Vivanco Ruiz MdM, Stunnenberg HG 1994 Characterization of the ligand-dependent transactivation domain of the thyroid hormone receptor. EMBO J **13**:3039–3049.
122. Durand B, Saunders M, Gaudon C, Roy B, Losson R, Chambon P 1994 Activation function 2 (AF-2) of retinoic acid receptor and 9-*cis*-retinoic acid receptor: Presence of a conserved autonomous constitutive activating domain and influence of the nature of the response element on AF-2 activity. EMBO J **13**:5370–5382.
123. Mitchell PJ, Tjian R 1989 Transcriptional regulation in mammalian cells by sequence-specific DNA binding proteins. Science **245**:371–378.
124. Halachmi S, Marden E, Martin G, MacKay H, Abbondanza C, Brown M 1994 Estrogen receptor-associated proteins: Possible mediators of hormone-induced transcription. Science **264**:1455–1458.
125. Cavailles V, Dauvois S, L'Horset F, Lopez G, Hoare S, Kushner PJ, Parker MG 1995 Nuclear factor RIP140 modulates transcriptional activities by the estrogen receptor. EMBO J **14**:3741–3751.
126. Le Douarin B, Zechel C, Garnier J-M, Lutz Y, Tora L, Pierrat B, Heery D, Gronenmyer H, Chambon P, Losson R 1995 The N-terminal part of TIF1, a putative mediator of the ligand dependent activation function (AF-2) of nuclear receptors, is fused to B-Raf in the oncogenic protein T18. EMBO J **14**:2020–2033.
127. Onate SA, Tsai SY, Tsai MJ, O'Malley BW 1995 Sequence and characterization of a co-activator for the steroid hormone receptor superfamily. Science **270**:1354–1357.
128. Lee JW, Ryan F, Swaffield JC, Johnston SA, Moore DD 1995 Interaction of thyroid-hormone receptor with a conserved transcriptional mediator. Nature **374**:91–94.
129. Kamei Y, Xu L, Heinzel T, Torchia J, Kurokawa R, Gloss B, Lin S-C, Heyman RA, Rose DW, Glass CK, Rosenfield MG 1996 A CBP-integrator complex mediates transcriptional activation and AP-1 inhibition by nuclear receptors. Cell **85**:403–414.
130. vom Baur E, Zechel C, Heery D, Heine MJ, Garnier JM, Vivat V, Le Dourarin B, Gronemeyer H, Chambon P, Losson R 1996 Differential ligand-dependent interactions between the AF-2 activating domain of nuclear receptors and the putative transcriptional intermediary factors mSUG1 and TIF1. EMBO J **15**: 110–124.
131. Wurtz J-M, Bourguet WJ-PR, Vivat V, Chambon P, Moras D, Gronemeyer H 1996 A canonical structure for the ligand-binding domain of nuclear receptors. Nature Struct Biol **3**:87–94.
132. Rosen ED, Beninghof EG, Koenig RJ 1993 Dimerization interfaces of thyroid hormone, retinoic acid, vitamin D, and retinoid X receptors. J Biol Chem **268**:11534–11541.
133. Whitfield GK, Hsieh JC, Nakajima S, MacDonald PN, Thompson PD, Jurutka PW, Haussler CA, Haussler MR 1995 A highly conserved region in the hormone-binding domain of the human vitamin D receptor contains residues vital for heterodimerization with retinoid X receptor and for transcriptional activation. Mol Endocrinol **9**:1166–1179.
134. Leid M, Kastner P, Lyons R, Nakshatri H, Saunders M, Zacharewski T, Chen JY, Staub A, Garnier JM, Mader S, Chambon P 1992 Purification, cloning, and RXR identity of the HeLa cell factor with which RAR or TR heterodimerizes to bind target sequences efficiently. Cell **68**:377–395.
135. Nakajima S, Hsieh J-C, MacDonald PN, Galligan MA, Haussler CA, Whittfield GK, Haussler MR 1994 The C-terminal region of the vitamin D receptor is essential to form a complex with a receptor auxiliary factor required for high affinity binding to the vitamin D-responsive element. Mol Endocrinol **8**:159–172.
136. Kurokawa R, DiRenzo J, Boehm M, Sugarman J, Gloss B, Rosenfeld MG, Heyman RA, Glass CK 1994 Regulation of retinoid signalling by receptor polarity and allosteric control of ligand binding. Nature **371**:528–531.
137. Forman BM, Umesono K, Chen J, Evans RM 1995 Unique response pathways are established by allosteric interactions among nuclear hormone receptors. Cell **81**:541–550.
138. Perlmann T, Jansson L 1995 A novel pathway for vitamin A signaling mediated by RXR heterodimerization with NGFI-B and NURR1. Genes Dev **9**:769–782.
139. Willy PJ, Umesone K, Ong ES, Evans RM, Heyman RA, Manglesdorf DJ 1995 LXR, a nuclear receptor that defines a distinct retinoid response pathway. Genes Dev **9**:1033–1045.
140. Forman BM, Goode E, Chen J, Oro AE, Bradley DJ, Perlmann T, Noonan DJ, Burka LT, McMorris T, Lamph WW, Evans RM, Weinberger C 1995 Identification of a nuclear receptor that is activated by farnesol metabolites. Cell **81**:687–693.
141. Schrader M, Bendik I, Becker-Andre M, Carlberg C 1993 Interaction between retinoic acid and vitamin D signaling pathways. J Biol Chem **268**:17830–17836.
142. Schrader M, Muller KM, Nayeri S, Kahlen J-P, Carlberg C 1994 Vitamin D_3-thyroid hormone receptor heterodimer polarity directs ligand sensitivity of transactivation. Nature **370**:382–386.
143. Schrader M, Muller KM, Carlberg C 1994 Specificity and flexibility of vitamin D signaling. J Biol Chem **269**:5501–5504.
144. Blanco JCG, Dey A, Leid M, Minucci S, Park B-K, Jurutka PW, Haussler MR, Ozato K 1995 Inhibition of ligand induced promoter occupancy in vivo by a dominant negative RXR. Genes Cells in press.
145. Schulman IG, Juguilon H, Evans RM 1996 Activation and repression by nuclear hormone receptors: Hormone modulates an equilibrium between active and repressive states. Mol Cell Biol **16**:3807–3813.
146. Baniahmad A, Leng X, Burris TP, Tsai SY, Tsai M-J, O'Malley BW 1995 The $\tau 4$ activation domain of the thyroid hormone receptor is required for release of a putative co-repressor(s) necessary for transcriptional silencing. Mol Cell BIol **15**:76–86.
147. Baniahmad A, Ha I, Reinberg D, Tsai S, Tsai M-J, O'Malley BW 1993 Interaction of human thyroid hormone receptor β with transcription factor TFIIB may mediate target gene derepression and activation by thyroid hormone. Proc Natl Acad Sci USA **90**:8832–8836.

148. Blanco JCG, Wang I-M, Tsai SY, Tsai M-J, O'Malley BW, Jurutka PW, Haussler MR, Ozato K 1995 Transcription factor TFIIB and the vitamin D receptor cooperatively activate ligand-dependent transcription. Proc Natl Acad Sci USA **92**:1535–1539.
149. MacDonald PN, Sherman DR, Dowd DR, Jefcoat SC, Delisle RK 1995 The vitamin D receptor interacts with general transcription factor IIB. J Biol Chem **270**:4748–4752.
150. Luisi BF, Freedman LP 1995 Dymer, dymer binding tight. Nature **375**:359–360.
151. Baker AR, McDonnell DP, Hughes M, Crisp TM, Mangelsdorf DJ, Haussler MR, Pike JW, Shine J, O'Malley BW 1988 Cloning and expression of full-length cDNA encoding the human vitamin D receptor. Proc Natl Acad Sci USA **85**:3294–3298.

CHAPTER 10

Nuclear Vitamin D Receptor: Structure–Function, Phosphorylation, and Control of Gene Transcription

MARK R. HAUSSLER, PETER W. JURUTKA, JUI-CHENG HSIEH, PAUL D. THOMPSON, CAROL A. HAUSSLER, SANFORD H. SELZNICK, LENORE S. REMUS, AND G. KERR WHITFIELD
Department of Biochemistry, College of Medicine, The University of Arizona, Tucson, Arizona

I. INTRODUCTION

A. Historical Aspects and Overview of VDR

The existence of the vitamin D receptor (VDR) was originally appreciated when it was observed that administration of radioactive vitamin D_3 at physiological doses to rachitic chickens elicited the selective nuclear localization of a bioactive metabolite of the parent vitamin in the target intestine [1]. The binding of this vitamin D metabolite in the nucleus was specific, saturable, and confined to the deoxynucleoprotein chromatin subfraction [1]. Subsequent salt extraction experiments resulted in the discovery of the chromosomal protein that comprises the vitamin D receptor [2]. The vitamin D metabolite associated with VDR was shown to be more polar than 25-hydroxyvitamin D_3 ($25OHD_3$) [3,4] and at least five times as potent and more rapidly acting than either vitamin D_3 or $25OHD_3$ [4]. These results set the stage for the chemical identification [5,6] of the VDR ligand as 1,25-dihydroxyvitamin D_3 [$1,25(OH)_2D_3$] (Fig. 1), a hormone produced in the kidney [7].

When isotopically labeled $1,25(OH)_2D_3$ could be produced in quantity, biochemical and pharmacological characterization of VDR was accomplished. VDR was found to be a 3.3–3.7 S protein (depending on species) that bound $1,25(OH)_2D_3$ with high affinity (K_d = 0.1–1 nM) [8,9] and displayed a pharmacological profile for binding various vitamin D metabolites and analogs that is consistent with their relative biological activities [10,11]. Avian intestinal VDR was next shown to be a DNA binding protein, and this property was exploited in the first purification of VDR [12]. An enriched avian VDR preparation, containing one major and several minor proteins of molecular mass 50,000–70,000 [12,13], was then utilized to generate monoclonal antibodies to the VDR that cross-reacted with all species of vitamin D receptor tested [14]. Finally, via monoclonal antibody-based screening of a λGT-11 expression vector library, a partial cDNA clone of avian VDR was obtained [15], revealing the presence of the classic $(Cys_2–Cys_2)_2$ zinc finger motif of the chromatin-bound steroid hormone receptors [16]. Independent biochemical evidence emphasized the importance of reduced cysteine thiol groups for DNA asociation [17] and implicated zinc in maintaining the DNA binding integrity of VDR [18]. Subsequently, full-length human [19], rat [20], and avian [21] VDRs were cloned and their sequences deduced, allowing for overexpression and functional studies of the type discussed in this chapter and elsewhere in this volume. It is notable that the cloning and classification of VDR as a DNA binding transcription factor activated by ligand not only validated the discovery of VDR as a chromosomal protein [2], but also verified the initial report of its uncharacterized hormonal ligand localized in nuclear chromatin [1].

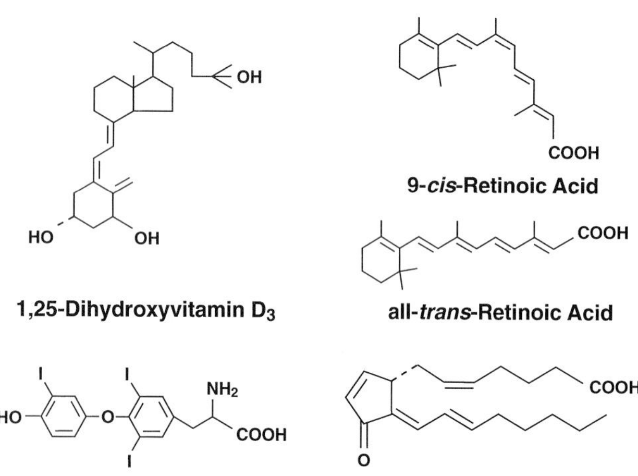

FIGURE 1 Selected lipophilic ligands of nutritional origin that bind to a subfamily of nuclear receptors. $1,25(OH)_2D_3$ occupies VDR; T_3 is a ligand for the nuclear thyroid hormone receptor; 9-*cis*-retinoic acid is a natural ligand for RXR but also binds to the retinoic acid receptor, for which all-*trans*-retinoic acid is a potent ligand; 15-deoxy-$\Delta^{12,14}$-PGJ_2 is an apparent ligand for one isoform of peroxisome proliferator-activated receptor [218,219].

Research on VDR in the 1990s includes a key demonstration by Sone *et al.* [22,23] that DNA binding by VDR requires the presence of a coreceptor, termed nuclear accessory factor (NAF). Utilizing baculovirus-expressed human VDR (hVDR), MacDonald *et al.* [24] provided independent evidence, as did Ross *et al.* [25] employing rat VDR, that such a NAF or receptor auxiliary factor (RAF) was essential for VDR binding to the vitamin D responsive element (VDRE) as a VDR–RAF heterodimer. With the discovery of retinoid X receptor β (RXRβ) and retinoid X receptor α (RXRα), originally as orphan nuclear receptors by Hamada *et al.* [26] and Mangelsdorf *et al.* [27], respectively, the coreceptor field developed rapidly. By late 1991, Yu *et al.* [28] provided the initial evidence that RXRβ functioned as a significant coregulator in transcription with the retinoic acid receptor (RAR) and the thyroid hormone receptor (TR), but only modestly with VDR. Leid *et al.* [29] also identified RXRβ as the heteropartner of RAR and TR, and Kliewer *et al.* [30] provided further evidence for RXR participation in RAR, TR, and VDR signaling. These observations led to the purification and identification of RXRβ as the HeLa cell RAF that cooperates with hVDR both in VDRE binding and in $1,25(OH)_2D_3$ stimulation of transcription [31]. Furthermore, it was demonstrated that VDR–RXR heterodimer action occurs on a VDRE in the natural context of the osteocalcin promoter in intact bone cells, and that 9-*cis*-retinoic acid (9-*cis*-RA) (Fig. 1), the natural ligand of the RXR partner [32], operates to suppress VDR-mediated stimulation of transcription by $1,25(OH)_2D_3$ [31].

MacDonald *et al.* [31] published the first evidence that the $1,25(OH)_2D_3$ hormone stimulated, and the 9-*cis*-RA hormone inhibited, the actual formation of a VDR–RXR heterocomplex on the VDRE. Therefore, preliminary and inferential evidence from studies of TR and RAR was extended to show conclusively that RXR is the functional partner of VDR in a binary heterocomplex (Fig. 2) that is switched positively by the vitamin D hormone and negatively by 9-*cis*-RA or a syn-

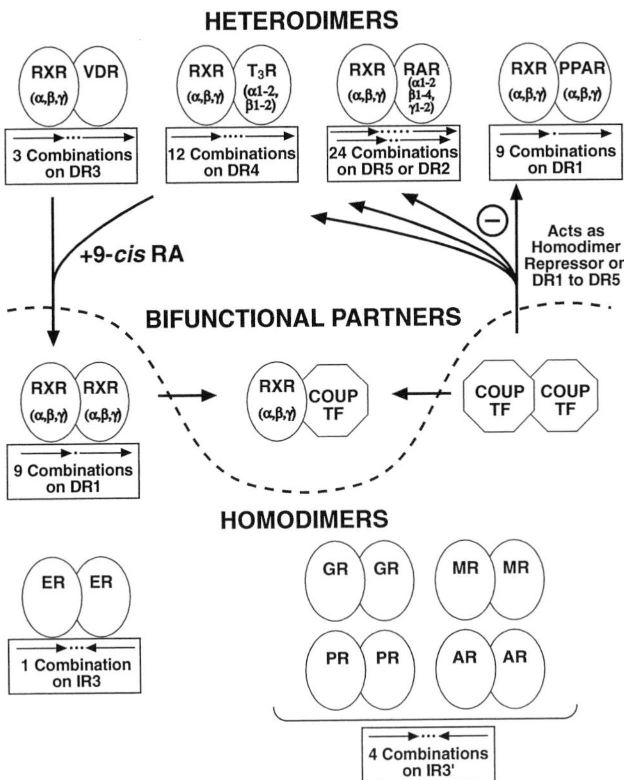

FIGURE 2 Dimerization and DNA binding by three distinct subclasses of nuclear receptors. T_3R is the thyroid hormone receptor. All other abbreviations are defined in text.

thetic RXR-specific ligand [31,33–38]. Apparently, 9-*cis*-RA attenuates $1,25(OH)_2D_3$ responsiveness by diverting RXRs away from VDR-mediated transcription toward RXR homodimers (Fig. 2) and other RXR-dependent transcriptional pathways [39,40]. Nevertheless, the RXR–VDR–$1,25(OH)_2D_3$ heterocomplex is the VDRE-bound and transcriptionally active species that effects the regulation of vitamin D controlled target genes. In addition to the discourse on VDR to follow, the reader is referred by the editors to Chapter 8 for a detailed discussion of the VDR protein and its gene, and to Chapter 9, which focuses on VDR–RXR heterodimerization, interaction with the VDRE, and VDR molecular modeling.

B. The VDR Subfamily of Nuclear Receptors and Their Ligands

In order to discuss VDR, it is necessary first to introduce it in the context of the nuclear receptor superfamily and appropriate cognate ligands. Space limitations prevent a comprehensive analysis of the topic with citations, but the reader is referred to excellent reviews by Mangelsdorf *et al.* [41] and Mangelsdorf and Evans [42] of the steroid/retinoid/thyroid hormone receptor superfamily and its actions to modulate transcription of target genes. A common feature of the nuclear receptors is that they are occupied by small, lipophilic ligands which readily penetrate the plasma membrane and diffuse into the nucleus. Figure 1 illustrates the structure of the VDR ligand, $1,25(OH)_2D_3$, along with other functionally related superfamily hormones such as triiodothyronine (T_3), two retinoic acid isomers, and the prostaglandin metabolite 15-deoxy-$\Delta^{12,14}$-PGJ_2. Analogous chemically to steroid hormones, $1,25(OH)_2D_3$ is derived from cholesterol. However, vitamin D can also be obtained from the diet and therefore is more appropriately classified with ligands of nutritional origin like thyroid hormone (from phenylalanine/tyrosine and iodine), the retinoids (from β-carotene/vitamin A), and 15-deoxy-$\Delta^{12,14}$-PGJ_2 (from linoleic/arachidonic acid). The diversity of this subfamily of nuclear receptor ligands (Fig. 1) indicates that the vitamin D_3 hormone participates in the regulation of a broad network of phenomena dependent upon essential nutrients from the environment. The fact that the $1,25(OH)_2D_3$ precursor, vitamin D_3, is also photobiosynthesized in skin in the presence of ultraviolet light renders $1,25(OH)_2D_3$ unique in that it is derived from two independent external sources.

The second characteristic property of the nuclear receptors, like many transcription factors, is their ability to form multimers that facilitate high affinity and specific DNA binding. Accordingly, this superfamily of lipophilic ligand receptors controls target genes by binding as dimers to specific hormone responsive elements [43]. On the basis of interactions between nuclear receptors and their natural responsive elements, as illustrated in Fig. 2, the superfamily can be organized [44,45] into the glucocorticoid receptor (GR) subfamily of classic steroid hormone receptors [including mineralocorticoid (MR), androgen (AR), progesterone (PR), and estrogen (ER) receptors] that homodimerize in response to ligand and bind to inverted repeat (IR) elements spaced by 3 bp in DNA, and the VDR subfamily [including TRs, RARs, and peroxisome proliferator-activated receptors (PPARs)] that heterodimerize with RXR and bind to direct repeat (DR) elements spaced by 1–5 bp [42]. Such heterodimerization in the latter subgroup, together with the existence of multiple isoforms of RXR, TR, RAR, and PPAR, leads to an extensive combinatorial repertoire for responsive elements and receptors of this subfamily occupied by nutritionally derived ligands (Fig. 2).

Two nuclear receptors do not fit well into either subfamily, namely, the RXR master heteropartner and the chicken ovalbumin upstream promoter-transcription factor (COUP-TF) orphan receptor (Fig. 2). RXR, after occupation by its natural 9-*cis*-RA ligand, forms an RXR homodimer capable of gene activation. Conversely, COUP-TF, which has no known cognate ligand, appears to function as a homodimeric repressor (Fig. 2) or complexes with RXR to generate a transcriptionally inactive heterocomplex [42].

Based upon the grouping of the nuclear receptors into subfamilies, it is clear that the VDR, which binds $1,25(OH)_2D_3$ as its hormonal ligand, is a bona fide member of the heterodimerizing ensemble of nuclear receptors that is occupied by a ligand of nutritional origin and associates with RXR isoforms in order to bind to DNA with high affinity. Specifically, VDR and RXR cooperate to interact with a hormone responsive element (Fig. 2) that is a direct repeat spaced by 3 bp (DR3). We might expect VDR to share many of the general functional features with close subfamily relatives like TR and RAR, to which it is most homologous in amino acid sequence, and with which it shares the distinction of depending on environmentally derived ligand precursors.

II. NUCLEAR VDR MEDIATES THE ACTIONS OF $1,25(OH)_2D_3$

A. Vitamin D Target Tissues

As detailed in Sections II and III of this volume, the classic target tissues for vitamin D are those involved in calcium translocation and skeletal remodeling, including intestine [46], kidney [47,48], and bone [49,50]. Systematic autoradiographic analysis of the retention of radiolabeled $1,25(OH)_2D_3$ reveals evidence for other vitamin D target tissues, including various areas of the brain, heart, pancreas, pituitary, skin, and parathyroid gland [51,52]. The presence of nuclear VDR in some of these same tissues was demonstrated by biochemical liganding experiments [53] and via immunocytochemical techniques in studies that extended the list of potential target tissues to include breast, thyroid, and adrenal glands [54,55].

Because it is known that osteoclasts in bone are derived from hematopoietic precursors, it is not surprising that vitamin D has been found to elicit proliferative/antiproliferative effects in the hematopoietic system, particularly in macrophages and monocytes (reviewed by Manolagas *et al.* [56]; see also Chapter 64). For example, $1,25(OH)_2D_3$ has been reported to mediate transcriptional repression of the interleukin-2 gene via the direct inhibition by VDR of NFATp/AP-1 complex formation [57]. Certain actions of $1,25(OH)_2D_3$ in the immune system, especially suppression of T-helper cells, suggest that analogs of vitamin D might be useful therapeutic agents in procedures such as tissue transplants [58] and in the treatment of autoimmune disorders [59] (Chapters 69 and 70). In addition, $1,25(OH)_2D_3$ is thought to play an important role in differentiation of bone (Chapters 18–21), skin (Chapter 25), and the immune system (Chapter 29). Several illustrations of this action have been reported, including differentiation of a human promyelocytic leukemia cell line (HL-60) into macrophage-like cells [60,61] and development of osteoclasts in bone from macrophage-like precursors (see Manolagas *et al.* [56], Chapter 21, and references therein). The prodifferentiation/antiproliferative actions of vitamin D analogs in skin cells already have been applied therapeutically to treat psoriasis [61] (Chapter 72).

B. Evidence that VDR Mediates $1,25(OH)_2D_3$ Action

Probably the most compelling argument proving that VDR is the obligatory mediator of vitamin D biological actions is the existence of natural mutations in the nuclear VDR which confer tissue insensitivity and clinical resistance to $1,25(OH)_2D_3$ [63], as considered in detail in Chapter 48. Target tissue insensitivity to $1,25(OH)_2D_3$, known as hereditary vitamin D-resistant rickets (HVDRR), is therefore caused by defects in the gene on chromosome 12 [64] coding for the VDR. Clinically significant HVDRR is an autosomal recessive disorder resulting in a phenotype characterized by severe bowing of the lower extremities, short stature, and often alopecia [65]. The pattern of serum chemistries in HVDRR includes frank hypocalcemia, secondary hyperparathyroidism, elevated alkaline phosphatase, variable hypophosphatemia, and markedly increased $1,25(OH)_2D_3$. The symptoms of HVDRR, with the exception of alopecia, mimic those of classic vitamin D-deficiency rickets, suggesting that VDR not only mediates all of the bone mineral homeostatic actions of vitamin D, but may also participate in the differentiation of hair follicles, *in utero*. VDR knock-out mice have been created [66], revealing apparently normal heterozygotes but severely affected homozygotes (VDR−/−), 90% of which die within 8–10 weeks. Surviving mice lose their hair and possess low bone mass, hypocalcemia, hypophosphatemia, as well as a 10-fold elevation of $1,25(OH)_2D_3$ coincident with extremely low $24,25(OH)_2D_3$. These parameters in the VDR knockout mouse mimic the phenotype of patients with HVDRR, confirming that VDR normally mediates all the functions of vitamin D, most prominently intestinal calcium and phosphate absorption, and the differentiation of certain cells in the skin. Both HVDRR in humans and VDR knock-out mice are particularly instructive in illuminating the physiological relevance of the $1,25(OH)_2D_3$–VDR complex, and, as detailed below (in Sections III,C and III,D), specific natural VDR mutations in HVDRR permit the elucidation of structure–function relationships within the receptor.

C. Target Genes and Vitamin D Responsive Elements

In some vitamin D target tissues, specific $1,25(OH)_2D_3$-regulated genes have been identified and cloned. Sequencing and promoter analysis of several of these genes have led to the identification of vitamin D responsive elements (VDREs), short DNA sequences to which the VDR binds and exerts its effects on transcription. Several positive VDREs described to date (Fig. 3) include those in the promoter regions of genes coding for osteocalcin [67–70] and osteopontin [71] expressed in bone osteoblasts, β_3 integrin in bone osteoclasts and blood macrophages [72], calbindin-D_{28K} from kidney [73], and 25-hydroxyvitamin D 24-hydroxylase (24-OHase) [74], a $1,25(OH)_2D_3$-catabolizing enzyme found in many vitamin D target tissues.

As detailed in Fig. 3, there is a reasonable amount of sequence variation between natural VDREs; however, a consensus positive VDRE can still be gleaned (consensus I). In addition, two laboratories have probed VDRE criteria by random selection of oligonucleotides that form complexes with VDR in combination with RXR (consensus II [75,76]). The results from random selection are in general agreement with the natural VDREs and define the VDRE as a direct repeat of two 6-base half-elements that resemble estrogen responsive element (ERE) half-sites, separated by a spacer of three nucleotides, denoted DR3. As depicted in Fig. 2, VDR occupies the 3′ half-site of positive DR3 VDREs, whereas the 5′ half-site is bound by the RXR heteropartner [77].

Examination of both VDRE consensus sequences (Fig. 3) suggests that the G at position 3 of the spacer is important in VDR binding, a deduction consistent with the finding [24] that this base is partially protected by RXR–VDR in methylation interference assays. However, interesting differences arise when comparing the most frequently encountered 3′ half-element bases in natural VDREs, namely, the GGGGCA composite which actually occurs in the human osteocalcin promoter, with the GGTTCA random consensus selection

TISSUE	GENE	RESPONSE ELEMENT		
Bone	rat osteocalcin	GGGTGA T	ATG	AGGACA C
	human osteocalcin	GGGTGA	ACG	GGGGCA
	mouse osteopontin	GGTTCA TTCC	CGA	GGTTCA TTCC
Most Targets	rat 24-OHase: proximal	AGGTGA	GTG	AGGGCG
	distal	GGTTCA	GCG	GGTGCG
Macrophages, Osteoclasts	chicken β3 integrin	GAGGCA	GAA	GGGAGA
Kidney	mouse calbindin-28K	GGGGGA	TGT	GAGGAG
	Consensus I	GGGTGA	NNG	GGGGCA
	Consensus II	A/G GGTCA	NNG	A/G GTTCA

FIGURE 3 Natural responsive elements from vitamin D-stimulated genes, arranged by target tissues. References for descriptions of each VDRE are given in the text. Two half-sites are indicated by arrows at top right. Black dots indicate guanine bases that are protected by VDR–RXR binding in methylation interference assays [24,67]. Small capital letters below sequences indicate nucleotide substitutions that abrogate VDRE activity [24,71]. Consensus VDREs were arrived at as follows: (I) visual inspection of known positive VDREs and (II) polymerase chain reaction (PCR) amplification of random oligonucleotides selected by VDR–RXR heterodimers [75,76].

for the 3′ half-element (Fig. 3). Clearly, GGTTCA represents a potent VDR binding site, a supposition bolstered by the fact that the osteopontin VDRE, a perfect DR3 of GGTTCA, is the highest affinity VDRE we have tested [38]. Intriguingly, Ts at both positions 3 and 4 in the 3′ VDR half-site (boxed in Fig. 3) occur infrequently in the balance of natural VDREs. The paucity of Ts in the 3′ half-element could be related to a need for VDREs of varying potency in regulated genes, or may even provide for a repertoire of different VDR conformations that could be induced by contact with distinct 3′ half-site core sequences [78]. This postulated range of VDR conformations might endow the receptor with the ability to recruit a variety of different coactivators or corepressors. Finally, the presence of two VDREs in the 24-hydroxylase gene promoter [79,80] intimates that the regulation of this catabolic enzyme is even more complex and may embody extra sensitivity in order to effect the rapid elimination of the $1,25(OH)_2D_3$ hormone. Irrespective of the above considerations, it is evident that the primary positive VDRE is a DR3 recognition site in DNA that directs RXR–VDR to the promoter region of $1,25(OH)_2D_3$-regulated genes, ultimately altering the functions of target cells as a result of transcriptional control of gene expression.

D. Comparison of VDR Across Species and with Other Members of the Nuclear Receptor Superfamily

To date VDRs have been cloned from four species, namely, human [19], rat [20], mouse [81], and chicken [15,21]. As depicted in Fig. 4, a schematic representation of the amino acid sequences portrays the significant similarities among these VDRs (rat VDR was chosen to represent the two rodent sequences, which display >96% sequence similarity to one another). Not surprisingly, the highest degree of conservation is seen in those regions of known functional significance (as detailed below in Section III). In addition to the zinc finger DNA binding motifs (identified as DBD in Fig. 4), these homologous areas include the T- and A-box helical region [82], the E1 segment [83], as well as a series of degenerately conserved heptad repeats of hydrophobic amino acids [84]. A striking feature that emerges from this comparison is the existence of a variable region of the receptor (residues 159–202 in hVDR) between the T- and A-box helical stretch and the E1 domain, which is much less conserved (39% between human and rat, and only 19% between human and avian VDRs). In addition, chicken VDR has an

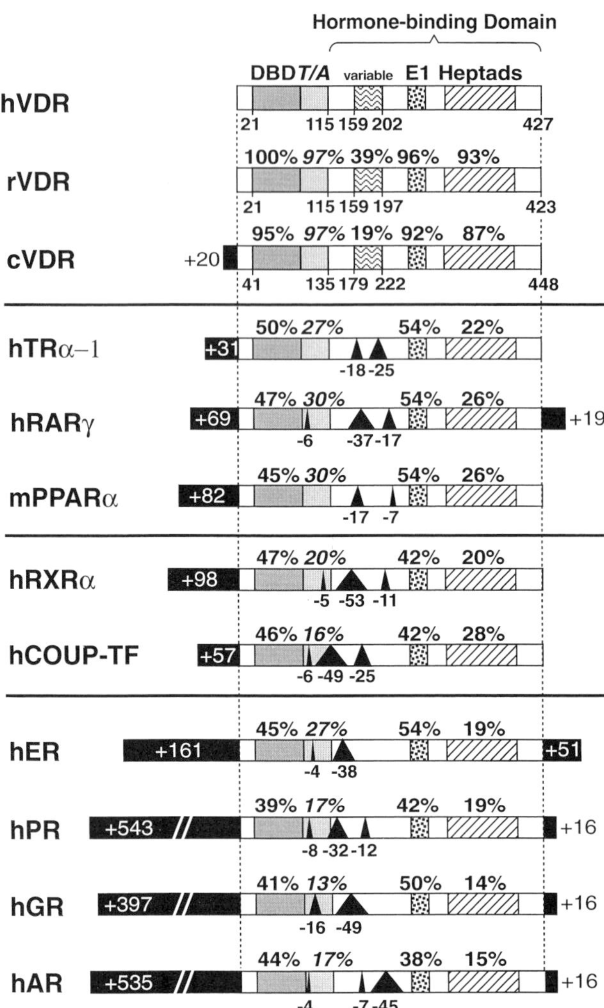

FIGURE 4 Schematic diagram of amino acid conservation among known VDRs and similarities to members of the nuclear receptor superfamily. Abbreviations for each receptor are defined in the text (h, human; r, rat; c, chicken; m, mouse). Areas of particularly high conservation are boxed and include (from left to right) the zinc finger DNA binding domain (DBD), the adjacent T- and A-box helical region (T/A), the conserved E1 region, and a series of partially conserved heptad repeats of hydrophobic amino acids (Heptads). Approximate percent identity in each region compared with human VDR is given above the appropriate box. A poorly conserved (variable) region in the central portion of each VDR is also delineated. Extensions at the N or C termini are shown in black boxes, and black triangles indicate central portions of each receptor that are absent as compared with the more extended hormone binding domain of human VDR. For both types of nonhomologies, the number of amino acids in the extension or missing portion is indicated with a + or − symbol, respectively.

N-terminal extension of 20 amino acids with no counterpart in other VDRs. Neither of these unique regions has been shown to subserve any major known function in the VDRs.

When VDRs are compared with other members of

the superfamily, it is evident that the categorization of nuclear receptors into two classes with distinct dimerization and DNA responsive element patterns (Fig. 2) is mirrored also in their respective primary structures (Fig. 4). Thus, VDR possesses the greatest similarity in amino acid sequence to members of its own subfamily, TR, RAR, and PPAR. The subclass of receptors to which VDR belongs possesses N-terminal extensions that are shorter than those found in the classic steroid receptors (Fig. 4), many of which possess transactivation domains in this region [85]. With the notable exception of VDR, both subclasses of receptors also contain modulatory phosphorylation sites near their respective N termini (detailed below in Section V,E).

Although hVDR has an extremely truncated N terminus, it does possess a hormone binding region which is more expansive than that found in other superfamily members. As illustrated in Fig. 4, all superfamily members are shown to possess significant internal deletions in the central portion of the molecules, thereby placing the "extra" amino acids in VDR between the T- and A-box helical region and the E1 domain, including the variable domain (Fig. 4). It should also be noted that the T-box amino acids [86] are partially deleted in RAR and COUP-TF, as well as in most of the classic steroid receptors (Fig. 4). From these comparisons, it can be expected that conserved domains likely mediate the important actions of VDR and that its structure–function relationships and mechanism may resemble those of TR and/or RAR.

III. FUNCTIONAL DOMAINS IN VDR

The major functional domains of hVDR participate in nuclear localization, ligand binding, heterodimerization, DNA binding, and transcriptional activation. These regions of the receptor are summarized in schematic form in Fig. 5 and discussed in detail below.

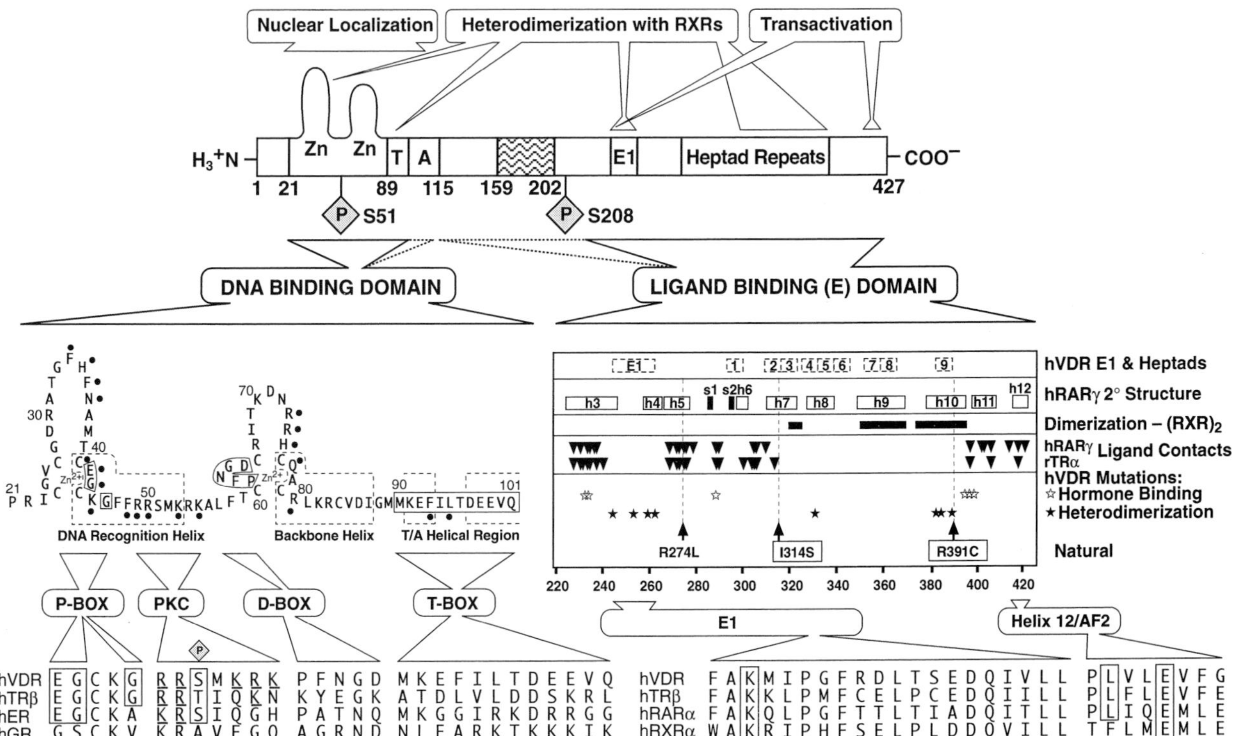

FIGURE 5 Schematic representation of hVDR, highlighting general functional domains for DNA and ligand binding, nuclear localization, heterodimerization with RXR, and transactivation, as well as the two major phosphorylation sites. Variable region 159–202 is unconserved among VDRs from various species, whereas other domains such as the zinc fingers, T- and A-box helical region, E-1, and heptad repeats are highly conserved among VDRs (Fig. 4). The DNA binding domain probably extends through residue 114, and the ligand binding region begins at amino acid 115 (see text). The dashed boxes in the zinc finger domain (lower left) refer to α helices in the hTRβ crystal structure; black dots refer to DNA contacts in this same structure [82]. The following residues are boxed in the detailed sequence data in the lower portion (from left to right): P-box amino acids EGG, PKC site Ser-51, E1 Lys-246 transactivation amino acid, and AF-2 amino acids Leu-417/Glu-420. Within the ligand binding domain (boxed at lower right), hRARγ secondary structural elements are α helices (h3, h4, etc.) and short β strands (s1, s2).

A. Nuclear Localization

Biochemical fractionation studies by Walters *et al.* [87] suggested that even predominantly unoccupied $1,25(OH)_2D_3$ receptors are localized (61–92%) to nuclear preparations from various vitamin D target tissues, including chick intestinal mucosa, parathyroid, kidney, testes, and pancreas; rat intestinal mucosa, kidney, and testes; as well as a mouse osteoblast-like bone cell line. However, these experiments also revealed that the nuclear/cytosol distribution depends on the ionic strength of the extraction buffer used in the subcellular fractionation procedure [87]. Nevertheless, immunocytochemical localization studies confirm that, *in situ*, VDR is primarily a nuclear protein even in the unoccupied state [54,55,88].

In spite of this evidence that VDR is a nuclear protein (reviewed by Haussler *et al.* [89]), little is known about the nature of nuclear localization signals (NLSs) within the VDR molecule or the process by which VDR is transferred into the nucleus. A number of NLSs in nuclear proteins have been identified, and most contain a short stretch of basic amino acids, Lys-Lys/Arg-x-Lys/Arg [where x = Lys (K), Arg (R), Pro (P), Val (V), or Ala (A)] [90]. One of the best characterized is PKKKRKV of the SV40 T-antigen [91,92], and this NLS sequence may represent a prototype for similar sequences in other nuclear proteins. Such a sequence of short basic amino acids has been identified in several of the nuclear receptors, including GR [93], PR [94], and AR [95].

Analyzing basic amino acid sequences present in hVDR, Luo *et al.* [96] selected a region between residues 70 and 111 to probe for the existence of a NLS capable of conferring nuclear retention. Three peptides, VDR(79–105), VDR(70–83), and VDR(100–111), were conjugated to fluorescein-labeled immunoglobulin G (IgG) and the chimeras were then microinjected into the cytoplasm of human osteosarcoma MG-63 cells. The VDR(79–105) construct was able to enter the nuclei as did an SV40 T-peptide, but the VDR(70–83) and VDR(100–111) conjugates, each of which contain only one of the basic clusters in VDR(79–105), remained in the cytosol. These results suggest that the basic residues at both ends of VDR(79–105) are equally necessary for nuclear accumulation. However, further characterization of this bipartite region is required to determine which individual basic residues actually constitute a VDR NLS(s).

By site-directed mutagenesis, Hsieh *et al.* [97] have identified a novel basic sequence between the two zinc fingers of hVDR that apparently represents a second NLS region. Preliminary data show that when positively charged residues between Arg-49 and Lys-55 (a seven residue stretch containing five basic amino acids; see Fig. 5 and Boulikas [98]) are mutated to nonbasic residues, a significant shift in VDR distribution results, favoring cytoplasmic retention as analyzed by Western blotting and immunocytochemistry [97]. These new findings raise the question of whether the region described by Hsieh *et al.* [97] between the zinc fingers and the dual region probed by Luo *et al.* [96], which extends from the second zinc finger into the T- and A-box helical region, function together or independently. Finally, as depicted in Fig. 5, the entire DNA binding domain, as a functional unit, may be required for nuclear transfer and optimal retention of VDR in the nucleus as a result of the general attraction of VDR for DNA [99].

B. Hormonal Ligand Binding and Overexpression

The precise ligand binding domain of VDR has not been defined, although carboxypeptidase cleavage studies of avian VDR suggest that it extends nearly to the C terminus [100]. Deletion analyses of hVDR [101] confirmed the significance of C-terminal receptor residues in $1,25(OH)_2D_3$ binding and established that the N-terminal boundary of ligand association is situated between residues 115 and 166. Thus, as illustrated in Figs. 4 and 5, a conservative estimate places the $1,25(OH)_2D_3$ binding region of hVDR as encompassing the entire section of the receptor C-terminal of the T- and A-box helical region. Actual ligand contact sites in the crystal structures of ligand-occupied hRARγ [102] and agonist-occupied rat TRα-1 [103] (Fig. 5) map to positions corresponding to hVDR residues between 220 and the C terminus, raising the possibility that hVDR amino acids spanning from positions 115 to 220 may not be integral to hormone binding. By analogy with studies of GR [104], however, this N-terminal extension of the ligand binding core could be required to maintain the structural integrity of the modular $1,25(OH)_2D_3$ binding pocket. To date, the few natural and artificial mutations we have studied in hVDR that compromise ligand binding are confined to a region C-terminal of residue 220 and virtually coincide with corresponding hormone contact sites in RAR and TR (Fig. 5).

One approach to understanding $1,25(OH)_2D_3$ ligand binding is to overexpress full-length hVDR, as well as fragments of the receptor, followed by biochemical and physical studies of hormone occupation. Overexpression of hVDR has been accomplished in *Escherichia coli* [105–107]. Using the pT7-7 vector, the full-length receptor was overproduced in our laboratory to a level of 15% of total protein, although it resided as an insoluble protein within inclusion bodies [107]. After solubilization and enrichment, the specific $1,25(OH)_2D_3$ binding activity of *E. coli*-expressed hVDR ranged from 2.0 to 3.4 nmol/mg protein [107]. Utilizing the baculovirus expression vector system, full-length hVDR has been overexpressed in Sf9 *Spodoptera frugiperda* fall army worm ovary cells [24]; the receptor was produced to a level of 0.5% of total soluble protein [3.2 nmol/mg protein $1,25(OH)_2D_3$ binding activity]. Rat VDR has also been overexpressed in the baculovirus system [108].

A potentially significant observation with respect to both purified baculovirus- and *E. coli*-expressed hVDRs is that ultrahigh affinity $1,25(OH)_2D_3$ ligand binding ($K_d \sim 0.1$ nM) requires the reintroduction of a factor(s) present in mammalian cell nuclear extracts [109]. In other words, purified expressed hVDR displays a K_d for ligand binding that is at least one order of magnitude higher (i.e., lower affinity) than that found for endogenous VDR in cellular extracts. Such binding kinetics are acceptable for most *in vitro* studies, but apparently, for hVDR to bind the ligand with native, *in vivo* high affinity, a stabilizing cofactor must be present.

One specific application of baculovirus-expressed hVDR has been to covalently label it with a chemical affinity probe comprised of a binding site-directed ligand analog, $1,25(OH)_2D_3$-3-deoxy-3β-bromoacetate, which reacts with cysteine residue side chains [110]. To complement and extend this finding, the three conserved cysteines in the ligand binding domain of hVDR (Cys-288, Cys-337, and Cys-369) were individually altered by site-directed mutagenesis [111]. Hormone binding results and transactivation data obtained from COS-7 monkey kidney epithelial cells transfected with hVDR expression plasmids indicated that mutating Cys-288 markedly reduced the $1,25(OH)_2D_3$ binding affinity of hVDR. Based on the published crystal structures of RARγ [102] and TRα [103], Cys-288 is likely to be configured in a β-strand region that forms part of the ligand binding pocket of hVDR (Fig. 5), supporting but not proving that this residue accounts for the propensity for chemical affinity labeling by the $1,25(OH)_2D_3$ analog. With additional overexpression of the full-length receptor and a minimal hormone binding fragment, it will be possible to begin to understand the structural conformation of VDR, which is not only the first step in elucidating the role of the $1,25(OH)_2D_3$ ligand, but also a prerequisite for eventually designing clinically effective analogs.

C. Heterodimerization with RXR

As mentioned previously, VDR heterodimerizes with its partner RXR and binds to a DR3 VDRE to exert an influence on transcription of target genes. Because heterodimerization with RXR is fundamental to VDR action, it is important to characterize the regions of hVDR that actually contact RXR as well as those that provide structural support for this interaction. As summarized in Fig. 5, four such regions have been delineated. First, in examining the activities of C-terminally truncated hVDRs, Nakajima *et al.* [112] located a specific RXR interaction surface between Lys-382 and Arg-402 (including heptad 9). Point mutations verified this domain and identified a second segment between Leu-325 and Leu-332 (heptad 4), as potentially supporting RXR heterodimerization.

Subsequent research [113] resulted in the characterization of a third, highly conserved region of hVDR (residues 244 to 263, known as E1; Fig. 5), that is critical for optimal interactions with RXR. Point mutations within E1 revealed that virtually all residues tested supported RXR contact [113]. Rosen *et al.* [114] also provided evidence that distinct E1 amino acids of VDR, as well as their corresponding residues in RAR and TR, are involved in heterodimerization with RXR. Moreover, employing small internal hVDR deletions to identify segments of VDR participating in heterodimerization with RXR and in transactivation, using yeast as an experimental system, Jin *et al.* [35] defined two regions of the hVDR ligand binding domain critical for dimerization. These regions, namely, residues 241–272 and 320–397, precisely encompass the E1 [113] and the heptad 4/9 [112] domains, respectively. Their identification, both in the yeast system and via point mutations and expression in transfected mammalian cells, lends strong support to a designation for these two stretches of hVDR as being highly significant in heterodimerization with RXR (Fig. 5).

A fourth region was identified by site-directed mutagenesis. Hsieh *et al.* [115] showed that Lys-91 and Glu-92, which are situated in the T box in hVDR (Fig. 5), apparently mediate heterodimerization between the VDR and RXR DNA binding domains on the VDRE scaffold. Asn-37 in the first zinc finger of hVDR, as well as Lys-91 and Glu-92 in the T box, have been proposed in a modeling study to contact residues in the second zinc finger of the RXR partner in determining selective association of VDR with a DR3 hormone responsive element in DNA [82]. Thus, as illustrated in Fig. 5, the schematic picture of hVDR that is developing not only maintains the modular concept of receptor organization with general DNA and hormone binding domains, but also illuminates the complexity of regions required for supporting heterodimerization.

Because many of the aforementioned conclusions concerning hVDR heterodimerization were derived from artificially generated mutants combined with VDR overexpression in heterologous systems, Whitfield *et al.* [116] sought to investigate a more natural setting where hVDR mutations actually elicit the phenotype of vitamin D-resistant rickets in patients, and where endogenous receptors can be probed in cultured fibroblasts. Most natural mutations found in hVDR are located in the zinc finger region, resulting in defective DNA binding and clinical vitamin D hormone resistance [63,65,117]. Whitfield *et al.* [116] investigated three natural mutations in the hormone binding domain of hVDR (R274L, I314S, and R391C) that confer the vitamin D-resistant phenotype, two of which significantly impair RXR heterodimerization (I314S and R391C). We have incorporated these three hVDR natural mutations into a hypothetical structural context (Fig. 5) based on the X-ray crystal structures of the ligand binding domains of hRARγ [102], rat TRα_1 [103], as well as unoccupied but homodimeric hRXRα [118].

Simply stated, the ligand binding domain of all nuclear receptors characterized to date consists of a sandwich of 12 α helices and several β strands organized in three dimensions around a lipophilic hormone binding pocket. When depicted linearly, as in Fig. 5, the dimerization face encompasses the region from helix 7 to helix 10, flanked by clusters of ligand contacts. The fact that this model applies to hVDR has been verified through site-directed mutagenesis in these regions (depicted as open and filled stars in Fig. 5) [111–113,119]. However, the striking finding is that the three hVDR natural mutations mentioned above can be explained in this proposed structural context. The R274L mutation, originally identified by O'Riordan and associates [120,121], behaves biochemically like a pure hormone binding mutation [116,121] and corresponds precisely to a ligand contact in both TR and RAR (Fig. 5). The mutation I314S, which endows hVDR with combined defects in both hormone retention and heterodimerization [116], lies within helix-7, at a presumed juncture of the ligand binding and heterodimerization activities of the receptor. Finally, R391C is positioned within the helix 10 dimerization surface, but not far removed from C-terminal ligand binding contacts (Fig. 5), consistent with the observed phenotype of a primary heterodimerization defect and a milder, secondary ligand retention deficiency [116]. Thus, all three mutations confer a phenotype that would have been predicted from their location within deduced functional domains. The primary conclusion from this work is that mutant hVDRs I314S and R391C (boxed in Fig. 5) establish a new subclass of clinically relevant, genetically altered receptors illustrating, *in vivo*, the fundamental importance of RXR heterodimerization in the physiological actions of hormone-occupied VDR [116].

D. DNA Binding Zinc Finger Region

The term "zinc finger" was first used to describe a 30-residue, repeated sequence motif found in *Xenopus* transcription factor IIIA [122], which now appears to be a ubiquitous structural motif for nucleic acid recognition. Well over 200 different cDNA sequences have been found to encode zinc finger motifs [123,124], including GAL4, protein kinase C (PKC), and the entire superfamily of nuclear hormone receptors [125]. Within the receptors that comprise the GR subfamily, there are two regions in the zinc finger DNA binding domain that play important roles in responsive element binding. The first region is known as the proximal or P box (Fig. 5), which confers target gene specificity for GR and ER [126–128]. For example, if the P box of GR is replaced with an ER P box, then the target gene specificity of GR will be changed to that of ER. The second region, known as the distal or D box (Fig. 5), mediates homodimerization of the GR subfamily of receptors on the DNA framework [126]. A third important region has been identified in mouse RXRβ, termed the T box (Fig. 5), which coincides with an α-helical structure just C-terminal of the zinc finger region [129]. It has been demonstrated that the T box mediates RXR homodimer binding to tandem repeats of estrogen responsive element half-sites [86,129]. Just C-terminal of the T box in NGF1-B, an RXR-related orphan receptor, lies an A box, which is required for the recognition of two A-T base pairs at the 5′ end of the single half-site DNA-binding element for the NGF1-B monomer [86].

The significance of the P, D, and T boxes in the VDR DNA-binding domain differs from that of the GR/ER subfamily (Fig. 5). As indicated earlier, GR and ER do not appear to contain a conserved T-box α-helical region, whereas the hVDR T box, specifically residues Lys-91 and Glu-92 [82,115], is engaged in heterodimerization with RXR on DNA. On the basis of the cocrystal structure of the RXR–TR heterodimer bound to a DR4 thyroid hormone responsive element (TRE), and modeling for VDR [82], it is now evident that the T- and A-box helical region (Fig. 5) of TR, VDR, and other subfamily members also contacts DNA, thereby significantly extending the DNA binding domain of the heterodimerizing nuclear receptors C-terminal of the zinc fingers. This hypothesis, in agreement with the original deletion analysis by McDonnell *et al.* [101], approximates the C-terminal boundary of the DNA binding domain of hVDR at amino acid 114.

Because VDR is not a homodimerizing receptor like GR and ER [38], it is not surprising that the D-box amino acids (Fig. 5) in hVDR play little or no role in specific DNA binding [115]. Further, results from gel mobility retardation analysis of VDRE binding and from transcriptional activity in transfected cells [115] reveal that the VDR P-box amino acids, EGG (boxed in Fig. 5), can be altered to GR P-box residues GSV without appreciable loss in VDR DNA binding or transcriptional activity utilizing a rat osteocalcin VDRE-linked reporter gene. This provocative finding suggests that the VDR P box alone does not confer target gene specificity as it does in the GR/ER subfamily of steroid hormone receptors. Instead, it appears that other amino acid–nucleotide contacts are also important, as is the likelihood that heterodimerization with RXR may be capable of overriding alterations of VDR P box residues to amino acids that are still compatible with an α-helical structure required to recognize the major groove of DNA. Interestingly, a similar observation has been made for TR [130,131], leading us to conclude that for both VDR and TR, the P and D boxes may only be significant as general structural elements in the two zinc fingers during the functioning of the receptors as heterodimers with RXR on cognate hormone responsive elements (HREs).

The solution of the cocrystal structure for the DNA binding domains of RXR–TR on a DR4 TRE [82] allows one to superimpose the key structural elements of TR on the VDR sequence (Fig. 5) in order to obtain an initial approximation of VDR structure–function (also covered in Chapter 9). VDR amino acids corresponding to TR residues that interact with DNA [82], which are designated with black dots in Fig. 5, segregate into groups on the C-terminal side of each zinc finger. In general, the α helix on the C-terminal side of finger 1 is involved in DNA base recognition, whereas the α helix C-terminal of finger 2 participates in DNA backbone contacts (Fig. 5). Because naturally occurring point mutations in putative DNA-interacting residues Lys-45, Arg-50, Arg-73, or Arg-80 of hVDR elicit the HVDRR phenotype of $1{,}25(OH)_2D_3$ resistance in patients [65], we would assert that there exists reasonable structural congruity between the VDR finger region and that of TR. Further support for this notion arises from the following predictions derived from Fig. 5: the D box in hVDR appears to be of little relevance in DNA binding while important DNA contacts occur in the T- and A-box helical region, and numerous DNA interaction sites exist besides the EG of the P box. As discussed above, precisely these observations have been reported for hVDR [115]. Ultimately, the elucidation of the definitive role of these various segments within the DNA binding region will require the determination of the X-ray crystal structure of RXR–VDR on a DR3 VDRE. A final point of interest concerning the DNA-binding domain of VDR is the unique presence of five basic amino acids in the intervening sequence between the two zinc fingers (Fig. 5). Four of these positively charged residues are hypothesized to make DNA contact (Fig. 5), but this region apparently is also crucial for nuclear localization of the receptor (detailed above in Section III,A) and, in addition, includes Ser-51, a site of hVDR phosphorylation by PKC (detailed below in Section V,C). Thus, the phosphorylation state of Ser-51 could control the DNA binding and nuclear localization capacity of VDR.

E. Transactivation and Interaction with Basal Transcription Factors

The precise mechanism of transcriptional regulation employed by the activated RXR–VDR heterodimer, once bound to DNA, is not well understood. Functional analyses of members of the nuclear receptor superfamily have demonstrated the presence of at least two major domains involved in receptor-mediated transcriptional stimulation. The N-terminal region of several nuclear receptors contains a constitutive activation domain referred to as AF-1 [132–134]. In addition, the entire ligand binding domain of various receptors can be fused to the GAL4 or glucocorticoid receptor DNA binding domain to produce a chimeric protein capable of activating transcription in response to the cognate ligand [135–137]. A C-terminal subdomain of this hormone-dependent activation function is known as AF-2 (Fig. 5, lower right) [136,138–140]. The AF-2 region of the thyroid hormone receptor has been shown to interact with the general transcription factor IIB (TFIIB), implying that association of nuclear receptors with the basal transcriptional machinery may serve as one means of achieving transcriptional control of hormone-stimulated genes [141]. Interestingly, the VDR has also been shown to interact physically with TFIIB using both glutathione-*S*-transferase (GST) fusion protein coprecipitation assays and the yeast two-hybrid system [142,143]. The precise domain in VDR responsible for this interaction is not well defined; nonetheless, we have shown that $1{,}25(OH)_2D_3$ induces the formation of an RXR–VDR–TFIIB ternary complex in solution (P. W. Jurutka, L. S. Remus, and M. R. Haussler, 1997, unpublished results), suggesting a functional role for the VDR–TFIIB association. This concept is consistent with the observation that VDR-mediated transcription is dramatically enhanced in transfected mouse embryonal carcinoma P19 cells by coexpression of additional TFIIB [142].

Functional evaluation of VDR by site-directed mutagenesis points to at least two regions of the receptor required for transcriptional activation (Fig. 5). One of these domains, known as E1 (residues 244 to 263), is highly conserved in the VDR subfamily and is critical for optimal interaction with RXR, which secondarily leads to competent transactivation in response to $1,25(OH)_2D_3$. However, embedded in this region is a pure or primary transactivation amino acid, Lys-246 (Fig. 5), apparently not involved in heterodimerization [113]. The second region (AF-2) resides near the C terminus, as confirmed by truncation of the last 25 amino acids in hVDR, which generates a transcriptionally inactive receptor that retains heterodimerization capacity and relatively high affinity ligand binding ability [112]. Comparison of this extreme C-terminal region of the receptor with other members of the nuclear receptor superfamily (Fig. 5, lower right) reveals a high degree of similarity, with hVDR indeed containing an AF-2-like domain. Further probing of this domain has identified Leu-417 and Glu-420 (boxed in Fig. 5) as essential for competent, hormone-dependent transcriptional activation [144]. These amino acids, located in a segment of VDR containing several hydrophobic and negative residues (Fig. 5), are presumed to form an α-helical structure analogous to helix 12 in TR, RAR, and RXR [102,103,118] and likely define the AF-2 function within the VDR.

Because mutation of Leu-417 or Glu-420 to alanine does not affect binding of VDR to hormone, RXR, or DNA to any significant degree, these amino acids are thought to participate directly in the mechanism of transcriptional activation following the recruitment of ligand-bound VDR–RXR heterodimers to the VDRE [144]. Such a mechanism might involve the interaction of VDR with basal factors like TFIIB as has been reported for TR [141] or with TATA binding protein (TBP)-associated factors (TAFs), thus facilitating the formation of an active transcriptional complex. However, the VDR AF-2 domain does not appear to mediate association with TFIIB because the wild-type receptor and Leu-417/Glu-420 mutant VDRs interact similarly with TFIIB, *in vitro* [144]. These observations suggest that Leu-417 and Glu-420 represent contact sites for the interaction of VDR with a novel coactivator protein(s) required for ligand-dependent transcriptional stimulation.

IV. THE FUNCTIONAL VDR IS A HETERODIMER WITH RXR

Precedents established by other subfamily members indicate that VDR functions primarily as a heterodimer with RXR, at least in the case of the stimulation of gene transcription. Studies with artificial and natural hVDR mutants, the latter eliciting the $1,25(OH)_2D_3$-resistant phenotype, confirm that RXR association is required for VDR to bind DNA and activate transcription. However, persistent questions regarding VDR include (1) the relative importance of VDR homodimers, (2) the role of $1,25(OH)_2D_3$ in RXR–VDR complex formation, and (3) the influence of the RXR ligand, 9-*cis*-RA.

A. $1,25(OH)_2D_3$ Ligand Enhances Heterodimer Formation

Among the classic steroid hormone receptors, it is known that ligand binding promotes the dissociation of auxiliary factors such as heat-shock proteins (hsps), resulting in the unmasking of nuclear localization and DNA binding domains [145], accompanied by ligand-induced homodimerization and HRE binding [146]. However, in the case of the nuclear receptors for thyroid hormone and the hormonally active metabolites of vitamins A and D, the role of ligand is less well defined.

The thyroid hormone receptor is generally accepted to form homodimers as well as heterodimers with RXR on TREs. Data suggest that the unliganded TR homodimer operates as a repressor of transcription, whereas thyroid hormone is proposed to dissociate these unoccupied homodimers to facilitate TR–RXR heterodimerization on the TRE and stimulate transcription [147,148]. In contrast, RAR does not appear to form homodimers on DR5 retinoic acid responsive elements (RAREs) [149], instead cooperating exclusively with RXR in RARE association and vitamin A metabolite-responsive transcription. When present in excess in gel mobility shift DNA binding assays, both TR and RAR display RXR heterodimeric association with their respective HREs in the absence of added cognate ligand. These *in vitro* studies are consistent with immunocytochemical data indicating that, unlike classic steroid hormone receptors which reside in the cytoplasm complexed with hsp90 and other proteins in their unoccupied state, unliganded TR, RAR, and VDR [54] exist in the nucleus in general association with DNA. These findings have led to the dogma that ligand is not required for TR, RAR, and VDR binding to target HREs. However, *in vivo* footprinting experiments [150,151] have prompted the conclusion that, in the case of RAR–RXR heterodimers, RAR ligands are required for RARE binding under physiological conditions, and similar studies of the rat osteocalcin promoter in intact rat osteosarcoma ROS 17/2.8 cells [152] suggest that ligand may be necessary for VDRE occupation by endogenous VDR.

Investigations of the molecular function of $1,25(OH)_2D_3$ in the interaction of VDR with DNA and hence the role of this ligand in the regulation of its specific hormone responsive genes, *in vitro*, have produced conflicting results and conclusions. While many groups report that VDRE binding requires RXR and is enhanced by ligand, interpretations vary as to whether ligand binding precedes heterodimer function [23], or whether DNA binding occurs initially in the absence of ligand [153]. Yet another model, presented by Cheskis and Freedman [34], proposes that VDR binding to the osteopontin-like DR3 leads to the formation of a homodimeric complex through a monomeric intermediate. They hypothesized that $1,25(OH)_2D_3$ increases the dissociation of the DNA bound homodimer by decreasing the rate of conversion of DNA bound monomer to homodimer, which is assumed to ultimately promote the formation of a DNA bound VDR–RXR heterodimer, again through a monomeric intermediate [34].

The relative importance of VDR associating with DNA as a homodimer is itself a controversial issue. Several studies involving gel mobility shift assays reveal that when physiological quantities of VDR are utilized, the receptor only binds efficiently to DR3 VDREs in the presence of purified RXRs or suitable nuclear extract sources of RXRs [23,24,31,38,72,74,154]. However, a study examining DNA binding of purified hVDR, *in vitro,* indicates that the receptor can bind in the absence of RXR to a DR3 consisting of an osteopontin-like half-site, AGTTCA. These experiments suggested that T–A base pairs at the third and fourth positions of each half-site are of primary importance for homodimeric hVDR recognition [155]. A comparative study of VDR binding to the mouse osteopontin and rat osteocalcin VDREs (Fig. 3) [76] yielded similar findings that VDR binds to the mouse osteopontin VDRE alone, whereas binding to the rat osteocalcin VDRE was dependent on the presence of RXR.

In probing the role of ligand in the association of VDR with DNA, the interpretation of gel mobility shift assays is complicated by the vast excess of expressed receptor often included in such reactions. Because of the sensitivity of this assay, even a minuscule fraction of receptor that actually binds to the labeled VDRE probe under these conditions can be visualized. In addition to employing supraphysiological receptor levels, a second problem often encountered in this *in vitro* assay is that subphysiological concentrations of salt (≤ 0.1 *M* KCl) have been used in some studies, allowing for the formation of relatively low affinity protein–DNA complexes. We therefore sought to devise an *in vitro* gel mobility shift assay that would more accurately reflect *in vivo* conditions, primarily by utilizing physiological salt concentrations (0.15 *M* KCl) and limited amounts of partially purified, baculovirus-expressed VDR and RXRs [38]. Employing this assay, we addressed some of the questions enumerated above.

Our studies reveal that when 20 ng of VDR ($\sim$10 n*M*) or 20 ng of VDR plus 20 ng of RXR are incubated with either the rat osteocalcin or mouse osteopontin VDREs, no DNA-bound homodimeric VDR species is apparent, but a VDRE-complexed VDR–RXR heterodimer occurs that is strikingly dependent on the presence of the $1,25(OH)_2D_3$ ligand [38]. Thus, at receptor levels approaching those in a typical target cell, a VDR ligand-dependent heterodimer with RXR is the preferred VDRE binding species. Only when VDR alone or VDR plus RXR levels are raised to 100 ng of each receptor with the mouse osteopontin VDRE [38], or 260 ng with the weaker rat osteocalcin VDRE [112], can faint homodimers of VDR bound to the DNA probe be visualized. In addition, at these greater amounts of receptors, neither the VDR homodimer nor the VDR–RXR heterocomplexes are modulated significantly by inclusion of $1,25(OH)_2D_3$ in the incubation mixture [38]. Therefore, we conclude that higher receptor levels, *in vitro,* generate artifactual VDR homodimers as well as attenuate the normal physiological ligand dependence of VDR–RXR binding to the VDRE. To explain seemingly ligand-independent VDR–RXR association with the VDRE, we postulate the existence of a minor subpopulation of VDR that is unstably activated in the absence of $1,25(OH)_2D_3$ [148] and therefore capable of heterodimerization to generate a positive gel mobility shift under conditions of vast receptor excess. In contrast, the physiologically relevant gel shift assay at or below 10 n*M* receptor levels and at 0.15 *M* KCl presumably reflects the *in vivo* events of ligand-triggered heterodimerization and DNA binding [150–152]. It also confirms and extends earlier *in vitro* data showing that $1,25(OH)_2D_3$ enhances VDR–RXR complex formation [23,31,74].

B. The RXR Ligand, 9-*cis*-Retinoic Acid, Attenuates $1,25(OH)_2D_3$ Action: An Allosteric Model for VDR–RXR Interaction

Conflicting experimental data also exist pertaining to the role played by the cognate ligand for RXR in the binding and transcriptional regulation of $1,25(OH)_2D_3$-responsive genes by the VDR–RXR heterodimer, including demonstration of synergistic action with $1,25(OH)_2D_3$ [156–159], negligible action [160], or an inhibitory effect [31,33,36]. These marked differences likely result from varying transfection and ligand addition protocols, as well as cell and species specificity.

Employing the physiological gel mobility shift procedure with biochemically defined components, clear evidence has been obtained that 9-*cis*-RA is a potent inhibitor of $1,25(OH)_2D_3$-enhanced VDR–RXR binding to VDREs such as osteopontin, with dramatic attenuation by the retinoid occurring at concentrations as low as 10^{-7} *M* [38]. Previous gel shift data also hinted at 9-*cis*-

RA inhibition [31,34], despite the higher concentrations of 9-*cis*-RA utilized in these studies. However, one puzzling finding indicated that the suppressive effect of 9-*cis*-RA seemed more pronounced *in vitro* than in transfected cells, where retinoid inhibition of $1,25(OH)_2D_3$-stimulated transcription is significant but 50% or less in magnitude [31]. This suggested that multiple pathways may exist for the assembly of the RXR–VDR heterocomplex, *in vivo*. To probe for distinct routes of assembly, we varied the order of addition of VDR, RXR, $1,25(OH)_2D_3$, and 9-*cis*-RA in the gel shift assay for VDRE binding [38]. The results showed that 9-*cis*-RA is a potent inhibitor of VDR–RXR heterodimerization on the VDRE in all situations excluding the circumstance when VDR alone is preincubated with $1,25(OH)_2D_3$ followed by addition of unliganded RXR [38].

To explain these data, we developed the model depicted in Fig. 6 which hypothesizes alternative allosteric pathways for the interaction of VDR–RXR with the VDRE. In pathway A of Fig. 6, $1,25(OH)_2D_3$ first occupies monomeric VDR, altering the conformation of the ligand binding domain such that it recruits RXR for heterodimerization and subsequent VDRE recognition. We postulate that VDR, previously occupied with $1,25(OH)_2D_3$, conformationally influences RXR in the resulting heterodimer such that it is resistant to liganding by 9-*cis*-RA, and is also transcriptionally activated as its AF-2/helix 12 domain is brought into the "closed" configuration [102]. Therefore, this action to suppress RXR ligand binding prevents 9-*cis*-RA from dissociating the RXR–VDR complex in order to divert RXR for retinoid signal transduction. In contrast, as illustrated in pathway B of Fig. 6, we propose that RXR exists in a different, 9-*cis*-RA-receptive, allosteric state in most other circumstances, such as when present as a monomer, in an apoheterodimer with unoccupied VDR, or even when the apoheterodimer is subsequently liganded with $1,25(OH)_2D_3$. This latter species of RXR–VDR–$1,25(OH)_2D_3$ (pathway B) is hypothesized to be fully competent in VDRE recognition, but the 9-*cis*-RA binding function of the RXR partner has not been conformationally repressed, rendering this form susceptible to dissociation by 9-*cis*-RA, which would then favor the formation of retinoid-occupied RXR homodimers. Therefore, unless VDR monomers are first occupied by $1,25(OH)_2D_3$ (pathway A), 9-*cis*-RA can operate to divert or dissociate RXR and direct it to form RXR homodimers (pathway B).

We speculate that the $1,25(OH)_2D_3$-liganded heterodimer in pathway A is active in transcriptional stimulation because both AF-2 domains are positioned in the

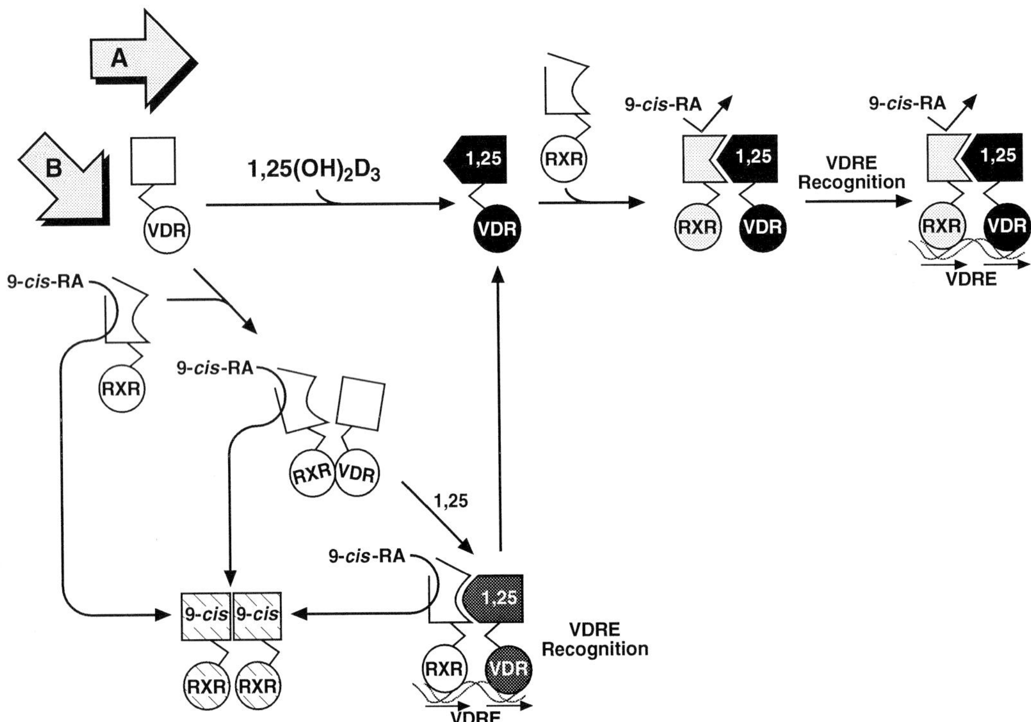

FIGURE 6 Allosteric pathways for VDR–RXR–$1,25(OH)_2D_3$ binding to DNA. In pathway A, $1,25(OH)_2D_3$-occupied VDR recruits RXR to form a transcriptionally active heterodimer on a VDRE. In pathway B, 9-*cis*-RA diverts/dissociates RXR from VDR to form an RXR homodimer.

sealed, efficacious configuration, whereas the analogous species in pathway B is relatively inactive, probably because the AF-2 function of the RXR partner is not conformed in the allosterically activated, closed position. The $1,25(OH)_2D_3$-occupied VDR–RXR dimer in pathway B has the advantage of flexible regulation because it is effectively a two-ligand switch. We propose that both pathways occur *in vivo* since 9-*cis*-RA blunting of $1,25(OH)_2D_3$ responsiveness in intact cells is significant, but incomplete, suggesting that at least two populations of RXR–VDR heterodimers exist. When our model is compared to those for RXR–RAR and RXR–TR [161], it is evident that VDR is closer in mechanism of action to the TR, where 9-*cis*-RA is likewise inhibitory, diverting RXR to a retinoid responsive pathway [162]. Also analogous is the fact that thyroid hormone occupation of TR attenuates 9-*cis*-RA binding to the RXR counterpart [161]. The action of RXR–RAR heterodimers, on the other hand, seems to be fundamentally different from that of RXR–VDR or RXR–TR in that RAR liganding by a retinoid facilitates RXR occupation by its retinoid ligand, resulting in cooperative stimulation of gene transcription by both of the vitamin A metabolites pictured in Fig. 1.

V. MULTISITE PHOSPHORYLATION OF VDR: MODULATION OF RECEPTOR ACTIVITY

A. Posttranslational Modification of Rodent, Avian, and Human VDR in Cell Culture

Several members of the steroid/thyroid receptor superfamily are known to be phosphorylated, including PR [163], GR [164], ER [165], TR [166], and VDR. The first demonstration of VDR phosphorylation [167], reported in 3T6 mouse embryonic fibroblasts, revealed a marked enhancement in phosphorylation of the endogenous murine receptor in response to the $1,25(OH)_2D_3$ ligand. The phosphorylated amino acid was subsequently identified as serine by Haussler *et al.* [89]. Employing organ culture experiments, ligand-dependent phosphorylation has also been reported for the avian VDR in the embryonic chick duodenum, a primary target tissue for $1,25(OH)_2D_3$. The incorporation of phosphate into the endogenous avian VDR appears to be a rapid event, taking place prior to significant $1,25(OH)_2D_3$-induced calbindin-D_{28K} mRNA accumulation [168]. Bone is another major site of action of $1,25(OH)_2D_3$, and the phosphorylation of VDR has been characterized biochemically in a rat osteoblast-like osteosarcoma cell line, ROS 17/2.8 [169]. These studies demonstrate that, like the mouse and chicken VDRs, the rat receptor is phosphorylated in the absence of the hormonal ligand and is rapidly ($\leq$30 min) hyperphosphorylated when ROS 17/2.8 cells are exposed to physiological concentrations of $1,25(OH)_2D_3$. The hyperphosphorylated receptor, similar to other phosphoproteins, exhibits a decrease in its electrophoretic mobility in denaturing polyacrylamide gels, a property that is reversed by treatment of the hyperphosphorylated VDR with alkaline phosphatase. In addition, the extent of phosphorylation induced in these cells by various concentrations of the natural $1,25(OH)_2D_3$ hormone or a noncalcemic analog, 22-oxa-$1,25(OH)_2D_3$, is positively correlated with the transcriptional activity of VDR when analyzed using a VDRE from the rat osteocalcin gene [169]. Thus, phosphorylation of rat VDR is a rapid event that temporally correlates with $1,25(OH)_2D_3$-induced transcription and precedes mRNA accumulation of osteocalcin, a bone matrix protein whose synthesis is transcriptionally regulated by $1,25(OH)_2D_3$.

McDonnell *et al.* [101] first showed that human VDR is also phosphorylated, although the magnitude of the ligand response of this posttranslational modification in hVDR was less pronounced than observed for mouse or rat VDR [167,169]. In addition, it was noted that hVDR was not detectably retarded in its mobility on sodium dodecyl sulfate (SDS)–polyacrylamide gel electrophoresis when phosphorylated as is rodent VDR [101,167,169,170]. Similar results were obtained in COS-7 monkey kidney epithelial cells [171] and ROS 17/2.8 cells [119] transfected with an expression vector coding for hVDR. To map the domain(s) in hVDR that is phosphorylated in intact mammalian cells, ROS 17/2.8 cells were transfected with internally deleted hVDR mutants followed by ortho[^{32}P]phosphate labeling. The results of these studies revealed that removal of hVDR residues in the N-terminal region of the hormone binding domain between Met-197 and Val-234 reduced the level of phosphorylation by the endogenous kinase(s) present in these osteoblast-like cells [119]. In agreement with these mutagenesis studies, peptide mapping experiments utilizing phosphorylated VDR immunoprecipitated from pig kidney (LLC-PK_1) cells localized the major site(s) of phosphorylation in the porcine receptor to a 23-kDa peptide fragment encompassing the hinge region and the N-terminal half of the hormone binding domain [172].

B. Phosphorylation of VDR by Casein Kinase II

An analysis of the deduced amino acid sequence of the rat VDR [20] for the presence of theoretical consensus recognition sites for several protein kinases revealed a number of minimal casein kinase-II (CK-II) consensus recognition sites of the type Ser/Thr-x-x-Asp/Glu-x-Asp/Glu [173], where either the +3 or the +5 position possesses an acidic residue. Most intriguing were two large clusters of potential CK-II sites within the N-terminal half of the hormone binding domain, and the rat VDR was subsequently found to be an efficient substrate for CK-II, *in vitro* [169]. The potential relationship between rat VDR phosphorylation in ROS 17/2.8 cells and that catalyzed by CK-II, *in vitro,* was addressed by examining the electrophoretic mobility of *in vivo* and *in vitro* phosphorylated receptor [169]. The results from these experiments demonstrated that the receptor, phosphorylated in response to $1{,}25(OH)_2D_3$ in intact cells, was retarded in its migration in denaturing polyacrylamide gels, as expected, relative to hypophosphorylated VDR control. However, this *in vivo* phosphorylated receptor species migrated slightly more rapidly in denaturing polyacrylamide gels than did VDR phosphorylated by CK-II, *in vitro,* suggesting a more extensive phosphorylation of VDR by this enzyme *in vitro* [169].

The sequence of hVDR [19] has also been examined for known kinase recognition sites, disclosing several consensus sites for CK-II which exhibit a pattern of distribution similar to that of rat VDR, although not all the sites are absolutely conserved. The possibility that the VDR from distinct species is phosphorylated by CK-II suggests this modification may be functionally significant. The ability of purified bovine CK-II to catalyze phosphorylation of full-length receptor produced from cloned hVDR cDNA by *in vitro* transcription/translation was tested [171], and the human receptor was found to serve as an *in vitro* substrate for CK-II similar to phosphorylation of rat VDR by this enzyme. Analysis of C-terminally truncated hVDR mutants showed that CK-II-catalyzed phosphorylation of hVDR, *in vitro,* is confined primarily to a region encompassing Ser-194 to Asp-232 [171]. As noted above, a major site of hVDR phosphorylation, *in vivo,* was initially localized between Met-197 and Val-234 [119]. This striking colocalization of *in vivo* and CK-II mediated *in vitro* phosphorylation of hVDR led us to hypothesize that a CK-II-like enzyme was the natural catalyst for this major posttranslational modification. Individual site-directed mutagenesis of five candidate Ser/Thr residues in this region demonstrated that replacement of Ser-208 (Fig. 5, top) with glycine or alanine selectivity abrogated CK-II phosphorylation of hVDR, both *in vitro* and *in vivo* [171]. Our studies showed that Ser-208 accounts for at least 60% of the phosphorylation of the receptor [171]. Subsequently, Hilliard *et al.* [174] confirmed Ser-208 as the predominant hVDR phosphorylation site in transfected COS-1 transformed monkey kidney cells utilizing phosphopeptide mapping and demonstrated that phosphorylation of this residue is enhanced by high concentrations of the $1{,}25(OH)_2D_3$ ligand. Therefore, two powerful and distinct methodologies have led to the elucidation of Ser-208 as the primary phosphorylated residue in hVDR.

C. Phosphorylation of Human VDR Serine-51 by Protein Kinase C

An important cellular kinase, protein kinase C (PKC), plays a crucial role in the modulation of multiple biochemical functions [175], and several studies [176,177] indicate that PKC could regulate VDR action in various cell lines. To investigate whether VDR is a substrate for PKC, VDR-containing immunocomplexes from ROS 17/2.8 cells were prepared and incubated in the presence of $[\gamma^{32}P]$ATP and the α, β, or γ isozymes of PKC. The results showed that only the β isozyme of PKC significantly phosphorylated immunoprecipitated rat VDR, *in vitro* [178]. Based on the deduced amino acid sequence, there are three potential PKC phosphorylation sites in the rat and human VDR, namely, Ser-51, Ser-119, and Ser-125. To examine whether any of these sites might actually be selectively phosphorylated by PKC-β, point mutant hVDRs were constructed by site-directed mutagenesis, expression plasmids were transfected into CV-1 monkey kidney cells, and cell extracts were tested for phosphorylation by this enzyme, *in vitro.* The results indicated that Ser-51 (Fig. 5, lower left) is the unique PKC-β-phosphorylation site in hVDR [178]. When CV-1 cells are transfected with wild-type or S51G mutant hVDR expression plasmids and treated with the PKC activator phorbol 12-myristate 13-acetate (PMA), an increase in ortho$[^{32}P]$phosphate incorporation into wild-type hVDR was observed, whereas the activator had no effect on phosphate incorporation in the S51G mutant hVDR [178]. These observations strongly suggest that PMA-induced phosphorylation of hVDR, *in vivo,* like the phosphorylation of hVDR *in vitro* by PKC-β, is localized to Ser-51.

As illustrated in Fig. 5 (lower left), Ser-51 in hVDR fits the consensus PKC phosphorylation criteria, $R/K_{1-3}X_{0-2}S/TX_{0-2}R/K_{1-3}$ [179]. Of particular significance is the conservation of a phosphorylatable residue at the position corresponding to Ser-51 in RAR, TR, and ER, but not in the GR group of receptors including PR, MR, and AR, which possess alanine at that location (Fig. 5, lower left). Moreover, as illustrated by underlining the basic amino acids immediately flanking serine or threonine in Fig. 5, the ER serine is not a PKC consensus site, indicating that the VDR subfamily of heterodimerizing receptors (including TR and RAR) could be a specific target for regulation by PKC.

To address the potential functional involvement of PKC phosphorylation in VDR action, additional mutations at the PKC site and its flanking basic residues were created, and the effects of these mutations on DNA binding, transactivation, and PKC-β phosphorylation were then examined in transfected cells [180]. Ser-51 phosphorylation was found to constitute up to 30% of total hVDR phosphorylation, *in vivo,* and was not enhanced by $1,25(OH)_2D_3$ ligand. This latter result is not surprising because Ser-51 is positioned in hVDR just C-terminal of the first zinc finger (Fig. 5). The positionally analogous threonine in TR occurs within an α helix that makes specific base contacts in the major groove of DNA during receptor–HRE recognition [82]. Therefore, whether VDR is DNA bound or free would seem to be more relevant than $1,25(OH)_2D_3$ liganding as a determinant of Ser-51 phosphorylation. On the basis of the studies outlined above [180], it was concluded that the basic residues surrounding Ser-51 are required for both PKC phosphorylation and VDRE binding of hVDR. However, alternation of Ser-51 to alanine reveals that phosphorylation of this residue is neither obligatory for DNA binding nor for transactivation [180]. Instead, VDR phosphorylation by PKC apparently renders the receptor incapable of localizing to the nucleus and binding to DNA (detailed below in Section V,F). Although some studies have reported that activators of PKC stimulate $1,25(OH)_2D_3$-mediated gene transcription [181], or that inhibitors of this kinase attenuate $1,25(OH)_2D_3$ induction of osteocalcin [182], these observations may be related to the composite effects of perturbing the PKC signal transduction pathway, such as stimulation of AP-1 activity, superimposed on the specific inhibitory action of PKC on VDR-mediated transcription.

D. Phosphorylation of VDR by Other Kinases (Protein Kinase A, etc.)

Although phosphorylation of Ser-51 and Ser-208 in hVDR accounts for approximately 90% of the total phosphorylation of the receptor in our systems, there is evidence for the existence of perhaps one or possibly two additional such postranslational modifications [183]. Another site of hVDR phosphorylation catalyzed by cAMP-dependent protein kinase (PKA) appears to reside between amino acids 133 and 201 [184], in a region containing a stretch of residues not conserved among species of VDR (Figs. 4 and 7). Although intact cell studies by others suggested that activators of PKA amplify $1,25(OH)_2D_3$-stimulated transcription [185,186], our direct probing of hVDR phosphorylation by PKA via coexpression of hVDR and the catalytic subunit of PKA in COS-7 cells revealed that receptor phosphorylation in this context diminishes its transcriptional activity [184]. Similar conclusions have been reached by Nakajima *et al.* [187], supporting PKA as initiating a negative regulatory loop in VDR action similar to that of PKC; however, instead of blunting nuclear localization/DNA binding, the PKA loop may suppress coactivator binding to VDR.

In an S51G/S208G double mutant hVDR, there exists a 5–10% residual phosphorylation that is enhanced significantly by the $1,25(OH)_2D_3$ hormone [183]. One possibility is that altering Ser-208 perturbs the structure of hVDR to uncover a cryptic hormone-dependent site of phosphorylation [174]. Alternatively, there may exist a quantitatively minor but novel site of hVDR phosphorylation that is highly responsive to ligand (Fig. 8). Ultimately, elucidation of the multiple phosphorylation sites in VDR and their responsiveness to ligand-enhanced modification will require a combination of peptide mapping and site-directed mutagenesis, with the possibility that even hormone-enhanced phosphorylation may be dependent on the expression of cell-specific kinases.

E. Phosphorylation of Other Nuclear Receptors and Comparison with VDR

Presently, direct phosphorylation of nuclear receptors by PKC seems unique to VDR. However, precedents exist for phosphorylation of nuclear receptors by both of the other kinases described thus far (CK-II and PKA), although the actual phosphorylation sites are not conserved. As depicted in Fig. 7, CK-II has been reported to phosphorylate chicken TRα-1 near the N terminus at serine-12 [188]. Another site fitting the criteria of a CK-II consensus has been located in the novel C-terminal extension found in the rat α-2 subtype of TR, as well as in the N-terminal regions of human ER, chicken PR, and mouse GR. Likewise, PKA has been shown to phosphorylate other members of the superfamily, notably chicken TRα-1 at two sites close to the above-mentioned CK-II site [166], and mouse RARα at a site within the conserved heptad region [189].

A kinase target with some similarity to a PKA site has been reported in the orphan receptor nur77 (also named NGFI-B) in the A-box region (Fig. 7). Although one group has claimed that PKA can phosphorylate this site [190], as well as a nearby site in the T box, *in vitro,* another group, without discounting PKA as a possibility, suggests that pp90^{rsk} or other kinases might be responsible for modification of the A-box site, *in vivo* [191].

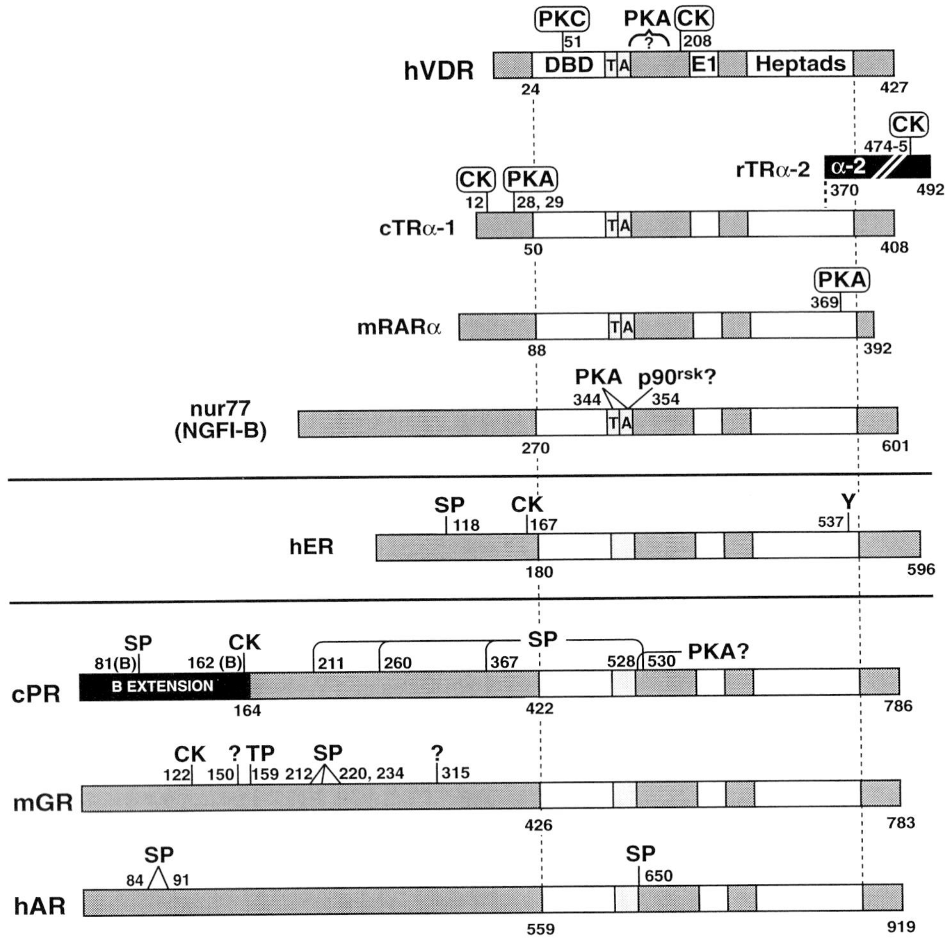

FIGURE 7 Known phosphorylation sites in receptors related to VDR. References are given in the text. Conserved functional domains (as in Fig. 4) are represented by white areas of the schematic diagram. SP and TP denote serines and threonines followed by prolines, which are sites for either the $p34^{cdc2}$ or the MAP kinases; PKA represents sites for cAMP-dependent protein kinase (or possibly $p90^{rsk}$ kinase, which has overlapping recognition criteria); CK signifies casein kinase-II sites; PKC denotes protein kinase C recognition sites; and a "?" indicates sites of receptor phosphorylation not fitting any known consensus recognition criteria. Outlining around the kinase designation indicates that the kinase identity has been confirmed by means other than consensus recognition criteria.

The most common type of phosphorylation site described so far for the superfamily of nuclear receptors, however, is a type not yet observed in VDR, namely, a serine or threonine followed immediately by a proline residue (SP or TP). This type of site potentially could be recognized by kinases such as proline-directed protein kinase [192] or mitogen-activated protein (MAP) kinase [193] and is commonly found in the N-terminal domain of nuclear receptors, especially the classic steroid receptors. Evidence for the actual *in vivo* phosphorylation of some of these sites has been obtained for the human ER [194,195], chicken PR [196,197], mouse GR [198], and human AR [199]. In addition to these numerous sites in the N termini, SP sites have been found just C-terminal of the DNA binding domain in human AR [199] and at a position slightly more C-terminal in chicken PR [196] (Fig. 7).

In addition to sites for CK-II, PKA, and the proline-directed kinases, there are two phosphorylation sites in mouse GR that do not fit the consensus recognition criteria for any known kinase. Clearly, this and other observations illustrate that nuclear receptor phosphorylation requires further characterization, both in terms of the specific kinases and amino acid recognition sites as well as with respect to functional significance of the phosphorylation event(s) for each receptor. To date, phosphorylation of steroid, retinoid, and thyroid hormone receptors has not been shown to be obligatory to primary signal transduction, and it appears instead to modulate such receptor actions as HRE

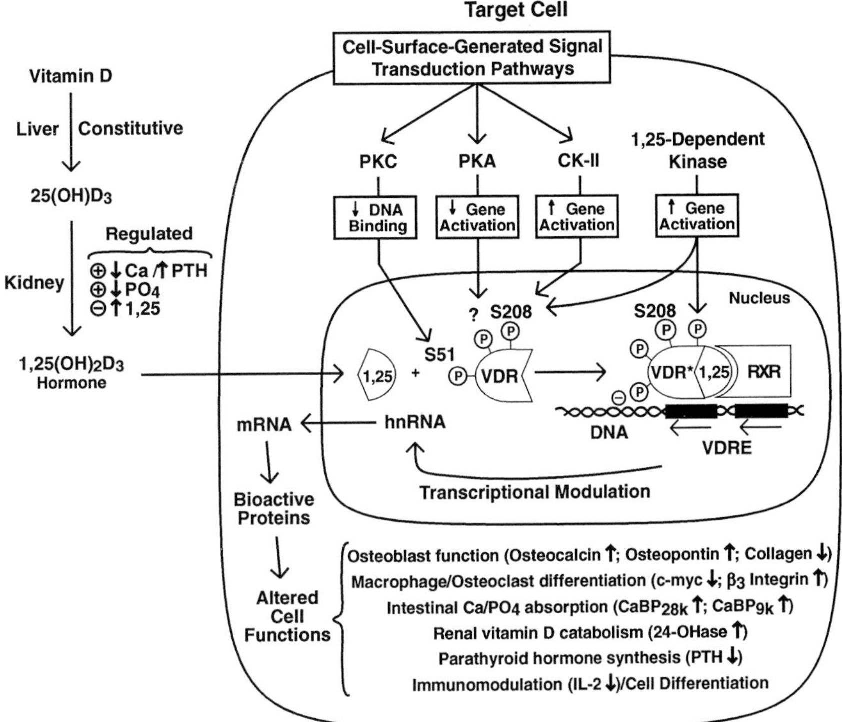

FIGURE 8 Integrative model of VDR regulation mediated by multisite receptor phosphorylation. The active $1,25(OH)_2D_3$ hormone, synthesized by two successive enzymatic hydroxylations of the vitamin D precursor, binds to its nuclear receptor (VDR) resulting in activation (VDR*) and high affinity interaction of the protein with the VDRE in combination with retinoid X receptor (RXR), thereby modulating the transcription of target genes like osteocalcin and leading to altered cell function. The receptor is shown to be a substrate for PKC (Ser-51), CK-II (Ser-208), and PKA (a site between residues 133 and 201). The activity of these kinases is regulated by cell surface-generated signal transduction pathways. The proposed function of each phosphorylation is shown in the box below each kinase. Alternatively, the receptor may also be phosphorylated by a putative $1,25(OH)_2D_3$-dependent protein kinase at Ser-208 and/or another hormone-stimulated site(s), possibly resulting from a ligand-induced conformational change in the VDR substrate.

binding and transactivation (reviewed by Jurutka *et al.* [200]).

F. Functional Implications of VDR Phosphorylation

As previously discussed, human VDR is likely phosphorylated by at least three and perhaps four distinct kinases. In the case of PKC-mediated phosphorylation, in studies utilizing site-directed mutagenesis we observed that replacement of Ser-51 with a negatively charged aspartic acid residue abolished VDRE binding, diminished nuclear localization, and reduced transactivation by 90% [180]. Independent experiments with purified, *E. coli*-expressed hVDR confirmed that phosphorylation of the receptor by PKC-β, *in vitro,* dramatically decreased its ability to bind to the VDRE in cooperation with RXR [180]. Thus, phosphorylation of VDR by PKC constitutes a negative feedback loop that may operate to limit VDR DNA-binding/nuclear localization. It is possible that PKC-mediated inhibition of hVDR activity could limit transactivation by $1,25(OH)_2D_3$ during certain periods of cell growth. Furthermore, $1,25(OH)_2D_3$ selectively stimulates transcription of the PKC-β gene in the human promyelocytic cell line HL-60 [201], and expression of the VDR gene is inhibited by activated PKC in NIH3T3 mouse embryo cells [177]. Taken together, these observations suggest the presence of a feedback loop in which stimulation of PKC-β gene expression by $1,25(OH)_2D_3$ can eventually result in PKC-mediated down-regulation of the VDR gene and in the inhibition of VDR protein activity by PKC-catalyzed phosphorylation. Under certain circumstances, phosphorylation by PKC could also be initiated by extracellularly generated signal transduction pathways. This regulatory interplay between the hormonal ligand and PKC pathways may be important in cellular growth control mechanisms.

Phosphorylation of hVDR at the other major site, namely, Ser-208, appears to be enhanced at high concentrations of the ligand [174] and to some degree at lower $1,25(OH)_2D_3$ levels with baculovirus-expressed hVDR [171], suggesting that modification of this residue may play a positive role in VDR action. However, our initial tests did not show any effect of phosphorylation at Ser-208 on either $1,25(OH)_2D_3$ binding or heterodimerization/DNA binding [200]. Further, transcriptional activation by $1,25(OH)_2D_3$, per se, is not dependent on Ser-208 phosphorylation because replacement of this residue with alanine does not compromise activity in transfected cells [174,200]. However, we have discovered that coexpression of CK-II further potentiates $1,25(OH)_2D_3$-stimulated transactivation via a mechanism absolutely dependent on Ser-208 phosphorylation [200]. Thus, it is reasoned that Ser-208 phosphorylation is a positive modulatory event which apparently maintains hVDR in an active conformational state for interaction with coactivators or components of the basal transcription apparatus.

Although less is known about PKA phosphorylation of VDR, our present data suggest that this modification acts, like PKC-mediated phosphorylation, to negatively modulate VDR action. This effect of PKA phosphorylation may be significant in terms of calcium homeostasis. Parathyroid hormone (PTH) functions to raise plasma calcium primarily through its action to stimulate bone resorption, utilizing cAMP and the PKA signal transduction pathway to mediate these effects in osteoblasts [202,203]. PTH also increases metabolism in the kidney of vitamin D to its active hormonal form, $1,25(OH)_2D_3$. This effect of PTH shifts calcium recovery from skeletal reserves in the acute situation to intestinal absorption as mediated by $1,25(OH)_2D_3$ in the chronic situation. However, since the increased levels of $1,25(OH)_2D_3$ can also act at the osteoblast to elicit the production of bone resorptive factors, it is possible that enhanced activity of PKA in osteoblasts may down-regulate the $1,25(OH)_2D_3$-mediated increase in bone resorptive factors through PKA-catalyzed phosphorylation and inhibition of VDR. This putative mechanism would provide a powerful cross talk between the PTH-stimulated PKA transduction pathway and VDR, and in this way would further facilitate the shift of calcium reclamation from bone to intestine. In addition, similar to the PKC-mediated down-regulation of the VDR gene, the PTH–cAMP system has been shown to suppress VDR mRNA accumulation in osteoblasts [204]. Therefore, the inhibition of VDR activity by PTH may operate at both the transcriptional and posttranslational level. Collectively, these observations suggest that although phosphorylation of VDR by PKC or CK-II may alter the activity of the receptor under various conditions of cell growth and differentiation, PKA phosphorylation of VDR may be more related to control of receptor activity in classic, physiological target tissues like bone in concert with other calcium regulators such as PTH.

The principal conclusions derived from phosphorylation studies of VDR, as summarized in Fig. 8, are that, at least with respect to rat osteocalcin VDRE–reporter constructs, phosphorylation of hVDR modulates receptor action via both negative (PKC and PKA) and positive (CK-II) effectors. It therefore can be envisioned that the nuclear functions of $1,25(OH)_2D_3$ are fine tuned by kinase cascades initiated at the cell surface, providing mechanisms for the cross talk control of vitamin D action with systemic hormones/growth factors and even with the state of cell differentiation. This modulation is likely superimposed on the primary signaling pathway of $1,25(OH)_2D_3$ action through VDR–RXR heterodimerization on the VDRE. It thus appears that the receptor-mediated genomic actions of $1,25(OH)_2D_3$ are orchestrated in an elegant fashion in what will likely turn out to be target cell-specific modulation by a myriad of phosphorylation events. When these intracellular regulatory forces are added to the fact that VDREs exist in composite enhancer elements with other *cis*-acting sequences such as AP-1 [67], gene control by the vitamin D hormone will ultimately be characterized as an extremely complex process controlled by multiple inputs.

VI. INTEGRATED HYPOTHESIS FOR THE MOLECULAR MECHANISM OF VDR ACTION

It is probable that hVDR forms an amphipathic helix (corresponding to helix 12 in the other receptors) surrounding glutamic acid-420 (Fig. 5) that is analogous to the AF-2 characterized for TR [205], RAR [102], RXR [140], and ER [138]. Although this AF-2 domain is capable of autonomously activating transcription [140], the reason that such activity is modest may be because the AF-2 region is proposed to operate in a ligand-dependent fashion, involving a structural rearrangement to reposition the AF-2 for both intramolecular and intermolecular protein–protein interactions. Specifically, based on the crystal structure of unoccupied RXR [118] and liganded RAR [102] and TR [103], helix 12/AF-2 appears to protrude outward from the more globular ensemble of helices 1 through 11 in the absence of ligand, such that it is unable to interact efficiently with a coactivator/transcription factor. Upon ligand binding, a conformational signal is then transmitted to helix 12, causing it to fold back on helix 11 and interface with the globular ligand binding domain. The pivoting of helix 12 seemingly accomplishes two feats that mediate ligand-activated transcription by the receptor: (1) closing of a "door" on the channel through which the lipophilic ligand enters the internal binding pocket of the receptor and (2) locking helix 12 into a stable conformation that facilitates its interaction with a coactivator/transcription factor. Ligand binding contacts on or near helix 12 (Fig. 5) probably are significant in maintaining this active positioning of helix 12, essentially trapping ligand in the binding pocket to sustain transactivation events.

Experiments with VDR [144] are in concert with insight into the function of AF-2 in other nuclear receptors, which is to recruit coactivators of the type of steroid receptor coactivator-1 (SRC-1) [206]. A number of candidate coactivators have been isolated in addition to SRC-1 [207–211], and, in several cases, interaction with nuclear receptors requires intact AF-2 core regions [208,209]. Moreover, AF-2 mutations create dominant negative receptors, as in hRARγ [102]. Indeed, Jurutka *et al.* [144] have observed that the VDR AF-2 mutants E420A and L417A exhibit dominant negative properties with respect to transcriptional activation. Such AF-2-altered receptors are inactive transcriptionally but can bind $1,25(OH)_2D_3$ ligand and heterodimerize normally on VDREs, consequently competing with wild-type VDR–RXR heterodimers for VDRE binding. These data argue that the AF-2 of VDR in an RXR–VDR heterodimer is absolutely required for the mediation of $1,25(OH)_2D_3$-activated transcription, not only for its intrinsic activation potential, but also because of its presumed role in stabilizing the retention of $1,25(OH)_2D_3$ ligand in the VDR binding pocket.

To further comprehend the molecular functioning of the RXR–VDR–$1,25(OH)_2D_3$ complex, it is important to investigate what part, if any, is played by the AF-2 domain (Fig. 5) of the RXR "silent" partner in this signal transduction pathway. To explore this phenomenon, AF-2-truncated mutants of RXRα or RXRβ were created and tested for their ability to function as dominant negative modulators of $1,25(OH)_2D_3$-stimulated transcription [150]. Because previous data with RXR–RAR control of gene expression seemed to indicate that the RXR AF-2 was dispensable [212], it was surprising to find that AF-2-truncated RXRs were potent dominant negative effectors of $1,25(OH)_2D_3$ action in transfected cells [150]. It is therefore concluded that although the RXR "silent" partner in VDR signaling apparently is not occupied by retinoid ligand (Fig. 6), its AF-2 still plays an active role in transcriptional stimulation. A similar conclusion also has been reached by two groups studying RXR–RAR action [148,151], with the use of RAR-specific ligands precluding ligand binding by the RXR partner. However, Shulman *et al.* [148] have introduced a caveat to the above theory, as they point out that AF-2-truncated RXRs in heterodimers become strong, constitutive binders of corepressors similar to the silencing mediators of retinoid and thyroid hormone receptors (SMRTs). Thus, an alternative explanation to an active coactivator-binding role for RXR AF-2 in heterodimers is that it plays a more passive role in excluding corepressors. In the latter scenario, truncation [150] or point mutation [148] of RXR AF-2 may generate spurious corepressor binding rather than preventing coactivator contact. Only additional research examining coactivator and corepressor associations of VDR–RXR heterodimers will resolve this issue.

To provide a working hypothesis for $1,25(OH)_2D_3$ action at the molecular level, we have developed the model illustrated in Fig. 9. It is based primarily on data from our laboratory and others studying $1,25(OH)_2D_3$ and VDR but also relies on assumed similarities between VDR, TR, and RAR. VDR is proposed to exist in target cell nuclei, perhaps very weakly associated with DNA, in a monomeric, inactive conformation with the C-terminal AF-2 domain extended away from the hormone binding cavity (see key at left in Fig. 9). In addition, VDR is probably bound to a corepressor analogous to SMRT [213] or N-CoR [214]. As suggested in Fig. 9, part of the action of ligand may be to drive off the corepressor. On binding with $1,25(OH)_2D_3$, VDR assumes an active phosphorylated conformation, with the AF-2 pivoted into correct position for both ligand retention and coactivator contact. Further, the hormone facilitates interaction of VDR with RXR through a stabilized heterodimerization interface. We propose that this event also repels a corepressor from RXR (Fig. 9). In addition, $1,25(OH)_2D_3$-occupied VDR may itself function as a kind of allosteric regulator of RXR, perhaps by conveying a conformational signal through the juxtaposed dimerization domains to induce the AF-2 of RXR into an active conformation for coactivator binding and/or corepressor exclusion, even in the absence of its 9-*cis*-RA ligand. As discussed above (Fig. 6), the joining of preliganded VDR and unliganded RXR apparently renders the RXR partner unresponsive to 9-*cis*-RA. Alternatively, if 9-*cis*-RA encounters RXR monomer first (Figs. 6 and 9), or binds to RXR that is complexed with VDR in an apoheterodimer (Fig. 6), the retinoid is able to divert the RXR to generate homodimers and effectively blunt $1,25(OH)_2D_3$-driven transcription.

In the primary activation pathway pictured in Fig. 9 (center), the RXR–VDR–$1,25(OH)_2D_3$ complex recognizes and targets the genes to be controlled through high affinity association with a VDRE in a gene pro-

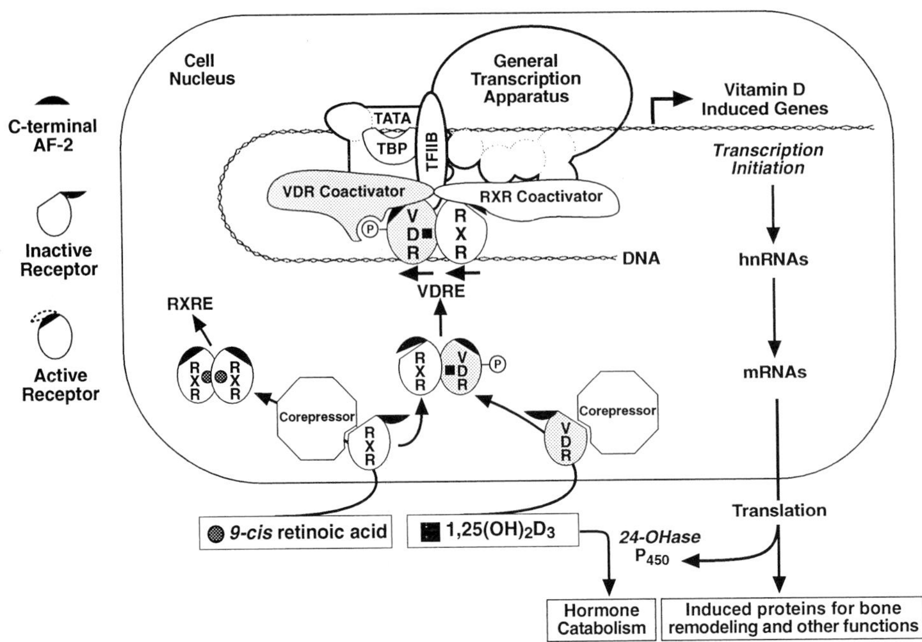

FIGURE 9 Model for transcriptional activation by $1,25(OH)_2D_3$ on the promoter of a target gene. VDR ligand binding is proposed to dissociate a corepressor and promote phosphorylation and RXR recruitment, including conformational activation of RXR. Thus, hormone-occupied and phosphorylated VDR recruits an unliganded RXR, forming a VDR–RXR heterodimer in which the occupied VDR conformationally influences RXR to become active in coactivator binding via AF-2, similar to the effect of ligand on VDR. The RXR–VDR–$1,25(OH)_2D_3$ heterocomplex possesses high affinity and selectivity for binding the DR3 VDRE, thereby targeting the gene whose transcription is to be controlled by vitamin D. DNA looping is facilitated by hypothetical coactivators bound to VDR (potentiated by phosphorylation) and to RXR that bridge to TBP and/or associated factors (TAFs, not shown). We propose that coactivator–TAF association, possibly mediated by a cointegrator analogous to CREB-binding protein (not pictured; see Jankneckht and Hunter [220]), constitutes one "arm" of the transcriptional activation pathway. The second "arm" is hypothesized to be a recruitment of TFIIB by VDR that is now positioned in the vicinity of the TATA box, an event that initiates the assembly of the RNA polymerase II preinitiation complex and elicits repeated rounds of transcriptional initiation of the regulated gene.

moter region. Coactivators that are presumed to bind to VDR (facilitated by phosphorylation) and RXR AF-2s then are postulated to link with TAFs/TBP, thereby looping out DNA 5′ of the TATA box. This series of events positions VDR such that it can independently recruit TFIIB to the promoter complex, commencing the assembly of the RNA polymerase II holoenzyme into the preinitiation complex. Precedents exist for transcription factors independently attracting TFIIB, such as hepatocyte nuclear factor-4 [215], as well as for a sequential, two-step pathway for activator-stimulated transcriptional initiation [216,217]. Using the latter model as an analogy, VDR would contact both TBP/TAFs (via coactivator bridges) and TFIIB in order to initiate RNA polymerase II holoenzyme assembly. The order of attachment of these two "arms" of activation has not been determined; however, in the case of acidic activators, it is known that attraction to the TATA element precedes interaction with components of the initiation complex [217]. Interestingly, the mechanism of $1,25(OH)_2D_3$ action depicted in Fig. 9 is not only essential for induction of bone remodeling and other vitamin D functions, but is also self-limiting via 24-hydroxylase induction. In addition, these actions of $1,25(OH)_2D_3$ would be blunted under conditions within a cell where 9-*cis*-RA concentrations dominate over those of $1,25(OH)_2D_3$, thus explaining an important facet of cross talk between the vitamin A and vitamin D systems.

In conclusion, the actions of the vitamin D hormone as mediated by VDR appear to involve a delicate balance of the control of gene expression, and the intricacies of this mechanism not only illuminate coregulation of cellular functions by vitamins D and A, but also unveil the potential participation of novel cofactors in the bridging/targeting of the VDR–RXR complex to the assembly and initiation of the transcription machine. The future directions of VDR research designed to test this model will no doubt involve characterizing the receptor physically through X-ray crystallography, biochemically with the aid of an *in vitro* transcription system, biologically via transgenic animal studies, and clinically through an evaluation of the role played by VDR variants in the etiology of skeletal and proliferative disorders.

Acknowledgments

The authors thank the other members of our laboratory, Hope Dang, Carlos Encinas, Michael Galligan, Anish Oza, Jack Price, Michelle Thatcher, and Cristina Velazquez, for their important contributions. This work was supported by National Institutes of Health grants to M.R.H., J.-C.H., and G.K.W.

References

1. Haussler MR, Myrtle JF, Norman AW 1968 The association of a metabolite of vitamin D_3 with intestinal mucosa chromatin *in vivo*. J Biol Chem **243**:4055–4064.
2. Haussler MR, Norman AW 1969 Chromosomal receptor for a vitamin D metabolite. Proc Natl Acad Sci USA **62**:155–162.
3. Myrtle JF, Haussler MR, Norman AW 1970 Evidence for the biologically active form of cholecalciferol in the intestine. J Biol Chem **245**:1190–1196.
4. Haussler MR, Boyce DW, Littledike ET, Rasmussen H 1971 A rapidly acting metabolite of vitamin D_3. Proc Natl Acad Sci USA **68**:177–181.
5. Holick MF, Schnoes HK, DeLuca HF, Suda T, Cousins RJ 1971 Isolation and identification of 1,25-dihydroxycholecalciferol, a metabolite of vitamin D active in the intestine. Biochemistry **10**:2799–2804.
6. Lawson DEM, Fraser DR, Kodicek E, Morris HR, Williams DH 1971 Identification of 1,25-dihydroxycholecalciferol, a new kidney hormone controlling calcium metabolism. Nature **230**:228–230.
7. Fraser DR, Kodicek E 1970 Unique biosynthesis by kidney of a biologically active vitamin D metabolite. Nature **228**:764–766.
8. Brumbaugh PF, Haussler MR 1974 1α,25-Dihydroxycholecalciferol receptors in intestine. I. Association of 1α,25-dihydroxycholecalciferol with intestinal mucosa chromatin. J Biol Chem **249**:1251–1257.
9. Brumbaugh PF, Haussler MR 1974 1α,25-Dihydroxycholecalciferol receptors in intestine. II. Temperature-dependent transfer of the hormone to chromatin via a specific cytosol receptor. J Biol Chem **249**:1258–1262.
10. Brumbaugh PF, Haussler MR 1973 1α,25-Dihydroxyvitamin D_3 receptor: Competitive binding of vitamin D analogs. Life Sci **13**:1737–1746.
11. Norman AW, Procsal DA, Okamura WH, Wing RM 1975 Structure–function studies of the interaction of the hormonally active form of vitamin D_3, 1α,25-dihydroxy-vitamin D_3, with the intestine. J Steroid Biochem **6**:461–467.
12. Pike JW, Haussler MR 1979 Purification of chicken intestinal receptor for 1,25-dihydroxyvitamin D. Proc Natl Acad Sci USA **76**:5485–5489.
13. Pike JW, Donaldson CA, Marion SL, Haussler MR 1982 Development of hybridomas secreting monoclonal antibodies to the chicken intestinal 1α,25-dihydroxyvitamin D_3 receptor. Proc Natl Acad Sci USA **79**:7719–7723.
14. Pike JW, Marion SL, Donaldson CA, Haussler MR 1983 Serum and monoclonal antibodies against the chick intestinal receptor for 1,25-dihydroxyvitamin D_3. J Biol Chem **258**:1289–1296.
15. McDonnell DP, Mangelsdorf DJ, Pike JW, Haussler MR, O'Malley BW 1987 Molecular cloning of complementary DNA encoding the avian receptor for vitamin D. Science **235**:1214–1217.
16. Evans RM 1988 The steroid and thyroid hormone receptor superfamily. Science **240**:889–895.
17. Pike JW 1981 Evidence for a reactive sulfhydryl in the DNA binding domain of the 1,25-dihydroxyvitamin D_3 receptor. Biochem Biophys Res Commun **100**:1713–1719.
18. Haussler MR, Mangelsdorf DJ, Yamaoka K, Allegretto EA, Komm BS, Terpening CM, McDonnell DP, Pike JW, O'Malley BW 1988 Molecular characterization and actions of the vitamin D hormone receptor. In: Ringold G (ed) Steroid Hormone Action. Alan R. Liss, New York, pp. 247–262.
19. Baker AR, McDonnell DP, Hughes M, Crisp TM, Mangelsdorf DJ, Haussler MR, Pike JW, Shine J, O'Malley BW 1988 Cloning and expression of full-length cDNA encoding human vitamin D receptor. Proc Natl Acad Sci USA **85**:3294–3298.
20. Burmester JK, Wiese RJ, Maeda N, DeLuca HF 1988 Structure and regulation of the rat 1,25-dihydroxyvitamin D_3 receptor. Proc Natl Acad Sci USA **85**:9499–9502.
21. Elaroussi MA, Prahl JM, DeLuca HF 1994 The avian vitamin D receptors: Primary structures and their origins. Proc Natl Acad Sci USA **91**:11596–11600.
22. Sone T, Ozono K, Pike JW 1991 A 55-kilodalton accessory factor facilitates vitamin D receptor DNA binding. Mol Endocrinol **5**:1578–1586.
23. Sone T, Kerner S, Pike JW 1991 Vitamin D receptor interaction with specific DNA: Association as a 1,25-dihydroxyvitamin D_3-modulated heterodimer. J Biol Chem **266**:23296–23305.
24. MacDonald PN, Haussler CA, Terpening CM, Galligan MA, Reeder MC, Whitfield GK, Haussler MR 1991 Baculovirus-mediated expression of the human vitamin D receptor: Functional characterization, vitamin D response element interactions, and evidence for a receptor auxiliary factor. J Biol Chem **266**:18808–18813.
25. Ross TK, Moss VE, Prahl JM, DeLuca HF 1992 A nuclear protein essential for binding of rat 1,25-dihydroxyvitamin D_3 receptor to its response elements. Proc Natl Acad Sci USA **89**:256–260.
26. Hamada K, Gleason SL, Levi B-Z, Hirschfeld S, Appella E, Ozato K 1989 H-2RIIBP, a member of the nuclear hormone receptor superfamily that binds to both the regulatory element of major histocompatibility class I genes and the estrogen response element. Proc Natl Acad Sci USA **86**:8289–8293.
27. Mangelsdorf DJ, Ong ES, Dyck JA, Evans RM 1990 Nuclear receptor that identifies a novel retinoic acid response pathway. Nature **345**:224–229.
28. Yu VC, Delsert C, Andersen B, Holloway JM, Devary OV, Näär AM, Kim SY, Boutin J-M, Glass CK, Rosenfeld MG 1991 RXRβ: A coregulator that enhances binding of retinoic acid, thyroid hormone, and vitamin D receptors to their cognate response elements. Cell **67**:1251–1266.
29. Leid M, Kastner P, Lyons R, Nakshatri H, Saunders M, Zacharewski T, Chen J-Y, Staub A, Garnier J-M, Mader S, Chambon P 1992 Purification, cloning, and RXR identity of the HeLa cell factor with which RAR or TR heterodimerizers to bind target sequences efficiently. Cell **68**:377–395.
30. Kliewer SA, Umesono K, Mangelsdorf DJ, Evans RM 1992 Retinoid X receptor interacts with nuclear receptors in retinoic

acid, thyroid hormone and vitamin D_3 signalling. Nature **355**:446–449.

31. MacDonald PN, Dowd DR, Nakajima S, Galligan MA, Reeder MC, Haussler CA, Ozato K, Haussler MR 1993 Retinoid X receptors stimulate and 9-*cis*-retinoic acid inhibits 1,25-dihydroxyvitamin D_3-activated expression of the rat osteocalcin gene. Mol Cell Biol **13**:5907–5917.
32. Heyman RA, Mangelsdorf DJ, Dyck JA, Stein RB, Eichele G, Evans RM, Thaller C 1992 9-*cis*-Retinoic acid is a high affinity ligand for the retinoid X receptor. Cell **68**:397–406.
33. Jin CH, Pike JW 1994 DNA binding site and coregulator requirements for 1,25-dihydroxyvitamin D_3-dependent activation. J Bone Miner Res **9**:(Suppl. 1):S160 (abstract).
34. Cheskis B, Freedman LP 1994 Ligand modulates the conversion of DNA-bound vitamin D_3 receptor (VDR) homodimers into VDR–retinoid X receptor heterodimers. Mol Cell Biol **14**:3329–3338.
35. Jin CH, Kerner SA, Hong MH, Pike JW 1996 Transcriptional activation and dimerization functions in the human vitamin D receptor. Mol Endocrinol **10**:945–957.
36. Lemon BD, Freedman LP 1996 Selective effects of ligands on vitamin D_3 receptor- and retinoid X receptor-mediated gene activation *in vivo*. Mol Cell Biol **16**:1006–1016.
37. Cheskis B, Freedman LP 1996 Modulation of nuclear receptor interactions by ligands: Kinetic analysis using surface plasmon resonance. Biochemistry **35**:3309–3318.
38. Thompson PD, Jurutka PW, Haussler CA, Whitfield GK, Haussler MR 1997 Heterodimeric DNA binding by the vitamin D receptor and retinoid X receptors is enhanced by 1,25-dihydroxyvitamin D_3 and inhibited by 9-*cis*-retinoic acid: Evidence for allosteric receptor interactions. J Biol Chem submitted.
39. MacDonald PN, Dowd DR, Haussler MR 1994 New insight into the structure and functions of the vitamin D receptor. Semin Nephrol **14**:101–118.
40. Haussler MR, Jurutka PW, Hsieh J-C, Thompson PD, Selznick SH, Haussler CA, Whitfield GK 1995 New understanding of the molecular mechanism of receptor-mediated genomic actions of the vitamin D hormone. Bone **17**:(Suppl.) 33S–38S.
41. Mangelsdorf DJ, Thummel C, Beato M, Herrlich P, Schütz G, Umesono K, Blumberg B, Kastner P, Mark M, Chambon P, Evans RM 1995 The nuclear receptor superfamily: The second decade. Cell **83**:835–839.
42. Mangelsdorf DJ, Evans RM 1995 The RXR heterodimers and orphan receptors. Cell **83**:841–850.
43. Truss M, Beato M 1993 Steroid hormone receptors: Interaction with deoxyribonucleic acid and transcription factors. Endocr Rev **14**:459–479.
44. Haussler M, Terpening C, Haussler C, MacDonald P, Hsieh J-C, Jones B, Jurutka P, Meyer J, Komm B, Galligan M, Selznick S, Whitfield GK 1991 A tale of two genes: Regulation of rat osteocalcin and chicken calbindin-D_{28K} by the vitamin D hormone and its phosphorylated nuclear receptor. In: Norman AW, Bouillon R, Thomasset M (eds) Vitamin D: Gene Regulation, Structure–Function Analysis and Clinical Application. de Gruyter, Berlin and New York, pp. 3–11.
45. Gronemeyer H 1992 Control of transcription activation by steroid hormone receptors. FASEB **6**:2524–2529.
46. Haussler MR 1986 Vitamin D receptors: Nature and function. Annu Rev Nutr **6**:527–562.
47. Colston KW, Feldman D 1979 Demonstration of a 1,25-dihydroxycholecalciferol cytoplasmic receptor-like binder in mouse kidney. J Clin Endocrinol Metab **49**:798–800.
48. Chandler JS, Pike JW, Haussler MR 1979 1,25-Dihydroxyvitamin D_3 receptors in rat kidney cytosol. Biochem Biophys Res Commun **90**:1057–1063.
49. Chen TL, Hirst MA, Feldman D 1979 A receptor-like binding macromolecule for 1 alpha, 25-dihydroxycholecalciferol in cultured mouse bone cells. J Biol Chem **254**:7491–7494.
50. Manolagas SC, Haussler MR, Deftos LJ 1980 1,25-Dihydroxyvitamin D_3 receptor-like macromolecule in rat osteogenic sarcoma cell lines. J Biol Chem **255**:4414–4417.
51. Stumpf WE, Sar M, Reid FA, Tanaka Y, DeLuca HF 1979 Target cells for 1,25-dihydroxyvitamin D_3 in intestinal tract, stomach, kidney, skin, pituitary, and parathyroid. Science **206**:1188–1190.
52. Bidmon H-J, Gutkowska J, Murakami R, Stumpf WE 1991 Vitamin D receptors in heart: Effects on atrial natriuretic factor. Experimentia **47**:958–962.
53. Pike JW, Goozé LL, Haussler MR 1980 Biochemical evidence for 1,25-dihydroxyvitamin D receptor macromolecules in parathyroid, pancreatic, pituitary, and placental tissues. Life Sci **26**:407–414.
54. Clemens TL, Garrett KP, Zhou X-Y, Pike JW, Haussler MR, Dempster DW 1988 Immunocytochemical localization of the 1,25-dihydroxyvitamin D_3 receptor in target cells. Endocrinology **122**:1224–1230.
55. Berger U, Wilson P, McClelland RA, Colston K, Haussler MR, Pike JW, Coombes RC 1988 Immunocytochemical detection of 1,25-dihydroxyvitamin D receptors in normal human tissues. J Clin Endocrinol Metab **67**:607–613.
56. Manolagas SC, Yu X-P, Girasole G, Bellido T 1994 Vitamin D and the hematolymphopoietic tissue: A 1994 update. Semin Nephrol **14**:129–143.
57. Alroy I, Towers TL, Freedman LP 1995 Transcriptional repression of the interleukin-2 gene by vitamin D_3: Direct inhibition of NFATp/AP-1 complex formation by a nuclear hormone receptor. Mol Cell Biol **15**:5789–5799.
58. Lemire JM, Archer DC, Khulkarni A, Ince A, Uskokovic MR, Stephkowski S 1992 The vitamin D_3 analogue 1,25-dihydroxy-Δ^{16}-cholecalciferol prolongs the survival of murine cardiac allografts. Transplantation **54**:762–763.
59. Lemire JM 1995 Immunomodulatory actions of 1,25-dihydroxyvitamin D_3. J Steroid Biochem Mol Biol **53**:599–602.
60. Reitsma PH, Rothberg PG, Astrin SM, Trial J, Bar-Shavit Z, Hall A, Teitelbaum SL, Kahn AJ 1983 Regulation of myc gene expression in HL-60 leukaemia cells by a vitamin D metabolite. Nature **306**:492–494.
61. Mangelsdorf DJ, Koeffler HP, Donaldson CA, Pike JW, Haussler MR 1984 1,25-Dihydroxyvitamin D_3-induced differentiation in a human promyelocytic leukemia cell line (HL-60): Receptor-mediated maturation to macrophage-like cells. J Cell Biol **98**:391–398.
62. Smith EL, Pincus SH, Donovan L, Holick MF 1988 A novel approach for the evaluation and treatment of psoriasis: Oral or topical use of 1,25-dihydroxyvitamin D_3 can be a safe and effective therapy for psoriasis. J Am Acad Dermatol **19**:516–528.
63. Feldman D, Malloy PJ 1990 Hereditary 1,25-dihydroxyvitamin D resistant rickets: Molecular basis and implications for the role of $1,25(OH)_2D_3$ in normal physiology. Mol Cell Endocrinol **72**:C57-C62.
64. Faraco JH, Morrison NA, Baker A, Shine J, Frossard PM 1989 ApaI dimorphism at the human vitamin D receptor gene locus. Nucleic Acids Res **17**:2150.
65. Rut AR, Hewison M, Kristjansson K, Luisi B, Hughes MR, O'Riordan JLH 1994 Two mutations causing vitamin D resistant rickets: Modelling on the basis of steroid hormone receptor DNA-binding domain crystal structures. Clin Endocrinol **41**:581–590.

66. Yoshizawa T, Handa Y, Uematsu Y, Sekine K, Takeda S, Yoshihara Y, Kawakami T, Sato H, Alioka K, Tanimoto K, Fukamizu A, Masushige S, Matsumoto T, Kato S 1996 Disruption of the vitamin D receptor (VDR) in the mouse. J Bone Miner Res **11**(Suppl 1):S124 (abstract).
67. Ozono K, Liao J, Kerner SA, Scott RA, Pike JW 1990 The vitamin D-responsive element in the human osteocalcin gene: Association with a nuclear proto-oncogene enhancer. J Biol Chem **265**:21881–21888.
68. DeMay MB, Gerardi JM, DeLuca HF, Kronenberg HM 1990 DNA sequences in the rat osteocalcin gene that bind the 1,25-dihydroxyvitamin D_3 receptor and confer responsiveness to 1,25-dihydroxyvitamin D_3. Proc Natl Acad Sci USA **87**:369–373.
69. Markose ER, Stein JL, Stein GS, Lian JB 1990 Vitamin D-mediated modifications in protein–DNA interactions at two promoter elements of the osteocalcin gene. Proc Natl Acad Sci USA **87**:1701–1705.
70. Terpening CM, Haussler CA, Jurutka PW, Galligan MA, Komm BS, Haussler MR 1991 The vitamin D-responsive element in the rat bone gla protein is an imperfect direct repeat that cooperates with other cis-elements in 1,25-dihydroxyvitamin D_3-mediated transciptional activation. Mol Endocrinol **5**:373–385.
71. Noda M, Vogel RL, Craig AM, Prahl J, DeLuca HF, Denhardt DT 1990 Identification of a DNA sequence responsible for binding of the 1,25-dihydroxyvitamin D_3 receptor and 1,25-dihydroxyvitamin D_3 enhancement of mouse secreted phosphoprotein 1 (*Spp-1* or osteopontin) gene expression. Proc Natl Acad Sci USA **87**:9995-9999.
72. Cao X, Ross FP, Zhang L, MacDonald PN, Chappel J, Teitelbaum SL 1993 Cloning of the promoter for the avian integrin β_3 subunit gene and its regulation by 1,25-dihydroxyvitamin D_3. J Biol Chem **268**:27371–27380.
73. Gill RK, Christakos S 1993 Identification of sequence elements in mouse calbindin-D_{28K} gene that confer 1,25-dihydroxyvitamin D_3- and butyrate-inducible responses. Proc Natl Acad Sci USA **90**:2984–2988.
74. Ohyama Y, Ozono K, Uchida M, Shinki T, Kato S, Suda T, Yamamoto O, Noshiro M, Kato Y 1994 Identification of a vitamin D-responsive element in the 5′ flanking region of the rat 25-hydroxyvitamin D_3 24-hydroxylase gene. J Biol Chem **269**:10545–10550.
75. Colnot S, Lambert M, Blin C, Thomasset M, Perret C 1995 Identification of DNA sequences that bind retinoid X receptor-$1,25(OH)_2D_3$-receptor heterodimers with high affinity. Mol Cell Endocrinol **113**:89–98.
76. Nishikawa J, Kitaura M, Matsumoto M, Imagawa M, Nishihara T 1994 Difference and similarity of DNA sequence recognized by VDR homodimer and VDR/RXR heterodimer. Nucleic Acids Res **22**:2902–2907.
77. Jin CH, Pike JW 1996 Human vitamin D receptor-dependent transactivation in *Saccharomyces cerevisiae* requires retinoid X receptor. Mol Endocrinol **10**:196–205.
78. Staal A, van Wijnen AJ, Birkenhäger JC, Pols HA, Prahl J, DeLuca H, Gaub M-P, Lian JB, Stein GS, van Leeuwen JPTM, Stein JL 1996 Distinct conformations of vitamin D receptor/retinoid X receptor-α heterodimers are specified by dinucleotide differences in the vitamin D-responsive elements for the osteocalcin and osteopontin genes. Mol Endocrinol **10**:1444–1456.
79. Jurutka PW, Hsieh J-C, Haussler MR 1994 Characterization of a new functional 1,25-dihydroxyvitamin D_3 responsive element in the promoter region of the rat 25-hydroxyvitamin D_3 24-hydroxylase gene. J Bone Miner Res **9**(Suppl. 1):S160 (abstract).
80. Pike JW, Kerner SA, Jin CH, Allegretto EA, Elgort M 1994 Direct activation of the human 25-hydroxyvitamin D_3-24-hydroxylase promoter by $1,25(OH)_2D_3$ and PMA: Identification of cis elements and transactivators. J Bone Miner Res **9**(Suppl. 1):S144 (abstract).
81. Kamei Y, Kawada T, Fukuwatari T, Ono T, Kato S, Sugimoto E 1995 Cloning and sequencing of the gene encoding the mouse vitamin D receptor. Gene **152**:281–282.
82. Rastinejad F, Perlmann T, Evans RM, Sigler PB 1995 Structural determinants of nuclear receptor assembly on DNA direct repeats. Nature **375**:203–211.
83. Lee JW, Gulick T, Moore DD 1992 Thyroid hormone receptor dimerization function maps to a conserved subregion of the ligand binding domain. Mol Endocrinol **6**:1867–1873.
84. Forman BM, Samuels HH 1990 Interactions among a subfamily of nuclear hormone receptors: The regulatory zipper model. Mol Endocrinol **4**:1293–1301.
85. Gronemeyer H 1991 Transcription activation by estrogen and progesterone receptors. Annu Rev Genet **25**:89–123.
86. Wilson TE, Paulsen RE, Padgett KA, Milbrandt J 1992 Participation of non-zinc finger residues in DNA binding by two nuclear orphan receptors. Science **256**:107–110.
87. Walters R, Hunziker W, Norman AW 1980 Unoccupied 1,25-dihydroxyvitamin D_3 receptors. J Biol Chem **255**:6799–6805.
88. Milde P, Merke J, Ritz E, Haussler MR, Rauterberg EW 1989 Immunohistochemical detection of 1,25-dihydroxyvitamin D_3 receptors and estrogen receptors by monoclonal antibodies: Comparison of four immunoperoxidase methods. J Histochem Cytochem **37**:1609–1617.
89. Haussler MR, Mangelsdorf DJ, Komm BS, Terpening CM, Yamaoka K, Allegreto EA, Baker AR, Shine J, McDonnell DP, Hughes M, Weigel NL, O'Malley BW, Pike JW 1988 Molecular biology of the vitamin D hormone. Recent Prog Horm Res **44**:263–305.
90. Chelsky D, Ralph R, Jonak G 1989 Sequence requirements for synthetic peptide-mediated translocation to the nucleus. Mol Cell Biol **9**:2487–2492.
91. Kalderon D, Roberts BL, Richardson WD, Smith AE 1984 A short amino acid sequence able to specify nuclear location. Cell **39**:499–509.
92. Lanford RE, Butel JS 1984 Construction and characterization of an SV40 mutant defective in nuclear transport of T antigen. Cell **37**:801–813.
93. Picard D, Yamamoto KR 1987 Two signals mediate hormone-dependent nuclear localization of the glucocorticoid receptor. EMBO J **6**:3333–3340.
94. Guiochon-Mantel A, Loosfelt H, Lescop P, Sar S, Atger M, Perrot-Applanat M, Milgrom E 1989 Mechanisms of nuclear localization of the progesterone receptor: Evidence for interaction between monomers. Cell **57**:1147–1154.
95. Simental JA, Sar M, Lane MV, French FS, Wilson EM 1991 Transcriptional activation and nuclear targeting signals of the human androgen receptor. J Biol Chem **266**:510–518.
96. Luo Z, Rouvinen J, Maenpaa PH 1994 A peptide C-terminal to the second Zn finger of human vitamin D receptor is able to specify nuclear localization. Eur J Biochem **223**:381–387.
97. Hsieh J-C, Shimizu Y, Minoshima S, Shimizu N, Galligan MA, Haussler CA, Haussler MR 1997 Identification of a nuclear localization signal between the zinc fingers of the human vitamin D receptor. J Cell Biochem submitted.
98. Boulikas T 1996 Nuclear import of protein kinases and cyclins. J Cell Biochem **60**:61–82.
99. Sackey FNA, Haché RJG, Reich T, Kwast-Welfeld J, Lefebvre YA 1996 Determinants of subcellular distribution of the glucocorticoid receptor. Mol Endocrinol **10**:1191–1205.

100. Allegretto EA, Pike JW, Haussler MR 1987 C-terminal proteolysis of the avian 1,25-dihydroxyvitamin D_3 receptor. Biochem Biophys Res Commun **147**:479–485.
101. McDonnell DP, Scott RA, Kerner SA, O'Malley BW, Pike JW 1989 Functional domains of the human vitamin D_3 receptor regulate osteocalcin gene expression. Mol Endocrinol **3**:635–644.
102. Renaud J-P, Rochel N, Ruff M, Vivat V, Chambon P, Gronemeyer H, Moras D 1995 Crystal structure of the RAR-γ ligand-binding domain bound to all-*trans*-retinoic acid. Nature **378**:681–689.
103. Wagner RL, Apriletti JW, McGrath ME, West BL, Baxter JD, Fletterick RJ 1995 A structural role for hormone in the thyroid hormone receptor. Nature **378**:690–697.
104. Xu M, Chakraborti PK, Garabedian MJ, Yamamoto KR, Simons SS Jr. 1996 Modular structure of glucocorticoid receptor domains is not equivalent to functional independence. J Biol Chem **271**:21430–21438.
105. Kumar R, Schaefer J, Weiben E 1992 The expression of milligram amounts of functional human 1,25-dihydroxyvitamin D_3 receptor in a bacterial expression system. Biochem Biophys Res Commun **189**:1417–1423.
106. Towers TL, Luisi BF, Asianov A, Freedman LP 1993 DNA target selectivity by the vitamin D_3 receptor: Mechanism of dimer binding to an asymmetric repeat element. Proc Natl Acad Sci USA **90**:6310–6314.
107. Hsieh J-C, Nakajima S, Galligan MA, Jurutka PW, Haussler CA, Whitfield GK, Haussler MR 1995 Receptor mediated genomic action of the $1,25(OH)_2D_3$ hormone: Expression of the human vitamin D receptor in *E. coli*. J Steroid Biochem Mol Biol **53**:583–594.
108. Ross TK, Prahl JM, DeLuca HF 1991 Overproduction of rat 1,25-dihydroxyvitamin D_3 receptor in insect cells using the baculovirus expression system. Proc Natl Acad Sci USA **88**:6555–6559.
109. Nakajima S, Hsieh J-C, MacDonald PN, Haussler CA, Galligan MA, Jurutka PW, Haussler MR 1993 Purified human vitamin D receptor overexpressed in *E. coli* and baculovirus systems does not bind 1,25-dihydroxyvitamin D_3 hormone efficiently unless supplemented with a rat liver nuclear extract. Biochem Biophys Res Commun **197**:478–485.
110. Ray R, Swamy N, MacDonald PN, Ray S, Haussler MR, Holick MF 1996 Affinity labeling of the 1α,25-dihydroxyvitamin D_3 receptor. J Biol Chem **271**:2012–2017.
111. Nakajima S, Hsieh J-C, Jurutka PW, Galligan MA, Haussler CA, Whitfield GK, Haussler MR 1996 Examination of the potential functional role of conserved cysteine residues in the hormone binding domain of the human 1,25-dihydroxyvitamin D_3 receptor. J Biol Chem **271**:5143–5149.
112. Nakajima S, Hsieh J-C, MacDonald PN, Galligan MA, Haussler CA, Whitfield GK, Haussler MR 1994 The C-terminal region of the vitamin D receptor is essential to form a complex with a receptor auxiliary factor required for high affinity binding to the vitamin D-responsive element. Mol Endocrinol **8**:159–172.
113. Whitfield GK, Hsieh J-C, Nakajima S, MacDonald PN, Thompson PD, Jurutka PW, Haussler CA, Haussler MR 1995 A highly conserved region in the hormone binding domain of the human vitamin D receptor contains residues vital for heterodimerization with retinoid X receptor and for transcriptional activation. Mol Endocrinol **9**:1166–1179.
114. Rosen ED, Beninghof EG, Koenig RJ 1993 Dimerization interfaces of thyroid hormone, retinoic acid, vitamin D, and retinoid X receptors. J Biol Chem **268**:11534–11541.
115. Hsieh J-C, Jurutka PW, Selznick SH, Reeder MC, Haussler CA, Whitfield GK, Haussler MR 1995 The T-box near the zinc fingers of the human vitamin D receptor is required for heterodimeric DNA binding and transactivation. Biochem Biophys Res Commun **215**:1–7.
116. Whitfield GK, Selznick SH, Haussler CA, Hsieh J-C, Galligan MA, Jurutka PW, Thompson PD, Lee SM, Zerwekh JE, Haussler MR 1996 Vitamin D receptors from patients with resistance to 1,25-dihydroxyvitamin D_3: Point mutations confer reduced transactivation in response to ligand and impaired interaction with the retinoid X receptor heterodimeric partner. Mol Endocrinol **10**:1617–1631.
117. Hughes MR, Malloy PJ, Kieback DG, Kesterson RA, Pike JW, Feldman D, O'Malley BW 1988 Point mutations in the human vitamin D receptor gene associated with hypocalcemic rickets. Science **242**:1702–1705.
118. Bourguet W, Ruff M, Chambon P, Gronemeyer H, Moras D 1995 Crystal structure of the ligand-binding domain of the human nuclear receptor RXR-α. Nature **375**:377–382.
119. Jones BB, Jurutka PW, Haussler CA, Haussler MR, Whitfield GK 1991 Vitamin D receptor phosphorylation in transfected ROS 17/2.8 cells is localized to the N-terminal region of the hormone-binding domain. Mol Endocrinol **5**:1137–1146.
120. Fraher LJ, Karmali R, Hinde FRJ, Hendy GN, Jani H, Nicholson L, Grant D, O'Riordan JLH 1986 Vitamin D-dependent rickets type II: Extreme end organ resistance to 1,25-dihydroxy vitamin D_3 in a patient without alopecia. Eur J Pediatr **145**:389–395.
121. Kristjansson K, Rut AR, Hewison M, O'Riordan JLH, Hughes MR 1993 Two mutations in the hormone binding domain of the vitamin D receptor cause tissue resistance to 1,25-dihydroxyvitamin D_3. J Clin Invest **92**:12–16.
122. Miller J, McLachlan AD, Klug A 1985 Repetitive zinc-binding domains in the protein transcription factor IIIA from *Xenopus* oocytes. EMBO J **4**:1609–1614.
123. Jacobs GH 1992 Determination of base recognition positions of zinc fingers from sequence analysis. EMBO J **11**:4507–4517.
124. Hoovers JMN, Mannens M, John R, Bliek J, van Heyningen V, Poryeus DJ, Leschot NJ, Westerveld A, Little PFR 1992 High-resolution localization of 69 potential human zinc finger protein genes: A number are clustered. Genomics **12**:254–263.
125. Klug A, Schwabe JWR 1995 Zinc fingers. FASEB J **9**:597–604.
126. Umesono K, Evans RM 1989 Determinants of target gene specificity for steroid/thyroid hormone receptors. Cell **57**:1139–1146.
127. Danielsen M, Hinck L, Ringold GM 1989 Two amino acids within the knuckle of the first zinc finger specify DNA response element activation by the glucocorticoid receptor. Cell **57**:1131–1138.
128. Mader S, Kumar V, de Verneuil H, Chambon P 1989 Three amino acids of the oestrogen receptor are essential to its ability to distinguish an oestrogen from a glucocorticoid-responsive element. Nature **338**:271–274.
129. Lee MS, Kliewer SA, Provencal J, Wright PE, Evans RM 1993 Structure of the retinoid X receptor α DNA binding domain: A helix required for homodimeric DNA binding. Science **260**:1117–1121.
130. Hartong R. Wang N, Kurokawa R, Lazar MA, Glass CK, Apriletti JW, Dillmann WH 1994 Delineation of three different thyroid hormone-response elements in promoter of rat sarcoplasmic reticulum Ca^{2+} ATPase gene. J Biol Chem **269**:13021–13029.
131. Nelson CC, Hendy SC, Faris JS, Romaniuk PJ 1996 Retinoid X receptor alters the determination of DNA binding specificity by the P-box amino acids of the thyroid hormone receptor. J Biol Chem **271**:19464–19474.
132. Gronemeyer H, Turcotte B, Quirin-Stricker C, Bocquel MT, Meyer ME, Krozowski Z, Jeltsch JM, Lerouge T, Garnier FM, Chambon P 1987 The chicken progesterone receptor: Sequence, expression and functional analysis. EMBO J **6**:3985–3994.
133. Nagpal S, Friant S, Nakshatri H, Chambon P 1993 RARs and

RXRs: Evidence for two autonomous transactivation domains (AF-1 and AF-2) and heterodimerization *in vivo*. EMBO J **12**:2349–2360.
134. Hadzic E, Desai-Yajnik V, Helmer E, Guo S, Wu S, Koudinova N, Casanova J, Raaka BM, Samuels HH 1995 A 10-amino acid sequence in the N-terminal A/B domain of thyroid hormone receptor α is essential for transcriptional activation and interaction with the general transcription factor TFIIB. Mol Cell Biol **15**:4507–4517.
135. Hollenberg SM, Evans RM 1988 Multiple and cooperative transactivation domains of the human glucocorticoid receptor. Cell **55**:899–906.
136. Webster NJG, Green S, Jin JR, Chambon P 1988 The hormone binding domains of the estrogen and glucocorticoid receptors contain an inducible transcription activation function. Cell **54**:199–207.
137. Thompson CC, Evans RM 1989 Trans-activation by thyroid hormone receptors: Functional parallels with steroid hormone receptors. Proc Natl Acad Sci USA **86**:3494–3498.
138. Danielian PS, White R, Lees JA, Parker MG 1992 Identification of a conserved region required for hormone dependent transcriptional activation by steroid hormone receptors. EMBO J **11**:1025–1033.
139. Saatcioglu F, Deng T, Karin M 1993 A novel cis element mediating ligand-independent activation by c-ErbA: Implications for hormonal regulation. Cell **75**:1095–1105.
140. Leng X, Blanco J, Tsai SY, Ozato K, O'Malley BW, Tsai M-J 1995 Mouse retinoid X receptor contains a separable ligand-binding and transactivation domain in its E region. Mol Cell Biol **15**:255–263.
141. Baniahmad A, Ha I, Reinberg D, Tsai S, Tsai M-J, O'Malley BW 1993 Interaction of human thyroid-hormone receptor-beta with transcription factor TFIIB may mediate target gene derepression and activation by thyroid-hormone. Proc Natl Acad Sci USA **90**:8832–8836.
142. Blanco JCG, Wang I-M, Tsai SY, Tsai MJ, O'Malley BW, Jurutka PW, Haussler MR, Ozato K 1995 Transcription factor TFIIB and the vitamin D receptor cooperatively activate ligand-dependent transcription. Proc Natl Acad Sci USA **92**:1535–1539.
143. MacDonald PN, Sherman DR, Dowd DR, Jefcoat SC Jr, DeLisle PK 1995 The vitamin D receptor interacts with general transcription factor IIB. J Biol Chem **270**:4748–4752.
144. Jurutka PW, Hsieh J-C, Remus LS, Whitfield GK, Haussler CA, Blanco JCG, Ozato K, Haussler MR 1997 Mutations in the 1,25-dihydroxyvitamin D_3 receptor identifying C-terminal amino acids required for transcriptional activation that are functionally dissociated from hormone binding, heterodimeric DNA binding and interaction with TFIIB. J Biol Chem. in press.
145. Beato M 1989 Gene regulation by steroid hormones. Cell **56**:335–344.
146. Wang H, Peters GA, Zeng X, Tang M, Ip W, Khan SA 1995 Yeast two-hybrid system demonstrates that estrogen receptor dimerization is ligand-dependent *in vivo*. J Biol Chem **270**:23322–23329.
147. Chin WW, Yen PM 1996 Editorial: T_3 or not T_3—The slings and arrows of outrageous TR function. Endocrinology **137**:387–389.
148. Schulman IG, Juguilon H, Evans RM 1996 Activation and repression by nuclear hormone receptors: Hormone modulates an equilibrium between active and repressive states. Mol Cell Biol **16**:3807–3813.
149. Perlmann T, Umesono K, Rangarajan PN, Forman BM, Evans RM 1996 Two distinct dimerization interfaces differentially modulate target gene specificity of nuclear hormone receptors. Mol Endocrinol **10**:958–966.
150. Blanco JCG, Dey A, Leid M, Minucci S, Park B-K, Jurutka PW, Haussler MR, Ozato K 1996 Inhibition of ligand induced promoter occupancy *in vivo* by a dominant negative RXR. Genes Cells **1**:209–211.
151. Chen J-Y, Clifford J, Zsui C, Starrett J, Tortolani D, Ostrowski J, Reczek PR, Chambon P, Gronemeyer H 1996 Two distinct actions of retinoid-receptor ligands. Nature **382**:819–822.
152. Breen EC, van Wijnen AJ, Lian JB, Stein GS, Stein JL 1994 In vivo occupancy of the vitamin D responsive element in the osteocalcin gene supports vitamin D-dependent transcriptional upregulation in intact cells. Proc Natl Acad Sci USA **91**:12902–12906.
153. Ross TK, Darwish HM, Moss VE, DeLuca HF 1993 Vitamin D-influenced gene expression via a ligand-independent, receptor–DNA complex intermediate. Proc Natl Acad Sci USA **90**:9257–9260.
154. Koszewski NJ, Lapuz MH, Russell J, Malluche HH 1994 Vitamin D receptor interactions with positive and negative DNA response elements: An interference footprint comparison. J Bone Miner Res **9**:(Suppl. 1):S290 (abstract).
155. Freedman LP, Arce V, Fernandez RP 1994 DNA sequences that act as high affinity targets for the vitamin D_3 receptor in the absence of the retinoid X receptor. Mol Endocrinol **8**:265–273.
156. Carlberg C, Bendik I, Wyss A, Meier E, Sturzenbecker LJ, Grippo JF, Hunziker W 1993 Two nuclear signalling pathways for vitamin D. Nature **361**:657–660.
157. Schräder M, Müller KM, Becker-André M, Carlberg C 1994 Response element selectivity for heterodimerization of vitamin D receptors with retinoic acid and retinoid X receptors. J Mol Endocrinol **12**:327–339.
158. Sasaki H, Harada H, Honda Y, Morino H, Suzawa M, Shimpo E, Katsumata T, Masuhiro Y, Matsuda K, Ebihara K, Ono T, Massusshige S, Kato S 1995 Transcriptional activity of a fluorinated vitamin D analog on VDR–RXR mediated gene expression. Biochemistry **34**:370–377.
159. Kato S, Sasaki H, Suzawa M, Masushige S, Tora L, Chambon P, Gronemeyer H 1995 Widely spaced, directly repeated PuGGTCA elements act as promiscuous enhancers for different classes of nuclear receptors. Mol Cell Biol **15**:5858–5867.
160. Ferrara J, McCuaig K, Hendy GN, Uskokovic M, White JH 1994 Highly potent transcriptional activation by 16-ene derivatives of 1,25-dihydroxyvitamin D_3. J Biol Chem **269**:2971–2981.
161. Forman BM, Umesono K, Chen J, Evans RM 1995 Unique response pathways are established by allosteric interactions among nuclear hormone receptors. Cell **81**:541–550.
162. Lehmann JM, Zhang X-K, Graupner G, Lee M-O, Hermann T, Hoffmann B, Pfahl M 1993 Formation of retinoid X receptor homodimers leads to repression of T_3 response: Hormonal cross talk by ligand-induced squelching. Mol Cell Biol **13**:7698–7707.
163. Logeat F, Le Cunff M, Pamphile R, Milgrom E 1985 The nuclear-bound form of the progesterone receptor is generated through a hormone-dependent phosphorylation. Biochem Biophys Res Commun **131**:421–427.
164. Singh VB, Moudgil VK 1985 Phosphorylation of rat liver glucocorticoid receptor. J Biol Chem **260**:3684–3690.
165. Migliaccio A, Rotondi A, Auricchio R 1984 Calmodulin-stimulated phosphorylation of 17β-estradiol receptor on tyrosine. Proc Natl Acad Sci USA **81**:5921–5925.
166. Goldberg Y, Glineur C, Gesquière J-C, Ricouart A, Sap J, Vennström B, Ghysdael J 1988 Activation of protein kinase C or cAMP-dependent protein kinase increases phosphorylation of the c-*erbA*-encoded thyroid hormone receptor and of the v-*erbA*-encoded protein. EMBO J **7**:2425–2433.
167. Pike JW, Sleator NM 1985 Hormone-dependent phosphorylation

of the 1,25-dihydroxyvitamin D_3 receptor in mouse fibroblasts. Biochem Biophys Res Commun **131**:378–385.

168. Brown TA, DeLuca HF 1990 Phosphorylation of the 1,25-dihydroxyvitamin D_3 receptor: A primary event in 1,25-dihydroxyvitamin D_3 action. J Biol Chem **265**:10025–10029.
169. Jurutka PW, Terpening CM, Haussler MR 1993 The 1,25-dihydroxy-vitamin D_3 receptor is phosphorylated in response to 1,25-dihydroxy-vitamin D_3 and 22-oxacalcitriol in rat osteoblasts, and by casein kinase II, *in vitro*. Biochemistry **32**:8184–8192.
170. Haussler MR, Terpening CM, Komm BS, Whitfield GK, Haussler CA 1988 Vitamin D hormone receptors: Structure, regulation and molecular function. In: Normal AW, Schaefer K, and Grigoleit HG, von Herrath D (eds) Vitamin D: Molecular, Cellular and Clinical Endocrinology. de Gruyter, Berlin and New York, pp. 205–214.
171. Jurutka PW, Hsieh J-C, MacDonald PN, Terpening CM, Haussler CA, Haussler MR, Whitfield GK 1993 Phosphorylation of serine 208 in the human vitamin D receptor: The predominant amino acid phosphorylated by casein kinase II, *in vitro*, and identification as a significant phosphorylation site in intact cells. J Biol Chem **268**:6791–6799.
172. Brown TA, DeLuca HF 1991 Sites of phosphorylation and photoaffinity labeling of the 1,25-dihydroxyvitamin D_3 receptor. Arch Biochem Biophys **286**:466–472.
173. Marin O, Meggio F, Marchiori F, Borin G, Pinna LA 1986 Site specificity of casein kinase-2 (TS) from rat liver cytosol: A study with model peptide substrates. Eur J Biochem **160**:239–244.
174. Hilliard GM, Cook RG, Weigel NL, Pike JW 1994 1,25-Dihydroxyvitamin D_3 modulates phosphorylation of serine 205 in the human vitamin D receptor: Site-directed mutagenesis of this residue promotes alternative phosphorylation. Biochemistry **33**:4300–4311.
175. Nishizuka Y 1989 The family of protein kinase C for signal transduction. JAMA **262**:1826–1833.
176. Yu X-P, Mocharia H, Hustmyer FG, Manolagas SC 1991 Vitamin D receptor expression in human lymphocytes. J Biol Chem **266**:7588–7595.
177. Krishnan AV, Feldman D 1991 Activation of protein kinase-C inhibits vitamin D receptor gene expression. Mol Endocrinol **5**:605–612.
178. Hsieh J-C, Jurutka PW, Galligan MA, Terpening CM, Haussler CA, Samuels DS, Shimizu Y, Shimizu N, Haussler MR 1991 Human vitamin D receptor is selectively phosphorylated by protein kinase C on serine 51, a residue crucial to its trans-activation function. Proc Natl Acad Sci USA **88**:9315–9319.
179. Kennelly PJ, Krebs EG 1991 Consensus sequences as substrate specificity determinants for protein kinases and protein phosphatases. J Biol Chem **266**:15555–15558.
180. Hsieh J-C, Jurutka PW, Nakajima S, Galligan MA, Haussler CA, Shimizu Y, Shimizu N, Whitfield GK, Haussler MR 1993 Phosphorylation of the human vitamin D receptor by protein kinase C: Biochemical and functional evaluation of the serine 51 recognition site. J Biol Chem **268**:15118–15126.
181. Chen ML, Boltz MA, Armbrecht HJ 1993 Effects of 1,25-dihydroxyvitamin D_3 and phorbol ester on 25-hydroxyvitamin D_3 24-hydroxylase cytochrome P450 messenger ribonucleic acid levels in primary cultures of rat renal cells. Endocrinology **132**:1782–1788.
182. van Leeuwen JP, Birkenhäger JC, van den Bemd GJ, Buurman CJ, Staal A, Bos MP, Pols HAP 1992 Evidence for the functional involvement of protein kinase C in the action of 1,25-dihydroxyvitamin D_3 in bone. J Biol Chem **267**:12562–12569.
183. Haussler MR, Jurutka PW, Hsieh J-C, Thompson PD, Selznick SH, Haussler CA, Whitfield GK 1994 Receptor mediated genomic actions of $1,25(OH)_2D_3$: Modulation by phosphorylation. In: Norman AW, Bouillon R, Thomasset M (eds) Vitamin D: A Pluripotent Steroid Hormone: Structural Studies, Molecular Endocrinology and Clinical Applications. de Gruyter, Berlin, pp. 209–216.
184. Jurutka PW, Hsieh J-C, Haussler MR 1993 Phosphorylation of the human 1,25-dihydroxyvitamin D_3 receptor by cAMP-dependent protein kinase, *in vitro*, and in transfected COS-7 cells. Biochem Biophys Res Commun **191**:1089–1096.
185. Darwish HM, Burmester JK, Moss VE, DeLuca HF 1993 Phosphorylation is involved in transcriptional activation by the 1,25-dihydroxyvitamin D_3 receptor. Biochim Biophys Acta **1167**:29–36.
186. Matkovits T, Christakos S 1995 Ligand occupancy is not required for vitamin D receptor and retinoid receptor-mediated transcriptional activation. Mol Endocrinol **9**:232–242.
187. Nakajima S, Yamagata M, Ozono K 1996 Effects of cyclic adenosine monophosphate and protein kinase-A on ligand-dependent transactivation via vitamin D receptor. J Bone Miner Res **11**:(Suppl. 1):S162 (abstract).
188. Glineur C, Bailly M, Ghysdael J 1989 The c-*erbA*α-encoded thyroid hormone receptor is phosphorylated in its amino terminal domain by casein kinase II. Oncogene **4**:1247–1254.
189. Rochette-Egly C, Oulad-Abdelghani M, Staub A, Pfister V, Scheuer I, Chambon P, Gaub M-P 1995 Phosphorylation of the retinoic acid receptor-α by protein kinase A. Mol Endocrinol **9**:860–871.
190. Hirata Y, Kiuchi K, Chen H-C, Milbrandt J, Guroff G 1993 The phosphorylation and DNA binding of the DNA-binding domain of the orphan nuclear receptor NGFI-B. J Biol Chem **268**:24808–24812.
191. Davis IJ, Hazel TG, Chen R-H, Blenis J, Lau LF 1993 Functional domains and phosphorylation of the orphan receptor Nur77. Mol Endocrinol **7**:953–964.
192. Vulliet PR, Hall FL, Mitchell JP, Hardie DG 1989 Identification of a novel proline-directed serine/threonine protein kinase in rat pheochromocytoma. J Biol Chem **264**:16292–16298.
193. Boulton TG, Nye SH, Robbins DJ, Ip NY, Radziejewska E, Morgenbesser SD, DePinho RA, Panayotatos N, Cobb NH, Yancopoulos GD 1991 ERKs: A family of protein-serine/threonine kinases that are activated and tyrosine phosphorylated in response to insulin and NGF. Cell **65**:663–675.
194. Ali S, Metzger D, Bornert J-M, Chambon P 1993 Modulation of transcriptional activation by ligand-dependent phosphorylation of the human oestrogen receptor A/B region. EMBO J **12**:1153–1160.
195. Le Goff P, Montano MM, Schodin DJ, Katzenellenbogen BS 1994 Phosphorylation of the human estrogen receptor: Identification of hormone-regulated sites and examination of their influence on transcriptional activity. J Biol Chem **269**:4458–4466.
196. Denner LA, Schrader WT, O'Malley BW, Weigel NL 1990 Hormonal regulation and identification of chicken progesterone receptor phosphorylation sites. J Biol Chem **265**:16548–16555.
197. Zhang Y, Beck CA, Poletti A, Edwards DP, Weigel NL 1994 Identification of phosphorylation sites unique to the B form of human progesterone receptor. In vitro phosphorylation by casein kinase II. J Biol Chem **269**:31034–31040.
198. Bodwell JE, Ortí E, Coull JM, Pappin DJC, Smith LI, Swift F 1991 Identification of phosphorylated sites in the mouse glucocorticoid receptor. J Biol Chem **266**:7549–7555.
199. Zhou Z-x, Kemppainen JA, Wilson EM 1995 Identification of three proline-directed phosphorylation sites in the human androgen receptor. Mol Endocrinol **9**:605–615.
200. Jurutka PW, Hsieh J-C, Nakajima S, Haussler CA, Whitfield

GK, Haussler MR 1996 Human vitamin D receptor phosphorylation by casein kinase II at Ser-208 potentiates transcriptional activation. Proc Natl Acad Sci USA **93**:3519–3524.

201. Obeid LM, Okazaki T, Karolak LA, Hannun YA 1990 Transcriptional regulation of protein kinase C by 1,25-dihydroxyvitamin D_3 in HL-60 cells. J Biol Chem **265**:2370–2374.
202. Partridge NC, Kemp BE, Veroni MC, Martin TJ 1981 Activation of adenosine 3′,5′-monophosphate-dependent protein kinase in normal and malignant bone cells by parathyroid hormone, prostaglandin E2 and prostacyclin. Endocrinology **108**:220–226.
203. Civitelli R, Hruska KA, Shen V, Avioli LV 1990 Cyclic AMP-dependent and calcium-dependent signals in parathyroid hormone function. Exp Gerentol **25**:223–231.
204. Reinhardt TA, Horst RL 1990 Parathyroid hormone down-regulates 1,25-dihydroxyvitamin D receptors (VDR) and VDR messenger ribonucleic acid *in vitro* and blocks homologous up-regulation of VDR *in vivo*. Endocrinology **127**:942–948.
205. Barettino D, Ruiz MdMV, Stunnenberg HG 1994 Characterization of the ligand-dependent transactivation domain of thyroid hormone receptor. EMBO J **13**:3039–3049.
206. Oñate SA, Tsai SY, Tsai M-J, O'Malley BW 1995 Sequence and characterization of a coactivator for the steroid hormone receptor superfamily. Science **270**:1354–1357.
207. Halachmi S, Marden E, Martin G, MacKay H, Abbondanza C, Brown M 1994 Estrogen receptor-associated proteins: Possible mediators of hormone-induced transcription. Science **264**:1455–1458.
208. Cavailles V, Dauvois S, L'Horset F, Lopez G, Hoare S, Kushner PJ, Parker MG 1995 Nuclear factor RIP140 modulates transcriptional activation by the estrogen receptor. EMBO J **14**:3741–3751.
209. Baniahmad C, Nawaz Z, Baniahmad A, Gleeson MAG, Tsai M-J, O'Malley BW 1995 Enhancement of human estrogen receptor activity by SPT6: A potential coactivator. Mol Endocrinol **9**:34–43.
210. Lee JW, Ryan F, Swaffield JC, Johnston SA, Moore DD 1995 Interaction of thyroid-hormone receptor with a conserved transcriptional mediator. Nature **374**:91–94.
211. Hong H, Kohli K, Trivedi A, Johnson DL, Stallcup MR 1996 GRIP1, a novel mouse protein that serves as a transcriptional coactivator in yeast for the hormone binding domains of steroid receptors. Proc Natl Acad Sci USA **93**:4948–4952.
212. Durand B, Saunders M, Gaudon C, Roy B, Losson R, Chambon P 1994 Activation function 2 (AF-2) of retinoic acid receptor and 9-*cis*-retinoic acid receptor: Presence of a conserved autonomous constitutive activating domain and influence of the nature of the response element of AF-2 activity. EMBO J **13**:5370–5382.
213. Chen JD, Evans RM 1995 A transcriptional co-repressor that interacts with nuclear hormone receptors. Nature **377**:454–457.
214. Hörlein AJ, Näär AM, Heinzel T, Torchia J, Gloss B, Kurokawa R, Ryan A, Kamei Y, Söderström M, Glass CK, Rosenfield MG 1995 Ligand-independent repression by the thyroid hormone receptor mediated by a nuclear receptor co-repressor. Nature **377**:397–404.
215. Malik S, Karanthanasis SK 1996 TFIIB-directed transcriptional activation by the orphan nuclear receptor hepatocyte nuclear factor 4. Mol Cell Biol **16**:1824–1831.
216. Struhl K 1996 Chromatin structure and RNA polymerase II connection: Implications for transcription. Cell **84**:179–182.
217. Stargell LA, Struhl K 1996 A new class of activation-defective TATA-binding protein mutants: Evidence for two steps of transcriptional activation *in vivo*. Mol Cell Biol **16**:4456–4464.
218. Forman BM, Tontonoz P, Chen J, Brun RP, Spiegelman BM, Evans RM 1995 15-Deoxy-$\Delta^{12,14}$-prostaglandin J_2 is a ligand for the adipocyte determination factor PPARγ. Cell **83**:803–812.
219. Kliewer SA, Lenhard JM, Willson TM, Patel I, Morris DC, Lehmann JM 1995 A prostaglandin J_2 metabolite binds peroxisome proliferator-activated receptor γ and promotes adipocyte differentiation. Cell **83**:813–819.
220. Jankneckht R, Hunter T 1996 A growing coactivator network. Nature **383**:22–23.

CHAPTER 11

Regulation of Vitamin D Receptor Abundance

ARUNA V. KRISHNAN AND DAVID FELDMAN
Division of Endocrinology, Gerontology and Metabolism, Stanford University School of Medicine, Stanford, California

I. INTRODUCTION

The actions of 1,25-dihydroxyvitamin D [1,25 $(OH)_2D$] are mediated through the nuclear vitamin D receptor (VDR) as described in detail in Chapters 8, 9 and 10, although rapid actions of this hormone at the membrane have been identified *in vitro* and postulated to be mediated via an alternative receptor mechanism (Chapter 15). Changes in the circulating levels of $1,25(OH)_2D$ are the most obvious mechanism to regulate target cell responses. However, it has become increasingly clear that the target cells themselves have evolved a complex ability to respond to the hormone, resulting in graded levels of response to a constant $1,25(OH)_2D$ signal. This variability in target cell response is intuitively determined by the abundance of VDR, although many other factors including posttranslational modifications of the VDR itself likely play significant roles. This chapter explores the variety of mechanisms and factors that regulate VDR abundance and thus modulate the amplitude of target cell responses to $1,25(OH)_2D$.

It seems clear from many studies that VDR is essential for most $1,25(OH)_2D$ actions. Although many other factors are required for optimal hormone response, such as coactivators and corepressors, the absence of a functional VDR due to genetic defects, or minor genetic abnormalities which disrupt the functionality of VDR (as described in Chapter 48), cause partial or total resistance to the action of the hormone. These genetic defects usually lead to major changes in target cell sensitivity to the hormone and affect all tissues of the body. More subtle modulations of VDR abundance, however, are mediated by physiological means, whereby the concentrations of VDR are adjusted to regulate or fine-tune the amplitude of the hormone response. Thus far, the physiological mechanisms appear to affect only the concentration of VDR and not the affinity of VDR for the hormone. These physiological mechanisms may be especially important in the modulation of tissue-specific sensitivity to $1,25(OH)_2D_3$, whereas the mutations appear to affect the expression of VDR in all tissues. In this chapter we discuss the physiological regulation of VDR by a variety of hormones and other growth and developmental signals.

This introduction emphasizes three important concepts that are repeated throughout the chapter: (1) VDR abundance determines the level of hormone response, and the abundance of VDR is regulated by many physiological signals; (2) major differences in VDR regulation exist among species and between various target tissues of the same species; and (3) several different cellular mechanisms are involved in VDR regulation.

1. VDR Abundance Modulates the Magnitude of Target Cell Response

The regulation of the abundance of VDR is an important mechanism that modulates cellular responsiveness to $1,25(OH)_2D_3$. Several studies have demonstrated that there is a strong correlation between VDR concentration and biological response in target cells [1–5]. Figure 1 depicts the results of a study in NIH-3T3 mouse fibroblasts [5] illustrating this important point. NIH-3T3 cells were transfected with a plasmid containing the vitamin D response element (VDRE) from the human osteocalcin gene promoter fused to the reporter gene chloramphenicol acetyltransferase (CAT). Treatment of these cells with forskolin (FSK), a stimulator of adenylate cyclase caused an 8-fold increase in VDR levels. Treatment with a phorbol diester, phorbol 12-myristate-13-acetate (PMA), which activates protein kinase C (PKC), suppressed VDR levels to about 30% of that seen in untreated cells. After pretreatment with FSK or PMA to regulate VDR abundance, the cells were exposed to graded concentrations of $1,25(OH)_2D_3$, and CAT activity was subsequently measured to determine hormone responsiveness. In the absence of any pretreatment (control), there was a concentration-dependent induction of CAT activity in response to $1,25(OH)_2D_3$ treatment. FSK pretreatment, which up-regulated VDR levels, caused a doubling of the magnitude of the response compared to control cells. In contrast, PMA treatment, which down-regulated VDR, resulted in a suppression of $1,25(OH)_2D_3$-induced CAT activity at all concentrations of $1,25(OH)_2D_3$. Thus, up- and down-regulation of VDR in these cells by FSK and PMA, respectively, resulted in a corresponding enhancement or attenuation of the amplitude of the $1,25(OH)_2D_3$-mediated response. It is interesting to note that the change (-fold) in the $1,25(OH)_2D_3$ response was less than the change observed at the VDR level, suggesting that other factors in the target cells may be limiting. However, it is clear that a change in the abundance of VDR is reflected in a parallel modulation of the amplitude of the target cell response to $1,25(OH)_2D_3$.

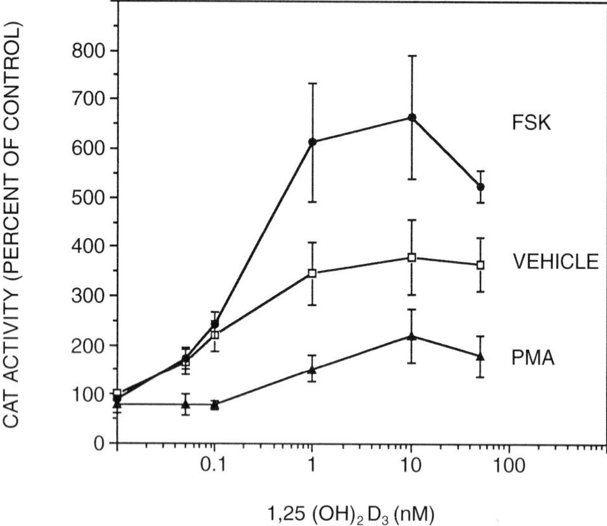

FIGURE 1 Changes in VDR levels predict the magnitude of $1,25(OH)_2D_3$ response. Confluent quiescent cultures of NIH-3T3 mouse fibroblasts stably transfected with a human osteocalcin promoter–CAT reporter plasmid were pretreated with serum-free basal medium containing dimethyl sulfoxide (vehicle), 50 μM forskolin (FSK), or 100 ng/ml PMA (PMA) for 4 hr. The cells were then washed and incubated with basal medium containing either vehicle (ethanol) or various doses of $1,25(OH)_2D_3$ (0.05–50 nM) for 24 hr, and CAT activity was assayed. CAT levels are expressed as a percentage of the control value, which was determined to be 43.5 ± 6.4 pmol/min/mg protein. Values are expressed as the mean ± SE, and there were three to six determinations for each concentration point. Reprinted with permission from Krishan and Feldman [5].

2. Tissue and Species Differences in VDR Regulation

VDR abundance may be regulated by $1,25(OH)_2D$ or other VDR ligands (referred to as homologous regulation) or by other hormones or growth factors that do not bind to the VDR (referred to as heterologous regulation). Both types of VDR regulation appear to occur in a tissue-specific as well as species-specific manner. For example, homologous up-regulation of VDR by $1,25(OH)_2D_3$ occurs to a greater extent in the kidneys of $1,25(OH)_2D_3$-treated rats than in the intestine [6], suggesting differential regulation among target organs in the same species. Treatment with glucocorticoids (heterologous regulation) caused a decline in mouse intestinal VDR, whereas rats treated with glucocorticoids showed a substantial increase in intestinal VDR levels [7], illustrating species-specific differences in VDR regulation.

Another factor that appears to influence VDR regulation in cultured cell systems is the state of differentiation of the cells under study, which is intimately related to the rate of proliferation of the cells. As VDR abundance has been shown to be closely associated with cellular proliferation rate [1,8] and as $1,25(OH)_2D_3$ has been shown to have antiproliferative and differentiating actions in a number of normal and malignant cells [9–12], the proliferation status of the cells under study may play an important role in understanding VDR regulation. Differences in VDR regulation by physiological stimuli have also been identified between normal and malignant cells [13,14].

3. Mechanisms Underlying VDR Regulation

The mechanisms underlying the regulation of VDR abundance include alterations in the rate of transcription of the VDR gene and/or the stability of VDR mRNA, which would be reflected as changes in VDR mRNA levels. Posttranslational effects can affect the synthesis of the VDR protein and/or its stability. Posttranslational modifications of the VDR (such as phosphorylation) could lead to changes in the VDR functionality [15]. In some experiments, such as those on the homologous up-regulation of VDR, there appears to be increased stabilization of the VDR due to ligand occupancy [16].

In this chapter, we discuss the various studies that have examined the homologous and heterologous regulation of VDR. We also analyze *in vitro* studies using cell culture systems that examine VDR changes in relation to cell growth and differentiation. Finally, we discuss the effects of growth, development, and aging of the organism on VDR concentrations and, therefore, $1,25(OH)_2D_3$ responsiveness. Regulation of VDR abundance is also discussed in detail in other chapters in specific settings such as in the parathyroid glands (Chapter 23), renal failure (Chapter 52), and hematological malignancy (Chapter 68).

II. HOMOLOGOUS REGULATION OF VDR

Ligands that bind to steroid hormone receptors are generally regarded as initiators of transcription; however, the hormone may also regulate the concentration of its own receptor (homologous regulation). $1,25(OH)_2D_3$ and other vitamin D metabolites have been shown to up-regulate VDR abundance, whereas many ligands down-regulate their own receptors. To prevent ever increasing levels of VDR due to $1,25(OH)_2D_3$ action, the VDR up-regulation must somehow trigger a "turnoff" of the system. This probably occurs by at least two mechanisms. First, $1,25(OH)_2D_3$ induces the enzyme 24-hydroxylase, which increases the metabolic conversion of the active hormone $1,25(OH)_2D_3$ into inactive metabolites. Second, the actions of $1,25(OH)_2D_3$ increase serum calcium which inhibits parathyroid hormone (PTH) production and turns off the renal synthesis of $1,25(OH)_2D$. $1,25(OH)_2D_3$ itself mediates both of these actions, by directly inhibiting PTH synthesis and 1α-hydroxylase activity while stimulating 24-hydroxylase. Thus, the effects of homologous up-regulation of VDR appear to be modulated by these mechanisms which prevent continual $1,25(OH)_2D_3$ action.

VDR regulation plays a key role in the synthesis of $1,25(OH)_2D$, acting in both a short feedback loop at the kidney and a long feedback loop to maintain blood levels. The proximal nephrons possess both 1α-hydroxylase activity and VDR and are thus a special case, being both the source of $1,25(OH)_2D$ production as well as a target for $1,25(OH)_2D$ action. Homologous up-regulation of VDR in these cells would enhance $1,25(OH)_2D$ actions, suppress 1α-hydroxylase activity, and block further production of $1,25(OH)_2D_3$. Because the kidneys synthesize $1,25(OH)_2D$ for systemic requirements, when the renal production of this hormone is favored (such as due to an increase in PTH levels or a decrease in serum phosphate), the VDR levels in these cells have to be suppressed or the synthesis would be inhibited. Iida *et al.* [17] tested this hypothesis in rat and quail models of enhanced renal production of $1,25(OH)_2D_3$ and found a down-regulation of VDR mRNA expression in kidney proximal tubules compared to other tissues such as duodenum. The mechanism underlying this regulation is not clear although PTH and serum calcium levels are postulated to play a role [18,19].

The first evidence for homologous up-regulation was provided by the study of Costa *et al.* [20] using LLC-PK_1 pig kidney cells and human skin fibroblasts. Treatment of the cells with $1,25(OH)_2D_3$ and other vitamin D metabolites caused a 2- to 5-fold increase in VDR levels as measured by ligand binding assays. This up-regulation was due to an increase in VDR abundance without any change in the affinity for the ligand or in its DNA binding properties. The order of potency among vitamin D metabolites to cause up-regulation was $1,25(OH)_2D_3 = 24,25(OH)_2D_3 > 1,24,25(OH)_3D_3 > 25OHD_3$. Costa *et al.* [20] further showed that this up-regulation required a functional VDR, as it did not occur in skin fibroblasts from patients with hereditary vitamin D-resistant rickets (HVDRR) with mutations in the VDR within the DNA binding domain but retaining normal hormone binding.

Early studies on the kinetics of $1,25(OH)_2D_3$ binding to mammary carcinoma cell cultures [21,22] also suggested homologous up-regulation of VDR in these cancer cells. Since then, a number of investigators have reported homologous up-regulation of VDR in a variety of cultured cell systems of both normal and cancer origin including rat and human osteosarcoma cells [16,23–26], human breast cancer cells [27], mouse fibroblasts [28,29], mouse adipose cells [30], human HL-60 promyelocytic leukemia cells [31,32], and human colon cancer cells [33]. Santiso-Mere *et al.* [34] reported that the levels of recombinantly expressed VDR protein produced in transiently transfected COS-1 monkey kidney cells increased after $1,25(OH)_2D_3$ treatment of the cells, and they also demonstrated a similar effect in a *Saccharomyces cerevisiae* expression system. Costa and Feldman [6] demonstrated homologous up-regulation *in vivo* in vitamin D-deficient rats. Their results suggested differential sensitivity between organs, with the up-regulation being more pronounced in the kidneys than in the intestine. Homologous up-regulation *in vivo* has also been demonstrated in rat and mouse intestine [35–37], cow colon [38], and rat parathyroid glands [39]. In contrast, Hulla *et al.* [40] found that the vitamin D sterols had no effect on VDR levels on CaCo-2 human colon cancer cells, whereas Kizaki *et al.* [41] found no change or a possible decrease, in total unoccupied VDR in HL-60 promyelocytic leukemic cells. According to a report by Song [42], VDR mRNA expression was not regulated by $1,25(OH)_2D_3$ in HOS-8603 human osteosarcoma cells, whereas it was decreased in a human megakaryoblastic leukemia cell line after $1,25(OH)_2D_3$ treatment.

The enzymes that metabolize $1,25(OH)_2D_3$ to less active or inactive compounds appear to occur in most $1,25(OH)_2D_3$ target tissues. These enzymes, especially the 24-hydroxylase, are induced by $1,25(OH)_2D_3$ in parallel with the up-regulation of VDR. This self-induced increase in catabolism may modulate cellular responsiveness to $1,25(OH)_2D_3$, including the ability to induce homologous up-regulation of VDR. Pols *et al.* [24,43] showed enhanced homologous up-regulation of VDR in UMR-106 rat osteosarcoma cells after inhibition of $1,25(OH)_2D_3$ metabolism by ketoconazole. Ketoconazole inhibits several cytochrome P450 enzymes including 24-hydroxylase [44] and 1-hydroxylase [45]. Pols *et al.* [24] found that ketoconazole increased the cellular concentration of $1,25(OH)_2D_3$ and, consequently, VDR occupancy by $1,25(OH)_2D_3$, and this effect augmented the homologous up-regulation of VDR. Studies by Reinhardt and Horst in rat osteosarcoma cells [25] and in human breast cancer cells [27] likewise suggested that self-induced catabolism of $1,25(OH)_2D_3$ was an important modulator of VDR occupancy, and treatment with ketoconazole or competitive inhibitors of $1,25(OH)_2D_3$ metabolism amplified the $1,25(OH)_2D_3$-mediated up-regulation of VDR and cellular responsiveness to $1,25(OH)_2D_3$.

In vitamin D toxicity, target tissues are exposed to high concentrations of 25-hydroxyvitamin D. At high concentrations, 25-hydroxyvitamin D can compete for binding on the VDR and produce biological effects in target cells similar to those of $1,25(OH)_2D$, including the homologous up-regulation of VDR. This leads to hypercalcemia and other pathophysiological changes (for detailed discussion, see Chapter 54).

Several *in vitro* and *in vivo* studies have attempted to investigate the mechanisms underlying the homologous up-regulation of VDR abundance. The results appear controversial, perhaps because of the differences in the model systems employed in these studies. Costa and Feldman [46], using a dense amino acid labeling technique, found that the up-regulation of VDR in $24,25(OH)_2D_3$-treated LLC-PK_1 pig kidney cells resulted predominantly from a prolongation of receptor half-life (4.3 ± 0.4 hr in control cells versus 8.9 ± 1.0 hr in treated cells) and from a smaller increase in the rate of receptor synthesis. Pan and Price [23], studying the phenomenon in ROS 17/2 rat osteosarcoma cells, did not find changes in VDR half-life due to $1,25(OH)_2D_3$ treatment. They concluded that mRNA synthesis was required to achieve the maximal up-regulation. In contrast, Arbour *et al.* [16], who studied VDR regulation in ROS 17/2.8 cells (derived from the parental cell line ROS 17/2), concluded that $1,25(OH)_2D_3$ treatment prolonged VDR half-life. They did not detect changes in VDR mRNA levels. A possible reason for these different findings may be that there are differences in characteristics of receptor regulation between the clonal osteosarcoma cell lines even though their properties are indicative of the osteoblastic phenotype. Santiso-Mere *et al.* [34] examined the expression of VDR in transiently transfected COS-1 cells and a reconstituted *Saccharomyces cerevisiae* system using heterologous promoters and found that $1,25(OH)_2D_3$ had a positive regulatory effect on VDR in both systems. Their results indicated that up-regulation by the ligand occurred at the level of protein turnover and that $1,25(OH)_2D_3$ treatment stabilized the VDR, perhaps by decreasing its degradation. Wiese *et al.* [47] similarly, found increases in the half-life of the VDR protein after $1,25(OH)_2D_3$ treatment of mouse fibroblasts and rat intestinal epithelial cells.

Studies that examined changes in VDR mRNA during homologous up-regulation also reveal a transcriptional and posttranscriptional component. Although no mRNA changes were seen in rat osteosarcoma cells [16], rat intestinal epithelial cells, and mouse fibroblasts [47], other investigators have found clear-cut increases in VDR mRNA on $1,25(OH)_2D_3$ treatment of other cells. Homologous up-regulation of VDR was associated with increased mRNA in ROS 17/2.8 cells [48], 3T3L1 mouse adipose cells [30], human osteosarcoma cells [26], and human colon cancer cells [33]. Mahonen *et al.* [26] also showed that $1,25(OH)_2D_3$ treatment of MG-63 human osteosarcoma cells produced an increase in VDR mRNA half-life. In *in vivo* studies using vitamin D-deficient rats, Huang *et al.* [49] observed that the administration of $1,25(OH)_2D_3$ did not change VDR mRNA levels in the intestine and kidney. However, Strom *et al.* [35] found that VDR mRNA levels increased substantially in the intestine of vitamin D-deficient rats 6 and 12 hr following intravenous administration of $1,25(OH)_2D_3$. Nakajima *et al.* [37] observed increases in duodenal VDR mRNA levels in mice treated with $1,25(OH)_2D_3$.

Thus, vitamin D may or may not increase VDR mRNA abundance, depending on the model system used, and it appears to affect VDR protein turnover by increasing the stability of the VDR protein. The effect on VDR protein stability may be due directly to the occupancy of the ligand binding site, causing a conformation that is less rapidly metabolized by degrading enzymes. This stabilization may also be due to posttranslational modifications of the VDR. As described in Chapter 10, $1,25(OH)_2D_3$-dependent phosphorylation of the VDR has been shown to occur. These modifications might increase the stability of VDR directly via conformational changes or indirectly by increasing its ability to dimerize with other accessory factors [such as the retinoid X receptor (RXR)], to bind DNA, or to affect any additional steps that stabilize the protein or diminish the availability of the VDR to the degradation system.

III. HETEROLOGOUS REGULATION OF VDR

This section deals with the regulation of VDR abundance by hormones other than $1,25(OH)_2D_3$ and includes compounds that activate specific intracellular second messenger pathways. The hormones are divided into two categories, those that bind to the steroid–thyroid–retinoid superfamily of intracellular receptors and the peptide hormones, which include growth factors and the calciotropic hormones PTH and PTH-related peptide (PTHrP) that act via membrane receptors.

A. Steroids and Retinoids

1. GLUCOCORTICOIDS

Of the steroid hormones that interact with the vitamin D endocrine system, glucocorticoids are of particular interest because of their apparent "anti-vitamin D" effects in several clinical settings (see Chapter 49). Exposure to increased levels of glucocorticoids results in osteopenia whether the source of the hormones is endogenous as in Cushing's syndrome or, more commonly, exogenous as in corticosteroid therapy. The cause of glucocorticoid-induced osteopenia is multifactorial, with important glucocorticoid actions on calcium metabolism at the levels of intestine and bone contributing to accelerated bone loss [50]. The basis of the apparent antagonism of glucocorticoids to vitamin D action is not entirely clear, but several possible mechanisms have been raised (reviewed in Chapter 49). In this section, we focus on the role of glucocorticoid regulation of VDR as one mechanism for the anti-vitamin D action of glucocorticoids.

The inhibition of intestinal calcium absorption has been shown to be a critical event in the development of glucocorticoid-induced osteopenia [51–53]. Because $1,25(OH)_2D_3$ is the major hormonal regulator of intestinal calcium absorption, several investigators have examined the possibility that glucocorticoids modulate $1,25(OH)_2D_3$ actions through the regulation of intestinal VDR, thus inhibiting calcium absorption and contributing to osteoporosis. It is in this area of glucocorticoid effects on VDR that species- and tissue-specific differences in VDR regulation have become very apparent (Table I).

Administration of glucocorticoids has been shown to cause decreased levels of intestinal VDR in mice by Hirst and Feldman [54]. The same investigators however, found that, in contrast to this effect in mice, rats

TABLE I VDR Regulation by Glucocorticoids[a]

	Mouse	Rat	Chick	Dog	Human
In vivo hormone treatment					
Intestine	↓ [54]	↑ [7,56] ↓ [55]		↑ [57]	
Kidney		↓ [56]			
Testis, heart, and lung		No change [56]			
Isolated monocytes					↓ [64]
In vitro hormone treatment					
Fetal osteoblastic cells	↓ [60]	↑ [4,59,62]			
Osteosarcoma cells		↑ [63]			↑ [26] ↓ [65]
Intestinal cytosol			No change [58]		

[a] ↑, Up-regulation; ↓, down-regulation.

treated with glucocorticoids showed a substantial increase in the number of intestinal VDR and that adrenalectomy led to a significant reduction in VDR concentration [7]. Chan and Atkins [55] studied the distribution of VDR in isolated rat jejunal villous and crypt cells and found that adrenalectomy caused an increase in VDR, whereas dexamethasone treatment led to a decrease in VDR levels. These changes were more pronounced in the crypt cells. The fact that this study was carried out in isolated cell populations may account for why the results are different from the observations of Hirst and Feldman [7] in rat intestine. Gensure *et al.* [56] reported that dexamethasone exerted tissue-specific effects on VDR levels in the rat, specifically, up-regulation in the intestine, down-regulation in the kidneys, and no effects on VDR in testis, heart, or lung. Korkor *et al.* [57] found increases in VDR concentrations in duodena of dogs after prednisone administration. Wilhelm and Norman [58] investigated the effects of a large molar excess of triamcinolone on the binding performance of VDR in a duodenal chromatin fraction of vitamin D-replete chickens and found that the glucocorticoid had no effect on VDR binding capacity or its affinity for $1,25(OH)_2D_3$. Thus, there appears to be a pronounced species difference in the response of intestinal VDR to glucorticoids; an increase occurs in the rat (most investigators) and dog, a decrease occurs in the mouse, and no change is seen in chickens (Table I).

In addition to VDR regulation in the intestine, a number of *in vitro* studies have investigated the regulation of VDR in bone, the other classic vitamin D target tissue. Manolagas *et al.* [59], in early studies, found that the addition of cortisol to cultures of fetal rat calvaria prevented the decline in VDR otherwise seen in these cultures. Chen *et al.* showed that dexamethasone produced a dose-dependent decrease in the concentration of VDR in primary cultures of mouse osteoblasts and a significant increase in VDR levels in cultures of rat osteoblasts [4,60–62]. Glucocorticoids have also been shown to increase VDR levels in rat osteosarcoma cells [63].

Because the experiments with mouse and rat models, or those utilizing cells derived from these species, produced opposite results (up-regulation in the rat and down-regulation in the mouse), studies using human cells or tissues have become particularly important. Nielsen *et al.* [64] studied the nuclear uptake of ^3H-$1,25(OH)_2D_3$ in freshly isolated monocytes from healthy human subjects before and after prednisone treatment and found an approximately 40% decrease in VDR binding after treatment. The observations of Mahonen *et al.* [26] revealed that glucocorticoid treatment caused an increase in VDR mRNA in MG-63 human osteosarcoma cells, reaching maximal levels by 96 hr. In the same cells, however, Godschalk *et al.* [65] found a decrease in VDR mRNA after a shorter exposure (<24 hr) to dexamethasone. Table I summarizes the results of the various *in vivo* and *in vitro* studies on the effect of glucocorticoids on VDR abundance in different species. Clearly, more studies with better systems that model the human intestine or bone are needed to ascertain the effect of glucocorticoid actions on VDR in humans.

As the few studies using human cell systems suggest, if glucocorticoids indeed act to lower VDR abundance (as has been shown in mouse cells and model systems), the sensitivity of target cells to respond to $1,25(OH)_2D_3$ would be decreased. This would provide a basis for the hypothesis that the anti-vitamin D action of glucocorticoids on calcium absorption in humans is due to the down-regulation of intestinal VDR. This hypothesis, although attractive as a contributing mechanism for glucocorticoid-induced osteopenia, remains unproved.

In addition to direct actions on VDR itself, glucocorticoids may modulate the actions of other hormones or stimuli that might regulate VDR and interact with the vitamin D system. For instance, some studies have demonstrated that glucocorticoids potentiate the action of PTH on bone cells [66–69]. Massaro *et al.* [70] have shown that glucocorticoids accelerate the developmental appearance of VDR in rat intestine (see Section VI below).

2. Estrogens

As steroid hormones that might interact with the vitamin D endocrine system, estrogens are of great interest because of the importance of the syndrome of postmenopausal osteoporosis. It has been repeatedly demonstrated that estrogen deficiency correlates with osteoporosis and that estrogen replacement therapy has a beneficial effect on bone (see Chapter 44). Direct estrogen actions on bone or bone marrow cells are likely, as estrogen receptors (ER) have been demonstrated in normal human osteoblast-like cells [71] and in osteosarcoma cells [72]. The effect of estrogens to inhibit cytokines in bone marrow cells has also been demonstrated [73]. Intestinal calcium absorption is known to be lower in postmenopausal women [74], and estrogen therapy has been shown to improve calcium balance in both normal postmenopausal women and in women with osteoporosis [75]. Since $1,25(OH)_2D_3$ is the principal hormonal regulator of intestinal calcium absorption, a tenable hypothesis is that estrogens might influence calcium absorption by regulating intestinal VDR levels. Support for this hypothesis is provided by the early study of Navickis *et al.* [76] showing that sex steroids modulate the ability of vitamin D to induce calcium binding protein in chick intestine.

Several investigators have attempted to study estrogen regulation of VDR in intestine as well as bone. The difficulties, as with glucocorticoid regulation of VDR, arise from a lack of good model systems for the human intestine and bone. Early reports on the effect of estrogens on VDR were from studies of the rat uterus. Walters [77] showed that injection of estradiol to ovariectomized rats increased the number of VDR in the uterus. Levy *et al.* [78] found that VDR was not present in the uteri of saline-treated immature rats and that estradiol administration to these rats caused an induction of VDR. The effect of estradiol directly on the intestine to regulate intestinal VDR is possible; ER have been found in rat intestinal cells [79], and an effect of estradiol on intestinal calcium transport has been demonstrated [80]. Chan *et al.* [81] studied the distribution of VDR in isolated rat jejunal cells in normal and ovariectomized rats and found that ovariectomy caused a selective reduction in VDR levels in villous cells. Duncan *et al.* [82] studied the regulation of VDR in ovariectomized rats and concluded that estradiol regulation of VDR was organ specific. In their study, administration of estradiol to ovariectomized rats increased VDR concentrations in the liver, whereas levels were decreased in the kidney. Estradiol treatment did not change intestinal VDR levels in castrated male rats. In unpublished studies, Hirst and Feldman found that aged ovariectomized rats exhibited decreased intestinal VDR compared to age-matched controls and that estradiol replacement caused a 25% rise in VDR. Thus, the decrease in VDR concentrations following ovariectomy and the increase observed after estradiol administration, although not a uniform finding, provide a reasonable model that may contribute to the reduction in calcium absorption that occurs after menopause.

There are fewer studies that have examined the regulation of VDR by estrogens in bone. Liel *et al.* [83] studied the effect of estradiol on VDR in ROS 17/2.8 rat osteosarcoma cells. VDR binding capacity doubled in the presence of 10 n*M* estradiol, whereas the affinity of VDR for $1,25(OH)_2D_3$ did not change. The effect of estradiol was blocked by the antiestrogen tamoxifen, indicating that the effect was specific to estradiol and was mediated by ER. In the presence of estradiol, maximal secretion of osteocalcin in response to $1,25(OH)_2D_3$ was 3.5-fold higher than in response to $1,25(OH)_2D_3$ alone. Up-regulation of VDR by estradiol was also observed in MG-63 human osteosarcoma cells by Mahonen and Maenpaa [26]. Estradiol produced a dose-dependent increase in VDR mRNA levels in these cells, and the mRNA levels were maximally elevated 72 hr after estradiol treatment, at which time they were 425% higher than control levels. The up-regulation appeared to occur through the induction of mRNA synthesis. However, these investigators did not detect an increase in VDR protein level as determined by metabolic labeling and immunoprecipitation using their antibody [26]. Ishibe *et al.* [84] also found increases in $1,25(OH)_2D_3$ binding and VDR mRNA levels after estrogen treatment of another human osteosarcoma cell line, OGA.

In conclusion, the effect of estradiol on VDR is complex and, like glucocorticoids, shows species and organ differences. It is attractive to hypothesize that one of the many actions of estradiol to prevent osteoporosis is an induction of intestinal VDR to maintain normal calcium absorption. Although some data from rats support this formulation, more conclusive data in humans are not yet available to prove this hypothesis.

3. Retinoids

Vitamin A and its naturally occurring metabolites retinol and retinoic acid, known largely for their role in visual and reproductive cycles and in the growth and development of epithelial tissues, are also necessary for normal bone metabolism. Both retinoic acid and $1,25(OH)_2D_3$ share a striking number of physiological effects on bone including their abilities to regulate cellular differentiation. Receptors for the retinoids all-*trans*-retinoic acid (ATRA) and 9-*cis*-retinoic acid (9-cis-RA), belong to the steroid–thyroid–retinoid receptor superfamily of proteins, like VDR. The retinoid X receptor (RXR), which binds 9-*cis*-RA, plays an important role in the VDR action pathway by virtue of its ability to heterodimerize with VDR and modulate target gene expression (Chapters 8–10). One level at which retinoids and vitamin D may modulate the action of one another is through the cross-regulation of the concentration of the receptor for the other hormone. It is therefore particularly interesting to examine the regulation of VDR by retinoids in vitamin D target cells and vice versa.

Petkovich *et al.* [14] showed that ATRA produced a dose-dependent increase in VDR levels as measured by ligand binding in ROS 17/2 rat osteosarcoma cells due to a selective increase in the number of VDR without affecting the affinity of VDR for $1,25(OH)_2D_3$. As opposed to the results in the malignant cells, Chen and Feldman [13] found that ATRA caused a reduction (up to one-third of control levels) in VDR abundance in primary cultures of rat osteoblasts while increasing the receptor level (up to 3-fold more than control) in mouse osteoblasts. The up-regulation observed in the mouse cells required protein and RNA synthesis. Mahonen and Maenpaa [26] reported up-regulation of VDR by ATRA in MG-63 human osteosarcoma cells. They observed an increase in VDR mRNA levels in these cells after ATRA treatment and showed that the increase was due to a prolongation of the half-life of VDR mRNA. The apparent half-life of VDR mRNA was increased by 11 and 6 hr by $1,25(OH)_2D_3$ and ATRA, respectively. ATRA did not affect VDR protein synthesis as measured by metabolic labeling of VDR in the presence of the hormones and followed by analysis after immunoprecipitation. Davoodi *et al.* [85] showed that ATRA caused VDR up-regulation in T47-D human breast cancer cells.

In conclusion, there are multiple interactions between $1,25(OH)_2D_3$ and retinoids, one of which is the regulation of each other's receptor. The significance of this receptor regulation in tissue responsiveness remains to be clarified. Among the hormones that bind the other members of the steroid–thyroid–retinoid receptor superfamily, triiodothyronine has been shown to increase VDR mRNA levels in MG-63 human osteosarcoma cells [86]. There is also a report of VDR up-regulation in T47-D human breast cancer cells by the progesterone analog R5020 [85].

B. Activators of Second Messenger Pathways

The different physiological stimuli that might regulate the intracellular concentrations of VDR include hormones [such as $1,25(OH)_2D_3$, steroid hormones, PTH], nutritional signals (such as dietary calcium), and alterations in the physiological state (such as growth and development, differentiation, aging, pregnancy, lactation). This results in a complex interplay of various signal transduction pathways within a target cell that might affect VDR levels. This section deals with the regulation of VDR in relation to the activation of intracellular second messenger systems. We summarize the *in vitro* studies on the roles of the protein kinase C (PKC) and protein kinase A (PKA) signal transduction pathways and intracellular calcium levels in the regulation of VDR abundance and function.

1. Protein Kinase C

While studying the effect of growth factor stimulation of cellular proliferation and VDR levels, Krishnan and Feldman [87] observed that basic fibroblast growth factor (bFGF), which activates calcium–phospholipid–dependent PKC, produced a significant down-regulation of VDR levels in NIH-3T3 mouse fibroblasts. They went on to examine the role of PKC in VDR down-regulation in these cells using phorbol esters. Potent phorbol esters such as PMA and phorbol 12,13-dibutyrate (PDBu), which activate PKC, caused down-regulation of VDR in a time- and dose-dependent manner. The down-regulation did not occur when an inactive phorbol ester that could not activate PKC was used or when PKC was desensitized after prolonged exposure to phorbol esters, nor in the presence of the PKC inhibitor staurosporine. Elevation of intracellular calcium levels in combination with the use of a synthetic diacylglycerol resulted in the down-regulation of VDR to a similar extent. This further confirmed the role of PKC in VDR down-regulation. Significant decreases in the steady-state levels of VDR mRNA and protein were observed. This study also demonstrated that the PKC-mediated inhibition of VDR gene expression could override the stimulation of VDR expression brought about by other mitogenic agents such as serum [87].

van Leeuwen *et al.* [88] studied the regulation of VDR in the osteoblast-like rat osteosarcoma cells UMR-106 and found that PMA exerted a bidirectional effect, namely, a moderate ($<30\%$) decrease at earlier time points after treatment (4 hr or less) and a 2-fold increase in receptor levels after 24 hr. VDR mRNA levels in PMA-treated UMR-106 cells recorded similar changes. They postulated that the increased VDR mRNA seen with longer treatment with PMA might be due to desensitization of PKC based on the fall in PDBu binding detected after 24 hr of treatment. However, Reinhardt and Horst [48] observed a time- and dose-dependent increase in VDR protein and mRNA levels in ROS 17/2.8 cells due to PMA treatment, which was also abolished by several PKC inhibitors. They found that PMA pretreatment caused a pronounced enhancement of $1,25(OH)_2D_3$-mediated up-regulation of VDR in these cells. Merke *et al.* [89] reported increased VDR levels in cultured bovine aortic endothelial cells due to PKC activation. Thus, there appear to be target cell-specific effects of PKC activation on VDR abundance (Table II).

The hormone $1,25(OH)_2D_3$ has been shown to increase phorbol ester binding in HL-60 cells [90], indicating PKC activation. Ways *et al.* [91] have demonstrated that $1,25(OH)_2D_3$-treated U937 human histiocytic lymphoma cells exhibited increased PKC-dependent phosphorylation of endogenous substrates. These observations together with the above-mentioned studies suggest a functional link between PKC activation, $1,25(OH)_2D_3$ action, and VDR regulation in target cells. In this connection, it is interesting to note that VDR has been shown to be a substrate for PKC *in vitro* and that phosphorylation of VDR at serine-51 by PKC alters its ability to modulate gene transcription [15].

2. Protein Kinase A

Several studies indicate that the other major intracellular signal transduction pathway, namely the cAMP-dependent PKA system, is also an important regulator of VDR abundance. Early indications that PKA might

TABLE II VDR Regulation in Cells of the Osteoblastic Phenotype

	MC3T3-E1 (clonal mouse osteogenic cells)	UMR-106 (clonal rat osteosarcoma)	ROS 17/2.8 (clonal rat osteosarcoma)
Differentiation state	Preosteoblastic	Preosteoblastic	Osteoblastic
PKA activation	Up-regulation [94]	Up-regulation [92–94]	Down-regulation [95]
PKC activation	Down-regulation [94]	Down-regulation [88,94]	Up-regulation [48]
PTH treatment	Up-regulation [94]	Up-regulation [92–94]	Down-regulation [19,95]

regulate VDR came from the study of Pols *et al.* [92] in UMR-106 osteosarcoma cells on the regulation of VDR by PTH. PTH caused a dose-dependent increase in VDR levels that was preceded by a stimulation of cAMP production by this hormone, suggesting a role for cAMP in PTH-mediated VDR up-regulation. Krishnan and Feldman [5] investigated the role of PKA activation in VDR regulation in NIH-3T3 cells and the relationship between the PKA and PKC pathways. They found that an elevation in intracellular cAMP levels produced by the adenylate cyclase activator forskolin (FSK) resulted in a substantial (8- to 10-fold) increase in VDR abundance as measured by ligand binding and Western blot analysis. FSK produced a rapid increase in VDR mRNA (by 1 hr), followed by a decline and a subsequent increase at 15 hr. FSK has been shown to produce significant increases in VDR mRNA in UMR-106 rat osteosarcoma cells [93,94] and in MC3T3-E1 mouse osteoblasts [94]. However, in ROS 17/2.8 rat osteosarcoma cells, FSK caused a down-regulation of VDR abundance and suppressed the homologous up-regulation of VDR [95].

Table II summarizes the results of the studies on VDR regulation in osteoblastic cells and demonstrates cell-specific effects due to the activation of PKC and PKA. FSK has been shown to increase VDR mRNA levels in a human megakaryoblastic leukemia cell line and in HOS-8603 human osteosarcoma cells [42]. Interestingly, simultaneous activation of PKC and PKA in NIH-3T3 cells abolished the PKA-mediated increases in VDR protein and mRNA [5]. Thus, these two major signal transduction pathways were mutually antagonistic in these cells at the level of VDR regulation. Regulation of VDR levels by PKC or PKA activation causes corresponding modulations in $1,25(OH)_2D_3$ responsiveness of these target cells (Fig. 1). van Leeuwen *et al.* [88], who demonstrated a down-regulation of VDR by PMA in UMR-106 cells, found that PKC activation inhibited a forskolin-stimulated rise in VDR levels at 4 hr. Krishnan *et al.* [94] also demonstrated this functional antagonism between the PKA and PKC pathways in MC3T3-E1 mouse osteoblasts and UMR-106-01 rat osteosarcoma cells.

The activation of PKC or PKA by specific agents or hormones appears to produce dramatic changes in VDR mRNA [5,87] that could be due to an increase in the rate of transcription [93]. Further studies are needed to distinguish between changes in transcriptional rate and mRNA stability. If the rate of transcription of the VDR gene is affected, one could speculate that the 5′ upstream region of the VDR gene might contain specific response elements such as a 12-*O*-tetradecanoylphorbol-13-acetate (TPA) response element (TRE) or a cAMP response element (CRE) as has been shown for other genes activated by phorbol ester or cAMP [96,97]. Interactions at the level of these response elements to regulate VDR gene expression would be possible, and differences in these interactions might contribute to the cell- and tissue-specific variations in VDR regulation.

3. Intracellular Calcium

The hormone $1,25(OH)_2D_3$ has been shown to produce an increase in intracellular calcium concentrations in several cultured cell systems [98–100]. Many hormonal stimuli that activate the PKC signal transduction pathway do so by increasing diacylglycerol concentration as well as intracellular calcium levels through the production of inositol 3-phosphate [101]. As both $1,25(OH)_2D_3$ and PKC have been shown to be major regulators of VDR abundance, it is likely that elevation of intracellular calcium may play a role in VDR regulation.

van Leeuwen *et al.* [102] studied the role of cAMP and calcium in an attempt to understand the mechanisms underlying the PTH-mediated regulation of VDR in UMR-106 cells. They found that agonists of the cAMP signal pathway produced an increase in $1,25(OH)_2D_3$ binding. Increasing the intracellular concentration by the calcium agonist Bay 8644 and the ionophore A23187 did not affect the basal $1,25(OH)_2D_3$ binding. However, calcium seemed to play a role in the cAMP-mediated up-regulation. The reduction in calcium levels elicited by the use of calcium channel blockers, such as nitrendipine and verapamil, or by the chelation of extracellular calcium levels with EGTA, reduced the cAMP-mediated increase in $1,25(OH)_2D_3$ binding. In a study of the activation of PKC and VDR regulation in NIH-3T3 mouse fibroblasts, Krishnan and Feldman [87] observed that an elevation in intracellular calcium levels caused by the calcium ionophore A23187, resulted in a time-dependent decrease in VDR levels. The simultaneous addition of a synthetic diacylglycerol, which would activate PKC, had an additive effect and caused further decreases in VDR levels. More studies are needed to obtain a clearer understanding of the role of calcium in VDR regulation and its interaction with other signal transduction pathways that regulate VDR. Figure 2 depicts the changes in VDR levels due to the activation of these second messenger systems.

C. Peptide Hormones

The peptide hormones that are known to play an important role in VDR regulation include the polypeptide growth factors and the calciotropic hormones PTH and PTHrP.

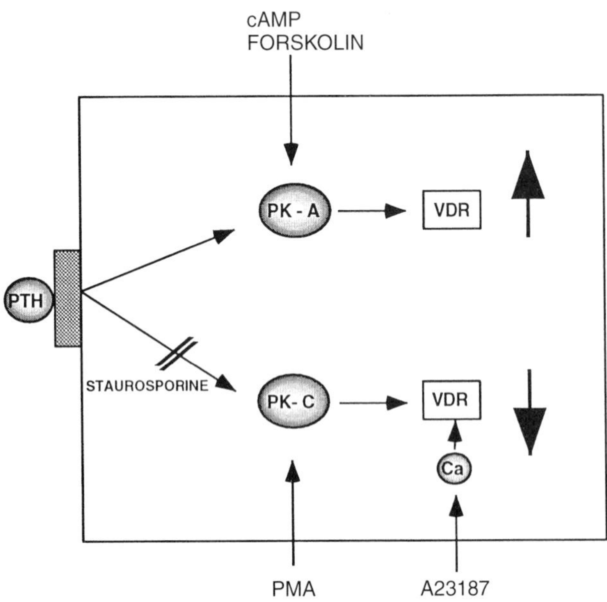

FIGURE 2 Regulation of VDR by PTH and activators of second messenger pathways. Stimulation of the cAMP-dependent PKA by agents such as dibutyryl cAMP or forskolin results in VDR up-regulation [5,91,92,94], and stimulation of PKC by the phorbol ester PMA causes VDR down-regulation [84,85,92]. Staurosporine, a PKC inhibitor, abolishes VDR down-regulation. Elevation in intracellular calcium levels produced by the ionophore A23187 also results in the down-regulation of VDR [84]. PTH activates multiple second messenger pathways, and the regulation of VDR by this hormone could represent the net effect of the various signaling systems [94].

1. Growth Factors and Serum

In addition to being regulated by the calciotropic hormones, the processes of bone formation and bone resorption are directed by the actions of several locally produced growth factors and cytokines. Insulin-like growth factor I (IGF-I) has been suggested to play a role in the coupling of bone formation to resorption [103]. Transforming growth factor-β (TGFβ) alters the proliferation, differentiation, and differentiated functions of osteoblasts [104–106] and osteoclasts [107]. Several growth factors including epidermal growth factor (EGF) have been shown to increase serum calcium *in vivo* and bone resorption *in vitro* [108,109]. $1,25(OH)_2D$ is the major hormonal regulator of calcium absorption, and it also plays an important role in bone metabolism. It is likely, therefore, that there are interactions between $1,25(OH)_2D$ actions and growth factor effects on bone and intestine. At the cellular level, growth factors act as potent mitogens, and some of these factors cause differentiation of specific cell types. $1,25(OH)_2D_3$ exhibits antiproliferative and differentiating properties in several cultured cell systems [12], and its receptor levels correlate with the rate of proliferation in many cells (for a detailed discussion see Section V below). Therefore, it is not surprising that growth factors influence VDR concentrations in target cells.

Studies on VDR in primary cultures of mouse bone cells by Chen and Feldman [1] led to the conclusion that the level of VDR in these cells was a regulated function related to the rate of cell division. Conditions that lowered the rate of cell division, such as treatment of cells with colcemid or reducing the serum concentration in the culture medium, caused a decrease in the VDR content of the cells. Addition of basic fibroblast growth factor (bFGF) to these cultures partially restored cell proliferation as well as VDR concentration. In primary cultures of mouse osteoblasts, Chen *et al.* [60] found that in the early log phase EGF blocked the inhibition of growth caused by dexamethasone and also reversed the dexamethasone-mediated decline in VDR levels.

Krishnan and Feldman [8] also observed a close correlation between VDR abundance and cell proliferation rate in NIH-3T3 mouse fibroblasts and in MCF-7 human breast cancer cells. They further demonstrated that serum and growth factors such as EGF, high concentrations of insulin, or IGF-I stimulated cell proliferation as well as increased VDR levels. Serum exerted the maximum effect, as it contained several growth factors. The growth factor-induced increase was seen at the level of VDR mRNA, which peaked at about 4 hr after serum addition and was superinduced in the presence of protein synthesis inhibitors. This study indicated that the increase in VDR gene expression occurred in the early phase of the cell cycle, suggesting that the VDR gene appeared to be one of a set of early genes activated when quiescent cells were stimulated by a mitogenic signal [8]. It was difficult to determine whether these mitogens had a direct effect on the VDR gene or whether their effects on VDR were secondary to an increase in the population of cells traversing the G_1 phase of the cell cycle. However, further studies by these investigators [87] showed that bFGF, which was a potent mitogen for NIH-3T3 fibroblasts, concurrently stimulated proliferation while dramatically decreasing VDR abundance in these cells due to the activation of PKC by this growth factor. This uncoupling of growth factor action on mitogenesis and VDR regulation indicates possible direct actions of some growth factors including serum to regulate VDR gene expression. The availability of the 5′ upstream sequence of the VDR gene would enable investigation of this question to be undertaken, and it is likely that a "serum response element" resides in the promoter region of the VDR gene, as has been shown for the c-*fos* gene promoter [110,111].

A number of other studies have demonstrated that growth factors are regulators of VDR abundance. Mezzetti *et al.* [112] cultured mammary explants from pregnant mice and showed that a combination of insulin, cortisol and prolactin caused the cells to express their differentiated function, namely the expression of the milk protein, casein, as well as to increase their VDR levels. TGFβ has been shown to cause an up-regulation in VDR levels in rat osteoblastic osteosarcoma cells [113,114]. This up-regulation was seen at the level of VDR mRNA [114]. However, in UMR-106 rat osteosarcoma cells, EGF inhibited PTH-mediated up-regulation of VDR and reduced the basal VDR levels as well [115]. EGF has been shown to increase intestinal calcium absorption in suckling rats [116]. Administration of EGF to rat pups influenced the changes in intestine associated with development including increased VDR concentration [117]. EGF increased the levels of VDR in the proximal segments of the small intestine and those of the vitamin D-dependent calcium binding protein calbindin-D_{9K} [117]. Additional changes in intestinal VDR, as related to growth and development, are discussed in detail below in Section VI.

The implications of VDR regulation by serum and growth factors are important in several spheres. On an experimental level, when VDR abundance is investigated in cell culture systems, the proliferation status of the cells under study as well as the time interval after medium (containing serum and growth factors) replenishment must be taken into account. The ability of cells to respond to $1,25(OH)_2D$ is limited by the level of VDR, which is influenced by the time interval between hormone and serum addition [118]. On a mechanistic level, it is not yet clear why changes in VDR may be a part of the early response of cells to serum and other mitogenic stimuli. Because $1,25(OH)_2D$ mediates the inhibition of growth and stimulates differentiation of many cells, perhaps an elevation in VDR levels before cell division provides a mechanism by which the $1,25(OH)_2D$ in the cellular environment might direct the dividing cells down a pathway toward cellular differentiation. Further studies in this area are necessary to test this hypothesis and shed light on the role of VDR regulation in the $1,25(OH)_2D$-mediated inhibition of cell growth.

2. PTH AND PTHrP

Parathyroid hormone (PTH) is a major calcium regulating hormone which modulates the synthesis of $1,25(OH)_2D$. As discussed in this section, PTH may act on selected vitamin D targets, such as bone and kidney cells, to affect $1,25(OH)_2D_3$ action by regulating VDR abundance. PTHrP, a hormone involved in the pathogenesis of humoral hypercalcemia of malignancy, is known to bind to and activate PTH receptors in target cells. PTH regulates the renal production of $1,25(OH)_2D$, and $1,25(OH)_2D$, in turn, has a feedback effect on PTH production and regulates VDR in the parathyroid glands. Chapters 23 and 52 include detailed discussions of VDR regulation in the parathyroid glands. Receptors for both PTH/PTHrP and $1,25(OH)_2D_3$ are found in osteoblasts and kidney. Several *in vivo* and *in vitro* studies provide evidence for the interaction between the two hormones in their target cells including osteoblasts [68,119,120].

The hormones PTH, PTHrP, and $1,25(OH)_2D_3$ can also interact at the level of receptor regulation. $1,25(OH)_2D_3$ has been shown to modulate the concentrations of PTH/PTHrP receptors in bone cells [120], and this section deals with the regulation of VDR abundance by PTH and PTHrP. At the cellular level, as depicted in Fig. 2, the biological actions of PTH are mediated through the activation of the two major signal transduction systems, PKA and PKC [121], both of which have been shown to be potent regulators of VDR abundance in target cells (see Section III,B above). In addition, PTH causes an increase in intracellular calcium levels in target cells [122,123], and calcium is another modulator of VDR levels (Section III,B). As these second messenger systems have differential effects on VDR regulation, several groups have studied the net effect of PTH signaling and have attempted to analyze the effects of PTH on VDR in terms of the individual signal transduction pathways activated by PTH.

Pols *et al.* [92] observed that PTH and PTHrP increased the VDR content in UMR-106 rat osteosarcoma cells, which was preceded by an increase in cAMP production in these cells. Further evaluation revealed that PTH caused a time- and dose-dependent increase in VDR mRNA in these cells, and this up-regulation was also elicited by activators of the cAMP–PKA pathway [93]. In contrast, Reinhardt and Horst [19] demonstrated that PTH and PTHrP produced a significant down-regulation of $1,25(OH)_2D_3$ binding and VDR mRNA levels in ROS 17/2.8 rat osteosarcoma cells. PTH also inhibited the $1,25(OH)_2D_3$-mediated up-regulation of VDR in these cells. Although both of these studies used rat osteosarcoma cells as models for the osteoblastic phenotype, there appears to be intrinsic differences between the cell lines. For example, in the degree of differentiation of the osteoblastic phenotype that they represent, the ROS 17/2.8 cells are osteoblastic whereas the UMR-106 cells are preosteoblastic [124–127]. Thus, cell-specific factors seem to differentially influence the regulation of VDR gene expression by hormones such as PTH as well as other activators as illustrated in Table II.

Reinhardt and Horst [19] expanded their *in vitro* observations in ROS 17/2.8 cells by performing *in vivo* experiments in which rats were infused for 5 days with $1,25(OH)_2D_3$. This treatment increased the level of expression of intestinal VDR, kidney VDR, and kidney 24-hydroxylase. Coinfusion of PTH along with $1,25(OH)_2D_3$ completely inhibited the $1,25(OH)_2D_3$-mediated increases in intestinal VDR and kidney 24-hydroxylase and caused more than a 50% reduction in the $1,25(OH)_2D_3$-mediated increase in kidney VDR levels. This study provides an explanation for the lack of up-regulation of VDR in the intestine and kidney due to increases in serum levels of $1,25(OH)_2D$ elicited by dietary calcium deficiency in rats in contrast to the response to exogenous administration of $1,25(OH)_2D_3$ [128]. In the case of dietary calcium deficiency, increases in plasma PTH would precede increases in $1,25(OH)_2D$, which would then inhibit $1,25(OH)_2D$-mediated VDR up-regulation in target tissues.

Another study performed in ROS 17/2.8 cells [95] attempted to elucidate the stucture–function relationship of PTH in the regulation of VDR. This study concluded that the first two amino acids of PTH appeared to be important in the stimulation of cAMP production and that PTH fragments containing this adenylate cyclase stimulating domain (PTH 1–34 and 1–31) were responsible for the PTH-mediated down-regulation of VDR in these cells. PTH fragments 3–34 or 13–34, which would retain PKC stimulating activity but not PKA activity, did not decrease the VDR content. However, Klaus *et al.* [129] showed an up-regulation in the VDR levels of subconfluent rat chondrocyte cultures by PTH as well as activators of PKC, and not by dibutyryl cAMP, suggesting that PKC activation plays a major role in PTH action to modulate VDR in these cells.

Both PTH and PTHrP have been shown by Krishnan *et al.* [94] to regulate VDR concentration in the mouse osteogenic cell line MC3T3-E1, established from newborn mouse calvaria [130]. These cells represent the preosteoblast phenotype and can be made to differentiate into osteoblasts in culture by a number of agents [130]. PTH/PTHrP treatment resulted in an up-regulation of VDR abundance in MC3T3-E1 cells as well as in UMR-106-01 rat osteosarcoma cells (Table II). Krishnan *et al.* [94] further concluded that the activation of PKA appeared to play a predominant role in the PTH-mediated VDR regulation because of the following three observations. First, inhibition of PKC by staurosporine did not further enhance the PTH-mediated up-regulation of VDR in MC3T3-E1 cells. Second, in UMR4-7 cells, a subclone of UMR-106-01 cells that expressed a mutant form of the type 1 regulatory subunit of PKA and, therefore, were cAMP resistant, VDR up-regulation by PTH was substantially abolished. Third, PTH pretreatment augmented a $1,25(OH)_2D_3$-mediated response in these cells, namely, the induction of the enzyme 24-hydroxylase. Thus, PTH may sensitize target cells to $1,25(OH)_2D$ through VDR up-regulation. van Leeuwen *et al.* [131] have observed that PTH treatment sensitized bone to $1,25(OH)_2D_3$-induced resorption. Although extrapolation from experiments in cultured cells must be made with caution, these data may provide a rational explanation for some clinical observations. An augmented response to $1,25(OH)_2D$ in hyperparathyroidism may contribute to hypercalcemia in these patients. Treatment of hypocalcemia in hypoparathyroid patients often requires higher than expected doses of $1,25(OH)_2D$ [132], possibly due to a relative resistance to $1,25(OH)_2D$ in the absence of PTH.

Because PTH is a hormone capable of activating multiple signaling pathways in target cells, the differential activation of these second messenger systems by PTH might provide a mechanism for tissue- and cell-specific variations in VDR regulation. However, the regulation appears to be even more complex and dependent on the state of differentiation of the cells under study. Activation of the same second messenger pathway has opposite effects on VDR regulation in different target cells exhibiting varying degrees of differentiation (Table II). For example, the activation of the cAMP pathway by PTH in MC3T3-E1 and UMR-106 cells with preosteoblastic features causes VDR up-regulation, whereas in ROS 17/2.8 cells, which exhibit a more mature osteoblastic phenotype, PKA activation by PTH decreases VDR abundance. This suggests that the differences in VDR gene regulation between target cells lies distal to the level of second messenger activation. In the case of PTH, for instance, one might speculate that the specificity may be conferred at the level of DNA binding proteins activated by the second messenger system, such as, for example, between the different isoforms of cAMP response element binding proteins (CREBs). Further studies to understand these effects require the availability of the VDR gene promoter sequences.

IV. VDR REGULATION BY DIETARY CALCIUM AND PHOSPHATE

Studies on the *in vivo* regulation of VDR and the role it plays in the physiological regulation of intestinal calcium transport must take into account several factors, particularly serum levels of $1,25(OH)_2D$, PTH, and calcium. These factors are, however, influenced by dietary intake of calcium and phosphate, so that changes in the diet may act indirectly through changes in the known regulators. For example, low levels of dietary calcium increase the serum $1,25(OH)_2D$ concentration [133] resulting from an increase in the production of PTH. Similarly, low phosphate diets are known to increase the endogenous production of $1,25(OH)_2D$, albeit through a PTH-independent mechanism [134]. Therefore, the concentrations of calcium and phosphate in the diet are important as indirect modulators of VDR abundance and vitamin D action.

Favus *et al.* [135] examined the homologous up-regulation of intestinal VDR *in vivo* during dietary calcium restriction in male Sherman rats. Their findings showed that a low calcium diet produced significant increases in duodenal and jejunal VDR concentrations. This up-regulation appeared to result primarily from posttranslational processes that decreased receptor degradation. In contrast, studies by Goff *et al.* [128] in Holtzman Sprague-Dawley rats revealed that increases in endogenously produced $1,25(OH)_2D$, induced by a low calcium diet, did not up-regulate duodenal VDR in these animals. This discrepancy might have resulted from the different strains of rats used. The Sherman rats used in the Favus *et al.* study had lower intestinal VDR levels, and the dietary calcium restriction in these animals achieved smaller increases in serum $1,25(OH)_2D$ levels than reported in other studies [128,133]. The Holtzman Sprague-Dawley rats used by Goff *et al.* had much higher levels of intestinal VDR, and up-regulation was not achieved by dietary calcium restriction. Although exogenously administered $1,25(OH)_2D_3$ achieved some degree of VDR up-regulation in the intestine (<2-fold), it was lower than that reported in the Favus *et al.* study (3-fold). To explain the lack of VDR up-regulation by endogenously produced $1,25(OH)_2D$ upon dietary calcium restriction, Goff *et al.* [128] raised the possibility that dietary calcium restriction might induce some other factor, such as PTH, that inhibited the $1,25(OH)_2D$-mediated increase in VDR. Reinhardt and Horst [19] validated this hypothesis by showing that coinfusion of PTH along with $1,25(OH)_2D_3$ inhibited the $1,25(OH)_2D_3$-mediated up-regulation of VDR in the intestine of male Holtzman rats. A study of VDR regulation in the chick [136] arrived at a similar conclusion, namely, that reduced dietary calcium diminished intestinal VDR levels despite increases in circulating $1,25(OH)_2D$. These data suggest that another factor, such as PTH, is a predominant down-regulator of VDR.

Other investigators [18,137,138] have examined the role of serum calcium in the homologous up-regulation of VDR. These studies showed that normal levels of serum calcium were necessary for $1,25(OH)_2D_3$-induced up-regulation of kidney VDR. An increase of serum calcium alone, in the absence of exogenous vitamin D, caused a 2-fold increase in kidney VDR [18]. $1,25(OH)_2D_3$ administration appeared to produce a rapid increase in kidney VDR, and the increase was independent of serum calcium or phosphate levels. However, adequate levels of serum calcium were necessary for a sustained effect by $1,25(OH)_2D_3$ [137]. In avian parathyroid glands, increases in dietary calcium alone produced a modest (60%) increase in VDR mRNA, and vitamin D treatment, when combined with a high level of dietary calcium, resulted in a substantial (6- to 8-fold) increase in VDR mRNA [139]. Brown *et al.* [39] also found that increases in dietary calcium produced a dose-dependent increase in VDR mRNA in rat parathyroid glands.

Dietary phosphate restriction has been shown to up-regulate intestinal VDR concentrations in rats [140] and chicks [136], probably due to an increase in the endogenous production of $1,25(OH)_2D$. Sriussadaporn *et al.* [141] have shown the dietary phosphate-mediated VDR regulation to be tissue specific. Acute restriction of phosphate in the diet caused a 3.5-fold increase in intestinal VDR, which followed an increase in VDR mRNA. Kidney and splenic monocyte/macrophage levels of VDR showed no change. Chronic phosphate restriction appeared to produce alterations in VDR metabolism that were consistent with the prolongation of receptor half-life [141].

In summary, decreases in dietary intake of calcium and phosphate appear to regulate VDR levels in target tissues owing, at least in part, to increases in endogenous synthesis of PTH and $1,25(OH)_2D$. The elevated circulating $1,25(OH)_2D$ levels appear to augment the biological responses to $1,25(OH)_2D$ by homologous up-regulation of VDR. The role of PTH to modulate this up-regulation is complex, and it depended on the cell system being studied, as a number of *in vitro* studies found that PTH either up-regulated VDR, and sensitized target cells to $1,25(OH)_2D_3$ actions, or caused a down-regulation of VDR and inhibited the homologous up-regulation (see Section III,C,2 for a discussion). However, a number of *in vivo* studies suggest that PTH attenuates the homologous up-regulation of VDR.

V. VDR LEVELS DURING CELL GROWTH AND DIFFERENTIATION

A number of investigators in carrying out *in vitro* studies on VDR regulation in cell culture model systems have observed that VDR abundance appears to be dependent on the proliferation status and/or degree of confluence of the cells in culture. Section III,C presents a detailed discussion of the correlation between VDR regulation and the rate of proliferation of the cells under study. In addition to its chemical role as a regulator of calcium homeostasis, $1,25(OH)_2D$ has also been shown to function as a modulator of cell growth and differentiation in a variety of normal and malignant cells (Chapter 64). Not surprisingly, therefore, in several cell culture systems, $1,25(OH)_2D$ exerts antiproliferative and/or differentiating effects that may be associated with changes in the abundance of VDR [1,11,33,142]. This section presents an analysis of studies on VDR abundance as it relates to cell growth and differentiation and attempts to interpret these changes in terms of vitamin D function in these cells.

The rate of cell division is usually linked to the state of cellular differentiation, and it is not always possible to distinguish whether alterations in VDR abundance reflect changes in the proliferation rate or the differentiation state. When nondividing peripheral blood lymphocytes, which lack VDR, are stimulated by mitogenic agents, they are blast transformed and respond to this complex signal, among many other changes, with the appearance of VDR [41,143]. Other examples of relatively undifferentiated, rapidly dividing cells with higher VDR levels compared to nondividing and/or more differentiated cells with lower levels of VDR, include HL-60 promyelocytic leukemia cells [10,41], malignant melanoma cells [9], rat incisors [144], human keratinocytes [11], and human colon carcinoma cells [33,142,145]. Chan and Atkins [55], in studying the distribution of VDR in rat jejunum, observed that the jejunal crypt cells, which appeared poorly differentiated but were rapidly dividing, had higher levels of VDR than the more differentiated, nondividing villous cells. Some of the differences in VDR regulation observed in different rat and mouse osteoblasts (Table II) may be the result of the different stages of differentiation of the cells under study. For example, ROS 25/1 cells are fibroblast-like, the UMR-106 and MC3T3-E1 cells are preosteoblastic, and ROS 17/2.8 cells exhibit a mature osteoblast phenotype [124–127,130].

In studies on human colon cancer cells [33,142,145] and human keratinocytes [11], $1,25(OH)_2D_3$ itself has been shown to be a powerful agent mediating the transition of cells from a rapidly proliferating, nondifferentiated phenotype with relatively high VDR levels to a less proliferating, more differentiated phenotype exhibiting relatively low VDR levels. Interestingly, the study by Zhao and Feldman [33] demonstrated that the more differentiated population of HT-29 human colon cancer cells, which still had half the abundance of VDR seen in the proliferating cell population, failed to exhibit biological responses to $1,25(OH)_2D_3$. These cells, in contrast to the proliferating population, did not exhibit homologous up-regulation of VDR, nor did they induce 24-hydroxylase in response to $1,25(OH)_2D_3$ treatment. The mechanism underlying the unresponsiveness to $1,25(OH)_2D_3$ in the more differentiated cells with diminished levels of VDR is not clear. It may be due to changes associated with the differentiated phenotype rather than merely due to decreased VDR abundance. This phenomenon, involving an altered phenotype after postproliferative maturation of cells, has been well described during the differentiation of osteoblasts in culture [146]. Although the pattern that emerges from the studies thus far reported indicates that the most differentiated cells have lower VDR abundance, this is not always the case. For example, in studies of cloned human skeletal muscle cells, the differentiated myotubes exhibited an equivalent abundance of VDR and equal responsiveness to $1,25(OH)_2D_3$ as the undifferentiated myoblasts [147].

In several cell culture systems, sodium butyrate treatment has been shown to cause hyperacetylation of histones [148], inhibition of cell growth, and alterations in the patterns of gene regulation by steroid and thyroid hormones as well as in the regulation of their receptors [149–151]. A few studies have investigated the effect of sodium butyrate on VDR abundance in cultured cells. Maiyar and Norman found that sodium butyrate increased VDR levels about 4-fold in primary cultures of chick renal epithelial cells and in a macrophage cell line, but VDR levels were not altered in other chick cell lines [152]. Costa and Feldman [153] found that sodium butyrate inhibited the proliferation of LLC-PK_1 pig kidney cells and caused a 50% decrease in VDR concentration in these cells as well. In sodium butyrate-treated cells, the induction of 24-hydroxylase by higher doses of $1,25(OH)_2D_3$ was diminished, and this paralleled the reduction in receptor abundance. At lower doses of the hormone, however, the treated cells appeared to be more active than the cells not exposed to sodium butyrate. Sodium butyrate treatment also abolished the homologous up-regulation of VDR in these cells. The understanding of the mechanism for this effect is not clear. However, the data from this experiment and others (lack of homologous up-regulation in differentiated HT-29 colon cancer cells [33], HVDRR fibroblasts that still bind hormone but fail to up-regulate [20], and the many cell types that do not exhibit homologous up-regulation [40–42] suggest that the occupancy of the VDR by ligand and the resultant stabilization do not provide a sufficient explanation for the phenomenon of homologous up-regulation.

Because VDR regulation appears to exhibit a correlation with the proliferation status of the cells under study, one would expect some differences in VDR regulation between malignant and nonmalignant cells. Malignant cells in culture do not exhibit the strict growth control mechanisms present in nonmalignant cells. Lee *et al.* [154], in their studies on retinoic acid regulation of VDR levels in rat bone cells, observed differences in response patterns between the tumorigenic ROS 17/2A and UMR-106M cells and nontumorigenic RCJ 1-20 cells, all of which exhibited features of the osteoblastic phenotype in culture. Petkovich *et al.* [14] showed that ATRA increased VDR levels in rat osteosarcoma cells, whereas Chen and Feldman [13] found that ATRA decreased the VDR content of primary cultures of rat calvarial osteoblast-like cells. In their studies on growth factor regulation of VDR levels, Krishnan and Feldman [8] observed differences between the magnitude of responses of the nonmalignant NIH-3T3 mouse fibroblasts and MCF-7 breast cancer cells. The magnitude of the growth factor-mediated increase in VDR was more pronounced in the NIH-3T3 cells. Also, in the MCF-7 cancer cells, unlike the fibroblasts, any single growth factor could produce effects on growth and VDR levels equivalent to serum, which contains many growth factors. Thus, there appear to be qualitative and quantitative differences in VDR regulation between malignant and nonmalignant cells in culture, and these may well be due to differences in the growth control mechanisms operating in these cells. In conclusion, the responses of a target cell to a hormone appear to be dependent on its state of differentiation at the time of hormone exposure, and changes in the differentiation state would affect different hormonal responses in different ways.

VI. VDR CHANGES DURING DEVELOPMENT AND AGING

Many investigators have observed that VDR levels change during various life stages. These changes in VDR levels determine when during development a target cell attains the ability to respond to $1,25(OH)_2D$. This section explores the studies evaluating VDR levels during development and aging and analyzes the role of VDR regulation in calcium homeostasis during various stages of life, such as pregnancy, parturition, lactation, menopause, and aging.

A. Changes in VDR During Neonatal Development

The VDR is detectable during neonatal development at different times in different tissues. Halloran and DeLuca [155] and Massaro *et al.* [70] showed that intestinal VDR gradually increased from low levels at 7–14 days postpartum to approximate adult levels at 21–28 days of age in the suckling rat. This gradual increase during the first few weeks of life explains the lack of responsiveness to $1,25(OH)_2D_3$ in the early postpartum period, and it is the determining factor in the initiation of active calcium absorption in response to $1,25(OH)_2D_3$. Interestingly, glucocorticoid administration during a critical window of time, on days 15–17, could induce VDR prematurely, whereas adrenalectomy delayed the rise in VDR [70]. This shows the important role of glucocorticoids in the developmental appearance of VDR. These findings, based on ligand binding assays, were subsequently confirmed by immunoblotting [156] and measuring mRNA levels [157]. In a more recent analysis of these phenomena, Lee *et al.* [158] concurred with the role of glucocorticoids to modulate the appearance of VDR mRNA but concluded that the glucocorticoid effect was not specific to VDR, but rather a general stimulation of differentiation. On the other hand, homologous regulation of VDR by $1,25(OH)_2D_3$ during the same critical postnatal period was found to be a VDR-specific effect.

Using immunocytochemistry, Johnson *et al.* [159] studied the ontogeny of the VDR in fetal rat bone. VDR epitopes were detected in mesenchyme that formed skeletal tissues on day 13 of gestation in the vertebral column and limbs, on day 17 in calvaria, and on day 19 within osteoblasts in the calvaria. The presence of VDR so early in development led the investigators to speculate that $1,25(OH)_2D_3$ may play a role in differentiating mesenchymal precursors into bone. The situation in bone is therefore much different than in intestine: the intestine is unresponsive to $1,25(OH)_2D_3$ until several weeks postpartum, while fetal rat bone cells are clearly responsive to $1,25(OH)_2D_3$ [4]. VDR was also detected in the skin of fetal limbs at all gestational ages [159].

The same authors examined VDR ontogeny in rat and mouse kidney [160] and noted that both species exhibited a similar pattern of development. VDR was detected as early as fetal day 15 in specific areas of the developing nephron, predominantly in the proximal and distal tubules. Calbindin-D_{28K} appeared later and was distributed differently than the VDR.

Berdal *et al.* [144] studied the distribution and stage-specific expression of VDR and calbindin genes in the developing rat incisor. Immunoreactivity for VDR was present in all progenitor cells and progressively decreased during the differentiation process. Homologous up-regulation was restricted to hard tissue forming cells, the ameloblasts and odontoblasts. Calbindin-D_{9K} mRNA was regulated by $1,25(OH)_2D_3$ in these cells. These changes suggest an important role for $1,25(OH)_2D_3$ in the hormonal control of tooth genes during development (see Chapter 27).

B. Tissue Distribution and VDR Changes During Life Stages in the Adult

1. Distribution of VDR in the Intestine

The VDR is not distributed equally along the adult rat intestine; rather, it is expressed at highest levels in the duodenum and gradually decreases in the distal areas of the small intestine [161]. Although most calcium absorption occurs in the proximal small intestine, VDR is present throughout the bowel including the colon [162]. At any cross-sectional level, the concentration is much higher in the crypts and decreases as the cells migrate up into the villi [55].

2. Effect of Pregnancy and Lactation on VDR Abundance

Breast appears to be a $1,25(OH)_2D_3$ target tissue, and the presence of VDR in the breast has been demonstrated [163,164]. VDR has also been found in most breast cancer samples that have been examined [165]. $1,25(OH)_2D_3$ increases alkaline phosphatase activity in breast cell cultures [166]. Changes in alkaline phosphatase activity have been shown to correlate with casein production [164], and presumably vitamin D is important in maintaining the calcium concentration of milk. Increases in VDR levels have been observed during pregnancy and lactation in the rat mammary gland [164] and cow colon [167]. At parturition, VDR levels decrease precipitously in cow colon, which has been considered to be a possible contributing factor in the development of milk fever, a condition of severe hypocalcemia in dairy cows [168]. However, a subsequent study has shown that colon VDR levels are not different in cows with milk fever and those that do not have milk fever [167].

3. VDR Levels in Hypercalciuria

In considering disease susceptibility due to genetically determined levels of intestinal VDR, Li *et al.* studied idiopathic hypercalciuria [169]. In these experiments, the investigators bred rats for hypercalciuria over 10 generations. Ultimately, they developed a strain of rats with substantially increased rates of urinary calcium excretion. There was a doubling of intestinal VDR in hypercalciuric rats compared to normal littermates. Presumably, the elevated levels of VDR led to increased calcium absorption and subsequent increases in urinary calcium excretion. The authors postulate that this is the first genetic disease to be described that is due to an increased concentration of VDR. It is not clear whether humans with idiopathic hypercalciuria and a greater tendency toward renal stone formation harbor increased levels of intestinal VDR.

4. VDR Changes in Aging and Osteoporosis

As osteoporosis is associated with the postmenopausal period and with decreased intestinal calcium absorption, an interesting hypothesis involves the effect of aging or menopause on intestinal VDR content as a contributing factor in the development of osteoporosis. Horst and colleagues have shown that VDR levels in rat intestine [170] and kidney [171] decrease with advancing age. Similarly, Liang *et al.* reported a 23% decrease in the expression of VDR mRNA in aged rats compared to adult rats [172]. A comparable decrease in calbindin-D_{9K} mRNA was also noted [172]. M. A. Hirst and D. Feldman (unpublished data) found that the concentration of rat intestinal VDR fell by approximately 25% in chronically ovariectomized, aged rats compared to age-matched controls. In an important study of human subjects, Ebeling *et al.* [173] determined VDR levels using an immunoradiometric assay of duodenal biopsy specimens from 35 female volunteers. They found a statistically significant decrease in VDR with age. Despite an age-related increase in PTH and $1,25(OH)_2D_3$ levels, the fractional calcium absorption did not change with age. Ebeling *et al.* concluded that there was an age-related impairment in intestinal responsiveness to $1,25(OH)_2D_3$ [173]. They speculated that this defect led to a compensatory rise in PTH levels and in $1,25(OH)_2D_3$ production to maintain normal calcium absorption and serum calcium levels at the expense of increased bone loss. The observations of McKane *et al.* [174] on the role of calcium intake in age-related increases in parathyroid function and bone resorption support this hypothesis.

Thus, the presence of VDR determines when in development target cells attain the ability to respond to $1,25(OH)_2D_3$. Regulation of VDR concentration presumably has physiological relevance during pregnancy, lactation, menopause, and aging and may contribute to disease susceptibility as exemplified by milk fever in cows and hypercalciuria and menopausal osteoporosis in humans.

VII. CONCLUDING COMMENTS

The regulation of VDR abundance plays an important role in determining the magnitude of the target cell response to $1,25(OH)_2D$. A variety of factors influence VDR levels in target cells including $1,25(OH)_2D$ itself, other vitamin D metabolites, hormones such as glucocorticoids, estrogens, retinoids, growth factors, activators of specific intracellular second messenger systems, and other growth and developmental signals. There appear to be pronounced species-specific as well as tissue- and cell-specific variations in VDR regulation. The mechanisms contributing to these variations are not entirely clear and could possibly be dependent on several factors. Such factors could include the state of differentiation of target cells, differences in the intracellular signaling pathways activated in these cells by the regulator, and differences in the interaction with the downstream nuclear proteins involved in the regulation of gene transcription in various target cells. Further studies are necessary to unravel the mechanisms that confer such specificities and the role they play in modulating $1,25(OH)_2D$ actions. The availability of the 5′ upstream sequences of the VDR gene will be helpful in understanding the mechanisms underlying these complex regulatory events. VDR regulation is closely related to the rate of cell proliferation, and the role this phenomenon plays in the antiproliferative and/or differentiation-promoting actions of $1,25(OH)_2D$ warrants further investigation. The VDR gene thus represents an interesting and important model to study the complex interactions between the various mechanisms that regulate gene transcription.

Acknowledgment

The authors thank T. Ross Eccleshall for critical reading of the manuscript.

References

1. Chen TL, Feldman D 1981 Regulation of 1,25-dihydroxyvitamin D_3 receptors in cultured mouse bone cells. Correlation of receptor concentration with the rate of cell division. J Biol Chem **256**:5561–5566.
2. Walters MR, Rosen DM, Norman AW, Luben RA 1982 1,25-Dihydroxyvitamin D receptors in an established bone cell line. Correlation with biochemical responses. J Biol Chem **257**:7481–7484.
3. Dokoh S, Donaldson CA, Haussler MR 1984 Influence of 1,25-dihydroxyvitamin D_3 on cultured osteogenic sarcoma cells: Correlation with the 1,25-dihydroxyvitamin D_3 receptor. Cancer Res **44**:2103–2109.
4. Chen TL, Hauschka PV, Cabrales S, Feldman D 1986 The effects of 1,25-dihydroxyvitamin D_3 and dexamethasone on rat osteoblast-like primary cell cultures: Receptor occupancy and functional expression patterns for three different bioresponses. Endocrinology **118**:250–259.
5. Krishnan AV, Feldman D 1992 Cyclic adenosine 3′,5′-monophosphate up-regulates 1,25-dihydroxyvitamin D_3 receptor gene expression and enhances hormone action. Mol Endocrinol **6**:198–206.
6. Costa EM, Feldman D 1986 Homologous up-regulation of the 1,25-dihydroxyvitamin D_3 receptor in rats. Biochem Biophys Res Commun **137**:742–747.
7. Hirst M, Feldman D 1982 Glucocorticoid regulation of $1,25(OH)_2$ vitamin D_3 receptors: Divergent effects on mouse and rat intestine. Endocrinology **111**:1400–1402.
8. Krishnan AV, Feldman D 1991 Stimulation of 1,25-dihydroxyvitamin D_3 receptor gene expression in cultured cells by serum and growth factors. J Bone Miner Res **6**:1099–1107.
9. Colston K, Colston MJ, Feldman D 1981 1,25-Dihydroxyvitamin D_3 and malignant melanoma: The presence of receptors and inhibition of cell growth in culture. Endocrinology **108**:1083–1086.
10. Tanaka H, Abe E, Miyaura C, Kuribayashi T, Konno K, Nishii Y, Suda T 1982 1,25-Dihydroxycholecalciferol and a human myeloid leukemia cell line (HL-60). Biochem J **204**:713–719.
11. Pillai S, Bikle DD, Elias PM 1988 1,25-Dihydroxyvitamin D production and receptor binding in human keratinocytes varies with differentiation. J Biol Chem **263**:5390–5395.
12. Pols HAP, Birkenhager JC, Foekens JA, van Leeuwen JPM 1990 Vitamin D: A modulator of cell proliferation and differentiation. J Steroid Biochem Mol Biol **37**:873–876.
13. Chen TL, Feldman D 1985 Retinoic acid modulation of $1,25(OH)_2D_3$ receptors and bioresponse in bone cells: Species differences between rat and mouse. Biochem Biophys Res Commun **132**:74–80.
14. Petkovich PM, Heersche JNM, Tinker DO, Jones G 1984 Retinoic acid stimulates 1,25-dihydroxyvitamin D_3 binding in rat osteosarcoma cells. J Biol Chem **259**:8274–8280.
15. Hseih J-C, Jurutka PW, Nakajima S, Galligan MA, Haussler CA, Shimizu Y, Shimizu N, Whitefield GK, Haussler MR 1993 Phosphorylation of the human vitamin D receptor by protein kinase C: Biochemical and functional evolution of the serine 51 recognition site. J Biol Chem **268**:15118–15126.
16. Arbour NC, Prahl JM, DeLuca HF 1993 Stabilization of the vitamin D receptor in rat osteosarcoma cells through the action of 1,25-dihydroxyvitamin D_3. Mol Endocrinol **7**:1307–1312.
17. Iida K, Shinki T, Yamaguchi A, DeLuca HF, Kurokawa K, Suda T 1995 A possible role of vitamin D receptors in regulating vitamin D activation in the kidney. Proc Natl Acad Sci USA **92**:6112–6116.
18. Sandgren M, DeLuca HF 1990 Serum calcium and vitamin D regulate 1,25-dihydroxyvitamin D_3 receptor concentration in rat kidney in vivo. Proc Natl Acad Sci USA **87**:4312–4314.
19. Reinhardt TA, Horst RL 1990 Parathyroid hormone down-regulates 1,25-dihydroxyvitamin D_3 receptors (VDR) and VDR mes-

senger ribonucleic acid in vitro and blocks homologous up-regulation in vivo. Endocrinology **127**:942–948.
20. Costa EM, Hirst MA, Feldman D 1985 Regulation of 1,25-dihydroxyvitamin D_3 receptors by vitamin D analogs in cultured mammalian cells. Endocrinology **117**:2203–2210.
21. Eisman JA, Sher E, Suva LJ, Frampton RJ, McLean FL 1984 1 alpha,25-Dihydroxyvitamin D_3 specifically induces its own metabolism in a human cancer cell line. Endocrinology **114**:1225–1231.
22. Sher E, Frampton RJ, Eisman JA 1985 Regulation of the 1,25-dihydroxyvitamin D_3 receptor by 1,25-dihydroxyvitamin D_3 in intact human cancer cells. Endocrinology **116**:971–977.
23. Pan LC, Price PA 1987 Ligand-dependent regulation of the 1,25-dihydroxyvitamin D_3 receptor in rat osteosarcoma cells. J Biol Chem **262**:4670–4675.
24. Pols HA, Schilte HP, Visser TJ, Birkenhager JC 1987 Effect of ketoconazole on metabolism and binding of 1,25-dihydroxyvitamin D_3 by intact rat osteogenic sarcoma cells. Biochim Biophys Acta **931**:115–119.
25. Reinhardt TA, Horst RL 1989 Ketoconazole inhibits self-induced metabolism of 1,25-dihydroxyvitamin D_3 and amplifies 1,25-dihydroxyvitamin D_3 receptor up-regulation in rat osteosarcoma cells. Arch Biochem Biophys **272**:459–465.
26. Mahonen A, Maenpaa PH 1994 Steroid hormone modulation of vitamin D receptor levels in human MG-63 osteosarcoma cells. Biochem Biophys Res Commun **205**:1179–1186.
27. Reinhardt TA, Horst RL 1989 Self-induction of 1,25-dihydroxyvitamin D_3 metabolism limits receptor occupancy and target tissue responsiveness. J Biol Chem **264**:15917–15921.
28. McDonnell DP, Mangelsdorf DJ, Pike JW, Haussler MR, O'Malley BW 1987 Molecular cloning of complementary DNA encoding the avian receptor for vitamin D. Science **235**:1214–1217.
29. Mangelsdorf DJ, Pike JW, Haussler MR 1987 Avian and mammalian receptor for 1,25-dihydroxyvitamin D_3: *In vitro* translation to characterize size and hormone-dependent regulation. Proc Natl Acad Sci USA **84**:354–358.
30. Kamei Y, Kawada T, Kazuki R, Ono T, Kato S, Sugimoto E 1993 Vitamin D receptor gene expression is up-regulated by 1,25-dihydroxyvitamin D_3 in 3T3-L1 preadipocytes. Biochem Biophys Res Commun **193**:948–955.
31. Lee Y, Inaba M, DeLuca HF, Mellon WS 1989 Immunological identification of 1,25-dihydroxyvitamin D_3 receptors in human promyelocytic leukemic cells (HL-60) during homologous regulation. J Biol Chem **264**:13701–13705.
32. Goto H, Chen KS, Prahl JM, DeLuca HF 1992 A single receptor identical with that from intestine/T47D cells mediates the actions of 1,25-dihydroxyvitamin D_3 in HL-60 cells. Biochim Biophys Acta **1132**:103–108.
33. Zhao X, Feldman D 1993 Regulation of vitamin D receptor abundance and responsiveness during differentiation of HT-29 human colon cancer cells. Endocrinology **132**:1808–1814.
34. Santiso-Mere D, Sone T, Hilliard GMI, Pike JW, McDonnell DP 1993 Positive regulation of the vitamin D receptor by its cognate ligand in heterologous expression systems. Mol Endocrinol **7**:833–839.
35. Strom M, Sandgren ME, Brown TA, DeLuca HF 1989 1,25-Dihydroxyvitamin D_3 up-regulates the 1,25-dihydroxyvitamin D_3 receptor in vivo. Proc Natl Acad Sci USA **86**:9770–9773.
36. Beckman MJ, Horst RL, Reinhardt TA, Beitz DC 1990 Up-regulation of the intestinal 1,25-dihydroxyvitamin D receptor during hypervitaminosis D: A comparison between vitamin D_2 and vitamin D_3. Biochem Biophys Res Commun **169**:910–915.
37. Nakajima S, Yamaoka K, Okada S, Pike JW, Seino Y, Haussler MR 1992 1,25-Dihydroxyvitamin D_3 does not up-regulate vitamin D receptor messenger ribonucleic acid levels in hypophosphatemic mice. Bone Miner **19**:201–213.
38. Naito Y, Goff JP, Horst RL, Reinhardt TA 1989 Effects of continuous administration of 1,25-dihydroxyvitamin D_3 on plasma minerals and unoccupied colon mucosal 1,25-dihydroxyvitamin D_3 receptor concentrations. J Dairy Sci **72**:2936–2941.
39. Brown AJ, Zhong M, Finch J, Ritter C, Slatopolsky E 1995 The roles of calcium and 1,25-dihydroxyvitamin D_3 in the regulation of vitamin D receptor expression by rat parathyroid glands. Endocrinology **136**:1419–1425.
40. Hulla W, Kallay E, Krugluger W, Peterlik M, Cross HS 1995 Growth control of human colon-adenocarcinoma-derived Caco-2 cells by vitamin D compounds and extracellular calcium in vitro: Relation to c-myc oncogene and vitamin-D-receptor expression. Int J Cancer **62**:711–716.
41. Kizaki M, Norman AW, Bishop JE, Lin CW, Karmakar A, Koeffler HP 1991 1,25-Dihydroxyvitamin D_3 receptor RNA: Expression in hematopoietic cells. Blood **77**:1238–1247.
42. Song L 1996 Demonstration of vitamin D receptor expression in a human megakaryoblastic leukemia cell line: Regulation of vitamin D receptor mRNA expression and responsiveness by forskolin. J Steroid Biochem Mol Biol **57**:265–274.
43. Pols HA, Birkenhager JC, Schilte JP, Visser TJ 1988 Evidence that the self-induced metabolism of 1,25-dihydroxyvitamin D_3 limits the homologous up-regulation of its receptor in rat osteosarcoma cells. Biochim Biophys Acta **970**:122–129.
44. Loose DS, Kan PB, Hirst MA, Marcus RA, Feldman D 1983 Ketaconazole blocks adrenal steroidogenesis by inhibiting cytochrome P450-dependent enzymes. J Clin Invest **71**:1495–1499.
45. Henry HL 1985 Effect of ketoconazole and miconazole on 25-hydroxyvitamin D_3 metabolism by cultured chick kidney cells. J Steroid Biochem **23**:991–994.
46. Costa EM, Feldman D 1987 Measurement of 1,25-dihydroxyvitamin D_3 receptor turnover by dense amino acid labeling: Changes during receptor up-regulation by vitamin D metabolites. Endocrinology **120**:1173–1178.
47. Wiese RJ, Uhland-Smith A, Ross TK, Prahl JM, DeLuca HF 1992 Up-regulation of the vitamin D receptor in response to 1,25-dihydroxyvitamin D_3 results from ligand-induced stabilization. J Biol Chem **267**:20082–20086.
48. Reinhardt TA, Horst RL 1994 Phorbol 12-myristate 13-acetate and 1,25-dihydroxyvitamin D_3 regulate 1,25-dihydroxyvitamin D_3 receptors synergistically in rat osteosarcoma cells. Mol Cell Endocrinol **101**:159–165.
49. Huang YC, Lee S, Gabrielides C, Pansini-Porta A, Bruns ME, Bruns DE, Miffin TE, Pike JW, Christakos S 1989 Effect of hormones and development on the expression of the rat 1,25-dihydroxyvitamin D_3 receptor gene. Comparison with calbindin gene expression. J Biol Chem **264**:17454–17461.
50. Feldman D, Krishnan AV 1987 Glucocorticoid effects on calcium metabolism and bone in the development of osteopenia. In: Christiansen C, Johansen JS, Riis BJ (eds) Osteoporosis (Proceedings of the International Symposium on Osteoporosis, Denmark, 1987), Vol. 2. Norhaven, Viborg, Denmark, pp. 1006–1013.
51. Kimberg DV, Baerg RD, Gershon E, Grandusius RT 1971 Effect of cortisone treatment on the active transport of calcium by the small intestine. J Clin Invest **50**:1309–1321.
52. Favus MJ, Walling MW, Kimberg DV 1973 Effects 1,25-dihydroxycholecalciferol on intestinal calcium transport in cortisone-treated rats. J Clin Invest **52**:1680–1685.
53. Hahn TJ 1980 Drug-induced disorders of vitamin D and mineral metabolism. Clin Endocrinol Metab **9**:107–127.
54. Hirst MA, Feldman D 1982 Glucocorticoids down-regulate the

number of 1,25-dihydroxyvitamin D_3 receptors in mouse intestine. Biochem Biophys Res Commun **105**:1590–1596.
55. Chan SD, Atkins D 1984 The temporal distribution of the 1 alpha,25-dihydroxycholecalciferol receptor in the rat jejunal villus. Clin Sci **67**:285–290.
56. Gensure RC, Fish JM, Walters MR 1993 Dexamethasone down-regulates vitamin D receptors in rat kidney, unmasking a high affinity binding site. Biochem Biophys Res Commun **195**:1139–1144.
57. Korkor AB, Kuchibotla J, Arrieh M, Gray RW, Gleason WAJ 1985 The effects of chronic prednisone administration on intestinal receptors for 1,25-dihydroxyvitamin D_3 in the dog. Endocrinology **117**:2267–2273.
58. Wilhelm F, Norman AW 1985 Influence of triamcinolone, estradiol-17β and testosterone on 1,25-dihydroxyvitamin D_3 binding performances to its chick intestinal receptor. J Steroid Biochem **23**:913–918.
59. Manolagas SC, Anderson DC, Lumb GA 1979 Glucocorticoids regulate the concentration of 1,25-dihydroxycholecalciferol receptors in bone. Nature **227**:314–315.
60. Chen TL, Cone CM, Morey-Holton E, Feldman D 1982 Glucocorticoid regulation of 1,25$(OH)_2$-vitamin D_3 receptors in cultured mouse bone cells. J Biol Chem **257**:13564–13569.
61. Chen TL, Cone CM, Feldman D 1983 Effects of 1 alpha,25-dihydroxyvitamin D_3 and glucocorticoids on the growth of rat and mouse osteoblast-like bone cells. Calcif Tissue Int **35**:806–811.
62. Chen TL, Cone CM, Morey-Holton E, Feldman D 1983 1,25-Dihydroxyvitamin D_3 receptors in cultured rat osteoblast-like cells. Glucocorticoid treatment increases receptor content. J Biol Chem **258**:4350–4355.
63. Manolagas SC, Abare J, Deftos LJ 1984 Glucocorticoids increase the 1,25$(OH)_2D_3$ receptor concentration in rat osteogenic sarcoma cells. Calcif Tissue Int **36**:153–157.
64. Nielsen HK, Eriksen EF, Storm T, Mosekilde L 1988 The effects of short-term, high-dose treatment with prednisone on the nuclear uptake of 1,25-dihydroxyvitamin D_3 in monocytes from normal human subjects. Metabolism **37**:109–114.
65. Godschalk M, Levy JR, Downs Jr RW 1992 Glucocorticoids decrease vitamin D receptor number and gene expression in human osteosarcoma cells. J Bone Miner Res **7**:21–27.
66. Wong GL, Luben RA, Cohn DV 1977 1,25-Dihydroxycholecalciferol and parathormone: Effects on isolated osteoclast-like and osteoblast-like cells. Science **197**:663–665.
67. Ng B, Hekkelman JW, Heersche JNM 1979 The effect of cortisol on the adenosine 3′,5′-monophosphate response to parathyroid hormone of bone in vitro. Endocrinology **104**:1130–1135.
68. Chen TL, Feldman D 1984 Modulation of PTH-stimulated cAMP in cultured rodent bone cells: The effect of 1,25$(OH)_2$vitamin D_3 and its interaction with glucocorticoids. Calcif Tissue Int **36**:580–585.
69. Catherwood BD 1985 1,25-Dihydroxycholecalciferol and glucocorticoid regulation of adenylate cyclase in an osteoblast-like cells. J Biol Chem **260**:736–743.
70. Massaro ER, Simpson RU, DeLuca HF 1983 Glucocorticoids and appearance of 1,25-dihydroxyvitamin D_3 receptor in rat intestine. Am J Physiol **244**:E230–E235.
71. Eriksen EF, Colvard DS, Berg NJ, Graham ML, Mann KG, Spelsberg TC, Riggs BL 1988 Evidence of estrogen receptors in normal human osteoblast-like cells. Science **241**:84–86.
72. Komm BS, Terpening CM, Benz DJ, Graeme KA, Gallegos A, Korc M, Greene GL, O'Malley BW, Haussler MR 1988 Estrogen binding, receptor mRNA, and biologic response in osteoblast-like osteosarcoma cells. Science **241**:81–84.
73. Manolagas SC, Jilka RL 1995 Bone marrow, cytokines, and bone remodelling. Emerging insights into the pathophysiology of osteoporosis. N Engl J Med **332**:305–311.
74. Gallagher JC, Riggs BL, Eisman J, Hamstra A, Arnaud SB, DeLuca HF 1979 Intestinal calcium absorption and serum vitamin D metabolites in normal subjects and osteoporotic patients: Effect of age and dietary calcium J Clin Invest **64**:729–736.
75. Gallagher JC, Riggs BL, DeLuca HF 1980 Effect of estrogen on calcium absorption and serum vitamin D metabolites in postmenopausal osteoporosis. J Clin Endocrinol Metab **51**:1359–1364.
76. Navickis RJ, Dial OK, Katzenellenbogen BS, Nalbandov AV 1979 Effects of gonadal hormones on calcium-binding protein in chick duodenum. Am J Physiol **237**:E409–E417.
77. Walters MR 1981 An estrogen-stimulated 1,25-dihydroxyvitamin D_3 receptor in the rat uterus. Biochem Biophys Res Commun **103**:721–726.
78. Levy J, Zuili I, Yankowitz N, Shany S 1984 Induction of cytosolic receptors for 1,25-dihydroxyvitamin D_3 in the immature rat uterus by oestradiol. J Endocrinol **100**:265–269.
79. Thomas ML, Xu X, Norfleet AM, Watson CS 1993 The presence of functional estrogen receptors in intestinal epithelial cells. Endocrinology **132**:426–430.
80. Arjmandi BH, Salih MA, Herbert DC, Sims SH, Kalu DN 1993 Evidence for estrogen receptor-linked calcium transport in the intestine. Bone Miner **21**:63–74.
81. Chan SDH, Chiu DKH, Atkins D 1984 Oophorectomy leads to a selective decrease in 1,25-dihydroxycholecalciferol receptors in rat jejunal villous cells. Clin Sci **66**:745–748.
82. Duncan WE, Glass AR, Wray HL 1991 Estrogen regulation of the nuclear 1,25-dihydroxyvitamin D_3 receptor in rat liver and kidney. Endocrinology **129**:2318–2324.
83. Liel Y, Kruas S, Levy J, Shany S 1992 Evidence that estrogens modulate activity and increase the number of 1,25-dihydroxyvitamin D_3 receptors in osteoblast-like cells (ROS 17/2.8). Endocrinology **130**:2597–2601.
84. Ishibe M, Nojima T, Ishibashi T, Koda T, Kaneda K, Rosier RN, Puzas JE 1995 17β-Estradiol increases the receptor number and modulates the actions of 1,25-dihydroxyvitamin D_3 in human osteosarcoma-derived osteoblast-like cells. Calcif Tissue Int **57**:430–435.
85. Davoodi F, Brenner RV, Evans SR, Schumaker LM, Nauta RJ, Buras RR 1995 Modulation of vitamin D receptor and estrogen receptor by 1,25$(OH)_2$-vitamin D_3 in T-47D human breast cancer cells. J Steroid Biochem Mol Biol **54**:147–153.
86. Mahonen A, Pirskanen A, Maenpaa PH 1991 Homologous and heterologous regulation of 1,25-dihydroxyvitamin D_3 receptor mRNA levels in human osteosarcoma cells. Biochim Biophys Acta **1088**:111–118.
87. Krishnan AV, Feldman D 1991 Activation of protein kinase-C inhibits vitamin D receptor gene expression. Mol Endocrinol **5**:605–612.
88. van Leeuwen JPTM, Birkenhager JC, Buurman CJ, van den Bemd GJCM, Pols HAP 1992 Bidirectional regulation of the 1,25-dihydroxyvitamin D_3 receptor by phorbol ester-activated protein kinase C in osteoblast-like cells: Interaction with adenosine 3′,5′-monophosphate-induced up-regulation of the 1,25-dihydroxyvitamin D_3 receptor. Endocrinology **130**:2259–2266.
89. Merke J, Milde P, Lewicka S, Hugel U, Klaus G, Mangelsdorf DJ, Haussler MR, Rauterberg EW, Ritz E 1989 Identification and regulation of 1,25-dihydroxyvitamin D_3 receptor activity and biosynthesis of 1,25-dihydroxyvitamin D_3. Studies in cultured bovine aortic endothelial cells and human dermal capillaries. J Clin Invest **83**:1903–1915.

90. Martell RE, Simpson RV, Taylor JM 1987 1,25-Dihydroxyvitamin D_3 regulation of phorbol ester receptors in HL-60 leukemia cells. J Biol Chem **262**:5570–5575.
91. Ways DK, Dodd RC, Bennett TE, Gray TK, Earp HS 1987 1,25-Dihydroxyvitamin D_3 enhances phorbol ester-stimulated differentiation and protein kinase C-dependent substrate phosphorylation activity in the U937 human monoblastoid cell. Endocrinology **121**:1654–1660.
92. Pols HAP, van Leeuwen JPTM, Schilte JP, Visser TJ, Birkenhager JC 1988 Heterologous up-regulation of the 1,25-dihydroxyvitamin D_3 receptor by parathyroid hormone (PTH) and PTH-like peptide in osteoblast-like cells. Biochem Biophys Res Commun **156**:588–594.
93. van Leeuwen JPTM, Birkenhager JC, Vin-van Wijngaarden T, van der Bemd GJCM, Pols HAP 1992 Regulation of 1,25-dihydroxyvitamin D_3 receptor gene expression by parathyroid hormone and cAMP-agonists. Biochem Biophys Res Commun **185**:881–886.
94. Krishnan AV, Cramer SD, Bringhurst RF, Feldman D 1995 Regulation of 1,25-dihydroxyvitamin D_3 receptors by parathyroid hormone in osteoblastic cells: Role of second messenger pathways. Endocrinology **136**:705–712.
95. Sriussadaporn S, Wong MS, Whitfield JF, Tembe V, Favus MJ 1995 Structure–function relationship of human parathyroid hormone in the regulation of vitamin D receptor gene expression in osteoblast-like cells (ROS 17/2.8). Endocrinology **136**:3735–3742.
96. Angel P, Imagawa M, Chiu R, Stein B, Imbra RJ, Rahnsdorf HJ, Jonat C, Herrlich P, Karin M 1987 Phorbol ester inducible genes contain a common 'cis' element recognized by a TPA-modulated 'trans'-acting factor. Cell **49**:729–739.
97. Montminty MR, Bilezikjian LM 1987 Binding of a nuclear protein to the cyclic-AMP response element of the somatostatin gene. Nature **328**:175–178.
98. Desai SS, Appel MC, Baran DT 1986 The differential effect of $1,25(OH)_2D_3$ on cytosolic calcium in two human cell lines (HL-60 and U-937). J Bone Miner Res **1**:497–501.
99. MacLaughlin JA, Cantley LC, Chawala SB, Holick MF 1987 $1,25(OH)_2D_3$ increases intracellular calcium in human keratinocytes by stimulating phosphatidylinositol turnover. J Bone Miner Res **2**(Suppl. 1):76A.
100. Civitelli R, Kim YS, Gunsten SL, Fujimori A, Huskey M, Avioli LV, Hruska K 1990 Nongenomic activation of the calcium message system by vitamin D metabolites in osteoblast-like cells. Endocrinology **127**:2253–2262.
101. Berridge MJ 1984 Inositol triphosphate and diacylglycerol as second messengers. Biochem J **220**:345–360.
102. van Leeuwen JP, Birkenhager JC, Scilte JP, Buurman CJ, Pols HA 1990 Role of calcium and cAMP in heterologous up-regulation of the 1,25-dihydroxyvitamin D_3 receptor in an osteoblast cell line. Cell Calcium **11**:281–289.
103. Canalis E, McCarthy TL, Centrella M 1989 The role of growth factors in skeletal modelling. Endocrinol Metab Clin North Am **18**:903–918.
104. Noda M, Rodan GA 1986 Type-beta transforming growth factor inhibits proliferation and expression of alkaline phosphatase in murine osteoblast-like cells. Biochem Biophys Res Commun **140**:56–65.
105. Centrella M, McCarthy TL, Canalis E 1987 Mitogenesis in fetal rat bone cells simultaneously exposed to type-beta transforming growth factor and other growth regulators. FASEB J **1**:312–317.
106. Robey PG, Young MF, Flanders KC, Roche NS, Kondiah P, Reddi AH, Termine JD, Sporn MB, Roberts MB 1987 Osteoblasts synthesize and respond to transforming growth factor-beta (TGF-β) *in vitro*. J Cell Biol **105**:457–463.
107. Shinar DM, Rodan GA 1990 Biphasic effects of transforming growth factor-β on the production of osteoclast-like cells in mouse bone marrow cultures: The role of prostaglandins in the generation of these cells. Endocrinology **126**:3153–3158.
108. Tashjian AHJ, Levine L 1978 Epidermal growth factor stimulates prostaglandin production and bone resorption in cultured mouse calvaria. Biochem Biophys Res Commun **85**:966–975.
109. Raisz LG, Simmons HA, Sandberg AL, Canalis E 1980 Direct stimulation of bone resorption by epidermal growth factor. Endocrinology **107**:270–273.
110. Treisman R 1986 Identification of a protein binding site that mediates transcriptional response of the c-*fos* gene to serum factors. Cell **46**:567–574.
111. Greenberg ME, Siegfried Z, Ziff EB 1987 Mutation of the c-*fos* gene dyad symmetry element inhibits serum inducibility of transcription in vivo and the nuclear regulatory factor binding in vitro. Mol Cell Biol **1**:1217–1225.
112. Mezzetti G, Barbiroli B, Oka T 1987 1,25-Dihydroxycholecalciferol receptor regulation in hormonally induced differentiation of mouse mammary glands in culture. Endocrinology **120**:2488–2493.
113. Schneider HG, Michelangeli VP, Frampton RJ, Grogan JL, Ikeda K, Martin TJ, Findlay DM 1992 Transforming growth factor-beta modulates receptor binding of calciotropic hormones in osteoblast-like cells. Endocrinology **131**:1383–1389.
114. Staal A, Birkenhager JC, Pols HA, Buurman CJ, Vink-van Wijngaarden T, Kleinekoort WM, van den Bemd GJ, van Leeuwen JP 1994 Transforming growth factor beta-induced dissociation between vitamin D receptor level and 1,25-dihydroxyvitamin D_3 action in osteoblast-like cells. Bone Miner **26**:27–42.
115. van Leeuwen JP, Pols HA, Schilte JP, Visser TJ, Birkenhager JC 1991 Modulation by epidermal growth factor of the basal $1,25(OH)_2D_3$ receptor level and the heterologous up-regulation of the $1,25(OH)_2D_3$ receptor in clonal osteoblast-like cells. Calcif Tissue Int **49**:35–42.
116. Oka Y, Ghishan FK, Greene HL, Orth DN 1983 Effect of mouse epidermal growth factor/urogastrone on the functional maturation of rat intestine. Endocrinology **112**:940–944.
117. Bruns DE, Krishnan AV, Feldman D, Gray RW, Christakos S, Hirsch GN, Bruns ME 1989 Epidermal growth factor increases intestinal calbindin-D_{9k} and 1,25-dihydroxyvitamin D receptors in neonatal rats. Endocrinology **125**:478–485.
118. Hirst M, Feldman D 1983 Regulation of $1,25(OH)_2D_3$ receptor content in cultured LLC-PK_1 kidney cells limits hormonal responsiveness. Biochem Biophys Res Commun **116**:121–127.
119. Titus L, Rubin JE, Nanes MS, Catherwood BD 1989 Glucocorticoid and 1,25-dihydroxyvitamin D modulate the degree of adenosine 3′,5′-monophosphate-dependent protein kinase isoenzyme I and II activation by parathyroid hormone in rat osteosarcoma cells. Endocrinology **125**:2806–2811.
120. Titus L, Jackson E, Nanes MS, Rubin JE, Catherwood BD 1991 1,25-Dihydroxyvitamin D reduces parathyroid hormone receptor number in ROS 17/2.8 cells and prevents the glucocorticoid-induced increase in these receptors: Relationship to adenylate cyclase activation. J Bone Miner Res **6**:631–637.
121. Fitzpatrick LA, Coleman DT, Bilezikian JP 1992 The target tissue actions of parathyroid hormone. In: Coe FL, Favus MJ (eds) Disorders of Bone and Mineral Metabolism. Raven, New York, pp. 123–148.
122. Babich M, Choi H, Johnson RM, King KL, Alford GE, Nissenson RA 1991 Thrombin and parathyroid hormone mobilize intracel-

lular calcium in rat osteosarcoma cells by distinct pathways. Endocrinology **129**:1463–1470.

123. Bidwell JP, Carter WB, Fryer MJ, Heath H 1991 Parathyroid hormone (PTH)-induced intracellular Ca^{2+} signalling in naive and PTH-desensitized osteoblast-like cells (ROS 17/2.8): Pharmacological characterization and evidence for synchronous oscillation of intracellular Ca^{2+}. Endocrinology **129**:2993–3000.
124. Majeska RJ, Rodan SB, Rodan GA 1980 Parathyroid hormone-responsive clonal lines from rat osteosarcoma. Endocrinology **107**:1494–1503.
125. Partridge NC, Alcorn D, Michelangeli VP, Ryan G, Martin TJ 1983 Morphological and biochemical characterization of four clonal osteogenic sarcoma cell lines of rat origin. Cancer Res **43**:4308–4314.
126. Williams GR, Bland R, Sheppard MC 1994 Characterization of thyroid hormone (T_3) receptors in three osteosarcoma cell lines of distinct osteoblast phenotype: Interactions among T_3, vitamin D_3, and retinoid signalling. Endocrinology **135**:2375–2385.
127. Williams GR, Bland R, Sheppard MC 1995 Retinoids modify regulation of endogenous gene expression by vitamin D_3 and thyroid hormone in three osteosarcoma cell lines. Endocrinology **136**:4304–4314.
128. Goff JP, Reinhardt TA, Beckman MJ, Horst RL 1990 Contrasting effects of exogenous 1,25-dihydroxyvitamin D [$1,25(OH)_2D$] versus endogenous $1,25(OH)_2D$ induced by dietary calcium restriction, on vitamin D receptors. Endocrinology **126**:1031–1035.
129. Klaus G, von Eichel B, May T, Hugel U, Mayer H, Ritz E, Mehls O 1994 Synergistic effects of parathyroid hormone and 1,25-dihydroxyvitamin D_3 on proliferation and vitamin D receptor expression of rat growth cartilage cells. Endocrinology **135**:1307–1315.
130. Sudo H, Kodama H, Amagai Y, Yamamoto S, Kasai S 1983 In vitro differentiation and calcification in a new clonal osteogenic cell line derived from newborn mouse calvaria. J Cell Biol **96**:191–198.
131. van Leeuwen JPTM, Birkenhager JC, Bos MP, van der Bemd GJCM, Hermann-Erlee MPM, Pols HAP 1992 Parathyroid hormone sensitizes long bones to the stimulation of bone resorption by 1,25-dihydroxyvitamin D_3. J Bone Miner Res **7**:303–309.
132. Chesney RA 1984 Resistance to 1,25-dihydroxyvitamin D in hypoparathyroidism: An enigma. NY State J Med **84**:226–227.
133. Langman CB, Favus MJ, Bushinsky DA, Coe FL 1985 Effects of dietary calcium restriction on 1,25-dihydroxyvitamin D_3 net synthesis by rat proximal tubules. J Lab Clin Med **106**:286–292.
134. Hughes MR, Brumbaugh PF, Haussler MR 1975 Regulation of serum 1α,25-dihydroxyvitamin D_3 by calcium and phosphate in the rat. Science **190**:578–580.
135. Favus MJ, Mangelsdorf DJ, Tembe V, Coe BJ, Haussler MR 1988 Evidence for in vivo upregulation of the intestinal vitamin D receptor during dietary calcium restriction in the rat. J Clin Invest **8**:218–224.
136. Meyer J, Fullmer CS, Wasserman RH, Komm B, Haussler MR 1992 Dietary restriction of calcium, phosphorus and vitamin D elicits differential regulation of the mRNAs for avian intestinal calbindin-D_{28K} and the 1,25-dihydroxyvitamin D_3 receptor. J Bone Miner Res **7**:441–448.
137. Uhland-Smith A, DeLuca HF 1992 Short-term regulation of the renal vitamin D receptor in rats by 1,25-dihydroxycholecalciferol is calcium insensitive. J Nutr **122**:2316–2321.
138. Uhland-Smith A, DeLuca HF 1993 The necessity for calcium for increased renal vitamin D receptor in response to 1,25-dihydroxyvitamin D. Biochim Biphys Acta **1176**:321–326.
139. Russell J, Bar A, Sherwood LM, Hurwitz S 1993 Interaction between calcium and 1,25-dihydroxyvitamin D_3 in the regulation of preproparathyroid hormone and vitamin D receptor messenger ribonucleic acid in avian parathyroids. Endocrinology **132**:2639–2644.
140. Goff JP, Horst RL, Reinhardt TA 1990 Dietary phosphorus restriction upregulates intestinal but not renal $1,25(OH)_2$vitamin D receptor concentration in the rat. J Bone Miner Res **5**(Suppl. 2):S269 (abstract).
141. Sriussadaporn S, Wong M, Pike JW, Favus MJ 1995 Tissue specificity and mechanism of vitamin D receptor up-regulation during dietary phosphorus restriction in the rat. J Bone Miner Res **10**:271–280.
142. Brehier A, Thomasset M 1988 Human colon cell line HT-29: Characterisation of 1,25-dihydroxyvitamin D_3 receptor and induction of differentiation by the hormone. J Steroid Biochem **29**:265–270.
143. Manolagas SC, Yu XP, Girasole G, Bellido T 1994 Vitamin D and the hematolymphopoietic tissue: A 1994 update. Semin Nephrol **14**:129–143.
144. Berdal A, Hotton D, Pike JW, Mathieu H, Dupret JM 1993 Cell- and stage-specific expression of vitamin D receptor and calbindin genes in rat incisor: Regulation by 1,25-dihydroxyvitamin D_3. Dev Biol **155**:172–179.
145. Harper KD, Iozzo RV, Haddad JG 1989 Receptors for and bioresponses to 1,25-dihydroxyvitamin D in a human colon carcinoma cell line (HT-29). Metabolism **38**:1062–1069.
146. Owen TA, Bortell R, Shalhoub V, Heinrichs A, Stein JL, Stein GS, Lian JB 1993 Postproliferative transcription of the rat osteocalcin gene is reflected by vitamin D-responsive developmental modifications in protein–DNA interactions at basal and enhancer promoter elements. Proc Natl Acad Sci USA **90**:1503–1507.
147. Costa EM, Blau HM, Feldman D 1986 1,25-Dihydroxyvitamin D_3 receptors and hormonal responses in cloned human skeletal muscle cells. Endocrinology **119**:2214–2220.
148. Candido EPM, Reeves R, Davie JR 1978 Sodium butyrate inhibits histone deacetylation in cultured cells. Cell **14**:105–113.
149. Samuels HH, Stanley F, Casanova J, Shao TC 1980 Thyroid hormone nuclear receptor levels are influenced by the acetylation of chromatin-associated proteins. J Biol Chem **255**:2499–2508.
150. Kruh J 1982 Effects of sodium butyrate, a new pharmacological agent, on cells in culture. Mol Cell Biochem **42**:65–82.
151. Stevens MS, Zarrinta A, Moore MR 1984 Associated effects of sodium butyrate on histone acetylation and estrogen receptor in the human breast cancer cell line MCF-7. Biochem Biophys Res Commun **119**:132–138.
152. Maiyar AC, Norman AW 1992 Effects of sodium butyrate on 1,25-dihydroxyvitamin D_3 receptor activity in primary chick kidney cells. Mol Cell Endocrinol **84**:99–107.
153. Costa EM, Feldman D 1987 Modulation of 1,25-dihydroxyvitamin D_3 receptor binding and action by sodium butyrate in cultured pig kidney cells (LLC-PK1). J Bone Miner Res **2**:151–159.
154. Lee K, Petkovich M, Heersche JNM 1988 The effects of sodium butyrate on the retinoic acid-induced changes in 1,25-dihydroxyvitamin D_3 receptors in tumorigenic and nontumorigenic bone derived cells. Endocrinology **122**:2399–2406.
155. Halloran BP, DeLuca HF 1981 Appearance of intestinal cytosolic receptor for 1,25-dihydroxyvitamin D_3 during neonatal development in the rat. J Biol Chem **256**:7338–7342.
156. Pierce EA, DeLuca HF 1988 Regulation of 1,25-dihydroxyvitamin D_3 receptor during neonatal development in the rat. Arch Biochem Biophys **261**:241–249.
157. Burmster JK, Wiese RJ, Maeda N, DeLuca HF 1988 Structure

and regulation of the rat 1,25-dihydroxyvitamin D_3 receptor. Proc Natl Acad Sci USA **85**:9499–9502.

158. Lee S, Szlachetka M, Christakos S 1991 Effect of glucocorticoids and 1,25-dihydroxyvitamin D_3 on the developmental expression of the rat intestinal vitamin D receptor gene. Endocrinology **129**:396–401.
159. Johnson JA, Grande JP, Roche PC, Kumar R 1996 Ontogeny of the 1,25-dihydroxyvitamin D_3 receptor in fetal rat bone. J Bone Miner Res **11**:56–61.
160. Johnson JA, Grande JP, Roche PC, Sweeney WE, Avner ED, Kumar R 1995 1,25-Dihydroxyvitamin D_3 receptor ontogenesis in fetal renal development. Am J Physiol **269**:F419–F428.
161. Feldman D, McCain TA, Hirst MA, Chen TL, Colston KW 1979 Characterization of a cytoplasmic receptor-like binder for 1,25-dihydroxycholecalciferol in rat intestinal mucosa. J Biol Chem **254**:10378–10384.
162. Hirst MA, Feldman D 1981 1,25-Dihydroxyvitamin D_3 receptors in mouse colon. J Steroid Biochem **14**:315–319.
163. Eisman JA, Martin TJ, MacIntyre I 1980 Presence of 1,25-dihydroxy vitamin D receptor in normal and abnormal breast tissue. Prog Biochem Pharmacol **17**:143–150.
164. Colston KW, Berger U, Wilson P, Hadcocks L, Naeem I, Earl HM, Coombes RC 1988 Mammary gland 1,25-dihydroxyvitamin D_3 receptor content during pregnancy and lactation. Mol Cell Endocrinol **60**:15–22.
165. Berger U, McClelland RA, Wilson P, Greene GL, Haussler MR, Pike JW, Colston K, Easton D, Coombes RC 1991 Immunocytochemical determination of estrogen receptor, progesterone receptor, and 1,25-dihydroxyvitamin D_3 receptor in breast cancer and relationship to prognosis. Cancer Res **51**:239–244.
166. Mulkins MA, Sussman HH 1987 1,25-Dihydroxyvitamin D_3 exerts opposite effects on the regulation of human embryonic and nonembryonic alkaline phosphatase isoenzymes. Endocrinology **120**:416–425.
167. Goff JP, Reinhardt TA, Horst RL 1995 Milk fever and dietary cation–anion balance effects on concentration of vitamin D receptor in tissue of periparturient dairy cows. J Dairy Sci **78**:2388–2394.
168. Goff JP, Reinhardt TA, Horst RL 1991 Enzymes and factors controlling vitamin D metabolism and action in normal and milk fever cows. J Dairy Sci **74**:4022–4032.
169. Li XQ, Tembe V, Horwitz GM, Bushinsky DA, Favus MJ 1993 Increased intestinal vitamin D receptor in genetic hypercalciuric rats. A cause of intestinal calcium hyperabsorption. J Clin Invest **91**:661–667.
170. Horst RL, Goff JP, Reinhardt TA 1990 Advancing age results in the reduction of intestinal and bone 1,25-dihydroxyvitamin D receptor. Endocrinology **126**:1053–1057.
171. Koszewski NJ, Reinhardt TA, Beitz DC, Horst RL 1990 Developmental changes in rat kidney 1,25-dihydroxyvitamin D receptor. Biochem Biophys Res Commun **170**:65–72.
172. Liang CT, Barnes J, Imanaka S, DeLuca HF 1994 Alterations in mRNA expression of duodenal 1,25-dihydroxyvitamin D_3 receptor and vitamin D-dependent calcium binding protein in aged Wistar rats. Exp Gerentol **29**:179–186.
173. Ebeling PR, Sandgren ME, DiMagno EP, Lane AW, DeLuca HF, Riggs BL 1992 Evidence of an age-related decrease in intestinal responsiveness to vitamin D: Relationship between serum 1,25-dihydroxyvitamin D_3 and intestinal vitamin D receptor concentrations in normal women. J Clin Endocrinol Metab **75**:176–182.
174. McKane WR, Khosla S, Egan KS, Robins SP, Burritt MF, Riggs BL 1996 Role of calcium intake in modulating age-related increases in parathyroid function and bone resorption. J Clin Endocrinol Metab **81**:1699–1703.

CHAPTER 12

Vitamin D Regulation of Type I Collagen Synthesis and Gene Expression in Bone

BARBARA E. KREAM Department of Medicine, The University of Connecticut Health Center, Farmington, Connecticut

ALEXANDER C. LICHTLER Department of Pediatrics, The University of Connecticut Health Center, Farmington, Connecticut

I. INTRODUCTION

Vitamin D has multiple functions in humans and animals [1,2]. It is best known as a nutrient that is required for adequate growth and mineralization of bone. Vitamin D, like parathyroid hormone (PTH), is also an important calcium-regulating hormone. PTH is primarily responsible for the acute physiological maintenance of serum calcium levels. In the presence of prolonged hypocalcemia, PTH increases the renal production of 1,25-dihydroxyvitamin D_3 [1,25$(OH)_2D_3$], the active metabolite of vitamin D. 1,25$(OH)_2D_3$, together with PTH, normalizes serum calcium levels by increasing intestinal calcium transport, bone resorption, and renal calcium reabsorption. Vitamin D deficiency leads to undermineralized bone in children (rickets) and adults (osteomalacia), a reduction in bone matrix formation [3] and mineralization [4], and an alteration in the pattern of collagen cross-linking [5]. Calcium deficiency also decreases bone formation and mineralization in rats [6]. Defective bone formation and mineralization in vitamin D-deficient rats can be largely corrected by the administration of calcium, suggesting that the trophic effect of vitamin D on the skeleton may be due to its ability to stimulate intestinal calcium absorption [7–9].

The discovery of high affinity 1,25$(OH)_2D_3$ receptors (VDRs) in cytosolic extracts of embryonic chick and fetal rat calvariae in the late 1970s suggested that 1,25$(OH)_2D_3$ had a direct effect on bone cell function

[10,11]. One of the hallmarks of $1{,}25(OH)_2D_3$ action in bone is its ability to stimulate osteoclastic bone resorption [12]. It does so by enhancing the formation of osteoclasts and the activity of differentiated osteoclasts [13] (also see Chapter 21). Because VDRs are found in osteoblasts but not differentiated osteoclasts, $1{,}25(OH)_2D_3$ is thought to increase osteoclast formation and activity by a primary effect on cells of the osteoblast lineage [14] (see also Chapter 20).

Type I collagen is one of the genetic targets of $1{,}25(OH)_2D_3$ action in osteoblastic cells. Type I collagen is the most abundant protein in the body and comprises at least 90% of the organic component of the bone matrix. The biochemistry, molecular biology, and hormonal regulation of collagen genes have been extensively reviewed [15–19]. Type I collagen is produced at high levels by differentiated skeletal osteoblasts and is required for the formation of the mineralized bone matrix (Chapter 20). Many other cell types synthesize and secrete type I collagen, although in lesser amounts. Each type I collagen molecule consists of three polypeptides; two are $\alpha1$(I) chains, and one is an $\alpha2$(I) chain. These polypeptides are encoded by separate genes (COL1A1 and COL1A2, respectively) that are expressed in a 2:1 ratio [20].

Collagen synthesis in bone is modulated by a variety of hormones, growth factors, and cytokines, some of which are produced locally by osteoblastic cells [21–23]. Insulin [24], insulin-like growth factor-I [25], and transforming growth factor-β [26] increase type I collagen synthesis, whereas PTH [27], interleukin-1 [28], and tumor necrosis factor [28] are inhibitory. Glucocorticoids [29] and prostaglandins [30] can be either stimulatory or inhibitory *in vitro* depending on the model and culture conditions. The effects of $1{,}25(OH)_2D_3$ on bone collagen synthesis and our current knowledge of the molecular mechanisms by which $1{,}25(OH)_2D_3$ alters COL1A1 expression are described in this chapter. Because of the critical role of type I collagen in maintaining the structure and function of the skeleton, we believe that it is important to elucidate mechanisms by which $1{,}25(OH)_2D_3$ regulates type I collagen expression.

II. REGULATION OF BONE COLLAGEN SYNTHESIS

Collagen synthesis in organ cultures of rodent calvariae and cell cultures can be assessed using several different assays [31,32]. In the most widely used assay, calvariae and cells are incubated with [^{3}H]proline for several hours prior to the end of culture. The incorporation of [^{3}H]proline into collagenase-digestible protein (CDP labeling) and noncollagen protein (NCP labeling) is then measured in extracts of the cultures using highly purified bacterial collagenase [33]. The percent collagen synthesis is calculated from the CDP and NCP values after correcting for the greater abundance of proline in collagen relative to noncollagen proteins [34]. Collagen production can also be determined by measuring the hydroxyproline content of cell or organ cultures, because hydroxyproline is virtually unique to collagens. These methods do not distinguish between different types of fibrillar collagen. However, the collagen synthesized by bone organ cultures and most osteoblastic cell cultures is largely type I (>95%) so that the CDP labeling value usually reflects type I collagen synthesis. If necessary, the production of different collagen types can be distinguished by ion-exchange chromatography and polyacrylamide gel electrophoresis of radiolabeled extracts of cell or organ cultures [35,36]. Finally, the use of specific cDNA probes in Northern blot assays can be used to assess collagen mRNA expression in bone models. As discussed below, measurement of the effect of $1{,}25(OH)_2D_3$ on collagen synthesis and mRNA levels has given comparable results using these different assays.

The hormone $1{,}25(OH)_2D_3$ inhibited collagen synthesis in organ cultures of 21-day fetal rat calvariae [37] and neonatal mouse calvariae [38] with little or no effect on noncollagen protein synthesis. Maximal inhibition of collagen synthesis by $1{,}25(OH)_2D_3$ in rat calvariae (~50%) occurred at 10 n*M* [37]. $1{,}24R{,}25(OH)_3D_3$ also inhibited collagen synthesis but was less potent than $1{,}25(OH)_2D_3$ [37]. $25OHD_3$ and $24R{,}25(OH)_2D_3$ did not alter collagen synthesis at concentrations below 100 n*M* [37,38]. Vitamin D metabolites inhibited collagen synthesis and stimulated resorption of fetal rat long bones with similar relative potencies that correlated with the binding affinity of the metabolites for the skeletal VDRs [39]. To determine the cell selectivity of the $1{,}25(OH)_2D_3$ inhibition of collagen synthesis, organ cultures of fetal rat calvariae were treated with $1{,}25(OH)_2D_3$ for 22 hr and then radiolabeled with [^{3}H]proline for the final 2 hr of culture. The central bone (mature osteoblasts) was dissected free of the periosteum (less mature osteoprogenitors and fibroblasts), and both compartments were analyzed separately for the incorporation of [^{3}H]proline. $1{,}25(OH)_2D_3$ decreased collagen synthesis in the central bone but not the periosteum, indicating selectivity of the $1{,}25(OH)_2D_3$ effect for mature osteoblasts [40,41]. With an *in vivo* protocol in which neonatal rats were given multiple injections of [^{3}H]proline to radiolabel newly synthesized bone matrix, 25 ng of $1{,}25(OH)_2D_3$ given on days 1, 3, and 5 inhibited bone matrix synthesis as assessed by histomorphometric analysis of autoradiographs of tibia and calvariae [42].

In addition, $1,25(OH)_2D_3$ inhibited collagen production in rat osteoblastic osteosarcoma ROS 17/2.8 cells [43], primary rat [44,45] and mouse osteoblastic cells [46], and an immortalized murine osteoblastic cell line called MMB-1 [47]. $1,25(OH)_2D_3$ had a greater inhibitory effect on type I collagen synthesis in primary murine osteoblastic cells during log phase growth than at confluence, perhaps because proliferative cells contained more VDR [48]. Likewise, $1,25(OH)_2D_3$ inhibition of collagen synthesis was greater in sparse cultures of murine MMB-1 cells, which had higher VDR levels than confluent MMB-1 cells [49]. $1,25(OH)_2D_3$ inhibition of collagen synthesis was equivalent in sparse and confluent rat primary osteoblastic cells [44], whereas VDR number did not change during growth of the cells [50]. Taken together, these data showed that the extent of inhibition of collagen synthesis by $1,25(OH)_2D_3$ was probably a function of the cellular quantity of VDRs, although other factors probably contribute as well. $1,25(OH)_2D_3$ inhibited collagen mRNA levels during the proliferative phase of long-term cultures of rat primary osteoblastic cells [51] and prevented the formation of mineralized bone nodules by these cultures [51,52]. These studies showed that $1,25(OH)_2D_3$ inhibits the differentiation of osteoprogenitors that form the nodules in primary rat osteoblastic cell cultures [52]. However, $1,25(OH)_2D_3$ inhibition of mineralized nodule formation may have been due in part to the regression of type I collagen synthesis in the cultures.

In contrast to the inhibitory effects described above, $1,25(OH)_2D_3$ caused a transient stimulation of collagen and noncollagen protein synthesis (about 2-fold) that peaked at 12–24 hr in the immortalized murine osteoblastic cell line MC3T3-E1 [53]. In this study, the percent collagen synthesized by the cultures (collagen relative to total protein synthesis) was not reported; as a result, it was not possible to determine the selectivity of the $1,25(OH)_2D_3$ effect for collagen synthesis. However, $1,25(OH)_2D_3$ was shown to decrease the percent collagen synthesis in MC3T3-E1 cells in subsequent reports [54,55]. $1,25(OH)_2D_3$ also increased collagen synthesis in the human osteoblastic osteosarcoma cell line MG-63 [56] and primary cultures of human osteoblastic cells [57]. MC3T3-E1 and MG-63 are thought to be preosteoblastic cells that undergo *in vitro* osteogenic differentiation with ascorbic acid. Moreover, $1,25(OH)_2D_3$ inhibited cell growth and increased osteocalcin expression and alkaline phosphatase activity in both cell lines. Collectively, these data suggest that $1,25(OH)_2D_3$ may act as a differentiating hormone in immature cells of the osteoblast lineage, perhaps, by decreasing cell replication, resulting in increased type I collagen expression. In contrast, $1,25(OH)_2D_3$ may inhibit the differentiation of more mature osteoblastic cells and decreases type I collagen expression in mature osteoblastic cells (Fig. 1).

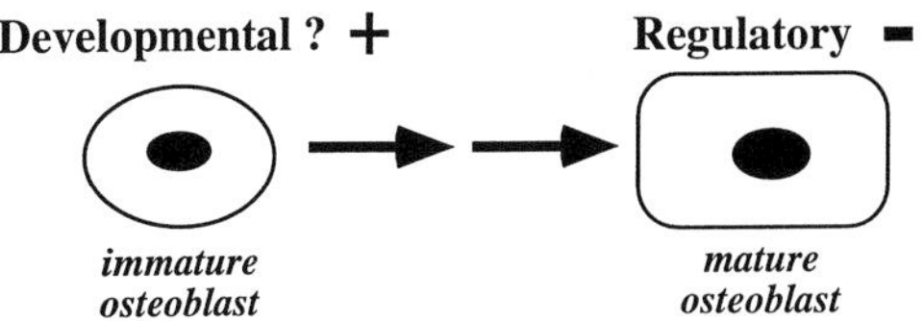

FIGURE 1 Differential effects of $1,25(OH)_2D_3$ on type I collagen expression in cells of the osteoblast lineage. $1,25(OH)_2D_3$ increases type I collagen expression in some osteoblastic cell lines. This may repesent an increase in the differentiation of immature cells of the osteoblast lineage (a development effect). In contrast, $1,25(OH)_2D_3$ inhibits differentiation of more mature osteoblastic cells and decreases type I collagen expression in mature osteoblastic cells (a regulatory effect).

III. MOLECULAR MECHANISMS OF REGULATION

With the development of assays to measure type I collagen mRNA levels (COL1A1 and COL1A2), it was shown that $1,25(OH)_2D_3$ repressed collagen synthesis in mature osteoblasts at a pretranslational level [40]. Measurements of procollagen mRNA activity by translation of total RNA in a reticulocyte lysate showed that $1,25(OH)_2D_3$ inhibited collagen mRNA in the osteoblast-rich central bone but not the periosteum of 21-day fetal rat calvariae [40]. $1,25(OH)_2D_3$ at 10 n*M* inhibited procollagen mRNA activity at 6 hr; maximal inhibition of about 50% occurred at 24 hr [40]. A single subcutaneous injection of $1,25(OH)_2D_3$ (1.6 ng/g body weight) also decreased procollagen mRNA activity in calvariae [40]. Subsequently, specific cDNA probes were used to show that $1,25(OH)_2D_3$ inhibited COL1A1 mRNA levels in ROS 17/2.8 cells [43] and in primary rat [45] and chick calvarial osteoblastic cells [58,59]. Nuclear run-on assays in ROS 17/2.8 cells demonstrated that $1,25(OH)_2D_3$ repressed COL1A1 and COL1A2 mRNA levels by a transcriptional mechanism [60]. $1,25(OH)_2D_3$ at 1 and 10 n*M* decreased the rate of COL1A1 and COL1A2 transcription by about 50%, similar to its effect on collagen synthesis and type I collagen mRNA levels, whereas actin and tubulin transcription were unaffected. $1,25(OH)_2D_3$ repressed COL1A1 and COL1A2 transcription as early as 4 hr, while maximal inhibition occurred at 24 hr [60].

DNA motifs that mediate stimulatory effects of $1,25(OH)_2D_3$ on gene expression have been well characterized for several genes; however, elements that mediate $1,25(OH)_2D_3$ repression of genes are only now being identified. This topic has been the subject of several reviews [61,62] and is discussed in Chapters 9 and 10. Vitamin D response elements (VDREs) that mediate $1,25(OH)_2D_3$ induction of target genes such as human [63] and rat [64] osteocalcin, mouse osteopontin [65], rat 24-hyroxylase [66], and rat calbindin-D_{9K} [67] contain two perfect or imperfect direct hexameric repeats of the consensus AGGTCA motif separated by three spacer nucleotides [61,62]. The consensus VDRE binds a heterodimer of the VDR and the retinoic acid X receptor (RXR) [68]. The negative VDRE in the avian PTH promoter is analogous to the consensus VDRE in that it contains two imperfect direct repeats separated by three spacer nucleotides and binds VDR and RXR [69]. In contrast, the negative VDRE in the human PTH gene contains a single AGGTTC motif; binding of the VDR to this site does not require RXR [70,71]. The negative VDRE of the parathyroid hormone-related peptide (PTHrP) gene contains two potential VDREs, one similar to the negative VDRE in the human PTH gene and another identical to the stimulatory VDRE; both motifs bind the VDR [72]. $1,25(OH)_2D_3$ repression of the interleukin-2 gene in Jurkat cells involves the displacement of an NFATp/AP-1 transcription factor complex with a VDR–RXR heterodimer at a NF-AT-1 promoter element [73].

To characterize the regions of the COL1A1 gene that are involved in its repression by $1,25(OH)_2D_3$, we produced a chimeric gene containing a fragment of the rat COL1A1 gene extending from −3518 to +116 bp fused to the chloramphenicol acetyltransferase (CAT) reporter gene termed ColCAT3.6 [74]. $1,25(OH)_2D_3$ inhibited ColCAT3.6 activity in transiently transfected ROS 17/2.8 cells by 50%, similar to its effect of the endogenous COL1A1 gene [74]. Subsequently, we produced a series of ColCAT constructs containing progressive 5′ promoter deletions of the COL1A1 promoter to map $1,25(OH)_2D_3$ response elements [75,76]. Analysis of multiple pools of stably transfected ROS 17/2.8 cells produced more consistent results than transient transfection assays. $1,25(OH)_2D_3$ inhibited a COL1A1 promoter fragment deleted to −2295 bp (ColCAT2.3) but did not affect a promoter fragment deleted to −1670 bp [76]. These experiments localized an inhibitory $1,25(OH)_2D_3$ element to a region of the COL1A1 promoter from −2295 to −1670 bp. Sequence analysis of the COL1A1 promoter revealed a site between −2240 and −2234 bp that had high sequence similarity to both the human and rat osteocalcin VDREs. We hypothesized that the VDR would bind to this motif and inhibit COL1A1 transcription. Electrophoretic mobility shift assays using VDR expressed in COS cells or by an adenovirus vector demonstrated that the VDR bound to this sequence *in vitro* [76]. However, deletion of the sequence between −2256 and −2216 bp from the ColCAT3.6 or ColCAT2.3 constructs did not affect the inhibitory effect of $1,25(OH)_2D_3$ on promoter activity [76]. Therefore, $1,25(OH)_2D_3$ does not repress COL1A1 transcription in ROS 17/2.8 cells solely via the −2240/−2234 bp site.

To determine the effect of $1,25(OH)_2D_3$ on COL1A1 promoter activity *in vivo*, we produced a series of transgenic mice lines harboring ColCAT constructs [77,78]. $1,25(OH)_2D_3$ at 10 n*M* inhibited ColCAT3.6 activity in organ cultures of 6- to 8-day-old transgenic mouse calvariae at 24–28 hr (A. Bedalov, R. Salvatori, and A. C. Lichtler, unpublished data). $1,25(OH)_2D_3$ also inhibited CAT mRNA as early as 3 hr; maximal inhibition of CAT mRNA (50%) was seen at 24 hr. The inhibition of CAT mRNA by $1,25(OH)_2D_3$ was not affected by cycloheximide, suggesting that new protein synthesis was not required for the effect. A series of COL1A1 promoter fragments deleted to −1719 bp were fully inhibited by $1,25(OH)_2D_3$; however, a COL1A1 promoter construct deleted to −1670 could not be analyzed because it did not have detectable baseline activity in transgenic mouse calvariae [79].

Taken together, the results in ROS 17/2.8 cells and transgenic calvariae described above suggest that the sequences necessary for $1,25(OH)_2D_3$ repression of the COL1A1 gene are located downstream of −1719 bp, or possibly within the −1719/−1670 bp region (Fig. 2). There are no good matches to consensus VDREs within this region, suggesting the following possibilities. First, $1,25(OH)_2D_3$ repression of COL1A1 may involve binding of the VDR to a site(s) distinct from known negative VDREs. As discussed above, negative VDREs have been identified in the human PTH [70,71], avian PTH [69], and rat PTHrP [72] promoters. Another possibility is that $1,25(OH)_2D_3$ inhibition of COL1A1 expression involves displacement of a stimulatory transcription factor(s) from its cognate DNA binding site, similar to the mechanism by which $1,25(OH)_2D_3$ inhibits the interleu-

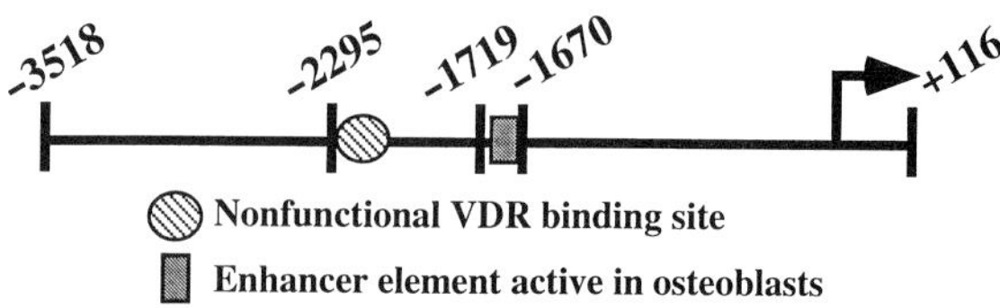

FIGURE 2 Functional analysis of the rat COL1A1 promoter. The rat COL1A1 promoter contains a strong enhancer element between −1683 and −1670 bp that confers high basal expression in osteoblasts. $1,25(OH)_2D_3$ inhibits COL1A1 expression in osteoblasts. The COL1A1 promoter contains a VDR binding site between −2240 and −2234 bp. However, mutational analysis demonstrated that this site is not functional in ROS 17/2.8 cells. Deletion of the COL1A1 promoter to −1719 bp did not eliminate $1,25(OH)_2D_3$ inhibition in transgenic mouse calvariae. However, deletion to −1670 bp blocked $1,25(OH)_2D_3$ inhibition in ROS 17/2.8 cells.

kin-2 gene [73]. It is also possible that $1,25(OH)_2D_3$ inhibition of COL1A1 involves interaction of the VDR with other transcription factors rather than binding of the VDR to DNA. Such a mechanism has been described for the inhibition of collagenase expression by glucocorticoids [80,81]. Finally, $1,25(OH)_2D_3$ repression of COL1A1 expression may be mediated by an alternative signal transduction pathway. In this regard, it has been suggested that some biological effects of $1,25(OH)_2D_3$ may be mediated by the protein kinase C (PKC) signaling pathway [62]. We have shown that stimulation of PKC with phorbol myristate acetate inhibits collagen synthesis in fetal rat calvariae [82] and ColCAT3.6 expression in transgenic mouse calvariae [83]. Therefore, one possibility is that $1,25(OH)_2D_3$ stimulation of the PKC pathway inhibits COL1A1 expression. Future experiments to identify $1,25(OH)_2D_3$ response elements in the COL1A1 gene will involve the analysis of constructs having site-directed mutations and internal deletions of the COL1A1 promoter downstream of −1719 bp in cultured osteoblastic cells and transgenic mice.

IV. CONCLUSIONS AND PERSPECTIVES

The effect of $1,25(OH)_2D_3$ (inhibitory or stimulatory) may depend in part on *in vitro* culture conditions such as cell density, the timing and concentration of $1,25(OH)_2D_3$ addition, the presence of ascorbic acid and the state of maturation of the model. Franceschi and co-workers proposed a model based on the premise that cells of the osteoblast lineage differ in their response to $1,25(OH)_2D_3$ depending on their state of maturation [84]. $1,25(OH)_2D_3$ stimulates osteoblast markers in immature osteoprogenitor cells (MC3T3-E1 and MG-63 cells) but inhibits these markers in mature osteoblasts (rodent calvarial organ cultures, primary rodent osteoblastic cell cultures, and ROS 17/2.8 cells) [84]. Such a model is consistent with the effects of $1,25(OH)_2D_3$ on bone remodeling during period of calcium and phosphate deficiency. When serum calcium and phosphate are low, PTH increases the synthesis of $1,25(OH)_2D_3$. Both hormones increase bone resorption to increase the supply of calcium and phosphate for soft tissues. During periods of mineral deficiency, it would be appropriate for $1,25(OH)_2D_3$ to repress collagen synthesis and inhibit the differentiation of late osteoprogenitors as a means of temporarily limiting new bone formation. Such an effect would prevent calcium and phosphate from being redeposited at sites of new osteoid formation. At the same time, $1,25(OH)_2D_3$ may stimulate the differentiation of early osteoprogenitors to form new osteoblasts that would initiate the phase of coupled formation [84]. *In vitro* experiments using cells at various stages of osteoblastic differentiation and *in vivo* experiments using *in situ* hybridization and histomorphometric analyses will be required to validate this model.

References

1. Raisz LG 1980 Direct effects of vitamin D and its metabolites on skeletal tissue. Clin Endocrinol Metab **9**:27–41.
2. DeLuca HF 1986 The metabolism and functions of vitamin D. Adv Exp Med Biol **196**:361–375.
3. Canas F, Brand JS, Neuman WF, Terepka AR 1969 Some effects of vitamin D_3 on collagen synthesis in rachitic chick cortical bone. Am J Physiol **216**:1092–1096.
4. Baylink D, Stauffer M, Wergedal J, Rich C 1970 Formation, mineralization and resorption of bone in vitamin D-deficient rats. J Clin Invest **49**:1122–1134.
5. Mechanic GL, Toverud SU, Ramp WK, Gonnerman WA 1975 The effect of vitamin D on the structural crosslinks and maturation of chick bone collagen. Biochim Biophys Acta **393**:419–425.
6. Stauffer M, Baylink D, Wergedal J, Rich C 1973 Decreased bone formation, mineralization, and enhanced resorption in calcium-deficient rats. Am J Physiol **225**:269–276.
7. Howard GA, Baylink DJ 1980 Matrix formation and osteoid maturation in vitamin D-deficient rats made normocalcemic by dietary means. Miner Electrolyte Metab **3**:44–50.
8. Underwood JL, DeLuca HF 1984 Vitamin D is not directly necessary for bone growth and mineralization. Am J Physiol **246**:E493–E498.
9. Weinstein RS, Underwood JL, Hutson MS, DeLuca HF 1984 Bone histomorphometry in vitamin D-deficient rats infused with calcium and phosphorus. Am J Physiol **246**:E499–E505.
10. Kream BE, Jose M, Yamada S, DeLuca HF 1977 A specific high-affinity binding macromolecule for 1,25-dihydroxyvitamin D_3 in fetal bone. Science **197**:1086–1088.
11. Manolagas SC, Taylor CM, Anderson DC 1979 Highly specific binding of 1,25-dihydroxycholecalciferol in bone. J Endocrinol **80**:35–39.
12. Raisz LG, Trummel CL, Schnoes HK, DeLuca HF 1972 1,25-Dihydroxycholecalciferol: A potent stimulator of bone resorption in tissue culture. Science **175**:768–769.
13. Suda T, Takahashi N, Abe E 1992 Role of vitamin D in bone resorption. J Cell Biochem **49**:53–58.
14. Suda T, Takahashi N, Martin TJ 1992 Modulation of osteoclast differentiation. Endocr Rev **13**:66–80.
15. Bornstein P, Sage H 1980 Structurally distinct collagen types. Annu Rev Biochem **49**:958–1003.
16. Raghow R, Thompson JP 1989 Molecular mechanisms of collagen gene expression. Mol Cell Biochem **86**:5–18.
17. Ramirez F, Di Liberto M 1990 Complex and diversified regulatory programs control the expression of vertebrate collagen genes. FASEB J **4**:1616–1623.
18. Vuorio E, de Crombrugghe B 1990 The family of collagen genes. Annu Rev Biochem **59**:837–872.
19. van der Rest M, Garrone R 1991 Collagen family of proteins. FASEB J **5**:2814–2823.
20. Vuurst J, Sobel ME, Martin GR 1985 Regulation of type I collagen synthesis: Total proα(I) and proα2(I) mRNAs are maintained in a 2:1 ratio under varying rates of collagen synthesis. Eur J Biochem **151**:449–453.

21. Raisz LG, Kream BE 1983 Regulation of bone formation. N Engl J Med **309**:29–35 and 83–89.
22. Raisz LG 1988 Bone metabolism and its hormonal regulation. Triangle **27**:5–10.
23. Canalis E, McCarthy TL, Centrella M 1991 Growth factors and cytokines in bone cell metabolism. Annu Rev Med **42**:17–24.
24. Kream BE, Smith MD, Canalis E, Raisz R 1985 Characterization of the effect of insulin on collagen synthesis in fetal rat bone. Endocrinology **116**:296–302.
25. Canalis E 1980 Effect of insulin-like growth factor-I on DNA and protein synthesis in cultured rat calvaria. J Clin Invest **66**:709–719.
26. Centrella M, Casinghino S, Ignotz R, McCarthy TL 1992 Multiple regulatory effects by transforming growth factor-beta on type I collagen levels in osteoblast-enriched cultures from fetal rat bone. Endocrinology **131**:2863–2872.
27. Kream BE, Rowe DW, Gworek SC, Raisz LG 1980 Parathyroid hormone alters collagen synthesis and procollagen mRNA levels in fetal rat calvaria. Proc Natl Acad Sci USA **77**:5654–5658.
28. Harrison JR, Vargas CJ, Petersen DN, Lorenzo JA, Kream BE 1990 Interleukin-1α and phorbol ester inhibition collagen synthesis by a transcriptional mechanism. Mol Endocrinol **4**:184–190.
29. Hodge BO, Kream BE 1988 Variable effects of dexamethasone on protein synthesis in clonal rat osteosarcoma cells. Endocrinology **122**:2127–2133.
30. Raisz LG, Fall PM 1990 Biphasic effects of prostaglandin E_2 on bone formation in cultured fetal rat calvariae: Interaction with cortisol. Endocrinology **126**:1654–1659.
31. Stern PH, Raisz LG 1979 Organ culture of bone. In: Simmons DJ, Kunin AS (eds) Skeletal Research: An Experimental Approach. Academic Press, New York, pp. 21–59.
32. Gronowicz G, Raisz LG 1996 bone formation assays. In: Bilezikian JP, Rodan GA, Raisz LG (eds) Principles of Bone Biology. Academic Press, San Diego, pp. 1253–1265.
33. Peterkofsky B, Diegelmann R 1971 Use of a mixture of proteinase-free collagenases for the specific assay of radioactive collagen in the presence of other hormones. Biochemistry **6**:988–994.
34. Diegelmann RF, Peterkofsky B 1972 Collagen biosynthesis during connective tissue development in chick embryo. Dev Biol **28**:443–453.
35. Sykes BD, Puddle B, Francis M, Smith R 1976 The estimation of two collagens from human dermis by interrupted gel electrophoresis. Biochem Biophys Res Commun **72**:1472–1480.
36. Sage H, Bornstein P 1982 Preparation and characterization of procollagens and procollagen–collagen intermediates. In: Cunningham LW, Frederiksen DW (eds) Methods in Enzymology, Vol. 82. Academic Press, New York, pp. 96–127.
37. Raisz LG, Maina DM, Gworek SC, Dietrich JW, Canalis EM 1978 Hormonal control of bone collagen synthesis in vitro: Inhibitory effect of 1-hydroxylated vitamin D metabolites. Endocrinology **102**:731–735.
38. Bringhurst FR, Potts JT Jr 1982 Effects of vitamin D metabolites and analogs on bone collagen synthesis in vitro. Calcif Tissue Int **34**:103–110.
39. Raisz LG, Kream BE, Smith MD, Simmons HA 1980 Comparison of the effects of vitamin D metabolites on collagen synthesis and resorption of fetal rat bone in organ culture. Calcif Tissue Int **32**:135–138.
40. Rowe DW, Kream BE 1982 Regulation of collagen synthesis in fetal rat calvaria by 1,25-dihydroxyvitamin D_3. J Biol Chem **257**:8009–8015.
41. Canalis E, Lian JB 1985 1,25-Dihydroxyvitamin D_3 effects on collagen and DNA synthesis in periosteum and periosteum-free calvaria. Bone **6**:457–460.
42. Hock JM, Kream BE, Raisz LG 1982 Autoradiographic study of the effect of 1,25-dihydroxyvitamin D_3 on bone matrix synthesis in vitamin D replete rats. Calcif Tissue Int **34**:347–351.
43. Kream BE, Rowe D, Smith MD, Maher V, Majeska R 1986 Hormonal regulation of collagen synthesis in a clonal rat osteosarcoma cell line. Endocrinology **119**:1922–1928.
44. Chen TL, Hauschka PV, Cabrales S, Feldman D 1986 The effects of 1,25-dihydroxyvitamin D_3 and dexamethasone on rat osteoblast-like primary cell cultures: Receptor occupancy and functional expression patterns for three different bioresponses. Endocrinology **118**:250–259.
45. Kim HT, Chen TL 1989 1,25-Dihydroxyvitamin D_3 interaction with dexamethasone and retinoic acid: Effects on procollagen messenger ribonucleic acid levels in rat osteoblast-like cells. Mol Endocrinol **3**:97–104.
46. Wong GL, Luben RA, Cohn DV 1977 1,25-Dihydroxycholecalciferol and parathormone: Effects on isolated osteoclast-like and osteoblast-like cells. Science **197**:663–665.
47. Rosen DM, Luben RA 1983 Multiple hormonal mechanisms for the control of collagen synthesis in an osteoblast-like cell line, MMB-1. Endocrinology **112**:992–999.
48. Chen TL, Li JM, Ye TV, Cone CM, Feldman D 1986 Hormonal responses to 1,25-dihydroxyvitamin D_3 in cultured mouse osteoblast-like cells—Modulation by changes in receptor level. J Cell Physiol **126**:21–28.
49. Walters MR, Rosen DM, Norman AW, Luben RA 1982 1,25-Dihydroxyvitamin D receptors in an established bone cell lines. Correlation with biochemical responses. J Biol Chem **257**:7481–7484.
50. Chen TL, Cone CM, Morey-Holton E, Feldman D 1983 1α-25-Dihydroxyvitamin D_3 receptors in cultured rat osteoblast-like cells. Glucocorticoid treatment increases receptor content. J Biol Chem **258:**4350–4355.
51. Owen TA, Aronow MS, Barone LM, Bettencourt B, Stein GS, Lian JB 1991 Pleiotropic effects of vitamin D on osteoblast gene expression are related to the proliferative and differentiated state of the bone cell phenotype: Dependency upon basal levels of gene expression, duration of exposure, and bone matrix competency in normal rat osteoblast cultures. Endocrinology **128**:1496–1504.
52. Ishida H, Bellows CG, Aubin JE, Heersche JNM 1993 Characterization of the $1,25(OH)_2D_3$-induced inhibition of bone nodule formation in long-term cultures of fetal rat calvaria cells. Endocrinology **132**:61–66.
53. Kurihara N, Ikeda K, Hakeda Y, Tsunoi M, Maeda N, Kumegawa M 1984 Effect of 1,25-dihydroxyvitamin D_3 on alkaline phosphatase activity and collagen synthesis in osteoblastic cells, clone MC3T3-E1. Biochem Biophys Res Commun **119**:767–771.
54. Kurihara N, Ishizuka S, Kiyoki M, Haketa Y, Ikeda K, Kumegawa M 1986 Effects of 1,25-dihydroxyvitamin D_3 on osteoblastic MC3T3-E1 cells. Endocrinology **118**:940–947.
55. Matsumoto T, Igarashi C, Takeuchi Y, Harada S, Kikuchi T, Yamato H, Ogata E 1991 Stimulation by 1,25-dihydroxyvitamin D_3 of in vitro mineralization induced by osteoblast-like MC3T3-E1 cells. Bone **12**:27–32.
56. Franceschi RT, Park K-Y, Romano PR 1988 Regulation of type I collagen synthesis by 1,25-dihydroxyvitamin D_3 in human osteosarcoma cells. J Biol Chem **263**:18938–18945.
57. Beresford JN, Gallagher JA, Russell RGG 1986 1,25-Dihydroxyvitamin D_3 and human bone-derived cells in vitro: Effects on alkaline phosphatase, type I collagen and proliferation. Endocrinology **119**:1776–1785.
58. Broess M, Riva A, Gerstenfeld LC 1995 Inhibitory effects of 1,25-$(OH)_2$ vitamin D_3 on collagen type I, osteopontin and osteocalcin gene expression in chicken embryo fibroblasts. J Cell Biochem **57**:440–451.

59. Gerstenfeld LC, Zurakowski D, Schaffer JL, Nichols DP, Toma CD, Broess M, Bruder SP, Caplan AI 1996 Variable hormone responsiveness of osteoblast populations isolated at different stages of embryogenesis and its relationship to the osteogenic lineage. Endocrinology **137**:3957–3968.
60. Harrison JR, Petersen DN, Lichter AC, Mador AT, Rowe DW, Kream BE 1989 1,25-Dihydroxyvitamin D3 inhibits transcription of type I collagen genes in the rat osteosarcoma cell line ROS 17–2.8. Endocrinology **125**:327–333.
61. Ozono K, Sone T, Pike W 1991 The genomic mechanism of action of 1,25-dihydroxyvitamin D_3. J Bone Miner Res **6**:1021–1027.
62. Darwish HM, DeLuca HF 1996 Recent advances in the molecular biology of vitamin D action. Prog Nucleic Acid Res Mol Biol **53**:321–344.
63. Kerner SA, Scott RA, Pike JW 1989 Sequence elements in the human osteocalcin gene confer basal activation and inducible response to hormonal vitamin D_3. Proc Natl Acad Sci USA **86**:4455–4459.
64. Demay MB, Gerardi JM, DeLuca HF, Kronenberg HM 1990 DNA sequences in the rat osteocalcin gene that bind the 1,25-dihydroxyvitamin D_3 receptor and confer responsiveness to 1,25-dihydroxyvitamin D_3. Proc Natl Acad Sci USA **87**:369–373.
65. Noda M, Vogel RL, Craig AM, Prahl J, DeLuca HF, Denhardt DT 1990 Identification of a DNA sequence responsible for binding of the 1,25-dihydroxyvitamin D_3 receptor and 1,25-dihydroxyvitamin D_3 enhancement of mouse secreted phosphoprotein 1 (Spp-1 or osteopontin) gene expression. Proc Natl Acad Sci USA **87**:9995–9999.
66. Ohyama Y, Ozono K, Uchida M, Shinki T, Kato S, Suda T, Yamamoto O, Noshiro M, Kato Y 1994 Identification of a vitamin D-responsive element in the 5′-flanking region of the rat 24-hydroxylase gene. J Biol Chem **269**:10545–10550.
67. Darwish HM, DeLuca HF 1992 1,25-Dihydroxyvitamin D_3-response element in the 5′-flanking region of the rat calbindin D-9k gene. Proc Natl Acad Sci USA **89**:603–607.
68. Towers TL, Luisi BF, Asianov A, Freedman LP 1993 DNA target selectivity by the vitamin D_3 receptor: Mechanism of dimer binding to an asymmetric repeat element. Proc Natl Acad Sci USA **90**:6310–6314.
69. Liu SM, Koszewski N, Lupez M, Malluche HH, Olivera A, Russell J 1996 Characterization of a response element in the 5′-flanking region of the avian (chicken) PTH gene that mediates negative regulation of gene transcription by 1,25-dihydroxyvitamin D_3 and binds the vitamin D_3 receptor. Mol Endocrinol **10**:206–215.
70. Demay MB, Kiernan MS, DeLuca HF, Kronenberg HM 1992 Sequences in the human parathyroid hormone gene that bind the 1,25-dihydroxyvitamin D_3 receptor and mediate transcriptional repression in response to 1,25-dihydroxyvitamin D_3. Proc Natl Acad Sci USA **89**:8097–8101.
71. Mackey SL, Heymont JL, Kronenberg HM, Demay MB 1996 Vitamin D receptor binding to the negative human parathyroid hormone vitamin D response element does not require the retinoid X receptor. Mol Endocrinol **10**:298–305.
72. Falzon M 1996 DNA sequences in the rat parathyroid hormone-related peptide gene responsible for 1,25-dihydroxyvitamin D_3-mediated transcriptional repression. Mol Endocrinol **10**:672–681.
73. Alroy I, Tower TL, Freedman LP 1995 Transcriptional repression of the interleukin-2 gene by vitamin D_3: Direct inhibition of NFATp/AP-1 complex formation by a nuclear hormone receptor. Mol Cell Biol **15**:5789–5799.
74. Lichtler A, Stover ML, Angilly J, Kream B, Rowe DW 1989 Isolation and characterization of the rat $\alpha 1(I)$ collagen promoter. Regulation by 1,25-dihydroxyvitamin D. J Biol Chem **264**:3072–3077.
75. Pavlin D, Lichtler AC, Bedalov A, Kream BE, Harrison JR, Thomas HF, Gronowicz GA, Clark SH, Woody CO, Rowe DW 1992 Differential utilization of regulatory domains within the $\alpha 1(I)$ collagen promoter in osseous and fibroblastic cells. J Cell Biol **116**:227–236.
76. Pavlin D, Bedalov A, Kronenberg MS, Kream BE, Rowe DW, Smith CL, Pike JW, Lichtler AC 1994 Analysis of regulatory domains in the COL1A1 gene responsible for 1,25-dihydroxyvitamin D_3-mediated transcriptional repression in osteoblastic cells. J Cell Biochem **56**:490–501.
77. Bedalov A, Salvatori R, Dodig M, Kronenberg MS, Kapural B, Bogdanovic Z, Kream BE, Woody CO, Clark SH, Mack K, Rowe DW, Lichtler AC 1995 Regulation of COL1A1 expression in type I collagen producing tissues: Identification of a 49 base pair region which is required for transgene expression in bone of transgenic mice. J Bone Miner Res **10**:1443–1451.
78. Dodig M, Kronenberg MS, Bedalov A, Kream BE, Gronowicz G, Clark S, Mack K, Liu Y, Maxon R, Pan ZZ, Upholt WH, Rowe DW, Lichtler AC 1996 Identification of a TAAT-containing motif required for high level expression of a COL1A1 promoter in differentiated osteoblasts of transgenic mice. J Biol Chem **271**:16422–16429.
79. Bodganovic Z, Bedalov A, Krebsbach PH, Woody CO, Clark SH, Thomas HF, Rowe DW, Kream BE, Lichtler AC 1993 Upstream regulatory elements necessary for expression of the rat COL1A1 promoter in transgenic mice. J Bone Miner Res **9**:285–292.
80. Schule R, Rangarajan P, Kliewer S, Ransone LJ, Bolado J, Yang N, Verma IM, Evans RM 1990 Functional antagonism between oncoprotein c-Jun and the glucocorticoid receptor. Cell **62**:1217–1226.
81. Jonat C, Rahmsdorg HJ, Park K-K, Cato ABC, Bebel S, Ponta H, Herrlich P 1990 Antitumor promotion and antiinflammation: Downregulation of AP-1 (Fos/Jun) activity by glucocorticoid hormone. Cell **62**:1189–1204.
82. Feyen JHM, Petersen DN, Kream BE 1988 Inhibition of bone collagen synthesis by the tumor promoter phorbol 12-myristate 13-acetate. J Bone Miner Res **3**:173–179.
83. Bogdanovic Z, Harrison JR, LaFrancis D, Bedalov A, Woody CO, Lichtler AC, Rowe DW, Kream BE 1993 Parathyroid hormone regulation of the $\alpha 1(I)$ collagen promoter and NF-IL6 expression in transgenic mouse calvariae. J Bone Miner Res **8**(Suppl. 1):S190.
84. Franceschi RT, Young J 1990 Regulation of alkaline phosphatase by 1,25-dihydroxyvitamin D_3 and ascorbic acid in bone-derived cells. J Bone Miner Res **5**:1157–1167.

Chapter 13

Calbindin-D_{28K}

Sylvia Christakos, Joshua D. Beck, and Sven Johan Hyllner
Department of Biochemistry and Molecular Biology, University of Medicine and Dentistry of New Jersey, New Jersey Medical School, Newark, New Jersey

I. INTRODUCTION AND GENERAL CONSIDERATIONS

In the two major target tissues of 1,25-dihydroxyvitamin D_3 [$1,25(OH)_2D_3$] action, intestine and kidney, one of the most pronounced effects of $1,25(OH)_2D_3$ known is the induction of the calcium binding protein, calbindin, the first identified target of $1,25(OH)_2D_3$ action. There are two major subclasses of calbindin: a protein of approximately 28,000 molecular weight (calbindin-D_{28K}), which will be discussed in this chapter, and a protein of approximately 9,000 molecular weight (calbindin-D_{9K}), which will be discussed in Chapter 14. Calbindin-D_{28K} is present in highest concentration in avian intestine and in avian and mammalian kidney, brain, and pancreas. Calbindin-D_{28K} has four functional high affinity calcium binding sites and is highly conserved in evolution. Calbindin-D_{9K} has two calcium binding domains, is present in highest concentration in mammalian intestine, and, unlike calbindin-D_{28K}, is not evolutionarily conserved but has been observed only in mammals. There is no amino acid sequence similarity between calbindin-D_{9K} and calbindin-D_{28K}.

The discussion that follows in this chapter regarding calbindin-D_{28K} reviews the chemistry, localization, proposed functional significance, and regulation of these calcium binding proteins. In addition, these chapters provide insight into the information obtained by studying these proteins concerning the multiple actions of the vitamin D endocrine system and the basic molecular mechanism of $1,25(OH)_2D_3$ action. Findings indicating that calbindins can be regulated by a number of different hormones and factors are also reviewed. The study of the molecular interactions of several members of the steroid hormone–retinoic acid family as well as the role of signal transduction pathways in the regulation of calbindin-D may be applicable to the regulation of other targets of $1,25(OH)_2D_3$ action. Elucidation of multiple factors and interactions regulating $1,25(OH)_2D_3$ target genes should result in novel insights related to tissue-specific molecular mechanisms involved in calcium homeostasis.

One of the most important findings in the vitamin D field has been the discovery by Wasserman and Taylor in 1966 of a 28,000 M_r vitamin D-dependent calcium binding protein in avian intestine [1]. Although previously known as the vitamin D-dependent calcium binding protein (CaBP), in 1985 it became officially known as calbindin-D_{28K} [2]. Initially identified in avian intestine (1), calbindin-D_{28K} has since been reported in many other tissues including kidney and bone, and in tissues that are not primary regulators of serum calcium such as pancreas, testes, and brain and in a variety of species [3–9] (see Christakos *et al.* [9] for review). The

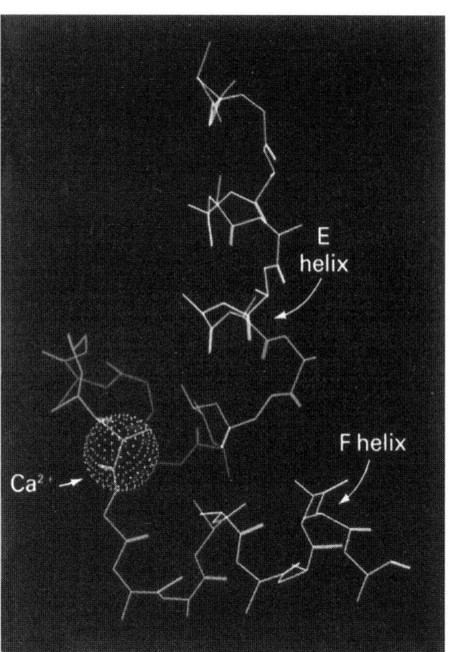

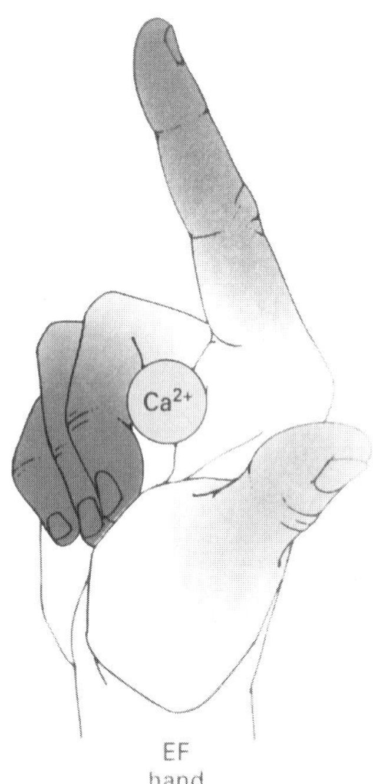

FIGURE 1 EF-hand structural motif (helix–loop–helix). The helices are represented by the extended forefinger and thumb. The clenched middle finger represents the loop that contains the oxygen ligands of the calcium ion. The EF hand is a recurring motif in calbindin and other calcium binding proteins. Reprinted with permission from Stryer L 1995 Biochemistry. Freeman, San Francisco, p. 1064.

importance of the discovery of calbindin is that key advances in our understanding of the diversity of the vitamin D endocrine system have been made through the study of the tissue distribution of calbindin-D_{28K} and its colocalization with the vitamin D receptor (VDR). In addition, the biosynthesis of calbindin has provided a model for studies that have resulted in an important basic understanding of the molecular mechanism of action of $1,25(OH)_2D_3$ in major target tissues such as intestine and kidney.

Chicken and mammalian calbindin-D_{28K} proteins contain 261 amino acid residues, have a molecular weight of approximately 28,000 (30,000 based on amino acid sequence and 28,000 based on migration on sodium dodecyl sulfate–polyacrylamide gels), and are blocked at the amino terminus [9–12]. The mammalian calbindin-D_{28K} sequences are 98% similar to one another and 79% similar to chicken calbindin-D_{28K} [9,11,12]. Calbindin-D_{28K} is highly conserved in evolution, suggesting an important, fundamental role for calbindin-D_{28K} in mediating intracellular calcium-dependent processes [9].

Calbindin-D_{28K} belongs to a family of high affinity calcium (Ca^{2+}) binding proteins ($K_d = 10^{-8}$–10^{-6} M) that contains over 200 members and is characterized by the EF-hand structural motif [13] (Fig. 1). The EF-hand domain is an octahedral structure consisting of two α helices separated by a 12-amino acid loop that contains side chain oxygens necessary for orienting the divalent calcium cation [14]. Calbindin-D_{28K} contains six EF hands; however, only four of these actively bind Ca^{2+} [15,16]. Other calcium binding proteins belonging to this family include calmodulin, parvalbumin, troponin C, calretinin, calcineurin, calpain, Spec I, myosin light chains, calbindin-D_{9K}, S100α, S100β, and recoverin [13,17]. Although the structure of calbindin-D_{28K} has yet to be elucidated by X-ray crystallography, circular dichroism experiments have shown that calbindin-D_{28K} contains approximately 30% α helix, 20.6% β sheet, and 51% random coil [18]. It is a heat-stable protein and is acidic, having a pI value of approximately 4.7 [19]. Calbindin-D_{28K} binds other cations in addition to calcium with reduced affinity: $Ca^{2+} > Cd^{2+} > Sr^{2+} > Mn^{2+} > Zn^{2+} > Ba^{2+} > Co^{2+} > Mg^{2+}$ [20].

II. LOCALIZATION AND PROPOSED FUNCTIONAL SIGNIFICANCE

A. Intestine

One of the most pronounced effects of $1,25(OH)_2D_3$ is increased synthesis of intestinal calbindin. Calbindin-D_{28K} has been localized primarily in the cytoplasm of absorptive cells in avian intestine [21], which supports its proposed role in intestinal calcium absorption [22–24]. A strong correlation has been observed between the level of calbindin and an increase in intestinal Ca^{2+} transport in the chick [25–27]. However, other investigators have reported that early actions of $1,25(OH)_2D_3$ on calcium uptake by brush-border membrane vesicles of vitamin D-deficient chicks may be independent of calbindin [28,29].

In the intestine, $1,25(OH)_2D_3$ effects the transfer of Ca^{2+} across the luminal brush-border membrane, the transfer of calcium through the cell interior, and active calcium extrusion from the basolateral membrane. A calbindin-based facilitated diffusion model has been proposed for the vitamin D-dependent movement of Ca^{2+} through the cytosolic compartment of the enterocyte [24] (also see Chapter 16 and Fig. 2). This model suggests binding of Ca^{2+} to calmodulin at the brush-border membrane, followed by Ca^{2+} binding to calbindin due to its higher affinity and greater concentration and translocation of calcium bound to calbindin to the basolateral membrane. At the basolateral membrane, calcium is actively transported out of the cell into the lamina propria via the plasma membrane calcium pump (PMCA). In addition to facilitated diffusion, it has been suggested that calbindin-D_{28K} may also act as a cytosolic buffer to prevent toxic levels of calcium from accumulating in the intestinal cell during vitamin-D mediated translocation of calcium [24,30]. Although most intestinal calbindin is localized in the cytosol, it should be noted that some calbindin has also been reported in small vesicles and lysosome-like structures [31] (also see Chapter 16) and in the nucleus [32,33]. It has been suggested that changes in the cellular localization of calbindin are dynamically involved in $1,25(OH)_2D_3$ dependent intestinal calcium transport [31,33].

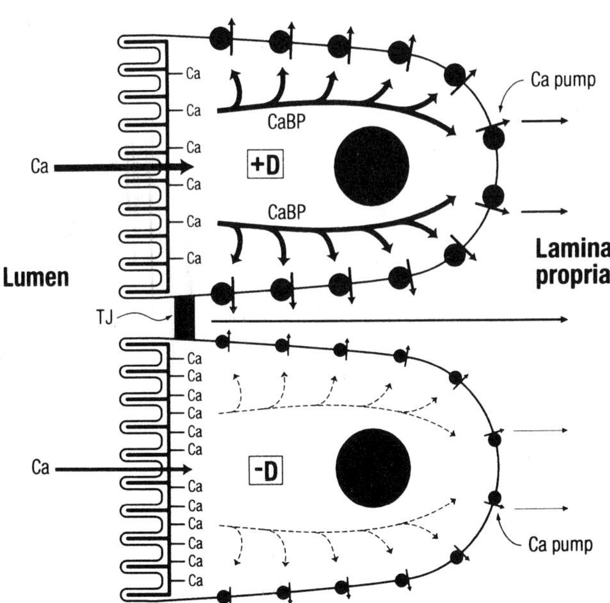

FIGURE 2 Calbindin-based facilitated diffusion model of vitamin D-dependent movement of Ca^{2+} through the cytosolic compartment of the enterocyte. Absorbed Ca^{2+} moves from the lumen to the brush-border membrane to calmodulin and to calbindin (CaBP). Ca^{2+} bound to calbindin is translocated to the basolateral membrane, where it is actively transported out of the cell into the lamina propria via the plasma membrane calcium pump. The process of calcium translocation through the enterocyte is increased in the vitamin D-replete animal by calbindin. Reprinted with permission from Wasserman and Fullmer [24].

B. Kidney

Immunocytochemical studies have reported the exclusive localization of calbindin-D_{28K} in the distal nephron (distal convoluted tubule and connecting tubule) in a variety of species [3,34–39] (see also Chapter 17). Renal calbindin is localized in the cytosol and the nucleus and is not associated with membranes or filamentous elements. Autoradiographic data indicated that the VDR is also predominantly localized in the distal nephron [40,41]. Although micropuncture data [42] as well as studies using a mouse distal convoluted tubule cell line [43] have indicated that vitamin D metabolites can enhance calcium transport in the distal nephron, little information is available concerning the exact role of vitamin D-inducible renal calbindin in this process.

In the kidney, calcium is reabsorbed by a gradient-dependent paracellular process in the proximal tubule and the thick ascending limb of the loop of Henle [44,45]. In the distal convoluted tubule (DCT), calcium reabsorption occurs via a transcellular, energy-dependent process [44,45]. It has been suggested that calbindin-D_{28K} acts to ferry Ca^{2+} across the cell, as in the intestine [44,46]. Following translocation through the cell, Ca^{2+} is actively extruded at the basolateral pole by the PMCA [45,46]. Similar to the calbindins and VDR, the PMCA has also been localized immunocytochemically exclusively to the DCT and the collecting duct [47]. Multiple isoforms of the Na^+,Ca^{2+} exchanger are also present in the distal tubule [48]. However, the interplay between the Na^+,Ca^{2+} exchanger and the PMCA in mediating calcium efflux from the distal tubule remains to be defined. Besides the suggestion of a role for calbindin in Ca^{2+} translocation in the renal cell, kinetic studies using apical membrane vesicles reconstituted with calbindin-D_{28K} revealed that this calcium binding protein may stimulate apical calcium entry which is mediated by dihydropyridine-sensitive calcium channels [49]. In addition, as in the intestine, it has been suggested that renal calbindin may act to lower cytosolic Ca^{2+} levels, preventing the accumulation of toxic levels of calcium [50].

C. Bone

Calbindin-D_{28K} is present in chondrocytes of growth plate cartilage in rats and chicks [51,52]. Although it is not clear from the immunocytochemical studies whether calbindin-D_{28K} is vitamin D dependent in chondrocytes, $1,25(OH)_2D_3$ receptors have been reported in developing chick bone, specifically in dividing chondrocytes [53]. It has been suggested that calbindin may be involved in the movement of calcium in the process of calcification in the chondrocyte [51]. Calbindin-D_{28K} has also been localized to osteoblasts and ameloblasts of rodent teeth, and it has been reported that calbindin-D_{28K} mRNA is induced by $1,25(OH)_2D_3$ in these cells [54,55]. It has also been suggested that elevated calbindin-D_{28K} may phenotypically characterize cells that are involved in calcium handling during mineralization [55].

D. Pancreas

The pancreas was the first nonclassic target tissue in which receptors for $1,25(OH)_2D_3$ were identified [56]. Although $1,25(OH)_2D_3$ has been reported to play a role in insulin secretion, the exact mechanisms remain unclear [57–59]. An early indication that the pancreas may be a target for $1,25(OH)_2D_3$ was the immunocytochemical study of Morrissey *et al.* [3] which localized calbindin-D_{28K} to the islet. In the chick, calbindin-D_{28K} is detected exclusively in insulin-producing β cells [60] and is responsive to vitamin D [61]. In the rat, however, calbindin-D_{28K} has been reported to be localized in α as well as β cells of the pancreas [62]. Because autoradiographic data have indicated that $1,25(OH)_2D_3$ receptors are localized only in rat β cells [63], and because insulin but not glucagon secretion is affected in vitamin D-deficient animals [57], studies in the rat suggest that β-cell calbindin may be regulated by $1,25(OH)_2D_3$ while non-β-cell calbindin may be independent of vitamin D. Calbindin-D_{28K} has also been identified in human pancreatic islet cells [64].

To determine a role for calbindin in β-cell function, calbindin-D_{28K} was transfected and overexpressed in the established β-cell line RIN1046–38 [65]. In cells transfected with calbindin, there was a marked induction [6- to 35-fold) in insulin secretion and insulin mRNA. Immunocytochemical studies also revealed a marked increase in insulin immunoreactivity in RIN cells overexpressing calbindin (Fig 3; see color insert). These studies suggest for the first time a direct role for calbindin in insulin secretion and biosynthesis.

E. Testes

Calbindin-D_{28K} has been reported in both chick and rat testes [5,6]. In chick and rat, immunocytochemical studies have revealed that calbindin-D_{28K} is present in spermatogonia and spermatocytes of the seminiferous tubules and some interstitial Leydig cells [5,6]. It has been reported that vitamin D-deficient chicks have significantly (3-fold) lower testicular calbindin levels than vitamin D-replete chicks [66]. As calbindin-D_{28K} as well as VDR (which is also present in seminiferous tubules) have been shown to correlate with testicular maturation [6,67], the involvement of calbindin-D_{28K} and vitamin D in spermatogenesis and steroidogenesis has been suggested [66].

F. Egg Shell Gland and Uterus

In the chick, calbindin-D_{28K} has been localized in the tubular gland cells of the shell gland [68], which are involved in calcium secretion during egg shell formation. Although receptors for $1,25(OH)_2D_3$ have been described in the chick shell gland [69], egg shell calbindin is induced by estradiol but is unaffected by $1,25(OH)_2D_3$ administration *in vivo* [70–72]. Because calbindin-D_{28K} mRNA levels fluctuated during the daily egg cycle, correlating with egg shell calcification, it was suggested that calcium flux through the egg shell gland could also modulate calbindin expression [73,74]. The direct involvement of calcium in the regulation of egg shell calbindin was shown in an *in vitro* study by Corradino [75].

Calbindin-D_{28K} is also found in the reproductive tissues of female mice (uterus, oviduct, and ovary) but is absent in those of the female rat [76]. Calbindin-D_{28K} is expressed in mouse oviduct epithelium and in mouse endometrial and glandular epithelium of the uterus. Estrogen has been reported to reduce levels of calbindin protein and mRNA in mouse uterus, whereas $1,25(OH)_2D_3$ has no effect on uterine calbindin [76]. It has been suggested that transepithelial calcium transport in mouse uterus and oviduct is facilitated by calbindin-D_{28K} [76]. Thus, although the proposed role for calbindin in these female reproductive tissues is similar to the proposed role of calbindin in intestine and kidney (facilitator of calcium diffusion), unlike intestinal and renal calbindin, calbindin in oviduct and uterus does not appear to be vitamin D dependent.

G. Nervous Tissue

Calbindin-D_{28K} is widely distributed throughout the brain of mammals, avians, reptiles, amphibians, fish, and mollusks [9]. It is present in most neuronal cell groups and fiber tracts and is localized in neuronal elements and some ependymal cells [8,77–79]. In brain, calbindin is not vitamin D dependent [9]. Neurons containing calbindin-D_{28K} are found in the cerebral cortex in layers 2–4, primarily in pyramidal neurons [8,77–79]. In the hippocampus, both basket cells and pyramidal neurons in CA1 stain positively for calbindin, as do granule cells and fibers in the dentate gyrus [8,77–79]. Purkinje cells of the cerebellum stain most intensely for calbindin-D_{28K} [8,77–79]. Calbindin immunoreactive cells are also observed in the hypothalamus, amygdala, pyriform region and thalamus [8,77–79].

In addition, specific neuronal sensory cells have been shown to contain calbindin-D_{28K} [80–88]. These cell populations include cochlear and vestibular hair cells in the inner ear [80–82], avian basilar papilla [83], cone but not rod photoreceptor cells of avian and mammalian retina [84–87], and conelike, modified photoreceptor cells (pinealocytes) of pineal transducers [88]. The presence of calbindin in specific cells of the sensory pathway suggests the possible involvement of calbindin in mechanisms of signal transduction.

It is also of interest that the mRNA for the calcium receptor, originally cloned from bovine parathyroid glands [89] and more recently from rat kidney [90] and rat thyroidal C cells [91], has been reported to be localized in brain in some of the same areas as calbindin (e.g., in Purkinje cells of the cerebellum) [92]. Since calbindin-D_{28K} is also localized in kidney and in calcitonin-containing ultimobranchial glands and thyroidal C cells [93], it will be of interest in future studies to determine whether there is an interrelationship between the calcium receptor and calbindin in mediating calcium-dependent functions in various systems.

In the nervous system it has been suggested that neuronal calbindin, by buffering calcium, can regulate intracellular calcium responses to physiological stimuli and can protect neurons against calcium-mediated neurotoxicity [94]. It has been demonstrated that introduction of exogenous calbindin into sensory neurons can modulate calcium signaling by decreasing the rate of rise of intracellular calcium and by changing the kinetics of decay of the calcium signal [95]. Using adenovirus as an expression vector, overexpression of calbindin in hippocampal neurons was reported to suppress posttetanic potentiation, possibly by restricting and destabilizing the evoked calcium signal [96]. Calbindin was also reported to play a role in the control of hypothalamic neuroendocrine neuronal firing patterns [97]. The whole cell patch clamp method was used to introduce calbindin into rat supraoptic neurons. Calbindin-D_{28K} suppressed Ca^{2+}-dependent depolarization afterpotentials and converted phasic into continuous firing [97]. As different firing patterns promote the release of different hypothalamic hormones, it was suggested that calbindin, by regulating firing patterns, may be involved in the control of hormone secretion from hypothalamic neuroendocrine neurons. The above studies [95–97] indicated directly by introduction of exogenous calbindin via the patch clamp method or by transfection and overexpression that calbindin is an important and effective regulator of calcium-dependent aspects of neuronal function. These studies also provide, for the first time, mechanisms whereby calbindin may act to protect against calcium-mediated neurotoxicity or to modulate hormone secretion from neuroendocrine neurons.

Correlative evidence between decreases in neuronal calbindin and neurodegeneration in studies of ischemic injury [98], seizure activity [99,100], and chronic neurodegeneration (Alzheimer's, Huntington's and Parkinson's diseases) [101–103] have been reported. It has been suggested that decreased calbindin levels may lead to a loss of calcium buffering or intracellular calcium homeostasis, which leads to cytotoxic events associated with neuronal damage and cell death. Thus, induction of calbindin expression, which may possibly prevent neuronal damage in neuropathies, has important therapeutic implications.

H. Calbindin-D_{28K} and Apoptosis

Calcium is thought to play an important regulatory function in apoptosis [104], but the precise mechanism(s) by which calcium promotes cell death is unknown. The first study suggesting that calbindin-D_{28K} plays a protective role in the process of apoptosis used subtraction analysis between the cDNA libraries of two human prostate cell lines [105]. One of the cell lines was androgen independent and the other one androgen dependent, and the results revealed that a hybrid calbindin-D_{28K} gene was specifically expressed in the hormone-independent cell line. Apoptosis has been observed in prostate cells on androgen depletion. Androgen deprivation of prostate cells triggers an influx of calcium ions into the cells, leading to an increase in intracellular calcium. It was suggested that calbindin-D_{28K} might buffer intracellular calcium and contribute to protection against apoptosis and thus androgen independence in the prostatic cell line.

It has been reported that stable transfection and overexpression of calbindin-D_{28K} in lymphocytes protects against apoptosis-induced by calcium ionophore, cAMP, and glucocorticoid [106]. A similar protective role for calbindin-D_{28K} has been observed in apoptosis susceptible cells in the central nervous system [107,108]. It will be of interest in future studies to determine in what other cells calbindin may protect against apoptotic cell death.

I. Calbindin-D_{28K} Enzyme Activation and Other Potential Targets

In addition to its role as an intracellular calcium buffer and in transepithelial calcium transport, there is *in vitro* evidence that calbindin may modulate the activities of calmodulin-sensitive enzymes such as calcium-dependent ATPase and phosphodiesterase [109]. There is also evidence for a calcium-dependent, specific interaction of calbindin-D_{28K} with intestinal alkaline phosphatase [110]. Although the physiological relevance of these findings is not known, these studies suggest that calbindin may possibly act as an enzyme regulator. In addition, studies in opossum kidney cells indicated that expression of calbindin in these cells resulted in increased phosphate transport which was associated with alterations in the actin cytoskeleton. Pollock and Santiesteban suggested a possible role for calbindin in cytoskeletal reorganization [111].

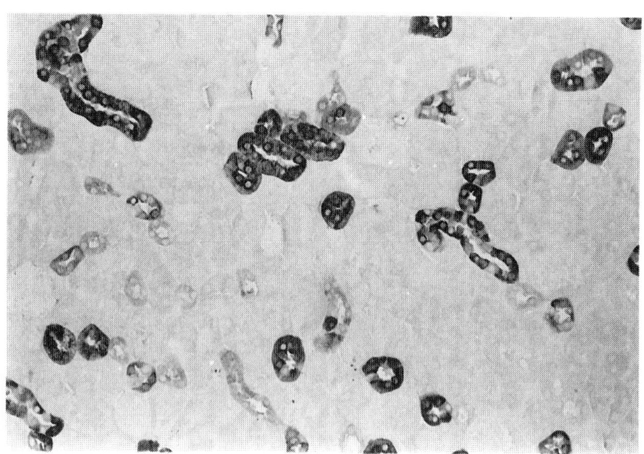

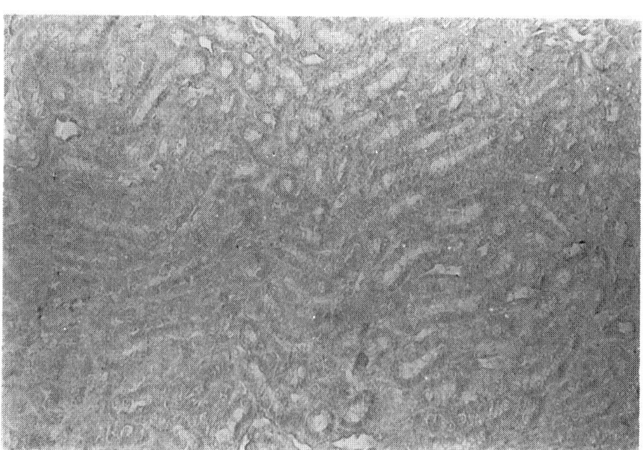

FIGURE 4 Wild-type and knock-out mouse kidney immunostained for calbindin-D_{28K}. Immunopositive cells were observed in the distal convoluted tubules (with a few cells also in connecting tubules and cortical collecting tubules) in wild-type mice (top). No calbindin-D_{28K} immunopositive cells were detected in mice homozygous for the targeted mutation (bottom).

J. Calbindin-D_{28K} Knock-out Mice

Calbindin-D_{28K} null mutant mice have been generated by gene targeting [112]. These mice are phenotypically normal and exhibit normal fecundity but do not express calbindin-D_{28K} as determined by immunohistochemical and Western blot analyses (112; S. Christakos, J. D. Beck, and K. Sooy unpublished; also see Fig. 4). Initial studies have indicated that synaptic calcium transients in cerebellar Purkinje cells are markedly altered in null mutant mice [112]. Future studies utilizing these "knock-out" mice represent an exciting area of further research that should provide new insight into the role of calbindin in insulin biosynthesis and secretion, in protection against calcium-mediated cell death, and in calcium reabsorption in the distal tubule of the nephron.

III. REGULATION OF CALBINDIN-D_{28K} GENE EXPRESSION

A. Genomic Organization of the Calbindin-D_{28K} Gene

The genomic organization of the chicken calbindin-D_{28K} gene has been elucidated [113,114] (Fig. 5) and a partial structure for the human gene was also reported [12,115]. The human gene is located on chromosome 8 [115,116] and is believed to consist of 11 exons, analogous to that demonstrated for the avian gene. Moreover, the total size of the gene is reported to be 18.5 kb in chicken. The protein coding region of the mouse gene shares 77% sequence similarity with the chicken gene [117]. However, no obvious sequence similarity exists between the mammalian and avian promoters except in the region of the TATA box [12,118].

B. Regulation by $1,25(OH)_2D_3$

It is well known that calbindin-D_{28K} in the avian intestine [1,119] and kidney [119,120], and in the mammalian kidney [121,122], is induced by $1,25(OH)_2D_3$. A similar $1,25(OH)_2D_3$ dependency of calbindin-D_{28K} has also been suggested in rat mineralized tissue [54]. The geno-

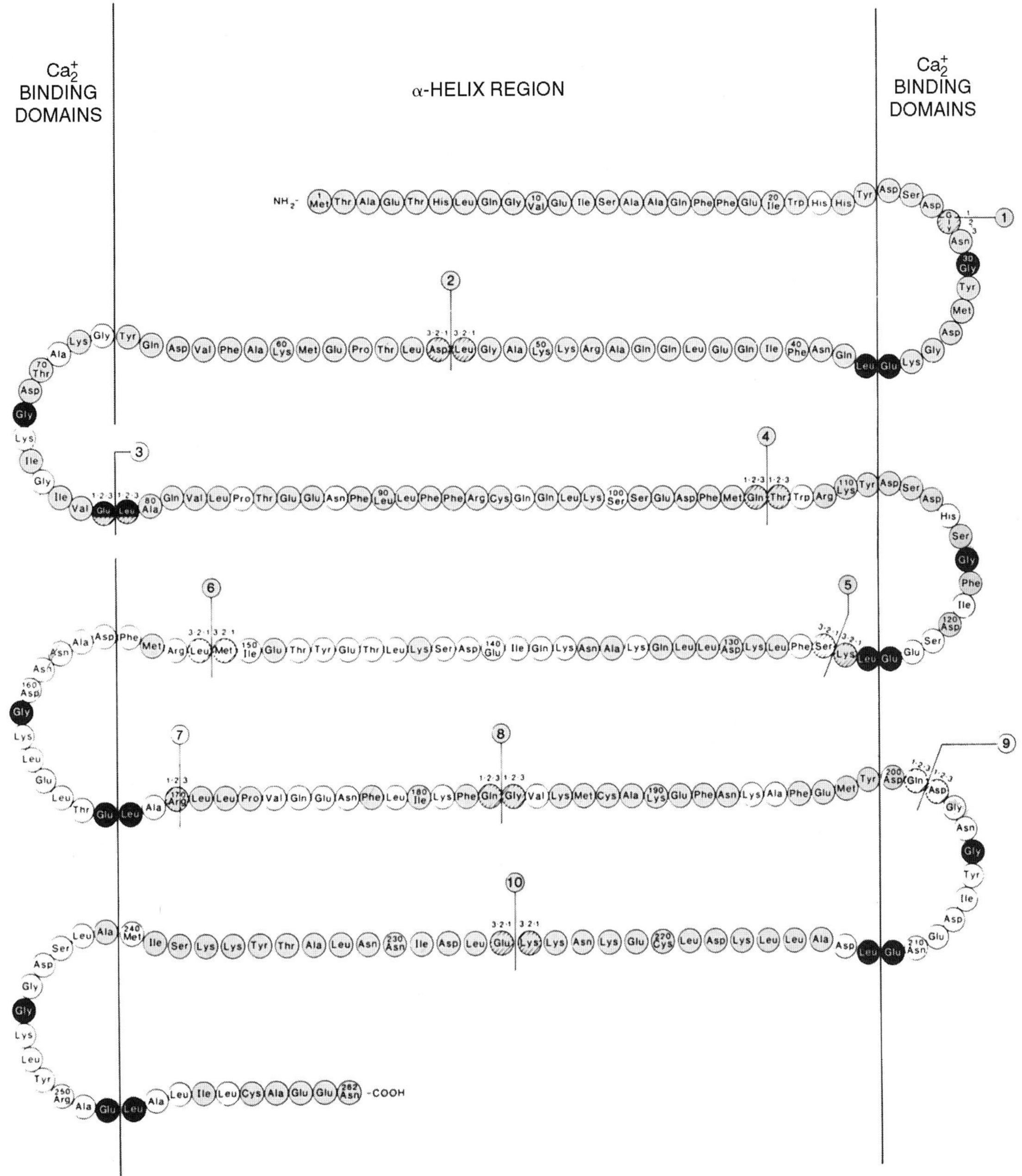

FIGURE 5 Position of intervening sequences within the structure of chicken calbindin-D_{28K}. Locations of introns are indicated by circled numbers. Numbers above amino acids indicate codon positions. Invariant Glu/Leu and Gly amino acids are indicated by black circles. Calcium binding domains are separated from the α-helix region by vertical lines. Reprinted with permission from Minghetti *et al.* [113].

mic response to $1,25(OH)_2D_3$ in general appears to involve the direct binding of a high affinity, low capacity receptor protein, the vitamin D receptor (VDR) to DNA sequences [123,124]. A limited number of vitamin D response elements (VDREs) have been identified within genes, including the human and rat osteocalcin VDREs [125,126] and the mouse osteopontin VDRE [127]. In chicken, a putative calbindin-D_{28K} VDRE was suggested after computer analysis of the promoter sequence [128], but only a 2-fold response to $1,25(OH)_2D_3$ was detected in primary kidney cells after transfection with a 2.1-kb segment of the 5' flanking region of the promoter [129]. A relatively inactive putative VDRE was also reported in the chicken calbindin-D_{28K} promoter by others [130,131].

Although sequence elements with homology to the rat osteocalcin VDRE that respond to $1,25(OH)_2D_3$ have been identified in the mouse calbindin-D_{28K} promoter, the response observed is modest [5-fold maximal induction in chloramphenicol acetyltransferase (CAT) activity] [118]. The modest response reflects previous *in vivo* findings that indicated that $1,25(OH)_2D_3$ induces the expression of the renal calbindin-D_{28K} gene by a small, rapid transcriptional stimulation (peak 1–3 hr; 3.5-fold induction as indicated by nuclear run-on assays) followed by a sustained accumulation of mRNA long after cessation of $1,25(OH)_2D_3$ treatment (at 12 hr >100-fold induction) [116,122]. Similar findings were reported for the *in vivo* induction of the chick intestinal calbindin-D_{28K} gene by $1,25(OH)_2D_3$ [132]. Thus, although the early increase in calbindin-D_{28K} mRNA is mediated via a transcriptional effect, *in vivo* studies and CAT activity indicate that there is only a modest $1,25(OH)_2D_3$ transcriptional induction of calbindin-D_{28K}. There is now increasing evidence that the large induction of calbindin-D_{28K} mRNA long after $1,25(OH)_2D_3$ treatment may be due primarily to post-transcriptional mechanisms [116,122,132–134]. Exactly how this action is exerted is not known, but one report suggests that $1,25(OH)_2D_3$ may regulate the expression of an intermediate protein which may be involved in calbindin-D_{28K} mRNA accumulation [134]. These studies, in both chicken and mouse, suggest that the mechanism of action of $1,25(OH)_2D_3$ on calbindin-D_{28K} regulation is more complicated than the conventional hormone receptor–transcriptional activation model which explains the regulation of osteocalcin and osteopontin genes by $1,25(OH)_2D_3$ [125–127].

C. Regulation by Other Steroids and Factors

Further studies in intestine and kidney have provided evidence that the calbindin-D_{28K} gene is not exclusively regulated by $1,25(OH)_2D_3$ and that other factors can modulate gene expression. It has been reported that glucocorticoids inhibit the levels of calbindin-D_{28K} mRNA and protein in intestine of vitamin D treated chicks, resulting in a comparable decrease in intestinal calcium absorption [135,136]. These findings suggest the involvement of the inhibition of intestinal calbindin in the clinically important hypocalcemic action of glucocorticoids. However, it should be noted that in organ cultures of embryonic chick intestine, low concentrations of dexamethasone were reported to enhance $1,25(OH)_2D_3$-dependent calbindin-D_{28K} biosynthesis [137].

The discrepancy between these results is not clearly understood. However, several reasons can be discerned. For example, the concentration of dexamethasone could be crucial for the response (low dose is stimulatory; high dose is inhibitory). It is also possible that the *in vivo* effects observed are not due to a primary action of glucocorticoids. Although a putative glucocorticoid-like response element has been found in the promoter of the chicken calbindin-D_{28K} gene [113], functional studies using the chromosomal gene for calbindin are needed to determine whether glucocorticoids have a direct effect on calbindin gene expression. Adding to the growing body of data which suggests that the regulation of calbindin-D_{28K} is more complex than previously thought are studies describing the modulation by calcium of intestinal and renal calbindin-D_{28K} gene expression [133,138–140]. Moreover, butyrate treatment of a rat pancreatic β-cell line elevates the levels of calbindin-D_{28K} and its mRNA in accord with the induction by butyrate of insulin content and secretion and β-cell differentiation, suggesting a role for calbindin-D_{28K} in these processes [59]. A butyrate-mediated induction of calbindin-D_{28K} had previously been observed in avian kidney cells [141], and 5′ flanking sequences have been identified that mediate butyrate stimulation of the mammalian calbindin-D_{28K} [118].

In other tissues where calbindin-D_{28K} is present in significant amounts, for example, in parts of the brain, its regulation appears to be very different from that in the intestine and kidney. In the central nervous system (CNS), $1,25(OH)_2D_3$ has no apparent effect on the levels of calbindin-D_{28K} [9]. Instead, a variety of different factors have been reported to be involved in regulating neuronal calbindin. Using rat hippocampal cultures, evidence has indicated that neurotropin 3 (NT-3) [142,143], brain-derived neurotropic factor (BDNF) [142–143], fibroblast growth factor (FGF) [142], and tumor necrosis factors (TNFs) [144,145] all can induce calbindin. It has been reported that neurotropic factors may protect against excitotoxic neuronal damage [146]. The induction of calbindin by those factors suggests a role for calbindin in the process of protection against cytotoxicity.

In addition, corticosterone administration *in vivo* has been reported to increase calbindin-D_{28K} expression in rat hippocampus [147,148]. The glucocorticoid response is specifically localized to the CA1 region [148]. Retinoic acid has also been reported to induce calbindin-D_{28K} protein and mRNA in medulloblastoma cells, which express a neuronal phenotype [149]. Furthermore, the content of calbindin-D_{28K} in cultured Purkinje cells can be increased by insulin-like growth factor I (IGF-I) [150]. Similarly IGF-I, as well as insulin, promoted the expression of calbindin-D_{28K} protein in cultured rat embryonic neuronal cells [151]. Thus, neuronal calbindin-D_{28K} can be regulated by steroids as well as by factors that affect signal transduction pathways. These different modes of activation may be important for cell-specific effects of calbindin-D_{28K}.

The molecular basis for tissue-specific calbindin-D_{28K} regulation is still not understood, but *in vivo* experiments using transgenic mice suggest that tissue specificity of calbindin-D_{28K} expression to some degree is controlled by separate elements on the promoter [152]. The transgenic mouse study demonstrated the importance of an *in vivo* system to investigate the role of sequence elements needed for tissue-specific gene expression and regulation [152].

As discussed previously, calbindin-D_{28K} is also present in the avian egg shell gland and in the reproductive tissues of female mice. In the avian egg shell gland, estradiol-17β induces calbindin-D_{28K} in *in vivo* experiments, whereas $1,25(OH)_2D_3$ seems to have no effect [70–72]. It should be noted, however, that under very specific conditions, an increase of calbindin-D_{28K} levels by $1,25(OH)_2D_3$ could be detected in *in vitro* experiments [75]. Estradiol-17β treatment further increased the response to $1,25(OH)_2D_3$ [75]. The discrepancy between the results obtained *in vivo* and *in vitro* indicates, similar to the results observed with dexamethasone treatment, that multiple factors may be involved in the observed *in vivo* effects. In the mouse, calbindin-D_{28K} gene expression was found to be down-regulated by estradiol in the uterus and oviduct [76,153] but up-regulated in the ovaries [153]. Multiple imperfect half-palindromic estrogen responsive elements, which are likely to mediate the estrogen responsiveness of the calbindin-D_{28K} gene by estradiol-17β, were present in two regions (−1075/−702 and −175/−78] of the promoter [153].

IV. CONCLUSION

We once viewed calbindin-D_{28K} as an exclusively vitamin D-dependent calcium binding protein. It is now evident that calbindin is not under the exclusive regulatory control of $1,25(OH)_2D_3$. Calbindin-D_{28K} is present in many different tissues and may serve many different functions. Accordingly, the regulation of calbindin-D_{28K} is varied and quite complex. Several steroids as well as a number of factors that affect signal transduction pathways are reported to be involved in the regulation of calbindin-D_{28K}. It is obvious that further research will be needed in order to understand fully the regulation of calbindin-D_{28K} and to clarify further its functions in the organism.

References

1. Wasserman RH, Taylor AN 1966 Vitamin D_3 induced calcium binding protein in chick intestinal mucosa. Science **152**:791–793.
2. Wasserman RH 1985 Nomenclature of the vitamin D induced calcium binding proteins. In: Norman AW, Schaefer K, Grigoleit HG, Herrath DV (eds) Vitamin D: Chemical, Biochemical and Clinical Update. de Gruyter, Berlin, pp. 321–322.
3. Morrissey RL, Bucci TJ, Richard B, Empson RN, Lufkin EG 1975 Calcium binding protein: Its cellular localization in jejunum, kidney and pancreas. Proc Soc Exp Biol Med **149**:56–60.
4. Christakos S, Norman AW 1978 A vitamin D-dependent calcium binding protein in bone tissue. Science **202**:70–71.
5. Inpanbutr N, Taylor AN 1992 Expression of calbindin-D_{28K} in developing and growing chick testes. Histochemistry **97**:335–339.
6. Kagi U, Chafouleas JG, Norman AW, Heizmann CW 1988 Developmental appearance of the Ca^{2+}-binding proteins parvalbumin, calbindin-D_{28K}, S100 proteins and calmodulin during testicular development in the rat. Cell Tissue Res **252**:359–365.
7. Taylor AN 1974 Chick brain calcium binding protein: Comparison with intestinal vitamin D-induced calcium binding protein. Arch Biochem Biophys **161**:100–108.
8. Jande SS, Maler L, Lawson DEM 1981 Immunohistochemical mapping of vitamin D-induced calcium binding protein in brain. Nature **294**:765–767.
9. Christakos S, Gabrielides C, Rhoten WB 1989 Vitamin D-dependent calcium binding proteins: Chemistry, distribution, functional considerations, and molecular biology. Endocr Rev **10**:84–107.
10. Fullmer CS, Wasserman RH 1987 Chicken intestinal 28-kilodalton calbindin-D: Complete amino acid sequence and structural considerations. Proc Natl Acad Sci USA **84**:4772–4776.
11. Hunziker W, Schrickel S 1988 Rat brain calbindin-D_{28K}: Six domain structure and extensive amino acid homology with chicken calbindin-D_{28K}. Mol Endocrinol **2**:465–473.
12. Parmentier M 1990 Structure of the human cDNAs and genes coding for calbindin-D_{28K} and calretinin. Adv Exp Med Biol **140**:21–25.
13. Van Eldik LJ, Zendegui JG, Marshak DR, Watterson DM 1982 Calcium binding proteins and the molecular basis of calcium action. Int Rev Cytol **77**:1–61.
14. Tufty RM, Kretsinger RH 1975 Troponin and parvalbumin calcium binding regions predicted in myosin light chain and T4 lysozyme. Science **187**:167–169.
15. Hunziker W 1986 The 28-kDa vitamin D-dependent calcium-binding protein has a six-domain structure. Proc Natl Acad Sci USA **83**:7578–8582.
16. Akerfeldt KS, Coyne AN, Wilk RR, Thulin E, Linse S 1996 Ca^{2+}-binding stoichiometry of calbindin-D_{28K} as assessed by spectroscopic analyses of synthetic peptide fragments. Biochemistry **35**:3662–3669.
17. Heizmann CW, Braun K 1992 Changes in Ca^{2+}-binding proteins in human neurodegenerative disorders. Trends Neurosci **15**:259–264.
18. Gross MD, Nelsestuen GL, Kumar R 1987 Observations on the binding of lanthanides and calcium to vitamin D-dependent chick intestinal calcium binding protein. J Biol Chem **262**:6539–6545.
19. Wasserman RH, Corradino RA, Taylor, AN 1968 Vitamin D dependent calcium binding protein: Purification and some properties. J Biol Chem **243**:3978–3986.
20. Ingersoll RJ, Wasserman RH 1971 Vitamin D_3 induced calcium binding protein: Binding characteristics, conformational effects and other properties. J Biol Chem **246**:2808–2814.
21. Taylor AN 1981 Immunohistochemical localization of vitamin D-induced calcium binding protein: Relocation of antigen during frozen section processing. J Histochem Cytochem **29**:65–73.
22. Feher JJ 1983 Facilitated calcium diffusion by intestinal calcium-binding protein. Am J Physiol **244**:C303–C307.

23. Bronner F, Pansu D, Stein WD 1986 An analysis of intestinal calcium transport across the rat intestine. Am J Physiol **150**: G561–G569.
24. Wasserman RH, Fullmer CS 1995 Vitamin D and intestinal calcium transport: Facts, speculations and hypotheses. J Nutr **125**:1971S–1979S.
25. Taylor AN, Wasserman RH 1969 Correlations between the vitamin D-induced calcium binding protein and intestinal absorption of calcium. Fed Proc **28**:1834–1838.
26. Feher JJ, Wasserman RH 1979 Calcium absorption and intestinal calcium-binding protein: Quantitative relationship. Am J Physiol **236**:E556–E561.
27. Corradino RA, Fullmer CS, Wasserman RH 1976 Embryonic chick intestine in organ culture: Stimulation of calcium transport by exogenous vitamin D-induced calcium-binding protein. Arch Biochem Biophys **174**:738–743.
28. Bikle DD, Munson S, Zolock DT 1983 Calcium flux across chick duodenal brush border membrane vesicles: Regulation by 1,25-dihydroxyvitamin D. Endocrinology **113**:2072–2080.
29. Bikle DD, Zolock DT, Munson S 1984 Differential response of duodenal epithelial cells to 1,25-dihydroxyvitamin D_3 according to position in the villus: A comparison of calcium uptake, calcium binding protein and alkaline phosphatase activity. Endocrinology **115**:2077–2084.
30. Murer H, Hildmann B 1981 Transcellular transport of calcium and inorganic phosphate in the small intestine epithelium. Am J Physiol **240**:G409–G416.
31. Nemere I, Leathers VL, Thompson BS, Luben RA, Norman AW 1991 Redistribution of calbindin-D_{28K} in chick intestine in response to calcium transport. Endocrinology **129**:2972–2984.
32. Thorens B, Roth J, Norman AW, Perrelet A, Orci L 1982 Immunocytochemical localization of vitamin D dependent calcium binding protein in chick duodenum. J Cell Biol **94**:115–122.
33. Taylor AN, Inpanbutr N 1988 Vitamin D mediated intestinal calcium transport: Ultrastructural distribution of calbindin-D_{28K} following calcitriol. In: Bronner F, Peterlik M (eds) Cellular Calcium and Phosphate Transport in Health and Disease. Alan R. Liss, New York, pp. 109–114.
34. Rhoten WB, Christakos S 1981 Immunocytochemical localization of vitamin D-dependent calcium binding protein in mammalian nephron. Endocrinology **109**:981–983.
35. Roth J, Thorens B, Hunziker W, Norman AW, Orci L 1981 Vitamin D-dependent calcium binding protein: Immunocytochemical localization in chick kidney. Science **214**:197–200.
36. Taylor AN, McIntosh JE, Bordeau JE 1982 Immunocytochemical localization of vitamin D-dependent calcium binding protein in renal tubules of rabbit, rat and chick. Kidney Int **21**:765–773.
37. Rhoten WB, Lubit B, Christakos S 1984 Avian and mammalian vitamin D-dependent calcium-binding protein in reptilian nephron. Gen Comp Endocrinol **55**:96–103.
38. Rhoten WB, Gona O, Christakos S 1986 Calcium binding protein [28,000 M_r calbindin-D_{28K}) in kidneys of the bullfrog *Rana catesbeiana* during metamorphosis. Anat Rec **186**:127–132.
39. Parmentier M, Ghysens M, Rypens F, Lawson DEM, Pasteels JL, Pochet R 1987 Calbindin in vertebrate classes: Immunohistochemical localization and Western blot analysis. Gen Comp Endocrinol **65**:399–407.
40. Stumpf WE, Sar M, Reid FA, Tanaka Y, DeLuca HF 1979 Target cells for 1,25-dihdyroxyvitamin D_3 in intestinal tract, stomach, skin, pituitary and parathyroid. Science **209**:1188–1190.
41. Stumpf WE, Sar M, Narbauitz R, Reid FA, DeLuca HF, Tanaka Y 1980 Cellular and subcellular localization of $1,25(OH)_2D_3$ in rat kidney: Comparison with localization of parathyroid hormone and estradiol. Proc Natl Acad Sci USA **77**:1149–1153.
42. Winaver J, Sylk DB, Robertson JS, Chen TC, Puschett JB 1980 Micropuncture study of the acute renal tubular transport effects of 25-hydroxyvitamin D_3 in the dog. Miner Electrolyte Metab **4**:178–188.
43. Friedman PA, Gesek FA 1993 Vitamin D accelerates PTH-dependent calcium transport in distal convoluted tubule cells. Am J Physiol **265**:F300–F308.
44. Bronner F 1989 Renal calcium transport: Mechanisms and regulation—an overview. Am J Physiol **257**:F707–F711.
45. Friedman PA, Gesek FA 1993 Calcium transport in renal epithelial cells. Am J Physiol **264**:F181–F198.
46. Johnson JA, Kumar R 1994 Renal and intestinal calcium transport: roles of vitamin D and vitamin D-dependent calcium binding proteins. Semin Nephrol **14**:119–128.
47. Borke JL, Caride A, Verma AK, Penniston JT, Kumar R 1989 Plasma membrane calcium pump and 28-kDa calcium binding protein in cells of rat kidney distal tubules. Am J Physiol **257**:F842–F849.
48. White KE, Gesek FA, Friedman PA 1996 Structural and functional analysis of Na^+/Ca^{2+} exchange in distal convoluted tubule cells. Am J Physiol **271**:F560–F570.
49. Bouhtiauy I, Lajeunesse D, Christakos S, Brunette MG 1994 Two vitamin D_3-dependent calcium binding proteins increase calcium reabsorption by different mechanisms I: Effect of $CaBP_{28k}$. Kidney Int **45**:461–468.
50. Koster HPG, Hartog A, Van Os CH, Bindels RJM 1995 Calbindin-D_{28K} facilitates cytosolic calcium diffusion without interfering with calcium signaling. Cell Calcium **18**:187–196.
51. Balmain, N, Brehier A, Cuisinier-Gleizes P, Mathieu H 1986 Evidence for the presence of calbindin-D_{28K} ($CaBP_{28K}$) in the tibial growth cartilages of rats. Cell Tissue Res **245**:331–335.
52. Zhou XY, Dempster DW, Marion SL, Pike JW, Haussler MR, Clemens TL 1986 Bone vitamin D-dependent calcium binding protein is localized in chondrocytes of growth plate cartilage. Calcif Tissue Int **38**:244–247.
53. Suda S, Takahashi N, Shinki T, Horiuchi N, Yamaguchi A, Yoshiki S, Enomato S, Suda T 1985 1,25-Dihydroxyvitamin D_3 receptors and their actions in embryonic chick. Calcif Tissue Int **37**:82–90.
54. Berdal A, Hotton, D, Pike JW, Mathieu H, Dupre J-M 1993 Cell and stage-specific expression of vitamin D receptor and calbindin genes in rat incisor: Regulation by 1,25-dihydroxyvitamin D_3. Dev Biol **155**:172–179.
55. Berdal A, Hotton D, Saffar JL, Thomasset M, Nanci A 1996 Calbindin-D_{9k} and calbindin-D_{28K} expression in rat mineralized tissues in vivo. J Bone Miner Res **11**:768–779.
56. Christakos S, Norman AW 1979 Evidence for a specific high affinity protein for 1,25-dihydroxyvitamin D_3 in chick kidney and pancreas. Biochem Biophys Res Commun **89**:56–63.
57. Norman AW, Frankel BJ, Heldt A,M Grodsky GM 1980 Vitamin D deficiency inhibits pancreatic secretion of insulin. Science **209**:823–825.
58. Kadowaki S, Norman AW 1985 Demonstration that the vitamin D metabolite $1,25(OH)_2D_3$ and not $24R,25(OH)_2$-vitamin D_3 is essential for normal insulin secretion in the perfused rat pancreas. Diabetes **34**:315–320.
59. Lee, S, Clark SA, Gill RK, Christakos S 1994 1,25-Dihydroxyvitamin D_3 and pancreatic β-cell function: Vitamin D receptors, gene expression and insulin secretion. Endocrinology **134**:1602–1610.
60. Roth J, Bonner-Weir S, Norman AW, Orci L 1982 Immunocytochemistry of vitamin D-dependent calcium binding protein in chick pancreas: Exclusive localization in β cells. Endocinology **111**:2216–2218.
61. Kadowaki S, Norman AW 1984 Pancreatic vitamin D-dependent

calcium binding protein: Biochemical properties and response to vitamin D. Arch Biochem Biophys **233**:228–236.

62. Pochet R, Pipeleers DG, Malaisse WJ 1987 Calbindin-D_{28K} Da: preferential localization in non-β islet cells of the rat pancreas. Biol Cell **61**:155–161.
63. Clark SA, Stumpf WE, Sar M, DeLuca HF, Tanaka Y 1980 Target cells for 1,25-dihydroxyvitamin D_3 in the pancreas. Cell Tissue Res **209**:515–520.
64. Johnson JA, Grande JP, Roche PC, Kumar R 1994 Immunohistochemical localization of the 1,25$(OH)_2D_3$ receptor and calbindin-D_{28K} in human and rat pancreas. Am J Physiol **267**:E356–E360.
65. Reddy D, Pollock AS, Clark S, Sooy K, Vasavada RC, Stewart AF, Honeyman T, Christakos S 1997 Transfection and overexpression of the calcium binding protein calbindin-D_{28K} results in a stimulatory effect on insulin synthesis in a rat B cell line (RIN 1046-38). Proc Natl Acad Sci USA **94**:1961–1966.
66. Inpanbutr N, Reiswig JD, Bacon WI, Slemons RD, Iacopino AM 1996 Effect of vitamin D on testicular $CaBP_{28k}$ expression and serum testosterone in chickens. Biol Reprod **54**:242–248.
67. Walters MR 1984 1,25-Dihydroxyvitamin D_3 receptors in seminiferous tubules of the rat testes increase at puberty. Endocrinology **114**:2167–2173.
68. Lippiello L, Wasserman RH 1975 Fluorescent antibody localization of the vitamin D dependent calcium binding protein in the oviduct of the laying hen. J Histochem Cytochem **23**:111–116.
69. Coty WA 1980 A specific high affinity binding protein for 1,25-dihydroxyvitamin D in chick oviduct shell gland. Biochem Biophys Res Commun **93**:285–292.
70. Bar A, Rosenberg J, Hurwitz S 1984 The lack of relationship between vitamin D_3 metabolites and calcium binding protein in the eggshell gland of laying birds. Comp Biochem Physiol **78**:75–79.
71. Nys Y, Mayel-Afshar S, Bouillon R, Van Bailin H, Lawson DEM 1989 Increases in calbindin-D_{28K} mRNA in the uterus of the domestic fowl induced by sexual maturity and shell formation. Gen Comp Endocrinol **76**:322–329.
72. Corradino RA, Smith CA, Krook LP, Fullmer CS 1993 Tissue specific regulation of shell gland calbindin-D_{28K} biosynthesis by estradiol in precociously matured, vitamin D-depleted chicks. Endocrinology **132**:193–198.
73. Bar A, Striem S, Vax E, Talpaz H, Hurwitz S 1992 Regulation of calbindin mRNA and calbindin turnover in intestine and shell gland of the chicken. Am J Physiol **262**:R800–R805.
74. Nys Y, Baker K, Lawson DEM 1992 Estrogen and a calcium flux dependent factor modulate calbindin gene expression in the uterus of the laying hen. Gen Comp Endocrinol **87**:87–94.
75. Corradino RA 1993 Calbindin-D_{28K} regulation in precociously matured chick egg shell gland *in vitro*. Gen Comp Endocrinol **91**:158–166.
76. Opperman LA, Saunders TJ, Bruns DE, Boyd JC, Mills SE, Bruns ME 1992 Estrogen inhibits calbindin-D_{28K} expression in mouse uterus. Endocrinology **130**:1728–1735.
77. Roth J, Baeten D, Norman AW, Garcia-Sergura LM 1981 Specific neurons in chick central nervous system stain with an antibody against intestinal vitamin D-dependent calcium binding protein. Brain Res **222**:452–457.
78. Feldman SC, Christakos S 1983 Vitamin D-dependent calcium binding protein in rat brain: Biochemical and immunocytochemical characterization. Endocrinology **112**:290–302.
79. Celio MR 1990 Calbindin and paravalbumin in the rat nervous system Neuroscience **35**:375–475.
80. Rabie A, Thomasset M, Legrand C 1983 Immunocytochemical detection of a calcium binding protein in the cochlear and vestibular hair cells of the rat. Cell Tissue Res **232**:691–696.
81. Sans A, Etchecopar B, Brehier A, Thomasset M 1986 Immunocytochemical detection of vitamin D-dependent calcium binding protein ($CaBP_{28k}$) in vestibular sensory hair cells and vestibular ganglion neurons of the cat. Brain Res **364**:190–194.
82. Legrand C, Brehier A, Clavel MC, Thomasset M, Rabie A 1988 Cholecalcin (28 kDa CaBP) in rat cochlea: Development in normal and hypothyroid animals: An immunocytochemical study. Brain Res **466**:121–129.
83. Oberholtzer JC, Buettger C, Summers MC, Matschinsky FM 1988 The 28 kDa calbindin-D is a major calcium binding protein in the basilar papilla of the chick. Proc Natl Acad Sci USA **85**:3387–3390.
84. Rabie A, Thomasset M, Parkes CO, Clavel MC 1985 Immunocytochemical detection of 28,000 MW calcium binding protein in horizontal cells of the rat retina. Cell Tissue Res **240**:493–496.
85. Schreiner DS, Jande SS, Lawson DEM 1985 Target cells of vitamin D in the retina. Acta Anat **121**:153–162.
86. Pasteels B, Parmentier M, Lawson DEM, Verstappen A, Pochet R 1987 Calcium binding protein immunoreactivity in pigeon retina. Invest Ophthalmol Vis Sci **28**:658–664.
87. Haley TL, Pochet R, Baizer L, Burton MD, Crabb JW, Parmentier M, Polans AS 1995 Calbindin-D_{28K} immunoreactivity of human cone cells varies with retinal position. Vis Neurosci **12**: 301–307.
88. Roman A, Brisson P, Pasteels B, Demol S, Pochet R, Collen JP 1988 Pineal–retinal molecular relationships: Immunocytochemical evidence of calbindin-D_{28K} in pineal transducers. Brain Res **442**:33–42.
89. Brown EM, Gamba G, Riccardi D, Lombardi M, Butters R, Kifor O, Sun A, Hediger MA, Lytton J, Hebert SC 1993 Cloning and characterization of an extracellular Ca^{2+} sensing receptor from bovine parathyroid. Nature **366**:575–580.
90. Riccardi D, Park J, Lee WS, Gamba G, Brown EM, Hebert SC 1995 Cloning and functional expression of a rat kidney extracellular calcium polyvalent cation sensing receptor. Proc Natl Acad Sci USA **92**:131–135.
91. Garrett JE, Tamir H, Kifor O, Simin RT, Rogers KV, Mithal A, Gagel RF, Brown EM 1995 Calcitonin-secreting cells of the thyroid express an extracellular calcium sensing receptor gene. Endocrinology **136**:5202–5211.
92. Rogers KV, Brown EM, Hebert SC 1994 Localization of calcium receptor mRNA in the adult rat central nervous system using *in situ* hybridization. Abstracts Soc Neurosci **20**:1061.
93. Taylor AN, Inpanbutr N, Deftos LJ 1987 Localization of calbindin-D_{28K} in calcitonin containing cells of chick ultimobranchial glands. Anat Rec **219**:86–90.
94. Mattson MP, Rychlick B, Chu C, Christakos S 1991 Evidence for calcium reducing and excitoprotective roles for the calcium binding protein calbindin-D_{28K} in cultured hippocampal neurons. Neuron **6**:41–51.
95. Chard PS, Bleakman D, Christakos S, Fullmer CS, Miller RJ 1993 Calcium buffering properties of calbindin-D_{28K} and parvalbumin in rat sensory neurones. J Physiol **472**:341–357.
96. Chard PS, Jordan J, Marcuccilli CJ, Miller RJ, Leiden JM, Roos RP, Ghadge GD 1995 Regulation of excitatory transmission at hippocampal synapses by calbindin-D_{28K}. Proc Natl Acad Sci USA **92**:5144–5148.
97. Li Z, Decavel C, Hatton GI 1995 Calbindin-D_{28K}: Role in determining intrinsically generated firing patterns in rat supraoptic neurones. J Physiol **488.3**:601–608.
98. Burke RE, Baimbridge KG 1993 Relative loss of the striatal striosome compartment, defined by calbindin-D_{28K} immunostaining, following developmental hypoxic–ischemic injury. Neuroscience **56**:305–315.

99. Baimbridge KG, Mody I, Miller JJ 1985 Reduction of rat hippocampal calcium binding protein following commissural, amygdala, septal, perforant path and olfactory bulb kindling. Epilepsia **26**:460–465.
100. Sonnenberg J, Frantz GD, Lee S, Heick A, Chu C, Tobin AJ, Christakos S 1991 Calcium binding protein (calbindin-D_{28K}) and glutamate decarboxylase gene expression after kindling induced seizures. Mol Brain Res **9**:179–190.
101. Iacopino AM, Christakos S 1990 Specific reduction of neuronal calcium binding protein (calbindin-D_{28K}) gene expression in aging and neurodegenerative diseases. Proc Natl Acad Sci USA **87**:4078–4082.
102. Chan-Palay V, Hochli M, Savaskan E, Hungerecker G 1993 Calbindin-D_{28K} and monoamine oxidase A immunoreactive neurons in the nucleus basalis of Meynert in senile dementia of the Alzheimer type and Parkinson's disease. Dementia **4**:1–15.
103. Ferrer I, Tunon T, Soriano E, Del Rio A, Iraizoz I, Fonseca M, Guionnet N 1993 Calbindin-D_{28K} immunoreactivity in the temporal neocortex in patients with Alzheimer's disease. Clin Neuropathol **12**:53–58.
104. Trump BF, Berezesky IK 1992 The role of cytosolic Ca^{2+} in cell injury, necrosis and apoptosis. Curr Opin Cell Biol **4**:227–232.
105. Liu AY, Abraham BA 1991 Subtractive cloning of a hybrid human endogenous retrovirus and calbindin gene in the prostate cell line PC3. Cancer Res **51**:4104–4110.
106. Dowd DR, MacDonald PN, Komm BS, Haussler MR, Miesfeld RL 1992 Stable expression of the calbindin-D_{28K} complementary DNA interferes with the apoptotic pathway in lymphocytes. Endocrinology **6**:1843–1848.
107. Ho B-K, Alexianu ME, Colom LV, Mohamed AH, Serrano F, Appel SH 1996 Expression of calbindin-D_{28K} in motoneuron hybrid cells after retroviral infection with calbindin-D_{28K} cDNA prevents amyotrophic lateral sclerosis IgG-mediated cytotoxicity. Proc Natl Acad Sci USA **93**:6796–6801.
108. Lee S, Wernyj R, Christakos S 1995 Evidence for a role for calbindin-D_{28K} in cell survival. J Bone Miner Res **10**(Suppl. 1), S493.
109. Reisner PD, Christakos S, Vanaman TC 1992 *In vitro* enzyme activation with calbindin-D_{28K} the vitamin D dependent 28kDa calcium binding protein. FASEB J **297**:127–131.
110. Norman AW, Leathers V 1982 Preparation of a photoaffinity probe for the vitamin D dependent intestinal calcium bindng protein: Evidence for a calcium-dependent, specific interaction with intestinal alkaline phosphatase. Biochem Biophys Res Commun **108**:220–226.
111. Pollock AS, Santiesteban HL 1995 Calbindin expression in renal tubular epithelial cells: Altered sodium phosphate co-transport in association with cytoskeletal rearrangement. J Biol Chem **270**:16291–16301.
112. Airaksinen MS, Eilers J, Garaschuk O, Thoenen H, Konnerth A, Meyer M 1997 Ataxia and altered dendritic calcium signaling in mice carrying a targeted null mutation of the calbindin-D_{28K} gene. Proc Natl Acad Sci USA **94**:1488–1493.
113. Minghetti PP, Cancela L, Fujisawa Y, Theofan G, Norman AW 1988 Molecular structure of the chicken vitamin D-induced calbindin-D_{28K} gene reveals eleven exons, six Ca^{2+}-binding domains, and numerous promoter regulatory elements. Mol Endocrinol **2**:355–367.
114. Wilson PW, Rogers J, Harding M, Pohl V, Pattyn G, Lawson DEM 1988 Structure of chick chromosomal genes for calbindin and calretinin. J Mol Biol **200**:615–625.
115. Parmentier M, De Vijlder JJM, Muir E, Szpirer C, Islam MQ, Geurts van Kessel A, Lawson DEM, Vassart G 1989 The human calbindin 27 kDa gene: Structure organization of the 5′ and 3′ regions, chromosomal assignment, and restriction fragment length polymorphism. Genomics **4**:309–319.
116. Varghese S, Deaven LL, Huang Y-C, Gill RK, Iacopino AM, Christakos S 1989 Transcriptional regulation and chromosomal assignment of the mammalian calbindin-D_{28K} gene. Mol Endocrinol **3**:495–502.
117. Wood TL, Kobayashi Y, Franz G, Varghese S, Christakos S, Tobin AJ 1988 Molecular cloning of mammalian 28,000 M_r vitamin D dependent calcium binding protein (calbindin-D_{28K}) in rodent kidney and brain. DNA **7**:585–594.
118. Gill RK, Christakos S 1993 Identification of sequence elements in mouse calbindin-D_{28K} gene that confer 1,25-dihydroxyvitamin D_3- and butyrate-inducible responses. Proc Natl Acad Sci USA **90**:2984–2988.
119. Christakos S, Norman AW 1980 Vitamin D-dependent calcium-binding protein synthesis by chick kidney and duodenal polysomes. Arch Biochem Biophys **203**:809–815.
120. Craviso GL, Garrett KP, Clemens TL 1987 1,25-Dihydroxyvitamin D_3 induces the synthesis of vitamin D-dependent calcium-binding protein in cultured chick kidney cells. Endocrinology **120**:894–902.
121. Pansini AR, Christakos S 1984 Vitamin D-dependent calcium binding protein in rat kidney: Purification and physicochemical and immunological characterization. J Biol Chem **259**:9735–9741.
122. Varghese S, Lee S, Huang Y-C, Christakos S 1988 Analysis of rat vitamin D-dependent calbindin-D_{28K} gene expression. J Biol Chem **263**:9776–9784.
123. Darwish HM, DeLuca HF 1993 Vitamin D regulated gene expression. Crit Rev Eukaryotic Gene Expression **3**:89–116.
124. Christakos S, Raval-Pandya M, Wernyj RP, Yang W 1996 Genomic mechanisms involved in the pleiotropic actions of 1,25-dihydroxyvitamin D_3. Biochem J **316**:361–371.
125. Kerner SA, Scott RA, Pike JW 1989 Sequence elements in the human osteocalcin gene confer basal activation and inducible response to hormonal vitamin D_3. Proc Natl Acad Sci USA **86**:4455–4459.
126. Demay MB, Gerardi JM, DeLuca HF, Kronenberg HM 1990 DNA sequences in the rat osteocalcin gene that bind the 1,25-dihydroxyvitamin D_3 receptor and confer responsiveness to 1,25-dihydroxyvitamin D_3. Proc Natl Acad Sci USA **87**:369–373.
127. Noda M, Vogel RL, Craig AM, Prahl J, DeLuca HF, Denhardt DT 1990 Identification of a DNA sequence responsible for binding of the 1,25-dihydroxyvitamin D_3 receptor and 1,25-dihydroxyvitamin D_3 enhancement of mouse secreted phosphoprotein 1 (Spp-1 or osteopontin) gene expression. Proc Natl Acad Sci USA **87**:9995–9999.
128. Minghetti PP, Gibbs PEM, Norman AW 1989 Computer analysis of 1,25-dihydroxyvitamin D_3-receptor regulated promoters: Identification of a candidate D_3-response element. Biochem Biophys Res Commun **162**:869–875.
129. Maiyar AC, Minghetti PP, Norman AW 1991 Transfection of avian vitamin D-dependent calbindin-D_{28K} 5′ flanking promoter sequence in primary chick kidney cells. Mol Cell Endocrinol **78**:127–135.
130. Ferrari S, Battini R, Pike WJ 1990 Functional analysis of the promoter region of the gene encoding chicken calbindin-D_{28K}. Adv Exp Med Biol **140**:21–25.
131. MacDonald PN, Whitfield GK, Haussler CA, Hocker AM, Haussler MR 1992 Evaluation of a putative vitamin D response element in the avian calcium binding protein gene. DNA Cell Biol **11**:377–383.
132. Theofan G, Nguyen AP, Norman AW 1986 Regulation of calbin-

din-D_{28K} gene expression by 1,25-dihydroxyvitamin D_3 is correlated to receptor occupancy. J. Biol Chem **261**:16943–16947.
133. Enomoto H, Hendy GN, Andrews GK, Clemens TL 1992 Regulation of avian calbindin-D_{28K} gene expression in primary chick kidney cells: Importance of posttranscriptional mechanisms and calcium ion concentration. Endocrinology **130**:3467–3474.
134. Meyer J, Galligan MA, Jones G, Komm BS, Haussler CA, Haussler MR 1995 1,25$(OH)_2D_3$-dependent regulation of calbindin-D_{28K} mRNA requires ongoing protein synthesis in chick duodenal organ culture. J Cell Biochem **58**:315–327.
135. Hall AK, Bishop JE, Norman AW 1987 Inhibitory and stimulatory effects of dexamethasone and 1,25-dihydroxyvitamin D_3 on chick intestinal calbindin-D_{28K} and its mRNA. Mol Cell Endocrinol **51**:25–31.
136. Tohmon M, Fukase M, Kishihara M, Kadowaki S, Fujita T 1988 Effect of glucocorticiod administration on intestinal, renal and cerebellar calbindin-D_{28K} in chicks. J Bone Miner Res **3**:325–331.
137. Corradino RA, Fullmer C 1991 Positive cotranscriptional regulation of intestinal calbindin-D_{28K} gene expression by 1,25-dihydroxyvitamin D_3 and glucocorticoids. Endocrinology **128**: 944–950.
138. Theofan G, Kong MW, Hall AK, Norman AW 1987 Expression of calbindin-D_{28K} mRNA as a function of altered serum calcium and phosphate levels in vitamin D-replete chick intestine. Mol Cell Endocrinol **54**:135–140.
139. Huang Y-C, Christakos S 1988 Modulation of rat calbindin-D_{28K} gene expression by 1,25-dihydroxyvitamin D_3 and dietary alteration. Mol Endocrinol **2**:928–935.
140. Bar A, Shani M, Fullmer CS, Brindak ME, Striem S 1990 Modulation of chick intestinal and renal calbindin gene expression by dietary vitamin D_3, 1,25-dihydroxyvitamin D_3, calcium and phosphorus. Mol Cell Endocrinol **72**:23–31.
141. Maiyar AC, Norman AW 1992 Effects of sodium butyrate on 1,25-dihydroxyvitamin D_3 receptor activity in primary chick kidney cells. Mol Cell Endocrinol **84**:99–107.
142. Collazo D, Takahashi H, McKay RDG 1992 Cellular targets and trophic functions of neurotrophin-3 in the developing rat hippocampus. Neuron **9**:643–656.
143. Ip NY, Li Y, Yancopoulos GD, Lindsay RM 1993 Cultured hippocampal neurons show responses to BDNF, NT-3 and NT-4 but not NGF. J Neurosci **13**:3394–3405.
144. Cheng B, Christakos S, Mattson MP 1994 Tumor necrosis factors protect neurons against excitotoxic/metabolic insults and promote maintenance of calcium homeostasis. Neuron **12**:139–153.
145. Mattson MP, Cheng B, Baldwin SA, Smith-Swintosky VL, Keller J, Geddes JW, Scheff SW, Christakos S 1995 Brain injury and tumor necrosis factors induce calbindin-D_{28K} in astrocytes: Evidence for a cytoprotective response. J Neurosci Res **42**:357–370.
146. Lindvall O, Kokaia Z, Bengzon J, Elmer E, Kokaia M 1994 Neurotrophins and brain insults. Trends Neurosci **17**:490–490
147. Iacopino AM, Christakos S 1990 Corticosterone regulates calbindin-D_{28K} mRNA and protein levels in rat hippocampus. J Biol Chem **265**:10177–10180.
148. Krugers HJ, Medema RM, Postema F, Korf J 1995 Region-specific alterations of calbindin-D_{28K} immunoreactivity in the rat hippocampus following adrenalectomy and corticosterone treatment. Brain Res **696**:89–96.
149. Wang Y-Z, Christakos S 1995 Retinoic acid regulates the expression of the calcium binding protein, calbindin-D_{28K}. Mol Endocrinol **9**:1510–1521.
150. Nieto-Bona MP, Busiguina S, Torres-Aleman I 1995 Insulin-like growth factor I is an afferent trophic signal that modulates calbindin-D_{28K} in adult Purkinje cells. J Neuosci Res **42**:371–376.
151. Yamaguchi T, Keino K, Fukuda J 1995 The effect of insulin and insulin-like growth factor-1 on the expression of calretinin and calbindin-D_{28K} in rat embryonic neurons in culture. Neurochem Int **26**:255–262.
152. Pavlou O, Ehlenfeldt R, Horn S, Orr HT 1996 Isolation, characterization and in vivo analysis of the murine calbindin-D_{28K} upstream regulatory region. Mol. Brain Res **36**:268–279.
153. Gill RK, Christakos S 1995 Regulation by estrogen through the 5′-flanking region of the mouse calbindin-D_{28K} gene. Mol Endocrinol **9**:319–326.

CHAPTER 14

Calbindin-D_{9K}

MONIQUE THOMASSET INSERM U.458, Alliée CNRS, Hôpital Robert Debré, Paris, France

I. INTRODUCTION

Calbindin-D_{9K} [1], previously named CaBP9k or ICaBP, was first reported by Wasserman's group in the rat intestinal mucosa as a calcium binding protein that was induced by vitamin D [2]. Calbindin-D_{9K}, like calbindin-D_{28K}, (see Chapter 13), belongs to the superfamily of EF-hand helix–loop–helix Ca^{2+} binding proteins which now contains 39 subfamilies and a total of more than 250 proteins, among them parvalbumin, calmodulin, and troponin C [3–5]. Calbindin-D_{9K} is a member of the S100 Ca^{2+}-binding protein subfamily; these are typically small, acidic proteins containing two Ca^{2+} binding sites [6]. Calbindin-D_{9K} is not closely related to calbindin-D_{28K}, a member of the calbindin subfamily, and there is no evidence to indicate that calbindin-D_{9K} arose evolutionarily from calbindin-D_{28K} [4,7]. Aside from its Ca^{2+} binding properties, several physiological functions have been postulated for calbindin-D_{9K}. Speculation on its function range from a pure Ca^{2+}-dependent regulatory function analogous to that of calmodulin to its role as a Ca^{2+} shuttle between the two cell membranes [8,9]. Several reviews of calbindin-D_{9K} were published prior to 1990 [10,11]. This chapter reviews more recent publications on the physical properties of the calbindin-D_{9K} protein, gene structure, tissue-specific distribution, and regulation by 1,25-dihydroxyvitamin D_3 [$1,25(OH)_2D_3$] and estrogens of calbindin-D_{9K} gene expression.

II. PHYSICAL CONSIDERATIONS

A. Isolation and Identification

Vitamin D supplementation of vitamin D-deficient rats increases the calcium binding activity of crude extracts of intestinal mucosa, as measured by competitive ion exchange and equilibrium dialysis [2]. Calbindin-D_{9K} was purified by gel filtration and DEAE-cellulose column chromatography in the presence or absence of Ca^{2+} ions and used to raise specific antisera in rabbits and to produce a radioimmunoassay [12,13]. Antisera to intestinal calbindin-D_{9K} from one species do not cross-react with calbindin-D_{9K} from other species in radioimmunoassays, but there is weak cross-reactivity by immunostaining. Antisera to calbindin-D_{9K} do not cross-react with calbindin-D_{28K} or S100 protein [13,14]. Calbindin-D_{9K} cDNA from the rat was cloned [15,16] using a library constructed from size-selected mRNA from the intestine of vitamin D-treated rats followed by screening by differential *in situ* hybridization using enriched cDNAs (made from RNA from intestines of vitamin D-replete rats) and unenriched probes (RNA from intestines of vitamin D-deficient animals). A cDNA encoding human calbindin-D_{9K} has also been reported [17,18]. The deduced amino acid sequence contains two helix–loop–helix structures of 29 amino acids, each binding one Ca^{2+} ion, as would be expected from the reviewed EF-hand concept [3] (see Chapter 13).

B. Protein Structure and Biological Activities

The primary structures of bovine, porcine, and murine intestinal calbindin-D_{9K} proteins were obtained by amino acid sequencing and deduced from the sequence of appropriate rat [15,16,19], bovine [1], porcine [20], and human [17,18] cDNA clones. The bovine protein, which is the prototype for these macromolecules, has 78 amino acid residues and a molecular weight of 8788. It is 87 and 81% homologous with the porcine and murine proteins, respectively [21].

Human calbindin-D_{9K} has 79 amino acids and a calculated molecular mass of 9015 daltons. It is 89% homologous with the bovine and porcine sequences, 78% with rat, and 77% with mouse calbindin-D_{9K}. The three-dimensional structures of the protein in the presence and absence of Ca^{2+} suggest that calbindin-D_{9K} responds to Ca^{2+} binding by limited conformational changes as compared to calmodulin and troponin C [22]. Calbindin-D_{9K} is heat stable and acidic (p$I \sim 4.7$), with two EF-hand regions that bind calcium with high affinity (K_a 10^6 M^{-1}). It also binds several other cations ($Ca^{2+} > Cd^{2+} > Sr^{2+} > Mn^{2+} > Zn^{2+} > Ba^{2+} > Co^{2+} > Mg^{2+}$), including Pb^{2+}.

III. TISSUE/CELLULAR DISTRIBUTION AND FUNCTIONAL SIGNIFICANCE

Calbindin-D_{9K} gene expression is characteristic of mammals. Its pattern of tissue-specific activity, as demonstrated by radioimmunoassay [13,23] (Table I), and cell specific distribution using immunostaining and *in situ* hybridization [24–43] (Table II) are characteristic. Most studies have been done on rats. Calbindin-D_{9K} is abundant in the rat duodenum, placenta, and uterus but is also present in the yolk sac, fallopian tube, lung, kidney, and mineralized tissues. Calbindin-D_{9K} gene expression has also been reported in the adult human gut [44–46], uterus [47], and in the duodenum, kidney, placenta, lung, and thymus of the 20-week human fetus [48].

TABLE I Tissue Distribution of Calbindin-D_{9k} [13] and 1,25-Dihydroxyvitamin D_3 Receptor (VDR) [23] in the Male Rat[a]

Tissue	Calbindin-D_{9K} (ng/mg soluble protein)	VDR (fmol/mg protein)
Esophagus	14 ± 4	—
Stomach	10 ± 16	10 ± 1.0
Duodenum	18,250 ± 1620	—
Jejunum	2930 ± 430	1152 ± 77
Ileum	430 ± 10	590 ± 39.7
Cecum	3080 ± 679	—
Colon	—	1005 ± 94.7
Kidney cortex	35 ± 3	224 ± 36.9
Submaxillary gland	33 ± 10	—
Thymus	10 ± 2	—
Liver	1.6 ± 0.4	0.4 ± 0.3
Pancreas	9.8 ± 2.7	—
Parathyroid gland	0	75 ± 6.4
Bone	209 ± 7	42 ± 7.7
Skin	1.0 ± 0.3	81 ± 31.1
Skeletal muscle	1.0 ± 0.3	0 ± 0
Blood (ng/ml plasma)	10 ± 2.8	—
Heart	4.5 ± 1.7	6 ± 0.9
Lung	384 ± 77	20 ± 3.1
Testis	6.5 ± 3.5	—
Spleen	—	5.2 ± 0.5
Cerebellum	4.0 ± 1.0	0 ± 0

[a] Calbindin-D_{9K} was determined by radioimmunoassay, and VDR by ligand binding.

A. Intestine

Calbindin-D_{9K} has been found in the intestine of most mammals tested to date, including the pig, cow, rat, and mouse (Table II). Calbindin-D_{9K} gene expression has been extensively studied in the rat intestine, in which the calbindin-D_{9K} protein accounts for 2% of the soluble protein in the duodenal mucosa (Fig. 1). The calbindin-D_{9K} protein and mRNA concentrations gradually decrease along the jejunum to the distal ileum, and they are almost absent from the large intestine, except for the cecum [13,49]. This distribution pattern correlates with active calcium transport [8] and the presence of vitamin D receptor (VDR) (Table II) [23,50]. Calbindin-D_{9K} and its mRNA are present in absorptive epithelial cells, but not in mucous or endocrine cells of the rat duodenum [25,27]. In the villus itself, the concentration of calbindin-D_{9K} increases from the crypt to the upper

TABLE II Cellular Distribution of Calbindin-D_{9K} Gene Expression in the Rat, Detected by Immunostaining and *in Situ* Hybridization[a]

Tissue	Species	Cellular localization	Ref.
Intestine	Rat	Enterocyte	24
			25
			6,27
	Pig	Enterocyte	28
	Mouse	Enterocyte	29
Kidney	Rat	DCT, loop of Henle, ClD	30
		(few)	31
	Mouse	DCT, CnT, ClD	29
Placenta	Rat	Endoderm	32
Yolk sac	Mice	Endoderm	33
Fetal placenta	Mouse	Epithelium	34
	Rat	Trophoblastic epithelium	26
Uterus (end of gestation)	Rat	Endometrial epithelium	35,36
Fallopian tube	Rat	Luminal epithelium	37
Cartilage	Rat	Mature chondrocytes	38
Bone	Rat	Osteoblasts, osteocytes	39,40
Tooth	Rat	Ameloblasts	40,41
Uterus	Rat	Myometrium: muscular fiber	36,42
	Rat	Endometrium: stroma	35
Lung	Rat	Epithelial alveolar cell	43

[a] The underlined tissues are involved in calcium transport. DCT, Distal convoluted tubule; ClD, collecting duct; CnT, connecting tubule.

part of the villus and correlates with increased alkaline phosphatase and decreased thymidine kinase activities [24]. Therefore, calbindin-D_{9K} is considered to be a marker of enterocyte differentiation. The difference between its concentration and the VDR distribution along the villus [51,52] suggests that vitamin D is not the only factor governing the expression of the calbindin-D_{9K} gene.

Duodenal calbindin-D_{9K} gene activity varies during development. Calbindin-D_{9K} is already detectable in the day-17 rat fetus, and its concentration increases before birth and reaches a maximum at weaning, which plateaus until the age of 10 weeks and then decreases in 30-week-old rats [13]. This profile parallels that of calcium absorption. There is still controversy concerning the presence or absence of VDR before birth in rats [53,54] which prevents any conclusion regarding the involvement of factors other than vitamin D.

Intestinal calbindin-D_{9K} gene activity is controlled by $1,25(OH)_2D_3$. Because vitamin D also induces the absorption of Ca^{2+} from the intestinal lumen, the possibility of a link was evident. However, the results of kinetic studies of duodenal calbindin-D_{9K} and Ca^{2+} transport induction in response to a single $1,25(OH)_2D_3$ injection into vitamin D-deficient rats during the first 2 hr, are controversial; they revealed that calcium absorption returned to the basal level, whereas the duodenal calbindin-D_{9K} content remained maximal [55,56]. These studies do not show a direct, single role of calbindin-D_{9K} in intestinal calcium transport. This major cytosolic protein could therefore act as a shuttle for calcium transfer from the luminal to laterobasal poles of the enterocyte, as suggested by the correlation between calbindin-D_{9K} and calcium pump activity [57]. Alternatively, calbindin-D_{9K} could act as an intracellular Ca^{2+} buffer to protect the cells from the toxic effects of increased cytoplasmic calcium concentrations. Calbindin-D_{9K} also increases ATP-dependent Ca^{2+} transport in the duodenal basolateral membrane vesicles [58,59] and can bind to the regulatory calmodulin binding domain of the plasma membrane Ca^{2+} pump (see Christakos *et al.* [10] and Gross and Kumar [11] for review and chapter herein).

B. Uterus

In the uterus, the calbindin-D_{9K} gene is expressed mainly in the myometrium and the endometrial stroma of nonpregnant rats [36,42]. In pregnant rats, the calbindin-D_{9K} gene is also expressed in the endometrial epithelium [35] (Fig. 1). In contrast to the case in intestine, the calbindin-D_{9K} gene is not under the control of $1,25(OH)_2D_3$ in the uterus, despite the presence of VDR in this tissue [60]. The uterine calbindin-D_{9K} gene is under the positive control of estradiol [36,42,61,62] and negative control of progesterone [36,62]. Changes in the myometrial calbindin-D_{9K} content parallel changes in the intracellular calcium-dependent uterine muscular function. In the rat, calbindin-D_{9K} concentration increases at the begining of gestation, when the uterine contractions are responsible for the distribution of the embryos; decreases at mid-gestation, which corresponds to a relaxation stage, and increases at the end of gestation, when the uterine contractions appear [63]. Similarly, human calbindin-D_{9K} concentration is elevated in the myometrium at the end of gestation and decreases after labor [47]. Calbindin-D_{9K} is also present in the luminal epithelial cells of the fallopian tube of mature rats and is under the control of estradiol [37].

C. Placenta

The calbindin-D_{9K} gene is very active in fetal placenta and to a lesser extent in the maternal placenta and yolk sac of the rat (Table II and Fig. 1). The concentrations

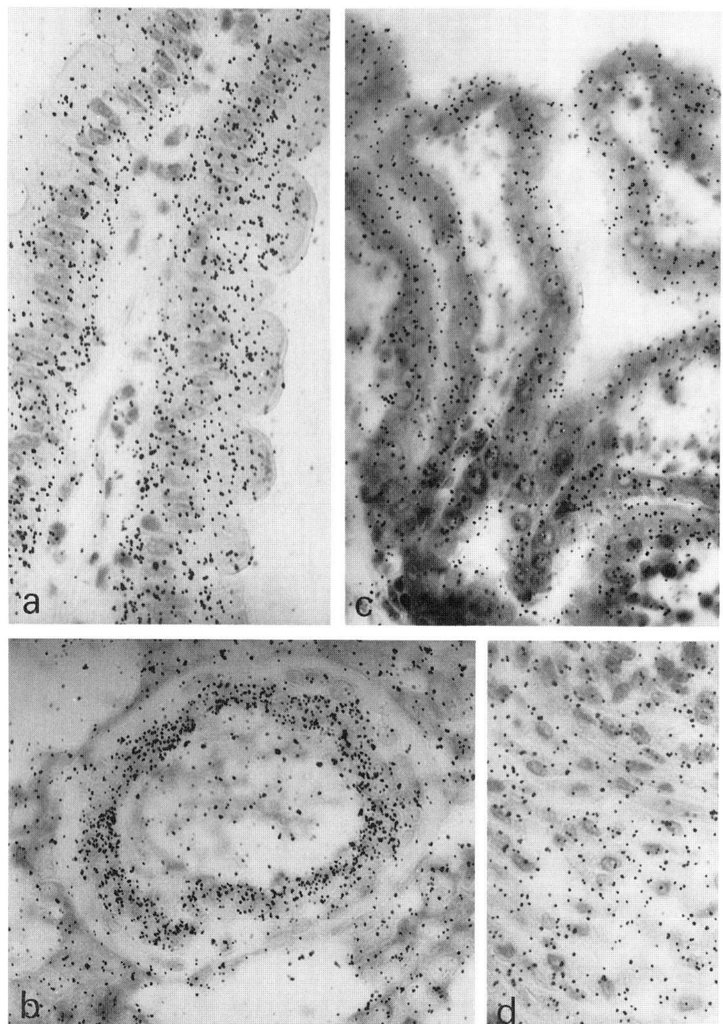

FIGURE 1 *In situ* hybridization of ^{3}H-labeled calbindin-D_{9K} mRNA to paraffin sections of several tissues from 21-day pregnant rats [26,27,35]. (a) Duodenal villus; (b) placenta: labyrinth zone showing intense labeling at the level of trophoblastic epithelium; (c) endometrium of the uterine horns: cells of the luminal epithelium; (d) myometrium of the uterine horn. Original magnification $\times$ 530. The corresponding tissue distribution of calbindin-D_{9K} is 20.0 $\pm$ 0.8, 1.2 $\pm$ 0.14, 0.1 $\pm$ 0.02, and 0.9 $\pm$ 0.007 μg/mg soluble protein in rat duodenum, fetal placenta, maternal placenta, and uterine horns, respectively, by radioimmunoassay [26,35].

of calbindin-D_{9K} protein and its mRNA increase at the end of gestation and parallel the increased calcium needs of the fetus. Hormonal regulation of calbindin-D_{9K} by 1,25$(OH)_2D_3$ in the mouse placenta and yolk sac has not been clearly demonstrated, as compared to that of calbindin-D_{9K} expression in mouse intestine and kidney [64].

D. Kidney

The calbindin-D_{9K} gene is expressed in mouse kidney [29,65], but it is not as active in rat kidney where the calbindin-D_{28K} gene is intensively expressed [13] (Tables I and II). Calbindin-D_{9K} is restricted to the basolateral membranes in the rat distal nephron [31] and its expression is regulated by 1,25$(OH)_2D_3$ in the mouse [29,65]. Calbindin-D_{9K} activates the Ca^{2+} pump in the rabbit basolateral membrane of both proximal and distal tubules [66]. Vitamin D depletion decreases both Ca^{2+} uptake by the luminal membrane of distal segments and ATP-dependent Ca^{2+} transport by the basolateral membranes of these same segments [67]. *In vivo*, 1,25$(OH)_2D_3$ may enhance Ca^{2+} reabsorption in the distal nephron through the synthesis of calbindin-D_{9K} and calbindin-D_{28K}.

E. Lung

The calbindin-D_{9K} gene is expressed in alveolar epithelial cells but not in alveolar macrophages of the lung in the rat. Neither vitamin D nor estradiol regulate its expression in this tissue [43]. Its function in lung remains unknown.

F. Mineralized Tissues

Calbindin-D_{9K} is specifically found in mature chondrocytes [38], osteoblasts and osteocytes [40] and in the ameloblasts of incisors and molars, but not in odontoblasts [38,40]. Calbindin-D_{9K} is 100-fold less abundant than calbindin-D_{28K} in dental tissues [40]. $1,25(OH)_2D_3$ appears to regulate calbindin-D_{9K} gene expression in bone and teeth [39,41,68,69]. Although calbindin-D_{9K} is strictly a cytosolic protein in soft tissue, it is restricted to the lateral edges of the longitudinal septa in the epiphyseal cartilage, the same area in which matrix vesicles are found, and in the matrix vesicles near the mineralizing front of trabecular and compact bone. Therefore, calbindin-D_{9K} could be necessary for mineral nucleation in the matrix vesicles of epiphyseal cartilage and bone. Several functions have been proposed for the protein in ameloblasts [40].

IV. GENE STRUCTURE

The rat intestinal calbindin-D_{9K} cDNA [70] (GenBank accession number X16635) was used to isolate calbindin-D_{9K} genomic clones from a rat genomic library. A total of 30 kbp overlapping the calbindin-D_{9K} gene was sequenced and analyzed. The gene is 2.5 kb long and contains three exons interrupted by two introns. The first exon contains the 5′ untranslated region (I); the second exon, the first EF hand (II); and the third exon, the second EF hand (III) and the 3′ untranslated region (Fig. 2). The two calcium binding sites are separated by a class o intron that interrupts the exon between two codons. The promoter region contains a TATA box 31 bp upstream from the single transcription start site [70,71].

The gene encoding human calbindin-D_{9K} has also been cloned and sequenced [17,18]. It lies on chromosome Xp22.2 and is not linked to the hypophosphatemic rickets gene on the X-chromosome ([18]; Oudet, ICGM Strasbourg, personal communication). The structure of the human gene is very similar to that of the rat gene. The promoter regions of the rat and human genes are also very similar, suggesting that important regulatory factors bind to that region [72].

Mapping of the DNase I-hypersensitive sites in nuclear chromatin isolated from several rat tissues revealed a cluster of DNase-I hypersensitive sites in the proximal promoter region of the active gene in the intestine and uterus [73]. The cis-acting elements involved in the intestine-specific expression of the rat calbindin-D_{9K} gene have been identified, and the trans-acting factors have been characterized in *ex vivo* and *in vitro* experiments [72]. Caudal homeobox-2 (Cdx-2), an intestinal-specific transcription factor, binds to the TATA box of the rat calbindin-D_{9K} gene combined with liver-specific (HNF-1, C/EBP, and HNF4) and ubiquitous transcription factors that bind to the proximal promoter region (Fig. 3) [72]. Besides the proximal region, a distal

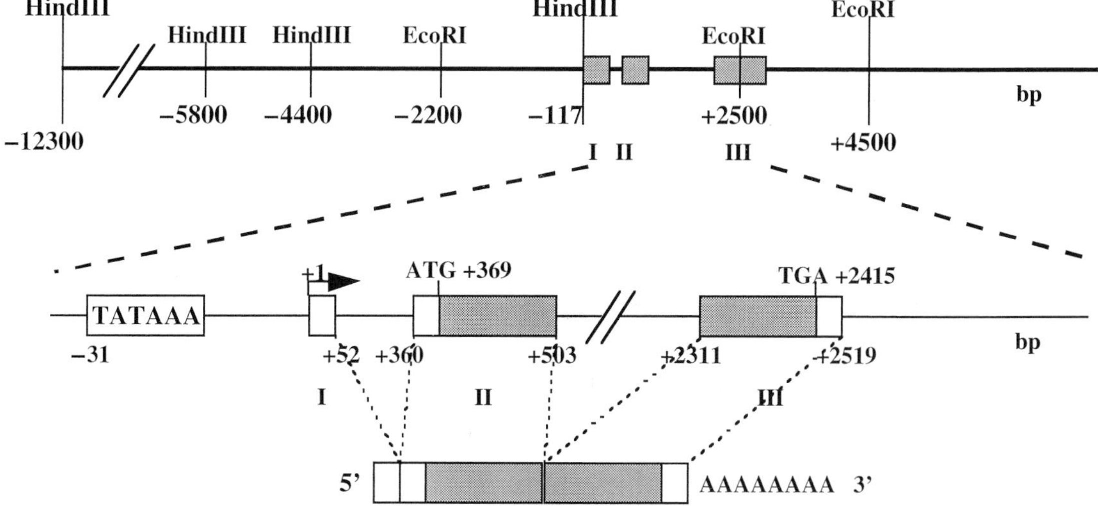

FIGURE 2 Sequence of the rat calbindin-D_{9K} gene [19,70]. Selected restriction sites, exons (numbered) and introns (—) are shown in the top line. The proximal promoter with the start site of transcription (+1) and the exons (rectangle) including the coding sequences (shaded) are shown below. Splicing (---) gives rise to a 405-nucleotide messenger RNA.

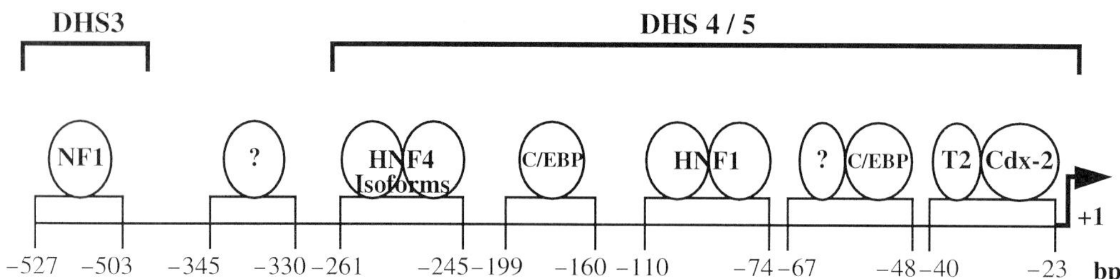

FIGURE 3 Identification of potential cis elements and corresponding bound trans-acting factors in the proximal promoter of the rat calbindin-D_{9K} gene in the intestine [70]. DNase I-hypersensitive (DHS) sites 3, 4, and 5 are noted at top [73].

region located 3.5 kb upstream of the site of transcription also seems to be necessary for expression of the rat calbindin-D_{9K} gene in the intestine, as suggested by DNase I-hypersensitive site analysis [73], cis- and trans-acting phenomena analysis [72], and investigations in transgenic mice [74,75].

V. REGULATION OF CALBINDIN-D_{9K} GENE EXPRESSION

A. Regulation by $1,25(OH)_2D_3$

Calbindin-D_{9K} synthesis is under the control of $1,25(OH)_2D_3$, the hormonally active metabolite of vitamin D_3, in the intestine of rats [13,49]. $1,25(OH)_2D_3$ rapidly induces calbindin-D_{9K} synthesis in rats fed a vitamin D-free diet for 5 weeks from weaning and in organ culture of fetal rat duodenum [76]. Run-on experiments [77] suggest that $1,25(OH)_2D_3$ controls the expression of the calbindin-D_{9K} gene at the transcriptional and posttranscriptional levels. $1,25(OH)_2D_3$ acts through its nuclear receptor (VDR), a member of the nuclear steroid/thyroid receptor superfamily which functions as a ligand-inducible transcription factor. The VDR binds to the vitamin D responsive element (VDRE) as a heterodimer with retinoid X receptor (RXR) [78–80] and see Chapters 8–10. A consensus binding site for the RXR–VDR heterodimer, RGGTCA NNR RGTTCA, has been characterized using PCR-mediated binding site selection [81,82]. No such consensus sequence has been found in the rat calbindin-D_{9K} promoter sequence. However, the responsiveness to vitamin D in the duodenum is contained in the largest 4400 bp construct, as shown by the transgenic mouse analysis [74]. A vitamin D response element (VDRE) has been proposed for the rat calbindin-D_{9K} gene using *in vitro* analysis [83]. It is clear that this element, GGGTGT NNN AAGCCC, located at −481 bp is different from the above selected consensus sequence. Calbindin-D_{9K} regulatory sequence constructs in rat osteosarcoma (ROS) 17/2.8 cells showed no increase in CAT activity in response to 10^{-8} *M* $1,25(OH)_2D_3$, whereas the rat osteocalcin gene VDRE was used as a positive control [84]. Therefore, the physiological relevance of this element remains to be demonstrated using transgenic mice. These results suggest that the transcriptional control of the rat calbindin-D_{9K} gene by $1,25(OH)_2D_3$ in the rat duodenum is complex, involving trans-acting factors and perhaps nonclassic pathways.

The capacity of the jejunum, ileum, and cecum in vitamin D calcium deficient rats to respond to exogenous $1,25(OH)_2D_3$, in terms of calbindin-D_{9K} protein and mRNA concentrations, is greater than that of the duodenum [49]. This hormonal sensitivity is pertinent since, up to now, the duodenum rather than the distal intestine has been considered to be the primary target for vitamin D. The phenomenon appears to occur as a result of calcium absorptive stress imposed by calcium deprivation and may have considerable relevance in physiological and clinical applications [85]. The molecular mechanisms underlying the response remain to be investigated. A sequence related to the VDRE has been detected in the human gene about 1.1 kb upstream of the promoter [17]. The physiological relevance of this element remains to be determined. Similar regulation of calbindin-D_{9K} gene expression by vitamin D in kidney, placenta, bone, and teeth remains to be demonstrated.

B. Regulation by Calcium

There is some evidence that dietary calcium modulates intestinal calbindin-D_{9K} gene expression independently of the renal $1,25(OH)_2D_3$-stimulated synthesis in the rat [8,86]. This has been verified *in vitro* [76]. Increasing the calcium concentration in the medium of fetal rat intestinal organ cultures causes an increase in the production of calbindin-D_{9K} mRNA in the presence and absence of $1,25(OH)_2D_3$. These data underline the complexity of the molecular mechanisms involved in the tissue-specific regulation of the expression of the calbindin-D_{9K} gene in the intestine.

C. Regulation by 17β-Estradiol

Unlike the case in intestine, calbindin-D_{9K} gene expression is under the control of estrogens in the rat uterus and appears to be independent of the vitamin D status of this tissue [36,42], despite the presence of VDR [60]. Estradiol is the major factor controlling calbindin-D_{9K} gene expression [61,62]. Calbindin-D_{9K} mRNA is undetectable in the uterus of ovariectomized rats. A single injection of 17(β-estradiol rapidly induces the expression of the calbindin-D_{9K} gene in this tissue. Estrogens control calbindin-D_{9K} gene expression, at least in part, at the transcriptional level. Similarly, under physiological conditions, calbindin-D_{9K} gene expression is maximum during the estrogen-dominated phase of the estrous cycle [62]. There is also *in vitro* evidence that estradiol acts via an imperfect estrogen responsive element (ERE) located at the border of the first exon and first intron [87,88]:

Rat calbindin-D_{9K}ERE	AGGTCAgggTGATCT
Vitellogenin consensus ERE	AGGTCAacgTGACCT

In vivo experiments using transgenic mice strongly suggest that this imperfect ERE is functional, since the shortest −117 bp calbindin-D_{9K} gene construct is only expressed in the uterine myometrium and is strictly estrogen-dependent, like that of the endogenous gene [74,75].

The human gene also contains a similar element at the border of the first exon/intron. The rat and human calbindin-D_{9K} EREs differ by two essential nucleotides in this region. The human sequence does not bind the estrogen receptor. It has been proposed that a two-nucleotide mutation causes the lack of gene expression in the human uterus [17]. However, controversial data have been reported for the human myometrium [47]. Tamoxifen and ICI 182780 were reported to have opposing effects on estrogen-induced calbindin-D_{9K} *in vivo* in the rat uterus and in primary cultures of myometrial cells. As for many uterine genes, tamoxifen, a mixed estrogen antagonist, acts as agonist, whereas ICI 182780, a pure estrogen antagonist, acts as an antagonist in the control of rat calbindin-D_{9K} gene expression [89]. Progesterone has a late negative effect on this estradiol-induced expression [62]. A promising approach to the problem of the tissue-specific genomic mechanism by which steroid hormones act is the use of overexpressed calbindin-D_{9K}- and calbindin-D_{9K}-deficient transgenic mice. Results from these studies should also clarify the physiological function(s) of calbindin-D_{9K}.

VI. CONCLUSION

Calbindin-D_{9K} belongs to the S100 Ca^{2+} binding protein subfamily of the troponin C superfamily of Ca^{2+} binding proteins. This small, acidic protein containing two Ca^{2+} binding sites seems to occur in most mammals. Calbindin-D_{9K} gene expression is tissue specific. Calbindin-D_{9K} is abundant in the duodenum (2% of soluble proteins), placenta, and uterus but is also present in the yolk sac, fallopian tube, lung, kidney, cartilage, and teeth. Aside from its Ca^{2+} binding property, no true physiological function has been clearly established. Calbindin-D_{9K} may be involved in active transport of calcium in the duodenum, kidney, and placenta, in contraction in the myometrium, and in mineralization in bone and teeth. A protective role against intracellular Ca^{2+} toxicity has also been postulated. Expression of the calbindin-D_{9K} gene is positively regulated in a tissue-specific manner at both transcriptional and posttranscriptional levels by $1,25(OH)_2D_3$ in the duodenum and kidney, and by estrogens in the uterus. The calbindin-D_{9K} gene has been cloned, and DNase I-hypersensitive sites have been described in the duodenum. Duodenum and uterus utilize distinct cis-acting elements to direct and regulate the expression of the rat calbindin-D_{9K} gene. Several trans-acting factors other than the VDR are probably involved in the intestine-specific calbindin-D_{9K} gene expression. The calbindin-D_{9K} gene contains an imperfect ERE which seems to be functional in transgenic mice. The calbindin-D_{9K} gene is thus an appropriate model to enhance our understanding of the specific genomic mechanisms by which the vitamin D_3 hormone acts.

References

1. Wasserman RH 1985 Nomenclature of the vitamin D induced calcium binding proteins. In: Norman AW, Schaefer K, Grigoleit HG, Herrath DV (eds) Vitamin D: Chemical, Biochemical and Clinical Update. de Gruyter, Berlin, pp. 321–322.
2. Kallfelz FA, Taylor AN, Wasserman RH 1967 Vitamin D-induced calcium-binding factor in rat intestinal mucosa. Proc Soc Exp Biol Med **125**:54–58.
3. Kawasaki H, Kretsinger RH 1994 Calcium-binding proteins. 1:EF-hands. Protein Profile **1**:343–517.
4. Perret C, Lomri N, Thomasset M 1988 Evolution of the EF-hand calcium-binding protein family: Evidence for exon shuffling and intron insertion. J Mol Evol **27**:351–364.
5. Thomasset M, L'Horset F, Blin C, Lambert M, Colnot S, Perret C 1996 Calbindin-D9k. In: Celio MR (ed) Guidebook to the Calcium-Binding Proteins. Oxford Univ. Press, London, pp. 157–159.
6. Schäfer BW 1996 S100 Proteins. In: Celio MR (ed) Guidebook to the Calcium-Binding Proteins. Oxford Univ. Press, London, pp. 132–134.
7. Kretsinger RH, Nakayama S 1993 Evolution of EF-hand calcium-

modulated proteins IV. Exon shuffling did not determine the domain compositions of the EF-hand proteins. J Mol Evol **36**:477–488.
8. Bronner F, Pansu D, Stein WD 1986 An analysis of intestinal calcium transport across the rat intestine. Am J Physiol **150**:G561–G569.
9. Wasserman RH, Fullmer CS 1995 Vitamin D and intestinal calcium transport: Facts, speculations and hypotheses. J Nutr **125**:1971S–1979S.
10. Christakos S, Gabrielides C, Rhoten WB 1989 Vitamin D-dependent calcium binding proteins: Chemistry, distribution, functional considerations, and molecular biology. Endocr Rev **10**:84–107.
11. Gross M, Kumar R 1990 Physiology and biochemistry of vitamin D-dependent calcium-binding protein. Am J Physiol **259**:F195–F209.
12. Marche P, Pradelles P, Gros C, Thomasset M 1977 Radioimmunoassay for a vitamin D dependent calcium-binding protein in rat duodenal mucosa. Biochem Biophys Res Commun **76**:1020–1026.
13. Thomasset M, Parkes CO, Cuisinier-Gleizes P 1982 Rat calcium-binding proteins: Distribution, development, and vitamin D dependence. Am J Physiol **243**:E483-E488.
14. Baudier J, Glassert N, Strid L, Brehier A, Thomasset M, Gerard D 1985 Purification, calcium-binding properties, and conformational studies on a 28-kDa cholecalcin-like protein from bovine brain. J Biol Chem **260**:10662–10670.
15. Desplan C, Thomasset M, Moukhtar M 1983 Synthesis, molecular cloning, and restriction analysis of DNA complementary to vitamin D-dependent calcium-binding mRNA from rat duodenum. J Biol Chem **258**:2762–2765.
16. Darwish HM, Krisinger J, Strom M, DeLuca HF 1987 Molecular cloning of the cDNA and chromosomal gene for vitamin D-dependent calcium-binding protein of rat intestine. Proc Natl Acad Sci USA **84**:6108–6111.
17. Jeung EB, Krinsinger J, Dann JL, Leung PCK 1992 Molecular cloning of the full-length cDNA encoding the human calbindin-D9k. FEBS Lett **307**:224–228.
18. Howard A, Legon S, Spurk NK, Walters JRF 1992 Molecular cloning and chromosomal assignment of human calbindin-D9k. Biochem Biophys Res Commun **185**:663–669.
19. Desplan C, Heidmann O, Lillie JW, Auffray C, Thomasset M 1983 Sequence of rat intestinal vitamin D-dependent calcium-binding protein derived from a cDNA clone. Evolutionary implications. J Biol Chem **258**:13502–13505.
20. Jeung EB, Krisinger J, Dann JL, Leung PCK 1992 Cloning of the porcine calbindin-D9k complementary deoxyribonucleic acid by anchored polymerase chain reaction technique. Biol Reprod **47**:503–508.
21. Forsen S, Drakenberg T, Johansson C, Linse S, Thulin E, Kordel J 1990 Protein engineering and structure/function relations in bovine calbindin-D9k. Adv Exp Med Biol **269**:37–42.
22. Skelton HJ, Kördel J, Akke M, Forsen S, Chazin WJ 1994 Signal transduction versus buffering activity in Ca^{2+}-binding proteins. Struct Biol **1**:239–244.
23. Sandgren ME, Bronnegard M, DeLuca HF 1991 Tissue distribution of the 1.25-dihydroxyvitamin D_3 receptor in the male rat. Biochem Biophys Res Commun **181**:611–616.
24. Marche P, Cassier P, Mathieu H 1980 Intestinal calcium-binding protein. A protein indicator of enterocyte maturation associated with the terminal web. Cell Tissue Res **212**:63–72.
25. Taylor AN, Gleason WA, Lankord GL 1984 Immunocytochemical localization of rat intestinal vitamin D-dependent calcium-binding protein. J Histochem Cytochem **32**:153–158.
26. Warembourg M, Perret C, Thomasset M 1986 Distribution of vitamin D-dependent calcium-binding protein messenger ribonucleic acid in rat placenta and duodenum. Endocrinology **119**:176–184.
27. Warembourg M, Tranchant O, Perret C, Desplan C, Thomasset M 1986 In situ detection of vitamin D-induced calcium-binding protein (9-kDa CaBP) messenger RNA in rat duodenum. J Histochem Cytochem **34**:277–280.
28. Arnold BM, Kovacs K, Murray TM 1976 Cellular localization of intestinal calcium-binding protein in pig duodenum. Digestion **14**:77–84.
29. Rhoten WB, Bruns ME, Christakos S 1985 Presence and localization of two vitamin D-dependent calcium-binding proteins in kidney of higher vertebrates. Endocrinology **117**:674–683.
30. Schreiner DS, Jande SS, Parkes CO, Lawson EM, Thomasset M 1983 Immunocytochemical demonstration of two vitamin D-dependent calcium-binding proteins in mammalian. Acta Anat **117**:1–14.
31. Bindels RJM, Timmermans JAH, Hartog A, Coers W, Van Os CH 1991 Calbindin-D9k and parvalbumin are exclusively located along basolateral membranes in rat distal nephron. J Am Soc Nephrol **2**:1122–1129.
32. Delorme AC, Cassier P, Geny B, Mathieu H 1983 Immunocytochemical localization of vitamin D-dependent calcium-binding protein in the yolk sac of the rat. Placenta **4**:263–270.
33. Bruns ME, Kleeman E, Millis SE, Bruns DE, Herr JC 1985 Immunocytochemical localization of vitamin D-dependent calcium-binding protein in mouse placenta and yolk sac. Anat Rec **213**:514–517.
34. Bruns ME, Vollmer S, Wallshein V, Bruns DE 1981 Vitamin D-dependent calcium-binding protein. Immunocytochemical studies and synthesis by placental tissue in vitro. J Biol Chem **256**:4649–4653.
35. Warembourg M, Perret C, Thomasset M 1987 Analysis and in situ detection of cholecalcin messenger RNA (9000 M_r CaBP) in the uterus of the pregnant rat. Cell Tissue Res **247**:51–57.
36. Bruns ME, Overpeck JG, Smith GC, Hirsch GN, Milis SE, Bruns DE 1988 Vitamin D-dependent calcium-binding protein in rat uterus: Differential effects of estrogen, tamoxifen, progesterone and pregnancy on accumulation and cellular localization. Endocrinology **122**:2371–2378.
37. Mathieu H, Mills SE, Burnett SH, Cloney DL, Bruns DE, Bruns ME 1989 The presence and estrogen control of immunoreactive calbindin-D9k in the fallopian tube of the rat. Endocrinology **125**:2745–2749.
38. Balmain N, Tisserand-Jochem E, Thomasset M, Cuisinier-Gleizes P, Mathieu H 1986 Vitamin D-dependent calcium binding protein (CaBP9k) in rat growth cartilage. Histochemistry **84**:161–168.
39. Balmain N 1991 Calbindin-D9k. A vitamin D-dependent calcium-binding protein in mineralized tissues. Clin Orthop Related Res **265**:276–290.
40. Berdal A, Hotton D, Saffar JL, Thomasset M, Nanci A 1996 Calbindin-D_{9k} and calbindin-D_{28k} expression in rat mineralized tissues in vivo. J Bone Miner Res **11**:768–779.
41. Berdal A, Hotton D, Pike JW, Mathieu H, Dupre J-M 1993 Cell and stage-specific expression of vitamin D receptor and calbindin genes in rat incisor: Regulation by 1,25-dihydroxyvitamin D_3. Dev Biol **155**:172–179.
42. Delorme AC, Danan JL, Acker MG, Ripoche MA, Mathieu H 1983 In rat uterus 17β-estradiol stimulates a calcium-binding protein similar to the duodenal vitamin D-dependent calcium-binding protein. Endocrinology **113**:1340–1347.
43. Dupret JM, L'Horset F, Perret C, Bernaudin JF, Thomasset M 1992 Calbindin-D9k gene expression in the lung of the rat. Absence of regulation by 1,25-dihydroxyvitamin D_3 and estrogen. Endocrinology **131**:2643–2648.

44. Alpers DH, Lee SW, Avioli LV 1972 Identification of two calcium-binding proteins in human small intestine. Gastroenterology **62**:559.
45. Staun M 1987 Distribution of the 10,000 molecular weight calcium-binding protein along the small and large intestine of man. Gut **28**:878–882.
46. Staun M, Jarnum S 1988 Measurement of the 10,000-molecular weight calcium-binding protein in small-intestinal biopsy specimens from patients with malabsorption syndromes. Scand J Gastroenterol **23**:827–832.
47. Miller E, Word RA, Goodall CA, Iacopino AM 1994 Calbindin-D9k gene expression in human myometrium during pregnancy and labor. J Clin Endocr Metab **79**:609–615.
48. Brun P, Dupret JM, Perret C, Thomasset M, Mathieu H 1987 Vitamin D-dependent calcium-binding proteins (CaBPs) in human fetuses: Comparative distribution of 9k CaBP mRNA and 28k CaBP during development. Pediatr Res **21**:362–367.
49. Perret C, Desplan C, Thomasset M 1985 Cholecalcin a 9-kDa cholecalciferol-induced calcium-binding protein) messenger RNA. Distribution and induction by calcitriol in the rat digestive tract. Eur J Biochem **150**:211–217.
50. Huang YC, Lee S, Stolz R, Gabrielides C, Pansini-Porta A, Bruns ME, Bruns DE, Miffin TE, Pike JW, Christakos S 1989 Effect of hormones and development on the expression of the rat 1,25-dihydroxyvitamin D_3 receptor gene. Comparison with calbindin gene expression. J Biol Chem **264**:17454–17461.
51. Takahashi N, Shinki T, Kawate N, Samjima K, Nishi Y, Suda T 1982 Distribution of ornithine decarboxylase activity induced by 1,25-dihydroxyvitamin D_3 in chick duodenal villus mucosa. Endocrinology **111**:1539–1545.
52. Chan SDH, Atkins D 1984 The temporal distribution of the 1,25-dihydroxyvitamin D_3 receptor in the rat jejunal villus. Clin Sci **67**:285–290.
53. Danan JL, Delorme AC, Cuisinier-Gleizes P 1989 Biochemical evidence for a cytoplasmic 1α,25-dihydroxyvitamin D_3, receptor-like protein in rat yolk sac. J Biol Chem **256**:4847–4850.
54. Kessler MA, Lamm L, Jarnagen K, DeLuca HF 1986 1,25-Dihydroxyvitamin D_3-stimulated messenger-RNAs in rat small intestine. Arch Biochem Biophys **251**:403–412.
55. Thomasset M, Cuisinier-Gleizes P, Mathieu H 1979 1,25-Dihydroxyvitamin D_3: Dynamics of the stimulation of duodenal calcium-binding protein, calcium transport and bone calcium mobilization in vitamin D- and calcium-deficient rats. FEBS Lett **107**:91–94.
56. Bishop CW, Kendrick NC, DeLuca HF 1983 Induction of calcium binding protein before 1,25-dihydroxyvitamin D_3 stimulation of duodenal calcium uptake. J Biol Chem **258**:1305–1310.
57. Kumar R 1995 Calcium transport in epithelial cells of the intestine and kidney. J Cell Biochem **57**:392–398.
58. Walters S 1989 Calbindin-D9k stimulates the calcium pump in rat enterocyte basolateral membranes. Am J Physiol **256**:G124–G128.
59. Walters JR, Howard A, Charpin MV, Gniecko KC, Brodin P, Thulin E, Forsen S 1990 Stimulation of intestinal basolateral membrane calcium pump activity by recombinant synthetic calbindin-D9k and specific mutants. Biochem Biophys Res Commun **170**:603–608.
60. Walters RW 1981 An estrogen stimulated 1,25-dihydroxyvitamin D_3 receptor in the rat uterus. Biophys Biochem Res Commun **103**:721–726.
61. L'Horset F, Perret C, Brehier A, Thomasset M 1990 17b-Estradiol stimulates the calbindin-D9k (CaBP9k) gene expression at the transcriptional and postranscriptional levels in the rat uterus. Endocrinology **127**:2891–2897.
62. L'Horset F, Blin C, Brehier A, Thomasset M, Perret C 1993 Estrogen-induced calbindin-D9k gene expression in the rat uterus during the estrous cycle: Late antagonistic effect of progesterone. Endocrinology **132**:489–495.
63. Mathieu CL, Burnett SH, Mills SE, Overpeck JG, Bruns MEB, Bruns DE 1989 Gestational changes in calbindin-D9k in rat uterus, yolk sac and placenta: Implications for maternal fetal calcium transport and uterine muscle function. Proc Natl Acad Sci USA **86**:3433.
64. Bruns ME, Kleeman E, Bruns DE 1986 Vitamin D-dependent calcium-binding protein of mouse yolk sac. Biochemical and immunochemical properties and response to 1,25-dihydroxycholecalciferol. J Biol Chem **261**:7485–7490.
65. Delorme AC, Danan JL, Mathieu H 1983 Biochemical evidence for the presence of two vitamin D-dependent calcium-binding proteins in mouse kidney. J Biol Chem **258**:1878–1884.
66. Bouhtiauy I, Lajeunesse D, Christakos S, Brunette MG 1994 Two vitamin D_3-dependent calcium binding proteins increase calcium reabsorption by different mechanisms II. Effect of CaBP9k. Kidney Int **45**:469–474.
67. Bouhtiauy I, Lajeunesse D, Brunette MG 1993 Effect of vitamin D depletion on calcium transport by the luminal and basolateral membranes of the proximal and distal nephron. Endocrinology **132**:115–120.
68. Balmain N, Berdal A, Hotton D, Cuisinier-Gleizes P, Mathieu H 1989 Calbindin-D9k immunolocalization and vitamin D-dependence in the bone of growing and adult rats. Histochemistry **92**:359–365.
69. Balmain N, Hauchecorne M, Pike WJ, Cuisinier-Gleizes P, Mathieu H 1993 Distribution and subcellular immunolocalization of 1,25-dihydroxyvitamin D_3 receptors in rat epiphyseal cartilage. Cell Mol Biol **39**:329–350.
70. Perret C, Lomri N, Gouhier N, Auffray C, Thomasset M 1988 The rat vitamin D-dependent calcium-binding protein (9-kDa CaBP) gene. Complete nucleotide sequence and structural organization. Eur J Biochem **172**:43–51.
71. Krisinger J, Darwish H, Maeda N, Deluca HF 1988 Structure and nucleotide sequence of the rat intestinal vitamin D-dependent calcium binding protein gene. Proc Natl Acad Sci USA **85**:8988–8992.
72. Lambert M, Colnot S, Suh Er, L'horset F, Blin C, Calliot ME, Raymondjean M, Thomasset M, Traber PG, Perret C 1996 Identification of the cis-acting elements and transcription factors required for the intestinal specific expression of the rat calbindin-D9k gene: The intestine specific transcription factor Cdx-2 binds to the TATA box. Eur J Biochem **236**:778–788.
73. Perret C, L'Horset F, Thomasset M 1991 DNase I-hypersensitive sites are associated, in a tissue-specific manner, with expression of the calbindin-D9k encoding gene. Gene **108**:227–235.
74. Romagnolo B, Cluzeaud F, Lambert M, Colnot S, Porteu A, Molina T, Thomasset M, Vandewalle A, Kahn A, Perret C 1996 Tissue specific and hormonal regulation of calbindin-D9k fusion genes in transgenic mice. J Biol Chem **271**:16820–16826.
75. Romagnolo B, Molina T, Leroy G, Blin C, Porteu A, Thomasset M, Vanderwalle A, Kahn A, Perret C 1996 Estradiol-dependent uterine leiomyomas in transgenic mice. J Clin Invest **98**:777–784.
76. Bréhier A, Thomasset M 1990 Stimulation of calbindin-D9k (CaBP9k) gene expression by calcium and 1,25-dihydroxycholecalciferol in fetal rat duodenal organ culture. Endocrinology **127**:580–587.
77. Dupret JM, Brun P, Perret C, Lomri N, Thomasset M, Cuisinier-Gleizes P 1987 Transcriptional and post-transcriptional regulation of vitamin D-dependent calcium-binding protein gene expression in the rat duodenum by 1,25-dihydroxyvitamin D_3. J Biol Chem **262**:16553–16557.

78. Ross TK, Moss VE, Prahl JM, DeLuca HF 1992 A nuclear protein essential for binding of rat 1,25-dihydroxyvitamin D_3 receptor to its response elements. Proc Natl Acad Sci USA **89**:256–260.
79. Schräder M, Nayeri S, Kahlen JP, Muller KM, Carlberg C 1995 Natural vitamin D_3 response element formed by inverted palindromes: Polarity-directed ligand sensitivity of vitamin D_3 receptor–retinold X receptor heterodimer-mediated transactivation. Mol Cell Biol **15**:1154–1161.
80. Sone T, Kerner SA, Pike JW 1991 Vitamin D receptor interaction with specific DNA. Association as a 1,25-dihydroxyvitamin D_3-modulated heterodimer. J Biol Chem **266**:23296–23305.
81. Nishikawa JI, Kitaura M, Matsumoto M, Imigawa M, Nishihara T 1994 Difference and similarity of DNA sequence recognized by VDR homodimer and VDR/RXR heterodimer. Nucleic Acids Res **22**:2902–2907.
82. Colnot S, Lambert M, Blin C, Thomasset M, Perret C 1995 Identification of DNA sequences that bind retinoid X receptor–1,25-dihydroxyvitamin D_3 receptor heterodimers with high affinity. Mol Cell Endocrinol **113**:89–98.
83. Darwish HM, DeLuca HF 1992 Identification of a 1,25-dihydroxyvitamin D_3 response element in the 5′ flanking region of the rat calbindin-D9k gene. Proc Natl Acad Sci USA **89**:603–607.
84. Thomasset M, L'Horset F, Blin C, Drittanti L, Lambert M, Colnot S, Brehier A, Perret C 1992 Tissue-specific regulation of calbindin-D9k gene expression by steroid hormones. In: Cohn DV et al. (eds) Calcium Regulating Hormones and Bone Metabolism. Amsterdam, Elsevier, pp. 83–91.
85. Thomasset M, Pointillart A, Cuisinier-Gleizes P, Guéguen L 1979 Effect of vitamin D or calcium deficiency on duodenal, jejunal and ileal calcium-binding protein and on plasma calcium and 25-dihydroxycholecalciferol levels in the growing pig. Ann Biol Anim Biochem Biophys **19**:769–773.
86. Freund T, Bronner F 1975 Regulation of intestinal calcium-binding protein by calcium intake in the rat. Am J Physiol **228**:861–869.
87. Darwish H, Krisinger J, Furlox D, Smith C, Murdoch FE, DeLuca HF 1991 An estrogen-responsive element mediates the transcriptional regulation of calbindin-D9k gene in rat uterus. J Biol Chem **266**:551–558.
88. L'Horset F, Blin C, Colnot S, Lambert M, Thomasset M, Perret C 1994 Calbindin-D9k gene expression in the uterus: Study of the two messenger ribonucleic acid species and analysis of an imperfect estrogen-responsive element. Endocrinology **134**:11–18.
89. Blin C, L'Horset F, Leclerc T, Lambert M, Colnot S, Thomasset M, Perret C 1995 Contrasting effects of tamoxifen and ICI 182 780 on estrogen-induced calbindin-D9k gene expression in the uterus and in primary culture of myometrial cells. J Steroid Biochem Mol Biol **55**:1–7.

CHAPTER 15

Rapid Biological Responses Mediated by 1α,25-Dihydroxyvitamin D$_3$: A Case Study of Transcaltachia (Rapid Hormonal Stimulation of Intestinal Calcium Transport)

ANTHONY W. NORMAN Department of Biochemistry and Division of Biomedical Sciences, University of California at Riverside, Riverside, California

I. INTRODUCTION TO VITAMIN D_3 AND $1\alpha,25(OH)_2D_3$

The seco-steroid 1α,25-dihydroxyvitamin D_3 [$1\alpha,25(OH)_2D_3$] is now known to initiate biological responses both via regulation of gene transcription [1–3] as well as via "rapid" pathways [4–6], some of which involve opening of voltage-gated Ca^{2+} channels [7] and involvement of protein kinase C (PKC) [8–12]. The objectives of this chapter are to provide an overview of the current status of the rapid or nongenomic responses mediated by $1\alpha,25(OH)_2D_3$ with respect to the current understanding of the signal transduction pathways involved and with consideration of the similarities and differences of agonist and antagonist ligands for the rapid responses in comparison to the nuclear receptor for $1\alpha,25(OH)_2D_3$. A major focus of the chapter consists of an in-depth consideration of the rapid response of transcaltachia or the rapid hormonal stimulation of intestinal Ca^{2+} transport.

The first issue to address is the problem of definition; what is implied by the term "rapid response"? For the purposes of this chapter, "rapid response" is used to describe a biological response to $1\alpha,25(OH)_2D_3$ which appears to occur too rapidly to be simply explained via interaction of the ligand with the nuclear receptor. Frequently these responses are generated within seconds to 1–2 min. However, as there is not yet available a detailed molecular mechanism to describe all the observed rapid responses to $1\alpha,25(OH)_2D_3$, it is not known with certainty whether all of them occur via signal transduction pathways which do not involve immediate involvement of $1\alpha,25(OH)_2D_3$ with its nuclear receptor.

A. Structure of $1\alpha,25(OH)_2D_3$

The hormone $1\alpha,25(OH)_2D_3$ is one of 37 known metabolites of vitamin D_3 [13,14]. The structure of $1\alpha,25(OH)_2D_3$ is presented in Fig. 1A. The structure and chemistry of vitamin D compounds are related to those of the classic steroid hormones, for example, the glucocorticoids, estrogens, androgens, progestins, and mineralocorticoids. However, there are three structural aspects of the vitamin D compounds that make them distinct from classic steroids. First, the vitamin D_3 compounds have a side chain comprising eight carbons whereas the classic steroids have no side chain or only a two-carbon side chain. Second, the 9,10 carbon–carbon bond is broken in this family of related compounds, and, accordingly, they are designated as being seco steroids. The term "seco" designates that one of the rings of the cyclopentanoperhydrophenanthrene ring structure of classic steroids has been broken. Third, the vitamin D_3 family of molecules are unusually conformationally flexible in contradistinction to the classic steroids (see below).

B. Conformational Flexibility

Breakage of the 9,10 carbon–carbon bond "releases" the A ring from being held in place adjacent to the B ring, and, accordingly, the cyclohexane-like A ring is free to rapidly interchange (many thousands of times per second) between a pair of chair–chair conformers. This has the consequence of changing the orientation

FIGURE 1 Structural aspects of $1\alpha,25(OH)_2D_3$ that contribute to its conformational flexibility. The reader is also referred to Chapter 57 of this volume for an additional discussion of this topic. (A) The location of the three structural aspects of vitamin D seco-steroids that contribute to the conformational flexibility of these molecules are indicated and further illustrated in B–E. (B) The cyclohexane-like A ring is free to rapidly interchange between a pair of chair–chair conformers, effectively equilibrating the key 1α- and 3β-hydroxyls between equatorial and axial orientations [15,16]. Thus, when the 1α-hydroxyl is axial, the 3β-hydroxyl is equatorial, and *vice versa*. (C) Rotational freedom about the 6,7 carbon–carbon single bond of the seco B ring allows conformations ranging from the more steroid like 6-s-cis conformation to the open and extended 6-s-trans form of the hormone [17]. See also part E. (D) The dynamic single bond rotation of the cholesterol-like side chain of the hormone is shown at center (i.e., rotations about the six single bonds are indicated by the curved arrows at left). Energy minimization calculations to define the most stable orientation of the side chain indicate that there are 394 conformers within 4 kcal/mol of the global minimum (lowest energy state). "Dots" indicate the position in three-dimensional space of the 25-hydroxyl group for the 394 minima; collectively, these dots define the "volume in space" that the side chain occupies. The center drawing illustrates a "top view" looking down on the C/D rings, whereas the right-hand structure presents an "edge-on" view of the C/D rings. A discussion of the consequences of the side-chain conformational mobility has been previously presented [18–20]. (E) Excursions of the 1α-hydroxyl of $1\alpha,25(OH)_2D_3$ in three-dimensional space as a consequence of the A- and B-ring conformational flexibility are illustrated. "Dots" indicate the position of the 1α-hydroxyl for approximately 100 conformers that are close to the global minimum for rotation about the 6,7 carbon–carbon bond. The three drawings at right provide different views of the $1\alpha,25(OH)_2D_3$ molecule (shown at left). Structure 1 is a top view looking down on the A, C, and D rings; structure 2 is an end view where the A ring is closest to the observer and the D ring and side chain are away from the observer; structure 3 is an in-plane view with the A ring at the bottom and the D ring and side chain at the top. Three clusters of dots are apparent. The circled set describes the excursions of the C-1 hydroxyl for the 6-s-trans conformer of $1\alpha,25(OH)_2D_3$. The two other clusters, indicated by an asterisk (*) and a double asterisk (**), represent the two families of 6-s-cis conformers. One family (*) has its C10(19) double bond under the C/D ring, whereas the other family (**) has its C-10(19) double bond above the C/D ring (see structure 2).

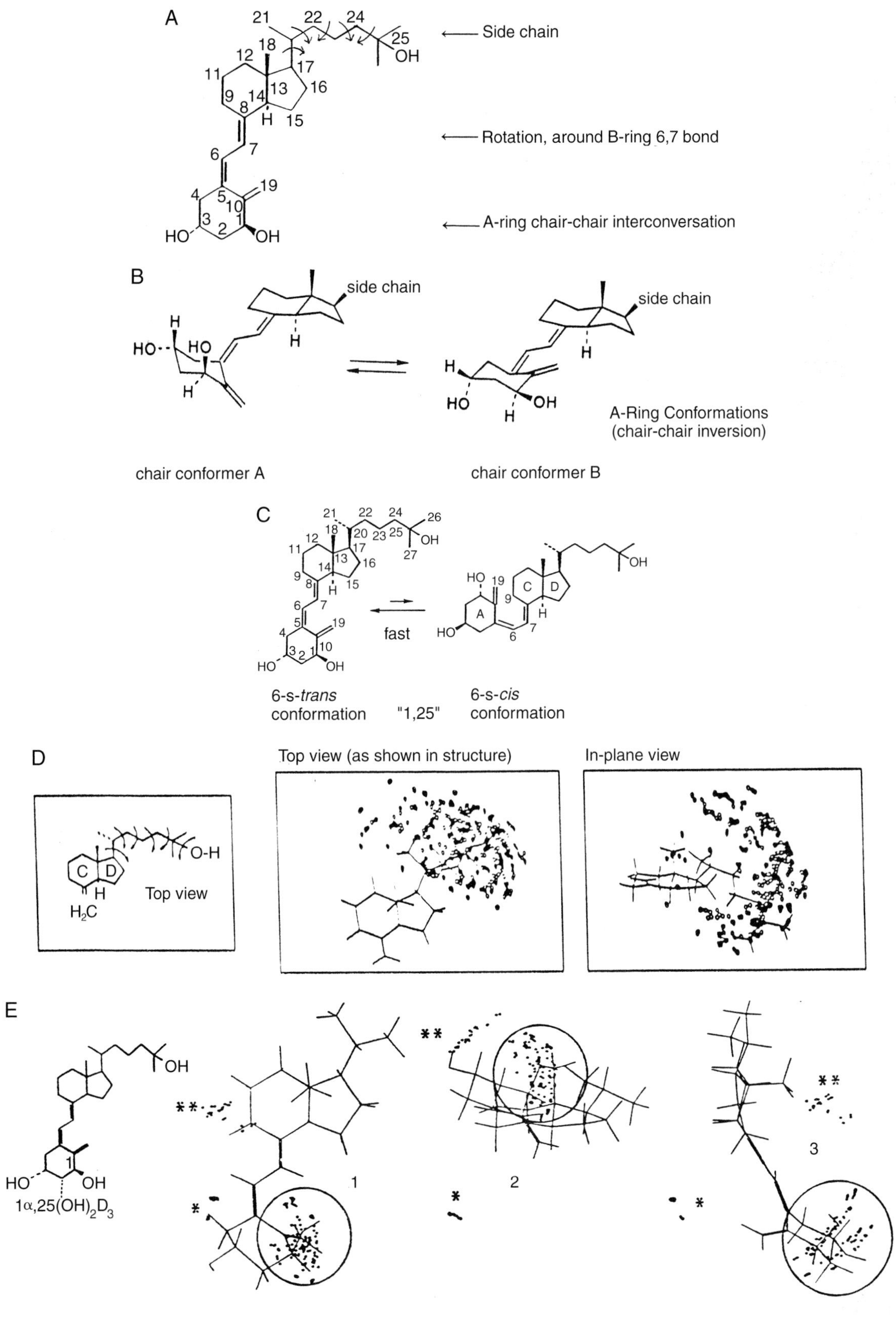

A
21 22 24
18 25
12 OH
11 17
9 14 13 16
8
H 15
6 7
19
4 5
10
3 1
HO 2 OH
Side chain
Rotation, around B-ring 6,7 bond
A-ring chair-chair interconversation
B
side chain
H
HO HO
H
H
side chain
H
H
HO H OH
A-Ring Conformations
(chair-chair inversion)
chair conformer A
chair conformer B
C
21 22 24 26
18 20 23 25
12 OH
11 17
13 27
14 16
9
8 H 15
6 7
5 19
4
3 2 1 10
HO OH
fast
HO 19
C D
9
A H
7
HO 6
OH
6-s-*trans*
conformation
"1,25"
6-s-*cis*
conformation
D
Top view (as shown in structure)
In-plane view
C D
O-H
H
H_2C
Top view
E
OH
1
HO OH
1α,25(OH)$_2$D$_3$
1
2
3

of the key 1α- and 3β-hydroxyls between equatorial and axial orientations [15,16] (see Fig. 1B). Rotational freedom about the 6,7 carbon–carbon bond of the seco B ring allows a continuum of conformations ranging from the more steroid like 6-s-cis conformation to the open and extended 6-s-trans form of the hormone [17]; these conformations are illustrated in Fig. 1C. The intact eight-carbon side chain of vitamin D_3 and related seco-steroids can easily assume numerous shapes and positions by virtue of rotation about its many single carbon–carbon bonds (see Fig. 1D). The dots in Fig. 1D indicate the position in three-dimensional space of the 25-hydroxyl group for some 394 readily identifiable side-chain conformations. A discussion of the consequences of the side-chain conformational mobility has been previously presented [18–20].

With respect to conformational flexibility of the vitamin D seco-steroids, it is generally accepted that the mobility is displayed in both an organic solvent as well as an aqueous environment similar to that encountered in biological systems. It is generally assumed for receptor–ligand interactions that the ligand is "frozen" out in a single shape or conformation which is dictated both by the structural constraints of the ligand and as well by the three-dimensional architecture of the peptide chains that create the ligand binding domain of the receptor. The X-ray crystallographic structures of the thyroid receptor [21] and the retinoic acid receptor (RAR) [22] have been determined, with their respective ligands present in the ligand binding domain (see Chapter 9). Both of the ligands for these receptors are also conformationally flexible, and the X-ray crystallographic structure indicated for each ligand that only one definitive conformer was present in the ligand binding domain. This clearly implies that steroid receptor proteins have the capability to "cap-

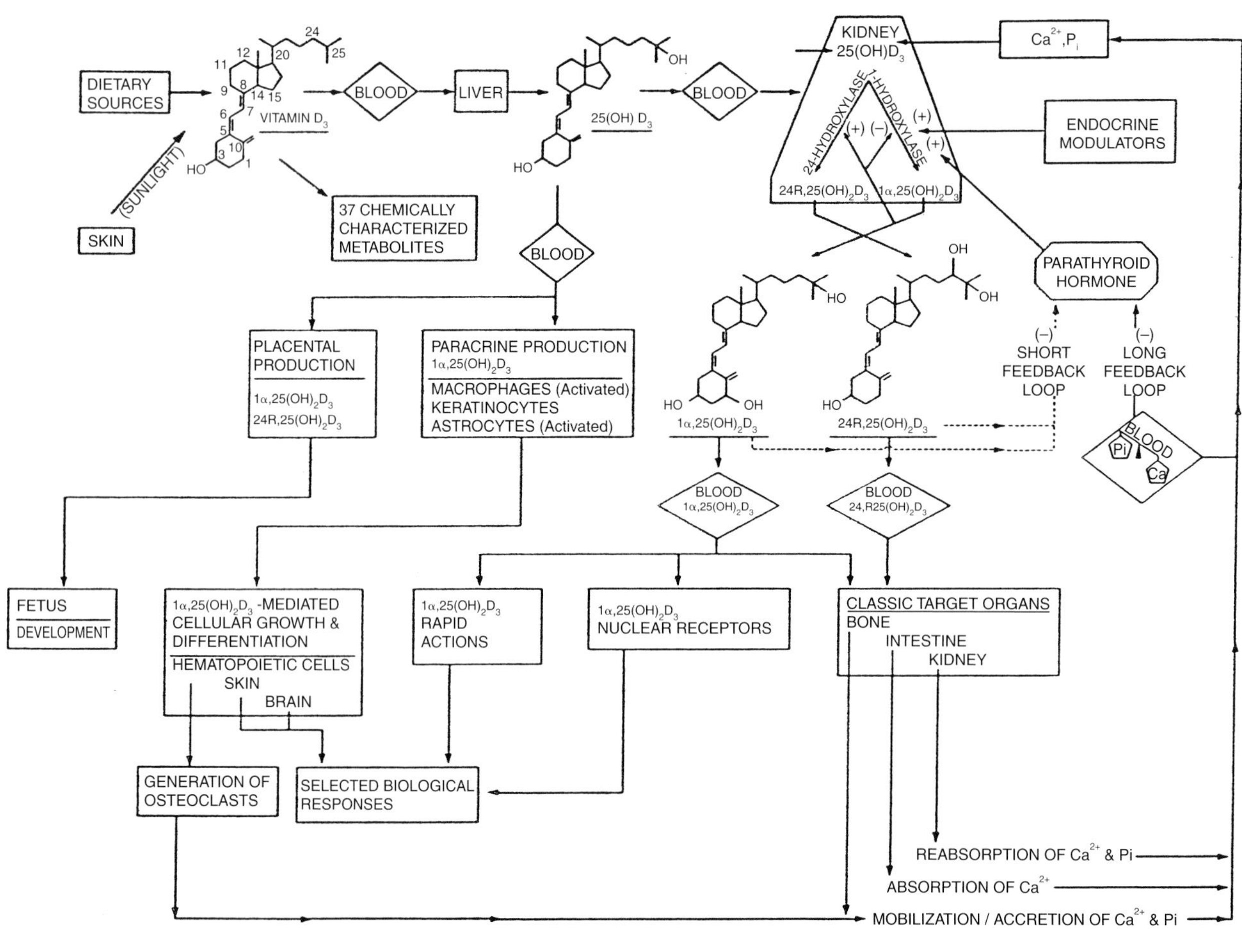

FIGURE 2 Summary of the vitamin D endocrine system. The parent vitamin D_3 is first converted in the liver to $25OHD_3$ and then in the kidney to one of two hormones, $1\alpha,25(OH)_2D_3$ or $24R,25(OH)_2D_3$. $1\alpha,25(OH)_2D_3$ produces biological responses in its target organs via interacting either with a nuclear receptor (VDR_{nuc}) or with a system that generates rapid responses and is believed to involve a membrane receptor (VDR_{mem}). The precise biological role(s) of $24R,25(OH)_2D_3$ is not yet defined, but several studies suggest the presence of receptors in chondrocytes [29,96] and bone [97].

ture" conformationally flexible ligands and choose a shape which accommodates to the reality of the ligand binding domain. Thus, it can be anticipated that receptors for $1\alpha,25(OH)_2D_3$ also will be able to bind one conformer of the population of conformers present in solution.

Certainly the most important polar functional group of $1\alpha,25(OH)_2D_3$ is the 1α-hydroxyl group; it is this group that makes the molecule distinct from 25-hydroxyvitamin D_3 ($25OHD_3$), which is the major circulating form of vitamin D_3 present in the blood. $1\alpha,25(OH)_2D_3$ binds some 333-fold more tightly to the nuclear receptor than does $25OHD_3$ [23]. Also, the orientation of the hydroxyl on C-1 is important to the nuclear receptor; thus, the nuclear receptor binds $1\alpha,25(OH)_2D_3$ 3000-fold better than $1\beta,25(OH)_2D_3$ [23]. Therefore, there is no question that receptors for $1\alpha,25(OH)_2D_3$ must be able to distinguish both the presence of this hydroxyl and its orientation (1α versus 1β) and as well to capture it to form a stable receptor–ligand complex. Figure 1E illustrates the extremely flexible nature of the 1α-hydroxyl with respect to its excursions in three-dimensional space.

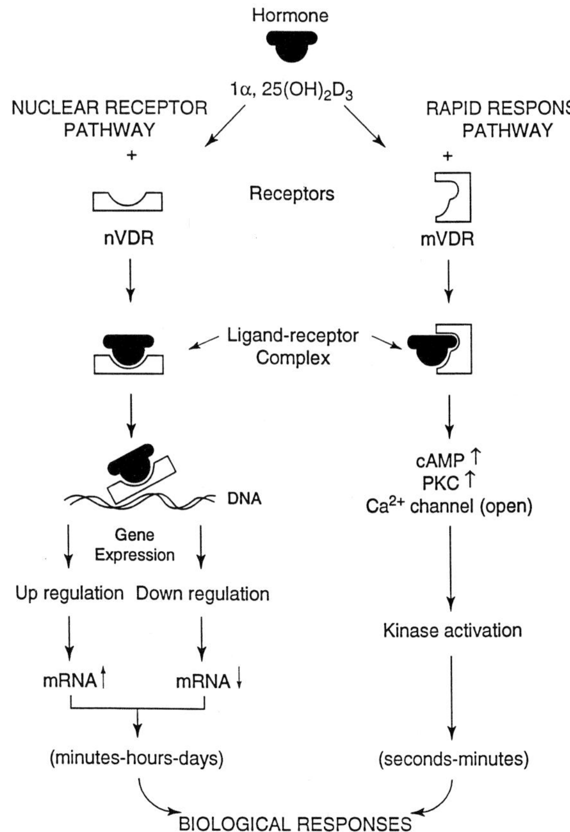

FIGURE 3 Pathways for generation of biological responses by $1\alpha,25(OH)_2D_3$. In the genomic pathway, occupancy of the nuclear receptor for $1\alpha,25(OH)_2D_3$ (nVDR) by a ligand leads to an up- or down-regulation of genes subject to hormone regulation. In the membrane-initiated pathway, occupancy of a membrane receptor for $1\alpha,25(OH)_2D_3$ (mVDR) by a ligand leads to activation of protein kinase C (PKC), adenylate cyclase, phospholipase C, and/or voltage-gated Ca^{2+} channel opening, which is coupled to generation of the end biological response(s).

TABLE I Tissue Distribution of Nuclear $1\alpha,25(OH)_2D_3$ Receptor (VDR_{nuc})[a]

Tissue	Ref.	Tissue	Ref.
Adipose	100	Muscle, embryonic	117
Adrenal	103	Muscle, smooth	118
Bone	105	Osteoblast	126
Bone marrow	108	Ovary	129
Brain	110,111	Pancreas β cell	104,106
Breast	115	Parathyroid	104,132
Cancer cells (many)	120–124	Parotid	102
Cartilage	125	Pituitary	104
Colon	128	Placenta	104
Eggshell gland	131	Prostate	107
Epididymis	119	Skin	113,114
Ganglion	134,135	Stomach	113
Hair follicle	101	Testis	119
Intestine	31	Thymus	127
Kidney	106	Thyroid	130
Liver (fetal)	109	Uterus	119
Lung	112	Yolk sac (bird)	133
Muscle, cardiac	116		

[a] Only one or two literature citations are provided documenting the presence of VDR_{nuc} in the various target tissues; in most instances many additional studies have been reported. Further references documenting the tissue distribution of nuclear receptors for $1\alpha,25(OH)_2D_3$ appear elsewhere [2,4,39,136].

II. VITAMIN D ENDOCRINE SYSTEM

A. General Description

The concept of the existence of the vitamin D endocrine system is now firmly established (see Fig. 2) [4,24–26]. This endocrine system includes four key components. First, the kidney functions as an endocrine gland to produce in a physiologically regulated manner the two seco-steroid hormones 1α,25-dihydroxyvitamin D_3 [$1\alpha,25(OH)_2D_3$] and $24R$,25-dihydroxyvitamin D_3 [$24R,25(OH)_2D_3$]. Second, target organs possess receptors for the hormones which, when occupied with ligand, initiate the signal transduction pathways leading to the appearance of biological responses. Third, a feedback loop regulates the secretion of parathyroid hormone (PTH) in response to the serum Ca^{2+} level. A prime function of PTH is to interact in the proximal tubule of the kidney to increase (or, in the case of falling PTH levels, decrease) the activity of the $25OHD_3$ 1α-hydroxylase, which, in turn, determines the output of $1\alpha,25(OH)_2D_3$ available to generate biological responses. A fourth key component in the operation of the vitamin D endocrine system is the plasma vitamin D binding protein (DBP), which carries vitamin D_3 and all its metabolites to the various target organs. The DBP is known to have a specific ligand binding domain for vitamin D-related ligands that is different in specificity from the ligand binding domain of the nuclear vitamin D receptor [23]. Many of these topics are reviewed in detail in other chapters of this book.

The seco-steroid $1\alpha,25(OH)_2D_3$ is now known to initiate biological responses via both regulation of gene transcription [1–3] as well as "rapid" membrane-initiated pathways [4–6], some of which involve opening of voltage-gated Ca^{2+} channels [7] and involvement of protein kinase C (PKC) [8–12] (see Fig. 3). The regulation of gene transcription by $1\alpha,25(OH)_2D_3$ is known to be mediated by interaction of this ligand with a nuclear receptor protein, termed the VDR_{nuc}, whereas the membrane-initiated responses are postulated to be mediated through interaction of the $1\alpha,25(OH)_2D_3$ with a protein receptor located on the external surface of the cell [27,28], referred to as the VDR_{mem}.

B. Proteins with Ligand Binding Domains for Vitamin D Steroids

Only a limited number of proteins have ligand binding domains for vitamin D_3 seco-steroids; these include the plasma transport protein or vitamin D-binding protein (DBP), the various cytochrome P450 hydroxylases that mediate the myriad metabolic transformations of vitamin D_3, and receptors for $1\alpha,25(OH)_2D_3$, including the classic nuclear receptor (VDR_{nuc}) and membrane receptors (VDR_{mem}), both of which are discussed later in this chapter. There is also a report from Seo *et al.* [29] describing a putative membrane receptor for $24R,25(OH)_2D_3$. An ongoing project in the laboratory of the author is to determine the shape of the vitamin D_3 metabolite ligand(s) for all these classes of proteins. As these proteins are not believed to be structurally related, it is not anticipated that there necessarily will be structural homology of the various ligand binding domains.

III. SIGNAL TRANSDUCTION PATHWAYS USED BY $1\alpha,25(OH)_2D_3$ TO GENERATE BIOLOGICAL RESPONSES

A. Introduction

Since the discovery of $1\alpha,25(OH)_2D_3$ in 1968 [30] and its VDR_{nuc} in 1969 [31], research has focused on defining how these agents collaborate to generate biological responses; of course, the principal focus was on understanding regulation of gene transcription [1]. However, starting in approximately 1983, a series of observations were reported which suggested that some of the biological responses generated by $1\alpha,25(OH)_2D_3$ occurred too rapidly to be simply explained via genomic events. Accordingly, there is now active research on how

TABLE II Distribution of Rapid Responses to $1\alpha,25(OH)_2D_3$[a]

Organ/cell/ system	Response studied	Ref.
Intestine	Rapid transport of intestinal Ca^{2+} (transcaltachia)	41,90,137
	CaCo-2 cells, PKC, G proteins	8
	Activation of PKC	9,137
Colon	PKC effects	56
Fibroblasts	Accumulation of cGMP near VDR_{nuc}	65,140
Kidney	PKC effects	
	Subcellular distribution and	138,139
	regulation of $25OHD_3$ 24-hydroxylase	54
Osteoblast	ROS 17/2.8 cells	
	Ca^{2+} channel opening	7
	Cl^- channel opening	141
	UMR-106 cells	
	Ca^{2+} channel opening by $24R,25(OH)_2D_3$	
Liver	Lipid metabolism	47,48
	Activation of PKC and MAP kinase	142
Muscle	PKC and Ca^{2+} effects	11,51–53
Promyelocytic leukemic cells	Aspects of cell differentiation	10,57
	PKC effects	12,143
Keratinocytes	Alter PKC subcellular distribution	144
Parathyroid cells	Phospholipid metabolism	145
	Cytosolic Ca^{2+}	146
Lipid bilayer	Activation of highly purified PKC	55

[a] The reader should compare the information here with the concepts illustrated in Figs. 2 and 3 which summarize the vitamin D endocrine system and signal transduction pathways utilized by $1\alpha,25(OH)_2D_3$ for generation of biological responses.

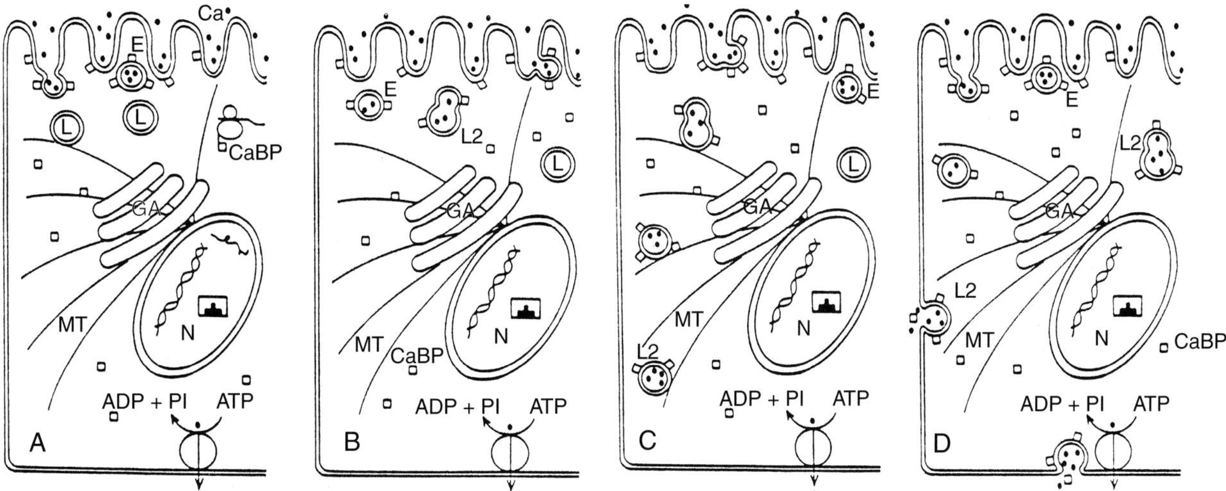

FIGURE 4 Schematic model of transcaltachia illustrating the various steps of Ca^{2+} translocation across intestinal epithelial cells via lysosome-like vesicles. (A) Formation of Ca^{2+}-containing endocytic vesicles (E) at the brush border membrane. (B) Fusion of endocytic vesicles with lysosomes (L). (C) Movement of lysosomes along microtubules (MT) of the epithelial cell. (D) Exocytosis of Ca^{2+}-bearing vesicles via fusion of the lysosomes with the basal lateral membrane. The exocytotic event is believed to be triggered via occupancy of a membrane receptor resident on the basal lateral membrane by the ligand $1\alpha,25(OH)_2D_3$.

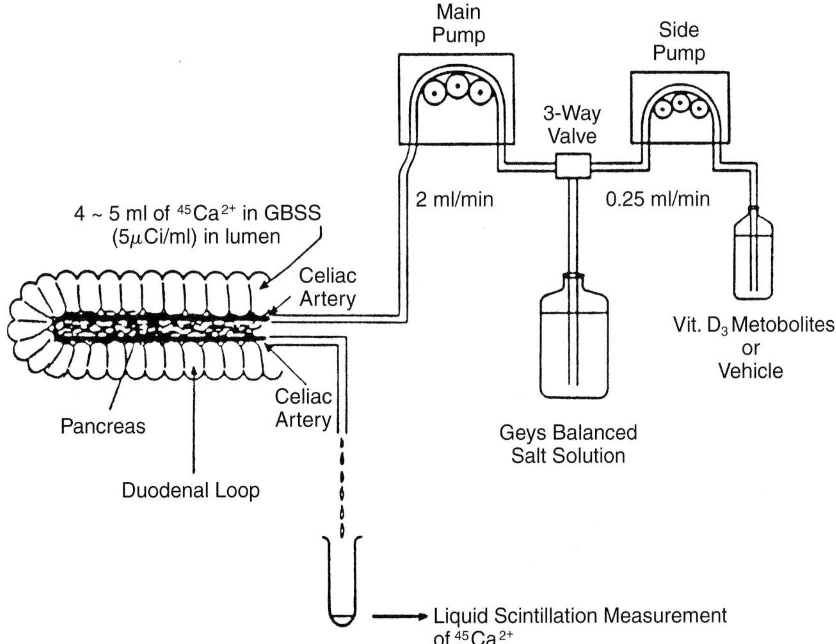

FIGURE 5 Schematic diagram of the chick duodenal perfusion system used to study transcaltachia. Both the celiac artery and vein are cannulated. The celiac artery is connected to a perfusion system with two "side" pumps that can supply nutrients, O_2 and CO_2, and a buffer system, and as well potential agonists [e.g., $1\alpha,25(OH)_2D_3$] and antagonists of transcaltachia. $^{45}Ca^{2+}$ is placed in the lumen of the intestine (duodenum). The effluent from the celiac vein is collected in sequential fractions, and the $^{45}Ca^{2+}$content is determined via liquid scintillation spectrometry. Details of this system are described elsewhere [17,81]. Typical data for transcaltachia are presented in Fig. 6.

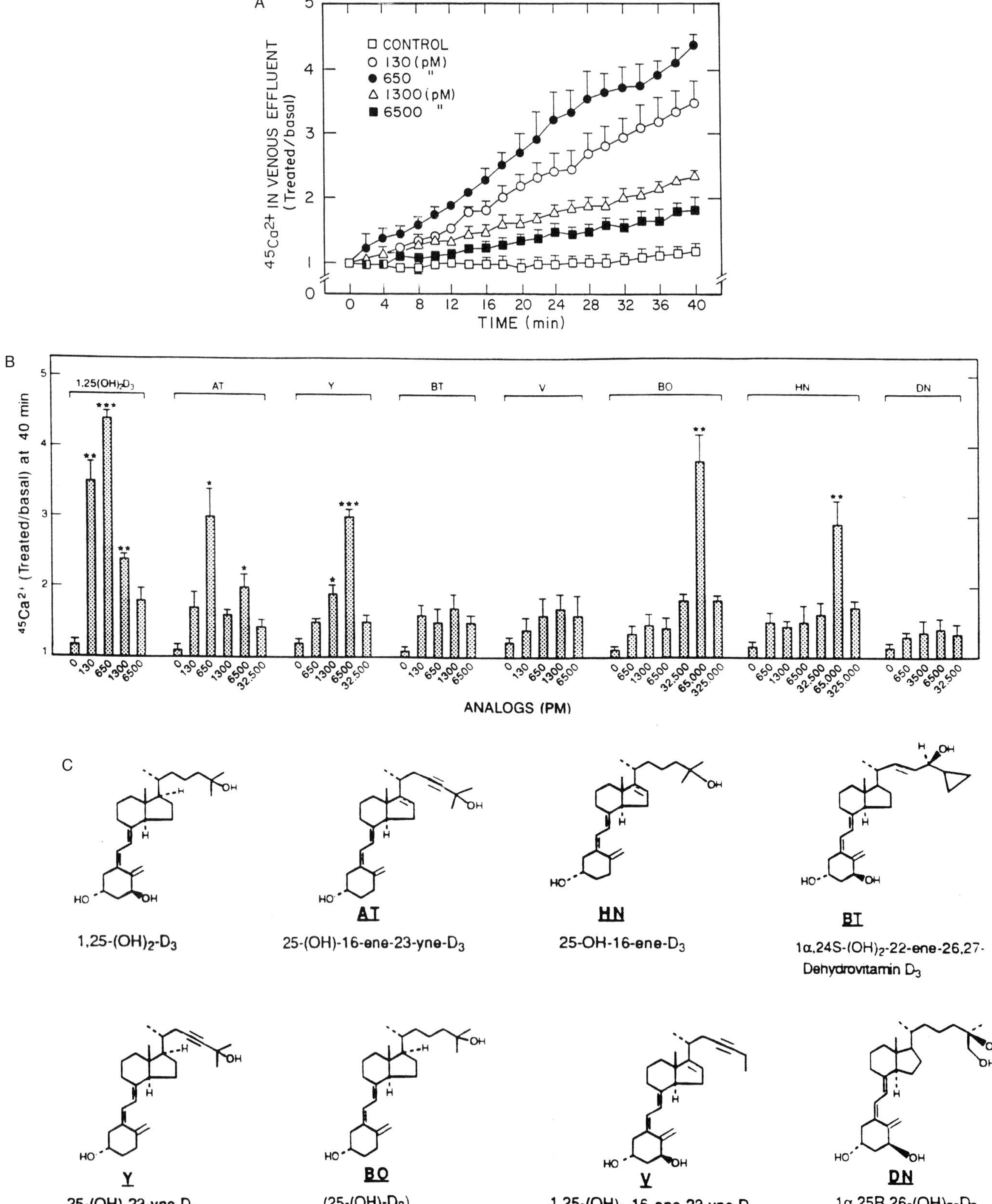

FIGURE 6 Effect of $1\alpha,25(OH)_2D_3$ on the appearance of $^{45}Ca^{2+}$ in the venous effluent of perfused duodena from vitamin D-replete chicks. (A) $1\alpha,25(OH)_2D_3$ dose–response study of transcaltachia. The rate of appearance of $^{45}Ca^{2+}$ data for 0–40 min are presented. The perfusion labeled "control" was exposed only to the vehicle. The ordinate reports the ratio of the rate of appearance of $^{45}Ca^{2+}$ from zero time to 40

$1\alpha,25(OH)_2D_3$ generates biological responses via both genomic and rapid pathways; a model is presented in Fig. 3.

B. Genomic Responses

Table I summarizes the target organs known to possess the VDR_{nuc}. The nuclear responses to $1\alpha,25(OH)_2D_3$ are generated in a manner homologous to that of classic steroid hormones, for example, glucocorticoids, progesterone, estradiol, testosterone, and aldosterone. In the general model, the hormone is produced in an endocrine gland in response to a physiological stimulus, then circulates in the blood bound to a protein carrier (the vitamin D binding protein or DBP), which delivers it to target tissues where the hormone enters the cell and interacts with a specific, high affinity intracellular receptor(s). The receptor–hormone complex then localizes in the nucleus, undergoes some type of "activation" perhaps involving phosphorylation [32–36] and binds to a hormone response element (HRE), specifically, a vitamin D response element (VDRE), on the DNA to modulate the expression of hormone-sensitive genes. The modulation of gene transcription results in either the induction or the repression of specific mRNAs, ultimately resulting in changes in protein expression needed to produce the required biological response [3,37,38]. High affinity receptors for $1\alpha,25(OH)_2D_3$ have been identified in at least 35 target tissues [2,4,39], and over 50 genes are known to be regulated by $1\alpha,25(OH)_2D_3$ [3]. The mechanism of the regulation of these genes may be direct, as described above, through VDREs, or indirect. Information is available for only a few of these genes.

C. Rapid/Nongenomic Responses

Many studies [40] suggest that not all of the actions of $1\alpha,25(OH)_2D_3$ can be explained by receptor–hormone interactions with the genome. Rapid actions have been shown to occur in intestine [41–46], liver [47,48], rat osteoblast ROS 17/2.8 cells [7,49,50], muscle [51–53], and parathyroid cells [54]. In addition, a number of reports have suggested that $1\alpha,25(OH)_2D_3$ either interacts directly with protein kinase C (PKC) [55] or mediates rapid changes in the intracellular location of PKC in a variety of cells [12,56–58]. Table II summarizes the organs or cell location where many "rapid" responses to $1\alpha,25(OH)_2D_3$ have been described. In comparison to our understanding of the interaction of $1\alpha,25(OH)_2D_3$ with the nuclear VDR and the plethora of details concerning regulation of gene transcription, it is clear that at the time of preparation of this review (early 1997) the field of nongenomic responses is only in its infancy.

As a consequence of many of these studies, it was originally proposed that some of the actions of $1\alpha,25(OH)_2D_3$ may be mediated at the cell membrane or by extranuclear subcellular components [41,59,60]. A candidate membrane receptor (VDR_{mem}) for $1\alpha,25(OH)_2D_3$ has been identified [27], and is discussed below.

Still other effects of $1\alpha,25(OH)_2D_3$ that do not appear to be mediated by the nuclear receptor are phosphoinositide breakdown [42], enzymatic activity in osteoblast-derived matrix vesicles [61], certain secretion events in osteoblasts [62], rapid changes in cytosolic Ca^{2+} levels in primary cultures of osteoblasts and osteosarcoma cells [7,63,64], and increases in cyclic guanosine monophosphate levels in fibroblasts [65]. Other steroid hormones, such as estrogen [66], progesterone [67–70], testosterone [71], glucocorticoids [72–74], and thyroid hormone[75,76], have similarly also been shown to have membrane effects [60] which result in the rapid onset of biological responses. An integrated model for the genomic and rapid response signal transduction pathways utilized by $1\alpha,25(OH)_2D_3$ for generation of biological responses is provided in Fig. 3.

IV. DESCRIPTION OF THE "RAPID" RESPONSE OF TRANSCALTACHIA

A particularly well-studied system in the laboratory of the author is the duodenum of the vitamin D-replete chick, where $1\alpha,25(OH)_2D_3$ mediates "the rapid hormonal stimulation of Ca^{2+} transport" which has been termed transcaltachia [41,77,78]. This process is only operative in the vitamin D-replete chick and is not ob-

min divided by the rate of appearance of $^{45}Ca^{2+}$ from −20 min to zero time. Each duodenum was filled with $^{45}Ca^{2+}$ (5 μCi) in Gey's balanced salt solution (GBSS) and was vascularly perfused (24°C) for the first 20 min with control medium (GBSS containing 0.125% bovine serum albumin and 0.5 μl ethanol) and then with $1\alpha,25(OH)_2D_3$ at the indicated concentration. See Zhou *et al.* [87] for experimental details. (B) Ratios of treated/basal data were obtained at the 40-min time point from the experiment in A and others and replotted as a histogram. Values are means ±SEM for three duodena. Statistical analysis was performed by using Duncan's multiple comparison test. Bars that share letters are not significantly different from one another ($p > 0.05$). Data were abstracted from Zhou *et al.* [87]. (C) Structures of $1\alpha,25(OH)_2D_3$ and analogs evaluated for transcaltachic activity (data presented in Fig. B).

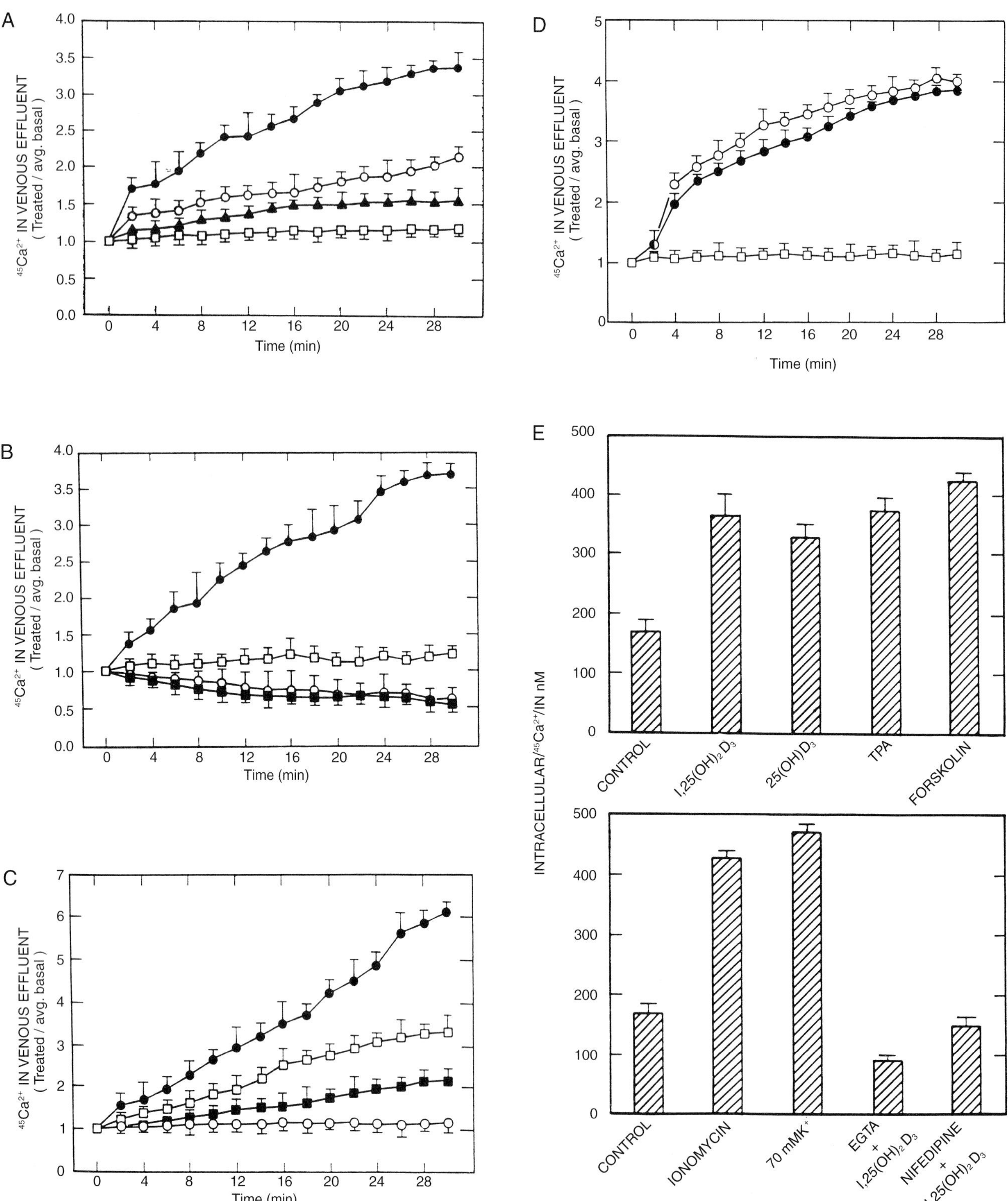

FIGURE 7 Stimulation of transcaltachia by drugs or experimental procedures that increase the intracellular concentration of Ca^{2+}. (A) Stimulation of transcaltachia by the dihydropyridine agonist BAY K-8644. During the treatment phase, the duodena were perfused with BAY K-8644 at 0.5 (▲), 1.0 (○), or 2.0 μM (●) or with control medium (□). Data were abstracted from de Boland *et al.* [94]. (B) Inhibition of transcaltachia by the Ca^{2+} channel blocker nifedipine. During the treatment phase, the duodena were vascularly perfused with 130 pM 1α,25(OH)$_2$D$_3$ (●), 130 pM 1α,25(OH)$_2$D$_3$, + 1.0 μM nifedipine (○), 0.05 μl/ml ethanol (□), 0.05 μl/ml ethanol + 1.0 μM nifedipine (■), or control medium (○). Values are means ± SD of three duodena for each treatment. Data were abstracted from de Boland *et al.* [94]. (C)

served in vitamin D-deficient chick intestine. Figure 4 presents a schematic model of the various steps associated with transcaltachia. In this model, Ca^{2+} is believed to be absorbed across the brush border membrane of the epithelial cell via an endocytic process (Fig. 4A). This would result in Ca^{2+}-containing endocytic vesicles traveling along microtubules and eventually undergoing fusion with lysosomes (Fig. 4B). The lysosome-like vesicles would then continue movement along microtubules (Fig. 4C), eventually leading to fusion with the basal lateral membrane (Fig. 4D). After stimulation by interaction of $1\alpha,25(OH)_2D_3$ with a putative membrane receptor which is coupled to opening of a Ca^{2+} channel, the fused lysosome-like vesicles undergo exocytosis, resulting in extrusion of the Ca^{2+} outside of the cell, adjacent or near to the vascular system. Norman, Nemere, and co-workers have published an extensive series of papers presenting a biochemical characterization of various aspects of the transcaltachic system for transepithelial Ca^{2+} transport; they include evidence for the process of vesicular transport Ca^{2+} [79–81], the involvement of microtubules [82], the involvement of endosomal and lysosomal vesicles [83,84], and, importantly, the dynamic involvement of calbindin-D_{28K} in Ca^{2+} translocation [85]. A number of other general properties have also been described [40,78,80,86].

The process of transcaltachia is studied in an *ex vivo* perfused duodenal system (Fig. 5) [17,81]. One important advantage of this system is that it allows the operator to study the effects of agonists and antagonists on either, or both, the brush border and basal lateral sides of the intestinal epithelial cell. In a typical experiment $^{45}Ca^{2+}$ is placed in the lumen of the intestine. Appropriate drugs may be placed in the lumen of the duodenum to study their immediate effects on the brush border membrane handling of Ca^{2+}. In addition, through use of perfusion pumps leading into the celiac artery, other drugs and agents can be delivered to the basal lateral surface of the epithelial cell. The absorbed and translocated $^{45}Ca^{2+}$ is collected in the perfusate exiting via the celiac vein. Thus the system permits the replication under controlled conditions of the translocation of Ca^{2+} from the lumen of the duodenum, across the epithelial cell, and appearance on the vascular side of the cell. In retrospect, probably the biggest advantage to the transcaltachic system has been the ability to study events occurring at the basal lateral side of the cell. After mounting of the duodenum in the system, the organ remains viable for 60–90 min; a typical experiment would study the absorption kinetics of $^{45}Ca^{2+}$ over 40 min.

Figure 6 presents results of a series of typical experiments comparing the potency of $1\alpha,25(OH)_2D_3$ and related analogs to stimulate transcaltachia. As clearly illustrated in Fig. 6A, the application of a physiological concentration of $1\alpha,25(OH)_2D_3$ to the basal lateral surface of the intestinal epithelial cell results in a very prompt (within 2 min) increase in the rate of appearance of $^{45}Ca^{2+}$ in the perfusate exiting via the celiac artery. This is $^{45}Ca^{2+}$ which has moved from the lumen of the duodenum, across the cell, and which has then exited the cell in response to the presence of $1\alpha,25(OH)_2D_3$ on the basal lateral surface of the cell. The data are expressed as the ratio of the amount of $^{45}Ca^{2+}$ which has appeared in a basal interval (from −20 to 0 min) divided by the amount of $^{45}Ca^{2+}$ which appears after application of $1\alpha,25(OH)_2D_3$ (or analog) starting at time zero for an additional 40 min. Thus in the untreated control, the ratio is 1.0. If the time-dependent ratios of $^{45}Ca^{2+}$ which are obtained at 40 min (Fig. 6A) are replotted as a function of the concentration of $1\alpha,25(OH)_2D_3$ (Fig. 6B), then the biphasic nature of the transcaltachic response is apparent.

Figure 6B also summarizes the dose–response results for evaluation of seven analogs of $1\alpha,25(OH)_2D_3$ to stimulate transcaltachia [87]; the structure of the analogs are presented in Fig. 6C. The analogs can be classified on the basis of their relative ability to stimulate transcaltachia. Analogs AT, Y, BO, and HN all were potent agonists of transcaltachia, whereas analogs BT, V, and DN were weak agonists of transcaltachia. In separate studies [23,87], these same analogs BT, V and HN were shown to be excellent competitors of ^3H-labeled $1,25(OH)_2D_3$ for binding to the nuclear receptor for $1\alpha,25(OH)_2D_3$ (VDR_{nuc}), whereas analogs AT, Y, BO, and DN were found to be weak competitors. These

Stimulation of transcaltachia by the carboxylic ionophore ionomycin. Duodena were perfused with ionomycin at 1.0 (■), 2.0 (□), or 4.0 μM (●) or with control medium (○). Data were abstracted from de Boland *et al.* [90]. (D) Stimulation of transcaltachia by depolarization of the intestinal epithelial cells via perfusion with 40 mM K^+. During the treatment phase, the duodena were perfused with 70 mM KCl (○), 70 mM KCl + 130 pM $1\alpha,25(OH)_2D_3$ (●), or control medium (□). Data were abstracted from de Boland *et al.* [90]. (E) Summary of the effects of agonists and antagonists that mediate changes in intestinal epithelial cell cytosolic $[Ca^{2+}]_i$. The intracellular concentration of Ca^{2+} was measured via loading the epithelial cells with Fura 2/AM and monitoring the changes in fluorescence as described by de Boland *et al.* [94]. The calculated cytosolic $[Ca^{2+}]_i$ values were determined 1 min after addition of $1\alpha,25(OH)_2D_3$ (130 pM), $25OHD_3$ (10 nM), TPA (100 mM), and forskolin (10 μM) (top graph) and in the presence of ionomycin (4 μM), KCl (70 mM), EGTA (5 mM) + $1\alpha,25(OH)_2D_3$ (130 pM), or nifedipine (1.0 μM) + $1\alpha,25(OH)_2D_3$ (130 pM) (bottom graph).

results suggested to Zhou *et al.* [87] that the membrane components present in the perfused duodenum which respond to the analogs of $1\alpha,25(OH)_2D_3$ have a ligand specificity different from that of the classic nuclear receptor.

Figure 7 presents a series of results emerging from a study of transcaltachia that implicate the importance of achieving an increase in intracellular Ca^{2+} concentrations to effect a transcaltachic response. As shown in Fig. 7A, when the intestine is perfused with BAY K-8644, a known activator of voltage-gated L type Ca^{2+} channels [88,89], there is duplication of the results achieved by perfusion with $1\alpha,25(OH)_2D_3$. BAY K-8644 would effect via opening of Ca^{2+} channels an increase in internal calcium ion concentration ($[Ca^{2+}]_i$). Further, the transcaltachic response that is stimulated by BAY K-8644 can be blocked by nifedipine (Fig. 7B), a known inhibitor of Ca^{2+} channels [90]. One consequence of the opening of a Ca^{2+} channel would be to increase the intracellular concentration of Ca^{2+}. As shown in Fig. 7C, perfusion with the carboxylic ionophore ionomycin stimulates transcaltachia. Insertion of ionomycin into membranes results in the creation of Ca^{2+} selective pores [91] that would allow entry of extracellular Ca^{2+} into the cell and a resultant increase in $[Ca^{2+}]_i$. When the duodenum is perfused with 70 m*M* KCl, which effects a depolarization of the basal lateral membrane of the epithelial cell, again there is an apparent stimulation of transcaltachia (Fig. 7D) [91]. Fig. 7E presents a summary of agonists and antagonists that have been shown capable of modulating both $[Ca^{2+}]_i$ and as well transcaltachia [91].

At approximately the same time that the results presented in Fig. 7 were obtained, Caffrey and Farach-Carson presented evidence from biophysical patch clamp studies that ROS 17/2.8 cells (a rat osteosarcoma osteoblastic cell line) also are capable of exhibiting the opening of voltage-gated L type Ca^{2+} channels in response to physiological concentrations of $1\alpha,25(OH)_2D_3$ [7]; these results are summarized in Fig. 8. As was the case with transcaltachia, both $1\alpha,25(OH)_2D_3$ and BAY K-8644 were found to function as agonists for the opening of Ca^{2+} channels. One advantage of these studies is the sophistication of the single cell "patch clamp" technique. As shown in Fig. 8D, these authors obtained evidence that $1\alpha,25(OH)_2D_3$ significantly increased the open time of the Ca^{2+} channels.

These "rapid" actions of $1\alpha,25(OH)_2D_3$ in ROS 17/2.8 cells were important in several respects; they clearly indicated that the rapid responses could be achieved in a second distinct target organ of $1\alpha,25(OH)_2D_3$, namely, the bone and intestine, and, in addition, they could occur in two species, the rat and the chick. This clearly suggested that the phenomenon of rapid responses achieved by $1\alpha,25(OH)_2D_3$ was likely to be a manifestation of the normal biological actions of the hormone $1\alpha,25(OH)_2D_3$. Certainly the summary of rapid responses of $1\alpha,25(OH)_2D_3$ presented in Table II is consistent with this view.

Table III presents a summary of all the known agonists of transcaltachia. Besides $1\alpha,25(OH)_2D_3$, the activator of adenylate cyclase forskolin, the activator of PKC 12-*O*-tetradecanoylphorbol-13-acetate (TPA), mastoparan, a G-protein activator, as well as the aforementioned BAY K-8644, ionomycin, and 70 m*M* KCl all were shown to stimulate transcaltachia. In one respect it is confusing that so many substances can achieve the same end result, namely, activation of transcaltachia. However a possible common theme to these agonists is that each has been shown to initiate signal transduction pathways which can result in an increase in $[Ca^{2+}]_i$. Clearly much additional work is required to sort out the relative importance of these signal transduction pathways to the transcaltachic process.

Table IV presents a summary of all the known antagonists of transcaltachia. Some antagonists are known to interfere with the early steps of transcaltachia (see Fig. 4), whereas others are known to be specific inhibitors of the signal transduction pathways that have been implicated in the transcaltachic process (see Table III). Thus, colchicine, which can depolymerize microtubles [41], would interfere with the movement of the lysosomal-like vesicles across the epithelial cell. Leupeptin blocks cathepsin B, which may be involved with the uptake steps of Ca^{2+} across the brush border of the cell [40,80]. Actinomycin D, the inhibitor of DNA-directed mRNA synthesis, when administered to the chicks 3 hr before duodenal perfusion, was without effect on the transcaltachia process; this result is consistent with the notion that the rapid response of transcaltachia is not immediately dependent on regulation of gene transcription by the VDR_{nuc}. In addition, as reported by Yoshimoto *et al.* [78], perfusion of the basal lateral surface of the epithelial cell with a Ca^{2+}-containing solution that was equivalent to the physiological state of hypercalcemia (high ionized Ca^{2+} concentration) resulted in a complete inhibition of the ability of $1\alpha,25(OH)_2D_3$ to stimulate transcaltachia. One interpretation is that high ionized Ca^{2+} concentrations are able to down-regulate the transcaltachic response; it is possible that under conditions of hypercalcemia there is a negative feedback regulation of the process of intestinal Ca^{2+} absorption.

The current view of this laboratory is that transcaltachia is initiated by $1\alpha,25(OH)_2D_3$ via a membrane-localized receptor, the VDR_{mem}, which is separate and distinct from the VDR_{nuc}. One basis for this postulate has been the ability to carry out a series of structure–

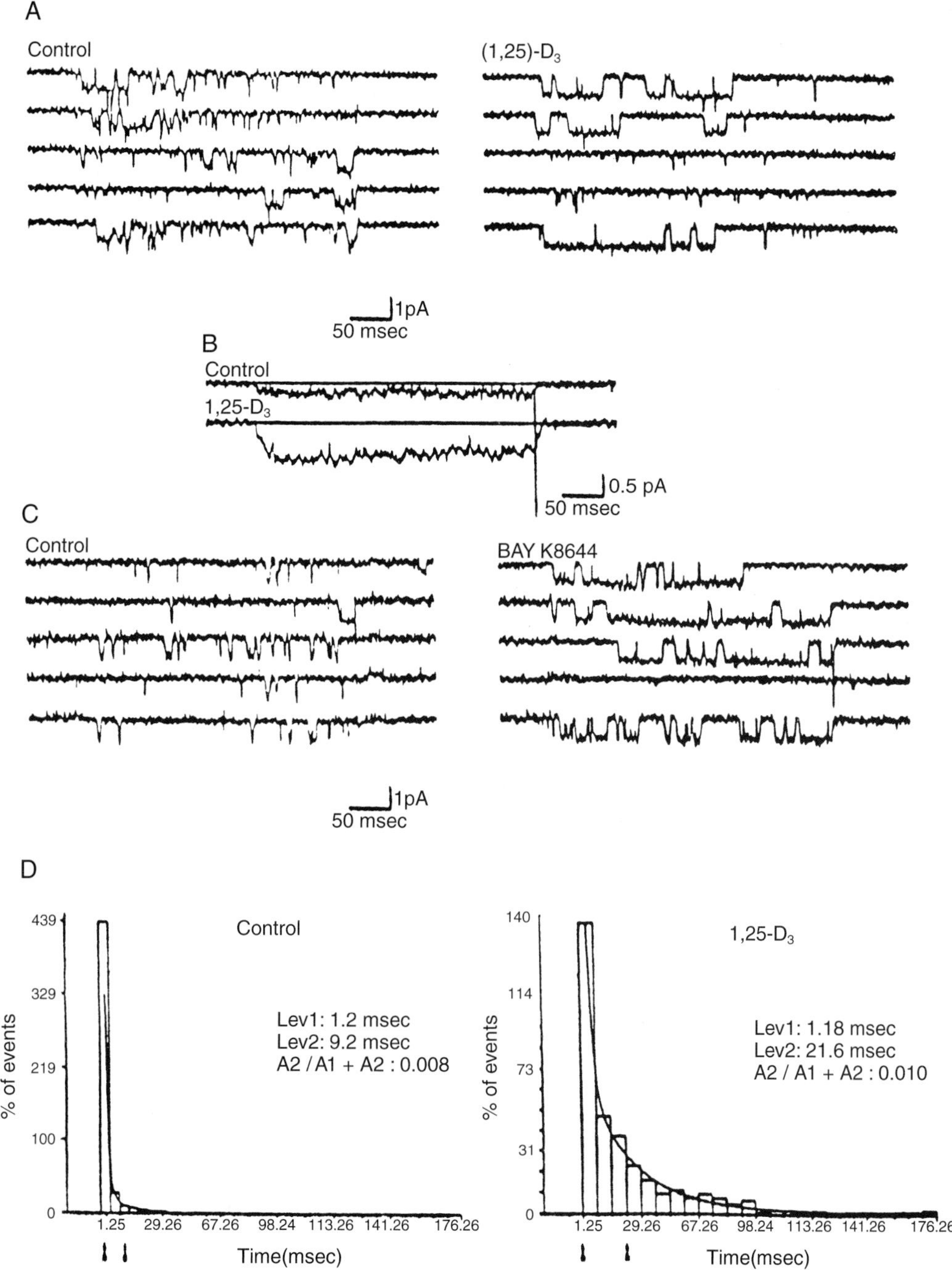

FIGURE 8 Demonstration that 1α,25$(OH)_2D_3$ and BAY K-8644 both modulate dihydropyridine-sensitive Ca^{2+} currents in clonal rat osteosarcoma cells (ROS 17/2.8 cells) as studied by single cell patch-clamping. Data are abstracted from Caffrey and Farach-Carson [7]. (A) Consecutive traces at a test potential of 0 mV, before (left) and after (right) of addition of 3 n*M* 1α,25$(OH)_2D_3$. Showing evidence of the prolongation of channel openings and increased incidence of traces with low opening probability (third and fourth traces at right) relative to control. All traces were made at a holding potential of −50 mV. (B) Illustration of the increase in averaged current in A (right) mediated by 1α,25$(OH)_2D_3$. (C) Consecutive traces at a test potential of −50 mV before (left) and after (right) addition of the dihydropyridine agonist BAY K-8644 (1 μ*M*). The effects are similar to those resulting from addition of 1α,25$(OH)_2D_3$ shown in A. (D) Summary of the increased channel open time resulting from exposure of a single ROS 17/2.8 cell to 1α,25$(OH)_2D_3$ (same data as in A and B). In the control (left), open times are fit by two exponentials with means of 1.2 and 9.2 msec. Following addition of 1α,25$(OH)_2D_3$ (right), the open times are described by exponentials with means of 1.15 and 21.5 msec.

TABLE III Activators of Transcaltachia

Compound	Process	Cell side for agonist presentation	Stimulation of transcaltachia	Ref.
$1{,}25(OH)_2D_3$	Interaction with VDR_{mem}	Basal lateral	Yes	41
$1{,}25(OH)_2D_3$	Ca^{2+} Absorption	Brush border	No	41
Forskolin	Adenylate cyclase	Basal lateral	Yes	90,137
BAY K-8644	Ca^{2+} channel agonist	Basal lateral	Yes	94
BAY K-8644	Ca^{2+} channel agonist	Brush border	No	94
Ionomycin	Carboxylic ionophore	Basal lateral	Yes	90
Phorbol ester (TPA)	Protein kinase C (PKC)	Basal lateral	Yes	90,137
Mastoparan	G-protein activator	Basal lateral	Yes	147
K^+, 70 mM	Depolarization of intestinal cell	Basal lateral	Yes	90

function studies utilizing analogs of 1α,$25(OH)_2D_3$ to evaluate the ligand specificities for VDR_{nuc} mediated events, such as binding to the chick intestinal VDR_{nuc} and induction of the calbindin-D_{28K}, and to compare these with the relative ability of the analogs to stimulate the "rapid" responses of transcaltachia [87,92] and to stimulate $^{45}Ca^{2+}$ uptake in ROS 17/2.8 cells [6]. One glimpse of such studies was presented above in Fig. 6.

TABLE IV Inhibitors of Transcaltachia

Compound	Process	Inhibition of transcaltachia	Ref.
Actinomycin D	Nuclear transcription	No	40
Leupeptin (luminal side)	Cathepsin B	Yes	40,80
Pepstatin	Cathepsin B, pepsin	No	40,80
Monensin	Golgi function	No	40,80
Iodoacetate	Sulfhydryl group	No	40
Colchicine	Depolymerize microtubules	Yes	41
Cytochalasin B	Antimicrofilaments	No	40,80
EGTA	Ca^{2+} chelator	Yes	90
Verapamil	Ca^{2+} channel antagonist	Yes	137
Nifedipine	Ca^{2+} channel antagonist	Yes	90,94
Staurosporine	Protein kinase C	Yes	137
NIP (synthetic peptide)	Inhibits protein kinase C	Yes	137
H7	Protein kinase C	Yes	137
U73122	Protein kinase C	Yes	147
Ca^{2+}	Hypercalcemia (ionized Ca^{2+})	Yes	78

Our collective results clearly suggest that the ligand binding domains of the VDR_{nuc} and VDR_{mem} recognize different structural features of the 1α,$25(OH)_2D_3$ molecule. Support for this conclusion is presented in Table V, which summarizes the membrane-mediated and genomic responses of the A-ring diasteromers of 1α,$25(OH)_2D_3$. The results clearly indicate that whereas 1α,$25(OH)_2D_3$ (with it 1α- and 3β-hydroxyls) is an optimal ligand for initiation of both genomic events and transcaltachia, there is a divergence in responsivity of these two receptor systems for the other three A-ring diastereomers. Thus although analog HH is 25% as potent as 1α,$25(OH)_2D_3$ for transcaltachia, it is only 0.2% as effective in binding to the VDR_{nuc}. This general conclusion is further emphasized by the results for analog HL or 1β,$25(OH)_2D_3$, which is inactive as an agonist for both VDR_{nuc}- and VDR_{mem}-mediated responses but is a potent inhibitor of transcaltachia [92]. 1β,$25(OH)_2D_3$ is the first analog known to be capable of functioning as an antagonist of 1α,$25(OH)_2D_3$.

Table VI reports further studies of the shape preferences for the VDR_{nuc} and VDR_{mem}. A comparison is made of the previtamin form of 1α,$25(OH)_2D_3$, as represented by analog HF or $1{,}25(OH)_2$-d_5-pre-D_3, which is locked in the 6-s-cis shape, with the unlocked 1α,$25(OH)_2D_3$ molecule, which can rapidly interchange between the 6-s-cis and 6-s-trans shapes (see Fig. 3B). The results (Table VI) clearly emphasize that the 6-s-cis locked analog HF is a full agonist of VDR_{mem}-mediated events (both $^{45}Ca^{2+}$ uptake in ROS 17/2.8 cells and stimulation of transcaltachia) but is only 1–2% as active as 1α,$25(OH)_2D_3$ in stimulating genomic responses [17].

A further test of the observation that the VDR_{mem} preferred 6-s-cis-locked analogs was made by a study of the four analogs JM, JN, JO, and JP, which are diaste-

TABLE V Comparison of the Membrane-Mediated and Genomic Responses of the A-Ring Diasteromers of $1\alpha,25(OH)_2D_3$[a]

Analog code	Hydroxyl orientation C-1	Hydroxyl orientation C-3	Genomic responses[b] RCI (%)	Genomic responses[b] ICA (%)	Genomic responses[b] Calbindin-D_{28K} induction (%)	Membrane transcaltachia(%)
$1\alpha,25(OH)_2D_3$	α	β	100	100	100	100
HJ	α	α	24	2.8	10	75
HH	β	α	0.2	—	1	25
HL	β	β	0.2	<0.1	1	0
C + HL					96%	~0%

[a] Data were abstracted from Norman *et al.* [99].
[b] RCI, Relative competitive index (i.e., relative binding to the VDR_{nuc}); ICA, relative ability to stimulate intestinal Ca^{2+} absorption.

reomers at C-9 and C-19 of the non-seco-steroid provitamins; their structures are provided in Fig. 9A. The experimental results are presented in Fig. 9B, 9C; the data document that the analog $1\alpha,25(OH)_2$-lumisterol (JN) is a full agonist and $1\alpha,25(OH)_2$-7-dehydrocholesterol (JM) is a partial agonist of transcaltachia [93]. In contrast Fig. 9C clearly documents that the four provitamin diastereomers (JM, JN, JO, and JP) collectively have only 0.01 to 1.4% of the activity of acting as an agonist for the VDR_{nuc} to mediate the genomic induction of osteocalcin in human osteosarcoma MG-63 cells. However, the 6-s-cis locked analogs JO and JP are not good transcaltachia agonists, showing that the 6-s-cis conformation is not the sole structural element leading to transcaltachia. Collectively these results reinforce the conclusion that the VDR_{mem} is fully responsive to some 6-s-cis locked analogs.

TABLE VI Relative Ability of a 6-s-cis Previtamin Analog of $1\alpha,25(OH)_2D_3$ to Stimulate the VDR_{mem} and the VDR_{nuc} Signal Transduction Systems[a]

Biological evaluation	Compound evaluated: $1\alpha,25(OH)_2D_3$ (%)	Compound evaluated: $1\alpha,25(OH)_2$-d_5-previtamin D_3 (HF) (%)
Genomic responses		
RCI for VDR_{nuc}	100	10 ± 6
Induction of osteocalcin (MG-63 cells)	100	2
Inhibition of proliferation (HL-60 cells)	100	1.5
Membrane initiated responses		
Transcaltachia (chick intestine)	100	100
$^{45}Ca^{2+}$ uptake (ROS 17/2.8 cells)	100	95

[a]The structure of the 6-s-cis-locked analog HF is presented in Fig. 9A. Data were abstracted from Norman *et al.* [17].

These results then raise the question of what is the activity of 6-s-trans locked analogs to act as agonists for the VDR_{mem} and VDR_{nuc}? These results are presented in Fig. 9BC, which shows the ability of 6-s-trans-locked 1α,25(OH)-dihydrotachysterol$_3$ (JB) and $1\alpha,25(OH)_2$-*trans*-isotachysterol$_3$ (JD) to interact with the VDR_{nuc} and mediate the induction of osteocalcin in MG-63 cells. Neither analog JB nor JD bound well to the VDR_{nuc} in steroid competition assays [relative competitive index (RCI) < 1%], and neither were more than 0.03% as active as $1\alpha,25(OH)_2D_3$ in stimulating osteocalcin induction. Our interpretation of this result is that the VDR_{nuc} is not responsive to analogs which are locked in a 6-s-trans shape; we are currently evaluating analogs JB and JD in VDR_{mem} responsive systems.

Table VII presents a summary of our current understanding of the physiological and biochemical properties of receptor proteins that possess ligand binding domains for vitamin D-related seco-steroids. We have concluded that the VDR_{nuc} and VDR_{mem} proteins each have the ability to recognize and optimally respond to structurally diverse ligands. Thus, one possible way in which the vitamin D endocrine system is able to achieve such diversity in its biological responses is through utilization of receptors and also transport proteins (DBP) that are differentially responsive to the myriad shapes available in the conformationally flexible $1\alpha,25(OH)_2D_3$.

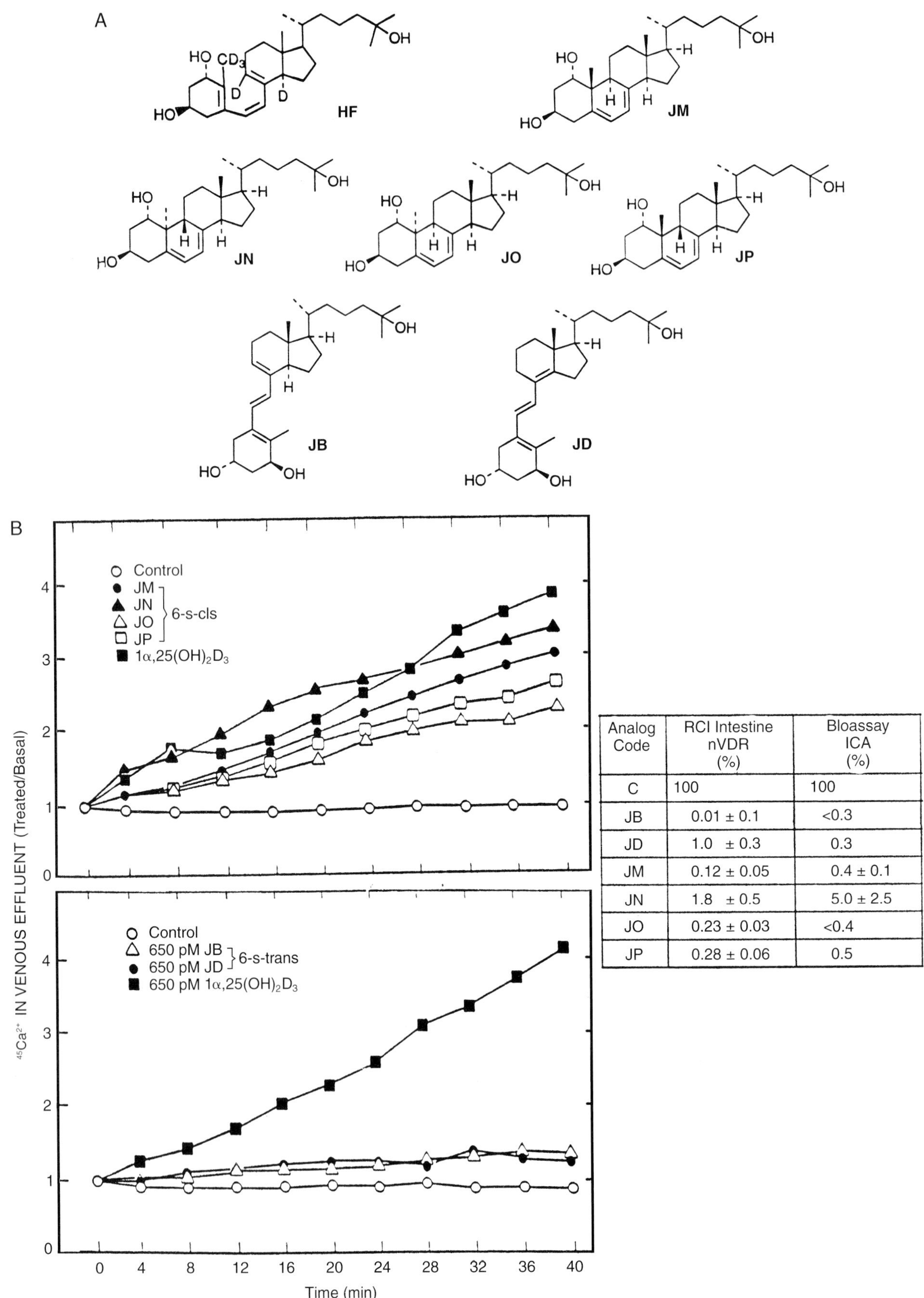

Analog Code	RCI Intestine nVDR (%)	Bioassay ICA (%)
C	100	100
JB	0.01 ± 0.1	<0.3
JD	1.0 ± 0.3	0.3
JM	0.12 ± 0.05	0.4 ± 0.1
JN	1.8 ± 0.5	5.0 ± 2.5
JO	0.23 ± 0.03	<0.4
JP	0.28 ± 0.06	0.5

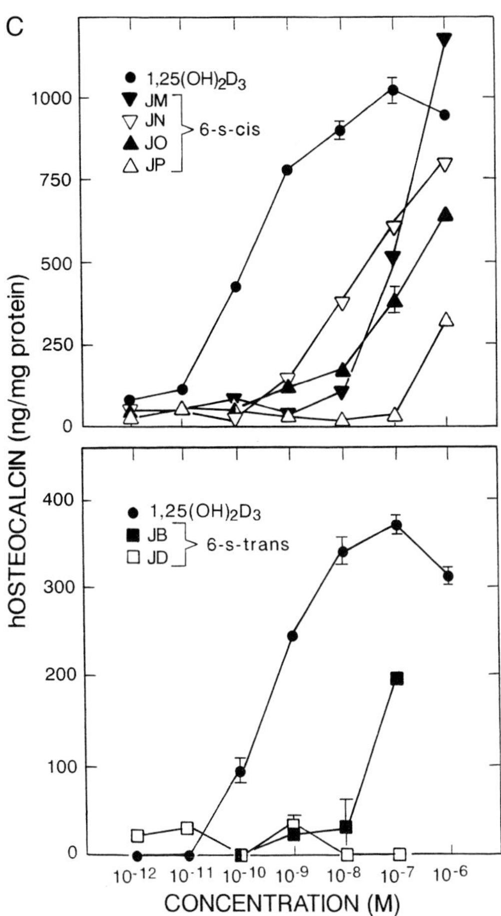

FIGURE 9 Comparison of genomic and rapid responses of 6-s-cis- versus 6-s-trans-locked analogs. (A) Structures of 6-s-cis- and 6-s-trans-locked analogs evaluated with respect to their genomic and rapid action responses. (B, top) Comparison of the abilities of $1\alpha,25(OH)_2D_3$, $1\alpha,25(OH)_2$-7-dehydrocholesterol (JM), $1\alpha,25(OH)_2$-lumisterol (JN), $1\alpha,25(OH)_2$-pyrocalciferol (JO), or $1\alpha,25(OH)_2$-isopyrocalciferol$_3$ (JP) to stimulate transcaltachia; also tabulated are values for the RCI or relative competitive index (relative binding to the VDR_{nuc}) and the ICA or relative ability to stimulate intestinal Ca^{2+} absorption. Results were abstracted from Norman *et al.* [98]. (B, bottom) Comparison of the abilities of 6-s-*trans*-locked analogs 1α,25(OH)-dihydrotachysterol$_3$ (JB) and $1\alpha,25(OH)_2$-*trans*-isotachysterol$_3$ (JD) to stimulate transcaltachia. Results were abstracted from Norman *et al.* [98]. (C, top) Comparison of the relative abilities of $1\alpha,25(OH)_2D_3$ and the provitamins JM, JN, JO, and JP to stimulate the genomic response of induction of osteocalcin in MG-63 cells. Results were abstracted from Norman *et al* [99]. (C, bottom) Ability of $1\alpha,25(OH)_2D_3$ and 6-s-trans locked JB and JD to interact with the VDR_{nuc} and mediate the induction of osteocalcin in MG-63 cells. Results were abstracted from Norman *et al.* [99].

V. MEMBRANE RECEPTOR (VDR_{mem}) FOR $1\alpha,25(OH)_2D_3$

A. Evidence for Existence of a VDR_{mem}

The concept of the existence of a membrane receptor for $1\alpha,25(OH)_2D_3$ has its origin in studies of the process of the $1\alpha,25(OH)_2D_3$-mediated response of transcaltachia, or the rapid hormonal stimulation of intestinal Ca^{2+} absorption in the perfused chick intestine. The most potent agonist is the natural hormone $1\alpha,25(OH)_2D_3$, which can stimulate the transport of $^{45}Ca^{2+}$ across the intestine within 1–2 min [41,90,94].

The current view of the author's laboratory is that transcaltachia is initiated by $1\alpha,25(OH)_2D_3$ via a membrane-localized receptor, the VDR_{mem}, which is separate and distinct from the VDR_{nuc}. One basis for this postulate has been the results of a series of structure–function studies with analogs of $1\alpha,25(OH)_2D_3$ comparing the potency of analogs in well-known VDR_{nuc} mediated responses, such as the induction of calbindin-D_{28K}, with the biological response of transcaltachia [17,87,92,93]. Also, a similar comparison of vitamin D analog potencies of nuclear responses as compared with rapid responses have been carried out in ROS 17/2.8 cells [6,95]. These results clearly suggest that the ligand binding domains of the VDR_{nuc} and VDR_{mem} recognize

TABLE VII $1\alpha,25(OH)_2D_3$ Binding Proteins for Vitamin D Seco-Steroids: A Status Report

	DBP	VDR_{nuc}	VDR_{mem}
Cellular location	Blood	Nucleus	Outer cell membrane
Mode of action	—	Gene transcription	PKC, Ca^{2+} channels
Molecular weight	40,000	51,000	~60,000
K_d for $1\alpha,25(OH)_2D_3$	5×10^{-7} *M*	$1\text{–}4 \times 10^{-10}$ *M*	$2\text{–}7 \times 10^{-10}$ *M*
Copies per cell	—	~5000	~50–500
Acceptable shape of ligand	Rigid side chain	Not 6-s-trans	6-s-cis
Specific antagonist	—	?	$1\beta,25(OH)_2D_3$

different structural features of the $1\alpha,25(OH)_2D_3$ molecule; more particularly it was concluded that the VDR_{mem} preferred some ligands with a 6-s-cis shape, whereas the VDR_{nuc} had little affinity for ligands of this shape.

B. Purification

An important verification of the concept of the existence of the proposed VDR_{mem} is the isolation and purification of such a protein. In 1994, the Riverside laboratory described a seven-step purification of a protein derived from the basal lateral membrane of chick intestinal mucosal cells that displays a 4500-fold enrichment in binding activity for $1\alpha,25(OH)_2D_3$ [27]. We believe that the purified protein represents the VDR_{mem}; typical purification results are presented in Table VIII. A functional correlation between VDR_{mem} and transcaltachia was observed in three experimental situations: (1) in circumstances of vitamin D deficiency, which suppresses transcaltachia, there resulted a reduced specific binding of $^3H\text{-}1,25(OH)_2D_3$ in those protein fractions which possess the VDR_{mem} obtained from vitamin D-replete chicks [27]; (2) the VDR_{mem} exhibited down-regulation of specific $^3H\text{-}1,25(OH)_2D_3$ binding following exposure *in vivo* to nonradioactive $1\alpha,25(OH)_2D_3$ [27]; and (3) the relative potencies of two 6-s-cis analogs, $1\alpha,25(OH)_2$-7-dehydrocholesterol (JM) and $1\alpha,25(OH)_2$-lumisterol (JN), to initiate transcaltachia and their ability to compete for binding to the VDR_{mem} were parallel [18,93]. Also shown in Table VIII is an evaluation of the extent of "carryover" of ^{35}S-labeled VDR_{nuc} which was added to the chick intestinal mucosal homogenate prior to isolation of the VDR_{mem}. Approximately 0.02% of the $^{35}S\text{-}VDR_{nuc}$ was found in step 5 of the isolation procedure for the VDR_{mem}. These results support the conclusion that the membrane binding activity for $^3H\text{-}1,25(OH)_2D_3$ is not due to contamination with the VDR_{nuc}.

TABLE VIII Purification Steps for the VDR_{mem} and Lack of Contamination by the VDR_{nuc}[a]

Purification step	Binding of $^3H\text{-}1\alpha,25(OH)_2D_3$ in HAP assay (fmol/mg protein	Carryover of $^{35}S\text{-}VDR_{nuc}$ (%)
1. Chick intestinal mucosa homogenate	0.3	100
2. (a) Nuclear pellet (800 *g*)		84
(b) Nuclear pellet supernatant		14
3. Basal lateral membranes (BLM) (20,000 *g* postnuclear pellet)	9	2.5
4. (a) Percoll gradient (fractions 18–20 yield BLM)	—	0.6
(b) TED-washed membranes	—	0.11
5. BLM solubilization in 10 m*M* CHAPSO and centrifugation (160,000 *G*, 60 min); supernatant	8	0.022
6. Mono-Q anion chromatography (fractions 15–17)	~1200	
7. Superose-12 chromatography (~60-kDa band)	~1350	

[a]Results were abstracted from Norman *et al.* [99]. An overall purification of 4500-fold was achieved. BLM, basal lateral membranes; TED, buffer containing Tris, EDTA, and dithiothreitol; CHAPSO, anonionic detergent; HAP, hydroxyl apatite.

VI. SUMMARY

The objectives of this chapter have been to provide an overview of the subject of rapid or nongenomic responses mediated by $1\alpha,25(OH)_2D_3$. It is apparent that there is a burgeoning list of reports describing new locations and new examples of "rapid" responses to this seco-steroid. One particularly well-studied system is the rapid hormonal stimulation of Ca^{2+} across the intestine, known as transcaltachia. It is now known that initiation of transcaltachia by $1\alpha,25(OH)_2D_3$ utilizes signal transduction pathways that are not known to be subserved by the nuclear receptor (VDR_{nuc}) for $1\alpha,25(OH)_2D_3$; these include participation of a membrane receptor (VDR_{mem}) that is likely coupled to a G-protein mediator which ultimately activates a Ca^{2+} channel, leading to

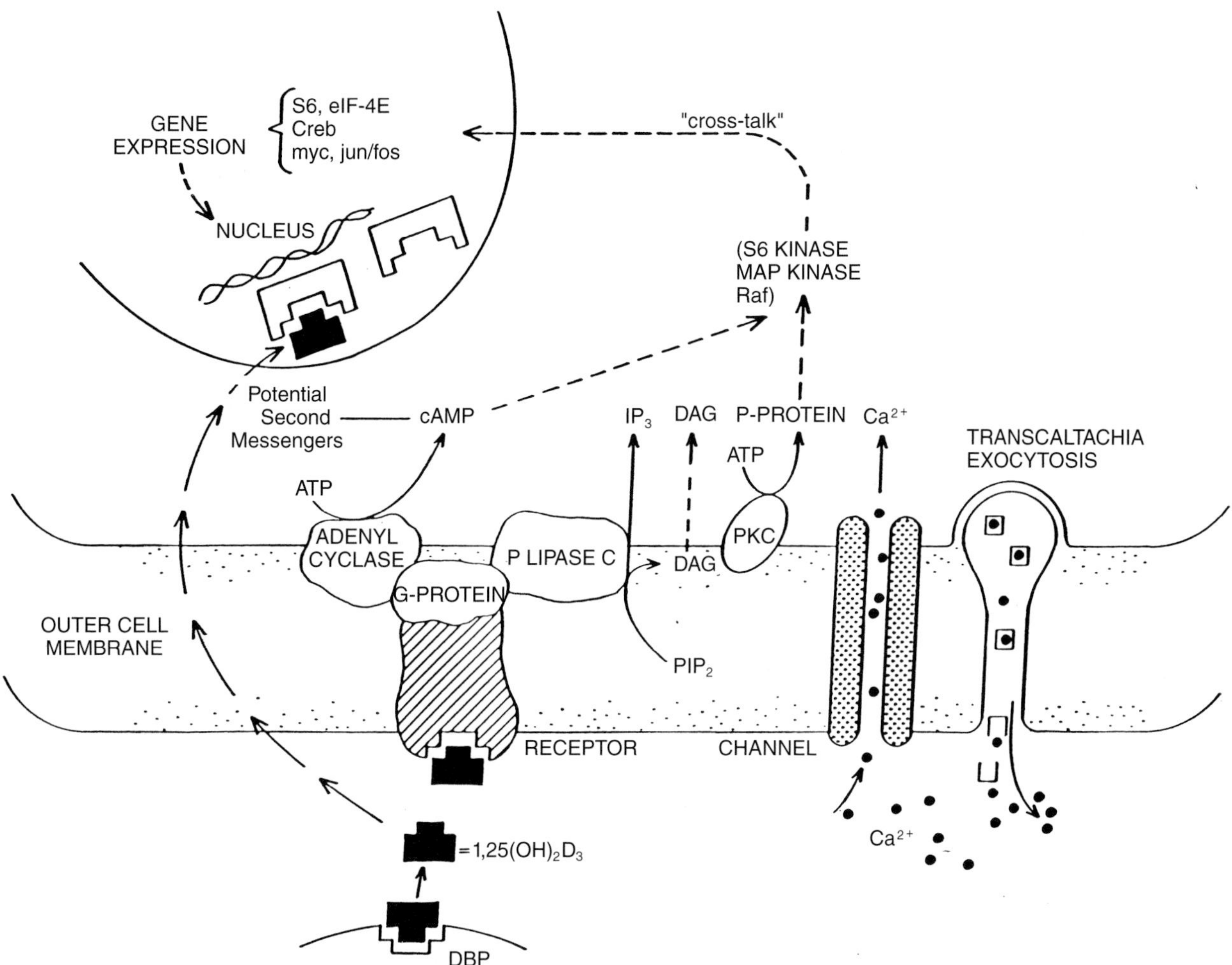

FIGURE 10 Schematic model for the integrated actions of the VDR_{nuc} and VDR_{mem} to elicit biological responses through the combined actions of their respective signal transduction pathways. Binding of $1\alpha,25(OH)_2D_3$ to the membrane surface receptor may result in an increase of one or more second messengers [cAMP, diacylglycerol (DAG), inositol 1,4,5-trisphosphate (IP_3), PKC, or intracellular Ca^{2+}], which, in some systems, may result in the transient opening of Ca^{2+} channels. The increased $[Ca^{2+}]_i$, possibly in collaboration with another second messenger, may, if the biological response is transcaltachia, initiate the exocytosis of the Ca^{2+}-bearing lysosomal vesicles. Alternatively, the activation of the second messengers may include PKC and the mitogen-activated protein (MAP) kinase, S6 kinase, or Raf or other kinases which engage in cross talk with the nucleus of the cell so as to initiate biological responses that are dependent on nuclear events.

exocytosis of lysosomal Ca^{2+}-bearing vesicles. Through the use of conformationally restricted analogs of $1\alpha,25(OH)_2D_3$, it has been concluded that the VDR_{nuc} and VDR_{mem} respond to differing shapes of the pluripotent $1\alpha,25(OH)_2D_3$. Studies are currently in progress to further describe the biochemical properties of the VDR_{mem} and to define the intimate details of the signal transduction pathways to which it is coupled.

Figure 10 presents a summary figure which attempts to integrate the signal transduction pathways which are subserved by the nuclear receptor (VDR_{nuc}) and membrane receptor (VDR_{mem}) for $1\alpha,25(OH)_2D_3$. This figure emphasizes the reality of the complexity of overlapping and interconnecting signal transduction pathways. An important question to answer is to understand whether the VDR_{mem} can "communicate" with the nucleus of the cell to modulate gene transcription. A second important question relates to whether there is specific cross talk between the VDR_{mem} and the VDR_{nuc}. It will be intriguing to monitor developments in this field in the future.

References

1. Minghetti PP, Norman AW 1988 $1,25(OH)_2$-vitamin D_3 receptors: Gene regulation and genetic circuitry. FASEB J **2**:3043–3053.

2. Lowe KE, Maiyar AC, Norman AW 1992 Vitamin D-mediated gene expression. Crit Rev Eukayotic Gene Expression **2**:65–109.
3. Hannah SS, Norman AW 1994 1α,25$(OH)_2$-vitamin D_3-regulated expression of the eukaryotic genome. Nutr Revi **52**:376–382.
4. Norman AW, Roth J, Orci L 1982 The vitamin D endocrine system: steroid metabolism, hormone receptors and biological response (calcium binding proteins). Endocr Rev **3**:331–366.
5. Norman AW, Nemere I, Zhou L-X, Bishop JE, Lowe KE, Maiyar AC, Collins ED, Taoka T, Sergeev I, Farach-Carson MC 1992 1,25$(OH)_2$-vitamin D_3, a steroid hormone that produces biologic effects via both genomic and nongenomic pathways. J Steroid Biochem Mol Biol **41**:231–240.
6. Farach-Carson MC, Sergeev IN, Norman AW 1991 Nongenomic actions of 1,25-dihydroxyvitamin D_3 in rat osteosarcoma cells: Structure–function studies using ligand analogs. Endocrinology **129**:1876–1884.
7. Caffrey JM, Farach-Carson MC 1989 Vitamin D_3 metabolites modulate dihydropyridine-sensitive calcium currents in clonal rat osteosarcoma cells. J Biol Chem **264**:20265–20274.
8. Khare S, Tien X-Y, Wilson D, Wali RK, Bissonnette BM, Scaglione-Sewell B, Sitrin MD, Brasitus TA 1994 The role of protein kinase-Cα in the activation of particulate guanylate cyclase by 1α,25-dihydroxyvitamin D_3 in CaCo-2 cells. Endocrinology **135**:277–283.
9. Bissonnette M, Tien X-Y, Niedziela SM, Hartmann SC, Frawley BP, Jr, Roy HK, Sitrin MD, Perlman RL, Brasitus TA 1994 1,25$(OH)_2$ vitamin D_3 activates PKC-α in Caco-2 cells: A mechanism to limit secosteroid-induced rise in $[Ca^{2+}]_i$. Am J Physiol Gastrointest Liver Physiol **267**:G465-G475.
10. Bhatia M, Kirkland JB, Meckling-Gill KA 1995 Monocytic differentiation of acute promyelocytic leukemia cells in response to 1,25-dihydroxyvitamin D_3 is independent of nuclear receptor binding. J Biol Chem **270**:15962–15965.
11. Vazquez G, de Boland AR 1996 Involvement of protein kinase C in the modulation of 1α,25-dihydroxy-vitamin D_3-induced $^{45}Ca^{2+}$ uptake in rat and chick cultured myoblasts. Biochim Biophys Acta Mol Cell Res **1310**:157–162.
12. Berry DM, Antochi R, Bhatia M, Meckling-Gill KA 1996 1,25-Dihydroxyvitamin D_3 stimulates expression and translocation of protein kinase Cα and Cδ via a nongenomic mechanism and rapidly induces phosphorylation of a 33-kDa protein in acute promyelocytic NB4 cells. J Biol Chem **271**:16090–16096.
13. Henry HL, Norman AW 1991 Metabolism of vitamin D. In: Coe FL, Favus MJ (eds) Disorders of Bone and Mineral Metabolism. Raven, New York, pp. 149–162.
14. Henry HL, Norman AW 1984 Vitamin D: Metabolism and biological action. Annu Rev Nutr **4**:493–520.
15. Wing RM, Okamura WH, Pirio MR, Sine SM, Norman AW 1974 Vitamin D_3: Conformations of vitamin D_3, 1α,25-dihydroxyvitamin D_3, and dihydrotachysterol$_3$. Science **186**:939–941.
16. Okamura WH, Norman AW, Wing RM 1974 Vitamin D: Concerning the relationship between molecular topology and biological function. Proc Natl Acad Sci USA **71**:4194–4197.
17. Norman AW, Okamura WH, Farach-Carson MC, Allewaert K, Branisteanu D, Nemere I, Muralidharan KR, Bouillon R 1993 Structure–function studies of 1,25-dihydroxyvitamin D_3 and the vitamin D endocrine system. 1,25-Dihydroxy-pentadeuterio-previtamin D_3 (as a 6-s-cis analog) stimulates nongenomic but not genomic biological responses. J Biol Chem **268**:13811–13819.
18. Okamura WH, Midland MM, Hammond MW, Rahman NA, Dormanen MC, Nemere I, Norman AW 1994 Conformation and related topological features of vitamin D: Structure–function relationships. In: Norman AW, Bouillon R, Thomasset M (eds) Vitamin D, A Pluripotent Steroid Hormone: Structural Studies, Molecular Endocrinology and Clinical Applications. de Gruyter, Berlin, pp. 12–20.
19. Midland MM, Plumet J, Okamura WH 1993 Effect of C20 stereochemistry on the conformational profile of the side chains of vitamin D analogs. Bioorg Med Chem Lett **3**:1799–1804.
20. Okamura WH, Palenzuela JA, Plumet J, Midland MM 1992 Vitamin D: Structure–function analyses and the design of analogs. J Cell Biochem **49**:10–18.
21. Wagner RL, Apriletti JW, McGrath ME, West BL, Baxter JD, Fletterick RJ 1995 A structural role for hormone in the thyroid hormone receptor. Nature **378**:690–697.
22. Bourguet W, Ruff M, Chambon P, Gronemeyer H, Moras D 1995 Crystal structure of the ligand-binding domain of the human nuclear receptor RXR-α. Nature **375**:377–382.
23. Bishop JE, Collins ED, Okamura WH, Norman AW 1994 Profile of ligand specificity of the vitamin D binding protein for 1α,25$(OH)_2$-vitamin D_3 and its analogs. J Bone Miner Res **9**:1277–1288.
24. Reichel H, Koeffler HP, Norman AW 1989 The role of the vitamin D endocrine system in health and disease. N Engl J Med **320**:980–991.
25. Norman AW, Hurwitz S 1993 The role of the vitamin D endocrine system in avian bone biology. J Nutr **123**:310–316.
26. Bouillon R, Okamura WH, Norman AW 1995 Structure–function relationships in the vitamin D endocrine system. Endocr Rev **16**:200–257.
27. Nemere I, Dormanen MC, Hammond MW, Okamura WH, Norman AW 1994 Identification of a specific binding protein for 1α,25-dihydroxyvitamin D_3 in basal-lateral membranes of chick intestinal epithelium and relationship to transcaltachia. J Biol Chem **269**:23750–23756.
28. Baran DT, Ray R, Sorensen AM, Honeyman T, Holick MF 1994 Binding characteristics of a membrane receptor that recognizes 1α,25-dihydroxyvitamin D_3 and its epimer, 1β,25-dihydroxyvitamin D_3. J Cell Biochem **56**:510–517.
29. Seo EG, Kato A, Norman AW 1996 Evidence for a 24*R*,25$(OH)_2$-vitamin D_3 receptor/binding protein in a membrane fraction isolated from a chick tibial fracture-healing callus. Biochem Biophys Res Commun **225**:203–208.
30. Haussler MR, Myrtle JF, Norman AW 1968 The association of a metabolite of vitamin D_3 with intestinal mucosa chromatin, *in vivo*. J Biol Chem **243**:4055–4064.
31. Haussler MR, Norman AW 1969 Chromosomal receptor for a vitamin D metabolite. Proc Natl Acad Sci USA **62**:155–162.
32. Ortí E, Bodwell JE, Munck A 1992 Phosphorylation of steroid hormone receptors. Endocr Rev **13**:105–128.
33. Darwish HM, Burmester JK, Moss VE, DeLuca HF 1993 Phosphorylation is involved in transcriptional activation by the 1,25-dihydroxyvitamin D_3 receptor. Biochim Biophys Acta Lipids Lipid Metab **1167**:29–36.
34. Jurutka PW, Hsieh J-C, Haussler MR 1993 Phosphorylation of the human 1,25-dihydroxyvitamin D_3 receptor by cAMP-dependent protein kinase, *in vitro*, and in transfected COS-7 cells. Biochem Biophys Res Commun **191**:1089–1096.
35. Hsieh J-C, Jurutka PW, Nakajima S, Galligan MA, Haussler CA, Shimizu Y, Shimizu N, Whitfield GK, Haussler MR 1993 Phosphorylation of the human vitamin D receptor by protein kinase C. Biochemical and functional evaluation of the serine 51 recognition site. J Biol Chem **268**:15118–15126.
36. Jurutka PW, Hsieh J-C, MacDonald PN, Terpening CM, Haussler CA, Haussler MR, Whitfield GK 1993 Phosphorylation of serine 208 in the human vitamin D receptor. J Biol Chem **268**:6791–6799.
37. Freedman LP, Arce V, Perez Fernandez R 1994 DNA sequences

that act as high affinity targets for the vitamin D_3 receptor in the absence of the retinoid X receptor. Mol Endocrinol **8**:265–273.

38. Liu M, Freedman LP 1994 Transcriptional synergism between the vitamin D_3 receptor and other nonreceptor transcription factors. Mol Endocrinol **8**:1593–1604.
39. Walters MR 1992 Newly identified actions of the vitamin D endocrine system. Endocr Rev **13**:719–764.
40. Nemere I, Norman AW 1987 Studies on the mode of action of calciferol. LII. Rapid action of 1,25-dihydroxyvitamin D_3 on calcium transport in perfused chick duodenum: Effect of inhibitors. J Bone Miner Res **2**:99–107.
41. Nemere I, Yoshimoto Y, Norman AW 1984 Studies on the mode of action of calciferol. LIV. Calcium transport in perfused duodena from normal chicks: Enhancement with 14 minutes of exposure to 1α,25-dihydroxyvitamin D_3. Endocrinology **115**:1476–1483.
42. Lieberherr M, Grosse B, Duchambon P, Drüeke T 1989 A functional cell surface type receptor is required for the early action of 1,25-dihydroxyvitamin D_3 on the phosphoinositide metabolism in rat enterocytes. J Biol Chem **264**:20403–20406.
43. Inoue M, Wakasugi M, Wakao R, Gan N, Tawata M, Nishii Y, Onaya T 1992 A synthetic analogue of vitamin D_3, 22-oxa-1,25-dihydroxy-vitamin D_3, stimulates the production of prostacyclin by vascular tissues. Life Sci **51**:1105–1112.
44. Wali RK, Baum CL, Sitrin MD, Brasitus TA 1990 $1,25(OH)_2$ vitamin D_3 stimulates membrane phosphoinositide turnover, activates protein kinase C, and increases cytosolic calcium in rat colonic epithelium. J Clin Invest **85**:1296–1303.
45. Wali RK, Baum CL, Sitrin MD, Bolt MJG, Dudeja PK, Brasitus TA 1992 Effect of vitamin D status on the rapid actions of 1,25-dihydroxycholecalciferol in rat colonic membranes. Am J Physiol **262**:G945–G953.
46. Bolt MJG, Bissonnette BM, Wali RK, Hartmann SC, Brasitus TA, Sitrin MD 1993 Characterization of phosphoinositide-specific phospholipase C in rat colonocyte membranes. Biochem J **292**:271–276.
47. Baran DT, Sorensen AM, Honeyman TW, Ray R, Holick MF 1989 Rapid actions of 1α,25-dihydroxyvitamin D_3 and calcium and phospholipids in isolated rat liver nuclei. FEBS Lett **259**:205–208.
48. Baran DT, Sorensen AM, Honeyman RW, Ray R, Holick MF 1990 1α,25-Dihydroxyvitamin D_3-induced increments in hepatocyte cytosolic calcium and lysophosphatidylinositol: Inhibition by pertussis toxin and 1β,25-dihydroxyvitamin D_3. J Bone Miner Res **5**:517–524.
49. Baran DT, Sorensen AM, Shalhoub V, Owen T, Oberdorf A, Stein G, Lian J 1991 1,25-Dihydroxyvitamin D_3 rapidly increases cytosolic calcium in clonal rat osteosarcoma cells lacking the vitamin D receptor. J Bone Miner Res **6**:1269–1275.
50. Civitelli R, Kim YS, Gunsten SL, Fujimori A, Huskey M, Avioli LV, Hruska KA 1990 Nongenomic activation of the calcium message system by vitamin D metabolites in osteoblast-like cells. Endocrinology **127**:2253–2262.
51. de Boland AR, Boland RL 1993 1,25-Dihydroxyvitamin D_3 induces arachidonate mobilization in embryonic chick myoblasts. Biochim Biophys Acta Mol Cell Res **1179**:98–104.
52. Morelli S, de Boland AR, Boland RL 1993 Generation of inositol phosphates, diacylglycerol and calcium fluxes in myoblasts treated with 1,25-dihydroxyvitamin D_3. Biochem J **289**:675–679.
53. Selles J, Boland RL 1991 Evidence on the participation of the 3′,5′-cyclic AMP pathway in the non-genomic action of 1,25-dihydroxy-vitamin D_3 in cardiac muscle. Mol Cell Endocrinol **82**:229–235.
54. Mandla S, Boneh A, Tenenhouse HS 1990 Evidence for protein kinase C involvement in the regulation of renal 25-hydroxyvitamin D_3-24-hydroxylase. Endocrinology **127**:2639–2647.
55. Slater SJ, Kelly MB, Taddeo FJ, Larkin JD, Yeager MD, McLane JA, Ho C, Stubbs CD 1995 Direct activation of protein kinase C by 1α,25-dihydroxyvitamin D_3. J Biol Chem **270**:6639–6643.
56. Bissonnette M, Wali RK, Hartmann SC, Niedziela SM, Roy HK, Tien X-Y, Sitrin MD, Brasitus TA 1995 1,25-Dihydroxyvitamin D_3 and 12-*O*-tetradecanoyl phorbol 13-acetate cause differential activation of Ca^{2+}-dependent and Ca^{2+}-independent isoforms of protein kinase C in rat colonocytes. J Clin Invest **95**:2215–2221.
57. Bhatia M, Kirkland JB, Meckling-Gill KA 1996 1,25-Dihydroxyvitamin D_3 primes acute promyelocytic cells for TPA-induced monocytic differentiation through both PKC and tyrosine phosphorylation cascades. Exp Cell Res **222**:61–69.
58. Sylvia VL, Schwartz Z, Ellis EB, Helm SH, Gomez R, Dean DD, Boyan BD 1996 Nongenomic regulation of protein kinase C isoforms by the vitamin D metabolites $1\alpha,25\text{-}(OH)_2D_3$ and $24R,25\text{-}(OH)_2D_3$. J Cell Physiol **167**:380–393.
59. Norman AW, Nemere I, Williams G, King M 1984 1α,25-Dihydroxyvitamin D_3 mediates biological responses both as a steroid hormone and as a membrane-active agent. Cohn DV, Potts JT Jr, Fujita T (eds) Endocrine Control of Bone and Calcium Metabolism, 1st Ed. Elsevier, Amsterdam, pp. 316–319.
60. Nemere I, Zhou L-X, Norman AW 1993 Nontranscriptional effects of steroid hormones. Receptor **3**:277–291.
61. Boyan BD, Schwartz Z, Bonewald L, Swain L 1989 Localization of $1,25\text{-}(OH)_2D_3$-responsive alkaline phosphatase in osteoblast-like cells (ROS 17/2.8, MG 63, and MC 3T3) and growth cartilage cells in culture. J Biol Chem **264**:11879–11886.
62. Meikle MC 1988 Hypercalcaemia of malignancy. Nature **336**:311
63. Cancela L, Nemere I, Norman AW 1988 $1\alpha,25(OH)_2$-vitamin D_3: A steroid hormone capable of producing pleiotropic receptor-mediated biological responses by both genomic and nongenomic mechanisms. J Steroid Biochem Mol Biol **30**:33–39.
64. Lieberherr M 1987 Effects of vitamin-D_3 metabolites on cytosolic free calcium in confluent mouse osteoblasts. J Biol Chem **262**:13168–13173.
65. Barsony J, Marx SJ 1991 Rapid accumulation of cyclic GMP near activated vitamin D receptors. Proc Natl Acad Sci USA **88**:1436–1440.
66. Morley P, Whitfield JF, Vanderhyden BC, Tsang BK, Schwartz JL 1992 A new, nongenomic estrogen action: The rapid release of intracellular calcium. Endocrinology **131**:1305–1312.
67. Mendoza C, Tesarik J 1993 A plasma-membrane progesterone receptor in human sperm is switched on by increasing intracellular free calcium. FEBS Lett **330**:57–60.
68. Aurell Wistrom C, Meizel S 1993 Evidence suggesting involvement of a unique human sperm steroid receptor/Cl^- channel complex in the progesterone-initiated acrosome reaction. Dev Biol **159**:679–690.
69. Blackmore PF, Beebe SJ, Danforth DR, Alexander N 1990 Progesterone and 17α-progesterone: Novel stimulators of calcium influx in human sperm. J Biol Chem **265**:1376–1380.
70. Majewska MD, Vaupel DB 1991 Steroid control of uterine motility via gamma-aminobutyric $acid_A$ receptors in the rabbit: A novel mechanism. J Endocrinol **131**:427–434.
71. Koenig H, Fan C-C, Goldstone AD, Lu CY, Trout JJ 1989 Polyamines mediate androgenic stimulation of calcium fluxes and membrane transport in rat myocytes. Circ Res **64**:415–426.
72. Rehberger P, Rexin M, Gehring U 1992 Heterotetrameric structure of the human progesterone receptor. Proc Natl Acad Sci USA **89**:8001–8005.
73. Gametchu B, Watson CS, Wu S 1993 Use of receptor antibodies

to demonstrate membrane glucocorticoid receptor in cells from human leukemic patients. FASEB J **7**:1283–1292.

74. Orchinik M, Murray TF, Moore FL 1991 A corticosteroid receptor in neuronal membranes. Science **252**:1848–1851.
75. Smith TJ, Davis FB, Davis PJ 1992 Stereochemical requirements for the modulation by retinoic acid of thyroid hormone activation of Ca^{2+}-ATPase and binding at the human erythrocyte membrane. Biochem J **284**:583–587.
76. Segal J 1990 Thyroid hormone action at the level of the plasma membrane. Thyroid **1**:83–87.
77. Yoshimoto Y, Norman AW 1986 Biological-activity of vitamin D metabolites and analogs–dose–response study of Ca-45 transport in an isolated chick duodenum perfusion system. J Steroid Biochem Mol Biol **25**:905–909.
78. Yoshimoto Y, Nemere I, Norman AW 1986 Hypercalcemia inhibits the "rapid" stimulatory effect on calcium transport in perfused duodena from normal chicks mediated in vitro by 1,25-dihydroxyvitamin D_3. Endocrinology **118**:2300–2304.
79. Nemere I, Norman AW 1989 1,25-Dihydroxyvitamin D_3-mediated vesicular calcium transport in intestine: Dose–response studies. Mol Cell Endocrinol **67**:47–53.
80. Nemere I, Leathers VL, Norman AW 1986 1,25-Dihydroxyvitamin D_3-mediated intestinal calcium transport: Biochemical identification of lysosomes containing calcium and calcium binding protein (calbindin-D_{28K}). J Biol Chem **261**:16106–16114.
81. Nemere I, Norman AW 1988 1,25-Dihydroxyvitamin D_3-mediated vesicular transport of calcium in intestine: Time course studies. Endocrinology **122**:2962–2969.
82. Nemere I, Norman AW 1990 Transcaltachia, vesicular calcium transport, and microtubule-associated calbindin-D_{28K}: Emerging views of 1,25-dihydroxyvitamin D_3-mediated intestinal calcium absorption. Miner Electrolyte Metab **16**:109–114.
83. Nemere I, Norman AW 1991 Redistribution of cathepsin B activity from the endosomal–lysosomal pathway in chick intestine within 3 min of calcium absorption. Mol Cell Endocrinol **78**:7–16.
84. Nemere I, Norman AW 1985 Effect of $1,25(OH)_2D_3$ on the subcellular redistribution of lysosomal hydrolases and potential relation to calcium transport. Norman AW, Schaefer K, Grigoleit H-G Herrath Dv (eds). Vitamin D: Chemical, Biochemical and Clinical Update. de Gruyter, Berlin, pp. 417–418.
85. Nemere I, Leathers VL, Thompson BS, Luben RA, Norman AW 1991 Redistribution of calbindin-D_{28K} in chick intestine in response to calcium transport. Endocrinology **129**:2972–2984.
86. Nemere I, Norman AW 1987 The rapid, hormonally stimulated transport of calcium (transcaltachia). J Bone Miner Res **2**:167–169.
87. Zhou L-X, Nemere I, Norman AW 1992 $1,25(OH)_2$-vitamin D_3 analog structure–function assessment of the rapid stimulation of intestinal calcium absorption. (transcaltachia). J Bone Miner Res **7**:457–463.
88. Greenberg DA, Carpenter CL, Cooper EC 1985 Stimulation of calcium uptake in PC12 cells by the dihydropyridine agonist BAY K 8644. J Neurochem **45**(No. 3):990–993.
89. Enyeart JJ, Hinkle PM 1984 The calcium agonist Bay K 8644 stimulates secretion from a pituitary cell line. Biochem Biophys **122**:991–996.
90. de Boland AR, Norman AW 1990 Influx of extracellular calcium mediates 1,25-dihydroxyvitamin D_3-dependent transcaltachia (the rapid stimulation of duodenal Ca^{2+} transport). Endocrinology **127**:2475–2480.
91. D'Emden MC, Wark JD 1989 Vitamin D-enhanced thyrotrophin release from rat pituitary cells: Effects of Ca^{2+}, dihydropyridines and ionomycin. J Endocrinol **121**:441–450.
92. Norman AW, Bouillon R, Farach-Carson MC, Bishop JE, Zhou L-X, Nemere I, Zhao J, Muralidharan KR, Okamura WH 1993 Demonstration that 1β,25-dihydroxyvitamin D_3 is an antagonist of the nongenomic but not genomic biological responses and biological profile of the three A-ring diastereomers of 1α,25-dihydroxyvitamin D_3. J Biol Chem **268**:20022–20030.
93. Dormanen MC, Bishop JE, Hammond MW, Okamura WH, Nemere I, Norman AW 1994 Nonnuclear effects of the steroid hormone $1\alpha,25(OH)_2$-vitamin D_3: Analogs are able to functionally differentiate between nuclear and membrane receptors. Biochem Biophys Res Commun **201**:394–401.
94. de Boland AR, Nemere I, Norman AW 1990 Ca^{2+}-channel agonist Bay K8644 mimics $1,25(OH)_2$-vitamin D_3 rapid enhancement of Ca^{2+} transport in chick perfused duodenum. Biochem Biophys Res Commun **166**:217–222.
95. Khoury R, Ridall AL, Norman AW, Farach-Carson MC 1994 Target gene activation by 1,25-dihydroxyvitamin D_3 in osteosarcoma cells is independent of calcium influx. Endocrinology **135**:2446–2453.
96. Corvol M, Ulmann A, Garabedian M 1980 Specific nuclear uptake of 24,25-dihydroxycholecalciferol, a vitamin D_3 metabolite biologically active in cartilage. FEBS Lett **116**:273–276.
97. Somjen D, Somjen GJ, Weisman Y, Binderman I 1982 Evidence for 24,25-dihydroxycholecalciferol receptors in long bones of newborn rats, and in chick embryo limb-bud cells. Personal communication.
98. Norman AW, Okamura WH, Hammond MW, Bishop JE, Dormanen MC, Bouillon R, van Baelen H, Ridal AL, Daane E, Khoury R, 1996 Comparison of 6-*s-cis* and 6-*s-trans* locked analogs of $1\alpha,25(OH)_2$-vitamin D_3 indicates that the 6-*s-cis* conformation is preferred by the membrane but not the nuclear receptor for $1\alpha,25(OH)_2$-vitamin D_3. Mol Endo in press.
99. Norman AW, Bishop JE, Collins ED, Seo E-G, Satchell DP, Dormanen MC, Zanello SB, Farach-Carson MC, Bouillon R, Okamura WH 1996 Differing shapes of 1α,25-dihydroxyvitamin D_3 function as ligands for the D-binding protein, nuclear receptor and membrane receptor: A status report. J Steroid Biochem Mol Biol **56**:13–22.
100. Kamei Y, Kawada T, Kazuki R, Ono T, Kato S, Sugimoto E 1993 Vitamin D receptor gene expression is up-regulated by 1,25-dihydroxyvitamin D_3 in 3T3-L1 preadipocytes. Biochem Biophys Res Commun **193**:948–955.
101. Reichrath J, Schilli M, Kerber A, Bahmer FA, Czarnetzki BM, Paus R 1994 Hair follicle expression of 1,25-dihydroxyvitamin D_3 receptors during the murine hair cycle. Brit J Dermatol **131**:477–482.
102. Peterfy C, Tenenhouse A 1982 Vitamin D receptors in isolated rat parotid and acinar cells. Biochim Biophys Acta **721**:158–163.
103. Clark SA, Stumpf WE, Bishop CW, DeLuca HF, Park DH, Joh TH 1986 The adrenal—A new target organ of the calciotropic hormone 1,25-dihydroxyvitamin-D_3. Cell Tissue Res **243**:299–302.
104. Pike JW, Gooze LL, Haussler MR 1980 Biochemical evidence for 1,25-dihydroxyvitamin D_3 receptor macromolecules in parathyroid, pancreatic, pituitary, and placental tissues. Life Sci **26**:407–414.
105. Kream BE, Jose M, Yamada S, DeLuca HF 1977 A specific high-affinity binding macromolecule for 1,25-dihydroxycholecalciferol in fetal bone. Science **197**:1086–1088.
106. Christakos S, Norman AW 1981 Studies on the mode of action of calciferol XXIX—Biochemical characterization of 1,25-dihydroxyvitamin D_3 receptors in chick pancreas and kidney cytosol. Endocrinology **108**:140–149.
107. Schleicher G, Bartke A, Bidmon HJ, Stumpf WE 1993 $1,25(OH)_2$ vitamin D_3 binding sites in male sex organs of the Siberian ham-

ster (*Phodopus sungorus*). An autoradiographic study. J Steroid Biochem Mol Biol **46**:331–335.

108. Yu X-P, Mocharla H, Hustmyer FG, Manolagas SC 1991 Vitamin D receptor expression in human lymphocytes. Signal requirements and characterization by western blots and DNA sequencing. J Biol Chem **266**:7588–7595.
109. Duncan WE, Glass AR, Wray HL 1991 Estrogen regulation of the nuclear 1,25-dihydroxyvitamin D_3 receptor in rat liver and kidney. Endocrinology **129**:2318–2324.
110. Findlay I, Ashcroft FM, Kelly RP, Rorsman P, Petersen OH, Trube G 1989 Calcium currents in insulin-secreting β-cells. Ann NY Acad Sci **560**:403–409.
111. Stumpf WE, Sar M, Clark SA, DeLuca HF 1982 Brain target sites for 1α,25-dihydroxyvitamin D_3. Science **215**:1403–1405.
112. Nguyen TM, Guillozo H, Marin L, Dufour ME, Tordet C, Pike JW, Garabedian M 1990 1,25-Dihydroxyvitamin D_3 receptors in rat lung during the perinatal period: Regulation and immunohistochemical localization. Endocrinology **127**:1755–1762.
113. Stumpf WE, Sar M, Reid FA, Tanaka Y, DeLuca HF 1979 Target cells for 1,25-dihydroxyvitamin D_3 in intestinal tract, stomach, kidney, skin, pituitary, and parathyroid. Science **206**:1188–1190.
114. Merke J, Milde P, Lewicka S, Hügel U, Klaus G, Mangelsdorf DJ, Haussler MR, Rauterberg EW, Ritz E 1989 Identification and regulation of 1,25-dihydroxyvitamin D_3 receptor activity and biosynthesis of 1,25-dihydroxyvitamin D_3. Studies in cultured bovine aortic endothelial cells and human dermal capillaries. J Clin Invest **83**:1903–1915.
115. Colston KW, Berger U, Wilson P, Hadcocks L, Naeem I, Earl HM, Coombes RC 1988 Mammary gland 1,25-dihydroxyvitamin D_3 receptor content during pregnancy and lactation. Mol Cell Endocrinol **60**:15–22.
116. Simpson RU, Thomas GA, Arnold AJ 1985 Identification of 1,25-dihydroxyvitamin D_3 receptors and activities in muscle. J Biol Chem **260**:8882–8891.
117. Boland R, Norman AW, Ritz E, Hasselbach W 1992 Presence of a 1,25-dihydroxyvitamin D_3 receptor in chick skeletal muscle myoblasts. Biochem Biophys Res Commun **128**:305–311.
118. Merke J, Hofmann W, Goldschmit D, Ritz E 1987 Demonstration of $1{,}25(OH)_2$ vitamin-D_3 receptors and actions in vascular smooth-muscle cells *in vitro*. Calcif Tissue Int **41**:112–114.
119. Walters MR 1982 1,25-Dihydroxyvitamin D receptors in rat testes, epididymis, and uterus. Fed Proc **41**:1165.
120. Hedlund TE, Moffatt KA, Miller GJ 1996 Stable expression of the nuclear vitamin D receptor in the human prostatic carcinoma cell line JCA-1: Evidence that the antiproliferative effects of 1α,25-dihydroxyvitamin D_3 are mediated exclusively through the genomic signaling pathway. Endocrinology **137**:1554–1561.
121. Shabahang M, Buras RR, Davoodi F, Schumaker LM, Nauta RJ, Evans SRT 1993 1,25-Dihydroxyvitamin D_3 receptor as a marker of human colon carcinoma cell line differentiation and growth inhibition. Cancer Res **53**:3712–3718.
122. Mangelsdorf DJ, Koeffler HP, Donaldson CA, Pike JW, Haussler MR 1984 1,25-Dihydroxyvitamin D_3-induced differentiation in a human promyelocytic leukemia cell line (HL-60): Receptor-mediated maturation to macrophage-like cells. J Cell Biol **98**:391–398.
123. Berger U, McClelland RA, Wilson P, Greene GL, Haussler MR, Pike JW, Colston K, Easton D, Coombes RC 1991 Immunocytochemical determination of estrogen receptor, progesterone receptor, and 1,25-dihydroxyvitamin D_3 receptor in breast cancer and relationship to prognosis. Cancer Res **51**:239–244.
124. Albert DM, Marcus DM, Gallo JP, O'Brien JM 1992 The antineoplastic effect of vitamin D in transgenic mice with retinoblastoma. Invest Ophthalmol Vis Sci **33**:2354–2364.
125. Klaus G, Merke J, Eing H, Hügel U, Milde P, Reichel H, Ritz E, Mehls O 1991 $1{,}25(OH)_2D_3$ receptor regulation and $1{,}25(OH)_2D_3$ effects in primary cultures of growth cartilage cells of the rat. Calcif Tissue Int **49**:340–348.
126. Walters MR, Rosen DM, Norman AW, Luben RA 1982 1,25-Dihydroxyvitamin D receptors in an established bone cell line: Correlation with biochemical responses. J Biol Chem **257**:7481–7484.
127. Stumpf WE, Downs TW 1987 Nuclear receptors for $1{,}25(OH)_2$ vitamin D_3 in thymus reticular cells studied by autoradiography. Histochemistry **87**:367–369.
128. Vandewalle B, Adenis A, Hornez L, Revillion F, Lefebvre J 1994 1,25-Dihydroxyvitamin D_3 receptors in normal and malignant human colorectal tissues. Cancer Lett **86**:67–73.
129. Dokoh S, Donaldson CA, Marion SL, Pike JW, Haussler MR 1983 The ovary: a target organ for 1α,25-dihydroxyvitamin D_3. Endocrinology **112**:200–206.
130. Berg JP, Torjesen PA, Haug E 1991 Calcitriol receptors in rat thyroid follicular cells (FRTL-5). Acta Endocrinol **125**:574–580.
131. Elaroussi MA, Prahl JM, DeLuca HF 1994 The avian vitamin D receptors: Primary structures and their origins. Proc Natl Acad Sci USA **91**:11596–11600.
132. Wecksler WR, Ross FP, Mason RS, Posen S, Norman AW 1980 Studies on the mode of action of calciferol. XXIV. Biochemical properties of the 1α,25-dihydroxyvitamin D_3 cytoplasmic receptors from human and chick parathyroid glands. Arch Biochem Biophys **201**:95–103.
133. Danan J, Delorme A, Cuisinier-Gleizes P 1981 Biochemical evidence for a cytoplasmic 1α,25-dihydroxyvitamin D_3 receptor-like protein in rat yolk sac. J Biol Chem **256**:4847–4850.
134. Johnson JA, Grande JP, Windebank AJ, Kumar R 1996 1,25-Dihydroxyvitamin D_3 receptors in developing dorsal root ganglia of fetal rats. Dev Brain Res **92**:120–124.
135. Stumpf WE, Clark SA, O'Brien LP, Reid FA 1988 $1{,}25(OH)_2$ vitamin D_3 sites of action in spinal cord and sensory ganglion. Anat Embryo **177**:307–310.
136. Pike JW 1991 Vitamin D_3 receptors: Structure and function in transcription. Annu Rev Nutr **11**:189–216.
137. de Boland AR, Norman AW 1990 Evidence for involvement of protein kinase C and cyclic adenosine 3′,5′-monophosphate-dependent protein kinase in the 1,25-dihydroxyvitamin D_3-mediated rapid stimulation of intestinal calcium transport (transcaltachia). Endocrinology **127**:39–45.
138. Simboli-Campbell M, Gagnon A, Franks DJ, Welsh J 1994 1,25-Dihydroxyvitamin D_3 translocates protein kinase C_β to nucleus and enhances plasma membrane association of protein kinase C_α in renal epithelial cells. J Biol Chem **269**:3257–3264.
139. Simboli-Campbell M, Franks DJ, Welsh JE 1992 $1{,}25(OH)_2D_3$ Increases membrane associated protein kinase C in MDBK cells. Cell Signal **4**:99–109.
140. Barsony J, Pike JW, DeLuca HF, Marx SJ 1990 Immunocytology with microwave-fixed fibroblasts shows 1α,25-dihydroxyvitamin D_3-dependent rapid and estrogen-dependent slow reorganization of vitamin D receptors. J Cell Biol **111**:2385–2395.
141. Zanello LP, Norman AW 1996 1a,$25(OH)_2$ Vitamin D_3-mediated stimulation of outward anionic currents in osteoblast-like ROS 17/2.8 cell. Biochem Biophys Res Commun **225**:551–556.
142. Beno DWA, Brady LM, Bissonnette M, Davis BH 1995 Protein kinase C and mitogen-activated protein kinase are required for 1,25-dihydroxyvitamin D_3-stimulated Egr induction. J Biol Chem **270**:3642–3647.
143. Biskobing DM, Rubin J 1993 1,25-Dihydroxyvitamin D_3 and

phorbol myristate acetate produce divergent phenotypes in a monomyelocytic cell line. Endocrinology **132**:862–866.

144. Yada Y, Ozeki T, Meguro S, Mori S, Nozawa Y 1989 Signal transduction in the onset of terminal keratinocyte differentiation induced by 1α,25-dihydroxyvitamin D_3: Role of protein kinase C translocation. Biochem Biophys Res Commun **163**:1517–1522.
145. Bourdeau A, Atmani F, Grosse B, Lieberherr M 1990 Rapid effects of 1,25-dihydroxyvitamin D_3 and extracellular Ca^{2+} on phospholipid metabolism in dispersed porcine parathyroid cells. Endocrinology **127**:2738–2743.
146. Sugimoto T, Ritter C, Ried I, Morrissey J, Slatopolsky E 1992 Effect of 1,25-dihydroxyvitamin D_3 on cytosolic calcium in dispersed parathyroid cells. Kidney Int **33**:850–854.
147. Zhou L-X, Norman AW 1996 Characterization of the second messengers involved in 1α,25-dihydroxyvitamin D_3 stimulated intestinal calcium absorption (transcaltachia). Endocrine **5**:47–50.

Section III

Target Organs and Actions

Vitamin D and the Intestinal Absorption of Calcium and Phosphorus

ROBERT H. WASSERMAN Department of Physiology, College of Veterinary Medicine, Cornell University, Ithaca, New York

I. INTRODUCTION

Calcium is required for many physiological processes, including proper nerve and muscle function, blood coagulation, as a second or third messenger in hormonal and neurotransmitter action, and in the mineralization of bones and teeth. Phosphorus is also multifunctional, participating in anabolic and catabolic reactions in the form of ATP, serving as a precursor in the synthesis of biologically important molecules (phospholipids, deoxyribonucleic acids, ribonucleic acids), as a contributor to the buffering capacity of body fluids and cells, and, with calcium, as a major constituent of the mineral of bones and teeth. The hard tissues contain 99 and 86% of the body's content of calcium and phosphorus, respectively. The physiology of calcium and phosphate is discussed in Chapters 31 and 32. These mineral ions gain access to the body by way of absorption from the gastrointestinal tract, and vitamin D is required in most vertebrate species for the optimization and regulation of the absorptive processes integral to this uptake.

Previous studies by investigators like E. Mellanby, R. Nicolaysen, E. Kodicek, A. Carlsson, and B. Lindquist established that an important site of vitamin D action was the intestinal tract and absorption of calcium and phosphorus. A number of uncertainties, as one might expect, surrounded these pioneer studies. Hypotheses were offered that vitamin D directly affected calcium absorption and secondarily phosphate absorption, or that the direct effect was on phosphate absorption, with calcium secondarily affected. No special physiological mechanisms for the absorption of these mineral ions were evident at that time. More recently, a considerable amount of research has been accomplished that has advanced our understanding of these transport processes and contributed to the possible application of newer information to clinical medicine.

This chapter describes some of the earlier as well as more current information on calcium and phosphorus absorption, emphasizing the role of vitamin D. The amount of material to be included was necessarily limited, and important contributions unfortunately are omitted. However, the reader is referred to various reviews [1–16] for more detailed discussions and contrasting views on the role of vitamin D on these specific transport processes.

II. THE VITAMIN D HORMONE

As discussed in detail in Section I of this volume vitamin D, from either dietary sources or endogenous synthesis by ultraviolet irradiation of the skin, is transformed into the most active hormonal form, 1,25-dihydroxyvitamin D [$1,25(OH)_2D$], by two sequential hydroxylation reactions [17]. The first, the 25-hydroxylation reaction, takes place in the liver, and the second, a 1α-hydroxylation reaction, occurs in the kidney. The renal synthesis of $1,25(OH)_2D_3$ is a direct function of circulating parathyroid hormone (PTH) level, which is in turn responsive to other feedback controlling mechanisms.

The derivative $1,25(OH)_2D_3$, with PTH and calcitonin, constitute the traditional calciotropic hormones that maintain serum calcium within narrow, "normal" limits through either direct or indirect effects on the intestine, kidney, and the skeleton. Each of the calciotropic hormones also influences phosphate absorption and metabolism with effects on the same organ systems. In addition to its homeostatic function, the vitamin D hormone also influences a number of other biological systems as discussed in the reviews by Bikle [18] and Walters [19] and extensively discussed in Section VIII of this volume.

In the intestine and kidney, $1,25(OH)_2D_3$ elicits effects on the genome as reflected by induced or enhanced synthesis of various proteins (see Chapters 8–10). Nongenomic actions are discussed in Chapter 15.

III. CALCIUM ABSORPTION

A. Overall Processes of Calcium Absorption

Analysis of the transport processes requires a consideration of the transmural thermodynamic factors that may influence transepithelial ion absorption. These parameters for calcium include an intracellular ionic concentration for Ca^{2+} of about 10^{-7} *M* and an intracellular electropotential difference of about -50 mV with respect to the lumen. These factors clearly indicate that luminal Ca^{2+}, usually present in the millimolar range, can enter the enterocyte by diffusing down an extremely steep electrochemical gradient. The transfer of Ca^{2+} out of the cell in either direction would require a considerable input of energy into the extrusion process.

Schachter and colleagues [20,21], using an *in vitro* everted gut sac technique, reported for the first time that rat intestine can actively transport calcium against a concentration gradient and that the active calcium transport was highly dependent on the vitamin D status of the donor animal. The Harrisons [22] also demonstrated that vitamin D increased the movement of Ca^{2+} across the *in vitro* everted gut sac of the rat but proposed that the steroid increased the diffusivity of Ca^{2+} across the membrane rather than acting in an active transport step.

The presence of an active Ca^{2+} transport mechanism in the duodenum was first demonstrated in the intact rat by determining the unidirectional Ca^{2+} fluxes across the duodenum and the transmural potential difference [23]. These data were subjected to analysis by the method of Ussing (see Wasserman and Taylor [1] for a fuller discussion and references), and it was evident that there was a net movement of Ca^{2+} against an electrochemical gradient. An active calcium transport system is also present in both the upper and lower segments of the rat small intestine [1,24,25]. A similar approach was employed with chick intestine *in situ.* Like the rat, the presence of active Ca^{2+} transport by chick duodenum was demonstrated. In addition, it was shown that the uphill transport of Ca^{2+} was enhanced by vitamin D repletion of vitamin D-deficient chicks [26], verifying the earlier findings of Schachter *et al.* [20,21]. Further, this same study suggested that vitamin D increased the nonsaturable diffusive movement of Ca^{2+} across the epithelium (see Section III,C) in addition to its effect on active calcium transport.

The relative absorption of calcium along the gastrointestinal tract of the rat was determined by Marcus and Lengemann [27], using a nonabsorbable indicator as the reference substance. After a single oral dose of tracer calcium, residual tracer was measured in the various segments at different times after the gavage. Most absorption was shown to occur in the ileum (62%), followed by the jejunum (23%) and then the duodenum (15%), with minor amounts absorbed from the stomach and large intestine. The major contributor to the amount absorbed was the residence time in the particular segment and, secondarily, the rate of absorption from that segment. The latter follows the following order: duodenum > jejunum > ileum.

B. Vitamin D and Transcellular Calcium Transport

The steps in the transcellular calcium transport path include the transfer of calcium across the brush border membrane into the enterocyte interior, movement through the cytosolic compartment, and finally its extrusion across the basolateral membrane. Part of the evidence for multiple effects of vitamin D on these transport steps comes from experiments such as that depicted in Fig. 1. The mucosal accumulation and absorption of calcium ($^{47}Ca^{2+}$) from ligated duodenal loops *in situ* were determined as a function of time after the vitamin D hormone was given to vitamin D-deficient chicks [28]. As shown in Fig. 1, $1,25(OH)_2D_3$ stimulated a relatively rapid (within 30 min) accumulation of ^{47}Ca by the mucosal tissue, and only later (3–4 hr after hormone) was there an increase in overall ^{47}Ca absorption. The initial vitamin D-dependent increase in Ca^{2+} entry into the tissue appears to be nongenomic. The transport reactions that followed with a 3- to 4-hr time lag include, first, movement of Ca^{2+} through the cytosolic compartment and, second, extrusion from the cell. The latter reactions, as discussed in Section III,B,2,3 below, are highly dependent on the genomic action of vitamin D.

1. CALCIUM ENTRY

The entry of Ca^{2+} from intestinal lumen across the brush border membrane into the enterocyte is a downhill, diffusional process [28]. The vitamin D hormone increases the rate of entry by a rapid, nongenomic process proposed to involve an alteration in the fluidity state of the microvillus membrane, as was inferred from vitamin D-dependent changes in phospholipid composition [29]. Direct measurements of microvillar fluidity by Bikle *et al.* [30], Putkey *et al.* [31], and Schedl *et al.*. [32] did not support the fluidity hypothesis. Brasitus *et al.* [33], however, using specific probes, did find an early effect of $1,25(OH)_2D_3$ on brush border fluidity, providing some support for the fluidity notion of Rasmussen *et al.* [29]. An increase in overall Ca^{2+} absorption was not discernible in the Brasitus study until 5 hr after $1,25(OH)_2D_3$, suggesting that the early modification of brush border membrane properties, which might increase Ca^{2+} entry, is not the limiting step in vitamin D-dependent calcium absorption.

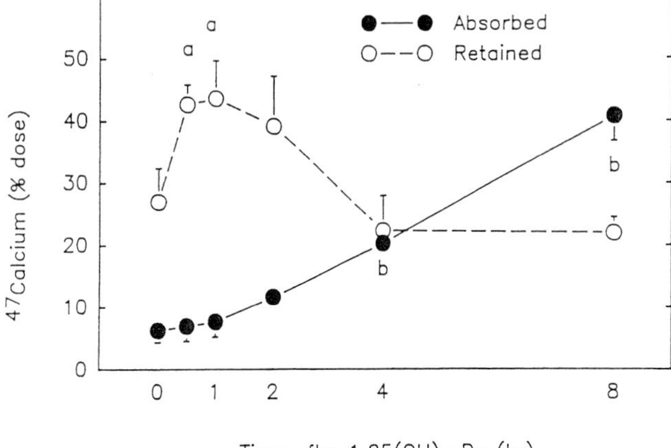

FIGURE 1 Intestinal absorption and tissue retention of luminally administered $^{47}Ca^{2+}$ in vitamin D-deficient chicks at varying times following a single dose of $1,25(OH)_2D_3$. Data are means ± SEM for six chicks. Lowercase letters represent statistically significant results: a denotes significant difference from the zero time control ($p < 0.05$) in $^{47}Ca^{2+}$ tissue retention, and b denotes significant difference ($p < 0.05$) in $^{47}Ca^{2+}$ absorption above the control value. From Fullmer *et al.* [47]; reprinted with permission.

In addition to changes in membrane properties, other vitamin D-dependent factors associated with the brush border membrane have been implicated in the calcium entry step. These include the brush border-bound calbindin-D_{28K} [34,35], the integral brush border membrane protein of Schachter and Kowarski [36] and Miller *et al.* [37], alkaline phosphatase, and calmodulin, the latter possibly acting via calmodulin-associated brush border proteins [38,39]. The exact role of these factors in the permeation of Ca^{2+} across the microvillar membrane has not been defined. Evidence of the presence of a mobile membrane carrier of Ca^{2+} has been provided by Wilson *et al.* [40], an observation that will require additional confirmation and information on its vitamin D dependency.

The entrance of calcium into the enterocyte, even in the absence of vitamin D, suggests the necessity of a mechanism to control Ca^{2+} influx because of the known toxicity of excessive amounts of intracellular calcium. A possible mechanism, as previously proposed [41], involves the binding of Ca^{2+} to elements within the brush border-terminal web region, wherein resides a relatively high concentration of the ubiquitous calcium-binding protein calmodulin [42–44]. Calmodulin in this region is bound to a variable extent to myosin I, a 110-kDa ATPase with mechanoenzyme properties that tethers F-actin filaments to the microvillar membrane [45]. Mooseker *et al.* [45] had speculated that the actin–myosin I–calmodulin complex, via a Ca^{2+}-dependent reaction, might elicit an effect on microvillar membrane permeability, which we considered might include the control of Ca^{2+} entry. It was proposed that the binding of Ca^{2+} to calmodulin–myosin I–actin would result in the closure of a membrane calcium "channel," and the opening of the Ca^{2+} channel would occur when Ca^{2+} dissociates from the complex, facilitated by a vitamin D-dependent mechanism.

At present, however, there is no direct evidence supporting this hypothesis. Nevertheless, a system that limits Ca^{2+} entry by a feedback loop seems entirely reasonable, especially when the extrusion of Ca^{2+} from the enterocyte, as occurs in vitamin D deficiency, is compromised.

2. INTRACELLULAR CALCIUM TRANSFER

In the vitamin D-deficient animal, luminal calcium on entering the enterocyte appears to localize in a region subjacent to the brush border membrane as shown by ion microscopy. Ion microscopy is a technique that allows one to visualize the tissue localization of specific isotopes within sections that have been specially prepared to minimize translocation. Isotopes of the same element (e.g., ^{40}Ca and ^{44}Ca) can be visualized separately and in sequence by this instrumentation (see Chandra and Morrison [46] for details of the method). The stable calcium nuclide of mass 44, $^{44}Ca^{2+}$, was used in our studies as the transported species in order to distinguish the transported species from residual tissue calcium (primarily $^{40}Ca^{2+}$).

In one study, $^{44}Ca^{2+}$ was injected into the lumen of a ligated duodenal segment of a vitamin D-deficient chick, and, after an absorption period of 10 min, a section of the intestinal tissue was frozen *in situ* and further processed for ion microscopic imaging [47]. These images showed that $^{44}Ca^{2+}$ can readily enter the mucosal cells even in the absence of $1,25(OH)_2D$ hormone to become sequestered within the apical region of the enterocyte (Fig. 2; see color insert). With time after $1,25(OH)_2D_3$ (~4 hr), the sequestered calcium moves away from apical regions, distributing throughout the cell and subsequently moving into the lamina propria to complete the vitamin D-stimulated absorption cycle.

Observations of a similar nature were made in 1970 by Sampson *et al.* [48], using rat intestine, the radionuclide ^{45}Ca, and autoradiography at the electron microscopic level. The labeled calcium, given orally to a rachitic rat, was observed to enter the cell and associate primarily with the microvillus of the enterocyte. Movement of the tracer away from this region took place only in vitamin D-replete animals. Thus, both the electron and ion microscopic studies indicate that the calcium entering the cell in the absence of vitamin D becomes largely sequestered in the microvillar and/or apical region of the intestine, and that vitamin D is requisite for enabling calcium to leave these regions to eventually be extruded from the cell. The vitamin D-dependent factors most likely to be involved in the latter process are the calbindins, the vitamin D-induced, high affinity calcium binding proteins.

The calbindins were first identified in chick intestine in 1966 [49] and then in rat intestine in 1967 [50]. Many of their properties and characteristics have since been defined, and the relative dependency of their synthesis on the vitamin D hormone from tissue to tissue was established [9,10,51,52]. The avian form has a molecular weight of about 30,000, binds four Ca^{2+} ions per molecule with high affinity, and is termed calbindin-D_{28K}. The mammalian intestinal form has a molecular weight of about 9700, binds two Ca^{2+} ions with high affinity, and is termed calbindin-D_{9K}. The calbindins are further discussed in Chapters 13 and 14. The high affinity binding sites of both proteins have an average association constant (K_a) of 2–4 $\times 10^6\ M^{-1}$ and a dissociation constant (K_d) of 2.5–5 $\times 10^{-7}\ M$, in a range which indicates that the calbindins could serve as effective intracellular Ca^{2+} buffers. Direct correlations between the mucosal concentration of calbindin and the efficiency of calcium absorption under a wide variety of experimental conditions have been demonstrated and support a role of the calbindins in vitamin D-mediated calcium absorption. As reviewed elsewhere [51,53], these include adaptation to either a low calcium or low phosphorus diet, growth, aging, and egg laying in hens. Cortisol, which inhibits calcium absorption, appears to do so by depressing calbindin synthesis [54]. There are situations, however, where noncorrelations exist. These noncorrelations can be interpreted in terms of the multifunctional role of $1,25(OH)_2D_3$ on calcium absorption and suggests that under certain circumstances a vitamin D-dependent reaction or factor other than the calbindins becomes limiting.

The suggestion that calbindin might facilitate the intracellular diffusional transfer of Ca^{2+} from the brush border region to the basolateral membrane gained support from the theoretical analysis of Kretsinger *et al.* [55]. Calbindin, largely a cytosolic soluble protein, was proposed to function simultaneously as an intracellular Ca^{2+} translocater as well as a Ca^{2+} buffering agent. Further support for this model came from the quantitative analysis of the Ca^{2+} absorptive system by Bronner *et al.* [7]. Their analysis indicated that the rate of transfer of Ca^{2+} through the cell interior was likely to be exceedingly slow and limiting without the presence of cytosolic calbindin. Indeed, the ingenious *in vitro* diffusion experiments of Feher and colleagues [56,57] clearly demonstrated that the presence of calbindin in the central compartment of a three-compartment diffusion cell enhances the rate of diffusion of Ca^{2+} through the cell when compared to serum albumin, a low affinity calcium binding protein.

Of considerable interest are the experiments of Koster *et al.* [58] which showed that the incorporation of another soluble Ca^{2+} ligand, BAPTA, into isolated rabbit cortical collecting tubules "fully mimicked" the stimulatory effect of calbindin-D_{28K} on transcellular Ca^{2+} transport. This study rather impressively suggests that any soluble Ca^{2+} chelator with sufficiently high affinity Ca^{2+} binding sites is capable of increasing the intracytosolic movement of Ca^{2+}. Koster *et al.* [58], using spectrofluorometry with the Ca^{2+} probe Fura 2, also demonstrated that calbindin-D_{28K} in the renal tubules does not dampen Ca^{2+} oscillations induced by a Na^+-free buffer. In contrast, BAPTA effectively dampened these oscillations. The difference, apparently, was due to the more sluggish on-rate of Ca^{2+} binding to calbindin as compared to BAPTA.

The movement of luminally derived Ca^{2+} through the cytosolic compartment within membrane-bound structures was proposed by Jande and Brewer in 1974 [59] and by Warner and Coleman in 1975 [60] on the basis of histological and electron probe analysis, respectively. More recently, Nemere and Norman [61] also implicated vesicular transport as a means by which Ca^{2+} moves through the cell, to be released from the enterocyte by exocytosis.

3. CALCIUM EXTRUSION

The transfer of Ca^{2+} from the cell interior to the extracellular space requires energy input in order to overcome the considerable electrochemical potential difference. It was estimated that 9.3 kcal of energy is required to move 1 mol of Ca^{2+} against this gradient. Processes potentially involved in this transfer are depicted in Fig. 3. Important roles are played by the plasma membrane calcium pump (PMCA) and a sodium–calcium (Na^+/Ca^{2+}) exchanger.

a. Plasma Membrane Calcium Pump Isolated basolateral vesicles were initially used to demonstrate the presence of an ATP-dependent uphill transporter in enterocyte membranes [62–64]. The transport capacity of the ATP-dependent Ca^{2+} uptake by rat basolateral vesicles correlated with the degree of Ca^{2+} absorption as a function of maturation and aging [65,66], intestinal segment [67], and crypt-to-villus axis [68]. The stimulatory effect of vitamin D or $1,25(OH)_2D_3$ on the ATP-dependent uptake of Ca^{2+} by isolated basolateral membranes was reported by several groups [62,68,69], and the results of one study [41] with isolated basolateral vesicles from vitamin D-deficient and vitamin D-replete chicks are shown in Fig. 4. Kinetic analysis of the data, using an inverse plotting procedure, indicated that $1,25(OH)_2D$ increases the V_{max} of the transport process by a factor of about 3, whereas the K_m was unaffected by treatment.

The question of interest was whether vitamin D indirectly affected Ca^{2+} pump activity by altering membrane composition or activating nascent intracellular PMCA proteins. Alternatively, was there a direct effect of the vitamin D hormone on PMCA synthesis? The approach to the latter question was the use of a monoclonal anti-

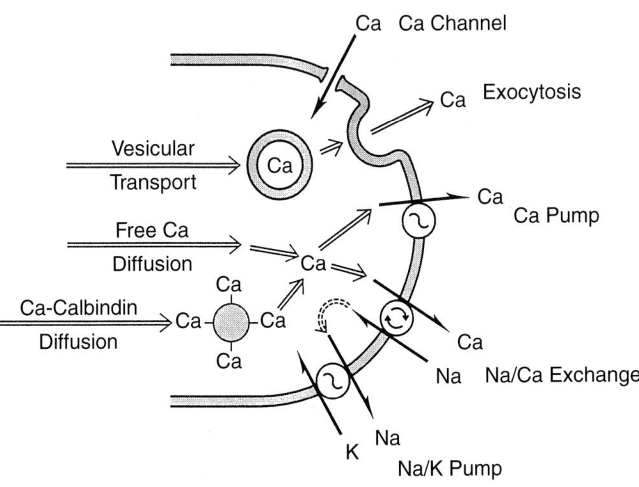

FIGURE 3 Calcium-related events at the basolateral membrane of the enterocyte. The Na^+/K^+ pump maintains a transmembrane gradient of Na^+ for the operation of the Na^+/Ca^{2+} exchanger. © J Nutr (125, 1971S–1979S), American Institute of Nutrition.

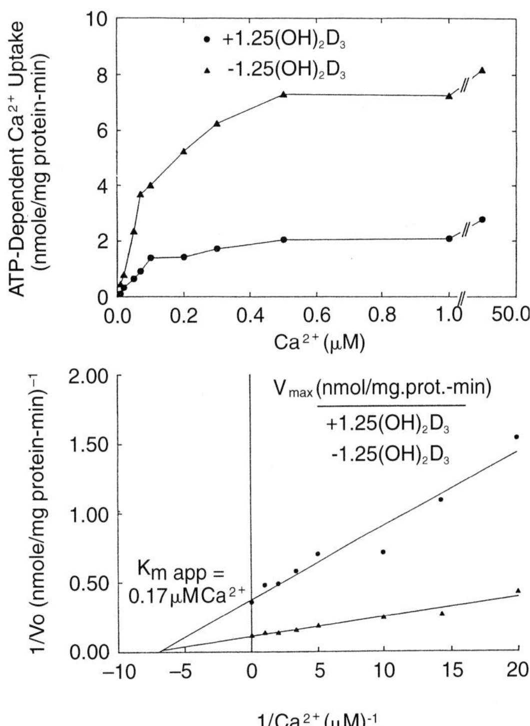

FIGURE 4 ATP-dependent uptake of Ca^{2+} by isolated chick basolateral membrane vesicles as a function of free Ca^{2+} concentrations. The free Ca^{2+} concentrations were monitored with a calcium-selective electrode (WPI, New Haven, CT). The basolateral membrane vesicles were from 3-week-old rachitic chicks or those given 500 ng $1,25(OH)_2D_3$ by intracardiac injection at 18 hr before membrane preparation. (Top) Difference between uptake with or without ATP. Initial rates were calculated from the linear part of the uptake curve. (Bottom) Double-reciprocal plot of the initial rates. The V_{max} values are ~2.8 and 8.2 nmol mg protein^{-1} min^{-1} for the rachitic and $1,25(OH)_2D_3$ chicks, respectively. The apparent K_m for both groups is 0.17 μmol/liter. © J Nutr (122, 662–671), American Institute of Nutrition.

body produced against the erythrocyte membrane calcium pump but that cross-reacts with avian PMCA [70]. Western analysis of basolateral preparations from chick intestine demonstrated that vitamin D increased the density of the PMCA band by a factor of about 2–3. Adaptation of vitamin D-sufficient chicks to diets deficient in calcium or phosphorus also resulted in an increase in the PMCA protein bands by a similar factor. Northern analysis revealed that $1,25(OH)_2D_3$ mediated an increase in the gene expression of chick intestinal PMCA [71], as was previously demonstrated by Zelinski *et al.* [72] for rat intestine.

More recently, it was determined that the increase in PMCA gene expression by $1,25(OH)_2D_3$ was due, at least in part, to a transcriptional event. In the latter study, nuclei were isolated from the intestines of vitamin D-deficient chicks given $1,25(OH)_2D_3$ at different times before the experiment [73]. Total mucosal RNA and nuclear RNA were assayed for PMCA RNA by a ribonuclease protection assay (RPA), and, as shown in Fig. 5, an increase in nuclear and total cellular PMCA RNA occurred at 1.5 hr post-$1,25(OH)_2D_3$ administration and peaked at 3 hr. As both nuclear and cellular RNA were elevated at 1.5 hr and later, it was necessary to prove that the nuclear RNA actually originated in the nucleus and was not a contaminant of cellular RNA. This was done by spiking intestinal mucosa with nuclei-free avian ovarian granulosa cell cytosol, a source of exogenous α-inhibin RNA, before isolation of nuclei. The RPA with the inhibin riboprobe showed only a minor amount of nuclear RNA that was derived from cytosolic RNA. Therefore, $1,25(OH)_2D_3$ mediation of PMCA synthesis does take place at the transcriptional level, although posttranscriptional modifications might also be occurring.

Calmodulin, the ubiquitous high affinity calcium binding protein, is known to stimulate the activity of the PMCA [74]. Several studies have shown that calbindin-D_{9K} and calbindin-D_{28K}, under certain conditions, can also stimulate the activity of the basolateral Ca^{2+} pump; others were unable to show this response to calbindin (refer to Wasserman *et al.* [75] for references and discussion). An example of the stimulatory effect of calbindin-D_{28K} on the ATP-dependent uptake of Ca^{2+} by isolated avian basolateral vesicles is depicted in Fig. 6. Also shown is the expected stimulatory effect of calmodulin. More recently, Timmermans *et al.* [76] reported that, in addition to calmodulin and calbindin-D_{28K}, another high affinity calcium binding protein, parvalbumin, stimulates the PMCA activity of rat isolated duodenal basolateral membrane vesicles.

The exact mechanism by which these high affinity calcium binding proteins affect PMCA activity is not known. One possibility is a direct interaction with the PMCA molecule, a calmodulin-like response. Another is a means of efficiently transferring Ca^{2+} to the calcium binding site of PMCA, analogous to an explanation given for the stimulatory effect of the calcium chelator, EGTA, on PMCA activity (see Carafoli [74] for a discussion of the EGTA effect).

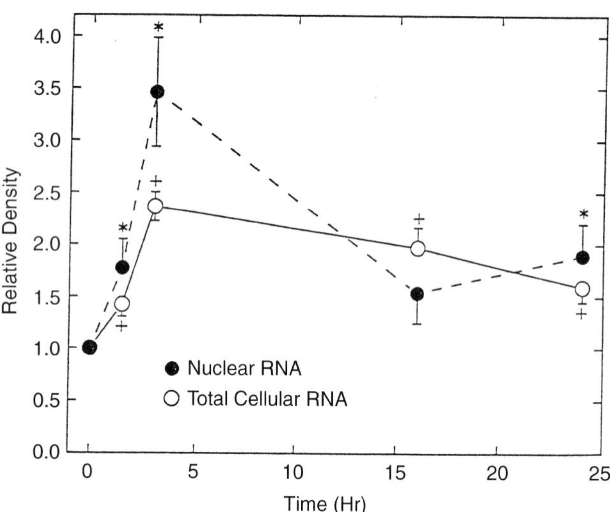

FIGURE 5 Ribonuclease protection analysis of duodenal PMCA RNA at various times after $1,25(OH)_2D_3$ dosage of vitamin D-deficient chicks. The $1,25(OH)_2D_3$ dosage was 0.5 μg per chick given by intracardiac injection. See text for experimental details. The data represent means ± SE of 20 experiments. The symbols (*, +) indicate a significant difference from zero time control values at $p < 0.05$. From Ref. [73], with permission.

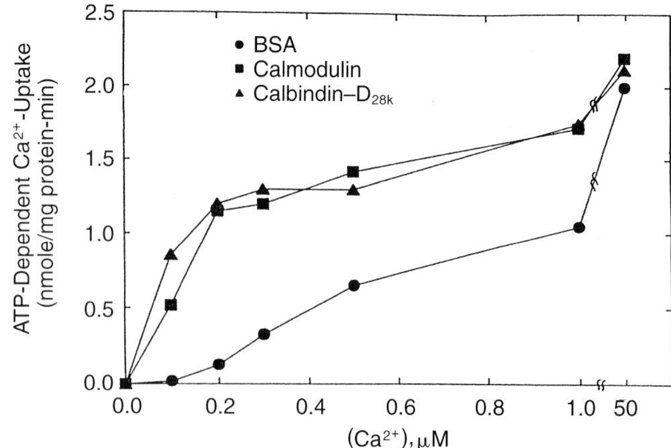

FIGURE 6 Effect of calmodulin and calbindin-D_{28K} on the ATP-dependent Ca^{2+} uptake by EGTA-washed basolateral membrane vesicles. The buffer contained 0.6 μmol/liter of either BSA, calmodulin, or calbindin-D_{28K}, and uptake was initiated after a preincubation period (15 min). © J Nutr (122, 662–671), American Institute of Nutrition.

b. Sodium–Calcium Exchange The extrusion of Ca^{2+} by the Na^+/Ca^{2+} exchanger is linked to the downhill movement of interstitial Na^+ into the cell. The exchanger is rheogenic, transferring 3 Na^+ per 1 Ca^{2+}, and therefore is responsive to the cytosolic negative electropotential with respect to the external fluid phase. The Ca^{2+} binding affinity of the Na^+/Ca^{2+} exchanger was reported to vary from 0.1 to 8 μM in different systems (see review by Wasserman and Fullmer, [41] for references), in a range appropriate for the extrusion of Ca^{2+} across the basolateral membrane. Ghijsen *et al.* [77] estimated that the Na^+/Ca^{2+} exchanger in rat intestine might account for 20% of the Ca^{2+} extrusion capacity of the basolateral membrane. They also reported that vitamin D did not affect exchanger activity.

4. Limiting Steps in Transcellular Calcium Transport

The multiple effects of vitamin D on the calcium transport processes of the enterocyte raises the question of the limiting step or steps in the calcium transport path. As shown in Fig. 1, calcium entry into the intestinal tissue is increased rapidly by the administration of $1,25(OH)_2D_3$ to vitamin D-deficient animals. Only later, and at a time that coincides with the synthesis of calbindin, is there an increase in overall calcium absorption. The limiting step in this situation is apparently not calcium entry but a process dependent on vitamin D-mediated protein synthesis.

Conditions can be developed experimentally in which the rate of entry of calcium across the brush border membrane can be limiting. This was previously shown in an experiment with vitamin D-deficient chicks partially repleted with vitamin D and a submaximal level of intestinal calbindin [77a]. Calcium absorption occurred at a relatively slow rate. The administration of $1,25(OH)_2D_3$ resulted in a rapid increase in Ca^{2+} uptake by the intestinal tissue and a rapid increase in overall calcium absorption without a significant effect on calbindin synthesis. The limiting step in this case was calcium entry, with the amount of calbindin already present being sufficient to allow the vitamin D hormone to increase overall Ca^{2+} absorption.

The *ex vivo* intestinal perfusion preparation of Nemere and Norman [61] might be another system in which the limiting step is calcium entry (see Chapter 15). The injection of $1,25(OH)_2D_3$ into the perfused vascular system of their preparation resulted in the rapid movement of radiocalcium from the intestinal lumen to the "blood" side of the preparation. Of significance was the finding that this rapid effect occurs only with intestine derived from vitamin D-replete animals (i.e., the vitamin D-dependent proteins are already present). Thus, this rapid effect, termed transcaltachia, could well be due to the known early effect of $1,25(OH)_2D_3$ on calcium entry, the step that might have become limiting in their *ex vivo* preparation.

5. Summary of the Transcellular Transport of Calcium

Figure 7 depicts the scheme by which vitamin D, through its hormonal form, is considered to enhance the intestinal absorption of Ca^{2+}. In vitamin D deficiency, Ca^{2+} is able to enter the cell at a basal rate, becoming sequestered intracellularly in the brush border terminal web region, as shown by ion microscopic images (Fig. 2). Relatively little Ca^{2+} is actually absorbed in the absence of vitamin D. After a dose of $1,25(OH)_2D_3$ is given to a deficient animal, there is a rather rapid (30 min) increase in Ca^{2+} entry, but the completion of the vitamin D-stimulated transcellular transfer and absorption cycle does not occur until some time later (3–4 hr) (Fig. 1). The 3- to 4-hr lag period is sufficient for the induction of the synthesis of calbindin [53] and other vitamin D-induced entities. Calbindin-D_{28K}, with its high affinity binding sites, can affect Ca^{2+} entry by two mechanisms: (1) decreasing the Ca^{2+} concentration immediately adjacent to the brush border to increase the lumen–intracellular Ca^{2+} gradient and (2)

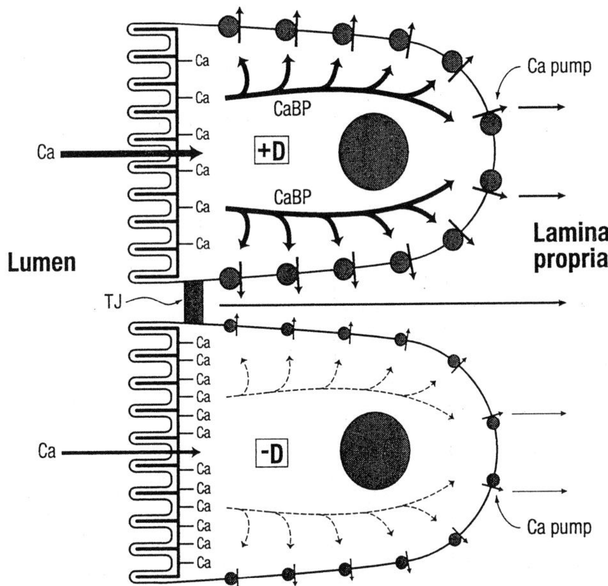

FIGURE 7 Transcellular transport model of Ca^{2+} absorption. Calcium entering the enterocyte is sequestered transiently by components of the brush border complex subjacent to the microvillar membrane. In vitamin D deficiency, Ca^{2+} diffuses slowly from this region to the basolateral membrane. After vitamin D repletion, trans-cytosolic movement is considerably increased, due most likely to the presence of calbindin (CaBP). The number of Ca^{2+} pump units is also increased by vitamin D. © J Nutr (125, 1971S–1979S), with permission.

releasing Ca^{2+} from sequestered sites. The transfer of Ca^{2+} through the cytosol is accelerated by the translocater function of calbindin and, at the same time, maintains intracellular Ca^{2+} concentrations at safe levels by its Ca^{2+} buffering capacity. The extrusion of Ca^{2+} from the enterocyte is achieved by the action of the plasma membrane calcium pump and the Na^+/Ca^{2+} exchanger, as depicted in Fig. 3. Calcium is made available to the Ca^{2+} pump by free Ca^{2+} diffusion and its release from calbindin-D_{28K}. The number of Ca^{2+} pump units are increased by the stimulation of pump synthesis by vitamin D.

The scheme as shown in Fig. 7 requires that the binding affinities along the transport path be appropriate for the site-to-site transfer of absorbed Ca^{2+}. This is indeed the case, as illustrated in Fig. 8 which shows a progressive decrease in the dissociation constants as Ca^{2+} moves from the brush border membrane, to the proposed calmodulin-associated sequestration site in the apical region, to the plasma membrane Ca^{2+} pump. These relative binding values would seem to make the scheme, as described, thermodynamically feasible.

C. Vitamin D and Calcium Absorption via the Paracellular Path

The curve describing the relationship between increasing luminal Ca^{2+} concentrations and Ca^{2+} absorption is biphasic, with the first segment reflecting a saturable process and the second segment, a nonsaturable process [1]. The saturable phase is considered to represent active transport, and the nonsaturable segment, movement by diffusion through the paracellular route. Net movement through the paracellular path occurs when the intraluminal concentration of Ca^{2+} is sufficiently high to overcome the energy barrier consisting of an electropotential difference between lumen and lamina propria of about 5–8 mV, lamina propria positive, and a plasma ionized Ca^{2+} concentration of about 1.25 m*M*. Previous estimates indicate that this occurs at a luminal concentration of calcium about 6 m*M* [1].

The earlier studies in which unidirectional fluxes of calcium were determined in the intact chick showed that vitamin D, in addition to enhancing active Ca^{2+} transport in the chick duodenum, also increased the nonsaturable phase of calcium absorption [26]. An increase in the reverse diffusional flow of Ca^{2+} from plasma to intestinal lumen by vitamin D was also observed [26,79], supporting the concept that vitamin D has a stimulatory effect on the diffusional process. Although the proposed stimulatory effect of vitamin D on paracellular calcium transport is somewhat controversial [80,81], support for this concept has come from the work of Toverud and Dostal [82], Nellans and Kimberg [83], and Karbach [84]. The paracellular permeability of CaCo-2 cells in monolayers is also increased by the addition of $1,25(OH)_2D_3$ to the culture medium [85]. The mechanism of this paracellular effect is not known, although Stenson *et al.* [86] have reported that activation of protein kinase C (PKC) increases paracellular permeability, and $1,25(OH)_2D_3$ activation of PKC has been shown in various systems [78,87,88].

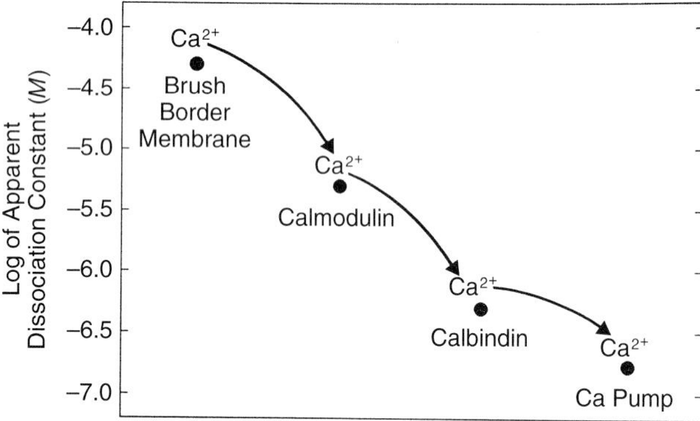

FIGURE 8 Dissociation constants of components of the transcellular Ca^{2+} transport path, illustrating the thermodynamically favorable downhill "gradient" of binding as Ca^{2+} moves through the intestinal cell. Dissociation constants for the brush border complex, calmodulin, calbindin-D_{28K}, and the Ca^{2+} pump are from Wilson and Lawson [128], Glenney and Glenney [129], Bredderman and Wasserman [130], and Wasserman *et al.* [75], respectively. © J Nutr (125, 1971S–1979S), American Institute of Nutrition.

IV. PHOSPHATE ABSORPTION

A. Overall Processes of Phosphate Absorption

Phosphate, like calcium, is also transferred across the intestinal epithelium by a multistep process. Here, though, the thermodynamic parameters have a different effect on the negatively charged phosphate ion as compared to the positively charged calcium ion. The −50 mV electropotential difference between the lumen and the cytosol, and a probable intracellular ionic phosphate concentration of the order of 1 m*M* [89], disallows downhill net diffusion of phosphate into the cell. An energy-dependent entry step, for which there is substantial evidence, is required. The extrusion step could occur by a diffusional process, which is thermodynamically feasible because of the factors referred to above; in addition, the electropotential difference between the lumen and the blood side of the intestine membrane is positive, about a 5 mV difference. The transfer of phosphate through the cell in a sequestered form has been proposed, as discussed later.

As mentioned for Ca^{2+}, the amount of absorption of phosphate from different segments of the intestinal tract is dependent on two main factors: the rate of absorption from the ligated segments and the residence time of the absorbable material in each segment. In the rat, Cramer [90] showed that the rate of absorption followed the following order: duodenum > jejunum > ileum ≫ colon > stomach. When the progress of ingesta along the tract is considered, the greatest amount of phosphate to be absorbed takes place in the ileum (38%), followed in order by the duodenum (29%), jejunum (25%), and colon (8%). Kayne *et al.* [91], using a more sophisticated compartmental analysis approach with the rat, obtained values "surprising similar" (their words) to those of Cramer [90]. The overall intestinal absorption of calcium follows a somewhat different pattern in which the following order was found: ileum > jejunum > duodenum > colon [27].

B. Vitamin D and Phosphate Absorption

The evidence for an active intestinal transport system for phosphate came from the *in vitro* everted gut sac experiments of Harrison and Harrison [92,93]. They clearly demonstrated that the intestinal transport of phosphate was energy dependent (inhibitable by metabolic poisons and anaerobiosis), that vitamin D increased phosphate transport, and that this process is Na^+ dependent.

The effect of vitamin D on phosphate absorption is further illustrated by studies done with vitamin D-deficient chicks, some given vitamin D and some, vehicle only [94]. Figure 9A shows the rapid rate at which [^{32}P]phosphate is transferred from a ligated intestinal segment *in situ* into mucosal tissue, and the pronounced effect of vitamin D on this process. The phosphate that accumulated in the mucosal tissue reached a steady-state value at about 5 min; the intramucosal content of phosphate was significantly enhanced by the vitamin (Fig. 9B). Last, Figure 9C shows a rather immediate, linear increase in the movement of ^{32}P from the mucosal tissue into the circulatory system in the vitamin D-treated chicks, whereas, in the absence of vitamin D, there was a delay of about 10 min before a linear relationship was seen. These results were interpreted as indicating that vitamin D exerts at least a dual effect on ^{32}P absorption (i.e., entry into the mucosal tissue and a subsequent effect on transport through and/or out of the tissue). In this same series of experiments, phosphate absorption during a 10-min period became saturated at intraluminal phosphate concentrations between 2 and

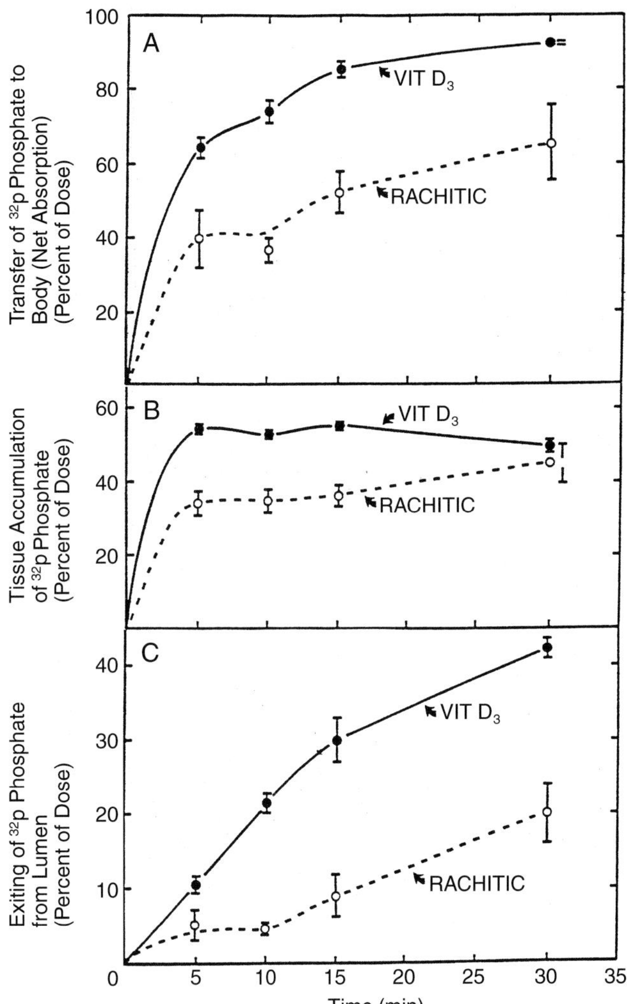

FIGURE 9 Effect of vitamin D_3 on the transfer of [^{32}P]phosphate from duodenal lumen *in situ*. (A) Transfer from lumen into the mucosa tissue, (B) accumulation in mucosal tissue, and (C) transfer from mucosal tissue into body (net absorption). The vitamin D-deficient chicks were given 500 IU vitamin D_3 48 hr before the experiment. Each point represents the mean ± SEM of five to six chicks. From Ref. [70]; with permission.

5 m*M*, and, with increasing luminal concentrations of phosphate, an apparent nonsaturable, diffusion-type process became evident.

C. Relation between Calcium and Phosphate Absorption

The active transport of phosphate by the everted gut sac preparation of Harrison and Harrison [92,93] seemed to be dependent on the presence of Ca^{2+}, which was thought to be required for the operation of the phosphate pump. This observation reflected back to the even earlier studies of Nicolaysen [95,96] in which vitamin D was shown to increase phosphate absorption by the rat but only if calcium were present in the diet. The calcium effect, according to Nicolaysen, was related to the relative unavailability of phosphate due to the formation of insoluble calcium phosphate complex. When calcium was removed from the intestinal tract by vitamin D-dependent absorption, phosphate was then available for absorption.

The differential and independent influence of vitamin D on calcium and phosphate absorption was later fairly well established by several reports. For example, vitamin D was shown to have a greater effect on calcium transport by duodenum than jejunum, and the reverse held for phosphate transport by rat intestinal segments *in vitro* [97] and in the intact chick [98]. Taylor [99] reported that phosphate transport by chick ileum *in vitro* is not dependent on the presence of the Ca^{2+} ion, and Lee *et al.* [100] showed that $1,25(OH)_2D_3$ specifically stimulated calcium absorption in rat colon without affecting phosphate absorption. These and other features of the phosphate transport system and vitamin D effects thereon have been the subject of several reviews [3,4,101–106].

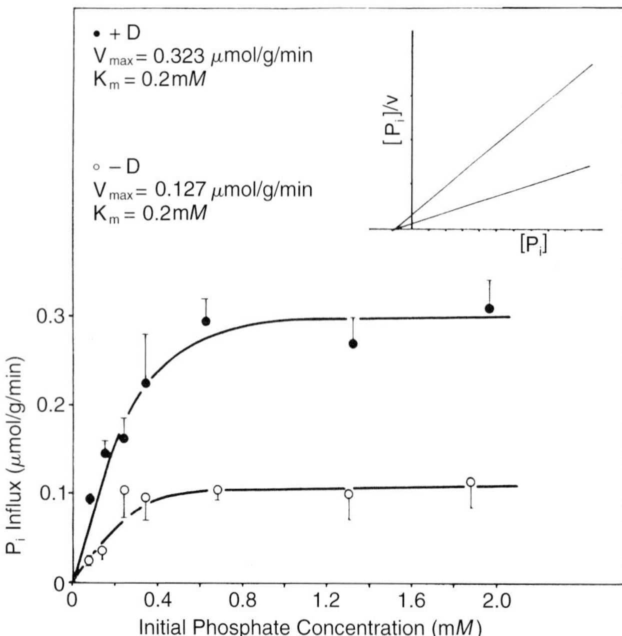

FIGURE 10 Effect of vitamin D on the initial rate of [^{32}P]phosphate uptake by chick jejunum *in vitro*, showing the dependence of rate of mucosal phosphate influx on extracellular concentration. The inset shows a linearized plot of the concentration dependency of the entry rate. Regression lines were calculated by least squares method. Statistically significant differences ($p < 0.025$) were found for each point of determination. The number of everted sacs ranged from 6 to 12 for an individual concentration in each experimental group, and measurements were made at 10 min. Results are presented as means ± SE. From Peterlik and Wasserman [107]; reprinted with permission.

D. Vitamin D and Transcellular Phosphate Transport

1. Phosphate Entry

The saturability of the influx step, using initial rates, was clearly demonstrated with the chick jejunum *in vitro*, as shown in Fig. 10 [107]. The apparent K_m of about 0.2 m*M* was similar to that reported by others and was unaffected by vitamin D. The V_{max}, in contrast, was increased by a factor of 2–3. The energy dependency of this influx step was demonstrated by the inhibitory effect of anaerobiosis.

The transfer of phosphate from the intestinal lumen into the enterocyte has a dependency on intraluminal Na^+, as first reported by Harrison and Harrison [92,93]. The Na^+ site of action was placed at the luminal pole of the intestinal cell by Berner *et al.* [108], suggesting that the entry of phosphate into the mucosa is coupled to the entry of Na^+, and that the asymmetrical distribution of Na^+ across this membrane due to the operation of the Na^+,K^+-ATPase provides the driving force for phosphate entry. The Na^+-dependent uptake of phosphate by rabbit mucosa was further shown to be stimulated by $1,25(OH)_2D_3$, whereas the Na^+-independent entry of phosphate was unaffected [109]. The uniqueness of vitamin D-mediated phosphate absorption was also supported by the observation that arsenate, an "analog" of the phosphate ion, inhibited phosphate absorption but only in the vitamin D-replete chick, suggesting that the vitamin D-dependent process was indeed different from that which occurs in the absence of vitamin D [94].

The Na^+ dependency of phosphate entry was immediately reminiscent of the Na^+ dependency of the absorption of a number of nonelectrolytes such as glucose and amino acids, where Na^+/substrate cotransporters have been identified. For phosphate, several isoforms and variants of renal sodium/phosphate (P_i) cotransporters have been identified, cloned, and sequenced [106]. An intestinal Na/P_i cotransporter was first identified in rabbit intestine by Peerce [110] as a 130-kDa brush border protein. Studies on the purified intestinal Na^+/P_i cotransporter led to the identification of a divalent phosphate allosteric regulatory site [111] and the observation that the sodium and the phosphate ions bound to the transporter were in an occluded state, that is, were relatively inaccessible for ionic exchange with ions in the ambient buffer mixture [112]. Occlusion of ions by P-type ion transporters (e.g., the Na^+,K^+-ATPase) has been taken to indicate that the transported ion is transiently enclosed within the hydrophobic, membrane spanning region of the protein (i.e., within a channel). In the deocclusion step, the fully loaded cotransporter released two Na^+ ions prior to the release of the phosphate ion.

The dependency of $1,25(OH)_2D_3$-mediated intestinal phosphate transport on protein synthesis was shown by the inhibitory effect of cycloheximide [113]. In a sense, this confirmed the report of Ferraro *et al.* [114] who had previously shown that the maintenance of the phosphate absorption system of the intact rat was dependent on continuous protein synthesis. Cycloheximide and actinomycin D also block the $1,25(OH)_2D_3$-stimulated uptake of phosphate by isolated chick renal cells [115]. In the latter study, the K_m of the uptake process in renal cells was also about 0.2 mM, the same as for the chick jejunal uptake process [107].

The evidence that the vitamin D-depdendent effect involves the synthesis of the Na^+/P_i cotransporter was obtained by the expression of poly(A)-rich RNA from duodenal mucosa in *Xenopus* oocytes [116]. The RNAs were derived from normal rabbits, rabbits given $1,25(OH)_2D_3$, or those fed a low phosphorus diet. Sodium-dependent phosphate uptake by the occytes was increased by a factor of about 4 to 5 when the RNA was from $1,25(OH)_2D_3$-treated rabbits or from rabbits fed the phosphate-deficient diet. These results suggest that $1,25(OH)_2D_3$ increases the expression of the intestinal Na^+/P_i cotransporter gene.

The first intestinal Na^+/P_i cotransporter to be cloned and expressed was from the flounder, and, in this species, the intestinal and the renal Na^+/P_i cotransporter were identical [117]. Contrary to this, a cDNA of a rat renal cotransporter did not hybridize with rat intestinal mRNA, which indicates a substantial molecular difference [118].

Vitamin D was also observed to decrease the voltage-sensitive Na^+ flux across chick jejunal brush border membrane vesicles and the acitivity of brush border Na^+/H^+ antiporter. This decrease in Na^+ permeability of the brush border membrane would better maintain the Na^+ gradient that is required for the uphill transport of phosphate by the Na/P_i cotransporter [104,119].

Karsenty *et al.* [120] reported a relatively rapid effect of $1,25(OH)_2D_3$ on phosphate uptake by isolated enterocytes possibly due to an increase in membrane fluidity. Kurnik and Hruska [121] also proposed that $1,25(OH)_2D_3$ enhances renal phosphate transport by altering membrane composition.

2. Intracellular Phosphate Transfer

Transport studies with *in vitro* intestinal preparations suggest that phosphate does not substantially exchange or mix with the intracellular phosphate pool as it moves through the cytosol of enterocytes [97,107]. Entrapment of phosphate in a vesicle or in a "transcellular channel" could account for the lack of extensive interchange of absorbed phosphate with residual cellular phosphate. Bearing on the vesicular transport proposal is the observation that cytochalasin B, a disrupter of microfilaments, inhibits the vitamin D-dependent transcellular phosphate transport by jejunal everted gut sacs, suggesting a role of microfilaments in the transport process [122]. Cytochalasin B, however, did not affect the influx of phosphate from the lumen into the jejunal tissue. The stimulation of actin synthesis by $1,25(OH)_2D_3$ in chick intestine is consistent with a role of the microfilament system in vitamin D-mediated ion transport [123], and the participation of lysosomes in intracellular phosphate movement has also been suggested [124]. The involvement of a soluble, high affinity cytosolic phosphate binding protein as part of the transport process was considered, but attempts to identify such a protein have proved unsuccessful thus far [94].

3. Phosphate Extrusion

A vitamin D-dependent Na^+/phosphate cotransporter was identified on the basolateral membrane of rat intestine that displayed kinetic properties different from those of the cotransporter of the brush border membrane [125,126]. A Na^+-dependent phosphate cotransporter has also been identified on the basolateral membranes of renal cells, along with an Na^+-independent carrier-mediated transporter and a anion exchanger system [127]. As proposed by Hammerman [127], the Na^+/phosphate transporter of renal basolateral membranes provides a means by which plasma phosphate can enter the renal cell to meet cellular requirements. The other phosphate transporters of the basal lateral membrane (i.e., a Na^+-independent carrier or channel) and an anion exchange mechanism could be involved in the transfer of phosphate from the renal cell to the interstitial fluid phase. It is reasonable to suppose that similar mechanisms are present in the basolateral membranes of intestinal cells to facilitate the extrusion of phosphate from the intestinal cell across the basolateral membrane into the lamina propria.

4. Summary of Intestinal Phosphate Absorption

Transmural phosphate transport is summarized pictorially in Fig. 11. Shown on the brush border membrane is the Na^+/P_i cotransporter that moves luminal phosphate into the cell uphill against a thermodynamic gradient. The energy for the reaction is provided by the downhill movement of Na^+ via the Na^+/P_i cotransporter into the cell. The Na^+ gradient that contributes the en-

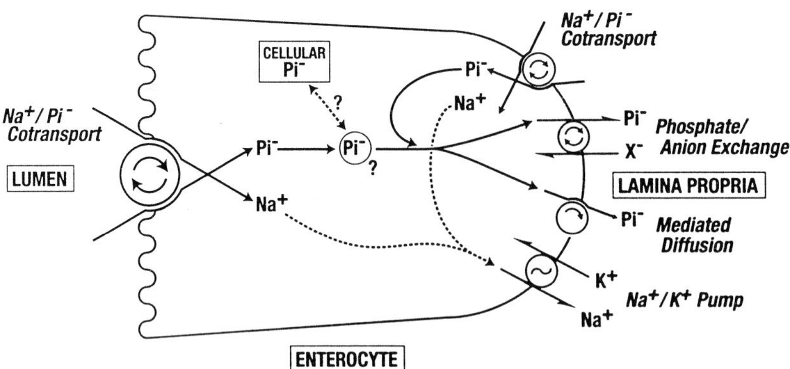

FIGURE 11 Model of intestinal phosphate transport. Phosphate (Pi^-), illustrated as the monovalent species, enters the enterocyte across the brush border membrane by Na^+-dependent secondary cotransport. An "in–out" Na^+ gradient is produced by the extrusion of cellular Na^+ by the basolateral Na^+,K^+-ATPase pump. The mechanism of transfer of Pi^- through the cytosol is not known. Depicted in the intracellular path is Pi^- being transported in a form that may minimally mix with cellular Pi^-, which is consistent with experimental observations. Extrusion of Pi^- across the basolateral membrane could be via mediated diffusion and/or a phosphate/anion exchange mechanism. With the assumption of vesicular transport, a exocytotic process at the basolateral membrane would be present. A Na^+/Pi^- cotransporter on basolateral membranes, with different properties than the brush border membrane cotransporter, has been identified. See text for discussion and references.

ergy for the secondary active phosphate transport is established and maintained by the basolateral ATP-dependent Na^+/K^+ pump. One action of $1,25(OH)_2D_3$ on phosphate absorption appears to be the stimulation of the synthesis of additional cotransporter units. A Na^+-independent entry of phosphate also occurs which apparently is not affected by vitamin D status. Much of the phosphate transported through the cell interior appears to be in a sequestered form, that is, in a state which does not substantially intermix with residual phosphate. Extrusion of phosphate across the basolateral membrane might occur by a facilitative-type diffusional mechanism or by an anion exchange mechanism, as suggested for the renal cell. Specific information on the extrusion of phosphate from the intestinal epithelium is sparse or lacking. The basolateral membrane has also been shown to contain a Na^+/P_i cotransporter which most likely functions to assure that adequate phosphate is made available to epithelial cells to meet their metabolic requirements.

References

1. Wasserman RH, Taylor AN 1969 Some aspects of the intestinal absorption of calcium, with special reference to vitamin D. In: Comar CL, Bronner F (eds) Mineral Metabolism, An Advanced Treatise, Vol 3. Academic Press, New York, pp. 321–403.
2. DeLuca HF 1980 William C. Rose lectureship in biochemistry and nutrition. Some new concepts emanating from a study of the metabolism and function of vitamin D. Nutr Rev **38**:169–182.
3. Murer H, Hildmann B 1981 Transcellular transport of calcium and inorganic phosphate in the small intestinal epithelium. Am J Physiol **240**:G409–G416.
4. Walling MW 1982 Regulation of intestinal calcium and inorganic phosphate absorption. In: Parsons JA (ed) Endocrinology of Calcium Metabolism. Raven Press, New York, pp. 87–101.
5. DeLuca HF 1985 Vitamin D-dependent calcium transport. Soc Gen Physiol Ser **39**:159–176.
6. Favus MJ 1985 Factors that influence absorption and secretion of calcium in the small intestine and colon. Am J Physiol **248**:G147–G157.
7. Bronner F, Pansu D, Stein WD 1986 An analysis of intestinal calcium transport across the rat intestine. Am J Physiol **250**:G561–G569.
8. van Os CH 1987 Transcellular calcium transport in intestinal and renal epithelial cells. Biochim Biophys Acta **906**:195–222.
9. Christakos S, Gabrielides C, Rhoten WB 1989 Vitamin D-dependent calcium binding proteins: Chemistry, distribution, functional considerations and molecular biology. Endocr Rev **10**:3–26.
10. Gross M, Kumar R 1990 Physiology and biochemistry of vitamin D-dependent calcium binding proteins. Am J Physiol **259**:F195–F209.
11. Norman AW 1990 Intestinal calcium absorption: A vitamin D-hormone-mediated adaptive response. Am J Clin Nutr **51**:290–300.
12. Nemere I, Norman AW 1991 Transport of calcium. Handb Physiol **4**:337–360.
13. Feher JJ, Fullmer CS, Wasserman RH 1992 The role of facilitated diffusion of calcium by calbindin in intestinal calcium absorption. Am J Physiol **262**:C517–C526.
14. Wasserman RH 1992 The intestinal calbindins: Their function, gene expression and modulation in genetic disease. In: Bronner F, Peterlik M (eds) Extra- and Intracellular Calcium and Phosphatase Regulation. CRC Press, Boca Raton, Florida, pp. 43–70.

15. Bindels RJM 1993 Calcium handling by the mammalian kidney. J Exp Biol **183**:89–104.
16. Johnson JA, Kumar R 1994 Renal and intestinal calcium transport: Roles of vitamin D and vitamin D-dependent calcium binding proteins. Semin Nephrol **14**:119–128.
17. Holick MF 1995 Noncalcemic actions of 1,25-dihydroxyvitamin D_3 and clinical applications. Bone **17**:107S–111S.
18. Bikle DD 1992 Clinical counterpoint: vitamin D: New actions, new analogs, new therapeutic potential. Endocr Rev **13**:765–784.
19. Walters MR 1992 Newly identified actions of the vitamin D endocrine system. Endroc Rev **13**:719–764.
20. Schachter D, Rosen S 1959 Active transport of Ca^{45} by the small intestine and its dependence on vitamin D. Am J Physiol **196**:357–362.
21. Schachter D, Dowdle EB, Schenker H 1960 Active transport of calcium by the small intestine of the rat. Am J Physiol **198**:263–268.
22. Harrison HC, Harrison HE 1970 Dibutyryl cyclic AMP, vitamin D and intestinal permeability to calcium. Endocrinology **86**:756–760.
23. Wasserman RH, Kallfelz FA, Comar CL 1961 Active transport of calcium by rat duodenum in vivo. Science **133**:883–884.
24. Krawitt EL, Schedl HP 1968 In vivo calcium transport by rat small intestine. Am J Physiol **214**:232–236.
25. Younoszai MK, Urban E, Schedl HP 1973 Vitamin D and intestinal calcium fluxes in vivo in the rat. Am J Physiol **225**:287–292.
26. Wasserman RH, Kallfelz FA 1962 Vitamin D_3 and the unidirectional calcium fluxes across the rachitic chick duodenum. Am J Physiol **203**:221–224.
27. Marcus CS, Lengemann FW 1962 Absorption of Ca45 and Sr85 from solid and liquid food at various levels of the alimentary tract of the rat. J Nutr **77**:155–160.
28. Fullmer CS 1992 Intestinal calcium absorption: Calcium entry. J Nutr **122**:644–650.
29. Rasmussen H, Matsumoto T, Fontaine O, Goodman DB 1982 Role of changes in membrane lipid structure in the action of 1,25-dihydroxyvitamin D_3. Fed Proc **41**:72–77.
30. Bikle DD, Whitney J, Munson S 1984 The relationship of membrane fluidity to calcium flux in chick intestinal brush border membranes. Endocrinology **114**:260–267.
31. Putkey JA, Spielvogel AM, Sauerheber RD, Sunlap CS, Norman AW 1982 Effects of essential fatty acid deficiency and spin label studies of enterocyte membrane lipid fluidity. Biochim Biophys Acta **688**:177–190.
32. Schedl HP, Ronnenberg W, Christensen KK, Hollis BW 1994 Vitamin D and enterocyte brush border membrane calcium transport and fluidity in the rat. Metabolism **43**:1093–1103.
33. Brasitus TA, Dudeja PK, Eby B, Lau K 1986 Correction by 1,25-dihydroxycholecalciferol of the abnormal fluidity and lipid composition of enterocyte brush border membranes in vitamin D-deprived rats. J Biol Chem **261**:16404–16409.
34. Feher JJ, Wasserman RH 1978 Evidence for a membrane-bound fraction of chick intestinal calcium-binding protein. Biochim Biophys Acta **540**:134–143.
35. Shimura F, Wasserman RH 1984 Membrane-associated vitamin D-induced calcium binding protein (CaBP): Quantification by a radio-immunoassay and evidence for a specific CaBP-binding protein in purified intestinal brush borders. Endocrinology **115**:1–9.
36. Schachter A, Kowarski S 1982 Isolation of the protein IMCal, a vitamin D-dependent membrane component of the intestinal transport mechanism for calcium. Fed Proc **41**:84–87.
37. Miller A, Ueng T-H, Bronner F 1979 Isolation of a vitamin D-dependent, calcium-binding protein from brush border of rat duodenal mucosa. FEBS Lett **103**:319–322.
38. Bikle DD, Munson S, Chafouleas J 1984 Calmodulin may mediate 1,25-dihydroxyvitamin D-stimulated intestinal calcium transport. FEBS Lett **174**:30–33.
39. Bikle DD, Munson S 1985 1,25-Dihydroxyvitamin D increases calmodulin binding to specific protein in the chick duodenal brush border membrane. J Clin Invest **76**:2313–2316.
40. Wilson HD, Schedl HP, Christensen K 1989 Calcium uptake by brush-border membrane vesicles from the rat intestine. Am J Physiol **257**:F446–F453.
41. Wasserman RH, Fullmer CS 1995 Vitamin D and intestinal calcium transport: Facts, speculations and hypotheses. J Nutr **125**:1971S–1979S.
42. Glenney JR Jr, Glenney P 1985 Comparison of Ca^{++}-regulated events in the intestinal brush border. J Cell Biol **100**:754–763.
43. Howe CL, Keller III TCS, Mooseker MS, Wasserman RH 1982 Analysis of cytoskeletal proteins and Ca^{2+}-dependent regulation of structure in intestinal brush borders from rachitic chicks. Proc Natl Acad Sci USA **79**:1134–1138.
44. Kaune R, Munson S, Bikle DD 1994 Regulation of calmodulin binding to the ATP extractable 110 kDa protein (myosin I) from chicken duodenal brush border by $1,25(OH)_2D_3$. Biochim Biophys Acta **1190**:329–336.
45. Mooseker MS, Wolenski JS, Coleman TR, Hayden SM, Cheney RD, Espreafico E, Heintzelman MB, Peterson MD 1991 Structural and functional dissection of a membrane-bound mechanoenzyme: Brush border myosin I. Curr Topics Membr **33**:31–55.
46. Chandra S, Morrison GH 1988 Ion microscopy in biology and medicine. In: Riordan JF, Vallee BL (eds) Methods in Enzymology, Vol. 158. Academic Press, Orlando, Florida, pp. 157–179.
47. Fullmer CS, Chandra S, Smith CA, Morrison GH, Wasserman RH 1996 Ion microscopic imaging of calcium during 1,25-dihydroxyvitamin D-mediated intestinal absorption. Histochem Cell Biol **106**:215–222.
48. Sampson HW, Matthews JL, Martin JH, Kunin AS 1970 An electron microscopic localization of calcium in the small intestine of normal, rachitic, and vitamin-D-treated rats. Calcif Tissue Res **5**:305–316.
49. Wasserman RH, Taylor AN 1960 Vitamin D_3-induced calcium-binding protein in chick intestinal mucosa. Science **152**:791–793.
50. Kallfelz FA, Taylor AN, Wasserman RH 1967 Vitamin D-induced calcium-binding factor in rat intestinal mucosa. Proc Soc Exp Biol Med **125**:54–58.
51. Wasserman RH, Fullmer CS 1982 Vitamin D-induced calcium-binding protein (CaBP). In: Cheung WY (ed) Calcium and Cell Function, Vol. 2. Academic Press, New York, pp. 175–216.
52. Feher JJ, Wasserman RH 1979 Calcium absorption and calcium-binding protein: Quantitative relationship. Am J Physiol **236**:E556–E561.
53. Wasserman RH, Fuller CS 1989 On the molecular mechanism of intestinal calcium transport. Adv Exp Med Biol **249**:45–65.
54. Feher JJ, Wasserman RH 1979 Intestinal calcium-binding protein and calcium absorption in cortisol-treated chicks: Effects of vitamin D_3 and 1,25-dihydroxyvitamin D_3. Endocrinology **104**:547–551.
55. Kretsinger RH, Mann JE, Simmonds JG 1982 Model of facilitated diffusion of calcium by the intestinal calcium binding protein. In: Norman AW, Schaefer K, von Herrath D, Grigoleit H-G (eds) Vitamin D: Chemical, Biochemical and Clinical Endocrinology of Calcium Metabolism. de Gruyter, Berlin, pp. 233–248.
56. Feher JJ 1983 Facilitated calcium diffusion by intestinal calcium-binding protein. Am J Physiol **244**:303–307.

57. Feher JJ, Fullmer CS, Fritzsch GK 1989 Comparison of the enhanced steady-state diffusion of calcium by calbindin-D_{9K} and calmodulin: Possible importance in intestinal calcium absorption. Cell Calcium **10**:189–203.
58. Koster HPG, Hartog A, Van Os CH, Bindels RJM 1995 Calbindin-D_{28k} facilitates cytosolic calcium diffusion without interfering with calcium signaling. Cell Calcium **18**:187–196.
59. Jande SS, Brewer LM 1974 Effects of vitamin D_3 on duodenal absorptive cells of chicks. An electron microscopic study. Z Anat Entwickl Gesch **144**:249–265.
60. Warner RR, Coleman JR 1975 Electron probe analysis of calcium transport by small intestine. J Cell Biol **64**:54–74.
61. Nemere I, Norman AW 1990 Transcaltachia, vesicular calcium transport, and microtubule-associated calbindin-D_{28K}: Emerging views of 1,25-dihydroxyvitamin D_3-mediated intestinal calcium absorption. Miner Electrolyte Metab **16**:109–114.
62. Freedman RA, Weiser MM, Isselbacher KJ 1977 Calcium translocation by Golgi and lateral-basal membrane vesicles from rat intestine: Decrease in vitamin D-deficient rats. Proc Natl Acad Sci USA **74**:3612–3616.
63. Ghijsen WEJM, De Jong MD, Van Os CH 1982 ATP-dependent calcium transport and its correlation with Ca^{2+}-ATPase activity in basolateral plasma membranes of rat duodenum. Biochim Biophys Acta **689**:327–336.
64. Hildmann B, Schmidt A, Murer H 1982 Ca^{++}-transport across basal-lateral plasma membranes from rat small intestinal epithelial cells. J Membr Biol **65**:55–62.
65. Armbrecht HJ, Boltz M, Strong R, Richardson A, Bruns MEH, Christakos S 1989 Expression of calbindin-D decreases with age in intestine and kidney. Endocrinology **125**:2950–2956.
66. Ghishan FK, Leonard D, Pietsch J 1988 Calcium transport by plasma membranes of enterocytes during development: Role of 1,25-$(OH)_2$ vitamin D_3. Pediatr Res **24**:338–341.
67. Van Corven EJJM, de Jong MD, van Os CH 1986 Enterocyte isolation procedure specifically effects ATP-dependent Ca^{2+}-transport in small intestinal plasma membranes. Cell Calcium **7**:89–99.
68. Walters JR, Weiser MM 1987 Calcium transport by rat duodenal villus and crypt basolateral membranes. Am J Physiol **252**:G170–G177.
69. Takito J, Shinki T, Sasaki T, Suda T 1990 Calcium uptake by brush-border and basolateral membrane vesicles in chick duodenum. Am J Physiol **258**:G16–G23.
70. Wasserman RH, Smith CA, Brindak ME, de Talamoni N, Fullmer CS, Penniston JT, Kumar R 1991 Vitamin D and mineral deficiencies increase the plasma membrane calcium pump of chicken intestine. Gastroenterology **102**:886–894.
71. Cai Q, Chandler JS, Wasserman RH, Kumar R, Penniston JT 1993 Vitamin D, and adaptation to dietary calcium and phosphate deficiencies increase intestinal plasma membrane calcium pump gene expression. Proc Natl Acad Sci USA **90**:1345–1349.
72. Zelinski JM, Sykes DE, Weiser MM 1991 The effect of vitamin D on rat intestinal plasma membrane Ca-pump mRNA. Biochem Biophys Res Commun **179**:749–755.
73. Pannabecker TL, Chandler JS, Wasserman RH 1995 Vitamin D-dependent transcriptional regulation of the intestinal plasma membrane calcium pump. Biochem Biophys Res Commun **213**:499–505.
74. Carafoli E 1991 Calcium pump of the plasma membrane. Physiol Rev **71**:129–153.
75. Wasserman RH, Chandler JS, Meyer SA, Smith CA, Brindak ME, Fullmer CS, Penniston JT, Kumar R 1992 Intestinal calcium transport and calcium extrusion processes at the basolateral membrane. J Nutr **122**:662–671.
76. Timmermans JAH, Bindels RJM, Van Os CH 1995 Stimulation of plasma membrane Ca^{2+} pump by calbindin-D_{28k} and calmodulin is additive in EGTA-free solutions. J Nutr **125**:1981S–1986S.
77. Ghijsen WEJM, De Jong MD, Van Os CH 1983 Kinetic properties of Na^+/Ca^{2+} exchange in basolateral plasma membranes of rat small intestine. Biochim Biophys Acta **730**:85–94.
77a. Wasserman RH, Brindak ME, Meyer SA, Fullmer CS 1982 Evidence for multiple effects of vitamin D on calcium absorption: Response of rachitic chicks, with and without partial vitamin D repletion, to 1,25-dihydroxyvitamin D_3. Proc Natl Acad Sci USA **79**:7939–7943.
78. De Boland AR, Norman AW 1990 Evidence for involvement of protein kinase C and cyclic adenosine 3′,5′-monophosphate-dependent protein kinase in the 1,25-dihydroxy-vitamin D_3-mediated rapid stimulation of intestinal calcium transport, (transcaltachia). Endocrinology **127**:39–45.
79. Wasserman RH, Taylor AN, Kallfelz FA 1966 Vitamin D and transfer of plasma calcium to intestinal lumen in chicks and rats. Am J Physiol **211**:419–423.
80. Pansu D, Bellaton C, Bronner F 1981 Effect of Ca intake on saturable and nonsaturable components of duodenal Ca transport. Am J Physiol **240**:32–37.
81. Pansu D, Bellaton C, Roche C, Bronner F 1983 Duodenal and ileal calcium absorption in the rat and effects of vitamin D. Am J Physiol **244**:695–700.
82. Toverud SU, Dostal LA 1986 Calcium absorption during development: Experimental studies of the rat small intestine. J Pediatr Gastroenterol Nutr **5**:688–695.
83. Nellans HN, Kimberg DV 1978 Cellular and paracellular calcium transport in rat ileum: Effects of dietary calcium. Am J Physiol **236**:E726–E737.
84. Karbach U 1992 Paracellular calcium transport across the small intestine. J Nutr **122**:672–677.
85. Chirayath MV, Gaidzik L, Graf J, Cross HS, Peterlik M 1994 Cellular and paracellular Ca^{++} transport in Caco-2 cell monolayers: Effect of vitamin D. Proc Int Conf Prog Bone Miner Res Vienna, Austria (abstract).
86. Stenson WF, Easom RA, Riehl TE, Turk J 1993 Regulation of paracellular permeability in Caco-2 cell monolayers by protein kinase C. Am J Physiol **265**:G955–G962.
87. Obeid LM, Okazaki T, Karolak LA, Hannun YA 1990 Transcriptional regulation of protein kinase C by 1,25-dihydroxyvitamin D_3 in HL-60 cells. J Biol Chem **265**:2370–2374.
88. Wali RK, Baum CL, Sitrin MD, Brasitus TA 1990 1,25$(OH)_2$ vitamin D_3 stimulates membrane phosphoinositide turnover, activates protein kinase C, and increases cytosolic calcium in rat colonic epithelium. J Clin Invest **85**:1296–1303.
89. Barac-Nieto M, Dowd TL, Gupta RK, Spitzer A 1991 Changes in NMR-visible kidney cell phosphate with age and diet: Relationship to phosphate transport. Am J Physiol **261**:F153–F162.
90. Cramer CF 1961 Progress and rate of absorption of radiophosphorus through the intestinal tract of rats. Can J Biochem Physiol **39**:499–503.
91. Kayne LH, D'Argenio DZ, Meyer JH, Hu MS, Jamgotchian N, Lee DB 1993 Analysis of segmental phosphate absorption in intact rats. A compartmental analysis approach. J Clin Invest **91**:915–922.
92. Harrison HE, Harrison HC 1961 Intestinal transport of phosphate: Action of vitamin D, calcium, and potassium. Am J Physiol **201**:1007–1012.
93. Harrison HE, Harrison HC 1963 Sodium, potassium, and intestinal transport of glucose, 1-tyrosine, phosphate and calcium. Am J Physiol **205**:107–111.
94. Wasserman RH, Taylor AN 1973 Intestinal absorption of phos-

phate in the chick: Effect of vitamin D_3 and other parameters. J Nutr **103**:586–599.

95. Nicolaysen R 1937 Studies upon the mode of action of vitamin D. II. The influence of vitamin D on the fecal output of endogenous Ca and P. Biochem J **31**:107–121.
96. Nicolaysen R 1937 Studies upon the mode of action of vitamin D. III. The influence of vitamin D on absorption of Ca and P. Biochem J **31**:122–129.
97. Kowarski S, Schachter D 1969 Effects of vitamin D on phosphate transport and incorporation into mucosal constituents of rat intestinal mucosa. J Biol Chem **244**:211–217.
98. Hurwitz S, Bar A 1972 Site of action of vitamin D. Am J Physiol **222**:761–767.
99. Taylor AN 1974 In vitro phosphate transport in chick ileum: Effect of cholecalciferol, calcium, sodium and metabolic inhibitors. J Nutr **104**:489–494.
100. Lee DBN, Walling MW, Gafter U, Silis V, Coburn JW 1980 Calcium and inorganic phosphate transport in rat colon. Dissociated response to 1,25-dihydroxyvitamin D_3. J Clin Invest **65**:1326–1331.
101. Wasserman RH, Taylor AN 1976 Gastrointestinal absorption of calcium and phosphorus. In: Aurbach GD (ed.) Handbook of Physiology, Section 7: Endocrinology, Vol. VII, Parathyroid Gland. Amer Physiol Soc, Washington, DC, pp. 137–155.
102. Lee DB 1986 Mechanisms and regulation of intestinal phosphate transport. Adv. Exp Med Biol **208**:207–212.
103. Armbrecht HJ 1990 Effect of age on calcium and phosphate absorption. Role of 1,25-dihydroxyvitamin D. Miner Electrolyte Metab **16**:159–166.
104. Cross HS, Debiec H, Peterlik M 1990 Mechanism and regulation of intestinal phosphate absorption. Miner Electrolyte Metab **16**:115–124.
105. Care AD 1994 The absorption of phosphate from the digestive tract of ruminant animals. Br Vet J **150**:197–205.
106. Murer H, Markovich D, Biber J 1994 Renal and small intestinal sodium-dependent symporters of phosphate and sulphate. J Exp Biol **196**:167–181.
107. Peterlik M, Wasserman RH 1978 Effect of vitamin D on transepithelial phosphate transport in chicken intestine. Am J Physiol **234**:E379–E388.
108. Berner W, Kinne R, Murer H 1976 Phosphate transport into brush-border membrane vesicles isolated from rat small intestine. Biochem J **160**:467–474.
109. Danisi G, Bonjour JP, Straub RW 1980 Regulation of Na-dependent phosphate influx across the mucosal border of duodenum by 1,25-dihydroxycholecalciferol. Pfluegers Arch **388**:227–232.
110. Peerce BE 1989 Identification of the intestinal Na-phosphate cotransporter. Am J Physiol **256**:G645–G652.
111. Peerce BE 1995 Effect of substrates and pH on the intestinal Na^+/phosphate cotransporters: Evidence for an intervesicular divalent phosphate allosteric regulatory site. Biochim Biophys Acta **1239**:1–10.
112. Peerce BE 1996 Simultaneous occlusion of Na^+ and phosphate by the intestinal brush border membrane Na^+/phosphate cotransporter. Kidney Int **49**:988–991.
113. Peterlik M, Wasserman RH 1980 Regulation by vitamin D of intestinal phosphate absorption. Horm Metab Res **12**:216–219.
114. Ferraro C, Ladizesky M, Cabrejas M, Montoreano R, Mautalen C 1976 Intestinal absorption of phosphate: Action of protein synthesis inhibitors and glucocorticoids. J Nutr **106**:1752–1756.
115. Liang CT, Barnes J, Balakir R, Cheng L, Sacktor B 1982 In vitro stimulation of phosphate uptake in isolated chick renal cells by 1,25-dihydroxycholecalciferol. Proc Natl Acad Sci USA **79**:3532–3536.
116. Yaegci A, Werner A, Murer H, Biber J 1992 Effect of rabbit duodenal mRNA on phosphate transport in *Xenopus laevis* oocytes: Dependence on 1,25-dihydroxy-vitamin-D_3. Pfluegers Arch **422**:211–216.
117. Kohn B, Herter P, Hulseweh B, Elger M, Hentschel H, Kinne RK, Werner A 1996 Na-P_i cotransport in flounder: Same transport system in kidney and intestine. Am J Physiol **270**:F937–F944.
118. Magagnin S, Werner A, Markovich D, Sorribas V, Stange G, Biber J, Murer H 1993 Expression cloning of human and rat renal cortex Na/P_i cotransport. Proc Natl Acad Sci USA **90**:5979–5983.
119. Fuchs R, Graf J, Peterlik M 1985 Effects of 1α-dihydroxycholecalciferol on sodium-ion translocation across chick intestinal brush-border membrane. Biochem J **230**:441–449.
120. Karsenty G, Lacour B, Ulmann A, Pierandréi E, Drúeke T 1985 Early effects of vitamin D metabolites on phosphate fluxes in isolated rat enterocytes. Am J Physiol **248**:G40–G45.
121. Kurnik BRC, Hruska KA 1985 Mechanism of stimulation of renal phosphate transport by 1,25-dihydroxycholecalciferol. Biochim Biophys Acta **817**:42–50.
122. Fuchs R, Peterlik M 1979 Vitamin D-induced transepithelial phosphate and calcium transport by chick jejunum. FEBS Lett **100**:357–359.
123. Wilson PW, Lawson DEM 1978 Incorporation of [^{3}H]leucine into an actin-like protein in response to 1,25-dihydroxycholecalciferol in chick intestinal brush borders. Biochem J **173**:627–631.
124. Nemere I 1996 Apparent nonnuclear regulation of intestinal phosphate transport: Effects of 1,25-dihydroxyvitamin D_3, 24,25-dihydroxyvitamin D_3, and 25-hydroxyvitamin D_3. Endocrinology **137**:2254–2262.
125. Ghishan FK, Kikuchi K, Arab N 1987 Phosphate transport by rat intestinal basolateral-membrane vesicles. Biochem J **243**:641–646.
126. Ghishan FK 1992 Phosphate transport by plasma membranes of enterocytes during development: Role of 1,25-dihydroxycholecalciferol. Am J Clin Nutr **55**:873–877.
127. Hammerman MR 1986 Phosphate transport across renal proximal tubular cell membranes. Am J Physiol **251**:F385–F398.
128. Wilson PW, Lawson DEM 1980 Calcium binding activity by chick intestinal brush-border membrane vesicles. Pfluegers Arch **389**:69–74.
129. Glenney JR, Jr, Glenney P 1985 Comparison of Ca^{++}-regulated events in the intestinal brush border. J Cell Biol **100**:754–763.
130. Bredderman PJ, Wasserman RH 1974 Chemical composition, affinity for calcium, and some related properties of the vitamin D dependent calcium-binding protein. Biochemistry **13**:1687–1694.

CHAPTER 17

Vitamin D and the Kidney

RAJIV KUMAR Division of Nephrology, Mayo Clinic and Foundation, Rochester, Minnesota

I. INTRODUCTION

The kidney has a unique function in mineral homeostasis and plays a vital role in the control of plasma calcium and phosphorus. While examining how the kidney controls calcium and phosphate homeostasis it is worthwhile to keep the following facts in mind.

1. In humans, in a 24-hr period, about 8 g of calcium are filtered at the glomerulus, and about 7.8 g are reabsorbed in the proximal and distal tubules and the loop of Henle [1–5]. This is carried out in a manner such that, under normal circumstances (i.e., in states of neutral calcium balance), the amount of calcium in the urine closely approximates that absorbed in the intestine. The mechanisms by which calcium is reabsorbed in the kidney are complex and yield several insights into cellular regulation of calcium transport.
2. The reabsorption of calcium in the kidney is controlled by several factors such as the filtered load of sodium, urine flow, and the activity of several hormones, most notably, parathyroid hormone (PTH), 1,25-dihydroxyvitamin D [1,25(OH)$_2$D], and calcitonin, in addition to others [1–5].
3. The kidney is the site of synthesis of 1,25(OH)$_2$D, the active and hormonal form of the vitamin D [6,7].
4. The kidney expresses several vitamin D-dependent proteins such as the plasma membrane calcium pump, the sodium–calcium exchanger, and the calbindins, some of which play a vital role in calcium transport.
5. The kidney expresses the vitamin D receptor (VDR).
6. The kidney expresses the 25-hydroxyvitamin D_3 and 1,25-dihydroxyvitamin D_3 24-hydroxylase (24-hydroxylase) and other 1,25-dihydroxyvitamin D and vitamin D analog metabolizing enzymes [6,7].
7. Finally, the kidney has an equally important role in the control of plasma phosphate and the filtration and reabsorption of phosphate. As in the case of calcium, the reabsorption and secretion of phosphate are under hormonal control, and many of the same hormones and factors involved in calcium regulation play a significant role in the regulation of phosphate reabsorption [8–10].

Given the complex role of the kidney in calcium and phosphate homeostasis, a brief overview of calcium and phosphate reabsorption in the kidney is in order. The details of calcium and phosphate homeostasis are discussed in Chapters 31 and 32, respectively.

A. Calcium Handling by the Kidney

About 55% of total plasma calcium is ultrafilterable [11]. The ultrafilterable calcium concentration is about 1.35 m*M* (5.4 mg/dl) and closely approximates the concentrations of calcium present in glomerular fluid [12–14]. The total amount of calcium filtered at the glomerulus in a 24 hour period is about 8000 mg. Approximately 98% of the filtered load of calcium is reabsorbed in the tubules. Thus, the amount of calcium excreted in the urine in a 24-hr period is about 150–200 mg [1,11].

In the proximal tubule, about 50–60 percent of the filtered load of calcium is reabsorbed [2,3,15]. The reabsorption of calcium is thought to occur as result of solvent drag and by a paracellular route, and it is sodium dependent. Volume expansion and a reduction of tubular sodium reabsorption inhibits calcium reabsorption, whereas volume contraction and an increase in sodium reabsorption enhance calcium reabsorption [2,16]. Inhibition of Na^+, K^+-ATPase activity and sodium reabsorption by ouabain reduce the amount of calcium reabsorbed, as does the substitution of sodium with lithium [16]. The concentration of calcium at the end of the proximal tubule is virtually similar to that in the glomerular fluid. Importantly, calcium reabsorption in the proximal tubule is not influenced by thiazide diuretics, by hormones such as PTH or $1,25(OH)_2D_3$, or by hydrogen ions [2,3,15,16]. As discussed later, some of vitamin D-dependent proteins (Section IV) such as the calbindins and the plasma membrane calcium pump are either not expressed in the proximal tubule or are expressed in low amounts when compared with the amounts expressed in the distal tubule.

Calcium reabsorption in the descending loop and the thin ascending limb of the loop of Henle is minimal. In the thick ascending limb of the loop of Henle about 20% of the filtered load of calcium is reabsorbed, whereas another 10–15% is reabsorbed in the distal tubule, with the remaining 5% being reabsorbed in the collecting ducts [2,3,15,16]. There are important distinctions between the factors influencing calcium reabsorption in the proximal tubule and the mechanisms of calcium reabsorption in the distal segments of the nephron. First, the movement of calcium in the distal nephron occurs against a concentration gradient (lumen relative to extracellular fluid). Second, the lumen of the tubule is electronegative and becomes progressively more so toward the end of the distal tubule. Third, calcium reabsorption can be dissociated from sodium reabsorption by thiazide diuretics which inhibit sodium reabsorption but enhance calcium reabsorption. Fourth, hydrogen ions inhibit calcium reabsorption in the distal tubule, whereas they have no effect on calcium reabsorption in the proximal tubule.

TABLE I Factors That Alter Renal Phosphate Excretion

Increase	Decrease
High phosphate diet	Low phosphate diet
Parathyroid hormone	Thyroparathyroidectomy
Calcitonin	Thyroxine
Chronic vitamin D	Acute vitamin D
Glucagon	Insulin
Glucocorticoids	Growth hormone
Volume expansion	Volume contraction
Increased pCO_2	Decreased pCO_2
Chronic acidosis	
Starvation	
Diuretics	
"Phosphatonin"	

B. Phosphate Handling by the Kidney

Virtually all inorganic phosphate in the serum is filtered by the glomerulus [8–10]. About 80% of filtered phosphorus is reabsorbed in the kidney, mostly in the proximal tubule. The amount of phosphorus reabsorbed in the proximal tubule is greatest in the first half of the proximal tubule and exceeds that of sodium. There is evidence for further phosphorus reabsorption in the pars recta. Little or no phosphorus reabsorption occurs in the loop of Henle or the distal tubule, although there is some debate about whether there is phosphorus reabsorption in the distal tubule. The reabsorption of phosphorus is sodium dependent and is mediated by a sodium–phosphate cotransporter. Some of the factors that are involved in phosphorus reabsorption are noted in Table I.

II. ROLE OF THE KIDNEY IN THE METABOLISM OF 25OHD

A. Formation of $1,25(OH)_2D$

The central role of the kidney in the formation of $1,25(OH)_2D$ was first noted by Fraser and Kodicek, who demonstrated that nephrectomy abolished the formation of $1,25(OH)_2D$ [17,18]. This was subsequently confirmed by others [19,20]. Nephrectomy greatly decreases circulating $1,25(OH)_2D$ concentration *in vivo* except during pregnancy, and in granulomatous diseases and lymphomas associated with the ectopic production of $1,25(OH)_2D$ [21–24]. The proximal tubule is the site of $1,25(OH)_2D$ synthesis in the kidney, and there is no evidence that cells of other tubular segments synthesize $1,25(OH)_2D$ [25]. There appears to be little intranephron segment heterogeneity, as all parts of the proximal tubule are capable of synthesizing the metabolite from its substrate, 25OHD. *In vitro*, chick renal epithelial cells in culture, mammalian nephron segments, and homogenates derived from avian and mammalian (mostly rodent) renal cells all appear to metabolize $25OHD_3$ to $1,25(OH)_2D_3$ [26–30]. The 25-hydroxyvitamin D_3 1α-hydroxylase is a multicomponent, cytochrome P450-containing enzyme in the mitochondria of renal proximal tubular cells [31–37]. The characteristics of the enzyme are described in Chapter 5. Table II summarizes some of the key factors known to regulate the activity of this enzyme *in vivo* and *in vitro* [38–76]. The major regulators appear to be PTH, inorganic phosphorus, and $1,25(OH)_2D$ itself.

B. Formation of $24R,25(OH)_2D_3$ and Other Metabolites of $25OHD_3$ in the Kidney

The kidney is the major, but not exclusive, site of the formation of $24R,25(OH)_2D_3$ [77–85]. This metabolite of $25OHD_3$ is formed in states of normocalcemia or normophosphatemia and is inducible by $1,25(OH)_2D_3$ [44,45,75]. Although some have suggested that it has certain unique properties and actions [86–88], others using C-24 fluorinated compounds that are incapable of undergoing transformation at C-24 have suggested that

TABLE II Factors Altering Serum $1,25(OH)_2D$ Levels or 25OHD 1α-Hydroxylase Activity[a]

Factor	Level or activity change of substance	Effect on $1,25(OH)_2D_3$ levels or 1α-hydroxylase activity[b]		Ref.
		Animals	Humans	
Parathyroid hormone	Increase	+	+	40,42,44,45,48,53,54,67,70
	Decrease	−	−	
Serum inorganic phosphorus	Increase	−	−	56,75
	Decrease	+	+	
$1,25(OH)_2D$	Increase	−	−	57
	Decrease	+	+	
Calcium (direct)	Increase	?	?	65
	Decrease	?	+	
Calcitonin	Increase	+,−,0	+	11,40,70
	Decrease	?	?	
Hydrogen ion	Increase	−	0	39,41,76
	Decrease	?	?	
Sex steroids	Increase	+	+	47,74
	Decrease	?	?	
Prolactin	Increase	+	0	38,49,60,73
	Decrease	?	?	
Growth hormone and insulin-like growth factor-1	Increase	+	0,−,+	50,55,63,64,65,72
	Decrease	?	?	
Glucocorticoids	Increase	−	−,0,+	46,51,52,58,59,71
	Decrease	?	?	
Thyroid hormone	Increase	?	−*	43,66,68
	Decrease	−	+	
Pregnancy	—	+	+*	61,62,69

[a] Modified from Kumar [6] with permission.

[b] +, Stimulation or increase; −, suppression or decrease; 0, no effect; ?, effect not known; *, effects may be secondary to changes in calcium, phosphorus, or parathyroid hormone.

TABLE III Metabolism of $25OHD_3$ by the Kidney

	↗ $24R,25(OH)_2D_3$ →	24-Keto-$25OHD_3$
$25OHD_3$ →	→ $25S,26(OH)_2D_3$	$25OHD_3$-lactone
	↘ $23S,25(OH)_2D_3$ →	23-Keto-$25OHD_3$

it has no unique properties [89–94]. The 24-hydroxylase is a multicomponent enzyme whose terminal cytochrome P450 has been structurally characterized [80,95–99] (see also Chapter 6). The precise ligand binding residues remain to be identified. The kidney is also capable of transforming $25OHD_3$ to several other compounds listed in Table III [100–120]. The specific physiological roles of these various metabolites are not known with certainty.

Several polar metabolites of $1,25(OH)_2D_3$ are formed in the liver, including calcitroic acid and glucuronide and sulfate conjugates of the hormone; these and small amounts of unchanged dihydroxylated and trihydroxylated metabolites of vitamin D are excreted in the urine [121–135]. In addition, many of the transformations that occur with $25OHD_3$ also occur in the case of $1,25(OH)_2D_3$ although, in all likelihood, they occur to a larger extent in the intestine than in the kidney.

III. EFFECTS OF VITAMIN D, 25OHD, AND $1,25(OH)_2D$ ON THE RENAL HANDLING OF CALCIUM AND PHOSPHORUS

Clinical studies have shown that vitamin D deficiency is associated with a low urine calcium concentration whereas vitamin D excess or intoxication is associated with hypercalciuria [1]. The concentrations of calcium and phosphorus in the urine in these *in vivo* situations, however, reflect the decreased or increased calcium absorption in the intestine, the presence of hypo- or hypercalcemia, and the presence of diminished or elevated concentrations of circulating PTH. Puschett *et al.* examined the effect of vitamin D_3 and $25OHD_3$ on the renal transport of phosphate, sodium, and calcium in parathyroidectomized dogs [136–138]. They showed that short-term infusions of vitamin D_3 and $25OHD_3$ were associated with decreases in the clearances of phosphate, calcium, and sodium relative to the clearance of inulin. These studies were interpreted as showing that vitamin D_3 and $25OHD_3$ enhance phosphate, calcium, and sodium reabsorption. A follow-up study by Puschett *et al.* showed that $1,25(OH)_2D_3$ had similar effects on the excretion of phosphate, calcium, and sodium in thyroparathyroidectomized (TPTX) dogs [139]. We performed similar studies examining the effects of $25OHD_3$ on renal bicarbonate and phosphate reabsorption [140]. Unlike the studies of Puschett *et al.*, we observed that phosphate and bicarbonate reabsorption increased only in intact animals and not in parathyroidectomized animals, suggesting the need for PTH. Our observations are similar to those of Popovtzer *et al.* [141].

Yamamoto *et al.* carried out perhaps the most comprehensive examination of the effects of $1,25(OH)_2D_3$ on the reabsorption of calcium [142]. Vitamin D-deficient rats, vitamin D-deficient rats supplemented with dietary calcium to normalize plasma calcium and PTH levels, and vitamin D-replete rats were examined following TPTX and the infusion of graded amounts of calcium. Figure 1 shows the relationship between the amount of calcium excreted in the urine and serum calcium concentrations in the three groups of animals. Urinary calcium excretion was lower in vitamin D-replete rats than in vitamin D-deficient rats, suggesting that vitamin D increased the efficiency of calcium absorption in the absence of PTH. In a second group of experiments, rats, treated in the manner noted above, were subjected to TPTX and infused with PTH. The results of this experiment show that a lower dose of PTH is needed to exhibit a comparable effect on renal calcium reabsorption in vitamin D-replete rats when compared to vitamin D-deficient rats (Fig. 2).

Winaver *et al.* used micropuncture to examine the sites along the nephron at which $25OHD_3$ exerted its antiphosphaturic and hypocalciuric effects [143]. They observed that the effects of $25OHD_3$ were not mediated at the level of the superficial proximal tubule but were likely due to effects in other segments of the nephron, most likely the distal tubule. These studies are similar to those of Sutton *et al.* and others who observed distal tubular effects of $25OHD_3$ in the dog [144]. Harris and Seely examined the effects of $1,25(OH)_2D_3$ on tubular calcium handling and concluded that the enhanced reabsorption of calcium occurred in distal tubular segments of the nephron [145]. All of these studies do not exclude an effect of the vitamin D analogs on a segment of the proximal tubule that is not accessible to micropuncture.

In vitro studies have shown that vitamin D deficiency is associated with decreased calcium uptake in membranes derived from the distal segments of the nephron [146]. This occurs in membranes of both luminal and basolateral origin. In addition, Bindels *et al.* have demonstrated that cultured rabbit connecting tubule cells

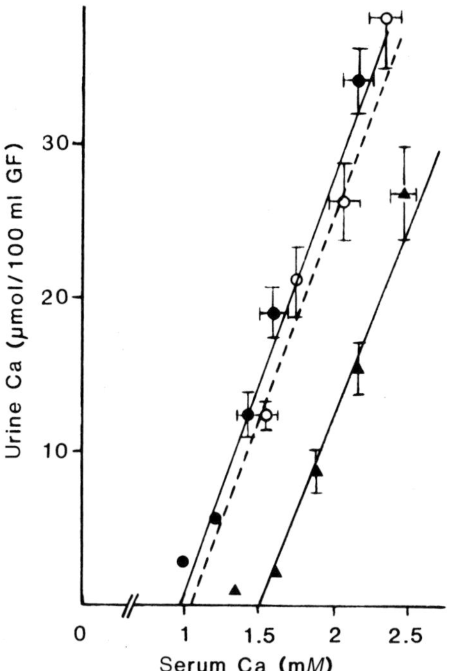

FIGURE 1 Relationships between urinary calcium excretion and serum calcium concentration in three groups of thyroparathyroidectomized (TPTX) rats. Serum concentration and urinary excretion of calcium were determined 16–19 hr after continuous infusion of an electrolyte solution containing 0–30 m*M* of $CaCl_2$. Each point represents the data pooled according to a continuous series of 0.25-m*M* changes in serum calcium concentration. Horizontal bars indicate standard error of mean serum calcium concentration, and vertical bars indicate standard error of mean urinary calcium excretion. The lines were derived from the regression analysis of the linear portion of data. Three groups of animals were studied: ●, group A rats fed vitamin D-deficient standard diet; ○, group B rats fed vitamin D-deficient diet containing high calcium and lactose; and ▲, group C rats fed vitamin D-replete standard diet. For any given serum calcium level, the urinary calcium excretion was significantly lower in vitamin D-replete rats (group C) than in vitamin D-deficient rats (groups A and B). Thus, the apparent serum calcium threshold determined as an intercept of the regression line on the serum calcium axis was higher in vitamin D-replete rats (~1.5 m*M*) than in vitamin D-deficient rats (~1.0 m*M*). There was no significant difference in the calcium threshold between group A and group B. Reprinted with permission from Yamamoto *et al.* [142].

show an increase in the transport of calcium when exposed to $1,25(OH)_2D_3$ [147].

A synthesis of the experimental results suggests that vitamin D metabolites have effects on the distal tubular reabsorption of calcium. In addition, it is possible that they may have effects on calcium reabsorption in portions of the proximal tubule not accessible to micropuncture. The effects of vitamin D metabolites on phosphate and sodium reabsorption are also in portions of the nephron other than the superficial proximal tubule.

IV. DISTRIBUTION AND REGULATION OF VITAMIN D-DEPENDENT PROTEINS IN THE KIDNEY

Several vitamin D-dependent proteins are expressed in the kidney, and many of these play a role in calcium transport. We have shown that the VDR, the calbindins-D, and the plasma membrane calcium pump are colocalized in the principal cells of the distal tubule and are very likely to act coordinately in the regulation of calcium transport in this segment of the nephron [5,15,148–152]. Other proteins, such as the 24-hydroxylase, are involved in the metabolism of various vitamin D metabolites. This section discusses pertinent information about the following vitamin D-dependent proteins: (1) the vitamin D receptor, (2) calbindins-D_{9K} and -D_{28K}, (3) the plasma membrane calcium pump, and (4) the 25-hydroxyvitamin D_3 24-hydroxylase cytochrome P450.

A. 1,25-Dihydroxyvitamin D_3 Receptor (VDR) in the Kidney

1. DISTRIBUTION

The VDR mediates many, if not all, of the effects of $1,25(OH)_2D_3$ in diverse organs [153–161]. The distribution of the VDR in the kidney has been assessed using a variety of techniques, including ligand binding assays with protein obtained from specific microdissected nephron segments, localization of radiolabeled $1,25(OH)_2D_3$ by autoradiography, and various antibody techniques [162–165]. Kidneys from adult animals and humans, and fetal kidneys from rodents, have been studied using one or the other of these techniques. Using protein from microdissected nephron segments, the VDR was found in the proximal and distal tubules [164,165]. With autoradiographic methods, following the administration of labeled ligand *in vivo*, silver grains were localized over distal tubule segments [163,164].

We have used sensitive polyclonal antibodies to localize the receptor in human and rat kidneys [152]. The antibodies were raised against highly purified antigen that was expressed in bacteria. The specificity of the antibodies was determined by absorption with purified antigen and by Western analysis, which showed that they detected a protein band with a molecular mass of approximately 50,000 daltons [152,166]. With these antibodies, we found that the VDR was present abundantly in the distal tubule and to a lesser extent in the proximal tubule (Fig. 3). Interestingly, not all cells in

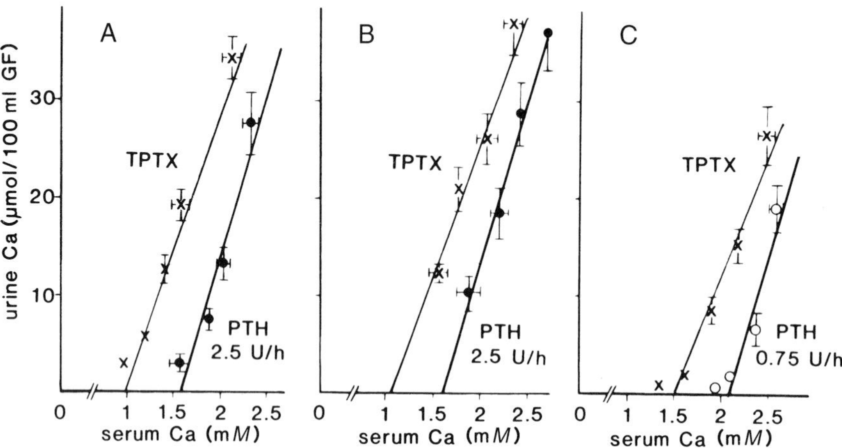

FIGURE 2 Effects of PTH infusion on the renal handling of calcium among three groups of TPTX rats. The data are presented as described in the legend to Fig. 1. PTH was delivered at 2.5 U/hr (●) to group A and B rats, and 0.75 U/hr (○) to group C rats. A, B, and C illustrate the results in group A, group B, and group C, respectively. The enhancement of calcium reabsorption by PTH is shown as shifts of the lines to the right in each group. A striking difference exists between vitamin D-deficient (groups A and B) and vitamin D-replete (group C) rats in the doses of PTH required to induce a comparable shift in the calcium threshold. x, data of TPTX rats in each group (from Fig. 1). Reprinted with permission from Yamamoto *et al.* [142].

the distal tubule expressed the receptor. Only those cells expressing calbindin-D_{28K} also expressed the VDR. As noted later, these cells also express the calcium pump. Acid-secreting cells do not express the VDR in significant amounts. Taken together, the results are consistent with the notion that the VDR is present in significant amounts in the distal tubule, where it regulates the amount and the activity of several vitamin D-dependent proteins such as the plasma membrane calcium pump and calbindin-D_{28K}. It is also present, although in lesser amounts, in the proximal tubule, where it regulates the activity of the 1α-hydroxylase and the 24-hydroxylase. Other work has suggested that VDR concentrations are low in proximal tubular cells when 1α-hydroxylase activity is enhanced [167].

We have shown that the VDR is present in cells of the developing rodent kidney [168] (Figs. 4 and 5) and in the cultured metanephros (Fig. 6). The VDR was detected as early as day 15 *post coitum* (p.c.) in the developing rat kidney *in vivo*. Significant amounts of the receptor were found in the metanephric mesenchyme as well as in the ureteric bud. As the kidney matured, the receptor was observed in the S-shaped and comma-shaped bodies and in the developing glomerulus, specifically in the parietal and visceral epithelial cells. The VDR staining in the latter cells persists in the adult kidney as well. Despite slightly different gestational periods, similar patterns of VDR distribution were found in the developing mouse kidney *in vivo*. Of great interest is the observation that calbindin-D_{28K} appears in the distal tubule only around day 18 p.c. This is when urine flow begins in the kidney. The VDR is also present in mouse metanephric cultures in the same distribution pattern as is found *in vivo*. By day 3 of culture, the pattern of expression of the VDR was similar to that seen in the mouse kidney *in vivo* at day 15. Calbindin-D_{28K} does not appear in the cultured metanephros. These results showing that the VDR appears well before the appearance of calbindin-D_{28K} suggests that the VDR may play a role in fetal renal development; however, the precise mechanisms underlying its role remain to be determined.

2. REGULATION OF VDR IN THE KIDNEY

The VDR is regulated by several factors in diverse tissues [169]. This general subject is discussed in detail in Chapter 11 and for the kidney and parathyroid gland in Chapters 52 and 23, respectively. Concentrations of the VDR in the kidney and parathyroid glands are mainly regulated by $1,25(OH)_2D_3$, PTH, and dietary calcium. Studies have shown that somewhat different results are obtained *in vivo* with respect to VDR abundance when $1,25(OH)_2D_3$ concentrations are altered by dietary manipulations when compared to the effects of intravenous administration of the hormone [170]. In Table IV are shown in the effects of exogenous administration of $1,25(OH)_2D_3$ on VDR concentrations in rats receiving a vitamin D-replete diet. $1,25(OH)_2D_3$ increases duodenal and kidney VDR concentrations. When endogenous $1,25(OH)_2D_3$ concentrations are in-

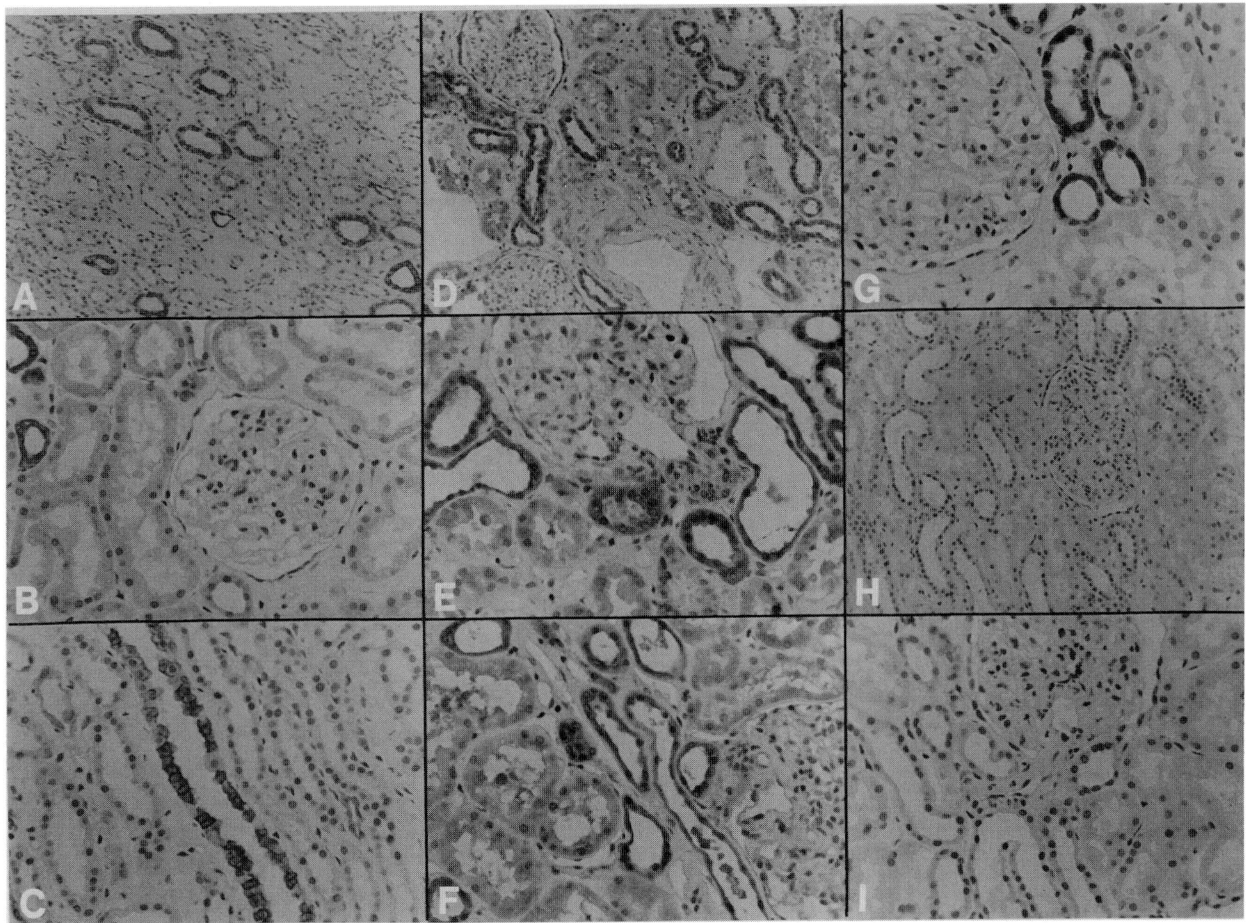

FIGURE 3 (A–C) Immunohistochemical detection of VDR in normal human kidney tissue with polyclonal anti-hVDR antibody 2-152. (D–F) Immunohistochemical detection of $25OHD_3$ 24-hydroxylase cytochrome P450 in human kidney. (G) Immunohistochemical detection of calbindin-D_{28K} in human kidney. (H,I) Staining of kidney tissue with preimmune serum. Magnifications: A, ×200; B, ×400; C, ×400; D, ×200; E, ×400; F, ×400. Reprinted with permission from Kumar *et al.* [152].

creased by adapting an animal to a low calcium diet (Table V), VDR concentrations in the duodenum and kidney do not increase. The difference appears to be due to increased levels of PTH elicted by the low calcium diet and decreased levels following $1,25(OH)_2D_3$ administration. PTH has been shown to down-regulate VDR in osteosarcoma cells as well as block homologous up-regulation of VDR in rats infused with $1,25(OH)_2D_3$ [171].

DeLuca and co-workers have examined the effect of dietary conditions and the administration of vitamin D or $1,25(OH)_2D_3$ on VDR concentrations in the kidney [172,173]. Vitamin D-deficient animals were fed either a low calcium or a normal calcium diet. They subsequently received either vitamin D_3 supplementation orally or $1,25(OH)_2D_3$ by osmotic minipump. In the presence of normal dietary calcium, the administration of vitamin D_3 or $1,25(OH)_2D_3$ resulted in an increase in renal VDR content. The administration of $1,25(OH)_2D_3$ to animals fed a vitamin D-deficient diet containing a high amount of calcium that resulted in a normalization of plasma calcium also resulted in an increase in the amount of the VDR found in the kidney. However, in the presence of a low calcium diet, neither vitamin D nor $1,25(OH)_2D_3$ had a significant effect on the concentrations of the VDR in the kidney. These results have also been confirmed by others who measured the mRNA concentrations for the VDR in kidneys of rats that had been administered $1,25(OH)_2D_3$ [174]. These studies are in agreement with the observations of Reinhardt and Horst [171] and suggest that increases in PTH caused by low dietary calcium are important modulators of renal VDR levels.

The results obtained from *in vivo* studies are consistent with those obtained following the administration of $1,25(OH)_2D_3$ to yeast cells transfected with a VDR construct [175]. There is now evidence that, in certain cells such as fibroblasts, $1,25(OH)_2D_3$ may have its pre-

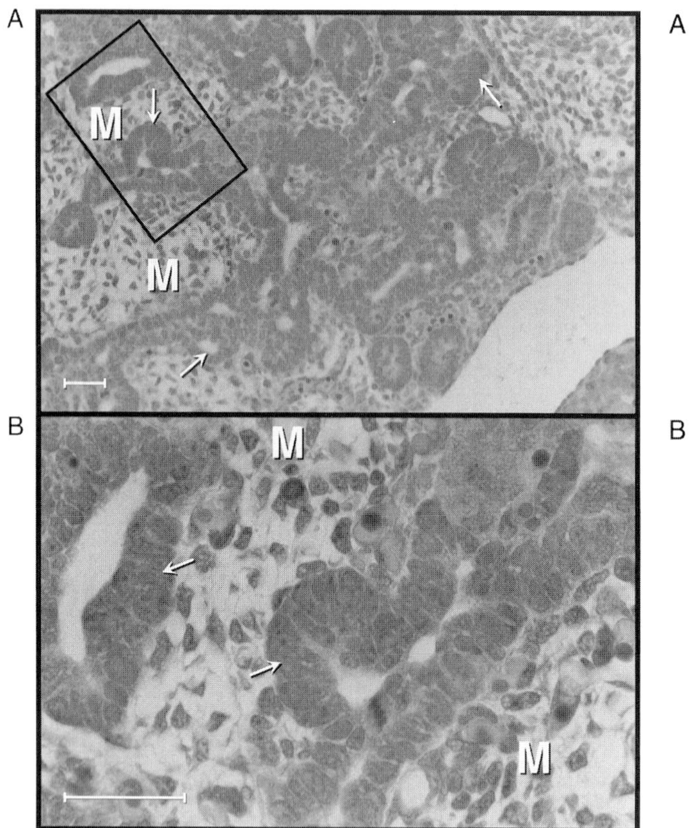

FIGURE 4 Vitamin D receptor (VDR) immunostaining in metanephros of rat fetus on gestational day 15. (B) Higher power view of boxed area in A. Branching ureteric buds (arrows) and mesenchyme (M) are indicated. Bars = 2.1 μm. Reprinted with permission from Johnson *et al.* [168].

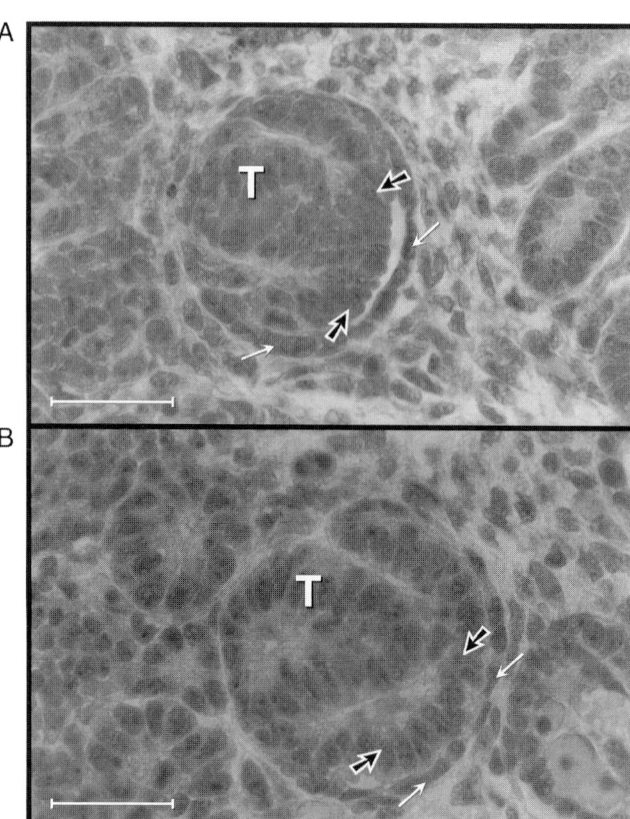

FIGURE 5 VDR (A) and calbindin-D_{28K} (B) immunostaining in metanephros of rat fetus on gestational day 17. Parietal epithelial cells (small arrows), visceral epithelial cells (open arrows), and tubule portion of developing comma-shaped body (T) are indicated. Bars = 2.1 μm. Reprinted with permission from Johnson *et al.* [168].

dominant effect to increase VDR abundance by ligand-induced stabilization of the protein [176]. Wiese *et al.* [176] showed that the addition of $1,25(OH)_2D_3$ to fibroblasts did not result in a significant increase in VDR mRNA concentrations but did cause increases in the amount of VDR protein, suggesting that the predominant effect of $1,25(OH)_2D_3$ in fibroblasts is the stabilization of preexisting VDR protein or the stabilization of preexisting vitamin D receptor mRNA.

Considerable work has also been carried out on the regulation of the VDR in parathyroid glands [177,180] (see also Chapters 23 and 52). The effects of different concentrations of calcium on the amount of VDR mRNA in the parathyroid glands of vitamin D_3-deficient chicks has been examined by Russell *et al.* [177]. These investigators showed that the administration of diets containing higher amounts of calcium resulted in an increase in VDR mRNA levels. Examination of interrelationships between $1,25(OH)_2D_3$ administration and dietary calcium clearly shows that the stimulatory effect of $1,25(OH)_2D_3$ on VDR mRNA occurs in the presence of high dietary calcium.

Renal failure is associated with alterations in the amount of the VDR in parathyroid glands [181–183]. Korkor showed that the amount of the VDR in the parathyroid glands of patients with chronic renal failure was substantially decreased when compared with patients with primary hyperparathyroidism or secondary hyperparathyroidism post kidney transplantation [181]. These studies have been confirmed by others in animal models. Additionally, Koyama *et al.* have shown that VDR content in the duodena of animals with uremia is diminished [183]. The administration of $1,25(OH)_2D_3$ does not increase the amount of the VDR protein but does increase the amount of VDR mRNA. There is also evidence now that uremic ultrafiltrates contain a factor that inhibits interaction of the VDR with the vitamin response elements in the DNA of target genes [184–186].

In addition to the VDR, several vitamin D-dependent proteins are expressed in the kidney. The pattern of

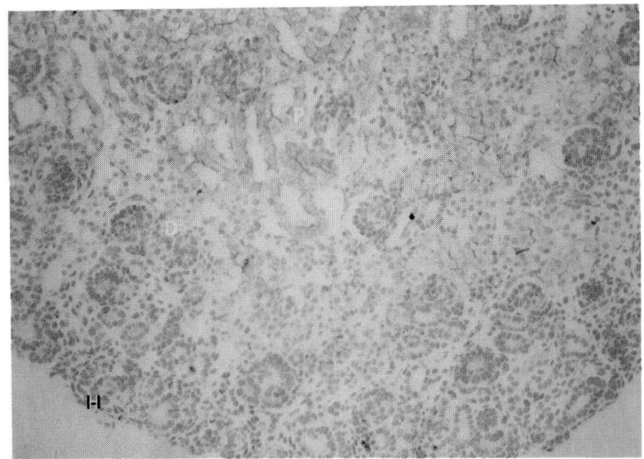

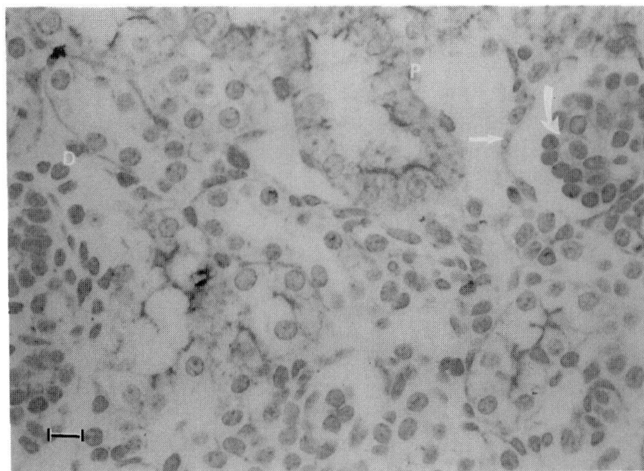

FIGURE 6 VDR immunostaining in mouse metanephric organ culture explant following 120 hr of incubation. Parietal epithelial cells (small arrow), visceral epithelial cells (large arrow), proximal tubule (P), and distal tubule (D) are indicated. Bars = 2.1 μm. Reprinted with permission from Johnson *et al.* [168].

regulation of these proteins is of great interest, in as much as it casts light on the different mechanisms by which calcium is transported in the kidney. Those proteins involved in calcium transport include calbindin-D_{28K}, calbindin-D_{9K}, and the plasma membrane calcium pump. Additionally, although not involved in the transport of calcium, the 24-hydroxylase is also expressed in the kidney. All of these proteins appear to be regulated by $1,25(OH)_2D_3$.

B. Calbindins-D

1. DISTRIBUTION

The calbindins-D are widely distributed in many tissues of the body and in a variety of species [14] (see also Chapters 13 and 14). Except in the brain, synthesis of the calbindins-D is dependent on vitamin D but not calcium or phosphorus. There are two forms of calbindin-D, namely, calbindin-D_{28K} and calbindin-D_{9K}, that are variably distributed in different tissues of the body [187–204]. The apparent molecular mass of the larger protein is about 30,000 daltons, whereas that of the smaller protein is 9,000 daltons [187]. The proteins are classic EF-hand proteins, the calbindin-D_{28K} having six EF-hand structures and the calbindin-D_{9K} having two such motifs [205–224]. The proteins bind calcium with high affinity and in different molar amounts. Calbindin D_{28K} binds 3–4 mol of calcium per mol of protein and calbindin D_{9K} 2 mol of calcium per mol of protein [187]. In the mouse kidney, both forms are present and are regulated by vitamin D. In other species, only calbindin-D_{28K} is expressed in the kidney.

Both calbindin-D_{9K} and calbindin-D_{28K} undergo conformational changes on binding to calcium. This phenomenon has been studied extensively by us in the case of calbindin-D_{28K}. We have shown that the protein undergoes a two-step change in conformation that is associated with binding to a high affinity site in EF hand 1 and a further change on calcium binding to sites in EF hands 4 and 5 [214,216]. The conformation change is probably what allows the protein to act as a modulator of the activity of the plasma membrane calcium pump (see Section IV,C below). In this manner its actions are analogous to those of calmodulin.

TABLE IV Plasma Calcium, Phosphorus, $1,25(OH)_2D_3$, and Unoccupied VDR Content of Duodenum and Kidney of Rats Receiving 36 ng $1,25(OH)_2D_3$/day for 7 Days and Control Rats[a]

	Plasma Ca^{2+} (mg/dl)	Plasma phosphorus (mg/dl)	Plasma $1,25(OH)_2D_3$ (pg/ml)	Duodenum VDR (fmol/mg protein)	Kidney VDR (fmol/mg protein)
Control	10.78 ± 0.11	8.23 ± 0.27	53 ± 6	351 ± 16	60 ± 5
$1,25(OH)_2D_3$-treated	13.40 ± 0.17**	7.75 ± 0.24	261 ± 0.17**	510 ± 21**	194 ± 23**

[a] Data are means ± SEM (n = 6). Asterisks (**) denote statistically significant differences in treated versus control values ($p < 0.001$). Reprinted from Goff *et al.* [170] with permission.

TABLE V Unoccupied VDR Content of Duodenal and Renal Tissue from Rats Fed a 1 or 0.02% Calcium Diet after 2, 7, 14, and 21 Days of Dietary Treatment[a]

Tissue	Diet	Day 2	Day 7	Day 14	Day 21
Duodenum	1% calcium	341 ± 26	197 ± 17	202 ± 17	259 ± 26
	0.02% calcium	365 ± 27	226 ± 28	221 ± 16	267 ± 28
Kidney	1% calcium	ND**	163 ± 11	165 ± 9	124 ± 8
	0.02% calcium	ND	120 ± 4*	131 ± 10*	77 ± 3*

[a] Values are mean ± SEM expressed as fmol ^3H-1,25$(OH)_2D_3$ bound per mg cytosol protein. *, $p < 0.05$; **, ND = not done. Reprinted from Goff *et al.* [70] with permission.

2. Regulation

Both calbindin-D_{28K} and calbindin-D_{9K} are regulated by 1,25$(OH)_2D$ in the kidney [218–224]. The calbindin effect on calcium transport in the distal tubule cell is probably mediated by increasing the transcellular movement of calcium and by increasing the activity of the plasma membrane calcium pump [225,226]. The proteins increase the maximal capacity (V_{max}) but not the affinity (K_m) of the plasma membrane calcium pump.

C. The Plasma Membrane Calcium Pump

1. Distribution

We raised monoclonal and polyclonal antibodies directed against the plasma membrane calcium pump and used them to examine the distribution of these proteins in the adult human kidney [15,148–151]. We found that epitopes for the Ca^{2+}, Mg^{2+}-ATPase (calcium pump) were expressed in the basolateral membrane of the distal tubular cells (Fig. 7). Similar patterns of expression were apparent in the rat [150] and rabbit [227] kidney. Interestingly, not all cells of the distal tubule stained positively for the calcium pump. Further investigation showed that cells expressing carbonic anhydrase and presumably involved in acid secretion did not express the plasma membrane calcium pump, whereas the other cells of the distal tubule did so. We found that the calbindin-D_{28K} was present in the same cells of the distal tubule as the plasma membrane calcium pump. Our studies on the distribution of the plasma membrane calcium pump in the kidney and its localization, predominantly in the distal tubule of the kidney, are supported by others who have shown that calcium ATPases are present mostly in the distal tubule [227]. In addition, it appears that the plasma membrane calcium pump is widely distributed in a large number of other calcium transporting tissues, many of which display vitamin D-dependent calcium transport (228–234). Table VI shows the distribution of the plasma membrane calcium pump in different tissues.

2. Regulation

In collaboration with Wasserman's group at Cornell University, we have shown that the plasma membrane calcium pump is regulated in the intestine by vitamin D [229,235,236]. We showed by Western analysis using monoclonal antibodies directed against the pump that shortly after the administration of 1,25$(OH)_2D_3$ to vitamin D-deficient chicks there is an increase in the amount of immunoreactive plasma membrane calcium pump in the cells of the duodenum, jejunum, and ileum [229]. This is associated with an increase in the amount of mRNA for the pump in the same segments of the intestine [235]. The increase occurs within 3–6 hr following the administration of 1,25$(OH)_2D_3$, and the effect is dose dependent. Furthermore, dietary calcium and phosphorus depletion are also associated with an increase in the amount of the pump expressed in the intestine. Thus, in the intestine, 1,25$(OH)_2D_3$ increases the synthesis of the plasma membrane calcium pump. In addition, 1,25$(OH)_2D_3$ increases the activity of the plasma membrane calcium pump in intestinal cell basolateral membranes. The mechanism by which this occurs is uncertain, although several possibilities arise. It could be a direct effect either of 1,25$(OH)_2D_3$ or stimulation via calbindin-D_{9K} or calbindin-D_{28K} [225,226]. Indeed, calmodulin is an important regulator of the plasma membrane calcium pump, and the calbindins-D have biophysical properties similar to those of calmodulin [187]. There is also evidence that calbindin-D_{28K} and calbindin-D_{9K} increase the activity of the calcium pump derived from red cell and intestinal cell membranes [237,238].

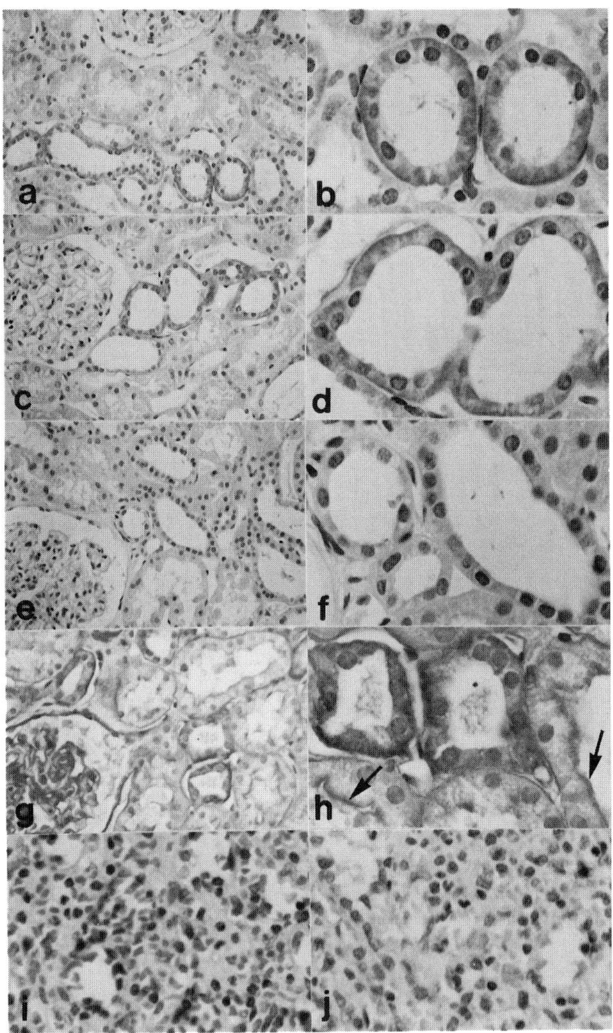

FIGURE 7 Immunoperoxidase localization of Ca^{2+},Mg^{2+}-ATPase within human kidney distal tubules and human spleen erythrocytes. Kidney was treated with (a,b) monoclonal antibody JA3 or (c,d) monoclonal antibody JA8 (e and f represent the negative control). Kidney was also treated with (g,h) double staining using the periodic acid–Schiff reagent (PAS) and JA3; arrows mark PAS-positive proximal tubule brush border. Spleen was treated with (i) monoclonal antibody JA3 (j represents the negative control). Magnifications: a, c, e, and g, ×200; b, d, f, h, i, and j, ×640. Reprinted with permission from Borke *et al.* [148].

In the kidney, Bouhtiauy *et al.* have examined the effects of vitamin D deficiency on the activity of the plasma membrane calcium pump [225,226]. They observed that vitamin D deficiency was associated with a decrease in plasma membrane calcium pump activity in the distal tubule of the kidney. These same authors also observed that calbindin-D_{28K} and calbindin-D_{9K} increased the activity of the renal basolateral membrane calcium pump. Whether $1,25(OH)_2D_3$ increases the synthesis of the plasma membrane calcium pump in the kidney is not known, although it is likely that mechanisms are similar to those noted in the case of the intestine.

In summary, it is likely that $1,25(OH)_2D_3$ increases calcium transport in the distal tubule of the kidney by increasing the synthesis of calbindin-D_{28K} and calbindin-D_{9K}, as well as the synthesis and activity plasma membrane calcium pump in distal tubules of the kidney. In addition, $1,25(OH)_2D_3$ via calbindin D-dependent mechanisms, also increases the activity of the plasma membrane calcium pump.

D. The 25-Hydroxyvitamin D_3 and 1,25-Dihydroxyvitamin D_3 24-Hydroxylase

The 24-hydroxylase enzyme is widely distributed in a number of renal and nonrenal tissues [44,77–87] (see also Chapter 6). Pioneering work by DeLuca's laboratory showed that the renal 24-hydroxylase was regulated by calcium and phosphorus such that elevated or normal calcium levels induced the 24-hydroxylase whereas low calcium levels inhibited it [74,77,79,80]. Similarly, elevated serum phosphorus concentrations increased the synthesis of the 24-hydroxylase enzyme, whereas low phosphorus diets decreased the activity of the enzyme [75]. Using antibodies against the 24-hydroxylase cytochrome P450 to examine the distribution of the enzyme in the human kidney, we found exceptionally high concentrations of the cytochrome P450 in distal tubular cells [152]. Lower amounts were found in the proximal tubule. Using enzymatic methods, several investigators have found 24-hydroxylase activity in kidney cells of the proximal tubule of the rat nephron. Iida *et al.* were unable to find 24-hydroxylase activity in microdissected distal tubule segment [239]. The reason for this apparent discrepancy between human and rat tissues is uncertain. Certainly, it would make biological sense for the 24-hydroxylase to be present in the distal tubule where other elements of the vitamin D responsive system are present. Other enzymes responsible for the transformation of 25-hydroxyvitamin D are also present in the kidney; they mediate the reactions shown in Table III.

V. CONCLUSION

The kidney plays a vital role in the conservation of calcium and phosphorus. Besides being the site of synthesis of $1,25(OH)_2D_3$, the kidney responds to the hormone by increasing the efficiency of calcium and phosphorus reabsorption. Elements of the calcium transport systems including calbindin-D_{28K}, calbindin-D_{9K}, and the plasma membrane calcium pump are present in the distal tubule, where vitamin D receptors are also found.

TABLE VI Distribution of Plasma Membrane Calcium Pump in Transporting Epithelia as Assessed by Immunohistochemistry

Tissue	Source	Cell type	Location in cell	Ref.
Kidney	Rat, human	Distal convoluted tubule, principal cell	Basolateral	15,148–151
Intestine	Rat, chick	Absorptive cell	Basolateral	228,229
Trophoblast	Rat, human	Syncytiotrophoblast	Basal	230
Choroid plexus	Cat Human	Choroid plexus Secretory cell	Apical	231
Shell gland	Chick	Principal cell	Apical	232
Bone	Human	Osteoblast	Not vectorially oriented	233
Bone	Chick	Osteoclast	Not vectorially oriented	234

References

1. Nordin BEC, Need AG, Morris HA (eds) 1993 Metabolic Bone and Stone Disease. Churchill Livingstone, London.
2. Friedman PA, Gesek FA 1995 Cellular calcium transport in renal epithelia: Measurement, mechanisms, and regulation. Physiol Rev **75**:429–471.
3. Borke JL, Penniston JT, Kumar R 1990 Recent advances in calcium transport by the kidney. Semin Nephrol **10**:15–23.
4. Johnson JA, Kumar R 1994 Vitamin D and renal calcium transport. Curr Opin Nephrol Hypertens **3**:424–429.
5. Kumar R 1995 Calcium transport in epithelial cells of the intestine and kidney. J Cell Biol **57**:392–398.
6. Kumar R 1984 The metabolism of 1,25-dihydroxyvitamin D_3. Physiol Rev **64**:478–504.
7. Kumar R 1991 Vitamin and calcium transport. Kidney Int **40**:1177–1189.
8. Berndt TJ, Knox FG 1992 Renal regulation of phosphate excretion. In: Seldin DW, Giebisch G (eds) The Kidney: Physiology and Pathology, 2nd Ed, Raven, New York, pp. 2511–2532.
9. Knox FG, Haramati A 1985 Renal regulation of phosphate excretion. In: Seldin DW, Giebish G (eds) The Kidney: Physiology and Pathology. Raven, New York, pp. 1351–1396.
10. Knox FG, Haas JA, Marchand GR, Youngberg SP 1977 Phosphate transport in superficial and deep nephrons in phosphate-loaded rats. Am J Physiol **233**:F150–F153.
11. Kumar R 1996 Calcium disorders. In: Kokko JP, Tannen RL (eds) Fluids and Electrolytes, 3rd Ed. Saunders, Philadelphia, Pennsylvania, pp. 391–419.
12. Neuman WF, Neuman MW 1958 The Chemical Dynamics of Bone Mineral. Univ. of Chicago Press, Chicago.
13. Moore EW 1970 Ionized calcium in normal serum, ultrafiltrates, and whole blood determined by ion-exchange electrodes. J Clin Invest **49**:318–334.
14. Raman A 1971 The calcium fractions of normal serum. Clin Chem **4**:141–146.
15. Kumar R, Penniston JT, Borke JL 1988 Ca^{2+}, Mg^{2+}-ATPase calcium pumps in the kidney. News Physiol Sci **3**:219–222.
16. Friedman PA 1988 Renal calcium transport: Sites and insights. New Physiol Sci **3**:17–21.
17. Lawson DE, Fraser DR, Kodicek E, Morris HR, Williams DH 1971 Identification of 1,25-dihydroxycholecalciferol, a new kidney hormone controlling calcium metabolism. Nature **230**: 228–230.
18. Fraser DK, Kodicek E 1970 Unique biosynthesis by kidney of a biological active vitamin D metabolite. Nature **228**:764–766.
19. Gray R, Boyle I, DeLuca HF 1971 Vitamin D metabolism: The role of kidney tissue. Science **172**:1232–1234.
20. Shultz TD, Fox J, Heath III H, Kumar R 1983 Do tissues other than the kidney produce 1,25-dihydroxyvitamin D_3 in vivo? A re-examination. Proc Natl Acad Sci USA **80**:1746–1750.
21. Weisman Y, Vargas A, Duckett G, Reister E, Root A 1978 Synthesis of 1,25-dihydroxyvitamin D in the nephrectomized pregnant rat. Endocrinology **103**:1992–1998.
22. Tanaka Y, Halloran B, Schnoes HK, DeLuca HF 1979 *In vitro* production of 1,25-dihydroxyvitamin D_3 by rat placental tissues. Proc Natl Acad Sci USA **76**:5033–5035.
23. Bell NH 1984 Sarcoidosis. In: Kumar R (ed) Vitamin D: Basic and Clinical Aspects. Martinus Nijhoff, Boston, pp. 581–589.
24. Golconda MS, Larson TS, Kolb L, Kumar R 1966 1,25-Dihydroxyvitamin D-mediated hypercalcemia in a renal transplant recipient. Mayo Clin Proc **71**:32–36.
25. Brunette MG, Chan M, Ferriere C, Roberts KD 1978 Site of $1,25(OH)_2$ vitamin D_3 synthesis in the kidney. Nature **276**: 287–289.
26. Henry HL 1984 Regulation of the synthesis of 1,25-dihydroxyvitamin D_3 and 24,25-dihydroxyvitamin D_3 in the kidney cell culture. In: Kumar R (ed) Vitamin D: Basic and Clinical Aspects. Martinus Nijhoff, Boston, pp. 152–174.
27. Turner RL 1984 Mammalian 25-hydroxyvitamin D-1α-hydroxylase: Measurement and regulation. In: Kumar R (ed) Vitamin D: Basic and Clinical Aspects. Martinus Nijhoff, Boston, pp. 175–196.
28. Rasmussen H, Wong M, Bikle D, Goodman DBP 1972 Hormonal control of the renal conversion of 25-hydroxycholecalciferol to 1,25-dihydroxycholecalciferol. J Clin Invest **51**:2502–2504.
29. Treschsel U, Bonjour JP, Fleisch H 1979 Regulation of the metabolism of 25-hydroxyvitamin D_3 in primary cultures of chick kidney cells. J Clin Invest **64**:206–217.
30. Henry HL 1979 Regulation of the hydroxylation of 25-hydroxyvitamin D_3 *in vivo* and in primary cultures of chick kidney cells. J Biol Chem **254**:2722–2729.
31. Ghazarian JG, DeLuca HF 1974 25-Hydroxycholecalciferol-1-hydroxylase: A specific requirement for NADPH and a hemoprotein component in chick kidney mitochondria. Arch Biochem Biophys **160**:63–72.
32. Ghazarian JG, Jefcoats CR, Knutson JC, Orme-Johnson WH, DeLuca HF 1974 Mitochondrial cytochrome P_{450}: A component of chick kidney 25-hydroxycholecalciferol-1α-hydroxylase. J Biol Chem **249**:3026–3033.

33. Pedersen JI, Ghazarian JG, Orme-Johnson NR, DeLuca HF 1976 Isolation of chick renal mitochondrial ferredoxin active in the 25-hydroxyvitamin D_3-1-hydroxylase system. J Biol Chem **251**:3933–3941.
34. Yoon PS, DeLuca HF 1980 Purification and properties of chick renal mitochondrial ferredoxin. Biochemistry **19**:2165–2171.
35. Yoon PS, Rawlings J, Orme-Johnson WH, DeLuca HF 1980 Renal mitochondrial ferredoxin active in 25-hydroxyvitamin D_3 1α-hydroxylase. Characterization of the iron-sulfur cluster using interprotein cluster transfer and electron paramagnetic resonance spectroscopy. Biochemistry **19**:2172–2176.
36. Okamoto Y, DeLuca HF 1994 Separation of two forms of chick 1,25-dihydroxyvitamin D_3 and 25-hydroxyvitamin D_3 24-hydroxylase. Proc Soc Exp Biol Med **205**:52–55.
37. Burgos-Trinidad M, Ismail R, Ettinger RA, Prahl JM, DeLuca HF 1992 Immunopurified 25-hydroxyvitamin D 1α-hydroxylase and 1,25-dihydroxyvitamin D 24-hydroxylase are closely related but distinct enzymes. J Biol Chem **267**:3498–3505.
38. Adams ND, Garthwaite TL, Gray RW, Hagen TC, Lemann J Jr 1979 The interrelationships among prolactin, 1,25-dihydroxyvitamin D, and parathyroid hormone in humans. J Clin Endocrinol Metab **49**:628–630.
39. Adams NR, Gray W, Lemann J Jr 1979 The calciuria of increased fixed acid production in humans: Evidence against a role for parathyroid hormone and 1,25-dihydroxyvitamin D. Calcif Tissue Int **28**:233–238.
40. Adams ND, Gray RW, Lemann J Jr 1979 The effects of oral $CaCO_3$ loading and dietary calcium deprivation on plasma 1,25-dihydroxyvitamin D concentrations in healthy adults. J Clin Endocrinol Metab **48**:1008–1016.
41. Bauveur B, Garabedian M, Fellot C, Mougin P, Balsan S 1977 The effect of induced metabolic acidosis on vitamin D_3 metabolism in rachitic chicks. Calcif Tissue Res **23**:121–124.
42. Bilezikian JP, Canfield RE, Jacobs TP, Polay JS, D'Adamo AP, Eisman JA, DeLuca HF 1978 Response of 1α-25-hydroxyvitamin D_3 to hypocalcemia in human subjects. N Engl J Med **299**:437–441.
43. Bouillon R, Muls E, De Moor P 1980 Influence of thyroid function on the serum concentration of 1,25-dihydroxyvitamin D_3. J Clin Endocrinol Metab **51**:793–797.
44. Boyle IT, Gray RW, DeLuca HF 1971 Regulation by calcium of *in vivo* synthesis of 1,25-dihydroxycholecalciferol and 21,24-dihydroxycholecalciferol. Proc Natl Acad Sci USA **68**:2131–2134.
45. Boyle IT, Gray RW, Omdahl JL, DeLuca HF 1972 Calcium control of the *in vivo* biosynthesis of 1,25-dihydroxyvitamin D_3: Nicolaysen's endogenous factor. In: Taylor S (ed) Proceeding of the International Symposium on Endocrinology. Heinemann, London, pp. 468–476.
46. Carre M, Ayigbede O, Miravet L, Rasmussen H 1974 The effect of prednisolone upon the metabolism and action of 25-hydroxy and 1,25-hydroxyvitamin D_3. Proc Natl Acad Sci USA **71**:2996–3000.
47. Castillo L, Tanaka Y, DeLuca F, Sunde ML 1977 The simulation of 25-hydroxyvitamin D_3-1α-hydroxylase by estrogen. Arch Biochem Biophys **179**:211–217.
48. Drezner MK, Neelon FA, Haussler R, McPherson HT, Lebovitz HE 1976 1,25-Dihydroxycholecalciferol deficiency: The probable cause of hypocalcemia and metabolic bone disease in pseudohypoparathyroidism. J Clin Endocrinol Metabl **42**:621–628.
49. Emmersten K, Melsen F, Mosekilde L, Lund BI, Lund BJ, Sorensen OH, Nielsen HE, Solling H, Hansen HH 1981 Altered vitamin D metabolism and bone remodeling in patients with medullary thyroid carcinoma and hypercalcitoninemia. Metab bone Dis Related Res **4**:17–23.
50. Eskildsen PC, Lund BJ, Sorensen OH, Lund BI, Bishop JE, Norman AW 1979 Acromegaly and vitamin D metabolism effect of bromocriptine treatment. J Clin Endocrinol Metab **49**:484–486.
51. Favus MM, Walling MW, Kimberg DV 1973 Effects of 1,25-dihydroxycholecalciferol on intestinal calcium transport in cortisone-treated rats. J Clin Invest **52**:1680–1685.
52. Feher JJ, Wasserman RH 1979 Intestinal calcium-binding protein and calcium absorption in cortisol treated chicks: Effects of vitamin D_3 and 1,25-dihydroxyvitamin D_3. Endocrinology **104**:547–551.
53. Fraser DR, Kodicek E 1973 Regulation of 25-hydroxycholecalciferol-1-hydroxylase activity in kidney by parathyroid hormone. Nature New Biol **241**:163–166.
54. Garabedian M, Holick MF, DeLuca HF, Boyle IT 1972 Control of 25-hydroxycholecalciferol metabolism by the parathyroid glands. Proc Natl Acad Sci USA **69**:1673–1676.
55. Gertner JM, Horst RL, Broadus AE, Rasmussen H, Genel M 1979 Parathyroid function and vitamin D metabolism during human growth hormone replacement. J Clin Endocrinol Metab **49**:185–188.
56. Gray RW, Wilz DR, Caldas AE, Lemann J Jr 1977 The importance of phosphate in regulating plasma 1,25-$(OH)_2$-vitamin D levels in human studies in healthy subjects, in calcium-stone formers and in patients with primary hyperparathyroidism. J Clin Endocrinol Metab **45**:299–306.
57. Kawashima H, Torikai S, Kurokawa K 1981 Calcitonin selectively stimulates 25-hydroxyvitamin D_3-1α-hydroxylase in proximal straight tubule of rat kidney. Nature **291**:327–329.
58. Kimberg DV, Baerg RD, Gershon E, Grandusius RT 1971 Effect of cortisone treatment on the active transport of calcium by the small intestine. J Clin Invest **50**:1309–1321.
59. Kumar R, Abboud CF, Riggs BL 1980 The effect of elevated prolactin levels on plasma 1,25-dihydroxyvitamin D and intestinal absorption of calcium. Mayo Clin Proc **55**:51–53.
60. Kumar R, Cohen WR, Epstein FH 1980 Vitamin D and calcium hormones in pregnancy. N Engl J Med **302**:1143–1145.
61. Kumar R, Cohen WR, Silva P, Epstein FH 1979 Elevated 1,25-dihydroxyvitamin D plasma levels in normal human pregnancy and lactation. J Clin Invest **63**:342–344.
62. Kumar R, Merimee TJ, Silva P, Epstein FH 1979 The effect of chronic growth hormone excess or deficiency on plasma 1,25-dihydroxyvitamin D levels in man. In: Norman AW, Schaefer K, von Herrath D, Grigoleit H-G, Coburn JW, DeLuca HF, Mawer EB, Suda T (eds) Proceedings of the Fourth Workshop on Vitamin D. de Gruyter, Elmsford, New York, pp. 1005–1009.
63. Menaa C, Vrtovsnik F, Friedlander G, Corvol M, Garabedian M 1995 Insulin-like growth factor I, a unique calcium-dependent stimulation of 1,25-dihydroxyvitamin D_3 production. Studies in cultured mouse kidney cells. J Biol Chem **270**:25461–25467.
64. Condamine L, Menaa C, Vrtovsnik F, Vztovsnik F (corrected to Vrtovsnik F), Friedlander G, Garabedian M 1994 Local action of phosphate depletion and insulin-like growth factor 1 on *in vitro* production of 1,25-dihydroxyvitamin D by cultured mammalian kidney cells. J Clin Invest **94**:1673–1679.
65. Lund BJ, Sorensen OH, Lund BI, Bishop JE, Norman AW 1980 Stimulation of 1,25-dihydroxyvitamin D production by parathyroid hormone and hypocalcemia in man. J Clin Endocrinol Metab **50**:480–484.
66. Mosekilde BJ, Melsen F, Lund BI, Lund BJ, Sorensen OH 1982 Serum levels of vitamin D metabolites and bone remodeling in hyperthyroidism. Metabolism **31**:126–132.

67. Omdahl JL, Gray RW, Boyle IT, Knutson J, DeLuca HF 1972 Regulation of metabolism of 25-hydroxycholecalciferol by kidney tissue *in vitro* by dietary calcium. Nature New Biol **237**:63–64.
68. Pahuja DN, DeLuca HF 1982 Thyroid hormone and vitamin D metabolism in the rat. Arch Biochem Biophys **213**:293–298.
69. Pike JW, Parker JB, Haussler MR, Boass A, Toverud SU 1979 Dynamic changes in circulating 1,25-dihydroxyvitamin D during reproduction in rats. Science **204**:1427–1429.
70. Rasmussen H, Wong M, Bikle D, Goodman DBP 1972 Hormonal control of the renal conversion of 25-hydroxycholecalciferol to 1,25-dihydroxycholecalciferol. J Clin Invest **51**:2502–2504.
71. Seeman E, Kumar R, Hunder GG, Scott M, Heath H, Riggs BL 1980 Production, degradation, and circulating levels of 1,25-dihydroxyvitamin D in health and in chronic glucocorticoid excess. J Clin Invest **66**:664–669.
72. Spanos E, Barrett D, MacIntyre I, Pike JN, Safillian EC, Haussler MR 1978 Effect of growth hormone on vitamin D metabolism. Nature **273**:2420–2433.
73. Spanos E, Colson KW, Evans A, Galante LS, MacAuley SJ, MacIntyre I 1976 Effect of prolactin on vitamin D metabolism. Nature **5**:153–167.
74. Tanaka Y, Castillo L, DeLuca HF 1976 Control of renal vitamin D hydroxylases in birds by sex hormones. Proc Natl Acad Sci USA **73**:2701–2705.
75. Tanaka Y, DeLuca HF 1973 The control of 25-hydroxyvitamin D metabolism by inorganic phosphorus. Arch Biochem Biophys **154**:566–574.
76. Weber H, Gray RW, Sommghly JH, Lemann J 1976 The lack of effect of chronic metabolic acidosis on 25-OH-vitamin D metabolism and serum parathyroid hormones in humans. J Clin Endocrinol Metab **43**:1047–1055.
77. Holick MF, Schnoes HK, DeLuca HF, Gray RW, Boytle IT, Suda T 1972 Isolation and identification of 24,25-dihydroxycholecalciferol: A metabolite of vitamin D_3 made in the kidney. Biochemistry **11**:4251–4255.
78. Tanaka Y, DeLuca HF, Ikekawa N, Morisaki M, Koizumi N 1975 Determination of stereochemical configuration of the 24-hydroxyl group of 24,25-dihydroxyvitamin D_3 and its biological importance. Biochem Biophys **170**:620–626.
79. Holick MF, Baxter LA, Schraufrogel PK, Tavela TE, DeLuca HF 1976 Metabolism and biological activity of 24,25-dihydroxyvitamin D_3 in the chick. J Biol Chem **251**:397–402.
80. Knutson JC, DeLuca HF 1974 25-Hydroxyvitamin D_3-24-hydroxylase: Subcellular location and properties. Biochemistry **13**:1543–1548.
81. Kumar R, Schnoes HK, DeLuca HF 1978 Rat intestinal 25-hydroxyvitamin D_3-and 1α,25-hydroxyvitamin D_3-24-hydroxylase. J Biol Chem **253**:3804–3809.
82. Garabedian M, DuBois MB, Corvol MT, Pezant E, Balsan S 1978 Vitamin D and cartilage. I. *In vitro* metabolism of 25-hydroxycholecalciferol by cartilage. Endocrinology **102**:1262–1268.
83. Kulkowski JA, Chan T, Martinez J, Ghazarian JG 1979 Modulation of 25-hydroxyvitamin D_3-24-hydroxylase by aminophylline: A cytochrome P-450 monoxygenase system. Biochem Biophys Res Commun **90**:50–57.
84. Tanaka Y, Lorenc RS, DeLuca HF 1975 The role of 1,25-dihydroxyvitamin D_3 and parathyroid hormone in the regulation of chick renal 25-hydroxyvitamin D_3-24-hydroxylase. Arch Biochem Biophys **171**:521–526.
85. DeLuca HF 1979 The vitamin D system in the regulation of calcium and phosphorus metabolism. Nutr Rev **37**:161–194.
86. Ornoy A, Goodwin D, Noff D, Edelstein S 1978 24,25-Dihydroxyvitamin D is a metabolite of vitamin D essential for bone formation. Nature **276**:517–519.
87. Henry HL, Taylor AN, Norman AW 1977 Response of chick parathyroid glands to the vitamin D metabolites 1,25-dihydroxyvitamin D_3 and 24,25-dihydroxyvitamin D_3. J Nutr **107**:1918–1926.
88. Henry HL, Norman AW 1978 Vitamin D: Two dihydroxylated metabolites are required for normal chicken egg hatchability. Science **201**:835–837.
89. Kobayashi Y, Taguchi T, Terada T, Oshida J, Morisaki M, Ikekawa N 1979 Synthesis of 24,24-difluoro- and 24ε-fluoro-25-hydroxyvitamin D_3. Tetrahedron Lett **22**:2023–2026.
90. Yamada S, Ohmori M, Takayama H 1979 Synthesis of 24,25-difluoro-25-hydroxyvitamin D_3. Tetrahedron Lett **21**:1859–1862.
91. Halloran BP, DeLuca HF, Barthell E, Yamada S, Ohmori M, Takayama H 1981 An examination of the importance of 24-hydroxylation to the function of vitamin D during early development. Endocrinology **108**:2067–2071.
92. Tanaka Y, DeLuca HF, Kobayashi Y, Taguchi T, Ikekawa N, Morisaki M 1979 Biological activity of 24,24-difluoro-25-hydroxyvitmin D_3. Effect of blocking of 24-hydroxylation on the functions of vitamin D. J Biol Chem **254**:7163–7167.
93. Mathews CHE, Parfitt AM, Brommage R, Jarnagin K, DeLuca HF 1983 Only 1,25-dihydroxyvitamin D_3 is needed for normal bone growth and mineralization in the rat. Am Soc Bone Miner Res S66 (abstract).
94. Tanaka Y, Pahuja DN, Wichmann JK, DeLuca HF, Kobayashi Y, Taguchi T, Ikekawa N 1982 25-Hydroxy-26,26,26,27,27,27-hexafluorovitamin D_3: Biological activity in the rat. Arch Biochem Biophys **218**:134–141.
95. Madhok TC, Schnoes HK, DeLuca HF 1977 Mechanism of 25-hydroxyvitamin D_3 24-hydroxylation: Incorporation of oxygen-18 into the 24-position of 25-hydroxyvitamin D_3. Biochemistry **16**:2142–2145.
96. Ohyama Y, Noshiro M, Eggertsen G, Gotch O, Kato Y, Bjorkhem I, Okuda K 1993 Structural characterization of the gene encoding rat 25-hydroxyvitamin D_3 24-hydroxylase. Biochemistry **32**:76–82.
97. Itoh S, Yoshimura T, Iemura O, Yamada E, Tsujikawa K, Kohama Y, Mimura T 1995 Molecular cloning of 25-hydroxyvitamin D_3 24-hydroxylase (Cyp-24) from mouse kidney: Its inducibility by vitamin D_3. Biochim Biophys Acta **1264**:26–28.
98. Labuda M, Lemieux N, Tihy F, Prinster C, Glorieux FH 1993 Human 25-hydroxyvitamin D 24-hydroxylase cytochrome P450 subunit maps to a different chromosomal location than that of pseudovitamin D-deficient rickets. J Bone Miner Res **8**:1397–1406.
99. Chen KS, DeLuca HF 1995 Cloning of the human 1α,25-hydroxyvitamin D_3 24-hydroxylase gene promoter and identification of two vitamin D-responsive elements. Biochim Biophys Acta **1263**:1–9.
100. Takasaki Y, Noriuchi N, Takahashi N, Abe E, Shinki T, Suda T, Yamada S, Takayama H, Norikawa H, Masumura T, Sugahara M 1980 Isolation and identification of 25-hydroxy-24-oxochole calciferol: A metabolite of 25-hydroxycholecalciferol. Biochem Biophys Res Commun **95**:177–181.
101. Takasaki Y, Suda T, Yamada S, Takayama H, Nishii Y 1981 Isolation, identification, and biological activity of 25-hydroxy-24-oxovitamin D_3: A new metabolite of vitamin D_3 generated by *in vitro* incubations with kidney homogenates. Biochemistry **20**:1681–1686.
102. Wickman JK, Schnoes HK, DeLuca HF 1981 23,24,25-Trihydroxyvitamin D_3, 24,25,26-trihydroxyvitamin D_3, and 23-dehy-

dro-25-hydroxyvitamin D_3: New *in vivo* metabolites of vitamin D_3. Biochemistry **20**:2350–2353.
103. Redel J, Bazely Y, Tanaka Y, DeLuca HF 1978 The absolute configuration of the natural 25,26-dihydroxycholecalciferol. FEBS Lett **94**:228–230.
104. Partridge JJ, Shiuey S-J, Chadha NK, Baggiolini EG, Blount JF, Uskokovic MR 1981 Synthesis and structure proof of a vitamin D_3 metabolite, 25(*S*),26-dihydroxycholecalciferol. J Am Chem Soc **103**:1253–1255.
105. Cesario M, Guilhem J, Pascard C, Redel J 1978 The absolute configuration of C-25 epimers of 25,26-dihydroxycholecalciferol by x-ray diffraction analysis. Tetrahedron Lett **12**:1097–1098.
106. Cesario M, Guilhem J, Pascard C, Redel J 1980 The absolute configuration of C-25 epimers of 25,26-dihydroxycholecalciferol by x-ray diffraction analysis (Erratum). Tetrahedron Lett **21**:1588.
107. Horst RL, Littledike ET 1980 25-Hydroxyvitamin D_3-26,23-lactone, demonstration of kidney-dependent synthesis in the pig and rat. Biochem Biophys Res Commun **93**:149–154.
108. Gray RW, Caldas AE, Weber JL, Ghazarian TG 1978 Biotransformations of 25-hydroxyvitamin D_3 by kidney microsomes. Biochem Biophys Res Commun **82**:121–128.
109. Tanaka Y, Shepard RA, DeLuca HF, Schnoes HK 1978 The 26-hydroxylation of 25-hydroxyvitamin D_3 *in vitro* by chick renal hemogenates. Biochem Biophys Res Commun **83**:7–13.
110. Horst RL 1979 25-Hydroxyvitamin D_3-26,23-lactone, a metabolite of vitamin D_3 that is five times more potent than 25-OH-D_3 in the rat plasma competitive protein binding radioassay. Biochem Biophys Res Commun **89**:286–293.
111. Wichmann JK, DeLuca HK, Schnoes HK, Horst RL, Shepard RM, Jorgensen NA 1979 25-Hydroxyvitamin D_3-26,23-lactone, a new *in vivo* metabolite of vitamin D. Biochemistry **18**:4775–4780.
112. Ishizuka S, Ishimoto S, Norman AW 1982 Metabolic pathway to 25-hydroxyvitamin D_3-26,23-lactone from 25-hydroxyvitamin D_3. FEBS Lett **138**:83–87.
113. Tanaka Y, Wichmann TK, Schnoes HK, DeLuca HF 1981 Isolation and identification of 23,25-dihydroxyvitamin D_3. Biochemistry **20**:3875–3879.
114. Horst RL, Pramanik BC, Reinhardt TA, Shiuey SJ, Partridge JJ, Uskokovic MR, Napoli JL 1982 Binding properties of 23*S*,25-dihydroxyvitamin D_3. Biochem Biophys Res Commun **106**:1006–1011.
115. Partridge JJ, Chadha NK, Shiuey SJ, Wovkulich PM, Uskokovic MR, Napoli JL, Horst RL 1982 Synthesis and structure proof of 23*S*,25-dihydroxycholecalciferol, a new *in vivo* vitamin D_3 metabolite. In: Norman AW, Schaefer K, v. Herrath D, Grigoleit H-G (eds) Vitamin D. Chemical, Biochemical and Clinical Endocrinology of Calcium Metabolism. de Gruyter, New York, pp. 1073–1078.
116. Morris DS, Williams DH, Norris AF 1981 Structure and synthesis of 25-hydroxycholecalciferol-26,23-lactone, a metabolite of vitamin D_3. J Chem Soc Chem Commun **9**:424–425.
117. Morris DS, Williams DH, Norris AF 1981 Structure and synthesis of 25-hydroxycholecalciferol-26,23-lactone, a metabolite of vitamin D. J Org Chem **46**:3422–3428.
118. Tanaka Y, Wichmann JK, Paaren HE, Schnoes HK, DeLuca HF 1980 Role of kidney tissue in the production of 25-hydroxyvitamin D_3-26,23-lactone. Proc Natl Acad Sci USA **77**:6411–6416.
119. Horst RL, Reinhardt TA, Napoli JL 1982 23-Keto-25-hydroxyvitamin D_3 and 23-keto-1,25-dihydroxyvitamin D_3 receptor. Biochem Biophys Res Commun **107**:1319–1325.
120. Horst RL, Reinhardt TA, Pramanik BC, Napoli JL 1982 23-Keto-25-dihydroxyvitamin D_3: A vitamin D_3 metabolite with high affinity for the 1,25-dihydroxyvitamin D-specific cytosol receptor. Biochemistry **22**:245–250.
121. Kumar R 1984 The metabolism of dihydroxylated vitamin D metabolites. In: Kumar R (ed) Vitamin D: Basic and Clinical Aspects. Martinus Nijhoff, Boston, pp. 69–90.
122. Frolik CA, DeLuca HF 1972 Metabolism of 1,25-dihydroxycholecalciferol in the rat. J Clin Invest **51**:2900–2906.
123. Frolik CA, DeLuca HF 1973 The stimulation of 1,25-dihydroxycholecalciferol treatment. J Clin Invest **52**:543–548.
124. Kumar R, Harnden DH, DeLuca HF 1976 Metabolism of 1,25-dihydroxyvitamin D_3: Evidence for side chain oxidation. Biochemistry **15**:2420–2423.
125. Harnden DH, Kumar R, Holick MF, DeLuca HF 1976 Side chain oxidation of 25-hydroxy[26,27-^{14}C]vitamin D_3 and 1,25-dihydroxy[26,27-^{14}C]vitamin D_3 *in vivo*. Science **193**:493–494.
126. Esvelt RP, Schnoes HK, DeLuca HF 1979 Isolation and characterization of 1α-hydroxy-23-carboxytetranorvitamin D: A major metabolite of 1,25-dihydroxyvitamin D_3. Biochemistry **18**:3977–3983.
127. Ohnuma N, Kruse JR, Popjak G, Norman AW 1982 Isolation and chemical characterization of two new vitamin D metabolites produced by intestine: 1,25-Dihydroxyvitamin D_3 and 1,25,26-trihydroxyvitamin D_3. J Biol Chem **257**:5097–5102.
128. Ohnuma N, Norman AW 1982 Identification of a new C-23 oxidation pathway of metabolism for 1,25-dihydroxyvitamin D_3 present in intestine. J Biol Chem **257**:8261–8271.
129. Reinhart TA, Napoli JL, Beitz DC, Littledike ET, Horst RL 1981 A new *in vivo* metabolite of vitamin D_3: 1,25,26-Trihydroxyvitamin D_3. Biochem Biophys Res Commun **99**:302–307.
130. Kumar R, Nagubandi S, Mattox VR, Londowski JM 1980 Enterohepatic physiology of 1,25-dihydroxyvitamin D_3. J Clin Invest **65**:277–284.
131. Kumar R, Nagubandi S, Londowski JM 1981 Production of a polar metabolite of 1,25-dihydroxyvitamin D_3 in rat liver perfusion system. Digest Dis Sci **26**:242–246.
132. Litwiller RD, Mattox VR, Jardine I, Kumar R 1982 Evidence for a monoglucuronide of 1,25-dihydroxyvitamin D_3 in rat bile. J Biol Chem **257**:2491–2494.
133. Gray RW, Caldas AE, Wilz DR, Lemann J Jr, Smith GA, DeLuca HF 1978 Metabolism and excretion of ^{3}H-1,25-$(OH)_2$-vitamin D_3 in healthy adults. J Clin Endocrinol Metab **46**:756–765.
134. Seeman E, Kumar R, Hunder GG, Scott M, Heath H, Riggs BL 1980 Production, degradation, and circulating levels of 1,25-dihydroxyvitamin D in health and in chronic glucocorticoid excess. J Clin Invest **66**:664–669.
135. Wiesner RH, Kumar R, Seeman E, Go VLW 1980 Enterohepatic physiology of 1,25-dihydroxyvitamin D_3 metabolites in normal man. J Lab Clin Med **96**:1094–1100.
136. Puschett JB, Beck WS Jr 1975 Parathyroid hormone and 25-hydroxy vitamin D_3: Synergistic and antagonistic effects on renal phosphate transport. Science **190**:473–475.
137. Puschett JB, Beck WS Jr, Jelonek A, Fernandez PC 1974 Study of the renal tubular interactions of thyrocalcitonin, cyclic adenosine 3′,5′-monophosphate, 25-hydroxycholecalciferol, and calcium ion. J Clin Invest **53**:756–767.
138. Puschett JB, Moranz J, Kurnick WS 1972 Evidence for a direct action of cholecalciferol and 25-hydroxycholecalciferol on the renal transport of phosphate, sodium, and calcium. J Clin Invest **51**:373–385.
139. Puschett JB, Fernandez PC, Boyle IT, Gray RW, Omdahl JL, DeLuca HF 1972 The acute renal tubular effects of 1,25-dihydroxycholecalciferol. Proc Soc Exp Biol Med **141**:379–384.
140. Siegfried D, Kumar R, Arruda J, Kurtzman N 1977 Influence of vitamin D on bicarbonate reabsorption. In: Massry SG, Ritz E

(ed) Advances in Experimental Medicine and Biology Phosphate Metabolism **81**:395–408. Heidelberg. Plenum, New York.

141. Popovtzer MM, Robinette JB, DeLuca HF, Holick MF 1974 The acute effect of 25-hydroxycholecalciferol on renal handling of phosphorus. J Clin Invest **53**:913–921.
142. Yamamoto M, Kawanobe Y, Takahashi H, Shimazawa E, Kimura S, Ogata E 1984 Vitamin D deficiency and renal calcium transport in the rat. J Clin Invest **74**:507–513.
143. Winaver J, Sylk DB, Robertson JS, Chen TC, Puschett JB 1980 Micropuncture study of the acute renal tubular transport effects of 25-hydroxyvitamin D_3 in the dog. Miner Electrolyte Metab **4**:178–188.
144. Sutton RAL, Wong NLM, Dirks JH 1975 25-Hydroxy vitamin D_3 (25(OH)D_3): enhancement of distal tubular calcium reabsorption in the dog. 8th Annu Meet Am Soc Nephrol 8.
145. Harris CA, Seely JF 1979 Effects of 1,25$(OH)_2D_3$ on renal handling of electrolytes in the thyroparathyroidectomized (TPTX) D-deficient rat. Clin Res **27**:417A.
146. Bouhtiauy I, Lajeunesse D, Brunette MG 1993 Effect of vitamin D depletion on calcium transport by the luminal and basolateral membranes of the proximal and distal nephron. Endocrinology **132**:115–120.
147. Bindels RJM, Hartog A, Timmermans JAH, Van Os Ch 1991 Active Ca^{2+} transport in primary cultures of rabbit kidney: Stimulation by 1,25-dihydroxyvitamin D_3 and PTH. Am J Physiol **261**:F799–F807.
148. Borke JL, Minami J, Verma A, Penniston JT, Kumar R 1987 Monoclonal antibodies to human erythrocyte membrane Ca^{++}-Mg^{++} adenosine triphosphatase pump recognize an epitope in the basolateral membrane of human kidney distal tubule cells. J Clin Invest **80**:1225–1231.
149. Borke JL, Minami J, Verma AK, Penniston JT, Kumar R 1988 Co-localization of erythrocyte Ca^{++}-Mg^{++} ATPase and vitamin D-dependent 28-kilodalton calcium binding protein. Kidney Int **34**:262–267.
150. Borke JL, Caride A, Verma AK, Penniston JT, Kumar R 1989 Plasma membrane calcium pump and 28 kDa calcium binding protein in cells of rat kidney distal tubules. Am J Physiol **257**:F842–F849.
151. Johnson JA, Kumar R 1994 Renal and intestinal calcium transport: Role of vitamin and vitamin D-dependent calcium binding proteins. Semin Nephrol **14**:119–128.
152. Kumar R, Schaefer J, Grande JP, Roche PC 1994 Immunolocalization of calcitriol receptor, 24-hydroxylase cytochrome P-450, and calbindin D_{28K} in the human kidney. Am J Physiol **266**:F477–F485.
153. Ozono K, Sone T, Pike JW 1991 The genomic mechanism of action of 1,25-dihydroxyvitamin D_3. J Bone Miner Res **6**:1021–1027.
154. Pike JW 1991 Vitamin D_3 receptors: Structure and function in transcription. Annu Rev Nutr **11**:189–216.
155. Hughes MR, Malloy PJ, O'Malley BW, Pike JW, Feldman D 1991 Genetic defects of 1,25-dihydroxyvitamin D_3 receptor. J Receptor Res **11**:699–716.
156. Feldman D, Malloy DJ 1990 Hereditary 1,25-dihydroxyvitamin D resistant rickets: Molecular basis and implications of the role of 1,25$(OH)_2D_3$ in normal physiology. Mol Cell Endocrinol **72**:C57–C62.
157. Malloy PJ, Hochberg Z, Tiosano D, Pike JW, Hughes MR, Feldman D 1990 The molecular basis of hereditary 1,25-dihydroxyvitamin D_3 resistant rickets in seven related families. J Clin Invest **86**:2071–2079.
158. Hughes M, Malloy P, Kieback D, McDonnel D, Feldman D, Pike JW, O'Malley B 1989 Human vitamin D receptor mutations: Identification of molecular defects in hypocalcemic vitamin D resistant rickets. Adv Exp Med Biol **255**:491–503.
159. Ritchie HH, Hughes MR, Thompson ET, Malloy PJ, Hochberg Z, Feldman D, Pike JW, O'Malley BW 1989 An ochre mutation in the vitamin D receptor gene causes hereditary 1,25-dihydroxyvitamin D_3-resistant rickets in three families. Proc Natl Acad Sci USA **86**:9783–9787.
160. Sone T, Scott RA, Hughes MR, Malloy PJ, Feldman D, O'Malley BW, Pike JW 1989 Mutant vitamin D receptors which confer hereditary resistance of 1,25-dihydroxyvitamin D_3 in humans are transcriptionally inactive *in vitro*. J Biol Chem **264**:20230–20234.
161. Hughes MR, Malloy PJ, Kieback DG, Kesterson RA, Pike JW, Feldman D, O'Malley BW 1988 Point mutations in the human vitamin D receptor gene associated with hypocalcemic rickets. Science **242**:1702–1705.
162. Stumpf WE, Sar M, Narbaitz R, Reid FA, DeLuca HF, Tanaka Y 1980 Cellular and subcellular localization of 1,25$(OH)_2$ vitamin D_3 rat kidney: Comparison with localization of parathyroid hormone and estradiol. Proc Natl Acad Sci USA **77**:1149–1153.
163. Stumpf WE, Sar M, Reid FA, Tanaka Y, DeLuca HF 1979 Target cells for 1,25-dihydroxyvitamin D_3 in intestinal tract, stomach, kidney, skin pituitary, and parathyroid. Science **206**:1188–1190.
164. Kawashima H, Torikai S, Kurokawa K 1981 Localization of 25-hydroxyvitamin D_3 1α-hydroxylase and 24-hydroxylase along the rat nephron. Proc Natl Acad Sci USA **78**:1199–1203.
165. Kawashima H, Kurokawa K 1982 Localization of receptors for 1,25-dihydroxyvitamin D_3 along the rat nephron. Direct evidence for presence of the receptors in both proximal and distal nephron. J Biol Chem **257**:13428–13432.
166. Solvsten H, Fogh K, Svendsen M, Kristensen P, Astrom A, Kumar R, Kragballe K 1996 Normal levels of the vitamin D receptor and its message in psoriatic skin. J Invest Dermatol Symp Proc **1**:28–32.
167. Iida K, Shinki T, Yamaguchi A, DeLuca HF, Kurokawa K, Suda T 1995 A possible role of vitamin D receptors in regulating vitamin D activation in the kidney. Proc Natl Acad Sci USA **92**:6112–6116, 1995.
168. Johnson JA, Grande JP, Roche PC, Sweeney WE Jr, Avner ED, Kumar R 1995 1α,25-Dihydroxyvitamin D_3 receptor ontogenesis in fetal renal development. Am J Physiol **269**:F419–F428.
169. Kumar R 1996 Abnormalities in the vitamin D receptor in uremia. Nephrol Dial Transplant **11**(Suppl. 3):6–10.
170. Goff JP, Reinhardt TA, Beckman MJ, Horst RL 1990 Contrasting effects of exogenous 1,25-dihydroxyvitamin D [1,25$(OH)_2$D] versus endogenous 1,25$(OH)_2$D, induced by dietary calcium restriction, on vitamin D receptors. Endocrinology **126**:1031–1035.
171. Reinhardt TA, Horst RL 1990 Parathyroid hormone down-regulates 1,25-dihydroxyvitamin D receptors (VDR) and VDR messenger ribonucleic acid *in vitro* and blocks homologous up-regulation of VDR *in vivo*. Endocrinology **127**:942–948.
172. Uhland-Smith A, DeLuca HF 1993 The necessity for calcium for increased renal vitamin D receptor to 1,25-dihydroxyvitamin D. Biochim Biophys Acta **1176**:321–326.
173. Sandgren ME, DeLuca HF 1990 Serum calcium and vitamin D regulation 1,25-dihydroxyvitamin D_3 receptor concentration in rat kidney *in vivo*. Proc Natl Acad Sci USA **87**:4312–4314.
174. Huang YC, Lee S, Stolz R, Gabrielides C, Pansini-Porta A, Bruns ME, Bruns DE, Miffin TE, Pike JW, Christakos S 1989 Effect of hormones and development on the expression of the rat 1,25-dihydroxyvitamin D_3 receptor gene. J Biol Chem **264**:17454–17461.
175. Santiso-Mere D, Sone T, Hilliar IV, GM, Pike JW, McDonnell DP 1993 Positive regulation of the vitamin D receptor by its

cognate ligand in heterologous expression systems. Mol Endocrinol **7**:833–839.
176. Wiese RJ, Uhland-Smith A, Ross TK, Prahl JM, DeLuca HF 1992 Up-regulation of the vitamin D receptor in response to 1,25-dihydroxyvitamin D_3 results from ligand-induced stabilization. J Biol Chem **267**:20082–20086.
177. Russell J, Bar A, Sherwood LM, Hurtwitz S 1993 Interaction between calcium and 1,25-dihydroxyvitamin D_3 in the regulation of preproparathyroid hormone and vitamin D receptor messenger ribonucleic acid in avian parathyroids. Endocrinology **132**:2639–2644.
178. Brown AJ, Zhong M, Finch J, Ritter C, Slatopolsky E 1995 The roles of calcium and 1,25-dihydroxyvitamin D_3 in the regulation of vitamin D receptor expression by rat parathyroid glands. Endocrinology **136**:1419–1425.
179. Naveh-Many T, Marx R, Keshet E, Pike JW, Silver J 1990 Regulation of 1,25-dihydroxyvitamin D_3 receptor gene expression by 1,25-dihydroxyvitamin D_3 in the parathyroid *in vivo.* J Clin Invest **86**:1968–1975.
180. Brown AJ, Berkoben M, Ritter CS, Slatopolsky E 1992 Binding and metabolism of 1,25-dihydroxyvitamin D_3 in cultured bovine parathyroid cells. Endocrinology **130**:276–281.
181. Korkor AB 1987 Reduced binding of [^{3}H]1,25-dihydroxyvitamin D_3 in the parathyroid glands of patients with renal failure. N Engl J Med **316**:1573–1577.
182. Brown AJ, Dusso A, Lopez-Hilker S, Lewis-Finch S, Grooms P, Slatopolsky E 1989 1,25-$(OH)_2D$ receptors are decreased in parathyroid glands from chronically uremic dogs. Kidney Int **35**:19–23.
183. Koyama H, Nishizawa Y, Inaba M, Hino M, Prahl JM, DeLuca HF, Morrii H 1994 Impaired homologous upregulation of vitamin D receptor in rats with chronic renal failure. Am J Physiol **266**:F706–F712.
184. Patel SR, Ke HQ, Hsu CH 1994 Regulation of calcitriol receptor and its mRNA in normal and renal failure rats. Kidney Int **45**:1020–1027.
185. Hsu CH, Patel SR, Vanholder R 1993 Mechanism of decreased intestinal calcitriol receptor concentration in renal failure. Am J Physiol **264**:F662–F669.
186. Patel SR, Ke H-Q, Vanholder R, Loenig RJ, Hsu CH 1995 Inhibition of calcitriol receptor binding to vitamin D response elements by uremic toxins. J Clin Invest **96**:50–59.
187. Gross M, Kumar R 1990 Physiology and biochemistry of vitamin D-dependent calcium binding proteins. Am J Physiol **259**:F195–F209.
188. Desplan C, Heidmann O, Lillie JW, Auffray C, Thomasset M 1983 Sequence of rat intestinal vitamin D-dependent calcium-binding protein derived from a cDNA clone. Evolutionary implications. J Biol Chem **258**:13502–13505.
189. Desplan C, Thomasset M, Moukhtar M 1983 Synthesis, molecular cloning, and restriction analysis of DNA complementary to vitamin D-dependent calcium-binding protein mRNA from rat duodenum. J Biol Chem **258**:2762–2765.
190. Fullmer CS, Wasserman RH 1987 Chicken intestinal 28-kilodalton calbindin-D: Complete amino acid sequence and structural consideration. Proc Natl Acad Sci USA **84**:4772–4776.
191. Hoffman T, Kawakami M, Hichman AJW, Harrison JE, Dorrington KJ 1979 The amino acid sequence of porcine intestinal calcium-binding protein. Can J Biochem **57**:737–748.
192. Horrocks WD Jr 1982 Lanthanide ion probes of bimolecular structure. In Eichhorn GL, Marzitti LG (eds) Advances in Inorganic Biochemistry. Elsevier, New York, pp. 201–261.
193. Hunziker W 1986 The 28-kDa vitamin D-dependent calcium-binding protein has a six domain structure. Proc Natl Acad Sci USA **83**:7578–7582.
194. Hunziker W, Schrickel S 1988 Rat brain calbindin D28: Six domain structure and extensive amino acid homology with chicken calbindin D28. Mol Endocrinol **2**:465–474.
195. Kumar R, Wieben R, Beecher SJ 1989 The molecular cloning of the cDNA for bovine vitamin D-dependent calcium binding protein: Structure of the full-length protein and evidence for homologies with other calcium-binding proteins of the troponin-C superfamily of proteins. Mol Endocrinol **3**:427–432.
196. Lomri N, Perret C, Gouhier N, Thomasset M 1989 Cloning and analysis of calbindin-D_{28k} cDNA in its expression in the central nervous system. Gene **80**:87–98.
197. Parmentier M, Lawson DEM, Vassart G 1987 Human 27 kDa calbindin complementary DNA sequence. Evolutionary and functional implications. Eur J Biochem **170**:207–215.
198. Tankagi T, Nojiri M, Koniski K, Maruyama K, Nonomura Y 1986 Amino acid sequence of vitamin D-dependent calcium-binding protein from bovine cerebellum. FEBS Lett **201**:41–45.
199. Tsarbopoulos A, Gross M, Kumar R, Jardines I 1989 Rapid identification of calbindin-D_{28k} cyanogen bromide peptide fragments by plasma desorption mass spectrometry. Biomed Environ Mass Spectrom **18**:387–393.
200. Wilson PW, Harding M, Lawson DEM 1985 Putative amino acid sequence of chick calcium-binding protein deduced from a complementary DNA sequence. Nucleic Acids Res **13**:18867–18881.
201. Yamakuni T, Kuwano R, Odani S, Miki N, Yamaguchi Y, Takahashi Y 1986 Nucleotide sequence of cDNA to mRNA for a cerebellar Ca-binding protein, spot 35 protein. Nucleic Acids Res **14**:6728 (abstract).
202. Darwish HM, Kirsinger J, Strom M, DeLuca HF 1987 Molecular cloning of the cDNA and chromosomal gene for vitamin D-dependent calcium-binding protein of rat intestine. Proc Natl Acad Sci USA **84**:6108–6111.
203. Fullmer CS, Wasserman RH 1981 The amino acid sequence of bovine intestinal calcium-binding protein. J Biol Chem **256**:5669–5674.
204. MacManus JP, Watson DC, Yaguci M 1986 The purification and complete amino acid sequence of the 9000-M_r Ca^{2+}-binding protein from rat placenta. Identity with the vitamin D-dependent intestinal Ca^{2+}-binding protein. Biochem J **225**:585–595.
205. Szebenyl DME, Abendorf SK, Moffat K 1981 Structure of vitamin D-dependent calcium-binding protein from bovine intestine. Nature **294**:327–332.
206. Elms TN, Taylor 1987 Calbindin-D_{28k} kappa localization in rat molars during odontogenesis. J Dent Res **66**:1431–1434.
207. Dorrington KJ, Kells PIC, Hitchman AJW, Harrison JE, Hoffmann T 1978 Spectroscopic studies on the binding of divalent cations to porcine intestinal calcium-binding protein. Can J Biochem **56**:492–499.
208. O'Neil JDJ, Dorrington KJ, Hofmann T 1984 Luminescence and circular-dichroism analysis of terbium binding by pig intestinal calcium-binding protein (relative mass = 9000). Can J Biochem Cell Biol **12**:434–442.
209. Chiba K, Ohyashiki T, Mokri T 1983 Quantitative analysis of calcium binding to porcine intestinal calcium-binding protein. J Biochem (Tokyo) **93**:487–493.
210. Vogel HJ, Drakenberg T, Farsen S, O'Neil JDJ, Hofmann T 1985 Structural differences in the two calcium bidning sites of the porcine intestinal calcium binding protein: A multinuclear NMR study. Biochemistry **24**:3870–3876.
211. Shelling, JG, Sykes B 1985 ^{1}H nuclear magnetic resonance study

of the two calcium-binding sites of porcine intestinal calcium-binding protein. J Biol Chem **260**:8342–8347.
212. Bredderman PJ, Wasserman RH 1974 Chemical composition, affinity for calcium, and some related properties of the vitamin D-dependent calcium-binding protein. Biochemistry **13**:1687–1694.
213. Fullmer CS, Edelstein S, Wasserman RH 1985 Lead-binding properties of intestinal calcium-binding proteins. J Biol Chem **260**:6816–6819.
214. Gross MD, Nelsestuen GL, Kumar R 1987 Observations on the mechanism of lanthanide/calcium binding to vitamin D-dependent chick intestinal calcium binding protein: Implications regarding calcium binding protein function. J Biol Chem **262**:6539–6545.
215. Gross MD, Kumar R, Hunziker W 1988 Expression *E. coli* of full-length and mutant rat brain calbindin D28: Comparison with the purified native protein. J Biol Chem **253**:11426–11432.
216. Veenstra T, Gross MD, Hunziker W, Kumar R 1995 Identification of metal binding sites in rat brain calcium-binding protein. J Biol Chem **270**:30353–30358.
217. Kumar R, Hunziker W, Gross M, Naylor S, Londowski JM, Schaefer J 1994 The high efficient production of full-length and mutant rat brain calcium-binding proteins (calbindins-D_{28k}) in a bacterial expression system. Arch Biochem Biophys **308**: 311–317.
218. Wasserman RH, Corradino RA, Fullmer CS, Taylor AV 1974 Some aspects of vitamin D action: Calcium absorption and the vitamin D-dependent calcium-binding protein, vitamins and hormones. Adv Res Appl **32**:299–323.
219. Christakos S, Brunette MG, Norman AW 1981 Localization of immunoreactive vitamin D-dependent calcium binding protein in chick nephron. Endocrinology **109**:322–324.
220. Taylor AN, Wasserman RH 1967 Vitamin D_3 induced calcium-binding protein: Partial purification, electrophoretic visualization and tissue distribution. Arch Biochem Biophys **119**:536–540.
221. Taylor AN, McIntosh JE, Bourdeau JE 1982 Immunocytochemical localization of vitamin D-dependent calcium-binding protein renal tubules of rabbit, rat, and chick. Kidney Int **21**:765–773.
222. Roth J, Thorens B, Hunziker W, Norman AW, Orci L 1981 Vitamin D-dependent calcium binding protein: Immunocytochemical localization in chick kidney. Science **214**:197–200.
223. Roth J, Brown D, Norman AW, Orci L 1982 Localization of the vitamin D-dependent calcium-binding protein in mammalian kidney. Am J Physiol **243**:F243–F252.
224. Thomasset M, Parkes CO, Cuininier-Gleizes P 1982 Rat calcium-binding proteins: Distribution, development and vitamin D dependence. Am J Physiol **243**:E483–E488.
225. Bouhtiauy I, Lajeunesse D, Christakos S, Brunette MG 1994 Two vitamin D_3-dependent calcium binding proteins increase calcium reabsorption by different mechanisms I. Effect of CaBP 28K. Kidney Int **45**:461–468.
226. Bouhtiauy I, Lajeunesse D, Christakos S, Brunette MG 1994 Two vitamin D_3-dependent calcium bidning proteins increase calcium reabsorption by different mechanisms II. Effect of CaBP 9K. Kidney Int **45**:469–474.
227. Doucet A, Katz AI 1982 High-affinity Ca,Mg-ATPase along the rabbit nephron. Am J Physiol **242**:F346–F352.
228. Borke JL, Caride A, Verma AK, Penniston JT, Kumar R 1990 Cellular and segmental distribution of Ca^{++}-pump epitopes in rat intestines. Pfluegers Arch **417**:120–122.
229. Wasserman RH, Smith CA, Brindak ME, de Talamoni N, Fullmer CS, Penniston JT, Kumar R 1992 Vitamin D and mineral deficiencies increase the PMCa pump of chicken intestine. Gastroenterology **102**:886–894.
230. Borke JL, Caride A, Verma AK, Kelley LK, Smith CH, Penniston JT, Kumar R 1989 Ca pump epitopes in placental trophoblast membranes. Am J Physiol **257**:C341–C346.
231. Borke JL, Caride AJM, Yaksh TL, Penniston JT, Kumar R 1989 Cerebrospinal fluid Ca homeostasis: Evidence for a $PMCa^{++}$-pump in mammalian choroid plexus. Brain Res **489**:355–360.
232. Wasserman RH, Smith CA, Smith CM, Brindak ME, Fullmer CS, Krook L, Penniston JT, Kumar R 1991 Immunohistochemical localization of a Ca pump and calbindin-D_{28k} in the oviduct of the laying hen. Histochemistry **96**:413–418.
233. Borke JL, Eriksen EF, Minami J, Keeting P, Mann KG, Penniston JT, Riggs BL, Kumar R 1988 Epitopes to the human erythrocyte Ca^{++},Mg^{++}-ATPase pump in human osteoblast-like cell PMs. J Clin Endocrinol Metab **67**:1299–1304.
234. Kumar R, Haugen JD, Medora M, Yamakawa K, Hruska KA 1993 Detection and partial sequencing of a $PMCa^{2+}$ pump in the chicken osteoclast. J Bone Miner Res **8**:S386.
235. Cai Q, Chandler JS, Wasserman RH, Kumar R, Penniston JT 1993 Vitamin D and adaptation to dietary calcium and phosphate deficiencies increase intestinal plasma membrane calcium pump gene expression. Proc Natl Acad Sci USA **90**:1345–1349.
236. Wasserman RH, Chandler JS, Meyer SA, Smith CA, Brindak ME, Fullmer CS, Penniston JT, Kumar R 1992 Intestinal calcium transport and calcium extrusion processes at the basolateral membrane. J Nutr **122**:662–671.
237. Morgan DW, Welton AF, Heick AE, Christakos S 1986 Specific *in vitro* activation of Ca,Mg-ATPase by vitamin D-dependent rat renal calcium binding protein (calbindin D_{28K}). Biochem Biophys Res Commun **138**:547–553.
238. Walters JR 1989 Calbindin-D9k stimulates the calcium pump in rat enterocyte basolateral membranes. Am J Physiol **256**:G124–G128.
239. Iida K, Taniguchi S, Kurokawa K 1993 Distribution of 1,25-dihydroxyvitamin D_3 receptor and 25-hydroxyvitamin D_3-24-hydroxylase mRNA expression along rat nephron segments. Biochem Biophys Res Commun **194**:659–664.

CHAPTER 18

Vitamin D and Bone Development

RENÉ ST-ARNAUD Genetics Unit, Shriners Hospital for Children and Departments of Surgery and Human Genetics, McGill University, Montréal, Québec, Canada

FRANCIS H. GLORIEUX Genetics Unit, Shriners Hospital for Children and Departments of Surgery and Pediatrics, McGill University, Montréal, Québec, Canada

I. INTRODUCTION

Vitamin D is a known regulator of osteoblastic differentiation and function. It is also involved in the regulation of osteoclastogenesis. Thus, vitamin D plays a key role in bone development. This chapter first reviews the two modes of ossification observed during embryogenesis and then discusses the role of vitamin D during bone development. The effects of vitamin D on bone formation secondary to the modulation of mineral homeostasis are briefly mentioned, as they are presented elsewhere in this book (see Chapters 41, 42, 46, 47, 48, and 54), while the direct effects of both 1,25-dihydroxyvitamin D [1,25(OH)$_2$D] and 24,25-dihydroxyvitamin D [24,25(OH)$_2$D] are discussed in more detail.

II. OVERVIEW OF SKELETAL DEVELOPMENT

Skeletal development is dependent on the properly regulated differentiation, function, and interaction of the component cell types of the skeleton, which include chondrocytes, the cartilage-forming cells; osteoblasts, the bone-forming cells; and osteoclasts, the bone-resorbing cells. Chondrocytes and osteoblasts share a common mesenchymal origin [1], whereas osteoclasts derive from hematopoetic lineages [2].

Two mechanisms of bone formation have classically been distinguished: intramembranous bone formation (flat bones) and endochondral bone formation (long bones) [3]. The essential difference between them is the presence or absence of a cartilaginous phase. Intramembranous bone formation occurs when mesenchymal precursor cells proliferate and produce a mucoprotein matrix, called osteoid, in which collagen fibers are embedded. The mesenchymal cells subsequently differentiate directly into osteoblasts that begin to deposit inorganic crystals of calcium phosphate on, between, and within the collagen fibers, forming an immature bone tissue called woven bone. Woven bone is characterized by irregular bundles of randomly oriented collagen fibers and an abundance of partially calcified immature new bone. At later stages, the woven bone is progressively remodeled to mature, lamellar bone.

Endochondral bone formation entails the conversion of a cartilaginous template into bone. Mesenchymal cells condense and differentiate into chondrocytes

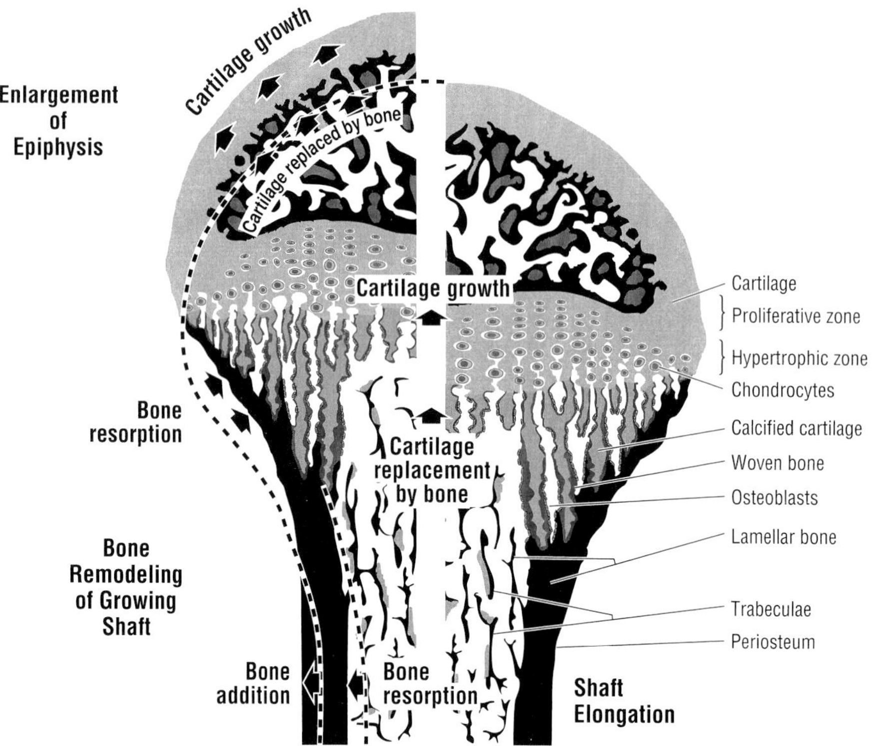

FIGURE 1 Endochondral ossification in long bones. Longitudinal bone growth occurs at the epiphyseal growth plate where chondrocytes appear in regular columns that initially proliferate and progressively hypertrophy. The mineralized cartilaginous matrix is resorbed by osteoclasts, and osteoblasts differentiate to deposit woven bone on top of the calcified cartilage remnants.

that secrete the cartilaginous matrix (Chapter 26). This embryonic cartilage is avascular, and, during its early development, a ring of woven bone is formed by intramembranous ossification in the future midshaft area. This calcified woven bone is then invaded by vascular tissue, and osteoblasts are recruited to replace the cartilage scaffold with bone matrix and excavate the hematopoietic bone marrow cavity. At the extremities of long bones (epiphyses), longitudinal growth occurs by a similar process of endochondral ossification at the growth plates (Fig. 1). Growth plate chondrocytes appear in regular columns that initially proliferate and progressively hypertrophy. The cartilaginous matrix becomes mineralized just below the hypertrophic zone of the growth plate and the chondrocytes then die. The calcified cartilage is resorbed by osteoclasts, and osteoblasts differentiate from mesenchymal cells that were brought with the invading vascular tissue. Woven bone is then deposited on top of the calcified cartilage remnants. Still lower in the growth plate, the mixture of woven bone and calcified cartilage is further remodeled and replaced by mature lamellar bone trabeculae.

III. MODULATION OF MINERAL HOMEOSTASIS BY VITAMIN D

An analysis of the effects of vitamin D on bone development are compounded by the well-established role of $1,25(OH)_2D$ in the regulation of mineral homeostasis. This section briefly mentions the abnormalities of bone development observed in the hypo- and hypervitaminosis D state. These conditions mostly represent indirect effects of vitamin D that are secondary to concomitant alterations in the concentrations of calcium and/or phosphate in the extracellular space.

A. Vitamin D Deficiency during Embryogenesis

The importance of vitamin D during embryogenesis was demonstrated in classic studies using the chicken as an experimental animal: embryonic chick development and normal egg hatchability both show essential requirements for vitamin D metabolites [4–6]. Some of the experiments in the chicken have highlighted a putative role for $24,25(OH)_2D$ in embryonic development and are detailed in Section IV,B,2 below. Studies dealing with the role of $1,25(OH)_2D$ are discussed here.

Vitamin D deficiency can be induced in chicken embryos by maintaining laying hens on diets containing $1,25(OH)_2D_3$ as the sole source of vitamin D [4–6]. The resulting eggs have very low hatchability [4,5] and show abnormal bone development [4]. The bones from the vitamin D-deficient embryos exhibit wide osteoid seams typical of rickets and osteomalacia (see below) [6,7]. Surprisingly, some of the impaired development can be rescued by direct injection of $1,25(OH)_2D_3$ into the eggs [4], suggesting that parts of the observed defects are caused by insufficient transfer of $1,25(OH)_2D_3$ from hen to egg. The injection of $1,25(OH)_2D_3$ also corrects the hypocalcemia in the vitamin D-deficient embryos, supporting the view that the changes in bone mineralization and growth are secondary to alterations in the circulating concentrations of calcium and/or phosphate. Experiments in which hypervitaminosis D has been induced in chicken embryos appear to confirm this hypothesis. Although high doses of $1,25(OH)_2D_3$ stimulate the rapid growth of abundant undermineralized trabeculae in chicken embryonic bone [8,9] (see also Section III,C below), this effect is not observed in "shell-less" cultured embryos [10]. Thus, the inability of the $1,25(OH)_2D_3$ to mobilize calcium from the shell in the shell-less cultured embryos prevented the bone defects [10].

Vitamin D-deficient rats can develop and reproduce [11]. Examination of bones from late-gestation fetuses from vitamin D-deficient mothers revealed a slight increase in the amount of unmineralized osteoid on trabecular bone surfaces; the fetal bones otherwise appeared normal [12]. This finding is surprising considering the severe skeletal changes apparent in the vitamin D-depleted mothers [13,14]. Plasma calcium and phosphate concentrations were virtually normal in the vitamin D-deficient pups [12]; this would suggest that calcium transport across the placenta to the fetus is independent of maternal plasma calcium concentrations [15,16] and is not impaired by the absence of vitamin D. The engineering of a mutant strain of mice deficient for the vitamin D receptor gene [16a] will permit unequivocal assessment of the role of $1,25(OH)_2D$-mediated signaling on embryonic bone development.

The effect of vitamin D deprivation is more evident after parturition. Circulating levels of calcium and phosphate steadily decline, and the characteristic skeletal changes associated with rickets and osteomalacia (see below) progressively appear [12,17].

B. Vitamin D Deficiency in Early Postnatal Development: Rickets and Osteomalacia

Separate chapters cover the pathophysiology of rickets in detail (see Chapters 41, 42, 47, and 48). Abnormalities at any step of vitamin D metabolism, from dietary deficiency through metabolic errors to end-organ resistance due to mutations in the vitamin D receptor (VDR), all lead to rickets and osteomalacia. In infants, the symptoms include failure to thrive, bone pain, deformity of the thorax, softening of the skull, and enlargement of wrists and ankles. Delayed diagnosis or treatment results in very severe deformities of the long bones and the spine, together with generalized muscle weakness simulating myopathy [18]. The characteristic bone lesions result mainly from inadequate mineralization. At the growth plate, calcium fails to be deposited in the matrix at the zone of provisional calcification; consequently, this zone fails to be resorbed. Since the cartilage continues proliferating but is not resorbed at a corresponding rate, it increases in width. There is a concomitant irregular proliferation of undermineralized bone trabeculae in the region directly underlying the growth plate cartilage.

Mineralization is also abnormal in the diaphysis, affecting both cortical and trabecular bone, a condition defined as osteomalacia. The mineralization defect is characterized by a widening of the osteoid seams, that is, an increase in the amount of unmineralized bone matrix located between the layer of osteoblasts and the fully mineralized bone.

In most instances, treatment with vitamin D metabolites cures the mineralization defects [18]. Although vitamin D itself or polar metabolites such as 25-hydroxyvitamin D (25OHD) can be used to treat nutritional rickets, the hereditary disease pseudovitamin D deficiency rickets (PDDR; see Chapter 47) requires administration of $1,25(OH)_2D$. These observations suggested that vitamin D is required for normal bone and cartilage mineralization; however, since the healing of the bone lesions is preceded by the normalization of blood concentrations of calcium and phosphate, it was rapidly appreciated that the main mechanism by which vitamin D prevents or heals rickets is through the restoration of normal circulating levels of calcium and phosphate. This hypothesis is supported by studies in patients with hereditary vitamin D-resistant rickets (HVDRR; Chapter 48). These individuals carry mutations in the VDR [19] that mediates the genomic actions of vitamin D. They therefore exhibit end-organ resistance to vitamin D and do not respond to treatment with vitamin D metabolites [18]. Alternative treatment has been devised in the form of intravenous calcium infusions with oral phosphate supplementation. Several studies based on this regimen have shown that normalization of the blood levels of calcium and phosphate can cure the rickets and osteomalacia in these patients [20–23]. This demonstrates that normal mineralization can be achieved in the absence of an intact VDR–effector system and suggests that the effects of vitamin D on biomineralization are indirect.

Studies in experimental animals support the clinical investigations. Growth plates isolated from rachitic animals mineralize when cultured in normal serum or solutions containing normal concentrations of calcium and phosphate [24]. Moreover, cartilage and bone mineralization was normal in vitamin D-deficient rats in which the circulating levels of phosphate and calcium were normalized by intravenous infusion [25].

Despite the above-mentioned results, extensive research has been conducted to evaluate the role of vitamin D metabolites on bone development, growth, and mineralization under conditions of normal calcium and phosphate concentrations. These studies utilized varied experimental protocols in a number of species and cell types. Overall, they support a direct role for vitamin D metabolites in bone development. The findings from these experiments are detailed in Section IV below.

C. Hypervitaminosis D

Vitamin D supplementation is indicated in a number of clinical conditions including nutritional deficiency, malabsorption, hypoparathyroidism, renal osteodystrophy, drug-induced osteomalacia, and refractory rickets [26] (see also Chapter 39). Treatment with vitamin D metabolites such as 25OHD or $1,25(OH)_2D$ is less frequently associated with intoxication than treatment with cholecalciferol and ergocalciferol [27]. This is most likely due to the longer metabolic half-life of the parent vitamin compared to its metabolites and the fact that it can be stored in adipose, hepatic, and muscle tissue for many months [28]. Chronic hypervitaminosis D leads to increased absorption of calcium and phosphate from the gut and concomitant resorption of mineral from bone. Consequently, hypercalcemia, hyperphosphatemia, calciuria, and phosphaturia develop. This condition results in metastatic calcification occurring principally in kidney (nephrocalcinosis) and large blood vessels [27]. Surprisingly, chronic vitamin D intoxication in patients does not always result in significant alterations in trabecular bone mass [27].

The skeletal changes associated with hypervitaminosis D are more readily evaluated in experimental animals. Whereas classic studies were conducted by administering large doses of vitamin D [29], identical results were obtained using 25OHD or $1,25(OH)_2D$. Chronic administration of physiological or pharmacological doses of $1,25(OH)_2D$ in vitamin D-replete rats leads to increased trabecular bone mass accompanied by marked osteoid accumulation [30–34]. Histomorphometric analysis of the bone from treated animals revealed that the hyperosteoidosis was due primarily to impaired bone mineralization [33,34]. Although similar responses are observed in rats and chicks [8,9,30–34], it is worth mentioning that significant species differences exist in the response to high doses of vitamin D or its active metabolites. For example, no increase in bone mass was observed in a patient that suffered vitamin D intoxication for 19 years [27]. Similarly, treatment with high doses of $1,25(OH)_2D_3$ in mice resulted in decreased matrix apposition rates with concomitant enhancement of the mineral apposition rates, resulting in a net reduction in the amount of osteoid [35]. The studies in mice also showed evidence of enhanced bone resorption in the form of an increased number of acid phosphatase-stained osteoclasts [35]. In experiments utilizing rats as the model species, osteoclastic resorption becomes predominant only in the later stages of the chronic treatment or when the animals are fed diets that are poor in phosphate or contain a large calcium/phosphate ratio [29].

Thus, both vitamin D deficiency and vitamin D intoxication lead to abnormal bone development, suggesting that vitamin D is required for normal bone and cartilage mineralization. As previously discussed, the observed defects implicate indirect effects of vitamin D related to impaired mineral homeostasis. However, experimental evidence also supports a direct role for vitamin D and its metabolites in the regulation of bone development. We now review those findings.

IV. DIRECT EFFECTS OF VITAMIN D METABOLITES ON BONE

A. $1,25(OH)_2D$

1. VDR Ontogeny during Bone Development

A direct role for vitamin D in bone development is first supported by the fact that bone cells express VDR. Several studies have documented the expression of the VDR in osteoblasts from neonatal or late gestational rats [36,37]. More recently, investigators have analyzed the expression pattern of the VDR throughout the course of appendicular and axial skeletal development in rats. Prior to the onset of ossification, the VDR was detected in condensing mesenchyme in the vertebral column and limbs at day 13 of gestation [38], supporting a role for $1,25(OH)_2D$ in the regulation of bone cell differentiation and the proliferation of preosteogenic mesenchyme. The VDR was first detected in osteoblasts at sites of endochondral ossification in the vertebral column on embryonic day 17; osteoblastic expression of the VDR was also observed at sites of intramembranous ossification in the calvaria at the same time. Expression of the VDR was maintained in bone forming cells until birth; similarly, the VDR was detected in proliferating and hypertrophic chondrocytes from embryonic day 17 onward [38]. A number of investigators have studied the VDR in fetal and neonatal calvarial cells from both mice and rats (see Chapter 12).

Interestingly, Johnson *et al.* also claimed to detect VDR in large, multinucleated cells reminiscent of osteoclasts in developing bone [38]. Earlier studies have reported the expression of the VDR in mononucleate osteoclastic precursors but failed to detect the protein in differentiated cells [39]. However, analysis based on the sensitive technique of reverse transcription coupled to the polymerase chain reaction (RT-PCR) has shown that the mRNA for the VDR is expressed in multinucleated osteoclasts [40]. More studies will be required to determine the functional significance of the detection of the VDR mRNA and protein in differentiated osteoclasts.

2. *In Vitro* Effects

Consistent with the detection of VDR expression in osteoblasts, a number of effects have been described when primary cultures of bone cells or established osteoblastic cell lines are treated with $1,25(OH)_2D$ *in vitro.* Vitamin D treatment inhibits type I collagen expression by osteoblasts [41–45]; this is a direct effect mediated at the level of gene transcription [46] (see also Chapter 12). The regulatory element within the alpha 1 chain of collagen type I [$\alpha_1(I)$] mediating this transcriptional inhibition is currently under investigation [47].

Contrary to its inhibitory effect on type I collagen expression, $1,25(OH)_2D_3$ enhances the expression of genes that are specific markers of osteoblastic differentiation. These include alkaline phosphatase [48–51], osteopontin [52,53], osteocalcin [54,55], and matrix-Gla protein [56–58] (see Chapter 12). Other responses of osteoblasts to vitamin D treatment *in vitro* include an increase in the number of receptor molecules for insulin-like growth factor I (IGF-I) [59]. Moreover, vitamin D treatment has been shown to induce the expression of VDR mRNA and protein in bone cells [60] (see Chapter 11). The stimulation of the expression of osteocalcin and osteopontin occurs at the transcriptional level, and the molecular mechanisms involved have been studied in considerable detail. Functional vitamin D response elements (VDREs) have been identified and characterized for the promoters of the osteocalcin [55,61] and osteopontin [62] genes. Overall, the direct effects of $1,25(OH)_2D$ on bone cells are consistent with the induction of differentiation of osteoblasts *in vitro* (see Chapter 20). Accordingly, the metabolite has been shown to stimulate mineralization in cultures of clonal osteoblast-like cells [63]. This effect appears to implicate genomic responses to vitamin D as the multifunctional protein calreticulin, which binds to the VDR DNA binding domain to inhibit its transcriptional activating function [64], perturbs the vitamin D response, and prevents mineralization in cultured osteoblasts [65].

A survey of the literature describing the effects of vitamin D on osteoblasts *in vitro* reveals discrepancies, with some groups reporting stimulatory effects on type I collagen synthesis [50,51] while the majority of investigators have observed decreased expression of collagen following vitamin D treatment [41–45] (discussed in detail in Chapter 12). Similarly, both increases [50,51] and decreases [49,57] in alkaline phosphatase expression have been reported in vitamin D-treated osteoblasts. Although some of these discrepancies could be due to clonal variations or duration of treatment, a significant factor involved in the response of osteoblasts to vitamin D appears to be the differentiated state of the osteoblast. Cultured bone cells undergo an ordered developmental sequence characterized by a precise temporal pattern of expression of cell growth- and differentiation-related genes leading to the progressive acquisition of the differentiated osteoblastic phenotype [66]. Owen *et al.* [67] have performed a careful study evaluating the effects of vitamin D treatment on gene expression in cultured osteoblasts as a function of the differentiated state of the bone cells. Their results show that vitamin D can both positively and negatively regulate the expression of osteoblastic phenotype markers. In early cultures characteristic of the preosteoblastic phenotype, vitamin D treatment inhibited type I collagen and alkaline phosphatase mRNA levels. In contrast, the expression of both genes was stimulated in late-stage cultures [67]. Similar variations were noted for the response of additional markers to vitamin D treatment, with matrix-Gla protein and osteopontin expression being strongly stimulated by the metabolite during the early proliferative period but exhibiting a blunted response at later stages [67]. Finally, osteocalcin was not expressed and did not respond to vitamin D exposure in early bone cell cultures but showed a strong induction by vitamin D in mid to late cultures [67]. These results can thus be used to reconcile the data obtained by different investigators and support the concept of pleiotropic direct effects of vitamin D on bone cells that are modulated as a function of the progressive growth and maturation of the osteoblast.

3. *In Vivo* Effects

As previously noted, the interpretation of studies aimed at determining the role of $1,25(OH)_2D$ in bone development *in vivo* has been complicated because vitamin D-deficient animals were invariably hypocalcemic. Several experimental manipulations have been devised to prevent severe hypocalcemia in vitamin D-deficient animals. These include high calcium-containing diets [68], continuous infusions with calcium and phosphate solutions [25,69], and supplementation with high calcium, vitamin D-free diets containing lactose [70–72].

Most of these studies concluded that the effects of vitamin D are indirect and mediated through the well-documented role of $1,25(OH)_2D$ on the regulation of mineral homeostasis, as evidenced by normal body growth rates and normal skeletal bone growth and mineralization in vitamin D-deficient but normocalcemic animals [25,68,69,71,72]. However, some of the experimental protocols used have provided evidence for direct actions of vitamin D metabolites on bone development. The effects ascribed to $24,25(OH)_2D$ are discussed below. An interesting finding related to $1,25(OH)_2D$ was obtained using heterotopic osteogenesis as an experimental system. Osteogenesis, defined as the frequency of appearance and amount of newly formed bone, was markedly decreased in normocalcemic, vitamin D-deficient rodents when bone formation was induced by intramuscular transplantation of living epithelium [73]. Interestingly, the reduced amount of newly formed heterotopic bone mineralized normally [73]. These data demonstrate that normal $1,25(OH)_2D$ concentrations are required for *de novo* bone formation.

B. $24,25(OH)_2D$

The high concentration of $24,25(OH)_2D_3$ in biological fluids as well as the hormonal regulation of its synthesis have long suggested that this metabolite of vitamin D may have revelant biological activity. However, data from experiments addressing this putative function have remained controversial.

1. *In Vitro* Effects

Studies using organ cultures or isolated cells have revealed that chondrocytes may represent specific target cells for the effects of $24,25(OH)_2D$ (see Chapter 26). The response of epiphyseal growth plate chondrocytes to vitamin D metabolites is a function of the zone of maturation from which the cells were originally derived. Thus, resting zone chondrocytes respond to $24,25(OH)_2D_3$, whereas chondrocytes from the growth zone are primarily responsive to $1,25(OH)_2D_3$. Measured responses of resting zone chondrocytes to $24,25(OH)_2D_3$ include the inhibition of cell proliferation [74], increased collagen and noncollagen protein synthesis [74], and regulation of plasma membrane and matrix vesicle enzyme activity, such as alkaline phosphatase, phospholipase A_2, and protein kinase C [75,76]. Moreover, it has been shown that chondrocytes from the resting and proliferative zones of the growth plate can hydroxylate $25OHD_3$ to form $24,25(OH)_2D_3$ and that this process can be regulated by dexamethasone or transforming growth factor-$\beta 1$ [77]. Overall, these observations suggest an autocrine regulation of chondrocytic differentiation and function in which $24,25(OH)_2D$ would induce immature resting zone chondrocytes to differentiate along the endochondral pathway to acquire a growth zone-like phenotype [78]. Whether these effects are relevant *in vivo* remains to be clarified.

Osteoclastogenesis may also be regulated by $24,25(OH)_2D$. A report using *in vitro* fusion of hemopoietic progenitor cells into osteoclast-like tartrate-resistant acid phosphatase-positive multinucleated cells has shown that treatment with $24,25(OH)_2D_3$ inhibited the $1,25(OH)_2D_3$-induced cell fusion [79]. The results suggest a specific inhibitory effect of $24,25(OH)_2$ on the formation and function of osteoclastic cells. Again, the physiological relevance of these observations remain unclear.

2. *In Vivo* Effects

As previously mentioned, the development of vitamin D-deficient chick embryos is impaired, and a large proportion of the eggs from the vitamin D-depleted hens fail to hatch [4–6]. Some of the observed defects can be corrected by direct injection of $1,25(OH)_2D_3$ into the vitamin D-deficient eggs [4]; however, this rescue is not complete [4], suggesting that additional vitamin D metabolites may be essential for embryonic development. Indeed, it has been shown that normal egg hatchability requires both $1,25(OH)_2D$ and $24,25(OH)_2D$ [5,6]. The unnatural epimer $24S,25(OH)_2D_3$ cannot substitute for $24,25(OH)_2D_3$ in this system [80], supporting the view that $24,25(OH)_2D$ may play a unique role in egg hatchability.

In mammals, the use of vitamin D analogs fluorinated at position 24 (thus preventing further hydroxylation at that position) has revealed that these compounds produce the same biological response as those resulting from 25OHD concerning intestinal calcium transport, the mobilization of calcium from bone, and the mineralization of vitamin D-deficient bone [81]. The mineralized bone is histologically indistinguishable whether it is healed by the fluoro compound or by 25OHD [82]. Furthermore, when animals supported with the fluorinated analog are allowed to breed, the pups that are born show no developmental abnormalities, and the mineral content of their bones is normal [83–85]. These results have been used to substantiate the notion that 24-hydroxylation does not play a significant role in bone growth and development in the rat.

However, it can be argued that the fluoro compound might in fact act as an agonist of $24,25(OH)_2D$ since the carbon at position 24 is indeed occupied. No data, however, support this view. Moreover, the conclusions derived from these studies rely on the certainty that the animals were totally vitamin D deficient, which may be disputed. The controversy is further fueled by studies that demonstrate that treatment with pharmacological doses of $24,25(OH)_2D_3$ increases bone mass in vitamin D-replete rats [86], rabbits [87,88], and dogs [89,90]. These experiments support a clear pharmacologic activity for $24,25(OH)_2D_3$, although they do not establish a putative physiological activity of the metabolite.

To address the putative physiological role of $24,25(OH)_2D_3$ during bone development, we have used the powerful technique of homologous recombination in embryonic stem (ES) cells [91,92] to engineer a strain of mice deficient in the 24-hydroxylase enzyme. The targeted mutation effectively deleted the heme binding domain of the cytochrome P450 enzyme, ensuring that the mutated allele could not produce a functional protein. Northern blot analysis of mRNA from kidneys of homozygotes and heterozygous control littermates demonstrated that the homozygotes do not express any detectable levels of 24-hydroxylase mRNA (not shown), showing that we have engineered a true null mutation [92a]. The survival of some 24-hydroxylase mutant animals to adulthood has allowed us to make some novel and fascinating observations concerning the loss of 24-hydroxylation during development. When adult mutant homozygous females are mated to heterozygous males, litters comprise an equal proportion of mutant homozygotes and control heterozygous littermates. Because of the homozygous genotype of the female, homozygous embryos are completely deprived of vitamin D metabolites hydroxylated at position 24 during development. Heterozygous littermates can synthesize those metabolites because they carry one functional allele of the 24-hydroxylase gene. Bone development is abnormal in homozygous mutants born of homozygous females (Fig. 2, bottom; see color insert). Histological examination of the bones from these animals revealed an accumulation of osteoid at sites of intramembranous ossification, particularly the calvaria and the exocortical (periosteal) surface of long bones (Fig. 2, bottom). Control heterozygote littermates showed normal bone structure (Fig. 2, top). These results show that a complete absence of vitamin D metabolites hydroxylated at position 24 during embryogenesis leads to abnormal bone structure and suggest a key role for $24,25(OH)_2D_3$ in the developmental regulation of intramembranous ossification. Interestingly, the growth plates from these mutant animals were normal (not shown), suggesting that $24,25(OH)_2D_3$ is not a major regulator of chondrocyte maturation *in vivo*.

We do not believe that the observed mineralization defect is due to the loss of the C-24 oxidation pathway and therefore a consequence of hypervitaminosis D. Treatment with high doses of vitamin D perturbs mineralization in rats [30–34] but has little or no consequence on bone structure in mice [35]. Moreover, the defects observed in rats treated with high doses of vitamin D were localized at the endosteal surface of long bones as well as in bone trabeculae [30–34], sites that are absolutely normal in 24-hydroxylase mutant animals (Fig. 2, bottom). It is also very unlikely that the observed phenotype is secondary to a perturbation in mineral homeostasis, as the growth plate is normal in homozygotes. Moreover, preliminary data reveal equivalent blood calcium levels in homozygotes compared to heterozygotes. The most likely hypothesis is that $24,25(OH)_2D_3$ is required for normal intramembranous ossification. It is interesting to note that chicken embryos from hens maintained on $1,25(OH)_2D_3$ as their sole source of vitamin D have defective mandibles [14], which are bones that form by intramembranous ossification.

Homozygous mutant animals born from heterozygous females do not exhibit the bone phenotype (not shown [92a]. We surmise that this is because the heterozygous female can supply the $24,25(OH)_2D_3$ across the fetoplacental barrier.

Although these data are highly suggestive of a role for $24,25(OH)_2D_3$ in bone development, caution is still necessary. Our findings are at odds with the findings in cartilage cells *in vitro* [74–78] since the growth plates appear normal in homozygous mutants. Also, the localized changes within bone remain difficult to explain. Finally, a potential role of 24-hydroxylase affecting $1,25(OH)_2D_3$ action rather than a role for $24,25(OH)_2D_3$ remains possible. Experiments attempting to rescue the mutant phenotype by exogenous administration of $24,25(OH)_2D_3$ to gestating homozygous mutant females are in progress and will help clarify the interpretation of the observed results.

3. A PUTATIVE RECEPTOR FOR $24,25(OH)_2D_3$?

Our results showing abnormal bone development in mice deficient for the 24-hydroxylase enzyme suggest the possibility that a specific receptor for $24,25(OH)_2D_3$ might exist that would mediate the effects of the vitamin D metabolite. Moreover, our data have identified probable sites of expression for the hypothetical receptor, namely, in bones that form via the process of intramembranous ossification such as the calvaria. Alternatively, 24,25$(OH)_2D_3$ might act in some manner via the VDR. Autoradiographic and binding saturation analyses support the possibility of a hypothetical 24,25$(OH)_2D_3$ receptor in the epiphyses of rat bone [93,94], in cultured growth plate chondrocytes [95], in chick embryo limbbud cultures [96], and in the parathyroid gland [97]. It is important to note here, however, that these studies were highly equivocal, and virtually no progress has been made in this area for over a decade. Thus, the existence of a 24,25$(OH)_2D_3$ receptor has yet to be established.

By analogy with the vitamin D receptor that binds 1,25$(OH)_2$D, we hypothesize that if a 24,25$(OH)_2$D receptor exists, it will be a member of the nuclear hormone receptor superfamily [98]. More than 150 members of this superfamily have now been identified [98]. Although it is possible that the 24,25$(OH)_2$D receptor has already been cloned as a so-called orphan receptor for which no ligand has yet been identified, it might be impractical to screen all the orphan receptors cloned to date for putative 24,25$(OH)_2$D binding activity. Moreover, many of the orphan receptors were not cloned from tissues that are likely to express the 24,25$(OH)_2$D receptor (i.e., the periosteum of long bones or the calvaria); in fact, many of the orphan receptors were cloned from *Caenorhabditis elegans, Xenopus,* or HeLa cells [98]. Thus, it is likely that the putative 24,25$(OH)_2$D receptor has not yet been identified, and a cloning strategy using cDNA derived from likely sites of expression appears to be the most straightforward way to identify and characterize the receptor at the molecular level. Such studies are in progress in our laboratory and should unravel novel molecular mechanisms implicated in the regulation of bone development by vitamin D metabolites.

References

1. Bruder SP, Fink DJ, Caplan AI 1994 Mesenchymal stem cells in bone development, bone repair, and skeletal regeneration therapy. J Cell Biochem **56**:283–294.
2. Suda T, Udagawa N, Nakamura I, Miyaura C, Takahashi N 1995 Modulation of osteoclast differentiation by local factors. Bone **17**:87S–91S.
3. Ham AW 1974 Histology. Lippincott, Philadelphia, Pennsylvania.
4. Sunde ML, Turk CM 1978 The essentiality of vitamin D metabolites for embryonic chick development. Science **200**:1067–1069.
5. Henry HL, Normal AW 1978 Vitamin D: Two dihydroxylated metabolites are required for normal chicken egg hatchability. Science **201**:835–837.
6. Ornoy A, Goodwin D, Noff D, Edelstein S 1978 24,25-Dihydroxyvitamin D is a metabolite of vitamin D essential for bone formation. Nature **276**:517–519.
7. Narbaitz R, Tsang CPW 1989 Vitamin D deficiency in the chick embryo. Effects on prehatching motility and on the growth and differentiation of bone, muscles and parathyroid glands. Calcif Tissue Int **44**:348–355.
8. Narbaitz R, Tolnai S 1978 Effects produced by the administration of high doses of 1,25-dihydroxycholecalciferol to the chick embryo. Calcif Tissue Res **26**:221–226.
9. Narbaitz R, Fragiskos B 1984 Hypervitaminosis D in the chick embryo: Comparative study of various vitamin D_3 metabolites. Calcif Tissue Int **36**:392–400.
10. Narbaitz R 1992 Effects of vitamins A, C, D, and K on bone growth mineralization, and resorption. In: Hall BK (ed) Bone, Vol. 4: Bone Metabolism and Mineralization. CRC Press, Boca Raton, Florida, pp. 141–170.
11. Halloran BP, DeLuca HF 1979 Vitamin D deficiency and reproduction in rats. Science **204**:73–74.
12. Miller, SC, Halloran BP, DeLuca HF, Lee WSS 1983 Studies on the role of vitamin D in early skeletal development, mineralization, and growth in rats. Calcif Tissue Int **35**:455–460.
13. Halloran BP, DeLuca HF 1980 Skeletal changes during pregnancy and lactation: The role of vitamin D. Endocrinology **107**:1923–1929.
14. Miller SC, Halloran BP, DeLuca HF, Lee WSS 1982 Role of vitamin D in maternal skeletal changes during pregnancy and lactation: A histomorphometric study. Calcif Tissue Int **34**:245–252.
15. Delivoria-Papdopoulos M, Battaglia FC, Bruns PD, Meschia G 1967 Total, protein bound, and ultra-filterable calcium in maternal and fetal plasmas. Am J Physiol **213**:363–366.
16. Twardock AR, Austin MK 1970 Calcium transfer in perfused guinea pig placenta. Am J Physiol **219**:540–545.

16a. Yoshizawa T, Handa Y, Uematsu Y, Sekine K, Takeda S, Yoshihara Y, Kawakami T, Sato H, Alioka K, Tanimot K, Fukamizu A, Masushige S, Matsumoto T, Kato S 1996 Disruption of the vitamin D receptor (VDR) in the mouse. J Bone Miner Res **11**(Suppl. 1):S124.

17. Halloran BP, DeLuca HF 1981 Effect of vitamin D deficiency on skeletal development during early growth in the rat. Arch Biochem Biophys **209**:7–14.
18. Glorieux FH 1991 Rickets. Raven, New York.
19. Hughes MR, Malloy PJ, Kieback DG, Kesterson RA, Pike JW, Feldman D, O'Malley BW 1988 Point mutations in the human vitamin D receptor gene associated with hypocalcemic rickets. Science **242**:1702–1705.
20. Balsan S, Garabedian M, Larchet M, Gorski A, Cournot G, Tau C, Bourdeau A, Slive C, Ricour C 1986 Long-term nocturnal calcium infusions can cure rickets and promote normal mineralization in hereditary resistance to 1,25-dihydroxyvitamin D. J Clin Invest **77**:1661–1667.
21. Weisman Y, Bab I, Gazit D, Spirer Z, Jaffe M, Hochberg Z 1987 Long-term intracaval calcium infusion therapy in end organ resistance to 1,25-dihydroxyvitamin D. Am J Med **83**:984–990.
22. Bliziotes M, Yergey AL, Nanes M, Muenzer J, Begley M, Vieira N, Kher K, Brandi M, Marx SJ 1988 Absent intestinal response to calciferols in hereditary resistance to 1,25-dihydroxyvitamin D: Documentation and effective therapy with high dose intravenous calcium infusions. J Clin Endocrinol Metab **66**:294–300.
23. Al-Aqeel A, Ozand P, Sobki S, Sewairi W, Marx S 1993 The combined use of intravenous and oral calcium for the treatment of vitamin D dependent rickets type II (VDDR II). Clin Endocrinol **39**:229–237.
24. Anderson HC, Cecil R, Sajdera SW 1975 Calcification of rachitic rat cartilage in vitro by extracellular matrix vesicles. Am J Pathol **79**:237–245.

25. Underwood JL, DeLuca HF 1984 Vitamin D is not directly necessary for bone growth mineralization. Am J Physiol **246**:E493–498.
26. Bikle DD 1993 Formulary of drugs commonly used in treatment of mineral disorders. In: Favus MJ, Christakos S, Gagel RF, Kleerekoper M, Langman CB, Shane E, Stewart AF (eds) Primers on the Metabolic Bone Diseases and Disorders of Mineral Metabolism, 2nd Ed. Raven, New York, pp. 414–416.
27. Allen SH, Shah JH 1992 Calcinosis and metastatic calcification due to vitamin D intoxication. Horm Res **37**:68–77.
28. Adams JS 1989 Vitamin D metabolite-mediated hypercalcemia. Endocrinol Metab Clin North Am **18**:765–778.
29. Harris LJ 1956 Vitamin D and bone. In: Bourne GH (ed) The Biochemistry and Physiology of Bone. Academic Press, New York, pp. 581–622.
30. Boyce RW, Weisbrode SE 1983 Effect of dietary calcium on the response of bone to $1,25(OH)_2D_3$. Lab Invest **48**:683–689.
31. Boyce RW, Weisbrode SE 1985 The histogenesis of hyperosteoidosis in $1,25(OH)_2D_3$-treated rats fed high levels of dietary calcium. Bone **6**:105–112.
32. Boyce RW, Weisbrode SE, Kindig O 1985 Ultrastructural development of hyperosteoidosis in $1,25(OH)_2D_3$-treated rats fed high levels of dietary calcium. Bone **6**:165–172.
33. Wronski TJ, Halloran BP, Bikle DD, Globus RK, Morey-Holton ER 1986 Chronic administration of $1,25(OH)_2D_3$: Increased bone but impaired mineralization. Endocrinology **119**:2580–2585.
34. Hock JM, Gunness-Hey M, Poser J, Olson H, Bell NH, Raisz LG 1986 Stimulation of undermineralized matrix formation by 1,25-dihydroxyvitamin D_3 in long bones of rats. Calcif Tissue Int **38**:79–86.
35. Marie PJ, Hott M, Garba MT 1985 Contrasting effects of 1,25-dihydroxyvitamin D_3 on bone matrix and mineral appositional rates in the mouse. Metabolism **34**:777–783.
36. Narbaitz R, Stumpf WE, Sar M, Huang S, DeLuca HF 1983 Autoradiographic localization of target cells for 1α,25-dihydroxyvitamin D_3 in bones from fetal rats. Calcif Tissue Int **35**:177–182.
37. Boivin G, Mesguich P, Pike JW, Bouillon R, Meunier PJ, Haussler MR, Dubois PM, Morel G 1987 Ultrastructural immunocytochemical localization of endogenous 1,25-dihydroxyvitamin D_3 and its receptor in osteoblasts and osteocytes from neonatal mouse and rat calvaria. Bone Miner **3**:125–136.
38. Johnson JA, Grande JP, Roche PC, Kumar R 1996 Ontogeny of the 1,25-dihydroxyvitamin D_3 receptor in fetal rat bone. J Bone Miner Res **11**:56–61.
39. Merke J, Klaus G, Hugel U, Waldherr R, Ritz E 1986 No 1,25-dihydroxyvitamin D_3 receptors on osteoclasts of calcium-deficient chicken despite demonstrable receptors on circulating monocytes. J Clin Invest **77**:312–314.
40. Mee AP, Hoyland JA, Braidman IP, Freemont AJ, Davies M, Mawer EB 1996 Demonstration of vitamin D receptor transcripts in actively resorbing osteoclasts in bone sections. Bone **18**:295–299.
41. Kream BE, Rowe DW, Gworek G, Raisz LG 1980 Parathyroid hormone alters collagen synthesis and procollagen mRNA levels in fetal rat calvaria. Proc Natl Acad Sci USA **77**:5654–5658.
42. Bringhurst FR, Potts JT Jr 1981 Bone collagen synthesis in vitro: Structure/activity relations among parathyroid hormone fragments and analogs. Endocrinology **108**:103–108.
43. Row DW, Kream BE 1982 Regulation of collagen synthesis in fetal rat calvaria by 1,25-dihydroxyvitamin D_3. J Biol Chem **257**:8009–8015.
44. Rosen DM, Luben RA 1983 Multiple hormonal mechanisms for the control of collagen synthesis in an osteoblast-like cell line, MMB-1. Endocrinology **112**:992–999.
45. Kream BE, Rowe DW, Smith MD, Maher V, Majeska R 1986 Hormonal regulation of collagen production in a clonal rat osteosarcoma cell line. Endocrinology **119**:1922–1928.
46. Harrison JR, Petersen DN, Lichtler AC, Mador AT, Rowe DW, Kream BE 1989 1,25-Dihydroxyvitamin D_3 inhibits transcription of type I collagen genes in the rat osteosarcoma cell line ROS 17/2.8. Endocrinology **125**:327–333.
47. Pavlin D, Bedalov A, Kronenberg MS, Kream BE, Rowe DW, Smith CL, Pike JW, Lichtler AC 1994 Analysis of regulatory regions in the COL1A1 gene responsible for 1,25-dihydroxyvitamin D_3-mediated transcriptional repression in osteoblastic cells. J Cell Biochem **56**:490–501.
48. Manolagas SC, Burton DW, Deftos LJ 1981 1,25-Dihydroxyvitamin D_3 stimulates the alkaline phosphatase activity of osteoblast-like cells. J Biol Chem **256**:7115–7117.
49. Majeska RJ, Rodan GA 1982 The effects of $1,25(OH)_2D_3$ on alkaline phosphatase in osteoblastic osteosarcoma cells. J Biol Chem **257**:13362–13365.
50. Kurihata N, Ikeda K, Hakeda Y, Tsunoi M, Maeda N, Kumegawa M 1984 Effect of 1,25-dihydroxyvitamin D_3 on alkaline phosphatase activity and collagen synthesis in osteoblastic cells, clone MC3T3-E1. Biochem Biophys Res Commun **119**:767–771.
51. Kurihara N, Ishizuka S, Kiyoki M, Haketa Y, Ikeda K, Kumegawa M 1986 Effects of 1,25-dihydroxyvitamin D_3 on osteoblastic MC3T3-E1 cells. Endocrinology **118**:940–947.
52. Prince CW, Butler WT 1987 1,25-Dihydroxyvitamin D_3 regulates the biosynthesis of osteopontin, a bone-derived cell attachment protein, in clonal osteoblast-like osteosarcoma cells. Collagen Related Res **7**:305–315.
53. Oldberg A, Jinskod-Hed B, Axelsson S, Heinegard D 1989 Regulation of bone sialoprotein mRNA by steroid hormones. J Cell Biol **109**:3183–3186.
54. Lian J, Stewart C, Puchacz E, Mackowiak S, Shalhoub V, Collart D, Zambeti G, Stein G 1989 Structure of the rat osteocalcin gene and regulation of vitamin D-dependent expression. Proc Natl Acad Sci USA **84**:1143–1147.
55. Demay MB, Gerardi JM, DeLuca HF, Kronenberg HM 1990 DNA sequences in the rat osteocalcin gene that bind the 1,25-dihydroxyvitamin D_3 receptor and confer responsiveness to 1,25-dihydroxyvitamin D_3. Proc Natl Acad Sci USA **87**:369–373.
56. Fraser JD, Otawara Y, Price PA 1988 1,25-Dihydroxyvitamin D_3 stimulates the synthesis of matrix γ-carboxyglutamic acid protein by osteosarcoma cells. J Biol Chem **263**:911–916.
57. Spiess YH, Price PA, Deftos JL, Manolagas SC 1986 Phenotype-associated changes in the effects of 1,25-dihydroxyvitamin D_3 on alkaline phosphatase and bone Gla-protein of rat osteoblastic cells. Endocrinology **118**:1340–1346.
58. Fraser JD, Price PA 1990 Induction of matrix Gla protein synthesis during prolonged 1,25-dihydroxyvitamin D_3 treatment of osteosarcoma cells. Calcif Tissue Int **46**:270–279.
59. Kurose H, Yamaoka K, Okada S, Nakajima S, Seino Y 1990 1,25-Dihydroxyvitamin D_3 [$1,25(OH)_2D_3$] increases insulin-like growth factor I (IGF-I) receptors in clonal osteoblastic cells. Study on interaction of IGF-I and $1,25(OH)_2D_3$. Endocrinology **126**:2088–2094.
60. Mahonen A, Pirskanen A, Keinänen R, Mäenpää PH 1990 Effect of $1,25(OH)_2D_3$ on its receptor mRNA levels and osteocalcin synthesis in human osteosarcoma cells. Biochim Biophys Acta **1048**:30–37.
61. Kerner SA, Scott RA, Pike JW 1989 Sequence elements in the human osteocalcin gene confer basal activation and inducible

response to hormonal vitamin D_3. Proc Natl Acad Sci USA **86**:4455–4459.
62. Noda M, Vogel RL, Craig AM, Prahl J, DeLuca HF, Denhardt DT 1990 Identification of 1,25-dihydroxyvitamin D_3 receptor and 1,25-dihydroxyvitamin D_3 enhancement of mouse secreted phosphoprotein 1 (*Spp*-1 or osteopontin) gene expression. Proc Natl Acad Sci USA **87**:9995–9999.
63. Matsumoto T, Igarashi C, Takeuchi Y, Harada S, Kikuchi T, Yamato Y, Ogata E 1991 Stimulation by 1,25-dihydroxyvitamin D_3 of in vitro mineralization induced by osteoblast-like MC3T3-E1 cells. Bone **12**:27–32.
64. Wheeler DG, Horsford J, Michalak M, White JH, Hendy GN 1995 Calreticulin inhibits vitamin D_3 signal transduction. Nucleic Acids Res **23**:3268–3274.
65. St-Arnaud R, Prud'homme J, Leung-Hagesteijn C, Dedhar S 1995 Constitutive expression of calreticulin in osteoblasts inhibits mineralization. J Cell Biol **131**:1351–1359.
66. Owen TA, Aronow M, Shalhoub V, Barone LM, Wilming L, Tassinari M, Kennedy MB, Pockwinse S, Lian JB, Stein G 1990 Progressive development of the rat osteoblast phenotype in vitro: Reciprocal relationships in expression of genes associated with osteoblast proliferation and differentiation during formation of the bone extracellular matrix. J Cell Physiol **143**:420–430.
67. Owen TA, Aronow MS, Barone LM, Bettencourt B, Stein GS, Lian JB 1991 Pleiotropic differentiated state of the bone cell phenotype: Dependency upon basal levels of gene expression, duration of exposure, and bone matrix competency in normal rat osteoblast cultures. Endocrinology **128**:1496–1504.
68. Clark SA, Boass A, Toverud SU 1987 Effect of high dietary contents of calcium and phosphorus on mineral metabolism and growth of vitamin D-deficient suckling and weaned rats. Bone Miner **2**:257–270.
69. Weinstein RS, Underwood JL, Hutson MS, DeLuca HF 1984 Bone histomorphometry in vitamin D-deficient rats infused with calcium and phosphorus. Am J Physiol **246**:E499–E505.
70. Howard GA, Baylink DJ 1980 Matrix formation and osteoid maturation in vitamin D-deficient rats made normocalcemic by dietary means. Miner Electrolyte Metab **3**:44–50.
71. Holtrop ME, Cox KA, Carnes DL, Holick MF 1986 Effects of serum calcium and phosphorus on skeletal mineralization in vitamin D-deficient rats. Am J Physiol **251**:E234–E240.
72. Kollenkirchen U, Fox J, Walters MR 1991 Normocalcemia without hyperparathyroidism in vitamin D-deficient rats. J Bone Miner Res **6**:273–278.
73. Dziedzic-Goclawska A, Toverud SU, Kaminski A, Boass A, Yamauchi M 1992 Decreased heterotopic osteogenesis in vitamin-D-deficient, but normocalcemic guinea pigs. Bone Miner **19**:127–143.
74. Schwartz Z, Schlader DL, Ramirez V, Kennedy MB, Boyan BD 1989 Effects of vitamin D metabolites on collagen production and cell proliferation of growth zone and resting zone cartilage cells in vitro. J Bone Miner Res **4**:199–207.
75. Schwartz Z, Boyan B 1988 The effects of vitamin D metabolites on phospholipase A_2 activity of growth zone and resting zone cartilage cells in vitro. Endocrinology **122**:2191–2198.
76. Boyan BD, Schwartz Z, Carnes DL Jr, Ramirez V 1988 The effects of vitamin D metabolites on the plasma and matrix vesicle membranes of growth and resting cartilage cells in vitro. Endocrinology **126**:2851–2860.
77. Schwartz Z, Brooks B, Swain L, Del Toro F, Norman A, Boyan BD 1992 Production of 1,25-dihydroxyvitamin D_3 and 24,25-dihydroxyvitamin D_3 by growth zone and resting zone chondrocytes is dependent on cell maturation and is regulated by hormones and growth factors. Endocrinology **130**:2495–2504.
78. Schwartz Z, Dean DD, Walton JK, Brooks BP, Boyan BD 1995 Treatment of resting zone chondrocytes with 24,25-dihydroxyvitamin D_3 [24,25-$(OH)_2D_3$] induces differentiation into a 1,25$(OH)_2D_3$-responsive phenotype characteristic of growth zone chondrocytes. Endocrinology **136**:402–411.
79. Yamato H, Okazaki R, Ishii T, Ogata E, Sato T, Kumegawa M, Akaogi K, Taniguchi N, Matsumoto T 1993 Effect of 24*R*,25-dihydroxyvitamin D_3 on the formation and function of osteoclastic cells. Calcif Tissue Int **52**:255–260.
80. Norman AW, Leathers V, and Bishop JE 1983 Normal egg hatchability requires the simultaneous administration to the hen of 1α,25-dihydroxycholecalciferol and 24*R*,25-dihydroxycholecalciferol. J Nutr **113**:2505–2515.
81. Tanaka Y, DeLuca HF, Kobayashi Y, Taguchi T, Ikekawa N, Morisaki M 1979 Biological activity of 24,24-difluoro-25-hydroxyvitamin D_3. Effects of blocking of 24-hydroxylation on the functions of vitamin D. J Biol Chem **254**:7163–7167.
82. Miller SC, Halloran BP, DeLuca HF, Yamada S, Takayama H, Jee WS 1981 Studies on the role of 24-hydroxylation of vitamin D in the mineralization of cartilage and bone of vitamin D-deficient rats. Calcif Tissue Int **33**:489–497.
83. Jarnagin K, Brommage R, DeLuca HF, Yamada S, Takayama H 1983 1- but not 24-hydroxylation of vitamin D is required for growth and reproduction in rats. Am J Physiol **244**:E290–E297.
84. Brommage R, Jarnagin K, DeLuca HF, Yamada S, Takayama H 1983 1- but not 24-hydroxylation of vitamin D is required for skeletal mineralization in rats. Am J Physiol **244**:E298–E304.
85. Parfitt AM, Mathews CHE, Brommage R, Jarnagin K, DeLuca HF 1984 Calcitriol but no other metabolite of vitamin D is essential for normal bone growth and development in the rat. J Clin Invest **73**:576–586.
86. Nakamura T, Kurokawa T, Orimo H 1988 Increase of bone volume in vitamin D-repleted rats by massive administration of 24*R*,25-$(OH)_2D_3$. Calcif Tissue Int **43**:235–243.
87. Nakamura T, Hirai T, Suzuki K, Orimo H 1992 Osteonal remodeling and mechanical properties of femoral cortex in rabbits treated with 24*R*,25-$(OH)_2D_3$. Calcif Tissue Int **50**:74–79.
88. Nakamura T, Suzuki K, Hirai T, Kurokawa T, Orimo H 1992 Increased bone volume and reduced bone turnover in vitamin D-replete rabbits by the administration of 24*R*,25-dihydroxyvitamin D_3. Bone **13**:229–236.
89. Nakamura T, Nagai Y, Yamato H, Suzuki K, Orimo H 1992 Regulation of bone turnover and prevention of bone atrophy in ovariectomized beagle dogs by the administration of 24*R*,25-dihydroxyvitamin D_3. Calcif Tissue Int **50**:221–227.
90. Yamaura M, Nakamura T, Nagai Y, Yoshihara A, Suzuki K 1993 Reduced mechanical competence of bone by ovariectomy and its preservation with 24*R*,25-dihydroxyvitamin D_3 administration in beagles. Calcif Tissue Int **52**:49–56.
91. Capecchi MR 1989 The new mouse genetics: Altering the genome by gene targeting. Trends Genet **5**:70–76.
92. Rossant J, Nagy A 1995 Genome engineering: The new mouse genetics. Nature Med **1**:592–595.
92a. St-Arnaud R, Arabian A, Glorieux FH 1996 Abnormal bone development in mice deficient for the vitamin D 24-hydroxylase gene. J Bone Miner Res **11**:S126 (abstract).
93. Sömjen D, Sömjen GJ, Weisman Y, Binderman I 1982 Evidence for 24,25-dihydroxycholecalciferol receptors in long bones of newborn rats. Biochem J **204**:31–36.
94. Fine N, Binderman I, Sömjen D, Earon Y, Edelstein S, Weisman Y 1985 Autoradiographic localization of 24*R*,25-dihydroxyvitamin D_3 in epiphyseal cartilage. Bone **6**:99–104.

95. Corvol MM, Ulmann A, Garabedian M 1980 Specific nuclear uptake of 24*R*,25-dihydroxycholecalciferol, a vitamin D_3 metabolite biologically active in cartilage. FEBS Lett **116**:273–276.
96. Sömjen D, Sömjen GJ, Harell A, Mechanic GL, Binderman I 1982 Partial characterization of a specific high affinity binding macromolecule for 24*R*,25-dihydroxyvitamin D_3 in differentiating skeletal mesenchyme. Biochem Biophys Res Commun **106**:644–651.
97. Merke J, Norman AW 1981 Studies on the mode of action of calciferol XXXII. Evidence for 24(*R*),25$(OH)_2$-vitamin D_3 receptor in the parathyroid gland of the rachitic chick. Biochem Biophys Res Commun **100**:551–558.
98. Mangelsdorf DJ, Thummel C, Beato M, Herrlich P, Schütz G, Umesono K, Blumberg B, Kastner P, Mark M, Chambon P, Evans RM 1995 The nuclear receptor superfamily: The second decade. Cell **83**:835–839.

CHAPTER 19

Vitamin D Metabolites and Bone

HIROYUKI TANAKA AND YOSHIKI SEINO
Department of Pediatrics, Okayama University Medical School, Okayama, Japan

I. INTRODUCTION

Vitamin D was originally discovered as an antirachitic factor. Thus the effects of vitamin D on bone are of particular importance. 1,25-Dihydroxyvitamin D_3 [$1,25(OH)_2D_3$] has been considered to be the most potent and perhaps the only metabolite that functions in bone and which produces the complete spectrum of functions that are observed for vitamin D. However, studies have suggested that vitamin D metabolites other than $1,25(OH)_2D_3$ may also play an important role. Thus, it has become exceedingly important to define, in bone, the actions not only of $1,25(OH)_2D_3$ but also of other metabolites. Thus far, two natural metabolites of vitamin D have been suggested to play important roles in bone, $24R,25(OH)_2D_3$ and $1,25R(OH)_2$-vitamin D_3-$26,23S$-lactone. The establishment of firm biological roles for these metabolites will prompt additional attention to their mechanisms of action in bone. It is possible that mechanistic insights will lead to the synthesis of new analogs useful in the treatment of various bone diseases.

II. $1,25(OH)_2D_3$ AND BONE

The hormone $1,25(OH)_2D_3$ stimulates both bone formation and bone resorption. However, it is unclear at present whether all of the actions of $1,25(OH)_2D_3$ are direct or indirect [1]. It is generally accepted that the bone resorbing cell, the mature osteoclast, does not contain the nuclear vitamin D receptor (VDR) for $1,25(OH)_2D_3$ [2]. This implies that vitamin D is unlikely to directly induce bone resorption by stimulating the activity of preexisting osteoclasts. Suda *et al.* [3,4] suggested that $1,25(OH)_2D_3$ is important for supporting two discrete steps in osteoclastogenesis: (1) to induce immature hematopoietic cells to differentiate into monocytes and (2) to define the commitment of the differentiating osteoclast through an essential mechanism involving the osteoblast. As a final component of the differentiation process, VDR disappears from the mature osteoclasts (these specific issues are discussed in Chapters 20 and 21).

The metabolite $1,25(OH)_2D_3$ is also known to exert several effects on the bone forming cell, the osteoblast.

It is clear that osteoblastic cells possess VDR [1,5,6]. Both estrogens and glucocorticoids may alter the number of VDR [7,8]. Moreover, VDR concentration is affected by the stage of the cell cycle [9]. There is considerable evidence to suggest a direct stimulatory effect of $1,25(OH)_2D_3$ on osteoblastic functions. $1,25(OH)_2D_3$ increases the level of mRNAs for alkaline phosphatase [10,11], osteocalcin [11–15], osteopontin [16,17], and matrix Gla protein [18]. The metabolite also enhances type 1 collagen synthesis. Because both anabolic and catabolic effects are observed, these differences in response may reflect the state of differentiation of the cells [19–21]. Finally, $1,25(OH)_2D_3$ can cause inhibition of proteoglycan synthesis and stimulation of its degradation [22]. The rat osteosarcoma cell line ROS 17/2.8 can respond to $1,25(OH)_2D_3$ to increase Ca^{2+} flux through a "nongenomic pathway" [23]. Intracellular pH changes can also be induced by $1,25(OH)_2D_3$ in the osteoblast [24].

The metabolite $1,25(OH)_2D_3$ also induces changes in growth factor and cytokine production that are important for bone formation. Thus, $1,25(OH)_2D_3$ induces the secretion of insulin-like growth factor binding protein [25] while suppressing the secretion of insulin-like growth factor type I (IGF-I) itself [26] and stimulating the deposition of transforming growth factor-β into bone [27].

The hormone $1,25(OH)_2D_3$ is produced under the action of renal 1α-hydroxylase. In special circumstances there is evidence of extrarenal production of $1,25(OH)_2D_3$ such as in alveolar macrophages and keratinocytes (see Chapter 55). However, there are no data to suggest the production of $1,25(OH)_2D_3$ in the osteoblast. Thus, it is unlikely that differences in local activity of $1,25(OH)_2D_3$ in bone plays a role. However, Lidor *et al.* [28] reported a decrease in bone levels of $1,25(OH)_2D_3$ in women with femoral neck fracture.

A number of studies have evaluated the clinical efficacy of $1,25(OH)_2D_3$ in the treatment of osteoporosis (see Chapter 44). These studies provide some insights into the effects of $1,25(OH)_2D_3$ on bone *in vivo*. Because of the heterogeneity in the pathophysiology of osteoporosis, it may not be possible to identify a single treatment modality that is beneficial for all cases. However, the rationale for the utilization of $1,25(OH)_2D_3$ in the treatment of osteoporosis is supported by the widespread decrease in the serum concentrations of $1,25(OH)_2D_3$ in elderly subjects [29]. In an extensive 3-year prospective, multicenter, single blind study involving 622 women who had one or more vertebral compression fractures, Tilyard *et al.* [30] demonstrated that women who received $1,25(OH)_2D_3$ had a significantly lower incidence of new vertebral fractures during the second and third years of treatment, compared with women who received only a calcium supplement. An important issue in this study is that there was no significant incidence of vitamin D side effects, such as hypercalcemia.

Gallagher and Goldgar [31] performed a separate randomized controlled study in which they reported that utilization of replacement doses of $1,25(OH)_2D_3$ resulted in significant increases in total body calcium and spine bone mineral density. A 5-year follow-up experiment with 1 μg/day of $1\alpha OHD_3$ confirmed the ability of the latter compound to maintain trabecular bone mass and to prevent vertebral fracture in postmenopausal Japanese women [32]. In Italy, an increase in bone mass after prolonged therapy with $1\alpha OHD_3$ was also demonstrated [33]. In contrast, several initial studies and an extensive double blind study with $1,25(OH)_2D_3$ were unable to demonstrate any beneficial effect on bone mass of postmenopausal women who received adequate calcium intake [34]. A possible interpretation for the controversy regarding the efficacy of $1,25(OH)_2D_3$ in osteoporosis treatment is a difference in the basal nutritional condition of the subjects. In the case of a subclinical deficiency of calcium or vitamin D, which is frequently observed in elderly subjects, combined treatment with vitamin D and calcium could reverse bone loss and reduce the rate of new hip fractures by 43%.

A provocative report correlates polymorphisms in the gene encoding the VDR with peak bone mass [35]. In that study, common allelic variations in the VDR gene predicted differences in bone mineral density and presumed to identify those individuals at risk for osteoporosis. Both supportive and unsupportive reports have followed by other investigators [36,37], and the subject is discussed in detail in Chapter 45. The finding of a link between polymorphisms in the VDR gene and rate of bone loss from the femoral neck in postmenopausal women (enhanced at low calcium intakes) suggests that intestinal calcium absorption may be a site of differential action by the VDR alleles. Dawson-Hughes *et al.* [38] reported a significant difference among VDR genotypes in calcium absorption at low calcium intakes.

These human studies suggest that $1,25(OH)_2D_3$ exerts positive effects on bone formation mainly by increasing intestinal calcium absorption. Moreover, DeLuca and co-workers [39,40] clearly demonstrated that normal bone formation might be achieved even in the vitamin D-deficient rat, if the supply of calcium and phosphorus was adequate. Therefore, the effect of $1,25(OH)_2D_3$ on bone *in vivo* is a combination of direct actions on bone cells and important indirect actions on the intestine to regulate the delivery of mineral to bone.

III. 24*R*,25(OH)$_2$D$_3$ AND BONE

Although it has been established that 1,25(OH)$_2$D$_3$ facilitates the intestinal absorption of calcium and phosphorus and enhances renal mineral reabsorption, the biological role played by 24*R*,25(OH)$_2$D$_3$ is a matter of continuing controversy. Attention was originally focused on this metabolite when it was learned that administration of 1,25(OH)$_2$D$_3$ alone appeared unable to produce the same biological effects as that of parental vitamin D$_3$ administration; this suggested that other vitamin D metabolites might also be required [41,42]. Circulating levels of 24*R*,25(OH)$_2$D$_3$ are at least 5-fold higher than those of 1,25(OH)$_2$D$_3$ (see Table I), whereas its affinity for VDR is 1/100 to 1/1000 lower than that of 1,25(OH)$_2$D$_3$ [43]. Osteoblasts contain 24-hydroxylase activity which in turn is regulated differently from that in kidney [43]. These observations suggest the possibility of a unique biological role for 24*R*,25(OH)$_2$D$_3$, particularly in bone and cartilage tissue.

Several reports have suggested that 24*R*,25(OH)$_2$D$_3$ is important and necessary for chondrocyte function, demonstrating that it increases the synthesis of DNA [44] and proteoglycan [45] and increases alkaline phosphatase activity [46,47] (see also Chapter 26). Takigawa *et al.* [48] demonstrated that 24*R*,25(OH)$_2$D$_3$ induced chondrocyte differentiation. A putative receptor for 24*R*,25(OH)$_2$D$_3$ was reported to be present in cartilage, and ^{3}H-labeled 24*R*,25(OH)$_2$D$_3$ was found by autoradiography to accumulate in epiphyseal cartilage [49,50]. However, there has been no follow-up on the latter studies for over a decade, calling into question the validity of these reports.

Interestingly, rats with elevated circulating levels of 1,25(OH)$_2$D$_3$ exhibit an increased rate of bone resorption, whereas rats with an increased level of 24*R*,25(OH)$_2$D$_3$ exhibit reduced rates of bone turnover [51]. In a rabbit model, administration of 24*R*,25(OH)$_2$D$_3$ resulted in increased bone volume and a reduction in bone turnover [52]. Nakamura *et al.* reported that 24*R*,25(OH)$_2$D$_3$ had bone forming ability (increase in bone mineral density and mechanical strength) in normal rats [53] and in ovariectomized dogs [54]. In a chick fracture repair model, radiolabeled 24*R*,25(OH)$_2$D$_3$ was found in the healing callus [55]. In addition, 24*R*,25(OH)$_2$D$_3$ was found to have a specific inhibitory effect on the formation and function of osteoclastic cells stimulated by 1,25(OH)$_2$D$_3$ or parathyroid hormone (PTH) [56]. It is interesting that 24*R*,25(OH)$_2$D$_3$ exhibited these effects on bone formation without inducing hypercalcemia even at the highest doses. The most recent information, based on data from a 24-hydroxylase knock-out model in the mouse, suggests that 24*R*,25(OH)$_2$D$_3$ may have a unique role in the development of intramembranous bone (see Chapter 18).

In the Hyp mouse, a model of human X-linked hypophosphatemic vitamin D-resistant rickets (XLH), it has been shown that a massive dose of 24*R*,25(OH)$_2$D$_3$ cured rachitic changes more effectively than 1,25(OH)$_2$D$_3$ and did so without hypercalcemia [57,58]. The most important difference in bone histology between the effect of the two metabolites was the number of osteoclasts. Osteoclast number was restored to normal in 24*R*,25(OH)$_2$D$_3$-treated bone, whereas 1,25(OH)$_2$D$_3$-treated bone showed higher values (Table II) [57,58]. Massive doses of 24*R*,25(OH)$_2$D$_3$ induced low levels of 1,25(OH)$_2$D$_3$ in this report. The mechanism underlying this effect has not been elucidated. Matsumoto *et al.* also suggested that 24*R*,25(OH)$_2$D$_3$ enhanced the degradation of 1,25(OH)$_2$D$_3$ in a calcium-deficient rat model [59]. Norman and co-workers [60,61] demonstrated that 24*R*,25(OH)$_2$D$_3$ acts as an allosteric effector and reduces the affinity of 1,25(OH)$_2$D$_3$ for its receptor.

TABLE I Serum Concentrations of Vitamin D Metabolites in Normal Individuals and Patients with Osteoporosis and Renal Failure

Subjects	*n*	Age	25OHD (ng/ml)	24,25(OH)$_2$D (ng/ml)	1,25(OH)$_2$D (pg/ml)	Calcitriol-lactone (pg/ml)
Normal	11	25–40	27.7 ± 5.9	2.53 ± 0.41	56.1 ± 12.9	131.3 ± 55.5
Elderly healthy women	15	60–80	18.5 ± 3.4	1.61 ± 0.46	46.8 ± 5.8	66.6 ± 9.6
Osteoporosis	23	60–88	12.2 ± 4.0	0.73 ± 0.45	31.1 ± 10.4	36.6 ± 16.1
Renal failure						
On hemodialysis	6	38–45	18.5 ± 5.7	0.29 ± 0.17	12.6 ± 3.5	19.9 ± 3.6
On conservative	4	38–44	14.6 ± 3.3	0.36 ± 0.15	18.9 ± 2.7	28.7 ± 4.7

Each metabolite was extracted by chloroform-methanol and was purified by LH-20 column chromatography following HPLC. Purified materials were analyzed by house-made radioimmunoassay. Reproduced from Ishizuka [72] with permission.

TABLE II Osteoclast Histomorphometry in the Hyp Mouse Treated with Vitamin D Metabolites[a]

	Normal mouse, vehicle	Hyp mouse		
		Vehicle	$24,25(OH)_2D_3$	$1,25(OH)_2D_3$
N.Oc/B.Pm (/10 cm)[b]	258 ± 65	70.4 ± 52	237 ± 71	346 ± 71
Oc.S/BS (%)	6.0 ± 1.0	1.8 ± 1.3	5.2 ± 1.1	7.6 ± 1.9

[a] Hyp mice were either treated with 1000 μg/kg of $24,25(OH)_2D_3$ or 0.1 μg/kg of $1,25(OH)_2D_3$. Mice were treated with vitamin D metabolites for 4 weeks [58].

[b] N.Oc/B.Pm (/10 cm): number of osteoclasts per 10 cm of bone tissue. Oc.S/BS%: fraction of total trabecular surface covered by osteoclasts.

$24R,25(OH)_2D_3$ may therefore act through a suppression of the effects of $1,25(OH)_2D_3$. One of the alternate possibilities is that $24R,25(OH)_2D_3$ may also exert effects on bone through suppression of PTH secretion. Clinical findings have suggested that $24R,25(OH)_2D_3$ may be a useful adjunct to standard therapy in XLH [62].

Collectively, these results suggest the possibility of unique biological actions of $24R,25(OH)_2D_3$ on normal bone biology. However, the distinction between biological effects of $24R,25(OH)_2D_3$ at normal circulating concentrations and pharmacological effects at high concentrations must be made. Nevertheless, if the differences between $1,25(OH)_2D_3$ and $24R,25(OH)_2D_3$ are confirmed, further chemistry focused on the $24R,25(OH)_2D_3$ molecule will be warranted.

IV. $1,25R(OH)_2$-VITAMIN D_3-26,23S-LACTONE AND BONE

Originally, it was proposed that $1,25R(OH)_2$-vitamin D_3-26,23*S*-lactone (calcitriol-lactone) was a metabolic excretion product of $1,25(OH)_2D_3$ [63,64]. Studies have revealed that the calcitriol-lactone may also be capable of generating unique biological actions in osteoblasts and osteoclasts.

The most unique feature of this metabolite is that it has four diastereoisomers, namely, 23*R*,25*S*-lactone, 23*S*,25*S*-lactone, 23*R*,25*R*-lactone, and calcitriol-lactone. The natural metabolite is the calcitriol-lactone, and it is produced in osteoblasts [65]. Although 23*R*,25*S*-lactone and 23*R*,25*R*-lactone merely mimic the action of $1,25(OH)_2D_3$, and increase serum calcium levels, 23S,25*R*-lactone (calcitriol-lactone) and 23*R*,25*S*-lactone decrease serum calcium levels [66]. The presence of the 26,23-lactone ring on $1,25(OH)_2D_3$ increased affinity for the vitamin D binding protein (DBP) by 2- to 3-fold, although DBP is unable to distinguish between the four diastereoisomers [67]. In contrast, although the VDR displays a reduced affinity for the four diastereoisomers, the receptor is capable of discriminating between each diastereoisomer. The natural calcitriol-lactone exhibits the lowest affinity for VDR [0.4–10% as effective as $1,25(OH)_2D_3$] (Table III) [67]. These findings suggest that the natural calcitriol-lactone might have a unique signal transduction pathway and unique biological functions.

In the chick system, the calcitriol-lactone was found to be only 1/30 as effective as $1,25(OH)_2D_3$ in promoting intestinal calcium absorption and was found to induce a significant decrease in serum calcium [66]. In a murine osteoblast culture system, the calcitriol-lactone was

TABLE III Ability of $1,25(OH)_2D_3$ and Four Diastereoisomers of $1,25(OH)_2D_3$-26,23-lactone to Compete with ^3H-$1,25(OH)_2D_3$ for Binding to the Chick Intestinal VDR[a]

Vitamin D_3 analog	RCI
$1,25(OH)_2D_3$	100.00
23*S*,25*S*-$1,25(OH)_2D_3$-26,23-lactone	7.90
23*R*,25*R*-$1,25(OH)_2D_3$-26,23-lactone	2.27
23*S*,25*R*-$1,25(OH)_2D_3$-26,23-lactone	0.17
23*R*,25*S*-$1,25(OH)_2D_3$-26,23-lactone	0.22
In vivo $1,25(OH)_2D_3$-26,23-lactone	0.17

[a] The ability of each diastereoisomer to complete with ^3H-$1,25(OH)_2D_3$ for binding to the chick intestinal VDR was quantitated by measuring the decrease in specific binding of ^3H-$1,25(OH)_2D_3$ when increasing concentrations of each diastereoisomer were incubated with a fixed concentration of ^3H-$1,25(OH)_2D_3$. The reciprocal of the percent of maximal binding of ^3H-$1,25(OH)_2D_3$ was then calculated and plotted as a function of the relative concentrations of each diastereoisomer and ^3H-$1,25(OH)_2D_3$. The slope of each plot was defined as the competitive index value. Relative competitive index (RCI) is a percentage of the competitive index value of each diastereoisomer relative to $1,25(OH)_2D_3$. Reprinted from Seino and Ishizuka [72] with permission.

shown to stimulate collagen production and to increase alkaline phosphatase activity [68]. In a bone organ culture study, bone resorption induced by $1,25(OH)_2D_3$ and PTH was inhibited by the calcitriol-lactone [69]. Moreover, an important finding was the demonstration of a selective ability of the natural calcitriol-lactone to inhibit $1,25(OH)_2D_3$-induced osteoclastogenesis [70].

There is a single report concerning the *in vivo* effects of calcitriol-lactone on bone formation. Shima *et al.* reported that the metabolite stimulated new bone formation *in vivo* by increasing matrix formation and mineral uptake in a system based on bone morphogenetic protein-induced ectopic bone formation [71].

The conclusion from these observations is that the calcitriol-lactone might act as a natural antagonist to $1,25(OH)_2D_3$-mediated bone resorption when used in concentrations in excess of those of $1,25(OH)_2D_3$. However, it remains to be determined whether the calcitriol-lactone has a functional physiological regulatory role in bone metabolism. A first approach to this end should be the measurement of serum concentrations of this metabolite in various abnormal metabolic bone diseases. As shown in Table III, patients with osteoporosis and renal failure showed significantly lower levels of calcitriol-lactone [72]. The significance of these findings is, at present, unclear. Further studies may delineate more accurately the physiological role of the calcitriol-lactone.

V. CONCLUSIONS

Available data have been summarized demonstrating that $24R,25(OH)_2D_3$ and calcitriol-lactone have effects on bone. It is unclear whether these effects are mediated through unique receptors. Alternatively, the activity on bone may be due to multiple interactions with factors in the classic vitamin D endocrine system, VDR, DBP, metabolizing enzymes, etc. Whatever the mechanism, important and somewhat unique patterns of activity have emerged when animal models or patients have been treated with these metabolites. Further study appears warranted to determine the mechanism of action and the potential therapeutic role of these metabolites in treating metabolic bone disease.

References

1. Mellon WS, Deluca HF 1980 A specific 1,25-dihydroxyvitamin D_3 binding macromolecule in chicken bone. J Biol Chem **255**: 4081–4086.
2. Merke J, Klaus G, Hugel U, Waldherr R, Ritz E 1986 No 1,25-dihydroxyvitamin D_3 receptors on osteoclasts of calcium-deficient chicken despite demonstrable receptors on circulating monocytes. J Clin Invest **77**:312–314.
3. Suda T, Takahashi N, Abe E 1992 Role of vitamin D in bone resorption. J Cell Biochem **49**:53–58.
4. Suda T, Shinki T, Takahashi N 1990 The role of vitamin D in bone and intestinal cell differentiation. Annu Rev Nutr **10**:195–211.
5. Walters MR, Rosen DM, Norman AW, Luben RA 1982 1,25-Dihydroxyvitamin D receptors in an established bone cell line: Correlation with biochemical responses. J Biol Chem **257**:7481–7484.
6. Jurutka PW, Terpening CM, Haussler MR 1993 The 1,25-dihydroxyvitamin D_3 receptor is phosphorylated in response to 1,25-dihydroxyvitamin D_3 and 22-oxacalcitriol in rat osteoblasts, and by casein kinase II, *in vitro*. Biochemistry **32**:8184–8192.
7. Liel Y, Klaus S, Levy J, Shany S 1992 Evidence that estrogens modulate activity and increase the number of 1,25-dihydroxyvitamin D receptors in osteoblast-like cells (ROS 17/2.8). Endocrinology **130**:2597–2601.
8. Godschalk M, Levy JR, Downs RW JR 1992 Glucocorticoids decrease vitamin D receptor number and gene expression in human osteosarcoma cells. J Bone Miner Res **7**:21–27.
9. Suzuki S, Koga M, Takaoka K, Ono K, Sato B 1993 Effects of retinoic acid on steroid and vitamin D_3 receptors in cultured mouse osteosarcoma cells Bone **14**:7–12.
10. Mulkins MA, Manolagas SC, Deftos LJ, Sussman HH 1983 1,25-Dihydroxyvitamin D_3 increases bone alkaline phosphatase isoenzyme levels in human osteogenic sarcoma cells. J Biol Chem **258**:6219–6225.
11. Marie PJ, Connes D, Hott M, Miravet L 1990 Comparative effects of a novel vitamin D analogue MC-903 and 1,25-dihydroxyvitamin D_3 on alkaline phosphatase activity, osteocalcin and DNA synthesis by human osteoblastic cells in culture. Bone **11**:171–179.
12. Theofan G, Price PA 1989 Bone Gla protein messenger ribonucleic acid is regulated by both 1,25-dihydroxyvitamin D_3 and 3′,5′-cyclic adenosine monophosphate in rat osteosarcoma cells. Mol Endocrinol **3**:36–43.
13. Demay MB, Kierman MS, DeLuca HF, Kronenberg HM 1992 Characterization of 1,25-dihydroxyvitamin D_3 receptor interactions with target sequence in the rat osteocalcin gene. Mol Endocrinol **6**:557–562.
14. Mahonen A, Pirskanen A, Keinanen R, Maenpaa PH 1990 Effect of $1,25(OH)_2D_3$ on its receptor mRNA levels and osteocalcin synthesis in human osteosarcoma cells. Biochim Biophys Acta **1048**:30–37.
15. Yoon K, Rutledge SJC, Buenaga RF, Rodan GA 1988 Characterization of the rat osteocalcin gene: Stimulation of promoter activity by 1,25-dihydroxyvitamin D_3. Biochemistry **27**:8521–8526.
16. McKee MD, Glimcher MJ, Nanci A 1992 High-resolution immunolocalization of osteopontin and osteocalcin in bone and cartilage during endochondral ossification in the chicken tibia. Anat Rec **234**:479–492.
17. Noda M, Vogel RL, Craig AM, Prahl J, DeLuca HF, Denhardt DT 1990 Identification of a DNA sequence responsible for binding of the 1,25-dihydroxyvitamin D_3 receptor and 1,25-dihydroxyvitamin D_3 enhancement of mouse secreted phosphoprotein 1 (SPP-1 or osteopontin) gene expression. Proc Natl Acad Sci USA **87**:9995–9999.
18. Fraser JD, Price PA 1990 Induction of matrix Gla protein synthesis during prolonged 1,25-dihydroxyvitamin D_3 treatment of osteosarcoma cells. Calcif Tissue Int **46**:270–279.
19. Kurihara N, Ishizuka S, Kiyoki M, Haketa Y, Ikeda K, Kumegawa M 1986 Effects of 1,25-dihydroxyvitamin D_3 on osteoblastic MC3T3-E1 cells. Endocrinology **118**:940–947.
20. Owen TA, Aronow MS, Barone LM, Bettencourt B, Stein GS,

Lian JB 1991 Pleiotropic effects of vitamin D on osteoblast gene expression are related to the proliferative and differentiated state of the bone cell phenotype: Dependency upon basal levels of gene expression, duration of exposure, and bone matrix competency in normal rat osteoblast cultures. Endocrinology **128**:1496–1504.

21. Pavlin D, Bedalov A, Kronenberg MS, Kream BE, Rowe DW, Smith CL, Pike JW, Lichtler AC 1994 Analysis of regulatory regions in the COL1A1 gene responsible for 1,25-dihydroxyvitamin D_3-mediated transcriptional repression in osteoblastic cells. J Cell Biochem **56**:490–501.
22. Takeuchi Y, Matsumoto T, Ogata E, Shishiba Y 1989 1,25-Dihydroxyvitamin D_3 inhibits synthesis and enhances degradation of proteoglycans in osteoblastic cells. J Biol Chem **264**:18407–18413.
23. Farach-Carson MC, Abe J, Nishii Y, Khoury R, Wright GC, Norman AW 1993 22-Oxacalcitriol: Dissection of $1,25(OH)_2D_3$ receptor mediated and Ca^{2+} entry-stimulating pathways. Am J Physiol **265**:F705–F711.
24. Jenis LG, Lian JB, Stein GS, Baran DT 1993 1α,25-Dihydroxyvitamin D_3-induced changes in intracellular pH in osteoblast-like cells modulate gene expression. J Cell Biochem **53**:234–239.
25. Moriwake T, Tanaka H, Kanzaki S, Higuchi J, Seino Y 1992 1,25-Dihydroxyvitamin D_3 stimulates the secretion of insulin-like growth factor binding protein 3 (IGFBP-3) by cultured human osteosarcoma cells. Endocrinology **130**:1071–1073.
26. Scharla SH, Strong DD, Mohan S, Baylink DJ, Linkhart TA 1991 1,25-Dihydroxyvitamin D_3 differentially regulates the production of insulin-like growth factor-1 (IGF-1) and IGF binding protein-4 in mouse osteoblasts. Endocrinology **129**:3139–3146.
27. Finkelman RD, Linkhart TA, Mohan S, Lau KH, Baylink DJ, Bell NH 1991 Vitamin D deficiency causes a selective reduction in deposition of transforming growth factor β in rat bone: Possible mechanism for impaired osteoinduction. Proc Natl Acad Sci USA **88**:3657–3660.
28. Lidor C, Sagiv P, Amdur B, Gepstein R, Otremski I, Hallel T, Edelstein S 1993 Decrease in bone levels of 1,25-dihydroxyvitamin D in women with subcapital fracture of the femur. Calcif Tissue Int **52**:146–148.
29. Chapuy MC, Arlot ME, Duboeuf, Brun J, Crouzet B, Arnaud S, Delmas PD, Meunier PJ 1992 Vitamin D_3 and calcium to prevent hip fractures in elderly women. N Engl J Med **327**:1637–1642.
30. Tilyard MW, Spears GFS, Thomson J, Dovey S 1992 Treatment of postmenopausal osteoporosis with calcitriol or calcium. N Engl J Med **326**:357–362.
31. Gallagher JC, Goldgar D 1990 Treatment of postmenopausal osteoporosis with high doses of synthetic calcitriol. A randomized controlled study. Ann Intern Med **113**:649–655.
32. Fujita T 1990 Studies of osteoporosis in Japan. Metabolism **39**:39–42.
33. Brandi ML 1993 New treatment strategies: Ipriflavone, strontium, vitamin D metabolites and analogs. Am J Med **95**:69S–74S.
34. Ott SM, Chesnut CH 1989 Calcitriol treatment is not effective in postmenopausal osteoporosis. Ann Intern Med **110**:267–274.
35. Morrison NA, Qi JC, Tokita A, Kelly PJ, Crofts L, Nguyen TV, Sambrook PN, Eisman JA 1994 Prediction of bone density from vitamin D receptor alleles. Nature **367**:284–287.
36. Garnero P, Borel O, Sornay-Rendu E, Delmas PD 1995 Vitamin D receptor gene polymorphisms do not predict bone turnover and bone mass in healthy premenopausal women. J Bone Miner Res **10**:1283–1288.
37. Ferrari S, Rizzoli R, Chevalley T, Slosman D, Eisman JA, Bonjour JP 1995 Vitamin-D-receptor-gene polymorphisms and change in lumbar-spine bone mineral density. Lancet **345**:423–424.
38. Dawson-Hughes B, Harris SS, Finneran S 1995 Calcium absorption on high and low calcium intakes in relation to vitamin D receptor genotype. J Clin Endocrinol Metab **80**:3657–3661.
39. Halloran BP, DeLuca HF 1981 Effect of vitamin D deficiency on skeletal development during early growth in the rat. Arch Biochem Biophys 209:7–14.
40. Weinstein RS, Underwood JL, Hutson MS, DeLuca HF 1984 Bone histomorphometry in vitamin D-deficient rats infused with calcium and phosphorus. Am J Physiol **246**:E499–E505.
41. Henry HL, Norman AW 1978 Vitamin D: Two dihydroxylated metabolites are required for normal chicken hatchability. Science **201**:835–837.
42. Norman AW, Leathera VL, Bishop JE 1983 Studies on the mode of action of calciferol: Normal egg hatchability requires the simultaneous administration to the hen of 1α,25-dihydroxyvitamin D_3 and 24*R*,25-dihydroxyvitamin D_3. J Nutr 113:2505–2525.
43. Nishimura A, Shinki T, Jin CH, Ohyama Y, Noshiro M, Okuda K, Suda T 1994 Regulation of messenger ribonucleic acid expression of 1α,25-dihydroxyvitamin D_3-24-hydroxylase in rat osteoblasts. Endocrinology **134**:1794–1799.
44. Somjen D, Binderman I, Weisman Y 1983 The effects of 24*R*,25-dihydroxycholecalciferol on ornithine decarboxylase activity and on DNA synthesis in the epiphysis and diaphysis of rat bone and in the duodenum. Biochem J **214**:293–298.
45. Blaugrund E, Edelstein S 1988 Response of rachitic cartilage cells to metabolites of vitamin D_3. Cell Biol Int Rep **12**:373–381.
46. Schwaltz Z, Boyan B 1988 The effects of vitamin D metabolites on phospholipase A_2 activity of growth zone and resting cartilage cells *in vitro*. Endocrinology **122**:2191–2198.
47. Hale LV, Kemick MLS, Wuthier RE 1986 Effects of vitamin D metabolites on the expression of alkaline phosphatase activity by epiphyseal hypertrophic chondrocytes in primary cell culture. J Bone Miner Res **1**:489–495.
48. Takigawa M, Enomoto M, Shirai E, Nishii Y, Suzuki F 1988 Differential effects of 1α,25-dihydroxycholecalciferol and 24*R*,25-dihydroxycholecalciferol on the proliferation and the differentiation phenotype of rabbit costal chondrocyte in culture. Endocrinology **122**:831–839.
49. Somjen D, Somjen GJ, Weisman Y, Binderman I 1982 Evidence for 24,25-dihydroxycholecalciferol receptors in long bones of newborn rats. Biochemistry **204**:31–36.
50. Fine N, Binderman I, Somjen D, Earon Y, Edelstein S, Weisman Y 1985 Autoradigraphic localization of 24*R*,25-dihydroxyvitamin D_3 in epiphyseal cartilage. Bone **6**:99–104.
51. Mortensen BM, Gautvik KM, Gordeladze JO 1993 Bone turnover in rats treated with 1,25-dihydroxyvitamin D_3, 25-hydroxyvitamin D_3 or 24,25-dihydroxyvitamin D_3. Biosci Rep **13**:27–39.
52. Nakamura T, Suzuki K, Hirai T, Kurokawa T, Orimo H 1992 Increased bone volume and reduced bone turnover in vitamin D-repleted rabbits by the administration of 24*R*,25-dihydroxyvitamin D_3. Bone **13**:229–236.
53. Nakamura T, Kurokawa T, Orimo H 1989 Increased mechanical strength of the vitamin D replete rat femur by the treatment with a large dose of $24R,25(OH)_2D_3$. Bone **10**:117–123.
54. Nakamura T, Nagai Y, Yamato H, Suzuki K, Orimo H 1992 Regulation of bone turnover and prevention of bone atrophy in ovariectomized beagle dogs by the administration of $24R,25(OH)_2D_3$. Calcif Tissue Int **50**:221–227.
55. Lidor C, Delel S, Hallel T, Edelstein S 1987 Levels of active metabolites of vitamin-D_3 in the callus of fracture repair in chicks. J Bone Joint Surg **69**:132–136.
56. Yamato H, Okazaki R, Ishii T, Ogata E, Sato T, Kumegawa M, Akaogi K, Taniguchi N, Matsumoto T 1993 Effect of 24*R*,25-dihydroxyvitamin D_3 on the formation and function of osteoclastic cells. Calcif Tissue Int **52**:255–260.

57. Yamate T, Tanaka H, Nagai Y, Yamato H, Taniguchi N, Nakamura T, Seino Y 1994 Bone-forming ability of 24*R*,25-dihydroxyvitamin D_3 in the hypophosphatemic mouse. J Bone Miner Res **9**:1967–1974.
58. Ono T, Tanaka H, Yamate T, Nagai Y, Nakamura T, Seino Y 1996 24*R*,25-Dihydroxyvitamin D_3 promotes bone formation without causing excessive resorption in hypophosphatemic mice. Endocrinology **137**:2633–2637.
59. Matsumoto T, Ikeda K, Yamato H, Morita K, Ezawa I, Fukushima M, Nishii Y, Ogata E 1988 Effect of 24*R*,25-dihydroxyvitamin D_3 on 1,25-dihydroxyvitamin D_3 metabolism in calcium-deficient rats. Biochem J **250**:671–677.
60. Wilhelm F, Ross FP, Norman AW 1986 Specific binding of 24*R*,25-dihydroxyvitamin D_3 is an allosteric effector of 1,25-dihydroxyvitamin D_3 binding. Arch Biochem Biophys **249**:88–94.
61. Wilhelm F, Norman AW 1985 24*R*,25-Dihydroxyvitamin D_3 regulates 1,25-dihydroxyvitamin D_3 binding to its chick intestinal receptor. Biochem Biophys Res Commun **126**:496–501.
62. Carpenter TO, Keller M, Schwartz D, Mitnick M, Smith C, Ellison A, Carey D, Comite F, Horst R, Travers R, Glorieux FH, Gundberg CM, Poole AR, Insogna KL 1996 24,25-Dihydroxyvitamin D supplementation corrects hyperparathyroidism and improves skeletal abnormalities in X-linked hypophosphatemic rickets—A clinical research center study. J Clin Endocrinol Metabl **81**:2381–2388.
63. Ishizuka S, Ishimoto S, Norman AW 1984 Isolation and identification of 1α,25-dihydroxy-24-oxo-vitamin D_3, 1α,25-dihydroxyvitamin D_3-26,23-lactone and 1α,24(*S*),25-trihydroxyvitamin D_3: *In vivo* metabolites of 1α,25-dihydroxyvitamin D_3. Biochemistry **23**:1473–1478.
64. Ishizuka S, Norman AW 1987 Metabolic pathways of 1α,25-dihydroxyvitamin D_3 to 1α,25-dihydroxyvitamin D_3-26,23-lactone: Stereo-retained and stereo-selective lactonization. J Biol Chem **262**:7165–7170.
65. Siu-Caldera ML, Zou L, Ehrlich MG, Schwartz ER, Ishizuka S, Reddy GS 1995 Human osteoblasts in culture metabolize both 1α,25-dihydroxyvitamin D_3 and its precursor 25-hydroxyvitamin D_3 into their respective lactones. Endocrinology **136**:4195–4203.
66. Ishizuka S, Ishimoto S, Norman AW 1984 Biological activity assessment of 1α,25-dihydroxyvitamin D_3-26,23-lactone in the rat. J Steroid Biochem **20**:611–615.
67. Wilhelm F, Mayer E, Norman AW 1984 Studies on the mode of action of calciferol. Biological activity assessment of 26,23-lactones of 1,25-dihydroxyvitamin D_3 and 25-hydroxyvitamin D_3 and their binding-properties to chick intestinal receptor and plasma vitamin-D binding-protein. Arch Biochem Biophys **233**:322–329.
68. Ishizuka S, Kiyoki M, Kurihara N, Hakeda Y, Ikeda K, Kumegawa M, Norman AW 1988 Effects of diastereoisomers of 1,25-dihydroxyvitamin D_3-26,23-lactone on alkaline phosphatase and collagen synthesis in osteoblastic cells. Mol Cell Endocrinol **55**:77–86.
69. Kiyoki M, Kurihara N, Ishizuka S, Ishii S, Hakeda Y, Kumegawa M, Norman AW 1985 Unique action for bone metabolism of 1α,25-dihydroxyvitamin D_3-26,23-lactone. Biochem Biophys Res Commun **127**:693–698.
70. Ishizuka S, Kurihara N, Haleda Y, Maeda N, Ikeda K, Kumegawa M, Norman AW 1988 1α,25-Dihydroxyvitamin D_3 [1α,25-$(OH)_2D_3$]-26,23-lactone inhibits 1,25-$(OH)_2D_3$-mediated fusion of mouse bone marrow mononuclear cells. Endocrinology **123**:781–786.
71. Shima M, Tanaka H, Norman AW, Yamaoka K, Yoshikawa H, Takaoka K, Ishizuka S, Seino Y 1990 23(*S*),25(*R*)-1,25-Dihydroxyvitamin D_3-26,23-lactone stimulates murine bone formation *in vivo*. Endocrinology **126**:832–836.
72. Seino Y, Ishizuka S 1992 23(*S*),25(*R*)-1α,25$(OH)_2D_3$-26,23-Lactone. Drugs Future **17**:655–659.

CHAPTER 20

Vitamin D and Osteoblasts

JANE E. AUBIN Department of Anatomy and Cell Biology, Faculty of Medicine, University of Toronto, Toronto, Ontario, Canada

JOHAN N. M. HEERSCHE Faculty of Dentistry, University of Toronto, Toronto, Ontario, Canada

I. INTRODUCTION

1,25-Dihydroxyvitamin D_3 [$1,25(OH)_2D_3$] affects the proliferation and differentiation of numerous cell lineages/cell types. For example, $1,25(OH)_2D_3$ promotes the differentiation of keratinocytes from epidermal precursors [1,2], monocytes–macrophages from myelopoietic progenitors/stem cells and osteoclasts from mononuclear precursors [3,4], and adipocytes in at least some models [5,6] (see Section II.D). It also induces hypertrophy of chondrocytes [7,8]. The ability of $1,25(OH)_2D_3$ to modulate differentiation pathways as well as its ability to alter cell expression through regulation of transcription or through potential genomic pathways have been reviewed [9–13], and many aspects are covered elsewhere in this volume. This chapter focuses on the ability of $1,25(OH)_2D_3$ to modulate osteoblast differentiation and activity *in vitro* and *in vivo*.

II. EFFECTS OF $1,25(OH)_2D_3$ ON OSTEOBLASTS *IN VITRO*

A. $1,25(OH)_2D_3$ Modifies the Osteoblast Differentiation Pathway *in Vitro* and Has Selective Effects Depending on the Relative Differentiation Status of the Cells

It has been known for some time that osteoblasts possess specific receptors for $1,25(OH)_2D_3$ [14–17], and several early studies of osteoblastic cells *in vitro* showed that $1,25(OH)_2D_3$ alters expression of osteoblast-associated genes, for example, increasing alkaline phosphatase activity [18–21] and stimulating expression of some of the noncollagenous proteins of bone such as osteopontin [22–24] and osteocalcin [25]. However, concomitant with stimulation of expression of some osteoblast phenotypic markers, there occurs inhibition of others, such as type I collagen [26,27] and bone sialoprotein [28,29]. Similarly, $1,25(OH)_2D_3$ was reported to decrease type I collagen and alkaline phosphatase mRNA and protein levels in rat organ cultures [30] and in rat osteosarcoma cells [31]. Whereas these studies were important for establishing the basis for further analysis, the observations were difficult to reconcile with any straightforward model of osteoblast functional status, although it was suggested that at least some of the variable effects might be dependent on the maturational state of the cells used. Many of the early studies relied on analysis of cell lines whose proliferation–differentiation coupling was aberrant, and few experiments were done under conditions in which a complete differentiation sequence could be followed. Better understanding of some of the apparent contradictions and discrepancies has come from several approaches including the use of primary cultures of osteoblastic cells undergoing a developmental sequence *in vitro*. With these models, it has become clearer that $1,25(OH)_2D_3$ can alter, in a biphasic manner, osteoblastic cells as they undergo a differentiation sequence from early proliferating cell to mature osteoblast.

One model that has been useful is that of rat calvaria cell populations. It has been well established that rat calvaria populations contain osteoprogenitor cells that, under appropriate culture conditions (i.e., addition of ascorbic acid and β-glycerophosphate to the culture medium), can differentiate into mature osteoblasts that are capable of forming a mineralized tissue resembling woven bone [32–35] (Fig. 1). *In vitro* the cells progress through a developmental sequence of proliferation and differentiation with stage-specific expression of growth- and bone-related genes; the sequence has been well-characterized by morphological [33,36], immunohistochemical [33,37,38], and molecular [38–40] approaches. The process has been categorized overall as a period of proliferation (~days 1–7), followed by multilayering and early nodule formation (~days 7–11) that is characterized by biosynthesis and organization of type I collagen and noncollagenous proteins of bone, growth of nodules (~days 11–15), and finally osteoid mineralization (>day 15) (reviewed by Aubin *et al.* [34] and Stein and Lian [35]). Of particular interest is the fact that multiple factors have been found to influence this proliferation–differentiation process, some with effects at multiple stages and often biphasic in nature [41,42]. Some of the most widely studied agents are the glucocorticoids, which stimulate nodule formation [43], increase the proliferative lifetime of the bone-forming precursor [44], and induce primitive osteoprogenitors to differentiate that do not do so in its absence [45].

Although some factors appear to be inhibitory throughout the proliferation–differentiation sequence that leads to formation of a bone nodule, for example, transforming growth factor-β (TGFβ) [46–50], others

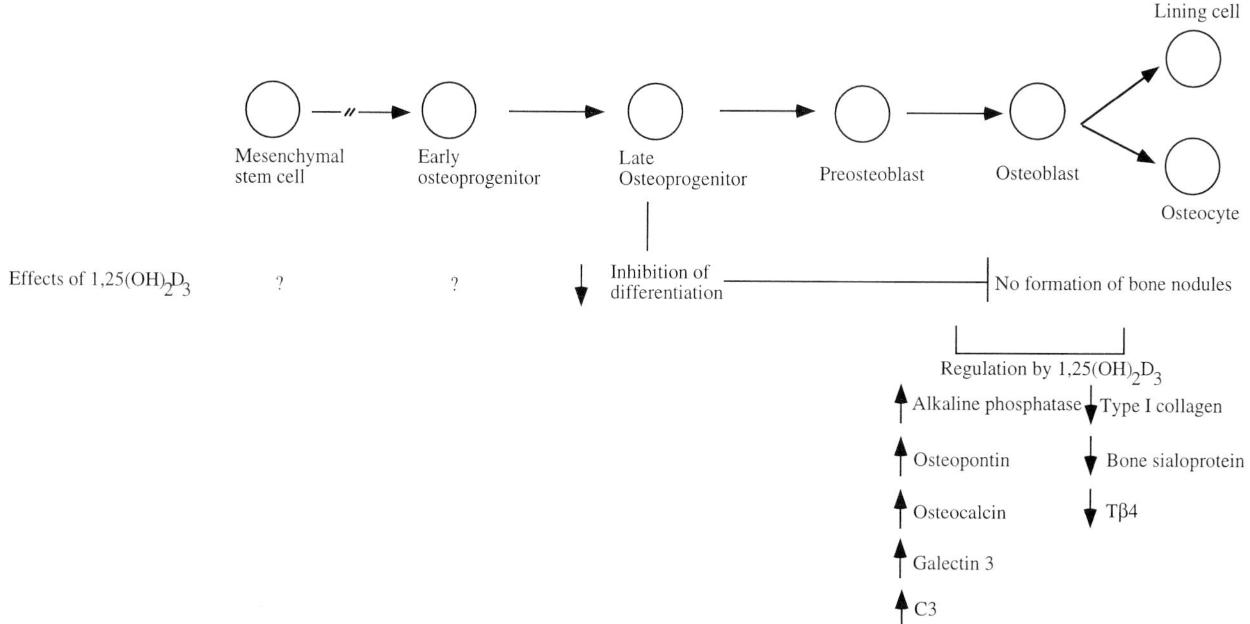

FIGURE 1 The hormone $1,25(OH)_2D_3$ has various effects on cells at different maturational stages during their differentiation from osteoprogenitors to mature osteoblasts, osteocytes, and lining cells. In rat calvaria populations, osteoprogenitor cells are inhibited from undergoing further differentiation, while some osteoblast-associated genes are up-regulated and others are down-regulated in more mature cells in the lineage. Tβ4, Thymosin β_4. C3, third component of complement.

such as interleukin-1 (IL-1) [51] and epidermal growth factor (EGF) [52] have biphasic effects that depend on the differentiation stage at which cells are treated; the latter stimulate nodule formation presumably through a mitogenic effect on proliferating progenitors or inhibit when present during particular sensitive differentiation windows. It is within the latter category—a hormone with biphasic effects that can modify (abrogate or stimulate) the normal developmental pathway or gene expression profiles associated with the normal developmental process—that $1,25(OH)_2D_3$ belongs. The ability to manipulate the *in vitro* model during precise windows of time corresponding to proliferative stages, early events, and late events in osteoblast differentiation, bone formation, and mineralization makes it a useful alternative to *in vivo* studies or most cell lines.

In the rat calvaria cell culture model, when $1,25(OH)_2D_3$ treatment is initiated prior to confluence during the proliferative phase, mRNA levels for alkaline phosphatase and type I collagen are suppressed, bone nodules do not form, and the up-regulation of osteopontin and osteocalcin mRNAs that is seen in the normal differentiation process does not occur [53]. In contrast to the inhibitory effects of $1,25(OH)_2D_3$ on osteoblast differentiation when added during the proliferative period, acute or continuous treatment of mature (postproliferative) cultures results in up-regulation of at least some osteoblast-associated genes, such as osteopontin and osteocalcin [53], consistent with other observations showing stimulation of the production of these two proteins by $1,25(OH)_2D_3$ in this and other osteoblastic models [22–25] (see also below). Together with the earlier studies reported above, these data suggested that $1,25(OH)_2D_3$ may stimulate mature osteoblasts while inhibiting early osteoid formation or the differentiation of osteoprogenitors.

Further evidence for this hypothesis was obtained by treating rat calvaria cell cultures either continuously or acutely with pulse treatment of $1,25(OH)_2D_3$ at particular stages during the differentiation process [54]. Continuous treatment of rat calvaria cells with $1,25(OH)_2D_3$ dose-dependently inhibited bone nodule formation, with half-maximal inhibition at 0.06 n*M*, while concomitantly stimulating rat calvaria cell growth. Notably, inhibition of bone nodule formation occurred when $1,25(OH)_2D_3$ was present during the proliferation phase and up to the early multilayering stage (e.g., up to day 11) even when it was present for as little as 48 hr [54]. Thus, $1,25(OH)_2D_3$ was inhibitory to osteoprogenitor differentiation during proliferation and the earlier stages of differentiation before visible nodule formation occurred, at a time when cell growth was stimulated, and the effect was not reversible on removal of $1,25(OH)_2D_3$ (Fig. 1). Two other vitamin D_3 metabolites, $24,25(OH)_2D_3$ and $1,24,25(OH)_3D_3$, had dose-dependent inhibitory effects similar to those of $1,25(OH)_2D_3$, with effectiveness in the sequence $1,25(OH)_2D_3 > 1,24,25(OH)_3D_3 > 24,25(OH)_2D_3$ [54], consistent with the biological efficacy of these metabolites in other systems where they have been compared (e.g., [55–58]).

Do other primary osteoblastic cell models behave similarly to the rat calvaria cell model with respect to $1,25(OH)_2D_3$ effects? Interestingly, although some features are the same, others are not, and these are beginning to lead to a consistent picture as more data accumulate on the relative differentiation status of the majority of cells present in each model. Similar to the rat calvaria culture results, addition of $1,25(OH)_2D_3$ to preconfluent/proliferating chick embryonic osteoblastic cultures over a 30-day period resulted in a 2- to 10-fold reduction in collagen, osteopontin, osteocalcin, and alkaline phosphatase content and mineral deposition [59]. However, in contrast to the rat calvaria system, acute treatments for 24 hr during the proliferative phase (day 5), or when cell proliferation had ceased and cultures had started differentiating, or late during the matrix mineralization phase, caused inhibition at the latter two time points but not during the proliferative period. Broess *et al.* [59] concluded that the inhibitory effects of $1,25(OH)_2D_3$ on the chick cells reflected effects on relatively mature stages and were independent of effects on proliferation, and that the inhibition was of genes associated with the differentiated cells such that matrix deposition and mineralization were inhibited. The authors hypothesized that the lack of correlation between the chick and rat systems at the mature osteoblast stage might reside in the fact that the cells are from bones of different developmental ages, namely, day 17 fetal chick versus day 21 fetal rat. Because the chick calvaria is more highly mineralized at this developmental age than is the rat calvaria at fetal day 21, the cells from the former might be more mature than the latter. Data from preliminary experiments in which they compared cultures of chick osteoblasts derived from early embryos (day 12) to osteoblasts from day 17 embryos support the conclusion that $1,25(OH)_2D_3$ blocks differentiation of the day 12 embryonic cells and up-regulates differentiation markers in more mature osteoblastic cells in a manner similar to the rat calvaria cultures [59]. These data suggest that although cells from both ages are able to undergo a maturational sequence and deposit a mineralizing matrix *in vitro*, some characteristics of the cells and the process in the two cases are different, and the age or maturational stage of the cells at isolation influences the effect of hormones including $1,25(OH)_2D_3$.

A number of other studies support the idea that the stage of cellular differentiation of different osteoblastic lines affects their responsiveness to $1,25(OH)_2D_3$.

Fraser *et al.* [60] found that different rat osteosarcoma lines were differentially responsive to $1,25(OH)_2D_3$ in terms of osteocalcin or matrix Gla protein; notably, prolonged $1,25(OH)_2D_3$ treatment of ROS 17/2.8 rat osteosarcoma cells resulted in down-regulation of osteocalcin and up-regulation of matrix Gla [61], supporting the notion that, at some differentiation stages, $1,25(OH)_2D_3$ can alter cell expression to a more mature phenotype. Franceschi and colleagues came to similar conclusions by treating the MG-63 human osteosarcoma line with $1,25(OH)_2D_3$; the $1,25(OH)_2D_3$ treatment promoted partial differentiation of the line [62–64].

Evidence that $1,25(OH)_2D_3$ may "push" relatively mature osteoblasts toward an even more mature state comes from human osteoblastic cells also. Innumerable studies have documented the ability of $1,25(OH)_2D_3$ to stimulate expression of osteogenic markers, most notably osteocalcin, which is often undetectable without $1,25(OH)_2D_3$ stimulation, in cells derived from human trabecular bone [21,65–69] (for reviews, see Rodan and Rodan [70], and Marie [71]). The accumulating evidence suggests that cultures derived from human trabecular bone fragments may be largely representative of more mature cells (see discussion in Marie [71]), consistent with the ability of $1,25(OH)_2D_3$ to stimulate genes such as osteopontin and osteocalcin in mature cells in rat calvaria cultures. However, it is also clear that, concomitant with stimulation of the expression of these markers, the expression of other markers such as bone sialoprotein is inhibited by $1,25(OH)_2D_3$ in several other cell systems (see above). This may be hard to reconcile with the suggestion that $1,25(OH)_2D_3$ acts synergistically with dexamethasone to stimulate expression of all osteogenic markers investigated, including alkaline phosphatase, bone sialoprotein, and osteocalcin, in human marrow-derived stromal cells [72]. The observation that sequential addition of $1,25(OH)_2D_3$ enhances the actions of dexamethasone in bone marrow cultures and is required for maximum osteocalcin and osteopontin mRNA expression [37,73], however, supports the view that $1,25(OH)_2D_3$ is up-regulating some markers of the mature osteoblast.

Rickard *et al.* [74] found that dexamethasone and $1,25(OH)_2D_3$ did not have the same effects on all osteoblast markers tested in early or more mature colonies and cultures of human bone marrow stromal-derived cells. In particular, $1,25(OH)_2D_3$ stimulated osteocalcin production in both younger and older cultures, and collagen type I especially in older cultures, but was largely without effect on other markers. Combinations of dexamethasone and $1,25(OH)_2D_3$ led to coexpression of osteoblast and adipocyte markers. Rickard *et al.* [74] discuss various alternative explanations for the discrepancies, including both species differences and developmental stage differences, but more work will be required to correlate the human and the rodent and avian data in the context of multilineage cells and maturational stage of restricted osteoblast progenitors (see also Section II.D).

The apparently relatively mature state of human bone-derived cells and the ability of $1,25(OH)_2D_3$ to induce the cells to a more mature state is also supported by other results. Monoclonal antibodies raised against $1,25(OH)_2D_3$-treated human trabecular bone osteoblast-like cells were largely directed against epitopes present in end-stage osteoblasts and osteocytes [75]. $1,25(OH)_2D_3$ used alone or in combination with TGFβ in primary cultures of human osteoblastic cells caused the largely spindle-shaped cells to become stellate [76]. When proliferation and expression of collagen type I, alkaline phosphatase, and osteocalcin were analyzed, the authors concluded that TGFβ stimulates matrix synthesis in these human cells and that $1,25(OH)_2D_3$ may push the cells to an end-stage phenotype (lower proliferation, increased osteocalcin). This is consistent with the ideas proposed for rodent and chick cells and other studies in human bone-derived cells in which $1,25(OH)_2D_3$ inhibited proliferation and increased alkaline phosphatase and osteocalcin expression [77]. However, the complexity of the results and interpretation is evident when alkaline phosphatase and osteocalcin were colocalized in another study of $1,25(OH)_2D_3$-treated human bone-derived cells. Not all cells that made osteocalcin in response to $1,25(OH)_2D_3$ (9%) synthesized alkaline phosphatase (24%) and vice versa, whereas a proportion produced both (12%). Thavarajah *et al.* concluded that during differentiation in response to $1,25(OH)_2D_3$ human cells ended up with heterogeneous phenotypes [77].

Heterogeneity of mature osteoblasts has also been documented in rat bones *in vivo* by *in situ* hybridization of osteocalcin in comparison to type I collagen [78] and bone sialoprotein in comparison to osteopontin [79,80], by immunohistochemistry of bone sialoprotein versus alkaline phosphatase in secretory osteoblasts [81], and by an analysis of osteocalcin expression in transgenic mice [82]. Heterogeneity has also been found in rat calvaria osteoblasts *in vitro* by both immunocytochemistry and polymerase chain reaction (PCR) analyses for all markers of mature osteoblasts including alkaline phosphatase, type I collagen, and most strikingly bone sialoprotein, osteocalcin, and osteopontin [37,38,83]. The heterogeneity is also seen before and after $1,25(OH)_2D_3$ treatment in nodules in rat calvaria cultures assessed by *in situ* hybridization [84]. Thus, heterogeneity is not a consequence of $1,25(OH)_2D_3$ treatment; rather, a possibly heterogeneous response to $1,25(OH)_2D_3$ must be taken into account in interpreting $1,25(OH)_2D_3$ effects on bone and osteoblastic cells.

The concept of a differentiating cell having different windows of responsiveness to the same physiological mediator underscores the need for rigorous determination of cell behavior throughout the differentiation and maturational sequence (see e.g., Aubin *et al.* [34,41] and Stein and Lian [35]). *In situ* hybridization and immunocytochemistry can be used to further elucidate whether it is specific subpopulations of cells within cultures and within nodules that respond to hormones [84,85]. As discussed above, in contrast to the inhibitory effects of $1,25(OH)_2D_3$ on osteoblast differentiation when the hormone is added during the proliferative period, acute or continuous treatment to postproliferative mature osteoblasts causes up-regulation of genes such as osteopontin and osteocalcin [53] (Fig. 1). Altered osteocalcin and osteopontin mRNA levels resulting from regulation by acute exposure to $1,25(OH)_2D_3$ were found by *in situ* hybridization analysis to be restricted to nodule-associated differentiated cells [84,85]. In analyses done with immunocytochemistry in differentiating bone marrow cell cultures, acute exposure to $1,25(OH)_2D_3$ was also found to up-regulate expression of osteocalcin and in addition galectin 3 (i.e., RCC455.4 positive cells), a member of an important class of lectin adhesion-related molecules and found in high abundance in cells also expressing osteocalcin [37]. However, in this case, the up-regulation was evident also in single cells and small groups of cells acutely treated early in the culture period prior to nodule formation; it was concluded that $1,25(OH)_2D_3$ was up-regulating expression in small groups of cells already mature when released from the bone [37]. Consistent with this interpretation, these soon disappeared *in vitro*, and new osteocalcin positive cells with the concomitant stimulation by $1,25(OH)_2D_3$ appeared later as nodule formation occurred and as seen in the *in situ* hybridization studies. $1,25(OH)_2D_3$ pulse treatment of nodule-associated cells was found to be accompanied by shape changes (cells become more elongated, flattened, polarized), and it was in the flat cells that osteocalcin was so markedly up-regulated [85]. On the other hand, as already outlined, expression of type I collagen and bone sialoprotein are inhibited by $1,25(OH)_2D_3$ at these cell maturational stages.

These changes in morphology and gene expression may relate to bone resorbing effects of $1,25(OH)_2D_3$ on bone surface lining osteoblasts [85] (see also Section III). Indeed, osteocalcin expression as determined in bone lining cells is observed by *in situ* hybridization at a significant level *in vivo* [78,86]. Both Owen *et al.* [53] and Broess *et al.* [59] also found that mineralization of osteoid was inhibited by $1,25(OH)_2D_3$ when it was added after deposition of the collagenous matrix, suggesting that the hormone can affect the ability of osteoblasts to mineralize their matrix, perhaps independently of an effect on matrix formation.

The effects of $1,25(OH)_2D_3$ on osteoblastic cells indirectly through nonadherent cells in the bone marrow must also be considered. $1,25(OH)_2D_3$ has been found to stimulate stromal cell proliferation and alkaline phosphatase activity in rat bone marrow cultures in a manner dependent on the nonadherent fraction of cells [87]. Although the authors believed they were stimulating the osteoblastic subpopulation of stromal cells, given that it appears that only a proportion of alkaline phosphatase positive colonies are osteoblastic at least under a given set of conditions [88], caution must be used in concluding the results are specific for osteoblastic cells. In another series of experiments, Long and colleagues have been isolating osteoblast precursors and osteoprogenitors from the nonadherent fraction of human bone marrow. Intriguingly, some nonadherent cells give rise to colonies with cells that express osteocalcin, osteonectin, and bone alkaline phosphatase and respond to $1,25(OH)_2D_3$ [89]. Clearly, the widespread presence of the vitamin D receptor (VDR) and its multiple activities in many cell types suggests that many effects of the hormone to alter osteoblast activity may also be mediated via other cell types.

B. $1,25(OH)_2D_3$ Regulates Genes Associated with Osteoblast Proliferation and Differentiation

As discussed above, $1,25(OH)_2D_3$ up-regulates or down-regulates a wide variety of genes associated with the osteoblast phenotype. What are the molecular mediators of these complex and sometimes biphasic biological effects? As discussed elsewhere in this volume, $1,25(OH)_2D_3$ activates multiple signaling pathways in osteoblasts, inducing rapid nongenomic and long-term genomic pathways. Nongenomic pathways involve lipid turnover, activation of Ca^{2+} channels, and elevation of intracellular Ca^{2+}, all of which occur within seconds of administration of the steroid *in vitro*.

Genomic pathways are mediated by the VDR, a member of the steroid receptor superfamily, and involve transcriptional regulation of target genes. Thus, in at least some cases, $1,25(OH)_2D_3$ effects may be relatively direct on the osteoblast-specific gene. For example, it is known that $1,25(OH)_2D_3$ following its association with the VDR regulates osteocalcin via transcriptional mechanisms; together with other transcription factors, the hormone and receptor form a complex that interacts in a sequence-specific manner with promoter regulatory elements [90–94] (see Chapter 8–10); other hormone response elements may interfere with this induction [95]. Some other factors may be effective through their ability to regulate VDR levels, for example, tumor necrosis factor-α (TNFα) [96,97], glucocorticoids [98], and TGF-β [99,100] (see Chapter 11). However, it is becoming clear that other transcription factors may also vary as a function of the differentiation status of osteoblasts [101–104], such that changing levels of these may also play a role in the ability of $1,25(OH)_2D_3$ to regulate responsive genes. This appears to be true, for example, of Fos–Jun family members, which vary as a function of the rat calvaria cell proliferation–differentiation cycle [105]. Interestingly, $1,25(OH)_2D_3$ has been shown to have differential effects on different members of the Fos–Jun family and in some cases to act via regulation of initiation and elongation of transcription (e.g., c-*fos*) versus at a posttranscriptional level distinct from mRNA stabilization (e.g., c-*jun*, *jun-B*) in MC3T3-El mouse osteoblast cells [106]. It may also be that one needs to consider another regulatory level, namely, the ability of $1,25(OH)_2D_3$ to modulate binding of other nonreceptor transcription factors to other regulatory sequences; for example, $1,25(OH)_2D_3$ up-regulates the homeodomain protein Msx-2 [107], which is known to be important in skeletal development [108,109] and down-regulates osteocalcin [101,110].

In addition to considering the ability of $1,25(OH)_2D_3$ to regulate genes such as osteocalcin that are associated with differentiated osteoblasts, it will be important to determine how $1,25(OH)_2D_3$ alters the proliferation of precursors and what role this plays in the ability of $1,25(OH)_2D_3$ to block or alter the differentiation status of osteoblastic cells. Observations in other cell types may be enlightening. It has been known for many years that $1,25(OH)_2D_3$ induces myeloid cell lines to differentiate into monocytes–macrophages [3,4]. To isolate target genes of the VDR that initiate the differentiation process, Freedman and colleagues [111] used probes prepared from $1,25(OH)_2D_3$-treated or untreated cells to survey a cDNA library from the myelomonocytic U937 cell line. One clone that differentially hybridized is the cyclin-dependent kinase (Cdk) inhibitor p21^{WAF1,CIP1}, a protein that inhibits the cell cycle by associating with cyclin–Cdk complexes and blocking their function. The p21 mRNA appeared very early (2 hr) after treating cells with $1,25(OH)_2D_3$, and the p21 promoter contains a vitamin D response element (VDRE). In studies to confirm that p21 was functionally involved in the $1,25(OH)_2D_3$ regulation of U937 cell differentiation, p21 was transiently overexpressed in these cells, and several monocyte–macrophage markers were up-regulated in the absence in $1,25(OH)_2D_3$. The related Cdk inhibitor p27 had similar activity, suggesting that alterations in cyclin–Cdk complexes induced by $1,25(OH)_2D_3$ can induce the terminal differentiation of this cell type [111].

The widespread presence of VDREs in a number of promoters studied to date means that many genes may be modulated directly by $1,25(OH)_2D_3$ in osteoblastic cells, as the examples already given indicate. It is striking that several genes, only recently identified as osteoblast products and whose functions are not yet fully elucidated in the osteoblast, are also regulated by $1,25(OH)_2D_3$. As mentioned above, galectin 3 (by mRNA levels or as immunolabeled with the monoclonal antibody RCC455.4) is coexpressed at high levels in cells expressing osteocalcin, and both proteins are stimulated by $1,25(OH)_2D_3$ [37]. Importantly, the regulation of galectin 3 expression appears different in osteoblastic cells compared to rat skin fibroblasts, suggesting cell type-specific regulation by hormones such as $1,25(OH)_2D_3$. For example, $1,25(OH)_2D_3$ had no significant effect on and dexamethasone enhanced galectin 3 expression in skin fibroblast cultures, whereas dexamethasone down-regulated and $1,25(OH)_2D_3$ up-regulated galectin 3 expression in rat calvaria and ROS17/2.8 cells. This was especially true at later time points in culture, that is, times corresponding to late stages in the osteogenic differentiation sequence [112].

Other unexpected molecules have been found to be products of osteoblastic cells and to be regulated by $1,25(OH)_2D_3$. For example, thymosin β_4 (Tβ_4), a 43-residue peptide member of a family of closely related peptides with unknown function but with actin-binding capacity, was found in a differential hybridization screen of osteoblastic cells and is expressed in ROS 17/2.8 and UMR-106 cells, and in neonatal and fetal calvaria of rat. Tβ_4 mRNA levels have been reported to be 8 to 10-fold higher in well-differentiated ROS 17/2.8 cells compared to the less differentiated ROS 25/1 line, and $1,25(OH)_2D_3$ down-regulates Tβ_4 mRNA expression [113]. In response to $1,25(OH)_2D_3$, the third component of complement (C3) is produced by both the stromal cell line ST2 and primary mouse osteoblastic cells. $1,25(OH)_2D_3$ also stimulates C3 expression *in vivo*, and C3 has been immunolocalized to periosteal and sutural cells in calvaria and in the tibial metaphyses [114]. An interesting series of experiments has suggested that C3 produced by stromal cells in response to $1,25(OH)_2D_3$ may be involved in the differentiation of osteoclasts from their precursors [115–118].

C. $1,25(OH)_2D_3$ Modifies Osteoblast Responsiveness to Other Hormones and Growth Factors That Influence Osteoblast Differentiation and Activity

Among genes regulated by $1,25(OH)_2D_3$ are those for other hormones, growth factors, or their receptors, suggesting that $1,25(OH)_2D_3$ may act in part by modifying responsiveness to other regulators of osteoblast differentiation and activity. Although it is beyond the scope of this chapter to discuss all of the hormone and cytokine networks with which $1,25(OH)_2D_3$ may interact, some examples may be interesting and informative. In MC3T3-E1 cells, which express large numbers of IL-4 receptors, $1,25(OH)_2D_3$ augments IL-4 binding, increases receptor abundance (B_{max}), and increases receptor mRNA levels, all of which could enhance IL-4 effects, which include increasing cell proliferation and inhibiting alkaline phosphatase activity in these cells [119]. MC3T3-E1 cells also express IL-1 receptor type 1 (by both mRNA and binding studies) and respond to IL-1 through the IL-1 type I receptor to induce IL-6. Accordingly, the ability of $1,25(OH)_2D_3$ to increase the type I receptor in MC3T3-E1 cells may modulate the effect of IL-1 to stimulate IL-6 production. Thus, $1,25(OH)_2D_3$ could be responsible, in part, for modulating IL-1 effects on the skeleton [120]. In other examples, $1,25(OH)_2D_3$ has been shown to increase expression of nerve growth factor (NGF) mRNA in ROS 17/2.8 cells which bind NGF to the apparently low affinity NGF receptor [121]. $1,25(OH)_2D_3$ down-regulates specific cell surface endothelin receptors in ROS 17/2.8 cells [122]. The opioid gene proenkephalin (PENK), which has been identified in osteoblastic cells and found to be down-regulated as osteoblasts mature *in vitro* and *in vivo*, is regulated by a variety of osteotropic hormones including $1,25(OH)_2D_3$ [123]. Finally, in human bone marrow stromal cell cultures, $1,25(OH)_2D_3$ appears to potentiate fluoride-mediated anabolic effects such as increased type I collagen production, alkaline phosphatase, and osteocalcin [124].

A related issue that should also be mentioned with regard to the effects of $1,25(OH)_2D_3$ on the development and function of skeletal cells *in vivo* and *in vitro* is the possible interdependence of the effects of $1,25(OH)_2D_3$, retinoids, and thyroid hormones as a consequence of the heterodimerization capabilities of their receptors with a common member of the family, the retinoid X receptor. Williams *et al.* [125] demonstrated, using a variety of osteoblastic cell lines representing different stages of osteoblastic development, that retinoids were required for expression of responses to $1,25(OH)_2D_3$ and triiodothyronine (T_3). The observations by us and others in the rat calvaria bone nodule-forming system that retinoids inhibit osteoprogenitor proliferation and differentiation [126] and that T_3 has variable effects, depending on the stage of differentiation of the target cells [127,128], suggest that receptor interactions do occur in this and other systems, although a variety of other explanations are possible. $1,25(OH)_2D_3$ is synergistic or antagonistic with retinoic acid with respect to levels of alkaline phosphatase, osteopontin, matrix Gla protein, and type I collagen in the immortalized rat preosteoblast line UMR-201-10B [129]. Evidence that TGFβ interacts with $1,25(OH)_2D_3$ and that the effects are either synergistic (with regard to alkaline phosphatase production) or antagonistic (with regard to osteocalcin production) further illustrate the importance of multiple interactions in studies evaluating the effects of the various vitamin D metabolites [76].

D. Does $1,25(OH)_2D_3$ Alter Commitment of Progenitor Cells Capable of Forming Fat and Bone?

As mentioned in the introduction (Section I), $1,25(OH)_2D_3$ is a known regulator of the differentiation and activity of cells of various lineages, including mesenchymal and hemopoietic cells. In some systems, $1,25(OH)_2D_3$ appears to act in synergy with other steroids, including glucocorticoids, to regulate osteoblast differentiation. While investigating the inhibitory effects of vitamin D_3 metabolites on bone nodule formation in primary fetal rat calvaria cells [54] discussed above, we noted an increased number of adipocyte foci in cultures that had been grown in the presence of $1,25(OH)_2D_3$ (Fig. 2). Dexamethasone also increases adipocyte number in rat calvaria cultures [5], consistent with earlier observations that glucocorticoids stimulate the preadipocytes derived from various sources to differentiate into mature adipocytes [130–134]. The reported effects of $1,25(OH)_2D_3$ on adipogenesis are more discrepant. Sato and Hiragun [135] reported that in the mouse preadipocyte cell lines 3T3L1 and ST13, $1,25(OH)_2D_3$ receptors were present in both cell lines at the preadipocyte cell stage, but not in mature adipocytes, and that $1,25(OH)_2D_3$ suppressed the differentiation of preadipocytes into adipocytes. In contrast, Ishida *et al.* [136] found that $1,25(OH)_2D_3$ decreased cell proliferation and $[^3H]$thymidine uptake in a dose-dependent manner in 3T3L1 cells and increased the number of lipid droplets in the cytoplasm without a significant stimulation in glyceraldehyde-3-phosphate dehydrogenase (G3PDH) activity. This study was in agreement with studies by Vu *et al.* [6] in the same cell line. It has also been reported, however, that $1,25(OH)_2D_3$ at lower concentration stimulated triglyceride accumulation in 3T3L1 cells, whereas at higher concentration it inhibited triglyceride accumulation and G3PDH activity [137].

Given the ability of $1,25(OH)_2D_3$ to inhibit adipogenesis when it is added together with dexamethasone in murine calvariae or a bone marrow-derived adipocytic clone [135,138,139], the stimulatory effects observed above when rat calvaria cultures are treated with both $1,25(OH)_2D_3$ and dexamethasone is puzzling. Whether

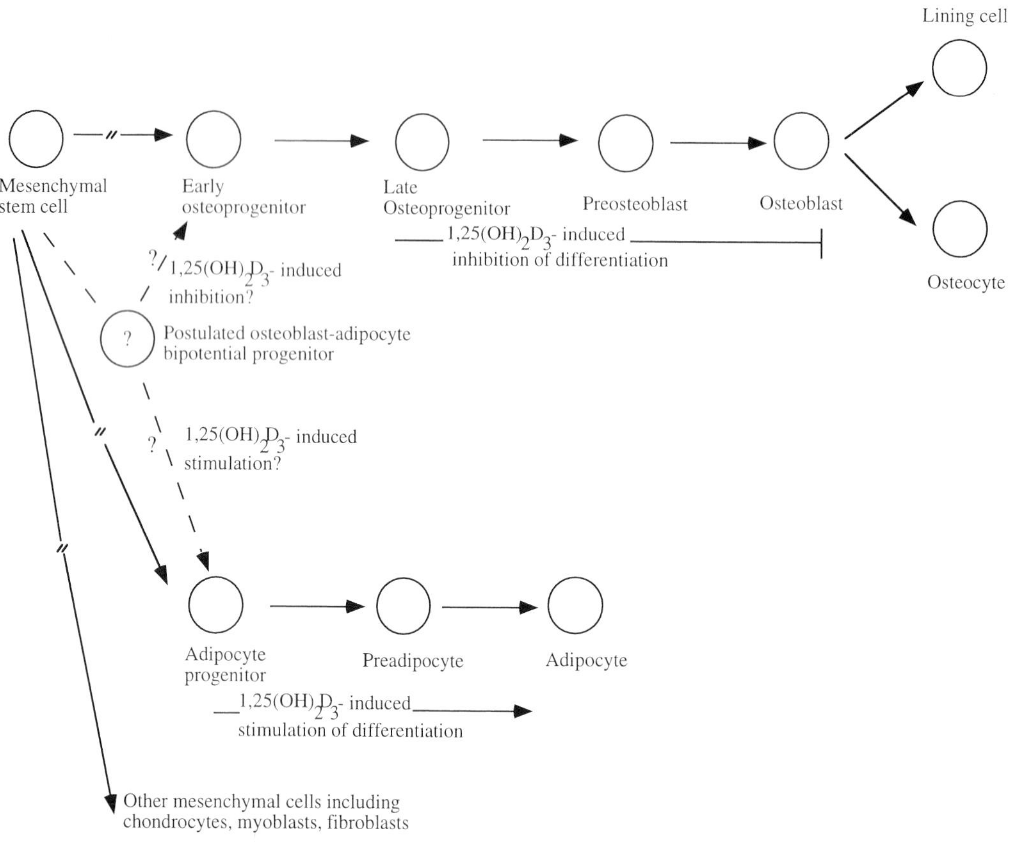

FIGURE 2 The hormone $1,25(OH)_2D_3$ has been proposed to alter the commitment or fate of putative bipotential osteoblast–adipocyte progenitors, stimulating commitment to adipocytes and differentiation of the adipocyte progenitors while inhibiting commitment to osteoblasts and differentiation of osteoprogenitors.

the effects of $1,25(OH)_2D_3$ and dexamethaxone in combination were additive or synergistic depended on the assay used for adipogenesis and the concentrations tested, but the effects were not less than additive in rat calvaria cultures [5]. $1\alpha OHD_3$ had a similar effect to $1,25(OH)_2D_3$ in stimulating adipocyte focus number and G3PDH activity, suggesting either that the $1\alpha OHD_3$ form of vitamin D_3 may be capable of acting itself or that cells within the differentiating rat calvaria cell cultures are capable of converting the $1\alpha OHD_3$ precursor to $1,25(OH)_2D_3$.

Understanding the discrepancies noted above will be important. Some, such as differences observed in mouse versus rat calvaria, may be species related, given the observation that concentrations of glucocortcoid that are stimulatory for osteogenesis in rat bone marrow cultures may be inhibitory in mouse marrow cultures [140], that osteogenesis in mouse marrow cultures may not be glucocorticoid dependent [140], and that mouse calvaria cells show a different dose–response profile to glucocorticoids as compared to rat calvaria cells for osteogenic stimulation [141]. Others may reflect differences in primary cultures versus established cell lines, or the presence and types of accessory cells present in heterogeneous primary cultures of calvaria or marrow stroma. Although the adipocyte progenitor cells present in rat calvaria may differ from those present in rat bone marrow stroma, the findings of Owen *et al.* [53] and Ishida *et al.* [54] and the apparent lack of stability of the adipocyte progenitor cells from rat calvaria cell cultures could explain the decrease in adipogenesis and increase in osteocalcin mRNA in secondary rat bone marrow stromal cell cultures seen by Beresford *et al.* [138]. However, it is also known that adipocytic cells of different tissues, in particular those in true adipose tissue versus those in bone marrow, exhibit heterogeneity in size, esterase activity, and collagen-associated reticulin (reviewed by Tavasolli [142]) and differ in several of their responses to hormones, for example, in their insulin dependence [143]. As adipocyte formation is regulated by several other hormones, such as growth hormone [144] and T_3 [145], it is probable that interactions between these hormones and $1,25(OH)_2D_3$ and glucocorticoids do occur.

It is striking that in $1,25(OH)_2D_3$-treated rat calvaria cell populations, concomitant with the inhibition of osteoprogenitor cell differentiation, a class of progenitor cells is stimulated by $1,25(OH)_2D_3$ to differentiate along the adipogenic pathway [5]. Given the similarities in effect between dexamethasone and $1,25(OH)_2D_3$ on adipogenesis but differences between the two in osteogenesis in the rat calvaria system, it is of interest to compare the adipogenic effects of $1,25(OH)_2D_3$ with those of dexamethasone. Although short-term pulses of either hormone during log phase growth stimulated the number of adipocyte foci, maximal stimulation by either hormone required the continual presence of hormone throughout the log and early multilayering phases of culture, a similar time window to that in which osteoprogenitors are inhibited by $1,25(OH)_2D_3$. Once induced, the adipocyte progenitor cells no longer required $1,25(OH)_2D_3$ or dexamethasone for terminal differentiation. However, $1,25(OH)_2D_3$ consistently stimulated more adipocyte formation than dexamethasone in rat calvaria cell cultures, and the adipocytes that did form were morphologically different (more compact, three-dimensional foci, with individual adipocytes being spindle shaped, refractile, and containing abundant liposomes) from those that formed in response to dexamethasone (less compact foci, and individual adipocytes containing fewer liposomes) [5].

Whether the cells stimulated by $1,25(OH)_2D_3$ or dexamethasone represent different classes of progenitor cells or differences between the actions of the two hormones on the same class of progenitor cells at different stages of differentiation will require further work. The fact that the effects of the hormones are additive or synergistic and that the morphology of adipocytes and the clusters of foci induced by the two agents are different is consistent with the former possibility. However, the proposed multilineage hierarchy in which both adipocytes and osteoblasts reside give multiple precursor cell points at which commitment and/or differentiation could be regulated [34,146], indicating that the latter possibility cannot be ruled out.

Our data on the effects of $1,25(OH)_2D_3$ and dexamethasone on adipocyte formation together with the data showing $1,25(OH)_2D_3$-induced inhibition of osteoblastic differentiation may have relevance to the bone and bone marrow changes that are observed during glucocorticoid-induced osteoporosis and postmenopausal osteoporosis, namely, a reduction in bone mass appearing with a concomitant increase in marrow adipose tissue [147,148]. In older humans, marrow changes from "red" or erythropoietic with few adipocytes to "yellow" as a result of increased numbers of adipocytes [149,150]. A significant enhancement of marrow fat volume correlating with a decrease in bone volume in samples from paraplegic compared to age-matched nonparaplegic patients has also been reported [151]. The correlation between bone loss and marrow fat gain is seen in animal models as well, for example, after ovariectomy [152,153], chronic glucocorticoid treatment [154,155], or immobilization [156]. The marrow of older mice contains fewer fibroblast colony-forming units (CFU-F) than that of younger mice [157]. In addition, fewer osteoprogenitor cells have been measured in assays of bone nodule formation *in vitro* in cells from aged rats [158]. This observation correlates with studies on the senescence-accelerated mouse (SAMP6), wherein osteoblastogenesis is decreased and adipogenesis is increased [159,160].

The fact that adipocytes and osteoblasts evolve from a common multipotential stem cell (references as above) has suggested to some that fat is forming instead of bone via a redirection of a bipotential cell totally or almost exclusively to the adipocyte pathway. In rodent stromal cultures, enhanced expression of adipocyte markers (aP2, lipid vacuoles) is paralleled by decreased expression of osteoblast markers (osteopontin, osteocalcin) and by a switch from fibrillar (types I and III) to basement membrane (type IV) collagens [137,161,162]. These findings are consistent with the suggestion that the commitment of the adipocyte lineage occurs at the expense of osteoblast numbers or function, although it does not disprove the possibility that the two occur independently without involvement of a common precursor. Bennett *et al.* [163] reported that adipocytes cultured from rat bone marrow revert to a proliferative phase in culture and differentiate to form bonelike tissue when placed in diffusion chambers. Several clonal cell lines from mouse [164] stromal cell cultures have been found to make fat and bone *in vitro* under appropriate culture conditions, often with the loss of osteoblast properties concomitant with the expression of adipocyte properties. Stromal adipocytes decrease their expression of osteopontin, a ligand for CD44 [165] which is also highly up-regulated in mature osteoblasts (see above), compared to the levels in preadipocytes [161]; hyaluronate, another ligand for CD44, increases [166]. On the other hand, Rickard *et al.* [174] identified the expression of osteoblast and adipocyte markers in the same CFU-F colonies grown from purified human marrow stromal cultures grown with dexamethasone or $1,25(OH)_2D_3$.

These observations support the idea that a bipotential osteoblast–adipocyte progenitor may exist (Fig. 2) and that $1,25(OH)_2D_3$ is one of the possible regulators of its commitment and differentiation. The difficulty remains, however, that in mixed stromal or calvaria cultures different kinds of progenitors may be present, each separately regulated by multiple factors including

$1,25(OH)_2D_3$. The number of such bipotential progenitors in such cultures may be too small to be detectable in the presence of larger numbers of monopotential adipocyte and osteoblast progenitors. In the case of clonal lines, undifferentiated/uncommitted progenitor cells may always exist in the colonies being analyzed. The difficulty of interpreting some of these experimental observations without detailed subclone analyses and rigorous assessment of all cells present has been discussed elsewhere [34]. Nevertheless, the possibility that bipotential osteoblast–adipocyte precursors exist either in the presence or absence of other mesenchymal precursors and the molecular mediators that elicit their commitment and differentiation in marrow or other tissues remains worthy of further analyses. With respect to regulation, Devchand *et al.* reported that a generalized inflammatory response is controlled by the direct binding of leukotriene B4 (LTB4) to the transcription factor peroxisome proliferator-activated receptor-α (PPARα) [167]. Thus, LTB4 acts like certain hypolipidemic drugs to activate a transcription factor that regulates expression of enzymes involved in lipid metabolism, adipogenesis, and eicosanoid synthesis and therefore inflammation. Antidiabetic agents, the thiazolidinediones, also bind to the steroid receptor-like peroxisome proliferator-activated receptors. These latter agents induce bone marrow stromal stem cell adipogenesis *in vivo* [168], possibly at the expense of bone [169], and *in vitro* [148]. It is intriguing to speculate that factors acting through this pathway may act as adipocytic agonists and osteogenic antagonists [148].

III. EFFECTS OF $1,25(OH)_2D_3$ ON BONE APPOSITION RATES *IN VIVO*

It is clear from the observations *in vitro* that osteoprogenitors, particularly early osteoprogenitors, are target cells for $1,25(OH)_2D_3$ and $24,25(OH)_2D_3$ action and that the effects of these metabolites on early osteoprogenitor proliferation and differentiation are inhibitory. It is equally clear that another group of target cells for these hormones are mature, functional osteoblasts, and that the effects of $1,25(OH)_2D_3$ on these are stimulatory with regard to osteocalcin and osteopontin secretion but inhibitory with respect to collagen and bone sialoprotein production. The *in vivo* consequence of this could be that $1,25(OH)_2D_3$-mediated increased deposition of osteopontin and osteocalcin by secretory osteoblasts approaching the end of their secretory lifetime may render the bone surface in those locations resorbable by osteoclasts, as osteoclasts require the bone matrix to be mineralized in order to be able to resorb it and simultaneously push the osteoblastic cells to differentiate into lining cells (see Section II.A). $1,25(OH)_2D_3$-induced inhibition of bone sialoprotein and collagen secretion by early secretory phase osteoblasts could reflect inhibition of matrix deposition by these cells, revert the phenotype to a presecretory cell, and result in the disappearance of osteoblasts from the trabecular bone surface observed *in vivo* in $1,25(OH)_2D_3$-repleted vitamin D-deficient rats [170] (see below).

With regard to the effects of vitamin D metabolites on bone apposition *in vivo*, the information available is limited. We have previously reported a series of investigations carried out for the purpose of clarifying the functions of vitamin D metabolites on the linear rate of bone mineral apposition (BMAR) in adult vitamin D-deficient rats [171]. These studies were carried out in young adult, previously vitamin D-replete rats, and the BMAR was determined by a sequential tetracycline-labeling method. Vitamin D restriction for 8 weeks resulted in a progressive fall of the serum concentration of 25OHD and $24,25(OH)_2D$ while serum $1,25(OH)_2D$ remained normal. The BMAR decreased by about 30%, and the amount of osteoid increased, suggesting that, in the absence of $24,25(OH)_2D_3$, $1,25(OH)_2D_3$ may inhibit mineralization of osteoid. This interpretation is compatible with the observations of Hock *et al.* [172] that pharmacological concentrations of $1,25(OH)_2D_3$ administered to intact rats increased the amount of osteoid and inhibited mineralization. Repletion of these rats with 200 ng/day of either $25(OH)D_3$ or $24,25(OH)_2D_3$ for the last 15 days of the 8-week period resulted in a rapid restoration of the BMAR, but treatment with 200 ng/day of $1,25(OH)_2D_3$ had no effect. We also observed in these experiments that $1,25(OH)_2D_3$-repleted animals exhibited increased osteoclastic activity and decreased numbers of osteoblasts on the trabecular bone surfaces, whereas the $24,25(OH)_2D_3$-repleted animals maintained osteoblastic activity on most of the trabecular bone surfaces without increased osteoclastic activity (C.S. Tam and J.N.M. Heersche, unpublished). This observation is similar to that of Ono *et al.* [173] who found that $24,25(OH)_2D_3$ normalized bone formation and resorption in rachitic hypophosphatemic mice, whereas $1,25(OH)_2D_3$ normalized bone formation but also induced excessive stimulation of bone resorption (see also Chapter 19).

The fact that all three metabolites of vitamin D tested were capable of reducing the mean osteoid seam width within 15 days of their administration in the vitamin D-restricted rats, whereas only $25OHD_3$ and $24,25(OH)_2D_3$ stimulated BMAR, points out the complexity of the process of bone formation. Interestingly, the observed effects of $24,25(OH)_2D_3$ and $25OHD_3$ on the mineralization lag time are in keeping with their effects in stimulating BMAR. However, $1,25(OH)_2D_3$

repletion reduced the lag time but did not affect the BMAR. It is possible that these differences can be accounted for by differences in the effects of these metabolites on organic matrix synthesis. Unfortunately, a systematic evaluation of the effects of vitamin D and its metabolites on organic matrix apposition has not been published to date.

A number of studies in various animals and humans have suggested that $24,25(OH)_2D_3$ is an active metabolite of vitamin D_3 which has a qualitatively different action than $1,25(OH)_2D_3$ [170,174–177] (also see Chapters 19 and 26). The effects on BMAR that we observed after dietary repletion with various metabolites of both vitamin D-restricted and vitamin D-sufficient animals support this suggestion. We would like to stress, however, that the BMAR, as determined by these data, may not reflect the overall effects of various metabolites of vitamin D on bone formation. Our observations on the effects of $1,25(OH)_2D_3$ and $24,25(OH)_2D_3$ on BMAR would suggest that although $24,25(OH)_2D_3$ is directly involved in stimulation of mineralization of the bone matrix, $1,25(OH)_2D_3$ is not. It is of interest to note, in this regard, that observations in mice deficient for the vitamin D 24-hydroxylase gene [178] also indicate that a primary role for $24,25(OH)_2D_3$ may be related to mineralization (see Chapter 18).

IV. CONCLUSIONS

The major challenge in the area of regulation of osteoblast differentiation and activity by vitamin D metabolites is to place findings made in the *in vitro* systems into an *in vivo* context. This remains difficult at present, partly because of the paucity of detailed information on the *in vivo* effects of vitamin D_3 on osteoblast numbers and activity under "normal" vitamin D-sufficient conditions. The complex interactions between, and interdependence of, the actions of the different metabolites with other endocrine and local regulatory systems discussed in some detail above add to the complexity of the issues and will make them more difficult to resolve.

References

1. Studzinski GP, McLane JA, Uslokovic MR 1993 Signalling pathways for vitamin D-induced differentiation: Implications for therapy of proliferative and neoplastic diseases. Crit Rev Eukaryotic Gene Expression **3**:279–312.
2. Bickle DD, Pillai S 1993 Vitamin D, calcium, and epidermal differentiation. Endocr Rev **14**:3–19.
3. Suda T, Takahashi N, Etsuko A 1992 Role of vitamin D in bone resorption. J Cell Biochem **49**:53–58.
4. Suda T, Takahashi N, Martin TJ 1995 Modulation of osteoclast differentiation: An update. Endocr Rev **4**:266–279.
5. Bellows CG, Wang Y-H, Heersche JNM, Aubin JE 1994 1,25-Dihydroxyvitamin D_3 stimulates adipocyte differentiation in cultures of fetal rat calvaria cells: Comparison with the effects of dexamethasone. Endocrinology **134**:2221–2229.
6. Vu D, Ong JM, Clemens TL, Kern PA 1996 1,25-Dihydroxyvitamin D induces lipoprotein lipase expression in 3T3-L1 cells in association with adipocyte differentiation. Endocrinology **137**:1540–1544.
7. Gerstenfeld LC, Kelly CM, von Deck M, Lian JB 1990 Effect of 1,25-Dihydroxyvitamin D_3 on induction of chondrocyte maturation in culture: Extracellular matrix gene expression and morphology. Endocrinology **126**:1599–1609.
8. Schwartz Z, Brooks B, Swain L, Del Toro F, Norman A, Boyan B 1992 Production of 1,25-dihydroxyvitamin D_3 and 24,25-dihydroxyvitamin D_3 by growth zone and resting zone chondrocytes is dependent on cell maturation and is regulated by hormones and growth factors. Endocrinology **130**:2495–2504.
9. Ozono K, Sone T, Pike JW 1991 The genomic mechanism of action of 1,25-dihydroxyvitamin D_3. J Bone Miner Res **6**:1021–1027.
10. Lowe KE, Maiyar AC, Norman AW 1992 Vitamin D-mediated gene expression. Crit Rev Eukaryotic Gene Expression **2**:65–109.
11. Walters MR 1992 Newly identified actions of the vitamin D endocrine system. Endocr Rev **13**:719–755.
12. Lian JB, Stein GS 1992 Transcriptional control of vitamin D-regulated proteins. J Cell Biochem **49**:37–45.
13. Darwish H, DeLuca HF 1993 Vitamin D-regulated gene expression. Crit Rev Eukaryotic Gene Expression **3**:89–116.
14. Manolagas SC, Haussler MR, Deftos LJ 1980 1,25-Dihydroxyvitamin D_3 receptor-like macromolecules in rat osteogenic sarcoma cell lines. J Biol Chem **255**:4414–4417.
15. Chen TL, Feldman D 1981 Regulation of 1,25-dihydroxyvitamin D_3 receptors in mouse bone cells. Correlation of receptor concentration with the rate of cell division. J Biol Chem **256**:5561–5566.
16. Narbaitz R, Stumpf WE, Sar M, Huong S, DeLuca HF 1983 Autoradiographic localization of target cells for 1,25-dihydroxyvitamin D_3 in bones from fetal rats. Calcif Tissue Int **35**:177–182.
17. Petkovich PM, Heersche JNM, Aubin JE, Grigoriadis AE, Jones G 1984 Retinoic acid-induced changes in 1,25-dihydroxyvitamin D_3 receptor levels in tumor and non-tumor cells derived from rat bone. J Natl Cancer Inst **78**:265–270.
18. Majeska RJ, Rodan GA 1982 The effect of $1,25(OH)_2D_3$ on alkaline phosphatase in osteoblastic osteosarcoma cells. J Biol Chem **257**:3362–3365.
19. Manolagas SC, Burton DW, Deftos LJ 1981 1,25-Dihydroxyvitamin D_3 stimulates the alkaline phosphatase activity of osteoblast-like cells. J Biol Chem **257**:7115–7117.
20. Kurihara N, Ikeda K, Haleta Y, Tsunoi M, Maeda N, Kumegawa M 1984 Effects of 1,25-dihydroxyvitamin D_3 on alkaline phosphatase activity and collagen synthesis in osteoblastic cells, clone MC3T3-E1. Biochem Biophys Res Commun **119**:767–771.
21. Beresford JN, Gallagher JA, Russell RGG 1986 1,25-Dihydroxyvitamin D_3 and human bone-derived cells in vitro: Effects on alkaline phosphatase, type I collagen and proliferation. Endocrinology **119**:1776–1785.
22. Prince CW, Butler WT 1987 1,25-Dihydroxyvitamin D_3 regulates the biosynthesis of osteopontin, a bone-derived cell attachment protein, in clonal osteoblast-like osteosarcoma cells. Collagen Related Res **7**:305–313.
23. Chang PL, Prince CW 1991 1,25-Dihydroxyvitamin D_3 stimulates synthesis and secretion of nonphosphorylated osteopontin (se-

creted phosphoprotein 1) in mouse JB6 epidermal cells. Cancer Res **51**:2144–2150.
24. Sodek J, Chen J, Nagata T, Kasugai S, Todescan R, Li J, Kim RH 1995 Regulation of osteopontin expression in osteoblasts. Ann NY Acad Sci **760**:223–241.
25. Price PA, Baukol SA 1980 1,25-Dihydroxyvitamin D_3 increases synthesis of the vitamin K-dependent bone protein by osteosarcoma cells. J Biol Chem **255**:11660–11663.
26. Raisz LG, Maina DM, Gworek SC, Dietrich JW, Canalis EM 1978 Hormonal control of bone collagen synthesis in vitro: Inhibitory effect of 1-hydroxylated vitamin D metabolites. Endocrinology **102**:731–735.
27. Kream BE, Rowe D, Smith MD, Maher V, Majeska R 1986 Hormonal regulation of collagen synthesis in a clonal rat osteosarcoma cell line. Endocrinology **119**:1922–1928.
28. Oldberg A, Jirskog-Hed B, Axelsson S, Heinegard D 1989 Regulation of bone sialoprotein mRNA by steroid hormones. J Cell Biol **109**:3183–3186.
29. Sodek J, Kim RH, Ogata Y, Li J, Yamauchi M, Zhang Q, Freedman LP 1995 Regulation of bone sialoprotein gene transcription by steroid hormones. Connect Tissue Res **32**:209–217.
30. Canalis E, Lian JB 1985 $1{,}25(OH)_2D_3$ Effects on collagen and DNA synthesis in periosteum and periosteum-free calvariae. Bone **6**:457–460.
31. Harrison JR, Peterson DN, Lichtler AC, Mador AT, Rowe DW, Kream BE 1989 1,25-Dihydroxyvitamin D_3 inhibits transcription of type I collagen genes in the rat osteosarcoma cell line ROS 17/2.8. Endocrinology **125**:327–333.
32. Nefussi J-R, Boy-Lefevre ML, Boulekbache H, Forest N 1985 Mineralization in vitro of matrix formed by osteoblasts isolated by collagenase digestion. Differentiation **29**:160–168.
33. Bellows CG, Aubin JE, Heersche JNM, Antosz ME 1986 Mineralized bone nodules formed in vitro from enzymatically released rat calvaria cell populations. Calcif Tissue Int **38**:143–154.
34. Aubin JE, Turksen K, Heersche JNM 1993 Osteoblastic cell lineage. In: Noda M (ed) Cellular and Molecular Biology of Bone. Academic Press, New York, pp. 1–45.
35. Stein GS, Lian JB 1993 Molecular mechanisms mediating proliferation/differentiation interrelationships during progressive development of the osteoblast lineage. Endocr Rev **14**:424–442.
36. Bjargava U, Bar-Lev M, Bellows CG, Aubin JE 1988 Ultrastructural analysis of bone nodules formed in vitro by isolated fetal rat calvaria cells. Bone **9**:155–163.
37. Malaval L, Modrowski D, Gupta AK, Aubin JE 1994 Cellular expression of bone related proteins during in vitro osteogenesis in rat bone marrow stromal cell cultures. J Cell Physiol **158**:555–572.
38. Liu F, Malaval L, Gupta AK, Aubin JE 1994 Simultaneous detection of multiple bone-related mRNAs and protein expression during osteoblast differentiation: Polymerase chain reaction and immunocytochemical studies at the single cell level. Dev Biol **166**:220–234.
39. Aronow MA, Gerstenfeld LC, Owen TA, Tassinari MS, Stein GS, Lian JB 1990 Factors that promote progressive development of the osteoblast phenotype in cultured fetal rat calvarial cells. J Cell Physiol **143**:213–221.
40. Owen TA, Aronow MA, Shaloub V, Barone LM, Wilming L, Tassinari MS, Kennedy MB, Pockwinse S, Lian JB, Stein GS 1990 Progressive development of the rat osteoblast phenotype in vitro: Reciprocal relationships in expression of genes associated with osteoblast proliferation and differentiation during formation of the bone extracellular matrix. J Cell Physiol **143**: 420–430.
41. Aubin JE, Bellows CG, Turksen K, Liu F, Heersche JNM 1992 Analysis of the osteoblast lineage and regulation of differentiation. In: Slavkin H, Price P (eds) Chemistry and Biology of Mineralized Tissues. Elsevier, Amsterdam, pp. 267–276.
42. Heersche JNM, Bellows CG, Aubin JE 1994 Cellular actions of parathyroid hormone on osteoblast and osteoclast differentiation. In: Bilezekian JP, Levine M, Marcus R (eds) The Parathyroids. Raven, New York, pp. 83–91.
43. Bellows CG, Aubin JE, Heersche JNM 1987 Physiological concentrations of glucocorticoids stimulate formation of bone nodules from isolated rat calvaria cells in vitro. Endocrinology **121**:1985–1992.
44. Bellows CG, Heersche JNM, Aubin JE 1990 Determination of the capacity for proliferation and differentiation of osteoprogenitors cells in the pesence and absence of dexamethasone. Dev Biol **140**:132–138.
45. Turksen K, Aubin JE 1991 Positive and negative immunoselection for enrichment of two classes of osteoprogenitor cells. J Cell Biol **114**:373–384.
46. Antosz ME, Bellows CG, Aubin JE 1987 Effects of transforming growth factor β and epidermal growth factor on cell proliferation and the formation of bone nodules in isolated fetal rat calvaria cells. J Cell Physiol **140**:386–395.
47. Gehron-Robey P, Young MF, Flanders KC, Roche NS, Reddi AH, Termine JD, Sporn MB, Roberts AB 1987 Osteoblasts synthesize and respond to transforming growth factor β (TGFβ) in vitro. J Cell Biol **105**:457–463.
48. Iwasaki M, Nakata K, Nakahara H, Nakase T, Kimura T, Kimata K, Caplan AL, Ono K 1993 Transforming growth factor beta 1 stimulates chondrogenesis and inhibits osteogenesis in high density culture of periosteum-derived cells. Endocrinology **132**:1602–1608.
49. Harris SE, Bonewald LF, Harris MA, Sabatini M, Dallas S, Feng J, Ghosh-Choudury N, Wozney J, Mundy GR 1994 Effects of transforming growth factor β on bone nodule formation and expression of bone morphogenetic protein 2, osteocalcin, osteopontin, alkaline phosphatase and type I collagen mRNA in long-term cultures of fetal rat calvarial osteoblasts. J Bone Miner Res **9**:855–863.
50. Breen EC, Ignotz R, McCabe L, Lian JB, Stein GS 1994 TGFβ alters growth and differentiation related gene expression in proliferating osteoblasts in vitro preventing development of the mature osteoblast phenotype. J Cell Physiol **160**:323–335.
51. Ellies LG, Aubin JE 1990 Temporal sequence of interleukin 1 alpha-mediated stimulation and inhibition of bone formation by isolated fetal rat calvaria cells in vitro. Cytokine **2**:430–437.
52. Antosz ME, Bellows CG, Aubin JE 1987 Biphasic effects of epidermal growth factor on the formation of bone nodules from isolated rat calvaria cells in vitro. J Bone Miner Res **13**:185–194.
53. Owen TA, Aronow MA, Barone LM, Bettencourt B, Stein G, Lian JB 1991 Pleiotropic effects of vitamin D on osteoblast gene expression are related to the proliferative and differentiated state of the bone cell phenotype: Dependency upon basal levels of gene expression, duration of exposure and bone matrix competency in normal rat osteoblast cultures. Endocrinology **128**:1496–1504.
54. Ishida H, Bellows CG, Aubin JE, Heersche JNM 1993 Characterization of the $1{,}25(OH)_2D_3$-induced inhibition of bone nodule formation in long-term cultures of fetal rat calvaria cells. Endocrinology **132**:61–66.
55. Spiess YH, Price PA, Deftos JL, Manolagas SC 1986 Phenotype-associated changes in the effects of 1,25-Dihydroxyvitamin D_3 on alkaline phosphatase and bone-GLA-protein of rat osteoblastic cells. Endocrinology **118**:1340–1346.

56. Chen TL, Cone CM, Feldman D 1983 Effects of 1,25-Dihydroxyvitamin D_3 and glucocorticoids on the growth of rat and mouse osteoblast-like bone cells. Calcif Tissue Int **35**:806–811.
57. Harrison JR, Clark NB 1986 Avian medullary bone in organ culture: Effects of vitamin D metabolites on collagen synthesis. Calcif Tissue Int **39**:35–43.
58. Bringhurst FR, Potts JT Jr 1982 Effects of vitamin D metabolites and analogs on bone collagen synthesis in vitro. Calcif Tissue Int **34**:103–110.
59. Broess M, Riva A, Gerstenfeld LC 1995 Inhibitory effects of $1,25(OH)_2$ vitamin D_3 on collagen type I, osteopontin, and osteocalcin gene expression in chicken osteoblasts. J Cell Biochem **57**:440–451.
60. Fraser JD, Otawara Y, Price PA 1988 1,25-Dihydroxyvitamin D_3 stimulates the synthesis of matrix γ-carboxyglutamic acid protein by osteosarcoma cells. J Biol Chem **263**:911–916.
61. Fraser JD, Price PA 1990 Induction of matrix gla protein synthesis during prolonged $1,25(OH)_2D_3$ treatment of osteosarcoma cells. Calcif Tissue Int **46**:270–279.
62. Franceschi RT, Young J 1990 Regulation of alkaline phosphatase by 1,25-dihydroxyvitamin D_3 and ascorbic acid in bone-derived cells. J Bone Miner Res **5**:1157–1167.
63. Franceschi RT, Romano PR, Park KY 1988 Regulation of type I collagen synthesis by 1,25-Dihydroxyvitamin D_3 in human osteosarcoma cells J Biol Chem **263**:18938–18945.
64. Matsumoto T, Igarashi C, Taksuchi Y, Harada S, Kikuchi T Yamato H, Ogata E 1991 Stimulation by 1,25-dihydroxyvitamin D_3 of in vitro mineralization induced by osteoblast-like MC3T3-E1 cells. Bone **12**:27–32.
65. Wergedal JE, Baylink DJ 1984 Characterization of cells isolated and cultured from human bone. Proc Soc Exp Biol Med **176**:60–69.
66. Auf'Mkolk B, Hauschka PV, Schwartz E 1985 Characterization of human bone cells in culture. Calcif Tissue Int **37**:228–235.
67. Robey PG, Termine JD 1985 Human bone cells in vitro. Calcif Tissue Int **37**:453–460.
68. Marie PJ, Lomri A, Sabbagh A, Basle M 1989 Culture and behavior of osteoblastic cells isolated from normal trabecular bone surfaces. In Vitro Cell Biol Dev.**25**:373–380.
69. Chavassieux PM, Chenu C, Valentin-Opran A, Merle B, Delmas PD, Hartmann DJ, Saez S, Meunier PJ 1990 Influence of experimental conditions on osteoblast activity in human primary bone cell cultures. J Bone Miner Res **5**:337–343.
70. Rodan GA, Rodan SB 1990 Expression of the osteoblastic phenotype. In: Peck WA (ed) Bone and Mineral Research, Vol. 2. Elsevier, Amsterdam, pp. 244–285.
71. Marie PJ 1994 Human osteoblastic cells: A potential tool to assess the etiology of pathologic bone formation. J Bone Miner Res **9**:1847–1850.
72. Beresford JN, Joyner CJ, Devlin C, Triffitt JT 1994 The effects of dexamethasone and 1,25-dihydroxyvitamin D_3 on osteogenic differentiation of human marrow stromal cells in vitro. Arch Oral Biol **39**:941–947.
73. Leboy RS, Beresford JN, Devlin C, Owen ME 1991 Dexamethasone induction of osteoblast mRNAs in rat marrow stromal cell cultures. J Cell Physiol **146**:370–378.
74. Rickard DJ, Kassem M, Hefferan TE, Sarkar G, Soelsberg TC, Riggs BL 1996 Isolation and characterization of osteoblast precursor cells from human bone marrow. J Bone Miner Res **11**:312–324.
75. Walsh S, Dodds RA, James IE, Bradbeer JN, Gowen MJ 1994 Monoclonal antibodies with selective reactivity against osteoblasts and osteocytes in human bone. J Bone Miner Res **9**:1687–1696.
76. Ingram RT, Bonde SK, Riggs BL, Fitzpatrick LA 1994 Effects of transforming growth factor beta (TGFβ) and 1,25-dihydroxyvitamin D_3 on the function, cytochemistry and morphology of normal human osteoblast-like cells. Differentiation **55**:153–163.
77. Thavarajah M, Evans DB, Kanis JA 1993 Differentiation of heterogeneous phenotypes in human osteoblast cultures in response to 1,25-dihydroxyvitamin D_3. Bone **14**:763–767.
78. Heersche JNM, Reimers SM, Wrana JL, Waye MMY, Gupta AK 1992 Changes in expression of alpha 1 type 1 collagen and osteocalcin mRNA in osteoblasts and odontoblasts at different stages of maturity as shown by in situ hybridization. Proc Finn Dent Soc **88**:173–1182.
79. Chen J, Shapiro HS, Sodek J 1992 Developmental expression of bone sialoprotein mRNA in rat mineralized connective tissues. J Bone Miner Res **7**:987–997.
80. Chen J, Singh K, Mukherjee B, Sodek J 1993 Developmental expression of osteopontin (OPN) mRNA in rat tissues: Evidence for a role for OPN in bone formation and resorption. Matrix **12**:113–123.
81. Bianco P, Riminucci M, Bonucci E, Termine JD, Robey PG 1993 Bone sialoprotein (BSP) secretion and osteoblast differentiation: Relationship to bromodeoxyuridine incorporation, alkaline phosphatase, and matrix deposition. J Histochem Cytochem **41**:183–191.
82. Pan LC, Grasser WA, McCurdy SP, Thompson DD 1995 A trangenic mouse model that distinguishes subsets of osteoblasts. J Bone Miner Res **10**(Suppl. 1):S149.
83. Liu F, Malaval L, Aubin JE 1997 The mature osteoblast phenotype is characterized by extensive plasticity. Exp Cell Res **232**:97–105.
84. Bellows CG, Reimers S, Heersche JNM 1995 Expression of mRNA for osteopontin, osteocalcin, bone sialoprotein and type I collagen at different stages of osteoblastic differentiation and its regulation by $1,25(OH)_2D_3$. J Bone Miner Res **10**(Suppl. 1):S414.
85. Pockwinse SM, Stein JL, Lian JB, Stein GS 1995 Developmental stage-specific cellular responses to vitamin D_3 and glucocorticoids during differentiation of the osteoblast phenotype: Interrelationships of morphology and gene expression by in situ hybridization. Exp Cell Res **216**:244–260.
86. Weinreb M, Shinar D, Rodan GA 1990 Different pattern of alkaline phosphatase, osteopontin, and osteocalcin expression in developing rat bone visualized by in situ hybridization. J Bone Miner Res **5**:831–842.
87. Rickard DJ, Kazhdan I, Leboy PS 1995 Importance of 1,25-dihydroxyvitamin D_3 and the nonadherent cells of marrow for osteoblast differentiation from rat marrow stromal cells. Bone **16**:671–678.
88. Herbertson A, Aubin JE 1995 Dexamethasone alters the subpopulation make-up of rat bone marrow stromal cell cultures. J Bone Miner Res **10**:285–294.
89. Long MW, Robinson JA, Ashcraft EA, Mann KG 1995 Regulation of human bone marrow-derived osteoprogenitor cells by osteogenic growth factors. J Clin Invest **95**:881–887.
90. Morrison NA, Shine J, Fragonas JC, Verkest V, McMenemy L, Eisman JA 1989 1,25-Dihydroxyvitamin D-responsive element and glucocorticoid repression in the osteocalcin gene. Science **246**:1158–1161.
91. Demay MB, Gerardi JM, DeLuca HF, Kronenberg HM 1990 DNA sequences in the rat osteocalcin gene that bind the 1,25-dihydroxyvitamin D_3 receptor and confer responsiveness to 1,25-dihydroxyvitamin D_3. Proc Natl Acad Sci USA **87**:369–373.
92. Markose ER, Stein JL, Stein GS, Lian JB 1990 Vitamin D-mediated modifications in protein–DNA interactions at two pro-

moter elements of the osteocalcin gene. Proc Natl Acad Sci USA **87**:1701–1705.
93. Kerner SA, Scott RA, Pike JW 1989 Sequence elements in the human osteocalcin gene confer basal activation and inducible response to hormonal vitamin D_3. Proc Natl Acad Sci USA **86**:4455–4459.
94. Terpening CM, Haussler CA, Jurutka PW, Galligan MA, Komm BS, Haussler MR 1991 The vitamin D-responsive element in the rat bone Gla protein gene is an imperfect direct repeat that cooperates with other cis-elements in 1,25-dihydroxyvitamin D_3-mediated transcriptional activation. Mol Endocrinol **5**:373–385.
95. MacDonald PN, Dowd DR, Nakajima S, Galligan MA, Reeder MC, Haussler CA, Ozato K, Haussler MR 1993 Retinoid X receptors stimulate and 9-*cis*-retinoic acid inhibits 1,25-dihydroxyvitamin D_3-activated expression of the rat osteocalcin gene. Mol Cell Biol **13**:5907–5917.
96. Nanes MS, Kuno H, Demay MB, Kurian M, Hendy GN, DeLuca HF, Catherwood BD, Titus L, Rubin J 1994 A single upstream element confers responsiveness to $1,25(OH)_2D_3$ and tumor necrosis factor-α in the rat osteocalcin gene. Endocrinology **134**:1113–1120.
97. Mayur N, Lewis S, Catherwood BD, Nanes MS 1993 Tumor necrosis factor α decreases 1,25-dihydroxyvitamin D_3 receptors in osteoblastic ROS 17/2.8 cells. J Bone Miner Res **8**:997–1003.
98. Godschalk M, Levy JR, Downs RW Jr 1992 Glucocorticoids decrease vitamin D receptor number and gene expression in human osteosarcoma cells. J Bone Miner Res **7**:21–27.
99. Pirskanen A, Jaaskelainen T, Maenpaa PH 1994 Effects of transforming growth factor β_1 on the regulation of osteocalcin synthesis in human MG-63 osteosarcoma cells. J Bone Miner Res **9**:1635–1642.
100. Staal A, Van Wijnen AJ, Desai RK, Pols HA, Birkenhager JC, DeLuca HF, Denhardt DT, Stein JL, Van Leeuwen JP, Stein GS, Lian JB 1996 Antagonistic effects of transforming growth factor-beta on vitamin D_3 enhancement of osteocalcin and osteopontin transcription: Reduced interactions of vitamin D receptor/retinoid X receptor complexes with vitamin D response elements. Endocrinolgy **137**:2001–2011.
101. Hoffmann HM, Catron KM, van Wijnen AJ, McCabe LR, Lian JB, Stein GC, Stein JL 1994 Transcriptional control of the tissue-specific developmentally regulated osteocalcin gene requires a binding motif for the MSX-family of homeodomain proteins. Proc Natl Acad Sci USA **91**:12887–12891.
102. Owen TA, Bortell R, Shalhoub V, Heinrichs A, Stein JL, Stein GS, Lian JB 1993 Postproliferative transcription of the rat osteocalcin gene is reflected by vitamin D-responsive developmental modifications in protein–DNA interactions at basal and enhancer promoter elements. Proc Natl Acad Sci USA **90**:1503–1507.
103. Heinrichs AAJ, Bortell R, Bourke M, Lian JB, Stein GS, Stein JL 1996 A proximal promoter binding protein contributes to developmental, tissue-restricted expression of the rat osteocalcin gene. J Cell Biochem **57**:90–100.
104. Ducy P, Karsenty G 1995 Two distinct osteoblast-specific *cis*-acting elements control expression of a mouse osteocalcin gene. Mol Cell Biol **15**:1858–1869.
105. McCabe LR, Kockx M, Lian JB, Stein JL, Stein CG 1995 Selective expression of fos- and jun-related genes during osteoblast proliferation and differentiation. Exp Cell Res **218**:255–262.
106. Candeliere GA, Prud'homme J, St Arnaud R 1991 Differential stimulation of Fos and Jun family members by calcitriol in osteoblastic cells. Mol Endocrinol **5**:1780–1788.
107. Hodgkinson JE, Davidson CI, Beresford J, Sharpe PT 1993 Expression of a human homeobox-containing gene is regulated by $1,25(OH)_2D_3$ in bone cells. Biochim Biophys Acta **1174**:11–16.
108. Jabs EW, Muller U, Li X, Ma L, Luo W, Haworth IS, Klisak I, Sparkes R, Warman ML, Mulliken JB, Snead ML, Maxson R 1993 A mutation in the homeodomain of the human MSX2 gene in a family affected with autosomal dominant craniosynostosis. Cell **785**:443–450.
109. Satokata I, Maas R 1994 *Msx1* deficient mice exhibit cleft palate and abnormalities of craniofacial and tooth development. Nature Genet **6**:348–356.
110. Towler DA, Bennett CD, Rodan GA 1994 Activity of the rat osteocalcin basal promoter in osteoblastic cells is dependent upon homedomain and CP1 binding motifs. Endocrinol **8**: 614–624.
111. Liu M, Lee M-H, Bommakanti M, Freedman LP 1996 Vitamin D_3 transcriptionally activates the p21 gene leading to the induced differentiation of the myelomonocytic cell line U937. Genes Dev **10**:142–152.
112. Aubin JE, Gupta AK, Bhargava U, Turksen K 1996 Expression and regulation of galectin 3 in rat osteoblastic cells. J Cell Physiol **169**:468–480.
113. Atkinson MJ, Freeman MW, Kronenberg HM 1990 Thymosin $\beta 4$ is expressed in ROS 17/2.8 osteosarcoma cells in a regulated manner. Mol Endocrinol **4**:69–74.
114. Jin CH, Sinki T, Hong MH, Sato T, Yamaguchi A, Ikeda T, Yamaguchi S, Abe E, Suda T 1992 1,25-Dihydroxyvitamin D_3 tissue-specifically regulates in vivo production of the third component of complement (C3) in bone. Endocrinology **131**:2468–2475.
115. Hong MH, Jin CH, Sato T, Ishimi Y, Abe E. Suda T 1991 Transcriptional regulation of the production of the third component of complement (C3) by 1α,25-dihydroxyvitamin D_3 in mouse marrow-derived stromal cells (ST2) and primary osteoblastic cells. Endocrinology **129**:2774–2779.
116. Sato T, Hong MH, Jin CH, Ishimi Y, Udagawa N, Shinki T, Abe E, Suda T 1991 The specific production of the third component of complement by osteoblastic cells treated with 1,25-dihydroxyvitamin D_3 FEBS Lett **285**:21–24.
117. Sato T, Abe S, Jin CH, Hong MH, Katagiri T, Kinoshita T, Amizuka N, Ozawa H, Suda T 1993 The biological roles of the third component of complement in osteoclast formation. Endocrinology **133**:397–404.
118. Sato T, Ono T, Tuan RS 1993 1,25-Dihydroxyvitamin D_3 stimulation of TGF-beta expression in chick embryonic calvarial bone. Differentiation **52**:139–150.
119. Lacey DL, Erdmann JM, Tan HL, Ohara J 1993 Murine osteoblast interleukin 4 receptor expression: Upregulation by 1,25-dihydroxyvitamin D_3. J Cell Biochem **53**:122–134.
120. Lacey DL, Grosso LE, Moser SA, Erdmann J, Tan HL, Pacifici R, Villareal DT 1993 IL-1-Induced murine osteoblast IL-6 production is mediated by the type I IL-1 receptor and is increased by 1,25-dihydroxyvitamin D_3. J Clin Invest **91**:1731–1742.
121. Jehan F, Naveilhan P, Neveu I, Harvie D, Dicou E, Brachet P, Wion D 1996 Regulation of NGF, BDNF, and LNGFR gene expression in ROS 17/2.8 cells. Mol Cell Endocrinol **116**: 149–156.
122. Nambi P, Lipshutz D, Prabhakar U 1995 Identification and characterization of endothelin receptors on rat osteoblastic cells: Down-regulation by 1,25-dihydroxy-vitamin D_3. Mol Pharmacol **47**:266–271.
123. Rosen H, Krichevsky A, Polsakiewicz RD, Benzakine S, Bar-Shavit Z 1995 Developmental regulation of proenkephalin gene expression in osteoblasts. Mol Endocrinol **9**:1621–1631.
124. Kassem M, Mosekilde L, Eriksen EF 1993 1,25-Dihydroxy-

vitamin D_3 potentiates fluoride-stimulated collagen type I production in cultures of human bone marrow stromal osteoblast-like cells. J Bone Miner Res **8**:1453–1458.

125. Williams GR, Bland R, Sheppard MC 1995 Retinoids modify regulation of endogenous gene expression by vitamin D_3 and thyroid hormone in three osteosarcoma cell lines. Endocrinology **136**:4304–4314.

126. Ohishi K, Nishikawa S, Nagata T, Yamauchi N, Shinohara H, Kido J, Ishida H 1995 Physiological concentrations of retinoic acid suppress the osteoblastic differentiation of fetal rat calvaria cells *in vitro*. Eur J Endocrinol **133**:335–341.

127. Ohishi K, Ishida H, Nagata T, Yamauchi N, Tsuzumi C, Nishikawa S, Wakano Y 1994 Thyroid hormone suppresses the differentiation of osteoprogenitor cells to osteoblasts, but enhances functional activities of mature osteoblasts in cultured rat calvaria cells. J Cell Physiol **161**:544–552.

128. Ishida H, Bellows CG, Aubin JE, Heersche JNM 1995 Tri-iodothyronine (T_3) and dexamethasone interact to modulate osteoprogenitor cell differentiation in fetal rat calvaria cell cultures. Bone **16**:545–549.

129. Choong PF, Martin TJ, Ng KW 1993 Effects of ascorbic acid, calcitriol, and retinoic acid on the differentiation of preosteoblasts J Orthopedic Res **11**:538–547.

130. Chapman AB, Knight DM, Ringold GM 1985 Glucocorticoid regulation of adipiocyte differentiation: Hormonal triggering of the developmental program and induction of a differentiation-dependent gene. J Cell Biol **101**:1227–1235.

131. Schwiek DR, Loffler G 1987 Glucocorticoid hormones contribute to the adipogenic activity of human serum. Endocrinology **120**:469–474.

132. Deryugina EI, Ratnikov BI, Bourdon MA, Muller-Sieburg CE 1994 Clonal analysis of primary marrow stroma: Functional heterogeneity in support of lymphoid and myeloid cell lines and identification of positive and negative regulators. Exp Hematol **22**:910–918.

133. Gimble JM, Youkhana K, Hua X, Bass H, Medina K, Sullivan MB, Greenberger J, Wang CS 1992 Adipogenesis in a myeloid supporting bone marrow stromal cell line. J Cell Biochem **50**:73–82.

134. Nuttall ME, Olivera DL, Gowen M 1994 Control of osteoblast/adipocyte differentiation of MG-63 cells. J Bone Miner Res **9**(Suppl. 1):A28.

135. Sato M, Hiragun A 1988 Demonstration of 1α,25-dihydroxyvitamin D_3 receptor-like molecule in ST 13 and 3T3 L1 preadipocytes and its inhibitory effects on preadiocyte differentiation. J Cell Physiol **135**:545–550.

136. Ishida Y, Taniguchi H, Baba S 1988 Possible involvement of 1α,25-dihydroxyvitamin D_3 in proliferation and differentiation of 3T3-L1 cells. Biochem Biophys Res Commun **151**:1122–1127.

137. Kawada T, Aoki N, Kamei Y, Maeshige K, Nishiu S, Sugimoto E 1990 Comparative investigation of vitamins and their analogues on terminal differentiation, from preadipocytes to adipocytes of 3T3-L1 cells. Comp Biochem Physiol **96A**:323–326.

138. Beresford JN, Bennet JH, Devlin C, Leboy PS, Owen ME 1992 Evidence for an inverse relationship between the differentiation of adipocytic and osteogenic cells in rat marrow stromal cell cultures. J Cell Sci **102**:341–351.

139. Shinome M, Shinski T, Takahashi N, Hasegawa K, Suda T 1992 1α,25-Dihydroxyvitamin D_3 modulation in lipid metabolism in established bone marrow-derived stromal cells, MC3T3–G2/PA6. J Cell Biochem **48**:424–430.

140. Falla N, Van Vlassalaer P, Bierkens J, Borremans B, Schoeters G, Van Gorp U 1993 Characterization of a 5-fluorouracil-enriched osteoprogenitor population of the murine bone marrow. Blood **82**:3580–3591.

141. Bellows CG, Ciaccia A, Heersche JNM 1996 Osteoprogenitor cells isolated from mouse and rat calvaria differ in their response to corticosterone and their density-dependent differentiation. J Bone Miner Res **11**(Suppl. 1):S259.

142. Tavasolli M 1989 Fatty involution in narrow and the role of adipose tissue in hemopoiesis. In: Tavasolli M (ed) Handbook of the Hemopoietic Microenvironment, Vol. 4. Humana Press, Clifton, New Jersey, pp. 157–187.

143. Tavassoli M 1984 Marrow adipose cells and hemopoiesis: An interpretative review. Exp Haemotol **12**:139–146.

144. Morikawa M, Nixon T, Green H 1982 Growth hormone and the adipose conversion of 3T3 cells. Cell **29**:783–789.

145. Gharbi-Chihi J, Grimaldi P, Torresani L, Ailhaud G 1981 Triiodothyronine and adipose conversion of Ob17 preadipocytes: Binding to high affinity sites and effects on fatty acid synthesizing and esterifying enzymes. J Receptor Res **2**:153–173.

146. Grigoriadis AE, Heersche JNM, Aubin JE 1990 Continuously growing bipotential and monopotential myogenic, adipogenic and chondrogenic subclones isolated from the multipotential LRCJ3.1 clonal cell line. Dev Biol **142**:313–318.

147. Meunier PJ, Aaron J, Edouard C, Vignon G 1971 Osteoporosis and the replacement of cell populations of the marrow by adipose tissue: A quantitative study of 84 iliac bone biopsies. Clin Orthopedic Related Res **80**:147–154.

148. Burkhardt R, Kettner G, Bohm W, Schmidtmeir M, Schlag R, Frisch B, Mallmann B, Eisenmenger W, Gilg TH 1987 Changes in trabecular bone, hematopoiesis and bone marrow vessels in aplastic anemia, primary osteoporosis and old age. Bone **8**:157–164.

149. Gimble JM, Robinson CE, Wu X, Kelly KA 1996 The function of adipocytes in bone marrow stroma: An update. Bone **19**:421–428.

150. Rozman C, Feliu E, Berga L, Reverter J-C, Climent C, Ferran M-J 1989 Age related variations of fat tissue fraction in normal human bone marrow depend both on size and number of adipocytes: A stereological study. Exp Hematol **17**:34–37.

151. Minaire P, Edouard C, Arlot M, Meunier PJ 1984 Marrow changes in paraplegic patients. Calcif Tissue Int **36**:338–340.

152. Wronski TJ, Smith JM, Jee WSS 1981 Variations in mineral apposition rate of trabecular bone within the beagle. Calcif Tissue Int **33**:583–586.

153. Martin RB, Chow BD, Lucas PA 1990 Bone marrow fat content in relation to bone remodeling and serum chemistry in intact and ovariectomized dogs. Calcif Tissue Int **46**:189–194.

154. Wang GW, Sweet D, Reger S, Thompson R 1977 Fat cell changes as a mechanism of avascular necrosis in the femoral head of cortisone-treated rabbits. J Bone Joint Surg **59A**:729–735.

155. Kwai K, Tamaki A, Hirohata K 1985 Steroid induced accumulation of lipid in the osteocytes of the rabbit femoral head. J Bone Joint Surg **67A**:755–763.

156. Minaire P, Meunier PJ, Edouard C, Bernard C, Courpron J, Bourret J 1974 Histological data on disuse osteoporosis: Comparison with biological data. Calcif Tissue Res **17**:57–73.

157. Jiang D, Fei RG, Pendergrass WR, Wolf NS 1992 An age-related reduction in the replicative capacity of two murine hematopoietic stroma cell types. Exp Hematol **20**:1216–1222.

158. Tsuji T, Hughes FJ, McCulloch CA, Melcher AH 1990 Effects of donor age on osteogenic cells of rat bone marrow in vitro. Mech Ageing Dev **51**:121–132.

159. Kajkenva O, Gubrij I, Hauser SP, Takahashi K, Jilka RL, Manolagas SC, Lipshitz DA 1995 Increased hematopoiesis accompanies reduced osteoblastogenesis in the senescence-accelerated mouse (SAM-P/6). J Bone Miner Res **10**(Suppl 1):T309.

160. Jika RL, Weinstein RS, Takahashi K, Parfitt AM, Manolagas SC 1996 Linkage of decreased bone mass with impaired osteoblastogenesis in a murine model of accelerated senescense. J Clin Invest **97**:1732–1740.
161. Dorheim MA, Sullivan M, Dandapani V, Wu X, Hudson J, Segarini PR, Rosen DM, Aulthouse AL, Gimble JM 1993 Osteoblastic gene expression during adipogenesis in homopoietic supporting murine bone marrow stromal cells. J Cell Physiol **154**:317–328.
162. Locklin RM, Williamson MC, Beresford JN, Triffitt JT, Owen ME 1995 In vitro effects of growth factors and dexamethasone on rat marrow stromal cells. Clin Orthop **313**:27–35.
163. Bennett JH, Joyner CJ, Triffit JT, Owen ME 1991 Adipocytic cells cultured from marrow have osteogenic capacity. J Cell Sci **99**:131–139.
164. Thompson DL, Nygaard S, Moore E 1996 The use of p53 knockout mice to generate bone marrow stromal cell lines representing various stages of osteoblast differentiation. J Bone Miner Res **1**(Suppl. 1):S108.
165. Weber GF, Ashkar S, Glimcher MJ, Cantor H 1996 Receptor–ligand interaction between CD44 and osteopontin (Eta-1). Science **271**:509–512.
166. Abumrad NA, El-Maghrabi MR, Amri E-Z, Grimaldi PA 1993 Cloning of rat adipocyte membrane protein implicated in binding or transport of long-chain fatty acids that is induced during preadipocyte differentiation: Homology with CD36. J Biol Chem **268**:17665–17668.
167. Devchand PR, Keller H, Peters JM, Vazquez M, Gonzalez FJ, Wahli W 1996 The PPARα-leukotriene B4 pathway to inflammation control. Nature **384**:39–43.
168. Williams GD, Deldar A, Jordan WH, Gries C, Long GG, Dimarchi RD 1993 Subchronic toxicity of the thiazolidinedione Tanabe-174 (LY282449) in the rat and dog. Diabetes **42**(Suppl.):186.
169. Jennerman C, Triantafillou J, Cowan D, Pennink BGA, Connolly KM, Morris DC 1995 Effects of thiazolidinediones on bone turnover in rat. J Bone Miner Res **10**(Suppl. 1):241.
170. Tam CS, Jones G, Heersche JNM 1981 The effect of vitamin D restriction on bone apposition in the rats and its dependence on parathyroid hormone. Endocrinology **101**:1448.
171. Tam CS, Heersche JNM, Jones G, Murray TM, Rasmussen H 1986 The effect of vitamin D on bone *in vivo*. Endocrinology **118**:2217–2224.
172. Hock JM, Gunnes-Hey M, Poser J, Olson H, Bell NH, Raisz LG 1986 Stimulation of undermineralized matrix formation by 1,25-dihydroxyvitamin D_3 in long bones of rats. Calcif Tissue Int **28**:79–86.
173. Ono T, Tanaka H, Yamate T, Nagai Y, Nakamura T, Seino Y 1996 24*R*,25-Dihydroxyvitamin D_3 promotes bone formation without causing excessive resorption in hypophosphatemic mice. Endocrinology **137**:2633–2637.
174. Ornoy AD, Goodwin D, Noff D, Edelstein S 1978 24,25-Dihydroxyvitamin D is a metabolite of vitamin D essential for bone formation. Nature **276**:5.
175. Endo H, Kiyoki M, Kawashima K, Naguchi T, Hashimoto Y 1980 Vitamin D_3 metabolites and PTH synergistically stimulate bone formation. Nature **286**:262.
176. Bordier P, Rasmussen H, Maria P, Miravet L, Gueris J, and Ryckewaert A 1978 Vitamin D metabolites and bone mineralization in man. J Clin Endocrinol Metab **46**:284.
177. Malluche HH, Massry SG 1980 The role of vitamin D metabolites in the management of bone in renal disease. Adv Exp Med Biol **128**:615.
178. St-Arnaud R, Arabian A, Glorieux FH 1996 Abnormal bone development in mice deficient for the vitamin D 24-hydroxylase gene. J Bone Miner Res **11**(Suppl. 1):S126.

CHAPTER 21

Vitamin D and Osteoclastogenesis

TATSUO SUDA AND NAOYUKI TAKAHASHI Department of Biochemistry, School of Dentistry, Showa University, Shinagawa-ku, Tokyo, Japan

I. INTRODUCTION

Throughout the course of vitamin D study in bone, emphasis has been placed on the mineralization process as the possible site of vitamin D action [1–3]. Indeed, osteoblasts are the major bone cells having vitamin D receptors (VDR) [4]. However, whether the action form of vitamin D_3, 1α,25-dihydroxyvitamin D_3 [$1\alpha,25(OH)_2D_3$], directly promotes osteoblastic bone mineralization is still a matter of controversy. It is widely believed that the administration of vitamin D to rachitic animals cures rickets and osteomalacia mainly through stimulation of intestinal absorption of calcium and phosphate [1–3].

It appears paradoxical, but vitamin D definitely mobilizes calcium from bone to the bloodstream. This important observation was made initially by Carlsson in 1952 [52], who provided convincing evidence that vitamin D functions in the process of calcium mobilization from calcified bone, making calcium available to the extracellular fluid on demand by the calcium homeostatic system. It is now well recognized that $1\alpha,25(OH)_2D_3$ is able to stimulate osteoclastic bone resorption, although the mechanism by which osteoclastogenesis is induced by the vitamin was not known until 1981.

Several lines of evidence have indicated that multinucleated osteoclasts, the principal bone-resorbing cells, are derived from immature bone marrow cells of the monocyte–macrophage lineage and are localized only in bone [6]. $1\alpha,25(OH)_2D_3$, when administered in picomolar to nanomolar doses to rachitic animals fed a vitamin D-deficient, low calcium diet, markedly increases the number of osteoclasts, then stimulates osteoclastic bone resorption. This suggests that $1\alpha,25(OH)_2D_3$ is involved in osteoclastic bone resorption in concert with parathyroid hormone (PTH).

In 1981, we reported that 1α,25$(OH)_2D_3$ induces differentiation of mouse myeloid leukemia cells (M1) into macrophages *in vitro* [7]. The degree of cell differentiation in various markers of macrophages induced by 10^{-8} *M* 1α,25$(OH)_2D_3$ was nearly equivalent to that induced by 10^{-6} *M* dexamethasone, the most potent known stimulator. Neither 25-hydroxyvitamin D_3 (25OHD$_3$) nor 24,25-dihydroxyvitamin D_3 [24,25$(OH)_2D_3$] showed potent inducing activity. Subsequently, we found that 1α,25$(OH)_2D_3$ induced not only differentiation but also activation and fusion of monocytes–macrophages directly and also by an indirect mechanism mediated by spleen cells [8]. From these results, we hypothesized that 1α,25$(OH)_2D_3$ could be involved in osteoclastogenesis by stimulating differentiation and fusion of osteoclast progenitors.

II. STRUCTURE AND FUNCTION OF OSTEOCLASTS

Osteoclasts have several unique ultrastructural characteristics (Fig. 1) [9–11]. The cells have a number of nuclei, abundant pleomorphic mitochondria, and a large number of vacuoles and lysosomes. The most characteristic morphological feature of osteoclasts is the presence of ruffled borders and clear zones. It is generally accepted that the clear zone, sometimes called the sealing zone, serves for the attachment of osteoclasts to the bone surface. Osteoclasts are also known to be positive for the presence of tartrate-resistant acid phosphatase (TRAP, a marker enzyme of osteoclasts), but the precise role of TRAP in osteoclasts is not known. Calcitonin inhibits bone resorption by directly acting on osteoclasts [12]. Numerous specific binding sites for calcitonin have been demonstrated by autoradiographic and biochemical analysis of disaggregated mammalian osteoclasts [13]. Thus, the expression of calcitonin receptors is one of the most reliable markers of osteoclasts [14]. Antibodies to vitronectin receptors strongly inhibited bone resorption by osteoclasts [15,16]. Thus, the vitronectin receptors on osteoclasts appeared to be involved in their function, possibly in the attachment of osteoclasts to the bone matrix.

The resorbing area under the ruffled border of osteoclasts is acidic, which favors dissolution of bone mineral. In bone-resorbing osteoclasts, hydrogen ions are provided by carbonic anhydrase II, which catalyzes the hydration of CO_2 to H_2CO_3 [17]. The proton pump of vacuolar H^+-ATPase exists in the ruffled border membranes of osteoclasts [18,19]. The transport of protons into Howship's lacunae appears to be mediated by this

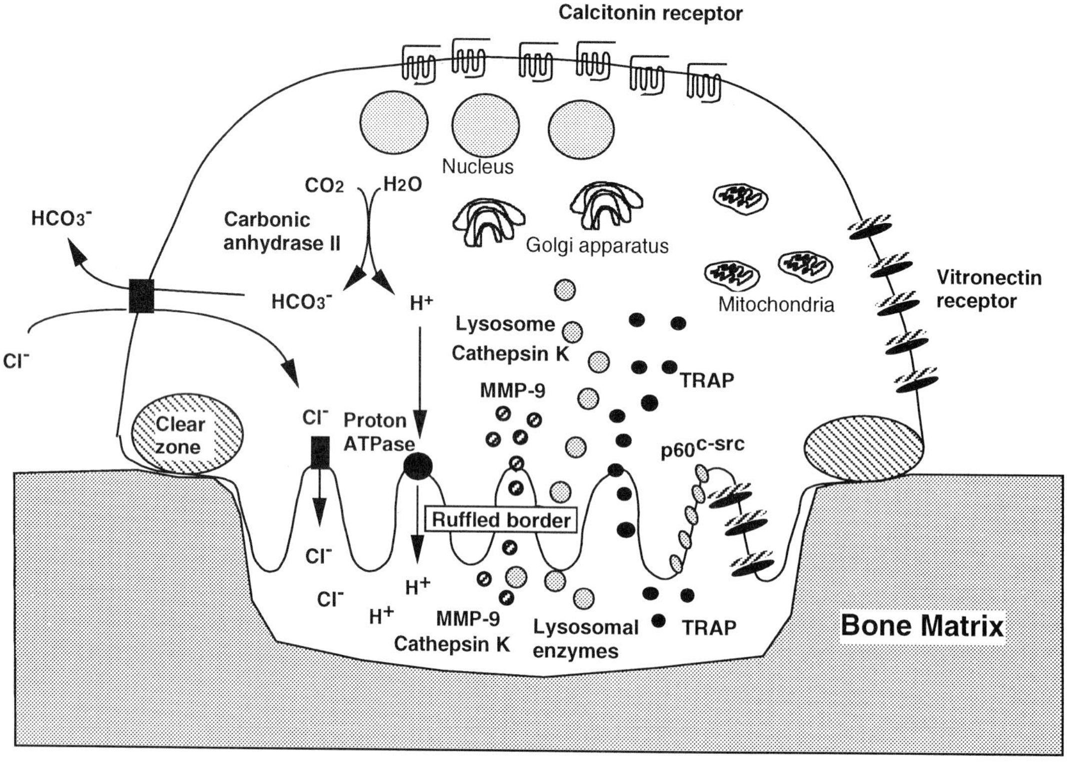

FIGURE 1 Schematic representation of typical structures and functions of osteoclasts. TRAP, Tartrate-resistant acid phosphatase; MMP, metalloproteinases.

vacuolar type H^+-ATPase. Secretion of hydrogen ions by osteoclasts generates an equal amount of cytoplasmic base equivalents, principally as HCO_3^-. Osteoclasts have a chloride/biocarbonate exchanger, which normalizes the intracellular pH when osteoclasts resorb bone [20]. Lysosomal enzymes of osteoclasts are also secreted into the resorbing area to degrade the organic matrix of bone. Cysteine proteinases such as cathepsin K (OC-2) [21,22] and metalloproteinases (MMPs) such as MMP-9 [23] have been reported to be present in osteoclasts, which degrade type I collagen of demineralized bone in an acidic environment.

III. OSTEOCLAST FORMATION IN MOUSE BONE MARROW CULTURES AND IN COCULTURES WITH OSTEOBLASTIC CELLS

Development of bone marrow culture systems has greatly contributed to our understanding of osteoclast differentiation. The formation of TRAP-positive multinucleated cells has been demonstrated in marrow cultures of several species such as feline, baboon, mouse, and humans [6,24]. $1\alpha,25(OH)_2D_3$ greatly promoted TRAP-positive multinucleated cell formation in those marrow cultures. In mouse marrow cultures, TRAP-positive multinucleated cells were formed only near the clusters of alkaline phosphatase (ALP)-positive stromal cells [25]. This suggested that osteoblastic cells are somehow involved in osteoclast differentiation.

To further examine the possible involvement of osteoblastic cells in osteoclast formation, we established an efficient coculture system to recruit osteoclasts [26]. When mouse spleen cells and primary osteoblastic cells were cocultured in the presence of $1\alpha,25(OH)_2D_3$, a number of osteoclast-like multinucleated cells were formed. These osteoclast-like cells were formed only when spleen cells and osteoblastic cells were cocultured in the presence of $1\alpha,25(OH)_2D_3$; when osteoblastic cells and spleen cells were cultured separately, no osteoclast-like cells were formed [26]. From these results, we hypothesized that osteoblastic cells are prerequisite to the differentiation of osteoclast progenitors into osteoclasts. This is the key point of our strategy for osteoclast differentiation, which is based on the pioneering hypothesis by Rodan and Martin [27]. Spleen cells represent osteoclast progenitors, in other words, "seeds," and osteoblastic cells represent supporting cells to provide a suitable microenvironment for osteoclast differentiation in bone, in other words, a "farm." This hypothesis has been named the seeds and farm theory.

To examine whether osteoclast-like multinucleated cells that were formed in mouse cocultures could resorb calcified tissues, cocultures were performed on dentine slices in the presence of $1\alpha,25(OH)_2D_3$. Numerous resorption pits were formed on the dentine slices, whereas control cultures without $1\alpha,25(OH)_2D_3$ formed no resorption pits. Thus, osteoclast formation in the mouse coculture system is strictly hormone dependent. The multinucleated cells formed in mouse cocultures had abundant calcitonin receptors, carbonic anhydrase II, vacuolar-type proton ATPase, and vitronectin receptors ($\alpha_v\beta_3$) [6]. Also, they expressed the protooncogene c-*src* protein ($p60^{c\text{-}src}$) [28]. Thus, we believe that the TRAP-positive multinucleated cells formed in the mouse coculture system are indeed genuine osteoclasts. Whether this *in vitro* assay for the formation of osteoclasts reflects the processes that occur *in vivo*, however, remains to be clarified.

IV. ESTABLISHED OSTEOBLASTIC STROMAL CELL LINES THAT HAVE THE CAPACITY TO SUPPORT OSTEOCLAST FORMATION

Subsequent experiments indicated that primary osteoblastic cells capable of inducing osteoclasts in mouse cocultures with spleen cells could be replaced by certain established stromal cell lines [29,30]. We first reported that two marrow-derived mouse stromal cell lines, MC3T3-G2/PA6 and ST2, had the capacity to support osteoclast formation in cocultures with spleen cells (Table I) [29]. MC3T3-G2/PA6 and ST2 cells have been classified as preadipocytes that support growth and differentiation of hemopoietic cells in cocultures with bone marrow cells. The osteoclast formation induced by $1\alpha,25(OH)_2D_3$ in cocultures with MC3T3-G2/PA6 or ST2 cells was markedly stimulated by simultaneous addition of dexamethasone [29]. Glucocorticoids per se had no effect on osteoclast formation in cocultures with primary osteoblastic cells. The steroids may have potentiated differentiation of these clonal stromal cells into cells that could support osteoclastogenesis in the presence of $1\alpha,25(OH)_2D_3$.

We also established a clonal stromal cell line (KS-4) from mouse calvaria on the basis of the ability to support osteoclast formation in cocultures with spleen cells [30]. Glucocorticoids were not required for osteoclast formation induced by $1\alpha,25(OH)_2D_3$ in cocultures with KS-4 cells, as in the case of the coculture with primary osteoblastic cells. The KS-4 cells showed several characteristics of the osteoblast phenotype. They synthesized mainly type I collagen, formed nodules in the absence of adding β-glycerophosphate in long-term cultures, and

TABLE I Comparison of Characteristics of Three Stromal Cell Lines in Supporting Osteoclast Formation

Characteristic	Stromal cell line								
	ST2			MC3T3-G2/PA6			KS-4		
Origin	Bone marrow			Calvaria			Calvaria		
Cell types	Preadipocytes			Preadipocytes			Osteoblasts		
Hormones	PTH	PGE_2	$1,25D_3$	PTH	PGE_2	$1,25D_3$	PTH	PGE_2	$1,25D_3$
Existence of hormone receptor[a]	−	+	+	−	+	+	+	+	+
Osteoclast formation-supporting activity[b]	−	+	+	−	+	+	+	+	+

[a] The presence or absence of receptors for parathyroid hormone (PTH), prostaglandin E_2 (PGE_2), and $1\alpha,25(OH)_2D_3$ ($1,25D_3$) on each cell line is represented by + or −.

[b] The presence or absence of osteoclast formation-supporting activity evaluated in cocultures of the respective cell line and mouse spleen cells in the presence of each hormone is represented by + or −.

increased ALP activity after they became confluent. In addition, PTH and prostaglandin E_2 (PGE_2) stimulated cyclic AMP production in KS-4 cells (Table I). These findings suggest that some osteoblast lineage cells retain the capacity to support osteoclast development in cocultures with spleen cells.

Another osteoblastic cell line, MC3T3-E1, and some stromal cell lines of nonbony origin such as ST13, BALB-3T3, and NIH-3T3 did not support osteoclast development in cocultures with spleen cells [6]. At present, it is difficult to identify the common phenotype in positive cell lines which support osteoclastogenesis in cocultures with spleen cells. However, it should be noted that some of the stromal cell lines which support osteoclast development are endowed with the ability to support hemopoiesis in cocultures with bone marrow cells. It appears that there is a common mechanism between hemopoiesis and osteoclastogenesis [6]. When spleen cells and primary osteoblastic cells were cultured with one another but separated by a membrane filter, no osteoclasts were formed even in the presence of $1\alpha,25(OH)_2D_3$ [26]. Similar findings were obtained in cocultures of spleen cells and clonal stromal cells which support osteoclast formation [29,30], suggesting the importance of cell-to-cell or cell-to-cell matrix recognition in osteoclast recruitment. Thus, it is concluded that the development of osteoclasts proceeds within a microenvironment provided by mesenchymal cells exemplified by osteoblastic cells or their precursors.

V. ORIGIN AND DIFFERENTIATION PROCESS OF OSTEOCLASTS

The conclusive evidence that osteoclasts are derived from hemopoietic cells came from three different *in vivo* investigations: parabiosis experiments [31], chick–quail chimera experiments [32], and the restoration of bone resorption in osteopetrotic mice by transplanting normal marrow cells or spleen cells [33]. Cocultures of mouse osteoblastic cells and spleen cells also provided evidence that osteoclast progenitors were present in spleen cell preparations but not in osteoblastic cell preparations. To determine which hemopoietic lineage cells are able to differentiate into osteoclasts, we developed a coculture system using ST2 cells [34]. When a relatively small number of mononuclear cells obtained from mouse bone marrow, spleen, thymus, or peripheral blood were cultured for 12 days on the ST2 cell layer, they formed single cell-derived colonies. Addition of $1\alpha,25(OH)_2D_3$ and dexamethasone induced TRAP-positive cells in some of the colonies (TRAP-positive colonies) formed on the ST2 cell layer. The bone marrow formed the largest number of TRAP-positive colonies, followed by peripheral blood, spleen, and thymus, in that order. The monocyte-depleted population prepared from peripheral blood mononuclear cells failed to form TRAP-positive colonies on the ST2 cell layer, whereas the monocyte-enriched population formed many TRAP-positive colonies. All of the colonies consisted of mainly cells positive for nonspecific esterase (NSE, a marker enzyme of monocytes–macrophages). Alveolar macrophages similarly formed TRAP-positive colonies in the presence of $1\alpha,25(OH)_2D_3$ and dexamethasone. The TRAP-positive cells formed from alveolar macrophages had calcitonin receptors and the ability to form resorption pits on dentine slices. These findings suggested that blood monocytes and some tissue macrophages are capable of differentiating into osteoclasts in the presence of stromal cells which support osteoclast development.

We also found that the precursor cells undergoing

osteoclast differentiation expressed macrophage-associated phenotypes. In the coculture system, the 6-day culture period could be separated into two phases: the first 4 days, in which proliferation of osteoclast progenitors primarily occurred, and the final 2 days, in which differentiation into osteoclasts was predominant [35]. The presence of hydroxyurea in the cocultures during the final 2 days completely inhibited proliferation of osteoclast progenitors, but their differentiation into osteoclasts proceeded even in the presence of hydroxyurea. Using this coculture system, chronological changes of macrophage-associated phenotypes such as expression of NSE and antigens to Mac-1, Mac-2, and F4/80 were examined in postmitotic osteoclast precursors during differentiation into osteoclasts induced by 1α,25$(OH)_2D_3$ added on day 4 (Fig. 2) [36]. Osteoclast differentiation was monitored by the expression of calcitonin receptors (CTRs) assessed by autoradiography using ^{125}I-labeled calcitonin. CTRs were first detected on small mononuclear cells within 12 hr after the addition of 1α,25$(OH)_2D_3$. The number of CTR-positive mononuclear cells attained a maximum at 24 hr and decreased thereafter. CTR-positive multinucleated cells were first detected at 24 hr, and reached a maximum at 48 hr. All CTR-positive cells showed strong TRAP activity. Most of the CTR-positive mononuclear cells which appeared at 12 hr were positive for NSE and antigens to Mac-1 and Mac-2, but they were negative for F4/80 antigen (Fig. 2). The proportion of CTR-positive cells expressing NSE and Mac-1 to total CTR-positive mononuclear cells decreased with the passage of time. Like authentic osteoclasts, CTR-positive multinucleated cells were negative for NSE and antigens to Mac-1 and F4/80, but positive to Mac-2. These findings indicate that postmitotic osteoclast precursors are mononuclear phagocytes with macrophage-associated phenotypes, some of which disappear rapidly during osteoclast differentiation (Fig. 2) [36].

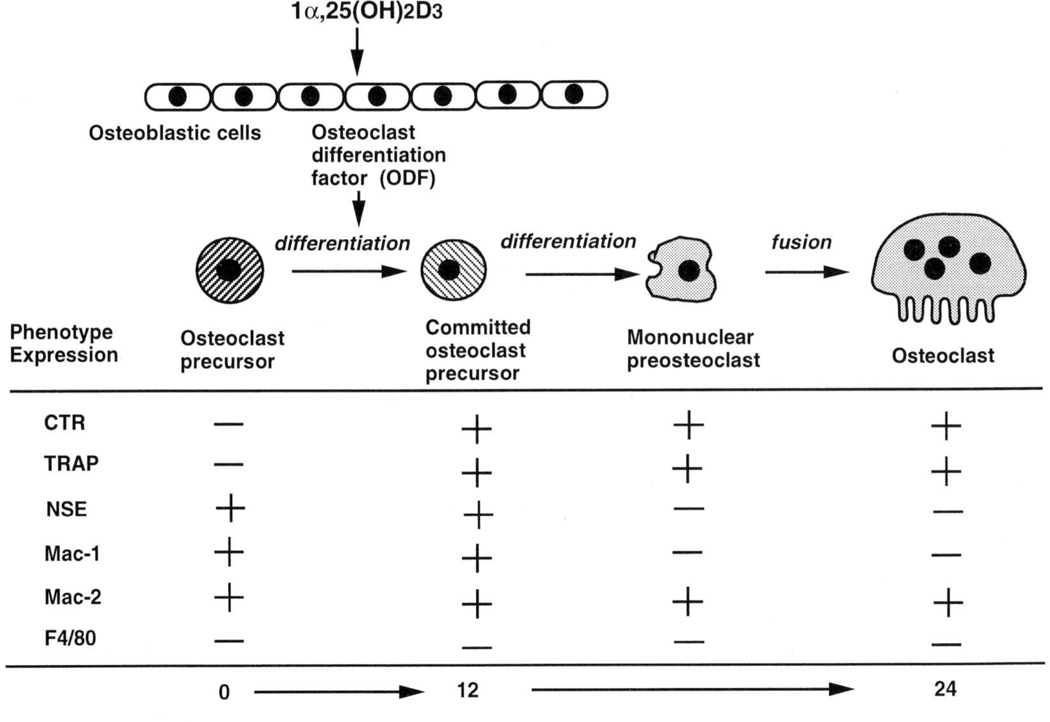

FIGURE 2 Expression of the osteoclast/macrophage-associated phenotype during differentiation of postmitotic osteoclast precursors into osteoclasts. Postmitotic osteoclast precursors appear to express macrophage-associated phenotypes such as nonspecific esterase (NSE), Mac-1, and Mac-2. When osteoblastic cells express osteoclast differentiation factor (ODF) in response to 1α,25$(OH)_2D_3$, postmitotic osteoclast precursors recognize ODF through cell-to-cell or cell-to-matrix interactions and begin to differentiate into osteoclasts. They express calcitonin receptors (CTR) and tartrate-resistant acid phosphatase (TRAP) almost simultaneously. In contrast, some of the macrophage-associated phenotype markers such as NSE and Mac-1 rapidly disappear in the precursor cells during differentiation. After the precursor cells are committed to differentiate into mononuclear preosteoclasts, they begin to fuse to form multinucleated cells. Some of the macrophage phenotype markers such as Mac-2 are still retained in mature osteoclasts.

VI. SIGNAL TRANSDUCTION OF BONE-RESORBING FACTORS FOR INDUCING OSTEOCLAST DEVELOPMENT

Evidence has indicated that various bone-resorbing hormones and cytokines including $1\alpha,25(OH)_2D_3$ commonly act directly on osteoblastic stromal cells for osteoclast development in cocultures of osteoblastic stromal cells and spleen cells [37]. The bone-resorbing hormones and cytokines could be classified into three categories in terms of their signal transduction mechanism: cAMP-mediated effects are shown by PTH, parathyroid hormone-related peptide (PTHrP), PGE_2, and interleukin-1 (IL-1); VDR-mediated effects are shown by $1\alpha,25(OH)_2D_3$; and gp130-mediated effects are caused by IL-6, IL-11, leukemia inhibitory factor (LIF), and oncostatin M (OSM) [37].

A. cAMP-Mediated Signal Transduction

Osteoclast formation induced by prostaglandins (PGs) in mouse marrow cultures was mediated by a mechanism involving cAMP [38]. The potency of PGs in inducing osteoclast formation was the highest for PGE_1 and PGE_2, followed by $PGF_{2\alpha}$ and 6-keto-$PGF_{1\alpha}$, in that order. Potency to induce osteoclastogenesis was highly correlated with the order of the potency to increase cAMP production in bone marrow cells. Addition of dibutyryl cAMP to mouse marrow cultures also induced osteoclast formation in a dose-dependent manner. Moreover, isobutylmethylxanthine (IBMX), a potent inhibitor of phosphodiesterase, potentiated the PGE_2-induced osteoclast formation. These findings indicate that the activity of PGs in inducing osteoclasts is mediated by a mechanism involving cAMP. Similarly, osteoclast formation induced by PTH in mouse marrow cultures appeared to be mediated mainly by cAMP [39]. IL-1 also stimulated osteoclast formation by a mechanism involving PGE_2 production both in mouse marrow cultures and in cocultures of osteoblastic cells and spleen cells [40]. Indomethacin completely blocked osteoclast formation induced by IL-1 but not by $1\alpha,25(OH)_2D_3$, PTH, and PGE_2. These findings indicate that osteoclast formation induced by IL-1, PTH, and PGE_2 is commonly mediated by a mechanism involving cAMP production.

We compared the abilities of three stromal cell lines (MC3T3-G2/PA6, ST2, and KS-4) to produce cAMP with their ability to support osteoclast formation in cocultures with spleen cells in response to PTH and PGE_2 [29,30]. These stromal cell lines appeared to possess receptors for PGE_2, since PGE_2 stimulated cAMP production in all three cell lines (Table I). Osteoclasts were similarly formed in cocultures of the respective stromal cells and spleen cells when they were treated with PGE_2. In contrast, the PTH-induced osteoclast formation occurred only in cocultures with KS-4 cells. PTH stimulated cAMP production in KS-4 cells but not in MC3T3-G2/PA6 and ST2 cells (Table I). These results suggest that cAMP-mediated signals of bone-resorbing agents are transduced into osteoblastic stromal cells [29,30].

B. gp130-Mediated Signal Transduction

More convincing evidence for the hypothesis that bone-resorbing agents act on osteoblastic cells but not osteoclast progenitors was obtained in IL-6-induced osteoclast formation [41,42]. IL-6 exerts its activity via a cell surface receptor that consists of two components: a ligand-binding glycoprotein gp80 (IL-6 receptor, IL-6R) and a non-ligand-binding but signal-transducing glycoprotein gp130 [43]. The genetically engineered human and mouse soluble IL-6R (sIL-6R), which lacked both the transmembrane and cytoplasmic domains, has been shown to transduce IL-6 signals through gp130 [44]. We examined the role of IL-6 and sIL-6R in osteoclast formation in the mouse coculture system [41]. Neither mouse recombinant IL-6 nor sIL-6R induced osteoclast formation when they were added separately to the cocultures. In contrast, simultaneous treatment of the coculture with IL-6 and sIL-6R strikingly induced osteoclast formation. Other cytokines such as IL-11, oncostatin M, and leukemia inhibitory factor, which exert their functions through gp130 as a common signal transducer, also stimulated osteoclast formation in the absence of soluble receptors of the respective cytokines [41]. This suggests that osteoblastic cells and/or osteoclast progenitors express receptors for IL-11, oncostatin M, and leukemia inhibitory factor, but not receptors for IL-6.

Using transgenic mice constitutively expressing human IL-6R, we showed that the expression of membrane-bound IL-6R in osteoblastic cells is indispensable for osteoclast formation induced by IL-6 in the coculture [42]. When osteoblastic cells obtained from the transgenic mice were cocultured with normal spleen cells from ddY mice, osteoclast formation was stimulated in response to human IL-6 without the addition of human sIL-6R. Murine IL-6 failed to induce osteoclast formation in the cocultures with osteoblastic cells constitutively expressing human IL-6R. In contrast, no osteoclasts were formed in response to human IL-6 in the cocultures of spleen cells from the transgenic mice and normal osteoblastic cells. It is known that human IL-6 binds to human IL-6R, but mouse IL-6 does not. These findings confirm the hypothesis that signals of gp130-mediated bone-resorbing agents are transduced into osteoblastic cells [37,42].

C. VDR-Mediated Signal Transduction

Using culture systems described above, we examined the role of vitamin D derivatives in osteoclast development. Neither $25OHD_3$ nor $24,25(OH)_2D_3$ at 10^{-8} *M* showed a significant stimulatory effect on osteoclast formation in our culture system. This indicates that osteoclast development induced by 10^{-8} *M* $1\alpha,25(OH)_2D_3$ occurs by a mechanism involving VDR. For understanding the role of $1\alpha,25(OH)_2D_3$ in osteoclast recruitment, it is important to know how signals mediated by VDR are transduced for osteoclast development. Kato and colleagues succeeded in producing VDR-deficient mice by targeted disruption of the gene encoding VDR [68]. Osteoblastic cells obtained from VDR (−/−) mice failed to support osteoclast development in cocultures with spleen cells obtained from wild-type mice in response to $1\alpha,25(OH)_2D_3$. In contrast, when spleen cells derived from VDR (−/−) mice were cocultured with osteoblastic cells obtained from wild-type mice, TRAP-positive osteoclasts were formed in response to $1\alpha,25(OH)_2D_3$ (S. Takeda, T. Matsumoto, and S. Kato, personal communication, 1996). These results strongly suggest that signals mediated by VDR are also transduced into osteoblastic cells to recruit osteoclasts.

The relationship among the three signaling pathways of bone-resorbing factors in osteoclasts during osteoclastogenesis is interesting. Girasole *et al.* [45] reported that PTH and $1\alpha,25(OH)_2D_3$ stimulated IL-11 production in mouse bone marrow cultures, and an antibody against IL-11 suppressed osteoclast formation in response to $1\alpha,25(OH)_2D_3$ and PTH. Romas *et al.* [46] reported that osteoblasts produced IL-11 in response to various bone-resorbing factors including IL-1, tumor necrosis factor-α (TNFα), PGE_2, PTH, and $1\alpha,25(OH)_2D_3$. Unstimulated osteoblastic cells did not release any detectable IL-11 into the culture medium. To examine whether gp130-mediated signal transduction is required for osteoclast formation by bone-resorbing factors which do not require gp130 in their signal transduction, we prepared a rat monoclonal anti-mouse gp130 antibody. The anti-gp130 antibody completely abolished osteoclast formation induced by IL-1 and partially abolished that induced by PGE_2, PTH, or $1\alpha,25(OH)_2D_3$. These results suggest that gp130 signals are somehow important for osteoclastogenesis induced by all bone-resorbing factors.

Figure 3 compares two possible signaling pathways of bone-resorbing factors in osteoblastic cells for osteoclast recruitment. In Fig. 3A, bone-resorbing factors were classified into three different categories: VDR-mediated [$1\alpha,25(OH)_2D_3$], gp130-mediated (cytokines), and cAMP-mediated factors (PTH, PGE_2, and IL-1). These bone-resorbing factors appeared to act independently on osteoblastic cells to induce osteoclast formation. In contrast, in Fig. 3B, IL-11 may be involved in a common pathway in osteoclastogenesis induced by all bone-resorbing factors. At present, the contribution of each pathway toward osteoclast recruitment from their progenitors *in vivo* is not known.

VII. PROTEINS PRODUCED BY OSTEOBLASTS IN RESPONSE TO $1\alpha,25(OH)_2D_3$

Evidence that the $1\alpha,25(OH)_2D_3$ receptor (VDR) is present in osteoblasts but not in osteoclasts indicates that the major target cells for $1\alpha,25(OH)_2D_3$ in bone are osteoblasts or preosteoblastic stromal cells. Indeed, $1\alpha,25(OH)_2D_3$ acts on cells of the osteoblast phenotype to produce several noncollagenous proteins (Table II) [47–61]. Bone tissue contains two vitamin K-dependent calcium-binding proteins, bone Gla protein (BGP, osteocalcin) [47] and matrix Gla protein (MGP) [48]. BGP is soluble in water and preferentially bound to the mineral phase [47], but MGP is insoluble and anchored to the matrix phase [48]. Price and Baukol [49] reported for the first time that osteoblast-like rat osteosarcoma cells, ROS 17/2.8, can produce BGP in response to $1\alpha,25(OH)_2D_3$. The promoter regions of the rat BGP gene [47] and human BGP gene [51] contain an element responsive to $1\alpha,25(OH)_2D_3$ (VDRE). The synthesis of MGP is also reported to be stimulated by $1\alpha,25(OH)_2D_3$ in another osteosarcoma cell line, UMR-106 [52]. Osteopontin (bone sialoprotein I) is a 44-kDa glycoprotein that contains the residues Gly-Arg-Gly-Asp-Ser, a sequence identical to the cell binding domain of fibronectin and several other cell-adhesion molecules [53]. An immunological examination revealed that osteopontin is present in bone matrix, osteoid, osteoblasts, and osteocytes [54]. The levels of osteopontin mRNA were markedly increased by $1\alpha,25(OH)_2D_3$ in osteoblastic cells such as ROS 17/2.8 [55] and MC3T3-E1 [56]. The promoter region of the human osteopontin gene has been reported to contain a sequence responsive to $1\alpha,25(OH)_2D_3$ [57]. Although possible roles of BGP and osteopontin in osteoclastic bone resorption have been suggested, the precise biological role(s) of those noncollagenous proteins in bone remains to be elucidated.

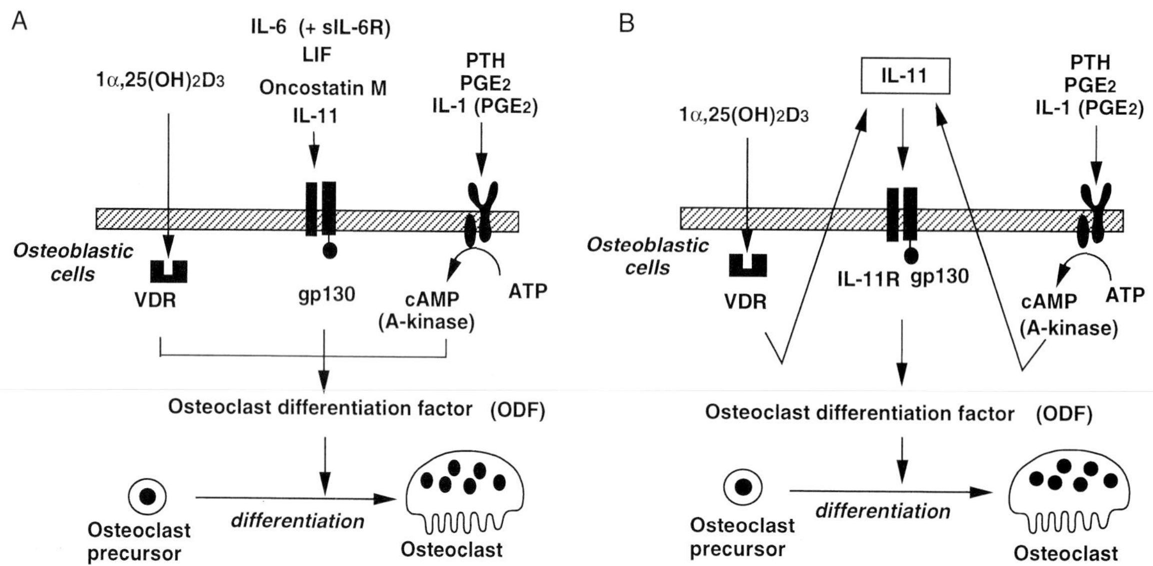

FIGURE 3 Two hypothetical concepts for osteoclast differentiation. (A) Three different signaling pathways mediated by $1\alpha,25(OH)_2D_3$ receptors (VDR), cAMP, and gp130 independently stimulate osteoclast differentiation. Target cells for these signal-inducing hormones and cytokines are osteoblastic cells. Osteoblastic cells express osteoclast differentiation factor (ODF) in response to these cytokines and hormones. The osteoclast precursors then recognize ODF and differentiate into osteoclasts. (B) IL-11 is a common key factor involving osteoclastogenesis induced by various bone-resorbing factors. Bone-resorbing factors and hormones stimulate IL-11 production by osteoblastic cells. IL-11 then acts on osteoblastic cells to express ODF. In this hypothesis, the signals mediated by gp130 regulate osteoclastogenesis induced by other bone-resorbing factors.

We have purified a 190-kDa protein from conditioned media of ST2 cell cultures treated with $1\alpha,25(OH)_2D_3$ and identified it as the third component of mouse complement (C3) [58]. Northern and Western blot analyses revealed that the production of C3 by ST2 and primary osteoblastic cells was strictly dependent on $1\alpha,25(OH)_2D_3$, but its production by hepatocytes was not [59]. Adding $1\alpha,25(OH)_2D_3$ together with antibodies against mouse C3 and its receptors to bone marrow cultures greatly inhibited the formation of TRAP-positive osteoclasts [58,60]. Adding C3 alone induced no TRAP-positive cell formation [58]. These results indicate the possibilities that C3 is one of the factors produced by osteoblasts in response to $1\alpha,25(OH)_2D_3$ and that C3 is somehow involved in determining the differentiation of bone marrow cells into osteoclasts in concert with other $1\alpha,25(OH)_2D_3$-dependent factors.

It is important to determine whether the vitamin D-dependent production of C3 by osteoblastic cells and marrow-derived stromal cells is biologically relevant to bone metabolism. The circulating level of C3 in mice is about 1.5 mg/ml, which is at least 500- to 1000-fold

TABLE II Proteins Produced by Osteoblastic Cells in Response to $1\alpha,25(OH)_2D_3$

Protein	Characteristics	Nucleotide sequences for vitamin D response elements	Ref.
Bone Gla protein (BGP, osteocalcin)	Water-soluble protein composed of 49 amino acid residues containing 3 Gla (γ-carboxylated glutamic acid) residues with calcium binding capacity	Rat: GGGTGAatgAGGACA Human: GGGTGAacgGGGGCA	47,49–51
Matrix Gla protein (MGP)	Water-insoluble protein composed of 78 amino acid residues containing 5 Gla residues	Not identified	48,52
Osteopontin (bone sialoprotein I)	44-kDa glycoprotein containing the specific binding sequence -Gly-Arg-Asp-	Mouse: GGTTGAatgGGTTCA	53–57
Third component of complement (C3)	190-kDa protein consisting of two polypeptide chains, α and β, of 116 and 65 kDa, respectively	Not identified	58–61

higher than the C3 concentration in the conditioned media of ST2 cells and primary osteoblastic cells treated with $1\alpha,25(OH)_2D_3$ for 3 days [59]. However, when $1\alpha,25(OH)_2D_3$ was administered to vitamin D-deficient mice, the calvarial level of C3 greatly increased [vitamin D-deficient mice, 0.18, 0.24 ng/mg protein (n = 2); $1\alpha,25(OH)_2D_3$-supplemented mice, 0.78, 0.90 ng/mg protein (n = 2)], whereas the level in serum was not changed [vitamin D-deficient mice, 1.4, 1.5 mg/ml (n = 2); $1\alpha,25(OH)_2D_3$-supplemented mice, 1.5, 1.8 mg/ml (n = 2)] [61]. Why bone C3 levels are not influenced by the extremely high circulating levels of C3 has yet to be elucidated.

VIII. REGULATION OF OSTEOCLAST FUNCTION BY OSTEOBLASTIC CELLS

Osteoclasts formed on plastic culture dishes cannot be detached by treatment with trypsin–EDTA and bacterial collagenase. However, we have established a method for obtaining a large number of osteoclasts with bone-resorbing activity from mouse cocultures [62]. Primary mouse osteoblastic cells and bone marrow cells were first cocultured on collagen gel-coated dishes in the presence of 10^{-8} M $1\alpha,25(OH)_2D_3$. After culture for 6 days, the plates were treated with bacterial collagenase to release all cells. Usually, 4–10 × 10^4 osteoclasts can be recovered from a 10-cm collagen gel-coated dish, and the purity of osteoclasts was 2–3% [63]. A reliable pit formation assay was also established to investigate the control mechanism of osteoclast function using the crude osteoclast preparation. Calcitonin, bafilomycin A1 (an inhibitor of vacuolar type proton ATPase), and bisphosphonates each inhibited the pit-forming activity of osteoclasts placed on dentine slices [62–64]. We examined the effect of $1\alpha,25(OH)_2D_3$ on pit-forming activity of osteoclasts placed on dentine slices using the crude osteoclast preparation, which was contaminated with many osteoblastic cells [63]. $1\alpha,25(OH)_2D_3$ added to the pit formation assay did not stimulate further pit-forming activity of osteoclasts [63].

It has been reported that the pit-forming activity of osteoclasts is stimulated by $1\alpha,25(OH)_2D_3$ [65] and PTH [66] through a mechanism involving osteoblastic cells. Many osteoblastic cells were present as contaminants in the crude osteoclast preparation. To determine the role of osteoblastic cells in the regulation of osteoclast function, we also developed a method for purifying functionally active osteoclasts from the coculture [67]. Cocultures were performed for 6 days on collagen gel-coated dishes in the presence of $1\alpha,25(OH)_2D_3$. The coculture was then treated with pronase to remove some osteoblastic cells, followed by treatment with collagenase to recover all the cells from the dishes. The cells were then placed on 30% Percoll solution and centrifuged at 250 g for 30 min. Multinucleated osteoclasts accumulating in the interface layer were recovered. The purity of osteoclasts in this preparation was 50–70% (purified osteoclasts). Interestingly, when the purified osteoclast preparation was cultured on dentine slices, osteoclasts formed only a few resorption pits [67]. The pit-forming activity of purified osteoclasts was strikingly increased by adding osteoblastic cells. These results indicate that osteoblastic cells stimulate not only osteoclast development but also osteoclast activity (Fig. 4). The pit-forming activity of purified osteoclasts was not affected by adding $1\alpha,25(OH)_2D_3$ (T. Suda and N. Takahashi, unpublished observation, 1996). This suggests that osteo-

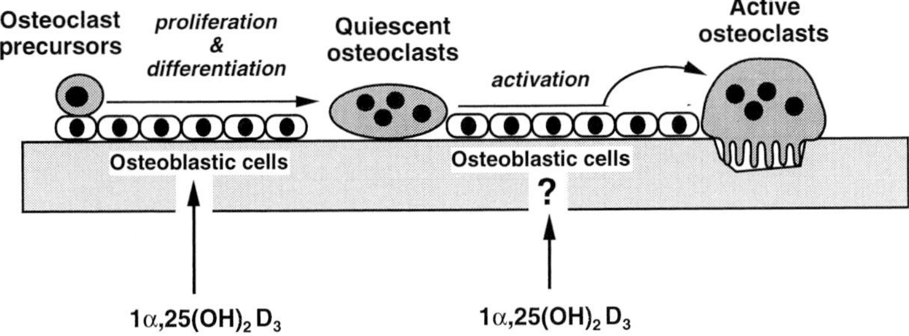

FIGURE 4 Schematic representation of the role of $1\alpha,25(OH)_2D_3$ in osteoclast differentiation and function. Osteoblastic cells play important roles in osteoclastic bone resorption in two different manners: stimulation of osteoclast differentiation and activation of osteoclast function. Both processes are essentially regulated by osteoblastic cells through a mechanism involving cell-to-cell or cell-to-matrix interactions. $1\alpha,25(OH)_2D_3$ stimulates the differentiation process of osteoclast precursors through a mechanism involving cell-to-cell interaction with osteoblastic cells. It is not clear how $1\alpha,25(OH)_2D_3$ stimulates osteoclast function. Osteoclasts may be activated by osteoblastic cells in response to $1\alpha,25(OH)_2D_3$ during their differentiation process.

clasts in the preparation had already been activated in response to 1α,25$(OH)_2D_3$ during cocultures with osteoblastic cells. Further studies are needed to elucidate the regulatory mechanism of 1α,25$(OH)_2D_3$ in osteoclast activation.

IX. CONCLUSION

The ability of 1α,25$(OH)_2D_3$ to induce osteoclast differentiation and activation is closely linked to its classic action in maintaining bone and calcium homeostasis. Osteoclast recruitment is induced by 1α,25$(OH)_2D_3$ through a mechanism involving osteoblasts or bone marrow-derived stromal cells. Like other osteotropic factors such as PTH, PGE_2, and IL-6, 1α,25$(OH)_2D_3$ acts on osteoblastic cells to induce differentiation of osteoclast progenitors into osteoclasts. Further studies are necessary to understand how osteoblasts and bone marrow-derived stromal cells form the microenvironment suitable for osteoclast formation and function as well as the multiplicity of pathways by which the interaction between these cell types promotes osteoclastogenesis.

References

1. Suda T, Takahashi N, Abe E 1992 The role of vitamin D in bone resorption. J Cell Biochem **49**:53–58.
2. Suda T, Shinki T, Takahashi N 1990 The role of vitamin D in bone and intestinal cell differentiation. Annu Rev Nutr **10**:195–211.
3. DeLuca HF 1988 The vitamin D story: A collaborative effort of basic science and clinical medicine. FASEB J **15**:61–79.
4. Berger U, Wilson P, McClelland RA, Colston M, Haussler MR, Pike JW, Coombers RC 1988 Immunocytochemical detection of 1α,25-dihydroxyvitamin D receptors in normal human tissues. J Clin Endorinol Metab **67**:607–618.
5. Carlsson A 1952 Tracer experiments on the effects of vitamin D on the skeletal metabolism of calcium and phosphorus. Acta Physiol Scand **26**:212–220.
6. Suda T, Takahashi N, Martin TJ 1992 Modulation of osteoclast differentiation. Endocr Rev **13**:66–80.
7. Abe E, Miyaura C, Sakagami H, Takeda M, Konno K, Yamazaki T, Yoshiki S, Suda T 1981 Differentiation of mouse myeloid leukemia cells induced by 1α,25-dihydroxyvitamin D_3. Proc Natl Acad Sci USA **78**:4990–4994.
8. Abe E, Miyaura C, Tanaka H, Shiina Y, Kuribayashi T, Suda S, Nishii Y, DeLuca HF, Suda T 1983 1α,25-Dihydroxyvitamin D_3 promotes fusion of alveolar macrophages both by a direct mechanism and by a spleen cell-mediated indirect mechanism. Proc Natl Acad Sci USA **80**:5583–5587.
9. Vaes G 1988 Cellular biology and biochemical mechanisms of bone resorption: A review of recent developments on the formation and mode of action of osteoclasts. Clin Orthop Related Res **231**:239–271.
10. Baron R 1989 Molecular mechanisms of bone resorption by the osteoclast. Anat Rec **234**:317–324.
11. Mundy GR, Roodman GD 1987 Osteoclast ontogeny and function. In: Peck WA (ed) Bone and Mineral Research, Vol. 5. Elsevier, Amsterdam, pp. 209–281.
12. Chambers TJ, Magnus CJ 1982 Calcitonin alters behaviour of isolated osteoclasts. J Pathol **136**:27–39.
13. Nicholson GC, Moseley JM, Sexton PM, Mendelsohn FAO, Martin TJ 1986 Abundant calcitonin receptors in isolated rat osteoclasts. Biochemical and autoradiographic characterization. J Clin Invest **78**:355–360.
14. Takahashi N, Akatsu T, Sasaki T, Nicholson GC, Moseley JM, Martin TJ, Suda T 1988 Induction of calcitonin receptors by 1α,25-dihydroxyvitamin D_3 in osteoclast-like multinucleated cells formed from mouse bone marrow cells. Endocrinology **123**:1504–1510.
15. Horton MA, Davies J 1989 Adhesion receptors in bone. J Bone Miner Res **4**:803–808.
16. Lakkakorpi PT, Horton MA, Helfrich MH, Karhukorpi EK, Väänänen HK 1991 Vitronectin receptor has a role in bone resorption but does not mediate tight sealing zone attachment of osteoclasts to the bone surface. J Cell Biol **115**:1179–1186.
17. Väänänen HK 1984 Immunohistochemical localization of carbonic anhydrase isoenzymes I and II in human bone, cartilage and giant cell tumor. Histochemistry **81**:485–487.
18. Blair HC, Teitelbaum SL, Ghiselli R, Gluck S 1989 Osteoclastic bone resorption by a polarized vacuolar proton pump. Science **245**:855–857.
19. Väänänen HK, Karhukorpi EK, Sundquist K, Wallmark B, Roininen I, Hentunen T, Tuukkanen J, Lakkakorpi P 1990 Evidence for the presence of a proton pump of the vacuolar H^+-ATPase type in the ruffled borders of osteoclasts. J Cell Biol **111**:1305–1311.
20. Blair HC, Teitelbaum SL, Tan HL, Koziol CM, Schlesinger PH 1991 Passive chloride permeability charge coupled to H^+-ATPase of avian osteoclast ruffled membrane. Am J Physiol **260**:C1315–C1324.
21. Tezuka K, Tezuka Y, Maejima A, Sato T, Nemoto K, Kamioka H, Hakeda Y, Kumegawa M 1994 Molecular cloning of a possible cysteine proteinase predominantly expressed in osteoclasts. J Biol Chem **269**:1106–1109.
22. Drake FH, Robert AD, James IE, Conner JR, Debouck CC, Richardson S, Lee-Rykaczewski E, Coleman L, Rieman D, Barthlow R, Hastings G, Gowen M 1996 Cathepsin K, but not cathepsins B, L, or S, is abundantly expressed in human osteoclasts. J Biol Chem **271**:12511–12516.
23. Tezuka K, Nemoto K, Tezuka Y, Sato T, Ikeda Y, Kobori M, Kawashima H, Eguchi H, Hakeda Y, Kumegawa M 1994 Identification of matrix metalloproteinase 9 in rabbit osteoclasts. J Biol Chem **269**:15006–15009.
24. Roodman GD, Ibbotson KM, MacDonald BR, Kuehl TJ, Mundy GR 1985 1,25-Dihydroxyvitamin D_3 causes formation of multinucleated cells with several osteoclast characteristics in cultures of primate marrow. Proc Natl Acad Sci USA **82**:8213–8217.
25. Takahashi N, Yamana H, Yoshiki S, Roodman GD, Mundy GR, Jones SJ, Boyde A, Suda T 1988 Osteoclast-like cell formation and its regulation by osteotropic hormones in mouse bone marrow cultures. Endocrinology **122**:1373–1382.
26. Takahashi N, Akatsu T, Udagawa N, Sasaki T, Yamaguchi A, Moseley JM, Martin TJ, Suda T 1988 Osteoblastic cells are involved in osteoclast formation. Endocrinology **123**:2600–2602.
27. Rodan GA, Martin TJ 1981 Role of osteoblasts in hormonal control of bone resorption—A hypothesis. Calcif Tissue Int **33**:349–351.
28. Tanaka S, Takahashi N, Udagawa N, Sasaki T, Fukui Y, Kurokawa T, Suda T 1992 Osteoclasts express high levels of $p60^{c\text{-}src}$, preferentially on ruffled border membranes. FEBS Lett **313**:85–89.
29. Udagawa N, Takahashi N, Akatsu T, Sasaki T, Yamaguchi A,

Kodama H, Martin TJ, Suda T 1989 The bone marrow-derived stromal cell lines MC3T3-G2/PA6 and ST2 support osteoclast-like cell differentiation in cocultures with mouse spleen cells. Endocrinology **125**:1805–1813.
30. Yamashita T, Asano K, Takahashi N, Akatsu T, Udagawa N, Sasaki T, Martin TJ, Suda T 1990 Cloning of an osteoblastic cell line involved in the formation of osteoclast-like cells. J Cell Physiol **145**:587–595.
31. Walker DG 1973 Osteopetrosis in mice cured by temporary parabiosis. Science **180**:875.
32. Kahn AJ, Simmons DJ 1975 Investigation of cell lineage in bone using a chimaera of chick and quail embryonic tissue. Nature **258**:325–327.
33. Walker DG 1975 Bone resorption restored in osteopetrotic mice by transplants of normal bone marrow and spleen cells. Science **190**:784–785.
34. Udagawa N, Takahashi N, Akatsu T, Tanaka H, Sasaki T, Nishihara T, Koga T, Martin TJ, Suda T 1990 Origin of osteoclasts: Mature monocytes and macrophages are capable of differentiating into osteoclasts under a suitable microenvironment prepared by bone marrow-derived stromal cells. Proc Natl Acad Sci USA **87**:7260–7264.
35. Tanaka S, Takahashi N, Udagawa N, Tamura T, Akatsu T, Stanley ER, Kurokawa T, Suda T 1993 Macrophage colony-stimulating factor is indispensable for both proliferation and differentiation of osteoclast progenitors. J Clin Invest **91**:257–263.
36. Takahashi N, Udagawa N, Tanaka S, Murakami H, Owan I, Tamura T, Suda T 1993 Postmitotic osteoclast precursors are mononuclear cells which express macrophage-associated phenotypes. Dev Biol **163**:212–221.
37. Suda T, Takahashi N, Martin TJ 1995 Modulation of osteoclast differentiation: Update 1995 Endoc Rev Monogr **4**:266–270.
38. Akatsu T, Takahashi N, Debari K, Morita I, Murota S, Nagata N, Takatani O, Suda T 1989 Prostaglandins promote osteoclast-like multinucleated cell formation by a mechanism involving cyclic adenosine 3′,5′-monophosphate in mouse marrow cell cultures. J Bone Miner Res **4**:29–35.
39. Akatsu T, Takahashi N, Udagawa N, Sato K, Nagata N, Moseley JM, Martin TJ, Suda T 1989 Parathyroid hormone (PTH)-related protein is a potent stimulator of osteoclast-like multinucleated cell formation to the same extent as PTH in mouse marrow cultures. Endocrinology **125**:20–27.
40. Akatsu T, Takahashi N, Udagawa N, Imamura K, Yamaguchi A, Sato K, Nagata N, Suda T 1991 Role of prostaglandins in interleukin 1-induced bone resorption in mice in vitro. J Bone Miner Res **6**:183–190.
41. Tamura T, Udagawa N, Takahashi N, Miyaura C, Tanaka S, Koishihara Y, Ohsugi Y, Kumaki K, Taga T, Kishimoto T, Suda T 1993 Soluble interleukin-6 receptor triggers osteoclast formation by interleukin-6. Proc Natl Acad Sci USA **90**:11924–11928.
42. Udagawa N, Takahashi N, Katagiri T, Tamura T, Wada S, Findlay DM, Martin TJ, Hirota H, Taga T, Kishimoto T, Suda T 1995 Interleukin (IL) 6 induction of osteoclast differentiation depends on IL-6 receptors expressed on osteoblastic cells but not on osteoclast progenitors. J Exp Med **182**:1461–1468.
43. Kishimoto T, Akira S, Taga T 1992 Interleukin 6 and its receptor: A paradigm for cytokines. Science **258**:593–597.
44. Saito T, Yasukawa K, Suzuki H, Futatsugi K, Fukunaga T, Yoko mizo C, Koishihara Y, Fukui H, Ohsugi Y, Yawata H, Kobayashi I, Hirano T, Taga T, Kishimoto T 1991 Preparation of soluble murine IL-6 receptor and anti-murine IL-6 receptor antibodies. J Immunol **147**:168–173.
45. Girasole G, Passeri G, Jilka RL, Manolagas SC 1994 Interleukin 11: A new cytokine critical for osteoclast development. J Clin Invest **93**:1516–1524.
46. Romas E, Udagawa N, Zhou H, Tamura T, Saito M, Taga T, Hilton DJ, Suda T, Ng KW, Martin TJ 1996 The role of gp130-mediated signals in osteoclast development: Regulation of interleukin 11 production by osteoblasts and distribution of its receptor in bone marrow cultures. J Exp Med **183**:2581–2591.
47. Price PA, Pose JW, Raman M 1976 Primary structure of γ-carboxyglutamic acid-containing protein from bovine bone. Proc Natl Acad Sci USA **73**:3374–3375.
48. Price PA, Williamson MK 1985 Primary structure of bovine matrix gla protein, a new vitamin K-dependent bone protein. J Biol Chem **260**:14971–14975.
49. Price PA, Baukol SA 1980 1,25-Dihydroxyvitamin D_3 increases synthesis of the vitamin K-dependent bone protein by osteosarcoma cells. J Biol Chem **255**:11660–11663.
50. Yoon K, Rutledge SJ, Buenaga RF, Rodan GA 1988 Characterization of the rat osteocalcin gene: Stimulation of promoter activity by 1,25-dihydroxyvitamin D_3: Biochemistry **27**:8521–8526.
51. Kerner SA, Scott BA, Pike JW 1989 Sequence elements in the human osteocalcin gene confer basal activation and inducible response to hormonal vitamin D_3. Proc Natl Acad Sci USA **86**:4455–4459.
52. Fraser JD, Otawara Y, Price PA 1988 1,25-Dihydroxyvitamin D_3 stimulates the synthesis of matrix γ-carboxyglutamic acid protein by osteosarcoma cells. J Biol Chem **263**:911–916.
53. Oldberg A, Franzen A, Heinegard D 1986 Cloning and sequence analysis of rat bone sialoprotein (osteopontin) cDNA reveals an Arg-Gly-Asp cell binding sequence. Proc Natl Acad Sci USA **83**:8819–8823.
54. Mark MP, Prince CW, Oosawa T, Gay S, Bronckers ALJJ, Butler WT 1987 Immunohistochemical demonstration of a 44-kD phosphoprotein in developing rat bones. J Histochem Cytochem **35**:707–715.
55. Yoon K, Buenaga R, Rodan GA 1987 Tissue developmental expression of rat osteopontin. Biochem Biophys Res Commun **148**:1129–1136.
56. Noda M, Yoon K, Prince CW, Butler WT, Rodan GA 1988 Transcriptional regulation of osteopontin production in rat osteosarcoma cells by type β transforming growth factor. J Biol Chem **263**:13916–13921.
57. Noda M, Vogel RL, Craig AM, Prahl J, DeLuca HF, Denhardt DT 1990 Identification of cDNA sequence responsible for binding of the 1,25-dihydroxyvitamin D_3 receptor and 1,25-dihydroxyvitamin D_3 enhancement of mouse secreted phosphoprotein 1 (SSP-1or osteopontin) gene expression. Proc Natl Acad Sci USA **91**:900–902.
58. Sato T, Hong MH, Jin CH, Ishimi Y, Udagawa N, Shinki T, Abe E, Suda T 1991 The specific production of the third component of complement by osteoblastic cells treated with 1α,25-dihydroxyvitamin D_3. FEBS Lett **285**:21–24.
59. Hong MH, Jin CH, Sato T, Ishimi Y, Abe E, Suda T 1991 Transcriptional regulation of the production of the third component of complement (C3) by 1α,25-dihydroxyvitamin D_3 in mouse marrow-derived stromal cells (ST2) and primary osteoblastic cells. Endocrinology **129**:2774–2779.
60. Sato T, Abe E, Jin CH, Hong MH, Katagiri T, Kinoshita T, Amizuka N, Ozawa H, Suda T 1993 The biological roles of the third component of complement in osteoclast formation. Endocrinology **133**:397–404.
61. Jin CH, Shinki T, Hong MH, Sato T, Yamaguchi A, Ikeda T, Yoshiki S, Abe E, Suda T 1992 1α,25-Dihydroxyvitamin D_3 tissue-specifically regulates in vivo production of the third component of complement (C3) in bone. Endocrinology **131**:2468–2475.

62. Akatsu T, Tamura T, Takahashi N, Udagawa N, Tanaka S, Sasaki T, Yamaguchi A, Nagata N, Suda T 1992 Preparation and characterization of a mouse osteoclast-like multinucleated cell population. J Bone Miner Res **7**:1297–1306.
63. Tamura T, Takahashi N, Akatsu T, Sasaki T, Udagawa N, Tanaka S, Suda T 1993 New resorption assay with mouse osteoclast-like multinucleated cells formed in vitro. J Bone Miner Res **8**:953–960.
64. Murakami H, Takahashi N, Sasaki T, Udagawa N, Tanaka S, Nakamura I, Zhang D, Barbier A, Suda T 1995 A possible mechanism of the specific action of bisphosphonates on osteoclasts: Tiludronate preferentially affects polarized osteoclasts having ruffled borders Bone **17**:137–144.
65. McSheehy PMJ, Chambers TJ 1987 1,25-Dihydroxyvitamin D_3 stimulates rat osteoblastic cells to release a soluble factor that increases osteoclastic bone resorption. J Clin Invest **80**:425–429.
66. McSheehy PMJ, Chambers TJ 1986 Osteoblastic cells mediate osteoclastic responsiveness to parathyroid hormone. Endocrinology **118**:824–828.
67. Jimi E, Nakamura I, Amano H, Taguchi Y, Turukai T, Tamura N, Takahashi N, Suda T 1996 Osteoclast function is activated by osteoblastic cells through a mechanism involving cell-to-cell contact. Endocrinology **137**:2187–2190.
68. Kato *et al.* 1996 J Bone Miner Res **11**:S124 (Abstract).

CHAPTER 22

1,25-Dihydroxyvitamin D_3 Interactions with Local Factors in Bone Remodeling

PAULA H. STERN Department of Molecular Pharmacology and Biological Chemistry, Northwestern University Medical School, Chicago, Illinois

I. INTRODUCTION

Our perspective on how the cellular functions of bone are regulated has changed markedly with the recognition that factors produced locally by bone cells modulate important bone cell activities. Furthermore, it is now known that the concentrations of these local factors, which include growth factors, cytokines, other bioactive peptides, and prostaglandins, are influenced by circulating hormones. This has led to the concept that many of the actions of systemic hormones on bone could be mediated through their effects on the production of local factors within the skeletal tissue. In addition, the systemic hormones can interact with local factors resulting in either enhancement or antagonism of responses. This chapter addresses the effects of the systemic hormone 1,25-dihydroxyvitamin D_3 [$1,25(OH)_2D_3$, calcitriol] on the production of local factors in bone cells and the potentially significant interactions of $1,25(OH)_2D_3$ with these local mediators.

II. VITAMIN D AND GROWTH FACTORS IN BONE

A. Insulin-like Growth Factors

1. $1,25(OH)_2D_3$ Effects on IGF Production in Bone

Insulin-like growth factor type I (IGF-I) is considered to be an important local mediator of bone growth, as it is produced by osteoblastic cells [1–3] and promotes bone cell proliferation and synthesis of bone matrix components [2,4,5]. The effect of $1,25(OH)_2D_3$ on IGF-I production in bone has been studied in several models, and, interestingly, diverse responses have been found in the different systems. Stimulatory effects were seen in cultures of biopsy-derived human bone cells, where $1,25(OH)_2D_3$ increased IGF-I production in a time- and dose-dependent manner [2]. In rat osteoblastic cells, the inhibitory effect of 100 n*M* dexamethasone on IGF-I production in rat osteoblastic cells was antagonized by 10 n*M* $1,25(OH)_2D_3$ [6]. In mouse MC3T3-E1 cells, however, $1,25(OH)_2D_3$ failed to increase IGF-I production, although IGF-I production could be stimulated by growth hormone [3]. Inhibitory effects on IGF-I production have been noted in several systems. In subconfluent clonal mouse osteoblasts, $1,25(OH)_2D_3$, over a 10 p*M* to 10 n*M* concentration range, inhibited IGF-I production [7]. In intact mouse calvaria, $1,25(OH)_2D_3$ inhibited basal release of IGF-I as well as that stimulated by parathyroid hormone (PTH) or transforming growth factor-β (TGFβ) [8]. IGF-II production, on the other hand, was not affected by $1,25(OH)_2D_3$ in cells derived from fetal rat calvariae, based on measurements of the peptide product and transcripts for prepro-IGF-II [9].

2. Interactions of IGF-I and $1,25(OH)_2D_3$ on Cell Proliferation and Early Response Genes

In clonal mouse osteoblasts, $1,25(OH)_2D_3$ inhibited the increase in [^{3}H]thymidine incorporation elicited by IGF-I [7]. IGF-I actions to increase proliferation of MG-63 cells and to increase expression of c-*jun* mRNA were decreased by 10 n*M* $1,25(OH)_2D_3$ [10] (Fig. 1A).

3. Interactions of IGF-I and $1,25(OH)_2D_3$ on Alkaline Phosphatase

Because alkaline phosphatase is a characteristic osteoblast marker and has been implicated in bone mineralization, the effects of $1,25(OH)_2D_3$ on this enzyme, especially in combination with growth factors, are of interest and have been studied in a number of models. The findings reveal mixed effects, with different studies showing either enhancement or antagonism of activity in response to combined treatment. Enhancement of alkaline phosphatase activity was seen with combined IGF-I and $1,25(OH)_2D_3$ treatment in normal human osteoblastic cells [2] and in mouse MC3T3-E1 clonal osteoblasts [11]. In the MC3T3-E1 cells, $25OHD_3$ or $24,25(OH)_2D_3$, at concentrations 1000–2000 times those of $1,25(OH)_2D_3$, also elicited synergistic effects on alkaline phosphatase. In human osteoblastic cells, growth hormone likewise produced synergistic effects on alkaline phosphatase when added together with $1,25(OH)_2D_3$ [2]. In contrast to the stimulatory effects

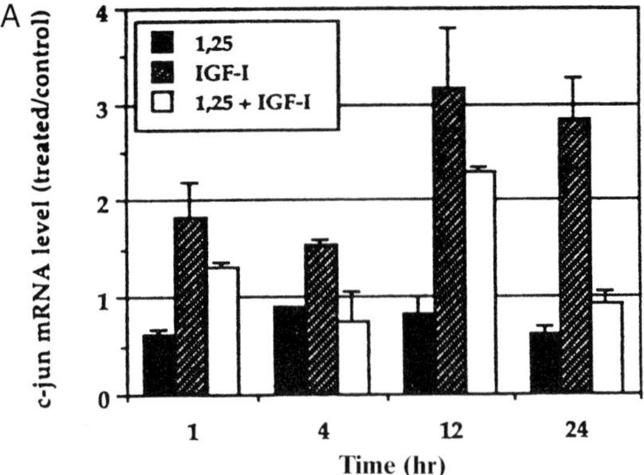

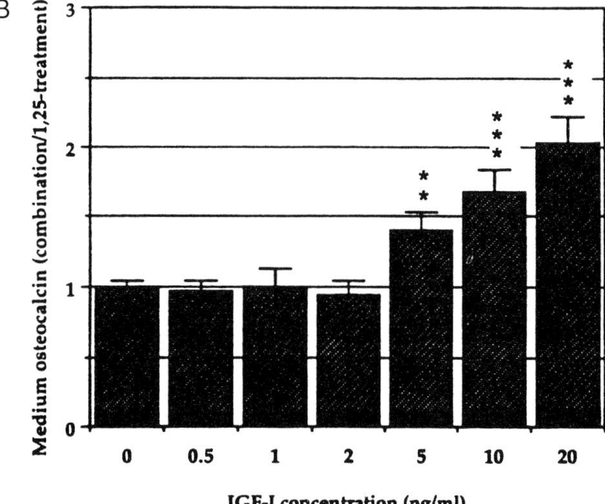

FIGURE 1 Interactions of $1,25(OH)_2D_3$ and IGF-I in MG-63 human osteosarcoma cells. (A) $1,25(OH)_2D_3$ inhibits the stimulatory effect of IGF-I on c-*jun* mRNA, and (B) IGF-I enhances $1,25(OH)_2D_3$-induced osteocalcin synthesis. ** = $p < 0.01$, *** = $p < 0.001$ vs. $1,25(OH)_2D_3$ alone. From Pirskanen *et al.* [10] with permission.

in these two models, IGF-I treatment of the MG-63 human osteosarcoma cell line resulted in small decreases in alkaline phosphatase activity and antagonism of the stimulatory effect of $1,25(OH)_2D_3$ on alkaline phosphatase [10].

4. Interactions of IGF-I and $1,25(OH)_2D_3$ on Osteocalcin

In several studies in which the interaction of $1,25(OH)_2D_3$ with IGF-I was investigated, enhanced osteocalcin protein and/or message was seen. The model systems included human osteoblastic cells [2], MG-63 human osteosarcoma cells [10] (Fig. 1B), and human OHS-4 cells [12]. In the MG-63 cells, 24-hr pretreatment with IGF-I prior to 48-hr cotreatment resulted in a greater effect than cotreatment alone [10].

5. INTERACTIONS OF IGF-I AND $1,25(OH)_2D_3$ ON COLLAGEN SYNTHESIS

The interactions of $1,25(OH)_2D_3$ and IGF-I on collagen synthesis have been less extensively studied, and the effects revealed thus far are less dramatic than effects on alkaline phosphatase and osteocalcin. In MC3T3-E1 cells treated with IGF-I and $1,25(OH)_2D_3$, there was less enhancement of collagen synthesis than of alkaline phosphatase [11]. In 21-day fetal rat calvaria, under conditions in which PTH treatment inhibited the stimulatory effects of IGF-I on collagen synthesis, $1,25(OH)_2D_3$ did not inhibit IGF-I effects on collagen [13].

6. MECHANISMS OF INTERACTIONS BETWEEN IGF-I AND $1,25(OH)_2D_3$

Enhancement or diminution of IGF-I or $1,25(OH)_2D_3$ effects by cotreatment could be effected by modulation by IGF-I of $1,25(OH)_2D_3$ receptors (VDR) or response elements, or alternatively by effects of $1,25(OH)_2D_3$ on IGF-I receptors or on the specific IGF-I binding proteins that promote or inhibit IGF responses in bone [7,14–16]. Kurose *et al.* [3] carried out studies to determine the mechanism of the synergistic effects of IGF-I and $1,25(OH)_2D_3$ on alkaline phosphatase in MC3T3-E1 cells. Treatment of the cells for 2–3 days with 50 p*M* $1,25(OH)_2D_3$ or 1000 times higher concentrations of $25OHD_3$ or $24,25\text{-}(OH)_2D_3$ increased the number of IGF-I receptors by 50%. The affinity of the receptors for IGF-I was not altered by the $1,25(OH)_2D_3$ treatment. VDR were not altered by IGF-I treatment.

$1,25(OH)_2D_3$ increased secretion of the inhibitory IGF binding protein IGFBP-4 in clonal mouse osteoblasts, potentially decreasing interaction of IGF-I with its receptors [7,17]. In MG-63 cells, $1,25(OH)_2D_3$ increased secretion of IGFBP-3 [15]. The IGF binding protein IGFBP-3 has been found to both promote [18,19] and antagonize [18–22] IGF-I effects. Although $1,25(OH)_2D_3$ treatment of rat osteoblastic cells increased IGF-binding protein measured by Western blot, the $1,25(OH)_2D_3$ treatment did not reverse the dexamethasone-induced decrease in the binding protein [23].

In other studies in MG-63 cells, gel mobility shift assays revealed that pretreatment with IGF-I for 48–96 hr decreased binding of the transcription factor AP-1 to its cognate response element [10]. It was noted by the authors that this finding was consistent with the enhancement of osteocalcin production by $1,25(OH)_2D_3$, since previous studies had suggested that AP-1 binding in the rat osteocalcin promoter can mediate transcription repression during certain stages of development [24].

B. Transforming Growth Factor-β

1. $1,25(OH)_2D_3$ EFFECTS ON TGFβ PRODUCTION IN BONE

Transforming growth factor-β is produced in bone by chondrocytes [25], osteoblasts [26,27], and osteoclasts [28]. Its actions are consistent with the possibility that it is an important regulator of bone remodeling. TGFβ has a range of effects on chondrocyte differentiation [27,29–33]. TGFβ stimulates bone formation [34–36] and inhibits resorption [37]. It is incorporated into bone matrix [38] and its release on resorption and conversion from a latent to an active form may result in both the termination of resorption and the activation of new osteogenesis. In a chick embryo model, treatment with $1,25(OH)_2D_3$ increased TGFβ expression [39]. In cultures of confluent chick chondrocytes, low concentrations of $1,25(OH)_2D_3$ (1–100 p*M*) decreased total TGFβ activity, whereas higher concentrations (10 n*M*) increased TGFβ activity. TGFβ2 mRNA was selectively increased, and mRNA for the β1 and β3 isoforms was decreased [25]. $1,25(OH)_2D_3$ likewise increased TGFβ in mouse calvaria, in osteoblasts derived from the calvaria [40], and in cells derived from human bone [41]. Bone from vitamin D-deficient rats had lower concentrations of TGFβ than tissue from normal animals, whereas IGF-I and IGF-II were unaffected [40]. The authors postulated that this deficiency in TGFβ was the cause of the weaker osteoinductive activity of bone matrix derived from vitamin D-deficient rats.

2. INTERACTIONS OF TGFβ AND $1,25(OH)_2D_3$ ON CELL PROLIFERATION

[^{3}H]Thymidine incorporation in cells derived from human trabecular bone is markedly stimulated by TGFβ [42]. This mitogenic effect is inhibited by $1,25(OH)_2D_3$ (Fig. 2A).

3. INTERACTIONS OF TGFβ AND $1,25(OH)_2D_3$ ON ALKALINE PHOSPHATASE

Transforming growth factor-β has synergistic effects with $1,25(OH)_2D_3$ on osteoblast alkaline phosphatase. This has been shown in MG-63 cells [43] and in cells isolated from human trabecular bone [42,44] (Fig. 2B). Synergistic effects on alkaline phosphatase are also seen in chondrocytes [33].

4. INTERACTIONS OF TGF-β AND $1,25(OH)_2D_3$ ON OSTEOCALCIN AND OSTEOPONTIN

Transforming growth factor-β inhibited $1,25(OH)_2D_3$-stimulated osteocalcin production in ROS 17/2.8 cells [45,46], MG-63 cells [43,46], and human osteoblastic bone-derived cells [42,44,47] (Fig. 2C). The inhib-

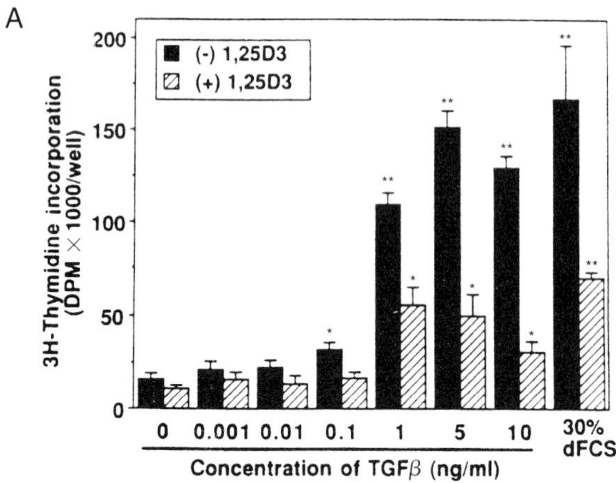

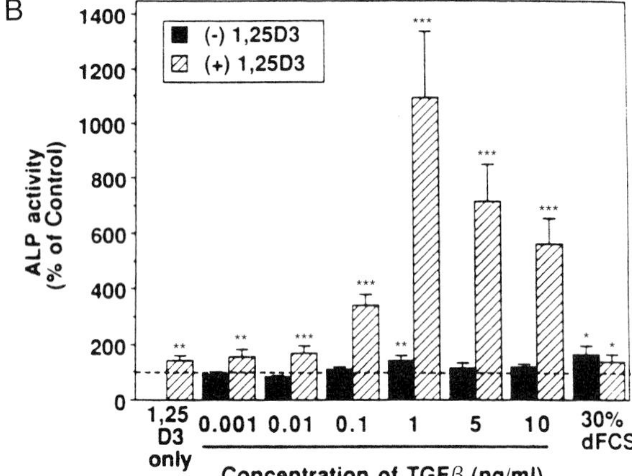

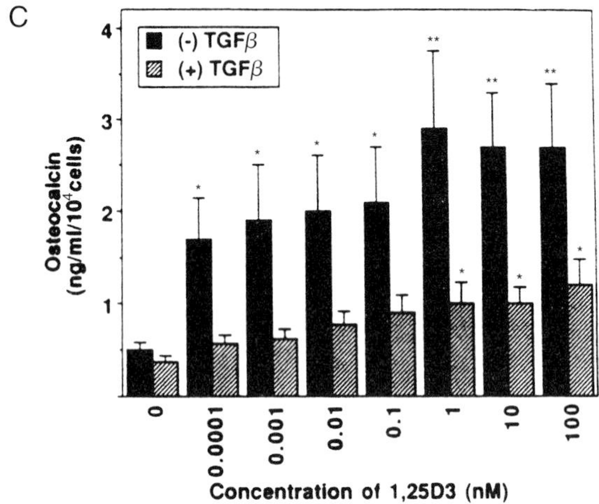

FIGURE 2 Interactions of $1,25(OH)_2D_3$ and TGFβ in cells derived from human trabecular bones. (A) $1,25(OH)_2D_3$ inhibits the mitogenic effect of TGFβ, * $p < 0.03$, *** $p < 0.001$ vs. control, (B) $1,25(OH)_2D_3$ and TGFβ have synergistic effects on alkaline phosphatase, * $p < 0.003$, *** $p < 0.005$ vs. control, and (C) TGFβ inhibits the stimulatory effect of $1,25(OH)_2D_3$ on osteocalcin synthesis * $p < 0.03$, ** $p < 0.008$ vs. control. From Ingram *et al.* [42] with permission.

itory effect is seen both at the level of mRNA and product. An interesting finding was that the TGFβ treatment did not inhibit the increase in osteocalcin elicited by the vitamin D analog DMC-903 [47].

Osteopontin mRNA was increased by both $1,25(OH)_2D_3$ and TGFβ in ROS 17/2.8 cells. However, in cells stimulated by $1,25(OH)_2D_3$, the addition of TGFβ did not elicit a further increase [48].

5. INTERACTIONS OF TGFβ AND $1,25(OH)_2D_3$ ON COLLAGEN

In MG-63 cells, $1,25(OH)_2D_3$ and TGFβ produced additive effects on collagen and fibronectin synthesis [43]. An enhanced effect on collagen was also seen in cells derived from human femoral bone [44]. In another study using human trabecular bone, addition of $1,25(OH)_2D_3$ did not significantly influence the stimulatory effects of TGFβ on procollagen type I peptide in osteoblast cultures derived from the tissues, or in medium from the cultures [42]. Although treatment with $1,25(OH)_2D_3$ enhanced TGFβ expression in the calcium-deficient chick embryo [39] (see Section II,B,1), and restored serum calcium and improved calvarial calcification, $1,25(OH)_2D_3$ failed to correct the altered collagen phenotype resulting from the calcium deficiency (i.e., a chondrogenic phenotype with production of type II collagen) [39]. Thus, there appeared to be a dissociation between TGFβ and this manifestation of the calcium-deficient phenotype.

The hormone $1,25(OH)_2D_3$ promotes synthesis of collagenase by osteoblastic cells, which may play a role in the removal of osteoid prior to resorption. In a model system employing mouse calvarial osteoblasts and type I collagen films, $1,25(OH)_2D_3$ promoted degradation of the collagen, and TGFβ, epidermal growth factor (EGF), platelet-derived growth factor (PDGF), and α- and β-fibroblast growth factor (FGF), but not IGF-I or IGF-II, inhibited this effect [49]. TGFβ was the most potent of the growth factors. Higher concentrations of each of the active growth factors elicited an increase in the tissue inhibitor of metalloproteinases (TIMP).

6. MECHANISMS OF INTERACTIONS OF TGFβ OR EGF WITH $1,25(OH)_2D_3$ IN SKELETAL CELLS

Enhanced responses to $1,25(OH)_2D_3$ with TGFβ treatment could derive from effects of TGFβ at the level of $1,25(OH)_2D_3$ receptors. A 3-day incubation of UMR-106-06 cells with 1 ng/ml TGFβ elicited a 130% increase in the number of $1,25(OH)_2D_3$ receptors, with no change in receptor affinity [50]. Effects of TGFβ to increase the number of VDR were also observed in both rat (UMR-106, ROS 17/2.8) and human (MG-63) osteoblastic cells [46]. In chondrocyte cultures, TGFβ increased the production of $1,25(OH)_2D_3$ and

24,25$(OH)_2D_3$ in cells derived from both resting zone and growth zone [51]. This autocrine regulation could modulate interactions of the two factors to regulate the functions and differentiation of the tissue. EGF treatment decreased the number of 1,25$(OH)_2D_3$ binding sites in UMR-106 cells without changing the affinity of the ligand for its receptor [52]. The up-regulation of the 1,25$(OH)_2D_3$ receptor by PTH was also inhibited by EGF.

7. Interactions of TGFβ and 1,25$(OH)_2D_3$ on Osteoclast Generation

Transforming growth factor-β elicited dose-dependent biphasic effects on the 1,25$(OH)_2D_3$-stimulated generation of osteoclast-like multinucleated giant cells in mouse bone marrow cultures, with lower concentrations (10–100 pg/ml) promoting the 1,25$(OH)_2D_3$ effect and higher concentrations inhibiting it [53]. Antibodies to TGFβ prevented the effect of 1,25$(OH)_2D_3$ to stimulate generation of osteoclast-like cells, providing evidence that TGFβ plays a role in the process.

C. Effect of 1,25$(OH)_2D_3$ on Vascular Endothelial Growth Factor Production in Bone

Vascular endothelial growth factor (VEGF) was first recognized as a promoter of vascular endothelial cell proliferation and angiogenesis [54]. Its potential importance to osteoblastic cell function derives from findings that VEGF increases alkaline phosphatase in bovine fetal calvarial osteoblastic cells [55] and that prostaglandins increase VEGF expression in osteoblastic cells [56]. In human osteoblastic cells derived from trabecular bone and in human SaOS-2 cells, expression of mRNA for this growth factor was increased by 1,25$(OH)_2D_3$, and there was increased VEGF activity in conditioned medium from the 1,25$(OH)_2D_3$-treated cells [57]. The increased VEGF secretion in response to 1,25$(OH)_2D_3$ was postulated to be partially mediated by phosphokinase C (PKC), based on its enhancement by phorbol esters and the inhibition of the 1,25$(OH)_2D_3$ effect by the PKC inhibitors H-7 and staurosporine.

D. Implications of Growth Factor Interactions for 1,25$(OH)_2D_3$ Effects on Skeletal Tissues

Vitamin D has effects on the growth and differentiation of many tissues, including bone [58]. As the growth factors described above likewise affect bone cell growth and the acquisition of phenotypic characteristics, it is conceivable that the effects of 1,25$(OH)_2D_3$ to modulate the production and action of these factors contributes to the anabolic actions of vitamin D. It is also important to note that many of the studies found that these effects and interactions are elicited at concentrations of 1,25$(OH)_2D_3$ that are within the physiological range.

With regard to mitogenic responses, it appears that 1,25$(OH)_2D_3$ generally inhibits the proliferative effects of growth factors, suggesting that it shifts cellular activity toward differentiation (Chapter 64). The interactions of growth factors and 1,25$(OH)_2D_3$ on osteoblast phenotypic responses (alkaline phosphatase, osteocalcin, collagen synthesis) appear to be different for IGF-I and TGFβ, although generally the effects of one or the other growth factor, and not both, have been investigated in a given study. The disparate effects in different studies could reflect differences in sensitivity due to the stages of differentiation manifested by each of the cell lines, differences in differentiation due to the extent of confluence, the presence or production of other local factors, as well as species or bone site differences (Chapter 20). Viewing this complex situation in a positive way, these differences in response raise the exciting possibility that once the contributions of the different conditions represented by each of the models are sorted out, they will lead to the understanding that there is a tightly regulated system for the local control of bone growth and differentiation. Certainly it would be expected that some of the observed differences reflect physiological regulatory factors that normally determine the direction and extent of the anabolic response.

III. VITAMIN D AND INFLAMMATORY CYTOKINES IN BONE

A. 1,25$(OH)_2D_3$ Effects on Cytokine Production in Bone and Their Potential Role

The production of several inflammatory cytokines is increased by 1,25$(OH)_2D_3$ in osteoblastic cells. 1,25$(OH)_2D_3$ increases interleukin-6 (IL-6) in osteoblastic cells. The response is enhanced in the presence of IL-1 and has been associated with increased numbers of IL-1 receptors [59]. The increase in IL-6 in osteoclast precursor cells in marrow may be a critical step in the differentiation of the osteoclast and may be important for the bone-resorbing effects of 1,25$(OH)_2D_3$. This differentiation pathway evoked by 1,25$(OH)_2D_3$ is permissive for the responses to other agents affecting the generation of osteoclasts and manipulations such as the effect of estrogen withdrawal to promote IL-6 production [60]. Production of IL-11 may also play an important role in the process of osteoclast differentiation. IL-11 bioactivity is increased in marrow cultures by $1,25(OH)_2D_3$ (Fig. 3). The effects of $1,25(OH)_2D_3$, as well as effects of PTH to stimulate osteoclast development in bone marrow/calvarial cell cocultures was inhibited by a neutralizing antibody to IL-11 [61]. The importance of cytokines in bone resorption stimulated by many agents will become clearer as new, selective pharmacological agents become available. A synthetic inhibitor of cytokine production, 5-(4-pyridyl)-6-(4-fluorophenyl-2,3-dihydroimidazo(2,1-*b*)thiazole (SK&F 86002), and other cytokine inhibitors inhibited the resorption elicited in fetal rat limb bones by $1,25(OH)_2D_3$, PTH, tumor necrosis factor-α (TNFα), and bacterial lipopolysaccharide [62].

B. Interactions of $1,25(OH)_2D_3$ and Inflammatory Cytokines in Bone

Several of the inflammatory cytokines inhibit the anabolic effects of $1,25(OH)_2D_3$ in osteoblastic cells. In cells isolated from human trabecular bone, granulocyte–macrophage colony-stimulating factor (GM-CSF), IL-1, and TNFα inhibited effects of $1,25(OH)_2D_3$ on alkaline phosphatase and osteocalcin [63–66]. Inhibitory effects of TNFα on alkaline phosphatase and collagen were also seen in cells isolated from fetal rat calvaria [67]. $1,25(OH)_2D_3$ stimulation of osteocalcin mRNA and protein production in ROS 17/2.8 cells was inhibited by both IL-1 and TNFα, and IL-6 decreased the $1,25(OH)_2D_3$ effect on osteocalcin mRNA but not on the protein [68]. The effects of TNFα and IL-6 were at the transcriptional level, whereas the effect of IL-1 occurred posttranscriptionally. TNFα decreased the number of VDR in ROS 17/2.8 cells; however, mRNA for the 4.4-kb VDR was not selectively decreased relative to cyclophilin mRNA, indicating that the inhibitory effect was subsequent to the synthesis of VDR message [69].

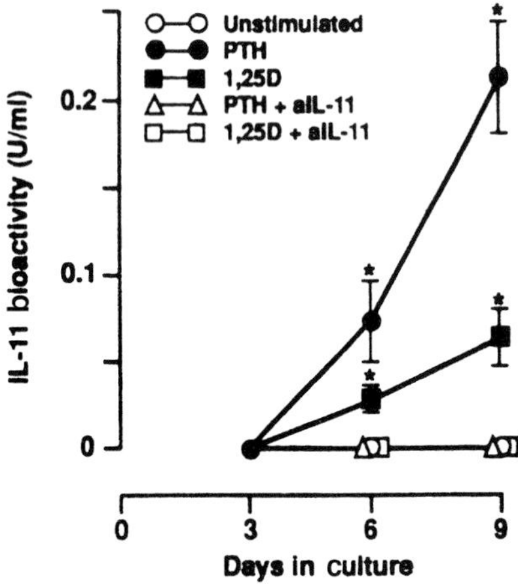

FIGURE 3 $1,25(OH)_2D_3$ and PTH increase IL-11 bioactivity in mouse bone marrow cultures. * $p < 0.05$ vs. unstimulated cultures. From Girasole *et al.* [61] with permission.

The interaction between $1,25(OH)_2D_3$ and TNFα on osteocalcin gene expression appears to involve a novel mechanism [70–72] (Fig. 4). It was determined that the inhibitory effect of TNFα was elicited by interaction with a 26-bp sequence that constitutes a vitamin D response element (VDRE), and that there was impaired formation of retinoid X receptor (RXR)–VDR–DNA complexes [71]. Further studies defined a critical sequence of a 5′ half-site GGGTGA sequence [72]. TNFα also decreased the $1,25(OH)_2D_3$-induced production of IGF-I and IGFBP-4 in MC3T3-E1 cells [17].

The hormone $1,25(OH)_2D_3$ stimulated the synthesis of IL-4 receptors in MC3T3-E1 osteoblastic cells, as shown by both binding studies (Fig. 5), in which the B_{max} for IL-4 was increased, and by an increase in mRNA for the IL-4 receptor [73]. The effect was seen with exposure times as short as 4 hr. Since IL-4 is an inhibitor of osteoclast generation [74] and bone resorption [75,76], it is conceivable that this effect on IL-4 receptors could play a regulatory role in modulating the resorptive effects of $1,25(OH)_2D_3$.

C. Implications of Cytokine Interactions for $1,25(OH)_2D_3$ Effects on Skeletal Tissues

Calcitriol thus stimulates the production of several inflammatory cytokines that promote osteoclast development. These cytokines also have inhibitory effects on the anabolic actions of $1,25(OH)_2D_3$. Implications of these findings are that the interactions with cytokines could play a significant role in the effects of vitamin D, especially excess amounts of the hormone, to promote bone resorption. The antagonism of anabolic effects of vitamin D could be a factor facilitating the bone loss elicited by elevated concentrations of inflammatory cytokines.

IV. INTERACTIONS OF $1,25(OH)_2D_3$ AND OTHER LOCALLY PRODUCED PEPTIDES IN BONE

A. Natriuretic Peptides

Production of the atrial natriuretic factor (ANF) analog C-type natriuretic peptide (CNP) is increased by

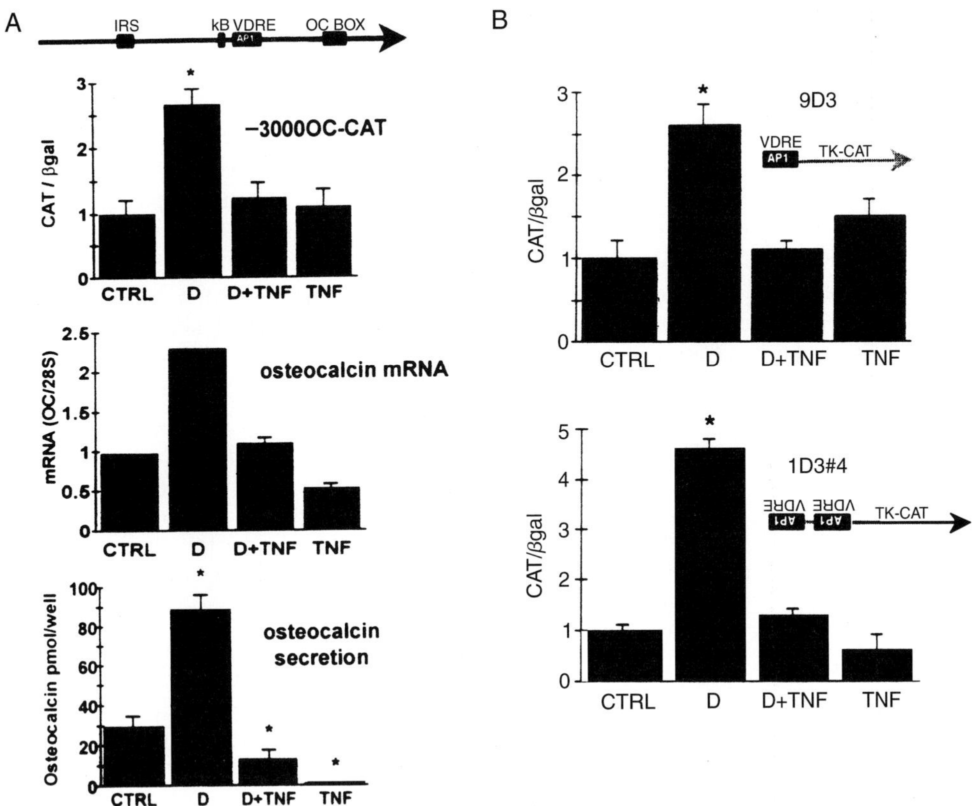

FIGURE 4 Regulation of the osteocalcin gene in ROS 17/2.8 rat osteosarcoma cells by $1,25(OH)_2D_3$ and TNFα. (A) TNFα inhibits $1,25(OH)_2D_3$-stimulated transcription of −3000-OC-CAT, as well as basal and $1,25(OH)_2D_3$-stimulated osteocalcin mRNA and osteocalcin secretion; (B) $1,25(OH)_2D_3$ stimulates chloramphenicol acetyltransferase (CAT) activity in cells transfected with a synthetic 26-bp VDRE in correct or reverse orientation and the inhibitory effect of TNFα on this stimulation. * $p < 0.05$ vs. all other groups. From Nanes *et al.*. [71] with permission.

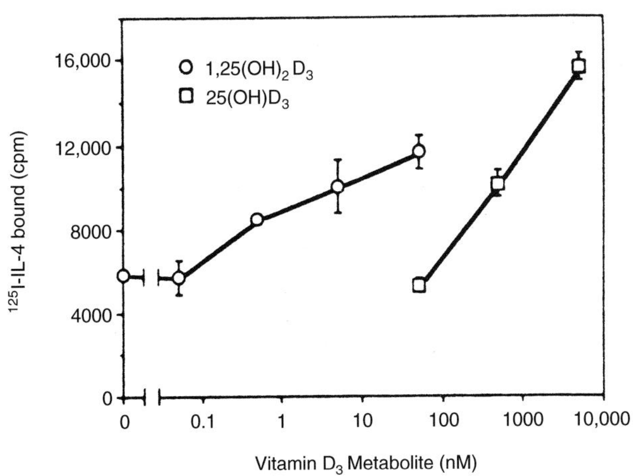

FIGURE 5 Dose-dependent effects of $1,25(OH)_2D_3$ and $25OHD_3$ on IL-4 binding in MC3T3-E1 mouse osteoblastic cells. From Lacey *et al.* [73] with permission.

$1,25(OH)_2D_3$ in osteoblastic cells [77]. The class of natriuretic peptides, including ANP and CNP, increase membrane-associated guanylate cyclase [78]. A synthetic rat ANF (Ile-ANF-26) stimulated cGMP in rat calvarial cells and in UMR-106-01 cells [79], and a synthetic human ANF (ANF 99–126) increased cGMP in calvarial cultures [80]. CNP stimulated osteoclastic activity [77]. Synthetic ANF did not affect collagen synthesis, and its effects on resorption in the organ culture systems in which it was studied were limited to a partial inhibition of the resorption elicited by prostaglandin E_2 (PGE_2) in fetal limb bones [80]. Thus, the effect of $1,25(OH)_2D_3$ on natriuretic factor synthesis seems unlikely to play a major role in the phenotypic effect of $1,25(OH)_2D_3$ in bone.

B. Endothelins

Endothelins, 21-amino acid peptides found in bone [81,82] affect many functions of bone cells. Treatment of ROS 17/2.8 cells with 10 nM 1,25(OH)$_2$D$_3$ for 14 hr decreased the expression of receptors of both ET$_A$ and ET$_B$ subtypes of endothelin receptors [83]. ET$_A$ receptors mediate effects of endothelin-1 to increase osteocalcin mRNA expression [84] and osteocalcin synthesis [83] in ROS 17/2.8 cells. Endothelin-1 also decreased the production of alkaline phosphatase in MC3T3-E1 cells [85,86]. Endothelin-1 treatment increased mRNA for type I procollagen in ROS 17/2.8 cells and chicken osteoblasts [87,88] and stimulated the synthesis of both collagen and noncollagenous proteins in mouse calvaria [89]. Effects elicited by stimulation of endothelin receptor production could contribute to anabolic effects of 1,25(OH)$_2$D$_3$ on bone.

V. EFFECTS OF 1,25(OH)$_2$D$_3$ ON PROSTAGLANDIN PRODUCTION IN BONE

Treatment with 1,25(OH)$_2$D$_3$ increases prostaglandin production in mouse bones [90,91] and in some but not all lines of osteoblastic cells [92,93]. Although species may be a factor in this response, it is not the sole factor. In a study of human osteoblastic cells obtained by bone biopsy, cells isolated from bones of women with postmenopausal osteoporosis and high bone turnover showed increased production of PGE$_2$ following 48-hr incubation with 10 nM 1,25(OH)$_2$D$_3$. In contrast, a decrease in PGE$_2$ production was observed in cells that had lower rates of proliferation [94]. In rat costochondral chondrocytes, cells derived from growth zone showed increased PGE$_2$ production in response to 1,25(OH)$_2$D$_3$, whereas cells from the resting zone did not [95]. Both genomic [96] and nongenomic [97] processes may be involved in the increased generation of the eicosanoids in skeletal tissues.

The importance of prostaglandin production for the stimulatory effect of 1,25(OH)$_2$D$_3$ on bone resorption seems to be dependent on the model in which it has been studied. In the neonatal mouse calvarial bone organ culture model, inhibitors of prostaglandin synthesis decrease the bone resorption produced by the 1,25(OH)$_2$D$_3$ [90,98]. However, this is not the case in all bone organ culture models. In fetal mouse radii and ulnae, indomethacin failed to inhibit the resorption elicited by 1,25(OH)$_2$D$_3$ [99]. The resorption elicited by 1,25(OH)$_2$D$_3$ in fetal rat limb bones was likewise unaffected by indomethacin, ibuprofen and naproxen [62]. A dissociation, on the basis of both time course and dose dependence, between PGE$_2$ production and resorption was described in mouse parietal bones, with resorption being seen at earlier times and lower 1,25(OH)$_2$D$_3$ concentrations [91]. This would be consistent with the existence of both PGE$_2$-dependent and PGE$_2$-independent mechanisms of 1,25(OH)$_2$D$_3$-mediated resorption in calvarial bones. Interactions at the level of precursor cells could be important for the prostaglandin-dependent resorptive actions of 1,25(OH)$_2$D$_3$. Stromal cells isolated from bone promoted osteoclastic differentiation from murine marrow; the differentiation was stimulated by PGE$_2$, and 1,25(OH)$_2$D$_3$ augmented the PGE$_2$ production [100]. Prostaglandin dependence was also shown for the stimulatory effect of phorbol myristate acetate on osteoclast formation in 1,25(OH)$_2$D$_3$-primed mouse embryonic calvarial cells [101]. 1,25(OH)$_2$D$_3$ promoted the PGE$_1$-induced inhibition of proliferation in promyelocytic cells and enhanced the PGE$_1$-induced differentiation, including a stimulation of the production of IL-1α, IL-6, and TNFα [102].

The stimulatory effect of 1,25(OH)$_2$D$_3$ on osteoclast generation in bone marrow cultures was enhanced in the presence of low (10–100 pg/ml) concentrations of TGFβ, and this enhancement was prevented by indomethacin [53]. The stimulatory effects of IL-11 on osteoclast generation from bone marrow was also blocked by indomethacin [61]. All of these findings point to a functionally important broad role for 1,25(OH)$_2$D$_3$/prostaglandin interactions in the promotion of osteoclastogenesis. Preincubation with 10 μM PGE$_2$ for 24 hr potentiated the bone-resorbing effect of 1,25(OH)$_2$D$_3$ in fetal mouse radii and ulnae as well as the increase in the number of cells staining positively for tartrate-resistant acid phosphatase [99]. A similar potentiation of 1,25(OH)$_2$D$_3$ effects was observed after pretreatment with PTH. This potentiation appeared to be related to prostaglandin production, as it was prevented by indomethacin [99].

Antagonistic interactions between 1,25(OH)$_2$D$_3$ and prostaglandins have been described for several processes in bone. PGE$_2$ inhibited 1,25(OH)$_2$D$_3$-stimulated osteocalcin production in human bone-derived cells and in the MG-63 human osteosarcoma cell line [65,103]. The signaling pathway for this effect was though a cyclic AMP mechanism. The effect of IL-1 to inhibit 1,25(OH)$_2$D$_3$-stimulated osteocalcin production in human osteoblastic cells was partially inhibited by indomethacin, indicating that the effect of this cytokine is mediated, in part, through a prostaglandin-dependent pathway [65]. Thus, whereas prostaglandins may enhance the effects of 1,25(OH)$_2$D$_3$ that lead to enhanced osteoclast generation, there is also evidence that, in osteoblastic cells, increased prostaglandin production can antagonize the actions of 1,25(OH)$_2$D$_3$.

Treatment with 1,25(OH)$_2$D$_3$ modulated prostaglandin signal transduction, inhibiting the cyclic AMP and inositol trisphosphate (IP_3) production elicited by PGE_2 in MC3T3-E1 cells [104] and the increase in cytosolic calcium and IP_3 production elicited by PGE_2 in UMR-106 cells [105]. Analysis of the mechanisms of these effects led to the interpretation that the effect on IP_3 production reflected an action at a GTP-binding protein level, whereas the site of action on cAMP occurred downstream of adenylate cyclase. These results suggest possibilities for further feedback interactions between $1,25(OH)_2D_3$ at the level of cross talk between signal transduction pathways.

VI. SUMMARY AND CONCLUSIONS

$1,25(OH)_2D_3$ stimulates the production, by bone cells, of local factors that affect bone remodeling. This finding suggests that these local factors, which include growth factors, inflammatory cytokines, and prostaglandins, could mediate some of the effects of vitamin D on bone. The local factors also interact with $1,25(OH)_2D_3$ to modulate bone cell activity. Mitogenic effects of IGF-I and TGFβ are inhibited by $1,25(OH)_2D_3$, whereas one or more phenotypic responses (e.g., alkaline phosphatase activity, osteocalcin production, and collagen synthesis) are commonly, although not universally, enhanced by combined treatment of osteoblastic cells with $1,25(OH)_2D_3$ and IGF-I or TGFβ. Inflammatory cytokines such as IL-1, TNFα, and IL-6 may mediate resorptive effects of $1,25(OH)_2D_3$ through effects on osteoclastogenesis, and they can inhibit the anabolic effects of $1,25(OH)_2D_3$. $1,25(OH)_2D_3$ stimulates prostaglandin production in some models, and there is evidence that prostaglandins function as mediators of certain of the interactions of $1,25(OH)_2D_3$ with growth factors and cytokines. Many of the interactions discussed vary in magnitude and even direction in different model systems. These differences indicate that the responses are sensitive to other, unidentified modulators. These additional modulators could regulate the timing and expression of the interactions of $1,25(OH)_2D_3$ with the growth factors, cytokines, and prostaglandins during normal bone remodeling.

Thus, there is already a sizable literature documenting the ability of $1,25(OH)_2D_3$ to affect the production of local factors in bone and to interact with these local factors to influence both anabolic and resorptive responses in bone cells. These effects could conceivably contribute to the physiological effects of vitamin D on bone as well as to responses that result from hypervitaminosis D. Future studies utilizing specific inhibitors of the local factors or genetically modified animals in which expression of these factors is amplified or prevented will provide further evidence to assess the importance of these effects in vitamin D action on bone.

References

1. Centrella M, Canalis E 1985 Local regulators of skeletal growth: A perspective. Endocr Rev **6**:544–551.
2. Chenu C, Valentin-Opran A, Chavassieux P, Saez S, Meunier PJ, Delmas PD 1990 Insulin like growth factor I hormonal regulation by growth hormone and by $1,25(OH)_2D_3$ and activity on human osteoblast-like cells in short-term cultures. Bone **11**:81–86.
3. Kurose H, Yamaoka K, Okada S, Nakajima S, Seino Y 1990 1,25-Dihydroxyvitamin D_3 [$1,25\text{-}(OH)_2D_3$] increases insulin-like growth factor I (IGF-I) receptors in clonal osteoblastic cells. Study on interaction of IGF-I and $1,25\text{-}(OH)_2D_3$. Endocrinology **126**:2088–2094.
4. Canalis E, Lian JB 1988 Effects of bone associated growth factors on DNA, collagen and osteocalcin synthesis in cultured fetal rat calvariae. Bone **9**:243–246.
5. Canalis E, McCarthy TL, Centrella M 1988 Growth factors and the regulation of bone remodeling. J Clin Invest **81**:277–281.
6. Chen TL, Mallory JB, Hintz RL 1991 Dexamethasone and $1,25(OH)_2$vitamin D_3 modulate the synthesis of insulin-like growth factor-1 in osteoblast-like cells. Calcif Tissue Int **48**:278–282.
7. Scharla SH, Strong DD, Mohan S, Baylink DJ, Linkhart TA 1991 1,25-Dihydroxyvitamin D_3 differentially regulates the production of insulin-like growth factor I (IGF-I) and IGF-binding protein-4 in mouse osteoblasts. Endocrinology **129:**3139–3146.
8. Linkhart TA, Keffer MJ 1991 Differential regulation of insulin-like growth factor-I (IGF-I) and IGF-II release from cultured neonatal mouse calvaria by parathyroid hormone, transforming growth factor-beta, and 1,25-dihydroxyvitamin D_3. Endocrinology **128:**1511–1518.
9. Canalis E, Centrella M, McCarthy TL 1991 Regulation of insulin-like growth factor-II production in bone cultures. Endocrinology **129:**2457–2462.
10. Pirskanen A, Jääskeläinen T, Mäenpää PH 1993 Insulin-like growth factor-I modulates steroid hormone effects on osteocalcin synthesis in human MG-63 osteosarcoma cells. Eur J Biochem **218:**883–891.
11. Kurose H, Seino Y, Yamaoka K, Tanaka H, Shima M, Yabuuchi H 1989 Cooperation of synthetic insulin-like growth factor I/somatomedin C and 1,25-dihydroxyvitamin D_3 on regulation of function in clonal osteoblastic cells. Bone Miner **5**:335–345.
12. Fournier B, Ferralli JM, Price PA, Schlaeppi JM 1993 Comparison of the effects of insulin-like growth factors-I and -II on the human osteosarcoma cell line OHS-4. J Endocrinol **136:**173–180.
13. Kream BE, Petersen DN, Raisz LG 1990 Parathyroid hormone blocks the stimulatory effect of insulin-like growth factor-I on collagen synthesis in cultured 21-day fetal rat calvariae. Bone **11:**411–415.
14. Andress DL, Birnbaum RS 1992 Human osteoblast-derived insulin-like growth factor (IGF) binding protein-5 stimulates osteoblast mitogenesis and potentiates IGF action. J Biol Chem **267**:22467–22472.
15. Moriwake T, Tanaka H, Kanzaki S, Higuchi J, Seino Y 1992 1,25-Dihydroxyvitamin D_3 stimulates the secretion of insulin-like growth factor binding protein 3 (IGFBP-3) by cultured human osteosarcoma cells. Endocrinology **130:**1071–1073.

16. Schmid Ch, Schläpfer I, Waldvogel M, Meier PJ, Schwander J, Böni-Schnetzler M, Zapf J, Froesch ER 1992 Differential regulation of insulin-like growth factor binding protein (IGFBP)-2 mRNA in liver and bone cells by insulin and retinoic acid in vitro. FEBS Lett **202**:205–209.
17. Scharla SH, Strong DD, Mohan S, Chevalley T, Linkhart TA 1994 Effect of tumor necrosis factor-α on the expression of insulin-like growth factor I and insulin-like growth factor binding protein 4 in mouse osteoblasts. Eur J Endocrinol **131**:293–301.
18. De Mellow JSM, Baxter RC 1988 Growth hormone-dependent insulin-like growth factor (IGF) binding protein both inhibits and potentiates IGF-I stimulated DNA synthesis in human skin fibroblasts. Biochem Biophys Res Commun **159**:199–204.
19. Conover CA, Powell DR 1991 Insulin-like growth factor (IGF)-binding protein-3 blocks IGF-I induced receptor down-regulation and cell desensitization in cultured bovine fibroblasts. Endocrinology **129**:710–716.
20. Kawaguchi H, Pilbeam CC, Raisz LG 1994 Anabolic effects of 3,3′5-triiodothyronine and triiodothyroacetic acid in cultured neonatal mouse parietal bones. Endocrinology **135**:971–976.
21. Samaras SE, Hammond JM 1995 Insulin-like growth factor binding protein-3 inhibits porcine granulosa cell function in vitro. Am J Physiol **268**:E1057–E1064.
22. Schmid Ch, Schläpfer I, Keller A, Waldvogel M, Froesch ER, Zapf J 1995 Effects of insulin-like growth factor (IGF) binding proteins (BPs)-3 and -6 on DNA synthesis of rat osteoblasts: Further evidence for a role of auto-/paracrine IGF-I but not IGF-II in stimulating osteoblast growth. Biochem Biophys Res Commun **212**:242–248.
23. Chen TL, Chang LY, Bates RL, Perlman AJ 1991 Dexamethasone and 1,25-dihydroxyvitamin D_3 modulation of insulin-like growth factor-binding proteins in rat osteoblast-like cell cultures. Endocrinology **128**:73–80.
24. Owen TA, Bortell R, Shalhoub V, Heinrichs A, Stein JL, Stein GS, Lian JB 1993 Postproliferative transcription of the rat osteocalcin gene is reflected by vitamin D-responsive developmental modifications in protein-DNA interactions at basal and enhancer promoter elements. Proc Natl Acad Sci USA **90**:1503–1507.
25. Farquharson C, Law AS, Seawright E, Burt DW, Whitehead CC 1996 The expression of transforming growth factor-β by cultured chick growth plate chondrocytes: Differential regulation by 1,25-dihydroxyvitamin D_3. J Endocrinol **149**:277–285.
26. Robey PG, Young MF, Flanders KC, Roche MS, Kandaih P, Reddi AH, Termine JD, Sporn MB, Roberts AB 1987 Osteoblasts synthesize and respond to transforming growth factor-type β (TGFβ) in vitro. J Cell Biol **105**:457–463.
27. Rosen DM, Stempien SA, Thompson AY, Seyedin SM 1988 Transforming growth factor-beta modulates the expression of osteoblast and chondroblast phenotypes in vitro. J Cell Physiol **134**:337–346.
28. Oursler MJ 1994 Osteoclast synthesis, secretion, and activation of latent transforming growth factor-β. J Bone Miner Res **9**:443–452.
29. Kato Y, Iwamoto M, Koike T, Suzuki F, Takano Y 1988 Terminal differentiation and calcification in rabbit chondrocyte cultures grown in centrifuge tubes: Regulation by transforming growth factor β and serum factors. Proc Natl Acad Sci USA **85**:9552–9556.
30. Rosier RN, O'Keefe RJ, Crabb ID, Puzas JE 1989 Transforming growth factor β: An autocrine regulator of chondrocytes. Connect Tissue Res **20**:295–301.
31. Kulyk WM, Rodgers BJ, Greer K, Kosher RA 1989 Promotion of embryonic chick limb cartilage differentiation by transforming growth factor-β. Dev Biol **135**:424–430.
32. Ballock RT, Heydemann A, Wakefield LM, Flanders KC, Roberts AB, Sporn MB 1993 TGF-β1 prevents hypertrophy of epiphyseal chondrocytes: Regulation of gene expression for cartilage matrix proteins and metalloproteases. Dev Biol **158**:414–429.
33. Schwartz Z, Bonewald LF, Caulfield K, Brooks B, Boyan BD 1993 Direct effects of transforming growth factor-β on chondrocytes are modulated by vitamin D metabolites in a cell maturation-specific manner. Endocrinology **132**:1544–1552.
34. Centrella M, McCarthy TL, Canalis E 1987 Transforming growth factor-β is a bifunctional regulator of replication and collagen synthesis in osteoblast-enriched cell cultures from rat bone. J Biol Chem **262**:2869–2874.
35. Noda M, Camilliere JJ 1989 In vivo stimulation of bone formation by transforming growth factor-β. Endocrinology **124**:2991–2994.
36. Joyce ME, Roberts AB, Sporn MB, Bolander ME 1990 Transforming growth factor-β and the initiation of chondrogenesis and osteogenesis in the rat femur. J Cell Biol **110**:2195–2207.
37. Pfeilschifter J, Seyedin SM, Mundy GR 1988 Transforming growth factor beta inhibits bone resorption in fetal rat long bone cultures. J Clin Invest **82**:680–685.
38. Canalis E, McCarthy TL, Centrella M 1988 Isolation of growth factors from adult bovine bone. Calcif Tissue Int **43**:346–351.
39. Sato T, Ono T, Tuan RS 1993 1,25-Dihydroxy vitamin D_3 stimulation of TGF-beta expression in chick embryonic calvarial bone. Differentiation **52**:139–150.
40. Finkelman RD, Linkhart TA, Mohan S, Lau KH, Baylink DJ, Bell NH 1991 Vitamin D deficiency causes a selective reduction in deposition of transforming growth factor beta in rat bone: Possible mechanism for impaired osteoinduction. Proc Natl Acad Sci USA **88**:3657–3660.
41. Merry K, Gowen M 1992 The transcriptional control of TGF-beta in human osteoblast-like cells is distinct from that of IL-1 beta. Cytokine **4**:171–179.
42. Ingram RT, Bonde SK, Riggs BL, Fitzpatrick LA 1994 Effects of transforming growth factor beta (TGFβ) and 1,25 dihydroxyvitamin D_3 on the function, cytochemistry and morphology of normal human osteoblast-like cells. Differentiation **55**:153–163.
43. Bonewald LF, Kester MB, Schwartz Z, Swain LD, Khare A, Johnson TL, Leach RJ, Boyan BD 1992 Effects of combining transforming growth factor beta and 1,25-dihydroxyvitamin D_3 on differentiation of a human osteosarcoma (MG-63). J Biol Chem **267**:8943–8949.
44. Wergedal JE, Matsuyama T, Strong DD 1992 Differentiation of normal human bone cells by transforming growth factor-beta and 1,25$(OH)_2$ vitamin D_3. Metab Clin Exp **41**:42–48.
45. Noda M 1989 Transcriptional regulation of osteocalcin production by transforming growth factor-beta in rat osteoblast-like cells. Endocrinology **124**:612–617.
46. Staal A, Birkenhager JC, Pols HA, Buurman CJ, Vink-van Wijngaarden T, Kleinekoort WM, van den Bemd GJ, van Leeuven JP 1994 Transforming growth factor beta-induced dissociation between vitamin D receptor level and 1,25-dihydroxyvitamin D_3 action in osteoblast- like cells. Bone Miner **26**:27–42.
47. Marie PJ, Connes D, Hott M, Miravet L 1990 Comparative effects of a novel vitamin D analogue MC-903 and 1,25-dihydroxyvitamin D_3 on alkaline phosphatase activity, osteocalcin and DNA synthesis by human osteoblastic cells in culture. Bone **11**:171–179.
48. Noda M, Yoon K, Prince CW, Butler WT, Rodan GA 1988 Transcriptional regulation of osteopontin production in rat osteosarcoma cells by type beta transforming growth factor. J Biol Chem **263**:13916–13821.
49. Meikle MC, McGarrity AM, Thomson BM, Reynolds JJ 1991

Bone-derived growth factors modulate collagenase and TIMP (tissue inhibitor of metalloproteinases) activity and type I collagen degradation by mouse calvarial osteoblasts. Bone Miner **12:**41–55.

50. Schneider HG, Michelangeli VP, Frampton RJ, Grogan JL, Ikeda K, Martin TJ, Findlay DM 1992 Transforming growth factor-beta modulates receptor binding of calciotropic hormones and G protein-mediated adenylate cyclase responses in osteoblast-like cells. Endocrinology **131**:1383–1389.
51. Schwartz Z, Brooks B, Swain L, Del Toro F, Norman A, Boyan B 1992 Production of 1,25-dihydroxyvitamin D_3 and 24,25-dihydroxyvitamin D_3 by growth zone and resting zone chondrocytes is dependent on cell maturation and is regulated by hormones and growth factors. Endocrinology **130**:2495–2504.
52. van Leeuwen JP, Pols HA, Schilte JP, Visser TJ, Birkenhager JC 1991 Modulation by epidermal growth factor of the basal $1,25(OH)_2D_3$ receptor level and the heterologous up-regulation of the $1,25(OH)_2D_3$ receptor in clonal osteoblast-like cells. Calcif Tissue Int **49**:35–42.
53. Shinar DM, Rodan G 1990 Biphasic effects of transforming growth factor-beta on the production of osteoclast-like cells in mouse bone marrow cultures: The role of prostaglandins in the generation of these cells. Endocrinology **126:**3153–3158.
54. Neufeld G, Tessler S, Gitay-Goren H, Cohen T, Levi BZ 1994 Vascular endothelial growth factor and its receptors. Prog Growth Factor Res **5**:89–97.
55. Midy V, Plout J 1994 Vasculotropin/vascular endothelial growth factor induces differentiation in cultured osteoblasts. Biochem Biophys Res Commun **199**:380–386.
56. Harada S, Nagy JA, Sullivan KA, Thomas KA, Endo N, Rodan GA, Rodan SB 1994 Induction of vascular endothelial growth factor expression by prostaglandin E_2 and E_1 in osteoblasts. J Clin Invest **93**:2490–2496.
57. Wang DS, Yamazaki K, Nohtomi K, Shizume Z, Ohsumi K, Shibuya M, Demura H, Sato K 1996 Increase of vascular endothelial growth factor mRNA expression by 1,25-dihydroxyvitamin D_3 in human osteoblast-like cells. J Bone Miner Res **11:**42–49.
58. Walters MR 1992 Newly identified actions of the vitamin D endocrine system. Endocr Rev **13:**719–764.
59. Lacey DL, Grosso LE, Moser SA, Erdmann J, Tan HL, Pacifici R, Willareal DT 1993 IL-1-induced murine osteoblast IL-6 production is mediated by the type 1 IL-1 receptor and is increased by 1,25-dihydroxyvitamin D. J Clin Invest **91**:1731–1742.
60. Passeri G, Girasole G, Jilka RL, Manolagas SC 1993 Increased interleukin-6 production by murine bone marrow and bone cells after estrogen withdrawal. Endocrinology **133:**822–828.
61. Girasole G, Passeri G, Jilka RL, Manolagas SC 1994 Interleukin-11: A new cytokine critical for osteoclast development. J Clin Invest **93:**1516–1524.
62. Votta BJ, Bertolini DR 1994 Cytokine suppressive anti-inflammatory compounds inhibit bone resorption in vitro. Bone **15**:533–538.
63. Gowen M, MacDonald BR, Russell RG 1988 Actions of recombinant human γ-interferon and tumor necrosis factor alpha on the proliferation and osteoblastic characteristics of human trabecular bone cells in vitro. Arth Rheum **31**:1500–1507.
64. Evans DB, Bunning RA, Russell RG 1989 The effects of recombinant human granulocyte–macrophage colony-stimulating factor (rhGM-CSF) on human osteoblast-like cells. Biochem Biophys Res Commun **160**:588–595.
65. Evans DB, Bunning RA, Russell RG 1990 The effects of recombinant human interleukin-1 beta on cellular proliferation and the production of prostaglandin E_2, plasminogen activator, osteocalcin and alkaline phosphatase by osteoblast-like cells derived from human bone. Biochem Biophys Res Commun **166**:208–216.
66. Evans DB, Thavarajah M, Kanis JA 1990 Involvement of prostaglandin E_2 in the inhibition of osteocalcin synthesis by human osteoblast-like cells in response to cytokines and systemic hormones. Biochem Biophys Res Commun **167**:194–202.
67. Centrella M, McCarthy TL, Canalis E 1988 Tumor necrosis factor-α inhibits collagen synthesis and alkaline phosphatase activity independently of its effect on deoxyribonucleic acid synthesis in osteoblast-enriched bone cell cultures. Endocrinology **123**:1442–1448.
68. Li YP, Stashenko P 1992 Proinflammatory cytokines tumor necrosis factor-α and IL-6 but not IL-1, down-regulate the osteocalcin gene promoter. J Immunol **148**:788–794.
69. Mayur N, Lewis S, Catherwood BD, Nanes MS 1993 Tumor necrosis factor alpha decreases 1,25-dihydroxyvitamin D_3 receptors in osteoblastic ROS 17/2.8 cells. J Bone Miner Res **8**:997–1003.
70. Nanes MS, Rubin J, Titus L, Hendy GN, Catherwood BD 1991 Tumor necrosis factor alpha inhibits 1,25-dihydroxyvitamin D_3-stimulated bone gla protein synthesis in rat osteosarcoma cells (ROS 17/2.8) by a pre translational mechanism. Endocrinology **128**:2577–2582.
71. Nanes MS, Kuno H, Demay MB, Kurian M, Hendy GN, DeLuca HF, Titus L, Rubin J 1994 A single up-stream element confers responsiveness of 1,25-dihydroxyvitamin D_3 and tumor necrosis factor-α on the rat osteocalcin gene. Endocrinology **134**:1113–1120.
72. Kuno H, Kurian SM, Hendy GN, White J, DeLuca HF, Evans CO, Nanes MS 1994 Inhibition of 1,25-dihydroxyvitamin D_3 stimulated osteocalcin gene transcription by tumor necrosis factor-alpha: Structural determinants within the vitamin D response element. Endocrinology **134**:2524–2531.
73. Lacey DL, Erdmann JM, Tan HL, Ohara J 1993 Murine osteoblast interleukin 4 receptor expression: Upregulation by 1,25-dihydroxyvitamin D_3. J Cell Biochem **53:**122–134.
74. Shioi A, Teitelbaum SL, Ross FP, Welgus HW, Suzuki H, Ohara J, Lacey DL 1991 Interleukin 4 inhibits murine osteoclast formation in vitro. J Cell Biochem **47**:272–277.
75. Watanabe K, Tanaka Y, Morimoto I, Eto S 1990 Interleukin 4 as a potent inhibitor of bone resorption. Biochem Biophys Res Commun **172**:1035–1041.
76. Watanabe K, Sato K, Kasono K, Nakano Y, Eto S 1991 Interleukin-4 inhibits hypercalcemia in parathyroid hormone related protein-infused normal mice and tumor-bearing nude mice in vivo. J Bone Miner Res **6**:815S.
77. Holliday LS, Dean AD, Greenwald JE, Glucks SL 1995 C-type natriuretic peptide increases bone resorption in 1,25-dihydroxyvitamin D_3-stimulated mouse bone marrow cultures. J Biol Chem **270**:18983–18989.
78. Waldman SA, Rapoport RM, Murad F 1984 Atrial natriuretic factor selectively activates particulate guanylate cyclase and elevates cyclic GMP in rat tissues. J Biol Chem **259**:14332–14334.
79. Fletcher AE, Allen EH, Casley DJ, Martin TJ 1986 Atrial natriuretic factor receptors and stimulation of cyclic GMP formation in normal and malignant osteoblasts. FEBS Lett **208**:263–268.
80. Vargas SJ, Holden SN, Fall PM, Raisz LG 1989 Effects of atrial natriuretic factor on cyclic nucleotides, bone resorption, collagen and deoxyribonucleic acid synthesis, and prostaglandin E2 production in fetal rat bone culture. Endocrinology **125**:2527–2531.
81. Sasaki T, Hong MH 1993 Endothelin-1 localization in bone cells and vascular endothelial cells in rat bone marrow. Anat Rec **237**:332–337.

82. Sasaki T, Hong MH 1993 Localization of endothelin-1 in the osteoclast. J Electron Microsc **42**:193–196.
83. Nambi P, Wu HL, Lipshutz D, Prabhakar U 1995 Identification and characterization of endothelin receptors on rat osteoblastic osteosarcoma cells: Down-regulation by 1,25-dihydroxy-vitamin D_3. Mol Pharmacol **47**:266–271.
84. Shioide M, Noda M 1993 Endothelin modulates osteopontin and osteocalcin messenger ribonucleic acid expression in rat osteoblastic osteosarcoma cells. J Cell Biochem **53**:176–180.
85. Takuwa Y, Ohue Y, Takuwa N, Yamashita K 1989 Endothelin-1 activates phospholipase C and mobilizes Ca^{2+} from extra- and intracellular pools in osteoblastic cells. **257**:E797–E803.
86. Takuwa Y, Masaki T, Yamashita K 1990 The effects of endothelin family peptides on cultured osteoblastic cells from rat calvariae. Biochem Biophys Res Comm **170**:998–1005.
87. del Puerto G, McAdams J, Mathews C, Pun K-K, Arnaud CD 1994 Endothelin is anabolic in cultured osteoblast-like cells. J Bone Miner Res **9**(Suppl):S363.
88. Khoury E, del Puerto G, McAdams J, Nyiredy K, Mathews C, Pun K-K, Arnaud C 1993 Anatomical and functional associations between endothelial cells and osteoblasts. J Bone Miner Res **8**(Suppl):S159.
89. Tatrai A, Foster S, Lakatos P, Shankar G, Stern PH 1992 Endothelin-1 actions on resorption, collagen and noncollagen protein synthesis, and phosphatidylinositol turnover in bone organ cultures. Endocrinology **131**:603–607.
90. Klaushofer K, Hoffmann O, Horandner H, Karasegh S, Fratzl-Zelman N, Leis HJ, Gleispach H, Koller K, Peterlik M 1989 Effect of auranofin on resorption, prostaglandin synthesis and ultrastructure of bone cells in cultured mouse calvaria. J Rheumatol **16:**749–756.
91. Klein-Nulend J, Pilbeam CC, Raisz LG 1991 Effect of 1,25-dihydroxyvitamin D_3 on prostaglandin E_2 production in cultured mouse parietal bones. J Bone Miner Res **6**:1339–1344.
92. Rodan SB, Weslowski G, Rodan GA 1986 Clonal differences in prostaglandin synthesis among osteosarcoma cell lines. J Bone Miner Res **1**:213–220.
93. Schwartz Z, Dennis R, Bonewald L, Swain L, Gomez R, Boyan BD 1992 Differential regulation of prostaglandin E_2 synthesis and phospholipase A_2 activity by 1,25-$(OH)_2D_3$ in three osteoblast-like cell lines (MC-3T3-E1, ROS 17/2.8, and MG-63). Bone **13:**51–58.
94. Marie PJ, Hott M, Launay JM, Graulet AM, Gueris J 1993 In vitro production of cytokines by bone surface-derived osteoblastic cells in normal and osteoporotic postmenopausal women: Relationship with cell proliferation. J Clin Endocrinol Metab **77**:824–830.
95. Schwartz Z, Swain LD, Kelly DW, Brook B, Boyan BD 1992 Regulation of prostaglandin E_2 production by vitamin D metabolites in growth zone and resting zone chondrocyte cultures is dependent on cell maturation. Bone **13**:395–401.
96. Raisz LG, Pilbeam CC, Fall PM 1993 Prostaglandins: Mechanisms of action and regulation of production in bone. Osteopor Int **3**(Suppl. 1):136–140.
97. Boyan BD, Dean DD, Sylvia VL, Schwartz Z 1994 Nongenomic regulation of extracellular matrix events by vitamin D metabolites. J Cell Biochem **56**:331–339.
98. Hara K, Akiyama Y, Tajima T, Shiraki M 1993 Menatetrenon inhibits bone resorption partly through inhibition of PGE_2 synthesis in vitro. J Bone Miner Res **8**:535–542.
99. van Leeuwen JP, Birkenhager JC, Bos MP, van der Bemd GJ, Herrmann-Erlee MP, Pols HA 1992 Parathyroid hormone sensitizes long bones to the stimulation of bone resorption by 1,25-dihydroxyvitamin D_3. J Bone Miner Res **7**:303–309.
100. Collins DA, Chambers TJ 1992 Prostaglandin E_2 promotes osteoclast formation in murine hematopoietic cultures through an action on hematopoitic cells. J Bone Miner Res **7**:555–561.
101. Amano S, Hanazawa S, Kawata Y, Nakada Y, Miyata Y, Kitamo S 1994 Phorbol myristate acetate stimulates osteoclast formation in 1α,25-dihydroxyvitamin D_3-primed mouse embryonic calvarial cells by a prostaglandin-dependent mechanism. J Bone Miner Res **9**:465–472.
102. Kawase T, Ogata S, Orikasa M, Burns DM 1995 1,25-Dihydroxyvitamin D_3 promotes prostaglandin E_1-induced differentiation of HL-60 cells. Calcif Tissue Int **57:**359–366.
103. Lajeunesse D, Kiebzak GM, Frondoza C, Sacktor B 1991 Regulation of osteocalcin secretion by human primary bone cells and by the osteosarcoma cell line MG-63. Bone Min **14**:237–250.
104. Tokuda H, Kotoyori J, Suzuki A, Oiso Y, Kozawa O 1993 Effects of vitamin D_3 signaling by prostaglandin E_2 in osteoblast-like cells. J Cell Biochem **52**:220–226.
105. Green J, Kleeman CR, Schotland S, Ye LH 1992 1,25$(OH)_2D_3$ blunts hormone-elevated cytosolic Ca^{2+} in osteoblast-like cells. Am J Physiol **263:**E1070–E1076.

CHAPTER 23

Vitamin D and the Parathyroid Glands

JUSTIN SILVER AND TALLY NAVEH-MANY
Minerva Center for Calcium and Bone Metabolism, Nephrology Services, Hadassah University Hospital, Hebrew University Hadassah Medical School, Jerusalem, Israel 91120

I. INTRODUCTION

Calcium homeostasis in the body is sensitively regulated by an interplay between the peptide hormone parathyroid hormone (PTH) and the sterol hormone 1,25-dihydroxyvitamin D_3 [1,25$(OH)_2D_3$]. The secretion of PTH is regulated by small changes in the extracellular fluid (ECF) calcium concentration, and the synthesis of 1,25$(OH)_2D_3$ is also regulated by calcium. PTH and 1,25$(OH)_2D_3$ both act to increase serum calcium. In addition, PTH acts on the renal tubules to increase 1,25$(OH)_2D_3$ synthesis directly. Serum calcium concentration is maintained by the action of PTH on bone and the kidney, and the action of 1,25$(OH)_2D_3$ on the intestine. The classic action of vitamin D in mineral homeostasis is to increase calcium absorption. However, 1,25$(OH)_2D_3$ has an additional action which is also central to calcium metabolism: it acts directly on the parathyroid glands to regulate PTH gene expression, thus completing an endocrinological feedback loop [1–3]. This chapter reviews that role.

II. THE PARATHYROID GLANDS

In the human, there are two pairs of parathyroid glands in the anterior cervical region. They are of endodermal origin, derived from the third and fourth pharyngeal pouches. In the rat, there is a single pair of glands embedded in the cranial part of the thyroid. The chief cell is the predominant cell in the human and the only cell in the rat. The second cell type is the oxyphil cell, which has an acidophilic cytoplasm and many mitochondria. Roth and Raisz [4,5] had suggested that the chief cells had a secretory cycle in which they changed from inactive to active cells. However, the change in cells from dark to light is probably artifactual due to problems with fixation. For example, Larsson *et al.* [6] used fixation by vascular perfusion, which reduced the number of light cells, and the cells had a much more uniform ultrastucture. The parathyroid cells have secretory granules that contain PTH, but these are limited in number [7,8]. There are relatively few mitoses in the parathyroid cells.

III. PARATHYROID HORMONE BIOSYNTHESIS

Parathyroid hormone, a protein of 84 amino acids, is synthesized as a larger precursor, preproparathyroid hormone [9,10]. PreproPTH has a 25-residue "pre" or signal sequence, and a 6-residue "pro" sequence. The signal sequence, along with the short pro sequence, functions to direct the protein into the secretory pathway. Like other signal sequences, the pre sequence binds to a signal recognition particle during protein synthesis. The signal recognition particle then delivers the nascent peptide chain to the rough endoplasmic reticulum, where it is threaded through a protein-lined aqueous pore [11]. During this transit, the signal sequence is cleaved off by a signal peptidase, and the pre sequence is rapidly degraded. Because the process of transport and cleavage occurs during protein synthesis, very little intact preproPTH is found within the parathyroid cell.

Mature PTH has a molecular mass of approximately 9600 daltons and is the only form that is secreted from the parathyroid cell. The amino acid sequence has been determined in several species, and there exists a high degree of identity among species, particularly in the amino-terminal region of the molecule [12]. The parathyroids synthesize another protein that is also secreted [13–15]. This protein, secretory protein I, is identical to chromogranin A isolated from the adrenal medulla, and it is present in other endocrine cells and neoplasms as well [13]. Its function is not known, but it is stored and secreted with PTH despite its differential transcriptional regulation relative to PTH. Drees and Hamilton [16] have isolated and characterized a 26-kDa N-terminal fragment of chromogranin A (CgA) secreted by bovine parathyroid glands. This fragment, when added to dispersed parathyroid cells in primary culture, inhibits the low calcium-stimulated secretion of both PTH and CgA, suggesting an autocrine or paracrine regulation of secretion [16].

IV. PTH SECRETION AND THE CALCIUM SENSOR

The major regulator of PTH secretion is the concentration of ionized calcium in blood and ECF. Increases in the level of calcium lead to a decrease in PTH secretion. The shape of the dose–response curve is sigmoidal. Such a curve can be defined by four parameters: the maximal secretory rate, the slope of the curve at its midpoint, the level of calcium at the midpoint (often called the set point), and the minimal secretory rate [17]. The parathyroid calcium sensor has been characterized and cloned and probably mediates the physiological responses of the cell to calcium [18]. The deduced amino acid sequence of the receptor suggests that it spans the plasma membrane seven times, like other receptors in the G-protein-linked receptor family [18]. A large extracellular domain, presumed to bind calcium, resembles similar domains in brain metabotropic glutamate receptors as well as bacterial periplasmic proteins designed to bind small ligands, including cations.

The most convincing proof of the identity of the parathyroid calcium sensor has been the finding that most patients with familial hyopcalciuric hypercalcemia, a disease of defective calcium sensing, have a variety of inactivating mutations in the calcium sensor gene [19]. Mice genetically engineered to have only one functioning copy of the calcium sensor gene also have the expected defects in parathyroid calcium sensing [20]. The findings in both patients with familial hypocalciuric hypercalcemia and in the genetically engineered mice establish a link between the parathyroid calcium sensor and control of PTH secretion [11].

V. REGULATION OF THE PTH GENE

The minute-to-minute regulation of PTH blood levels can be explained by the calcium sensor and amplification of this regulation by intracellular degradation of stored hormone. Over a longer time frame, the parathyroid cell regulates the expression of the PTH gene to allow the cell to roughly match the production of PTH to the demand. Regulation of PTH biosynthesis has been studied in intact animals and in primary parathyroid cells in culture. In intact animals, it may be difficult to separate direct effects of a manipulation from indirect responses of the animal [11]. There is no parathyroid cell line, and thus the use of primary cells in culture complements the *in vivo* studies. Primary cells are unstable, however, and rapidly lose their characteristics of parathyroid cells in culture. Despite these experimental limitations, much has been learned about the regulation of the PTH gene.

VI. THE PTH GENE

Complementary DNA encoding human [21,22], bovine [23,24], rat [25], pig [25] chicken [26,27], and dog [28] PTH have all been cloned. Corresponding genomic DNA has also been cloned from human [22], bovine [29], and rat [30]. The genes all have two introns or

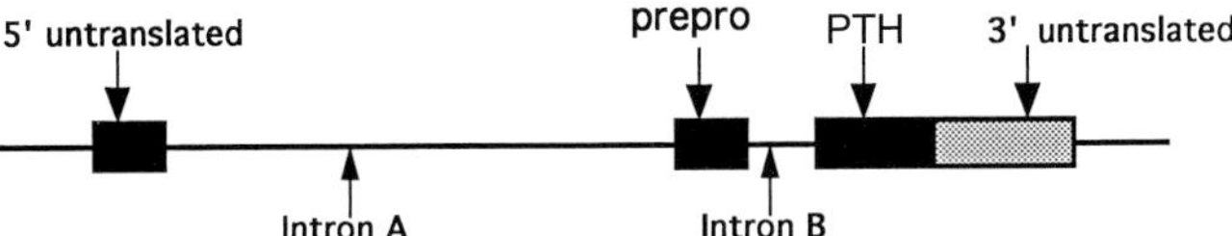

FIGURE 1 Schematic diagram of the PTH gene structure. Exons are indicated by the rectangles, and regions of the gene that code for preproPTH are indicated. Modified after Kronenberg *et al.* [31].

intervening sequences and three exons [31] (Fig. 1). The primary RNA transcript consists of RNA transcribed from both the introns and exons, and then the RNA sequences derived from the introns are spliced out. The product of this RNA processing, which represents the exons, is the mature PTH mRNA, which is then translated into preproPTH. The first intron separates the 5′-untranslated region of the mRNA from the rest of the gene, and the second intron separates most of the sequence encoding the precursor-specific "prepro" region from exon 3, which encodes the mature PTH and the 3′-untranslated region (UTR) (Fig. 1). The three exons that result are thus roughly divided into functional domains. There is considerable identity among the mammalian PTH genes, which is reflected in an 85% identity between the human and bovine proteins and 75% identity between the human and rat protein. There is less identity in the 3′-noncoding region. The two homologous TATA sequences flanking the human PTH gene direct the synthesis of two human PTH gene transcripts both in normal parathyroid glands and in parathyroid adenomas [32]. The termination codon immediately following the codon for glutamine at position 84 of PTH indicates that there are no additional precursors of PTH with peptide extensions at the carboxyl position.

VII. PROMOTER SEQUENCES

The regions upstream of the transcribed structural gene determine the tissue specificity and contain the regulatory sequences for the gene. For PTH, this analysis has been hampered by the lack of a parathyroid cell line. Rupp *et al.* analyzed the human PTH promoter region up to position −805 and identified a number of consensus sequences by computer analysis [33]. These included a sequence resembling the canonical cAMP responsive element 5′-TGACGTCA-3′ at position −81 with a single residue deviation. This element was fused to a reporter gene (chloramphenicol acetyltransferase, CAT) and then transfected into different cell lines. Pharmacological agents that increase cAMP led to an increased expression of the CAT gene, suggesting a functional role for the cAMP-responsive element (CRE). The role of this possible CRE in the context of the PTH gene in the parathyroid gland remains to be established.

Demay *et al.* [34] identified DNA squences in the human PTH gene that bind the $1,25(OH)_2D_3$ receptor (VDR). Nuclear extracts containing the VDR were examined for binding to sequences in the 5′-flanking region of the human PTH gene. A 25-bp oligonucleotide containing the sequences from −125 to −101 from the start of exon 1 bound nuclear proteins that were recognized by monoclonal antibodies against the VDR. The sequences in this region contained a single copy of a motif (AGGTTCA) that is homologous to the motifs repeated in the up-regulatory VDR response element (VDRE) of osteocalcin. When placed upstream of a heterologous viral promoter, the sequences contained in this 25-bp oligonucleotide mediated transcriptional repression in response to $1,25(OH)_2D_3$ in GH4C1 rat pituitary cells but not in ROS 17/2.8 rat osteosarcoma cells. Therefore, this down-regulatory VDRE differs from up-regulatory VDREs both in sequence composition and in the requirement for particular cellular factors other than the VDR for repressing PTH transcription [34]. Further work is needed to demonstrate that this negative VDRE functions in the context of the PTH gene promoter and to establish whether other VDREs control PTH gene expression. Farrow *et al.* [35,36] have identified DNA sequences upstream of the bovine PTH gene that bind the VDR. Liu *et al.* [37] have identified such sequences in the chicken PTH gene and demonstrated their functionality after transfection into the opossum kidney OK cell line.

VIII. REGULATION OF PTH GENE EXPRESSION

A. 1,25-Dihydroxyvitamin D

Parathyroid hormone is intimately involved in the homeostasis of normal serum concentrations of calcium and phosphate, which, in turn, regulate the synthesis and secretion of PTH. A major control mechanism in PTH synthesis occurs at the level of gene expression. $1,25(OH)_2D_3$ is also important in the maintenance of normal mineral metabolism, and there is a well defined feedback loop between $1,25(OH)_2D$ and PTH.

Parathyroid hormone increases the renal synthesis of $1,25(OH)_2D$. $1,25(OH)_2D$ then increases blood calcium largely by increasing the efficiency of intestinal calcium absorption. $1,25(OH)_2D$ also potently decreases the transcription of the PTH gene. This action was first demonstrated *in vitro* in bovine parathyroid cells in primary culture, wherein $1,25(OH)_2D_3$ led to a marked decrease in PTH mRNA levels [38,39] and a consequent decrease in PTH secretion [40–42]. Earlier studies of the effect of $1,25(OH)_2D_3$ on PTH secretion were negative because insufficient time was allowed for the effect of $1,25(OH)_2D_3$ on PTH synthesis to become manifest [43,44]; studies at longer time intervals have shown an effect [45]. Cantley *et al.* correlated the effect of $1,25(OH)_2D_3$ on PTH secretion and PTH mRNA levels in primary cultures of bovine parathyroid cells [40]. In short-term incubations (30–120 min) $1,25(OH)_2D_3$ had no effect on PTH secretion, but in long term incubations (24–96 hr) there was a dose-dependent decrease in PTH mRNA levels and PTH secretion. This study and that of Chan *et al.* [42] confirm that the $1,25(OH)_2D_3$ effect on PTH gene expression is reflected in a decrease in PTH synthesis and subsequent secretion.

The physiological relevance of these findings was established by *in vivo* studies in rats [46]. Rats injected with amounts of $1,25(OH)_2D_3$ that did not increase serum calcium exhibited a marked decrease in PTH mRNA levels, reaching less than 4% of control at 48 hr (Fig. 2). This effect was shown to be transcriptional both in *in vivo* studies in rats [46] and in *in vitro* studies with primary cultures of bovine parathyroid cells [47]. When 684 bp of the 5′-flanking region of the human PTH gene was linked to a reporter gene and transfected into a rat pituitary cell line (GH4C1), gene expression was lowered by $1,25(OH)_2D_3$ [48]. These studies suggest that $1,25(OH)_2D_3$ decreases PTH transcription by acting on the 5′-flanking region of the PTH gene, probably at least partly through interactions with the VDRE noted above.

A further level at which $1,25(OH)_2D_3$ might regulate PTH gene expression may be at the level of the VDR.

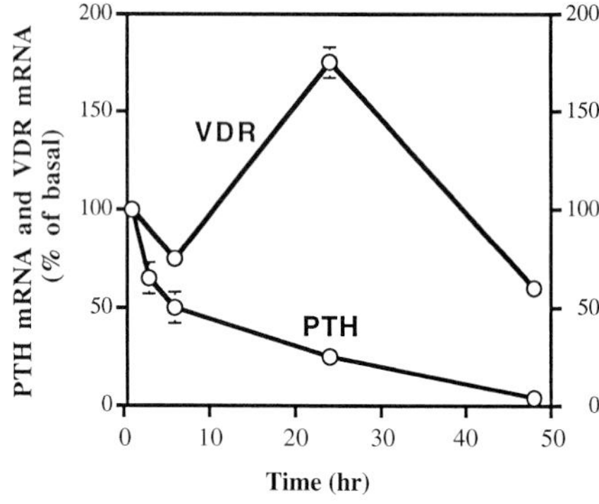

FIGURE 2 Time course for the effect of $1,25(OH)_2D_3$ on mRNA levels for PTH and the VDR in rat thyroparathyroid glands. Rats were injected with a single dose of 100 pmol $1,25(OH)_2D_3$ at 0 hr. The data represent means ± SE for four rats. Reproduced from *The Journal of Clinical Investigation*, 1990, Vol. 86, pp. 1968–1975 [53], by copyright permission of The American Society for Clinical Investigation.

$1,25(OH)_2D_3$ acts on its target tissues by binding to the VDR, which in turn regulates the transcription of genes with the appropriate recognition sequences. Changes in the concentration of the VDR in the $1,25(OH)_2D_3$ target sites could allow a modulation of the $1,25(OH)_2D_3$ effect, with an increase in receptor concentration leading to an amplification of its effect and a decrease in receptor concentration dampening the $1,25(OH)_2D_3$ effect (see Chapter 11). Ligand- and cation-dependent up-regulation of the VDR has been shown *in vivo* in rat intestine [49,50] and *in vitro* in a number of systems [51,52].

Naveh-Many *et al.* [53] injected $1,25(OH)_2D_3$ into rats and measured the levels of the VDR mRNA and PTH mRNA in the thyroparathyroid tissue. They showed that $1,25(OH)_2D_3$, in physiologically relevant doses, led to an increase in VDR mRNA levels in the parathyroid glands in contrast to the decrease in PTH mRNA levels (Fig. 2). This increase in VDR mRNA occurred after a time lag of 6 hr, and a dose–response study showed a peak at 25 pmol. The localization of the VDR mRNA to the parathyroids was demonstrated by *in situ* hybridization studies of the thyroparathyroid glands and duodenum (Fig. 3). VDR mRNA was localized to the parathyroids in the same concentration as in the duodenum, the $1,25(OH)_2D$ classic target organ. Weanling rats fed a diet deficient in calcium were markedly hypocalcemic at 3 weeks and had very high serum $1,25(OH)_2D$ levels. Despite the chronically high serum $1,25(OH)_2D$ levels there was no increase in VDR mRNA levels, and furthermore PTH mRNA levels did not fall and in fact were markedly increased in the parathyroid glands. The low calcium may have prevented the increase in parathyroid VDR levels, and this may partially explain the lack of PTH mRNA suppression. Whatever the mechanism, the lack of suppression of PTH synthesis in the setting of hypocalcemia and increased serum $1,25(OH)_2D_3$ is crucial physiologically, because it allows an increase in both PTH and $1,25(OH)_2D_3$ at a time of chronic hypocalcemic stress.

Russell *et al.* [54] studied the parathyroids of chicks with vitamin D deficiency and confirmed that $1,25(OH)_2D_3$ regulates PTH and VDR gene expression in the avian parathyroid gland. The chicks in this study were fed a vitamin-D deficient diet from hatching for 21 days and had established secondary hyperparathyroidism. These hypocalcemic chicks were then fed a diet with different calcium contents (0.5, 1.0 and 1.6%) for 6 days. The serum calcium levels were all still low (5, 6, and 7 mg/dl) with the expected inverse relationship between PTH mRNA and serum calcium. There was also a direct relationship between serum calcium and VDR mRNA levels. This result suggests either that VDR mRNA was not up-regulated in the setting of

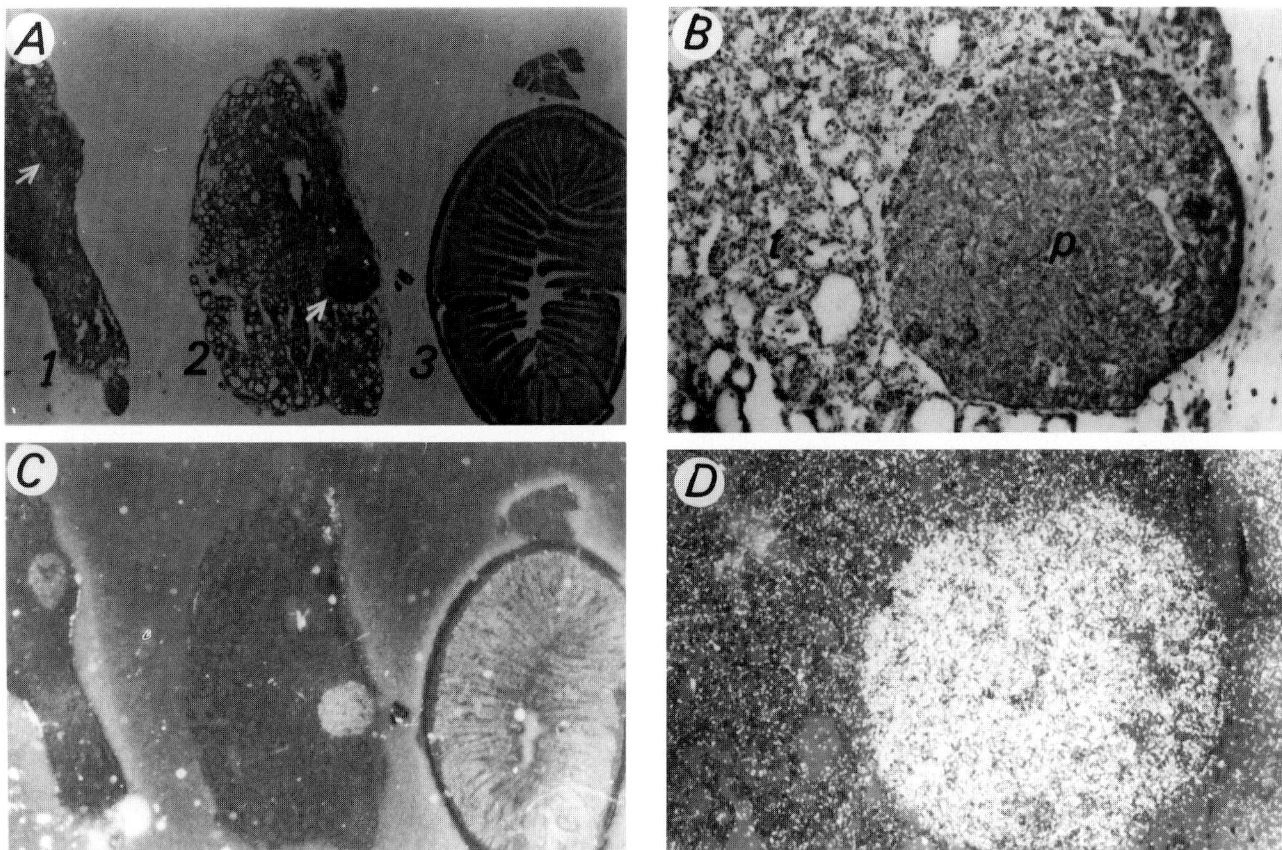

FIGURE 3 *In situ* hybridization of thyroparathyroid and duodenum sections with VDR. (A1) Thyroparathyroid tissue from a control rat. (A2) Thyroparathyroid tissue from a $1,25(OH)_2D_3$-treated rat (100 pmol at 24 hr). (A3) Duodenum from the $1,25(OH)_2D_3$-treated rat. White arrows point to the parathyroid glands. (B) Higher power view of A2 showing the parathyroid gland (p) and thyroid follicle (t). A and B were photographed under bright-field illumination. (C, D) Dark-field illumination of the same sections in A and B. Hybridization was with an antisense VDR probe. After 4 days of autoradiographic exposure, sections were stained with Giemsa stain and photographed. Magnification in B is 7 times the magnification of A. Reproduced from *The Journal of Clinical Investigation,* 1995, Vol. 96, pp. 1786–1793 [94], by copyright permission of The American Society for Clinical Investigation.

secondary hyperparathyroidism or that calcium directly regulates the VDR gene. Brown *et al.* [55] studied vitamin D-deficient rats and confirmed that $1,25(OH)_2D_3$ up-regulated the parathyroid VDR mRNA and that in secondary hyperparathyroidism with hypocalcemia the PTH mRNA was up-regulated without change in the VDR mRNA [53].

All these studies show that $1,25(OH)_2D_3$ increases the expression of the VDR gene in the parathyroid gland, which would result in increased VDR protein synthesis and increased binding of $1,25(OH)_2D_3$. This ligand-dependent receptor up-regulation would lead to an amplified effect of $1,25(OH)_2D_3$ on the PTH gene and might help explain the dramatic effect of $1,25(OH)_2D_3$ on the PTH gene (see Fig. 4).

The therapeutic use of $1,25(OH)_2D_3$ is limited by its hypercalcemic effect. As a result a number of calcitriol analogs have been synthesized that are biologically active but are less hypercalcemic than $1,25(OH)_2D_3$ (see Section VII of this volume). These analogs usually involve modifications of the $1,25(OH)_2D_3$ side chain, such as replacing atoms in 22-oxa-$1,25(OH)_2D_3$, which is the chemical modification in oxacalcitriol [56], or adding a cyclopropyl group at the end of the side chain in calcipotriol [57,58]. Brown *et al.* showed that oxacalcitriol *in vitro* decreased PTH secretion from primary cultures of bovine parathyroid cells with a similar dose–response relationship to that of calcitriol [59]. *In vivo* the injection of both vitamin D compounds led to a decrease in rat parathyroid PTH mRNA levels [59]. However, detailed *in vivo* dose–response studies showed that calcitriol is the most effective analog for decreasing PTH mRNA levels, even at doses which do not cause hypercalcemia [60]. Oxacalcitriol and calcipotriol are less effective for decreasing PTH RNA levels but have a wider dose range at which they do not cause hypercalcemia; this property might be useful clinically (see Chapters 52 and 73). The marked activity of calcitriol analogs *in vitro* as compared to their modest hypercalcemic actions *in vivo* probably reflects their rapid clearance from the circulation [61] (Chapter 59).

The ability of calcitriol to decrease PTH gene transcription is used therapeutically in the management of patients with chronic renal failure. They are treated with calcitriol in order to prevent the secondary hyperparathyroidism of chronic renal failure. The most effective dosing regime to suppress the elevated serum PTH levels without causing hypercalcemia has been shown to involve administration of calcitriol in single oral or intravenous doses three times a week [62–64] (Chapters 52 and 73). The poor response in some patients who do not respond may well result from poor control of serum phosphate [63,65], decreased vitamin D receptor concentration [66], or tertiary hyperparathyroidism with monoclonal parathyroid tumors [67].

Retinoid X receptors (RXRs) are involved in $1,25(OH)_2D_3$-mediated transcriptional events. MacDonald *et al.* [68] showed that all-*trans*- or 9-cis-retinoic acid suppressed the release of PTH from bovine parathyroid cell cultures. Both retinoids were remarkably potent, with significant decreases evident at 10^{-10} *M* and a maximally suppressive effect (~65%) at 10^{-7} *M*. All-*trans*-Retinol was considerably less potent in this system. The effect was not evident until 12 hr, suggesting that retinoids did not affect the rapid secretion of preexisting PTH stores. PreproPTH mRNA levels were also suppressed by retinoic acid, and the retinoid potencies were similar to those observed in the secretion studies. Combined treatment with 10^{-6} *M* retinoic acid and 10^{-8} *M* $1,25(OH)_2D_3$ more effectively decreased PTH secretion and preproPTH mRNA than did either compound alone. These data indicate that retinoic acid (1) elicits a bioresponse in bovine parathyroid cells; (2) attenuates PTH expression at the protein and mRNA levels, and (3) acts independently of $1,25(OH)_2D_3$ in the control of PTH expression. MacDonald *et al.* [68] suggested that, because all-*trans*- and 9-*cis*-retinoic acid are equipotent, the suppressive effect may be mediated through the retinoic acid receptors (RARs) and not through the RXRs. This is based on the facts that retinoic acid binds selectively to RAR and 9-*cis*-retinoic acid represents a panagonist capable of binding both RAR and RXR.

Liu *et al.* [37] characterized a VDRE in the 5′-flanking region of the chicken PTH gene that mediated negative regulation of gene transcription by $1,25(OH)_2D_3$ and bound the VDR. Using gel mobility shift assays they showed that the chicken VDRE, incubated with partially purified nuclear extract from dog intestine, showed two bound complexes. These complexes were competed for by an excess of unlabeled VDRE or monoclonal antibody specific for the VDR protein. In addition, they demonstrated similar protein–DNA complexes when the chicken VDRE was incubated with a mixture of purified preparations of recombinant VDR and RXRα proteins [37]. By themselves, neither of the recombinant proteins were able to bind the VDRE significantly. This study suggests that RXR is necessary for the binding of the VDR complex to the VDRE. Mackey *et al.* [69] studied the VDRE of the human PTH gene [34] in protein–DNA binding studies. They showed that with bovine parathyroid nuclear extracts, the VDR bound the VDRE independently of the RXR. In GH4C1 nuclear extracts, there were two VDR-containing complexes, one lacking and one containing RXR. The human PTH VDRE mediates transcriptional repression in GH4C1 cells but not in ROS 17/2.8 cells. In ROS 17/2.8 nuclear extracts, a single VDR-dependent complex was observed that contained RXR. In contrast, when the up-regulatory rat osteocalcin VDRE was used as a probe, only VDR–RXR-containing complexes were generated using nuclear extracts from bovine parathyroids and GH4C1 and ROS 17/2.8 cells.

From these studies, it is clear that retinoic acid acts to decrease PTH gene expression. The study by Liu *et al.* [37] using the chicken VDRE shows that RXR is involved in this action, but the Mackey *et al.* study [69] suggests that this may not apply for the human VDRE. Therefore, the parathyroid nuclear proteins involved in the binding of the VDR to the PTH down-regulatory VDRE need to be isolated to provide a more comprehensive understanding of how $1,25(OH)_2D_3$ decreases PTH gene transcription.

B. Calcium

The interrelationships of the effect of calcium and $1,25(OH)_2D_3$ on the PTH gene and parathyroid cell are very interesting. This is of particular relevance to the *in vivo* situation where a low serum calcium, as a result of dietary calcium deficiency for instance, leads to a marked increase in serum $1,25(OH)_2D_3$ levels, which would be expected to decrease PTH gene transcription. However, this dietary hypocalcemia is associated with a marked increase in PTH mRNA and serum PTH levels. The effect of calcium on the parathyroid gland and in particular the interrelationship between hypocalcemia and the lack of effect of $1,25(OH)_2D_3$ is reviewed in this section.

1. *In Vitro* Studies

A remarkable characteristic of the parathyroid gland is its sensitivity to small changes in serum calcium, which leads to large changes in PTH secretion. This calcium sensing is also expressed at the levels of PTH gene expression and parathyroid cell proliferation. The results of *in vitro* and *in vivo* experiments agree that calcium regulates PTH mRNA levels, but the data differ in important ways. *In vitro* studies with bovine parathyroid cells in primary culture showed that calcium regulated PTH mRNA levels [70,71], with an effect mainly of high calcium to decrease PTH mRNA. These effects were most pronounced after more prolonged incubations such as 72 hr. The parathyroid calcium sensor is markedly decreased in cells in culture by 24 hr, so it is difficult to interpret the data at 72 hr [19,72].

2. *In Vivo* Studies

Naveh-Many *et al.* [73] studied rats *in vivo*. They showed that small decreases in serum calcium from 2.6 to 2.1 mmol/liter led to large increases in PTH mRNA levels, reaching 3-fold over that of controls at 1 and 6 hr. On the other hand, a high serum calcium had no effect on PTH mRNA levels even at concentrations as high as 6.0 mmol/liter. Interestingly, in the same thyroparathyroid tissue RNA extracts, calcium had no effect on the expression of the calcitonin gene. Thus, although high calcium is a secretagogue for calcitonin it does not regulate calcitonin gene expression (see Chapter 24). Yamamoto *et al.* [74] also studied the effect of calcium on PTH mRNA levels in rats. They showed that hypocalcemia induced by a calcitonin infusion for 48 hr led to a 7-fold increase in PTH mRNA levels. Rats made hypercalcemic (2.9–3.4 m*M*) for 48 hr had the same PTH mRNA levels as controls which had received a normal calcium infusion (2.5 m*M*) but modestly lower levels than those which had received a calcium-free infusion. In further studies, Naveh-Many *et al.* [75] transplanted Walker carcinosarcoma 256 cells into rats, which led to serum calcium levels of 18 mg/dl at day 10. There was no change in PTH mRNA levels in the rats with marked chronic hypercalcemia [75]. The differences between *in vivo* and *in vitro* results probably reflects the instability of the *in vitro* system, but it is also difficult to eliminate the possibility that *in vivo* effects are influenced by indirect actions caused by a high or low serum calcium. Nevertheless, the physiological conclusion is that common causes of hypercalcemia *in vivo* do not importantly decrease PTH mRNA levels; these results emphasize that the parathyroid gland is geared to respond to hypocalcemia and not hypercalcemia.

3. Mechanisms of Regulation of PTH mRNA by Calcium

The mechanism whereby calcium regulates PTH gene expression is particularly interesting. Changes in extracellular calcium are monitored by the calcium sensor, which then regulates PTH secretion [18]. It is not known what mechanism transduces the message of changes in extracellular calcium that lead to changes in PTH mRNA. Okazaki *et al.* have identified a negative calcium regulatory element (nCaRE) in the atrial natiuretic peptide gene, with a homologous sequence in the PTH gene [76]. They identified a redox factor protein (ref1), which was known to activate several transcription factors via alterations of their redox state, that bound a nCaRE, and the level of ref1 mRNA and protein were elevated by an increase in extracellular calcium concentration [77]. They suggested that ref1 had transcription repressor activity in addition to its function as a transcriptional auxiliary protein [77]. Because no parathyroid cell line is available, these studies were performed in nonparathyroid cell lines, so their relevance to physiological PTH gene regulation remains to be established.

Preliminary studies have been reported by Moallem *et al.* which indicate that the effect of Ca^{2+} on PTH mRNA levels *in vivo* occurs with no detectable change in PTH gene transcription rate [78]. These results suggest that the effect of calcium *in vivo* may entail posttranscriptional mechanisms. Such mechanisms could well involve proteins that bind directly to PTH mRNA. To identify such proteins they used ultraviolet crosslinking and RNA gel mobility shift assays and demonstrated that cytoplasmic proteins bind to the PTH mRNA 3′ untranslated region (3′-UTR). RNA fragments spanning the full-length rat PTH cDNA incubated with thyroparathyroid cytoplasmic proteins revealed three bands that bound specifically to the mRNA. RNA corresponding to the 5′-UTR and the coding region of the gene did not bind any proteins. An RNA fragment representing the 3′-UTR bound parathyroid proteins with the same pattern as the full-length mRNA; this region, therefore, contains the protein binding sequence. Protein–RNA binding with parathyroid proteins from hypocalcemic rats was increased as compared to control and decreased from hypophosphatemic rats (see later). These studies show that regions in the 3′-UTR of the PTH mRNA specifically bind cytosolic proteins and may be involved in the posttranscriptional increase in PTH gene expression induced by hypocalcemia.

In vitro studies by Hawa *et al.* [79] have also suggested a posttranscriptional effect of calcium on PTH gene expression. They incubated bovine parathyroid cells for 48 hr in 0.4 m*M* calcium. This did not increase PTH mRNA levels as compared to controls but did increase the membrane-bound polysomal content of PTH mRNA 2-fold. Actinomycin D reduced PTH mRNA levels by about 50% at 48 hr in cells incubated in 0.4 and 1.0 m*M* calcium [79]. However, actinomycin D did not prevent the rise in polysomal PTH mRNA induced by low calcium. Vadher *et al.* [80] have shown that the 3′-

untranslated region is of major importance in mediating translational regulation of PTH synthesis by cytosolic regulatory proteins. 5′-UTR or 3′-UTR constructs of PTH mRNA fused to a luciferase reporter were synthesized and then translated with wheat germ lysate with or without parathyroid cell cytosol [80]. The addition of cytosol inhibited translation from the 5′- and 3′-UTR PTH mRNA . These *in vitro* studies did not demonstrate the changes in PTH mRNA that are so marked in the *in vivo* studies, but they do show a mechanism of translational regulation of the PTH mRNA and a role for the UTR in this process. Together, these results suggest that calcium regulates PTH mRNA by posttranscriptional mechanisms, which might involve binding of calcium-sensitive proteins to the 5′- and 3′-UTRs. Transcriptional regulation of the PTH gene by calcium in parathyroid cells remains possible, but direct evidence for such regulation has not yet been obtained.

A pulse of $1,25(OH)_2D_3$ decreases PTH gene transcription, but in chronic hypocalcemia there is a marked increase in $1,25(OH)_2D_3$ levels which paradoxically do not decrease PTH mRNA levels (Fig. 4). Calreticulin binds to the sequence KXGFFKR, found in steroid hormone receptors, resulting in altered transcription of steroid responsive genes *in vitro* [81,82]. Wheeler *et al.* [83] showed that calreticulin prevented the action of $1,25(OH)_2D_3$ to increase osteocalcin gene transcription *in vitro.* Sela *et al.* [84] have presented preliminary results on the role of calreticulin and the VDR in preventing the action of chronic high $1,25(OH)_2D_3$ levels on the PTH gene *in vivo.* Normal rats given $1,25(OH)_2D_3$ by continuous minipump for 7 days (50 pmol/day) had no change in their parathyroid gland VDR levels but had decreased nuclear calreticulin and PTH mRNA levels. Normal rats given a single injection of $1,25(OH)_2D_3$ (50 pmol) alone or after a minipump infusion for 7 days (50 pmol/day) also had decreased PTH mRNA and calreticulin but had increased VDR levels. Therefore, the response to $1,25(OH)_2D_3$ is associated with a decrease in calreticulin. The increase in VDR may amplify the effect of $1,25(OH)_2D_3$ on the PTH gene. In a further experiment, rats were fed diets

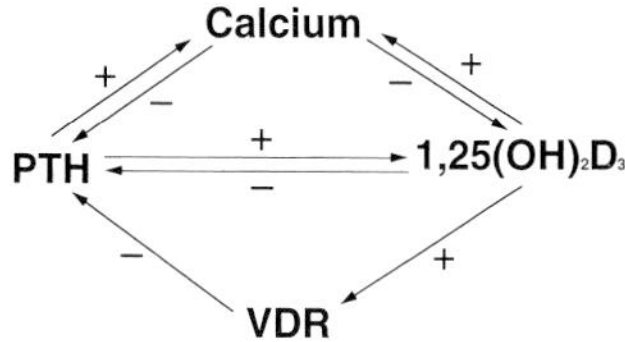

FIGURE 4 Interrelationships of calcium, $1,25(OH)_2D_3$, VDR, and PTH. *In vivo,* rats with chronic hypocalcemia have high PTH mRNA and serum PTH levels despite the increased serum $1,25(OH)_2D_3$ levels, which are not effective in decreasing PTH gene transcription.

with a normal vitamin D content and a low calcium content, leading to hypocalcemia and a marked increased serum $1,25(OH)_2D_3$. These hypocalcemic rats, with increased serum $1,25(OH)_2D_3$, had increased PTH mRNA and calreticulin with decreased VDR levels. Therefore, under these conditions, where $1,25(OH)_2D_3$ increased calreticulin and decreased the VDR in the parathyroid glands, there was no effect on PTH mRNA. In rats with chronic hypocalcemia there are enhanced levels of calreticulin which dominate over the effect of $1,25(OH)_2D_3$ on the PTH gene.

In summary, elevated VDR levels together with decreased levels of calreticulin correlate with an effect of $1,25(OH)_2D_3$ on the PTH gene, and increased calreticulin correlates with a lack of an effect. These *in vivo* results demonstrate that the ratio of calreticulin to the VDR determines the response of the PTH gene to circulating or pulsed $1,25(OH)_2D_3$. They suggest that high circulating levels of $1,25(OH)_2D_3$ in dietary induced hypocalcemia do not decrease PTH gene transcription, because the increased levels of calreticulin in the parathyroid gland together with decreased VDR may prevent the binding of the VDR, or the VDR–RXR complex, to its VDRE in the PTH gene [84] (Fig. 4).

4. SECONDARY HYPERPARATHYROIDISM AND PARATHYROID CELL PROLIFERATION

Chronic changes in the physiological milieu often lead to both changes in parathyroid cell proliferation and PTH gene regulation, as discussed by Silver and Kronenberg [11]. In such complicated settings, the regulation of PTH gene expression may well be controlled by mechanisms that differ from those in nonproliferating cells. Further, the effects of change in cell number and activity of individual cells can be complicated and difficult to dissect. Nevertheless, such chronic changes represent commonly observed clinical circumstances that require examination. Secondary parathyroid hyperplasia is a complication of chronic renal disease [85,86] or vitamin D deficiency, and it may lead to disabling skeletal complications. The expression and regulation of the PTH gene has been studied in two models of secondary hyperparathyroidism: (1) rats with experimental uremia due to 5/6 nephrectomy and (2) rats with nutritional secondary hyperparathyroidism due to diets deficient in vitamin D and/or calcium.

Rats with 5/6 nephrectomy had higher serum creatinine levels and also appreciably higher levels of parathyroid gland PTH mRNA [87]. Their PTH mRNA levels decreased after single injections of $1,25(OH)_2D_3$, a response similar to that of normal rats [87]. Interestingly, the secondary hyperparathyroidism is characterized by an increase in parathyroid gland PTH mRNA but not in VDR mRNA [87]. This suggests that in 5/6 nephrectomy rats there is relatively less VDR mRNA per parathyroid cell, or a relative down-regulation of the VDR, as has been reported in VDR binding studies [88–91]. Fukagawa *et al.* [92] also studied 5/6 nephrectomized rats and confirmed that $1,25(OH)_2D_3$ decreased PTH mRNA levels, as did the $1,25(OH)_2D_3$ analog 22-oxacalcitriol.

The second model of experimental secondary hyperparathyroidism studied was that due to dietary deficiency of vitamin D (−D) and/or calcium (−Ca), as compared to normal vitamin D (ND) and normal calcium (NCa) [93]. These dietary regimes were selected to mimic the secondary hyperparathyroidism in which the stimuli for the production of hyperparathyroidism are the low serum levels of $1,25(OH)_2D_3$ and ionized calcium. Weanling rats were maintained on the diets for 3 weeks and then studied. Rats on diets deficient in both vitamin D and calcium (−D, −Ca) exhibited a 10-fold increase in PTH mRNA as compared to controls (ND, NCa) together with much lower serum calcium levels and also lower serum $1,25(OH)_2D_3$ levels. Calcium deficiency alone (−Ca, ND) led to a 5-fold increase in PTH mRNA levels, whereas a diet deficient in vitamin D alone (−D, NCa) led to a 2-fold increase in PTH mRNA levels.

Because renal failure and prolonged changes in blood calcium and $1,25(OH)_2D_3$ can affect both parathyroid cell number and the activity of each parathyroid cell, the change in both these parameters must be assessed in each model in order to understand the various mechanisms of secondary hyperparathyroidism. Parathyroid cell number was determined in thyroparathyroid tissue of normal rats and −D, −Ca rats. To do this, the tissue was enzymatically digested into an isolated cell population, which was then passed through a flow cytometer [fluorescence-activated cell sorter (FACS)] and separated by size into two peaks. The first peak of smaller cells contained parathyroid cells as determined by the presence of PTH mRNA, and the second peak contained thyroid follicular cells and calcitonin-producing cells that hybridized positively for thyroglobulin mRNA and calcitonin mRNA but not PTH mRNA. There was a 1.6-fold increase in cells from the −D, −Ca rats than from the normal rats, and a 10-fold increase in PTH mRNA. Therefore, this model of secondary hyperparathyroidism is characterized by increased gene expression per cell, together with a smaller increase in cell number.

Further studies by Naveh-Many *et al.* [94] have clearly demonstrated that hypocalcemia is a stimulus for parathyroid cell proliferation. They studied parathyroid cell proliferation by staining for proliferating cell nuclear antigen (PCNA) and found that a low calcium diet led to increased levels of PTH mRNA and a 10-fold increase in parathyroid cell proliferation (Fig. 5) [94]. The secondary hyperparathyroidism of 5/6 nephrectomized rats was characterized by an increase in both PTH mRNA levels and PCNA-positive parathyroid cells. Therefore, both hypocalcemia and uremia induce parathyroid cell proliferation *in vivo*. The effect of $1,25(OH)_2D_3$ on parathyroid cell proliferation was also studied in this dietary model of secondary hyperparathyroidism. $1,25(OH)_2D_3$ at a dose (25 pmol for 3 days) that lowered PTH mRNA levels had no effect on the number of PCNA-positive cells. However, *in vitro* in bovine parathyroid cells using pharmacological doses of $1,25(OH)_2D_3$ [95] and *in vivo* in rats with experimental uremia, $1,25(OH)_2D_3$ decreased the amount of $[^3H]$thymidine incorporated into the parathyroid [96]. These findings emphasize the importance of a normal calcium in the prevention of parathyroid cell hyperplasia and indicate that the role of $1,25(OH)_2D_3$ is as yet not clear.

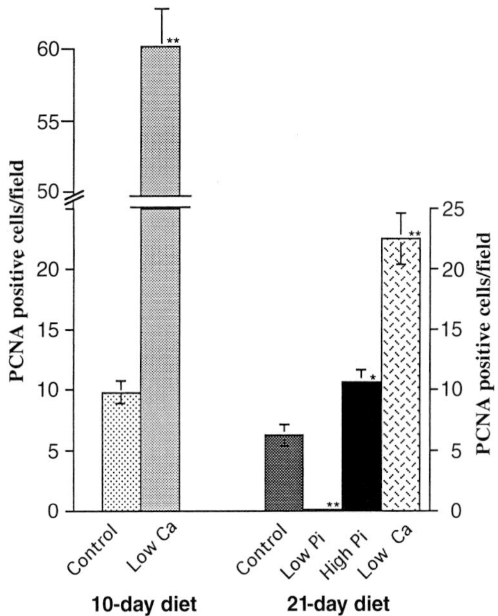

FIGURE 5 Dietary phosphate and calcium regulate parathyroid cell proliferation. Weanling rats were fed different diets for 10 or 21 days and the number of proliferating cells determined by PCNA staining. The diets were either control, low calcium (0.02%), low phosphate (0.02%), or high phosphate (1.2%). The results are expressed as PCNA positive cells per microscope field, as means ± SE for four different rats. Statistically significant differences compared to rats fed the control diet were noted: *, $p < 0.05$; **, $p < 0.01$. Reproduced from *The Journal of Clinical Investigation,* 1995, Vol. 96, pp. 1786–1793 [94], by copyright permission of The American Society for Clinical Investigation.

A further mechanism by which calcium might regulate parathyroid cell number is through induction of apoptosis. This has been studied in the parathyroid glands of hypocalcemic rats as well as rats with experimental uremia fed different diets [94]. Apoptosis was determined by the TUNEL method, which detects nuclear DNA fragmentation *in situ*. In no situation were apoptotic cells detected in the parathyroid glands. However, apoptosis in the parathyroid glands needs to be explored in other models using different methodologies.

Wernerson and co-workers have studied parathyroid cell number in dietary induced secondary hyperparathyroidism (ND, −Ca) using stereoscopic electron microscopy and showed that the cells are markedly hypertrophic without an increase in cell number [97,98]. Thus, in models of secondary parathyroid enlargement such as this one, parathyroid cell hypertrophy can precede the development of parathyroid cell hyperplasia. These experimental findings are relevant to the management of patients with secondary hyperparathyroidism. Increased transcription of the PTH gene is readily reversible, but the reversibility of an increased number of parathyroid cells by accelerating cell death has not yet been demonstrated.

C. Phosphate

Serum phosphate levels, like serum calcium, lead to large changes in serum $1,25(OH)_2D_3$. A low serum phosphate increases serum $1,25(OH)_2D_3$, and a high serum phosphate decreases $1,25(OH)_2D_3$. However, in contrast to the effect of hypocalcemia, which increases serum PTH despite the high $1,25(OH)_2D_3$, dietary induced hypophosphatemia, which lowers both phosphate and $1,25(OH)_2D_3$, cause a decrease in serum PTH. This implies a simple additive effect, although it may not be so simple. Kilav *et al.* [99] showed *in vivo* that the effect of hypophosphatemia on the PTH gene is posttranscriptional. Thus, it may be expected that the transcriptional effect of the high $1,25(OH)_2D_3$ would exert a dominant effect on the PTH gene. Therefore, there must be a mechanism at play preventing the action of the chronically elevated serum $1,25(OH)_2D_3$ acting on the PTH gene.

Parathyroid hormone regulates serum phosphate concentration through its effect on the kidney to decrease phosphate reabsorption. In moderate renal failure there is a decrease in phosphate clearance and an increase in serum phosphate, which becomes an important problem in severe renal failure. This has always been considered to be central to the pathogenesis of secondary hyperparathyroidism, but it has been difficult to separate the effects of hyperphosphatemia from those of the attendant hypocalcemia and decrease in serum $1,25(OH)_2D_3$ levels. In the 1970s, Slatopolsky and Bricker showed in dogs with experimental chronic renal failure that dietary phosphate restriction prevented secondary hyperparathyroidism [100]. Clinical studies [101] have demonstrated that phosphate restriction in patients with chronic renal insufficiency is effective in preventing the increase in serum PTH levels [65,101–104]. The mechanism of this effect was not clear, although at least part of it was considered to be due to changes in serum $1,25(OH)_2D_3$ concentrations. *In vitro* [105,106] and *in vivo* [101,107] phosphate directly regulated the production of $1,25(OH)_2D_3$. A raised serum phosphate decreases serum $1,25(OH)_2D_3$ levels which then leads to decreased calcium absorption from the diet and eventually a low serum calcium levels. Also, the raised phosphate complexes calcium, which is then deposited in bone and soft tissues and, thereby, decreases serum calcium, and also prevents the mobilization of calcium from bone.

A number of careful clinical and experimental studies suggested, however, that the effect of phosphate on serum PTH levels was independent of changes in both serum calcium and $1,25(OH)_2D_3$ levels. Lopez-Hilker *et al.* have shown in dogs with experimental chronic renal failure that phosphate restriction corrected their secondary hyperparathyroidism independent of changes in serum calcium and $1,25(OH)_2D_3$ levels [108]. This was done by placing the uremic dogs on diets deficient in both calcium and phosphate. This led to lower levels of serum phosphate and calcium, with no increase in the low levels of serum $1,25(OH)_2D_3$. Despite this, there was a 70% decrease in PTH levels. The study by Lopez-Hilker *et al.* on the effect of a low phosphate diet on serum PTH levels suggested that phosphate had an effect on the parathyroid cell through a mechanism independent of its effect on serum $1,25(OH)_2D_3$ and calcium levels [108]. Yi *et al.* [109] studied rats in early chronic renal failure. They showed that with a normal phosphate diet (0.6%) there was secondary hyperparathyroidism, with an increase in serum PTH and PTH mRNA levels and no change in serum calcium, phosphate, or $1,25(OH)_2D_3$ levels. However, rats fed a low phosphate diet, either 0.3 or 0.1%, had no evidence of secondary hyperparathyroidism [109]. At least in the rats fed the 0.3% phosphate diet there was no increase in serum calcium or $1,25(OH)_2D_3$. Therefore, phosphate plays a central role in the pathogenesis of secondary hyperparathyroidism, both through its effect on serum $1,25(OH)_2D_3$ and calcium levels and also independently.

Kilav *et al.* [110] have studied expression of the PTH gene in hypophosphatemic rats and showed that phosphate regulated the gene independent of its effects on serum calcium and $1,25(OH)_2D_3$. Weanling rats were fed diets with low, normal, or high phosphate contents for 3 weeks. The low phosphate diet led to hypophosphatemia, hypercalcemia, and increased serum $1,25(OH)_2D_3$ together with decreased PTH mRNA levels (25% of controls) and a similar decrease in serum PTH levels. A high phosphate diet led to increased PTH mRNA levels. *In situ* hybridization for PTH mRNA showed that hypophosphatemia decreased PTH mRNA in all the parathyroid cells. To separate the effect of low phosphate from changes in calcium and vitamin D, rats were fed diets to maintain them vitamin D deficient and normocalcemic despite the hypophosphatemia. Hypophosphatemic, normocalcemic rats with normal serum $1,25(OH)_2D_3$ levels still had decreased PTH mRNAs (Fig. 6). Nuclear transcript run-on assays showed that the effect of low phosphate was posttranscriptional, unlike the predominantly transcriptional effect of $1,25(OH)_2D_3$ on the PTH gene. They, therefore, demonstrated that dietary phosphate regulates the parathyroid gland through a mechanism which remains to be defined but is clearly independent of changes in serum calcium and $1,25(OH)_2D_3$ (Fig. 7) [110].

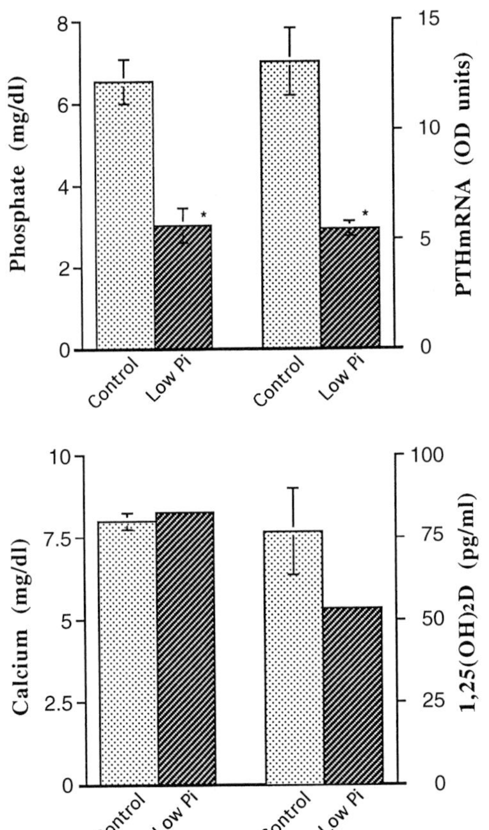

FIGURE 6 Effect of dietary phosphate on PTH mRNA levels in second generation vitamin D-deficient rats fed a vitamin D-deficient diet with both a low phosphate and low calcium content for 1 day. Levels of serum phosphate and PTH mRNA (top) and serum calcium and $1,25(OH)_2D$ (bottom) were measured in rats fed a vitamin D-deficient, normal phosphate, normal calcium diet (control group) or a vitamin D-deficient, low phosphate, low calcium diet for 1 day. The results are means ± SE for five rats in each group; *, $p \leq 0.01$ versus control. Data for $1,25(OH)_2D$ represent pooled serum from two or three rats, with three samples for control rats and two samples for low phosphate rats. Reproduced from *The Journal of Clinical Investigation,* 1995, Vol. 96, pp. 327–333 [110], by copyright permission of The American Society for Clinical Investigation.

There is now evidence that reveals a direct effect of phosphate on the parathyroid gland. Almaden *et al.* [111] showed that high phosphate directly stimulated PTH secretion from whole rat parathyroid glands in culture and PTH mRNA levels in human parathyroid tissue in culture [112]. Slatopolsky *et al.* [113] and Nielsen *et al.* [114] showed similar results for PTH secretion from rat parathyroid glands maintained in primary culture. These results indicate that phosphate regulates the parathyroid directly. In these *in vitro* studies, parathyroid slices or tissue were incubated, rather than isolated cells, suggesting either that the sensing mechanism for phosphate is damaged during the preparation of the isolated cells or that the intact gland structure is important to the phosphate response.

One of the effects of either prolonged hypocalcemia or experimental chronic renal failure on the parathyroid is to increase parathyroid cell proliferation [94]. Hypophosphatemic rats have a marked decrease in parathyroid cell proliferation. This effect of hypophosphatemia occurred in both normal and uremic rats. Rats fed a high dietary phosphate had an increase in parathryoid cell proliferation, particularly in those rats with experimental chronic renal failure [94]. Therefore, phosphate regulates the parathyroid gland at a number of levels, namely, PTH gene expression, parathyroid cell proliferation, and serum PTH levels. However, the *in vivo* studies reported here utilized diets that led to very low serum phosphate levels, which may have no direct relevance to possible direct effects of high phosphate in renal failure. It is necessary to separate nonspecific effects of very low phosphate from true physiological regulation.

The effect of calcium on PTH secretion from dispersed bovine parathyroid cells occurs within seconds [115,116]. However, the effect of phosphate *in vitro* in parathyroid glands in culture requires about 4 hr [111,113,114] before any change in PTH secretion occurs. The reason for this delay is not clear. It may reflect

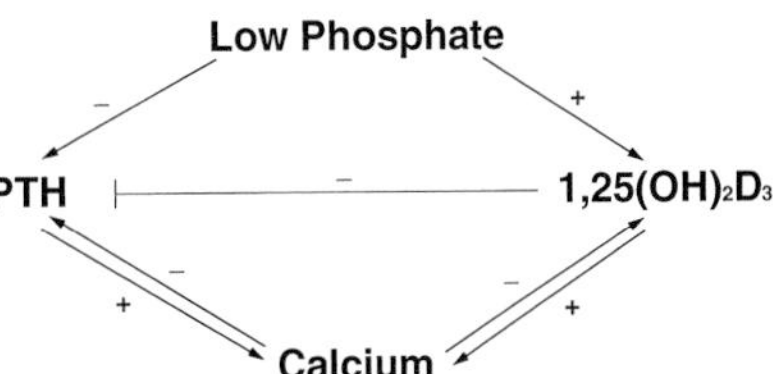

FIGURE 7 Interrelationships among low phosphate, PTH, $1,25(OH)_2D$, and calcium. *In vivo,* rats with chronic hypophosphatemia have increased serum $1,25(OH)_2D$ levels, which do not decrease PTH gene transcription. The decreased PTH mRNA levels in chronic hypophosphatemia are posttranscriptional despite the elevated serum $1,25(OH)_2D$ levels.

a delay in signal transduction or PTH synthesis. What is clear from the preliminary studies of Naveh-Many *et al.* [117] is that at least the effect on PTH gene expression correlates with a decrease in PTH mRNA protein binding. This compares to the effect of hypocalcemia, which increases this binding. These results indicate that the final pathway of effects of low phosphate and low calcium on PTH mRNA share a common mechanism.

IX. SUMMARY

The PTH gene is regulated by a number of factors. $1,25(OH)_2D_3$ acts on the PTH gene to decrease its transcription, and this action is used in the management of patients with chronic renal failure. Mechanisms of transcriptional down-regulation are less well defined than mechanisms of transcriptional up-regulation. It is clear that the effect of retinoic acid is additive to that of $1,25(OH)_2D_3$ to down-regulate PTH gene expression, but the role of RXR in the binding of the VDR to the VDRE of the PTH gene remains controversial. A down-regulatory VDRE has been defined in the PTH gene promoter, which requires other cellular factors for its function. What is clear is the powerful action of $1,25(OH)_2D_3$ *in vivo* in experimental animals and also in patients to decrease PTH gene expression through transcriptional mechanisms. What is now only beginning to be clarified is why in *in vivo* situations of very high serum $1,25(OH)_2D_3$ levels there is no inhibitory activity on the PTH gene. In the case of chronic hypocalcemia, preliminary studies suggest that calreticulin may play a role in preventing the $1,25(OH)_2D_3$ action on the PTH gene. Calcium and phosphate also determine cell proliferation in the parathyroid gland, an effect largely independent of $1,25(OH)_2D_3$ *in vivo*.

In diseases such as chronic renal failure secondary hyperparathyroidism involves abnormalities in PTH secretion and synthesis. An understanding of how the parathyroid gland is regulated at each level will help in devising a rational therapy for the management of such conditions. The effect of $1,25(OH)_2D_3$ to decrease PTH gene transcription and serum PTH levels has led to the rational use of $1,25(OH)_2D_3$ in the management of patients with renal failure. This therapy represents a fine example of advances in basic science and clinical investigation being effectively applied to patient well being.

Acknowledgments

The studies reported in this chapter were performed in the Minerva Center for Calcium and Bone Metabolism, and they were supported in part by the Israel Academy of Sciences and Humanities, US–Israel Binational Science Foundation, the German–Israel Foundation (GIF), and the National Institutes of Health.

References

1. Silver J 1992 Regulation of parathyroid hormone synthesis and secretion. In: Coe FL, Favus MJ (eds) Disorders of Bone and Mineral Metabolism. Raven, New York, pp. 83–106.
2. Naveh-Many T, Silver J 1996 Parathyroid hormone synthesis, secretion and action. In: Coe FL, Favus MJ, Pak C, Parks J, Preminger G (eds) Kidney Stones: Medical and Surgical Management. Raven, New York, pp. 175–199.
3. Silver J, Moallem E, Epstein E, Kilav R, Naveh-Many T 1994 New aspects in the control of parathyroid hormone secretion. Curr Opin Nephrol Hypertens **3**:379–385.
4. Roth SI, Raisz LG 1966 The course and reversibility of the calcium effect on the ultrastructure of the rat parathyroid gland in organ culture. Lab Invest **15**:1187–1211.
5. Shannon WA, Roth SI 1974 An ultrastructural study of acid phosphatase activity in normal adenomatous and hyperplastic (chief cell type) human parathyroid glands. Am J Pathol **77**:493–501.
6. Larsson H-O, Lorentzon R, Boquist L 1984 Structure of the parathyroid glands as revealed by different methods of fixation. A quantitative light- and electron-microscopic study in untreated Mongolian gerbils. Cell Tissue Res **235**:51–58.
7. Roth SI, Raisz LG 1964 Effect of calcium concentration on the ultrastructure of rat parathyroid in organ culture. Lab Invest **13**:331–345.
8. Setoguti T, Inoue Y, Kato K 1981 Electron-microscopic studies on the relationship between the frequency of parathyroid storage granules and serum calcium levels in the rat. Cell Tissue Res **219**:457–467.
9. Potts JTJ, Kronenberg HM, Habener JF, Rich A 1980 Biosynthesis of parathyroid hormone. Ann NY Acad Sci **343**:38–55.
10. Habener JF 1981 Regulation of parathyroid hormone secretion and biosynthesis. Annu Rev Physiol **43**:211–223.
11. Silver J, Kronenberg HM 1996 Parathyroid hormone—Molecular biology and regulation. In: Bilezikian JB, Raisz LG, Rodan GA (eds) Principles of Bone Biology. Academic Press, San Diego.
12. Habener JF, Potts JT Jr 1990 Fundamental considerations in the physiology, biology, and biochemistry of parathyroid hormone. In: Avioli LV, Krane SM (eds) Metabolic Bone Disease, 2nd Ed. Saunders, Philadelphia, Pennsylvania, pp. 69–130.
13. Cohn DV, Kumarasamy R, Ramp WK 1986 Intracellular processing and secretion of parathyroid gland proteins. Vitam Horm **43**:283–316.
14. Majzoub JA, Dee PC, Habener JF 1982 Cellular and cell-free processing of parathyroid secretory proteins. J Biol Chem **257**:3581–3588.
15. Cohn DV, Morrissey JJ, Shofstall RE, Chu LL 1982 Cosecretion of secretory protein-I and parathormone by dispersed bovine parathyroid cells. Endocrinology **110**:625–630.
16. Drees BM, Hamilton JW 1992 Pancreastatin and bovine parathyroid cell secretion. Bone Miner **17**:335–346.
17. Brown EM 1983 Four-parameter model of the sigmoidal relationship between parathyroid hormone release and extracellular calcium concentration in normal and abnormal parathyroid tissue. J Clin Endocrinol Metab **56**:572–581.
18. Brown EM, Gamba G, Riccardi R, Lombardi M, Butters R,

Kifor O, Sun A, Hediger MA, Lytton J, Hebert J 1993 Cloning and characterization of an extracellular Ca^{2+}-sensing receptor from bovine parathyroid. Nature **366**:575–580.
19. Mithal A, Kifor O, Kifor I, Vassilev P, Butters R, Krapcho K, Simin R, Fuller F, Hebert SC, Brown EM 1995 The reduced responsiveness of cultured bovine parathyroid cells to extracellular Ca^{2+} is associated with marked reduction in the expression of extracellular Ca^{2+}-sensing receptor messenger ribonucleic acid and protein. Endocrinology **136**:3087–3092.
20. Ho C, Conner DA, Pollak MR, Ladd DJ, Kifor O, Warren HB, Brown EM, Seidman JG, Seidman CE 1995 A mouse model of human familial hypocalciuric hypercalcemia and neonatal severe hyperparathyroidism. Nat Genet **11**:389–394.
21. Hendy GN, Kronenberg HM, Potts JT Jr, Rich A 1981 Nucleotide sequence of cloned cDNAs encoding human preproparathyroid hormone. Proc Natl Acad Sci USA **78**:7365–7369.
22. Vasicek TJ, McDevitt BE, Freeman MW, Fennick BJ, Hendy GN, Potts JT Jr, Rich A, Kronenberg HM 1983 Nucleotide sequence of the human parathyroid hormone gene. Proc Natl Acad Sci USA **80**:2127–2131.
23. Kronenberg HM, McDevitt BE, Majzoub JA, Nathans J, Sharp PA, Potts JT Jr, Rich A 1979 Cloning and nucleotide sequence of DNA coding for bovine preproparathyroid hormone. Proc Natl Acad Sci USA **76**:4981–4985.
24. Weaver CA, Gordon DF, Kemper B 1982 Nucleotide sequence of bovine parathyroid hormone messenger RNA. Mol Cell Endocrinol **28**:411–424.
25. Schmelzer HJ, Gross G, Widera G, Mayer H 1987 Nucleotide sequence of a full-length cDNA clone encoding preproparathyroid hormone from pig and rat. Nucleic Acids Res **15**:6740.
26. Khosla S, Demay M, Pines M, Hurwitz S, Potts JT Jr, Kronenberg HM 1988 Nucleotide sequence of cloned cDNAs encoding chicken preproparathyroid hormone. J Bone Miner Res **3**:689–698.
27. Russell J, Sherwood LM 1989 Nucleotide sequence of the DNA complementary to avian (chicken) preproparathyroid hormone mRNA and the deduced sequence of the hormone precursor. Mol Endocrinol **3**:325–331.
28. Rosol TJ, Steinmeyer CL, McCauley LK, Grone A, DeWille JW, Capen CC 1995 Sequences of the cDNAs encoding canine parathyroid hormone-related protein and parathyroid hormone. Gene **160**:241–243.
29. Weaver CA, Gordon DF, Kissil MS, Mead DA, Kemper B 1984 Isolation and complete nucleotide sequence of the gene for bovine parathyroid hormone. Gene **28**:319–329.
30. Heinrich G, Kronenberg HM, Potts JTJ, Habener JF 1984 Gene encoding parathyroid hormone. Nucleotide sequence of the rat gene and deduced amino acid sequence of rat preproparathyroid hormone. J Biol Chem **259**:3320–3329.
31. Kronenberg HM, Igarashi T, Freeman MW, Okazaki T, Brand SJ, Wiren KM, Potts JT Jr 1986 Structure and expression of the human parathyroid hormone gene. Recent Prog Horm Res **42**:641–663.
32. Igarashi T, Okazaki T, Potter H, Gaz R, Kronenberg HM 1986 Cell-specific expression of the human parathyroid hormone gene in rat pituitary cells. Mol Cell Biol **6**:1830–1833.
33. Rupp E, Mayer H, Wingender E 1990 The promoter of the human parathyroid hormone gene contains a functional cyclic AMP-response element. Nucleic Acids Res **18**:5677–5683.
34. Demay MB, Kiernan MS, DeLuca HF, Kronenberg HM 1992 Sequences in the human parathyroid hormone gene that bind the 1,25-dihydroxyvitamin D_3 receptor and mediate transcriptional repression in response to 1,25-dihydroxyvitamin D_3. Proc Natl Acad Sci USA **89**:8097–8101.
35. Farrow SM, Hawa NS, Karmali R, Hewison M, Walters JC, O'Riordan JL 1990 Binding of the receptor for 1,25-dihydroxyvitamin D_3 to the 5′-flanking region of the bovine parathyroid hormone gene. J Endocrinol **126**:355–359.
36. Hawa NS, O'Riordan JL, Farrow SM 1994 Binding of 1,25-dihydroxyvitamin D_3 receptors to the 5′-flanking region of the bovine parathyroid hormone gene. J Endocrinol **142**:53–60.
37. Liu SM, Koszewski N, Lupez M, Malluche HH, Olivera A, Russell J 1996 Characterization of a response element in the 5′-flanking region of the avian (chicken) parathyroid hormone gene that mediates negative regulation of gene transcription by 1,25-dihydroxyvitamin D_3 and binds the vitamin D_3 receptor. Mol Endocrinol **10**:206–215.
38. Silver J, Russell J, Sherwood LM 1985 Regulation by vitamin D metabolites of messenger ribonucleic acid for preproparathyroid hormone in isolated bovine parathyroid cells. Proc Natl Acad Sci USA **82**:4270–4273.
39. Russell J, Silver J, Sherwood LM 1984 The effects of calcium and vitamin D metabolites on cytoplasmic mRNA coding for pre-proparathyroid hormone in isolated parathyroid cells. Trans Assoc Am Physicians **97**:296–303.
40. Cantley LK, Russell J, Lettieri D, Sherwood LM 1985 1,25-Dihydroxyvitamin D_3 suppresses parathyroid hormone secretion from bovine parathyroid cells in tissue culture. Endocrinology **117**:2114–2119.
41. Karmali R, Farrow S, Hewison M, Barker S, O'Riordan JL 1989 Effects of 1,25-dihydroxyvitamin D_3 and cortisol on bovine and human parathyroid cells. J Endocrinol **123**:137–142.
42. Chan YL, McKay C, Dye E, Slatopolsky E 1986 The effect of 1,25 dihydroxycholecalciferol on parathyroid hormone secretion by monolayer cultures of bovine parathyroid cells. Calcif Tissue Int **38**:27–32.
43. Golden P, Greenwalt A, Martin K, Bellorin-Font E, Mazey R, Klahr S, Slatopolsky E 1980 Lack of a direct effect of 1,25-dihydroxycholecalciferol on parathyroid hormone secretion by normal bovine parathyroid glands. Endocrinology **107**:602–607.
44. Chertow BS, Baker GR, Henry HL, Norman AW 1980 Effects of vitamin D metabolites on bovine parathyroid hormone release in vitro. Am J Physiol **238**:E384-E388.
45. Au WY 1984 Inhibition by 1,25-dihydroxycholecalciferol of hormonal secretion of rat parathyroid gland in organ culture. Calcif Tissue Int **36**:384–391.
46. Silver J, Naveh-Many T, Mayer H, Schmelzer HJ, Popovtzer MM 1986 Regulation by vitamin D metabolites of parathyroid hormone gene transcription *in vivo* in the rat. J Clin Invest **78**:1296–1301.
47. Russell J, Lettieri D, Sherwood LM 1986 Suppression by $1,25(OH)_2D_3$ of transcription of the pre-proparathyroid hormone gene. Endocrinology **119**:2864–2866.
48. Okazaki T, Igarashi T, Kronenberg HM 1988 5′-Flanking region of the parathyroid hormone gene mediates negative regulation by $1,25\text{-}(OH)_2$ vitamin D_3. J Biol Chem **263**:2203–2208.
49. Costa EM, Feldman D 1986 Homologous up-regulation of the $1,25(OH)_2$ vitamin D_3 receptor in rats. Biochem Biophys Res **137**:742–747.
50. Favus MJ, Mangelsdorf DJ, Tembe V, Coe BJ, Haussler MR 1988 Evidence for *in vivo* upregulation of the intestinal vitamin D receptor during dietary calcium restriction in the rat. J Clin Invest **82**:218–224.
51. Huang Y, Lee S, Stolz R, Gabrielides C, Pansini-Porta A, Bruns ME, Bruns DE, Miffins TE, Pike JW, Christakos S 1989 Effect of hormones and development on the expression of the rat 1,25-dihydroxyvitamin D_3 receptor gene. J Biol Chem **264**:17454–17461.

52. Sandgren ME, DeLuca HF 1990 Serum calcium and vitamin D regulate 1,25-dihydroxyvitamin D_3 receptor concentration in rat kidney in vivo. Proc Natl Acad Sci USA **87**:4312–4314.
53. Naveh-Many T, Marx R, Keshet E, Pike JW, Silver J 1990 Regulation of 1,25-dihydroxyvitamin D_3 receptor gene expression by 1,25-dihydroxyvitamin D_3 in the parathyroid in vivo. J Clin Invest **86**:1968–1975.
54. Russell J, Bar A, Sherwood LM, Hurwitz S 1993 Interaction between calcium and 1,25-dihydroxyvitamin D_3 in the regulation of preproparathyroid hormone and vitamin D receptor messenger ribonucleic acid in avian parathyroids. Endocrinology **132**:2639–2644.
55. Brown AJ, Zhong M, Finch J, Ritter C, Slatopolsky E 1995 The roles of calcium and 1,25-dihydroxyvitamin D_3 in the regulation of vitamin D receptor expression by rat parathyroid glands. Endocrinology **136**:1419–1425.
56. Nishii Y, Abe J, Mori T, Brown AJ, Dusso AS, Finch J, Lopez-Hilker S, Morrissey J, Slatopolsky E 1991 The noncalcemic analogue of vitamin D, 22-oxacalcitriol, suppresses parathyroid hormone synthesis and secretion. Contrib Nephrol **91**:123–128.
57. Kissmeyer AM, Binderup L 1991 Calcipotriol (MC 903): Pharmacokinetics in rats and biological activities of metabolites. A comparative study with $1{,}25(OH)_2D_3$. Biochem Pharmacol **41**:1601–1606.
58. Evans DB, Thavarajah M, Binderup L, Kanis JA 1991 Actions of calcipotriol (MC 903), a novel vitamin D_3 analog, on human bone-derived cells: Comparison with 1,25-dihydroxyvitamin D_3. J Bone Miner Res **6**:1307–1315.
59. Brown AJ, Ritter CR, Finch JL, Morrissey J, Martin KJ, Murayama E, Nishii Y, Slatopolsky E 1989 The noncalcemic analogue of vitamin D, 22-oxacalcitriol, suppresses parathyroid hormone synthesis and secretion. J Clin Invest **84**:728–732.
60. Naveh-Many T, Silver J 1993 Effects of calcitriol, 22-oxacalcitriol and calcipotriol on serum calcium and parathyroid hormone gene expression. Endocrinology **133**:2724–2728.
61. Bouillon R, Allewaert K, Xiang DZ, Tan BK, van-Baelen H 1991 Vitamin D analogs with low affinity for the vitamin D binding protein: Enhanced in vitro and decreased *in vivo* activity. J Bone Miner Res **6**:1051–1057.
62. Slatopolsky E, Weerts C, Thielan J, Horst R, Harter H, Martin KJ 1984 Marked suppression of secondary hyperparathyroidism by intravenous administration of 1,25-dihydroxy-cholecalciferol in uremic patients. J Clin Invest **74**:2136–2143.
63. Quarles LD, Yohay DA, Carroll BA, Spritzer CE, Minda SA, Bartholomay D, Lobaugh BA 1994 Prospective trial of pulse oral versus intravenous calcitriol treatment of hyperparathyroidism in ESRD. Kidney Int **45**:1710–1721.
64. Caravaca F, Cubero JJ, Jimenez F, Lopez JM, Aparicio A, Cid MC, Pizarro JL, Liso J, Santos I 1995 Effect of the mode of calcitriol administration on PTH-ionized calcium relationship in uraemic patients with secondary hyperparathyroidism. Nephrol Dial Transplant **10**:665–670.
65. Aparicio M, Combe C, Lafage MH, De Precigout V, Potaux L, Bouchet JL 1994 In advanced renal failure, dietary phosphorus restriction reverses hyperparathyroidism independent of the levels of calcitriol. Nephron **63**:122–123.
66. Fukuda N, Tanaka H, Tominaga Y, Fukagawa M, Kurokawa K, Seino Y 1993 Decreased 1,25-dihydroxyvitamin D_3 receptor density is associated with a more severe form of parathyroid hyperplasia in chronic uremic patients. J Clin Invest **92**:1436–1443.
67. Arnold A, Brown MF, Urena P, Gaz RD, Sarfati E, Drueke TB 1995 Monoclonality of parathyroid tumors in chronic renal failure and in primary parathyroid hyperplasia. J Clin Invest **95**:2047–2053.
68. MacDonald PN, Ritter C, Brown AJ, Slatopolsky E 1994 Retinoic acid suppresses parathyroid hormone (PTH) secretion and preproPTH mRNA levels in bovine parathyroid cell culture. J Clin Invest **93**:725–730.
69. Mackey SL, Heymont JL, Kronenberg HM, Demay MB 1996 Vitamin D binding to the negative human parathyroid hormone vitamin D response element does not require the retinoid X receptor. Mol Endocrinol **10**:298–305.
70. Russell J, Lettieri D, Sherwood LM 1983 Direct regulation by calcium of cytoplasmic messenger ribonucleic acid coding for pre-proparathyroid hormone in isolated bovine parathyroid cells. J Clin Invest **72**:1851–1855.
71. Brookman JJ, Farrow SM, Nicholson L, O'Riordan JL, Hendy GN 1986 Regulation by calcium of parathyroid hormone mRNA in cultured parathyroid tissue. J Bone Miner Res **1**:529–537.
72. Brown AJ, Zhong M, Ritter C, Brown EM, Slatopolsky E 1995 Loss of calcium responsiveness in cultured bovine parathyroid cells is associated with decreased calcium receptor expression. Biochem Biophys Res Commun **212**:861–867.
73. Naveh-Many T, Friedlaender MM, Mayer H, Silver J 1989 Calcium regulates parathyroid hormone messenger ribonucleic acid (mRNA), but not calcitonin mRNA *in vivo* in the rat. Dominant role of 1,25-dihydroxyvitamin D. Endocrinology **125**:275–280.
74. Yamamoto M, Igarashi T, Muramatsu M, Fukagawa M, Motokura T, Ogata E 1989 Hypocalcemia increases and hypercalcemia decreases the steady-state level of parathyroid hormone messenger RNA in the rat. J Clin Invest **83**:1053–1056.
75. Naveh-Many T, Raue F, Grauer A, Silver J 1992 Regulation of calcitonin gene expression by hypocalcemia, hypercalcemia, and vitamin D in the rat. J Bone Miner Res **7**:1233–1237.
76. Okazaki T, Ando K, Igarashi T, Ogata E, Fujita T 1992 Conserved mechanism of negative gene regulation by extracellular calcium. Parathyroid hormone gene versus atrial natriuretic peptide. J Clin Invest **89**:1268-1273.
77. Okazaki T, Chung U, Nishishita T, Ebisu S, Usuda S, Mishiro S, Xanthoudakis S, Igarashi T, Ogata E 1994 A redox factor protein, ref1, is involved in negative gene regulation by extracellular calcium. J Biol Chem **269**:27855–27862.
78. Moallem E, Silver J, Naveh-Many T 1995 Post-transcriptional regulation of PTH gene expression by hypocalcemia due to protein binding to the PTH mRNA 3′UTR. J Bone Miner Res **10**:S142 (abstract)
79. Hawa NS, O'Riordan JL, Farrow SM 1993 Post-transcriptional regulation of bovine parathyroid hormone synthesis. J Mol Endocrinol **10**:43–49.
80. Vadher S, Hawa NS, O'Riordan JLH, Farrow SM 1996 Translation of parathyroid hormone gene expression and RNA : protein interactions. J Bone Miner Res **11**:746–753.
81. Burns K, Duggan B, Atkinson EA, Famulski KS, Nemer M, Bleakley RC, Michalak M 1994 Modulation of gene expression by calreticulin binding to the glucocorticoid receptor. Nature **367**:476–480.
82. Dedhar S, Rennie PS, Shago M, Leung-Hagestein C-Y, Yang H, Hilmus J, Hawley RG, Bruchovsky N, Cheng H, Matusik RJ, Giguere V 1994 Inhibition of nuclear hormone receptor activity by calreticulin. Nature **367**:480–483.
83. Wheeler DG, Horsford J, Michalak M, White JH, Hendy GN 1995 Calreticulin inhibits vitamin D_3 signal transduction. Nucleic Acids Res **23**:3268–3274.
84. Sela A, Silver J, Naveh-Many T 1996 Chronic hypocalcemia increases PTH mRNA despite high $1{,}25(OH)_2D$ levels: Roles

of calreticulin and the vitamin D receptor. J Bone Miner Res **11**:(abstract).
85. Castleman B, Mallory TB 1932 The pathology of the parathyroid gland in hyperparathyroidism. Am J Pathol **11**:1–72.
86. Castleman B, Mallory TB 1937 Parathyroid hyperplasia in chronic renal insufficiency. Am J Pathol **13**:553–574.
87. Shvil Y, Naveh-Many T, Barach P, Silver J 1990 Regulation of parathyroid cell gene expression in experimental uremia. J Am Soc Nephrol **1**:99–104.
88. Brown AJ, Dusso A, Lopez-Hilker S, Lewis-Finch J, Grooms P, Slatopolsky E 1989 1,25-$(OH)_2D$ receptors are decreased in parathyroid glands from chronically uremic dogs. Kidney Int **35**:19–23.
89. Korkor AB 1987 Reduced binding of [^3H]1,25-dihydroxyvitamin D_3 in the parathyroid glands of patients with renal failure. N Engl J Med **316**:1573–1577.
90. Merke J, Hugel U, Zlotkowski A, Szabo A, Bommer J, Mall G, Ritz E 1987 Diminished parathyroid 1,25$(OH)_2D_3$ receptors in experimental uremia. Kidney Int **32**:350–353.
91. Denda M, Finch J, Brown AJ, Nishii Y, Kubodera N, Slatopolsky E 1996 1,25-Dihydroxyvitamin D_3 and 22-oxacalcitriol prevent the decrease in vitamin D receptor content in the parathyroid glands of uremic rats. Kidney Int **50**:34–39.
92. Fukagawa M, Kaname S, Igarashi T, Ogata E, Kurokawa K 1991 Regulation of parathyroid hormone synthesis in chronic renal failure in rats. Kidney Int **39**:874–881.
93. Naveh-Many T, Silver J 1990 Regulation of parathyroid hormone gene expression by hypocalcemia, hypercalcemia, and vitamin D in the rat. J Clin Invest **86**:1313–1319.
94. Naveh-Many T, Rahamimov R, Livni N, Silver J 1995 Parathyroid cell proliferation in normal and chronic renal failure rats: The effects of calcium, phosphate and vitamin D. J Clin Invest **96**:1786–1793.
95. Nygren P, Larsson R, Johansson H, Ljunghall S, Rastad J, Akerstrom G 1988 1,25$(OH)_2D_3$ inhibits hormone secretion and proliferation but not functional dedifferentiation of cultured bovine parathyroid cells. Calcif Tissue Int **43**:213–218.
96. Szabo A, Merke J, Beier E, Mall G, Ritz E 1989 1,25$(OH)_2$ vitamin D_3 inhibits parathyroid cell proliferation in experimental uremia. Kidney Int **35**:1049–1056.
97. Wernerson A, Svensson O, Reinholt FP 1989 Parathyroid cell number and size in hypercalcemic rats: A stereologic study employing modern unbiased estimators. J Bone Miner Res **4**:705–713.
98. Svensson O, Wernerson A, Reinholt FP 1988 Effect of calcium depletion on the rat parathyroids. Bone Miner **3**:259–269.
99. Kilav R, Silver J, Biber J, Murer H, Naveh-Many T 1995 Coordinate regulation of the rat renal parathyroid hormone receptor mRNA and the Na-Pi cotransporter mRNA and protein. Am J Physiol **268**:F1017-F1022.
100. Slatopolsky E, Bricker NS 1973 The role of phosphorus restriction in the prevention of secondary hyperparathyroidism in chronic renal disease. Kidney Int **4**:141–145.
101. Portale AA, Booth BE, Halloran BP, Morris RCJ 1984 Effect of dietary phosphorus on circulating concentrations of 1,25-dihydroxyvitamin D and immunoreactive parathyroid hormone in children with moderate renal insufficiency. J Clin Invest **73**:1580–1589.
102. Lucas PA, Brown RC, Woodhead JS, Coles GA 1986 1,25-Dihydroxycholecalciferol and parathyroid hormone in advanced chronic renal failure: Effects of simultaneous protein and phosphorus restriction. Clin Nephrol **25**:7–10.
103. Lafage MH, Combe C, Fournier A, Aparicio M 1992 Ketodiet, physiological calcium intake and native vitamin D improve renal osteodystrophy. Kidney Int **42**:1217–1225.
104. Combe C, Aparicio M 1994 Phosphorus and protein restriction and parathyroid function in chronic renal failure. Kidney Int **46**:1381–1386.
105. Tanaka Y, DeLuca HF 1973 The control of vitamin D by inorganic phosphorus. Arch Biochem Biophys **154**:566–570.
106. Condamine L, Vztovsnik F, Friedlander G, Menaa C, Garabedian M 1994 Local action of phosphate depletion and insulin-like growth factor 1 on in vitro production of 1,25-dihydroxyvitamin D by cultured mammalian kidney cells. J Clin Invest **94**:1673–1679.
107. Portale AA, Halloran BP, Curtis Morris J 1989 Physiologic regulation of the serum concentration of 1,25-dihydroxyvitamin D by phosphorus in normal men. J Clin Invest **83**:1494–1499.
108. Lopez-Hilker S, Dusso AS, Rapp NS, Martin KJ, Slatopolsky E 1990 Phosphorus restriction reverses hyperparathyroidism in uremia independent of changes in calcium and calcitriol. Am J Physiol **259**:F432-F437.
109. Yi H, Fukagawa M, Yamato H, Kumagai M, Watanabe T, Kurokawa K 1995 Prevention of enhanced parathyroid hormone secretion, synthesis and hyperplasia by mild dietary phosphorus restriction in early chronic renal failure in rats: Possible direct role of phosphorus. Nephron **70**:242–248.
110. Kilav R, Silver J, Naveh-Many T 1995 Parathyroid hormone gene expression in hypophosphatemic rats. J Clin Invest **96**:327–333.
111. Almaden Y, Canalejo A, Hernandez A, Ballesteros E, Garcia-Navarro S, Torres A, Rodriguez M 1996 Direct effect of phosphorus on parathyroid hormone secretion from whole rat parathyroid glands in vitro. J Bone Miner Res **11**:970–976.
112. Almaden Y, Hernandez A, Torregrosa V, Campistol J, Torres A, Rodriguez MS 1995 High phosphorus directly stimulates PTH secretion by human parathyroid tissue. J Am Soc Nephrol **6**:957 (abstract)
113. Slatopolsky E, Finch J, Denda M, Ritter C, Zhong A, Dusso A, MacDonald P, Brown AJ 1996 Phosphate restriction prevents parathyroid cell growth in uremic rats. High phosphate directly stimulates PTH secretion in vitro. J Clin Invest **97**:2534–2540.
114. Nielsen PK, Feldt-Rasmusen U, Olgaard K 1996 A direct effect of phosphate on PTH release from bovine parathyroid tissue slices but not from dispersed parathyroid cells. Nephrol Dial Transplant (in press)
115. Nemeth EF, Scarpa A 1987 Rapid mobilization of cellular Ca^{2+} in bovine parathyroid cells evoked by extracellular divalent cations. Evidence for a cell surface calcium receptor. J Biol Chem **262**:5188–5196.
116. Brown EM 1991 Extracellular Ca^{2+} sensing, regulation of parathyroid cell function, and role of Ca^{2+} and other ions as extracellular (first) messengers. Physiol Rev **71**:371–411.
117. Naveh-Many T, Kilav R, Moallem E, Silver J 1994 Post-transcriptional regulation of the PTH gene by calcium and phosphate in vivo. J Bone Miner Res **9**: S338 (abstract)

CHAPTER 24

Thyroidal C Cells and Medullary Thyroid Carcinoma as Target Tissues for Vitamin D Action

ROBERT F. GAGEL AND SARA PELEG
Section of Endocrinology, University of Texas M.D. Anderson Cancer Center, Houston, Texas

I. INTRODUCTION

The parafollicular or C cells of the thyroid gland are a unique subset of cells within the thyroid gland that have their embryological origin in the neural crest. These cells are located adjacent to the thyroid follicle and represent less than 0.1% of the total cells within the thyroid gland. These cells differ from the more common thyroid follicular cells in several important ways. First, the C cells are derived from neural crest tissue and have many of the characteristics of neuroendocrine cells, including the following: amine uptake and decarboxylation; the presence of secretory granules that are released directly into blood vessels rather than into the thyroid follicle; and the production of several neuroendocrine-type peptides including somatostatin, bombesin, chromogranin A, and its predominant secretory product, calcitonin (CT). The parafollicular or C cells do not concentrate iodine and do not produce thyroglobulin or thyroid hormone. The low abundance of C cells within the mammalian thyroid gland makes it difficult to identify these cells without specific immunohistochemical staining techniques for secretory products such as CT or chromogranin A.

II. ORIGIN OF THYROIDAL C CELLS AND THEIR FUNCTION IN HEALTH AND DISEASE

A. Embryonic Development of Thyroidal C Cells

In a series of elegant experiments, Le Douarin and colleagues [1,2] tracked the migration of quail neural crest-derived cells, with a distinctive nuclear shape, in a chicken model of neural crest development. They showed migration of cells from the neural crest to the ultimobranchial body, a discrete structure in fish and birds composed predominantly of CT-producing cells. The migration is thought to be similar in mammalian species, although the C cells are dispersed throughout the thyroid gland. The regulatory signals that determine the migratory pattern have not been elucidated, although findings suggest a role for the glial cell-derived neurotropic factor (GDNF) signaling system in this pathway. GDNF and its receptor complex, Ret and GDNFR-α, form a signaling system important for migration of neural crest cells to form the neural network responsible for normal gastrointestinal motility. The GDNF signaling system includes several components (Fig. 1), namely, the ligand GDNF [3] and a multicomponent receptor composed of an extracellular protein named GDNF receptor-α (GDNFR-α) [4] and the Ret tyrosine kinase receptor [3,5]. The peptide GDNF is a member of the cysteine knot family of growth factors, which includes transforming growth factor-β (TGFβ), vascular endothelial growth factor (VEGF), several bone matrix proteins, the platelet-derived growth factors, and the pituitary glycoprotein hormones. The expression of this receptor complex in the C cell and evidence that mutant Ret receptor causes transformation of the C cell make it reasonable to hypothesize a role for this signaling system in the migration and differentiation of C cells.

B. C Cells in Normal Adult Thyroid Gland: Morphology and Function

In the normal mammalian thyroid gland the C cells are distributed throughout the thyroid gland, although in humans the highest concentration is located along a central cephalad–caudal axis in each lobe of the thyroid gland at the junction of the upper one-third and lower two-thirds of the thyroid gland. The C cells have morphological features of neuroendocrine cells with secretory granules that empty into adjacent capillaries.

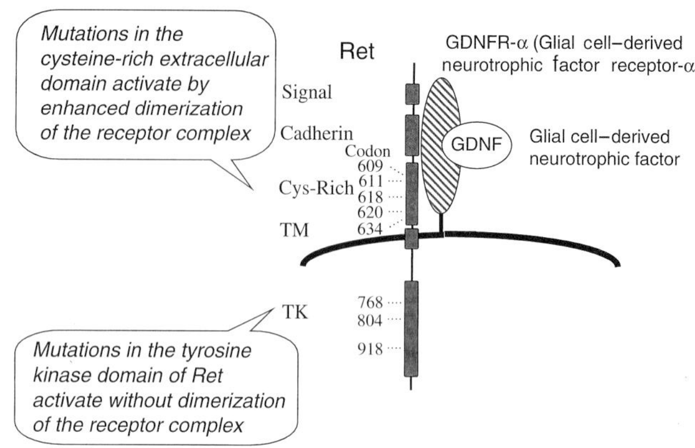

FIGURE 1 The Ret/glial cell-derived neurotropic factor receptor-α (GDNFR-α) signaling system. Glial cell-derived neurotropic factor (GDNF) signals through a novel type of receptor system that combines a classic tyrosine kinase receptor (Ret) and an extracellular protein (GDNFR-α). Mutations in a cysteine-rich (Cys-Rich) portion of the Ret receptor cause multiple endocrine neoplasia type 2A and familial medullary thyroid carcinoma. Mutations of the intracellular tyrosine kinase (TK) cause multiple endocrine neoplasia type 2B. Other abbreviations: TM, transmembrane portion of Ret; Cadherin, a cadherin-like extracellular domain of Ret; and Signal, the signal peptide for Ret.

C. Neoplasia of the C Cell

Medullary thyroid carcinoma (MTC) is a malignant neoplasm of the C cells. Hereditary MTC, an autosomal dominant form of thyroid cancer, accounts for 25% of all medullary thyroid carcinoma. Medullary thyroid carcinoma occurs in combination with tumors of the adrenal medulla and parathyroid tumors in a disorder called multiple endocrine neoplasia type 2 (MEN 2) [6,7] or by itself in familial medullary thyroid carcinoma (FMTC) [8]. Mapping studies initiated in the early 1980s localized the causative gene to centromeric chromosome 10 and led to the identification in 1993 of *c-ret* protooncogene mutations as the cause for this neoplasm [9,10]. The most common mutations, accounting for greater than 85% of all mutations found in the hereditary form of MCT, convert highly conserved cysteines in the extracellular domain of the Ret tyrosine kinase receptor to another amino acid (Fig. 1). A less common mutation, associated with a rarer form of MEN 2 (MEN 2B), affects the intracellular tyrosine kinase portion of the Ret receptor [11]. Several lines of evidence support a causative role for these mutations in the genesis of MCT, including genotype/phenotype correlation [12] and the demonstration that transfection of mutant Ret cDNA into NIH-3T3 cells causes transformation [13,14]. These studies have identified two different mechanisms of activation: dimerization of the receptor in the absence of ligand is found with mutations of the cysteine-rich extracellular domain, and mutations of the tyrosine kinase domain cause activation in the absence of either ligand or dimerization (Fig. 1).

Sporadic MTC accounts for approximately 75% of all medullary thyroid carcinoma. A somatic mutation of codon 918, the same mutation which causes MEN 2B when present as a germ line mutation, is found in approximately 25% of all sporadic MTC (Fig. 1) [15].

III. HORMONAL PRODUCTION BY NORMAL OR TRANSFORMED C CELLS

A. Calcitonin Gene Expression by the C Cell

The CT gene has six exons that encode two peptides, CT and calcitonin gene-related peptide (CGRP) (Fig. 2). The primary RNA transcript of the CT gene is processed to produce an mRNA encoding CT by splicing together exons 1 through 4 of the CT gene and utilizing the polyadenylation site immediately downstream of exon 4 [16]. In the C cell, 98–99% of the primary CT gene transcript is processed in this way. The CT gene is also expressed in a variety of neuronal cells; in these cell types 95% of the primary transcript is processed to exclude exon 4 [17,18]. The resultant mRNA contains sequence from exons 1, 2, 3, 5, and 6 and yields CGRP (Fig. 2). Very little CGRP is produced by the normal C cell, although processing of the primary transcript may be altered significantly in MTC, with up to 50% production of CGRP. The mechanism by which the alternative splice is regulated is not understood, although several regulatory elements that contribute to enhanced recognition and inclusion of exon 4 have been identified [19–23]. The most clearly defined of these is an intron element located approximately 250 nucleotides downstream of the CT exon 4, whose function is essential for CT exon 4 polyadenylation. This novel splicing element contains a 5′ splice site-like sequence [24,25] that interacts with U1 small nuclear ribonucleoproteins (snRNPs), pyrimidine tract binding protein, and accessory splicing factor (ASF) to form a complex which facilitates polyadenylation of CT exon 4 [26].

In thyroidal C cells, the primary transcript is processed to include exon 4 and produce an mRNA encoding calcitonin (CT).
A(n)
Intronic regulatory element necessary for CT-specific splicing
A
A
Introns
1
2
3
4
5
Exons
1
2
3
4
5
6
A(n)
In neuronal cells, the primary transcript is processed to exclude exon 4 and produce an mRNA encoding calcitonin gene–related peptide.

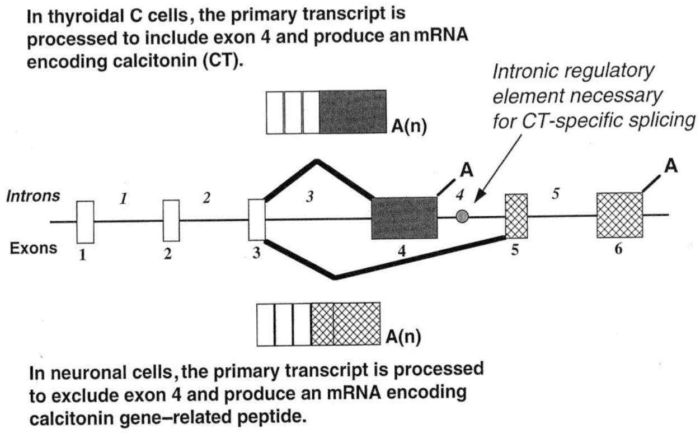

FIGURE 2 Alternative RNA processing of the calcitonin gene. The primary transcript of the calcitonin gene can be processed to produce mRNA encoding either calcitonin (CT) or calcitonin gene-related peptide (CGRP). A regulatory element located downstream of exon 4 contains a pyrimidine tract and pseudo 5′ splice site and is required for inclusion and polyadenylation of exon 4. Mutation(s) of this element results in a skip splice with production of mRNA encoding CGRP. Other elements within or preceding exon 4 have been implicated in regulation of this splice choice, but they have not been characterized. A(n), polyA tail; A, polyadenylation site.

B. Transcriptional Regulation of Calcitonin Gene Expression in the C Cell

Calcitonin gene expression is limited to the thyroidal C cell and certain neuronal cell types. Transcription of this gene is enhanced by glucocorticoids or butyrate or by activation of protein kinase A or C and is inhibited by the active metabolite of vitamin D, 1,25-dihydroxyvitamin D_3 [$1,25(OH)_2D_3$]. Two regions of the CT promoter have been implicated in the transcriptional regulation of the CT gene. The first is located approximately 1000 nucleotides upstream and consists of three E-box-like (CANNTG) sequences, termed the neuroendocrine-specific enhancer (Fig. 3). A second regulatory region located approximately 250 nucleotides upstream of the transcription start site is responsible for cAMP-mediated transcriptional enhancement of the CT gene (Fig. 4).

IV. REGULATION OF CALCITONIN GENE EXPRESSION BY VITAMIN D

Although early studies of the effect of vitamin D on CT production produced conflicting results, more recent studies have defined an inhibitory role of vitamin D on CT gene expression. The studies of Raue and co-workers [27] provided the first clear demonstration that systemic $1,25(OH)_2D_3$ administration causes a reversible decrease in intrathyroidal CT content. Naveh-Many and

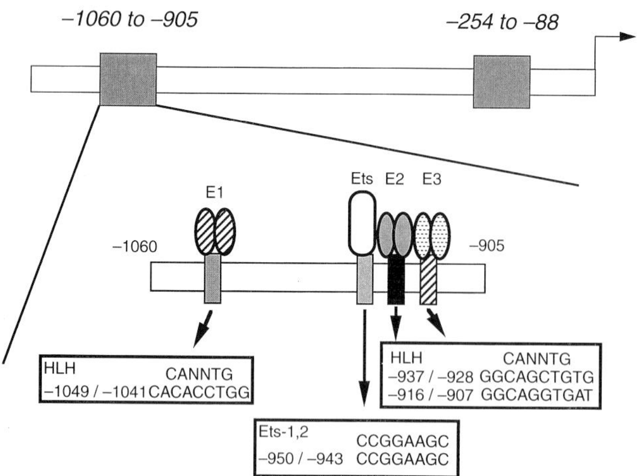

FIGURE 3 Structure of the calcitonin neuroendocrine-specific enhancer. The neuroendocrine-specific enhancer of the CT gene is located between −1060 and −905 nucleotides upstream of the transcription start site. The locations of transcriptionally functional elements of the CT neuroendocrine-specific enhancer are shown. The elements E1, E2, and E3 are all CANNTG motifs, potential binding sites for transcription factors from the helix–loop–helix (HLH) family of transcription factors. The elements E2 and E3 are absolutely essential for basal and cell-specific transcription, whereas the E1 and Ets contribute only to maximal activity of the enhancer.

Silver [28] showed subsequently that doses of $1,25(OH)_2D_3$ which had no effect on the serum calcium concentration lowered CT mRNA to 10% of control within 6 hr of treatment by inhibition of transcription. Although these authors suggest a regulatory loop between CT and $1,25(OH)_2D_3$, the concentration of $1,25(OH)_2D_3$ required for these effects suggests a pharmacological rather than a physiological effect.

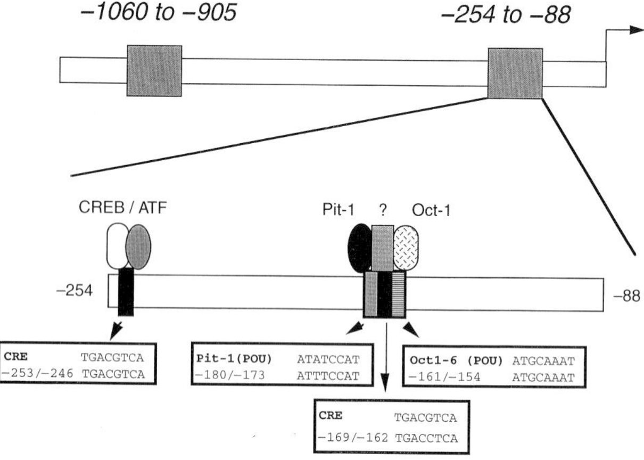

FIGURE 4 Structure of the cAMP responsive enhancer of the calcitonin gene. The cAMP response element is located between −254 and −88 nucelotides upstream from the transcription start site. Several elements are located within this complex transcriptional unit. The CRE, a cAMP responsive element of this enhancer, is a binding site for CRE binding protein (CREB) and ATF. O/CREL/O is a complex element containing a transcriptionally active CRE-like motif that does not bind CREB or ATF. The motifs flanking CREL are homologous to binding sites of the POU/homeobox transcription factors Pit-1 and Oct-1, but only the downstream Oct-1-like site is transcriptionally functional in TT cells. O/CREL/O = Pit-1 ? Oct-1.

The exact mechanism by which $1,25(OH)_2D_3$ suppresses CT gene transcription in normal C cells is not known. However, the cells contain detectable levels of the vitamin D receptor (VDR) [29,30], and the inhibition is induced preferentially by the VDR-activating metabolite $1,25(OH)_2D_3$ and not by other metabolites of the vitamin D endocrine system [29]. These findings suggest that repression of the gene is mediated by ligand-dependent transcriptional effects of the VDR and has enabled mapping of the DNA sequences responsible for this effect.

A. Development of a Model System in Which to Study the Effect of $1,25(OH)_2D_3$ on CT Transcription

To understand the molecular mechanism(s) that leads to repression of CT gene transcription, it was necessary to identify a cell culture system in which calcitonin gene expression was down-regulated by the sterol. Several transformed rat or human C-cell lines were examined; one human cell line (TT), with a likely hereditary defect because of the presence of a codon 634 cysteine to tryptophan mutation [31,32] (Fig. 1), expressed the calcitonin gene at a high level and was regulated by vitamin D [30].

The TT cell line produces several polypeptide hormones including CT, CGRP, somatostatin, and parathyroid hormone-related protein [33–40]. In the TT cell line, vitamin D down-regulates transcription of the somatostatin and CT genes [30,40,41] and reverses enhanced expression of the CT gene produced by glucocorticoids [42]. Transcription of the CT gene by the TT cell line is enhanced by glucocorticoids [33,43,44], butyrate, or substances that activate protein kinase A or C [45–51]. Although the regulation of CT gene expression in this cell line mimics that observed in the normal C cells, there exists the possibility that the transformed phenotype affects the response to sterols.

V. MECHANISM FOR DOWN-REGULATION OF CALCITONIN GENE TRANSCRIPTION BY VITAMIN D

The calcitonin gene is expressed constitutively in the C cell. As a first step toward an understanding of transcriptional regulation, it was important to analyze the mechanisms that regulate constitutive expression of the gene. Analysis of the CT promoter from either human or rat species in MTC cell lines revealed almost identical transcriptional regulatory elements [49,52–54]. In this section we describe the results obtained by analysis of the promoter of the human gene in the TT cell line.

A. Cloning and Characterization of Cis-Acting Elements in the CT Promoter

The calcitonin I (CALC I) gene is expressed almost exclusively in thyroidal C cells, in a scattered population of neuroendocrine cells in the lung, prostate, and pituitary, and in selected neuronal cells. Cloning of the CT promoter [55] and detailed analysis of its functional transcription elements revealed that it contained a constitutive neuroendocrine-specific component [54] and another second messenger-regulated component [51].

1. The Neuroendocrine-Specific Enhancer

The neuroendocrine specific enhancer of the CT/CGRP gene has been studied extensively in rat and human species [52–54]. It is a complex enhancer that contains a cluster of elements located approximately 1 kb upstream from the transcription start site (Fig. 3). The full-length enhancer functions constitutively and in a cell-restricted manner; it is active only in CT-expressing neuroendocrine cell lines derived from MTC or small cell lung carcinoma, or in a CT-negative neuronal cell line. Detailed dissection of the enhancer by deletions and point mutations showed that it contained several elements: three CANNTG motifs (E boxes) and an Ets-like motif (Fig. 3). The E boxes, termed E1, E2, and E3, resemble binding sites for helix–loop–helix proteins [53,54], transcription factors that regulate cell lineage and differentiation in a variety of cell types. The Ets-like responsive element (CCGGAAGC) is a potential binding site for one or more of the cellular homologs of the viral E26 oncogene. The Ets element of the CT promoter can function autonomously as a constitutive enhancer, but it does not function in a cell-specific manner. On the other hand, E2, one of the three E boxes, functions autonomously as a neuroendocrine enhancer element [54]. There has been an attempt to link the activity of this element to the mammalian achete-scute homolog (MASH-1, hASH-1), a helix–loop–helix transcription factor that is important for development of a fraction of neuronal and neuroendocrine cellular components. The MASH-1 gene product is expressed in TT cells; however, convincing evidence for its role in the neuroendocrine-specific transactivation of the CT/CGRP gene is not available [56]. Furthermore, disruption of this gene in animal models does not affect the expression of CGRP in neuronal cells [57].

2. The cAMP-Induced Enhancer

Transcription of the CT gene, like many others, is regulated by second messengers such as diacylglycerol or cAMP. A search for a promoter element that responds to these signals in the calcitonin gene identified a complex cAMP responsive enhancer located proximal to the transcription start site (Fig. 4). It contains a perfect cAMP response element or CRE that functions as a binding site for cAMP response element binding protein (CREB)/ATF family members. A second element that responds to cAMP contains three overlapping motifs: a CRE-like motif flanked by two homeo/POU domain-like binding sites (O/CREL/O) (Fig. 4).

Mutation or deletion of the ATF/CREB binding site reduces cAMP-induced transcription in transformed C cells but does not abolish it, indicating that the composite cAMP responsive element can function without the CREB/ATF element (Fig. 5). Mutation of the composite octamer/CRE motif, however, seems to diminish the activity of the CREB/ATF binding site [51]. The function of these cAMP-responsive elements is cell specific to some extent: a construct containing the CREB binding site is functional in a cAMP-independent manner in HeLa cells, whereas there is little expression in a CT-negative MTC cell line [51]. The composite O/CREL/O motif is functional only in the CT-positive MTC cell lines [51].

The neuroendocrine-specific enhancer is responsible for basal transcription. Basal expression of a construct

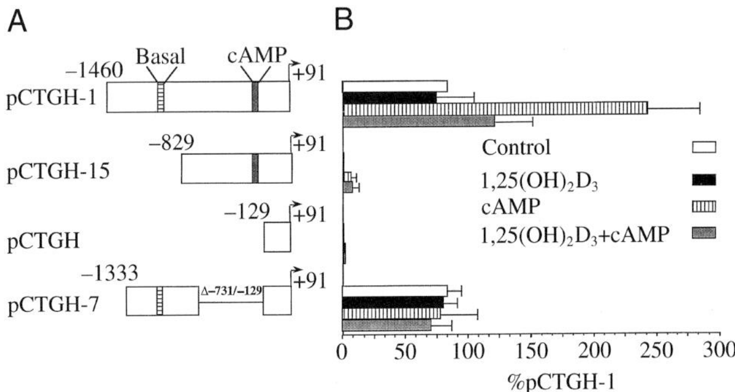

FIGURE 5 Requirement for both the cAMP-induced and neuroendocrine-specific elements for $1,25(OH)_2D_3$ inhibition of CT gene transcription. Constructs containing either the basal enhancer, the cAMP-induced enhancer, or both (A) were transfected into the human medullary thyroid carcinoma TT cell line. Transiently transfected cells were treated with cAMP, $1,25(OH)_2D_3$, or both. Reporter gene expression (growth hormone) was measured as an indicator of transcriptional activity of these constructs (B). Note that basal transcription is similar with or without the cAMP-induced enhancer, but the cAMP-induced transcription is significantly enhanced by the presence of the basal enhancer. Modified from Peleg *et al.* [29].

containing this sequence and the cAMP-induced enhancer causes expression no greater than that observed in a construct containing the neuroendocrine-specific enhancer by itself (Fig. 5). However, the magnitude of the response stimulated by cAMP is greatly enhanced by the combination of enhancers when compared with either enhancer alone (Fig. 5). There is little effect of cAMP on the neuroendocrine-specific enhancer, whereas cAMP analogs stimulate a severalfold response in cells transfected with a construct containing the cAMP-induced enhancer, but to a level that is far less than that observed basally with the neuroendocrine-specific enhancer and less than 5% of the level observed when cAMP analog is added to cells expressing both enhancers (Fig. 5). These relationships suggest the two enhancers function synergistically, but it is the basal enhancer that modulates the activity of the proximal cAMP-induced enhancer, and not vice versa.

B. A Complex Negative VDRE That Interacts with Both the Neuroendocrine-Specific and cAMP Enhancers

Initial examination of the response of the CT promoter construct to $1,25(OH)_2D_3$ showed that only cAMP-induced transcription was repressed (Fig. 5). These results suggested that the hormone acts directly on the cAMP-induced enhancer. However, when the two enhancers were examined separately, neither the basal enhancer nor the cAMP-induced enhancer were repressed by $1,25(OH)_2D_3$ (Fig. 5). These results suggested the possibility that a vitamin D-induced protein (perhaps the VDR) inhibited the synergism between the two elements rather than acting directly on either. Because such a factor must interact physically with DNA adjacent to one of these enhancers or with one or more of the transacting transcription factors that modulate their activities, it was necessary to map a putative negative response by mutating individual enhancer elements.

In the cAMP response element we have identified three major components: the CRE, the composite octamer/CRE-like motif, and the proximal promoter. Internal deletion that removed only the CRE reduced cAMP-induced transcription by one-half but had no effect on the repression by $1,25(OH)_2D_3$. A replacement of the proximal CT promoter with the thymidine kinase promoter showed that the basal enhancer, cAMP-induced transcription, and $1,25(OH)_2D_3$ repression of the latter were all unaffected [51,54]. These experiments suggested that in the cAMP-induced enhancer the element important for $1,25(OH)_2D_3$ action was the O/CREL/O motif.

A similar mapping study was performed within the upstream neuroendocrine-specific enhancer: individual elements were removed, and the effect of the deletion or mutation on basal transcription, cAMP-induced transcription, and repression by $1,25(OH)_2D_3$ was examined (Fig. 6). These mapping experiments showed that E1 did not act synergistically with cAMP-induced transcription and did not contribute to its repression. The E2 element, which was essential for the neuroendocrine-specific constitutive transcription, also contributed to cAMP-induced transcription but not to $1,25(OH)_2D_3$-induced repression. On the other hand, a nucleic acid sequence containing E3 was essential for basal transcription, cAMP-induced transcription, and repression by $1,25(OH)_2D_3$ (Fig. 6).

C. Proposed Mechanism for Repression of CT Gene Transcription by $1,25(OH)_2D_3$

From the mapping experiments described above, we propose a model that involves at least three separate components: a $1,25(OH)_2D_3$-induced factor (perhaps VDR), the E3 binding protein (probably a helix–loop–helix transcription factor), and the O/CREL/O binding protein(s). In our model, the interaction between E3 and O/CREL/O synergizes cAMP-induced transcription. The $1,25(OH)_2D_3$-induced factor must bind at a DNA site that interferes with the interaction between E3 and O/CREL/O, thus preventing the synergism be-

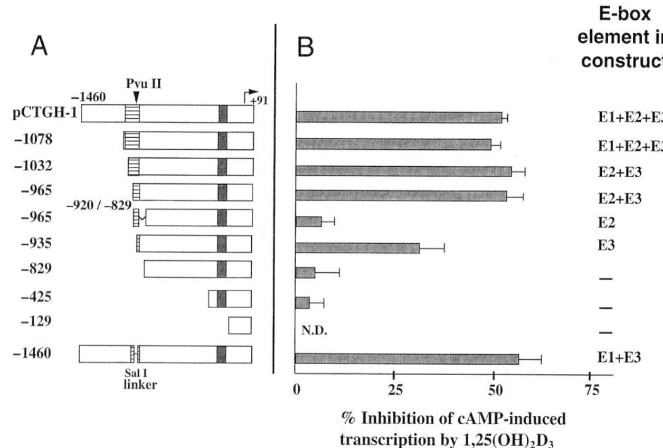

FIGURE 6 Mapping the negative VDRE of the calcitonin gene. To identify DNA sequences that are required for down-regulation of CT gene transcription, studies were focused on the region surrounding the basal enhancer. Constructs containing deletions of individual elements (A) were transfected individually into TT cells, and the response of cells to cAMP or a combination of cAMP and $1,25(OH)_2D_3$ was examined. (B) The E3 element of the neuroendocrine-specific enhancer was required for $1,25(OH)_2D_3$-induced repression. Modified from Peleg *et al.* [29].

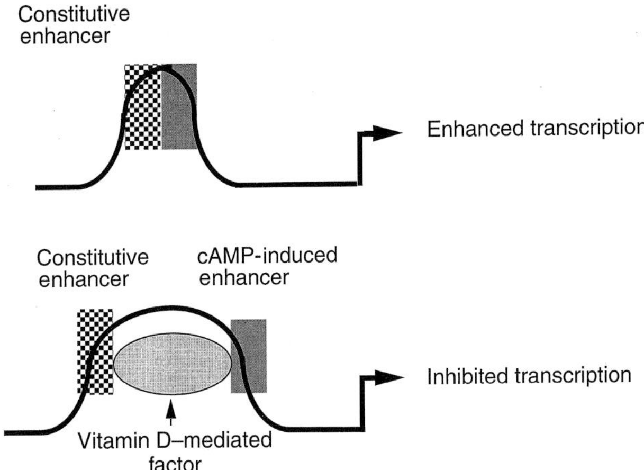

FIGURE 7 Model for $1,25(OH)_2D_3$-induced repression of CT gene transcription. Interaction between two distant enhancers, the neuroendocrine-specific and cAMP-induced enhancers, are necessary for maximal cAMP-induced transcription. A $1,25(OH)_2D_3$-induced factor may interfere with the synergistic interaction by binding directly to a DNA sequence overlapping the E3 motif or by interacting with the upstream and downstream enhancer binding proteins.

tween the two (Fig. 7). In the search for a putative binding site for VDR in the vicinity of O/CREL/O or E3, one sequence was found that exhibits sequence similarity with the functional VDRE motifs of the osteocalcin, osteopontin, and 24-hydroxylase genes (Table I). This sequence overlaps the binding site for the E3 binding protein(s), suggesting that binding of VDR at this site may have a direct effect on E3 function. In our studies, however, we were unable to identify significant binding of VDR at this site [51,54]. It is possible that E3 binding proteins are required for anchorage of VDR into this weak binding site, but proving that hypothesis will require purification and characterization of these

TABLE I Sequence Comparison of Vitamin D Responsive Elements

Promoter[a]	DNA sequence
r24-OH (distal)	GGTTCA GCG GGTGCG
r24-OH (proximal)	AGGTGA GTG AGGGCG
rOC	GGGTGA ATG AGGACA
hOC	GGGTGA ACG GGGGCA
mOP	GGTTCA CGA GGTTCA
Calcitonin (−916/−900)	GCAGGTGA TGG ATGGCA

[a] Shown are VDREs from the rat 24-hydroxylase promoter (r24-OH), the rat and the human osteocalcin promoters (rOC and hOC), and the mouse osteopontin promoter (mOP). The calcitonin sequence implicated in the $1,25(OH)_2D_3$ effect overlaps with the E3 motif (underlined). This sequence resembles the 24-hydroxylase, osteocalcin, and osteopontin gene response elements, known binding sites for VDR–retinoid X receptor (RXR) heterodimers.

proteins. Nevertheless, these studies define a complex and interesting transcriptional unit and a potentially novel mechanism by which $1,25(OH)_2D_3$ down-regulates calcitonin gene transcription.

VI. EFFECTS OF VITAMIN D ON CELL GROWTH

Through their interaction with the VDR and retinoic acid receptor systems, vitamin D and its analogs have been shown to affect differentiation and growth in a variety of cell types (see Chapter 64). There is a limited and conflicting literature regarding effects in the C cell. Studies in the TT cell line have demonstrated either no effect of $1,25(OH)_2D_3$ on cell growth [30] or stimulation of cell growth [58]. In the studies in which $1,25(OH)_2D_3$ stimulated thymidine incorporation and cell growth, c-*myc* antisense DNA oligomers abolished the proliferative effect of $1,25(OH)_2D_3$ but not the effect on inhibition of CT gene expression [59]. These results suggest the potential for an independent effect of $1,25(OH)_2D_3$ in this cell line, although more investigation is needed.

VII. SUMMARY AND FUTURE DIRECTIONS

The primary effect of $1,25(OH)_2D_3$ in the C cell is to inhibit transcription of the calcitonin gene. The available evidence supports a model in which the VDR interacts with a partial VDRE and interferes with the positive function of several transcription factors, one of which is likely to be a CREB variant. Further definition of this model will require the identification of the specific transcription factors that interact with the upstream enhancer to regulate calcitonin gene transcription.

Although these studies point to a clearly defined pathway for vitamin D action in the C cell, it is less clear whether this pathway has regulatory significance in normal physiology, in part because the exact role of calcitonin in normal physiology has not been defined. It is also unclear at this point whether vitamin D analogs will have any role in the treatment of medullary thyroid carcinoma. The studies reported in this chapter describe contradictory effects of $1,25(OH)_2D_3$ on cell growth in human and rat model systems. Further studies will be required to clarify these issues.

References

1. Le Douarin N, Le Lievre C 1970 Demonstration de l'origine neurales des cellules a calcitonine du corps ultimobranchial chez l'embryon de poulet. Compt Rend **270**:2857–2860.

2. Le Douarin N 1982 The neural crest. In: Developmental and Cell Biology Series, Vol. 12. Cambridge Univ. Press, Cambridge, pp. 1–259.
3. Trupp M, Ryden, M, Jornvall H, Funakoshi H, Timmusk T, Arenas E, Ibanez CF 1995 Peripheral expression and biological activities of GDNF, a new neurotrophic factor for avian and mammalian peripheral neurons. J Cell Biol **130**:137–148.
4. Jing S, Wen D, Yu Y, Holst PL, Luo Y, Fang M, Tamir R, Antonio L, Hu Z, Cupples R, Louis JC, Hu S, Altrock BW, Fox GM 1996 GDNF-induced activation of the Ret protein tyrosine kinase is mediated by GDNFR-α, a novel receptor for GDNF. Cell **85**:1113–1124.
5. Durbec P, Marcos-Gutierrez CV, Kilkenny C, Suvanto P, Smith D, Ponder B, Costantini F, Saarma M, Sarlola H, Pachnis V 1996 GDNF signaling through the ret receptor tyrosine kinase. Nature **381**:789–793.
6. Sipple JH 1961 The association of pheochromocytoma with carcinoma of the thyroid gland. Am J Med **31**:163–166.
7. Steiner AL, Goodman AD, Powers SR 1968 Study of a kindred with pheochromocytoma, medullary carcinoma, hyperparathyroidism and Cushing's disease: Multiple endocrine neoplasia, type 2. Medicine **47**:371–409.
8. Farndon JR, Leight GS, Dilley WG, Baylin SB, Smallridge RC, Harrison TS, Wells SA Jr 1986 Familial medullary thyroid carcinoma without associated endocrinopathies: A distinct clinical entity. Br J Surg **73**:278–281.
9. Donis-Keller H, Shenshen D, Chi D, Carlson KM, Toshima K, Lairmore TC, Howe JR, Moley JF, Goodfellow P, Wells SA Jr 1993 Mutations in the RET proto-oncogene are associated with MEN 2A and FMTC. Hum Mol Genet **2**:851–856.
10. Mulligan LM, Kwok JBJ, Healey CS, Elsdon MJ, Eng C, Gardner E, Love DR, Mole SE, Moore JK, Papi L, Ponder MA, Telenius H, Tunnacliffe A, Ponder BAJ 1993 Germline mutations of the RET proto-oncogene in multiple endocrine neoplasia type 2A (MEN 2A). Nature **363**:458–460.
11. Hofstra RMW, Landsvater RM, Ceccherini I, Stulp RP, Stelwagen T, Luo Y, Pasini B, Höppener JWM, van AmStel HKP, Romeo G, Lips CJM, Buys CHCM 1994 A mutation in the RET proto-onocogene associated with multiple endocrine neoplasia type 2B and sporadic medullary thyroid carcinoma. Nature **367**:376.
12. Mulligan LM, Marsh DJ, Robinson BG, Lenoir G, Schuffenecker I, Zedenius J, Lips CJM, Gagel RF, Takai SI, Noll WW, Fink M, Raue F, LaCroix A, Thibodeau SN, Frilling A, Ponder BAJ, Eng C 1995 Genotype–phenotype correlation in multiple endocrine neoplasia type 2: Report of the international RET mutation consortium. J Int Med **238**:343–346.
13. Asai N, Iwashita T, Matsuyama M, Takahashi M 1995 Mechanisms of activation of the ret proto-oncogene by multiple endocrine neoplasia 2A mutations. Mol Cell Biol **15**(3):1613–1619.
14. Santoro M, Carlomagno F, Romano A, Bottaro DP, Dathan NA, Grieco M, Fusco A, Vecchio G, Matoskova B, Kraus MH, Di Fiiore PP 1995 Activation of RET as a dominant transforming gene by germline mutations of MEN 2A and MEN 2B. Science **267**:381–383.
15. Wohllk N, Cote GJ, Bugalho MMJ, Ordonez N, Evans DB, Goepfert H, Khorana S, Schultz PS, Richards CS, Gagel RF 1996 Relevance of RET proto-oncogene mutations in sporadic medullary thyroid carcinoma. J Clin Endocrinol Metab **81**:3740–3745.
16. Amara SG, Jonas V, Rosenfeld MG, Ong ES, Evans RM 1982 Alternative RNA processing in calcitonin gene expression generates mRNAs encoding different polypeptide products. Nature **298**:240–244.
17. Leff SE, Evans RM, Rosenfeld MG 1987 Splice commitment dictates neuron-specific alternative RNA processing in calcitonin/CGRP gene expression. Cell **48**:517–524.
18. Crenshaw EB, Russo AF, Swanson LW, Rosenfeld MG 1987 Neuron-specific alternative RNA processing in transgenic mice expressing a metallothionein–calcitonin fusion gene. Cell **49**:389–398.
19. Yeakley JM, Hedjran F, Morfin JP, Merillat N, Rosenfeld MG, Emeson RB 1993 Control of calcitonin/calcitonin gene-related peptide pre-mRNA processing by constitutive intron and exon elements. Mol Cell Biol **13**:5999–6011.
20. Emeson RB, Hedjran F, Yeakley JM, Guise JW, Rosenfeld MG 1989 Alternative production of calcitonin and CGRP mRNA is regulated at the calcitonin-specific splice acceptor. Nature **341**:76–80.
21. Cote GJ, Stolow DT, Peleg S, Berget SM, Gagel RF 1992 Identification of exon sequences and an exon binding protein involved in alternative RNA splicing of calcitonin/CGRP. Nucleic Acids Res **20**:2361–2366.
22. Adema GJ, Bovenberg RA, Jansz HS, Baas PD 1988 Unusual branch point selection involved in splicing of the alternatively processed calcitonin/CGRP-I pre-mRNA. Nucleic Acids Res **16**:9513–9526.
23. van Oers CC, Adema GJ, Zandberg H, Moen TC, Baas PD 1994 Two different sequence elements within exon 4 are necessary for calcitonin-specific splicing of the human calcitonin/calcitonin gene-related peptide I pre-mRNA. Mol Cell Biol **14**:951–960.
24. Lou H, Yang Y, Cote GJ, Berget SM, Gagel RF 1995 An intron enhancer containing a 5′ splice site sequence in the human calcitonin/calcitonin gene-related peptide gene. Mol Cell Biol **15**:7135–7142.
25. Lou H, Cote GJ, Gagel RF 1994 The calcitonin exon and its flanking intronic sequences are sufficient for the regulation of human calcitonin/calcitonin gene-related peptide alternative RNA splicing. Mol Endocrinol **8**:1618–1626.
26. Lou H, Gagel RF, Berget SM 1996 An intron enhancer recognized by splicing factors activates polyadenylation. Genes Dev **10**:208–219.
27. Raue F, Deutschle I, Küntzel C, Ziegler R 1984 Reversible diminished calcitonin secretion in the rat during chronic hypercalcemia. Endocrinology **115**:2362–2367.
28. Naveh-Many T, Silver J 1988 Regulation of calcitonin gene transcription by vitamin D metabolites in vivo in the rat. J Clin Invest **81**:270–273.
29. Peleg S, Abruzzese RV, Cooper CW, Gagel RF 1993 Down-regulation of calcitonin gene transcription by vitamin D requires two widely separated enhancer sequences. Mol Endocrinol **7**:999–1008.
30. Cote GJ, Rogers DG, Huang ES, Gagel RF 1987 The effect of 1,25-dihydroxyvitamin D_3 treatment on calcitonin and calcitonin gene-related peptide mRNA levels in cultured human thyroid C-cells. Biochem Biophys Res Commun **149**:239–243.
31. Cooley LD, Elder FB, Knuth A, Gagel RF 1995 Cytogenetic characterization of three human and three rat medullary thyroid carcinoma cell lines. Cancer Genet Cytogenet **80**:138–149.
32. Carlomagno F, Salvatore D, Santoro M, de Franciscis V, Quadro L, Panariello L, Colantuoni V, Fusco A 1995 Point mutation of the RET proto-oncogene in the TT human medullary thyroid carcinoma cell line. Biochem Biophys Res Commun **207**:1022–1028.
33. Cote GJ, Gagel RF 1986 Dexamethasone differentially affects the levels of calcitonin and calcitonin gene-related peptide mRNAs expressed in a human medullary thyroid carcinoma cell line. J Biol Chem **261**:15524–15528.
34. Cote GJ, Palmer WN, Leonhart K, Leong SS, Gagel RF 1986

The regulation of somatostatin production in human medullary thyroid carcinoma cells by dexamethasone. J Biol Chem **261**:12930–12935.

35. Cote GJ, Gould JA, Huang SC, Gagel RF 1987 Studies of short-term secretion of peptides produced by alternative RNA processing. Mol Cell Endocrinol **53**:211–219.
36. Gagel RF, Palmer WN, Leonhart K, Chan L, Leong SS 1986 Somatostatin production by a human medullary thyroid carcinoma cell line. Endocrinology **118**:1643–1651.
37. Berger CL, de Bustros A, Roos BA, Leong SS, Mendelsohn G, Gesell MS, Baylin SB 1984 Human medullary thyroid carcinoma in culture provides a model relating growth dynamics, endocrine cell differentiation, and tumor progression. J Clin Endocrinol Metab **59**:338–343.
38. Nelkin BD, Chen KY, de Bustros AC, Roos BA, Baylin SB 1989 Changes in calcitonin gene RNA processing during growth of a human medullary thyroid carcinoma cell line. Cancer Res **49**:6949–6952.
39. Leong SS, Horoszewicz JS, Shimaoka K, Friedman M, Kawinski E, Song MJ, Ziegel R, Chu TM, Baylin S, Mirand EA 1981 A new cell line for the study of human medullary thyroid carcinoma. In: Advances in Thyroid Neoplasia. Itaha, Rome, Italy, pp. 95–108.
40. Ikeda K, Lu C, Weir EC, Mangin M, Broadus AE 1989 Transcriptional regulation of the parathyroid hormone-related peptide gene by glucocorticoids and vitamin D in a human C-cell line. J Biol Chem **264**:15743–15746.
41. Rogers DG, Cote GJ, Huang ES, Gagel RF 1989 1,25-Dihydroxyvitamin D_3 silences 3′,5′-cyclic adenosine monophosphate enhancement of somatostatin gene transcription in human thyroid C cells. Mol Endocrinol **3**:547–551.
42. Lazaretti CM, Grauer A, Raue F, Ziegler R 1990 1,25-Dihydroxyvitamin D_3 suppresses dexamethasone effects on calcitonin secretion. Mol Cell Endocrinol **71**:13–18.
43. Russo AF, Nelson C, Roos BA, Rosenfeld MG 1988 Differential regulation of the coexpressed calcitonin/alpha-CGRP and beta-CGRP neuroendocrine genes. J Biol Chem **263**:5–8.
44. Muszynski M, Birnbaum RS, Roos BA 1983 Glucocorticoids stimulate the production of preprocalcitonin-derived secretory peptides by a rat medullary thyroid carcinoma cell line. J Biol Chem **258**:11678–11683.
45. Nakagawa T, Nelkin BD, Baylin SB, de Bustros A 1988 Transcriptional and posttranscriptional modulation of calcitonin gene expression by sodium *n*-butyrate in cultured human medullary thyroid carcinoma. Cancer Res **48**:2096–2100.
46. de Bustros A, Baylin SB, Berger CL, Roos BA, Leong SS, Nelkin BD 1985 Phorbol esters increase calcitonin gene transcription and decrease c-*myc* mRNA levels in cultured human medullary thyroid carcinoma. J Biol Chem **260**:98–104.
47. Nelkin BD, de Bustros AC, Mabry M, Baylin SB 1989 The molecular biology of medullary thyroid carcinoma. A model for cancer development and progression. JAMA **261**:3130–3135.
48. Nelkin BD, Borges M, Mabry M, Baylin SB 1990 Transcription factor levels in medullary thyroid carcinoma cells differentiated by Harvey *ras* oncogene: c-*jun* is increased. Biochem Biophys Res Commun **170**:140–146.
49. Peleg S, Cote GJ, Abruzzese RV, Gagel RF 1989 Transcription regulation of the human calcitonin gene: A progress report. Henry Ford Hosp Med J **37**:194–197.
50. Nelkin BD, Rosenfeld KI, de Bustros A, Leong SS, Roos BA, Baylin SB 1984 Structure and expression of a gene encoding human calcitonin and calcitonin gene related peptide. Biochem Biophys Res Commun **123**:648–655.
51. Monla YT, Peleg S, Gagel RF 1995 Cell type-specific regulation of transcription by cyclic adenosine 3′,5′-monophosphate-responsive elements within the calcitonin promoter. Mol Endocrinol **9**:784–793.
52. de Bustros A, Lee RY, Compton D, Tsong TY, Baylin SB, Nelkin BD 1990 Differential utilization of calcitonin gene regulatory DNA sequences in cultured lines of medullary thyroid carcinoma and small-cell lung carcinoma. Mol Cell Biol **10**:1773–1778.
53. Fredman LS, Leff SE, Klein ES, Crenshaw III EB, Yeakley J, Rosenfeld MG 1990 A tissue specific enhancer in the rat calcitonin/CGRP gene is active in both neural and endocrine cell types. Mol Endocrinol **4**:497–504.
54. Peleg S, Abruzzese RV, Cote GJ, Gagel RF 1990 Transcription of the human calcitonin gene is mediated by a C cell-specific enhancer containing E-box-like elements. Mol Endocrinol **4**:1750–1757.
55. Cote GJ, Abruzzese RV, Lips CJM, Gagel RF 1990 Transfection of calcitonin gene regulatory elements into a cell culture model of the C-cell. J Bone Miner Res **5**:165–171.
56. Ball DW, Azzoli CG, Baylin SB, Chi D, Dou S, Donis-Keller H, Cumaraswamy A, Borges M, Nelkin BD 1993 Identification of a human achaete-scute homolog highly expressed in neuroendocrine tumors. Proc Natl Acad Sci USA **90**:5648–5652.
57. Guillemot F 1995 Analysis of the role of basic helix–loop–helix transcription factors in the development of neural lineages in the mouse. Biol Cell **84**:3–6.
58. Lazaretti-Castro M, Grauer A, Baier R, Vieira JG, Ziegler R, Raue F 1995 1,25-Dihydroxyvitamin D_3 stimulates growth and inhibits calcitonin secretion in a human C cell carcinoma cell line. Braz J Med Biol Res **28**:1013–1018.
59. Grauer A, Baier R, Ziegler R, Raue F 1995 Crucial role of c-*myc* in $1,25(OH)_2D_3$ control of C-cell-carcinoma proliferation. Biochem Biophys Res Commun **213**:922–927.

CHAPTER 25

Vitamin D and Skin

DANIEL D. BIKLE Endocrine Research Unit, VA Medical Center, UCSF Medical Services, San Francisco, California

I. INTRODUCTION

It has become clear that 1,25-dihydroxyvitamin D [$1,25(OH)_2D$] and possibly other vitamin D metabolites have functions that extend beyond those of regulating bone mineralization and intestinal calcium transport. The skin is one such tissue where such a broader role is being intensively explored. Besides producing vitamin D, epidermal cells (keratinocytes) make $1,25(OH)_2D$ [1], contain $1,25(OH)_2D$ receptors (VDR) [2–4], and respond to $1,25(OH)_2D$ with changes in proliferation and differentiation [3,5,6]. Calcium is an important modulator of these pathways. Calcium decreases $1,25(OH)_2D$ production and regulates the effects of $1,25(OH)_2D$ on proliferation and differentiation [7]. Calcium is itself an important regulator of keratinocyte proliferation and differentiation [8,9], effects which are in turn modulated by $1,25(OH)_2D$. Exploring the production and action of D and its metabolites in the epidermis should not only increase our understanding of the mechanisms by which this hormone family regulates keratinocyte function but also lead to advances in our ability to treat diseases of disordered epidermal differentiation including psoriasis and squamous cell carcinoma. The use of calcipotriol for the treatment of psoriasis is the first example of this therapeutic potential.

II. CUTANEOUS PRODUCTION OF $1,25(OH)_2D$

A. Metabolism of 25OHD and Identification of the Products

Keratinocytes are not only capable of producing vitamin D_3 from endogenous sources of 7-dehydrocholesterol (7-DHC) in a regulated fashion (Chapter 3), they are also capable of producing a variety of vitamin D metabolites including $1,25(OH)_2D$, $24,25(OH)_2D$, and $1,24,25(OH)_3D$ [1,10,11] from exogenous sources of 25OHD (Fig. 1). Production of 25OHD from vitamin D_3 does not seem to take place in keratinocytes [12], so the complete pathway from 7-DHC to $1,25(OH)_2D$ is not found in this or any other cell. The processes by which $1,25(OH)_2D$ is produced and catabolized are tightly regulated and coupled to the differentiation of these cells.

Extrarenal production of $1,25(OH)_2D$ has been clearly demonstrated in both anephric humans [13,14] and pigs [15], although the tissue(s) responsible for the circulating levels of $1,25(OH)_2D$ in anephric animals has not been established (Chapter 55). The epidermis is likely to contribute as human keratinocytes rapidly and extensively convert 25OHD to $1,25(OH)_2D$. Peak levels of $1,25(OH)_2D$ are reached in the cell within

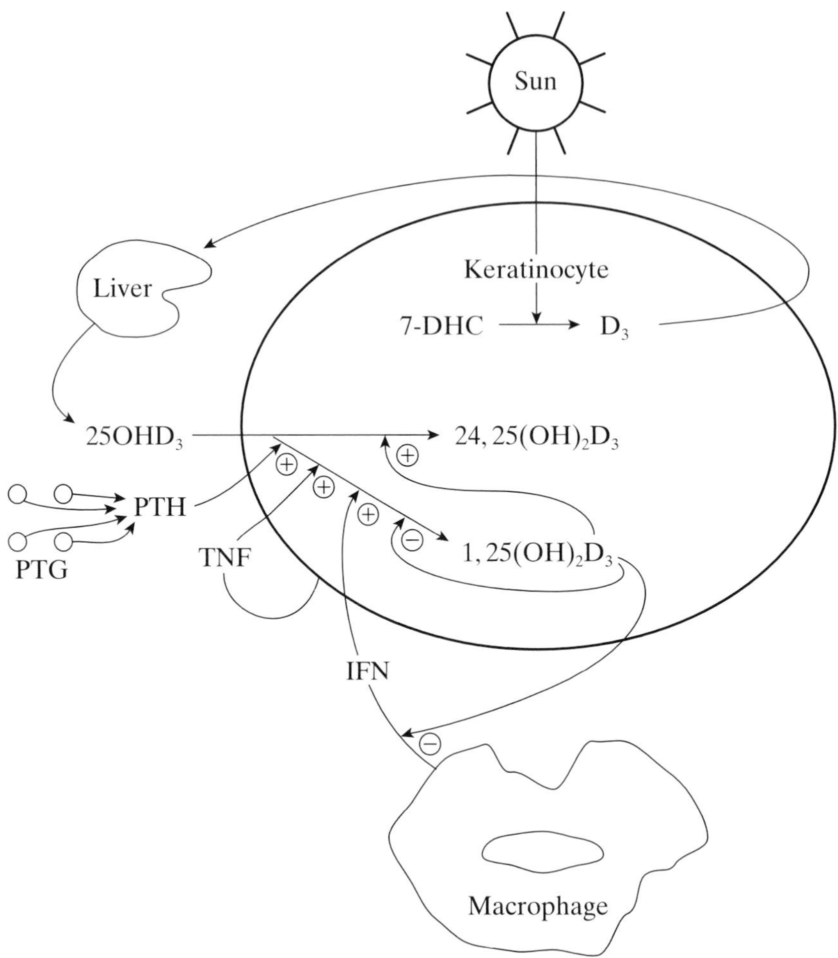

FIGURE 1 Regulation of $1,25(OH)_2D$ production in the keratinocyte. 7-Dehydrocholesterol (7-DHC) is converted to vitamin D_3 by a photochemical reaction. The vitamin D_3 produced is transported out of the keratinocyte to the liver, where it is converted to $25OHD_3$. $25OHD_3$ then enters the keratinocyte, where it is metabolized either to $24,25(OH)_2D_3$ or to $1,25(OH)_2D_3$. Parathyroid hormone (PTH) secreted by the parathyroid gland (PTG) stimulates the production of $1,25(OH)_2D_3$, as does tumor necrosis factor-α (TNF) secreted by keratinocytes and interferon-γ (IFN) secreted by macrophages. $1,25(OH)_2D_3$ inhibits its own production, promotes its metabolism by inducing the 24-hydroxylase [also responsible for $24,25(OH)_2D_3$ production], and decreases IFN secretion by macrophages.

1 hr after adding 25OHD. By 1 hr $1,25(OH)_2D$ is the main metabolite observed; however, other metabolites appear with continued incubation, many of which represent degradation products of $1,25(OH)_2D$ [10]. The apparent K_m for the enzyme (25OHD 1α-hydroxylase) metabolizing 25OHD to $1,25(OH)_2D$ is estimated to be 5×10^{-8} M, a value lower than that estimated for the kidney. The production of $1,25(OH)_2D$ by isolated keratinocytes in culture has been confirmed using intact pig skins perfused with 25OHD [16]. However, when renal production of $1,25(OH)_2D$ is normal the circulating levels of $1,25(OH)_2D$ are sufficient to limit the contribution from epidermal production.

B. Hormonal Regulation

Both the formation and catabolism of $1,25(OH)_2D$ are under hormonal control. Parathyroid hormone (PTH) (optimal concentration 20 ng/ml) exerts a modest stimulation of $1,25(OH)_2D$ production, whereas the phosphodiesterase inhibitor isobutylmethylxanthine (IBMX) inhibits its degradation [10]. In combination, PTH and IBMX markedly increase the amount of $1,25(OH)_2D$ that accumulates within the keratinocyte following the addition of 25OHD. These effects are not reproduced by cAMP or its membrane-permeable derivatives, suggesting that the actions of PTH and

IBMX may be operating through a mechanism independent of cAMP. The effects of PTH and IBMX are maximal after a 4-hr incubation of cells with these agents before adding 25OHD; that is, the effects are not immediate. In renal cells PTH directs a more acute stimulation of $1,25(OH)_2D$ production [17], and cAMP appears to play a second messenger role [18]. Thus, the regulation of $1,25(OH)_2D$ production in keratinocytes by PTH and cAMP differs from their regulation of $1,25(OH)_2D$ production in renal cells. The mechanism by which PTH stimulates $1,25(OH)_2D$ production in keratinocytes is unclear; PTH receptors have been difficult to detect in normal keratinocytes, and PTH fails to stimulate adenylate cyclase activity in these cells.

The hormone $1,25(OH)_2D$ negatively regulates its own levels within the cell. This negative feedback loop is similar to that observed in the kidney, but it differs from that seen in the macrophage, which lacks this feedback loop. Incubation of keratinocytes with exogenous $1,25(OH)_2D$ inhibits $1,25(OH)_2D$ production and stimulates $1,25(OH)_2D$ catabolism in part through induction of the enzyme 25OHD 24-hydroxylase that converts 25OHD and $1,25(OH)_2D$ to $24,25(OH)_2D$ and $1,24,25(OH)_3D$, respectively. $1,25(OH)_2D$ actually induces a number of enzymatic reactions involved with 25OHD and $1,25(OH)_2D$ catabolism, but most of these have not been unequivocally identified. Although inhibition of the 1-hydroxylase occurs within 4 hr of the addition of $1,25(OH)_2D$ (IC_{50} 10^{-11} *M*), induction of the 24-hydroxylase requires more time (within 16 hr) and higher concentrations of $1,25(OH)_2D$ (EC_{50} 2.3×10^{-11}) [10,19]. The concentrations required to regulate 25OHD metabolism in keratinocytes are free concentrations and are independent of vitamin D binding proteins (DBPs). Addition of serum (which contains DBP) or albumin reduces the free fraction of $1,25(OH)_2D$ and increases the apparent EC_{50} for total $1,25(OH)_2D$. However, using direct measurements of the free fraction of $1,25(OH)_2D$ [19], we showed that the EC_{50} for the free concentration was not altered by serum or albumin.

An important difference in the regulation of 25OHD metabolism by $1,25(OH)_2D$ between keratinocytes and renal cells is that the concentration of $1,25(OH)_2D$ required to inhibit the 1-hydroxylase and induce the 24-hydroxylase in renal cells appears to be several orders of magnitude greater than that required for comparable effects in keratinocytes [10,20–22]. Thus, $1,25(OH)_2D$ production by keratinocytes is exquisitely sensitive to exogenous $1,25(OH)_2D$. This difference in sensitivity to feedback inhibition by $1,25(OH)_2D$ between keratinocytes and renal cells may account for the observation that following acute nephrectomy extrarenal production of $1,25(OH)_2D$ is very low [23,24]; only with time after renal production has ceased does extrarenal production emerge. A similar observation was made in pig skins perfused with 25OHD: the amount of $1,25(OH)_2D$ produced was initially low but increased after 4–8 hr of perfusion [16].

C. Effects of Differentiation

In Section III.D we review the role of $1,25(OH)_2D$ in promoting differentiation of the keratinocyte. In this section, the ability of differentiation to regulate $1,25(OH)_2D$ production is considered. In the presence of adequate levels of calcium, keratinocytes progress from rapidly proliferating cells to cells capable of making cornified envelopes, one of the most distinctive features of epidermal terminal differentiation. The cornified envelope is formed by the cross-linking of precursor molecules such as involucrin and loricrin into an insoluble, durable sheet by the membrane-bound enzyme transglutaminase. As the cells differentiate in culture there is a successive increase in involucrin, transglutaminase, and cornified envelope formation [4,25]. 25OHD 1-hydroxylase and 24-hydroxylase also change with differentiation [4] (Fig. 2). Preceding the rise in transglutaminase and involucrin is a rise in 25OHD 1-hydroxylase activity; the 1-hydroxylase activity, transglutaminase activity, and involucrin then fall as cornified envelopes appear; the appearance of cornified envelopes and the fall in 1-hydroxylase activity coincides with a rise in 24-hydroxylase activity [4]. Growing the cells in 0.1 m*M* calcium, which retards differentiation [25], permits the cells to maintain higher 1-hydroxylase activity than when they are grown in 1.2 m*M* calcium [7], although acute changes in calcium have little effect on $1,25(OH)_2D$ production [10].

D. Regulation by Cytokines

Both tumor necrosis factor-α (TNFα) and interferon-γ (IFN-γ) bind to and promote the differentiation of keratinocytes [26,27]. Both cytokines regulate $1,25(OH)_2D$ production by these cells in a manner consistent with their effects on differentiation [7,28]. Unlike PTH and $1,25(OH)_2D$, these cytokines must be incubated with the keratinocytes for at least 1 day (not hours) before their effects on $1,25(OH)_2D$ production are observed. These cells are exquisitely sensitive to IFN-γ, with maximal stimulation of $1,25(OH)_2D$ production at concentrations less than 10 p*M*. Higher con-

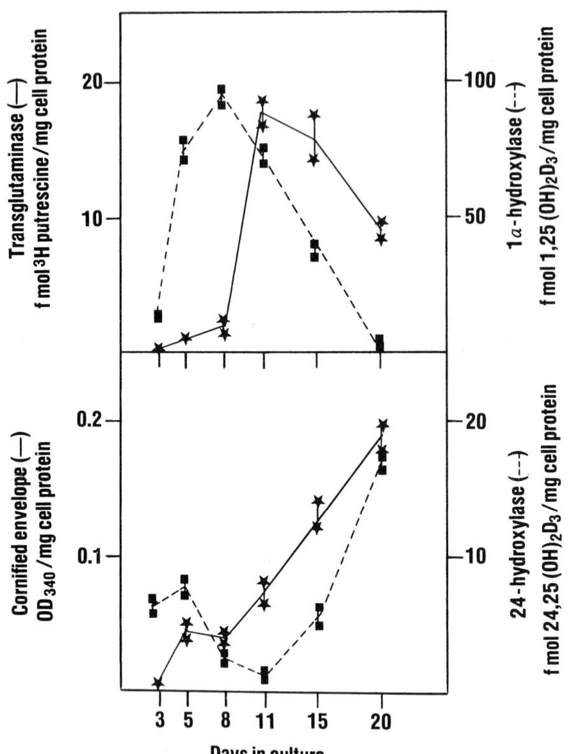

FIGURE 2 Change in 1- and 24-hydroxylase activities in comparison to transglutaminase activity and cornified envelope formation as keratinocytes grow and differentiate in culture. The cells were plated on day 1, and the various measurements were made on the days indicated. Production of $1{,}25(OH)_2D$ rises ahead of transglutaminase activity as the cells approach confluence, then falls along with transglutaminase activity as cornified envelope formation and $24{,}25(OH)_2D$ production ensue. Modified from Pillai *et al.* [4] with permission from the J Biol Chem.

centrations are inhibitory, but such concentrations also profoundly inhibit the proliferation of these cells and limit their ability to differentiate. Keratinocytes grown in 0.1 m*M* calcium are more sensitive to IFN-γ than cells grown in 1.2 m*M* calcium [7]. Serum markedly reduces the potency of IFN-γ in this system, for reasons which are obscure. TNFα stimulates $1{,}25(OH)_2D$ production and transglutaminase activity in preconfluent cells [28], and it can reverse the inhibition seen with the higher concentrations of IFN-γ. The effects of TNFα and IFN-γ are not additive at the lower and stimulatory concentrations of IFN-γ. When TNFα is added after the cells have reached confluence, a time after which $1{,}25(OH)_2D$ production (and transglutaminase activity) has peaked, TNFα inhibits $1{,}25(OH)_2D$ production (and transglutaminase activity) even as it stimulates cornified envelope formation. Although IFN-γ is not made in keratinocytes, TNFα is, and its production is stimulated by ultraviolet light [29] and barrier disruption [30]. Thus, environmental perturbations could enhance $1{,}25(OH)_2D$ production in the skin, leading to accelerated repair.

E. $1{,}25(OH)_2D$ Production by Transformed Keratinocytes

Keratinocytes from squamous cell carcinomas (SCC) do not differentiate normally in response to calcium [31] or $1{,}25(OH)_2D$ [32] despite having genes for the differentiation markers which can be induced by serum [33]. Nevertheless, these cells produce $1{,}25(OH)_2D$ [and $24{,}25(OH)_2D$], and in some cases the rates of production are comparable to those of normal keratinocytes [32]. Furthermore, the SCC lines respond to exogenous $1{,}25(OH)_2D$ with a reduction in $1{,}25(OH)_2D$ production and an increase in $24{,}25(OH)_2D$ production, although in some cases the sensitivity of the SCC line to $1{,}25(OH)_2D$ is less than normal [32]. The levels of the VDR mRNA and protein in SCC are comparable to those in normal keratinocytes [33], suggesting that the reason why $1{,}25(OH)_2D$ can regulate 25OHD metabolism but not differentiation in SCC lies in other transcription factors required for calcium and $1{,}25(OH)_2D$ regulation of the differentiation pathway.

F. Clinical Implications

The finding of $1{,}25(OH)_2D$ production by keratinocytes indicates that the skin is at least one source for extrarenal production of this important metabolite. The kidney is the major source, but in anephric individuals the circulating level of $1{,}25(OH)_2D$ may fall sufficiently low such that the cutaneous production of $1{,}25(OH)_2D$ is no longer inhibited and its degradation is no longer induced. Thus, the epidermis may provide $1{,}25(OH)_2D$ to the body in patients with decreased or absent renal function, accounting for the normalization of $1{,}25(OH)_2D$ levels when such individuals are provided adequate amounts of 25OHD [34]. Although a variety of squamous cell carcinomas are associated with hypercalcemia, this appears to be due to their elaboration of a parathyroid hormone like protein and not due to uncontrolled $1{,}25(OH)_2D$ production. Most likely the $1{,}25(OH)_2D$ produced by keratinocytes serves an autocrine or paracrine function, regulating the proliferation and differentiation of these cells. The ability of the epidermis to produce $1{,}25(OH)_2D$ from 25OHD and to catabolize the vitamin D metabolites quickly offers several possibilities for the topical administration of vitamin D compounds in the treatment of skin disorders.

III. REGULATION OF KERATINOCYTE DIFFERENTIATION

A. Microanatomy of the Epidermis

The epidermis is composed of four layers of cells at different stages of differentiation (Fig. 3). The basal layer (stratum basale) rests on the basal lamina separating the dermis and epidermis. These cells proliferate, providing the cells for the upper differentiating layers. They are large, columnar cells forming intercellular attachments with adjacent cells through desmosomes. An asymmetric distribution of integrins on their lateral and basal surface may also regulate their attachment to the basal lamina and adjacent cells [35–37]. They contain an extensive keratin network comprising principally keratins K5 (58 kDa) and K14 (50 kDa) [38]. By a process which we are only beginning to understand cells migrate upward from this basal layer, acquiring the characteristics of a fully differentiated corneocyte, which is eventually sloughed off.

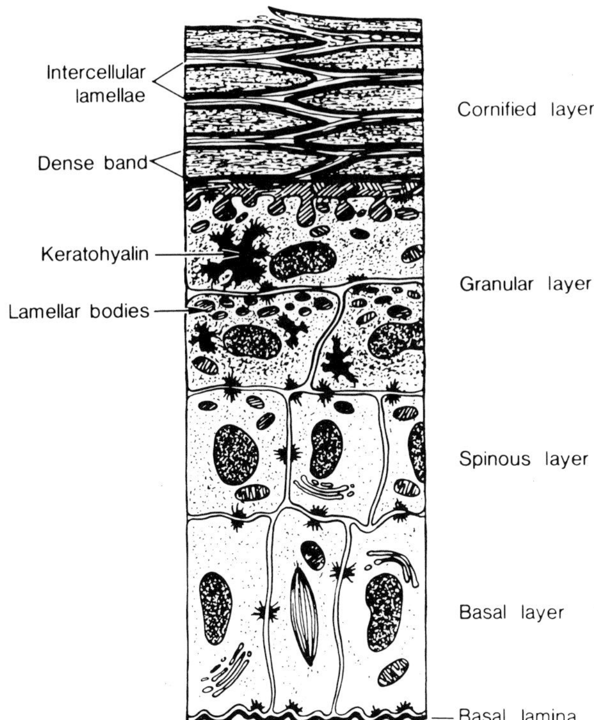

FIGURE 3 Four principal layers of the epidermis. The basal layer rests on the basal lamina separating the epidermis from the dermis. The cornified layer provides the major barrier to the outside environment. The basal layer provides the cells which differentiate as they pass through the different layers to become corneocytes. Each layer has its distinct appearance, function, and differentiation markers as described in the text.

The layer above the basal cells is the spinous layer (stratum spinosum). These cells initiate the production of the keratins K1 and K10, which are the keratins characteristic of the more differentiated layers of the epidermis [39]. Cornified envelope precursors such as involucrin [40] also appear in the spinous layer as does the enzyme transglutaminase, responsible for the ε-(γ-glutamyl)lysine cross-linking of these substrates into the insoluble cornified envelope [41]. The keratinocyte contains both a soluble (tissue, TG-C, or type II) and a membrane-bound (particulate, TG-K, or type I) form of transglutaminase. It is the membrane-bound form that correlates with differentiation and is thought to be responsible for the formation of the cornified envelope [41].

The granular layer (stratum granulosum), lying above the spinous layer, is characterized by electron-dense keratohyalin granules. These are of two types [42]. The larger of the two granules contains profilaggrin, the precursor of filaggrin, a protein thought to facilitate the aggregation of keratin filaments [43]. The smaller granule contains loricrin, a major component of the cornified envelope [44]. The granular layer also contains lamellar bodies, lipid-filled structures that fuse with the plasma membrane, divesting their contents into the extracellular space where the lipid contributes to the permeability barrier of skin [45]. As the cells pass from the granular layer to the cornified layer (stratum corneum) they undergo destruction of their organelles with further maturation of the cornified envelope into an insoluble, highly resistant structure surrounding the keratin–filaggrin complex and linked to the extracellular lipid milieu [46].

B. Regulators of Growth and Differentiation—General

The ability to grow keratinocytes in culture in a manner that permits at least partial differentiation has made it possible to study the regulation of this process [47]. Although this chapter emphasizes the roles of calcium and $1,25(OH)_2D$ in keratinocyte differentiation, a number of hormones, cytokines, and ions are involved.

1. VITAMIN A

Vitamin A and its metabolites and analogs have long been known to influence epidermal development. Vitamin A deficiency induces squamous metaplasia, causing even normal keratinizing epithelia (e.g., the epidermis) to become hyperkeratotic [48]. In contrast, pharmacological doses of retinoids can induce a mucous metaplasia in what is otherwise a keratinizing epithelium [49]. In culture, it has long been appreciated that retinoids block the terminal differentiation of keratinocytes [50]. Retinoids block the appearance of the suprabasal keratins (K1 and K10) while enhancing the appearance of keratins characteristic of undifferentiated cells [51–53]. Retinoids decrease cornified envelope formation [54] by decreasing type I (or type K) transglutaminase [55] and substrate (involucrin, loricrin) [55,56] levels as well as the expression of filaggrin [57]. These effects are opposite to and antagonize the prodifferentiation actions of $1,25(OH)_2D_3$.

Like the actions of $1,25(OH)_2D_3$, many of the actions of retinoic acid (RA) are mediated through changes in gene expression. Two members of the retinoic acid receptor family (RARα and RARγ), whose structures are homologous to steroid hormone receptors (including the VDR), have been identified in keratinocytes [58–60]. It was found that epidermal differentiation was blocked in a transgenic mouse with a dominant negative form of RAR in the skin [61]. The RARs are found in transformed as well as normal keratinocytes [62,63], although their expression in SCC may not be normal [62]. Certain retinoic acid metabolites (e.g., 9-*cis*-retinoic acid) also bind to another family of receptors, the retinoid X receptors (RXRs). The major member of this family in skin is RXRα [60,64]. It is likely that RA and $1,25(OH)_2D_3$ interact at the genomic level in their control of keratinocyte differentiation [65–67]. Furthermore, since RARs and VDR each can form heterodimers with RXR [68–71], both proteins may compete for the RXR pool within a given cell. Thus, the antagonism that exists between retinoic acid and $1,25(OH)_2D$ on keratinocyte differentiation may have an explanation at the molecular level.

2. Cytokines and Peptide Growth Factors

Keratinocytes produce an array of cytokines and growth factors, many of which have autocrine activity, and respond to still other cytokines and growth factors produced by stromal cells in the dermis [72,73]. Transforming growth factor-α (TGFα) is produced by the keratinocyte and acts through the epidermal growth factor (EGF) receptor to stimulate proliferation and migration [74]. Transforming growth factors-β1 and -β2 (TGFβ1, TGFβ2) are also produced by keratinocytes [75,76], but they inhibit proliferation [77]. TGFβ exerts a number of effects on keratinocytes including reducing the differentiation markers K1 and filaggrin [53], increasing fibronectin, laminin β1, and α1 (IV) collagen [53,78,79], increasing type II transglutaminase without altering type I transglutaminase [80], and decreasing c-*myc* expression [81]. The c-*myc* gene has been shown to have a response element for TGFβ [81], but whether the other actions of TGFβ are mediated through a similar response element in other genes remains to be demonstrated. Even though TGFα and TGFβ exert opposite effects on keratinocyte proliferation, neither promotes keratinocyte differentiation. Basic fibroblast growth factor (bFGF) is produced by keratinocytes [82] and stimulates their proliferation [83]. A related molecule, keratinocyte growth factor (KGF), is produced by stromal fibroblasts but stimulates the proliferation of keratinocytes [84].

The ability of EGF (or TGFα), bFGF, and KGF to stimulate keratinocyte proliferation is markedly enhanced by the presence of high concentrations of insulin, indicating that insulin-like growth factors (IGF-1 and -2) are also keratinocyte mitogens [78,83,85,86]. Receptors for IGF-1 on keratinocytes have been found [87]. TNFα is produced by keratinocytes [73] and promotes their differentiation with only a modest antiproliferative effect [28]. IFN-γ, on the other hand, is not made by keratinocytes, but it markedly inhibits their proliferation with little effect on differentiation [7,88]. IFN-γ increases class II antigens (HLA-DR) [26,88] and the intercellular adhesion molecule (ICAM-1) [89] of these cells. Keratinocytes produce platelet-derived growth factors (PDGF) but do not have the PDGF receptor [73]. Parathyroid hormone related peptide (PTHrP) is also produced by keratinocytes [90], but, like PDGF, receptors for PTHrP on normal keratinocytes have been difficult to demonstrate [91,92]. Nevertheless, PTHrP and/or its C-terminal fragments have been reported to inhibit or stimulate keratinocyte proliferation [93,94], and a 7–34 PTH antagonist has been claimed to promote hair growth by reversing the antiproliferative effects of PTHrP on the epidermis [93]. Interleukins 1 (α and β) [95,96], 3 [97], 6 [98], and 8 [99] are all produced by keratinocytes, as are the colony stimulating factors (CSF), granulocyte–macrophage (GM)-CSF [100], G-CSF, and M-CSF [101]. The roles the interleukins and colony-stimulating factors play in keratinocyte proliferation, migration, and differentiation are uncertain, but acting in a coordinated fashion they probably regulate important aspects of wound healing, inflammation, and cell growth [102] in the epidermis.

C. Calcium-Regulated Differentiation

Calcium is the best studied prodifferentiating agent for keratinocytes (Fig. 4). *In vivo*, a calcium gradient exists in the epidermis such that in the basal and spinous layers calcium is found primarily intracellularly and in low amounts, but in the upper granular layers calcium accumulates in large amounts in the intercellular matrix [103]. This gradient of calcium may provide the driving force for differentiation in intact epidermis. However,

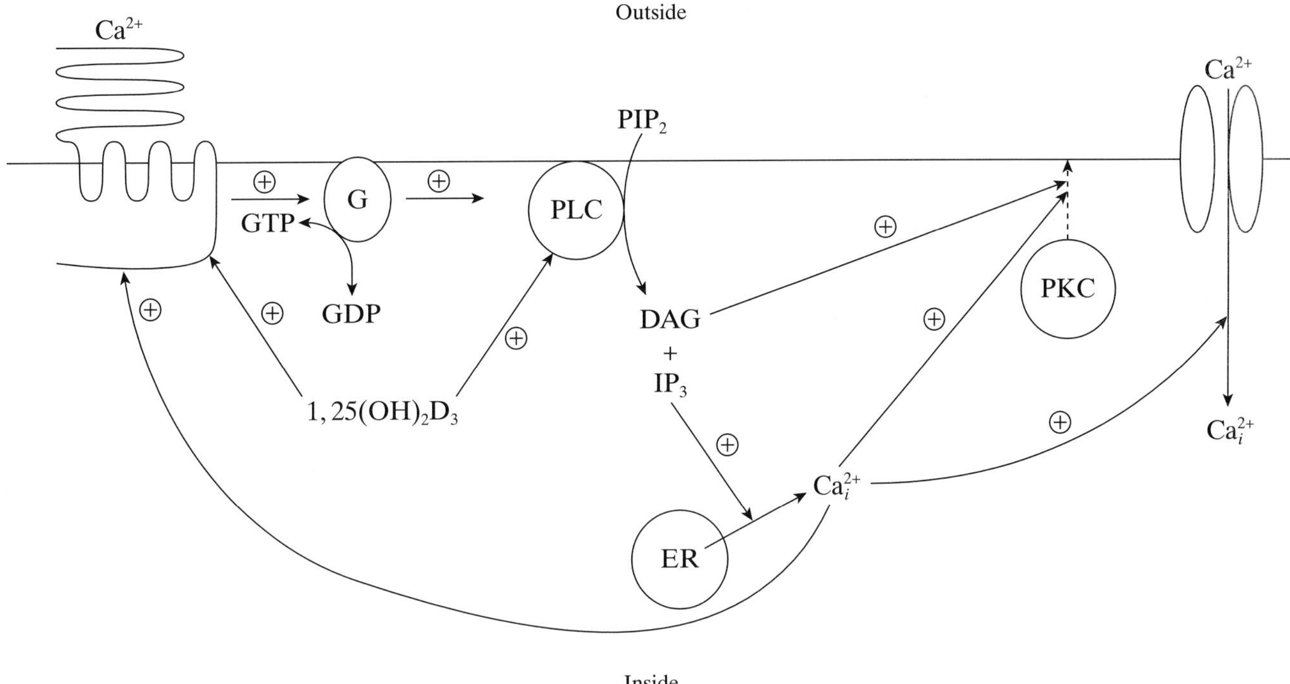

FIGURE 4 Calcium sensing mechanism of the keratinocyte. The keratinocyte contains a calcium receptor that increases intracellular calcium by activating phospholipase C (PLC) via a G-protein-coupled process. $1,25(OH)_2D$ enhances this process by increasing the number of calcium receptors and inducing PLC. PLC hydrolyzes phosphatidylinositol bisphosphate (PIP_2) to inositol trisphosphate (IP_3) and diacylglycerol (DAG). IP_3 releases calcium (Ca_i^{2+}) from intracellular stores such as the endoplasmic reticulum (ER). In combination with DAG, $[Ca^{2+}]_i$ stimulates the translocation (and activation) of the classic isozymes of protein kinase C (primarily PKC-α in keratinocytes) to the membrane. $[Ca^{2+}]_i$ also opens the nonspecific cation channel, which is the main calcium-carrying channel in keratinocytes. $[Ca^{2+}]_i$ along with $1,25(OH)_2D$ may also increase the number of calcium receptors in the keratinocyte.

most of the information regarding calcium-induced differentiation comes from *in vitro* studies with cultured keratinocytes. In low calcium-containing medium, keratinocytes proliferate readily but differentiate slowly if at all and remain as a monolayer in culture. On switching the cells to higher calcium concentrations (referred to as the calcium switch), keratinocytes undergo a coordinated set of responses at both the genomic and nongenomic levels that eventuates in a stratified culture in which the cells contain many of the features of the differentiated epithelium. For reasons that are not apparent, mouse keratinocytes are more sensitive to the antiproliferative effects of calcium and require lower concentrations of calcium for differentiation than do human keratinocytes; however, the qualitative effects of calcium on human and murine cells are comparable.

1. Changes Induced by the Calcium Switch

Within minutes to hours of the calcium switch, morphological changes are apparent, with rapid development of cell to cell contact [8], desmosome formation [104], and a realignment of actin and keratin bundles near the cell membrane at the point of intercellular contacts [105]. Desmoplakin (a component of desmosomes), fodrin (an actin and calmodulin binding spectrin-like protein), and calmodulin are redistributed to the membrane shortly after the calcium switch by a mechanism that is blocked by cytochalasin, an agent which disrupts microfilament reorganization [105–107]. These effects do not appear to be under genomic control, although this has not been tested rigorously. Within hours to days of the calcium switch the cells begin to make involucrin [9,55,108], loricrin [56], transglutaminase [9,55,108], keratins K1 and K10 [109], and filaggrin [109], and they start to form cornified envelopes [9,109]. As evidenced by a rise in mRNA levels for these proteins following the calcium switch [56,108,109], these effects of calcium represent genomic actions. Calcium response regions have been identified in the involucrin [110] and K1 [111] genes. The redistribution of integrin isoforms within days following the calcium switch [35,37,112] may participate in the mechanism by which cells begin to stratify. Calcium-induced increases in TGFβ [75,113] may contribute to the decrease in proliferation that accompanies the calcium switch, but, as discussed above, TGFβ does not stimulate differentiation.

2. ROLE OF INTRACELLULAR CALCIUM

The mechanisms by which calcium exerts its effects on keratinocyte differentiation are multiple. The intracellular free calcium ion concentration ($[Ca^{2+}]_i$) increases as keratinocytes differentiate, correlating closely with their ability to form cornified envelopes [6]. Raising the extracellular calcium concentration ($[Ca^{2+}]_o$) increases $[Ca^{2+}]_i$ [6,114–118]. This response is saturable, reaching a maximum at 2 m*M* extracellular Ca^{2+} in human foreskin keratinocytes [6]. The response of $[Ca^{2+}]_i$ to $[Ca^{2+}]_o$ is multiphasic and changes with differentiation. In undifferentiated keratinocytes and in transformed keratinocytes that are unable to differentiate, the switch to higher $[Ca^{2+}]_o$ results in an initial spike of $[Ca^{2+}]_i$ that is followed by a plateau level which persists as long as the $[Ca^{2+}]_o$ remains elevated [114,116,117]. As the cells differentiate, this acute response to $[Ca^{2+}]_o$ is lost [116,117]. Lanthanum, which blocks calcium entry, blocks this response to $[Ca^{2+}]_o$, indicating that much of the rise in $[Ca^{2+}]_i$ following a change in $[Ca^{2+}]_o$ is dependent on calcium entry [119], an increase of which has been shown following the calcium switch [117,120]. Calcium channel blockers do not prevent the rise in calcium uptake [120]. The prolonged increase in $[Ca^{2+}]_i$ after elevation of $[Ca^{2+}]_o$ stands in contrast to the response of $[Ca^{2+}]_i$ to ATP [118,121,122]. ATP increases $[Ca^{2+}]_i$ acutely and transiently at all stages of differentiation. The pool from which internal Ca^{2+} is released by ATP is intracellular and is rapidly depleted by a single dose of ATP, although ionomycin can mobilize additional intracellular calcium. As ATP inhibits rather than promotes keratinocyte differentiation [123], the sustained increase in $[Ca^{2+}]_i$ following a $[Ca^{2+}]_o$ increase appears to be essential for the differentiation process.

3. THE CALCIUM RECEPTOR

The acute response of the keratinocyte to calcium resembles that of the parathyroid cell [124], which senses $[Ca^{2+}]_o$ via a seven-transmembrane domain, GTP binding protein-coupled calcium receptor (CaR) [125,126]. This receptor is not limited to the parathyroid gland, and we have identified the same structure in the keratinocyte [117]. As mentioned above, keratinocytes lose their ability to sense calcium with differentiation. This change in calcium responsiveness is reflected by the loss of CaR mRNA levels in normal keratinocytes grown under low calcium conditions, with no changes seen in transformed keratinocytes [117], which do not lose this aspect of calcium responsiveness. $1,25(OH)_2D$ increases the CaR mRNA levels and prevents their decrease with time [127]. Furthermore, $1,25(OH)_2D$ potentiates the ability of these cells to respond to $[Ca^{2+}]_o$ with a rise in $[Ca^{2+}]_i$ [127]. Thus, the CaR appears to be important in mediating the internal Ca^{2+} response of keratinocytes to $[Ca^{2+}]_o$, and it provides a mechanism by which $1,25(OH)_2D_3$ can regulate calcium-induced epidermal differentiation.

4. PHOSPHOINOSITIDE METABOLISM

The calcium switch stimulates phosphoinositide metabolism, which potentially could provide additional second messengers for mediating its effects on the keratinocyte [128–131]. As for the response of $[Ca^{2+}]_i$ to $[Ca^{2+}]_o$, the rise in inositol trisphosphate (IP_3) and diacylglycerol (DAG) is both immediate and prolonged following the calcium switch. Other agents such as ATP raise IP_3 levels at least as effectively as calcium and yet do not stimulate differentiation [123]. Just as the rise in $[Ca^{2+}]_i$ after ATP is transient, so is the rise in IP_3. Conceivably, the prolonged rise in $[Ca^{2+}]_i$ and IP_3 after increases in $[Ca^{2+}]_o$ compared to the transient effects of ATP contributes to the ability of $[Ca^{2+}]_o$ and not ATP to stimulate differentiation.

5. PROTEIN KINASE C

The rise in DAG and $[Ca^{2+}]_i$ following the calcium switch provides a means to link the abilities of both calcium and phorbol esters to stimulate keratinocyte differentiation, and that link is protein kinase C (PKC). Phorbol esters, which bind to and activate PKC, are well-known tumor promoters in skin [132–135]. However, the initial effects of phorbol esters *in vitro are* to promote differentiation in cells grown in low calcium [132–134,136–142], effects which are potentiated by calcium [139]. Phorbol esters stimulate PKC, and PKC inhibitors block the ability of both calcium and phorbol esters to promote differentiation [140,141]. Nevertheless, differences between the acute effects of phorbol esters and calcium on differentiation are clear. Phorbol esters block rather than promote calcium induction of K1 and K10 (early differentiation markers), in contrast to their synergistic effects with calcium on the later differentiation markers such as involucrin, loricrin and filaggrin [142; D. C. Ng and D. D. Bikle, unpublished]. Calcium does not duplicate all the changes in protein phosphorylation caused by phorbol esters [130,143,144], and phorbol esters, unlike calcium, do not activate the phospholipase C (PLC) pathway [128,145]. Calcium activation of the PLC pathway, by increasing the amount of DAG, an endogenous activator of PKC, is a likely means by which calcium stimulates PKC [144], whereas phorbol esters tend to reduce hormonal activation of PLC and the increase in $[Ca^{2+}]_i$ at least in other cells [146,147]. Thus, it seems certain that although calcium-regulated differentiation involves the PKC pathway, the effects of calcium are not solely due to PKC activation and are not mimicked in their entirety by phorbol esters.

Regardless of the fact that PKC activation cannot explain all of the effects of calcium on keratinocyte differentiation, it clearly plays a major role. However, the study of PKC in differentiation is complicated by the large number of isozymes of PKC in the epidermis, most of which are separate gene products and under different modes of regulation and distribution within the epidermis. Mouse keratinocytes contain PKC-α, -δ, -ε, -ζ, and -η [148]. Transformed mouse keratinocytes lose their responsiveness to calcium but retain the same complement of isozymes [148,149]. Human keratinocytes contain the same set of isozymes as mouse keratinocytes [150,151]. HaCaT cells (an immortalized human keratinocyte cell line) contain PKC-α, -δ, -ε, and -ζ (no PKC-β or -η) [152]. PKC-α and -β are classic PKC isozymes and are activated by calcium, phorbol esters, and DAG. PKC-δ, -ε, and -η are novel PKCs that are activated by phorbol esters and DAG, like the classic PKCs, but are not activated by calcium. PKC-ζ is an atypical PKC that does not respond to calcium or phorbol esters. The cDNAs for these isozymes have been sequenced, and the structural basis for these differences is known [153]. During the first 48 hr after the calcium switch in mouse keratinocytes, the translocation of PKC-α from the cytosol to the membrane paralleled in time the induction of loricrin and profilaggrin better than did the changes in the other isozymes [154]. Down-regulation of PKC-α by bryostatin-1 and 12-deoxyphorbol 13-phenylacetate (DPP) also correlated with the inhibition of calcium-induced loricrin and profilaggrin expression better than did the effects of these phorbol esters on the other isozymes [154,155]. Thus, PKC-α may be the major isozyme associated with calcium-induced differentiation.

The actual mechanism by which PKC induces differentiation is not clear [156–159], although a mechanism involving transcription factors in the Fos and Jun families acting on their AP-1 sites in the promoter regions of the genes involved with differentiation seems likely. PKC activation leads to a rapid increase in expression of c-*fos* and c-*jun* [121,122,160] and increased binding to AP-1 sites as assessed by gel retardation studies [161]. However, c-Fos and c-Jun are not the only transcription factors capable of binding to AP-1 sites. Other members of the Fos and Jun families (Fra-1, Fra-2, Jun B, Jun D) have been found in keratinocytes [162] and are differentially distributed throughout the epidermis. The AP-1 site from the involucrin gene was found to bind Fra-1, Jun B, and Jun D on gel retardation analysis [163]. A dominant negative mutant of c-*jun* [70], which blocks c-*jun/fos*-regulated transcriptional activity in AP-1 regulated genes such as prolactin, markedly stimulated transcriptional activity of involucrin gene constructs (D. C. Ng and D. D. Bikle, unpublished), suggesting that some members of the Fos–Jun family are playing an inhibitory role rather than a stimulatory role in involucrin gene transcription. As both keratin 1 and involucrin contain an AP-1 site within their calcium responsive regions, a role for members of the Fos and Jun family in calcium-induced differentiation appears likely.

6. Calmodulin

Calmodulin also participates in the response of the cell to calcium. Calmodulin levels rise after the calcium switch under circumstances in which proliferation is increased [164,165], and calmodulin antagonists block proliferation [165–167]. Calmodulin and calcium may activate calmodulin/calcium-dependent kinases that could phosphorylate and thus activate critical transcription factors [168]. Calmodulin moves to the membrane along with fodrin [107] and desmoplakin [106] (calmodulin binding proteins) following the calcium switch. Calmodulin antagonists block the development of desmosome formation [136] and cell to cell contacts. At the membrane calmodulin may participate not only in the formation of intercellular bridges but in the regulation of calcium influx and efflux as well.

D. 1,25(OH)$_2$D-Regulated Differentiation

The observation that 1,25(OH)$_2$D induces keratinocyte differentiation was first made by Hosomi *et al.* [3] and provided a rationale for the previous and unexpected finding of 1,25(OH)$_2$D receptors in the skin [2]. As discussed earlier, 1,25(OH)$_2$D is likely to be an autocrine or paracrine factor for epidermal differentiation since it is produced by the keratinocyte, but under normal circumstances keratinocyte production of 1,25(OH)$_2$D does not appear to contribute to circulating levels [1,10]. The receptors for and the production of 1,25(OH)$_2$D vary with differentiation [4,169,170] in a manner that suggests feedback regulation; both are reduced in the later stages of differentiation. 1,25(OH)$_2$D increases involucrin, transglutaminase activity, and cornified envelope formation at subnanomolar concentrations in preconfluent keratinocytes [3,5,6,32,171]. At these concentrations 1,25(OH)$_2$D has been found to promote proliferation in some studies [172–174], although the antiproliferative actions are most frequently observed, especially when concentrations above 10^{-9} *M* are employed. The mechanisms underlying the proproliferative actions are not known. The antiproliferative effects are accompanied by a reduction in the mRNA levels for c-*myc* [175], and the rise in involucrin and transglutaminase is accompanied by a rise in their respective mRNAs [108]. Other likely examples of genomic regulation by $1,25(OH)_2D$ include modulation of the skin calcium binding protein [176], reciprocal effects on the 25OHD 1- and 24-hydroxylases [10], decreases in parathyroid hormone-related peptide (PTHrP) [177] and EGF receptor gene expression [175], and increases in TGFβ [178] and plasminogen activator inhibitor [179] gene expression.

Where examined, these effects of $1,25(OH)_2D$ can be reproduced by 25OHD [11,32], presumably because of endogenous conversion of 25OHD to $1,25(OH)_2D$, but are not observed with the biologically inactive 1β isomer of $1,25(OH)_2D$ [5] [the natural isomer is $1\alpha,25(OH)_2D$]. On the other hand, 1αOHD has little effect on differentiation [180], indicating that keratinocytes are unable to convert 1αOHD to $1,25(OH)_2D$ and confirming that they are missing the vitamin D 25-hydroxylase [12], the enzyme that converts vitamin D to 25OHD and 1αOHD to $1\alpha,25(OH)_2D$.

The mechanisms by which $1,25(OH)_2D$ alters keratinocyte differentiation are not fully elucidated (Fig. 5). The $1,25(OH)_2D$ receptor (VDR) is critical for the genomic actions of $1,25(OH)_2D$; its role, if any, in potential nongenomic actions of $1,25(OH)_2D$ remains to be determined. An acute increase in $[Ca^{2+}]_i$ associated with an acute increase in phosphoinositide turnover (producing a rise in both IP_3 and DAG) has been observed in several studies [181–185]. Not all investigators (including ourselves) have been able to reproduce these acute effects of $1,25(OH)_2D$ [117], although a gradual rise in $[Ca^{2+}]_i$ and cornified envelope formation is observed [6]. Furthermore, vitamin D metabolites that do not induce differentiation [1αOHD and $24,25(OH)_2D$] have been reported to increase both $[Ca^{2+}]_i$ and IP_3 acutely [183]. The rise in $[Ca^{2+}]_i$, IP_3, and DAG is accompanied by translocation of PKC

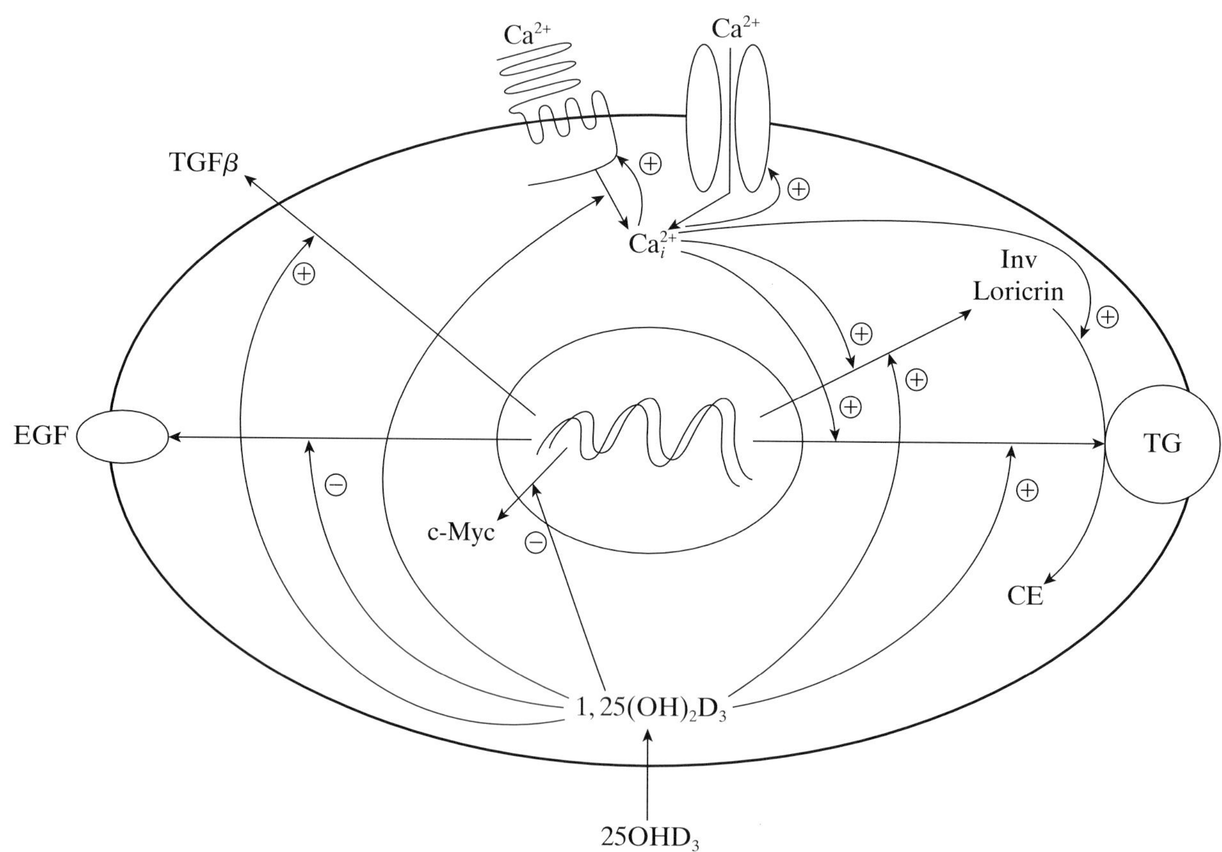

FIGURE 5 Interactions of $1,25(OH)_2D$ and calcium in the regulation of keratinocyte proliferation and differentiation. As described in the legend to Fig. 4, both the calcium receptor and the nonspecific cation channel regulate $[Ca^{2+}]_i$, and $1,25(OH)_2D$ enhances the rise in $[Ca^{2+}]_i$ in part by increasing calcium receptors and the PLC pathway. $1,25(OH)_2D$ at least at higher concentrations inhibits proliferation in part by reducing c-Myc and epidermal growth factor (EGF) receptor production while stimulating TGFβ production. $1,25(OH)_2D$ in combination with calcium stimulate differentiation by inducing the substrates [e.g., involucrin (Inv) and loricrin] for cornified envelope (CE) formation as well as the enzyme transglutaminase (TG), which cross-links these substrates into the CE. Calcium further activates TG once it is produced.

to the membrane [183]. Down-regulation of PKC and inhibition of its activity have been reported to block the ability of 1,25$(OH)_2$D to stimulate cornified envelope formation [183]. However, the role of PKC in mediating or interacting with 1,25$(OH)_2$D in its effects on keratinocyte differentiation remains virtually unexplored.

Calcium and 1,25$(OH)_2$D interact in their ability to inhibit proliferation and stimulate involucrin and transglutaminase gene expression [108]. The higher the $[Ca^{2+}]_o$, the more sensitive is the keratinocyte to the antiproliferative effect of 1,25$(OH)_2$D (and vice versa) [171]. The interaction on gene expression is more complex. Both calcium [in the absence of 1,25$(OH)_2$D] and 1,25$(OH)_2$D (at 0.03 mM Ca^{2+}) raise the mRNA levels for involucrin and transglutaminase in a dose-dependent fashion. The stimulation is synergistic at intermediate concentrations of calcium (0.1 mM) and 1,25$(OH)_2$D (10^{-10} M), but inhibition is observed in combination at higher concentrations. The synergism is more apparent at earlier times after the calcium switch (4 hr) than later (24–72 hr), when increased turnover of the mRNA by the higher combined concentrations of calcium and 1,25$(OH)_2$D becomes dominant. Reductions in K1/K10 keratins, involucrin, transglutaminase, and filaggrin were similarly observed in immunolocalization studies of keratinocytes grown at the air–liquid interface when treated with high concentrations of 1,25$(OH)_2$D (10^{-6} M) in the presence of calcium [186]. The molecular mechanisms underlying this complex, dose- and time-dependent interaction between calcium and 1,25$(OH)_2$D on keratinocyte differentiation are being actively investigated; they may also underlie some of the side effects such as facial irritation when 1,25$(OH)_2$D analogs are used topically for the management of psoriasis.

These actions of 1,25$(OH)_2$D defined primarily at the level of tissue culture have physiological implications. Rats maintained in the dark on a vitamin D-deficient diet that is otherwise normal demonstrate striking changes in gross appearance including matted hair, patchy alopecia, scaling and erythema, and a variable but significant defect in the barrier to transcutaneous water loss [187]. These changes are not due to gross alterations in cutaneous sterologenesis [188], but rather suggest a more subtle abnormality in differentiation. Histologically, the skin from such rats is thinner with fewer granular cell layers and increased nuclei in the basal cell layer [189]. Such changes are similar to the changes observed in psoriasis. In fact, the antiproliferative, prodifferentiating effects of 1,25$(OH)_2$D and its analogs have been put to clinical use in the management of this disease, as discussed more fully in Chapter 72.

References

1. Bikle DD, Nemanic MK, Whitney JO, Ellas PW 1986 Neonatal human foreskin keratinocytes produce 1,25-dihydroxyvitamin D_3. Biochemistry **25**:1545–1548.
2. Stumpf WE, Sar M, Reid FA, Tanaka Y, DeLuca HF 1979 Target cells for 1,25-dihydroxyvitamin D_3 in intestinal tract, stomach, kidney, skin, pituitary and parathyroid. Science **206**:1188–1191.
3. Hosomi J, Hosoi J, Abe E, Suda T, Kuroki T 1983 Regulation of terminal differentiation of cultured mouse epidermal cells by 1α,25-dihydroxyvitamin D_3. Endocrinology **113**:1950–1957.
4. Pillai S, Bikle DD, Elias PM 1988 1,25-Dihydroxyvitamin D production and receptor binding in human keratinocytes varies with differentiation. J Biol Chem **263**:5390–5395.
5. Smith EL, Walworth NC, Holick MF 1986 Effect of 1α,25-dihydroxyvitamin D_3 on the morphologic and biochemical differentiation of cultured human epidermal keratinocytes grown in serum-free conditions. J Invest Dermatol **86**:709–714.
6. Pillai S, Bikle DD 1991 Role of intracellular free calcium in the cornified envelope formation of keratinocytes: Differences in the mode of action of extracellular calcium and 1,25-dihydroxyvitamin D. J Cell Physiol **146**:94–100.
7. Bikle DD, Pillai S, Gee E, Hincenbergs M 1988 Regulation of 1,25-dihydroxyvitamin D production in human keratinocytes by interferon-gamma. Endocrinology **124**:655–660.
8. Hennings H, Michael D, Cheng C, Steinert P, Holbrook K, Yuspa SH 1980 Calcium regulation of growth and differentiation of mouse epidermal cells in culture. Cell **19**:245–254.
9. Pillai S, Bikle DD, Mancianti ML, Cline P, Hincenbergs M 1990 Calcium regulation of growth and differentiation of normal human keratinocytes: Modulation of differentiation competence by growth and extracellular calcium. J Cell Physiol **143**:294–302.
10. Bikle DD, Nemanic MK, Gee EA, Elias P 1986 1,25-dihydroxyvitamin D_3 production by human keratinocytes: Kinetics and regulation. J Clin Invest **78**:557–566.
11. Matsumoto K, Azuma Y, Kiyoki M, Okumura H, Hashimoto K, Yoshikawa K 1991 Involvement of endogenously produced 1,25-dihydroxyvitamin D_3 in the growth and differentiation of human keratinocytes. Biochim Biophys Acta **1092**:311–318.
12. McLaughlin JA, Castonguay A, Holick MF 1991 Cultured human keratinocytes cannot metabolize vitamin D_3 to 25-hydroxyvitamin D_3. FEBS Lett **282**:409–411.
13. Barbour GL, Coburn JW, Slatoposky E, Norman AW, Horst RL 1981 Hypercalcemia in an anephric patient with sarcoidosis: Evidence for extrarenal generation of 1,25-dihydroxyvitamin D_3. N Engl J Med **305**:440–443.
14. Lambert PW, Stern PH, Avioli RC, Brackett NC, Turner RT, Green A, Fu IY, Bell NH 1982 Evidence for extrarenal production of 1α,25-dihydroxyvitamin D in man. J Clin Invest **69**:722–725.
15. Littledike ET, Horst RL 1982 Metabolism of vitamin D_3 in nephrectomized pigs given pharmacological amounts of vitamin D_3. Endocrinology **111**:2008–2013.
16. Bikle DD, Halloran BP, Riviere JE 1994 Production of 1,25-dihydroxyvitamin D_3 by perfused pig skin. J Invest Dermatol **102**:796–798.
17. Rasmussen H, Wong M, Bikle DD, Goodman DB 1972 Hormonal control of the renal conversion of 25-hydroxycholecalciferol to 1,25-dihydroxycholecalciferol. J Clin. Invest **51**:2502–2504.
18. Rost CR, Bikle DD, Kaplan RA 1981 In vitro stimulation of 25-hydroxycholecalciferol 1 alpha-hydroxylation by parathyroid hormone in chick kidney: Evidence for a role for adenosine 3′, 5′-monophosphate. Endocrinology **108**:1002–1006.

19. Bikle DD, Gee E 1989 Free and not total $1,25(OH)_2D$ regulates 25 hydroxyvitamin D metabolism by keratinocytes. Endocrinology **124**:649–654.
20. Henry HL 1979 Response of chick kidney cell cultures to 1,25-dihydroxyvitamin D_3. In: Norman AW, Schaefer K, Herrath DV, Grigoleit HG, Coburn JW, DeLuca HF, Mawer EB, Suda T (eds) Vitamin D: Basic Research and Its Clinical Application. de Gruyter, Berlin and New York, pp. 467–474.
21. Spanos E, Barrett DD, Chong KT, MacIntyre I 1978 Effect of oestrogen and 1,25-dihydroxycholecalciferol metabolism in primary chick kidney cell cultures. Biochem J **174**:231–236.
22. Trechsel U, Bonjour J-P, Fleisch H 1979 Regulation of the metabolism of 25-hydroxyvitamin D_3 in primary cultures of chick kidney cells. J Clin Invest **64**:206–217.
23. Reeve L, Tanaka Y, DeLuca HF 1983 Studies on the site of 1,25-dihydroxyvitamin D_3 synthesis in vivo. J Biol Chem **258**:3615–3617.
24. Shultz TD, Fox J, Heath III H, Kumar R 1983 Do tissues other than the kidney produce 1,25-dihydroxyvitamin D_3 in vivo? A re-examination. Proc Natl Acad Sci USA **80**:1746–1750.
25. Pillai S, Bikle DD, Hincenbergs M, Elias PM 1988 Biochemical and morphological characterization of growth and differentiation of normal human neonatal keratinocytes in a serum-free medium. J Cell Physiol **134**:229–239.
26. Morhenn VB, Wood GS Gamma interferon-induced expression of class II major histocompatibility complex antigens by human keratinocytes: Effects of conditions of culture. In: Milstone LM, Edelson RL (eds) Endocrine, Metabolic and Immunologic Functions of Keratinocytes. Annals of the New York Academy of Sciences, New York, pp. 321–330.
27. Pillai S, Bikle DD, Essalu TE, Aggarwal BB, Elias PM 1989 Binding and biological effects of tumor necrosis factor-alpha on cultured human neonatal foreskin keratinocytes. J Clin Invest **83**:816–821.
28. Bikle DD, Pillai S, Gee EA, Hincenbergs M 1991 Tumor necrosis factor-alpha regulation of 1,25-dihydroxyvitamin D production by human keratinocytes. Endocrinology **129**:33–38.
29. Trefzer U, Brockhaus M, Lotscher H, Parlow F, Budnik A, Grewe M, Christoph H, Kapp A, Schoff E, Luger TA, Krutmann J 1993 The 55-KD tumor necrosis factor receptor is regulated by tumor necrosis factor-alpha and by ultraviolet B irradiation. J Clin Invest **92**:462–470.
30. Wood LC, Elias PM, Sequeira-Martin SM, Grunfeld C, Feingold KR 1994 Occlusion lowers cytokine mRNA levels in essential fatty acid-deficient and normal mouse epidermis, but not after acute barrier disruption. J Invest Dermatol **103**:834–838.
31. Rheinwald JG, Beckett MA 1980 Defective terminal differentiation in culture as a consistent and selectable character of malignant human keratinocytes. Cell **22**:629–632.
32. Bikle DD, Pillai S, Gee E 1991 Squamous carcinoma cell lines produce 1,25-dihydroxyvitamin D_3, but fail to respond to its prodifferentiating effect. J Invest Dermatol **97**:435–441.
33. Ratnam AV, Bikle DD, Su M-J, Pillai S 1996 Squamous carcinoma cell lines fail to respond to the prodifferentiation actions of 1,25-dihydroxyvitamin D despite normal levels of the vitamin D receptor. J Invest Dermatol **106**:522–525.
34. Halloran BP, Schaefer P, Lifshcitz M, Levens M, Goldsmith RS 1984 Plasma vitamin D metabolite concentration in chronic renal failure: Effect of oral administration of 25-hydroxyvitamin D_3. J Clin Endocrinol Metab **59**:1063–1069.
35. Marchisio PC, Bondanza S, Cremona O, Cancedda R, DeLuca M 1991 Polarized expression of integrin receptors ($\alpha_6\beta_4$, $\alpha_2\beta_1$, $\alpha_3\beta_1$,and $\alpha_v\beta_5$) and their relationship with the cytoskeleton and basement membrane matrix in cultured human keratinocytes. J Cell Biol **112**:761–773.
36. Peltonen J, Larjava J, Jaakkola S, Gralnick J, Akiyama SK, Yamada SS, Yamada KM, Uitto J 1989 Localization of integrin receptors for fibronectin, collagen and laminin in human skin: Variable expression in basal and squamous cell carcinomas. J Clin Invest **84**:1916–1923.
37. Guo M, Kim LT, Akiyama SK, Gralnick HR, Yamada KM, Grinnell F 1991 Altered processing of integrin receptors during keratinocyte activation. Exp Cell Res **195**:315–322.
38. Moll R, Franke WW, Schiller DL, Geiger B, Krepler R 1982 The catalog of human cytokeratins: Patterns of expression in normal epithelia, tumors and cultured cells. Cell **31**:11–24.
39. Eichner R, Sun TT, Aebi U 1986 The role of keratin subfamilies and keratin pairs in the formation of human epidermal intermediate filaments. J Cell Biol **102**:1767–1777.
40. Warhol MJ, Roth J, Lucocq JM, Pinkus GS, Rice RH 1985 Immuno-ultrastructural localization of involucrin in squamous epithelium and cultured keratinocytes. J Histochem Cytochem **33**:141–149.
41. Thacher SM, Rice RH 1985 Keratinocyte-specific transglutaminase of cultured human epidermal cells: Relation to cross-linked envelope formation and terminal differentiation. Cell **40**:685–695.
42. Steven AC, Bisher ME, Roop DR, Steinert PM 1990 Biosynthetic pathways of filaggrin and loricrin—Two major proteins expressed by terminally differentiated epidermal keratinocytes. J Struct Biol **104**:150–162.
43. Dale BA, Resing KA, Lonsdale-Eccles JD 1985 A keratin filament associated protein. Ann NY Acad Sci **455**:330–342.
44. Mehrel T, Hohl D, Rothnagel JA, Longley MA, Bundman D, Cheng C, Lichti U, Bisher ME, Steven AC, Steinert PM, Yuspa SH, Roop DR 1990 Identification of a major keratinocyte cell envelope protein, loricin. Cell **61**:1103–1112.
45. Elias PM, Menon GK, Grayson S, Brown BE 1988 Membrane structural alterations in murine stratum corneum: Relationship to the localization of polar lipids and phospholipases. J Invest Dermatol **91**:3–10.
46. Hohl D 1990 Cornified cell envelope. Dermatologica **180**:201–211.
47. Rheinwald JG, Green H 1975 Serial cultivation of strains of human epidermal keratinocytes: The formation of keratinizing colonies from single cells. Cell **6**:331–344.
48. Wolbach SB, Howe PR 1925 Tissue changes following deprivation of fat-soluble A vitamin. J Exp Med **43**:753–777.
49. Fell HB, Mellanby E 1953 Metaplasia produced in cultures of chick ectoderm by high vitamin A. J Physiol **119**:470–488.
50. Wolbach SB, Howe PR 1933 Epithelial repair in recovery from vitamin A deficiency. J Exp Med **57**:511–526.
51. Fuchs E, Green H 1981 Regulation of terminal differentiation of cultured human keratinocytes by vitamin A. Cell **25**:617–625.
52. Gilfix BM, Eckert RL 1981 Coordinate control by vitamin A of keratin gene expression in human keratinocytes. J Biol Chem **260**:14026–14029.
53. Choi Y, Fuchs E 1990 TGF-β and retinoic acid: Regulators of growth and modifiers of differentiation in human epidermal cells. Cell Regul **1**:791–809.
54. Kline PR, Rice RH 1983 Modulation of involucrin and envelope competence in human keratinocytes by hydrocortisone, retinyl acetate and growth arrest. Cancer Res **43**:3203–3207.
55. Rubin AL, Parenteau NL, Rice RH 1989 Coordination of keratinocyte programming in human SCC-13 carcinoma and normal epidermal cells. J Cell Physiol **138**:208–214.
56. Hohl D, Lichti U, Breitkreutz D, Steinert PM, Roop DR 1991

Transcription of the human loricrin gene in vitro is induced by calcium and cell density and suppressed by retinoic acid. J Invest Dermatol **96**:414–418.

57. Fleckman P, Dale BA, Holbrook KA 1985 Profilaggrin, a high molecular-weight precursor of filaggrin in human epidermis and cultured keratinocytes. J Invest Dermatol **85**:507–512.
58. Vollberg TM Sr, Nervi C, George MD, Fujimoto W, Krust A, Jetten AM 1992 Retinoic acid receptors as regulators of human epidermal keratinocyte differentiation. Mol Endocrinol **6**:667–676.
59. Kastner P, Krust A, Mendelsohn C, Garnier JM, Zelent A, Leroy P, Staub A, Chambon P 1990 Murine isoforms of retinoic acid receptor gamma with specific patterns of expression. Proc Natl Acad Sci USA **87**:2700–2704.
60. Elder JT, Astrom A, Pettersson U, Tavakkol A, Krust A, Kastner P, Chambon P, Voorhees JJ 1992 Retinoic acid receptors and binding proteins in human skin. J Invest Dermatol **98**:36S–41S.
61. Saitou M, Sugai S, Tanaka T, Shimouchi K, Fuchs E, Narumiya S, Kakizuka A 1995 Inhibition of skin development by targeted expression of a dominant-negative retinoic acid receptor. Nature **34**:159–162.
62. Hu L, Crowe DL, Rheinwald JG, Chambon P, Gudas LJ 1991 Abnormal expression of retinoic acid receptors and keratin 19 by human oral and epidermal squamous cell carcinoma cell lines. Cancer Res **51**:3972–3981.
63. Finzi E, Blake MJ, Celano P, Skouge J, Diwan R 1992 Cellular localization of retinoic acid receptor-gamma expression in normal and neoplastic skin. Am J Pathol **140**(6):1463–1471.
64. Torma H, Rollman O, Vahlquist A 1993 Detection of mRNA transcripts for retinoic acid, vitamin D_3, and thyroid hormone (c-erb-A) nuclear receptors in human skin using reverse transcription and polymerase chain reaction. Acta Dermatol Venereol **73**:102–107.
65. Schule R, Umesono K, Mangelsdorf DJ, Bolado J, Pike JW, Evans RM 1990 Jun–Fos and receptors for vitamin A and D recognize a common response element in the human osteocalcin gene. Cell **61**:497–504.
66. Xie Z, Bikle DD 1997 Cloning of the human phospholipase C-α_1 promoter and identification of a DR6-type vitamin D response element. J Biol Chem **272**:6573–6577.
67. Li X-Y, Xiao J-H, Feng X, Qin L, Voorhees JJ 1997 Retinoid X receptor-specific ligands synergistically upregulate 1,25-dihydroxyvitamin D_3-dependent transcription in epidermal keratinocytes in vitro and in vivo. J Invest Dermatol **108**:506–512.
68. Kliewer SA, Umesono K, Mangelsdorf DJ, Evans RM 1992 Retinoid X receptor interacts with nuclear receptors in retinoic acid, thyroid hormone and vitamin D_3 signalling. Nature **355**:446–449.
69. Yu VC, Delsert C, Andersen B, Holloway JM, Devary OV, Näär AM, Kim SY, Boutin J-M, Glass CK, Rosenfeld MG 1991 RXRβ: A coregulator that enhances binding of retinoic acid, thyroid hormone, and vitamin D receptors to their cognate response elements. Cell **67**:1251–1266.
70. Forman BM, Casanova J, Raaka BM, Ghysdael J, Samuels HH 1992 Half-site spacing and orientation determines whether thyroid hormone and retinoic acid receptors and related factors bind to DNA response elements as monomers, homodimers, or heterodimers. Mol Endocrinol **6**:429–442.
71. Glass CK, Devary OV, Rosenfeld MG 1990 Multiple cell type-specific proteins differentially regulate target sequence recognition by the α retinoic acid receptor. Cell **63**:729–738.
72. McKay IA, Leigh IM 1991 Epidermal cytokines and their roles in cutaneous wound healing. Br J Dermatol **124**:513–518.
73. Ansel J, Perry P, Brown J, Damm D, Phan T, Hart C, Luger T, Hefeneider S 1990 Cytokine modulation of keratinocyte cytokines. J Invest Dermatol **94**:101S–107S.
74. Coffey RJ Jr, Derynck R, Wilcox JN, Bringman TS, Goustin AS, Moses HL, Pittelkow MR 1987 Production and auto-induction of transforming growth factor-α in human keratinocytes. Nature **328**:817–820.
75. Glick AB, Danielpour DD, Morgan D, Sporn MB, Yuspa SH 1990 Induction and autocrine receptor binding of transforming growth factor-B2 during terminal differentiation of primary mouse keratinocytes. Mol Endocrinol **4**:46–52.
76. Coffey RJ, Sipes NJ, Bascom CC, Graves-Deal R, Pennington CY, Weissman BE, Moses HL 1988 Growth modulation of mouse keratinocytes by transforming growth factors. Cancer Res **48**:1596–1602.
77. Shipley GD, Pittelkow MR, Willie JJ, Scott RE, Moses HL 1986 Reversible inhibition of normal human prokeratinocyte proliferation by type-B transforming growth factor-growth inhibitor in serum free medium. Cancer Res **46**:2068–2071.
78. DeLapp NW, Dieckman BS 1990 Effect of basic fibroblast growth factor (bFGF) and insulin-like growth factors type I (IGF-I) and type II (IGF-II) on adult human keratinocyte growth and fibronectin secretion. J Invest Dermatol **94**:777–780.
79. Vollberg TM, George MD, Jetten AM 1991 Induction of extracellular matrix gene expression in normal human keratinocytes by transforming growth factor B is altered by cellular differentiation. Exp Cell Res **193**:93–100.
80. George MD, Vollberg TM, Floyd E, Stein JP, Jetten AM 1990 Regulation of transglutaminase type II by transforming growth factor-β1 in normal and transformed human epidermal keratinocytes. J Biol Chem **265**:11098–11104.
81. Pietenpol JA, Holt JT, Stein RW, Moses HL 1990 Transforming growth factor β1 suppression of c-myc gene transcription: Role in inhibition of keratinocyte proliferation. Proc Natl Acad Sci USA **87**:3758–3762.
82. Halaban R, Langdon R, Birchall N, Cuono C, Baird A, Scott G, Moellmann G, McGuire J 1988 Basic fibroblast growth factor from human keratinocytes is a natural mitogen for melanocytes. J Cell Biol **107**:1611–1619.
83. O'Keefe EJ, Chiu ML, Payne RE 1988 Stimulation of growth of keratinocytes by bFGF. J Invest Dermatol **90**:767–769.
84. Bottaro DP, Rubin JS, Ron D, Finch PW, Florio C, Aaronson SA 1990 Characterization of the receptor for keratinocyte growth factor. J Biol Chem **265**:12767–12770.
85. Krane JF, Murphy DP, Carter DM, Krueger JG 1991 Synergistic effects of epidermal growth factor (EGF) and insulin-like growth factor I/somatomedin C (IGF-I) on keratinocyte proliferation may be mediated by IGF-I transmodulation of the EGF receptor. J Invest Dermatol **96**:419–424.
86. Rubin JS, Osada H, Finch PW, Taylor WG, Rudikoff S, Aaronson SA 1989 Purification and characterization of a newly identified growth factor specific for epithelial cells. Proc Natl Acad Sci USA **86**:802–806.
87. Misra P, Nickoloff BJ, Morhenn VB, Hintz RL, Rosenfeld RG 1986 Characterization of insulin-like growth factor-I/somatomedin-C receptors on human keratinocyte monolayers. J Invest Dermatol **87**:264–267.
88. Nickoloff BJ 1987 Binding of ^{125}I-gamma interferon to cultured human keratinocytes. J Invest Dermatol **89**:132–135.
88. Nickoloff BJ, Griffiths CEM, Barker JNWN 1990 The role of adhesion molecule, chemotactic factors and cytokines in inflammatory and neoplastic skin disease—1990 update. J Invest Dermatol **94**:151S–157S.
90. Merendino JJ Jr, Insogna KL, Milstone LM, Broadus AE, Stew-

art AF 1986 Parathyroid hormone-like protein from cultured human keratinocytes. Science **231**:388–390.

91. Hanafin NM, Chen TC, Heinrich G, Segre GV, Holick MF 1995 Cultured human fibroblasts and not cultured human keratinocytes express a PTH/PTHrP receptor mRNA. J Invest Dermatol **105**:133–137.
92. Orloff JJ, Kats Y, Urena P, Schipani E, Vasavada RC, Philbrick WM, Behal A, Abou-Samra AB, Segre GV, Juppner H 1995 Further evidence for a novel receptor for amino-terminal parathyroid hormone-related protein on keratinocytes and squamous carcinoma cell lines. Endocrinology **136**:3016–3023.
93. Holick MF, Ray S, Chen TC, Tian X, Persons KS 1994 A parathyroid hormone antagonist stimulates epidermal proliferation and hair growth in mice. Proc Natl Acad Sci USA **91**:8014–8016.
94. Whitfield JR, Isaacs RJ, Jouishomme H, MacLean S, Chakravarthy BR, Morley P, Barisoni D, Regalia E, Armato U 1996 C-terminal fragment of parathyroid hormone-related protein, PTHrP-(107–111), stimulates membrane-associated protein kinase C activity and modulates the proliferation of human and murine skin keratinocytes. J Cell Physiol **166**:1–11.
95. Luger TA, Stadler BM, Luger BM, Sztein MB, Schmidt JA, Hawley-Nelson P, Grabner G, Oppenheim JJ 1983 Characteristics of an epidermal cell thymocyte-activating factor (ETAF) produced by human epidermal cells and a human squamous cell carcinoma line. J Invest Dermatol **81**:187–193.
96. Kupper TS, Ballard DW, Chau AO, McGuire JS, Flood PM, Horowicz MC, Langdon R, Lightfoot L, Gubler U 1986 Human keratinocytes contain mRNA indistinguishable from monocyte interleukin 1α and β mRNA. J Exp Med **164**:2095–2100.
97. Luger TA, Kock A, Kirnbauer R, Scharwarz T, Ansel JC 1988 Keratinocyte-derived interleukin 3. In: Milstone LM, Edelson RL (eds) Endocrine, Metabolic, and Immunologic Functions of Keratinocytes. Annals of the New York Academy of Sciences, New York, pp. 253–261.
98. Grossman RM, Krueger J, Yourish D, Granelli-Piperno A, Murphy DP, May LT, Kupper TS, Sehgal PB, Gottlieb AB 1989 Interleukin 6 is expressed in high levels in psoriatic skin and stimulates proliferation of cultured human keratinocytes. Proc Natl Acad Sci USA **86**:6367–6371.
99. Larsenk CG, Anderson AO, Oppenheim JJ, Matsushima K 1989 Production of interleukin-8 by human dermal fibroblasts and keratinocytes in response to interleukin-1 or tumor necrosis factor. Immunology **68**:31–36.
100. Kupper TS, Lee F, Coleman D, Chodakewitz J, Flood P, Horowitz M 1988 Keratinocyte derived T-cell growth factor (KTGF) is identical to granulocyte macrophage colony stimulation factor (GM-CSF). J Invest Dermatol **91**:185–188.
101. Chodakewitz JA, Lacy J, Edwards SE, Birchall N, Coleman DL 1990 Macrophage colony-stimulating factor production by murine and human keratinocytes. J Immunol **144**:2190–2196.
102. Kupper TS, Horowitz M, Birchall N, Mizutani H, Coleman D, McGuire J, Flood P, Dower S, Lee F 1988 Hematopoietic, lymphopoietic, and proinflammatory cytokines produced by human and murine keratinocytes. In: Milstone LM, Edelson RL (eds) Endocrine, Metabolic and Immunologic Functions of Keratinocytes. Annals of the New York Academy of Sciences, New York, pp. 262–270.
103. Menon GK, Grayson S, Elias P 1985 Ionic calcium reservoirs in mammalian epidermis: Ultrastructural localization by ion-captured cytochemistry. J Invest Dermatol **84**:508–512.
104. Hennings H, Holbrook KA 1983 Calcium regulation of cell–cell contact and differentiation of epidermal cells in culture. An ultrastructural study. Exp Cell Res **143**:127–142.
105. Zamnsky GB, Nguyen U, Chou I-N 1991 An immunofluorescence study of the calcium-induced coordinated reorganization of microfilaments and microtubules in cultured human epidermal keratinocytes. J Invest Dermatol **97**:985–994.
106. Inohara S, Tatsumi Y, Cho H, Tanaka Y, Sagami S 1990 Actin filament and desmosome formation in cultured human keratinocytes. Arch Dermatol Res **282**:210–212.
107. Yoneda K, Fujimoto T, Imamura S, Ogawa K 1990 Fodrin is localized in the cytoplasm of keratinocytes cultured in low calcium medium: Immunoelectron microscopic study. Acta Histochem Cytochem **23**:139–148.
108. Su M-J, Bikle DD, Mancianti ML, Pillai S 1994 1,25-Dihydroxyvitamin D_3 potentiates the keratinocyte response to calcium. J Biol Chem **269**:14723–14729.
109. Yuspa SH, Kilkenny AE, Steinert PM, Roop DR 1989 Expression of murine epidermal differentiation markers is tightly regulated by restricted extracellular calcium concentrations in vitro. J Cell Biol **109**:1207–1217.
110. Ng DC, Su M-J, Kim R, Bikle DD 1996 Regulation of involucrin gene expression by calcium in normal human keratinocytes. Front Biosci **1**:16–24.
111. Huff CA, Yuspa SH, Rosenthal D 1993 Identification of control elements 3′ to the human keratin 1 gene that regulate cell type and differentiation-specific expression. J Biol Chem **268**:377–384.
112. Ryynanen J, Jaakkola S, Engvall E, Peltonen J, Uitto J 1991 Expression of B4 integrins in human skin: Comparison of epidermal distribution with B1-integrin epitopes, and modulation by calcium and vitamin D_3 in cultured keratinocytes. J Invest Dermatol **97**:562–567.
113. Glick AB, Sporn MB, Yuspa SH 1991 Altered regulation of TGF-β1 and TGF-α in primary keratinocytes and papillomas expressing v-Ha-ras. Mol Carcinogen **4**:210–219.
114. Hennings H, Kruszewski FH, Yuspa SH, Tucker RW 1989 Intracellular calcium alterations in response to increased external calcium in normal and neoplastic keratinocytes. Carcinogenesis **10**:777–780.
115. Sharpe GR, Gillespie JI, Greenwell JF 1989 An increase in intracellular free calcium is an early event during differentiation of cultured human keratinocytes. FEBS Lett **254**:25–28.
116. Kruszewski FH, Hennings H, Tucker RW, Yuspa SH 1991 Differences in the regulation of intracellular calcium in normal and neoplastic keratinocytes are not caused by ras gene mutations. Cancer Res **51**:4206–4212.
117. Bikle DD, Ratnam AV, Mauro T, Harris J, Pillai S 1996 Changes in calcium responsiveness and handling during keratinocyte differentiation: Potential role of the calcium receptor. J Clin Invest **97**:1085–1093.
118. Pillai S, Bikle DD, Mancianti M-L, Hincenbergs M 1991 Uncoupling of the calcium-sensing mechanism and differentiation in squamous carcinoma cell lines. Exp Cell Res **192**:567–573.
119. Pillai S, Bikle DD 1992 Lanthanum influx into cultured human keratinocytes: Effect on calcium flux and terminal differentiation. J Cell Physiol **151**:623–629.
120. Reiss M, Lipsey L, Zhou Z-L 1991 Extracellular calcium-dependent regulation of transmembrane calcium fluxes in murine keratinocytes. J Cell Physiol **147**:281–291.
121. Holladay K, Fujiki H, Bowden GT 1992 Okadaic acid induces the expression of both early and secondary response genes in mouse keratinocytes. Mol Carcinogen **5**:16–24.
122. Bollag WB, Xiong Y, Ducote J, Harmon CS 1994 Regulation of fos–lacZ fusion gene expression in primary mouse epidermal keratinocytes isolated from transgenic mice. Biochem J **300**:263–270.
123. Pillai S, Bikle DD 1992 ATP stimulates phosphoinositide metabolism, mobilizes intracellular calcium and inhibits terminal differ-

entiation of human epidermal keratinocytes. J Clin Invest **90**:42–51.
124. Nemeth EF, Scarpa A 1987 Rapid mobilization of cellular calcium in bovine parathyroid cells evoked by extracellular divalent cations. J Biol Chem **262**:5188–5196.
125. Brown EM, Gamba G, Riccardi D, Lombardi M, Butters R, Kifor O, Sun A, Hediger MA, Lytton J, Hebert SC 1993 Cloning and characterization of an extracellular Ca^{2+}-sensing receptor from bovine parathyroid. Nature **366**:575–580.
126. Garrett JE, Capuano IV, Hammerland LG, Hung BCP, Brown EM, Hebert SC, Nemeth EF, Fuller F 1995 Molecular cloning and functional expression of human parathyroid calcium receptor cDNAs. J Biol Chem **270**:12919–12925.
127. Ratnam AV, Cho J-K, Bikle DD 1996 1,25-dihydroxyvitamin D_3 enhances the calcium response of keratinocytes. J Invest Dermatol **106**:910.
128. Jaken S, Yuspa SH 1988 Early signs for keratinocyte differentiation: Role of Ca^{2+}-mediated inositol lipid metabolism in normal and neoplastic epidermal cells. Carcinogenesis **9**:1033–1038.
129. Tang W, Ziboh VA, Isseroff R, Martinez D 1988 Turnover of inositol phospholipids in cultured murine keratinocytes: Possible involvement of inositol triphosphate in cellular differentiation. J Invest Dermatol **90**:37–43.
130. Moscat J, Flemming TP, Molloy CJ, Lopez-Barahona M, Aaronson SA 1989 The calcium signal for Balb 1 MK keratinocyte terminal differentiation induces sustained alterations in phosphoinositide metabolism without detectable protein kinase C activation. J Biol Chem **264**:11228–11235.
131. Lee E, Yuspa SH 1991 Aluminum fluoride stimulates inositol phosphate metabolism and inhibits expression of differentiation markers in mouse keratinocytes. J Cell Physiol **148**:106–115.
132. Hawley-Nelson P, Stanley JR, Schmidt J, Gullino M, Yuspa SH 1982 The tumor promoter 12-*O*-tetradecanoylphorbol-13-acetate accelerates keratinocyte differentiation and stimulates growth of an unidentified cell type in cultured human epidermis. Exp Cell Res **137**:155–167.
133. Yuspa SH, Ben T, Hennings H, Lichti U 1982 Divergent responses in epidermal basal cells exposed to the promoter 12-*O*-tetradecanoylphorbol-13-acetate. Cancer Res **42**:2344–2349.
134. Kitajima Y, Inoue S, Nagao S, Nagata K, Yaoita H, Nozawa Y 1988 Biphasic effects of 12-*O*-tetradecanoylphorbol-13-acetate on the cell morphology of low calcium-grown human epidermal carcinoma cells: Involvement of translocation and down regulation of protein kinase C. Cancer Res **48**:964–970.
135. Rice R, Rong X, Chakravarty R 1988 Suppression of keratinocyte differentiation in SSC-9 human squamous carcinoma cells by benzo[*a*]pyrene, 12-*O*-tetradecanoylphorbol-13-acetate and hydroxyurea. Carcinogenesis **9**:1885–1890.
136. Sheu H, Kitajima Y, Yaoita H 1989 Involvement of protein kinase C in translocation of desmoplakins from cytosol to plasma membrane during desmosome formation in human squamous cell carcinoma cells grown in low to normal calcium concentration. Exp Cell Res **185**:176–190.
137. Stanwell C, Denning MF, Rutberg SE, Cheng C, Yuspa SH, Dlugosz AA 1996 Staurosporine induces a sequential program of mouse keratinocyte terminal differentiation through activation of PKC isozymes. J Invest Dermatol **106**:482–489.
138. Nagao S, Kitajima Y, Nagata K, Inoue S, Yaoita H, Nozawa Y 1989 Correlation between cell–cell contact formation and activation of protein kinase C in a human squamous cell carcinoma cell line. J Invest Dermatol **92**:175–178.
139. Yuspa SH, Ben T, Hennings H 1983 The induction of epidermal transglutaminase and terminal differentiation by tumor promoters in cultured epidermal cells. Carcinogenesis **4**:1413–1418.
140. Dlugosz AA, Yuspa SH 1994 Protein kinase C regulates keratinocyte transglutaminase (TGK) gene expression in cultured primary mouse epidermal keratinocytes induced to terminally differentiate by calcium. J Invest Dermatol **102**:409–414.
141. Matsui MS, Illarda I, Wang N, DeLeo VA 1993 Protein kinase C agonist and antagonist effects in normal human epidermal keratinocytes. Exp Dermatol **2**:247–256.
142. Dlugosz AA, Yuspa SH 1993 Coordinate changes in gene expression which mark the spinous to granular cell transition in epidermis are regulated by protein kinase C. J Cell Biol **120**:217–225.
143. Filvaroff E, Stern D, Dotto G 1990 Tyrosine phosphorylation is an early and specific event involved in primary keratinocyte differentiation. Mol Cell Biol **10**:1164–1173.
144. Chakravarthy BR, Isaacs RJ, Morley P, Durkin JP, Whitfield JF 1995 Stimulation of protein kinase C during Ca^{2+}-induced keratinocyte differentiation. Selective blockade of MARCKS phosphorylation by calmodulin. J Biol Chem **270**:1362–1368.
145. Punnonen K, Denning M, Lee E, Li L, Rhee SG, Yuspa SH 1993 Keratinocyte differentiation is associated with changes in the expression and regulation of phospholipase C isoenzymes. J Invest Dermatol **101**:719–726.
146. Hepler JR, Earp HS, Harden TK 1988 Long-term phorbol ester treatment down-regulates protein kinase C and sensitizes the phosphoinositide signaling pathway to hormone and growth factor stimulation. Evidence for a role of protein kinase C in agonist-induced desensitization. J Biol Chem **263**:7610–7619.
147. Kojima I, Shibata H, Ogata E 1986 Phorbol ester inhibits angiotensin-induced activation of phospholipase C in adrenal glomerulosa cells. Its implication in the sustained action of angiotensin. Biochem J **237**:253–258.
148. Dlugosz AA, Mischak H, Mushinski JF, Yuspa SH 1992 Transcripts encoding protein kinase C-alpha, -delta, -epsilon, -zeta, and -eta are expressed in basal and differentiating mouse keratinocytes in vitro and exhibit quantitative changes in neoplastic cells. Mol Carcinogen **5**:286–292.
149. Denning MF, Dlugosz AA, Howett MK, Yuspa SH 1993 Expression of an oncogenic rasHa gene in murine keratinocytes induces tyrosine phosphorylation and reduced activity of protein kinase C delta. J Biol Chem **268**:26079–26081.
150. Reynolds NJ, Baldassare JJ, Henderson PA, Shuler JL, Ballas LM, Burns DJ, Moomaw CR, Fisher CJ 1994 Translocation and downregulation of protein kinase C isoenzymes-alpha and -epsilon by phorbol ester and bryostatin-1 in human keratinocytes and fibroblasts. J Invest Dermatol **103**:364–369.
151. Fisher GJ, Tavakkol A, Leach K, Burns D, Basta P, Loomis C, Griffiths CE, Cooper KD, Reynolds NJ, Elder JT 1993 Differential expression of protein kinase C isoenzymes in normal and psoriatic adult human skin: Reduced expression of protein kinase C-beta II in psoriasis. J Invest Dermatol **101**:553–559.
152. Geiges D, Marks F, Gschwendt M 1995 Loss of protein kinase C delta from human HaCaT keratinocytes upon ras transfection is mediated by TGF alpha. Exp Cell Res **219**:299–303.
153. Goodnight J, Mischak H, Mushinski JF 1994 Selective involvement of protein kinase C isozymes in differentiation and neoplastic transformation. Adv Cancer Res **64**:159–209.
154. Denning MF, Dlugosz AA, Williams EK, Szallasi Z, Blumberg PM, Yuspa SH 1995 Specific protein kinase C isozymes mediate the induction of keratinocyte differentiation markers by calcium. Cell Growth Differ **6**:149–157.
155. Szallasi Z, Kosa K, Smith CB, Dlugosz AA, Williams EK, Yuspa SH, Blumberg PM 1995 Differential regulation by anti-tumor-promoting 12-deoxyphorbol-13-phenylacetate reveals distinct roles of the classical and novel protein kinase C isozymes in

biological responses of primary mouse keratinocytes. Mol Pharmacol **47**:258–265.

156. Downward J, Waterfield MD, Parker PJ 1985 Autophosphorylation and protein kinase C phosphorylation of the epidermal growth factor receptor. Effect on tyrosine kinase activity and ligand binding affinity. J Biol Chem **260**:14538–14546.
157. Schlessinger J 1986 Allosteric regulation of the epidermal growth factor receptor kinase. J Cell Biol **103**:2067–2072.
158. Kolch W, Heidecker G, Kochs G, Hummel R, Vahidi H, Mischak H, Finkenzeller G, Marme D, Rapp UR 1993 Protein kinase C alpha activates RAF-1 by direct phosphorylation. Nature **364**:249–252.
159. Goode N, Hughes K, Woodgett JR, Parker PJ 1992 Differential regulation of glycogen synthase kinase-3 beta by protein kinase C isotypes. J Biol Chem **267**:16878–16882.
160. Yaar M, Gilani A, DiBenedetto PJ, Harkness DD, Gilchrest BA 1993 Gene modulation accompanying differentiation of normal versus malignant keratinocytes. Exp Cell Res **206**:235–243.
161. Stanwell C, Denning MF, Rutberg SE, Cheng C, Yuspa SH, Dlugosz AA 1996 Staurosporine induces a sequential program of mouse keratinocyte terminal differentiation through activation of PKC isozymes. J Invest Dermatol **106**:482–489.
162. Welter JF, Eckert RL 1995 Differential expression of the fos and jun family members c-fos, fosB, Fra-1, Fra-2, c-jun, junB and junD during human epidermal keratinocyte differentiation. Oncogene **11**:2681–2687.
163. Welter JF, Crish JF, Agarwal C, Eckert RL 1995 Fos-related antigen (Fra-1), junB, and junD activate human involucrin promoter transcription by binding to proximal and distal AP1 sites to mediate phorbol ester effects on promoter activity. J Biol Chem **270**:12614–12622.
164. Fairley JA, Marcelo CL, Hogan VA, Voorhees JJ 1985 Increased calmodulin levels in psoriasis and low Ca^{++} regulated mouse epidermal keratinocyte cultures. J Invest Dermatol **84**:195–198.
165. Al-Ani AM, Messenger AG, Lawry J, Bleehen SS, MacNeil S 1988 Calcium/calmodulin regulation of the proliferation of human epidermal keratinocytes, dermal fibroblasts and mouse B16 melanoma cells in culture. Br J Dermatol **119**:295–306.
166. Eichelberg D, Fuchs A 1988 Calmodulin-antagonism inhibits human keratinocyte proliferation. Arch Dermatol Res **280**:323–324.
167. Grief F, Soroff H, Albers K, Taichman L 1985 The effect of trifluoperazine, a calmodulin antagonist, on the growth of normal and malignant epidermal keratinocytes in culture. J Cell Biol **100**:1455–1465.
168. Sheng M, Thompson MA, Greenberg ME 1991 CREB: A Ca^{2+}-regulated transcription factor phosphorylated by calmodulin-dependent kinases. Science **252**:1427–1430.
169. Horiuchi N, Clemens TL, Schiller AL, Holick MF 1985 Detection and developmental changes of the 1,25-$(OH)_2$-D_3 concentration in mouse skin and intestine. J Invest Dermatol **84**:461–464.
170. Merke J, Schwittay D, Furstenberger G, Gross M, Marks F, Ritz E 1985 Demonstration and characterization of 1,25-dihydroxyvitamin D_3 receptors in basal cells of epidermis of neonatal and adult mice. Calcif Tissue Int **37**:257–267.
171. McLane JA, Katz M, Abdelkader N 1990 Effect of 1,25-dihydroxyvitamin D_3 on human keratinocytes grown under different culture conditions. In Vitro Cell Dev Biol **26**:379–387.
172. Bollag WB, Ducote J, Harmon CS 1995 Biphasic-effect of 1,25-dihydroxyvitamin D_3 on primary mouse epidermal keratinocyte proliferation. J Cell Physiol **163**:248–256.
173. Itin PH, Pittelkow MR, Kumar R 1994 Effects of vitamin D metabolites on proliferation and differentiation of cultured human epidermal keratinocytes grown in serum-free or defined culture medium. Endocrinology **135**:1793–1798.
174. Gniadecki R 1996 Stimulation versus inhibition of keratinocyte growth by 1,25-dihydroxyvitamin D_3: Dependence on cell culture conditions. J Invest Dermatol **106**:510–516.
175. Matsumoto K, Hashimoto K, Nishida Y, Hashiro M, Yoshikawa K 1990 Growth-inhibitor effects of 1,25-dihydroxyvitamin D_3 on normal human keratinocytes cultured in serum-free medium. Biochem Biophys Res Commun **166**:916–923.
176. Rabin-Rizk M, Pavlovitch JH 1988 Effect of vitamin D deficiency and 1,25-dihydroxycholecalciferol treatment on epidermal calcium-binding protein (ECaBP) RNA activity. Mol Cell Endocrinol **60**:145–149.
177. Kremer R, Karaplis AC, Henderson J, Gulliver W, Banville D, Hendy GN, Goltzman D 1991 Regulation of parathyroid hormone-like peptide in cultured normal human keratinocytes. Effect of growth factors and 1,25-dihydroxyvitamin D_3 on gene expression and secretion. J Clin Invest **87**:884–893.
178. Kim HJ, Abdelkader N, Katz M, McLane JA 1992 1,25-Dihydroxy-vitamin-D_3 enhances antiproliferative effect and transcription of TGF-beta1 on human keratinocytes in culture. J Cell Physiol **151**:579–587.
179. Iwasha M, Hashimoto K, Hashiro M, Yoshikawa K 1991 Production of plasminogen activator inhibitor-1 (PA-1) by human keratinocytes is induced by 1,25-dihydroxyvitamin D_3. J Invest Dermatol **96**:558.
180. Kragballe K, Wildfang IL 1990 Calcipotriol (MC 903), a novel vitamin D_3 analogue, stimulates terminal differentiation and inhibits proliferation of cultured human keratinocytes. Arch Dermatol Res **282**:164–167.
181. Tang W, Ziboh VA, Isseroff RR, Martinez D 1987 Novel regulatory action of 1,25-dihydroxyvitamin D on the metabolism of polyphosphoinositides in murine epidermal keratinocytes. J Cell Physiol **132**:131–136.
182. McLaughlin JA, Cantley LC, Holick MF 1990 1,25$(OH)_2D_3$ increased calcium and phosphatidylinositol metabolism in differentiating cultured human keratinocytes. J Nutr Biochem **1**:81–87.
183. Yada Y, Ozeki T, Meguro S, Mori S, Nozawa Y 1989 Signal transduction in the onset of terminal keratinocyte differentiation induced by 1α,25-dihydroxyvitamin D_3: Role of protein kinase C translocation. Biochem Biophys Res Commun **163**:1517–1522.
184. Tang W, Ziboh VA 1991 Agonist/inositol trisphosphate-induced release of calcium from murine keratinocytes: A possible link with keratinocyte differentiation. J Invest Dermatol **96**:134–138.
185. Bittiner B, Bleehen SS, MacNeil S 1991 1α,25$(OH)_2$ vitamin D_3 increased intracellular calcium in human keratinocytes. Br J Dermatol **124**:230–235.
186. Regnier M, Darmon M 1991 1,25-Dihydroxyvitamin D_3 stimulates specifically the last steps of epidermal differentiation of cultured human keratinocytes. Differentiation **47**:173–188.
187. Pillai S, Bikle DD, Elias PM 1988 Vitamin D and epidermal differentiation: Evidence for a role of endogenously produced vitamin D metabolites in keratinocyte differentiation. Skin Pharmacol **1**:149–160.
188. Feingold KR, Williams ML, Pillai S, Menon GK, Bikle DD, Elias PM 1987 The effect of vitamin D status on cutaneous sterologenesis in vivo and in vitro. Biochim Biophys Acta **930**:193–200.
189. Pavlovitch JH, Galoppin L, Rizk M, Didierjean L, Balsan S 1984 Alterations in rat epidermis provoked by chronic vitamin D deficiency. Am J Physiol **247**:E228–E233.

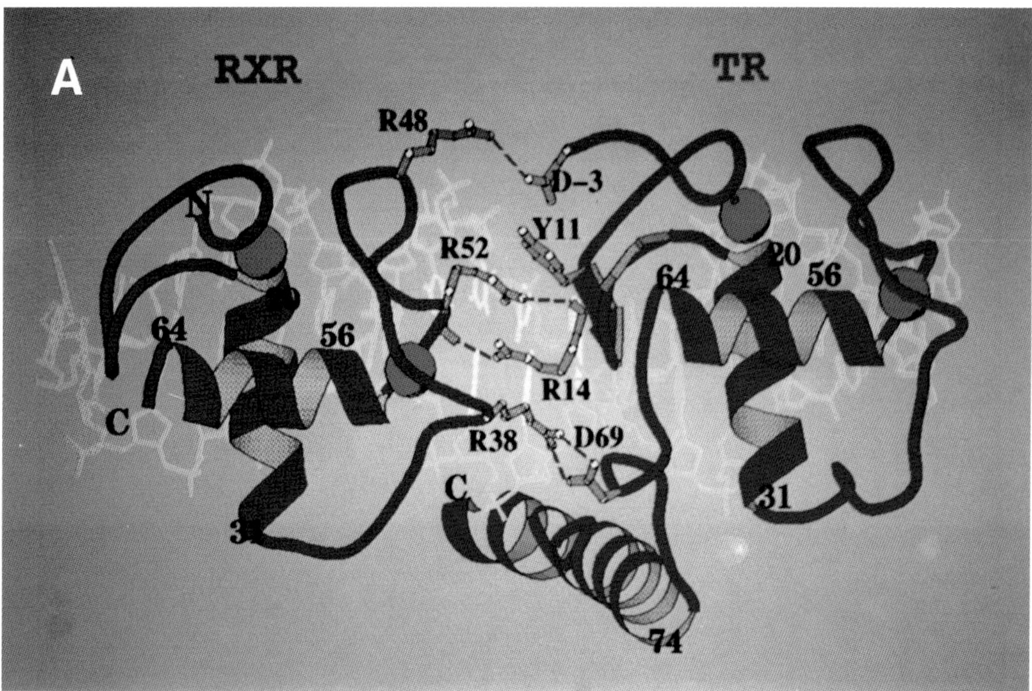

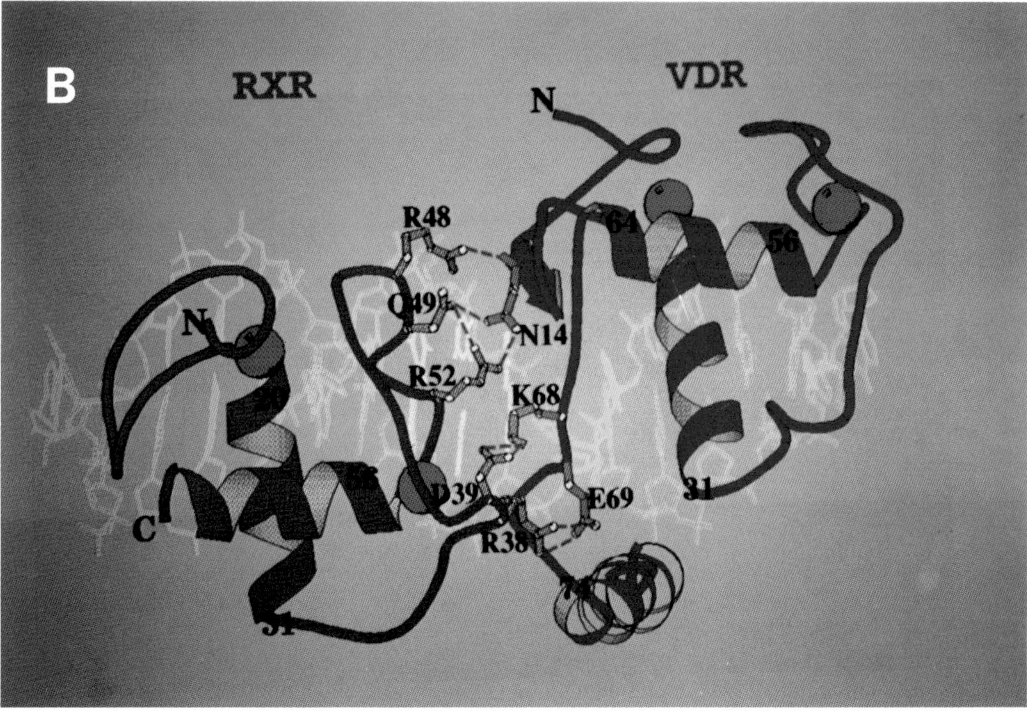

FIGURE 9-4 Asymmetric dimer interfaces of the RXR–TR DBDs (A) and modeled RXR–VDR DBDs (B) bound to DR4 and DR3 elements, respectively. Yellow base pairs are the two hexameric half-sites; white base pairs are spacers; green balls are zinc ions; TR is blue, RXR is red, and VDR is purple. Dotted lines depict hydrogen bonds between subunit DBDs. Only those residues participating in the dimer interfaces are shown. Amino acid numbering takes the first coordinating Cys as 1. For VDR, this is residue 24. See text for details. From Rastinjead *et al.* [19].

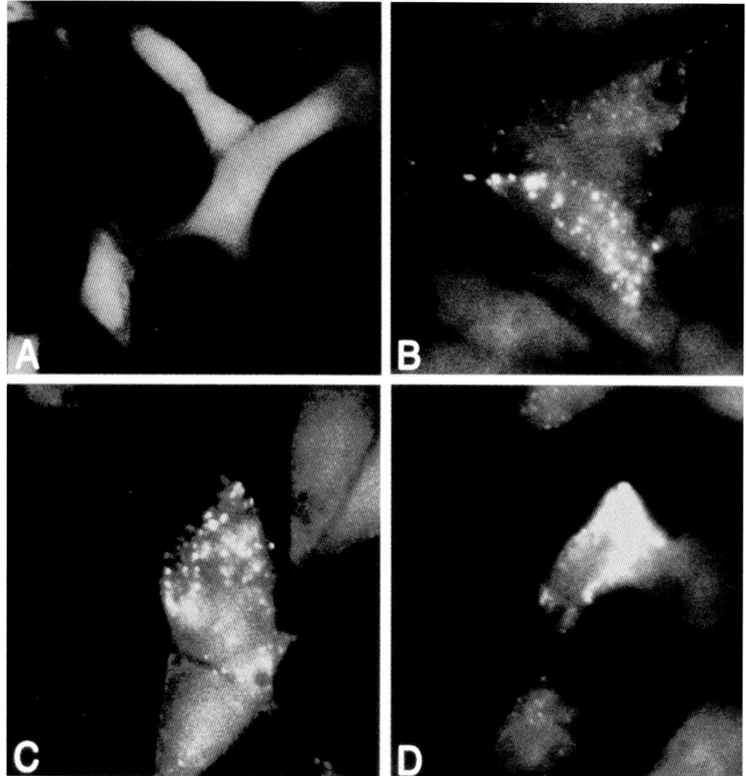

FIGURE 13-3 Analysis of insulin distribution by immunocytochemistry with anti-insulin antibody in a rat β-cell line (RIN1046-38) overexpressing calbindin. (A) Vector-transfected clone V9. (B–D) Calbindin-D_{28K}-transfected clone C2. Note the marked increase in insulin immunoreactivity in RIN cells overexpressing calbindin. Approximately 20% of calbindin-transfected cells demonstrated many insulin granules, which are not observed in the control vector-transfected clones.

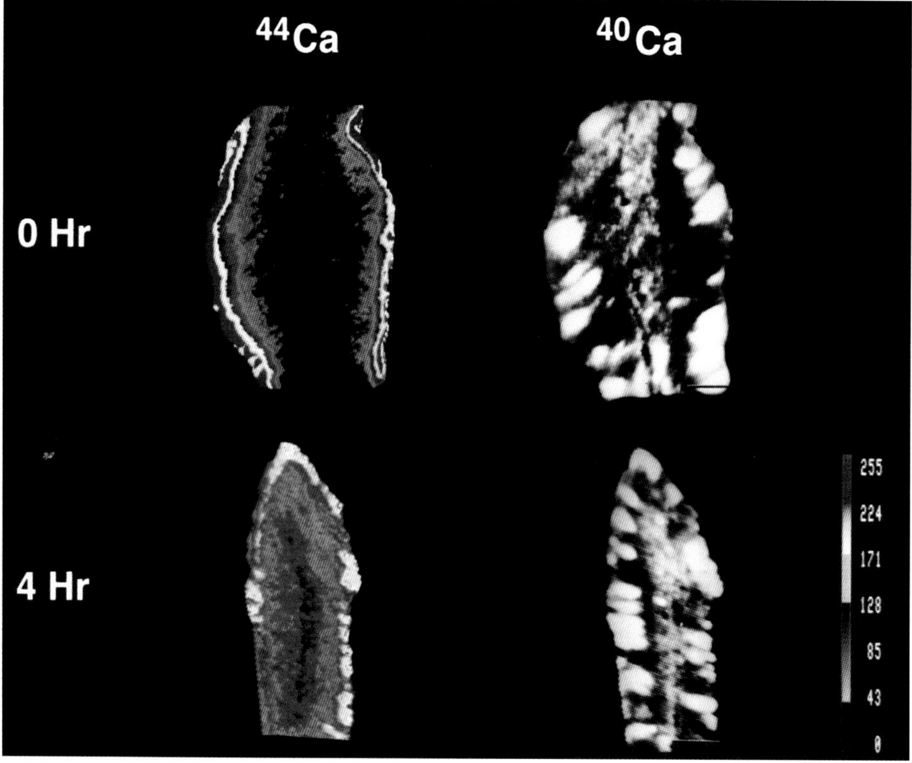

FIGURE 16-2 Ion microscopy images of luminally administered ^{44}Ca in intestinal tissue of vitamin D-deficient chicks either given vehicle (0 hr) or at 4 hr after a single dose of $1,25(OH)_2D_3$. The images are presented in pseudocolor, with a color-coded bar included for estimation of relative intensities. The corresponding ^{40}Ca images of the same sections are shown at right. The high density structures in the ^{40}Ca images are goblet cells, known to have a high concentration of calcium. From Fullmer *et al.* [47]; reprinted with permission.

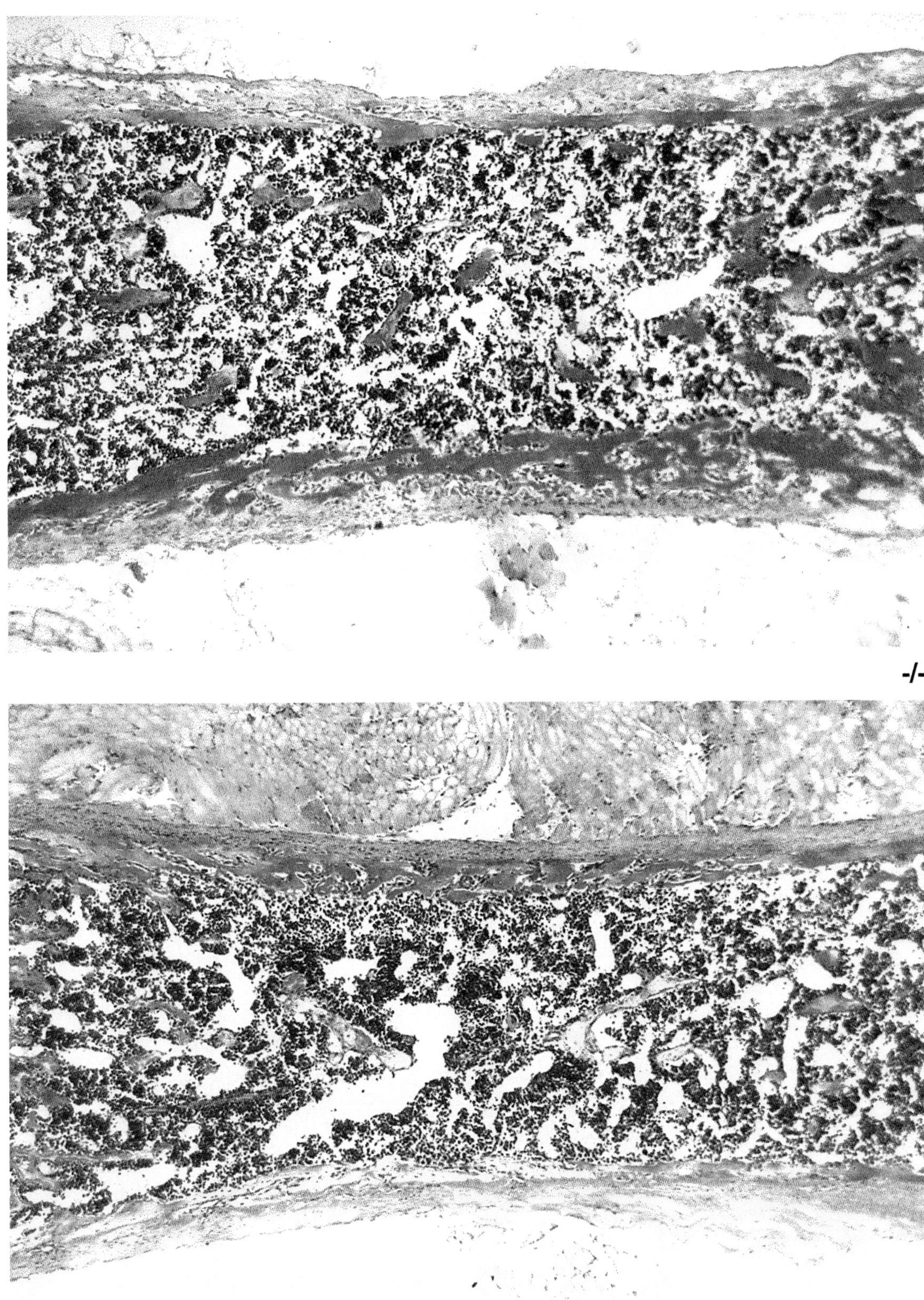

FIGURE 18-2 Abnormal intramembranous ossification in mice deficient for the vitamin D 24-hydroxylase gene, as shown in longitudinal sections through the femur of 5-day-old neonates. (Bottom) Homozygous mutant (–/–) sample; (top) heterozygote (+/–) control littermate. Note the abnormal cortex with abundant unmineralized osteoid (stained in red) at the exocortical surface of the bone from mutant mice.

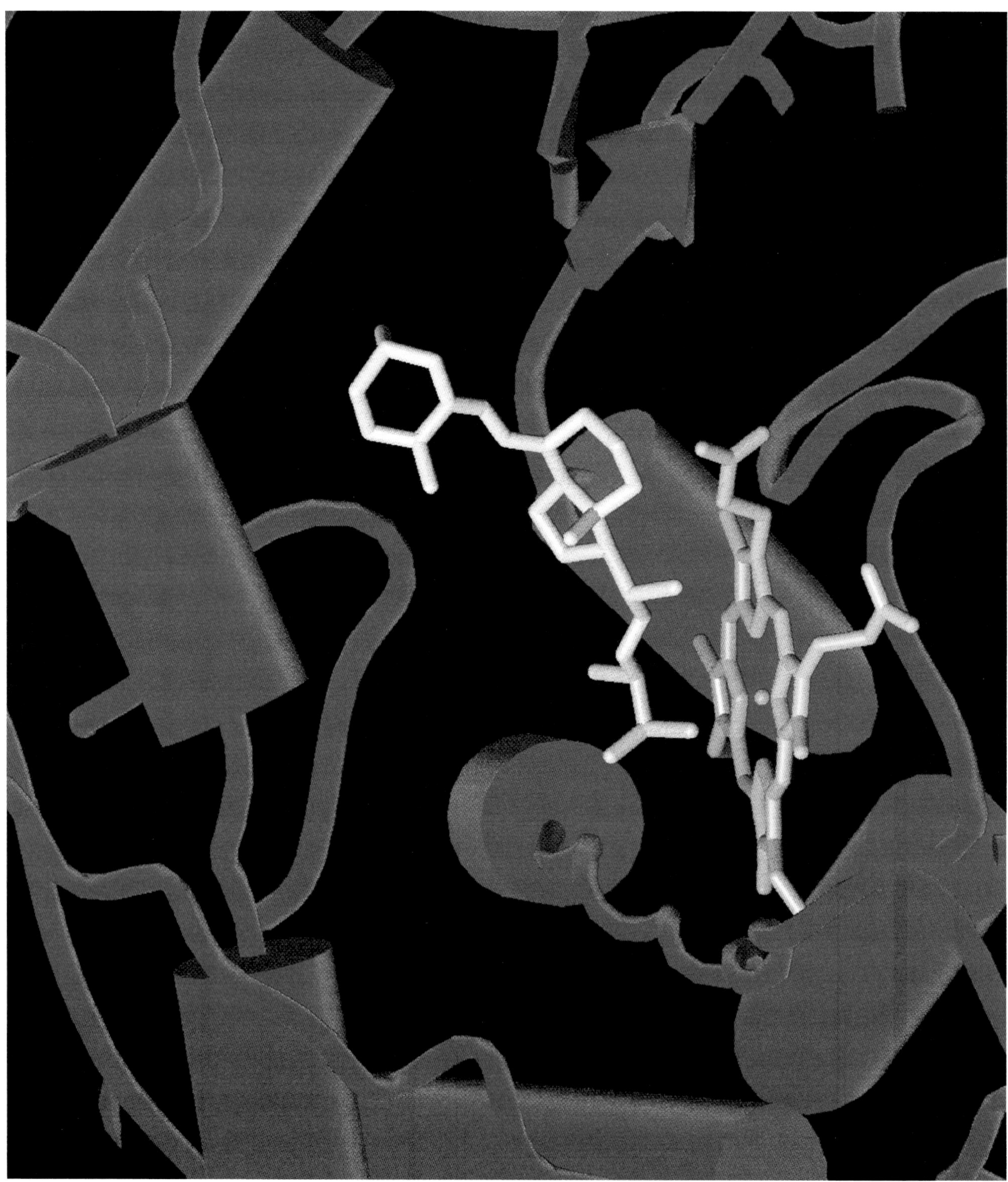

FIGURE 58-8 Molecular modeling of the vitamin D-related cytochrome P450, CYP27. An unobstructed view of the active site of CYP27 shows a possible vitamin D_2 binding orientation above the heme prosthetic group. Residues 120–160, which normally block the view of the binding site, are not displayed in the figure. The heme is anchored between the I helix and a β-sheet structure by interactions with Cys-476, Arg-405, and possibly Arg-474, Arg-158, and Trp-154. The model was derived [108] from prokaryotic P450 crystal structures for CAM, TERP, BM-3, and EryF and is currently being used to guide site-directed mutagenesis experiments and vitamin D docking studies. From D. Prosser, D. Weaver, and G. Jones (unpublished results).

CHAPTER 26

Cartilage and Vitamin D: Genomic and Nongenomic Regulation by $1,25(OH)_2D_3$ and $24,25(OH)_2D_3$

BARBARA D. BOYAN,[1,2,3] DAVID D. DEAN,[1] VICTOR L. SYLVIA,[1] AND ZVI SCHWARTZ[1,2,4]

Departments of [1]Orthopedics, [2]Periodontics, and [3]Biochemistry, The University of Texas Health Science Center at San Antonio, San Antonio, Texas, and [4]Hebrew University Hadassah Faculty of Dental Medicine, Jerusalem, Israel

I. CHONDROGENESIS AND ENDOCHONDRAL OSSIFICATION *IN VIVO*

A. The Chondrocyte Lineage

Cartilage is a group of tissues produced by chondrocytes that share a set of specific characteristics. These include an extracellular matrix consisting of predominantly type II collagen and proteoglycan, often in the form of proteoglycan aggregate [1–3]. Moreover, the glycosaminoglycan side chains on the proteoglycan core protein are highly sulfated in the mature tissue [2–4]. Cartilage is also distinguished by its relative lack of vascularity [5–10].

Embryologically, cartilage forms from mesoderm and

provides the structural framework for a number of other tissues. The best understood of these is bone. Mesenchymal stem cells differentiate into chondroblasts, which produce and maintain the proteoglycan-rich type II collagen matrix [11–14]. Eventually, the chondrocytes undergo hypertrophy and mineralize their extracellular matrix. Once this has occurred, the tissue is resorbed by osteoclasts, accompanied by vascular invasion of the mineralized regions. Osteoprogenitor cells migrate to the calcified cartilage scaffold and form bone [15,16]. Ultimately the cartilage within the newly forming bone also resorbs, leaving a marrow cavity [15–18]. In addition, cartilage at the ends of the bones further differentiates along two divergent pathways. At the juncture of the newly formed bone and the terminal cartilage, the cells develop what will eventually become the growth plate. In the areas that will become the joints, the cells form what will eventually be articular cartilage.

Within the developing embryo, there are many variations on this theme. For example, the mandibular condyle develops a secondary cartilage with a layer of pluripotential cells at the outer surface of the cartilage. In postfetal life, these cells can respond to physical forces on the cartilage, enabling considerable modeling to occur. Other cartilages develop with a fibroblastic component, resulting in a broad range of structural properties.

In general, chondrocytes go through a series of maturational stages as they undergo their lineage cascade. In the simplest terms, mesenchymal stem cells are induced to become osteochondral progenitor cells and, finally, chondroprogenitor cells. Once the cells are committed to the chondrogenic pathway, their further maturation is limited to this lineage [19–22]. The ability of committed chondrocytes to proliferate is considerably reduced. However, the chondrocytes do continue to undergo mitosis, and, because of the physical properties of the tissue, they can be seen to do so as clonal foci [23]. Once mitosis is complete, the cells produce and maintain an extensive extracellular matrix. In hyaline-type cartilages, the cells are relatively far apart, but they are interconnected through fibrils in the extracellular matrix that form a meshwork, enabling the individual chondrocytes to sense events throughout the tissue.

Most of the cells in hyaline cartilages, like the articular cartilages at the ends of long bones, are in a relatively stable state of maturation. However, the cells continue to progress through the lineage, although at a very slow rate. For example, at the interface between articular cartilage and the adjacent subchondral bone, there is a region of cartilage similar to the growth plate in which chondrocytes undergo hypertrophy and mineralize their extracellular matrix. Calcification of the cartilage in this region, termed the "tidemark," continues throughout the life of the individual, or until the articular cartilage is lost from the joint surface [1].

At the other end of the spectrum is fracture callus and induced bone formation, in which cells traverse the entire lineage relatively rapidly [24–27]. As demonstrated by Reddi *et al.* [28,29] and Urist *et al.* [30], when demineralized bone powder is implanted into mesenchymal tissues that would otherwise not support bone formation, mesenchymal cells migrate to the implant and undergo proliferation. By the third day postimplantation, these cells begin to acquire a chondroblast phenotype, and by day 7 they have formed cartilage. By day 11 the cartilage is calcified. Although these events do not occur with the spatial orientation normally associated with growth plate cartilage, they retain their temporal organization, with most of the population of cells undergoing transition from maturation state to maturation state in synchrony. The transition from mesenchymal cells to chondrocytes has been observed *in vitro* following treatment of neonatal rat muscle cells [31] or chick limb bud cells [32] with bone morphogenetic protein (BMP).

The growth plate, itself, represents a specialized situation in which the terminal differentiation of the chondrocyte occurs in a linear array so that chondrocytes appear as columns of cells traversing the lineage in clearly demarcated zones of maturation [33,34] (Fig. 1). At one end of the growth plate is the resting zone, also called the reserve zone, in which the cells exhibit a hyaline cartilage-like phenotype. The length of time a cell may remain in the resting zone is unknown. Hormonal signals stimulate the cells to undergo a proliferative burst, after which they remain in a prehypertrophic state for varying times. Studies using knock-out mice for a number of proteins believed to play a role in bone formation have shown that the lesions are frequently expressed most notably in cells of the prehypertrophic zone of cartilage [35,36]. These postproliferative cells are undergoing a major shift in phenotype. Not only do they begin to change shape, but they also modify their extracellular matrix. It is during the prehypertrophic maturation state that they synthesize the proteins that will make this possible.

Once hypertrophy begins, the chondrocytes appear to be in lockstep for their final differentiation along the endochondral lineage [10,37–40]. Hypertrophic chondrocytes increase markedly in size and must make major changes in the composition of their extracellular matrix to accommodate this [41]. In addition, they must prepare their matrix for mineral deposition. It is perhaps in this aspect of their maturation that the greatest amount of controversy has arisen [33,34]. As the cells reach the lowest regions of the growth plate, they cease to make

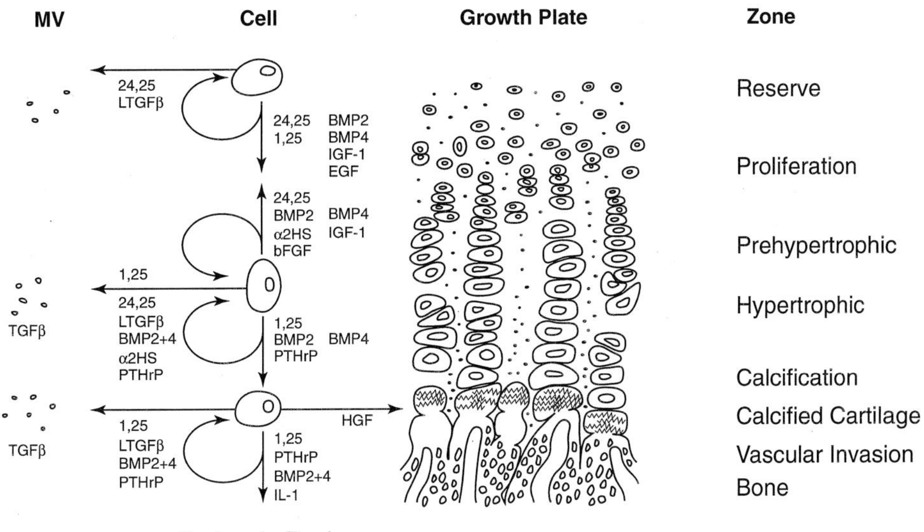

FIGURE 1 Schematic drawing of the growth plate showing the local factors and hormones that regulate the chondrocytes at various points in their differentiation. Cells synthesize and secrete a number of cytokines and growth factors, as well as $1,25(OH)_2D_3$ (1,25) and $24,25(OH)_2D_3$ (24,25), as they mature in the endochondral lineage. Although some of these factors have been demonstrated *in vivo* by immunohistochemistry, others have been inferred from cell culture data in the literature. This list is not intended to be complete, but rather emphasizes the important role of local factors, including vitamin D metabolites, in conveying messages between cells both up and down the growth plate, as well as between cells and their extracellular matrix vesicles (MV). Our own studies have shown that resting zone cells synthesize latent transforming growth factor-β (LTGFβ) *in vivo* and *in vitro* and incorporate it into their extracellular matrix. MV isolated from growth zone chondrocyte cultures (prehypertrophic and hypertrophic zones) can activate LTGFβ when they are incubated with $1,25(OH)_2D_3$ but not $24,25(OH)_2D_3$. In contrast, MV produced by resting zone cells do not activate LTGFβ when treated with either metabolite. BMP, Bone morphogenetic protein; IGF-1, insulin-like growth factor-1; α2HS-glycoprotein; bFGF, basic fibroblast growth factor; PTHrP, parathyroid hormone-related peptide; IL-1, interleukin-1; HGF, hepatocyte growth factor.

cartilage-specific proteins, such as proteoglycan aggregate and type II collagen [3,39,42–44]. They synthesize type X collagen at this time [12–14,45–47]. However, they also begin to make proteins normally associated with bone and tooth mineralization such as osteocalcin, osteopontin, and type I collagen [48–50]. It is important to note that these cells are not osteoblasts, just as odontoblasts are not osteoblasts. They are chondrocytes that are mineralizing their matrix, and, to do so, they must make a matrix that can support hydroxyapatite formation.

Once the cells accomplish this goal, they die by apoptosis [51]. There have been reports that sister cells may survive the apoptotic event, proceeding on to form osteoblasts [51,52]. Other investigators have reported that, in the mandibular condyle, some of the cells "transdifferentiate" into osteoblasts. Whether either is the case is difficult to prove, as pluripotential cells are present and may undergo differentiation into osteoblasts at this time. This is particularly true in cell culture. Depending on culture conditions, cells may adapt or redifferentiate. Thus, chondrocytes may make type I collagen in culture as an adaptation to the presence of serum or to the process of attachment and spreading. In spite of this, these cells will retain many of the features of their *in vivo* chondrocyte phenotype.

Although there is considerable overlap in the articular cartilage and growth plate chondrocyte phenotypes, the cells do exhibit differences, depending on their geographical location within the tissue and the presence of pathology such as osteoarthritis. Unfortunately, the nature of cell culture has led to a confusing array of findings concerning chondrocyte cell biology. *In vivo*, articular chondrocytes and other hyaline cartilage cells exist individually or as isolated clones within the extracellular matrix. These cells tend to have a rounded morphology that can be retained in culture through growth in agarose or on type I collagen gels [53–56]. Neither of these materials are physiologically identical to the type II collagen and proteoglycan matrix the cells experience *in vivo* and, consequently, may impose artifacts on the model. In contrast, growth plate chondrocytes tend to exist in columns and are much more closely packed than the articular cartilage cells. Their phenotype *in vivo* is distinctly different. They produce a different matrix, they produce extracellular matrix vesicles in much greater numbers, and they are actively undergoing extensive shape changes. Most investigators do not grow these cells in three-dimensional cultures [57–65], so direct comparisons with studies on articular chondrocytes are often difficult.

In this chapter, we focus on growth plate chondrocytes since this tissue has been most extensively studied with respect to vitamin D. In general, the discussion refers to growth plates typical of postfetal development. Two animal models, chickens and rats, are most often used. Major differences exist between them, and neither model is a precise analog for human endochondral ossification. Most *in vivo* studies have used the tibial growth plate, although by no means exclusively. The tibia is an example of a primary growth plate. In contrast, the mandibular condyle is a secondary growth plate and, in humans, remains open even in adults. Thus, direct comparisons with the tibial growth plate may not be appropriate. The increased use of cell culture has led investigators to develop the costochondral cartilage of rabbits and rats as a growth plate model. Because these tissues are not exposed to the compressive forces experienced by tibial growth plate, there may be biochemical differences that could affect the generalization of results obtained with these models.

A number of studies have used chicken tissue as the model system, and the chick limb bud is a frequently used culture system [66,67]. At the time it is placed in culture, the limb bud contains pluripotential mesenchymal cells together with committed osteochondroprogenitor cells. These cells continue along their lineage cascade in culture. In postfetal life, chick tibial growth plate is used because of the extreme sensitivity of chickens to vitamin D [68]. Unfortunately, the complex interdigitation of bone and hypertrophic cartilage can make interpretation of results more difficult; nonetheless, this has been an extremely productive model.

The rachitic rat has also been of considerable value to investigators [41,69–71]. To maintain the rat in a rachitic state it is necessary to make it hypophosphatemic as well as vitamin D deficient. In addition, the rat growth plate remains open throughout life, making direct comparisons with the human growth plate more complex. Even so, the rat costochondral cartilage has been an exceptionally valuable model for understanding the role of maturational state on cellular response to vitamin D [64,65].

B. Cartilage Proteins

The purpose of this section is to acquaint the reader with those proteins which give cartilage its unique character. A schematic drawing of a generalized growth plate is provided for reference (Fig. 1).

1. Collagen

The structural integrity of the growth plate is largely dependent on its collagenous matrix. Types II, IX, X, and XI collagens are present, with each zone containing a unique mixture of the various collagen types at different concentrations [72–75]. The amount of collagen in each zone also varies. In the upper part of the growth plate, collagen represents as much as 70% of the dry weight of the tissue. This drops to a value of 23–25% in the hypertrophic cell zone [76,77] owing to the decrease in the amount of extracellular matrix per unit volume. To accommodate the high degree of remodeling of the matrix that occurs in the growth plate, collagen turnover is relatively rapid [41].

The growth plate in long bones like the tibia and femur is bounded by the bony end plates of the metaphysis and epiphysis and the fibrous perichondral structures along the margins. By itself, this architectural design would be unable to withstand the high osmotic pressures transmitted through the growth plate or maintain the shape of the tissue. Thus, an important function of collagen is to help provide a structural network that can impart strength to the tissue. Both class 1, or fibril forming, collagens and class 2, or helical, collagens are found in the growth plate. In addition, other factors, including a variety of noncollagenous proteins, contribute to the maintenance of shape and overall integrity of the tissue [78–81].

a. Type II Collagen Approximately 70% of the total collagen found in the growth plate is type II. It forms narrow, cross-striated fibrils that are heavily glycosylated and virtually unique to cartilaginous tissues. This collagen type predominates in hyaline cartilages and in growth plate may constitute at least half of the tissue dry weight [72,74,75]. In many respects, it is very similar to type I collagen, but the fibrils formed by type II collagen are much thinner and do not pack with the density of type I collagen [73].

b. Type IX Collagen Type IX collagen represents 10–12% of the total collagen present in growth plate [72]. It is a fibril-associated collagen with an interrupted helix formed from three genetically distinct chains [74,82,83]. The molecule contains seven regions along its length: three helices are separated by four nonhelical domains. It has been postulated that this structure allows type IX collagen, as it is positioned on the type II collagen molecule through lysine-derived cross-links, to project a large amino-terminal globular domain from the surface of the fibril into the surrounding matrix [74,84,85]. These projections, then, could be possible attachment sites for proteoglycans or adhesive proteins and serve to generate and maintain the three-dimensional structure of the matrix required for mineralization. In addition, type IX may also be responsible for maintaining a tight collagen network [86].

c. Type X Collagen Type X collagen, a relatively short molecule, is synthesized virtually only by hypertrophic chondrocytes [45,46,87,88], and it may account for up to 45% of the total collagen synthesized by these cells [12,45,72,89]. During embryonic development, deposition of type X precedes calcification per se [47]. The collagen seems to exist as a fine filamentous macromolecular aggregate, which is distinct from the small-diameter fibrils produced by type II collagen, and its role in the growth plate is not clear.

d. Type XI Collagen The concentration of type XI collagen in the growth plate has been surprisingly high, approaching 30% of the level found for type II collagen [72]. Interactions with proteoglycans have been demonstrated, and a role for type XI in regulating fibril formation like type V collagen has been postulated [90].

2. Proteoglycans

Proteoglycans are required to maintain the elastic properties of growth plate cartilage. In addition, they modulate cellular differentiation and may regulate mineralization. The load-bearing properties of the growth plate are largely due to the osmotic properties of the proteoglycan aggregates, which are confined within the collagen network [91,92]. There is a close relationship between collagen and proteoglycan aggregates [93]. Hyaluronic acid spans the spaces between collagen fibers and stabilizes the proteoglycan subunits, preventing their loss during mechanical stress.

Proteoglycans are characterized by their content of covalently attached glycosaminoglycan (GAG) chains to a central protein core. The mammalian growth plate contains a high proportion of chondroitin 4-sulfate. Small proteoglycans (e.g., biglycan and decorin) are also present in growth plate, and these may play a role in regulating collagen fibril formation [94]. Evidence has also shown that proteoglycans are involved in binding of growth factors in the matrix [95,96].

There is a close relationship between the amount and size of proteoglycan aggregate and the calcification of cartilage [43,44]. Prior to calcification, approximately 70–80% of the proteoglycans are found in aggregate form [97]. As the growth plate matures, the proteoglycan aggregate is degraded [2,3]. At the same time, there is degradation of the glycosaminoglycan side chains, and, finally, the core protein is degraded. A number of theories have been proposed with respect to this process, including the hypotheses that proteoglycan aggregate sterically prevents crystal growth and that sulfated glycosaminoglycans bind Ca^{2+}, thereby reducing the pool available for crystal formation and growth [98].

3. Noncollagenous Proteins in the Growth Plate

Although the classification and determination of function of various noncollagenous proteins in the growth plate are still incomplete, the existence of these molecules is of great interest [78,80]. One of the most interesting new molecules to be described in the growth plate is anchorin CII [99], a 34-kDa glycoprotein that was originally isolated from purified membranes of embryonic chick chondrocytes [100]. The name of the protein is derived from its ability to anchor type II collagen in the chondrocyte cell membrane [100,101] and to facilitate attachment of the cells to their substrate. Chondronectin is another protein produced by chondrocytes that mediates the binding of these cells to type II collagen [102,103]. This 150-kDa protein does not bind to collagen directly, but does so in the presence of chondroitin sulfate.

Additional proteins, such as the cartilage matrix protein originally described by Fife, and collagen peptides have been reported in cartilage [78,99]. Osteonectin has been found at calcifying sites in growth cartilage [104,105]. Chondrocalcin, also known as the C-propeptide of type II collagen, may also play a role in the growth plate [106–108]. Chondrocalcin is a calcium binding protein that is distributed with proteoglycan in regions undergoing active calcification. Various reviews have reported on the presence of several proteins, including thrombospondin, fibromodulin, a 148-kDa protein, cartilage oligomeric high molecular weight protein (COMP), and a 21-kDa protein in cartilage [78–80]. The roles these proteins may play are not yet known.

C. Changes Occurring during Cartilage Maturation

The combinations of collagens, proteoglycans, and noncollagenous proteins described above are part of an assembly required for normal growth, differentiation, and mineralization of this tissue. As the growth plate matures, the relative distribution and composition of many of these proteins changes, in some cases dramatically [33,34]. How the chondrocyte accomplishes this is of considerable interest; the amount of matrix relative to the cell is quite large, and, once it has been synthesized by the cell, management of the matrix must occur by predetermined mechanisms. As shown in Fig. 1, growth plate chondrocytes produce factors that may act on the cell in an autocrine manner, on neighboring cells both above and below via paracrine mechanisms, and directly on matrix vesicles in the extracellular matrix.

Throughout growth plate maturation, the chondrocytes continue to synthesize matrix components. In the resting zone, the matrix produced by the cells is predominantly large proteoglycan aggregate and type II collagen. These cells produce extracellular matrix vesicles in small numbers and incorporate them into the matrix as it is synthesized. The matrix vesicles are enriched in alkaline phosphatase specific activity in comparison with the plasma membrane of the cells [64]. In addition, they contain matrix metalloproteinases capable of metabolizing proteoglycan [109]. As described in greater detail in Section V of this chapter, the activity of the matrix vesicles is regulated by the cells through signaling molecules secreted by them [110]. Thus, the chondrocyte can modulate the expanse of extracellular matrix in the resting zone though activation of matrix processing enzymes found in the matrix vesicles [109]. In addition, several of the matrix metalloproteinases are stored in the matrix in latent form. These are activated via a cascade of enzyme actions, but the regulatory mechanisms involved are not well understood [111–113].

The nature of the signals that induce resting zone cells to enter into proliferation is also not understood. Controversy still exists as to whether the supply of cells in the proliferating cell zone comes from the resting zone or from pluripotential cells that migrate into the tissue. Once proliferation is instituted, the cells undergo a set number of divisions and then enter a prehypertrophic state. The columns of cells that are the hallmark of growth plate anatomy are the result of the directional forces applied to the tissue. When growth plate cartilage is cultured under gravity-free conditions, as in spaceflight, the architecture of the tissue loses some of its distinctive orientation [114].

Studies using transgenic and knock-out mice have showcased the importance of the prehypertrophic state [35,36]. During this phase of growth plate maturation, the cells begin to produce large numbers of matrix vesicles. The levels of alkaline phosphatase specific activity are higher than seen in resting zone matrix vesicles [64], as is the amount of matrix metalloproteinases [109]. The extracellular matrix is predominantly type II collagen and proteoglycan aggregate, but the cells also begin to produce type X collagen [12–14,47]. These cells also synthesize transforming growth factor-β (TGFβ) [115], BMP-2 [116], cartilage-derived growth factors [117], and parathyroid hormone-related peptide (PTHrP) [118]. Thus, disruption of the regulation of this tissue has profound consequences for the ability of the growth plate chondrocyte to undergo hypertrophy.

As the cells begin to hypertrophy, the extracellular matrix undergoes extensive and rapid remodeling. Calcium transport from the cells to the matrix is increased, and ultimately hydroxyapatite crystal formation occurs along the inner leaflet of the matrix vesicle membrane. These 20 to 50-nm-diameter extracellular organelles first appear granular and then swell, and finally, as the crystal grows along the membrane surface, membrane integrity is lost [119]. At the transition from the lower hypertrophic zone to calcified cartilage, the crystals in the erupted matrix vesicles serve as foci for crystal multiplication and growth throughout the matrix. It is important to note that matrix vesicles are in intimate contact with the collagen fibrils [120], such that the mineral deposition occurring in association with the matrix vesicle membranes can contribute to the collagen-associated mineralization also occurring at the base of the hypertrophic cartilage.

The crystal formation within the matrix vesicle is accompanied by a series of biochemical changes. Throughout the hypertrophic cartilage, alkaline phosphatase specific activity continues to increase [121]. At the same time phospholipase A_2 activity is also increasing [122]. There is a change in the lipid composition of the matrix as well. The amount of calcium–phospholipid–phosphate complexes (CPLX) increases [123], as does the content of phosphatidylserine [124], the major phospholipid present in CPLX [125]. The content of lysophospholipids also increases [71], presumably due to the action of phospholipase A_2. The relative content of proteolipids, matrix vesicle membrane proteins associated with initial mineral formation, increases [126], indicating that as the non-nucleational lipids are metabolized, those lipids incorporated into the mineral phase of the tissue remain. Once calcification begins, the increases in alkaline phosphatase and phospholipase A_2 specific activities and CPLX content are no longer seen [127].

Matrix proteins are also modulated during hypertrophy. To allow for the marked increase in chondrocyte size, there is an extensive retailoring of the matrix. Proteoglycanases degrade the proteoglycan aggregate [44,128] and chondroitinases remove the glycosaminoglycan side chains [43,129,130]. In addition, the cells cease synthesizing type II collagen and begin to secrete type X collagen into the matrix [45,46,87,88]. At the base of the hypertrophic zone, the chondrocytes synthesize and secrete type I collagen, osteonectin, and osteopontin [48–50].

Serum factors are also found in the matrix of the calcified cartilage. Evidence indicates that the chondrocytes may synthesize and secrete some of the proteins themselves. For example, growth plate chondrocytes have been shown to produce α2-HS-glycoprotein [131], hypothesized to be a chemoattractant for osteoclast progenitor cells [132]. The chondrocytes also make hepatocyte growth factor, which promotes vascular invasion [133].

D. Regulation by Vitamin D *in Vivo*

Vitamin D has long been associated with cartilage and bone metabolism. Administration of vitamin D has been used effectively for over 60 years in the treatment of rickets and osteomalacia, as well as for treatment of several metabolic bone diseases, such as renal osteodystrophy. The effect of vitamin D deficiency on cartilage has been extensively studied. The primary disturbance caused by vitamin D deficiency is a failure of maturation of prehypertrophic cartilage and reduced mineralization of hypertrophic cartilage matrix [134].

In rickets, the proliferating cartilage zone usually does not differ much from the normal in extent or in arrangement of its cells. However, the prehypertrophic and hypertrophic zones are significantly increased, there is abnormal invasion of metaphyseal blood vessels, and calcification along the longitudinal septa of the cartilage columns is defective. Figure 2 diagrams the changes seen in epiphyseal cartilage in chicks raised for 3 weeks in the absence of ultraviolet light and fed a diet deficient in vitamin D. Crystallographic studies suggest that the mineral crystals which do form in rachitic rats are less mature than those of normal controls [135].

The pathology associated with rickets may reflect lesions in a number of aspects of growth plate development. 1,25-Dihydroxyvitamin D_3 [$1,25(OH)_2D_3$] may regulate the differentiation of postproliferative prehypertrophic cells, such that they do not enter into hyper-

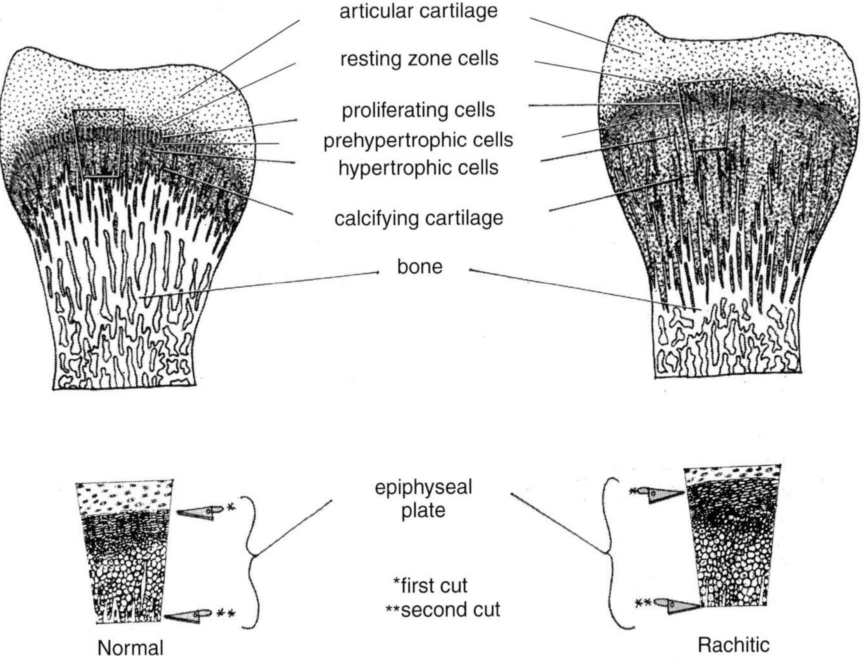

FIGURE 2 Schematic drawing showing the cellular organization of both the normal and rachitic chick growth plate. Cornish Rock hens were maintained from birth either under normal lighting conditions and chicken diet containing normal levels of calcium and phosphate or in the absence of ultraviolet light and with chicken diet containing low calcium and phosphate. At the age of 3 weeks, the animals were sacrificed, and histologic sections of the long bone epiphyses were prepared, stained with hematoxylin and eosin, and examined. Rachitic cartilage exhibited expanded prehypertrophic and hypertrophic cell zones. The amount of metaphyseal bone was reduced, and there was a widening of the epiphysis. Note the interdigitation of metaphyseal bone spicules with the hypertrophic cartilage, a feature of both normal and rachitic chick growth plates. Enlarged views of the epiphyseal plate appear at bottom.

trophy. There are considerable data demonstrating that a mineralization defect exists. This may be due in part to a failure of the cells to transport calcium effectively. The distribution of mitochondrial mineral granules in the growth plates of rachitic rats (low phosphate, vitamin D-deficient diet) appears to be modified, as revealed by electron microscopic observation [136]. In control rats, intramitochondrial mineral granules were found in maturing and hypertrophic chondrocytes and not in the chondrocytes at the zone of provisional calcification. In the rachitic rats, intramitochondrial mineral granules were reduced in number and found only in some hypertrophic chondrocytes. The gradient of these intramitochondrial mineral granules was restored by the addition of vitamin D or phosphate [136,137].

The mineralization defect may also result in part from altered lipid metabolism of the growth plate. Early studies by Irving and Wuthier [138] showed that there was a marked increase in acidic lipids in the rachitic rat growth plate when compared to normal tissue. When the rats were made vitamin D replete, the acidic lipid content was decreased. That acidic phospholipid metabolism is regulated by vitamin D was noted by others, including Creuss and Clark [139] and Howell and coworkers [140]. These studies led to the hypothesis that phospholipase A_2 is regulated by vitamin D in growth plate. Indeed, Wuthier [122] showed that treatment of rachitic chicks with vitamin D led to an increase in growth plate enzyme activity. Similarly, Boyan and Ritter [141] showed that the relative content of chick growth plate matrix vesicle proteolipids, which are resistant to phospholipase A_2 activity, also increased with vitamin D treatment. Subsequent studies by Boskey and Wientroub [71] confirmed the hypothesis, showing that treatment of rachitic chicks with vitamin D metabolites resulted in an increase in lysophospholipid content. Moreover, they showed that the effects of the hormones were directed to specific classes of phospholipids.

Failure of rachitic cartilage to calcify may be a direct consequence of a change in the regulation of matrix vesicle enzymes including alkaline phosphatase and the matrix metalloproteinases. Several investigators have shown that expression of alkaline phosphatase mRNA is dependent on $1,25(OH)_2D_3$ [49,142]. Thus, in a rachitic animal, one would expect matrix vesicles that are deficient in this important enzyme to be produced by the chondrocytes. This is in fact the case. Although the precise role of alkaline phosphatase in calcification is not known, its activity is directly correlated with the ability of a tissue to calcify. Under normal conditions, alkaline phosphatase activity increases through the hypertrophic zone [121]. In rachitic animals, activity is markedly suppressed. Restoration of vitamin D causes a rapid increase in enzyme activity. While it is certainly possible that this is due to new gene transcription, it is also possible that preexisting matrix vesicle alkaline phosphatase is activated by the hormone [143]. Other matrix vesicle enzymes are also regulated by vitamin D. Carbonic anhydrase, which plays a role in maintaining an alkaline pH in the region of crystal growth, is also regulated by the hormone [144]. Studies have suggested that both matrix metalloproteinase activity and protein kinase C (PKC) activity of matrix vesicles are also regulated by vitamin D metabolites [111–113,145].

Our studies have shown that proteoglycan degradation is regulated by vitamin D [112,113]. However, the hormone also modulates growth plate maturation and calcification by regulating proteoglycan synthesis. In vitamin D deficiency in the chick, smaller aggregating proteoglycans are synthesized [146,147]. Restoration of calcium results in the production of larger proteoglycans. These effects, which were shown to be specific for growth cartilages, have been confirmed by others [148]. In healing rickets in the rat, there is a substantial decrease in proteoglycan binding to hyaluronic acid [149]. This suggests that proteoglycan cleavage, which often occurs in the part of the molecule adjacent to the hyaluronic acid binding region and likely mediated by stromelysin [128], is increased in the presence of vitamin D and retarded in rickets.

Vitamin D deficiency can also alter the rate at which cartilage is resorbed. If resorption of calcified cartilage does not occur, negative feedback through the growth plate to the prehypertrophic and proliferative zones may result. Thus, the absence of hydroxyapatite crystals is important. However, equally important is the absence of molecules that coat the hydroxyapatite crystals and serve as attachment sites for the osteoclasts which will remove the calcified cartilage prior to bone formation. In addition, $1,25(OH)_2D_3$ regulates osteoclast formation, and these cells are needed for proper removal of the calcified cartilage once it forms.

Table I provides a summary overview of the actions of vitamin D in each cell zone of the growth plate. There has been some controversy over the role of various vitamin D metabolites in the regulation of growth plate cartilage; however, some of this controversy has resulted from the selection of animal model. For example, the antirachitic activity of vitamin D_2 is much less than that of vitamin D_3 in the chicken and in a New World monkey, but there is no difference between them in the rat [150]. In addition, *in vivo* studies have been complicated by the ability of the animal to further metabolize the hormone in the liver and kidney. More recently, it has been shown that various tissues, including growth plate cartilage [151–155], have the ability to metabolize 25-hydroxyvitamin D_3 ($25OHD_3$) locally.

The hormone $1,25(OH)_2D_3$ has been found to be

TABLE I The Role of Vitamin D in the Growth Plate

Growth plate zone	Effect of vitamin D
Resting	Metabolite-specific regulation of cell differentiation and maturation $1,25(OH)_2D_3$ modulates cell proliferation *in vitro* $24,25(OH)_2D_3$ regulates matrix production, matrix vesicle composition and enzyme activity, and cell maturation to a $1,25(OH)_2D_3$-responsive phenotype
Proliferation	No effects reported
Prehypertrophic	Induction of differentiation and maturation Alteration of acidic lipid content Alteration of activity of enzymes both in the cell and in matrix vesicles, e.g., alkaline phosphatase, phospholipase A_2, carbonic anhydrase, matrix metalloproteinase Regulation of proteoglycan aggregate size Regulation of glycosaminoglycan chain sulfation
Hypertrophic	Induction of differentiaion and maturation Modulation of enzyme activity both in the cell and in matrix vesicles, e.g., alkaline phosphatase, phospholipase A_2, carbonic anhydrase, matrix metalloproteinase Regulation of proteoglycan aggregate size Regulation of glycosaminoglycan chain sulfation Induction of mineralization Regulation of mineral crystal maturation as well as number Alteration in phospholipid metabolism and content Alteration in matrix vesicle proteolipid content Alteration in calcium levels, thereby regulating calcification
Replacement of cartilage by bone	Regulation of metaphyseal blood vessel invasion Regulation of longitudinal septa calcification Regulation of osteoclast formation and activation Induction of the removal of calcified cartilage

essential for normal differentiation of cartilage and the prevention of rickets. However, calcification of rachitic cartilage may occur, even in the absence of $1,25(OH)_2D_3$, if the calcium and phosphate content of the extracellular fluid are normal [156]. This can be achieved in rachitic chicks either by $1,25(OH)_2D_3$ or even through administration of a high calcium and phosphate diet [134,157,158]. One of the most elegant experiments demonstrating the importance of $1,25(OH)_2D_3$ in maintaining calcium homeostatis was done using rachitic rats born of rachitic rat mothers [156]. By infusing these rats with calcium, the investigators were able to "heal" the rickets. Similar observations were made by Underwood and DeLuca [159].

Studies such as these suggest that maintenance of the supply of calcium and phosphate is the primary function of vitamin D metabolites in cartilage. It is certainly possible that enough redundancy exists in the growth plate that administration of $1,25(OH)_2D_3$ to rachitic rats can lead to rapid calcification of the matrix. In fact, Howell *et al.* [160,161] demonstrated that rachitic rat cartilage contains a nucleator capable of promoting calcification *in vitro*, suggesting that if the calcium and phosphate content of the extracellular fluid were sufficiently high, mineralization *in vivo* will occur. Once calcification takes place, the resorption of the calcified cartilage can occur, and it may well be that this is critical to the flow of messages up the growth plate to the prehypertrophic zone.

More recent studies have indicated that "healing" of rickets by administration of $1,25(OH)_2D_3$ may only be a part of the vitamin D story. $24,25(OH)_2D_3$ may also play an important role, which is most evident in the resting zone. This has led to the hypothesis that $24,25(OH)_2D_3$ is involved in promoting the differentiation of the resting zone chondrocyte into a more mature phenotype [162]. Local injection of $24,25(OH)_2D_3$ into the upper tibial growth plate [163] or systemic injection of $24,25(OH)_2D_3$ [164] can heal rickets. It is unlikely that this is due to hydroxylation of $24,25(OH)_2D_3$ to $1,24,25(OH)_3D_3$ and subsequent actions of $1,24,25(OH)_3D_3$ on Ca^{2+} release, as local injection of $1,25(OH)_2D_3$ is relatively ineffective [163]. Rather, it is more likely that $24,25(OH)_2D_3$ acts directly on chondrocytes in the upper growth plate to promote their differentiation along the endochondral lineage. When $24,25(OH)_2D_3$ is injected with $1,25(OH)_2D_3$ into fracture callus, bone repair occurs more rapidly than when $1,25(OH)_2D_3$ is injected alone [165]. The role of $24,25(OH)_2D_3$ in growth plate cartilage development is discussed in greater detail in Section III.

Only a few studies on the effects of hypervitaminosis D on cartilage and bone in growing animals have been published. Wider epiphyses were reported with excess metaphyseal bone, causing "hypervitaminosis D rickets." On the other hand, when pregnant rats were treated with high doses of vitamin D_2 during days 10–21 of gestation, the fetuses had short bones, mainly due to short diaphyses. The cartilaginous epiphyses were of normal length, but they contained less calcium and atypical chondrocytes [166,167].

II. MODELS FOR STUDYING VITAMIN D-DEPENDENT CHONDROGENESIS *IN VITRO*

Endochondral bone formation involves a developmental cascade of cell differentiation and maturation, culminating in mineralization of the extracellular matrix, first of chondrocytes and later of osteoblasts. This is a very complex cascade of events, which is highly regulated by both hormones and local factors. To understand this cascade, especially its regulation, a variety of *in vitro* models have been developed, each of which has provided clues to the regulatory mechanisms involved.

A. Embryonic Bone Formation

During embryonic bone formation, mesenchymal cells in the limb bud anlage differentiate into cartilage. As the cartilage matures, the cells at the periphery calcify their matrix. Osteogenesis occurs on the calcified cartilage scaffold, and finally, the cartilage core is replaced with marrow [168]. Chick limb bud cultures are frequently used to study this process [15,16,169,170]. The fetal limb bud grows in length by forming an embryonic growth plate [15]. Soskolne *et al.* [171] have developed a fetal mouse long bone organ culture model that permits assessment of the bone modeling process in mammalian species as well.

B. Postfetal Bone Growth and Development

Endochondral ossification in postfetal life is comparable in many ways to its embryological counterpart. One of the most notable differences is that during long bone growth the resting zone serves as a major source of cells for the growth plate. Thus, the cells that proliferate and hypertrophy have been committed chondrocytes for extended periods of time, depending on the age of the individual. The exceptions to this are during bone wound healing involving formation of a callus or during bone induction. In these instances, mesenchymal cells become chondrocytes and synthesize cartilage matrix. Within a relatively short time, the cells calcify their matrix, thus supporting subsequent osteogenesis. Bone healing following fracture [172] and heterotopic bone induction [28] have both been exploited as models for understanding the regulatory events associated with endochondral ossification.

Regardless of the model used to study postfetal endochondral bone formation, the histological and cytological results are comparable. This is the case whether the tissue examined is the growth plate in chickens [68,173], rats [64], rabbits [174], or bovines [175]. It is also true of heterotopic bone resulting from implantation of demineralized bone powder [29,176] or isolated bone-inductive components [177,178], either in the superficial fascia of the thorax or in the thigh muscle of rats or mice [179]. This implies that there is a cascade of regulatory factors that operate in the development of calcified cartilage. In addition, it suggests that, because the events are constant among models, at least part of the regulation is a function of cells, independent of mechanical forces. Imposition of mechanical stresses on the cascade results in the regulation of tissue geometry.

C. Cell Culture Models

1. Three-dimensional Models

Scientists have begun to dissect the specific events in the regulatory cascade by examining the cells in culture. The models that have been developed are as diverse as the laboratories. Micromass cultures [57–59] have been useful for recapitulating the development of the growth plate *in vitro*. Under these conditions, the cells form multilamellar structures with the chondrocytes at the base of the mass, mineralizing their matrix. The same series of events can be observed in cultures of growth plate chondrocytes that are incubated over relatively long periods in medium containing ascorbate [180]. The cells form nodules that have a histological appearance reminiscent of growth plate morphology. At the ultrastructural level, the cells appear identical to those commonly observed in long bones. At the top of the nodule, cells are flattened and relatively undifferentiated in appearance. As the microscopist progresses more deeply into the nodule, the cells become more rounded, and matrix vesicles are observed in the extracellular matrix. At the base of the nodule, matrix vesicles contain initial crystals of hydroxyapatite, and mineral deposits appear in the interterritorial matrix. Finally, the cells are encased in highly mineralized lacunae. Similar results have been obtained using three-dimensional collagen gels [55].

Three-dimensional cultures (nodules, micromass, gels) have been used to examine the role of $1,25(OH)_2D_3$ [174] and other hormones on endochondral development. The results provide an important part of the picture, but unless the studies are designed to follow the cultures over time, they present only a composite picture of numerous cell types. This is not substantially different from organ culture models or *in vivo* studies.

2. SHORT-TERM MONOLAYER CULTURES

An alternative approach is to study the cells in short-term culture following separation of the cells on the basis of their maturational status in the original tissue. One way of doing this is by countercurrent centrifugal elutriation [68]. This technique separates the cells on the basis of density, working on the hypothesis that there is a direct relationship between cell density and cell size and presuming that the most immature cells will be the smallest. Using alkaline phosphatase activity as a parameter of cell differentiation, O'Keefe *et al.* [68] have shown that the method does produce cell populations of increasing enzyme activity. Because alkaline phosphatase activity *in situ* increases as the cells differentiate and calcify their matrix, it is inferred that the cell separation by density also separates cells of differing maturation. These cultures have been used to study the short-term response of chondrocytes to growth factors like transforming growth factor β (TGFβ). Other gradient systems, like Ficoll, have been tried with varying success [175]. Unfortunately, it has proved difficult to remove the Ficoll from the cells following separation.

3. LONG-TERM MONOLAYER CULTURES

Monolayer culture systems have been established using epiphyseal growth plate chondrocytes from a variety of animals, including rabbits [181,182], chicks [183,184], rats [59,185,186], and bovines [175]. Because the growth plate is a mineralizing tissue, it is sometimes difficult to obtain cells. One strategy has been to use rachitic animals, since the growth plate is more pliable in its non-mineralized state [59,185]. However, these cells may be compromised as a result of the hormone deficiency *in vivo* and may not reflect the normal chondrocyte phenotype. Other models have used mesenchymal cells that developed into chondrocytes in culture [187,188]. Although clonal cell lines have been established [189–192], their use has suffered from attendant artifact typical of cell lines, whether or not they are transformed.

a. Chick Growth Plate Chondrocytes For long-term cultures, some of the best results have been obtained using cells isolated from slices of the chick long bone growth plate [183,184,193]. Because the chondrocytes in the growth plate are aligned in columns, with the least mature cells at the top and the most mature cells at the bottom, it is possible to separate cell phenotypes by sharp dissection of the tissue. Cells are then released from the tissue by collagenase digestion and cultured for varying periods of time. There is some problem of cross-contamination of cells because the chick growth plate is irregular and is highly interdigitated with bone spicules, raising the possibility of bone cell contamination. To overcome this, rat growth cartilage [57] and bovine growth plate [175] are often used, as the cartilage and bone are sharply demarcated, facilitating dissection.

b. Costochondral Chondrocytes Among the most completely described cell culture models for examining regulation are the rat [64,143] and rabbit [174,181,182,194] costochondral cartilage systems. In the rat model, the resting zone (reserve zone) cartilage is separated from the adjacent bone and proliferative cartilage by sharp dissection. Similarly, the prehypertrophic and upper hypertrophic zones (growth zone) are separated from the proliferative zone and calcified cartilage. Cells in the reserve zone and growth zone are cultured separately, ensuring a clean demarcation of cell maturation. Although chondrocytes have a well-known tendency to lose expression of phenotypic markers during long-term culture and subculture [56], these cells retain their ability to synthesize cartilage-specific proteoglycan and type II collagen through fourth passage [195]. Furthermore, they respond in distinctly different ways to a number of regulatory agents. These experiments are described in detail below.

III. ROLES OF $24,25(OH)_2D_3$ AND $1,25(OH)_2D_3$ IN CARTILAGE

Numerous studies have shown that $1,25(OH)_2D_3$ regulates the terminal differentiation of hypertrophic cartilage. In states of vitamin D deficiency, such as rickets, the hypertrophic cell zone of the growth greatly increases in size [41,164]. Restoration of $1,25(OH)_2D_3$ results in mineralization of the extracellular matrix and resolution of the rachitic state. Because infusion of calcium also heals the rickets [156,196], it has been hypothesized that the primary function of $1,25(OH)_2D_3$ is to maintain the concentration of extracellular Ca^{2+}.

Confusion over the role of $24,25(OH)_2D_3$ in the growth plate has stemmed in large part from studies published in the 1980s that addressed the role of vitamin D metabolites in the treatment of rickets using a series of novel compounds [197,199]. One vitamin D analog incorporated fluoride in place of the hydroxyl group on C-24. Because this fluorinated analog appeared to have no effect, the investigators concluded that $24,25(OH)_2D_3$ did not possess bioactivity other than as a mechanism to direct excess $25\text{-}OHD_3$ to be excreted. However, more recent studies (described later in this section) seem to demonstrate a distinct role for $24,25(OH)_2D_3$ in the early stages of endochondral maturation.

Growth plate cartilage appears to be a target organ for $24,25(OH)_2D_3$. Fine *et al.* [200] demonstrated that rabiolabeled $24,25(OH)_2D_3$ could be found in epiphyseal cartilage. Similarly, when vitamin D-replete rats were injected with radiolabeled $25\text{-}OHD_3$, radiolabeled $24,25(OH)_2D_3$ was concentrated in growth plate cartilage, but not articular cartilage [201]. Finally, injection of radiolabeled $24,25(OH)_2D_3$ into vitamin D-replete rats also resulted in enhanced uptake of the metabolite by the growth plate [201]. Receptors for $24,25(OH)_2D_3$ have been suggested to be present by autoradiographic studies in growth plate cartilage [200,202,203] but never definitively identified.

It is possible that the growth plate gives the impression of being a target tissue for $24,25(OH)_2D_3$ because it is one of the few tissues in the body in which cells are found together at essentially the same maturation state in their lineage cascade. Thus, other cells may be sensitive to this vitamin D metabolite, but, because of the models used, the $24,25(OH)_2D_3$-responsive phenotype is not present in high enough density to rise above the non-$24,25(OH)_2D_3$ noise. Support for this hypothesis comes from two sources. First, ROS 17/2.8 cells increase alkaline phosphatase specific activity in response to $24,25(OH)_2D_3$ [204]. Second, St. Arnaud and colleagues demonstrated that in the 24-hydroxylase knockout mouse (see Chapter 18), the predominant effect noted was in osteoblasts found producing unmineralized osteoid on the endosteal surface of cortical bone [205]. This does not, however, support an unequivocal role for $24,25(OH)_2D_3$, as the effects of this gene knock out could be due to altered degradation of $1,25(OH)_2D_3$.

It appears that regulation of endochondral bone formation involves both $1,25(OH)_2D_3$ and $24,25(OH)_2D_3$. Studies using fetal mouse long bone organ cultures have shown that maintenance of normal hydration, hypertrophy, columnar arrangement, and preservation of mineral in growth plate cartilage requires the presence of both $1,25(OH)_2D_3$ and $24,25(OH)_2D_3$ at physiological concentrations [181]. Similarly, both metabolites are required for maximal production of chondrocalcin (C-propeptide of type II collagen) and calcification of rat [206] and bovine [207] growth plate chondrocyte cultures. Again, however, the effects of $24,25(OH)_2D_3$ administration may alter $1,25(OH)_2D_3$ metabolism.

In vivo studies of rachitic rats in which normal plasma levels of $1,25(OH)_2D_3$ were retained during the first week on a rachitogenic diet showed that mild rachitic lesions occurred anyway [208], suggesting that vitamin D metabolites other than $1,25(OH)_2D_3$ might be involved. Studies of this type do not rule out the fact that low calcium and phosphate availability might be the critical factor. Thus, until the advent of cell culture models, it was difficult to identify specific roles for each of the vitamin D metabolites. It is becoming increasingly evident that $24,25(OH)_2D_3$ and $1,25(OH)_2D_3$ exert independent effects on growth plate chondrocytes. For example, $1,25(OH)_2D_3$ stimulates the proliferation of rabbit costal growth plate cells but inhibits proteoglycan synthesis [174]. In contrast, $24,25(OH)_2D_3$ has no effect on proliferation of these cells but stimulates proteoglycan synthesis. The relevance of these culture systems to animal biology remains to be determined, however.

Both $1,25(OH)_2D_3$ and $24,25(OH)_2D_3$ exert different effects on growth plate cartilage, possibly by acting on specific target cells. Autoradiographic evidence using radiolabeled ligands [200] suggests that receptors for both metabolites exist in the growth plate. These studies indicate that subpopulations of cells were responsive to either one metabolite or the other. Putative receptors for $24,25(OH)_2D_3$ were localized to the proliferating cells in the growth plate, whereas those for $1,25(OH)_2D_3$ were found in osteoprogenitor cells and osteoblasts. *In vitro* studies have demonstrated that growth plate chondrocytes exhibit differential responsiveness to $1,25(OH)_2D_3$ and $24,25(OH)_2D_3$. When rabbit growth plate organ cultures were treated with $24,25(OH)_2D_3$, calcium granules were seen in the cytoplasm, but matrix vesicles were devoid of crystals [181]. In contrast, $1,25(OH)_2D_3$ promoted mineral deposition in the matrix. Thus, the target cells for both metabolites may be different, implying a distinct biological function for each.

There is a growing body of evidence supporting the hypothesis that $1,25(OH)_2D_3$ and $24,25(OH)_2D_3$ act on different target cells in endochondral bone formation. Using embryonic chick limb bud cells as a model, Boskey *et al.* [67] found that the less mature cells are affected by $24,25(OH)_2D_3$, whereas the more mature cells are affected by $1,25(OH)_2D_3$. When chick embryo cartilage cells were examined for evidence of $1,25(OH)_2D_3$ receptors, the $1,25(OH)_2D_3$ receptor was found only in growth cartilage cells and not in resting cartilage [209]. Increases in calcification of growth cartilage are paralleled by an increase in $1,25(OH)_2D_3$ receptor levels. The differential response is seen in rats as well. Rat growth plate chondrocytes contain a $1,25(OH)_2D_3$-dependent calcium binding protein found in the duodenum [210,211]. This protein is localized in the growth zone of the rat growth plate and is found in the cytoplasm of prehypertrophic cells and in the cytoplasmic processes of hypertrophic chondrocytes. Similarly, the protein is found in the growth zone of the chick embryo growth plate [211]. In contrast to this specific effect of $1,25(OH)_2D$, studies using embryonic chick chondrocytes have demonstrated that $24,25(OH)_2D_3$, but not $1,25(OH)_2D_3$, induces increases in cell proliferation and protein synthesis [143,212].

Cell culture studies have shown that the response of rat costochondral chondrocytes to vitamin D_3 metabolites depends on the zone of maturation from which the cells were originally derived [213]. Resting zone chondrocytes respond primarily to $24,25(OH)_2D_3$, whereas growth zone chondrocytes respond primarily to $1,25(OH)_2D_3$. In addition, treatment of resting zone chondrocytes with $24,25(OH)_2D_3$ for 36 hr caused a change in maturation state, indicating that this hormone has a very specific role in chondrocyte differentiation [162]. The response of each cell to its metabolite includes differential regulation of plasma membrane and matrix vesicle enzyme activity [214] and fluidity [215], collagen and noncollagen protein synthesis and cell proliferation [195], calcium flux [216], phospholipid metabolism [217,218], and production of vitamin D metabolites [155]. These studies are discussed in greater detail in Section IV.

IV. RAPID ACTIONS OF VITAMIN D AND NONGENOMIC MECHANISMS

A. Definitions and Models for Studying Rapid Actions

In addition to $1,25(OH)_2D_3$ receptors (VDR), putative receptors for $24,25(OH)_2D_3$ have been identified in cartilage [200,202,203]; therefore, it is probable that many of the effects are via classic VDR pathways, involving changes in gene transcription and mRNA stabilization [49,145,195,196]. This is clearly the case for $1,25(OH)_2D_3$. Chick limb bud mesenchymal cells [219], chick sternal chondrocytes [49], and chick epiphyseal chondrocytes [220] all exhibit $1,25(OH)_2D_3$-dependent synthesis of proteins such as alkaline phosphatase and osteocalcin. Although less is known concerning the action of $24,25(OH)_2D_3$, its effects on various markers of chondrocyte differentiation strongly suggest that genomic regulatory mechanisms are involved [185,195]. This is supported by studies using inhibitors of gene transcription and translation [145].

New evidence has suggested that at least some of the effects of vitamin D in cartilage, as well as effects in bone, occur via nongenomic mechanisms. This subject is discussed in detail in Chapter 15. We focus here on nongenomic events in cartilage. A nongenomic action of the hormone is one that involves neither new gene transcription nor protein synthesis. Although many rapid effects of vitamin D metabolites may be nongenomic, time course of action is not an *a priori* proof, as new protein synthesis may occur even in very short times. Examples of nongenomic actions include changes in membrane fluidity [215], turnover of phospholipids [217,218], changes in calcium flux [216,221,222], and activation of protein kinase C [145]. Even these actions, however, may be downstream from a vitamin D-dependent nuclear event, whether or not the VDR is involved. Thus, the only proofs for nongenomic mechanisms are those which examine responses in the absence of DNA, RNA, or protein synthesis. To definitively prove physiological relevance of rapid and nongenomic actions of vitamin D, it is necessary to demonstrate that they occur *in vivo*. Few such experiments have been done. However, new molecular techniques are making experiments of this kind more possible.

The rat costochondral chondrocyte culture model has been particularly useful for studying the mechanisms of $1,25(OH)_2D_3$ and $24,25(OH)_2D_3$ action. Both resting zone and growth zone chondrocytes produce extracellular matrix vesicles in culture; however, the phospholipid composition [64] and enzyme activity [213,214] of these matrix vesicles depend on the cell of origin. Not only do matrix vesicles produced by growth zone chondrocytes differ from those produced by resting zone chondrocytes, but each type of matrix vesicle differs from the plasma membrane of the cell from which it was derived. Not surprisingly, the basal membrane fluidity of matrix vesicles is distinct from that of the plasma membrane [215], reflecting differences in structure and chemical composition.

Because matrix vesicles are external to the cells in the extracellular matrix, they can be isolated from cultures without disrupting the cell membranes. Following a brief trypsin digestion, matrix vesicles can be collected by differential centrifugation. They are right-side out since no homogenization is necessary, and they are free of DNA, RNA, ribosomes, or contamination from intracellular membranes such as Golgi or endoplasmic reticulum. The advantages of such a system for examining direct, nongenomic effects of lipophilic hormones are obvious [110]. One can examine the genomic response of cells to $1{,}25(OH)_2D_3$ and $24{,}25(OH)_2D_3$ by incubating cultures with hormone and then studying isolated matrix vesicles in comparison with plasma membranes prepared by differential centrifugation of cellular homogenates. By varying the time course and using transcription and translation inhibitors, one can begin to sort out those responses which require genomic events. One can also isolate matrix vesicles and plasma membranes from naive cultures and incubate the membrane fractions with hormone *in vitro*, thereby obtaining definitive evidence of a nongenomic response. The caveat to the latter aproach, however, is that this becomes a purely *in vitro* approach.

Our early experiments demonstrate the power of the model to elucidate nongenomic events [64,213,223]. When rat costochondral growth zone chondrocytes were incubated with $1{,}25(OH)_2D_3$, alkaline phosphatase specific activity was increased, whereas $24{,}25(OH)_2D_3$ had no effect. The $1{,}25(OH)_2D_3$-dependent response was found to be targeted to the matrix vesicles as there was no increase in plasma membrane alkaline phosphatase specific activity. When examining cell layer alkaline phosphatase specific activity in resting zone chondrocyte cultures exposed to $1{,}25(OH)_2D_3$ or $24{,}25(OH)_2D_3$, no apparent effect of either hormone was observed. However, matrix vesicles isolated from the $24{,}25(OH)_2D_3$-treated cultures did exhibit a dose-dependent increase in enzyme activity, whereas isolated plasma membranes did not. This suggests that the effects of $24{,}25(OH)_2D_3$ were targeted exclusively to the matrix vesicles produced by resting zone chondrocytes and that it was necessary to isolate them to detect the hormone-specific effect.

It should be noted, parenthetically, that Hale *et al.* [220] found that $24{,}25(OH)_2D_3$ caused an increase in alkaline phosphatase activity in serum-free cultures of chick growth plate chondrocytes at low doses, whereas at higher concentrations enzyme activity was inhibited. Although species differences may be involved, it is more likely that the effects on chick cells represent an initial response in the absence of other serum factors, whereas in the rat costochondral cultures effects of hormone are examined in a background of serum. Indeed, when examining the combined effects of TGFβ and $24{,}25(OH)_2D_3$ on resting zone chondrocytes, a synergistic increase in matrix vesicle alkaline phosphatase was observed (B. D. Boyan, D. D. Dean, V. L. Sylvia, and Z. Schwartz, unpublished data, 1995). This was of sufficient magnitude to be detectable in the cell layer [115,224]. In contrast, the combined effect of $1{,}25(OH)_2D_3$ and TGFβ on growth zone cell cultures was comparable to the effect of TGFβ alone.

Phospholipase A_2 specific activity was also assayed in membrane fractions isolated from rat costochondral chondrocyte cultures [214]. $1{,}25(OH)_2D_3$ stimulated phospholipase A_2 specific activity in matrix vesicles isolated from growth zone chondrocyte cultures but had no effect on the plasma membrane enzyme activity; in contrast, enzyme activity in both membrane fractions from resting zone chondrocyte cultures was unaffected. $24{,}25(OH)_2D_3$ inhibited phospholipase A_2 specific activity in matrix vesicles isolated from resting zone chondrocyte cultures but had no effect on plasma membrane enzyme activity; enzyme activity in both of the membrane fractions from growth zone cultures was unaffected. Once again, to observe the $24{,}25(OH)_2D_3$-dependent effect on resting zone cultures, it was necessary to examine isolated matrix vesicles.

Because matrix vesicles are already external to the cells, these data suggested that either the chondrocytes were making new matrix vesicles in response to hormone or that the hormones were acting directly on the matrix vesicles in the matrix. To test this, matrix vesicles were isolated from cultures of growth zone and resting zone chondrocytes that had no prior exposure to hormone, and then incubated directly with either $1{,}25(OH)_2D_3$ or $24{,}25(OH)_2D_3$ [143]. $1{,}25(OH)_2D_3$ stimulated both alkaline phosphatase and phospholipase A_2 specific activities in matrix vesicles isolated from growth zone chondrocytes but had no effect on either enzyme in matrix vesicles isolated from resting zone cell cultures. In contrast, $24{,}25(OH)_2D_3$ stimulated alkaline phosphatase and inhibited phospholipase A_2 specific activities in matrix vesicles from resting zone cell cultures, but it had no effect on either enzyme in matrix vesicles isolated from growth zone chondrocyte cultures. These experiments clearly demonstrated that vitamin D metabolites were able to mediate specific effects in the absence of genomic machinery. Moreover, the direct effects were essentially identical to those observed in the intact cultures [225]. Thus, it is possible that $1{,}25(OH)_2D_3$ and $24{,}25(OH)_2D_3$ might operate even *in vivo*, at least in part, through nongenomic pathways.

Subsequent studies showed that $1{,}25(OH)_2D_3$ and $24{,}25(OH)_2D_3$ elicited a variety of rapid effects on resting zone and growth zone chondrocytes. These effects were found to be cell maturation specific. In general, $1{,}25(OH)_2D_3$ exerted its effects on growth zone cells, whereas $24{,}25(OH)_2D_3$ exerted its effects on resting zone cells, although in some instances one or both metabolites would have an effect on its nontarget cell population. Even when this occurred, however, the effects were cell maturation specific. Moreover, the rapid effects of the two metabolites were frequently membrane specific, causing one response in the plasma membrane and another in the matrix vesicle membrane.

B. Membrane Fluidity

The simplest explanation for how vitamin D metabolites could regulate resting zone chondrocytes in a differential manner is based on the concept that these lipophilic molecules could alter the fluidity of the membrane in much the same way as is seen in artificial membranes treated with cholesterol [218,226]. The difference in charge density between $1,25(OH)_2D_3$ and $24,25(OH)_2D_3$ would automatically make the metabolites assume different habits in the membrane, with different consequences to its fluid mosaic structure. Furthermore, plasma membranes from resting zone cells differ from those of growth zone cells with respect to their lipid composition and enzyme activity [64]. Matrix vesicles produced by these cells differ from their parent plasma membranes and from one another [64]. Not surprisingly, basal fluidity of these membrane preparations differ as well [215].

When resting zone or growth zone chondrocyte cultures are treated with $1,25(OH)_2D_3$ or $24,25(OH)_2D_3$, plasma membrane fluidity is changed in a cell maturation-specific manner [215]. $1,25(OH)_2D_3$ increased the fluidity of the growth zone cell membrane but had no effect on resting zone cell membrane fluidity. $24,25(OH)_2D_3$ decreased the fluidity of the growth zone cell membrane but increased the fluidity of resting zone cell membranes. Similar effects are seen when these metabolites are incubated directly with matrix vesicles isolated from naive cultures.

C. Phospholipid Metabolism

A change in membrane fluidity may result from the physical presence of the vitamin D metabolite, or it may be due to rapid retailoring of membrane phospholipids. As described above, phospholipase A_2 activity is differentially regulated by $1,25(OH)_2D_3$ and $24,25(OH)_2D_3$ [214]. In these experiments, phospholipase A_2 was measured as a function of the release of radiolabeled arachidonic acid from prelabeled phosphatidylethanolamine. The effect of the vitamin D metabolites on general incorporation and release of radiolabeled arachidonic acid is also cell maturation specific [217,227]. Within 5 min, $1,25(OH)_2D_3$ and $24,25(OH)_2D_3$ begin to modulate arachidonic acid turnover by the cells. $1,25(OH)_2D_3$ stimulated arachidonic acid turnover in growth zone chondrocytes, but had no effect on resting zone chondrocytes. In contrast, $24,25(OH)_2D_3$ stimulated arachidonic acid turnover in resting zone chondrocytes but inhibited turnover in growth zone chondrocytes. Interestingly, when purified phospholipase A_2 was incubated directly with the vitamin D metabolites, $1,25(OH)_2D_3$ stimulated enzyme activity, whereas $24,25(OH)_2D_3$ inhibited enzyme activity [227]. These results suggest that at least part of the effect of the metabolites on cell membranes may be related to changes in phospholipase A_2 activity and fatty acid turnover.

It is important to note that arachidonic acid incorporation is regulated as well. This suggests that the vitamin D metabolites modulate the activity of phospholipid synthesis. Moreover, this effect is rapid, indicating a change in the phospholipid composition of the membrane. Rasmussen and colleagues [228] showed that $1,25(OH)_2D_3$ alters kidney membrane phospholipids. Other investigators have shown similar effects in intestinal epithelia [229]. In the costochondral chondrocyte model discussed here, these vitamin D metabolites also cause a shift in the relative distribution of cellular and matrix vesicle membrane phospholipids [217,227].

As a consequence of altered arachidonic acid release, $1,25(OH)_2D_3$ and $24,25(OH)_2D_3$ regulate prostaglandin production by growth zone and resting zone chondrocytes [230]. $1,25(OH)_2D_3$ stimulates production of prostaglandins E_1 and E_2 (PGE_1 and PGE_2) from growth zone cells. Basal production of these prostaglandins is higher in resting zone cell cultures, and when the cultures are treated with $24,25(OH)_2D_3$, production is reduced to levels somewhat similar to those seen in the growth zone cell cultures.

D. Calcium Flux

One of the most rapid effects of $1,25(OH)_2D_3$ and $24,25(OH)_2D_3$ on the costochondral chondrocytes is to alter Ca^{2+} flux [216,221]. Using uptake and release of $^{45}Ca^{2+}$ as an indicator, a series of studies showed that the effects of these metabolites were observed within 1 min of exposure. The effects of the metabolites were complex. Although there was crossover in target cell specificity, the kinetics of the response were cell maturation specific.

E. Membrane Signaling

One of the most exciting aspects of how $1,25(OH)_2D_3$ and $24,25(OH)_2D_3$ exert their differential effects on chondrocytes has arisen from application of classic peptide membrane receptor technology to the study of lipophilic hormones. These studies suggest that several vitamin D metabolites work through signal transduction mechanisms involving protein kinase C (PKC). When the costochondral chondrocytes were treated with $1,25(OH)_2D_3$ and $24,25(OH)_2D_3$ and PKC activity measured, it was found that $1,25(OH)_2D_3$ activated enzyme activity exclusively in growth zone cells, whereas $24,25(OH)_2D_3$ activated enzyme activity only in resting zone cells [145]. Furthermore, there were distinct differences in the kinetics of action and the requirement for genomic expression. $1,25(OH)_2D_3$ exerted its effect without mRNA expression or protein synthesis, whereas $24,25(OH)_2D_3$ required both.

Studies using plasma membranes and matrix vesicles isolated from the chondrocyte cultures indicated that at least part of the response was due to direct effects of $1,25(OH)_2D_3$ and $24,25(OH)_2D_3$ on the cellular and matrix vesicle membranes [231]. When isolated plasma membranes were incubated directly with the vitamin D metabolites, $1,25(OH)_2D_3$ stimulated PKC-α in growth zone cell membranes, and $24,25(OH)_2D_3$ stimulated PKC-α in resting zone cell membranes. However, when isolated matrix vesicles were incubated directly with the metabolites, $1,25(OH)_2D_3$ and $24,25(OH)_2D_3$ inhibited PKC-ζ; the effect was specific to the extracellular organelles produced by the target cell cultures. This differential distribution of PKC isoforms suggests an explanation for how $1,25(OH)_2D_3$ and $24,25(OH)_2D_3$ may exert one set of effects on the cells and another on the matrix vesicles. Other studies have indicated that incorporation of PKC-ζ into matrix vesicles is under genetic control but regulation of preexisting matrix vesicle PKC-ζ is under nongenomic control [231]

F. Membrane Receptors

The experiments described above indicate that the direct effects of vitamin D metabolites are likely to occur on the membrane; however, it remains unclear whether specific membrane receptors are involved. Nemere and co-workers [232,233] have identified a putative $1,25(OH)_2D_3$ receptor in membranes from chick intestinal epithelium; Baran *et al.* [234–236] have evidence supporting a similar membrane-bound receptor in bone cells; and Seo and Norman report the possibility of a $24,25(OH)_2D_3$ membrane receptor in chick fracture callus [237]. Studies using $1,25(OH)_2D_3$ analogs that have modified A rings and/or CD-ring side chains have provided definitive evidence of stereospecific interactions of the compounds with the membrane, suggesting specificity between cellular and matrix vesicle membranes [238]. The present data concerning nongenomic effects of vitamin D metabolites support three potential hypotheses, summarized as follows: (1) a specific protein receptor does not exist, and specificity is conferred by changes in the fluid mosaic microstructure of the membrane [239]; (2) specific membrane receptors exist that are unique and distinct from the classic VDR [233,240]; or (3) classic VDRs interact with the hormone as it passes through the membrane, resulting in subsequent nongenomic and/or genomic actions [241]. Whether any of these hypotheses is correct remains to be determined.

G. Mechanism of Action

The precise mechanisms by which vitamin D metabolites regulate chondrocytes via rapid effects on cell membranes and matrix vesicles are not clear. Various data, summarized below, suggest that the mechanism for each metabolite is different. Events at the cell membrane may be nongenomic, but it is likely that they will elicit a genomic response via VDR-independent signal transduction cascades. Furthermore, in intact cells, rapid effects on the plasma membrane may be rapidly down-regulated. In contrast, rapid effects on matrix vesicles may involve similar mechanisms; however, down-regulation is not possible, nor is there a downstream modulation of a genomic response.

A proposal for the mechanism of rapid action of $1,25(OH)_2D_3$ on growth zone chondrocytes is presented in Fig. 3. The effect of $1,25(OH)_2D_3$ on growth zone cells acting via a hypothetical membrane receptor is to increase phospholipase A_2 (PA_2) activity in the membrane [214]. This results in increased fatty acid turnover [217] and, ultimately, to a change in membrane fluidity [215]. This fluidity change can alter membrane enzyme activity and calcium flux [216,221]. $1,25(OH)_2D_3$ also increases phospholipase C (PLC) activity, which can act on phosphatidylinositol 4,5-bisphosphate (PIP_2) to increase the production of inositol trisphosphate (IP_3) and diacylglycerol (DAG) [242]. IP_3 stimulates release of Ca^{2+} from the endoplasmic reticulum (ER) and into the cell from the extracellular fluid. DAG activates protein kinase C (PKC), which is the main second messenger of $1,25(OH)_2D_3$ in these cells [145,242]. Stimulation of arachidonic acid release increases local concentrations of this fatty acid. Arachidonic acid by itself can stimulate PKC activity [243]. In addition, production of arachidonic acid is the rate-limiting step in prostaglandin production. Increasing arachidonic acid release results in an increase in PGE_2 production, which serves as an autocrine regulator of the chondrocytes [230], acting through protein kinase A (PKA) via G-protein activation of adenylate cyclase (AC) and increased cyclic AMP (cAMP). Tyrosine kinases do not appear to play a role in this process [242]. Once PKC is activated, it can activate a signal transduction cascade, by phosphorylation of serine and threonine residues, leading to mitogen-activated protein kinase (MAPK) and phosphorylation of AP-1 sites on the relevant gene promoters.

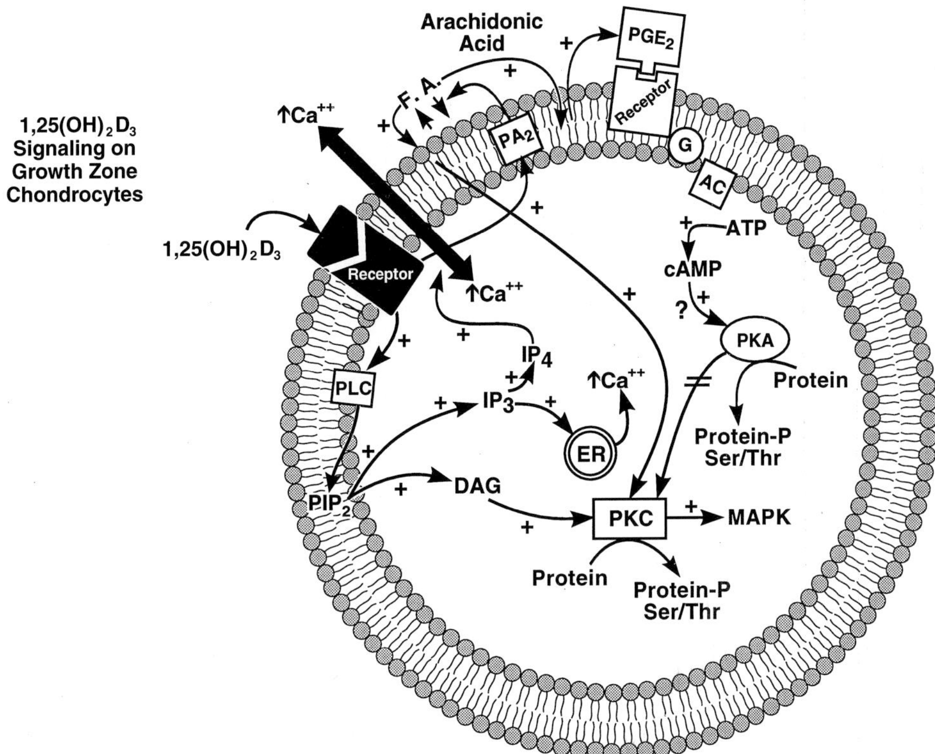

FIGURE 3 Proposed mechanism for the rapid action of $1,25(OH)_2D_3$ on growth zone chondrocytes. In growth zone cells, $1,25(OH)_2D_3$, acting via a hypothetical membrane receptor, increases phospholipase A_2 (PA_2) activity in the membrane. This results in increased fatty acid turnover and changes in membrane fluidity. Changes in membrane fluidity can alter the activity of enzymes in the membrane and also calcium flux. $1,25(OH)_2D_3$ also increases phospholipase C (PLC), which can act on phosphatidylinositol 4,5-bisphosphate (PIP_2) to increase the production of inositol trisphosphate (IP_3) and diacylglycerol (DAG). IP_3 can stimulate the release of calcium from the endoplasmic reticulum (ER) and into the cell. DAG can activate protein kinase C (PKC), which is the main second messenger of $1,25(OH)_2D_3$ in these cells. Stimulation of arachidonic acid release activates PKC. Increased production of arachidonic acid also increases the production of prostaglandins, such as prostaglandin E_2 (PGE_2), which are potent regulators of chondrocytes. PGE_2 activates the G-protein pathway, stimulating adenylate cyclase (AC) activity and increasing protein kinase A (PKA) activity. Once PKC is activated, it can activate a signal transduction cascade, by phosphorylation of serine and threonine residues, leading to mitogen-activated protein kinase (MAPK) activation, phosphorylation of AP-1, and increased transcription from relevant gene promoters. This hypothetical model remains to be proved.

The mechanism of $24,25(OH)_2D_3$ action on resting zone cells is schematically shown in Fig. 4. It is predicated on the observation that $24,25(OH)_2D_3$, again acting via a hypothetical membrane receptor, inhibits PA_2 [214] activity, an effect that will change fatty acid turnover [217,227], release of arachidonic acid [217,227], and production of PGE_2 [230]. There is a resultant change in membrane fluidity [215] and Ca^{2+} flux [216,221]. $24,25(OH)_2D_3$ increases DAG production [242], stimulating PKC activity [145,231,242]. Tyrosine kinase and phospholipase C are not involved in this process, suggesting that phospholipase D is responsible. Inhibition of phospholipase A_2 results in increased PKC activity. Moreover, inhibition of prostaglandin E_2 production increases PKC. In contrast, addition of prostaglandin E_2 inhibits PKC through PKA. PKC may change cell behavior through phosphorylation of different proteins, such as stromelysin-1 [111].

V. PHYSIOLOGIC RELEVANCE OF NONGENOMIC REGULATION OF MATRIX VESICLES

Although the cell can down-regulate undesired nongenomic effects at the plasma membrane, this is more difficult in the matrix. To control events in the matrix, the cell may modulate the rate and nature of matrix vesicles produced by two mechanisms [110]. Initially,

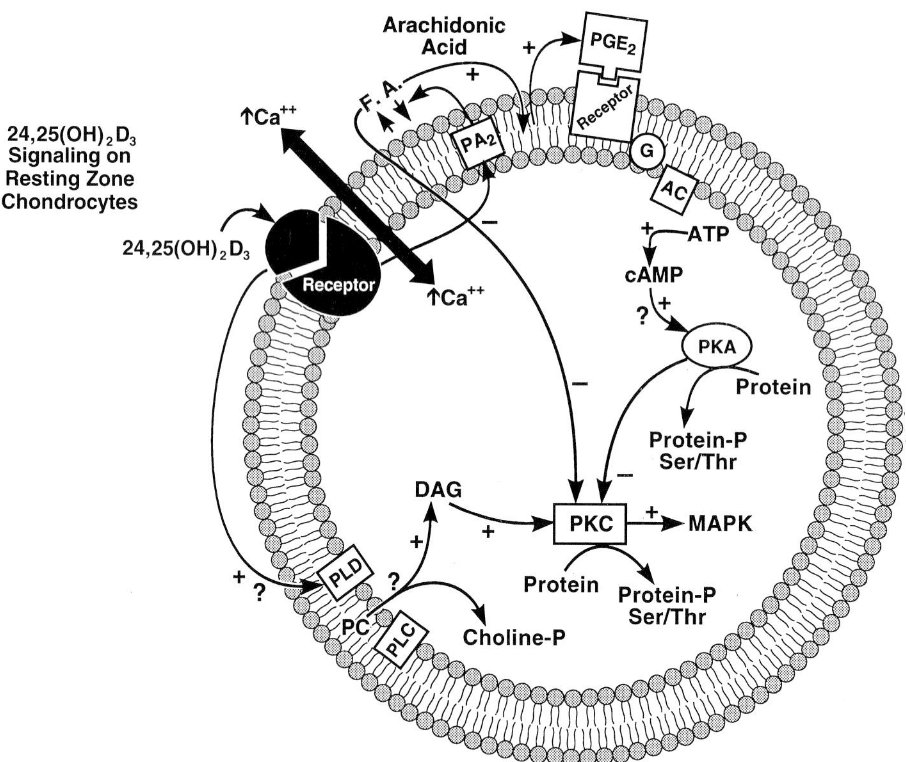

FIGURE 4 Proposed mechanism for the rapid action of $24,25(OH)_2D_3$ on resting zone chondrocytes. In resting zone chondrocytes, $24,25(OH)_2D_3$, acting via a hypothetical receptor, inhibits phospholipase A_2 (PA_2) activity. This results in changes in fatty acid turnover, release of arachidonic acid, and production of prostaglandin E_2 (PGE_2). There is a resultant change in membrane fluidity and calcium flux. $24,25(OH)_2D_3$ increases diacylglycerol (DAG) production, which activates protein kinase C (PKC). Tyrosine kinase and phospholipase C (PLC) are not involved in this process, suggesting that phospholipase D is responsible. Inhibition of PA_2 results in increased PKC activity. Moreover, inhibition of PGE_2 production increases PKC. In contrast, addition of PGE_2 inhibits PKC through PKA. PKC may then change cell behavior through phosphorylation of different proteins. This hypothetical model remains to be proved.

matrix vesicles are produced under genomic control. $1,25(OH)_2D_3$ and perhaps $24,25(OH)_2D_3$, regulate the composition of matrix vesicles through new gene transcription, protein synthesis, and, finally, membrane synthesis. Once matrix vesicles are released into the extracellular matrix, the cells may regulate their maturation through secretion of vitamin D metabolites that act on the matrix vesicle through nongenomic mechanisms.

If this is the case, then it is obligatory that the cells produce vitamin D metabolites and that this production be regulated by growth factors and hormones. In fact, chondrocytes have been found to produce both ^3H-$1,25(OH)_2D_3$ and ^3H-$24,25(OH)_2D_3$ when incubated with ^3H-$25(OH)D_3$ [155]. Moreover, production of vitamin D metabolites is regulated by $1,25(OH)_2D_3$, $24,25(OH)_2D_3$, TGFβ, and dexamethasone in a cell maturation-specific manner. It is likely that systemic $1,25(OH)_2D_3$ and $24,25(OH)_2D_3$ also play roles in this process by providing the conditioning background for cell response. Locally secreted metabolites would permit fine-tuning of the matrix vesicles.

Matrix vesicles may have multiple functions in the matrix. Those in the lower hypertrophic cell zone of cartilage or in the osteoid of bone are probably involved in matrix calcification [244]. In addition, matrix vesicles also appear to be involved in matrix maturation, as they contain matrix-processing enzymes that degrade proteoglycans [109,112,113,245]. We have shown that matrix vesicles can reverse the inhibition of calcification caused by proteoglycan aggregate *in vitro* [246], suggesting that the proteinases in matrix vesicles are active in a functional capacity as well.

In addition to their role in calcification, it now seems evident that matrix vesicles may play an important role in activation of growth factors present in the extracellular matrix [247]. When chondrocytes are incubated with vitamin D metabolites, only growth zone cells incubated with $1,25(OH)_2D_3$ show a decrease in the amount of latent TGFβ found in the conditioned medium. This decrease is further correlated with an increase in active TGFβ when isolated matrix vesicles are incubated directly with latent TGFβ and $1,25(OH)_2D_3$ *in vitro*. These observations suggest that nongenomic regulation of matrix vesicles can result in changes in local growth factor activation. This is a particularly attractive hypothesis in cartilage, where activation of latent growth factor by local decreases in pH (as occurs in osteoclasts) has not been reported.

These observations led to the following hypothesis for nongenomic regulation of events in the extracellular matrix (Fig. 5). Chondrocytes produce matrix vesicles under hormonal and growth factor regulation. At the same time, vitamin D metabolites are synthesized by the cells and secreted in response to regulatory factors like $1,25(OH)_2D_3$, $24,25(OH)_2D_3$, TGFβ, or corticosteroids [155]. These factors diffuse into the matrix and interact directly with the plasma membrane. In addition, they also interact with the matrix vesicle membrane, where their effects initiate a cascade of biochemical events that lead to maturation of the matrix vesicle, hydroxyapatite crystal formation, degeneration of the integrity of the matrix vesicle membrane, and eventual release of active proteases. The proteases then degrade proteoglycan aggregates in the vicinity of the matrix vesicle, facilitating extracellular matrix calcification. In addition, they may activate latent growth factors that can then act on the cell in an autocrine manner or on adjacent cells via paracrine interactions.

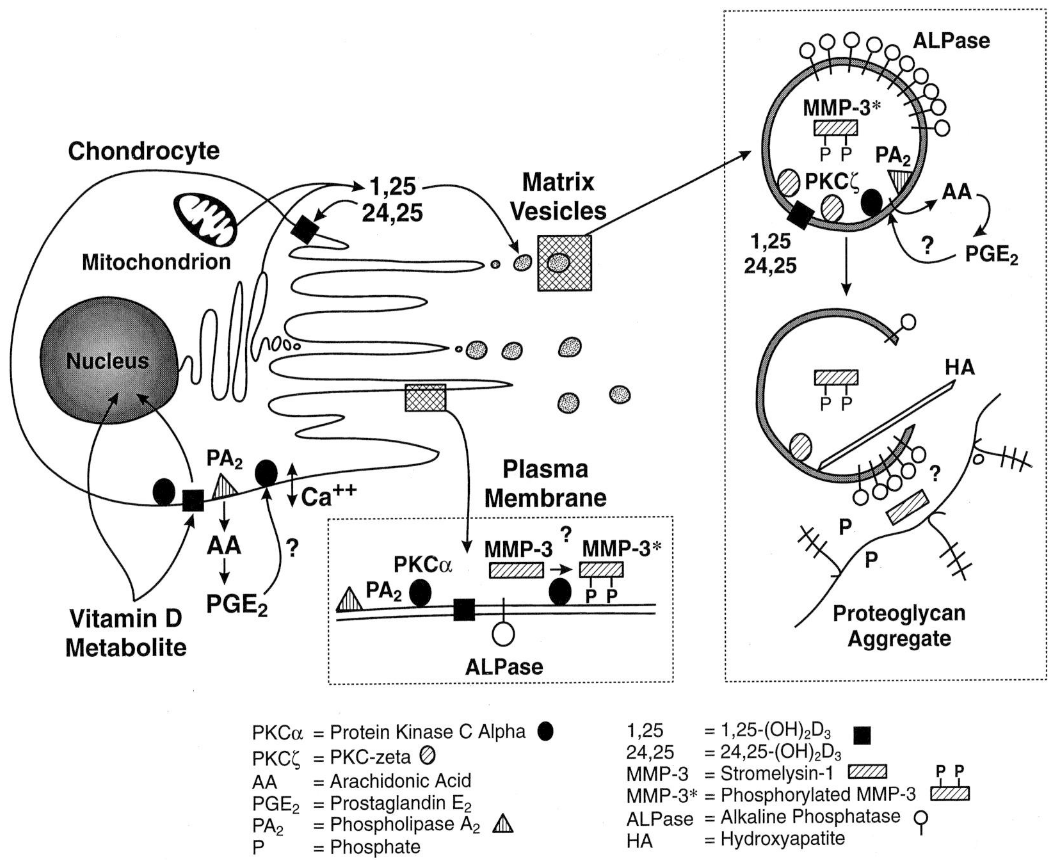

FIGURE 5 Proposed mechanism for the nongenomic regulation of matrix vesicles in the extracellular matrix. Chondrocytes produce matrix vesicles under hormonal and growth factor control. Systemic $1,25(OH)_2D_3$ or $24,25(OH)_2D_3$ interact with classic receptors or stimulate rapid membrane-mediated signal transduction pathways, resulting in new gene expression and, ultimately, new matrix vesicle production. Rapid membrane responses include release of arachidonic acid and prostaglandin E_2 (PGE_2) production as well as altered calcium flux. At the same time, vitamin D metabolites are synthesized by the cells. Although evidence for the 1α- and 24-hydroxylases exists, it is not yet known whether these enzymes are localized to mitochondria or endoplasmic reticulum. The vitamin D metabolites are secreted into the matrix and interact directly with the plasma membrane, causing PKCα-dependent phosphorylation of matrix metalloproteinase-3 (MMP-3). In addition, they also interact with the membrane of preexisting matrix vesicles, where they initiate a cascade of events leading to matrix vesicle maturation, hydroxyapatite crystal formation, and, in matrix vesicles produced by growth zone chondrocytes, loss of matrix vesicle membrane integrity through stimulation of phospholipases. Once this occurs, active proteinases such as MMP-3 are released. The proteinases degrade proteoglycan aggregates, facilitating matrix calcification. In addition, they may activate latent growth factors. See the text for additional details.

Further evidence that $1,25(OH)_2D_3$ and $24,25(OH)_2D_3$ can directly affect proteoglycan degradation and matrix calcification via nongenomic effects on matrix vesicles is now available. When growth zone chondrocytes are treated with $1,25(OH)_2D_3$, there is an increase in matrix vesicle matrix metalloproteinase (MMP) activity [112,113]. Analysis of the direct effect of $1,25(OH)_2D_3$ on isolated membrane fractions indicates that plasma membrane-associated PKC is increased, resulting in the PKC-α-dependent phosphorylation of MMP-3 [111]. In this state, MMP-3 is then packaged into matrix vesicles and released into the matrix. However, no MMP activity is detectable in isolated matrix vesicles unless membrane integrity is lost [109,245]. When treated with $1,25(OH)_2D_3$, isolated matrix vesicles contain increased phospholipase A_2, which destabilizes the matrix vesicle membrane, releasing the MMP into the matrix. At the same time, alkaline phosphatase in the matrix vesicles is elevated in response to $1,25(OH)_2D_3$. We hypothesize that alkaline phosphatase then dephosphorylates MMP-3, resulting in increased MMP activity. As PKC-ζ activity in the matrix vesicle is decreased by $1,25(OH)_2D_3$ treatment, this isoform of PKC has no effect on matrix vesicle MMP-3, leading to the hypothesis that the enzyme is then fully active in the matrix [111].

When isolated matrix vesicles are incubated in gelatin gels in the presence of proteoglycan, the inhibition of crystal formation normally associated with proteoglycan is lost. Furthermore, treatment of the matrix vesicles with $1,25(OH)_2D_3$ causes an increase in the rate and extent of new crystal formation (B. D. Boyan, A. L. Boskey, and Z. Schwartz, unpublished data, 1996).

VI. SUMMARY

This chapter has shown that cartilage, much like other tissues, is really a family of tissues spanning a broad spectrum of cell maturation states. In growth plate, a subset of the cartilage phenotype, chondrocytes can be seen at distinct states of maturation in a linear array. Using a variety of *in vivo* and *in vitro* assays, investigators have been able to show that the growth plate is sensitive to vitamin D regulation, with $24,25(OH)_2D_3$ affecting less mature cells, particularly those of the resting zone, and $1,25(OH)_2D_3$ modulating activities in the growth zone (prehypertrophic and upper hypertrophic) cartilage. Both metabolites exert their effects through genomic mechanisms. However, some of the responses of the cell may involve nongenomic mechanisms. Rapid cell membrane-mediated events may result in secondary genomic responses via protein phosphorylation cascades and MAP kinase; in matrix vesicles, rapid membrane effects may be termed nongenomic because no gene expression or protein synthesis are possible. This is relevant to *in vivo* regulation of endochondral ossification, since chondrocytes produce and secrete $1,25(OH)_2D_3$ or $24,25(OH)_2D_3$, which then may interact with the matrix vesicles, resulting in modulation of the activity of this extracellular organelle. The consequences of this include activation of latent growth factors, degradation of matrix proteoglycans, and calcium phosphate deposition. Thus, $1,25(OH)_2D_3$ and $24,25(OH)_2D_3$ regulate chondrocyte proliferation, metabolism, differentiation, and maturation, as well as events in the extracellular matrix. The effects are cell maturation-dependent, and organelle specific, and may involve both VDR-dependent and VDR-independent genomic as well as nongenomic mechanisms.

References

1. Caplan AI, Boyan BD 1994 Endochondral bone formation: The lineage cascade. In: Hall BK. (ed) Bone 8. Mechanisms of Bone Development and Growth. CRC Press, Boca Raton, Florida, pp. 1–46.
2. Hardingham TE, Fosang AJ, Dudhia J 1992 Aggrecan, the chondroitin sulfate/keratan sulfate proteoglycan from cartilage. In: Kuettner KE, Schleyerbach R, Peyron JG, Hascall VC (eds) Articular Cartilage and Osteoarthritis. Raven, New York, pp. 5–20.
3. Weitzhandler M, Carrino DA, Caplan AI 1988 Proteoglycans synthesized during the cartilage-to-bone transition in developing chick embryos. Bone **9**:225–233.
4. Mourao PAS 1988 Distribution of chondroitin-4-sulfate and chondroitin-6-sulfate in human articular and growth cartilage. Arth Rheum **31**:1028–1033.
5. Drushel RF, Pechak DG, Caplan AI 1985 The anatomy, ultrastructure and fluid dynamics of the developing vasculature of the embryonic chick wing bud. Cell Differ **16**:13–28.
6. Caplan AI 1985 The vasculature and limb development. Cell Differ **16**:1–11.
7. Jargiello DM, Caplan AI 1983 The establishment of vascular-derived microenvironments in the developing chick wing. Dev Biol **97**:364–374.
8. Drushel RF, Caplan AI 1988 Extravascular fluid dynamics of the embryonic chick wing bud. Dev Biol **126**:7–18.
9. Schenk RK, Wiener J, Spiro D 1968 Fine structural aspects of vascular invasion of the tibial epiphyseal plate of growing rats. Acta Anat **69**:1–17.
10. Hunziker EB, Herrmann W, Schenk RK, Mueller M, Moor H 1984 Cartilage ultrastructure after high pressure freezing, freeze substitution, and low temperature embedding. I. Chondrocyte ultrastructure: Implications for the theories of mineralization and vascular invasion. J Cell Biol **98**:267–276.
11. Solursh M, Ahrens PB, Reiter RS 1978 A tissue culture analysis of the steps in limb chondrogenesis. In Vitro **14**:51–61.
12. Schmid TM, Linsenmayer TF 1985 Immunohistochemical localization of short chain cartilage collagen (type X) in avian tissues. J Cell Biol **100**:598–605.
13. Stocum DL, Davis RM, Leger M, Conrad HE 1979 Development of the tibiotarsus in the chick embryo: Biosynthetic activities of

histologically distinct regions. J Embryol Exp Morphol **54**: 155–170.
14. Pacifici M, Golden EB, Iwamoto M, Adams SL 1991 Retinoic acid treatment induces type X collagen gene expression in cultured chick chondrocytes. Exp Cell Res **195**:38–46.
15. Pechak DG, Kujawa MJ, Caplan AI. 1986 Morphological and histological events during first bone formation in embryonic chick limbs. Bone **7**:441–458.
16. Pechak DG, Kujawa MJ, Caplan AI 1986 Morphology of bone development and bone remodeling in embryonic chick limbs. Bone **7**:459–472.
17. Caplan AI, Pechak DG 1987 The cellular and molecular embryology of bone formation. In: Peck WA. (ed), Bone and Mineral Research. Elsevier, New York; pp. 117–183.
18. Caplan AI 1988 Bone development. Cell and Molecular Biology of Vertebrate Hard Tissues. Wiley, Chichester, pp. 3–21.
19. Caplan AI 1991 Mesenchymal stem cells. J Orthop Res **9**:641–650.
20. Beresford JN 1989 Osteogenic stem cells and the stromal system of bone and marrow. Clin Orthop Related Res **240**:270–280.
21. Owen M 1985 Lineage of osteogenic cells and their relationship to the stromal system. Bone Miner Res **3**:1–25.
22. Fridenstein AJ 1990 Osteogenic stem cells in the bone marrow. In: Peck WA. (ed) Bone and Mineral Research. Elsevier, New York, pp. 243–272.
23. Janners MY, Searls RL 1970 Changes in the rate of cellular proliferation during the differentiation of cartilage and muscle in mesenchyme of the embryonic chick wing. Dev Biol **23**:136–165.
24. Brighton CT 1984 The biology of fracture repair. In: Murray JA. (ed) American Academy of Orthopaedic Surgeons Instructional Course Lectures. Mosby, St. Louis, Missouri, pp. 60–82.
25. Cornell CN, Lane JM 1992 Newest factors in fracture healing. Clin Orthop Related Res **277**:297–311.
26. Nemeth GG, Bolander ME, Martin GR 1988 Growth factors and their role in wound and fracture healing. Prog Clin Biol Res **266**:1–17.
27. Joyce ME, Terek RM, Jingushi S, Bolander ME 1990 Role of transforming growth factor-β in fracture repair. Ann NY Acad Sci **593**:107–123.
28. Reddi AH, Wientroub S, Muthukumaran N 1987 Biologic principles of bone induction. Orthop Clin North Am **18**:207–212.
29. Reddi AH, Huggins CB 1972 Biochemical sequences in the transformation of normal fibroblasts in adolescent rats. Proc Natl Acad Sci USA **69**:1601–1605.
30. Urist MR, Delange RJ, Finerman GA 1983 Bone cell differentiation and growth factors. Science **220**:680–686.
31. Boyan BD, Schwartz Z, Swain LD, Khare AG, Heckman JD, Ramirez V, Peters P, Carnes DL Jr 1992 Initial effects of partially purified bone morphogenetic protein on the expression of glycosaminoglycan, collagen, and alkaline phosphatase in nonunion cell cultures. Clin Orthop **278**:286–304.
32. Chen P, Carrington JL, Hammonds RG, Reddi AH 1991 Stimulation of chondrogenesis in limb bud mesoderm cells by recombinant human bone morphogenetic protein 2B (BMP-2B) and modulation by transforming growth factor beta 1 and beta 2. Exp Cell Res **195**:509–515.
33. Howell DS, Dean DD 1992 Biology, chemistry and biochemistry of the mammalian growth plate, In: Coe FL, Favus MJ. (eds) Disorders of Bone and Mineral Metabolism, Raven, New York, pp. 313–353.
34. Poole AR 1991 The growth plate: Cellular physiology, cartilage assembly and mineralization. In: Hall BK, Newman S. (eds) Cartilage: Molecular Aspects. CRC Press, Boca Raton, Florida, pp. 179–211.
35. Reddi AH 1994 Bone and cartilage differentiation. Curr Opin Genet Dev **4**:737–744.
36. Lanske B, Karaplis AC, Lee K, Luz A, Vortkamp A, Pirro A, Karperien M, Defize LHK, Ho C, Mulligan RC, Abou-Samra AB, Juppner H, Segre GV, Kronenberg HM 1996 PTH/PTHrP receptor in early development and indian hedgehog-regulated bone growth. Science **273**:663–666.
37. Hunziker EB, Schenk RK, Cruz-Orive LM 1987 Quantitation of chondrocyte performance in growth plate cartilage during longitudinal bone growth. J Bone Joint Surg **69A**:162–173.
38. Carlson CS, Hilley HD, Henrickson CK 1985 Ultrastructure of normal epiphyseal cartilage of the articular-epiphyseal cartilage complex in growing swine. Am J Vet Res **46**:306–313.
39. Howell DS, Carlson L 1968 Alterations in the composition of growth cartilage septa during calcification studied by microscopic x-ray elemental analysis. Exp Cell Res **51**:185–195.
40. Buckwalter JA, Mower D, Ungar R 1986 Morphometric analysis of chondrocyte hypertrophy. J Bone Joint Surg **68A**:243–255.
41. Dean DD, Muniz OE, Berman I, Pita JC, Carreno MR, Woessner JF Jr, Howell DS 1985 Localization of collagenase in the growth plate of rachitic rats. J Clin Invest **76**:716–722.
42. Poole AR, Pidoux I, Rosenberg LC 1982 Role of proteoglycans in endochondral ossification: immunofluorescent localization of link protein and proteoglycan monomer in bovine fetal epiphyseal growth plate. J Cell Biol **92**:249–260.
43. Buckwalter JA 1983 Proteoglycan structure in calcifying cartilage. Clin Orthop Related Res **172**:207–232.
44. Buckwalter JA, Rosenberg LC, Ungar R 1987 Changes in proteoglycan aggregates during cartilage mineralization. Calcif Tissue Int **41**:228–236.
45. Grant WT, Sussman MD, Balian G 1985 A disulfide-bonded short chain collagen synthesized by degenerative and calcifying zones of bovine growth plate cartilage. J Biol Chem **260**:3798–3803.
46. Kielty CM, Kwan APL, Holmes DF, Schor SL, Grant ME 1985 Type X collagen, a product of hypertrophic chondrocytes. Biochem J **227**:545–554.
47. Schmid TM, Linsenmayer TF 1985 Developmental acquisition of type X collagen in the embryonic chick tibiotarsus. Dev Biol **107**:373–381.
48. Gerstenfeld LC, Kelly CM, Von Deck M, Lian JB 1990 Effect of 1,25-dihydroxyvitamin D_3 on induction of chondrocyte maturation in culture: Extracellular matrix gene expression and morphology. Endocrinology **126**:1599–1609.
49. Gerstenfeld LC, Kelly CM, Von Deck M, Lian JB 1990 Comparative morphological and biochemical analysis of hypertrophic, non-hypertrophic and $1,25(OH)_2D_3$-treated non-hypertrophic chondrocytes. Connect Tissue Res **24**:29–39.
50. Quarto R, Dozin B, Bonaldo P, Cancedda R, Colombatti A 1993 Type IV collagen expression is upregulated in the early events of chondrocyte differentiation. Development **117**:245–251.
51. Roach HI, Erenpreisa J, Aigner T 1995 Osteogenic differentiation of hypertrophic chondrocytes involves asymmetric cell divisions and apoptosis. J Cell Biol **131**:483–494.
52. Roach HI 1992 Trans-differentiation of hypertrophic chondrocytes into cells capable of producing a mineralized bone matrix. Bone Miner **19**:1–20.
53. Aydelotte MB, Raiss RX, Caterson B, Kuettner KE 1992 Influence of interleukin-1 on the morphology and proteoglycan metabolism of cultured bovine articular chondrocytes. Connect Tissue Res **28**:143–159.
54. Hauselmann HJ, Aydelotte MB, Schumacher BL, Kuettner KE, Gitelis SH, Thonar EJ 1992 Synthesis and turnover of proteogly-

cans by human and bovine adult articular chondrocytes cultured in alginate beads. Matrix **12**:116–129.
55. McClure J, Bates GP, Rowston H, Grant ME 1988 A comparison of the morphological, histochemical and biochemical features of embryonic chick sternal chondrocytes *in vivo* with chondrocytes cultured in three-dimensional collagen gels. Bone Miner **3**:235–247.
56. Benya PD, Shaffer JD 1982 Dedifferentiated chondrocytes reexpress the differentiated collagen phenotype when cultured in agarose gels. Cell **30**:215–224.
57. Suzuki F, Takase T, Takigawa M, Uchida A, Shimomura S 1981 Simulation of the initial stage of endochondral ossification: *In vitro* sequential culture of growth cartilage cells and bone marrow cells. Proc Natl Acad Sci USA **78**:2368–2372.
58. Ali SY 1985 Calcification of cartilage. In: Hall BK. (ed) Cartilage Structure, Function and Biochemistry. Academic Press, New York, pp. 343–378.
59. Vaananen HK, Morris DC, Anderson HC 1983 Calcification of cartilage matrix in chondrocyte cultures derived from rachitic rat growth plate cartilage. Metab Bone Dis Related Res **5**:87–92.
60. Hancock RH, Schwartz Z, Swain LD, Ramirez V, Boyan BD 1990 Effect of dexamethasone on cartilage cell differentiation. J Dent Res **69**:293
61. Schwartz Z, Hancock RH, Dean DD, Brooks BP, Gomez R, Boskey AL, Balian G, Boyan BD 1995 Dexamethasone promotes von Kossa-positive nodule formation and increased alkaline phosphatase activity in costochondral chondrocyte cultures. Endocrine **3**:351–360.
62. Nakahara H, Dennis JE, Bruder SP, Haynesworth SE, Lennon DP, Caplan AI 1991 In vitro differentiation of bone and hypertrophic cartilage from periosteal-derived cells. Exp Cell Res **195**:492–503.
63. Nakahara H, Watanabe K, Sugrue SP, Olsen BR, Caplan AI 1990 Temporal and spatial distribution of type XII collagen in high cell density culture of periosteal-derived cells. Dev Biol **142**:481–485.
64. Boyan BD, Schwartz Z, Swain LD, Carnes DL Jr, Zislis T 1988 Differential expression of phenotype by resting zone and growth region costochondral chondrocytes in vitro. Bone **9**:185–194.
65. Boyan BD, Schwartz Z, Swain LD 1992 *In vitro* studies on the regulation of endochondral ossification by vitamin D. Crit Rev Oral Biol Med **3**:15–30.
66. Chen P, Carrington JL, Paralkar VM, Pierce GF, Reddi AH 1992 Chick limb bud mesodermal cell chondrogenesis: Inhibition by isoforms of platelet-derived growth factor and reversal by recombinant bone morphogenetic protein. Exp Cell Res **200**:110–117.
67. Boskey AL, Stiner D, Doty SB, Binderman I 1991 Requirement of vitamin C for cartilage calcification in a differentiating chick limb-bud mesenchymal cell culture. Bone **12**:277–282.
68. O'Keefe RJ, Crabb ID, Puzas JE, Rosier RN 1989 Countercurrent centrifugal elutriation: High-resolution method for the separation of growth-plate chondrocytes. J Bone Joint Surg **71**(A): 607–620.
69. Dean DD, Muniz OE, Woessner JF Jr, Howell DS 1990 Production of collagenase and tissue inhibitor of metalloproteinases (TIMP) by rat growth plates in culture. Matrix **10**:320–330.
70. Atkin I, Dean DD, Muniz OE, Agundez A, Castiglione G, Cohen G, Howell DS, Ornoy A 1992 Enhancement of osteoinduction by vitamin D metabolites in rachitic host rats. J Bone Miner Res **7**:863–875.
71. Boskey AL, Wientroub S 1986 The effect of vitamin D deficiency on rat bone lipid composition. Bone **7**:277–281.
72. Wu JJ, Eyre DR, Dean DD, Howell DS 1989 Quantitative studies on the collagen composition of the growth plate in the rachitic rat. Trans Orthop Res Soc **14**:58 (Abstract).
73. Miller EJ 1985 The structure of fibril-forming collagens. Ann NY Acad Sci **460**:1–13.
74. Mayne R, Irwin MH 1986 Collagen types in cartilage. In: Kuettner K, Schleyerbach R, Hascall VC. (eds) Articular Cartilage Biochemistry. Raven, New York, pp. 23–38.
75. Eyre DR 1987 Collagens of the disc. In: Ghosh P. (ed) The Biology of the Invertebral Disc. CRC Press, Boca Raton, Florida, pp. 171–188.
76. Howell DS, Delchamps EE, Reimer W 1960 A profile of electrolytes in the cartilaginous plate of growing ribs. J Clin Invest **39**:919–929.
77. Wuthier RE 1971 Zonal analysis of electrolytes in epiphyseal cartilage and bone of normal and rachitic chickens and pigs. Calcif Tissue Res **8**:24–35.
78. Heinegard D, Oldberg A 1989 Structure and biology of cartilage and bone matrix noncollagenous macromolecules. FASEB J **3**:2042–2051.
79. Miller RR, McDevitt CA 1988 Thrombospondin is present in articular cartilage and is synthesized by articular chondrocytes. Biochem Biophys Res Commun **153**:708–714.
80. Boskey AL 1989 Noncollagenous matrix proteins and their role in mineralization: A mini-review. Bone Miner **6**:111–123.
81. Mechanic GL, Banes AJ, Henmi M, Yamauchi M 1985 Possible collagen structural control of mineralization. In: Butler WT. (ed) The Chemistry and Biology of Mineralized Tissues. Ebsco Media, Birmingham, U.K., pp. 98–106.
82. van der Rest M, Mayne R 1987 Type IX collagen. In: Mayne R, Burgeson RE (eds.) Structure and Function of Collagen Types. Academic Press, New York; pp. 195–221.
83. Noro A, Kimata K, Oike Y, Shinomura T, Maeda N, Yano S, Takahashi N, Suzuki S 1983 Isolation and characterization of a third proteoglycan (PG-Lt) from chick embryo cartilage which contains a disulfide-bonded collagenous polypeptide. J Biol Chem **258**:9323–9331.
84. Eyre DR, Apone S, Wu JJ 1987 Collagen type IX: Evidence for covalent linkages to type II collagen in cartilage. FEBS Lett **220**:337–341.
85. van der Rest M, Mayne R 1988 Type IX collagen proteoglycan from cartilage is covalently cross-linked to type II collagen. J Biol Chem **263**:1615–1618.
86. Bruckner P, Vaughan L, Winterhalter KH 1985 Type IX collagen from sternal cartilage of chicken embryo contains covalently bound glycosaminoglycans. Proc Natl Acad Sci USA **82**:2608–2612.
87. Schmid TM, Conrad HE 1982 A unique low molecular weight collagen secreted by cultured chick embryo chondrocytes. J Biol Chem **257**:12444–12450.
88. Jimenez SA, Yankowski R, Reginato AM 1986 Quantitative analysis of type X collagen biosynthesis by embryonic chick sternal cartilage. Biochem J **233**:357–367.
89. Ninomiya Y, Gordon M, van der Rest M, Schmid T, Linsenmayer TF, Olsen BR 1986 The developmentally regulated type X collagen gene contains a long open reading frame without introns. J Biol Chem **261**:5041–5050.
90. Poole AR 1986 Proteoglycans in health and disease: Structures and functions. Biochem J **236**:1–14.
91. Mow VC, Mak AF, Lai WM, Rosenberg LC, Tang LH 1984 Viscoelastic properties of proteoglycan subunits and aggregates in varying solution concentrations. J Biomech **17**:325–338.
92. Hardingham TE, Muir H, Kwan MK, Lai WM, Mow VC 1987 Viscoelastic properties of proteoglycan solutions with varying proportions present as aggregates. J Orthop Res **5**:36–46.

93. Hunziker EB, Schenk RK 1987 Structural organization of proteoglycans in cartilage. In: Wight TN, Mecham RP (eds) Biology of Proteoglycans. Academic Press, New York, pp. 155–185.
94. Bianco P, Fisher LW, Young MF 1989 Differential expression of small proteoglycans I and II (biglycan and decorin) in human developing bone and non-bone tissues as revealed by immunolocalization and in situ hybridization. J Bone Miner Res **4:**S321 (Abstract).
95. Andres JL, Stanley K, Cheifetz S, Massague J 1989 Membrane-anchored and soluble forms of betaglycan, a polymorphic proteoglycan that binds transforming growth factor-beta. J Cell Biol **109**:3137–3145.
96. Ruohola JK, Valve EM, Vainikka S, Alitalo K, Harkonen PL 1995 Androgen and fibroblast growth factor (FGF) regulation of FGF receptors in S115 mouse mammary tumor cells. Endocrinology **136**:2179–2188.
97. Pita JC, Muller FJ, Howell DS 1979 Structural changes of sulfated proteoglycans of rat growth cartilage during endochondral calcification. Proceedings of the Fourth International Symposium on Glycoconjugate Research. Academic Press, New York; pp. 743–746.
98. Boskey AL. 1992 Mineral–matrix interactions in bone and cartilage. Clin Orthop Related Res **281**:244–274.
99. Von der Mark K, Mollenhauer J, Pfaffle M 1986 Role of anchorin CII in the interaction of chondrocytes with extracellular collagen. In: Kuettner KE, Schleyerbach R, Hascall VC (eds.) Articular Cartilage Biochemistry. Raven, New York; pp. 125–141.
100. Mollenhauer J, Bee JA, Lizarbe MA, Von der Mark K 1984 Role of anchorin CII, a 31,000 molecular weight membrane protein, in the interaction of chondrocytes with type II collagen. J Cell Biol **98**:1572–1579.
101. Von der Mark K, Mollenhauer J, Muller PK, Pfaffle M 1985 Anchorin CII: a type II collagen-binding glycoprotein from chondrocyte membranes. Ann NY Acad Sci **460**:214–223.
102. Hewitt AT, Kleinman HK, Pennypacker JP, Martin GR 1980 Identification of an adhesion factor for chondrocytes. Proc Natl Acad Sci USA **77**:385–388.
103. Hewitt AT, Varner HH, Silver MH, Dessau W, Wilkes CM, Martin GR 1982 The isolation and partial characterization of chondronectin, an attachment factor for chondrocytes. J Biol Chem **257**:2330–2334.
104. Romberg RW, Werness PG, Lollar P, Riggs BL, Mann KG 1985 Isolation and characterization of native adult osteonectin. J Biol Chem **260**:2728–2736.
105. Leboy PS, Shapiro IM, Uschmann BD, Oshima O, Lin D 1988 Gene expression in mineralizing chick epiphyseal cartilage. J Biol Chem **263**:8515–8520.
106. Choi HU, Tang LH, Johnson TL, Pal S, Rosenberg LC, Reiner A, Poole AR 1983 Isolation and characterization of a 35,000 molecular weight subunit fetal cartilage matrix protein. J Biol Chem **258**:655–661.
107. Poole AR, Pidoux I, Reiner A, Rosenberg LC, Hollister D, Murray L, Rimoin D 1988 Kniest dysplasia is characterized by an abnormal processing of the C-propeptide of type II cartilage collagen resulting in imperfect fibril assembly. J Clin Invest **81**:579–589.
108. Poole AR, Pidoux I, Reiner A, Choi H, Rosenberg LC 1984 Association of an extracellular protein (chondrocalcin) with the calcification of cartilage in endochondral bone formation. J Cell Biol **98**:54–65.
109. Dean DD, Schwartz Z, Muniz OE, Gomez R, Swain LD, Howell DS, Boyan BD 1992 Matrix vesicles are enriched in metalloproteinases that degrade proteoglycans. Calcif Tissue Int **50**: 342–349.
110. Boyan BD, Dean DD, Sylvia VL, Schwartz Z 1994 Nongenomic regulation of extracellular matrix events by vitamin D metabolites. J Cell Biochem **56**:331–339.
111. Schmitz JP, Schwartz Z, Sylvia VL, Dean DD, Calderon F, Boyan BD 1996 Vitamin D_3 regulation of stromelysin-1 (MMP-3) in chondrocyte cultures is mediated by protein kinase C. J Cell Physiol **168**:570–579.
112. Dean DD, Schwartz Z, Schmitz JP, Muniz OE, Lu Y, Calderon FJ, Howell DS, Boyan BD 1996 Vitamin D regulation of metalloproteinase activity in matrix vesicles. Connect Tissue Res **35**:331–336.
113. Dean DD, Boyan BD, Muniz OE, Howell DS, Schwartz Z 1996 Vitamin D metabolites regulate matrix vesicle metalloproteinase content in a cell maturation-dependent manner. Calcif Tissue Int **59**:109–116.
114. Klement BJ, Spooner BS 1994 Pre-metatarsal skeletal development in tissue culture at unit and microgravity. J Exp Zool **269**:230–241.
115. Schwartz Z, Bonewald LF, Caulfield K, Brooks BP, Boyan BD 1993 Direct effects of transforming growth factor-β on chondrocytes are modulated by vitamin D metabolites in a cell maturation-specific manner. Endocrinology **132**:1544–1552.
116. Lyons KM, Pelton RW, Hogan BL 1989 Patterns of expression of murine Vgr-1 and BMP-2a RNA suggest that transforming growth factor-beta-like genes coordinately regulate aspects of embryonic development. Genes Dev **3**:1657–1668.
117. Chang SC, Hoang B, Thomas JT, Vukicevic S, Luyten FP, Ryba NJ, Kozak CA, Reddi AH, Moos M Jr 1994 Cartilage-derived morphogenetic proteins. New members of the transforming growth factor β superfamily predominantly expressed in long bones during human embryonic development. J Biol Chem **269**:28227–28234.
118. Amizuka N, Warshawsky H, Henderson JE, Goltzman D, Karaplis AC 1994 Parathyroid hormone-related peptide-depleted mice show abnormal epiphyseal cartilage development and altered endochondral bone formation. J Cell Biol **126**:1611–1623.
119. Schwartz Z, Sela J, Ramirez V, Amir D, Boyan BD 1989 Changes in extracellular matrix vesicles during healing of rat tibial bone: A morphometric and biochemical study. Bone **10**:53–60.
120. Wu LN, Genge BR, Lloyd GC, Wuthier RE 1991 Collagen-binding proteins in collagenase-released matrix vesicles from cartilage. Interaction between matrix vesicle proteins and different types of collagen. J Biol Chem **266**:1195–1203.
121. Wuthier RE, Register TC 1985 Role of alkaline phosphatase, a polyfunctional enzyme, in mineralizing tissues, In: Butler WT. (eds) The Chemistry and Biology of Mineralized Tissues. Ebsco Media Inc., Birmingham, Alabama, pp. 113–124.
122. Wuthier RE 1973 The role of phospholipids in biological calcification: Distribution of phospholipase activity in calcifying epiphyseal cartilage. Clin Orthop Related Res **90**:191–200.
123. Boskey AL, Reddi AH 1983 Changes in lipids during matrix-induced endochondral bone formation. Calcif Tissue Int **35**: 549–554.
124. Boskey AL, Posner AS, Lane JM, Goldberg MR, Cordella DM 1980 Distribution of lipids associated with mineralization in the bovine epiphyseal growth plate. Arch Biochem Biophys **199**:305–311.
125. Boskey AL, Posner AS 1976 Extraction of a calcium–phospholipid–phosphate complex from bone. Calcif Tissue Int **19**:273–283.
126. Boyan-Salyers B 1981 A role for proteolipid in membrane-initiated calcification. In: Veis A. (ed) The Chemistry and Biology of Mineralized Connective Tissues. Elsevier/North-Holland, New York, pp. 539–542.

127. Boskey AL 1981 Current concepts of the physiology and biochemistry of calcification. Clin Orthop Related Res **157**:225–255.
128. Buckwalter JA, Ehrlich MG, Armstrong AL, Mankin HJ 1987 Electron microscopic analysis of articular cartilage proteoglycan degradation by growth plate enzymes. J Orthop Res **5**:128–132.
129. Ingmar B, Wasteson A 1979 Sequential degradation of a chondroitin sulphate trisaccharide by lysosomal enzymes from embryonic-chick epiphyseal cartilage. Biochem J **179**:7–13.
130. Amado R, Ingmar B, Lindahl U, Wasteson A 1974 Depolymerisation and desulphation of chondroitin sulphate by enzymes from embryonic chick cartilage. FEBS Lett **39**:49–52.
131. Yang F, Schwartz Z, Swain LD, Lee CC, Bowman BH, Boyan BD 1991 Alpha-2-HS-glycoprotein: Expression in chondrocytes and augmentation of alkaline phosphatase and phospholipase A_2 activity. Bone **12**:7–15.
132. Malone JD, Teitelbaum SL, Griffin GL, Senior RM, Kahn AJ 1982 Recruitment of osteoclast precursors by purified bone matrix constituents. J Cell Biol **92**:227–230.
133. Grumbles RM, Roos BA, Brandi ML, Howard GA, Howell DS 1994 Rachitic growth plate regulation by $1,25(OH)_2D_3$: A putative role for hepatocyte growth factor in angiogenesis. J Bone Miner Res **9**(Suppl 1):S373.
134. Edelstein S, Ornoy A 1979 24,25-Dihydroxyvitamin D: The preferred metabolite for bone. In: Norman AW, Schaefer K, Herrath DV, Grigoleit HG, Coburn JW, DeLuca HF (eds), Vitamin D: Basic Research and Its Clinical Application. de Gruyter, Berlin, pp. 381–389.
135. Muller SA, Posner AS, Firschein HE 1966 Effect of vitamin D deficiency on the crystal chemistry of bone mineral. Proc Soc Exp Biol Med **121**:844–846.
136. Mathews JL, Martin JH, Sampson HW, Kunin AS, Roan JH 1970 Mitochondrial granules in the normal and rachitic epiphysis. Calcif Tissue Res **5**:91–99.
137. Brighton CT, Hunt RM 1978 The role of mitochondria in growth plate calcification as demonstrated in a rachitic model. J Bone Joint Surg Am **60A**:630–639.
138. Irving JT, Wuthier RE 1968 Histochemistry and biochemistry of calcification with special reference to the role of lipids. Clin Orthop Related Res **56**:237–260.
139. Cruess R, Clark I 1965 Alterations in the lipids of bone caused by hypervitaminosis A & D. Biochem J **96**:262–265.
140. Howell DS, Marquez JF, Pita JC 1965 The nature of phospholipids in normal and rachitic costochondral plates. Arth Rheum **8**:1039–1046.
141. Boyan BD, Ritter NM 1984 Proteolipid–lipid relationships in normal and vitamin D-deficient chick cartilage. Calcif Tissue Int **36**:332–337.
142. Kyeyune-Nyombi E, Lau KH, Baylink DJ, Strong DD 1991 1,25-Dihydroxyvitamin D_3 stimulates bone alkaline phosphatase gene transcription and mRNA stabilization in human bone cells. Arch Biochem Biophys **291**:316–325.
143. Schwartz Z, Schlader DL, Swain LD, Boyan BD 1988 Direct effects of 1,25-dihydroxyvitamin D_3 and 24,25-dihydroxyvitamin D_3 on growth zone and resting zone chondrocyte membrane alkaline phosphatase and phospholipase-A_2 specific activities. Endocrinology **123**:2878–2884.
144. Gay CV, Anderson RE, Leach RM 1985 Activities and distribution of alkaline phosphatase and carbonic anhydrase in the tibial dyschondroplastic lesion and associated growth plate of chicks. Avian Dis **29**:812–821.
145. Sylvia VL, Schwartz Z, Schuman L, Morgan RT, Mackey S, Gomez R, Boyan BD 1993 Maturation-dependent regulation of protein kinase C activity by vitamin D_3 metabolites in chondrocyte cultures. J Cell Physiol **157**:271–278.
146. Roughley PJ, Dickson I 1980 Factors influencing proteoglycan size in rachitic-chick growth cartilage. Biochem J **185**:33–39.
147. Roughley PJ, Dickson IR 1986 A comparison of proteoglycan from chick cartilage of different types and a study of the effect of vitamin D on proteoglycan structure. Connect Tissue Res **14**:187–197.
148. Carrino DA, Lidor C, Edelstein S, Caplan AI 1989 Proteoglycan synthesis in vitamin D-deficient cartilage: recovery from vitamin D-deficiency. Connect Tissue Res **19**:135–147.
149. Reinholt FP, Engfeldt B, Heinegard D, Hjerpe A 1985 Proteoglycans and glycosaminoglycans of epiphyseal cartilage in florid and healing low phosphate, vitamin D deficiency rickets. Collagen Related Res **5**:55–64.
150. Frame B, Parfitt AM 1978 Osteomalacia: Current concepts. Ann Intern Med **89**:966–982.
151. Howard GA, Turner RT, Sherrard DJ, Baylink DJ 1981 Human bone cells in culture metabolize 25-hydroxyvitamin D_3 to 1,25-dihydroxyvitamin D_3 and 24,25-dihydroxyvitamin D_3. J Biol Chem **256**:7738–7740.
152. Turner RT, Puzas JE, Forte MD, Lester GE, Gray TK, Howard GA, Baylink DJ 1980 *In vitro* synthesis of 1-alpha,25dihydroxycholecalciferol and 24,25-dihydroxycholecalciferol by isolated calvarial cells. Proc Natl Acad Sci USA **77**:5720–5724.
153. Reichel H, Koeffler HP, Bishop JE, Norman AW 1987 25-Hydroxyvitamin D_3 metabolism by lipopolysaccharide-stimulated normal human macrophages. J Clin Endocrinol Metab **64**:1–9.
154. Reichel H, Koeffler HP, Norman AW 1987 25-Hydroxyvitamin D_3 metabolism by human T-lymphotropic virus-transformed lymphocytes. J Clin Endocrinol Metab **65**:519–525.
155. Schwartz Z, Brooks BP, Swain LD, Del Toro F, Norman AW, Boyan BD 1992 Production of 1,25-dihydroxyvitamin D_3 and 24,25-dihydroxyvitamin D_3 by growth zone and resting zone chondrocytes is dependent on cell maturation and is regulated by hormones and growth factors. Endocrinology **130**:2495–2504.
156. Holtrop ME, Cox KA, Carnes DL, Holick MF 1986 Effects of serum calcium and phosphorus on skeletal mineralization in vitamin D-deficient rats. Am J Physiol **251**:E234-E240.
157. Dekel S, Ornoy A, Sekeles E, Noff D, Edelstein S 1979 Contrasting effects on bone formation and on fracture healing of cholecalciferol and of 1α-hydroxycholecalciferol. Calcif Tissue Int **28**:245–251.
158. Ornoy A, Sekeles E, Cohen R, Edelstein S 1979 The role of vitamin D metaboliltes in calcification of chicken epiphyseal cartilage. In: Norman AW, Schaefer K, Harrath DV, Grigoleit HG, Coburn JW, DeLuca HF (eds) Vitamin D: Basic Research and Its Clinical Application. de Gruyter, Berlin, pp. 363–367.
159. Underwood JL, DeLuca HF 1984 Vitamin D is not directly necessary for bone growth and mineralization. Am J Physiol **246**:E493–E498.
160. Howell DS, Pita JC, Alvarez J 1976 Possible role of extracellular matrix vesicles in initial calcification of healing rachitic cartilage. Fed Proc **35**:122–126.
161. Howell DS, Blanco LN, Pita JC 1978 Further characterization of a nucleational agent in hypertrophic cell extracellular cartilage fluid. Metab Bone Dis Related Res **1**:155–160.
162. Schwartz Z, Dean DD, Walton JK, Brooks BP, Boyan BD 1995 Treatment of resting zone chondrocytes with 24,25-dihydroxyvitamin D_3 [$24,25(OH)_2D_3$] induces differentiation into a $1,25(OH)_2D_3$-responsive phenotype characteristic of growth zone chondrocytes. Endocrinology **136**:402–411.
163. Lidor C, Atkin I, Ornoy A, Dekel S, Edelstein S 1987 Healing of rachitic lesions in chicks by 24*R*,25-dihydroxycholecalciferol administered locally into bone. J Bone Miner Res **2**:91–98.
164. Atkin I, Pita JC, Ornoy A, Agundez A, Castiglione G, Howell

DS 1985 Effects of vitamin D metabolites on healing of low phosphate, vitamin D-deficient induced rickets in rats. Bone **6**:113–123.

165. Seo EG, Einhorn TA, Norman AW 1996 24*R*-25Dihydroxyvitamin D_3: An essential vitamin D_3 metabolite for both normal bone integrity and healing of tibial fracture in chicks. J Bone Miner Res **11**(Suppl 1):S422 (abstract).
166. Ornoy A, Menczel J, Nebel L 1968 Alterations in the mineral composition and metabolism of rat fetuses and their placentas induced by maternal hypervitaminosis D_2. Isr J Med Sci **4**:827–832.
167. Ornoy A, Nebel L, Menczel Y 1969 Impaired osteogenesis of fetal long bones induced by maternal hypervitaminosis D_2. Arch Pathol **87**:563–571.
168. Brighton CT 1984 The growth plate. Orthop Clin North Am **15**:571–595.
169. Somjen D, Somjen GJ, Harell A, Mechanic GL, Binderman I 1982 Partial characterization of a specific high affinity binding macromolecule for 24,25-dihydroxyvitamin D_3 in differentiating skeletal mesenchyme. Biochem Biophys Res Commun **106**: 644–651.
170. Bruder SP, Caplan A 1989 Cellular and molecular events during embryonic bone development. Connect Tissue Res **20**:65–71.
171. Soskolne WA, Schwartz Z, Ornoy A 1986 The development of fetal mice long bones in vitro: An assay of bone modeling. Bone **7**:41–48.
172. McKibbin B 1978 The biology of fracture healing in long bone. J Bone Joint Surg **60**(B):150–162.
173. Wuthier RE 1982 A review of the primary mechanism of endochondral calcification with special emphasis on the role of cells, mitochondria and matrix vesicles. Clin Orthop Related Res **169**:219–242.
174. Takigawa M, Enomoto M, Shirai E, Nishii Y, Sazuki F 1988 Differential effects of 1 alpha-25-dihydroxycholeclciferol and 24*R*,25-dihydroxycholecalciferol on the proliferation and the differentiated phenotype of rabbit costal chondrocytes in culture. Endocrinology **122**:831–839.
175. Ray RL, Ehrlich MG, Mankin HJ 1982 Isolation of proliferating chondrocytes from bovine growth-plate cartilage by rate-zonal centrifugation in a Ficoll density gradient. J Bone Joint Surg **64**(A):1221–1224.
176. Urist MR 1965 Bone: Formation by autoinduction. Science **150**:893–899.
177. Urist MR, Huo YK, Brownell AG, Hohl WM, Buyske J, Lietze A, Tempst P, Hunkapiller M, Delange RJ 1984 Purification of bovine bone morphogenetic protein by hydroxyapatite chromatography. Proc Natl Acad Sci USA **81**:371–375.
178. Urist MR 1981 New bone formation induced in post-fetal life by bone morphogenetic protein. In: Becker R, (ed) Mechanisms of Growth Control. Thomas, Springfield, Illinois, pp. 406–434.
179. Urist MR, Chang JJ, Lietze A, Huo YK, Brownell AG, Delange RJ 1987 Preparation and bioassay of bone morphogenetic protein and polypeptide fragments. In: Barnes D, Sirbasku DA (eds) Methods Enzymology, Vol. 146. Academic Press, Orlando, Florida, pp. 294–312.
180. Hsu HHT, Anderson HC 1978 Calcification of isolated matrix vesicles and reconstituted vesicles from fetal bovine cartilage. Proc Natl Acad Sci USA **75**:3805–3808.
181. Plachot JJ, DuBois MB, Halpern S, Cournot-Witmer G, Garabedian M, Balsan S 1982 *In vitro* action of 1,25-dihydroxycholecalciferol and 24,25-dihydroxycholecalciferol on matrix organization and mineral distribution in rabbit growth plate. Metab Bone Dis Related Res **4**:135–142.
182. Corvol MT, Dumontier MF, Garabedian M, Rappaport R 1978 Vitamin D and cartilage. II. Biological activity of 25-hydroxycholecalciferol and 24,25- and 1,25-dihydroxycholecalciferol in cultured growth plate chondrocytes. Endocrinology **102**:1269–1274.
183. Burch WM, Lopez-Claros M, Uskokovic MR, Drezner MK 1988 1,25-Dihydroxyvitamin D_3 stimulates avian and mammalian cartilage growth *in vitro*. J Bone Miner Res **3**:87–91.
184. Capusso O, Gionti E, Pontarelli G, Ambesi-Impionbato FS, Nitsch L, Tajana G, Cancedda R 1982 The culture of chick embryo chondrocytes and the control of their differentiated functions *in vitro*. Exp Cell Res **142**:197.
185. Levy J, Shimshoni Z, Somjen D, Berger E, Fine N, Silbermann M, Binderman I 1988 Rat epiphyseal cells in culture: Responsiveness to bone-seeking hormones. In Vitro Cell Dev Biol **24**:620–624.
186. Thyberg J, Moskalewski S. 1979 Bone formation in cartilage produced by transplanted epiphyseal chondrocytes. Cell Tissue Res **204**:77–94.
187. Grigoriadis AE, Heersche JN, Aubin JE 1988 Differentiation of muscle, fat, cartilage, and bone from progenitor cells present in a bone-derived clonal cell population: Effect of dexamethasone. J Cell Biol **106**:2139–2151.
188. Binderman I, Somjen D 1984 24,25-Dihydroxycholecalciferol induces the growth of chick cartilage in vitro. Endocrinology **115**:430–432.
189. Engel FE, Swain LD, Dean DD, Boyan BD 1994 Nodule formation and calcification of mandibular condylar cartilage cell cultures mimic in vivo ultrastructure. Acta Anat **150**:136–149.
190. Engel FE, Swain LD, Miller P, Khare AG, Boyan BD 1990 Ultrastructure of mandibular condylar cartilage cells in culture. J Dent Res **69**:206.
191. Grigoriadis AE, Aubin JE, Heersche JN 1989 Effects of dexamethasone and vitamin D_3 on cartilage differentiation in a clonal chondrogenic cell population. Endocrinology **125**:2103–2110.
192. Spencer CA, Palmer TN, Mason RM 1990 Intermediary metabolism in the swarm rat chondrosarcoma chondrocyte. Biochem J **265**:911–914.
193. Matsumoto H, Silverton SF, Debolt K, Shapiro IM 1991 Superoxide dismutase and catalase activities in the growth cartilage: relationship between oxidoreductase activity and chondrocyte maturation. J Bone Miner Res **6**:569–574.
194. Kato Y, Nasu N, Takase T, Suzuki F 1978 Demonstration of somatomedin activity of "multiplication-stimulating activity" in rabbit costal chondrocytes in culture. J Biochem (Tokyo) **84**:1001–1004.
195. Schwartz Z, Schlader DL, Ramirez V, Kennedy MB, Boyan BD 1989 Effects of vitamin D metabolites on collagen production and cell proliferation of growth zone and resting zone cartilage cells in vitro. J Bone Miner Res **4**:199–207.
196. Weinstein RS, Underwood JL, Hutson MS, DeLuca HF 1984 Bone histomorphometry in vitamin D-deficient rat infused with calcium and phosphorus. Am J Physiol **246**:E499-E505.
197. Miller SC, Halloran BP, DeLuca HF, Yamada S, Takayama H, Jee WS 1981 Studies on the role of 24-hydroxylation of vitamin D in the mineralization of cartilage and bone of vitamin D-deficient rats. Calcif Tissue Int **33**:489–497.
198. Sai H, Takatsuto S, Ikekawa N, Tanaka Y, Smith C, DeLuca HF 1984 Synthesis and biological activity of (22*E*,24*R*)- and (22*E*,24*S*)-1 alpha,24-dihydroxy-22-dehydrovitamin D_3. Chem Pharm Bull **32**:3866–3872.
199. Brommage R, DeLuca HF 1985 Evidence that 1,25-dihydroxyvitamin D_3 is the physiologically active metabolite of vitamin D_3. Endocr Rev **6**:491–511.
200. Fine N, Binderman I, Somjen D, Earon Y, Edelstein S, Weisman

Y 1985 Autoradiographic localization of 24*R*,25-dihydroxyvitamin D_3 in epiphyseal cartilage. Bone **6**:99–104.

201. Seo EG, Schwartz Z, Dean DD, Norman AW, Boyan BD 1996 Preferential accumulation in vivo of 24*R*,25-dihydroxyvitamin D_3 in growth plate cartilage of rats. Endocrine in press.
202. Corvol MT, Ulmann A, Garabedian M 1980 Specific nuclear uptake of 24,25-dihydroxycholecaliferol, a vitamin D_3 metabolite biologically active in cartilage. FEBS Lett **116**:273–276.
203. Balmain N, Hauchecorne M, Pike JW, Cuisinier-Gleizes P, Matlieu H 1993 Distribution and subcellular immunolocalization of 24,25-dihydroxyvitamin D_3 receptors in rat epiphyseal cartilage. Cell Mol Biol **39**:339–350.
204. Boyan BD, Schwartz Z, Bonewald LF, Swain LD 1989 Localization of $1,25(OH)_2D_3$ responsive alkaline phosphatase in osteoblast-like cells (ROS 17/2.8, MG 63, and MC 3T3) and growth cartilage cells in culture. J Biol Chem **264**:11879–11886.
205. St-Arnaud R, Arabian A, Glorieux FH 1996 Abnormal bone development in mice deficient for the vitamin D 24-hydroxylase gene. J Bone Miner Res **11:**S126 (abstract).
206. Hinek A, Poole AR 1988 The influence of vitamin D metabolites on the calcification of cartilage matrix and the C-propeptide of type II collagen (chondrocalcin). J Bone Miner Res **3**:421–429.
207. Hinek A, Reiner A, Poole AR 1987 The calcification of cartilage matrix in chondrocyte culture. Studies of the C-propeptide of type II collagen (chondrocalcin). J Cell Biol **104**:1435–1441.
208. Harrison JE, Hitchman AJ, Jones G, Tam CS, Heersche JN 1982 Plasma vitamin D metabolite levels in phosphorus deficient rats during the development of vitamin D deficient rickets. Metab Clin Exp **31**:1121–1127.
209. Suda S, Takahashi N, Shinki T, Horiuchi N, Yamaguchi S, Yoshiki S, Enomoto S, Suda T 1985 1α,25,-Dihydroxyvitamin D_3 receptors and their action in embryonic chick chondrocytes. Calcif Tissue Int **37**:82–90.
210. Balmain N, Tisserand-Jochem E, Thomasset M, Cuisinier-Gleizes P, Mathieu H 1986 Vitamin D-dependent calcium-binding protein (CaBP-9k) in rat growth cartilage. Histochemistry **84**:161–168.
211. Zhou XY, Dempster DW, Marion SL, Pike JW, Haussler MR, Clemens TL 1986 Bone vitamin D-dependent calcium-binding protein is localized in chondrocytes of growth plate cartilage. Calcif Tissue Int **38**:244–247.
212. Somjen D, Kaye AM, Binderman I 1984 24,25-Dihydroxyvitamin D stimulates creatine kinase BB activity in chick cartilage cells in culture. FEBS Lett **167**:281–284.
213. Boyan BD, Schwartz Z, Carnes DL Jr, Ramirez V 1988 The effects of vitamin D metabolites on the plasma and matrix vesicle membranes of growth and resting cartilage cells in vitro. Endocrinology **122**:2851–2860.
214. Schwartz Z, Boyan BD 1988 The effects of vitamin D metabolites on phospholipase A_2 activity of growth zone and resting zone cartilage cells *in vitro*. Endocrinology **122**:2191–2198.
215. Swain LD, Schwartz Z, Caulfield K, Brooks BP, Boyan BD 1993 Nongenomic regulation of chondrocyte membrane fluidity by 1,25-$(OH)_2D_3$ and 24,25-$(OH)_2D_3$ is dependent on cell maturation. Bone **14**:609–617.
216. Langston GG, Swain LD, Schwartz Z, Del Toro F, Gomez R, Boyan BD 1990 Effect of $1,25(OH)_2D_3$ and $24,25(OH)_2D_3$ on calcium ion fluxes in costochondral chondrocyte cultures. Calcif Tissue Int **47**:230–236.
217. Schwartz Z, Swain LD, Ramirez V, Boyan BD 1990 Regulation of arachidonic acid turnover by 1,25-$(OH)_2D_3$ and 24,25-$(OH)_2D_3$ in growth zone and resting zone chondrocyte cultures. Biochim Biophys Acta **1027**:278–286.
218. Kimelberg HK 1975 Alterations in phospholipid-dependent (Na^+,K^+)-ATPase activity due to lipid fluidity. Effects of cholesterol and Mg^{2+}. *Biochim Biophys Acta* **413**:143–156.
219. Boskey AL, Stiner D, Doty SB, Binderman I, Leboy PS 1992 Studies of mineralization in tissue culture: Optimal conditions for cartilage calcification. Bone Miner **16**:11–36.
220. Hale LV, Kemick ML, Wuthier RE 1986 Effect of vitamin D metabolites on the expression of alkaline phosphatase activity by epiphyseal hypertrophic chondrocytes in primary cell culture. J Bone Miner Res **1**:489–495.
221. Schwartz Z, Langston GG, Swain LD, Boyan BD 1991 Inhibition of 1,25-$(OH)_2D_3$- and 24,25-$(OH)_2D_3$-dependent stimulation of alkaline phosphatase activity by A23187 suggests a role for calcium in the mechanism of vitamin D regulation of chondrocyte cultures. J Bone Miner Res **6**:709–718.
222. Farach-Carson MC, Abe J, Nishii Y, Khoury R, Wright GC, Norman AW 1993 22-Oxacalcitriol: Dissection of $1,25(OH)_2D_3$ receptor-mediated and Ca^{2+} entry-stimulating pathways. Am J Physiol **265**:F705-F711.
223. Schwartz Z, Knight G, Swain LD, Boyan BD 1988 Localization of vitamin D_3-responsive alkaline phosphatase in cultured chondrocytes. J Biol Chem **263**:6023–6026.
224. Schwartz Z, Sylvia VL, Dean DD, Boyan BD 1996 The synergistic effect of TGFβ and 24,25-$(OH)_2D_3$ on resting zone chondrocytes is metabolite-specific and mediated by PKC. Connect Tissue Res **35**:101–106.
225. Boyan BD, Sylvia VL, Dean DD, Schwartz Z 1994 Nongenomic effects of vitamin D. In: Norman AW, Bouillon R, Thomasset M (eds), Vitamin D: A Pluripotent Steroid Hormone: Structural Studies, Molecular Endocrinology and Clinical Applications. de Gruyter, Berlin, pp. 333–340.
226. Shinitzky M, Inbar M 1974 Difference in microviscosity induced by different cholesterol levels in the surface membrane lipid layer of normal lymphocytes and malignant lymphoma cells. J Mol Biol **85**:603–615.
227. Swain LD, Schwartz Z, Boyan BD 1992 1,25-$(OH)_2D_3$ and 24,25-$(OH)_2D_3$ regulation of arachidonic acid turnover in chondrocyte cultures is cell maturation-specific and may involve direct effects on phospholipase A_2. Biochim Biophys Acta **1136**:45–51.
228. Rasmussen H, Matsumoto T, Fontaine O, Goodman DB 1982 Role of changes in membrane lipid structure in the action of 1,25-dihydroxyvitamin D_3. Fed Proc **41**:72–77.
229. O'Doherty PJA 1979 1,25-Dihydroxyvitamin D_3 increases the activity of the intestinal phosphatidylcholine deacylation reacylation cycle. Lipids **14**:75–77.
230. Schwartz Z, Swain LD, Kelly DW, Brooks BP, Boyan BD 1992 Regulation of prostaglandin E_2 production by vitamin D metabolites in growth zone and resting zone chondrocyte cultures is dependent on cell maturation. Bone **13**:395–401.
231. Sylvia VL, Schwartz Z, Ellis EB, Helm SH, Gomez R, Dean DD, Boyan BD 1996 Nongenomic regulation of protein kinase C isoforms by the vitamin D metabolites 1α,25-$(OH)_2D_3$ and 24*R*,25-$(OH)_2D_3$. J Cell Physiol **167**:380–393.
232. Nemere I, Norman AW 1991 Steroid hormone actions at the plasma membrane: Induced calcium uptake and exocytotic events. Mol Cell Endocrinol **80**:C165-C169 (review).
233. Norman AW, Dormanen MC, Okamura WH, Hammond M, Nemere I 1994 Non nuclear actions of $1\alpha,25(OH)_2D_3$ and 24*R*,25-$(OH)_2D_3$ in mediating intestinal calcium transport: The use of analogs to study membrane receptors for vitamin D metabolites and to determine receptor ligand conformational preferences, In: Norman AW, Bouillon R, Thomasset M (eds) Vitamin D: A Pluripotent Steroid Hormone. de Gruyter, Berlin, pp. 324–332.
234. Baran DT, Sorenson AM, Shalhoub V, Owen T, Oberdorf A, Stein G, Lian JB 1991 1 Alpha,25-dihydroxyvitamin D_3 rapidly

increases cytosolic calcium in clonal rat osteosarcoma cells lacking the vitamin D receptor. J Bone Miner Res **6**:1269–1275.

235. Baran DT, Ray R, Sorenson AM, Honeyman T, Holick MF 1994 Binding characteristics of a membrane receptor that recognizes 1 alpha,25-dihydroxyvitamin D_3 and its epimer, 1 beta,25-dihydroxyvitamin D_3. J Cell Biochem **56**:510–517.
236. Baran DT, Sorenson AM, Shalhoub V, Owen T, Stein G, Lian JB 1992 The rapid nongenomic actions of 1 alpha,25-dihydroxyvitamin D_3 modulate the hormone-induced increments in osteocalcin gene transcription in osteoblast-like cells. J Cell Biochem **50**:124–129.
237. Seo EG, Norman AW 1996 Evidence for the existence of a specific 24*R*,25$(OH)_2$-vitamin D_3 receptor in a chick callus membrane fraction. J Bone Miner Res **11**(Suppl 1):S316 (abstract).
238. Greising DM, Schwartz Z, White MC, Posner GH, Sylvia VL, Dean DD, Boyan BD 1996 Vitamin D_3 analogues with low affinity regulate chondrocyte metabolism. J Bone Miner Res **11**(Suppl. 1):S313 (abstract).
239. Scheetz MP 1993 Glycoprotein motility and dynamic domains in fluid plasma membranes. Annu Rev Biophys Biomol Struct **22**:417–431.
240. Zhou LX, Nemere I, Norman AW 1992 1,25-Dihydroxyvitamin D_3 analog structure–function assessment of the rapid stimulation of intestinal calcium absorption (transcaltachia). J Bone Miner Res **7**:457–463.
241. Kim YS, Dedhar S, Hruska K 1994 Binding of the occupied vitamin D receptor (VDR) to extranuclear sites. A potential mechanism of nongenomic actions of 1,25-$(OH)_2D_3$, In: Norman AW, Bouillon R, Thomasset M (eds), Vitamin D: A Pluripotent Steroid Hormone: Structural Studies, Molecular Endocrinology and Clinical Applications. de Gruyter, Berlin, pp. 341–344.
242. Helm SH, Sylvia VL, Harmon T, Dean DD, Boyan BD, Schwartz Z 1996 24,25-$(OH)_2D_3$ regulates protein kinase C through two specific phospholipid-dependent mechanisms. J Cell Physiol **169**:509–521.
243. Luo T, Luo Y, Vallano ML 1995 Arachidonic acid, but not sodium nitroprusside, stimulates presynaptic protein kinase C and phosphorylation of GAP-43 in rat hippocampal slices and synaptosomes. J Neurochem **64**:1808–1818.
244. Sela J, Schwartz Z, Swain LD, Boyan BD 1992 The role of matrix vesicles in calcification. In: Bonucci E (ed) Calcification in Biological Systems. CRC Press, Boca Raton, Florida, pp. 73–105.
245. Dean DD, Schwartz Z, Muniz OE, Gomez R, Swain LD, Howell DS, Boyan BD 1992 Matrix vesicles contain metalloproteinases that degrade proteoglycans. Bone Miner **17**:172–176.
246. Boskey AL, Boyan BD, Doty SB, Feliciano A, Greer K, Weiland D, Swain LD, Schwartz Z 1992 Studies of matrix vesicle-induced mineralization in a gelatin gel. Bone Miner **17**:257–262.
247. Boyan BD, Schwartz Z, Park-Snyder S, Dean DD, Yang F, Twardzik D, Bonewald LF 1994 Latent transforming growth factor-β is produced by chondrocytes and activated by extracellular matrix vesicles upon exposure to 1,25$(OH)_2D_3$. J Biol Chem **269**:28374–28381.

Vitamin D Action on Tooth Development and Biomineralization

ARIANE BERDAL Laboratoire de Biologie-Odontologie, Faculté de Chirurgie Dentaire, Université Paris VII, Institut Biomédical des Cordeliers, Paris, France

I. INTRODUCTION

One puzzling aspect of the biological effects of hormones is their tissue specificity and their differential actions that depend on the stage of cell differentiation. Cells devoted to bone formation have been shown to express the nuclear vitamin D receptor (VDR) gene and thus to be potentially under the control of the hormone 1,25-dihydroxyvitamin D [1,25$(OH)_2$D] throughout their life cycle [1]. Several target genes have been identified that play a role in bone formation: initial proliferation stage, c-*fos* [2]; in the later deposition and biomineralization of bone matrix, type I collagen [3]; osteocalcin [4]; osteopontin [5], and bone sialoprotein [6]; and calcium handling, calbindin-D_{9K} and calbindin-D_{28K} [7]. The orofacial skeleton provides a unique opportunity to jointly investigate different systems of elaboration of mineralized tissues *in vivo*. Dentin and cementum share many features with bone tissues [8–10]. Enamel is the only mineralized tissue of epithelial origin and shows the largest hydroxyapatite crystals of the mammalian body. Its matrix contains a set of unique proteins, the cDNAs for which the first to be isolated were from dental libraries (for review, see Snead [11]). VDR and vitamin D–dependent proteins important for the differentiation of cells involved in hard tissue formation and mineralization, notably the calbindins, were systematically identified in dental and bone tissues using various epitheliomesenchymal systems of the orofacial area (Fig. 1).

II. ODONTOGENESIS

A. Dental Morphogenesis and Differentiation

Odontogenesis follows the basic features of epitheliomesenchymal organ development involving a sequence of cell–cell and cell–matrix interactions leading to (1)

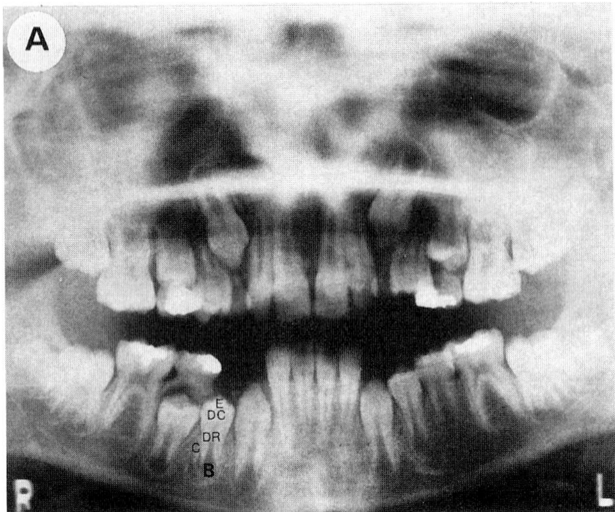

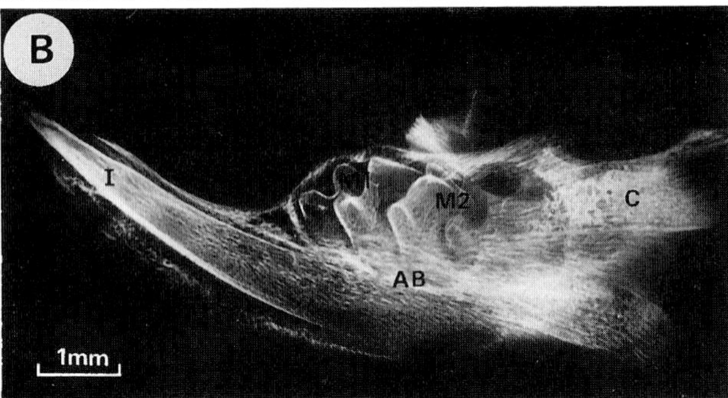

FIGURE 1 (A) Panoramic radiograph showing the features of mixed dentition in a 10-year-old boy. The formation of human teeth from the migration of neural crest cells until the eruption of the last tooth may last 20 years. In unusual cases, the third molar may even erupt in elderly people under full dentures. Human dentition is formed of lacteal teeth which include two incisors, one canine, and two molars. Their histogenesis begins when the primary palate is coalescing (first trimester of gestation), and they are approximately set up at 2 years of age. These lacteal teeth (in children from 6 to 12 years old) are replaced sequentially by permanent teeth. This dynamic period is called mixed dentition. The second permanent mandibular molar forms underneath the second lacteal molar, which radicular resorption has advanced. There, the three mineralized dental tissues may be observed: highly mineralized enamel (E), which covers the crown; dentin, which may be coronal (DC) or radicular (DR); and the cementum (C) covering the forming root. Teeth form inside lacunae surrounded by alveolar bone (B). Courtesy of E. Klingler. (B) Microradiograph of 4-day-old Sprague-Dawley rat mandible. Several biomineralization processes are illustrated. There is one continuously growing incisor (I) extending from the incisal area to the mandible gonodion. Three limited growth molars are similar to human teeth. Four days postnatally, the biomineralization stage is reached in the first (M1) and second (M2) molars, while the third one is still involved in morphogenesis. Two types of bone are present: alveolar bone (AB), which is dependent on the presence of teeth; and basal bone, which is independent of dental development. Calcified cartilage (C) also contributes to mandibular growth.

morphogenesis, that is, acquisition of distinctive morphotypes such as incisors, canines, premolars, and molars; (2) cell differentiation of ameloblast, odontoblast, and cementoblast lineages; and (3) matrix secretion and biomineralization. Tooth formation is part of the development of the branchial arches, which provide the structures of the face and neck [12–15]. Sequential and reciprocal intercellular communications between epithelium and mesenchyme lead to the progression of dental morphogenesis and differentiation. Epithelial and mesenchymal cells follow programmed cell proliferation, condensation, differentiation, and apoptosis [16]. Factors involved include transcription factors (DNA binding proteins containing conserved homeoboxes, paired boxes, and zinc finger encoding domains), growth factors (e.g., epidermal growth factor (EGF), fibroblast growth factors (FGFs), HGF, insulin-like growth factors (IGFs), nerve growth factor (NGF), transforming growth factor-β (TGFβ) and related bone morphogenic proteins (BMPs), platelet-derived growth factors (PDGFs), growth hormone (GH), and vitamins A and D and their receptors [17]). *In vitro* investigation, such as addition of diffusible factors (EGF, GH, IGF-1, acidic and basic FGFs [15], and more recently HGF [18]), has stressed their importance in tooth development. Antisense strategies have illustrated the predominant role of EGF [19,20]. Epitheliomesenchymal interactions provide a useful system to investigate the signaling pathways of BMPs and transcription factors [21]. Terminal differentiation of ameloblasts [22], odontoblasts [14], and cementoblasts [23] is characterized by cytological and functional modifications that are spatially ordered depending on tooth morphodifferentiation patterns.

B. Distinctive Characteristics of Dentin and Cementum Compared to Bone

Odontogenic cells, when overtly differentiated, are involved in the formation of specific mineralized tissues. Dentin and dentinogenesis share many basic structures and mechanisms with ectomesenchyme-derived bone and cementum [24]. Two types of cementum coexist: acellular and cellular. Their relative proportion follows a reverse gradient along the root. Cementoblasts that elaborate cellular cementum are phenotypically very similar to osteoblasts [25,26]. However, the primary role of cementum is the anchorage of tooth in bone, a function that may depend on several factors that will control cell migration and adhesion. Some are common with bone, such as osteopontin, bone sialoprotein, various growth factors, and matrix components, whereas at least one is tissue-specific, namely a collagenous cementum-derived attachment protein (CAP) isolated from various species [10].

Odontoblasts are arranged in a layer that specializes in secreting and metabolizing an ordered extracellular matrix and controlling the extracellular environment permissive for hydroxyapatite formation. Odontoblasts elaborate a set of matrix proteins, mainly type I collagen, that are secreted at specific levels of the secretory pole. Extracellularly, these proteins follow a programmed self-assembly evidenced by thickening of collagen fibrils and distribution of polyanions in the predentin [9]. Some components may also be metabolized [24]. Abruptly, at the mineralization front, ordered nucleation and growth of hydroxyapatite crystals occur with their c axis arranged parallel to collagen fibrils. To understand the basic mechanisms of dentin formation and biomineralization, one needs to isolate the extracellular matrix components [24,27] and to characterize their temporospatial pattern of expression and tissue distribution. These proteins may be divided into two categories, one made of dentin-specific species and the other made of dentin/bone/cementum species. For the latter, only distinctive dentin characters are discussed in this chapter [8,9,24,28].

Much attention in the dental field has been focused on dentin. Dentin phosphoproteins, the main noncollagenous proteins of dentin, are characterized by their high level of phosphorylation (45–50% of phosphoserine residues) [27]. They have been proposed to contribute to biomineralization, as they appear to be exported to the mineralization front [24,28]. Their chemical properties include polyanionicity, low affinity binding to ionized calcium, and binding to collagen fibrils. Several *in vitro* experiments support the notion that they may play a dual role, depending on their binding type I collagen, on the initiation and control of the final shape of hydroxyapatite crystals [8]. Another protein, dentin matrix protein 1, appears also to be characterized by a high level of serine residues [29]. Apart from these acidic phosphorylated proteins, a sialoprotein appears to be specific to dentin matrix (5–8% of noncollagenous proteins). Similarly to bone sialoproteins, this dentin sialoprotein is rich in glutamine, aspartic acid, serine, and glycine [30]. It also has a high carbohydrate content [31]. Interestingly, dentin phosphoprotein and dentin sialoprotein are cleavage products expressed from a single transcript coded by a gene on chromosome 4 [32]. Being acidic, these three dentin-specific proteins are thought to bind to a collagen scaffolding and cooperate in the regulation of hydroxyapatite formation through their high calcium binding properties [27]. The same may be true for the bone noncollagenous proteins present in dentin (osteocalcin, bone sialoprotein, osteopontin, osteonectin, and proteoglycans) [27].

Biomineralization requires the availability of calcium and phosphate in the extracellular compartment. Calcium transfer from the blood compartment may use paracellular and/or transcellular pathways [24]. Whatever the balance between these two pathways, it appears that odontoblasts are instrumental in concentrating calcium near the predentin. Calcium-ATPase [33], Na^+/Ca^{2+} exchanger [24], and calcium channels [34] allow the transit through the odontoblast membrane. Subcellular fractionation and functional studies have illustrated the respective role of mitochondria and microsomes in storage and release of calcium into the cytosol [24]. Finally, it has been shown that odontoblasts, osteoblasts, and ameloblasts share a selective expression of vitamin D–dependent calbindins-D [7] and parvalbumin in mineralized tissues [35,36].

C. The Enamel, a Unique Epithelium-Derived Mineralized Tissue

Ameloblasts produce enamel, which constitutes a unique example of epithelium-derived mineralized tissue. This most superficial hypermineralized tooth barrier contains a packed network of hydroxyapatite crystals assembled in a prismatic structure [37]. Enamel is almost devoid of organic matrix in its mature form [38]. A precise sequence of events appears to determine enamel morphogenesis and biomineralization. During the secretion stage, enamel matrix is synthesized and exported while biomineralization is initiated. When the full thickness of acellular matrix is reached, the maturation stage involves the selective proteolysis of matrix proteins and the completion of hydroxyapatite crystal growth.

Much attention has been devoted to the identification of enamel proteins [11,38–42]. The major one, amelogenin, which constitutes 90% of the enamel proteins, appears to be polymorphic because there are two distinct transcribed genes on the X and Y chromosomes [43]. Five different mutations of the chromosome X amelogenin gene were reported in an enamel-specific heterogenous genetic disorder: X-linked amelogenesis imperfecta [44]. Several *in vitro* inhibitors of amelogenin expression confirmed its functional relevance [45,46]. After the completion of hydroxyapatite crystal growth, amelogenin is nearly absent from enamel, except in pathological conditions, notably in a porcine model of vitamin D pseudodeficiency rickets [47]. Other isolated enamel proteins include tuftelin [40] and amelin/ameloblastin/sheathlin [11,41,42,48]. Interestingly, the gene that controls the latter is located on human chromosome 4, close to the locus identified for an autosomal form of amelogenesis imperfecta in a Swedish kindred [49].

III. VITAMIN D AND OROFACIAL DEVELOPMENT AND BIOMINERALIZATION

Vitamin D is intimately involved in tooth mineralization. Various experimental studies on rodents, dogs, and pigs (for review, see Berdal *et al.* [50]) have shown that hypovitaminosis D induces defects in dentinogenesis. Enamel dysplasia and hypomaturation were also observed. In contrast, hypervitaminosis D secondary to injections of $1,25(OH)_2D_3$ resulted in accelerated apposition of dentin and hypocellularity of the dental pulp [51]. In humans, dental defects have been associated with simple and pseudo-vitamin D deficiency and familial hypophosphatemia. In the latter, the major defect is the formation of globular dentin, whereas in the former conditions, enamel hypoplasia is the characteristic abnormality [52] (see also Chapters 46–48). Studies in the rat have confirmed the impact of vitamin D deficiency on enamel mineralization [53] (Fig. 2). Previous studies suggested that dental cells contain VDR [54–57]. However, the reported absence of VDR in ameloblasts and odontoblasts was in conflict with the concept of similarity between bone and dental mineralization and with the expression in teeth of numerous proteins [14,15,22,58–66] shown to be vitamin D–dependent in other systems [50,67–70].

A. Correspondence between VDR Expression Pattern and Tissue-Specific Disturbances Caused by Impaired Vitamin D Bioactivation

Tooth germs, as hair follicles, are characterized by their epitheliomesenchymal duality and cooperativity that determine their respective cell fate and activity [15]. One frequently reported feature of hereditary vitamin D–resistant rickets (HVDRR, see Chapter 48) caused by VDR mutations, is the association of alopecia and dental dysmorphogenesis (Fig. 3). Interestingly, during the period when morphogenesis and cell determination occur, both epithelial and mesenchymal cells express VDR [55] (Fig. 4). Furthermore, the characteristics of the dentition of these patients correspond to the anatomical distribution of epithelial vitamin D target cells [1] (Fig. 4). Dysmorphogenesis, with peculiar waving of the enamodentinal junction that can be discerned by plain radiographs (Fig. 3), is evident exclusively at the crown level, where epithelial cells express VDR mRNA [1] (Fig. 4) and protein [55,58]. In contrast, root morphology is normal (Fig. 3), and the epithelial root sheath appears to be devoid of VDR mRNA (Fig. 4). Such observations are also valid in the case of nutritional vitamin D deficiency rickets. Experimental vitamin D deficiency in rodents also results in major defects in crown morphogenesis [53]. Furthermore, such hypomineralization of dental tissues results in early tooth decay.

As observed for the hair follicle, where terminal differentiation is also based on epitheliomesenchymal interactions, both tissue components show differential developmental patterns. Whereas in the hair follicle [71] (Fig. 4) vitamin D target cells were first mesenchymal (fibroblasts of the dermal papilla) and then epithelial ones (outer sheath keratinocytes), the reverse situation is true in teeth. In teeth, modulation of VDR expression characterized both ectomesenchymal and epithelial cells. A transient VDR mRNA level down-regulation was observed after terminal differentiation, first in ameloblasts and thereafter in crown and root odontoblasts. Furthermore, when ameloblasts and odontoblasts were actively secreting their respective matrix, accessory cells located immediately in contact with them appeared to contain large amounts of VDR transcripts [1] as described previously by VDR radioautographic studies [59–62]. Finally (Fig. 4), the developmental down-regulation of odontoblasts after the deposition of mantle dentine contrasts with the hormonal up-regulation of VDR transcripts [1] and protein [58] induced by a single injection of $1,25(OH)_2D_3$. Because VDR functions as a dimer in cooperation with retinoid X receptor (RXR) [72], the final quantity of these receptor proteins in cells may participate in several potential pathways with distinctive biological effects depending on dental cell stage. Indeed, retinoid receptors have been shown to be present in developing teeth, exhibiting distinct developmental patterns (for review, see Mark *et al.* [73]). However, as attention was mainly focused on the initial stages of dental development, the pattern of RXR expression needs to be reevaluated during the formation of mineralized tissues.

Up- and down-regulation of VDR mRNA around the critical period of terminal differentiation of ameloblasts and odontoblasts is particularly interesting. Concomitant epitheliomesenchymal disturbances of tissue organization are characteristic of vitamin D deficient rat molars [53]. These data suggest that vitamin D may influence the occurrence of epitheliomesenchymal interactions leading to morphogenesis and cell differentiation by acting on the genomic expression of some regulatory proteins in epithelial and also in ectomesenchymal cells. A reverse scheme of action may also be suggested, from ectomesenchyme to epithelial components. Sev-

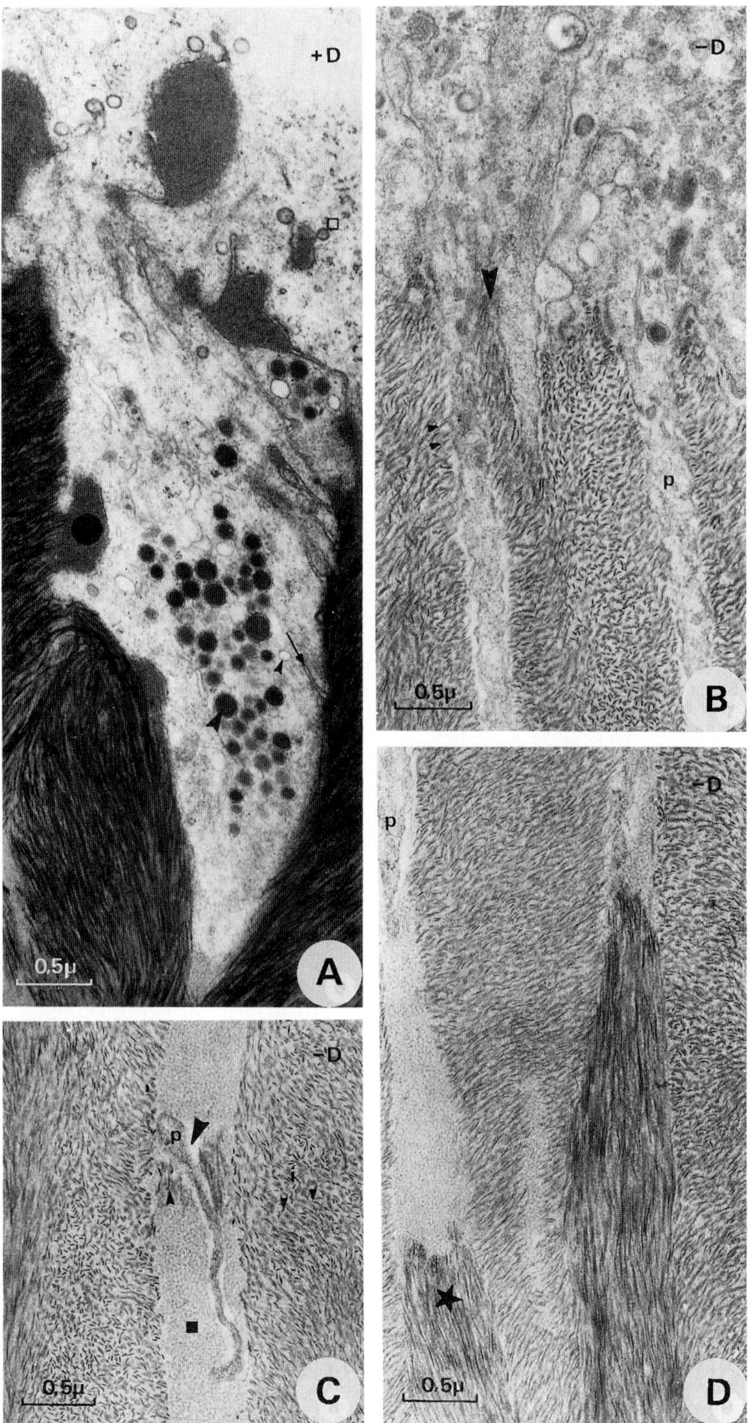

FIGURE 2 Secretion stage of amelogenesis in control and vitamin D–deficient rat molars. (A) Tomes process and the external prismatic layer of enamel in a control rat molar. In the apical pole, a specialized process secretes the interprismatic and intraprismatic enamel matrix. Magnification: × 20,000. Secretory granules (arrowhead) and membrane infoldings (arrow) are shown. (B, C, D) Corresponding area in vitamin D–deficient rat molar. The Tomes process appears elongated (p). Stippled material (square) is present in the intraprismatic area as evidence for the inhibition of enamel biomineralization (star).

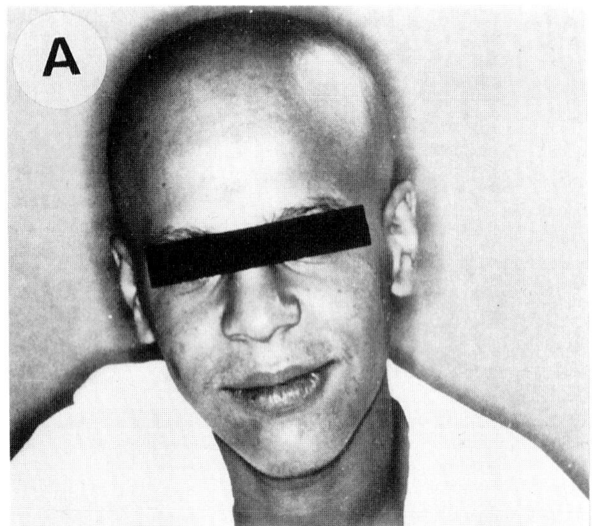

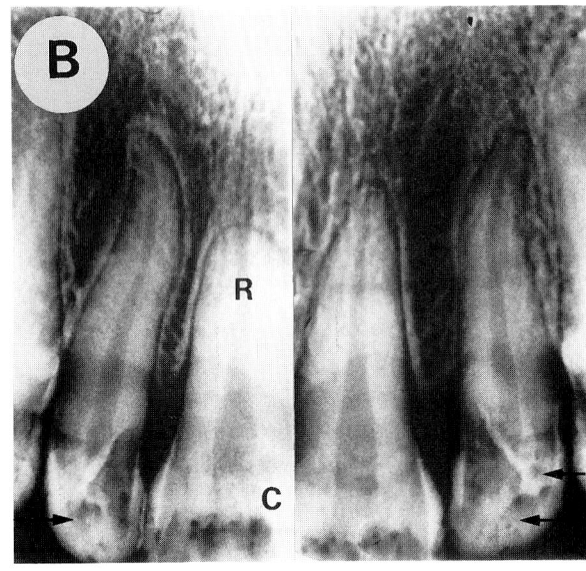

FIGURE 3 Clinical features of pathological vitamin D status. (A) Photograph of a 19-year-old male patient with hereditary vitamin D-resistant rickets (HVDRR) and alopecia. Cultured skin fibroblasts showed a near absence of functional VDR (A. Bréhier, personnal communication). (B) Radiographs of the upper incisors of the same patient showing abnormal morphology of the crown (C) and normal aspect of the root (R). Mineralization defects are particularly evident in the crown (arrows).

eral proteins that play a part in epitheliomesenchymal communications in odontogenesis are vitamin D-dependent in other systems. One interesting instance is Msx-2 [74], although its regulation by vitamin D is still unclear [75,76]. Expression of Msx-2 alternates in epithelium and mesenchyme depending on the stage of odontogenesis, and it is induced by cell interactions [77] that are mediated at least in part by BMP4 [21]. Therefore, Msx-2 is thought to be a key element in the transduction of growth factor–mediated epitheliomesenchymal interactions [78]. Its decrease in vitamin D deficiency states might contribute to the associated epitheliomesenchymal disturbances.

Other regulatory molecules may be involved in these early stages of development (c-*fos*, c-*myc*, TGFβs, receptors for EGF, NGF, etc.) [50]. Intercellular communication involves diffusible factors such as growth factors [78]. However, other kinds of interactions such as cell–cell contacts and cell–matrix cooperation have been described. Such is the case for fibronectin (for review, see Ruch *et al.* [14]). Fibronectin interacts with a membrane protein unrelated to the integrin family that has been purified from odontoblastic cells. The fibronectin gene contains a vitamin D response element (VDRE) sequence [50]. Odontoblasts from vitamin D–deficient rats are characterized by an irregular absence of immunoreactive fibronectin in the predentine (Fig. 5) along with a disorganized odontoblastic apical pole. Fibronectin has been clearly shown to participate in the elongation and organization of the odontoblast body through the organization of its cytoskeleton [14]. Fibronectin is also a good candidate for a vitamin D–mediated control of odontoblast polarization and apical organization.

Vitamin D deficiency is associated with major disturbances of enamel and dentin formation [47,52,53] (Figs. 2, 3, and 5). In vitamin D deficiency, predentin is widened and its mineralization front is irregular. Accompanying matrix disturbances are observed, such as abnormal collagen fibrils in predentin and decreased immunoreactive phosphoproteins and osteocalcin in dentin [28]. In odontoblasts from vitamin D–deficient rats, the depletion of osteocalcin contrasts with the presence of other phosphoproteins. These data suggest that osteocalcin is vitamin D dependent in dentin, as has been shown for bone. Enamel biomineralization also appears to be inhibited in vitamin D deficiency (Fig. 2). Furthermore, enamel structure is selectively affected. The secretion stage allows orderly deposition of various layers of enamel. The internal aprismatic layer is characterized by the parallel distribution of crystals, which are grossly perpendicular to the enamel–dentine junction. This is followed by the main constituent of enamel, the internal prismatic layer, where prisms grow to their largest size. Prisms are individualized by different crystal orientations in interprismatic and intraprismatic zones, corresponding to the secretion areas of the apically specialized ameloblasts. Thereafter, the size of the prisms is considerably reduced in the external prismatic layer. The secretion stage ends with the elaboration of an external aprismatic layer. In samples from vitamin D–deficient rats, the continuity between the internal aprismatic and external prismatic layers suggests that the step of internal prismatic layer formation is selectively

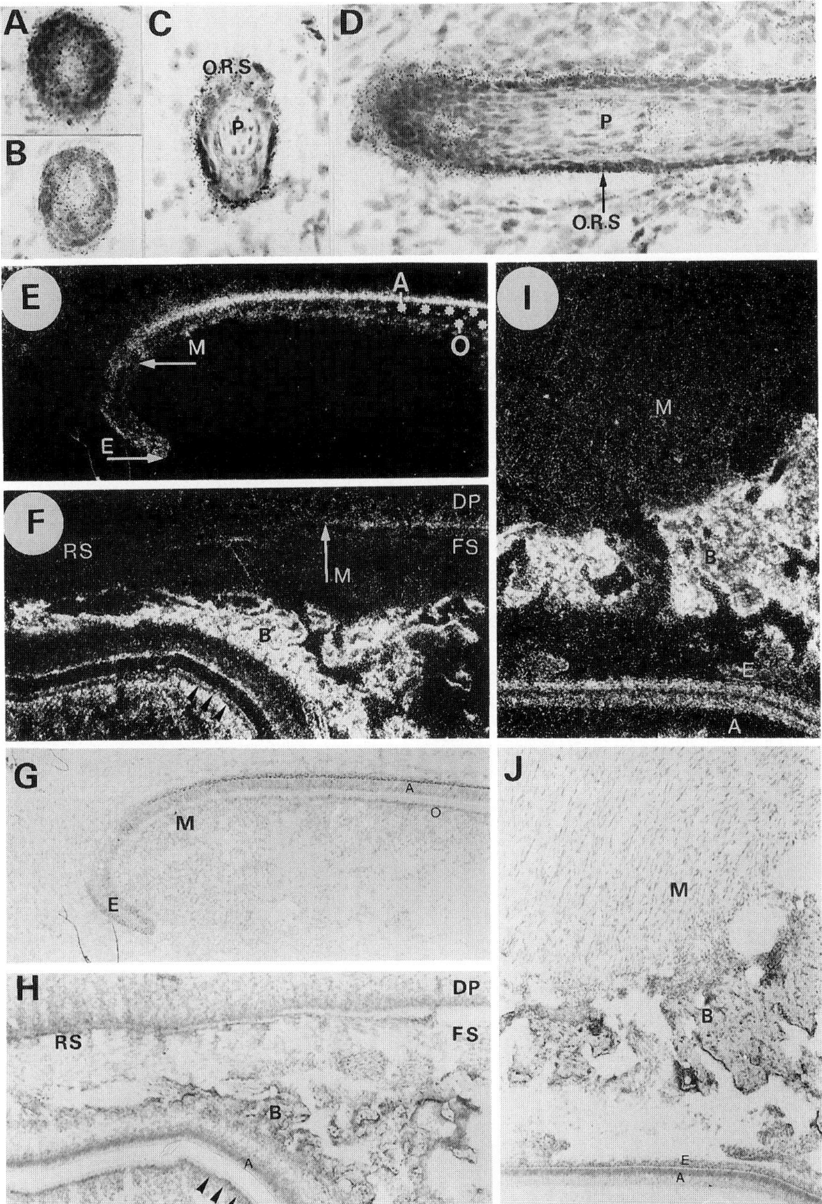

FIGURE 4 Vitamin D receptor (VDR) mRNA distribution in epitheliomesenchymal systems and mineralized tissues. Transverse and longitudinal sections from the base of a hair follicle (A–D) show the changing pattern of VDR mRNAs in the epithelium (ORS, outer root sheath) and mesenchyme (P, fibroblasts of the dermal papilla). Longitudinal sections of forming rat incisor illustrate the early stages of hard tissue formation in the crown-analog part (E, G) and root-analog part (F, H). In the crown-analog part, epithelial labeling (E) begins at an earlier stage than in the ectomesenchyme (M). Asterisks illustrate the stage when VDR mRNA expression reaches a nadir in both epithelial and mesenchymal cells. A, ameloblasts; O, odontoblasts. In the root-analog part, epithelial cells of the root sheath (RS) are devoid of VDR mRNA, although it is present in ectomesenchymal cells, its expression starting before odontoblast differentiation (M) in the root dental papilla (DP) as shown in the crown-analog part. FS, follicular sac; B, alveolar bone. When biomineralization progresses, in the molar (F) and in the incisor (I, J), the VDR mRNA concentration increases in cells devoted to tissue formation. Levels in odontoblasts (F, H; triple arrowhead) are less than subodontoblastic cells. Ameloblasts (A) in their secretion stage (not shown) and maturation stage, as well as adjoining epithelial cells (E), appear to contain VDR mRNA. Bone cells (B) are also reactive, but the dental follicle is not. M, jaw muscle.

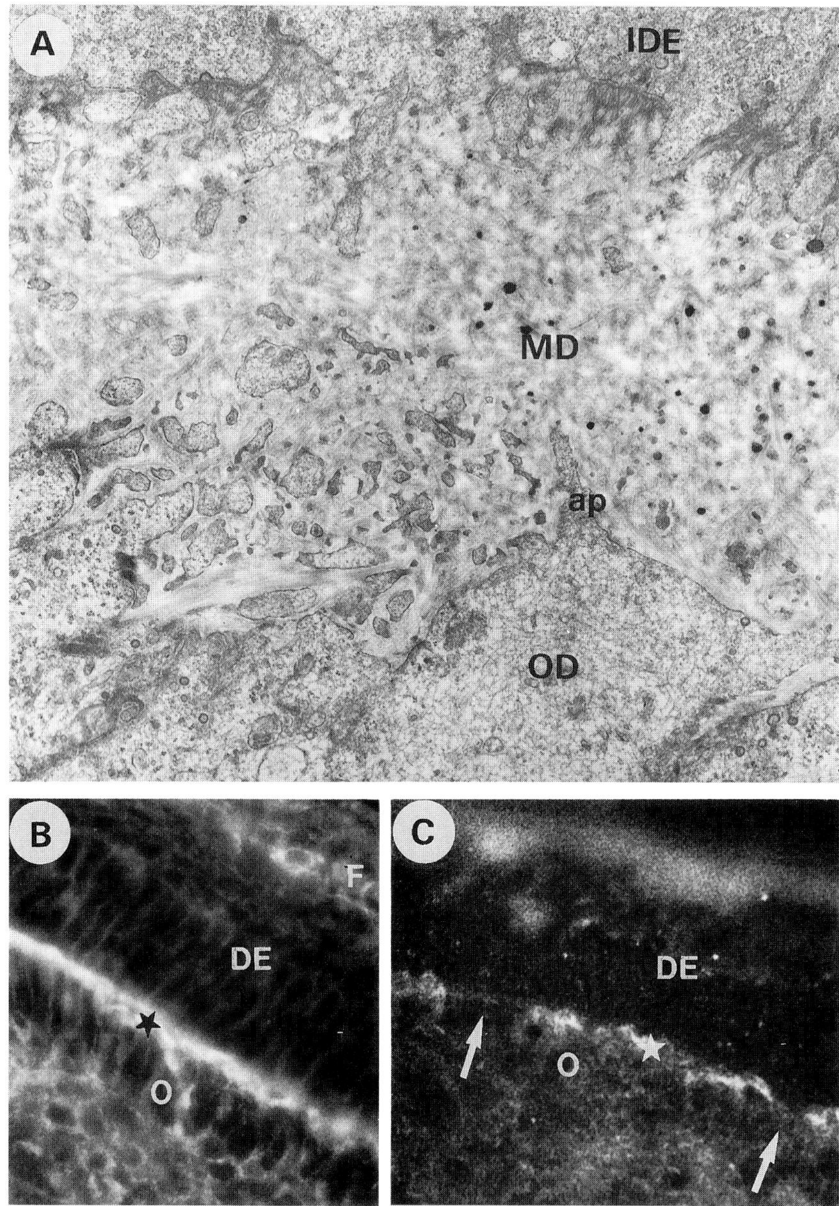

FIGURE 5 Odontoblasts in a vitamin D–deficient 3-day-old rat tooth. (A) Irregular apical pole (ap) of odontoblasts that form mantle–dentine (MD). Magnification: × 17,500. (B) Immunolabeling of fibronectin (star) in an incisor of a control rat showing the apical redistribution of the protein adjoining polarizing odontoblasts (O). DE, Dental epithelium; F, follicular sac. Magnification: × 300. (C) Immunolabeling of fibronectin (star) is irregular (arrow) in odontoblasts of the corresponding area in vitamin D–deficient rat incisors. Magnification: × 30.

withdrawn. This would suggest that vitamin D is a major determinant for this central step of amelogenesis.

These structural, ultrastructural, and immunochemical data were obtained in rat molars. Biochemical investigation could not be done in the same system in view of the small size of the samples. Thus, we have used the continously growing mandibular incisor as a developmental model, which is already well described morphologically [37,38]. This system shares many features with the hair follicle, where the life cycles of epithelial and mesenchymal cells are aligned along the longitudinal axis of the structures (Fig. 4). In the proximal end, continuous proliferation is followed by terminal differentiation of several cell lineages. In the same sample, the processes of amelogenesis and crown dentinogenesis may be observed on the labial aspect. On the lingual aspect of the tooth, the deposition of an acellular cementum occurs in parallel to root dentinogenesis. Therefore, by using the surrounding alveolar bone as a reference tissue, it is possible to obtain an overview of the biogenesis of enamel, acellular cementum, dentin, and alveolar bone. Target organs for $1,25(OH)_2D_3$ (i.e., duodenum, kidney, and bone) were selected as internal controls for the investigation of vitamin D pathways [54]. Some studies were also carried out on growth-limited teeth (rodent molars and human teeth) in order to confirm the data obtained in the continously growing organs.

B. Developmental and Hormonal Control of Calbindin Expression in Tooth and Bone Cells

Calbindins belong to the superfamily of intracellular proteins that bind calcium with high affinity by conserved EF-hand binding sites [79]. Two proteins, calbindin-D_{9K} and calbindin-D_{28K}, are encoded by separate genes [80] (see also Chapters 13 and 14). Their tissue distribution is divergent. The developmental pattern and subcellular distribution of calbindins have been systematically analyzed in mineralized tissues [7,81]. These calcium binding proteins appear restricted to the cells directly involved in the deposition of mineralizing matrices, namely ameloblasts, odontoblasts, osteoblasts, and chondrocytes. Cementoblasts from the rat incisor (acellular cementum) did not show a significant amount of either calbindin-D_{9K} or calbindin-D_{28K}. Mineralized tissues including bone, dentin, and enamel and cartilage (for review, see Berdal *et al.* [7]) contain both proteins but show distinct ratios of calbindin-D_{28K} and calbindin-D_{9K}. In dental cells, calbindin-D_{28K} appears 100-fold more abundant than calbindin-D_{9K}, which is almost absent in odontoblasts [7]. In dentin and bone, terminal differentiation is associated with a consistently present expression of calbindins in mineralized tissues. In enamel, the fluctuations of calbindins are finely tuned, in accord with the proposed variations of transcellular calcium transport [56]. The most striking features in epithelium are (1) the biphasic aspect of steady-state mRNA levels over the ameloblast life cycle and (2) the cocycling of concentrations of both calbindins in the cytosol and the nucleus of ruffle- and smooth-ended ameloblasts.

Because vitamin D response elements are present in the promoter regions of both calbindin genes [5,72], and because both proteins are present in significant amounts in ameloblasts and odontoblasts, they have been proposed as markers of the genomic action of $1,25(OH)_2D_3$. This view is supported by the observation that in vitamin D–deficient rats following Northern blot analyses, a single injection of $1,25(OH)_2D_3$, shows changes in the steady-state levels of both calbindins in enamel and of calbindin-D_{28K} only in the dental mesenchyme. The concomittant increase of VDR mRNA [1] and protein [58] suggests that variations of calbindins are likely related to the genomic action of $1,25(OH)_2D_3$.

C. Comparative Analysis of Gene Expression in Rat Mandibular Mineralized Tissues—Matrix Proteins and Molecules Involved in Calcium and Phosphate Handling

Most of the molecules shown to be vitamin D dependent in other systems and expressed in differentiated cells of dental tissues are also present in bone cells. As for calbindins, comparisons of their developmental pattern of expression throughout the different stages of formation of mineralized tissues were also demonstrated by *in situ* hybridization and immunolabelling [1,7,54,56,57] jointly in the mandibular incisor and its adjoining alveolar bone. Figure 6 integrates the data from other laboratories as well as our own concerning the cells involved directly in the deposition of extracellular matrix (i.e., ameloblasts, odontoblasts, and osteoblasts). Biphasic expression patterns concerning proteins involved in biomineralization matrix proteins characterize epithelial cells, whereas ectomesenchymal cells evenly express matrix proteins when they are terminally differentiated.

The initiation of expression of matrix proteins in ameloblasts and odontoblasts shows epitheliomesenchymal switching similar to that involving ones of growth factors and homeobox proteins in earlier development. Epitheliomesenchymal confusion in the expression of enamel and dentin matrix proteins characterizes the initial steps of enamel and dentin formation. This transitory step is prolonged in vitamin D–deficient rats [28,53].

During amelogenesis, two phases of mRNA expression are framed with clear-cut modifications during differentiation and transition (Fig. 6). Furthermore, the developmental pattern of detected proteins shows variations depending on the cyclic modulation of the maturation stage of ameloblasts as shown for calbindin-D_{9K}, calbindin-D_{28K} [82], amelogenin [37], H^+-ATPase [83], alkaline phosphatase [84] and calcium pump [33] activities, and EGF receptor [85]. Common posttranscriptional and even posttranslational systems of regulation might be involved.

Finally, the comparative investigation of VDR and

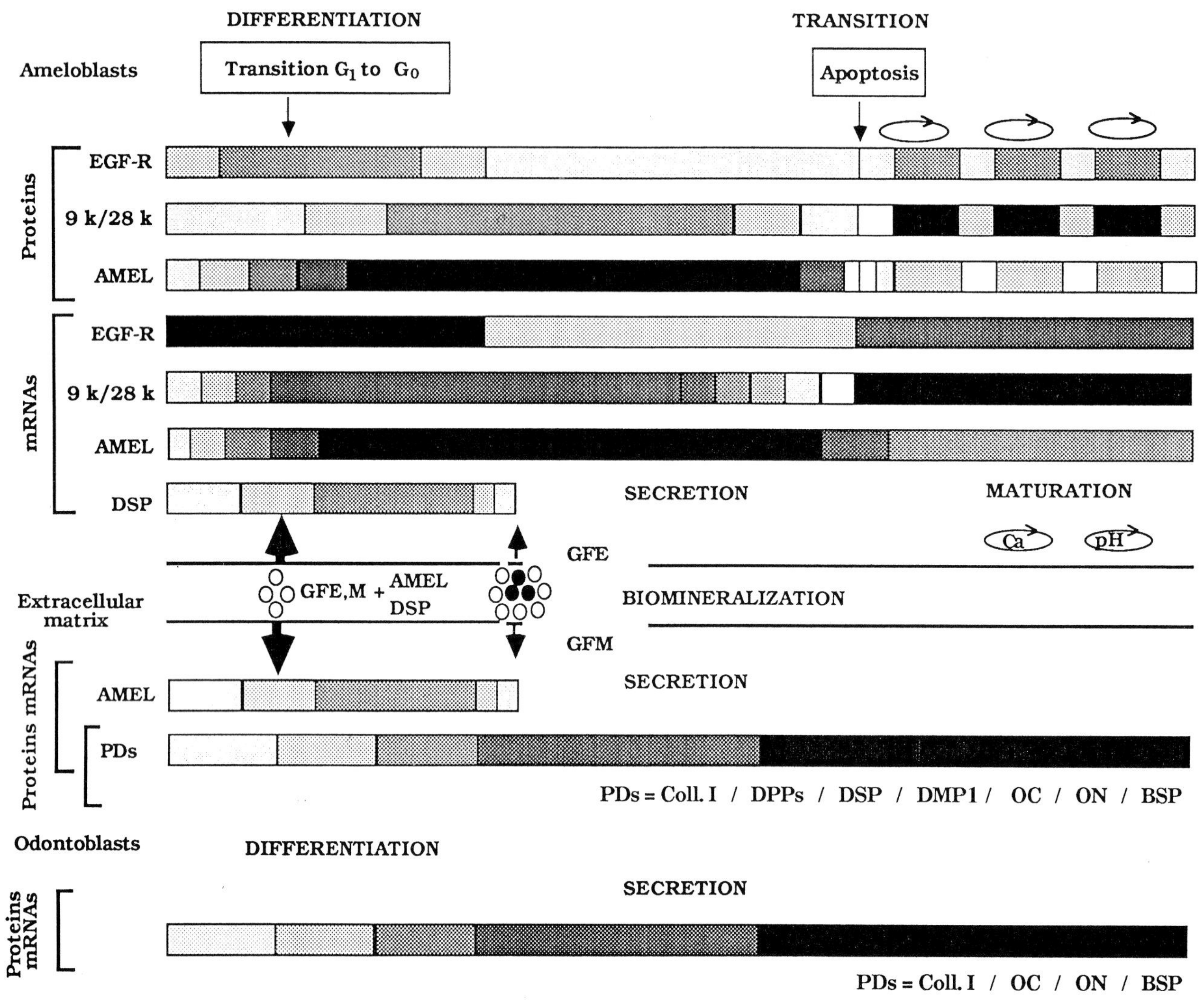

FIGURE 6 Schematic representation of relative quantities of mRNA and protein in cells of mineralized tissues. Life cycles of ameloblasts, odontoblasts, and osteoblasts are shown. EGF-R, receptor for EGF; 9k, calbindin-D_{9k}; 28k, calbindin-D_{28k}; AMEL, amelogenin; DSP, dentin sialoprotein; PDs, proteins of dentin; Coll. I, type I collagen; DPPs, dentin phosphoproteins; DMP1, dentin matrix protein I; OC, osteocalcin; ON, osteonectin; BSP, bone sialoprotein; GFE, growth factor from epithelium; GFM, growth factor from mesenchyme. Data were compiled from various sources [1,8,22–28] and from Hotton *et al.* (unpublished results). Terminal differentiation of ameloblasts and odontoblasts is not a single step but includes polarization, elongation, and matrix secretion. Morphologically, differentiation appears to be a complex progressive acquisition of cell and extracellular features for both enamel and dentin, although some phenotypic traits are shared between ameloblasts and odontoblasts. When enamel and dentin are mineralized, dentin protein expression is stopped in ameloblasts (Hotton *et al.*, unpublished results).

vitamin D–dependent gene expression in various mineralized tissues showed that apart from the cells directly in contact with the forming matrix, other cells may contribute to their vitamin D–controlled formation. The cells adjoining secreting cells (stratum intermedium, subodontoblastic cells, osteoprogenitor bone cells) have in common active synthesis of alkaline phosphatase and high levels of vitamin D receptor mRNA [1] and protein [55,84]. Developmentally controlled expression of proteins involved in biomineralization (i.e., tissue-specific matrix proteins and molecules devoted to calcium and phosphate handling) has been proposed to depend on various transcription factors. For instance, Msx-2 has been suggested to bind promoter regions and control the expression of one of the most classic target genes of vitamin D, osteocalcin [75,76]. Therefore, the differential intervention of transcription factors (genetically programmed Msx-2 and hormonally controlled vitamin D receptor) might cooperate in the regulation of the expression of genes important for skeletal histogenesis.

Interrelations between genetically controlled and hormonally induced pathways may contribute to the site-specific deformities in the craniofacial skeleton (e.g., dental crown and not root, frontal and parietal bones) that characterize rickets.

IV. CONCLUSION

Odontogenesis, an important developmental event of the craniofacial area, may serve as an exemplary system to investigate signaling pathways of the morphogenesis of the skeleton derived from the neural crest. The sequence of events occurring from the initiation event in the odontogenic placode visualized as a local thickening [15] to the erupted tooth, and its integration in the orofacial environment, has been studied. Much attention has been and is still devoted to the premineralization phases: morphogenesis and cytodifferentiation [15]. Some data concerning growth factors, bone morphogenetic proteins, and homeobox genes have been obtained initially in tooth germs [21,86]. At later stages, the orofacial area contains at least four different modalities of biomineralization, excluding mineralized cartilage. In contrast to bone shape, which also depends on mechanical forces and movements [87], tooth final shape is mostly determined by genetically and hormonally controlled sequential gene expression [1,16]. The last specificity of tooth versus bone is the absence of remodeling, which allows retrospective studies of developmental defects.

Vitamin D appears to be important for the final optimal morphogenesis of the skeleton and, notably, the teeth. Our data suggest that control of the expression of matrix proteins in teeth would share VDR–RXR activation pathways, as shown in bone. Furthermore, the anatomical specificities of gene activity for proteins involved in skeletogenesis (e.g., type I collagen, alkaline phosphatase) may be modulated through signaling pathways (Msx-2/BMP4) that are also under the control of vitamin D. The spatial and temporal coordinated process of odontogenesis may provide a good model system to investigate the relationships between genetically controlled and hormonally induced expression of proteins important for biomineralization.

Acknowledgments

The collaboration of D. Hotton, J.-L. Davideau, P. Papagerakis, and I. Bailleul-Forestier and the friendly scientific help of A. Nanci are gratefully acknowledged.

References

1. Davideau JI, Papagerakis P, Hotton D, Lezot F, Berdal A 1996 *In situ* investigation of vitamin D receptor, alkaline phosphatase, osteocalcin gene expression in oro-facial mineralized tissues. Endocrinology **137**:3577–3585.
2. St Arnaud R, Candeliere A 1994 A novel vitamin D responsive element in the murine c-*fos* promoter. In: Norman AW, Bouillon M, Thomasset M (eds) Vitamin D: A Pluripotent Steroid Hormone: Structural Studies, Molecular Endocrinology and Clinical Applications. de Gruyter, Berlin, pp. 286–287.
3. Lichtler A, Stowler ML, Angilly J, Kream B, Rowe DW 1989 Isolation and characterization of the rat alpha I (I) collagen promoter. Regulation by 1,25-dihydroxyvitamin D_3. J Biol Chem **261**:3072–3077.
4. Staal A, Van Wijnen AJ, Desai RK, Pols HAP, Birkenhager JC, DeLuca F, Denhardt DT, Stein JL, Van Leeuven JPTM, Stein S, Lian JB 1996 Antagonistic effects of transforming growth factor-β on vitamin D_3 enhancement of osteocalcin and osteopontin transcription: Reduced interactions of vitamin D receptor/retinoid X receptor complexes with vitamin D response elements. Endocrinology **137**:262–271.
5. Carlberg C 1995 Mechanisms of nuclear signalling by vitamin D_3 interplay with retinoid and thyroid hormone signalling. Eur J Biochem **231**:517–527.
6. Sodek J, Kim RH, Ocata JLI, Yamauchi M, Zhang Q, Freedman LP 1995 Regulation of bone sialoprotein gene transcription by steroid hormones. Connect Tissue Res **32**:209–217.
7. Berdal A, Hotton D, Saffar JL, Thomasset M, Nanci A 1996 Calbindin-D_{9k} and calbindin-D_{28k} expression in rat mineralized tissues *in vivo.* J Bone Miner Res **11**:768–779.
8. Butler WT, Ritchie H 1995 The nature and functional significance of dentin extracellular matrix proteins. Int J Dev Biol **39**:169–179.
9. Goldberg M, Septier D, Lecolle S, Chardin H, Quintana MA, Acevedo AC, Gafni G, Dillouya D, Vermelin L, Thonemann B, Schmalz G, Bissila-Mapahou P, Carreau JP 1995 Dental mineralization. Int J Dev Biol **39**:93–110.
10. Wu D, Ikezawa K, Parker T, Saito M, Narayanan AS 1996 Characterization of a collagenous cementum derived attachment protein. J Bone Miner Res **11**:686–692.
11. Snead ML 1996 Enamel biology logodaedaly: Getting to the root of the problem, or "Who's on first ...." J Bone Miner Res **11**:899–904.
12. Le Douarin NM, Ziller C, Couly GF 1993 Patterning of neural crest derivatives in the avian embryo: *In vivo* and *in vitro* studies. Dev Biol **159**:24–49.
13. Lumsden AGS 1988 Spatial organization of the epithelium and the role of neural crest cells in the initiation of the mammalian tooth germ. Development **103**:155–169.
14. Ruch JV, Lesot H, Bégue-Kirn C 1995 Odontoblast differentiation. Int J Dev Biol **39**:51–68.
15. Thesleff I, Vaahtokari A, Kettuner P, Aberg T 1995 Epithelial mesenchymal signalling during tooth development. Connect Tissue Res **32**:915–921.
16. Vaahtokari A, Aberg T, Thesleff I 1996 Apoptosis in the developing tooth: Association with an embryonic signaling center and suppression by EGF and FGF-4. Development **122**:121–129.
17. Slavkin HC 1991 Molecular determinants during dental morphogenesis and cytodifferentiation: A review. J Craniofacial Genet Dev Biol **11**:338–349.
18. Tabata MJ, Kim K, Liu JG, Yamashita K, Matsumura T, Kato J, Iwamoto M, Wakisaka S, Matsumoto K, Nakamura T, Kumegawa M, Kurisu K 1996 Hepatocyte growth factor involved in the

morphogenesis of tooth germ in murine molars. Development **122**:1243–1251.
19. Kronmiller JE, Upholt WB, Kollar EJ 1991 EGF antisense oligodeoxynucleotides block murine odontogenesis *in vitro*. Dev Biol **147**:485–488.
20. Shum I, Sakakura Y, Bringas JP, Luo W, Snead ML, Mayo M, Crohin C, Millar S, Werb Z, Buckley S, Hall FL, Warburton D, Slavkin HC 1993 EGF abrogation induced *fusili*-form dysmorphogenesis of Meckel's cartilage during embryonic mouse mandibular morphogenesis *in vitro*. Development **118**:913–917.
21. Vainio S, Karavanova I, Jowet A, Thesleff I 1993 Identification of BMP-4 as a signal mediating secondary induction between epithelial and mesenchymal tissues during early tooth development. Cell **75**:45–58.
22. Zeichner-David M, Diekwisch T, Fincham A, Lau E, MacDougall M, Moradian-Olkak J, Simmer J, Snead M, Slavkin HC 1995 Control of ameloblast differentiation. Int J Dev Biol **39**:69–92.
23. Thomas HF 1995 Root formation. Int J Dev Biol **39**:231–237.
24. Linde A, Lundgren T 1995 From serum to the mineral phase. The role of the odontoblast in calcium transport and mineral formation. Int J Dev Biol **39**:213–222.
25. Bronckers ALJJ, Farach-Carson MC, Vanwaren M, Butler WT 1994 Immunolocalization of osteopontin, osteocalcin and dentin sialoprotein during dental root formation and early cementogenesis in the rat. J Bone Miner Res **9**:833–841.
26. MacNeil RL, Sheng N, Strayhorn C, Fisher LW, Somerman MJ 1994 Bone sialoprotein is localized to the root surface during cementogenesis. J Bone Miner Res **9**:1597–1606.
27. Butler WT 1995 Dentin matrix proteins and dentinogenesis. Connect Tissue Res **33**:59–65.
28. Berdal A, Gorter de Vries I, Hotton D, Cuisinier-Gleizes P, Mathieu H 1991 The cellular and extracellular distribution of osteocalcin and dentin phosphoprotein in teeth of vitamin D–deficient rats. J Biol Bucc **19**:45–63.
29. George A, Sabsay B, Simonian PAL, Veis A 1993 Characterization of a novel dentin matrix acid phosphoprotein. J Biol Chem **268**:12624–12630.
30. Ritchie HH, Hou H, Veis A, Butler WT 1994 Cloning and sequence determination of rat dentin sialoprotein, a novel dentin protein. J Biol Chem **269**:3698–3702.
31. Ritchie HH, Pinero G, Hou H, Butler WT 1995 Molecular analysis of rat dentin sialoprotein (dsp). Connect Tissue **23**:73–79.
32. MacDougall M, Simmons D, Luan X, Nydegger J, Feng J, Gu TT 1997 Dentin phosphoprotein and dentin sialoprotein are cleavage products expressed from a single transcript coded by a gene on human chromosome 4. J Biol Chem **272**:836–842.
33. Borke JM, Zaki AE, Einsenmann DR, Mednieks Ml 1995 Localization of plasma membrane Ca^{2+} pump mRNA and protein in human ameloblasts by *in situ* hybridization and immunohistochemistry. Connect Tissue Res **33**:139–144
34. Seux D, Joffre A, Fosset M, Magloire H 1994 Immunohistochemical localization of L-type calcium channels in the developing first molar of the rat during odontoblast differentiation. Arch Oral Biol **39**:167–170.
35. Davideau JI, Celio MR, Hotton D, Berdal A 1993 Developmental pattern and subcellular localization of parvalbumin in the rat tooth germ. Arch Oral Biol **38**:707–715.
36. Toury R, Belqasmi F, Hauchecorne M, Leguellec D, Heizman C, Balmain N 1995 Localization of the Ca^{2+}-binding α-parvalbumin and its mRNA in epiphyseal plate cartilage and bone of growing rats. Bone **17**:121–130.
37. Nanci A, Smith CE 1992 Development and calcification of enamel. In: Bonucci E (ed) Calcification in Biological Systems. CRC Press, Boca Raton, Florida, pp. 313–343.
38. Robinson C, Kirkham J, Brookes SJ, Bonass WA, Shore RC 1995 The chemistry of enamel development. Int J Dev Biol **39**:145–152.
39. Fukae M, Tanabe T 1987 Nonamelogenin components of porcine enamel in the protein fraction free from the enamel crystals. Calcif Tissue Int **40**:286–293.
40. Deutsch D, Palmon Z, Dafni L, Catalano-Sherman J, Young MF, Fisher LW 1995 The enamelin (tuftelin) gene. Int J Dev Biol **39**:135–143.
41. Krebsbach PH, Lee SK, Matsuki Y, Kozak Ca. Yamada RM, Yamada Y 1996 Full length sequence, localization and chromosomal mapping of ameloblastin. A novel tooth-specific gene. J. Biol Chem **271**:4431–4435.
42. Hu CC, Fukae M, Uchida T, Qian Q, Zhang CH, Tyu OH, Tanabe T, Yamakoshi Y, Murakami C, Dohi N, Shimizu M, Shimmer JP, Sheathlin 1997 Cloning, cDNA/polypeptide sequences, and immunolocalization of porcine enamel sheath proteins. In press.
43. Sasaki S, Shimokawa H 1995 The amelogenin gene. Int J Dev Biol **39**:127–133.
44. Lagerström L, Niklas-Dahl LM, Iselius L, Backman B, Pettersson U 1990 Mapping of the gene for X-linked amelogenesis impertecta by linkage analysis. Am J Hum Genet **46**:120–125.
45. Diekwisch T, David S, Bringas P, Santos V, Slavkin HC 1993 Antisense inhibition of AMEL translation demonstrates supramolecular controls for enamel HAP crystal growth during embryonic mouse molar development. Development **117**:471–482.
46. Lyngstadaas SP, Risnes S, Sproat BS, Thrane PS, Prydz HP 1995 A synthetic, chemically modified ribozyme eliminates amelogenin, the major translation product in developing mouse enamel *in vivo*. EMBO J **14**:5224–5229.
47. Limeback H, Schlumbohm C, Sen A, Nikiforuk G 1992 Effects of hypocalcemia/hypophosphatemia on porcine bone and dental hard tissues in an inherited form of type 1 pseudo-vitamin D deficiency rickets. J Dent Res **71**:346–352.
48. Cerny R, Slaby I, Hammarstrom L, Wurtz T 1996 A novel gene expressed in rat ameloblasts codes for proteins with cell binding domains. J Bone Miner Res **11**:883–891.
49. Forsman K, Lind L, Bäckman B, Westermarke N, Holmgren G 1994 Localization of a gene for autosomal dominant amelogenesis imperfecta (ADAI) to chromosome 4q Hum Mol Genet **3**:1621–1625.
50. Berdal A, Papagerakis P, Hotton D, Bailleul-Forestier I, Davideau JL 1995 Ameloblasts and odontoblasts, target-cells for 1,25-dihydroxyvitamin D_3: A review. Int J Dev Biol **39**:257–262.
51. Pitaru S, Blauschild N, Noff D, Edelstein S 1982 The effects of toxic doses of 1,25-dihydroxycholecalciferol on dental tissues in the rat. Arch Oral Biol **27**:915–923.
52. Nikiforuk G, Fraser D 1981 Etiology of enamel hypoplasia and interglobular dentin: The roles of hypocalcemia and hypophosphatemia. Metab Bone Dis Related Res **4**:17–22.
53. Berdal A, Balmain N, Cuisinier-Gleizes P, Mathieu H 1987 Histology and microradiography of early post-natal molar tooth development in vitamin-D deficient rats. Arch Oral Biol **32**:493–498.
54. Berdal A, Balmain N, Brehier A, Hotton D, Cuisinier-Gleizes P, Mathieu H 1989 Immunological characterization, developmental pattern and vitamin D–dependency of calbindin-D_{28k} in rat teeth ameloblasts. Differentiation **40**:27–35.
55. Bailleul-Forestier I, Davideau JL, Papagerakis P, Noble I, Nessmann C, Peuchmaur M, Berdal A 1996 Immunolocalization of vitamin D receptor and calbindin-D_{28k} in human tooth germ. Pediatr Res **39**:636–642.
56. Berdal A, Hotton D, Kamyab S, Cuisinier-Gleizes P, Mathieu H 1991 Subcellular co-localization and co-variations of two vitamin D–dependent calcium-binding proteins in rat ameloblasts. Arch Oral Biol **10**:715–725.

57. Berdal A, Balmain N, Thomasset M, Brehier A, Hotton D, Cuisinier-Gleizes P, Malhieu H 1989 Calbindins-D_{9k} Da and $_{28k}$ Da and enamel secretion in vitamin D deficient and control rats. Connect Tissue Res **22**:161–165.
58. Berdal A, Hotton D, Pike JW, Mathieu H, Dupret JM 1993 Cell- and stage-specific expression of vitamin D receptor and calbindin-D genes in rat incisor: Regulation by 1,25 dihydroxyvitamin D_3. Dev Biol **155**:172–179.
59. Clark SA, Dame MC, Kim YS, Stumpf WE, DeLuca HF 1985 1,25-dihydroxyvitamin D_3 in teeth of rats and humans: Receptors and nuclear localization. Anat Rec **212**:250–254.
60. Kim YS, Stumpf WE, Clark SA, Sar M, DeLuca HF 1983 Target-cells for 1,25 dihydroxyvitamin D_3 in developing rat incisor teeth. J Dent Res **62**:58–59.
61. Kim YS, Clark SA, Stumpf WE, DeLuca HF 1985 Nuclear uptake of 1,25 dihydroxy vitamin D_3 in developing rodent teeth: An autoradiographic study. Anat Rec **212**:301–306.
62. Midgett RJ, Friedman CM, Schneider PE 1982 Localization of tritiated 1,25 dihydroxyvitamin D_3 in foetal rat mandible, kidney and intestine. Acta Vitaminol Enzymol **4**:233–236.
63. Helder M, Bronckers A, Wöltgens J 1993 Dissimilar expression patterns for the extracellular matrix proteins osteopontin (opn) and collagen type 1 in dental tissues and alveolar bone of the neonatal rat. Matrix **13**:415–425.
64. Mitsiadis TA, Luukko K 1995 Neurotrophins in odontogenesis. Int J Dev Biol **39**:195–202.
65. Byers MR, Mecifi KB, Iadarola MJ 1995 Focal c-*fos* expression in developing rat molars: Correlations with subsequent intradental and epithelial sensory innervation. Int Dev Biol **39**:181–189.
66. D'Souza RN, Happonen RP, Flanders KS, Butler WT 1990 Temporal and spatial patterns of transforming growth factor-1 expression in developing rat molars. Arch Oral Biol **35**:957–965.
67. Hirning U, Schmid D, Schulz WA, Rettenberger R, Hameister H 1992 A comparative analysis of N-*myc* and c-*myc* expression and cellular proliferation in mouse organogenesis. Mech Dev **33**:119–126.
68. MacKenzie A, Ferguson MWJ, Sharpe PT 1992 Expression patterns of the homeobox gene, Hox 8, in the mouse embryo suggest a role in specifying tooth initiation and shape. Development **115**:403–420.
69. Pavlin D, Lichtler AC, Bedalov A, Kream BE, Harrison JR, Thomas HF, Gronowics GA, Clark SH, Woody CO, Rowe DW 1992 Differential utilization of regulatory domains within the α1(I) collagen promoter in osseous and fibroblastic cells. J Cell Biol **116**:227–236.
70. Andujar MB, Couble P, Couble ML, Magloire H 1991 Differential expression of type I and III collagen genes during tooth development. Development **111**:691–698.
71. Paus R, Jung M, Schilli M, Kerber A, Egly C, Chambon P, Bahmer FA, Reichhrat J 1994 Hair follicle expression of 1,25-dihydroxyvitamin D_3 receptor (VDR) and retinoid-X receptor-α (RXRα). In: Norman AW, Bouillon R, Thomasset M (eds) Vitamin D: A Pluripotent Steroid Hormone: Structural Studies, Molecular Endocrinology and Clinical Applications. de Gruyter, Berlin, pp. 617–618.
72. Lowe KE, Maiyar AC, Norman AW 1992 Vitamin D-mediated gene expression. Crit Rev Eukariotic Gene Expression **2**:65–109.
73. Mark M, Lohnes D, Mendelsohn C, Dupé V, Vonesch JL, Kastner P, Rijli F, Bloch-Zupan A, Chambon P 1995 Roles of retinoid acid receptors and of Hox genes in the patterning of the teeth and of the jaw skeleton. Int J Dev Biol **39**:111–121.
74. Hodgkinson JE, Davidson CL, Beresford J, Sharpe PT 1993 Expression of human homeobox-containing gene is regulated by $1,25(OH)_2D_3$ in bone cells. Biochim Biophys Acta **1174**:11–16.
75. Towler DA, Rutledge SJ, Rodan GA 1994 Msx-2/Hox 8.1: A transcriptional regulator of the rat osteocalcin promoter. Mol Endocrinol **8**:1484–1493.
76. Hoffman HM, Catron KM, Van Wijnen AJ, McCabe LR, Lian JB, Stein GS, Stein JL 1934 Transcriptional control of the tissue-specific, developmentally regulated osteocalcin gene requires a binding motif for the *Msx* family of homeodomain proteins. Proc Natl Acad Sci USA **91**:12887–12891.
77. Jowett AK, Vainio S, Ferguson MWJ, Sharpe PT, Thesleff I 1993 Epithelial mesenchymal interactions are required for *msx-1* and *msx-2* gene expression in the developing murine molar tooth. Development **117**:461–470.
78. Thesleff I, Vaahtokari A, Vaino S, Jowett A 1996 Molecular mechanisms of cell and tissue interactions during early tooth development. Anat Rec **245**:151–161.
79. Christakos S, Gabrieledes C, Rhoten WB 1989 Vitamin D-dependent calcium binding proteins: Chemistry, distribution, functional considerations, and molecular biology. Endocr Rev **10**:3–26.
80. Perret C, Lomri N, Thomasset M 1988 Evolution of the EF-hand calcium binding protein family: Evidence for exon shuffling and intron insertion. J Mol Evol **27**:351–364.
81. Hotton D, Davideau JL, Bernaudin JF, Berdal A 1995 *In situ* hybridization of calbindin-D_{28k} transcripts in undecalcified sections of the rat continuously erupting incisor. Connect Tissue Res **32**:137–143.
82. Berdal A, Nanci A, Smith CE, Ahlualia JP, Thomasset M, Cuisinier-Gleizes P, Mathieu H 1991 Differential expression of calbindin-D_{28k} Da in rat incisor ameloblasts throughout enamel development. Anat Rec **230**:149–163.
83. Lin HM, Nakamura H, Noda T, Ozawa H 1994 Localization of H^+-ATPase and carbonic anhydrase II in ameloblasts at maturation. Calcif Tissue Int **55**:38–45.
84. Gomez S, Boyde A 1994 Correlated alkaline phosphatase histochemistry and quantitative backscattered electron imaging in the study of rat incisor ameloblasts and enamel mineralization. Microsc Res Tech **29**:29–36.
85. Davideau JL, Sahlberg D, Blin C, Papagerakis P, Thesleff I, Berdal A 1995 Differential expression of the full-length and secreted truncated forms of EGF receptor during formation of dental tissues. Int J Dev Biol **39**:605–615.
86. Satokata I, Maas R 1994 *Mxs-1* deficient mice exhibit cleft palate and abnormalities of cranio-facial and tooth development. Nature Genet **7**:348–356.
87. Buckwalter JA, Glimcher MJ, Cooper RR, Recker R 1995 Bone biology part II: Formation, modeling, remodeling, and regulation of cell function. J Bone Joint Surg **77**:1276–1289.

CHAPTER 28

Vitamin D in Pregnancy, the Fetoplacental Unit, and Lactation

ANTHONY D. CARE Institute of Biological Sciences, University of Wales, Aberystwyth, UK

I. INTRODUCTION

The human fetus accumulates a total of 25–30 g of calcium by term, with most of this accretion of calcium by the fetal skeleton taking place in the third trimester, when the growth rate is highest. The rate of calcium accretion then reaches 150 mg per day per kilogram fetal body weight. The control of transplacental calcium flux appears to rest primarily with the fetus itself, which is intimately involved with the control of fetal calcium homeostasis. In all mammalian fetuses examined during this period, the fetal plasma calcium and phosphate concentrations exceed those in the mother, presumably to facilitate the deposition of hydroxyapatite on the developing skeletal matrix.

To supply these demands of the fetus for calcium, marked adjustments in maternal calcium regulation are required to preserve overall homeostasis. These involve major changes in the metabolism of vitamin D. Calcitriol [1,25-dihydroxyvitamin D_3, $1,25(OH)_2D_3$], the most biologically active metabolite of vitamin D, also plays an important role in the transfer of calcium ions across the placenta. The genomic mechanism of vitamin D action is well established; a proposed nongenomic mechanism remains largely hypothetical. However, the importance of vitamin D, relative to other factors that regulate the control of the transfer of calcium ions across the placenta, is species dependent. In the cow, for example, lactation imposes an even greater demand for calcium than pregnancy, and the clinical condition of milk fever may result. However, this demand is usually satisfied, largely by way of an increase in the rate of synthesis of calcitriol and its subsequent calcium-mobilizing effects via increased absorption from the digestive tract and resorption of bone. In addition to its role in calcium homeostasis, it is now also clear that calcitriol has more general effects on cell growth and differentiation.

II. VITAMIN D DURING PREGNANCY

A. Plasma Concentrations and Placental Transport of Vitamin D Metabolites

It has long been known that fetal rickets can occur in humans when the mothers have osteomalacia [1,2]. This evidence that placental transfer of vitamin D and/or its biologically active metabolites can take place from mother to fetus is supported by both human [3,4] and animal data, for example, in sheep [5], calf [6], and rat [7]. It is generally accepted that the concentrations of 25-hydroxyvitamin D (25OHD) in fetal plasma are lower than the corresponding levels in maternal plasma in a wide variety of species, including human [8], sheep [9], cow [6], pig [10], and rat [11]. In humans, there is a significant relationship between these two concentrations, which probably signifies dependence of the fetus on its mother for the supply of 25OHD, the principal form of vitamin D in the circulation of both mother and fetus. This is the situation that might have been predicted for a nutrient like vitamin D or its prohormone equivalent, 25OHD.

With the exception of the sheep [12], fetal plasma levels of calcitriol are lower than the corresponding maternal concentrations in human and other species studied [13]. However, a positive correlation between these two concentrations is often difficult to demonstrate [14], as might be expected of a hormone whose secretion is independently regulated in mother and fetus. In the human [4], maternal to fetal transfer of calcitriol has been demonstrated, as it has in the sheep [5]. However, the extent of this transfer is relatively small because the metabolic clearance rate of calcitriol in pregnant sheep was found not to be different from that in non-pregnant animals [15]. Fetal to maternal passage of labeled calcitriol has also been demonstrated in the sheep, in which it was found to localize to the maternal intestinal mucosa [16].

Not more than 1% of vitamin D and its metabolites exists in plasma in the free state, the rest being bound to vitamin D binding protein (DBP). The concentration of this protein must therefore be taken into account when assessing the significance of a change in the circulating level of calcitriol. For example, the concentration of calcitriol in human plasma increases during pregnancy [17], but the plasma concentration of DBP also increases during this period [12,18]. Nevertheless, toward the end of human pregnancy there is an increase in the free calcitriol index [19], and at term the concentration of free calcitriol in plasmas of mother and baby are correlated [20].

As might have been expected from the metabolism of 25OHD, the plasma concentration of $24,25(OH)_2D$ falls during the last quarter of human pregnancy [21] but the fetal levels are generally lower than those in the maternal circulation [9], despite the likelihood of placental transfer of this metabolite. The biological significance of these concentrations of $24,25(OH)_2D$ remains undefined.

B. Placental Production of Calcitriol

Placental production of calcitriol has been clearly demonstrated in the rat [22,22a], rabbit [23], and human [24,25]. 1α-Hydroxylase activity is mostly localized to decidual cells of the human placenta [23]. Its absence was demonstrated in decidual cells harvested from patients with pseudo-vitamin D deficiency [26], showing clearly that the renal and placental enzymes are encoded by the same gene (see Chapter 47). However, in the sheep, bilateral fetal nephrectomy has been shown to reduce the fetal plasma concentration, indicating that the placental contribution to the normal circulating level of calcitriol would seem to be limited and to be insufficient to maintain the normal fetal plasma calcium concentration [27,28].

The control of the placental production of calcitriol is unclear, as it is in the fetal kidney. The fetal plasma concentrations of calcium and phosphate are both relatively high, and that of parathyroid hormone (PTH) is low. That is, the three major determinants of calcitriol production are in an inhibitory mode for the production of calcitriol. It is possible that parathyroid hormone-related protein (PTHrP), produced in the placenta itself, may stimulate the synthesis of the placental 1α-hydroxylase, although this hypothesis remains to be substantiated. The placenta may represent a site for the metabolism of calcitriol as well as for its production, and this may account in part for the greater metabolic clearance rate of calcitriol in the fetal lamb compared with the neonate [29].

C. Maternal Metabolism of Vitamin D during Pregnancy

During pregnancy, maternal calcium homeostasis must be adjusted to supply the additional requirements of the fetus for calcium. This involves increased conversion of 25OHD to $1,25(OH)_2D$ to respond to the tendency toward hypocalcemia and hypophosphatemia that is associated with the transfer of these ions to the fetus, and the ensuing rise in the circulating concentration of PTH. Indeed, it has been confirmed using rats [30] and sheep [29] that the elevated maternal plasma concentration of calcitriol during pregnancy is in fact the result of an increased production rate rather than a decreased rate of metabolic clearance. This increase in the concentration of calcitriol leads to a rise in the abundance of the vitamin D receptor (VDR) in the intestine in the pregnant animal, particularly the rat [31], and to an increase in the efficiency of the intestinal absorption of both calcium and phosphate ions [32–34]. Heaney and Skillman [32] demonstrated a doubling of the maternal intestinal absorption rate by the end of the second trimester of human pregnancy, which persisted until term. The increase in absorption of calcium probably involves an increase in the expression of the gene for the plasma membrane calcium pump in the enterocyte [35]. Also involved is an increase in the concentration of intestinal calbindin-D_{9K} [36]. In the rat, an increase in the intestinal mucosal content of calbindin-D_{9K} has been shown to take place toward the last part of pregnancy [37], when the demands of the fetal skeleton for calcium are rising rapidly.

Depending on the dietary supply of calcium and phosphorus, net mobilization of bone mineral, stimulated by PTH and calcitriol, may be required to meet the needs of the fetus in various species, including the human [38], rat [39], and particularly the sheep because of the relatively high degree of fetal skeletal mineralization in the latter species at birth [40]. In the Middle East, there is a comparatively high incidence of osteomalacia among Bedouin women and low plasma levels of 25OHD in mothers and in cord blood [41]. Although congenital rickets has been reported in association with maternal vitamin D deficiency [42], it is comparatively rare.

Adequate maternal vitamin D stores are not a requirement for normal development of the fetal skeleton in all species. For example, when pregnant rats were maintained on a long-term vitamin D-deficient diet, sufficient to reduce their plasma concentrations of calcitriol to undetectable levels, the offspring had normal skeletal mineralization at birth [43] and the fetuses were able to maintain normal transplacental gradients of calcium and phosphate [44]. This implies that the pregnant rat can deplete its skeleton sufficiently to provide the additional calcium required by the fetuses and transport it across the placenta without the involvement of calcitriol. A possible role of the calcitriol produced by the fetal kidney in this mineral transfer has yet to be addressed in detail. However, the fertility of the vitamin D-depleted rats was impaired [45]. This may reflect the wide distribution of VDR in ovarian follicles in rats, where it may mediate the action of calcitriol in this tissue.

D. Roles of Calcitriol in the Fetus

In the fetal rat, the concentration of calcitriol-dependent calbindin-D_{9K} in the intestinal mucosa increases toward the end of gestation, the concentration being highest in the duodenum and decreasing toward the ileocecal junction, which corresponds to the adult pattern [47]. Presumably, this prepares the fetus for the critical neonatal period during which the newborn changes from placental transfer to intestinal absorption of calcium. Calcitriol may also aid in the reabsorption of calcium, derived from urinary and tracheal fluid, which is excreted into the amnion and recycled as the fetus ingests amniotic fluid.

Calcitriol and its receptor also play an important part in cellular growth and differentiation [48,49] (see also Chapter 64). The VDR has been shown to be present in specific areas of the kidney of the developing mouse and rat early in gestation [50], and labeled calcitriol has been shown to be concentrated in the nuclei of the distal renal tubules in chick embryos, indicating that these cells are a probable target for calcitriol [51]. The VDR has also been shown to appear very early in the development of the fetal rat skeleton, suggesting that, in concert with calcitriol, it may play a role in the differentiation of mesenchymal precursors into bone and related tissue [52]. In addition, subnormal concentrations of calcitriol in human fetal blood are associated with small size for gestational age and low content of bone mineral [106]. Similarly, neonatal rats from vitamin D-deficient mothers show reduced growth of longitudinal bones [107].

III. CALCITRIOL AND THE FETOPLACENTAL UNIT

Despite markedly different placentation from that of humans, the sheep has nevertheless been used for most of the *in vivo* physiological studies on the fetus and on the placental transfer of nutrients to the fetus. This is also true of placental calcium transport and the effect of calcitriol on it. The ruminant placenta, of which the sheep is a prime example, consists of two distinct parts: a relatively flat, loose apposition of uterine epithelium to fetal trophectodermal epithelium in the interplacentomal regions, and a greater area of convoluted, intimate apposition of the two epithelia with villous microarchitecture in the 70 or so placentomes.

Calcitriol can probably stimulate the maternal to fetal transfer of calcium ions by one or more of the following processes: (1) increasing maternal calcemia, (2) stimulating the uptake of calcium into the trophectodermal cells by a nongenomic mechanism, (3) increasing the synthesis of calbindin-D_{9K} in the trophectodermal cells, and (4) stimulating the calcium pump in the basal (fetal-facing) plasma membrane of the trophoblasts.

A. Control of Calcium Concentration in Uterine Secretions (Uterine Milk)

A calcium ion electrode, implanted within the lumen of a uterine horn, has been used to demonstrate that the normal calcium concentration in the secretions of the endometrium and uterine glands in the nonpregnant ewe is similar to that in plasma but can be increased severalfold by the administration of progesterone [53]. This finding has been confirmed in pregnant ewes [54] using either the ligated uterine horn technique [55] or a microdialysis probe [56]. Furthermore, using the latter technique, it was shown that throughout pregnancy the calcium concentration in the uterine secretions was increased in response to induced maternal hypercalcemia [54].

It has also been demonstrated that the uptake of calcium across the apical surface of the ovine trophectodermal epithelium is increased as the calcium concentration to which that surface is exposed is increased [57] (see Section III.B). Similarly, with 2- to 4-day cultured human syncytiotrophoblast-like cells, Bax *et al.* [58] demonstrated a marked increase in cytosolic ionized calcium concentration in response to elevated extracellular calcium ion concentration.

It was earlier shown in both pregnant sheep [59] and guinea pigs [60] that the administration of 1α-hydroxycholecalciferol (1αOHD, a biologically active analog of calcitriol) to the sheep and calcitriol to the guinea pigs for 12 and 10 days, respectively, led to increases in the transfer of calcium to the fetuses. However, in both studies the plasma calcium concentrations in the pregnant animals were significantly increased by administration of the sterol. Thus, it is possible that at least part of this stimulatory effect of calcitriol on placental calcium transport may be accounted for by an increase in the calcium concentration in the uterine secretions and thus increased uptake of calcium ions by the trophectodermal cells in apposition to the uterine epithelium.

In pregnant rats, maternal hypocalcemia caused by thyroparathyroidectomy (TPTX) induced fetal hypocalcemia 6–9 days later, during the last 3 days of gestation. These changes in both maternal and fetal calcemia could be prevented by treating the mother with calcitriol after TPTX [61]. These results indicate the possibility of a relationship between maternal calcemia and the placental transfer of calcium to the fetus.

B. Calcium Uptake by Trophectodermal Cells of the Chorion

An *in vitro* technique has been developed with ovine tissue in which the uptake of $^{45}Ca^{2+}$ into disks of the placental chorion is measured. The maternal-facing surface, lined with trophectodermal cells, was exposed to incubation fluid containing the $^{45}Ca^{2+}$ under controlled conditions. It was demonstrated that the uptake of $^{45}Ca^{2+}$ was maximal after 15 min of incubation and was increased in association with a corresponding increase in the calcium concentration in the incubation fluid [57]. This entry of calcium could be reduced by the addition of calcium L-type channel blockers (e.g., 50 μM diltiazem).

Thus, prolonged maternal hypercalcemia leading to an increase in the calcium concentration in uterine secretions would be expected to lead to an enhanced uptake of calcium by the trophectodermal cells of the ovine chorion. Further work has demonstrated that this uptake can be rapidly increased by the addition of 600 pM calcitriol to the incubation fluid in contact with the maternal-facing membrane of the chorion from fetal lambs bilaterally nephrectomized 6–21 days before [62]. This effect could be seen within 20–30 min of the addition of the calcitriol and may represent an example of a nongenomic action of calcitriol [63] through the triggering of calcium influx via voltage-gated Ca^{2+} channels, as observed in isolated chick duodenum [64]. Furthermore, it has been demonstrated that the uptake of calcium ions by chorionic disks prepared from the placentas of twins, one member of which had been previously nephrectomized, was less than that by disks from their intact siblings [62].

Similarly, in ovine placental perfusion studies, using the method of Weatherley *et al.* [65], simultaneous perfusions of the placentas of twins, one member of each of which had been previously nephrectomized to reduce the circulating level of calcitriol in that fetus, showed that the placenta of the intact twins transported calcium more effectively than that of their nephrectomized siblings [66]. In these perfusion experiments, the putative action of calcitriol could be on events following entry of calcium into the trophectoderm (see Sections III.C, III.D) as well as on this process alone. However, the addition of calcitriol to the medium used to perfuse the placenta of a previously nephrectomized fetal lamb resulted in a rapid increase (1–2 hr) in the rate of placental transfer of calcium from the ewe. In the sheep, VDR have been detected in cotyledonary tissue from the placenta, and nuclear targeting of labeled calcitriol has been demonstrated [67]. This supports the above finding that calcitriol stimulates placental calcium transfer in the sheep.

Perfusion studies of the placentas of fetal rats, decapitated *in utero* 2 days earlier to induce fetal hypocalcemia, resulted in a reduction in the plasma clearance of $^{45}Ca^{2+}$ by the placenta. This could be rapidly although only partially restored by administration of calcitriol to the perfusion fluid [68]. It is concluded that calcitriol might play only a facilitatory role in placental calcium transport in the rat.

Transport of calcium across the microvillus membrane of the human placenta is also time and concentration dependent. The concentration dependence data for calcium transport can be accounted for by a Michaelis–Menten equation having two saturable sites and diffusion [69]. However, in contrast to the findings with ovine chorion, calcium entry was unaffected by the addition of calcium channel blockers. No studies have been made of the effects of added calcitriol on the entry of calcium using either human [69] or rat [70] placental tissue. If such studies were carried out, however, they would probably not lead to a positive finding, since previous nephrectomy of the fetus in sheep is necessary before enhanced uptake of calcium by the chorion in response to added calcitriol can be demonstrated [62]. Presumably, sufficient unoccupied VDR must be present before an effect of added calcitriol can be manifested.

It should be noted that in human hereditary vitamin D resistant rickets (vitamin D-dependent rickets type II; see Chapter 48), an autosomal recessive disease caused by mutations in the VDR, the affected baby is born free of any bone disease, and rickets develops only after birth [71]. Thus, as in the rat, vitamin D action in humans may not be essential for the placental transport of sufficient calcium to the fetus for adequate skeletal development as long as the mother is normocalcemic during the latter half of pregnancy.

C. Placental Synthesis of Calbindin-D_{9K}

In addition to uptake of calcium across the microvillus membrane of the human syncytiotrophoblast, there is evidence for a calcium binding process that shows saturation at micromolar concentrations [69]. This binding process is likely to play a vital role in the maintenance of a low intracellular concentration of calcium and may involve the human placental calcium binding protein described by Tuan [72].

Until relatively recently it was unclear whether calbindin-D_{9K} was present in human intestine. Human upper small intestinal mucosa has been shown to contain the mRNA for calbindin-D_{9K}; it could not be found in human placenta, however [73]. The mechanism by which calcium is carried across the human syncytiotrophoblast remains unknown.

In the sheep, goat, and cow, calbindin-D_{9K} has been demonstrated to be present in the uninucleate cells of the fetal trophectodermal epithelium in the interplacentomal regions but is absent in the maternal interplacentomal uterine epithelium [74]. Further studies also showed calbindin-D_{9K} to be present in the epithelial cells of the uterine glands, suggestive of a role in transcellular calcium transport [108]. It has not yet been unequivocally proved that this ovine placental calbindin-D_{9K} is vitamin D-dependent.

In the rat and mouse, although vitamin D-dependent calbindin-D_{9K} is found, as expected, in the duodenal mucosa in high concentration, the protein is also found in uterine myometrium, where its abundance is controlled by the stimulatory action of estrogen and the inhibitory action of progesterone [75,76]. It may be involved in the control of uterine contraction in these rodents [77]. Calbindin-D_{9K} is also present in the intraplacental yolk sac in the rat, and its levels rise in late gestation when fetal accumulation of calcium is maximal, suggesting a role for this calbindin in late gestation [77–79]. Studies by Bruns *et al.* [80] indicate that calcitriol can increase the synthesis of calbindin-D_{9K} in the cultured yolk sac of the mouse. However, the concentration of calbindin-D_{9K} in mouse placenta and yolk sac *in vivo* are not increased by the administration of calcitriol to the mother [75], possibly because there is only minimal transport of calcitriol from the mother to the fetal rat [81]. Perhaps a response to added calcitriol might depend on the action of estrogen to induce the development of VDR, which have been shown to be present in the yolk sac and placenta of the rat [82].

As in duodenal cells, the mechanism by which calbindin-D_{9K} is involved in calcium transport across the trophoblast is unknown, (see Chapters 14 and 16). During this process the protein may protect the cell from a dangerous increase in the intracellular calcium concentration. A popular hypothesis with some experimental support [83] is that each molecule of calbindin-D_{9K} binds two calcium ions near the apical border of the cell and increases the flux of this bound calcium across the cell by a process of facilitated diffusion. At the basolateral border, the calcium ions are removed from the calbindin by virtue of the superior binding affinity (10^{-9} *M* relative to 10^{-7} *M*) of the Ca^{2+}-adenosine triphosphatase, the activity of which regulates the Ca^{2+} pump in that plasma membrane. By analogy with the rat enterocyte [84], it is possible that calbindin-D_{9K} might also stimulate the placental calcium pump itself as well as the transport of calcium ions to it. The successful genetic ablation of this seemingly important gene should reveal much regarding its essentiality for calcium transport.

D. Stimulation of the Placental Calcium Pump

Extrusion of calcium ions from the basolateral border of the trophoblast into the fetal extracellular fluid is a process that requires energy. A high efficiency Ca^{2+} pump has been located in the basal (fetal-facing) layers of the human and rat placenta [85,86]. This mechanism is dependent on the ambient concentration of magnesium ions and has a calcium ion sensitivity in the nanomolar range, which is somewhat similar to the calcium ion pump in the erythrocyte [87]. So far, effects of calcitriol on the activity of this calcium pump in the placenta have not been investigated. However, in the chicken duodenal mucosa, it has been found that the gene expression of the plasma membrane calcium pump is increased by exogenous calcitriol or by a rise in endogenous calcitriol induced by either a low calcium or a low phosphorus diet [35,88]. These effects take some hours to become apparent; thus, if this result is demonstrated in the placenta, most of the rapid stimulatory effects of calcitriol on placental calcium transport that have been described above are unlikely to stem from increased synthesis of the placental calcium pump. It seems probable, although not yet proved unequivocally, that the placental calcium pump is activated by PTHrP, the active moiety being a sequence in the mid-region [89,90].

IV. VITAMIN D AND LACTATION

A. Skeletal Changes during Lactation

It has been established that PTH is not required for normal milk secretion or milk composition or indeed lactation-associated bone loss in normocalcemic rats [91]. However, the same group showed that thyroparathyroidectomized lactating rats fed a high calcium, normal phosphorus diet, to maintain normocalcemia, underwent significant loss of bone in association with lactation. However, this bone loss was not significantly different between the experimental group and their intact (non TPTX) lactating controls. Calcitriol was not likely to have been responsible for this difference. Although not measured here, the plasma concentrations of calcitriol in parathyroidectomized lactating rats, fed a diet adequate in calcium, would be expected to be depressed rather than elevated, as is normally found in lactation [92,93]. Moreover, experiments with vitamin D-deficient rats have shown that the efficiency of intestinal absorption of calcium is still increased during lactation despite significantly lower levels of calcitriol in plasma and intestinal mucosa [94]. It is thought that PTHrP secreted from the mammary gland into the blood, as well as the milk [95], rather than calcitriol, is probably the factor responsible for the similar loss of bone mineral in the two groups. PTHrP shares with PTH the same receptor in bone [96].

The normal increase in the circulating concentration of calcitriol that occurs during lactation leads to raised efficiency of calcium and phosphate absorption from the digestive tract, but this is not usually sufficient to compensate for the drain of lactation. Thus, in the cow, rat, and human, some skeletal loss of mineral can be demonstrated during lactation [39,97,109]. The extent depends on the species, milk yield, age, and breed.

B. Vitamin D and Milk Composition

The vitamin D content of bovine milk is comparatively low. It is not substantially increased by the ingestion of natural foodstuffs but is more affected by the exposure of the animal to sunlight. Thus, the vitamin D activity in bovine milk in the summer can be nine times higher than in the winter [98]. As in the cow, massive oral doses of vitamin D administered to women are required to increase significantly the vitamin D content of the milk [99], as measured by its biological activity.

The control of calcium secretion into milk is still unclear. Earlier reports using rats indicated that calcitriol localizes in normal mammary tissue of lactating rats where VDR has been identified [100]. The total calcium concentration in the milk of lactating vitamin D-deficient rats is increased 6 hr after the intraperitoneal administration of calcitriol. In fact, this increase took place before there was a significant increase in the serum calcium concentration. [101]. However, it now seems clear that in ruminant animals calcitriol supplementation does not increase the calcium concentration in milk; indeed, in rats, vitamin D has been shown not to be essential for the maintenance of a normal calcium concentration in the milk [102]. The administration of calcitriol as a glycoside in the dried leaves of *Solanum glaucophyllum* has been shown to increase the calcium concentration in the colostrum of cows but not in the subsequent milk [103]. Presumably, this is a reflection of the leaky junctions between the secretory cells during the secretion of colostrum and the hypercalcemia caused by the ingestion of the calcitriol glycoside. Similarly, calcitriol was found to increase the uptake of calcium by murine mammary explants, cultured in the presence of lactogenic hormones at the initiation of lactogenesis. This was a rapid change that was unaffected by the presence of cycloheximide and thus may be mediated by a nongenomic mechanism [104]. In addition to this effect on the calcium content of mammary secretions before lactation has been fully established, Mezzetti *et al.* [105] showed that calcitriol also augmented milk protein synthesis induced by lactogenic hormones.

V. CONCLUSIONS

Vitamin D through its most active metabolite calcitriol, plays an important role to maintain calcium homeostasis in most species during pregnancy and lactation. Its effects are primarily mediated on intestinal calcium absorption and bone resorption. The importance of calcitriol in the fetus also depends on the species. For example, in the rat it is not an essential factor for fetal calcium homeostasis, whereas in the sheep there is significant impairment in fetal calcium homeostasis if the circulating level of calcitriol is not maintained.

The placental transfer of calcium from mother to fetus can be stimulated by calcitriol at four possible levels. At the first stage, the maintenance of an adequate plasma calcium concentration in the mother, regulated by calcitriol, is important since the uptake by the maternal-facing membrane of the trophoblast cells is dependent on the concentration of calcium presented to it. The second stage has been demonstrated in the sheep, in which this calcium uptake is increased by calcitriol. In some species (e.g., the sheep), the third stage is represented by vitamin D-dependent calbindin-D_{9K}, which may be involved in the transport of the absorbed calcium to the fetal-facing membrane of the trophoblast. In the fourth stage, the calcium is transferred to a high affinity calcium pump which transfers calcium to the exterior of the cell from which it diffuses into the fetal circulation. By analogy with a comparable system in the duodenal mucosa, it is possible that calcitriol may be involved in stimulation of the synthesis of this calcium pump. Apart from calcium homeostasis, calcitriol plays important roles in cell differentiation and growth in the fetus. Finally, although calcitriol may be important in stimulating the uptake of calcium by developing mammary cells, it appears to play a small part in the determination of milk secretion or composition once lactation has been established.

References

1. Maxwell JP, Miles LM 1925 Osteomalacia in China. J Obstet Gynecol Br Empire **32**:433–438.
2. Ford JA, Davidson DC, McIntosh WB, Fyfe WM, Dunningan MG 1973 Neonatal rickets in Asian immigrant population. Br Med J **3**:211–212.
3. Hillman LS, Haddad JG 1974 Human perinatal vitamin D metabolism. 1,25-Hydroxyvitamin D in maternal and cord blood. J Pediatr **84**:742–749.
4. Ron M, Levitz M, Chuba M, Dancis J 1984 Transfer of 25-hydroxyvitamin D_3 and 1,25-dihydroxyvitamin D_3 across the perfused human placenta. Am J Obstet Gynecol **148**:370–374.
5. Ross R, Care AD, Taylor CM, Pelc B, Sommerville BA 1979 The transplacental movement of metabolites of vitamin D in the sheep. In:Norman AW (ed) Vitamin D: Basic Research and Its Clinical Application. de Gruyter, Berlin, pp. 341–344.
6. Goff JP, Horst RL, Littledike ET 1982 Effect of the maternal vitamin D status at parturition on the vitamin D status of the neonatal calf. J Nutr **112**:1387–1393.
7. Haddad JG, Boisseau V, Avioli LV 1971 Placental transfer of vitamin D_3 and 25-hydroxycholecalciferol in the rat. J Lab Clin Med **77**:908–915.
8. Weisman Y, Occhipinti M, Knox G, Reiter E, Root A 1978 Concentrations of 24,25-dihydroxyvitamin D and 25-hydroxyvitamin D in paired maternal–cord sera. Am J Obstet Gynecol **130**:704–707.
9. Paulson SK, DeLuca HF, Battaglia F 1987 Plasma levels of vitamin D metabolites in fetal and pregnant ewes. Proc Soc Exp Biol Med **185**:267–271.
10. Care AD, Ross R, Pickard DW, Weatherley AJ, Garel JM, Manning RM, Allgrove J, Papapoulos S, O'Riordan JLH 1982 Calcium homeostasis in the fetal pig. J Dev Physiol **4**:85–106.
11. Gertner JM, Glassman MS, Coustan DR, Goodman DBP 1980 Feomaternal vitamin D relationships at term. J Pediatr **97**:637–640.
12. Abbas SK, Care AD, Van Baelen, Bouillon R 1987 Plasma vitamin D-binding protein and free 1,25-dihydroxyvitamin D_3 index in pregnant ewes and their fetuses in the last month of gestation. J Endocrinol **115**:7–12.
13. Bouillon R, Van Assche FA, Van Baelen H, Heyns W, De Moor P 1981 Influence of the vitamin D-binding protein on the serum concentration of 1,25-dihydroxyvitamin D_3. J Clin Invest **67**:589–596.
14. Steichen JJ, Tsang R, Gratton T, Hamstra A, DeLuca HF 1980 Vitamin D homeostasis in the perinatal period: 1,25-Dihydroxyvitamin D in maternal, cord and neonatal blood. N Engl J Med **302**:315–319.
15. Ross R, Dorsey J, Ellis K 1990 Progressive increases in 1,25-dihydroxyvitamin D_3 production rate in multiple ovine pregnancies are independent of changes in the metabolic clearance rate. Pediatr Res 27:192A.
16. Ross R, Florer J, Chen M, Halbert K 1987 Uptake of intrafetally administered ^{3}H-1,25 dihydroxyvitamin D_3 by the maternal intestine. Pediatr Res **21**:220A.
17. Kumar R, Cohen WR, Silva P, Epstein FH 1979 Elevated 1,25-dihydroxyvitamin D plasma levels in normal human pregnancy and lactation. J Clin Invest **63**:342–344.
18. Haddad JG, Walgate J 1976 Radioimmunoassay of the binding protein for vitamin D and its metabolites in human serum: Concentrations in normal subjects and patients with disorders of mineral metabolism. J Clin Invest **58**:1217–1222.
19. Bikle DD, Gee E, Halloran BP, Haddad JG 1984 Free 1,25-dihydroxyvitamin D levels in serum from normal subjects, pregnant subjects and subjects with liver disease. J Clin Invest **74**:1966–1971.
20. Bouillon R, Van Assche FA, Van Baelen H 1980 Vitamin D metabolites and their transport protein in maternal and cord serum. Calcif Tissue Int **Suppl.**:31–32.
21. Reiter EO, Braustein GD, Vargas A, Root AW 1979 Changes in 25-hydroxyvitamin D and 24,25-dihydroxyvitamin D during pregnancy. Am J Obstet Gynecol **135**:227–229.

22. Weisman Y, Vargas A, Duckett G, Reiter E, Root AW 1978 Synthesis of 1,25-dihydroxyvitamin D in the nephrectomized pregnant rat. Endocrinology **103**:1992–1996.
22a. Tanaka Y, Halloran B, Schnoes HK, DeLuca HF 1979 In vitro production of 1,25-dihydroxyvitamin D_3 by rat placental tissue. Proc Natl Acad Sci USA **76**:5033–5035.
23. Delvin EE, Gilbert M, Pere MC, Garel JM 1988 In vivo metabolism of calcitriol in the pregnant rabbit doe. J Dev Physiol **10**:451–459.
24. Weisman Y, Harell A, Edelstein S, David M, Spirer Z, Golander A 1979 1,25-Dihydroxyvitamin D_3 and 24,25-dihydroxyvitamin D in vitro synthesis by human decidua and placenta. Nature **281**:317–319.
25. Whitsett JA, Ho M, Tsang RC, Norman EJ, Adams KG 1981 Synthesis of 1,25-dihydroxyvitamin D_3 by human placenta in vitro. J Clin Endocrinol Metab **53**:484–488.
26. Glorieux FH, Arabian A, Delvin A, Delvin EE 1995 Pseudovitamin D deficiency: absence of 25 hydroxyvitamin D 1α hydroxylase activity in human placental decidual cells. J Clin Endoc Metab **80**:2255–2257.
27. Ross R, Care AD, Robinson JS, Pickard DW, Weatherley AJ 1980 Perinatal 1,25-dihydroxycholecalciferol in the sheep and its role in the maintenance of the transplacental calcium gradient. J Endocrinol **87**:17P–18P.
28. Moore ES, Langman CB, Favus MJ, Coe FL 1985 Roleof fetal 1,25-dihydroxyvitamin D production in intrauterine phosphorus and calcium homeostasis. Pediatr Res **19**:566–569.
29. Ross R, Dorsey J 1991 Postnatal changes in plasma 1,25-dihydroxyvitamin D_3 in sheep: Role of altered clearance. Am J Physiol **261**:E635–641.
30. Paulson SK, Ford KK, Langman CB 1990 Pregnancy does not alter the metabolic clearance of 1,25-dihydroxyvitamin D in rats. Am J Physiol **258**:E158–E162.
31. Horst RL, Reinhardt TA 1988 Changes in intestinal 1,25-dihydroxyvitamin D receptors during aging, gestation and pregnancy in rats (abstract). In: Norman AW, Schaeffer K, Grigoleit H-G, von Herrath D(eds) Vitamin D: Molecular, Cellular and Clinical Endocrinology. de Gruyter, New York, p. 47.
32. Heany RD, Skillman TG 1971 Calcium metabolism in normal human pregnancy. J Clin Endocrinol Metab **33**:661–670.
33. Swaminathan R, Sommerville BS, Care AD 1977 The effect of dietary calcium on the activity of 25-hydroxycholecalciferol-1-hydroxylase and calcium absorption in vitamin D-replete chicks. Br J Nutr **38**:47–54.
34. Friedlander EJ, Henry HL, Norman AW 1977 Studies on the mode of action of cholecalciferol. J Biol Chem **252**:8677–8683.
35. Cai Q, Chandler JS, Wasserman RH, Kumar R, Penniston JT 1993 Vitamin D and adaptation to dietary calcium and phosphate deficiencies increase intestinal plasma membrane calcium pump gene expression. Proc Natl Acad Sci USA **90**:1345–1349.
36. Wasserman RH 1992 In: Bronner F, Peterlik M (eds) Extra- and Intracellular Calcium and Phosphate Regulation. CRC Press, Boca Raton, Florida, Chap. 5.
37. Bruns ME, Fausto A, Avioli LV 1978 Placental calcium binding protein in rats. Apparent identity with vitamin D-dependent calcium binding protein from rat intestine. J Biol Chem **253**:3186–3190.
38. Lamke B, Brundin J, Moberg P 1977 Changes in bone mineral content during pregnancy and lactation. Acta Obstet Gynecol Scand **56**:217–219.
39. Miller SC, Halloran BP, DeLuca HF 1982 Role of vitamin D in maternal skeletal changes during pregnancy and lactation; a histomorphometric study. Calcif Tissue Int **34**:245–252.
40. Braithwaite GD, Glascock RF, Riazuddin SH 1970 Calcium metabolism in pregnant ewes. Br J Nutr **24**:661–670.
41. Biale Y, Shany S, Levi M, Shainkin-Kestenbaum R, Berlyne GM 1979 25-Hydroxycholecalciferol levels in Bedouin women in labor and in cord blood of their infants. Am J Clin Nutr **32**:2380–2382.
42. Moncrieff M, Fadahunsi TO 1974 Congenital rickets due to maternal vitamin D deficiency. Arch Dis Child **49**:810–811.
43. Halloran BP, DeLuca HF 1981 Effect of vitamin D deficiency on skeletal development during early growth in the rat. Arch Biochem Biophys **209**:7–14.
44. Brommage R, DeLuca HF 1984 Placental transport of calcium and phosphorus is not regulated by vitamin D. Am J Physiol **246**:F526–F529.
45. Halloran BP, DeLuca HF 1979 Vitamin D deficiency and reproduction in rats. Science **204**:73–74.
46. Johnson JA, Grande JP, Roche PC, Kumar R 1966 Immunohistochemical detection and distribution of the 1,25-dihydroxyvitamin D_3 receptor in rat reproductive tissues. Histochem Cell Biol **105**:7–15.
47. Delorme AC, Marche PG, Arel JM 1979 Vitamin D-dependent calcium-binding protein changes during gestation, prenatal and postnatal development in rats. J Dev Physiol **1**:181–194.
48. Itin PH, Pittelkon MR, Kumar R 1994 Effects of vitamin D metabolites on proliferation and differentiation of cultured human epidermal keratinocytes grown in serum-free or defined culture medium. Endocrinology **135**:1793–1798.
49. Koh E, Morimoto S, Fukuo K, Itoh K, Hironaka T, Shiraishi T, Onishi T, Kumahara Y 1988 1,25-Dihydroxyvitamin D_3 binds specifically to rat vascular smooth muscle cells and stimulates their proliferation in vitro. Life Sci **42**:215–223.
50. Johnson JA, Grande JP, Roche PC, Sweeney WE, Avner ED, Kumar R 1995 1α,25-Dihydroxyvitamin D_3 receptor ontogenesis in fetal renal development. Am J Physiol **269**:F419–F428.
51. Narbaitz R, Stumpf WE, Sar M, DeLuca HF 1982 The distal nephron in the chick embryo as a target tissue for 1α,25-dihydroxycholecalciferol. Acta Anat **112**:208–216.
52. Johnson JA, Grande JP, Roche PC, Kumar R 1996. Ontogeny of the 1,25-dihydroxyvitamin D_3 receptor in fetal rat bone. J Bone Miner Res **11**:1–6.
53. Ward JPT, Watson PF, Noakes DE 1989 Chronic in situ monitoring of the free calcium ion concentration in the uterine tubes and horn of the sheep. Comp Biochem Physiol **94A**:765–769.
54. Care AD, Abbas SK, Jones GV, Robinson J, Wooding FBP 1996 Studies of the calcium ion concentration in ovine uterine secretions. J Physiol **494**:30P–31P.
55. Bazer FW, Roberts RM, Basha SMM, Zavy MT, Caton D, Barron DH 1979 Method for obtaining ovine uterine secretions from unilaterally pregnant ewes. J Anim Sci **49**:1522–1527.
56. Nordenvall M, Ulmsten U, Ungerstedt U 1989 Influence of progesterone on the sodium and potassium concentrations of rat uterine fluid investigated by microdialysis. Gynecol Obstet Invest **28**:73–77.
57. Jones GV, Taylor KR, Morgan G, Wooding FRC, Care AD 1996 Aspects of calcium transport by ovine placenta: Studies based on the inter-placentomal region of the chorion. Placenta in press.
58. Bax C, Bax B, Bain M, Zaidi M 1994 Ca^{2+} channels in human term trophoblast cells in vitro. A study using the Ca^{2+}-sensitive dye Fura 2. Trophoblast Res **8**:573–580.
59. Durand D, Braithwaite GD, Barlet JP 1983 The effect of 1α-hydroxycholecalciferol on the placental transfer of calcium and phosphate in sheep. Br J Nutr **49**:475–480.
60. Durand D, Barlet JP, Braithwaite GD 1983 The influence of

1,25-dihydroxycholecalciferol on the mineral content of foetal guinea pigs. Reprod Nutr Dev **23**(2A):235–244.
61. Garel JM, Gilbert M, Besuard P 1981 Fetal growth and 1,25-dihydroxyvitamin D_3 injections into thyroparathyroidectomized pregnant rats. Reprod Nutr Dev **21**(6A):961–968.
62. Jones GV, Care AD, Morgan G, Wooding FBP 1996 Calcitriol stimulates calcium uptake by ovine chorion. J Endocrinol **148**(Suppl.):P179.
63. Walters MR 1992 Newly identified actions of the vitamin D endocrine system. Endocr Rev **13**:719–764.
64. De Boland AR, Norman AW 1990 Transport of calcium across the dually perfused placenta of the rat. J Physiol **420**:295–312.
65. Weatherley AJ, Ross R, Pickard DW, Care AD 1983 The transfer of calcium during perfusion of the placenta in intact and thyroparathyroidectomized sheep. Placenta **4**:271–278.
66. Care AD 1991 The placental transfer of calcium. J Dev Physiol **15**:303–307.
67. Ross R, Halbert K, Tsang RC 1989 Determination of the production and metabolic clearance rates of 1,25-dihydroxyvitamin D_3 in the pregnant sheep and its chronically catheterized fetus by primed infusion technique. Pediatr Res **26**:633–638.
68. Robinson NR, Sibley CP, Mughal MZ, Boyd RDH 1989 Fetal control of calcium transport across the rat placenta. Pediatr Res **26**:109–115.
69. Kamath SG, Kelley LK, Friedman AF, Smith CH 1992 Transport and binding in calcium uptake by microvillous membrane of human placenta. Am J Physiol **262**:C789–C794.
70. Stulc J, Stulcova B, Svihovec J 1990 Transport of calcium across the dually perfused placenta of the rat. J Physiol **420**:295–312.
71. Hochberg Z, Weisman Y 1995 Calcitriol-resistant rickets due to vitamin D receptor defects. Trends Endocrinol Metab **6**:216–220.
72. Tuan RS 1985 Ca^{2+}-binding protein of the human placenta. Biochem J **227**:317–326.
73. Howard A, Legon S, Spurr NK, Walters JRF 1992 Molecular cloning and chromosomal assignment of human calbindin-D_{9k}. Biochem Biophys Res Commun **185**:663–669.
74. Wooding FBP, Bruns ME 1991 Distribution of 9kD calcium binding protein in ruminant placenta. J Anat **176**:249.
75. Delorme AC, Dana JL, Acker MG, Ripoche MA, Mattieu CL 1983 In rat uterus 17β-estadiol stimulates a calcium binding protein similar to duodenal vitamin D-dependent calcium binding protein. Endocrinology **113**:1340–1347.
76. Bruns ME, Overpeck JG, Smith GC, Hirsch GN, Mills SE, Bruns DE 1988 Vitamin D-dependent calcium binding protein in rat uterus: Differential effects of estrogen, tamoxifen, progesterone and pregnancy on accumulation and cellular localization. Endocrinology **122**:2371–2378.
77. Mattieu CL, Burnett SH, Mills SE, Overpeck JG, Bruns DE, Gruns ME 1989 Gestational changes in calbindin-D_{9k} in rat uterus, yolk sac and placenta: Implications for maternal fetal calcium transport and uterine muscle function. Proc Natl Acad Sci USA **86**:3433–3437.
78. Mattieu CL, Burnett SH, Mills SE, Bruns DE, Bruns ME 1988 Gestational appearance of 9kD calbindin in uterus, yolk sac and placenta: Implications for maternal: fetal calcium transport. J Bone Miner Res **3**(Suppl. 1):S151 (abstract 331).
79. Glazier JD, Atkinson DE, Thornburg KL, Sharpe PT, Edwards D, Boyd RDH, Sibley CP 1992 Gestational changes in Ca^{2+} transport across the rat placenta and mRNA for calbindin9K and Ca^{2+}-ATPase. Am J Physiol **263**:R930–R935.
80. Bruns ME, Kleeman E, Bruns DE 1986 Vitamin D-dependent calcium binding protein in mouse yolk sac: Biochemical and immunochemical properties and responses to 1,25-dihydroxycholecalciferol. J Biol Chem **261**:7485–7490.
81. Noff D, Edelstein S 1978 Vitamin D and its hydroxylated metabolites in the rat. Horm Res **9**:292–300.
82. Pike JW, Gooze LL, Haussler MR 1980 Biochemical evidence for 1,25-dihydroxyvitamin D receptor macromolecules in parathyroid, pancreatic, pituitary and placental tissues. Life Sci **26**:407–414.
83. Feher JJ 1983 Facilitated calcium diffusion by intestinal calcium-binding protein. Am J Physiol **244**:C303-C307.
84. Walters JRF 1989 Calbindin-D9k stimulates the calcium pump in rat enterocyte basolateral membranes. Am J Physiol **256**:G124–G128.
85. Fisher GJ, Kelley LK, Smith CH 1987 ATP-dependent calcium transport across basal plasma membranes of human placental trophoblast. Am J Physiol **252**:C38–C46.
86. Borke JL, Caride A, Verna AK, Kelley LK, Smith CH, Penniston JT, Kumar R 1989 Calcium pump epitopes in placental trophoblast basal plasma membranes. Am J Physiol **257**:C341–C346.
87. Penniston JT 1982 Plasma membrane Ca^{2+}-pumping ATPases. Ann NY Acad Sci **402**:296–303.
88. Wasserman RH, Smith CA, Brindak ME, de Talamoni N, Fullmer CS, Penniston JT, Kumar R 1992 Vitamin D and mineral deficiencies increase the plasma membrane calcium pump of chicken intestine. Gastroenterology **102**:886–894.
89. Abbas SK, Pickard DW, Rodda CP, Heath JA, Hammonds RG, Wood WI, Caple IW, Martin TJ, Care AD 1989 Stimulation of ovine placental calcium transport by purified natural and recombinant parathyroid hormone-related protein (PTHrP) preparations. Q J Exp Physiol **74**:549–552.
90. Care AD, Abbas SK, Pickard DW, Barri M, Drinkhill M, Findlay JBC, White IR, Caple IW 1990 Stimulation of ovine placental transport of calcium and magnesium by mid-molecule fragments of human parathyroid hormone-related protein. Exp Physiol **75**:605–608.
91. Garner SC, Boass A, Toverud SU 1990 Parathyroid hormone is not required for normal milk composition or secretion or lactation-associated bone loss in normocalcemic rats. J Bone Miner Res **5**:69–75.
92. Boass A, Toverud SU, McCain TA, Pike JW, Haussler MR 1977 Elevated serum levels of 1α,25-dihydroxycholecalciferol in lactating rats. Nature **267**:630–632.
93. Halloran BP, Barthell ED, DeLuca HF 1979 Vitamin D metabolism during pregnancy and lactation in the rat. Proc Natl Acad Sci USA **76**:5549–5553.
94. Boass A, Toverud SU, Pike JW, Haussler MR 1981 Calcium metabolism during lactation: Enhanced intestinal calcium absorption in vitamin D-deprived, hypocalcemic rats. Endocrinology **109**:900–905.
95. Ratcliffe WA, Thompson GE, Care AD, Peaker M 1992 Production of parathyroid hormone-related protein by the mammary gland of the goat. J Endocrinol **133**:87–93.
96. Juppner H, Abou-Samra AB, Uneo S, Gu WX, Potts JT, Segre GV 1988 The parathyroid hormone-like peptide associated with humoral hypercalcemia of malignancy and parathyroid hormone bind to the same receptor on the plasma membrane of ROS 17/2.8 cells. J Biol Chem **263**:8557–8560.
97. Atkinson PJ, West RR 1970 Loss of skeletal calcium in lactating women. J Obstet Gynecol **77**:555–560.
98. Bechtel HE, Hoppert CA 1936 A study of the seasonal variation of vitamin D in normal cow's milk. J Nutr **11**:537–549.
99. Polskin LJ, Kramer B, Sobel AE 1945 Secretion of vitamin D in milk of women fed fish oil. J Nutr **30**:451–466.
100. Fry JM, Curnow DH, Gutteridge DH, Retallack RW 1980 Vitamin D in lactation. I. The localization, specific binding and biolog-

ical effect of 1,25-dihydroxyvitamin D_3 in mammary tissue of lactating rats. Life Sci **27**:1255–1260.

101. Fry JM, Curnow DH, Retallack RW, Gutteridge DH 1980 Significance of 1,25 dihydroxyvitamin D "receptors" in normal and malignant breast tissue. Lancet **1**:1308–1309.
102. Toverud SU 1963 In: Wasserman RH (ed) The Transfer of Calcium and Strontium across Biological Membranes. Academic Press, New York, p. 341.
103. Roux R, Davicco MJ, Carrillo BJ, Barlet JP 1979 *Solanum glaucophyllum* in pregnant cows. Effect on colostrum mineral composition and plasma calcium and phosphorus levels in dams and newborn calves. Ann Biol Anim Biochem Biophys **19**(1A):91–101.
104. Mezzetti MG, Monti MG, Pernecco-Casolo L, Piccinini G, Moruzzi MS 1988 1,25-Dihydroxycholecalciferol-dependent calcium uptake by mouse mammary gland in culture. Endocrinology **122**:389–394.
105. Mezzetti MG, Barbiroli B, Oka T 1985 1,25-Dihydroxycholecalciferol in hormonally induced differentiation of mouse mammary gland in culture. In: Norman AW, Schaefer K, Grigoleit HG, von Herrath D (eds) Vitamin D: A Chemical, Biochemical and Clinical Update. de Gruyter, Berlin, p. 233.
106. Namgung R, Tsang RC, Specker BL, Sierra RI, Ho ML 1993 Reduced serum osteocalcin and 1,25-dihydroxyvitamin D concentrations and low bone mineral content in small for gestation age infants: evidence of decreased bone formation rates. J Pediatr **112**:269–275.
107. Miller SC, Halloran BP, DeLuca HF, Jee WSS 1983 Studies on the role of vitamin D in early skeletal development, mineralization and growth in rats. Calcif Tissue Int **33**:455–460.
108. Wooding FBP, Morgan G, Jones GV, Care AD 1996 Calcium transport and the localisation of calbindin-D_{9K} in the ruminant placenta during the second half of pregnancy. Cell Tissue Res **285**:477–489.
109. Ramberg CF, Johnson EK, Fargo RD, Kronfeld DS 1984 Calcium homeostasis in cows, with special reference to parturient hypocalcemia. Am J Physiol **245:**R698–R704.

CHAPTER 29

Immunomodulatory and Cell Differentiation Effects of Vitamin D

MARTIN HEWISON Department of Medicine, Queen Elizabeth Medical Centre, The University of Birmingham, Birmingham, UK

JEFFREY L. H. O'RIORDAN Bone and Mineral Centre, University College, London Medical School, London, UK

I. INTRODUCTION

The 1980s and 1990s have witnessed a considerable expansion in the endocrinological impact of vitamin D. Studies *in vitro* and *in vivo* have highlighted potential therapeutic applications for vitamin D that extend beyond its classic role as a regulator of calcium homeostasis. The active form of vitamin D, 1,25-dihydroxyvitamin D_3 [$1,25(OH_2D_3)$], has been shown to have antiproliferative and immunosuppressive properties, which may be further enhanced by the development of new and more potent vitamin D analogs. The search for nonclassic targets for $1,25(OH)_2D_3$ gained pace following studies in the late 1970s and early 1980s that suggested a role for vitamin D as an immunomodulatory agent. Several clinical reports linking vitamin D with diseases such as tuberculosis and sarcoidosis actually preceded these studies, but a specific interaction between vitamin D and the immune system was only proposed after the following observations were made: (1) ectopic production of $1,25(OH)_2D_3$ by activated macrophages, (2) the presence of high affinity receptors for $1,25(OH)_2D_3$ (vitamin D receptors, VDR) in leukocytes, and (3) antiproliferative and differentiative effects of $1,25(OH)_2D_3$ on monocytes and T cells.

This chapter reviews the principal immunomodulatory effects of vitamin D and provides an introduction to later chapters that detail the clinical importance of the interaction of vitamin D with the immune system. Section II of this chapter outlines key features of the cellular immune system, with subsequent sections detailing specific areas of vitamin D modulation. These include the synthesis of $1,25(OH)_2D_3$ by macrophages, its role in directing hematopoiesis, and the control of T-cell function and proliferation by $1,25(OH)_2D_3$.

II. CELL BIOLOGY OF THE IMMUNE SYSTEM

Key features of the interaction of vitamin D with the immune system are shown in Fig. 1. The schematic representation also outlines fundamental aspects of immune cell biology, with macrophages acting as the central cell type. Macrophages are large, phagocytic cells that exhibit both antibody-independent and antibody-dependent cytotoxic activities. Almost all bacteria will eventually be killed by phagocytic cells, principally via oxygen-dependent pathways that give rise to toxic hydroxyl radicals, superoxide anions, and hydrogen peroxide. Phagocytosis of bacteria is enhanced by specialized receptors present on the macrophage (e.g., Fc receptors and CR3b) that recognize immunoglobulin- or complement-coated microorganisms. Macrophages (together with dendritic cells and Langerhans cells) are also able to process foreign proteins and thus act as antigen presenting cells (APCs). Processed antigen, in conjunction with major histocompatibility complex (MHC), is recognized by T cells; class I MHC is recognized by cytotoxic T cells (T_C or CD8+ cells) and class II MHC by helper T cells (T_H or CD4+ cells). The former are active in lysing viral-infected cells, whereas the latter are active in stimulating antibody (immunoglobulin, Ig) production by B cells. APCs may also interact directly with B cells to produce antibodies, but the presence of an MHC molecule is also required. In humans, the MHC is known as HLA. The genes for the HLA complex are found on chromosome 6 and consists of regions that encode both class I (regions A, B, and C) and class II (region D) MHC.

Another important feature of the immune system is the role of cytokines in directing specific cellular responses. For example, interleukin 1 (IL-1) production by macrophages is crucial to the stimulation of resting T cells in response to infection. Activated T cells are then able to produce IL-2, which further promotes the proliferation of the T-cell pool and expression of other cytokines (such as IL-4 and IL-10) which, in turn, direct the nature of the T-cell response. An important cytokine produced by activated T cells is interferon-γ (IFN-γ), which acts in a suppressive fashion to limit further macrophage activity.

Other cells that are active in the immune system but are not shown in Fig. 1 include natural killer (NK) cells, which are leukocytes capable of recognizing membrane

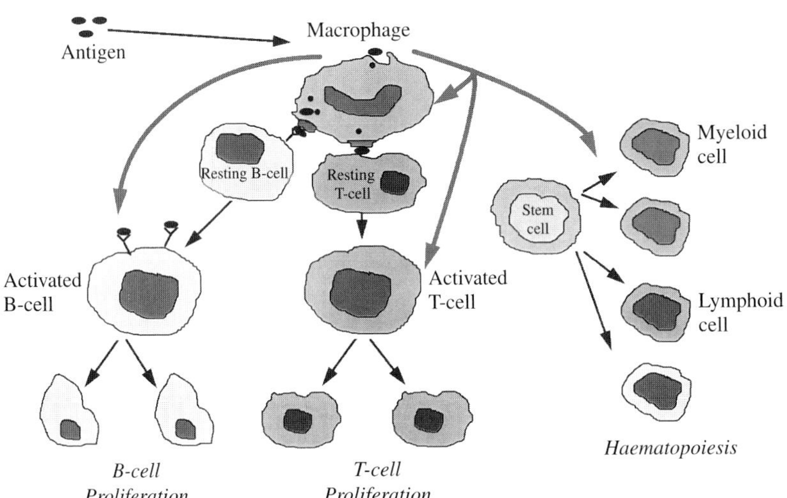

FIGURE 1 Schematic representation of the key features of the interaction of vitamin D with cellular immune responses. Macrophages act as an extrarenal source of $1,25(OH)_2D_3$, which is then able to modulate other cells within the immune system. In particular, $1,25(OH)_2D_3$ is able to maintain macrophage populations by stimulating the differentiation of myeloid stem cells toward a macrophage phenotype (hematopoiesis). Other important targets for locally produced $1,25(OH)_2D_3$ include activated T cells and B cells. $1,25(OH)_2D_3$ decreases T-cell proliferation, promotes specific T-helper cell responses, and down-regulates B-cell functions such as immunoglobulin production. The principal pathways for immune cell development, activation, and interaction are shown as thin arrows, whereas the targets for vitamin D action are shown as heavy arrows.

changes in virally infected cells. NK cells are a target for interferons such as IFN-γ, which is released by virus-infected cells as well as the aforementioned T cells. Immunity to protozoa and worms is thought to be mediated via eosinophils, which are phagocyte-like cells but appear to act by releasing their contents outside the cell in order to combat microorganisms that are too large to phagocytose. Eosinophils counteract the effects of mast cells and basophils, which themselves release allergenic IgE-mediated compounds as part of a mechanism to counteract parasitic infection.

Development of all of the cells described above occurs as a result of the maturation of specific lineages within the hematopoietic system. Pluripotent stem cells undergo initial proliferation and differentiation steps to give rise to myeloid (monocytes, macrophages, neutrophils, etc.) or lymphoid (T-cell, B-cell) progenitors. Further differentiation steps involving cell–cell interactions and cytokine production lead to the maturation of a specific cell type. The putative role of vitamin D as a modulator of myeloid and lymphoid development is detailed later in this chapter and in Fig. 3.

From the brief outline above, it is clear that maintenance of a normal immune response requires a complex series of cell to cell and chemical-mediated interactions. It is therefore likely that both systemic and localized factors will act as immunomodulatory agents. Prime candidates for this role are the steroid hormones which may have both endocrine and paracrine activity. Studies *in vitro* have indicated that vitamin D in particular may influence the activities of a wide variety of immune cell types.

III. EXTRARENAL SYNTHESIS OF $1,25(OH)_2D_3$ BY MACROPHAGES

Clinical studies in the first half of the twentieth century described occasional hypercalcemia in patients with the granulomatous disease sarcoidosis. This disorder occurs in response to an, as yet, unknown antigenic challenge that leads to dysfunction of circulating T cells and overactivity of B cells. The resulting inflammatory disease may occur in many sites including lymph nodes, peripheral blood, as well as tissues such as liver, kidney, eye, and brain. In particular, there appears to be increased T_H cell activity in pulmonary fluid.

A. Sarcoidosis

Support for a link between abnormal vitamin D metabolism and the hypercalcemia of sarcoidosis came from studies which described seasonal variations in this phenomenon that correlate with the amount of sunlight [1]. Papapoulos *et al.* later showed that the episodes of hypercalcemia associated with sarcoidosis were due to abnormally high circulating levels of $1,25(OH)_2D_3$ (Fig. 2) [2]. Raised serum $1,25(OH)_2D_3$ was also observed in an anephric patient with sarcoidosis, suggesting that ectopic production of $1,25(OH)_2D_3$ contributes to the hypercalcemia associated with this disorder [3]. Further studies showed that the 25-hydroxyvitamin D_3 1α-hydroxylase (1α-hydroxylase) was present in sarcoid lymph node homogenates [4], and in macrophages isolated from sarcoid pulmonary fluid [5]. Characterization of this activity indicated that the sarcoid macrophage had a similar affinity (K_m) for substrate (25-hydroxyvitamin D_3, $25OHD_3$) as renal cells [6–8]. This suggested possible competition with the classic endocrine source of $1,25(OH)_2D_3$. Importantly though, addition of exogenous $1,25(OH)_2D_3$ did not appear to inhibit macrophage

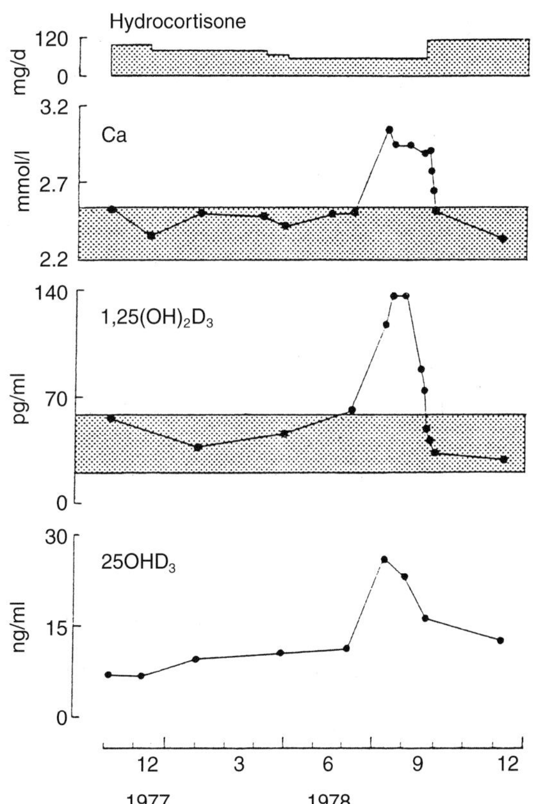

FIGURE 2 Changes in serum $1,25(OH)_2D_3$ and $25OHD_3$ during an episode of hypercalcemia in a patient with sarcoidosis. Serum calcium and vitamin D metabolite levels in the patient were normal during the first half of the year, indicating adequate control of mineral homeostasis. During summer months circulating $25OHD_3$ levels increased, resulting in abnormally high levels of $1,25(OH)_2D_3$ and, as a consequence, hypercalcemia. The patient remained on steroid therapy throughout the study. From Papapoulos SE, *et al.* 1,25Dihydroxycholecalciferol in the pathogenesis of the hypercalcaemia of sarcoidosis, 1, 627–630, © The Lancet Ltd, 1979.

1α-hydroxylase activity nor stimulate an accompanying 24-hydroxylase. This contrasts with the situation in renal epithelial cells in which 24-hydroxylase may compete with 1α-hydroxylase, resulting in the synthesis of less potent vitamin D metabolites. Indeed, clinical studies identified sarcoid patients whose serum $1,25(OH)_2D_3$ levels remained supranormal despite clear hypercalcemia. These findings indicated that metabolic control of 1α-hydroxylase activity in the macrophage was different from that in renal epithelial cells. More recent studies have shown that increased calcium uptake by sarcoid macrophages (using the calcium ionophore, A23187) further stimulates 1α-hydroxylase activity [9]. This appears to be mediated via increased arachidonic acid synthesis, through the 5-lipoxygenase pathway. Table I summarizes some of the differences between renal and immune 1α-hydroxylase, by contrasting the effect of known regulators of the two tissue types.

Stimulators of macrophage immune activity, such as IFN-γ, increased $1,25(OH)_2D_3$ production in sarcoid macrophages and were subsequently shown to induce 1α-hydroxylase activity in macrophages from the peripheral blood of normal subjects [10–12]. The observation that macrophage 1α-hydroxylase was inducible following infection strengthened the proposal that extrarenal $1,25(OH)_2D_3$ production was part of a normal immune response, with overproduction of $1,25(OH)_2D_3$ resulting from an inflammatory disorder such as sarcoidosis. Further studies have shown that synovial fluid from patients with inflammatory rheumatoid arthritis has relatively high levels of 1α-hydroxylase activity [13,14]. This might contribute to the osteoarticular arthritis associated with this disease, although the precise impact of raised localized levels of $1,25(OH)_2D_3$ has yet to be fully defined.

TABLE I Comparison of Effects of Modulators of 1α-Hydroxylase Activity on Renal Cells and Macrophages *In Vitro*

Modulator	Kidney	Macrophage
$1,25(OH)_2D_3$	Inhibition	No effect
PTH	Stimulation	No effect
Dexamethasone	Inhibition	Inhibition
Ketoconazole (cytochrome P450 inhibitor)	Inhibition	Inhibition
Forskolin (cAMP activator)	Stimulation	No effect
Phorbol esters (PKC activator)	Inhibition	Stimulation
IFN-γ	No effect	Stimulation
A23187 (calcium ionophore)	No effect	Stimulation

B. Lymphoid 1α-Hydroxylase Activity

Ectopic nonrenal 1α-hydroxylase activity may not be restricted to activated macrophages. Several groups have investigated the dysfunction of calcium homeostasis and vitamin D metabolism that can occur in patients with Hodgkin's or non-Hodgkin's lymphomas [15,16]. As in the case of sarcoidosis, patients with Hodgkin's disease may have seasonal episodes of hypercalcemia that coincide with elevation of serum $1,25(OH)_2D_3$ levels [17]. This rise in circulating $1,25(OH)_2D_3$ appears to occur in the presence of only moderate increases in serum $25OHD_3$, further emphasizing abnormal control of 1α-hydroxylase activity. Reports of $1,25(OH)_2D_3$ production in lymphoma homogenates suggest that this may be the origin of hypercalcaemia in Hodgkin's patients [18]. The precise cell type associated with lymphoma 1α-hydroxylase remains unclear. Activated macrophages are not commonly associated with lymph nodes, and so it is probable that $1,25(OH)_2D_3$ synthesis is due to a more lymphoid cell; studies using virus-transformed T cells have indicated that lymphocytes may be able to produce $1,25(OH)_2D_3$ [19].

A more comprehensive discussion of the mechanisms controlling extrarenal synthesis of $1,25(OH)_2D_3$ and its clinical impact is presented in Chapter 56. However, it is clear that our current understanding of both renal and immune 1α-hydroxylases is far from complete. It is to be hoped that the eventual cloning of the renal 1α-hydroxylase gene will enable more detailed analysis of its macrophage counterpart. Specifically, it will be necessary to determine whether the enzymes in different tissues result from of identical gene products. In the meantime, it is evident that immune production of $1,25(OH)_2D_3$ may reach significant levels not only in tissue microenvironments, such as lymph nodes, but in the circulation as a whole. This raises the question of identifying targets for $1,25(OH)_2D_3$ within the immune system and defining functional responses to this hormone. The remainder of this chapter gives an overview of our current understanding of these aspects of vitamin D immunobiology.

IV. MACROPHAGE RESPONSES TO $1,25(OH)_2D_3$

Apart from being a source of 1α-hydroxylase activity, macrophages are also potential targets for the immunomodulatory actions of $1,25(OH)_2D_3$. Various groups have reported enhanced antimicrobial action of macrophages following treatment with $1,25(OH)_2D_3$. This includes increased macrophage chemotaxis [20] and increased phagocytic killing of bacteria [21]. The latter effect did not appear to be entirely due to stimulation of a macrophage respiratory burst, although other groups have shown that $1,25(OH)_2D_3$ increases hydrogen peroxide production by macrophages [22]. Macrophage function may also be enhanced by the ability of $1,25(OH)_2D_3$ to induce fusion of macrophages and increase expression of F_c receptors, which are instrumental in recognizing antibody-coated bacteria [23]. The resulting multinucleate cells showed some of the characteristics of osteoclasts (e.g., tartrate-resistant acid phosphatase activity), but the precise relationship between monocyte–macrophage differentiation and the origin of bone-resorbing cells has yet to be fully defined (Chapter 22).

A. Inflammatory Disease

Vitamin D-induced cytotoxic responses may have important clinical consequences. The relationship between ectopic vitamin D metabolism and the granulomatous disorder sarcoidosis has already been described (see Section III,A above and Chapter 56), although the precise function of locally produced $1,25(OH)_2D_3$ remains unclear. Polymerase chain reaction (PCR) studies have identified DNA for *Mycobacterium tuberculosis* in granulomatous tissue from sarcoid patients, suggesting a link between sarcoidosis and tuberculosis (TB) [24]. As in sarcoidosis, hypercalcemia is an occasional feature of patients with TB. Rook and colleagues have shown that $1,25(OH)_2D_3$ increases the killing of *M. tuberculosis* by cultured human monocytes [21], but a question remains as to whether this is the principal function of locally produced $1,25(OH)_2D_3$ in maintaining immunity to TB and sarcoidosis. Indeed, it is perhaps more likely that excessive $1,25(OH)_2D_3$ synthesis by macrophages contributes to the pathophysiology of these diseases by influencing T_H cell function and the release of necrosis-promoting cytokines [25]. This aspect of vitamin D immunomodulation is discussed in more detail in later sections of this chapter.

Various studies have examined the effects of vitamin D on monocytes from patients with acquired immune deficiency syndrome (AIDS). Circulating levels of $1,25(OH)_2D_3$ have been shown to be subnormal in patients infected with the human immunodeficiency virus (HIV); decreased serum $1,25(OH)_2D_3$ levels in these patients appeared to correlate with survival [26]. In a similar fashion to effects in sarcoidosis and TB, $1,25(OH)_2D_3$ has been shown to enhance bacterial killing and monocyte migration in cells from AIDS patients [20]. However, regulation of the expression of HIV itself by vitamin D remains uncertain. Different studies have reported that $1,25(OH)_2D_3$ (10^{-9} *M* for 24 hr) inhibits HIV infection of cultured monocytes [27] but enhances replication of the virus in the same type of cells [28].

B. Accessory Cell Function

Modulation of macrophage cytotoxicity by $1,25(OH)_2D_3$ is complemented by its effects on antigen presentation, also known as accessory cell function. T_H cells recognize foreign antigen in association with class II MHC on monocytes. This stimulates the expansion of the T-cell pool, allowing the development of memory cells and enabling cooperation with B cells to induce antibody production. Human MHC class II antigens (HLA-D) are found predominantly on monocytes–macrophages, B cells and some activated T cells. Vitamin D down-regulates expression of all three MHC class II groups (HLA-DR, -DP, and -DQ) on monocytes [29–31]. The overall effect of this is to reduce antigen-dependent T-cell proliferation [30]. Modulation of HLA-D expression may be part of the mechanism by which $1,25(OH)_2D_3$ has been shown to influence HIV infection of monocytes, although this may also be due to parallel modulation of another cell surface antigen, CD4 [30].

Another important aspect of this immunosuppressive action of $1,25(OH)_2D_3$ is its role as a modulator of autoimmune disease. Particular HLA antigens are known to show association with autoimmune diseases so that an individual carrying the antigen is more likely to contract the disease. Examples include rheumatoid arthritis, systemic lupus erythematosus (SLE), and juvenile (type I) diabetes. The effects of $1,25(OH)_2D_3$ on HLA-D expression in these diseases have yet to be fully defined, but studies using animal models have highlighted the therapeutic potential of vitamin D in the treatment of autoimmune disease. A more detailed discussion of vitamin D, T-cell function, and autoimmune disease is given below in Section VII and in other chapters (see Chapter 70).

As part of their study of vitamin D and MHC class II expression, Xu *et al.* [32] proposed that $1,25(OH)_2D_3$ stimulates classic macrophage cytotoxicity, whereas cytokines such as IL-4 enhance accessory cell function. However, the impact of $1,25(OH)_2D_3$ on accessory cell function may extend beyond its effects on macrophages. Studies from our group using dendritic cells purified from human tonsillar tissue have demonstrated constitutive expression of VDR [33]. Dendritic cells are potent

APCs, stimulating both T_H and T_C cell responses. Expression of VDR in these cells would therefore suggest that $1,25(OH)_2D_3$ plays a role in the induction of specific T-cell responses, possibly in tissue sites such as lymph nodes and tonsils. This effect would occur prior to, or in conjunction with, the antiproliferative activity of $1,25(OH)_2D_3$ toward T cells that have already been activated. In other words, it is possible that in certain tissues vitamin D may direct the expansion of specific T-cell pools, while preventing overproliferation of these cells. This apparent biphasic response is discussed in further detail in Sections VI and VII of this chapter.

Vitamin D may thus influence both antibody-independent and antibody-dependent macrophage activity, thereby influencing responses to a wide range of antigen challenges. However, two further important macrophage responses to $1,25(OH)_2D_3$ are the control of macrophage development through hematopoietic differentiation of myeloid precursors and the regulation of cytokine production by macrophages. The former represents a potential target for the use of $1,25(OH)_2D_3$ analogs in the treatment of leukemias, and the latter highlights a possible role for $1,25(OH)_2D_3$ in directing T-cell responses to infection or autoimmunity.

V. VITAMIN D AND MONOCYTE DEVELOPMENT

A. Antiproliferative/Differentiative Effects of $1,25(OH)_2D_3$

One of the earliest functional responses linking vitamin D with the immune system was the ability of $1,25(OH)_2D_3$ to stimulate the differentiation of macrophage precursor cells, namely, monocytes. Studies by several groups using both mouse and human cell lines showed that $1,25(OH)_2D_3$ inhibits the proliferation of these cells in a dose-dependent fashion and stimulates their differentiation toward a more mature, macrophage-like phenotype [34–38]. An important aspect of the cells studied is that they express high affinity VDR, a feature which is observed in relatively mature (but not fully differentiated) cells within the myelomonocytic lineage. Examples of such cell lines include U937, HL-60, K562, and THP1, which were cloned from patients with myeloid leukemias, although there are phenotypic and functional differences between each of the cell lines. As a result, investigation of these cells has not only revealed important aspects of hormonal control of hematopoiesis (see Fig. 3), but has also fueled interest in the use of vitamin D as an antileukemic agent. For example, Munker and colleagues examined the effects

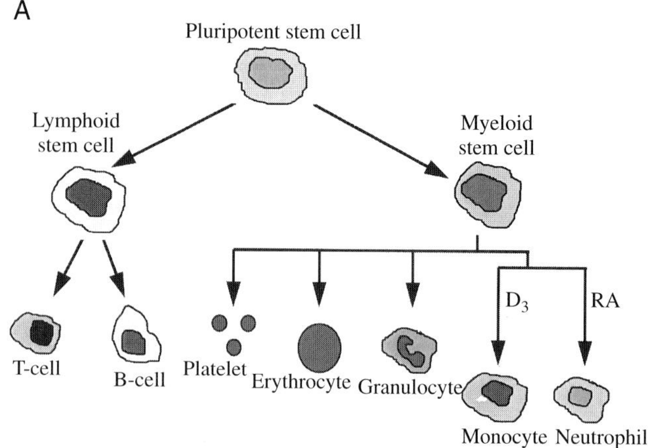

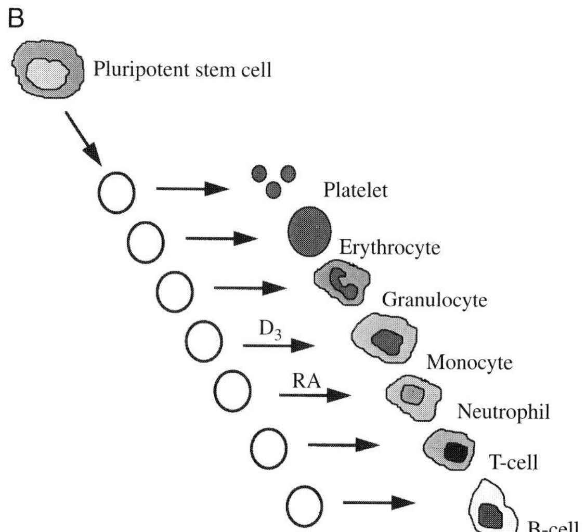

FIGURE 3 Schematic representation of (A) conventional and (B) sequential models for hematopoiesis. The conventional model highlights the commitment of stem cells to either a lymphoid or myeloid lineage followed by further differentiation to a specifc cell type. The sequential model proposes that there is a preferred order to the development of immune cells. At each stage of hematopoietic development cells may be committed to an individual pathway or progress to the next available differentiation option. Successful maturation toward a particular lineage is dependent on a variety of signals. The influence of $1,25(OH)_2D_3$ in differentiation of monocytes is indicated and contrasts with the induction of neutrophils by retinoids.

of different vitamin D metabolites on leukemic monocytes and granulocyte–macrophage colony-forming cells (GM-CFCs) from normal subjects [39]. They observed that the concentration of $1,25(OH)_2D_3$ which produced 50% inhibition of cell proliferation varied between 10^{-9} and 10^{-10} M for the different cell lines, with relatively mature monocytes being more responsive. Further experiments showed that a novel fluorinated analog of $1,25(OH)_2D_3$ (24,25-difluoro-1,25-dihydroxyvitamin D_3) was a more potent inhibitor of HL-60 proliferation than $1,25(OH)_2D_3$. However, both the analog and $1,25(OH)_2D_3$ were without effect on GM-CFCs from normal patients.

B. VDR Expression and Gene Regulation

Apart from highlighting the differences between normal and neoplastic cells, the above type of study raises several important points regarding the mechanism by which $1,25(OH)_2D_3$ regulates monocyte function. Variable monocyte responses to $1,25(OH)_2D_3$ may, in part, be attributable to differences in VDR expression. VDR expression has been assessed in a variety of human myelomonocytic cell lines using ^3H-$1,25(OH)_2D_3$ binding assays. Numbers of receptors vary from 500 to 6000 per cell, and this correlates well with VDR mRNA levels in the cell lines [40]. VDR expression does not appear to be regulated by the presence of ligand but may be modulated by other cell differentiation agents such as phorbol esters [41]. It is clear that the key effects of $1,25(OH)_2D_3$ on monocytes are mediated via VDR, with hormone–VDR complexes binding to DNA response elements to promote or suppress the transcription of target genes (see Chapters 8 and 9). Potential DNA targets for vitamin D within the immune system include genes associated with cell cycle progression. Elegant studies by Studzinski and colleagues using HL-60 cells showed that $1,25(OH)_2D_3$ blocked the progression of cells from the G_1 to S phase of the cell cycle. This suggested that the induction of cell differentiation in HL-60 cells by $1,25(OH)_2D_3$ is preceded by regulation of early replicative cellular events (i.e., DNA synthesis and cell proliferation) [42]. Consequently, there has been considerable interest in the genes associated with this part of the cell cycle, such as the oncogene c-*myc*. Karmali *et al.* reported that c-*myc* mRNA expression by U937 cells decreased within 4 hr following treatment with 10^{-8} *M* $1,25(OH)_2D_3$, prior to any changes in cell differentiation or proliferation [43]. Similar findings have been reported using HL-60 cells [44,45]. Other genes that are known to be down-regulated as part of myeloid cell growth arrest include stathmin [46] and the transferrin receptor [47]. Effects on other oncogenes such as c-*fos* and c-*fms* are less clear and are discussed in more detail in Chapter 64.

Vitamin D also regulates the expression of genes associated with cell differentiation. Changes in the phenotype of hematopoietic cells, such as monocytes and neutrophils, are accompanied by modulation of characteristic antigens that are found on the surface of these cells. Cell surface antigens associated with monocyte differentiation include CD14, which is up-regulated by $1,25(OH)_2D_3$, whereas neutrophil maturation is characterized by CD15 expression [48]. Characterization of these antigens has allowed phenotypic analysis of the differential effects of vitamin D and retinoids on hematopoietic development (this is discussed in greater detail below in Section V,D). Phenotypic variation may also occur within the monocyte lineage itself. Studies from our laboratory using monocyte precursor cells (U937) have shown that $1,25(OH)_2D_3$ induces different cell surface antigens to differentiation agents such as phorbol esters [49]. Other monocytic genes that are stimulated by $1,25(OH)_2D_3$ include those encoding a wide range of proteins associated with cell adhesion, and these are discussed in more detail in the next section.

C. New Models for Studying the Role of VDR in Monocytes

To study further the role of VDR in regulating expression of the above genes we have developed stable transfectant variants of U937 that either over- or under-express VDR. Increased VDR expression following transfection of a sense VDR cDNA expression vector resulted in an equivalent rise in the sensitivity of the cells to $1,25(OH)_2D_3$; inhibition of c-*myc* and stimulation of CD14 expression was much greater than in controls, and there was almost complete inhibition of proliferation after 48 hr of treatment with 10^{-8} *M* $1,25(OH)_2D_3$ [50]. The VDR transfectants also showed strong homotypic adhesion following treatment with $1,25(OH)_2D_3$. Subsequent experiments indicated that this was associated with increased expression of CD18, CD11a, CD11b, and CD11c cell surface antigens, which act as subunits to form the dimeric $\beta 2$-integrin cell adhesion molecules. Functional investigations indicated that vitamin D-induced cell–cell adhesion was mediated via the CD18/CD11a dimer, which binds to intercellular adhesion molecule I (ICAM-I) on other cells. Increased CD11b expression is now frequently used as a marker of $1,25(OH)_2D_3$-induced cell maturation whether in monocytes or neutrophils [51].

The cell–cell adhesion observed in sense VDR transfectants showed some similarity to the granulomata observed in inflammatory disorders such as sarcoidosis. It would therefore appear that the effects of $1,25(OH)_2D_3$ on monocytes may extend beyond regulation of cell proliferation and differentiation. Indeed, further studies using U937 cells transfected with antisense VDR cDNA have raised the possibility of a link between decreased VDR expression and apoptosis of monocytes [52]. Antisense RNA produced by the transfected expression vector only decreased VDR expression by 50%, but this was sufficient to render the cells resistant to the antiproliferative effects of

$1,25(OH)_2D_3$. The overall phenotype of the U937 cells was unaffected by decreased VDR expression, but the antisense transfectants showed a rate of proliferation that was almost twice that of controls; cell cycle analysis showed that a high proportion of the antisense cells were in S phase (i.e., that of DNA synthesis). Stimulation of the expression vector further increased antisense mRNA expression and enhanced cell cycle changes in the antisense tranfectants. The resulting cells showed evidence of chromatin condensation and DNA fragmentation characteristic of programmed cell death or apoptosis.

The data suggest a link between decreased VDR expression and disruption of normal cell cycling, leading ultimately to monocyte apoptosis. A possible explanation for this would be that efficient regulation of VDR expression is required to maintain control of specific target genes associated with the equilibrium between cell proliferation and cell death. Candidate genes would include the oncogenes c-*myc* and c-*fos*, both of which are abnormally regulated in the VDR antisense transfectants. The notion that $1,25(OH)_2D_3$–VDR interaction may play a role in controlling monocyte survival is not entirely new. Previous studies have shown that $1,25(OH)_2D_3$ is able to promote macrophage survival in inflammatory environments by inducing heat-shock protein synthesis [53]. More recent investigations have indicated that the regulation of monocyte differentiation and apoptosis by $1,25(OH)_2D_3$ is dependent on interaction with retinoids and their receptors. This aspect of VDR function within the immune system is outlined in the next section.

D. Interaction of VDR with Other Steroid/Thyroid Receptors

An important aspect of VDR function relating to control of monocyte development concerns the cooperativity between $1,25(OH)_2D_3$ and other differentiating agents. Other members of the steroid–thyroid receptor superfamily are able to participate in similar differentiation mechanisms to $1,25(OH)_2D_3$–VDR. Retinoids such as all-*trans*-retinoic acid (RA) and 9-*cis*-retinoic acid (9-*cis*-RA) exhibit a synergistic effect on $1,25(OH)_2D_3$-induced monocyte differentiation [54]. This is not entirely surprising in view of the fact that both VDR and receptors for RA (retinoic acid receptor, RAR) each function as a heterodimer, with a common partner, namely, the receptor for 9-*cis*-RA (retinoid X receptor, RXR) (see Chapter 9).

U937 cells have a relatively committed phenotype in that they differentiate toward monocytes regardless of the stimulus they receive. However, less mature cells such as HL-60 cells may develop toward a different phenotype depending on the activating agent. Vitamin D induces the formation of monocytes from HL-60 cells, whereas RA and 9-*cis*-RA induce neutrophil formation [55,56]. Thus $1,25(OH)_2D_3$ and retinoids, utilizing a common heterodimer partner (RXR), stimulate differentiation and maturation toward different phenotypes. Bunce *et al.*, through the use of combinations of retinoids and $1,25(OH)_2D_3$, were able to optimize neutrophil or monocyte formation from HL-60 cells [51]. For example, RA or 9-*cis*-RA at a relatively low dose (10^{-8} *M*) reduced the dose of $1,25(OH)_2D_3$ required to promote maximum monocyte differentiation from 10^{-7} to 5×10^{-9} *M*. Similarly, maximum neutrophil differentiation was obtained with a combination of 5×10^{-7} *M* RA or 9-*cis*-RA and 10^{-14} *M* $1,25(OH)_2D_3$. The authors also examined HL-60 cells that were initially stimulated toward neutrophil differentiation and then exposed to optimal stimuli for monocyte differentiation [57]. The result of this was that although the treatments decreased HL-60 cell proliferation, there was failure to differentiate into either a monocyte or a neutrophil. In fact, the cells appeared to be undetermined as to their choice of differentiation pathway and underwent apoptosis. These data therefore strengthen the putative link between VDR signal pathways and the control of programmed cell death.

The immediate conclusion from these findings is that certain amounts of retinoids and vitamin D promote specific interactions between nuclear receptors so as to activate genes and cellular properties that result in differentiation toward neutrophils or to monocytes. An alternative proposal is that $1,25(OH)_2D_3$ and retinoids do not differentiate cells to a particular lineage, but rather they reflect the different conditions that are required for survival, proliferation, and maturation of cells that have otherwise been committed to either of these two pathways. This putative role for $1,25(OH)_2D_3$ as a "survival factor" could be applied to either conventional or sequential models of hematopoiesis (see Fig. 3). In the conventional model, pluripotent stem cells initially give rise to myeloid and lymphoid sublineages. Further differentiation branches develop according to the different factors that are available within the immune microenvironment. The sequential model proposes that both myeloid and lymphoid cells develop as part of a single pathway, in which there are ordered stages beginning with erythrocytes and ending with B cells. At each stage a particular cell may be committed to an individual pathway of maturation, or it may progress to the next available differentiation option. A variety of genes encoding transcription factors are thought to be involved in determining the lineage options of a particular cell, and these are switched on and off by immune environment factors. Similarly, once a cell has committed itself to a particular pathway, other factors may be required to ensure the successful maturation of the cell; in the absence of these survival factors the cell will undergo apoptosis. In view of the studies described above, it is interesting to speculate on the possibility that a key aspect of the normal immune function of vitamin D is to act as a hematopoietic survival factor.

E. Vitamin D Analogs and Monocyte Proliferation

A considerable wealth of information now exists concerning the differentiation effects of analogs related to $1,25(OH)_2D_3$ [58]. Much of the data has been derived from analysis of antiproliferative/differentiative effects on HL-60 and U937 cells. Studies of the naturally occurring vitamin D metabolite 24,25-dihydroxyvitamin D_3 [$24,25(OH)_2D_3$] demonstrated similar antiproliferative effects to $1,25(OH)_2D_3$ but at concentrations that were 100 times higher than $1,25(OH)_2D_3$; the mechanism of action of $24,25(OH)_2D_3$ appears to be mediated via low specificity binding to VDR [59].

Further work with monocytic cell lines has described the antiproliferative effects of almost 300 different vitamin D analogs. The concentration at which these compounds produce 50% inhibition of cell proliferation (ED_{50}) varies considerably, and many are more potent than $1,25(OH)_2D_3$ itself [58]. This, coupled with low calciotropic activity, has highlighted the potential of certain analogs as antiproliferative agents. It is now possible to alter the conformation of the A, C, and D rings of the vitamin D molecule, but variations in the side chain have produced the most promising analogs so far. These include fluorinated analogs of $1,25(OH)_2D_3$ and double bond modifications. Other groups have reported potent effects with relatively subtle changes in the methyl group at C-20 (20-epi analogs) [59–62]. The ED_{50} values of these compounds vary) considerably (from 0.01 to 10,000 nM), but many have an improved calciotropic index, indicating their potential usefulness as therapeutic agents. The precise mechanism for improved action of specific analogs has yet to be fully defined, as all vitamin D metabolites appear to act via a single receptor (VDR). However, different responses to analogs and $1,25(OH)_2D_3$ are likely to be due, at least in part, to variations in analog metabolism. Further details of these analogs can be found in Chapters 58–63.

It is interesting to note that novel studies of HL-60 cells grown in serum-free cultures have shown that the effects of $1,25(OH)_2D_3$ and analogs on differentiation of HL-60 cells are significantly enhanced by various anti-inflammatory agents. The concentrations of these agents (e.g., indomethacin and Depo-Provera) were relatively low and were without effect by themselves. Preliminary Western blot data suggest that this effect is mediated via a single intracellular target: an isoenzyme of the aldoketo- reductase family [63,64]. The substrate and product of this novel enzyme have yet to be determined, but the data suggest that, by using agents such as indomethacin, it may be possible to reduce the therapeutic doses of $1,25(OH)_2D_3$ required for effective treatment of myeloid leukemias.

There is considerable literature documenting *in vitro* antileukemic effects of vitamin D, but *in vivo* responses are less well documented. However, studies by different groups using mice inoculated with a variety of myeloid leukemia cells have all shown that $1,25(OH)_2D_3$ (or an analog) increases survival rates [65–67]. Doses of $1,25(OH)_2D_3$ (administered intraperitoneally) were approximately 1 μg/kg/day, although the analog $1,25(OH)_2$-16-ene-23-yne-D_3 was effective at a much higher level (80 μg/kg/day). Further aspects of this work are discussed in Chapters 64–73.

VI. VITAMIN D AND T-LYMPHOCYTE FUNCTION

A. The Thymus and Lymphocyte Development

A key observation linking vitamin D with the immune system was the detection of VDR in lymphocytes. The development of T lymphocytes (T cells) and B lymphocytes (B cells) takes place in the thymus, with VDR being expressed in medullary thymocytes but not in the less mature cortical thymocytes [68]. This is in contrast to the glucocorticoid receptor (GR), which is expressed only in cortical thymocytes. Further studies showed that $1,25(OH)_2D_3$ inhibited proliferation of cytokine-activated medullary thymocytes, which are resistant to glucocorticoids. A particular feature of glucocorticoid regulation of the thymus is involution due to apoptosis of cortical cells [69]. In view of the differential expression of GR and VDR in this part of the thymus, it is interesting to speculate on the different roles for these receptors, and their signal pathways, in regulating programmed cell death.

B. Regulation of T-Cell Proliferation by $1,25(OH)_2D_3$

VDR expression is lost once cells leave the thymus and enter the circulation as T or B cells. However,

Bhalla *et al.*, and Provvedini *et al.* showed that VDR expression is reinduced in T-cell cultures which have been activated to proliferate by treatment with mitogens [70,71]. Lectins such as phytohemmaglutinin (PHA) induce VDR expression within 24 hr of treatment, coinciding with the transition of T cells from a resting (G_0) phase of the cell cycle into early G_1 phase (Fig. 4) [72]. Similar effects are also observed in T cells activated with antigens, such as the anti-CD3 T-cell antigen receptor. As with the monocytic cells described above, the numbers of VDR expressed by activated T cells are of the order of several thousand per cell, with K_d values being approximately 10^{-10} *M*.

Subsequent studies showed that $1{,}25(OH)_2D_3$ is a potent inhibitor of T-cell proliferation [73,74]. Fluorescence-activated cell sorting (FACS) showed that $1{,}25(OH)_2D_3$ blocked the transition of cells from early G_1 phase (G_{1a}) to late G_1 phase (G_{1b}); $1{,}25(OH)_2D_3$ did not affect movement from G_0 (resting) to G_{1a} or from G_{1b} to S phase [75]. As a result, $1{,}25(OH)_2D_3$ had no effect on expression of the receptor for IL-2, which is an early G_1 event, but inhibited expression of the transferrin receptor (a late G_1 event) (see Fig. 4). Much of this work has been carried out in T cells isolated from peripheral blood, although this represents a heterogeneous population of cells with varying degrees of inherent activation.

Analysis of specific subsets of T cells with more uniform patterns of proliferation can be achieved using lymphoid tissue. When centrifuged on bovine serum albumin density gradients, tonsillar T cells separate into three different populations according to density. Low density cells represent activated, proliferating T cells, whereas high density T cells correspond to resting T cells. Using the latter it has been possible to examine the effect of different T-cell activators on the induction of VDR expression and responsiveness to $1{,}25(OH)_2D_3$. Studies from our laboratory using different stimulators showed that the response to $1{,}25(OH)_2D_3$ is proportional to the rate of cell proliferation which, in turn, correlates with the number of VDR per cell [76]. Agents such as phorbol 12-myristate 13-acetate (PMA), anti-CD3 antibody, and the calcium ionophore A23187 produce only a small rise in T-cell proliferation. Accompanying induction of VDR expression is also relatively modest, and cells remain insensitive to $1{,}25(OH)_2D_3$. However, the lectin PHA and a combination of PMA and the anti-CD3 antibody were potent stimulators of T-cell proliferation. The resulting VDR numbers were 2- to 3-fold higher than with PMA, A23187, or anti-CD3 alone, suggesting that a threshold level of VDR per cell is required for $1{,}25(OH)_2D_3$ to be effective. The data also indicate that induction of VDR expression is not specific for a particular signal transduction pathway but clearly involves both protein kinase C (PMA) and phospholipase C (anti-CD3, PHA).

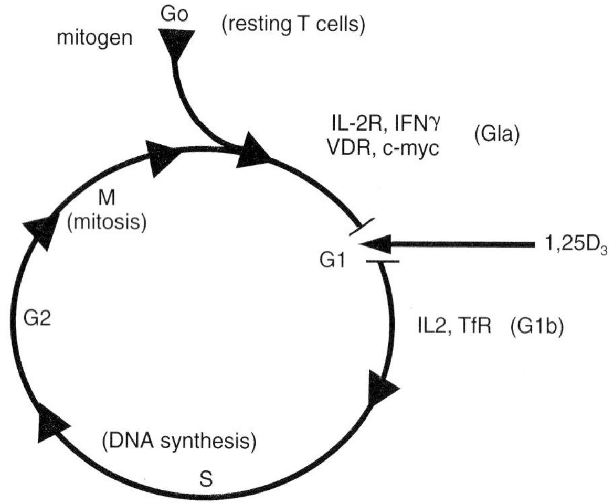

FIGURE 4 Impact of $1{,}25(OH)_2D_3$ on cell cycle progression of T cells. Resting (nonactivated) T cells in the G_0 phase of the cell cycle are induced to proliferate following a mitogenic challenge. On entering early G_1 phase (G_{1a}) cells express genes such as interleukin-2 receptor (IL-2R), interferon-γ (IFN-γ), vitamin D receptor (VDR), and c-*myc*. Late G_1 (G_{1b}) genes include interleukin-2 (IL-2) and transferrin receptor (TfR). Inhibition of cell cycle progression by $1{,}25(OH)_2D_3$ takes place after early G_1 phase events, so that progression to S phase is blocked.

C. Regulation of T-Cell Gene Expression by $1{,}25(OH)_2D_3$

Activation of T cells is associated with the regulation of a large number of cellular genes. For example, transferrin receptor expression occurs at a late stage of the G_1 phase and reflects the general requirement of all proliferating eukaryotic cells for delivery of iron. Another early T-cell response is induction of c-*myc* transcription. Expression of this oncogene (as with VDR) is low or undetectable in resting T cells, but it increases rapidly following activation and parallels the rate of proliferation and level of VDR expression [76]. Inhibition of c-*myc* expression by $1{,}25(OH)_2D_3$ in activated T cells is similar to that observed in proliferating monocytes and, again, reflects a concomitant decrease in proliferation.

During the early G_1 phase, induction of IL-2 receptor (IL-2R) expression enables the cell to respond to IL-2 and thus promote long-term growth of the T-cell population. Transcription of IL-2 mRNA is induced immediately after IL-2R, and, initially, this appeared to be a target for the T-cell effects of $1{,}25(OH)_2D_3$. However, although $1{,}25(OH)_2D_3$ rapidly down-regulates IL-2 expression in T cells, there was no abrogation of this response following treatment with exogenous cytokine [77,78]. Indeed, it is likely that vitamin D modulates a much more complex series of cytokine regulatory events than was previously thought. For example, it has been reported that $1,25(OH)_2D_3$ stimulates IL-1 expression by monocytes–macrophages [79–81]. IL-1 is a potent stimulator of T-cell activation from the G_0 to G_1 phase. This would therefore suggest that vitamin D plays a role in the initial activation of T-cell proliferation and then inhibits further T-cell proliferation at a later stage.

A biphasic T-cell response to $1,25(OH)_2D_3$ is postulated earlier in this chapter (Section IV,B) with reference to the expression of VDR on dendritic cells [33]. Other groups have reported short-term rises in IL-2, IFN-γ, and transferrin receptor mRNA accumulation, prior to the aforementioned down-regulation associated with decreased T-cell proliferation [1]. This indicates that the cytokine-regulatory effects of $1,25(OH)_2D_3$ may not be purely related to the antiproliferative effects of this hormone.

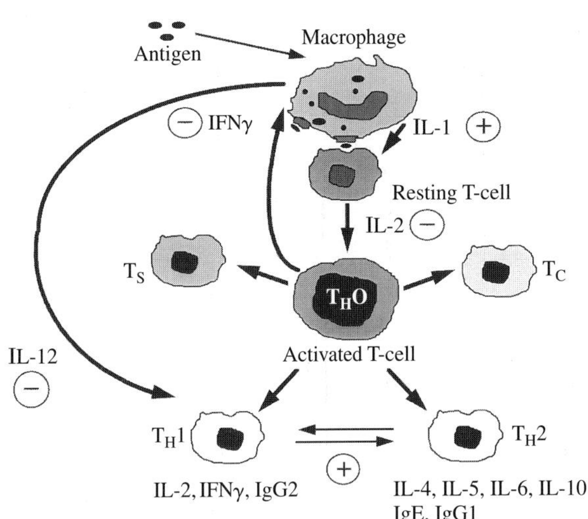

FIGURE 5 The hormone $1,25(OH)_2D_3$ modulates the cytokine profile of T_H subgroups. Activation of resting T cells results in the induction of specific T-cell subgroups. T-suppressor cells (T_S), cytotoxic T cells (T_C), and T-helper cells (T_H). T_H cells can be further divided according to their cytokine profile and the specific immunoglobulins associated with their activities. Cytokines induced by $1,25(OH)_2D_3$ are highlighted with a + symbol. Cytokines down-regulated by $1,25(OH)_2D_3$ are identified by a − symbol.

VII. ROLE OF CYTOKINES IN MEDIATING THE IMMUNOSUPPRESSIVE EFFECTS OF $1,25(OH)_2D_3$

Developments linking cytokine expression with specific T-cell functions are now discussed. The consequence of an antigenic challenge is that T and B cells acquire specific receptors for the antigen. If there is a repeat infection with the agent, the cells are immediately induced to proliferate in conjunction with accessory cell presentation. Lymphocyte proliferation essentially produces two type of cells: effector cells involved with counteracting the infection, and memory cells that express receptors for antigen without direct exposure to the agent. The latter form part of a large pool of "virgin" lymphocytes that circulate throughout the body, ensuring vigilance against repeat infection. As already described, a key feature in the function of these lymphocytes is activation by antigen and the regulation of subsequent clonal expansion by antiproliferative agents. It is clear from studies described in this chapter that vitamin D may act as a potent modulator of T-cell proliferation, although this may not be purely an antiproliferative effect. Indeed, analyses of T-cell subgroup activation suggest that $1,25(OH)_2D_3$ may have further, more subtle, effects on these cells.

A. Regulation of T-Cell Subgroups

The effector cells produced by antigen activation of T cells fall into several categories with distinct and important functions (Fig. 5). The differential activation of T_C and T_H cells by MHC class I and MHC class II accessory cells is highlighted earlier in this chapter (Section IV). T_H cells in general cooperate with lymphocytes; stimulating the maturation of T_C cells and cooperating with B cells to produce antibody. Other cells known as suppressor T cells (T_S) have been demonstrated in models of immune tolerance. The precise function of these cells has not been fully defined, but they appear to play a role in normal immune responses by limiting lymphocyte activity. Studies *in vitro* and *in vivo* have shown that $1,25(OH)_2D_3$ enhances T_S cell availability, further highlighting the apparent immunosuppressive nature of this hormone [82,83].

The preferential target for $1,25(OH)_2D_3$ appears to be T_H cells. It is now evident that this population of cells can itself be divided into two further subgroups which develop either cell-mediated or antibody (humoral) immunity. These sub-groups are known as T_H1 and T_H2 and are associated with distinct cytokine profiles. T_H1 cells produce IL-2 and IFN-γ and mediate cellular responses, such as macrophage and natural killer cell activation in response to intracellular and viral pathogens. However, development of T_H1 cells appears to be central to the adverse immune responses observed with autoimmune disease and transplantation rejection. T_H2 cells secrete IL-4, IL-5, IL-6, and IL-10 and provide more efficient help for B-cell activation and antibody production. In particular, T_H2 responses involve development of IgE and IgG_1 immunoglobulin isotypes. T_H2 cells may help to counteract the detrimental effects of T_H1 responses by modulating T_H1 cell activity. It has also been shown that IL-12 production by pathogen-infected macrophages is a potent inducer of T_H1 activity (i.e., cell-mediated responses) [84].

Development of one T-cell subgroup versus another allows the immune system to select the most appropriate response to a particular infection. Preferential development of a T_H cell type occurs immediately after exposure to infection and appears to be very sensitive to early regulation of cytokine levels. The established anti-IL-2 effect of $1,25(OH)_2D_3$ suggested that vitamin D could play a role in directing T_H cell responses, and there is much current interest in this aspect of vitamin D immunoregulation. Indeed, one of the initial immune functions proposed for $1,25(OH)_2D_3$ was suppression of T-helper cell activity [85]. Further studies have shown that $1,25(OH)_2D_3$ inhibits IL-2 and IFN-γ secretion, by T cells, and IL-12 secretion by monocytes [86–88]. No effects were observed with IL-4 secretion, suggesting that $1,25(OH)_2D_3$ promotes T_H2 responses by down-regulating T_H1 activity.

B. Vitamin D and Autoimmune Responses

It is now clear that control of T_H subgroups plays an important role in the development of several autoimmune diseases including arthritis, SLE (lupus), and type I (autoimmune/juvenile) diabetes. This has suggested new therapeutic uses for vitamin D that extend beyond its antileukemic potential. Studies *in vivo* using the NOD (nonobese diabetic) mouse model of type I diabetes have shown that $1,25(OH)_2D_3$ can prevent the development of clinical diabetes in these animals, without the induction of generalized immunosuppression. Specific effects included the induction of suppressor T cells, which are normally depleted in these animals [83]. More detailed aspects of the immunosuppressive actions of vitamin D and autoimmune disease are outlined in Chapter 70. This chapter and chapters in Section VII of this book also detail the potential of new vitamin D analogs as treatment for autoimmune disorders. For example, studies have shown that $1,25(OH)_2$-16-ene-D_3 may be as much as 100-fold more potent than $1,25(OH)_2D_3$ in down-regulating T_H1 responses and preventing autoimmune disease [88].

Other studies have highlighted the potential contribution of T_H1 to T_H2 switching in the development of disorders other than autoimmune disease. T_H2 responses may be inappropriate for some infections and lead to pathological conditions. In HIV infection, for example, a T_H1 to T_H2 shift has been proposed as part of the mechanism by which CD4+ (or T_H cell) levels are depleted [89]. Regulation of steroid hormone levels may play an important role in directing these changes, although a specific contribution by $1,25(OH)_2D_3$ has yet to be examined. There is also evidence for T_H2 activity in tuberculosis, resulting in cell-mediated damage at inflammatory sites. Again, subtle endocrinological changes may be pivotal to a deleterious T-cell shift [25]. Abnormal regulation of macrophage 1α-hydroxylase activity has been linked with episodic hypercalcemia in TB. This suggests that overproduction of $1,25(OH)_2D_3$ may influence the development of TB by directing T-cell and cytokine function at inflammatory sites.

Selection of T_H responses may play an important role in the pathogenesis of other disorders. Psoriasis is characterized by epidermal hyperproliferation, but the disorder has a substantial immune component that may serve as a target for the actions of $1,25(OH)_2D_3$. There is increasing evidence to suggest that T_H cells play a central role in maintaining the psoriatic process via induction of local inflammatory cytokines [90,91]. Vitamin D analogs are now widely used in the treatment of this skin disorder [92]. Their principal effect is in regulating overproliferation of keratinocytes, but it seems likely that vitamin D may also influence the cytokine- and cell-mediated immune dysfunction known to contribute to this disease. This aspect of vitamin D immunoregulation is discussed in further detail in Chapter 25.

VIII. REGULATION OF B-CELL FUNCTION BY $1,25(OH)_2D_3$

Both $1,25(OH)_2D_3$ and analogs have been shown to inhibit IgG_{2a} responses characteristic of T_H1 cell activity [87], although, at present, there is no specific evidence for vitamin D-induced up-regulation of T_H2-mediated B-cell activity. Studies *in vitro* on the interaction of $1,25(OH)_2D_3$ with B cells have shown that the hormone suppresses the development of Ig-secreting B cells following pokeweed mitogen stimulation [93]. Further reports confirmed the antiproliferative effect of $1,25(OH)_2D_3$ and described inhibitory effects on Ig production as well [94]. However, using peripheral blood mononuclear cells, Lemire and colleagues showed that the B-cell suppressive effects of $1,25(OH)_2D_3$ are likely to be mediated via inhibition of T-helper cells [85]. VDR expression by B cells is not so clearly characterized as for T cells, but (in common with other groups) we have routinely used transformed B-cell lines to characterize VDR. Although these cell lines are rapidly proliferating, expression of VDR is normally low. This can be resolved by treatment with mitogens such as PMA, which induce detectable levels of VDR and confer responses to

$1,25(OH)_2D_3$. However, *in vivo,* with a mixed cell population the likely scenario remains that vitamin D effects on B cells are mediated by T cells and monocytes–macrophages through a variety of mechanisms described above.

IX. CONCLUSIONS

It is now clear that the immunomodulatory effects of $1,25(OH)_2D_3$ are perhaps the best example of the pluripotent nature of this hormone. Vitamin D effects are highlighted by subtle changes in the differentiation of specific cell types and the promotion of specific cell responses. The ability of immune cells to synthesize $1,25(OH)_2D_3$ indicates a possible paracrine/autocrine mode of action for this hormone. Locally produced hormone may influence a wide range of immune functions including cytotoxic macrophage activity. However, the principal action of $1,25(OH)_2D_3$ is to act as an immunosuppressive agent, down-regulating the activity of T cell and B cells. In part this is achieved by antiproliferative effects on T cells, which prevent overexpansion of lymphocyte populations. It now seems likely that one of the key actions of $1,25(OH)_2D_3$ is to direct the type of T-cell response by modulating the cytokine profiles of T_H cells. The immune activity of vitamin D may, to some extent, be dependent on the availability of $1,25(OH)_2D_3$-synthesizing cells such as macrophages. Thus, the second key immune action of $1,25(OH)_2D_3$, is to facilitate the development of these cells through the hematopoietic system by promoting differentiation of the myeloid lineage, and the survival of monocytes.

Coordination of all the immune actions of vitamin D appears to be disrupted during the development of inflammatory disorders such as sarcoidosis, tuberculosis, and arthritis. Overproduction of $1,25(OH)_2D_3$ not only results in hypercalcemic complications but may also contribute to the pathogenesis of inflammatory disease. In particular, excessive hormone may produce inappropriate T_H2 responses which then contribute to tissue damage. Thus, although there is much interest in the use of $1,25(OH)_2D_3$ in autoimmune disease, future studies are also likely to address its detrimental effects in inflammatory disease. The precise cause of abnormal regulation of immune $1,25(OH)_2D_3$ production remains unclear. The eventual cloning of the gene for 1α-hydroxylase will reveal much concerning differences between macrophage 1α-hydroxylase and its kidney counterpart.

Several other questions are likely to be addressed by future studies. There may be crosstalk between vitamin D signals and other steroidogenic pathways, though the contribution of these effects to monocyte development remains to be determined. Expression of VDR appears to be central to all of the immune effects of vitamin D and, as described in Section V, may also contribute to novel actions such as apoptosis. However, the impact on the immune system of expression of abnormal VDR has yet to be clearly defined. Patients with VDR gene mutations due to hereditary vitamin D-resistant rickets (HVDRR) express defective VDR and show resistance to $1,25(OH)_2D_3$; the basis of this disorder and the clinical problems associated it are described detail in Chapter 49. Although patients with HVDRR appear to be susceptible to infection, there have been relatively few reports of specific immune deficiencies associated with the disease [95]. However, these have been supported by studies of vitamin D deficiency that also describe associated immune dysfunction [96].

Some of the questions regarding the role of VDR in mediating immune responses to $1,25(OH)_2D_3$ *in vivo* may be answered by consideration of nonclassical vitamin D pathways. Studies have identified a specific $1,25(OH)_2D_3$ binding protein in cell membranes, suggesting possible rapid, nongenomic actions for vitamin D [97] (see Chapter 15). Functional differences between this receptor and VDR [98] may help to clarify some of the immune effects of vitamin D, specifically the mechanism for decreased calciotropic action of $1,25(OH)_2D_3$ analogs. In turn, it is to be hoped that this will lead to an even wider range of vitamin D-related compounds for use in the treatment of immune disorders.

References

1. Taylor RL, Lynch HJ Jr, Wysor WG Jr 1963 Seasonal influence of sunlight on hypercalcaemia of sarcoidosis. Am J Med **34**:221–227.
2. Papapoulos SE, Clemens TL, Fraher LJ, Lewin IG, Sandler LM, O'Riordan JLH 1979 1,25-Dihydroxycholecalciferol in the pathogenesis of the hypercalcaemia of sarcoidosis. Lancet **1**:627–630.
3. Barbour GL, Coburn JW, Slatopolsky E, Norman AW, Horst RL 1981 Hypercalcemia in an anephric patient with sarcoidosis: Evidence for extra-renal generation of 1,25-dihydroxyvitamin D. N Engl J Med **305**:440–443.
4. Mason RS, Frankel T, Chan YL, Lissner D, Posen S 1984 Vitamin D conversion by sarcoid lymph node homogenate. Ann Intern Med **100**:59–61.
5. Adams JS, Sharma OP, Gacad MA, Singer FR 1983 Metabolism of 25-hydroxyvitamin D_3 by cultured pulmonary alveolar macrophages. J Clin Invest **72**:1856–1860.
6. Adams JS, Singer FR, Gacad MA, Sharma OP, Hayes MJ, Vouros P, Holick MF 1985. Isolation and structural identification of 1,25-dihydroxyvitamin D_3 produced by cultured alveolar macrophages in sarcoidosis. J Clin Endocrinol Metab **60**:960–966.
7. Adams JS, Gacad MA 1985 Characterization of 1-alpha-hydroxylation of vitamin D_3 sterols by cultured alveolar macrophages from patients with sarcoidosis. J Exp Med **161**:755–765.
8. Reichel H, Koeffler HP, Bishop JE, Norman AW 1987 25-Hy-

droxyvitamin D_3 metabolism by lipopolysaccharide-stimulated normal human macrophages. J Clin Endocrinol Metab **64**:1–9.
9. Adams JS, Gacad MA, Diz M, Nadler JL 1990 A role for endogenous arachidonate metabolites in the regulated expression of the 25-hydroxyvitamin D-1-hydroxylation reaction in cultured alveolar macrophages from patients with sarcoidosis. J Clin Endocrinol Metab **70**:595–600.
10. Koeffler HP, Reichel H, Bishop JE, Norman AW 1985 Gamma interferon stimulates production of 1,25-dihydroxyvitamin D_3 by normal human macrophage. Biochem Biophys Res Commununún **127**:596–603.
11. Reichel H, Koeffler HP, Barbers R, Norman AW 1987 Regulation of 1,25-dihydroxyvitamin D_3 production by cultured alveolar macrophages from normal human donors and from patients with pulmonary sarcoidosis. J Clin Endocrinol Metab **65**:1201–1209.
12. Reichel H, Koeffler HP, Norman AW 1987 Synthesis in vitro of 1,25-dihydroxyvitamin D_3 and 24,25-dihydroxyvitamin D_3 by interferon gamma-stimulated normal human bone marrow and alveolar macrophages. J Biol Chem **262**:10931–10937.
13. Hayes ME, Denton J, Freemont AJ, Mawer EB 1989 Synthesis of the active metabolite of vitamin D, $1,25(OH)_2D_3$, by synovial fluid macrophages in arthritic diseases. Ann Rheum Dis **48**:723–729.
14. Mawer EB, Hayes ME, Still PE, Davies M, Lumb GA, Palit J, Holt PJL 1991 Evidence for nonrenal synthesis of 1,25-dihydroxyvitamin D in patients with inflammatory arthritis. J Bone Miner Res **6**:733–739.
15. Davies M, Hayes ME, Mawer EB, Lumb GA 1985 Abnormal vitamin-D metabolism in Hodgkin's lymphoma. Lancet **1**:1186–1188.
16. Rosenthal N, Insogna KL, Godsall JW, Smaldone L, Waldron JA, Stewart AF 1985 Elevations in circulating 1,25-dihydroxyvitamin D in 3 patients with lymphoma-associated hypercalcemia. J Clin Endocrinol Metab **60**:29–33.
17. Karmali R, Barker S, Hewison M, Fraher L, Katz DR, O'Riordan JLH 1990 Intermittent hypercalcemia and vitamin D sensitivity in Hodgkin's disease. Postgrad Med J **66**:757–760.
18. Mudde AH, van den Berg H, Boshuis PG, Breedveld FC, Markusse HM, Kluin PM, Bijvoet OLM, Papapoulos SE 1987 Ectopic production of 1,25-dihydroxyvitamin D by B-cell lymphoma as a cause of hypercalcaemia. Cancer **59**:1543–1546.
19. Reichel H, Koeffler HP, Norman AW 1987 25-Hydroxyvitamin D_3 metabolism by human lymphotropic-T virus-transformed lymphocytes. J Clin Endocrinol Metab **65**:519–526.
20. Girasole G, Wang JM, Pedrazzoni M, Pioloi G, Balota C, Passeri M, Lazzarin A, Ridolfo A, Mantovani A 1990 Augmentation of monocyte chemotaxis by 1α,25-dihydroxyvitamin D_3: Stimulation of defective migration of AIDS patients. J Immunol **145**:2459–2464.
21. Rook GAW, Taverne J, Leveton C, Steele J 1987 The role of gamma-interferon, vitamin-D_3 metabolites and tumor necrosis factor in the pathogenesis of tuberculosis. Immunology **62**:229–234.
22. Cohen MS, Mesler DE, Snipes RG, Gray TK 1986 1,25-Dihydroxyvitamin D_3 activates secretion of hydrogen peroxide by human monocytes. J Immunol **136**:1049–1053.
23. Abe E, Shiina Y, Miyaura C, Tanaka H, Hayashi T, Kanegasaki S, Saito M, Nishii Y, Deluca HF, Suda T 1984 Activation and fusion induced by 1-alpha,25-dihydroxyvitamin D_3 and their relation in alveolar macrophages. Proc Natl Acad Sci USA **81**:7112–7116.
24. Saboor SA, Johnson NMI, McFadden J 1992 Detection of mycobacterial DNA in sarcoidosis and tuberculosis with polymerase chain reaction. Lancet **339**:1012–1015.
25. Rook GAW, Hernandezpandor R 1994 T-cell helper types and endocrines in the regulation of tissue-damaging mechanisms in tuberculosis. Immunobiology **191**:478–492.
26. Haug C, Muller F, Aukhurst P, Froland SS 1994 Subnormal serum concentration of 1,25-dihydroxyvitamin D_3 in human immunodeficiency virus infection: Correlation with degree of immune deficiency and survival. J Infect Dis **169**:889–893.
27. Connor RI, Rigby WFC 1991 1-alpha,25-dihydroxyvitamin-D_3 inhibits productive infection of human monocytes by HIV-1. Biochem Biophys Res Commun **176**:852–859.
28. Skolnik PR, Jahn B, Wang MZ, Rota TR, Hirsch MS, Krane SM 1991 Enhancement of human immunodeficiency virus 1 replication in monocytes by 1,25-dihydroxycholecalciferol. Proc Natl Acad Sci USA **88**:6632–6636.
29. Rigby WFC, Waugh M, Graziano RF 1990 Regulation of human monocyte HLA-DR and CD4 antigen expression, and antigen presentation by 1,25-dihydroxyvitamin-D_3. Blood **76**:189–197.
30. Tokuda N, Mizuki N 1992 1,25-dihydroxyvitamin D_3 down-regulation of HLA-DR on human peripheral blood monocytes. Immunology **75**:349–354.
31. Rigby WFC, Waugh MG 1992 Decreased accessory cell function and co-stimulatory activity by 1,25-dihydroxyvitamin D_3-treated monocytes. Arth Rheum **35**:110–119.
32. Xu H, Soruri A, Gieseler RKH, Peters JH 1993 1,25-Dihydroxyvitamin D_3 exerts opposing effects to IL-4 on MHC class II antigen expression, accessory activity and phagocytosis of human monocytes. Scand J Immunol **38**:535–540.
33. Brennan A, Katz DR, Nunn JD, Barker S, Hewison M, Fraher LJ, O'Riordan JLH 1987 Dendritic cells from human tissues express receptors for the immunoregulatory vitamin-D_3 metabolite, dihydroxycholecalciferol. Immunology **61**:457–461.
34. Abe E, Miyaura C, Sakagami K, Takeda M, Konno K, Yamazaki T, Yoshiki S, Suda T 1981 Differentiation of mouse myeloid leukemia cells induced by 1α,25-dihydroxyvitamin D_3. Proc Natl Acad Sci USA **78**:4990–4994.
35. Koeffler HP, Amatruda T, Ikekawa N, Kobayashi Y, Deluca HF 1984 Induction of macrophage differentiation of human normal and leukemic myeloid stem cells by 1,25-dihydroxyvitamin D_3 and its fluorinated analogs. Cancer Res **44**:5624–5628.
36. Koeffler HP, Hirj K, Itri L 1985 1,25-Dihydroxy vitamin D_3: In vivo and in vitro effects on human preleukemic and leukemic cells. Cancer Treat Rep **69**:1399–1407.
37. Mangelsdorf DJ, Koeffler HP, Donaldson CA, Pike JW, Haussler MR 1984 1,25-Dihydroxyvitamin D_3-induced differentiation in a human promyelocytic leukemia cell line (HL-60): Receptor-mediated maturation to macrophage-like cells. J Cell Biol **98**:391–398.
38. Amento EP, Bhalla AK, Kurnick JT, Kradin RL, Clemens TL, Holick SA, Holick MF, Krane SMI 1984 1-alpha,25-dihydroxyvitamin D_3 induces maturation of the human monocyte cell line U937, and, in association with a factor from human T-lymphocytes, augments production of the monokine, mononuclear cell factor. J Clin Invest **73**:731–739.
39. Munker R, Norman AW, Koeffler HP 1986 Vitamin D compounds: Effect on clonal proliferation and differentiation of human myeloid cells. J Clin Invest **78**:424–430.
40. Kizaki M, Norman AW, Bishop JE, Lin CW, Karmakar A, Koeffler HP 1991 1,25-Dihydroxyvitamin D_3 receptor RNA expression in hematopoietic cells. Blood **77**:1238–1247.
41. Hewison M, Barker S, Brennan A, Nathan J, Katz DR, O'Riordan JLH 1989 Autocrine regulation of 1,25-dihydroxycholecalciferol metabolism in myelomonocytic cells. Immunology **68**:247–252.
42. Studzinski GP, Bhandal AK, Brelvi ZS 1985 Cell cycle sensitivity

of HL-60 cells to the differentiation-inducing effects of 1-alpha,25-dihydroxyvitamin D_3. Cancer Res **45**:3898–3905.
43. Karmali R, Bhalla A, Farrow SM, Lydyard P, O'Riordan JLH 1989 1,25DHCC regulates oncogene expression in U937 cells. J Mol Endocrinol **3**:43–50.
44. Reitsma PH, Rothberg PG, Astrin SM, Trial J, Barshavit Z, Hall A, Teitelbaum SL, Kahn AJ 1983 Regulation of myc gene expression in HL-60 leukemia cells by a vitamin D metabolite. Nature **306**:492–494.
45. Collins SJ 1987 The HL60 promyelocytic leukaemia cell line. Proliferation, differentiation and cellular oncogene expression. Blood **70**:1233–1244.
46. Johnson EWB, Jones NA, Rowlands DC, Williams A, Guest SS, Brown G 1995 Down-regulation but not phosphorylation of stathmin is associated with induction of HL-60 cell growth arrest and differentiation by physiological agents. FEBS Lett **364**:309–313.
47. Gatter KC, Brown G, Trowbridge IS, Woolston RE, Mason DY 1983 Transferrin receptors in human tissues: Their distribution and possible clinical relevance. J Clin Pathol **36**:539–545.
48. Fischer, AG, Bunce CM, Toksoz D, Stone PCW, Brown G 1982 Studies of human myeloid antigens using monoclonal antibodies and variant cell lines from the promyeloid cell line HL60. Clin Exp Immunol **50**:374–381.
49. Hewison M, Brennan A, Singh-Ranger R, Walters JC, Katz DR, O'Riordan JLH 1992 The comparative role of 1,25-dihydroxycholecalciferol and phorbol esters in the differentiation of the U937 cell line. Immunology **77**:304–311.
50. Hewison M, Dabrowski M, Faulkner L, Hughson E, Vadher S, Rut A, Brickell PM, O'Riordan JLH, Katz DR 1994 Transfection of vitamin D receptor cDNA into the monoblastoid cell line U937: The role of vitamin D_3 in homotypic macrophage adhesion. J Immunol **153**:5709–5719.
51. Bunce CM, Wallington LA, Harrison P, Williams GR, Brown G 1995 Treatment of HL60 cells with various combinations of retinoids and 1α,25-dihydroxyvitamin D_3 results in differentiation towards neutrophils or monocytes or a failure to differentiate and apoptosis. Leukemia **9**:410–418.
52. Hewison M, Dabrowski M, Vadher S, Faulkner L, Cockerill FJ, Brickell PM, O'Riordan JLH, Katz DR 1996 Antisense inhibition of vitamin D receptor expression induces apoptosis in monoblastoid U937 cells. J Immunol **156**:4391–4400.
53. Polla BS, Healy AM, Amento EP, Krane SM 1986 1,25-Dihydroxyvitamin D_3 maintains adherence of human monocytes and protects them from thermal injury. J Clin Invest **77**:1332–1339.
54. Taimi M, Chateau MT, Cabane S, Marti J 1991 Synergistic effect of retinoic acid and 1,25-dihydroxyvitamin D_3 on the differentiation of the human monocytic cell line U937. Leuk Res **15**:1145–1152.
55. Brown G, Bunce CM, Rowlands DC, Williams GR 1994 all-*trans*-Retinoic acid and 1-alpha,25-dihydroxyvitamin D_3 cooperate to promote differentiation of the human promyeloid leukemia cell line HL-60 to monocytes. Leukemia **8**:806–815.
56. Miyaura C, Abe E, Suda T, Kuroki T 1985 Alternative differentiation of human promyelocytic leukemia cells (HL-60) induced selectively by retinoic acid and 1α,25-dihydroxyvitamin D_3. Cancer Res **45**:4244–4248.
57. Wallington LA, Bunce CM, Durham J, Brown G 1995 Particular combinations of signals, by retinoic acid and 1-alpha,25-dihydroxyvitamin D_3, promote apoptosis of HL-60 cells. Leukemia **9**:1185–1190.
58. Bouillon R, Okamura WH, Norman AW 1995 Structure–function relationships in the vitamin D endocrine system. Endocr Rev **16**:200–257.
59. Hewison M, Barker S, Brennan A, Cifelli AT, Katz DR, O'Riordan JLH 1987 24,25-$(OH)_2D_3$ stimulates differentiation of the monoblastic cell-line U937. Bone **8**:264- 275.
60. Zhou JY, Norman AW, Lubbert M, Collins ED, Uskokovic MR, Koeffler HP 1989 Novel vitamin D analogs that modulate leukemic cell growth and differentiation with little effect on either intestinal calcium absorption or bone calcium mobilization. Blood **74**:82–93.
61. Jung SJ, Lee YY, Pakkala S, Devos S, Elstner E, Norman AW, Green J, Uskokovic M, Koeffler HP 1994 1,25$(OH)_2$-16-ene-vitamin-D_3 is a potent antileukemic agent with low potential to cause hypercalcemia. Leuk Res **18**:453–463.
62. Elstner E, Lee YY, Hashiya M, Pakkala S, Binderup L, Norman AW, Okamura WH, Koeffler HP 1994 1-alpha,25-dihydroxy-20-epi-vitamin D_3: An extraordinarily potent inhibitor of leukemic cell growth in vitro. Blood **84**:1960–1967.
63. Bunce CM, French PJ, Durham J, Stockley RA, Michell RH, Brown G 1994 Indomethacin potentiates the induction of HL60 differentiation to neutrophils by retinoic acid and granulocyte colony-stimulating factor, and to monocytes by vitamin D_3. Leukemia **8**:595–604.
64. Bunce CM, Mountford JC, French PJ, Mole DJ, Durham J, Michell RH, Brown G 1995 Promotion of myeloid differentiation by anti-inflammatory agents by steroids and by retinoic acid involves a single intracellular target, an enzyme of the aldoketoreductase family. Biochim Biophys Acta **1311**:189–198.
65. Honma Y, Hozumi M, Abe E, Konno K, Fukushima M, Hata S, Nishii Y, DeLuca HF, Suda T 1983 1α,25-Dihydroxyvitamin D_3 and 1α-hydroxyvitamin D_3 prolong survival time of mice inoculated with myeloid leukaemia cells. Cell Biol **80**:201–204.
66. Potter GK, Mohamed AN, Dracapoli NC, Groshen SL, Shen RL, Moore MA 1985 Action of 1,25$(OH)_2D_3$ in nude mice bearing transplantable human myelogenous leukaemic cell lines. Exp Hematol **13**:722–732.
67. Kaukabe T, Honma Y, Hozumi M, Suda T, Nishii Y 1987 Control of proliferating potential of myeloid leukaemia cells during long-term treatment with vitamin D_3 analogues and other differentiation inducers in combination with anti-leukemic drugs: *In vitro* and *in vivo*. Cancer Res **47**:567–572.
68. Rigby WFC 1988 The immunobiology of vitamin D. Immunol Today **9**:54–58.
69. Thompson EB 1994 Apoptosis and steroid hormones. Mol Endocrinol **8**:665–673.
70. Bhalla AK, Amento EP, Clemens TL, Holick MF, Krane SM 1983 Specific high affinity receptors for 1,25-dihydroxyvitamin D_3 in human peripheral blood mononuclear cells: Presence in monocytes and induction in lymphocytes following activation. J Clin Endocrinol Metab **57**:1308–1310.
71. Provvedini DM, Tsoukas CD, Deftos LJ, Manolagas SC 1983 1,25-Dihydroxyvitamin D_3 receptors in human leukocytes. Science **221**:1181–1182.
72. Yu X-P, Mocharia H, Hustmeyer FG, Manolagas SC 1991 Vitamin D receptor expression in human lymphocytes. J Biol Chem **266**:7588–7595.
73. Bhalla AK, Amento EP, Serog B, Glimcher LH 1984 1,25-Dihydroxyvitamin D_3 inhibits antigen-induced T cell activation. J Immunol **133**:1748–1754.
74. Nunn JD, Katz DR, Barker S, Fraher LJ, Hewison M, Hendy GN, O'Riordan JLH 1986 Regulation of human tonsillar T-cell proliferation by the active metabolite of vitamin-D_3. Immunology **59**:479–484.
75. Rigby WFC, Noelle RJ, Krause K, Fanger MW 1985 The effects of 1,25-dihydroxyvitamin D_3 on human T-lymphocyte activation and proliferation: A cell-cycle analysis. J Immunol **135**:2279–2286.

76. Karmali R, Hewison M, Rayment N, Farrow SM, Brennan A, Katz DR, O'Riordan JLH 1991 $1,25(OH)_2D_3$ regulates c-myc messenger-RNA levels in tonsillar T-lymphocytes. Immunology **74**:589–593.
77. Rigby WFC, Denome S, Fanger MW 1987 Regulation of lymphokine production and human lymphocyte T activation by 1,25-dihydroxyvitamin-D_3—Specific inhibition at the level of messenger RNA. J Clin Invest **79**:1659–1664.
78. Rigby WFC, Hamilton BJ, Waugh MG 1990 1,25-Dihydroxyvitamin-D_3 modulates the effects of interleukin-2 independent of IL-2 receptor binding. Cell Immunol **125**:396–414.
79. Bhalla AK, Amento EP, Krane SM 1985 An enhancement of interleukin-1 production from monocytes by 1,25-dihydroxyvitamin D_3. Br J Rheumatol **24**:108–109.
80. Bhalla AK, Amento EP, Krane SMI 1985 Effects of 1,25-dihydroxyvitamin D_3 on interleukin-1 production by the human monocyte cell line U937 and peripheral blood monocytes. Br J Rheumatol **24**:143–146.
81. Fagan DL, Prehn JL, Adams JS, Jordan SC 1991 The human myelomonocytic cell line U937 as a model for studying alterations in steroid-induced monokine gene expression: Marked enhancement of lipopolysaccharide-stimulated interleukin-1-beta messenger RNA levels by 1,25-dihydroxyvitamin D_3. Mol Endocrinol **5**:179–186.
82. Meehan MA, Kerman RH, Lemire JM 1992 1,25-Dihydroxyvitamin D_3 enhances the generation of non-specific suppressor cells while inhibiting the induction of cytotoxic cells in human MLR. Cell Immunol **140**:400–409.
83. Mathieu C, Waer M, Laureys J, Tutgeerts O, Bouillon R 1994 Prevention of type 1 diabetes in NOD mice by 1,25-dihydroxyvitamin D_3. Diabetologia **37**:552–558.
84. Hsieh C-S, Macatonia SE, Tripp CS, Wolf SF, O'Garra A, Murphy KM 1993 Development of T_H1 CD4+ T cells through IL-12 produced by *Listeria*-induced macrophages. Science **260**:547–549.
85. Lemire JM, Adams JS, Kermaniarab V, Bakke AC, Sakai R, Jordan SC 1985 1,25-Dihydroxyvitamin D_3 suppresses human T-helper inducer lymphocyte activity in vitro. J Immunol **134**:3032–3035.
86. Reichel H, Koeffler HP, Tobler A, Norman AW 1987 1-alpha,25-dihydroxyvitamin D_3 inhibits gamma-interferon synthesis by normal human peripheral blood lymphocytes. Proc Natl Acad Sci USA **84**:3385–3389.
87. Lemire JM, Archer DC, Beck L, Spiegelberg HL 1995 Immunosuppressive actions of 1,25-dihydroxyvitamin D_3: Preferential inhibition of Th-1 functions. J Nutr **125**:1704–1708.
88. Lemire JM 1995 Immunomodulatory actions of 1,25-dihydroxyvitamin D_3. J Steroid Biochem Mol Biol **53**:599–602.
89. Clerici M, Shearer GM 1993 A T_H1 to T_H2 switch is a critical step in the etiology of HIV infection. Immunol Today **14**:107–110.
90. Bouillon R, Garmyn M, Verstuyf A, Segaert S, Casteels K, Mathieu C 1995 Paracrine role for calcitriol in the immune system and skin creates new therapeutic possibilities for vitamin D analogs. Eur J Endocrinol **133**:7–16.
91. Gerritsen MJ, Rulo JF, Van Vlijmen-Willems I, Van Erp PE, van der Kerkhof PC 1993. Topical treatment of psoriatic plaques with 1,25-dihydroxyvitamin D_3: A cell biological study. Br J Dermatol **128**:666–673.
92. Kragballe K 1992 Vitamin D analogues in the treatment of psoriasis. J Cell Biochem **49**:46–52.
93. Shiozawa K, Shiozawa S, Shimizu S, Fujita T 1985 1α,25-Dihydroxyvitamin D_3 inhibits pokeweed mitogen-stimulated human B-cell activation: An analysis using serum-free culture conditions. Immunology **56**:161–168.
94. Iho S, Takahashi T, Kura F, Sugiyama H, Hoshino T 1986 The effect of 1,25-dihydroxyvitamin D_3 on in vitro immunoglobulin production in human B-cells. J Immunol **136**:4427–4431.
95. Walka MM, Daumling S, Hadorn HB, Kruse K, Belohradsky BH 1991 Vitamin D dependent rickets type II with myelofibrosis and immune dysfunction. Eur J Pediatr **150**:665–668.
96. Kolb E, Grun E 1996 D-vitamins and the immune system. Praktische Tierarzt **77**:11–15.
97. Dormanen MC, Bishop JE, Hammond MW, Okamura WH, Nemere I, Norman AW 1994 Non-nuclear effects of the steroid hormone 1-alpha,$25(OH)_2$vitamin D_3 analogs are able to functionally differentiate between nuclear and membrane receptors. Biochem Biophys Res Commun **201**:394–401.
98. Nemere I, Dormanen MC, Hammond MW, Okamura WH, Norman AW 1994 Identification of a specific binding protein for 1-alpha,25-dihydroxyvitamin D_3 in basal lateral membranes of chick intestinal epithelium and relationship to transcaltachia. J Biol Chem **269**:23750–23756.

Other Vitamin D Target Tissues: Vitamin D Actions in Cardiovascular Tissue and Muscle, Endocrine and Reproductive Tissues, and Liver and Lung

MARIAN R. WALTERS Department of Physiology, Tulane University School of Medicine, New Orleans, Louisiana

I. INTRODUCTION

As reviewed previously [1,2], there are numerous vitamin D target tissues other than the classic Ca^{2+} translocating 1,25-dihydroxyvitamin D_3 [$1,25(OH)_2D_3$] targets intestine, kidney, and bone. Recognition of the ubiquitous distribution of vitamin D targets began in the late 1970s when vitamin D receptor (VDR)-like binding sites were observed in unexpected tissues such as pancreas [3,4], placenta [4], and testis [5]. Subsequent studies from many laboratories employed biochemical approaches, including Scatchard analysis (Fig. 1), sucrose density gradient ultracentrifugation (Fig. 2), and DNA-cellulose chromatography of ^{3}H-1,25(OH)$_2$D$_3$ binding sites, the technique of ^{3}H-1,25(OH)$_2$D$_3$ autoradiographic localization (Fig. 3), and more recently immunocytochemical and Northern analysis, to provide convincing evidence of detectable VDR sites in numerous tissues (reviewed by Walters [1,2]).

In classic vitamin D targets, 1,25(OH)$_2$D$_3$ effects often include induction of one or more calcium binding proteins (CaBPs), particularly calbindin-D$_{9K}$ or -D$_{28K}$ [6] (see also Chapters 13 and 14). Although the functions of these proteins are not completely understood, the parallels between their induction and known 1,25(OH)$_2$D$_3$ effects in the classic targets strongly suggest their involvement therein [1,6] (Chapter 16). However, extensive studies have demonstrated that there is not an obligatory relationship between VDR effects and induction of calbindin-D$_{9K}$ or calbindin-D$_{28K}$, both because the presence/induction of these CaBPs is not de-

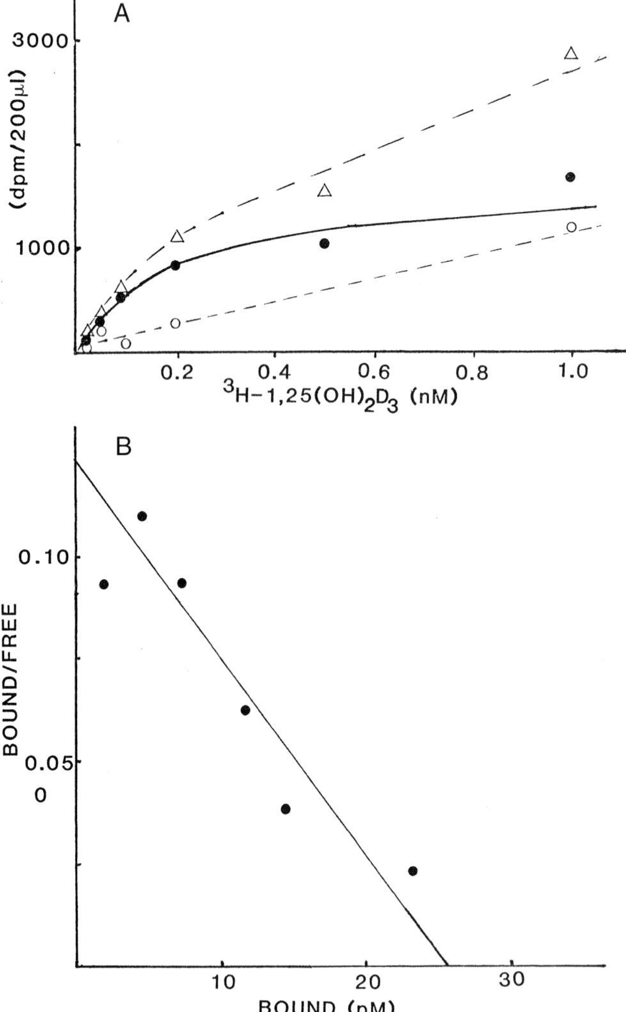

FIGURE 1 Saturation (A) and Scatchard (B) analysis of ^{3}H-1,25(OH)$_2$D$_3$ binding to chromatin preparations of adrenal gland of normal rats. Tissue was homogenized (20%, w/v) in TEDMo buffer containing Trasylol and phenylmethylsulfonyl fluoride (PMSF), and nuclear pellets were washed three times in TED-Triton. The crude chromatin preparation was resuspended in TEDMo. Aliquots (200 μl) were incubated at 4°C overnight with the indicated concentrations of ^{3}H-1,25(OH)$_2$D$_3$ in the presence (nonspecific binding) or absence (total binding) of 1μM 1,25(OH)$_2$D$_3$. Separation of bound and free hormone was accomplished by the hydroxyapatite assay. △, Total ^{3}H-1,25(OH)$_2$D$_3$ binding; ○, nonspecific binding; ●, specific binding (total minus nonspecific). A k_d value of 0.24 ± 0.8 nM (n = 4) was obtained. From J. Carron, J. Fox, and M. Walters, unpublished data, 1992.

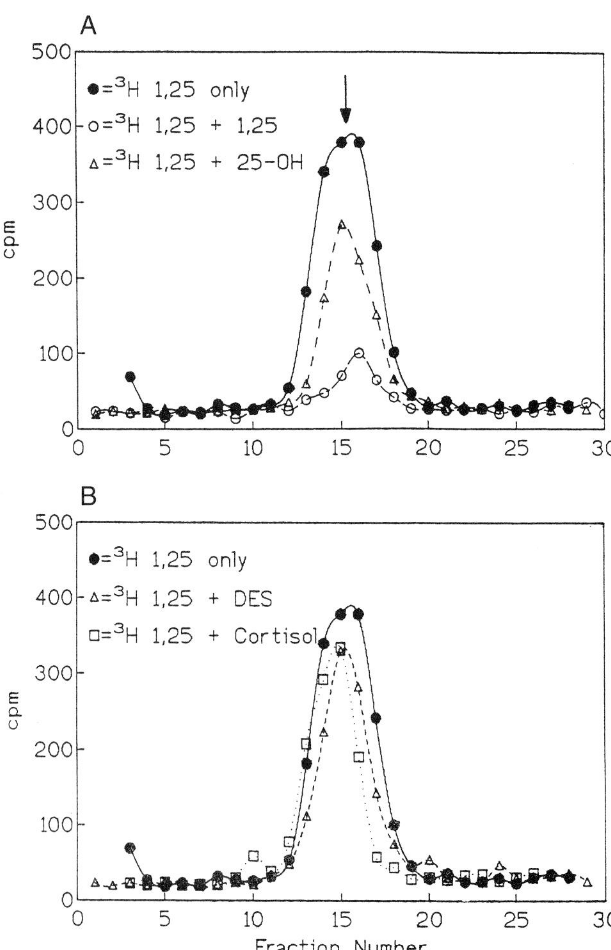

FIGURE 2 Sucrose density gradient analysis of ^{3}H-1,25(OH)$_2$D$_3$ binding in rat adrenal gland. Chromatin preparations incubated with 1.0 nM ^{3}H-1,25(OH)$_2$D$_3$ alone or with 50 nM 25OHD$_3$, 1 μM 1,25(OH)$_2$D$_3$, diethylstilbestrol, or cortisol were subjected to sucrose density gradient analysis followed by hydroxyapatite assay on each fraction. Arrow indicates the location of the VDR peak detected in kidney in similar experiments. From J. Carron, J. Fox, and M. Walters, unpublished data, 1992.

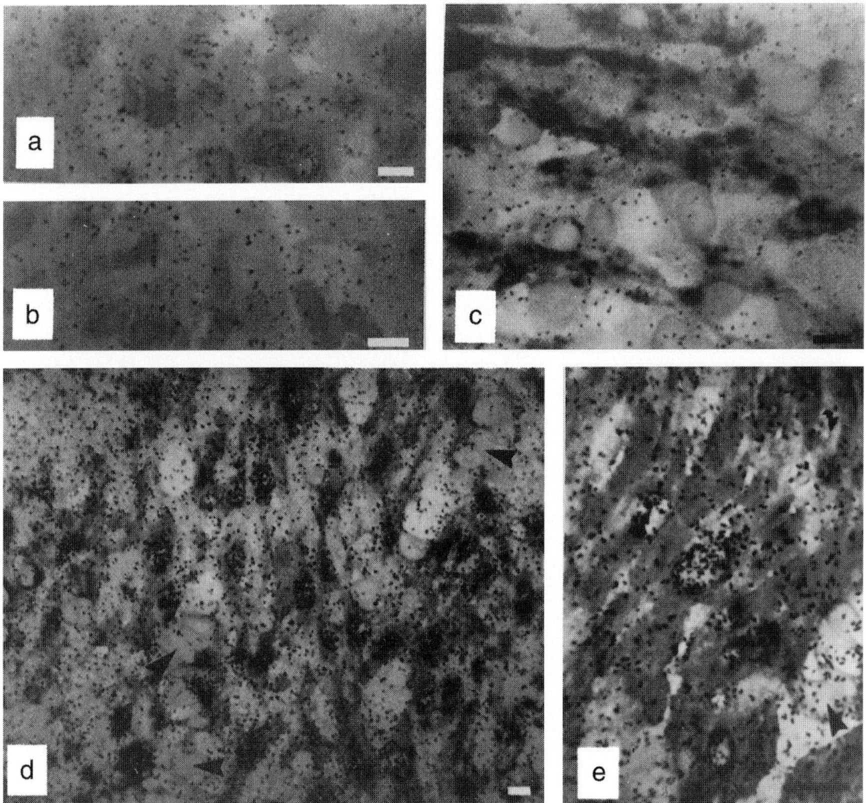

FIGURE 3 Autoradiography of ^3H-1,25(OH)$_2$D$_3$ in the heart of mice. (a) Autoradiograms showing nuclear concentration of ^3H-1,25(OH)$_2$D$_3$-related silver grains over many cardiomyocytes of the right atrium. (b) Competition controls from mice that had received unlabeled 1,25(OH)$_2$D$_3$ prior to the tracer show no nuclear silver grain concentration over cardiomyocytes of the right atrium. (c) In ANF positive cardiomyocytes in the left and right ventricles, binding sites for ^3H-1,25(OH)$_2$D$_3$ are absent. (d) In the posterior right atrium, more than 90% of the ANF-positive cardiomyocytes exhibit nuclear 1,25(OH)$_2$D$_3$ binding sites. (e) Only a few cell clusters (two to four ANF positive cells) in the left atrium concentrate 1,25(OH)$_2$D$_3$ in their nuclei. In both atria 1,25(OH)$_2$D$_3$ binding is not detectable in endothelial cells (arrowheads in d and e). Bar, 15 μm. Reprinted with permission from Bidmon *et al.* [33].

pendent on vitamin D in many tissues (e.g., brain, uterus, and lung) and because their induction does not occur in all 1,25(OH)$_2$D$_3$ targets (Fig. 4) [7,8] (see also Chapters 13, 14, and 16).

Functional studies of 1,25(OH)$_2$D$_3$ effects following demonstration of VDR sites therein have led to elucidation of a variety of roles for the vitamin D hormone, for example, regulation of growth and differentiation in hematolymphopoietic cells (Chapter 29), skin (Chapter 25), and mammary gland/breast cancer (Chapter 65), involvement in nongenomic signaling pathways in skeletal muscle cells and hepatocytes (see below), and tissue specific effects in pancreas (see below and Chapter 70), as well as a variety of other targets. The broad range of differing effects described in vitamin D target tissues to date likely reflects the pleiotropic nature of the effects of 1,25(OH)$_2$D$_3$ and the VDR in all targets tissues, including the classic Ca^{2+} translocating tissues.

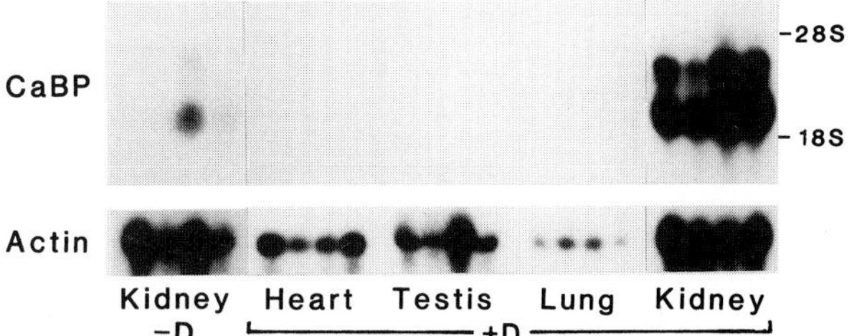

FIGURE 4 Northern analysis of calbindin-D$_{28K}$ expression in selected tissues from vitamin D-deficient (−D) or 1,25(OH)$_2$D$_3$-stimulated (100 ng/day for 7 days, +D) rats. Aliquots of total RNA (heart and lung, 20 μg; testis and kidney, 40 μg) from individual rats were separated by electrophoresis in the presence of formaldehyde, vacuum-blotted to Zeta-probe membrane (Bio-Rad), and incubated with ^{32}P-labeled cDNA to rat calbindin-D$_{28K}$ or mouse β-actin. Overexposure did not reveal additional bands in nonkidney tissues. Reprinted with permission from Walters and Hunziker [8].

This chapter addresses the effects of vitamin D, principally via the hormonal form $1,25(OH)_2D_3$, in a variety of targets not covered elsewhere in this volume. Each section of the chapter provides historical and/or clinical observations, outlines the studies leading to definition of the tissue/organ as a vitamin D target, and discusses subsequent studies of specific actions of $1,25(OH)_2D_3$ in each target. In retrospect, it is clear that much of the functional information either describes rapid, possibly nongenomic, effects of the vitamin D hormone (Chapter 15) and related signal transduction events or involves more long-term effects on functions such as gene expression and growth. Thus, more extensive details of some of the studies are given in a few instances to illustrate the evolution of these concepts with respect to $1,25(OH)_2D_3$ action.

II. CARDIOVASCULAR TARGETS

A. Heart

Although not a definitive correlate of the presence and actions of VDR in cardiac muscle cells, cardiac lesions resulting at least in part from excess calcium accumulation have been described after excessive vitamin D exposure in numerous species [9–14] (Chapter 54). Similar soft tissue calcification (calcinosis) can also occur in uremic patients given vitamin D therapy, depending on the dose [15,16]. In the most extensively studied model, dosing rats with $1–5 \times 10^5$ IU vitamin D results within a few days in myocardial lesions associated with Ca^{2+} deposits throughout necrotic and nonnecrotic tissue and significant alterations in biochemical, enzymatic, and contractile function 17–21]. Such soft tissue calcifications are likely due to the hypercalcemia state induced by excessive vitamin D.

Other experimental and clinical studies demonstrate beneficial effects of more modest vitamin D levels on cardiac function. Thus, vitamin D-deficient rats treated with 10^3 IU vitamin D demonstrated improved cardiac contractility in response to cardiac glycosides [22]. Moreover, treatment of uremic and/or hemodialysis patients with $1\alpha OHD_3$ or $25OHD_3$ results in improved left ventricular function [23–25]. These observations, particularly those that derive from clinical observations, probably reflect in part the extensive parathyroid hormone (PTH) effects, or their reversal, on the heart and on mechanisms regulating calcium therein [23–29].

More specific studies of $1,25(OH)_2D_3$ function in cardiac muscle followed the identification of ^3H-$1,25(OH)_2D_3$ binding sites in cultured rat cardiac muscle cells [30,31] and rat heart [32]. Subsequent autoradiographic and biochemical studies provided evidence for a general distribution of VDR throughout the heart, with some concentration in the atria (Fig. 3) [1,33].

As reviewed previously [1], an initial series of studies of $1,25(OH)_2D_3$ effects in the heart (and some other targets) focused on the hypothesis that the VDR functions to regulate intracellular Ca^{2+} homeostasis and/or Ca^{2+} regulatory functions within the cells. Tests of this hypothesis demonstrated rapid (significant within 15 min), specific, and cycloheximide-inhibitable effects of $1,25(OH)_2D_3$ on $^{45}Ca^{2+}$ uptake in cultured adult rat ventricular cardiac muscle cells, with no effects on cell growth or morphology [31]. Mechanistic studies of a similar phenomenon in chick cardiac ventricular slices indicated that $^{45}Ca^{2+}$ uptake could be prevented by the calcium channel blockers nifedipine and verapamil and did not correlate with changes in lipid biosynthesis, although there was an attendant stimulation of ^{32}P incorporation into microsomal proteins of 43 and 55 kDa [34]. Additionally, in cultured chick cardiac muscle cells, a $1,25(OH)_2D_3$ effect on $^{45}Ca^{2+}$ uptake was accompanied by stimulation of cAMP levels. Involvement of G protein and cAMP activation in these effects was suggested from observations that these $1,25(OH)_2D_3$ effects on $^{45}Ca^{2+}$ uptake and cAMP levels were abolished both by the adenylate cyclase inhibitor SQ22536 and by the G-protein inhibitor GDP-β-S [35]. Finally, although the relationship is not yet clear, a report demonstrates that sphingomyelin hydrolysis is stimulated in rat hearts after 1 hr of $1,25(OH)_2D_3$ exposure, which was interpreted as implicating protein kinase C (PKC) in these early $1,25(OH)_2D_3$ actions [36].

The effect of $1,25(OH)_2D_3$, or conversely vitamin D deficiency, in the heart has been studied extensively in the rat, but it remains problematic because of the complication of attendant alterations in plasma and tissue calcium levels. As previously reviewed [37], a particularly extensive series of functional studies demonstrated that vitamin D deficiency results in increased cardiac contractile function in this model. In these studies, participation of hypocalcemia and hyperparathyroidism in this effect was thought to have been ruled out, and it was shown that changes in the calcium slow channel and calcium uptake by sarcoplasmic reticulum were not involved in this vitamin D effect [37]. However, in similar studies in the chick, alterations in cardiac contractility and in the number of slow calcium channels correlated with the degree of hypocalcemia, rather than vitamin D deficiency [38].

On gross examination, the heart of a vitamin D-deficient rat has a flacid, enlarged appearance. This change is reflected in an increased heart weight/body weight ratio [39] (Table I). A similar effect is not observed in the kidney (Table I), and here again the role of altered calcium levels is unclear [39] (Table I). Further examina-

TABLE I Effect of Vitamin D Deficiency on Heart and Kidney Weights in the Rat[a]

Animal status	Ratio (× 1000)	
	Kidney/body weight (n)	Heart/body weight (n)
−D, low Ca	3.19 ± 0.11 (6)	2.79 ± 0.01 (6)**
−D, normal Ca	3.16 ± 0.01 (6)	2.61 ± 0.09 (6)
−D, normal Ca + 7 days, 1,25 $(OH)_2D_3$	3.01 ± 0.15 (6)	2.42 ± 0.08 (5)

[a] Rats were raised on the indicated vitamin D-deficient (−D) diets or treated with 100 ng $1,25(OH)_2D_3$ daily for 7 days prior to sacrifice and determination of the indicated weights. Data represent means ± SEM for five to six individual rats. $^{**}p < 0.01$ versus $1,25(OH)_2D_3$-treated group. From S. A. Antrobus and M. R. Walters, unpublished observations, 1988.

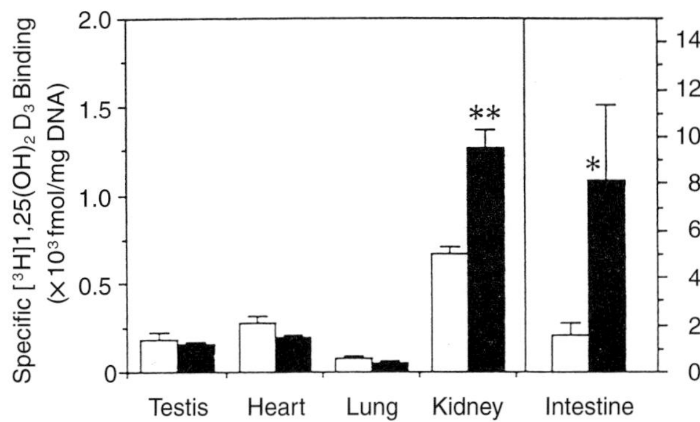

FIGURE 5 Effect of $1,25(OH)_2D_3$ injection on VDR levels in several tissues in the rat. Vitamin D-deficient rats were injected daily for 7 days with 100 ng $1,25(OH)_2D_3$ (solid bars) or vehicle control (open bars). VDR levels were determined by incubation with 1.0 nM ^3H-$1,25(OH)_2D_3$ ± 1 μM $1,25(OH)_2D_3$ and are expressed as means ± SEM for six individual rats. *, $p < 0.05$, **, $p < 0.005$ versus control by Student's one-tailed t-test. Reprinted with permission from Gensure *et al.* [45].

tion of the cardiomegaly in vitamin D-deficient rats showed decreased myofibrillar space, increased extracellular space, smaller but more numerous myocytes typical of hyperplasia, and increased levels of myocardial collagen, c-Myc, and myosin isozyme VI [39–41]. The latter observation is particularly interesting in light of the vitamin D-linked effects on cardiac contractility discussed above. Complementing the observations of cardiac hypertrophy in vitamin D deficiency is a report that $1,25(OH)_2D_3$ antagonizes endothelin-induced hypertrophy in neonatal rat cardiac myocytes [42]. Finally, in a similar model of cultured rat ventricular myocytes, $1,25(OH)_2D_3$ blocked cell maturation, apparently by a PKC-related mechanism [43].

Metabolic and biochemical consequences of vitamin D deficiency and/or $1,25(OH)_2D_3$ treatment have also been examined in the heart. ATP, creatine phosphate, and intracellular pH levels are similar in chicks raised on vitamin D-deficient or vitamin D-supplemented diets [38]. In rats, citrate content and the activity of a number of metabolic enzymes (NAD^+-dependent isocitrate dehydrogenase, citrate synthase, glyceraldehyde-3-phosphate dehydrogenase, 3-phosphoglycerate kinase, and acylphosphatase) increased with vitamin D dietary supplementation in vitamin D-deficient rats treated with $1,25(OH)_2D_3$ [44]. Exceptions included acid phosphatase, which decreased with these treatments, and cytochrome-c oxidase, which did not change [44]. A number of other vitamin D-associated biochemical parameters have been investigated in rat heart with negative results. Thus, $1,25(OH)_2D_3$ treatment in the rat does not alter cardiac levels of VDR (Fig. 5) [45], small calcium binding proteins [7] or calbindin-D_{28K} mRNA (Fig. 4) [7,8], or calmodulin binding proteins (Fig. 6) [46]. Consistent with VDR localization in atrial myocytes producing atrial natriuretic factor (ANF), $1,25(OH)_2D_3$ inhibits cardiac ANF release and gene expression [47,48], and this effect does not seem to be altered by plasma calcium levels. Finally, although not strictly a cardiac effect, cardiac graft rejection is reduced after treatment of animal models with $1,25(OH)_2D_3$ analogs [49,50].

Collectively, these observations indicate that $1,25(OH)_2D_3$ treatment exerts specific effects on a broad range of biochemical and functional events in cardiac muscle and endocrine elements in the atria. These effects are independent of effects on classically vitamin D-regulated phenomena in these tissues.

B. Vascular Smooth Muscle

In vascular smooth muscle, as in the heart (see above and Chapter 54), hypervitaminosis D results in cellular damage due to calcinosis in several species [10–15,51–53]. In fact, hypervitaminosis D treatments with cholesterol or nicotine supplementation have been explored as a model of experimental atherosclerosis [54,56]. Additional evidence of possible functional links between the vitamin D endocrine system and smooth muscle comes from the controversial hypothesis that altered levels of calcium regulatory hormones contribute to the development of salt-dependent hypertension [1,57–61]. Based in part on these experimental observations of possible deleterious effects of vitamin D treatment, VDR sites were described in cultured vascular smooth muscle cells by Scatchard and sucrose density gradient analysis [62–64] and by autoradiographic localization [65].

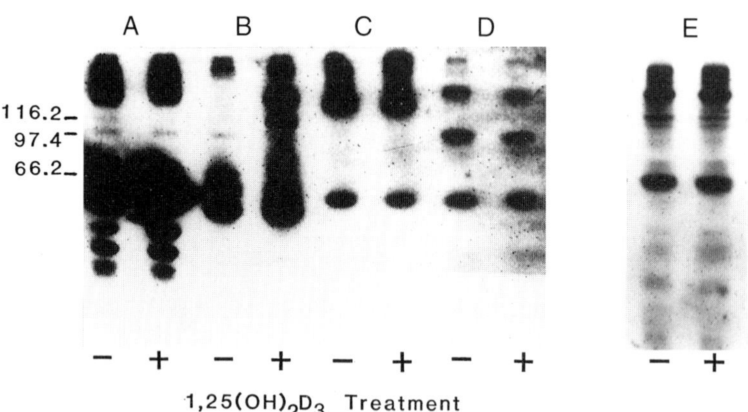

FIGURE 6 Tissue specificity of $1,25(OH)_2D_3$ induction of calmodulin binding proteins in the rat. After sacrifice of rats treated daily for 7 days with vehicle (−) or 100 ng/day $1,25(OH)_2D_3$ (+), ^{125}I-calmodulin binding was compared in cytosol preparations of brain (set A), kidney (set B), testis (set C), heart (set D), and intestinal mucosa (set E, separate experiment). Reprinted with permission from Antrobus and Walters [46].

In vitamin D-deficient rats, increased contractility of isolated aortic rings after norepinephrine exposure was reversed by improved calcium levels, suggesting that the effect was not a direct effect of vitamin D withdrawal [37]. Conversely, $1,25(OH)_2D_3$ treatment in normotensive or hypertensive rats resulted in calcium-independent enhancement of contractile force generation in mesenteric resistance arteries [66]. This effect was accompanied by elevated blood pressure *in vivo* and developed over a 3- to 7-day period of $1,25(OH)_2D_3$ treatment, suggesting that the $1,25(OH)_2D_3$ effect occurs via a genomic mechanism [66]. Similarly, although there is one report [67] of rapid effects of $1,25(OH)_2D_3$ to increase Ca^{2+} channel activity (5–10 min) and intracellular Ca^{2+} concentrations ($[Ca^{2+}]_i$, 3–4 min), most $1,25(OH)_2D_3$ effects on vascular smooth muscle described to date are likely to be genomic in nature. Thus, even vitamin D effects on prostacyclin synthesis, $^{45}Ca^{2+}$ uptake, and Ca^{2+}-ATPase activity require 10 hr to 7 days of treatment and are inhibited by cycloheximide [68–71]. The $1,25(OH)_2D_3$-induced growth effects—usually but not always [72,73] growth inhibition [64,73,74]—are accompanied by parallel changes in [^{3}H]thymidine incorporation. Other genes reported to be regulated by $1,25(OH)_2D_3$ in these systems include tropoelastin [75], *fos* and *jun*, [76], and the heavy and light chains of myosin [77].

III. SKELETAL MUSCLE

Clinically, vitamin D insufficiency associated with osteomalacia or chronic renal failure is accompanied by skeletal muscle myopathies (reviewed by Boland [78]). The myopathy improves on vitamin D treatment in humans [79,80] and in experimental models [79–83]. Some of the defect, and the improvement with therapy, may be due to changes in circulating levels of calcium and phosphate. However, direct effects on muscle may be part of the picture. Early functional studies correlated muscle changes with alterations in function and morphology of the sarcoplasmic reticulum, troponin C concentration, and impaired ATP and creatine phosphate production [78]. Thus, direct vitamin D-dependent biochemical and functional defects in skeletal muscle function may be a significant treatable cause of myopathy in a variety of clinical conditions.

Finding nuclear VDR by biochemical techniques in G8 and chick embryo skeletal muscle cells [30,84] and by autoradiography in skeletal muscle of several species [85,86] has strengthened the concept that vitamin D may exert direct effects on skeletal muscle function. Costa *et al.* demonstrated the presence of VDR in cultured human skeletal muscle cells and documented $1,25(OH)_2D_3$ induction of 24-hydroxylase activity [86a].

In a particularly extensive series of studies, the team of de Boland and Boland have used chick skeletal muscle myoblasts and soleus muscle to delineate potential nongenomic pathways related to the actions of $1,25(OH)_2D_3$ [87]. As modeled in Fig. 7, their studies have shown that phospholipases (C, A_2, D), adenylate cyclase, and protein kinase C are rapidly activated by $1,25(OH)_2D_3$ [87]. These $1,25(OH)_2D_3$-induced changes, as well as G-protein activation, lead to a rapid increase in cellular calcium uptake into skeletal muscle, an effect mimicked by the Ca^{2+} channel activator BAY

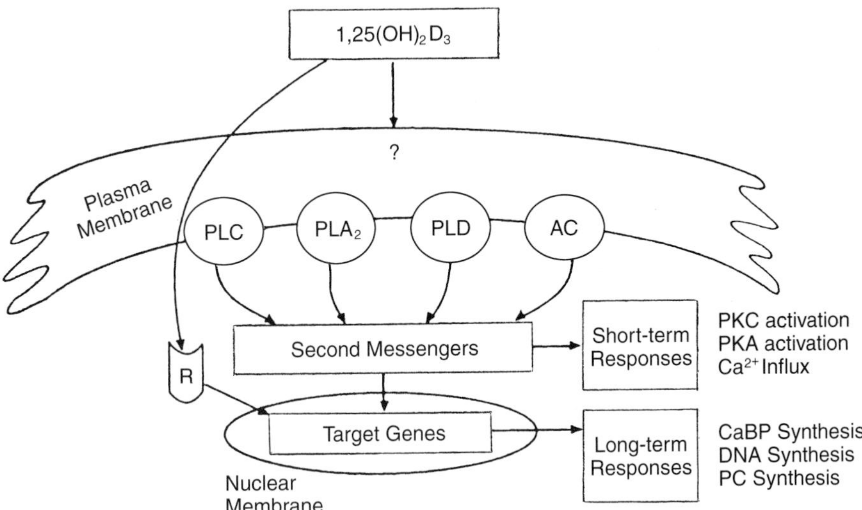

FIGURE 7 Integrated nongenomic and genomic modes of action of 1,25(OH)$_2$D$_3$ in target muscle cells. PLC, PLA$_2$, PLD, phospholipases C, A$_2$, and D, respectively: AC, adenylate cyclase; R, intracellular 1,25(OH)$_2$D$_3$ receptor. The question mark refers to an as yet hypothetical plasma membrane-bound 1,25(OH)$_2$D$_3$ binding protein. Reprinted by permission of the publisher from A. R. de Boland and R. L. Boland, Non-genomic signal transduction pathway of vitamin D in muscle, *Cellular Signalling* Vol. 6, pp. 717–724 [87]. Copyright 1994 by Elsevier Science Inc.

K8644 and inhibited by Ca^{2+} channel antagonists (Fig. 8) [87]. Other events in the rapid effects of 1,25(OH)$_2$D$_3$ in this sytem include cAMP-dependent phosphorylation of microsomal proteins of 28–29, 55, and 250 kDa. There is also a rapid calmodulin redistribution to microsomal membranes and binding to a 28-kDa microsomal protein by a mechanism that does not require elevated $[Ca^{2+}]_i$ [87].

Collectively, these observations have been interpreted as follows: 1,25(OH)$_2$D$_3$ interaction with an as yet unidentified membrane receptor activates a host of membrane signaling pathways, resulting in phospholipid turnover and cAMP and PKC activation (Fig. 7) [87]. In particular, there follows cAMP-dependent phosphorylation of a 28-kDa membrane protein, which causes calmodulin redistribution to the membrane and subsequently activates Ca^{2+} channels, putatively voltage-regulated, by an unknown mechanism to cause rapid Ca^{2+} entry into the cells [87]. PKC may be involved either through attenuation of the cAMP-dependent pathway (cross talk) or after phospholipase-generated diacylglycerol interacts with the elevated calcium (induced by cAMP-dependent pathways) to maximize a PKC signal for subsequent events [87]. Although many questions remain unanswered in this system [87], the efforts of this team have provided one of the most well-studied models of potential nongenomic 1,25(OH)$_2$D$_3$ effects (see Chapter 15).

In efforts to also elucidate long-term, genomic 1,25(OH)$_2$D$_3$ effects on skeletal muscle, 1,25(OH)$_2$D$_3$ was shown to exert varying effects on DNA synthesis [30,88]. This observation has been interpreted as representing first an early effect to stimulate myoblast proliferation accompanied by an increased c-*myc*/c-*fos*

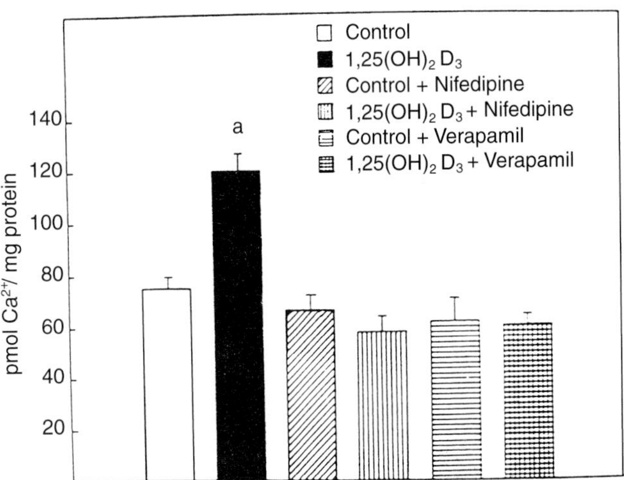

FIGURE 8 Rapid stimulation of muscle calcium uptake by 1,25(OH)$_2$D$_3$ and its suppression by calcium channel antagonists. Soleus muscle obtained from vitamin D-deficient chicks was used. $^{45}Ca^{2+}$ uptake (2 min) by the tissue was measured after treatment with 1,25(OH)$_2$D$_3$ for 3 min in the absence and presence of verapamil or nifedipine (50 μM). a, $p < 0.001$. Reprinted with permission from Boland R 1986 Role of vitamin D in skeletal muscle function. Endocrinology Review **7**:434–448; © The Endocrine Society.

mRNA ratio. There followed a later myoblast differentiation effect that was accompanied by increased synthesis of myofibrillar and microsomal proteins [88]. Stimulation of proliferation in this system is accompanied by elevated synthesis of proteins (^{3}H]leucine incorporation) of a variety of sizes and subcellular locales, including in particular calmodulin and a 110-kDa cytoskeletal calmodulin binding protein [89,90]. Thus, $1,25(OH)_2D_3$ affects structural and functional elements in skeletal muscle, and these $1,25(OH)_2D_3$ effects involve genomic mechanisms as well as apparent nongenomic actions (Fig. 7).

IV. REPRODUCTIVE TISSUES

Despite the essential importance of vitamin D as a regulator of mineral metabolism and availability, the question of whether vitamin D is required for normal reproduction has not been fully answered. Vitamin D-deficient rat mothers produce rather normal offspring, although there is some evidence for reductions in number and size of the pups [91,92]. In male rats, vitamin D deficiency results in disrupted spermatogenesis and reduced, but not abolished, reproductive potency [93–95]. Whether the rather modest reproductive defects in both male and female rats result from the hypocalcemia secondary to vitamin D deficiency or are due to more direct effects of the vitamin D hormones on the tissues remains unresolved.

A. Female

The presence of VDR in various mammary gland and breast cancer models has been described previously from a number of laboratories [e.g., 96–102] as reviewed earlier [1]. The presence of VDR in these tissues has created significant interest, particularly in light of observations that $1,25(OH)_2D_3$ inhibits breast cancer cell growth [103–106]. Much current interest is focused on developing less calcemic $1,25(OH)_2D_3$ analogs as possible therapeutic tools for breast cancer, either alone or in combination with other drugs (Chapter 65).

In the rat uterus, low levels of VDR have been reported using a variety of techniques [107–110], and the levels of uterine VDR are up-regulated by estrogen [107,108]. There are significant levels of calbindin-D_{9K} in the uterus; however, like the VDR, the uterine calbindin-D_{9K} is up-regulated by estrogen but not by $1,25(OH)_2D_3$ (reviewed by Christakos *et al.* [6]; see also Chapter 14). A rather novel study has also reported evidence for estrogen inhibition of uterine calbindin-D_{28K} expression, which may explain its absence in uteri of mature animals [111]. As in other reproductive tissues, the functions of VDR in the uterus and whether they relate exclusively to placental Ca^{2+} and phosphate transfer (see below) are not completely defined. However, $1,25(OH)_2D_3$ or $24,25(OH)_2D_3$ treatment alters sex steroid effects on alkaline phosphatase in cultured myometrial cells [112], and intraluminal $1,25(OH)_2D_3$ administration *in vivo* induces endometrial decidualization [113].

In the avian egg shell gland (sometimes termed the avian uterus), several features of the vitamin D endocrine system are similar to those in the mammalian uterus. Consistent with its role in eggshell mineralization, this tissue contains significant levels of VDR that are estrogen inducible [114–116]. Moreover, the chicken calbindin-D_{28K} is present in this tissue and is regulated by estrogen, but not vitamin D, treatment [117].

The appreciation that the placenta is a vitamin D target tissue has developed from studies of the regulation of placental–fetal Ca^{2+} and phosphate transfer (Chapter 28). Briefly, VDR have been defined in placental tissues of a number of species [4,118–120]. Additionally, the mammalian intestinal calbindin-D_{9K} is present in the placenta, but, similar to the uterus and lung, placental calbindin-D_{9K} is not regulated by $1,25(OH)_2D_3$ (Chapter 14). However, the vitamin D endocrine system affects a variety of other enzymatic and hormonal functions in the placenta (Chapter 28).

Finally, there is some evidence for VDR in ovarian models [110,121]. In Chinese hamster ovary (CHO) cell cultures, for example, $1,25(OH)_2D_3$ inhibited cell growth and protein synthesis, demonstrating the potential for effects of the vitamin D endocrine system on ovarian function [121].

B. Male

A number of groups have established the presence of VDR in testis, and several of these studies define the likely VDR cellular localization as the Sertoli cells [122–128]. Consistent with Sertoli cell localization, the testis VDR levels increase with sexual maturation [124,129] and increase with hormonal stimulation or decrease with disruption of Sertoli cell function [93,130].

Despite the many reports of testicular VDR, the hormonal functions of vitamin D in the testis are not well understood. Several reports suggest that spermatogenesis is disrupted in vitamin D deficiency [93,94], but, again, a role of altered Ca^{2+} underlying this effect has not been fully ruled out. Several studies also suggest possible $1,25(OH)_2D_3$ effects on Leydig (interstitial) cell function, including improved cAMP response to luteinizing hormone stimulation in uremic rats [131] and antagonism of interferon-γ-induced expression of HLA class II antigens [132]. The latter observation implies Sertoli-to-Leydig cell communication because of the apparent lack of VDR in the interstitial cells (see above). Other scattered reports of the effects of the vitamin D endocrine system on Sertoli cell function include altered follicle stimulating hormone function [133] and a preliminary report of effects on phosphate transport [134]. Furthermore, extensive biochemical and molecular approaches have provided no evidence for the presence of the vitamin D-related calbindin-D_{28K} in the rat testis (Fig. 4) [7,8]. There is, in addition, no evidence for $1,25(OH)_2D_3$ regulation of VDR levels (Fig. 5) or calmodulin binding proteins (Fig. 6) in this tissue.

Complementing the incomplete information on $1,25(OH)_2D_3$ effects on testis *in vivo*, its hormonal effects have been studied in mouse TM_4 Sertoli-derived cells [127]. In this system, a 2-hr exposure to $1,25(OH)_2D_3$ increased $^{45}Ca^{2+}$ uptake into the cells by a mechanism dependent on protein synthesis [127]. This $1,25(OH)_2D_3$ effect at 2 hr was accompanied by detectable increases in protein content and DNA synthesis, which were sustained at 24 hr and were not accompanied by changes in cell number or morphology [127]. Subsequent studies demonstrated that $1,25(OH)_2D_3$ treatment in the cells results in short-term VDR down-regulation, not due to VDR masking by hormone occupancy, as well as increased estrogen receptor levels (V.L. Akerstrom and M. R. Walters, unpublished observations, 1991).

Low levels of apparent VDR have also been demonstrated in epididymis [124,135] and vas deferens [135]. Moreover, despite reports of the apparent absence of VDR in the rat prostate [125], there is evidence for VDR in mouse prostate [135] and in a variety of prostatic cancer cell lines [136,137]. The latter observations and evidence for $1,25(OH)_2D_3$ inhibition of prostate cell growth *in vitro* has led to significant interest in $1,25(OH)_2D_3$ and its analogs as chemotherapeutic agents in prostate cancer (Chapter 66).

V. OTHER ENDOCRINE TISSUES

A. Adrenal Glands

Studies of vitamin D action in the adrenal glands are rather sketchy. Evidence for possible vitamin D effects on adrenal catecholamine ratios (see below), possible relationships between vitamin D and hypertension (see Section II above), and analogies between $1,25(OH)_2D_3$ regulation of chromogranin gene expression in the parathyroid and the role of chromogranin in adrenal catecholamine storage (J. Carron and J. Fox, personal communication, 1996) led several groups to test for adrenal VDR sites. Autoradiographic localization studies provided evidence for ^3H-$1,25(OH)_2D_3$ concentration in nuclei of adrenal medulla cells, both with and without immunocytochemical colocalization of phenylethanolamine-*N*-methyltransferase (PNMT) [138]. A separate biochemical study provided evidence for VDR sites in rat adrenal glands by both Scatchard and sucrose density gradient analysis (Figs. 1 and 2).

Although difficult to interpret because of the possible indirect role of hypo/hypercalcemia, there are scattered reports of the effects of vitamin D status on adrenal levels of dopamine [139,140] and epinephrine and norepinephrine [139,141] via effects on the PNMT enzyme [141]. Moreover, a brief report indicated that $1,25(OH)_2D_3$ increased tyrosine hydroxylase mRNA levels in cultured bovine adrenal medulla cells, possibly through an osteocalcin-type vitamin D response element found in the tyrosine hydroxylase promoter region [142]. Collectively, these observations seem to suggest that the vitamin D endocrine system may affect adrenal catecholamine metabolism at multiple biosynthetic steps. Whether this action might relate to the association of hypertension with hyperparathyroidism and elevated $1,25(OH)_2D_3$ levels is a highly speculative concept.

B. Pancreas

Clinical evidence for interrelationships between pancreatic function and the vitamin D endocrine system include evidence of bone disease in some diabetics, as well as impaired insulin secretion in vitamin D-deficient states (reviewed by Walters [1]; see also Chapter 70). An important advance in understanding these phenomena and the ubiquitous nature of VDR tissue distribution was the demonstration of the presence of VDR in the chick and rat pancreas [3,4]. Similarly important was the parallel demonstration of the presence of calbindin-D_{28K} in the chick pancreas [143]. Subsequent studies localized both the pancreatic VDR [144] and calbindin-D_{28K} (reviewed by Christakos *et al.* [6]; see also Chapter 13) to the β cells. Interestingly, in this vitamin D target tissue, calbindin-D_{28K} levels are regulated by $1,25(OH)_2D_3$ [145–147] (Chapter 13).

Since the initial observations were made, an extensive literature has developed describing $1,25(OH)_2D_3$ effects on insulin secretion and possible mechanisms thereof, including impaired glucose tolerance and insulin secretion in vitamin D deficiency, and altered vitamin D metabolism in diabetes (reviewed by Walters [1]; also see Chapter 70). As these topics are extensively reviewed elsewhere, they are not discussed further here.

C. Pituitary

Owing in part to interest in its possible interaction with the PTH–vitamin D axis, the pituitary was one of the early "nonclassical" tissues tested for the presence of VDR. These efforts resulted in parallel definition of pituitary VDR by several groups using biochemical [4,148] and autoradiographic [149] approaches. In subsequent studies comparing ^{3}H-1,25(OH)$_2$D$_3$ autoradiographic localization to immunocytochemical definition of pituitary cell types, nuclear concentration of ^{3}H-1,25(OH)$_2$D$_3$ was found in pituitary thyrotropes, with only weak localization in restricted subpopulations of lactotropes, gonadotropes, and somatotropes [150]. Conversely, VDR were identified by biochemical studies in cultured rat GH pituitary cells, which produce prolactin and growth hormone [151,152].

The predominance of pituitary VDR in the thyrotropes correlates with *in vivo* studies demonstrating that 1,25(OH)$_2$D$_3$ enhances thyrotropin-releasing hormone (TRH)-induced thyrotropin (TSH) secretion [153,154]. Similarly, the possible localization of low levels of VDR in prolactin-producing pituitary cells is supported by an in vivo study indicating 1,25(OH)$_2$D$_3$ enhancement of TRH-stimulated prolactin secretin [155].

There are numerous *in vitro* studies of the effects of 1,25(OH)$_2$D$_3$ on TRH-stimulated pituitary cell TSH and prolactin secretion and the mechanism thereof. TRH-stimulated TSH release has been predominantly studied in primary cultures of rat pituitary cells. Similar to the above *in vivo* studies, TRH-stimulated TSH release is enhanced by 1,25(OH)$_2$D$_3$ in this system, with no effect on basal TSH release nor on agonist-induced prolactin or growth hormone release [156–159]. This 1,25(OH)$_2$D$_3$ effect was linked to alterations in Ca^{2+} dependent TRH signal transduction [158]. Protracted 1,25(OH)$_2$D$_3$ exposure (8–24 hr) was required, and the effect was blocked by inhibitors of transcription, suggesting regulation at the level of gene expression [159,160].

In the rat GH pituitary cell system, 1,25(OH)$_2$D$_3$ stimulates Ca^{2+}-dependent synthesis of prolactin, but not growth hormone, and elevates prolactin mRNA levels [161]. Moreover, pulse labeling with [^{35}S]methionine demonstrated 1,25(OH)$_2$D$_3$-stimulated synthesis of three proteins of 12.5, 17.5, and 18 kDa [152]. Similar to the effect on stimulation of TSH release (see above), probing the mechanism of this 1,25(OH)$_2$D$_3$ effect on prolactin synthesis with Ca^{2+} channel blockers and activators demonstrated the presence, but not the mechanism, of a link between 1,25(OH)$_2$D$_3$ enhancement of the TRH stimulation and Ca^{2+} signaling by TRH [162]. Moreover, although 1,25(OH)$_2$D$_3$ did not acutely alter [Ca^{2+}]$_i$, protracted 1,25(OH)$_2$D$_3$ treatment (24–48 hr) enhanced the TRH-stimulated [Ca^{2+}]$_i$ spike by a mechanism that required influx of extracellular Ca^{2+}. This effect did not seem to involve inositol trisphosphate (IP$_3$) production or PKC activation, nor alterations in TRH receptor levels or in the TRH-affected intracellular Ca^{2+} pool (reviewed by Tashjian [163]). This protracted 1,25(OH)$_2$D$_3$ effect on TRH-induced Ca^{2+} entry involved modulation of both voltage-operated Ca^{2+} channels and Na$^+$/Ca^{2+} exchange mechanisms and was blocked by protein synthesis inhibitors [164]. However, subsequent studies in GH cells have demonstrated a 1,25(OH)$_2$D$_3$ enhancement of (1) TRH-stimulated IP$_3$ production [165,166], (2) Ca^{2+} release from an IP$_3$-sensitive intracellular Ca^{2+} pool [165], and (3) TRH receptor levels [167]. 1,25(OH)$_2$D$_3$ also enhanced bombesin stimulation of prolactin secretion by a mechanism reminiscent of the TRH effect [168]. Collectively, these observations on 1,25(OH)$_2$D$_3$ effects on pituitary cells clearly demonstrate 1,25(OH)$_2$D$_3$ modulation of agonist-induced hormone secretion via partially defined mechanisms regulating the agonist-induced signal transduction pathways.

The hormone 1,25(OH)$_2$D$_3$ also interferes with growth hormone induction by triiodothyronine (T$_3$) [169,170] and retinoic acid (RA) [170] in rat GH pituitary cells. Modulation of the T$_3$ effect includes 1,25(OH)$_2$D$_3$ down-regulation of T$_3$ receptor binding and mRNA levels [169]. Molecular studies demonstrated that the effect of vitamin D on T$_3$ and RA regulation of growth hormone gene expression may result from transcriptional interference by VDR blocking binding of the thyroid receptor or RA receptor to a common hormone response element [170]. Finally, the presence of a potent 1,25(OH)$_2$D$_3$ responsive enhancer element in the 5′-promoter region of the *Pit-1* gene suggests the hypothesis that 1,25(OH)$_2$D$_3$ plays a developmental role in pituitary cell organogenesis [171]. No evidence exists, however, to suggest that 1,25(OH)$_2$D$_3$ modulates endogenous expression of *Pit-1*.

Thus, 1,25(OH)$_2$D$_3$ exerts clear genomic and possibly nongenomic effects on production and gene expression for several pituitary hormones in *in vitro* models. Although VDR are apparently predominantly present in the thyrotropes, it is not yet clear whether the vitamin D hormonal effects are restricted to this cell type *in vivo*.

D. Thyroid

It is well known that vitamin D-induced hypercalcemia alters thyroid parafollicular (C cell) structure and calcitonin content [171–176], but appreciation that these effects are mediated in part by direct $1,25(OH)_2D_3$ actions on the C cells has emerged more recently (Chapter 24). In part, this development stems from biochemical studies demonstrating VDR in human medullary thyroid carcinoma C cells [177] and in FRTL-5 thyroid follicular cells [178,179]. Autoradiographic studies, in contrast, found nuclear ^3H-$1,25(OH)_2D_3$ localization in thyroid follicular cells, but not C cells [180]. The resultant question of whether VDR are present in normal C cells has been difficult to resolve, in part because of the low abundance of this cell type in the normal thyroid.

Although information on the effect of vitamin D treatment on thyroid calcitonin content (or circulating plasma levels) *in vivo* has been ambiguous [171–176,181,182], at least in part due to complications from hypercalcemia (reviewed by Naveh-Many and Silver [183]), studies of calcitonin gene regulation have established that $1,25(OH)_2D_3$ suppresses calcitonin mRNA levels in a variety of models [184–187] (see also Chapter 24). This inhibition is exerted at the transcriptional level [183,186,187] and likely results from VDR interaction with a negative response element in the 5′-upstream region of the calcitonin gene, which in turn exerts a prominent inhibitory influence on a downstream cAMP-induced enhancer element [187]. Other $1,25(OH)_2D_3$ effects in C-cell carcinoma cells include the stimulation of growth and DNA synthesis [188,189] accompanied by increased c-*myc* gene expression [188]. Moreover, $1,25(OH)_2D_3$ decreased transcription of PTH-related peptide mRNA [190] and induced a specific increase in somatostatin mRNA levels [191,192]; however, reminiscent of the effect on calcitonin gene expression (see above), $1,25(OH)_2D_3$ inhibited cAMP enhancement of somatostatin mRNA levels [190,192]. Thus, in contrast to exerting negative feedback on PTH gene expression (Chapter 23), $1,25(OH)_2D_3$ also may alter production of calcitonin, the hormone that has the potential of opposing its hypercalcemic effect (Chapter 24).

Consistent with evidence for VDR localization in the thyroid follicular cells (see above), $1,25(OH)_2D_3$ has been shown to inhibit cAMP-mediated TSH effects on follicular cell function, resulting in reductions in growth and iodide uptake, by reducing the TSH receptor number [193–196]. Attenuation of the cAMP response in this system involves both increases in the inhibitory G-protein-2α [196] and a selective increase of protein kinase A regulatory subunit type II β [197]. $1,25(OH)_2D_3$ decreases the number of α_1-adrenergic receptors on follicular cells [198] and increases ATP-evoked, but not noradrenaline-evoked [198], elevations in intracellular calcium levels [199].

VI. OTHER TISSUES

A. Lung

Biochemical studies have demonstrated VDR in lung of adult [1] and fetal [200,201] rats. Finding these putative VDR sites in the lung was rather exciting, in part because of parallel observations of a protein resembling the mammalian intestinal calbindin-D_{9K} in the rat lung [7,202]. Moreover, $1,25(OH)_2D_3$ treatment does not alter VDR levels in this tissue (Fig. 5). However, although this CaBP proved to be identical to the intestinal calbindin-D_{9K} [202], it is not regulated by $1,25(OH)_2D_3$ in the rat lung [7,202]. However, observations of abnormal alveolar and connective tissue development in pups born to vitamin D-deficient mothers suggested the possible importance of vitamin D to other functions in lung [203].

Subsequent immunohistochemical studies localized the VDR in fetal lung to type II epithelial pneumocytes [201]. Studies focusing on vitamin D function in these cells have demonstrated effects of $1,25(OH)_2D_3$ on their maturation, including stimulation of phospholipid synthesis and surfactant release [204,205], acceleration of maturation-dependent glycogenolytic processes [205], and stimulation of DNA synthesis coincident with a shift toward an increased proportion of cells in the S and G_2 cell cycle phases [206]. Thus, $1,25(OH)_2D_3$ seems to play important roles in lung epithelial cell growth and the development of functions characteristic of a mature phenotype.

B. Liver

Biochemical studies have demonstrated VDR in the liver of fish [207] and mammals [1,208]. Moreover, estrogen treatment induces a steroid- and organ-specific increase in VDR levels in the rat liver [209].

Treatment with $1,25(OH)_2D_3$ stimulates hepatic regeneration after partial hepatectomy through actions at many levels, although some of the described effects relate at least in part to altered calcemic status [210–212]. Effects of $1,25(OH)_2D_3$ treament include stimulation of the synthesis or activity of several enzymes, including stimulation of type 1 cAMP-dependent protein kinase activity [213] through a selective effect on the catalytic subunit [214] and stimulation of the DNA replication enzymes ribonucleotide reductase [215] and DNA polymerase α [212,216]. $1,25(OH)_2D_3$ also stimulates epidermal growth factor receptor levels by a posttranslational mechanism that is accompanied by increases in receptor autophosphorylation and tyrosine kinase activity [211]. Thus, $1,25(OH)_2D_3$-regulated effects in this system clearly stimulate growth-promoting pathways [217].

The hormone $1,25(OH)_2D_3$, either directly or through its calcemic effects, also has effects on liver in the absence of the growth promotion that follows hepatectomy. Vitamin D deficiency significantly alters hepatic function [218–220]. Moreover, $1,25(OH)_2D_3$ treatment attenuates hepatic cellular damage induced by bromobenzene, and it increases the bromobenzene-induced rise in circulating aspartate and alanine aminotransferases, and sorbitol dehydrogenase [218]. The hormone also stimulates the activity of isocitrate lyase and malate synthase, key enzymes in the glyoxylate cycle in the liver [220,221], increases the activity of the NADPH-dependent cytosolic triiodothyronine binding protein in rat liver and in dRLh rat hepatoma-derived cells [222], and enhances transferrin synthesis in primary cultures of rat hepatocytes [223].

Hepatocyte models have also been used to explore rapid, putatively membrane receptor-mediated, effects of $1,25(OH)_2D_3$. In cultured rat hepatocytes and in isolated rat liver nuclei, $1,25(OH)_2D_3$ rapidly (within 5 min) increased $[Ca^{2+}]_i$ by a mechanism that was dependent on phospholipase A_2-dependent turnover of phosphatidylinositol [224–227], including in particular production of the deacylation product lysophosphatidylinositol [224,225]. This effect is also dependent on cell alkalinization through activation of the Na^+–H^+ antiport [224,226]. The observation that the $1,25(OH)_2D_3$-induced increase in $[Ca^{2+}]_i$, but not the generation of lysophosphatidylinositol, was blocked by pertussis toxin (Fig. 9) was interpreted as reflecting G-protein involvement in transduction of the lysophosphatidylinositol signal for the increment in $[Ca^{2+}]_i$ [228]. Moreover, the membrane-selective epimer $1\beta,25(OH)_2D_3$ inhibited these $1\alpha,25(OH)_2D_3$ effects in hepatocytes [228], but not in isolated nuclei [225], supporting the hypothesis that the cellular $1,25(OH)_2D_3$ effects are mediated by an as yet uncharacterized membrane receptor specific for $1\alpha,25(OH)_2D_3$ (see Chapter 15).

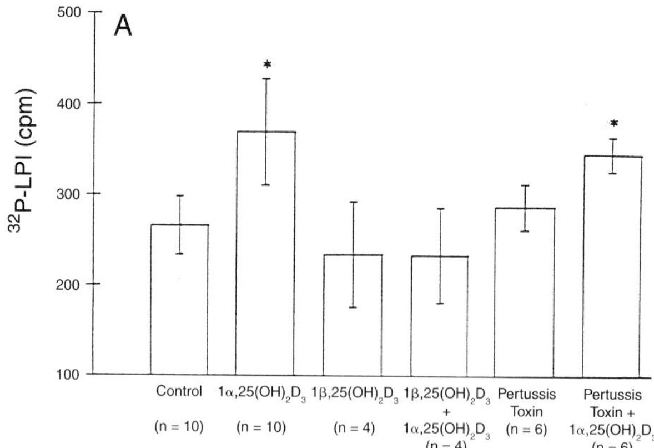

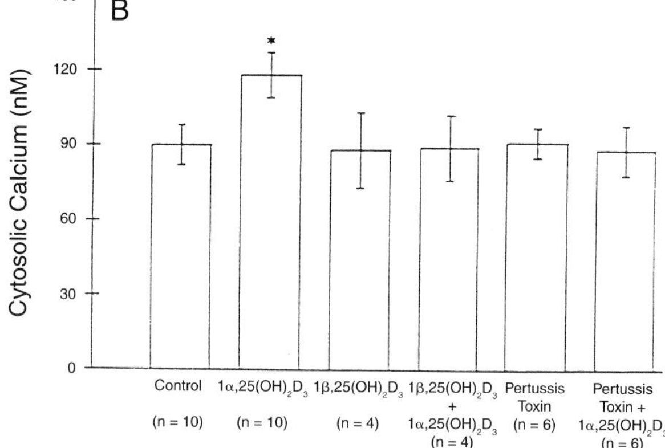

FIGURE 9 [^{32}P] Lysophosphatidylinositol production (A) and cytosolic calcium (B) in hepatocytes as affected by $1\alpha,25(OH)_2D_3$ (20 n*M*), $1\beta,25(OH)_2D_3$ (20 n*M*), and pertussis toxin (500 ng/ml). Values represent means ± SE (A) or SD (B). Numbers in parenthesis indicate the number of observations in each group. *This group differs from the other groups with $p < 0.05$ by Duncan's test for multiple comparisons. Reprinted with permission from Baran *et al.* [228].

C. Brain

An emerging area of interest is that of the effects of vitamin D in the brain and other areas of the central nervous system in both physiological and pathophysiological states. As these concepts are treated in detail elsewhere (Chapter 71), they are only briefly discussed here.

For technical reasons, the principal studies of VDR localization in brain have employed ^3H-$1,25(OH)_2D_3$ autoradiography. As reviewed previously [1], these studies have defined VDR-like sites in neurons in a wide range of specific brain nuclei that regulate variously sensory, motor, autonomic, and endocrine systems [229–231], as well as in spinal cord and sensory ganglia [232]. VDR have also been identified in choroid plexus of several species [233,234], a site of particular interest because of an unresolved issue of the degree to which vitamin D metabolites and the humoral factors they regulate cross the blood–brain barrier [235–238]. $1,25(OH)_2D_3$ has also been shown to affect certain enzymatic activities, for example, choline acetyltransferase [239] and creatinine kinase [240], in some specific brain nuclei. The vitamin D-related calbindin-D_{28K} is

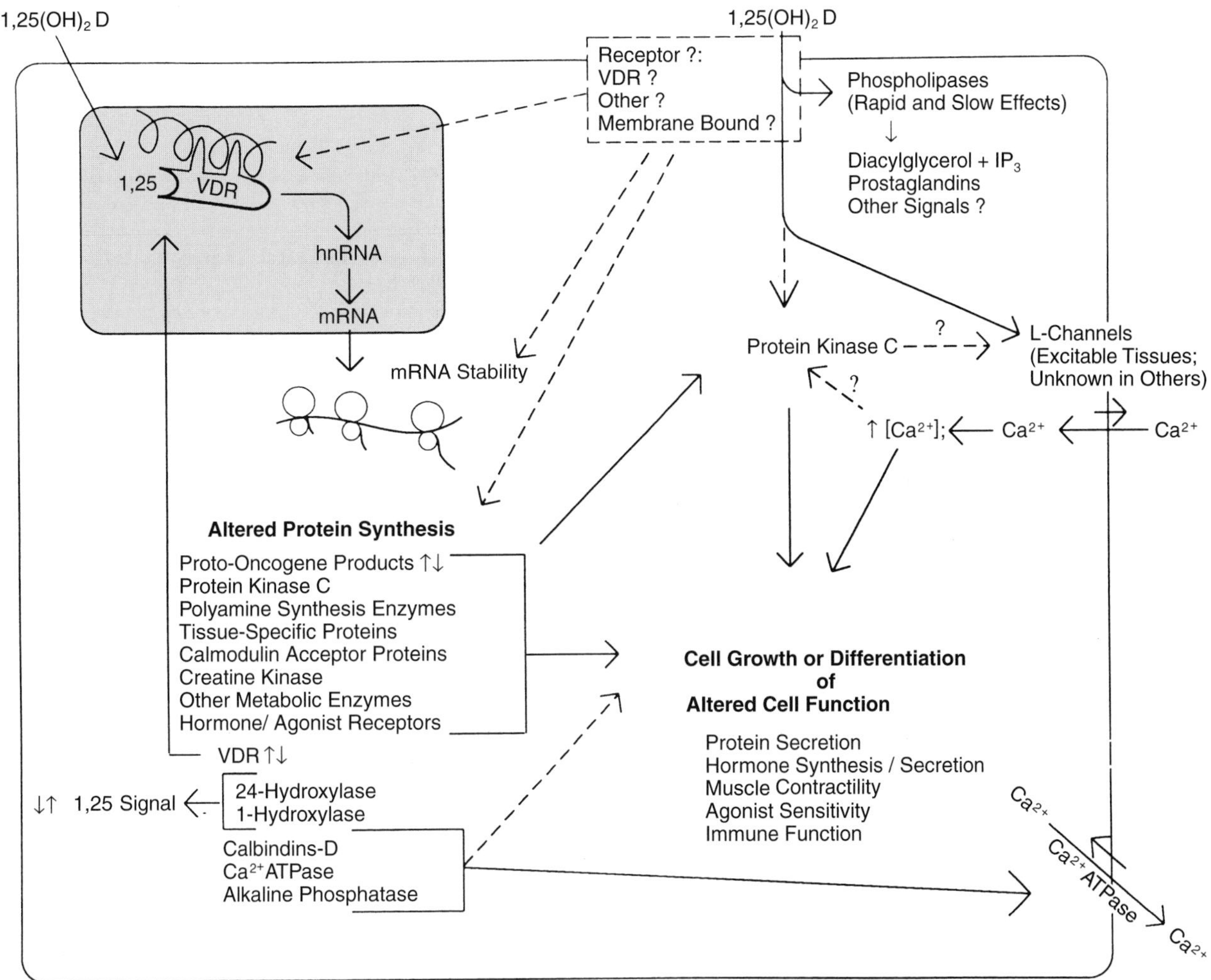

FIGURE 10 Generalized model of the mechanisms of the newly identified actions of $1,25(OH)_2D_3$. (Left) $1,25(OH)_2D_3$ interacts with its nuclear receptor (a zinc-finger nuclear transcription factor) and alters mRNA and protein synthesis. This action results in induction of numerous general regulatory and tissue-specific proteins that in turn alter cell function. (Right) $1,25(OH)_2D_3$ also alters signal transduction pathways involving extracellular Ca^{2+}, phospholipids, and PKC. Details of these effects are unknown (dashed lines), including the nature of the membrane binding protein(s) (if any) involved and the precise sequence of events that follow activation, and the pathways differ across systems. Thus, some putative interactions among the signals are omitted. Nevertheless, $1,25(OH)_2D_3$ induction of this set of signal transduction elements in many cases requires protein (and RNA?) synthesis and may mediate or augment the VDR-mediated regulatory effects of $1,25(OH)_2D_3$ on cell function. Reprinted with permission from Walters MR 1992 Newly identified actions of the Vitamin D endocrine system. Endocrinology Review **13**:719–764; © The Endocrine Society.

prominent in cerebellar Purkinje cells; however, these cells do not contain VDR, and the calbindin-D_{28K} therein is not affected by $1,25(OH)_2D_3$ treatment [6] (see also Chapter 13). The most interesting information on vitamin D and the brain relates to the observed effects of $1,25(OH)_2D_3$ on clinically important functions (reviewed by Walters [1]; see also Chapter 71), for example, hippocampal seizure threshold levels [241], or possible alterations in its response pathways in disease states such as Alzheimer's disease [242]. In this context, demonstration that $1,25(OH)_2D_3$ induces nerve growth factor protein and mRNA in several brain-derived models, including rat brain *in vivo* [243,244], has led to the suggestion that vitamin D analogs may have potential application in neurodegenerative disorders (Chapter 71).

VII. CONCLUSION

The 1980s and 1990s have provided very exciting developments in studies of vitamin D action. In particular, appreciation of the unexpectedly extensive tissue distribution of VDR has led to the concept that the hormonal effects of $1,25(OH)_2D_3$ are similarly extensive with respect both to the breadth of the targets involved and to the remarkable diversity of $1,25(OH)_2D_3$ effects within those tissues.

Thus, in this time span the concept of the vitamin D endocrine system has shifted away from that of a hormone system only regulating plasma Ca^{2+} and phosphate homeostasis through actions in a rather limited set of tissues. Now we recognize that this hormonal system exhibits extensive and pleiotropic actions throughout the body (Fig. 10). The multitude of $1,25(OH)_2D_3$ effects has in turn led to the recognition that analogs of this versatile hormone may have significant therapeutic potential in a broad range of disease states (see Section VIII of this volume).

Acknowledgments

Jeff Carron and John Fox are gratefully acknowledged for important contributions to the unpublished data included herein. Debbie Olavarrieta and Bridget Harbor provided the expert secretarial assistance necessary for this chapter. Studies from the author's laboratory were supported by National Institutes of Health Grants DK-31847 and DK-43846.

References

1. Walters MR 1992 Newly identified actions of the vitamin D endocrine system. Endocr Rev **13**:719–764.
2. Walters MR 1995 Newly identified actions of the vitamin D endocrine system: Update 1995. Endocr Rev Monogr **4**:47–56.
3. Christakos S, Norman AW 1979 Studies on the mode of action of calciferol XVIII. Evidence for a specific high affinity binding protein for 1,25-dihydroxyvitamin D_3 in chick kidney and pancreas. Biochem Biophys Res Commun **89**:56–63.
4. Pike JW, Gooze LL, Haussler MR 1980 Biochemical evidence for 1,25-dihydroxyvitamin D receptor macromolecules in parathyroid, pancreatic, pituitary, and placental tissues. Life Sci **26**:407–414.
5. Kream BE, Yamada S, Schnoes HK, DeLuca HF 1977 Specific cytosol-binding protein for 1,25-dihydroxyvitamin D_3 in rat intestine. J Biol Chem **252**:4501–4505.
6. Christakos S, Gabrielides C, Rhoten WB 1989 Vitamin D-dependent calcium binding proteins: Chemistry, distribution, functional considerations, and molecular biology. Endocr Rev **10**:3–26.
7. Walters MR, Bruns ME, Carter RM, Riggle PC 1991 Vitamin D-independence of small calcium binding proteins in nonclassical target tissues. Am J Physiol **260**:E794–E800.
8. Walters MR 1992 Absence of calbindin-D28 expression in nonclassical 1,25-dihydroxyvitamin D targets: Analysis by polymerase chain reaction. J Bone Miner Res **7**:1461–1466.
9. Seelig MS 1967 Vitamin D and cardiovascular, renal, and brain damage in infancy and childhood. Ann NY Acad Sci **147**:539–584.
10. Mullen PA, Bedford PGC, Ingram PL 1979 An investigation of the toxicity of 1α-hydroxycholecalciferol to calves. Res Vet Sci **27**:275–279.
11. Gilka F, Sugden EA 1981 Experimental ectopic mineralization in young boars I. The role of supplemental vitamin D in the etiology. Can J Comp Med **45**:266–270.
12. Long GG 1984 Acute toxicosis in swine associated with excessive dietary intake of vitamin D. J Am Vet Med Assoc **184**:164–170.
13. Harrington DD 1982 Acute vitamin D_2 (ergocalciferol) toxicosis in horses: Case report and experimental studies. J Am Vet Med Assoc **180**:867–873.
14. Simpson RU, Weishaar RE 1988 Involvement of 1,25-dihydroxyvitamin D_3 in regulating myocardial calcium metabolism: Physiological and pathological actions. Cell Calcium **9**:285–292.
15. Milliner DS, Zinsmeister AR, Lieberman E, Landing B 1990 Soft tissue calcification in pediatric patients with end-stage renal disease. Kidney Int **38**:931–936.
16. Saupe J, Hirschberg R, Hofer W, Bosaller W, von Herrath D, Pauls A, Schaefer K 1986 Effect of long-term administration of 1,25-dihydroxyvitamin D_3 and 1α-hydroxyvitamin D_3 on the calcium content of the aorta, heart and kidney of normal and uremic rats. Nephron **43**:290–292.
17. Wrzolkowa T, Zydowo M 1980 Ultrastructural studies on the vitamin D induced heart lesions in the rat. J Mol Cell Cardiol **12**:1117–1133.
18. Wrzolkowa T, Rudzinska-Kisiel T, Klosowska B 1991 Calcium content of serum and myocardium in vitamin D-induced cardionecrosis. Bone Miner **13**:111–121.
19. Takeo S, Tanonaka R, Tanonaka K, Miyake K, Hisayama H, Ueda N, Kawakami K, Tsumura H, Katsushika S, Taniguchi Y 1991 Alterations in cardiac function and subcellular membrane activities after hypervitaminosis D_3. Mol Cell Biochem **107**:169–183.
20. Wrzolek MA 1985 The effect of zinc on vitamin D_3-induced cardiac necrosis. J Mol Cell Cardiol **17**:109–117.
21. Swierczynski J, Nagel G, Zydowo MM 1987 Calcium content in some organs of rats treated with a toxic calciol dose. Pharmacology **34**:57–60.
22. Bazzani C, Arletti R, Bertolini A 1983 Vitamin D deficiency reduces the inotropic effect of ouabain. Acta Vitaminol Enzymol **5**:147–151.
23. Coratelli P, Petrarulo F, Buongiorno E, Giannattasio M, Antonelli G, Amerio A 1984 Improvement in left ventricular function during treatment of hemodialysis patients with 25-OHD_3. Contrib Nephrol **41**:433–437.
24. Coratelli P, Buongiorno E, Petrarulo F, Corciulo R, Giannattasio M, Passavanti G, Antonelli G 1989 Pathogenetic aspects of uremic cardiomyopathy. Miner Electrolyte Metab **15**:246–253.
25. McGonigle RJS, Fowler MB, Timmis AB, Weston MJ, Parsons V 1984 Uremic cardiomyopathy: Potential role of vitamin D and parathyroid hormone. Nephron **36**:94–100.
26. Jahn H, Schmitt R, Schohn D, Olier P 1984 Aspects of the myocardial function in chronic renal failure. Contrib Nephrol **41**:240–250.
27. Massry SG 1984 Parathyroid hormone and uremic myocardiopathy. Contrib Nephrol **41**:231–239.
28. Gafter U, Battler A, Eldar M, Zevin D, Neufeld HN, Levi J 1985 Effect of hyperparathyroidism on cardiac function in patients with end-stage renal disease. Nephron **41**:30–33.
29. Symons C, Fortune F, Greenbaum RA, Dandona P 1985 Cardiac hypertrophy, hypertrophic cardiomyopathy, and hyperparathyroidism—An association. Br Heart J **54**:539–542.
30. Simpson RU, Thomas GA, Arnold AJ 1985 Identification of

1,25-dihydroxyvitamin D_3 receptors and activities in muscle. J Biol Chem **260**:8882–8891.

31. Walters MR, Ilenchuk TT, Claycomb WC 1987 1,25-Dihydroxyvitamin D_3 stimulates $^{45}Ca^{2+}$ uptake by cultured adult rat ventricular cardiac muscle cells. J Biol Chem **262**:2536–2541.
32. Walters MR, Wicker DC, Riggle PC 1986 1,25-Dihydroxyvitamin D_3 receptors identified in the rat heart. J Mol Cell Cardiol **18**:67–72.
33. Bidmon H-J, Gutkowska J, Murakami R, Stumpf WE 1991 Vitamin D receptors in heart: Effects on atrial natriuretic factor. Experientia **47**:958–962.
34. Selles J, Boland R 1991 Rapid stimulation of calcium uptake and protein phosphorylation in isolated cardiac muscle by 1,25-dihydroxyvitamin D_3. Mol Cell Endocrinol **77**:67–73.
35. Selles J, Bellido T, Boland R 1994 Modulation of calcium uptake in cultured cardiac muscle cells by 1,25-dihydroxyvitamin D_3. J Mol Cell Cardiol **26**:1593–1599.
36. Kotsiuruba AB, Tuhanova AV, Bukhanevich OM, Tarakanov SS 1995 Mechanisms of the early effect of biologically active hydroxysterols: Calcitriol and ecdysterone. Identification of sphingomyelin as the effector mechanism of the early effect. Ukr Biokhim Zh **67**:53–58.
37. Weishaar RE, Simpson RU 1989 The involvement of the endocrine system in regulating cardiovascular function: Emphasis on vitamin D_3. Endocr Rev **10**:351–365.
38. Hochhauser E, Kushnir T, Navon G, Rehavi M, Barak J, Edelstein S, Vidne B 1993 The effect of vitamin D_3 deficiency on the isolated chick heart: Hemodynamic, P-31 NMR and membrane studies. J Mol Cell Cardiol **25**:93–102.
39. Weishaar RE, Kim S-N, Saunders DE, Simpson RU 1990 Involvement of vitamin D_3 with cardiovascular function. III. Effects on physical and morphological properties. Am J Physiol **258**:E134–E142.
40. O'Connell TD, Simpson RU 1995 1,25-Dihydroxyvitamin D_3 regulation of myocardial growth and c-myc levels in the rat heart. Biochem Biophys Res Commun **213**:59–65.
41. O'Connell TD, Weishaar RE, Simpson RU 1994 Regulation of myosin isozyme expression by vitamin D_3 deficiency and 1,25-dihydroxyvitamin D_3 in the rat heart. Endocrinology **134**:899–905.
42. Wu J, Garami M, Cheng T, Gardner DG 1996 $1,25(OH)_2$ vitamin D_3 and retinoic acid antagonize endothelin-stimulated hypertrophy of neonatal rat cardiac myocytes. J Clin Invest **97**:1577–1588.
43. O'Connell TD, Giacherio DA, Jarvis AK, Simpson RU 1995 Inhibition of cardiac myocyte maturation by 1,25-dihydroxyvitamin D_3. Endocrinology **136**:482–488.
44. Stio M, Lunghi B, Lantomasi T, Vincenzini MT, Treves C 1994 Effect of vitamin D deficiency and 1,25-dihydroxyvitamin D_3 on rat heart metabolism. J Mol Cell Cardiol **26**:1421–1428.
45. Gensure RC, Riggle PC, Antrobus SD, Walters MR 1991 Evidence for two classes of 1,25-dihydroxyvitamin D_3 binding sites in classical vs. nonclassical target tissues. Biochem Biophys Res Commun **180**:867–873.
46. Antrobus SD, Walters MR 1992 1,25-Dihydroxyvitamin D effects in the kidney: Induction of calmodulin binding proteins. Biochem Biophys Res Commun **185**:636–640.
47. Wong NLM, Halabe A, Wong EFC, Sutton RAL 1991 The effect of calcitriol on atrial natriuretic factor release from isolated atrium. Metabolism **40**:135–138.
48. Wu J, Garami M, Cao L, Li Q, Gardner DG 1995 $1,25(OH)_2D_3$ suppresses expression and secretion of atrial natriuretic peptide from cardiac myocytes. Am J Physiol **268**:E1108–E1113.
49. Lemire JM, Archer DC, Khulkarni A 1992 Prolongation of the survival of murine cardiac allografts by the vitamin D_3 analogue 1,25-dihydroxy-delta^{16}-cholecalciferol. Transplantation **54**:762–763.
50. Johnsson C, Tufveson G 1994 MC 1288—A vitamin D analogue with immunosuppressive effects on heart and small bowel grafts. Transplant Int **7**:392–397.
51. Taura S, Taura M, Imai H, Kummerow FA, Tokuyasu K, Cho SB 1978 Ultrastructure of cardiovascular lesions induced by hypervitaminosis D and its withdrawal. Paroi Arterille **4**:245–259.
52. Seydewitz V, Staubesand J 1983 Calcification of the arterial wall in the aging process. Electron microscopic and biochemical studies of a model of vitamin D_3 overdose. Aktuelle Gerontol **13**:115–118.
53. Cannon EP, Williams BJ 1990 Raised vascular calcium in an animal model: Effects on aortic function. Cardiovasc Res **24**:47–52.
54. Henrion D, Chillon J-M, Hoffman M 1991 Decrease in endothelium-dependent relaxation in the mesenteric arterial bed following vascular calcium overload produced by vitamin D_3 and nicotine in rats. Life Sci **49**:575–586.
55. Kamio A, Kummerow FA, Imai H 1977 Degeneration of aortic smooth muscle cells in swine fed excess vitamin D_3. Arch Pathol Lab Med **101**:378–381.
56. Kunitomo M, Takaoka K, Matsumoto J, Iwai H, Bando Y 1983 Experimental induction of atherosclerosis in guinea pigs fed a cholesterol and vitamin D_2-rich diet. Nippon Yakurigaku Zasshi **81**:275–283.
57. Resnick LM, Muller FB, Laragh JH 1986 Calcium regulating hormones in essential hypertension: Relation to plasma renin activity and sodium metabolism. Ann Intern Med **105**:649–654.
58. Resnick LM 1990 Calciotropic hormones in human and experimental hypertension. Am J Hypertens **3**:171S–178S.
59. McCarron DA 1985 Is calcium more important than sodium in the pathogenesis of essential hypertension? Hypertension **7**:607–620.
60. Lau K, Eby B 1995 The role of calcium in genetic hypertension. Hypertension **7**:657–673.
61. Bukoski RD, Kremer D 1991 Calcium-regulating hormones in hypertension: Vascular actions. Am J Clin Nutr **54**:220S–226S.
62. Kawashima H 1987 Receptor for 1,25-dihydroxyvitamin D in a vascular smooth muscle cell line derived from rat aorta. Biochem Biophys Res Commun **146**:1–6.
63. Koh E, Morimoto S, Fukuo K, Itoh K, Hironaka T, Shiraishi T, Onishi T, Kumahara Y 1987 1,25-Dihydroxyvitamin D_3 binds specifically to rat vascular smooth muscle cells and stimulates their proliferation in vitro. Life Sci **42**:215–223.
64. Merke J, Milde P, Lewicka S, Hugel U, Klaus G, Mangelsdorf DJ, Haussler MR, Rauterberg EW, Ritz E 1989 Identification and regulation of 1,25-dihydroxyvitamin D_3 receptor activity and biosynthesis of 1,25-dihydroxyvitamin D_3. Studies in cultured bovine aortic endothelial cells and human dermal capillaries. J Clin Invest **83**:1903–1915.
65. Stumpf WE 1990 Steroid hormones and the cardiovascular system: Direct actions of estradiol, progesterone, testosterone, gluco- and mineralcorticoids, and soltriol [vitamin D] on central nervous regulatory and peripheral tissues. Experientia **46**:13–25.
66. Bukoski RD, Xue H 1993 On the vascular inotropic action of $1,25(OH)_2$ vitamin D_3. Am J Hypertens **6**:388–396.
67. Shan J, Resnick LM, Lewanczuk RZ, Karpinski E, Li B, Pang PKT 1993 1,25-Dihydroxyvitamin D as a cardiovascular hormone. Effects on calcium current and cytosolic free calcium in vascular smooth muscle cells. Am J Hypertens **6**:983–988.
68. Inoue M, Wakasugi M, Wakao R, Gan N, Tawata M, Nishii Y, Onaya T 1992 A synthetic analogue of vitamin D_3, 22-oxa-1,25-

dihydroxy-vitamin D_3, stimulates the production of prostacyclin by vascular tissues. Life Sci **51**:1105–1112.
69. Inoue T, Kawashima H 1988 1,25-Dihydroxyvitamin D_3 stimulates $^{45}Ca^{2+}$-uptake by cultured vascular smooth muscle cells derived from rat aorta. Biochem Biophys Res Commun **152**:1388–1394.
70. Bukoski RD, Xue H, McCarron DA 1987 Effect of $1{,}25(OH)_2$ vitamin D_3 and ionized Ca^{2+} on ^{45}Ca uptake by primary cultures of aortic myocytes of spontaneously hypertensive and Wistar Kyoto normotensive rats. Biochem Biophys Res Commun **146**:1330–1335.
71. Kawashima H 1988 1,25-Dihydroxyvitamin D_3 stimulates Ca-ATPase in a vascular smooth muscle cell line. Biochem Biophys Res Commun **150**:1138–1143.
72. Koh E, Morimoto S, Nabata T, Takamoto S, Kitano S, Ogihara T 1990 Effects of 1,25-dihydroxyvitamin D_3 on the synthesis of DNA and glycosaminoglycans in rat aortic smooth muscle cells in vitro. Life Sci **46**:1545–1551.
73. Mitsuhashi T, Morris RC, Ives HE 1991 1,25-Dihydroxyvitamin D_3 modulates growth of vascular smooth muscle cells. J Clin Invest **87**:1889–1895.
74. MacCarthy EP, Yamashita W, Hsu A, Ooi BS 1989 1,25-Dihydroxyvitamin D_3 and rat vascular smooth muscle cell growth. Hypertension **13**:954–959.
75. Hinek A, Botney MD, Mecham RP, Parks WC 1991 Inhibition of tropoelastin expression by 1,25-dihydroxyvitamin D_3. Connect Tissue Res **26**(3):155–166.
76. Tu-Yu AH, Morris RC, Ives HE 1993 Differential modulation of fos and jun gene expression by 1,25-dihydroxyvitamin D_3. Biochem Biophys Res Commun **193**:161–166.
77. Ishibashi K, Evans A, Shingu T, Bian K, Bukoski RD 1995 Differential expression and effect of 1,25-dihydroxyvitamin D_3 on myosin in arterial tree of rats. Am J Physiol **269**:C443–C450.
78. Boland R 1986 Role of vitamin D in skeletal muscle function. Endocr Rev **7**:434–448.
79. Rimaniol JM, Authier FJ, Chariot P 1994 Muscle weakness in intensive care patients: Initial manifestation of vitamin D deficiency. Intensive Care Med **20**:591–592.
80. Matthews C, Heimberg K-W, Ritz E, Agostini B, Fritzsche J, Hasselbach W 1977 Effect of 1,25-dihydroxycholecalciferol on impaired calcium transport by the sarcoplasmic reticulum in experimental uremia. Nephrol Int **11**:227–235.
81. Sjostrom M, Lorentzon R, Larsson SE, Holmlund D 1978 The influence of 1,25-dihydroxycholecalciferol on the ultrastructural organization of skeletal muscle fibres. Morphometric analyses on vitamin D deficient or calcium deficient growing rats. Med Biol **56**:209–215.
82. Rodman JS, Baker T 1978 Changes in the kinetics of muscle contraction in vitamin D-depleted rats. Kidney Int **13**:189–193.
83. Davie MWJ, Chalmers TM, Hunter JO, Pelc B, Kodicek E 1976 1-Alphahydroxycholecalciferol in chronic renal failure. Studies of the effect of oral doses. Ann Intern Med **84**:281–285.
84. Boland R, Norman A, Ritz E, Hasselbach W 1985 Presence of a 1,25-dihydroxyvitamin D_3 receptor in chick skeletal muscle myoblasts. Biochem Biophys Res Commun **128**:305–311.
85. Stumpf WE, Sar M, O'Brien LP, Morin J 1988 Pyloric gastrin-producing cells and pyloric sphincter muscle cells are nuclear targets for ^3H-$1{,}25(OH)_2$ vitamin D_3. Studies by autoradiography and immunohistochemistry. Histochemistry **89**:447–450.
86. Bidmon HJ, Stumpf WE 1995 1,25-Dihydroxyvitamin D_3 binding sites in the eye and associated tissues of the green lizard *Anolis carolinensis*. Histochem J **27**:516–523.
86a. Costa EM, Blau HM, Feldman D 1986 1,25-Dihydroxyvitamin D_3 receptors and hormonal responses in cloned human skeletal muscle cells. Endocrinology **119**:2214–2220.
87. de Boland AR, Boland RL 1994 Non-genomic signal transduction pathway of vitamin D in muscle. Cell Signal **6**:717–724.
88. Drittanti L, de Boland AR, Boland R 1989 Modulation of DNA synthesis in cultured muscle cells by 1,25-dihydroxyvitamin D_3. Biochim Biophys Acta **1014**:112–119.
89. Drittanti L, de Boland AR, Boland R 1990 Stimulation of calmodulin synthesis in proliferating myoblasts by 1,25-dihydroxyvitamin D_3. Mol Cell Endocrinol **74**:143–153.
90. Brunner A, de Boland AR 1990 1,25-Dihydroxyvitamin D_3 affects the synthesis, phosphorylation and in vitro calmodulin binding of myoblast cytoskeletal proteins. Z Naturforsch **45c**:1156–1160.
91. Kwiecinski GG, Petrie GI, DeLuca HF 1989 1,25-Dihydroxyvitamin D_3 restores fertility of vitamin D-deficient female rats. Am J Physiol **256**:E483–E487.
92. Halloran BP 1989 Is 1,25-dihydroxyvitamin D_3 required for reproduction? Proc Soc Exp Biol Med **191**:227–232.
93. Osmundsen BC, Huang HFS, Anderson MB, Christakos SC, Walters MR 1989 FSH stimulation of 1,25-dihydroxyvitamin D receptors in the testis. J Steroid Biochem **34**:339–343.
94. Sood S, Reghunandanan R, Reghunandanan V, Marya RK, Singh PI 1995 Effect of vitamin D repletion on testicular function in vitamin D-deficient rats. Ann Nutr Metab **39**:95–98.
95. Kwiecinski GG, Petrie GI, DeLuca HF 1989 Vitamin D is necessary for reproductive functions of the male rat. J Nutr **119**:741–744.
96. Eisman JA, MacIntyre I, Martin TJ, Moseley JM 1979 1,25-Dihydroxyvitamin-D receptors in breast cancer cells. Lancet 1335–1336.
97. Eisman JA, Suva LJ, Martin TJ 1986 Significance of 1,25-dihydroxyvitamin D_3 receptor in primary breast cancers. Cancer Res **46**:5406–5408.
98. Colston K, Hirst M, Feldman D 1980 Organ distribution of the cytoplasmic 1,25-dihydroxycholecalciferol receptor in various mouse tissues. Endocrinology **107**:1916–1922.
99. Berger U, McClelland RA, Wilson P, Greene GL, Haussler MR, Pike JW, Colston K, Easton D, Coombes RC 1991 Immunocytochemical determination of estrogen receptor, progesterone receptor, and 1,25-dihydroxyvitamin D_3 receptor in breast cancer and relationship to prognosis. Cancer Res **51**:239–244.
100. Fry JM, Curnow DH, Gutteridge DH, Retallack RW 1980 Vitamin D in lactation. I. The localization, specific binding and biological effect of 1,25-dihydroxyvitamin D_3 in mammary tissue of lactating rats. Life Sci **27**:1255–1263.
101. Narbaitz R, Sar M, Stumpf WE, Huang S, DeLuca HF 1981 1,25-Dihydroxyvitamin D_3 target cells in rat mammary gland. Horm Res **15**:263–269.
102. Gross M, Kost SB, Ennis B, Stumpf W, Kumar R 1986 Effect of 1,25-dihydroxyvitamin D_3 on mouse mammary tumor (GR) cells: Evidence for receptors, cellular uptake, inhibition of growth and alteration in morphology at physiologic concentrations of hormone. J Bone Miner Res **1**:457–467.
103. Frampton RJ, Osmond SA, Eisman JA 1983 Inhibition of human cancer cell growth by 1,25-dihydroxyvitamin D_3 metabolites. Cancer Res **43**:4443–4447.
104. Simpson RU, Arnold AJ 1986 Calcium antagonizes 1,25-dihydroxyvitamin D_3 inhibition of breast cancer cell proliferation. Endocrinology **119**:2284–2289.
105. Eisman JA, Sutherland RL, McMenemy ML, Fragonas J-C, Musgrove EA, Pang GYN 1988 Effects of 1,25-dihydroxyvitamin D_3 on cell-cycle kinetics of T47D human breast cancer cells. J Cell Physiol **138**:611–616.

106. Colston KW, Berger U, Coombes RC 1989 Possible role for vitamin D in controlling breast cancer cell proliferation. Lancet **1**:188–191.
107. Walters MR 1981 An estrogen-stimulated 1,25-dihydroxyvitamin D_3 receptor in rat uterus. Biochem Biophys Res Commun **103**:721–726.
108. Levy J, Zuili I, Yankowitz N, Shany S 1984 Induction of cytosolic receptors for 1α,25-dihydroxyvitamin D_3 in the immature rat uterus by oestradiol. J Endocrinol **100**:265–269.
109. Acker GM, Gross B, Lieberherr M 1991 Interactions of calcitriol and sex steroids in rat myometrial cells. In: Norman AW, Bouillon R, Thomasset M (eds) Vitamin D: Gene Regulation, Structure–Function Analysis and Clinical Application. de Gruyter, Berlin, pp. 675–676.
110. Stumpf WE, Denny ME 1989 Vitamin D (soltriol), light, and reproduction. Am J Obstet Gynecol **161**:1375–1384.
111. Opperman LA, Saunders TJ, Bruns DE, Boyd JC, Mills SE, Bruns ME 1992 Estrogen inhibits calbindin-D_{28K} expression in mouse uterus. Endocrinology **130**:1728–1735.
112. Lieberherr M, Acker GM, Grosse B, Pesty A, Balsan S 1984 Rat endometrial cells in primary culture: Effects and interaction of sex hormones and vitamin D_3 metabolites on alkaline phosphatase. Endocrinology **115**:824–829.
113. Halhali A, Acker GM, Garabiedian M 1991 1,25-Dihydroxyvitamin D_3 induces in vivo the decidualization of rat endometrial cells. J Reprod Fertil **91**:59–64.
114. Coty WA 1980 A specific, high affinity binding protein for 1α,25-dihydroxyvitamin D in the chick oviduct shell gland. Biochem Biophys Res Commun **93**:285–292.
115. Takahashi N, Abe E, Tanabe R, Suda T 1980 A high-affinity cytosol binding protein for 1α,25-dihydroxycholecalciferol in the uterus of Japanese quail. Biochem J **190**:513–518.
116. Coty WA, Scordato JC, McConkey CL 1982 Estrogen regulates 1,25-dihydroxyvitamin D receptor levels in the chick oviduct shell gland. In: Norman AW, Schaefer K, Herrath DV, Grigoleit H-G (eds) Vitamin D, Chemical Biochemical and Clinical Endocrinology of Calcium Metabolism. de Gruyter, pp. 121–123.
117. Navickis RJ, Katzenellenbogen BS, Nalbandov AV 1979 Effects of the sex steroid hormones and vitamin D_3 on calcium-binding proteins in the chick shell gland. Biol Reprod **21**:1153–1162.
118. Stumpf WE, Sar M, Narbaitz R, Huang S, DeLuca HF 1983 Autoradiographic localization of 1,25-dihydroxyvitamin D_3 in rat placenta and yolk sac. Horm Res **18**:215–220.
119. Ross R, Florer J, Halbert K, McIntyre L 1989 Characterization of 1,25-dihydroxyvitamin D_3 receptors and in vivo targeting of $[^3H]$-1,25$(OH)_2D_3$ in the sheep placenta. Placenta **10**:553–567.
120. Saunders TJ, Opperman LA, Gorman J, Bruns DE, Bruns ME 1990 Tissue distribution of vitamin D receptor (VDR) mRNA in mice and rats and its control during reproduction. J Bone Miner Res **5**:S263 (abstract).
121. Dokoh S, Donaldson CA, Marion SL, Pike JW, Haussler MR 1983 The ovary: A target organ for 1,25-dihydroxyvitamin D_3. Endocrinology **112**:200–206.
122. Merke J, Kreusser W, Bier B, Ritz E 1983 Demonstration and characterisation of a testicular receptor for 1,25-dihydroxycholecalciferol in the rat. Eur J Biochem **130**:303–308.
123. Levy FO, Eikvar L, Jutte NHPM, Hansson V 1984 Properties and compartmentalization of the testicular receptor for 1,25-dihydroxyvitamin D_3. Ann NY Acad Sci **438**:591–593.
124. Walters MR 1984 1,25-Dihydroxyvitamin D_3 receptors in the seminiferous tubules of the rat testis increase at puberty. Endocrinology **114**:2167–2174.
125. Stumpf WE, Sar M, Chen K, Morin J, DeLuca HF 1987 Sertoli cells in the testis and epithelium ductuli efferentes are targets for ^3H-1,25$(OH)_2$ vitamin D_3. Cell Tissue Res **247**:453–455.
126. Habib FK, Maddy SQ, Gelly KJ 1990 Characterisation of receptors for 1,25-dihydroxyvitamin D_3 in the human testis. J Steroid Biochem **35**:195–199.
127. Akerstrom VL, Walters MR 1992 Physiological effects of 1,25-dihydroxyvitamin D_3 in the TM_4 Sertoli cell line. Am J Physiol **262**:E884–E890.
128. Merke J, Hugel U, Ritz E 1985 Nuclear testicular 1,25-dihydroxyvitamin D_3 receptors in Sertoli cells and seminiferous tubules of adult rodents. Biochem Biophys Res Commun **127**:303–309.
129. Levy FO, Eikvar L, Jutte NHPM, Cervenka J, Yoganathan T, Hansson V 1985 Appearance of the rat testicular receptor for calcitriol (1,25-dihydroxyvitamin D_3) during development. J Steroid Biochem **23**:51–56.
130. Levy FO, Eikvar L, Froysa A, Yoganathan T, Cervenka J, Hansson V 1986 Testicular receptos for 1,25-dihydroxyvitamin D_3: Cellular localization and changes during experimental crytorchidism. In: Stefanini M (ed) Molecular and Cellular Endocrinology of the Testis. Elsevier, Amsterdam, pp. 107–112.
131. Kreusser W, Mader H, Haag WD, Ritz E 1982 Diminished response of ovarian cAMP to luteinizing hormone in experimental uremia. Kidney Int **22**:272–279.
132. Tokuda N, Mano T, Levy RB 1990 1,25-Dihydroxyvitamin D_3 antagonizes interferon-gamma-induced expression of class II major histocompatibility antigens on thyroid follicular and testicular Leydig cells. Endocrinology **127**:1419–1427.
133. Majumdar SS, Bartke A, Stumpf WE 1994 Vitamin D modulates the effects of follicle-stimulating hormone on Sertoli cell function and testicular growth in Siberian hamsters. Life Sci **55**:1479–1486.
134. Eliaschewitz FG, Goldberg AC, Borelli A, Quintao ECR 1982 Effect of 1,25$(OH)_2D_3$ on P transport in rat seminiferous tubules. Clin Res **30**:391A.
135. Schleicher G, Privette TH, Stumpf WE 1989 Distribution of soltriol [1,25$(OH)_2$-vitamin D_3] binding sites in male sex organs of the mouse: An autoradiographic study. J Histochem Cytochem **37**:1083–1086.
136. Miller GJ, Stapleton GE, Ferraral JA, Lucia MS, Pfister S, Hedlund TE, Upadhya P 1992 The human prostatic carcinoma cell line LN-CaP expresses biologically active, specific receptors for 1 alpha,25-dihydroxyvitamin D_3. Cancer Res **52**:515–519.
137. Skowronski RJ, Peehl DM, Feldman D 1993 Vitamin D and prostate cancer: 1,25-Dihydroxyvitamin D_3 receptors and actions in human prostate cancer cell lines. Endocrinology **132**:1952–1960.
138. Clark SA, Stumpf WE, Bishop CW, DeLuca HF, Park DH, Joh TH 1986 The adrenal: A new target organ of the calciotropic hormone 1,25-dihydroxyvitamin D_3. Cell Tissue Res **243**:299–302.
139. Baksi SN, Hughes MJ 1984 Alteration of adrenal catecholamine levels in the rat after dietary calcium and vitamin D deficiencies. J Auton Nerv Syst **11**:393–396.
140. Brion F, Dupuis Y 1980 Calcium and monoamine regulation: Role of vitamin D nutrition. Can J Physiol Pharmacol **58**:1431–1444.
141. Brion F, Parvez S, Parvez H, Marnay-Gulat C, Raoul Y 1978 Effects of cortisol on adrenal phenylethanolamine-*N*-methyltransferase: Antagonistic effects of vitamin D in hypophysectomized rats fed a vitamin D free diet. Can J Physiol Pharmacol **56**:1017–1021.
142. Puchacz E, Goc A, Stumpf WE, Bidmon HJ, Stachowiak EK, Stachowiak MW 1991 Vitamin D regulates expression of tyrosine hydroxylase gene in adrenal chromaffin cells. Soc Neurosci **17**:391.8 (abstract).

143. Christakos S, Friedlander EJ, Frandsen BR, Norman AW 1979 Studies on the mode of action of calciferol. XIII. Development of a radioimmunoassay for vitamin D-dependent chick intestinal calcium-binding protein and tissue distribution. Endocrinology **104**:1495–1501.
144. Narbaitz R, Stumpf WE, Sar S 1981 The role of autoradiographic and immunocytochemical techniques in the clarification of sites of metabolism and action of vitamin D. J Histochem Cytochem **29**:91–100.
145. Kadowaki S, Norman AW 1984 Pancreatic vitamin D-dependent calcium binding protein: Biochemical properties and response to vitamin D. Arch Biochem Biophys **233**:228–236.
146. Hall AK, Norman AW 1991 Acute actions of 1,25-dihydroxyvitamin D_3 upon chick pancreatic calbindin-D_{28K}. Biochem Biophys Res Commun **176**:1057–1061.
147. Bourlon P-M, Faure-Dussert A, Billaudel B, Sutter BCJ 1996 Relationship between calbindin-D_{28K} levels in the A and B cells of the rat endocrine pancreas and the secretion of insulin and glucagon: Influence of vitamin D_3 deficiency and 1,25-dihydroxyvitamin D_3. J Endocrinol **148**:223–232.
148. Gelbard HA, Stern PH, U'Prichard DC 1980 1α,25-Dihydroxyvitamin D_3 nuclear receptors in the pituitary. Science **209**:1247–1249.
149. Stumpf WE, Sar M, Reid FA, Tanaka Y, DeLuca HF 1979 Target cells for 1,25-dihydroxyvitamin D_3 in intestinal tract, stomach, kidney, skin, pituitary, and parathyroid. Science **206**:1188–1190.
150. Stumpf WE, Sar M, O'Brien LP 1987 Vitamin D sites of action in the pituitary studied by combined autoradiography–immunohistochemistry. Histochemistry **88**:11–16.
151. Haug E, Gautvik KM 1985 Demonstration and characterization of a 1 alpha,25-$(OH)_2D_3$ receptor-like macromolecule in cultured rat pituitary cells. J Steroid Biochem **23**:625–635.
152. Murdoch GH, Rosenfeld MG 1981 Regulation of pituitary function and prolactin production in the GH_4 cell line by vitamin D. J Biol Chem **256**:4050–4055.
153. Tornquist K, Lamberg-Allardt C 1987 1,25-Dihydroxycholecalciferol on the TRH-induced TSH release in rats. Acta Endocrinol **114**:55–59.
154. Smith MA, McHenry C, Oslapas R, Hofmann C, Hessel P, Paloyan E 1989 Altered TSH levels associated with increased serum 1,25-dihydroxyvitamin D_3: A possible link between thyroid and parathyroid disease. Surgery **106**:987–991.
155. Tornquist K 1987 Effect of 1,25-dihydroxyvitamin D_3 on rat pituitary prolactin release. Acta Endocrinol **116**:459–464.
156. Tornquist K, Lamberg-Allardt C 1987 Effect of 1,25-dihydroxyvitamin D_3 on TSH secretion from rat pituitary cells in culture. Acta Endocrinol **114**:357–361.
157. d'Emden MC, Wark JD 1987 1,25-Dihydroxyvitamin D_3 enhances thyrotropin releasing hormone induced thyrotropin secretion in normal pituitary cells. Endocrinology **12**:1192–1194.
158. d'Emden MC, Wark JD 1989 Vitamin D-enhanced thyrotrophin release from rat pituitary cells: Effects of Ca^{2+}, dihydropyridines and ionomycin. J Endocrinol **121**:441–450.
159. d'Emden MC, Wark JD 1991 Culture requirements for optimal expression of 1α,25-dihydroxyvitamin D_3-enhanced thyrotropin secretion. In Vitro Cell Dev Biol **27A**:197–204.
160. d'Emden MC, Wark JD 1989 Effects of tri-iodothyronine, cortisol and transcriptional inhibitors on vitamin D_3-enhanced thyrotrophin secretion by rat pituitary cells in vitro. J Endocrinol **121**:451–458.
161. Wark JD, Tashjian AH Jr 1983 Regulation of prolactin mRNA by 1,25-dihydroxyvitamin D_3 in GH_4C_1 cells. J Biol Chem **258**:12118–12121.
162. Wark JD, Gurtler V 1988 Vitamin D-induction of secretory responses in rat pituitary tumour (GH_4C_1) cells. J Endocrinol **117**:293–298.
163. Tashjian AH 1990 Hormonal regulation of cytosolic free calcium and its functional consequences: The GH-cell model. Environ Health Perspect **84**:27–30.
164. Tornquist K, Tashjian AH 1989 Dual actions of 1,25-dihydroxycholecalciferol on intracellular Ca^{2+} in GH_4C_1 cells: Evidence for effects on voltage-operated Ca^{2+} channels and Na^+/Ca^{2+} exchange. Endocrinology **124**:2765–2776.
165. Tornquist K 1992 Pretreatment with 1,25-dihydroxycholecalciferol enhances thyrotropin-releasing hormone- and inositol 1,4,5-trisphosphate-induced release of sequestered Ca^{2+} in permeabilized GH_4C_1 pituitary cells. Endocrinology **131**:1677–1681.
166. Sornes G, Haug E, Torjesen PA 1993 Calcitriol attenuates the thyrotropin-releasing hormone-stimulated inositol phosphate production in clonal rat pituitary (GH_4C_1) cells. Mol Cell Endocrinol **93**:149–156.
167. Atley LM, Lefroy N, Wark JD 1995 1,25-Dihydroxyvitamin D_3-induced upregulation of the thyrotropin-releasing hormone receptor in clonal rat pituitary GH_3 cells. J Endocrinol **147**:397–404.
168. Tornquist K 1991 1,25-Dihydroxycholecalciferol enhances both the bombesin-induced transient in intracellular free Ca^{2+} and the bombesin-induced secretion of prolactin in GH_4C_1 pituitary cells. Endocrinology **128**:2175–2182.
169. Kaji H, Hinkle PM 1989 Attenuation of thyroid hormone action by 1,25-dihydroxyvitamin D_3 in pituitary cells. Endocrinology **124**:930–936.
170. Garcia-Villalba P, Jimenez-Lara AM, Aranda A 1996 Vitamin D interferes with transactivation of the growth hormone gene by thyroid hormone and retinoic acid. Mol Cell Biol **16**:318–327.
171. Rhodes SJ, Chen R, DiMattia GE, Scully KM, Kalla KA, Lin SC, Yu VC, Rosenfeld MG 1993 A tissue-specific enhancer confers Pit-1-dependent morphogen inducibility and autoregulation on the Pit-1 gene. Genes Dev **7**:913–932.
172. Ericson LE 1968 Degranulation of the parafollicular cells of the rat thyroid by vitamin-D_2-induced hypercalcemia. J Ultrastruct Res **24**:145–149.
173. Young DM, Capen CC 1970 Thyrocalcitonin content of thyroid glands from cows with vitamin D-induced hypercalcemia. Endocrinology **86**:1463–1466.
174. Roszkiewicz J 1974 Histologic and histochemical studies on C cells in the thyroid gland in white rats under conditions of chronic hypercalcemia induced with vitamin D_3. Folia Morphol **33**:177–189.
175. Thurston V, Williams ED 1982 Experimental induction of C cell tumours in thyroid by increased dietary content of vitamin D_3. Acta Endocrinol **100**:41–45.
176. Rix E, Raue F, Deutschle I, Ziegler R 1984 Different effects of hypercalcemic state induced by Walker tumor (HWCS 256) and 1,25$(OH)_2D_3$ intoxication on rat thyroid C cells. An ultrastructural, immunocytochemical, and biochemical study. Histochemistry **80**:503–508.
177. Freake HC, MacIntyre I 1982 Specific binding of 1,25-dihydroxycholecalciferol in human medullary thyroid carcinoma. Biochem J **206**:181–184.
178. Berg JP, Torjesen PA, Haug E 1991 Calcitriol receptors in rat thyroid follicular cells (FRTL-5). Acta Endocrinol **125**:574–580.
179. Lamberg-Allardt C, Valtonen E, Polojarvi M, Stewen P 1991 Characterization of a 1,25-dihydroxyvitamin D_3 receptor in FRTL-5 cells. Mol Cell Endocrinol **81**:25–31.
180. Stumpf WE, O'Brien LP 1987 Autoradiographic studies with ^{3}H-1,25-dihydroxyvitamin D_3 in thyroid and associated tissues of the neck region. Histochemistry **87**:53–58.

181. Okada H, Saito E, Matsukawa K 1991 Ultrastructure of thyroid C cells in sheep treated with vitamin D_3. J Vet Med Sci **53**:921–927.
182. Segond N, Legendre B, Tahri EH, Besnard P, Jullienne A, Moukhtar MS, Garel JM 1985 Increased level of calcitonin mRNA after 1,25-dihydroxyvitamin D_3 injection in the rat. FEBS Lett **184**:268–272.
183. Naveh-Many T, Silver J 1988 Regulation of calcitonin gene transcription by vitamin D metabolites in vivo in rat. J Clin Invest **81**:270–273.
184. Cote GJ, Rogers DG, Huang ES, Gagel RF 1987 The effect of 1,25-dihydroxyvitamin D_3 treatment on calcitonin and calcitonin gene-related peptide mRNA levels in cultured human thyroid C-cells. Biochem Biophys Res Commun **149**:239–243.
185. Collignon H, Laborie C, Tahri EH, el M'Selmi A, Garel JM 1992 Effects of dexamethasone, calcium and 1,25-dihydroxycholecalciferol on calcitonin and calcitonin gene-related peptide mRNA levels from the CA-77 C cell line. Thyroid **2**:361–365.
186. Silver J, Naveh-Many T 1993 Calcitonin gene regulation in vivo. Horm Metab Res **25**:470–472.
187. Peleg S, Abruzzese RV, Cooper CW, Gagel RF 1993 Down-regulation of calcitonin gene transcription by vitamin D requires two widely separated enhancer sequences. Mol Endocrinol **7**:999–1008.
188. Baier R, Grauer A, Lazaretti-Castro M, Ziegler R, Raue F 1994 Differential effects of 1,25-dihydroxyvitamin D_3 on cell proliferation and calcitonin gene expression. Endocrinology **135**:2006–2011.
189. Lazaretti-Castro M, Grauer A, Baier R, Vieira JG, Ziegler R, Raue F 1995 1,25-Dihydroxyvitamin D_3 stimulates growth and inhibits calcitonin secretion in a human C cell carcinoma cell line. Braz J Med Biol Res **28**:1013–1018.
190. Ikeda K, Lu C, Weir EC, Mangin M, Broadus AE 1989 Transcriptional regulation of the parathyroid hormone-regulated peptide gene by glucocorticoids and vitamin D in a human C-cell line. J Biol Chem **264**:15743–15746.
191. Rogers DG, Cote GJ, Huang ES, Gagel RF 1989 1,25-Dihydroxyvitamin D_3 silences 3′,5′-cyclic adenosine monophosphate enhancement of somatostatin gene transcription in human thyroid C cells. Mol Endocrinol **3**:547–551.
192. Odum L, Bundgaard JR, Monstein HJ 1995 Vitamin D_3 effects on basal and cAMP modulated expression of cholecystokinin and somatostatin genes in a rat medullary thyroid carcinoma cell line (CA-77). Neuropeptides **29**:45–51.
193. Berg JP, Sornes G, Torjesen PA, Haug E 1991 Cholecalciferol metabolites attenuate cAMP production in rat thyroid cells (FRTL5). Mol Cell Endocrinol **76**:201–206.
194. Ongphiphadhanakul B, Ebner SA, Fang SL, Lombardi A, Baran DT, Braverman LE 1992 1,25-Dihydroxycholecalciferol modulates ^{3}H-thymidine incorporation in FRTL5 cells. J Cell Biochem **49**:304–309.
195. Lamberg-Allardt C, Valtonen E 1992 1,25-Dihydroxycholecalciferol attenuates thyrotropin stimulated iodine accumulation in rat thyroid follicular FRTL-5 cells by reducing iodide porter number. Biochem Biophys Res Commun **182**:1435–1439.
196. Berg JP, Sandvik JA, Ree AH, Sornes G, Bjoro T, Torjesen PA, Gordeladze JO, Haug E 1994 1,25-Dihydroxyvitamin D_3 attenuates adenylyl cyclase activity in rat thyroid cells: Reduction of thyrotropin receptor number and increase in guanine nucleotide-binding protein G_i-2 alpha. Endocrinology **135**:595–602.
197. Berg JP, Ree AH, Sandvik JA, Tasken K, Landmark BF, Torjesen PA, Haug E 1994 1,25-Dihydroxyvitamin D_3 alters the effect of cAMP in thyroid cells by increasing the regulatory subunit type II beta of the cAMP-dependent protein kinase. J Biol Chem **269**:32233–32238.
198. Tornquist K, Stewen P, Lamberg-Allardt C 1992 Regulatory effect of 1,25-dihydroxycholecalciferol on calcium fluxes in thyroid FRTL-5 cells. Mol Cell Endocrinol **86**:21–27.
199. Ahlstrom M, Tornquist K, Lamberg-Allardt C 1995 1,25-Dihydroxyvitamin D_3 reduces the number of alpha 1-adrenergic receptors in FRTL-5 rat thyroid cells. Mol Cell Endocrinol **108**:143–148.
200. Nguyen M, Guillozo H, Garabedian M, Balsan S 1987 Lung as a possible additional target organ for vitamin D during fetal life in the rat. Biol Neonate **52**:232–240.
201. Nguyen TM, Guillozo H, Marin L, Dufour ME, Tordet C, Pike JW, Garabedian M 1990 1,25-Dihydroxyvitamin D_3 receptors in rat lung during the perinatal period: Regulation and immunohistochemical localization. Endocrinology **127**:1755–1762.
202. Dupret JM, L'Horset F, Perret C, Bernaudin JF, Thomasset M 1992 Calbindin-D_{9K} gene expression in the lung of the rat. Absence of regulation by 1,25-dihydroxyvitamin D_3 and estrogen. Endocrinology **131**:2643–2648.
203. Gaultier C, Harf A, Balmain N, Cuisinier-Gleizes P, Mathieu H 1984 Lung mechanics in rachitic rats. Am Rev Respir Dis **130**:11–16.
204. Marin L, Dufour ME, Tordet C, Nguyen M 1990 $1,25(OH)_2D_3$ stimulates phospholipid biosynthesis and surfactant release in fetal rat lung explants. Biol Neonate **57**:257–260.
205. Marin L, Dufour ME, Nguyen TM, Tordet C, Garabedian M 1993 Maturational changes induced by 1 alpha,25-dihydroxyvitamin D_3 in type II cells from fetal rat lung explants. Am J Physiol **265**:L45–L52.
206. Edelson JD, Chan S, Jassal D, Post M, Tanswell AK 1994 Vitamin D stimulates DNA synthesis in alveolar type-II cells. Biochim Biophys Acta **1211**:159–166.
207. Sundell K, Bishop JE, Bjornsson BT, Horst RL, Hollis BW, Norman AW 1991 Organ distribution of a high affinity receptor for $1,25(OH)_2$ vitamin D_3 in the Altlantic cod (*Gadus morhua*), a marine teleost. In: Norman AW, Bouillon R, Thomasset M (eds) Vitamin D: Gene Regulation, Structure–Function Analysis and Clinical Application. de Gruyter, Berlin, pp. 114–115.
208. Duncan WE, Whitehead D, Wray HL 1988 A 1,25-dihydroxyvitamin D_3 receptor-like protein in mammalian and avian liver nuclei. Endocrinology **122**:2584–2589.
209. Duncan WE, Glass AR, Wray HL 1991 Estrogen regulation of the nuclear 1,25-dihydroxyvitamin D_3 receptor in rat liver and kidney. Endocrinology **129**:2318–2324.
210. Rixon RH, MacManus JP, Whitfield JF 1979 The control of liver regeneration by calcitonin, parathyroid hormone and 1 alpha,25-dihydroxycholecalciferol. Mol Cell Endocrinol **15**:79–89.
211. Ethier C, Goupil D, Demers C, Hendy GN, Gascon-Barre M 1993 Hypocalcemia, regardless of the vitamin D status, decreases epidermal growth factor receptor density and autophosphorylation in rat livers. Endocrinology **133**:780–792.
212. Sikorska M, De Belle I, Whitfield JF, Walker PR 1989 Regulation of the synthesis of DNA polymerase-α in regenerating liver by calcium and 1,25-dihydroxyvitamin D_3. Cell Biol **67**:345–351.
213. Sikorska M, Whitfield JF, Rixon RH 1983 The effects of thyroparathyroidectomy and 1,25-dihydroxyvitamin D_3 on changes in the activities of some cytoplasmic and nuclear protein kinases during liver regeneration. J Cell Physiol **115**:297–304.
214. Sikorska M, Whitfield JF 1985 The regulatory and catalytic subunits of rat liver cyclic AMP-dependent protein kinases respond differently to thyroparathyroidectomy and 1 alpha,25-dihydroxyvitamin D_3. Biochem Biophys Res Commun **129**:766–772.
215. Youdale T, Whitfield JF, Rixon RH 1985 1 alpha,25-Dihydroxyvitamin D_3 enables regenerating liver cells to make functional

ribonucleotide reductase subunits and replicate DNA in thyroparathyroidectomized rats. Can J Biochem Cell Biol **63**:319–324.

216. Rixon RH, Isaacs RJ, Whitfield JF 1989 Control of DNA polymerase-alpha activity in regenerating rat liver by calcium and 1 alpha,25$(OH)_2D_3$. J Cell Physiol **139**:354–360.
217. Ethier C, Kestekian R, Beaulieu C, Dube C, Havrankova J, Gascon-Barre M 1990 Vitamin D depletion retards the normal regeneration process after partial hepatectomy in the rat. Endocrinology **126**:2947–2959.
218. Haddad P, Coulombe PA, Gascon-Barre M 1987 Influence of the vitamin D hormonal status on the hepatic response to bromobenzene. J Pharmacol Exp Ther **242**:354–363.
219. Pahuja DN, Deshpande UR, Soman CS, Nadkarni GD 1989 Altered hepatic function in vitamin D-deprived rats. J Hepatol **9**:209–216.
220. Davis WL, Matthews JL, Goodman DBP 1989 Glyoxylate cycle in the rat liver: Effect of vitamin D_3 treatment. FASEB J **3**:1651–1655.
221. Davis WL, Jones RG, Farmer GR, Dickerson T, Cortinas E, Cooper OJ, Crawford L, Goodman DB 1990 Identification of glyoxylate cycle enzymes in chick liver—the effect of vitamin D_3: Cytochemistry and biochemistry. Anat Rec **227**:271–284.
222. Hashizume K, Suzuki S, Ichikawa K, Takeda T, Kobayashi M 1991 Effect of active vitamin D_3 on the levels of NADPH-dependent cytosolic 3,5,3′-triiodo-L-thyronine-binding protein. Biochem Biophys Res Commun **177**:388–394.
223. Taniguchi K, Morimoto S, Itoh K, Morita R, Fukuc K, Imanaka S, Ogihara T 1990 Stimulatory effect of 1,25-dihydroxyvitamin D_3 on transferrin synthesis in primary cultures of adult rat hepatocytes. Biochem Int **22**:37–44.
224. Baran DT, Kelly AM 1988 Lysophosphatidylinositol: A potential mediator of 1,25-dihydroxyvitamin D-induced increments in hepatocyte cytosolic calcium. Endocrinology **122**:930–934.
225. Baran DT, Sorensen AM, Honeyman TW, Ray R, Holick MF 1989 Rapid actions of 1α-25-dihydroxyvitamin D_3 on Ca^{2+} and phospholipids in isolated rat liver nuclei. FEBS Lett **259**:205–208.
226. Rockwell JC, Sorensen AM, Baran DT 1993 Na^+/H^+ exchange and PLA_2 activity act interdependently to mediate the rapid effects of 1 alpha,25-dihydroxyvitamin D_3. Steroids **58**:491–494.
227. Baran DT, Sorensen AM, Honeyman TW 1988 Rapid action of 1,25-dihydroxyvitamin D_3 on hepatocyte phospholipids. J Bone Miner Res **3**:593–600.
228. Baran DT, Sorensen AM, Honeyman TW, Ray R, Holick MF 1990 1α,25-Dihydroxyvitamin D_3-induced increments in hepatocyte cytosolic calcium and lysophosphatidylinositol: Inhibition by pertusis toxin and 1β,25-dihydroxyvitamin D_3. J Bone Miner Res **5**:517–521.
229. Stumpf WE, O'Brien LP 1987 1,25$(OH)_2$ vitamin D_3 sites of action in the brain. An autoradiographic study. Histochemistry **87**:393–406.
230. Musiol IM, Stumpf WE, Bidmon HJ, Heiss C, Mayerhofer A, Bartke A 1992 Vitamin D nuclear binding to neurons of the septal, substriatal and amygdaloid area in the Siberian hamster (*Phodopus sungorus*) brain. Neuroscience **48**:841–848.
231. Bidmon HJ, Stumpf WE 1994 Distribution of target cells for 1,25-dihydroxyvitamin D_3 in the brain of the yellow bellied turtle *Trachemys scripta*. Brain Res **640**:277–285.
232. Stumpf WE, Clark SA, O'Brien LP, Reid FA 1988 1,25$(OH)_2$ vitamin D_3 sites of action in spinal cord and sensory ganglion. Anat Embryol **177**:307–310.
233. Bidmon H-J, Mayerhofer A, Heiss C, Bartke A, Stumpf WE 1991 Vitamin D (soltriol) receptors in the choroid plexus and ependyma: Their species-specific presence. Mol Cell Neurosci **2**:145–156.
234. Walters MR, Fischette T, Fetzer C, May B, Riggle PC, Tibaldo-Bongiorno M, Christakos S 1992 Specific 1,25-dihydroxyvitamin D_3 binding sites in choroid plexus. Eur J Pharmacol **213**:309–311.
235. Gascon-Barre M, Huet P-M 1983 Apparent [^{3}H]1,25-dihydroxyvitamin D_3 uptake by canine and rodent brain. Am J Physiol **244**:E266–E271.
236. Pardridge WM, Sakiyama R, Coty WA 1985 Restricted transport of vitamin D and A derivatives through the rat blood–brain barrier. J Neurochem **44**:1138–1141.
237. Luine VN, Sonnenberg J, Christakos S 1987 Vitamin D: Is the brain a target? Steroids **49**:133–153.
238. Murphy VA, Smith QR, Rapoport SI 1988 Regulation of brain and cerebrospinal fluid calcium by brain barrier membranes following vitamin D-related chronic hypo- and hypercalcemia in rats. J Neurochem **51**:1777–1782.
239. Sonnenberg J, Luine VN, Krey LC, Christakos S 1986 1,25-Dihydroxyvitamin D_3 treatment results in increased choline acetyltransferase activity in specific brain nuclei. Endocrinology **118**:1433–1439.
240. Binderman I, Harel S, Earon Y, Tomer A, Weisman Y, Kaye AM, Somjen D 1988 Acute stimulation of creatine kinase activity by vitamin D metabolites in the developing cerebellum. Biochim Biophys Acta **972**:9–16.
241. Siegel A, Malkowitz L, Moskovits MJ, Christakos S 1984 Administration of 1,25-dihydroxyvitamin D_3 results in the elevation of hippocampal seizure threshold levels in rats. Brain Res **298**:125–129.
242. Sutherland MK, Somerville MJ, Yoong LK, Bergeron C, Haussler MR, McLachlan DR 1992 Reduction of vitamin D hormone receptor mRNA levels in Alzeheimer as compared to Huntington hippocampus; Correlation with calbindin-28k mRNA levels. Brain Res Mol Brain Res **13**:239–250.
243. Saporito MS, Brown ER, Hartpence KC, Wilcox HM, Vaught JL, Carswell S 1994 Chronic 1,25-dihydroxyvitamin D_3-mediated induction of nerve growth factor mRNA and protein in L929 fibroblasts and in adult rat brain. Brain Res **633**:189–196.
244. Neveu I, Naveilhan P, Jehan F, Baudet C, Wion D, DeLuca HF, Brachet P 1994 1,25-Dihydroxyvitamin D_3 regulates the synthesis of nerve growth factor in primary cultures of glial cells. Brain Res Mol Brain Res **24**:70–76.

Section IV

Physiology

CHAPTER 31

Vitamin D: Role in the Calcium Economy

ROBERT P. HEANEY Creighton University, Omaha, Nebraska

I. INTRODUCTION

Vitamin D functions in many body systems, but perhaps the best attested of its actions—and certainly the one most clearly associated with human disease—is its role in transferring calcium (and phosphorus) from ingested food into the body fluids. Calcium, like most divalent cations, is only partially absorbed from the chyme as it travels through the small intestine. This situation creates an opportunity for regulation of absorption, with room both to increase and to decrease calcium extraction efficiency in response to physiological controls.

Details of the many cellular and tissue effects of vitamin D, and of the absorptive process itself, are covered in other chapters in this volume. Here I summarize mainly the meaning and importance of the vitamin D-mediated transfer process from gut to blood and outline how it fits into the maintenance of the calcium economy. My frame of reference will be the integrated functioning of the intact organism.

II. OVERVIEW OF THE CALCIUM ECONOMY

A. Body Calcium Compartments

Body calcium in an adult human amounts to about 15–20 g (0.375–0.5 mol) per kilogram body weight. This calcium exists in three quite distinct divisions (or compartments). They are distinct because (1) movement of calcium atoms between them is both limited and regulated and (2) they can and do vary in magnitude independently of one another.

The first and most obvious compartment is the calcium in the bones and teeth. Here calcium exists as inorganic mineral crystals, arranged in an imperfect apatite lattice with variable stoichiometry, and embedded in a dense protein matrix. Although cells (osteocytes) ramify throughout bony tissue, the intercellular bony material itself lacks appreciable free water. As a result there is very limited exchange of calcium ions between the bone and the circulating body fluids. Isotopic exchange with tracers injected into the blood is confined to the surface layer of crystals in the bone situated along vascular channels and spaces, and to still incompletely mineralized new forming sites. Taken together the exchangeable bone calcium moieties amount to only about 25 mmol, or ~0.1% of total skeletal calcium [1]. Moreover, the insolubility of bone mineral is such that, even when there is exchange, there is virtually never net transfer out of bone into the body fluids. Net transfer normally requires formation or resorption of bone tissue.[1]

A second, biologically critical compartment is cell calcium. Here calcium serves as a ubiquitous second messenger, linking signals from outside the cell to the mechanisms constituting the response of the cell. Although cytosolic free calcium ion concentrations are typically on the order of 1×10^{-5} mmol [2], total cell calcium is on the order of 0.5–2 mmol/kg tissue [1]. Most of this quantity (~99.99%) is bound to specialized calcium storage proteins (e.g., sequestrin, parvalbumin, calbindin) and located in storage vesicles, typically specialized units of the endoplasmic reticulum. Because the calcium ion is of just the right radius to fit neatly into folds of a peptide chain, and because calcium is capable of forming up to 12 (typically 6–8) coordination bonds with oxygen atoms in the side chains of amino acids projecting off the peptide spine, calcium stabilizes the tertiary structure of many catalytic molecules, thereby activating them. Cytosolic [Ca^{2+}] must be kept very low to prevent constant, uncontrolled activity of the many cell functions in which calcium plays a messenger or activation role. On the other hand, substantial intracellular stores of calcium are necessary because the binding avidity of cytosolic proteins for calcium is so high that, typically, the free path of a calcium ion in the cytosol is only a tiny fraction of a cell diameter [2]. If reservoirs were not diffusely distributed throughout the cytosol, and if extracellular calcium were the only source for this critical second messenger function, activation would be limited to the zone immediately beneath the plasma membrane.

Because the cell membrane is relatively impermeable to calcium ions, and because the cytosolic Ca^{2+} compartment is so tiny, tracer exchange between extracellular calcium and cell stores of calcium is surprisingly slow, typically requiring hours or days to come into tracer equilibrium [1]. Further, cell calcium levels are typically unaffected by calcium concentrations in the extracellular fluids (ECF).

The third and smallest division of body calcium is the calcium in the circulating blood and the ECF bathing all the body tissues. This compartment contains typically about 0.4 mmol calcium per kilogram. Ionized calcium concentration in these fluids is ordinarily about 1.25 mmol/liter (5 mg/dL). [Ca^{2+}] is regulated across the higher vertebrate orders with the same exquisite precision as are, for example, the concentrations of sodium and potassium. Departures from this level produce well-studied, significant effects on interneuronal signal transmission and on muscular excitability. ECF [Ca^{2+}] is also important for supra-cellular protein activations such as those in the coagulation cascade.

Into and out of this compartment passes all the calcium entering and leaving the body from the outside, as well as that entering and leaving bone. These fluxes are summarized schematically in Fig. 1. Together they involve daily quantities amounting to 35–50% of the size of the entire compartment in healthy adults, and to several times that compartment size in infants. Without tight regulation, ECF [Ca^{2+}] would oscillate between

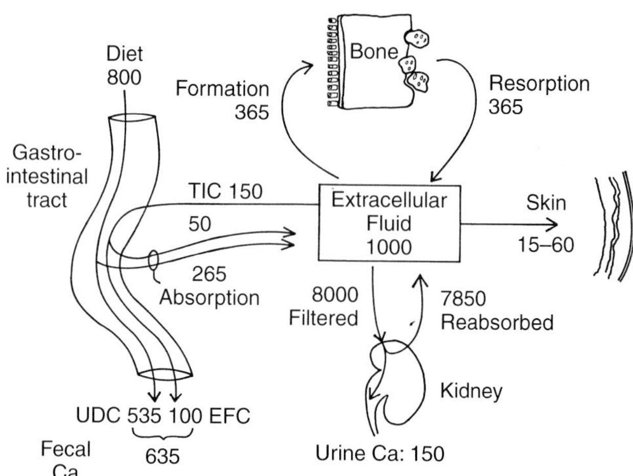

FIGURE 1 Schematic representation of the principal inputs and outputs of the extracellular fluid compartment. TIC, Total digestive juice calcium; UDC, unabsorbed dietary calcium; EFC, endogenous fecal calcium. Units are millimoles Ca^{2+} per day. Copyright Robert P. Heaney, 1986, 1996. Reprinted with permission.

[1] One possible exception is the calcium carbonate of bone. Carbonate substitutes poorly for phosphate in the apatite lattice, and it is generally considered that, because of different valences and ionic radii for the two anions, carbonate is confined to crystal surfaces. Generally it is assumed that the carbonate is more or less uniformly distributed throughout the bony material. However, bone carbonate is substantially more labile than is bone calcium generally, and it is likely therefore that the carbonate is situated even more superficially than generally presumed, that is, primarily on anatomic bone surfaces, rather than diffusely on crystal surfaces generally. Calcium carbonate might, thus, be a kind of "icing" on the underlying mineralized matrix, sensitive to pH and pCO_2 in the extracellular fluid. If this is the case, a limited amount of net movement of calcium into and out of bone would be possible without involving cell-mediated formation or resorption of bone tissue.

possibly fatal extremes of hypo- and hypercalcemia as the organism goes from fasting to feeding.

B. Regulation of [Ca^{2+}] in Extracellular Fluids

A detailed description of the processes involved in regulating ECF [Ca^{2+}] is beyond the purpose of this chapter. To situate vitamin D in this system it is necessary only to note that regulation occurs both by controlling the renal excretory threshold for calcium and by regulating calcium fluxes into and out of the ECF. Two of the transfers illustrated in Fig. 1 drive the system and are essentially unregulated. The other transfers are at least partially responsive and constitute the basis of the regulatory control of ECF [Ca^{2+}]. The unregulated, driving transfers are (1) mineralization of newly deposited bone matrix at bone forming sites and (2) daily obligatory loss out of the body. Vitamin D and its metabolites play critical roles in the two regulated transfers into the ECF, namely, from ingested food and from bone resorption.

1. Driving Transfers

a. Mineralization of Bone Mineralization of bone constitutes an unregulated drain because it is a passive phenomenon, lagging several days, and even weeks, behind the osteoblastic cellular activity that initiates the process. The growing mineral deposits extract calcium and phosphorus from blood flowing past the mineralizing site at a decreasing exponential rate for up to 40 weeks after the matrix has been deposited. The only way, in theory, to stop the process is to shut down blood flow to bone. The magnitude of this mineralization drain varies with skeletal remodeling activity and with body size. It amounts to about 0.15 mmol/kg body weight in healthy mature adults. It is much higher during growth, of course, and curiously seems to rise once again after midlife [3,4].

b. Obligatory Loss Obligatory loss consists of a combination of dermal loss and the fixed components of urinary and endogenous fecal calcium excretion. Dermal losses consist not just of sweat calcium but of the calcium contained in shed skin, hair, and nails. As has been noted above, all cells contain substantial amounts of calcium, and their loss from body surfaces inexorably takes calcium with them. Dermal losses have been difficult to quantify but are estimated to be at least 0.4 mmol/day and more likely closer to 1.5 mmol/day [5]. Much higher losses have been reported with vigorous physical activity [5a].

The fixed component of endogenous fecal and urinary calcium losses is somewhat more complicated [6]. On average, 3.5–4.0 mmol calcium enters the gut each day from endogenous sources, principally as a component of digestive secretions, but also as the calcium contained in shed mucosal cells (which turn over once every 4–5 days). The precise quantity varies directly with body size and with the amount of food consumed. For reasons not understood, the amount of calcium entering the gut in the digestive secretions varies directly also with phosphorus intake [6]. In any event, this endogenous calcium mixes with food calcium, and much of it is subject to absorption, just as is the calcium of food. However, as noted above, calcium absorption is always incomplete. Gross absorption averages about 25–30% in healthy adults, and net absorption about 10%. Hence, much of the secreted calcium is lost in the feces. Moreover, some of the endogenous calcium secretion enters the intestinal stream effectively so low in the gut as to be essentially unreclaimable (e.g., the calcium in colon mucus and in shed colon cells). On a normal diet this distal component averages about 0.6 mmol [6]. Typical endogenous fecal calcium values average in the range of 3 mmol/day. Since absorption efficiency in adults is essentially never above 60%, even on a very low calcium diet (see below), there is an irreducible minimum loss of endogenous calcium through the gut averaging close to 2 mmol/day. If absorption is not greater than this figure, the gut becomes a net excretory organ for calcium.

The third obligatory loss is through the kidney. It is generally held that renal calcium excretion is controllable, but this is only partly true. Parathyroid hormone (PTH) certainly regulates tubular calcium reabsorption. However, there is a fixed limit to what that mechanism can accomplish. This limit is itself a function of other variables that are outside the regulatory loop. Best studied of these factors is the renal excretion of sodium, dietary intake of protein, and absorbed dietary phosphorus. Sodium [7–9] and protein [10,11] increase urinary calcium loss; phosphorus decreases it [12].

Because sodium and calcium compete for the same reabsorption mechanism in the proximal tubule, the two ions influence the excretion of one another [13]. On average, urine calcium increases by 0.5 to 1.5 mmol for every 100 mmol of sodium excreted [7,8]. Similarly, urine calcium rises by about 0.25 mmol for every 10 g of protein ingested [10]. The result, for an adult woman ingesting the RDA for protein and the median sodium intake for North Americans, is a level of obligatory urinary loss amounting to about 2 mmol/day. Reducing sodium intake would certainly reduce this obligatory loss. Nevertheless, such voluntary dietary change is clearly not a part of any physiological regulatory loop. Thus, to the extent that sodium intake influences obligatory calcium loss, it constitutes a demand to which the calcium homeostatic system must respond.

Given typical adult diets in Europe and North America, the sum of these obligatory losses through skin, gut, and kidney is about 5 ± 1 mmol/day, or about one-fifth of the total calcium in the ECF. To offset these losses (plus the demands of bone mineralization), the organism regulates countervailing inputs into the ECF from food and bone. Vitamin D plays a role in both transfers. The first source is dependent on both the presence of food in the upper GI tract and the presence of sufficient calcium in that food. Because neither condition can be guaranteed, the osseous response is the more reliable and is, in fact, the first line of defense against hypocalcemia.[2]

2. Response of the System to Demand

a. PTH Mediation of Response Briefly, a fall in ECF [Ca^{2+}] evokes a prompt rise in PTH release. PTH acts in a classic negative feedback loop to raise the ECF [Ca^{2+}], thereby closing the loop and reducing PTH release. The mechanisms by which PTH raises ECF [Ca^{2+}] illustrate well the complexities of the calcium economy. These mechanisms include (1) increasing renal phosphate clearance, thereby lowering ECF phosphate levels; (2) increasing renal tubular reabsorption of calcium, thereby allowing system inputs to elevate ECF [Ca^{2+}]; (3) both activating new bone resorption loci and augmenting osteoclast work at existing resorption loci; and (4) increasing the activity of the renal 1α-hydroxylase, thereby increasing serum levels of 1,25-dihydroxyvitamin D [1,25$(OH)_2$D] and augmenting absorption efficiency for ingested food calcium.

These four effects reinforce one another in important ways. The earliest effects, occurring within minutes, are a decrease in renal tubular phosphate reabsorption and the resulting fall in serum phosphate. The latter immediately augments existing osteoclastic bone resorption [14] and increases activity of the renal 1α-hydroxylase [15]. The elevated production of 1,25$(OH)_2$D leads to increased intestinal absorption, elevating ECF [Ca^{+2}] and thereby closing the feedback loop. 1,25$(OH)_2$D also suppresses PTH release in its own right. Finally, 1,25$(OH)_2$D is a necessary condition for efficient osteoclast work. In this latter role, it is not known whether variations of 1,25$(OH)_2$D in the physiologic range produce corresponding alterations in bone resorption, and it is a difficult question to study because of the tight regulation of the various components of the system. Nevertheless, it is well established that resorptive work is severely impaired in vitamin D deficiency states (see below), and that 1,25$(OH)_2$D in supraphysiological doses is capable of causing substantial increases in bone resorption. Finally, all of the components of the intestinal calcium transport system, including vitamin D receptors and calbindin, are also found in the distal tubule of the nephron [16]. It is possible, though not yet firmly established, that 1,25$(OH)_2$D may thus enhance recovery of filtered calcium and contribute to the PTH effect of elevating the renal calcium threshold.

An example of the integrated response of the system to demand is afforded by what occurs during antler formation in deer in the spring [17]. The calcium demands of antler formation exceed the calcium available in the nutrient-poor late winter foliage. Other things being equal, this drain would lower ECF [Ca^{2+}], but the parathyroid glands, sensing minute reductions in ECF [Ca^{2+}], respond by increasing secretion of PTH, which in turn activates numerous remodeling loci throughout the skeleton of the deer. Because remodeling is asynchronous, with resorption preceding formation at each remodeling locus, the extra burst of newly activated remodeling sites provides a temporary surplus of calcium, which the animal promptly uses to mineralize matrix in the forming antler. A few weeks later, when the skeletal remodeling loci reach their own formation phase, the calcium content of the ingested foliage will be higher, and the "mineral debt" created by antler formation will be repaid from ingested greens.

b. Vitamin D Deficiency As already noted, vitamin D is essential for this mobilization of calcium from the skeletal calcium reserves. This is well illustrated in the course of nutritional rickets observed in the children of China and Northern Europe before the discovery of vitamin D. Typically, solar exposure produced some vitamin D during summer months—enough to allow reasonably normal mineralization of the hypertrophic cartilage zone beneath the growth plate. Then, during the 8 months or so when solar vitamin D was lacking, the rachitic process resumed. The result, visible on X-ray, was a series of bands parallel with the growth plate,

[2] Because it is beyond the scope of this chapter, which is concerned mainly with calcium transfers, I shall not develop further what is actually an even more fundamental mechanism by which PTH regulates ECF [Ca^{2+}], that is, the control of the renal calcium threshold. Suffice it to say here only that in the shift from a lower to higher PTH level renal calcium losses are temporarily curtailed until calcium inputs from bone and gut succeed in elevating [Ca^{2+}]. Then, as the filtered load at the glomerulus rises, calcium excretion returns to its previous level (although renal calcium clearance is now reduced).

with densely mineralized bone tissue alternating with undermineralized layers (Fig. 2). One might have thought that, during the winter, PTH-mediated bone resorption would have attacked bone in the dense layers, making its calcium available to new forming sites (as occurs with antler formation in deer). In the absence of vitamin D, however, the bone resorptive process is severely impaired. As a result the rachitic child is unable to mobilize its own calcium reserves adequately.

c. Feast and Famine The diet of hominids was high in calcium [18], just as is the diet of contemporary deer and other higher mammals. Foods available to contemporary hunter-gatherers exhibit an annual mean calcium nutrient density of 70–80 mg/100 kcal. For individuals of contemporary body size, doing the work of a hunter-gatherer, that value translates to calcium intakes in the range of 2000–3000 mg/day. However, as noted, the environment could not be relied on to supply calcium-rich food continuously. Periods of fasting, famine, or drought would undoubtedly have threatened hypocalcemia. This fact underscores the importance both of bone as a calcium reserve and of the vitamin D–parathyroid hormone control system, with its ability to release calcium rapidly from bone.[3]

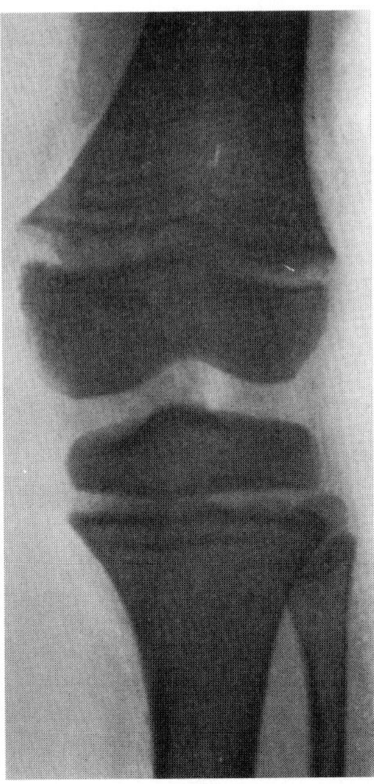

FIGURE 2 Radiograph of the knee in a child with vitamin D deficiency rickets. The alternating bands of high and low density reflect annual periods of greater and lesser vitamin D availability. Reprinted from *A Text-book of X-Ray Diagnosis,* Vol. 6, 4th edition, 1971, edited by S. C. Shanks and P. Kerley. HK Lewis & Co., Ltd., Toronto. Chapter XLIII, Metabolic and endocrine-induced bone disease, by C. J. Hodson, p. 661, Fig. 774.

Contemporary humans in industrialized nations have the same paleolithic physiology as our hominid ancestors. We experience, however, a few crucial differences in external conditions. One is a generally lower exposure to sunlight. Rickets was endemic in Northern Europe in the nineteenth century, partly because of latitude, partly because of air pollution, and partly because of child labor. Routine use of vitamin D supplementation today has all but eliminated that problem in children, but adults, and particularly the elderly, are often vitamin D insufficient. (The consequences of this low vitamin D exposure are explored elsewhere in this volume.) The other difference is a much reduced nutrient density for calcium in our ingested food. The importance of this latter departure from primitive conditions lies in the fact that it limits the efficacy of vitamin D in augmenting input from the intestine. This limitation is not widely recognized, and rarely is its quantitative impact adequately appreciated. Section III,B of this chapter (below) develops this important component of the calcium economy in greater detail.

3. INDEPENDENCE OF PTH EFFECTOR MECHANISMS

The PTH regulatory system is unique in that the feedback loop operates through three independent effector mechanisms, already described (elevated renal calcium threshold, elevated intestinal calcium absorption, and elevated bone resorption). This seeming redundancy underscores the physiological importance of maintaining constant ECF [Ca^{2+}]. The independence of response of the PTH effectors constitutes the substratum for still incompletely explored, but interesting, differences in the way the organism adapts to deficiency and surfeit of calcium.

The feedback loop of calcium regulation is typically closed by some variable combination of all three effects. For example, when calcium intake falls, there is an obvious limit to what the absorptive mechanism can yield, forcing higher PTH levels and correspondingly greater renal calcium retention and enhanced net transfer from bone. This instance of quasi-imbalance of the three mechanisms arises from an extrinsic stress, but relative differences in the intrinsic responsiveness of those ef-

[3] Technically calcium is never actually "released" from bone. (See, however, note 1, above.) Rather a volume of bone tissue is resorbed. In the process its calcium is released into the ECF. Thus, transfer of calcium out of bone always means some removal of bone tissue. Bone, however, is a very rich source of calcium. A single cubic centimeter of bone has as much calcium as is contained in the entire circulating blood of an adult human.

fector organs can also have important consequences for bone mass [19]. Slightly lower intrinsic responsiveness of osteoclastic resorption to PTH, other things being equal, leads to a slightly higher PTH level, which, in turn, drives all three effector mechanisms slightly harder. The result is that the intestinal absorptive and renal recovery mechanisms for calcium operate at higher efficiency, and bone resorption, lower. This leads to more bone, and is probably a large part of the explanation for the higher bone mass in blacks.

Although there has been some inconsistency in the data reported to date in this regard, the bulk of the evidence points to relative refractoriness to PTH-stimulated bone resorption in African Americans [20–23]. For example, Dawson-Hughes *et al.* [24] showed that blacks exhibited a greater response to dietary calcium reduction than did whites, with larger increments of both PTH and $1,25(OH)_2D$, indicating a black–white difference in bony response. Clear evidence in this regard also comes from the calcium tracer studies of Abrams and colleagues, showing higher calcium absorption and retention in adolescent black females than in whites at the same pubertal stage [23]. There seem also to be differences in vitamin D metabolism in the two ethnic groups, with blacks showing typically lower serum 25OHD levels and higher $1,25(OH)_2D$ levels than whites [20,21,24]. The relative importance of these two metabolites to the calcium economy, and of differences in their concentrations in ethnic groups, is still unclear (see below).

III. CALCIUM ABSORPTIVE INPUT

Chapter 16 describes the mechanisms of calcium transfer from the intestinal lumen. Here I describe quantitative aspects of that transfer as a part of the integrated calcium economy, focusing as well on the factors determining the magnitude of the input from the gut into the ECF.

A. Location and Timing of Absorption in the Gut

As noted in Chapter 16, there is a gradient of concentrations of vitamin D receptors and of calbindin in mucosa along the gut, with highest levels in the duodenum and lowest in the colon mucosa. Accordingly, the avidity (or rate) of active absorption is highest in the duodenum. It is sometimes said that absorption itself is highest there, but this is not correct. That conclusion is based on studies of isolated loops or gut sacs, where movement of the chyme along the intestine cannot occur. Absorbed quantity is the product of absorption rate and residence time, and residence time of the chyme in the duodenum is very brief. Only at very low calcium intakes (or test loads), and with maximal $1,25(OH)_2D$-stimulated active transport, will it be true that most of the calcium absorbed will be from the duodenum. At more usual intakes, the much longer residence time in the jejunum and ileum means that most of the quantity absorbed occurs from the lower small intestine. The importance of length of exposure to the absorptive surface is reflected in the finding that absorptive efficiency varies directly with mouth-to-cecum transit time [25].

Absorption does not occur from the healthy stomach, and thus the beginning of absorption is delayed until gastric emptying begins. This, in turn, is dependent on the character of the ingested meal or other calcium source. Emptying tends to be most rapid with small fluid ingestates, and it is slower with solid food and with fat. In healthy individuals ingesting light meals (such as would commonly be employed to test absorption efficiency), calcium absorption is nearly complete by 5 hr after ingestion [26]. Figure 3 presents data on the time course of absorption, using the ratio of the time-dependent apparent absorption fraction to its ultimate value in the individual being tested. As Fig. 3 shows, absorption has reached better than 80% of its ultimate value by 3 hr after ingestion, and 96% by 7 hr. There is then only a very gradual approach to completion over the next 20 hr. This latter component probably reflects

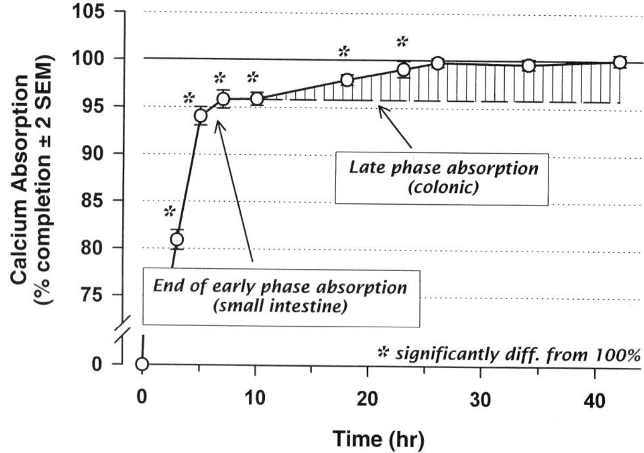

FIGURE 3 Time course of absorption (derived from Barger-Lux *et al.* [26]). The data plot the percentage completion of absorption (derived from expressing the double-isotope absorption fraction as a ratio of its value at any given time to its final value after 24 hr). Completion of absorption is thus expressed as a value of 100%. As the curve shows, absorption is about 94% complete by 5 hr after oral ingestion. The remaining 6% occurs more slowly and may be presumed to reflect absorption from the colon and/or from ileocecal reflux. Copyright Robert P. Heaney, 1966. Reprinted with permission.

a small amount of colonic absorption (or, alternatively, cecal–ileal reflux, with delayed ileal absorption). It should be stressed that the percentage values in Fig. 3 refer to the quantity absorbed, not the quantity ingested. Thus, with typically only 25–30% of a load absorbed (see below), the 4–5% colonic component represents absorption of only about 1% of the ingested load.

B. Absorption as a Function of Intake

It has long been recognized that absorption efficiency varies inversely with intake. Figure 4 illustrates this relationship with data obtained from healthy, middle-aged women in whom absorption fraction was measured under controlled metabolic ward conditions and plotted as a function of their usual ingested intakes [27]. The best fit regression line through the data shows the expected rise in absorption fraction at low calcium intakes. (Note, however, that even at the lowest intakes, predicted mean absorption efficiency is only ~45%.)

The higher efficiency at low intakes is traditionally attributed to adaptation, specifically to higher production of $1,25(OH)_2D$, with a corresponding increase in active calcium absorption. Although that explanation is undoubtedly correct, it is substantially incomplete. This is shown by the data in Fig. 5, which plots nonadaptive absorption fraction as a function of a broad range of calcium load sizes. These studies were performed in women assigned randomly on any given morning to intake loads spanning a 30-fold range, from 0.4 to 12.5 mmol [28]. Clearly, an inverse relationship is present,

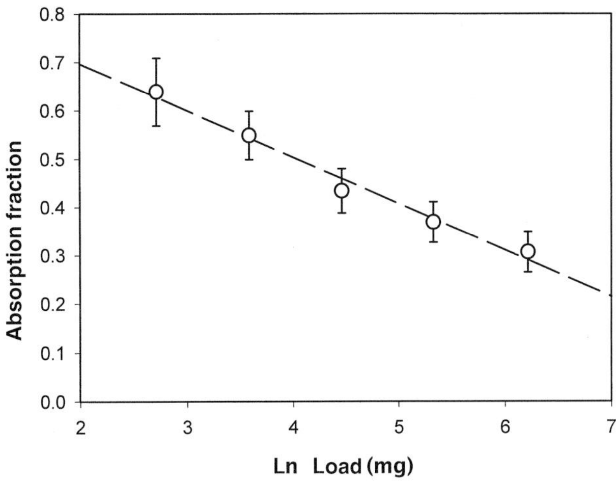

FIGURE 5 Fractional absorption in nonadapted healthy women for loads ranging from 0.4 to 12.5 mmol Ca. Error bars are ±2 SEM. The dashed line is the least squares regression through the actual data ($N = 75$), derived from Heaney *et al.* [28]. The units of the horizontal axis are natural logarithms of load size (in mmol Ca). 1 mmole equals 40 mg of calcium. Copyright Robert P. Heaney, 1996. Reprinted with permission.

just as in the data of Fig. 4. Equally clearly, it cannot be due to adaptation, as the test load was the first exposure these women had to the intake level concerned. Figure 6 plots these two sets of data together, and shows that, although both exhibit an inverse relationship between absorption and intake, there is in fact a difference between them, with the adapted women absorbing more at low intakes than the nonadapted women (as would be predicted). The zone between the two lines is a quan-

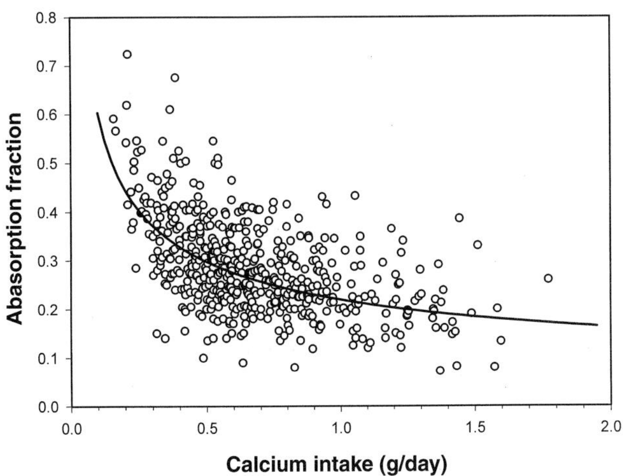

FIGURE 4 Fractional absorption plotted as a function of usual calcium intake (in g/day) in 525 studies in healthy, middle-aged women [27]. The solid line is the least squares regression line derived from a log–log fit. Copyright Robert P. Heaney, 1989. Reprinted with permission.

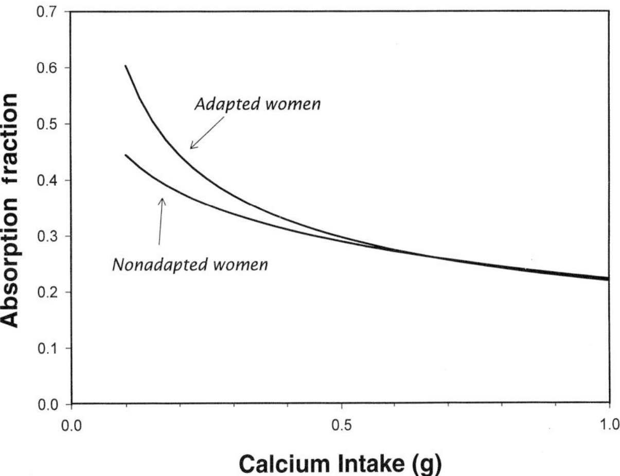

FIGURE 6 Combination of the regression lines from Figs. 4 and 5, showing the extent of the difference produced by adaptation to the various intakes. Copyright Robert P. Heaney, 1996. Reprinted with permission.

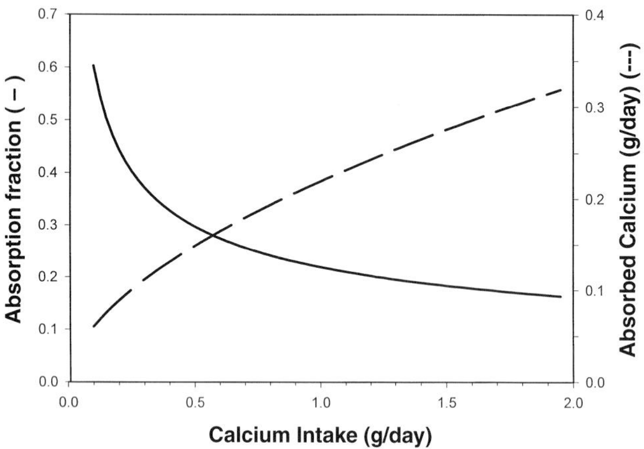

FIGURE 7 Fractional absorption and mass absorption for the 525 studies of Fig. 4. The left axis and solid line represent fractional absorption, and the right axis and dashed line denote mass absorption. Copyright Robert P. Heaney, 1996. Reprinted with permission.

titative expression of the PTH–vitamin D-mediated adaptation to the lower intake.[4]

The most likely explanation for the inverse relationship observed under both sets of conditions is that calcium transfer, whether active or passive, is a slow, inefficient process, with only a limited number of carrier molecules or pores available at any given instant. In the brief interval between exit of a bolus of food from the stomach and the time it reaches the colon, only so many calcium ions can use the available transport. If the number of ions reaching the absorptive site is small, then by numerical necessity the fraction absorbed will be larger than when the number of ions is large.

Absorption fraction (or efficiency) is thus a potentially misleading measure (at least if we stop there). It is, however, a necessary starting point because it is the primary datum available from most studies of absorptive physiology. Figure 7 presents the regression line from Fig. 4 and adds a second line representing the actual quantity of calcium absorbed in these same women (i.e., the product of absorption fraction and intake). This variable is obviously the nutritionally relevant one because, in offsetting obligatory losses (or special demands such as antler building or fetal skeletal development), it is a quantity of calcium (not a fraction) that is needed to balance the drains created by the quantity of calcium leaving the ECF.

Figure 7 also illustrates another important aspect of this input to the calcium economy. At low intakes, absorption is quantitatively low, despite being relatively more efficient. A moment's reflection suffices to show that a large fraction of a small number is, of necessity, a smaller number still. Thus, absorbing even a large fraction of a small intake cannot produce much calcium. The result is that, in the range of intakes commonly encountered in contemporary, industrialized humans, absorptive adaptation (via vitamin D) mitigates the problem created by a low intake, but it does not counterbalance it. A concrete example, employing realistic numbers, will help illustrate this point, and will show additionally how optimal operation of the vitamin D hormonal system is dependent on—and in fact presumes—the kinds of high calcium intakes found among hunter-gatherer humans and high primates (in whom the system evolved).

Contrast how two individuals are able respond to the increased obligatory loss occasioned by regular daily ingestion of an additional 100 mmol sodium (approximately the sodium contained in a single fast-food chicken dinner). Assume that one individual is ingesting 5 mmol Ca/day (1 mmol = 40 mgm) (corresponding to the lower quintile of calcium intakes in U.S. women [29]), and the other, 40 mmol (approximately the NIH Consensus Conference recommendation [30] for estrogen-deprived, postmenopausal women). Using data from the curve in Fig. 4, the individual with the lower intake absorbs at an efficiency of 44.5% prior to the extra sodium load, and the individual with the higher intake absorbs at 17.8%. (The first, therefore, has a gross absorbed quantity from the diet of 2.2 mmol/day, and the second, 7.1 mmol/day.) The increase in obligatory urinary loss occasioned by the increase in sodium intake will be about 1 mmol/day (see above). To offset this loss, the first individual, with the low intake, would have to increase her absorbed quantity to 3.2 mmol/day, which means increasing her already high absorption efficiency a factor of nearly 1.5 (from 44.5 to 64.5%). In contrast, the individual with the high intake needs to increase only from 17.5 to 20.3%. These calculations are summarized in Table I. The adjustment for the individual with a low calcium is substantially more than most adults can accomplish, whereas the second is easily accommodated. The first individual must, therefore, get the needed calcium from bone, whereas the second easily gets it from her food.

Thus, although vitamin D plays a critical role in increasing absorptive efficiency in response either to increased losses from the body or to decreased intake, it must be stressed that there is little room in which the PTH–vitamin D endocrine system can operate when intakes are already low. That does not mean that ECF

[4] As Fig. 6 shows, most of the difference occurs at intakes below 500 mg (12.5 mmol). However, the two sets of observations were performed in different groups of women, and even the nonadapted set must have had some basal level of $1,25(OH)_2D$-mediated adaptive absorption. Hence, the true extent of adaptive absorption is undoubtedly somewhat greater than indicated solely by the difference between the two lines in Fig. 6.

TABLE I Two Responses to Increased Obligatory Loss

Parameter	NIH optimal[c] Basal	NIH optimal[c] +100 mmol Na[a]	NHANES 20th percentile[d] Basal	NHANES 20th percentile[d] +100 mmol Na[a]
Calcium intake (mmol/day)	40	40	5	5
Absorption fraction	0.178	0.203[b]	0.445	0.645[b]
Absorbed calcium (mmol/day)	7.1	8.1	2.2	3.2

[a] The extra sodium causes an increase in obligatory calcium loss of ~1 mmol.

[b] The increase in absorption fraction required to produce an extra millimole of absorbed calcium. 1 mmole of calcium = 40 mgm.

[c] National Institutes of Health Consensus Development Conferences on Optimal Calcium Intake [30].

[d] National Health and Nutrition Examination Survey–II [29].

$[Ca^{2+}]$ regulation suffers. The bony calcium reserves are vast, essentially limitless. They will readily be drawn on to support ECF $[Ca^{2+}]$, using the well-studied mechanisms already described. Naturally, if this drain continues, bone mass will inevitably decline. At high calcium intakes, such as prevailed during hominid evolution, the vitamin D hormonal system helps maintain not only ECF $[Ca^{2+}]$, but total body calcium as well; at low intakes, only the ECF is protected.

C. Partition of Absorption between Active and Passive Mechanisms

As described in Chapter 16, absorption occurs both by vitamin D-mediated active transport across the mucosal cells and by passive diffusion around the cells. Is it possible to partition absorption between the active, cellular process and the passive, paracellular process? To a limited extent, the answer is yes. As should be clear from the foregoing, both passive and active absorption exhibit an inverse relation to intake or load. Absorption will be high by either mechanism, even approaching 100%, if the load is sufficiently small. However, small loads are not nutritionally relevant, no matter how they are (or are not) absorbed, so this discussion is confined to consideration of loads or intakes in the range of (or above) the RDA.

The right end of the regression line in Fig. 4 represents an absorption fraction of approximately 0.15. In work published earlier from our laboratory [31], we extended intakes well above the 2 g upper limit of Fig. 4, to as high as 8.0 g (200 mmol) Ca/day. Absorption fraction in that study also averaged about 0.15 at these very high intakes, or approximately what we observed at 50 mmol in Fig. 4. The essentially linear character of this absorption across such a broad range of intakes probably reflects the fact that, if passive absorption is due largely to solvent drag, the quantity of calcium absorbed will be a linear function of luminal calcium concentration.

It is likely, at habitual intakes in the range of 40–50 mmol Ca/day, that active absorption would be minimal, and it is a virtual certainty that this would be so at supraphysiological intakes of 200 mmol. Studies in patients with end-stage renal disease, with limited ability to synthesize $1,25(OH)_2D$, also report absorption fractions in the range of 10–20% [32,33]. Taken together these data indicate that passive absorption is able to extract about 10–15% of the calcium in the ingested food at nutritionally relevant loads. Variation around that level will presumably relate to interindividual variations in mucosal mass and in intestinal transit time.

What would an absorption efficiency of 15% mean for the calcium economy if passive absorption were the only means for extracting calcium from the diet, as, for example, must be the case with hereditary vitamin D-resistant rickets (HVDRR) type II? Assume an intake at the RDA (i.e., 20 mmol/day). As noted above, digestive secretions add 3.5 mmol to the stream, 0.6 mmol of that total too low to be absorbed. Calculation from these data easily shows that, at a 15% absorption efficiency, the gut will be a net excretory organ (if only barely: ~0.1 mmol/day net loss). No calcium gain into the body could occur at such an intake. Hence a 15% background absorption figure without vitamin D seems entirely compatible with development of severe rickets, either as found in HVDRR or in typical nutritional vitamin D deficiency.

IV. PHYSIOLOGICAL SOURCES OF VITAMIN D ACTIVITY

Thus far I have spoken of vitamin D only generically. As discussed extensively elsewhere in this volume, native vitamin D (cholecalciferol) has very little biological activity in its own right. It is converted in the body to a number of hydroxylated metabolites, the most important of which for the purposes of this chapter are 25-hydroxyvitamin D (25OHD) and $1,25(OH)_2D$. It is well established that cholecalciferol is 25-hydroxylated in the liver. The reaction is loosely controlled by circulating levels of $1,25(OH)_2D$ as well as by limited end product inhibition exerted by 25OHD itself. For the most part, however, circulating 25OHD levels are driven mainly by the circulating levels of the precursor, cholecalciferol. In unpublished data from my laboratory (1997), serum 25OHD rises by approximately 4–6 ng/ml for every 1000 IU of oral cholecalciferol after 8 weeks of daily oral administration. This relationship is consistent with published data from studies of high dose vitamin D treatment and cases of vitamin D intoxication [34,35].

The metabolite 1,25$(OH)_2$D is produced in the kidney, as already described, under control by serum PTH and changes in inorganic phosphate renal handling (see Chapters 32 and 46). It is generally considered to be the biologically active form of the vitamin, responsible for its full spectrum of actions; and it is considered also that the precursors (cholecalciferol and 25OHD) exert activity in their own right only under conditions of intoxication. This conclusion is based mainly on studies of binding kinetics of the various metabolites with the vitamin D receptor (VDR). If binding affinity for 1,25$(OH)_2$D is taken as 1.0, reported values for 25OHD are in the range of 1×10^{-3} to 1×10^{-4}, and for native vitamin D, 1×10^{-6} or lower [36–38]. Accordingly, it is reasoned that normal serum levels of the precursor compounds are too low to exert significant action.

However, because serum levels of 25OHD are typically three orders of magnitude higher than those of 1,25$(OH)_2$D, it is not clear, *a priori,* that 25OHD would be without effect under normal conditions. Protein binding in serum complicates interpretation of these relationships. It is generally presumed that total serum levels are physiologically less meaningful than free serum levels of a metabolite. As discussed in Chapter 7, the vitamin D family of compounds is carried in serum complexed to a vitamin D binding carrier protein (an α-globulin designated DBP, or Gc) with a gradient of affinities: highest for 25OHD, lower for native vitamin D, and lower still for 1,25$(OH)_2$D. However, binding to the carrier protein cannot be interpreted without reference to other factors, usually not readily determined in any given situation. These include the relative affinities of the binding protein and the receptor as well as the numbers of both. Available data on binding at the intact organism level are not sufficient to resolve this issue of hormone dynamics. Such theoretical considerations aside, however, there is a substantial body of clinical data indicating that 25OHD exerts appreciable biological activity at normal physiological concentrations.

Several studies show a surprisingly strong correlation between serum 25OHD and intestinal calcium absorption efficiency in intact humans [25,39–42]. If 25OHD were acting only as a precursor, one should have expected no correlation at all. Where 1,25$(OH)_2$D levels were measured in these studies, the correlation of 1,25$(OH)_2$D with absorption fraction was usually weaker than for 25OHD, or nonsignificant altogether. Clinically, it is widely recognized that 1,25$(OH)_2$D levels are commonly normal in sporadic nutritional rickets, whereas 25OHD levels are invariably low. Moreover, Colodro *et al.* [42], in a dose–response study for both metabolites, using calcium absorption in healthy human adults as the end point, reported a molar potency for administered 25OHD relative to 1,25$(OH)_2$D of 1 : 125, not the less than 1 : 2000–1 : 4000 figure predicted from *in vitro* receptor binding studies. Barger-Lux *et al.* [43], in a similar study, found a nearly identical potency ratio (1 : 100). In this study the rise in absorption fraction occurred without a detectable change in 1,25$(OH)_2$D level. Taken together these studies provide evidence that some fraction of circulating vitamin D-like activity can be attributed to 25OHD.[5]

This conclusion raises important questions about the interactions of the two principal hydroxylated vitamin D metabolites in the intact organism. To begin with, there can be no question about the potency of 1,25$(OH)_2$D itself. This very active metabolite produces a strong enhancement of absorption efficiency when given to intact humans [42–45]. In addition, whereas Norman and colleagues [46,47] have described a rapid response, nongenomic action of the D vitamin metabolites (termed transcaltachia), occupancy of the receptor by 1,25$(OH)_2$D seems to be required [47]. Also, patients with HVDRR lack functional vitamin D receptors and are not able to absorb calcium efficiently despite normal to high circulating levels of both 1,25$(OH)_2$D and 25OHD [48,49]. Thus, vitamin D-mediated absorption requires both a functioning receptor and some combination of 1,25$(OH)_2$D and 25OHD. Given the short serum half-life of 1,25$(OH)_2$D and the much longer half-life of 25OHD, and given the apparent potency of 25OHD in its own right, it may be that there is a floor or background level of absorption in normal individuals that is determined in part by 25OHD, and that 1,25$(OH)_2$D produces a quick acting fine-tuning of the system, in part via the rapid response transcaltachia pathway.

V. OPTIMAL VITAMIN D STATUS

Optimal vitamin D status can be defined as the daily intake or production of the vitamin (and/or the serum

[5] Some of the apparent action of 25OHD in oral dosing studies may be due to a direct effect of the metabolite on the mucosal cell (a first-pass effect). During absorption of 25OHD, mucosal exposure to the agent would be substantially higher than would occur from exposure to plasma 25OHD.

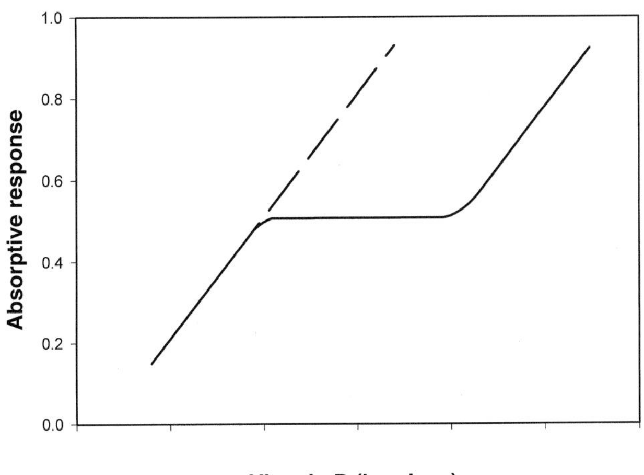

FIGURE 8 Schematic diagram of the likely relationship between absorptive performance and vitamin D dose. The dashed line represents the curve in the absence of physiological controls (such as might be found in isolated gut loop studies), whereas the solid line represents the likely intake relationship in organisms with functioning control systems. Copyright Robert P. Heaney, 1966. Reprinted with permission.

level of 25OHD) which is sufficient so that its availability does not limit any of the metabolic functions dependent on the vitamin. This notion of limit is illustrated in Fig. 8. A large body of data indicates that vitamin D-mediated absorption of calcium follows a curve such as the one presented in Fig. 8, rising with intake at levels below the requirement, then flat through a range of sufficiency, then rising again at pharmacological (or toxic) intakes. It is likely that the absorptive response, taken in isolation, would be a continuous smooth function of vitamin D dose, such as indicated by the dashed line in Fig. 8. However, at the whole organism level, once calcium absorption is optimal, other factors alter the response to vitamin D exposure. For example, at levels of calcium absorption higher than the body needs, PTH levels would drop and $1,25(OH)_2D$ synthesis would fall. Thus, despite a rising solar or oral dose of the precursor vitamin D, absorption would plateau. When the dose becomes sufficiently high, however, system controls are saturated and bypassed, and absorption begins to rise once again.[6] Intakes below the plateau are clearly insufficient, since they limit absorption anterior to any physiological controls. Intake levels above the plateau, in contrast, represent toxicity, that is, the overriding of physiological controls. (The subject of vitamin D toxicity is discussed in Chapter 54.)

[6] This second rise in absorption has never been reported for purely solar sources of vitamin D; however, it is well recognized as a component of vitamin D intoxication.

Optimal status could thus be defined as an intake or production of the vitamin sufficient to get an individual to the plateau. Optimal position on the plateau would depend on the relative risks of toxicity and deficiency, which would be predicted to touch on the tails of the population distributions of requirements. Defining such a level can be approached operationally in at least two ways. One is by determining the level of vitamin D intake at which calcium absorption does not change further on giving extra vitamin D at doses within the physiological range. A basically similar approach can be taken to the parathyroid response evoked by inadequate absorbed calcium intake, that is, to circulating PTH levels and to PTH-mediated bone resorption. Both are known to rise in the face of vitamin D insufficiency.

Using either the absorptive response to supplemental vitamin D [39,50] or such indices of vitamin D status as seasonal variation in serum immunoreactive PTH and bone remodeling [51–56], one can derive a level of 25OHD located at the theoretical threshold of the plateau in Fig. 8. Both approaches yield serum values of approximately 32 ng/ml (80 nmol/liter). This value is well above the bottom end of the range currently considered normal. However, not surprisingly, it is well below the mean for individuals with levels of sun exposure such as must have prevailed during hominid evolution [57].

As vitamin D is not found in appreciable concentrations in most of the items in the food supply (either primitive or modern), maintenance of optimal vitamin D status requires either sun exposure or, in high northern or southern latitudes, some degree of supplementation/fortification. Such a conclusion has often been uncongenial to traditional nutritionists, who have maintained that humans ought to be able to get all of the nutrients they need from a well-balanced diet. However, it must be stressed that vitamin D is an accidental nutrient, included with the other vitamins by mistake at an early stage of the development of nutritional science. Whatever the merit of the position held by traditional nutritionists, it cannot apply to this essential compound. Now that we understand the situation, we must see that it is certainly no more unnatural to sustain vitamin D status in high latitudes by supplementation than it is to sustain body temperature there by clothing or shelter.

Although more work clearly needs to be done to define the bottom end of the acceptable normal range with precision, for now the prudent course would seem to be to aim for a vitamin D intake sufficient to produce a serum 25OHD level of at least 32 ng/ml (see also Chapter 3). Levels below that point carry a risk of bone loss. A level of 32 ng/ml is certainly safe, as healthy college-age adults (with often generous sun exposure) commonly have levels two to three times that high.

VI. SUMMARY

The best attested function of vitamin D is the facilitation of transfer of calcium (and phosphorus) into the extracellular fluid from ingested food and from bone. In this capacity, vitamin D functions as a part of a control system that operates to maintain constancy of the calcium ion concentrations in the extracellular fluid against the demands of obligatory excretory losses and skeletal mineralization. In both transfers vitamin D works in concert with PTH. Quantitative analysis of the inputs and drains of the calcium economy reveals that, at contemporary calcium intakes, vitamin D-mediated absorptive enhancement only partially mitigates the impact of low calcium intake or large calcium losses. However, at intakes closer to those prevailing during hominid evolution, minor shifts in vitamin D-mediated absorption are fully adequate to compensate for stresses on the calcium economy. Although $1,25(OH)_2D$ is clearly the most potent form of the vitamin, 25OHD exerts significant vitamin D-like activity in its own right at physiological serum levels. Optimal vitamin D status is operationally defined as a level of vitamin D intake (or production) high enough to ensure that the vitamin D-mediated transfers are not limited by vitamin D availability. Available data point to a value for serum 25OHD of about 32 ng/ml (80 nmol/liter) as the bottom end of the optimal range.

References

1. Heaney RP 1963 Evaluation and interpretation of calcium kinetic data in man. Clin Orthop **31**:153–183.
2. Clapham DE 1995 Calcium signaling. Cell **80**:259–268.
3. Eastell R, Delmas PD, Hodgson SF, Eriksen EF, Mann KG, Riggs BL 1988 Bone formation rate in older normal women: Concurrent assessment with bone histomorphometry, calcium kinetics, and biochemical markers. J Clin Endocrinol Metab **67**:741–748.
4. Chapuy MC, Schott AM, Garnero P, Hans D, Delmas PD, Meunier PJ, EPIDOS Study Group 1996 Healthy elderly French women living at home have secondary hyperparathyroidism and high bone turnover in winter. J Clin Endocrinol Metab **81**:1129–1133.
5. Charles P 1989 Metabolic bone disease evaluated by a combined calcium balance and tracer kinetic study. Dan Med Bull **36**:463–479.

5a. Klesges RC, Ward KD, Shelton ML, Applegate WB, Cantler ED, Palmieri GMA, Harmon K, Davis J 1996 Changes in both mineral content in male athletes. Mechanisms of action and intervention effects. JAMA **276**:226–230.

6. Heaney RP, Recker RR 1994 Determinants of endogenous fecal calcium in healthy women. J Bone Miner Res **9**:1621–1627.
7. Nordin BEC, Need AG, Morris HA, Horowitz M 1993 The nature and significance of the relationship between urinary sodium and urinary calcium in women. J Nutr **123**:1615–1622.
8. Itoh R, Suyama Y 1996 Sodium excretion in relation to calcium and hydroxyproline excretion in a healthy Japanese population. Am J Clin Nutr **63**:735–740.
9. Devine A, Criddle RA, Dick IM, Kerr DA, Prince RL 1995 A longitudinal study of the effect of sodium and calcium intakes on regional bone density in postmenopausal women. Am J Clin Nutr **62**:740–745.
10. Heaney RP, Recker RR 1982 Effects of nitrogen, phosphorus, and caffeine on calcium balance in women. J Lab Clin Med **99**:46–55.
11. Johnson NE, Alcantara EN, Linkswiler H 1970 Effect of level of protein intake on urinary and fecal calcium and calcium retention of young adult males. J Nutr **100**:1425–1430.
12. Parfitt AM, Higgins BA, Nassim JR, Collins JA, Hilb A 1964 Metabolic studies in patients with hypercalciuria. Clin Sci **27**:463–482.
13. Walser M 1961 Calcium clearance as a function of sodium clearance in the dog. Am J Physiol **200**:769–773.
14. Raisz LG 1965 Bone resorption in tissue culture. Factors influencing the response of parathyroid hormone. J Clin Invest **44**:103–116.
15. Portale AA, Halloran BP, Morris RC Jr 1987 Dietary intake of phosphorus modulates the circardian rhythm in serum concentration of phosphorus. J Clin Invest **80**:1147–1154.
16. Feldman D, Malloy PJ, Gross C 1996 Vitamin D: Metabolism and Action. In: Marcus R, Feldman D, Kelsey J (eds) Osteoporosis. Academic Press, San Diego, pp. 205–225.
17. Banks WJ Jr, Epling GP, Kainer RA, Davis RW 1968 Antler growth and osteoporosis. Anat Rec **162**:387–398.
18. Eaton SB, Nelson DA 1991 Calcium in evolutionary perspective. Am J Clin Nutr **54**:281S–287S.
19. Heaney RP 1965 A unified concept of osteoporosis. Am J Med **39**:877–880.
20. Bell NH, Greene A, Epstein S, Oexmann MJ, Shaw S, Shary J 1985 Evidence for alteration of the vitamin D-endocrine system in blacks. J Clin Invest **76**:470–473.
21. Bell NH, Stern PH, Pauslon SK 1985 Tight regulation of circulating 1α,25-dihydroxyvitamin D in black children. N Engl J Med **313**:1418.
22. Weinstein RS, Bell NH 1988 Diminished rates of bone formation in normal black adults. N Engl J Med **319**:1698–1701.
23. Abrams SA, O'Brien KO, Liang LK, Stuff JE 1995 Differences in calcium absorption and kinetics between black and white girls aged 5–16 years. J Bone Miner Res **10**:829–833.
24. Dawson-Hughes B, Harris S, Kramich C, Dallal G, Rasmussen HM 1993 Calcium retention and hormone levels in black and white women on high- and low-calcium diets. J Bone Miner Res **8**:779–787.
25. Barger-Lux MJ, Heaney RP, Lanspa SJ, Healy JC, DeLuca HF 1995 An investigation of sources of variation in calcium absorption efficiency. J Clin Endocrinol Metab **80**:406–411.
26. Barger-Lux MJ, Heaney RP, Recker RR 1989 Time course of calcium absorption in humans: Evidence for a colonic component. Calcif Tissue Int **44**:308–311.
27. Heaney RP, Recker RR, Stegman MR, Moy AJ 1989 Calcium absorption in women: Relationships to calcium intake, estrogen status, and age. J Bone Miner Res **4**:469–475.
28. Heaney RP, Weaver CM, Fitzsimmons ML 1990 The influence of calcium load on absorption fraction. J Bone Miner Res **11**:1135–1138.
29. Carroll MD, Abraham S, Dresser CM 1983 Dietary intake source data: US, 1976–80. Vital and Health Statistics, Serv. 11-NO. 231,

DHHS. Publ. No. (PHS) 83-PHS. U.S. Government Printing Office, Washington, D.C.

30. NIH Consensus Conference: Optimal calcium intake 1994 JAMA **272**:1942–1948.
31. Heaney RP, Saville PD, Recker RR 1975 Calcium absorption as a function of calcium intake. J Lab Clin Med **85**:881–890.
32. Recker RR, Saville PD 1971 Calcium absorption in renal failure: Its relationship to blood urea nitrogen, dietary calcium intake, time on dialysis, and other variables. J Lab Clin Med **78**:380–388.
33. Coburn JW, Koppel MH, Brickman AS, Massry SG 1973 Study of intestinal absorption of calcium in patients with renal failure. Kidney Int **3**:264–272.
34. Whyte MP, Haddad JG Jr, Walters DD, Stamp TCB 1979 Vitamin D bioavailability: Serum 25-hydroxyvitamin D levels in man after oral, subcutaneous, intramuscular, and intravenous vitamin D administration. J Clin Endocrinol Metab **48**:906–911.
35. Byrne PM, Freaney R, McKenna MJ 1995 Vitamin D supplementation in the elderly: Review of safety and effectiveness of different regimes. Calcif Tissue Int **56**:518–520.
36. Hughes MR, Baylink DJ, Jones PG, Haussler MR 1976 Radioligand receptor assay for 25-hydroxyvitamin D_2/D_3 and 1α,25-dihydroxyvitamin D_2/D_3. J Clin Invest **58**:61–70.
37. Brumbaugh PF, Haussler MR 1973 1,25-Dihydroxyvitamin D_3 receptor: Competitive binding of vitamin D analogs. Life Sci **13**:1737–1746.
38. DeLuca HF 1983 The vitamin D–calcium axis. In: Rubin RP, Weiss GB, Putney Jr JW (eds) Calcium in Biological Systems. Plenum, New York, pp. 491–511.
39. Francis RM, Peacock M, Storer JH, Davies AEJ, Brown WB, Nordin BEC 1983 Calcium malabsorption in the elderly: The effect of treatment with oral 25-hydroxyvitamin D_3. Eur J Clin Invest **13**:391–396.
40. Bell NH, Epstein S, Shary J, Greene V, Oexmann MJ, Shaw S 1988 Evidence of probable role for 25-hydroxyvitamin D in the regulation of human calcium metabolism. J Bone Miner Res **3**:489–495.
41. Reasner CA, Dunn JF, Fetchick D, Liel Y, Hollis BW, Epstein S, Shary J, Mundy GR, Bell NH 1990 Alteration of vitamin D metabolism in Mexican-Americans [Letter to the Editor]. J Bone Miner Res **5**:793–794.
42. Colodro IH, Brickman AS, Coburn JW, Osborn TW, Norman AW 1978 Effect of 25-hydroxy-vitamin D_3 on intestinal absorption of calcium in normal man and patients with renal failure. Metabolism **27**:745–753.
43. Barger-Lux MJ, Heaney RP, Dowell S, Bierman J 1996 Relative molar potency of 25-hydroxyvitamin D indicates a major role in calcium absorption. J Bone Miner Res **11**:S423.
44. Gallagher JC, Jerpbak CM, Jee WSS, Johnson KA, DeLuca HF, Riggs BL 1982 1,25-Dihydroxyvitamin D_3: Short- and long-term effects on bone and calcium metabolism in patients with postmenopausal osteoporosis. Proc Natl Acad Sci USA **79**:3325–3329.
45. Dawson-Hughes B, Harris SS, Finneran S, Rasmussen HM 1995 Calcium absorption responses to calcitriol in black and white premenopausal women. J Clin Endocrinol Metab **80**:3068–3072.
46. Norman AW 1990 Intestinal calcium absorption: A vitamin D-hormone-mediated adaptive response. Am J Clin Nutr **51**:290–300.
47. Norman AW, Nemere I, Zhou L-X, Bishop JE, Lowe KE, Maiyar AC, Collins ED, Taoka T, Sergeev I, Farach-Carson MC 1992 $1,25(OH)_2$-vitamin D_3, a steroid hormone that produces biologic effects via both genomic and nongenomic pathways. Steroid Biochem Mol Biol **41**:231–240.
48. al-Aqeel A, Ozand P, Sobki S, Sewairi W, Marx S 1993 The combined use of intravenous and oral calcium for the treatment of vitamin D dependent rickets type II (VDDRII). Clin Endocrinol **39**:229–237.
49. Simonin G, Chabrol B, Moulene E, Bollini G, Strouc S, Mattei JF, Giraud F 1992 Vitamin D-resistant rickets type II: Apropos of 2 cases. Pediatrie-Bucur **47**:817–820.
50. Krall EA, Dawson-Hughes B 1991 Relation of fractional ^{47}Ca retention to season and rates of bone loss in healthy postmenopausal women. J Bone Miner Res **6**:1323–1329.
51. McKenna JM, Freaney R, Meade A, Muldowney FP 1985 Hypovitaminosis D and elevated serum alkaline phosphatase in elderly Irish people. Am J Clin Nutr **41**:101–109.
52. Rosen CJ, Morrison A, Zhou H, Storm D, Hunter SJ, Musgrave K, Chen T, Wei W, Holick MF 1994 Elderly women in northern New England exhibit seasonal changes in bone mineral density and calciotropic hormones. Bone Miner **25**:83–92.
53. Dawson-Hughes B, Harris SS, Dallal GE 1997 Plasma 25-hydroxyvitamin D, season, and serum parathyroid hormone levels in healthy elderly men and women. Am J Clin Nutr **65**:67–71.
54. Salamone LM, Dallal GE, Zantos D, Makrauer F, Dawson-Hughes B 1993 Contributions of vitamin D intake and seasonal sunlight exposure to plasma 25-hydroxyvitamin D concentration in elderly women. Am J Clin Nutr **58**:80–86.
55. Dawson-Hughes B, Dallal GE, Krall EA, Harris S, Sokoll LJ, Falconer G 1991 Effect of vitamin D supplementation on wintertime and overall bone loss in healthy postmenopausal women. Ann Intern Med **115**:505–512.
56. Krall EA, Sahyoun N, Tannenbaum S, Dallal GE, Dawson-Hughes B 1989 Effect of vitamin D intake on seasonal variations in parathyroid hormone secretion in postmenopausal women. N Engl J Med **321**:1777–1783.
57. Matsuoka LY, Wortsman J, Hollis BW 1990 Suntanning and cutaneous synthesis of vitamin D_3. J Lab Clin Med **116**:87–90.

CHAPTER 32

Regulation of Phosphate Transport

KEITH HRUSKA Renal Division, Washington University School of Medicine, Barnes-Jewish Hospital, St. Louis, Missouri

ANANDARUP GUPTA Renal Division, Barnes-Jewish Hospital, St. Louis, Missouri

JEAN-PHILIPPE BONJOUR Division of Clinical Pathophysiology, WHO Collaborating Center for Osteoporosis and Bone Disease, Department of Internal Medicine, University Hospital, Geneva, Switzerland

JOSEPH CAVERSAZIO Division of Clinical Pathophysiology, WHO Collaborating Center for Osteoporosis and Bone Disease, Department of Internal Medicine, University Hospital, Geneva, Switzerland

I. INTRODUCTION

Phosphate balance involves absorption of dietary phosphorus by the intestine, its distribution in body fluids and tissues, especially bone, and its excretion largely by the kidney. Scientific advances in the 1990s have improved our understanding of phosphate balance. We are now able to discuss at the cellular and molecular level phosphate handling in the kidney and bone, and sections discussing the molecular basis of osteoblastic and osteoclastic phosphate transport are contained in this chapter. Advances in our understanding of the regulation of phosphate have kept pace with the rapid rate of scientific progress in general, and this is apparent in the discussions of transport processes. As a result, a much clearer role for calcitriol in the regulation of phosphate transport can be provided than previously possible. Finally, the prospect of a new and exciting chapter in phosphate homeostasis is on the horizon owing to the recent discovery of the PEX gene for X-linked hypophosphatemia.

The physiological concentration of serum phosphate ranges from 2.8 to 4.5 mg/dl (0.9–1.45 m*M*) in adults [1]. There is a diurnal variation of 0.6 to 1.0 mg/dl, with the nadir occurring between 8 and 11 A.M. Ingestion of meals rich in carbohydrate decreases serum phosphate concentrations as a result of movement of phosphate from the extracellular to the intracellular space. In the extracellular fluid, phosphorus is present predominantly in the inorganic form. In serum, phosphate exists mainly as the free ion (HPO_4^{2-} and $H_2PO_4^-$ at a ratio of 4:1), and only a small fraction (15%) is protein bound [2,3]. Approximately 80 to 85% of the total phosphorus is present in the skeleton. Phosphate is a critical, basic component of hydroxyapatite, the mineral phase of bone. The rest of the phosphorus is widely distributed throughout the body in the form of organic phosphate compounds. Phosphorus plays a fundamental role in seven aspects of cellular metabolism. The energy required for many cellular reactions including biosynthesis derives from hydrolysis of adenosine triphosphate (ATP). Organic phosphate is an important component of phospholipids in cell membranes. Its concentration influences the activity of several metabolic pathways such as ammoniagenesis, glycolysis, gluconeogenesis, and the formation of 1,25-dihydroxyvitamin D [$1,25(OH)_2D$ from 25-hydroxyvitamin D (25OHD). Changes in serum phosphorus may also influence the dissociation of oxygen from hemoglobin through its regulation of the concentrations of 2,3-diphosphoglycerate.

II. GASTROINTESTINAL ABSORPTION OF PHOSPHORUS

Approximately 1 g of phosphorus is ingested daily in an average diet in the United States. About 300 mg is excreted in the stool, and a net of 800 mg is absorbed (Fig. 1). Most of the phosphorus is absorbed in the duodenum and jejunum, with minimal absorption occurring in the ileum [4]. Phosphorus transport in proximal segments of the small intestine appears to involve both passive and active components and to be under the influence of vitamin D (see Chapter 16). The movement of phosphorus from the intestinal lumen to the blood requires (1) transport across the luminal brush-border membrane of the intestine, (2) transport through the cytoplasm, and (3) transport across the basolateral plasma membrane of the epithelium.

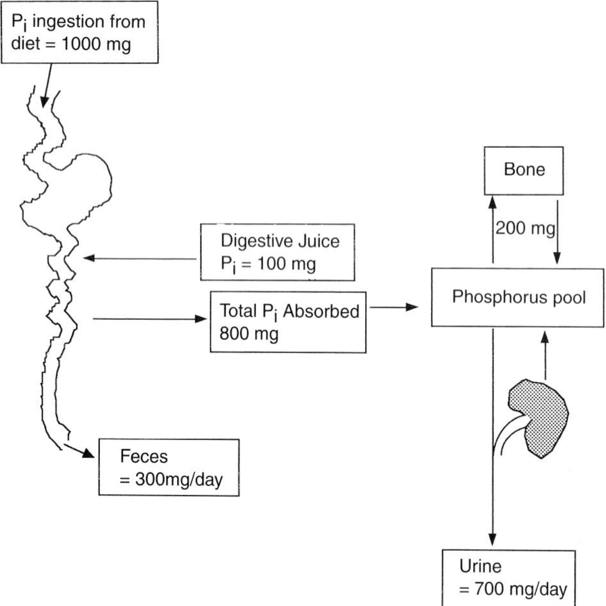

FIGURE 1 Overview of phosphorus (P_i) metabolism, showing pathways involved in phosphorus balance and turnover rates in major tissue compartments.

A. Luminal Brush-Border Transport and Effect of $1,25(OH)_2D_3$

The mechanism of transport across the intestinal brush-border epithelial membrane involves a sodium–phosphate cotransport system as suggested by Berner *et al.* [5]. The role of $1,25(OH)_2D$ and $25OHD_3$ in this transport system has been studied by several investigators [6–11]. Murer and Hildmann have shown that *in vivo* administration of $1,25(OH)_2D$ to rabbits affects *in vitro* uptake of phosphorus by intestinal brush-border membrane vesicles [7]. Conditions that affect $1,25(OH)_2D_3$ production have similar effects on duodenal and overall intestinal phosphate absorption [11]. Reduction of endogenous $1,25(OH)_2D_3$ decreases uptake of phosphorus by approximately 65% in brush-border membrane vesicles [7,10]. The action of $1,25(OH)_2D_3$ is on the maximal velocity (V_{max}) of transport, and it does not affect the affinity of inorganic phosphate (P_i) for the transporter. This uptake increased 3-fold after injection of high doses of $1,25(OH)_2D_3$. These studies are in agreement with those of Fuchs and Peterlik, who found similar results utilizing chick intestinal epithelial brush-border membrane vesicles [4].

The molecular basis for the sodium-dependent phosphate transport by the intestinal brush border has not been discovered at this writing (early 1997). The finding that intestinal brush-border phosphate transport is normal in the murine homolog (Hyp mouse) of X-linked hypophosphatemia indicates that the genetic control of intestinal transport is separate from the gene responsible for renal brush-border phosphate transport [12]. These findings are in agreement with studies demonstrating the tissue distribution of the renal phosphate cotransport proteins.

B. Transcellular Movement of Phosphorus

The second component of transcellular intestinal phosphorus transport involves the movement of phosphorus from the luminal to the basolateral membrane. Although little is known about the cellular events that mediate this transcellular process, evidence suggests a role for the microtubule system of intestinal cells [4]. Microtubules may be important in conveying phosphorus from the brush-border membrane to the basolateral membrane and may be involved in the extrusion of phosphorus at the basolateral membrane of the epithelial cell. 1,25$(OH)_2D_3$ stimulates the production of the 28- and 9-kDa calbindins that have been proposed as participating in transcellular movement of calcium. Whether they also participate in phosphate transport is unknown. This subject is discussed in detail in Chapters 13, 14, and 16 of this book.

C. Phosphate Exit at Basolateral Membrane

Little is known about the mechanisms of phosphorus extrusion at the basolateral membrane of intestinal epithelial cells. The electrochemical gradient for phosphorus favors movement from the intracellular to the extracellular compartment because the interior of the cell is electrically negative compared with the basolateral external surface [7]. Phosphate may exit the cell via an anion exchange mechanism [7], although there have been reports of an ATP- and Na^+-dependent transport process [10,13].

D. Regulation of Intestinal Phosphate Transport

The events that lead to stimulation of membrane uptake of phosphorus by vitamin D are of major interest but have not been completely elucidated. Even at the level of the renal brush border where the molecular identity of the transporter is known, the studies examining the transcriptional effects of 1,25$(OH)_2D_3$ are just in progress at this writing. Studies of phosphorus accumulation by rat intestinal brush-border vesiscles have demonstrated that at physiological pH phosphorus uptake is dependent on luminal sodium but is unaffected by the transmembrane potential, a finding that is consistent with an electroneutral cotransport of sodium and monovalent phosphate ($H_2PO_4^-$); at lower pH values (pH 6.0) the increased uptake is independent of intraluminal sodium concentration and appears to involve transfer of net negative charge. Vitamin D seems to stimulate phosphorus absorption by two mechanisms: a calcium-dependent duodenal process and a calcium-independent jejunal system. In the rat the major effect of 1,25$(OH)_2D_3$ on phosphorus absorption appears to involve increased active transport. Studies using perfused duodenal loops of normal chicks demonstrate that 1,25$(OH)_2D_3$ and some analogs of vitamin D stimulate P_i absorption within 2–8 min, suggesting that 1,25$(OH)_2D_3$ may exert nongenomic actions on P_i transport analogous to actions on Ca transport [14]. Parathyroid hormone (PTH) may influence phosphorus absorption in the gut but presumably does so indirectly by stimulating the renal synthesis of 1,25$(OH)_2D_3$ [11]. Low phosphorus diets increase the absorption of phosphorus from the intestine by stimulating the formation of 1,25$(OH)_2D_3$ [15], and by direct effects independent of 1,25$(OH)_2D_3$ [16]. In animals fed a high phosphorus diet, suppression of 1,25$(OH)_2D_3$ may result in decreased intestinal absorption of phosphorus.

III. RENAL REABSORPTION OF PHOSPHORUS

Most (90 to 95%) of the inorganic phosphorus in plasma (serum is not a physiological compartment) is ultrafilterable at the level of the glomerulus. At physiological levels of plasma phosphorus approximately 7 g of phosphorus is filtered daily by the kidney, of which 80 to 90% is reabsorbed by the renal tubules and the remainder are excreted in the urine (about 700 mg on a 1-g phosphorus diet) (Fig. 1) [17]. Micropuncture studies have demonstrated that 60 to 70% of the filtered phosphorus is reabsorbed in the proximal tubule (Fig. 2). However, there is also evidence that a significant amount of filtered phosphorus is reabsorbed in distal

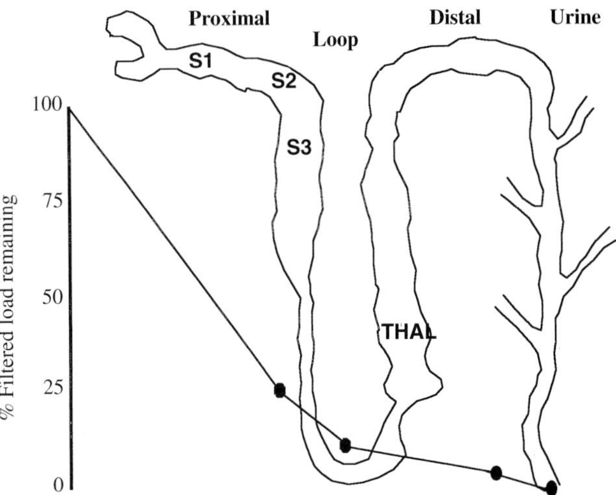

FIGURE 2 Profile of phosphate reabsorption along the nephron derived from micropuncture data. S_1, S_2, and S_3 are segments of the proximal tubule, and THAL stands for the thick ascending loop of Henle.

segments of the nephron [18,19]. When plasma phosphorus levels increase and the filtered load of phosphorus increases, the capacity to reabsorb phosphorus also increases. However, a maximum rate of transport for phosphorus reabsorption (TmPi) is obtained usually at plasma phosphorus concentrations of 6 mg/dl.

There is a direct correlation between T_m phosphorus values and glomerular filtration rate (GFR) even when the latter is varied over a broad range. Micropuncture studies suggest two different mechanisms responsible for phosphorus reabsorption in the proximal tubule. In the first third of the proximal tubule, in which only 10 to 15% of the filtered sodium and fluid is reabsorbed, the ratio of tubular fluid (TF) phosphorus to plasma ultrafilterable phosphorus (UF) falls to values of approximately 0.6. This indicates that the first third of the proximal tubule accounts for approximately 50% of the total amount of phosphorus reabsorbed in this segment of the nephron. In the last two-thirds of the proximal tubule the reabsorption of phosphorus parallels the movement of salt and water. In the remaining 70% of the pars convoluta, the TF/UF phosphorus ratio remains at a value of 0.6 to 0.7, whereas fluid reabsorption increases to approximately 60 to 70% of the filtered load. Thus, in the last two-thirds of the proximal tubule, the TF/UF phosphorus reabsorption is directly proportional to sodium and fluid reabsorption. A significant amount of phorphorus, perhaps on the order of 20 to 30%, is reabsorbed beyond the portion of the proximal tubule that is accessible to micropuncture. There is little phosphorus transport within the loop of Henle [20], with most distal transport occurring in the distal convoluted tubule. In this segment, approximately 15% of filtered phosphorus is reabsorbed under baseline conditions in animals subjected to parathyroidectomy [18,19], but the value falls to about 6% after administration of large doses of PTH. The collecting duct is a potential site for distal nephron reabsorption of phosphorus [21–23]. Transport in this nephron segment may explain the discrepancy between the amount of phosphorus delivered to the late distal tubule in micropuncture studies and the considerably smaller amount of phosphorus that appears in the final urine of the same kidney. Phosphorus transport in the cortical collecting tubule is independent of regulation by PTH. This is in agreement with the absence of PTH-dependent adenylate cyclase in the cortical collecting tubule [23].

A. Comparison of Superficial and Deep Nephron Transport

The contribution of superficial and deep nephrons to phosphorus homeostasis differs. Nephron heterogeneity in phosphorus handling has been evaluated in a number of conditions by puncture of the papillary tip and of the superficial early distal tubule, with the recorded fractional delivery representing deep and superficial nephron function, respectively. Using this technique, Haramati *et al.* [24,25] have demonstrated that in thyroparathyroidectomized (TPTX) rats fed a normal phosphorus diet, phosphorus reabsorption is greater in deep than in superficial nephrons, and this heterogeneous handling of phosphorus can be mitigated by both PTH infusion [25] and a low phosphorus diet [24]. Studies in TPTX rats fed a normal phosphorus diet have presented *in vivo* measurements of maximum tubular reabsorptive capacities for phosphorus of deep and superficial nephrons. Deep nephron values (5.05 pmol/ml/min) were significantly greater than the values obtained in superficial nephrons (3.38 pmol/ml/min). These results suggest that the juxtamedullary nephrons are more responsive to body phosphorus requirements than are the superficial nephrons. Also, microinjection of phosphorus tracer into thin ascending and descending limbs of Henle's loop reveal that only 80% of phosphorus was recovered in the urine, whereas 88 to 100% of phosphorus was recovered when the tracer was injected into the late superficial distal tubule. It was concluded that a significant amount of phosphorus must be reabsorbed by juxtamedullary distal tubules and/or by segments connecting the juxtamedullary distal tubules to the collecting ducts to account for the discrepancy between the results of superficial nephron injection and injection into the juxtamedullary ascending limb of Henle's loop. These data seem to support an increased reabsorptive capacity for phosphorus in deep as opposed to superficial nephrons.

In summary, phosphorus transport occurs in the distal nephron, particularly in the distal convoluted tubule and cortical collecting tubular system. This transport may be considerable under certain experimental conditions, but the importance of the terminal nephron system in day to day phosphorus homeostasis remains to be defined. It is also evident from data obtained from various micropuncture and microinjection studies that juxtamedullary and superficial nephrons have different capacities for phosphorus transport. The increased responsiveness of the deep nephrons to phosphorus intake suggests a key regulatory role for this system in phosphorus homeostasis.

B. Cellular Mechanisms of Phosphate Reabsorption in the Kidney

The apical membrane of renal tubular cells is the initial barrier across which phosphorus and other solutes

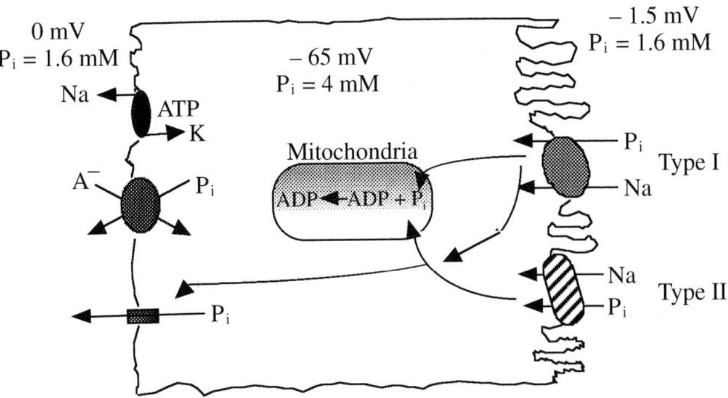

FIGURE 3 Phosphate transport in the proximal convoluted tubule. A^- represents an unspecified anion (Pi, OH^-).

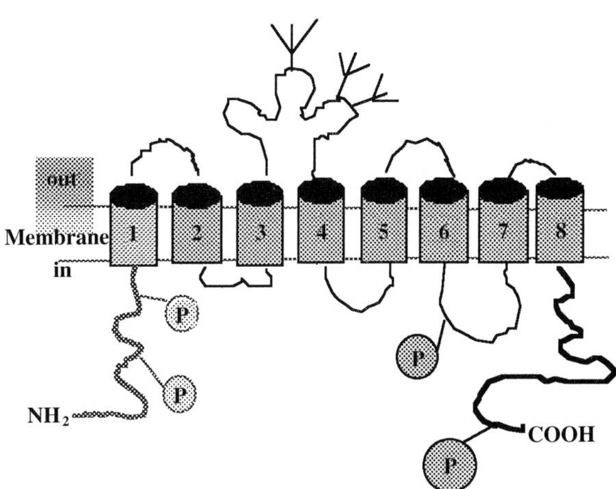

FIGURE 4 Structure of type II Na^+-dependent phosphate cotransporter (NaPi-2, -3, -4, and -6).

present in the tubular fluid must pass to be transported into the peritubular capillary network (Fig. 3). Because the electrical charge of the cell interior is negative compared to the exterior, and phosphorus concentrations are higher in the cytosol, phosphorus must move against an electrochemical gradient into the cell interior, whereas, at the antiluminal membrane, the transport of phosphorus into the peritubular capillary is favored by the high intracellular phosphorus concentration and the electronegativity of the cell interior. Studies with apical membrane vesicles have demonstrated cotransport of Na^+ with phosphate across the brush-border membrane, whereas the transport of phosphorus across the basolateral membrane is independent of that of Na^+ [26]. The apical membrane Na^+–phosphate cotransport protein (NaPi) energizes the uphill transport of phosphate across the brush-border membrane by the movement of Na^+ down its electrochemical gradient. The latter gradient is established and maintained by active extrusion of Na^+ across the basolateral cell membrane into the peritubular capillary through the action of Na^+, K^+-ATPase (Fig. 3) [27].

Two families of Na^+–phosphate cotransport proteins of the proximal tubule (Type I and Type II) have been cloned using *Xenopus* oocyte expression strategies [28–30]. The DNA clones encode 80- to 90-kDa proteins that reconstitute Na^+-dependent concentrative or "uphill" transport of phosphate on complementary RNA injection in oocytes or transfection of Sf9 insect cells [28,31]. The Type II NaPi proteins (NaPi-2, -3, -4, and -6) are similar among several species including humans (Fig. 4) [28,29]. Nephron localization of NaPi proteins have been limited to the proximal tubule of superficial and deep nephrons (greatest in the latter, concordant with physiological studies) [32]. At the time of writing, the nature of the apical phosphate cotransport protein of the distal nephron, and of the enterocyte cotransporter, remains to be described. The osteoblast phosphate cotransporter has been identified recently as a homolog of the gibbon ape leukemia virus receptor (Glvr-1) [33,34]. Immunolocalization studies in renal epithelial cells demonstrated apical membrane and subapical membrane vesicle staining [32], suggesting that a functional pool of transporters is available for insertion or retrieval from the brush-border membrane itself. This has been shown to be a major mechanism of P_i transport regulation [29,35,36] in response to acute changes in P_i and PTH.

The role of the Type I (NaPi-1) family of cotransporters in physiological regulation has not been elucidated. The NaPi-1 transporter family may be constitutive and represent a basic "housekeeping" function. The Type II family (NaPi-2, -6) is up-regulated by chronic feeding of low P_i diets [37,38]. Both the level of mRNA and the amount of transport protein in the apical membrane are increased chronically, suggesting a transcriptional event (Fig. 5). Acutely, there appears to be an early stimulation of protein insertion into the apical membrane that precedes increased transcription (Fig. 5). The NaPi-2 protein is also regulated by PTH, mainly through removal of transport protein from the apical membrane, and chronically there is also a decrease in the mRNA [36].

Studies of phosphorus exit across the basolateral membrane suggest that it is accompanied by the net transfer of a negative charge and occurs down a favorable electrochemical gradient via sodium-independent mechanisms [39].

C. Factors That Affect the Urinary Excretion of Phosphorus

Several factors are known to affect the urinary excretion of phosphorus. Of the multiplicity of factors that

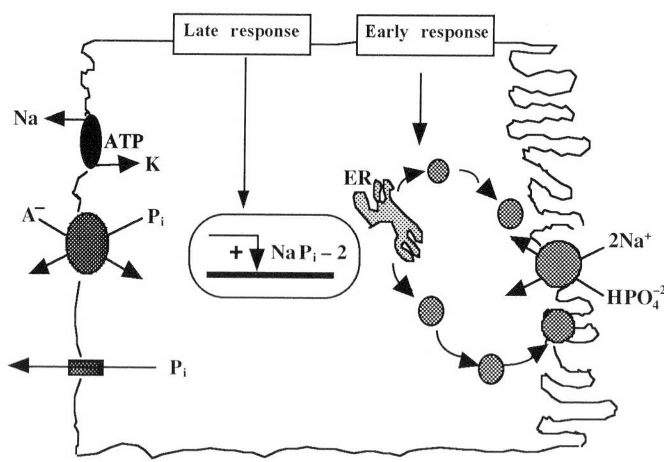

FIGURE 5 Biphasic response of proximal tubular phosphate transport to reduced delivery of phosphate.

regulate phosphate transport in the kidney (Table I), the most important are phosphate supply and PTH. The focus of this chapter, $1,25(OH)_2D_3$, affects P_i transport, but it may not be an important determinant of steady-state TmPi/GFR.

1. VITAMIN D

Controversy still surrounds the regulatory role of vitamin D in renal phosphorus handling. Several studies have demonstrated that the chronic administration of $1,25(OH)_2D_3$ to parathyroidectomized animals is phosphaturic [40–42]. Conversely, other investigators have reported that vitamin D acutely stimulates proximal tubular phosphorus transport in both parathyroidectomized and vitamin D-depleted rats [43]. A unifying interpretation of these studies was hampered by the fact that the dosages of vitamin D administered and the status of the serum calcium, phosphorus, and PTH levels varied considerably from study to study.

TABLE I Factors Regulating Renal Sodium–Phosphate Cotransport

I. Main factors influencing steady-state TmPi/GFR
   A. Parathyroid hormone
   B. P_i supply
      1. High
      2. Low
   C. IGF-I
   D. Thyroid hormone
II. Other factors that acutely affect TmPi/GFR (indirectly or pharmacologically)
   A. $1,\alpha 25(OH)_2D_3$
   B. Calcitonin
   C. Growth hormone
   D. Insulin
   E. Atrial natriuretic factor
   F. Glucocorticoids
   G. Metabolic acidosis
   H. Fasting
III. Unknown
   A. Stanniocalcin

Liang and co-workers [44] administered $1,25(OH)_2D_3$ to vitamin D-deficient chicks and subsequently examined the transport characteristics of isolated renal tubular cells. Three hours following the *in vivo* administration of $1,25(OH)_2D_3$ phosphorus uptake by the cells was significantly increased, whereas 17 hr after the administration of vitamin D metabolite phosphorus uptake was reduced. The serum phosphorus concentration, however, was significantly increased at the 17-hr time point, and administration of phosphorus to vitamin D-depleted animals so that their serum phosphorus reached a level comparable to that of the 17-hr vitamin D-replete group resulted in a similar decrease in phosphorus uptake [44]. In response to *in vitro* preincubation with as little as 0.01 p*M* of $1,25(OH)_2D_3$, renal cells isolated from vitamin D-deficient chicks demonstrated a specific increase in sodium-dependent phosphorus uptake, which was blocked by pretreatment with actinomycin D. The stimulatory effect was relatively specific for $1,25(OH)_2D_3$, and kinetic analysis indicated that the V_{max} of the phosphorus transport system was increased, whereas the affinity of the system for phosphorus was unaffected [44].

Kurnik and Hruska [45] also examined the relationship between vitamin D and renal phosphorus excretion in normocalcemic, normophosphatemic weanling rats fed a vitamin D-deficient diet. The animals were mildly vitamin D deficient [92 pg/ml of $1,25(OH)_2D$ versus

169 pg/ml in controls] but had no evidence of secondary hyperparathyroidism. Clearance studies performed in the basal, partially vitamin D-deficient state, showed an increase in both absolute and fractional phosphorus excretion compared to controls. Animals repleted with $1,25(OH)_2D_3$ and maintained on diets designed to protect against the development of hyperphosphatemia demonstrated a significant decrease in urinary phosphorus excretion. Other animals were similarly repleted with vitamin D but did not receive dietary adjustment, and in that group both the serum phosphorus and the urinary phosphorus excretion levels increased significantly. A third group was fed a normal diet and received smaller doses of $1,25(OH)_2D_3$ (15 pmol/g body weight) for shorter periods of time; although this dose had no effect on the serum phosphorus concentration, the phosphaturia was completely resolved. Studies on brush-border membrane vesicles prepared from these animals revealed that in the partially vitamin D-deficient state, sodium-dependent phosphorus uptake was significantly reduced compared with control animals. Animals repleted with vitamin D and fed a control diet had a greater sodium-dependent phosphorus uptake than both vitamin D-depleted animals and vitamin D-repleted animals not maintained on control diets.

Regulation of the Types I and II NaPi families of proteins by $1,25(OH)_2D_3$ has not been studied extensively. Preliminary studies of the 5′ promoter region of the NaPi-1 and -2 genes suggest that $1,25(OH)_2D_3$ may play an important role in transactivation of both genes. Thus, relatively decreased levels of $1,25(OH)_2D_3$ synthesis in X-linked hypophosphatemia (see Chapter 46) could contribute to the underexpression of the NaPi proteins in this disorder [46], although other studies support the concept that the loss of the capacity to adjust to a low P_i supply is the hallmark of X-linked hypophosphatemia [47–49] (see below for additional discussion of this issue). The administration of $1,25(OH)_2D_3$ in X-linked hypophosphatemia does not stimulate P_i reabsorption beyond that expected from an inhibition of PTH secretion [48].

The results of this series of studies [40–45] have left the understanding of the role of $1,25(OH)_2D_3$ in the regulation of renal P_i transport as an unresolved issue. Long-term administration of vitamin D represents a complex situation wherein phosphaturia may occur secondary to changes in the filtered load of phosphorus, to the distribution of phosphorus within the body, or to the intracellular phosphorus activity. These effects could serve to mute actions related to acute stimulation of tubular phosphate reabsorption. Furthermore, vitamin D deficiency does not prevent complete reabsorption of phosphate in the face of low P_i supply [50].

2. Alterations in Dietary Phosphorus Intake

The mechanism by which the kidney conserves phosphorus when dietary phosphorus is reduced or increased continues to be intriguing. The original discovery was that adaptation of renal tubular phosphate transport to dietary P_i supply is not mediated by factors generally thought to be important in the regulation of renal P_i handling [51–53]. Early micropuncture studies suggested that the most striking adaptive increase in phosphorus transport occurs in the proximal tubule, but later studies [24,54,55] suggested that the entire nephron participates in the reduction of phosphorus excretion during dietary phosphorus deprivation. Isolated perfused tubules obtained from rabbits fed a normal or low phosphorus diet differ in their capacity to reabsorb phosphate. When obtained from normal animals, the proximal convoluted tubule is capable of reabsorbing 7.2 ± 0.8 pmol/ml/min, whereas tubules obtained from phosphorus-deprived animals reabsorb 11.1 ± 1.3 pmol/ml/min. Conversely, animals fed a diet high in phosphorus show reduced phosphorus reabsorption when the proximal tubules are perfused *in vitro* (2.7 ± 2.6 pmol/ml/min).

Judging from renal brush-border membrane preparations, the effect of reduced dietary phosphorus to stimulate renal phosphorus transport appeared to be intrinsic to the renal tubular epithelium, occurring specifically at the brush-border membrane Na^+-phosphate cotransporter [15,56]. Considerable evidence has accrued from studies performed in cell lines isolated from mammalian kidneys, indicating that the adaptation to phosphate supply by the sodium–phosphate cotransporter is biphasic [57–59]. These studies demonstrated that incubation of cells in low phosphate medium results in a 2-fold increase in Na^+-independent phosphate cotransport. The first phase of adaptation is observed rapidly (within 10 min) and is characterized by an increase in the V_{max} of the transporter. This initial phase is independent of new protein synthesis. A slower phase resulting in a doubling of the phosphate transport rate, again through an increase in V_{max}, occurs over several hours (maximal at ~15 hr) and is inhibited by blocking new protein synthesis. These studies have been interpreted to indicate that the immediate response to reduced phosphate availability is the insertion of new transport units into the brush-border membrane from an intracellular store (see Fig. 5). Second, through gene transcription and increased NaPi protein synthesis, additional units are produced and inserted into the brush-border membrane (Fig. 5). The reverse of these processes is produced by high phosphorus availability and by PTH.

Cloning of the genes encoding the Type II family of NaPi phosphate cotransport proteins has enabled further confirmation of these results. As discussed above, chronic feeding of low phosphorus diets increases steady-state mRNA levels of NaPi-6 in rabbits and NaPi-2 in rats [37,38]. Acute P_i deprivation does not affect NaPi-2 mRNA levels, compatible with a protein synthesis-independent mechanism related to insertion of new transport proteins from subapical vesicles into the brush-border membrane [35,60].

3. Effects of PTH on Phosphorus Reabsorption by the Kidney

Parathyroidectomy decreases urinary phosphorus excretion, whereas administration of PTH rapidly increases phosphorus excretion [61,62]. Micropuncture studies indicate that PTH inhibits phosphorus transport in the proximal tubule [55,63–65] and probably in segments of the nephron located beyond the proximal tubule [18]. The TF/UF phosphorus ratio reaches a value of 0.6 by the S_2 segment of the proximal tubule, and, once achieved, this equilibrium ratio is maintained along the accessible portion of the proximal tubule. Within 6 to 24 hr of parathyroidectomy, the proximal TF/UF phosporus falls to a value of 0.2 to 0.4, indicating an increase in phosphorus reabsorption [64–67]. Tubular fluid phosphorus falls progressively with continuous fluid absorption along the length of the tubule, so that by the end of the proximal tubule the reabsorption of phosphorus is 70 to 85% of the filtered load, resulting in decreased phosphorus delivery to distal segments of the nephron. Because decreased delivery of phosphorus out of the proximal tubule complicates the evaluation of any distal effects of PTH on phosphorus excretion, maneuvers have been designed to increase phosphorus delivery to the distal nephron to study distal effects of parathyroidectomy on phosphorus reabsorption (e.g., phosphorus loading by intravenous infusion [55,68,69]). In the non-phosphorus-loaded, acutely parathyroidectomized animal, virtually all the distal load of phosphorus is reabsorbed by the distal nephron, reducing urinary phosphorus excretion to very low levels. In the phosphorus-loaded animal, the distal reabsorption of phosphorus increases until saturation is approached and urinary phosphorus excretion begins to rise [70,71]. Acute administration of PTH to phosphorus-loaded, parathyroidectomized dogs sharply lowers the distal reabsorption. These experiments indicate that PTH inhibits reabsorption of phosphorus in the distal as well as in the proximal nephron.

Administration of PTH *in vivo* results in decreased rates of Na^+-dependent phosphorus transport in brush-border membrane vesicles isolated from the kidneys of treated rats [72–74], and parathyroidectomy increases P_i transport in vesicles from these animals [56]. The uptakes of D-glucose and Na^+ were not affected by administration of PTH. Intravenous infusion of dibutryl cyclic AMP also decreased Na^+-dependent phosphorus uptake in isolated brush-border vesicles, but neither PTH nor dibutyrl cyclic AMP decreased phosphate transport when added directly to membrane vesicles [72]. These observations suggested that the effects of PTH on renal phosphate transport were mediated through altered functional characteristics (decreased V_{max}) of the renal brush-border membrane Na^+-dependent phosphate transporter [72,73]

Measurements of renal reabsorption of phosphorus, *in vivo,* and calculations of kinetic parameters of Na^+-dependent phosphorus transport in membrane vesicles isolated from the renal brush-border membranes of normal dogs and rats, thyroparathyroidectomized animals, animals fed a low phosphorus diet, and animals receiving human growth hormone (GH) have been performed (Table II) [56,73–77]. The latter three groups of dogs had greater baseline values for absolute tubular reabsorption of phosphorus compared with normal dogs. Na^+-dependent phosphate transport in brush-border membrane vesicles isolated from kidneys of these dogs were significantly increased compared with transport in brush-border vesicles from kidneys of normal dogs. Administration of PTH significantly decreased the apparent V_{max} for Na^+-dependent phosphorus transport in brush-border membrane vesicles isolated from kidneys of each of the four groups of dogs (Table II). The apparent K_m (intrinsic binding affinity) for Na^+-dependent phosphorus transport was not significantly changed by experimental maneuvers. Absolute tubular reabsorption of phosphorus measured *in vivo* was decreased by administration of PTH in each group of dogs and rats with the exception of the animals fed a low phosphorus diet [73–76]. Thus, alterations in phosphorus reabsorption measured *in vivo* were paralleled by alterations in Na^+-dependent phosphorus transport in isolated membrane vesicles, and the administration of PTH *in vivo* resulted in altered transport characteristics of the isolated brush-border membranes.

The cloning of the NaPi cotransporter genes has further elucidated the mechanisms of PTH action on phosphate transport. Because the phosphaturic effect of PTH can be reproduced by analogs of cyclic AMP, the intracellular mechanism of phosphate transport regulation is thought to involve the cyclic AMP/protein kinase A signal pathway. However, the NaPi transport proteins are not characterized by a protein kinase A-mediated phosphorylation site [78]. Phosphorylation of brush-border membrane proteins, *in vitro,* occurs in parallel with inhibition of sodium–phosphate cotransport [73], and these may be involved in endocytic traffic of NaPi-2 transporters. Parathyroidectomy of rats causes a 2- to 3-fold increase in the NaPi-2 protein content of brush-

TABLE II Effect of Parathyroid Hormone on Kinetic Parameters of Phosphate Uptake in Brush-Border Membrane Vesicles[a]

Animals	K_m (μM)	V_{max} (pmol P_i/20 sec/mg protein)
Normal	51 ± 2	1585 ± 20
Normal + PTH	45 ± 3	892 ± 68
Thyroparathyroidectomized	43 ± 4	2490 ± 139
Thyroparathyroidectomized + PTH	55 ± 8	1259 ± 68
Phosphate depleted	39 ± 6	3364 ± 75
Phosphate depleted + PTH	36 ± 2	2680 ± 69
GH treated	52 ± 1	2124 ± 99
GH treated + PTH	48 ± 4	1388 ± 104

[a] Adapted from S. Klahr, *et al.*, Renal effects of parathyroid hormone and calcitonin. In *Renal Endocrinology* (M. Dunn, ed.), p. 269. Williams & Wilkins, Baltimore, 1983.

border membrane vesicles [79]. Immunocytochemistry reveals the increase in protein exclusively in apical brush-border membranes of proximal tubules. PTH treatment of parathyroidectomized rats for 2 hr decreased protein levels in the brush-border membrane and decreased the abundance of NaPi-2 specific mRNA by 31% [79]. Parathyroidectomy did not affect NaPi-2 mRNA levels. The effects of PTH were apparent within 2 hr of administration and indicate that PTH regulation of NaPi is determined in large part by changes in the expression of NaPi-2 protein in the renal brush-border membranes by a mechanism that results in endocytic withdrawal of NaPi-2 into a cytoplasmic pool [79].

4. Effects of Changes in Acid–Base Balance on Phosphate Excretion

The effect of acid–base status on the renal excretion and transport of phosphate is complex. Acute respiratory acidosis increases and acute respiratory alkalosis decreases phosphate excretion [80]. These effects occur independent of PTH, and plasma or luminal bicarbonate levels, but they may be mediated by changes in plasma phosphate [80].

Acute metabolic acidosis has minimal effects on phosphate excretion; however, the phosphaturic effect of PTH is blunted. Acute metabolic alkalosis causes an increase in phosphate excretion independently of PTH [81–85]. This effect is due, in part, to volume expansion produced by the infusion of bicarbonate [82,83]. Chronic acidosis increases phosphate excretion, again independent of PTH or changes in ionized Ca^{2+} [85–88]. The effect appears to be directly on the sodium-dependent phosphate transport mechanism [89]. Chronic alkalosis decreases phosphate excretion, probably by the same mechanism as acidosis, operating in the opposite direction [81,90]. The effects of acid–base perturbations are complex and depend on antecedent dietary intake, the chronicity of the change, and whether the change affects luminal or intracellular pH, or both.

5. Adrenal Hormones

Administration of of pharmacological amounts of cortisol leads to phosphaturia. Acute adrenalectomy diminishes GFR and increases the reabsorption of phosphorus in the proximal tubule. Frick and Durasin [91] concluded that glucocorticoid hormones could play an important role in the regulation of fractional reabsorption of phosphorus. The current notion of an important role for cellular metabolism in regulating the proximal tubular reabsorption of phosphorus is relevant to these observations. An effect of glucocorticoid hormones in altering carbohydrate metabolism within the proximal tubular cells could underlie the effects described.

6. Growth Hormone

An increase in serum phosphorus and a rise in renal phosphorus transport are characteristics of growth hormone excess either during the period of rapid growth in the child, during acromegaly, or during exogenous growth hormone administration to experimental animals. Hammerman *et al.* examined this phenomenon in canine brush-border membrane vesicle preparations [92] and demonstrated that growth hormone treatment *in vivo* resulted in an increased sodium-dependent phosphorus transport.

7. Insulin-Like Growth Factor

The effect of growth hormone on renal tubular P_i transport is relatively slow to develop, raising the possibility that its actions are mediated by induction of a secondary factor. Growth hormone increases production of the insulin-like growth factors (IGFs). IGF-I directly stimulates proximal tubular P_i transport and $1,25(OH)_2D_3$ production [77,93,94]. These studies have suggested that some of the actions of growth hormone on renal tubular P_i transport are mediated by systemic and locally produced IGF-I. IGF may be a critical modulator of P_i transport and $1,25(OH)_2D_3$ production during growth [95].

8. Stanniocalcin

Stanniocalcin is a newly recognized calcium-regulating hormone found in serum and the kidney [96]. Its name derives from its synthesis by the corpuscles of Stannius, endocrine glands found in the kidneys of bony fishes. In mammals, stanniocalcin lowers calcium transport and increases phosphate reabsorption. The bony fishes use this action to increase phosphate deposition into bone and scales. The role of stanniocalcin in human physiology is unstudied and unknown, as is regulation of its secretion.

9. Phosphatonin

Phosphatonin is a substance that was given a name before having been identified [97] (see Chapter 46). This substance circulates in the plasma of patients with oncogenic osteomalacia (tumor-induced osteomalacia) [98–102], and it causes hypophosphatemia, phosphate wasting, and osteomalacia. Oncogenic osteomalacia is rare, and it is found in patients with mesenchymal tumors [98,99,102]. The role of phosphatonin in physiology is unknown, but the discovery of the molecular basis of X-linked hypophosphatemia (XLH) as an extracellular neutral endopeptidase (PEX) has given rise to the speculation that the role of PEX is to neutralize phosphatonin (Fig. 6). Thus, according to this formulation of the etiology of XLH, the defective PEX gene results in high circulating levels of phosphatonin and inhibition of NaPi-3 and -7 levels. Phosphatonin must be specific for the renal transporter, NaPi-3 or -7, because intestinal and osteoblast transporters function normally in Hyp mice. Much of this scenario is still speculative, but, if confirmed, it will establish the discovery of a new major regulator of phosphate homeostasis and the probable humoral basis of XLH. There is no evidence to date that the humoral factor of XLH is in fact phosphatonin (see Chapter 46 for further discussion).

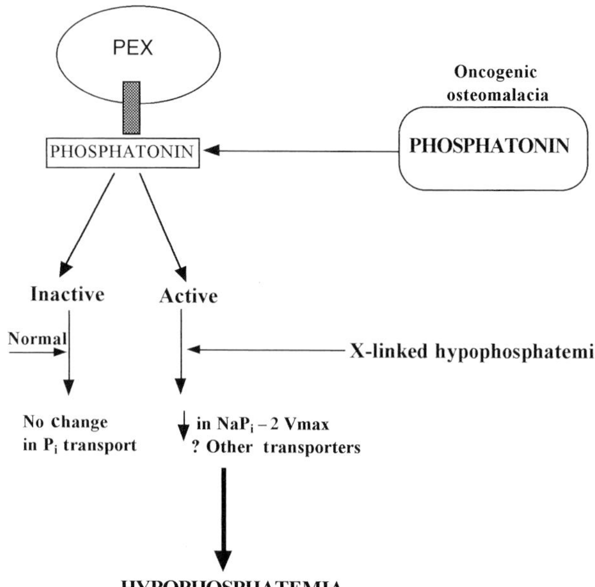

FIGURE 6 Conceptual diagram of a new phosphate regulatory system. The discovery that the genetic defect in X-linked hypophosphatemia is in a gene (PEX) encoding an endopeptidase gives even greater interest to findings in oncogenic osteomalacia, wherein a circulating substance that inhibits phosphate transport has been described. The concept is that PEX normally inactivates phosphatonin unless it is overproduced, as in oncogenic osteomalacia, or unless PEX is defective.

IV. BONE REMODELING AND PHOSPHATE TRANSPORT

The main physiological function of bone is to act as a mechanical support. This function is achieved by mineralization of an extracellular matrix formed by specialized cells. Inorganic phosphate is an essential element for the physiological functioning of bone cells, not only because it is an integral component of apatite crystals, but also because it can affect the production rate of bone matrix and regulate bone resorption [103,104]. Until relatively recently, little was known about the transport of phosphorus in bone cells. Recent advances have led to the description of phosphate transport as an exciting component of bone cell physiology that offers opportunity for regulation, leading to modulation of bone remodeling [103,104].

A. Osteoclast Function and Phosphate Transport

Osteoclasts are polarized cells involved in bone resorption. They are exposed to high ambient concentrations of inorganic phosphate during the active process of bone resorption. Osteoclasts possess specific phosphate transport systems including a sodium-dependent P_i cotransporter whose activity is dependent on ATP production and Na^+,K^+-ATPase activity to maintain the inwardly directed Na^+ gradient driving force for the concentration P_i cotransport [104]. When osteoclasts are isolated *in vitro* and allowed to attach to bone particles, there is stimulation of P_i transport, without an increment of transport protein synthesis [104]. The stimulatory effect of bone particles on P_i transport was inhibitable by peptides containing the Arg-Gly-Asp cell adhesion motif, an effect that implicates integrins and cell matrix interaction in the regulation of P_i transport. Western analysis of osteoclast lysates utilizing antibodies to the NaPi-2 family of proteins detects two proteins of 100 and 105 kDa. Immunofluorescence studies of osteoclasts cultured on glass coverslips reveals the proteins to be present in discrete cytosolic vesicles. Following attachment to bone, these vesicles fuse with the basolateral membrane of osteoclasts coordinate with the stimulation of phosphate transport. Immunofluorescence-based confocal microscopy reveals colocalization of the osteoclast sodium–phosphate cotransporter with the Na^+,H^+ antiporter on the basolateral membrane. The osteoclast isoform of the NaPi-2 protein family exhibits an apparent molecular mass greater than the 80–90 kDa of the analog in the renal tubule. Analogous to the renal tubule, vesicular traffic regulates membrane transporter protein levels in acute regulatory responses of phosphate transport [104]. In addition, agents that specifically inhibit the renal tubule phosphate transporter also inhibit osteoclast phosphate transport and bone resorption [105]. Phosphate transport is essential to osteoclast function as it provides the substrate for ATP synthesis, which is required to support the high rates of proton transport involved in bone resorption. At the time of this writing, whether $1,25(OH)_2D_3$ affects Na^+-dependent P_i transport in osteoclast precursors remains to be determined.

B. Phosphate Transport and Osteogenic Cell Function

The osteogenic cells osteoblasts and chondrocytes do not exhibit a phosphate transporter with sequence similarity to the NaPi-1 or NaPi-2 protein family [103]. A new sodium–phosphate cotransport system, which is expressed in several tissues including bone marrow, has been identified [33]. This transport system also serves as a membrane receptor for some retroviruses. It is a candidate protein for the Na^+-dependent P_i contransporter of osteogenic cells [34].

Analysis of Na^+-dependent P_i transport in osteoblasts reveals that phosphate transport in these cells differs significantly from that of renal epithelial cells. Affinity constants for P_i are significantly higher (300–500 μM compared to 50–100 μM for renal epithelial cells). The affinity of the transporter for Na^+ is similar to that of epithelial cells (40 mM). Phosphate transport activity of osteoblasts is stimulated by an acidic extracellular pH in contradistinction to the renal epithelial transporter. Regulation of osteoblast phosphate transport also differs from that of renal epithelial cells. PTH stimulates phosphate transport in osteoclast-like cells through the cAMP signal transduction pathway [106] (Fig. 7). This is directionally opposite to the effect of PTH in the renal epithelium. IGF-1 is also a selective stimulator of P_i transport in osteoblasts and chondrocytes [107] (Fig. 7). Fluoride, another agent that, at low doses, stimulates bone formation, also increases P_i transport in osteoblasts [108]. The effect of fluoride may be mediated by the formation of fluoroaluminate [103].

1. Matrix Vesicles

Biological mineralization is a complex process involving different types of tissues. Primary mineralization is regulated by chondrocytes, osteoblasts, and odontoblasts and is observed in epiphyseal cartilage, embryonic

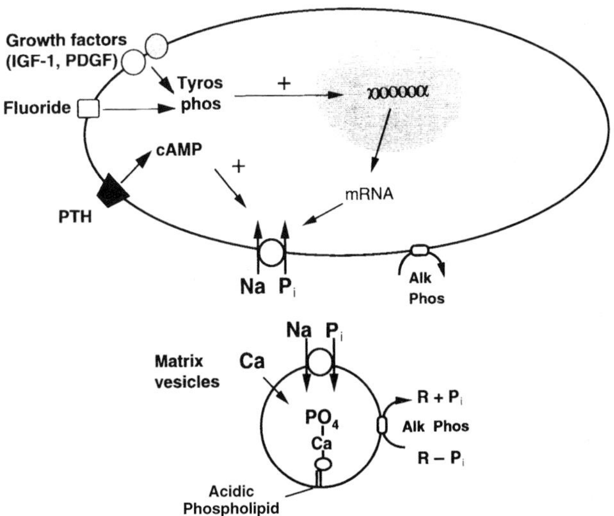

FIGURE 7 Schematic representation of regulatory factors and their signal transduction mechanisms known to stimulate Na^+-coupled P_i transport in osteoblast-like cells. As mentioned in the text, experimental evidence has been obtained suggesting parallel regulation of P_i transport activity in the plasma membrane of bone-forming cells and in their derived matrix vesicles. With the notion that the P_i transport activity in matrix vesicles is important for accumulation of mineral ions, it is hypothesized that the regulation of P_i transport activity in matrix vesicles by osteogenic cells might play a significant role in the control of processes involved in the calcification of the bone matrix. Tyros phos = Tyrosine phosphorylation; Alk Phos = Alkaline phosphatase; R − Pi = Phosphorylated regulatory protein.

bone, postnatal ossification, and the development of predentin. The mineralization process mediated by osteoblasts is not fully understood. Two general but not mutually exclusive theories exist on how calcification is initiated in skeletal tissues: (1) collagen nucleation and (2) matrix vesicles [109]. The first theory holds that the collagen fibril is the major site of crystal nucleation such that calcium phosphate crystals are deposited in the whole region of these fibrils. The alternative theory involves matrix vesicle production by osteogenic cells. When matrix vesicles are calcified in the extracellular matrix, they serve as nucleation sites for mineralization.

Matrix vesicles are small organelles with a diameter of 100–200 nm that have been observed in mineralizing tissues in cultures of osteogenic cells often in contact with initial mineral crystals. Matrix vesicles arise from bone-forming cells by a process of budding from elongated tubular extensions that project from the plasma membrane of these cells [110]. Matrix vesicles may participate in the uptake of mineral ions dependent on the presence of labile forms of mineral within the vesicles that serve as a trap for newly entering mineral ions [111]. Calcium enters matrix vesicles by a protease-sensitive carrier that may be related to the annexins [112]. Matrix vesicles have a similar Na^+-dependent P_i transport system to osteoblasts and chondrocytes (Fig. 7) [113]. Formation of phosphate–calcium–phospholipid complexes in the vesicular space may maintain a favorable P_i gradient, allowing the continuation of phosphate transfer into matrix vesicles. In this system, sodium only facilitates translocation of P_i across the membrane of matrix vesicles, because a mechanism of maintaining the sodium gradient is not present as in the plasma membrane of cells.

Studies [114] have shown that P_i transport activity is low in matrix vesicles released during the proliferative phase of osteoblast differentiation but is significantly increased during osteoblast differentiation, peaking at the time of bone matrix formation. This observation suggests the existence of specific mechanisms for enhanced phosphate transport activity in matrix vesicles released during the formation of the collagenous matrix. It may be explained by enrichment of matrix vesicles with phosphate carriers during the differentiation process. Because the activity of P_i transport in osteogenic cells is regulated by calciotropic hormones and growth factors, regulation of transport may represent a mechanism by which osteogenic cells regulate the calcification of their extracellular matrix (Fig. 7).

2. Regulation by $1,25(OH)_2D_3$

At the time of this writing, regulation of Na^+-dependent P_i transport in osteogenic cells by $1,25(OH)_2D_3$ has been reported only in UMR-106 rat osteogenic sarcoma cells. Surprisingly, $1,25(OH)_2D_3$ decreased Na^+-dependent P_i uptake after 8 hr of incubation in UMR-106 cells, and decreased intracellular P_i levels [115]. Whether these results would also be observed in normal osteoblasts remains to be shown. The findings are surprising in view of the effects of $1,25(OH)_2D_3$ to correct vitamin D-deficient osteomalacia. Because osteomalacia is a defect of mineralization, one would naturally predict that $1,25(OH)_2D_3$ deficiency would increase P_i transport, as the latter is potentially involved in mineralization, and that $1,25(OH)_2D_3$ administration would stimulate P_i transport. Much still needs to be done in this area.

V. CLINICAL DISEASE STATES ASSOCIATED WITH ABNORMAL PHOSPHATE TRANSPORT AND VITAMIN D

A. Hypophosphatemia

Hypophosphatemia refers to serum phosphorus concentrations of less than 2.5 mg/dl. The main determinant of plasma P_i concentration in human subjects and experimental animals at steady state is the renal tubular reabsorptive capacity as assessed by the TmPi/GFR [116]. Tubular reabsorptive capacity for phosphate varies physiologically, and as a result plasma phosphate concentrations vary. The best example of this physiological variation is the high TmPi/GFR seen in growing children and animals associated with higher plasma P_i levels than in older individuals. The urinary excretion of P_i corresponds, at steady state, to the net entry of P_i into the extracellular compartment. In other words, the sum of differences between phosphate influx and efflux at the intestinal, skeletal, and soft tissues will determine urinary excretion at a set TmPi/GFR. When the differences are small and TmPi/GFR is normal, hypophosphatemia will result. This may be physiological depending on the cause. The major causes of hypophosphatemia are listed in Table III. For the purposes of this chapter, discussion is limited to disorders that affect both P_i transport and as a consequence $1,25(OH)_2D_3$ synthesis (see also Chapter 46).

1. X-Linked Hypophosphatemia

The disease of X-linked hypophosphatemia (XLH) is discussed extensively in chapter 46 of this book, but because of the insights it provides into current concepts of P_i transport and $1,25(OH)_2D_3$ action it is necessary to discuss it here briefly. This is an X-linked dominant disorder characterized by hypophosphatemia, decreased reabsorption of phosphorus by the renal tubule,

TABLE III Causes of Hypophosphatemia

I. Decreased TmPi/GFR
- A. Primary hyperparathyroidism
- B. Secondary hyperparathyroidism
- C. Renal tubular defects
- D. Diuretic phase of acute tubular necrosis
- E. Postobstructive diuresis
- F. After renal transplantation
- G. ECF volume expansion
- H. Familial
  - 1. X-linked hypophosphatemia
  - 2. McCune-Albright Syndrome

II. Decrease in gastrointestinal absorption of phosphorus
- A. Malabsorption
- B. Malnutrition–starvation
- C. Administration of phosphate binders
- D. Abnormalities of vitamin D metabolism
  - 1. Vitamin D deficiency rickets
  - 2. Familial
    - a. Vitamin D-resistant rickets
    - b. X-linked hypophosphatemia

III. Miscellaneous causes/translocation of phosphorus
- A. Diabetes mellitus: during treatment for ketoacidosis
- B. Severe respiratory alkalosis
- C. Recovery phase of malnutrition
- D. Alcohol withdrawal
- E. Toxic shock syndrome
- F. Leukemia, lymphoma
- G. Severe burns

decreased absorption of calcium and phosphorus from the gastrointestinal tract, and varying degrees of rickets and/or osteomalacia. The roles of alterations in vitamin D sensitivity and/or synthesis in the defective transport of phosphorus observed in both the gut and the kidney have been investigated extensively in the murine homolog of XLH, the Hyp mouse [117]. Tenehouse and Scriver found that administration of small doses of $1,25(OH)_2D_3$ to normal animals significantly increased plasma calcium, plasma phosphorus, and fractional calcium excretion without a change in fractional phosphorus excretion or phosphorus transport as measured in proximal tubular brush-border membrane vesicles [118]. However, there was no response to a similar dose of $1,25(OH)_2D_3$ in the Hyp mouse. At a 5-fold higher dose of $1,25(OH)_2D_3$ the Hyp mice did show increased plasma calcium and fractional calcium excretion as well as plasma phosphorus levels, but again there was no change in fractional phosphorus excretion or in sodium-dependent phosphorus transport in brush-border membrane vesicles. Although $1,25(OH)_2D_3$ increased phosphorus transport in the intestine, there was no defect of phosphorus transport in the intestine of untreated Hyp mice. The authors concluded that $1,25(OH)_2D_3$ treatment influenced phosphorus homeostasis in Hyp mice by pharmacological stimulation of phosphorus absorption from the gastrointestinal tract. The defect in renal phosphorus reabsorption remained unchanged despite high levels of vitamin D, as did adaptation to low P_i supply [47–49]. These observations are consistent with studies performed in humans with XLH and in whom the renal reabsorptive defect persists despite correction of growth by administration of vitamin D and oral phosphate supplements [119,120].

There is, however, evidence of defective or altered metabolism of vitamin D_3 in Hyp mice. Meyer *et al.* found that on a normal diet plasma $1,25(OH)_2D_3$ levels were the same in Hyp and normal mice [121], although $25OHD_3$ levels were reduced in Hyp mice. Because hypophosphatemia increases plasma $1,25(OH)_2D_3$ levels, hypophosphatemic mice were resistant to the stimulatory effect of hypophosphatemia. To test this possibility, these authors fed the animals a low phosphorus diet and found a paradoxical reduction in $1,25(OH)_2D_3$ levels in hypophosphatemic mice. They concluded that Hyp mice have a defective control system for plasma $1,25(OH)_2D_3$ that is unresponsive to a low phosphate diet stimulus. The issue is further complicated by the studies of Beamer *et al.* [122], who examined the effect of various preparations of vitamin D_3 on intestinal transport of phosphorus. The Hyp mice responded to 1α-dihydroxyvitamin D_3 but not to $1,25(OH)_2D_3$, suggesting that intestinal phosphorus transport is not genetically absent in this model but rather does not respond to normal endogenous levels of $1,25(OH)_2D_3$. Thus, mice with familial hypophosphatemia appear to have impaired metabolism of vitamin D, but this impairment does not cause the renal phosphate transport defect. There is also evidence suggesting that the decreased tubular reabsorption of phosphorus is not due to increased PTH levels [122]. It seems that a component of renal phosphorus transport, which may be PTH independent, may be abnormal and responsible for the increased phosphaturia. Tenenhouse *et al.* [123] have shown that tissue phosphate levels were normal in familial Hyp mice, whereas these levels tended to be low in animals that have hypophosphatemia secondary to reduction in dietary phosphorus. It would seem, therefore, that XLH is a selective disorder of the transepithelial transport of phosphate in the kidney. However, there is no correlation between the degree of hypophosphatemia and the severity of bone disease. Moreover, alterations in vitamin D metabolism cannot be explained solely by the presence of hypophosphatemia. Therefore, some undefined pathological mechanism may involved both abnormal phosphate transport and renal 1-hydroxylase function.

Tenenhouse *et al.* [124] have extended their studies in Hyp mice to demonstrate that the mRNA levels for the NaPi-2 cotransport protein are reduced by 50% in the renal cortex, similar to the reduction in apical membrane vesicle NaPi-2 protein levels. This demonstrates a significant underexpression of the proximal tubule Na^+-dependent phosphate transport protein in Hyp mice. However, the mapping of NaPi-2 and NaPi-1 to chromosomes 5 and 6, respectively [125,126], indicates that these genes are not candidates for the genetic defects leading to XLH or Hyp. One possibility is that XLH and Hyp are produced by defective transcription of the NaPi-2 and NaPi-3 genes, respectively [127]. Another possibility is that the pathogenesis of XLH and the defect in Hyp mice is linked to abnormal levels of a circulating hormonal factor. Results of the cross-perfusion studies of Meyer *et al.* [128] and the renal cross-transplantation studies of Nesbitt *et al.* [129] have been interpreted to indicate the presence of a circulating factor. Furthermore, in the latter study [129] the authors were unable to demonstrate transmission of the Hyp renal defect by transplanting a Hyp kidney into a normal recipient. Likewise, transplantation of a normal kidney into a Hyp recipient factor to correct the Hyp phenotypic abnormalities. These studies, in addition to those by Cai *et al.* [98] in oncogenic osteomalacia, suggest that the presence of a circulating humoral factor could produce the Hyp and XLH phenotypes.

The circulating humoral factor hypothesis for the etiology of XLH (see Fig. 6) has gained considerable support from the discovery of the defective gene, PEX, through positional cloning [130]. The PEX gene product encodes a neutral endopeptidase of the same family as the endothelin converting enzyme and is characterized by zinc regulation and a single transmembrane-spanning domain. The model proposed by this result suggests that a circulating phosphaturic factor is normally catabolized by the PEX gene product. When PEX is mutated, the phosphaturic factor levels are increased and hypophosphatemia results.

An interesting aspect of the Hyp phenotype is the presence of a normal Na^+-dependent phosphate cotransport in osteoblasts [116,131]. However, the osteoblasts of Hyp mice have been shown to be defective in models of endochondral bone formation [132,133]. Defective mineralization is an important aspect of osteomalacia, and mineralization is controlled by calcium, phosphate, and bone matrix proteins. We demonstrated defective phosphorylation of a key bone matrix protein, osteopontin, in Hyp mice and abnormal low activity of a protein kinase (casein kinase-II like activity) responsible for osteopontin phosphorylation [127,134]. Thus, multiple proteins appear to be defective in the Hyp phenotype. This is compatible with multiple targets of the PEX gene such as phosphatonin and an inhibitor of casein kinase-II, which achieve greater activity when PEX is defective [98,127–129]. The mouse PEX gene has now been cloned. Interestingly, its expression was only detected in bone and cultured osteoblasts and was greatly decreased in Hyp mice [134a].

From a therapeutic point of view, the combination of neutral phosphate and $1,25(OH)_2D_3$ has led to an improvement in the bone disease of some patients with XLH as well as in the Hyp mice [135,136]. The administration of phosphorus in XLH is usually divided into four doses, with the total amount ranging between 1 and 4 g per day. Pharmacological doses of $1,25(OH)_2D_3$ of the order of 1 to 3 μg per day may be necessary to correct the skeletal alterations. $1,25(OH)_2D_3$ does not correct the increased fractional excretion of phosphate. The enthusiasm for this regimen is tempered by a high incidence of nephrocalcinosis and occasional renal failure [135–137].

2. Hereditary Hypocalcemic Rickets

The condition of hereditary hypocalcemia rickets is characterized by hypocalcemia, hypophosphatemia, elevated levels of serum alkaline phosphatase, and, sometimes, generalized aminoaciduria and severe bone lesions. The hypophosphatemia is primarily due to secondary hyperparathyroidism induced by chronic hypocalcemia. Currently, two main forms of vitamin D-resistant rickets have been characterized, and the serum concentration of $1,25(OH)_2D$ serves to differentiate the two types. The first one (type I) is an inborn error in conversion of 25OHD to $1,25(OH)_2D_3$ due to deficiency of the renal 1-hydroxylase enzyme [138]. The condition is also known as pseudo-vitamin D deficiency rickets (PDDR) and is discussed in detail in Chapter 47. PDDR was formerly treated with very large doses of vitamins D_2 and D_3 (100 to 300 times the normal requirement of physiological doses); however, it responds well to daily administration 0.5 to 1.0 μg per day of $1,25(OH)_2D_3$.

The second form of heritable hypocalcemic rickets, hereditary vitamin D-resistant rickets (HVDRR), also known as vitamin D-resistant rickets type II, is characterized by end-organ resistance to $1,25(OH)_2D$. This condition is discussed in detail in Chapter 48. Plasma levels of $1,25(OH)_2D$ are elevated, as expected if there is resistance of the target tissue to $1,25(OH)_2D$. Cellular defects found in patients with HVDRR are heterogenous, providing, in part, an explanation for the different clinical manifestations of this disorder. Numerous studies [139–147] have demonstrated that HVDRR is a genetic disease affecting the vitamin D receptor (VDR). Defects in the hormone binding domain [140,141] and the DNA binding domain [142,143] have been defined. In addition, rare cases of HVDRR have been studied where no abnormality in the coding region of the VDR has been found [144], suggesting a defect elsewhere in the hormone action pathway. The pathogenic basis of

HVDRR as a genetic defect in the VDR has similarities to genetic diseases involving receptors of other members of the steroid–thyroid receptor superfamily such as resistance to thyroid hormone, androgens, and estrogens [145–147]. However, much remains to be learned from the genetic analysis of this disease. The treatment of HVDRR may require large pharmacological doses of calcium that bypass the defects in the VDR and normalize bone [143].

3. ONCOGENIC OSTEOMALACIA

The rare syndrome of oncogenic or tumor-induced osteomalacia is characterized by hypophosphatemia associated with tumors. It was described initially in association with mesenchymal tumors; however, more recent reports have emphasized the association of this syndrome with malignant tumors as well [99,100]. Other characteristics of this syndrome are increased phosphate excretion, inappropriately low plasma $1,25(OH)_2D$ concentrations, and osteomalacia. All of the biochemical and pathological abnormalities disappear when the tumor is resected. The tumors associated with this syndrome are thought to secrete a substance that inhibits the renal tubular reabsorption of phosphate and suppresses 25-hydroxycholecalciferol 1-hydroxylase activity. Whether this factor interacts directly with renal tubular cells is unknown.

Studies by Cai *et al.* [98] have investigated the ability of conditioned media from cultures of sclerosing hemangioma cells to inhibit sodium-dependent phosphate transport. The cell culture was established from a tumor removed from a patient with osteogenic osteomalacia. They found that the medium inhibited sodium-dependent phosphate transport, without increasing cellular concentrations of cyclic AMP. The medium had PTH-like immunoreactivity but not PTH-related protein (PTHrP) immunoreactivity, but the action of the tumor medium was not blocked by a PTH antagonist. The plasma $1,25(OH)_2D$ concentrations are inappropriately low in patients with oncogenic osteomalacia despite the presence of hypophosphatemia [148–150], which usually increases plasma $1,25(OH)_2D$ concentrations by stimulating the renal 1-hydroxylase in a PTH-independent manner [151] (see also Chapter 46). Besides the hypophosphatemia, deficient production of $1,25(OH)_2D$ is a factor contributing to the pathogenesis of the osteomalacia in these patients. Miyauchi *et al.* [101] have reported that 1-hydroxylase activity of cultured renal tubular cells was decreased by incubating the cells with tumor extracts. This supports the concept that the tumor extracts contain a substance that inhibits the formation of $1,25(OH)_2D$ in the proximal tubule.

The varied studies of oncogenic osteomalacia and some studies associated with XLH rickets support the possibility that a hormone primarily responsible for the regulation of renal phosphate reabsorption is abnormally produced in these conditions. Econs and Drezner [97] have termed this putative factor "phosphatonin." The similarity between oncogenic osteomalacia and XLH raises the possibility that phosphatonin may be the factor normally degraded by the PEX gene product that causes abnormal phosphate transport in XLH and Hyp mice.

B. Hyperphosphatemia

Hyperphosphatemia is said to occur when the plasma phosphorus concentration exceeds 5 mg/dl in adults. Hyperphosphatemia may be physiological (i.e., growth related increase in TmPi/GFR) and in growing children and adolescents serum levels of phosphorus of up to 6 mg/dl may be physiological. The most frequent cause of hyperphosphatemia is decreased excretion of phosphorus in the urine as a consequence of a fall in GFR. However, increases in plasma phosphorus can also occur as a result of increased entry into the extracellular fluid due to excessive intake of phosphorus or increased release of phosphorus from tissue breakdown. The major causes of hyperphosphatemia are listed in Table IV.

1. DECREASED EXCRETION OF PHOSPHORUS IN URINE WITH DECREASED RENAL FUNCTION

In progressive renal failure, phosphorus homeostasis is maintained by a progressive increase in phosphorus excretion per nephron [152,153] (also see Chapter 52). As a consequence of this increased phosphorus excretion, it is unusual to see marked hyperphosphatemia until GFR values fall below 25 ml/min [154]. Under physiological conditions with a GFR of 120 ml/min, a fractional excretion of 5 to 15% of the filtered load of phosphorus is adequate to maintain phosphorus homeostasis. However, as renal insufficiency progresses and the number of nephrons decreases, fractional excretion of phosphorus may increase to as high as 60 to 80% of the filtered load. This progressive phosphaturia per nephron, as renal disease progresses, serves to maintain the concentration of phosphorus within normal limits in plasma (see Fig. 3). However, when the number of nephrons is greatly diminished, if the dietary intake of phosphorus remains constant, phosphorus homeostasis can no longer be maintained and hyperphosphatemia develops. This usually occurs when the GFR falls below 25 ml/min. As hyperphosphatemia develops, the filtered load of phosphorus per nephron increases, phosphorus excretion rises, and phosphorus balance is reestablished but at higher concentrations of serum phosphorus. Hyperphosphatemia is a usual finding in patients with far-

TABLE IV Causes of Hyperphosphatemia[a]

I. Increased renal tubular TmPi/GFR or absence of glomerular filtration
- A. Renal insufficiency (decreased GFR)
  - 1. Chronic
  - 2. Acute
- B. Hypoparathyroidism
- C. Pseudohypoparathyroidism
  - 1. Type I
  - 2. Type II
- D. Abnormal circulating PTH
- E. Acromegaly
- F. Tumoral calcinosis
- G. Administration of bisphosphonates

II. Increased entrance of phosphorus
- A. Neoplastic diseases
  - 1. Leukemia
  - 2. Lymphoma
- B. Increased catabolism
- C. Respiratory acidosis

III. Administration of phosphate salts or vitamin D
- A. Pharmacological administration of vitamin D metabolites
- B. Ingestion and/or administration of phosphate salts

IV. Miscellaneous
- A. Cortical hyperostosis
- B. Intermittent hyperphosphatemia
- C. Artifacts

[a] Adapted from E. Slatopolsky, Pathophysiology of calcium, magnesium, and phosphorus. In *The Kidney and Body Fluids and Disease* (S. Klahr, ed.), p. 269. Plenum, New York, 1983.

advanced renal insufficiency unless phosphorus intake in the diet has been decreased through dietary manipulations or unless the patient is receiving phosphate binders such as calcium carbonate or aluminum-containing salts that will decrease the absorption of phosphate from the gastrointestinal tract [155]. A deficiency of $1,25(OH)_2D_3$ develops early in the course of chronic renal failure, and the role this deficiency plays in permitting some of the adaptive changes in P_i transport that occur is unclear.

In patients with acute renal failure, hyperphosphatemia is a usual finding [156]. The degree of hyperphosphatemia in patients with acute renal failure varies considerably. It is quite marked in patients with renal insufficiency secondary to severe trauma or known traumatic rhabdomyolysis, as frequently occurs in patients ingesting large amounts of alcohol or in heroin addicts [157]. The degree of hyperphosphatemia depends on the amount of phosphorus released from damaged tissue, as phosphorus intake and decreased GFR, to less than 2 ml/min, is constant across different forms of oliguric acute renal failure. In most patients with acute renal failure, hyperphosphatemia is transitory, and serum phosphorus values return toward normal as renal function improves. However, in some patients infection or tissue destruction due to many causes may maintain relatively high serum phosphorus values even during the recovery phase of renal function.

2. TUMORAL CALCINOSIS

Although the etiology of tumoral calcinosis is not completely characterized, its pathogenesis is probably related to a primary increase in phosphorus reabsorption by the kidney [158] (also see Chapter 46). This condition, which is seen more frequently in young blacks, is characterized by hyperphosphatemia, ectopic calcification around large joints, normal levels of circulating immunoreactive PTH, and a normal response to administration of exogenous PTH [159,160]. The extensive calcification of soft tissues observed in patients with this condition is most likely due to an elevated phosphorus–calcium product in blood. Despite the development of hyperphosphatemia, patients with tumoral calcinosis do not develop secondary hyperparathyroidism. This may be due to the fact that circulating levels of $1,25(OH)_2D_3$ remain normal in these patients despite hyperphosphatemia. These normal levels of $1,25(OH)_2D_3$ maintain normal gastrointestinal absorption of calcium. This, combined with the decreased urinary calcium observed in these patients, may serve to maintain normal serum calcium values and prevent the development of secondary hyperparathyroidism.

3. ADMINISTRATION OF VITAMIN D OR ITS METABOLITES

Administration of vitamin D_3 or its metabolites, particularly $1,25(OH)_2D_3$, may result in increases in serum phosphorus, particularly in uremic patients. These compounds very likely may result in hyperphosphatemia in uremic individuals by increasing phosphorus absorption from the gut and perhaps by potentiating the effect of PTH on the skeleton with increased release of phosphorus from bone. Decreased renal function limits the compensatory mechanism of the kidney to excrete the increased load of phosphate entering the extracellular space. In addition to elevating serum phosphorus, vitamin D metabolites may result in hypercalcemia. An increase in the phosphorus–calcium product may result in tissue deposition of calcium, particularly in the kidney, leading to further renal functional deterioration.

References

1. Levine BS, Kleeman CR 1994 Hypophosphatemia and hyperphosphatemia: Clinical and pathophysiologic aspects. In: Maxwell MH, Kleeman CR (eds) Clinical Disorders of Fluid and Electrolyte Metabolism. McGraw–Hill, New York, p. 1045.

2. Hopkins T, Howard JE, Eisenberg H 1952 Ultrafiltration studies on calcium and phosphorus in human serum. Bull Johns Hopkins Hospital **91**:1–21.
3. Walser M 1961 Ion association. VI. Interactions between calcium, magnesium, inorganic phosphate, citrate and protein in normal human plasma. J Clin Invest **40**:723–730.
4. Fuchs R, Peterlik M 1980 Intestinal phosphate transport. In: Massry SG, Ritz E, Jahn H (eds) Phosphate and Minerals in Health and Disease. Plenum, New York, p. 381.
5. Berner W, Kinne R, Murer H 1976 Phosphate transport into brush border membrane vesicles isolated from rat small intestine. Biochem J **160**:467–474.
6. Matsumoto T, Fontaine O, Rasmussen H 1980 Effect of $1{,}25(OH)_2$ vitamin D_3 on phosphate uptake into chick intestinal brush border membrane vesicles. Biochim Biophys Acta **599**:13–23.
7. Murer H, Hildmann B 1981 Transcellular transport of calcium and inorganic phosphate in the small intestinal epithelium. Am J Physiol **240**:G409–G416.
8. Norman AW 1978 Calcium and phosphorus absorption. In: Lawson DEM (ed) Vitamin D. Academic Press, New York, p. 93.
9. Peterlik M, Wasserman RH 1978 Effect of vitamin D on transepithelial phosphate transport in chick intestine. Am J Physiol **234**:E379–E388.
10. Ghishan FK 1992 Phosphate transport by plasma membranes of enterocytes during development: Role of 1,25-dihydroxycholecalciferol. Am J Clin Nutr **55**:873–877.
11. Rizzoli R, Fleisch H, Bonjour J-P 1977 Role of 1,25-dihydroxyvitamin D_3 on intestinal phosphate absorption in rats with a normal vitamin D supply. J Clin Invest **60**:639–647.
12. Nakagawa N, Ghishan FK 1993 Sodium-phosphate transport in the kidney and intestine of the hypophosphatemic mouse. Pediatr Nephrol **7**:815–818.
13. Kikuchi K, Ghishan FK 1987 Phosphate transport by basolateral plasma membranes of human small intestine. Gastroenterology **93**:106–113.
14. Nemere I 1996 Apparent nonnuclear regulation of intestinal phosphate transport: Effects of 1,25-dihydroxyvitamin D_3, 24,25-dihydroxyvitamin D_3, and 25-hydroxyvitamin D_3. Endocrinology **137**:2254–2261.
15. Caverzasio J, Danisi G, Straub RW, Murer H, Bonjour J-P 1987 Adaptation of phosphate transport to low phosphate diet in renal and intestinal brush border membrane vesicles: Influence of sodium and pH. Pfluegers Arch **409**:333–336.
16. Lee DBN, Walling MW, Brautbar N 1986 Intestinal phosphate absorption: Influence of vitamin D and non-vitamin D factors. Am J Physiol **250**:G369–G373.
17. Knox FG, Osswald H, Marchand GR, Spielman WS, Haas JA, Berndt T, Youngberg SP 1977 Phosphate transport along the nephron. Am J Physiol **233**:F261–F268.
18. Pastoriza-Munoz E, Colindres RE, Lassiter WE, Lechene C 1978 Effect of parathyroid hormone on phosphate reabsorption in rat distal convolution. Am J Physiol **235**:F321–F330.
19. Amiel C, Kuntziger H, Richet G 1970 Micropuncture study of handling of phosphate by proximal and distal nephron in normal and parathyroidectomized rat. Evidence for distal reabsorption. Pfluegers Arch **317**:93–109.
20. Jamison RL, Arrascue JF 1980 Calcium and phosphate reabsorption by the loop of Henle. Miner Electrolyte Metab **4**:97.
21. Peraino RA, Suki WN 1980 Phosphate transport by isolated rabbit cortical collecting tubule. Am J Physiol **238**:F358–F362.
22. Shareghi GR, Agus ZS 1982 Phosphate transport in the light segment of the rabbit cortical collecting tubule. Am J Physiol **242**:F379–F384.
23. Chabardes D, Imbert M, Clique A, Montegut M, Morel F 1975 PTH sensitive adenyl cyclase activity in different segments of the rabbit nephron. Pfluegers Arch (Eur J Physiol) **354**:229–239.
24. Haramati A, Haas JA, Knox FG 1983 Adaptation of deep and superficial nephrons to changes in dietary phosphate intake. Am J Physiol **244**:F265–F269.
25. Haramati A, Hass JA, Knox FG 1984 Nephron heterogeneity of phosphate reabsorption: Effect of parathyroid hormone. Am J Physiol **246**:F155–F158.
26. Hoffman N, Thees M, Kinne F 1976 Phosphate transport by isolated renal brush border vesicles. Pfluegers Arch (Eur J Physiol) **362**:147–156.
27. Sacktor B 1977 Transport in membrane vesicles isolated from the mammalian kidney and intestine. In: Sanadi R (ed) Current Topics in Bioenergetics. Academic Press, New York, p. 39.
28. Magagnin S, Werner A, Markovich D, Sorribas V, Stange G, Biber J, Murer H 1993 Expression cloning of human and rat renal cortex Na/P_i cotransport. Proc Natl Acad Sci USA **90**:5979–5983.
29. Murer H, Biber J 1996 Molecular mechanisms of renal apical Na phosphate cotransport. Ann Rev Physiol **58**:607–618.
30. Werner A, Moore ML, Mantei N, Biber J, Semenza G, Murer H 1991 Cloning and expression of cDNA for a Na/P_i cotransport system of kidney cortex. Proc Natl Acad Sci USA **88**:9608–9612.
31. Fucentese M, Murer H, Biber J 1994 Expression of rat renal Na/cotransport of phosphate and sulfate in Sf9 insect cells. J Am Soc Nephrol **5**:862.
32. Custer M, Lotscher M, Biber J, Murer H, Kaissling B 1994 Expression of Na-Pi cotransport in rat kidney: Localization by RT-PCR and immunohistochemistry. Am J Physiol **266**:F767.
33. Kavanaugh MP, Miller DG, Zhang W, Law W, Kozak SL, Kabat D, Miller AD 1994 Cell-surface receptors for gibbon ape leukemia virus and amphotropic murine retroviruses are inducible sodium-dependent phosphate symporters. Proc Natl Acad Sci USA **91**:7071–7075.
34. Caverzasio J, Bonjour J-P, Fleisch H 1982 Tubular handling of Pi in young growing and adult rats. Am J Physiol **242**:F705–F710.
35. Murer H, Werner A, Reshkin S, Wuarin F, Biber J 1991 Cellular mechanisms in proximal tubular reabsorption in inorganic phosphate. Am J Physiol **260**:C885–C899.
36. Beck N 1981 Effect of metabolic acidosis on renal response to parathyroid hormone in phosphorus-deprived rats. Am J Physiol **241**:F23–F27.
37. Verri T, Markovich D, Perego C, Norbis F, Stange G, Sorribas V, Biber J, Murer H 1995 Cloning of a rabbit renal Na/P_i-cotransporter which is regulated by dietary phosphate. Am J Physiol **268**:F626–F636.
38. Werner A, Kempson SA, Biber J, Murer H 1994 Increase of Na/P_i-cotransport encoding mRNA in response to low P_i diet in rat kidney cortex. J Biol Chem **269**:6637–6639.
39. Schwab SJ, Hammerman MR 1984 Mechanisms of phosphate exit across the basolateral membrane of the renal proximal tubule cell. Clin Res **32**:535.
40. Bonjour JP, Preston C, Fleisch H 1977 Effect of 1,25-dihydroxyvitamin D_3 on renal handling of P_i in thyroparathyroidectomized rats. J Clin Invest **60**:1419–1482.
41. Muhlbauer RC, Bonjour JP, Fleisch H 1981 Tubular handling of Pi: Localization of effects of $1{,}25(OH)_2D_3$ and dietary Pi in TPTX rats. Am J Physiol **241**:F123–F128.
42. Stoll R, Kinne R, Murer H, Fleisch H, Bonjour JP 1979 Phosphate transport by rat renal brush border membrane vesicles: Influence of dietary phosphate, thyroparathyroidectomy, and 1,25-dihydroxyvitamin D_3. Pfluegers Arch (Eur J Physiol) **380**:47–52.
43. Gekle DJ, Stroder J, Rostock D 1971 The effect of vitamin D on renal inorganic phosphate reabsorption on normal rats,

parathyroidectomized rats, and rats with rickets. Pediatr Res **5**:40.

44. Liang CT, Barnes J, Cheng L, Balakir R, Sacktor B 1982 Effects of $1,25(OH)_2D_3$ administered in vivo on phosphate uptake by isolated chick renal cells. Am J Physiol **242**:C312–C318.
45. Kurnik BR, Hruska KA 1984 Effects of 1,25-dihydroxycholecalciferol on phosphate transport in vitamin D-deprived rats. Am J Physiol **247**:F177–F184.
46. Taketani Y, Miyamoto K, Chikamori M, Tanaka K, Yamamoto H, Tatsumi S, Morita K, Takeda E 1996 Characterization of 5′ flanking region of the human renal phosphate transporter (NPT-1) gene. J Bone Miner Res **11**:S164.
47. Insogna KL, Broadus AE, Gertner JM 1983 Impaired phosphorus conservation and 1,25dihydroxyvitamin D generation during phosphorus deprivation in familial hypophosphatemic rickets. J Clin Invest **71**:1562–1569.
48. Muhlbauer RC, Bonjour J-P, Fleisch H 1982 Abnormal tubular adaptation to dietary Pi restriction in X-linked hypophosphatemic mice. Am J Physiol **242**:F353–F359.
49. Thornton SW, Tenenhouse HS, Martel J, Bockian RW, Meyer MH, Meyer RA Jr 1994 X-linked hypophosphatemic *Gy* mice: Renal tubular maximum for phosphate vs. brush-border transport after low-P diet. Am J Physiol **266**:F309–F315.
50. Steele TH, Engle JE, Tanaka Y, Lorenc RS, Dudgeon KL, DeLuca HF 1975 Phosphatemic action of 1,25-dihydroxyvitamin D_3. Am J Physiol **229**:489–495.
51. Trohler U, Bonjour J-P, Fleisch H 1976 Renal tubular adaptation to dietary phosphorus. Nature **261**:145–146.
52. Trohler U, Bonjour J-P, Fleisch H 1976 Inorganic phosphate homeostasis. Renal adaptation to the dietary intake in intact and thyroparathyroidectomized rats. J Clin Invest **57**:264–273.
53. Steele TH, DeLuca HF 1976 Influence of dietary phosphorus on renal phosphate reabsorption in the parathyroidectomized rat. J Clin Invest **57**:867–874.
54. Muhlbauer RC, Bonjour J-P, Fleisch H 1977 Tubular localization of adaptation to dietary phosphate in rats. Am J Physiol **233**:F342–F348.
55. Muhlbauer RC, Bonjour J-P, Fleisch H 1981 Tubular handling of Pi: Localization of effects of $1,25(OH)_2D_3$ and dietary Pi in TPTX rats. Am J Physiol **241**:F123–F128.
56. Stoll R, Kinne R, Murer H, Fleisch H, Bonjour J-P 1979 Phosphate transport by rat renal brush border membrane vesicles: Influence of dietary phosphate, thyroparathyroidectomy, and 1,25-dihydroxyvitamin D_3. Pfluegers Arch **380**:47–52.
57. Biber J, Murer H 1985 Na–P_i cotransport in LLC-PK_1 cells: Fast adaptive response to P_i deprivation. Am J Physiol **249**:C430–C434.
58. Brown CD, Bodmer B, Biber J, Murer H 1984 Sodium-dependent phosphate transport by apical membrane vesicles from a cultured renal epithelial cell line (LLC-PK_1). Biochem Biophys Acta **769**:471–478.
59. Caverzasio J, Brown CD, Biber J, Bonjour JP, Murer H 1985 Adaptation of phosphate transport in phosphate-deprived LLC-PK_1 cells. Am J Physiol **248**:F122–F127.
60. Levi M, Lotscher M, Sorribas V, Custer M, Arar M, Kaissling B, Murer H, Biber J 1994 Cellular mechanisms of acute and chronic adaptation of rat renal P(i) transporter to alterations in dietary P(i). Am J Physiol **267**:F900–F908.
61. Beutner EH, Munson PL 1960 Time course of urinary excretion of inorganic phosphate by rats after parathyroidectomy and after injection of parathyroid extract. Endocrinology **66**:610–616.
62. Pullman TN, Lavender AR, Aho I, Rasmussen H 1960 Direct renal action of a purified parathyroid extract. Endocrinology **67**:570–582.
63. Agus ZS, Gardner LB, Beck LH, Goldberg M 1973 Effects of parathyroid hormone on renal tubular reabsorption of calcium, sodium and phosphate. Am J Physiol **224**:1143–1148.
64. Ullrich KJ, Rumrich G, Kloss S 1977 Phosphate transport in the proximal convolution of the rat kidney. I. Tubular heterogeneity, effect of parathyroid hormone in acute and chronic parathyroidectomized animals and effect of phosphate diet. Pfluegers Arch **372**:269–274.
65. Muhlbauer RC, Bonjour J-P, Fleisch H 1980 Chronic thyroparathyroidectomy and tubular handling of phosphate: Increased reabsorption in late but not in early proximal tubule. Pfluegers Arch **388**:185–189.
66. Beck LH, Goldberg M 1973 Effects of acetazolamide and parathyroidectomy on renal transport of sodium, calcium and phosphate. Am J Physiol **224**:1136–1142.
67. Beck LH, Goldberg M 1974 Mechanism of the blunted phosphaturia in saline-loaded thyroparathyroidectomized dogs. Kidney Int **6**:18–23.
68. Amiel C, Kuntziger H, Richet G 1970 Micropuncture study of handling of phosphate by proximal and distal nephron in normal and parathyroidectomized rat. Evidence for distal reabsorption. Pfluegers Arch (Eur J Physiol) **317**:93–109.
69. Goldfarb S, Beck LH, Agus ZS, Goldberg M 1978 Dissociation of tubular sites of action of saline, PTH and DbCAMP on renal phosphate reabsorption. Nephron **21**:221–229.
70. Knox FG, Preiss J, Kim JK, Dousa TP 1977 Mechanism of resistance to the phosphaturic effect of the parathyroid hormone in the hamster. J Clin Invest **59**:675–683.
71. Le Grimellec C, Roinel N, Morel F 1974 Simultaneous Mg, Ca, P, K and Cl analysis in rat tubular fluid IV. During acute phosphate plasma loading. Pfluegers Arch (Eur J Physiol) **346**:189–204.
72. Evers C, Murer H, Kinne R 1978 Effect of parathyrin on the transport properties of isolated renal brush-border vesicles. Biochem J **172**:49–56.
73. Hammerman MR, Hruska KA 1982 Cyclic AMP-dependent protein phosphorylation in canine renal brush-border membrane vesicles is associated with decreased Pi transport. J Biol Chem **257**:992–999.
74. Hruska KA, Hammerman MR 1981 Parathyroid hormone inhibition of phosphate transport in renal brush border vesicles from phosphate-depleted dogs. Biochim Biophys Acta **645**:351–356.
75. Gloor HJ, Bonjour J-P, Caverzasio J, Fleisch H 1979 Resistance to the phosphaturic and calcemic actions of parathyroid hormone during phosphate depletion. J Clin Invest **63**:371–377.
76. Caverzasio J, Montessuit C, Bonjour J-P 1990 Stimulatory effect of insulin-like growth factor-1 on renal Pi transport and plasma 1,25-dihydroxyvitamin D_3. Endocrinology **127**:453–459.
77. Caverzasio J, Bonjour J-P 1989 Insulin-like growth factor I stimulates Na-dependent Pi transport in cultured kidney cells. Am J Physiol **257**:F712–F717.
78. Murer H, Biber J 1996 Molecular mechanisms of renal apical Na/phosphate cotransport. Annu Rev Physiol **58**:607–618.
79. Kempson SA, Lotscher M, Kaissling B, Biber J, Murer H, Levi M 1995 Parathyroid hormone action on phosphate transporter mRNA and protein in rat renal proximal tubules. Am J Physiol **268**:F784–F791.
80. Hoppe A, Metler M, Berndt TJ, Knox FG, Angielski S 1982 Effect of respiratory alkalosis on renal phosphate excretion. Am J Physiol **243**:F471–F475.
81. Fulop M, Brazeau P 1968 The phosphaturic effect of sodium bicarbonate and acetazolamide in dogs. J Clin Invest **47**:983–991.
82. Quamme GA 1984 Urinary alkalinization may not result in an increase in urinary phosphate excretion. Kidney Int **25**:152 (abstract).

83. Mercado A, Slatopolsky E, Klahr S 1975 On the mechanisms responsible for the phosphaturia of bicarbonate administration. J Clin Invest **56**:1386–1395.
84. Puschett JB, Goldberg M 1969 The relationship between the renal handling of phosphate and bicarbonate in man. J Lab Clin Med **73**:956–969.
85. Kuntziger HE, Amiel C, Couette S, Coureau C 1980 Localization of parathyroid-hormone-independent sodium bicarbonate inhibition of tubular phosphate reabsorption. Kidney Int **17**:749–755.
86. Cuche JL, Ott CE, Marchand GR, Diaz-Buxo JA, Knox FG 1976 Intrarenal calcium in phosphate handling. Am J Physiol **230**:790–796.
87. Pitts RF, Alexander RS 1944 The renal reabsorptive mechanism for inorganic phosphate in normal and acidotic dogs. Am J Physiol **142**:648–662.
88. Guntupalli J, Eby B, Lau K 1982 Mechanism for the phosphaturia of NH_4Cl: Dependence on acidemia but not on diet PO_4 or PTH. Am J Physiol **242**:F552–F560.
89. Kempson SA 1982 Effect of metabolic acidosis on renal brush border membrane adaptation to low phosphorus diet. Kidney Int **22**:225–233.
90. Quamme GA, Mizgala CL, Wong NLM, Whiting SJ 1985 Effects of intraluminal pH and dietary phosphate on phosphate transport in the proximal convoluted tubule. Am J Physiol **249**:F759–F768.
91. Frick A, Durasin I 1980 Proximal tubular reabsorption of inorganic phosphate in adrenalectomized rats. Pfluegers Arch (Eur J Physiol) **385**:189–192.
92. Hammerman MR, Karl IE, Kruska KA 1980 Regulation of canine renal vesicle P_i transport by growth hormone and parathyroid hormone. Biochim Biophys Acta **603**:322–335.
93. Caverzasio J, Bonjour J-P 1996 Characteristics and regulation of P_i transport in osteogenic cells for bone metabolism. Kidney Int **49**:975–980.
94. Caverzasio J, Murer H, Fleisch H, Bonjour J-P 1982 Phosphate transport in brush border membrane vesicles isolated from renal cortex of young growing and adult rats. Pfluegers Arch **394**:217–221.
95. Caverzasio J, Bonjour J-P 1985 Expression of chronic thyroparathyroidectomy on phosphate transport in whole kidney and proximal luminal membranes during phosphate deprivation. Pfluegers Arch **405**:395–399.
96. Olsen HS, Cepeda MA, Zhang Q-Q, Rosen CA, Vozzolo BL, Wagner GF 1996 Human stanniocalcin: A possible hormonal regulator of mineral metabolism. Proc Natl Acad Sci USA **93**:1792–1796.
97. Econs MJ, Drezner MK 1994 Tumor-induced osteomalacia: Unveiling a new hormone. N Engl J Med **330**:1645–1649.
98. Cai Q, Hodgson SF, Kao PC, Lennon VA, Klee GG, Zinsmiester AR, Kumar R 1994 Brief report: Inhibition of renal phosphate transport by a tumor product in a patient with oncogenic osteomalacia. N Engl J Med **330**:1645–1649.
99. Parker MS, Klein I, Haussler MR, Mintz DH 1981 Tumor-induced osteomalacia: Evidence of a surgically correctable alteration in vitamin D metabolism JAMA **245**:492–493.
100. Rowe PSN, Ong ACM, Cockerill FJ, Goulding JN, Hewison M 1996 Candidate 56 and 58 kDa protein(s) responsible for mediating the renal defects in oncogenic hypophosphatemic osteomalacia. Bone **18**:159–169.
101. Miyauchi A, Fukase M, Tsutsumi M, Fujita T 1988 Hemangiopericytoma-induced osteomalacia: Tumor transplantation in nude mice causes hypophosphatemia and tumor extracts inhibit renal 25-hydroxyvitamin D 1-hydroxylase activity. J Clin Endocrinol Metab **67**:46–53.
102. Wilkins GE, Granleese S, Hegele RG, Holden J, Anderson DW, Bondy GP 1995 Oncogenic osteomalacia: Evidence for a humoral phosphaturic factor. J Clin Endocrinol Metab **80**:1628–1634.
103. Caverzasio J, Montessuit C, Bonjour J-P 1996 Functional role of P_i transport in osteogenic cells. News Physiol Sci **11**:119–125.
104. Gupta A, Miyauchi A, Fujimori A, Hruska K 1996 Phosphate transport in osteoclasts: A functional and immunochemical characterization. Kidney Int **49**:968–974.
105. Gupta A, Guo X-L, Alvarez UM, Hruska KA 1997 Regulation of sodium-dependent phosphate transport in osteoclasts. J Clin Invest (in press).
106. Ecarot B, Caverzasio J, Desbarats M, Bonjour J-P, Glorieux FH 1994 Phosphate transport by osteoblasts from X-linked hypophosphatemic mice. Am J Physiol **266**:E33–E38.
107. Montessuit C, Bonjour JP, Caverzasio J 1994 P_i transport regulation by chicken growth plate chondrocytes. Am J Physiol **267**:E24–E31.
108. Selz T, Caverzasio J, Bonjour JP 1990 Sodium fluoride stimulates P_i transport in osteoblast-like cells. Am J Physiol **260**:E833–E838.
109. William DC, Frolik CA 1991 Physiological and pharmacological regulation of biological calcification. Int Rev Cytol **126**:195–292.
110. Anderson HC 1985 Matrix vesicle calcification: Review and update. In: Peck WA (ed) Bone and Mineral Research, 3rd Ed. Elsevier, Amsterdam, pp. 109–150.
111. Felix R, Herrmann W, Fleisch H 1978 Stimulation of precipitation of calcium phosphate by matrix vesicles. Biochem J **170**:681–691.
112. Genge BR, Cao X, Wu LN, Buzzi WR, Showman RW, Arsenault AL, Ishikawa Y, Wuthier RE 1992 Establishment of the primary structure of the major lipid-dependent Ca^{2+} binding proteins of chicken growth plate cartilage matrix vesicles: Identity with anchorin CII (annexin V) and annexin II. J Bone Miner Res **7**:807–819.
113. Montessuit C, Caverzasio J, Bonjour JP 1991 Characterization of a P_i transport system in cartilage matrix vesicles. Potential role in the calcification process. J Biol Chem **266**:17791–17797.
114. Quarles LD, Yohay DA, Lever LW, Caton R, Wenstrup RJ 1992 Distinct proliferative and differentiated stages of murine MC3T3-E1 cells in culture: An in vitro model of osteoblast development. J Bone Miner Res **7**:683–692.
115. Green J, Luong KV, Kleeman CR, Ye LH, Chaimovitz C 1993 1,25-Dihydroxyvitamin D_3 inhibits Na^+-dependent phosphate transport in osteoblast cells. Am J Physiol **264**:C287–C295.
116. Bijvoet OL, Morgan DB, Fourman P 1969 The assessment of phosphate reabsorption. Clin Chim Acta **26**:15–24.
117. Eicher EM, Southard JL, Scriver CR, Glorieux FH 1976 Hypophosphatemia: Mouse model for human familial hypophosphatemic (vitamin D-resistant) rickets. Proc Natl Acad Sci USA **73**:4667–4671.
118. Tenenhouse HS, Scriver CR 1981 Effect of 1,25-dihydroxyvitamin D_3 on phosphate homeostasis in the X-linked hypophosphatemic (Hyp) mouse. Endocrinology **109**:658–660.
119. Brickman AS, Coburn JW, Kurokawa K, Bethune JE, Harrison HE, Norman AW 1973 Actions of 1,25-dihydroxycholecalciferol in patients with hypophosphatemic, vitamin D-resistant rickets. N Engl J Med **289**:495–498.
120. Russell RG, Smith R, Preston C, Walton RJ, Woods CG, Henderson RG, Norman AW 1975 The effect of 1,25-dihydroxycholecalciferol on renal tubular reabsorption of phosphate, intestinal absorption of calcium and bone histology in hypophosphataemic renal tubular rickets. Clin Sci Mol Med **48**:177–186.
121. Meyer RA Jr, Gray RW, Meyer MH 1980 Abnormal vitamin D metabolism and the X-linked hypophosphatemic mouse. Endocrinology **107**:1577–1581.
122. Beamer WG, Wilson MD, DeLuca HF 1980 Successful treatment

of genetically hypophosphatemic mice by 1-alpha-hydroxy vitamin D_3 but not 1,25-dihydroxy vitamin D_3. Endocrinology **106**:1949–1955.
123. Tenenhouse HS, Scriver CR, McInnes RR, Glorieux FH 1978 Renal handling of phosphate *in vivo* and *in vitro* by the X-linked hypophosphatemic male mouse: Evidence for a defect in the brush border membrane. Kidney Int **14**:236–244.
124. Tenenhouse HS, Werner A, Biber J, Ma S, Martel J, Roy S, Murer H 1994 Renal Na^+-phosphate cotransport in murine X-linked hypophosphatemic rickets: Molecular characterization. J Clin Invest **93**:671–676.
125. Kos CH, Lemieux N, Tihy F, Biber J, Murer H, Econs MJ, Tenenhouse HS 1993 The renal specific Na^+-phosphate cotransporter cDNA maps to human chromosome 5q35. J Am Soc Nephrol **4**:816 (abstract).
126. Chong SS, Kristjansson K, Zoghbi HY, Hughes MR 1993 Molecular cloning of the cDNA encoding a human renal sodium phosphate transport protein and its assignment to chromosome 6p21.3-p23. Genomics **18**:355–359.
127. Hruska KA, Rifas L, Cheng S-L, Gupta A, Halstead L, Avioli L 1995 X-linked hypophosphatemic rickets and the murine Hyp homologue. Am J Physiol **268**:F357–F362.
128. Meyer RA Jr, Tenenhouse HS, Meyer M, Klugerman AH 1989 The renal phosphate transport defect in normal mice parabiosed to X-linked hypophosphatemic mice persists after parathyroidectomy. J Bone Miner Res **4**:523–532.
129. Nesbitt T, Coffman TM, Griffiths R, Drezner MK 1992 Cross-transplantation of kidneys in normal and Hyp mice. Evidence that the Hyp mouse phenotype is unrelated to an intrinsic renal defect. J Clin Invest **89**:1453–1459.
130. The Hyp Consortium 1995 A gene (PEX) with homologies to endopeptidases is mutated in patients with X-linked hypophosphatemic rickets. Nature Genet **11**:130–136.
131. Rifas L, Dawson LL, Halstead LH, Roberts M, Avioli LV 1995 Phosphate transport in osteoblasts from normal and X-linked hypophosphatemic mice. Calcif Tissue Int **54**:505–510.
132. Ecarot B, Glorieux FH, Desbarats M, Travers R, Labelle L 1992 Defective bone formation by Hyp mouse bone cells transplanted into normal mice: Evidence in favor of an intrinsic osteoblast defect. J Bone Miner Res **7**:215–220.
133. Ecarot B, Glorieux FH, Desbarats M, Travers R, Labelle L 1992 Effect of dietary phosphate deprivation and supplementation of recipient mice on bone formation by transplanted cells from normal and X-linked hypophosphatemic mice. J Bone Miner Res **7**:523–530.
134. Rifas L, Avioli LV, Cheng SL 1994 $1,25(OH)_2D_3$ Corrects underphosphorylation of osteopontin in the *Hyp*/Y mouse osteoblast. In: Bouillon R, Normon AW, Thomasset M (eds) Vitamin D, A Pluripotent Steroid Hormone: Structural Studies, Molecular Endocrinology and Clinical Applications. de Gruyter, New York, p. 704.
134a. Du L, Desbarats M, Viel J, Glorieux FH, Cawthorn C, Ecarot B 1996 cDNA cloning of the murine Pex gene implicated in X-linked hypophosphatemia and evidence for expression in bone. Genomics **36**:22–28.
135. Glorieux FH, Marie PJ, Pettifor JM, Delvin EE 1980 Bone response to phosphate salts, ergocalciferol, and calcitriol in hypophosphatemic vitamin D-resistant rickets. N Engl J Med **303**:1023–1031.
136. Verge CF, Lam A, Simpson JM, Cowell CT, Howard NJ, Silink M 1991 Effect of therapy in X-linked hypophosphatemic rickets. N Engl J Med **325**:1875–1877.
137. Friedman NE, Lobaugh B, Drezner MK 1993 Effects of calcitriol and phosphorus therapy on the growth of patients with X-linked hypophosphatemia. J Clin Endocrinol Metab **7**:839–844.
138. Fraser D, Kooh SW, Kind HP, Holick MF, Tanaka Y, DeLuca HF 1973 Pathogenesis of hereditary vitamin-D-dependent rickets. An inborn error of vitamin D metabolism involving defective conversion of 25-hydroxyvitamin D to 1alpha,25-dihydroxyvitamin D. N Engl J Med **289**:817–822.
139. Liberman UA, Eil C, Marx SJ 1983 Resistance of 1,25-dihydroxyvitamin D. Associated with heterogeneous defects in cultured skin fibroblasts. J Clin Invest **71**:192–200.
140. Feldman D, Chen T, Cone C, Hirst M, Shani S, Benderli A, Hochberg Z 1982 Vitamin D resistant rickets with alopecia: Cultured skin fibroblasts exhibit defective cytoplasmic receptors and unresponsiveness to $1,25(OH)_2D_3$. J Clin Endocrinol Metab **55**:1020–1022.
141. Chen TL, Hirst MA, Cone CM, Hochberg Z, Tietze H-U, Feldman D 1984 1,25-Dihydroxyvitamin D resistance, rickets and alopecia: Analysis of receptors and bioresponse in cultured fibroblasts from patients and parents. J Clin Endocrinol Metab **59**:383–388.
142. Malloy PJ, Hochberg Z, Pike JW, Feldman D 1989 Abnormal binding of vitamin D receptors to deoxyribonucleic acid in a kindred with vitamin D-dependent rickets, type II. J Clin Endocrinol Metab **68**:263–269.
143. Hochberg Z, Weisman Y 1995 Calcitriol-resistant rickets due to vitamin D receptor defects. Trends Endocrinol Metab **6**:216–220.
144. Hewison M, Rut AR, Kristjansson K, Walker RE, Dillon MJ, Hughes MR, O'Riordan JL 1993 Tissue resistance to 1,25-dihydroxyvitamin D without a mutation of the vitamin D receptor gene. Clin Endocrinol **39**:663–670.
145. Refetoff S, Weiss RE, Usala SJ 1993 The syndromes of resistance to thyroid hormone. Endocr Rev **14**:348–399.
146. McPhaul MJ, Marcelli M, Zoppi S, Griffin JE, Wilson JD 1993 Genetic basis of endocrine disease. 4. The spectrum of mutations in the androgen receptor gene that causes androgen resistance. J Clin Endocrinol Metab **76**:17–23.
147. Smith EP, Boyd J, Frank GR, Takahashi H, Cohen RM, Specker B, Williams TC, Lubahn DB, Korach KS 1994 Estrogen resistance caused by a mutation in the estrogen receptor gene in a man. N Engl J Med **331**:1088–1089.
148. Betro MG, Pain RW 1972 Hypophosphatemia and hyperphosphatemia in a hospital population. Br Med J **1**:273–276.
149. Ryback RS, Eckardt MJ, Pautler CP 1980 Clinical relationships between serum phosphorus and other blood chemistry values in alcoholics. Arch Intern Med **140**:673–677.
150. Seldin DW, Tarail R 1950 The metabolism of glucose and electrolytes in diabetic acidosis. J Clin Invest **29**:552–565.
151. Ribovich ML, DeLuca HF 1978 Effect of dietary calcium and phosphorus on intestinal calcium absorption and vitamin D metabolism. Arch Biochem Biophys **188**:145–156.
152. Slatopolsky E, Gradowska K, Kashemsant C, Keltner R, Manley C, Bricker NS 1966 The control of phosphate excretion in uremia. J Clin Invest **45**:672–677.
153. Slatopolsky E, Robson AM, Elkan I, Bricker NS 1968 Control of phosphate excretion in uremic man. J Clin Invest **47**:1865–1874.
154. Goldman R, Bassett SH 1954 Phosphorus excretion in renal failure. J Clin Invest **33**:1623.
155. Rutherford E, Mercado A, Hruska K, Harter H, Mason N, Sparks R, Klahr S, Slatopolsky E 1973 An evaluation of a new and effective phosphate binding agent. Trans Am Soc Artif Intern Organs **19**:446–449.
156. Massry SG, Arieff AI, Coburn JW, Palmieri G, Kleeman CR

1974 Divalent ion metabolism in patients with acute renal failure: Studies on the mechanism of hypocalcemia. Kidney Int **5**:437–445.
157. Koffler A, Friedler RM, Massry SG 1976 Acute renal failure due to nontraumatic rhabdomyolysis. Ann Intern Med **85**:23–28.
158. Mitnick PD, Goldbarb S, Slatopolsky E, Lemann J Jr, Gray RW, Agus ZS 1980 Calcium and phosphate metabolism in tumoral calcinosis. Ann Intern Med **92**:482–487.
159. Lufkin EG, Wilson DM, Smith LH, Bill NJ, DeLuca HF, Dousa TP, Knox FG 1980 Phosphorus excretion in tumoral calcinosis: Response to parathyroid hormone and acetazolamide. J Clin Endocrinol Metab **50**:648–653.
160. Zerwekh JE, Sanders LA, Townsend J, Pak CY 1980 Tumoral calcinosis: Evidence for concurrent defects in renal tubular phosphorus transport and in 1alpha,25-dihydroxycholecalciferol synthesis. Calcif Tissue Int **32**:1–6.

CHAPTER 33

Racial, Geographic, Genetic, and Body Habitus Effects on Vitamin D Metabolism

DEVASHIS MITRA AND NORMAN H. BELL
Department of Medicine, Medical University of South Carolina, Department of Veterans Affairs Medical Center, Charleston, South Carolina

I. INTRODUCTION

To understand why changes in vitamin D metabolism occur in blacks, Asian Indians, and Pakistanis, it is helpful to briefly review vitamin D metabolism and the vitamin D endocrine system. These subjects are discussed in greater detail in Chapter 2 of this book. An outline of races and populations at risk for developing rickets and osteomalacia and factors involved are shown in Table I.

Vitamin D_3 is synthesized in the skin from 7-dehydrocholesterol. 7-Dehydrocholesterol is formed from previtamin D_3 by absorption of one photon of ultraviolet light, and the conversion of previtamin D_3 to vitamin D_3 is regulated by body heat, a process that is temperature dependent and requires a period of several days [1,2]. Vitamin D_3 is transported from dermal capillaries in association with vitamin D binding protein (DBP) to the liver where it undergoes 25-hydroxylation and forms 25-hydroxyvitamin D (25OHD), a conversion that is modulated by vitamin D 25-hydroxylase (25-hydroxylase) [3,4]. Conversion of 25OHD to 1,25-dihydroxyvitamin D [$1,25(OH)_2D$] takes place in the proximal tubule of the kidney by action of 25-OHD 1α-hydroxylase (1α-hydroxylase), an enzyme that is stimulated by parathyroid hormone [5–7] and by growth hormone, through stimulation of insulin-like growth factor I [8,9] and is inhibited by calcium [10,11] and inorganic phosphorus [12,13]. In states of vitamin D excess, 25OHD is converted to 24,25-dihydroxyvitamin D [$24,25(OH)_2D$] by 25-OHD 24-hydroxylase (24-hydroxylase), an enzyme induced by $1,25(OH)_2D$ and present in the kidney and

TABLE I Causes and Consequences of Vitamin D Deficiency in Various Races and Populations

Population	Increased skin pigment	Diminished exposure to sunshine	Inadequate intake of vitamin D	Rickets	Osteomalacia
Asian Indians	+	+	+	+	+
Blacks	+	+	+	+	−
Caucasians	−	−	+	+	−
Chinese	−	−	+	+	+
Egyptians	+	+	+	+	−
Hispanics	+	−	−	−	−
Jordanians	+	+	+	+	−
Libyans	+	+	+	+	−
Moroccans	+	+	+	+	−
Pakistanis	+	+	+	+	+
Polynesians	+	−	−	−	−
Saudi Arabians	+	+	+	+	−

[a] Newborn infants of any race or population are prone to develop rickets when breast fed and kept indoors. Asians and Pakistanis are at risk to develop vitamin D deficiency and osteomalacia, particularly when they reside away from the equator. Despite knowledge of prevention of these diseases, they are widespread throughout the world. + = occurs; − = does not occur.

other organs [14–16]. In states of vitamin D deficiency, less $24,25(OH)_2D$ is produced. Both $1,25(OH)_2D$ and $24,25(OH)_2D$ undergo further hydroxylation to form $1,24,25(OH)_3D$ which follows a degradative pathway to calcitroic acid, a pathway similar to that of formation of bile acids from cholesterol. By inducing 24-hydroxylase, $1,25(OH)_2D$ regulates its rate of degradation as well as that of 25OHD. In rats, $1,25(OH)_2D$ increases the metabolic clearance rate of 25OHD [17], and in human subjects, administration of $1,25(OH)_2D_3$ prevents the increase in serum 25OHD produced by administration of vitamin D [18].

A negative feedback control system exists for regulation of calcium metabolism that involves the parathyroids, skeleton, kidney, and intestine (Fig. 1) Parathyroid hormone (PTH) together with $1,25(OH)_2D$ maintains the serum calcium within a very narrow range by causing osteoclastic bone resorption [19], by enhancing the tubular reabsorption of calcium [20], and by enhancing intestinal absorption of calcium, a biochemical event that is mediated by $1,25(OH)_2D$ and the vitamin D receptor (VDR) [21,22]. Secretion of PTH in turn is inhibited by both calcium and $1,25(OH)_2D$; inhibition by calcium is mediated by a calcium sensing receptor [23] and $1,25(OH)_2D$ by a VDR [24,25]. $1,25(OH)_2D$ also modulates secretion of PTH by up-regulating the VDR in parathyroid tissue [26].

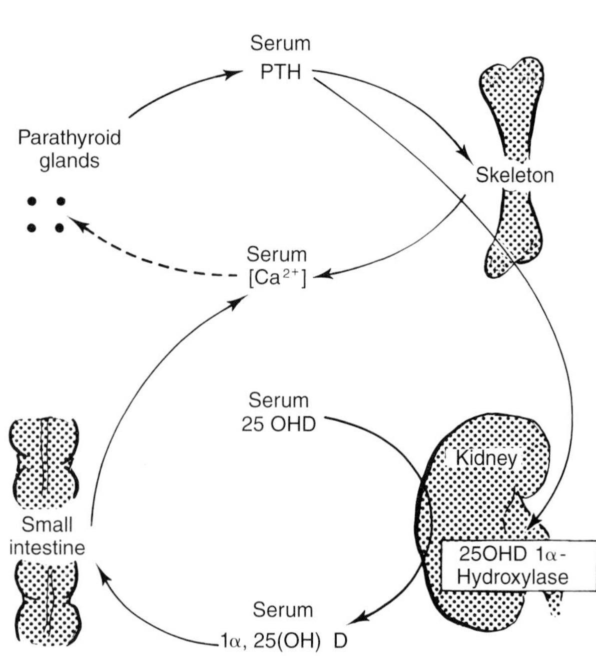

FIGURE 1 Feedback regulation of serum ionized calcium (Ca^{2+}) by serum PTH. The hormone stimulates bone resorption directly (top) and intestinal absorption of calcium indirectly by modulating renal synthesis of $1,25(OH)_2D$ (bottom). Reprinted with permsision (J SC Med Assoc 78:21–26, 1982).

II. EFFECTS OF RACE AND GEOGRAPHY

A. Blacks

1. VITAMIN D ENDOCRINE SYSTEM

In blacks who live in areas away from the equator, the vitamin D endocrine system is altered as a consequence of diminished exposure to sunlight. Thus, one study demonstrated that serum 25OHD was 29 ± 4 pg/ml in blacks in Zaire and 9 ± 1 pg/ml in blacks in Belgium and that declines in serum 25OHD in blacks moving from Zaire to Belgium took place over a period of several years [27]. Reduction in serum 25OHD in blacks is attributed to increased skin pigment and diminished dermal production of vitamin D_3 [28]. Melanin absorbs ultraviolet light and prevents formation of previtamin D_3. Most but not all studies in the United States report reductions in serum 25OHD in African-Americans compared to Caucasians that are on the order of 50% and that occur in infants and children as well as adults [29–40].

In blacks, alteration of the vitamin D endocrine system is characterized by decreased serum 25OHD, mild secondary hyperparathyroidism, increased serum $1,25(OH)_2D$ and urinary cyclic adenosine 3′,5′-monophosphate, and decreased urinary calcium [29–40]. These changes are reversed by treatment with $25OHD_3$ [39]. Thus, reduction of urinary calcium is attributed to a low serum 25OHD and secondary hyperparathyroidism [39]. As a result of reduced urinary calcium, the incidence of calcium-containing kidney stones in blacks in South Africa is reduced [41].

Seasonal variation in serum 25OHD and $1,25(OH)_2D$ occurs in both blacks and Caucasians. In the northern hemisphere, serum values for the two metabolites are higher in summer than in winter, and serum 25OHD is lower and serum $1,25(OH)_2D$ is higher in black than in Caucasian men and women regardless of season [42]. The variation results from differences in duration of exposure to and intensity of sunlight.

Increased secretion of PTH in blacks is underscored by studies of dynamics of secretion showing an exaggerated stimulation of secretion of the hormone in response to hypocalcemia and diminished suppression of secretion of the hormone in response to hypercalcemia compared to white men and women [37]. Further, postmortem studies in the United States show that parathyroid glands are significantly larger in black than in white subjects [43]. Thus, in blacks moderate hypertrophy of the parathyroid glands is associated with long-term increased secretion of parathyroid hormone.

Reduction of serum 25OHD in blacks could occur because of either (1) decreased synthesis of vitamin D_3 and decreased production rate of 25OHD or (2) increased metabolic clearance as a consequence of elevated serum $1,25(OH)_2D$ and induction of 24-hydroxylase or both. Preliminary results show that V_{max} of the 24-hydroxylase enzyme in cultured skin fibroblasts, a measure of enzymatic response to $1,25(OH)_2D_3$, is modestly higher in blacks than in Caucasians so that reduction in serum 25OHD was found to probably result from a diminished production rate and not an increased metabolic rate of the metabolite [44]. In these studies, low values for serum 25OHD were associated with low values for serum vitamin D [44]. Thus, decreased availability of substrate is another factor to account for lower 25OHD values. There was wide variation in the production rate of 25OHD in both black and Caucasian men and women. It is not clear whether genetic heterogeneity for 25-hydroxylase itself or nonspecific hydroxylation by other enzymes accounts for the wide variation.

The bulk of the vitamin D and 25OHD pool is transmitted from mother to fetus before birth. Hence, it is important that pregnant women receive dietary supplementation. This is as critical as postnatal supplements to newborns. Black infants, especially when breast fed, are at risk for developing nutritional rickets [45–49]. Because 25OHD is stored in milk and provides a source for 25OHD in infants, low maternal serum 25OHD concentrations result in low concentrations of 25OHD in milk. However, as the concentrations of vitamin D and 25OHD in milk even from Caucasian women are also low (39 and 310 pg/ml, respectively), they provide only a modest portion of nutritional needs for infants, and other sources are necessary [49]. A total of 400 IU (10 μg) of vitamin D per day is required for maintenance of a normal serum 25OHD in both premature and full-term healthy infants [50,51]; preterm babies fed human milk should receive a daily supplement of 1000 IU/day (see Chapter 34). It is clear that in well infants, there is no maturational delay in the appearance of 25-hydroxylase.

2. BONE MASS

Beginning in childhood, bone mass is higher in African-Americans than in Caucasians [52–59]. Multivariate analysis has demonstrated that bone mineral density of the midradius, lumbar spine, trochanter, and femoral neck are higher in black than in white boys and girls in the age range of 7 to 17 years [53]. Bone mineral density of the lumbar spine, trochanter, and femoral neck are higher in African-American than in white premenopausal and postmenopausal women [54–58] and are higher in African-American than in white men in the United States [59–61].

Whereas one study attributed the greater accumulation of bone mass during childhood and adolescence in black than white girls to higher intestinal absorption of calcium [40], another study attributed the racial difference to lower urinary calcium excretion in black as compared to white children [36]. Whether these differences represent heterogeneity in the populations or differences in methodology is not known. In the former study, calcium accretion, determined by radiocalcium kinetic analysis, was higher in black than in white girls. [40].

The skeleton undergoes continuous remodeling that begins with activation of osteoclasts, resorption of small areas called Howship's lacunae, and attraction of osteoblasts or bone-forming cells which repair the resorption site. In older individuals, however, repair is never complete, and this results in age-related bone loss. In general, high rates of skeletal remodeling are associated with high rates of bone loss and low rates of remodeling with low rates of bone loss. Available evidence indicates that the rate of skeletal remodeling in blacks is reduced, and this could contribute to or be responsible for the greater bone mass (see Section II,A,3) [62]. In addition, serum 17β-estradiol, a major determinant of growth hormone secretion in both men and women, and growth hormone secretion are 50% higher in black than in white men [60]. In these studies, bone mineral density of the total body, forearm, trochanter, and femoral neck were significantly higher in the black than in the white men. Thus, because estrogen diminishes skeletal remodeling and growth hormone stimulates skeletal growth, racial differences in serum 17β-estradiol and growth hormone levels in men could be a contributing factor to differences in bone mineral density [60]. In contrast, no racial difference in growth hormone secretion or serum 17β-estradiol was found in women, despite a higher bone mineral density of the total body and hip in the black women [58]. Thus, the biochemical mechanisms for increased bone mineral density in African-Americans is not established.

The incidence of osteoporosis and atraumatic fractures is lower in African-Americans than in Caucasian subjects in the United States. Hip fracture rates for African-American women are 40 to 60% lower than those of Caucasian women in the United States [63–71]. The lower incidence of fractures in black than in white men and women is attributed in part to the racial difference in bone mass. Low body weight is a risk factor for hip fracture in both black and white women [69] as is previous stroke, use of aids in walking, and alcohol consumption [70]. A shorter hip axis length may contribute to the lower incidence of fractures in black women [71]. In addition, the incidence of obesity is twice as high in black women than in white women, and because body weight is a major determinant of bone mineral density, body weight is a contributing factor to the higher bone mineral density in black women. Also, by cushioning falls, fat helps prevent fractures.

In South Africa, the incidences of fractures of the lumbar spine and femoral neck are lower in black than in Caucasian subjects [72,73]. No racial difference in bone mineral content of the forearm was found in black and white children [74]. Bone mineral density of the femoral neck is higher in black than in white women, but there is no difference in bone mineral density of the forearm or lumbar spine [75]. Whereas greater bone mineral density of the femoral neck may account for the lower incidence of femoral fractures in blacks, other factors must account for the racial difference in vertebral fractures.

3. Bone Histomorphometry

Histomorphometric analysis of biopsies of the iliac crest after double-tetracycline labeling shows that the bone formation rate is lower by about two-thirds in African-American compared to Caucasian subjects [62]. Static measurements are not different. This is not inconsistent with differences in bone mineral density, as the error of measurement is 10% and racial differences in bone mineral density are of the order 5 to 10%. Markers of skeletal remodeling support the histomorphometric findings. For example, serum osteocalcin, a marker of bone formation, is lower in black than in Caucasian subjects [29,38,56], and urinary hydroxyproline, a marker of bone resorption, is lower in black than in white women both before and after menopause [38,56]. Urinary excretion of N-telopeptides of type I collagen, a highly specific marker of bone resorption, is lower in black than in white men [76]. As noted, decreases in the rate of skeletal remodeling could contribute to the racial difference in bone mineral density. Findings in patients with thyroid disease support this concept. Bone mineral density varies inversely with rates of skeletal remodeling. In patients with hypothyroidism, for example, the rate of skeletal remodeling is reduced [77]. Further, inhibition of bone resorption with the bisphosphonate alendronate increases bone mineral density at both the lumber spine and hip [78,79].

In contrast to the findings in blacks in the United States, histomorphometric studies without double-tetracycline labeling from South Africa demonstrated thicker trabecular bone, greater osteoid volume, surface, and thickness, and greater erosion surfaces in blacks compared to Caucasians [80]. These results were interpreted to mean that the greater values for osteoid volume and erosion in blacks reflected higher rates of bone turnover and that trabecular bone in blacks was renewed more frequently and thus was less prone to fatigue failure and spontaneous fracture [80]. The reason for the differences in skeletal histomorphometry in the present [80] compared to the previous study showing decreased skeletal remodeling [62] is not apparent, but differences in study populations and methodology could be responsible.

B. Asian Indians and Pakistanis

1. Vitamin D Endocrine System

It is well documented that Asian Indians and Pakistanis who immigrate to Great Britain are prone to develop vitamin D deficiency [81–87], rickets [81,83,84,88–96], and osteomalacia [81,94–98]. Compared to values in Caucasians, serum 25OHD is reduced [81–98], and vitamin D deficiency often leads to rickets in neonates [84,88], infants [89], as well as children and adolescents [81,90–96] and to osteomalacia in adults [81,94–98]. Reductions in serum 25OHD are associated with secondary hyperparathyroidism [81,86,96], and in one study increases in serum immunoreactive intact PTH were shown to be a highly sensitive indicator of the presence of occult osteomalacia [99]. Vitamin D deficiency, rickets, and osteomalacia can be treated and prevented by vitamin D in daily doses of 400 to 3000 IU [84,96,98,100,101].

In immigrant Asian Indians, [102] and Pakistanis, vitamin D deficiency is attributed to increased skin pigmentation and diminished dermal production of vitamin D_3 [28], diminished exposure to sunlight in more northern climates [86,103], reduced intake of vitamin D [83,90,94], and consumption of nonfortified Chapputi flour [83,90] and vegetarian diets [82,85,98]. Studies show that serum vitamin D_3 was below the limit of detection (1 ng/ml) during the summer in Asian Indians living in Charleston, South Carolina, and reductions were associated with alterations of the vitamin D endocrine system similar to those found in blacks: increases in serum immunoreactive intact PTH and $1,25(OH)_2D$ and decreases in urinary calcium excretion [102]. However, increases in serum vitamin D after exposure to ultraviolet light are comparable in Indians, Pakistanis, and Caucasians [103]. Values for serum 25OHD are higher in Pakistanis living closer to the equator in Pakistan and are comparable to values in Caucasians living in Britain [86,104]. One case–control study showed that vegetarian diets, particularly those without meat, eggs, or fish, and not intake of Chapputi flour, contribute to vitamin D deficiency [105]. However, serum 25OHD also was lower in Asian adults and children who did not consume a vegetarian diet than in Caucasians [82,86]. Thus, consumption of a vegetarian diet, diminished vitamin D intake, increased skin pigment, limited exposure to sunlight, and possibly consumption of unfortified Chapputi flour are all contributing factors for vitamin D deficiency in Asians.

Vitamin D deficiency and rickets occur in Pakistani infants, particularly those who are breast fed, in Karachi [106]. The rickets is attributed to maternal vitamin D deficiency and lack of supplementation of vitamin D.

2. Bone Mass

There is no difference in bone mass of Asian Indians and Caucasians when body mass index and the area over which the X-ray beam is projected is taken into account. Whereas earlier studies demonstrated lower bone densities in Japanese and Chinese, who are smaller compared to Caucasians, the studies did not take body mass index or size into consideration [107]. There was no difference in the incidence of hip fracture in Asian Indians and Caucasians living in England [108].

3. Bone Histomorphometry

Biopsies of the iliac crest in Asian Indians showed increases in osteoid volume in patients with clinical and biochemical changes of osteomalacia [94]. However, only static histomorphometric measurements were performed. Dynamic measurements after double-tetracycline labeling also should be performed to demonstrate lack of mineralization, an essential requirement for the diagnosis of osteomalacia.

C. Chinese

Infantile rickets frequently occurs in China, and the incidence is highest in infants age 2 to 4 months and in breast-fed infants [109,110]. It sometimes occurs in infants who have a normal serum 25OHD, indicating that other factors may play a role in the pathogenesis of the bone disease in these infants [111]. However, normal serum 25OHD values may be found in rachitic patients who had received a dose of vitamin D or have had significant exposure to sunlight before blood sampling. As expected, serum 25OHD is higher in spring, summer, and fall than in winter and is higher in maternal than cord blood. There is a high correlation between concentrations of serum 25OHD in cord and maternal blood [111]. In a prospective study in two northern and two southern Chinese cities, X rays, cord blood, and 6-month blood samples from term infants given 100, 200, or 400 IU vitamin D daily were obtained [112]. Half of them were studied in the fall and the other half in the spring. It was found that serum 25OHD from cord blood was lower in the north than in the south; wrist ossification was less likely to be present in the northern than in the southern infants and was more likely to be present in infants born in the fall who had a higher serum 25OHD. Serum 25OHD was lower in infants at 6 months of age and was higher with increasing doses of $25OHD_3$. None of the children had rickets at 6 months of age [112]. In another study in breast-fed infants, it was found that although increased facial exposure to sunshine could increase serum 25OHD in some infants, supplementation with vitamin D is the only means to prevent vitamin D-deficient rickets in the general population [113].

D. Hispanics

Compared to Caucasians, Mexican-American men and women have decreases in serum vitamin D and 25OHD and increases in serum $1,25(OH)_2D$ and serum immunoreactive PTH [114]. However, unlike blacks [29], no differences were found in urinary calcium, osteocalcin, and urinary cyclic adenosine 3′,5′-monophosphate. Reduction in serum vitamin D and 25OHD in Mexican-Americans is attributed to increased skin pigment. The incidence of fractures of the lumbar spine and hip is lower in Mexican-Americans than in Caucasians [115,116].

E. Polynesians

In Polynesians, serum 25OHD is reduced compared to Caucasians, but reduction is not associated with changes in serum immunoreactive PTH, $1,25(OH)_2D$, osteocalcin, alkaline phosphatase, or calcitonin, or urinary calcium/creatinine and urinary hydroxyproline/creatinine ratios [117]. Reduction in serum 25OHD is attributed to increased skin pigment. Bone mineral density of the forearm, lumbar spine, and femoral neck is higher in Polynesian than in Caucasian women [117,118]. The increases persist after adjustment for body mass index [117].

F. Saudi Arabians

In Saudi Arabia, serum 25OHD is often low in Saudi women because of lack of exposure to sunlight since women are veiled and stay inside [119,120]. Serum 25OHD is very low and vitamin D deficiency is very common in Saudi women who are pregnant, and serum 25OHD is low in cord blood [120]. Serum 25OHD also is low in Saudi men as well as in Jordanian, Egyptian, and other men living in Saudi Arabia. This is because of consumption of diets that are inadequate in vitamin D and avoidance of exposure to sunlight by remaining indoors [121]. Serum 25OHD is lower in elderly Saudi men and women with fractures of the femoral neck than in age-matched controls and is associated with avoidance of exposure to sunlight [122]. Thus, vitamin D deficiency is very common in Saudis of all ages and can cause rickets and osteoporosis with fractures.

G. Other Groups

Reductions in serum 25OHD and $24,25(OH)_2D$, increases in serum $1,25(OH)_2D$, and biochemical and clinical rickets occur in Libyan infants who are breast fed. The mechanism is a consequence of inadequate maternal exposure to sunlight and inadequate maternal intake of vitamin D [123]. Compared to Caucasians, reductions in serum 25OHD and increases in serum immunoreactive PTH occur in children of Turkish and Moroccan immigrants to the Netherlands who have dark skin [124]. In contrast, dark-skinned Bedouins of the Negev Desert of Israel do not have decreased serum 25OHD compared to Jewish men and women, presumably because of the longer duration of exposure to sunshine and greater intensity of sunlight [125].

As in Chinese infants, the concentration of serum 25OHD in similar populations can vary as a consequence of geographical location. For example, serum values are low in subjects in southern Argentinian (Ushuaia) and are higher in subjects in northern Argentina (Buenos Aires) [126].

III. EFFECTS OF DIET

Decreases in serum 25OHD and increases in serum immunoreactive PTH were found in Finland in Caucasian subjects who were strict vegetarians with reduced intake of vitamin D [105]. Serum 25OHD was reduced, and there was a high prevalence of rickets in infants on macrobiotic diets (unpolished rice, pulses, vegetables with high fiber content, seaweeds, fermented foods, nuts, seeds, and fruits) in the Netherlands [127]. Thus, the vitamin D content of strict vegetarian and macrobiotic diets is inadequate, and the diets need to be fortified to prevent vitamin D deficiency.

IV. EFFECTS OF BODY HABITUS

A. Vitamin D Endocrine System

Vitamin D metabolism is altered by obesity. Obese individuals have decreased serum 25OHD values as a result of low serum vitamin D [128–133]. Reduction of serum vitamin D presumably occurs because the vitamin is fat soluble and is stored in fat [134,135]. As a consequence of the low serum 25OHD, there are increases in serum immunoreactive PTH, serum $1,25(OH)_2D$, and urinary cyclic adenosine 3′,5′-monophosphate and decreases in urinary calcium [131,133,136]. Decreases in serum 25OHD and secondary hyperparathyroidism in obese subjects are corrected by weight loss [128,137]. Further, alteration of the vitamin D endocrine system in obese subjects is corrected by administration of $25OHD_3$ [133]. There are increases in serum 25OHD and urinary calcium and decreases in serum $1,25(OH)_2D$ and urinary cyclic adenosine 3′,5′-monophosphate. After treatment, the changes return to baseline values [133]. Some degree of vitamin D depletion apparently is required for these changes to occur, as administration of $25OHD_3$ to nonobese subjects increases serum 25OHD, decreases serum $1,25(OH)_2D$, and does not alter urinary calcium [133].

As noted already, blacks living in the northern hemisphere show alteration of the vitamin D endocrine system as a consequence of increased skin pigment so that dermal production of vitamin D_3 is impaired [27]. However, obesity in blacks does not result in diminished circulating vitamin D, the serum vitamin D may be reduced in blacks as a consequence of diminished dermal production.

B. Bone Mineral Density

Bone mineral density of the lumbar spine and hip is higher in obese than nonobese women, both black and white [57,138]. In postmenopausal women, multivariate analysis showed that body weight is a major determinant of bone mineral density of the lumbar spine, trochanter, and femoral neck [140,141]. Further, radiographic measurements of the metacarpal cortical area demonstrated that skeletal mass is greater in obese than nonobese individuals [139]. Other studies show that fat mass is an important determinant of total body bone mineral density in premenopausal and postmenopausal women [57,140,141].

C. Bone Histomorphometry

Despite the low serum 25OHD and mild secondary hyperparathyroidism that occurs in obese subjects, bone histomorphometry is usually normal. In a study in 24 grossly obese individuals in whom serum 25OHD was reduced, one patient had mild osteomalacia and secondary hyperparathyroidism and a second patient had evidence of increased bone turnover [142].

Partial or total biliopancreatic bypass is sometimes used to treat morbid obesity. In 41 patients in which bone histomorphometric analysis was performed 1 to 5 years after this procedure, all had a normal serum 25OHD and 30 of them (73%) had impaired bone mineralization, diminished bone formation, and increased bone resorption [143]. Thus, histomorphometric changes of osteomalacia are common after biliopancreatic bypass for treatment of obesity despite the absence of vitamin D deficiency. The cause of bone disease under these circumstances is not known.

V. EFFECTS OF EXERCISE

In adult and adolescent males, weight lifting and other weight-bearing exercises resulted in increased bone mass of the lower extremities [144], lumbar spine, trochanter, and femoral neck [144–147]. Exercise was found to increase total body calcium and bone mineral density of the lumbar spine in older individuals [148–150], and muscle strength of the back correlated with bone mineral density of the lumbar spine and midradius in elderly men [150]. In college women, bone mineral density of the lumbar spine was increased in tennis players but not in swimmers [151], indicating the necessity that the exercise be weight bearing.

In men who regularly engage in muscle-building exercise for at least 1 year, serum osteocalcin, an index of bone formation, serum $1,25(OH)_2D$, and urinary cyclic adenosine 3′,5′-monophosphate were higher than in men who were sedentary [152]. Further, serum $1,25(OH)_2D$ did not correlate with serum osteocalcin. The results were interpreted to mean that exercise may increase bone formation and that increases in serum $1,25(OH)_2D$ may provide the means for increased calcium absorption and increased bone mineral density.

VI. SUMMARY

In summary, compared to Caucasians, vitamin D metabolism in blacks, Asian Indians and Pakistanis, and other dark-skinned peoples is altered when they live away from the equator because adequate dermal production of vitamin D_3 is prevented by increased skin pigment. As a consequence, reductions in serum 25OHD and urinary calcium and secondary hyperparathyroidism occur. Further, vitamin D-deficiency rickets sometimes occurs in infants of blacks, Asian Indians, and Chinese, and vitamin D-deficiency rickets and osteomalacia often occur in Asian Indians and Pakistanis without vitamin D supplementation. Vitamin D deficiency can contribute to the development of osteoporosis and fractures in Saudi Arabians, as it does in Caucasians (see Chapter 43). Hispanic men and women with increased skin pigment have reductions in serum vitamin D and 25OHD, but no increased incidence of vitamin D-deficiency rickets has been reported. Serum 25OHD is reduced in Polynesians as a consequence of increased skin pigment, but there is no alteration of the vitamin D endocrine system. Macrobiotic and vegetarian diets are deficient in vitamin D, and infants on macrobiotic diets are prone to develop rickets. In obese men and women, serum vitamin D is markedly reduced, probably because vitamin D is fat soluble and is stored in fat. As a consequence, serum 25OHD and urinary calcium are reduced and secondary hyperparathyroidism occurs. However, skeletal changes of osteomalacia are only rarely present, and osteomalacia is not a clinical problem. Thus, despite knowledge of the necessity for supplementary dietary vitamin D and adequate exposure to sunshine, particularly during pregnancy, childhood, and adolescence, vitamin D deficiency commonly occurs in people from many races throughout the world in both developed and undeveloped countries, even those near the equator.

Weight lifting and muscle-building exercise are associated with increases in serum $1,25(OH)_2D$, serum osteocalcin (an index of bone formation), urinary cyclic adenosine 3′,5′-monophosphate (an index of PTH secretion), and bone density. Thus, vigorous exercise may lead to increases in serum $1,25(OH)_2D$, calcium absorption, and increased bone mass and may help to prevent osteoporosis.

References

1. Holick MF, MacLaughlin JA, Doppelt SH 1981 Regulation of cutaneous previtamin D photosynthesis in man: Skin pigment is not an essential regulator. Science **211**:590–593.
2. Holick MF 1981 The cutaneous synthesis of previtamin D_3: A unique photo-endocrine system. J Invest Dermatol **77**:51–58.
3. Ponchon C, DeLuca HF 1969 The role of the liver in the metabolism of vitamin D. J Clin Invest **48**:1273–1279.
4. Masumoto O, Ohyama Y, Okuda K 1988 Purification and characterization of vitamin D_3 25-hydroxylase from rat liver. J Biol Chem **263**:14256–14260.
5. Garabedian M, Holick MF, DeLuca HF, Boyle IT 1972 Control of 25-hydroxycholecalciferol metabolism by parathyroid glands. Proc Natl Acad Sci USA **69**:1673–1676.
6. Henry HL 1989 Parathyroid hormone modulation of 25-hydrovitamin D_3 metabolism by cultured chick kidney cells is mimicked and enhanced by forskolin. Endocrinology **116**:503–510.
7. Horiuchi N, Suda T, Takahashi H, Shimazawa E, Ogata E 1977 In vivo evidence for the intermediary role of 3′,5′-cyclic AMP in parathyroid hormone-mediated stimulation of 1,25-dihydroxyvitamin D_3 synthesis in rats. Endocrinology **101**:969–974.
8. Caverzasio J, Montessuit C, Bonjour JP 1990 Stimulatory effect of insulin-like growth factor-I on renal Pi transport and plasma 1,25-dihydroxyvitamin D_3. Endocrinology **127**:453–459.
9. Nesbitt T, Drezner MK 1993 Insulin-like growth factor-I regulation of renal 25-hydroxyvitamin D-1-hydroxylase activity. Endocrinology **132**:133–138.
10. Boyle IT, Graw RW, DeLuca HF 1971 Regulation by calcium of in vitro synthesis of 1,25-dihydroxycholecalciferol. Proc Natl Acad Sci USA **68**:2131–2134.
11. Bushinsky DA, Riera G, Mavus MJ, Coe FL 1985 Evidence that blood ionized calcium can regulate serum $1,25(OH)_2D_3$ independently of PTH and phosphorus in the rat. J Clin Invest **76**:1599–1604.
12. Tanaka Y, DeLuca HF 1973 The control of 25-hydroxyvitamin D metabolism by inorganic phosphorus. Arch Biochem Biophys **154**:566–574.
13. Portale AA, Halloran BP, Murphy MM, Morris RC Jr 1986 Oral intake of phosphorus can determine the serum concentration of 1,25-dihydroxyvitamin D by determining its production rate in humans. J Clin Invest **77**:7–12.
14. Henry HL 1979 Regulation of the hydroxylation of 25-hydroxyvitamin D_3 in vivo and in primary cultures of chick kidney cells. J Biol Chem **254**:2722–2729.
15. Trechsel U, Bonjour J-P, Fleisch H 1978 Regulation of the metabolism of 25-hydroxyvitamin D in primary cultures of chick kidney cells. J Clin Invest **64**:206–217.
16. Henry HL 1992 Vitamin D hydroxylases. J Cell Biochem **49**:4–9.
17. Halloran BP, Bikle DD, Levens MJ, Castro ME, Globu RK, Holton E 1986 Chronic 1,25-dihydroxyvitamin D_3 administration in the rat reduces the serum concentration of 25-hydroxyvitamin D by increasing metabolic clearance rate. J Clin Invest **78**: 622–628.
18. Bell NH, Shaw S, Turner RT 1984 Evidence that 1,25-dihydroxyvitamin D_3 inhibits the hepatic production of 25-hydroxyvitamin D in man. J Clin Invest **74**:1540–1544.
19. McSheely PM, Bibby NJ 1985 Osteoblastic cells mediate osteoclastic responsiveness to parathyroid hormone. Endocrinology **118**:824–828.
20. Agus ZS, Gardner LB, Beck LH, Goldberg M 1973 Effects of parathyroid hormone on renal tubular reabsorption of calcium, sodium and phosphate. Am J Physiol **224**:1143–1148.
21. Favus MJ 1985 Factors that influence absorption and secretion of calcium in the small intestine and colon. Am J Physiol **248**:G147–G157.
22. Bikle DD, Zolock DT, Munson S 1984 Differential response of duodenal epithelial cells to 1,25-dihydroxyvitamin D_3 according to position in the villus: A comparison of calcium uptake, calcium-binding protein, and alkaline phosphatase activity. Endocrinology **115**:2077–2084.
23. Brown EM, Gamba G, Riccardi D, Lombardi M, Butters R, Kifor O, Sun A, Hediger MA, Lytton J, Hebert SC 1993 Cloning and characterization of an extracellular Ca^{2+}-sensing receptor from bovine parathyroid. Nature **366**:575–580.
24. Naveh-Many T, Friedlaender MM, Mayer H, Silver J 1989 Calcium regulates parathyroid hormone messenger ribonucleic acid (mRNA), but not calcitonin mRNA in vivo in the rat. Dominant role of 1,25-dihydroxyvitamin D. Endocrinology **125**:275–280.
25. Silver J, Russell J, Sherwood LM 1985 Regulation by vitamin D metabolites of messenger ribonucleic acid for preproparathyroid hormone in isolated bovine parathyroid cells. Proc Natl Acad Sci USA **82**:4270–4273.
26. Naveh-Many T, Marx R, Keshet E, Pike JW, Silver J 1990 Regulation of 1,25-dihydroxyvitamin D_3 receptor gene expression by 1,25-dihydroxyvitamin D_3 in the parathyroid in vivo. J Clin Invest **86**:1968–1975.
27. M'Buyamba-Kabangu JR, Fagard R, Lijnen P, Bouillon R, Lissens W, Amery A 1987 Calcium, vitamin D-endocrine system

and parathyroid hormone in black and white males. Calcif Tissue Int **41**:70–74.

28. Clemens TL, Henderson SL, Adams JS, Holick MF 1982 Increased skin pigment reduces the capacity of skin to synthesize vitamin D_3. Lancet **1**:74–76.
29. Bell NH, Greene A, Epstein S, Oexmann MJ, Shaw S, Shary J 1985 Evidence for alteration of the vitamin D-endocrine system in blacks. J Clin Invest **76**:470–473.
30. Hollis BW, Pittard III WB 1984 Evaluation of the total fetomaternal vitamin D relationships at term: Evidence for racial differences. J Clin Endocrinol Metab **59**:652–657.
31. Luckey MM, Meier DE, Clemens TL 1986 PTH and vitamin D metabolites in premenopausal white and black women. J Bone Miner Res **1**(Suppl. 1):381.
32. Meier DE, Luckey MM, Wallenstein S, Clemens TL, Orwoll ES, Waslien CI 1991 Calcium, vitamin D and parathyroid hormone status in young white and black women: Association with racial differences in bone mass. J Clin Endocrinol Metab **72**:703–710.
33. Brickman AS, Nydy MD, Griffiths RF, von Hungen K, Tuck ML 1993 Racial differences in platelet cytosolic calcium and calciotropic hormones in normotensive subjects. Am J Hypertens **6**:46–51.
34. Dawson-Hughes B, Harris S, Kramich C, Dallal G, Rasmussen HM 1993 Calcium retention and hormone levels in black and white women on high and low calcium diets. J Bone Miner **8**:779–787.
35. Katz BS, Jackson GJ, Hollis BW, Bell NH 1992 Diagnostic criteria of vitamin D deficiency. Endocrinologist **3**:248–253.
36. Bell NH, Yergey AL, Vieira NE, Oexmann MJ, Shary JR 1993 Demonstration of a difference in urinary calcium, not calcium absorption, in black and white adolescents. J Bone Miner Res **8**:1111–1115.
37. Fuleihan GE-H, Gundberg CM, Gleason R, Brown EM, Stromski ME, Grant FD, Conlin PR 1994 Racial differences in parathyroid hormone dynamics. J Clin Endocrinol Metab **79**:1642–1647.
38. Kleerekoper M, Nelson DA, Peterson EL, Flynn MJ, Pawluszka AS, Jacobsen G, Wilson P 1994 Reference data for bone mass, calciotropic hormones, and biochemical markers of bone remodeling in older (55–75) postmenopausal white and black women. J Bone Miner Res **8**:1267–1276.
39. Bell NH 1995 25-Hydroxyvitamin D_3 reverses alteration of the vitamin D-endocrine system in blacks. Am J Med **99**:597–599.
40. Abrams SA, O'Brien KO, Liang LK, Stuff JE 1995 Differences in calcium absorption and kinetics between black and white girls aged 5–16 years. J Bone Miner Res **10**:829–833.
41. Katz BS, Jackson GJ, Hollis BW, Bell NH 1993 Diagnostic criteria of vitamin D deficiency. Endocrinologist **3**:248–253.
42. Modlin M 1967 Urinary calcium in normal adults and in patients with renal stones: An interracial study. Invest Urol **5**:49–57.
43. Ghandur-Mnaymneh L, Cassady J, Hajanpour MA, Reiss E 1986 The parathyroid gland in health and disease. Am J Pathol **125**:292–299.
44. Awumey E, Hollis BW, Bell NH 1996 Low serum 25-hydroxyvitamin D in blacks results from decreased production rate and not increased metabolic clearance rate. J Bone Miner Res **11**:S165 (abstract).
45. Bachrach S, Fisher J, Parks JS 1980 An outbreak of vitamin D deficiency rickets in a susceptible population. Pediatrics **64**:277–283.
46. Edidin DV, Levitsky LL, Schey W, Dumbovic M, Campos A 1980 Resurgence of nutritional rickets associated with breastfeeding and special dietary practices. Pediatrics **65**:232–235.
47. Kruger DM, Lyne ED, Kleerekoper M 1987 Vitamin D deficiency rickets. A report of three cases. Clin Orthop **224**:277–283.
48. Key LL 1992 Vitamin D deficiency rickets. Trends Endocrinol Metab **2**:81–85.
49. Chang YT, Germain-Lee EL, Doran TF, Migeon CJ, Levine MA, Berkovitz GD 1992 Hypocalcemia in nonwhite breast-fed infants. Clin Pediat **31**:695–698.
50. Hollis BW, Roos BA, Draper HH, Lambert PW 1981 Vitamin D and its metabolites in human and bovine milk. J Nutr **111**:1240–1248.
51. Pittard WB, Geddes KM, Husley TC, Hollis BW 1991 How much vitamin D for neonates? Am J Dis Child **145**:1147–1149.
52. Li J-Y, Specker BL, Ho ML, Tsang RC 1989 Bone mineral content in black and white children 1 to 6 years of age. Early appearance of race and sex differences. Am J Dis Child **143**:1346–1349.
53. Bell NH, Shary J, Stevens J, Garza M, Gordon L, Edwards J 1991 Demonstration that bone mass is greater in black than in white children. J Bone Miner Res **6**:719–723.
54. Cohn SH, Abesamis C, Yasumara S, Aloia JF, Zanai I, Ellis KJ 1977 Comparative skeletal mass and radial bone mineral content in black and white women. Metabolism **26**:171–178.
55. Liel Y, Edwards J, Shary J, Spicer KM, Gordon L, Bell NH 1988 The effects of race and body habitus on bone mineral density of the radius, hip and spine in premenopausal women. J Clin Endocrinol Metab **66**:1247–1250.
56. Meier De, Luckey MM, Wallenstein S, Lapinski RH, Catherwood B 1992 Racial differences in pre- and postmenopausal bone homeostasis: Association with bone density. J Bone Miner Res **7**:1181–1189.
57. DeSimone DP, Stevens J, Edwards J, Shary J, Gordon L, Bell NH 1989 Influence of body habitus and race on bone mineral density of the midradius, hip and spine in aging women. J Bone Miner Res **4**:827–830.
58. Wright NM, Papadea N, Willi S, Veldhuis JD, Pandey JP, Key LL, Bell NH 1996 Demonstration of a lack of racial difference in secretion of growth hormone despite a racial difference in bone mineral density in premenopausal women—A clinical research center study. J Clin Endocrinol Metab **81**:1023–1026.
59. Bell NH, Gordon L, Stevens J, Shary J 1995 Demonstration that bone mineral density of the lumbar spine, trochanter and femoral neck is higher in black than in white young men. Calcif Tissue Int **56**:11–13.
60. Wright NM, Renault J, Willi S, Veldhuis JD, Pandey JP, Gordon L, Key LL, Bell NH 1995 Greater secretion of growth hormone in black than in white men: Possible factor in greater bone mineral density—A clinical research center study. J Clin Endocrinol Metab **80**:2291–2297.
61. Nelson DA, Jacobsen G, Barondess DA, Parfitt AM 1995 Ethnic differences in regional bone density, hip axis length, and lifestyle variables among healthy black and white men. J Bone Miner Res **10**:782–787.
62. Weinstein RS, Bell NH 1988 Diminished rates of bone formation in normal black adults. N Engl J Med **319**:1698–1701.
63. Gyepes M, Melliaz HZ, Katz I 1962 The low incidence of fracture of the hip in the Negro. JAMA **181**:1073–1074.
64. Bollet AJ, Engh G, Parson W 1965 Epidemiology of osteoporosis. Arch Intern Med **116**:191–194.
65. Farmer ME, White LR, Brody JA, Bailey KR 1984 Race and sex differences in hip fracture incidence. Am J Public Health **74**:1374–1380.
66. Silverman SL, Madison RE 1988 Decreased incidence of hip fracture in hispanics, Asians and Blacks: California hospital discharge data. Am J Public Health **78**:1482–1483.

67. Kellie SE, Brody JA 1990 Sex-specific and race-specific hip fracture rates. Am J Public Health **80**:326–328.
68. Griffin MR, Ray WA, Fought RL, Melton III LJ 1992 Black–white differences in fracture rates. Am J Epidemiol **136**:1378–1385.
69. Pruzansky ME, Turano M, Luckey M, Senie R 1989 Low body weight is a risk factor for hip fracture in both black and white women. J Orthop Res **7**:192–197.
70. Grisso AJ, Kelsey JL, Strom BL, O'Brien LA, Maislin G, La Pann K, Samelson L, Hoffman S 1994 Risk factors for hip fracture in black women. The Northeast Hip Fracture Study Group. N Engl J Med **330**:1555–1559.
71. Cummings SR, Cauley JA, Palermo L, Ross PD, Wasnich RD, Black D, Faulkner KG 1994 Racial differences in hip axis lengths might explain racial differences in rates of hip fracture. Osteoporosis Int **4**:226–229.
72. Solomon L 1968 Osteoporosis and fracture of the femoral neck in the South African Bantu. J Bone Joint Surg **50**:2–5.
73. Solomon L 1979 Bone density in aging Caucasian and African populations. Lancet **3**:1326–1329.
74. Patel DN, Pettifor JM, Becker PJ, Grieve C, Leschner K 1993 The effect of ethnicity on appendicular bone mass in white, coloured and Indian schoolchildren. S Afr Med J **83**:847–853.
75. Daniels ED, Pettifor JM, Schnitzler CM, Russel SW, Paten DN 1995 Ethnic differences in bone density in female South African nurses. J Bone Miner Res **10**:359–367.
76. Jackson G, Hollis BW, Eyre DR, Baylink DJ, Bell NH 1994 Effects of race and calcium intake on bone markers and calcium metabolism in young adult men. J Bone Miner Res **9**:S185 (abstract).
77. Eriksen EF, Mosekilde L, Melsen F 1986 Kinetics of trabecular bone resorption and formation in hypothyroidism: Evidence for a positive balance per remodeling cycle. Bone **7**:101–108.
78. Chestnut CH, McClung MR, Ensrud KE, Bell NH, Genant HK, Harris ST, Singer FR, Stock JL, Yood RA, Delmas PD, Pryor-Tillotson S, Santora AC 1995 Alendronate treatment of the postmenopausal osteoporotic woman: Effect of multiple dosages on bone mass and bone remodeling. Am J Med **99**:144–152.
79. Liberman UA, Weiss SR, Broll J, Minne HW, Quan H, Bell NH, Rodriquez-Portales J, Downs RW, Dequecker J, Favus M, Capizzi T, Santora II AC, Lombardi A, Shah RV, Hirsch LJ, Karpf DB 1995 Effect of three years treatment with oral alendronate on fracture incidence in women with postmenopausal osteoporosis. N Engl J Med **33**:1437–1443.
80. Schnitzler CM, Pettifor JM, Mesqita JM, Bird MD, Schnaid E, Smyth AE 1990 Histomorphometry of iliac crest bone in 346 normal black and white South African adults. Bone Miner **10**:183–199.
81. Preece MA, Ford JA, McIntosh WB, Dunnigan MG, Tomlinson S, O'Riordan JLH 1973 Vitamin D deficiency among Asian immigrants to Britain. Lancet **1**:907–910.
82. Dent CE, Gupta MM 1975 Plasma 25-hydroxyvitamin-D levels during pregnancy in Caucasians and in vegetarian and non-vegetarian Asians. Lancet **2**:1057–1060.
83. Hunt SP, O'Riordan JLH, Windo J, Truswell AS 1976 Vitamin D status in different sub-groups of British Asians. Br Med J **2**:1351–1354.
84. Heckmatt JZ, Peacock M, Davies AEJ, McMurray J, Isherwood DM 1979 Plasma 25-hydroxyvitamin D in pregnant Asian women and their babies. Lancet **2**:546–548.
85. Brooke OG, Brown IRF, Cleeve HJW, Sood A 1981 Observations on the vitamin D state of pregnant Asian women in London. Br J Obstet Gynacol **88**:18–26.
86. Ellis G, Woodhead JS, Cooke WT 1977 Serum 25-hydroxyvitamin-D concentrations in adolescent boys. Lancet **1**:825–828.
87. O'Hare AE, Uttley WS, Belton NR, Westwood A, Levin SD, Anderson F 1984 Persisting vitamin D deficiency in the Asian adolescent. Arch Dis Child **59**:766–770.
88. Ford JA, Davidson DC, McIntosh WB, Fyfe WM, Dunnigan MG 1973 Neonatal rickets in Asian immigrant population. Br Med J **3**:211–212.
89. Arneil GC, Crosbie JC 1963 Infantile rickets returns to Glasgow. Lancet **2**:423–425.
90. Goel KM, Sweet EM, Logan RW, Warren JM, Arneil GC, Shanks RA 1976 Florid and subclinical rickets among immigrant children in Glasgow. Lancet **1**:1141–1145.
91. Henderson JB, Dunnigan MG, McIntosh WB, Abdul-Motaal AA, Gettinby G, Glekin BM 1987 The importance of limited exposure to UV radiation and dietary factors in the aetiology of Asian rickets: A risk factor model. Q J Med **63**:413–425.
92. Dent CE, Rowe DJF, Round JM, Stamp TCB 1973 Effect of chapattis and ultraviolet irradiation on nutritional rickets in an Indian immigrant. Lancet **1**:1282–1284.
93. Wills MR, Day RC, Phillips JB, Bateman EC 1972 Phytic acid and nutritional rickets in immigrants. Lancet **1**:771–773.
94. Holmes AM, Enoch BA, Taylor JL, Jones ME 1973 Occult rickets and osteomalacia amongst the Asian immigrant population. Q J Med **42**:125–149.
95. Ford JA, Colhoun EM, McIntosh WB, Dunnigan MG 1972 Rickets and osteomalacia in the Glasgow Pakistani community, 1961–71. Br Med J **1**:677–679.
96. Preece MA, Tomlinson S, Ribot CA, Pietrek J, Korn HT, Davies DM, Ford JA, Dunnigan MG, O'Riordan JLH 1975 Studies of vitamin D deficiency in man. Q J Med **44**:575–589.
97. Finch PJ, Ang L, Colston KW, Nisbet J, Maxwell JD 1992 Blunted seasonal variation in serum 25-hydroxyvitamin D and increased risk of osteomalacia in vegetarian London Asians. Eur J Clin Nutr **46**:509–515.
98. Henderson JB, Dunnigan MG, McIntosh WB, Motaal AA, Hole D 1990 Asian osteomalacia is determined by dietary factors when exposure to ultraviolet radiation is restricted: A risk factor model. Q J Med **76**:923–933.
99. Ashby JP, Newman DJ, Rinsler MG 1989 Is intact PTH a sensitive biochemical market of deranged calcium homeostasis in vitamin D deficiency? Ann Clin Biochem **26**:24–327.
100. Pietrek J, Preece MA, Windo J, O'Riordan J, Dunnigan MG, McIntosh WB, Ford JA 1976 Prevention of vitamin-D deficiency in Asians. Lancet **1**:1145–1148.
101. Dunnigan MG, McIntosh WB, Sutherland GR, Gardee R, Glekin B, Ford JA, Robertson I 1981 Policy for prevention of Asian rickets in Britain: A preliminary assessment of the Glasgow rickets campaign. Br Med J **1**:357–360.
102. Mitra DA, Hollis BW, Bell NH 1996 Demonstration that the vitamin D-endocrine system of Asian Indians in South Carolina is altered. J Bone Miner Res **11**:S255 (abstract).
103. Lo CW, Paris PW, Holick MF 1986 Indian and Pakistani immigrants have the same capacity as Caucasians to produce vitamin D in response to ultraviolet irradiation. Am J Clin Nutr **44**:683–685.
104. Rashid A, Mohammed T, Stephens WP, Warrington S, Berry JL, Mawer EB 1983 Vitamin D state of Asians living in Pakistan. Br Med J **1**:182–184.
105. Lamberg B, Allardt C, Karkkainen M, Seppanen R, Bistrom H 1993 Low serum 25-hydroxyvitamin D concentrations and secondary hyperparathyroidism in middle-aged strict vegetarians. Am J Clin Nutr **58**:684–689.
106. Ahmed I, Atiq M, Iqbal J, Khurshid M, Whittaker P 1995 Vitamin

D deficiency rickets in breast-fed infants presenting with hypocalcemic seizures. Acta Paediatr **84**:941–942.

107. Cundy C, Cornish J, Evans MC, Gamble G, Stapleton J, Reid Ir 1995 Sources of interracial variation in bone mineral density. J Bone Miner Res **10**:368–373.
108. Parker M, Anand JK, Myles JW, Lodwick R 1992 Proximal femoral fractures: Prevalance in different racial groups. Eur J Epidemiol **8**:730–732.
109. Fong JC, Huong YM, Shen KY, Dewen PL 1980 Prevention and treatment of rickets (report). Zhengzhou, Henan, People's Republic of China: Henan College of Medicine publication (Chinese) 1–10.
110. Xiao L 1982 Survey of infant rickets. J Chinese Pediatr (Chinese) **20**:63–64.
111. Danhui Z, Quingbing Z, Yan X 1990 Serum 25OHD levels in maternal and chord blood in Beijing, China. Acta Paediatr Scand **79**:1240–1241.
112. Specker BL, Ho ML, Oestreich A, Yin T-A, Shui Q-M, Chen X-C, Tsang RC 1992 Prospective study of vitamin D supplementation and rickets in China. J Pediatr **120**:733–739.
113. Specker BL, Valanis B, Hertzberg V, Edwards N, Tsang RC 1985 Sunshine exposure and serum 25-hydroxyvitamin D concentrations in breast-fed infants in Beijing, China. J Pediatr **107**:928–931.
114. Reasner III CA, Dunn JF, Fetchik DA, Leil Y, Hollis BW, Epstein S, Shary J, Mundy GR, Bell NH 1990 Alteration of vitamin D metabolism in Mexican-Americans. J Bone Miner Res **5**:13–17.
115. Bauer RL, Diehl AK, Barton SA, Brender JA, Deyo RA 1986 Risk of postmenopausal hip fracture in Mexican-American women. Am J Public Health **76**:1020–1021.
116. Bauer RL, Deyo RA 1987 Low risk vertebral fracture in Mexican-American women. Arch Intern Med **147**:1437–1439.
117. Reid IR, Cullen S, Schooler BA, Livingstone NE, Evans MC 1990 Calciotropic hormone levels in Polynesians, evidence against their role in inter-racial differences in bone mass. J Clin Endocrinol Metab **70**:1452–1456.
118. Reid IR, Mackey M, Ibbertson HK 1986 Bone mineral content in Polynesian and white New Zealand women. Br Med J **2**:1457–1458.
119. Fonseca V, Tongia R, El-Hazmi M, Abu-Aisha H 1984 Exposure to sunlight and vitamin D deficiency in Saudi Arabian women. Postgrad Med J **60**:589–591.
120. Serinius F, Elidrissy A, Dandona P 1984 Vitamin D nutrition in pregnant women at term and in newly born babies in Saudi Arabia. J Clin Pathol **37**:444–447.
121. Sedrani SH 1984 Low 25-hydroxyvitamin D and normal serum calcium concentrations in Saudi Arabia: Riyadh region. Ann Nutr Metab **28**:181–185.
122. Al-Arabi KM, Wahab A, Elidrissy TH, Sedrani SH 1984 Is avoidance of sunlight a cause of fractures of the femoral neck in elderly Saudis? Trop Geogr Med **36**:273–279.
123. Elzouki AY, Markestad T, Elgarrah M, Elhoni N, Aksnes L 1989 Serum concentrations of vitamin D metabolites in rachitic Libyan children. J Pediatr Gastroenterol Nutr **9**:507–512.
124. Meulmeester JF, van den Berg H, Wedel M, Boshuis PG, Hulshof KFAM, Luyken R 1990 Vitamin D status, parathyroid hormone and sunlight in Turkish, Moroccan and Caucasian children in The Netherlands. Eur J Clin Nutr **44**:461–470.
125. Shany S, Hirsh J, Berlyne GM 1976 25-Hydroxycholecalciferol levels in Bedouins in the Negev. Am J Clin Nutr **29**:1104–1107.
126. Ladizesky M, Lu Z, Oliveri B, San Roman N, Diaz S, Holick MF, Mautalen C 1995 Solar ultraviolet B radiation and photoproduction of vitamin D_3 in central and southern parts of Argentina. J Bone Miner Res **10**:545–548.
127. Dagnelie PC, Vergote FJVRA, van Staveren WA, van den Berg H, Dingjan PG, Hautvast JGAJ 1990 High prevalence of rickets in infants on macrobiotic diets. Am J Clin Nutr **51**:202–208.
128. Teitelbaum SL, Halverson JD, Bates M, Wise L, Haddad JG 1977 Abnormalities of circulating 25-OH vitamin D after jejunal bypass for obesity: Evidence of an adaptive response. Ann Intern Med **86**:289–293.
129. Compston JE, Vedi S, Ledger JE, Webb A, Gazet JC, Pilkington TR 1981 Vitamin D status and bone histomorphometry. Am J Clin Nutr **34**:2359–2363.
130. Rickers H, Christiansen C, Balslev I, Rodbro P 1984 Impairment of vitamin D metabolism and bone mineral content after intestinal bypass surgery. Scand J Gastroenterol **19**:184–189.
131. Bell NH, Epstein S, Greene A, Shary J, Oexmann MJ, Shaw S 1985 Evidence for alteration of the vitamin D-endocrine system in obese subjects. J Clin Invest **76**:370–373.
132. Liel Y, Ulmer E, Shary J, Hollis BW, Bell NH 1988 Low circulating vitamin D in obesity. Calcif Tissue Int **43**:199–201.
133. Bell NH, Epstein S, Shary J, Greene V, Oexmann MJ, Shaw S 1988 Evidence of a probable role for 25-hydroxyvitamin D in the regulation of calcium metabolism in man. J Bone Miner Res **3**:489–495.
134. Rosenstreich SJ, Rich C, Volwiler W 1971 Deposition in and release of vitamin D_3 from body fat: Evidence for a storage site in the rat. J Clin Invest **50**:679–687.
135. Mawer EB, Backhouse J, Holman CA, Lumb GA, Stanbury SW 1972 The distribution and storage of vitamin D and its metabolites in human tissues. Clin Sci **43**:414–431.
136. Epstein S, Bell NH, Shary J, Shaw S, Greene A, Oexmann MJ 1986 Evidence that obesity does not influence the vitamin D-endocrine system in blacks. J Bone Miner Res **1**:181–184.
137. Atkinson RL, Dahms WT, Bray GA, Schwartz AA 1978 Parathyroid hormone levels in obesity: Effects of intestinal bypass surgery. Miner Electrolyte Metab **1**:315–320.
138. Liel Y, Edwards J, Shary J, Spicer KM, Gordon L, Bell NH 1988 The effects of race and body habitus on bone mineral density of the radius, hip, and spine in premenopausal women. J Clin Endocrinol Metab **66**:1247–1250.
139. Dalen N, Hallberg D, Lamke B 1975 Bone mass in obese subjects. Acta Med Scand **197**:353–355.
140. Reid IR, Ames R, Evans MC, Sharpe S, Gamble F, France JT, Lim TMT, Cundy TF 1992 Determinants of total body and regional bone mineral density in normal postmenopausal women—A key role for fat mass. J Clin Endocrinol Metab **75**:45–51.
141. Reid IR, Plank LD, Evans MC 1992 Fat mass is an important determinant of whole body bone density in premenopausal women but not in men. J Clin Endocrinol Metab **75**:779–782.
142. Steiniche T, Vesterby A, Eriksen EF, Melsen F 1986 A histomorphometric determination of iliac bone structure and remodeling in obese subjects. Bone **7**:77–82.
143. Compston JE, Verdi S, Gianetta E, Watson G, Civalleri D, Scopinaro N 1984 Bone histomorphometry and vitamin D status after biliopancreatic bypass for obesity. Gastroenterology **87**:350–356.
144. Nilsson BE, Westlin NE 1971 Bone density in athletes. Clin Orthop Related Res **77**:179–182.
145. Granbed H, Jonson R, Hansson T 1987 The loads on the lumbar spine during extreme weight lifting. Spine **12**:146–149.
146. Colletti LA, Edwards J, Gordon L, Shary J, Bell NH 1989 The effects of muscle-building exercise on bone mineral density of the radius, spine, and hip in young men. Calcif Tissue Int **45**:12–14.

147. Conroy BP, Kraemer WJ, Maresh CM, Fleck SJ, Stone MH, Fry AC, Miller PD, Kalsky GP 1993 Bone mineral density in elite junior Olympic weightlifters. Med Sci Sports Exercise **25**:1103–1109.
148. Aloia JF, Cohn SH, Ostuni JA, Cane R, Ellis K 1978 Prevention of involutional bone loss by exercise. Ann Intern Med **89**:356–358.
149. Krolner B, Toft B, Pors Nielson KS, Tondevold E 1983 Physical exercise as prophylaxis against involutional vertebral bone loss: A controlled trial. Clin Sci **64**:541–546.
150. Bevier W, Pyka G, Kozac K, Newhall K, Wiswell R, Marcus R 1988 Aerobic capacity, muscle strength and bone density in elderly men and women. J Bone Miner Res **3**:S215.
151. Jacobson PC, Beaver W, Grubb SA, Taft TN, Talmadge RV 1984 Bone density in women: College athletes and older athletic women. J Orthop Res **2**:328–332.
152. Bell NH, Godsen RN, Henry DP, Shary J, Epstein S 1988 The effects of muscle building exercise on vitamin D and mineral metabolism. J Bone Miner Res **3**:369–373.

CHAPTER 34

Perinatal Vitamin D Actions

NICHOLAS J. BISHOP McGill University and Genetics Unit, Shriners' Hospitals for Children Canadian Unit, Montréal, Québec, Canada

BERNARD L. SALLE Department of Neonatology, Hôpital Édouard Herriot, Lyon, France

I. INTRODUCTION

The climactic transition from fetal to extrauterine life abruptly halts the provision of both mineral substrates and the placental factors that had participated in the regulation of skeletal maturation *in utero*. The response of the newborn infant to these sudden changes depends in part on the reserves of calcium, phosphate, and vitamin D laid down during pregnancy. Although the role of vitamin D in the pregnant mother is dealt with elsewhere in this volume (see Chapter 28), a summary of the clinical consequences of deficiency or excess is appropriate in considering the status of the neonate at birth and during the perinatal period. The events of the last trimester are particularly important in establishing good nutritional reserves while maintaining rapid growth, and the period covered by this review thus spans the last 3 months of pregnancy and the first month of extrauterine life for infants born at term: for those born preterm, the postnatal period covered is the first 3 months. In addition, because of the potential for metabolic bone disease arising as a result of inadequate calcium and phosphate intake in preterm infants, vitamin D metabolism in this population is considered in the light of mineral substrate provision.

II. THE LAST TRIMESTER OF PREGNANCY

A. Malnutrition

In countries where vitamin D supplementation of table milk is routine, vitamin D deficiency is unlikely to arise during pregnancy except in recent immigrants with chronic dietary insufficiency of calcium, vitamin D, and other essential nutrients, in groups avoiding dairy products for cultural or dietary reasons (e.g., cow's milk protein intolerance), and where sunlight exposure is negligible. In malnourished populations, evidence for vitamin D deficiency causing osteomalacia in the mother and abnormal skeletal metabolism in the fetus and infant is strong [1–3]. Infants of severely malnourished mothers may be born with rickets and suffer fractures in the neonatal period. Although much of the literature relating to such infants was written early in the twentieth century, recent reports from European centers indicate the need for continued vigilance, particularly in immigrant or refugee populations. Four infants were born with craniotabes to immigrant mothers with osteomalacia in Berlin [2]. The infants exhibited typical biochemical and radiological changes of rickets, which responded well to vitamin D therapy. Observational studies suggest that radiographic bone density is reduced in both malnourished mothers and their infants [3], and this reduction can be ameliorated by calcium supplementation during pregnancy.

B. Vitamin D Deficiency

Vitamin D nutrition in pregnancy was investigated by Brooke and colleagues in 115 Asian (Pakistani, Hindu Indian, and East African Asian) women living in London, and in 50 of their newborn infants [4]. Maternal serum 25-hydroxyvitamin D (25OHD) concentration at the beginning of the last trimester was 20.2 nmol/liter (8.1 ng/ml), falling to 16.0 nmol/liter (6.4 ng/ml) after delivery. Postpartum, 36% of the women and 32% of the infants had undetectable 25OHD concentrations (less than 3 nmol/liter or 1.2 ng/ml). Alkaline phosphatase bone isoenzyme was elevated (compared with appropriate well-nourished controls) in 20% of the women postpartum, and in 50% of the infants. Five infants developed symptomatic hypocalcemia.

C. Vitamin D Supplementation

In a double blind study of supplementary vitamin D (1000 IU per day) administered during the third trimester to Asian women living in London, Maxwell and colleagues showed increased maternal weight gain and a 50% reduction in the numbers of infants classified as "growth retarded" (born weighing less than 2500 g at term), which closely approached significance at the 5% level [5]. Infants in the control group also had larger fontanelles, suggesting delayed ossification. Further follow-up of the cohort was reported to age 1 year, with the observers still blinded to the original randomization. There was increasing divergence of the groups in terms of both weight and length, so that by 1 year of age the infants whose mothers had received the supplemental vitamin D during the third trimester were on average 0.4 kg heavier and 1.6 cm longer than the control group [6].

Thus, maternal malnutrition with coexisting vitamin D deficiency can result in metabolic bone disease and disturbed calcium and vitamin D metabolism in the neonate. Vitamin D supplementation in the malnourished mother results in improved growth of the child both in terms of birth weight and also subsequent linear growth during infancy. The neonatal metabolic bone disease resulting from maternal malnutrition is amenable, at least in the short term, to standard treatment with vitamin D and calcium supplements. There are no long-term data on the outcome for infants treated in this way.

In well-nourished mothers who do not receive vitamin D supplementation during pregnancy, the situation is less clear. Delvin and colleagues studied the effect of vitamin D supplementation from the end of the first trimester in a group of French mothers [7]. They reported differences (comparing supplemented with unsupplemented controls) in vitamin D metabolite levels at birth in both mothers and their infants. In addition, the postnatal fall in plasma calcium in the infants of unsupplemented mothers was more likely to be associated with symptomatic hypocalcemia. In contrast with the studies of malnourished mothers detailed above, there were no differences in maternal blood calcium or inorganic phosphate concentrations between groups, and the authors did not report any evidence of active rickets in the infants. There were no differences in birth weight or length between the groups (Delvin *et al.*, unpublished data, 1986).

Where vitamin D supplementation of dairy products is not undertaken, specific recommendations are required for pregnant mothers. In the study by Marya *et al.*, vitamin D (600,000 IU) given twice during the seventh and eighth months of pregnancy to Hindu women living in India had a greater effect on infant birth weight and cord blood levels of calcium, inorganic phosphate, and alkaline phosphatase activity than a daily supplement of 1200 IU per day throughout the third trimester; however, there were no data regarding the compliance of the group taking the daily supplement [8].

If daily vitamin D supplementation can be undertaken, this should be 400 IU per day. Where presentation for antenatal care is delayed, 1000 IU per day during the last 3 months of pregnancy should be given. If only monthly dosing is feasible, this should be 100,000 IU during each of the last 3 months of pregnancy.

III. THE NORMAL TERM INFANT

Stearns *et al.* in the 1930s showed that the rate of linear growth and weight gain in normal infants was related to vitamin D intake. Infants supplemented with 340 as opposed to 135 IU of vitamin D grew more rapidly in both weight and length [9]. Over the years, there was a tendency to increase the amount of vitamin D given to infants so that by the 1950s some infants were receiving over 2000 IU per day. Contemporaneously, a number of cases of "idiopathic hypercalcemia" were reported. Hypervitaminosis D resulted in hypercalcemia with polyuria leading to dehydration and its typical consequences, which were documented in some case reports. Of more concern, a recent report suggests that some infants receiving intermittent high dose vitamin D prophylaxis may go on to develop nephrocalcinosis [10].

Normal term infants born to vitamin D-sufficient mothers have plasma levels of total vitamin D metabolites that correlate closely with those of their mothers [11]. A number of studies have reported that total levels of 25OHD and 1,25-dihydroxyvitamin D [$1,25(OH)_2D$] are decreased in cord blood compared with maternal blood at the time of delivery [11–13]; unbound (free) metabolite levels of 25OHD, $24,25(OH)_2D$, and $25,26(OH)_2D$ have been reported as higher in infants' blood, with free $1,25(OH)_2D$ levels being equal [11]. The first report of $1,25(OH)_2D$ levels in infants born at term indicated that the initially low levels in cord blood rose to normal adult values by 24 hr of age [12]. Longitudinal measurement of vitamin D metabolites in the serum of breast-fed infants (not receiving vitamin D supplements) who were born to vitamin D-replete mothers suggested that depletion of vitamin D stores occurs within 8 weeks of delivery in the majority [11].

In 1963 the American Academy of Pediatrics recommended that, for infants, daily intakes of vitamin D be restricted to 400 IU from all sources [14]. These recommendations are still regarded as appropriate by most pediatricians, although a more recent report from the U.S. National Research Council Food and Nutrition Board suggested that the recommended dietary allowance (RDA) should be 300 IU per day to age 6 months [15] and the United Kingdom's Department of Health recommends 340 IU per day to age 6 months [16].

Sufficient vitamin D is available in normally reconstituted infant formulas to meet these recommendations. However, the vitamin D content of human milk is low. Unless the infant is exposed (face and hands) to sunlight for 10 min each day, there is a good case for providing an oral supplement of vitamin D of up to 400 IU per day. The effect of vitamin D supplementation on bone mineralization in wholly breast fed-fed infants was investigated by Greer *et al.* [17]. Infants who received 400 IU per day of vitamin D had higher bone mineral content and serum 25OHD levels at age 12 weeks than those not supplemented. The effect on growth and bone mineralization beyond this period remains unknown. There is no evidence that increasing the level of vitamin D supplementation beyond 400 IU per day influences either linear growth or bone mineral accretion.

IV. THE TERM GROWTH-RETARDED INFANT

There is no direct evidence that vitamin D metabolism is altered in the infant who is growth-retarded (small for gestational age) due to uteroplacental factors rather than maternal malnutrition. Reduced bone mineral content and reduced serum $1,25(OH)_2D$ and osteocalcin levels have been documented for growth retarded as opposed to term infants, but no difference in 25OHD status was recorded [18]. The authors suggested that reduced uteroplacental blood flow resulted in reduced fetal production of $1,25(OH)_2D$ and hence lower osteocalcin and reduced bone mineral accretion, but reduced transfer of all nutrients including minerals is also likely to have contributed.

V. THE PREMATURE INFANT

A. Early Neonatal Hypocalcemia

Early neonatal hypocalcemia is a common event occurring in up to 75% preterm infants, chiefly those born with very low birth weight (under 1500 g) [19]. It is usually of short duration and does not express itself clinically in the majority of infants. Immaturity of the vitamin D activation pathway has been suggested as a major underlying factor either alone or in combination with other abnormalities, particularly transient hypoparathyroidism, hypercalcitoninemia, and end-organ resistance to hormonal effects [20]. However, it has been clearly shown that there is an appropriate secretion of parathyroid hormone (PTH) in response to this hypocalcemic stimulus [21,22]. This increase in serum immunoreactive PTH concentration appeared within the initial 24 hr after birth, with levels of both intact PTH (1–84) and the carboxyl-terminal fragment (cPTH) following the same trend [23]. This physiological response to a hypocalcemic stimulus is substantiated by the observation that the increment in immunoreactive PTH levels was blunted when premature infants received calcium infusion; this calcium load buffered the postnatal depression of serum calcium. By day 10 serum levels of PTH (1–84) and cPTH return to euparathyroid values [23,24].

B. Vitamin D in the Neonate

In preterm as in full-term newborns, both total and free 25OHD cord blood levels were lower than in maternal blood and were correlated to those of the mothers [25–27]. Bouillon *et al.* [28] reported a positive correlation between maternal and cord serum concentrations of both total and free $1,25(OH)_2D$ in premature babies; others found that only those of free $1,25(OH)_2D$ were correlated [25]. This discrepancy could be due to the vitamin D depletion state of the subjects studied [23]. Cord and maternal blood vitamin D binding protein (DBP) levels were also positively related. Hirsfeld and Lunell [29] have excluded the possibility of a placental transfer of this protein by DBP polymorphism analysis; the most likely explanation for this fetomaternal relationship would therefore be common fetal and maternal factors affecting its synthesis.

Longitudinal trends in total and free 25OHD and $1,25(OH)_2D$ estimates for preterm infants in the first month of life are shown in Figs. 1 and 2. Many reports have clearly shown that in premature infants, after 28 weeks of gestation, activation of vitamin D is operative as early as 24 hr after birth [21,30–33]. In European countries where dairy products are not enriched with vitamin D, average levels of 25OHD in cord blood are lower than those in North America [21]. Vitamin D supplementation (from 500 to 2000 IU/day) in French preterm infants just after birth improved vitamin D nutritional status as evidenced by rising plasma 25OHD levels. In addition, the administration of vitamin D resulted in an increase in the circulating concentration of $1,25(OH)_2D$ (Fig. 1). By 5 days of age, the plasma levels of $1,25(OH)_2D$ were well above the range observed in reference adolescent groups [31–33]. This sharp elevation was probably linked to hypocalcemia and the concomitant elevated PTH levels. Substrate concentration is a rate limiting factor in the synthesis of $1,25(OH)_2D$ in the presence of hypocalcemia, and thus a strong positive correlation between serum 25OHD and $1,25(OH)_2D$ concentrations was observed during the first 10 days of life over a wide range of 25OHD levels [21, 33]. This was well illustrated by the report of Glorieux *et al.* [21] of twin preterm babies in a controlled study of vitamin D supplementation. Their levels of 25OHD at birth were identical and much higher than those measured in other preterm newborns. In the protocol of early supplementation, one of the twins was assigned to the vitamin D-supplemented group. $1,25(OH)_2D$ increased in both infants in a similar fashion and paralleled the average increase recorded in the supplemented group. This observation emphasizes the importance of maternal 25OHD adequacy during pregnancy and indicates the potential for limitation of $1,25(OH)_2D$ production.

Serum bone Gla protein (BGP) values are high at birth (15 ± 3 ng/ml); maternal and cord serum BGP levels were not correlated [34,35]. During the first month of life, serum BGP increases and parallels the changes in $1,25(OH)_2D$ but without sustained correlation. These results indicate that serum BGP does not reflect changes in serum $1,25(OH)_2D$ but rather probably the overall rate of bone formation or growth at the tissue level.

C. Postnatal Vitamin D Supplementation

After the first week of life, in premature infants who received vitamin D, plasma 25OHD remained constant; $1,25(OH)_2D$ concentration increased up to day 30 with no further change until the end of the first 3 months [33]. The levels of $1,25(OH)_2D$ were more than 2 or 3

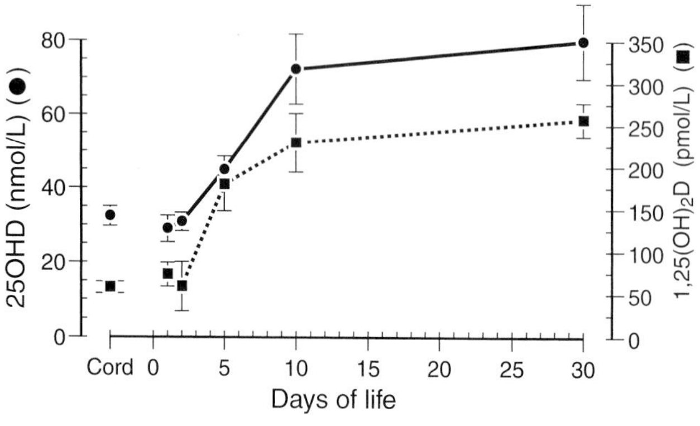

FIGURE 1 Newborn serum total 25OHD and $1,25(OH)_2D$ levels as a function of age. Values are means ± SEM. From Delvin *et al.* [23] with permission.

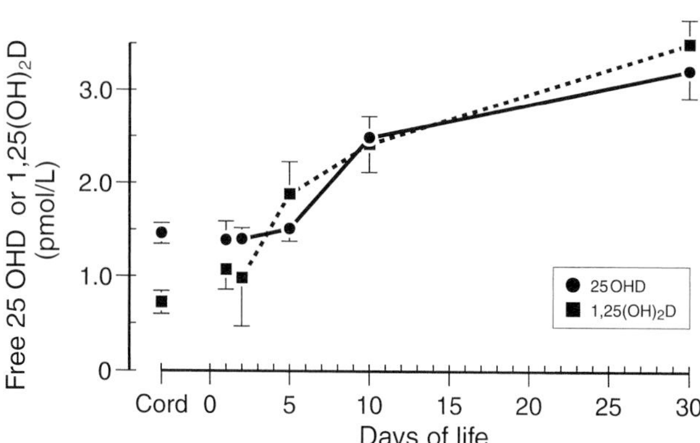

FIGURE 2 Newborn serum free 25OHD and $1,25(OH)_2D$ levels as a function of age. Values are means ± SEM. From Delvin *et al.* [23] with permission.

times higher than those seen in older children. During this time, there was no significant correlation between vitamin D metabolite concentration and serum calcium and phosphorus levels or calcium and phosphorus intake.

The high levels of plasma $1,25(OH)_2D$ beyond the neonatal period may represent a compensatory effect to ensure calcium and phosphorus absorption from the diet at a time where bone demineralization may occur. Osteopenia is seen commonly in premature infants, particularly in those who have received prolonged periods of parenteral feeding or who received a diet insufficient in calcium and phosphate (European formula or human milk). There is now a widespread agreement that deficiency of mineral substrate and not intake and metabolism of vitamin D is the principal etiological factor of osteopenia in low birth weight infants [36–38].

The vitamin D requirements of low birth weight infants are influenced by the body stores at birth, which in turn are related to the length of gestation and maternal stores. These factors should be taken into consideration when deciding on the policy concerning vitamin D supplementation in each country. The American Association of Pediatrics recommended that daily intake should be at least 400 IU independently of the vitamin D content of low birth weight formula [39]. The European Society of Pediatric Gastroenterology and Nutrition recommended that when low birth weight infants are fed human milk they should receive a vitamin D supplement of 1000 IU per day. Formula-fed infants should also be supplemented with vitamin D in order to achieve the same intake as babies receiving breast milk. It is recommended that the maximum vitamin D content of a formula should not exceed 3 μg/dl (120 IU/dl) [40].

VI. INFANTS OF DIABETIC MOTHERS

Hypocalcemia in infants of diabetic mothers (IDM) has been the subject of a number of studies, and several pathogenic factors have been suggested in this metabolic disorder including hypoparathyroidism, hyperphosphatemia, hypomagnesemia, and defective vitamin D metabolism [41,42]. Hypocalcemia of IDM shares some features with the early neonatal hypocalcemia (ENH) observed in premature babies. Indeed, it appears during the very early hours of life and shows little further change after 24 hr of age. However, it tends to be more severe than ENH and to persist for a longer time. No consistent abnormality in vitamin D metabolism has been observed in IDM, similar to the observations of ENH of premature infants [43]. Neither Salle *et al.* [43] nor Noguchi *et al.* [44] detected any major impairment in PTH responsiveness in IDM. Thus, the pathogenesis of hypocalcemia of IDM remains unclear. Possibly, the increased fetal body size of these infants may be responsible for increased calcium needs; the whole body bone mineral content of these babies measured by dual energy X-ray absorptiometry corresponds to that of a newborn baby of the same weight (Lapillonne A, Braillon PM, Delmas P, Salle BL, unpublished data, 1997).

VII. LATE NEONATAL HYPOCALCEMIA

Late neonatal hypocalcemia (LNH) is less frequent than ENH and usually brought to attention by the clinical manifestations of tetany and convulsions [45,46]. These are observed from the third or fourth day up to the end of the first month of life. LNH generally affects term infants, but it is also observed in premature infants and IDM in the form of prolonged and severe ENH (see above). A now uncommon cause is the feeding of "doorstep" (unmodified) cow's milk; the ensuing excessive phosphate intake suppresses production of $1,25(OH)_2D$.

In the presence of hypocalcemia, serum PTH levels remain inappropriately low (normal values). Relative hypoparathyroidism thus appears to be the main abnormality but is transient and not due to the absence or hypoplasia of the parathyroid glands as found in the Di George syndrome [47]. Serum $1,25(OH)_2D$ remains normal to moderately elevated, corresponding to what one would expect in response to the low serum PTH levels. This contrasts with the sharp increase in serum $1,25(OH)_2D$ levels observed in infants with ENH [47]. Severe hypocalcemia occurs in perinatal asphyxia when hypoxemia and acidosis persist, despite raised serum PTH levels, suggesting possible end-organ resistance [48] in the context of generally sick or failing cells.

Mild LNH requires only a watchful eye, but symptomatic and persistent LNH may require more aggressive intervention. The therapeutic management of LNH based on the pathophysiological findings would be to apply the treatment strategy of hypoparathyroidism. The active form of vitamin D [$1,25(OH)_2D$] may be effective. In most cases the treatment can be discontinued after a few days without relapse of hypocalcemia. In the case of symptomatic hypocalcemia with convulsions, intravenous calcium is recommended (1–2 mmol/kg over 30–60 min, preferably by a central line), within the context of managing the underlying clinical situation.

VIII. SUMMARY

Maternal vitamin D intake during the last trimester of pregnancy significantly influences neonatal vitamin

D stores and metabolism, and may influence growth in infancy. There is no impairment of neonatal vitamin D metabolism consequent on "immaturity," whatever the gestational age or birth weight of the infant. All mothers should receive an adequate vitamin D intake during the last trimester of pregnancy, and all infants should receive vitamin D in their diet, either as a supplement when the infant is completely breast-fed or as part of a modified cow's milk-derived formula. There is no place for the use of active vitamin D metabolites in the routine care of healthy infants.

References

1. Coutinho M de L, Dormandy TL, Yudkin S 1968 Maternal malabsorption presenting as rickets. Lancet **1**:1048–1052.
2. Park W, Paust H, Kaufmann HJ, Offermann G 1987 Osteomalacia of the mother—rickets of the newborn. Eur J Pediat **146**:292–293.
3. Krishnamachari KAVR, Iyengar L 1975 Effect of maternal malnutrition on the bone density of neonates. Am J Clin Nutr **28**: 482–486.
4. Brooke OG, Brown IRF, Cleeve HJW, Sood A 1980 Observations on the vitamin D state of pregnant Asian women in London. Br J Obstet Gynaecol **88**:18–26.
5. Maxwell JD, Ang L, Brooke OG, Brown IR 1981 Vitamin D supplements enhance weight gain and nutritional status in pregnant Asians. Br J Obstet Gynaecol **88**:987–991.
6. Brooke OG, Butters F, Wood C 1981 Intrauterine vitamin D nutrition and postnatal growth in Asian infants. Br Med J Clin Res **283**:1024.
7. Delvin EE, Salle BL, Glorieux FH, Adeleine P, David LS 1986 Vitamin D supplementation during pregnancy: effect on neonatal calcium homeostasis. J Pediat **109**:328–334.
8. Marya RK, Rathee S, Lata V, Mudgil S 1981 Effects of vitamin D supplementation in pregnancy. Gynecol Obstet Invest**12**:155–161.
9. Stearns G, Jeans PC, Vandecar V 1936 The effect of vitamin D on linear growth in infancy. J Pediatr **9**:1–10.
10. Misselwitz J, Hesse V, Markestad T 1990 Nephrocalcinosis, hypercalciuria and elevated serum levels of 1,25-dihydroxyvitamin D in children. Possible link to vitamin D toxicity. Acta Paediatr Scand **79**:637–643.
11. Hoogenboezem T, Degenhart J, Munick Keizer-Schrama SM, Bouillon R, Grose WF, Hackeng WH, Visser HK 1989 Vitamin D metabolism in breast-fed infants and their mothers. Pediatr Res **25**:623–627.
12. Steichen JJ, Tsang RC, Gratton TL, Hamstra A, DeLuca HF 1980 Vitamin D homeostasis in the perinatal period: 1,25-Dihydroxyvitamin D in maternal, cord, and neonatal blood. N Engl J Med **302**:315–319.
13. Delvin EE, Salle BS, Glorieux FH 1991 Vitamin D and calcium homeostasis in pregnancy: Feto–maternal relationships. In: Glorieux FH (ed) Rickets. Vevey/Raven, New York, pp. 91–105.
14. American Academy of Pediatrics Committee on Nutrition 1963 The prophylactic requirement and the toxicity of vitamin D. Pediatrics **31**:512.
15. National Research Council Food and Nutrition Board 1989 Recommended Dietary Allowances. 10th Ed. National Academy Press, Washington, D.C., pp. 78–114.
16. HMSO Report Number 41 1991 Dietary and nutrient reference values for food energy and nutrients for the United Kingdom. United Kingdom Department of Health, London.
17. Greer FR, Searcy JE, Levin RS, Steichen JJ, Asch PS, Tsang RC 1981 Bone mineral content and serum 25-hydroxyvitamin D concentration in breast-fed infants with and without supplemental vitamin D. J Pediatr **98**:696–701.
18. Namgung R, Tsang RC, Specker BL, Sierra RI, Ho ML 1993 Reduced serum osteocalcin and 1,25-dihydroxyvitamin D concentrations and low bone mineral content in small for gestational age infants: Evidence of decreased bone formation rates. J Pediatr **122**:269–275.
19. Rösli A, Fanconi A 1973 Neonatal hypocalcemia. 'Early type' in low birth weight newborns. Helv Paediatr Acta **28**:443–457.
20. Tsang RC, Light IJ, Sutherland JM, Kleinman LI 1973 Possible pathogenetic factors in neonatal hypocalcemia of prematurity. The role of gestation, hyperphosphatemia, hypomagnesemia, urinary calcium loss, and parathormone responsiveness. J Pediatr **82**:423–429.
21. Glorieux FH, Salle BL, Delvin EE, David LS 1981 Vitamin D metabolism in preterm infants: Serum calcitriol values during the first five days of life. J Pediatr **99**:640–643.
22. David L, Salle BL, Chopard JP, Frederich A 1976 Parathyroid function in low birth weight newborns during the first 48 hours of life. In: Stern F-H (ed) Symposium on Intensive Care of the Newborn. Masson, New York, pp. 107–117.
23. Delvin E, Salle BL, Glorieux FH 1996 Calciotropic hormones in the perinatal period in preterm infants. Submitted for publication.
24. David L, Salle BL, Putet G, Grafmeyer D 1981 Serum immunoreactive calcitonin in low birth weight infants. Description of early changes; effect of intravenous calcium infusion: relationships with early changes in serum calcium, phosphorus, magnesium, parathyroid hormone, and gastrin levels. Pediatr Res **15**:803–808.
25. Delvin EE, Salle BL, Glorieux FH, David LS 1988 Vitamin D metabolism in preterm infants: Effect of a calcium load. Biol Neonate **53**:321–326.
26. Delvin E, Glorieux F, Salle B, David L, Varenne J 1982 Control of vitamin D metabolism in preterm infants: Fetomaternal relationships. Arch Dis Child **57**:754–757.
27. Hillman LS, Haddad JG 1975 Perinatal vitamin D metabolism II. Serial 25-hydroxyvitamin D concentrations in sera of term and preterm infants. J Pediatr **86**:928–935.
28. Bouillon R, van Assche FA, van Baelen H, Heyns W, De Moor P 1981 Influence of the vitamin D-binding protein on the serum concentration of 1,25-dihydroxyvitamin D_3: Significance of the free 1,25-dihydroxyvitamin D_3 concentration. J Clin Invest **67**:589–596.
29. Hirsfeld J, Lunell O 1963 Serum protein syntesis in foetus: Haptoglobins and group-specific components. Nature **196**:1220–1222.
30. Salle BL, Glorieux FH, Delvin EE, David LS, Meunier G 1983 Vitamin D metabolism in preterm infants. Serial serum calcitriol values during the first four days of life. Acta Paediatr Scand **72**:203–206.
31. Markestad T, Elzouki A, Legrain M, Ulstein M, Asknes L 1984 Serum concentration of vitamin D metabolites in maternal and umbilical cord blood of Libyan and Norwegian women. Hum Nutr Clin Nutr **38**:55–62.
32. Schilling R, Haschke F, Schatten C, Schmid M, Woloszczuk W, Steffan I, Schuster E 1990 High total and free 1,25-dihydroxyvitamin D concentrations in serum of premature infants. Acta Paediatr Scand **79**:36–40.
33. Salle BL, Senterre J, Glorieux FH, Delvin EE, Putet G 1987 Vitamin D metabolism in preterm infants. Biol Neonate **52**:119–130.
34. Delmas PD, Glorieux FH, Delvin EE, Salle BL, Melki I 1987

Perinatal serum bone Gla-protein and vitamin D metabolites in preterm and full term neonates. J Clin Endocrinol Metab **65**:588–591.

35. Pittard WBD, Geddes KM, Hulsey TC, Hollis BW 1992 Osteocalcin, skeletal alkaline phosphatase, and bone mineral content in very low birth weight infants: A longitudinal assessment. Pediatr Res **31**:181–185.
36. Tsang RC, Demarini S 1995 Rickets and calcium and phosphorus requirements in very low birth weight infants. Monatsschr Kinderheilkd **43**:S125–S129.
37. Shaw NJ, Bishop NJ 1995 Mineral accretion in growing bones—A framework for the future? Arch Dis Child **72**:177–179.
38. Lapillonne A, Glorieux FH, Salle BL, Braillon PM, Chambon M, Rigo J, Putet G, Senterre J 1994 Mineral balance and whole body bone mineral content in very low-birth-weight infants. Acta Paediatr **405**(Suppl.):117–122.
39. American Academy of Pediatrics, Committee on Nutrition 1977 Nutritional needs of low birth weight infants. Pediatrics **60**:519–530.
40. European Society of Paediatric Gastroenterology and Nutrition, Committee on Nutrition of the Preterm Infant 1987 Nutrition and feeding of preterm infants. Acta Paediatr Scand **336**(Suppl.):6.
41. Tsang RC, Chen IW, Friedman FA, Gigger M, Steichen J, Koffler H, Fenton L, Brown D, Pramanik A, Keenan W, Strub R, Joyce T 1975 Parathyroid function in infants of diabetic mothers. J Pediatr **86**:399–404.
42. Bergman L, Kjellmer I, Seltam U 1974 Calcitonin and parathyroid hormone. Relation to early neonatal hypocalcemia in infants of diabetic mothers. Biol Neonate **24**:151–160.
43. Salle BL, David L, Glorieux FH, Delvin EE, Louis JJ, Troncy G 1982 Hypocalcemia in infants of diabetic mothers. Studies on circulating calciotropic hormone concentrations. Acta Paediatr Scand **71**:573–577.
44. Noguchi A, Eren M, Tsang R 1980 Parathyroid hormone in hypocalcemia and normocalcemic infants of diabetic mothers. J Pediatr **97**:112–114.
45. Balsan S, Alizon M 1968 L'hypoparathyroïdie transitoire idiopathique du nourrisson. Arch Fr Pediatr **25**:1151–1170.
46. Fanconi A, Prader A 1967 Transient congenital idiopathic hypoparathyroidism. Helv Paediatr Acta **22**:342–359.
47. David L, Salle BL, Varenne P, Glorieux FH, Delvin EE 1981 Treatment of late neonatal hypocalcemia with 1α-hydroxycholecalciferol (1αOHCC). Pediatr Res **15**:1222 (abstract).
48. Schedewie HK, Odell WD, Fisher DA, Krutzik SR, Dodge M, Cousins L, Fiser WP 1979 Parathormone and perinatal calcium homeostasis. Pediatr Res **13**:1–6.

CHAPTER 35

Vitamin D Metabolism: The Effects of Aging

BERNARD P. HALLORAN Department of Medicine, University of California, San Francisco, and Divisions of Endocrinology and Geriatrics, Veterans Affairs Medical Center, San Francisco, California

ANTHONY A. PORTALE Departments of Medicine and Pediatrics, University of California, San Francisco, San Francisco, California

I. INTRODUCTION

Aging in the context of this chapter refers to postmaturational aging and not to growth and development. Postmaturational aging is a complex process beginning with senescence at the cellular level. Cells undergo intrinsic aging. All cells gradually change phenotypically and eventually lose the ability to proliferate as they age. Tissues and organs also age, with consequent changes in metabolism and function. And of course organisms age, resulting in diminished health and performance. Aging is heterogeneous. Differences in genetic makeup and lifestyle (diet, activity level, environment) influence the progress of senescence. Some people appear to age more rapidly than others. Disease confounds the aging process. Are the changes observed in an aging population a consequence of true aging or the cumulative effect of chronic disease? The changes that occur during aging are often subtle, but their cumulative effect is dramatic. Predictably aging influences vitamin D metabolism, and vitamin D influences the aging process. This chapter deals with how aging affects cutaneous production, dietary availability, metabolism, and action of vitamin D; and how vitamin D influences the progression of aging. An attempt is made to separate the effects of common age-related diseases from the effects of aging per se. Each individual ages differently depending on the disease burden.

II. CUTANEOUS PRODUCTION OF VITAMIN D

The production of vitamin D in the skin is described in Chapter 3. Skin function begins to deteriorate during the third decade of life [1]. Epidermal thinning begins around age 20, and tissue loss continues with advancing age [2]. Skin elasticity, keratinocyte number, and cell turnover rate decrease with aging [3–6]. The pattern in gene expression in cultured keratinocytes also changes with donor age [7,8]. Barrier function is compromised, and total lipid content (including cholesterol) in the stratum corneum is decreased in aged animals [9]. Skin blood flow falls by nearly 40% between 20 and 70 years of age, thus reducing dermal clearance of vitamin D [10].

Not surprisingly, cutaneous production of vitamin D decreases with advancing age [11,12]. Approximately 80% of the vitamin D formed in the skin is produced in the epidermis, and the amount of precursor to vitamin D_3, 7-dehydrocholesterol, is decreased in the epidermis of elderly subjects. Despite a decrease in the number of melanocytes in the skin (normally melanin in the melanocyte acts as a natural sunscreen to reduce production of vitamin D) conversion of 7-dehydrocholesterol to previtamin D_3 in human skin samples exposed to ultraviolet radiation is decreased by as much as 2-fold in elderly subjects. Holick *et al.* [13] report that whole body exposure to one minimal erythemal dose of ultraviolet B radiation can increase serum vitamin D to a maximum of 78.1 nmol/liter in young subjects but to a maximum of only 20.8 nmol/liter in elderly subjects 68–80 years of age. The decrease in vitamin D synthesis is presumably a consequence of the decrease in substrate (7-dehydrocholesterol) and reduced dermal clearance of vitamin D.

The diminished ability to produce vitamin D in the skin of older subjects is often aggravated by changes in lifestyle. Many older people are homebound or hesitant to venture out-of-doors, and when exposed to the sun frequently wear potent sunscreens to reduce the risk of skin cancer. Sunlight that has passed through a glass window pane will not produce any vitamin D because of the UV absorbance by the glass [14]. Furthermore, chronic use of sunscreens can dramatically reduce serum vitamin D levels [15]. The elderly, especially those that are homebound or confined to a nursing home, are, therefore, at increased risk of becoming vitamin D deficient, both as a result of diminished efficiency in cutaneous production and reduced effective solar exposure. Gloth *et al.* [16] report that of 244 elderly men and women over the age of 65 and deprived of direct sunlight exposure (homebound elderly and nursing home residents) 54% of the community dwellers and 38% of nursing home residents have serum 25-hydroxyvitamin D (25OHD) levels below 25 n*M* (normal range 25–137 n*M*, 10–55 ng/ml). The subject of vitamin D insufficiency in elderly subjects in relationship to osteoporosis is also discussed in Chapter 43.

III. DIETARY VITAMIN D AND INTESTINAL ABSORPTION

Dietary intake of vitamin D decreases slightly or remains unchanged with advancing age. The United States Department of Agriculture (USDA) has established and recently reaffirmed that the adult recommended dietary allowance (RDA) for vitamin D is 5.0 μg/day or 200 IU/day [17]. The major sources of vitamin D are milk, milk products, butter, fortified margarine, eggs, and some fatty fish. Although calculations of dietary vitamin D intake are somewhat unreliable (the vitamin D content of many foods is not available), data from the USDA National Food Consumption Survey indicates that the usual adult (all ages) dietary intake in the United States is 1.25–1.75 μg/day, less than about one-third of the RDA [18]. This does not take into account those foods for which data on vitamin D content are not available and thus probably underestimates the actual vitamin D intake. The most recent National Health and Nutrition Examination Survey, NHANES III, did not measure vitamin D intake, and no other large-scale United States survey data are available.

Because of differences in dietary supplementation and food preferences, dietary vitamin D varies over a broad range. Krall *et al.* [19] in a study of 333 women observed vitamin D intakes ranging from 0.5 to more than 40 μg/day. With advancing age, protein, fat, and total caloric intake decrease [20,21]. Dietary calcium also decreases (NHANES III), and vitamin D intake is tightly coupled to total dietary calcium [22]. Thus, one would predict that vitamin D intake decreases with age. However, vitamin D supplementation of the normal diet can account for as much as 50% of the total intake of the vitamin [22,23], and supplementation in at least some populations increases with age. Sowers *et al.* [23] report that in 373 women ranging in age from 20 to 80 years the intake of vitamin D from supplements increased from 104 ± 18 IU/day in 25–40 year olds to 202 ± 43 IU/day in 60–75 year olds. Although direct calculation of vitamin D intake in elderly populations suggests that intake decreases with age in some elderly [24], the predominance of evidence suggests that in most elderly there is no change [19,22,23,25].

Intestinal absorption of vitamin D does not appear to change with aging. Vitamin D is absorbed in the proximal intestine, and in the absence of disease associated with intestinal malabsorption, most studies, but not all [26], support the idea that vitamin D absorption remains normal in the elderly [27–29]. Clemens *et al.* [27] compared serum vitamin D_2 levels in young adults and 25 chronically institutionalized but otherwise healthy elderly adults (mean age 72) after administration of 50,000 IU of vitamin D_2. In the absence of gastrointestinal disease, they found no evidence of impaired vitamin D absorption. However, studies to examine whether aging disrupts enterohepatic recirculation of vitamin D have not been performed. The effects of gastrointestinal and hepatobiliary disease on vitamin D absorption are discussed in Chapter 51.

Despite normal dietary intake and normal or near normal intestinal absorption, vitamin D status in elderly populations, as judged by serum concentrations of either vitamin D or 25OHD, deteriorates with advancing age [16,22,30–37]. The serum concentration of 25OHD in the elderly, a reflection of serum vitamin D levels (see Section IV below), is related inversely to age and directly to sun exposure. The age-related decrease in serum 25OHD, in the absence of disease, relfects both a decrease in cutaneous production (all elderly) and reduced dietary intake (some elderly). In the presence of disease, malabsorption can play a major role.

In a study of 433 postmenopausal women, Need *et al.* [35] showed that serum 25OHD decreased with age and was positively correlated to hours of sunlight. Van der Wielan *et al.* [31], in a study of 824 elderly people from 11 European countries, report that 36% of men and 47% of women had serum 25OHD levels below 12 ng/ml (the lower range of normal). Serum 25OHD concentrations were directly related to hours of ultraviolet light exposure and factors of physical health status. A study by Fardellone *et al.* [38] further exemplifies the importance of sun exposure and health status, and it demonstrates the heterogeneity of aging populations as well. These investigators studied a group of chronically institutionalized elderly and observed a mean serum 25OHD level of only 3.7 ng/ml. Eighty-five percent of the subjects had serum 25OHD levels below 5 ng/ml, and 98% had levels below 10 ng/ml (normal range 10–55 ng/ml). Clearly, many older people are vitamin D insufficient or frankly deficient. Disease and medication use frequently exacerbate the problem.

Mild hyperparathyroidism, increased bone turnover, and diminished bone density are associated with moderate vitamin D insufficiency in the elderly. In elderly populations serum 25OHD has been observed to be inversely correlated with both the serum concentration of parathyroid hormone (PTH) and markers of bone turnover [19,36,38–41]. Supplementation with vitamin D increases serum 25OHD, in most cases decreases serum PTH, and normalizes bone turnover [42–44]. Ooms *et al.* [40], in a study of 330 healthy women over the age of 70, report that serum 25OHD in 65% of the subjects was less than 12 ng/ml. 25-Hydroxyvitamin D was negatively correlated to serum PTH but only at levels of 25OHD below 10 ng/ml. Bone mineral density was positively correlated with serum 25OHD but only below 12 ng/ml. Brazier *et al.* [39] compared vitamin D-sufficient and -insufficient elderly men and women before and after treatment of the insufficient group with 800 IU/day of vitamin D. In the vitamin D-insufficient subjects, serum PTH and markers of bone turnover were increased prior to treatment but normalized to levels found in age-matched vitamin D-sufficient subjects after treatment.

The data suggest that dietary supplementation of elderly men and women with poor or marginal vitamin D status may be beneficial in reducing bone loss associated with aging. Indeed, Dawson-Hughes *et al.* [45] report that vitamin D supplementation (10 μg/day) to postmenopausal women with an otherwise normal mean intake of 5 μg/day can reduce wintertime bone loss. Further studies by these investigators [46] in 247 healthy ambulatory postmenopausal women consuming an average of 2.5 μg/day of vitamin D showed that supplementation with 17.5 μg/day could reduce the rate of loss of bone mineral density in the femoral neck year-round. Interestingly, the anabolic effects of vitamin D treatment, although associated with an increase in serum 25OHD, are not usually accompanied by an increase in serum 1,25-dihydroxyvitamin D [$1,25(OH)_2D$] unless the patients are frankly vitamin D deficient [41]. The report by Barger-Lux *et al.* [47] demonstrating a positive correlation between serum 25OHD [but not $1,25(OH)_2D$] and calcium absorption in healthy premenopausal women treated with 25OHD is consistent with this idea. These data suggest that 25OHD may be acting directly on the intestine to stimulate calcium absorption (see Chapter 31).

Not all individuals benefit from vitamin D supplementation. Orwoll *et al.* [48] studied normal healthy men ranging in age from 30 to 87 years. Administration of both calcium (1000 mg/day) and vitamin D (25 μg/day) did not affect the normal rate of bone loss. Importantly, however, mean basal dietary calcium and vitamin D intakes in this population were 1159 mg/day and approximately 9 μg/day, respectively. These data are consistent with the findings of Ooms *et al.* [42] and suggest that elderly populations with adequate dietary calcium and adequate vitamin D status are not likely to improve their mineral balance with additional vitamin D.

IV. SYNTHESIS OF 25-HYDROXYVITAMIN D

Synthesis of 25OHD does not appear to be influenced by aging. The primary site for synthesis of 25OHD is the liver (see Chapter 4). Severe liver disease can lead to a reduction in the synthesis of 25OHD and the serum concentrations of 25OHD, vitamin D binding protein (DBP), and total (but not free) $1,25(OH)_2D$ [49,50], and elderly patients with inadequate hepatic function may have a reduction in vitamin D 25-hydroxylase reserve. In the absence of hepatic disease, however, activity of the vitamin D 25-hydroxylase appears to be normal in the elderly. Aksnes *et al.* [30] examined the relationships among the serum concentrations of vitamin D, 25OHD, and DBP in healthy adults ranging in age from 22 to 96 years. They observed decreases in vitamin D, 25OHD, and DBP but no change or a small increase in the molar ratio of 25OHD to vitamin D. This suggests that hepatic hydroxylation at the 25 position of vitamin D, in the absence of disease, is not impaired by aging. Matsuoka *et al.* [15] studied the response of serum $25OHD_2$ to an oral load of vitamin D_2 in young and elderly subjects. The increase in serum 25OHD over time was similar between age groups, providing evidence that both intestinal absorption and 25-hydroxylation are normal in healthy elderly people.

V. METABOLISM OF 1,25-DIHYDROXYVITAMIN D

A clear consensus of the effects of aging on the synthesis and metabolism of $1,25(OH)_2D$ is slowly emerging. The picture is made complex by the heterogeneity of the aging process and the independent effects of age-related changes in kidney function, acid–base balance, sex steroid levels, growth hormone status, and other factors on $1,25(OH)_2D$ metabolism (see Chapter 5).

A. Serum Concentration of $1,25(OH)_2D$

Whether the serum concentration of $1,25(OH)_2D$ changes with age has been controversial. With advancing age serum concentrations of $1,25(OH)_2D$ have been reported in men to either decrease [51,52] or remain unchanged [53–55]; in women, to decrease [51,56–58], remain unchanged [54], to increase and then decrease [52], or to decrease and then increase [59]; and in mixed populations to either decrease [60] or remain unchanged [27,61–63]. Results in animal studies are similar. Using the rat as an animal model of human aging, serum $1,25(OH)_2D$ levels have been reported to be decreased [64–67], unchanged [68,69], and increased [70] in aged animals. These confusing results stem in part from differences in the definition of "old" and differences in the sex, lifestyle, and the presence of underlying disease/medication use in the populations studied. In many early studies data obtained from elderly subjects residing in nursing homes were compared with data obtained in young hospital staff. More recent studies support the results of early reports, but by better defining the populations we are now in a position to reconcile earlier discrepancies.

The definition of "old" in animal studies of aging is critical. The most popular animal model for aging studies has been the Fischer 344 (F344) rat. The mean life span of the F344 rat is 2 years. Animals are sexually mature at approximately 3 months. Serum $1,25(OH)_2D$ is elevated during growth in the rat and reaches mature adult levels by 3–4 months of age. Many "aging" studies in rats have compared growing or not yet mature animals to older adults. The results of these studies, which clearly show that serum $1,25(OH)_2D$ is lower in adults than in growing animals, have been interpreted to indicate that serum $1,25(OH)_2D$ falls with age. Indeed serum $1,25(OH)_2D$ does fall with age if animals 3 weeks old are compared to animals older than 3 months. These studies, however, do not address whether serum levels change during postmaturational aging (3 months to the time of death).

However, even when young adult animals (4–6 months) are compared to old adults (20–28 months) discrepancies in data still exist. These differences can be accounted for, at least in part, by the heterogeneity of the aging process and the small number of animals normally sampled in animal aging studies. For example, loss of renal function is a major cause of death in the F344 rat. The primary source of serum $1,25(OH)_2D$ is the kidney, and loss of renal function can dramatically reduce serum levels (see Chapter 52). Predictably, serum $1,25(OH)_2D$ will be low in aged animals with renal insufficiency. Thus, comparing relatively small numbers of adult and randomly selected aged F344 rats may or may not indicate that serum $1,25(OH)_2D$ decreases with age depending on the renal status of the aged population. If the serum concentrations of $1,25(OH)_2D$ in aged (24-month-old) male F344 rats with normal or near normal glomerular filtration rates (GFR) and adult male F344 rats are compared, the levels are not significantly different [69]. Interestingly, there is a trend in aging male F344 rats between the ages of 12 and 24 months with good kidney function for the serum concentration of $1,25(OH)_2D$ to increase. The trend does not reach significance but may explain why some investigators have observed an increase in serum $1,25(OH)_2D$ in the rat with advancing age [70].

In humans, renal function also deteriorates with postmaturational aging [71–74]. The rate of deterioration varies from subject to subject, and some older people maintain normal or near normal GFR well into their eighth decade of life. The low serum concentrations of $1,25(OH)_2D$ reported in many elderly populations reflect in large part the age-related fall in GFR [54,55,63]. In elderly individuals with moderate to severe renal insufficiency (GFR <70 ml/min), renal synthesis and serum concentrations of $1,25(OH)_2D$ are low. In healthy elderly men in whom renal function is normal or near normal, serum $1,25(OH)_2D$ levels are not significantly different from those in young men [55] (Fig. 1). Moreover, serum DBP levels in men do not change with age [60], suggesting that free levels of $1,25(OH)_2D$ do not change. In elderly men, there is general agreement that if the population is healthy (GFR >50–70 ml/min, free of chronic diseases) and active serum $1,25(OH)_2D$ is not different than in young men. If the population is ill or inactive, serum $1,25(OH)_2D$ is likely to be low.

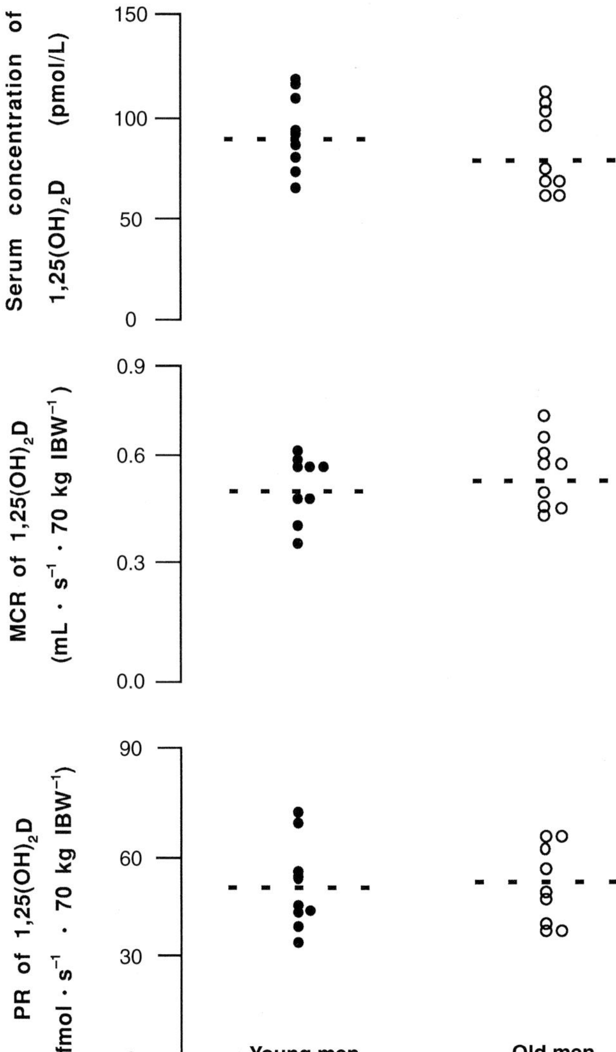

FIGURE 1 Serum concentration, metabolic clearance rate (MCR), and production rate (PR) of $1,25(OH)_2D$ in healthy young and elderly men after ingesting a constant normal diet for 9 days. The mean of each population is indicated by a dashed line. IBW, ideal body weight. From Halloran BP, Portale AA, Lonergan ET, and Morris RC. Production and metabolic clearance of $1,25(OH)_2D$ in men: Effect of advancing age. *J Clin Endocrinol Metab.* Vol. 70;318–323; 1990; © The Endocrine Society.

Women are more difficult to evaluate because of problems in establishing a baseline in young women with normal menstrual cycles [serum $1,25(OH)_2D$ has been reported to change during the menstrual cycle] and because of the confounding effects of menopause. The majority of data suggest that serum total $1,25(OH)_2D$ levels in healthy women remain the same or tend to gradually increase through the eighth decade [54,75,76]. Menopause has little effect on the serum concentration of $1,25(OH)_2D$ [59,77,78]. Prince *et al.* [59] report, however, that there may be a transitory fall in total serum $1,25(OH)_2D$ 5–15 years after menopause. This decrement is followed by a gradual rise in serum levels of $1,25(OH)_2D$ that correlate with rising serum PTH concentrations.

The serum concentration of DBP has been reported to be decreased [60,79,80], normal [30,36], or increased [59] in postmenopausal women. In a cross-sectional study of 655 women ranging in age from 35 to 90, Prince *et al.* [59] report that circulating DBP levels increase after the menopause, then fall transiently, and finally increase progressively with age, resulting in a continual fall in the serum free $1,25(OH)_2D$ index beginning 10–15 years after menopause. Despite these convincing data, other studies do not support these findings [79]. Thus, it is not clear whether the serum concentration of free $1,25(OH)_2D$ changes with age in women.

B. Kinetics of $1,25(OH)_2D$ Metabolism

The serum concentration of $1,25(OH)_2D$ is determined by its production (PR) and metabolic clearance rates (MCR). Clearance of $1,25(OH)_2D$ is increased by vitamin D deficiency [81], chronic exogenous administration of $1,25(OH)_2D$ [82], and in the rat [81] and pig [83], by depletion of dietary calcium. The MCR of $1,25(OH)_2D$ is unaffected by changes in dietary phosphorus [83,84], PTH administration [85], and glucocorticoid excess [86]. Loss of renal function decreases, the MCR of $1,25(OH)_2D$ in the rat [87].

With advancing age, the metabolic clearance of $1,25(OH)_2D$ gradually increases in the F344 male rat with normal or near normal renal function [69] (Fig. 2). When compared to 6-month-old animals, clearance is 24% ($p < 0.10$) higher at 12 months, 30% ($p < 0.05$) higher at 18 months, and 57% higher ($p < 0.01$) at 24 months of age. Production of $1,25(OH)_2D$ also increases with age in the F344 male rat, and by 24 months of age it is 91% ($p < 0.01$) higher than at 6 months (Fig. 3). The

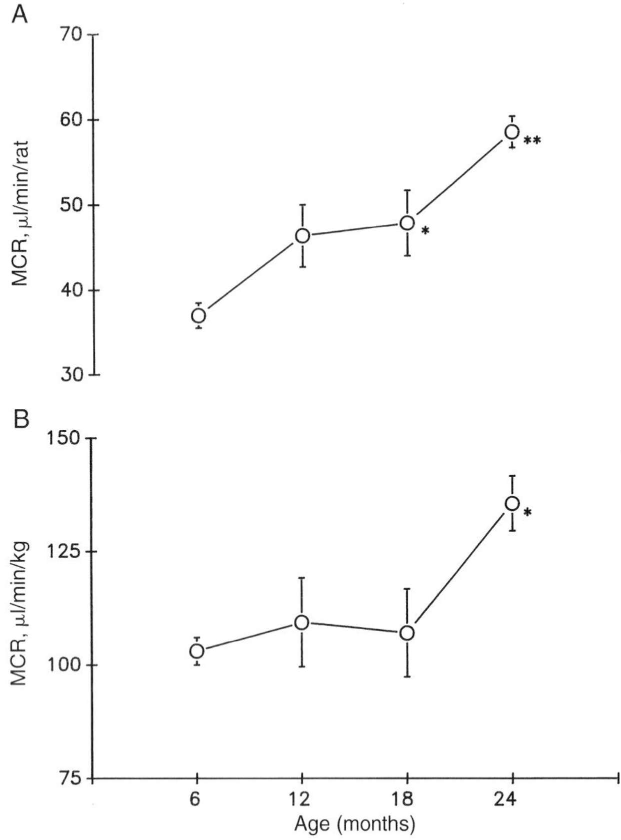

FIGURE 2 Metabolic clearance rate of $1,25(OH)_2D$ expressed per rat (A) and per kilogram body weight (B) in 6-, 12-, 18-, and 24-month-old Fischer 344 rats (means ± SE). $*p < 0.05$ and $**p < 0.01$ when compared to 6-month-old animals, using Dunnett's test. From Wada *et al.* [69].

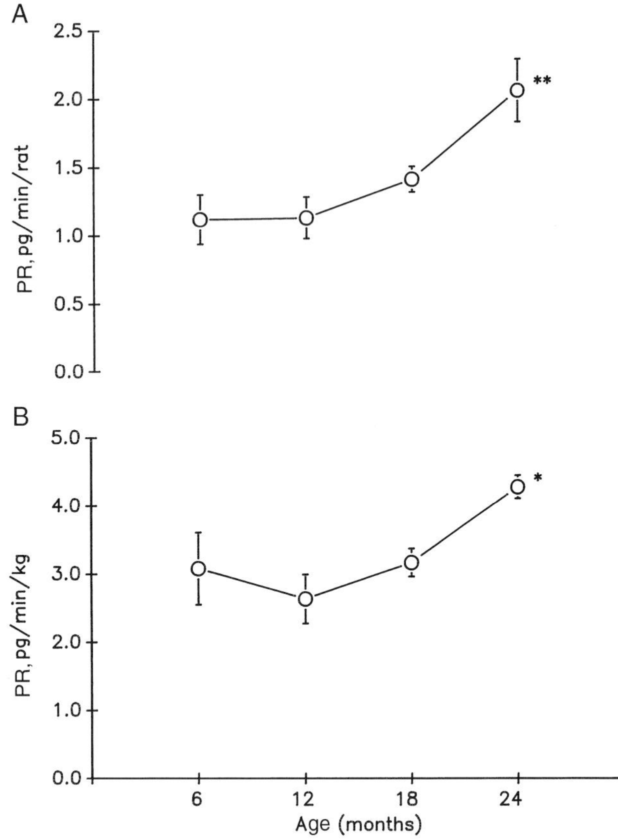

FIGURE 3 Production rate of $1,25(OH)_2D$, expressed per rat (A) and per kilogram body weight (B), in 6-, 12-, 18-, and 24-month-old Fischer 344 rats (means ± SE). $*p < 0.05$ and $**p < 0.01$ when compared to 6-month-old animals, using Dunnett's test. From Wada *et al.* [69].

increase in $1,25(OH)_2D$ production correlates positively with an increase in serum PTH, presumably reflecting the normal stimulatory effect of PTH on 1-hydroxylase activity (Fig. 4). These data demonstrate that both production and clearance of 1,25(OH)D can increase during postmaturational aging in the rat. They also suggest that the kidney (in animals with normal GFR) remains sensitive to PTH.

The kinetics of $1,25(OH)_2D$ metabolism in healthy men [55] and women [75] remain unchanged with advancing age. In a group of healthy young (34 ± 5 years, $n = 9$) and elderly (72 ± 5 years, $n = 9$) men with normal or near normal GFR (creatinine clearance in elderly was 1.61 ± 0.34 ml sec^{-1} 1.73 m^{-2}; clearance in young controls was 1.87 ± 0.18 ml sec^{-1} 1.73 m^{-2}), we have shown, under strictly controlled metabolic conditions, that the production and metabolic clearance rates for $1,25(OH)_2D$, as well as the serum concentration, are not different [55] (Fig. 1). Serum $1,25(OH)_2D$ correlates with production but not clearance in this population. There is also no evidence that substrate (25OHD) is any more or less limiting in the conversion of 25OHD to $1,25(OH)_2D$ in the elderly than in the young.

C. Synthesis of $1,25(OH)_2D$: Trophic Factors

1. SEX STEROIDS

Serum total and free testosterone levels decrease with advancing age in men [88,89], and testosterone treatment has been shown to increase modestly both serum total and free $1,25(OH)_2D$ levels in hypogonadal men [90]. However, treatment of healthy elderly men with testosterone for 3 months did not change otherwise normal serum $1,25(OH)_2D$ levels (E. S. Orwoll and B. P. Halloran, unpublished data, 1991). These data suggest that the decline in serum testosterone may reduce tonic stimulation of the 1-hydroxylase, but if this does occur other factors must come into play to compensate for this loss because serum $1,25(OH)_2D$ remains unchanged.

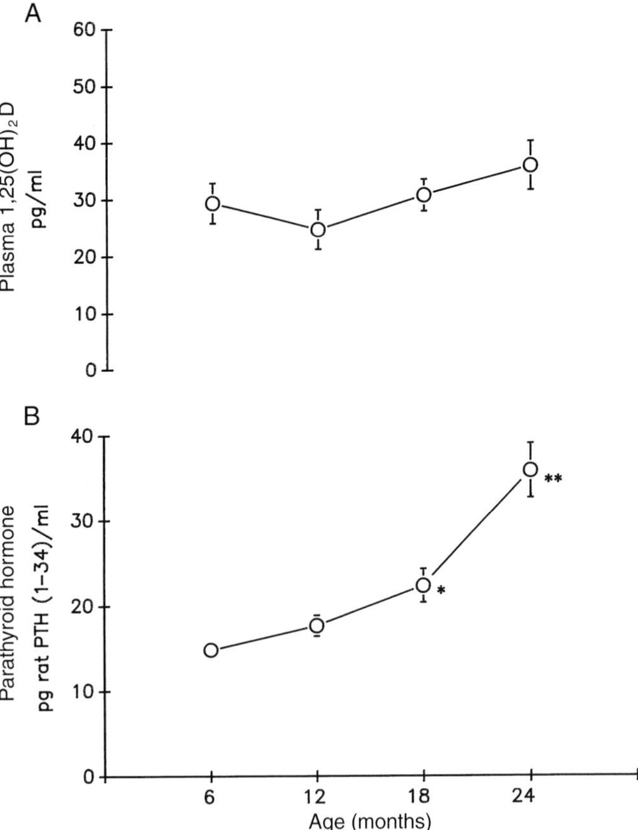

FIGURE 4 Plasma concentrations of $1,25(OH)_2D$ (A) and parathyroid hormone (PTH) (B) in 6-, 12-, 18-, and 24-month-old Fischer 344 rats (means ± SE). $*p < 0.05$ and $**p < 0.01$ when compared to 6-month-old animals, using Dunnett's test. From Wada *et al.* [69].

Estrogen deficiency leads to a decrease in serum $1,25(OH)_2D$ regardless of age [59,91,92]. Estrogen replacement in postmenopausal women can increase both total and free serum $1,25(OH)_2D$ [79,91,93–97], suggesting that menopause and the accompanying estrogen deficiency may remove an important trophic factor for the maintenance of serum $1,25(OH)_2D$ in aging women.

2. Calcium and Phosphorus

The morning fasting serum concentration of Ca^{2+} in elderly men and women has been reported to be either normal [52,54,98], decreased [36,54,98,99], or increased [100,101]. Serum Ca^{2+} has been shown to influence directly 1-hydroxylase activity [102], but studies to determine whether the response of serum $1,25(OH)_2D$ to direct modulation by Ca^{2+} changes with age have not been done.

There is general agreement as to the changes in serum phosphorus with advancing age. Serum phosphorus decreases in men and remains unchanged in women [54,56]. As phosphorus directly regulates 1-hydroxylase activity, it is of interest to determine whether there is a change in responsiveness of serum $1,25(OH)_2D$ to changes in serum phosphorus. Villa *et al.* [103] report that older women retain the capacity to respond to restriction of dietary phosphorus. These investigators studied the effects of aluminum hydroxide administration on serum PTH and $1,25(OH)_2D$ in postmenopausal women. Treatment with aluminum hydroxide lowered serum phosphorus and increased serum $1,25(OH)_2D$ without affecting serum Ca^{2+} or PTH.

To examine the effects of aging on the serum concentrations of Ca^{2+} and phosphorus, and on phosphorus regulation of $1,25(OH)_2D$, we studied healthy young and elderly men with normal or near normal renal function on a constant whole food diet containing normal levels of calcium and phosphorus [104,105]. Whole blood Ca^{2+} concentrations, both morning fasting levels and 24-hr means, in young and elderly men were not different (Table I). In contrast, concentrations of serum phosphorus, both morning fasting levels and 24-hr means, were significantly lower in the elderly men. Because low serum phosphorus stimulates $1,25(OH)_2D$ production, mild hypophosphatemia in the elderly would be expected to increase serum levels of $1,25(OH)_2D$. That serum $1,25(OH)_2D$ is not increased in this population suggests that sensitivity to changes in serum phosphorus may diminish with age.

To determine whether aging influences the relationship between the serum concentrations of phosphorus and $1,25(OH)_2D$, we varied dietary phosphorus within its normal range and measured the effect on serum $1,25(OH)_2D$. Restriction of dietary phosphorus from 2300 to 625 mg/day increased serum $1,25(OH)_2D$ by 47% in each age group. At each dietary level of phosphorus, serum $1,25(OH)_2D$ levels in the young and elderly men were virtually identical but serum phosphorus levels in the elderly were consistently lower than in the young men (Table II). The 24-hr mean serum concentration of phosphorus and the morning fasting serum

TABLE I Whole Blood Ionized Ca^{2+} in Healthy Young and Elderly Men After 9 Days on a Constant Whole Food Diet[a]

	Young men (39 ± 1 years, $n = 13$)	Elderly men (74 ± 2 years, $n = 9$)
Whole blood Ca^{2+} (mg/dl)		
Morning fasting	4.84 ± 0.03	4.84 ± 0.04
24-hr mean	4.77 ± 0.04	4.80 ± 0.04

Arterialized fresh whole blood was analyzed with a calcium ion electrode.

[a] Values are means ± SE. From Portale *et al.* [105].

TABLE II Serum Concentrations of Phosphorus (24-hr mean) and $1,25(OH)_2D$ in Healthy Young and Elderly Men Consuming Constant Whole Food Diets Containing Various Amounts of Phosphorus[a]

	Diet P (mg/day/70 kg)		
	625	1500	2300
Serum P (mg/dl)			
Young men	3.7 ± 0.1	4.2 ± 0.1	4.3 ± 0.1
Elderly men	3.2 ± 0.2	3.7 ± 0.2	3.8 ± 0.2
Serum $1,25(OH)_2D$ (pg/ml)			
Young men	43 ± 2	36 ± 2	29 ± 2
Elderly men	43 ± 2	33 ± 2	31 ± 2

[a] Values are means ± SE. The young men were aged 29 ± 2 years (n = 9), and the elderly men were aged 71 ± 1 years (n = 7). From Portale *et al.* [104].

$1,25(OH)_2D$ level varied inversely ($r = 0.92, p < 0.0001$) (Fig. 5). These data suggest that the magnitude of the response in serum $1,25(OH)_2D$ to changes in serum phosphorus is unaffected by aging. However, for a given level of serum phosphorus, serum $1,25(OH)_2D$ is lower in the elderly. The slope of the regression of serum $1,25(OH)_2D$ onto serum phosphorus was not different for young and elderly men, but the intercept was lower in the elderly group (Fig. 5). Thus, the normal relationship between the serum concentrations of phosphorus and $1,25(OH)_2D$ is altered with advancing age, even in healthy individuals with normal or near normal glomerular filtration rates.

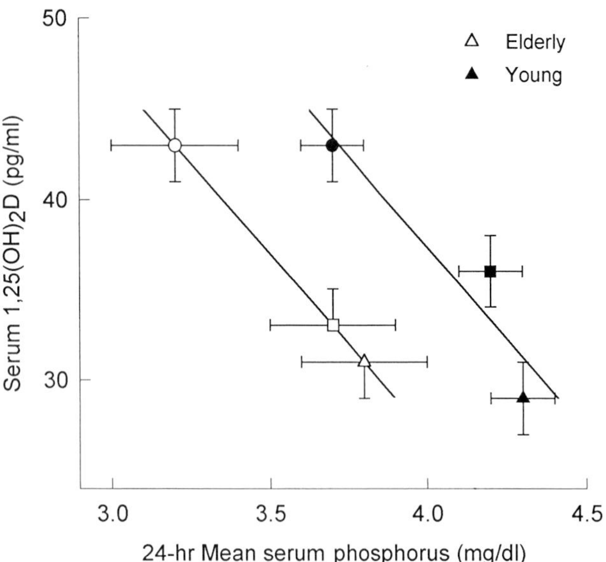

FIGURE 5 Relationship between concentrations of serum $1,25(OH)_2D$ and 24-hr mean serum phosphorus when dietary phosphorus was normal (squares), then supplemented (triangles), and then restricted (circles) in healthy elderly and young men. Data represent means ± SE of serum $1,25(OH)_2D$ and 24-hr serum phosphorus at steady state. Multiple linear regression analysis indicates that the intercept is significantly lower in elderly men (p < 0.001). From Portale *et al.* [104].

3. PARATHYROID HORMONE

The serum concentration of parathyroid hormone (PTH) increases progressively with advancing age in men and women [52–55,74,75,98,106–109]. As early as the fifth decade [98] serum PTH is increased in normal healthy men and by age 70 is 2- to 3-fold higher than in young men (30–40 yr old) [55,98]. The age-related increase in serum PTH is linked in part to diminishing renal function, but even in healthy elderly individuals, in whom GFR is greater than 70 ml/min, serum PTH is increased [105].

In men serum PTH increases with aging, but $1,25(OH)_2D$ does not change, suggesting that the ability of the kidney to respond to PTH may decrease with advancing age [57,62]. Indeed, Slovik *et al.* [62] found that intravenous infusion of human PTH(1–34) can induce an increase in serum $1,25(OH)_2D$ in healthy young subjects but not in elderly patients with osteoporosis. Tsai *et al.* [57] report that the increase in serum $1,25(OH)_2D$ induced by infusion of bovine PTH(1–34) is blunted in elderly postmenopausal women with mild to moderate renal insufficiency when compared to that in healthy young women. These studies, however, involved elderly patients with osteoporosis or mild to moderate renal insufficiency. Thus, they do not permit separation of the effects of aging from other conditions that may impair the ability of the kidney to respond to PTH.

To determine whether aging in normal subjects without osteoporosis or renal insufficiency influences the ability of the kidney to respond to PTH, we studied the renal response to infusion of human PTH(1–34) for 24 hr in 16 healthy men, 9 young (39 ± 1 years) and 7 elderly (70 ± 1 years), free of any conditions known to influence mineral metabolism and in whom the glomerular filtration rate (GFR) was greater than 70 ml/min/1.73 m^2 [110]. Basal concentrations of blood ionized calcium, serum $1,25(OH)_2D$, and urinary calcium and phosphorus were similar in both age groups, but basal serum PTH (+148%), plasma cAMP (+44%), nephrogenous cAMP (NcAMP) (+56%), and fractional excretion of phosphorus (FEPi) (+44%) were higher in the elderly. PTH infusion increased serum $1,25(OH)_2D$ to the same maximum level in the young (47 ± 4 pg/ml) and elderly men (44 ± 5 pg/ml), but the response in the elderly was delayed (Fig. 6). Urinary cAMP, NcAMP, and FEPi increased, but neither the time course nor increment were significantly different be-

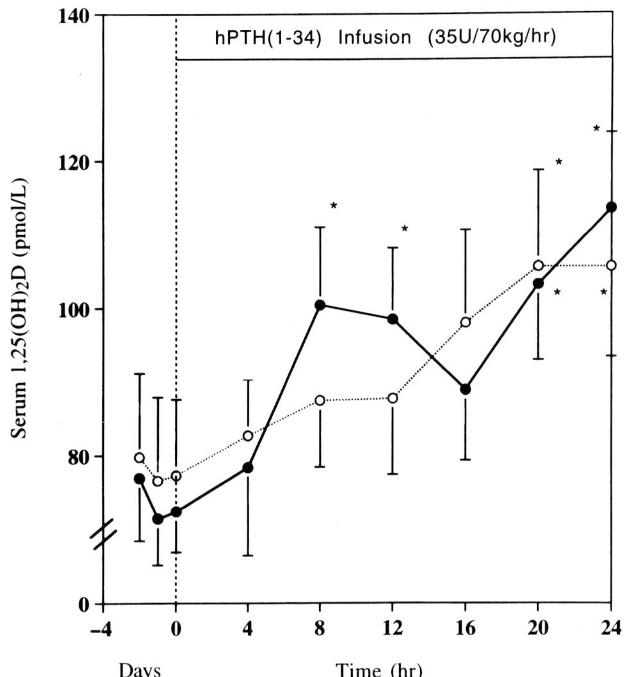

FIGURE 6 Effect of PTH infusion on the serum concentration of $1,25(OH)_2D$ in young (solid line) and elderly men (dotted line). Values are given for the 2 days before PTH administration and every 4 hr during a 24-hr infusion. Asterisks denote significant differences ($p < 0.05$) from baseline (i.e., days 0, −1, and −2), using two-way repeated measures ANOVA. From Halloran BP, Portale AA, Lonergan ET, and Morris RC. Production and metabolic clearance of $1,25(OH)_2D$ in men: Effect of advancing age. *J Clin Endocrinol Metab.* Vol. 70;318–323; 1990; © The Endocrine Society.

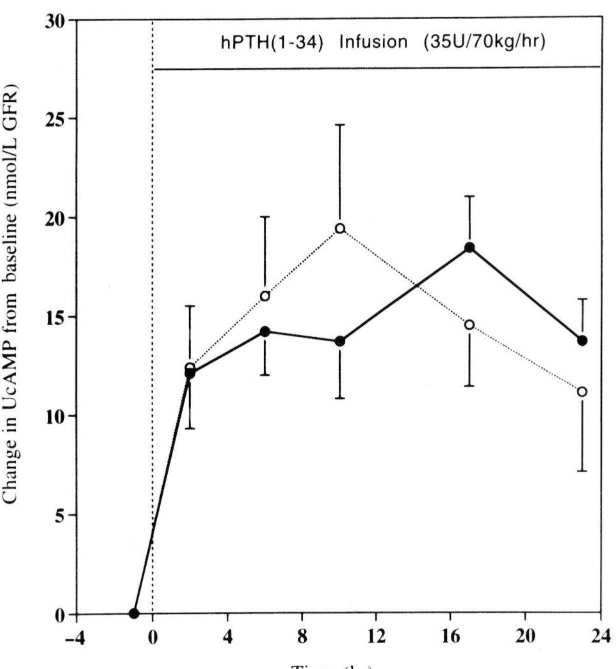

FIGURE 7 Effect of PTH infusion on the change in urinary cAMP from baseline in young (solid line) and elderly men (dotted line). Urinary cAMP increased significantly ($p < 0.05$) in both age groups during PTH infusion, and although the absolute value was significantly higher in the elderly at all time points (data not shown), the magnitude of the change was not significantly different between young and elderly men (by two-way repeated measures ANOVA). From Halloran BP, Portale AA, Lonergan ET, and Morris RC. Production and metabolic clearance of $1,25(OH)_2D$ in men: Effect of advancing age. *J Clin Endocrinol Metab.* Vol. 70;318–323; 1990; © The Endocrine Society.

tween age groups (Figs. 7–9). Tubular absorptive maximum for phosphorus (TmPi)/GFR ratios decreased in response to PTH to the same extent in both age groups. These results demonstrate that although the increment in serum $1,25(OH)_2D$ induced by PTH in the elderly may be delayed relative to that in the young, maximum renal responsiveness to PTH in healthy elderly men is not impaired.

These studies leave open the question as to how basal serum PTH can be elevated in the elderly without causing an increase in serum $1,25(OH)_2D$. The answer to this question and the reason for the altered relationship between the serum concentrations of phosphorus and $1,25(OH)_2D$ may lie with growth hormone (GH) and insulin-like growth factor (IGF). Growth hormone and IGF-I both stimulate 1-hydroxylase activity as well as mediate the response of serum $1,25(OH)_2D$ to dietary phosphorus restriction [111–114]. With advancing age the serum concentrations of GH and IGF-I decrease [115,116], and serum $1,25(OH)_2D$ has been shown to increase in elderly subjects treated with growth hormone [117,118]. These observations suggest that the age-related decrease in serum GH and IGF may reduce tonic stimulation of the 1-hydroxylase. In response, the serum concentration of $1,25(OH)_2D$ would be predicted to decrease. This would predictably reduce calcium absorption in the intestine, thereby producing hypocalcemia. PTH would be expected to increase in response to the hypocalcemia. Accordingly $1,25(OH)_2D$ synthesis would increase through the anabolic actions of PTH, thus offsetting the normal stimulus provided by GH/IGF. The mild hyperparathyroidism observed in elderly men and the consequent hypophosphatemia are consistent with this hypothesis. Of course, other anabolic factors besides GH and IGF may also play a role.

VI. TISSUE RESPONSIVENESS AND THE $1,25(OH)_2D$ RECEPTOR

Active intestinal calcium absorption is an excellent marker of tissue responsiveness to vitamin D. Intestinal calcium absorption has been reported to decrease with

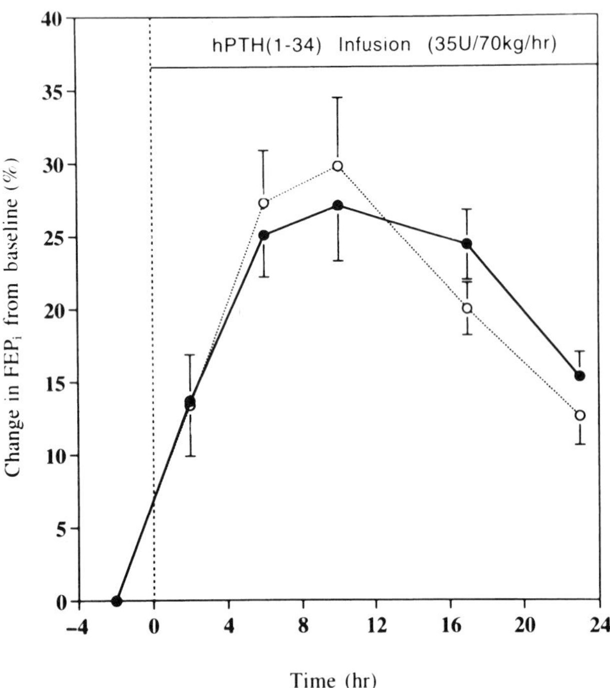

FIGURE 8 Effect of PTH infusion on the change in fractional excretion of phosphorus (FEPi) in young (solid line) and elderly men (dotted line). FEPi increased significantly ($p < 0.05$) in both age groups during PTH infusion, and although the absolute value was significantly higher ($p < 0.05$) in the elderly at all time points (data not shown), the magnitude of the change was not significantly different between young and elderly men (by two-way repeated measures ANOVA). From Halloran BP, Portale AA, Lonergan ET, and Morris RC. Production and metabolic clearance of 1,25(OH)$_2$D in men: Effect of advancing age. *J Clin Endocrinol Metab.* Vol. 70;318–323; 1990; © The Endocrine Society.

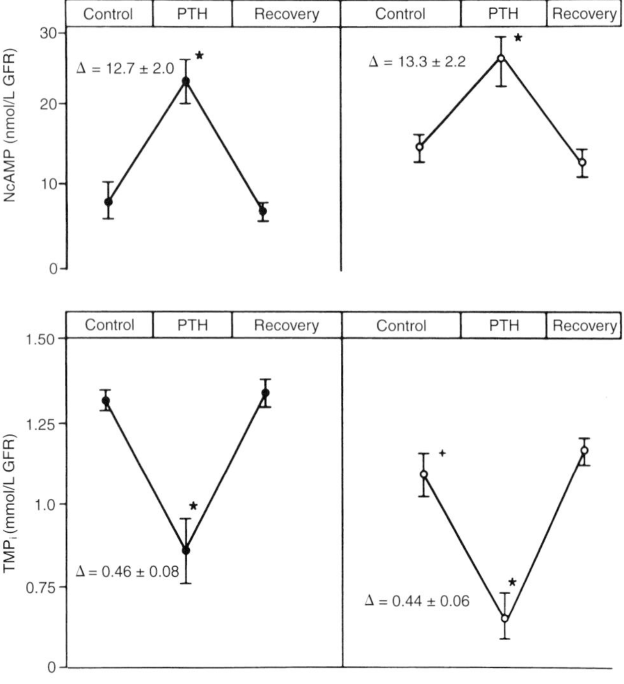

FIGURE 9 Effect of PTH infusion and recovery on NcAMP and tubular reabsorptive maximum for phosphorus (TmPi) in young and elderly men. Values represent measurements taken between 0700 and 0900 hr. Differences are indicated between the control and PTH periods. Using two-way ANOVA, asterisks denote a significant ($p < 0.05$) difference from the control period; plus signs denote a significant difference from the control period in young men. From Halloran BP, Portale AA, Lonergan ET, and Morris RC. Production and metabolic clearance of 1,25(OH)$_2$D in men: Effect of advancing age. *J Clin Endocrinol Metab.* Vol. 70;318–323; 1990; © The Endocrine Society.

postmaturational aging, suggesting that aging maybe accompanied by intestinal (and other end organ) resistance to 1,25(OH)$_2$D [119–122]. In animal studies, Liang *et al.* [67] reported that calcium uptake into duodenal cells declines with donor age but that the decline can be accounted for by decreased serum levels of 1,25(OH)$_2$D in the older animals. Response to *in vivo* administration of 1,25(OH)$_2$D in adult and aged animals was similar. Nevertheless, vitamin D receptor (VDR) protein [123] and mRNA [124] is decreased by 22 and 23%, respectively, in old when compared to young adult rats. VDR binding affinity for 1,25(OH)$_2$D is not affected by aging.

In humans, Eastell *et al.* [75] measured serum 1,25(OH)$_2$D and intestinal calcium absorption in women 26 to 88 years of age. Calcium absorption did not change with age, but serum 1,25(OH)$_2$D gradually increased. Eastell *et al.* [75] concluded that true calcium absorption in healthy elderly women must be resistant to 1,25(OH)$_2$D. To assess the relationship among calcium absorption, serum 1,25(OH)$_2$D, and VDR concentration in the intestinal mucosa, Ebeling *et al.* [76] studied 44 healthy women between the ages of 20 and 87. Over the age range of the population, the serum concentration of 1,25(OH)$_2$D increased by approximately 27%, VDR concentration decreased by 20% (affinity did not change), and intestinal calcium absorption did not change. Although these data suggest that intestinal responsiveness to 1,25(OH)$_2$D declines with age, it is not clear that changes in serum or VDR concentrations of 20–30% have a significant effect on absorption. In a separate study, Ebeling *et al.* [125] report that the responses of serum 1,25(OH)$_2$D and intestinal calcium absorption to restriction of dietary calcium are the same in young (30 ± 1 years) and elderly women (74 ± 1 years). These data support the contention that intestinal responsiveness to 1,25(OH)$_2$D is not affected by age. Collectively the animal and human data suggest that VDR concentration in the intestinal mucosa decreases by approximately 25% during postmaturational aging, but the putative decline in tissue responsiveness to 1,25(OH)$_2$D is either too modest to be detected or does not occur. Interestingly, the response of calcium absorption to $1,25(OH)_2D$ is lower in black than in white women, but the significance of this for aging has not been investigated [126].

Parallel studies in bone comparing responsiveness to $1,25(OH)_2D$ and VDR concentration in bones from young and old adult donors have not been done. Horst *et al.* [127] studied young (4–5 weeks) and adult rats (15–18 months) and found lower receptor concentrations in both intestine and bone in adult animals, but the studies did not address what happens to VDR concentrations during postmaturational aging.

In the rat kidney the concentration of the vitamin D-dependent protein calbindin-D_{28K} decreases with advancing age, but the decline has been linked to lower serum $1,25(OH)_2D$ levels in the older animals [101]. In cultured renal cells from adult (10–12 months) and old rats (20–24 months), Chen *et al.* [101] report that calbindin-D_{28K} induction by $1,25(OH)_2D$ is 51% greater in cells from old than from adult animals. VDR concentrations were not measured.

VII. CONCLUSION

A complete and accurate description of the effects of aging on vitamin D metabolism cannot yet be made. Many aspects have not been adequately investigated, and much of the existing data remain controversial. The following is an attempt to integrate what we think we know combined with a bit of speculation.

In the absence of disease and in a free-living population consuming a normal Western diet, vitamin D metabolism begins to change around midlife. Dietary calcium and cutaneous production of vitamin D decline, serum growth hormone and IGF-I begin to decrease, and renal function begins to deteriorate. As a consequence of these changes the serum concentration of 25OHD decreases, calcium absorption in the intestine diminishes (diet calcium and VDR concentration decrease), and calcium bioavailability to meet serum demands declines. This stimulates PTH secretion, but because of declining renal function and loss of the normal trophic effects of testosterone, estrogen, and GH/IGF on 1-hydroxylase activity, production of $1,25(OH)_2D$ either decreases, remains the same, or increases modestly depending on the individual. The mild hyperparathyroidism associated with aging stimulates bone turnover. Coupled with the modest decrease in serum 25OHD and decline in intestinal calcium absorption, this shifts the source of calcium for maintenance of the serum pool away from the diet and toward the bone. This subtle shift in mineral balance aggravates the gradual loss of bone associated with aging and contributes to the eventual development of senile osteoporosis. The impact of disease and medication use complicates this enormously. Further studies are clearly needed to better define the effects of aging per se and the confounding effects of disease burden on mineral homeostasis and vitamin D metabolism.

References

1. Cerimele D, Celleno L, Serri F 1990 Physiological changes in ageing skin. Br J Dermatol **122**(Suppl. 35):13–20.
2. Lavker RM 1979 Structural alterations in exposed and unexposed aged skin. J Invest Dermatol **73**:59–66.
3. Goodson WH, Hunt TK 1979 Wound healing and aging. J Invest Dermatol **73**:88–91.
4. Daly CH, Odland GF 1979 Age-related changes in the mechanical properties of skin. J Invest Dermatol **73**:84–89.
5. Grove GL, Kligman AM 1983 Age-associated changes in human epidermal cell renewal. J Gerontol **38**:137–142.
6. Escoffier C, DeRegal J, Rochefort A 1989 Age-related mechanical properties of human skin. J Invest Dermatol **93**:353–360.
7. Gilchrest BA, Garmyn M, Yaar M 1994 Aging and photoaging affect gene expression in cultured human keratinocytes. Arch Dermatol **130**:82–86.
8. Puschel HU, Chang J, Muller PK, Brinckman J 1995 Attachment of intrinsically and extrinsically aged fibroblasts on collagen and fibronectin. J Photochem Photobiol **27**:39–46.
9. Ghadially R, Brown BE, Sequeira-Martin SM, Feingold KR, Elias PM 1995 The aged epidermal permeability barrier. J Clin Invest **95**:2281–2290.
10. Tsuchida Y 1993 The effect of aging and arteriosclerosis on human skin blood flow. J Dermatol Sci **5**:175–181.
11. MacLaughlin J, Holich MF 1985 Aging decreases the capacity of human skin to produce vitamin D. J Clin Invest **76**:1536–1538.
12. Webb AR, Kline L, Holick MF 1988 Influence of season and latitude on the cutaneous synthesis of vitamin D_3. J Clin Endocrinol Metab **67**:373–378.
13. Holick MF, Matsuoka LY, Wortsman J 1989 Age, vitamin D and solar ultraviolet. Lancet **2**:1104–1105.
14. Holick MF 1994 Vitamin D—New horizons for the 21st century. Am J Clin Nutr **60**:619–630.
15. Matsuoka LY, Wortsman J, Hanifan N, Holick MF 1988 Chronic sunscreen use decreases circulating concentrations of 25-hydroxyvitamin D. Arch Dermatol **124**:1802–1804.
16. Gloth FM, Gundberg CM, Hollis BW, Haddad JG, Tobin JD 1995 Vitamin D deficiency in homebound elderly persons. J Am Med Assoc **274**:1683–1686.
17. Food and Nutrition Board, Commission of Life Sciences, National Research Council 1989 Recommended Daily Allowances, 10th Ed., National Academic Press, Washington, D.C., pp. 92–98.
18. U.S. Department of Agriculture 1983 Nationwide Food Consumption Survey for 1977–78, Report No. I01. Consumer Nutrition Division, Human Nutrition Information Service, U.S. Department of Agriculture, Hyattsville, Maryland, p. 126.
19. Krall EA, Sahyoun N, Tannenbaum S, Dallal GE, Dawson-Hughs B 1989 Effect of vitamin D intake on seasonal variations in parathyroid hormone secretion in postmenopausal women. N Engl J Med **321**:1777–1783.
20. Schock NW 1972 Energy metabolism caloric intake and physical activity of the aging. In: Carson LA (ed) Nutrition in Old Age. Almqvist and Wiksell, Uppsala, Sweden.

21. Elahi VK, Elahi D, Andres R, Tobin JD, Butler MG, Norris AH 1983 A longitudinal study of nutritional intake in men. J Gerontol **38**:162–180.
22. Omdahl JL, Garry PJ, Hunsaker LA, Hunt WC, Goodwin JS 1982 Nutritional status in a healthy elderly population: Vitamin D. Am J Clin Nutr **36**:1225–1233.
23. Sowers MR, Wallace RB, Hollis BW, Lemke JH 1986 Parameters related to 25OHD levels in a population-based study of women. Am J Clin Nutr **43**:621–628.
24. Delvin EE, Imbach A, Copti M 1988 Vitamin D nutritional status and related biochemical indices in an autonomous elderly population. Am J Clin Nutr **48**:373–378.
25. Payette H, Gray-Donald K 1991 Dietary intake and biochemical indicies of nutritional status in an elderly population with estimates of the precision of the 7-day food record. Am J Clin Nutr **54**:478–488.
26. Barragry JM, France MW, Corless P, Gupta SP, Switala S, Cohen RD 1978 Intestinal cholecalciferol absorption in elderly and young adults. Clin Sci Mol Med **55**:213–220.
27. Clemens TL, Zhou XY, Myles M, Endres D, Lindsay R 1986 Serum vitamin D_3 and vitamin D_2 concentrations and absorption of vitamin D_2 in elderly subjects. J Clin Endocrinol Metab **63**:656–660.
28. Hollander D, Tarnawski H 1984 Influence of aging on vitamin D absorption and unstirred water layer dimensions in the rat. J Lab Clin Med **103**:462–469.
29. Fleming BB, Barrows CH 1982 The influence of aging on intestinal absorption of vitamins A and D by the rat. Exp Gerontol **17**:115–120.
30. Aksnes L, Rodland O, Aarskog D 1988 Serum levels of vitamin D_3 and 25-hydroxyvitamin D_3 in elderly and young adults. Bone Miner **3**:351–357.
31. Van der Wielen RP, Lowik MR, van den Berg H, de Groot LC, Haller J, Moreiras O, van Staveren WA 1995 Serum vitamin D concentrations among elderly people in Europe. Lancet **346**:207–210.
32. McMurtry CT, Young SE, Downs RW, Adler RA 1992 Mild vitamin D deficiency and secondary hyperparathyroidism in nursing home patients receiving adequate dietary vitamin D. J Am Geriatr Soc **40**:343–347.
33. Weisman Y, Schen RJ, Eisenberg Z, Edelstein S, Harell A 1981 Inadequate status and impaired metabolism of vitamin D in the elderly. Isr J Med Sci **17**:19–21.
34. Guillemant J, Oberlin F, Bourgeois P, Guillemant S 1994 Age-related effect of a single oral dose of calcium on parathyroid function: Relationship with vitamin status. Am J Clin Nutr **60**:403–407.
35. Need AG, Morris HA, Horowitz M, Nordin C 1993 Effects of skin thickness, age, body fat and sunlight on serum 25-hydroxyvitamin D. Am J Clin Nutr **58**:882–885.
36. Quesada JM, Coopmans W, Ruiz B, Aljama P, Jans I, Bouillon R 1992 Influence of vitamin D on parathyroid function in the elderly. J Clin Endocrinol Metab **75**:494–501.
37. McKenna MJ 1992 Differences in vitamin D status between countries in young adults and the elderly. Am J Med **93**:69–77.
38. Fardellone P, Sebert JL, Garabedian M, Bellony R, Maamer M, Agbomson F, Brasier M 1995 Prevalance and biological consequences of vitamin D deficiency in elderly institutionalized subjects. Rev Rhum Engl Ed **62**:576–581.
39. Brazier M, Kamel S, Maamer M, Agbomson F, Elesper I, Garabedian M, Desmet G, Sebert JL 1995 Markers of bone remodeling in the elderly subject: Effects of vitamin D insufficiency and its correction. J Bone Miner Res **10**:1753–1761.
40. Ooms ME, Lipps P, Roos JC, van der Vvijgh WJ, Popp-Snijders C, Bezemer PD, Bouter LM 1995 Vitamin D status and sex hormone binding globulin: Determinants of bone turnover and bone mineral density in elderly women. J Bone Miner Res **10**:1177–1184.
41. Guillemant S, Guillemant J, Feteanu D, Sebag-Lanoe R 1989 Effect of vitamin D_3 administration on serum 25-hydroxyvitamin D_3, 1,25-dihydroxyvitamin D and osteocalcin in vitamin D deficient elderly people. J Steroid Biochem **33**:1155–1159.
42. Ooms ME, Roos JC, Bezemer PD, van der Vvijgh WJ, Bouter LM, Lipps P 1995 Prevention of bone loss by vitamin D supplementation in elderly women: A randomized double-blind trial. J Clin Endocrinol Metab **80**:1052–1058.
43. Himmelstein S, Clemens TL, Rubin A, Lindsay R 1990 Vitamin D supplementation in elderly nursing home residents increases 25OHD but not $1,25(OH)_2D$. Am J Clin Nutr **52**:701–706.
44. Lips P, Wiersinga A, van Ginkel FC 1988 The effect of vitamin D supplementation on vitamin D status and parathyroid function in elderly subjects. J Clin Endocrinol Metab **67**:644–650.
45. Dawson-Hughes B, Dallal GE, Krall EA, Harris S, Sokoll LJ, Falconor G 1991 Effect of vitamin D supplementation on wintertime overall bone loss in healthy postmenopausal women. Ann Intern Med **115**:505–512.
46. Dawson-Hughes B, Harris SS, Krall EA, Dallal GE, Falconer G, Green CL 1995 Rates of bone loss in postmenopausal women randomly assigned to one of two doses of vitamin D. Am J Clin Nutr **61**:1140–1145.
47. Barger-Lux MJ, Heaney RP, Lanspa SJ, Healy JC, DeLuca HF 1995 An investigation of sources of variation in calcium absorption efficiency. J Clin Endocrinol Metab **80**:406–411.
48. Orwoll ES, Oviatt SK, McClung MR, Deftos LJ, Sexton G 1990 The rate of bone mineral loss in normal men and the effects of calcium and cholecalciferol supplementation. Ann Intern Med **112**:29–34.
49. Bouillon R, Auwerx J, Dekeyser L, Fevery J, Lisssns W, De Moor P 1984 Serum vitamin D metabolites and their binding protein in patients with liver cirrhosis. J Clin Endocrinol Metab **59**:86–89.
50. Bikle DD, Gee E, Halloran B, Haddad JG 1984 Free 1,25-dihydroxyvitamin D levels in serum from normal subjects, pregnant subjects and subjects with liver disease. J Clin Invest **74**:1966–1977.
51. Manolagas S, Culler FL, Howard JE, Brickman AS, Deftos LJ 1983 The cytoreceptor assay for $1,25(OH)_2D$ and its application to clinical studies. J Clin Endocrinol Metab **56**:751–760.
52. Epstein S, Bryce G, Hinman JW 1986 The influence of age on bone mineral regulating hormones. Bone **7**:421–425.
53. Orwoll ES, Meier D 1986 Alterations in calcium, vitamin D and PTH physiology in normal men with aging. J Clin Endocrinol Metab **63**:1262–1269.
54. Sherman SS, Hollis BW, Tobin JD 1990 Vitamin D status and related parameters in a healthy population: The effects of age, sex and season. J Clin Endocrinol Metab **71**:405–413.
55. Halloran BP, Portale AA, Lonergan ET, Morris RC 1990 Production and metabolic clearance of $1,25(OH)_2D$ in men: Effect of advancing age. J Clin Endocrinol Metab **70**:318–323.
56. Gallagher C, Riggs BL, Eisman JA, Hamstra A, Arnaud SB, DeLuca HF 1979 Intestinal calcium absorption and vitamin D metabolites in normal subjects and osteoporotic patients. J Clin Invest **64**:729–736.
57. Tsai KS, Health H, Kumar R, Riggs BL 1984 Impaired vitamin D metabolism with aging in women. J Clin Invest **73**:1668–1672.
58. Buchanan JR, Myers CA, Greer RB 1988 Effect of declining renal function on bone density in aging women. Calcif Tissue Int **43**:1–6.

59. Prince RL, Dick I, Garcia WP, Retallack RW 1990 The effects of the menopause on calcitriol and PTH: Responses to a low dietary calcium stress test. J Clin Endocrinol Metab **70**:1119–1123.
60. Fujisawa Y, Kida K, Matsuda H 1984 Role of change in vitamin D metabolism with age on calcium and phosphorus metabolism in normal human subjects. J Clin Endocrinol Metab **59**:719–726.
61. Dokoh S, Morita R, Fukunaga M, Yamamoto I, Torizuka K 1978 Competitive protein binding assay for $1,25(OH)_2D$ in human plasma. Endocrinol Jpn **25**:431–436.
62. Slovik SM, Adams JS, Neer RM, Holick MF, Potts JT 1981 Deficient production of $1,25(OH)_2D$ in elderly osteoporotic patients. N Engl J Med **305**:372–374.
63. Lund B, Sorensen OH, Lund B, Agner E 1982 Serum $1,25(OH)_2D$ in normal subjects and in patients with postmenopausal osteopenia. Horm Metab Res **14**:271–274.
64. Armbrecht HJ, Strong R, Boltz M, Rocco D, Wood WG, Richardson A 1988 A modulation of age-related changes in serum $1,25(OH)_2D$ and PTH by dietary restriction in F344 rats. J Nutr **118**:1360–1365.
65. Fox J, Mathew B 1991 Heterogeneous response to PTH in aging rats: Evidence for skeletal PTH resistance. Am J Physiol **260**:E933–E937.
66. Gray RW, Gambert SR 1982 Effect of age on plasma $1,25(OH)_2D$ in the rat. Age **5**:54–56.
67. Liang CT, Barns J, Takamoto S, Sacktor B 1989 Effect of age on calcium uptake in isolated duodenal cells: Role of $1,25(OH)_2D$. Endocrinology **124**:2830–2836.
68. Armbrecht HJ, Forte L, Halloran BP 1984 Effect of age and dietary calcium on renal 25(OH)D metabolism, serum $1,25(OH)_2D$ and PTH. Am J Physiol **246**:E266–E270.
69. Wada L, Daly R, Kern D, Halloran B 1992 Kinetics of 1,25-dihydroxyvitamin D metabolism in the aging rat. Am J Physiol **262**:E906–E910.
70. Theuns HM, van der Vijgh WJF, Hackeng WHL, Bekker H, Barto R, Lips P, Roholl PJM, Netelenbos JC, Knook DL 1991 Parathyroid hormone and vitamin D metabolites throughout life of two inbred rat strains. In: Norman AW, Bouillon R, Thomasset M (eds) Vitamin D. de Gruyter, New York, pp. 769–770.
71. Schock NW 1945 Inulin, diodrast and urea clearance studies on aged human subjects. Fed Proc **4**:65–71.
72. Rowe JW, Anders R, Tobin JR, Norris AH, Schock NW The effect of age on creatinine clearance in man: A cross-sectional and longitudinal study. J Gerontol **31**:155–163.
73. Lindeman RD, Tobin J, Schock NW 1985 Longitudinal studies on the rate of decline in renal function with age. J Am Geriatr Soc **33**:278–285.
74. Marcus R, Madvig P, Young G 1984 Age related changes in PTH and PTH action in normal humans. J Clin Endocrinol Metab **58**:223–230.
75. Eastell R, Yergey AL, Vierira NE, Cedel SE, Kumar R, Riggs BL 1991 Interrelationship among vitamin D metabolism, true calcium absorption, parathyroid function and age in women: Evidence of an age-related intestinal resistance to $1,25(OH)_2D$. J Bone Miner Res **6**:125–132.
76. Ebeling PR, Sandgren ME, DiMagno EP, Lane AW, DeLuca HF, Riggs BL 1992 Evidence of an age-related decrease in intestinal responsiveness to vitamin D: Relationship between serum 1,25-dihydroxyvitamin D and intestinal vitamin D receptor concentrations in normal women. J Clin Endocrinol Metab **75**:176–182.
77. Hartwell D, Riis BJ, Christiansen C 1990 Changes in vitamin D metabolism during natural and medical menopause. J Clin Endocrinol Metab **71**:127–132.
78. Falch JA, Oftebro H, Haug E 1987 Early postmenopausal bone loss is not associated with a decrease in circulating 25OHD, $1,25(OH)_2D$ or vitamin D binding protein. J Clin Endocrinol Metab **64**:836–841.
79. Bikle DD, Halloran BP, Harris ST, Portale AA 1992 Progestin antagonism of estrogen stimulated $1,25(OH)_2D$ levels. J Clin Endocrinol Metab **75**:519–524.
80. Dick IM, Prince RL, Kelly JJ, Ho KK 1995 Oestrogen effects on calcitriol levels in postmenopausal women: A comparison of oral verses transdermal administration. Clin Endocrinol **43**:219–224.
81. Jongen MJ, Bishop JE, Cade C, Norman AW 1987 Effect of dietary calcium, phosphate and vitamin D deprivation on the pharmacokinetics of $1,25(OH)_2D$ in the rat. Horm Metab Res **19**:481–485.
82. Halloran BP, Castro ME 1989 Vitamin D kinetics in vivo: Effect of $1,25(OH)_2D$ administration. Am J Physiol **256**:E686–E691.
83. Fox J, Ross R 1985 Effect of low phosphorus and low calcium diets on the production and metabolic clearance rates of $1,25(OH)_2D$ in pigs. J Endocrinol **105**:169–173.
84. Portale AA, Halloran BP, Murphy MM, Morris RC Jr 1986 Oral intake of phosphorus can determine the serum concentration of $1,25(OH)D_2$ by determining its production rate in humans. J Clin Invest **77**:7–12.
85. Young EW, Hsu CH, Patel S, Simpson RU, Komanicky P 1987 Metabolic degradation and synthesis of calcitriol in spontaneously hypertensive rat. Am J Physiol **252**:E778–E782.
86. Seeman E, Kumar R, Hunder GG, Scott M, Heath H, Riggs BL 1980 Production, degradation and circulating levels of $1,25(OH)_2D$ in health and in chronic glucocorticoid excess. J Clin Invest **66**:664–669.
87. Hsu CH, Patel S, Young EW, Simpson RU 1987 Production and degradation of calcitriol in renal failure rats. Am J Physiol **253**:F1015–F1019.
88. Gray A, Feldman HA, McKinlay JB, Longccope C 1991 Age, disease and changing sex hormone levels in middle-aged men: Results of the Massachusetts male aging study. J Clin Endocrinol Metab **73**:1016–1025.
89. Vermeulen A 1991 Androgens in the aging male. J Clin Endocrinol Metab **73**:221–224.
90. Hagenfeldt Y, Linde K, Sjoberg HE, Zumkeller W, Arver S 1992 Testosterone increases serum $1,25(OH)_2D$ and IGF-I in hypogonadal men. Int J Androl **15**:93–102.
91. Boucher A, D'Amour P, Hamel L 1989 Estrogen replacement decreases the set-point of PTH stimulation by calcium in normal postmenopausal women. J Clin Endocrinol Metab **68**:831–836.
92. Scharla SH, Minne HW, Waibel TS 1990 Bone mass reduction after estrogen deprivation by long-acting gonadotropin-releasing hormone agonists and its relation to pretreatment serum concentrations of $1,25(OH)_2D$. J Clin Endocrinol Metab **70**:1055–1061.
93. Marcus R, Villa ML, Cheema M, Cheema C, Newhall K, Holloway L 1992 Effects of conjugated estrogen on the calcitriol response to parathyroid hormone in postmenopausal women. J Clin Endocrinol Metab **74**:413–418.
94. Packer E, Holloway L, Newhall K, Kanwar G, Butterfield K, Marcus R 1990 Effects of estrogen on daylong circulating calcium, phosphorus, 1,25-dihydroxyvitamin D and parathyroid hormone in postmenopausal women. J Bone Miner Res **5**:877–884.
95. Stock JL, Coderre JA, Mallette LE 1985 Effects of a short course of estrogen on mineral metabolism in postmenopausal women. J Clin Endocrinol Metab **61**:595–600.
96. Gallagher C, Riggs BL, DeLuca HF 1980 Effect of estrogen on

calcium absorption and vitamin D metabolites in postmenopausal osteoporosis. J Clin Endocrinol Metab **51**:1359–1364.
97. Selby PL, Peacock M, Barkworth SA, Brown WB, Taylor GA 1985 Early effects of ethinyloestradiol and norethisterone treatment in postmenopausal women. Clin Sci **69**:265–271.
98. Minisola S, Pacitti M, Scarda A 1993 Serum ionized calcium, parathyroid hormone and related variables: Effect of age and sex. J Bone Miner **23**:183–193.
99. Wiske PS, Epstein S, Bell NH, Queener SF, Edmondson J, Johnston CC 1979 Increases in immunoreactive PTH with age. N Engl J Med **300**:1419–1421.
100. Endres DB, Morgan CH, Garry PJ, Omdahl JL 1987 Age-related changes in serum immunoreactive parathyroid hormone and its biological action in healthy men and women. J Clin Endocrinol Metab **65**:724–731.
101. Chen ML, Boltz M, Christakos S, Armbrecht HJ 1992 Age-related alterations in calbindin-D_{28K} induction by $1,25(OH)_2D$ in primary cultures of rat renal tubule cells. Endocrinology **130**:3295–3300.
102. Hulter HN, Halloran BP, Toto RD, Peterson JC 1985 Long term control of calcitriol concentration in dog and man: The dominant role of plasma calcium concentration in experimental hyperparathyroidism. J Clin Invest **76**:695–702.
103. Villa ML, Packer E, Cheema M, Holloway L, Marcus R 1991 Effects of aluminum hydroxide on the parathyroid-vitamin D axis of postmenopausal women. J Clin Endocrinol Metab **73**:1256–1261.
104. Portale AA, Halloran BP, Morris RC, Lonergan ET 1996 Effect of aging on the metabolism of phosphorus and 1,25-dihydroxyvitamin D in healthy men. Am J Physiol **270**:E483–E490.
105. Portale AA, Lonergan ET, Tanney DM, Halloran BP 1997 Aging alters calcium regulation of serum concentration of parathyroid hormone in healthy men. Am J Physiol **272** (part 1):E139–E146.
106. Imanaka S, Onishi T, Morimoto S, Takamoto S, Kohno H, Kumahara Y 1985 Comparison of renal responses to synthetic human PTH(1–34) administration in normal young and elderly male subjects. J Calcif Tissue Int **37**:357–362.
107. Young G, Marcus R, Minkoff JR, Kim LY, Segre GV 1987 Age-related rise in parathyroid hormone in man: The use of intact and midmolecule antisera to distinguish hormone secretion from retention. J Bone Miner Res **2**:367–373.
108. Forero MS, Klein RF, Nissenson RA, Nelson K, Heath III H, Arnaud CD, Riggs BL 1987 Effect of age on circulating immunoreactive and bioactive parathyroid hormone levels in women. J Bone Miner Res **2**:363–368.
109. Ledger GA, Burritt MF, Kao PC, O'Fallon WM, Riggs BL, Khosla S 1994 Abnormalities of PTH secretion in elderly women that are reversible by short term therapy with $1,25(OH)_2D$. J Clin Endocrinol Metab **79**:211–216.
110. Halloran BP, Lonergan ET, Portale AA 1996 Aging and renal responsiveness to parathyroid hormone in healthy men. J Clin Endocrinol Metab **81**:2192–2197.
111. Gray RW, Garthwaite TL 1985 Activation of renal $1,25(OH)_2D$ synthesis by phosphate deprivation: Evidence for a role for growth hormone. Endocrinology **116**:189–193.
112. Gray RW 1987 Evidence that somatomedians mediate the effect of hypophosphatemia to increase serum $1,25(OH)_2D$ levels in rats. Endocrinology **121**:504–509.
113. Halloran BP, Spencer EM 1988 Dietary phosphorus and 1,25-dihydroxyvitamin D metabolism: Influence of insulin-like growth factor I. Endocrinology **123**:1225–1230.
114. Menaa C, Vrtovsnik F, Friedlander G, Corvol M, Garabedian M 1995 Insulin-like growth factor I, a unique calcium-dependent stimulator of 1,25-dihydroxyvitamin D production. J Biol Chem **270**:25461–25467.
115. Florini JR, Prinz PN, Vitiello MV, Hintz RL 1985 Somatomedin C levels in healthy young and old men: Relationship to peak and 24-hour integrated levels of growth hormone. J Gerontol **40**:2–7.
116. Rudman D 1985 Growth hormone, body composition and aging. J Am Geriatr Soc **33**:800–806.
117. Marcus R, Butterfield G, Holloway L, Gilland L, Baylink D, Hintz R, Sherman B 1990 Effects of short term administration of recombinant human growth hormone to elderly people. J Clin Endocrinol Metab **70**:519–523.
118. Lieberman SA, Holloway L, Marcus R, Hoffman AR 1994 Interactions of growth hormone and parathyroid hormone in renal phosphate, calcium and calcitriol metabolism and bone remodeling in postmenopausal women. J Bone Miner Res **9**:1723–1728.
119. Francis RM, Peacock M, Taylor GA, Storer JH, Nordin BEC 1984. Calcium malabsorption in elderly women with vertebral fractures: Evidence for resistance to the actions of vitamin D metabolites on the bowl. Clin Sci **66**:103–107.
120. Avioli LV, McDonald JE, Lee SW 1965 The influence of age on the intestinal absorption of Ca in women and its relation to Ca absorption in postmenopausal osteoporosis. J Clin Invest **44**:1960–1967.
121. Bullammore JR, Gallagher JC, Wilkinson JR, Nordin BEC 1970 Effect of age on calcium absorption. Lancet **2**:535–537.
122. Heaney RP, Reckler RR, Stegman MR, Moy AJ 1989 Calcium absorption in women: Relationships to calcium intake, estrogen status, and age. J Bone Miner Res **4**:469–475.
123. Takamoto S, Seino Y, Sacktor B, Liang CT 1990 Effect of age on duodenal 1,25-dihydroxyvitamin D receptors in Wistar rats. Biochim Biophys Acta **1034**:22–28.
124. Liang CT, Barns J, Imanaka S, DeLuca HF 1994 Alterations in mRNA expression of duodenal 1,25-dihydroxyvitamin D receptor and vitamin D dependent calcium binding protein in aged Wistar rats. Exp Gerontol **29**:179–186.
125. Ebeling PR, Yergey AL, Vierira NE, Burritt MF, O'Fallon WM, Kumar R, Riggs BL 1994 Influence of age on effects of endogenous $(1,25(OH)_2D$ on calcium absorption in normal women. Calc Tissue Res **55**:330–334.
126. Dawson-Hughes B, Harris SS, Finnerman S, Rasmussen HM 1995 Calcium absorption responsiveness to calcitriol in black and white premenopausal women. J Clin Endocrinol Metab **80**:3068–3074.
127. Horst RL, Goff JP, Reinhardt TA 1990 Advancing age results in the reduction of intestinal and bone 1,25-dihydroxyvitamin D receptor. Endocrinology **126**:1053–1057.

Section V

Diagnosis and Management

CHAPTER 36

Approach to the Patient with Metabolic Bone Disease

MICHAEL P. WHYTE Division of Bone and Mineral Diseases, Washington University School of Medicine at Barnes-Jewish Hospital, St. Louis, Missouri, and Metabolic Research Unit, Shriners Hospital for Children, St. Louis, Missouri

I. INTRODUCTION

Metabolic bone disease traditionally encompasses a great number and variety of conditions [1–3]. In fact, the list is now rapidly growing as the molecular basis of inherited skeletal syndromes and bone dysplasias is being elucidated using DNA technology [4,5]. Although these medical problems are often rare, several are epidemic in many regions of the world (e.g., osteoporosis and vitamin D deficiency rickets). Some can be life-threatening (e.g., severe forms of osteopetrosis and osteogenesis imperfecta); others can be incidental findings (e.g., Paget bone disease and fibrous dysplasia). Any age may be affected. Cumulatively, the number of patients with clinically important metabolic bone disease is significant [1–3].

Diagnosis and treatment of metabolic bone disease in patients of all ages can be both intriguing and satisfying, but there are numerous challenges. The physician now has at his/her disposal techniques to image the skeleton and to measure bone mass, assays for many of the factors that affect mineral and skeletal homeostasis as well as for a considerable number of biochemical markers of bone turnover, and quantitative and qualitative histopathological methods to directly examine osseous tissue [6]. Furthermore, molecular tests for genetic disorders of the skeleton are becoming increasingly available [4]. Initial and follow-up patient evaluation are often helped greatly by correct use of these tools. Nevertheless, for this medical discipline, clinical skill is more important than ever. Judicious selection from among these advances in technology and circumspect interpretation of the information they provide comes from experience with patients. Successful treatment of metabolic bone disease often requires multidisciplinary medical skills that may need to be especially broad-based when deformity or other structural problems of the skeleton are present. Significantly, a variety of potent hormones and drugs that affect mineral homeostasis and/or alter bone remodeling are available to clinicians [3]. This pharmaceutical armamentarium must be used knowledgeably to care for patients effectively yet safely.

This chapter emphasizes a number of considerations for the approach to the patient with metabolic bone disease, particularly those with disturbances in vitamin D homeostasis. Subsequent chapters in this section of the book discuss in detail: bone historphometry (Chapter 37), measurement of the vitamin D metabolites (Chapter 38), the pharmacology and therapeutic use of vitamin D preparations (Chapter 39), and the radiology of rickets and osteomalacia (Chapter 40).

II. DIAGNOSTIC EVALUATION

Patients with metabolic bone disease are often challenging to physicians because many nutritional, environmental, genetic, pharmacological, and toxic factors can adversely affect mineral metabolism and impact the skeleton [2,3]. This book is testimony to the number of diversity of internal and external perturbations that can disturb the biosynthesis or bioactivation of vitamin D.

Patient age is yet another challenge. There are several reasons. In infants, children, and adolescents, complications of metabolic bone disease can be especially severe and complex, because bone growth and modeling are occurring in addition to skeletal remodeling. All three physiological processes can be disturbed in pediatric patients with possibly long lasting consequences on final growth and shape of various bones. The result is novel physical and radiological findings compared to adults. Age conditions the pathogenesis and clinical manifestations of these disorders and provides a guide to the etiology. However, diagnoses can be missed or delayed if the physician is unaware that the reference ranges for some of the biochemical parameters of mineral homeostasis and for all of the markers of skeletal turnover are different for infants and children compared to adults [2,3]. The elderly are challenging, because they are especially likely to have metabolic bone disease with multifactorial etiology and pathogenesis [1–3,7].

Broad-based medical knowledge is necessary to fully understand the relationships between mineral and skeletal homeostasis [8]. Clinicians who encounter patients with metabolic bone disease routinely need some of the skills of the endocrinologist, nutritionist, nephrologist, geneticist, and often the pediatrician or gerontologist. Familiarity with skeletal radiology and pathology is required. Furthermore, if there is significant bony deformity, a working understanding of orthopedics, rheumatology, and rehabilitation medicine is helpful. Metabolic bone disease is a subspecialty that is remarkable for the many disciplines of medicine that contribute to the comprehensive evaluation and effective care of patients.

Diagnosis of metabolic bone disease must begin with acquisition of all of the important information obtainable from the medical history and from a thorough physical examination. The medical history and physical examination are then supplemented with helpful and cost-effective choices from the ever expanding menu of biochemical tests as well as use of relevant radiological and histological studies. For the majority of patients, histological assessment before therapy or during follow-up is not necessary. However, most will require some biochemical and radiological investigation [2,3].

The importance of the medical history for evaluation of metabolic bone disease cannot be overemphasized. Foremost in diagnosis is the orderly accumulation of information from patients [9]. The data help to guide the physical examination and subsequent laboratory studies. Examples are endless. For deficiency of vitamin D_3 historical details alone often explain how the patient has come to require supplementation. Lack of sunlight exposure together with failure to consume foods fortified with vitamin D, use of certain anticonvulsants, or a family history of similar symptoms exemplify critical information that will be obtained by talking with the patient. For the physician skilled in metabolic bone disease, it is not unusual for the physical examination and laboratory testing to merely confirm the diagnosis indicated by the medical history.

Physical examination of the patient with metabolic disease may show findings for or against the diagnosis suspected from the medical history, but the evaluation is also important because it can reveal deformity or other structural problems of the skeleton that need attention.

Radiological investigation of metabolic bone disease may appropriately be minimal or extensive, but it should be directed by the complete medical history and physical examination. Not uncommonly, the diagnosis is established from characteristic radiographic findings (e.g., Paget bone disease, fibrous dysplasia) [10–12]. If not, X-ray studies often provide an important starting point for differential diagnosis (e.g., osteopenia, rickets) or support the diagnostic impression that must be confirmed by additional tests (e.g., pseudohypoparathyroidism, mastocytosis, hyperparathyroidism). However, radiological studies are also important because they may help to assess the severity and evolution of the disorder. In addition, these procedures can reveal skeletal complications not detected by physical examination.

Laboratory procedures are crucial to characterize any disturbance of mineral homeostasis. Furthermore, biochemical testing provides quantitative information to help guide the intensity of medical treatment and to follow the patient's response to therapy. Proper selection of laboratory studies at the time of diagnosis is important to establish the baseline data. Most metabolic bone diseases will be treated medically. If laboratory investigation is incomplete when pharmacological or nutritional intervention begins, an opportunity for diagnosis or for evaluating therapy may be lost. In fact, freezing away some patient serum and an aliquot of urine before medication is prescribed is sometimes worthwhile. This effort provides a means for retrospection. Additionally, in some select circumstances, access to pretreatment specimens can be a tactic for saving money. Expensive and low-yield testing can be held in abeyance pending results of a brief trial of therapy (e.g., suspected environmental vitamin D deficiency rickets). Among the biochemical tests for metabolic bone disease there is now some redundancy (e.g., markers of skeletal turnover) [2,3,6]. Furthermore, many assays are suited only for research purposes where groups of patients are studied. The utility of a single measurement of many of the markers of bone turnover for an individual patient is often limited, because of technical or physiological variability leading to just modest correlation with pathophysiological processes.

Histological study of the skeleton is essential in relatively limited clinical circumstances, though it has considerable research importance. Here, too, consideration of bone biopsy must recognize that an opportunity for diagnosis and/or baseline assessment may be lost after therapy begins [13].

TABLE I Some Potentially Adverse Influences on Skeletal Homeostasis

Genetic	Medical disorders
Ethnic background	Acromegaly
Lifestyle	Anorexia nervosa/bulimia
Inactivity (immobilization)	Cushing syndrome
Smoking	Cystic fibrosis
Nutritional	Early menopause
Alcohol abuse	Fanconi syndrome
Consistently high protein intake	Gastrointestinal disease
Low dietary calcium intake	Glycogen storage disease
Milk intolerance	Hemochromatosis
Vegetarian diet	Hemolytic anemia
Drugs	Hepatobiliary disease
Anticonvulsants	Homocystinuria
Chemotherapy	Hyperparathyroidism
Cyclosporin A	Hypogonadism
Diuretics producing calciuria	Lymphoproliferative disease
Glucocorticoids	Mastocytosis
Gonadotropin-releasing hormone (GnRH) agonists or antagonists	Multiple myeloma
Heparin	Osteogenesis imperfecta
Lithium	Pancreatic insufficiency
Thyroid replacement therapy	Prolactinoma
Vitamin A or D	Renal failure (transplantation)
	Rheumatologic disorders
	Secondary amenorrhea
	Thyrotoxicosis
	Turner syndrome
	Type I diabetes

A. Medical History

The complete, accurate, and detailed medical history and thorough physical examination are critical elements for diagnosis, effective and safe patient management, and sound clinical investigation [9].

> The more resources we have, and the more complex they are, the greater are the demands upon our clinical skill. These resources are calls upon judgment and not substitutes for it. Do not, therefore, scorn clinical examination; learn it sufficiently to get from it all it holds, and gain in it the confidence it merits.
>
> Sir F.M.R. Walshe (1881–1973)
> *Canadian Medical Association Journal* **67**:395, 1952.

The detailed medical history helps assure that the many internal derangements and adverse external factors that can affect the patient with metabolic bone disease will be uncovered (Table I). A questionnaire completed by the patient may serve as a useful beginning, but it is not a satisfactory substitute for a medical history. Only by talking with his/her patient will the physician sense how knowledgeable this individual might be and the value of their historical and current information. The medical history should be a narrative description of the clinical problem.

As discussed below, all of the principal elements of the medical history are potentially important for patients with metabolic bone disease.

1. CHIEF COMPLAINT

The chief complaint (or CC) may readily lead to the diagnosis, for example, pain from hip fracture due to osteoporosis or increasing head size due to Paget bone disease. Often, however, the patient's major concern is more subtle. For example, the above conditions may manifest with gradual loss of height or leg deformity, respectively. These less dramatic problems must be noted not only because they may be pointing to a diagnosis, but also because they can provide a means for assessing the efficacy of medical therapy; for example, when vitamin D deficiency causing weakness from myopathy or bone pain from osteomalacia is effectively treated, these difficulties should resolve. Furthermore, no matter how mild or severe the CC, this is the principal worry that the patient with metabolic bone disease comes to the physician to correct.

With disturbances in vitamin D homeostasis, there may be one or more chief complaints that can be metabolic or skeletal in origin (Table II). This aspect of the medical history is sometimes particularly challenging, because one problem can be emphasized from among

TABLE II Vitamin D Deficiency: Age-Dependent Signs and Symptoms

Metabolic	Skeletal
Hypocalcemia (see Table IV)	Bone tenderness
Muscle	Cranial sutures widened
Asthenia	Craniotabes (skull asymmetry)
Potbelly with lumbar lordosis	Dystocia
Proximal myopathy	Flared wrists and ankles
Waddling gait	Fracture
Dental	Frontal bossing
Caries	Harrison's groove
Delayed eruption	Hypotonia
Enamel defects	Kyphosis
	Lax ligaments
	Limb deformity
	Listlessness
	Low back pain
	Pneumonia
	Rachitic rosary
	Rib deformity → respiratory compromise
	Short stature
	Sternal indention or protrusion
	"String-of-pearls" deformity in hands

this myriad collection of symptoms or signs or because complaints may seem vague or excessive.

2. HISTORY OF PRESENT ILLNESS

Most metabolic bone diseases (including those caused by disturbances in vitamin D homeostasis) are chronic conditions. The detailed history of present illness (or HPI) may be lengthy. Nevertheless, the time invested is crucial because this effort provides the most important historical data. Attention to the details also demonstrates to the patient that the doctor understands and cares about their illness, and this effort is necessary to secure the patient's confidence needed for effective treatment.

From the HPI the physician should obtain an understanding of the temporal evolution of the disorder. This information may be essential for accurate and complete diagnosis. Clues predating the symptoms of vitamin D deficiency would be necessary to fully uncover the pathogenesis and etiology. An understanding of past therapeutic attempts may help reveal factors that are masking a diagnosis. Furthermore, the outcome of previous therapy (successful or unsuccessful) may help to guide future treatment.

At the outset, it is useful for the sake of time and effort, organization, and completeness to tell the patient that this information will be most valuable if obtained in historical sequence. Patients often require some guidance at the beginning of this effort, but most will then describe their medical history in this helpful way. Tactfully diverting them from excessive or extraneous detail is usually quite possible, because they now realize that the physician is both concerned and wishes to achieve this important objective.

Nutritional factors could be considered in the HPI. Both mineral metabolism and vitamin D homeostasis are influenced by diet in countries that fortify foods with vitamin D. Strict vegetarians (vegans), who avoid all foods derived from animals, and nearly as restrictive lactovegetarians, will not benefit from the safety net of vitamin D supplementation of milk products in the United States. Avoidance of dairy products may also lead to a calcium-deficient diet (Table III).

When the HPI is described, not only will critical clues to etiology and pathogenesis emerge, but the physician may also gain important insight for treatment, and here may be a glimpse of the patient's prognosis. Have the manifestations of the disorder been lifelong (suggesting a congenital or genetic problem), or has there been the development of recent symptomatology that should prompt very different diagnostic considerations and interventions? Have complications been substantial and likely to remain so if medical therapy cannot effect a cure? Has the patient been compliant with regard to recommendations for diagnostic study or medical care; if not, will pharmacological therapy be safe?

Only by discussing the patient's illness in detail is the physician likely to learn that previous medical records, radiographs, etc., are available to help avoid expensive duplication of effort, and perhaps to help address important diagnostic and prognostic issues.

Finally, by carefully documenting this aspect of a metabolic bone disease, the physician is providing the basis for sound clinical research.

3. PAST MEDICAL HISTORY

A considerable number and variety of perturbations can cause metabolic bone disease or can influence the

TABLE III Vegetarians

Vegans (strict vegetarians who do not consume dairy products or eggs)
Buddhists (some sects)
Zen (Chinese)
Ch´an (Japanese)
Muslims (some Islamic sects)
Ethnic groups of African, Hispanic, American Indian, Jewish, or Oriental descent (who may have lactose intolerance)
Yoga
Seventh Day Adventists
International Society for Krishna Consciousness
Zen Macrobiotic Movement

outcome of medical treatment by impacting on mineral and skeletal homeostasis (Table I). The detailed past medical history (or PMH) will help to disclose these disturbances or factors that can obscure a diagnosis.

In the PMH, previous studies may be revealed that could prove useful for assessing the metabolic bone disease. Radiographs (e.g., chest X rays, intravenous pyelogram) taken years ago for other purposes could show whether osteopenia or rachitic change is new or old. Routine biochemical test reports found in the patient's medical record might help to date the onset of vitamin D deficiency by documenting when serum alkaline phosphatase activity began to rise.

Drug history must be carefully assessed here and, if relevant, perhaps included instead in the HPI. Many pharmaceuticals can adversely affect the skeleton and disturb mineral or vitamin D homeostasis (e.g., glucocorticoids, certain diuretics, and anticonvulsants) (Chapter 50). A seemingly incidental problem like acne, if overlooked, might fail to reveal tetracycline or retinoid exposure. However, this concern also applies to some over-the-counter preparations (e.g., vitamin A, calcium supplements, antacids). In large amounts, each of these nonprescription items can have important effects on mineral or skeletal homeostasis and result in illness. Nevertheless, they can be overlooked when eliciting a drug history ("Pardon me, doctor. You asked me what *medications* I was taking"). Many patients will not consider vitamins, mineral supplements, or antacids in this category (Table I). Furthermore, some medications confound interpretation of biochemical studies aimed at metabolic bone disease, because they alter mineral homeostasis (e.g., diuretics that increase or decrease urine calcium levels, and pharmaceuticals that elevate serum alkaline phosphatase activity from the liver). Fortunately, in the United States, pharmacological doses of vitamin D require a prescription. Nevertheless, vitamin D intoxication has occurred from excesses inadvertently added to milk and other industrial errors that would have gone unrecognized were it not for a detailed medical history.

The PMH may also help to predict the nature and frequency of recurrent illness. Depending on the patient's prior medical problems, will use of vitamin D_2, with its long biological effect, be safer or more risky than a shorter-acting metabolite? Such information might help to forestall the abrupt development of hypocalcemia if therapy is suddenly compromised, or prevent a prolonged episode of hypercalcemia if dosing becomes excessive.

4. Social History

Patient compliance for medical treatment, especially for chronic disorders, is often imperfect. Furthermore, medical care is expensive. Recognition that a patient will be uncooperative or has health insurance problems or financial difficulties that could affect his/her ability or willingness to undergo diagnostic testing or influence compliance for therapy or follow-up may be disclosed in the social history (or SH). This information may be necessary to formulate not only an effective but also a safe treatment plan, particularly when the disease is severe and/or requires potent medication. The various pharmaceutical preparations of vitamin D, such as 1,25-dihydroxyvitamin D_3 [$1,25(OH)_2D_3$], 1α-hydroxyvitamin D_3 ($1\alpha OHD_3$), dihydrotachysterol (DHT), $25OHD_3$, and vitamin D_2, have very different potencies, biological half-lives, and price (Chapters 39 and 56). What will the issues of long-term treatment and drug cost mean for patient compliance and/or follow-up? The medical literature is scarred with cases of renal failure from vitamin D intoxication when patient monitoring was inadequate.

Because vitamin D_3 is produced naturally by exposure to UV light in sunshine and because nutrition also importantly influences mineral homeostasis, appreciation of climate, clothing, skin pigmentation, and diet may be important. Several religious, ethnic, and other populations have vegetarian members who will not consume dairy products for ethical, spiritual, or health reasons (Table III). Recognition in the SH that the patient belongs to one of these groups may disclose a significant contribution to metabolic bone disease.

Physical factors (e.g., exercise and work activities) often impact patients with metabolic bone disease. Indeed, until more effective regimens are developed to restore diminished skeletal mass, much of what the clinician can do for a child or adult with osteoporosis comes from cautioning them against potentially traumatic pursuits at play or at work. Prevention of falls and proscription against heavy lifting for patients with pediatric or adult forms of osteoporosis are important for diminishing fracture and halting progressive spinal deformity. This straightforward advice, often guided by history-taking, can even help correct vertebral deformities in osteoporotic children who, unlike adults, are capable of reconstituting their spinal anatomy (Fig. 1).

5. Family History

The family history (or FH) is often important when dealing with metabolic bone disease, because many of these conditions are inherited [1–5]. A correct diagnosis may be revealed by study of kindred members, familial benign (hypocalciuric) hypercalcemia or X-linked hypophosphatemia are good examples. Furthermore, significant benefit may come from screening studies to identify and then to treat or to counsel other affected relatives. Inborn errors of vitamin D bioactivation or

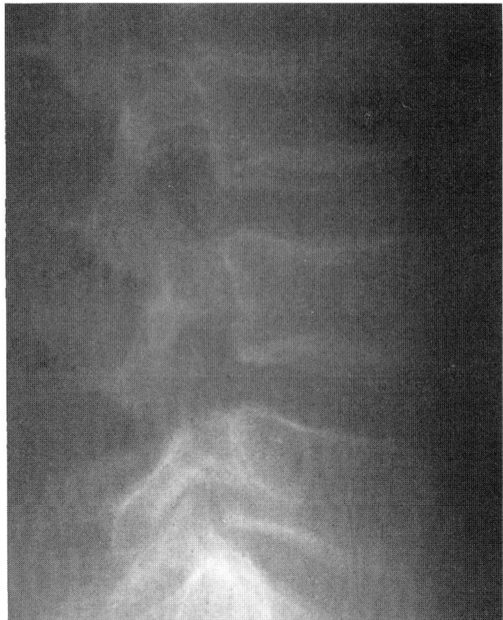
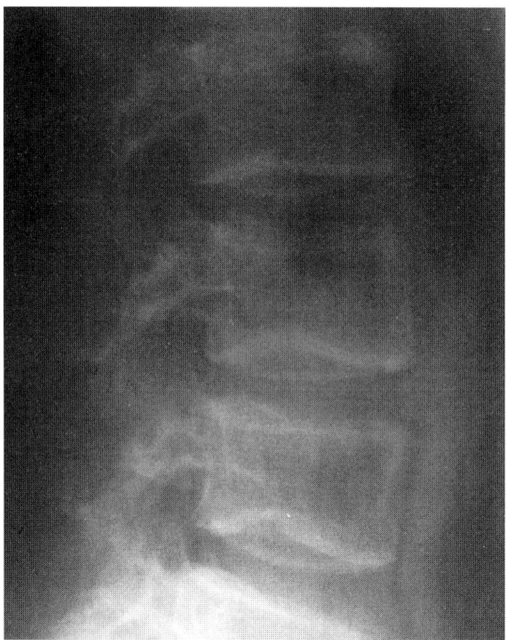

FIGURE 1 Considerable reconstitution of vertebrae (shown here, L_3 to L_5) has occurred between ages 14 (left) and 16 (right) years in a boy with idiopathic osteoporosis. He was counseled against lifting and to avoid falls and stopped participating in traumatic exercises in physical education classes. No pharmacological intervention was attempted.

resistance are rare, but they may be uncovered in the FH.

To report that the FH is "negative" without establishing the value of this information is misleading. Obviously, it is important to know if the patient is adopted and cannot give relevant data. However, an understanding of the size of a kindred is helpful before dismissing the possibility that a heritable disorder has been transmitted. The patient who is the only child of only children, or from a disrupted family, is not as likely to disclose a heritable disorder as will be the patient from a large cohesive kindred. The benefit from this effort can exceed other more costly attempts to reach a diagnosis; medical records from similarly affected family members may be an important guide to prognosis and treatment.

6. Review of Systems

Metabolic bone disease can engender a considerable variety of symptoms. This is especially true for disorders that disturb vitamin D homeostasis and lower extracellular calcium and phosphate levels (Table II). A careful "review of systems" may uncover a sufficient number of these problems so that a diagnosis becomes clearer, or a new or additional condition is suspected. Furthermore, this effort provides a baseline from which to judge the impact of subsequent medical therapy. Symptoms that persist after a course of treatment may need further investigation if they are not expected sequelae of the underlying metabolic problem.

B. Physical Examination

Many clinical signs as well as significant skeletal deformities can accompany or result from metabolic bone disease. Furthermore, not all of these disorders manifest themselves with overt disturbances of hormone or mineral homeostasis (e.g., postmenopausal osteoporosis). Accordingly, physical examination is especially important. Discovery as well as successful treatment of metabolic bone disease often depends on this skill.

Occasionally, diagnosis of a metabolic bone disease rests with the identification of one physical finding, for example, blue or gray sclerae (osteogenesis imperfecta), café-au-lait spots (McCune-Albright syndrome), or tumor (oncogenic rickets/osteomalacia). Other unusual signs, such as premature loss of deciduous teeth (hypophosphatasia), hallux valgus (fibrodysplasia ossificans progressiva), and brachydactyly (pseudohypoparathyroidism), are important clues to additional disorders. For some metabolic bone diseases, a constellation of physical findings leads to the diagnosis. Paget bone disease of bone, in many patients, features an enlarging calvarium with bulging temporal arteries, deafness, asymmetrical bowing of the limbs, and localized areas of skeletal pain and warmth. Postmenopausal osteoporosis causes loss of vertebral height manifested by reduced stature, kyphosis or a gibbus (dowager's hump), a protuberant abdomen (that the patient may confuse with obesity), ribs lowered toward the pelvic brim, paravertebral muscle spasm, and thin skin (McConkey's sign) [7]. Unless physical abnormalities individually or in combination are correctly identified, a diagnosis may be missed. Furthermore, these findings should focus attention on structures of concern, perhaps requiring treatment.

With disturbances in vitamin D homeostasis, a plethora of physical findings can develop (Table II). Patient age determines which may be encountered. Low levels or ineffective action of vitamin D can be especially harmful for infants and for children. As discussed below, distinctive physical findings occur in the pediatric age group that are not observed in adults.

Rickets disturbs the most actively growing bones. Because the skull is enlarging especially quickly at birth, craniotabes (flattened posterior skull) is characteristic of congenital disease. A rachitic rosary (enlargement of the costochondral junctions) can appear during the first year of life, when the rib cage forms rapidly. During infancy or childhood, there may be flared wrists and ankles from metaphyseal widening, Harrison's groove (rib cage ridging from diaphragmatic pull producing a horizontal depression along the lower border of the chest at costal insertions of the diaphragm), and bony tenderness. Although weight bearing typically bows especially the lower limbs, at the time of the adolescent growth spurt, knock-knee deformity may occur. The patient can also have myopathy with reduced muscle strength and tone, a waddling gait, lax ligaments, indentation of the sternum from the force exerted by the diaphragm and intercostal muscles, delayed eruption of permanent teeth, and enamel defects. In infants, floppiness and hypotonia are characteristic. Rachitic infants and young children commonly are listless and irritable. Bone pain can also occur from fracture and deformity.

Hypocalcemia (Chapter 56) can result from vitamin D deficiency (Chapters 42 and 43), pseudodeficiency (Chapter 47), or resistance (Chapter 48) [14]. Hence, it is important that not only the symptoms of hypocalcemia are elicited during the medical history, but the physical signs of latent or overt tetany are recognized during the physical examination (Table IV). Hypocalcemia enhances neuromuscular excitability. This disturbance may lead to varying degrees of tetany, which usually presents with numbness and tingling around the mouth and in the fingertips, and can be followed by muscle spasm in the extremities, face, and elsewhere. Symptoms and signs may be particularly striking when reductions in levels of extracellular ionized calcium are severe or when hypocalcemia occurs rapidly. Typically, there is carpopedal spasm manifest as adduction of the thumb, metacarpophalangeal joint flexion, and interphalangeal joint extension. Latent tetany can be unmasked by eliciting Chvostek's sign or Trousseau's sign. Chvostek's sign is a spasm of the ipsilateral muscles of the face on tapping the facial nerve near its exit from the skull in the region of the parotid gland (just anterior to the ear lobe, below the zygomatic arch). A positive Chvostek's sign ranges from twitching of the lip at the corner of the mouth to spasm of all of the facial muscles on the stimulated side. Slightly positive responses occur in as many as 10 to 15% of normal adults. Trousseau's sign is provoked when a sphygmomanometer is inflated on the arm above the systolic blood pressure for up to 3 min [9]. Positive responses consist of carpal spasm with resolution occurring 5 to 10 sec after the cuff is deflated (i.e., relaxation is not immediate). Both Chvostek's and Trousseau's signs can be absent, however, even in severe hypocalcemia. They appear to be a reflection of the rapidity of change in serum calcium levels. Profound hypocalcemia can also manifest with epileptic seizures, mental status changes, and stridor from laryngeal muscle spasm. Chronic hypocalcemia causes cataracts, dermopathy (Fig. 2), and basal ganglia calcification [14].

Growth rate is an important parameter to follow in infants and children with metabolic bone disease, especially rickets. Improvement or correction of short stat-

TABLE IV Signs and Symptoms of Hypocalcemia[a]

Nervous system
Increased irritability with latent or overt tetany
Seizures
Mental status change, retardation
Basal ganglia calcification
Cardiovascular
Prolonged ST interval with arrhythmia
Cardiomyopathy with congestive heart failure
Hypotension
Other
Papilledema
Lenticular cataracts
Intestinal malabsorption
Dysplastic teeth
Rickets/osteomalacia
Integument changes
Joint contractures
Vertebral ligament calcification

[a] Reprinted with permission from M. P. Whyte. Hypocalcemia, in *Metabolic Bone and Stone Disease,* 3rd Ed. B. E. C. Nordin, A. G. Need, and H. A. Morris, eds., pp. 147–162. Churchill Livingstone, Edinburgh, 1993.

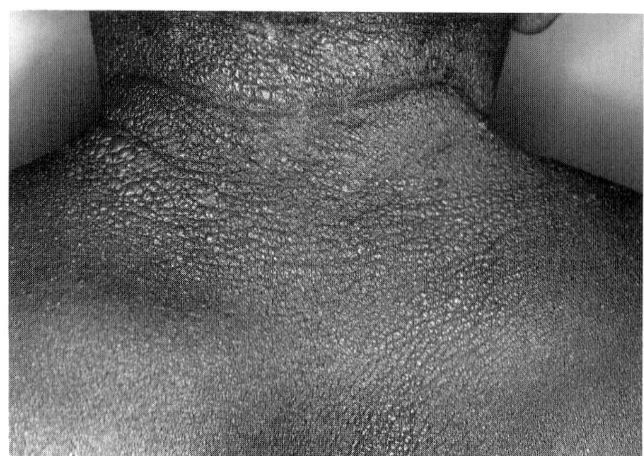

FIGURE 2 Hyperkeratotic dermatosis (shown here, posterior neck) recurs in a 19-year-old man with pseudohypoparathyroidism, who is only intermittently compliant with medical therapy. When he stops treatment with calcium and vitamin D, he becomes markedly hypocalcemic, and the hyperkeratotic lesions reappear.

ure is a major benefit of effective treatment. With it should also come reduction or resolution of skeletal deformity if there is sufficient time for growth before physes close after puberty. Height and length are determined with stadiometers with the patient in bare feet. Weight should also be carefully assessed (and controlled if excessive). With obesity or inordinate weight gain in girls during late childhood and adolescence, there may be transient improvement in stature that does not necessarily reflect a response to medical therapy. Instead, the physician can mistake the influence of excess weight on precipitating puberty for efficacy of treatment. Here, the growth spurt has merely occurred earlier, but physes will fuse soon after menarche, negating any improvement in ultimate adult stature.

Skeletal deformation can cause much of the morbidity of metabolic bone disease. Bowing of the lower limbs predisposes the patient to osteoarthritis especially in the knees. Prevention, control, or correction of deformities is an important goal of patient care. Without a complete physical examination, important clinical problems may go unnoticed. Something as inexpensive as a shoe lift can be of considerable benefit, but the correct size and placement must come from accurate evaluation of leg-length inequality if iatrogenic problems are to be avoided.

For children with rickets, measurements of the upper and lower segment lengths and calculation of their ratio, as well as determination of arm span and height, will help to quantitate skeletal deformity. With rickets (e.g., X-linked hypophosphatemia), some time may pass before the metabolic bone disease is controlled medically. Accordingly, clinical quantitation, photography, gait analysis, and even videotaping of skeletal deformity may help assess progression or document response to therapy.

A "metabolic myopathy" is a prominent clinical feature of vitamin D deficiency and tumor-induced rickets or osteomalacia. Proximal muscle weakness of the limbs is suspected from a history of difficulty rising from sitting position, negotiating stairs, or combing hair, but must be confirmed by physical examination. Gower's sign is an excellent way to detect this problem in children who are observed getting up from a seated position on the floor. If they must push up with their hands on their thighs to achieve upright posture, this is a positive test. Other routine assessments of muscle strength should be performed [9].

In infants and children with rachitic disease, skull shape should be determined and calvarial growth followed by plotting head circumference on standard charts. Early closure of cranial sutures is not uncommon in these disorders [15]. Premature union of sagittal sutures causes a dolichocephalic skull and is very common in X-linked hypophosphatemia, but usually of only cosmetic concern. In hypophosphatasia, however, there can be either functional or true premature fusion of multiple cranial sutures, leading to a scafalocephalic skull, sometimes with raised intracranial pressure.

Dystocia of the pelvis with too narrow a birth canal following vitamin D deficiency during childhood was a major cause of puerperal mortality in mothers in the early twentieth century. This deformity should be searched for in pregnant women with a history of rickets during late childhood or adolescence.

Alopecia is a distinctive clinical finding in some patients with hereditary vitamin D-resistant rickets (vitamin D-dependent rickets, type II) (Chapter 48). However, it is also a manifestation of vitamin D-deficiency rickets accompanying malnutrition and some forms of metaphyseal dysplasia that can be confused clinically and radiographically with rickets [5].

The benign tumors that cause oncogenic rickets or osteomalacia (tumor-induced osteomalacia) are often at least palpable if not visible, but they may be no more than pea-size and are sometimes hidden (Chapter 46). Typically, they are found subcutaneously, but anywhere on the body. However, some have occurred intravaginally or in the nasopharynx, and some lie within the skeleton. Because extirpation of these lesions is curative, especially thorough physical examination for a candidate neoplasm is essential when this disorder is suspected. If the physician cannot find the tumor, patients should be instructed concerning periodic searches for subcutaneous masses. Physical examination yearly is warranted in hopes that previously undetectable lesions will appear.

C. Laboratory Testing

The medical history and physical examination help guide further assessment of the patient with metabolic bone disease by laboratory methods. Beginning the evaluation without a history and physical examination, but instead with biochemical or radiological tests, is unconscionable.

Of some consternation for many physicians who only occasionally encounter patients with metabolic bone disease is the often bewildering array of expensive assays for factors that condition mineral or skeletal homeostasis and for markers of skeletal apposition or resorption (Table V). Although some biochemical testing is necessary for effective diagnosis or treatment of metabolic bone disease, especially disturbances of vitamin D homeostasis, a few studies are usually all that are needed. Appropriate radiological procedures are relatively limited for skeletal disease but often give critical information. Histopathological study is indicated in fewer cases but may provide definitive findings. However, it is expensive and wasteful to cover the diagnostic "waterfront" by ordering a "bone battery" of biochemical tests or a series of low-yield radiological studies.

D. Radiological Examination

1. X-Ray Images

Radiographs of the skeleton chosen selectively are often an important part of the diagnosis and follow-up of patients with metabolic bone disease (Chapter 40) [10–12,16]. However, the "skeletal survey" that examines all bones is an expensive, relatively insensitive, and laborious procedure with a not trivial exposure to X-irradiation. Visualization of the entire skeleton is often indicated to fully assess a bone dysplasia, but it is rarely necessary for evaluation of metabolic bone disease. Instead the "metabolic bone survey" generally provides the necessary radiographic information for diagnosis. This is a study of the appendicular as well as the axial skeleton and therefore delineates both cortical and trabecular bone as well as "red" and "yellow" marrow spaces. The necessary films are an anteroposterior view of the pelvis and a knee, posteroanterior view of a hand and wrist and chest, and a lateral view of the skull and thoracolumbar spine. Chapter 40 as well as several comprehensive texts describe the radiographic findings of the metabolic bone diseases and help to distinguish them from the skeletal dysplasias [10–12,16].

When there is rickets, anteroposterior radiographs of the knees and posteroanterior radiographs of the hands are used to precisely document the presence and degree of physeal and metaphyseal change and to evaluate the response to therapy. Long cassette films of the lower extremities, taken while the patient is standing, help to quantitate bowing or knock-knee deformity. Radiographic studies can also provide clues to the particular etiology or pathogenesis of rickets. For those disorders that cause rickets by disturbing vitamin D homeostasis and result in secondary hyperparathyroidism, subperiosteal bone resorption and osteopenia may be noted in addition to growth plate widening and irregular metaphyses (Fig. 3). These findings contrast to most cases of X-linked hypophosphatemia where the skeleton typically has normal or sometimes increased radiodensity and evidence of hyperparathyroidism is generally absent. In hypophosphatasia, there are often peculiar "tongues" of radiolucency that project from the physes into the metaphyses. Recent-onset rachitic disease will symmetrically widen growth plates, whereas long-standing rachitic disease with bony deformity changes the mechanical forces acting on the physes, which in turn become asymmetrically widened. Accordingly, chronic rachitic disease may seem especially difficult to diagnose with X rays (Fig. 4). Similarly, the rapidity of resolution of rachitic changes on X-ray images may be of diagnostic significance. With vitamin D deficiency rickets from lack of sunlight exposure, radiographic improvement can occur rapidly (a few weeks) following a single pharmacological dose of vitamin D. Other forms of rickets, especially those from renal phosphate wasting, generally take longer (several months or more) to improve or correct with appropriate medical therapy.

TABLE V Biochemical Markers of Bone Remodeling

Marker
Resorption (osteoclast products)
Tartrate-resistant acid phosphatase (serum)
Hydroxyproline (urine)
Hydroxylysine glycosides (urine)
Collagen cross-links (urine and serum)
Total pyridinolines (Pyr and/or Dpy)
Free pyridinolines (Pyr and/or Dpy)
Cross-linked N- and C-telopeptides
Formation (osteoblast products)
Propeptides of type I collagen (serum)
C-propeptide
N-propeptide
Osteocalcin (serum)
Alkaline phosphatase (serum)
Total activity
Bone-specific enzyme

Pyr = pyridinoline; Dpy = deoxypyridinoline.

2. Bone Scintigraphy

Bone scintigraphy is an excellent tool to uncover abnormalities of the skeleton, but it does not establish

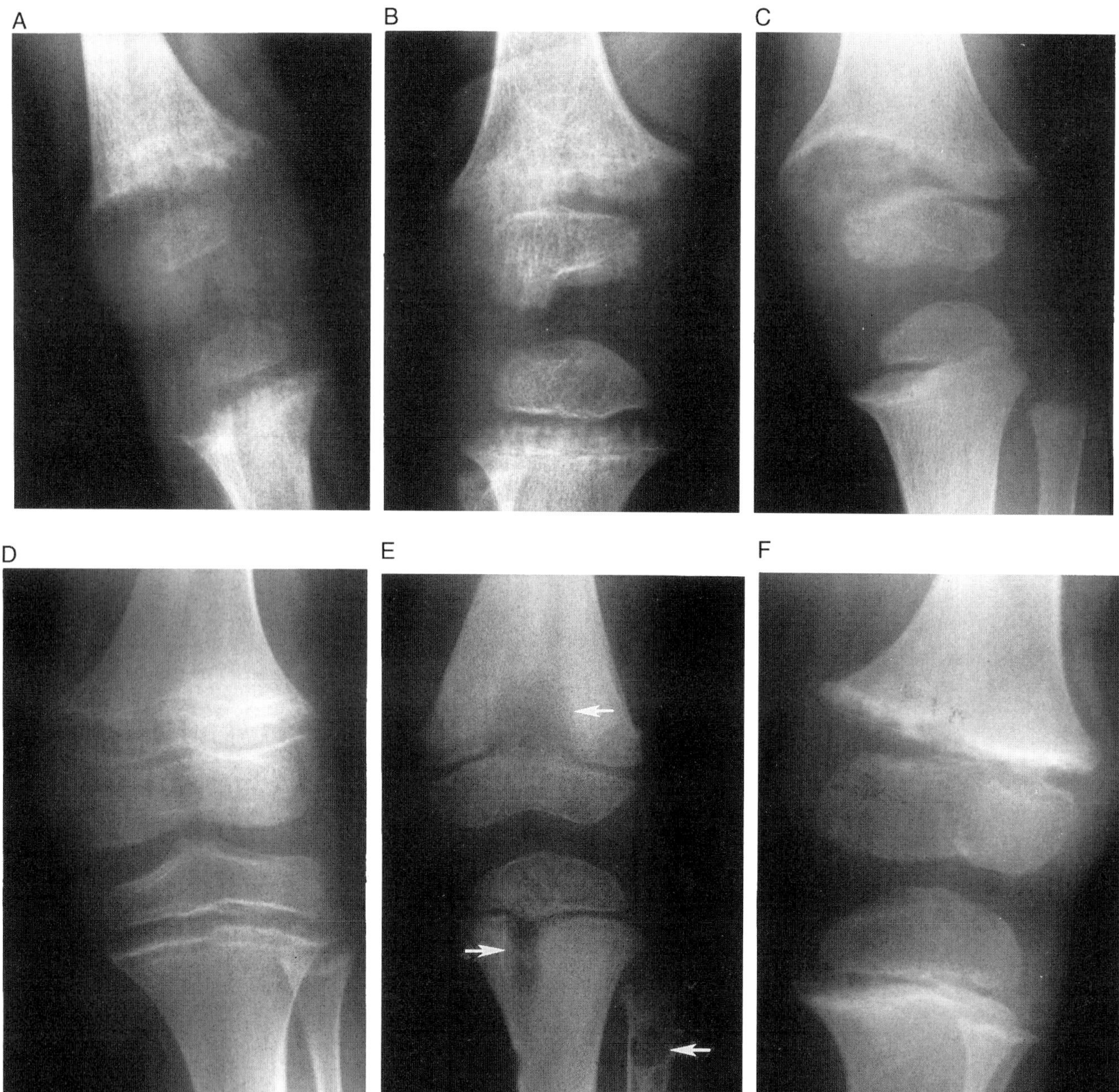

FIGURE 3 Characteristic changes of rickets (growth plate widening and frayed ends of metaphyses) are present in anteroposterior radiographs of the knees of five boys each newly diagnosed with a different type of rachitic disease (A–E). However, additional information is apparent in the films concerning the pathogenesis and/or etiology. (A) A 1-year-old boy and (B) a 5 5/12-year-old boy have osteopenia consistent with documented secondary hyperparathyroidism from "nutritional" rickets and anticonvulsant-induced rickets, respectively. (C) A 2 8/12-year-old boy has normal bone mass and no secondary hyperparathyroidism, consistent with X-linked hypophosphatemia. (D) A 10 3/12-year-old boy has symmetrical widening of the growth plates seen best in the proximal tibia and no long bone deformity that together suggest recent-onset disease in keeping with suspected tumor-induced rickets. (E) A 10 3/12-year-old boy has characteristic "tongues" of radiolucency (arrows), that project from physes into metaphyses, found in the childhood form of hypophosphatasia. However, not all disorders that cause metaphyseal irregularity are forms of rickets. (F) A 3 10/12-year-old boy with normal biochemical studies and iliac crest histology following "tetracycline labeling" has metaphyseal dysplasia.

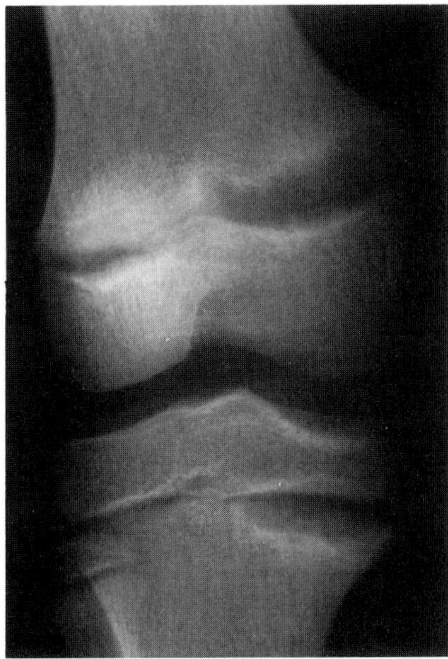

FIGURE 4 A 9 3/12-year-old girl, who is poorly compliant for medical therapy for X-linked hypophosphatemia, has lower limb bowing that asymmetrically deforms the growth plates in her knees.

a diagnosis [17]. Enhanced radioisotope uptake occurs in areas of increased blood flow to the skeleton, excess osteoid, and/or particularly active bone formation. A cost-effective means to assess the patient is to order the bone scan and request X rays of the "hot spots" to provide a diagnosis or to guide further study. Bone scanning is generally not necessary in children with rachitic disease. However, a search for a skeletal source of tumoral rickets by this technique is indicated when physical examination fails to yield an obvious lesion. In adults, bone scanning can also disclose complications of osteomalacia including "true" as well as "false" (pseudo) fractures.

3. Bone Densitometry

Bone mass quantitation is now a widely available tool for the clinician and researcher [6]. Dual energy X-ray absorptiometry (DEXA) and quantitative computed tomography (QCT) usually give reliable assessments of skeletal mass or bone mineral density (BMD). Unfortunately, however, there is currently promulgation of the notion that DEXA can diagnose osteoporosis. DEXA, however, does not provide a diagnosis. In fact, each of the principal categories of metabolic bone disease (see below), namely, osteoporosis, osteomalacia, and osteitis fibrosa cystica, can manifest with low BMD. The presence of osteopenia demonstrable by these techniques is merely a point of departure for differential diagnosis. For those rare conditions associated with increased BMD, these tools are similarly helpful.

Densitometry has some important technical caveats. When using DEXA for children, it is crucial to understand that the tecnique generates a so-called areal (two-dimensional) rather than volumetric (three-dimensional) assessment of BMD. Accordingly, values for BMD are importantly dependent on body size. Children will appear to have lower BMD on DEXA compared to adults. Small stature individuals, who have otherwise normal skeletons, will seem to be osteopenic compared to taller subjects. Hence, the pediatric age group or individuals who are small for reasons other than metabolic bone disease may be incorrectly diagnosed with osteopenia if this technical phenomenon with DEXA goes unrecognized or uncorrected. QCT can measure BMD and focus on either cortical or trabecular bone with the advantage that it provides a volumetric measurement at the expense of a higher level of X-irradiation. QCT can also evaluate the anatomy of the skeleton and detect extracellular calcification [17].

4. Other Radiological Procedures

Magnetic resonance imaging (MRI) is particularly useful for marrow space examination, including delineation of ischemic necrosis of bone. New applications of MRI promise assessment of trabecular bone microanatomy [17]. Ultrasound studies may provide qualitative as well as quantitative information concerning the skeleton [6].

E. Biochemical Investigation

Understandably, circulating calcium levels are closely scrutinized in patients with metabolic bone disease, especially those suspected of having rickets or osteomalacia (Chapters 31 and 56). However, extracellular phosphate levels may be equally if not more important when there is defective skeletal mineralization (Chapters 32 and 46). Chronic hypophosphatemia accompanies a number of conditions that in turn can cause rickets or osteomalacia (Table VI). Accordingly, hypophosphatemia is an especially important finding. Clinicians must recognize that the pediatric age group normally has higher fasting blood phosphate levels compared to adults. Serum phosphate levels should be assayed with fasting blood specimens, because food (depending on content) can acutely raise or lower the level. Assay of phosphate levels in a urine specimen can help to distinguish whether hypophosphatemia is due to renal phosphate wasting or to dietary deficiency. More detailed tests of renal phosphate handling [e.g., tubular

TABLE VI Causes of Chronic Hypophosphatemia

Decreased intestinal absorption
- Antacid abuse
- Vitamin D deficiency
- Malabsorption
- Starvation, alcohol abuse

Increased urinary losses
- Renal tubular defects
- X-linked hypophosphatemia (XLH)
- Oncogenic osteomalacia (tumor-induced osteomalacia)
- Abnormalities of vitamin D metabolism
  - Vitamin D deficiency
  - Vitamin D-dependent rickets
- Hyperparathyroidism
- Renal transplantation
- Alcohol abuse
- Poorly controlled diabetes mellitus
- Metabolic or respiratory acidosis
- Drugs: calcitonin, diuretics, glucocorticoids, bicarbonate
- Respiratory alkalosis
- Extracellular fluid volume expansion
- Severe burns

absorptive maximum for phosphorus/glomerular filtration rate (TmP/GFR) ratios] are necessary for definitive assessment [2,3] (Chapter 32).

Disturbances in vitamin D stores or bioactivation commonly result in hypocalcemia, secondary hyperparathyroidism, and, consequently, hypophosphatemia. Assay of the serum levels of the major active vitamin D metabolites 25OHD and $1,25(OH)_2D$ is essential to detect disturbances in vitamin D stores or in vitamin D bioactivation, respectively. However, the important effects of season on serum 25OHD levels should be considered (Chapters 3 and 33). Furthermore, 25OHD is transported bound to a vitamin D-binding protein (DBP) in the circulation. Accordingly, hypoproteinemia should be recognized before interpreting a serum 25OHD concentration.

Low levels of serum 25OHD generally indicate vitamin D deficiency, but this biochemical finding is merely a starting point for several diagnostic considerations. Furthermore, a considerable number of disorders can cause rickets or osteomalacia (Table VII). Assay of serum $1,25(OH)_2D$ levels is most helpful in exploring hypercalcemia, but this effort is also necessary to characterize inborn errors of vitamin D bioactivation (Chapter 47). Finding a low serum $1,25(OH)_2D$ level also helps in the investigation of oncogenic osteomalacia where concentrations can be low. Serum $1,25(OH)_2D$ levels should be interpreted with pediatric or adult reference ranges and with fasting serum phosphate levels in mind. Hypophosphatemia without physiological evaluation in $1,25(OH)_2D$ levels points to a disturbance in $1,25(OH)_2D$ biosynthesis (Chapter 46).

F. Histopathological Assessment

Throughout most of the twentieth century, clinicians have diagnosed three major types of metabolic bone disease based on the principal histopathological features and the mineral-to-protein ratio of osseous tissue, namely osteoporosis, osteomalacia, and hyperparathyroidism (osteitis fibrosis cystica). Biopsy of the iliac crest, followed by modern laboratory techniques—including preparation of nondecalcified specimens and histomorphometric analysis—readily samples a representative part of the skeleton and can distinguish among these types of metabolic bone disease [13] (Chapter 37). Histological examination of the iliac crest should follow *in vivo* "labeling" of the patient's skeleton by ingestion of two 3-day courses of oxytetracycline or demeclocycline [6,13]. However, documentation that one of these three types of bone disease is present merely provides another starting point for differential diagnosis, as many disorders can cause these types of metabolic bone disease [2,3,13].

Rachitic disease is diagnosed technically by radiological studies of the physes of the wrist and knees together with biochemical testing; hence, bone biopsy is not routinely needed. In adults with osteomalacia, however, growth plate changes on X-ray imaging are "lost" as guideposts by which to judge the diagnosis as well as the efficacy of treatment. Thus, bone biopsy may be especially helpful in adult patients in appropriate situations.

III. TREATMENT

Treatment of metabolic bone disease can range from the simple (e.g., exposure to sunlight for environmental vitamin D deficiency rickets) to the very complex (e.g., bone marrow transplantation for malignant forms of osteopetrosis). When there is a disturbance in vitamin D homeostasis, pharmacological therapy is usually necessary, but additional approaches may be required when there is skeletal deformity. Patients who require vitamin D sterols can benefit greatly from treatment; however, medical care must be skilled (Chapter 39).

Rickets or osteomalacia can occur from vitamin D deficiency or from impaired vitamin D bioactivation. Either disturbance will decrease calcium absorption from the gastrointestinal tract, leading to variable degrees of hypocalcemia, secondary hyperparathyroidism, and hypophosphatemia. The hypocalcemia and hypophosphatemia engenders the defective skeletal mineralization (Chapter 41).

Fortunately, it is possible to prevent, cure, or control most aberrations in vitamin D homeostasis. Vitamin

TABLE VII Causes of Rickets or Osteomalacia

I. Vitamin D deficiency
  A. Deficient endogenous synthesis
    1. Inadequate sunshine
    2. Other factors, e.g., genetic, aging, pigmentation
  B. Dietary
    1. Classic "nutritional"
    2. Fat-phobic
II. Gastrointestinal
  A. Intestinal
    1. Small-bowel diseases with malabsorption, e.g., celiac disease (gluten-sensitive enteropathy)
  B. Hepatobiliary
    1. Cirrhosis
    2. Biliary fistula
    3. Biliary atresia
  C. Pancreatic
    1. Chronic pancreatic insufficiency
III. Disorders of vitamin D bioactivation
  A. Hereditary
    1. Vitamin D dependency, type I (pseudovitamin D deficiency)
    2. Vitamin D dependency, type II (hereditary vitamin D-resistant rickets)
  B. Acquired
    1. Anticonvulsant therapy
    2. Renal insufficiency (see below)
IV. Acidosis
  A. Distal renal tubular acidosis (classic, type I)
    1. Primary (specific etiology not determined)
      a. Sporadic
      b. Familial
    2. Secondary
      a. Galactosemia
      b. Hereditary fructose intolerance with nephrocalcinosis
      c. Fabry's disease
    3. Hypergammaglobulinemic states
    4. Medullary sponge kidney
    5. Post renal transplantation
  B. Acquired
    1. Ureterosigmoidostomy
    2. Drug-induced
      a. Chronic acetazolamide use
      b. Chronic ammonium chloride use
V. Chronic renal failure
VI. Phosphate depletion
  A. Dietary
    1. Low phosphate intake
    2. Total parenteral nutrition
    3. Aluminum hydroxide antacid abuse (or other nonabsorbable hydroxides)
  B. Impaired renal tubular (? intestinal) phosphate reabsorption
    1. Hereditary
      a. X-linked hypophosphatemic rickets
      b. Adult-onset vitamin D-resistant hypophosphatemic osteomalacia
    2. Acquired
      a. Sporadic hypophosphatemic osteomalacia (phosphate diabetes)
      b. Tumor-associated rickets and osteomalacia
      c. Neurofibromatosis
      d. McCune-Albright syndrome
VII. General renal tubular disorders (Fanconi syndrome)
  A. Primary renal
    1. Idiopathic
      a. Sporadic
      b. Familial
    2. Associated with systemic metabolic process
      a. Cystinosis
      b. Glycogenosis
      c. Lowe's syndrome
  B. Systemic disorder with associated renal disease
    1. Hereditary
      a. Inborn errors
        (i) Wilson's disease
        (ii) Tyrosinemia
    2. Acquired
      a. Multiple myeloma
      b. Nephrotic syndrome
      c. Transplanted kidney
    3. Intoxications
      a. Cadmium
      b. Lead
      c. Outdated tetracycline
VIII. Primary mineralization defects
  A. Hereditary
    1. Hypophosphatasia
  B. Acquired
    1. Bisphosphonate intoxication
    2. Fluorosis
IX. States of rapid bone formation with or without a relative defect in bone resorption
  A. Postoperative hypoparathyroidism with osteitis fibrosa cystica
  B. Osteopetrosis ("osteopetrorickets")
X. Defective matrix synthesis
  A. Fibrogenesis imperfecta ossium
XI. Miscellaneous
  A. Magnesium-dependent
  B. Steroid-sensitive
  C. Axial osteomalacia

D deficiency stemming from socioeconomic factors has relative uniformity of etiology and pathogenesis, considerable frequency, and a long history for mankind that make prevention or treatment well understood and usually straightforward (Chapters 42 and 43). Inborn errors of vitamin D bioactivation may be relatively easy to control, merely by providing a "replacement" dose of the missing vitamin D metabolite such as $1,25(OH)_2D_3$ for pseudovitamin D deficiency rickets (vitamin D-dependent rickets, type I) (Chapter 47), or extremely difficult to treat, as in some patients with hereditary vitamin D resistant rickets (vitamin D-dependent rickets, type II) whose resistance to $1,25(OH)_2D$ in some cases is so great that it must be bypassed by providing calcium intravenously (Chapter 48). For other clinical situations, especially hepatobiliary, gastrointestinal, or pancreatic disease that lead to vitamin deficiency, the precise regimen and duration of treatment will vary greatly from patient to patient and likely depend, in part, on any measure of control of the primary disorder (Chapter 51). Hence, it is apparent that an accurate diagnosis, understanding of pathogenesis, and familiarity with the pharmacological armamentarium is essential for successful therapy that may need to be adjusted from time to time.

There is now a considerable number of pharmaceuticals for treatment of metabolic bone disease. Included in the United States are the three major biological forms of vitamin D [vitamin D, 25OHD, and $1,25(OH)_2D$] as well as several preparations that provide inorganic phosphate and innumerable forms of calcium supplementation (Chapters 39 and 56). Hence, repletion of vitamin D reserves or bypassing steps in vitamin D bioactivation is feasible. Furthermore, direct (intravenous) administration of calcium is possible when theremight be insufficient or delayed action of a vitamin D sterol on the gastrointestinal tract. It is essential, however, to recognize that these drugs have different durations of biological storage and potency. Additionally, the lag time between onset of action and longevity of their effects vary considerably (Chapter 39). Finally, the price of these pharmaceuticals differ greatly (Chapter 56). An accurate diagnosis from among the many disorders that cause rickets or osteomalacia is essential for therapeutic efficacy, safety, and economy.

An understanding of the pathophysiology of the disturbance in vitamin D homeostasis is required for effective and safe therapy. Use of $1,25(OH)_2D_3$, when there is depletion of body stores of vitamin D, merely circumvents but does not correct the basic problem. Because many disorders disturb vitamin D homeostasis and because their severity can differ markedly from patient to patient, each condition and patient must be treated individually.

From the above it should be apparent that follow-up of medical treatment is essential for many reasons. Only some generalities and guidelines can be discussed here. For infants, children, and adolescents with rickets, close monitoring is especially important because they are growing. Not only will biochemical and skeletal dynamics change in response to treatment, but there is increasing body size to contend with as well. Alterations in patient weight, a growth spurt, or healing of rachitic disease (alone, or altogether) will likely require changes in dosage. With greater weight, higher doses of vitamin D sterols and/or mineral supplements will likely be required to maintain control of a given metabolic bone disease. A growth spurt, however, can undo such success if medical therapy does not increase to keep up with the greater skeletal demands. Conversely, healing of rickets or osteomalacia may require a reduction in dosage, because the skeletal "sump" is now fully mineralized. It should be clear that follow-up, especially of the pediatric age group of patients with metabolic bone disease, is crucial.

When treating chronic forms of rickets, input from medical and surgical (orthopedic) subspecialties must occur with significant interaction. The child who is not responding to pharmacological therapy may benefit from bracing, epiphysiodesis (physeal stapling), or osteotomy to help correct lower limb deformity. However, these procedures usually reflect failure of medical therapy that can be avoided if dosing, compliance, and follow-up of the treatment regimen are satisfactory. For some patients, particularly children and adolescents, pill counts to assess compliance can provide important information. If deformities are not obviously undergoing correction, at least yearly orthopedic evaluation (perhaps biannually during the growth spurt) is advisable.

Not all types of rickets or osteomalacia manifest with low serum levels of 25OHD or $1,25(OH)_2D$. Hypophosphatemia, with or without hypocalcemia, can lead to defective bone mineralization. Although generally not regarded as disturbances in vitamin D homeostasis, many of these disorders respond to treatment with $1,25(OH)_2D_3$ and mineral supplements [2,3]. Here again, the etiology and pathogenesis of the specific condition must be understood for successful medical management (Chapter 46).

The most common form of heritable rickets, X-linked hypophosphatemia (XLH), is transmitted as an X-linked dominant trait (Chapter 46). The pathogenesis of XLH includes a renal tubular defect engendering loss of phosphate by the kidney [3]. Despite hypophosphatemia, serum $1,25(OH)_2D$ levels are paradoxically normal instead of elevated. More complex proximal renal tubular defects that also feature renal phosphate wasting (Fanconi syndrome) can be due to other inborn errors of metabolism, heritable disorders, or certain drugs or toxins including heavy metal poisoning (Table VI). Developmental hypophosphatemic rickets is also a common complication in McCune-Albright syndrome and is characteristic of oncogenic rickets. Treatment with $1,25(OH)_2D_3$ and phosphate is used for these disorders. Normalization of blood phosphate levels occurs only transiently with phosphate supplementation in conditions characterized by renal phosphate wasting. Clinical improvement can be substantial (Fig. 5), even with-

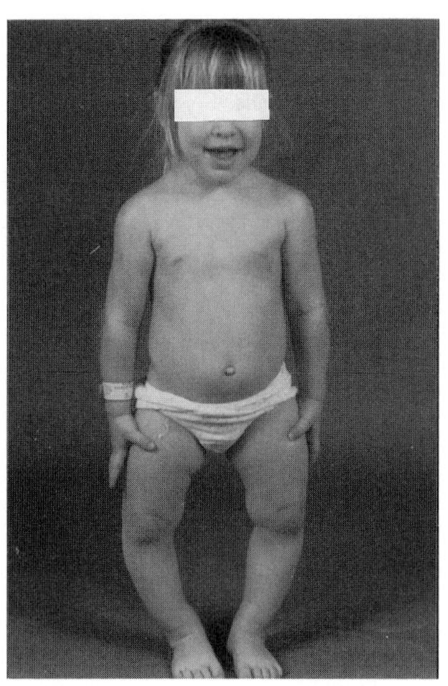
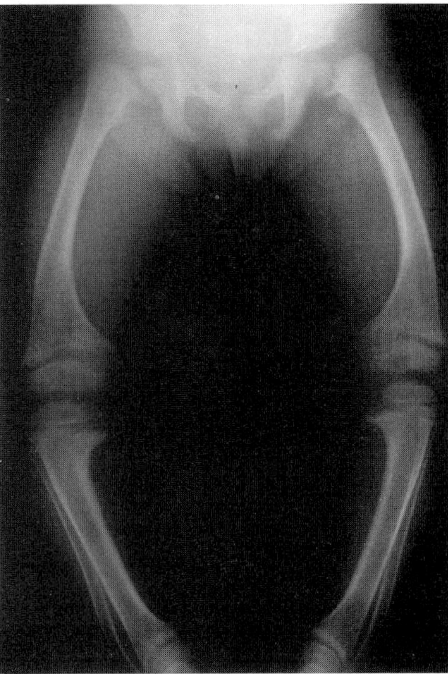
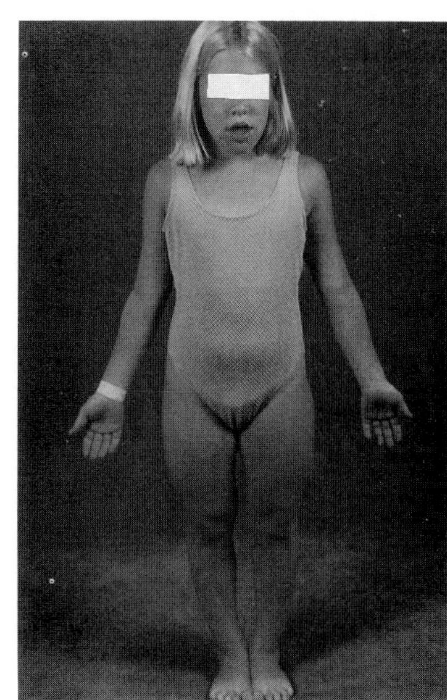

FIGURE 5 In X-linked hypophosphatemia (XLH), the extremes of lower limb deformity in children that can be corrected by pharmacological therapy alone are not well delineated. Accordingly, close cooperation between the medical and orthopedic disciplines is essential. At age 3 years (left), a girl with XLH has severe bowing deformity of the lower extremities that can be quantitated with standing, long-cassette radiographs (middle). Compliance for $1,25(OH)_2D_3$ and inorganic phosphate supplementation therapy was excellent, and her deformity is remarkably improved by age 10 years (right) without osteotomy, epiphysiodesis, or lower extremity bracing.

out correction of hypophosphatemia detected on fasting blood specimens. This biochemical finding should not be considered the objective of therapy, but rather correction of deformity and adequate growth rate.

IV. SUMMARY

The complex interaction of the many endogenous and exogenous factors that impact on vitamin D homeostasis discussed throughout this text explain why patients with disturbances in vitamin D homeostasis are especially challenging. Nevertheless, they typically benefit greatly from the time and effort of a knowledgeable physician. Demonstration of concern for and commitment to the patient by a complete medical history and thorough physical examination helps to win the confidence and trust of the patient. This rapport will likely be essential for effective management of what will often be a chronic disorder. Physical examination is also essential for diagnosis and to uncover structural problems. Information gathered by the medical history and physical examination will be the guide to the myriad of biochemical, radiological, and other technologies that are often important to establish the etiology and pathogenesis and to set the stage for subsequent treatment. Effective therapies are often available for derangements of vitamin D homeostasis, but the pharmaceuticals vary greatly in potency, duration of effect, and cost. Once the proper clinical foundation is in place, the physician can then be gratified by a patient that he/she has helped.

Acknowledgment

Darlene Harmon provided expert secretarial help. Supported by Grant 15958 from the Shriners Hospitals for Children.

References

1. Avioli LV, Krane SM (eds) 1997 Metabolic Bone Disease. Academic Press, San Diego, in press.
2. Coe FL, Favus MJ 1992 Disorders of Bone and Mineral Metabolism. Raven, New York.
3. Favus MJ (ed) 1996 Primer on Metabolic Bone Disease and Disorders of Mineral Metabolism, 3rd Ed. Raven, New York.
4. Scriver CR, Beaudet AL, Sly WS, Valle D (eds) 1995 The Metabolic and Molecular Bases of Inherited Disease. McGraw-Hill, New York.
5. McKusick V 1996 Online Mendelian Inheritance in Man. Monograph online, available from the National Library of Medicine, Bethesda, Maryland.
6. Tovey FI, Stamp TCB (eds) 1995 The Measurement of Metabolic Bone Disease, Vol. 1. Parthenon, London.

7. Marcus R, Feldman D, Kelsey J 1996 Osteoporosis. Academic Press, San Diego.
8. Bilezikian JP, Raisz LG, Rodan GA 1996 Principles of Bone Biology. Academic Press, San Diego.
9. DeGowen and DeGowen 1994 Diagnostic Examination, 6th Ed. McGraw-Hill Health Professionals Division, New York.
10. Resnick D, Niwayama G 1995 Diagnosis of Bone and Joint Disorders, 3rd Ed. Saunders, Philadelphia, Pennsylvania.
11. Edeiken J, Dalinka MK, Karasack D (eds) 1990 Edeiken's Roentgen Diagnosis of Diseases of Bone, 4th Ed. Williams & Wilkins, Baltimore, Maryland.
12. Taybi H, Lachman R 1996 Radiology of Syndromes, Metabolic Disorders, and Skeletal Dysplasias, 4th Ed. Mosby, St. Louis, Missouri.
13. Ravell PA 1985 Pathology of Bone. Springer-Verlag, Berlin.
14. Whyte MP 1993 Hypocalcemia. In: Nordin BEC, Need AG, Morris HA 1993 Metabolic Bone and Stone Disease, 3rd Ed. Churchill Livingstone, Edinburgh, pp. 147–162.
15. Reilly RJ, Leeming JM, Fraser D 1964 Craniosynostosis in the rachitic spectrum. J Pediatr **64**:396–405.
16. Greenfield GB 1986 Radiology of Bone Diseases, 4th Ed. Lippincott, Philadelphia, Pennsylvania.
17. Resnick D (ed) 1996 Bone and Joint Imaging, 2nd Ed. Saunders, Philadelphia, Pennsylvania.

CHAPTER 37

Bone Histomorphometry

JULIET COMPSTON University of Cambridge School of Clinical Medicine, Cambridge, United Kingdom

I. INTRODUCTION

Bone histomorphometry describes the quantitative assessment of bone remodeling, modeling, and structure. It provides information that is not currently available from other investigative approaches, for example, bone densitometry and biochemical markers of bone turnover, and enables a more precise characterization of disease states and their response to treatment than can be obtained from qualitative examination of bone histology (also see Chapter 41). In the last two decades there have been significant advances in histomorphometric techniques, most notably the use of computerized rather than manual techniques and the development of sophisticated approaches to the assessment of bone microarchitecture. The application of these techniques has been particularly valuable in determining the cellular pathophysiology of different forms of bone diseases and in defining the mechanisms by which drugs affect bone.

II. BONE BIOPSY

A. Procedure

The iliac crest is the preferred site for bone biopsy in patients with metabolic bone disease. Most investigators favor the transverse approach, in which a biopsy containing two cortices and intervening cancellous bone is obtained (Fig. 1), in contrast to vertical biopsies, which contain only one cortical plate. In growing individuals, only the transverse approach should be considered, because of the presence of the growth plate along the top of the crest. A number of specially designed trephines are commercially available; ideally for bone histomorphometry, the internal diameter of the specimen should be at least 6 mm.

In most cases, the biopsy is performed as an outpatient procedure. For a transiliac biopsy, the patient lies in the supine position, and the specimen is obtained approximately 2 cm below and behind the anterior superior iliac spine. Most operators use a mild sedative, such as midazolam, and some also routinely administer an analgesic by mouth or injection before the procedure. However, if care is taken to ensure adequate local anesthesia during the biopsy, the latter measure is not generally necessary. The biopsy should be performed under sterile conditions.

The area around the anterior superior iliac spine is infiltrated with local anesthetic, and the inner and outer periosteum are then anesthetized. The author uses a small trocar and stilette which is driven with a weighted instrument until it lies just in the outer cortex; local anesthetic is infiltrated under the periosteum, and the trocar and stilette are then advanced through the bone

FIGURE 1 Section of transiliac biopsy obtained with an 8 mm internal diameter trephine. The biopsy contains inner and outer cortical plates and intervening cancellous bone.

to the inner cortex, where the procedure is repeated. Other investigators anesthetize the inner cortex by introducing a needle through the skin from opposite the biopsy site. A small skin incision is then made, and a hollow cannula with a serrated edge is introduced and placed firmly on the outer periosteum. A smaller, hollow cannula is then inserted through the larger cannula, and the biopsy is obtained by advancing the serrated edge of this cannula through the iliac crest until it has reached the outer surface of the inner cortex. The cannula is then withdrawn after rotation through 360° (to ensure that the core of bone has been freed from adjoining tissues), and the biopsy is removed from the cannula using a metal rod. The incision is then sutured and dressed, and the patient is instructed to lie on the side of the biopsy to apply pressure to the site and reduce the risk of bruising. Ideally, this position should be maintained for 2 hr, and thereafter the patient should be advised to rest for 24 hr.

B. Adverse Effects

Bone biopsy is safe and generally well tolerated, although there is often some discomfort after the procedure, usually lasting between 24 and 48 hr. The morbidity is low and mainly due to hematoma, which may occasionally be extensive; this is most likely to occur in obese subjects or in patients with bleeding diatheses. Other reported complications include infection, transient femoral nerve palsy, avulsion of the superior ramus of the iliac crest, fracture of the iliac crest, and osteomyelitis. It should be stressed, however, that these are extremely rare, and, overall, the incidence of all complications is less than 1% [1].

C. Indications for Bone Biopsy

In clinical practice bone biopsy is most often performed to exclude or confirm a diagnosis of osteomalacia and to characterize the different forms of renal osteodystrophy. In patients with suspected osteomalacia, the diagnosis may be evident from biochemical and/or radiological abnormalities, but in the absence of these, bone biopsy is required. Chronic renal failure is associated with several types of bone disease, as discussed in Chapter 52; accurate diagnosis is essential to establish the correct treatment. Bone biopsy is also helpful in a number of other, rarer forms of metabolic bone disease, for example, fibrogenesis imperfecta ossium and hypophosphatemic osteomalacia. For diagnostic purposes, qualitative assessment of bone by an experienced histopathologist is sufficient, and histomorphometry is not required.

Although bone biopsy is a valuable research tool in metabolic bone disease, it is of little value diagnostically, mainly because of the heterogeneity of bone loss and the relatively weak correlations between bone mass in iliac crest biopsies and clinically relevant sites such as the spine and femoral neck.

III. HISTOMORPHOMETRY

A. Theoretical Considerations

Histological sections of bone provide a two-dimensional representation of a three-dimensional structure; extrapolation from one to the other requires the application of stereological formulas that are based on the assumptions that sampling is random and unbiased and, for most applications, that the structure is isotropic (i.e., evenly dispersed and randomly orientated in space) [2]. Although these conditions cannot be strictly fulfilled in the case of bone histomorphometry, conversion of histomorphometric indices to three-dimensional units is used by some investigators, whereas others express their data as two-dimensional values. The absolute values generated by these different approaches clearly differ, but in practical terms either is acceptable provided that consistency is maintained.

The requirement, for stereological purposes, for bone to be isotropic is clearly not fulfilled in bone in which the macro- and microarchitecture are primarily determined by mechanical forces. However, random and unbiased sampling can be achieved using vertical sections, as first described by Baddeley *et al.* [3] and subsequently applied to bone by Vesterby *et al.* [4]; the vertical axis of the sections is kept parallel to the axis of the cycloid test system, in which the test lines are defined in relation to the axis (sine-weighted).

B. Methodology

The use of manual techniques using grids and graticules has now been almost entirely replaced by interactive computerized systems. These are faster, more operator-friendly, and possess the ability to perform complex measurements that could not be achieved manually. A number of systems are now available, both commercial and in-house. An example of the latter is shown in Fig. 2. Sections are viewed directly through a binocular microscope with a drawing attachment and direct measurements are made using a digitizing tablet, cursor and light-emitting diode (LED) point light source. The system is developed around a personal computer; for structural analysis of cancellous bone, a frame-store board is incorporated to capture images from a charge-coupled device (CCD) television camera on a video monitor [5,6].

C. Terminology

The nomenclature applied to bone histomorphometry has been standardized [7] in an attempt to clarify the cumbersome and sometimes unintelligible terminology

FIGURE 2 Image analysis system used for bone histomorphometry. The image of the biopsy section is captured by a CCD television camera onto a video monitor.

used previously. The revised system expresses all data in terms of the source (the structure on which the measurement is made) the measurement, and the referent. The recommended format (including punctuation) is source-measurement/referent, although because only one source is usually used in a particular study, it is unnecessary to specify this once it has been defined. Area or perimeter (or volume and surface) are used as referents for most measurements (Table I).

Histomorphometric data may be described in two-dimensional or three-dimensional terms; the system used should be consistent within studies. Primary measurements are referred to as area, perimeter, and width (two-dimensional nomenclature) or volume, surface, and thickness (three-dimensional nomenclature) (Table II). It should be noted that absolute area and perimeter measured on conventional histological sections have no three-dimensional equivalent, but, if three-dimensional nomenclature is adopted, these are referred to as volume and surface although the absolute values are identical. Cortical width and thickness are also numerically equal; however, in other situations, for example, trabecular width, conversion to thickness requires division of width by $4/\pi$ (1.273) for isotropic structures or 1.2 for human iliac cancellous bone [2,8]. The units recommended for the revised nomenclature are micrometer and millimeter for length and day and year for time; surface/surface and volume/volume ratios are expressed as percentages whereas volume/surface ratios are expressed in mm^3/mm^2. Some of the more commonly used derived histomorphometric indices are shown in Table III.

Most histomorphometric studies have been confined to cancellous bone which, because of its high surface to volume ratio, exhibits greater remodeling activity than cortical bone. Although cortical bone predominates throughout the skeleton and is a major determinant of bone strength and fracture risk, it is largely ignored by histomorphometrists. The application of histomorphometric techniques to cortical bone was first described by Frost [9] and has been revived more recently by a detailed analysis of the bone remodeling cycle in human cortical bone [10].

TABLE I Referents Commonly Used in Bone Histomorphometry[a]

Referent (3D/2D)	Abbreviation (3D/2D)
Bone surface/perimeter	BS/B.Pm
Bone volume/area	BV/B.Ar
Tissue volume/area	TV/T.Ar
Core volume/area	CV/C.Ar
Osteoid surface/perimeter	OS/O.Pm
Eroded surface/perimeter	ES/E.Pm
Mineralized surface/perimeter	Md.S/Md.Pm
Osteoblast surface/perimeter	Ob.S/Ob.Pm
Osteoclast surface/perimeter	Oc.S/Oc.Pm

[a] 3D, Three dimensions; 2D, two dimensions. Adapted from Parfitt *et al.* [7], with permission.

TABLE II Primary Histomorphometric Indices of Bone Remodeling

Name	Abbreviation	Units
Bone area	B.Ar/T.Ar	%
Osteoid area	O.Ar/T.Ar or O.Ar/B.Ar	%
Osteoid perimeter	O.Pm/B.Pm	%
Osteoblast perimeter	Ob.Pm/B.Pm	%
Osteoid width	O.Wi	μm
Interstitial width	It.Wi	μm
Trabecular width	Tb.Wi	μm
Eroded perimeter	E.Pm/B.Pm	%
Osteoclast perimeter	Oc.Pm/B.Pm	%
Mineralizing surface	Md.Pm/B.Pm	%
Mineral apposition rate	MAR	μm/day
Wall width	W.Wi	μm
Erosion depth	E.De	μm
Erosion length	E.Le	μm
Erosion area	E.Ar	μm^2
Cavity number	N.CV./B.Pm or N.Cv./T.Ar	No./mm or no./mm^2
Quiescent perimeter	Q.Pm	%
Reversal perimeter	Rv.Pm	%

Adapted from Parfitt *et al.* [7], with permission.

D. Limitations of Bone Histomorphometry

Certain limitations of bone histomorphometry should be recognized. Some of these are inherent in the restrictions imposed by a single biopsy site and disease heterogeneity, whereas others reflect imperfections in measurement techniques and difficulties in identification of some of the key processes in remodeling; at least some of those in the latter categories may eventually be overcome by methodological improvements in the future.

A number of studies have documented the large measurement variance associated with bone histomorphometry, which arises from a number of sources including intra- and interobserver variation, sampling variation, and methodological factors [11–13]. Intra- and interobserver variation reflect the subjective approach to identification of many of the histological features assessed,

TABLE III Derived Histomorphometric Indices of Bone Remodeling

Name	Abbreviation	Units
Adjusted apposition rate	Aj.AR	μm/day
Bone formation rate	BFR/B.Pm	$\mu m^2/\mu m$/day
	BFR/B.Ar	%/year
Erosion Rate	ER	μm/day
Mineralization lag time	Mlt	day(s)
Osteoid maturation time	Omt	day(s)
Formation period	FP	day(s)
Active formation period	FP(a+)	day(s)
Erosion period	EP	day(s)
Reversal period	Rv.P	day(s)
Quiescent period	QP	day(s)
Remodeling period	Rm.P	day(s)
Total period	Tt.P	day(s)
Activation frequency	Ac.f	/year
Trabecular separation[a]	Tb.Sp	μm or mm
Trabecular number[a]	Tb.N	/mm

[a] May also be measured directly. Adapted from Parfitt *et al.* [7], with permission.

for example, resorption cavities and osteoid seams. Methodological factors include the criteria used for corticomedullary differentiation, which are often arbitrary, the staining method used, and the magnification at which measurements are made. The technique used for quantitation may also affect the values obtained, as a result of different sampling procedures and variations in the number of sampling points utilized. Many of these sources of variance can be minimized by the standardization of staining, corticomedullary delineation, and magnification and the employment of criteria for identification of osteoid seams, resorption cavities, and newly formed bone structural units.

The important issue of how closely bone remodeling in the iliac crest resembles that at other skeletal sites has not been fully resolved. Some metabolic bone disorders, for example, osteomalacia and most forms of renal osteodystrophy, appear to affect the whole skeleton, and in such cases an iliac crest biopsy is representative. In osteoporosis, however, there is clear evidence of disease heterogeneity, and it is well documented that bone volume in iliac crest biopsies is a poor indicator of bone loss elsewhere in the skeleton [14]. There is also some evidence for variations in bone turnover at different skeletal sites [15]; however, the demonstration of clear abnormalities of bone remodeling in iliac crest bone obtained from patients with osteoporosis indicates that changes responsible for the disease process are reflected, at least to some extent, in bone from this site. Similarly, changes in bone turnover and remodeling balance have been shown in patients with treated osteoporosis and have generally been consistent with the observed changes in bone mass at sites such as the spine and hip.

Finally, current histomorphometric techniques are seriously limited by the lack of reliable markers for activation and resorption. Dynamic indices related to these processes are at present calculated from bone formation rates, based on the assumptions that bone resorption and formation are coupled temporally and spatially and that bone remodeling is in a steady state; however, these are unlikely to be tenable in many cases of untreated or treated osteoporosis [16].

IV. ASSESSMENT OF MINERALIZATION

A. *In vivo* Tetracycline Labeling

The administration of two time-spaced doses of a tetracycline derivative prior to bone biopsy enables measurement and calculation of dynamic indices of bone formation and, by extrapolation, bone resorption [17]. Such information is not currently obtainable by other means, and tetracycline labeling should therefore be performed whenever possible. Various regimes have been described; most involve a 10-day gap between the two doses, bone biopsy being performed 3–5 days after the last dose. The regime used by the author is as follows:

Days 1 and 2	150 mg demeclocycline twice daily
10 days	No demeclocycline
Days 13 and 14	150 mg demeclocycline twice daily

The bone biopsy is performed 3–5 days after day 14.

Adverse effects of demeclocycline and related compounds are rare but include diarrhea and other gastrointestinal symptoms. Occasionally, skin rashes occur; these are sometimes severe and may exhibit photosensitivity.

There is evidence that different tetracycline compounds differ with respect to their uptake by mineralizing bone. Parfitt *et al.* [18] reported that demeclocycline labeling resulted in a greater surface extent of fluorescence than oxytetracycline, significantly affecting values for dynamic indices of bone formation. These differences should be borne in mind when comparing data between centers and, in particular, when using control data obtained from other sources.

B. Measurement of Osteoid

Osteoid may be distinguished from mineralized bone by several staining procedures, including von Kossa, toluidine blue, Goldners trichrome, and solochrome cyanin. Primary measurements of osteoid include area, perimeter, and seam width; assessment of dynamic indices of mineralization, for example, mineral apposition rate, mineralization lag time, and osteoid maturation rate, requires double tetracycline labeling prior to biopsy [7]. Calcification fronts along osteoid seams can be demonstrated by staining with toluidine blue (Fig. 3; see color insert), Sudan black B, or thionin, but the surface extent assessed in this way does not always correlate well with the mineralizing perimeter as measured using tetracycline uptake [19].

Osteoid seam width may be measured directly or calculated from osteoid area and perimeter. However, when relatively small amounts of osteoid are present, the latter approach is inaccurate because the value for osteoid area, expressed as a percentage of bone area, is influenced not only by the seam width but also by the mineralized bone area [20]. Direct assessment can be made using an eyepiece micrometer and calculating the mean of several equidistant measurements of width along each seam.

Measurements of osteoid, in particular its perimeter, are strongly influenced by the magnification used. At high magnifications, it becomes difficult to distinguish osteoid seams from the thin endosteal membrane covering the quiescent bone surface, and for this reason it is preferable to use defined criteria, for example, all seams less than one lamella (3 μm) are excluded. Osteoid measurements may also be affected by the staining procedure used, and poor differentiation of osteoid from mineralized bone in images used for semiautomatic techniques may reduce the accuracy of measurements; thus, values obtained using image analysis are generally lower than those generated by manual measurements [12]. Finally, the delineation of the corticomedullary junction may affect the values obtained for primary measurements of osteoid, as osteoid amount tends to be greater in the cortico-endosteal region than in pure cancellous bone.

C. Dynamic Indices of Mineralization

The administration of time-spaced tetracycline labels prior to biopsy enables calculation of the dynamic indices of matrix formation and mineralization that are central to the histomorphometric diagnosis of osteomalacia. Tetracycline binds to calcium and becomes permanently incorporated into the mineralization fronts at sites of active mineralization (Fig. 4; see color insert). In this respect, the timing of the biopsy in relation to the labeling regime is crucial, with a period of 3 to 5 days after the last label enabling deposition of a sufficiently thick layer of new mineral to retain the tetracycline label [21]. The primary histomorphometric indices of bone remodeling are listed in Table II and the derived indices in Table III. Normative histomorphometric data from the author's laboratory are shown for females (Table IV) and males (Table V).

TABLE IV Normative Histomorphometric Data in Females[a]

Parameter	Age range: 19–30 years (n = 5)	31–40 years (n = 6)[b]	41–50 years (n = 6)[c]	51–60 years (n = 10)[d]	61–80 years (n = 6)
BV/TV (%)	25.9(3.1)	27.7(5.5)	29.6(2.1)	23.9(4.5)	19.8(3.9)
OV/BV (%)	2.4(1.4)	2.7(2.4)	2.3(1.3)	3.1(1.9)	4.7(1.9)
OS/BS (%)	13.1(5.9)	17.7(11.5)	14.5(7.8)	21.2(11.3)	35.0(12.1)
ES/BS (%)	2.15(0.36)	1.84(0.92)	1.78(1.03)	1.76(0.83)	1.66(0.66)
Md.S/BS (%)	9.4(1.8)	8.1(3.4)	8.1(4.2)	13.0(6.7)	14.8(8.1)
O.Th (μm)	5.4(1.9)	3.9(1.6)	5.8(1.1)	5.8(1.6)	5.8(2.3)
W.Th (μm)	45.7(4.9)	51.2(6.6)	47.5(4.9)	36.1(2.9)	32.5(3.6)
MAR (μm/day)	0.59(0.06)	0.60(0.12)	0.61(0.09)	0.61(0.10)	0.54(0.07)
Mlt (days)	12.2(6.2)	16.5(11.8)	20.7(8.7)	21.2(19.6)	29.6(13.5)
BFR ($\mu m^3/\mu m^2$/day)	0.056(0.013)	0.060(0.022)	0.051(0.031)	0.084(0.045)	0.081(0.043)

[a] Results are expressed as means, with SD values in parentheses. Data from Vedi *et al.* [66,67]. Md.S/BS was calculated as the double plus half the single tetracycline-labeled surface.

[b] n = 4 for dynamic variables.

[c] n = 5 for dynamic variables.

[d] n = 9 for dynamic variables.

TABLE V Normative Histomorphometric Data in Males[a]

Parameter	19–30 years (n = 3)	31–40 years (n = 6)	41–50 years (n = 3)	51–60 years (n = 6)	61–80 years (n = 6)
BV/TV (%)	31.3(6.4)	22.2(3.9)	26.9(7.1)	23.0(5.5)	21.4(2.6)
OV/BV (%)	1.6(0.8)	4.1(1.6)	2.8(0.9)	3.1(0.9)	5.6(3.6)
OS/BS (%)	10.0(5.3)	28.3(7.8)	26.3(6.0)	20.0(7.0)	34.8(15.7)
ES/BS (%)	2.84(1.27)	1.69(0.62)	1.68(0.32)	1.77(0.68)	1.91(0.42)
Md.S/BS (%)	9.9(4.0)	13.5(8.4)	8.7(7.0)	8.8(1.5)	9.2(5.1)
O.Th (μm)	8.3(3.0)	6.1(1.3)	6.9(2.9)	6.2(2.0)	6.6(2.8)
W.Th (μm)	49.7(9.6)	45.9(4.4)	42.8(4.0)	36.9(2.0)	33.4(3.3)
MAR (μm/day)	0.67(0.07)	0.60(0.10)	0.55(0.10)	0.57(0.13)	0.53(0.05)
Mlt (days)	10.5(1.9)	28.5(14.6)	36.1(6.2)	34.7(40.8)	49.1(28.1)
BFR ($\mu m^3/\mu m^2$/day	0.066(0.006)	0.079(0.049)	0.047(0.005)	0.051(0.015)	0.045(0.023)

[a] Results are expressed as means with SD values in parentheses. Data from Vedi *et al.* [66,67]. Md.S/BS was calculated as the double plus half the single tetracycline-labeled surface.

1. MINERAL APPOSITION RATE

The mineral apposition rate (MAR) is calculated as the distance between two time-spaced tetracycline labels divided by the time between the administration of the labels. Measurements are made from the midpoint of each label at approximately equidistant points along the labeled surface, and the interlabel period is calculated as the number of days between the midpoints of the two labeling periods [22]. MAR is used in the calculation of many derived indices of bone formation, and accurate measurement is thus of key importance. In the absence of tetracycline uptake, mineral apposition rate and derived indices should be treated as missing data; in biopsies in which only single labels can be detected, the finite lower limit of 0.3 μm/day for mineral apposition rate should be used for the calculation of derived indices [23].

2. ADJUSTED APPOSITION RATE

The adjusted apposition rate (Aj.AR) represents the mineral apposition rate averaged over the osteoid surface. In the absence of a mineralization defect the apposition of matrix and mineral, whilst not synchronous, can be assumed to occur at the same rate and under such circumstances the adjusted apposition rate is equivalent to the osteoid or matrix apposition rate. It is calculated as follows:

$$\text{Aj.AR} = \text{MAR} \times \text{Md.Pm/O.Pm}.$$

From the above formula it is clear that Aj.AR is usually less than MAR and cannot exceed it.

3. MINERALIZATION LAG TIME AND OSTEOID MATURATION TIME

The mineralization lag time (Mlt) is the interval between deposition and mineralization of a given amount of osteoid, averaged over the life span of the osteoid seam. It is calculated as

$$\text{Mlt} = \text{O.Wi/Aj.AR}.$$

The osteoid maturation time (Omt) is the period between the deposition and onset of mineralization of a given amount of osteoid and results from processes such as collagen cross-linking that are necessary before mineralization can proceed. In humans it is usually shorter than Mlt and never exceeds it. It is calculated as follows:

$$\text{Omt} = \text{O.Wi/MAR}.$$

V. HISTOLOGICAL DIAGNOSIS OF OSTEOMALACIA

A. Generalized Osteomalacia

Osteomalacia is essentially a histological diagnosis, although biochemical and radiological abnormalities may enable a firm diagnosis to be made without the necessity for histological examination of bone. Nonetheless, osteomalacia may exist in the absence of biochemical and radiological abnormalities [24], and in such cases bone biopsy is the only certain means by which the diagnosis can be established.

The cardinal feature of osteomalacia is defective mineralization, which results in accumulation of osteoid with an increase in the width of osteoid seams (Fig. 5; see color insert) and a reduction in the surface extent of osteoid showing tetracycline labeling; there is often an increase in the width of individual tetracycline labels, and the distance between double labels is reduced or undetectable. In severe cases, tetracycline labeling may be absent. Increased bone turnover and erosion depth, due to secondary hyperparathyroidism, are present in the earlier stages of hypocalcemic osteomalacia but become less apparent as the mineralized bone surface becomes covered with thick osteoid seams and, therefore, inaccessible to osteoclastic resorption. In such cases tunneling resorption may be apparent, and the irregular outline of mineralized bone beneath the thick osteoid seams provides evidence of previous resorption. Paratrabecular fibrosis is also seen in severe cases. In contrast, histological evidence of secondary hyperparathyroidism is absent in untreated hypophosphatemic osteomalacia.

In histomorphometric terms, osteomalacia is defined as an increased osteoid seam width and a prolonged mineralization lag time [21]. The criteria for abnormality in these indices depend on the source of the reference data, which may vary as a result of both geographical and methodological factors, but in most centers a mean osteoid seam width greater than 12.5 μm and a mineralization lag time in excess of 100 days would be regarded as abnormal. Although the mineral apposition rate is also reduced in osteomalacia, this is not a specific feature because low mineral apposition rates may also result from a reduction in matrix apposition rate as occurs, for example, in postmenopausal osteoporosis, osteogenesis imperfecta, and some forms of secondary osteoporosis. Similarly, the mineralization lag time will be increased in the presence of reduced matrix apposition rate and is therefore not, by itself, pathognomonic of osteomalacia. The distinguishing feature of osteomalacia is that the mineralization lag time is prolonged relative to the adjusted apposition rate, whereas in osteoporosis the reverse is true; this phenomenon accounts for the increase in osteoid seam width that is specific to osteomalacia.

B. Focal Osteomalacia

Focal osteomalacia has been described in patients taking bisphosphonate therapy and is characterized by the focal distribution of abnormally thick osteoid seams with impaired mineralization [25,26]. In such cases the osteoid area and perimeter may be normal, and, because some osteoid seams are of normal width and exhibit normal mineralization, the abnormality may only be detected by examination of the distribution, within single biopsies, of values for osteoid seam width. The significance of these histological changes in terms of fracture risk has not been established; they do not appear, however, to be associated with clinical symptoms or biochemical abnormalities.

VI. ASSESSMENT OF BONE TURNOVER

Bone turnover describes the tissue level of bone resorption and formation, a key determinant of which is activation frequency, that is, the probability that a new remodeling cycle will be initiated at any point of the bone surface. The uptake of tetracycline derivatives at sites of actively forming bone enables the rate of mineralization of osteoid seams and the surface extent of bone formation to be assessed and a number of derived indices, including activation frequency, to be calculated. Tetracycline labeling in bone is detected by fluorescence microscopy under blue light.

A. Mineralizing Perimeter

The extent of bone perimeter or surface that exhibits tetracycline fluorescence is an important primary measurement from which many dynamic indices are derived. When a double tetracycline label has been administered, both double and single labels will be seen; this reflects the labeling escape error, caused by initiation of mineralization before the first label or its termination between administration of the two labels [27], and probably also the switch from an active to resting state in a minority of osteoid seams (the so-called on/off phenomenon) [2]. Because of the former, the extent of double labeled surface underestimates the actively mineralizing surface, and the double plus half the single label is therefore used to estimate the mineralizing perimeter. In cases where only single labels can be detected, it has been suggested that the mineralizing perimeter should be expressed as half the single-labeled perimeter [23].

In the absence of tetracycline administration prior to biopsy, the osteoid perimeter may provide some indication of bone turnover, with increased osteoid perimeter being characteristic of high turnover states. An increase in the extent of perimeter occupied by resorption cavities does not necessarily imply increased bone turnover, however, as these may not represent active resorption but rather reflect failure of formation to occur in previously resorbed cavities.

B. Bone Formation Rates

Bone formation rates are usually expressed in terms of the bone perimeter or area. In the former case, either the osteoid perimeter may be used as the referent (adjusted apposition rate) or the total bone perimeter (tissue-based bone formation rate). They are calculated as follows:

$$\text{Adjusted apposition rate} = \text{MAR} \times \text{Md.Pm/O.Pm},$$

$$\text{Bone formation rate (tissue-based) (BFR/BS)} = \text{MAR} \times \text{Md.Pm/B.Pm}.$$

The bone formation rate/bone area (BFR/B.Ar) is calculated as

$$\text{BFR/B.Ar} = \text{MAR} \times \text{Md.Pm} \times (\text{B.Pm/B.Ar}) \times 100.$$

C. Bone Resorption Rates

Because of the lack of markers of active resorption analogous to the use of tetracycline to identify actively forming bone, bone resorption rates can only be calculated indirectly, from bone formation rates, based on the assumptions discussed earlier. Erosion rate (ER) is expressed in micrometers per day and calculated as follows:

$$\text{ER} = \text{E.De/EP}.$$

Calculation of the erosion period (EP) is shown below.

D. Remodeling Periods

The average duration of a single remodeling cycle is described as the remodeling period, which can be further divided into quiescent, erosion, reversal, and formation remodeling periods [2]. The formation period (FP) can be further divided into the active [FP(a+)] and inactive formation period [FP(a−)] [28]; the latter is a measure of the "off-time," which accounts for the discrepancy between osteoid and mineralizing perimeter after correction for label escape [27]. Formation periods are calculated as follows:

$$\text{FP} = \text{W.Wi/Aj.AR}$$

$$\text{FP(a+)} = \text{W.Wi/MAR},$$

$$\text{FP(a−)} = \text{FP} - \text{FP(a+)},$$

where W.Wi (wall width) is the mean width of completed bone structural units. Quiescent, erosion, and reversal periods are calculated as follows:

$$\text{QP} = \text{Q.Pm/B.Pm} \times \text{FP},$$

$$\text{EP} = \text{E.Pm/B.Pm} \times \text{FP},$$

$$\text{Rv.P} = \text{Rv.Pm/B.Pm} \times \text{FP}.$$

where Rv.Pm = E.Pm − osteoclastic perimeter; Q.Pm = B.Pm. − (O.Pm) + E.Pm).

The mean time between initiation of two successive remodeling cycles at the same site is defined as the total period and calculated as follows:

$$\text{Tt.P} = \text{Rm.P} + \text{QP}.$$

E. Activation Frequency

Activation frequency (Ac.f) is a key determinant of bone mass in the adult skeleton, and increased activation frequency, resulting in high bone turnover, forms quantitatively the most important mechanism of bone loss in osteoporosis [29]. At present, however, there are no *in situ* markers of activation, and hence direct assessment of activation frequency cannot be made. Rather, it is calculated as the frequency with which a given site on the bone surface undergoes new remodeling, as follows:

$$\text{Ac.f} = 1/\text{Tt.P} \quad \text{or} \quad (\text{BFR/B.Pm})/\text{W.Wi}.$$

These formulas define activation frequency as the reciprocal of the time taken from the initiation of one remodeling cycle to initiation of a new one at that site, thus implying its dependence on the life span of previously activated units. Because activation may occur at any site on the bone surface, however, there seems no *a priori* reason why this should be so, particularly in nonsteady states [16]. In addition, the use of indices of bone formation to calculate activation frequency relies on assumptions about the coupling of bone resorption and formation that are unlikely to be tenable in many disease states.

VII. ASSESSMENT OF REMODELING BALANCE

A. Bone Formation

Within individual remodeling units, the amount of bone formed is termed the wall width [30]; this is measured as the mean width of completed bone structural units that are identified under polarized light (Fig. 6; see color insert) or by stains such as toluidine blue or thionin [31], which demonstrate the cement line. Completed structural units are identified by the absence of resorption lacunae or osteoid. There is a large variation in reported values for wall width in both normal and osteoporotic subjects [16], reflecting differences in sampling procedures and difficulties in accurate identification of the cement line, which forms the base of the original resorption cavity.

Investigation of the effect of a disease or its treatment on wall width (and calculated dynamic indices for which wall width is required) necessitates differentiation of those units formed during the period of observation from those formed prior to this time. This can only be achieved by identification of uncompleted bone structural units, which have a covering of osteoid and hence can be presumed to represent current or recent remodeling activity. Reconstruction of these forming sites can then be achieved [32]; however, the number of such units that can be identified in any one biopsy is likely to be extremely small and variance correspondingly high.

B. Bone Resorption

There are several problems associated with accurate assessment of the amount of bone resorbed during each remodeling cycle. The identification of resorption cavities is often difficult and always to some extent subjective; in addition, it is difficult to identify those cavities in which resorption has been completed. The use of polarized light microscopy to demonstrate cutoff lamellae at the edges of the cavity assists recognition [33] (Fig. 7; see color insert), as does the presence of osteoclast-like cells within the cavity. More precise identification of osteoclasts can be achieved by histochemical techniques that demonstrate the presence of tartrate-resistant acid phosphatase, although this is not specific to osteoclasts [34,35]. Finally, it is not usually possible to identify those cavities that have resulted in trabecular perforation.

Indirect assessment of erosion depth was first reported by Courpron *et al.* [36], based on the postulate that the interstitial width, (i.e., the distance between two bone structural units on opposite sides of a trabecula) is inversely proportional to erosion depth. In this model, interstitial width is calculated as the difference between the trabecular width and twice the wall width. However, the relationship between interstitial width and erosion depth is not a simple inverse one, as it is influenced by concomitant changes in wall width and trabecular width [37,38].

A new approach to the direct measurement of erosion depth was reported by Eriksen *et al.* [39]. In this method, the number of lamellae eroded beneath the bone surface is counted, and cavities are characterized according to the presence of osteoclasts, mononuclear cells, and preosteoblastic cells, these being specifically associated with increasing stages of completion of the resorptive phase. This approach depends critically upon the accurate identification, on morphological grounds, of these different cell types within resorption cavities; even in high quality histological sections this can be difficult, and in the author's hands 24% of resorption cavities were excluded from measurement because of failure to classify by cell type or inability to define and count the eroded lamellae. This method has not been widely adopted by other groups, although some have used a simplified approach in which the eroded lamellae are counted without subdivision of cavities by cell type [40]. The latter technique provides an estimate of the mean depth of cavities in all stages of completion and thus underestimates the final resorption depth.

The computerized method developed by Garrahan *et al.* [41] involves reconstruction of the eroded bone surface by a curve-fitting technique (cubic spline) and provides measurements of mean and maximum erosion depth together with the area, surface extent, and number of cavities (Fig. 8; see color insert). Because all resorption cavities are included in the measurements, the mean values for mean and maximum depth and for area considerably underestimate the final resorption depth and area. This method may also be performed using interactive reconstruction of the eroded bone surface [42]. Another modification is to include for measurement only those cavities that contain a thin layer of osteoid, ensuring that the final resorption depth has been achieved [42]; however, the number of such cavities that can be identified in a single biopsy is usually extremely small. Interestingly, the values reported using this approach in normal human biopsies were approximately 17% lower than those obtained in the same biopsies by the technique of counting eroded lamellae and were more consistent with reported age-related changes in trabecular width [43,44].

Further work is thus required to improve existing techniques for the measurement of erosion depth. The approaches most widely used at present measure the depth of all resorption cavities and therefore underestimate the completed erosion depth, whereas measurements made using Eriksen's method probably overestimate the true value [45]. Although these limitations preclude accurate assessment of remodeling balance at present, measurement of erosion depth has generated valuable information about the pathophysiology of bone loss in untreated and treated disease states.

VIII. ASSESSMENT OF BONE STRUCTURE

The importance of cortical and cancellous bone structure as a determinant of bone strength is well established, and there has been increasing interest in the quantitative assessment of bone microarchitecture. Changes in bone structure have important implications not only for bone strength and fracture risk but also for the timing and efficacy of treatment in osteoporosis. In particular, the potential for anabolic agents to restore cancellous bone architecture in patients with advanced bone loss is of considerable interest.

A. Structural Determinants of Bone Strength

Structural determinants of the mechanical strength of bone include cortical width and porosity and, in cancellous bone, trabecular size, shape, connectivity, and anisotropy. Early approaches to the quantitative assessment of cancellous bone structure were based on direct or indirect measurements of trabecular width, separation, and number [46–48]. Direct measurements of trabecular width can be made using an eyepiece graticule or grid but nowadays are most commonly performed using computerized techniques [49,50]. These measurements provide information not only about the mean trabecular width but also about the distributions of trabecular width within individual biopsies. Calculation of trabecular width from area and perimeter measurements may also be performed [51], this approach being based on the assumption that the width of measured structures is small relative to their length. Calculation of trabecular separation and number from trabecular width and bone area is based on a parallel plate model [51] that may often be inappropriate in cancellous bone. Nevertheless, these approaches have generated useful information and have stimulated the development of more sophisticated techniques for analysis of bone structure.

Because all of the structural determinants of cancellous bone strength are three-dimensional characteristics, their assessment on conventional histological sections provides only indirect information about these qualities, and three-dimensional images are required for direct measurement of connectivity, anisotropy, and trabecular size and shape [52]. Although further studies are required to examine the relationship between structural indices obtained using two- and three-dimensional approaches, there are several lines of evidence to support the contention that measurements from two-dimensional sections are representative of three-dimensional structure [53,54].

B. Two-dimensional Approaches

1. Strut Analysis

Strut analysis is based on the definition of nodes and termini and the topological classification of trabeculae and struts. Garrahan *et al.* [5] have described a semiautomated procedure in which the binary image of a section is skeletonized and the different strut types are classified as shown in Fig. 9 (see color insert). The total length of each strut type may be expressed as a percentage of the total strut length or in absolute terms. Node-to-node and node-to-loop strut lengths are positively related to connectivity, whereas node-to-terminus and terminus-to-terminus strut lengths are inversely related. Because of the edge effect, termini may be created artifactually, whereas node-to-node and node-to-loop struts are true indices of connectedness, although their number may be underestimated. Termini may also result from sectioning through a trabecular window in a connected structure.

2. Star Volume

The star volume is defined as the mean volume of solid material or empty space that can be seen unobscured from a point of measurement chosen at random inside the material [55]. Its assessment in histological sections of bone was first reported by Vesterby [56] using the vertical section technique [3] and a cycloid test system. The method may be applied to measurement both of trabecular width (trabecular star volume) and trabecular separation (marrow space star volume) and theoretically provides an unbiased stereological approach to these indices. The method involves the generation of intercepts from random sampling points, with the cubed length of the intercepts being used in the calculation of star volume (Fig. 10; see color insert). The values generated by marrow star volume measurements are significantly influenced by biopsy size, particularly in poorly interconnected cancellous bone in which a large proportion of the measured intercepts may hit the boundary rather than bone, resulting in underestimation of star volume [57].

3. Trabecular Bone Pattern Factor

Another method is based on the concept that patterns or structures can be defined by the relationship between convex and concave surfaces [58], with convexity indicating poor connectivity and concavity reflecting structural integrity. Using a computer-based system, convexity and concavity are assessed by measurement of the bone perimeter before and after computer-based dilatation of the trabecular surface (Fig. 11; see color insert); whereas thickening of convex structures increases their perimeter, the reverse applies to concave structures. The values obtained for trabecular bone pattern factor, which is calculated as the difference between perimeter measurements before and after dilatation divided by the corresponding difference in area, may be significantly influenced by the computer-based smoothing technique used, the degree of dilatation employed, and the magnification at which the measurements are performed.

4. FRACTAL ANALYSIS

Fractal objects are characterized by scale invariance or self-similarity over a wide range of magnifications so that any one piece of a fractal, if magnified sufficiently, resembles the intact object [59]. Fractal analysis has been described in a number of biological systems including the bronchial tree and vascular networks; in bone, it has been applied to radiological images and histological sections of bone [60]. The value obtained for the fractal dimension is critically dependent on the magnification used for measurement, and the relationship between fractal dimension and connectivity in cancellous bone has not been established. Multidirectional fractal analysis can be used to assess structural anisotropy [61].

C. Three-dimensional Approaches

A number of techniques have been used to generate three-dimensional images of bone. These include reconstruction of serial sections, scanning and stereo microscopy, volumetric, high resolution, and microcomputed tomography, and magnetic resonance imaging [62,63]. The potential for imaging techniques such as magnetic resonance imaging and computed tomography to provide information about bone structure *in vivo* is an important and active area of current research. At present, such application of these approaches is restricted by limited resolution, partial volume effects, and noise. Nevertheless, such approaches enable direct assessment of connectivity using the Euler number, a topological property based on the number of holes and number of connected components in an object [64]. This measurement can also be obtained using the ConnEuler method in which projections are made through parallel thin section pairs, or dissectors, spaced approximately 10–40 μm apart [65].

IX. FUTURE DEVELOPMENTS

Despite the current limitations of bone histomorphometry, its value in defining pathophysiological processes responsible for disease and its unique ability to demonstrate the mechanisms by which drugs affect bone are increasingly recognized. The rapid advances that have occurred in molecular biological techniques and in our knowledge of bone cell biology should soon enable better identification of the processes of activation and resorption *in situ*, leading to a better understanding of mechanisms of bone loss and bone gain. Finally, rapid advances in the *in vivo* assessment of cancellous bone architecture will provide new insights into structural determinants of bone strength and the ability of anabolic skeletal agents to restore bone architecture in patients with advanced bone loss.

References

1. Rao DS 1983 Practical approach to bone biopsy. In Recker R (ed) Bone Histomorphometry: Techniques and Interpretation. CRC Press, Boca Raton, Florida, pp. 3–11.
2. Parfitt AM 1983 The physiological and clinical significance of bone histomorphometric data, In Recker R (ed) Bone Histomorphometry: Techniques and Interpretations. CRC Press, Boca Raton, Florida, pp. 143–224.
3. Baddeley AJ, Gundersen HJG, Cruz Orive LM 1986 Estimation of surface area from vertical sections. J Microsc **142**:259–276.
4. Vesterby A, Kragstrup J, Gundersen HJG, Melsen F 1987 Unbiased stereologic estimation of surface density in bone using vertical sections. Bone **8**:13–17.
5. Garrahan NJ, Mellish RWE, Compston JE 1986 A new method for the analysis of two-dimensional trabecular bone structure in human iliac crest biopsies. J Microsc **142**:341–349.
6. Compston JE, Garrahan NJ, Croucher PI, Yamaguchi K 1993 Quantitative analysis of trabecular bone structure. Bone **14**:187–192.
7. Parfitt AM, Drezner MK, Glorieux FH, Kanis JA, Malluche H, Meunier PJ, Ott SM, Recker RR 1987 Bone histomorphometry. Standardization of nomenclature, symbols and units. J Bone Miner Res **2**:595–610.
8. Schwartz MP, Recker RR 1981 Comparison of surface density and volume of human iliac trabecular bone measured directly and by applied sterology. Calcif Tissue Int **33**:561–565.
9. Frost HM 1963 Mean formation time of human osteons. Can J Biochem Physiol **41**:1307–1310.
10. Agerbaek MO, Eriksen EF, Kragstrup J, Mosekilde LE, Melsen F 1991 A reconstruction of the remodelling cycle in normal human cortical iliac bone. Bone Miner **12**:101–112.
11. de Vernejoul MC, Kuntz D, Miravet L, Goutalier D, Ryckewaert A 1981 Histomorphometric reproducibility in normal patients. Calcif Tissue Int **33**:369–374.
12. Chavassieux PM, Arlot ME, Meunier PJ 1985 Intermethod variation in bone histomorphometry: Comparison between manual and computerised methods applied to iliac bone biopsies. Bone **6**:211–219.
13. Wright CDP, Vedi S, Garrahan NJ, Stanton M, Duffy SW, Compston JE 1992 Combined inter-observer and inter-method variation in bone histomorphometry. Bone **13**:205–208.
14. Compston JE, Crowe JP, Wells IP, Horton LWL, Hirst D, Merrett AL, Woodhead JS, Williams R 1980 Vitamin D prophylaxis and osteomalacia in chronic cholestatic liver disease. Dig Dis **25**: 28–32.
15. Eventov I, Frisch B, Cohen Z, Hammel I 1991 Osteopenia, hematopoiesis, and bone remodelling in iliac crest and femoral biopsies: A prospective study of 102 cases of femoral neck fractures. Bone **12**:1–6.
16. Compston JE, Croucher PI 1991 Histomorphometric assessment of trabecular bone remodelling in osteoporosis. Bone Miner **14**:91–102.
17. Frost HM 1969 Tetracycline-based histological analysis of bone remodelling. Calcif Tissue Int **3**:211–237.

18. Parfitt AM, Foldes J, Villanueva AR, Shih MS 1991 Difference in length between demethylchlortetracycline and oxytetracycline: Implications for the interpretation of bone histomorphometric data. Calcif Tissue Int **48**:74–77.
19. Compston JE, Vedi S, Webb A 1985 Relationship between toluidine blue-stained calcification fronts and tetracycline labelled surfaces in normal human iliac crest biopsies. Calcif Tissue Int **37**:32–35.
20. Vedi S, Compston JE 1984 Direct and indirect measurements of osteoid seam width in human iliac crest biopsies. Metab Bone Dis Related Res **5**:269–274.
21. Parfitt AM 1990 Osteomalacia and related disorders. In: Krane SM (ed) Metabolic Bone Disease, 2 Ed., Grune & Stratton, New York.
22. Frost HM 1983 Bone histomorphometry: Analysis of trabecular bone dynamics. In Recker R (ed) Bone Histomorphometry: Techniques and Interpretations. CRC Press, Boca Raton, Florida, pp. 109–131.
23. Foldes J, Shih M-S, Parfitt AM 1990 Frequency distributions of tetracycline-based measurements: Implications for the interpretation of bone formation indices in the absence of double-labelled surfaces. J Bone Miner Res **5**:1063–1067.
24. Peach H, Compston JE, Vedi S, Horton LWL 1982 The value of plasma calcium, phosphate and alkaline phosphatase in the diagnosis of histological osteomalacia. J Clin Pathol **35**:625–630.
25. Boyce BF, Fogelman I, Ralston S, Johnston E, Ralston S, Boyle IT 1984 Focal osteomalacia due to low-dose diphosphonate therapy in Paget's disease. Lancet **1**:821–824.
26. Adamson BB, Gallacher SJ, Byars J, Ralston SH, Boyle IT, Boyce BF 1993 Mineralisation defects with pamidronate therapy for Paget's disease. Lancet **342**:1459–1460.
27. Frost HM 1983 Bone histomorphometry: Choice of marking agent and labelling schedule, In Recker R (ed) Bone Histomorphometry: Techniques and Interpretations. CRC Press, Boca Raton, Florida, pp. 37–51.
28. Arlot M, Edouard C, Meunier PJ, Neer RM, Reeve J 1984 Impaired osteoblast function in osteoporosis: Comparison between calcium balance and dynamic histomorphometry. Br Med J **289**:517–520.
29. Frost HM 1985 The pathomechanics of osteoporosis. Clin Orthop Related Res **200**:198–225.
30. Lips P, Courpron P, Meunier PJ 1978 Mean wall thickness of trabecular bone packets in the human iliac crest: Changes with age. Calcif Tissue Res **26**:13–17.
31. Derkz P, Birkenhäger-Frenkel DH 1995 A thionin stain for visualizing bone cells, mineralizing fronts and cement lines in undecalcified bone sections. Biotech Histochem **70**:70–74.
32. Steiniche T, Eriksen EF, Kudsk H, Mosekilde L, Melsen F 1992 Reconstruction of the formative site in trabecular bone by a new, quick, and easy method. Bone **13**:147–152.
33. Vedi S, Tighe JR, Compston JE 1984 Measurement of total resorption surface in iliac crest trabecular bone in man. Metab Bone Dis Related Res **5**:275–280.
34. Burstone MS 1959 Histochemical demonstration of acid phosphatase activity in osteoclasts. J Histochem Cytochem **7**:39–41.
35. Evans RA, Dunstan CR, Baylink DJ 1979 Histochemical identification of osteoclasts in undecalcified sections of human bone. Miner Electrolyte Metab **2**:179–185.
36. Courpron P, Lepine P, Arlot M, Lips P, Meunier PJ 1980 Mechanisms underlying the reduction with age of the mean wall thickness of trabecular basic structure unit (BSU) in human iliac bone. In Jee WSS, Parfitt AM (eds) Bone histomorphometry, 3rd International Workshop. Armour Montagu, Paris, pp. 323–329.
37. Croucher PI, Mellish RWE, Vedi S, Garrahan NJ, Compston JE 1989 The relationship between resorption depth and mean interstitial bone thickness: Age-related changes in man. Calcif Tissue Int **45**:15–19.
38. Parfitt AM, Foldes J 1991 The ambiguity of interstitial bone thickness: A new approach to the mechanism of trabecular thinning. Bone **12**:119–122.
39. Eriksen EF, Gunderson HJG, Melsen F, Mosekilde L 1984 Reconstruction of the resorptive site in iliac trabecular bone; a kinetic model for bone resorption in 20 normal individuals. Metab Bone Dis Related Res **5**:235–242.
40. Palle S, Chappard D, Vico L, Riffat G, Alexandre C 1989 Evaluation of the osteoclastic population in iliac crest biopsies from 36 normal subjects: A histoenzymologic and histomorphometric study. J Bone Miner Res **4**:501–506.
41. Garrahan NJ, Croucher PI, Compston JE 1990 A computerised technique for the quantitative assessment of resorption cavities in trabecular bone. Bone **11**:241–246.
42. Cohen-Solal ME, Shih M-S, Lundy MW, Parfitt AM 1991 A new method for measuring cancellous bone erosion depth: Application to the cellular mechanisms of bone loss in postmenopausal osteoporosis. J Bone Miner Res **6**:1331–1338.
43. Weinstein RS, Hutson MS 1987 Decreased trabecular width and increased trabecular spacing contribute to bone loss with ageing. Bone **8**:137–142.
44. Mellish RWE, Garrahan NJ, Compston JE 1989 Age-related changes in trabecular width and spacing in human iliac crest biopsies. Bone Miner **6**:331–338.
45. Parfitt AM 1991 Bone remodelling in type 1 osteoporosis (letter). J Bone Miner Res **6**:95–97.
46. Wakamatsu E, Sissons HA 1969 The cancellous bone of the iliac crest. Calcif Tissue Res **4**:147–161.
47. Whitehouse WJ 1974 The quantitative morphology of anisotropic trabecular bone. J Microsc **101**:153–168.
48. Aaron JE, Makins NB, Sagreiya K 1987 The microanatomy of trabecular bone loss in normal aging men and women. Clin Orthop Related Res **215**:260–271.
49. Clermonts ECGM, Birkenhäger-Frenkel DH 1985 Software for bone histomorphometry by means of a digitizer. Comput Math Prog Biomed **21**:185–194.
50. Garrahan NJ, Mellish RWE, Vedi S, Compston JE 1987 Measurement of mean trabecular plate thickness by a new computerized method. Bone **8**:227–230.
51. Parfitt AM, Mathews CHE, Villanueva AR, Kleerekoper M, Frame B, Rao DS 1983 Relationship between surface, volume and thickness of iliac trabecular bone in aging and in osteoporosis: Implications for the microanatomic and cellular mechanism of bone loss. J Clin Invest **72**:1396–1409.
52. Compston JE 1994 Connectivity of cancellous bone: Assessment and mechanical implications. Bone **15**:463–466.
53. Feldkamp LA, Goldstein SA, Parfitt AM, Jesion G, Kleerekoper M 1989 The direct examination of bone architecture in vitro by computed tomography. Bone **4**:3–11.
54. Odgäard A, Gundersen HJG 1993 Quantification of connectivity in cancellous bone with special emphasis on 3-D reconstruction. Bone **14**:173–182.
55. Serra J 1982 Image Analysis and Mathematical Morphology. Academic Press, London.
56. Vesterby A 1990 Star volume of marrow space and trabeculae in iliac crest: Sampling procedure and correlation to star volume of first lumbar vertebra. Bone **11**:149–155.
57. Croucher PI, Garrahan NJ, Compston JE 1996 Assessment of cancellous bone structure: Comparison of strut analysis, trabecular bone pattern factor and marrow space star volume. J Bone Miner Res **11**:955–961.

58. Hahn M, Vogel M, Pompesius-Kempa M, Delling G 1992 Trabecular bone pattern factor—A new parameter for simple quantification of bone microarchitecture. Bone **13**:327–330.
59. Mandelbrot BB 1977 Fractals: Form, chance and dimension. Freeman, San Francisco.
60. Weinstein RS, Majumdar S, Genant HK 1992 Fractal geometry applied to the architecture of cancellous bone biopsy specimens. Bone **13**:A38.
61. Jacquet G, Ohley WJ, Mont MA, Siffert R, Schmukler R 1990 Measurement of bone structure by fractal dimension. Proc Ann Conf IEEE/EMBS **12**:1402–1403.
62. Majumdar S, Genant HK. 1995 A review of the recent advances in magnetic resonance imaging in the assessment of osteoporosis. Osteoporosis Int **5**:79–92.
63. Genant HK, Engelke K, Fuerst T, Glüer C-C, Grampp S, Harris ST, Jergas M, Lang T, Lu Y, Majumdar S, Mathur A, Takada M 1996 Noninvasive assessment of bone mineral and structure: State of the art. J Bone Miner Res **11**:707–730.
64. De Hoff RT, Aigeltinger EH, Craig KR 1972 Experimental determination of the topological properties of three-dimensional microstructures. J Microsc **95**:69–91.
65. Gundersen HJG, Boyce RW, Nyengaard JR, Odgäard A 1993 The ConnEulor: Unbiased estimation of connectivity using physical dissectors under projection. Bone **14**:217–222.
66. Vedi S, Compston JE, Webb A, Tighe JR 1982 Histomorphometric analysis of bone biopsies from the iliac crest of normal British subjects. Metab Bone Dis Related Res **4**:231–236.
67. Vedi S, Compston JE, Webb A, Tighe JR 1983 Histomorphometric analysis of dynamic parameters of trabecular bone formation in the iliac crest of normal British subjects. Metab Bone Related Res **5**:69–74.

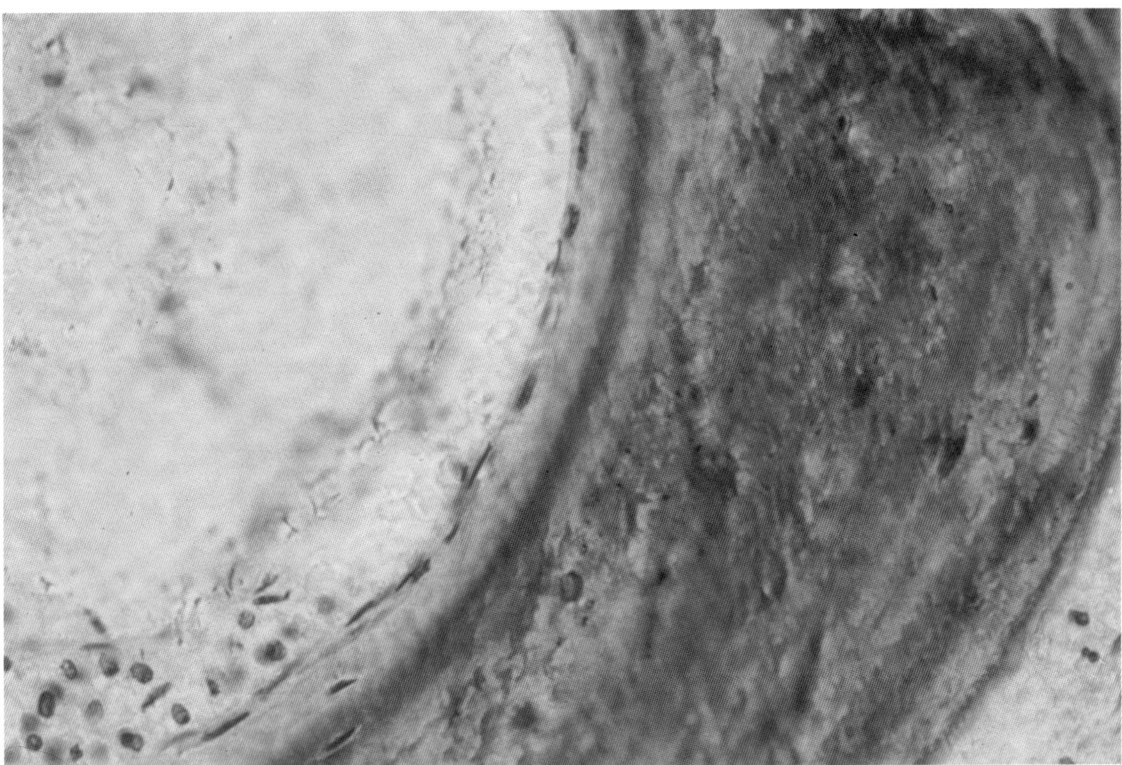

FIGURE 3 Toluidine blue-stained section of iliac crest cancellous bone showing mineralized bone (purple/blue) and osteoid (pale blue). The calcification front can be seen as a dark blue line at the interface of the osteoid and mineralized bone.

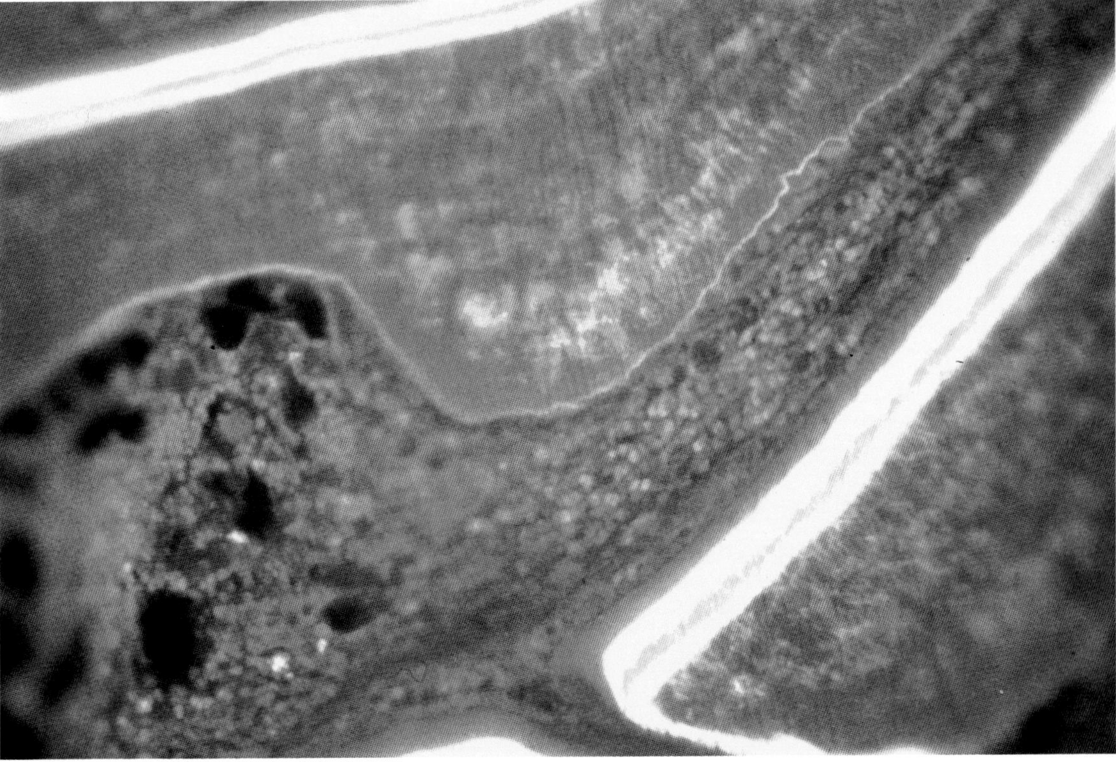

FIGURE 4 Unstained section of iliac crest cancellous bone viewed by fluorescence microscopy. The tetracycline labels are seen as double yellow fluorescent bands.

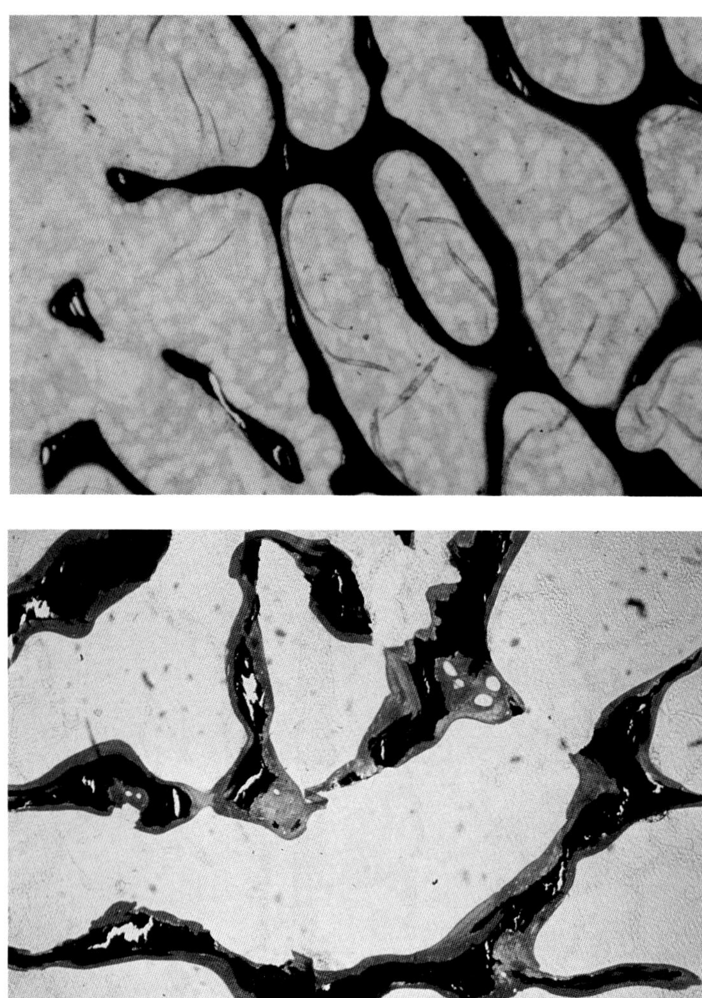

FIGURE 5 (Top) Section of iliac crest biopsy stained by the von Kossa technique to demonstrate mineralized bone (black) and osteoid (pink) in a normal subject. (Bottom) Section of iliac crest biopsy stained by the von Kossa technique to show osteoid accumulation in a woman with severe privational osteomalacia.

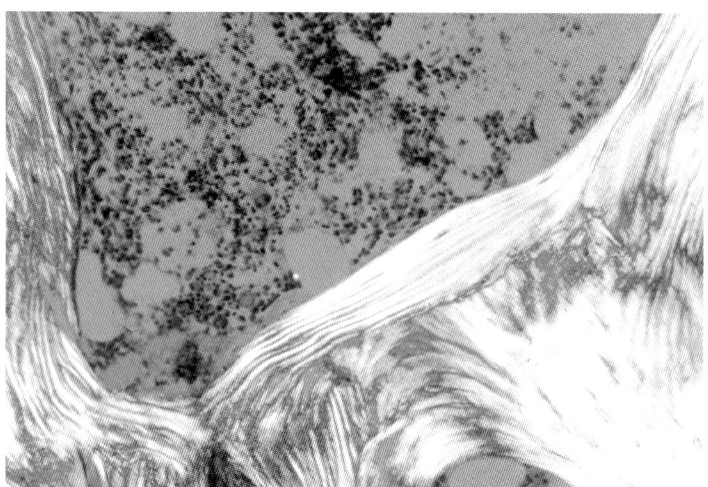

FIGURE 6 Section of iliac crest biopsy viewed under polarized light to show a completed bone structural unit bounded by the cement line and mineralized bone surface.

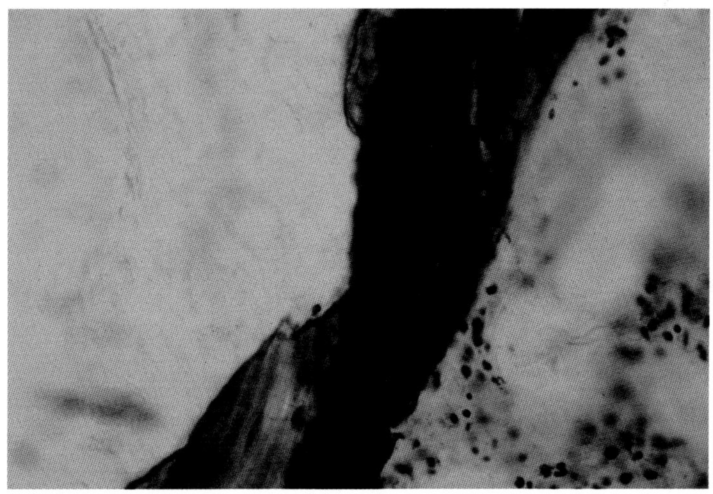

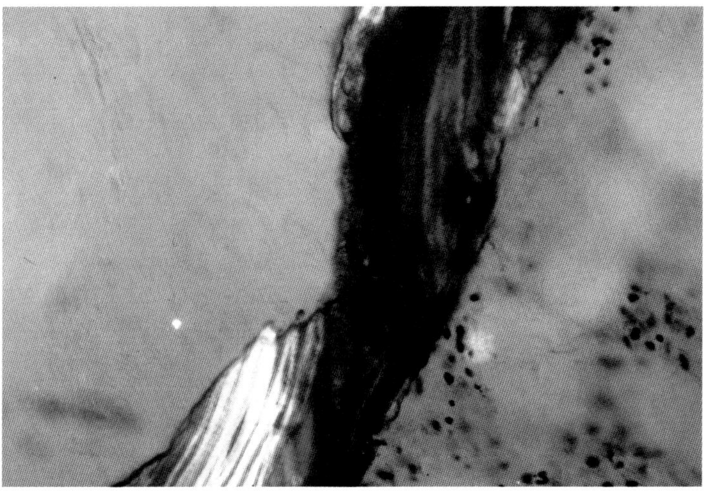

FIGURE 7 (Top) Resorption cavity in cancellous iliac crest bone stained by toluidine blue. (Bottom) Same resorption cavity viewed under polarized light. Note the cutoff collagen lamellae at the edges of the cavity.

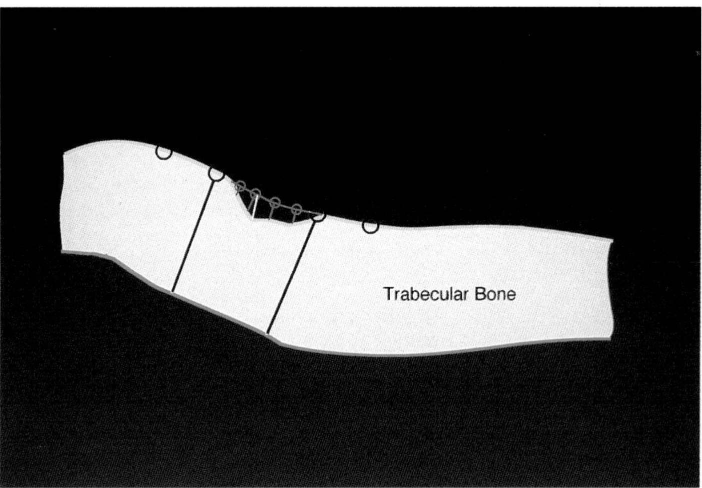

FIGURE 8 Measurement of erosion depth, with reconstruction of the mineralized surface using a cubic spline. Interactive graphics are superimposed over the image using the screen cursor controlled from a digitizer tablet. The two end points of the cavity are identified using the cursor, and two circles are drawn on the screen, centered on each end point. The positions of the intersections between the circles and the trabecular surface are also entered using the cursor. A smooth continuous curve (cubic spline) is generated to pass through the four defined points. Mean erosion depth is measured as the mean distance from four approximately equidistant points (green lines). The trabecular width on either side of the cavity is also measure (red lines).

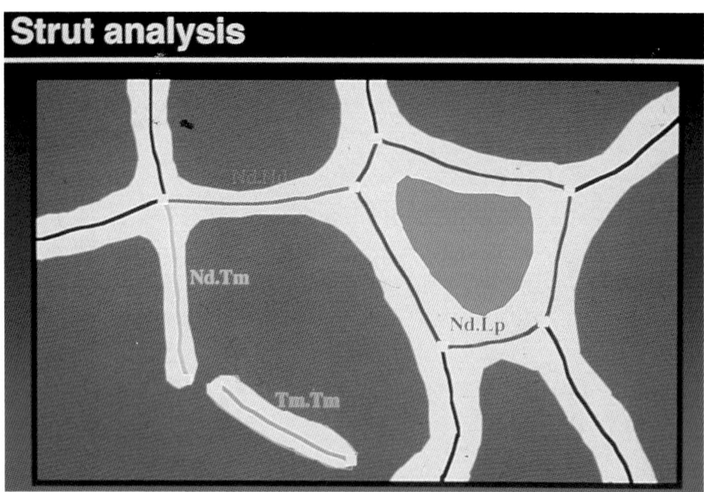

FIGURE 9 Diagrammatic representation of different strut types in strut analysis. Cancellous bone is shown in gray. Lines represent the skeletonized axis of the original bone profile. Squares represent nodes (Nd) and termini (Tm). Terminus-to-terminus (Tm.Tm), node-to-loop (Nd.Lp), and node-to-terminus (Nd.Tm) strut types are illustrated. Reprinted from Croucher *et al.* [57] with permission.

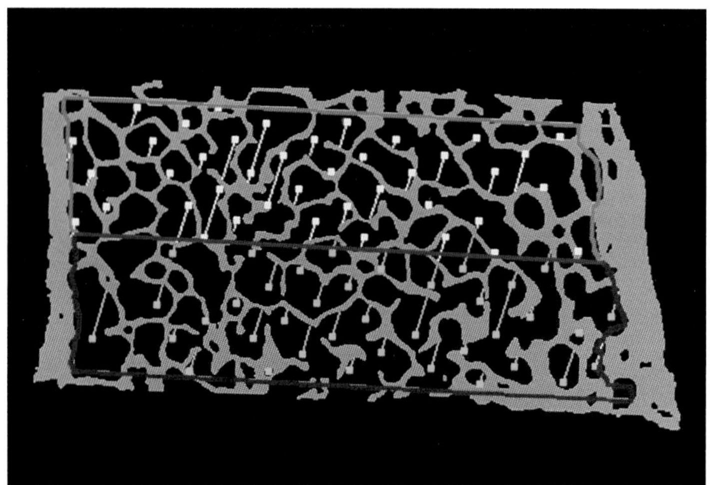

FIGURE 10 Binary image used in the assessment of marrow space star volume. The upper active region is defined by the green and blue lines and the lower region by red and blue lines. Blue and yellow squares represent grid points hitting the marrow space. Blue and yellow lines represent grid lines intercepting trabeculae and the edge of the active region. The white arrow shows the direction of the vertical axis. C, Cortex; CM, corticomedullary delineation. Reprinted from Croucher *et al.* [57] with permission.

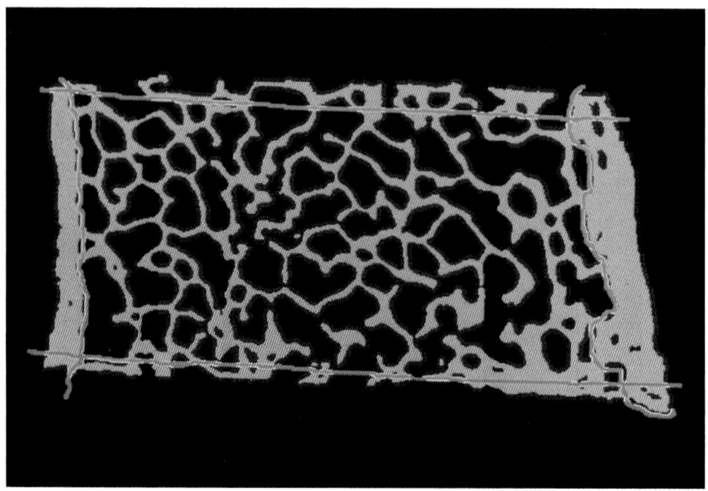

FIGURE 11 Binary image of cancellous bone used in the assessment of trabecular bone pattern factor. The active region is shown by the green line. The binary image is in gray and the dilated area is shown in red. Reprinted from Croucher *et al.* [57] with permission.

CHAPTER 38

Detection of Vitamin D and Its Major Metabolites

BRUCE W. HOLLIS Departments of Pediatrics, Biochemistry, and Molecular Biology, Medical University of South Carolina, Charleston, South Carolina

I. INTRODUCTION

Vitamin D is a 9,10-seco steroid and is treated as such in the numbering of its carbon skeleton (Fig. 1). Vitamin D occurs in two distinct forms: vitamin D_2 and vitamin D_3. As shown in Fig. 1, vitamin D_3 is a 27-carbon derivative of cholesterol; vitamin D_2 is a 28-carbon molecule derived from the plant sterol ergosterol. Besides containing an extra methyl group, vitamin D_2 differs from vitamin D_3 in that it contains a double bond between carbons 22 and 23. The most important aspects of vitamin D chemistry center on its *cis*-triene structure. This unique *cis*-triene structure makes vitamin D and related metabolites susceptible to oxidation, ultraviolet (UV) light-induced conformational changes, heat-induced conformational changes, and attack by free radicals. As a rule, the majority of these transformation products have lower biological activity than vitamin D. It is important to note that, in humans, vitamin D_2 and D_3 provide equal potency, and in this chapter the term vitamin D refers to both compounds.

Metabolic activation of vitamin D is achieved through hydroxylation reactions at both carbon 25 of the side chain and, subsequently, carbon 1 of the A ring. Metabolic inactivation of vitamin D takes place primarily through a series of oxidative reactions at carbons 23,24, and 26 of the side chain of the molecule. These metabolic activations and inactivations are well characterized (discussed in Section I of this book) and result in a plethora of vitamin D metabolites (Fig. 2). Of the compounds shown in Fig. 2, only four, vitamin D, 25-hydroxyvitamin D (25OHD), 24,25-dihydroxyvitamin D [$24,25(OH)_2D$], and 1,25-dihydroxyvitamin D [$1,25(OH)_2D$] have been extensively quantitated, and only two of those, namely, 25OHD and $1,25(OH)_2D$, provide any clinically relevant information. However, the quantitation of vitamin D and $24,25(OH)_2D$ can provide important information in a research environment. Thus, this chapter addresses the quantitation of

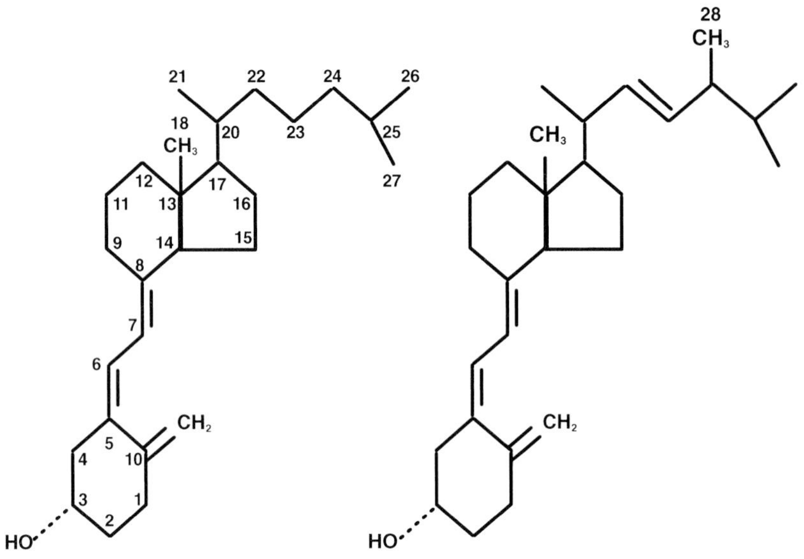

FIGURE 1 Molecular structures of vitamins D_3 and D_2.

these four important vitamin D compounds. Further, it is not the intent of this chapter to address the detailed history of vitamin D metabolite analysis, as this can be obtained from previous reviews [1–3]. Rather, the intent of this chapter is to describe how we currently measure vitamin D and its major metabolites in our laboratory, as well as to discuss the appropriate clinical judgments in the selection of a given compound for analysis.

The first semiquantitative assay for vitamin D was a bioassay based on the rat-line test [4]. This assay was cumbersome, expensive, and relatively inaccurate. Real progress in vitamin D analysis was not achieved until the advent of high specific activity ^3H-labeled vitamin D_3 compounds [5]. The introduction of these tracers led to the development of competitive protein binding assays (CPBA) for vitamin D and 25OHD [6,7]. A short time later CPBA for 24,25$(OH)_2$D and radioreceptor assays (RRA) for 1,25$(OH)_2$D were introduced [8,9]. In the late 1970s, high-performance liquid chromatographic (HPLC) analytical procedures for vitamin D and 25OHD were described [10,11]. Subsequently, radioimmunoassay (RIA) techniques began to appear as a means to quantitate 25OHD and 1,25$(OH)_2$D [12,13]. Finally, the most recent advances in antirachitic sterol analysis have included RIA coupled with ^{125}I-labeled tracers that require little or no chromatographic treatment of the sample [14,15].

The assays for vitamin D and its major metabolites that we currently utilize in our laboratory are described in this chapter. These assays are all stand-alone types of assays as opposed to the multiple-metabolite assays described in years past [16,17]. We chose to do this because seldom in a clinical situation does one require a battery of vitamin D metabolite values. Further, many assays, especially for 25OHD and 1,25$(OH)_2$D, have been optimized as single metabolite procedures [14,15].

II. DETECTION OF VITAMIN D

A. Background

Vitamin D, the parent compound, is by far the most lipophilic of the antirachitic sterols, and for this reason it is the most difficult to quantitate (Table I). The first serious attempt to quantitate vitamin D was performed in 1971 utilizing CPBA [7]. This initial study grossly overestimated the actual amount of circulating vitamin D because of insufficient sample prepurification prior to CPBA. It was later shown that vitamin D could be assessed by CPBA, but only following extensive chromatographic purification of the organic extract, including HPLC [18,19]. The first valid determination of circulating vitamin D was achieved in 1978 by utilizing direct UV detection following a two-step HPLC purification procedure [10]. A short time later, valid CPBA were introduced for the quantitation of circulating vitamin D [18,19]. However, these procedures were cumbersome, as they required extensive sample prepurification prior to CPBA, including HPLC.

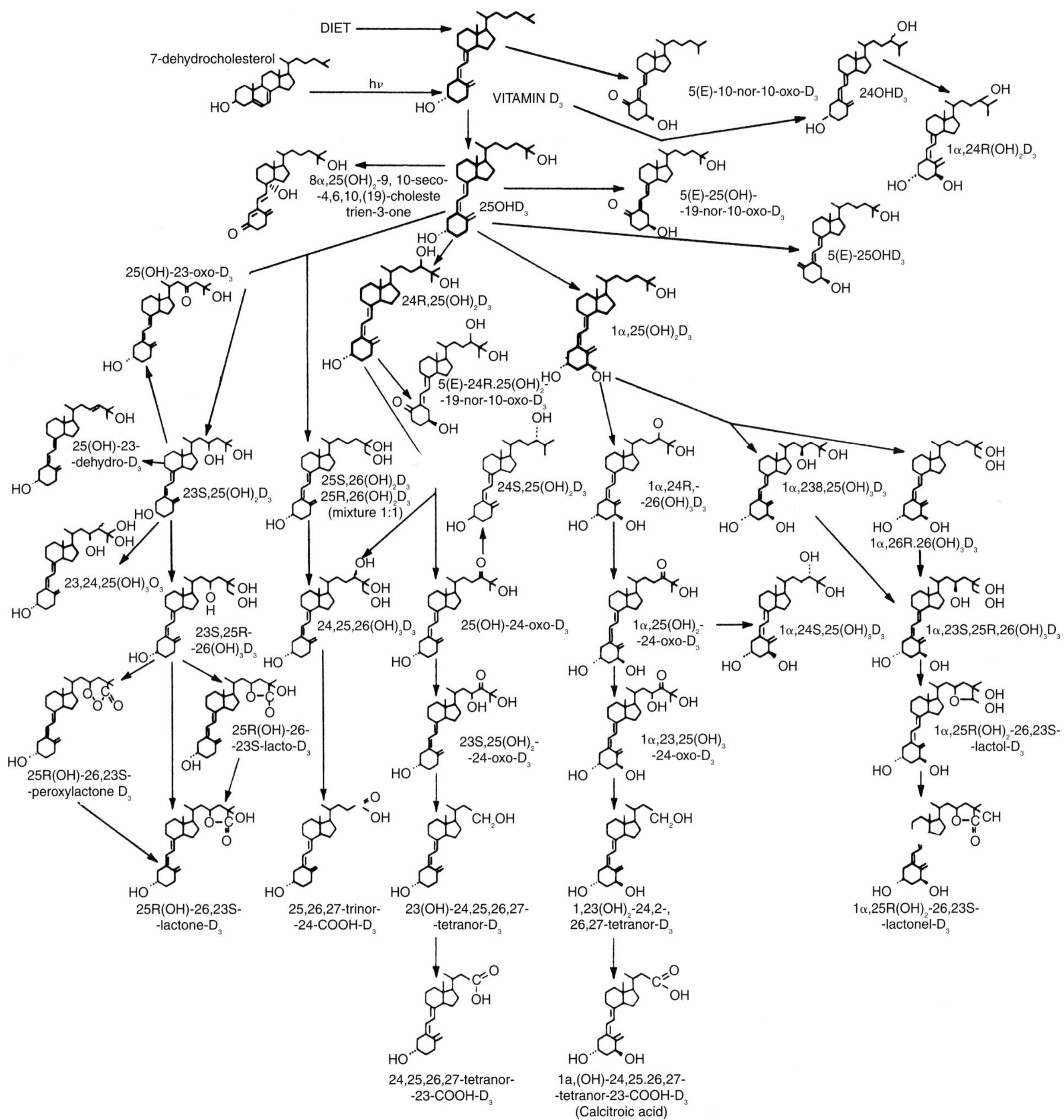

FIGURE 2 Summary of metabolic transformations of vitamin D_3. From Bouillon R, Okamura WH, Norman AW, Structure–function relationships in the vitamin D endocrine system. *Endocrine Review* 16;200–257; 1995. © The Endocrine Society.

Vitamin D is also difficult to quantitate because it is the only antirachitic sterol that cannot be extracted from aqueous media utilizing solid-phase extraction techniques [20]. Therefore, unlike its more polar metabolites, vitamin D must be extracted from serum or plasma using liquid–liquid organic extraction techniques. Many of the initial studies used Bligh and Dyer-type total lipid extraction to extract vitamin D from serum samples [7,10,18]. However, these types of extractions remove an extraordinary amount of lipid from the plasma sample. We therefore utilized a more selective organic extraction procedure incorporating methanol–hexane [21]. This extraction method coupled with open cartridge silica chromatography and direct UV quantitation of vitamins D_2 and D_3 following nonaqueous reversed phase LPLC provides an accurate, convenient method to measure circulating vitamin D. This method is described here in detail.

TABLE I Significant Methods for the Estimation of Vitamin D in Human Serum[a]

Detection method	Extraction	Preliminary chromatography	Ref.	Normal circulating levels[b]
CPBA	Methanol–chloroform	Silicic acid	Belsey *et al.* [7]	24–40 ng/ml
HPLC	Methanol–chloroform	Preparative HPLC	Jones [10]	2.2 ± 1.1 ng/ml
CPBA	Ether–methylene chloride	Sephadex LH-20, preparative HPLC	Horst *et al.* [19]	—
CPBA	Methanol–methylene chloride	Lipidex-5000, preparative HPLC	Hollis *et al.* [18]	2.3 ± 1.1 ng/ml
HPLC	Methanol–hexane	Silica cartridges, preparative HPLC	Liel *et al.* [21]	9.1 ± 1.0 ng/ml (normal), 1.3 ± 0.1 ng/ml (obese)

[a] As noted in the text, the Belsey method overestimated the circulating vitamin D levels because of insufficient prepurification prior to competitive protein binding assay (CPBA).
[b] ng/ml × 2.6 = nmol/liter.

B. Methodology

1. Sample Extraction

A 0.5- to 1-ml of sample serum or plasma is placed into a 13 × 100 mm borosilicate glass culture tube containing 1000 cpm of ^{3}H-vitamin D_3 in 25 μl of ethanol to monitor recovery of the endogenous compound through the extraction and chromatographic procedures. Following a 15-min incubation with the tracer, 2 plasma volumes of HPLC-grade methanol are added to each sample. The sample is then vortex-mixed for 1 min followed by the addition of 3 plasma volumes of HPLC-grade hexane. Each tube is capped and vortex-mixed for an additional 1 min followed by centrifugation at 1000 *g* for 10 min. The hexane layer is removed into another 13 × 100 mm culture tube, and the aqueous layer is reextracted in the same fashion. The hexane layers are combined and dried in a heated water bath, 55°C, under N_2. The lipid residue is then resuspended in 1 ml of HPLC-grade methylene chloride and capped.

2. SILICA CARTRIDGE CHROMATOGRAPHY

Silica Bond-Elut cartridges (500 mg) and a Vac-Elut cartridge rack were obtained from Varian Instruments (Harbor City, CA). The silica cartridges are washed in order with 5 ml HPLC-grade methanol, 5 ml HPLC-grade isopropanol, and 10 ml HPLC-grade methylene chloride. The sample, in 1 ml of methylene chloride, is then applied to the cartridge and eluted through the cartridge under vacuum into waste. This initial step is followed by 3 ml of 0.2% isopropanol in methylene chloride (discard) and 8 ml of the same solvent (vitamin D). The 8-ml fraction contains vitamins D_2 and D_3 and is subsequently dried in a heated water bath, 55°C, under N_2. The elution profile of vitamin D from the silica cartridge is displayed in Fig. 3. The silica cartridges can be cleaned and regenerated by washing with methanol, isopropanol, and methylene chloride and reused many times.

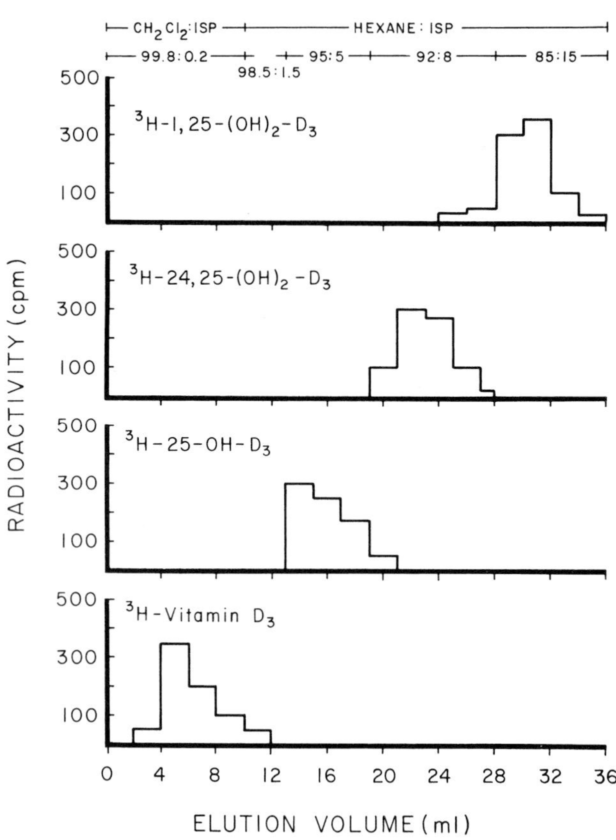

FIGURE 3 Elution profiles of radioactive vitamin D_3 and its metabolites chromatographed on a silica Bond-Elut (500 mg) cartridge.

3. Preparative Normal-Phase High-Performance Liquid Chromatography

High-performance liquid chromatography can be performed on any available HPLC system. This normal-phase HPLC step is performed in our laboratory using a 0.4 × 25 cm Zorbax-Sil column packed with 5 μm silica, but any equivalent column could be utilized. The mobile phase comprises hexane–methylene chloride–isopropanol (49.5:49.5:0.5, v/v) at a flow rate of 2 ml/min. The sample residue from the silica cartridge is dissolved in 150 μl of the mobile phase and injected onto the HPLC column that had been previously calibrated with 10 ng of standard vitamin D_3. The elution of vitamins D_2 and D_3 (they coelute on this system) can be seen in Fig. 4A. The vitamin D fraction is collected in a 12 × 75 mm glass culture tube and dried under N_2 at 55°C.

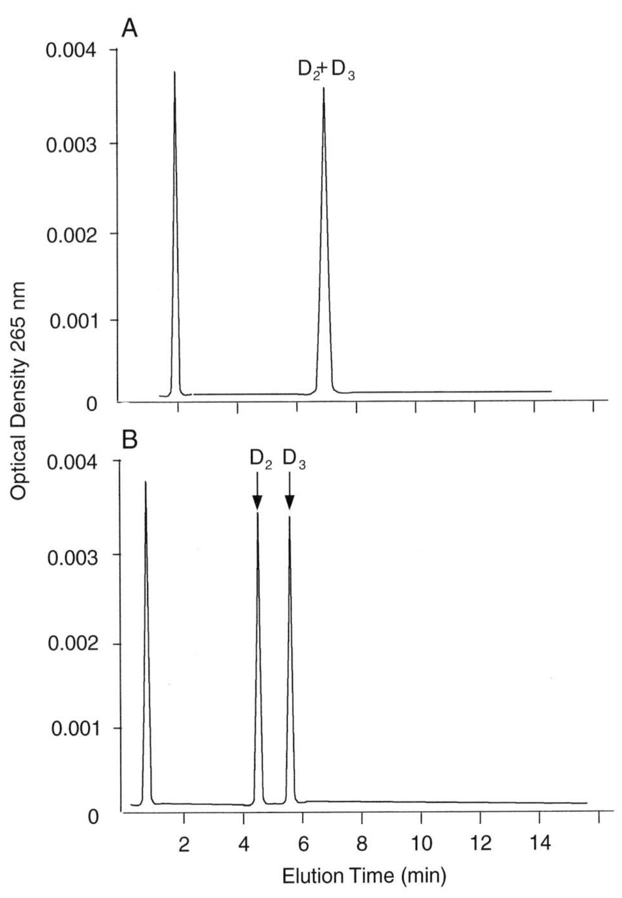

FIGURE 4 High-performance liquid chromatographic profiles of standard vitamins D_2 and D_3 on normal-phase (A) and reversed-phase (B) systems. Column calibration was achieved by injecting 10 ng of each compound and monitoring optical density at 265 nm. Column conditions are described in the text.

4. Quantitative Reversed-Phase High-Performance Liquid Chromatography

The final quantitative step is performed using non-aqueous reversed-phase HPLC. The column is a Vydac TP-201-54, 5 μm, wide pore, non-end-capped octadecylsilane (ODS) silica material, 0.4 × 25 cm. This particular column must be used for this procedure to work. The mobile phase comprises acetonitrile–methylene chloride (65:35, v/v) utilizing a flow rate of 1.2 ml/min. This system provides clear resolution of vitamins D_2 and D_3 (Fig. 4B). This system is calibrated with varying amounts of vitamin D_2 and D_3 (1–100 ng). The sample residue from normal-phase HPLC is dissolved in 15 μl methylene chloride followed by the addition of 135 μl of acetonitrile and injected onto the HPLC. After elution final quantitation of vitamin D_2 and D_3 is by direct UV monitoring of 265 nm. The vitamin D_3 portion is collected, dried under N_2, and subjected to liquid scintillation counting in order to determine the final recovery of endogenous vitamin D_3 from the sample. Calculations are then performed and the results reported in as nanograms vitamin D_2 and/or D_3 per milliliter. A flow diagram of the entire procedure is displayed in Fig. 5.

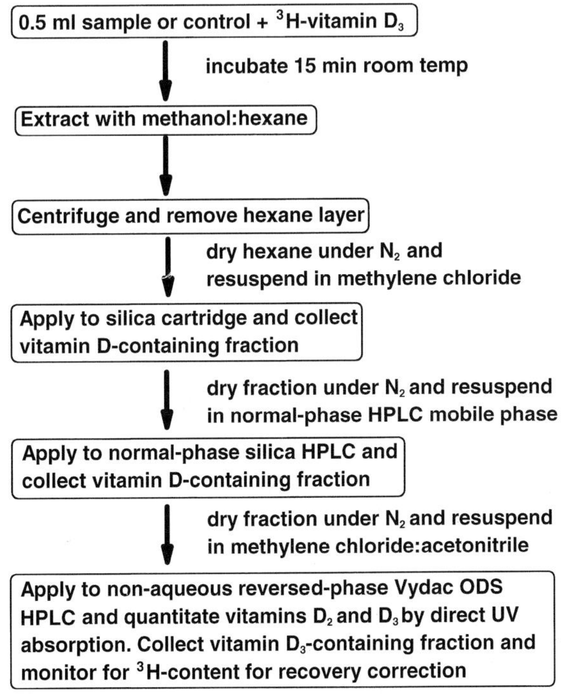

FIGURE 5 Flow diagram of the HPLC–UV assay for the quantitation of vitamins D_2 and D_3.

III. DETECTION OF 25OHD

A. Background

One of the major factors responsible for the explosion of knowledge related to vitamin D metabolism was the introduction of valid CPBA for 25OHD in the early 1970s [6,7] (Table II). One of these assays in particular gained widespread use owing to its relative simplicity and, as a result, has been cited nearly 1000 times [6]. The first assays utilized the vitamin D-binding protein (DBP) from rat serum as a specific binding agent. These assays all contained some type of organic extraction coupled with sample prepurification by column chromatography, all used ^{3}H-25OHD$_3$ as a tracer, and all required individual sample recovery estimates to account for endogenous losses of 25OHD during the extraction and purification procedures. A nonchromatographic assay for circulating 25OHD was introduced in 1974 [22], but it was never widely accepted because of its nonspecificity and susceptibility to serum lipid interference.

Various CPBA for 25OHD dominated the literature until 1977 when the first valid direct UV quantitative HPLC assay was introduced [11]. 25OHD circulates in the nanogram per milliliter (nanomole/liter) range and thus could be directly quantitated by UV detection following its separation by normal-phase HPLC. Also, HPLC detection provided the advantage of being able to individually quantitate 25OHD$_2$ and 25OHD$_3$. The disadvantages of HPLC quantitation methods are their requirements for expensive equipment and large sample size, cumbersome nature, and the technical expertise to perform this type of analysis. However, HPLC analysis for 25OHD is currently frequently used in research environments, including our own, and has provided a great deal of significant information.

As the clinical demand for circulating 25OHD analysis increased, it was clear that simpler, rapid yet valid assay procedures would be required. To this point, all valid assays required liquid–liquid organic extraction, some sort of chromatographic prepurification, and evaporation of the organic solvents, hardly practical for a clinical chemistry laboratory. Thus, in 1985, the first valid RIA for assessing circulating 25OHD was introduced [13]. This RIA eliminated the need for sample prepurification prior to assay and had no requirement for organic solvent evaporation. However, the method was still based on the use of ^{3}H-25OHD$_3$ as a tracer. This final shortcoming was solved in 1993 when an ^{125}I-labeled tracer was developed and incorporated into the RIA for 25OHD [14]. This assay has become the method of choice for assessing 25OHD status and has become the first test for vitamin D approved for clinical diagnosis by the U.S. Food and Drug Administration (FDA). This RIA along with an HPLC-based procedure routinely used in our laboratory for research purposes is described here.

B. HPLC Methodology

1. SAMPLE EXTRACTION

A 0.5-ml sample of serum or plasma is placed into a 12 × 75 mm borosilicate glass culture tube containing 1000 cpm of ^{3}H-25OHD$_3$ in 25 μl of ethanol to monitor recovery of endogenous compound through the extraction and chromatographic procedures. Following a 15-min incubation with the tracer, 1 plasma volume of HPLC-grade acetonitrile is added to each sample. The sample is then vortex-mixed for 1 min followed by centrifugation at 1000 g for 10 min. The supernatant is removed into another 12 × 75 mm culture tube, and 1 plasma volume of 0.4M K$_2$HPO$_4$, pH 10.4, is added.

TABLE II Significant Methods for the Estimation of 25OHD in Human Serum

Detection method	Extraction	Preliminary chromatography	Ref.	Normal circulating levels[a]
CPBA	Methanol–chloroform	Silicic acid	Belsey *et al.* [7]	18–36 ng/ml
CPBA	Ether	Silicic acid	Haddad and Chyu [6]	27.3 ± 11.8 ng/ml
CPBA	Ethanol	None	Belsey *et al.* [22]	20–100 ng/ml
HPLC	Methanol–chloroform	Sephadex LH-20	Eisman *et al.* [11]	31.9 ± 1.7 ng/ml
RIA	Acetonitrile	None	Hollis and Napoli [13]	25.5 ± 11.8 ng/ml
RIA	Acetonitrile	None	Hollis *et al.* [14]	9.9–41.5 ng/ml

[a] ng/ml × 2.4 = nmol/liter.

2. Solid-Phase Extraction Chomatography

C_{18} silica Sep-Pak cartridges (500 mg) and a Sep-Pak rack were obtained from Waters Associates (Milford, MA). The C_{18} cartridges are washed in order with 5 ml HPLC-grade isopropanol and 5 ml HPLC-grade methanol. The sample is applied to the cartridge and eluted through the cartridge under vacuum into waste. This initial step is followed by 5 ml of 30% water in methanol (discard) and 3 ml of acetonitrile (25OHD). The acetonitrile fraction is dried in a heated water bath, 55°C, under N_2. The lipid residue is then suspended in 1 ml of 1.5% isopropanol in hexane and capped. The C_{18} cartridges can be cleaned and regenerated by washing with 2 ml of methanol and reused many times.

3. Silica Cartridge Chromatography

Silica Bond-Elut cartridges (500 mg) and a Vac-Elut cartridge rack were obtained from Varian Instruments. The silica cartridges are washed in order with 5 ml HPLC-grade methanol, 5 ml HPLC-grade isopropanol, and 5 ml HPLC-grade hexane. The sample, in 1 ml of 1.5% isopropanol in hexane, is then applied to the cartridge and eluted through the cartridge under vacuum into waste. This initial elution is followed by 4 ml of 1.5% isopropanol in hexane (discard) and 6 ml of 5% isopropanol in hexane (25OHD). The 6-ml fraction contains $25OHD_2$ and $25OHD_3$ and is subsequently dried in a heated water bath, 55°C, under N_2. The elution profile of $25OHD_3$ from the silica cartridge is displayed in Fig. 3. The silica cartridges can be cleaned and reused many times.

4. Quantitative Normal-Phase High-Performance Liquid Chromatography

The final quantitative step is performed using normal-phase HPLC with a 0.4 × 25 cm Zorbax-Sil column packed with 5 μm spherical silica. The mobile phase is composed of hexane–methylene chloride–isopropanol (50:50:2.5, v/v) at a flow rate of 2 ml/min. The sample residue from the silica cartridge is dissolved in 150 μl of mobile phase and injected onto the HPLC column previously calibrated with varying amounts of $25OHD_2$ and $25OHD_3$ (1–100 ng). This HPLC system provides clear resolution of $25OHD_2$ and $25OHD_3$ (Fig. 6A). After elution, final quantitation of $25OHD_2$ and $25OHD_3$ is by direct UV monitoring at 265 nm. The $25OHD_3$ portion is collected, dried under N_2, and subjected to liquid scintillation counting in order to determine the final endogenous recovery of $25OHD_2$ and $25OHD_3$ from the sample. Calculations are then performed and the results reported as nanograms $25OHD_2$ and/or $25OHD_3$ per milliliter. A flow diagram of the entire procedure is displayed in Fig. 7.

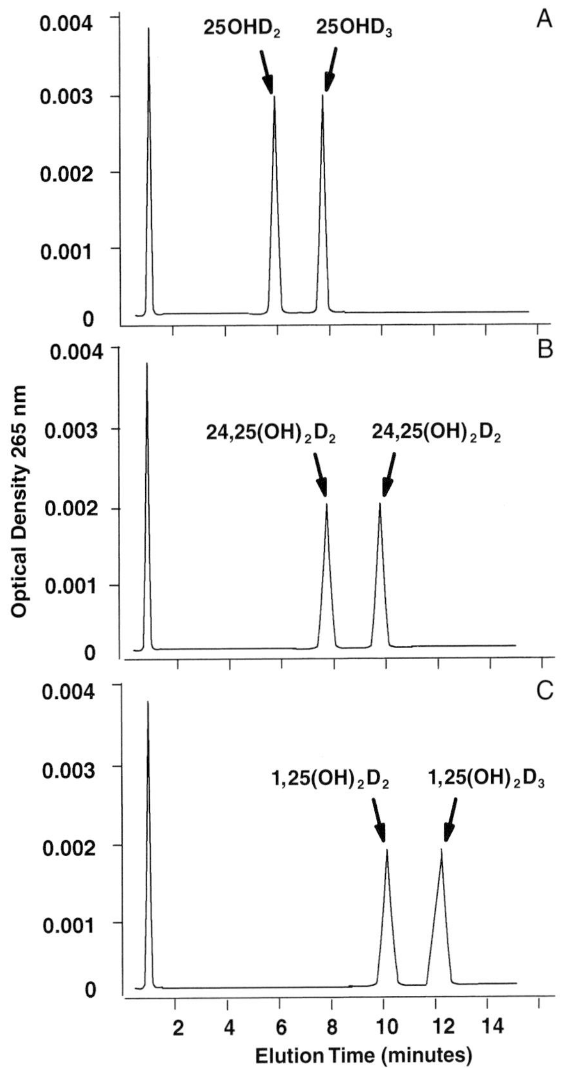

FIGURE 6 High-performance liquid chromatographic profiles of standard vitamin D metabolites. (A) $25OHD_2$ and $25OHD_3$; (B) $24,25(OH)_2D_2$ and $24,25(OH)_2D_3$; (C) $1,25(OH)_2D_2$ and $1,25(OH)_2D_3$. The HPLC was performed on a normal-phase Zorbax-Sil column, and column calibration was achieved by injecting 10 ng of each compound and monitoring optical density at 265 nm. Column conditions are described in the text.

C. RIA Methodology

1. Preparation of Assay Calibrators

One of the goals of the RIA procedure for 25OHD was to eliminate the need for individual sample recov-

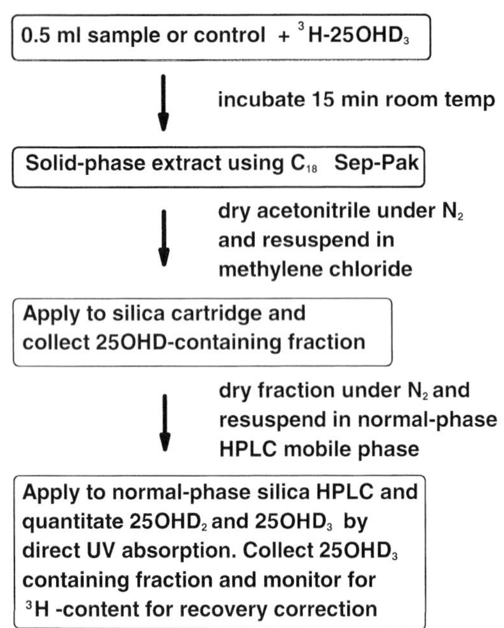

FIGURE 7 Flow diagram of the HPLC–UV assay for the quantitation of $25OHD_2$ and $25OHD_3$.

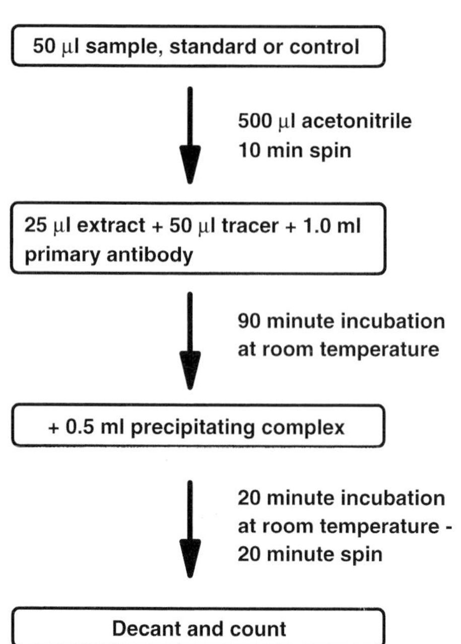

FIGURE 8 Flow diagram of the direct RIA for the quantitation of 25OHD.

ery. Another goal was to obtain FDA approval for clinical use of this procedure in the United States. Both of these goals placing $25OHD_3$ in a human serum-based set of assay calibrators. To prepare these calibrators, human serum was "stripped" free of vitamin D metabolites by treatment with activated charcoal. Absence of endogenous 25OHD in the stripped sera was confirmed by direct UV detection of 25OHD in serum following HPLC as described in Section III,B above. Subsequently, crystalline $25OHD_3$ dissolved in absolute ethanol was added to the stripped sera to yield calibrators at concentrations of 0, 5, 12, 40, 100 ng/ml.

2. SAMPLE AND CALIBRATOR EXTRACTION

To extract 25OHD from calibrators and samples, 0.5 ml of acetonitrile is placed into a 12 × 75 mm borosilicate glass tube after which 50 μl of sample of calibrator is dropped through the acetonitrile. After vortex-mixing, the tubes are centrifuged (1000 g, 4°C, 5 min) and 25 μl of supernatant transferred to 12 × 75 mm borosilicate glass tubes and placed on ice.

3. RADIOIMMUNOASSAY

The assay tubes are 12 × 75 mm borosilicate glass tubes containing 25 μl of acetonitrile-extracted calibrators or samples. To each tube add ^{125}I-25OHD derivative (50,000 cpm in 50 μl 1 : 1 ethanol–10 mM phosphate buffer, pH 7.4) that was synthesized as previously described [14]. Then added to each tube 1.0 ml of primary antibody diluted 1 : 15,000 in sodium phosphate buffer (50 mM, pH 7.4, containing 0.1% swine skin gelatin). Nonspecific binding is estimated using the above buffer minus the antibody. Vortex-mix the contents of the tubes and incubate them for 90 min at 20–25°C. Following this period, add 0.5 ml of a second antibody precipitating complex to each tube, vortex-mix, incubate at 20–25°C for 20 min, and centrifuge (20°C, 2000 g, 20 min). Discard the supernatant and bound the tubes in a gamma well counting system. 25OHD values are calculated directly from the standard curve by the counting system using a smooth-spline method of calculation. The entire 25OHD RIA procedure is displayed in Fig. 8.

4. COMMENTS ON THE 25OHD RIA

This ^{125}I-based RIA is similar to an RIA we introduced several years ago that used ^{3}H-$25OHD_3$ as a tracer [13]. In both cases, antisera were raised against the synthetic vitamin D analog 23,24,25,26,27-pentanor vitamin D-C(22)-carboxylic acid. The syntheses of this analog and its ^{125}I-labeled counterpart have been described in detail [13,14]. Coupling this compound to bovine serum albumin allowed us to generate antibodies that cross-reacted equally with most vitamin D_2 and D_3 metabolites (Table III). The structures for vitamins D_2 and D_3 differ only with respect to their side chains (Fig. 1). Because the analog retained the intact structure of vitamin D only up to carbon 22, the structural differences between vitamins D_2 and D_3 were not involved in the antibody recognition, and antibodies directed against this analog could not discriminate with respect

TABLE III Cross-reactivity of Various Vitamin D Compounds with 25OHD Antiserum and ^{125}I-labeled Vitamin D Derivative[a]

Steroid	Cross-reactivity (%)[b]
Vitamin D_2	0.8
Vitamin D_3	0.8
DHT	<0.1
25OHD_2	100
25OHD_3	100
25OHD_3-26,23-lactone	100
24,25$(OH)_2D_2$	100
24,25$(OH)_2D_3$	100
25,26$(OH)_2D_2$	100
25,26$(OH)_2D_3$	100
1,25$(OH)_2D_2$	2.5
1,25$(OH)_2D_3$	2.5

[a] From Hollis *et al. Clin Chem* 39:529–533.
[b] Ability to displace 50% of the ^{125}I tracer from the 25OHD antiserum diluted 15,000-fold.

to side chain metabolism of vitamin D. The antibody, however, was specific for the open B-ring *cis*-triene structure containing a 3β-hydroxyl group that is inherent in all vitamin D compounds.

Many vitamin D metabolites other than 25OHD are present in the circulation; however, they contribute only a small percentage (6–7%) to the overall assessment of nutritional vitamin D status as compared with 25OHD [23]. This fact is supported by the comparison of the 25OHD RIA with the UV quantitative HPLC assay for 25OHD described earlier in Section III,B on a variety of human serum samples (Fig. 9). Further, the present ^{125}I-based RIA was shown to identify vitamin D deficiency in biliary atresia patients as well as vitamin D toxicity in hypoparathyroid patients who were receiving massive vitamin D therapy for the maintenance of plasma calcium (Table IV).

IV. DETECTION OF 24,25(OH$_2$D

A. Background

Next to 25OHD, 24,25$(OH)_2$ D is quantitatively the most abundant circulating vitamin D metabolite, and, as a result, interest in its circulating levels have persisted. However, to this day the biological function(s) of 24,25$(OH)_2D_3$, if any, remains unresolved (discussed in Chapter 19). The first assay for 24,25$(OH)_2$D was first

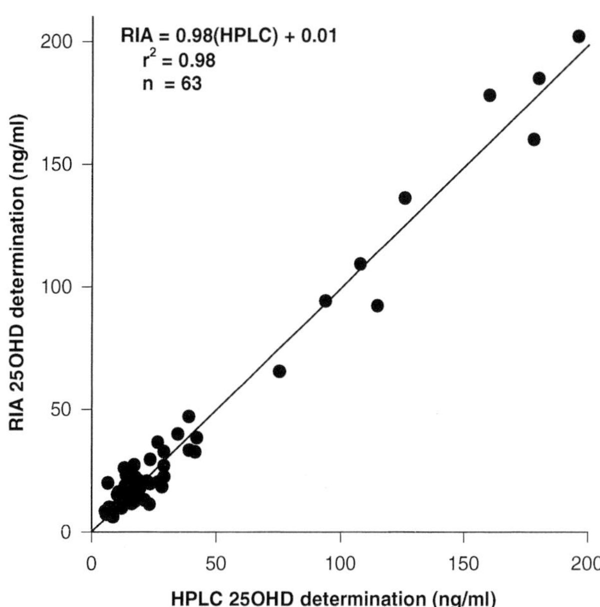

FIGURE 9 25OHD values obtained by the 25OHD RIA (*y* axis) and by direct UV quantitation of 25OHD following HPLC (*x* axis).

reported in 1977 and used CPBA in conjunction with sample prepurification on Sephadex LH-20 [8] (Table V). However, it was soon discovered that more extensive sample prepurification was required prior to 24,25$(OH)_2$D quantitation owing to substances that interfered in the 24,25$(OH)_2$D CPBA [24]. To further complicate matters, the quantitation of 24,25$(OH)_2$D is especially difficult when both the vitamin D_2 and D_3 forms are present in the circulation [16]. When both forms of 24,25$(OH)_2$D are present, it is extremely hard to remove other vitamin D metabolites that coelute with 24,25$(OH)_2D_2$ and 24,25$(OH)_2D_3$ on HPLC separation [16]. Further, once 24,25$(OH)_2D_2$ and 24,25$(OH)_2D_3$ are adequately separated and ready for CPBA, varying affinities of the two metabolites for the DBP require

TABLE IV Concentration of 25OHD as Determined by ^{125}I-RIA in Various Physiological States[a]

Subject type	*n*	Mean (ng/ml)[b]	Range (ng/ml)[b]
Normal[c]	36	25.7	9.9–41.5
Biliary atresia	12	6.3	4.3–8.3
Vitamin D therapy[d]	8	145	92–202

[a] From Hollis *et al. Clin Chem* 39:529–533.
[b] ng/ml × 2.496 = nmol/liter.
[c] Samples from subjects in Minnesota in October.
[d] Samples from subjects with hypoparathyroidism or pseudohypoparathyroidism receiving pharmacological doses of vitamin D_2,

TABLE V Significant Methods for the Estimation of $24,25(OH)_2D$ in Human Serum

Detection method	Extraction	Preliminary chromatography	Ref.	Normal circulating levels[a]
CPBA	Methanol–methylene chloride	Sephadex LH-20	Haddad *et al.* [8]	3.7 ± 0.2 ng/ml
CPBA	Methanol–methylene chloride	Sephadex LH-20, preparative HPLC	Shepard *et al.* [24]	3.5 ± 1.4 ng/ml
HPLC	Methanol–methylene chloride	Sephadex LH-20, preparative HPLC	Dreyer and Goodman [25]	2.4 ± 1.1 ng/ml
RIA	Methanol–methylene chloride	Sephadex LH-20, preparative HPLC	Hummer and Christiansen [26]	0.1–4.0 ng/ml
CPBA	Solid phase $C_{18}OH$	Silica cartridges	Wei *et al.* [27]	3.1 ± 0.7 ng/ml

[a] ng/ml × 2.4 = nmol/liter

standard curves to be constructed for final quantitation [28].

A report published in 1994 questions the requirement of HPLC prepurification of the serum sample prior to CPBA [27]. However, we are firm believers that in order to perform a valid assay for $24,25(OH)_2D$, one has to incorporate HPLC prepurification into the assay protocol. We have also chosen to do the final quantitation of $24,25(OH)_2D$ by RIA instead of CPBA. The RIA was chosen because the antibody used is cospecific for the vitamin D_2 and D_3 forms, and thus only one compound, $24,25(OH)_2D_3$, is required to construct the standard curve (Table III). This procedure is described here in detail.

B. Methodology

1. Sample Extraction

A 0.5-ml sample of serum or plasma is placed into a 12 × 75 mm borosilicate glass culture tube containing 1000 cpm of ^3H-$24,25(OH)_2D_3$ in 25 μl of ethanol to monitor recovery of endogenous compound through the extraction and chromatographic procedures. Following a 15-min incubation with the tracer, 1 plasma volume of HPLC-grade acetonitrile is added to each sample. The sample is then vortex-mixed for 1 min followed by centrifugation at 1000 *g* for 10 min. The supernatant is removed into another 12 × 75 mm culture tube, and 1 plasma volume of distilled water is added.

2. Solid-Phase Extraction Chromatography

C_{18} silica Bond-Elut cartridges (500 mg) and a Vac-Elut cartridge rack were obtained from Varian Instruments. The C_{18} cartridges are washed in order with 5 ml HPLC-grade isopropanol and 5 ml HPLC-grade methanol. The sample is then applied to the cartridge and eluted through the cartridge under vacuum into waste. This initial step is followed by 5 ml of 40% water in methanol (discard), 5 ml of 1% methylene chloride in hexane (discard), and 5 ml of 5% isopropanol in hexane [$24,25(OH)_2D$]. The final fraction is dried in a heated water bath, 55°C, under N_2. The lipid residue is then resuspended in 150 μl 5% isopropanol in hexane and capped.

3. Preparative Normal-Phase High-Performance Liquid Chromatography

The normal-phase HPLC step is performed using a 0.4 × 25 cm Zorbax-Sil column packed with 5 μm silica. The mobile phase is composed of hexane–methylene chloride–isopropanol (80:15:3.5, v/v) at a flow rate of 2 ml/min. The sample residue from the C_{18} silica cartridge is injected onto the HPLC column that had been previously calibrated with 10 ng of $24,25(OH)_2D_2$ and $24,25(OH)_2D_3$. The elution of these metabolites can be seen in Fig. 6B. The fractions containing $24,25(OH)_2D_2$ and $24,25(OH)_2D_3$ are collected individually in 12 × 75 mm glass tubes and dried under N_2 at 55°C. The residue is then redissolved in 500 μl absolute ethanol and capped.

4. Radioimmunoassay

The assay tubes are 12 × 75 mm borosilicate glass tubes containing 25 μl of the HPLC-purified extracts in ethanol. The standards for this RIA, which are $24,25(OH)_2D_3$, are placed in 12 × 75 mm tubes in 25 μl ethanol at concentrations between 0 and 200 pg/tube. To each tube add ^{125}I-25OHD derivative (50,000 cpm in 50 μl 1:1 ethanol–10 m*M* phosphate buffer, pH 7.4) or ^3H-$25OHD_3$ (5000 cpm in 25 μl ethanol). Then add to each tube 1.0 ml of primary antibody diluted 1:15,000 in sodium phosphate buffer (50 m*M*, pH 7.4, containing 0.1% swine skin gelatin). Nonspecific binding is estimated using the above buffer minus the antibody. Vortex-mix the contents of the tubes and incubate them for 90 min at 20–25°C. Following this period, add 0.5 ml of a second antibody precipitating complex to each tube if ^{125}I tracer was used 0.2 ml of 0.1 *M* borate buffer containing 1.0% norit A charcoal and 0.1% dextran T-70 if ^{3}H-25(OH)D_3 was used, vortex-mix, incubate at 20–25°C for 20 min and centrifuge (20°C, 2000 *g*, 20 min). In the case of ^{125}I tracer, discard the supernatant and count the tubes in a gamma well counting system. In the case of ^{3}H-25OHD_3, remove the supernatant into vials, add scintillation fluid, and monitor for radioactive content in a scintillation counter. 24,25$(OH)_2D$ values are calculated from the standard curve in picograms per tube. To convert this value to nanograms per milliliter, correct for dilution used as well as final recovery of ^{3}H-24,25$(OH)_2D_3$ added at the beginning of the sample extraction procedure. The entire 24,25$(OH)_2D$ RIA procedure is displayed in Fig. 10.

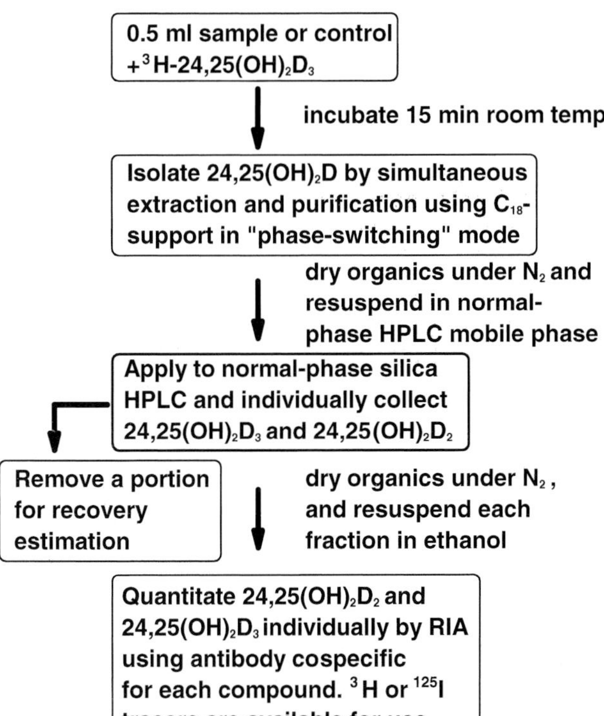

FIGURE 10 Flow diagram of the HPLC–RIA assay for the quantitation of 24,25$(OH)_2D_2$ and 24,25$(OH)_2D_3$.

V. DETECTION OF 1,25$(OH)_2D$

A. Background

Of all the steroid hormones, 1,25$(OH)_2D$ represented the most difficult challenge to the analytical biochemist with respect to quantitation. 1,25$(OH)_2D$ circulates at low picogram per milliliter concentrations (too low for direct UV quantitation), is highly lipophilic, and is relatively unstable, and its precursor, 25OHD, circulates at concentrations in excess of 10^3 to 10^4 times that of 1,25$(OH)_2D$. The first RRA for 1,25$(OH)_2D$ was introduced in 1974 [9] (Table VI). Although this initial assay was extremely cumbersome, it did provide invaluable information with respect to vitamin D homeostasis. This initial RRA required a 20-ml serum sample, which was extracted using Bligh–Dyer organic extraction. The extract had to be purified by three successive laborious chromatographic systems (there was no HPLC at the time), and chickens had to be sacrificed and vitamin D receptor (VDR) harvested from their intestines at the time of the RRA. By 1976, the volume requirement for this RRA had been reduced to a 5-ml sample and sample prepurification had been modified to include HPLC [29]. However, the sample still had to be extracted using a modified Bligh–Dyer extraction, and then prepurified on Sephadex LH-20, and chicken intestinal VDR was still utilized as a binding agent.

In 1978, the first RIA for 1,25$(OH)_2D$ was introduced [12]. Although it was an advantage not to have to isolate the intestinal VDR as a binding agent, this RIA was relatively nonspecific, so the cumbersome sample preparative steps were still required. Because of the extreme technical nature of these assays, and the cost of HPLC systems, few laboratories could afford to measure circulating 1,25$(OH)_2D$. Further, because these early techniques were so cumbersome, commercial laboratories did not offer 1,25$(OH)_2D$ determinations as a clinical service.

This all changed in 1984 with the introduction of a radically new concept for the determination of circulating 1,25$(OH)_2D$ [30]. This new RRA utilized solid-phase extraction of 1,25$(OH)_2D$ from serum along with silica cartridge purification of 1,25$(OH)_2D$. As a result, the need for HPLC sample prepurification was eliminated. Also, this assay utilized VDR isolated from calf thymus, which proved to be quite stable and thus had to be prepared only periodically. Further, the volume requirement was reduced to 1 ml of serum or plasma. This assay opened the way for any laboratory to measure circulating 1,25$(OH)_2D$. This procedure also resulted in the production of the first commercial kit for 1,25$(OH)_2D$ measurement. This RRA was further simplified in 1986 by decreasing the required chromatographic purification steps [31]. Through the mid-1990s, no new advances were reported with respect to the quantitation of circulating 1,25$(OH)_2D$.

As good as the calf thymus RRA for 1,25$(OH)_2D$ was, it still possessed two serious shortcomings. First, VDR had to be isolated from thymus glands, which was

TABLE VI Significant Methods for the Estimation of 1,25$(OH)_2$D in Human Serum

Detection method	Extraction	Preliminary chromatography	Ref.	Normal circulating levels[a]
RRA	Methanol–chloroform	Silicic acid, Sephadex LH-20, celite	Brumbaugh *et al.* [9]	39 ± 8 pg/ml
RRA	Methanol–methylene chloride	Sephadex LH-20, preparative HPLC	Eisman *et al.* [29]	29 ± 2 pg/ml
RIA	Methanol–chloroform	Sephadex LH-20, preparative HPLC	Clemens *et al.* [12]	35 pg/ml
RRA	Solid phase C_{18}OH	Silica cartridge	Reinhardt *et al.* [30]	37.4 ± 2.2 pg/ml
RRA	Solid phase C_{18}OH	None	Hollis [31]	28.2 ± 11.3 pg/ml
RIA	Solid phase C_{18}OH/silica	None	Hollis *et al.* [15]	32.2 ± 8.5 pg/ml

[a] pg/ml × 2.4 = pmol/liter.

still a difficult technique. Second, because the VDR is so specific for its ligand, only ^{3}H-1,25$(OH)_2D_3$ could be used as a tracer, eliminating the possibility of using a ^{125}I-based tracer. This is a major handicap, especially for the commercial laboratory. As a result, we have developed and reported in 1996 the first significant advance in 1,25$(OH)_2$D quantification in a decade [15]. This new RIA incorporates an ^{125}I-tracer as well as standards in an equivalent serum matrix so individual sample recoveries are no longer required. We describe this new RIA for 1,25$(OH)_2$D along with the standard RRA.

B. RRA Methodology

1. Preparation of Calf Thymus VDR

Frozen or fresh tissue is processed for VDR as follows (all steps are carried out at 4°C). Thymus tissue is minced with a meat grinder and homogenized (20% w/v) in a buffer containing 50 m*M* K_2HPO_4, 5 m*M* dithiothreitol, 1 m*M* EDTA, and 400 m*M* KCl, pH 7.5. The tissue is homogenized using five, 30-sec bursts of a Polytron PT-20 tissue disrupter at a maximum power setting. The homogenate is then centrifuged for 15 min at 20,000 *g* to remove large particles. The resulting supernatant is centrifuged at 100,000 *g* for 1 hr, and the "cytosol" (actually a high salt extract that includes nucleus VDR) is collected minus the floating lipid layer. The VDR is then precipitated by the slow addition of solid $(NH_4)_2SO_4$ to 35% saturation. The cytosol–$(NH_4)_2SO_4$ mixture is stirred for 30 min while maintaining the temperature at 4°C. The mixture is then divided into 15-ml centrifuge tubes and centrifuged at 20,000 *g* for 20 min. The supernatant is discarded and tubes allowed to drain for 5 min. The precipitated VDR is lyophilized and stored under inert gas at −70°C. VDR prepared in this manner is stable for up to 60 hr at room temperature. As an alternative, calf thymus VDR preparations can also be purchased from commercial sources (INCSTAR Corporation, Stillwater, MN).

2. Sample Extraction

A 1.0 ml sample of serum or plasma is placed into a 12 × 75 mm borosilicate glass culture tube containing 700 cpm of ^{3}H-1,25$(OH)_2D_3$ in 25 μl of ethanol to monitor recovery of endogenous compound through the extraction and chromatographic procedures. Following a 15-min incubation with the tracer, 1 ml of HPLC-grade acetonitrile is added to each sample. The sample is then vortex-mixed for 1 min, followed by centrifugation at 1000 *g* for 10 min. The supernatant is removed into another 12 × 75 mm culture tube, and 1 vol of 0.4 *M* K_2HPO_4, pH 10.4, is added followed by vortex-mixing.

3. Solid-Phase Extraction and Purification Chromatography

C_{18}OH silica Bond-Elut cartridges (500 mg) and a Vac-Elut cartridge rack were obtained from Varian Instruments. The C_{18}OH cartridges are washed in order with 5 ml HPLC-grade methylene chloride, 5 ml HPLC-grade isopropanol, and 5 ml HPLC-grade methanol. The sample is applied to the cartridge and eluted through the cartridge under vacuum into waste. This initial step is followed by 5 ml of 30% water in methanol (discard), 5 ml of 10% methylene chloride in hexane (discard), 5 ml of 1% isopropanol in hexane (discard), and 5 ml of 3% isopropanol in hexane [1,25$(OH)_2$D] (Fig. 11). This final fraction is dried in a heated water bath, 55°C, under N_2. The residue is then suspended in 200 μl absolute ethanol and capped.

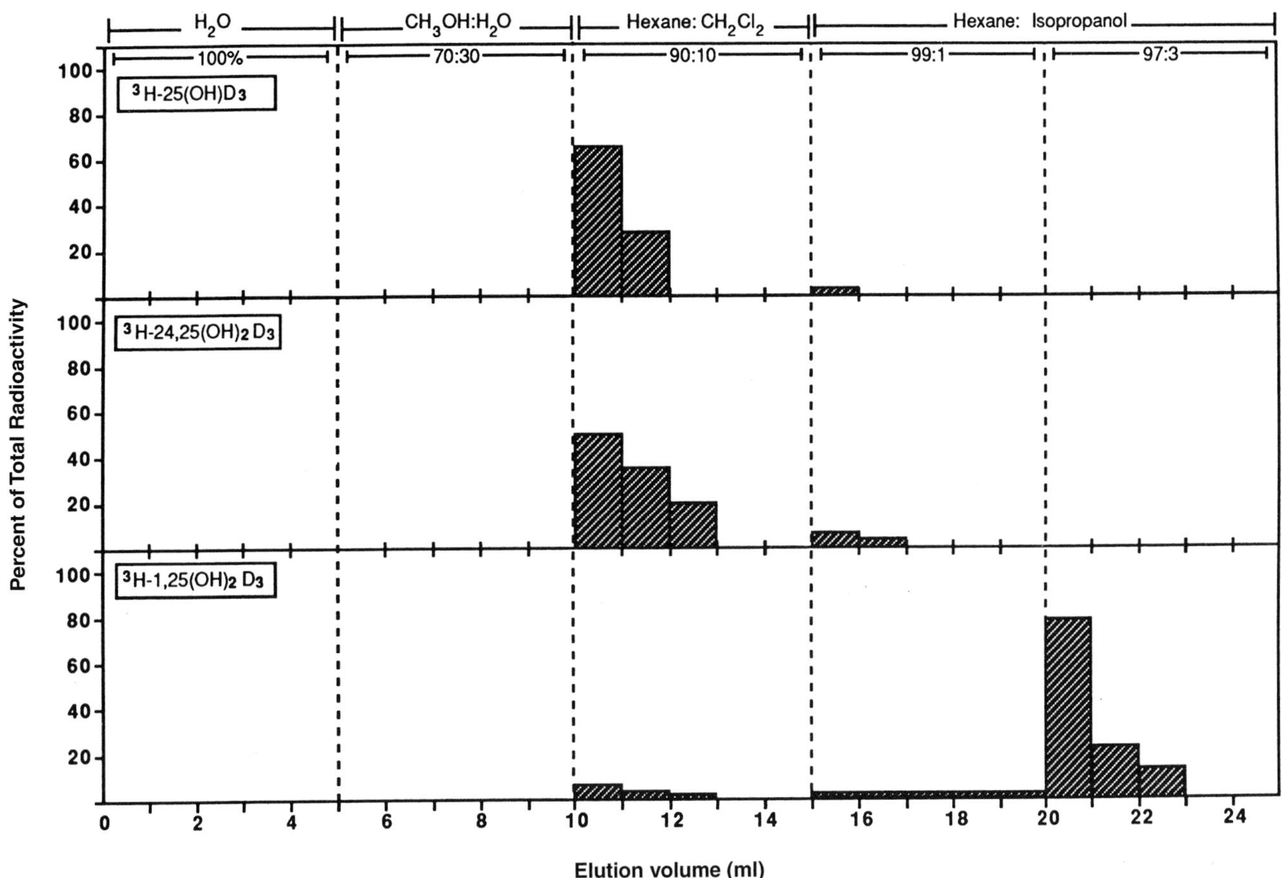

FIGURE 11 Elution of ^{3}H-vitamin D_3 and its metabolites from a $C_{18}OH$ Bond-Elut cartridge. From Hollis BW, *Clin Chem* 31:1815–1819.

4. Radioreceptor Assay

Prior to assay, the VDR-containing pellet is reconstituted to its original volume with assay buffer. The assay buffer contains 50 mM K_2HPO_4, 5 mM dithiothreitol, 1.0 mM EDTA, and 150 mM KCl at pH 7.5. The receptor pellet is dissolved by gentle stirring on ice using a magnetic stir bar. The receptor solution is allowed to mix for 20–30 min. Typically, a small portion of the pellet resists solubilization and is removed by centrifugation at 3000 g for 10 min. The receptor solution is then diluted 1:3–1:9 with assay buffer and kept on ice until use. The correct dilution of receptor used in the assay is determined for each new batch of receptor. At the appropriate dilution for assay use, specific binding in the absence of unlabeled $1,25(OH)_2D$ is 1600–2000 cpm; nonspecific binding is 200–300 cpm. These results assume a specific activity of 130 Ci/mmol for $^3H\text{-}1,25(OH)_2D_3$ and a 40% counting efficiency for tritium.

The assay tubes are 12 × 75 mm borosilicate glass tubes containing 50 μl of $C_{18}OH$-purified extracts in ethanol. The standards for the assay, $1,25(OH)_2D_3$, are placed in 12 × 75 mm tubes in 50 μl ethanol at concentrations between 1 and 15 pg/tube. Nonspecific binding is estimated by adding 1 ng/tube of $1,25(OH)_2D_3$. To each tube add 0.5 ml of reconstituted thymus cytosol, vortex-mix, and incubate for 1 hr at 15–20°C. Following this initial incubation, each tube receives $^3H\text{-}1,25(OH)_2D_3$ (5000 cpm in 50 μl ethanol) and the incubation proceeds for an additional 1 hr at 15–20°C. Finally, place the assay tubes in an ice bath and add 0.2 ml of 0.1 M borate buffer containing 1.0% norit A charcoal and 0.1% dextran T-70, vortex-mix, incubate 20 min, and centrifuge (4°C, 2000 g, 10 min). Remove the supernatant into vials, add scintillation fluid, and monitor for radioactive content in a scintillation counter. $1,25(OH)_2D$ values are calculated from the standard curve in picograms per tube. To convert this value to picograms per milliter, correct for dilution used as well as final recovery of $^3H\text{-}1,25(OH)_2D_3$ added at the beginning of the sample extraction procedure. The $1,25(OH)_2D$ RRA procedure is displayed in Fig. 12.

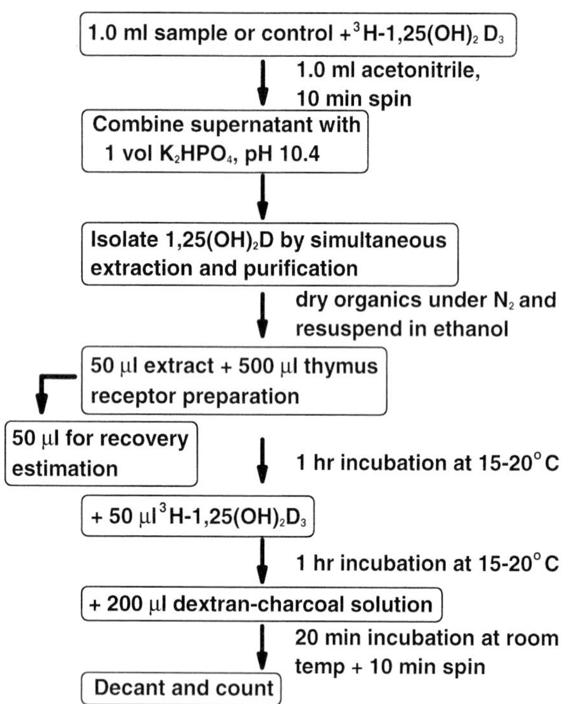

FIGURE 12 Flow diagram of the RRA for the quantitation of 1,25(OH)$_2$D.

C. RIA Methodology

1. Preparation of Assay Calibrators

As was described for the 25OHD RIA (Section III,C), one of the goals of this RIA procedure was to eliminate the need for individual sample recovery. To prepare the assay calibrators, human serum was stripped free of vitamin D metabolites. The absence of endogenous 1,25(OH)$_2$D in the stripped sera was confirmed by RRA for 1,25(OH)$_2$D as previously described in Section V,B. Subsequently, crystalline 1,25(OH)$_2$D$_3$ dissolved in absolute ethanol was added to the stripped sera to yield calibrators at concentrations of 0, 5, 15, 30, 60, and 160 pg/ml.

2. Sample and Calibrator Extraction and Pretreatment

The 1,25(OH)$_2$D is extracted from calibrators and samples as follows. First, 0.5 ml of serum of plasma is placed into a 12 × 75 mm borosilicate glass culture tube; 0.5 ml of HPLC-grade acetonitrile is added and vortex-mixed for 1 min followed by centrifugation at 1000 *g* for 10 min. The supernatant is removed into another 12 × 75 mm culture tube, and 0.5 ml of a 25 mg/ml solution of sodium *m*-periodate is added. The samples are incubated for 30 min at room temperature.

3. Solid-Phase Extraction and Silica Purification Chromatography

C$_{18}$OH silica, silica Bond-Elut cartridges (500 mg), and a Vac-Elut cartridge rack were obtained from Varian Instruments. The supernatant–sodium periodate mix is applied to a C$_{18}$OH cartridge that had been prewashed successively with 5-ml isopropanol and 5-ml methanol. Next, the cartridge is successively washed with 5 ml of 30% water in methanol (discard), 5 ml of 10% methylene chloride in hexane (discard), and 5 ml of 1% isopropanol in hexane (discard). The C$_{18}$OH cartridge is then placed into a silica cartridge previously washed successively with 5 ml methanol, 5 ml isopropanol, and 5 ml 20% isopropanol in hexane, and 1,25(OH)$_2$D is eluted onto the silica cartridge with 5 ml 8% isopropanol in hexane (discard). The C$_{18}$OH cartridge is removed from the silica cartridge, and the silica cartridge is eluted with 2 ml of 8% isopropanol in hexane (discard) and 5 ml of 20% isopropanol in hexane [1,25(OH)$_2$D]. Each C$_{18}$-OH cartridge can be regenerated for reuse by washing with 2 ml of methanol. The silica cartridges can be reused without any further washing steps. This final fraction is dried in a heated water bath, 55°C, under N$_2$. The residue is then suspended in 50 μl absolute ethanol and capped.

4. Radioimmunoassay

The assay tubes are 12 × 75 mm borosilicate glass tubes containing 20 μl of the ethanol-reconstituted extracted calibrators or samples. To each tube add ^{125}I-1,25(OH)$_2$D derivative (50,000 cpm in 50 μl 1:1 ethanol–10 m*M* phosphate buffer, pH 7.4) that was synthesized as previously described [14]. Then add to each tube 0.25 ml of primary antibody diluted 1:200,000 in sodium phosphate buffer [50 m*M*, pH 6.2, containing 0.1% swine-skin gelatin and 0.35% polyvinyl alcohol (M_r 13,000–23,000)]. Nonspecific binding is estimated by using the above buffer without the antibody. Vortex-mix the contents of the tubes, incubate them for 2 hr at 20–25°C, add 0.5 ml of second antibody precipitating complex, incubate at 20–25°C for 20 minutes, and then centrifuge (20°C, 2000 *g*, 20 min). Discard the supernatant and count the tubes in a gamma well counting system. 1,25(OH)$_2$D values are calculated directly from standard curve by the counting system using a smooth-spline method of calculation. The entire 1,25(OH)$_2$D RIA procedure is displayed in Fig. 13.

5. Comments on the 1,25(OH)$_2$D RIA

Of the procedures developed for determining 1,25(OH)$_2$D status in humans, only a few RRA [30,31] have been able to quantify circulating 1,25(OH)$_2$D without using HPLC for sample prepurification. Many RIA have been published and validated for the quantification

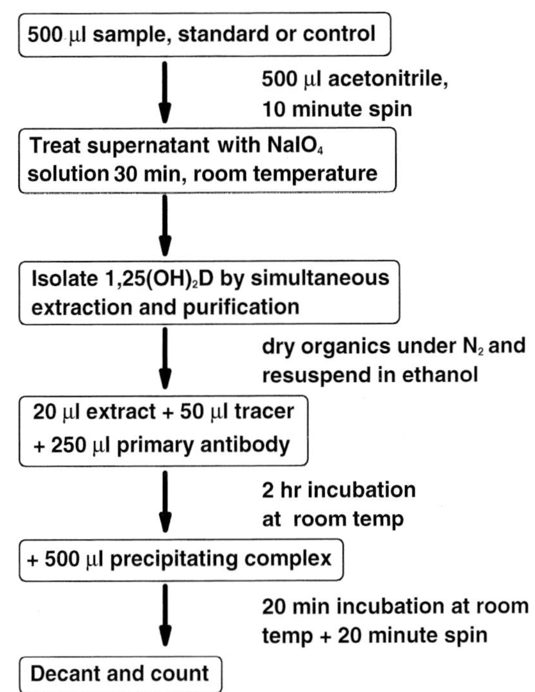

FIGURE 13 Flow diagram of the RIA for the quantitation of $1,25(OH)_2D$.

of $1,25(OH)_2D$, but all have included HPLC steps for sample prepurification [12,32,33]. Development of an RIA for quantification of circulating $1,25(OH)_2D$ has been hampered from the beginning by the relatively poor specificity of the antibodies that have been generated. To date, the best antibodies toward $1,25(OH)_2D$ have, at best, a cross-reactivity with the non-1-hydroxylated metabolites of vitamin D of approximately 1%. In comparison, the VDR used in the RRA has a cross-reactivity of approximately 0.01% with these more abundant metabolites [30]. Given that the non-1-hydroxylated metabolites circulate at concentrations over 1000 times greater than that of $1,25(OH)_2D$, the magnitude of the problem becomes clear. However, the VDR is so specific that any attempt to introduce a radionuclide such as ^{125}I into $1,25(OH)_2D_3$ erodes the binding between this steroid hormone and the VDR. Therefore, if one wishes to develop ^{125}I-based assays to quantify $1,25(OH)_2D$, RIA is the only choice.

The antibody we chose to use in the development of this RIA has been characterized previously [34]. Its relatively low cross-reactivity to non-1-hydroxylated vitamin D metabolites and high titer made it a good candidate. Still, the present antibody cross-reacts by 1–2% with the more abundant vitamin D metabolites [34]. This substantial cross-reactivity meant that some chromatographic prepurification of $1,25(OH)_2D$ before RIA was essential. We wished to avoid HPLC because of its cost and cumbersome nature. Therefore, we chose a simplified chromatographic procedure that had been incorporated into the RRA for $1,25(OH)_2D$ (Fig. 11). However, when we utilized this purification procedure with the RIA, the "apparent" circulating $1,25(OH)_2D$ concentrations were approximately 50% greater than those determined with the RRA. Further investigation revealed that the vitamin D metabolite most responsible for this overestimation was $25,26(OH)_2D$ (Table VII).

Efforts to resolve the $1,25(OH)_2D$ metabolite by chromatographic means short of HPLC failed. We therefore instituted a novel sample pretreatment with sodium periodate. Instituting this pretreatment step converted all of the $24,25(OH)_2D_3$ and $25,26(OH)_2D_3$ to their respective aldehyde and ketone forms, which are easily removed by the current chromatographic scheme. We also incorporated an extra silica column into the procedure to assure all contaminating substances were removed. Use of this purification procedure reduced the cross-reactivity of non-1-hydroxylated vitamin D compounds to insignificant levels Table VIII). Figure 14 demonstrates the necessity of sample pretreatment. Failure to use the pretreatment step resulted in a 40% overestimation of the actual concentration of circulating $1,25(OH)_2D$ in normal human subjects. We validated this new RIA against the standard RRA with excellent results (Figs. 15 and 16).

The concentrations of $1,25(OH)_2D$ as determined in serum from various groups of healthy and pathological

TABLE VII $1,25(OH)_2D$ Measured in a Serum Sample after Addition of Exogenous $25OHD_3$, $24,25(OH)_2D_3$, or $25,26(OH)_2D_3$ With and Without Sample Pretreatment with Sodium Periodate[a]

		$1,25(OH)_2D$ measured (pg/ml)[b]	
Steroid added	Amount added (ng/ml)	Treated	Not treated
None	—	25.5	30.2
$25OHD_3$	20	28.0	29.2
	50	28.0	28.2
	200	31.6	33.9
$24,25(OH)_2D_3$	5	25.6	29.8
	20	24.2	41.3
	100	30.4	57.5
$25,26(OH)_2D_3$	5	26.2	95.0
	20	30.7	126.0
	100	29.4	340.0

[a] From Hollis *et al. Clin Chem* 42:586–592.
[b] Determined in duplicate by radioimmunoassay with or without sodium periodate treatment following chromatographic isolation described in text. pg/ml × 2.4 = pmol/liter.

TABLE VIII Cross-reactivity of Various Vitamin D Compounds with $1,25(OH)_2D$ Antiserum and ^{125}I-labeled 1-Hydroxylated Tracer Following Sample Treatment and Purification[a]

Steroid	Cross-reactivity (%)[b]
Vitamin D_3	<0.001
$25(OH)D_3$	<0.001
$24,25(OH)_2D_3$	0.005
$25,26(OH)_2D_3$	<0.001
$1,25(OH)_2D_2$	73.1
$1,25(OH)_2D_3$	100
$1,25(OH)_2D_3$-26,23-lactone	13.1
$1,24,25(OH)_3D_3$	25.1
$1,25,26(OH)_3D_3$	23.2

[a] From B. W. Hollis, *et al. Clin Chem* 42:586–592.
[b] Ability to displace 50% of the ^{125}I tracer from the $1,25(OH)_2D$ antiserum diluted 1:150,000.

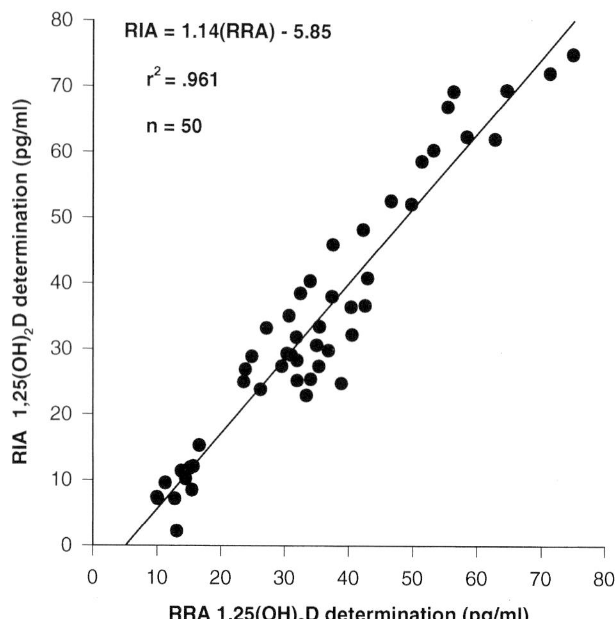

FIGURE 15 Comparison of circulating $1,25(OH)_2D$ measured by RIA and RRA. From Hollis *et al. Clin Chem* 42:586–592.

subjects (Fig. 16) agree well with values reported in previous studies [30,31]. It is very important to include pathological samples such as those from subjects with biliary atresia and vitamin D toxicity in any assay validation procedure for circulating $1,25(OH)_2D$. This importance was underlined in our previous report on an unpublished, unvalidated, but commercially available ^{125}I-based RIA for $1,25(OH)_2D$ that involves the immunoextraction of $1,25(OH)_2D$ from serum samples and is marketed by IDS Ltd. (Tyne and Wear, UK) [35]. The basis of this kit is selective immunoextraction of $1,25(OH)_2D$ from serum or plasma with a specific monoclonal antibody bound to a solid support. This antibody is directed toward the 1α-hydroxylated A ring of

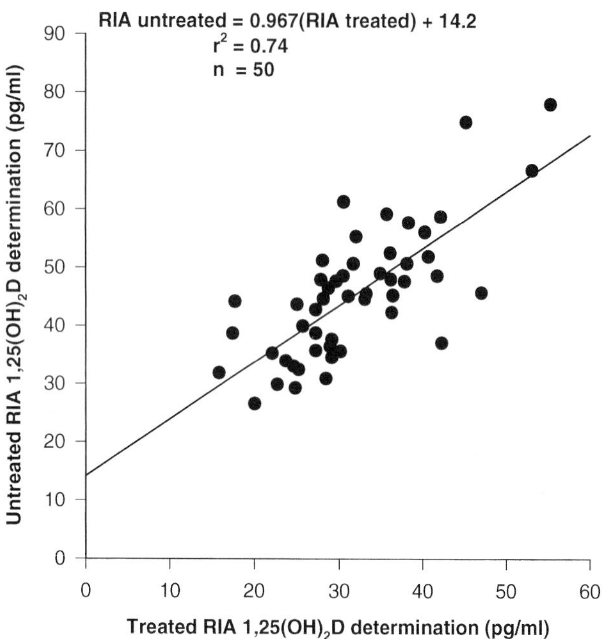

FIGURE 14 Comparison of circulating $1,25(OH)_2D$ measured by RIA with and without sample pretreatment with sodium *m*-periodate. From Hollis *et al. Clin Chem* 42:586–592.

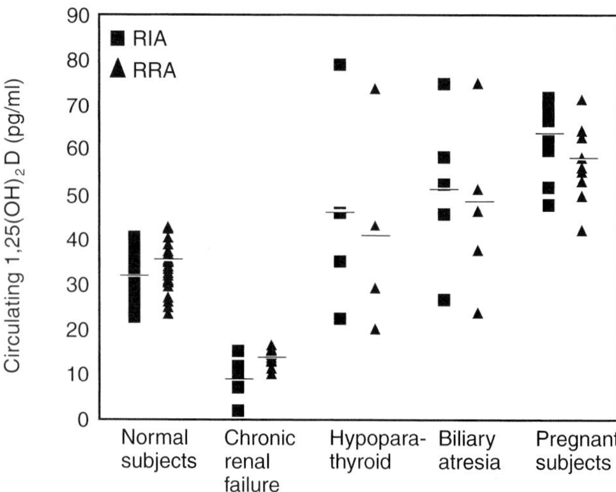

FIGURE 16 Circulating $1,25(OH)_2D$ as determined by the RIA (squares) or RRA (triangles) on a variety of clinical samples. The same samples were compared in each assay. Horizontal lines denote means.

$1,25(OH)_2D$ [36]. We concluded that this immunoextraction procedure was highly specific for the 1α-hydroxylated forms of vitamin D [35]. However, there was a serious flaw in the assumptions made when this kit was designed: $1,25(OH)_2D$ was the *only* significant 1α-hydroxylated vitamin D metabolite that circulates. Many other 1α-hydroxylated metabolites exist in the circulation, including $1,24,25(OH)_3D_3$, $1,25,26(OH)_3D_3$, $1,25(OH)_2D_3$-26,23-lactone, 1,25 $(OH)_2$-24-oxo-D_3, calcitroic acid, and probably various water-soluble, side-chain conjugates. Some of these compounds are bioactive, but most are not, and this assay cannot distinguish among them (Fig. 17). Compare what is measured by the IDS RIA (Fig. 17) versus what is measured by the RIA described in this text (Fig. 18). Further, compare how the IDS RIA for $1,25(OH)_2D$ performs outside of normal or chronic renal failure human samples (Fig. 19). This assay appears to be inadequate when presented with selected pathological samples.

We have specifically investigated the effects of $1,25(OH)_2D_3$-26,23-lactone on the "apparent" $1,25(OH)_2D$ levels using the immunoextraction technique and found it to interfere on an equal molar basis compared with $1,25(OH)_2D_3$ (Fig. 20). We also know that $1,25(OH)_2D_3$-26,23-lactone is a significant *in vivo* metabolite in a variety of clinical samples, and we find its concentration to be 0–30% of the respective $1,25(OH)_2D$ concentration. Further, using the RIA based on immunoextraction, we have found "apparent" levels of $1,25(OH)_2D$ to be grossly higher than the actual concentration in vitamin D-intoxicated subjects, hypoparathyroid subjects receiving vitamin D therapy, and biliary atresia patients (Fig. 19). We have also observed some normal samples that displayed 100% elevation from the actual levels. What this assay is recognizing in these samples remains unknown, but it is undoubtedly some 1α-hydroxylated metabolite, probably a catabolic product. It is important to note that the RIA described in this chapter, which is based on classic separation procedures, appears to escape this problem of detecting inactive 1α-hydroxylated vitamin D metabolites (Figs. 16 and 20).

VI. CLINICAL INTERPRETATION AND RELEVANCE OF ANTIRACHITIC STEROL MEASUREMENTS

A. Vitamin D

The quantitation of circulating vitamin D is essentially of no clinical importance. The parent compound is a poor indicator of nutritional status because of its short circulating half-life. The circulating levels of vitamin D are also difficult to interpret because the levels are greatly affected by short-term sun exposure and dietary intake of vitamin D [38,39]. Vitamin D has proved to be useful in assessing intestinal lipid absorptive capacity associated with fat malabsorption syndromes [40,41]. However, this use is more of a research application as opposed to an application used in a clinical diagnosis. Table IX lists a variety of clinical condi-

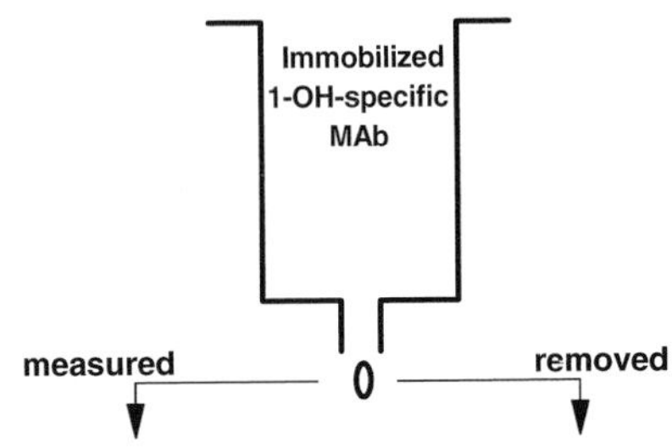

1) 100% $1,25(OH)_2D$
2) 100% $1,24,25(OH)_3D$
3) 100% $1,25,26(OH)_3D$
4) 100% $1,23,25(OH)_3D$
5) 100% $1,25(OH)_2D_3$-26,23-lactone
6) 100% Calcitroic acid
7) 100% water-soluble conjugates of calcitroic acid

1) all non-1-hydroxylated metabolites

FIGURE 17 Graphic description of the vitamin D metabolites assayed as "apparent" circulating $1,25(OH)_2D$ by utilizing immunoextraction with a 1-hydroxy-specific monoclonal antibody (MAb) in conjunction with RIA.

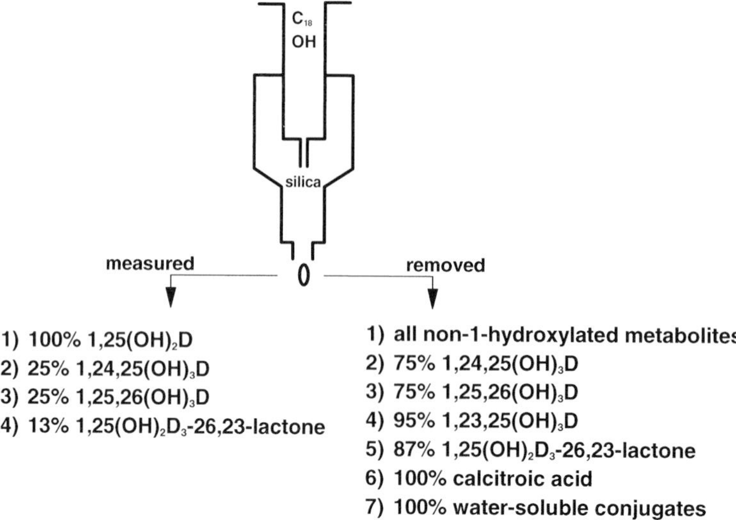

FIGURE 18 Graphic description of the vitamin D metabolites assayed as "apparent" circulating $1,25(OH)_2D$ by utilizing the extraction and purification procedure outlined in Section V,C of the text in conjunction with RIA.

tions for which the circulating levels of vitamin D have been defined.

B. 25OHD

Nutritional vitamin D status is defined by the amount of circulating 25OHD [42]. The assessment of circulating 25OHD is thus an important measurement to the clinician. Subnormal circulating levels of 25OHD usually result from inadequate vitamin D intake and/or insufficient sunlight exposure. This combination of events usually puts elderly patients at risk, especially if they are homebound, of developing vitamin D deficiency and ensuing secondary hyperparathyroidism [43]. This in turn has been shown to result in an increased incidence of hip fractures in the elderly (see Chapter 43) [44]. Other conditions that contribute to nutritional vitamin D deficiency include nephrotic syndrome, chronic renal disease (Chapter 52), cirrhosis, and malabsorption syndromes such as biliary atresia (Chapter 51) (Table IX).

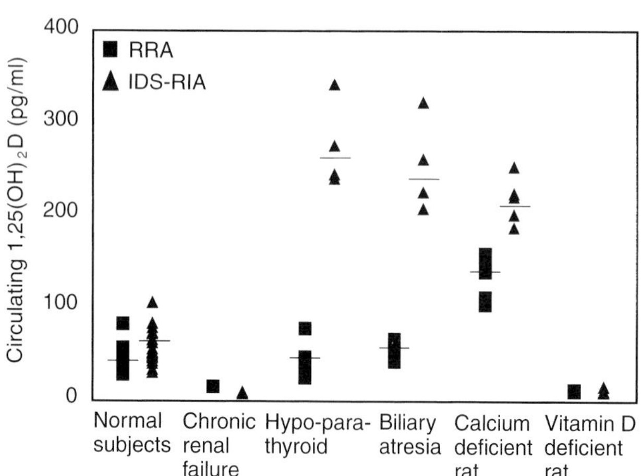

FIGURE 19 Circulating $1,25(OH)_2D$ as determined by the RRA (squares) or IDS immunoextraction RIA (triangles) on a variety of clinical samples. The same samples were compared in each assay. Horizontal lines denote means.

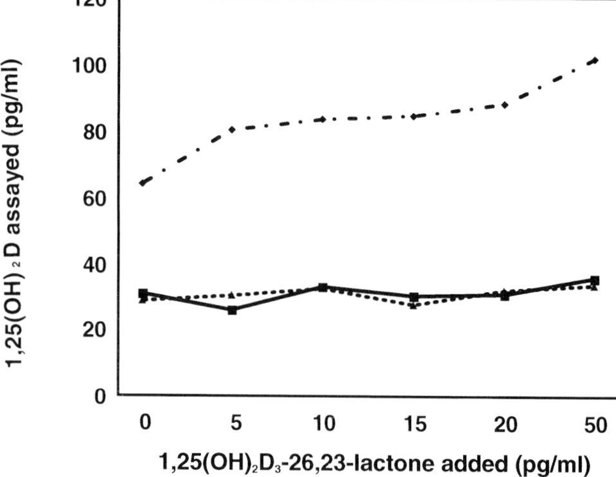

FIGURE 20 Effect of exogenously added $1,25(OH)_2D$-26,23-lactone on the "apparent" serum concentration of $1,25(OH)_2D$. Concentrations were assessed by RRA (squares), RIA as described in the text (triangles), and IDS immunoextraction RIA (diamonds). From Hollis BW. *Clin Chem* 41:1313–1314.

TABLE IX Relative Circulating Concentrations of Vitamin D, 25OHD, $24,25(OH)_2D$, and $1,25(OH)_2D$ in Various Disease States

Condition	Vitamin D[a]	25(OH)D[b]	$24,25(OH)_2D$[c]	$1,25(OH)_2D$[d]
Nutritional deficiency	Decreased	Decreased	Normal	Increased followed by decrease
Hypoparathyroidism	Normal	Normal	Normal	Decreased
Pseudohypoparathyroidism	Normal	Normal	Normal	Decreased
Hyperparathyroidism	Normal	Normal	Normal	Decreased
Tumor-induced osteomalacia	Normal	Normal	Normal	Decreased
Vitamin D-dependent rickets, type I	Normal	Normal	Normal	Decreased
Vitamin D-dependent rickets, type II	Normal	Normal	Normal	Increased
Sarcoidosis	Normal	Normal	Normal	Increased during hypercalcemia
Renal failure	Normal or decreased	Normal or decreased	Decreased	Decreased
Nephrotic syndrome	Decreased	Decreased	Decreased	Decreased
Hypervitaminosis D	Increased	Increased	Increased	Normal or decreased
Cirrhosis	Normal or decreased	Normal or decreased	Normal or decreased	Normal or increased
Tuberculosis	Normal	Normal	Normal	Increased during hypercalcemia
Hodgkin's disease	Normal	Normal	Normal	Increased during hypercalcemia
Lymphoma	Normal	Normal	Normal	Increased during hypercalcemia
Wegener's granulomatosis	Normal	Normal	Normal	Increased during hypercalcemia
X-linked hypophosphatemic rickets	Normal	Normal	Normal	Decreased or normal

[a] Normal range is 0–30 ng/ml (0–78 nmol/liter) and is extremely variable with respect to sunlight exposure and dietary intake.
[b] Normal range is 15–60 ng/ml (37–150 nmol/liter) and is related to season, latitude, and diet.
[c] Normal range is 0.5–4 ng/ml (1.2–9.6 nmol/liter) and is directly related to circulating 25OHD.
[d] Normal range is 20–60 pg/ml (48–144 pmol/liter).

Vitamin D intoxication, though rare, still occurs and is most accurately diagnosed by determining circulating 25OHD (Chapter 54). Thus, from a clinical standpoint, the determination of circulating 25OHD is the most frequently requested antirachitic sterol measurement.

C. $24,25(OH)_2D$

At the present time there does not appear to be a compelling reason to measure circulating $24,25(OH)_2D$ in a clinical setting. Even in a research setting, the determination of $24,25(OH)_2D$ is of questionable value as evidenced by the decreased usage of this assay in the literature since the early 1990s.

D. $1,25(OH)_2D$

Circulating $1,25(OH)_2D$ is diagnostic for several clinical conditions, including vitamin D-dependent rickets types I (Chapter 47) and II (Chapter 48), hypercalcemia associated with sarcoidosis, and other hypercalcemic disorders causing increased $1,25(OH)_2D$ levels (Chapters 54 and 55). These other disorders include tuberculosis, fungal infections, Hodgkin's disease, lymphoma, and Wegener's granulomatosis. In all other clinical conditions involving the vitamin D endocrine system, including hypoparathyroidism, hyperparathyroidism, and chronic renal failure, the assay of $1,25(OH)_2D$ is a confirmatory test. It is also important to remember that circulating $1,25(OH)_2D$ provides essentially no information with respect to the patient's nutritional vitamin D status. Thus, circulating $1,25(OH)_2D$ should not be used as an indicator for hypo- or hypervitaminosis D when nutritional factors are suspected (Table IX).

References

1. Seamark DA, Trafford DHJ, Makin HLJ 1981 The estimation of vitamin D and its metabolites in human plasma. J Steroid Biochem **14**:111–123.
2. Porteous CE, Coldwell RD, Trafford DJH, Makin HLJ 1987 Recent developments in the measurement of vitamin D and its metabolites in human body fluids. J Steroid Biochem **28**:785–801.
3. Jones G, Trafford DJH, Makin HLJ, Hollis BW 1992 Vitamin D: Cholecalciferol, ergocalciferol, and hydroxylated metabolites. In: DeLeenheer AP, Lambert WE, Nelis HJ (eds) Modern Chromatographic Analysis of Vitamins. Dekker, New York, pp. 73–151.
4. McCollum EV, Simmonds N, Shipley PG, Park EA 1922 Studies

on experimental rickets. XVI. A delicate biological test for calcium-depositing substances. J Biol Chem **51**:41–49.

5. Suda T, DeLuca HF, Hallick RB 1971 Synthesis of [26,27-^{3}H]-25-hydroxycholecalciferol. Anal Biochem **43**:139–146.
6. Haddad JG, Chyu KJ 1971 Competitive protein-binding radioassay for 25-hydroxycholecalciferol. J Clin Endocrinol Metab **33**:992–995.
7. Belsey R, DeLuca HF, Potts JT 1971 Competitive binding assay for vitamin D and 25-OH vitamin D. J Clin Endocrinol Metab **33**:554–557.
8. Haddad JG, Min C, Mendelsohn M, Slatopolsky E, Hahn TJ 1977 Competitive protein-binding radioassay of 24,25-dihydroxyvitamin D in sera from normal and anephric subjects. Arch Biochem Biophys **182**:390–395.
9. Brumbaugh PF, Haussler DH, Bursac DM, Haussler MR 1974 Filter assay for 1,25-dihydroxyvitamin D_3. Utilization of the hormones target tissue chromatin receptor. Biochemistry **13**:4091–4097.
10. Jones G 1978 Assay of vitamins D_2 and D_3, and 25-hydroxyvitamins D_2 and D_3 in human plasma by high-performance liquid chromatography. Clin Chem **24**:287–298.
11. Eisman JA, Shepard RM, DeLuca HF 1977 Determination of 25-hydroxyvitamin D_2 and 25-hydroxyvitamin D_3 in human plasma using high-pressure liquid chromatography. Anal Biochem **80**:298–305.
12. Clemens TL, Hendy GN, Graham RF, Baggiolini EG, Uskokovic MR, O'Riordan JLH 1978 A radioimmunoassay for 1,25-dihydroxycholecaliferol. Clin Sci Mol Med **54**:329–332.
13. Hollis BW, Napoli JL 1985 Improved radioimmunoassay for vitamin D and its use in assessing vitamin D status. Clin Chem **31**:1815–1819.
14. Hollis BW, Kamerud JQ, Selvaag SR, Lorenz JD, Napoli JL 1993 Determination of vitamin D status by radioimmunoassay with an ^{125}I-labeled tracer. Clin Chem **39**:529–533.
15. Hollis BW, Kamerud JQ, Kurkowski A, Beaulieu J, Napoli JL 1996 Quantification of circulating 1,25-dihydroxyvitamin D by radioimmunoassay with an ^{125}I-labeled tracer. Clin Chem **42**:586–592.
16. Horst RL, Littledike ET, Riley JL, Napoli JL 1981 Quantitation of vitamin D and its metabolites and their plasma concentrations in five species of animals. Anal Biochem **116**:189–203.
17. Lambert PW, DeOreo PB, Hollis BW, Fu IY, Ginsberg DJ, Roos BA 1981 Concurrent measurement of plasma levels of vitamin D_3 and five of its metabolites in normal humans, chronic renal failure patients, and anephric subjects. J Lab Clin Med **98**:536–548.
18. Hollis BW, Roos BA, Lambert PW 1981 Vitamin D in plasma: Quantitation by a nonequilibrium ligand binding assay. Steroids **37**:609–613.
19. Horst RL, Reinhardt TA, Beitz DC, Littledike ET 1981 A sensitive competitive protein binding assay for vitamin D in plasma. Steroids **37**:581–591.
20. Rhodes CJ, Claridge PA, Traffold DJH, Makin KLJ 1983 An evaluation of the use of Sep-Pak C_{18} cartridges for the extraction of vitamin D_3 and some of its metabolites from plasma and urine. J Steroid Biochem **19**:1349–1354.
21. Liel Y, Ulmer E, Shary J, Hollis BW, Bell NH 1988 Low circulating vitamin D in obesity. Calcif Tissue Int **43**:199–201.
22. Belsey RE, DeLuca HF, Potts JT 1974 A rapid assay for 25-OH-vitamin D_3 without preparative chromatography. J Clin Endocrinol Metab **38**:1046–1051.
23. Hollis BW, Pittard WB 1984 Evaluation of the total fetomaternal vitamin D relationships at term: Evidence for racial differences. J Clin Endocrinol Metab **59**:652–657.
24. Shepard RM, Horst RL, Hamstra AJ, DeLuca HF 1979 Determination of vitamin D and its metabolites in plasma from normal and anephric man. Biochem J **182**:55–69.
25. Dreyer BE, Goodman DBP 1981 A simple direct spectrophotometric assay for 24,25-dihydroxyvitamin D_3. Anal Biochem **114**:37–41.
26. Hummer L, Christiansen C 1984 A sensitive and selective radioimmunoassay for serum 24,25-dihydroxycholecaliferol in man. Clin Endocrinol **21**:71–79.
27. Wei S, Tanaka H, Kubo T, Ichikawa M, Seino Y 1994 A multiple assay for vitamin D metabolites without high-performance liquid chromatography. Anal Biochem **222**:359–365.
28. Jones G, Byrnes B, Palma F, Segev D, Mazur Y 1980 Displacement potency of vitamin D_2 analogs in competitive protein-binding assays for 25-hydroxyvitamin D_3, 24,25-dihydroxyvitamin D_3 and 1,25-dihydroxyvitamin D_3 J Clin Endocrinol Metab **50**:773–775.
29. Eisman JA, Hamstra AJ, Kream BE, DeLuca HF 1976 A sensitive, precise, and convenient method for determination of 1,25-dihydroxyvitamin D in human plasma. Arch Biochem Biophys **176**:235–243.
30. Reinhardt TA, Horst RL, Orf JW, Hollis BW 1984 A microassay for 1,25-dihydroxyvitamin D not requiring high performance liquid chromatography: Application to clinical studies. J Clin Endocrinol Metab **58**:91–98.
31. Hollis BW 1986 Assay of circulating 1,25-dihydroxyvitamin D involving a novel single-cartridge extraction and purification procedure. Clin Chem **32**:2060–2063.
32. Bouillon R, De Moor P, Baggiolini EG, Uskokovic MR 1980 A radioimmunoassay for 1,25-dihydroxycholecalciferol. Clin Chem **26**:562–567.
33. Gray TK, McAdoo T, Pool D, Lester GE, Williams ME, Jones G 1981 A modified radioimmunoassay for 1,25-dihydroxycholecalciferol. Clin Chem **27**:458–463.
34. Fraher LJ, Adami S, Clemens TL, Jones G, O'Riordan JLH 1983 Radioimmunoassay of 1,25-dihydroxyvitamin D_2: Studies on the metabolism of vitamin D_2 in man. Clin Endocrinol **18**:151–165.
35. Hollis BW 1995 1,25-Dihydroxyvitamin D_3-26,23-lactone interferes in determination of 1,25-dihydroxyvitamin D by RIA after immunoextraction. Clin Chem **41**:1313–1314.
36. Mawer EB, Berry JL, Bessone J, Shany S, White A 1985 Selection of high-affinity and high specificity monoclonal antibodies for 1,25-dihydroxyvitamin D. Steroids **46**:741–754.
37. Bouillon R, Okamura WH, Norman AW 1995 Structure–function relationships in the vitamin D endocrine system. Endocr Rev **16**:200–257.
38. Adams JA, Clemens TL, Parrish JA, Holick MF 1982 Vitamin D synthesis and metabolism after ultraviolet irradiation of normal and vitamin D-deficient subjects. N Engl J Med **306**:722–725.
39. Hollis BW, Lowery JW, Pittard WB, Guy DG, Hansen JW 1996 Effect of age on the intestinal absorption of vitamin D_3-palmitate and nonesterified vitamin D_2 in the term human infant. **81**:1385–1388.
40. Heubi JE, Hollis BW, Tsang RC 1990 Bone disease in chronic childhood cholestasis II. Better absorption of 25(OH)D than vitamin D in extrahepatic biliary atresia. Pediatr Res **27**:26–31.
41. Argao EA, Heubi JE, Hollis BW, Tsang RC 1992 *d*-alpha-tocopheryl polyethylene glycol-1000 succinate enhances the absorption of vitamin D in chronic cholestatic liver disease of infancy and childhood. Pediatr Res **31**:146–150.
42. Haddad JG, Stamp TCB 1974 Circulating 25-hydroxyvitamin D in man. Am J Med **57**:57–62.
43. Gloth MF, Gundberg CM, Hollis BW, Haddad JG, Tobin JD 19 Vitamin D deficiency in homebound elderly persons. JAMA **274**:1683–1686.
44. Chapuy MC, Arlot ME, DuBoeuf F, Brun J, Crouzet B, Arnaud S. Meunier P 1992 Vitamin D_3 and calcium to prevent hip fractures in elderly women. N Engl J Med **327**:1637–1642.

CHAPTER 39

Pharmacology of Vitamin D Preparations

NEIL A. BRESLAU AND JOSEPH E. ZERWEKH
University of Texas Southwestern Medical Center at Dallas, Center for Mineral Metabolism and Clinical Research and Baylor University Medical Center, Dallas, Texas

I. HISTORICAL FACTORS

It has been nearly two centuries since the first insight into the potential cause of rickets was made by Sniadecki in 1822 [1]. He observed that while rickets was prevalent in Warsaw, a highly industrialized city, children living in nearby rural areas were essentially free of this bone disease. This led him to advocate exposure of affected individuals to the sun as an efficient method for the prevention and cure of rickets. Although unknown to him at that time, his observations were probably responsible for the first clinical application of vitamin D therapy in the treatment of a specific disease. Nearly 100 years later, Huldschinsky [2] exposed four rachitic children to radiation from a mercury arc lamp and demonstrated by X-ray analysis that rickets was cured after 4 months of therapy. This observation was confirmed two years later by Hess and Unger [3] in seven rachitic children exposed to sunshine. Shortly thereafter, Hess and Weinstock [4] and Steenbock and Black [5] exposed a variety of foods and other substances including lettuce, vegetable oils, and rat feed to ultraviolet irradiation and found that this simple procedure imparted antirachitic activity to most of the food products. Steenbock [6] concluded that the prevention and cure of rickets in children might be accomplished simply by irradiating food substances. This concept led to the addition of provitamin D to milk and its subsequent irradiation to impart antirachitic activity. Ultimately, this form of vitamin D therapy led to the eradication of rickets as a significant health problem in the United States and other countries. When the structure of vitamin D became known [7,8], it was chemically synthesized and added directly to milk, making the irradiation of milk obsolete.

The chemical synthesis of vitamin D also opened another chapter in the pharmacology of vitamin D, namely, the discovery that vitamin D undergoes metabolic alteration to a physiologically active form, and that a number of chemically synthesized analogs of vitamin D and its metabolites retain antirachitic activity. Currently, there are some seven different preparations of vitamin D and its metabolites that have been approved for use in disorders of mineral metabolism either in the United States or abroad.

II. NATURAL SOURCES OF VITAMIN D

As in the rest of this text, the term "vitamin D" encompasses both vitamin D_2 and vitamin D_3 (where appropriate, we refer to the individual forms of the vitamin). Vitamin D is naturally obtained from dietary sources and from the endogenous synthesis of a precursor substance in the skin. The chemical structure, metabolism, and conversion of 7-dehydrocholesterol to previtamin D_3 and its isomerization to vitamin D_3 are extensively reviewed in Chapters 2 and 3 and are not considered here. Rather, it should be mentioned that the cutaneous production of vitamin D_3 is extremely efficient in meeting the vitamin D requirement of normal human subjects. Brief, casual exposure of the face, arms, and hands to sunlight is equivalent to ingesting 200 IU[1] of vitamin D, and repeated total-body exposure (causing mild erythema) raises plasma 25-hydroxyvitamin D (25OHD) concentrations as much as the long-term ingestion of 10,000 IU (250 μg) of vitamin D per day [9].

Dietary vitamin D is absorbed in the upper part of the small intestine in much the same way as are other fat-soluble compounds. It enters the circulation primarily through the thoracic duct in the chylomicron fraction, after which it associates with an α-globulin fraction in the blood. The dietary contribution to total body stores of vitamin D is probably minimal in view of the fact that very few natural foods contain much vitamin D. The few exceptions to this observation are fatty fish and fish oils, eggs, liver, and milk, although naturally occurring levels of vitamin D in milk are highly variable depending on the season.

Table 1 summarizes the vitamin D content of some common food products. From Table I it is apparent that it would be difficult to meet the recommended dietary allowance (RDA) of 400 IU for growing children and 200 IU for adults [10], based on naturally occurring levels. For this reason, many countries have adopted a policy of fortifying foods with vitamin D, a policy that has been applied to milk in the United States since the 1930s. Because of this fortification, dietary vitamin D intake of healthy adults is in the range of 4–6 μg/day (160–240 IU/day) in the United States [11,12]. In addition to milk, infant formula, and breakfast cereals, vitamin D is also added to baked goods, pasta, rice dishes, fats and oils, sauces, and nonalcoholic beverages. These food products contain either vitamin D_2 or D_3 because of fortification with these forms of vitamin D or from the UV light-mediated conversion of either ergosterol or 7-dehydrocholesterol to the natural vitamin. Unfortunately, in the early 1990s it was disclosed that milk and infant formula preparations rarely contain the amount of vitamin D stated on the label and may be either underfortified or overfortified [13,14]. In addition, errors in the labeling of the stock solutions used for milk fortification have been documented [15], pointing to the need for better regulation and testing procedures for this type of food fortification.

[1] One international unit (IU) = 0.025 μg or 25 ng.

TABLE I Vitamin D Content of Selected Food Products

Food category and example	Vitamin D content (IU)
Cereals, ready-to-eat	
Bran flakes—1 oz	50
Corn flakes—1 oz	40
Oat bran—1 oz	50
Rice Krispies—1 oz	50
Cheese and cheese products	
Cream cheese—1 oz	2
Swiss—1 oz	1
Cheese spread	3
Puddings	
Various types, instant and regular—1/2 cup	30–50[a]
Eggs	
Whole, fresh/frozen—1 large	25
Yolk, fresh—1	25
Fish oils	
Cod liver oil—1 T	34
Grain products	
Rice—1/2 cup	4
Stuffing—1/2 cup	9
Meats	
Bacon bits—0.25 oz	4
Luncheon	7–20
Milk	
Evaporated, low fat canned—8 fl oz	100[b]
Skim, canned—1 fl oz	25[b]
Whole—8 fl oz	100[b]
Salad dressings	
Ranch, from mix—1 T	5
Buttermilk farm style, from mix—1 T	3
Spreads	
Margarine, stick—1 t	21
Margarine, tube—1 t	21
Miracle Whip	1
Vegetables	
Broccoli, frozen with cheese sauce—1/2 cup	2
Green beans, frozen—1/2 cup	2
Peas, frozen with cheese sauce—1/2 cup	1

[a] Prepared with vitamin D-fortified milk.
[b] Fortified with vitamin D.
From: Pennington JAT, 1994. In Bowes and Church's Food Values of Portions Commonly Used—16th Edition, JB Lippincott, Philadelphia, pp. 411–414.

Interestingly, the physiologically active form of vitamin D_3, 1α,25-dihydroxyvitamin D_3 [$1,25(OH)_2D_3$], has been identified as a water-soluble glycoside in two wild plants, *Solanum malacoxylon* [16] and *Cestrum diurnum* [17]. The presence of this calcemic vitamin D metabolite in these plants is believed to be responsible for the pathological calcification that occurs in grazing animals ingesting these plants. Affected areas include the Argentinian pampas.

III. PHARMACOKINETICS AND METABOLISM OF VARIOUS VITAMIN D PREPARATIONS AND ANALOGS

The term vitamin D is a general one used to describe a number of chemically related compounds represented by both naturally occurring as well as organically synthesized examples. Most share common antirachitic properties but differ in the way they are produced and the rapidity with which they bring about metabolic effects. For the purposes of this discussion, we limit the consideration to seven different pharmacological preparations. They are vitamins D_2 and D_3, dihydrotachysterol, $25OHD_3$, $1,25(OH)_2D_3$, 1α-hydroxycholecalciferol (1α-OHD_3), and calcipotriol. Clinical use, dosage forms, and cost are further discussed in Chapter 56.

A. Vitamin D_2 (Ergocalciferol, Calciferol)

Ergocalciferol (vitamin D_2), (calciferol) was isolated as a pure crystalline substance by Windaus *et al.* [7] and Askew *et al.* [18] in the 1930s, making it one of the oldest of the known vitamin D substances. It was prepared from the irradiation of ergosterol, and the same method is still used today for mass production. Although this form of vitamin D is essentially a plant vitamin, it continues to be of extreme biological value in human metabolism because of its use as a food additive. It is also widely used in animal nutrition. Vitamin D_2 appears to be able to substitute for vitamin D_3 in every metabolic step in humans. Thus, discussion of its pharmacokinetics is covered in the next section with cholecalciferol. Vitamin D_2 continues to be the most frequently used form of vitamin D in pediatric and adult multivitamin preparations, and it is also the most commonly used pure form of vitamin D. One milligram of vitamin D_2 is equal to 40,000 IU of vitamin D activity. The vitamin is usually marketed in oil capsules or tablets each containing 50,000 IU (1.25 mg), as an oil injection of 500,000 IU/ml (12.5 mg), and as an over-the-counter oral preparation containing 8000 IU/ml (0.2 mg).

All vitamin D compounds appear to be sensitive to oxidation and are inactivated by moist air within a few days. Vitamin D_2 has a shelf life of about 9 months when stored in an amber container under refrigeration.

B. Vitamin D_3 (Cholecalciferol)

Cholecalciferol (vitamin D_3) represents the animal counterpart of vitamin D_2. As previously mentioned, it was first isolated and obtained in crystalline form by Schenk [8] from the numerous irradiation products of 7-dehydrocholesterol. This was followed by the identification of this vitamin in pure form from natural tuna liver oil [19]. 7-Dehydrocholesterol is found principally in the skin of animals, and it is the sun's action on this precursor that results in the cutaneous production of vitamin D_3. Another important source of vitamin D_3 is the diet. Like vitamin D_2, dietary vitamin D_3 is absorbed in the jejunum and upper ileum via the intestinal lymphatics.

Peak serum vitamin D concentrations occur approximately 12 hr after a single oral dose and return to basal levels within 72 hr [20]. However, in patients in whom chronic ingestion of large doses of vitamin D_2 was necessary, the time course of vitamin D disappearance from serum was prolonged to several months to a year [21]. Similar observations have been reported [22] using radioactive cholecalciferol. These studies suggest that under conditions of excessive administration or ingestion, body fat stores of vitamin D increase and provide a sustained release of this vitamin D metabolite for further metabolic activation. Although the pharmacological use of vitamin D_2 or D_3 has diminished in recent years, the above-mentioned sustained release of these vitamin D preparations has obvious clinical implications. For example, if vitamin D toxicity occurs, it may persist for months. Moreover, the prolonged activity of vitamin D preparations should be considered in the follow-up of patients shifted from treatment with traditional high doses of vitamin D to physiological doses of the more active vitamin D metabolites or analogs.

Vitamin D is excreted in bile principally as more polar metabolites. Increased concentrations of vitamin D or of 25OHD in the liver also cause an increased biliary excretion of these substances, and an increased hepatic formation and elimination of more polar metabolites of the vitamin.

Cholecalciferol is not currently available as a pharmacological agent. Because of the ease in producing ergocalciferol (vitamin D_2), this form of vitamin D has replaced the use of vitamin D_3. However, vitamin D_3 is manufactured and used for the fortification of various food products. Vitamin D_3 appears to have a shelf life slightly longer than that of vitamin D_2, demonstrating negligible deterioration after 1 year of storage in evacuated amber containers under refrigeration.

C. Dihydrotachysterol (DHT, Intensol)

Dihydrotachysterol (DHT, Intensol) is a structural analog of vitamin D_3. It was first isolated in 1930 by Holtz and Schreiber [23] and is of great historical interest because of its early successful use in the long-term management of patients with idiopathic or postsurgical hypoparathyroidism [24,25]. If the A ring of vitamin D_3 is rotated by 180°, the C-3 β-OH group simulates the C-1 α-OH group of the active dihydroxy metabolite of vitamin D_3. Such compounds are therefore called pseudo-1α- hydroxyvitamin D analogs. Early preparations of DHT varied significantly with respect to their purity and biological potency. Thus, the older preparations have been slowly discontinued and replaced with newer preparations containing crystalline DHT.

For many years few, if any, assays were available for measuring DHT in serum, and thus little was known regarding its pharmacokinetics. In 1988, Taylor *et al.* [26] developed a reliable method for quantitating DHT in serum and provided the first pharmacokinetic profile for this vitamin D analog. In normal subjects, oral administration of 0.4–0.8 mg DHT raised serum DHT values within 1 hr following administration. Peak concentrations occurred at 4 hr after DHT administration. This was followed by a decline in serum DHT levels by approximately 70% between 4 and 24 hr. The mean level 4 hr after DHT administration on days 2, 5, and 8 of administration was comparable to the 4-hr level on day 1, and after 24 hr it was negligible. Thus, serum DHT levels do not appear to be influenced by repeated short-term dosing. The half-life of this vitamin D preparation is also much shorter than that of vitamin D_3, with estimates ranging between 7 and 8 hr [27]. This relatively short half-life is probably responsible for the reduced toxicity of this compound [28].

Like vitamin D, DHT undergoes extensive metabolism. More than seven metabolites have been isolated and identified in rats [29]. For many years, it was felt that DHT only required one metabolic activation in the liver to 25-hydroxydihydrotachysterol (25OHDHT) before it became biologically active [30]. However, it was suggested that 1-hydroxylation of 25OHDHT also occurs, producing two isomers, 1α,25- and 1β,25$(OH)_2$DHT [31]. This was subsequently shown to be true not only in the rat [32] but also for humans [33]. Preliminary studies [34] suggested that 1,25$(OH)_2$DHT was produced *in vivo* in anephric human, and extensive *in vitro* studies using isolated perfused rat kidneys were unable to demonstrate the 1-hydroxylation of 25OHDHT, implying that the enzyme responsible for this conversion is not the PTH-controlled renal 1α-hydroxylase, and, indeed, is probably located at an extrarenal site. The formation of these 1,25-dihydroxylated metabolites casts doubt on the concept that the pseudo 1α-hydroxy compound DHT is effective in humans, requiring only 25-hydroxylation for activity. Previous studies have shown that 1α,25$(OH)_2DHT_3$, and presumably the same metabolite of DHT_2, is more active in a reconstructed cell system than $25OHDHT_3$, but only one-tenth as active as 1,25$(OH)_2D_3$, suggesting that the 1α,25-dihydroxylated compound may be the active metabolite of DHT.

Current preparations of DHT include tablets at 0.125, 0.2, and 0.4 mg dosages as well as a liquid formulation at a concentration of 0.2 mg/ml. Shelf-life and storage conditions are similar to those for vitamin D_2 and D_3.

D. 25-Hydroxycholecalciferol (Calcifediol, 25-Hydroxyvitamin D_3, Calderol)

25-Hydroxyvitamin D_3 (25-hydroxycholecalciferol, calcifediol, Calderol) and 25-hydroxyvitamin D_2 hold important central roles in the metabolic pathways of vitamin D. In humans, 25-hydroxyvitamin D is the major circulating metabolite of vitamin D and is considered to be the best indicator of a patient's vitamin D stores. The circulating concentration of 25OHD is directly proportional to an individual's UV light exposure as well as dietary and supplement contributions [35]. Early studies suggested that 25-hydroxyvitamin D_3 had approximately three times the antirachitic activity of vitamin D_3 on a weight basis. However, since these studies were performed in animals with intact kidneys, in whom production of the active form of vitamin D probably occurred, it was not possible to ascribe a specific biological action to $25OHD_3$. *In vitro* studies have demonstrated that $25OHD_3$ has approximately 1/500 to 1/1000 the affinity of 1,25$(OH)_2D_3$ for the intestinal receptor. However, it should also be remembered that the circulating concentration of $25OHD_3$ is approximately three orders of magnitude greater than that of 1,25$(OH)_2D_3$, thus leaving the possibility of some residual biologic action for $25OHD_3$.

The possibility that $25OHD_3$ could have specific biological actions at pharmacological doses has been demonstrated. Early studies by Teitelbaum *et al.* [36] in azotemic patients and more recently by Rutherford *et al.* [37] and Colodro *et al.* [38] in patients with advanced renal failure, or who were anephric, have clearly demonstrated that $25OHD_3$ elicited a pronounced biological effect, namely, an improvement in intestinal calcium absorption. In an analogous fashion, in patients with pseudovitamin D deficiency (PDDR, see Chapter 47), who have an inherited defect in 1α-hydroxylase activities, adequate control of the bone disease can be achieved with large doses of vitamin D that cause an up to 10-fold increase in 25OHD circulating levels, without significant change in 1α,25$(OH)_2D$ concentrations. Fournier *et al.* [39] demonstrated a better therapeutic effect from 25OHD$_3$ with respect to bone mineralization when compared with either 1α-OHD$_3$ or 1,25$(OH)_2D_3$ in patients with advanced renal failure and osteomalacia. Still to be resolved is whether this positive effect on bone mineralization following 25OHD$_3$ administration is solely the result of this metabolite or whether it results from the production of more polar metabolites, including 1,25$(OH)_2D_3$, produced at extrarenal sites containing 1α-hydroxylase activity.

25-Hydroxyvitamin D_3 is rapidly absorbed from the intestinal tract. Peak serum 25OHD$_3$ levels occur 4 to 8 hr after oral dosing, and the incremental increase is linear over a dose range of 1 to 10 μg [40]. Studies have suggested that the $t_{1/2}$ for 25OHD$_3$ is on the order of 3 weeks [40,41]. In anephric subjects, this value may be increased to as high as 7 weeks because of the lack of further metabolism by renal tissue [41]. Currently, 25OHD$_3$ is supplied under the name Calderol (calcifediol capsules from Organon) in 20 and 50 μg capsules. Storage conditions and shelf life are similar to those described for vitamin D_3 above.

E. 1α,25-Dihydroxycholecalciferol (Calcitriol, 1α,25-Dihydroxyvitamin D_3, Rocaltrol, Calcijex)

1,25-Dihydroxyvitamin D_3 (1α,25-dihydroxycholecalciferol, calcitriol, Rocaltrol, Calcijex) is well recognized as the natural hormonal form of vitamin D_3. For over two decades, experimental studies have shown it to be the most powerful weight-for-weight metabolite of vitamin D for stimulating intestinal calcium absorption and mobilizing calcium from bone. Because this vitamin D metabolite requires no further metabolism for biological activity, it is effective in both hypoparathyroid and anephric subjects. It is also the compound of choice for lifelong replacement therapy in patients with pseudovitamin D deficiency rickets and osteomalacia (see Chapter 47). In contrast to vitamin D_3 and 25OHD$_3$, calcitriol elicits an increase in intestinal calcium absorption as rapidly as 2 to 6 hr after the oral administration of 0.5 μg. This is consistent with the observation that this agent does not require any further metabolic activation before the induction of its metabolic effects.

Pharmacokinetic studies have demonstrated that peak serum concentrations of 1,25$(OH)_2D_3$ occur at 4 to 8 hr after a single oral dose and return to baseline values by 24 hr [42–44]. If 1,25$(OH)_2D_3$ is given in divided doses more frequently during the day, then baseline values tend to increase. However, the biological effects of calcitriol are dependent on the total dose administered rather than on the frequency of dosing. This difference between dosing regimens and the magnitude of the biological response is due to differences between the serum half-life of calcitriol and the biological half-life. Early studies that examined the plasma disappearance of 1,25$(OH)_2D_3$ suggested a serum half-life on the order of minutes to a few hours [42–45]. On the other hand, the biological effects of 1,25$(OH)_2D_3$ have been shown to persist for as long as 7 to 8 days in human studies [46,47]. Such observations make it unlikely that twice-daily dosing would affect intestinal calcium absorption differently than would a single daily dose.

Another area of considerable interest regarding the pharmacology of 1,25$(OH)_2D_3$, is the intravenous administration of this drug and its potential for reducing secondary hyperparathyroidism of renal failure. Slatopolsky and co-workers [48,49] have demonstrated that intravenous administration of calcitriol raises the serum level of this vitamin D metabolite some 4-fold higher than that observed with oral dosing, followed by a decline to values found with oral dosing at 6 to 8 hr post-dosing. The cause for this difference in serum levels is dependent not only on the efficiency of intestinal absorption of calcitriol when given orally, but also on metabolism of 1,25$(OH)_2D_3$ to inactive metabolites in the gut [50]. Thus, oral administration would raise intestinal calcium absorption by acting from the mucosal side, but it would raise blood levels only minimally by 6 to 8 hr. On the other hand, intravenous administration has been demonstrated to lower immunoreactive parathyroid hormone (PTH) levels with little or no calcemic activity in renal failure patients. The implication of these results is that attainment of certain peripheral effects of oral 1,25$(OH)_2D_3$ may be limited by the enhanced local effect on intestinal calcium absorption, leading to hypercalcemia. The hyperabsorption of calcium limits the dose of 1,25$(OH)_2D_3$ that can be administered. The quantity of 1,25$(OH)_2D_3$ available to peripheral tissues (e.g., parathyroid glands) may be further reduced by intestinal degradation of the vitamin D metabolite. Intravenous administration of 1,25$(OH)_2D_3$, on the other hand, may allow greater delivery to peripheral tissues and allow for expression of specific biological effects at these sites.

The principal route of excretion of 1,25$(OH)_2D_3$ is via the bile. This is probably true for the majority of the vitamin D metabolites after some type of structural modification resulting in metabolites of greater solubility in bile. For $1,25(OH)_2D_3$, this seems to occur through metabolism of the side chain [51,52]. Following administration of isotopically labeled $1,25(OH)_2D_3$, approximately 10–15% of the total administered dose appeared in the urine as a glucuronide [45], whereas the remainder of the dose (up to 65%) appeared in the feces. The total quantity of radioactivity recovered in the urine and feces by 6 days after labeling (65%) indicated that most of the radioactivity is excreted during this time. This metabolic clearance pathway, combined with the metabolism of $1,25(OH)_2D_3$ from side chain cleavage, accounts for nearly all of the administered dose of calcitriol. These results indicate that the body can rapidly remove from plasma amounts of $1,25(OH)_2D_3$ that are in excess of the normal quantities circulating in the plasma.

Currently, there are two pharmacological preparations of $1,25(OH)_2D_3$ available for clinical use. Rocaltrol (Roche) is available as soft gelatin capsules in doses of 0.25 and 0.5 μg. This product should be protected from light and heat. Calcijex (calcitriol injection) (Abbott) is synthetically manufactured calcitriol and is available as a sterile, isotonic, clear aqueous solution for intravenous injection. Calcijex is available in 1 ml ampoules at a concentration of 1 or 2 μg/ml. Although Calcijex should be protected from light, it can be stored at room temperature.

F. 1α-Hydroxycholecalciferol (Alfacalcidiol, 1α-Hydroxyvitamin D_3)

1α-Hydroxyvitamin D_3 (1αOHD_3), 1α-hydroxycholecalciferol, alfacalcidiol) is not a naturally occurring vitamin D metabolite. However, its chemical synthesis was reported in 1973 [53]. Several studies documented its antirachitic activity in animals as well as its need for conversion to $1,25(OH)_2D_3$ prior to eliciting a biological response [54,55]. Early studies in normal humans and patients with chronic renal failure suggested that pharmacological differences existed between 1αOHD_3 and $1,25(OH)_2D_3$ and that these differences were probably the result of the requirement for metabolic conversion to $1,25(OH)_2D_3$ [56].

Several studies are available in which measurements of circulating $1,25(OH)_2D_3$ have been performed following oral administration of 1αOHD_3. Lund *et al.* [57] studied the effect of a single oral dose of 2 μg of 1αOHD_3 on the serum $1,25(OH)_2D_3$ concentration in three elderly subjects. Although large individual variations were observed, peak concentrations of $1,25(OH)_2D_3$ were observed after 4–12 hr. The mean rise in serum $1,25(OH)_2D_3$ of 8 pg/ml was seen 24 hr after the administration of the dose. Kimura *et al.* [58] studied six renal failure patients maintained on hemodialysis and found that multiple dosing of 1αOHD_3 or $1,25(OH)_2D_3$ gave higher basal levels of circulating $1,25(OH)_2D_3$, resulting in a higher area under the curve at 24 hr (AUC_{24}, 409 $\pm$ 50 pg/ml in single dose versus 822 $\pm$ 64 pg/ml in multiple dosing). No significant difference in maximum increase over basal level, or in $t_{1/2}$, were observed, although the value of 36 to 43 hr reported by Kimura is considerably longer than the $t_{1/2}$ of minutes observed for $1,25(OH)_2D_3$ discussed above. Seino *et al.* [59] gave a 4 μg oral dose of 1αOHD_3 to 20 healthy men and reported that peak serum $1,25(OH)_2D_3$ concentrations were seen at 10.5 hr. Masuda *et al.* [60] found that a 5 μg dose of 1αOHD_3, given orally to healthy males, resulted in a gradual increase in serum $1,25(OH)_2D_3$ levels. Peak concentration of serum $1,25(OH)_2D_3$ was reached between 8 to 10 hr after administration, and the concentrations of $1,25(OH)_2D_3$ returned to baseline by 48 hr. Joffe *et al.* [61] observed a peak increase in serum $1,25(OH)_2D_3$ at 12 hr following an oral dose of approximately 5.6 μg (80 ng/kg body weight) to 8 patients on peritoneal dialysis. Although they were unable to calculate $t_{1/2}$ for orally administered 1αOHD_3, they did calculate a mean $t_{1/2}$ of 109 hr for intravenous administration. Taken together, these pharmacological results support clinical observations which indicate that 1αOHD_3 is less effective than $1,25(OH)_2D_3$ at lower doses, has a slower onset of maximal effect, and may have a more prolonged duration of action after therapy is discontinued. Although widely used in Europe and Japan, this vitamin D analog is currently not approved for clinical use in the United States.

G. Calcipotriol (Calcipotriene, Dovonex, MC903)

Calcipotriol (calcipotriene, Dovonex, MC903) is a 1,24-dihydroxyvitamin D_3 analog with an altered side chain that is discussed in detail in Chapter 61. In several *in vitro* models, calcipotriol and calcitriol markedly inhibited cell proliferation and enhanced cell differentiation over a range of concentrations from approximately 10^{-10} to 10^{-6} mol/liter. Calcipotriol binds to intestinal calcitriol receptors with affinity similar to that of calcitriol but is 100 to 200 times less potent than calcitriol in its effect on *in vivo* calcium metabolism in rats. The reduced calcemic activity of calcipotriol compared with calcitriol is largely due to pharmacokinetic differences. In liver homogenates from three different species, calcipotriol is transformed into two major metabolites [62]. The metabolic pathway involves oxidation at carbon 24 in the side chain and constitutes a deactivation pathway for calcipotriol. The enhanced catabolism of calcipotriol may be related to its low affinity for the serum vitamin D binding protein (DBP), which is about 1/30 that of calcitriol [63]. The metabolites of calcipotriol appear to have much weaker effects on cell proliferation and differentiation than those exerted by calcipotriol. After a single intravenous dose to rats, the half-life of calcipotriol was observed to be 4 min, and the serum concentration curve was more than 100 times lower than for $1,25(OH)_2D_3$, which agrees well with the relative calcemic potency [64].

Little is known regarding the transdermal absorption of calcipotriol. In one human study, less than 1% of calcipotriol was systemically absorbed after a single application of ^{3}H-calcipotriol ointment at 50 μg/g (0.3 to 1.7 g) to psoriatic lesions on the backs of four patients [65]. This relatively low systemic bioavailability, when coupled with its rapid metabolism and deactivation, minimize any systemic effects.

Calcipotriol is approved in several countries for the topical treatment of plaque psoriasis (see Chapter 72). In the United States it is prepared under the name Dovonex Ointment (Westwood-Squibb). It is supplied as a topical ointment at a concentration of 0.005% and is available in 30, 60, and 100 g size aluminum tubes. It should be stored at controlled room temperature and protected from freezing.

IV. APPROPRIATE CLINICAL USAGE OF VITAMIN D PREPARATIONS

The major therapeutic uses of vitamin D may be divided into four categories [66]: (1) prophylaxis and cure of nutritional rickets, (2) treatment of metabolic disorders of vitamin D metabolism resulting in rickets or osteomalacia, (3) treatment of hypoparathyroidism, and (4) prevention and treatment of osteoporosis.

A. Nutritional Rickets

Nutritional rickets (see also Chapters 41–43) may result from inadequate exposure to sunlight or deficiency of dietary vitamin D. The condition is rare in the United States and other countries where food fortification with the vitamin is practiced. Infants and children receiving adequate amounts of vitamin D-fortified food do not require additional vitamin D. However, breast-fed infants or those fed unfortified formula should receive 400 IU of vitamin D (ergocalciferol) daily as a supplement.

The curative dose of vitamin D for the treatment of fully developed rickets is larger than the prophylactic dose. A dose of 1000 IU daily will normalize serum calcium and phosphate concentrations in approximately 10 days, and radiographic evidence of healing will occur within about 3 weeks. However, a daily dose of 3000 to 4000 IU often is prescribed for more rapid healing. The same doses apply to children with rickets or adults with osteomalacia.

There are several advantages in using the parent vitamin D compound in nutritional rickets. Most of the administered drug accumulates in body fat, and that in the circulation must be activated by both 25- and 1α-hydroxylation. This storage characteristic is useful when daily drug administration is not practical or when parenteral therapy is required. Use of the parent vitamin D compounds also allows normal regulation of mineral metabolism by the parathyroid–renal axis. Moreover, these drugs are the least expensive vitamin D supplements.

Despite adequate sunlight and nutrition, certain disorders may lead to reduced availability of vitamin D. Gastrointestinal and hepatobiliary disorders may lead to reduced absorption of vitamin D (see Chapter 51). There have been concerns that an impaired enterohepatic circulation of biologically active vitamin D sterols could imperil sterol stores from any source [67]. Treatment to correct deficiency states due to malabsorption can include sunlight, parenteral vitamin D, or exaggerated oral doses of vitamin D. Guaranteed delivery of vitamin D is provided by intramuscular injections of vitamin D in oil preparations, and this can be done with injections of 2.5–12.5 mg (100,000–500,000 IU) every 6 months.

In patients with the nephrotic syndrome and heavy protein losses into urine, both 25OHD and vitamin D binding protein (DBP) appear in the urine in greater than normal amounts. In those cases with marginal vitamin D levels, such a condition can lead to depletion of stores and, on occasion, the development of vitamin D deficiency and osteomalacia [68]. Strikingly low plasma 25OHD levels have been observed in such patients, with a negative correlation between the amount of proteinuria and the plasma 25OHD concentration [69]. Such patients should be treated with vitamin D or $25OHD_3$ (Calderol).

B. Disordered Vitamin D Metabolism Resulting in Rickets/Osteomalacia

Various disorders are characterized by abnormalities in the synthesis of 25OHD or $1,25(OH)_2D$ or in the tissue response to $1,25(OH)_2D$. Reduced availability of 25OHD may be encountered in patients with advanced liver disease (see Chapter 51) and in patients taking anticonvulsants (see Chapter 50). Considerable attention has been given to hepatic disorders because plasma 25OHD levels, as well as DBP levels in plasma, are usually lower in such patients [70]. Deficiency of vitamin D is not easily attributable to hepatic cirrhosis; 25-hydroxylation may be adequate in severe, advanced cases, and the free 25OHD available to tissues may be normal. Yet, osteomalacia can occasionally be encountered in liver disease and is probably due to deficient sunlight exposure and poor dietary vitamin D intake or absorption.

An important interaction has been demonstrated between vitamin D and phenytoin or phenobarbital. Rickets and osteomalacia have been reported in patients receiving chronic anticonvulsant therapy. Plasma concentrations of 25OHD are decreased in patients receiving these drugs, and it was proposed that phenytoin and phenobarbital accelerate the metabolism of vitamin D to inactive products [71]. However, concentrations of calcitriol in plasma remain normal in patients receiving anticonvulsant therapy [72].

Calcifediol ($25OHD_3$, Calderol) is a useful metabolite for treating patients with hepatic osteodystrophy or anticonvulsant-induced osteomalacia. It differs from vitamin D insofar as a smaller proportion is stored in body fat and it bypasses hepatic 25-hydroxylation. Thus, the onset and offset of action are faster, and it is effective when there is impairment of hepatic 25-hydroxylation.

Reduced availability of $1,25(OH)_2D$ occurs in various renal tubular disorders, both acquired and genetic. Pseudovitamin D deficiency rickets is an autosomal recessive disease caused by an inborn error of vitamin D metabolism involving defective conversion of 25OHD to calcitriol [73] (see also Chapter 47). The disease has all the clinical and biochemical characteristics of vitamin D deficiency rickets. Although these patients require large doses of vitamin D for treatment, they respond to small doses (0.5–2 μg/day) of 1,25 $(OH)_2D_3$ [73,73a]. Hypophosphatemic rickets or osteomalacia (either the more common familial X-linked form or the less common sporadic form) appears to be a result of a defect in renal tubular phosphate transport, leading to phosphate wasting and hypophosphatemia (see Chapter 46). In these patients, serum 25OHD is normal, but serum $1,25(OH)_2D$ is within the normal range but inappropriately low for the concurrent level of serum phosphate [74,75]. Combined therapy with $1,25(OH)_2D_3$ and oral phosphate is superior to vitamin D in the management of these disorders.

Hereditary $1,25(OH)_2D_3$ resistance (also called hereditary vitamin D-resistant rickets and vitamin D-dependent rickets type II) is an autosomal recessive disorder that is characterized by hypocalcemia, osteomalacia and rickets, and in some cases complete alopecia. Studies of skin fibroblasts from these patients have identified mutations in the calcitriol receptor that lead either to defective hormone binding or defective binding of the hormone-receptor complex to DNA (see Chapter 48). Most affected children are completely unresponsive to massive doses of vitamin D or calcitriol, and they may require prolonged treatment with parenteral calcium.

Renal osteodystrophy (renal rickets) is associated with chronic renal failure and is characterized by decreased conversion of 25OHD to calcitriol. This topic, which is discussed in Chapters 52 and 73, is briefly reviewed here. Phosphate retention decreases serum calcium concentration leading to secondary hyperparathyroidism. In addition, calcitriol deficiency impairs intestinal calcium absorption and mobilization from bone. Hypocalcemia commonly results, although in some patients, prolonged and severe hyperparathyroidism eventually may lead to hypercalcemia. Aluminum deposition in bone also may play a role in the genesis of the skeletal disease. Pathologically, the bone lesions are typical of hyperparathyroidism (osteitis fibrosa), deficiency of vitamin D (osteomalacia), or a mixture of both.

In patients with chronic renal failure who are not receiving dialysis, emphasis has been on the prevention and treatment of hyperphosphatemia with phosphate binders and calcium supplementation. This can be accomplished by the oral administration of calcium carbonate, combined with dietary phosphate restriction [76]. The use of calcitrol in predialysis patients as well as in patients who are undergoing dialysis is now routine. The compound raises intestinal calcium absorption and serum calcium levels, lowers the concentration of PTH, and helps to maintain bone mineralization and growth in children. Intravenous calcitriol may be particularly effective in suppressing secondary hyperparathyroidism [77]. DHT and $1\alpha OHD_3$ can also be used effectively, as renal hydroxylation is not required for their activity. Although $25OHD_3$ also may be effective, high doses must be used.

C. Hypoparathyroidism and Pseudohypoparathyroidism

Hypoparathyroidism (PTH deficiency) and pseudohypoparathyroidism (PTH resistance) are characterized by hypocalcemia and hyperphosphatemia. $1,25(OH)_2D$ is inadequately synthesized in these disorders because of the lack of sufficient PTH stimulus and high phosphate levels [78]. Some subjects with mild or incomplete hypoparathyroidism may not require vitamin D therapy, and they can be managed with large supplements of calcium. When therapy with vitamin D is required, there has been an increasing trend toward utilizing $1,25(OH)_2D_3$, because more rapid effects are seen and because exaggerated stores of less potent forms can lead to intoxication periods of greater duration. Provision of vitamin D treatment as $1,25(OH)_2D_3$ usually requires 0.5–2.0 μg/day, along with a daily calcium supplement of 1.0 to 2.0 g in divided doses. Patients with PTH deficiency have defective calcium reclamation at the renal tubule. High renal calcium excretion may occur even before a normal serum calcium concentration is attained. To minimize the risk of kidney stones, serum calcium should be raised only to low normal levels, adequate hydration should be encouraged, and a thiazide diuretic should be added if urinary calcium is greater than 300 mg/day. The doses of calcitriol and calcium required for management of patients with pseudohypoparathyroidism are usually lower than those required for hypoparathyroidism, reflecting incomplete resistance to the action of PTH in pseudohypoparathyroidism [79,80]. Treatment of hypoparathyroidism is also discussed in Chapter 56.

D. Osteoporosis

Modest supplementation with vitamin D (400–800 IU per day) may improve intestinal calcium absorption, suppress bone remodeling, and improve bone mass in individuals with marginal or deficient vitamin D status. Such a regimen caused an impressive reduction in hip fractures (43%) in elderly persons living in France [81]. The rationale for the study was the known decline of serum 25OHD levels with age, which has been attributed to reduced solar exposure, impaired efficiency of vitamin D synthesis in the skin, and diminished intake of vitamin D-containing foods. It has been estimated that up to 60% of patients with hip fractures may have vitamin D deficiency, although usually not enough to cause frank osteomalacia [81].

The use of calcitriol to treat osteoporosis is distinct from assuring vitamin D nutritional adequacy. Here, the rationale is directly to suppress parathyroid function and reduce bone turnover. The results of major clinical trials of calcitriol therapy in women with osteoporosis have been conflicting [82–84]. Patients treated with calcitriol need to be monitored closely for development of hypercalciuria and hypercalcemia. The use of vitamin D metabolites to treat various forms of osteoporosis is discussed further in Chapters 43, 44, and 49.

E. Use of Vitamin D in Psoriasis

Clinical trials suggest that calcitriol may become an important agent for the treatment of psoriasis [85,86]. As such nontraditional uses of vitamin D are discovered, it will become important to develop noncalcemic analogs of calcitriol that achieve effects on cellular differentiation without the risk of hypercalcemia. One such example is calcipotriol, which was discussed earlier. Chapter 72 provides more detail on the use of vitamin D preparations in psoriasis.

V. VITAMIN D INTOXICATION

Physiological routes and sources of vitamin D are not known to cause hypervitaminosis D or intoxication (see also Chapter 54). Pharmacological doses are required, and these doses are on the order of 1–5 mg (40,000–200,000 IU) daily for weeks to months in humans [72]. Some time is required for conversion of vitamin D to 25OHD, but shorter periods to intoxication are expected if 25OHD is the sterol administered. The potent agent $1,25(OH)_2D_3$ can produce intoxication very quickly.

Studies have shown that vitamin D intoxication can occur in the absence of renal tissue and 1α-hydroxylase activity. It is generally agreed that a variety of metabolites, in high titer, can stimulate target tissues responsible for calcium translocation and can cause hypercalcemia. Under these circumstances, intoxication occurs at high 25OHD levels (200–1200 ng/ml). Serum concentrations of PTH and calcitriol are both suppressed.

In adults, hypervitaminosis D usually results from overtreatment of hypoparathyroidism or from faddist use of excessive doses. Improper fortification of milk by a dairy has also been reported to cause vitamin D toxicity [13,14]. The initial signs and symptoms of vitamin D toxicity are those associated with hypercalcemia. If uncorrected, this imposes a risk of nephrocalcinosis, nephrolithiasis, and/or often irreversible decrease in glomerular filtration.

Treatment of hypervitaminosis D consists of immediate withdrawal of the vitamin, a low calcium diet, administration of glucocorticoids, and vigorous fluid support. With this regimen, the serum calcium falls to normal, and calcium in soft tissues tends to be mobilized. Conspicuous improvement in renal function occurs unless renal damage has been severe. Actual reduction of sterol stores requires prolonged abstinence (months) from pharmacological vitamin D preparations. Shorter periods (days) are required after $1,25(OH)_2D$ toxicity.

References

1. Sniadecki J cited by Mozolowski W 1939 Jerdrzej Sniadecki (1768–1838) on the cure of rickets. Nature **143**:121.

2. Huldschinsky K 1919 Heilun von rachitis durch kunstliche honensonne. Curing rickets by artificial UV-radiation. Deut Med Wochenschr **45**:712–713.
3. Hess AF, Unger LF 1921 Cure of infantile rickets by sunlight. JAMA **39**:77–82.
4. Hess AF, Weinstock M 1924 Antirachitic properties imparted to inert fluids and green vegetables by ultraviolet irradiation. J Biol Chem **62**:301–313.
5. Steenbock H, Black A 1924 The induction of growth-promoting and calcifying properties in a ration by exposure to ultraviolet light. J Biol Chem **61**:408–422.
6. Steenbock H 1924 The induction of growth-promoting and calcifying properties in a ration exposed to light. Science **60**:224–225.
7. Windaus A, Linsert O, Luttringhaus A, Weidlich G 1932 Uber das krystallisierte vitamin D_2. Ann der Chem **492**:226–241.
8. Schenk F 1937 Uber das kristallisierte vitamin D_3. Naturwissenschaften **25**:159.
9. Stamp TCB, Haddad JG, Twigg CA 1977 Comparison of oral 25-hydroxycholecalciferol, vitamin D, and ultraviolet light as determinants of circulating 25-hydroxyvitamin D. Lancet **1**:1341–1343.
10. Subcommittee on the Tenth Edition of the Recommended Dietary Allowances; Food and Nutrition Board, Commission on Life Sciences, National Research Council. 1989 Recommended Dietary Allowances, 10th Ed., National Academy Press, Washington, D.C.
11. Haddad JG, Hahn TJ 1973 Natural and synthetic sources of circulating 25-hydroxyvitamin D in man. Nature **255**:515–517.
12. Haddad JG, Stamp TCB 1974 Circulating 25-hydroxyvitamin D in man. Am J Med **57**:57–62.
13. Holick MF, Shao Q, Liu WW, Chen TC 1992 The vitamin D content of fortified milk and infant formula. N Engl J Med **326**:1178–1181.
14. Chen TC, Heath III, H, Holick MF 1993 An update on the vitamin D content of fortified milk from the United States and Canada N Engl J Med **329**:1507.
15. Jacobs CH, Holick MF, Shao Q, Chen TC, Holm IA, Kolodny JM, El-Hajj Fuleihan G, Seely EW 1992 Hypervitaminosis D associated with drinking milk. N Engl J Med **326:** 1173–1177.
16. Wasserman RH, Henion JD, Haussler MR, McCain TA 1976 Calcinogenic factor in *Solanum malacoxylon*; evidence that it is 1,25-dihydroxyvitamin D_3-glycoside. Science **194**:853–855.
17. Hughes MR, McCain TA, Chang SY, Haussler MR, Villareale M, Wasserman RH 1977. Presence of 1,25-dihydroxyvitamin D_3-glycoside in the calcinogenic plant *Cestrum diurnum*. Nature **268**:347–349.
18. Askew FA, Bourdillon RB, Bruce HM, Callow RK, Philpot JL, Webster TA 1932 Crystalline vitamin D. Proc R Soc Lond Ser B **109**:488–506.
19. Brockmann H, Busse A 1938 Kristallisier des vitamin D austhundfischleberol. Naturwissenschaften **26**:122–123.
20. Holick MF 1989 Vitamin D: Biosynthesis, metabolism, and mode of action. In: DeGroot LJ, Besser GM, Cahill GF Jr, et al. (eds) Endocrinology, 2nd Ed., Vol. 2. Saunders, Philadelphia, Pennsylvania, pp. 902–926.
21. Aksnes L, Aarskog D 1980 Vitamin D metabolites in serum from hypoparathyroid patients treated with vitamin D_2 and 1α-hydroxyvitamin D_3. J Clin Endocrinol Metab **51**:823–829.
22. Mawer EB, Backhouse J, Holman CA, Lumb GA, Stanbury SW 1972 The distribution and storage of vitamin D and its metabolites in human tissues. Clin Sci **43**:413–431.
23. Holtz F, Schreiber E 1930 Einige weitere physiologische Erfahrungen uber das bestrahlte Ergosterin und seine Umwandlungsprodukte. Z Physiol Chem **191**:1–22.
24. Albright F, Reifenstein EC 1948 The Parathyroid Glands and Metabolic Bone Disease. Williams & Wilkins, Baltimore, Maryland.
25. Holtz F 1933 Die Behandlung der postoperativen tetanie. Arch Klin Chirugie **177**:32–33.
26. Taylor A, Bikle DD, Norman ME 1988 Serum dihydrotachysterol levels and biological action in normal man. J Clin Endocrinol Metab **67**:198–202.
27. Koytchev R, Alken R-G, Vagaday M, Kunter U, Kirkov V 1994 Differences in the bioavailability of dihydrotachysterol preparations. Eur J Clin Pharmacol **47**:81–84.
28. Stanbury SW, Mawer EB 1978 Physiological aspects of vitamin D metabolism. In: Lawson DEM (ed) Vitamin D. Academic Press, London, pp. 303–341.
29. Porteous C, Trafford DJH, Makin HLJ, Cunningham J, Jones G 1988 Use of mass spectrometry in the identification of in vivo and in vitro metabolites of $\text{dihydrotachysterol}_3$ in the rat. Biomed Environ Mass Spectrom **16**:87–92.
30. Jones G, Edwards N, Vriezen D, Porteous C, Trafford DJ, Cunningham J, Makin HL 1988 Isolation and identification of seven metabolites of $\text{25-hydroxydihydrotachysterol}_3$ formed in the isolated perfused rat kidney: A model for the study of side chain metabolism of vitamin D. Biochemistry **27**:7070–7079.
31. Bosch R, Versluis C, Terlouw JK, Thijssen JHH, Duursma SA 1985 Isolation and identification of 25-hydroxydihydrotachy-sterol_2: 1α,25-dihydroxydihydrotachysterol_2 and 1β,25-dihydroxydihydrotachysterol_2. J Steroid Biochem Mol Biol **23**:223–228.
32. Qaw F, Calverley MJ, Schroeder NJ, Trafford DJH, Makin HLJ, Jones G 1993 *In vivo* metabolism of the vitamin D analogue, dihydrotachysterol. Evidence for formation of 1α,25- and 1β,25-dihydroxydihydrotachysterol metabolites and studies of their biological activity. J Biol Chem **268**:282–292.
33. Schroeder NJ, Trafford DJH, Cunningham J, Jones G, Makin HLJ 1994 *In vivo* $\text{dihydrotachysterol}_2$ metabolism in normal man: 1α- and 1β-hydroxylation of $\text{25-hydroxydihydrotachysterol}_2$ and effects on plasma parathyroid hormone and 1α,25-dihydroxyvitamin D_3 concentrations. J Clin Endocrinol Metab **78**:1481–1487.
34. Qaw F, Schroeder NJ, Calverley MJ, Trafford DJH, Makin HLJ, Jones G 1992 The metabolism of dihydrotachysterols: Renal side chain and non-renal nuclear hydroxylations *in vivo* and *in vitro*. J Steroid Biochem Mol Biol **41**:859–870.
35. Haddad JG, Hahn TJ 1973 Natural and synthetic sources of circulating 25-hydroxyvitamin D in man. Nature **244**:515–516.
36. Teitelbaum SL, Bone JM, Stein PM, Gilden JJ, Bates M, Boisseau VC, Avioli LV 1976 Calciferol in chronic renal insufficiency; skeletal response. JAMA **235**:164–167.
37. Rutherford WE, Blondin J, Hruska K, Kopelman R, Klahr S, Slatopolsky E 1975 Effect of 25-hydroxycholecalciferol on calcium absorption in chronic renal disease. Kidney Int **8**:320–324.
38. Colodro IH, Brickman AS, Coburn JW, Osborn TW, Norman AW 1978 Effect of 25-hydroxyvitamin D_3 on intestinal absorption of calcium in normal man and patients with renal failure. Metabolism **27**:745–753.
39. Fournier A, Bordier P, Gueris J, Sebert JL, Marie P, Ferriere C, Bedrossian J, DeLuca HF 1979 Comparison of 1α-hydroxycholecalciferol and 25-hydroxycholecalciferol in the treatment of renal osteodystrophy: Greater effect of 25-hydroxycholecalciferol on bone mineralization. Kidney Int **15**:196–204.
40. Haddad JG Jr, Rojanasathit S 1976 Acute administration of 25-hydroxycholecalciferol in man J Clin Endocrinol Metab **42**:284–290.
41. Gray RW, Weber HP, Dominquez JH, Lemann J Jr 1974 The metabolism of vitamin D_3 and 25-hydroxyvitamin D_3 in normal and anephric humans. J Clin Endocrinol Metab **39**:1045–1056.
42. Mawer EB, Backhouse J, Davies M, Hill LF 1976 Metabolic fate

of administered 1,25-dihydroxycholecalciferol in controls and in patients with hypoparathyroidism. Lancet **1(7971)**:1203–1206.

43. Mason RS, Lissner D, Posen S 1980 Blood concentrations of dihydroxylated vitamin D metabolites after an oral dose. Br Med J **280**:449–450.
44. Levine BS, Singer FR, Bryce GF, Mallon JP, Miller ON, Coburn JW 1985 Pharmacokinetics and biologic effects of calcitriol in normal humans. J Lab Clin Med **105**:239–246.
45. Gray RW, Caldas AE, Wilz DR, Lemann J Jr, Smith GA, DeLuca HF 1978 Metabolism and excretion of ^{3}H-1,25-$(OH)_2$-vitamin D_3 in healthy adults. J Clin Endocrinol Metab **46**:756–765.
46. Brickman AS, Coburn JW, Massry SG, Norman AW 1974 1,25-Dihydroxyvitamin D_3 in normal men and patients with renal failure. Ann Intern Med **80**:161–168.
47. Rosen JF, Fleischman AR, Finberg L, Kisman J, DeLuca HF 1977 1,25-Dihydroxycholecalciferol: Its use in the long-term management of idiopathic hypoparathyroidism in children. J Clin Endocrinol Metab **45**:457–468.
48. Slatopolsky E, Weerts C, Thielan J, Horst R, Harter H, Martin KJ 1984 Marked suppression of secondary hyperparathyroidism by intravenous administration of 1,25-dihydroxycholecalciferol in uremic patients. J Clin Invest **74**:2136–2143.
49. Delmez JA, Tindira C, Grooms P, Dusso A, Windus DW, Slatopolsky E 1989 Parathyroid hormone suppression by intravenous 1,25-dihydroxyvitamin D: A role for increased sensitivity to calcium. J Clin Invest **83**:1349–1355.
50. Napoli JL, Premanik BC, Royal PM, Reinhardt TA, Horst RA 1983 Intestinal synthesis of 24-keto 1,25-dihydroxyvitamin D_3. J Biol Chem **258**:2100–2107.
51. Harnden D, Kumar R, Holick MF, DeLuca HF 1976 Side chain metabolism of 25-hydroxy-[26,27-^{14}C]vitamin D_3 and 1,25-dihydroxy-[26,27-^{14}C]vitamin D_3 *in vivo*. Science **193**:493–494.
52. Onisko BL, Esvelt RP, Schnoes HK, DeLuca HF 1980 Excretion of metabolites of 1α,25-dihydroxyvitamin D_3 in rat bile. Arch Biochem Biophys **205**:175–179.
53. Barton DHR, Hesse RH, Pechet MM, Rizzardo E 1973 A convenient synthesis of 1-hydroxyvitamin D_3. J Am Chem Soc **95**:2748–2749.
54. Haussler MR, Zerwekh JE, Hesse RH, Rizzardo E, Pechet MM 1973 Biological activity of 1α-hydroxycholecalciferol, a synthetic analog of the hormonal form of vitamin D_3. Proc Natl Acad Sci USA **70**:2248–2252.
55. Zerwekh JE, Brumbaugh PF, Haussler DH, Cork DJ, Haussler MR 1974 1α-Hydroxyvitamin D_3; an analog of vitamin D which apparently acts by metabolism to 1α,25-dihydroxyvitamin D_3. Biochemistry **13**:4097–4102.
56. Brickman AS, Coburn JW, Friedman GR, Okamura WH, Massry SG, Norman AW 1976. Comparison of effects of 1α-hydroxyvitamin D_3 and 1,25-dihydroxyvitamin D_3 in man. J Clin Invest **57**:1540–1547.
57. Lund B, Lund B, Sorenson OH 1979 Measurement of circulating 1,25-dihydroxyvitamin D in man; changes in serum concentration during treatment with 1α-hydroxycholecalciferol. Acta Endocrinol **91**:338–350.
58. Kimura Y, Nakayama M, Kuriyama S, Watanabe S, Kawaguchi Y, Sakai O 1991 Pharmacokinetics of active vitamin D_3, 1α-hydroxyvitamin D_3 and 1α,25-dihydroxyvitamin D_3 in patients on chronic hemodialysis. Clin Nephrol **35**:72–77.
59. Seino Y, Tanaka H, Yamaoka K, Yabuuchi H 1987 Circulating 1α,25-dihydroxyvitamin D levels after a single dose of 1α,25-dihydroxyvitamin D_3 or 1α-hydroxyvitamin D_3 in normal men. Bone Miner **2**:479–485.
60. Masuda S, Okano T, Matsuoka S, Kobayashi T 1990 Changes in the serum levels of 1,25$(OH)_2D_3$, PTH and other parameters after oral administration of 1α-$(OH)D_3$ to normal subjects. Renal bone disease, parathyroid hormone, and vitamin D. Singapore, p 73 (abstract).
61. Joffe P, Cintin C, Ladefoged SD, Rasmussen SN 1994 Pharmacokinetics of 1α-hydroxycholecalciferol after intraperitoneal, intravenous and oral administration in patients undergoing peritoneal dialysis. Clin Nephrol **41**:364–369.
62. Sorenson H, Binderup L Calverley MJ, Hoffmeyer L, Andersen NR 1990 In vitro metabolism of calcipotriol (MC 903), a vitamin D analog. Biochem Pharmacol **39**:391–403.
63. Bouillon R, Allewaert K, Xiang DZ, Vandewalle M, De Clerq P 1989 Structure–function analysis of vitamin D analogs; difference in binding to the vitamin D receptor and vitamin D binding serum protein. J Bone Miner Res **4**(Suppl 1):301 (abstract).
64. Kissmeyer AM, Binderup L 1991 Calcipotriol (MC 903); pharmacokinetics in rats and biological activities of metabolites; a comparative study with 1,25-$(OH)_2$-D_3. Biochem Pharmacol **41**:1601–1606.
65. Murdoch D, Clissold SP 1992 Calcipotriol: A review of its pharmacological properties and therapeutic use in psoriasis vulgaris. Drugs **43**:415–429.
66. Marcus R 1996 Agents affecting calcification and bone turnover: Calcium, phosphate, parathyroid hormone, vitamin D, calcitonin, and other compounds. In: Hardman JG, Limbird LE, Molinoff PG, Ruddon RW, Gilman AG (eds) Goodman and Gilman's The Pharmacological Basis of Therapeutics, 9th Ed. McGraw-Hill, New York, pp. 1519–1546.
67. Kumar R 1983 Hepatic and intestinal osteodystrophy and the hepatobiliary metabolism of vitamin D. Ann Intern Med **98**:662–663.
68. Malluche HH, Goldstein DA, Massry SG 1979 Osteomalacia and hyperparathyroid bone disease in patients with nephrotic syndrome. J Clin Invest **63**:494–500.
69. Goldstein DA, Oda Y, Kurokawa K, Massry SG 1977 Blood levels of 25OHD in nephrotic syndrome. Ann Intern Med **87**:664–667.
70. Haddad JG Jr 1992 Clinical aspects of measurements of plasma vitamin D sterols and the vitamin D binding protein. In: Coe FL, Favus MJ (eds) Disorders of Bone and Mineral Metabolism. Raven, New York, pp. 195–216.
71. Hahn TJ, Hendin BA, Scharp CR, Haddad JG Jr 1972 Effect of chronic anticonvulsant therapy on serum 25-hydroxycholecalciferol levels in adults. N Engl J Med **287**:900–904.
72. Jubiz W, Haussler MR, McCain TA, Tolman KG 1977 Plasma 1,25-dihydroxyvitamin D levels in patients receiving anticonvulsant drugs. J Clin Endocrinol Metab **44**:617–621.
73. Fraser D, Kooh SW, Kind HP, Holick MF, Tanaka Y, DeLuca HF 1973 Pathogenesis of hereditary vitamin D-dependent rickets: An inborn error of vitamin D metabolism involving defective conversion of 25-hydroxyvitamin D to 1α,25-dihydroxyvitamin D. N Engl J Med **289**:817–822.

73a. Delvin EE, Glorieux FH, Marie PS, Pettifor JM 1981 Vitamin D dependency: Replacement therapy with calcitriol. J Pediatr **99**:26–34.

74. Drezner MK, Lyles KW, Haussler MR, Harrelson JM 1980 Evaluation of a role for 1,25$(OH)_2D$ in the pathogenesis and treatment of X-linked hypophospatemic rickets and osteomalacia. J Clin Invest **66**:1020–1032.
75. Miyanchi A, Fukase M, Tsutsumi M, Fujita T 1988 Hemangiopericytoma-induced osteomalacia: Tumor transplantation in nude mice causes hypophosphatemia and tumor extracts inhibit renal 25-hydroxyvitamin D-1-hydroxylase activity. J Clin Endocrinol Metab **67**:46–53.
76. Coburn JW, Salusky IB 1989 Control of serum phosphorus in uremia. N Engl J Med **320**:1140–1142,

77. Andrew DK, Norris KC, Coburn JW, Slatspolsky EA, Sherard DJ 1989 Intravenous calcitriol in the treatment of refractory osteitis fibrosa of chronic renal failure. N Engl J Med **321**:274–279.
78. Breslau NA, Winstock R 1988 Regulation of 1,25-$(OH)_2D$ synthesis in hypoparathyroidism and pseudohypoparathyroidism. Am J Physiol **255** (Endocrinol Metab **18**):E730-E736.
79. Okano K, Furukawa Y, Hirotoshi M, Fujita T 1982 Comparative efficacy of various vitamin D metabolites in the treatment of various types of hypoparathyroidism. J Clin Endocrinol Metab **55**:238–243.
80. Breslau NA 1989 Pseudohypoparathyroidism: Current concepts. Am J Med Sci **246**:130–140.
81. Chapuy MC, Arlot ME, Duboeuf F, et al. 1992 Vitamin D_3 and calcium to prevent hip fractures in elderly women. N Engl J Med **327**:1637.
82. Ott SM, Chestnut III CH 1989 Calcitriol treatment is not effective in postmenopausal osteoporosis. Ann Intern Med **110**:267–274.
83. Gallagher JC, Goldgar D 1990 Treatment of postmenopausal osteoporosis with high doses of synthetic calcitriol: A randomized controlled study. Ann Intern Med **113**:649–362.
84. Tilyard MW, Spears GFS, Thomson J, et al. 1992 Treatment of postmenopausal osteoporosis with calcitriol or calcium. N Engl J Med **326**:357–362.
85. Holick MR 1993 Active vitamin D compounds and analogues: A new therapeutic era for dermatology in the 21st century. Mayo Clin Proc **68**:925–927.
86. Kragballe K 1992 Vitamin D analogues in the treatment of psoriasis. J Cell Biochem **49**:46–52.

CHAPTER 40

Radiology of Rickets and Osteomalacia

JUDITH E. ADAMS
Department of Diagnostic Radiology, The University, Manchester, England, United Kingdom

I. INTRODUCTION AND HISTORICAL ASPECTS

Bone resorption is a one-stage process, with osteoclasts resorbing mineral and osteoid together. In contrast, bone formation occurs in two stages: osteoblasts lay down osteoid, which subsequently becomes mineralized. The mineralization of bone matrix depends on the presence of adequate supplies of not only vitamin D, in the form of its active metabolite 1,25-dihydroxyvitamin D [$1,25(OH)_2D$], but also calcium, phosphorus, and alkaline phosphatase, and on a normal pH prevailing in the body environment. If there is a deficiency of these substances for any reason, or if there is severe systemic acidosis, then mineralization of bone will be defective. There will be a qualitative abnormality of bone, with reduction in the mineral to osteoid ratio, resulting in rickets in children and osteomalacia in adults. In the immature skeleton, the radiographic abnormalities predominate at the growing ends of the bones, where endochondral ossification is taking place, giving the classic appearances of rickets. When the skeleton reaches maturity and the process of endochondral ossification has ceased, the defective mineralization of osteoid is evident radiographically as Looser's zones (pseudofractures, Milkman's fractures), which are pathognomonic of osteomalacia. Rickets and osteomalacia are therefore synonymous and represent the same disease process, but are the manifestation in either the growing or the mature skeleton. A large number of different diseases can cause the same radiological abnormalities of rickets and osteomalacia [1–3].

Rickets appeared when people began to live in cities during the industrial revolution. The first descriptions of the condition are attributed to Daniel Whistler, who in 1692, while a student at Merton College Oxford, wrote a thesis entitled, "Inaugural Medical Disputation in the Disease of English Children which is popularly termed Rickets" for his doctorate at the University of Leyden [4]. Glisson in 1650 wrote a treatise for the Royal College of Physicians on, "De Rachitide sive morbo puerili qui vulgo The Rickets dicitur" [5,6]. In 1666 a postmortem examination was reported by John Locke of a child with rickets who died of pneumonia

[7]. The disease was well known in Europe at this time and was regarded as "the English disease." The origin of the term "rickets" is still debated [8,9].

Fish oil was recognized as a popular cure for "chronic rheumatism" (osteomalacia) and limb deformities (rickets) in children [10]. Robert Darbey, a house surgeon and apothecary to the Manchester Infirmary, was quoted by Percival in 1783 [11] as saying that 50–60 gallons of cod-liver oil were issued every year to patients, usually for the treatment of chronic rheumatism [6]. In 1890 Palm [12] published an essay on the distribution of rickets around the world and noted that it was common in cities where people were deprived of sunlight. He suggested that rickets could be treated with sunlight but did not proceed to perform the experiment. Others have observed that the disease is more common in cities [13].

There was much confusion between conditions that had similar clinical features but different courses of progression and responses to therapies of the day [14,15]. It was the microscopic studies of Pommer in 1885 [16] which distinguished, for the first time, between rickets, achondroplasia, and osteogenesis imperfecta. In that year also Röntgen discovered X rays, and it then became possible to display the radiographic features of rickets and osteomalacia.

The unraveling of the structure and function of vitamin D and its metabolites during the twentieth century has elucidated the causes for confusion which existed in the past as to the etiology of rickets and the variable response to treatment (see Chapter 1).

Vitamin D deficiency may occur as a consequence of simple nutritional deficiency (diet, lack of sunlight; see Chapters 42 and 43), due to malabsorption states, chronic liver disease which affects hydroxylation at the 25 position (see Chapter 51), and chronic renal disease in which the active metabolite $1,25(OH)_2D$ is not produced (see Chapter 52). Consequently, a large variety of diseases may result in vitamin D deficiency [17,18] (see Chapter 36). The radiological features of all will be similar, being those of rickets and osteomalacia. This similarity of radiological features but variations in response to treatment contributed to some of the early confusion [2,19,20]. Rickets due to nutritional deprivation was cured by ultraviolet light or physiological doses of vitamin D (400 IU per day), but that associated with chronic renal disease was not, except if very large pharmacological doses (up to 300,000 IU per day) were used. This lead to the terms "refractory rickets" and "vitamin D resistant rickets" being used for these conditions. In these groups were included the diseases that caused the clinical and radiological features of rickets but were related to phosphate, not vitamin D, deficiency, such as x-linked hypophosphatemia (see Chapter 46) and genetic diseases involving defects in 1-hydroxylase (Chapter 47) and the vitamin D receptor (Chapter 48).

II. VITAMIN D DEFICIENCY

The hormone $1,25(OH)_2D$ plays an important role in calcium homeostasis by its actions principally on the bone and intestine, but also on the kidney and parathyroid glands. It promotes the intestinal absorption of calcium and phosphorus from the intestine. On the bone it has two actions; one is to mobilize calcium and phosphorus from the skeleton as required, and the other is to promote maturation and mineralization of the osteoid matrix. Deficiency of vitamin D results in rickets in children and osteomalacia in adults [21]. There are known to be seasonal variations in vitamin D status, with plasma levels being lower in the winter months [22] (see also Chapters 3 and 33). The pathophysiology, clinical descriptions and treatments are discussed elsewhere in this volume.

A. Rickets

In the growing skeleton the effect of vitamin D deficiency and consequent defective mineralization of osteoid is seen at the growing ends of the bones. In the early phase there is widening of the growth plate, which is the translucent (unmineralized) gap between the mineralized metaphysis and epiphysis [23,24].

As the changes become more severe, there is "cupping" of the metaphysis with irregular and poor mineralization (Figs. 1–3). There is some expansion in width of the metaphysis which results in the apparent soft tissue swelling around the ends of the long bones affected. This produces the expansion at the anterior ends of the ribs referred to as a "rachitic rosary" (Fig. 1C). There is often a thin "ghost-like" rim of mineralization at the periphery of the metaphysis, since this mineralization occurs by membranous ossification at the periosteum. The margin of outline of the epiphysis also appears indistinct as endochondral ossification at this site is also defective.

The changes are most pronounced at the sites of bone which are growing most actively. These sites, in sequence, are around the knee, the wrist (particularly the ulna, Fig. 2), the anterior ends of the middle ribs, the proximal femur, and the distal tibia. It is these anatomical sites that should be radiographed if rickets is suspected. As rachitic bone is soft and bends, additional features are bowing of the legs (genu valgum) or knock-knees (genu varum), deformity of the hips (coxa valga or, more usually, coxa vara), indrawing of the ribs at

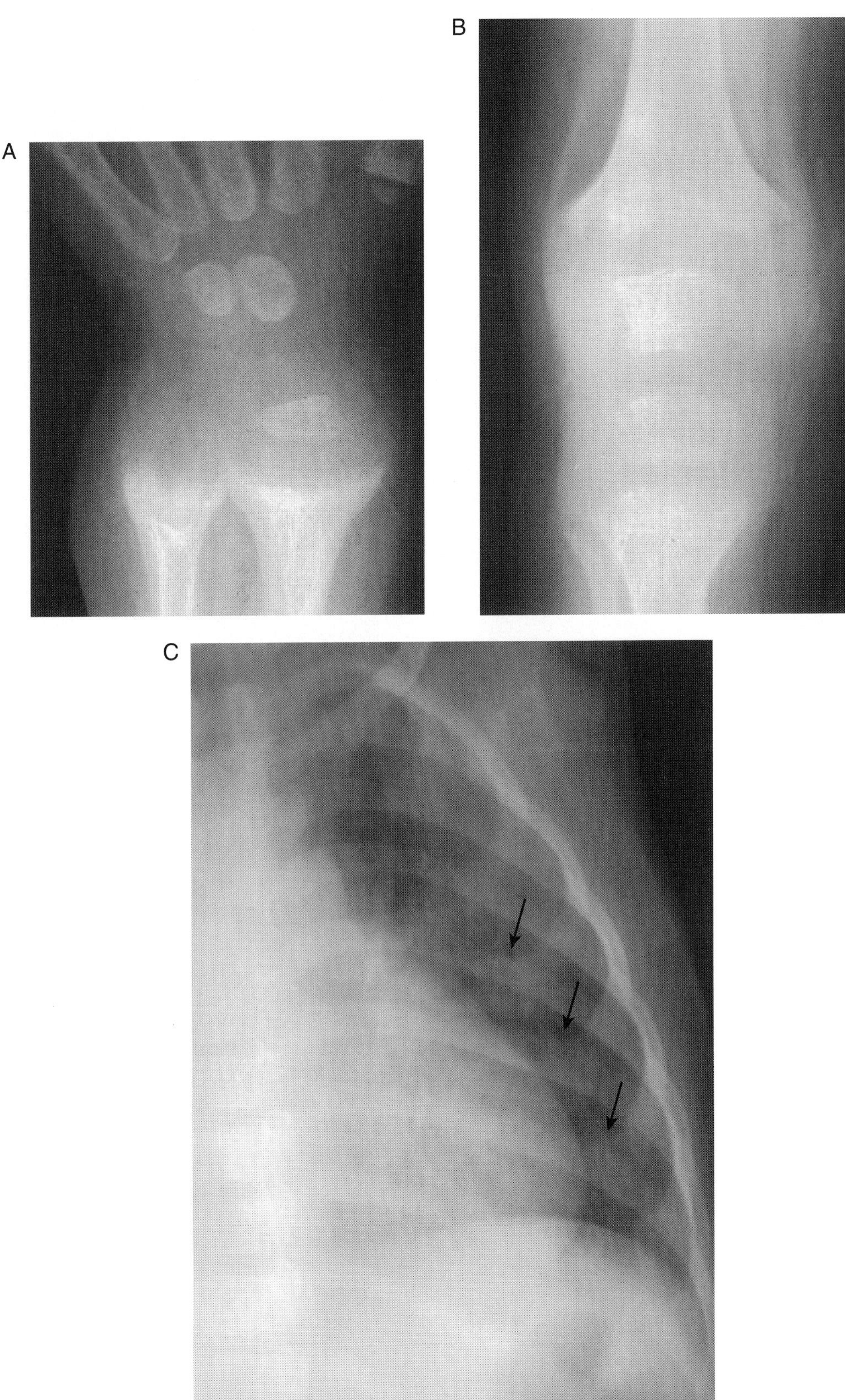

FIGURE 1 Nutritional vitamin D deficiency rickets in a young child. (A) Wrist showing widening of the growth plate with cupping and expansion of the metaphysis, which is poorly mineralized. The expansion results in soft tissue swelling around the ends of the bones. The cortical "tunneling" (subcortical erosion) and hazy trabecular pattern indicate secondary hyperparathyroidism. (B) Knee showing similar changes at the metaphyses. (C) Bulbous expansion of the anterior ends of the ribs (arrows) known as a "rachitic rosary."

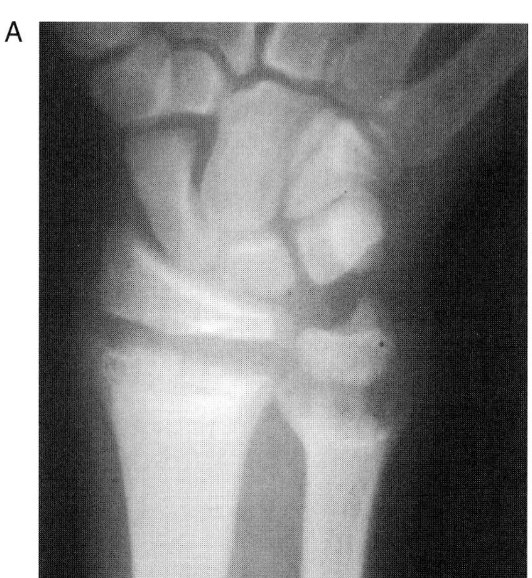

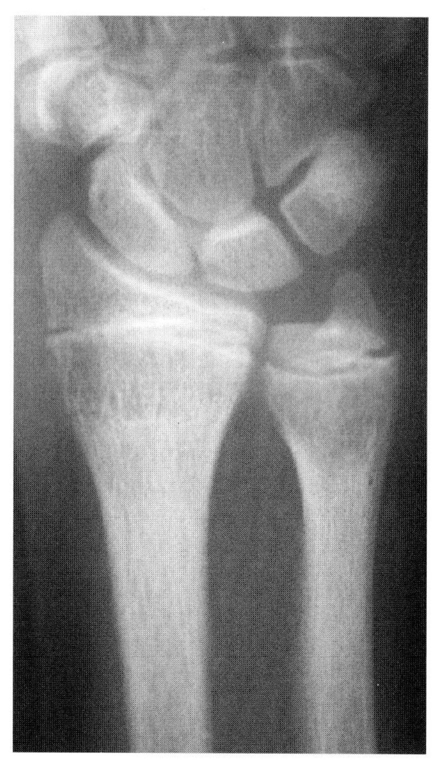

FIGURE 2 Nutritional rickets in an adolescent at puberty. (A) Wrist before treatment. The growth plate is increased in width, and there is some cupping of the metaphyses of the distal radius and ulna with irregular and deficient mineralization. Note that the ulnar plate is more severely affected than the radial plate. This is the consequence of the ulna growing in length exclusively from its distal end, the proximal end forming the olecranon. This is the most sensitive site for assessing rickets activity. (B) Wrist 4 months after treatment with vitamin D. There is healing of the rickets, although the distal segments of the radius and ulna are still reduced in density. This indicates the stage of bone development at which the rickets occurred and the amount of growth that has taken place since then. With time and remodeling the appearances will become normal.

the insertion of the diaphragm (Harrison's sulcus), and protrusio acetabulae and triradiate deformity of the pelvis. The latter can result in problems with parturition at subsequent pregnancies. Involvement of the bones of the thorax and respiratory tract (larynx and trachea) can result in stridor and respiratory distress [25,26]. Paradoxically, in very severe rickets, where little growth is taking place (i.e., owing to nutritional deprivation or chronic ill health), the features of rickets may not be evident at the growth plate level [27]. In mild vitamin D deficiency, the radiographic features of rickets may only become apparent during the growth spurt associated with puberty, and then the changes are most prominent at the knee.

In the rickets of prematurity there may be little abnormality at the metaphysis, as no skeletal growth is taking place in the premature neonate. However, the bones are osteopenic and prone to fracture (Fig. 4).

If the vitamin D deficiency is treated appropriately, then the radiographic abnormalities of rickets will heal over about 2–3 months (Figs. 2 and 3). The radiographic features of healing will lag behind the improvement in biochemical parameters (2 weeks) and clinical symptoms. With treatment the unmineralized osteoid of the growth plate of the metaphysis and epiphysis will mineralize. This section of abnormal bone may be visible for a period of time and gives some indication as to the age of onset and duration of the period of rickets (Fig. 2). Eventually, this zone becomes indistinguishable from the normal bone with time and remodeling. The zone of provisional calcification that was present at the onset of the disturbance to endochondral ossification may remain (Harris growth arrest line) (Fig. 13A) as a marker of the age of skeletal maturation at which the rickets occurred [28]. However, this is not specific for rickets and can result from any condition (i.e., a period of ill health, lead poisoning) that inhibits normal endochondral ossification. There is evidence of retarded growth and development in rickets, but this is more marked when the vitamin D deficiency is associated with chronic diseases that reduce calorie intake, general well-being, and activity (i.e., malabsorption, chronic renal disease)

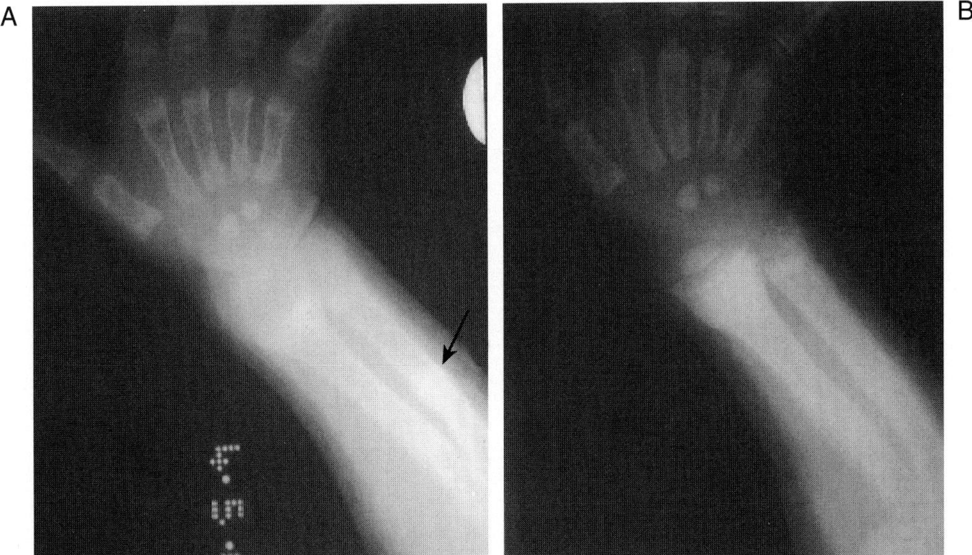

FIGURE 3 Severe nutritional rickets in a young child. (A) Forearm before treatment. There are expansion and cupping of the metaphysis, which is poorly mineralized. A Looser's zone is present in the mid shaft of the ulna (arrow). Extensive erosions and cortical tunneling indicate secondary hyperparathyroidism. (B) Forearm after several months of treatment with vitamin D, which resulted in some healing. There is mineralization of the metaphysis of the radius; healing of the metaphysis of the ulna is still incomplete. The ulna fracture has healed, and the bone texture is more normal, although there is still evidence of hyperparathyroidism. With time and remodeling the shape of the distal radius will become normal.

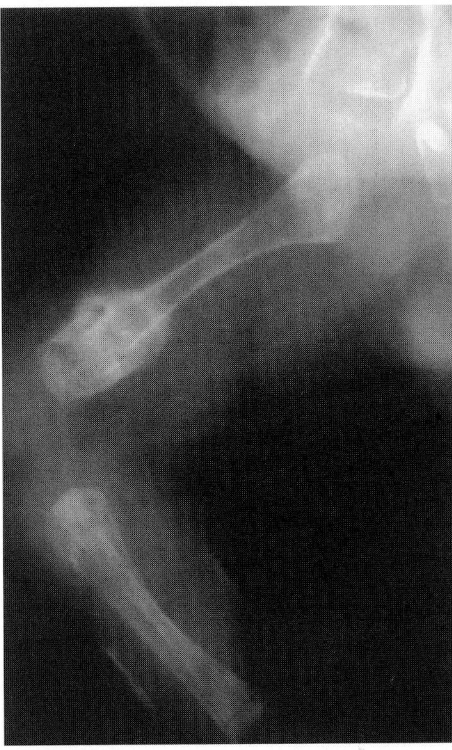

FIGURE 4 Rickets of prematurity. Although there is little abnormality of the metaphyses (because of poor growth), the bones are osteopenic, and there is a fracture of the distal femoral shaft. Profuse callus indicates that the patient is receiving vitamin D and calcium.

than with simple nutritional vitamin D deficiency (J. E. Adams, personal observation).

With vitamin D deficiency, there is associated hypocalcemia. To maintain calcium homeostasis, the parathyroid glands are stimulated to secrete parathyroid hormone (PTH). This results in another important feature of vitamin D deficiency rickets [29]. Evidence of secondary hyperparathyroidism, with increased osteoclastic resorption, is always evident histologically (see Chapter 37), although it may not always be revealed radiographically [30–32].

B. Osteomalacia

At skeletal maturity the epiphysis fuses to the metaphysis and longitudinal bone growth ceases. However, bone turnover continues throughout life to maintain the tensile integrity of the skeleton. Vitamin D deficiency in the adult skeleton results in osteomalacia, the pathognomonic feature of which is the Looser's zone (pseudofracture, Milkman's fracture) [33–37]. Looser's zones are translucent areas in the bone that are composed of unmineralized osteoid. They appear as translucent lines in the bone that are perpendicular to the bone cortex, do not extend across the entire bone shaft, and characteristically have a sclerotic margin (Fig. 5) [38]. They can occur in any bone but typically are found in

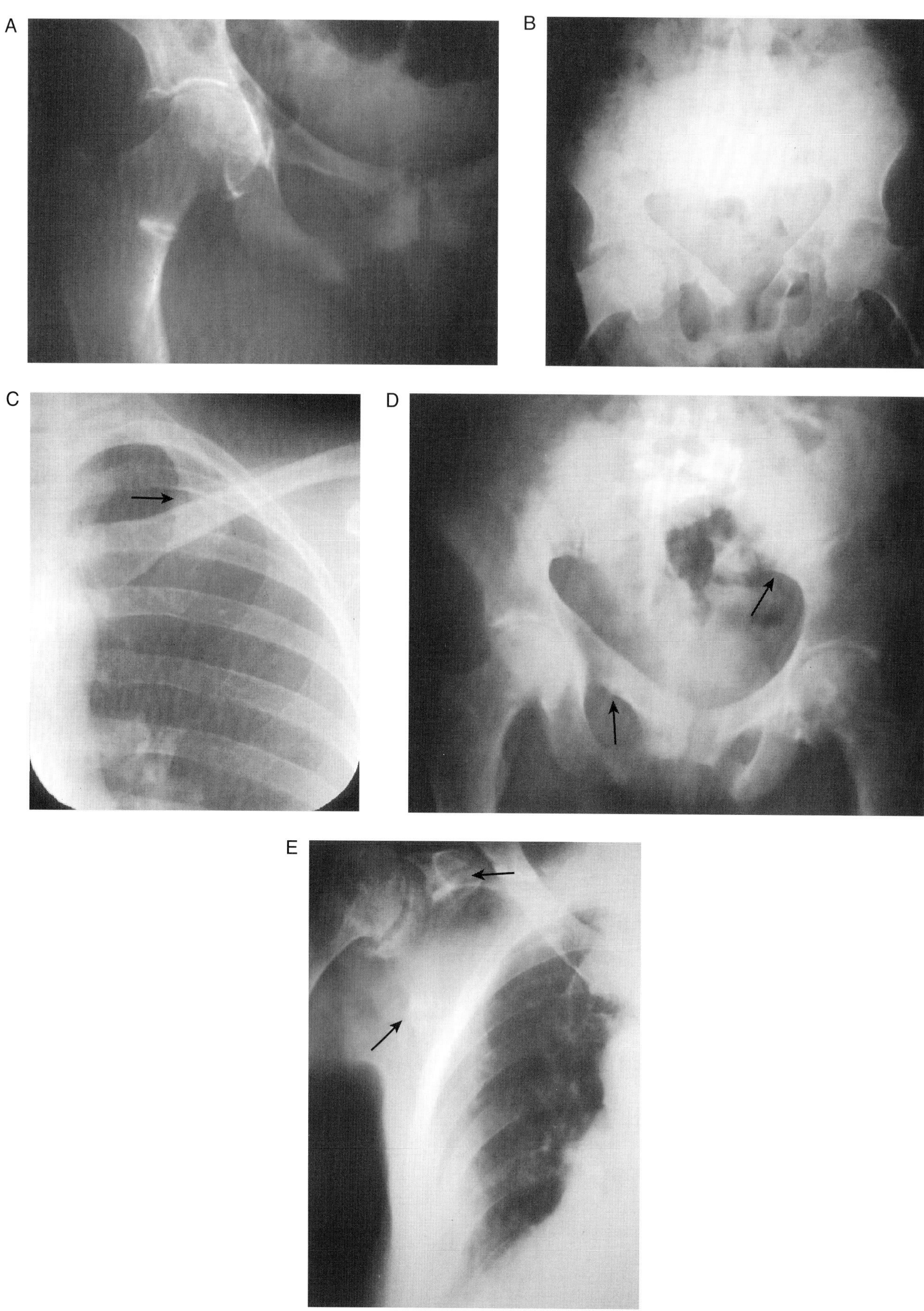

A
B
C
D
E

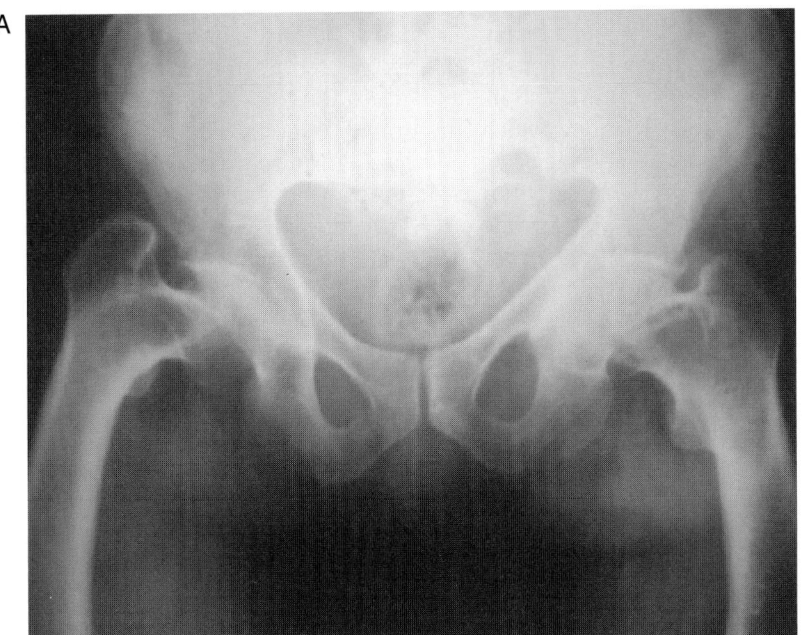

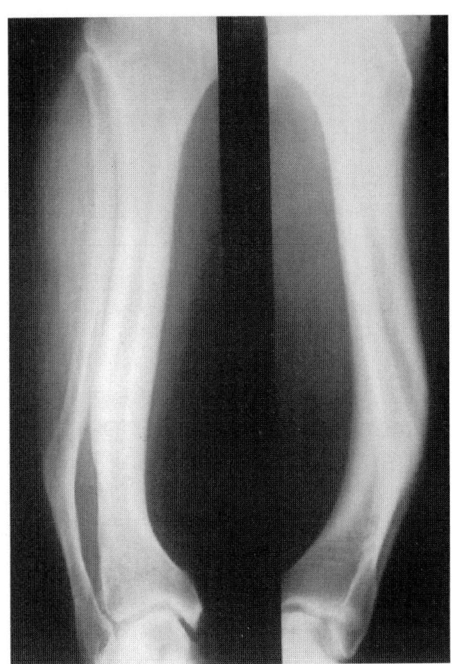

FIGURE 6 Rickets in childhood. (A) Pelvis showing stigmata of past rickets with a triradiate deformity of the pelvis and varus deformity of the femoral necks. (B) Bowing of tibia and fibula with evidence of Harris growth arrest lines in the distal shaft of the tibia.

the medial portion of the femoral neck, the pubic rami, the lateral border of the scapula, and the ribs. They may involve the first and second ribs, in which traumatic fractures are uncommon and are usually associated with severe trauma. Other less common sites are the metatarsals and metacarpals, the base of the acromium, and the ilium (Fig. 5D). Again, traumatic fractures of the ilium require very severe trauma, and a lack of history of such severe injury should alert one to the fact that a fracture of the ilium may be related to vitamin D deficiency.

The etiology of why Looser's zones occur in the anatomical sites which they do has been much debated [39,40]. At one time, it was thought that they were in the sites of vascular channels, but this theory has been discarded. Their position is most likely to be related to sites of stress in the skeleton. Looser's zones must be differentiated from insufficiency fractures that occur in osteoporotic bone, particularly in the pubic rami, sacral ala, and calcaneum [41–43]. Insufficiency fractures consist of multiple microfractures and often have florid callus formation, which differentiates them from Looser's zones. Incremental fractures occur in Paget's disease of bone and resemble Looser's zones in appearance (translucent zone, suggesting incomplete fracture, with sclerotic margin), but these tend to occur on the outer (convex) cortex of the bone involved. The typical features of Paget's disease (sclerotic, disordered trabecular pattern, enlarged bone) serve as distinguishing radiographic features.

Complete fractures can occur through Looser's zones. As in rickets, osteomalacic bone is soft and bends. This is evident by protrusio acetabulae, in which the femoral head deforms the acetabular margin so that the normal teardrop outline of the acetabulum is lost (Fig. 5A). There may be bowing of the long bones of the legs

FIGURE 5 Osteomalacia. (A) Nutritional osteomalacia in an Asian. The right hip shows a classic Looser's zone in the medial cortex of the femoral neck. The sclerotic margins of the translucent area are characteristic. There is also some protrusio acetabulae due to softening of the bone. (B) Pelvis in untreated celiac disease. Bilateral protrusio acetabulae and triradiate deformity of the pelvis are apparent. There are Looser's zones through both superior and inferior pubic rami, with complete fracture through that in the left superior pubic ramus. (C) Nutritional osteomalacia. In the chest there is a Looser's zone through the entire width of the left first rib (arrow). Traumatic fractures through the first and second rib normally occur only with severe trauma. (D) Malabsorption osteomalacia. In the pelvis there is an extensive Looser's zone through the left ilium and the right pubic ramus (arrows). There is evidence of pinning of a previous left hip fracture. Severe trauma is required to fracture the ilium. (E) Malabsorption osteomalacia. In the chest there are Looser's zones in the lateral border of the scapula and at the base of the acromium (arrows).

and triradiate deformity of the pelvis (Fig. 6), particularly if the cause of the vitamin D deficiency has persisted since childhood and has been inadequately treated or untreated.

As in rickets, secondary hyperparathyroidism is present and can be manifested radiographically as subperiosteal erosions, particularly in the phalanges but at other sites also (sacroiliac joints, symphysis pubis, proximal tibiae, outer ends of the clavicle, and "pepper-pot" skull), depending on the intensity of the hyperparathyroidism (Fig. 7). There can also be cortical tunneling and a "hazy" trabecular pattern. There may be generalized osteopenia, and vertebral bodies may have concave end plates. This is due to softening of the malacic bone, which is deformed by the cartilagenous intervertebral disc ("codfish" deformity). The etiology of this deformation is different from that which results in end plate irregularity in osteoporosis, in which microfractures occur owing to the bone being brittle rather than soft.

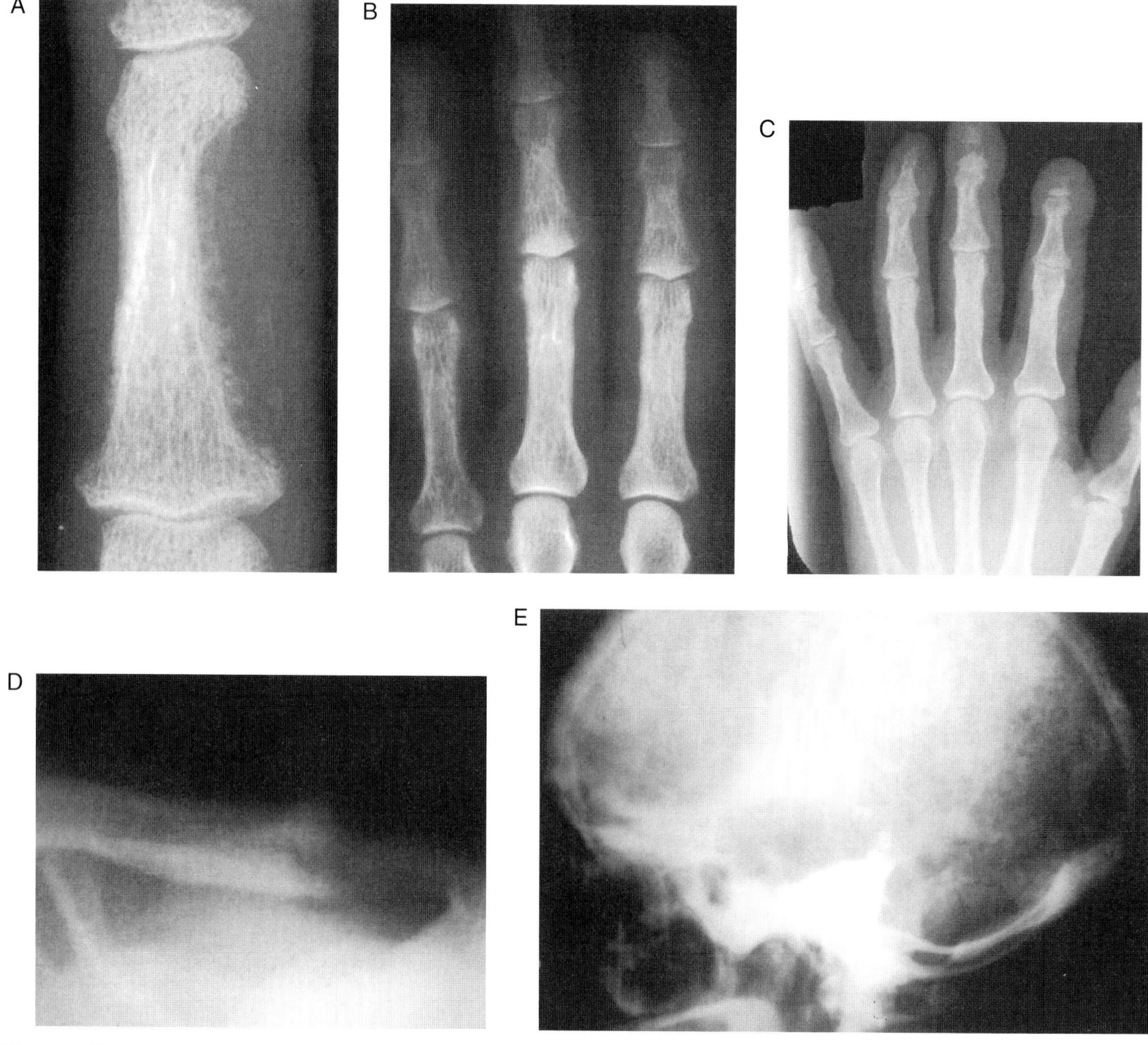

FIGURE 7 Hyperparathyroid erosions. (A) Extensive subperiosteal erosions along the radial border of the middle phalanx of the third finger. There is adjacent metastatic vascular calcification in the digital artery indicating phosphate retention and renal osteodystrophy. (B) Intracortical tunneling of the cortex of the phalanges, with subperiosteal erosions in the radial cortex of the middle phalanx of the index finger. (C) Acroosteolysis or resorption of the distal phalanges resulting in pseudoclubbing and subperiosteal erosions in the radial cortex of the middle phalanges of the third and fourth fingers (renal osteodystrophy). (D) Erosions in the outer end of the clavicle. (E) Lateral skull showing erosions in skull vault giving pepper-pot appearance.

C. Secondary Hyperparathyroidism

The most sensitive site for the radiographic features of hyperparathyroidism is the radial sides of the middle phalanges of the second and third fingers. There are characteristic erosions along the cortical surface of these bones (Fig. 7A,B). If there is florid hyperparathyroidism, then cortical erosions may be seen more widely and involve the distal phalanges (Fig. 7C), the outer ends of the clavicle (Fig. 7D), the symphysis pubis, the sacroiliac joints, the upper medial cortex of the tibiae, and the skull vault (pepper-pot skull) (Fig. 7E). The erosions of hyperparathyroidism in the sacroiliac joints tend to involve the iliac margin of the joint, in contrast to the involvement of both joint surfaces in inflammatory and erosive arthritides (Fig. 12A). The erosions of hyperparathyroidism in the skull vault (pepper-pot skull) must be differentiated from the "granularity" of the parietal region of the skull vault, which may be a variant of normal. In the former, the margins of the erosions are indistinct; in the latter they are distinct and corticated. Loss of the lamina dura around the teeth does occur but is not specific to hyperparathyroidism, occurring also in other conditions such as Paget's disease of bone and dental infection.

With intense hyperparathyroidism, there is an increase in cortical "tunneling" due to bone resorption and an indistinct, "hazy" trabecular pattern. Erosions may occur along the growth plate and result in displacement of the epiphysis from the metaphysis of the shaft of the bone. This is most likely to occur in association with chronic renal disease (see Section III below), since the intensity of the hyperparathyroidism secondary to vitamin D deficiency is related to the duration and severity of the hypocalcemia which acts as the stimulant.

Although the changes of rickets occur predominantly at the growth plates, Looser's zones may also be present in the juvenile skeleton (Fig. 3).

III. RENAL OSTEODYSTROPHY

The bone disease that occurs in chronic renal impairment, namely, renal osteodystrophy or uremic (azotemic) osteodystrophy, is complex and multifactorial [44–46]. There is a combination of vitamin D deficiency, which results in rickets and osteomalacia, and hypocalcemia [47,48]. The latter induces severe secondary hyperparathyroidism that stimulates osteoclastic resorption of bone [49–51]. This results radiographically in subperiosteal erosions, most frequently identified along the radial aspect of the middle phalanx of the second and third fingers (Fig. 7). Because the stimulus to secondary hyperparathyroidism has often been intense and present for a long time, the skeletal manifestations are often more extensive than in primary hyperparathyroidism,

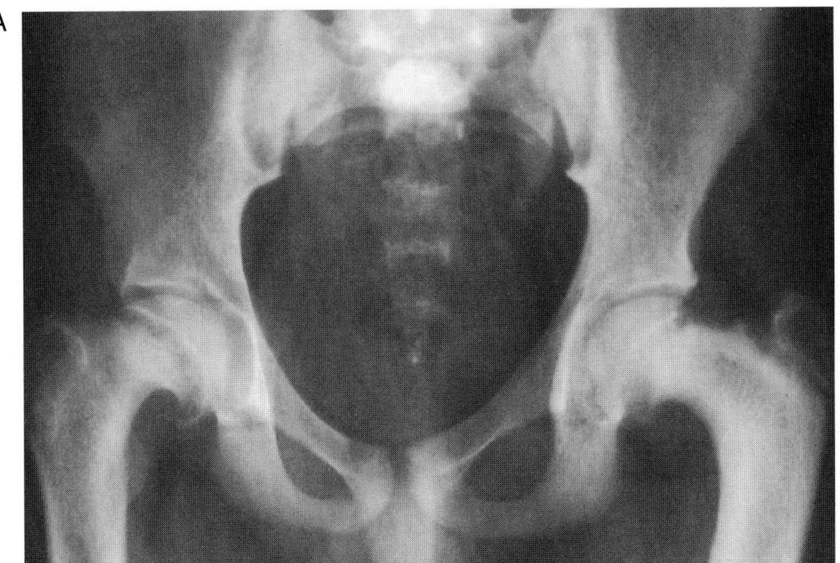

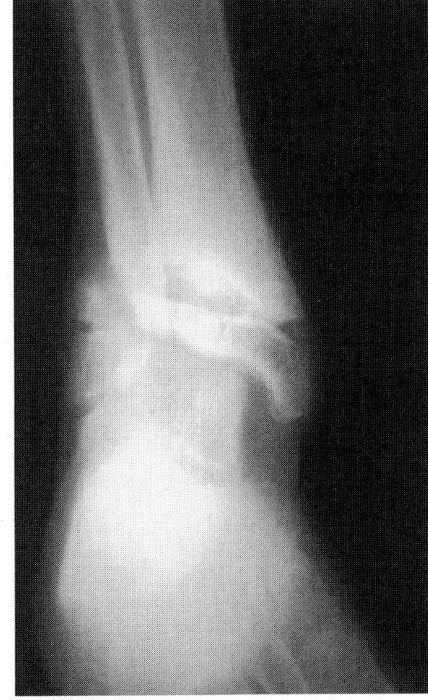

FIGURE 8 Renal osteodystrophy in an adolescent. (A) Pelvis showing rickets (widened growth plate at femoral metaphyses; varus deformity of femoral necks) and hyperparathryoidism (bone sclerosis and erosions in sacroiliac joints and along femoral metaphyses) resulting in slipped upper femoral epiphysis at right. (B) Anteroposterior view of ankle showing extensive erosions of the metaphyses due to severe secondary hyperparathyroidism.

which is now diagnosed early, and frequently, in asymptomatic patients.

A. Hyperparathyroidism

1. Subperiosteal Erosions

Subperiosteal erosions occur most commonly in the middle phalanges but are also present in the distal phalangeal tufts, causing acroosteolysis and the clinical sign of "pseudoclubbing" (Figs. 7A–C). Other sites of erosions include the outer end of the clavicle (Fig. 7D), the medial aspect of the proximal portion of the tibia, humerus, and femur, and the superior and inferior borders of the ribs [52].

Erosions may also occur adjacent to joints, and consequent damage to the articular subchondral bone can cause symptomatic arthritis [53]. These erosions may simulate those that occur in rheumatoid arthritis (RA) but tend to be a little further from the joint margin and are not generally associated with joint narrowing, periarticular osteopenia, or soft tissue swelling, which are early features of RA. Joints that can be affected include the acromioclavicular, sternoclavicular, sacroiliac, and the symphysis pubis (Fig. 12A). In the hand, the distal interphalangeal joints and ulnar aspect of the metacarpophalangeal joints can be involved [54]. Subperiosteal erosions of the phalanges are diagnostic of hyperparathyroidism. In children, erosions occur in the region of the growth plate, causing radiographic abnormalities which may be mistaken for rickets and which can result in slipped epiphysis and deformity [55,56] (Fig. 8).

If the hyperparathyroidism is treated successfully erosions will fill in, and the cortex will revert to its normal appearance. However, if there has been severe resorption of the distal phalangeal tufts, these cannot reconstitute to their normal shape and may remain shortened and stubby (Fig. 7C).

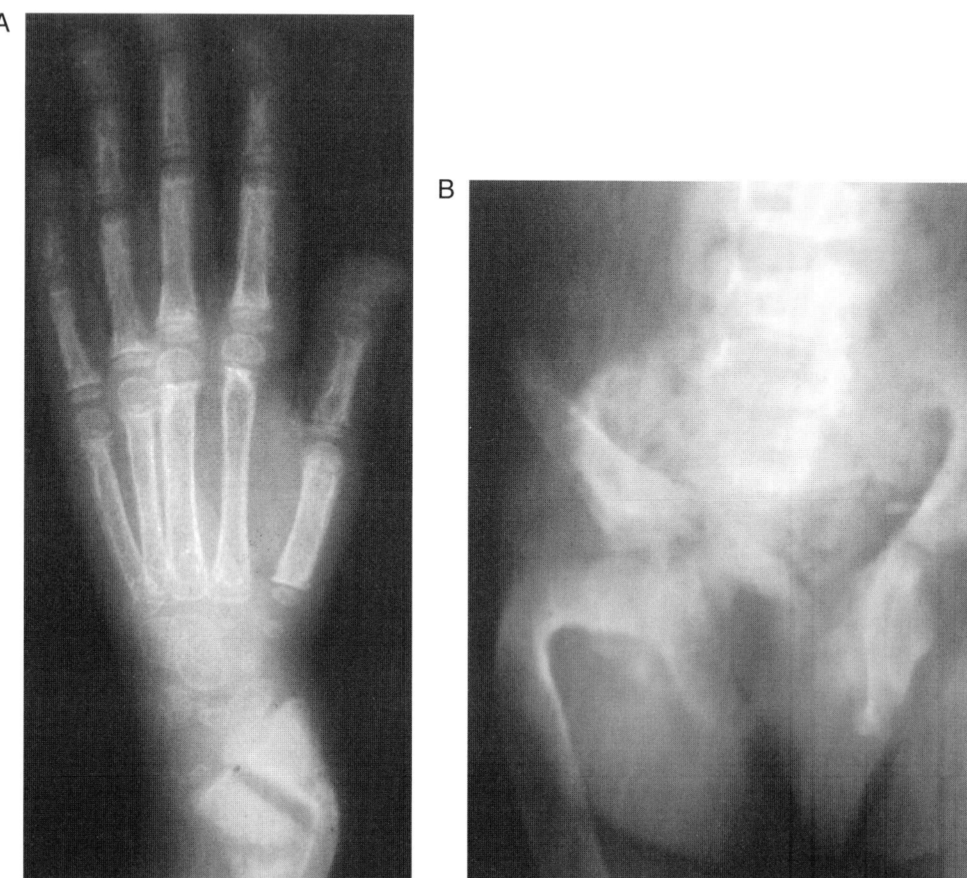

FIGURE 9 Severe renal osteodystrophy in a young boy. (A) Hand showing Looser's zones through distal ulna with bending of the radius and extensive bone resorption caused by hyperparathyroidism. (B) Pelvis showing evidence of rickets (soft bone) with triradiate deformity of the pelvis and varus deformity of the femoral necks. Cyst in the left ilium and erosions in the proximal left femur are related to hyperparathyroidism.

Bone resorption can occur in the regions of insertion of tendons and ligaments, particularly the trochanters, the ischial and humeral tuberosities, the inferior aspect of the calcaneum, and around the elbow. Excessive bone resorption in the skull vault causes the mottled texture of alternating areas of lucency and sclerosis referred to as a "salt and pepper" or pepper-pot skull (Fig. 7E).

2. Intracortical Bone Resorption

Intracortical bone resorption is caused by osteoclastic resorption of Haversian canals within the cortex of the bones. Radiographically this causes linear translucencies within the cortex (Figs. 7B and 9). This feature is not specific for hyperparathyroidism and can be found in other disorders in which bone turnover is increased (e.g., Paget's disease of bone).

3. Osteosclerosis

Osteosclerosis occurs uncommonly in primary hyperparathyroidism but is a common feature of disease secondary to chronic renal disease. Radiographically the bones appear increased in density (Fig. 10). This affects particularly the axial skeleton. In the vertebral bodies the end plates are preferentially involved, giving bands of dense bone adjacent to the end plates with a central band of lower, normal bone density. These alternating bands of normal and sclerotic bone give a striped pattern described as a "rugger jersey" spine (Fig. 10A). The osteosclerosis may be more generalized. It may result from excessive accumulation of poorly mineralized osteoid, which would appear more dense radiographically than normal bone. It was also suggested that it results from an exaggerated osteoblastic response following bone resorption [57].

4. Brown Tumors

Brown tumors represent cavities within the bone in which there has been excessive osteoclastic resorption. Histologically they are filled with fibrous tissue and osteoclasts, and may undergo necrosis and liquifaction. Radiographically they appear as low density cysts within

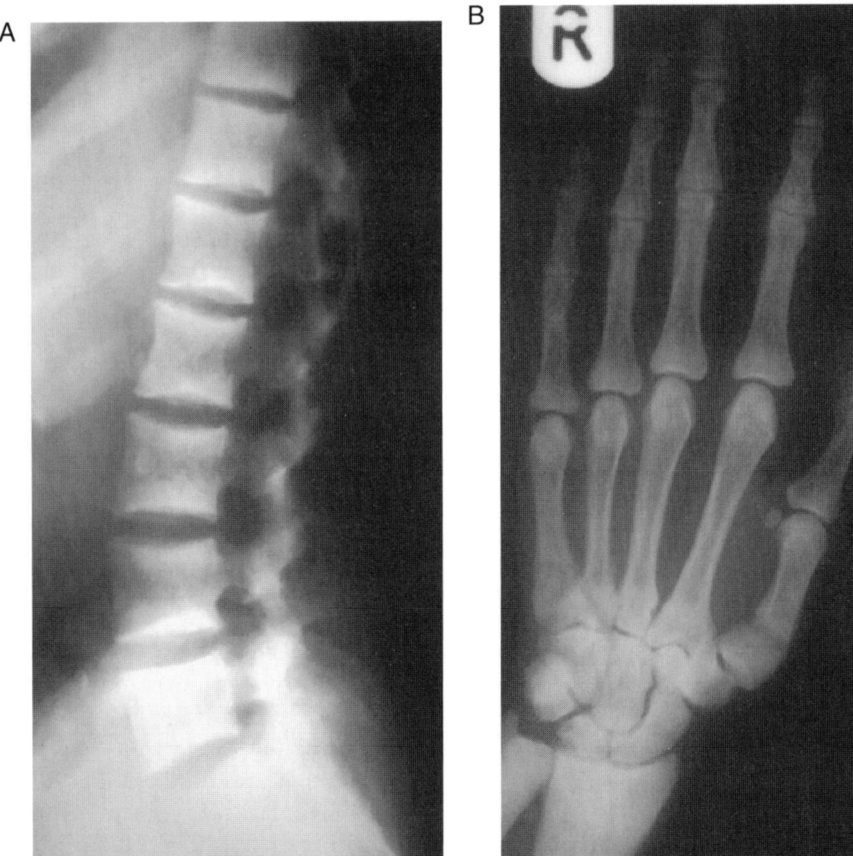

FIGURE 10 Renal osteodystrophy. (A) Bone sclerosis of the lateral lumbar spine showing end plate sclerosis, giving a rugger jersey appearance of stripes. (B) Hand radiograph showing Looser's zone at the base of the first metacarpal indicating osteomalacia and generalized osteosclerosis due to hyperparathyroidism.

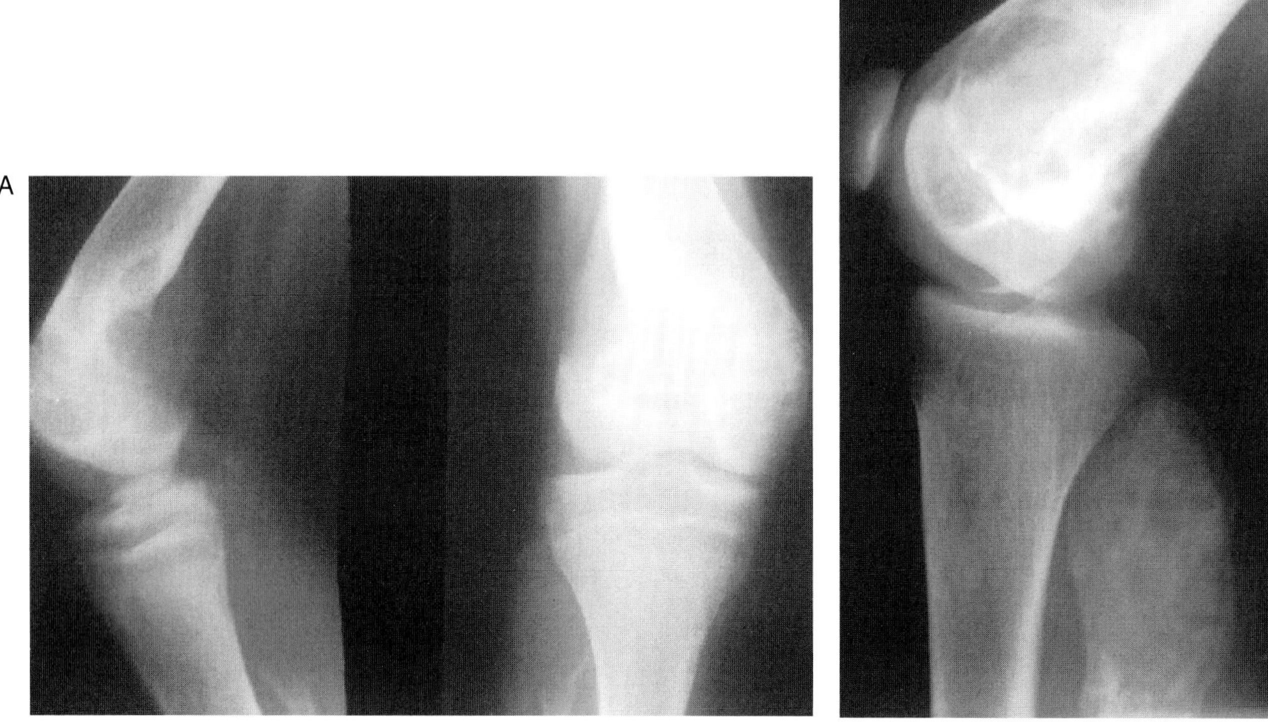

FIGURE 11 Brown tumors in renal osteodystrophy. (A) Right knee (anteroposterior and lateral views) showing well-defined lytic areas in the distal femur and the proximal fibula. The lesions are causing expansion of the bone with cortical destruction and a soft tissue mass, radiological features of aggressive bone lesions. (B) Following parathryoidectomy and regression of the hyperparathyroidism, the destructive lesions have filled in with woven bone and so have increased in density, and the cortex has reconstituted although there is still expansion of the fibula.

the bone. They can occur anywhere in the skeleton and may cause expansion of bones (Fig. 11). They constitute the osteitis fibrosa cystica of hyperparathyroidism first described by Von Recklinghausen. When appropriate treatment is given these bone cysts will fill with woven bone and increase in density (Fig. 11B).

5. Osteoporosis

With excessive bone resorption, combined with defective mineralization, the bones can appear osteopenic in some patients.

6. Periosteal Reaction

Periosteal new bone formation may be observed in up to 25% of patients with renal osteodystrophy. It occurs most frequently in those with severe bone disease and is thought to be a manifestation of intense hyperparathyroidism. It occurs in the metatarsals, femur, and pelvis and less commonly in the humerus, radius, ulna, tibia, metacarpals, and phalanges [58,59].

B. Metastatic Calcification

With the reduced glomerular function of chronic renal failure there is phosphate retention [60]. This results in an increase in the calcium × phosphate product, and, as a consequence, amorphous calcium phosphate is precipitated in organs and soft tissues [61]. This metastatic calcification can occur in the eyes and skin, causing symptoms of "gritty," sore eyes and itching. Other organs in which calcium is deposited include the heart, stomach, kidneys, lung, and skeletal muscle, where it is rarely detected radiographically. However, radiographic evidence of metastatic calcification is seen in arteries and around the joints (Fig. 12).

The periarticular calcification is more common around the large joints (hip, shoulder) but also occurs around small joints. The calcification can involve the joint capsule or tendons but more usually lies in the bursae adjacent to joints and bony protuberances (ischial tuberosity). These metastatic deposits of calcium

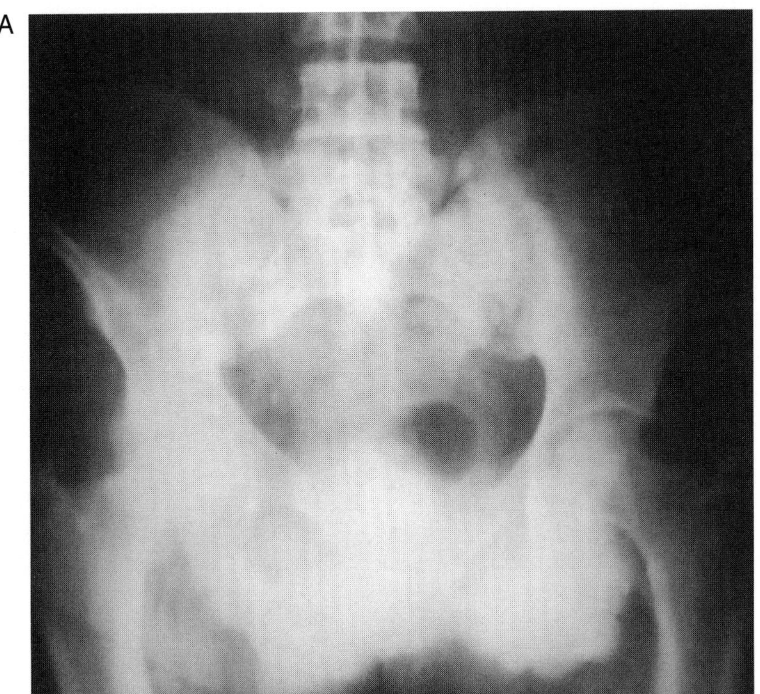

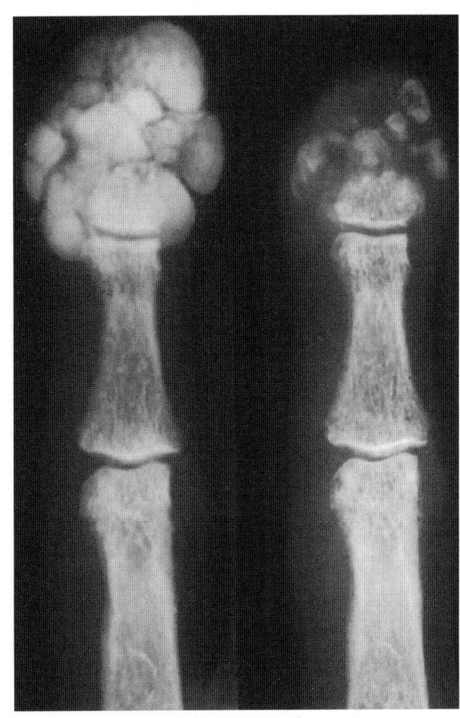

FIGURE 12 Metastatic calcification in renal osteodystrophy. (A) Pelvis showing extensive soft tissue calcification around the ischia and left hip. Widening of the sacroiliac joints indicates erosions of hyperparathyroidism. (B) Fingers showing (left) extensive metastatic calcification around the distal phalanx and subperiosteal erosions along the radial cortex of the middle phalanx. (Right) Following therapy with oral phosphate binder there is reduction in the extent of the metastatic soft tissue calcification. Note the acroosteolysis due to hyperparathyroidism.

will cause swelling and may be painful. They may increase rapidly in size and are sometimes erroneously diagnosed as tumors on initial clinical examination. These calcific masses can regress with appropriate treatment (Fig. 12B). Initially the masses may liquify, and apparent fluid levels are seen on radiographs taken with the patient in the erect position. These fluid levels are the result of the interface between serum and serum plus mineral. There can be complete regression of periarticular calcific masses with appropriate treatment; vascular calcification rarely regresses [62].

C. Aluminum Toxicity

Aluminum toxicity may occur in patients with chronic renal disease due to excessive ingestion of aluminum hydroxide taken to reduce serum phosphate. It also occurs in those patients on hemodialysis in which the dialysate water contains excessive amounts of aluminum [63–65]. The control of aluminum content in dialysate water in recent years has reduced the prevalence of this disorder [66]. Aluminum accumulates at the bone–osteoid interface and inhibits mineralization. This results in rickets and osteomalacia, but the bone also becomes "adynamic," with reduced osteoid production and turnover. Radiographically this results in reduced bone density and easy fracture [67].

IV. RENAL TUBULAR DEFECTS AND HYPOPHOSPHATEMIA

Inorganic phosphate, glucose, and amino acids are absorbed in the proximal renal tubule; concentration and acidification of the urine in exchange for a fixed base occur at the distal renal tubule. Disorders of the renal tubules may involve either the proximal or distal tubule or both. They will result in a spectrum of biochemical disturbances that may result in loss of phosphate, glucose, or amino acids alone, or in combination, with additional defects in urine acidification and concentration. These defects may be inherited and present from birth (de Toni-Fanconi syndrome, cystinosis, X-linked hypophosphatemia or XLH) [68] or acquired later in life (e.g., Wilson's disease, hereditary tyrosinemia, neurofibromatosis, mesenychymal tumors) by tubular function being interfered with either by crystal deposition (e.g., copper, in Wilson's disease) or a humoral substance, such as is produced by tumors in

A

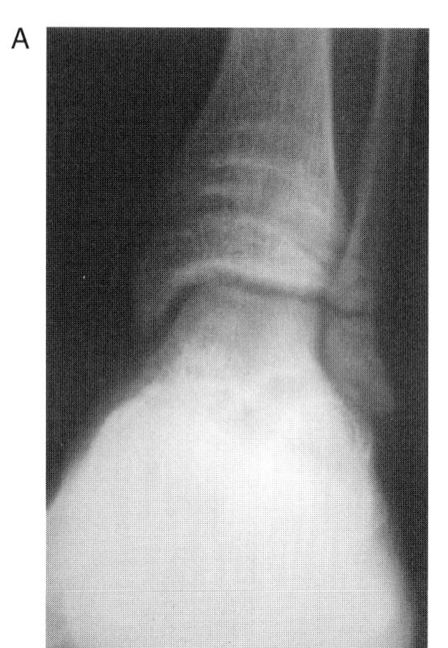

B

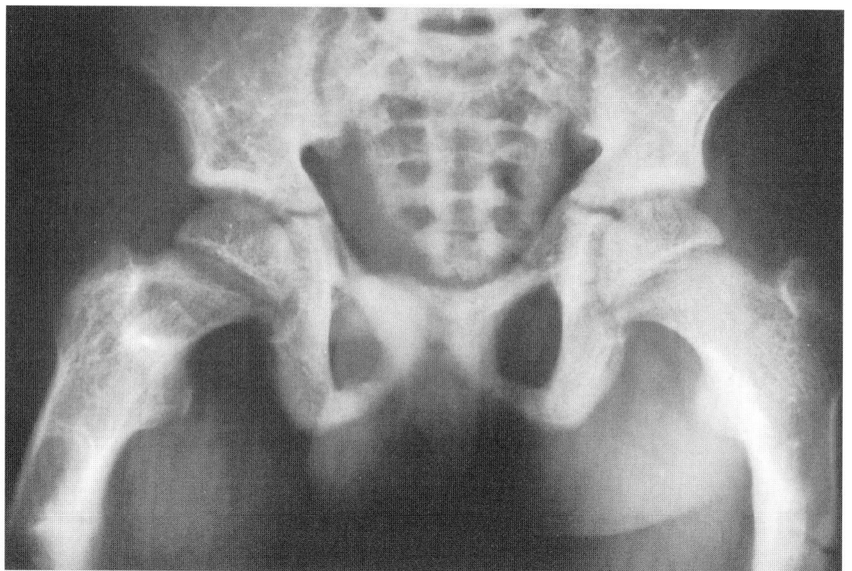

C

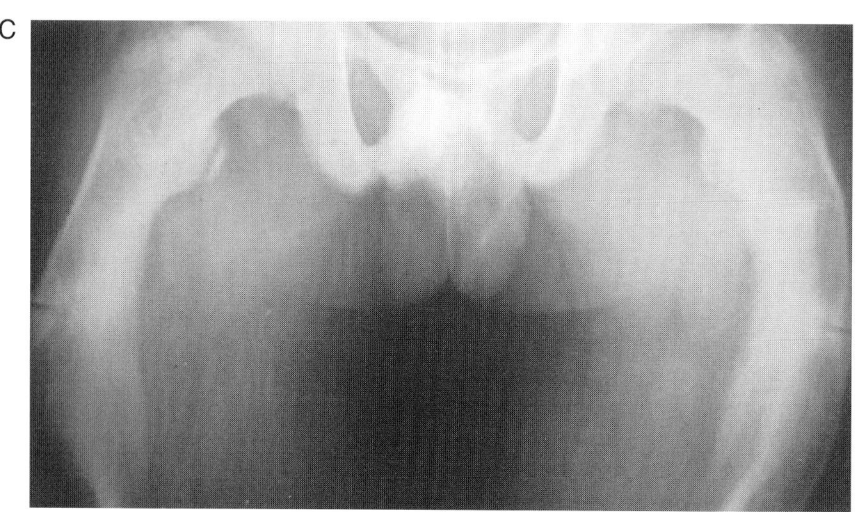

D

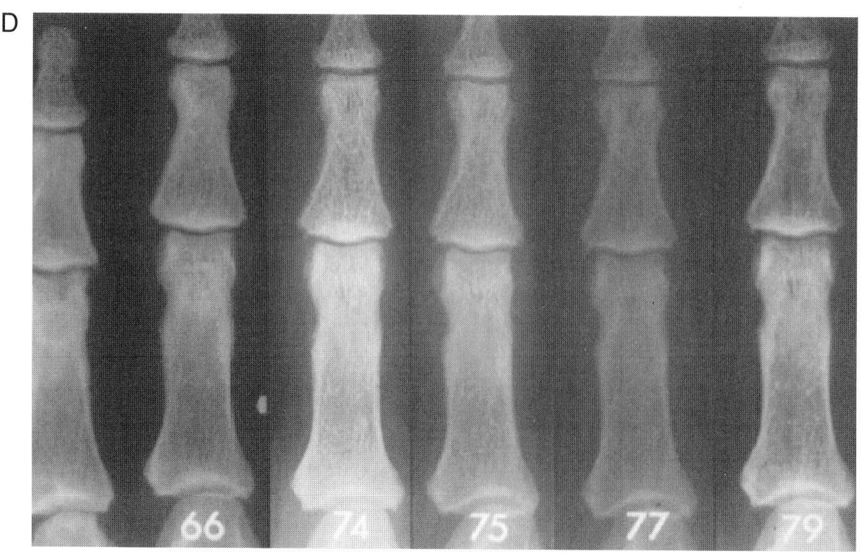

tumor-induced osteomalacia (TIO) also known as "oncogenic" rickets [69]. It is the renal tubular disorders that cause phosphaturia which result in rickets and osteomalacia [70]. As the serum calcium is generally normal in these diseases, secondary hyperparathyroidism does not occur (see Chapter 46).

A. X-linked Hypophosphatemia

The genetic disorder XLH is transmitted as an X-linked dominant trait. Sporadic cases through spontaneous mutations also occur. The incidence is approximately 1 in 25,000, and XLH is now the most common of genetically induced rickets [71–73]. The pathophysiology of XLH and its mode of treatment are discussed in detail in Chapter 46.

The disease is characterized by lifelong phosphaturia, hypophosphatemia, and rickets and osteomalacia. Rickets becomes clinically evident around 12 to 18 months of age. The radiological features of XLH are characteristic.

1. Rickets

There is widening of the growth plate and defective mineralization of the metaphysis; however the metaphyseal margin tends to be less indistinct than in nutritional rickets, and there is less expansion in the width of the metaphysis (Figs. 13A,B). Changes are most marked at the knee, wrist, ankle, and proximal femur. Healing does occur with appropriate treatment. The growth plates fuse normally at skeletal maturation. The bones are often short and undertubulated (shaft wide in relation to bone length), with bowing of the femur and tibia (Fig. 13C).

2. Osteomalacia

After skeletal maturation Looser's zones persist in patients with XLH. These tend to occur in sites different from those in nutritional osteomalacia, and are often present in the outer cortex of the bowed femur (Fig. 13C), although they occur along the medial cortex of the shaft also [74]. Looser's zones in the ribs and pelvis are rare. They may heal with appropriate treatment, but those that have been present for many years will persist and are presumably filled with fibrous tissue (Fig. 15D).

In the untreated patient there is no evidence of hyperparathyroidism as there is no hypocalcemia, in contrast to vitamin D deficiency osteomalacia. However, treatment with large doses of oral phosphate for long periods may induce secondary hyperparathyroidism, which may exceptionally be evident by subperiosteal erosions in the hand [75] (Fig. 13D).

3. Osteosclerosis

Although there is defective mineralization of osteoid in XLH, the bones are commonly increased in density, with a coarse and prominent trabecular pattern [76] (Fig. 13). This is a characteristic feature of the disease and is not related to treatment with vitamin D and phosphate supplements, as it can be present in those who have not received treatment. This bone sclerosis has been shown to involve the petrous bone and structures of the inner ear and may be responsible for the hydropic cochlear pattern of deafness that these patients can develop in later life [77].

4. Abnormalities of Bone Modeling

In XLH the bones are often short, with widening of the shaft. The ribs are broad and tend to slope downward more than normal, causing a bell-shaped chest (Fig. 14A). There can be broadening of the distal end of the ulna (Fig. 14B), and often marked bowing of the femur and tibia [78].

5. Extraskeletal Ossification

X-linked hypophosphatemia is characterized by an enthesopathy (inflammation in the junctional area between bone and tendon insertion) that heals by ossification of ligament and tendon insertions to bone in many affected patients [79,80]. This results in new bone formation around the pelvis and spine, with the changes resembling ankylosing spondylitis (Fig. 15). There can be complete ankylosis of the spine, which limits move-

FIGURE 13 Familial X-linked hypophosphatemic rickets (XLH). (A) Anteroposterior view of ankle showing widened growth plate and Harris growth arrest lines. The appearance of the metaphysis is different from that in nutritional rickets (see Fig. 1). (B) Pelvis showing rachitic changes at the proximal femoral metaphyses, dense bones with a coarse trabecular pattern, and bowing of the femoral shafts with bilateral Looser's zones (through medial cortex at right and outer cortex at left). (C) Femora showing bones that are bowed, shortened, and undertubulated with broad shafts. There are Looser's zones through the outer cortex of each femur. (D) Radiographic series of fingers. As the serum calcium is usually normal in XLH, there is no secondary hyperparathyroidism. However, large doses of oral phosphate may induce secondary hyperparathryoidism. This occurred in a patient in whom subperiosteal erosions are present in the phalanges in 1974 and 1975. The erosions subsequently healed with adjustments in therapy.

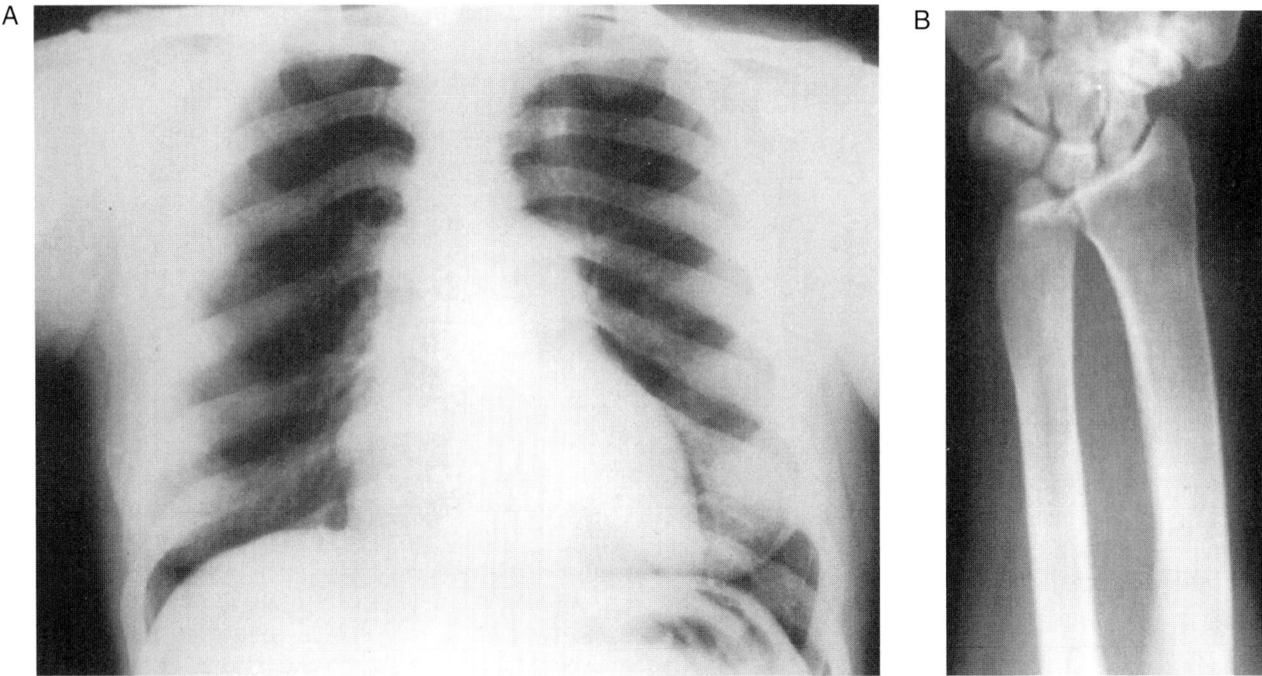

FIGURE 14 Modeling deformities in familial X-linked hypophosphatemic rickets (XLH). (A) Chest radiograph showing that the ribs are broad and slope downward more than normal, giving a rather bell-shaped chest. (B) Wrist showing some bulbous expansion of the distal shaft of the ulna.

ments. As there is no inflammatory arthritis, the sacroiliac joints are normal, an important radiographic feature that serves to differentiate this condition from ankylosing spondylitis (Figs. 16A,B). Ossification can occur in the interosseous membrane of the forearm, forming a synchondrosis between the radius and ulna, and in the leg between the tibia and fibula (Figs. 15C,D). Separate small ossicles can occur around the joints of the hands (Fig. 15B); there is also ossification at tendon insertions in the hands, causing "whiskering" of bone margins.

Rarely, spinal cord compression may be caused by a combination of ossification of the ligamentum flavum, thickening of the laminae, and hyperostosis around the apophyseal joints [81] (Fig. 16). It is the ossification of the ligamentum flavum that causes the most significant narrowing of the spinal canal [82]. This occurs most commonly in the thoracic spine and generally involves two or three adjacent vertebral segments. Patients may be asymptomatic even with severe spinal canal narrowing, and acute cord compression can be precipitated by quite minor trauma. It is important to be aware of this rare, but recognized, complication of the disease since surgical decompression by laminectomy is curative. The extent of ossification cannot be predicted by the degree of paraspinal or extraskeletal ossification at other sites. Computerized tomography (CT), with its cross-sectional depiction of anatomy, is a useful imaging technique for demonstrating the extent of intraspinal ossification (Figs. 16c–f). Extraskeletal ossification is uncommon in patients with XLH under 40 years of age.

6. Osteoarthritis

The deformity and bowing of the long bones in XLH cause altered stress through joints, predisposing to degenerative joint disease. This is particularly prominent in the knee joint.

The extent to which the radiographic abnormalities discussed in Section IV,A,1–6 are present varies between affected individuals [83]. In some, all the features are present and so are diagnostic of XLH. In others, there may only be minor abnormalities, and the condition may be overlooked [84].

B. Tumor-Induced ("Oncogenic") Rickets/Osteomalacia

Tumor induced osteomalcia (TIO) or "oncogenic" rickets and osteomalacia were first reported in 1947 [69,85]. There is hypophosphatemia, and the clinical and radiographic features of rickets or osteomalacia can precede identification of the causative tumor by long

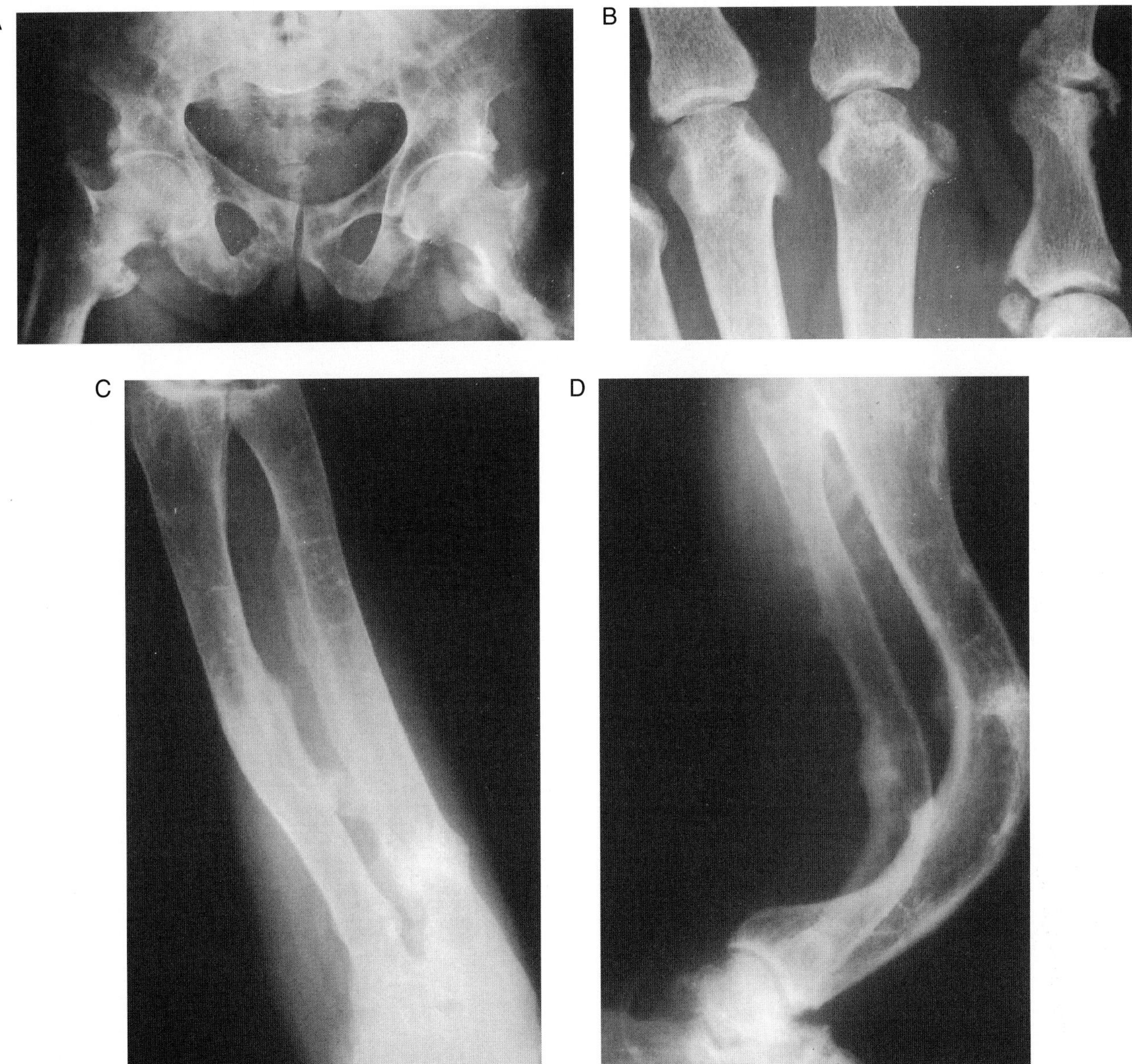

FIGURE 15 Enthesopathy and extraskeletal ossification in XLH. (A) Anteroposterior view of pelvis showing deformity of rickets, in the past, and chronic Looser's zone in the left femoral shaft. New bone formation is apparent around hip joints and at lesser trochanters. (B) Hands showing ossicles and bony outgrowths related to heads of the metacarpals and joints. (C) Forearm showing synostosis between the radius and ulna due to ossification of the interosseous ligament. (D) Lower leg showing deformity due to bowing and chronic Looser's zones with ossification of the interosseous membrane. Such severe deformity reflects inadequate treatment during childhood.

periods (1–16 years). The tumors are usually small, benign, and vascular in origin (hemangiopericytoma) [86], but some may be malignant [87] (Fig. 17). Rickets and osteomalacia will heal with surgical removal of the tumor [88,89]. Often the tumors are extremely small and elude detection for many years (Fig. 19). It is important that the patient is vigilant about self-examination and reports any small palpable lump or skin lesion. More sophisticated imaging (CT, magnetic resonance imaging or MRI) may be helpful in localizing more deep-seated lesions [90]. The condition is discussed in detail in Chapter 46.

V. VITAMIN D INTOXICATION

In times gone by vitamin D was advocated in the treatment of a great variety of conditions, including tuberculosis (especially lupus vulgaris), sarcoidosis, rheumatoid arthritis, hay fever, chilblains, and asthma.

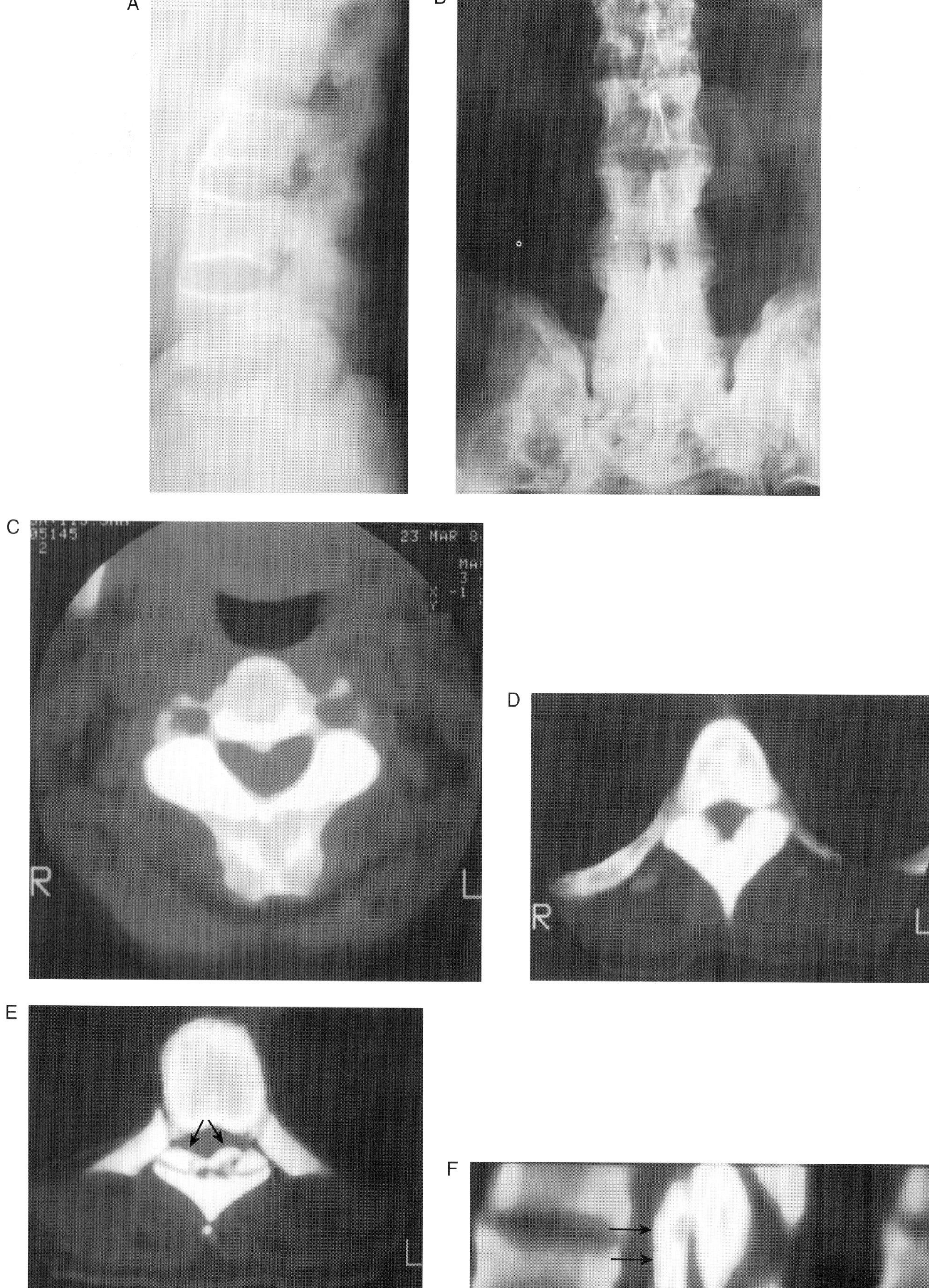
A
B
C
05145
2
23 MAR 8
R
L
D
R
L
E
L
F

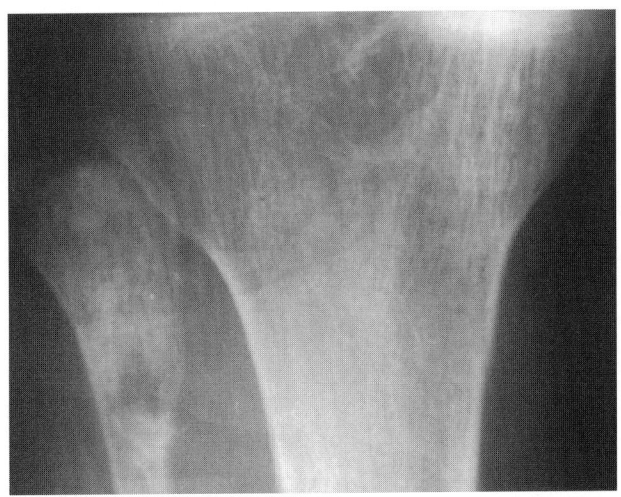

FIGURE 17 Tumor-induced or oncogenic osteomalacia in a 46-year-old woman who 18 months previously had presented with hypophosphatemic osteomalacia. There is a lytic lesion in the proximal fibula. This was a fibrosarcoma. The osteomalacia healed following surgical excision of the tumor.

This treatment had no beneficial effect in these conditions, and its use was eventually abandoned, but not before many cases of vitamin D intoxication had been described [91]. Although vitamin D intoxication has become less common with the advent of $1,25(OH)_2D_3$ and other active metabolites, the new era of vitamin D usage to treat cancer, psoriasis and immunological disease (see Section VIII of this volume) may see a resurgence of interest in vitamin D intoxication. Clinically, the symptoms are of fatigue, malaise, weakness, thirst and polyuria, anorexia, nausea, and vomiting due to hypercalcemia (see Chapter 54). The hypercalcemia results in hypercalciuria, nephrocalcinosis, renal impairment, and hypertension. Metastatic calcification and bone sclerosis also occur [92,93] (Fig. 18).

VI. TECHNICAL ASPECTS OF IMAGING

A. Plain Radiographs

Despite tremendous developments and expansion in the imaging techniques available since the 1970s, plain radiographs remain the principal imaging method in the radiographic diagnosis of metabolic bone disorders, including those involving vitamin D deficiency, rickets, and osteomalacia. When radiographing the hands and feet, image quality must be optimized by using fine grain, single-sided emulsion film and a fine X-ray focal spot (0.6 mm or less). Meticulous attention to detail of the radiographic technique used will enhance the diagnostic features present in the hand radiograph, such as the subperiosteal erosions and intracortical tunneling of hyperparathyroidism. Magnification techniques, either optical or radiographic, can further enhance identification of such diagnostic features of metabolic bone disease. High-resolution radiographs of the torso regions of the body are generally precluded because of the high radiation doses required.

B. Nuclear Medicine

In the imaging technique of the skeleton referred to as nuclear medicine, ^{99m}Tc-labeled phosphate compounds are administered intravenously [94,95]. They are incorporated into the skeleton, particularly in sites that have either increased vascularity or increased new bone formation. Such areas of increased uptake are evident as "hot spots" on the scan, which is performed, using a gamma camera, 2 hr after administration of the radionuclide. This bone scanning technique is very sensitive to disease in bone, but not specific, in that numerous pathologies may cause hot spots, including infection, Paget's disease of bone, metastases, and degenerative joint disease (hyperostosis). Radiographs of the relevant anatomical site will help to differentiate these various pathologies.

The radionuclide bone scan is sensitive to detecting Looser's zones that may not be evident radiographically [96,97] (Fig. 19). The areas of increased uptake of label may be bilateral and symmetrical and be present in anatomical sites typical for Looser's zones (femoral necks, ribs, pubic rami). If there is associated secondary hyperparathyroidism, there will be generalized increase in uptake of the radionuclide by the skeleton ("super scan"), with elevation of the bone/soft tissue ratio. There will be poor renal uptake of the radionuclide if

FIGURE 16 Changes in XLH. (A) Lateral lumbar spine showing ossification of the paraspinal ligaments and at apophyseal joints. The appearances resemble those of ankylosing spondylitis but can be differentiated from it by the normal aspect of the sacroiliac joints. (B) Anteroposterior view of lumbar spine showing ossification of paraspinal ligaments resulting in ankylosis but normal sacroiliac joints. (C–F) Narrowing of the spinal canal may be caused by various factors. (C) Thickening of the laminae. In a computerized tomography (CT) section through the cervical spine the laminae are increased in width 3-fold. (D) Hyperostosis of apophyseal joints causing trefoil deformity of the thoracic spinal canal at this level (CT scan). (E) Ossification of the ligamentum flavum (arrows) lying anterior to the laminae, severely narrowing the thoracic spinal canal (CT scan). (F) Saggital reformation of thin (3 mm) contiguous CT sections showing ossification of the ligamentum flavum, severely reducing the anteroposterior diameter of the spinal canal. C, E, and F reprinted from Adams and Davies [82] with permission.

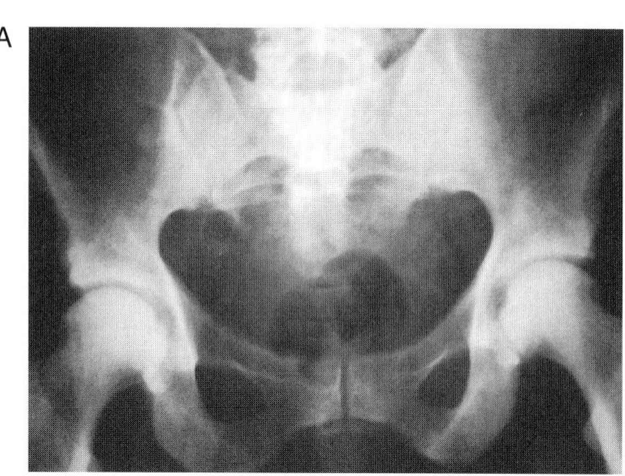

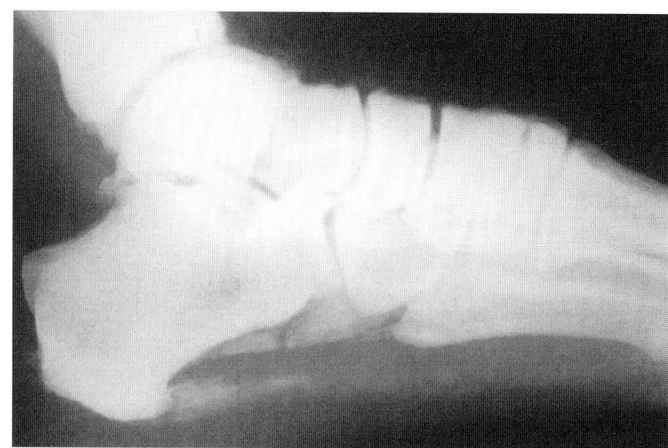

FIGURE 18 Vitamin D intoxication with associated renal impairment. (A) Pelvis showing bone sclerosis, calcification of vessels in the pelvis, and periosteal reaction around the pelvic brim. (B) Lateral foot showing bone sclerosis and calcification of the ligaments in the foot. There was calcification of other ligaments (iliolumbar) and of the falx and tentorium.

the cause of the osteomalacia and vitamin D deficiency is chronic renal disease. With appropriate treatment of the osteomalacia, the Looser's zones will heal, and the hot spots will not be present on subsequent radionuclide scanning. Radionuclide scanning offers a method of monitoring response of osteomalacia to therapy [98]; however, clinical symptoms, biochemical parameters, and plain radiographs often suffice for this purpose.

Radionuclide scanning in children is of limited value in metabolic bone disorders, because there is high uptake in the normal metaphysis of the growing skeleton. In addition, the examination carries a significant radiation dose to the bone marrow.

C. Other Imaging Techniques

Ultrasonography (US), computerized tomography (CT), magnetic resonance imaging (MRI), and angiography generally play little part in the diagnosis and management of rickets and osteomalacia. The exception to this is their application to the identification and localization of tumors that cause oncogenic rickets. Such tumors are often very small and may be deep seated, so they can be extremely difficult to identify (see Chapter 46).

D. Bone Mineral Densitometry

Methods of bone densitometry play an important role in diagnosis of patients with osteoporosis and monitoring the efficacy of treatment [99,100]. The methods available include single energy X-ray absorptiometry (SXA) for forearm measurements, dual energy X-ray absorptiometry (DXA) for measurements in the lumbar spine, femoral neck, and whole body, quantitative computerized tomography (QCT) for measuring cortical and trabecular bone separately in the lumbar spine and forearm, and broadband ultrasound attenuation (BUA), used to make measurements in the calcaneum. In disease, and treatment, these techniques can provide complementary information because they measure different types of bone in different sites of the skeleton. QCT is unique amongst methods in providing a true volumetric density and can measure cortical and trabecular bone density separately. SXA measures integral (cortical and trabecular) bone in the forearm, and DXA measures integral bone in the proximal femur, the lumbar spine, and the whole body. In vitamin D deficiency and rickets or osteomalacia, there may be osteopenia. If there is secondary hyperparathyroidism the forearm cortical bone mineral density (BMD) measurement might show the most marked reduction [101]. If the vitamin D deficiency osteomalacia is treated appropriately, there is very rapid increase (+25% or more) in BMD (2–4 weeks) on serial bone densitometry (see Chapter 47).

VII. CONCLUSIONS

Mineralization of bone matrix depends on the presence of adequate supplies of not only vitamin D, in the form of its active metabolite $1,25(OH)_2D$, but also calcium and phosphorus and the presence of alkaline phosphatase and a normal pH. If there is deficiency of these substances for any reason, or if there is severe acidosis, then mineralization of bone will be defective. This results in rickets in childhood and osteomalacia in adults. Radiographically, rickets is evident by bone

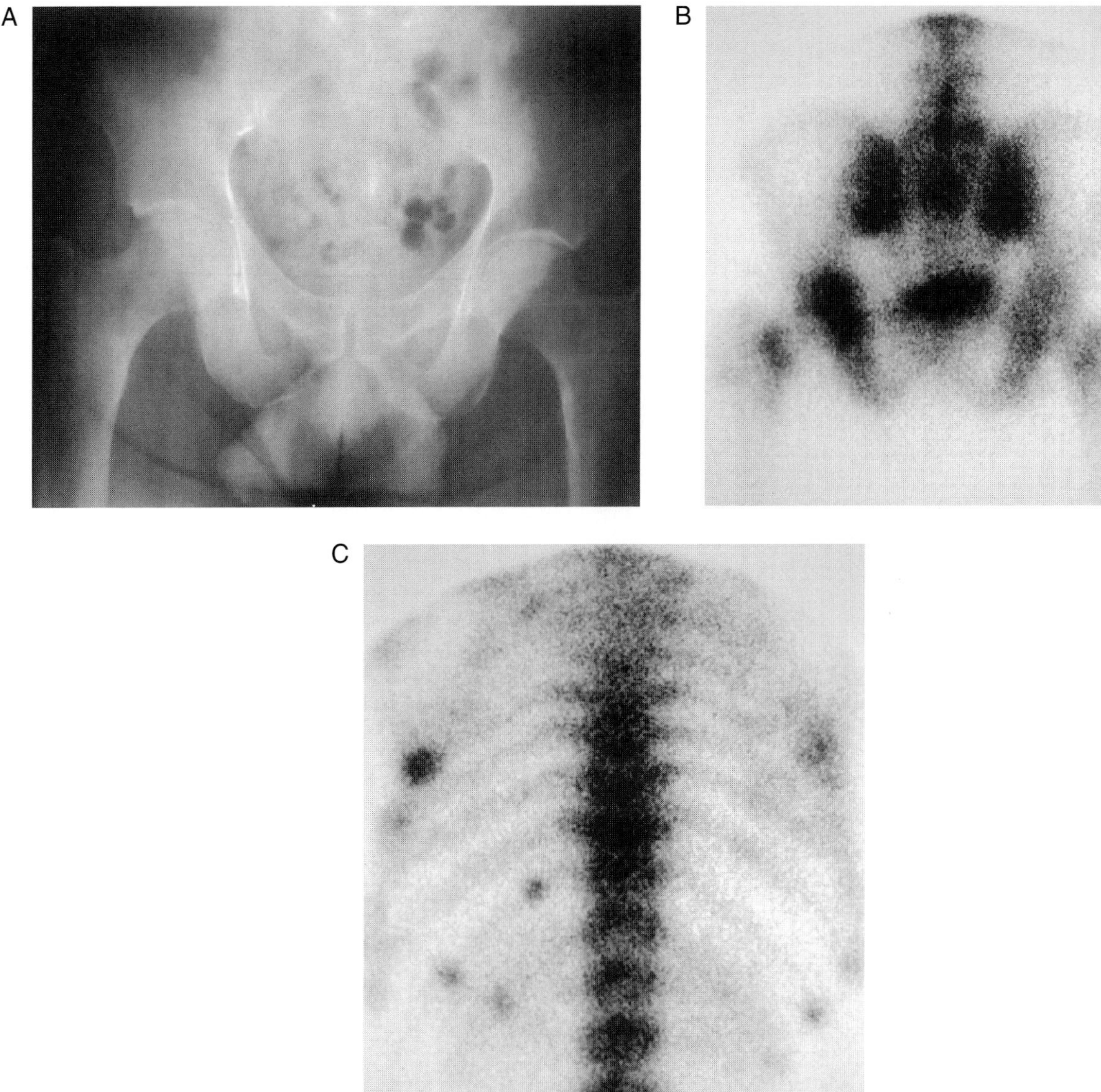

FIGURE 19 Acquired hypophosphatemic osteomalacia. A 40-year-old man presented with low back and buttock pain. He was thought to have ankylosing spondylitis. (A) Pelvic radiograph, which was considered to be normal at the original hospital of referral. (B) Radionuclide scan of pelvis showing increased uptake in both femoral necks. (C) Radionuclide scan of upper torso showing multiple hot spots in ribs. The distribution of the hot spots suggests Looser's zones. A radionuclide scan may be more sensitive than a radiograph for identifying Looser's zones. The patient was found to have hypophosphatemia presumed to be tumor-induced, but no tumor has yet been identified after 9 years, despite careful clinical examination and use of other imaging techniques.

deformity caused by softening and metaphyseal abnormalities where endochondral ossification is defective. The pathognomonic feature of osteomalacia is the Looser's zone (pseudofracture).

Many different diseases that result in rickets and osteomalacia (vitamin D deficiency, hypophosphatemia, hypophosphatasia, and acidemia) may therefore have similar radiographic appearances. There may be distinguishing features, such as bone sclerosis and extraskeletal ossification in XLH. In conditions in which there is hypocalcemia (vitamin D deficiency), this acts as a stimulus to secondary hyperparathyroidism, features of which can be identified radiographically (bone erosions). The bone disease of chronic renal impairment (renal osteodystrophy) is complex, being a combination of rickets and osteomalacia [$1{,}25(OH)_2D$ deficiency], hyperparathyroidism (bone erosions, sclerosis), and metastatic calcification (due to phosphate retention).

Plain radiographs remain the most important imaging technique for the diagnosis of metabolic bone disease; radionuclide scans may be more sensitive for identifying Looser's zones. Other techniques (ultrasound, computed tomograph, magnetic resonance imaging) have a role in identifying tumors that induce TIO or hypophosphatemic ("oncogenic") osteomalacia.

References

1. Harrison HE, Harrison HC 1975 Rickets then and now. J Pediatr **87**:1144 .
2. Pitt MJ 1991 Rickets and osteomalacia are still around. Radiol Clin North Am **29**:97–118.
3. Turner ML, Dalinka MK 1979 Osteomalacia: Uncommon causes. A J Roentgenol **133:**539–540.
4. Smerdon GT 1950 Daniel Whistler and the English Disease. A translation and biographical note. J Hist Med **5**:397–415.
5. Glendening L 1942 Source Book of Medical History. Dover, New York, pp. 268–269.
6. Fourman P, Royer P, 1960 Vitamin D. In: Calcium Metabolism and the Bone. Blackwell, Oxford and Edinburgh, pp 104–129.
7. Dewhurst K 1962 Postmortem examination in case of rickets performed by John Locke. Br Med J **2**:1466.
8. Hunter R 1972 Rickets, ruckets, rekets, or rackets? Lancet **1:**1176–1177.
9. LeVay D 1975 On the derivation of the name "rickets." Proc R Soc Med **68**:46–50.
10. Schutte D 1824 Beobachtungen uber den Nutzen des Berger Leberthrans. Arch Med Erfahrung **2**:79–92.
11. Percival T 1783 Observations on the medicinal uses of the oleum jecoris aselli or cod liver oil, in the chronic rheumatism and other painful disorders. Lond Clin Med J **3**:392–401.
12. Palm TA 1890 The geographical distribution and aetiology of rickets. Practitioner **45**:270–279.
13. Morse JL 1900 The frequency of rickets in infancy in Boston and vicinity. JAMA **34**:724.
14. Parsons LG 1927 The bone changes occurring in renal and coeliac infantilism, and their relationship to rickets. Part I. Renal rickets. Arch Dis Child **2**:1–25.
15. Parsons LG 1927 The bone changes occurring in renal and coeliac infantilism and their relationship to rickets. Part II. Coeliac rickets. Arch Dis Child **2**:198–211.
16. Pommer G 1885 Untersuchungen ober osteomalacie und rachitis. Leipzig.
17. Dent CE 1970 Rickets (and osteomalacia), nutritional and metabolic (1919–1969). Proc R Soc Med **63:**401–408.
18. Hutchison FN, Bell NH 1992 Osteomalacia and rickets. Semin Nephrol **12:**127–145.
19. Pitt MJ 1981 Rachitic and osteomalacia syndromes. Radiol Clin North Am **19**:581–598.
20. Smith R 1972 The pathophysiology and management of rickets. Orthop Clin North Am **3**:601–621.
21. Stamp TCB, Exton-Smith AN, Richens A 1976 Classical rickets and osteomalacia in Britain. Lancet **2**:308.
22. Stamp TCB, Round JM 1974 Seasonal changes in human plasma levels of 25-hydroxyvitamin D. Nature **247**:563–565.
23. Oestreich AE, Ahmad BS 1993 The periphysis and its effect on the metaphysis. II Application to rickets and other abnormalities. Skel Radiol **22:**115–119.
24. Steinbach HL, Noetzli M 1964 Roentgen appearance of the skeleton in osteomalacia and rickets. A J Roentgenol **91**:955–972.
25. Frankel A, Gruber B, Schey WL 1994 Rickets presenting as stridor and apnea. Ann Otol Rhinol Laryngol **103:**905–972.
26. Glasgow JFT, Thomas PS 1977 Rachitic respiratory distress in small preterm infants. Arch Dis Child **52:**268–273.
27. Park EA 1954 The influence of severe illness on rickets. Arch Dis Child **29:**369–380.
28. Harris HA 1933 Rickets. In: Bone Growth in Health and Disease. Oxford Medical Publications, Oxford University Press, London, p. 87.
29. Davies DR, Dent CE, Willcox A 1956 Hyperparathyroidism and steatorrhoea. B Med J **2**:1133–1137.
30. Mankin HJ 1974 Rickets, osteomalacia, and renal osteodystrophy—Part I. J Bone Joint Surg (Am) **56:**101–128.
31. Mankin HJ 1974 Rickets, osteomalacia, and renal osteodystrophy—Part II. J Bone Joint Surg (Am) **56:**352–386.
32. Mankin HJ 1990 Rickets, osteomalacia, and renal osteodystrophy. Orthop Clin North Am **21**:81–96.
33. Looser E 1908 Uber spatrachitis und die Beziehungen zwischen rachitis und osteomalacie. Mitt Grenzgeb Med Chir **18**:678–744.
34. Looser E 1920 Uber spatrachitis und osteomalacie Klinishe rontgenologische und pathologischanatomische Untersuchungen. Drsch Z Chir **152**:210–357.
35. Looser E 1920 Uber pathologische Formen von Infraktionen und Callusbildungen bei Rachitis und osteomalakie und anderen knocken Erkrankungen. Zbl Chir **47**:1470–1474.
36. Milkman LA 1930 Pseudofractures (hunger osteopathy, late rickets, osteomalacia). Am J Roentgenol **24**:29–37.
37. Milkman LA 1934 Multiple spontaneous idiopathic symmetrical fracture. Am J Roentgenol **32**:622–634.
38. Camp JD, McCullough JAL 1941 Pseudofractures in diseases affecting the skeletal system. Radiology **36**:651–661.
39. LeMay M, Blunt JW 1949 A factor determining the location of pseudofractures in osteomalacia. J Clin Invest **28**:521–525.
40. Steinbach HL, Kolb FO, Gilfillan R 1954 A mechanism of the production of pseudofractures in osteomalacia (Milkman's syndrome). Radiology **62**:388–394.
41. Cooper KL 1994 Insufficiency stress fractures. Curr Prob Diagn Radiol **23**:29–68.
42. McKenna MJ, Kleerekoper M, Ellis BI, Rao DS, Parfitt AM, Frame B 1987 Atypical insufficiency fractures confused with Looser zones of osteomalacia. Bone **8**:71–78.

43. North KAK 1966 Multiple stress fractures simulating osteomalacia. Am J Roentgenol **97**:672–675.
44. Pitt MJ 1995 Rickets and osteomalacia. In: Resnick D (ed) Diagnosis of Bone and Joint Disorders. Saunders, Philadelphia, Pennsylvania, pp 1885–1920.
45. Shapiro R 1972 Radiologic aspects of renal osteodystrophy. Radiol Clin North Am **10**:557–568.
46. Sundaram M 1989 Renal osteodystrophy. Skeletal Radiol **18**:415–426.
47. McCarthy JT, Kumar R 1986 Behavior of the vitamin D endocrine system in the development of renal osteodystrophy. Semin Nephrol **6**:21–30.
48. Peacock M 1978 Renal bone disease. Practitioner **220**:913–918.
49. Massry SG, Ritz E 1978 The pathogenesis of secondary hyperparathyroidism of renal failure: Is there a controversy? Arch Intern Med **138**:853–856.
50. Norfray J, Calenoff L, DelGreco F, Krumlovsky F 1975 Renal osteodystrophy in patients on hemodialysis as reflected in the bony pelvis. Am J Roentgenol **125**:352–358.
51. Parfitt AM 1972 Renal osteodystrophy. Orthop Clin North Am **3**:681–698.
52. Resnick D, Niwayama G 1976 Subchondral resorption of bone in renal osteodystrophy. Radiology **118**:315–321.
53. Andresen J, Nielsen HE 1981 Juxta-articular erosions and calcifications in patients with chronic renal failure. Acta Radiol Diagn **22**:709–713.
54. Resnick DL 1974 Erosive arthritis of the hand and wrist in hyperparathyroidism. Radiology **110**:263–269.
55. Goldman AB, Lane JM, Salvati E 1978 Slipped capital femoral epiphyses complicating renal osteodystrophy: A report of three cases. Radiology **126**:333–339.
56. Mehls O, Ritz E, Krempien B, Gilli G, Link K, Willich E, Scharer K 1975 Slipped epiphyses in renal osteodystrophy. Arch Dis Child **50**:545–554.
57. Wolf HL, Deubo JV 1958 Osteosclerosis in chronic renal disease. Am J Med Sci **235**:33–42.
58. Meema HE, Oreopoulos DG, Rabinovich S, Husdan H, Rapoport A 1974 Periosteal new bone formation (periosteal neostosis) in renal osteodystrophy. Radiology **110**:513–522.
59. Ritchie WGM, Winney RJ, Davison AM, Robson JS 1975 Periosteal new bone formation developing during haemodialysis for chronic renal failure. Br J Radiol **48**:656–661.
60. Slatopolsky E, Rutherford WE, Hruska K, Martin K, Klahr S 1978 How important is phosphate in the pathogenesis of renal osteodystrophy? Arch Intern Med **138**:848–852.
61. Parfitt AM 1969 Soft-tissue calcification in uremia. Arch Intern Med **124**:544–556.
62. Verberckmoes R, Bouillon R, Krempien B 1975 Disappearance of vascular calcification during treatment of renal osteodystrophy—two patients treated with high doses of vitamin D and aluminium hydroxide. Ann Intern Med **82**:529–533.
63. Andress DL, Maloney NA, Coburn JW, Endres DB, Sherrard DJ 1987 Osteomalacia and aplastic bone disease in aluminum-related osteodystrophy. J Clin Endocrinol Metab **65**:11–16.
64. Kriegshauser JS, Swee RG, McCarthy JT, Hauser MF 1987 Aluminium toxicity in patients undergoing dialysis: Radiographic findings and prediction of bone biopsy results. Radiology **164**:399–403.
65. Ward MK, Feest TG, Ellis HA, Parkinson IS, Kerr DNS 1978 Osteomalacic dialysis osteodystrophy: Evidence for a water-borne aetiological agent, probably aluminum. Lancet **4**:841–845.
66. Smith GD, Winney RJ, McLean A, Robson JS 1987 Aluminum-related osteomalacia: Response to reverse osmosis water treatment. Kidney Int **32**:96.
67. Sundaram M, Dessner D, Ballal S 1991 Solitary spontaneous cervical and large bone fracture in aluminum osteodystrophy. Skel Radiol **20**:91–94.
68. Schulman JD, Schneider JA 1976 Cystinosis and the Fanconi syndrome. Pediatr Clin North Am **23**:779–793.
69. Ryan EA, Reiss E 1984 Oncogenous osteomalacia: Review of the world literature of 42 cases. Am J Med **77**:501–512.
70. Clarke BL, Wynne AG, Wilson DM, Fitzpatrick LA 1995 Osteomalacia associated with adult Fanconi's syndrome: Clinical and diagnostic features. Clin Endocrinol **43**:479–490.
71. Hanna JD, Niimi K, Chan JCM 1991 X-linked hypophosphatemia. Genetic and clinical correlates. Am J Dis Child **145**:865–870.
72. Walton J 1976 Familial hypophosphatemic rickets: A delineation of its subdivisions and pathogenesis. Clin Pediatr **15**:1007–1012.
73. Weisman Y, Hochberg Z 1994 Genetic rickets and osteomalacia. Curr Ther Endocrinol Metab **5**:492–495.
74. Milgram JW, Compere CL 1981 Hypophosphatemic vitamin D refractory osteomalacia with bilateral femoral pseudofractures. Clin Orthop **160**:78–85.
75. Rivkees SA, el-Hajj-Fuleihan G, Brown EM, Crawford JD 1992 Tertiary hyperparathyroidism during high phosphate therapy of familial hypophosphatemic rickets. J Clin Endocrinol Metab **76**:1514–1518.
76. Steinbach HL, Kolb FO, Crane JT 1959 Unusual roentgen manifestations of osteomalacia. A J Roentgenol **82**:875–886.
77. O'Malley SP, Adams JE, Davies M, Ramsden RT 1988 The petrous temporal bone and deafness in X-linked hypophosphatemic osteomalacia. Clin Radiol **39**:528–530.
78. McAlister WH, Kim GS, Whyte MP 1987 Tibial bowing exacerbated by partial premature epiphyseal closure in sex-linked hypophosphatemic rickets. Radiology **162**:461–463.
79. Burnstein MI, Lawson JP, Kottamasu SR, Ellis BI, Micho J 1989 The enthesopathic changes of hypophosphatemic osteomalacia in adults: Radiologic findings. Am J Roentgenol **153**:785–790.
80. Polisson RP, Martinez S, Khoury M, Harrell RM, Lyles KW, Friedman N, Harrelson JM, Reisner E, Drezner MK 1985 Calcification of entheses associated with X-linked hypophosphatemic osteomalacia. N Engl J Med **313**:1–6.
81. Cartwright DW, Latham SC, Masel JP, Yelland JDN 1979 Spinal canal stenosis in adults with hypophosphatemic vitamin D-resistant rickets. Aust New Zealand J Med **9**:705–708.
82. Adams JE, Davies M 1986 Intra-spinal new bone formation and spinal cord compression in familial hypophosphataemic vitamin D resistant osteomalacia. Q J Med **61**:1117–1129.
83. Hardy DC, Murphy WA, Siegel BA, Reid IR, Whyte MP 1989 X-linked hypophosphatemia in adults: Prevalence of skeletal radiographic and scintigraphic features. Radiology **171**:403–414.
84. Econs MJ, Samsa GP, Monger M, Drezner ML, Fuessner JR 1994 X-linked hypophosphatemic rickets: A disease often unknown to affected patients. Bone Miner **24**:17–24.
85. Prader A, Illig R, Uehlinger RE, 1959 Rachitis infolge Knochentumors. Helv Paediatr Acta **14**:554–565.
86. Renton P, Shaw DG 1976 Hypophosphatemic osteomalacia secondary to vascular tumors of bone and soft tissue. Skeletal Radiol **1**:21–24.
87. Taylor HC, Fallon MD, Velasco ME 1984 Oncogenic osteomalacia and inappropriate antidiuretic hormone secretion due to oat-cell carcinoma. Ann Intern Med **101**:786–788.
88. Linovitz RJ, Resnick D, Keissling P, Kondon JJ, Sehler B, Nejdl RJ, Rowe JH, Deftos LJ 1976 Tumor-induced osteomalacia and rickets: A surgically curable syndrome, report of two cases. J Bone Joint Surg (Am) **58**:419–423.
89. Pollack JA, Schiller AL, Crawford JD 1973: Rickets and myopa-

thy cured by removal of a nonossifying fibroma of bone. Pediatrics **52**:363–371.

90. Leicht E, Kramann B, Seitz G, Trentz O, Remberger K 1993 Oncogenic osteomalacia: Imaging studies. Bildgebung **60**:13–17.
91. Anning ST, Dawson J, Dolby DE, Ingram JT 1948 The toxic effects of calciferol. Q J Med **17**:203–228.
92. Harris PWR 1969 An unusual case of calcinosis due to vitamin D intoxication. Guy's Hospital Rev **118**:553–541.
93. Irnell L 1969 Metastatic calcification of soft tissue on overdosage of vitamin D. Acta Med Scand **185**:147–152.
94. Alazraki N 1995 Radionuclide Techniques. In: Resnick, D. (Ed) Diagnosis of Bone and Joint Disorders. Saunders, Philadelphia, Pennsylvania; pp 430–474.
95. Murray IPC 1994 Nuclear medicine in disorders of bone and joints: Growth and Metabolic Disorders. In: Murray IPC and Ell PJ (eds). Nuclear Medicine in Clinical Diagnosis and Treatment Churchill Livingstone, London, 1035–1040.
96. Lee HK, Sung WW, Solodnik P, Shimshi M 1995 Bone scan in tumor-induced osteomalacia. J Nucl Med **36**:247–249.
97. Rosenthall L, Kaye M 1976 Observations in the mechanism of ^{99m}Tc-labelled phosphate complex uptake in metabolic bone disease. Semin Nucl Med **6**:59–67.
98. Wu YW, Seto H, Shimizu M, Kageyama M, Watanabe N, Kakishita M 1995 Postgastrectomy osteomalacia with pseudofractures assessed by repeated bone scintigraphy. Ann Nucl Med **9**:29–32.
99. Adams JE 1992 Osteoporosis and bone mineral densitometry. Curr Opin Radiol **4**:11–19.
100. Adams JE 1995 Quantitative measurements in osteoporosis. In: Tovey FI and Stamp TCB (eds). The measurement of metabolic bone disease. Parthenon Publishing Group, pp 107–142.
101. Wishart J, Horowitz M, Need A, Nordin BE 1990 Relationship between forearm and vertebral mineral density in postmenopausal women with primary hyperparathyroidism. Arch Intern Med **150**:1329–1331.

SECTION VI

Disorders of the Vitamin D Endocrine System

Chapter 41

Vitamin D and the Pathogenesis of Rickets and Osteomalacia

A. Michael Parfitt Division of Endocrinology and Center for Osteoporosis and Metabolic Bone Disease, University of Arkansas for Medical Sciences, Little Rock, Arkansas

I. INTRODUCTION

The vitamin D field has become so diverse and so complex that many forget how it all started—it was the study of rickets that led to the discovery of vitamin D. Despite the multiplicity of effects on nontraditional target tissues, the principal function of vitamin D and its derivatives, in humans and most other mammals, is still to facilitate the processes and mechanisms that are necessary to prevent rickets and its adult counterpart osteomalacia. These diseases are both consequences of defective mineralization, but within different tissues; the mineralization of growth plate cartilage and of bone have many features in common, but there are also important differences. Vitamin D deficiency may be broadly classified as extrinsic, due to some combination of nutritional deficiency and inadequate exposure to sunlight, and intrinsic, due to some combination of impaired absorption and accelerated catabolism of vitamin D metabolites. The relative importance of these mechanisms may be different in different parts of the world, and different in children and adults. In both rickets and osteomalacia, there may be hypophosphatemia, hypocalcemia, and secondary hyperparathyroidism, but their temporal relationships to one another and to the events in bone may be different. A major unsolved problem is whether changes in the composition of the blood are sufficient to account for the effects of vitamin D deficiency on bone, or whether one or more of the metabolites of vitamin D has actions on skeletal cells that promote mineral deposition.

II. MECHANISMS OF MINERALIZATION

The process whereby ions in solution are transformed into a solid phase falls within the domain of physical chemistry, but skeletal mineralization is also a biological process that is controlled with regard to its location, timing, rate, and relationship to cells and to extracellular connective tissue matrices. A comprehensive theory of mineralization must be consistent with the laws of chemistry and physics but must also account for its morphological features. Disregard of these features led early students of bone, such as Franklin McLean, to believe that the matrix became mineralized as soon as it was formed, and that the presence of any unmineralized matrix was pathological [1]. This belief matched the notion that biological mineralization was nothing more than the precipitation, within the appropriate matrix, of crystals from a supersaturated solution, and that only the composition of the solution determined whether mineralization occurred [2]. The invariable existence of a significant amount of unmineralized bone matrix, or osteoid, in mammalian bone was first demonstrated by Lacroix and students in dogs and cats [3], and soon after confirmed in human subjects by Frost and Villanueva [4].

A. The Contexts for Mineralization—Bone Growth, Modeling, and Remodeling

The development and growth of the bones and the processes of intramembranous and endochondral ossification are described in Chapters 18 and 26, but some aspects especially pertinent to the understanding of rickets are summarized here. Modeling and remodeling of bone both involve formation by osteoblasts and resorption by osteoclasts, but the spatial and temporal organization of these processes is different [5,6]. In modeling, bone is laid down at one location and after a few weeks or months is removed at another, with the location being defined in relation to the changing shape and structure of the bone (relative location), not in relation to a fixed reference point (absolute location). As bones grow in length, bone formed at the junction between the growth plate and the metaphysis is resorbed at the junction between the metaphysis and the diaphysis. As bones grow in width, bone formed beneath the periosteum is resorbed at the endosteum. At these various relative locations, resorption and formation continue without interruption for extended periods. In remodeling, bone is first resorbed and after a few weeks or months is replaced at the same absolute location. Unlike modeling, remodeling is a cyclical process. At each surface location, resorption and formation continue only for a relatively brief time, interrupted by prolonged periods of quiescence.

Between the epiphysis and metaphysis of a growing long bone lies the cartilaginous growth plate, which in the proximal tibia of a young rat is about 600 μm thick, consisting of columns of chondrocytes, embedded in a honeycomb of matrix [7,8]. The sequence of layers known as resting, proliferative, and hypertrophic zones, and the zones of provisional calcification and vascular invasion, represent a sequence of events that occur in the same absolute location, while the epiphysis retreats and the metaphysis approaches. A typical chondrocyte arises from division of a stem cell in the resting zone, divides once or twice more [9], and increases in volume about 10-fold, all the while synthesizing new matrix [8]. After about 40 hr, corresponding to a growth rate of about 350 μm/day, the longitudinal cartilaginous septa become mineralized. Contrary to previous belief, this requires that the chondrocytes do not die, but remain visible until mineralization in their vicinity is complete [8,10]. Ingrowth of capillaries brings macrophages and chondroclasts from the circulation, which resorb the transverse unmineralized septa and about two-thirds of the mineralized septa, respectively, and osteoblasts from the adjacent bone marrow, which deposit loosely textured bone on the surface of the remaining one-third of the mineralized cartilage septa, to form the primary spongiosa [7,8]. Continued minimodeling [11] replaces the cartilage cores with bone to form the secondary spongiosa, and the trabeculae are progressively resorbed so that after a few weeks the metaphysis has been replaced by diaphyseal marrow [12]. As growth slows, the life span of the secondary spongiosa continues to lengthen, and when longitudinal growth ceases, the most recently formed spongiosa becomes the permanent metaphysis.

Bone remodeling is a replacement mechanism that is a minor component of turnover in the growing skeleton but the major component in the adult skeleton. For reasons that are poorly understood, once bone has reached a certain age, it becomes less able to carry out its functions, whether mechanical or metabolic, and must be renewed [13]. The instrument of bone remodeling is the basic multicellular unit (BMU), a unique temporary anatomic structure consisting of a team of osteoclasts in front, a team of osteoclasts behind, and associated blood vessels, nerves, and connective tissue, with a life span of approximately 6 months [14]. Each BMU originates at a particular place at a particular time and travels toward its target, either through the bone (osteonal remodeling) or across the surface of bone (hemiosteonal remodeling), continues beyond its target for a variable distance, and terminates when its supply of precursor cells is interrupted. During its progression, each BMU creates and leaves behind successive cycles of resorption followed by formation, with each cycle being slightly out of step with the one before and representing an individual remodeling transaction, which takes about 4 months to complete. The number of new cycles initiated per unit time (activation frequency) is the product of the number of new BMUs originated, which is equivalent to the number of new remodeling projects, and the average distance traveled by each BMU [13]. The time interval between successive cycles of remodeling at the same location is about 3 years in axial cancellous bone, and 10 to 20 years at sites where bone turnover is lower. The end result of the work carried out by the BMU is the formation of a new bone structural unit (BSU), either an osteon in cortical bone or a hemiosteon in cancellous bone.

B. Morphological and Biochemical Aspects of Mineralization

Microscopic examination of undecalcified sections of bone obtained after double tetracycline labeling (Chapter 37) allows the process of mineralization to be observed *in situ,* with preservation of its spatial relationships to the bone and the cells and introduction of the dimension of time [15,16]. Tetracycline labeling has been applied much less frequently to the study of mineralization in cartilage than in bone, because the rate of advance of the mineralization front and the consequent distance between the two labels is driven mainly by the rate of longitudinal growth, which reflects the rate of chondrocyte proliferation [17]. However, two important features apply to mineralization in both tissues: spatial localization and a measurable time delay between the synthesis of matrix and the deposition of mineral within it. In cartilage, the label is an aggregate of discrete patches, each corresponding to a single longitudinal septum, that form a band about 50 μm in width, extending all the way across the growth plate [18], and the average time delay is about 24 hr. In bone, the label is continuous, more sharply demarcated, and only 2 to 3 μm in width, and the average time delay is about 2 weeks. In cartilage the delay may reflect changes in gene expression in the chondrocytes [19], but in bone the delay reflects the need for extracellular changes in the matrix, collectively referred to as maturation, to occur before mineralization can begin. These changes include completion of cross-linking between collagen fibrils [20,21] and the development of precise orientation, conformation, and aggregation of a variety of noncollagenous proteins and proteoglycans [22,23].

Bone mineral consists of Ca^{2+}, PO_4^{3-}, OH^-, and CO_3^{2-} ions, arranged in space in accordance with the crystal lattice structure of hydroxyapatite [24–27]. The composition is indeterminate because some of the constituent ions can be replaced by other ions of similar radius, and at some lattice points calcium ions can be missing altogether [27]. The mineralizing potential of extracellular fluid (ECF) depends on the free ionic activity, or effective concentration (denoted *a*), of Ca^{2+}, HPO_4^{2-}, and H^+ ions. The activity coefficients relating effective to actual ionic concentrations depend mainly on pH, temperature, and total ionic strength, which are all fairly constant in ECF, but the coefficients are lower and more variable for divalent than for univalent ions [24]. For both Ca and P, ionic concentrations differ from the total concentrations normally measured because of protein binding and ion complexing, which are also affected by pH. The often calculated total plasma calcium $\times$ phosphate product, although meaningless in terms of physical chemistry, bears a rough empirical relationship to the true thermodynamic activity product $[aCa^{2+}]\ [aHPO_4^{2-}]$.

Mineralization is a phase transformation, not a chemical reaction [26], but it is more likely that the complex structural order of hydroxyapatite is attained in steps rather than all at once [25]. At sites of mineralization, the successive addition of Ca^{2+} and HPO_4^{2-} ions present in ECF, and simultaneous removal of protons, generates a series of compounds beginning with secondary calcium phosphate or brushite ($CaHPO_4 \cdot 2H_2O$), the first solid phase to be formed, and ending with hydroxyapatite $[Ca_{10}(PO_4)_6(OH)_2]$. The relevant activity products in ECF are in the region of metastability with respect to bone mineral, being undersaturated with respect to brushite but supersaturated with respect to hydroxyapatite [25]. Mechanisms to accomplish initial mineral deposition include concentration gradients between mineralizing and nonmineralizing sites maintained by cells and by the ion binding and releasing properties of a variety of macromolecules synthesized by cells [23–25], sequestration and subsequent release of calcium by mitochondria [28], and heterogeneous nucleation by outside agents or substances [26]. Alkaline phosphatase is essential for normal mineralization [29], but its function remains unknown. Mechanisms to restrain the growth of hydroxyapatite crystals include the precise spatial relationships between mineral and matrix [23], the presence at critical locations of chelators of calcium [23] and inhibitors of mineralization such as pyrophosphate [30] and albumin [31], and the cellular and biochemical characteristics of the quiescent bone surface, which is the site of reversible mineral exchange with systemic ECF [27].

Within this general framework, two types of mineralization can be recognized [32]. In growth plate cartilage and woven bone, which are temporary structures destined soon to be removed, the matrix is loosely textured and the collagen fibrils are small, immature, and disordered. Mineral is deposited in the form of approximately spherical clusters of randomly oriented crystals of varying size. The clusters, termed calcospherulites, are spatially associated with matrix vesicles, which are small membrane-bound particles that are derived by an unknown mechanism from chondrocytes. The vesicles are abundant and appear to be the only structures available for nucleation [8]. Additional, more active roles in promoting mineralization that have been proposed [8,33] must be reconciled with their distribution, with highest density in the resting and hypertrophic zones and lowest density in the proliferative and calcifying zones [34]. In lamellar bone, which is invariably formed in apposition to an existing surface, the matrix is compact in texture, matrix vesicles are infrequent or absent, and the collagen fibrils are long and highly ordered. The mineral crystals are aligned with their long axis parallel to the collagen fibrils and are initially deposited within the hole zones by heterogeneous nucleation, but longer and wider crystals are subsequently formed on and between the fibrils [26,32]. The differences between these two types of mineralization explain why rickets and osteomalacia can under some circumstances vary independently in their severity and response to treatment [35].

III. PROCESSES LEADING TO ACCUMULATION OF UNMINERALIZED TISSUE

A. Growth Plate Cartilage, and Osteoid in the Growing Skeleton

The width of the epiphyseal growth plate depends on the rate of longitudinal growth, which is determined by the rate of new chondrocyte production [9] and the life span for completion of maturation prior to initial calcification. For example, in 5-week-old rats a width of about 600 μm corresponds to a growth rate of 330 μm/day and a life span of about 1.8 days [8], and in 10-week-old rats a width of 350 μm corresponds to a growth rate of 180 μm/day and a life span of about 1.9 days [36]. These and other data indicate that although longitudinal growth slows progressively with increasing age, growth plate life span remains approximately constant. In experimental rickets in rats, even though longitudinal growth is reduced 3-fold, growth plate width increases about 4-fold in 6 weeks of by about 27 μm/day because of an increase in life span of at least 12-fold [36]; in severe rickets, in the absence of treatment, growth plate life span is limited only by the age of the animal. Evidently, the characteristic increase in growth plate occurs despite a reduction in the rate of growth and is due entirely to a profound delay in mineralization, the pathogenesis of which is discussed later. There is also structural disorganization of the metaphysis, partly due to mechanical effects [37] and partly to a profoundly altered pattern of vascular invasion [38].

During the transformation of calcified cartilage to primary and then secondary spongiosa, there is extensive deposition of osteoid. The kinetics of its production, life span, and mineralization have never been studied by tetracycline labeling, but accumulation of osteoid contributes to the microscopic characteristics of the rachitic metaphysis. A possible source of confusion must be addressed at this point. The term "rickets" is commonly applied to the totality of skeletal abnormalities associated with defective mineralization in the growing skeleton, but it is more accurate to restrict the term to changes in the growth plate and adjacent metaphysis. In the vertebral bodies and ilium, cancellous bone tissue occupies all the available space enclosed by the cortices and does not undergo removal, but is remodeled in a manner similar to the adult skeleton. During both intramembranous ossification [39] and long bone growth, osteoid is formed beneath the periosteum and mineralized bone is removed on the inner surface. When mineralization is defective, the accumulation of osteoid at sites other than the growing metaphysis should be referred to as osteomalacia, not as rickets. Thus, impaired mineralization leads to both rickets and osteomalacia in the growing skeleton but only to osteomalacia in the mature skeleton.

The kinetics of osteoid production and mineralization in the growing skeleton have been studied most thoroughly in the rat tibia; several important observations were made [40]. First, although osteoid seams are generally thinner in rats than in larger animals, they are invariably present at sites of bone formation; because of the temporal separation between matrix apposition and mineralization, there is also spatial separation. Second, in 3-week-old rats, osteoid is found underneath the entire circumference of the periosteum; because its extent cannot change except as a result of growth, a significant increase in osteoid accumulation can occur only if the thickness of the seam increases. This contrasts with the mature skeleton, in which osteoid almost always increases in surface extent before it increases in thickness. Third, analogous to the regulation of growth plate width, osteoid thickness depends on the rate of matrix apposition, which corresponds to the rate of longitudinal growth, and on the delay before the onset of mineralization that is imposed by osteoid maturation, known as the mineralization lag time (Mlt), which corresponds to the growth plate life span. Like growth plate width, osteoid thickness declines with increasing age because of a decline in matrix apposition rate with a relatively constant lag time.

B. Life History of Individual Osteoid Seams in the Adult Skeleton

The formation of each new BSU begins at the cement surface, a thin layer of lowly mineralized collagen-poor but glycoprotein-rich connective tissue [41] that is laid down on the floor of the resorption cavity at the end of the reversal phase of each remodeling cycle, represented in two-dimensional histological sections by the cement line, which remains in the same location, separating new bone from old. The boundary between mineralized and unmineralized bone, referred to as the osteoid–bone interface, is the location of the mineralization front, which normally moves away from the cement line during bone formation. Its rate of advance must be distinguished from the rate at which mineralization proceeds after it is initiated [42]. In an individual moiety of bone matrix, mineral accumulation as a function of time is a continuous process that is conveniently subdivided into two stages. There is an early rapid increase to about 75–80% of maximum within the first few days, referred to as primary mineralization. It involves multiplication in the number of crystals, occurs close to the osteoblast, and may be influenced by its function. A much slower increase to about 95% of maximum or more, over many months or even years, is referred to as secondary mineralization. It involves slow growth in the size of crystals, with displacement of water, occurs remote in both time and space from the osteoblast, and is presumably governed entirely by physicochemical factors [42].

A team of osteoblasts assembles on the cement surface and begins to deposit a layer of bone matrix referred to as an osteoid seam, which in standard histological sections appears in cortical bone appear as a ring, and in cancellous bone as a crescent tapering at each end. Each seam has a measurable life span, during which characteristic changes occur in the morphological features and function of the osteoblasts, and in the thickness of the seam (Fig. 1). Matrix apposition is most rapid (2.0 to 3.0 μm/day) at the outset, and the seam reaches a maximum thickness of approximately 15 to 20 μm after about 10–15 days, just before mineralization begins. Mineral apposition is also most rapid at the outset (1.0 to 1.5 μm/day), and thereafter is always more rapid than matrix apposition so that osteoid seam thickness declines. Both matrix and mineral apposition progressively slow with time, as the osteoblasts become flatter and more extended in shape. About 80 days after the onset of mineralization when the bone surface has returned to its previous location about 50 μm from the cement line, matrix synthesis stops. The osteoid seam thickness has by now fallen to about 6 μm, and for a further 30 days mineral apposition continues at a progressively declining rate that is too slow for tetracycline fixation to occur [43]. Eventually, the osteoid seam disappears, because all the new matrix has become mineralized. The osteoblasts have now completed their histological transformation into lining cells, and construction of the new BSU at that cross-sectional location is finished.

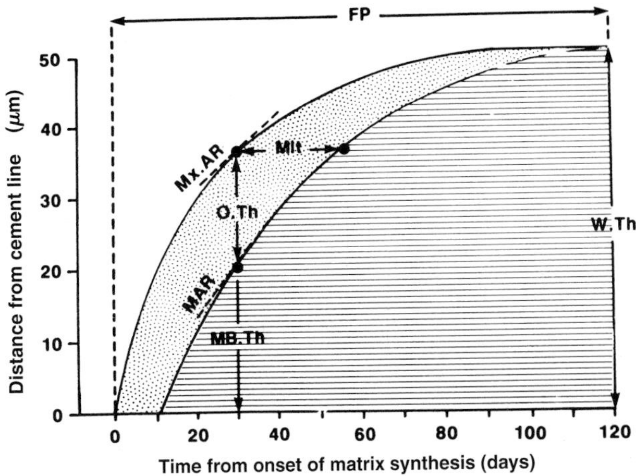

FIGURE 1 Model of bone formation, with growth curves for matrix apposition above and mineral apposition below, showing distances from the cement line as functions of time at a single cross-sectional location of a representative basic metabolic unit (BMU). At any distance from the cement line, the horizontal distance between the lines is the instantaneous mineralization lag time (Mlt) at that distance. At any time, the vertical distance between the lines is the instantaneous osteoid thickness (O.Th) at that time. At any point the slopes of the lines (tangents) represent instantaneous apposition rates for matrix (Mx.AR) or mineral (MAR). For example, at $t = 30$ days, the instantaneous values are 20 μm for mineralized bone thickness (MB.Th), 16 μm for osteoid thickness, 0.5 μm/day for matrix osteoid apposition rate, and 0.8 μm/day for mineral apposition rate; the matrix deposited at that time will have a mineralization lag time of 26 days. Formation period (FP) is counted from the onset of matrix synthesis to the completion of mineralization, and in this example is 120 days, at which time completed wall thickness (W.Th) equals 50 μm. It is evident that the total area between the curves is given by FP × mean O.Th and by W.Th × mean Mlt, so that these expressions are equal. Furthermore, it follows that O.Th = Mlt × Mx.AR. Reprinted from Parfitt [15], in *Chemistry and Biology of Mineralized Tissues,* 1992, pp. 465–474, with kind permission from Elsevier Science.

A key quantity in understanding the mechanisms of osteoid accumulation and the pathogenesis of osteomalacia is the mineralization lag time (Mlt); this was defined earlier for the rat, but its method of calculation and significance are somewhat different in the adult human skeleton. In the rat, periosteal bone formation is continuous, the entire bone-forming surface is labeled with tetracycline, there is no need to distinguish between instantaneous and mean values, and the best estimate of the matrix apposition rate is the mineral apposition rate (MAR) so that Mlt = O.Th/MAR (where O.Th is osteoid thickness). In humans, bone formation is cyclical, tetracycline fixation does not occur during terminal mineralization, and it is important to distinguish between instantaneous and mean values. Osteoid thickness at any distance from the cement line is the product of the instantaneous matrix apposition rate (Mx.AR) and the instantaneous Mlt (Fig. 1). Instantaneous values are provided only by complete remodeling sequence reconstruction [44], and in practice only mean values are obtained. The best estimate of the mean matrix apposition rate is the mineral apposition rate averaged over the entire osteoid surface, referred to as the adjusted apposition rate (Aj.AR), so that mean Mlt = mean O.Th/Aj.AR [42].

In the rat Mlt is identical with the osteoid maturation time (Omt), but in humans this may be true only for the initial Mlt, which is usually about 10 days (Fig. 1). The cause of the subsequent increase in Mlt to about 30 days, which has no counterpart in the rat, is unknown. One possibility is that the time required for matrix maturation increases with the age of the osteoblast, and changes in matrix apposition rate and lag time together determine the progress of mineralization (Fig. 2A). In this case, lag time would be an independent variable that remained identical with maturation time, and the changes in mineral apposition rate would follow automatically. Alternatively, the rates of matrix and mineral apposition could be separately and independently regulated as functions of osteoblast age (Fig. 2B). For example, there could be a decline in the supply of mineral, since the net inward calcium flux characteristic of osteoblasts must at some point change to the outward calcium gradient without net flux characteristic of lining cells [27]. In this case, lag time would not be an independent variable but would progressively exceed maturation time. Fortunately, this uncertainty does not detract from the usefulness of Mlt in the understanding of histomorphometric data and of the mechanisms of osteoid accumulation [15,40].

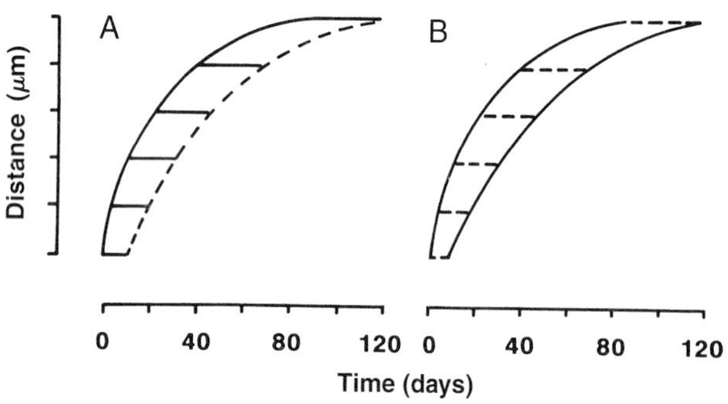

FIGURE 2 Two models of the relationship between matrix apposition, mineral apposition, and mineralization lag time. (A) Matrix apposition rate and mineralization lag time are separately and independently regulated as functions of osteoblast age, and the rate of mineral apposition changes as an automatic consequence. (B) Rates of matrix and mineral apposition are separately and independently regulated as functions of time, and the lag time changes as an automatic consequence. In both cases, the genuine independent variables are depicted by solid lines and the automatically determined variables by dashed lines. Reprinted from Parfitt [15], in *Chemistry and Biology of Mineralized Tissues,* 1992, pp. 465–474, with kind permission from Elsevier Science.

C. Osteoid Indices and the Recognition of Impaired Mineralization

Osteoid accumulation is assessed by three independent measurements (Chapter 37) that are related as follows [45] and defined below and in Fig. 1:

$$\text{OV/BV}\ (\%) = \text{O.Th}\ (\text{mm}) \times \text{OS/BS}\ (\%) \times \text{BS/BV}\ (\text{mm}^2/\text{mm}^3). \quad (1)$$

A fall in trabecular thickness, as occurs to a modest extent in aging and in osteoporosis, will increase BS/BV and to increase OV/BV even if surface and width are unchanged; for the most accurate interpretation, OV/BV should be corrected to the expected trabecular thickness for age and sex. Each of the three indices is related differently to the underlying kinetic determinants. Osteoid thickness has already been discussed and is given by

$$\text{O.Th}\ (\mu\text{m}) = \text{Aj.AR}\ (\mu\text{m/day}) \times \text{Mlt}\ (\text{days}). \quad (2)$$

Osteoid surface per unit of bone surface (OS/BS) is determined entirely by the mean osteoid seam life span or formation period (FP) and by the average frequency with which new osteoid appears at any point on the bone surface, which in the steady state is the same as the frequency of remodeling activation (Ac.f):

$$\text{OS/BS}\ (\%) = \text{FP (years)} \times \text{Ac.f}\ (\text{year}^{-1}) \times 100. \quad (3)$$

Osteoid volume is determined entirely by the fractional rate of bone turnover, which is the same as the volume-based bone formation rate (BFR/BV), and by the mean life span of an individual moiety of osteoid, which is the same as the mineralization lag time:

$$\text{OV/BV}\ (\%) = \text{BFR/BV}\ (\%/\text{year}) \times \text{Mlt (years)}. \quad (4)$$

Because in the steady state bone turnover is determined entirely by the frequency of remodeling activation and the surface to volume ratio, and because FP is inversely proportional to Aj.AR [42], each of the three static indices of osteoid accumulation is determined by a different pair of the same three kinetic indices [42]. Notably, osteoid volume is independent of matrix (or mineral) apposition rate, which in the steady state affects surface and thickness equally in opposite directions.

Although a reduction in mineral apposition rate is frequently taken to indicate defective mineralization, it is evident from Fig. 1 that matrix and mineral apposition are closely coupled and that the mean mineral apposition rate can never exceed the mean matrix apposition rate. Consequently, a reduction in the mean rate of matrix apposition inevitably leads to, and is much the most common cause of, a reduction in the mean rate of mineral apposition. Both in normal subjects and in patients with any metabolic bone disease except osteomalacia, there is a significant positive correlation between mean osteoid thickness and mean adjusted apposition rate, with broadly similar slopes (b) and intercepts (a) of the regression lines [45]. Although such a relationship is to be expected, it has an unanticipated consequence for the interpretation of the mineralization lag time, as we can write

$$\text{O.Th} = b(\text{Aj.AR}) + a. \tag{5}$$

If this is combined with Eq. (2), we obtain

$$\text{Mlt} = b + a/\text{Aj.AR} \tag{6}$$

Because of this relationship, which defines a rectangular hyperbola, when the matrix apposition rate falls, the mineralization lag time increases [16]. Another way of arriving at the same conclusion is to consider the effect of prolongation of FP, which is also an inevitable consequence of a reduction in matrix apposition rate. From Fig. 1 it is clear that

$$\text{FP} \times \text{mean O.Th} = \text{W.Th} \times \text{mean Mlt}. \tag{7}$$

Because O.Th has a minimum value [the intercept in Eq. (5)] and W.Th (the mean thickness of a completed BSU) is effectively constant in the short term, an increase in FP must be accompanied by an increase in Mlt. It follows from this reasoning that neither a reduction in Aj.AR nor an increase in Mlt indicate that mineralization is defective unless they are accompanied by an increase in O.Th.

IV. EVOLUTION OF VITAMIN D RELATED BONE DISEASE

A. Histological Evolution, and the Kinetic Definition of Osteomalacia

In most patients, osteomalacia is preceded for many years by clinically silent secondary hyperparathyroidism that accelerates the irreversible age-related loss of cortical bone [46]. Exposition of this concept is aided by using the term "hypovitaminosis D osteopathy" (HVO) to encompass the totality of osseous complications of deficiency of altered metabolism of vitamin D [16]. There is usually no relationship between osteoid thickness and osteoid surface, but in HVO there is a hyperbolic relationship between these variables [16,45] (Fig. 3A). This indicates that osteoid surface increases first and that osteoid thickness increases only slightly until OS/BS exceeds 70%, after which further increases in osteoid volume are due mainly to increasing thickness. In the same patients, osteoid thickness shows a more complex relationship to adjusted apposition rate (Aj.AR; Fig. 3B). When Aj.AR is above 0.1 μm/day, there is the usual positive relationship between these variables; below 0.1 μm/day further decrements in Aj.AR are associated not with a fall in osteoid thickness as in all other situations, but with a progressive increase, limited only by the normal thickness of new matrix or W.Th [47]. This reversal is the cardinal kinetic characteristic of defective mineralization; a similar hyperbolic relationship is found between osteoid thickness and the fraction of osteoid surface undergoing mineralization [48].

On the basis of these relationships, the author defines osteomalacia by a combination of mean mineralization lag time more than 100 days and mean osteoid thickness above the upper 95% confidence limit predicted by the

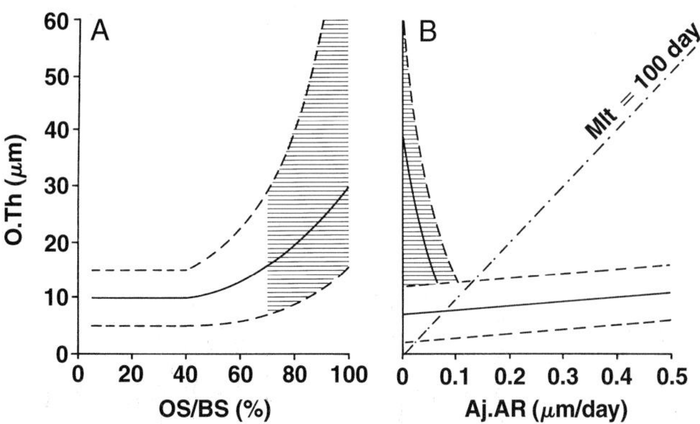

FIGURE 3 Relationship of osteoid thickness (O.Th) to osteoid surface (OS/BS) (A) and adjusted apposition rate (Aj.AR) (B). In normal subjects and in patients with osteoporosis there is no relationship between osteoid thickness and surface (A), but there is a significant positive relationship between osteoid thickness and adjusted apposition rate (B). During the development of osteomalacia there is a direct hyperbolic relationship between osteoid thickness and surface (A) and an inverse hyperbolic relationship between osteoid thickness and adjusted apposition rate (B); the usual ranges of values in established osteomalacia are shaded. The oblique line through the origin in (B) corresponds to a mineralization lag time of 100 days. Reprinted with permission from Parfitt AM. The physiologic and pathogenetic significance of bone histomorphometric data. In: Coe FL, Favus MJ (eds): Disorders of Bone and Mineral Metabolism. Raven, New York. 1992.

regression of osteoid thickness on Aj.AR in normal and osteoporotic subjects (Fig. 3B), or, more simply, above an absolute value of 12.5 μm (corrected for section obliquity). Patients with HVO who do not meet these criteria have increased volume and surface but not thickness of osteoid, increased bone formation rate, the normal positive relationship between O.Th and Aj.AR, and increased osteoclast indices [16], resembling in every respect the histological features of primary hyperparathyroidism. This analysis identifies the earliest stage of HVO (HVOi), when osteoid accumulation is due mainly to increased frequency of remodeling activation and bone turnover, before the emergence of a significant mineralization defect, as being due to secondary hyperparathyroidism (Chapter 43). Patients with HVO who meet the criteria for defective mineralization are further subdivided into those who retain some tetracycline double labels (HVOii) and those with no double labels (HVOiii).

A different perspective on the fundamental nature of osteomalacia can be gained from the model of osteoid seam life span (Fig. 4). In every other condition, all matrix formed eventually mineralizes; the slopes of the curves representing matrix and mineral apposition, although initially divergent, eventually converge, and the loop formed by these curves ultimately closes. In contrast, in untreated osteomalacia some matrix remains permanently unmineralized, the slopes of matrix and mineral apposition remain divergent, and the loop never closes. The model also illuminates the difference between HVOii, in which the earliest formed matrix becomes mineralized but the later formed matrix does not, and HVOiii, in which none of the matrix formed becomes mineralized. As all patients with HVOiii at the time of biopsy have likely been through the stage of HVOii, they show a mixture of the two types of osteoid seam depicted in Fig. 4. Both thickness and volume of osteoid are significantly greater in HVOiii than in HVOii [16], but even in the most severe cases individual values for mean osteoid thickness fall within the reference range for mean wall thickness [47].

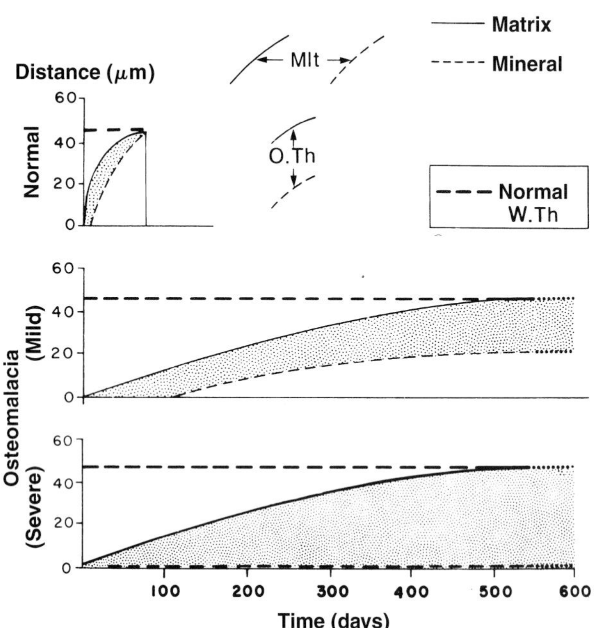

FIGURE 4 Kinetics of matrix and mineral apposition in osteomalacia. Evolution of bone formation at a single cross-sectional location. Each graph is constructed in a manner similar to Fig. 1 but with altered time scale. Mlt, Mineralization lag time; O.Th, osteoid thickness; W.Th, wall thickness. Note that in mild osteomalacia (HVOii) mineralization is delayed in onset, retarded in rate, and premature in termination, whereas in severe osteomalacia (HVOiii) no mineralization occurs at all. Reprinted from Parfitt AM. Bone-forming cells in clinical conditions. In: Hall BK (ed.), *Bone: A Treatise, Vol. 1, The Osteoblast and Osteocyte.* p. 395. Copyright CRC Press, Boca Raton, FL. © 1990.

B. Biochemical Evolution of Rickets and Osteomalacia

The cardinal metabolic consequence of vitamin D deficiency is reduced net intestinal absorption of calcium [16,48] (see also Chapter 16). Fecal calcium excretion is close to and can even exceed dietary intake, but urinary calcium is low and calcium balance rarely more negative than -100 mg/day [49]. There is an equimolar deficit in net absorption of inorganic phosphate, but the relative change is much smaller. According to the usual interpretation, calcium malabsorption leads in sequence to a fall in plasma calcium, secondary hyperparathyroidism, reduced renal tubular reabsorption of phosphate, hypophosphatemia, and reduction in calcium $\times$ phosphate product, which falls even further with the advent of more severe hypocalcemia. Eventually, deposition of mineral in osteoid is impaired because the supply of the relevant ions is reduced, and the alkaline phosphatase activity then rises. This traditional scheme requires considerable modification with regard to the differences related to age of onset, the order in which the changes occur, their pathophysiology, and their diagnostic significance.

Many years ago, three stages were recognized in simple vitamin D deficiency in infants [50]. In stage I there was hypocalcemia and normal plasma phosphate, in stage II plasma calcium rose to normal, plasma phosphate fell below normal, and alkaline phosphatase increased modestly, and in stage III plasma calcium and phosphate were both reduced, with no further change in alkaline phosphatase. However, in current practice there are many exceptions to this sequence [51] (Chapter 42). Early hypocalcemia is rarely observed in older children [51,52] but is quite common during adolescence [53], although very uncommon in adults [54] (Table

TABLE I Biochemical Evolution of Hypovitaminosis D Osteopathy (HVO)[a]

	Normal[b] (n = 23)	HVOi (n = 26)	HVOii (n = 11)	HVOiii (n = 28)
Age (years)	60.3 ± 1.3	57.2 ± 2.1	50.5 ± 6.3	58.1 ± 2.4
Plasma calcidiol (ng/ml)	23.7 ± 3.0	6.0 ± 0.5	6.8 ± 0.9(10)	4.1 ± 0.6(18)*
Plasma calcitriol (pg/ml)	40.8 ± 6.9	46.0 ± 4.7(11)	39.1 ± 3.7(8)	21.7 ± 3.2(10)†
Plasma calcium[c] (mg/dl)	9.64 ± 0.08	9.12 ± 0.11†	7.95 ± 0.39‡	8.02 ± 0.17
Plasma phosphate (mg/dl)	3.47 ± 0.08	3.36 ± 0.11	2.91 ± 0.23*	2.64 ± 0.13
Plasma Ca × P [$(mg/dl)^2$]	33.5 ± 0.7	30.7 ± 1.1*	23.3 ± 1.6‡	21.2 ± 1.2
Alkaline phosphatase (IU)	82.8 ± 3.9	132 ± 7.2‡	201 ± 31.1	284 ± 24.2*
NcAMP[d] (nmol/dl GF)	2.04 ± 0.26	4.01 ± 0.34	6.62 ± 0.79	5.94 ± 0.61

[a] Stages are defined in the text. Number of analyses are shown in parentheses when less than number of subjects. Data are means ± SE. Significance levels are shown for differences in mean values from the column immediately to the left: *, $p < 0.05$; †, $p < 0.01$; ‡, $p < 0.001$. Data reprinted from Parfitt [16].
[b] Volunteers for bone biopsy.
[c] Corrected for albumin.
[d] N cAMP, Nephrogenous cyclic AMP, which is given in units of nanomoles per deciliter of glomerular filtrate.

I). The ability to release calcium from bone may be compromised during periods of rapid growth, but in infants the abnormality is associated with a delayed increase in serum parathyroid hormone (PTH) [55], normal skeletal and renal tubular responsiveness to exogenous PTH [50], and spontaneous improvement; in adolescents and adults there is an appropriate increase in PTH levels but impaired renal tubular as well as skeletal responsiveness to PTH [53,56], and the abnormality persists in the absence of treatment. Further discussion of this acquired form of pseudohypoparathyroidism is beyond the scope of this chapter, except for two points pertinent to subsequent discussion. First, plasma calcium is determined by the homeostatic system at quiescent bone surfaces [27]. This system is independent of remodeling and is regulated jointly by PTH and one or more metabolites of vitamin D [24]. Second, deficiency of vitamin D causes hypocalcemia mainly by loss of its effects on bone.

In most adults with HVOi, the mean plasma calcium is slightly reduced, but the individual values are almost always normal (Table I). PTH secretion is increased as shown both by radioimmunoassay [55,57] and by excretion of nephrogenous cyclic AMP (NcAMP) (Table I). Although mean tubular absorptive maximum for phosphorus divided by the glomerular filtration rate (TmPi/GFR) and plasma phosphate are both slightly reduced, individual values are usually normal. Twenty-four hour urinary calcium excretion and fasting urinary calcium/creatinine are often but not invariably reduced, and a moderate elevation of alkaline phosphatase is the most consistent abnormality. In extrinsic vitamin D depletion the plasma calcidiol level at which abnormal mineral metabolism can first be detected in an individual is usually below 5 ng/ml [48,53], but in subjects with values between 5 and 10 ng/ml there is a slight but statistically significant depression of mean plasma calcium and phosphate and urinary calcium and elevation of PTH [57], and most patients with histologically verified HVOi have calcidiol values in this range (Table I). In intrinsic vitamin D depletion, the complete biochemical, histological, and bone densitometric syndrome of HVOi can occur at plasma calcidiol levels between 10 and 20 ng/ml [54], presumably because there is an independent mechanism for calcium malabsorption and consequent secondary hyperparathyroidism that is unrelated to vitamin D but accounts for the normal mean level of plasma calcitriol.

The progression of HVO through stages ii and iii is similar to the progression of infantile rickets through Stages II and III, but it occurs over a much longer time scale and differs somewhat in detail. In general all the biochemical abnormalities become more severe. Plasma and urinary calcium, TmPi/GFR, and plasma phosphate levels become lower, and PTH, NcAMP, and alkaline phosphatase levels higher (Table I), but there are many individual exceptions. As in rickets, patients with severe hyperparathyroidism may suffer impaired tubular reabsorption of bicarbonate and amino acids as well as phosphate, resembling proximal renal tubular acidosis or the Fanconi syndrome [16,48], except for increased rather than decreased tubular reabsorption of calcium. Hypophosphatemia is adequately explained by increased PTH secretion without the need to postulate an additional effect of vitamin D metabolite deficiency. Indeed, for the same increase in NcAMP, TmPi/GFR is higher in secondary than in primary hyperparathyroidism because of the independent effect of plasma calcium to decrease phosphate reabsorption [54,58]. The mean calcidiol level is not significantly lower in HVOii than in HVOi, but it does fall further in HVOiii (Table I). In contrast, the calcitriol levels can be normal in stage ii and do not become consistently subnormal until stage iii. Others have also found both normal and subnormal calcitriol levels in osteomalacia, although not classifying their cases in the same manner [16].

In summary, in HVOi there are characteristically no symptoms until a fracture occurs, which is why the existence of this intermediary stage was unrecognized for so long. The only biochemical abnormality that would be revealed by routine screening is a raised plasma alkaline phosphatase. Both fasting and 24-hr urinary calcium excretion are usually reduced. Skeletal radiographs are either normal or show only nonspecific osteopenia, but bone densitometry reveals that age-related loss of bone is accelerated, especially appendicular cortical bone but also axial trabecular bone, with a corresponding increase in fracture risk [54]. Plasma calcidiol is usually but not invariably low. There is both biochemical and histological evidence of secondary hyperparathyroidism and of increased bone turnover. Defective mineralization is either absent or no more severe than in primary hyperparathyroidism. Paradoxically, the deficit in forearm bone density is greater in HVOi than in HVOii and HVOiii despite less severe hyperparathyroidism. This can only be explained by slower progression and consequently longer duration of accelerated cortical bone loss and increased fracture risk [16]. It is likely that some patients remain arrested indefinitely at this stage, a few progress to more severe hyperparathyroidism with radiographic osteitis fibrosa, but others eventually develop the complete clinical, biochemical, radiographic, and histological syndrome of osteomalacia. However, it may reasonably be assumed that all patients in stages ii or iii at the time of diagnosis traveled earlier through stage i.

V. ASPECTS OF VITAMIN D METABOLISM RELEVANT TO RICKETS AND OSTEOMALACIA

Vitamin D metabolism can be affected at one of six levels [16]. Identification of the level is important in planning treatment, although the summation of independent factors at several levels may be needed to produce clinical effects, and some diseases affect more than one level. Each level is associated with a characteristic profile of vitamin D metabolite concentrations in blood, but these must be interpreted with caution because changes in vitamin D binding protein (DBP) can alter total concentrations without altering free concentrations [16] (Chapter 7). Body stores of vitamin D, located mainly in fat and to a lesser extent in muscle, are derived either from the photochemical production in skin of cholecalciferol or from dietary intake and intestinal absorption of either chole- or ergocalciferol. Although the former source is more physiological [59], with current lifestyles the latter is equally important. The distinction was made earlier between extrinsic vitamin D depletion, due to some combination of reduced skin synthesis and reduced intake, and intrinsic depletion, due to intestinal malabsorption of vitamin D, augmented by an additional mechanism for increased fecal loss of vitamin D [16].

Two infrequently emphasized features of normal vitamin D metabolism are relevant to the pathogenesis of rickets and osteomalacia: first, its wastefulness in normal circumstances [53] and, second, its susceptibility to disruption by calcium deficiency [60]. Metabolic pathways leading from calciferol to calcitriol have been extensively investigated but are preferentially followed only when body stores are greatly depleted. Normally, about 70% of the daily supply of both calciferol and calcidiol is converted to more polar metabolites of low or absent biological activity that undergo biliary and eventually fecal excretion [53], so that only about 10% of available calciferol is used for calcitriol production. The proportion following the alternate pathways can decrease to very low levels when necessary, but it increases to 90% for calciferol and 99% for calcidiol in vitamin D treated hypoparathyroidism [61]. Despite their quantitative importance in overall vitamin D economy, little is known about either the metabolites formed or their mechanism of production, and even less about how distribution between different pathways is regulated.

Because of the usual wide margin of safety, malabsorption of dietary vitamin D is rarely of sufficient severity to be the sole mechanism responsible for vitamin D depletion. The first additional mechanism to be proposed was interruption of a conservative enterohepatic circulation of calcidiol [16]. This proposal accounted for depletion of vitamin D of dermal as well as dietary origin, but the magnitude of this pathway in human subjects, if it occurs at all, is much too small to fulfill its postulated role [53,62] (Chapter 51). It now seems much more likely that the additional mechanism is accelerated catabolism of calcidiol in the liver initiated by calcium deprivation [63], whether due to dietary deficiency or intestinal malabsorption. It is because of calcium deprivation that plasma calcitriol levels are increased in the earlier stages of intrinsic HVOi [64], as they are also in patients treated with anticonvulsants [53], although this is not evident from Table I, which reports only data from patients who had bone biopsy. Accelerated calcidiol catabolism is mediated by secondary hyperparathyroidism, either as a direct effect of increased circulating PTH [60,63] or as an indirect effect due to increased serum levels of calcitriol [60,65]. This mechanism explains the occurrence of vitamin D deficiency in geographic regions with high sun exposure but low calcium intake [60], and it contributes to vitamin D deficiency in a variety of gastrointestinal disorders (Chapter 51).

The hepatic 25-hydroxylation of calciferol to calcidiol provides the principal transport form of vitamin D and an additional component of body stores, located mainly in muscle [16]. This process is impaired in cirrhosis of the liver, but rarely to a level that causes osteomalacia (unless there is also malabsorption, as in biliary cirrhosis) because the liver has such a large reserve capacity [53]. Significant calcidiol deficiency that is not due to depletion of its precursor is most commonly the result of increased catabolism to biologically inactive metabolites from drug-induced enzyme induction (Chapter 50) or from stimulation of existing enzymes by calcitriol or PTH [65] (Chapter 51). Loss of calcidiol bound to protein (both DBP and albumin) occurs in the nephrotic syndrome and leads to secondary hyperparathyroidism and osteoid accumulation in the absence of impaired renal function [16]. A similar mechanism operates during peritoneal dialysis, and urinary loss of calcidiol is also increased in patients with biliary cirrhosis (Chapter 51).

Calcitriol deficiency with normal body stores of vitamin D is most commonly the result of chronic renal failure (Chapter 52) but can also be due to a genetic defect in renal 1α-hydroxylation, referred to as pseudo vitamin D deficiency rickets (PDDR), hereditary hypocalcemia or vitamin D dependency type I (Chapter 47). Plasma calcitriol levels are reduced by magnesium depletion [16], but osteomalacia as a consequence has not been demonstrated. Calcitriol synthesis is impaired by deficiency of PTH (Chapter 50), but it is doubtful whether this causes osteomalacia, possibly because bone turnover is so low. However, there is one adequately documented case of osteomalacia due to pseudohypoparathyroidism with secondary hyperparathyroidism [66]. Very low plasma calcitriol levels are found during prolonged total parenteral nutrition, but they have not been clearly related to the presence or type of metabolic bone disease [16]. Finally, calcitriol may be ineffective because of one of several kinds of defect in its receptors (VDR), referred to as hereditary vitamin D-resistant rickets (HVDRR) or vitamin D dependency rickets type II (Chapter 48).

VI. VITAMIN D AND THE PATHOGENESIS OF IMPAIRED MINERALIZATION

The role of vitamin D in sustaining normal mineralization has given rise to two related controversies. First, is calcitriol the only metabolite of physiological importance, other than as a precursor [67], or must some other metabolite also be considered [68]? Second, is the action of vitamin D on bone mediated solely by changes in the calcium and phosphate concentrations in ECF [67], or does it influence mineralization more directly [49,53]? In both cases the contestants have often failed to recognize the difference between an essential function that confers an absolute requirement and a contributory function that confers only a relative requirement. Concerning the first controversy, calcitriol alone can correct the clinical, biochemical, radiographic, and histological effects of vitamin D deficiency both in humans [69,70] and in the rat [71,72]. Claims to the contrary [68,73] reflect the inability of intermittent oral administration to sustain an adequate blood level [71]. Although other metabolites appear not to be essential, 24-hydroxylase knock-out mice have defective mineralization of periosteal bone but not of growth plate cartilage, suggesting a role for 24-hydroxycalcidiol [24,25$(OH)_2D_3$] in the former process [74] (Chapter 18).

Concerning the second controversy, there is evidence that the effects of vitamin D deficiency in humans can be corrected by giving enough calcium and phosphate intravenously [75], and that defects in calcitriol receptor binding can also be bypassed by providing sufficient mineral substrate [76,77,77a]. The strength of the evidence is examined later in Section VI,B. In completely vitamin D-deficient rats, maintenance of normal plasma calcium and phosphate levels maintained normal growth, as well as normal growth plate and bone mineralization [78–79]. In clinical studies, the determination that radiographic and histological abnormalities are due to defective mineralization rather than to secondary hyperparathyroidism is not as straightforward as is often assumed, but if the conclusions of these studies are taken at face value [75–77a], then vitamin D is clearly not essential for mineralization. Nevertheless, one or more metabolites of vitamin D could have direct actions on those cells that contribute to the process in normal circumstances [16,53,73].

A. The Effects of Vitamin D Are Not Mediated Solely by Circulating Calcitriol

In patients with histologically verified osteomalacia or with radiographically unambiguous rickets, plasma calcitriol concentrations can be within the appropriate reference ranges [48,80–83] (Table I). The levels are indeed inappropriately low for the degrees of PTH hypersecretion and hypophosphatemia [82], as is indicated by the very high levels attained during the early stages of treatment with vitamin D [70,83,84], but the lack of target cell responses to a concentration of calcitriol that is normally adequate requires explanation. In adults with osteomalacia, both biochemical and histological indices of vitamin D depletion appear to correlate better with either the sum of calcidiol (in ng/ml) and calcitriol (in pg/ml) concentrations, or with calcidiol alone, than with calcitriol alone [48,54]. This suggests that calcidiol, or some other metabolite for which calcidiol is a precursor, such as 24-hydroxycalcidiol, might function as an agonist for calcitriol, or might have its own effects mediated by a different receptor. In infants with untreated rickets, the plasma Ca $\times$ P product correlated with the plasma concentration of calcitriol and not calcidiol, although a higher than normal calcitriol level was needed to maintain a normal product [84]. In this study the plasma level of 24-hydroxycalcidiol was often undetectable; this compound might function in a permissive manner, so that a fall in its concentration below a critical level would increase the need for calcitriol, but without dose-related effects above the critical level.

Higher than normal calcitriol levels could be needed to correct hypocalcemia and to restore normal mineralization when the calcitriol responsive cells are separated from the mineralized bone by a much wider than normal layer of uncalcified osteoid through which the mineral ions must travel [16,53] (Fig. 5). However, no similar reason is evident for the failure of intestinal mucosal cells to accomplish normal calcium transport, at least in patients who do not have an independent cause for impaired calcium absorption, such as intestinal disease (Chapter 51) or anticonvulsant administration [85]. Theories that ascribe all manifestations of rickets and osteomalacia to deficiency of circulating calcitriol alone may be able to account for its persistence, but they have much greater difficulty accounting for its initiation. At the onset of HVO in patients with extrinsic vitamin D deficiency, what causes calcium malabsorption and a small but significant fall in plasma calcium when plasma calcitriol is maintained at a normal (or even increased) level by secondary hyperparathyroidism [86,87]? A similar argument applies to the increased vitamin D requirement of primary hyperparathyroidism, due to accelerated calcidiol catabolism in the presence of increased plasma calcitriol levels [88].

During the evolution of HVO there is an early fall in plasma concentrations of both calcidiol [54,80,87,89] (Table I) and 24-hydroxycalcidiol [73,81,84]. Calcium absorption and retention in bone are increased in humans by pharmacological doses of 24-hydroxycalcidiol [68], but it is not known whether such effects occur at physiological blood levels. Calcidiol binds to intestinal receptors for calcitriol, but with approximately 500- to 1000-fold lower affinity (Chapter 8), although calcidiol is only 100 times less effective than calcitriol in promoting bone resorption *in vitro* [90]. Seemingly, these differences in activity could be offset by the much higher total plasma concentration of calcidiol, but there is only a 10-fold difference in free concentrations, based on estimates of the association constants for binding to the same circulating protein [91]. Consequently, a fall in plasma calcidiol level below normal could not significantly modify total receptor occupancy in the target cells that respond to circulating calcitriol, although some more complex effect on receptor function remains possible [92].

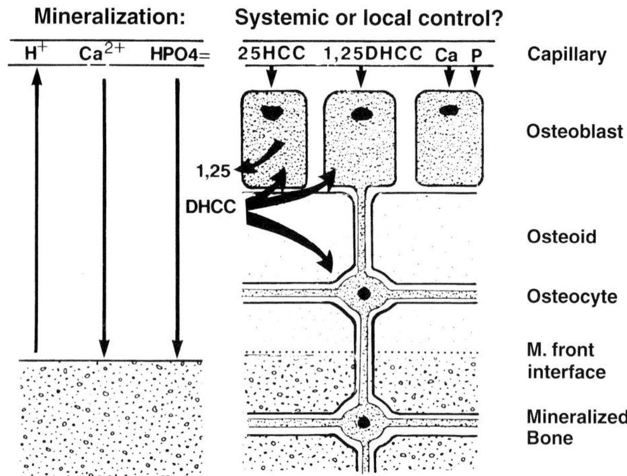

FIGURE 5 Biochemical and morphological approaches to mineralization. At left are shown the directional movements of ions between a blood vessel above and the bone below without reference to intervening structures. These structures are depicted diagrammatically at right. 25HCC, 25-Hydroxycholecalciferol (calcidiol); 1,25DHCC, 1,25-dihydroxycholecalciferol (calcitriol); M., mineralization; interface, boundary between osteoid and mineralized bone. The osteoblasts and osteocytes can be influenced by circulating levels of calcium, phosphate, and calcitriol, and also by locally produced calcitriol, either autocrine or paracrine. Reprinted from Parfitt [15]. In Chemistry and Biology of Mineralized Tissues, 1992, pp. 465–474, with kind permission from Elsevier Science.

A more promising approach to the clinical paradox is the possibility that one or more dihydroxylated metabolites are produced locally in target tissues, as is strongly suggested by studies with isolated bone and intestinal cells [93,94], by the *in vivo* intestinal response to a pharmacological oral dose of calcidiol [95], and by the *in vitro* effects of calcidiol to increase bone resorption [96] and intestinal calcium transport [97] in isolated tissues. If bone cell and intestinal cell 1α-hydroxylases were less influenced by PTH and phosphate than is the renal

1α-hydroxylase, as is generally the case for extrarenal calcitriol production [98], local production of calcitriol would be more substrate dependent than circulating calcitriol and would be impaired sooner by a fall in plasma calcidiol concentration below normal. This mechanism would also account for the improvement in bone mineralization and intestinal calcium absorption brought about by calcidiol administration in patients with chronic renal failure [99], and for the much greater relative therapeutic potency of dihydrotachysterol in hypoparathyroidism than in osteomalacia [61]. Similar considerations would apply to local production of 24-hydroxycalcidiol [94]. It seems reasonable to speculate that circulating calcitriol is most important for the regulation of calcium homeostasis, locally produced calcitriol and 24-hydroxycalcidiol are most important for the regulation of bone remodeling and mineralization, and both circulating and locally produced metabolites are important for the regulation of calcium absorption.

B. Evidence for Direct as Well as Indirect Effects of Vitamin D on Bone

It has been known since the early 1920s that active infantile rickets is associated with low plasma phosphate (less than 3.0 mg/dl) and low total plasma Ca × P product [less than 30 $(mg/dl)^2$] and healing rickets with increases in these values [100,101]. Freshly harvested rachitic rat growth plate cartilage will mineralize in normal rat serum and in aqueous solutions with the same pH and Ca × P product [102,103]. Calcification occurs in the same region as the cartilage as *in vivo,* and fails to occur if the viability of the tissue is compromised. The relationship of rickets to the total plasma Ca × P product has been confirmed many times [16,48], and the relationship of the product to the thermodynamic activity product for various solid phases has been analyzed [103]. The same relationship holds in rickets complicating osteopetrosis, in which low plasma levels of calcium and/or phosphate are due to increased mineral sequestration in bone [104]. The relationship is disturbed in patients with renal failure [105], possibly due to the presence of mineralization inhibitors, such as magnesium [106], or the effect of metabolic acidosis on the activity of PO_4^{3-} [103]. The data clearly established the importance of plasma composition, which is disturbed indirectly by vitamin D deficiency [103], but do not rule out an additional direct local effect.

In older children and adults the relationship between plasma composition and the state of mineralization is less consistent [16,53,73], being clearly evident in some series of patients [107–109] but not in others [110–112]. There are significant correlations between plasma phosphate and adjusted apposition rate and between plasma calcium and mean osteoid thickness [113], but their magnitude is too low for useful prediction in individual patients. Calculation of an ion product more clearly related to the physical chemistry of bone mineral may remove some discrepancies [103], but many remain. A more serious flaw in this line of reasoning is that single measurements in the fasting state, as in normal clinical practice, do not adequately represent body fluid composition because of the substantial circadian variation [27,114]. Nevertheless, it seems unlikely that such variation could account for the absence of osteomalacia in some patients with a degree of persistent hypophosphatemia that in other patients would be regarded as a sufficient explanation for their osteomalacia [113,115]. Even in the rat, a species in which mineralization probably depends more closely on plasma composition than in humans, healing of rickets can be detected radiographically in response to vitamin D administration while the Ca × P product is still subnormal [72].

It was previously demonstrated that the early effects of vitamin D deficiency in adults are due entirely to secondary hyperparathyroidism in the presence of normal bone mineralization; the correction of such effects by calcium has no bearing on the issue in question. Radiographic evidence for impaired mineralization is often ambiguous. Looser's zones can occur in the absence of osteomalacia and heal spontaneously [116]. Metaphyseal erosions in the absence of increased growth plate width are due to osteitis fibrosa, but they have often been attributed erroneously to rickets [117]. Problems in the histological recognition of defective mineralization were outlined earlier and are discussed in more detail elsewhere [16]. In the previously cited clinical studies [75–77], the authors demonstrated little awareness of the difficulties just mentioned, and the evidence that the lesions corrected by mineral administration alone were the result of defective mineralization rather than of secondary hyperparathyroidism was inconclusive. Of six cases included in the four reports, three had metaphyseal erosions with normal growth plate width, wrongly reported as rickets [76,77], three had qualitative bone histology only [75,76], and two had no bone biopsy at all [76,77]. Only in one case was there convincing histological evidence of osteomalacia, which healed with prolonged calcium infusion [77a].

Studies in the rat have provided stronger evidence that mineral alone can be effective *in vivo* in the absence of vitamin D. Restoration of normocalcemia in vitamin D-deficient rats by a high calcium diet corrected both abnormal bone enzyme activity [118] and defective osteoid maturation [119]. Complete vitamin D deficiency was not demonstrated in these experiments, but in 25-day-old rats weaned from vitamin D-deficient mothers, and maintained without access to ultraviolet light or dietary vitamin D, plasma levels of calcitriol were undetectable [78]. Calcium chloride and buffered sodium phosphate infused into separate jugular veins, in amounts sufficient to maintain the same plasma calcium and phosphate levels as in vitamin D-replete rats, maintained normal growth plate width and normal tetracycline based indices of mineralization [79]. This experiment demonstrated conclusively that vitamin D was not essential for mineralization but did not exclude a contributory role for vitamin D under normal conditions. First, as previously discussed, there is only a very approximate relationship between total plasma levels and thermodynamic activity products. More importantly, the plasma levels were measured only once every 3 days at an unspecified time; because of circadian variation [114] the mean levels could have been higher in the infused than in the vitamin D-replete animals. Osteoid surface and volume were significantly lower in the infused rats [79], consistent with oversuppression of PTH secretion. Consequently, it remains possible that higher mean plasma levels of calcium and phosphate are needed to sustain normal mineralization in the absence than in the presence of vitamin D.

The persistence in early osteomalacia of some doubly labeled surfaces with normal or only moderately reduced rates of mineral apposition [16] indicates that mineralization can proceed at the beginning of the osteoid seam life span, although it ceases prematurely (Fig. 4). Mineralizing and nonmineralizing osteoid seams are often close together, sometimes even in direct continuity, and they are exposed to the same microcirculation, so that the difference between them cannot be explained in terms of chemical change alone. However, at doubly labeled seams a higher proportion of the surface is lined by osteoblasts [47,120], suggesting that these cells, possibly in conjunction with the osteocytes derived from them lying within the osteoid [121], are able to promote mineralization in the face of a moderate reduction in plasma ion product, but do so for a shorter time than normal in vitamin D depletion. When this function is lost, mineralization ceases even though matrix apposition continues slowly and the osteoid seam gets progressively thicker. In more severe osteomalacia, osteoblasts are fewer or absent altogether [122], mineralization never begins, and double labels are not found (Fig. 4). A similar relationship is observed during treatment: the recovery of mineralization in response to calcitriol administration, indicated by double labeling, occurs preferentially at surfaces where new osteoblasts have appeared [123,124].

The bone histological data in patients with osteomalacia strongly suggest that deficiency of calcitriol (and possibly also other metabolites) impairs some function of the osteoblast that favors mineralization. This proposal is consistent with the presence in osteoblasts of calcitriol receptors (VDR) [125], the autoradiographic localization of labeled calcitriol in osteoblast nuclei [126], the stimulation by calcitriol of the *in vitro* production by osteoblasts of alkaline phosphatase [127] and osteocalcin [128], the *in vivo* enhancement by calcitriol of mineral apposition rate in young mice [129], and the morphological changes induced by calcitriol in the cells lining quiescent bone surfaces [130] that are of osteoblast lineage [27]. A local cellular effect of one or more vitamin D metabolites would also account for the abnormalities in collagen cross-linking and other changes in bone matrix maturation and composition [40,131–133], and the changes in intermediary metabolism in cartilage cells [134,135], that have been found in vitamin D deficiency, although it is less clear that these are the result of a direct rather than an indirect effect of vitamin D on osteoblast and chondrocyte function. Finally, the proposal also accounts for the greater ability of vitamin D than calcium carbonate to improve bone mineralization in chronic renal failure, despite equivalent changes in total Ca $\times$ P product [136], on the assumption that bone cells can make calcitriol from its precursor.

The usual approach to mineralization has been to study the physical chemistry of the solution in which the ions originate, and the events taking place in bone, and to largely ignore what happens in between [15]. But the circulating ions have to traverse a rather complex pathway before they arrive at the site of mineralization (Fig. 5). Having left the capillary and diffused through marrow connective tissue, they must pass through a layer of osteoblasts and a layer of osteoid before they can reach the site of mineral desposition. Osteoblasts on the surface, osteocytes within the osteoid, and osteocytes within mineralized bone are joined by a communicating network of cellular processes within the canaliculi. Very little is known about how mineral ions actually travel through this complex structure, but it would be surprising if cellular transport mechanisms of some kind were not involved in the movement of ions from the extracellular fluid to the site of mineral deposition. Indeed, it seems likely that calcitriol could stimulate the inward transport of calcium and/or phosphate ions through or between cells at sites of mineralization [24], consistent with its known effects on the cells of the intestinal mucosa (Chapter 16) and the renal tubule (Chapter 17).

The osteoblast thus influences mineralization in two ways, by its effects on matrix maturation [136] and by its effects on mineral transport [53]. Furthermore, calcium and phosphate ions must be regarded not just as substrates for apatite formation but as part of the environment of the cell that is involved in their transport, as osteoblast function is influenced by the circulating and presumably also local levels of calcium [137] and phosphate [53,113]. The concept that mineralization normally depends both on the availability of substrate ions via the circulation and on the activity of osteoblasts and chondrocytes, and that vitamin D influences both of these processes, although by no means rigorously established, enables all the apparently conflicting data, laboratory and clinical, to be reconciled. The concept has the additional merit of providing a basis for unifying the pathogenesis of all major forms of osteomalacia, as hereditary or acquired defects in phosphate transport across the renal tubular epithelium, whether intrinsic or due to humoral factors, could plausibly be accompanied by similar defects in transport across the quasi-epithelium that covers all bone surfaces [24,138].

References

1. McLean FC, Urist MR 1955 Bone: An Introduction to the Physiology of Skeletal Tissue. Univ. of Chicago Press, Chicago.
2. Robison R 1932 The Significance of Phosphoric Esters in Metabolism. New York Univ. Press, New York.
3. Lacroix P 1960 ^{45}Ca autoradiography in the study of bone tissue. In: Rodahl K, Nicholson JT, Brown EM (eds) Bone as a Tissue. McGraw-Hill, New York, pp. 262–279.
4. Frost HM, Villanueva AR 1960 Observations on osteoid seams. Henry Ford Hospital Med Bull **8**:212–219.
5. Parfitt AM 1990 Bone-forming cells in clinical conditions. In: Hall BK (ed) Bone: A Treatise, Volume 1, The Osteoblast and Osteocyte. Telford Press, Caldwell, New Jersey, pp. 351–429.
6. Parfitt AM 1995 Problems in the application of *in vitro* systems to the study of human bone remodeling. Calcif Tissue Int **56**(Suppl. 1):S5–S7.
7. Brighton CT 1978 Structure and function of the growth plate. Clin Orthop Related Res **136**:22–32.
8. Schenk RK, Hunziker EB 1991 Growth plate: Histophysiology, cell and matrix turnover. In: Glorieux FH (ed) Rickets, Nestle Nutrition Workshop Series, Vol. 21. Raven, New York, pp. 63–78.
9. Kember NF, Kirkwood, JK 1991 Cell kinetics and the study of longitudinal bone growth: A perspective. In: Dixon AD, Sarnat BG, Hoyte DA (eds) Fundamentals of Bone Growth: Methodology and Applications. CRC Press, Boca Raton, Florida, pp. 153–162.
10. Boskey AL, Doty SB, Stiner D, Binderman I 1996 Viable cells are a requirement for *in vitro* cartilage calcification. Calcif Tissue Int **58**:177–185.
11. Frost HM 1990 Structural adaptations to mechanical usage. 1. Redefining Wolff's law: The bone remodeling problem. Anat Rec **226**:403–413.
12. Turner RT 1994 Cancellous bone turnover in growing rats: Time-dependent changes in association between calcein label and osteoblasts. J Bone Miner Res **9**:1419–1424.
13. Parfitt AM 1996 Skeletal heterogeneity and the purposes of bone remodeling, Implications for the understanding of osteoporosis. In: Marcus R, Feldman D, Kelsey J (eds) Osteoporosis. Academic Press, San Diego, pp. 315–329.
14. Parfitt AM 1994 Osteonal and hemi-osteonal remodeling: The spatial and temporal framework for signal traffic in adult human bone. J Cell Biochem **55**:273–286.
15. Parfitt AM 1992 Human bone mineralization studied by *in vivo* tetracycline labeling: Application to the pathophysiology of osteomalacia. In Slavkin M, Price P (eds) Chemistry and Biology of Mineralized Tissues. Excerpta Medica, Amsterdam, pp. 465–474.
16. Parfitt AM Osteomalacia and related disorders. In: Avioli LV, Krane SM (eds) Metabolic Bone Disease and Clinically Related Disorders, 3rd Ed. Academic Press, San Diego, in press.
17. Hansson LI, Menander-Sellman K, Stenstrom A, Thorngren KG 1972 Rate of normal longitudinal bone growth in the rat. Calcif Tissue Res **10**:238–251.
18. Thorngren KG, Hansson LI, Menander-Sellman K, Stenstrom A 1973 Effect of hypophysectomy on longitudinal bone growth in the rat. Calcif Tissue Res **11**:281–300.
19. Caplan AI, Boyan BD 1994 Endochondral bone formation: The lineage cascade. In Hall BK (ed) Bone: A Treatise, Volume 8, Mechanisms of Bone Development and Growth. CRC Press, Boca Raton, Florida, pp. 1–46.
20. Yamauchi M, Chandler GS, Katz, EP 1992 Collagen cross-linking and mineralization. In: Slavkin M, Price P (eds) Chemistry and Biology of Mineralized Tissues. Excerpta Medica, Amsterdam, pp. 39–46.
21. Gerstenfeld LC, Riva A, Hodgens K, Eyre DR, Landis WJ 1993 Post-translational control of collagen fibrillogenesis in mineralizing cultures of chick osteoblasts. J Bone Miner Res **8**:1031–1043.
22. Boskey AL 1990 Mineral–matrix interactions in bone and cartilage. Clin Orthop Related Res **281**:244–274.
23. Robey PG, Boskey AL 1996 The biochemistry of bone. In: Marcus R, Feldman D, Kelsey J (eds) Osteoporosis. Academic Press, San Diego, pp. 95–183.
24. Parfitt AM, Kleerekoper M 1980 The divalent ion homeostatic system: Physiology and metabolism of calcium, phosphorus, magnesium and bone. In: Maxwell M, Kleeman CR (eds) Clinical Disorders of Fluid and Electrolyte Metabolism, 3rd Ed. McGraw-Hill, New York, pp. 269–398.
25. Neuman WF 1980 Bone material and calcification mechanisms. In: Urist MR (ed) Fundamental and Clinical Bone Physiology. Lippincott, Philadelphia, Pennsylvania, pp. 83–107.
26. Glimcher MJ 1992 The nature of the mineral component of bone and the mechanism of calcification. In: Coe FL, Favus MJ (eds) Disorders of Bone and Mineral Metabolism. Raven, New York, pp. 265–286.
27. Parfitt AM 1993 Calcium homeostasis. In: Mundy GR, Martin TJ (eds) Handbook of Experimental Pharmacology, Volume 107, Physiology and Pharmacology of Bone. Springer-Verlag, Heidelberg, pp. 1–65.
28. Brighton CT, Hunt RM 1978 The role of mitochondria in growth plate calcification as demonstrated in a rachitic model. J Bone Joint Surg **60-A**:630–639.
29. Fedde KN, Weinstein RS, Waymire KG, Mac Gregor GR, Whyte MP 1996 Alkaline phosphatase knock-out mice recapitulate the skeletal defects of infantile hypophosphatasia. J Bone Miner Res **11**(Suppl.):S140.
30. Alcock NW, Shils ME 1968 Association of inorganic pyrophosphatase activity with normal calcification of rat costal cartilage *in vivo*. Biochem J **112**:505–510.
31. Meyer JL, Fleisch H 1984 Determination of calcium phosphate ihibitor activity; critical assessment of the methodology. Miner Electrolyte Metab **10**:249–258.
32. Christoffersen J, Landis WJ 1991 A contribution with review to the description of mineralization of bone and other calcified tissues *in vivo*. Anat Rec **230**:435–450.
33. Anderson HC, Morris DC 1993 Mineralization. In: Mundy GR,

Martin TJ (eds) Handbook of Experimental Pharmacology, Volume 107, Physiology and Pharmacology of Bone. Springer-Verlag, Heidelberg, pp. 267–298.

34. Reinholt, Engfeldt B, Hjerpe A, Jansson K 1982 Stereological studies on the epiphyseal growth plate with special reference to the distribution of matrix vesicles. J Ultrastruct Res **80**:270–279.
35. Glorieux FH, Marie PJ, Pettifor JM, Delvin EE 1980 Bone response to phosphate salts, ergocalciferol and calcitriol in hypophosphatemic vitamin D resistant rickets. N Engl J Med **303**:1023–1031.
36. Dodds GS, Cameron HC 1934 Studies on experimental rickets in rats. I. Structural modifications of the epiphyseal cartilages in the tibia and other bones. Am J Anat **55**:135–165.
37. Park EA 1939 Observations on the pathology of rickets with particular reference to the changes at the cartilage–shaft junctions of the growing bones. Bull NY Acad Med **15**:495–543.
38. Bicknell F, Prescott F 1953 The Vitamins in Medicine. Heinemann, London, pp. 529–533.
39. Bailie JM, Irving JT 1955 Development and heating of rickets in intramembranous bone. Acta Med Sci **152**(Suppl. 306):1–14.
40. Baylink D, Stauffer M, Wergedal J, Rich C 1970 Formation, mineralization and resorption of bone in vitamin D-deficient rats. J Clin Invest **49**:1122–1134.
41. Schaffler MB, Burr DB, Frederickson RG 1987 Morphology of the osteonal cement line in human bone. Anat Rec **217**:223–228.
42. Parfitt AM 1992 The physiologic and pathogenetic significance of bone histomorphometric data. In: Coe FL, Favus MJ (eds) Disorders of Bone and Mineral Metabolism. Raven, New York, pp. 475–489.
43. Parfitt AM, Foldes J, Villanueva AR, Shih MS 1991 The difference in label length between demethychlortetracycline and oxytetracycline: Implications for the interpretation of bone histomorphometry. Calcif Tissue Int **48**:74–77.
44. Eriksen EF 1986 Normal and pathological remodeling of human trabecular bone: Three-dimensional reconstruction of the remodeling sequence in normals and in metabolic bone disease. Endocr Rev **4**:379–408.
45. Parfitt AM 1984 The cellular mechanisms of osteoid accumulation in metabolic bone disease. In: Mineral Metabolism Research in Italy, Vol. 4. Wichtig Editore, Milano, pp. 3–9.
46. Parfitt AM 1986 Accelerated cortical bone loss: Primary and secondary hyperparathyroidism. In: Uhthoff H (ed) Current Concepts of Bone Fragility. New York, Springer-Verlag, pp. 279–285.
47. Meunier PJ, Van Linthoudt D, Edouard C, Charhon S, Arlot M 1981 Histological analysis of the mechanisms underlying pathogenesis and healing of osteomalacia. Proceedings, Sixteenth European Calcified Tissue Symposium. Abstracts **33**:771–774.
48. Peacock M 1993 Osteomalacia and rickets. In: Nordin BEC, Need AG, Morris HA (eds), Metabolic Bone and Stone Disease, 3rd Ed. Churchill Livingstone, London, pp. 83–118.
49. Stanbury SW 1980 Vitamin D and calcium metabolism. In: Norman AW (ed) Vitamin D. Molecular Biology and Clinical Nutrition. Dekker, New York, pp. 251–319.
50. Fraser D, Kooh SW, Scriver CR 1967 Hyperparathyroidism as the cause of hyperaminoaciduria and phosphaturia in human vitamin D deficiency. Pediatr Res **1**:425–435.
51. David L 1991 Common vitamin D-deficiency rickets. In: Glorieux FH (ed) Rickets, Nestle Nutrition Workshop Series, Vol 21. Raven, New York, pp. 107–122.
52. Vainsel M, Manderlier T, Corvilain J, Vis HL 1974 Study of the secondary hyperparathyroidism in vitamin D deficiency rickets. I. Aspects of mineral metabolism. Biomedicine **21**:368–371.
53. Stanbury SW, Mawer EB 1990 Metabolic Disturbances in Acquired Osteomalacia. In: Cohen RD, Lewis B, Alberti KGMM, Denman AM (eds) The Metabolic and Molecular Basis of Acquired Disease. Bailliere Tindall, London, pp. 1717–1782.
54. Rao DS, Vinnanueva A, Mathews M, Pumo B, Frame B, Kleerekoper M, Parfitt AM 1983 Histologic evolution of vitamin depletion in patients with intestinal malabsorption or dietary deficiency. In: Frame B, Potts JT Jr (eds) Clinical Disorders of Bone and Mineral Metabolism. Excerpta Medica, Amsterdam, pp. 224–226.
55. Arnaud CD 1991 Parathyroid hormone and its role in the pathophysiology of the common forms of rickets and osteomalacia. In: Glorieux FH (ed) Rickets, Nestle Nutrition Workshop Series, Vol. 21. Raven, New York, pp. 47–61.
56. Rao DS, Parfitt AM, Kleerekoper M, Pumo BS, Frame B 1985. Dissociation between the effects of endogenous parathyroid hormone on cAMP generation and on phosphate reabsorption in hypocalcemia due to vitamin D depletion: An acquired disorder resembling pseudohypoparathyroidism type II. J Clin Endocrinol Metab **61**:285–290.
57. Brazier M, Kamel S, Maamer M, Agbomson F, Elesper I, Garabedian M, Desmet G, Sebert JL 1995 Markers of bone remodeling in the elderly subject: Effects of vitamin D insufficiency and its correction. J Bone Miner Res **10**:1753–1761.
58. Parfitt AM, Kleerekoper M, Cruz C 1986 Reduced phosphate reabsorption unrelated to parathyroid hormone after renal transplantation: Implications for the pathogenesis of hyperparathyroidism in chronic renal failure. Miner Electrolyte Metab **12**:356–362.
59. Fraser DR 1983 The physiological economy of vitamin D. Lancet **1**:969–972.
60. Fraser DR 1991 Physiology of vitamin D and calcium homeostasis. In: Glorieux FH (ed) Rickets, Nestle Nutrition Workshop Series, Vol 21. Raven, New York, pp. 23–34.
61. Parfitt AM 1978 Adult hypoparathyroidism: Treatment with calcifediol. Arch Intern Med **138**:874–881.
62. Clements MR, Chalmers TM, Fraser DR 1984 Enterohepatic circulation of vitamin D: A reappraisal of the hypothesis. Lancet **1**:1376–1379.
63. Clements MR, Johnson L, Fraser DR 1987 A new mechanism for induced vitamin D deficiency in calcium deprivation. Nature **325**:62–65.
64. Bisballe S, Eriksen EF, Melsen F, Mosekilde L, Sorensen OH, Hessov I 1991 Osteopenia and osteomalacia after gastrectomy: Interrelations between biochemical markers of bone remodelling, vitamin D metabolites, and bone histomorphometry. Gut **32**:1303–1307.
65. Clements MR, Davies M, Hayes ME, Hickey CD, Lumb GA, Mawer EB, Adams PH 1992 The role of 1,25-dihydroxyvitamin D in the mechanism of acquired vitamin D deficiency. Clin Endocrinol **37**:17–27.
66. Epstein S, Meunier PJ, Lambert PW, Stern PH, Bell NH 1993 1,25-Dihydroxyvitamin D_3 corrects osteomalacia in hypoparathyroidism and pseudohypoparathyroidism. Acata Endocrinol **103**:241–247.
67. Brommage R, DeLuca HF 1985 Evidence that 1,25-dihydroxyvitamin D_3 is the physiologically active metabolite of vitamin D_3. Endocr Rev **6**:491–511.
68. Norman AW, Roth J, Orci L 1982 The vitamin D endocrine system: Steroid metabolism, hormone receptors, and biological response (calcium binding proteins). Endocr Rev **3**:331–366.
69. Nagant de Deuxchaisnes C, Rombouts-Lindemans C, Huaux JP, Withofs H, Meesrsseman F 1979 Healing of vitamin D-deficient

osteomalacia by the administration of $1,25(OH)_2D_3$. In: MacIntyre I, Szelke M (eds) Molecular Endocrinology, pp. 375–404.

70. Papapoulos SE, Clements TL, Fraher LJ, Gleed J, O'Riordan JLH 1980 Metabolites of vitamin D_3 in human vitamin deficiency: Effect of vitamin D_3 or 1,25-dihydroxycholecalciferol. Lancet **2**:612–615.
71. Parfitt AM, Mathews CHE, Brommage R, Jarnagin K, DeLuca HF 1984 Calcitriol but no other metabolite of vitamin D is essential for normal bone growth and development in the rat. J Clin Invest **73**:576–586.
72. Lund B, Charles P, Egsmose C, Lund BJ, Melson F, Mosekilde L, Storm T, Søndergard H, Thode J, Sørensen OH 1985 Changes in vitamin D metabolites and bone histology in rats during recovery from rickets. Calcif Tissue Int **37**:478–483.
73. Rasmussen H, Baron R, Broadus A, DeFronzo R, Lang R, Horst R 1980 $1,25(OH)_2D_3$ is not the only D metabolite involved in the pathogenesis of osteomalacia. Am J Med **69**:360–368.
74. St. Arnaud R, Arabian A, Glorieux FH 1996 Abnormal bone development in mice deficient for the vitamin D 24-hydroxylase gene. J Bone Miner Res **11**:S216.
75. Popovtzer MM, Mathay R, Alfrey AC, Block M, Beck P, Miles J, Reeve EB 1973 Vitamin D deficiency osteomalacia—Healing of the bone disease in the absence of vitamin D with intravenous calcium and phosphorus infusions. In: Frame B, Parfitt AM, Duncan H (eds) Clinical Aspects of Metabolic Bone Disease. Excerpta Medica, Amsterdam, pp. 382–387.
76. Sakati N, Woodhouse NJY, Niles N, Harfi H, de Grange DA, Marx S 1986 Hereditary resistance to 1,25-dihydroxyvitamin D: Clinical and radiological improvement during high-dose oral calcium therapy. Horm Res **24**:280–287.
77. Weisman Y, Bab I, Gazit D, Spirer Z, Jaffe M, Hochberg Z 1987. Long-term intracaval calcium infusion therapy in end-organ resistance to 1,25-dihydroxyvitamin D. Am J Med **83**:984–990.
77a. Balsan S, Garabedian M, Larchet M, Gorski AM, Cournot G, Tau C, Bourdeau A, Silve C, Ricour C 1986 Long-term nocturnal calcium infusions can cure rickets and promote normal mineralization in hereditary resistance to 1,25-dihydroxyvitamin D. J Clin Invest **77**:1661–1667.
78. Underwood JR, DeLuca H 1984 Vitamin D is not directly necessary for bone growth and mineralization. Am J Physiol **246**:493–498.
79. Weinstein RS, Underwood JL, Hutson MS, DeLuca HF 1984 Bone histomorphometry in vitamin D-deficient rats infused with calcium and phosphorus. Am J Physiol **246**:E499–E505.
80. Compston JE, Vedi S, Merrett AL, Clemens TL, O'Riordan JLH, Woodhead JS 1981 Privational and malabsorption metabolic bone disease: Plasma vitamin D metabolite concentrations and their relationship to quantitative bone histology. Metab Bone Dis Related Res **3**:165–170.
81. Kashiwa H, Nishi Y, Usui T, Seino Y 1981 A case of rickets with normal serum level of $1,25\text{-}(OH)_2D$ and low 25-OHD. Hiroshima J Med Sci **30**:61–63.
82. Chesney RW, Zimmerman J, Hamstra A, DeYuca MF, Muzesr RB 1981 Vitamin D metabolite concentrations in vitamin D deficiency. Are calcitriol levels normal? Am J Dis Children **135**:1025–1028.
83. Markestad T, Halvorsen S, Halvorsen KS, et al. 1984 Plasma concentrations of vitamin D metabolites before and during treatment of vitamin D deficiency rickets in children. Acta Paediatr Scand **73**:225–231.
84. Stanbury SW, Taylor CM, Lumb GA, Mawer B, Berry J, Hann J, Wallace J 1981 Formation of vitamin D metabolites following correction of human vitamin D deficiency. Miner Electrolyte Metab **5**:212–227.
85. Wahl TO, Gobuty AH, Lukert BP 1981 Long-term anticonvulsant therapy and intestinal calcium absorption. Clin Pharmacol Ther **30**:506–516.
86. Demiaux B, Arlot ME, Chapuy M-C, Meunier PJ, Delmas PD 1992 Serum osteocalcin is increased in patients with osteomalacia: Correlations with biochemical and histomorphometric findings. J Clin Endocrinol Metab **74**:1146–1151.
87. Stanbury SW 1981 Vitamin D and hyperparathyroidism. J R Coll Phys Lond **15**:205–217.
88. Clements MR, Davies M, Fraser DR, Lumb GA, Mawer EB, Adams PH 1987 Metabolic inactivation of vitamin D is enhanced in primary hyperparathyroidism. Clin Sci **73**:659–664.
89. Stamp TCB, Walker PG, Perry W, Jenkins MW 1980 Nutritional osteomalacia and late rickets in greater London 1974–1979: Clinical and metabolic studies in 45 patients. Clin Endocrinol Metab **9**:81–105.
90. Raisz LG, Trummel CL, Holick MF, DeLuca HF 1972 1,25-Dihydroxycholecalciferol: A potent stimulator of bone resorption in tissue culture. Science **175**:768–869.
91. Bouillon R, Van Baelen H 1981 Transport of vitamin D: Significance of free and total concentrations of the vitamin D metabolites. Calcif Tissue Int **33**:451–453.
92. Wilhelm F, Norman AW 1984 Cooperativity in the binding of 1,25-dihydroxyvitamin D_3 to the chick intestinal receptor. FEBS Lett **170**:239–242.
93. Howard GA, Turner RT, Sherrard DJ, Baylink DJ 1981 Human bone cells in culture metabolize 25-hydroxyvitamin D_3 to 1,25-dihydroxyvitamin D_3 and 24,25-dihydroxyvitamin D_3. J Biol Chem **256**:7738–7740.
94. Puzas JE, Turner RT, Howard GA, Baylink DJ 1983 Cells isolated from embryonic intestine synthesize 1,25-dihydroxyvitamin-D_3 and 24,25-dihydroxyvitamin-D_3 in culture. Endocrinology **112**:378–380.
95. McDonald GB, Lau K-HW, Schy AL, Wergedal JE, Baylink DJ 1985 Intestinal metabolism and portal venous transport of $1,25(OH)_2D_3$, $25(OH)D_3$, and vitamin D_3 in the rat. Am J Physiol **248**:G633–G638.
96. Trummel CL, Raisz LG, Blunt JW, DeLuca HF 1969 25-Hydroxycholecalciferol: Stimulation of bone resorption in tissue culture. Science **163**:1450–1451.
97. Olson EB, DeLuca HF 1969 25-Hydroxycholecalciferol: Direct effect on calcium transport. Science **165**:405–407.
98. Parfitt AM, Gallagher JC, Heaney RP, *et al.* 1982 Vitamin D and bone health in the elderly. Am J Clin Nutr **36**:1014–1031.
99. Eastwood JB, Stamp TCB, DeWardener HE, Bordier PJ, Arnaud CD 1976 The effect of 25-hydroxy vitamin D_3 in the osteomalacia of chronic renal failure. Clin Sci Mol Med **52**:499–508.
100. Howland J, Kramer B 1921 Calcium and phosphorus in the serum in relation to rickets. Am J Dis Chldren **22**:105–119.
101. Howland J, Kramer B 1923 A study of the calcium and inorganic phosphorus of the serum in relation to rickets and tetany. Monatschrift fur Kinderheilkd **25**:279–293.
102. Shipley PG, Kramer B, Howland J 1926 Studies upon calcification in vitro. Biochem J **20**:379–387.
103. Nordin BEC, Smith DA 1967 Pathogenesis and treatment of osteomalacia. In: Hioco DJ (ed) L'Osteomalacie. Masson & Cie, Paris, pp. 374–399.
104. Kaplan FS, August CS, Fallon MD, Gannon F, Haddad JG 1993 Osteopetrorickets—The paradox of plenty. Pathophysiology and treatment. Clin Orthop Related Res **294**:64–78.

105. Stanbury SW 1962 Osteomalacia. Schweiz Med Wochenschr **29**:883–892.
106. Yendt ER, Connor TB, Howard JE 1955 In vitro calcification of rachitic rat cartilage in normal and pathological human sera with some observations on the pathogenesis of renal rickets. Bull Johns Hopkins Hospital **96/97**:1–19.
107. Bordier PH, Hioco D, Roquier, Hepner GW, Thompson GR 1969 Effects of intravenous vitamin D on bone and phosphate metabolism in osteomalacia. Calcif Tissue Res **4**:78–83.
108. Bordier P, Pechet MM, Hesse R, Marie P, Rasmussen H 1974 Response of adult patients with osteomalacia to treatment with crystalline 1α-hydroxy vitamin D_3. N Engl J Med **291**:866–871.
109. Bordier P, Rasmussen H, Marie P, Miravet L, Gueris J, Ryckwaert A 1978 Vitamin D metabolites and bone mineralization in man. J Clin Endocrinol Metab **46**:284–294.
110. Compston JE, Horton LWL, Thompson RPH 1979 Treatment of osteomalacia associated with primary biliary cirrhosis with parenteral vitamin D_2 or oral 25-hydroxyvitamin D_3. Gut **20**:133–136.
111. Compston JE, Crowe JP, Horton LWL 1979 Treatment of osteomalacia associated with primary biliary cirrhosis with oral 1-alpha-hydroxy vitamin D_3. Br Med J 309–312.
112. Compston JE, Horton LWL, Laker MF, Merrett AL, Woodhead JS, Gazet J-C, Pilkington TRE 1980 Treatment of bone disease after jejunoileal bypass for obesity with oral 1α-hydroxyvitamin D_3. Gut **21**:669–674.
113. Parfitt AM, Villanueva AR 1982 Hypophosphatemia and osteoblast function in human bone disease. In: Massry SG, Letteri JM, Ritz E (eds) Proceedings, 5th International Workshop on Phosphate and Other Minerals. Regulation of Phosphate and Mineral Metabolism. Adv Exp Med Biol **151**:209–216.
114. Markowitz ME, Rosen JF, Laxminarayan S, Mizruchi M 1984 Circadian rhythms of blood minerals during adolescence. Pediatr Res **18**:456–462.
115. De Vernejoul MC, Marie PJ, Miravet L, Ryckewaert A 1983 Chronic hypophosphatemia without osteomalacia. In: Frame B, Potts JT Jr (eds) Clinical Disorders of Bone and Mineral Metabolism. Excerpta Medica, Amsterdam, pp. 232–236.
116. McKenna MJ, Kleerekoper M, Ellis BI, Dao BS, Parfitt AM, Frame B 1987 Atypical insufficiency fractures confused with Looser zones of osteomalacia. Bone **8**:71–78.
117. Parfitt AM 1977 The clinical and radiographic manifestations of renal osteodystrophy. In: David DG (ed) Perspectives in Hypertension and Nephrology: Calcium Metabolism in Renal Failure and Nephrolithiasis. Wiley, New York, pp. 145–195.
118. Wergedal JE, Baylink JE 1971 Factors affecting bone enzymatic activity in vitamin D-deficient rats. Am J Physiol **220**:406–409.
119. Howard GA, Baylink DJ 1980 Matrix formation and osteoid maturation in vitamin D-deficient rats made normocalcemic by dietary means. Miner Electrolyte Metab **3**:44–50.
120. Sebert JL, Meunier PJ 1984. Role physiopathologique de la vitamine D et de ses metabolites dans l'osteomalacie. In: Bouillon R, Boudailliez B, Marie A, et al. (eds) Vitamine D et Maladies des Os et du Metabolisme Mineral. Masson, Paris, pp. 109–145.
121. Bordier PJ, Marie P, Miravet L, et al. 1976 Morphological and morphometrical characteristics of the mineralization front. A vitamin D regulated sequence of the bone remodeling. In: Meunier PJ (ed) Bone Histomorphometry. Second International Workshop. Armour Montagu, Paris, pp. 335–354
122. Weinstein RS 1992 Clinical use of bone biopsy. In: Coe FL, Favus MJ (eds) Disorders of Bone and Mineral Metabolism. Raven, New York, pp. 455–474.
123. Meunier PJ, Edouard C, Arlot M, et al. 1979 Effects of 1,25-dihydroxyvitamin D on bone mineralization. In: MacIntyre I, Szelke M (eds) Molecular Endocrinology. Elsevier/North-Holland, Amsterdam, pp. 283–292.
124. Marie PJ, Glorieux FH 1981 Histomorphometric study of bone remodeling in hypophosphatemic vitamin D-resistant rickets. Metab Bone Dis Related Res **3**:31–38.
125. Manolagas SC, Haussler MR, Deftos LJ 1980 1,25-Dihydroxyvitamin D_3 receptor-like macromolecule in rat osteogenic sarcoma cell lines. J Biol Chem **255**:4414–4417.
126. Stumpf WE, Ssar M, DeLuca HF 1981 Sites of action of $1,25(OH)_2$ vitamin D_3 identified by thaw-mount autoradiography. In: Cohn DV, Talmage RV, Matthews JL (eds) Hormonal Control of Calcium Metabolism. Excerpta Medica, Amsterdam, pp. 222–229.
127. Manolagas SC, Burton DW, Deftos LJ 1981 1,25-Dihydroxyvitamin D_3 stimulates the alkaline phosphatase activity of osteoblast-like cells. J Biol Chem **256**:7115–7117.
128. Lian JB, Coutts M, Canalis E 1985 Studies of hormonal regulation of osteocalcin synthesis in cultured fetal rat calvariae. J Biol Chem **260**:8706–8710.
129. Marie PJ, Hott M, Garba M-T 1985 Contrasting effects of 1,25-dihydroxyvitamin D_3 on bone matrix and mineral appositional rates in the mouse. Metabolism **34**:777–783.
130. Krempien B, Klimpel F 1980 Action of 1,25-dihydroxycholecalciferol on cartilage mineralization and on endosteal lining cells of bone. Virch Arch [A] **388**:335–347.
131. Stern PH 1980 The D vitamins and bone. Pharmacol Rev **32**:47–80.
132. Dickson IR, Roughley PJ 1993 The effects of vitamin D deficiency on proteoglycan and hyaluronate constituents of chick bone. Biochim Biophys Acta **1181**:15–22.
133. Tulpule PG, Patwardhan VN 1954 Mode of action of vitamin D. The effect of vitamin D deficiency on the rate of anaerobic glycolysis and pyruvate oxidation by epiphyseal cartilage. Biochem J **58**:61–65.
134. Klein GL, Simmons DJ 1993 Nutritional rickets: Thoughts about pathogenesis. Ann Med **25**:379–384.
135. Eastwood JB, Bordier PJ, Clarkson EM, Tun Chot S, De Wardener HE 1973 The contrasting effects on bone histology of vitamin D and of calcium carbonate in the osteomalacia of chronic renal failure. Clin Sci Mol Med **47**:23–42.
136. Marsh ME, Munne AM, Vogel JJ, Cui Y, Franceschi RT 1995 Mineralization of bone-like extracellular matrix in the absence of functional osteoblasts. J Bone Miner Res **10**:1635–1643.
137. Stauffer M, Baylink D, Wergedal J, Rich C 1975 Decreased bone formation, mineralization, and enhanced resorption in calcium-deficient rats. Am J Physiol **225**:269–276.
138. Ecarot B, Glorieux FH, Desbarats M, Travers R, Labelle L 1992 Defective bone formation by Hyp mouse bone cells transplanted into normal mice: Evidence in favor of an intrinsic osteoblast defect. J Bone Miner Res **7**:215–220.

Chapter 42

Vitamin D Deficiency and Nutritional Rickets in Children

JOHN M. PETTIFOR AND ELVIS D. DANIELS
MRC Mineral Metabolism Research Unit, Department of Paediatrics, University of the Witwatersrand and Baragwanath Hospital, Soweto, Gauteng, South Africa

I. INTRODUCTION

Rickets is a clinical syndrome that presents in children as a result of a failure of or delay in mineralization of the growth plate of growing bones. There are numerous different causes, the majority of which can be grouped into three major categories: those that primarily result in a failure to maintain normal calcium homeostasis; those that primarily affect phosphate homeostasis; and those that directly inhibit the mineralization process. Globally, rickets due to nutritional causes (which fall into the calciopenic group) remains the most frequent form of the disease seen. However, in a number of developed countries, such as the United States, the genetic forms of hypophosphatemic rickets are now probably more prevalent than the nutritional causes outside the neonatal period, as a result of the fortification of foods with vitamin D and the use of vitamin D supplements in at-risk groups.

II. HISTORICAL PERSPECTIVE

Although nutritional rickets is often considered to be a consequence of rapid industrialization, descriptions of rickets have been attributed to both Homer (900 B.C.) and Soranus Ephesius (A.D. 130). More recently, attention was drawn to rickets by Daniel Whistler in 1645, and later Francis Glisson (1650) provided a classic description of the disease. It was described as a disease that occurred in young children, produced severe deformities, and was often fatal. The condition, which was known in Europe as "the English disease," was more common in the cities than in rural areas. Prior to the industrial revolution, it was associated with affluence, as the children of well-to-do families were often completely covered by clothing and were kept indoors [1]. With the migration of large numbers of people from rural to urban areas at the time of the industrial revolution, the disease became associated with poverty and overcrowding in the developing urban slums.

A number of studies in the late nineteenth and early twentieth centuries documented the almost universal prevalence of rickets in young children in cities in northern Europe (for example, Glasgow) [2]. However, with the realization of the importance of ultraviolet light in preventing nutritional rickets and the discovery and isolation of vitamin D in the first quarter of the twentieth century (see Chapter 1), programs were introduced to prevent vitamin D deficiency.

In the United Kingdom, a number of foods were fortified with vitamin D during the Second World War. This led to a rapid reduction in the number of children diagnosed with rickets, but in the following years the incidence of idiopathic hypercalcemia rose in infants due to uncontrolled fortification of various foods (especially milk and cereals) leading to daily vitamin D intakes of 100 μg or more [2]. As a result, the fortification of foods and the use of vitamin D supplements fell into disrepute, and the prevalence of vitamin D deficiency and nutritional rickets has increased, particularly among the immigrant Asian population.

In the United States, the universal fortification of milk with vitamin D at 400 IU/quart from the 1930s has almost eradicated nutritional rickets except in families who exclude milk from their diets [3]. However, as had occurred in the United Kingdom, a few cases of vitamin D toxicity have been reported as a result of the lack of adequate monitoring of the fortification process [4]. In central Europe, rickets has been effectively prevented in infants and young children by the intermittent administration (every 3 to 5 months for the first 2 years of life) of high doses of vitamin D ("stosstherapie") [5].

III. THE EPIDEMIOLOGY OF VITAMIN D DEFICIENCY AND NUTRITIONAL RICKETS

Vitamin D deficiency is a prerequisite for the development of nutritional rickets in the majority of children. Thus, the disease is typically associated with a lack of ultraviolet light exposure or dietary vitamin D. As commonly ingested foods are generally deficient in vitamin D (the exceptions being oily fishes or fortified foods), the normal diet contributes little to the vitamin D status of an individual [6], thus, adequate skin exposure to ultraviolet radiation is essential for the prevention of rickets in most situations [7] (see Chapter 3). Consequently, rickets occurs most frequently in infants before they are able to walk and get outside, in children living in countries at the extremes of latitude, or in communities in which social norms prevent adequate sunlight exposure through excessive skin coverage by clothes or through the practice of purdah.

Vitamin D deficiency rickets is most prevalent in children under 2 years of age, with a peak incidence between 3 and 18 months [8,9]. The disease is uncommon in infants under 3 months of age, because 25-hydroxyvitamin D (25OHD) readily crosses the placenta [10,11], thus providing the newborn infant with some protection against vitamin D deficiency (see Chapter 28). Because 25OHD is not the major storage form of vitamin D and has a turnover time of 3 to 4 weeks, serum levels fall rapidly after birth unless additional sources of vitamin D are obtained by the young infant [12]. Neonatal or congenital rickets has been described in infants born to mothers who are themselves vitamin D deficient [13–15]. In a number of studies, vitamin D deficiency rickets has been noted to occur more commonly in boys than girls [16–18], but the mechanism for this remains unclear. It has been suggested that vitamin D deficiency rickets might be an hereditary disease, which manifests itself only under adverse circumstances [17,19]. In a study of infants with rickets and their parents [19], urinary excretion of α-amino acids was increased in one-third of the infants long after the rickets had healed, many of the parents had increased amino acid and phosphorus excretion, and a good correlation was found between the excretion of individual amino acids by an infant and its parents. The authors suggested that these findings indicate a genetic factor playing a role in predisposing a child to rickets, however, the mode of inheritance is unclear.

In the early literature, breast feeding was reported to be protective against rickets [20]. More recently, however, it has been described as a risk factor for the development of rickets [3,21–23]. In recent years, specifically designed breast milk substitutes have replaced natural cow's milk as the major source of nutrients for the non-breast-fed infant. This alteration in feeding patterns may account for the apparent change in risk associated with breast-feeding for several reasons; first, breast milk substitutes are fortified with vitamin D at 400 IU/liter, whereas natural cow's milk contains little vitamin D [24]; second, the calcium:phosphorus ratio in breast milk substitutes (ratio ~2 : 1) is more appropriate than that in cow's milk (ratio ~1 : 1) for optimizing intestinal calcium absorption; and third, breast milk usually contains only small quantities of vitamin D or its metabolites [24,25]. However, there is evidence that vitamin D metabolites may cross into breast milk from the mother in sufficient quantities to maintain normal serum concentrations of 25OHD in the suckling infant if the mother receives vitamin D supplements in high doses (~2000 IU/day) [12,26].

In the breast-fed infant not receiving vitamin D supplements, the maintenance of an adequate vitamin D status is dependent mainly on the infant's exposure to ultraviolet light [27,28]. Specker and co-workers [27,28] have shown a marked seasonal variation in serum 25OHD concentrations in breast-fed infants, which is dependent on the time spent outdoors and on the extent of skin exposed to sunlight. They have estimated that an infant in Cincinnati (latitude 39°09′ N) requires to be outdoors for either 20 min a week in a diaper only or for 2 hr a week fully clothed but without a hat to maintain normal circulating concentrations of 25OHD [27].

Seasonal variations in serum 25OHD concentrations have also been documented in a number of countries in older children and adults [29,30], and these variations appear to correlate with the amount of ultraviolet light reaching the earth [31]. These observations highlight the importance of the photobiosynthesis of vitamin D_3 in the skin to prevent vitamin D deficiency and thus rickets in many populations in the world. In a number of counties such as Turkey [32], Saudi Arabia [23], India [33], China [15], Algeria [30], Iran [8], Kuwait [34], Nigeria [35], and in others in the tropics and subtropics [36], however, rickets remains a problem despite generally good daily hours of sunshine. A number of factors contribute to the persistence of the problem in these areas; these include overcrowding and poverty, purdah, lack of access to sunlight, a lack of vitamin D-fortified foods or regular vitamin D supplements, and diets that are low in calcium and high in inhibitors of calcium absorption.

In the United States, despite the almost complete eradication of vitamin D deficiency among Caucasian children, several studies have highlighted the resurgence of the problem in specific groups [3,21,22,37,38], namely vegans and children on macrobiotic diets, children who are breast-fed for prolonged periods, and black children. It is suggested that the combination of the generally low dietary calcium intakes associated with vegetarian diets, low dietary vitamin D intakes because of the lack of dairy products, and decreased vitamin D_3 formation in dark skin contribute to an increased risk for vitamin D deficiency in these groups.

A similar pattern has also been documented in a number of European countries [39–42]. Perhaps the most intensively investigated community has been the Asian population in Great Britain because of the high prevalence of vitamin D deficiency and "Asian rickets" in children of all ages [43] and adults [44]. The age distribution of Asian children with rickets is described as being biphasic, with one peak in the classic age group of vitamin D deficiency (9–36 months) and the other related to the pubertal growth spurt [2,45]. Since the initial descriptions of the resurgence of rickets in the United Kingdom in the early 1960s, numerous studies have been undertaken to determine why Asians are predisposed to the problem when other immigrants such as West Indians are not. Among the hypotheses put forward are simple vitamin D deficiency due to dark skin and lack of skin surface exposed to sunlight [46,47], low calcium diets associated with vegetarianism [48], and impaired intestinal calcium absorption associated with high phytate diets [49]. A unifying hypothesis, proposed by Clements [39], suggests that in a situation of relative vitamin D insufficiency, the low dietary calcium and high phytate content of the typical Asian vegetarian diet leads to mild secondary hyperparathyroidism and a resultant increase in the catabolism of vitamin D. The progressive decline in vitamin D status culminates in the development of rickets (see Section VIII of this chapter).

IV. CLINICAL PRESENTATION

The majority of clinical signs in children with rickets results from the effects of vitamin D deficiency on the mineralization process at the growth plate or on calcium homeostasis. Fraser and co-workers [18] have described three stages in the progression of vitamin D deficiency. Stage I is characterized by hypocalcemia, with clinical signs related to the presence of hypocalcemia; in stage II, the clinical features of impaired bone mineralization become apparent; and in stage III signs of both hypocalcemia and severe rickets are present. This division of the progression of vitamin D deficiency rickets is conceptually useful, but there is considerable clinical overlap between the various stages.

The early clinical manifestations of vitamin D deficiency (stage I) are related to hypocalcemia and are more commonly seen in young infants (less than 6 months of age). They may present with convulsions [50], apneic episodes [51], or tetany with no clinical signs of rickets. Few children present clinically in stage I, as the majority who later present with rickets pass through this phase without developing symptomatic hypocalcemia. Pseudotumor cerebri and cataracts, probably due to hypocalcemia, have been reported in a young infant with rickets [52]. It has been suggested that symptomatic hypocalcemia in infants with vitamin D deficiency might be precipitated by an acute illness [53], in which there is a release of intracellular phosphate [54].

As the deficiency progresses, the classic features of rickets become apparent. Typically the infant or young child presents with a delay in motor milestones, hypotonia, and progressive deformities of the long bones. The deformities are most noticeable at the distal forearm with enlargement of the wrist and bowing of the distal radius and ulna, and in the legs with progressive lateral bowing of the femur and tibia. The site and type of deformity are dependent on the age of the child and the weight bearing patterns in the limbs. Thus, in the small infant deformities of the forearms and anterior bowing of the distal tibias are more common, whereas in the toddler who has started to walk an exaggeration of the normal physiological bowing of the legs (genu varum) is characteristic. In the older child valgus deformities of the legs or a windswept deformity (valgus deformity of one leg and varus deformity of the other) may be apparent. The characteristic feature in the ribs is enlargement of the costochondral junctions leading to visible beading along the anterolateral aspects of the chest (the rachitic rosary). In the infant or young child with severe rickets, the muscular pull of the diaphragmatic attachments to the lower ribs results in the development of the Harrison's sulcus (Fig. 1). The negative intrapleural pressure associated with breathing may result in narrowing of the lateral diameter of the chest (the violin case deformity) with consequent severe respiratory embarrassment. Increased sweating has also been described in young infants and probably relates to the increased work of breathing due to the decreased compliance associated with the excessively malleable ribs. In premature infants with rickets (which may also be due to dietary phosphorus deficiency), fractures of the ribs may be the first clinical sign to draw attention to the problem [55].

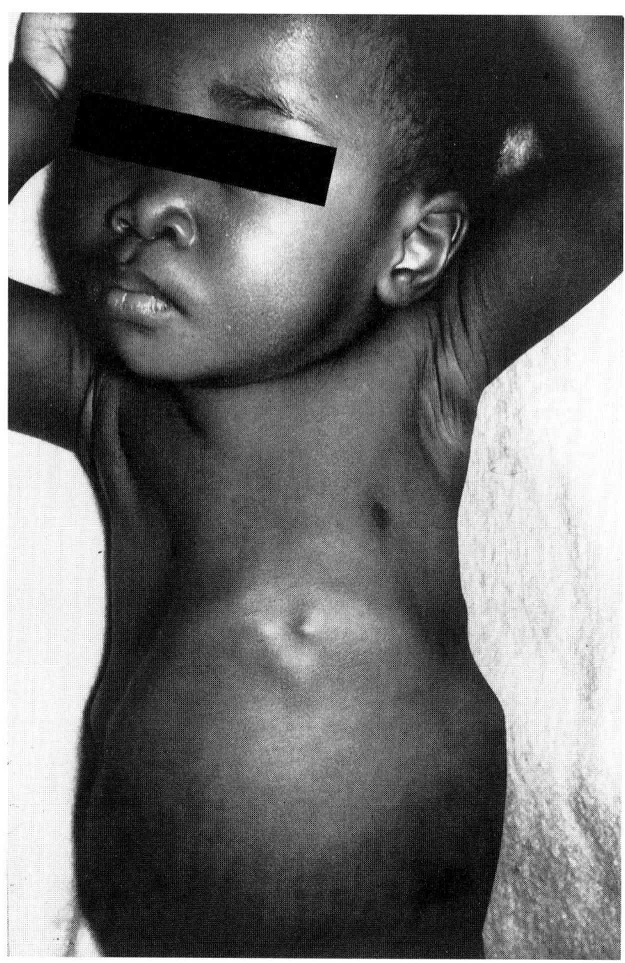

FIGURE 1 Young infant with vitamin D deficiency rickets, presenting with respiratory distress. The child shows the characteristic deformities of the chest associated with severe rickets. The lateral diameter of the chest is reduced and bilateral Harrison's sulci are present. The abdomen has a protuberant appearance.

Other skeletal abnormalities include a delay in the closure of the fontanelles, parietal and frontal bossing, and the presence of craniotabes [56]. Although craniotabes is often considered to be pathognomic of rickets, it may occur in normal young infants [57,58]. Craniosynostosis, involving the coronal or multiple sutures, has been described in approximately 25% of patients, who were followed up after having suffered from vitamin D deficiency rickets [59]. The condition appears to be related to the degree of severity of the rickets and thus to the severity of the mineralization defect, and inversely to the age of onset of the rickets.

A delay in tooth eruption is a feature of rickets in the young child, and enamel hypoplasia of teeth may occur if rickets develops prior to the completion of enamel deposition. The latter has been reported in the primary dentition of infants born to mothers who are vitamin D deficient [60], and it is seen in the secondary dentition of children who have suffered from rickets during early childhood.

Hypotonia, decreased activity, and a protruberant abdomen are characteristic features of advanced vitamin D deficiency rickets in the infant and young child. These signs are probably analogous to the proximal muscle weakness described in vitamin D-deficient adolescents and adults [61]. In this situation deep tendon reflexes are retained and may be brisk. The pathogenesis of the myopathy is thought to be due primarily to vitamin D deficiency, rather than hypophosphatemia (see Chapter 32).

Infants and young children with rickets are prone to an increased number and severity of infections [32]. Although the increase in respiratory infections may be explained on the thoracic cage abnormalities (softening of the ribs, enlarged costochondral junctions, and decreased thoracic movement due to muscle weakness), other reasons for the increase in diarrheal disease must be sought. The now well-documented role of 1,25-dihydroxyvitamin D [1,25$(OH)_2$D] in modulating immune function [62] may contribute to the observed increase in infections (see Chapter 29). Impaired phagocytosis [63] and neutrophil motility [64] have been described in children with vitamin D deficiency rickets.

A possibly associated abnormality is anemia, thrombocytopenia, leukocytosis, myelocytosis, erythroblastosis, myeloid metaplasia, and hepatosplenomegaly (von Jacksch-Luzet syndrome), which has been described in infants with rickets [65,66]. Although the exact pathogenetic mechanisms for this syndrome are unclear, vitamin D deficiency has been implicated on the basis of clinical observations that vitamin D therapy cures the condition and experimental evidence showing that 1,25$(OH)_2$D has antiproliferative activity on myeloid leukemia cell lines [67] (see Chapter 68).

V. BIOCHEMICAL ABNORMALITIES

The hallmark of vitamin D deficiency is a low circulating level of 25OHD. In children, a normal range of approximately 10 to 50 ng/ml (25–125 nmol/liter) has been found in the majority of studies [27,68,69] conducted in communities in which vitamin D deficiency rickets is uncommon. However, the normal range is dependent on the vitamin D and calcium contents of the diet and on the ultraviolet light exposure of the skin. In a number of studies a marked seasonal variation in levels has been recorded [28,29,70], reflecting in part the seasonal changes in the amount of ultraviolet light reaching the earth. In countries at high latitude where foods are not vitamin D fortified, serum 25OHD concentrations in some "normal" children may be in the range documented in symptomatic children with vitamin D deficiency [29,43]. Thus, the development of symptoms depends on the duration and severity of low 25OHD concentrations and on the ability of the kidney to achieve adequate 1,25$(OH)_2$D concentrations in the face of decreased substrate for the gastrointestinal tract to maintain calcium absorption at a level appropriate to meet the demands of the growing child. In children with 25OHD concentrations within the normal reference range, there is no correlation between serum 25OHD and 1,25$(OH)_2$D concentrations. However, once 25OHD levels fall below about 10 ng/ml (25 nmol/liter) 1,25$(OH)_2$D concentrations correlate with those of 25OHD [71,72]. In the majority of studies in which 25OHD values have been measured in children with rickets, concentrations have been found to be less than 4–5 ng/ml (10–12.5 nmol/liter) in most patients [43,73,74], although other workers have found higher values [75–77].

The classic biochemical changes in vitamin D-deficient children who have radiological changes of rickets are a combination of hypocalcemia, hypophosphatemia, and elevated alkaline phosphatase and parathyroid hormone (PTH) concentrations. In the early phase of vitamin D deficiency before the development of radiological signs (stage I), hypocalcemia may be the only biochemical abnormality [18]. Acute illness may precipitate hypocalcemia in the vitamin D-depleted infant through the sudden increase in serum phosphorus concentrations [54]. As the disease progresses, secondary hyperparathyroidism in response to the hypocalemia induces a partial correction of the low serum calcium concentration, which may return to levels within the normal range, and increases phosphate excretion by the kidney, resulting in hypophosphatemia (stage II) [78]. At this stage serum alkaline phosphatase concentrations are usually elevated, and other renal manifestations of secondary hyperparathyroidism, such as increased cyclic AMP excretion, generalized aminoaciduria, impaired acid excretion, and decreased urinary calcium excretion, are found [79]. In stage III of the disease, the radiological features are more severe, hypocalcemia once again becomes apparent, and alkaline phosphatase concentrations rise further [18].

The elegant studies conducted by Fraser and coworkers [18], before the availability of immunoassays for the measurement of serum PTH concentrations, suggested that in stage I vitamin D deficiency serum concentrations of PTH are normal, as serum phosphorus values and urinary amino acid excretion are within the normal range. Their patients only had radiological evidence of calvarial demineralization without other bone changes of rickets. More recent data support this conclusion, as normal PTH concentrations have been reported in the early hypocalcemic phase of symptomatic vitamin D deficiency [50]. However, Kruse [80] found elevated PTH values and increased urinary cyclic AMP excretion in children with stage I rickets. This discrepancy can possibly be explained by the fact that the patients in the latter study might represent a slightly later stage of vitamin D deficiency than those in the other studies as the children were selected on the presence of radiological changes.

Evidence of end-organ resistance to PTH has been found in young children with both mild and more severe radiological rickets [80,81]. In the study by Kruse [80] the children with mild rickets remained normophosphatemic and had normal renal handling of phosphate (tubular maximum for phosphate—TmP/GFR) despite elevated PTH concentrations and increased urinary cyclic AMP excretion. Similar indirect evidence of PTH resistance (hypocalcemia, normophosphatemia, and a decrease in the phosphate excretion index) was noted by Taitz *et al.* [81] in infants with more severe radiological rickets. Resistance to PTH has also been described in hypocalcemic adolescents with mild rickets [82]. Usually however, as the severity of the rickets increases (stages II and III), so PTH values rise further and renal hyporesponsiveness is overcome [80]. Thus, hypophosphatemia and a decrease in TmP/GFR become hallmarks of the disease.

Markers of bone turnover are typically elevated in nutritional rickets in response to the development of secondary hyperparathyroidism. Urinary hydroxyproline excretion may be within the normal range in stage I rickets but is elevated in patients with radiological rickets [80]. One report describes an increase in serum concentrations of pyridinoline cross-linked carboxyterminal telopeptide of type 1 collagen in children with rickets [83]. Similarly serum alkaline phosphatase values may be normal in stage I of vitamin D deficiency, but rise with the degree of severity of the radiological changes. Of all the readily available biochemical tests that might be deranged in nutritional rickets, alkaline phosphatase has been used most frequently as a screening test. However, although it is elevated in the vast majority of children with radiological changes, it lacks specificity [36,58,84]. Further, the degree of elevation of serum concentrations does not necessarily correlate with the radiological severity of the bone disease [36]. Whether the measurement of bone-specific alkaline phosphatase in patients with suspected rickets will be of greater sensitivity and specificity is unclear at present.

Osteocalcin is a noncollagenous bone matrix protein that binds to hydroxyapatite and is secreted by osteoblasts during mineralization [85]. Serum concentrations are higher in children than adults and peak during the pubertal growth spurt [86]. In the few children with untreated vitamin D deficiency rickets, in whom serum osteocalcin concentrations have been measured, values have been reported to be low or normal and rise rapidly with therapy to supranormal concentrations [87]. A Nigerian study [83] of 12 rachitic children found slightly elevated serum osteocalcin concentrations compared to values in age-matched controls; however, it was suggested that the children might have suffered from dietary calcium deficiency rather than vitamin D deficiency.

In patients with vitamin D deficiency, serum $1,25(OH)_2D$ concentrations have been reported to be low, normal, or even elevated [71,74,77,80,88,89], whereas $24,25(OH)_2D$ values are low or undetectable [71,74,77,89,90]. Kruse [80] found that $1,25(OH)_2D$ values were higher in children with stage II rickets than in those with either stage I or stage III rickets. The finding of normal or elevated levels of $1,25(OH)_2D$ in vitamin D deficiency rickets has led some researchers to conclude that other vitamin D metabolites, such as $24,25(OH)_2D$, are necessary for the maintenance of normal calcium homeostasis [88,91,92]. Others have suggested that although $1,25(OH)_2D$ concentrations are within the normal range, they are inappropriately low for the degree of hyperparathyroidism [80,89]. As discussed below, the latter hypothesis is more likely.

A possible pathophysiological progression of vitamin D deficiency rickets in children may be described as follows [80,93]. As the child becomes progressively vitamin D depleted, a stage is reached when the serum 25OHD concentration falls below that required to maintain a serum $1,25(OH)_2D$ level necessary for normal calcium homeostasis. The resultant hypocalcemia (stage I rickets) leads to secondary hyperparathyroidism, which through the stimulation of 1α-hydroxlase, increases $1,25(OH)_2D$ production despite falling 25OHD concentrations. In concert with PTH, $1,25(OH)_2D$ increases bone resorption and intestinal calcium absorption, thus returning serum calcium concentrations toward normal (stage II rickets). The presence of hypophosphatemia at this stage is probably responsible for the mineralization defect and the development of radiological rickets.

It is during this phase that serum $1,25(OH)_2D$ concentrations may be relatively elevated [80]. A possible explanation for the failure of the elevated $1,25(OH)_2D$ levels to reduce the hyperparathyroidism and heal the bone disease at this stage is that they are not high enough to meet the increased calcium requirements associated with the generalized mineralization defect and increased bone turnover. Support for this hypothesis comes from data which show that $1,25(OH)_2D$ concentrations rise to considerably higher levels (3–5 times normal) during the healing process even when only small doses of vitamin D are provided [71,80] and that intestinal calcium absorption may reach approximately 80% of dietary calcium intake during this phase [71].

As 25OHD concentrations fall further, $1,25(OH)_2D$ levels once again fall, despite persistent hyperparathyroidism, because of the lack of substrate. Hypocalcemia again becomes apparent as intestinal calcium absorption falls and calcium mobilization from bone decreases due to the lack of $1,25(OH)_2D$, which has a permissive action on bone resorption by PTH [94,95]. The combination of both hypocalcemia and hypophosphatemia increases the severity of the bone disease (stage III).

VI. RADIOLOGICAL CHANGES

The typical radiological changes associated with vitamin D deficiency rickets have been well described and are discussed and shown in Chapter 40. Stage I rickets characteristically shows few radiological signs; although demineralization of the calvarium and loss of definition of the skull sutures have been described [18], these signs are difficult to quantitate. The changes of rickets are best visualized at the growth plate of rapidly growing bones. Thus, in the upper limbs, the distal ulna is the site which may show best the early signs of impaired mineralization. In the older child, the metaphyses above and below the knees become more useful.

The early signs of rickets include widening of the epiphyseal plate and a loss of definition of the provisional zone of calcification at the metaphysis [96]. As the disease progresses, the disorganization of the growth plate becomes more apparent with cupping, splaying, spur formation, and stippling [54,97] (Fig. 2). The appearance of epiphyses may be delayed, or they may appear small, osteopenic, and ill-defined.

The shafts of the long bones show features of both hyperparathyroidism and osteomalacia. Osteopenia is a characteristic feature that in the so-called atrophic form of the disease may be very severe [36]. The cortices become thin and may show periosteal new bone formation, although this is more frequently seen during healing. The trabecular pattern is reduced and appears coarse. Deformities of the shafts of the long bones are typically present, and in severe rickets, pathological fractures and Looser's zones may be noted. In vitamin D deficiency rickets, features of hyperparathyroidism, such as subperiosteal erosions, are uncommon. However, loss of the lamina dura around the teeth is frequently seen.

Enlargement and splaying of costochondral junctions on the lateral radiographs of the chest have been used as a sign of rickets. However, in one study mild changes were found to be unreliable, as their presence did not correlate with serum 25OHD concentrations or with other features of rickets at the distal radius and ulna [98].

Rickets during adolescence may be difficult to detect using the conventional radiographic sites of the wrist and knees as the epiphyseal plates narrow and epiphyses fuse. A radiograph of the pelvis may be useful in this situation, as the secondary iliac and ischial ossification centers may be abnormally wide [99]. These centers appear at puberty and normally unite with the rest of the bone between 15 and 25 years of age.

The sign of early healing of rickets is described as broadened bands of increased density replacing the normal sharp metaphyseal lines (Fig. 2). The demarcation of the broad bands on the diaphyseal side of the shaft may be poorly defined [96]. Healing in more severe cases of rickets may first appear as bands of mineralization

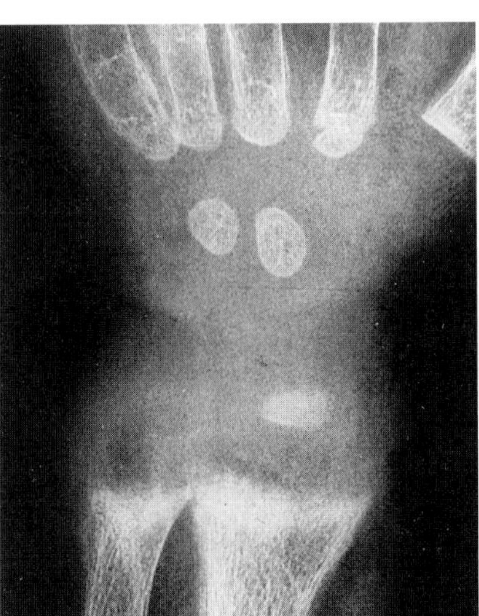
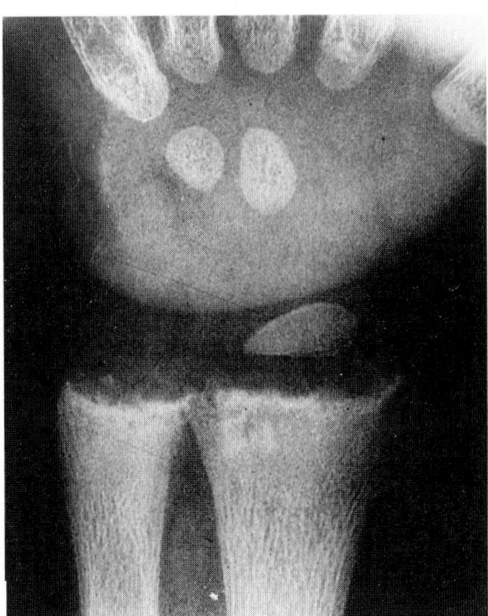
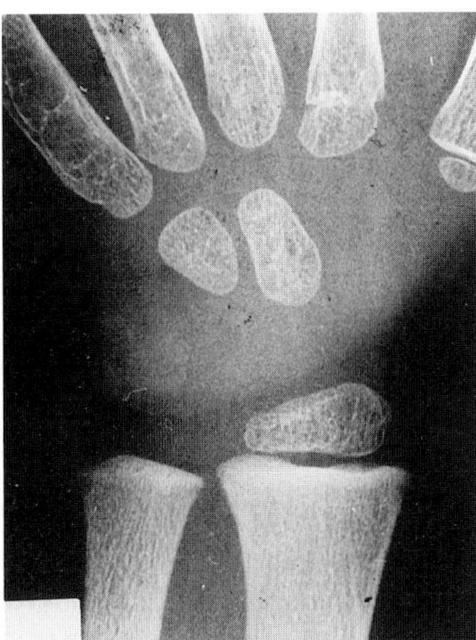

FIGURE 2 Radiographic features of vitamin D deficiency rickets at the wrist. (Left) Untreated vitamin D deficiency showing underdevelopment of the epiphyses, widening of the epiphyseal plates, splaying and irregularity of the metaphyses, and loss of the provisional zones of calcification. The shafts show coarsening of the trabecular pattern and loss of the normal cortical definition. (Middle) Response after 3 months of vitamin D therapy. The metaphyses show clear signs of healing with dense bands of calcification at the distal ends of the metaphyses, narrowing of the epiphyseal plates, and more clearly defined epiphyses. The trabecular pattern still appears coarse but shows improvement. (Right) Six months after starting vitamin D therapy. The radiographic changes of rickets have disappeared. The epiphyses, epiphyseal plates, metaphyses, and trabecular structure are normal. Reprinted with permission from Pettifor JM. Calcium, Phosphorus, and Vitamin D. In: Ballabriga A, Brusner O, Dobbing J, Gracey M, Senterre J (eds): *Clinical Nutrition of the Young Child.* Raven, New York. 1991.

occurring distal to and separated from the irregular and frayed metaphyses. There is then gradual filling in of the demineralized area proximal to the initial band of mineralization with remodeling and the development of a normal trabecular pattern. Periosteal new bone formation may be seen, which gradually becomes incorporated into the cortices of the long bones.

VII. TREATMENT AND PREVENTION

A. Treatment

Vitamin D deficiency rickets can be effectively treated by oral administration of small doses of either vitamin D_2 or vitamin D_3, provided there is no evidence of gastrointestinal malabsorption. Stanbury *et al.* [71] showed that an oral vitamin D dose of between 200 and 450 IU/day produced a rise in serum $1,25(OH)_2D$ concentrations to normal values within 1–3 days. The latter climbed to reach a peak some five times the normal mean after 1 to 3 weeks, despite serum 25OHD values remaining less than 10 ng/ml (25 μmol/liter). Spontaneous improvement in the biochemical features of rickets has been reported to occur in children with biochemical abnormalities during the summer months, associated with a rise in serum 25OHD values due presumably to increased ultraviolet light exposure [73].

More generally, however, doses of vitamin D between 5000 and 15,000 IU/day for 3 to 4 weeks are used in the management of rickets. Normalization of serum calcium and phosphorus concentrations occur within 1 and 3 weeks [80], although serum alkaline phosphatase concentrations and urinary hydroxyproline excretion remain elevated for several months. Despite the return to normal of serum PTH, calcium, and phosphorus values within 3 weeks, serum $1,25(OH)_2D$ concentrations may remain elevated for up to 10 weeks [77,80]. Serum $24,25(OH)_2D$ values, which are often undetectable in the untreated patient, rise with the progressive increase in serum 25OHD concentrations during treatment [77]. Lower doses of vitamin D (1000–2000 IU/day) do produce healing, but the response is less rapid.

In Central Europe, a single dose of 600,000 IU vitamin D (either orally or intramuscularly) has been found to be effective, resulting in a rapid improvement in biochemical abnormalities within a few days and radiological evidence of healing within 2 weeks [54,100]. A sustained drop in serum alkaline phosphatase is seen within 6 to 12 weeks [100]. Single dose therapy has an advantage over smaller daily doses as it avoids the problem of compliance, which was thought to be responsible for the lack of response in 40% of children with vitamin D deficiency rickets in a study conducted in Kuwait [101].

Besides ensuring an adequate vitamin D intake, the calcium content of the diet should be optimized (between 600 and 1000 mg/day) during the initial stages of management. This is particularly true for children who are on vegetarian or low calcium containing diets and for those who are severely hypocalcemic. In symptomatic patients, a single dose of calcium gluconate (1–2 ml/kg of a 10% solution) may be given slowly intravenously and the diet supplemented with 10% calcium gluconate (5 ml/kg/day in divided doses).

B. Prevention

As discussed in Section III of this chapter, vitamin D deficiency rickets remains a problem in a number of at-risk groups despite readily available methods of preventing the disease. A number of studies in several countries have been conducted prospectively in breast-fed infants to assess vitamin D status. Several have shown a fall in serum 25OHD concentrations in those infants who were not vitamin D supplemented, to levels in the vitamin D-deficient range [12,102,103], although this is not a universal finding [104,105]. Thus, although some nutritionists believe that it is unnecessary to supplement breast-fed infants with vitamin D if the social and environmental conditions are favorable [106], it is probably prudent to recommend a daily supplement of 400 IU for all breast-fed infants unless an adequate exposure to sunlight can be ensured [107–109]. In a prospective study conducted in China on infants from birth to 6 months of age, it was concluded that supplemental vitamin D at a dose of 400 IU/day produced more normal circulating 25OHD concentrations than did either 100 or 200 IU/day [110]; however, in a small number of infants even 400 IU/day did not maintain 25OHD levels above 11 ng/ml (27.5 nmol/liter). Nevertheless, no radiological evidence of rickets was found in any of the infants in the three groups at 6 months of age. The use of 400 IU vitamin D daily to prevent vitamin D deficiency in at-risk infants is supported by a study from Turkey, in which it was found that no rickets occurred in infants receiving the supplement compared to a prevalence of 3.8% in those that did not [32]. Infants fed milk formulas or cow's milk fortified with vitamin D do not require vitamin D supplements, as their intake of milk generally provides sufficient vitamin D to prevent deficiency [104].

As discussed earlier, high single dose therapy (stosstherapie) has been used with success in the treatment of vitamin D deficiency rickets in a number of countries. A similar dose has also been used on a regular intermittent basis of every 3 to 5 months for the first 18 months of life as a means of prevention of vitamin D deficiency.

Few data are available on the efficacy of such prophylaxis. However, in a study to assess the effect of these high doses of vitamin D (600,000 IU) on calcium and vitamin D metabolism in infants [5], it was found that serum 25OHD concentrations reached very high levels 2 weeks after each administration, but that these had returned to normal prior to the next dose. $1,25(OH)_2D$ generally remained within the normal range; however, 34% of infants were hypercalcemic at some stage during the study. These results led the authors to conclude that the dosage regimen as used during the study was excessive and unsafe [5].

Following these results, a study to assess the efficacy of a single dose of vitamin D (600,000 IU or 15 mg) at 15 days of life, compared to 200,000 IU (5 mg) at birth or 100,000 IU (2.5 mg) at birth and every 3 months for 9 months was undertaken [111]. Two weeks after the initial administration, 28 of 30 infants in the 15-mg group had serum 25OHD concentrations above the upper limit of normal (mean ± SD for the group, 307 ± 160 nmol/liter) compared to 58% (150 ± 55 nmol/liter) in the 5-mg group and 23% (92 ± 42 nmol/liter) in the 2.5-mg group. At 6 months of age, 50% of the infants who had received 15 mg at birth still had elevated 25OHD concentrations, whereas in the 5-mg group none had elevated levels. In the group receiving 2.5 mg every 3 months, serum 25OHD values were in the normal range on each occasion prior to receiving the next dose. Although hypercalcemia was not detected in any of the infants, serum calcium concentrations were higher in the 15-mg group 2 weeks after receiving the dose than in the other two groups. The authors concluded that intermittent doses of 15 mg vitamin D during the first year of life are excessive, and that 5 mg every 6 months or even better 2.5 mg every 3 months is more suitable for the prevention of vitamin D deficiency in at-risk infants.

Although the untargeted fortification of foods other than milk and infant milk formulas has been used in the past as a means of addressing the high prevalence of vitamin D deficiency in countries where the risk of vitamin D deficiency is high, the problems experienced in the United Kingdom after the Second World War has led to it falling into disfavor (see Section II of this chapter). More recently, the use of targeted food fortification has been studied in the Asian community in Great Britain as a means of reducing the high prevalence of vitamin D deficiency in both adults and children in that community [112]. In a small pilot study, it was found that the fortification of chapatti flour at a level of 6000 IU/kg produced a sustained and significant rise in serum 25OHD concentrations to values within the normal range over a 6-month period comparable to that achieved by a weekly dose of 3000 IU vitamin D. Over the 6-month period serum calcium and phosphorus values rose, and the number of subjects with biochemical abnormalities suggestive of rickets fell. The authors conclude that fortification of chapatti flour is an inexpensive and effective method of preventing vitamin D deficiency in the Asian community in Britain, and it has an advantage over daily or intermittent vitamin D supplementation, as the compliance utilizing the latter form of prevention is often poor. Nevertheless, food fortification remains an emotional public issue. In the United States not only have there been isolated reports of vitamin D toxicity related to inadequate monitoring of the fortification process, but underfortification is also a problem. Holick [7] reports that in one study less than 30% of milk samples from all sections of the Unted States and in British Columbia contained the specified amount, and that 14 to 21% of skim milk samples contained no detectable vitamin D.

VIII. DIETARY CALCIUM DEFICIENCY

Conventional wisdom would have it that nutritional rickets is primarily due to vitamin D deficiency, although dietary calcium intake modulates the severity and rapidity of onset of the disease [113,114]. However, since the 1970s evidence has been accummulating that implicates low dietary calcium intakes as a cause of rickets in the face of serum 25OHD concentrations within the normal reference range.

Isolated case reports of rickets developing in infants and toddlers, who were placed on very low calcium diets, have been published [115–117]. Their clinical and biochemical presentations were very similar to those of infants with vitamin D deficiency, but in three of five infants serum 25OHD and $1,25(OH)_2D$ values were reported to be greater than 9 ng/ml (22.5 nmol/liter) and 118 pg/ml (295 pmol/liter), respectively. In none of the five infants was a therapeutic trial of calcium supplementation of the diet alone tried; however, the clinical and biochemical presentation suggested to the authors that dietary calcium deficiency was the primary factor responsible for the development of rickets.

More convincing evidence of dietary calcium deficiency as a cause for rickets in children comes from studies in South Africa and Nigeria, where the staple diets of children are characteristically low in calcium because of the lack of readily available dairy products and the low calcium content of the cereals [maize (corn), cassava, yam, and plantain] [34,85,118–120]. Dietary calcium intakes have been estimated to be between 90 and 300 mg/day in children suffering from rickets compared to between 200 and 500 mg/day in age-matched controls [34,118]. The children typically come from rural areas of both countries and present with signs and symptoms of rickets between the ages of 4 and 15 years in South Africa [120] and between 1 and 9 years in Nigeria [34,85]. In the South African series, half the children presented with knock-knees, whereas the others presented with either bowlegs or windswept deformities (Fig. 3). Unlike vitamin D deficiency, symptoms of muscle weakness are characteristically absent in dietary calcium deficiency.

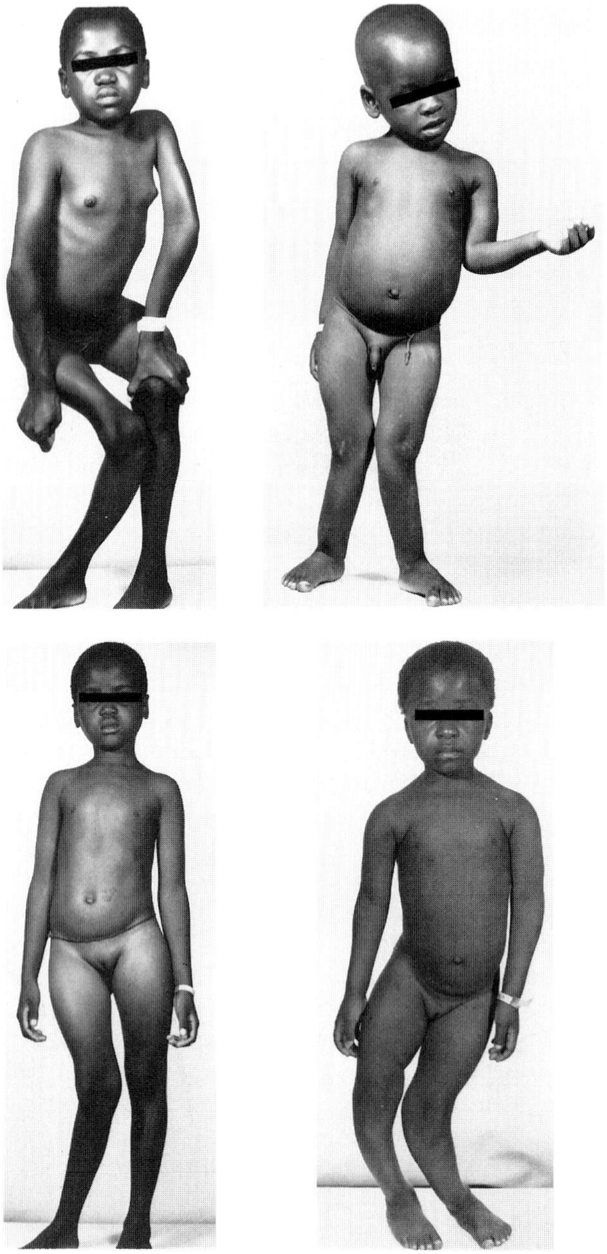

FIGURE 3 Clinical presentation of children with dietary calcium deficiency. The deformities are typically more severe in the legs, with a predominance of knock-knees or windswept deformities. Upper limb deformities are usually mild if present at all. Reprinted with permission from Pettifor [120].

Radiologically, the features are typical of calciopenic rickets, with osteopenia and features of hyperparathyroidism being frequent findings (Fig. 4). The severity of the metaphyseal changes is variable. Older children (teenagers) may have no radiological changes of rickets despite features of osteomalacia on iliac crest bone biopsy [121]. Younger children may show evidence of only minor degrees of impaired endochondral calcification, whereas in others the metaphyseal changes may be quite marked.

The biochemical features are similar to those of other causes of calciopenic rickets. Hypocalcemia, low urinary

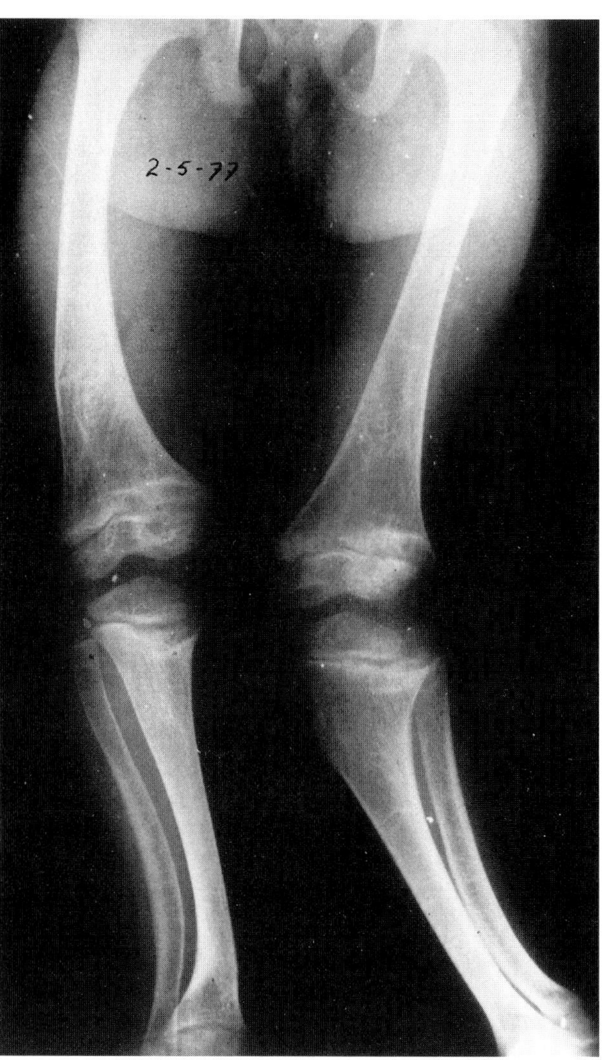

FIGURE 4 Radiographic features of dietary calcium deficiency rickets in the lower limbs of a child. The long bones are osteopenic with deformities characteristic of long-standing rickets. The metaphyses show evidence of impaired mineralization and growth arrest lines.

calcium excretion, and elevated serum PTH and alkaline phosphatase concentrations are characteristic, whereas serum phosphorus values are variable and often within the reference range for age [34,119,120]. Serum 25OHD values are normal (mean 16.4 and 14.4 ng/ml in the South African and Nigerian children, respectively), and $1,25(OH)_2D$ concentrations are elevated [85,122]. In a Nigerian study [34] and in the South African children (E.D. Daniels and J.M. Pettifor, unpublished data, 1997), serum osteocalcin levels are similar to those of nonrachitic controls in the majority of patients, although another report from Nigeria found slightly higher levels in rachitic patients than controls [83]. The finding of normophosphatemia and normal renal handling of phosphorus (TmP/GFR) suggests that a peripheral resistance to PTH might be prevalent in this form of rickets.

Iliac crest bone biopies reveal evidence of osteomalacia and hyperparathyroidism in those children who have radiological features of rickets [123], whereas in the teenagers without radiological changes, but with lower limb deformities, the histological picture varies from that of decreased bone volume, through features of hyperparathyroidism, to frank osteomalacia and hyperparathyroidism [121].

In both the Nigerian and South African studies, clinical, biochemical, and radiological healing has been achieved through increasing the calcium intake of the children to between 800 and 1500 mg/day without the administration of vitamin D supplements [34,124]. In the majority of the South African children, orthopedic corrective surgery has been necessary to correct the deformities of the legs once biochemical and radiological healing has occurred.

The data available from epidemiological studies conducted in a rural area in which a number of the affected children live suggest that asymptomatic dietary calcium deficiency is prevalent in schoolchildren living in the area. Some 13% of children between the ages of 7 and 12 years were hypocalcemic, 41.5% had elevated alkaline phosphatase concentrations, and 76% had low urinary calcium excretion [125]. It is unclear whether these children have long-term sequelae as a result of the poor calcium intakes. However, studies do indicate that asymptomatic children with biochemical abnormalities living in the rural community have lower appendicular bone mass than those with normal biochemistry [118] and that children in the community as a whole have lower appendicular bone mass than their urban peers (J.M. Pettifor, unpublished data, 1997).

Although dietary calcium intakes in children with biochemical changes suggestive of dietary calcium deficiency are very low, it is unclear what role the high phytate and oxalate content of the diet plays in aggravating the symptoms. Nevertheless, biochemical improvement can be achieved by supplementing the children on the rural diet with 500 mg calcium daily [126].

IX. THE PATHOGENETIC SPECTRUM OF NUTRITIONAL RICKETS

Nutritional rickets has been viewed for some time as being due to an inadequate supply of vitamin D through either an inadequate dietary intake or insufficient skin exposure to ultraviolet radiation, or more recently due to low dietary calcium intake in the face of a normal vitamin D status. However, these pathogenetic concepts are too simplistic. Early studies by Mellanby [127] had shown the effect of cereals in exacerbating the clinical development of vitamin D deficiency rickets in dogs. Studies in baboons have confirmed these findings [116].

The resurgence of rickets and osteomalacia in the Asian community in Great Britain provided the impetus for detailed studies into the pathogenesis of vitamin D deficiency and bone disease in that community. Although vitamin D deficiency as assessed by circulating 25OHD concentrations is the hallmark of the disease in Asians [43,128,129], the mechanisms for the low vitamin D status and the high prevalence of rickets were unclear. It is apparent that the majority of Asians in Britain do not spend less time outdoors than their Caucasian counterparts [130]. Further, although the Asians have darker skins than Caucasians, which might reduce the amount of vitamin D formed in response to sunlight exposure, West Indians living in Britain have even darker skins, yet very few cases of rickets have been described in this ethnic group [39]. Within the Asian community, studies have highlighted that risk factors for the disease include living at high latitude, Hindu religion, immigration from East Africa, vegetarianism, high fiber diets, and the consumption of chapatti [44,48,49]. The association with vegetarianism, high fiber diets, and cereals of high extraction suggests that dietary factors play a role. Support for this comes from two studies that have documented healing of rickets on removing chapattis from the diet [131,132], although this is not a universal finding [47].

Since the mid-1980s, research has shown that both high fiber diets and intestinal malabsorption reduce the serum half-life of 25OHD by approximately one-third [133,134]. Further, experiments in rats have demonstrated that an elevation in serum $1,25(OH)_2D$ concentrations, either by exogenous administration or endogenously through a low calcium diet, increases the metabolic clearance rate of 25OHD without altering its rate of production [135–137]. The fall in serum 25OHD levels could be accounted for by an increase in polar metabolites appearing in the feces. Similar findings have been reported from studies in humans [138,139]. Conversely, increasing the calcium content of the diet has been shown to increase serum 25OHD and decrease serum $1,25(OH)_2D$ concentrations [140]. Thus, these studies convincingly show that dietary calcium and phytate contents influence the catabolism of 25OHD through altering serum $1,25(OH)_2D$ concentrations.

In light of the above studies, Clements [39] has proposed that the low dietary calcium and high phytate diet of the Asian population in Britain increases vitamin D catabolism and thus vitamin D requirements. In the face of a marginal vitamin D status due to living at high latitude and the low dietary vitamin D content of the diet, the increased catabolism is sufficient to precipitate vitamin D deficiency and clinical rickets and osteomalacia.

Thus, nutritional rickets has a spectrum of pathogenetic mechanisms ranging from pure vitamin D deficiency associated with adequate calcium intakes, as might occur in the breast-fed infant, at one end of the spectrum, to pure dietary calcium deficiency with an adequate vitamin D status, as documented in Nigerian and South African rural children, at the other end [141]. In between these two extremes lies the situation exemplified by the Asian community in Britain, where both poor calcium intakes or absorption and marginal vitamin D status combine to lead to frank vitamin D deficiency and rickets (Fig. 5). It is likely that the high prevalence of rickets in vegetarian or immigrant children reported from the United States [21,22], Norway [40], Holland [41,42], and a number of tropical and subtropical countries [36] might be due to a mechanism similar to that in the Asian community, whereas osteo-

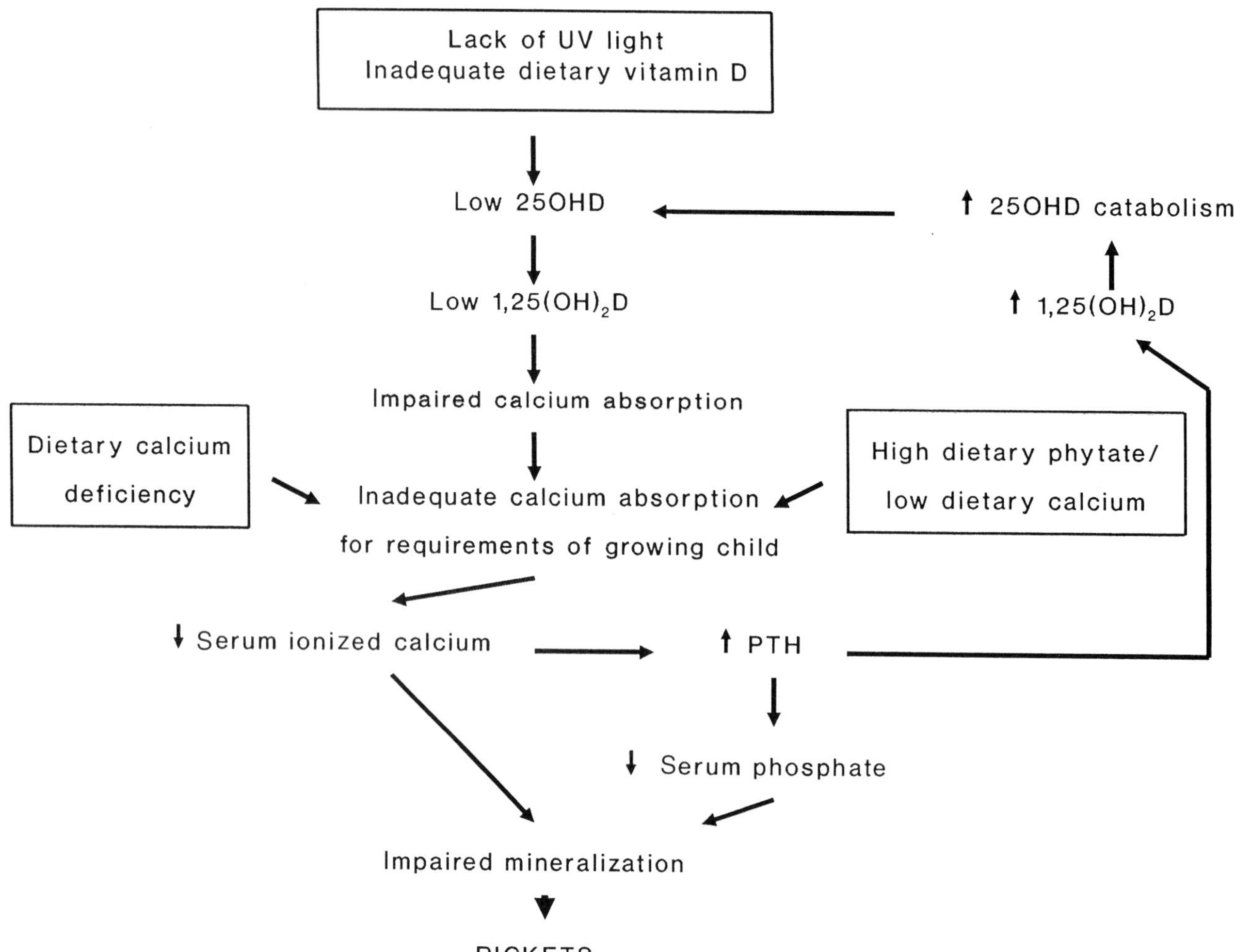

FIGURE 5 Spectrum of nutritional rickets. At either ends of the pathogenetic spectrum are vitamin D deficiency and dietary calcium deficiency. In between lie combinations in varying degrees of relative vitamin D insufficiency and decreased dietary calcium content or bioavailability.

malacia in Bedouin adults in the Middle East reflects mainly dietary calcium deficiency [142].

X. SUMMARY

Nutritional rickets in children became a major public health problem during the industrial revolution in Europe. Despite readily available and inexpensive preventive measures, the disease remains a problem in many tropical and subtropical countries. Further, it appears to be increasing in North America and Europe in certain cultural and religious groups.

The disease is most prevalent in breast-fed infants and young toddlers, who do not receive adequate sunlight exposure, vitamin D supplements, or vitamin D-fortified foods. Other at-risk groups include children who are on vegetarian diets, which are high in phytate and low in calcium content, and who due to social custom are not exposed to sufficient sunlight. Rickets due to dietary calcium deficiency in the face of a normal vitamin D status has also been described in otherwise well children in several African countries.

Clinically, nutritional rickets presents with progressive bone deformities, which may be accompanied by signs and symptoms of hypocalcemia. Hypotonia, delayed motor milestones, and an increase in infections may also occur. The biochemical hallmark of vitamin D deficiency is a low circulating concentration of 25OHD. Serum values of $1,25(OH)_2D$ may be low, normal, or even elevated. Hypophosphatemia and elevated alkaline phosphatase values are characteristic, whereas hypocalcemia is typical in the early and late phases of the disease. The majority of the biochemical markers of bone turnover are increased in association with secondary hyperparathyroidism. Dietary calcium deficiency is associated with normal 25OHD and elevated $1,25(OH)_2D$ concentrations and hypocalcemia.

Vitamin D deficiency is effectively treated by either small daily oral doses of vitamin D or by a single oral or intramuscular dose of 600,000 IU vitamin D. Prevention of vitamin D deficiency can be ensured by regular skin exposure to sunlight, by providing vitamin D supplements in at-risk groups, or by the vitamin D fortification of foods such as milk.

References

1. Stephen JML 1975 Epidemiological and dietary aspects of rickets and osteomalacia. Proc Nutr Soc **34**:131–138.
2. Arneil GC 1975 Nutritional rickets in children in Glasgow. Proc Nutr Soc **34**:101–109.
3. Rudolf M, Arulanantham K, Greenstein RM 1980 Unsuspected nutritional rickets. Pediatrics **66**:72–76.
4. Jacobus CH, Holick MF, Shao Q, Chen TC, Holm IA, Kolodny JM, Fuleihan GEH, Seeley EW 1992 Hypervitaminosis D associated with drinking milk. N Engl J Med **326**:1173–1177.
5. Markestad T, Hesse V, Siebenhuner M, Jahreis G, Aksnes L, Plenert W, Aarskog D 1987 Intermittent high-dose vitamin D prophylaxis during infancy: Effect on vitamin D metabolites, calcium, and phosphorus. Am J Clin Nutr **46**:652–658.
6. Portale AA, Halloran BP, Harris ST, Bikle DD, Morris RC 1992 Metabolic acidosis reverses the increase in serum $1,25(OH)_2D$ in phosphorus-restricted normal men. Am J Physiol **263**:E1164–E1170.
7. Holick MF 1995 Environmental factors that influence the cutaneous production of vitamin D. Am J Clin Nutr **61**:638S-645S.
8. Salimpour R 1975 Rickets in Tehran. Arch Dis Children **50**:63–66.
9. Al Hag AI, Karrar ZA 1995 Nutritional vitamin D deficiency rickets in Sudanese children. Ann Trop Paediatr **15**:69–76.
10. Hillman LS, Haddad JG 1974 Human perinatal vitamin D metabolism I: 25-Hydroxyvitamin D in maternal and cord blood. J Pediatr **84**:742–749.
11. Gertner JM, Glassman MS, Coutsan DR, Goodman DBP 1980 Fetomaternal vitamin D relationships at term. J Pediatr **97**:637–640.
12. Rothberg AD, Pettifor JM, Cohen DF, Sonnendecker EWW, Ross FP 1982 Maternal–infant vitamin D relationships during breast-feeding. J Pediatr **101**:500–503.
13. Maxwell JP 1934 Further studies in adult rickets (osteomalacia) and foetal rickets. Proc R Soc Med **28**:265–300.
14. Moncrieff M, Fadahunsi TO 1974 Congenital rickets due to maternal vitamin D deficiency. Arch Dis Children **49**:810–811.
15. Zhou H 1991 Rickets in China. In: Glorieux FH (ed) Rickets. Raven, New York, pp. 253–261.
16. Chen Y 1994 Prematurity as a predictor of rickets in Shanghai infants. Public Health **108**:333–339.
17. Lapatsanis P, Deliyanni V, Doxiadis S 1968 Vitamin D deficiency rickets in Greece. J Pediatr **73**:195–202.
18. Fraser D, Kooh SW, Scriver CR 1967 Hyperparathyroidism as the cause of hyperaminoaciduria and phosphaturia in human vitamin D deficiency. Pediatr Res **1**:425–435.
19. Doxiadis S, Angelis C, Karatzas P, Vrettos C, Lapatsanis P 1976 Genetic aspects of nutritional rickets. Arch Dis Children **51**:83–90.
20. Dancaster CP, Jackson WPU 1961 Studies in rickets in the Cape Peninsula II. Aetiology. S Afr Med J **35**:890–894.
21. Edidin DV, Levitsky LL, Schey W, Dumbuvic N, Campos A 1980 Resurgence of nutritional rickets associated with breast-feeding and special dietary practices. Pediatrics **65**:232–235.
22. Bachrach S, Fisher J, Parks JS 1979 An outbreak of vitamin D deficiency rickets in a susceptible population. Pediatrics **64**:871–877.
23. Elidrissy ATH, Sedrani SH, Lawson DEM 1984 Vitamin D deficiency in mothers of rachitic infants. Calcif Tissue Int **36**:266–268.
24. Hollis BW, Roos BA, Draper HH, Lambert PW 1981 Vitamin D and its metabolites in human and bovine milk. J Nutr **111**:1240–1248.
25. Specker BL, Tsang RC, Hollis BW 1985 Effect of race and diet on human-milk vitamin D and 25-hydroxyvitiamin D. Am J Dis Children **139**:1134–1137.
26. Ala-Houhala M, Koskinen T, Terho A, Kiovula T, Visakorpi J 1986 Maternal compared with infant vitamin D supplementation. Arch Dis Children **61**:1159–1163.
27. Specker BL, Valanis B, Hertzberg V, Edwards N, Tsang RC 1985 Sunshine exposure and serum 25-hydroxyvitamin D concentrations in exclusively breast-fed infants. J Pediatr **107**:372–376.

28. Specker BL, Tsang RC 1987 Cyclical serum 25-hydroxyvitamin D concentrations paralleling sunshine exposure in exclusively breast-fed infants. J Pediatr **110**:744–747.
29. Olivieri MB, Ladizesky M, Mautalen CA, Alonso A, Martinez L 1993 Seasonal variations of 25-hydroxyvitamin D and parathyroid hormone in Ushuaia (Argentina), the southernmost city of the world. Bone Miner **20**:99–108.
30. Garabedian M, Ben-Mekhbi H 1991 Is vitamin D-deficiency rickets a public health problem in France and Algeria? In: Glorieux FH (ed) Rickets. Raven, New York, pp. 215–221.
31. Ladizesky M, Lu Z, Oliveri B, Roman NS, Diaz S, Holick MF, Mautalen C 1995 Solar ultraviolet B radiation and photoproduction of vitamin D_3 in central and southern areas of Argentina. J Bone Miner Res **10**:545–549.
32. Beser E, Cakmakci T 1994 Factors affecting the morbidity of vitamin D deficiency rickets and primary protection. East Afr Med J **71**:358–362.
33. Ghai OP, Koul PB 1991 Rickets in India. In: Glorieux FH (ed) Rickets. Raven, New York, pp. 247–252.
34. Okonofua F, Gill DS, Alabi ZO, Thomas M, Bell JL, Dandona P 1991 Rickets in Nigerian children: A consequence of calcium malnutrition. Metabolism **40**:209–213.
35. Laditan AAO, Adeniyi A 1975 Rickets in Nigerian children: Response to vitamin D. J Trop Med Hyg **78**:206–209.
36. Bhattacharyya AK 1992 Nutritional rickets in the tropics. In: Simopoulos AP (ed) Nutritional Triggers for Health and in Disease. Karger, Basel, pp. 140–197.
37. Finberg L 1979 Human choice, vegetable deficiencies, and vegetarian rickets. Am J Dis Children **133**:129
38. Dwyer JT, Dietz WH, Hass G, Suskind R 1979 Risk of nutritional rickets among vegetarian children. Am J Dis Children **133**:134–140.
39. Clements MR 1989 The problem of rickets in UK Asians. J Hum Nutr Diet **2**:105–116.
40. Hellebostad M, Markestad T, Halvorsen KS 1985 Vitamin D deficiency rickets and vitamin B_{12} deficiency in vegetarian children. Acta Paediatr Scand **74**:191–195.
41. Meulmeester JF, van den Berg H, Wedel M, Boshuis PG, Hulshof KFAM, Luyken R 1990 Vitamin D status, parathyroid hormone and sunlight in Turkish, Moroccan and Caucasian children in The Netherlands. Eur J Clin Nutr **44**:461–470.
42. Dagnelie PC, Vergote FJVRA, van Staveren WA, van den Berg H, Dingjan PG, Hautvast JGAJ 1990 High prevalence of rickets in infants on macrobiotic diets. Am J Clin Nutr **51**:202–208.
43. Ford JA, McIntosh WB, Butterfield R, Preece MA, Pietrek J, Arrow-Smith WA, Arthurton MW, Turner W, O'Riordan JLH, Dunnigan MG 1976 Clinical and subclinical vitamin D deficiency in Bradford children. Arch Dis Child **51**:939–943.
44. Finch PJ, Ang L, Eastwood JB, Maxwell JD 1992 Clinical and histological spectrum of osteomalacia among Asians in South London. Q J Med **83**:439–448.
45. Moncrieff MW, Lunt HRW, Arthur LJH 1973 Nutritional rickets at puberty. Arch Dis Child **48**:221–224.
46. Hodgkin P, Kay GH, Hine PM, Lumb GA, Stanbury SW 1973 Vitamin-D deficiency in Asians at home and in Britain. Lancet **2**:167–172.
47. Dent CE, Round JM, Rowe DJF, Stamp TCB 1973 Effect of chapattis and ultraviolet irradiation on nutritional rickets in an Indian immigrant. Lancet **1**:1282–1284.
48. Henderson JB, Dunnigan MG, McIntosh WB, Abdul-Motaal A, Hole D 1990 Asian osteomalacia is determined by dietary factors when exposure to ultraviolet radiation is restricted: A risk factor model. Q J Med **76**:923–933.
49. Henderson JB, Dunnigan MG, McIntosh WB, Abdul-Motaal AA, Gettinby G, Glekin BM 1987 The importance of limited exposure to ultraviolet radiation and dietary factors in the aetiology of Asian rickets: A risk-factor model. Q J Med **63**:413–425.
50. Bonnici F 1978 Functional hypoparathyroidism in infantile hypocalcaemic stage I vitamin D deficiency rickets. S Afr Med J **54**:611–612.
51. Buchanan N, Pettifor JM, Cane RD, Bill PLA 1978 Infantile apnoea due to profound hypocalcaemia associated with vitamin D deficiency. S Afr Med J **53**:766–767.
52. Hochman HI, Mejlszenkier JD 1977 Cataracts and pseudotumor cerebri in an infant with vitamin D-deficiency rickets. J Pediatr **90**:252–254.
53. Park EA 1954 The influence of severe illness on rickets. Arch Dis Children **29**:369–380.
54. Harrison HE, Harrison HC 1979 Rickets and osteomalacia. In: Disorders of Calcium and Phosphate Metabolism in Childhood and Adolescence. Saunders, Philadelphia, Pennsylvania, pp. 141–256.
55. Geggel RL, Pereira GR, Spackman TJ 1978 Fractured ribs: Unusual presentation of rickets in premature infants. J Pediatr **93**:680–682.
56. Opie WH, Muller CJB, Kamfer H 1975 The diagnosis of vitamin D deficiency rickets. Pediatr Radiol **3**:105–110.
57. Pettifor JM, Pentopoulos M, Moodley GP, Isdale JM, Ross FP 1984 Is craniotabes a pathognomonic sign of rickets in 3-month-old infants? S Afr Med J **65**:549–551.
58. Dancaster CP, Jackson WPU 1960 Studies in rickets in the Cape Peninsula 1. Cranial softening in a coloured population and its relationship to the radiological and biochemical changes of rickets. S Afr Med J **34**:776–780.
59. Reilly BJ, Leeming JM, Fraser D 1964 Craniosynostosis in the rachitic spectrum. J Pediatr **64**:396–405.
60. Purvis RJ, Barrie MWJ, MacKay GS, Wilkinson EM, Cockburn F, Belton NR, Forfar JO 1973 Enamel hypoplasia of the teeth associated with neonatal tetany: A manifestation of maternal vitamin-D deficiency. Lancet **2**:811–814.
61. Schott GD, Wills MR 1976 Muscle weakness in osteomalacia. Lancet **1**:626–629.
62. Manolagas SC, Yu XP, Girasole G, Bellido T 1994 Vitamin D and the hematolymphopoietic tissue: A 1994 update. Semin Nephrol **14**:129–143.
63. Stroder J, Kasal P 1970 Evaluation of phagocytosis in rickets. Acta Paediatr Scand **59**:288–292.
64. Lorente F, Fontan G, Jara P, Casas C, Garcia-Rodriguez MC, Ojeda JA 1976 Defective neutrophil motility in hypovitaminosis D rickets. Acta Paediatr Scand **65**:695–699.
65. Yetgin S, Ozoylu S 1982 Myeloid metaplasia in vitamin D deficiency rickets. Scand J Haematol **28**:180–185.
66. David L 1991 Common vitamin D-deficiency rickets. In: Glorieux FH (ed) Rickets. Raven, New York, pp. 107–122.
67. Suda T 1987 Cellular mechanisms of fusion of hemopoietic cells induced by 1α,25-dihydroxyvitamin D_3. In: Cohn DV, Martin TJ, Meunier PJ (eds) Calcium Regulation and Bone Metabolism: Basic and Clinical Aspects. Excerpta Medica, Amsterdam, pp. 363–370.
68. Haddad JG, Chyu KJ 1971 Competitive protein binding radioassay for 25-hydroxycholecalciferol. J Clin Endocrinol **33**:992–995.
69. Pettifor JM, Ross FP, Moodley GP, Margo G 1978 Serum calcium, magnesium, phosphorus, alkaline phosphatase and 25-hydroxyvitamin D concentrations in a paediatric population. S Afr Med J **53**:751–754.
70. McLaughlin M, Raggatt PR, Fairney A, Brown DJ, Lester E,

Wills MR 1974 Seasonal variation in serum 25-hydroxycholecalciferol in healthy people. Lancet **1**:536–538.

71. Stanbury SW, Taylor CM, Lumb GA, Mawer EB, Berry J, Hann J, Wallace J 1981 Formation of vitamin D metabolites following correction of human vitamin D deficiency: Observations in patients with nutritional osteomalacia. Miner Electrolyte Metab **5**:212–227.
72. Mawer EB, Backhouse J, Hill LF, Lumb GA, De Silva P, Taylor CM, Stanbury SW 1975 Vitamin D metabolism and parathyroid function in man. Clin Sci Mol Med **48**:349–365.
73. Gupta MM, Round JM, Stamp TCB 1974 Spontaneous cure of vitamin-D deficency in Asians during summer in Britain. Lancet **1**:586–588.
74. Garabedian M, Vainsel M, Mallet E, Guillozo H, Toppet M, Grimberg R, Nguyen TM, Balsan S 1983 Circulating vitamin D metabolite concentrations in children with nutritional rickets. J Pediatr **103**:381–386.
75. Arnaud SB, Stickler GB, Haworth JC 1976 Serum 25-hydroxyvitamin D in infantile rickets. Pediatrics **57**:221–225.
76. Goel KM, Sweet EM, Logan RW, Warren JM, Arneil GC, Shanks RA 1976 Florid and subclinical rickets among immigrant children in Glasgow. Lancet **1**:1141–1145.
77. Markestad T, Halvorsen S, Seeger Halvorsen K, Aksnes L, Aarskog D 1984 Plasma concentrations of vitamin D metabolites before and during treatment of vitamin D deficiency rickets in children. Acta Paediatr Scand **73**:225–231.
78. Taitz LS, de Lacy CD 1962 Parathyroid function in vitamin D deficiency rickets 1. Phosphorus excretion index in vitamin D deficiency rickets in South African Bantu infants. Pediatrics **30**:875–883.
79. Muldowney FP, Freaney R, McGeeney D 1968 Renal tubular acidosis and amino-aciduria in osteomalacia of dietary or intestinal origin. Q J Med **37**:517–539.
80. Kruse K 1995 Pathophysiology of calcium metabolism in children with vitamin D-deficiency rickets. J Pediatr **126**:736–741.
81. Taitz LS, de Lacy CD 1962 Parathyroid function in vitamin D deficiency rickets II. The relationship of parathyroid function to bone changes and incidence of tetany in vitamin D deficiency rickets in South African Bantu infants. Pediatrics **30**:884–892.
82. Stanbury SW, Torkington P, Lumb GA, Adams PH, De Silva P, Taylor CM 1975 Asian rickets and osteomalacia: Patterns of parathyroid response in vitamin D deficiency. Proc Nutr Soc **34**:111–117.
83. Scariano JK, Walter EA, Glew RH, Hollis BW, Henry A, Ocheke I, Isichei CO 1995 Serum levels of the pyridinoline crosslinked carboxyterminal telopeptide of type I collagen (ICTP) and osteocalcin in rachitic children in Nigeria. Clin Biochem **28**:541–545.
84. Editorial 1971 Diagnosis of nutritional rickets. Lancet **2**:28–29.
85. Oginni LM, Worsfold M, Oyelami OA, Sharp CA, Powell DE, Davie MWJ 1996 Etiology of rickets in Nigerian children. J Pediatr **128**:692–694.
86. Cole DEC, Carpenter TO, Gundberg CM 1985 Serum osteocalcin concentrations in children with metabolic bone disease. J Pediatr **106**:770–776.
87. Greig F, Casas J, Castells S 1989 Changes in plasma osteocalcin concentrations during treatment of rickets. J Pediatr **114**:820–823.
88. Eastwood JB, de Wardener HE, Gray RW, Lemann JR Jr 1979 Normal plasma-1,25-$(OH)_2$-vitamin-D concentrations in nutritional osteomalacia. Lancet **1**:1377–1378.
89. Chesney RW, Zimmerman J, Hamstra A, DeLuca HF, Mazess RB 1981 Vitamin D metabolite concentrations in vitamin D deficiency. Am J Dis Children **135**:1025–1028.
90. NGuyen TM, Guillozo H, Garabedian M, Mallet E, Balsan S 1979 Serum concentrations of 24,25-dihydroxyvitamin D in normal children and in children with rickets. Pediatr Res **13**:973–976.
91. Rosen JF, Chesney RW 1983 Circulating calcitriol concentrations in health and disease. J Pediatr **103**:1–17.
92. Rasmussen H, Baron R, Broadus A, DeFronzo R, Lang R, Horst R 1980 1,25$(OH)_2D_3$ is not the only D metabolite involved in the pathogenesis of osteomalacia. Am J Med **69**:360–368.
93. Arnaud CD 1991 Parathyroid hormone and its role in the pathophysiology of the common forms of rickets and osteomalacia. In: Glorieux FH (ed) Rickets. Raven, New York, pp. 47–61.
94. Rasmussen H, De Luca H, Arnaud CD, Hawker C, Von Stedingk M 1963 The relationship between vitamin D and parathyroid hormone. J Clin Invest **42**:1940–1946.
95. Arnaud CD, Rasmussen H, Anast C 1966 Further studies on the interrelationship between parathyroid hormone and vitamin D. J Clin Invest **45**:1955–1964.
96. Richards IDG, Sweet EM, Arneil GC 1968 Infantile rickets persists in Glasgow. Lancet **1**:803–805.
97. Pettifor JM 1991 Calcium, phosphorus, and vitamin D. In: Ballabriga A, Brusner O, Dobbing J, Gracey M, Senterre J (eds) Clinical Nutrition of the Young Child. Raven, New York, pp. 497–516.
98. Pettifor JM, Isdale JM, Sahakian J, Hansen JDL 1980 Diagnosis of subclinical rickets. Arch Dis Children **55**:155–157.
99. Hunter GJ, Schneidau A, Hunter JV, Chapman M 1984 Rickets in adolescence. Clin Radiol **35**:419–421.
100. Shah BR, Finberg L 1994 Single-day therapy for nutritional vitamin D-deficiency rickets: A preferred method. J Pediatr **125**:487–490.
101. Lubani MM, Al-Shab TS, Al-Saleh QA, Sharda DC, Quattawi SA, Ahmed SAH, Moussa MA, Reavey PC 1989 Vitamin-D-deficiency rickets in Kuwait: The prevalence of a preventable disease. Ann Trop Paediatr **3**:134–139.
102. Greer FR, Searcy JE, Levin RS, Steichen JJ, Steichen-Asche PS, Tsang RC 1982 Bone mineral content and serum 25-hydroxyvitamin D concentrations in breast-fed infants with or without supplemental vitamin D: One-year follow-up. J Pediatr **100**:919–922.
103. Ala-Houhala M 1985 25-Hydroxyvitamin D levels during breastfeeding with or without maternal or infantile supplementation of vitamin D. J Pediatr Gastroenterol Nutr **4**:220–226.
104. Roberts CC, Chan GM, Folland D, Rayburn C, Jackson R 1981 Adequate bone mineralization in breast-fed infants. J Pediatr **99**:192–196.
105. Birkbeck JA, Scott HF 1980 25-Hydroxycholecalciferol serum levels in breast-fed infants. Arch Dis Children **50**:691–695.
106. Committee on Nutrition 1980 On the feeding of supplemental foods to infants. Pediatrics **65**:1178–1181.
107. Finberg L 1981 Human milk feeding and vitamin D supplementation—1981. J Pediatr **99**:228–229.
108. Ozsoylu S, Hansaoglu A 1981 25-Hydroxycholecalciferol levels in breast-fed infants. Arch Dis Children **56**:318
109. Editorial 1973 The need for vitamin-D supplements. Lancet **1**:1097–1098.
110. Specker BL, Ho ML, Oestreich A, Yin T, Shui Q, Chen X, Tsang RC 1992 Prospective study of vitamin D supplementation and rickets in China. J Pediatr **120**:733–739.
111. Zeghoud F, Ben-Mekhbi H, Djeghri N, Garabedian M 1994 Vitamin D prophylaxis during infancy: Comparison of the long-term effects of three intermittent doses (15, 5, or 2.5 mg) on 25-hydroxyvitamin D concentrations. Am J Clin Nutr **60**:393–396.
112. Pietrek J, Preece MA, Windo J, O'Riordan JLH, Dunnigan MG,

McIntosh WB, Ford JA 1976 Prevention of vitamin-D deficiency in Asians. Lancet **1**:1145–1148.
113. Walker ARP 1953 Does a low intake of calcium cause or promote the development of rickets? Am J Clin Nutr **3**:114–120.
114. Irwin MI, Kienholz EW 1973 A conspectus of research on calcium requirements of man. J Nutr **103**:1020–1095.
115. Maltz HE, Fish MB, Holliday MA 1970 Calcium deficiency rickets and the renal response to calcium infusion. Pediatrics **46**:865–870.
116. Sly MR, van der Walt WH, Du Bruyn D, Pettifor JM, Marie PJ 1984 Exacerbation of rickets and osteomalacia by maize: A study of bone histomorphometry and composition in young baboons. Calcif Tissue Int **36**:370–379.
117. Proesman W, Legius E, Eggermont E 1988 Rickets due to calcium deficiency. In: Proceedings of the Symposium on Clinical Disorders of Bone and Mineral Metabolism. Mary Ann Liebert, New York, p. 15.
118. Eyberg C, Pettifor JM, Moodley G 1986 Dietary calcium intake in rural black South African children. The relationship between calcium intake and calcium nutritional status. Hum Nutr Clin Nutr **40C**:69–74.
119. Bhimma R, Pettifor JM, Coovadia HM, Moodley M, Adhikari M 1995 Rickets in black children beyond infancy in Natal. S Afr Med J **85**:668–672.
120. Pettifor JM 1991 Dietary calcium deficiency. In: Glorieux FH (ed) Rickets. Raven, New York, pp. 123–143.
121. Schnitzler CM, Pettifor JM, Patel D, Mesquita JM, Moodley GP, Zachen D 1994 Metabolic bone disease in black teenagers with genu valgum or varum without radiologic rickets: A bone histomorphometric study. J Bone Miner Res **9**:479–486.
122. Pettifor JM, Ross FP, Travers R, Glorieux FH, DeLuca HF 1981 Dietary calcium deficiency: A syndrome associated with bone deformities and elevated serum 1,25-dihydroxyvitamin D concentrations. Metab Bone Related Res **2**:301–305.
123. Marie PJ, Pettifor JM, Ross FP, Glorieux FH 1982 Histological osteomalacia due to dietary calcium deficiency in children. N Engl J Med **307**:584–588.
124. Pettifor JM, Ross P, Wang J, Moodley G, Couper-Smith J 1978 Rickets in children of rural origin in South Africa: Is low dietary calcium a factor? J Pediatr **92**:320–324.
125. Pettifor JM, Ross FP, Moodley GP, Shuenyane E 1979 Calcium deficiency in rural black children in South Africa—A comparison between rural and urban communities. Am J Clin Nutr **32**:2477–2483.
126. Pettifor JM, Ross FP, Moodley GP, Shuenyane E 1981 The effect of dietary calcium supplementation on serum calcium, phosphorus and alkaline phosphatase concentrations in a rural black population. Am J Clin Nutr **34**:2187–2191.
127. Mellanby E 1919 An experimental investigation on rickets. Lancet **1**:407–412.
128. Iqbal SJ, Kaddam I, Wassif W, Nichol F, Walls J 1994 Continuing clinically severe vitamin D deficiency in Asians in the UK (Leicester). Postgrad Med J **70**:708–714.
129. Preece MA, Ford JA, McIntosh WB, Dunnigan MG, Tomlinson S, O'Riordan JLH 1973 Vitamin-D deficiency among Asian immigrants to Britain. Lancet **1**:907–910.
130. Dunnigan MG, McIntosh WB, Ford JA 1976 Rickets in Asian immigrants. Lancet **1**:1346
131. Wills MR, Day RC, Phillips JB, Bateman EC 1972 Phytic acid and nutritional rickets in immigrants. Lancet **1**:771–773.
132. Ford JA, Colhoun EM, McIntosh WB, Dunnigan MG 1972 Biochemical response of late rickets and osteomalacia to a chupatty-free diet. Br Med J **3**:446–447.
133. Batchelor AJ, Compston JE 1983 Reduced plasma half-life of radio-labelled 25-hydroxyvitamin D_3 in subjects receiving a high-fibre diet. Br J Nutr **49**:213–216.
134. Batchelor AJ, Watson G, Compston JE 1982 Changes in plasma half-life and clearance of ^{3}H-25-hydroxyvitamin D_3 in patients with intestinal malabsorption. Gut **23**:1068–1071.
135. Clements MR, Johnson L, Fraser DR 1987 A new mechanism for induced vitamin D deficiency in calcium deprivation. Nature **325**:62–65.
136. Halloran BP, Castro ME 1989 Vitamin D kinetics in vivo: Effect of 1,25-dihydroxyvitamin D administration. Am J Physiol **256**:E686–E691.
137. Halloran BP, Bikle DD, Levens MJ, Castro ME, Globus RK, Holton E 1986 Chronic 1,25-dihydroxyvitamin D_3 administration in the rat reduces serum concentration of 25-hydroxyvitamin D by increasing metabolic clearance rate. J Clin Invest **78**:622–628.
138. Clements MR, Davies M, Fraser DR, Lumb GA, Mawer B, Adams PH 1987 Metabolic inactivation of vitamin D is enhanced in primary hyperparathyroidism. Clin Sci **73**:659–664.
139. Clements MR, Davies M, Hayes ME, Hickey CD, Lumb GA, Mawer EB, Adams PH 1992 The role of 1,25-dihydroxyvitamin D in the mechanism of acquired vitamin D deficiency. Clin Endocrinol **37**:17–27.
140. Berlin T, Bjorkhem I 1988 Effect of calcium intake on serum levels of 25-hydroxyvitamin D_3. Eur J Clin Invest **18**:52–55.
141. Pettifor JM 1994 Privational rickets: A modern perspective. J R Soc Med **87**:723–725.
142. Shany S, Hirsh J, Berlyne GM 1976 25-Hydroxycholecalciferol levels in Bedouins in the Negev. Am J Clin Nutr **29**:1104–1107.

CHAPTER 43

Vitamin D Insufficiency in Adults and the Elderly

MARIE-CLAIRE CHAPUY AND PIERRE J. MEUNIER
INSERM Unit 403 and Department of Rheumatology and Bone Disease, Edouard Herriot Hospital, Lyon, France

I. INTRODUCTION

In contrast to the extensive attention paid to vitamin D as a component of nutritional health in infants and children, for whom vitamin D deficiency was synonymous with rickets, the relationship of vitamin D to the nutritional health of adults or elderly subjects had been largely ignored for a long time. Studies on the vitamin D status of the elderly began only in the mid-1970s when assays for serum 25-hydroxyvitamin D (25OHD) became available. The initial studies were carried out in Europe, and it was not until 1982 that studies on the nutritional status of the elderly living in the United States included data on the vitamin D status of the aged population [1].

With the major changes in demography that occurred since the 1960s leading to increased life expectancy of the population, vitamin D insufficiency represents a timely and common health problem in older people. Vitamin D insufficiency is frequently associated with abnormal bone metabolism including an increase in bone turnover and bone loss particularly in cortical bone. This leads to an increased risk of fracture, especially of hip fracture, in these subjects with vitamin D insufficiency. The purpose of this chapter is to define this state of vitamin D insufficiency, discuss its prevalence among adults and older subjects, and analyze its causes and its consequences on bone and muscle. At the end of the chapter, the utility of increasing vitamin D intake, not only in the elderly but also in adults, is discussed.

II. DEFINITION OF VITAMIN D DEFICIENCY AND INSUFFICIENCY

At the outset of this chapter it is essential that the definition of vitamin D "insufficiency" be clarified. Some confusion exists in the literature between the terms vitamin D "insufficiency" and "deficiency" or depletion. The term "insufficient" is defined as lacking in something necessary for completeness, and "deplete" is defined as empty [2]. "Repletion," on the other hand, connotes fullness. In this chapter we define a subject with reduced vitamin D as vitamin D insufficient, and a subject severely lacking in vitamin D as vitamin D deplete or deficient.

Peacock *et al.* have proposed other definitions that are based on serum 25OHD concentrations [3]. According to these authors, vitamin D deficiency occurs with serum 25OHD levels ranging from 0 to 4 ng/ml (0–10 nmol/liter) with evident secondary hyperparathyroidism and malabsorption of calcium, leading to osteomalacia and proximal myopathy. In vitamin D insufficiency, that is, 25OHD levels ranging from 4 to 20 ng/ml (10–50 nmol/liter), there is mild hyperparathyroidism, suboptimal calcium absorption, and reduced bone density. In vitamin D sufficiency the 25OHD levels range from 20 to 80 ng/ml (50 to 200 nmol/liter), and there is no disturbance in calcium homeostasis and bone metabolism. Peacock *et al.* have further defined vitamin D sufficiency by examining the serum 1,25-dihydroxyvitamin D [$1,25(OH)_2D$] response to treatment with $25OHD_3$ in various groups of patients and healthy subjects. Vitamin D sufficiency [i.e., no change in serum $1,25(OH)_2D$ levels] occurred at a serum 25OHD level in excess of 20 ng/ml (50 nmol/liter), which is above the lower limit of the normal range for 25OHD values.

Other investigators have used the parathyroid hormone (PTH) level as an index of vitamin D repletion, on the basis of the rationale that vitamin D insufficiency results in decreased calcium absorption, a subtle decline in blood ionized calcium, and consequently an increase in PTH values [4–6]. For Gloth and co-workers [7,8] and Webb *et al.* [9], the lowest acceptable limit for 25OHD levels was 37 nmol/liter (15 ng/ml). However, serum 25OHD concentration varies with country, season, and sunshine exposure. The mean values of serum 25OHD are much higher in the United States than in northwestern European countries. This results in different thresholds for diagnosing insufficiency versus sufficiency. As noted above, substrate-dependent synthesis is another way to diagnose a vitamin D deficient state; when vitamin D therapy results in an increase in the serum $1,25(OH)_2D$ concentration, vitamin D insufficiency is probable [3,10]. This approach might be complicated by an increase in PTH values secondary to low 25OHD values.

Also, the comparability of the cutoffs proposed by different authors may vary because different methods have been used for 25OHD assays (see Chapter 38). As the diagnosis of hypervitaminosis D is based on a blood test, whether very low or "undetectable" levels of vitamin D represent deficiency or depletion will depend in part on the sensitivity of the assay method. Assays using extraction and purification give lower results than those without a preparative chromatographic step. In the future, new radioimmunoassays, which give a higher correlation with the high-performance liquid chromatography (HPLC) method [11], certainly will be used more frequently and will allow better comparison of the data from different studies.

On the other hand, because the major biological effects of vitamin D are mediated by $1,25(OH)_2D$, many researchers would probably agree that vitamin D deficiency should entail $1,25(OH)_2D$ deficiency. However, low $1,25(OH)_2D$ may not always be associated with low 25OHD, and normal values of $1,25(OH)_2D$ can exist in the presence of low 25OHD levels. The determination of circulating $1,25(OH)_2D$ concentrations has very limited clinical utility except in the diagnosis of some conditions involving the vitamin D endocrine system such as vitamin D-dependent rickets types I (PDDR) and II (HVDRR) and some hypercalcemic states caused by increased levels of $1,25(OH)_2D$. Even if $1,25(OH)_2D$ is the active form of vitamin D, its measurement provides little information in many disorders and no information at all on the nutritional status of vitamin D [12].

Returning to the question of terminology, it appears that the term "deficiency" used in many studies or reports is heterogeneous and includes several different states. We believe that the term "insufficiency" is well-suited to define the state of hypovitaminosis D that induces other abnormalities of bone metabolism and changes in biochemical indices, but this is not yet accepted by all investigators.

III. DETERMINANTS OF VITAMIN D INSUFFICIENCY

The vitamin D status of a subject is derived mainly from cutaneous synthesis initiated by solar irradiation of the skin and also from dietary intake. A reduction of one or both sources unavoidably leads to vitamin D insufficiency. Some of these issues are also discussed in Chapter 35.

A. Sunlight Deprivation

The body stores of vitamin D are mainly dependent on the cutaneous synthesis of vitamin D_3, but this synthesis is in turn dependent on several factors such as the duration of sunlight exposure, the latitude of the country, the season, the time of day, and atmospheric conditions (see Chapter 3). Holick found that in Boston (42° N), exposure to sunlight on cloudless days between the months of November and February for up to 5 hr did not result in any significant production of vitamin D [13]. At high latitudes, the stores of vitamin D are mainly generated during the summer months. However, the penetration of effective ultraviolet rays (285 to 310 nm) into the cutaneous layers is modified by the type of clothing, the blockage of effective rays by window glass, and the capacity of skin to produce vitamin D.

Generally, on going outdoors healthy elderly subjects take protective action to reduce sunlight exposure either by use of clothing and sunscreens or simply by just avoiding direct sun. Recommendations for reducing the risk of skin cancer have heightened the situation. Aging has a dramatic impact on the skin: after the age of 20, skin thickness decreases linearly with age, and the capacity to produce previtamin D_3 is reduced. The increase in 25OHD values after the same amount of simulated sunlight was 3 times higher in young subjects aged 22–30 years than in elderly subjects aged 62–80 years [14]. This is due in part to the age-related decline in skin thickness [15]. Even in young adults, the seasonal variation of 25OHD concentrations might be explained by sun deprivation during winter months, when hypovitaminosis D is more frequent. Elderly institutionalized people often are unable to go outdoors because many are infirm or sick; thus, their vitamin D status depends exclusively on vitamin D supplied by food.

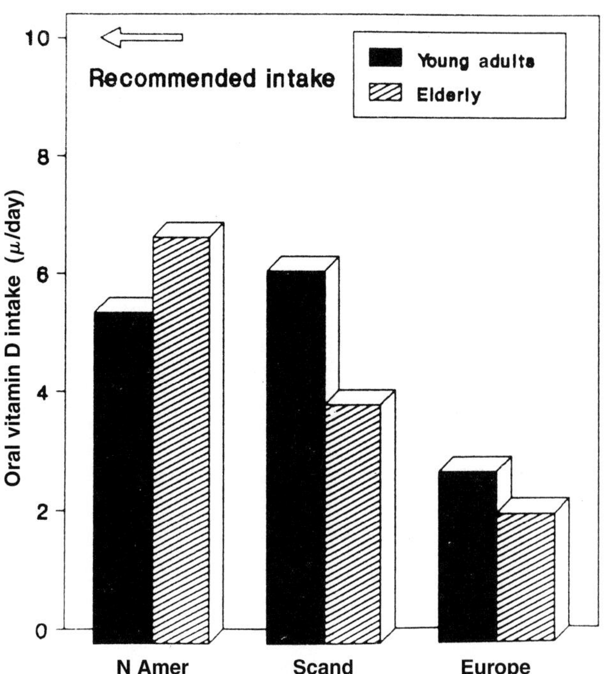

FIGURE 1 Mean of the average vitamin D intake in young adults and elderly from studies collated according to geographic region. The recommended intake is 10 μg/day. From McKenna [16] with permission.

B. Low Vitamin D Intake

The natural sources of dietary vitamin D are fatty fishes, fish liver oil, and to a lesser extent eggs. Regional differences in vitamin D intake are important. Mean vitamin D intake per day is lower in Europe compared with both North America and Scandinavia. In North America, milk is fortified with 400 IU/quart of either vitamin D_2 or vitamin D_3, but studies have revealed that upward of 70% of all milk samples tested did not contain this amount. In Scandinavia (as in Japan), people have as a part of their diet a substantial amount of fish (salmon, mackerel, herring). In the review by McKenna [16], the mean vitamin D intake was found to be 2.5 ± 1.3 μg/day in Europe, 6.2 ± 2.4 μg/day in North America, and 5.2 ± 2.0 μg/day in Scandinavia. In all studies combined no differences were found between the vitamin D intake of young adults and the elderly (Fig. 1).

The primacy of oral intake over sunlight exposure in maintaining vitamin D stores during the wintertime, especially in elderly subjects residing in Western Europe, and even in North American healthy postmenopausal women, has been demonstrated [6]. The current U.S. recommended dietary allowance of vitamin D for adults is 200 IU or 5 μg. From many studies, it appears that the need is greater and may approach 400 IU or 10 μg per day in order to maintain adequate 25OHD concentrations in adults living in high latitude countries. It is likely that the true vitamin D requirement, in the absence of any exposure to sunlight, is closer to 600 IU or 15 μg/day [13]. In elderly people who are confined indoors and who get no direct sunlight exposure the requirement is increased. In addition, some studies, but not all [17,18], found that intestinal vitamin D absorption was diminished in the elderly by as much as 40% compared to a younger population [19,20]. We have concluded that an oral vitamin D supplementation with 800 IU or 20 μg given every day is safe and has proved its efficacy to maintain adequate 25OHD concentrations [21].

C. Medical Causes

Vitamin D insufficiency also may be due to malabsorption of vitamin D (Chapter 51) or to drugs known to alter the metabolism of vitamin D (Chapter 50). Vitamin D is a fat-soluble vitamin, and disorders that lead to steatorrhea are associated with vitamin D insufficiency. This is seen in subjects with hepatobiliary and gastrointestinal disorders such as celiac disease, cystic fibrosis, chronic pancreatis, and partial or total gastrectomy. In such subjects the enteropatic circulation of 25OHD may be also disordered [22].

Vitamin D insufficiency may occur in patients taking anticonvulsant therapy such as phenytoins, phenobarbital, and glutethimide (see Chapter 50). In these patients, the 25OHD concentrations are decreased; however, the mechanisms involved in this reduction are uncertain. One possibility is that anticonvulsants induce hepatic microsomal enzymes which metabolize vitamin D to biologically inactive degradation products. Another possible mechanism may be drug interference in the synthesis of $1,25(OH)_2D$.

IV. CONSEQUENCES OF LOW VITAMIN D STATUS

A. Biochemical Consequences: Increased Serum PTH

As the synthesis of the active metabolite is substrate dependent, vitamin D insufficiency will lead to a small decrease in $1,25(OH)_2D$ levels. The lowest value of serum 25OHD necessary to maintain the normal synthesis of $1,25(OH)_2D$ is not known. In clinical studies, this threshold value was found to be lower than 12 ng/ml (30 nmol/liter) by Lips *et al.* [5] and 15 ng/ml (37.5 nmol/liter) by Bouillon *et al.* [10]. The subsequent subtle decrease in $1,25(OH)_2D$ levels leads to decreased serum calcium and induces an increase in serum PTH levels. This mechanism was suggested by Riggs and Melton as a pathophysiological factor for "type II osteoporosis" [23] (Fig. 2). This mild increase in serum PTH with age in subjects living either at home or in an institution has been reported in many studies [24–27]. It has been found to be related to renal dysfunction, age, and/or increasing years since menopause and to a fall in 25OHD levels [4,5,28] (Fig. 3). Renal insufficiency denoted by declining glomerular filtration rate appears to be a cause [29] but not the main cause of the age-related rise in serum PTH [30]. Vitamin D insufficiency seems to be a significant determinant of senile secondary hyperparathyroidism.

We have found that the main determinants for PTH values in free-living elderly women were, at equal weight, age and 25OHD values and, to a lesser degree, creatinine clearance [31]. When all these variables were adjusted in a multiple regression model, age and 25OHD were still significant predictors of PTH but not creatinine clearance [31]. Ooms *et al.* [32] have found that only 25OHD values below a threshold of 10 ng/ml (25 nmol/liter) were related to PTH. Below this threshold PTH increased 14.1% for every 4 ng/ml (10 nmol/liter) drop in serum 25OHD. Notwithstanding the fact that in several studies vitamin D only accounted for less than 10% of the variance in PTH, it must be emphasized that senile secondary hyperparathyroidism can be reversed by vitamin D supplements [4,5,21,33]. However, the association between 25OHD and PTH is likely to be influenced by the levels of calcium intake, as calcium supplementation was also demonstrated to be able to lower PTH levels in elderly subjects [34].

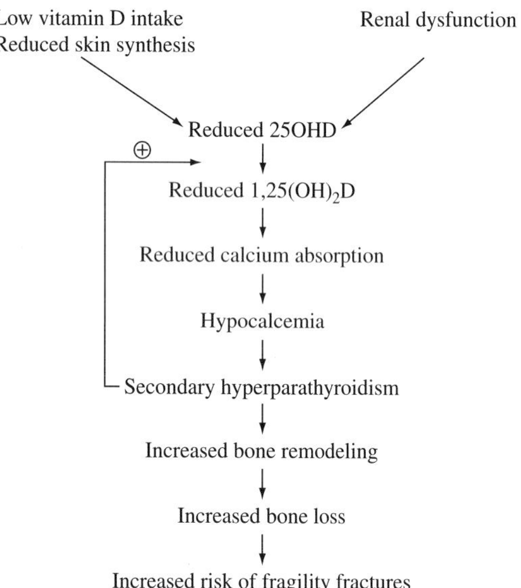

FIGURE 2 Pathophysiological mechanism of senile secondary hyperparathyroidism. Although increased PTH will stimulate $1,25(OH)_2D$ synthesis to correct hypocalcemia, this action is not efficient due to the decrease in the substrate (25OHD) concentration for 1α-hydroxylase activity.

B. Impact on Bone

Severe and prolonged deficiency in vitamin D is associated with osteomalacia characterized by defective mineralized bone and reduced bone strength (Chapter

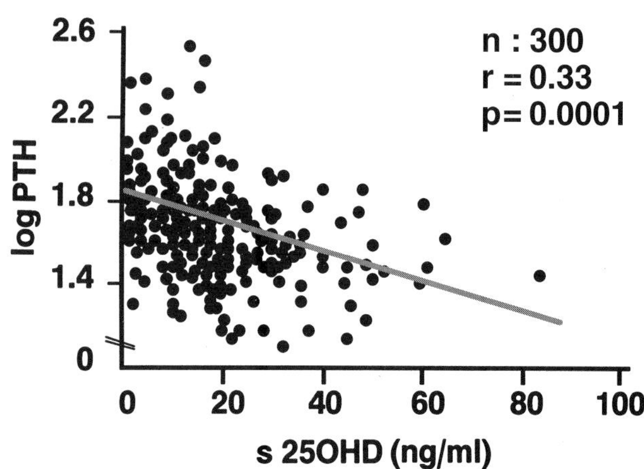

FIGURE 3 Inverse relationship between serum 25OHD and log PTH values in 300 elderly women living in institutions.

41). On the other hand, vitamin D insufficiency that is not severe enough to cause osteomalacia may nevertheless contribute to hip fracture risk in the elderly by decreasing calcium absorption and increasing PTH secretion, leading to increased bone turnover and bone loss particularly in cortical bone (Fig. 2) [1,23,35]. This concept seems to be confirmed by the higher PTH levels found in patients with hip fractures than in matched controls [36,37]. It is also supported by histomorphometric studies that have shown an increase in resorption parameter values (i.e., reduced thickness of iliac cortices and increased number of osteoclasts per square millimeter of bone section) in patients with vertebral fractures [35,36]. These bone changes have been confirmed by Okano *et al.* [38]. On the other hand, overt osteomalacia, as determined by both increased thickness of osteoid seams and decrease in the calcification rate measured through tetracycline double labeling, is very rare and has been found in less than 10% of patients with hip fracture [35,39]. The histological picture of the first stage of vitamin D-deficient osteopathy (vitamin D insufficiency) is caused by secondary hyperparathyroidism and cannot be distinguished from hyperparathyroid bone disease on histological grounds [40] (also see Chapter 41).

Several studies have shown a relationship between femoral neck bone mineral density (BMD), vitamin D insufficiency, and secondary hyperparathyroidism not only in elderly [32,41,42] but also in middle-aged women [43,44]. Martinez *et al.* found an association between low femoral BMD and low 25OHD levels in normal women older than 65 [45]. For Villareal *et al.* there was a relationship between vertebral bone density assessed by quantitative computerized tomography and PTH values only in subjects with low 25OHD values (≤15 ng/ml or 38 nmol/liter) [46]. Ooms *et al.* [32] found similar results for the BMD of the hip, the best fit being obtained with a threshold value of 25OHD at 12 ng/ml (30 nmol/liter). The femoral BMD was 5% higher for every 4 ng/ml (10 nmol/liter) increase in 25OHD up to the threshold. In this study, serum PTH was negatively related to BMD at all measurements sites, with the correlation coefficient ranging from −0.19 for the distal radius to −0.27 for the left femoral neck. In contrast, at a serum 25OHD level of 4 ng/ml (10 nmol/liter), the BMD of the femoral neck was reduced by 9.3%, which is 0.6 SD below the average BMD for an adequate vitamin D status (i.e., above 12 ng/ml or (30 nmol/liter). According to the data of Cummings *et al.* [47] and Ooms *et al.* [32], a 25OHD level of 4 ng/ml (10 nmol/liter) results in a relative risk of hip fracture of 1.8. In a German population, Scharla *et al.* found a borderline significant positive correlation between femoral neck BMD and 25OHD values in men older than 70 years ($r = 0.34, p \leq 0.03$). In women between 50 and 70 years ($r = 0.36$, $p \leq 0.02$), there was no association between vertebral BMD and 25OHD values [48].

In a cross-sectional study in the United Kingdom, a positive relationship between serum 25OHD values and BMD of the lumbar spine, the femoral neck, and greater trochanter was observed in a group of 138 middle-aged women volunteers (45–65 years) [43]. In a prospective study, Dawson-Hughes *et al.* [33] have shown that vitamin D insufficiency noted in winter contributes to spinal bone loss in healthy postmenopausal women (mean age 61 years), which can be reduced by daily supplements with 10 μg of vitamin D.

Several studies have shown correlations between bone biochemical markers and the kinetic and histomorphometric evaluation of bone formation and resorption, and it has been possible to demonstrate the impact of 25OHD insufficiency on bone status using the biochemical markers of bone turnover. Brazier, Kamel, and coworkers [49,50] have reported a 2- to 3-fold elevation of resorption estimated by excretion of pyridinoline crosslinks in elderly subjects with vitamin D insufficiency and secondary hyperparathyroidism as compared with vitamin D-sufficient elderly. In free-living healthy elderly women, we have found that the mean values of bone alkaline phosphatase, osteocalcin, and collagen C-telopeptide (crosslaps) were significantly increased as compared with the results obtained in young women [31] (Table I). For these markers, we found significant positive correlations with PTH values and negative correlations with hip BMD. In these healthy ambulatory women, we did not find a correlation between 25OHD

TABLE I Comparison of Biochemical Markers of Bone Remodeling in Elderly and Young Women during the Winter

Marker	Elderly women living at home (EPIDOS study, $n = 405$)	Young women (OFELY study, $n = 54$)
Bone alkaline phosphatase (μg/ml)	15.2 ± 6.2 (a)	8.5 ± 2.6
Osteocalcin (ng/ml)	24.9 ± 9.6 (a)	14.9 ± 4.4
Urinary calcium (mmol/mmol creatinine)	0.36 ± 0.22 (b)	0.26 ± 0.15
Urinary hydroxyproline (μmol/mmol creatinine)	29 ± 12 (a)	19 ± 7
Crosslaps[b] (μg/mmol creatinine)	311 ± 168 (a)	186 ± 108

[a] Data are expressed as means ± SD. Lowercase letters show results of statistical analyses: (a), significantly different from young women at $p = 0.0001$; (b), significantly different from young women at $p = 0.001$. From Chapuy *et al.* [31].

[b] Collagen C-telopeptide.

and PTH with BMD values as reported in some studies [43,45,46], but other researchers have also been unable to show a significant correlation between BMD and PTH values [42,51].

Thus, if an increase in PTH levels secondary to vitamin D insufficiency is one certain cause of age-related bone loss and fragility, this suggests that other unidentified factors may play a major role in the increase in bone turnover. Ooms *et al.* [52] found that low serum 25OHD concentrations are associated with higher PTH and osteocalcin and lower BMD of the hip, but also that when high sex hormone binding globulin (inverse measure of estrogen activity) is combined with vitamin D insufficiency, the secondary hyperparathyroidism is more severe. This suggests that low estrogen activity causes decreased sensitivity of the gut to $1,25(OH)_2D$ [53], leading to higher serum PTH levels and increasing the impact of vitamin D insufficiency. The observation, in several studies, of a seasonal variation in bone mineral density in normal subjects provides indirect evidence that relatively small changes in vitamin D status may have significant effects on bone mass. The demonstration that late wintertime bone loss could be prevented by small increases in vitamin D intake [33] provides a powerful argument in favor of the hypothesis that vitamin D status contributes to increased bone turnover and cortical bone loss.

C. Other Effects: Muscle Weakness

Hypovitaminosis D with or without osteomalacia has been associated with muscle weakness, limb pain, and impaired muscle function. This might be explained by the fact that vitamin D receptors (VDR) exist in muscle, or it may be secondary to the mineral abnormality. Birge and Haddad have shown that 25OHD directly influences intracellular accumulation of phosphate by muscle and offered this as an important role in the maintenance of muscle metabolism and function [54]. A syndrome of hyperesthesia has been described in association with hypovitaminosis D in five homebound elderly subjects by Gloth *et al.* [55], but no systematic studies of pain and muscle strength have been performed in an older population before and after treatment of vitamin D insufficiency [56]. Pain and weakness can lead to functional disability that may prevent a person from venturing outdoors, and this in turn exacerbates a poor vitamin D status by decreasing exposure to sunlight. Corless *et al.*, in a multicenter study, were unable to demonstrate an effect of vitamin D supplements on the functional activity levels in elderly subjects [57]. On the other hand, Sorensen *et al.* have shown that treatment with 1α-hydroxyvitamin D (1αOHD) was followed by a reduction in the time to dress [58]. This functional disability associated with vitamin D insufficiency may increase the incidence of falls and consequently the risk of hip fracture. Until relatively recently, studies on the prevention of senile osteoporosis and fractures with vitamin D supplements have failed to mention the incidence of falls [21,59,60].

V. PREVALENCE OF VITAMIN D INSUFFICIENCY

In analyzing the prevalence of vitamin D insufficiency in populations, it is necessary to divide the subjects according to age, country of residence, and season in which the study was performed (see also Chapter 33). To facilitate the analysis of vitamin D status, countries were grouped according to geographic regions: North America, Scandinavia, Europe, and others. Elderly subjects were subdivided into healthy elderly persons living at home and institutionalized subjects. In addition, the seasonal variations in serum 25OHD levels were taken into account. To analyze the values reported in the literature, we have largely relied on the review by McKenna [16] in which 117 studies of vitamin D status from 27 regions published between 1971 and 1990 were analyzed. In some groups of subjects, and especially in healthy elderly, the number of studies is very small.

A. Vitamin D Insufficiency in Adults (30–70 years)

In North America where dairy products are fortified with vitamin D, vitamin D insufficiency in young adults is rare [61–63]. In the study by Sherman *et al.* [27], less than 1% of men and 4% of women had 25OHD values lower than 14 ng/ml (35 nmol/liter) and these low values were found in May–June, when mean values are at their nadir. In Scandinavian countries, where vitamin D intake is high, a vitamin D insufficiency was noted in 4 to 9% of young adults during winter and in up to 5% during summer [64]. In Ireland [65], Italy [66], and England [16], 20 to 40% of young adults have hypovitaminosis D during the winter (Fig. 4). In a cross-sectional survey in a large Swiss population (3276 subjects aged 25–74 years) between November 1988 and June 1989, 6% of subjects had 25OHD values lower than 8 ng/ml (20 nmol/liter), and 34% had values lower than 15 ng/ml (37.5 nmol/liter) [67]. In subjects younger than 65 years, there was a small but clear cyclical seasonal variation with a nadir in February (median 41 nmol/liter i.e., 16.4 ng/ml) and a zenith in June (median 55 nmol/liter,

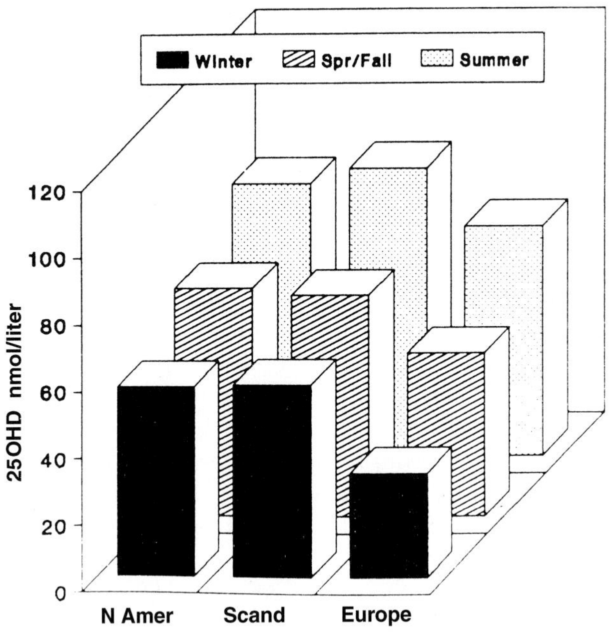

FIGURE 4 Mean of average serum 25OHD levels in young adults reported during the winter, spring and fall combined, and summer from studies collated according to geographic region. From McKenna [16] with permission.

i.e., 22 ng/ml). In contrast with previous studies which reported more important differences between young and elderly adults, in this population-based cohort, there was only a small and nonsignificant decrease in 25OHD values between the ages of 24 and 75 years. This may be because the upper age limit was only 75 years.

In a study on a large adult French population from the SU.VI.MAX project, 25OHD and PTH status was measured in 1584 healthy adults living in 16 large towns between November 1994 and April 1995 (772 men aged 45–60 years and 812 women aged 35–60 years) [68] (Table II). [The SUVIMAX (SUpplementation with VItamins Minerals and AntioXydants) project is a large international epidemiologic study to assess the effects of vitamins, minerals, and antioxydants in the prevention of global and specific mortality and morbidity due to cardiovascular diseases, cancers, and cataracts.] In this adult population, 14% of subjects had 25OHD lower than 12 ng/ml (30 nmol/liter). There was no age effect, but a sex difference between 25OHD levels was apparent (vitamin D insufficiency was present in 15.5% of women and only in 12.4% of men; $p \leq 0.04$). Major differences were observed between different regions within the country. In addition, there was a significant negative correlation between 25OHD and PTH values for the whole population ($r = 0.20$, $p \leq 0.001$). The mean PTH levels were 48.2 ± 22.0 pg/ml for subjects with 25OHD concentrations at or below 30 ng/ml and 37.9 ± 14.2 pg/ml for the others ($p \leq 0.01$). This study demonstrates that vitamin D insufficiency is present in a substantial percentage of the general adult urban population in France, a finding that has been confirmed by two other studies [69,70].

In low latitude countries, hypovitaminosis D is also present in adults. For example, in Saudi Arabia up to 40% of native inhabitants and immigrants from Africa have hypovitaminosis D in winter [71]. In healthy young adult Japanese, the frequency of hypovitaminosis D

TABLE II Prevalance of Vitamin D Insufficiency in the General Adult Population Living in 16 Large French Towns[a]

Region	n	Mean serum 25OHD (nmol/liter)[b]	Vitamin D insufficiency[c] (% hypovitaminosis D)	Mean serum PTH (pg/ml)[d]
North	200	43 ± 21	29	42 ± 15
Center	99	49 ± 25	27	40 ± 15
Northeast	199	52 ± 26	18	42 ± 16
Northwest	300	58 ± 29	14	38 ± 17
Paris	98	59 ± 25	13	46 ± 24
Rhône-Alpes	200	62 ± 27	9	40 ± 15
Mediterranean coast	299	68 ± 27	7	35 ± 13
South	89	81 ± 27	6	40 ± 11
Southwest	100	94 ± 38	0	37 ± 11

[a] A total of 1584 subjects were studied: 772 males from 45 to 60 years old and 812 females from 35 to 60 years old. From Chapuy *et al.* [68].
[b] To convert nmol/liter to ng/ml 25OHD, divide by 2.5.
[c] Vitamin D insufficiency is defined as a serum 25OHD level below 30 nmol/liter.
[d] PTH normal range is 55 pg/ml.

reaches 5% in males and 25% in females [72]. Actually, it appears that the vitamin D status of the young adult population needs to receive more attention, especially in winter, in countries where foods are not fortified. In the studies referenced by the McKenna analysis [16], the time of the year at which the measurements was performed is generally recorded, so as to permit a comparison of results with reference to the season of study [62,73,74]. In the studies performed earlier, the season and therefore the individual exposure to sunlight are not always reported.

B. Vitamin D Insufficiency in the Elderly (70–90 years)

Many studies, primarily those from Europe, have indicated that low levels of vitamin D may be more prevalent in older persons even if they do not appear related to the aging process per se (Fig. 5). In the healthy elderly population in North America and Scandinavia nearly 25% of subjects had low values in winter but less than 5% have low levels throughout the year [17,75–79]. So, in these countries, even if the vitamin D intake is adequate and equal or greater than the recommended dietary allowance the fraction of the vitamin D pool due to sunlight exposure is very important. As was found in normal adults, in elderly subjects there is a marked

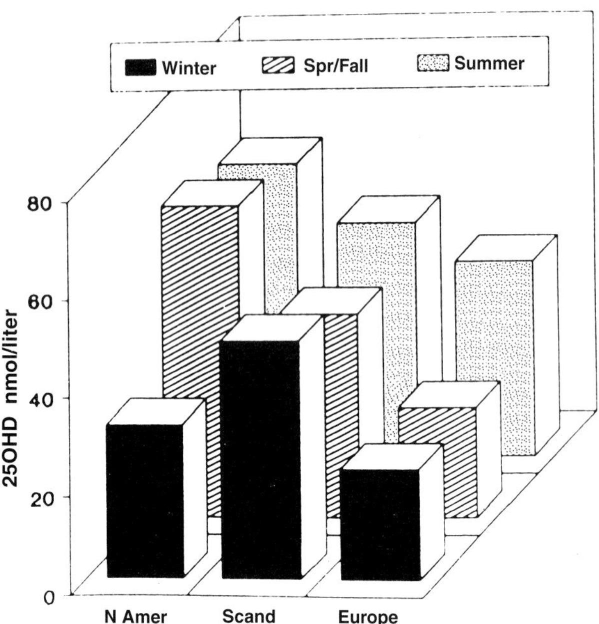

FIGURE 5 Mean of average serum 25OHD levels in healthy elderly reported during winter, spring and fall combined, and summer from studies collated according to geographic region. From McKenna [16] with permission.

seasonal variation in 25OHD levels. In Europe, the frequency of vitamin D insufficiency in winter ranges from 8 to 60% [10,16,73,80–83].

The Euronut Seneca study has evaluated the 25OHD concentrations in 824 free-living elderly people from 16 towns (latitudes between 35° and 61°N) in 11 European countries between December 1988 and March 1989 [84]. Town-specific mean 25OHD concentrations range from 10 to 24 ng/ml (25–59 nmol/liter) for men and from 8 to 19 ng/ml (21–48 nmol/liter) for women. Overall 36% of men and 47% of women had 25OHD concentrations below 12 ng/ml (30 nmol/liter), the lowest concentrations being found surprisingly in southern European towns in France, Spain, and Italy. The low 25OHD values in this study were largely explained by attitudes toward sunlight exposure. We have studied [31] the vitamin D status of 440 healthy free-living elderly women aged 75–90 years during winter and living in five French cities whose latitude varies from 49°9′ to 43°6′ N. The mean 25OHD level was not different among the five cities. However, in all of the cities the mean 25OHD level was significantly lower than the mean level found in young healthy women from the OFELY cohort study that were recruited during the same winter period (42.5 ± 25 versus 62.0 ± 40.0 nmol/liter). (OFELY is a prospective study of the bone-loss determinants in women aged 30–95 years randomly selected from a large insurance company in Lyon, France.) Nevertheless, the mean values of the free-living elderly women were significantly higher ($p = 0.03$) than the winter values obtained in 59 institutionalized elderly women (42.5 ± 25 versus 15.5 ± 6.5 nmol/liter) (Table III). Among the institutionalized elderly women, 39% exhibited vitamin D insufficiency (25OHD values ≤ 12 ng/ml, i.e., 30 nmol/liter) and only 16% had normal 25OHD values greater than or equal to 25 ng/ml (62.5 nmol/liter), the mean value of young women (Table II). Vitamin D insufficiency was associated in these healthy women with biochemical indices of secondary hyperparathyroidism and increased bone turnover.

In comparison to the few studies of free-living people, the vitamin D status of homebound subjects has been extensively studied. In the review by McKenna [16], the prevalence of vitamin D insufficiency in the institutionalized elderly population varies from 3 to 28% in North America. This prevalence reaches 54% in the study by Gloth *et al.* [85], who compared homebound elderly in community versus nursing homes. Despite apparently adequate vitamin D intake (200 to 400 IU/day or 5 to 10 μg/day), several studies have shown that between 30 and 50% of older homebound subjects have low vitamin D status, with 25OHD levels at or below 10 ng/ml (25 nmol/liter) [86–88]. In Europe, vitamin D insufficiency in elderly homebound subjects ranged

TABLE III Comparison of Biochemical Values in Elderly and Young Women during the Winter[a]

	Elderly women living at home (EPIDOS study)	Elderly women in institutions	Young women (OFELY study)
Number of subjects	440	59	54
Age (years)	80 ± 3	87 ± 6 (a)	34 ± 5 (a)
Calcium intake (mg/day)	569 ± 338	—	810 ± 280 (b)
Serum calcium (mmol/liter)	2.35 ± 0.07	2.22 ± 0.09 (a)	2.35 ± 0.07
25OHD (nmol/liter)	42.5 ± 25.0	15.5 ± 6.5 (c)	62.0 ± 40.0 (b)
PTH (pg/ml)	63 ± 28	76 ± 49 (d)	43 ± 15 (c)

[a] Data are expressed as means ± SD. Lowercase letter show results of statistical analyses: (a), $p < 0.0001$ compared to EPIDOS and OFELY studies; (b), $p < 0.0001$ compared to EPIDOS study; (c), $p < 0.001$ compared to EPIDOS study; (d), $p < 0.05$ compared to EPIDOS study. From Chapuy *et al.* [31].

from 70 to 100% [16]. The European 25OHD values in institutionalized subjects are significantly lower than both Scandinavian and North American values (Fig. 6). Only a few studies were conducted during specific seasons, but one would not expect much seasonal change as these elderly subjects have little or no exposure to sunlight [89]. The mean European values, based on year-round determinations of 25OHD and calculated from values reported in 16 European studies [16], is lower than 8 ng/ml (20 nmol/liter).

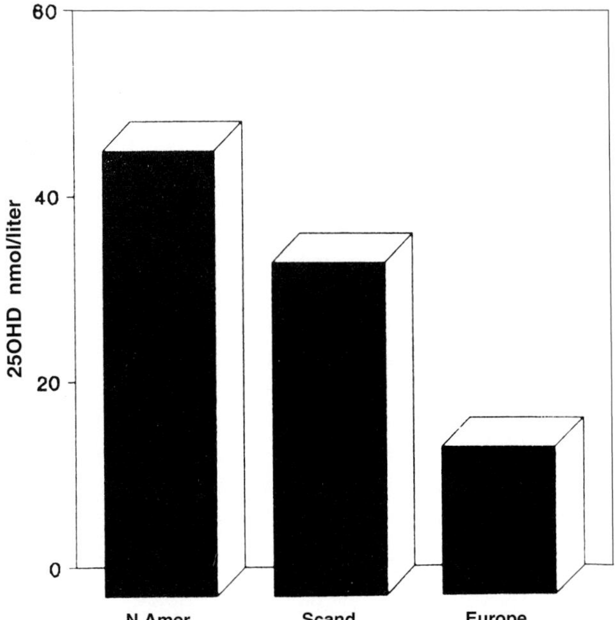

FIGURE 6 Mean of average serum 25OHD levels in institutionalized elderly reported throughout the year from studies collated according to geographic region. From McKenna [16] with permission.

In our DECALYOS study, we have measured the 25OHD values at baseline in 280 very elderly women (84 ± 6 years) who were ambulatory but lived in nursing homes. (The DECALYOS study was the prospective study undertaken to determine the effects of vitamin D and calcium supplements on hip fracture incidence [21] in Lyon, France.) The mean 25OHD values were 14 ± 10 ng/ml (35 ± 25 nmol/liter), and 44% of subjects had 25OHD values lower than or equal to 12 ng/ml (30 nmol/liter). In the Dutch study of Ooms *et al.* [52], 65% of the 330 healthy women, residents of home apartments for the elderly, had 25OHD values below 12 ng/ml (30 nmol/liter), and in 34% the 25OHD level was below 8 ng/ml (20 nmol/liter). In winter 83% and in summer 50% of subjects had 25OHD levels that were below 12 ng/ml (30 nmol/liter). The median levels of 25OHD were significantly higher for inhabitants of apartments for the elderly than for residents of homes for elderly (11.6 and 8.8 pg/ml, respectively, (i.e., 29 and 22 nmol/liter; $p < 0.0001$).

VI. PREVENTIVE MEASURES: CORRECTION OF LOW VITAMIN D STATUS

The effect on bone turnover of vitamin D insufficiency and secondary hyperparathyroidism have suggested that correction of subclinical hypovitaminosis D in the elderly may have a beneficial effect on the secretion of PTH, bone loss, and consequently the risk of fracture in elderly subjects.

A. Effects on Secondary Hyperparathyroidism

In the late 1980s, in a 6-month trial of supplementation with vitamin D_2 (800 IU or 20 μg/day) and calcium (1 g/day) given to elderly subjects with vitamin D insufficiency living in an institution, we were able to reduce PTH concentrations by more than 30% with normalization of 25OHD concentrations [4]. With 400 IU/day of vitamin D_3, Ooms *et al.* restored to normal 25OHD concentrations and decreased the mean PTH level by about 6 to 15% in an elderly Dutch population [32]. These two studies demonstrated that supplementation with low doses of vitamin D (400 to 800 IU/day or 10 to 20 μg/day) leads to adequate improvement of vitamin D status and parathyroid function. There does not appear to be a need for a higher dose, as has been suggested by others [1]. In contrast, vitamin D supplementation of nursing home residents in the United States having 25OHD levels of 16 ng/ml or greater (≥40 nmol/liter), did not significantly decrease PTH levels [90].

B. Effects on Bone Mass

By increasing the vitamin D intake from 100 to 500 IU daily, Dawson-Hughes *et al.* [33] were able to significantly reduce the late wintertime bone loss and to improve the net bone density of the spine. Among normal middle-aged women, Khaw *et al.* found a significant direct relationship between serum 25OHD and bone mass of the spine and femoral neck, with an inverse correlation between bone density and PTH concentration, which is consistent with effects of small changes in vitamin D status on bone mass [43]. The use of 15,000 IU or 375 μg of oral vitamin D_2 weekly has also been shown to reduce metacarpal bone loss in normal women aged 65 to 74 years living in the community [91].

In our DECALYOS study [21], the elderly women treated daily for 18 months with 800 IU or 20 μg of vitamin D_3 and 1.2 g of calcium showed an increase of 2.7% in the BMD of the proximal femur. During the same period, the women in the placebo group showed a decrease of femoral BMD equal to 4.6%. In the Dutch study of Ooms *et al.*, a daily treatment of elderly women (mean calcium intake 859 mg/daily) with 400 IU or 10 μg of vitamin D_3 over a 2-year period was associated with an increase in femoral neck BMD equal to 2.3% in comparison with the placebo group. In these women at baseline, hip BMD was positively correlated with the serum 25OHD concentration below a threshold level (serum 25OHD ≤ 12 ng/ml or 30 mmol/liter [32].

C. Effects on Fracture Rate

The effects of vitamin D supplements on bone mass suggested that correction of vitamin D insufficiency in the elderly with low doses of vitamin D may have beneficial effects on fracture incidence [92]. This was the aim of several studies. In our DECALYOS study, 3270 healthy ambulatory women (mean age 84 ± 6 years) living in a nursing home received daily either 800 IU or 20 μg of vitamin D_3 plus 1.2 g of elemental calcium or a double placebo. After 18 months of follow up, analysis showed that there was a 43% reduction in hip fractures ($p \leq 0.05$) and a 32% reduction in all nonvertebral fractures ($p = 0.015$) in the treated group. The results of the intention-to-treat analysis were also similar: 80 hip fractures in the vitamin D_3–calcium group and 110 in the placebo group (27% reduction; $p < 0.01$). At the same time, serum PTH decreased by 46% ($p < 0.001$) and the serum 25OHD level increased by 160% ($p < 0.001$) without change in the $1,25(OH)_2D$ levels. After a further 18 months of treatment for 1404 women, the beneficial effects of the treatment on nonvertebral fractures was confirmed. At the end of 36 months of follow up in the intention-to-treat analysis, there were 17.2% fewer nonvertebral fractures (255 versus 308, $p < 0.02$) and 23% fewer hip fractures (137 versus 178, $p < 0.02$) in the vitamin D_3–calcium group (Fig. 7). The probability of hip fractures was decreased (odds ratio 0.73, CI 0.62 to 0.84), as was that of all nonvertebral fractures (odds ratio 0.72, CI 0.60 to 0.84) [21,59] (Fig. 8). This study has pointed out the importance of vitamin D insufficiency as a major determinant of senile secondary hyperparathyroidism and bone loss. However, the study did not permit the elucidation of the relative importance of calcium versus vitamin D. Also, the possible effect of vitamin D supplements on the incidence of falls was not studied.

In the study of Heikinheimo *et al.* [93], 799 elderly men and women, living either in residential care or in their own home, were followed for 2 to 5 years. In those treated with an annual injection of 150,000 to 300,000 IU of vitamin D_2, there was a significant reduction in the number of fractures in the upper limbs and ribs but no significant reduction in hip fractures. The reduction in hip fracture incidence reached 22% (9.4% in the control group and 7.3% in the intervention group), which is similar to the reduction found in our DECALYOS study (23%). One reason for the lack of significance was possibly the smaller sample size. The prevention of fracture with vitamin D alone found by Heikinheimo might suggest that the results obtained with the association of calcium and vitamin D would be primarily due to vitamin D [93,94]. In a prospective double blind trial of 2578 men and women (mean age 80 ± 6 years) living either at home or in an institution, Lips *et al.* studied the effect of a daily supplement with 400 IU of vitamin D_3 [95]. After 1 year of vitamin D supplementation in a subgroup of women [32], the PTH level was reduced by only 6% from baseline, and after 2 years the BMD of the femoral neck had increased 2.2% in the vitamin D group compared to the placebo group. Nevertheless, the vitamin D supplementation did not decrease the incidence of hip fracture after a maximum treatment period of 3.5 years.

The differences observed in the results of the French and the Dutch studies may be explained by the lower dietary calcium intake in France than in the Netherlands and the use of calcium in the French study. In addition,

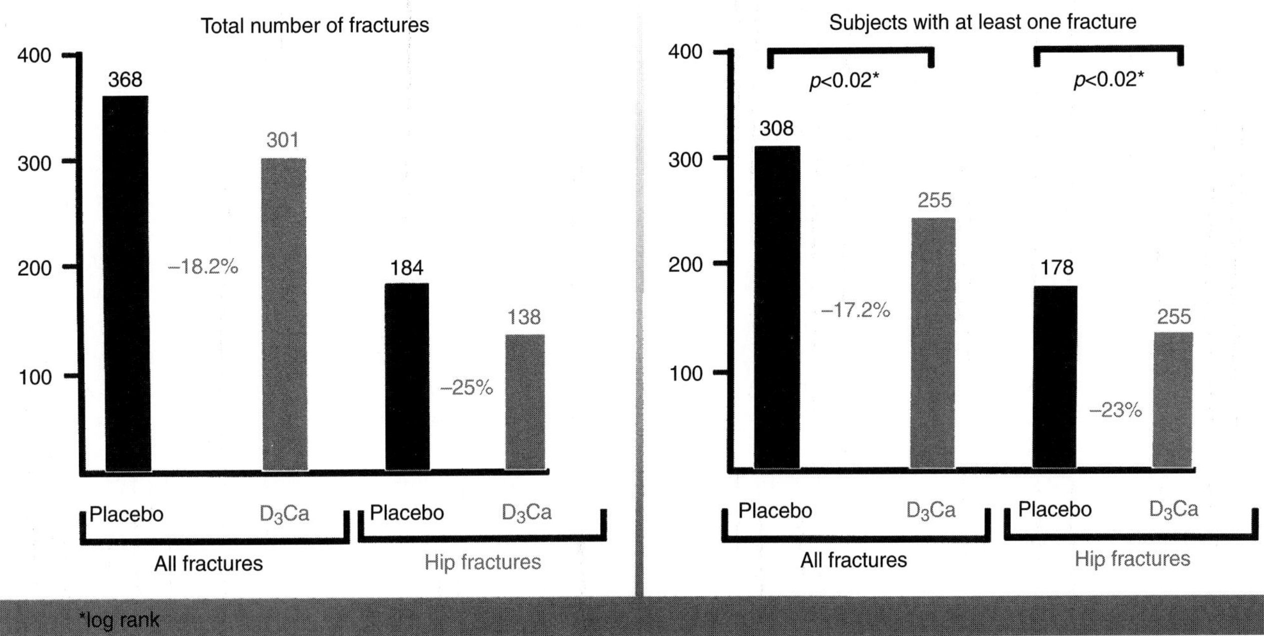

FIGURE 7 Effects of vitamin D_3 and calcium supplementation during 3 years on numbers of fractures in elderly women (DECALYOS study). One group received placebo, and the other group (D_3Ca) received 800 IU of vitamin D_3 plus 1.2 g elemental calcium.

the participants in the French study were 4 years older, less active at baseline, and were all residents of nursing homes. Hip fracture incidence was lower in the Dutch study (29 per 1000 versus 40 per 1000), so its power may have been insufficient to demonstrate significant differences in the fracture rates. Nevertheless, the increase in 25OHD levels and more importantly the reduction in PTH secretion were much lower in the Dutch study, raising the possibility that the daily dose of 400 IU of vitamin D_3 was suboptimal (−39% in our study versus −6% in the Dutch study after 1 year of treatment). These results indicate that a larger treatment effect was obtained in the French study [59].

In addition to these three prospective studies on the preventive effect of vitamin D on hip fracture incidence, Banstam and Kanis in a preliminary retrospective report from the MEDOS (MEDOS: *Med*iterranean *Os*teoporosis Study) study did not find that the use of vitamin

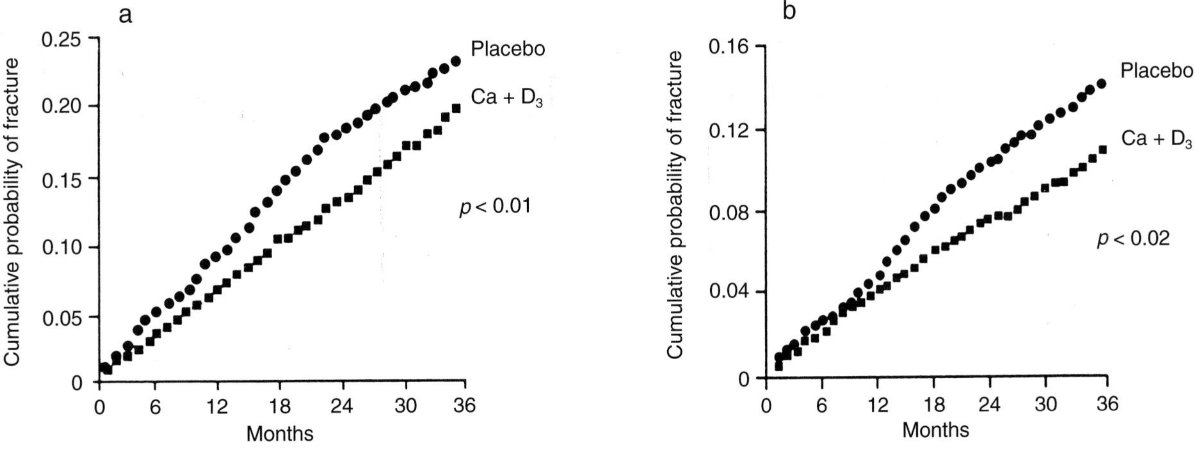

FIGURE 8 Cumulative probability of hip fractures (a) and all fractures (b) in the placebo and vitamin D_3-calcium groups (DECALYOS study). From Meunier *et al.* [41] with permission.

D was associated with a significant decrease in the risk of hip fracture, as was the use of calcium, estrogen, or calcitonins [96]. When the data were reanalyzed including not all the hip fractures but only the low energy fractures, the use of vitamin D supplement was associated with a 26% (but nonsignificant) decrease in the risk of hip fracture [82]. In addition, the risk reduction was influenced by age and body mass index (BMI). For women 80 years or older, the reduction in hip fracture risk for vitamin D users was 37% ($p = 0.04$), and for these women with a BMI less than 20 kg/m^2 the use of vitamin D was associated with a marked and significant reduction in hip fracture risk of 55% ($p = 0.01$). In this study, vitamin D was taken for a time ranging from 1 to 20 years, and the doses used were not known.

From all these studies, it appears that vitamin D supplements are undoubtedly useful in the prevention of hip fracture, but there are still two critical questions that need to be answered. What is the optimal dose, and who should receive supplements?

VII. CONCLUSIONS

There is now convincing evidence that there are several forms of low vitamin D states that induce different forms of bone disease. Severe and prolonged "deficiency" of vitamin D, which is no longer very common, is associated with osteomalacia. In contrast, less severe vitamin D depletion, which might be called vitamin D "insufficiency," is very common in elderly subjects living in institutions or at home but also appears not to be rare in healthy adults, particularly in winter. A progressive state of vitamin D insufficiency per se does not imply that bone disease is present but does identify a high risk status [1]. Vitamin D insufficiency will ultimately lead to a state of vitamin D deficiency. However, before overt clinical symptoms of osteomalacia became apparent, vitamin D insufficiency may give rise to problems at the bone level as a consequence of secondary hyperparathyroidism, resulting in increased bone turnover, bone loss, and risk of fractures. The observation of seasonal variation in bone mineral density in normal subjects provides evidence that relatively small changes in vitamin D status may have significant effects on bone mass.

In agreement with several authors [3–5,7–9], we propose that vitamin D insufficiency might be defined by a 25OHD level equal to or under 12 ng/ml (30 nmol/liter), a value usually found in sunlight-deprived subjects. Sometimes an elevation of serum PTH is observed with 25OHD levels greater than 12 ng/ml (30 nmol/liter), suggesting that the lower limit of the normal range for serum 25OHD level which initiates a PTH response could probably be higher: 14.8 ng/ml (37 nmol/liter) for Gloth and Tobin [8] and Webb *et al.* [9] and 20 ng/ml (50 nmol/liter) for Peacock *et al.* [3]. In many countries, natural dietary sources do not readily permit an intake of the recommended amounts of vitamin D; thus, the available mode of prevention of vitamin D insufficiency might be increased exposure to sunlight, fortification of foodstuffs, and oral or injectable vitamin D supplementation.

For young adults, adequate exposure to summer sunlight represents a means to make up for the wintertime depletion in vitamin D, but in some cases an oral supplementation with 400 IU/day during winter might be necessary. Fortification of dairy products with vitamin D is very common in North America but not in Europe and especially not in France, where vitamin D is considered to be a drug. This fortification, despite problems with manufacturing practices and consistency, is a means to provide a minimal supply for the general population. Supplementation should interest the health authorities in countries where such procedures are not yet employed.

Because the prevalence of vitamin D insufficiency is very high among elderly subjects, an increase in vitamin D intake becomes essential, especially for those living in an institution. Promoting sun exposure is difficult to realize, and placing ultraviolet lamps in the living room would probably not be efficient enough. Supplementation by vitamin D preparations seems to be the best solution. This supplementation should be at least 400 IU/day in the healthy elderly to increase 25OHD levels, but this dose was not sufficient for prevention of hip fractures in elderly subjects with a mean calcium intake of 600–1000 mg/day [95]. However, 800 IU of vitamin D_3 with 1.2 g of calcium was able to decrease by 25% the hip fractures incidence of institutionalized elderly women in 3 years and to maintain normal 25OHD and PTH concentrations. As it appears that vitamin D supplementation is effective and safe in preventing vitamin D insufficiency, daily low dose treatment (400–800 IU/daily) may be the best regimen. However, an intermittent high dose (100,000 IU) given orally or by injection every 6 months may be an effective alternative [83]. Very high single doses, like 600,000 IU twice a year, should be used very cautiously because of the risk of inducing transient hypervitaminosis D and side effects due to hypercalcemia, including increased calcification of vascular atheromatous plaques.

References

1. Parfitt MA, Gallagher JC, Heaney RP, Johnston CC, Neer R, Whedon GD 1982 Vitamin D and bone health in the elderly. Am J Clin Nutr **36:**1014–1031.

2. Walters MR, Kollenkirchen U, Fox J 1992 What is vitamin D deficiency? Proc Soc Exp Biol Med **199**:385–393.
3. Peacock M, Selby PL, Francis RM, Brown WB, Hordon L 1985 Vitamin D deficiency, insufficiency, sufficiency and intoxication. What do they mean? In: Norman A, Schaefer K, Grigoletti MG, Herrath DV (eds) Sixth Workshop on Vitamin D. de Gruyter, Berlin and New York, pp. 569–570.
4. Chapuy MC, Chapuy P, Meunier PJ 1987 Effect of calcium and vitamin D supplements on calcium metabolism in the elderly. Am J Clin Nutr **46**:324–328.
5. Lips P, Wierzinga A, Van Ginkel FC, Jongen MJM, Netelenbos JC 1988 The effect of vitamin D supplementation on vitamin D status and parathyroid function in elderly subjects. J Clin Endocrinol Metab **67**:644–50.
6. Krall EA, Sahyoun N, Tannebaum S, Dallal GE, Dawson-Hughes B 1989 Effect of vitamin D intake on seasonal variations in parathyroid hormone secretion in postmenopausal women. N Engl J Med **321**:1777–1783.
7. Gloth FM, Gundberg CM, Hollis BW, Haddad JG, Tobin JD 1995 Vitamin D deficiency in homebound elderly persons. JAMA **274**:1683–1686.
8. Gloth FM, Tobin JD 1995 Vitamin D deficiency in older people. J Am Geriatr Soc **43**:822–828.
9. Webb AR, Pilbeam C, Hanafin N, Holick MF 1990 An evaluation of the relative contributions of exposure to sunlight and of diet to the circulating concentrations of 25-hydroxyvitamin D in an elderly nursing home population in Boston. Am J Clin Nutr **51**:1075–1081.
10. Bouillon RA, Auwerx JD, Lissens WD, Pelemans WK 1987 Vitamin D status in the elderly, seasonal substrate deficiency causes 1,25-dihydroxycholecalciferol deficiency. Am J Clin Nutr **45**:755–763.
11. Lips P, Chapuy MC, Dawson-Hughes B, Pols HAP 1995 International comparison of serum 25-hydroxyvitamin D measurements. J Bone Miner Res **10**:S496.
12. Hollis B 1996 Assessment of vitamin D nutritional and hormonal status: What to measure and how to do it. Calcif Tissue Int **58**:4–5.
13. Holick MF 1994 McCollum Award lecture: Vitamin D—New horizons for the 21st century. Am J Clin Nutr **60**:619–630.
14. Holíck MF, Matsuoka LY, Wortsman J 1989 Age, vitamin D and solar ultraviolet. Lancet **1**:1104–1105.
15. Need AG, Morris HA, Horowitz M, Nordin BEC 1993 Effects of skin thickness, age, body fat, and sunlight on serum 25-hydroxyvitamin D. Am J Cl Nutr **58**:882–885.
16. McKenna MJ 1992 Differences in vitamin D status between countries in young adults and in elderly. Am J Med **93**:69–77.
17. Clemens TL, Zhou X, Myles M, Endres D, Lindsay R 1986 Serum vitamin D_2 and vitamin D_3 metabolite concentrations and absorption of vitamin D_2 in elderly subjects. J Clin Endocrinol Metab **63**:656–660.
18. Holick MF 1986 Vitamin D requirements for the elderly. Clin Nutr **5**:121–129.
19. Weisman Y, Schen RJ, Einsenberg Z, Edelstein S, Harell A 1981 Inadequate status and impaired metabolism of vitamin D in elderly. J Med Sci **17**:19–21.
20. Baragry JM, France MW, Corles D, Gupta SP, Switala S, Boucher BJ, Cohen RD 1978 Intestinal cholecalciferol absorption in the elderly and in younger adults. Clin Sci Mol Med **55**:213–220.
21. Chapuy MC, Arlot ME, Duboeuf F, Brun J, Crouzet B, Arnaud S, Delmas PD, Meunier PJ 1992 Vitamin D_3 and calcium to prevent hip fractures in elderly women. N Engl J Med **237**:1637–1642.
22. Mundy GR 1989 In Calcium Homeostasis: Hypercalcemia and Hypocalcemia. Martin Dunitz, London, pp. 185–186.
23. Riggs BL, Melton LJ 1983 Evidence of two distinct syndromes of involutional osteoporosis. Am J Med **75**:899–901.
24. Chapuy MC, Durr F, Chapuy P 1983 Age-related changes in parathyroid hormone and 25-hydroxycholecalciferol levels. J Gerontol **38**:19–22.
25. Forero MS, Klein RF, Nissenson RA, Nelson K, Heath H, Arnaud CD, Riggs BL 1987 Effect of age on circulating immunoreactive and bioactive parathyroid hormone levels in women. J Bone Miner Res **2**:363–366.
26. Epstein S, Bryce G, Hinman JW, Miller ON, Riggs BL, Hui SL, Johnston CC 1986 The influence of age on bone mineral regulating hormones. Bone **7**:421–425.
27. Sherman SS, Hollis BW, Tobin JD 1990 Vitamin D status and related parameters in a healthy population: The effects of age, sex and season. J Clin Endocrinol Metab **71**:405–413.
28. Prince RL, Dick I, Devine A, Price RI, Gutteridge DH, Kerr D, Criddle A, Garcia-Webb P, St John A 1995 The effects of menopause and age on calciotropic hormones: A cross sectional study on healthy women aged 35 to 90. J Bone Miner Res **10**:835–842.
29. Freaney R, McBrinn Y, Mc Kenna MJ 1993 Secondary hyperparathyroidism in elderly people: Combined effect of renal insufficiency and vitamin D deficiency. Am J Clin Nutr **58**:187–191.
30. Young G, Marcus R, Minkoff JR, Kim LY, Segre GV 1987 Age-related rise in parathyroid hormone in men: The use of intact and mid molecule antisera to distinguish hormone secretion from retention. J Bone Miner Res **2**:367–374.
31. Chapuy MC, Schott AM, Garnero P, Hans D, Delmas PD, Meunier PJ 1996 Healthy elderly French women living at home have secondary hyperparathyroidism and high bone turnover in winter. J Clin Endocrinol Metab **81**:1129–1133.
32. Ooms ME, Roos JC, Bezemer PD, Van Der Vijch WJF, Bouter LM, Lips P 1995 Prevention of bone loss by vitamin D supplementation in elderly women: A randomized double blind trial. J Clin Endocrinol Metab **80**:1052–1058.
33. Dawson-Hughes B, Dallal GE, Krall EA, Harris S, Sokoll LJ, Falconer G 1991 Effect of vitamin D supplementation on overall bone loss in healthy postmenopausal women. Ann Intern Med **115**:505–512.
34. Kochersberger G, Westlund R, Lyles KW 1990 Calcium supplementation lowers serum parathyroid hormone levels in elderly subjects. J Gerontol **45**:M159–M162.
35. Lips P, Netelendos C, Jongen MJM, Van Ginkel FC, Altuis AL, Vanschaik CL, Vandervijch WJF, Vermeiden JPW, Van Der Meer C 1982 Histomorphometric profile and vitamin D status in patients with femoral neck fracture. Metab Bone Dis Related Res **4**:85–93.
36. Johnston CC, Norton J, Khairi MRA, Kernek C, Edouard C, Meunier PJ 1985 Heterogeneity of fracture syndromes in postmenopausal women. J Clin Endocrinol Metab **61**:551–556.
37. Compston JE, Silver AC, Croucher PI, Brown RC, Woodhead JS 1989 Elevated serum intact parathyroid hormone levels in elderly patients with hip fracture. Clin Endocrinol **31**:667–672.
38. Okano T, Yamamoto K, Hagino H, Hishimoto H 1992 Iliac bone histomorphometry in Japanese women with hip fracture. In: Bone Morphometry, Sixth International Congress, Lexington 4–9 October 1992, Abstract 99.
39. Benhamou CL, Chappard D, Gauvain JB 1991 Hyperparathyroidism in proximal femur fractures: Biological and histomorphometric study in 21 patients over 75 years old. Clin Rheumatol **10**:144–160.
40. Rao DS, Villanueva A, Mathews SM 1983 Histologic evaluation of vitamin D depletion in patients with intestinal malabsorption or dietary deficiency. In: Frame B, Potts J Jr (eds) Clinical Disor-

ders of Bone and Mineral Metabolism. Excerpta Medica, Amsterdam, pp. 224–226.
41. Meunier PJ, Chapuy MC, Arlot ME, Delmas PD, Duboeuf F 1994 Can we stop bone loss and prevent hip fractures in the elderly? Osteoporosis Int **4**, Suppl 1:S71–S76.
42. Rosen CJ, Morrison A, Zhou H, Storm D, Hunter SJ, Musgrave K, Chen T, Wei W, Holick MF 1994 Elderly women in northern New England exhibit seasonal changes in bone mineral density and calciotropic hormones. Bone Miner **2**:83–92.
43. Khaw KT, Sheyd MJ, Compston J 1992 Bone density, parathyroid hormone and 25-hydroxyvitamin D concentrations in middle-aged women. Br Med J **305**:273–277.
44. Lukert B, Higgins J, Stoskopf M 1992 Menopausal bone loss is partially regulated by dietary intake of vitamin D. Calcif Tissue Int **51**:173–179.
45. Martinez ME, Delcampo MJ, Sanchez-Cabezudo MJ, Garcia JA, Sanchez-Calvin MT, Torrijos A, Coya J 1994 Relations between calcidiol serum levels and bone mineral density in postmenopausal women with low bone density. Calcif Tissue Int **55**:253–256.
46. Villareal DT, Civitelli R, Chines A, Avioli LV 1991 Subclinical vitamin D deficiency in postmenopausal women with low vertebral bone mass. J Clin Endocrinol Metab **72**:628–634.
47. Cummings SR, Black DM, Nevitt MC, Browner W, Cauley J, Ensrud K, Genant HK, Palermo L, Scott J, Vogt TM 1993 Bone density at various sites for prediction of hip fractures. Lancet **341**:72–75.
48. Scharla SH, Scheidt-Nave C, Leidig G, Seibel M, Ziegler R 1994 Association between serum 25-hydroxyvitamin D and bone mineral density in a normal population sample in Germany. In: Norman AW, Bouillon R, Thomasset M (eds) Vitamin D: A Pluripotent Steroid Hormone: Structural Studies, Molecular Endocrinology and Clinical Applications, de Gruyter, Berlin, p. 863.
49. Brazier M, Kamel S, Maamer M, Agbomson F, Elseper I, Garabedian M, Desmet G, Sebert JL 1995 Markers of bone remodeling in the elderly subjects: Effects of vitamin D insufficiency and its correction. J Bone Miner Res **10**:1753–1761.
50. Kamel S, Brazier M, Picar C, Boitte F, Samson L, Desmet G, Sebert JL 1994 Urinary excretion of pyridinoline crosslinks measured by immunoassay and HPLC techniques in normal subjects and elderly patients with vitamin D deficiency. Bone Miner **26**:197–208.
51. Dawson Hughes B, Harris SS, Krall EA, Dallal GE, Falconer G, Green CL 1995 Roles of bone loss in postmenopausal women randomly assigned to one or two dosages of vitamin D. Am J Clin Nutr **61**:1140–1145.
52. Ooms ME, Lips P, Roos JC, Vandervijch WJF, Popp-Snijdeers C, Bezemer PD, Bouter LM 1995 Vitamin D status and sex hormone binding globulin: Determinants of bone turnover and bone mineral density in elderly women. J Bone Miner Res **10**:1177–1184.
53. Gennari C, Agnusdei D, Nardi P, Civitelli R 1990 Estrogen preserves a normal intestinal response to 1,25-dihydroxyvitamin D_3 in oophorectomized women. J Clin Endocrinol Metab **71**:1288–1293.
54. Birge SJ, Haddad JC 1975 25-Hydroxycholecalciferol stimulation of muscle metabolism. J Clin Invest **56**:1100–1107.
55. Gloth FM, Lindsay JM, Zelesnick LB, Greenough WB 1991 Can vitamin D deficiency produce an unusual pain syndrome. Arch Intern Med **151**:1662–1664.
56. Gloth FM, Smith CE, Hollis BW, Tobin JD 1995 Functionnal improvement with vitamin D replenishment in a cohort of frail vitamin D deficient older people. J Am Geriatr **43**:1269–1271.
57. Corless D, Dawson D, Fraser F, Ellis M, Evans SJ, Perry JD, Reisner C, Silver CP, Beer M, Boucher BJ 1985 Do vitamin D supplements improve the physical capabilities of elderly hospital patients. Age Aging **14**:76–84.
58. Sorensen OH, Lund BI, Saltin B 1979 Myopathy in bone loss ageing: Improvement by treatment with 1α-hydroxycholecalciferol and calcium. Clin Sci **56**:157–161.
59. Chapuy MC, Arlot ME, Delmas PD, Meunier PJ 1994 Effect of calcium and cholecalciferol treatment for three years on hip fractures in elderly women. Br Med J **308**:1081–1082.
60. Tilyard MW, Spears GFS, Thomson J, Dovey S 1992 Treatment of postmenopausal osteoporosis with calcitriol or calcium. N Engl J Med **326**:357–362.
61. Haddad JG, Hahn TJ 1973 Natural and synthetic sources of circulating 25-hydroxyvitamin D in man. Nature **244**:515–517.
62. Styrd RP, Gilbertson TJ, Brunden MN 1979 A seasonal variation study of 25-hydroxyvitamin D_3 serum levels in normal humans. J Clin Endocrinol Metab **48**:771–775.
63. Chesney RW, Rosen JF, Hamstra AJ, Smith C, Mahaffey K, DeLuca HF 1981 Absence of seasonal variation in serum concentrations of 1,25-dihydroxyvitamin D despite a rise in 25-hydroxyvitamin D in summer. J Clin Endocrinol Metab **53**:139–142.
64. Lund B, Sorenson OH 1979 Measurement of 25-hydroxyvitamin D in serum and its relation to sunshine, age and vitamin D intake in the Danish population. Scand J Clin Lab Invest **39**:23–30.
65. Vik T, Try K, Stromme JH 1980 The vitamin D of man at 70° north. Scand J Clin Lab Invest **40**:227–232.
66. Benucci A, Tommasi M, Fantappie B, Scardigli S, Ottanelli S, Pratesi E, Romano S 1993 Serum 25-hydroxyvitamin D levels in normal subjects: Seasonal variation and relationship with parathyroid hormone and osteocalcin. J Nucl Biol Med **37**:77–82.
67. Burnand B, Slovtskis D, Gianoli F, Cornuz J, Rickenback M, Paccaud F, Burckhardt P 1992 Serum 25-hydroxyvitamin D : distribution and determinants in the Swiss population. Am J Clin Nutr **56**:537–542.
68. Chapuy MC, Preciozi P, Maamer M, Arnaud S, Galan P, Hercberg S, Meunier PJ 1996 Prevalence of hypovitaminosis D in the femoral adult french population. Osteoporosis Int **6**(Suppl. 1):abstract PSU151.
69. Ribot C, Pouilles JM, Tremollieres F 1996 Vitamin D status in a large cohort of early postmenopausal French women. Osteoporosis Int **6**(Suppl. 1):abstract PSU218.
70. Baudoin C, Aquino D, Charlier C, Cohen Solal M, Gueris J, Devernejoul MC 1996 Vitamin D deficiency in postmenopausal women in region of Paris. Osteoporosis Int **6**(Suppl. 1):abstract PSU281.
71. Sedrani SH, Elidrisy AWTH, El Arabi KM 1983 Sunlight and vitamin D status in normal Saudi subjects. Am J Clin Nutr **38**:129–132.
72. Takeuchi A, Okano T, Ishida Y, Kovayashi T 1995 Dietary vitamin D intake and vitamin D nutritional status in the healthy young and institutionalized elderly people. In: Burckhardt P Heaney RP (eds), Nutritional Aspects of Osteoporosis. Ares Serono Symposia Publications, Rome, pp. 351–355.
73. Stamp TCB, Round JM 1974 Seasonal changes in human plasma levels of 25-hydroxyvitamin. Nature **247**:563–565.
74. Poskitt EME, Cole TJ, Lawson DEM 1979 Diet, sunlight and 25-hydroxyvitamin D in healthy children and adults. Br Med J **1**:221–223.
75. Delvin EE, Imbach A, Copti M 1988 Vitamin D nutritional status and related biochemical indices in an autonomous elderly population. Am J Clin Nutr **48**:373–378.
76. Von Knorring J, Slätis P, Weber TH, Helenius T 1982 Serum levels of 25-hydroxyvitamin D, 24,25-dihydroxyvitamin D and parathyroid hormone in patients with femoral neck fracture in southern Finland. Clin Endocrinol **17**:189–194.
77. Omdahl JL, Garry PJ, Hunsaker LA, Hunt WC, Goodwin JS

1982 Nutritional status in a healthy elderly population: Vitamin D. Am J Clin Nutr **36**:1125–1133.
78. Lukert BP, Carey M, McCarty B, Tiemann S, Goodnight L, Helm M, Hassanein R, Stevenson S, Stoskopf M, Doolan L 1987 Influence of nutritional factors on calcium regulating hormone and bone mass. Calcif Tissue Int **40**:119–125.
79. Schmidt-Gayk H, Goosen J, Lendle F, Seidel D 1977 Serum 25-hydroxycholecalciferol in myocardial infarction. Atherosclerosis **26**:55–58.
80. Rapin CH, Lagier R, Boivin GF, Jung A, McGee W 1982 Biochemical findings in blood of aged patients with femoral neck fractures: A contribution to the detection of occult osteomalacia. Calcif Tissue Int **34**:465–469.
81. Newton HMV, Sheltawy M, Hay AWM, Morgan B 1985 The relations between vitamin D_2 and D_3 in elderly women in Great Britain. Am J Clin Nutr **41**:760–764.
82. Ranstam J, Kanis JA 1995 Influence of age and body mass on the effects of vitamin D on hip fractures risk. Osteoporosis Int **5**:450–454.
83. Byrne PM, Preatney R, McKenna MJ 1995 Vitamin D supplementation in the elderly: Review of safety and effectiveness of different regimes. Calcif Tissues Int **56**:518–520.
84. Van Der Wielen RP, Lowik MRH, Van Der Berg H, Degroot L, Haller J, Moreras O, Vanstaveren WA 1995 Serum 25OHD concentrations among elderly people in Europe. Lancet **346**:207–210.
85. Gloth MD, Tobin JD, Smith CE, Hollis BW, Gunberg CM 1993 The prevalence of vitamin D deficiency in home-bound elderly: Community vs nursing home. J Am Geriatr Soc **41**(Suppl.):SA11.
86. Gloth FM, Tobin JD, Shermann SS, Hollis BW 1991 Is the recommended daily allowance for vitamin D too low in the homebound elderly? J Am Geriatr Soc **39**:137–141.
87. McMurtry CT, Yound SE, Downs RW, Adler RA 1992 Mild vitamin D deficiency and secondary hyperparathyroidism in nursing home patients receiving adequate dietary vitamin D. J Am Geriatr Soc **40**:343–347.
88. O'Dowd KJ, Clemens TL, Kelsey JL, Lindsay R 1993 Exogenous calciferol (vitamin D) and vitamin D endocrine status among elderly nursing home residents in the New York City area. J Am Geriatr Soc **41**:414–421.
89. Davies M, Mawer EB, Hann JT, Taylor JL 1986 Seasonal changes in the biochemical indices of vitamin D deficiency in the elderly: A comparison of people in residential homes, long stay ward and attending a day hospital. Age Aging **15**:77–83.
90. Himmelstein S, Clemens TL, Rubin A, Lindsay R 1990 Vitamin D supplementation in elderly nursing home residents increases 25OHD but not 1,25$(OH)_2$D. Am J Clin Nutr **52**:701–706.
91. Nordin BEC, Baker MR, Horsman A, Peacock M 1985 A prospective trial on the effect of vitamin D supplementation on metacarpal bone loss in elderly women. Am J Clin Nutr **42**:470–747.
92. Compston JE 1995 The role of vitamin D and calcium supplementation in the prevention of osteoporotic fractures in the elderly. Clin Endocrinol **43**:393–405.
93. Heikinheimo RJ, Inkovaara JA, Harjv EJ, Haavisto MV, Kaarela RH, Kajata JM, Kokko AML, Kokko LA, Rajala SA 1992 Annual injections of vitamin D and fractures of aged bone. Calcif Tissue Int **51**:105–110.
94. Torgeson D, Campbell M 1994 Vitamin D alone may be helpful. Br Med J **309**:193 (letter).
95. Lips P, Graafmans WC, Ooms ME, Bezemer PD, Bouter LM 1996 Vitamin D supplementation and fracture incidence in elderly persons. Ann Intern Med **124**:400–406.
96. Kanis JA, Johnell O, Gullberg B, Allander E, Dilsen G, Gennari C, Lopez-Vaz AA, Lyritis JP, Mazuoli G, Miravet L, Passeri M, Cano RP, Rapado A, Ribot C 1992 Evidence for the efficacy of bone active drugs in the prevention of hip fracture. Br Med J **305**:1124–1128.

CHAPTER 44

Vitamin D and Osteoporosis

RICHARD EASTELL Department of Human Metabolism and Clinical Biochemistry, University of Sheffield, Sheffield, England, UK

B. LAWRENCE RIGGS Department of Endocrinology and Metabolism, Mayo Clinic and Foundation, Rochester, Minnesota

I. INTRODUCTION

Involutional osteoporosis begins in middle life and becomes progressively more common with advancing age. It is clinically manifest as fractures. Riggs and Melton [1] have proposed a model for the pathophysiology of osteoporosis that includes two types of osteoporosis. In type I osteoporosis, the major fractures are those of the forearm and vertebra, and postmenopausal women under the age of 75 years are affected. In type II osteoporosis, the major fractures are those of the proximal femur and vertebra, and men and women over the age of 70 years are affected. Type II osteoporosis is believed to be a result of age-related changes in calcium homeostasis that affect both men and women, and these changes are considered first. The effects of age are also considered in Chapters 35 and 43.

II. EFFECT OF AGE ON LEVELS OF 25OHD

A. Overview

Most studies have shown that plasma levels of 25-hydroxyvitamin D (25OHD) decrease with age by about 50% in both men and women [2–15] (Fig. 1). Lips [16] has proposed that the lower limit of the reference range for 25OHD in the summer be taken as 30 nmol/liter (12 ng/ml), on the basis of the following arguments. (1) This is the lower limit of the reference range established for the summer, as determined in blood donors. (2) This is the level of 25OHD above which 1,25-dihydroxyvitamin D [$1,25(OH)_2D$] is not substrate dependent. (3) This is the level of 25OHD above which bone mineral density (BMD) does not correlate with 25OHD (Fig. 2). (4)

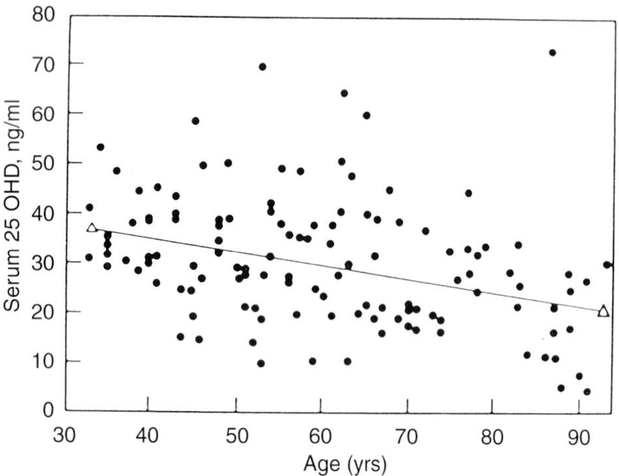

FIGURE 1 Influence of age on the serum concentration of 25OHD in a sample of healthy women. From Tsai *et al.* [14], with permission. (For conversion of 25OHD, ng/ml × 2.50 = nmol liter.)

This is the level of 25OHD above which further treatment with vitamin D will not suppress PTH any further.

On the basis of the 30 nmol/liter 25OHD lower limit, then 25 to 50% of the elderly are vitamin D deficient [17,18]. This figure is as high as 75% for the housebound elderly [17]. In Europe, latitude is not a major determinant of 25OHD status. Indeed, people living in southern Europe have lower levels of 25OHD mainly as a result of their attitudes to sunshine exposure, such as the use of long-sleeved clothes [18].

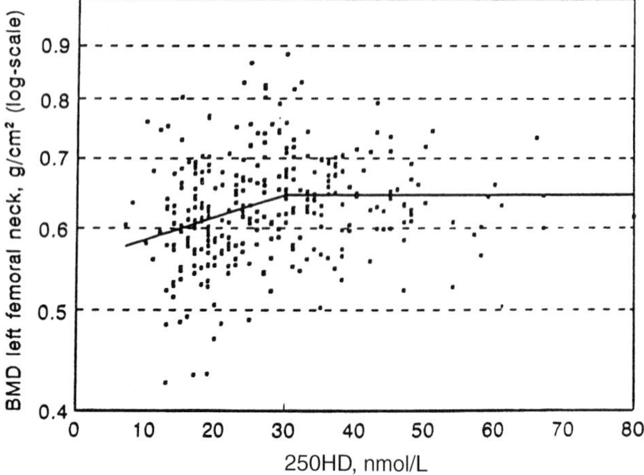

FIGURE 2 Relationship between bone mineral density (BMD) of the left femoral neck with serum 25OHD ($p = 0.001$), adjusted for serum SHBG, age, age at menopause, and body weight. Note the critical value of 25OHD of 30 nmol/liter, below which it is related to BMD. From Ooms *et al.* [137], with permission. (For conversion of 25OHD, nmol/liter ÷ 2.5 = ng/ml.)

B. Mechanisms for the Age-Related Decrease in 25OHD

There are several possible mechanisms for the age-related decrease in plasma 25OHD, and these are considered in turn.

1. Dietary Intake of Vitamin D

Most elderly subjects do not consume the recommended dietary allowance for vitamin D, but this is also true for younger subjects [19]. McKenna [19] found no evidence for differences in vitamin D intake between young adults and the elderly in any geographic region, although average intake varied greatly between countries.

2. Vitamin D Absorption and 25-Hydroxylation

Evidence relating to the effect of age on vitamin D absorption is conflicting. Barraguy *et al.* [4] administered ^{3}H-labeled vitamin D to young and elderly subjects and found that absorption was greater in young subjects (13.2 versus 7.6% in 6 hr). However, administration of vitamin D_2 results in similar increments in plasma 25OHD in young and elderly subjects [20,21]. One explanation for these apparently conflicting results is that vitamin D absorption might be decreased in the elderly but that 25-hydroxylase activity could be increased, possibly as a result of low levels of $1{,}25(OH)_2D$ in the elderly [22].

3. Effect of Age on Production of Vitamin D in Skin

Exposure to sunshine is often less in the elderly than in the young [3,7]. It has been suggested that widespread use of sunscreens to prevent skin cancer may make the elderly more susceptible to vitamin D deficiency [23], although it is not clear whether sunscreen use increases with age.

Evidence regarding the effect of age on the capacity of human skin to produce vitamin D remains controversial. MacLaughlin and Holick reported a decrease in 7-dehydrocholesterol (provitamin D) in skin biopsy specimens with aging [24]. Furthermore, conversion of 7-dehydrocholesterol by ultraviolet light was decreased in skin from elderly subjects. Exposure of subjects to ultraviolet light results in an increase in circulating levels of 25OHD, and this response has been reported to be the same in young and old subjects [25] or 4-fold greater in young subjects [26]. It has been suggested that the age-related decrease in skin fold thickness might be related to declining levels of 25OHD [27].

4. EFFECT OF AGE ON CLEARANCE OF 25OHD

The effect of age on the metabolic clearance rate of 25OHD has not been studied directly, but the half-life of plasma 25OHD after administration of ^{3}H-labeled vitamin D was reported as 21 to 27 days [4] and did not increase with age. Clemens *et al.* reported similar findings [20].

III. EFFECT OF AGE ON LEVELS OF 1,25(OH)$_2$D

A. Overview

Plasma levels of 1,25(OH)$_2$D probably decrease with age in most populations, especially after age 65 [6,8,13,20,28–32]. However, many studies have reported unchanged or increasing levels over a wide age range in studies of both men and women [10,12,33–36], whereas others have reported increasing levels from age 35 up to age 65, followed by a decrease [37,38]. Aksnes *et al.* found that 1,25(OH)$_2$D showed no age-related decrease in active healthy subjects living at home but decreased significantly in hospitalized geriatric patients not receiving vitamin D [2]. These conflicting findings suggest that more than one mechanism could account for the age-related changes in 1,25(OH)$_2$D, and that the importance of these mechanisms could vary in different populations.

B. Mechanisms for Age-Related Changes in 1,25(OH)$_2$D

The age-related decline in levels of 1,25(OH)$_2$D could be due to a change in the level of vitamin D binding protein (DBP), 25OHD substrate deficiency, estrogen deficiency, decreased renal 1α-hydroxylase activity, or increased metabolic clearance of 1,25(OH)$_2$D. These possible mechanisms are considered in turn.

1. CHANGES IN VITAMIN D-BINDING PROTEIN

Interpretation of 1,25(OH)$_2$D levels is complicated by the effect of vitamin D binding protein (DBP), which may also modulate its action. Less than 1% of circulating 1,25(OH)$_2$D is free, and the non-DBP-bound fraction may correspond better to its biological action under certain circumstances [39]. Changes in DBP can cause profound changes in plasma 1,25(OH)$_2$D levels, such as the increase found in pregnancy. Calculations of "free 1,25(OH)$_2$D" using measured concentrations of DBP and total 1,25(OH)$_2$D are likely to improve assessment of the non-DBP-bound fraction. However, a change in DBP concentration is unlikely to account for a decrease in 1,25(OH)$_2$D with aging, because DBP concentrations have been reported to show no significant change with age [2,29], or a minimal decrease [13] or even an increase, and when a free calcitriol index is calculated this clearly decreases with age in postmenopausal women [40] (Fig. 3).

2. 25OHD SUBSTRATE DEFICIENCY

A marked reduction in the level of 25OHD is likely to be a rate-limiting factor for 1,25(OH)$_2$D production [16]. However, in most elderly populations the level of 25OHD is probably not rate limiting. This is likely to account for the different effects of vitamin D supplementation in the elderly, with increased levels of 1,25(OH)$_2$D reported in some elderly populations [17,41] but not in others [42,43]. Lips *et al.* [17] found that the increase in 1,25(OH)$_2$D after vitamin D supplementation was inversely related to the initial 25OHD concentration. Some studies have shown a decrease in 1,25(OH)$_2$D with aging even though the level of 25OHD did not decline [6,32].

3. ESTROGEN DEFICIENCY AND 1,25(OH)$_2$D

The contribution of menopause and estrogen deficiency to the age-related decline in 1,25(OH)$_2$D is uncertain. Sowers *et al.* [31] found that the age-related decrease in 1,25(OH)$_2$D was largely accounted for by menopausal status, but other studies have found that the menopause has no effect on calculated free 1,25(OH)$_2$D [44,45]. Similarly, some studies have shown that postmenopausal estrogen administration increases total 1,25(OH)$_2$D [46,47], as well as calculated free 1,25(OH)$_2$D [46], whereas others have found that estrogen administration results in a transient increase in 1,25(OH)$_2$D and DBP, with no change in calculated free 1,25(OH)$_2$D [48].

4. RENAL 1α-HYDROXYLASE ACTIVITY

In subjects without marked 25OHD substrate deficiency, the most likely cause for the decrease in plasma 1,25(OH)$_2$D levels with aging is decreased renal lα-hydroxylase activity. Renal lα-hydroxylase activity is increased by parathyroid hormone (PTH), growth hormone, hypocalcemia, and hypophosphatemia and is decreased by 1,25(OH)$_2$D [49–51]. 1α-Hydroxylase activity is also affected by the amount of functioning renal tissue [52]. In early chronic renal failure, the glomerular filtration rate (GFR) determined by radioisotopic diethylenetriaminepentaacetic acid (DTPA) clearance is positively correlated with 1,25(OH)$_2$D [52], and these changes are first detectable when renal function is only slightly impaired (GFR $<$70 ml/min).

There are at least three possible reasons for reduced renal lα-hydroxylase activity with aging. First, aging is

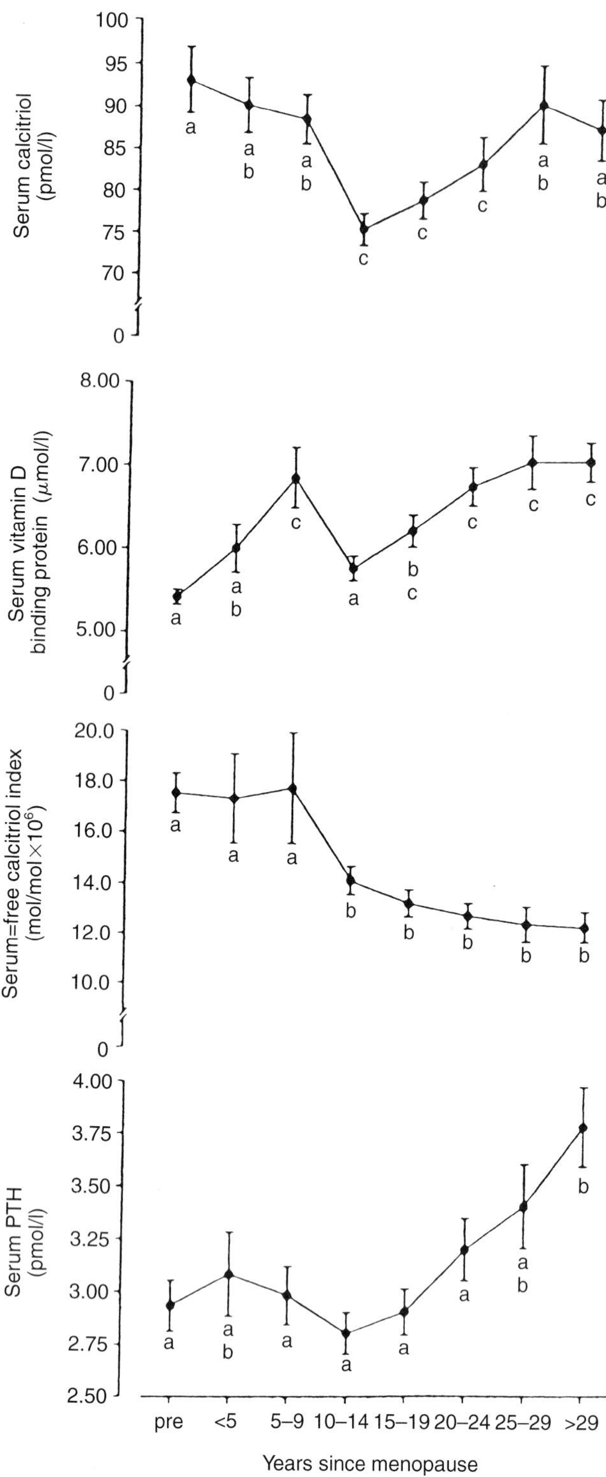

FIGURE 3 Serum $1,25(OH)_2D$ (calcitriol), vitamin D_3 binding protein, free calcitriol index, and intact parathyroid hormone (PTH) in groups of women differing in time since menopause. Results are expressed as means ± SE. Results with different letters are significantly different by Duncan's test ($p < 0.05$). From Prince *et al.* [40], with permission. (For conversion of calcitriol, pmol/liter ÷ 2.4 = pg/ml.)

associated with a decline in renal mass, and an association between impaired renal function and $1,25(OH)_2D$ levels with normal aging have been reported in some studies [32,53] but not in others [36,54,55]. Second, the rise in $1,25(OH)_2D$ in response to PTH infusion is blunted with aging [14,32,56]. This refractoriness to PTH with aging does not appear to be due to estrogen deprivation [45,47]. Studies of $1,25(OH)_2D$ production in renal slices from rats of different ages have indicated that the refractoriness to PTH may be a specific defect, and that the $1,25(OH)_2D$ response to calcitonin is preserved [57]. Finally, growth hormone stimulates renal $1,25(OH)_2D$ production [51,58,59], and growth hormone secretion decreases with age [60,61]. In one cross-sectional study [13] serum insulin-like growth factor type I (IGF-I) was found to be the most important determinant of the fall in $1,25(OH)_2D$ with age. Although 1α-hydroxylase activity is decreased by rising phosphate concentrations [49], this is not likely to account for an age-related decline in $1,25(OH)_2D$ because serum phosphate tends to decline with age [29,36,62–64].

5. CLEARANCE OF $1,25(OH)_2D$

Studies in rats indicate that the age-related decrease in $1,25(OH)_2D$ may be due to increased metabolic clearance of $1,25(OH)_2D$ [65,66]. However, Eastell *et al.* found that the clearance of infused ^3H-$1,25(OH)_2D$ tended to decrease with age in healthy women (age range 26 to 88 years) and that this was associated with an age-related increase in serum levels of $1,25(OH)_2D$ [37]. In another study, Halloran *et al.* found that clearance of $1,25(OH)_2D$ did not vary with age in men, but in this study subjects were selected so as not to show the usual age-related decrease in creatinine clearance [67].

IV. EFFECT OF AGE ON CALCIUM INTAKE AND ABSORPTION

A. Overview

Net calcium absorption is determined by both dietary calcium intake and the efficiency of calcium absorption. These issues are discussed in detail in Chapters 16 and 31. Cross-sectional studies in the United States have shown that calcium intake declines by about 10% in men and women between 35 and 75 years of age [68]. This decrease is due to a reduction in overall caloric intake, and calcium density of the diet does not decline with age [68].

Calcium is absorbed from the intestine by an active $1,25(OH)_2D$-dependent process and by passive vitamin

D-independent mechanisms [69]. The vitamin D-dependent mechanisms are saturated at low intake, and differences in fractional calcium absorption due to variations in the level of $1,25(OH)_2D$ may only be evident when calcium intake is low [69].

Several studies have shown that the efficiency of intestinal calcium absorption decreases with age. This decrease is found by calcium balance studies [70,71], by jejunal perfusion studies [72], and by radiocalcium absorption tests [6,70,73–75]. Nevertheless, there is some evidence that the actual amount of calcium absorbed from the habitual diet might not vary markedly with age. Eastell *et al.* [37] found that when true fractional calcium absorption (TFCA) was measured by tracing the habitual diet over 24 hr, true calcium absorption (TFCA multiplied by dietary calcium) did not decline with age. Ebeling *et al.* [35] reported similar findings employing the same technique using stable isotopes of calcium. In a longitudinal study, Heaney *et al.* [70] measured fractional calcium absorption by balance and double-tracer methods from a diet constructed to match the current dietary calcium intake of each subject. Although fractional calcium absorption clearly decreased in women who went through menopause between studies, fractional absorption decreased only slightly (about 0.2% per year) in women who experienced no change in estrogen status. Furthermore, in this study, calcium intake increased slightly with aging, and therefore the decline in net calcium absorption is likely to have been minimal.

The differences in the studies described above may relate to methodological factors. The interpretation and relevance of these tests depend on the test calcium load and the length of time over which measurements are made. Calcium absorption tests that use small fixed amounts of calcium carrier measure predominantly active calcium absorption. It is likely that active, but not passive, calcium absorption decreases with age, and in general the lower the amount of fixed calcium carrier, the greater is the apparent age-related decrease in calcium absorption. Using the small intestine perfusion technique, Ireland and Fordtran [72] found that calcium absorption in elderly subjects on a high calcium diet (2000 mg/day) was not different from that in younger subjects, but it was lower in elderly subjects when both groups were studied on a low calcium diet (300 mg/day). Restriction of measurements to the early time period after radiotracer ingestion (<6 hr) may also overestimate the effect of age on calcium absorption. At this time, part of the ingested dose is still in the large intestine, and isotopic equilibrium is not achieved until about 24 hr [76]. Colonic absorption probably accounts for about 5% of calcium absorption in healthy subjects, and this proportion is likely to increase in subjects in whom absorptive efficiency is low [76]. Calcium absorption tests that limit blood sampling to within 1 hr of tracer ingestion [33,77] may also overestimate the effect of aging because gastric emptying may be considerably delayed in the elderly [78]. Calcium absorption tests in which calcium tracer is administered without food are also likely to exaggerate the effect of age on calcium absorption. Gastric acidity is markedly reduced in a substantial proportion of elderly subjects [79]. In subjects with achlorhydria Recker found that calcium absorption was very low when administered while fasting but was normal in the presence of food [80].

In summary, there is good evidence to suggest that calcium absorption is somewhat less efficient in the elderly. Elderly subjects show poor adaptation to variations in calcium intake, and calcium balance in the elderly is likely to be more sensitive to calcium intake than in young adults. However, the apparent effect of age may be exaggerated when samples are collected in the early time period after tracer ingestion, when the carrier dose is low, and when calcium is administered without food.

B. Mechanisms for Decreased Calcium Absorption with Aging

1. Decreased $1,25(OH)_2D$

When subjects with widely varying levels of $1,25(OH)_2D$ are studied, calcium absorption and $1,25(OH)_2D$ are highly correlated. This correlation is greatest when subjects are fed low calcium meals [69]. However, in population studies, a decrease in the level of $1,25(OH)_2D$ with age is not a universal finding (see Section III.A), and several studies have not found a relationship between calcium absorption and $1,25(OH)_2D$ [33,37,81].

2. Intestinal $1,25(OH)_2D$ Resistance

Another factor that could be related to a decrease in calcium absorption with age is a decrease in intestinal responsiveness to $1,25(OH)_2D$. In rat studies, calcium uptake by isolated duodenal cells, the number of intestinal receptors for $1,25(OH)_2D$, the calbindin response to $1,25(OH)_2D$, and effect of $1,25(OH)_2D$ therapy on calcium absorption have been reported to decline with age, but there is disagreement about the relative importance of low serum $1,25(OH)_2D$, and $1,25(OH)_2D$ resistance in this model [82–86]. In human studies, there are two main lines of evidence supporting a possible role of intestinal $1,25(OH)_2D$ resistance. First, some studies have shown an age-related increase in the level of $1,25(OH)_2D$ in association with unchanged calcium absorption [33,87]. Second, direct measurements of tissue $1,25(OH)_2D$ receptor content in the duodenal mucosa have shown an age-related decrease of about 30% between age 20 and age 80 [33].

However, the response to an increase in endogenous $1,25(OH)_2D$ induced by a low calcium diet was not affected by age [35]. This study measured the important variable, true fractional calcium absorption. However, this measures both active and passive absorption. It is possible that had a lower calcium carrier been used, so as to trace active transport, that a blunted response to $1,25(OH)_2D$ could have been observed.

An important observation is that common allelic variants in the gene encoding the vitamin D receptor might predict up to 75% of the genetic effect on bone mineral density in healthy adults (see Chapter 45). Women with the low bone mineral density genotype (*BB*) also appear to have higher bone turnover than women with the *bb* genotype. These differences could relate to decreased vitamin D receptor expression in the *BB* genotype, with consequent effects on intestinal calcium absorption, PTH secretion, and bone mineralization.

3. Estrogen Deficiency and Calcium Absorption

Menopause is likely to account for a substantial proportion of the age-related decrease in calcium absorption in women. In a longitudinal study, Heaney *et al.* [70] found that the age-related decline in fractional absorption was considerably larger in women who went through menopause between studies, compared to subjects who had no change in estrogen status. Estrogen therapy increases intestinal calcium absorption [88], and studies indicate that estrogen may have a direct effect on the uptake of calcium by duodenal cells [89].

4. Calcium Absorption and the Aging Stomach

Atrophic gastritis and reduced gastric acid production is common among the elderly, with an estimated prevalence of 20 to 50% in the sixth and seventh decades of life [79]. Calcium bioavailability is probably not affected by high gastric pH per se, although this is controversial [80,90–92]. Although atrophic gastritis in the elderly is rarely severe enough to cause pernicious anemia, patients with pernicious anemia do appear to have reduced bone mineral density [91] and are at increased risk of osteoporotic fracture [93] even though calcium absorption from food is usually unimpaired [80,91]. It has been suggested that bone loss in pernicious anemia could be due to decreased secretion of a bone-stimulating factor produced by the gastric mucosa [93]. These findings raise the intriguing possibility that changes in the aging stomach could be associated with bone loss independently of any effect on calcium absorption. Partial gastrectomy is also associated with bone mineral loss, and it has been suggested that this could be related to reduced postprandial calcitonin secretion [94].

5. Which Form of Vitamin D Is the Major Determinant of Calcium Absorption?

The most potent natural metabolite of vitamin D is $1,25(OH)_2D$, and its potency ratio to 25OHD is 2000 : 1. However, the circulating level of 25OHD is almost 1000 times higher than that of $1,25(OH)_2D$. It has been calculated by Barger-Lux *et al.* [95] that 25OHD contributes about 25% of the circulating vitamin D activity. When Colodro *et al.* [96] tested the effect on calcium absorption in renal failure of different metabolites of vitamin D, it was calculated that the potency ratio in humans may be closer to 400 : 1. This would indicate that 25OHD contributes up to 90% of the circulating vitamin D activity. This would explain why several groups have found stronger relationships between 25OHD and calcium absorption fraction than between $1,25(OH)_2D$ and calcium absorption fraction [95,97].

V. AGE AND PARATHYROID HORMONE

A. Effect of Age on Parathyroid Function

The principal immunoreactive form of PTH in the plasma is heterogeneous C-terminal fragments, especially when renal function is impaired. Several cross-sectional studies using fragment-detecting assays showed increases in the concentration of PTH with age [12,13,29,63,98–103]. These studies are difficult to interpret, however, because renal function declines with age, resulting in increased retention of C-terminal fragments. However, many more recent studies have confirmed an age-related increase in intact PTH in women which is probably progressive after menopause [13,36,77,87, 102,104–108], although a few studies have not shown this [109,110].

In concurrent studies, the age-related increase in carboxy-terminal PTH is about 4-fold higher than the increase in intact PTH [13]. The effect of age on the level of intact PTH is modest, with average reported increases of about 10% per decade after menopause (Fig. 3). The rise in intact PTH is probably similar in men [36].

Bioactive PTH measured using an adenylate cyclase bioassay also appears to increase with age [111]. Moreover, there is evidence of increased PTH activity with age, based on excretion of cyclic AMP [12,100–102] and a decrease in the theoretical renal phosphate threshold, TmP/GFR [98,100,102], between the second and the ninth decade of life. This decrease in TmP/GFR is the likely cause of the decreased serum phosphorus discussed above.

It may be important to consider the pattern of PTH release when considering the effect of age on PTH secretion. PTH secretion shows a marked circadian rhythm, and the form of this rhythm may alter with aging [112]. Pulsatility of PTH secretion has also been shown in several studies using frequent sampling techniques [113–115], and it has been postulated that the pattern of pulsatility may govern the biological action of PTH. The pulse amplitude of PTH secretion may be markedly reduced in osteoporosis [113]. The effect of aging on the pattern of PTH release has not been studied, but this is clearly of some interest.

B. Mechanisms for Increased Intact PTH with Aging

Theoretically, increased PTH levels could be due to hypocalcemia, a greater parathyroid cell mass (parathyroid hyperplasia), a shift in the set point of calcium stimulated PTH secretion, or decreased PTH clearance. $1,25(OH)_2D$ can alter PTH secretion by altering the set point [116–119] as well as the maximum secretory capacity [120].

It is possible that subtle hypocalcemia is an important cause of age-related changes in PTH secretion. Total and ionized calcium in serum decrease slightly with age in most studies (see earlier). Dawson-Hughes *et al.* showed that healthy postmenopausal women with levels of serum ionized calcium in the lowest quintile had significantly higher levels of intact PTH than subjects with ionized calcium in the highest quintile [121]. A similar inverse relationship between ionized calcium and intact PTH has been shown by other investigators [13,122]. In rats, however, the age-related increase in PTH does not appear to be related to low plasma calcium levels [66,123].

Aging could also be associated with an increase in the set point for suppression of PTH secretion by calcium (implying higher PTH secretion at any given concentration of ionized calcium). A shift in set point could be a primary effect of aging on the sensitivity of the parathyroid gland to calcium, or it could be secondary to other factors such as low $1,25(OH_2D$ or decreased parathyroid responsiveness to $1,25(OH)_2D$. However, Landin-Wilhelmsen *et al.* [8] reported a positive correlation between $1,25(OH)_2D$ and PTH, and this would argue against a decrease in $1,25(OH)_2D$ as a cause of the increase in PTH. They did report a negative correlation between PTH and 25OHD (as did Hegarty *et al.* [124]), and this again raises the question of the relative contribution of 25OHD to the biological activity of vitamin D (see Section IV.B.5). Studies in rats have shown that PTH secretion is higher at any given level of calcium in aged animals *in vivo* [125,126] and *in vitro* [127].

In elderly human subjects, the set point does not appear to change with aging, but basal secretion and maximum secretory capacity are increased [120]. This suggests that the rise in PTH with aging is due to parathyroid hyperplasia, possibly secondary to chronic hypocalcemia, rather than an altered set point. Furthermore, in this study the magnitude of the decrease in basal and maximal PTH secretion after $1,25(OH)_2D$ therapy was similar in young and elderly subjects, suggesting that parathyroid responsiveness to $1,25(OH)_2D$ is preserved in the elderly. A number of studies have shown a significant inverse relationship between levels of intact PTH and $1,25(OH)_2D$ in elderly subjects [13,43], but this may be secondary to impaired calcium absorption rather than an effect of low $1,25(OH)_2D$ itself. Treatment of elderly subjects with vitamin D results in a reduction in the level of PTH [17,128,129].

A possible relationship between declining renal function and an increase in intact PTH has been explored in several studies. Some studies have shown that the level of intact PTH is independently related to declining renal function with age [55,107]. In contrast, other studies in rats [66,123] and humans [13,54] have shown that declining renal function is not independently associated with the age-related increase in intact PTH.

Finally, there is no evidence for a change in the clearance of intact PTH with age in humans. Studies in rats suggested that a decline in PTH clearance cannot account for high levels of PTH with advancing age [123].

VI. INTERRELATIONSHIPS BETWEEN AGE-RELATED BONE LOSS AND PTH AND VITAMIN D METABOLISM

The age-related increase in PTH may play a pivotal role in age-related bone loss [130]. In this model, decreased net intestinal calcium absorption [resulting from dietary calcium deficiency and reduced levels of $1,25(OH)_2D$] results in mild hypocalcemia and secondary hyperparathyroidism. PTH excess results in increased bone resorption and net bone loss. Although it is clear that PTH and vitamin D metabolism alter with aging, the relationship between these changes and bone loss is not certain. Inappropriately elevated PTH in primary hyperparathyroidism is associated with reduced bone density [131,132], but it is uncertain whether mild asymptomatic hyperparathyroidism is associated with an increased risk of fracture or progressive bone loss [133,134].

A number of studies have shown that the age-related increase in PTH or decrease in 25OHD and $1,25(OH)_2D$ are not independently predictive of increased bone turnover [108], bone mineral density [14,36,54,105], or bone loss [135]. Indeed, there is no conclusive evidence that bone turnover does increase progressively after menopause or in men.

In contrast to the above findings, some studies have shown a relationship between bone mineral density and the level of $1,25(OH)_2D$ [31,136], PTH, and 25OHD [136]. Ooms *et al.* [137] reported that 25OHD levels did correlate with hip BMD below the threshold of 30 nmol/liter [137] (Fig. 2). In one study, PTH did relate to rates of bone loss from the forearm in premenopausal women [138].

It is likely that differences in net calcium absorption account for only a small proportion (possibly about 10%) of the variance in bone density in the elderly population [139].

Pharmacological doses of calcium may reduce the rate of bone loss in elderly subjects [140,141], especially in subjects with a low habitual calcium intake. However, most of the variance in the rate of bone loss is likely to be independent of intestinal calcium supply. Orwoll *et al.* found that healthy men (30 to 87 years old) show substantial bone loss in the radius (1.0%/year) and spine (2.3%/year) which is not altered by calcium supplementation (1000 mg/day) or vitamin D (25 mg/day) [142]. Devine *et al.* reported that calcium absorption measured using the single-tracer stable strontium method did not correlate with bone density in women more than 10 years past menopause [81].

A model to describe the interrelationships between bone loss, PTH, and vitamin D metabolism is shown in Fig. 4.

VII. VITAMIN D STATUS IN ESTABLISHED OSTEOPOROSIS

A. Changes in Vitamin D Status in Hip Fracture (Type II Osteoporosis)

1. 25OHD

The plasma levels of 25OHD in patients with type II osteoporosis have been reported to be low [143,144] or normal [145].

2. Parathyroid Hormone

The level of PTH is increased in hip fracture cases, for example by 30% [144,146–148].

3. $1,25(OH)_2D$

It can be difficult to evaluate vitamin D status in hip fracture patients because the trauma itself lowers calcitriol levels [149]. The $1,25(OH)_2D$ levels may be increased after hip fracture as a result of recovery from recent osteomalacia. This was concluded from a Finnish study of hip fracture patients with high levels of this sterol and a U.K. study in which hip fracture patients had normal levels but those with mineralization defects had increased levels of $1,25(OH)_2D$ [148].

B. Changes in Vitamin D Status in Vertebral Fracture (Type I Osteoporosis)

1. 25OHD

The plasma levels of 25OHD in patients with type I osteoporosis have been reported to be normal [88,150]

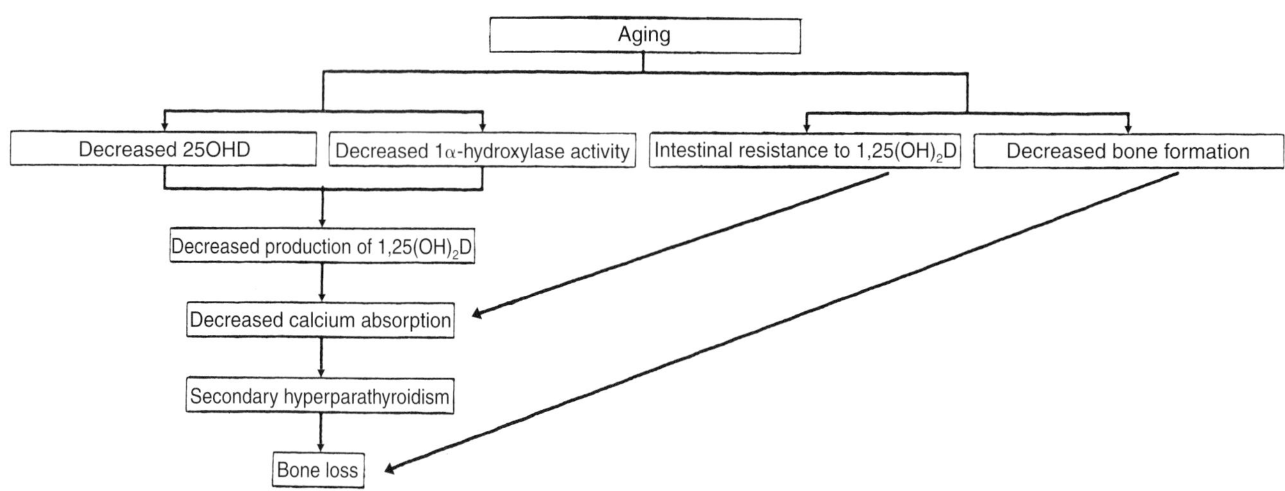

FIGURE 4 Model for the proposed changes in calcium homeostasis and bone turnover with age.

or low [151], or even high [9,152]. These reports may be difficult to interpret because not all controls were stated to be season matched [150,151] and because some cases had been already treated with vitamin D [153].

2. $1,25(OH)_2D$

A number of reports have described low levels of $1,25(OH)_2D$, with values 18 to 80% lower than controls [6,9,48,152–156]. In these studies no relationship was found between the hormone levels and calcium absorption. There were relationships described between the hormone and creatinine clearance. There have been some studies that have reported normal levels of $1,25(OH)_2D$ in type I osteoporosis [33,53,71,157,158]. It is not clear why these authors found no decrease. It could be that their cases had a mild form of the vertebral fracture syndrome.

Why is calcium absorption decreased in type I osteoporosis? There are several possibilities. (1) intestinal resistance to the action of $1,25(OH)_2D$, (2) renal 1α-hydroxylase defect, and (3) functional suppression of renal 1α-hydroxylase by a decrease in PTH secretion. Mechanism 3 is the most likely for the following reasons: the calcium absorptive response to calcitriol therapy was as expected [6,159,160]; PTH correlates negatively with bone resorption; the increase in plasma $1,25(OH)_2D$ as a result of PTH infusion was normal [152,156]; and oestrogen reverses the decreased levels of $1,25(OH)_2D$ and increases fractional calcium absorption [88].

3. Calcium Absorption

A decrease in fractional calcium absorption of between 20 and 30% was reported in type I osteoporosis [6,150,157,160–162]. In one study there was no decrease in calcium absorption [74]. The decrease in calcium absorption is likely to be a result of the decrease in levels of $1,25(OH)_2D$ (see Section VII.B.2).

4. Parathyroid Hormone

There is evidence for decreased PTH secretion [163,164]. According to the model shown in Fig. 5 this decrease is secondary to increased net bone resorption.

VIII. SUMMARY OF CHANGES IN VITAMIN D WITH AGING

Thus, with age there are a number of changes in vitamin D and its actions, as shown in Fig. 4.

1. There is a decrease in 25OHD that probably results from decreased UV light exposure, decreased effect of UV light on the skin synthesis of vitamin D, and decreased absorption of vitamin D from the diet.
2. There is a decrease in $1,25(OH)_2D$ after the age of 65 years that is partly a result of the decrease in the substrate, 25OHD, and partly a result of the decrease of renal 1α-hydroxylase activity.
3. There is a decrease in active calcium absorption that results from the decrease in vitamin D metabolites [25OHD and $1,25(OH)_2D$], intestinal resistance to the action of vitamin D, and estrogen deficiency.
4. There is an increase in PTH secretion.

There are further changes in patients with osteoporosis-related fractures, and these indicate different pathogenetic factors for these fracture syndromes, supporting the concept of type I and type II osteoporosis.

1. In type I osteoporosis (Fig. 5), there is a decrease in both calcium absorption and $1,25(OH)_2D$, but the PTH levels are lower than expected, indicating that the primary defect is likely to be an increase in net bone resorption.
2. In type II osteoporosis, there is an increase in PTH and a decrease in 25OHD that is greater than these changes that occur in all subjects with aging. This supports the importance of borderline vitamin D deficiency as an important cause of hip fracture.

IX. TREATMENT OF ESTABLISHED OSTEOPOROSIS WITH VITAMIN D

A. Rationale and General Principles

A distinction must be made between the use of physiological replacement dosages of vitamin D to treat nutritional deficiency and the use of therapeutic dosages of vitamin D or use of the natural vitamin D metabolites to treat osteoporosis when vitamin D stores are normal. Physiological replacement of vitamin D can be achieved with small dosages of 1000 IU per day or less. Because vitamin D plays a key role in the regulation of calcium metabolism and in the maintenance of bone mass, deficiency states should always be searched for and, when present, corrected with relatively small dosages of vitamin D.

As reviewed earlier in this chapter (Section VII), there is substantial evidence that many patients with both postmenopausal (type I) and age-related (type II) osteoporosis have an impairment in the metabolism of vitamin D to its physiologically active metabolite, $1,25(OH)_2D_3$, which contributes to their negative calcium metabolism and bone loss. This impairment can

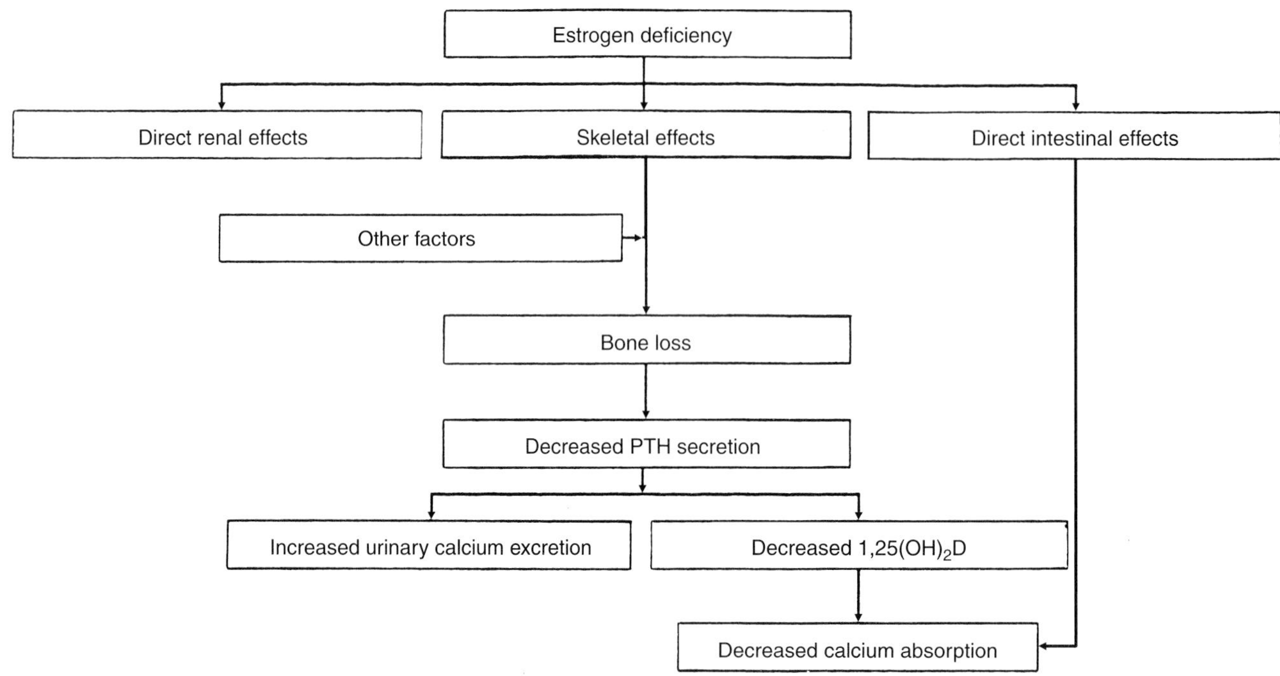

FIGURE 5 Model for the proposed changes in calcium homeostasis and bone turnover in type I osteoporosis.

be overcome by using large dosages of vitamin D (10,000 to 25,000 IU per day) or by small dosages of $1,25(OH)_2D_3$. Because both vitamin D and its main circulating metabolite 25OHD are stored in muscle and fat and over time, large amounts can be retained. Because of these large stores, severe hypercalcemia and hypercalciuria due to vitamin D intoxication can occur relatively suddenly when pharmacological dosages of vitamin D are used long-term, and, when it does occur, it may last for weeks or even months. In contrast, $1,25(OH)_2D_3$ does not have significant long-term storage, and if hypercalcemia or hypercalciuria occur, they rapidly resolve over a few days. Because of this and because physiological or near physiological dosages are effective in increasing calcium absorption [165], $1,25(OH)_2D_3$ is preferred over pharmacological dosages of vitamin D in the treatment of osteoporosis.

Calcium absorption is impaired in both postmenopausal and age-related osteoporosis and can be corrected by therapy with $1,25(OH)_2D_3$. If increasing calcium absorption were the only effect of $1,25(OH)_2D_3$ in the treatment of osteoporosis, large dosages of oral calcium might effectively substitute for it. However, as reviewed elsewhere in this book (Section III), $1,25(OH)_2D_3$ also enhances renal calcium conservation, increases the differentiation of bone cells, and acts directly on muscle.

Finally, from a therapeutic standpoint, it should be recognized that the effects of $1,25(OH)_2D_3$ on calcium metabolism and bone turnover are triphasic. In dosages below 0.5 μg per day, the drug fails to increase calcium absorption in many subjects [165], and thus these dosages should be considered to be incompletely effective. In dosages above 0.75 μg per day, some subjects will have increases in urine and serum calcium levels, particularly if the intake of calcium is not restricted, and there also may be evidence of increased bone resorption mediated by stimulation of osteoclasts by $1,25(OH)_2D_3$. In dosages above 1.0 μg per day, these adverse effects are much more common. The most favorable results from $1,25(OH)_2D_3$ treatment of established osteoporosis have been achieved using dosages in the intermediate range of 0.5 to 0.75 μg per day. Thus, the optimal therapeutic range appears to be quite narrow and may vary among patients. This restricted therapeutic range constitutes one of the major limitations to treatment using this drug. Differences in dosages employed in different clinical trials may explain some of the inconsistent and conflicting therapeutic results reported in the scientific literature.

B. Correction of Nutritional Vitamin D Deficiency

Although vitamin D deficiency classically is associated with osteomalacia, milder deficiency states can lead to osteoporosis [166]. As reviewed elsewhere in this book (Chapter 43), states of mild vitamin D deficiency may be more common than has been believed previously, particularly among the elderly who often have poorer diets and have decreased solar exposure. This is particularly likely to occur in those countries that do not supplement dairy products with vitamin D, in countries with more northerly latitudes, and during the winter months. Chapuy *et al.* [167] randomly treated 3270 elderly French women who were confined to either nursing homes or apartments for the elderly with either 800 IU of vitamin D and 500 mg of elemental calcium per day or double placebo. After 18 months, the numbers of hip fractures were 43% lower and nonvertebral fractures were 32% lower, and serum parathyroid hormone was 44% lower, all statistically significant changes. Heinkinheimo *et al.* [168] treated or followed without treatment a group of 621 elderly (>85 years of age) free-living or institutionalized men and women with annual intramuscular injections of ergocalciferol over a 4-year period. The treatment group had a slightly but significantly lower (16%) fracture rate than did the control group (22%). Ooms *et al.* [169] treated 348 elderly Dutch women with 400 IU per day of vitamin D or placebo and found that there were small but significant decreases in serum PTH and increases in femoral neck density. Finally, Lips *et al.* [170] treated 2578 elderly (>80 years of age) women and men who were living independently or in nursing homes with 400 IU of placebo. Over 3.5 years of observation, there were no differences in the number of hip or total fractures. The major difference between this study and that of the French study was that the subjects in the Chapuy study had lower levels of serum 25OHD and received a higher dosage of vitamin D and a calcium supplement.

From these and other studies, it appears that many elderly persons have varying degrees of vitamin D deficiency and that correction of this with small replacement dosages is beneficial. Lips *et al.* [171] found that an average of 300 IU per day of vitamin D was sufficient to correct vitamin D deficiency in elderly women. However, because of evidence that the elderly may absorb vitamin D less well than the young and because low dosages are safe and inexpensive, it would seem prudent to ensure that elderly individuals take 800 units per day of vitamin D and have a calcium intake of 1000 mg per day or more.

C. Treatment with $1,25(OH)_2D_3$

Despite many years of clinical trials, $1,25(OH)_2D_3$ still has not been approved for the treatment of established osteoporosis in the United States. It is approved, however, in the United Kingdom, Australia, Italy, Japan, New Zealand, as well as in 16 other countries. However, $1,25(OH)_2D_3$ has been approved in most countries for treatment of hypocalcemia and renal osteodystrophy and thus is available to physicians for "off label" usage.

1. Efficacy in Postmenopausal (Type I) Osteoporosis

Only randomized, prospective clinical trials with objective end points are reviewed here. The earliest evaluation of $1,25(OH)_2D_3$ in postmenopausal osteoporotic women was reported by Gallagher *et al.* [172], who compared the effect of 6 months of treatment with 0.5 μg per day of $1,25(OH)_2D_3$ or placebo. As compared with either baseline measurements or with changes in the placebo group, they found that $1,25(OH)_2D_3$ significantly increased calcium absorption, decreased bone resorption rate, and improved calcium balance. The most systematic study reported thus far was a large three-center study comparing the effects of $1,25(OH)_2D_3$ on the rate of change in bone density in postmenopausal osteoporotic women. All patients had their calcium intake adjusted to 1000 mg per day. Unfortunately, the protocol required that the dosage of $1,25(OH)_2D_3$ be increased until toxicity occurred and then reduced to a dosage that did not produce hypercalcemia or hypercalciuria. In the centers of Aloia *et al.* [173] and Gallagher *et al.* [174] there were significant increases in bone density, whereas in the center of Ott and Chesnut [175] there were no significant differences from the placebo group. However, Ott and Chesnut enrolled patients with milder disease (the mean number of vertebral fractures in their study was 1.2 versus 2.9 and 4.1 in the other two studies), and they had reduced the mean daily dosage of $1,25(OH)_2D_3$ to a much lower dosage (0.43 μg versus 0.80 and 0.67 μg in the other two centers). Subsequently, Ott and Chesnut [175] reanalyzed their data and found that those subjects who received 0.5 μg per day or more responded by increases in bone density whereas those with lower dosages did not.

Orimo *et al.* [176] found in 80 postmenopausal osteoporotic Japanese women that one year of treatment with 1α-hydroxyvitamin D_3 (1αOHD$_3$) increased bone density by 1.8% at the lumbar spine and by 4.6% at the femoral trochanter as compared with the placebo group, whereas there was no significant change at the femoral neck. Christiansen *et al.* [177] failed to demonstrate a significant effect of a small dosage of 0.25 μg per day of $1,25(OH)_2D_3$ on retarding bone loss in perimenopausal normal women. However, Need *et al.* [178] found that treatment of postmenopausal osteoporosis for 15 months with a dosage of 0.25 μg per day combined with a calcium supplement of 1000 mg per day reduced urine hydroxyproline excretion and increased bone density as compared with a control group.

Some studies have reported that $1,25(OH)_2D_3$ treatment decreases the occurrence of vertebral fractures. In a two-center double-blind study over 3 years, $1,25(OH)_2D_3$ resulted in a 65% reduction in vertebral fractures, although only the first year was placebo controlled [179]. Tilyard *et al.* [180] compared the effect of $1,25(OH)_2D_3$ treatment in postmenopausal osteoporotic women with a control group receiving only calcium supplementation. At the end of 3 years they found a significant, 63%, decrease in vertebral fractures compared to the control group. However, the data were only significant when the results of the first year and subjects having more than five fractures were excluded. Also, inexplicably, the vertebral fracture rate increased in the group receiving calcium supplementation. Finally, Orimo *et al.* [176] found that treatment of 80 Japanese women with $1\alpha OHD_3$ for 1 year reduced significantly the vertebral fracture rate by 83% as compared with placebo.

Thus far, there has been no controlled study of the effect of $1,25(OH)_2D_3$ in the treatment of elderly women with age-related (type II) osteoporosis.

2. SAFETY

Safety remains a concern because of the relatively narrow therapeutic range for $1,25(OH)_2D_3$. With dosages higher than 0.75 μg per day, hypercalciuria and mild hypercalcemia may occur in a minority of patients, particularly in those with unrestricted calcium intake. Thus far, significant deterioration of renal function has not been reported in subjects receiving recommended dosages, and the occurrence of kidney stones is rare. With the recommended dosages of 0.5 to 0.75 μg per day and with calcium intake restricted to 1000 mg per day, the drug appears to be quite safe. Thus, Tilyard *et al.* [180] found no evidence of hypercalcemia, hypercalciuria, or nephrocalcinosis after 528 patient-years of observation. Nonetheless, monitoring of serum and urine calcium values no less than yearly is recommended.

X. CONCLUSIONS

Although data in the literature are conflicting, the large majority of studies indicate that $1,25(OH)_2D_3$ is moderately effective in reducing bone loss and, possibly, also in reducing vertebral fracture occurrence in women with type I osteoporosis. Efficacy seems to be dose related, with the best results occurring in patients treated with 0.5 to 0.75 μg $1,25(OH)_2D_3$ per day. The drug appears to be quite safe in this dosage range, although restriction of calcium intake and monitoring serum and urine values for calcium are recommended. The osteoporotic women that are most likely to benefit from $1,25(OH)_2D_3$ are those with impaired calcium absorption. Unfortunately, it is not practical to measure calcium absorption directly. However, Riggs and Nelson [160] correlated calcium absorption and urinary calcium excretion and found that most osteoporotic women who had intestinal calcium malabsorption on a normal calcium diet had values for urine calcium excretion below 100 mg per day.

Many of the elderly patients with type II osteoporosis are vitamin D deficient and should be treated with small dosages of vitamin D. For the remainder, treatment with small dosages of 0.25 μg $1,25(OH)_2D_3$ per day combined with 1000 mg per day of elemental calcium seems rational because of the documented decrease in serum $1,25(OH)_2D_3$ levels in many patients. However, the value and safety of this drug in these elderly patients clearly need better documentation.

References

1. Riggs BL, Melton III LJ 1986 Involutional osteoporosis. N Engl J Med **314**:1676–1686.
2. Aksnes L, Rodland O, Odegaard OR, Bakke KJ, Aarskog D 1989 Serum levels of vitamin D metabolites in the elderly. Acta Endocrinol **127**:27–33.
3. Baker MR, Peacock M, Nordin BEC 1980 The decline in vitamin D status with age. Age Ageing **9**:249–252.
4. Barraguy JM, France MW, Corless D, Gupta SP, Switala S, Boucher BJ, Cohen RD 1978 Intestinal cholecalciferol absorption in the elderly and in younger adults. Clin Sci Mol Biol **55**:213–220.
5. Chapuy MC, Durr F, Chapuy P 1983 Age-related changes in parathyroid hormone and 25 hydroxycholecalciferol levels. J Gerontol **38**:19–22.
6. Gallagher JC, Riggs BL, Eisman J, Hamstra A, Arnaud SB, DeLuca HF 1979 Intestinal calcium absorption and serum vitamin D metabolites in normal subjects and osteoporotic patients. J Clin Invest **64**:729–736.
7. Guggenheim K, Kravitz M, Tal R, Kaufmann NA 1979 Biochemical parameters of vitamin D nutrition in old people in Jerusalem. Nutr Metab **23**:172–178.
8. Landin-Wilhelmsen K, Wilhalmsen L, Wilske J, Lappas G, Rosen T, Lindstedt G, Lundberg PA, Bengtsson BA 1995 Sunlight increases serum 24(OH) vitamin D concentration whereas $1,25(OH)_2D_3$ is unaffected. Eur J Clin Nutr **49**:400–407.
9. Lore F, Di Cairano G, Signorini AM, Caniggia A 1981 Serum levels of 25-hydroxyvitamin D in postmenopausal osteoporosis. Calcif Tissue Int **33**:467–470.
10. Lund B, Sorensen OH 1979 Measurement of 25-hydroxyvitamin D in serum and its relation to sunshine, age and vitamin D. Scand J Clin Lab Invest **39**:23–30.
11. Omdahl JL, Garry PJ, Hunsaker LA, Hunt WC, Goodwin JS 1982 Nutritional status in a healthy elderly population: Vitamin D. Am J Clin Nutr **36**:1225–1233.
12. Orwoll ES, Meier DE 1986 Alterations in calcium, vitamin D, and parathyroid hormone physiology in normal men with aging: Relationship to the development of senile osteopenia. J Clin Endocrinol Metab **63**:1262–1269.

13. Quesada JM, Coopmans W, Ruiz B, Aljama P, Jans I, Bouillon R 1992 Influence of vitamin-D on parathyroid function in the elderly. J Clin Endocrinol Metab **75**:494–501.
14. Tsai K, Wahner HW, Offord KP, Melton LJI, Kumar R, Riggs BL 1987 Effect of aging on vitamin D stores and bone density in women. Calcif Tissue Int **40**:241–243.
15. Weisman Y, Schen RJ, Eisenberg Z, Edelstein S, Harell A 1981 Inadequate status and impaired metabolism of vitamin D in the elderly. Isr J Med Sci **17**:19–21.
16. Lips P 1996 Vitamin D deficiency and osteoporosis: The role of vitamin D deficiency and treatment with vitamin D and analogues in the prevention of osteoporosis-related fractures. Eur J Clin Invest **26**:436–442.
17. Lips P, Wiersinga A, Vanginkel FC, Jongen MJM, Netelenbos JC, Hackeng WHL, Delmas PD, van der Vijgh WJF 1988 The effect of vitamin D supplementation on vitamin D status and parathyroid function in elderly subjects. J Clin Endocrinol Metab **67**:644–650.
18. Van Der Wielaen RPJ, Lowik MRH, Van Der Berg H, de Groot LCPGM, Haller J, Moreiral O, Van Staveren WA 1995 Serum vitamin D concentrations among elderly people in Europe. Lancet **346**:207–210.
19. McKenna MJ 1992 Differences in vitamin-D status between countries in young adults and the elderly. Am J Med **93**:69–77.
20. Clemens TL, Zhou X, Myles M, Endres D, Lindsay R 1986 Serum vitamin D_2 and vitamin D_3 metabolite concentrations and absorption of vitamin D_2 in elderly subjects. J Clin Endocrinol Metab **63**:656–660.
21. Somerville PJ, Lien JWK, Kaye M 1977 The calcium and vitamin D status in an elderly female population and their response to administered supplemental vitamin D_3. J Gerontol **32**:659–663.
22. Bell NH 1985 Vitamin D–endocrine system. J Clin Invest **76**:1–6.
23. Mutsuoka LY, Wortsman J, Hanifan N, Holick MF 1988 Chronic sunscreen use decreases circulating concentrations of 25-hydroxyvitamin D. Arch Dermatol **124**:1802–1804.
24. MacLaughlin J, Holick MF 1985 Aging decreases the capacity of human skin to produce vitamin D_3. J Clin Invest **76**:1536–1538.
25. Davie M, Lawson DEM 1980 Assessment of plasma 25-hydroxyvitamin D response to ultraviolet irradiation over a controlled area in young and elderly subjects. Clin Sci **58**:235–242.
26. Holick MF, Matsuoka LY, Wortsman J 1989 Age, vitamin D and solar ultraviolet. Lancet **2**:1104–1105.
27. Need AG, Morris HA, Horowitz M, Nordin BEC 1992 Skinfold thickness, age and vitamin D. J Bone Miner Res **7**:165.
28. Dandona P, Menon RK, Shenoy R, Houlder S, Thomas M, Mallinson WJW 1986 Low 1,25-dihydroxyvitamin D, secondary hyperparathyroidism, and normal osteocalcin in elderly subjects. J Clin Endocrinol Metab **63**:459–462.
29. Fujisawa Y, Kida K, Matsuda H 1984 Role of change in vitamin D metabolism with age in calcium and phosphorus metabolism in normal human subjects. J Clin Endocrinol Metab **59**:719–726.
30. Manolagas SC, Howard J, Culler F, Catherwood BD, Deftos LJ: 1982; Cytoreceptor assay for $1,25(OH)_2D$: A simple, rapid, and reliable test for clinical application. Clin Res **30**:527A (abstract).
31. Sowers MR, Wallace RB, Hollis BW 1990 The relationship of 1,25-dihydroxyvitamin D and radial bone mass. Bone Miner **10**:139–148.
32. Tsai K, Heath H, III., Kumar R, Riggs BL 1984 Impaired vitamin D metabolism with aging in women. Possible role in pathogenesis of senile osteoporosis. J Clin Invest **73**:1668–1672.
33. Ebeling PR, Sandgren ME, Dimagno EP, Lane AW, DeLuca HF, Riggs BL 1992 Evidence of an age-related decrease in intestinal responsiveness to vitamin D: Relationship between serum 1,25-dihydroxyvitamin D_3 and intestinal vitamin D receptor concentrations in normal women. J Clin Endocrinol Metab **75**:176–182.
34. Dokoh S, Morita R, Fukunaga M, Yamamoto I, Torizuka K 1978 Competitive protein binding assay for 1,25-dihydroxyvitamin D in human plasma. Endocrinol Jpn **25**:431–436.
35. Ebeling PR, Yergey AL, Vieira NE, Burritt MF, O'Fallon WM, Kumar R, Riggs BL 1994 Influence of age on effects of endogenous 1,25-dihydroxyvitamin D on calcium absorption in normal women. Calcif Tissue Int **55**:330–334.
36. Sherman SS, Hollis BW, Tobin JD 1990 Vitamin D status and related parameters in a healthy population: The effects of age, sex and season. J Clin Endocrinol Metab **71** (No. 2):405–413.
37. Eastell R, Yergey AL, Vieira NE, Cedel SL, Kumar R, Riggs BL 1991 Interrelationship among vitamin D metabolism, true calcium absorption, parathyroid function, and age in women: Evidence of an age-related intestinal resistance to 1,25-dihydroxyvitamin D action. J Bone Miner Res **6** (No. 2):125–132.
38. Epstein S, Bryce G, Hinman JW, Miller ON, Riggs BL, Hui SL, Johnston CC 1986 The influence of age on bone mineral regulating hormones. Bone **7**:421–425.
39. Vargas S, Bouillon R, Van Baelen H, Raisz LG 1990 Effects of vitamin D-binding protein on bone resorption stimulated by 1,25-dihydroxyvitamin D_3. Calcif Tissue Int **47**:164–168.
40. Prince RL, Dick I, Devine A, Price RI, Gutteridge DH, Kerr D, Criddle A, Garcia-Webb P, St John A 1995 The effects of menopause and age on calciotropic hormones: A cross-sectional study of 655 healthy women aged 35 to 90. J Bone Miner Res **10**:835–842.
41. Bouillon R, Auwerx J, Dekeyser L, Fevery J, Lissens W, DeMoor P 1984 Serum vitamin D metabolites and their binding protein in patients with liver cirrhosis. J Clin Endocrinol Metab **59**:86–89.
42. Fraher LJ, Caveney AN, Hodsman AB 1990 Vitamin D status of elderly institutionalized Canadians: effect of season and oral vitamin D_2 on circulating $1,2(OH)_2D$. J Bone Miner Res **5**:133.
43. Himmelstein S, Clemens TL, Rubin A, Lindsay R 1990 Vitamin D supplementation in elderly nursing-home residents increases 25(OH)D but not $1,25(OH)_2D$. Am J Clin Nutr **52**:701–706.
44. Falch JA, Oftebro H, Hang E 1987 Early postmenopausal bone loss is not associated with a decrease in circulating levels of 25-hydroxyvitamin D, 1,25-dihydroxyvitamin D, or vitamin D-binding protein. J Clin Endocrinol Metab **64**:836–841.
45. Prince RL, Dick I, Garcia-Webb P, Retallack RW 1990 The effects of the menopause on calcitriol and parathyroid hormone: Responses to a low dietary calcium stress test. J Clin Endocrinol Metab **70**:1119–1123.
46. Cheema C, Grant BF, Marcus R 1989 Effects of estrogen on circulating "free" and total 1,25-dihydroxyvitamin D and on the parathyroid–vitamin D axis in postmenopausal women. J Clin Invest **83**:537–542.
47. Marcus R, Villa ML, Cheema M, Cheema C, Newhall K, Holloway L 1992 Effects of conjugated estrogen on the calcitriol response to parathyroid hormone in postmenopausal women. J Clin Endocrinol Metab **74**:413–418.
48. Hartwell D, Hassager C, Overgaard K, Riis BJ, Podenphant J, Christiansen C 1990 Vitamin D metabolism in osteoporotic women during treatment with estrogen, an anabolic steroid, or calcitonin. Acta Endocrinol **122**:715–721.
49. Kumar R, Cahan DH, Madias NE, Harrington JT, Kurtin P, Dawson-Hughes BF 1991 Vitamin-D and calcium transport. Kidney Int **40**:1177–1189.
50. Reichel H, Koeffler P, Norman AW 1989 The role of the vitamin D endocrine system in health and disease. N Engl J Med **320**:980–991.
51. Spencer EM, Halloran B 1992 The dependence of PTH's action

on the renal 1-hydroxylase on insulin-like growth factor-1. J Bone Miner Res **7**:172.
52. St John A, Thomas MB, Davies CP, Mullan B, Dick I, Hutchison B, Van Der Schaff A, Prince RL 1992 Determinants of intact parathyroid hormone and free 1,25-dihydroxyvitamin D levels in mild and moderate renal failure. Nephron **61**:422–427.
53. Francis RM, Peacock M, Barkworth SA 1984 Renal impairment and its effects on calcium metabolism in elderly women. Age Ageing **13**:14–20.
54. Sherman SS, Tobin JD, Hollis BW, Gundberg CM, Roy TA, Plato CC 1992 Biochemical parameters associated with low bone density in healthy men and women. J Bone Miner Res **7**:1123–1130.
55. Webb AR, Pilbeam C, Hanafin N, Holick MF 1990 An evaluation of the relative contributions of exposure to sunlight and of diet to the circulating concentrations of 25-hydroxyvitamin D in an elderly nursing-home population in Boston. Am J Clin Nutr **51**:1075–1081.
56. Slovik DM, Adams JS, Neer RM, Holick MF, Potts JT 1981 Deficient production of 1,25-dihydroxyvitamin D in elderly osteoporotic patients. N Engl J Med **305**:372–374.
57. Armbrecht HJ, Wongsurawat N, Paschal R 1987 Effect of age on renal responsiveness to parathyroid hormone and calcitonin in rats. J Endocrinol **114**:173–178.
58. Bouillon R 1991 Growth hormone and bone. Horm Res **36**:49–55.
59. Spencer EM, Tobiassen O 1981 The mechanism of the action of growth hormone on vitamin D metabolism in the rat. Endocrinology **108**:1064–1070.
60. Rudman D, Rao UMP 1991 The hypothalamic-growth hormone–somatomedin C axis: The effect of aging. In: Morley JE (ed) Endocrinology and Metabolism in the Elderly. Blackwell, New York, pp. 35–55.
61. Zadik Z, Chalew SA, McCarter RJ, Meistas M, Kowarski AA 1985 The influence of age on the 24-hour integrated concentration of growth hormone in normal individuals. J Clin Endocrinol Metab **60**:513–516.
62. Tietz NW, Shuey DF, Wekstein DR 1992 Laboratory values in fit aging individuals—Sexagenarians through centenarians. Clin Chem **38**:1167–1185.
63. Wiske PS, Epstein S, Bell NH, Queener SF, Edmondson J, Johnston CC 1979 Increases in immunoreactive parathyroid hormone with age. N Engl J Med **300**:1419–1421.
64. Yendt ER, Cohanim M, Rosenberg GM 1986 Reduced serum calcium and inorganic phosphate levels in normal elderly women. J Gerontol **41**:325–330.
65. Johnson JA, Goff JP, Beitz RL, Horst RL, Reinhardt TA 1992 Expression of 1,25-hydroxyvitamin D-24-hydroxylase changes with aging. J Bone Miner Res **7**:150.
66. Wada L, Daly R, Kern D, Halloran B 1992 Kinetics of 1,25-dihydroxyvitamin-D metabolism in the aging rat. Am J Physiol **262**:E906–E910.
67. Halloran BP, Portale AA, Lonergan ET, Morris RC Jr 1990 Production and metabolic clearance of 1,25-dihydroxyvitamin D in men: Effect of advancing age. J Clin Endocrinol Metab **70** (No 2):318–323.
68. Carroll MD, Abraham S, Dresser CM 1983 Dietary intake source data: United States, 1976–80. Vital and health statistics. DHHS Publ **231**:83–1681.
69. Sheikh MS, Schiller LR, Fordtran JS 1990 In vivo intestinal absorption of calcium in humans. Miner Electrolyte Metab **16**:130–146.
70. Heaney RP, Recker RR, Stegman MR, Moy AJ 1989 Calcium absorption in women: Relationships to calcium intake, estrogen status, and age. J Bone Miner Res **4**:469–475.
71. Nordin BEC, Wilkinson R, Marshall DH, Gallagher JC, Williams A, Peacock M 1976 Calcium absorption in the elderly. Calcif Tissue Int **21**:442–451.
72. Ireland P, Fordtran JS 1973 Effect of dietary calcium and age on jejunal calcium absorption in humans studied by intestinal perfusion. J Clin Invest **52**:2672–2681.
73. Alevizaki CC, Ikkos DG, Singhelakis P 1973 Progressive decrease of true intestinal calcium absorption with age in normal man. J Nucl Med **14**:760–762.
74. Avioli LV, McDonald JE, Lee SW 1965 The influence of age on the intestinal absorption of ^{47}Ca in women and its relation to ^{47}Ca absorption in postmenopausal osteoporosis. J Clin Invest **44**:1960–1967.
75. Bullamore JR, Gallagher JC, Wilkinson R, Nordin BEC, Marshall DH 1970 Effect of age on calcium absorption. Lancet **2**:535–537.
76. Barger-Lux MJ, Heaney RP, Recker RR 1989 Time course of calcium absorption in humans: Evidence for a colonic component. Calcif Tissue Int **44**:308–311.
77. Chan EL, Lau E, Shek CC, MacDonald D, Wood J, Leung PC, Swaminathan R 1992 Age-related changes in bone density, serum parathyroid hormone, calcium absorption and other indexes of bone metabolism in Chinese women. Clin Endocrinol **36**:375–381.
78. Evans MA, Triggs EJ, Cheung M, Creasey H, Paull PD 1981 Gastric emptying rate in the elderly: Implications for drug therapy. J Am Geriatr Soc **29**:201–205.
79. Russell RM 1992 Changes in gastrointestinal function attributed to aging. Am J Clin Nutr **55**:1203S–1207S.
80. Recker RR 1985 Calcium absorption and achlorhydria. N Engl J Med **80**:70–73.
81. Devine A, Prince RL, Kerr DA, Dick IM, Criddle A, Kent GN, Price RI, Webb PG 1993 Correlates of intestinal calcium absorption in women 10 years past the menopause. Calcif Tissue Int **52**:358–360.
82. Armbrecht HJ, Bolz M, Strong R, Richardson A, Bruns MEH, Christakos S 1989 Expression of calbindin D decreases with age in intestine and kidney. Endocrinology **125**:2950–2956.
83. Horst RL, Goff JP, Reinhardt TA 1990 Advancing age results in reduction of intestinal and bone 1,25-dihydroxyvitamin D receptor. Endocrinology **126**:1053–1057.
84. Liang CT, Barnes J, Takamato S, Sacktor B 1989 Effect of age on calcium uptake in isolated duodenum cells: Role of 1,25-dihydroxyvitamin D_3. Endocrinology **124**:2830–2836.
85. Liang CT, Barnes J, Sacktor B, Takamoto S 1991 Alterations of duodenal vitamin D-dependent calcium-binding protein-content and calcium-uptake in brush-border membrane vesicles in aged Wistar rats: Role of 1,25-dihydroxyvitamin D_3. Endocrinology **128**:1780–1784.
86. Takamoto S, Seino Y, Sacktor B, Liang CT 1990 Effect of age on duodenal 1,25-dihydroxyvitamin D_3 receptors in Wistar rats. Biochim Biophys Acta **1034**:22–28.
87. Lissner L, Bengtsson C, Hansson T 1991 Bone mineral content in relation to lactation history in pre-and postmenopausal women. Calcif Tissue Int **48**:319–325.
88. Gallagher JC, Riggs BL, DeLuca HF 1980 Effect of estrogen on calcium absorption and serum vitamin D metabolism in postmenopausal osteoporosis. J Clin Endocrinol Metab **51**:1359–1364.
89. Arjmandi BH, Salih MA, Herbert DC, Sims SH, Kalu DN 1993 Evidence for estrogen receptor-linked calcium transport in the intestine. Bone Miner **21**:63–74.

90. Bo-Linn GW, Davis GR, Baddrus DJ, Morawski SG, Ana CS, Fordtran JS 1984 An evaluation of the importance of gastric acid secretion in the absorption of dietary calcium. J Clin Invest **73**:640–647.
91. Eastell R, Vieira NE, Yergey AL, Wahner HW, Silverstein MN, Kumar R, Riggs BL 1992 Pernicious anaemia as a risk factor for osteoporosis. Clin Sci **82**:681–685.
92. Knox TA, Kassarjian Z, Dawson-Hughes B, Golner BB, Dallal GE, Arora S, Russell RM 1991 Calcium absorption in elderly subjects on high-fiber and low-fiber diets—Effect of gastric acidity. Am J Clin Nutr **53**:1480–1486.
93. Goerss JB, Kim CH, Atkinson EJ, Eastell R, Ofallon WM, Melton LJ 1992 Risk of fractures in patients with pernicious anemia. J Bone Miner Res **7**:573–579.
94. Filipponi P, Gregorio F, Cristallini S, Mannarelli C, Blass A, Scarponi AM, Vespasiani G 1990 Partial gastrectomy and mineral metabolism: Effects on gastrin calcitonin release. Bone Miner **11**:199–208.
95. Barger-Lux MJ, Heaney RP, Lanspa SJ, Healy JC, DeLuca HF 1995 An investigation of sources of variation in calcium absorption efficiency. J Clin Endocrinol Metab **80**:406–411.
96. Colodro IH, Brickman AS, Coburn JW, Osborn TW, Norman AW 1978 Effect of 25-hydroxy-vitamin D_3 on intestinal absorption of calcium in normal man and patients with renal failure. Metabolism **27**:745–753.
97. Francis RM, Peacock M, Storer JH, Davies AEJ, Brown WB, Nordin BEC 1983 Calcium malabsorption in the elderly: The effect of treatment with oral 25-hydroxyvitamin D_3. Eur J Clin Invest **13**:391–396.
98. Berlyne GM, Ben-Ari J, Kushelevsky A, Idelman A, Galinsky D, Hirsch M, Shainkin R, Yagil R, Zlotnik M 1975 The aetiology of senile osteoporosis: Secondary hyperparathyroidism due to renal failure. Q J Med **64**:505–521.
99. Delmas PD, Stenner D, Wahner H, Mann KG, Riggs BL 1983 Increase in serum bone gamma-carboxyglutamic acid protein with aging in women. J Clin Invest **71**:1316–1321.
100. Insogna KL, Lewis AM, Lipinski BA, Byrant C, Baran DT 1981 Effect of age on serum immunoreactive parathyroid hormone and its biological effects. J Clin Endocrinol Metab **53**:1072–1075.
101. Kotowicz MA, Melton LJ, Cedel SL, O'Fallon WM, Riggs BL 1990 Effect of age on variables relating to calcium and phosphorus-metabolism in women. J Bone Miner Res **5**:345–352.
102. Marcus R, Madvig P, Young G 1984 Age-related changes in parathyroid hormone and parathyroid hormone action in normal humans. J Clin Endocrinol Metab **58**:223–230.
103. Van Breukelen FJM, Bijvoet OLM, Van Oosterom AT 1979 Inhibition of osteolytic bone lesions by (3-amino-1-hydroxypropylidene)-1,1-bisphosphonate (APD). Lancet **1**:803–805.
104. Eastell R, Peel NFA, Hannon RA, Blumsohn A, Price A, Colwell A, Russell RGG 1993 The effect of age on bone collagen turnover as assessed by pyridinium crosslinks and procollagen I C-terminal peptide. Osteoporosis Int **3**:S100–S101.
105. Flicker L, Lichtenstein M, Colman P, Buirski G, Kaymakei B, Hopper JL, Wark JD 1992 The effect of ageing on intact PTH and bone density in women. J Am Geriatr Soc **40**:1135–1138.
106. Klee GG, Preissner CM, Schryver PG, Taylor RL, Kao PC 1992 Multisite immunochemiluminometric assay for simultaneously measuring whole-molecule and amino-terminal fragments of human parathyrin. Clin Chem **38**:628–635.
107. Yendt ER, Cohanim M, Jarzylo S, Jones G, Rosenberg G 1991 Bone mass is related to creatinine clearance in normal elderly women. J Bone Miner Res **6**:1043–1050.
108. Young G, Marcus R, Minkoff JR, Kim LY, Sergre GV 1987 Age-related rise in parathyroid hormone in man—The use of intact and midmolecule antisera to distinguish hormone secretion from retention. Bone Miner Res **2**:367–374.
109. Bouillon R, Coopmans W, Degroote DEH, Radoux D, Ellard PH 1990 Immunoradiometric assay of parathyrin with polyclonal and monoclonal region-specific antibodies. Clin Chem **36**:271–276.
110. Sokoll LJ, Morrow FD, Quirbach DM, Dawson-Hughes B 1988 Intact parathyrin in postmenopausal women. Clin Chem **34**:407–410.
111. Forero MS, Klein RF, Nissenson RA, Nelson K, Heath III H, Arnaud CD, Riggs BL 1987 Effect of age on circulating immunoreactive and bioactive parathyroid hormone levels in women. J Bone Miner Res **2**:363–366.
112. Eastell R, Simmons PS, Colwell A, Assiri AMA, Burritt MF, Russell RGG, Riggs BL 1992 Nyctohemeral changes in bone turnover assessed by serum bone Gla-protein concentration and urinary deoxypyridinoline excretion—Effects of growth and ageing. Clin Sci **83**:375–382.
113. Harms HM, Prank K, Brosa U, Schlinke E, Neubauer O, Brabant G, Hesch RD 1992 Classification of dynamical diseases by new mathematical tools: Application of multi-dimensional phase space analyses to the pulsatile secretion of parathyroid hormone. Eur J Clin Invest **22**:371–377.
114. Kitamura N, Shigeno C, Shiomi K 1990 Episodic fluctuation in serum intact parathyroid hormone concentration in men. J Clin Endocrinol Metab **70**:252–263.
115. Samuels MH, Veldhuis J, Cawley C, Urban RJ, Luther M, Bauer R, Mundy G 1993 Pulsatile secretion of parathyroid hormone in normal young subjects—Assessment by deconvolution analysis. J Clin Endocrinol Metab **77**:399–403.
116. Cantley LK, Russell J, Lettieri D, Sherwood LM 1985 1,25-Dihydroxyvitamin D_3 suppresses parathyroid hormone secretion from bovine parathyroid cells in tissue culture. Endocrinology **117**:2114–2119.
117. Gruson M, Demignon J, Del Pinto Montes J, Miravet L 1982 Comparative effects of some hydroxylated vitamin D metabolites on parathyrin secretion by dispersed rat parathyroid cells in vitro. Steroids **40**:275–285.
118. Russell J, Lettieri D, Sherwood LM 1986 Suppression by $1,25(OH)_2D_3$ of transcription of the pre-proparathyroid hormone gene. Endocrinology **119**:2864–2866.
119. Silver J, Naveh-Many T, Mayer H, Schmelzer HJ, Popoutzer MM 1986 Regulation by vitamin D metabolites of parathyroid hormone gene transcription in vivo in the rat. J Clin Invest **78**:1296–1301.
120. Ledger GA, Burritt MF, Kao PC, O'Fallon WM, Riggs BL, Khosla S 1994 Abnormalities of parathyroid hormone secretion in elderly women that are reversible by short-term therapy with 1,25-dihydroxyvitamin D_3. J Clin Endocrinol Metab **79**:211–216.
121. Dawson-Hughes B, Harris S, Dallal GE 1991 Serum ionized calcium, as well as phosphorus and parathyroid hormone, is associated with the plasma 1,25-dihydroxyvitamin D_3 concentration in normal postmenopausal women. J Bone Miner Res **6**:461–468.
122. Sorva A, Risteli J, Risteli L, Valimaki M, Tilvis R 1991 Effects of vitamin-D and calcium on markers of bone metabolism in geriatric patients with low serum 25-hydroxyvitamin-D levels. Calcif Tissue Int **49**:S88–S89.
123. Fox J, Mathew MB 1991 Heterogeneous response to PTH in aging rats—Evidence for skeletal PTH resistance. Am J Physiol **260**:E933–E937.
124. Hegarty V, Woodhouse P, Khaw KT 1994 Seasonal variation in 25-hydroxyvitamin D and parathyroid hormone concentrations in healthy elderly women. Age Ageing **23**:478–482.

125. Fox J 1991 Regulation of parathyroid-hormone secretion by plasma calcium in aging rats. Am J Physiol **260**:E220–E225.
126. Uden P, Halloran B, Daly R, Duh QY, Clark O 1992 Set-point for parathyroid hormone release increases with postmaturational aging in the rat. Endocrinology **131**:2251–2256.
127. Wongsurawat N, Armbrecht HJ 1987 Comparison of calcium effect on in vitro calcitonin and parathyroid hormone release by young and aged thyroparathyroid glands. Exp Gerontol **22**:263–269.
128. Chapuy MC, Chapuy P, Meunier PJ 1987 Calcium and vitamin D supplements: Effects on calcium-metabolism in elderly people. Am J Clin Nutr **46**:324–328.
129. Chapuy MC, Arlot ME, Duboeuf F, Brun J, Crouzet B, Arnaud S, Delmas PD, Meunier PJ 1992 Vitamin-D_3 and calcium to prevent hip fractures in elderly women. N Engl J Med **327**:1637–1642.
130. Gallagher JC 1992 Pathophysiology of osteoporosis. Semin Nephrol **12**:109–115.
131. Eastell R, Kennedy NSJ, Smith MA, Tothill P, Edwards CRW 1986 Changes in total body calcium following surgery for primary hyperparathyroidism. Bone **7**:269–272.
132. Martin P, Bergmann P, Gillet C, Fuss M, Kinnaert P, Corvilain J, van Geertruyden J 1986 Partially reversible osteopenia after surgery for primary hyperparathyroidism. Arch Intern Med **146**:689–691.
133. Heath HI 1991 Clinical spectrum of primary hyperparathyroidism: Evolution with changes in medical practice and technology. J Bone Miner Res **6**:63–70.
134. Parfitt MA, Rao DS, Kleerekoper M 1991 Asymptomatic primary hyperparathyroidism discovered by multichannel biochemical screening: Clinical course and considerations bearing on the need for surgical intervention. J Bone Miner Res **6**:97–101.
135. Kleerekoper M, Wilson P, Peterson E, Nelson DA 1992 Acute parathyroid hormone response to intramuscular calcitonin may predict femoral neck bone loss in older postmenopausal women. J Bone Miner Res **7**:S312 (abstract).
136. Martinez ME, Del Campo MT, Sanchez-Cabezudo MJ 1994 Relations between calcidiol serum levels and bone mineral density in postmenopausal women with low bone density. Calcif Tissue Int **55**:253–256.
137. Ooms MEW, Lips P, Roos JC, van der Vijgh WJF, Popp-Snijders C, Bezemer PD, Louter LM 1995 Vitamin D status and sex hormone binding globulin: Determinants of bone turnover and bone mineral density in elderly women. J Bone Miner Res **10**:1177–1184.
138. Lukert B, Higgins J, Stoskopf M 1992 Menopausal bone loss is partially regulated by dietary intake of vitamin-D. Calcif Tissue Int **51**:173–179.
139. Avioli LV, Heaney RP 1991 Calcium intake and bone health. Calcif Tissue Int **48**:221–223.
140. Cumming RC 1990 Calcium intake and bone mass: A quantitative review of the evidence. Calcif Tissue Int **47**:194–201.
141. Dawson-Hughes B, Dallal GE, Drall EA, Sadowski L, Sahyoun N, Tannerbaum S 1990 A controlled trial of the effect of calcium supplementation on bone density in postmenopausal women. N Engl J Med **323**(No. 13):878–883.
142. Orwoll EW, Oviatt SK, McClung MR, Deftos LJ, Sexton G 1990 The rate of bone mineral loss in normal men and the effects of calcium and cholecalciferol supplementation. Ann Intern Med **112**:29–34.
143. Punnonen R, Salmi J, Tuimala R, Jarvinen M, Pystynen P 1986 Vitamin D deficiency in women with femoral neck fracture. Maturitas **8**:291–295.
144. Benhamou CL, Tourlier D, Gauvain JB, Picaper G, Audran M, Jallet P 1995 Calciotropic hormones in elderly people with and without hip fracture. Osteoporosis Int **5**:103–107.
145. Hordon LD, Peacock M 1987 Vitamin D metabolism in women with femoral neck fracture. Bone Miner **2**:413–426.
146. Compston JE, Silver AC, Croucher PI, Brown RC, Woodhead J 1989 Elevated serum intact parathyroid hormone levels in elderly patients with hip fracture. Clin Endocrinol **31**:667–672.
147. Johnston C, Norton J, Khairi MRA 1985 Heterogeneity of fracture syndromes in postmenopausal women. J Clin Endocrinol Metab **61**:551–565.
148. Harju E, Punnonen R, Tuimala R, Salmi J, Paronen I 1989 Vitamin D and calcitonin treatment in patients with femoral neck fracture: A prospective controlled clinical study. J Int Med Res **17**:226–242.
149. Lips P, Bouillon R, Jongen MJM, van Ginkel FC, van der Vijgh WJF, Netelenbos JC 1985 The effect of trauma on serum concentrations of vitamin D metabolites in patients with hip fracture. Bone **6**:63–67.
150. Nordin BEC, Peacock M, Crilly RG, Marshall DH 1979 Calcium absorption and plasma $1,25(OH)_2D$ levels in post-menopausal osteoporosis. In: Norman WE, Schaeffer K, Herrath D, Grigoleit HG, Coburn JW, DeLuca HF, *et al.*(eds) Vitamin D, Basic Research and Its Clinical Application: Proceedings of the Fourth Workshop on Vitamin D. de Gruyter, Berlin, pp. 99–106.
151. Bishop JE, Norman AW, Coburn JW, Roberts PA, Henry HL 1980 Studies on the metabolism of calciferol XVI. Determination of the concentration of 25-hydroxyvitamin D, 24,25-dihydroxyvitamin D and 1,25-dihydroxyvitamin D in a single two-milliliter plasma sample. Miner Electrolyte Metab **3**:181–189.
152. Sorensen OH, Lumholtz B, Lund B, Hjelmstrand IL, Mosekilde L, Melsen F, Bishop JE, Norman AW 1982 Acute effects of parathyroid hormone on vitamin D metabolism in patients with the bone loss of aging. J Clin Endocrinol Metab **54**:1258–1261.
153. Lund B, Sorensen OH, Agner E 1982 Serum 1,25-dihydroxyvitamin D in normal subjects and in patients with postmenopausal osteopenia. Influence of age, renal function and oestrogen therapy. Horm Metab Res **14**:271–274.
154. Aloia JF, Cohn SH, Vaswani A, Yeh JK, Yuen K, Ellis K 1985 Risk factors for postmenopausal osteoporosis. Am J Med **78**:95–100.
155. Orimo H, Shiraki M 1979 Role of calcium regulating hormones in the pathogenesis of senile osteoporosis. Endocrinol Jpn **1**:1–6.
156. Riggs BL, Hamstra A, DeLuca HF 1981 Assessement of 25-hydroxyvitamin D 1-alpha hydroxylase reserve in postmenopausal osteoporosis by administration of parathyroid extract. J Clin Endocrinol Metab **53**:833–835.
157. Nordin BEC, Peacock M, Crilly RG, Francis RM, Speed R, Barkworth S 1981 Summation of risk factors in osteoporosis. In: DeLuca HF, Frost HM, Jee WSS, Johnston C Jr, Parfitt AM (eds) Recent Advances in Pathogenesis and Treatment. Univ. Park Press, Baltimore, Maryland, pp. 359–367.
158. Haussler MR, Donaldson CA, Allegretto EA, Marion SL, Mangelsdorf J, Kelly MA, *et al.* 1984 New actions of 1,25-dihydroxyvitamin D_3: Possible due to the pathogenesis of postmenopausal osteoporosis. In: Christiansen C, Arnaud CD, Nordin BEC, Parfitt AM, Peck WA, Riggs BL (eds) Copenhagen International Symposium on Osteoporosis. Oskopress, pp. 725–736.
159. Gallagher JC, Riggs BL 1978 Current concepts in nutrition. Nutrition and bone disease. N Engl J Med **298**:193–195.
160. Riggs BL, Nelson KI 1985 Effect of long term treatment with calcitriol on calcium absorption and mineral metabolism in postmenopausal osteoporosis. J Clin Endocrinol Metab **61**:457–461.
161. Caniggia A, Gennari C, Bianchi V, Guideri R 1963 Intestinal

absorption of ^{45}Ca in senile osteoporosis. Acta Med Scand **173**:613–617.

162. Riggs BL, Kelly PJ, Kinney VR, Scholz DA, Bianco AJ 1967 Calcium deficiency and osteoporosis. Observations in 166 patients and critical review of the literature. J Bone Joint Surg **49A**:915–924.
163. Riggs BL, Arnaud CD, Jowsey J, Goldsmith RS, Kelly PJ 1973 Parathyroid function in primary osteoporosis. J Clin Invest **52**:181–184.
164. Riggs BL, Melton III LJ, Wahner HW 1983 Heterogeneity of involutional osteoporosis: Evidence for two distinct osteoporosis syndrome. In: Frame B, Potts JT (eds) Clinical Disorders of Bone and Mineral Metabolism. Excerpta Medica, Amsterdam, pp. 337–341.
165. Klein RG, Arnaud SB, Gallagher JC, DeLuca HF, Riggs BL 1977 Intestinal calcium absorption in exogenous hypercortisonism. J Clin Invest **60**:253–259.
166. Parfitt AM, Gallagher JC, Heaney RP, Johnston CC, Neer R, Whedon GD 1982 Vitamin D and bone health in the elderly. Am J Clin Nutr **36**:1014–1031.
167. Chapuy MC, Arlot ME, Duboeuf F, Brun J, Crouzet B, Arnaud S, Delmas PD, Meunier PJ 1992 Vitamin D_3 and calcium to prevent hip fractures in elderly women. N Engl J Med **327**:1637–1642.
168. Heikinheimo RJ, Inkovaara JA, Harju EJ, Haavisto MV, Kaarela RH, Kataja JM, Kokko AML, Kolho LA, Rajala SA 1992 Annual injection of vitamin-D and fractures of aged bones. Calcif Tissue Int **51**:105–110.
169. Ooms ME, Roos JC, Bezemer PD, van der Vijgh WJF, Bouter LM, Lips P 1995 Prevention of bone loss by vitamin D supplementation in elderly women: A randomised double-blind trial. J Clin Endocrinol Metab **80**:1052–1058.
170. Lips P, Graafmans WC, Ooms ME, Bezemer PD, Bouter LM 1996 Vitamin D supplementation and fracture incidence in elderly persons. A randomised, placebo-controlled clinical trial. Ann Intern Med **124**:400–406.
171. Lips P, van Ginkel FC, Jongen MJM, Rubertus F, van der Vijgh WJF, Netelenbos JC 1987 Determinants of vitamin D status in patients with hip fracture and in elderly control subjects. Am J Clin Nutr **46**:1005–1010.
172. Gallagher JC, Jerpbak CM, Jee WSS, Johnson KA, DeLuca HF, Riggs BL 1982 1,25-Dihydroxyvitamin D_3: Short- and long-term effects on bone and calcium metabolism in patients with postmenopausal osteoporosis. Proc Natl Acad Sci USA **79**:3325–3329.
173. Aloia JF, Vaswani A, Yeh JK, Ellis K, Yasumura A, Cohn SH 1988 Calcitriol in the treatment of postmenopausal osteoporosis. Am J Med **84**:401–408.
174. Gallagher JC, Riggs BL 1990 Action of 1,25-dihydroxyvitamin D on calcium balance and bone turnover and its effect on vertebral fracture rate. Metabolism **39**(Suppl. 1):30–34.
175. Ott SM, Chesnut CH 1989 Calcitriol treatment is not effective in postmenopausal osteoporosis. Ann Intern Med **110**:267–274.
176. Orimo H, Shiraki M, Hayashi Y, Hoshino T, Onaya T, Miyazaki S, Kurosawa H, Nakamura T, Ogawa N 1994 Effects of 1-hydroxyvitamin D_3 on lumbar bone mineral density and vertebral fractures in patients with postmenopausal osteoporosis. Calcif Tissue Int **54**:370–376.
177. Christiansen C, Christiansen MS, Rodbro P, Hagen C, Transbol I 1981 Effect of 1,25-dihydroxyvitamin D_3 in itself or combined with hormone treatment in preventing postmenopausal osteoporosis. Eur J Clin Invest **11**:305–309.
178. Need AG, Nordin BEC, Horwitz M, Morris HA 1990 Calcium and calcitriol therapy in osteoporotic postmenopausal women with impaired calcium absorption. Metabolism **39**:53–54.
179. Gallagher JC, Riggs BL, Recker RR, Goldgar D 1989 The effect of calcitriol on patients with postmenopausal osteoporosis with special reference to fracture frequency. Proc Soc Exp Biol Med **191**:287–292.
180. Tilyard MW, Spears GFS, Thomson J, Dovey S 1992 Treatment of Postmenopausal Osteoporosis with Calcitriol or Calcium. N Engl J Med **326**:357–362.

CHAPTER 45

Vitamin D Receptor Gene Variants and Osteoporosis: A Contributor to the Polygenic Control of Bone Density

NIGEL MORRISON Genomics Research Centre, Griffith University, Gold Coast 9726, Queensland, Australia

I. INTRODUCTION: GENETIC CLUES TO OSTEOPOROSIS

Bone mineral density (BMD) is a measure of the quantity of calcium in the bone matrix. BMD increases during early adulthood (see Fig. 1) and reaches an adult value that remains approximately constant until menopause where in women BMD undergoes a progressive decrease. At any stage during the life cycle, BMD accretion, maintenance, or breakdown could be affected by genetics, and such genetic influences could

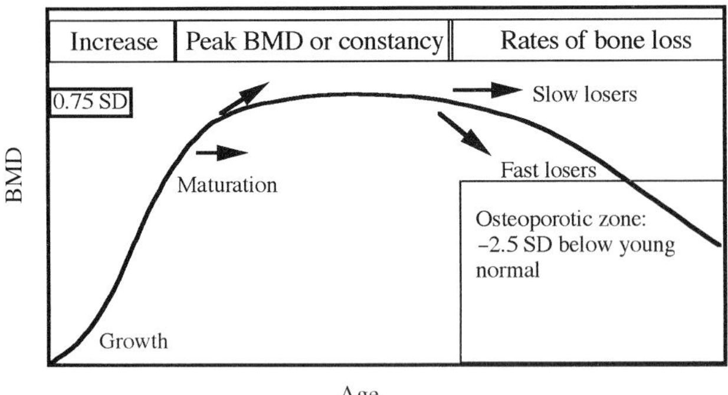

FIGURE 1 Relationship between bone mineral density (BMD) and age. BMD accumulates during growth, reaches approximate constancy in early adulthood, and then, in women, declines after menopause. Men suffer a similar but much less pronounced decline. The points indicated are positions in the life cycle where genes could influence the acquisition of bone density either by inefficient utilization of dietary calcium resources or by inappropriate maturation in terms of BMD. The arrows represent alternatives at these points where genes could have effects. During adulthood environmental and other nonspecific causes such as therapy for other disease, lifestyle, diet, and other environmental effects can alter BMD. After menopause there is a general decline in BMD, primarily resulting from estrogen deficiency. The rate of bone loss in this period is a critical parameter. At these three stages genes could influence bone metabolism and bone density. These could be the same genes acting at each time point in the same way, or the same genes acting differently, or indeed different genes responsible for the genetic control at different stages of the life cycle. It seems likely that the same genes would act throughout life but that their rank order or importance might alter through time. The effect of VDR needs to be examined at each of these time points. The box at lower right represents the area 2.5SD below the young normal mean, arbitrarily set as osteoporosis. The thick small box at left represents 0.75SD around the mean of young normal, to put the maximal effect of the VDR into biological perspective.

lead to increased risk of bone-related diseases. Osteoporosis is an age-related bone disease typified by reduction in BMD and a reduction in the strength of bone, leading to fracture with minimal trauma [1–3]. Osteoporosis is defined as a level of BMD 2.5 standard deviations (SD) below the mean BMD of young normal subjects. BMD values are used widely to diagnose osteoporosis and are measured by quantitative dual energy X-ray absorptiometry (DEXA) techniques. Although osteoporosis is a complex disease with a multifactorial etiology involving diet, life style, and other factors [3], there is a strong genetic component to risk, indicated by the high heritability of BMD values in twin and family studies [4–13].

Osteoporosis in women results primarily from a postmenopausal decrease in bone density as a result of a loss of estrogen coupled with a steady decrease as a result of the aging process. Estimates suggest that for those over 60 years of age the remaining lifetime risk of osteoporotic fracture is more than 50% in women and more than 25% in men [14]. The high prevalence of osteoporotic fracture makes an impractical definition of osteoporosis for the purpose of genetic studies. For genetic studies early fracture would be a more useful syndrome, but, apart from cases with mutations in type I collagen which overlap the phenotype of osteogenesis imperfecta, little progress has been made in this direction. BMD is the surrogate for fracture, and even this variable is strongly age dependent, making family studies problematical. With a normally distributed variable such as BMD and a high disease prevalence, the genetic basis of osteoporosis is built into the normal variation of our genome and resides in numerous polymorphic genes that have functionally distinct alleles. A relationship between BMD and the vitamin D receptor (VDR) gene has stimulated an expansion of interest in the search for genes related to osteoporosis. Although the study of the VDR gene has lead to controversy, it has generated a field of investigation and provides a framework for identifying other genes involved in the regulation of BMD and for determining the utility of explicit genetic information in both the prognosis of osteoporosis risk and response to therapy in established osteoporosis.

The VDR protein is a transcriptional regulator that controls the expression of a large number of target genes. A major role of the vitamin D endocrine system is the control of intestinal calcium absorption and calcium homeostasis [15]. In adults, vitamin D deficiency causes osteomalacia, and in children rickets, both characterized by poorly mineralized bone containing excess osteoid (unmineralized bone matrix) as a result of inadequate intestinal calcium absorption. Despite the fact that rickets and osteoporosis have little in common as disease entities, they both involve calcium metabolism. Point mutations in the VDR gene [16,17] lead to inherited rickets (see Chapter 48). Carriers of such mutations are ostensibly normal, although BMD values have not been reported, perhaps as a result of the small numbers and young age of such subjects. In contrast to osteomalacia, osteoporotic bone appears normally mineralized, although serum markers of bone turnover may be elevated. An imbalance in the ratio of bone formation by the osteoblast and bone resorption by the osteoclast, in time, will lead to a steady decrease in the content of normally mineralized bone. The VDR gene [18] is a prime candidate for involvement in the development of osteoporosis as a result of its central position in the control of calcium homeostasis.

II. GENETIC AND NONGENETIC EFFECTS ON BONE MINERAL DENSITY

Osteoporosis is not a genetic disease; rather, it is a complex multifactorial disease that is influenced by genetics. Osteoporosis is characterized by fracture and risk of fracture is related to postural instability, muscle strength, and BMD as well as other smaller contributors [14,19]. The genetic analysis of osteoporosis has as yet only seriously considered genetic effects on BMD; the other risk factors have been largely ignored so that currently the genetics of osteoporosis is synonymous with the genetics of BMD. This stems from the fact that BMD is easy to measure and is a reproducible quantitative trait.

Bone mineral density is normally distributed around a mean for any particular population. The population mean varies in different racial and ethnic groups and even with geographic distribution within an ethnic group. Mean BMD is higher in males than females, with BMD increasing during adolescent years, maintaining approximate constancy during early adulthood, and declining with age. In addition BMD in cross-sectional studies is strongly related to height, weight, and muscle strength: BMD is related to body size and load bearing and can be altered through diet, exercise, and fitness. Environmental influences on BMD have been quantitated by epidemiological studies. Smoking has a negative effect on BMD, whereas dietary calcium intake can affect the acquisition or peak BMD in the critical phase during development in juveniles. BMD can be affected by numerous drugs, therapies, and behavioral parameters. Nongenetic causes of variance in BMD need to be quantitated or controlled, for example, controlled mathematically by regression or in model design by exclusion. Uneven distributions of nongenetic causes of BMD variation could profoundly alter the ability of a study to detect the effect of a particular gene on BMD. Regression models for examining such correlated variables in genetic studies exist in other fields but have not yet been formally applied to bone, and this remains a problem for developing meaningful studies in this field.

A. Family Studies

Bone fracture is the unequivocal end point of disease, yet the influence of accidental events on bone fracture reduces its effectiveness as a phenotype for genetic studies. BMD has replaced fracture as the trait of choice in identifying genes involved in disease. Whole genome screens in families in which low BMD appears to segregate have not yielded positive linkage results, probably because of genetic heterogeneity. Studying single families large enough to generate significant linkages (defined by the logarithm of the odds of linkage or LOD scores) would seem the best strategy for family studies. Severe mutations of type I collagen cause osteogenesis imperfecta, so it is likely that milder mutations of type I collagen could account for some families segregating low BMD phenotypes [20,21] and that collagen mutations will characterize a distinct subcategory of osteoporotic patients. Sequence analysis of 26 patients with low BMD and a positive family history of osteoporosis revealed 5 with mutations in type I collagen [21]. Type I collagen is obviously an osteoporosis-related gene that requires further examination.

B. Twin Studies

The BMD values of an identical or monozygotic (MZ) twin pair are extremely similar. Twin studies rely on measuring the within pair difference of a trait in MZ twins and comparing them to the difference in a group of nonidentical or dizygotic (DZ) twins. A genetically determined trait will show higher similarity in MZ twin pairs than DZ pairs, because MZ twins are genetically identical whereas DZ twin pairs on average share only 50% of their genes (for any particular pair this proportion is not known). With a reasonable number of twins, the 50% sharing of genes in DZ twins will be approximated, and the comparison of MZ and DZ twins measures the relative genetic and environmental contributions to the trait.

Early twin studies used comparisons of forearm [5] and metacarpal BMD to identify genetic influences on BMD. Later studies confirmed that genetics regulates BMD at a number of skeletal sites using twins [6,7], mother–daughter pairs [10,12,22], and family studies [4]. An epidemiological study from Rancho Bernardo in the United States found a strong relationship between a family history of osteoporosis and decreased BMD [13]. Heritabilities are different at different skeletal sites within the same study group, suggesting that genetic effects are not constant throughout the skeleton.

As the study of genetic factors in osteoporosis increases in complexity, the search for candidate genes will have to be increased to include several classes of genes that may be involved: (1) genes that influence peak bone density, detectable in early adulthood; (2) genes related to the rate at which BMD is lost after menopause; and (3) genes that influence the quality of bone regardless of its density and hence its resistance to fracture under stress. Phenotypically different patient types may be required for each class of gene; however, the sib pair approach coupled with whole genome scanning would currently seem feasible for each particular phenotype.

Can we expect genetic effects to be of similar magnitude throughout life? Will the genes that dictate growth and development of the skeleton during early life be the same genes that regulate the gradual loss of BMD in later years? BMD is related to anthropomorphic variables, such as height, weight, and muscle strength. Therefore, genes that determine body size will be counted as regulators of BMD, even if the effect is indirect. Although genes that act indirectly on BMD will be of interest, it is not obvious how these could be used to develop new therapies for osteoporosis. It seems inherently likely that the rank order of significance for osteoporosis of various genes involved in skeletal growth, bone formation, and turnover will change during the aging process. To address this possibility, the magnitude of the effect of specific genes must be measured in twin studies across all age ranges.

C. Genetic Regulation of Bone Turnover

Bone is a dynamic organ that undergoes cycles of formation and breakdown mediated by osteoblasts and osteoclasts (see Chapters 18 and 26). It seems likely that the genetic control of BMD operates on both sides of this homeostatic mechanism. Twin studies measuring levels of serum osteocalcin [22] (a protein produced in the osteoblast) and serum markers of type I collagen synthesis and degradation [23] suggest that the bone turnover rate is genetically influenced.

D. The VDR Gene and Alleles

The structure of the VDR gene is very similar to other members of the steroid/thyroid hormone gene family, most of which have 10 exons (see Chapter 8). The VDR gene was initially described as having 9 exons but actually comprises at least 10 exons and spans some 60 kb, essentially in agreement with Pike *et al.* [24] except for the additional first exons (N. Morrison, L. Crofts, J. Eisman, unpublished; see Fig. 2). To avoid confusion the exons are numbered according to Pike *et al.* [24] with the addition of two exons designated −1 (see Chapter 8). The VDR gene maps to chromosome 12q14 [25]. The region in which the VDR gene maps is poorly characterized and contains other genes of interest to bone health. Labuda *et al.* [26] reported no recombination between the VDR gene locus and the putative 1-hydroxylase gene (involved in vitamin D metabolism) nor the type II collagen gene, which also maps nearby (see Chapter 47). The second exon of the VDR gene contains the initiation codon, and there are a number of alternative first exons. The last exon (exon 9) is the largest, and it contains a long 3′-untranslated region that harbors substantial sequence polymorphism in the noncoding region [27], possibly as a result of mutations

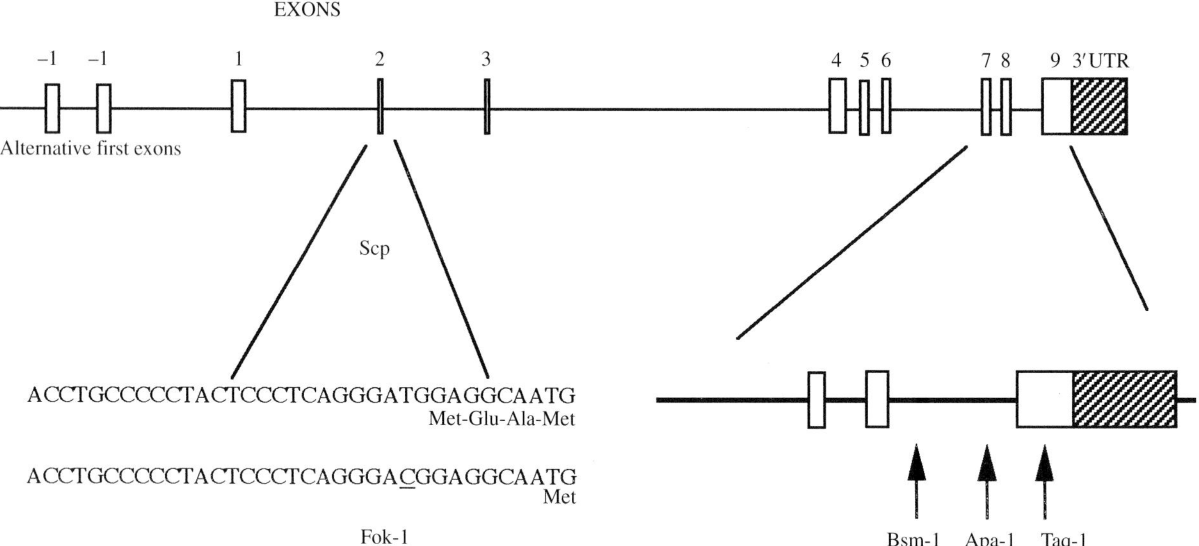

FIGURE 2 Schematic diagram of the vitamin D receptor gene and the position of various polymorphisms. The Bsm-1 and Apa-1 RFLP are in the last intron. The Taq-1 RFLP is a synonymous codon mutation in exon 9, whereas the start codon polymorphism (Scp) maps some 50 kbp 5′ in exon 2. Alternative first exons are present and are indicated generally with a −1.

in this part of the gene having little deleterious effect. The VDR gene promoter has not been described yet and should provide a wealth of information once cloned.

III. THE VDR GENE AND ALLELES: THE SYDNEY DATA

Further analysis of the VDR gene for polymorphic variation needs to be done. A very limited set of allelic variants are available for analysis, and all present markers in the VDR locus are single diallelic restriction fragment length polymorphisms (RFLP). The fact that these are used for studies is not a reflection of their inherent genetic interest, rather that there are limited tools. Bsm-1 and Apa-1 RFLP reside in the last intron (Fig. 2), whereas a Taq-1 RFLP resides in exon 9, making a synonymous codon change. Arbitrary nomenclature of the genotypes, with no inference as to function, is that the presence of the restriction site is designated with a lowercase letter and the absence of the restriction site with an uppercase letter. This, incidentally, merely reflected the position that a polymerase chain reaction (PCR) fragment or Southern blot detection of the genotype appeared on electrophoresis. The uppercase genotype was "up" on the gel and the lowercase was "down" on the gel in relation to molecular size. Hence, genotype *BB* represents homozygous absence of the Bsm-1 restriction site, and *TT* represents homozygous absence of the Taq-1 restriction site. A strong linkage disequilibrium has been observed between Bsm-1 and Taq-1 such that the *B* allele is mostly associated with the Taq-1 *t* allele. In nearly all studies to date the *BB* genotype is equivalent to the *tt* genotype. Different classifications in different papers has lead to some confusion; however, it can be generally taken that genotypes for these markers are either *BBtt* or *bbTT*, in homozygous form. The Apa-1 RFLP resides between the Bsm-1 and Taq-1 RFLP and is not as strongly related so that the combinations of Bsm-1, Apa-1, and Taq-1 form haplotypes that can become complex in different populations.

A. VDR Gene Alleles and Osteocalcin

The first indication that VDR alleles might be in linkage with functional effects on bone came from a study of osteocalcin. The osteocalcin gene maps to chromosome 1, so the only genetic connection between these genes is the vitamin D response element (VDRE) in the osteocalcin promoter, which is induced by 1,25-dihydroxyvitamin D_3 [1,25$(OH)_2D_3$] through the VDR [28,29]. Mean osteocalcin levels were significantly influenced by VDR genotypes in a cohort of unrelated subjects [30]. Higher osteocalcin levels were found in the homozygous genotype *BB* (absence of a polymorphic Bsm-1 site), and a codominant effect was seen in comparing heterozygotes (*Bb*) and the alternative homozygotes (*bb*).

B. VDR Gene Alleles and Bone Density in Twins

At any particular locus, a pair of twins can either have the same or a different genotype; this is termed allele sharing. If the gene regulates the trait in question the difference between the twins of a pair will be greater, on average, if the twins have different genotypes. In twin pairs where the genotypes are the same, the within pair difference is expected to be smaller. The within pair differences in BMD of 55 DZ twin pairs was very similar in twin pairs with the same VDR genotypes but greater in those twin pairs where genotype was not the same. This finding provided significant evidence for the hypothesis that BMD is influenced by VDR genotype [27].

However, many deficiencies in the original paper describing this effect have since come to be recognized. The magnitude of the effect seen initially in a single twin study [27] varied according to the skeletal site, with a maximal effect at the Ward's triangle in the femur and a marginally significant effect at the neck of femur. This variability illustrates the statistical problem of multiple comparisons of the same genotypic data against numerous skeletal sites. If the problem of multiple comparisons is taken into account, some of the genetic association studies in the bone field would be of marginal significance regardless of the reported p values. The entire concept of the level of significance necessary for candidate gene analysis in complex diseases has undergone a theoretical reappraisal [31]. Had this always been taken into account, less emphasis would have been placed on the magnitude of effect of the locus. Under the concept of the whole genome significance level (see Section V) the first study [27] might have only served as a tentative indication of the VDR gene locus being related to bone density.

C. VDR Gene Alleles and Bone Density in Postmenopausal Women

A second study in the original paper was directed at examining a series of essentially normal postmenopausal women drawn from the local population of Sydney [27]. The goal of the study was to assess whether a cross-sectional effect of VDR genotype on BMD could be seen in the population. A cross-sectional effect is unusual in any genetic study unless the genetic marker being tested (in this case a polymorphism) directly causes the genetic effect, or the marker is linked in strong linkage disequilibrium to such a causative gene. An example of such a cross-sectional effect, and a similar amount of controversy, relates to the role that a tryptophan to arginine mutation in the β_3-adrenergic receptor gene has in obesity [32].

The polymorphisms in the VDR gene [27] are not obvious candidates for a causative mutation, since one (Taq-1) is a coding region synonymous mutation, and the others (Bsm-1 *Bb*, and Apa-1 *Aa*) are intronic. The only reason that these RFLP were used is that, being immediately adjacent to the longest exon, they produce good hybridization signals in Southern blot analyses, whereas the other smaller exons produce less reliable signals and are more difficult to analyze. The rest of the VDR gene is likely to be equally polymorphic but has not been adequately investigated. This exemplifies the poor state of understanding of VDR gene polymorphisms. A new polymorphism at the start site of the VDR gene is discussed in Section VIII at the end of this chapter.

Simple diallelic RFLP have limited power to differentiate genotypes. However a number of closely spaced RFLPs can be grouped into haplotypes to increase the genetic information content [33], dependent on the linkage disequilibrium relationship between the RFLP [34]. The disequilibrium between the RFLP markers used produces four common haplotypes that can be treated individually as alleles. If the disequilibrium extends any distance in the gene, then it is possible that one haplotype exists which harbors a higher frequency of an unknown causative mutation that may alter BMD. If this is the case then differences in mean BMD in each particular genotype may be detected, derived from a low frequency "effect" mutation that is more common in one haplotype than others. In the study of 311 normal unrelated women from Sydney a cross-sectional effect was observed with a rather strong effect on the bone density of the lumbar spine and a much weaker effect on the neck of the femur [27]. The genotypes could be ranked as $BB < Bb < bb$ in relation to mean BMD using multivariate statistical analysis to model the effects of genotype, age, height, and weight. As far as a polygenic disease model goes, any gene locus that explains more than about 1% of the phenotypic variance is considered significant. In the cohort of unrelated women, the VDR gene allele effect was strong, with about 5% of the phenotypic variance explained by the genetic marker. Although significant, this is much less than the effect in the original twin study and argued against a large effect of the locus. These two studies were included in the original paper, but the discrepancy between the quite different magnitudes of effect in the two studies was not discussed. In retrospect, the magnitude of the VDR effect was disproportionately based on the single twin study, and this was an overinterpretation of the data.

IV. THE VDR GENE AND ALLELES: OTHER DATA

A second independent twin study [35] derived from London, focusing on postmenopausal twins, reproduced the VDR gene effect. The London twin study detected certain effects of the VDR and not others. In particular the cross-sectional effect of VDR was quite strong in the London study, with the means of each genotype group differing by 66% of the standard deviation. In contrast, within pair relationships were not significantly influenced by the VDR genotype status unless estrogen replacement status was taken into account. The differences in means of genotypes were also consistent with a dominant effect in the London data rather than co-dominant. Spector *et al.* [35] related the relative risk of fracture due to BMD to the cross-sectional effect of the VDR locus. The bone density differences suggested that VDR might confer about a 1.6-fold increased risk of osteoporotic fracture. A locus such as VDR which potentially influences BMD may have a strong effect on BMD which is translated to a rather modest effect on the relative risk of fracture. Even at a locus-specific effect of about 1SD, about a 2- to 3-fold relative risk of fracture would be expected [36]. This influences the power of fracture-based studies to detect an effect of a locus.

Studies of the potential relationship of VDR gene variation to fracture risk have not used a reasonable figure of genotype relative risk in power calculations. At least 150 each of fracture cases and age-matched controls are required to have 90% power of detecting an allele-related relative risk of 1.6-fold at a nominal p value of 0.05. If a more conservative genome-wide approach to statistical significance is taken, much higher numbers are required. The largest study so far [37] with 72 fracture cases compared to 70 controls did not have the power to refute a genotype relative risk of less than 2-fold. Those studies that have addressed fracture have taken the inflated concept of the magnitude of the VDR gene effect and are therefore completely underpowered to determine whether VDR gene alleles indeed confer a modestly increased relative risk of fracture. The problem of the power of studies to detect the more reasonable weak effects of VDR has led to what may be false-negative studies. The difficulty stems from the initial paper creating an inflated expectation of the magnitude of the genotype effect.

A second independent twin study from Indiana [38], formulated in much the same manner, did not find a relationship of VDR genotype to BMD. It is important to clarify what such a negative study means. The power to detect an effect of more moderate and reasonable magnitude in such a study is very low, and such a study cannot refute VDR having an effect at 5% of the phenotypic variance. However, this and other negative studies consistently refute VDR having an unreasonably large effect. This study clearly indicated that the VDR gene effect was not of the magnitude suggested in the original paper. It is now widely accepted that VDR RFLP are not related to large differences in bone density as originally hypothesized.

Previous twin studies show that the genetic effects on BMD are not constant throughout the skeleton. We should not expect any particular gene to have the same effect on all bones. Just how much of the genetic effect in terms of variance is explained by any locus will vary from a major influence on BMD at some skeletal sites and a minor influence at others. The VDR alleles had a weak effect at the neck of the femur, a predominantly cortical bone site, and a stronger effect at the lumbar spine, a predominantly trabecular bone site. The importance of this observation is that a single gene involved in BMD could have a wide range of effects depending on the skeletal site analyzed.

A. Difficulties in Replicating Association Studies in Complex Diseases

We should not be surprised if genotype–phenotype relationships in complex multifactorial conditions are not replicated consistently across all studies. The nature of the detection of such effects is such that false-negative and false-positive studies can both occur. False-negative studies do not tend to be followed up. True-positive studies tend to be replicated in adequately powered studies; however, even true-positive results can have a long waiting time before replication.

The theoretical statistics concerning the correlation of BMD at different sites and the multiple comparisons made in the analysis have not been formally addressed in this emerging field. These issues could profoundly alter the power to detect linkage in candidate gene association studies and in genome screening strategies to identify BMD-related genes. Clearly the amount of the phenotypic variance explained by a BMD-related gene will be less in a population study where age and other nongenetic variables are not controlled. Estimates of the component of the variance in BMD that is explained by genetics are problematic, owing to the increasing influence of environmental effects with time. Add to that the variable effects of genetics across the skeleton and it becomes apparent that estimates of the contribution of a gene to an effect in a complex multifactorial trait must be context specific. Indeed, using gene–environment models, and where gene effects change with age, the amount of the variance attributable to the locus in question can vary substantially, indicating that this question is model dependent [39].

B. Positive Studies

An example of the model dependency of this interpretation is found in Riggs *et al.* [40] who reported that the VDR effect was present in young subjects from Minnesota but declined with age. This study also found a cross-sectional effect, with about 15% difference (more than 1SD) between genotypes in BMD of younger subjects attributable to the VDR gene. Fleet *et al.* [41] found similar significant effects in American black and white premenopausal women of similar magnitude. American transmenopausal women from Omaha [42] are also reported to have BMD influenced by VDR gene alleles, with about a 10% difference in bone density at the spine between VDR genotypes. The Omaha result is similar in direction and magnitude to the result from the 311 Sydney women. A cross-sectional study of Japanese women [43] showed the same direction and similar magnitude of the VDR gene effect, but with differing allele frequencies. A second Japanese study [44] showed a significant difference in bone density between genotypes of one standard deviation, and a third Japanese study [45] again showed consistent significant effects. In an Austrian study of hospital controls, the VDR alleles were significantly related to BMD assessed by quantitative computerized tomography (QCT), which predominantly quantifies the trabecular bone [46]. The significance of the effect was dependent on exclusion of subjects with thyroid disorders and other pathological processes related to bone.

In 180 postmenopausal Italian women [47] VDR genotype was related to bone density, with the *bbaaTT* genotype having 11% higher BMD than the *BBAAtt* genotype, a result very similar to that from Australia and Omaha. Moreover, in this group the *BBAAtt* genotype was of significantly increased prevalence among osteoporotic women, and the *bbaaTT* genotype was significantly increased in normal women. A similar 12% difference between *BB* (low BMD) and *bb* (high BMD) was demonstrated in 127 Brazilian premenopausal women from a highly outbred and genetically mixed population [48], supporting the contention of Fleet *et al.* [41] that the VDR gene has similar effects regardless of racial group. In 470 premenopausal women from Pittsburgh [49], a modest but significant effect was seen, with the BMD of the *bb* genotype being 2.9% lower than that of the *BB* genotype (or 23% of a standard deviation). In an extension of this study Salamone *et al.* [50] described complex relationships between VDR genotype and environmentally influenced variables, such that BMD was determined by an interaction between VDR genotype and calcium intake and physical activity index.

Taken together, these studies have generally demonstrated statistically significant effects of the VDR locus on bone-related parameters in several ethnic groups. However, an effect can be statistically significant and still be clinically meaningless. It is not currently known whether the VDR gene alleles, as they currently are understood, will be clinically useful. After full analysis of the locus and identification of other genes which impact on bone density, it may be possible to use discrete genetic information in a clinical setting.

C. Negative and Equivocal Studies

A comprehensive French study of 268 postmenopausal women did not find an effect of VDR alleles on either BMD or serum markers of bone turnover [51]. This study started with 1039 women and used rigorous exclusion criteria to eliminate nongenetic sources of variation including the unusual practice of excluding those taking oral contraceptives (193 women), resulting in the final 268 women. The study was well powered to detect an effect of the VDR, and it did not find an effect on any parameter related to bone. In a comprehensive and large study of Danish patients, Jorgensen *et al.* [52] found no relationship between any VDR genotypes defined by RFLP and bone mineral density in 599 premenopausal women or bone loss in 136 postmenopausal women. In addition, no differences in genotype frequencies were observed in 125 women with low bone mass. These two studies are convincing examples of negative studies and add to the controversy about the VDR gene in that similar populations from Europe have conflicting results. These studies did not explore whether complex gene–environment relationships may exist; however, they do show that VDR alleles, as currently tested, have little effect in these populations. These studies clearly demonstrate that the VDR effect is complex and variable in different populations and that it is impossible to draw any sweeping generalizations about the role of this gene in the multivariate problem of bone density and osteoporosis.

In a similar large-scale population study from the Netherlands a strong effect was initially reported on 800 subjects of a community-based epidemiology study [53]. The magnitude of the effect decreased on sampling more people (902 women and 880 men) but was still quite large, with the difference between extremes of genotype being 50% of an SD in women and 30% of an SD in men [54]. Once again it was possible to detect an effect cross-sectionally in a large group of unrelated subjects. In the case of the Dutch, the cross-sectional effect was in the reverse direction to that observed in Australia, London, Japan, the United States, and Brazil but in the same direction as reported by Houston *et al.* [55] for a population from Aberdeen in Scotland. In other words, in the Dutch the *BB* genotype had the higher BMD, rather than the lower BMD. In the same group of Dutch subjects, the same VDR gene alleles that were related to higher BMD were strongly related to increased relative risk of osteoarthritis [56], a result confirmed by Keen *et al.* [57] from London and exhibiting the complexity of the relationship of VDR gene alleles to bone physiology.

Taken together, these studies have exhibited differences in the relative magnitude of effect and the relative positioning of mean BMD values for each particular genotype with respect to the population mean of BMD. The "direction of effect" is a simple description of a codominant effect with, for instance, the *BB* genotype having the lower BMD. A reverse direction with respect would suggest that the *BB* genotype had the higher BMD. This represents the "phase" of relationship between the RFLP markers and a putative causative mutation or polymorphism. In some populations, the putative mutation is more frequently associated with one haplotype, and in another population the relationship is reversed so that the alternative haplotype carries the putative mutation at higher frequency.

Considering that a cross-sectional effect is unexpected to begin with a reversal of phase or the "direction of effect" of RFLP-defined haplotypes on a quantitative trait in different populations is not surprising, as the putative causative polymorphism or mutations may exist at different frequencies in different populations and may be associated with different haplotypes as a result of the ancestry of the population. Highly admixed groups may have a balance of differing associations, making it impossible to detect such cross-sectional association effects, yet significant differences were observed in Brazilians as described [48]. However, if genetic variability in the VDR gene locus is related to quantitative trait differences in BMD, then it should still be possible to detect that effect in an outbred population using a nuclear family linkage strategy, given adequate power in the study to detect a locus contributing a reasonable amount of the phenotypic variance and adequate measures of the inheritance of the locus by descent. A quantitative trait linkage strategy does not presuppose any direction of effect and is based on allele sharing within family members. In this manner, it should still be possible to measure the contribution to the variance that the VDR gene locus has to BMD in population such as the French, in which there is convincing evidence that the VDR alleles, as defined by RFLP, are of no clinical significance.

V. STATISTICAL CONSIDERATIONS FOR STUDIES ON CANDIDATE GENES

The statistical problems that face candidate gene analysis are quite formidable. In the bone field, multiple physiological variables are usually available, and these are or will be compared to genotypes derived from many different target genes in addition to the VDR. The genome contains approximately 100,000 genes. One argument is that candidate genes will be tested until a relationship has been found and that the significance levels should be adjusted *a priori* to reflect the possibility that all genes are potential candidates. This is referred to as the whole genome significance level.

A formal analysis of the power of candidate gene association studies related to quantitative traits under whole genome significance is not yet available. A useful inference can be drawn from an analysis of dichotomous traits [58] in which the whole genome significance level for testing all genes has been estimated to be $p = 5 \times 10^{-8}$. This consideration tells us that many false-positive results will occur in the BMD field under nominal significance values ($p = 0.05$) as more and more candidate genes are studied. The findings of the VDR field appear to exceed the expected rate of false positives even if the first study is completely discounted (1 in 20 studies under the null hypothesis of no relationship). The probability in a similar linkage study that a false-positive study is fortuitously replicated by a second false-positive study is approximately 0.02 [59]. The problem of underpowered studies is the risk of the reverse: false rejection of a true effect.

Several factors argue against the concept of random result distributions around a mean of zero effect as an explanation for VDR genotype studies. Sufficient positive studies have been published from a number of independent laboratories to feel moderately comfortable that the VDR gene has a complex gene–environment effect on bone density and on other complex physiological processes related to the vitamin D endocrine system. Indeed a potent argument has been the demonstration of relationships between the RFLP in the VDR gene and other diseases and physiological traits that the VDR gene is supected of being involved in. If there really is a difference in VDR gene functionality related to the alleles and haplotypes in the populations, then it should be expected that these would influence other nonbone traits and conditions because of the pleiotropic nature of the VDR.

Although a bias generally exists toward the publication of positive studies in science, this has probably not occurred in such a controversial topic as the role of the VDR gene; thus, by examining abstract and full publications a balanced proportion of the available information is probably apparent. Given the inadequate analysis of the full sequence variation of the locus and surrounding genes, variations according to ethnic group, geographic effects on environmental factors such as vitamin D deficiency, nongenetic factors, and bone subcompartments, conclusions regarding the magnitude of the VDR effect will require a careful metaanalysis.

VI. CLINICAL SIGNIFICANCE OF VDR GENE ALLELES

A. VDR Gene Alleles and Osteoporotic Fracture Studies

Melhus *et al.* [37] reported a lack of an effect of the VDR gene on BMD or fracture incidence. However, the study demonstrated an increase in the frequency of the *BB* genotype in osteoporotic ($n = 70$) compared with normal elderly controls ($n = 76$) from 9 to 20%, a 2.2-fold increase in the expected direction, whereas the frequency of the *bb* genotype decreased from 45 to 39% among the osteoporotics. Although these authors concluded that there was no relationship between VDR genotype and osteoporosis the increase was in the expected direction and was of a magnitude similar to that predicted by Spector *et al.* [35]. The Melhus data were not statistically significant, possibly because the study lacked power [60], as in other studies with low subject numbers such as that by Looney *et al.* [61] where 41 cases of osteoporosis were compared with 23 controls without detecting an effect of the VDR locus. Unfortunately, a single gene which influences BMD may not have much impact on the clinically relevant event of fracture. Fracture depends on many variables apart from BMD, and although a gene that predicts a proportion of the variance in BMD will certainly influence fracture risk, large studies are required to see a difference in fracture risk between genotypes.

In addition, studies of single genes related to osteoporotic fracture risk will need to be confined to particular fracture sites in the skeleton and within discrete age ranges to reduce the phenotypic complexity, for instance, by accepting only atraumatic spine fracture into the study. Such comprehensive analysis has not yet been done but may be possible by retrospectively genotyping large numbers of subjects from multicenter clinical trials.

B. VDR Gene Alleles and the Rate of Bone Loss

The rate of loss of BMD after menopause is an important variable related to osteoporosis. Twin studies have suggested that the rate of loss is genetically influenced [62]. Yamagata *et al.* [43] observed a strong cross-sectional effect of VDR genotypes on bone density in Japanese and a significant effect of VDR alleles on the rate of loss of BMD, such that the *BB* genotype had both lower BMD and a greater rate of loss. A similar significant effect of VDR allele on the rate of change of bone density was seen in elderly calcium-supplemented Swiss patients [63], where a significant interaction between the rate of loss and calcium intake was described.

In contrast, an English long-term follow-up epidemiological study did not observe differences in the rate of bone loss according to the VDR genotype [64] in 195 perimenopausal women. The rate of loss in the younger English subjects was generally low to begin with, with the 95% confidence intervals being −1.32 to −0.88% at the spine and −1.08 and −0.45% at the hip. In this study the *tt* genotype consisted of 28 subjects, requiring a difference of about 1SD in the rate of loss to be reliably detected in a study of this size. This indicates that the power of a study to detect a difference is dependent on the total number of subjects in the least numerous genotype. If the more numerous genotypes are considered, being *TT* ($n = 71$) and *Tt* ($n = 96$), this comparison has the power to detect an effect of about 0.5SD in the rate of loss. If we hypothesize, for instance, that the rate of bone loss might differ between genotypes by a moderate level of 1.2SD, a study of perimenopausal females in which bone loss is generally low (such as the English study) would require more than 4000 patients to reliably detect such a difference. To avoid such large numbers, studies on rates of loss need to examine populations in which the mean rate of loss is actually clinically significant. Conditions in which bone loss is accelerated are the obvious model for determining whether a candidate gene has a relationship to bone loss.

A comprehensive American study of 299 healthy postmenopausal women who had been followed longitudinally confirmed that the VDR alleles are significantly related to the rate of loss of BMD [65]. The group was derived from a study of the effect of calcium supplementation in preventing bone loss after menopause. The genotype with the highest rate of loss (*BB*) exhibited a significant positive effect of calcium supplementation, suggesting that the accelerated loss in this genotype resulted from insufficient calcium in the diet, again suggesting a disturbance of calcium balance in the *BB* genotype. Calcium supplementation did not have a significant effect in the genotype with the lowest rate of loss (*bb*), suggesting that these subjects are in adequate calcium balance. Such data are strongly suggestive of a gene–environment interaction such that the effect of the VDR gene variation is dependent on the dietary calcium level. This proposition has physiological appeal and is logical. Even if there is a VDR allele effect in the population, the calcium intake history of the subjects may alter the ability of a study to detect the effect. Again, the Danish study [52] did not find such an effect. In agreement with the Danish study, Garnero *et al.* [51] found no effect of VDR gene alleles on bone loss in a study capable of detecting a difference of 0.42SD. An interaction effect with calcium intake was not observed.

In rheumatoid arthritis, bone loss can be a severe side effect of glucocorticoid therapy. The rapid bone loss that occurs in this disease may provide a model for understanding the slower progress of bone loss in osteoporosis. Gough *et al.* [66] reported that bone loss in rheumatoid patients ($n = 107$) was influenced by VDR genotype, such that the *BBtt* genotype group lost significantly more bone over the period of the study. The rate of loss in the *BBtt* group was −5.4% per annum compared with −2.9% in the *bbTT* genotype group. Given that there are now five positive and two negative studies we might cautiously conclude that VDR gene variation is capable of influencing the rate of bone loss. These data suggest the reasonable proposition that the same gene can influence peak bone density and the rate of bone loss.

C. VDR Gene Alleles and Response to Vitamin D Based Therapy

The response of BMD to therapy is influenced by genetics. In a study of 120 Japanese osteoporotic women [67], VDR genotypes differed in response of BMD to treatment with 1α-hydroxyvitamin D_3, an effective therapeutic used widely in Japan. The distribution of haplotypes is different in Japan in comparison to Europeans. Insufficient homozygous *BBAAtt* genotypes are present in Japan, so that the most useful comparisons are made between *BbAaTt* (heterozygotes) and *bbaaTT*. Those genotypes (*BbAATt* and *BbAaTt*) with lower BMD did not respond effectively to the standard dose of 1,α-hydroxyvitamin D_3, with a continued loss of BMD of 1.3 and 0.3% per annum, whereas the genotype *bbaaTT* responded significantly to the therapy, with a 2.2% increase in BMD per annum. These data are consistent with the concept that nonresponding VDR genotypes represent partial resistance to the action of vitamin D and its analogs. Clearly, clinical trials of different doses of vitamin D analogs are needed to determine the effective doses for the unresponsive genotypes.

D. Effect of VDR Alleles on Physiological Parameters

Significant effects of VDR alleles on physiological parameters were seen in a study comparing subjects of the alternative homozygote genotypes (*BB* versus *bb*) [68] treated with exogenous $1,25(OH)_2D_3$. The *BB* genotype group (associated with lower bone density) had a higher baseline parathyroid hormone (PTH) compared to the *bb* genotype group that was suppressed to a greater extent by calcitriol stimulation, despite the fact that this genotype group had higher osteocalcin at baseline. Urinary excretion of calcium and hydroxyproline rose significantly in response to $1,25(OH)_2D_3$ only in *BB* genotype.

The VDR gene polymorphism has been related to PTH activity in a number of other studies. In normal healthy male Swiss subjects, Ferrari *et al.* [69] selected individuals homozygous for alternate haplotypes (defined by Bsm-1 and Apa-1 RFLP as *BBAA* or as *bbaa*) for studies of dietary manipulation on parameters of bone and calcium physiology. Taking $1,25(OH)_2D_3$ levels into account, the *BBAA* genotype group had a significantly higher serum PTH concentration over wide ranges of dietary calcium. The results are compatible with a mild resistance at the level of the gut to the action of $1,25(OH)_2D_3$, leading to higher PTH levels regardless of the dietary calcium intake. This concept is compatible with the Japanese results [67] demonstrating resistance to 1α-hydroxyvitamin D therapy in a clinical trial of BMD improvement. Combining all the data on VDR effects, it seems reasonable to make the simple conclusion that the characteristics of the *BB* group are compatible with a partial resistance to the action of calcitriol. If so, intestinal calcium absorption should be related to VDR allelic variation, with the *BB* genotype having less efficient calcium absorption. More potent analogs of calcitriol may overcome such partial resistance by stimulating calcium absorption and suppressing bone turnover.

E. Intestinal Calcium Absorption and the VDR Gene

The relationship between VDR gene alleles and physiological parameters appears to be dependent on environmental factors and gene–environment interactions. A careful study of calcium absorption and VDR gene alleles from the United States demonstrated different calcium absorption in subjects with different VDR genotypes [70]. The data suggested that the calcium absorption of the *BBAA* genotype group was significantly lower than that of the alternate homozygous haplotype (*bbaa*). The deficit in calcium absorption could lead to higher PTH levels, increased bone turnover, and therefore accelerated bone loss in conditions of insufficient calcium balance. Although the genotype difference was not great, a small deficit in calcium balance over time could have a significant effect on BMD. A similar study of strontium absorption (a surrogate measure of calcium absorption activity) in 70 postmenopausal Italian women [47] demonstrated significantly lower strontium absorption in subjects with the *BBAA* genotype in comparison to both the heterozygote *BbAaTt* and the alternate homozygote *bbaa*. In young women, calcium supplementation during growth increases bone density. Lloyd *et al.* [71] showed that teenage women with the *bb* genotype accumulated significantly greater BMD during calcium supplementation than the *BB* genotype, again consistent with other studies showing an effect on intestinal calcium absorption. However, a study of twins by Peacock *et al.* [72] using radiocalcium absorption failed to show any effect of VDR alleles on intestinal calcium absorption. In the same study alleles of the PTH receptor gene were implicated in regulating serum levels of $1,25(OH)_2D_3$.

A study in Australia [73] of serum calcium and VDR allelic variations described a significant interaction effect between VDR alleles and serum 25-hydroxyvitamin D (25OHD), such that the genetic effect on serum calcium was strongest when serum 25OHD was lowest and such that increasing the serum 25OHD resulted in no genetic effect on serum calcium levels. Because 25OHD is the precursor of $1,25(OH)_2D_3$, this interaction effect seems entirely reasonable. When ambient 25OHD is low, then genetic differences in the VDR gene should be more apparent, if these differences relate to the functionality of the receptor. An unresolved factor in this matter is that the 1-hydroxylase gene that converts 25OHD to $1,25(OH)_2D_3$ maps very close to the VDR gene and is possibly in linkage disequilibrium with it.

Clearly all the genes that are near to the VDR locus need to be mapped and analyzed to resolve whether these effects are generated by VDR or other linked genes.

F. VDR Gene and Rickets

Genetic rickets (i.e., hereditary vitamin D-resistant rickets or HVDRR) due to complete homozygous defects in the VDR gene can be treated by intravenous calcium, circumventing the block in intestinal calcium absorption (see Chapter 48). Heterozygote carriers of the rickets mutations are reported to be normal, but no comprehensive evaluation of their bone density or intestinal calcium absorption has been reported, probably because of the low number of such identified carriers. If VDR effects on bone density can be generated by partial resistance at the level of the gut, as suggested above, then such carriers of defective VDR genes might be expected to have low bone density in comparison to age-matched normals. Likewise, the VDR gene knockout mouse as a heterozygous carrier would be expected to have low BMD.

G. VDR Gene Alleles and Bone Volume

Kovalinka *et al.* [74] described quantitative computerized tomography (QCT) studies on 100 prepubertal girls from Los Angeles in which bone density was significantly related to VDR gene alleles, with a 2.5% maximal effect of the VDR gene. Although VDR allelic variation appeared to influence bone density, it had no influence on bone size (measured as cross-sectional area and cortical bone volume). In a similar study of 90 prepubertal girls from Maine (United States), Matkovic *et al.* [75] found that the VDR gene alleles were related to insulin-like growth factor type 1 (IGF-1) levels, the rate of bone density increase in the forearm, and bone area. The effect was also influenced by the genotype of the estrogen receptor gene, suggesting the possibility of gene–gene interactions [76]. Although interpretation of such complex and derived physiological variables is difficult, these studies tend to support a role of VDR gene variation in determining parameters of skeletal maturation in young girls.

In a study of 110 premenopausal women from the United States, Gallagher *et al.* [77] showed that subjects with the *bb* genotype had 8% higher bone mass than the *BB* genotype, although the significance of the effect was dependent on not adjusting for weight effects on bone density. However, a significant effect on the rate of bone loss was observed. Barger-Lux *et al.* [78] also speculated that the VDR gene effect on bone density was through a more direct effect on body size.

H. Other Diseases to Which VDR Alleles Have Been Related

The VDR gene is a candidate for involvement in many pathogenic processes and so is now being tested in a range of diseases unrelated to the bone field, including prostate cancer, breast cancer, psoriasis, susceptibility to tuberculosis, and rheumatoid arthritis. As long as studies have the power to detect genes that confer a relatively small increase in relative risk of disease, we could expect the VDR to be related to a number of disorders.

In a study of 254 patients with hyperparathyroidism, Carling *et al.* [79] found a highly significant overrepresentation of the *bbaaTT* genotype in sporadic hyperparathyroid patients compared to normal controls. Subsequently, the same authors demonstrated that the *bbaaTT* genotype was related to a higher sensitivity of PTH release to exogenous calcium [80]. Studies on the genetic epidemiology of prostate cancer incidence have demonstrated a strong effect of the VDR gene as a risk factor gene in two different populations [81,82].

VII. POTENTIAL MOLECULAR GENETICS AND MOLECULAR BIOLOGY OF THE VDR EFFECT

A. Linkage Studies and the VDR Locus

Studies to date have not measured the real contribution of the VDR locus to the variance in bone density. The studies have concentrated on testing the hypothesis of "difference by state," in that the VDR RFLP confer differences in mean parameters in an unrelated population. For this to occur the RFLP have to be in strong linkage disequilibrium to a mutation that confers an effect. The method "identity by state" does not take into account from whom the gene was inherited. The method assumes that a particular sequence variant has the same effect regardless of the genetic background, ethnic group, or particular extended haplotype that the variant resides on. Identity by state is valid if the sequence change is causative of a phenotypic change. If the sequence variant under consideration is in linkage disequilibrium with an unmapped causative change, that confers a phenotypic change, then the method identity by state is an approximation of the actual situation.

The method "identity by descent" [83] is the alternative more sophisticated test, and this requires quantitative trait analysis in small or extended families that have been adequately phenotyped for the trait of interest, in this case BMD. With identity by descent the maternal and paternal VDR alleles are differentiated by arrays of highly polymorphic microsatellite markers to form haplotypes. In informative families it is possible to distinguish the alleles inherited from the parents in the offspring siblings. Generally, if a gene is related to a trait of interest, then those siblings who have identical genotypes (by descent) will be more similar to one another for the trait. Conversely, those siblings who have inherited different alleles from their parents will tend to have greater differences in the trait of interest. Siblings can have 2, 1, or 0 alleles identical by descent from their parents. Linkage is indicated by a significant negative regression between the within sib variance (the square of the within sibling pair difference) and the number of alleles identical by descent. The slope of the regression relationship is related to the amount of phenotypic variance explained by the gene locus of interest. This method needs to be applied to the VDR locus using highly polymorphic markers within and flanking the VDR gene, to quantitate the contribution of the VDR lows to the population variance in BMD.

B. Single Mutation or Infinite Alleles of Incremental Effect?

An "effect mutation" or polymorphism is a sequence change that has some causative alteration in VDR gene function that results in an alteration in physiology. In the case of the VDR gene, it is unknown whether there is a single rare mutation or multiple common functionally significant mutations of the VDR gene, each of differing magnitude and each in a different location in the gene. Data on ethnic differences could indicate differences in linkage disequilibrium between markers and possible effect loci mutations. If this is so, then studying populations derived from mixed ancestry should detect variable effects dependent on the origin of the progenitor populations.

Theoretically, a gene can vary almost infinitely, with four possible alleles at each nucleotide and combinations of these variants until the gene is diverged totally in function. Obviously there are only a constrained number of common variants of any gene in any population. It is possible through evolution and genetic drift that numerous additive incrementally small effects can accumulate as a result of sequence variation. In this manner a large number of genetic differences could contribute to an effect such as observed in the VDR. Alternatively, a single functional polymorphism (or mutation depending on definition) could exist in the population as a result of positive selection (so that the frequency will increase over time) or as a result of selection in the past which fixed the mutation at a certain frequency in the population. Selection could also favor the heterozygote, which has the effect of fixing the two alternate alleles in the population in a balanced equilibrium. A combination of multiple incremental effects and a single strong functional polymorphism could also occur, making it very difficult to unravel the relative contributions of these gene variants.

In considering the relationship between VDR gene alleles and disease, if we dichotomize bone density into "low" and "normal" groups we can approximate the power of studies to detect a modest effect of the VDR locus. Spector *et al.* [35] predicted that the difference in bone density seen in their twin study would translate into a 1.6-fold relative risk of osteoporosis in the *ttBB* genotype. The likelihood is that the RFLP are in linkage disequilibrium at an unknown rate, with some other mutation. The estimation of the position of such an effect polymorphism would require a multipoint mapping approach in a sufficiently powered study to estimate the potential recombination fraction from the known markers and the putative effect locus. In the situation of multiple small additive polymorphic variants, it would probably be impractical to differentiate their effects.

C. What Are the Likely Molecular Mechanisms of the VDR Allele Effect?

The VDR alleles must have some functional effect on the receptor or its gene to generate a phenotype, Numerous possibilities can be put forward to explain how a simple polymorphism could have an effect. Obtaining experimental support for this is another matter altogether. Although it is unappealing, something as simple as codon usage could have a small cumulative effect. For instance, the Taq-1 RFLP changes an isoleucine codon from the preferred to the least preferred codon. The 3′-untranslated region (3′UTR) of the VDR gene contains numerous sequence variants, including a polymorphic poly(A) tract. Such changes could have subtle effects on mRNA stability. A significant effect of alleles of the 3′UTR was observed in highly artificial minigene constructs where the 3′UTR was placed downstream of the luciferase gene and tested in transfection experiments [27]. Similarly, differential mRNA levels have only been reported in the original paper [27] and await confirmation in different studies. Speculation in this area needs to be replaced with firm experimental evidence from a number of different laboratories.

Mutations or functional polymorphisms in the VDR gene that might explain such effects have not been identified. The strength of disequilibrium between such mutations and known RFLP will vary in different test populations. The VDR gene RFLP, although at quite different frequencies in Japanese [43] and African Americans [41], showed the same direction of effect as in some Caucasian populations but not the Dutch [54]. These data suggest that there is no simple rule for the relationship between RFLP markers and genetic effect. This reflects the paucity of information regarding the extent of genetic variation in the VDR gene itself rather than providing insight into the mechanisms whereby an effect could be generated. An exhaustive analysis of the VDR locus to identify polymorphisms with functional consequences is required.

VIII. NEWLY DISCOVERED POLYMORPHISM IN EXON 2

A. High Frequency Polymorphism of the VDR Gene at the Start Codon

A remarkable mutation in the VDR gene was first observed some years ago in a Japanese sequencing experiment [17] and subsequently in cDNA clones from a cell line [84] and has been verified as a frequent polymorphic mutation in major ethnic groups [85,86] (N. Morrison, unpublished). The mutation results from T to C transition in the start codon of the VDR gene and can be detected by using the Fok-1 restriction enzyme [86] (Fig. 2). Although many genes have alternative start condons, this is one of a few rare examples of a polymorphic start codon occurring frequently in the normal population. The VDR gene has two ATG codons only 3 amino acids apart. The first ATG codon is conserved in comparison to the rat VDR sequence and is preceded immediately by a consensus Kozak sequence, suggesting that this may be the primary start codon. Surprisingly, in Japanese [85], Mexican-Americans [86], and Caucasians (N. Morrison, unpublished) the ACG codon is more frequent than the ATG codon, and the frequencies are similar between ethnic groups. Although initiation can occur on non-ATG codons, this has only been demonstrated for a very small number of genes and not for ACG codons.

There are several unknown facts about the human VDR start codons. First, it is not established which of these two codons, or both, are used. It is not known if either start codon can be used in the same cell; if this occurred then two protein species differing by 3 amino acids would be produced in the same cell. If the first ATG codon is used predominantly then polymorphism of the codon would result in a VDR protein truncated by 3 amino acids (MEA or Met-Glu-Ala). These two variant proteins were produced in cell culture and tested in assays of VDR action with the result that they were indistinguishable [84]. The fact that these experiments did not detect a difference in behavior does not mean that such differences do not exist in the human. Such short *in vitro* transfection experiments may not have addressed the relevant biochemical processes over the relevant time scale of bone and calcium homeostasis and may not have addressed the relevant physiological processes in physiologically meaningful receptor content or dose response.

In contrast to most members of the zinc finger transcription factor family, the VDR protein has a very short stretch of about 20 amino acids prior to the first zinc finger (Fig. 3). This so-called A region in other receptors contains transactivation functions. The function of the A region of the VDR is not known. In the context of such a short A region, a 3 amino acid truncation of the VDR protein may well have some effect. Because the VDR can heterodimerize with numerous partner proteins, it is possible that such a variant could generate a selective change in the response of a particular signaling pathway. To emphasize this possibility Miyamoto *et al.* [85] used the 24-hydroxylase promoter [which is a very strong responder to $1,25(OH)_2D_3$ induction] as the target for transfection experiments in COS-7 cells and found that the ACG variant was twice as potent a transactivator in comparison to the VDR containing the additional 3 amino acids. The VDR A region in question is proline rich and acidic. When bound on DNA, the leader portion of the protein must be in close proximity to the

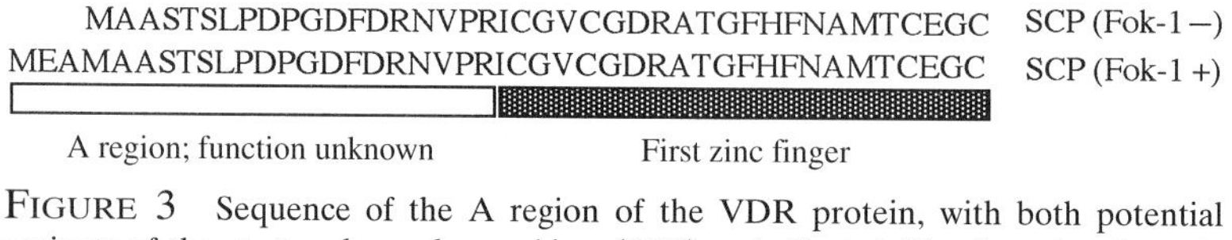

```
   MAASTSLPDPGDFDRNVPRICGVCGDRATGFHFNAMTCEGC   SCP (Fok-1 −)
MEAMAASTSLPDPGDFDRNVPRICGVCGDRATGFHFNAMTCEGC   SCP (Fok-1 +)
A region; function unknown        First zinc finger
```

FIGURE 3 Sequence of the A region of the VDR protein, with both potential variants of the start codon polymorphism (SCP) as indicated. The first zinc finger is shown to illustrate the short protein sequence that exists before the finger. The 3 amino acids missing are indicated.

DNA, and the effect of the start codon polymorphism (which removes a negative charge from the leader) on DNA binding is not known.

With respect to bone density, the Japanese study suggests that the presence of the ATG sequence is associated with lower bone density. Miyamoto *et al.* [85] examined 110 healthy premenopausal women, finding 16 ATG/ATG homozygotes, 64 ATG/ACG heterozygotes, and 30 ACG/ACG homozygotes, or allele frequencies of 0.43 for ATG and 0.56 for the ACG allele. These alleles are substantially more frequent than the Bsm-1, Apa-1, and Taq-1 RFLP, which are less frequent in Japanese than in Caucasians. The bone density of the ATG/ATG homozygotes was 10.7% lower than that of the ACG homozygotes. In a smaller study (56 subjects) Minamitani *et al.* [87] appeared to find the reverse result, with the ATG codon being associated with marginally higher bone density.

In 100 postmenopausal Mexican-Americans, Gross *et al.* [86] found a strong effect of the start codon variant, with the ACG genotype conferring increased BMD in a codominant manner. The effect was large, with a 12.8% or 1SD difference at the lumbar spine. If such an effect holds in other studies this would qualify as a major gene effect. Perhaps not surprisingly, the effect was not observed at the femoral neck. Twin studies demonstrate a lower impact of genetic effects at the femoral neck as apposed to the lumbar spine. The reason for this is unclear but may relate to the effect of weight bearing load having a greater impact at the neck of femur.

When the rate of bone loss was considered over a 2-year period of the study a significant effect of the VDR gene variants was observed at the femoral neck (−4.7 versus −0.5% for ATG versus ACG variants) but not the lumbar spine [86]. The effect was dramatic, with subjects having the *ff* genotype (the presence of the ATG codon) losing 5% over the 2-year period whereas mean bone loss in the heterozygote and the alternative homozygote groups was close to zero, suggestive of a dominant effect. No significant effect of the VDR gene variants on the rate of loss of BMD at the lumbar spine was observed. These data should stimulate an expansion of activity into the possible role of the start codon variant as a mechanism of generating genetic differences in BMD.

Why should such a protein-altering polymorphism exist in the population? Generally, in the absence of selection, gene frequencies remain in equilibrium over time. If a particular mutation or variant arises that has a selective advantage in a particular environment the frequency of this variant will increase over time until it dominates the population. Effects capable of balancing such an increase are possible reversals of such genetic advantage during the life cycle so that a gene which has a favorable effect early in life may lead to a deleterious effect in later life. An alternative is heterozygote advantage. It is possible that the heterozygote of the start codon polymorphism is functionally more adaptive than the alternative homozygotes. Such an effect fixes alternative forms of a gene in the population in a balanced equilibrium, dependent on the extent of the heterozygote advantage. It is easy to speculate about a heterozygote advantage in a complex signaling pathway such as the VDR, where heterodimer partners vary, response elements have different profiles of induction and affinity for receptors, and possible homodimers of VDR could form on certain response elements. We await the unraveling of the functional significance of this polymorphism with anticipation.

The existence of these variant forms of the VDR gene should change the emphasis of this controversial field from whether or not there is an effect of the VDR to focus on the physiological significance of this particular mutation and to extend the search for more amino acid polymorphisms. Adequately powered studies of this mutation in a number of cohorts are needed to determine if this coding region mutation has an effect on the vitamin D endocrine system and on BMD.

B. Single Mutation or Additive Effects?

A long chain of potential events could be altered by genetic variation in any transcription factor gene (see Fig. 4). The simplest explanation of the relationship between VDR gene variants and BMD is through partial resistance, which implies an inability of the VDR to respond normally to physiological levels of calcitriol. Elevated calcitriol concentrations presumably reflect a physiological drive to increase ligand concentration to normalize calcitriol signaling. Partial resistance at the level of the VDR could be generated by amino acid polymorphism or mutations that intrinsically alter the function of the receptor protein or reduce the amount of the receptor protein in target cells. Regulatory mutations that affect the protein content could reside in the promoter or in sequence motifs controlling mRNA stability and translation. Extensive sequence variation exists in the 3′-untranslated region of exon 9 (3′UTR) of the VDR gene which may relate to mRNA stability [88]. The VDR gene is about 60 kb long and has several alternative first exons. The RFLP (Bsm-1, Apa-1, and Taq-1) that have been used to date are in the 3′ end of the gene. In addition a polymorphic poly(A) length sequence variant and other RFLP exist in the long 3′UTR. Equally frequent polymorphisms probably exist

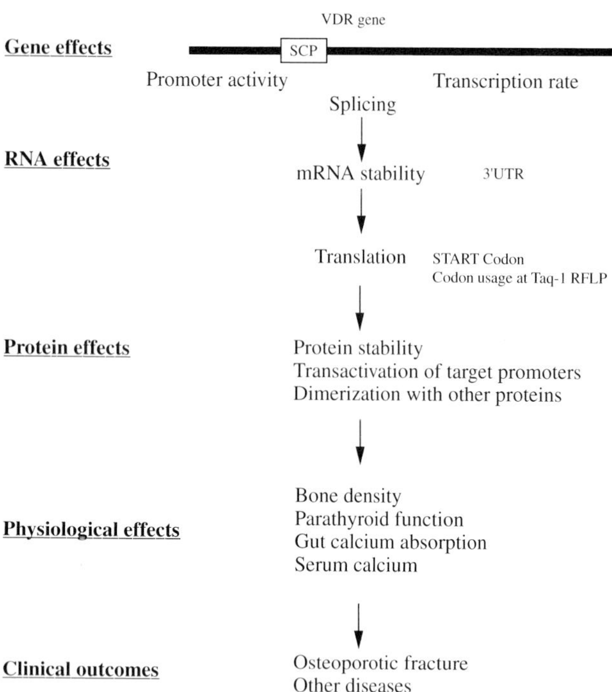

FIGURE 4 Potential chain of effects stemming from functional variation in the VDR gene. This hypothetical diagram shows key elements in a chain of effects that need to be verified conclusively before the VDR gene hypothesis can have validity. At present research is focused on the clinical and physiological end of the spectrum. More research is needed at the basic molecular and genetic end of the spectrum. The start codon polymorphism (SCP) is indicated as the only validated frequent mutation of the VDR gene that alters the protein sequence.

across the entire gene, and the relationships of these different RFLP need to be established.

It could be that several different mutations exist in various populations and that these all have small and additive effects on gene function. A gross abnormality causing a major change in phenotype is not to be expected. There is no reason to discount the possibility that multiple polymorphisms exist that have additive effects on VDR function. Such polymorphisms could reside anywhere in the gene, each conferring a slightly different genotype-mean BMD trait. If this is the case, the rate of disequilibrium between multiple polymorphic alleles would dictate the detectability of the VDR effect and its mean magnitude in different ethnic groups. If a single causative mutation exists, such as the start codon polymorphism might represent, then identification of that site will render all other polymorphic sequences irrelevant as a means of detecting VDR effects. If multiple mutations or functional sequence variants exist, then the situation will remain complex.

C. The Next Step in the Genetics of Osteoporosis and Bone Mineral Density

Clearly the concept that VDR is the major gene in osteoporosis has been rejected, and the effects of VDR gene variation can now be placed in the wider context of the polygenic control of bone density. This does not exclude a major role for VDR gene alleles in other disorders in which the vitamin D endocrine system is suspected of playing a role. The analysis of complex diseases and quantitative traits in the human has now become feasible for many investigators. In the case of bone density, a large number of gene effects will be discovered, all of which could be potentially useful as prognostic agents and as therapeutic targets. Perhaps the greatest utility in the bone field will come from the identification of genes which influence response to therapy or the rate of bone loss. There are consistent data that support a role for the VDR gene variants having an effect on response to vitamin D therapy. Understanding the basis of the effect of a gene on BMD will require physiological analysis of the relevant bone-related parameters. Identification of a new gene involved in the BMD trait should lead to new rounds of drug development and to genetic diagnosis coupled with choice of therapy directed by genetic information. Finally, as the pace of genetic analysis of disease increases, we can look forward to a comprehensive analysis of the relative risk conferred by VDR gene variants in a large number of conditions and diseases.

IX. SUMMARY AND CONCLUSIONS

Osteoporosis is a late developing multifactorial disease that has a strong genetic component. As a regulator of bone and calcium homeostasis and as a potent agent in cellular differentiation, the VDR, its ligands, and the vitamin D endocrine system in general are good candidates for involvement in metabolic processes that lead to osteoporosis. Genetic research in osteoporosis has two overlapping objectives. The first is to quantitate the involvement of candidate genes in the development of osteoporosis, and the second is to identify major genes that determine susceptibility to osteoporosis.

The VDR gene falls into the category of a candidate gene, and the controversy concerning claims about the role of this gene in osteoporosis highlights the underlying difficulties of genetic research in this disease. The actual magnitude of involvement of the gene in the disease would seem to be the source of the controversy rather than whether the VDR gene has a role to play in osteoporosis or in determining bone density. Although the upper limit of the genetic effect attributable to the VDR gene was originally claimed to be 75%, this is clearly incorrect. Any claim for a large effect was only weakly tenable to begin with and has been widely dismissed. However, sufficient studies from many laboratories have detected an effect of VDR gene variation on bone-related clinical parameters to indicate that variation in the VDR gene has a role to play in osteoporosis and in several other disorders.

The VDR gene and variations therein should now be viewed in the context of the polygenic nature of the regulation of bone density, where it is likely that many genes contribute to the control of bone density. The experiences with this locus and the development of substantial patient DNA resources by numerous investigators should mean that widespread testing of other candidate genes will be rapid. However, identification of mutations in the VDR gene sequence has been surprisingly lacking, with most focus addressed to a few simple restriction fragment length polymorphisms. The lack of sophistication in genetic analysis has hindered a balanced appraisal of the role of this gene. However, a widespread and interesting polymorphism in the VDR gene coding sequence involving the start codon of the VDR transcript may provide some answers to this controversial topic when it is fully analyzed.

References

1. Cooper C, Camplon G, Melton LJ 1992 Hip fracture in the elderly: A world wide projection. Ostoeporosis Int **2**:285–289.
2. Kroger H, Tuppurainen M, Honkanen R, Alhave E, Saankosi S 1994 Bone mineral density and risk factors of osteoporosis: A population based study of 1600 perimenopausal women. Calif Tissue Int **55**:1–7.
3. Cummings SR, Nevitt MC, Browner WS, Stone K, Fox KM, Ensrud KE, Cauley JC, Black D, Vogt TM 1995 Risk factors for hip fracture in white women. N Engl J Med **332**:767–773.
4. Smith D, Nance W, Won Kang K, Christian J, Johnston CJ 1973 Genetic factors in determining bone mass. J Clin Invest **52**:2800–2808.
5. Moller M, Horsman A, Harvald B, Hauge M, Henningsen K, Nordin BFC 1978 Metacarpal morphometry in monozygotic and dizygotic twins. Calcif Tissue Int **25**:197–201.
6. Dequecker J, Nijs J, Verstaeten A, Geusens P, Gevers G 1987 Genetic determinants of bone mineral content at the spine and the radius: A twin study. Bone **8**:207–209.
7. Pocock NA, Eisman JA, Hopper JL, Yeats MG, Sambrook PN, Eberl S 1987 Genetic determinants of bone mass in adults: A twin study. J Clin Invest **80**:706–710.
8. Evans RA, Marel GM, Lancaster EK, Kos S, Evans M, Wong SYP 1988 Bone mass is low in relatives of osteoporotic patients. Ann Intern Med **109**:870–873.
9. Kelly PJ, Hopper JL, Macaskill GT, Pocock NA, Sambrook PN, Eisman JA 1991 Genetic factors in bone turnover. J Clin Endocrinol Metab **72**:808–813.
10. Lutz J 1986 Bone mineral, serum calcium, and dietary intakes of mother/daughter pairs. Am J Clin Nutr **44**:99–106.
11. McKay HA, Bailey DA, Wilkinson AA, Houston CS 1994 Familial comparison of bone mineral density at the proximal femur and lumbar spine. Bone Miner **24**:95–110.
12. Seeman E, Hopper JL, Bach LA, Cooper ME, Parkinson E, McKay J, Jerums G 1989 Reduced bone mass in daughters of women with osteoporosis. N Engl J Med **320**:554–558.
13. Soroko SB, Barret-Conner E, Edelstein SL, Kritz-Silverstein D 1994 Family history of osteoporosis and bone mineral density at the axial skeleton: The Rancho Bernardo study. J Bone Miner Res **9**:761–769.
14. Nguyen TV, Sambrook PN, Kelly PJ, Lord SR, Freund J, Eisman JA 1993 Prediction of osteoporotic fractures by postural instability and bone density. Br Med J **307**:111–115.
15. Fraser D 1995 Vitamin D. Lancet **345**:104–107.
16. Hughes MR, Malloy PJ, Kieback DG, Kesterson RA, Pike JW, Feldman D, O'Malley BW 1988 Point mutations in the human vitamin D receptor gene associated with hypocalcaemia. Science **242**:1702–1705.
17. Saijo T, Ito M, Takeda E, Huq AHMM, Naito E, Yokota I, Sone T, Pike W, Kuroda Y 1991 A unique mutation in the vitamin D receptor gene in three Japanese patients with vitamin D-dependent rickets type II: Utility of single strand conformation polymorphism analysis for heterozygous carrier detection.Am J Hum Genet **49**:668–673.
18. Baker AR, McDonnell DP, Hughes M, Crisp TM, Mangelsdorf DJ, Haussler MR, Pike JW, Shine J, O'Malley BW 1988 Cloning and expression of full-length cDNA encoding human vitamin D receptor. Proc Natl Acad Sci USA **85**:3294–3298.
19. Hui SL, Selmenda CW, Johnston CCJ 1989 Baseline measurement of bone mass predicts fracture in white women. Ann Inter Med **111**:355–361.
20. Spotila LD, Constantinou CD, Sercda L, Ganguly A, Riggs BL, Prockop DJ 1991 Mutation in a gene for type I procollagen (COL1A2) in a woman with postmenopausal osteoporosis: Evidence for phenotypic and genotypic overlap with mild osteogenesis imperfecta. Proc Natl Acad Sci USA **88**:5423–5427.
21. Spotila LD, Colige A, Serada L, Constaninou-Deltas CD, Whyte MP, Riggs BL, Shaker JL, Spector TD, Hume E, Olsen N, Attic M, Tenenhouse A, Shane E, Brincy W, Prockop DJ 1994 Mutation analysis of coding sequences for type 1 procollagen in individuals with low bone density. J Bone Miner Res **9**:923–932.
22. Tylacsky FA, Bortz AD, Hancock RL, Anderson JJB 1989 Familial resemblance of radial bone mass between premenopausal mothers and their college-age daughters. Calcif Tissue Int **45**:265–272.
23. Tokita A, Kelly PJ, Nguygen TV, Qi J-C, Morrison NA, Ristell L, Risteli J, Sambrook PN, Eisman JA 1994 Genetic determinants of type I collagen synthesis and degradation: Further evidence for genetic regulation of bone turnover. J Clin Endocrinol Metab **78**:1461–1466.
24. Pike JW, McDonnell DP, Scott RA, Kerner SA, Kesterson RA, O'Malley B 1989 The vitamin D_3 receptor and its chromosomal gene. In Gustafson JA (ed) The Steroid/Thyroid Hormone Receptor Family and Gene Regulation. Birkhauser Verlag, Basel, pp. 147–159.
25. Faraco J, Morrison NA, Baker A, Shine J, Frossard PM 1990 Apal dimorphism at the vitamin D receptor locus. Nucleic Acids Res **17**:2150.
26. Labuda M, Ross MV, Fujiwara TM, Morgan K, Ledbetter D, Hughes MR, Glorieux FH 1991 Two hereditary defects related to vitamin D metabolism map to the same region of human chromosome 12q. Cytogenet Cell Genet **58**: (abstract).
27. Morrison NA, Qi JC, Tokita A, Kelly PJ, Crofts L, Sambrook P,

Eisman JA 1994 Prediction of bone density by vitamin D receptor alleles. Nature **367**:284–287.

28. Kerner S, Scott RA, Pike JW 1989 Sequence elements in the human osteocalcin gene confer basal activation and inducible response to hormonal vitamin D_3. Proc Natl Acad Sci USA **86**:4455–4459.
29. Morrison NA, Shine J, Fragonas JC, Verkest V, McMenemy ML, Eisman JA 1989 Dihydroxyvitamin D_3 responsive element and glucocorticoid repression in the osteocalcin gene. Science **246**:1158–1161.
30. Morrison NA, Yeoman R, Kelly PJ, Eisman JA 1992 Contribution of trans-acting factor alleles to normal physiological variability: Vitamin D receptor gene polymorphisms and circulating osteocalcin. Proc Natl Acad Sci USA **89**:6665–6669.
31. Lander E, Kurgylak L 1995 Genetic dissection of complex traits: Guidelines for interpreting and reporting linkage findings. Nature Genet **11**:241–247.
32. Mauriege P, Bouchard C 1996 Trp64Arg mutation of the β3-adrenoreceptor gene of doubtful significance for obesity and insulin resistance. Lancet **348**:698–699.
33. Uitterlinden AG, Pols HAP, van Daele PLA, Algra D, Hofman A, Birkenhåger, JC van Leeuwen JPTM 1995 Vitamin D receptor genotype is associated with bone mineral density in humans. Calcif Tissue Int **56**:419–512.
34. Thompson EA, Deeb S, Walker D, Motulsky AG 1998 The detection of linkage disequilibrium between closely linked markers: RFLPs at the AI-CIII apolipoprotein genes. Am J Hum Genet **42**:113–124.
35. Spector TD, Keen RW, Arden NNK, Major PJ, Baker JR, Morrison NA, Kelly PJ, Sambrook PN, Lanchbury JS, Eisman JA 1994 Vitamin D receptor gene (VDR) alleles and bone density in postmenopausal women; a UK twin study. Br Med J **310**:1357–1360.
36. Slemenda CW, Johnston CC, Hui SL 1996 Assessing fracture risk. In: Marcus R, Feldman D, Kelsey J (eds) Osteoporosis. Academic Press, San Diego, pp. 623–633.
37. Melhus H, Kindmark A, Amer S, Wilen B, Lindh E, Ljunghall S 1994 Vitamin D receptor genotypes in osteoporosis. Lancet **344**:949.
38. Hustmeyer FG, Peacock M, Hui S, Johnston CC, Christian JC 1994 Bone mineral density in relation to polymorphisms at the vitamin D receptor locus. J Clin Invest **94**:2130–2134.
39. Falconer DS, Mackay TFC 1996 Introduction to quantitative genetics. Longmans, Harlow, U.K., p. 131.
40. Riggs BL, Nguyen TV, Melton III LJ, Morrison NA, O'Fallon WM, Kelly PJ, Egan KS, Sambrook PN, Muhs JM, Eisman JA 1995 Contribution of vitamin D receptor gene alleles to the determination of bone mineral density in normal and osteoporotic women. J Bone Miner Res **10**:991–996.
41. Fleet JC, Hams SS, Wood RJ, Dawson-Hughes B 1995 The Bsm-1 vitamin D receptor restriction fragment length polymorphism (BB) predicts low bone density in premenopausal black and white women. J Bone Miner Res **10**:991–996.
42. Recker RR, Rogers J, Witte S, Johnson ML, Lappe JM, Kimmel DB 1996 Vitamin D receptor (VDR) genotype influence on bone mass in transmenopausal women. J Bone Miner Res **11**(Suppl. 1):S211 (abstract).
43. Yamagata Z, Miyamura T, Iijima S, Asaka A, Sasaki M, Kato J, Koizumi K 1994 Vitamin D receptor gene polymorphism and bone mineral density in healthy Japanese women. Lancet **344**:1027.
44. Shiraki M, Eguchi H, Aoki C, Shirki Y 1995 Can allelic variations in vitamin D receptor gene predict bone densities and serum osteocalcin level in Japanese women? Bone **16**(Suppl. 1):S84 (abstract).
45. Tokita A, Matsumoto H, Morrison NA, Tawa T, Miura Y, Fukamauchi K, Mitsuhashi N, Irimoto M, Yamamori S, Miura M, Watanabe T, Kuwabara Y, Yabuta K, Eisman JA 1996 Vitamin D receptor alleles, bone mineral density and turnover in premenopausal Japanese women. J Bone Miner Res **11**:1003–1009.
46. Gunzer C, Morrison NA, Eisman JA 1995 Genetic effect of the vitamin D receptor gene on bone density by QCT in continental Europeans. Bone **16**(Suppl. 1):84 (abstract).
47. Gennari L, Besherini L, Masi L, Gonelli S, Cepollaro C, Martini S, Mansari R, Montagani A, Brandi ML 1996 Vitamin D and estrogen receptor genotypes in postmenopausal women: Influences on BMD and intestinal calcium absorption. J Bone Miner Res **11**(Suppl. 1):S464 (abstract).
48. Lazaretti-Castro M, Duarte MA, Gilberto J, Vieira H 1996 Vitamin D receptor (VDR) alleles and bone mineral density (BMD) in a normal premenopausal Brazilian population. J Bone Miner Res **11**(Suppl. 1):S209 (abstract).
49. Salamone LM, Ferell R, Black DM, Palermo L, Epstein RS, Petro N, Steadman N, Kuller LH, Cauley JA 1996 The association between vitamin D receptor gene polymorphisms and bone mineral density at the spine, hip and total body in postmenopausal women. Osteoporosis Int **6**:63–68 (erratum **6**:187–188).
50. Salamone LM, Glynn NW, Black DM, Ferrel RE, Palermo L, Epstein RS, Kuller LH, Cauley JA 1996 Determinants of premenopausal bone mineral density: The interplay of genetic and lifestyle factors. J Bone Miner Res **11**:1557–1565.
51. Garnero P, Borel O, Sornay-Rendu E, Arlot MA, Delmas PD 1996 Vitamin D receptor gene polymorphisms are not related to bone turnover, rate of bone loss and bone mass in postmenopausal women: The OFELY study. J Bone Miner Res **11**:827–834.
52. Jorgensen HL, Scholler JC, Bjuring M, Hassager C, Christiansen C 1996 Relationship of common allelic variation at vitamin D receptor locus to bone mineral density and post menopausal bone loss: Cross sectional and longitudinal population study. Br Med J **313**:586–590.
53. Uitterlinden AG, Pols HAP, Burger H, Huang Q, van Daele PLA, van Duijn CM, Hofman A, Birkenhåger JC, van Leeuwen JPTM 1995 Association of the vitamin D receptor gene polymorphism and bone mineral density: Improved genetic resoluton by direct haplotyping. J Bone Miner Res **10**(Suppl. 1):S161.
54. Uitterlinden AG, Pols HAP, Burger H, Huang Q, van Daele PLA, van Diujn CM, Hofman A, Birkenhåger JC, van Leeuwen JPTM 1996 A large scale population based study of the association of vitamin D receptor polymorphisms with bone mineral density. J Bone Miner Res **11**:1241–1248.
55. Houston LA, Grant SFA, Reid DM, Ralston SH 1996 Vitamin D polymorphism, bone mineral density and osteoporotic fracture: Studies in a UK population. Bone **18**:249–252.
56. Uitterlinden AG, Burger H, Huang Q, van Duijn CM, Hofman A, Birkenhåger JC, van Leeuwen JPTM, Pols HAP 1996 Vitamin D receptor genotype is strongly associated with osteophytes in osteoarthritis and weakly with bone mineral density. J Bone Miner Res **11**(Suppl. 1):S116 (abstract).
57. Keen RW, Griffiths GO, Hart DJ, Nandra D, Lanchbury JS, Spector TD 1996 Polymorphism of the vitamin D receptor as an explanation for the inverse relationship between osteoporosis and osteoarthritis. J Bone Miner Res **11**(Suppl. 1):S213 (abstract).
58. Risch N, Merikangas K 1996 The future of genetic studies of complex human disorders. Science **273**:1516–1517.
59. Suarez BK, Hampe CL, Van Eerdewegh PV Problems of replicating linkage claims in psychiatry. In Geshon ES, Cloninger CR (eds) Genetic Approaches to Mental Disorders. American Psychopathological Press.

60 Nguyen TV, Kelly PJ, Morrison NA, Sambrook PN, Eisman JA

1994 Vitamin D receptor genotypes in osteoporosis. Lancet **344**:1580–1581.

61. Looney JE, Yoon HK, Fischer M, Farley SM, Wergedal JE, Baylink DJ 1995 Lack of evidence for an increased prevalence of the BB vitamin D receptor genotype in severely osteoporotic women. J Clin Endocrinol Metab **80**:2158–2162.
62. Kelly PJ, Nguyen T, Pocock N, Hopper J, Sambrook PN, Eisman JA 1993 Genetic determination of changes in bone density with age: A twin study. J Bone Miner Res **8**:11–17.
63. Ferrari SL, Rizzoll R, Chevalley T, Slosman D, Eisman JA, Bonjour JP 1995 Vitamin D receptor gene polymorphisms and the rate of change of lumbar spine bone mineral density in elderly men and women. Lancet **345**:423–424.
64. Keen RW, Major PJ, Lanchbury JS, Spector TD 1995 Vitamin D receptor gene polymorphism and bone loss. Lancet **345**:990.
65. Krall EA, Parry P, Lichter JB, Dawson-Hughes B 1995 Vitamin D receptor alleles and rates of bone loss: Influences of years since menopause and calcium intake. J Bone Miner Res **10**:978–984.
66. Gough A, Devlin J, Betteridge J, Franklyn J, Carke S, Nguyen T, Morrison NA, Eisman JA, Sambrook PN, Emery P 1995 Altered representation of vitamin D receptor gene alleles in early rheumatoid arthritis and their possible effect on bone turnover. Arth Rheum **38**:S217 (abstract).
67. Matsuyama T, Ishii S, Tokita A, Yabuta K, Yamamori S, Morrison NA, Eisman JA 1995 VDR gene polymorphisms and vitamin D analog treatment in Japanese. Lancet **345**:1238–1239.
68. Howard G, Nguyen T, Morrison N, Watanabe T, Sambrook PN, Eisman JA, Kelly PJ 1995 Genetic influences on bone density: Physiological correlates of vitamin D receptor gene alleles in premenopausal women. J Clin Endocrinol Metab **80**:2800–2805.
69. Ferrari SL, Rizzoli R, Eisman JA, Bonjour JP 1996 Relationship between vitamin D receptor gene polymorphisms and calcium and phosphate metabolism in young healthy males. J Bone Miner Res **11**(Suppl. 1):S477 (abstract).
70. Dawson-Hughes B, Harris SS, Finneran S 1995 Calcium absorption on high and low calcium intakes in relation to vitamin D receptor genotype. J Bone Miner Res **10**:978–984.
71. Lloyd T, Rollings N, Chinchilli V, Johnson ML 1996 Vitamin D receptor (VDR) gene polymorphism and the effect of calcium supplementation on bone mass acquisition in teenage women. J Bone Miner Res **11**(Suppl. 1):S466 (abstract).
72. Peacock M, Hustmyer FG, Johnston CC, Christian J 1996 Heritability of bone mass, bone turnover, calcium regulation and calcium absorption in relation to RFLPs at the vitamin D and PTH receptor genes. Bone **16**(Suppl. 1):S83 (abstract).
73. Stein MS, Scherer SC, O'Brien ML, Chick P, DiCarlantoinio M, Walton SL, Flicker L, Zajac JD, Wark JD 1996 Vitamin D receptor genotype and calcium levels in institutionalised elderly. J Bone Miner Res **11**(Suppl. 1):S211 (abstract).
74. Kovalinka A, Sainz J, Loro ML, Sayre J, Roe T, Gilanz V 1996 Relationship between vitamin D receptor alleles and the volume and density of cortical bone in children. J Bone Miner Res **11**(Suppl. 1):S208 (abstract).
75. Matkovic V, Ilich JZ, Klisovic D, Skugor M, Badenhop N, An B, Vereault D, Rosen CJ and Young AP 1996 Vitamin D receptor gene (VDR) polymorphism differentiates inslulin-like growth factor (IGF-1) and body composition during growth. J Bone Miner Res **11**(Suppl. 1):S209 (abstract).
76. Klisovic D, Ilich JZ, Skugor M, Young AP, Matkovic V 1996 Different distribution of growth related gain in weight, lean body mass, and bone area according to vitamin D receptor gene (VDR) polymorphism within distinct estrogen receptor genotype. J Bone Miner Res **11**(Suppl. 1):S209 (abstract).
77. Gallagher JC, Kinyamu HK, Knezetic JA, Ryschon KL 1996 Vitamin D receptor gene alleles and bone density in premenopausal women. J Bone Miner Res **11**(Suppl. 1):S210 (abstract).
78. Barger-Lux MJ, Heaney RP, Hayes J, DeLuca HF, Johnson ML, Gong G 1995 Vitamin D receptor gene polymorphism, bone mass, body size, and vitamin D receptor density. Calcif Tissue Int **57**:161–162.
79. Carling T, Kindmark A, Hellman P, Lundgren E, Ljunghall S, Rastad J, Akerstom G, Melhus H 1995 Vitamin D receptor genotypes in primary hyperparathyroidism. Nature Med **1**:1309–1311.
80. Carling T, Ridefelt P, Kindmark A, Hellman P, Rastad J Akerstom G, Melhus H 1995 Impact of vitamin D receptor alleles on parathyroid function in hyperthyroidism. J Bone Miner Res **11**:S115 (abstract).
81. Taylor JA, Hirvonen A, Watson M, Pittman G, Mohler JL, Bell DA 1996 Association of prostate cancer with vitamin D receptor gene polymorphism. Cancer Res **56**:4108–4110.
82. Ingles SA, Ross RK, Yu MC, Irvine RA, Pera GL, Haile RW, Coetzee GA 1997 Association of prostate cancer risk with vitamin D receptor and androgen receptor genetic polymorphisms. J Natl Cancer Inst **89**:166–170.
83. Lander E, Schork NJ 1994 Genetic dissection of complex traits. Science **265**:2037–2048.
84. Sturzenbecker L, Scardavolle B, Kratzeisen C, Katz M, Arbazua P, McLane J 1944 Isolation and analysis of cDNA encoding a naturally occuring truncated form of the human vitamin D receptor. Ninth Workshop on Vitamin D. Orlando, Florida, Abstract 25, pp. 112.
85. Miyamoto K, Takentani Y, Tohna E, Nishisho T 1996 A novel polymorphism in the vitamin D receptor gene and bone mineral density: A study of vitamin D receptor expression and function in COS-7 cells. J Bone Miner Res **11**(Suppl. 1):S116 (abstract).
86. Gross C, Eccleshall TR, Malloy PJ, Villa ML, Marcus R, Feldman D 1996 The presence of a polymorphism at the translation initiation site of the vitamin D receptor gene is associated with low bone mineral density in postmenopausal Mexican-American women. J Bone Miner Res **11**:1850–1855.
87. Minamitani K, Takahashi Y, Minagawa M, Someya T, Watanabe T, Yasuda T, Niimi H 1996 Exon 2 polymorphism in the human vitamin D receptor gene is a predictor of peak bone density. J Bone Miner Res **11**(Suppl. 1):S207 (abstract).
88. Crofts LA, Morrison NA, Dudman N, Eisman JA 1996 Differential expression of VDR gene alleles. J Bone Miner Res **11**(Suppl. 1):208 (abstract).

CHAPTER 46

Clinical Disorders of Phosphate Homeostasis

MARC K. DREZNER Departments of Medicine and Cell Biology and Sarah W. Stedman Nutrition Center, Duke University Medical Center, Durham, North Carolina

I. INTRODUCTION

Extensive studies since the early 1970s have established that phosphate homeostasis and vitamin D metabolism are reciprocally regulated. As discussed in Chapters 16, 17, and 32, calcitriol promotes phosphate absorption from the intestine, mobilization from bone, and reabsorption in the renal tubule [1]. In turn, phosphate depletion or hypophosphatemia stimulates renal production of 1,25-dihydroxyvitamin D [$1,25(OH)_2D$, calcitriol], whereas phosphate overload or hyperphosphatemia inhibits renal 25-hydroxyvitamin D (25OHD) 1α-hydroxylase activity [2].

Because phosphorus is one of the most abundant constituents of all tissues, disturbances in phosphate homeostasis can affect almost any organ system [3]. Indeed, a deficiency or excess of this mineral can have profound effects on a variety of tissues, which include consequences of hypophosphatemia, such as osteomalacia, rickets, red cell dysfunction, rhabdomyolysis, metabolic acidosis, and cardiomyopathy, and of hyperphosphatemia, such as soft tissue calcification, hypocalcemia, tetany, and secondary hyperparathyroidism. In many cases, the interrelationship between phosphate homeostasis and vitamin D metabolism precludes establishing if the consequences of hypo- or hyperphosphatemia are singularly related to this abnormality or are modified by changes in calcitriol production. Occasionally, however, discrimination between these possibilities has been achieved by evaluation of the therapeutic response to phosphate supplementation or depletion. Such studies indicate that few of the phosphate homeostatic disorders respond to alterations in phosphate alone, but do regress on coincident modification of the vitamin D status. The detailed control mechanisms that regulate phosphate homeostasis and vitamin D metabolism in intestine and kidney are reviewed in Chapters 16 and 17 and overall physiology in Chapter 32. Following is a summary of important elements of the phosphate homeostatic schema and the regulation of vitamin D metabolism that pertain to an understanding of the diseases to be described in this chapter.

A. Regulation of Phosphate Homeostasis

The kidney is the major arbiter of extracellular phosphate homeostasis and plays a key role in bone mineralization and growth. Most of the filtered phosphorus is reabsorbed in the proximal tubule, with approximately 60% of the filtered load reclaimed in the proximal convoluted tubule and 15–20% in the proximal straight tubule.

In addition, a small but variable portion (<10%) of filtered phosphorus is reabsorbed in more distal segments of the nephron.

Transepithelial phosphorus transport is effectively unidirectional and includes uptake at the brush border membrane of the renal tubule cell, translocation across the cell, and efflux at the basolateral membrane. Phosphorus uptake at the apical cell surface is the rate-limiting step in this process and the major site of regulation [4]. The uptake at this site is mediated by Na^+-dependent phosphate (P_i) transporters that reside in the brush border membrane and depend on the Na^+,K^+-ATPase to maintain the Na^+ gradient that drives the transport system. Two kinetically distinct Na^+–P_i cotransport systems have been identified in the brush border membrane: a high capacity, low affinity system in the proximal convoluted tubule and a low capacity, high affinity system in the proximal convoluted and straight tubules. This topological arrangement in series permits highly efficient reabsorption of phosphate in the proximal tubule.

Two renal-specific Na^+–P_i cotransporters have been identified and designated NPT1 [5] and NPT2 [6]. The genes for these transporters have been mapped to chromosomes 6p22 and 5q35, respectively. NPT1 transcripts are uniformly expressed in all segments of the proximal nephron, whereas NPT2 expression is highest in the S_1 segments of the proximal tubule. Both NPT1 and NPT2 mediate high affinity Na^+–P_i cotransport [7]. However, their pH profiles are significantly different, with that of NPT2 bearing closer resemblance to physiological pH. Moreover, NPT1 exhibits a broader substrate specificity and can induce a Cl^- conductance that is inhibited by Cl^- channel blockers and organic anions.

The hormonal and metabolic regulation of renal phosphate transport primarily involves alteration in the availability of NPT2 transporters (Fig. 1) [8]. In this regard, the parathyroid hormone (PTH)-dependent inhibition of Na^+–P_i cotransport depends on internalization of the cell surface NPT2 protein [9]. Likewise, the acute phase of the adaptation to phosphate restriction is associated with an increase in NPT2 protein, which is rapidly reversed by a high P_i diet [10]. In addition, chronic P_i deprivation is associated with an increase in renal abundance of both NPT2 mRNA and protein. Not surprisingly, this mechanism is also central to modulation of the abnormal renal phosphate transport underlying a large number of phosphate homeostatic disorders. In this regard, studies in *hyp* and *gy* mice, the murine homologs of the prototypic disease X-linked hypophosphatemia (XLH), indicate that abnormal renal tubule P_i transport is associated with a corresponding decrease in the renal abundance of NPT2 protein and mRNA, as well as diminished NPT2 gene transcription. Similarly, it is possible that NPT2 is a candidate gene responsible for hereditary hypophosphatemic rickets with hypercalciuria (HHRH).

B. Phosphate-Dependent Modulation of Vitamin D Metabolism

The production of $1,25(OH)_2D$, the active metabolite of vitamin D, is under stringent control (see Chapters 2 and 5). Indeed, 1α-hydroxylation represents the most important regulatory mechanism in the metabolism of vitamin D [11]. In normal adults, serum $1,25(OH)_2D$ concentrations change little in response to repeated dosing with vitamin D, and they remain normal, or even decline, in vitamin D intoxication. Phosphorus is one of the three major factors that regulate the activity of the enzyme [12]. In this regard, in rats, mice, and humans,

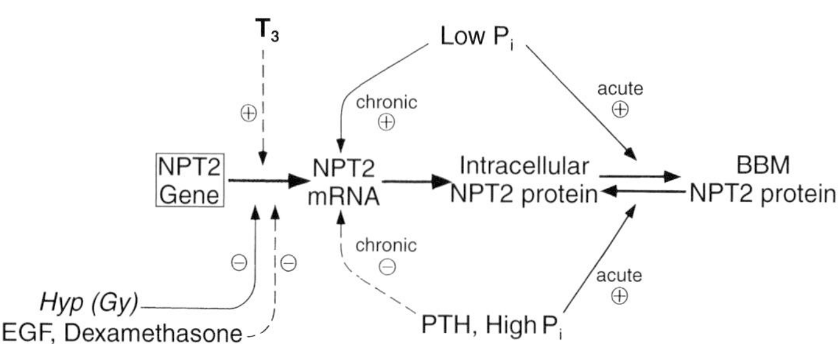

FIGURE 1 Hormonal and metabolic regulation of NPT2 gene expression and protein production in the kidney. Solid lines represent pathways of regulation that are supported by available data, and dashed lines depict instances where direct evidence for a regulatory mechanism is not available. +, Stimulation; −, inhibition; BBM, brush border membrane; EGF, epidermal growth factor; NPT2, Na^+-dependent phosphate cotransporter; *Hyp* (*Gy*), murine homologs of X-linked hypophosphatemic rickets (XLH); T_3, triiodothyronine. Adapted from Tenenhouse [8].

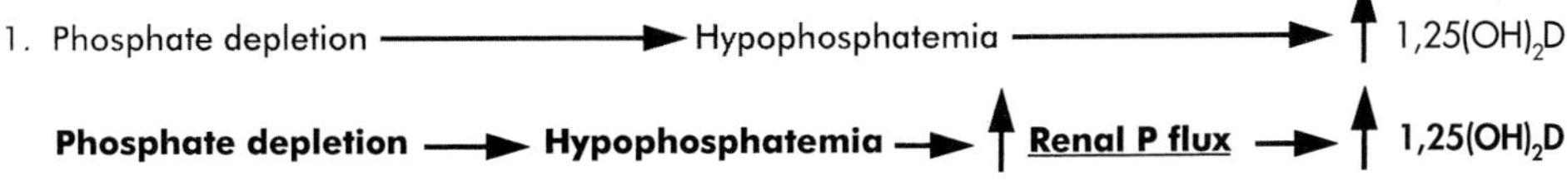

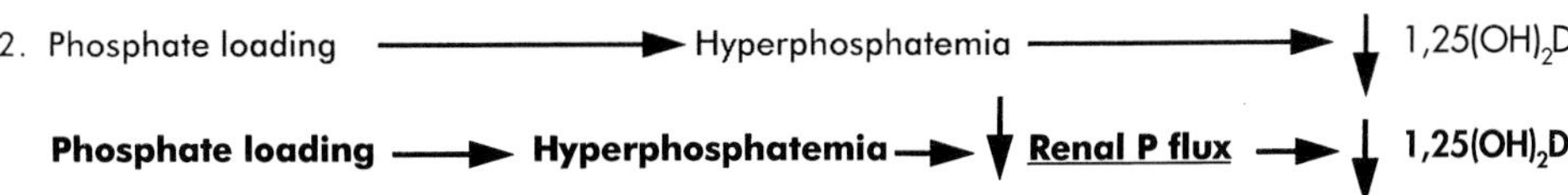

FIGURE 2 The proposed mechanism for phosphate regulation of renal 1,25(OH)$_2$D production presupposes that hypophosphatemia or hyperphosphatemia are the proximate trigger points that influence this process. Alternatively the effects of such deviations in phosphate homeostasis may be mediated by compensatory changes in the mechanisms of renal phosphate transport, which occurs in the cells dedicated to production of 1,25(OH)$_2$D. ↑, Increased; ↓, decreased.

phosphate depletion and resultant hypophosphatemia stimulate 1α-hydroxylase activity and increase the serum 1,25(OH)$_2$D concentration, whereas phosphate loading and consequent hyperphosphatemia inhibit formation of this metabolite. The mechanism whereby phosphorus modulates this adaptive effect remains unknown. Fukase *et al.* [13] reported that phosphorus may have a direct effect on the kidney, and in accord with several studies indicate that the effect is, in fact, independent of PTH. In contrast, alternative evidence suggests that phosphorus regulation of 1,25(OH)$_2$D production may depend on growth hormone [14] or insulin-like growth factor I [15].

Regardless of the proposed mechanism, the potential role of phosphorus in regulating vitamin D metabolism is central to understanding and appropriately treating many of the clinical disorders of phosphate homeostasis. As will become evident in the remainder of this chapter, however, the aberrant regulation of vitamin D metabolism encountered in many of these diseases is considered paradoxical. Thus, virtually all such disorders, in which abnormal renal phosphate transport underlies errant phosphate homeostasis, manifest hypophosphatemia or hyperphosphatemia that is perplexingly associated with decreased or increased serum 1,25(OH)$_2$D levels, respectively. These observations suggest that the effect of phosphorus on 1α-hydroxylase activity may, in fact, occur indirectly and secondary to alterations in renal phosphate transport systems (Fig. 2). In this regard, phosphate depletion and hypophosphatemia and phosphate loading and hyperphosphatemia are associated with compensatory changes in renal phosphate transport that may mediate changes in 1,25(OH)$_2$D production. In accord, the prevailing serum calcitriol levels in normals and patients with disorders of renal phosphate transport, namely, XLH, tumor-induced osteomalacia (TIO), and tumoral calcinosis, display a highly significant positive correlation with the renal tubular maximum for the reabsorption of phosphate per liter of glomerular filtrate (TmP/GFR) [16], supporting the possibility that renal tubular reabsorption of phosphate may be a major determinant of renal 1,25(OH)$_2$D production (Fig. 3). Of course, these data do not establish that alterations in 1α-hydroxylase activity in response to phosphate depletion or loading are, in fact, dependent on renal phosphate transport. However, the observa-

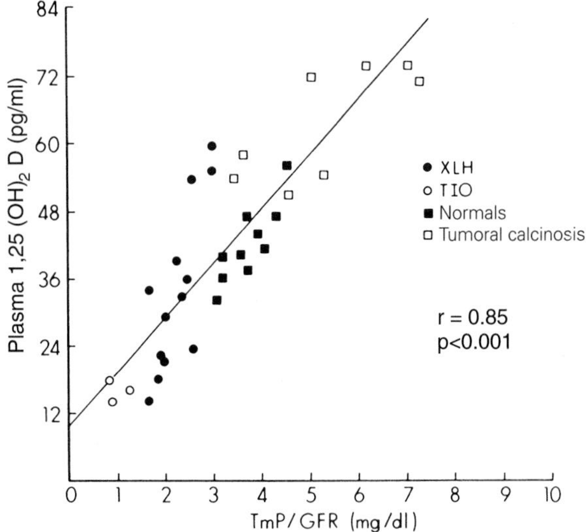

FIGURE 3 Correlation between renal TmP/GFR and plasma 1,25(OH)$_2$D levels in normals and patients with XLH, TIO, and tumoral calcinosis. Significant linear correlation is evident, suggesting a potential relationship between renal phosphate transport and the prevailing plasma levels of active vitamin D. From Drezner MK. Understanding the pathogenesis of X-linked hypophosphatemic rickets: A requisite for successful therapy. In: Zackson DA (ed.). *Cases in Metabolic Bone Disease.* Triclinica Publishing, New York. 1987.

tions that phosphate depletion does not increase $1,25(OH)_2D$ production in mice subject to treatment with phosphonoformic acid, which precludes a compensatory alteration in TmP/GFR, establishes a central role of the renal phosphate transport system in modulating phosphorus-mediated effects on 1α-hydroxylase activity under all conditions (Fig. 4).

Although these data indicate that the renal production of $1,25(OH)_2D$ and the renal tubular reabsorption of phosphate are linked, the precise mechanism(s) underlying this association remains uncertain. Indeed, it is unclear whether such factors as the topological localization of the aberrant phosphate transport along the renal tubule and the relationship to sites of 1α-hydroxylase activity play a role in the regulatory process. However, it is clear that understanding the pathophysiology of, and defining appropriate treatment regimens for, many clinical disorders of phosphate homeostasis require investigation of the vitamin D regulatory system.

C. Disorders of Phosphate Homeostasis

The variety of diseases, therapeutic agents, and physiological states that affect phosphate homeostasis are numerous and reflect a diverse pathophysiology. In-

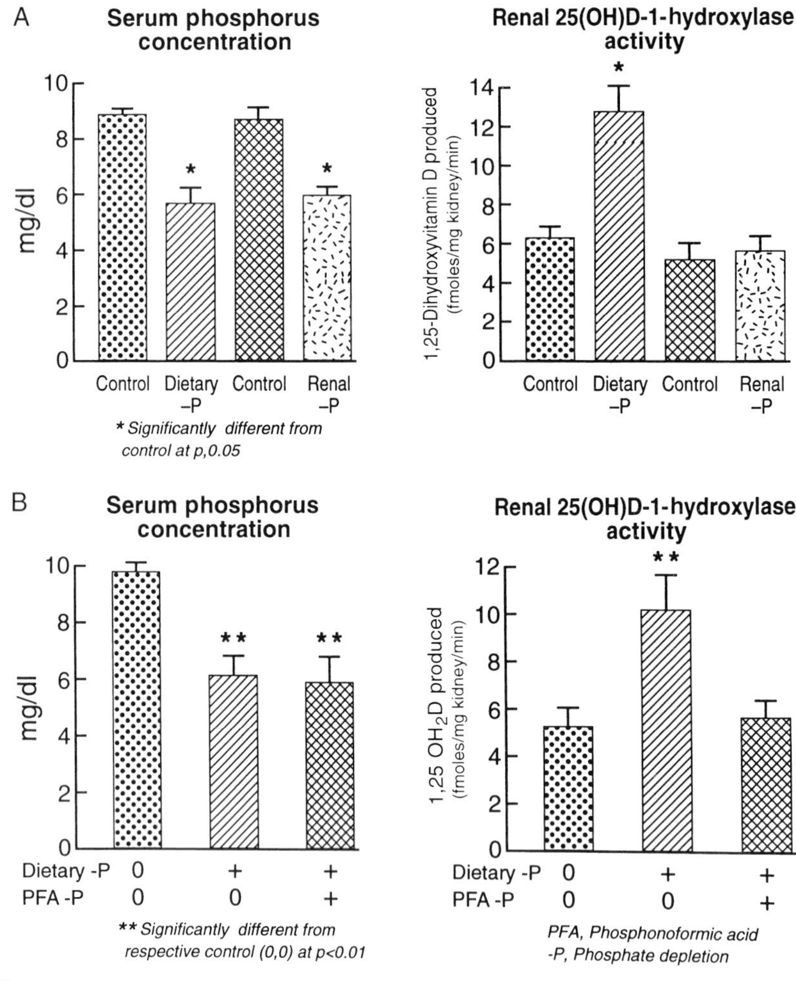

FIGURE 4 Although hypophosphatemia resulting from phosphate restriction (−P) increases renal 25(OH)D 1α-hydroxylase activity, hypophosphatemia of similar magnitude secondary to phosphonoformic acid (PFA)-mediated renal phosphate wasting has no effect on enzyme function. These data suggest that hypophosphatemia is not the proximal trigger mechanism for $1,25(OH)_2D$ production. (B) In similar experiments the effects of phosphate depletion and consequent hypophosphatemia on renal 25OHD 1α-hydroxylase activity are blocked by the coincident administration of phosphonoformic acid, which prevents the expected compensatory change in renal phosphate flux attendant on the depletion. These observations reaffirm the potentially important role of renal phosphate transport in modulating the effects of phosphate on $1,25(OH)_2D$ production.

deed, rational choice of an appropriate treatment for many of these disorders depends on determining the precise cause for the abnormality. In general, defects in phosphate homeostasis result from impaired renal tubular phosphate reabsorption and a consequent change in TmP/GFR, an altered phosphate load due to varied intake, or abnormal gastrointestinal absorption or translocation of phosphorus between the extracellular fluid and tissues (Table I).

The disorders of renal phosphate transport are the most common of these diseases and have been intensively studied. Indeed, investigation of these conditions has provided new insight into the reciprocal regulation of phosphate homeostasis and vitamin D metabolism. In the remainder of this chapter, several clinical states that represent disorders of phosphate homeostasis will be discussed and the potential role of the vitamin D endocrine system in their pathogenesis and phenotypic presentation highlighted.

TABLE I Disorders of Phosphate Homeostasis

Abnormal renal phosphate transport
Hypophosphatemic syndromes
X-Linked hypophosphatemia
Tumor-induced osteomalacia
Hereditary hypophosphatemic rickets with hypercalciuria
Fanconi's syndrome, type I
Familial idiopathic
Cystinosis (Lignac-Fanconi disease)
Lowe's syndrome
Glycogen storage disease
Wilson's disease
Galactosemia
Tyrosinemia
Hereditary fructose intolerance
Fanconi's syndrome, type II
Autosomal dominant hypophosphatemic rickets
Hyperphosphatemic syndromes
Tumor calcinosis
Altered phosphate load
Hypophosphatemic syndromes
Decreased phosphate availability
Phosphate deprivation
Gastrointestinal malabsorption
Transcellular shift of phosphate
Alkalosis
Glucose administration
Combined mechanisms
Alcoholism
Burns
Nutritional recovery syndrome
Diabetic ketoacidosis
Hyperphosphatemic syndromes
Vitamin D intoxication
Rhabdomyolysis
Cytotoxic therapy
Malignant hyperthermia

II. DISORDERS OF RENAL PHOSPHATE TRANSPORT

A. X-Linked Hypophosphatemia (XLH)

X-Linked hypophosphatemia (XLH) is the prototypic renal phosphate wasting disorder, characterized in general by progressively severe skeletal abnormalities and growth retardation. The syndrome occurs as an X-linked dominant disorder with complete penetrance of a renal tubular abnormality resulting in phosphate wasting and consequent hypophosphatemia (Table II) [17]. Clinical expression of the disease is widely variable, ranging from a mild abnormality, the apparent isolated occurrence of hypophosphatemia, to severe rickets and/or osteomalacia [18]. In children, the most common clinically evident manifestations include short stature and limb deformities. This height deficiency is more evident in the lower extremities, as they represent the fastest growing segment before puberty. In contrast, upper segment growth is generally less affected. The majority of children with the disease exhibit enlargement of the wrists and/or knees secondary to rickets as well as bowing of the lower extremities. Additional signs of the disease may include late dentition, tooth abscesses secondary to poor mineralization of the interglobular dentine, and premature cranial synostosis. Many of these features do not become apparent until age 6 to 12 months or older [19]. In spite of marked variability in the clinical presentation, bone biopsies in affected children and adults invariably reveal low turnover osteomalacia without osteopenia. The severity of the bone disorder has no relationship to sex, the extent of the biochemical abnormalities, or the severity of the clinical disability.

In untreated youths and adults with XLH, serum 25OHD levels are normal, and the concentration of $1,25(OH)_2D$ is in the low–normal range [20–22]. The paradoxical occurrence of hypophosphatemia and normal serum calcitriol levels is due to aberrant regulation of renal 25OHD 1α-hydroxylase activity due most likely to the abnormal phosphate transport. Indeed, studies in *hyp* and *gy* mice, the murine homologs of the human disease, have established that defective regulation is confined to the enzyme localized in the proximal convoluted tubule, the site of the abnormal phosphate transport [23–26].

TABLE II Abnormalities of Phosphate Homeostasis and Vitamin D Metabolism in Disorders of Renal Phosphate Transport[a]

	XLH	TIO	HHRH	FS I	FS II	ADHR	TC
Phosphate homeostasis							
Serum P_i	↓	↓	↓	↓	↓	↓	↑
TmP/GFR	↓	↓	↓	↓	↓	↓	↑
Gastrointestinal P_i absorption	↓	↓	↑	↓	↑	↓	↑
Vitamin D metabolism							
Serum 25OHD	N	N	N	N	N	N	N
Serum $1,25(OH)_2D$	(↓)	↓	↑	(↓)	↑	(↓)	↑

[a] XLH, X-Linked hypophosphatemia; TIO, tumor-induced osteomalacia; HHRH, hereditary hypophosphatemic rickets with hypercalciuria; FS I, Fanconi's syndrome, type I; FS II, Fanconi's syndrome, type II; ADHR, autosomal dominant hypophosphatemic rickets; TC, tumoral calcinosis. The effect of each abnormality is indicated by symbols: N, normal; ↓, decreased; ↑, increased; (↓), decreased relative to the serum phosphorus concentration. Modified from Econs *et al.* [17].

1. PATHOPHYSIOLOGY

Investigators generally agree that the primary inborn error in XLH results in an expressed abnormality of the renal proximal tubule that impairs P_i reabsorption. This defect has been indirectly identified in affected patients and directly demonstrated in the brush border membranes of the proximal nephron in *hyp* mice. Until relatively recently, whether this renal abnormality is primary or secondary to the elaboration of a humoral factor has been controversial. In this regard, demonstration that renal tubule cells from *hyp* mice maintained in primary culture exhibit a persistent defect in renal P_i transport [27,28], likely due to decreased expression of the Na^+–phosphate cotransporter (NPT2) mRNA and immunoreactive protein [29–31], supported the presence of a primary renal abnormality. In contrast, transfer of the defect in renal P_i transport to normal and/or parathyroidectomized normal mice parabiosed to *hyp* mice implicated a humoral factor in the pathogenesis of the disease [32,33]. Other studies, however, have provided compelling evidence that the defect in renal P_i transport in XLH is secondary to the effects of a circulating hormone or metabolic factor. In this regard, immortalized cell cultures from the renal tubules of *hyp* and *gy* mice exhibit normal Na^+–phosphate transport, suggesting that the paradoxical effects observed in primary cultures may represent the effects of impressed memory and not an intrinsic abnormality [34,35]. Moreover, the report that cross-transplantation of kidneys in normal and *hyp* mice results in neither transfer of the mutant phenotype nor its correction, unequivocally established the humoral basis for XLH [36]. Subsequent efforts, which resulted in localization of the gene encoding the Na^+–phosphate cotransporter to chromosome 5, further substantiated the conclusion that the renal defect in brush border membrane phosphate transport is not intrinsic to the kidney in XLH [37].

Although these data establish the presence of a humoral abnormality in XLH, the identity of the putative factor and the spectrum of its activity have not been definitively elucidated. However, several investigators have identified the presence of and partially characterized phosphaturic factors (inhibitors of Na^+-dependent phosphate transport) in patients with tumor-induced osteomalacia [38,39] (see below) and in patients with end-stage renal disease [40]. Whether any one of these factors is increased in patients with XLH remains unknown. Regardless, preliminary reports suggest production of a phosphaturic factor by *hyp* mouse osteoblasts, marrow mesenchymal cells, and hepatocytes maintained in culture. These studies argue that a circulating factor, phosphatonin, may play an important role in the pathophysiological cascade responsible for XLH.

2. GENETIC DEFECT

Efforts to better understand XLH have included attempts to identify the genetic defect underlying this disease. In 1986 Read *et al.* [41] and Machler *et al.* [42] reported linkage of the DNA probes DXS41 and DXS43, which had been previously mapped to Xp22.31-p21.3, to the *HYP* gene locus. In subsequent studies Thakker *et al.* [43] and Albersten *et al.* [44] reported linkage to the *HYP* locus of additional polymorphic DNA, DXS197, and DXS207 and, using multipoint mapping techniques, determined the most likely order of the markers as Xpter-DXS85-(DXS43/DXS197)-*HYP*-DXS41-Xcen and Xpter-DXS43-*HYP*-(DXS207/DXS41)-Xcen, respectively. The relatively small number of informative pedigrees available for these studies prevented definitive determination of the genetic map along the Xp22-p21 region of the X chromosome and only allowed identification of flanking markers for the *HYP* locus 20 centimorgans (cM) apart. More recently, the independent and collaborative efforts of the HYP consortium resulted in the study of some 13 multigenerational pedigrees and consequent refined mapping of the Xp22.1-p21 region of the X chromosome, identification of tightly linked flanking markers for the *HYP* locus, construction of a yeast artificial chromosome (YAC) contig spanning the *HYP* gene region, and eventual cloning and identification of the disease gene as PEX, a phosphate-regulating gene with homologies to endopeptidases located on the X chromosome. In brief, these studies ascertained a locus order on Xp22.1 of

Xcen-DXS451-(DXS41/DXS92)-DXS274-DXS1052-DXS1683-*HYP*-DXS7474-DXS365-(DXS443/DXS3424)-DXS257-(GLR/DXS43)-DXS315-Xtel.

Moreover, the physical distance between the flanking markers, DXS1683 and DXS7474, was determined as 350 kb and their location on a single YAC ascertained. Subsequently, a cosmid contig spanning the *HYP* gene region was constructed and efforts directed at discovery of deletions within the *HYP* region. Identification of several such deletions permitted characterization of cDNA clones that mapped to cosmid fragments in the vicinity of the deletions. Database searches with these cDNAs detected homologies at the peptide level to a family of endopeptidase genes that includes neutral endopeptidase (NEP), endothelin-converting enzyme-1 (ECE-1), and the Kell antigen. These efforts clearly established PEX as the candidate gene responsible for XLH [45–50].

Subsequent studies unequivocally indicate that deactivating mutations of the PEX gene underlie the phenotypic expression of XLH. In this regard, extensive mutational analysis of 90 families with X-linked hypophosphatemia reveal a range of defects in the PEX gene that include the following: nonsense mutations, frameshifts, missense, splice donor, and acceptor mutations, insertions, and deletions. Moreover, the mutations involve almost the entire length of the gene and in one-third of cases represent deletions of more than 104 bp, consistent only with diminished activity of the gene product. Unfortunately, the precise role that this gene and its product play in the regulation of phosphate homeostasis and the pathogenesis of the disease remains unknown.

Nevertheless, several groups have reported seminal observations that have begun to better define a potential function of the gene in XLH. In this regard, preliminary studies have identified bone (specifically osteoblasts), lung, and liver as potential sites of gene expression [51,52]. Such localization indicates that abnormalities of bone in this disease, as well as production of the putative phosphatonin, may be linked to the presence of PEX in bone and bone and liver, respectively. Further studies to define the regulation of gene expression will no doubt provide essential information and appropriate model systems for continued investigation of the PEX gene function.

3. PATHOGENESIS

In spite of the remarkable advances that have been made in understanding the genetic abnormality and pathophysiology of XLH, the detailed pathogenetic mechanism underlying this disease remains unknown. At present it is tempting to speculate that the PEX gene product acts directly or indirectly on a phosphaturic factor that regulates renal phosphate handling. Given the available data, the PEX gene product, which is a putative cell membrane-bound enzyme, may function normally to inactivate "phosphatonin," a presumed phosphaturic hormone. Indeed, precedent for such activity exists, as neutral endopeptidases inactivate enkephalins and atrial natriuretic peptide [53,54]. However, the data from parabiotic studies of normal and *hyp* mice argue strongly that extracellular degradation of the phosphaturic factor does not occur. Indeed, such activity would preclude transfer of the *hyp* mouse phenotype to parabiosed normals. Alternatively, the PEX gene product may function intracellularly to inactivate phosphatonin. In this regard, Jalal *et al.* [55] reported the internalization of neutral endopeptidase and a potential role for this enzyme in intracellular metabolism. Less likely, the PEX gene product may enzymatically activate a protein that suppresses production of phosphatonin. Although this is consistent with all previous data, it is a complex process and requires production of PEX, phosphatonin, and the suppressor protein in the same cell in order to accommodate the data from the parabiotic studies. Nevertheless, in accord with this possibility, Mari *et al.* [56] have reported that the neutral endopeptidase on human T cells may be involved in the production of lymphokines through the processing of an activating factor at the surface of the lymphocyte.

In any of these cases, a defect in the PEX gene will result in overproduction and circulation of phosphatonin and consequent inhibition of renal Na^+–phosphate transport, the likely scenario in the pathogenesis of XLH. Although such overproduction of phosphatonin is a favored hypothesis based on available data, including identification and isolation of such factors in patients with tumor-induced osteomalacia and in cell cultures from *hyp* mice, it is possible that XLH results from the inability of mutant PEX to activate a phosphate-conserving hormone. However, the only known phosphate-conserving hormone, stanniocalcin, is synthesized in active form within the kidney and has little known bioactivity in humans. These features strongly mitigate against a role for stanniocalcin in the pathogenesis of XLH.

The coexistence of osteoblast defects in X-linked hypophosphatemia further confounds understanding the pathophysiology of this disorder. Elegant experiments that documented the abnormal mineralization of periostea and osteoblasts of *hyp* mice following transplantation into the muscle of normal mice provide clear evidence for intrinsic defects in the bone of mutants [57]. Indeed, proof of specific osteoblast abnormalities has been provided by studies that show decreased phosphorylation of osteopontin and increased osteocalcin levels in cells from *hyp* mice [58]. On the basis of these observations, it is tempting to speculate that coordinate PEX expression and phosphatonin production in osteoblasts may impart innate functional abnormalities to these cells in X-linked hypophosphatemia [59,60]. In fact, the phosphatonin, likely processed by the PEX gene product, may have multiple activities beyond regulation of renal phosphate transport, which may include modulation of vitamin D metabolism and osteoid mineralization, defects that characterize XLH. In concert with this possibility Siddhanti and Quarles [59] have provided evidence that PEX expression correlates temporally with osteoblast-mediated mineralization *in vitro*.

In any case, it is evident that further information is requisite to enhance our understanding of the pathogenesis of XLH and, in turn, regulation of phosphate homeostasis, osteoblast function, vitamin D metabolism, and osteoid mineralization. Such data are not only critical to understanding the pathogenesis of XLH and the regulation of mineral homeostasis, but they may have significant impact on determination of optimal treatment strategies for many of the vitamin D-resistant diseases. Further studies may also reveal that the PEX gene product plays a central role in regulation of phosphate homeostasis. In this regard, alteration of PEX gene transcription may occur in response to a phosphate or vitamin D signal and mediate the changes in renal phosphate reabsorption that occur secondary to changes in phosphate availability such as diminished gastrointestinal absorption.

4. TREATMENT

In past years, physicians employed pharmacological doses of vitamin D as the cornerstone for treatment of XLH. However, long-term observations indicate that this therapy fails to cure the disease and poses the serious problem of recurrent vitamin D intoxication and renal damage. More recently, choice of therapy for this disease has been remarkably influenced by the increased understanding of the pathophysiological factors that affect phenotypic expression of the disorder. Thus, current treatment strategies for children directly address the combined calcitriol and phosphorus deficiency characteristic of the disease. Generally, the regimen includes a period of titration to achieve a maximum dose of calcitriol (1–3 μg/day in two divided doses) and phosphorus (1–4 g/day in four to five divided doses) [61,62]. Such combined therapy often improves growth velocity, normalizes lower extremity deformities, and induces healing of the attendant bone disease. Refractoriness to the growth-promoting effects of treatment, however, is often encountered, particularly in youths presenting at or below the fifth percentile in height [63]. For that reason the use of recombinant growth hormone as an additional treatment component has been advocated. Definite positive effects have been observed in young patients with XLH [64].

Of course, treatment involves a significant risk of toxicity that is generally expressed as abnormalities of calcium homeostasis, most notably secondary hyperparathyroidism that may become autonomous and require surgery. Detrimental effects on renal function secondary to abnormalities such as nephrocalcinosis are also possible. Thus, frequent monitoring of the phosphate and calcitriol dosage during growth is mandatory. Therapy in adults is reserved for episodes of intractable bone pain and refractory nonunion bone fractures.

B. Tumor-Induced Osteomalacia (TIO)

Since 1947 there have been reports of approximately 102 patients in whom rickets and/or osteomalacia has been induced by various types of tumors [65]. With time the name of the syndrome has varied and included oncogenic osteomalacia and tumor-induced osteomalacia. In at least 54 cases a tumor has been clearly documented as causing the rickets/osteomalacia, since the metabolic disturbances improved or completely disappeared on removal of the tumor. In the remainder of cases, patients had inoperable lesions (and investigators could not determine the effects of tumor removal on the syndrome), or surgery did not result in complete resolution of the evident abnormalities during the period of observation.

Affected patients generally present with bone and muscle pain, muscle weakness, rickets/osteomalacia, and occasionally recurrent fractures of the long bones. Additional symptoms common to younger patients are fatigue, gait disturbances, slow growth, and bowing of the lower extremities. Biochemistries include hypophosphatemia secondary to renal phosphate wasting and to a lesser extent gastrointestinal malabsorption of phosphate. The resultant negative phosphorus balance is associated with normal serum levels of calcium and 25OHD. In contrast, the serum $1,25(OH)_2D$ level is overtly low in 19 of 23 patients in whom measurements have been made, despite the presence of hypophosphatemia (Table II). Aminoaciduria, most frequently glycinuria, and glucosuria are occasionally present. Radiographic abnormalities manifest as generalized osteopenia, pseudofractures, and coarsened trabeculae, as well as widened epiphyseal plates in children. The duration of symptoms before diagnosis ranges from 2.5 months to 19 years with an average exceeding 2.5 years. The age at diagnosis is generally the sixth decade, with a range of 7–74 years. Approximately 18% of the patients are younger than 20 years at presentation.

The large majority of patients with TIO harbor tumors of mesenchymal origin that include primitive-appearing, mixed connective tissue lesions, osteoblastomas, nonossifying fibromas, and ossifying fibromas. However, the frequent occurrence of Looser zones in the radiographs of moribund patients with carcinomas of epidermal and endodermal derivation indicates that the disease may be secondary to a variety of tumor types. Indeed, the observations of tumor-induced osteomalacia concurrent with breast carcinoma, prostate carcinoma [66–68], oat cell carcinoma [69], small cell carcinoma [70], multiple myeloma, and chronic lymphocytic leukemia [71] support this conclusion. In addition, the occurrence of osteomalacia in patients with widespread fibrous dysplasia of bone, neurofibromatosis, and linear nevus sebaceous syndrome could also be tumor induced. Although proof of a causal relationship in these disorders has been precluded in general by an inability to surgically excise the multiplicity of lesions, in one case of fibrous dysplasia removal of virtually all of the abnormal bone did result in appropriate biochemical and radiographic improvement [72].

Regardless of the tumor cell type, the lesions that cause the syndrome are often small, difficult to locate, and present in obscure areas that include the nasopharynx, a sinus, the popliteal region, and the suprapatellar area. In any case, a careful and thorough examination is necessary to document/exclude the presence of such a tumor.

1. PATHOPHYSIOLOGY

The relatively infrequent occurrence of the disorder has confounded attempts to determine the pathophysiological basis for TIO. Nevertheless, most investigators agree that tumor production of a humoral factor(s) that may affect multiple functions of the proximal renal tubule, particularly phosphate reabsorption, is the probable pathogenesis of the syndrome. This possibility is supported by (1) the presence of phosphaturic activity in tumor extracts from 3 of 4 patients with TIO [73–75]; (2) the absence of parathyroid hormone and calcitonin from these extracts and the apparent cyclic AMP-independent action of the extracts; (3) the occurrence of hypophosphatemia and increased urinary phosphate excretion in heterotransplanted tumor-bearing athymic nude mice [76]; (4) the demonstration that extracts of the heterotransplanted tumor inhibit renal 25-hydroxyvitamin D 1α-hydroxylase activity in cultured murine kidney cells [76]; and (5) the coincidence of aminoaciduria and glycosuria with renal phosphate wasting in some affected subjects, indicative of complex alterations in proximal renal tubular function [77].

Indeed, partial purification of "phosphatonin" from a cell culture derived from a hemangiosceroma causing tumor-induced osteomalacia has reaffirmed this possibility [38]. These studies reveal that the putative phosphatonin is a peptide of 8–25 kDa which does not alter glucose or alanine transport but inhibits sodium-dependent phosphate transport in a cyclic AMP-independent fashion. Moreover, the activity of the phosphatonin is not blocked by a PTH receptor antagonist. However, other studies that document the presence in various disease states of additional phosphate transport inhibitors indicate that the tumor-induced osteomalacia syndrome is heterogeneous and that "phosphatonin" may be a family of hormones or metabolic factors. In fact, the complexity of the syndrome may reach greater proportions because additional observations indicate that some mesenchymal tumors from affected subjects do not secrete phosphaturic factors into the culture medium (M. K. Drezner, T. Nesbitt, unpublished observations). Thus, the pathogenesis of the disorder may be more complicated than is currently appreciated.

In any case, abnormal vitamin D metabolism is a complementary factor that likely contributes to the pathogenesis of the TIO. Several observations support this possibility including (1) the decreased circulating $1,25(OH)_2D$ level observed in patients who manifest the characteristic syndrome; (2) rapid normalization of the serum calcitriol concentration after surgical removal of the coincident tumor and in association with resolution of the biochemical abnormalities of the syndrome; (3) diminished renal 25OHD 1α-hydroxylase activity in heterotransplanted tumor-bearing athymic nude mice and in kidney cell cultures exposed to tumor extracts; and (4) the obligatory dependence on calcitriol, in combination with phosphorus, to effect healing of the osteomalacia in patients who are medically managed. Whether the abnormal vitamin D metabolism is solely the result of the aberrant renal phosphate transport or is independently caused by the activity of the tumor-produced circulating factor remains unknown. Indeed, an innate heterogeneity of the syndrome cannot be excluded. In such a case, the defective regulation of 25OHD 1α-hydroxylase may variably result from abnormal phosphate transport confined to a specific locus in the renal tubule (perhaps the S_1 segment), the singular effects of the circulating factor, or the combined actions of the circulating factor and deranged phosphate transport. Regardless, an interplay between abnormal phosphate homeostasis and defective vitamin D metabolism likely collectively contributes to the phenotypic expression of the disorder in the majority of the patients.

In contrast to these observations, patients with TIO secondary to hematogenous malignancy manifest abnormalities of the syndrome due to a distinctly different mechanism. In these subjects the nephropathy induced with light chain proteinuria or other immunoglobulin derivatives results in the decreased renal tubular reabsorption of phosphate characteristic of the disease. To date at least 15 patients have been reported who potentially manifest this form of the disorder [71]. In many instances, however, the diagnosis of tumor-induced osteomalacia was not considered. Nevertheless, at least in some cases of this syndrome, renal tubular damage may be mediated by tissue deposition of light chains or of some other immunoglobulin derivative with similar toxic effects on the kidney. Thus, light-chain nephropathy must be considered a possible mechanism for the TIO syndrome.

2. Pathogenesis

The tumor-induced osteomalacia syndrome has all of the classic biochemical and radiological criteria of the hypophosphatemic osteomalacias. Although it is certain that in the majority of cases expression of the disease is due to ectopic production of a hormonal factor, the identity of the normally produced hormone, its tissue of origin, and the chromosomal site coding its synthesis remain unknown. Identification of the primary genetic defect underlying XLH surprisingly shed little light on the nature of the ectopic hormone. In this regard, the similarity between this genetic disease and TIO dictated a common cause for the disorders. However, recognition that the PEX gene likely codes for a membrane-bound neutral endopeptidase makes this unlikely. Rather, it appears that the putative phosphatonin, likely inactivated by PEX under normal conditions, may be the hormone responsible for TIO. If so, normal production of this factor occurs in a variety of cells including osteoblasts, marrow osteogenic stromal cells, and hepatocytes. Of course, the unbridled production of this factor in a genetic disease, such as autosomal dominant hypophosphatemia, is possible and, under such circumstances, localizes the chromosomal site coding for the factor. Further studies will be required to discern this possibility.

As mentioned above, the full spectrum of the ectopic hormone activity in TIO remains ill defined. Thus, a potential role for the factor in modulating vitamin D metabolism and directing osteoid mineralization is certainly possible, in addition to the well-known function regarding regulation of renal phosphate transport. Whatever the spectrum of activity, however, studies by several groups further confound complete understanding of the pathogenesis of TIO. In this regard, examination of tumors from affected patients reflect high level expression of the PEX gene and its product (M. K. Drezner, M. Econs, unpublished observations). Whether this represents a compensatory homeostatic response to the prevailing hypophosphatemia has not been determined. However, the production of phosphatonin in cells that harbor the PEX gene under normal conditions suggests that ectopic phosphatonin production in TIO is in some fashion dependent on the presence of the PEX gene product. Further studies in this regard will produce important information regarding the pathophysiology of TIO.

3. Treatment

The first and foremost treatment of TIO is complete resection of the tumor. However, recurrence of mesenchymal tumors, such as giant cell tumors of bone, or inability to resect completely certain malignancies, such as prostatic carcinoma, has resulted in development of alternative therapeutic intervention for the syndrome. Administration of $1,25(OH)_2D_3$ alone or in combination with phosphorus supplementation has served as effective therapy for TIO [18,77]. Doses of calcitrol required range from 1.5 to 3.0 μg/day, whereas those of phosphorus are 2–4 g/day. Although little information is available regarding the long-term consequences of such treatment, the high doses of medicine required raise the possibility that nephrolithiasis, nephrocalcinosis, and hypercalcemia may frequently complicate the therapeutic course. Indeed, hypercalcemia secondary to parathyroid hyperfunction has been documented in at least five treated subjects. All of these patients received phosphorus as part of a combination regimen, which may have stimulated PTH secretion and led to parathyroid autonomy. Thus, careful assessment of parathyroid function, serum and urinary calcium levels, and renal function are essential to ensure safe and efficacious therapy.

C. Hereditary Hypophosphatemic Rickets with Hypercalciuria (HHRH)

This rare genetic disease is marked by hypophosphatemic rickets with hypercalciuria [78]. The cardinal biochemical features of hereditary hypophosphatemic rickets with hypercalciuria (HHRH) include hypophosphatemia due to increased renal phosphate clearance and normocalcemia. In contrast to other diseases in which renal phosphate transport is limited, patients with HHRH exhibit increased $1,25(OH)_2D_3$ production (Table II). The resultant elevated serum calcitriol levels enhance gastrointestinal calcium absorption, which in turn increases the filtered renal calcium load and inhibits parathyroid secretion. These events cause the characteristic hypercalciuria observed in affected patients.

The clinical expression of the disease is heterogeneous. In general, children become symptomatic between the ages of 6 months and 7 years. Initial symptoms consist of bone pain or deformities of the lower limbs (or both), which progressively interfere with gait and physical activity. The bone deformities vary from genu vara or genu valga to anterior external bowing of the femur and coxa vara. Additional features at presentation include short stature with disproportionately short lower limbs, muscle weakness, and radiological signs of rickets or osteomalacia (or both). These various symptoms and signs may exist separately or in combination and may be present in a mild or severe form.

A large number of apparently unaffected relatives of patients with HHRH exhibit an additional mode of disease expression. These subjects, although without evidence of bone disease, manifest hypercalciuria, most evident in postprandial periods, as well as a pattern of biochemical abnormalities similar to those of children with rickets and osteomalacia. Quantitatively, however, the abnormalities are milder, and the relevant biochemical values are intermediate between those observed in family members with HHRH and those in normal relatives. The absence of bone disease in these patients may be explained by relatively mild hypophosphatemia compared to the severe phosphate depletion evidenced in patients with the full spectrum of the disorder [79].

To date three unrelated kindreds with HHRH have been described: an extended family of Bedouin origin that includes 13 patients with HHRH and 42 with hypercalciuria; a smaller kindred of oriental Jewish origin with 5 affected members; and a family of Yemenite Jewish origin that includes 2 patients with HHRH and 2 with hypercalciuria [80]. However, a phenotypically similar disorder, namely, childhood idiopathic hypercalciuria with bone lesions (rickets) and stunted linear growth, has been independently recognized. The similarity of this syndrome to HHRH suggests that they may be one and the same pathological condition. Moreover, several patients with a sporadic occurrence of HHRH have been recognized. Studies are generally incomplete, however, and the presence of hypercalciuria in relatives has not been excluded.

1. Pathophysiology

Liberman and associates [78–81] have presented data which indicate that the primary inborn error underlying HHRH is an expressed abnormality in the renal proximal tubule which impairs phosphate reabsorption. They propose that this pivotal defect in turn stimulates renal 25(OH)D 1α-hydroxylase, thus promoting the production of $1,25(OH)_2D$ and increasing its serum and tissue levels. As a result intestinal calcium and phosphorus absorption is augmented and the renal filtered calcium load consequently increased. The enhanced intestinal calcium absorption also suppresses parathyroid function. In addition, prolonged hypophosphatemia diminishes osteoid mineralization, resulting in rickets and/or osteomalacia.

The proposal that abnormal phosphate transport results in increased calcitriol production remains untested. Indeed, the elevation of $1,25(OH)_2D$ in patients with HHRH is a unique phenotypic manifestation of the disease that distinguishes it from other disorders in which abnormal phosphate transport is likewise manifest (see Table II). However, Davidai *et al.* [82] observed that a murine homolog of the hypophosphatemic disorders, the *gy* mouse, exhibits elevated serum $1,25(OH)_2D$ levels similar to patients with HHRH. In this animal model renal 25(OH)D 1α-hydroxylase is normally responsive to hormonal/metabolic modulation. Moreover, in the baseline state the *gy* mice exhibit increased enzyme activity secondary to normally maintained phosphate-regulated function (an absent/mutated action in *hyp* mice; see Section IIA). These observations provide evidence that abnormal phosphate transport may be associated with biochemically diverse disease. Most likely, such heterogeneity results from disease at variable anatomical sites along the proximal convoluted tubule that uniformly impairs phosphate transport but not 25OHD 1α-hydroxylase activity.

2. Genetic Defect

Although X-linked transmission of the disease has been ruled out, the mode of genetic transmission for HHRH/hypercalciuria remains uncertain. Studies to date indicate that the disease(s) may be passed by an autosomal codominant mechanism with partial expression occurring in heterozygous subjects. In this case, as with benign familial hypercalcemia, a gene dosage effect is evident and thereby differentiates the phenotypic expression of disease in homozygous and heterozygous individuals. Alternatively, an autosomal recessive mechanism acting through the additive effects of two genes, each having two alleles (one normal and one defective), would explain the genetic passage of the disease. In such a case full expression of HHRH requires defective alleles at both genetic loci, whereas hypercalciuria occurs with a single homozygous defect. Although this mechanism may be operative in inbred families, the documentation of disease in families without inbreeding makes such an inheritance pattern unlikely.

3. Treatment

In accord with the hypothesis that a singular defect in renal phosphate transport underlies HHRH, patients have been treated successfully with high dose phosphorus (1–2.5 g/day in five divided doses) alone. Within several weeks after initiation of therapy, bone pain disappears, and muscular strength improves substantially. Moreover, the majority of treated subjects exhibit accelerated linear growth, and radiological signs of rickets are completely absent within 4–9 months. Concordantly serum phosphorus values increase toward normal, the $1{,}25(OH)_2D$ concentration decreases, and alkaline phosphatase activity declines to normal. Despite this favorable response, limited investigation indicates that the osteomalacia component of the bone disease does not exhibit normalization. Further studies will be necessary, therefore, to determine if phosphorus alone will be sufficient treatment for this rare disorder.

In any case, use of phosphorus in patients with this disease does not result in the same spectrum of complications encountered with its use in other disorders. Most notably, nephrocalcinosis, a common complication in treated patients with XLH, occurs infrequently in subjects with HHRH. In fact, the rare occurrence of this complication is associated with a history of vitamin D intoxication prior to initiation of treatment with phosphorus. Similarly, the development of secondary hyperparathyroidism in treated patients with HHRH has not been reported, although expectation of this complication is high because oral administration of phosphate does diminish the circulating calcium concentration and, in turn, stimulates parathyroid function.

D. Fanconi's Syndrome (FS)

Rickets and osteomalacia are frequently associated with Fanconi's syndrome (FS), a disorder characterized by hyperphosphaturia and consequent hypophosphatemia, hyperaminoaciduria, renal glycosuria, albuminuria, and proximal renal tubular acidosis [83–86]. Damage to the renal proximal tubule, seconary to genetic disease or environmental toxins, represents the common underlying mechanism of this disease. Resultant dysfunction causes renal wasting of those substances primarily reabsorbed at the proximal tubule. The associated bone disease in this disorder is likely secondary to hypophosphatemia and/or acidosis, abnormalities which occur in association with aberrantly (Fanconi's syndrome, type I) or normally regulated (Fanconi's syndrome, type II) vitamin D metabolism.

1. Fanconi's Syndrome, Type I

The type I disease resembles in many respects the more common genetic disease XLH (Table II). In this regard, occurrence of abnormal bone mineralization appears dependent on the prevailing renal phosphate wasting and resultant hypophosphatemia. Indeed, disease subtypes in which isolated wasting of amino acids, glucose, or potassium occur are not asociated with rickets and/or osteomalacia. Further, in the majority of patients studied, affected subjects exhibit abnormal vitamin D metabolism, characterized by serum $1{,}25(OH)_2D$ levels that are overtly decreased or abnormally low relative to the prevailing serum phosphorus concentration [87]. Although the aberrantly regulated calcitriol biosynthesis may be due to the abnormal renal phosphate transport, as suspected in patients with XLH, proximal tubule damage and acidosis may play important roles.

A notable difference between this syndrome and XLH is a prevailing acidosis, which may contribute to the bone disease. In this regard, several studies indicate that acidosis may exert multiple deleterious effects on bone. Such negative sequelae may be related to the loss of bone calcium that occurs secondary to calcium release for use in buffering. Alternatively, several investigators have reported that acidosis may impair bone mineralization secondary to direct inhibition of renal 25OHD 1α-hydroxylase activity. Others dispute these findings and claim that acidosis does not cause rickets or osteomalacia in the absence of hypophosphatemia. Indeed, hypophosphatemia and abnormally regulated vitamin D metabolism are the most likely primary factors underlying rickets and osteomalacia in this form of the disease.

2. Fanconi's Syndrome, Type II

Tieder *et al.* [88] have described two siblings (from a consanguineous mating) who presented with classic characteristics of Fanconi's syndrome, including renal phosphate wasting, glycosuria, generalized aminoaciduria and increased urinary uric acid excretion. However, these patients had appropriately elevated (relative to the decreased serum phosphorus concentration) serum $1{,}25(OH)_2D$ levels and consequent hypercalciuria (Table II). Moreover, treatment with phosphate reduced the serum calcitriol in these patients into the normal range and normalized the urinary calcium excretion. In many regards, this syndrome resembles HHRH and represents a variant of Fanconi's syndrome, referred to as type II disease. The bone disease in affected subjects is likely due to the effects of hypophosphatemia. In any case, the existence of this variant form of disease is probably the result of renal damage to a unique segment of the proximal tubule. Further studies will be necessary to confirm this possibility.

3. TREATMENT

Ideal treatment of the bone disease in FS is correction of the pathophysiological defect influencing proximal renal tubular function. In many cases, however, the primary abnormality remains unknown. Moreover, efforts to decrease tissue levels of causal toxic metabolites by dietary means (such as in fructose intolerance) or pharmacological means (such as in cystinosis and Wilson's syndrome) have met with variable success. Indeed, no evidence exists that indicates if the proximal tubule damage is reversible on relief of an acute toxicity. Thus, for the most part therapy of this disorder must be directed at raising the serum phosphorus concentration, replacing calcitriol (in type I disease), and reversing an associated acidosis. However, use of phosphorus and calcitriol in FS has been limited. In general, such replacement therapy leads to substantial improvement or resolution of the bone disease [89]. Unfortunately, growth and developmental abnormalities, more likely associated with the underlying genetic or acquired disease, remain substantially impaired. More efficacious therapy, therefore, is dependent on future research into the causes of the multiple disorders that underlie this syndrome.

E. Autosomal Dominant Hypophosphatemic Rickets (ADHR)

Several studies have documented an autosomal dominant inheritance of a hypophosphatemic disorder similar to XLH [90]. The phenotypic manifestations of the disorder, which is called autosomal dominant hypophosphatemic rickets (ADHR), include lower extremity deformities and rickets/osteomalacia. Indeed, affected patients display biochemical and radiographic abnormalities indistinguishable from those of individuals with XLH. These include hypophosphatemia secondary to renal phosphate wasting and normal levels of PTH and 25OHD, as well as inappropriately normal (relative to the serum phosphorus concentration) $1,25(OH)_2D$ (see Table II). However, long-term studies indicate that a few of the affected female patients exhibit delayed penetrance of clinically apparent disease and an increased tendency for bone fracture, uncommon occurrences in XLH. These individuals present in the second through the fourth decade with weakness and bone pain. Limited information is available regarding other aspects of the disease. However, recent studies have localized the disease gene to chromosome 12p.

An apparent *forme fruste* of this disease, (autosomal dominant) hypophosphatemic bone disease, has many of the characteristics of XLH and ADHR, but reports indicate that affected children with autosomal dominant transmission of the disease display no evidence of rickets [91,92]. As this syndrome is described in only a few small kindreds and radiographically evident rickets is not universal in children with familial hypophosphatemia, these families may have ADHR. Further observations will be necessary to discriminate this possibility.

F. Tumoral Calcinosis (TC)

Tumoral calcinosis (TC) is a rare genetic disease characterized by periarticular cystic and solid tumorous calcifications. Biochemical markers of the disorder include hyperphosphatemia and an elevated serum $1,25(OH)_2D$ concentration (see Table II). Using these criteria, evidence has been presented for autosomal recessive inheritance of this syndrome. However, an abnormality of dentition, marked by short bulbous roots, pulp stones, and radicular dentin deposited in swirls, is a phenotypic marker of the disease that is variably expressed [93]. Thus, this disorder may have multiple *formes frustes* that could complicate genetic analysis. Indeed, using the dental lesion as well as the more classic biochemical and clinical hallmarks of the disease, an autosomal dominant pattern of transmission has been documented.

The hyperphosphatemia characteristic of the disease results from an increase in capacity of renal tubular phosphate reabsorption secondary to an unknown defect. Hypocalcemia is not a consequence of this abnormality, however, and the serum PTH concentration is normal. Moreover, the phosphaturic and urinary cAMP responses to PTH are not disturbed. Thus, the defect does not represent renal insensitivity to a hormone, or hypoparathyroidism. Rather, the basis of the disease is probably an innate or hormone/metabolic factor-mediated abnormality of the renal tubule that enhances phosphate reabsorption. Interestingly, $1,25(OH)_2D$ levels are elevated despite hyperphosphatemia, underlining the fact that it is TmP/GFR rather than the serum phosphorus concentration that controls 25OHD 1α-hydroxylase activity (see Fig. 3). Undoubtedly, the calcific tumors result from the elevated calcium–phosphorus product. The observation that long-term phosphorus depletion alone or in association with administration of acetazolamide, a phosphaturic agent, leads to resolution of the tumor masses supports this possibility.

An acquired form of TC is rarely seen in patients with end-stage renal failure. Affected patients manifest hyperphosphatemia in association with either (1) an inappropriately elevated calcitriol level for the degree of renal failure, hyperparathyroidism, or hyperphosphatemia or (2) long-term treatment with calcium carbonate, calcitriol, or high calcium-content dialysates. Calcific tumors again likely result from an elevated calcium–phosphorus product. Indeed, complete remission of the tumors occurs on treatment with vinpocetine, a mineral scavenger drug.

III. DISORDERS RELATED TO AN ALTERED PHOSPHATE LOAD

A. Decreased Phosphate Load

1. Phosphate Deprivation

Hypophosphatemia and phosphate depletion due to inadequate dietary intake are rare. With a decline in ingested phosphate the renal TmP increases and urinary phosphate excretion decreases [94]. In addition, gastrointestinal phosphate secretion gradually lessens. However, severe dietary deprivation (less than 100 mg/day) leads to a prolonged period of negative phosphate balance and total body depletion. Affected females may display hypophosphatemia (1.5 to 2.5 mg/dl or 0.48 to 0.81 mmol/liter); in interesting contrast, males generally do not manifest a decreased serum phosphate concentration in response to dietary deprivation. Nevertheless, attempts to maintain phosphate homeostasis in both sexes include suppression of the serum PTH concentration and increased $1,25(OH)_2D_3$ production. Thus, hypercalciuria may be associated with the syndrome. Whether the efficiency of calcitriol responsiveness or differential effects on end organs, such as the gastrointestinal tract or bone, underlies the noted gender difference remains unclear.

Total starvation does not cause hypophosphatemia. The catabolic effects of total food deprivation result in the release of phosphate from intracellular stores, which compensates for the negative phosphorus balance. However, refeeding of the starved person will result in hypophosphatemia if phosphate deprivation is maintained.

2. Gastrointestinal Malabsorption

Gastrointestinal absorption of phosphorus may be decreased with the use of aluminum- or magnesium-containing antacids; prolonged use of these drugs in large amounts has been associated with hypophosphatemia and a negative phosphorus balance [95]. Long-term reduction of the serum phosphorus concentration owing to chronic, excessive use of antacids leads to frank osteomalacia and myopathy. The osteomalacia results as a direct consequence of the phosphate depletion and in spite of normal vitamin D stores and an increased serum $1,25(OH)_2D$ concentration that occurs in response to the hypophosphatemia.

In contrast, the pathophysiology of variably occurring osteomalacia secondary to gastrointestinal diseases that cause steatorrhea or rapid transit time (e.g., Crohn's disease, postgastrectomy states, and intestinal fistulas) is significantly different [96]. In this case, mild to moderate hypophosphatemia occurs due to vitamin D malabsorption/deficiency and resultant secondary hyperparathyroidism and renal phosphate wasting. Further, the relation between the metabolic bone disease, vitamin D deficiency, and hypophosphatemia is complex and likely involves the influence of phosphopenia and vitamin D-dependent calciopenia on bone mineralization. However, the impact of vitamin D deficiency may be overriding, as osteomalacia may occur in the absence of hypophosphatemia. Regardless, in the presence of hypophosphatemia (and/or phosphate depletion), an elevated serum calcitriol level is part of this syndrome and has no apparent effect on the evolution of the bone disease.

B. Transcellular Shift

In a large proportion of clinically important cases of hypophosphatemia, a sudden shift of phosphorus from the extracellular to the intracellular compartment is responsible for the decline of the serum phosphate concentration. This ion movement occurs in response to naturally occurring disturbances and after the administration of certain compounds.

1. Alkalosis

Alkalosis secondary to intense hyperventilation may depress serum phosphate levels to less than 1 mg/dl [97]. A similar degree of alkalemia owing to excess bicarbonate also causes hypophosphatemia, but of a much lesser magnitude (2.5 to 3.5 mg/dl). The disparity between the effects of a respiratory and metabolic alkalosis is related to the more pronounced intracellular alkalosis that occurs during hyperventilation. The phosphate shift to the intracellular compartment results from the utilization attendant on glucose phosphorylation, a process stimulated by a pH-dependent activation of phosphofructokinase.

2. Glucose Administration

The administration of glucose and insulin often results in moderate hypophosphatemia [98]. Endogenous or exogenous insulin increases the cellular uptake not only of glucose, but also of phosphorus. The most responsive cells are those of the liver and skeletal muscle. The decline of the serum phosphate concentration generally does not exceed 0.5 mg/dl. A lesser decrease is manifest in patients with type II diabetes mellitus and insulin resistance or those with a disease causing a diminished skeletal mass. The administration of fructose and glycerol similarly reduces the serum phosphorus concentration. In contrast to glucose, fructose administration may be associated with more pronounced hypophosphatemia; the more striking effect is due to unregulated uptake by the liver.

C. Combined Mechanisms

There are special clinical situations in which altered phosphate and consequent hypophosphatemia result from both a transcellular shift of phosphorus and phosphate deprivation or renal phosphate wasting. These disorders represent some of the more common and profound causes of decreased serum phosphorus concentrations.

1. Alcoholism

Alcoholic patients frequently enter the hospital with hypophosphatemia. However, many do not exhibit a decreased serum phosphate concentration until several days have elapsed and refeeding has begun. The multiple factors that underlie the hypophosphatemia include poor dietary intake, use of phosphate binders to treat gastritis, excessive urinary losses of phosphorus, and shift of phosphorus from the extracellular to the intracellular compartment, owing to glucose administration and/or hyperventilation that occurs in patients with cirrhosis or during alcohol withdrawal [98]. Moreover, many alcoholic patients are hypomagnesemic, which potentiates renal phosphate wasting by an unclear mechanism.

2. Burns

Within several days after sustaining an extensive burn, patients often manifest severe hypophosphatemia. The initial insult induces a transient retention of salt and water. When the fluid is mobilized, significant urinary phosphorus loss ensues. Coupled with the shift of phosphorus to the intracellular compartment, which occurs secondary to hyperventilation, and the anabolic state, profound hypophosphatemia may result.

3. Nutritional Recovery Syndrome

Refeeding of starved individuals or maintaining nutritional support by parenteral nutrition or by tube feeding, without adequate phosphorus supplementation, may also cause hypophosphatemia [99]. A prerequisite for the decreased serum phosphate concentration in affected patients is that their cells must be capable of an anabolic response. As new proteins are synthesized and glucose is transported intracellularly, phosphate demand depletes reserves. Several days are generally required after the initiation of refeeding in order to establish an anabolic condition. In patients receiving total parenteral nutrition, serum phosphate levels may be further depressed if sepsis supervenes and a respiratory alkalosis develops.

4. Diabetic Ketoacidosis

Poor control of blood glucose and consequent glycosuria, polyuria, and ketoacidosis invariably cause renal phosphate wasting [98]. The concomitant volume contraction may yield a normal serum phosphate concentration. However, with insulin therapy, the administration of fluids, and correction of the acidosis, serum and urine phosphate fall precipitously. The resultant hypophosphatemia may contribute to insulin resistance and slow the resolution of the ketoacidosis. Hence, the administration of phosphate may improve the capacity to metabolize glucose and facilitate recovery.

D. Increased Phosphate Load

1. Vitamin D Intoxication

An increase of the phosphate load from exogenous sources generally does not cause hyperphosphatemia because the excessive phosphorus is excreted by the kidney. However, an increased serum phosphate concentration may occur in vitamin D intoxication when the gastrointestinal absorption of phosphate is markedly enhanced. Increased phosphate mobilization from bone and a reduction of GFR, secondary to hypercalcemia and/or nephrocalcinosis, may also contribute to the evolution of the hyperphosphatemia.

The chronic ingestion of large doses of vitamin D, in excess of 100,000 IU/day, is generally required to cause intoxication. Suspected hypervitaminosis D may be investigated with specific assays, which can document excessive amounts of vitamin D and its metabolites in the circulation (see Chapter 38).

2. Rhabdomyolysis

Because muscle contains a large amount of phosphate, necrosis of muscle tissue may acutely increase the endogenous phosphate load and result in hyperphosphatemia. Such muscle necrosis (rhabdomyolysis) may complicate heat stroke, acute arterial occlusion, hyperosmolar nonketotic coma, trauma, toxic agents such as ethanol and heroin, and idiopathic paroxysmal myoglobinuria [100,101]. Muscle biopsy often reveals myolytic denervation, and as a consequence acute renal failure caused by myoglobin excretion frequently complicates the clinical presentation and contributes to the hyperphosphatemia. However, an elevated serum phosphate concentration may precede evidence of renal failure, or occur in its complete absence when rhabdomyolysis is present. The diagnosis is confirmed by elevated serum creatine phosphokinase, uric acid, and lactate dehydrogenase concentrations and the demonstration of heme-positive urine in the absence of red blood cells. Therapy is directed at the underlying disorder with maintenance of the extracellular volume to avoid volume depletion and alkalinization of the urine to prevent uric acid accumulation and consequent acute tubular necrosis.

3. Cytotoxic Therapy

Cytotoxic therapy often causes cell destruction and liberation of phosphorus into the circulation [102]. The lysis of tumor cells begins within 1 to 2 days after initiating treatment, and it is followed quickly by an elevation of the serum phosphate concentration. Hyperphosphatemia supervenes, however, only when the treated malignancies have a larger tumor burden, rapid cell turnover, and substantial intracellular phosphorus content. Such malignancies include lymphoblastic leukemia, various types of lymphoma, and acute myeloproliferative syndromes. Hyperkalemia and hyperuricemia also occur. Indeed, uric acid nephropathy may cause renal insufficiency that predisposes to further phosphate retention and worsening of the hyperphosphatemia.

4. Malignant Hyperthermia

Malignant hyperthermia is a rare familial syndrome characterized by an abrupt rise in body temperature during the course of anesthesia [103]. The disease appears to be autosomal dominant in transmission, and an elevated serum creatine phosphokinase concentration is found in otherwise normal family members. Hyperphosphatemia results from shifts of phosphate from muscle cells to the extracellular pool. A high mortality rate accompanies the syndrome.

E. Clinical Signs and Symptoms of Abnormal Serum Phosphorus in Diseases Caused by an Altered Phosphate Load

As related above, a wide variety of diseases and syndromes with varying clinical manifestations have the characteristic biochemical abnormalities of hyperphosphatemia or hypophosphatemia. A unique complex of disturbances often is directly related to the abnormal phosphate homeostasis. The recognition of these signs and symptoms may lead to appropriate biochemical testing, the diagnosis of an unsuspected disease, and initiation of lifesaving or curative treatment.

1. Hypophosphatemia

A low serum phosphorus level is associated with symptoms only if there is concomitant phosphate depletion. The presence of phosphate deficiency, however, may cause widespread disturbances. This is not surprising, as severe hypophosphatemia causes two critical abnormalities that impact on virtually all organ systems. First, a deficiency of 2,3-diphosphoglycerate (2,3-DPG) occurs in red cells, which is associated with an increased affinity of hemoglobin for oxygen and, therefore, tissue hypoxia. Second, there is a decline of tissue ATP content and a concomitant decrease in the availability of energy-rich phosphate compounds that are essential for cell function [104,105]. The major clinical syndromes resulting from these abnormalities include nervous system dysfunction, anorexia, nausea, vomiting, ileus, muscle weakness, cardiomyopathy, respiratory insufficiency, hemolytic anemia, and impaired leukocyte and platelet function. Additionally, phosphate deficiency causes osteomalacia and bone pain, clinical sequelae that are probably independent of the aforementioned abnormalities.

Central nervous system dysfunction has been well characterized in severe hypophosphatemia, especially in patients receiving total parenteral nutrition for diseases causing severe weight loss. A sequence of symptoms compatible with a metabolic encephalopathy usually begins one or more weeks after the initiation of therapy with solutions that contain glucose and amino acids but lack adequate phosphorus supplementation to prevent hypophosphatemia. The onset of dysfunction is marked by irritability, muscle weakness, numbness, and paresthesias, with progression to dysarthria, confusion, obtundation, coma, and death [106]. These patients have a profoundly diminished red cell 2,3-DPG content. Both biochemical abnormalities and clinical symptoms improve as patients receive phosphorus supplementation. Peripheral neuropathies, Guillain-Barre-like paralysis, hyporeflexia, intention tremor, and ballismus have also been described with hypophosphatemia and phosphate depletion.

The effects of hypophosphatemia on muscle depend on the severity and chronicity of the deficiency. Chronic phosphorus deficiency results in a proximal myopathy with striking atrophy and weakness. Osteomalacia frequently accompanies the myopathy, so patients complain of pain in weight-bearing bones. Normal values for serum creatine phosphokinase and aldolase activities are characteristically present. Rhabdomyolysis does not occur with chronic phosphate depletion.

In contrast, acute hypophosphatemia can lead to rhabdomyolysis with muscle weakness and pain. Most cases occur in chronic alcoholics or patients receiving total parenteral nutrition. In both groups of patients, muscle pain, swelling, and stiffness occur 3 to 8 days after the initiation of therapies that do not contain adequate amounts of phosphorus. Muscle paralysis and diaphragmatic failure may occur. Studies of muscle tissue from chronically phosphate-depleted dogs made acutely hypophosphatemic show a decrement in cellular content of phosphorus, ATP, and adenosine diphosphate (ADP). Rhabdomyolysis occurred in these muscles fibers. The laboratory findings in patients with hypophosphatemic myopathy and with rhabdomyolysis include elevated serum creatine phosphokinase levels; however, serum phosphate levels may become normal if enough necrosis has occurred with subsequent phosphorus release. Also, renal failure and hypocalcemia can be associated with the syndrome.

Myocardial performance can be abnormal at serum phosphate levels of 0.7 to 1.4 mg/dl. This occurs when ATP depletion causes impairment of the actin–myosin interaction, the calcium pump of the sarcoplasm, and the sodium–potassium pump of the cell membrane [107]. The net result is reduced stroke work and cardiac output, which may progress to congestive heart failure. These problems are reversible with phosphate replacement.

Respiratory failure can occur owing to failure of diaphragmatic contraction in hypophosphatemic patients. When serum phosphate levels are raised, diaphragmatic contractility improves. The postulated mechanism for the respiratory failure is muscle weakness secondary to inadequate levels of ATP and decreased glycolysis as the result of phosphate depletion [108].

Two disturbances of red cell function may occur secondary to phosphorus deficiency. First, as intraerythrocyte ATP production is decreased, the erythrocyte cell membrane becomes rigid, which can cause hemolysis [109]. This is rare and is usually seen in septic, uremic, acidotic, or alcoholic patients, when serum phosphate levels are less than 0.5 mg/dl. Second, the limited production of 2,3-DPG causes a leftward shift of the oxyhemoglobin curve and impairs the release of oxygen to peripheral tissues. Such a consequence of chronic hypophosphatemia has been documented in children with XLH and proposed as one factor underlying retarded statural growth [110].

Leukocyte dysfunction, which complicates phosphate deficiency, includes decreased chemotaxis, phagocytosis, and bactericidal activity [111]. These abnormalities increase the host susceptibility to infection. The mechanism by which hypophosphatemia impairs the various activities of the leukocyte probably is related to impairment of ATP synthesis. Decreased availability of energy impairs microtubules that regulate the mechanical properties of leukocytes and limit the rate of synthesis of organic phosphate compounds that are necessary for endocytosis.

Abnormal platelet survival, causing thrombocytopenia, profuse gastrointestinal bleeding, and cutaneous bleeding, has also been described in association with phosphate depletion in animal studies. Despite these abnormalities, there is little evidence that hypophosphatemia is a primary cause of hemorrhage in humans.

Perhaps the most consistent abnormalities associated with phosphate depletion are those on bone. Acute phosphate depletion induces dissolution of apatite crystals from the osseous matrix. This effect may be due to $1,25(OH)_2D_3$, which is increased in response to phosphate depletion in both animals and humans. More prolonged hypophosphatemia leads to rickets and osteomalacia. This complication is a common feature of phosphate depletion. However, the ultimate cause is variable. Although simple phosphate depletion alone may underlie the genesis of the abnormal mineralization, in many disorders the defect is secondary to phosphate depletion and commensurate $1,25(OH)_2D_3$ deficiency. Thus, treatment of this complication may often require combination therapy, phosphate supplements and calcitriol.

2. HYPERPHOSPHATEMIA

Hypocalcemia and consequent tetany are the most serious clinical sequelae of hyperphosphatemia [112]. The decreased serum calcium concentration results from the deposition of calcium phosphate salts in soft tissue, a process that may lead to symptomatic ectopic calcification. The dystrophic calcification is frequently seen in acute and chronic renal failure, hypoparathyroidism, pseudohypoparathyroidism, and tumoral calcinosis. Indeed, deposition of calcium phosphate complexes in the kidney may predipose to acute renal failure. When the calcium–phosphate product exceeds 70, the probability that soft tissue calcification will occur increases sharply. In addition, local factors such as tissue pH and injury (e.g., necrotic or hypoxic tissue) may predispose to precipitation of the calcium phosphate salts. In chronic renal failure, calcification occurs in arteries, muscle tissue, periarticular spaces, the myocardial conduction system, lungs, and the kidney. Affected patients may also have ocular calcification, causing the "red eye" syndrome of uremia, and subcutaneous calcification, which also plays a role in uremic pruritus. On the other hand, a predisposition to calcification of periarticular surfaces of the hips, elbows, shoulders, and other large joints occurs in tumoral calcinosis.

In some disease states, hyperphosphatemia may also play an important role in the development of secondary hyperparathyroidism [113]. A decrement in the serum calcium concentration secondary to hyperphosphatemia stimulates the release of PTH. Furthermore, hyperphosphatemia decreases the activity of 25OHD 1α-hydroxylase. The consequent diminished production of 1,25-$(OH)_2D_3$ impairs the gastrointestinal absorption of calcium and induces skeletal resistance to PTH, influences that augment the development of hyperparathyroidism.

Thus, hyperphosphatemia triggers a cascade of events that impact on calcium homeostasis at multiple sites. The prevention of secondary hyperparathyroidism, metabolic bone disease, and soft tissue and vascular calcification in affected patients, therefore, depends on ultimately controlling the serum phosphate concentration.

F. Treatment of Abnormal Serum Phorphorus in Diseases Caused by an Abnormal Phosphate Load

Treatment of the myriad of diseases that characteristically display hyperphosphatemia or hypophosphatemia depends on determining the mechanism underlying their pathogenesis. The cause can almost always be ascertained by assessment of the clinical setting, determination of renal function, measurement of urinary phosphate excretion, and analysis of arterial carbon dioxide tension and pH. Therapy is aimed at correcting both the serum phosphate concentration and associated complications.

The treatment of phosphate depletion depends on replacing body phosphorus stores. Preventive measures, however, will preclude the onset of phosphate depletion. Thus, appropriate monitoring of patients taking large doses of aluminum-containing antacids and provision of phosphorus intravenously to patients with diabetic ketoacidosis will preserve phosphate stores. Alternatively, treatment of established phosphate depletion may require 2.5 to 3.7 g of phosphate daily, preferably administered orally in four equally divided doses. This can be done by giving K-Phos Neutral tablets, which contain 250 mg of elemental phosphorus per tablet. If oral therapy is not tolerated and the serum phosphate shows a downward trend, approaching dangerous levels ($<$1.2 mg/dl), intravenous phosphate supplementation at a dose of 10 mg/kg body weight/day may be administered. Such therapy should be discontinued when the serum phosphate level reaches values of $>$2.0 mg/dl. However, therapy is not required for many of the conditions resulting in phosphate depletion. Only when the consequences of severe depletion are manifest need treatment be initiated.

Theoretically, elevated serum phosphate levels may be reduced by decreasing the TmP, increasing the GFR, or diminishing the phosphate load (exogenous or endogenous). There are no generally available pharmacological means of acutely altering the GFR or reducing the TmP. However, chronic use of drugs such as acetazolamide, which decreases TmP and induces phosphaturia, is effective as ancillary treatment of disorders such as tumoral calcinosis. Nevertheless, regulation of hyperphosphatemia is most often achieved by reducing the renal phosphate load. In tumoral calcinosis and chronic renal failure, such an effect is obtained by restricting the dietary phosphate intake or by administering phosphate binders such as calcium carbonate or aluminum hydroxide. Alternative strategies for management of load-dependent hyperphosphatemia include the administration of intravenous calcium or intravenous glucose and insulin. The consequence of such intervention is sequestration of phosphate in bone or soft tissues. Dialysis can also be used for acute management of load-dependent disorders or chronic maintenance of phosphate overload such as those that complicate chronic renal failure.

References

1. Reichel H, Koeffler HP, Norman AW 1989 The role of the vitamin D endocrine system in health and disease. N Engl J Med **320**:980–991.
2. Fraser DR 1980 Regulation of metabolism of vitamin D. Physiol Rev **60**:551–613.
3. Yanagawa N, Nakhoul F, Kurokawa K, Lee DBN 1994 Physiology of phosphorus metabolism. In: Narins RG (ed) Clinical Disorders of Fluid and Electrolyte Metabolism, 5th Ed. McGraw-Hill, New York, pp. 307–371.
4. Berndt TJ, Knox FG 1992 Renal regulation of phosphate excretion. In: Seldin DW, Giebisch G (eds) The Kidney, Physiology and Pathophysiology. Raven, New York, pp. 2511–2532.
5. Werner A, Moore ML, Mantei N, Biber J, Semenza G, Murer H 1991 Cloning and expression of cDNA for a Na/Pi cotransport system of kidney cortex. Proc Natl Acad Sci USA **88**:9608–9612.
6. Chong SS, Kristjansson K, Zoghbi HY, Hughes MR 1993 Molecular cloning of the cDNA encoding a human renal sodium phosphate transport protein and its assignment to chromosome 6p21.3–p23. Genomics **18**:355–359.
7. Custer M, Lotscher M, Biber J, Murer H, Kaissling B 1994 Expression of Na–Pi cotransport in rat kidney: Localization by RT-PCR and immunohistochemistry. Am J Physiol **266**:F767–F774.
8. Tenenhouse HS 1997 Cellular and molecular mechanisms of renal phosphate transport. J Bone Miner Res in press.
9. Kempson SA, Lotscher M, Kaissling B, Biber J, Murer H, Levi M 1995 Parathyroid hormone action on phosphate transporter

mRNA and protein in rat renal proximal tubules. Am J Physiol **268**:F784–F791.

10. Levi M, Lotscher M, Sorribas V, Custer M, Arar M, Kaissling B, Murer H, Biber J 1994 Cellular mechanisms of acute and chronic adaptation of rat renal phosphate transporter to alterations in dietary phosphate. Am J Physiol **267**:F900–F908.
11. Bell NH 1988 Vitamin D metabolism in health and disease. Henry Ford Hospital Med J **36**:40–52.
12. Tanaka Y, DeLuca HF 1973 The control of 25-hydroxyvitamin D metabolism by inorganic phosphorus. Arch Biochem Biophys **154**:566–574.
13. Fukase M, Birge SJ, Rifas L, Avioli LV, Chase LR 1982 Regulation of 25-hydroxyvitamin D_3 1-hydroxylase in serum-free monolayer culture of mouse kidney. Endocrinology **110**:1073–1075.
14. Gray RW, Garthwaite TL 1985 Activation of renal 1,25-dihydroxyvitamin D_3 synthesis by phosphate deprivation: Evidence for a role for growth hormone. Endocrinology **116**:189–193.
15. Gray RW, Lemann J Jr 1985 Vitamin D metabolism: The renal–pituitary axis. In: Norman AW, Schaefer K, Grigoleit H-G, Herrath DV (eds) Vitamin D—Chemical, Biochemical and Clinical Update. de Gruyter, Berlin, pp. 75–82.
16. Drezner MK 1987 Understanding the pathogenesis of X-linked hypophosphatemic rickets: A requisite for successful therapy. In: Zackson DA (ed) A CPC Series: Cases in Metabolic Bone Disease. Triclinica Communications, New York, pp. 1–11.
17. Econs MJ, Drezner MK 1992 Bone disease resulting from inherited disorders of renal tubule transport and vitamin D metabolism. In: Favus MJ, Coe FL (eds) Disorders of Bone and Mineral Metabolism. Raven, New York, pp. 935–950.
18. Lobaugh B, Burch WM Jr, Drezner MK 1984 Abnormalities of vitamin D metabolism and action in the vitamin D resistant rachitic and osteomalacic diseases. In: Kumar R (ed) Vitamin D: Basic and Clinical Aspects. Nijhoff, Boston, pp. 665–720.
19. Harrison HE, Harrison HC, Lifshitz F, Johnson AD 1966 Growth disturbance in hereditary hypophosphatemia. Am J Dis Children **112**:290–297.
20. Haddad JG, Chyu KJ, Hahn TJ, Stamp TCB 1973 Serum concentrations of 25-hydroxyvitamin D in sex linked hypophosphatemic vitamin D-resistant rickets. J Lab Clin Med **81**:22–27.
21. Delvin EE, Glorieux FH 1981 Serum 1,25-dihydroxyvitamin D concentration in hypophosphatemic vitamin D resistant rickets. Calcif Tissue Int **33**:173–175.
22. Lyles KW, Clark AG, Drezner MK 1982 Serum 1,25-dihydroxyvitamin D levels in subjects with X-linked hypophosphatemic rickets and osteomalacia. Calcif Tissue Int **34**:125–130.
23. Lobaugh B, Drezner MK 1983 Abnormal regulation of renal 25-hydroxyvitamin D levels in the X-linked hypophosphatemic mouse. J Clin Invest **71**:400–403.
24. Nesbitt T, Drezner MK, Lobaugh B 1986 Abnormal parathyroid hormone stimulation of renal 25-hydroxyvitamin D-1α-hydroxylase activity in the hypophosphatemic mouse: Evidence for a generalized defect of vitamin D metabolism. J Clin Invest **77**: 181–187.
25. Nesbitt T, Lobaugh B, Drezner MK 1987 Calcitonin stimulation of renal 25-hydroxyvitamin D-1α-hydroxylase activity in hypophosphatemic mice: Evidence that the regulation of calcitriol production is not universally abnormal in X-linked hypophosphatemia. J Clin Invest **75**:15–19.
26. Nesbitt T, Drezner MK 1990 Abnormal parathyroid hormone-related peptide stimulation of renal 25-hydroxyvitamin D-1α-hydroxylase activity in *hyp*-mice: Evidence for a generalized defect of enzyme activity in the proximal convoluted tubule. Endocrinology **127**:843–848.
27. Bell CL, Tenenhouse HS, Scriver CR 1988 Primary cultures of renal epithelial cells from X-linked hypophosphatemic (Hyp) mice express defects in phosphate transport and vitamin D metabolism. Am J Hum Genet **43**:293–303.
28. Dobre CV, Alvarez UM, Hruska KA 1990 Primary culture of hypophosphatemic proximal tubule cells express defective adaptation to phosphate. J Bone Miner Res **5**:S205 (abstract).
29. Collins JF, Ghishan FK 1994 Molecular cloning, functional expression, tissue distribution, and in situ hybridization of the renal sodium phosphate (Na^+/Pi) transporter in the control and hypophosphatemic mouse. FASEB J **8**:862–868.
30. Tenenhouse HS, Werner A, Biber J, Ma S, Martel J, Roy S, Murer H 1994 Renal Na^+–phosphate cotransport in murine X-linked hypophosphatemic rickets: Molecular characterization. J Clin Invest **93**:671–676.
31. Tenenhouse HS, Martel J, Biber J, Murer H 1995 Effect of Pi restriction on renal Na^+–Pi cotransporter mRNA and immunoreactive protein in X-linked Hyp mice. Am J Physiol **268**:F1062–F1069.
32. Meyer RA Jr, Meyer MH, Gray RW 1989 Parabiosis suggests a humoral factor is involved in X-linked hypophosphatemia in mice. J Bone Miner Res **4**:493–500.
33. Meyer RA Jr, Tenenhouse HS, Meyer MH, Klugerman AH 1989 The renal phosphate transport defect in normal mice parabiosed to X-linked hypophosphatemic mice persists after parathyroidectomy. J Bone Miner Res **4**:523–532.
34. Nesbitt T, Econs MJ, Byun JK, Martel J, Tenenhouse HS, Drezner MK 1995 Phosphate transport in immortalized cell cultures from the renal proximal tubule of normal and *hyp*-mice: Evidence that the *HYP* gene locus product is an extrarenal factor. J Bone Miner Res **10**:1327–1333.
35. Nesbitt T, Byun JK, Drezner MK 1996 Normal phosphate (Pi) transport in cells from the S_2 and S_3 segments of *hyp*-mouse proximal renal tubules. Endocrinology **137**:943–948.
36. Nesbitt T, Coffman TM, Griffiths R, Drezner MK 1992 Cross-transplantation of kidneys in normal and *hyp*-mice: Evidence that the *hyp*-mouse phenotype is unrelated to an intrinsic renal defect. J Clin Invest **89**:1453–1459.
37. Kos CH, Tihy F, Econs MJ, Murer H, Lemieux N, Tenenhouse HS 1994 Localization of a renal sodium phosphate cotransporter gene to human chromosome 5q35. Genomics **19**:176–177.
38. Cai Q, Hodgson SF, Kao PC, Lennon VA, Klee GG, Zinsmiester AR, Kumar R 1994 Brief report: Inhibition of renal phosphate transport by a tumor product in a patient with oncogenic osteomalacia. N Engl J Med **330**:1645–1649.
39. Wilkins GE, Granleese S, Hegele RG, Holden J, Anderson DW, Bondy GP 1995 Oncogenic osteomalacia: Evidence for a humoral phosphaturic factor. J Clin Endocrinol Metab **80**:1628–1634.
40. Kumar R, Haugen JD, Wieben ED, Londowski JM, Cai Q 1995 Inhibitors of renal epithelial phosphate transport in tumor-induced osteomalacia and uremia. Proc Assoc Am Physicians **107**:296–305.
41. Read AP, Thakker RV, Davies KE, Mountford RC, Brenton DP, Davies M, Glorieux F, Harris R, Hendy GN, King A, McGlade S, Peacock CJ, Smith R, O'Riordan JLH 1986 Mapping of human X-linked hypophosphatemic rickets by multilocus linkage analysis. Hum Genet **73**:267–270.
42 Machler M, Frey D, Gai A, Orth U, Wienker TF, Fanconi A, Schmid W 1986 X-Linked dominant hypophosphatemia is closely linked to DNA markers DXS41 and DXS43 at Xp22. Hum Genet **73**:271–275.
43. Thakker RV, Read AP, Davies KE, Whyte MP, Webber R, Glorieux F, Davies M, Mountford RC, Harris R, King A, Kim GS, Fraser D, Kooh SW, O'Riordan JLH 1987 Bridging markers

defining the map position of X-linked hypophosphatemic rickets. J Med Genet **24**:756–760.
44. Albersten HM, Ahrens P, Frey D, Machler M, Kruse TA 1987 Close linkage between X-linked hypophosphatemia and DXS207 defined by the DNA probe pPA4B. Ninth Workshop on Human Gene Mapping #401, p. 317 (abstract).
45. Econs MJ, Fain PR, Norman M, Speer MC, Pericak-Vance MA, Becker PA, Barker DF, Taylor A, Drezner MK 1993 Flanking markers define the X-linked hypophosphatemic rickets gene locus. J Bone Miner Res **8**:1149–1152.
46. Econs MJ, Francis F, Rowe PSN, Speer MC, O'Riordan J, Lehrach H, Becker PA 1994 Dinucleotide repeat polymorphism at the DXS1683 locus. Hum Mol Genet **3**:680.
47. Econs MJ, Rowe PSN, Francis F, Barker DF, Speer MC, Norman M, Fain PR, Weissenbach J, Read A, Davies KE, Becker PA, Lehrach H, O'Riordan J, Drezner MK 1994 Fine structure mapping of the human X-linked hypophosphatemic rickets gene locus. J Clin Endocrinol Metab **79**:1351–1354.
48. Francis F, Rowe PSN, Econs MJ, See CG, Benham F, O'Riordan JLH, Drezner MK, Hamvas RMJ, Lehrach H 1994 A YAC contig spanning the hypophosphatemic rickets gene candidate region. Genomics **21**:229–237.
49. The HYP Consortium: Lab 1: Francis F, Hennig S, Korn B, Reinhardt R, de Jong P, Poustka A, Lehrach H; Lab 2: Rowe PSN, Goulding JN, Summerfield T, Mountford R, Read AP, Popowska E, Pronicka E, Davies KE, O'Riordan JLH; Lab 3: Econs MJ, Nesbitt T, Drezner MK; Lab 4: Oudet C, Hanauer A; Lab 5: Strom T, Meindl A, Lorenz B, Cagnoli M, Mohnike KL, Murken J, Meitinger T 1995 Positional cloning of PEX: A phosphate regulating gene with homologies to endopeptidases is depleted in patients with X-linked hypophosphatemic rickets. Nature Genet **11**:130–136.
50. Rowe PSN, Goulding JN, Econs MJ, Francis F, Lehrach H, Read A, Mountford J, Oudet C, Hanauer A, Summerfield T, Meitinger T, Strom A, Drezner MK, Davies KE, O'Riordan JLH 1997 The gene for X-linked hypophosphatemic rickets maps to a 200–300 kb region in Xp22.1–Xp22.2, and is located on a single YAC containing a putative vitamin D response element (VDRE). Hum Genet in press.
51. Du L, Desbarats M, Viel J Glorieux FH, Cawthorn C, Ecarot B 1996 cDNA cloning of the murine Pex gene implicated in X-linked hypophosphatemia and evidence for expression in bone. Genomics **36**:22–28.
52. Tenenhouse HS, Woumounou Y, Goodyer C 1996 Pattern of Pex mRNA expression in human fetal tissues. J Bone Miner Res **11**:S137 (abstract).
53. Florentin D, Sassi A, Roques BP 1984 A highly sensitive fluorimetric assay for "enkephalinase," a neutral metalloendopeptidase that releases Tyr-Gly-Gly from enkephalins. Anal Biochem **141**:62–69.
54. Koehn JA, Norman JA, Jones BN, LeSueur L, Sakane Y, Ghai RD 1987 Degradation of atrial natriuretic factor by kidney cortex membranes. J Biol Chem **262**:11623–11627.
55. Jalal F, Lemay G, Zollinger M, Berthelot A, Boileau G, Crine P 1991 Neutral endopeptidase, a major brush border protein of the kidney proximal nephron, is directly targeted to the apical domain when expressed in Madin-Darby kidney cells. J Biol Chem **266**:19826–19857.
56. Mari B, Checler F, Ponzio G, Peyron JF, Manie S, Farahifar D, Rossi B, Anberger P, Jurkat T 1992 T cells express a functional neutral endopeptidase activity (CALLA) involved in T cell activation. EMBO J **11**:3875–3885.
57. Marie PJ, Glorieux FH 1983 Relation between hypomineralized periosteocytic lesions and bone mineralization in vitamin D resistant rickets. Calcif Tissue Int **35**:443–448.
58. Rifas L, Gupta A, Cheng S-L, Halstead LR, Hruska KA, Avioli LV 1996 Reduced casein kinase II activity is associated with underphosphorylated osteopontin in *HYP*/Y mice osteoblasts. J Bone Miner Res **11**:S253 (abstract).
59. Siddhanti SR, Quarles LD 1996 Identification of PEX in a human bone cDNA library: Further evidence for osteoblastic involvement in X-linked hypophosphatemic rickets. J Am Soc Nephrol **7**:A2744 (abstract).
60. Lajeunesse D, Aubin R 1996 Demonstration of a primary defect in osteogenic stromal stem cells from the hyp mouse: A cause for hypophosphatemia? J Bone Miner Res **11**:S136 (abstract).
61. Friedman NE, Drezner MK 1991 Genetic osteomalacia. In: Bardin CW (ed) Current Therapy in Endocrinology and Metabolism, 4th Ed. BC Decker, Philadelphia, Pennsylvania, pp. 421–428.
62. Glorieux FH, Marie PJ, Pettifor JM, Delvin EE 1980 Bone response to phosphate salts, ergocalciferol and calcitriol in hypophosphatemic vitamin D resistant rickets. N Engl J Med **303**:1023–1031.
63. Friedman NE, Lobaugh B, Drezner MK 1993 Effects of calcitriol and phosphorus therapy on the growth of patients with X-linked hypophosphatemia. J Clin Endocrinol Metab **76**:839–844.
64. Saggerve G, Baronelli GI, Butelloni S, Perri G 1995 Long-term growth hormone treatment in children with renal hypophosphatemic rickets: Effects on growth, mineral metabolism and bone density. J Pediatr **127**:395–402.
65. Drezner MK 1996 Tumor-induced rickets and osteomalacia. In: Favus MJ (ed) Primer on the Metabolic Bone Diseases and Disorders of Mineral Metabolism, 3rd Ed. Lippincott-Raven, Philadelphia, Pennsylvania, pp. 319–325.
66. Lyles KW, Berry WR, Haussler M, Harrelson JM, Drezner MK 1980 Hypophosphatemic osteomalacia: Association with prostatic carcinoma. Ann Intern Med **93**:275–278.
67. Murphy P, Wright G, Rai GS 1985 Hypophosphatemic osteomalacia induced with prostatic carcinoma. Br Med J **290**:1945.
68. Hosking DJ, Chamberlain MJ, Whortland-Webb WR 1975 Osteomalacia and carcinoma of prostate with major redistribution of skeletal calcium. Br J Radiol **48**:451–456.
69. Taylor HC, Fallon MD, Velasco ME 1984 Oncogenic osteomalacia and inappropriate antidiuretic hormone secretion due to oat-cell carcinoma. Ann Intern Med **101**:786–788.
70. Shaker JL, Brickner RC, Divgi AB, Raff H, Findling JW 1995 Case report: Renal phosphate wasting, syndrome of inappropriate antidiuretic hormone and ectopic corticotropin production in small cell carcinoma. Am J Med Sci **310**:38–41.
71. McClure J, Smith PS 1987 Oncogenic osteomalacia. J Clin Pathol **40**:446–453.
72. Saville PD, Nassim JR, Stevenson FH 1955 Osteomalacia in von Recklinghausen's neurofibromatosis: metabolic study of a case. Br Med J **1**:1311–1313.
73. Aschinberg LC, Soloman LM, Zeis PM, Justice P, Rosenthal IM 1977 Vitamin D-resistant rickets induced with epidermal nevus syndrome: Demonstration of a phosphaturic substance in the dermal lesions. J Pediat **91**:56–60.
74. Yoshikawa S, Nakamura T, Takagi M, Imamura T, Okano K, Sasaki S 1977 Benign osteoblastoma as a cause of osteomalacia. A report of two cases. J Bone Joint Surg **59B**:279–289.
75. Lau K, Strom MC, Goldberg M, Goldfarb S, Gray RW, Lemann R Jr, Agus ZS 1979 Evidence for a humoral phosphaturic factor in oncogenic hypophosphatemic osteomalacia. Clin Res **27**:421A (abstract).
76. Miyauchi A, Fukase M, Tsutsumi M, Fujita T 1988 Heman giopericytoma-induced osteomalacia: Tumor transplantation in

nude mice causes hypophosphatemia and tumor extracts inhibit renal 25-hydroxyvitamin D-1-hydroxylase activity. J Clin Endocrinol Metab **67**:46–53.

77. Drezner MK, Feinglos MN 1977 Osteomalacia due to 1,25-dihydroxycholecalciferol deficiency. Association with a giant cell tumor of bone. J Clin Invest **60**:1046–1053.
78. Tieder M, Modai D, Samuel R, Arie R, Halabe A, Bab I, Gabizon D, Liberman UA 1985 Hereditary hypophosphatemic rickets with hypercalciuria. N Engl J Med **312**:611–617.
79. Tieder M, Modai D, Shaked U, Samuel R, Arie R, Halabe A, Maor J, Weissgarten J, Averbukh Z, Cohen N, Liberman UA 1987 "Idiopathic" hypercalciuria and hereditary hypophosphatemic rickets: Two phenotypical expressions of a common genetic defect. N Engl J Med **316**:125–129.
80. Tieder M, Arie R, Bab I, Maor J, Liberman UA 1992 A new kindred with hereditary hypophosphatemic rickets with hypercalciuria: Implications for correct diagnosis and treatment. Nephron **62**:176–181.
81. Liberman UA 1988 Inborn errors in vitamin D metabolism—Their contribution to the understanding of vitamin D metabolism. In: Norman AW, Schaefer K, Grigoleit H-G, Herrath DV (eds) Vitamin D—Molecular, Cellular and Clinical Endocrinology. de Gruyter, Berlin, pp. 935–947.
82. Davidai GA, Nesbitt T, Drezner MK 1991 Variable phosphate-mediated regulation of vitamin D metabolism in the murine hypophosphatemic rachitic/osteomalacic disorders. Endocrinology **128**:1270–1276.
83. Chan JCM, Alon U 1985 Tubular disorders of acid–base and phosphate metabolism. Nephron **40**:257–279.
84. Chesney RW 1990 Fanconi syndrome and renal tubular acidosis. In: Favus MJ (ed) Primer on Metabolic Bone Diseases and Disorders of Mineral Metabolism. American Society for Bone and Mineral Research, Kelseyville, California, pp. 190–194.
85. De Toni G 1933 Remarks on the relations between renal rickets (renal dwarfism) and renal diabetes. Acta Paediatr Scand **16**:479–484.
86. McCune DJ, Mason HH, Clarke HT 1943 Intractable hypophosphatemic rickets with renal glycosuria and acidosis (the Fanconi syndrome). Am J Dis Children **65**:81–146.
87. Chesney RW, Kaplan BS, Phelps, M, DeLuca HF 1984 Renal tubular acidosis does not alter circulating values of calcitriol. J Pediatr **104**:51–55.
88. Tieder M, Arie R, Modai D, Samuel R, Weissgarten J, Liberman UA 1988 Elevated serum 1,25-dihydroxyvitamin D concentrations in siblings with primary Fanconi's syndrome. N Engl J Med **319**:845–849.
89. Schneider JA, Schulman JD 1983 Cystinosis. In: Stanbury JB, Wyngaarden JB, Fredrickson DS, Goldstein JL, Brown MS (eds) The Metabolic Basis of Inherited Disease, 5th Ed. McGraw-Hill, New York, pp. 1844–1866.
90. Harrison HE, Harrison HC 1979 Rickets and osteomalacia. In: Harrison HE, Harrison HC (eds) Disorders of Calcium and Phosphate Metabolism in Childhood and Adolescence. Saunders, Philadelphia, Pennsylvania, pp. 141–256.
91. Scriver CR, MacDonald W, Reade T, Glorieux FH, Nogrady B 1977 Hypophosphatemic nonrachitic bone disease: An entity distinct from X-linked hypophosphatemia in the renal defect, bone involvement and inheritance. Am J Med Genet **1**:101–117.
92. Scriver CR, Reade T, Halal F, Costa T, Cole DEC 1981 Autosomal hypophosphatemic bone disease responds to $1,25(OH)_2D_3$. Arch Dis Children **56**:203–207.
93. Lyles KW, Burkes EJ, Ellis GJ, Lucas KJ, Dolan EA, Drezner MK 1985 Genetic transmission of tumoral calcinosis: Autosomal dominant with variable clinical impressivity. J Clin Endocrinol Metab **60**:1093–1097.
94. Levine BS, Ho LD, Pasiecznik K, Coburn JV 1986 Renal adaptation to phosphorus deprivation. J Bone Miner Res **1**:33–40.
95. Lotz M, Zisman E, Bartter FC 1968 Evidence for a phosphorus-depletion syndrome in man. N Engl J Med **278**:409–415.
96. Baker LRI, Acknll P, Cattell WR, Stamp TC, Watson L 1974 Iatrogenic osteomalacia and myopathy due to phosphate depletion. Br Med J **3**:150–152.
97. Mostellar ME, Tuttle EP 1964 Effects of alkalosis on plasma concentration and urinary exretion of inorganic phosphate in man. J Clin Invest **43**:138–145.
98. Knochel JP 1977 The pathophysiology and clinical characteristics of severe hypophosphatemia. Arch Intern Med **137**:203–220.
99. Sheldon GF, Grzyb S 1975 Phosphate depletion and repletion: Relation to parenteral nutrition and oxygen transport. Ann Surg **182**:683–689.
100. Grossman RA, Hamilton RW, Morse BM, Penn AS, Goldberg M 1974 Nontraumatic rhabdomyolysis and acute renal failure. N Engl J Med **291**:807–811.
101. Koffler A, Friedler RM, Massry SG 1976 Acute renal failure due to nontraumatic rhabdomyolysis. Ann Intern Med **85**:23–28.
102. Zusman J, Brown DM, Nesbitt ME 1973 Hyperphosphatemia, hyperphosphaturia and hypocalcemia in acute lymphoblastic leukemia. N Engl J Med **289**:1335–1337.
103. Denborough MA, Forster JFA, Hudson MC, Carter NG, Zapf P 1970 Biochemical changes in malignant hyperpyrexia. Lancet **1**:1137–1138.
104. Duhm J 1971 2,3-DPG-induced displacements of the oxyhemoglobin dissociation curve of blood: Mechanisms and consequences. Adv Exp Biol Med **37A**:179–186.
105. Travis SF, Sugerman HJ, Ruberg RL, Dudrick SJ, Delivoria-Papadopoulos M, Miller LD, Oski FA 1977 Alterations of red cell glycolytic intermediates and oxygen transport as a consequence of hypophosphatemia in patients receiving intravenous hyperalimentation. N Engl J Med **297**:901–904.
106. Parfitt AM, Kleerekoper M 1980 Clinical disorders of calcium, phosphorus and magnesium metabolism. In: Maxwell MH, Kleeman CR (eds) Clinical Disorders of Fluid and Electrolyte Metabolism. McGraw-Hill, New York, pp. 947–1151.
107. O'Connor LR, Wheeler WS, Bethune JE 1977 Effect of hypophosphatemia on myocardial performance in man. N Engl J Med **297**:901–903.
108. Newman JH, Neff TA, Ziporen P 1977 Acute respiratory failure associated with hypophosphatemia. N Engl J Med **296**:1101–1103.
109. Klock JC, Williams HE, Mentzer WK 1974 Hemolytic anemia and somatic cell dysfunction in severe hypophosphatemia. Arch Intern Med **134**:360–364.
110. Glorieux FH, Scriver CR, Reade TM, Goldman H, Roseborough A 1972 Use of phosphate and vitamin D to prevent dwarfism and rickets in X linked hypophosphatemia. N Engl J Med **281**:481–487.
111. Craddock PR, Yawota Y, Van Santen L, Gilberstadt S, Silivis S, Jacob HS 1974 Acquired phagocyte dysfunction: A complication of the hypophosphatemia of parenteral hyperalimentation. N Engl J Med **290**:1403–1407.
112. Herbert LA, Lemann J, Peterson JR, Lennon EJ 1966 Studies of the mechanism by which phosphate infusion lowers serum calcium concentration. J Clin Invest **45**:1886–1891.
113. Sinha TK, Allen DO, Queener SF, Bell NH, Larson S, McClintock R 1977 Effects of acetazolamide on the renal excretion of phosphate in hyperparathyroidism and pseudohypoparathyroidism. J Lab Clin Med **89**:1188–1197.

CHAPTER 47

Vitamin D Pseudodeficiency

FRANCIS H. GLORIEUX Genetics Unit, Shriners Hospital for Children and Departments of Surgery and Pediatrics, McGill University, Montréal, Québec, Canada

RENÉ ST-ARNAUD Genetics Unit, Shriners Hospital for Children and Departments of Surgery and Human Genetics, McGill University, Montréal, Québec, Canada

I. INTRODUCTION

Following the description by Albright *et al.* in 1937 [1] of "rickets resistant to vitamin D," a number of observations were published [2,3] which indicated that there was a variant of resistant rickets which differed from the classic hypophosphatemic type (X-linked hypophosphatemic rickets, or XLH; see Chapter 46) by its clinical and biological symptoms and response to therapy. It was indeed shown by Prader *et al.* [4] that this form of rickets had an early onset (within the first year of life), contrary to the XLH type. The disease symptoms also included the development of profound hypocalcemia, tooth enamel hypoplasia, and a response to daily administration of large amounts of vitamin D. In view of the latter, the term "vitamin D dependency" was proposed to describe the new syndrome [5]. In 1973, when calcitriol [$1,25(OH)_2D_3$] became available as a therapeutic agent, it was demonstrated that this rare form of rickets responded to physiological amounts of calcitriol [6]. It was then recognized that this disease was due to an inborn error of metabolism involving the defective conversion of calcidiol ($25OHD_3$) to calcitriol (Fig. 1). For this reason, we feel it more appropriate to return to the original terminology of Prader and use the term pseudovitamin D-deficiency rickets (PDDR) to describe this form of rickets.

In 1978, another inborn error of vitamin D metabolism was recognized in which a clinical picture of pseudovitamin D-deficiency developed despite high circulating concentrations of endogenously produced calcitriol [7]. In some pedigrees the phenotype is compounded by the presence of complete alopecia [8]. This second variety of pseudodeficiency has been termed vitamin D dependency type II, pseudovitamin D-deficiency type II, calcitriol-resistant rickets, hypocalcemic vitamin D-resistant rickets, and hereditary 1,25-dihydroxyvitamin D-resistant rickets (HVDRR). The latter term is favored by Malloy, Pike, and Feldman, who discuss it in detail in Chapter 48. HVDRR is caused by a spectrum of mutations affecting the vitamin D receptor (VDR) in target tissues causing true resistance to calcitriol action. The human VDR, a 50-kDa protein, belongs to the steroid–thyroid–retinoic acid receptor superfamily

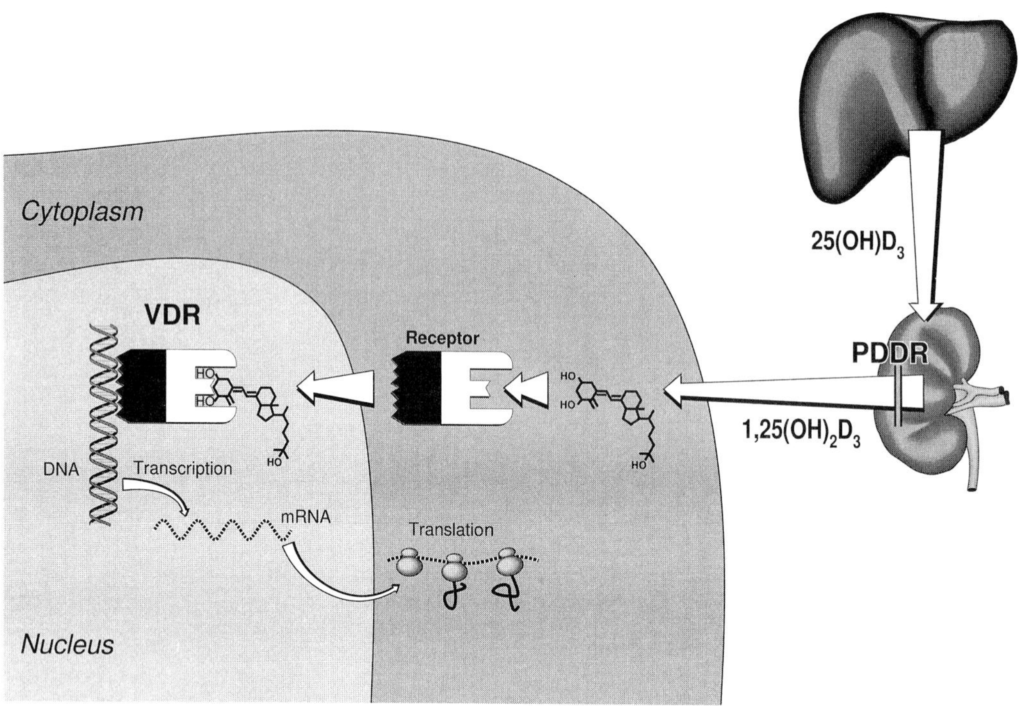

FIGURE 1 Schematic representation of the main steps of the vitamin D biosynthetic pathway, where genetic aberrations may lead to rickets and osteomalacia. The renal defect in PDDR is indicated by the break in the $1,25(OH)_2D_3$ arrow arising in the kidney. The mutation leads to insufficient synthesis of $1,25(OH)_2D_3$. The left part of the figure represents a target cell where schematic coupling of the ligand to its receptor (VDR) takes place in the cytosol or, more likely, in the nucleus. The complex then binds to DNA. Various mutations affecting either of the two VDR domains cause hereditary vitamin D-resistant rickets (HVDRR) (see Chapter 48].

of genes [9]. It comprises at least two functional domains, a ligand binding domain and a DNA binding domain (Fig. 1). Mutations affecting both have been found in HVDRR families (see Chapter 48).

II. CLINICAL MANIFESTATIONS

The clinical course of PDDR is similar to that of nutritional rickets due to simple vitamin D deficiency. The patients are healthy at birth. The first symptoms usually appear within the first year of life. Hypotonia, muscle weakness (proximal myopathy), and growth retardation are common manifestations. Motor problems translate into regression in head control and ability to stand. In some patients, the initial event is convulsions or tetany. Pathological fractures may occur. A history of adequate mineral and vitamin D intake, without evidence of intestinal malabsorption, is a constant finding. Infant death by hypocalcemia or pulmonary infections was not infrequent in the past when the diagnosis was either missed (confused with a neurological or respiratory condition) or made too late.

Physical examination reveals a small, hypotonic child with features similar to those found in patients with vitamin D deficiency rickets. There is a wide anterior fontanel with frontal bossing and frequent craniotabes (easy depression of the softened parietooccipital area). Tooth eruption is delayed, and erupted teeth show evidence of enamel hypoplasia. A rachitic rosary is either visible or palpable. In the appendicular skeleton, thickening of the metaphyseal areas is more evident in the wrists and ankles, and there is a variable degree of deformity (bowing) of long bone diaphyses. The Chvostek sign (twitching of the upper lip on light finger tapping of the facial nerve) reflects nerve irritability, a consequence of a rapid drop in serum calcium.

Radiological examination of the skeleton reveals diffuse osteopenia and the classic metaphyseal changes of vitamin D deficiency. There is fraying, cupping, widening, and fuzziness of the zone of provisional calcification immediately under the growth plate. These changes are seen better and detected earlier in the most active growth plates, namely, the distal ulna and femur and the proximal and distal tibia. Changes in the diaphyses may not be evident when metaphyseal changes are first detected. However, they will appear a few weeks later as rarefaction, coarse trabeculation, cortical thinning, and subperiosteal erosion (see Chapter 40). The latter is an expression of the increased resorption induced by secondary hyperparathyroidism.

III. BIOCHEMICAL FINDINGS

Hypocalcemia is the cardinal feature in PDDR. Serum calcium concentration will drop below 2 mmol/liter (8 mg/dl). This, particularly if the decrease is rapid, may give rise to tetany and convulsions, which may occur prior to any radiological evidence of rickets. Persistent hypocalcemia triggers secondary hyperparathyroidism and hyperaminoaciduria [10]. Urinary calcium content is low, whereas fecal calcium is high, reflecting impaired intestinal calcium absorption. Increased urinary cAMP is not a consistent finding, and normal values have been observed in PDDR patients with high circulating parathyroid hormone (PTH) levels [11].

Serum phosphate concentration may be normal or low. Hypophosphatemia, when present, is usually of a lesser degree than in XLH. It is the result of both impairment of intestinal absorption and increased urinary loss induced by secondary hyperparathyroidism. Serum alkaline phosphatase activity is consistently elevated (over 300 IU/liter). Its increase often precedes the appearance of clinical symptoms. The calcemic response to PTH is usually but not necessarily absent [12].

Studies of circulating vitamin D metabolites have provided a key insight into the pathogenesis of PDDR. Serum levels of 25OHD are normal in untreated patients and elevated in patients receiving large daily amounts of vitamin D [11]. These results indicate that intestinal absorption of vitamin D and its hydroxylation in the liver are not impaired in PDDR. Circulating levels of $1,25(OH)_2D$ are low in untreated patients [11–13]. This is evident immediately after birth, months before any clinical evidence of rickets develops. Even when patients are treated with large doses of vitamin D, causing major increases in the circulating levels of calcidiol, calcitriol levels do not reach the normal range (Fig. 2). This clearly identifies defective activity of the 25OHD 1α-hydroxylase enzyme as the basic abnormality in PDDR and differentiates it from HVDRR (see Chapter 48). Although $1,25(OH)_2D$ serum levels are low, they are not undetectable. This finding, coupled with the observation that serum levels of $1,25(OH)_2D$ are positively correlated to the serum concentrations of 25OHD in PDDR patients (either untreated or treated with large amounts of vitamin D) suggests that the renal 1α-hydroxylase is not totally absent in PDDR. Thus, the mutation probably affects the structural integrity of the enzyme, resulting in a modification of its kinetics [11]. Balsan and associates [14] have reported on normal

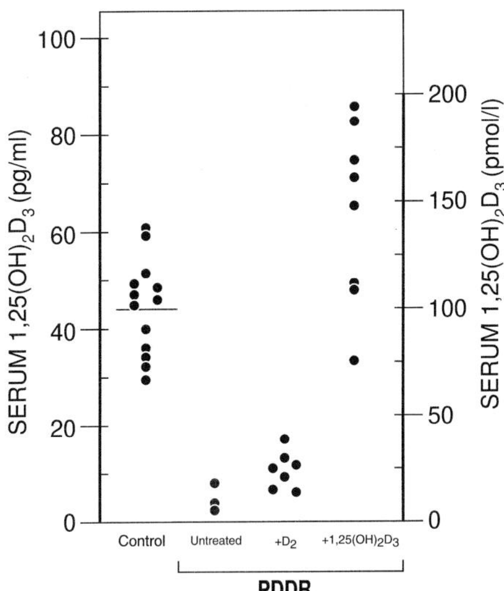

FIGURE 2 Serum calcitriol concentrations in control children and in PDDR patients either untreated or treated with high doses of vitamin D ($+D_2$) or calcitriol [$1,25(OH)_2D_3$]. The data scatter in the latter group reflects both dosage and the variable length of time between drug administration and blood sampling.

calcitriol levels in untreated PDDR patients. Such values, however, should be considered as inappropriate in the face of rickets, hypocalcemia, and secondary hyperparathyroidism. These differences may also reflect genetic heterogeneity among PDDR pedigrees that will only be resolved at the molecular level.

Circulating levels of $24,25(OH)_2D$ are normal in PDDR patients and are highly correlated with those of 25OHD, indicating a fully functional 24-hydroxylase enzyme [15,16]. These findings, as well as the observation that modulation of the expression of the 25OHD 24-hydroxylase is regulated independently from that of the 1α-hydroxylase [17], strongly suggest that the two renal hydroxylases are distinct gene products (see Chapters 4 and 5).

IV. PLACENTA STUDIES

In 1979, Weisman *et al.* [18] demonstrated that, besides the mammalian kidney, human placenta decidua was a major site of $1,25(OH)_2D$ synthesis. This was further substantiated by Delvin *et al.* [19], who also demonstrated that the involved enzyme was regulated by feedback mechanisms [20]. Over the years, in the cohort of our patients successfully treated with replacement therapy (see below), several have reached adulthood and became pregnant. At delivery, decidual cells were harvested from the placentas of these PDDR patients and were studied to evaluate their ability to hy-

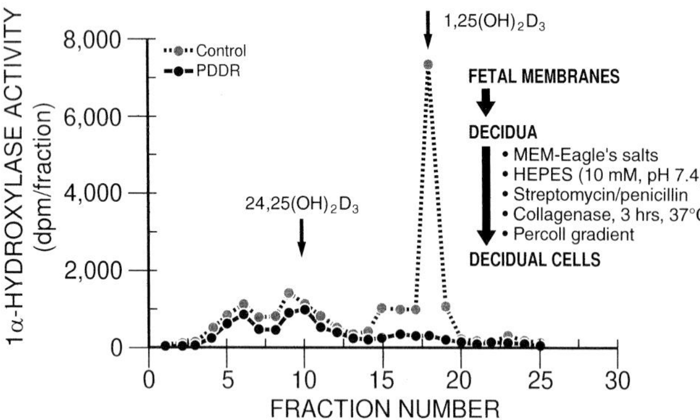

FIGURE 3 High-performance liquid chromatography elution patterns of the radioactivity extracted from medium and cells isolated from the decidua of one control and one PDDR patient. The elution positions of the two vitamin D dihydroxylated metabolites are indicated.

droxylate 25OHD at the 1α position. As shown in Fig. 3, we demonstrated that decidual cells from women with PDDR lack that function, making them likely targets for the mutation [21]. The physiological importance of this defect particularly with regard to fetal development is unclear. Replacement therapy with calcitriol was maintained throughout pregnancy, and patients had uneventful pregnancies and delivered healthy normocalcemic babies. The placenta thus represents a unique, albeit rare, source of mutant cells for further characterization of the PDDR mutation.

V. GENETICS STUDIES

Pseudovitamin D-deficiency rickets is inherited as a simple autosomal recessive trait [5]. No phenotypic abnormalities have been observed in presumed obligate heterozygotes [10]. Although quite rare, PDDR is present with unusual frequency in a subset of the French-Canadian population [22]. With the cooperation of the several large families under our care, we set out to map the PDDR locus by using DNA markers and linkage analysis to approach the primary defect in PDDR. The studied kindreds included 17 affected individuals and 59 healthy relatives, of whom 17 were obligate heterozygotes (having affected progeny). It was found that the mutated gene was linked to polymorphic RFLP (restriction fragment length polymorphism) markers in the region of band 14 of the long arm of chromosome 12 [12q14] [23]. Multipoint linkage analysis and studies of haplotypes (groups of tightly linked markers segregating together over the generations) and recombinants, strongly suggest the localization of the PDDR locus between COL2A1 (coding for the α1 chain of type II collagen) and a cluster of three anonymous probes (D12S14, D12S17, and D12S6] which segregate as a three-marker haplotype (Fig. 4). Linkage disequilibrium (i.e., combinations of closely linked genes occurring more often than expected with random distribution) has been observed between the PDDR locus and the three-marker haplotype in the group of kindreds studied [24]. The finding supports the notion of a founder effect that had taken place in the second half of the seventeenth century (about 12 generations ago). This is consistent with the present-day prevalence of 1 in 2400 births and carrier rate of 1 in 26 individuals in northeastern Quebec [25]. The localization of PDDR between flanking markers now allows for the prediction of the carrier status in healthy individuals from pedigrees where the disease is highly prevalent, and who seek genetic counseling (Fig. 5). The accuracy of such carrier detections could exceed 99% in families informative for the markers used [23]. Similar analyses have allowed us to detect affected newborns within the first week of life.

The VDR gene has also been assigned to chromosome 12 by Southern blot analysis of a panel of human–Chinese hamster cell hybrid DNAs. Using *in situ* hybridization, the VDR was found to map to the same 12q12–14 region where PDDR was localized [27]. Because the VDR cDNA exhibits an *Apa*I dimorphism [26], it was used as a RFLP marker in linkage analyses of samples from 21 of our PDDR families. The PDDR and VDR loci are located in close proximity to the markers COL2A1, ELA (elastase), and D12S15 (Fig. 5). It is likely that the genetic distance between the two genes involved in the control of vitamin D activity is in the range of a few centimorgans, which, in physical terms, may correspond to 1–10 megabases. We find, at present, no specific reason for this proximity, but its functional significance may be established in the future [27].

VI. MOLECULAR DEFECT

Although it is clear from clinical observations that PDDR is caused by defective activity of the renal 25OHD 1α hydroxylase leading to insufficient synthesis of calcitriol [6], the precise target of the mutation is not known. It could be either the enzyme itself leading to structural changes or, alternatively, a modulator of the activity of a normal enzyme. The 1α-hydroxylase is composed of three distinct moieties: cytochrome P450, ferredoxin, and ferredoxin reductase (reviewed by Ghazarian [28] (see also Chapter 5). *A priori*, each of the subunits could be affected by the mutation. However, the two components of the electron-transfer system are shared with other P450 enzymes and are encoded by single

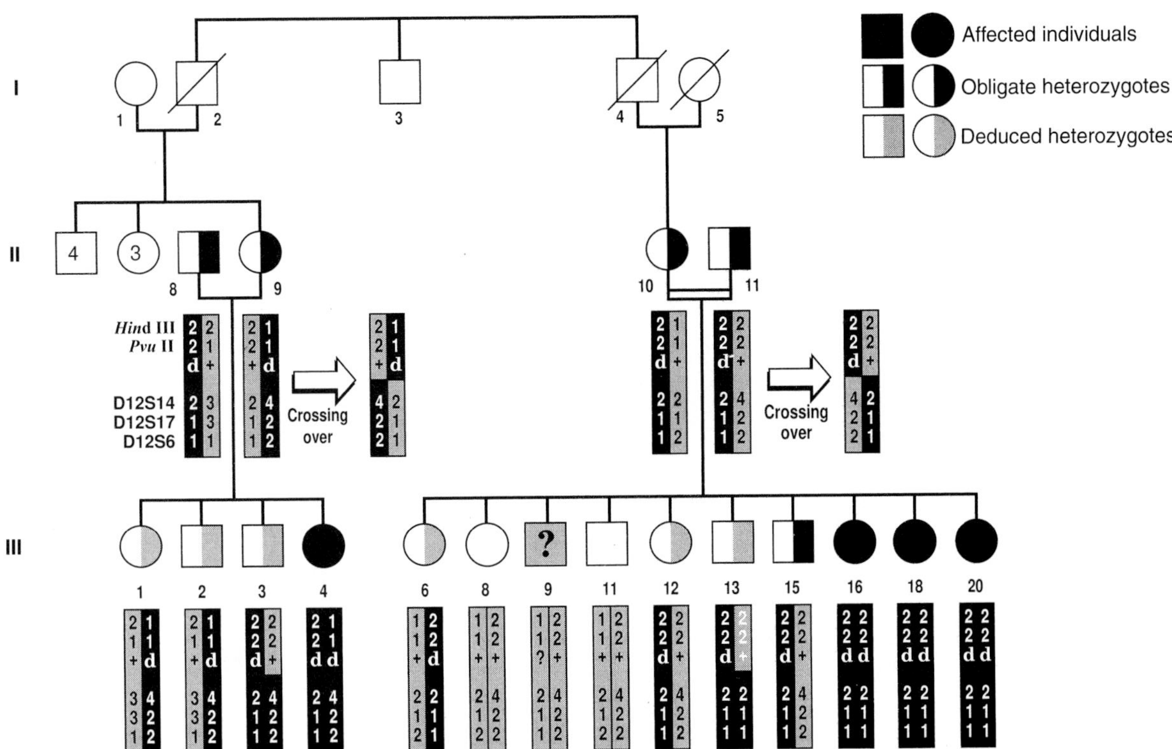

FIGURE 4 Representative PDDR family with two kindreds in which the disease is expressed. Under each genotyped subject, alleles are listed for the loci COL2A1 (*Hind*III and *Pvu*II, upper part of the box) and D12S14, D12S17, and D12S6 (lower part of the box). d, PDDR allele; +, normal allele. Black boxes designate chromosomes with the mutation. In individuals 9 and 11 in generation II, crossing-over was deduced from analysis of the progeny. Carrier status could be established for several individuals in generation III. The carrier status of individual III 9 is ambiguous (?) because his maternal haplotype is the result of a recombination with no precise point of crossing over. Recombination could have occurred anywhere between the COL2A1 and D12S6 loci.

copy genes, expressed in all steroidogenic tissues, that are located on chromosomes 11 and 17, respectively [29,30]. Because PDDR patients have no evident deficiencies in the activity of other cytochrome P450 systems, the P450 specific component of 1α-hydroxylase remains the most likely target of the PDDR mutation (Fig. 6).

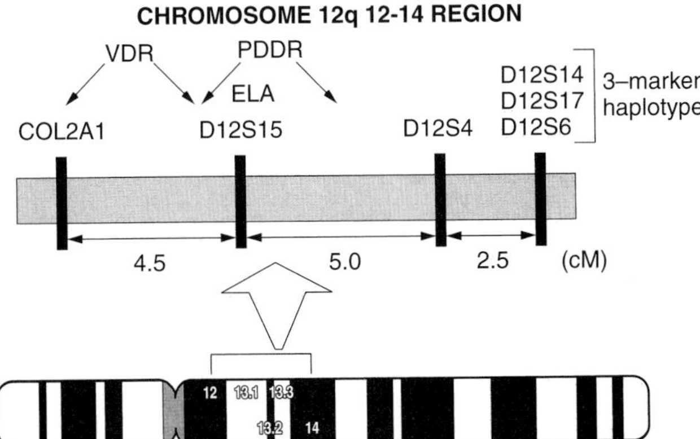

FIGURE 5 Schematic representation of the relevant part of the long arm of human chromosome 12 where the loci for PDDR and VDR have been located by linkage analysis. The various adjacent markers are indicated (see text for definitions). Genetic map distances between markers are measured in centimorgans (cM).

To understand the basis of the PDDR genotype, it will be necessary to clone and characterize a cDNA corresponding to the 1α-hydroxylase gene. Under the assumption that coding sequences for the 1α- and 24-hydroxylase genes are similar, we have used a probe centered on the heme binding domain of the rat 24-hydroxylase cDNA to screen a cDNA library from vitamin D-deficient rat kidney, under reduced stringency conditions. This procedure allowed us to identify a 2.4-kb full-length clone that codes for a protein with a predicted mass of 55 kDa. Amino acid sequence similarity with the 24-hydroxylase enzyme was 78% within the heme binding domain, but clear divergence was seen outside of this region, giving an overall sequence similarity of 26%. Transfection of embryonal carcinoma cells with the sense putative 1α-hydroxylase expression vector led to cells that produced a vitamin D metabolite that comigrated on high-performance liquid chromatography (HPLC) with the 1α,25$(OH)_2D_3$ standard using two different solvent systems. The compound was also authenticated using two different radioreceptor assays

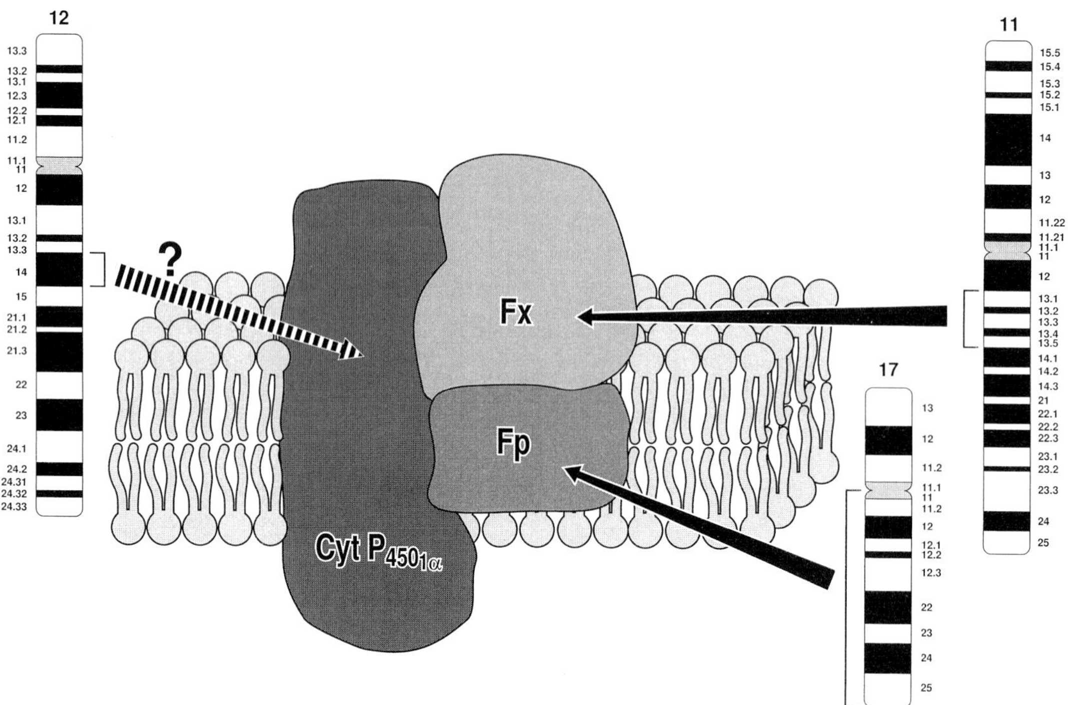

FIGURE 6 Schematic representation of the three subunits that form the 25OHD 1α-hydroxylase enzyme, with their respective genomic locations. The latter is not, as yet, established for the cytochrome P450 subunit. Its putative location on chromosome 12 is based on the linkage studies in PDDR (see text).

[31]. This rat cDNA probe will allow us to clone the human homolog, which will then be mapped to its genomic location. The hypothesis to be tested is that it will colocalize with the PDDR locus. If not the case, then the PDDR mutation is probably not affecting the structure of the enzyme but rather a factor controlling its activity.

Because we have demonstrated a founder effect for PDDR in northeastern Quebec, it is likely that all affected patients from that region will carry the same mutation. However, heterogeneity (in the sense of different loci being involved) cannot be excluded in PDDR. Once the mutation has been characterized in the Canadian patients, the study of PDDR pedigrees from other ethnic groups will help ascertain whether other mutations have occurred in the same gene and whether this may explain the variations observed in the biochemical phenotype [14].

VII. Treatment

Vitamin D_2, given at an appropriate daily dose, can be used to treat PDDR. The biochemical and clinical abnormalities regress and normal linear growth is restored. The dose of vitamin D_2 to heal the bone disease may be as high as 2.5 mg (100,000 IU) per day. This dose can be reduced by half or more to a maintenance dose, probably for the lifetime of the patient (Table I). Under such treatment, circulating levels of 25OHD increase sharply, with only minor changes in the levels of $1,25(OH)_2D$ (Fig. 2). It is likely that massive concentrations of 25OHD are able to bind to VDR and induce the response of the target organs to normalize calcium homeostasis. However, because such therapy leads to progressive accumulation of vitamin D in fat and muscle tissues, adjustment in case of overdose is difficult and slow to come into effect. Furthermore, the therapeutic doses are close to the toxic doses and place the patient at risk for nephrocalcinosis and impaired renal function.

There have been reports on the use of $25OHD_3$ as

TABLE I Vitamin D Dosage Requirements of PDDR Patients

Compound	Dosage (μg/day)	
	To heal rickets	Maintenance
Vitamin D	1000–2500	500–1250
$25OHD_3$	250–1000	100–500
$1\alpha OHD_3$	2–5	1–2
$1,25(OH)_2D_3$	1–3	0.25–1

a therapeutic agent in PDDR [32]. The doses used are smaller than those of vitamin D (Table I) and induce a similar response. The action of $25OHD_3$ is likely to be similar to the one of vitamin D itself, by maintaining high serum concentrations of 25OHD. The low availability and high cost of such a preparation have discouraged its widespread use as a long-term therapy for PDDR.

The treatment of choice is replacement therapy with calcitriol. Before this compound became available from commercial sources, several investigators used the monohydroxylated analog $1\alpha OHD_3$, which requires only liver hydroxylation at the 25 position (a step not affected by the PDDR mutation) to fully mimic $1,25(OH)_2D_3$ [33]. The response is rapid with healing of rickets in 7–9 weeks, requiring a daily dosage of 2–5 μg. The maintenance dose is about half the initial dose (Table I]. Withdrawal induces a reappearance of symptoms within 3 weeks. Thus, long-term compliance is a more important consideration than in the case of vitamin D treatment. On a weight basis, $1\alpha OHD_3$ is about half as potent as $1,25(OH)_2D$ [33], nullifying any possible economic advantage in favor of the monohydroxylated form. The reason for this difference in potency has not been investigated but may be related to a difference in intestinal absorption or to a variable degree of 25-hydroxylation of $1\alpha OHD_3$.

Replacement therapy with calcitriol results in rapid and complete correction of the abnormal phenotype, eliminating hypocalcemia, secondary hyperparathyroidism, and radiographic evidence of rickets (Figs. 7 and 8). The restoration of bone mineral content is equally rapid (Fig. 7), and histological evidence of healing has been documented [34]. Severe tooth enamel hypoplasia is a common complication of PDDR, which is only partially corrected if treatment, as is usually the case, is started around 12–15 months of age when permanent tooth enamel has already started to develop (Fig. 9).

The calcitriol regimen calls for an initial dose of 1–3 μg/day continued until bone is healed and is followed by a maintenance dose of 0.25–1 μg/day (Table I] to be continued probably throughout life. An important

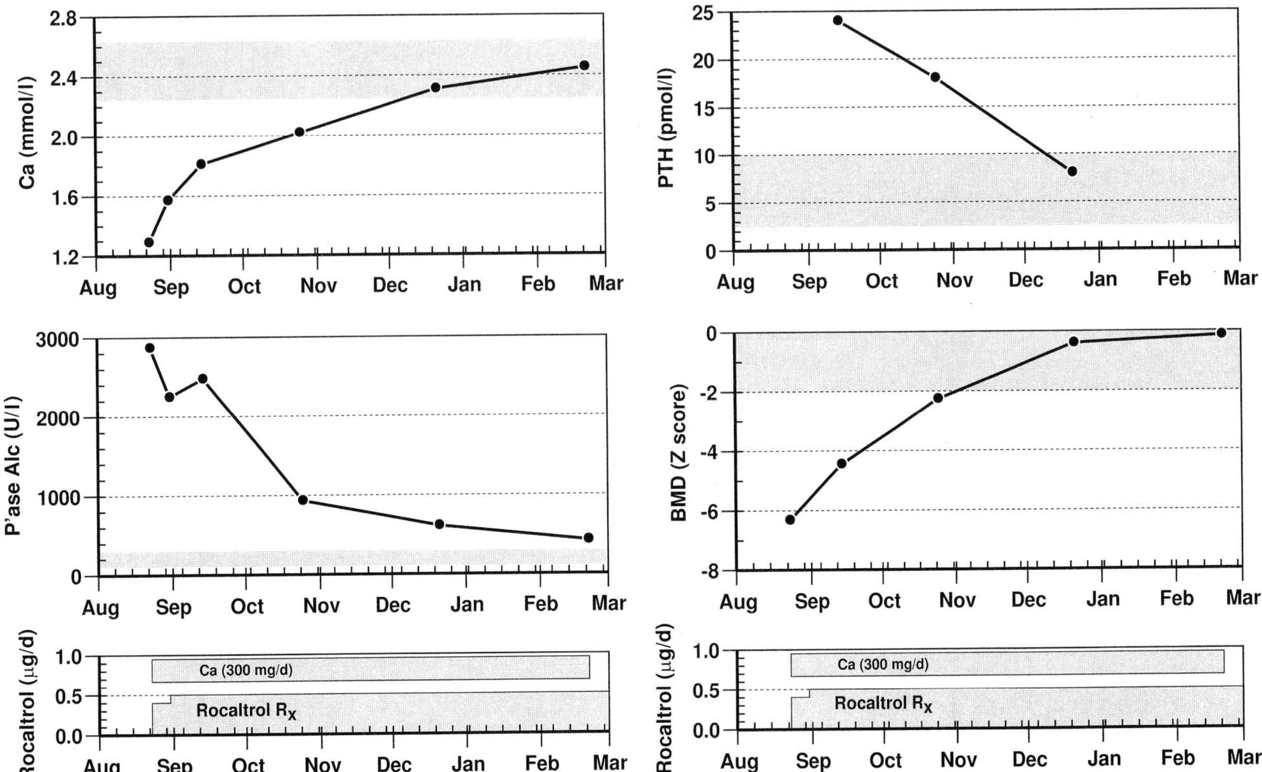

FIGURE 7 Biochemical response to treatment in a 16-month-old boy with PDDR treated with calcitriol (Rocaltrol) and calcium supplements. There was a progressive correction of the hypocalcemia (Ca) and secondary hyperparathyroidism (PTH) with concomittant decrease in alkaline phosphatase activity (P'ase Alc). Correction of the osteopenia followed the same pattern [BMD is bone mineral density of the lumbar spine by dual energy X-ray absorptiometry (DXA); the *Z* score is based on standard deviations from the mean BMD in age-matched controls].

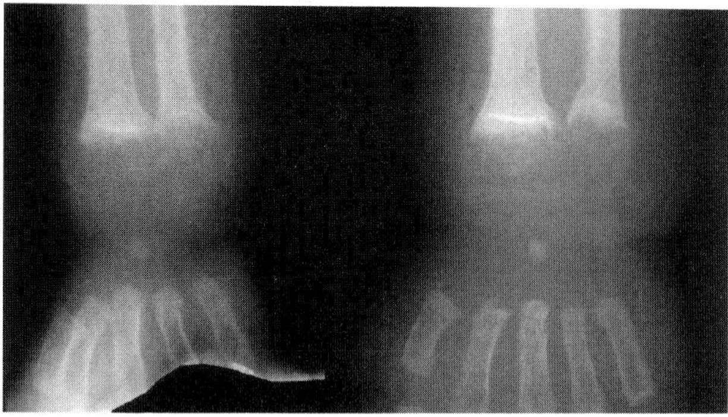

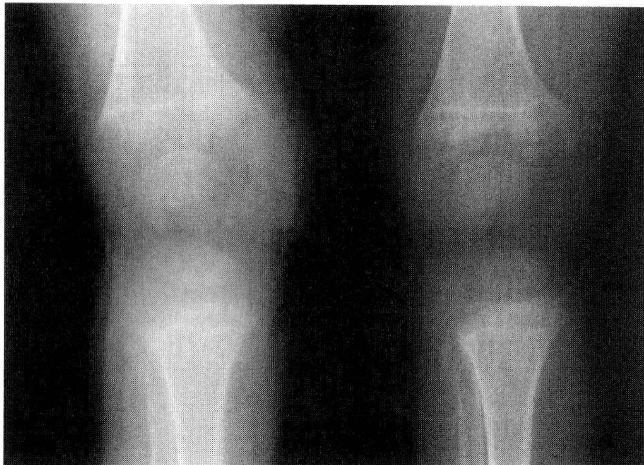

FIGURE 8 Radiographs of the right wrist (upper panel) and knee (lower panel) of a patient with PDDR (same as in Fig. 7). (Left) Before treatment; (right) after only 3 weeks of treatment, healing of rickets is well under way.

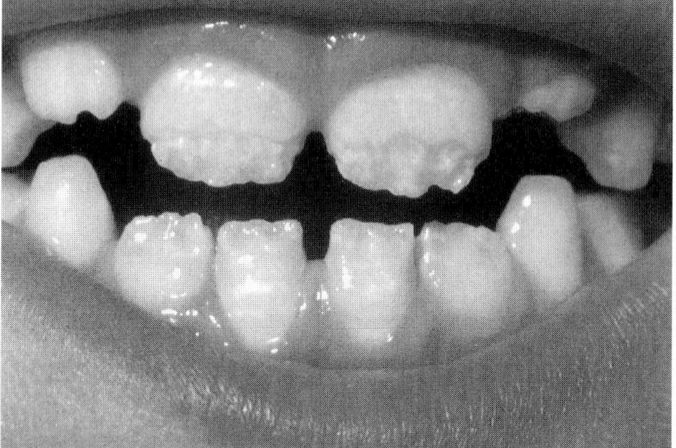

FIGURE 9 Permanent incisors of a 9-year-old patient with PDDR in whom calcitriol treatment was initiated at age 14 months. The part of the enamel that was formed before treatment remains hypoplastic. Subsequent to treatment normal enamel was produced.

component of treatment is to ensure adequate calcium intake during the bone healing phase. Dietary sources are supplemented to ensure a daily supply of around 1 g of elemental calcium. Needs are monitored by frequent (bimonthly) assessment of urinary calcium excretion. The latter can easily be assessed by measuring calcium and creatinine in an aliquot of the second void of the morning by a fasting patient. Normal values for the calcium/creatinine ratio are <0.35 (mg/mg) or <1.1 (mmol/mmol).

In the untreated state, calciuria is very low. It will stay low as long as bone is rapidly remineralizing. An increase in calciuria is the most sensitive index of efficient therapy. Hypercalciuria (which precedes hypercalcemia by weeks) calls for reducing calcitriol progressively to the maintenance dosage. Once the latter is established, assessment of calciuria every 3 months is sufficient to keep control of the treatment. Requirements have been remarkably stable in our cohort of 32 patients treated for up to 17 years.

During normal pregnancy, calcitriol circulating levels steadily increase to about twice the control values [35]. This adaptation to the specific needs of pregnancy can be mimicked in pregnant patients with PDDR by increasing the daily calcitriol dose during the second half of pregnancy. In seven such patients, we increased the calcitriol dose by 50–100% of the maintenance dose. All women gave birth to normal (obligate heterozygote) babies. The maintenance dose was progressively reestablished after delivery (F. H. Glorieux, unpublished data).

Hypercalciuria is not infrequent during treatment with calcitriol, particularly during the first year of administration, because changes in urinary calcium excretion are used to adjust the daily calcitriol dose. High levels of calcium excretion may amplify the pattern of calcium deposition in the normal kidney [36] and generate echodense images of the renal pyramids. This is referred to as nephrocalcinosis [37] and has been detected in several patients with PDDR treated with either vitamin D_2 or calcitriol [38]. As it may reflect a potential nephrotoxic effect of calcitriol we have now included in our treatment protocol an annual renal ultrasonography study and evaluation of the creatinine clearance. A positive ultrasound was observed in 10 of 20 patients with PDDR treated for a mean of 8.1 years. The intensity of the images did not change with time. Two patients have shown a decrease in creatinine clearance. However, both had a history of vitamin D intoxication prior to calcitriol therapy [39]. Thus, duration of administration and dosage of the compounds used for treatment will influence the development of renal medullary changes. Frequent renal imaging and assessment of renal function are therefore essential.

VIII. CONCLUSION

Pseudovitamin D-deficiency rickets is a rare condition inherited as an autosomal recessive trait that results in an inadequate synthesis of calcitriol that compromises intestinal calcium absorption and bone mineralization. The majority of the cases described are part of large kindreds from northeastern Quebec in Canada. Extensive genetic studies of those families have allowed an assignment of the PDDR locus to the long arm of chromosome 12 in close vicinity to the VDR gene. The exact nature of the PDDR mutation is still unknown, but it is most probably related to structural alteration of the cytochrome P450 subunit of the 25OHD 1α hydroxylase enzyme. Replacement therapy with small daily doses of calcitriol is the treatment of choice in PDDR. It is highly efficient, removing this condition from the list of lethal mutations. It should probably be continued throughout life. Because of the potential nephrotoxicity of this treatment, regular monitoring of kidney function is mandatory. PDDR was the first described inborn error of vitamin D metabolism. Through the complete unraveling of its molecular defect, it will continue to further our understanding of vitamin D biology.

References

1. Albright F, Butler AM, Bloomberg E 1937 Rickets resistant to vitamin D therapy. Am J Dis Children **54**:529–547.
2. Royer P 1960. Etude sur les rachitismes vitamino-résistants hypophosphatémiques idiopathiques. Acta Clin Belg **15**:499–517.
3. Fraser D, Salter RB (1958) The diagnosis and management of various types of rickets. Pediatr Clin North Am **5**:417–441.
4. Prader A, Illig R, Heierli E 1961 Eine besondere form des primäre vitamin D-resistenten rachitis mit hypocalcämie und autosomal-dominanten Erbgang: Die hereditäre PseudoMangelrachitis. Helv Paediatr Acta **16**:452–468.
5. Scriver CR 1970 Vitamin D dependency. Pediatrics **45**:361–363.
6. Fraser D, Kooh SW, Kind HP, Hollick MF, Tanaka Y, DeLuca HF 1973. Pathogenesis of hereditary vitamin D dependent rickets. An inborn error of vitamin D metabolism involving defective conversion of 25-hydroxyvitamin D to 1α,25-dihydroxyvitamin D. N Engl J Med **289**:817–822.
7. Marx SJ, Spiegel AM, Brown EM, Gardner DG, Downs RW, Attie M, Hamstra AJ, DeLuca HF 1978 A familial syndrome of decrease in sensitivity to 1,25-dihydroxyvitamin D, J Clin Endocrinol Metab **47**:1303–1310.
8. Rosen JF, Fleischman AR, Finberg L, Hamstra AJ, DeLuca HF 1979 Rickets with alopecia: An inborn error of vitamin D metabolism. J Pediatr **94**:729–735.
9. Baker AR, McDonnell DP, Hughes M, Crisp TM, Mangelsdorf DJ, Haussler MR, Pike JW, Shine J, O'Malley BW 1988 Cloning and expression of full length cDNA encoding human vitamin D receptor. Proc Natl Acad Sci USA **85**:3294–3298.
10. Arnaud C, Maijer R, Reade TM, Scriver CR, Whelan DT 1970 Vitamin D dependency: An inherited postnatal syndrome with secondary hyperparathyroidism. Pediatrics **46**:871–880.
11. Delvin EE, Glorieux FH, Marie PJ, Pettifor JM 1981 Vitamin D-dependency: Replacement therapy with calcitriol J Pediatr **99**:26–34.
12. Rosen JF, Finberg L 1972 Vitamin D-dependent rickets: Actions of parathyroid hormone and 25-hydroxycholecalciferol. Pediatr Res **6**:552–562.
13. Scriver CR, Reade TM, Hamstra AJ, DeLuca HF 1978 Serum 1,25-dihydroxyvitamin D levels in normal subjects and in patients with hereditary rickets or bone disease. N Engl J Med **299**:976–979.
14. Balsan S Garabedian M, Sorgniard R, Holick MF, DeLuca HF 1975 1,25-Dihydroxyvitamin D_3 and $1,\alpha$-hydroxyvitamin D_3 in children: Biologic and therapeutic effects in nutritional rickets and different types of vitamin D resistance. Pediatr Res **9**:593–599.
15. Glorieux FH, Delvin EE 1991 Pseudo-vitamin D deficiency rickets. *In*: Vitamin D: Regulation, Structure–Function Analysis and Clinical Application. de Gruyter, Berlin, and New York, pp. 238–245.
16. Mandla S, Jones G, Tenenhouse HS 1992 Normal 24-hydroxylation of vitamin D metabolites in patients with vitamin D-dependency rickets type I. Structural implications for vitamin D hydroxylases. J Clin Endocrinol Metab **74**:814–820.
17. Arabian A, Grover J, Barré MG, Delvin EE 1993 Rat kidney 25-hydroxyvitamin D_3 1α- and 24-hydroxylases: Evidence for two distinct gene products. J Steroid Biochem Mol Biol **45**:513–516.
18. Weisman Y, Harell A, Edelstein S, David M, Spirer Z, Golander A 1979 1α,25-Dihydroxyvitamin D_3 and 24,25-dihydroxyvitamin D_3 in vitro synthesis by human decidua and placenta. Nature **281**:317–319.
19. Delvin EE, Arabian A, Glorieux FH, Mamer OA 1985 *In vitro* metabolism of 25-hydroxycholecalciferol by isolated cells from human decidua. J Clin Endocrinol Metab **601**:880–885.
20. Delvin EE, Arabian A 1987 Kinetics and regulation of 25-hydroxycholecalciferol 1α-hydroxylase from cells isolated from human term decidua. Eur J Biochem **163**:659–662.
21. Glorieux FH, Arabian A, Delvin EE 1995 Pseudo-vitamin D deficiency: Absence of 25-hydroxyvitamin D 1α-hydroxylase activity in human placenta decidual cells. J Clin Endocrinol Metab **80**:2255–2258.
22. Bouchard G, Laberge C, Scriver CR, Glorieux F, Declos M, Bergeron L, Larochelle J, Mortezai S 1984 Étude démographique et généalogique de deux maladies héréditaires au Saguenay. Cahiers Québécois de Démographie **13**:117–137.
23. Labuda M, Morgan K, Glorieux FH 1990 Mapping autosomal recessive vitamin D dependency type I to chromosome 12q14 by linkage analysis. Am J Hum Genet **47**:28–36.
24. Labuda M, Labuda D, Korab-Laskowska M, Cole DEC, Zietkiewicz E, Weissenbach J, Popwska E, Pronicka E, Root AW Glorieux FH 1996 Linkage disequilibrium analysis in young populations: Pseudovitamin D deficiency rickets (PDDR) and the founder effect in French Canadians. Am J Hum Genet **59**:633–643.
25. De Braekeleer M. 1991 Hereditary disorders in Saguenay-Lac-St-Jean (Québec, Canada). Hum Hered **41**:141–146.
26. Faraco JH, Morrison NA, Baker A, Shine J, Frossard PM 1989 ApaI dimorphism at the human vitamin D receptor gene locus Nucleic Acids Res **17**:2150.
27. Labuda M, Fujiwara TM, Ross MV, Morgan K, Garcia-Heras J, Ledbetter DH, Hughes MR, Glorieux FH, 1992 Two hereditary defects related to vitamin D metabolism map to the same region of human chromosome 12q13–14. J Bone Miner Res **7**:1447–1453.
28. Ghazarian JG 1990 The renal mitochondrial hydroxylases of the vitamin D_3 endocrine complex: How are they regulated at the molecular level? J Bone Miner **5**:897–903.
29. Morel Y, Picado-Leonard J, Wu DA, Chang CY, Mohandas TK, Chung BC, Miller WL 1988 Assignment of the functional gene

for human adrenodoxin to chromosome 11q13→qter and of adrenodoxin pseudogenes to chromosome 20cen→q13.1 Am J Hum Genet **43**:52–59.

30. Solish SB, Picado-Leonard J, Morel Y, Kuhn RW, Mohandas TK, Hanukogly I, Miller WL 1988 Human adrenodoxin reductase: Two mRNAs encoded by a single gene on chromosome 17cen→q25 are expressed in steroidogenic tissues. Proc Natl Acad Sci USA **85**:7104–7108.
31. St-Arnaud R, Moir JM, Messerlian M, Glorieux FH 1996 Molecular cloning and characterization of a cDNA for vitamin D 1α-hydroxylase. J Bone Miner Res **11**(1):S124.
32. Balsan S, Garabedian M, Lieberherr M, Gueris J, Ulmann A 1979 Serum 1,25-dihydroxyvitamin D concentrations in two different types of pseudo-deficiency rickets. *In* Norman AW, Schaefer K, Herrath DV, Grigoleit H-G, Coburn JW, DeLuca HF, Mawer EB, Suda T (eds) Vitamin D: Basic Research and Its Clinical Application. de Gruyter, Berlin and New York, pp. 1143–1149.
33. Reade TM, Scriver CR, Glorieux FH, Nogrady B, Delvin E, Poirier R, Holick MF, DeLuca HF 1975 Response to crystalline 1α-hydroxyvitamin D_3 in vitamin D dependency. Pediatr Res **9**:593–599.
34. Delvin EE, Glorieux FH, Marie PJ, Pettifor JM 1991 Vitamin D-dependency: Replacement therapy with calcitriol. J Pediatr **99**:26–34, 1981.
35. Delvin EE, Salle BL, Glorieux FH, Adeleine P, David LS 1986 Vitamin D supplementation during pregnancy: Effect on neonatal calcium homeostasis. J Pediatr **109**:328–334.
36. Anderson L, McDonald JR 1946 The origin, frequency and significance of microscopic calculi in the kidney. Surg Gynecol Obstet **82**:275–282.
37. Alon U, Brewer WH, Chan JCM 1983. Nephrocalcinosis: Detection by ultrasonography. Pediatrics **71**:970–973.
38. Goodyear PR, Kronick JB, Jequier S, Reade TM, Seriver CR 1987 Nephrocalcinosis and its relationship to treatment of hereditary rickets. J Pediatr **11**:700–704.
39. Glorieux FH 1990 Calcitriol treatment in vitamin D-dependent and vitamin D-resistant rickets. Metabolism **39**:(Supp 1):10–12.

CHAPTER 48

Hereditary 1,25-Dihydroxyvitamin D Resistant Rickets

PETER J. MALLOY Division of Endocrinology, Gerontology, and Metabolism, Stanford University School of Medicine, Stanford, California

J. WESLEY PIKE Department of Molecular and Cellular Physiology, University of Cincinnati, Cincinnati, Ohio

DAVID FELDMAN Division of Endocrinology, Gerontology, and Metabolism, Stanford University School of Medicine, Stanford, California

I. INTRODUCTION

Hereditary vitamin D-resistant rickets (HVDRR), also known as vitamin D-dependent rickets, type II, is a rare autosomal recessive disease that is due to target organ resistance to 1,25-dihydroxyvitamin D [$1,25(OH)_2D$]. The intracellular response to $1,25(OH)_2D$, the active form of vitamin D, is initiated through binding of the hormone to specific vitamin D receptors (VDR) located in the nucleus of target cells. The VDR is a member of the steroid–thyroid–retinoid receptor gene superfamily and acts in turn as a transcription factor to regulate the expression of specific target genes, which ultimately determines the bioactivity of the hormone. In HVDRR, target organ resistance to vitamin D has been shown to be caused by mutations in the VDR. In this chapter, we describe the clinical syndrome as well as discuss the genetic defects in the VDR that render the receptor nonfunctional or of reduced function and thus define the underlying molecular basis for HVDRR.

In 1937, Albright *et al.* [1] described a patient with normocalcemic, hypophosphatemic rickets who required abnormally high doses of vitamin D as therapy for the disease. The authors suggested that the cause of the condition was end-organ resistance to vitamin D, and thus the concept of hormone resistance emerged. The disease they described appears to be what is now known as X-linked hypophosphatemic rickets (XLH, described in Chapter 46). Twenty-four years later, Prader *et al.* [2] described two patients with hypocalcemic, hypophosphatemic rickets who responded to high doses of vitamin D. This form of rickets appeared to be due to an inborn error in the conversion of vitamin D to the hormonally active form 1,25$(OH)_2$D. This disease (which is described in Chapter 47) is referred to by a number of names including vitamin D dependency rickets, pseudovitamin D deficiency rickets (PDDR), hypocalcemic vitamin D-resistant rickets, and vitamin D-dependent rickets type I (VDDR-I). In 1978, Brooks *et al.* [3] described a patient with rickets who had hypocalcemia, hypophosphatemia, secondary hyperparathyroidism, and markedly increased serum 1,25$(OH)_2$D levels. Since the patient had high circulating levels of 1,25$(OH)_2$D, they postulated that the rickets was due to impaired responsiveness of target organs to 1,25$(OH)_2$D. They termed the disease vitamin D-dependent rickets type II (VDDR-II) in order to distinguish it from VDDR I. Marx *et al.* [4] reported a similar syndrome in two children later the same year. Since these initial studies, there have been many other reports of patients with apparent target tissue resistance to 1,25$(OH)_2$D, and a number of different terms have been used to describe this syndrome. In addition to VDDR-II, the disease has been called pseudovitamin D deficiency type II (PDDR II), calcitriol-resistant rickets (CRR), hypocalcemic vitamin D-resistant rickets (HVDRR), and hereditary 1,25-dihydroxyvitamin D-resistant rickets (HVDRR). The latter term provides a better description of the disease, as it is based on a genetic defect and patients are actually resistant to and are not dependent on 1,25$(OH)_2$D. In this chapter, we refer to this disease as HVDRR, a more accurate depiction of the disease.

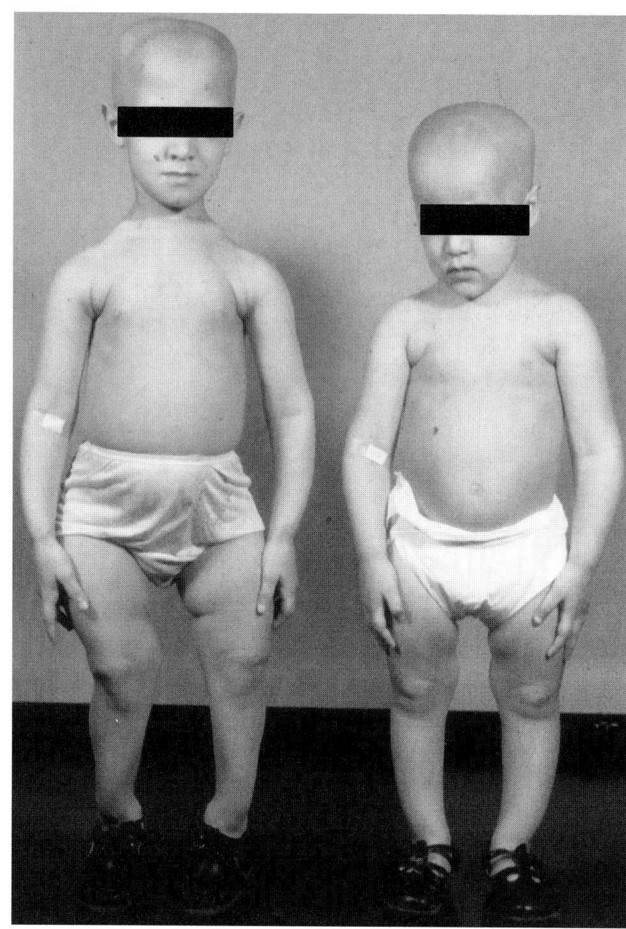

FIGURE 1 Two siblings with HVDRR and alopecia. The subjects are from the F3 family as detailed in Table I. Reprinted with permission from *The Journal of Pediatrics,* JF Rosen, AR Fleischman, L Fineberg, A Hamstra, and HF DeLuca. Rickets with alopecia: An inborn error of vitamin D metabolism. 1979;94:729–735.

II. THE CLINICAL SYNDROME

A. Differential Diagnosis

The HVDRR syndrome is manifested by a constellation of features caused by a generalized resistance to 1,25$(OH)_2$D [3–45]. The major feature of HVDRR is rickets (Fig. 1), which is often severe and generally displayed early in life, usually within months of birth. Individuals with HVDRR suffer from bone pain, muscle weakness, hypotonia, and occasionally convulsions. Children are often growth retarded and develop severe dental caries or exhibit enamel hypoplasia of the teeth [6,9,10,14,16,34,37]. Some infants have died from pneumonia due to severe rickets of the chest wall and inadequate respiratory movement [9,16,29]. Biochemically, patients with HVDRR exhibit hypocalcemia, hypophosphatemia, elevated serum alkaline phosphatase activity, and elevated serum parathyroid hormone (PTH) levels. These clinical and biochemical findings are also common to PDDR as described in Chapter 47. HVDRR is clearly distinct from PDDR, which is believed to be caused by a genetic defect in the 25-hydroxyvitamin D (25OHD) 1α-hydroxylase, the renal enzyme that converts 25OHD to 1,25$(OH)_2$D. The primary attribute that differentiates these two syndromes is the circulating level of serum 1,25$(OH)_2$D. In PDDR, the serum 1,25$(OH)_2$D level is depressed, whereas in HVDRR the level is elevated. Patients with PDDR can be effectively treated with physiological doses of calcitriol, which circumvents the enzyme deficiency and restores the circulating 1,25$(OH)_2$D to normal. In contrast, most individuals with HVDRR are resistant to supraphysiological doses of calcitriol or its analogs.

B. Pathophysiology

Among the many biological processes attributed to vitamin D, maintenance of calcium homeostasis is most apparent. The hormone 1,25$(OH)_2$D is essential for promoting calcium and phosphate transport across the small intestine and into the circulation, which is necessary for the normal mineralization of bone. It is now well established that the biological actions of 1,25(OH)$_2$D are mediated by the VDR, a nuclear transcription factor that regulates gene expression in 1,25(OH)$_2$D-responsive cells. Because vitamin D is involved in the translocation of calcium and phosphate, conditions that adversely affect the 1,25(OH)$_2$D action pathway can lead to a decrease in mineral transport, creating a condition of hypocalcemia and hypophosphatemia. The calcium and phosphate deficiency prevents normal bone mineralization, leading to osteomalacia and rickets (see Chapter 41). Rickets is a condition that results from an inability to mineralize bone sufficiently, and the poorly formed bone that is made is brittle and porous. In HVDRR, 1,25(OH)$_2$D target organs such as the intestine become resistant to the hormone; therefore, the intestine is less efficient in promoting calcium and phosphate transport into the circulation. The underlying cause of this hormone resistance is the result of changes in the VDR that render the receptor nonfunctional or less active.

As described in detail in Section V, HVDRR is caused by mutations in the VDR gene. The mutated VDR is unable or less able to respond to 1,25(OH)$_2$D, and therefore its capacity to mediate the actions of 1,25(OH)$_2$D to promote calcium absorption is impaired; this results in the HVDRR syndrome. HVDRR follows an autosomal recessive pattern of inheritance, affecting both sexes and multiple children in the same family. Parents of individuals with HVDRR are presumed heterozygotes; nevertheless, they are asymptomatic and show no obvious evidence of bone disease. This recessive genetic disorder has usually been found in families where there is a high incidence of consanguinity. HVDRR is characterized by hypocalcemia and the early onset of rickets, generally observed within the first year of life. As a consequence of the hypocalcemia, the children develop secondary hyperparathyroidism. The excess PTH contributes to the hypophosphatemia and stimulates increased renal 1α-hydroxylase enzyme activity, resulting in elevated serum 1,25(OH)$_2$D levels. The high circulating 1,25(OH)$_2$D levels in the face of hypocalcemia demonstrate hormone resistance and provide evidence for HVDRR. These findings exclude diseases of vitamin D deficiency, including PDDR. Interestingly, in some cases of HVDRR, a spontaneous healing of the rickets has been noted between 8 and 15 years of age. The mechanism for this clinical improvement is still unknown, and the details are discussed later in the chapter.

C. Alopecia

Another feature that is uniquely found in HVDRR is alopecia totalis (Fig. 1). The majority of HVDRR patients have sparse body hair, and some have total scalp and body alopecia [23,26,46]. Children with severe alopecia often lack eyebrows and in some cases eyelashes. Hair loss usually occurs during the first year of life. An analysis of HVDRR patients has shown that there is some correlation between the severity of rickets and the presence of alopecia [46]. In addition, patients with alopecia generally have more severe resistance to calcitriol than those without alopecia, In families with a prior history of the disease, the absence of body hair in newborns provides initial diagnostic evidence for HVDRR.

D. Other Aspects of HVDRR

As mentioned above and discussed extensively in this volume, in addition to maintaining calcium homeostasis, 1,25(OH)$_2$D regulates a number of biological processes in many tissues [47–52]. Despite the many tissue responses shown to be regulated by 1,25(OH)$_2$D, children with HVDRR appear relatively normal except for the constellation of features that relate to their calcium deficiency and alopecia. The VDR has been demonstrated in endocrine glands such as pituitary, pancreas, parathyroid, gonads, and placenta, and 1,25(OH)$_2$D has been shown to regulate hormone synthesis and secretion in several of these organs [47–52]. However, Hochberg *et al.* [53] examined insulin, thyrotropin (TSH), prolactin (PRL), growth hormone (GH), and testosterone levels in sera from patients with HVDRR and found no abnormalities in hormone secretion. A study by Even *et al.* [54] showed that urinary cAMP and excretion of potassium, phosphorus, and bicarbonate were normal in HVDRR patients treated with PTH. However, PTH failed to decrease urinary calcium and sodium excretion in these patients to the extent found in the control patients. This suggests that 1,25(OH)$_2$D may selectively modulate the renal response to PTH and faciliate the PTH-induced reabsorption of calcium and sodium [54]. The VDR has also been found in hematolymphopoietic cells, and 1,25(OH)$_2$D has been shown to regulate cell differentiation and the production of interleukins and cytokines [55]. Although minor aberrations have been noted in the fungicidal activity of neutrophils from HVDRR patients [56], the patients do not exhibit any clinically apparent immunological defects. In the light of the diverse actions of 1,25(OH)$_2$D demonstrated in many nonosteogenic tissues *in vitro*, the absence of related findings in children with HVDRR suggests that the lack of 1,25(OH)$_2$D responses in these nonosteogenic tissues is compensated by other mechanisms in such a way that abnormalities are not manifested.

TABLE I Compilation of HVDRR Cases

Family[a]	Patient name/description[b]	Ethnic origin	Consanguinity[c]	Age at onset	Sex	Total affected	Alopecia[d]	$1,25(OH)_2D$ binding	VDR mutation	Ref.
F1	IIB, patient 1, 1a		0	20 months	M	2	0	+		4,13,24,30
	IIC, patient 2, 1b			5 months	F					4,13,30
F2	Patient		0	15 years	F	1	0			3
F3	Patient 1, patient 2a, 2a		+	1 year	F	2	+	+		6,13,19,30,38
	Patient 2, patient 2b, 2b		+	1 year	F			+		6,13,19,24
F4	Patient		0	2 years	F	1	0	+		7,18
F5	Patient, patient 3, 3, kindred 3, P3		+	10 months	F	1	+	+	Arg80Gln	9,19,24,30,86
F6	K.N.	Japanese	+	15 months	F	1	+	–		11,28
F7	Patient		0	18 months	M	1	+			10
F8	Patient		?	45 years	M	1				8
F9	M.A., kindred 6, patient 6, 6	Arab	0	1 year	F	1	+	–		12,19,30
F10	Patient		0	12 years	F	1	0			14
F11	I.H., A1, patient 2, I.K., case 1	Arab	+	1 year	M	2	+	–	Tyr295stop	15,22,23,107
	R.K., patient 1, A2, case 2		+	<1 year	F	1	+	–	Tyr295stop	22,23,26,107
F12	Patient 4		+			1	+	±		19,20
F13	Patient 5		+	5 months		1	+	–		19,20
F14	Patient		+	2 years	M	1	0			21
F15	Patient A, patient 5		+	19 months	F	4	+	–		16
F16	Patient B, patient 4		+	8 months	F	2	+	±		16
F17	B patient		?			1	+	–		22
F18	S.H., patient 3, case 3, C1	Arab	0	6 months	M	2	+	–	Tyr295stop	22,23,26,64
	R.H., patient 4, case 4, C2		0	1 month	M		+	–	Tyr295stop	22,23,26,64
F19	D1	Haitian	+		F	3	+	+	Arg73Gln	25,63
	D2		+		F		+	+	Arg73Gln	25,63
F20	Kindred 7, patient 7, 7, P7					2	+	+	Arg 80Gln	24,30,38,86
F21	I.S., patient	Kuwait	+	1 year	M	1	0	–	Arg274Leu	29,88,90
F22	Patient	Hispanic	0	1 year	M	1	+	±		28
F23	Patient	Saudi	+		M	1	+	+	Gly46Asp	32,97
F24	Patient 1, N.D.	Arab	+	4 months	M	1	+	–		27,35
F25	Patient 1	Japanese	+	2 months	F	2	+	+		36
F26	Patient 2	Japanese	0	2 months	F	1	+	+	Arg50Gln	36,87
F27	Patient 3	Japanese	0	2 months	M	1	+	+		36
F28	Patient 2, M.T.	Persian-Jewish	+	9 months	M	1	+	–		27, 35
F29	Patient, line 10	Saudi	+	13 months	M	2	+	–	Tyr295stop	37,38,91
F30	Line 15	Saudi						–	Tyr295stop	91

continues

III. MECHANISM OF $1,25(OH)_2D$ ACTION

Classic vitamin D actions are mediated by the hormone $1,25(OH)_2D$, which binds with high affinity to the ligand binding domain (LBD) of the VDR in target cells as detailed in Chapter 8. $1,25(OH)_2D$ binding most likely induces a conformational change in the VDR, forming an activated hormone–receptor complex. The hormone-activated receptor is then able to interact with specific nucleotide sequences termed vitamin D response elements (VDREs). VDREs are generally located proximal to the core promoter sequences in $1,25(OH)_2D$ responsive genes. Recognition and binding of the VDR to a VDRE also require participation of retinoid X receptor (RXR), a multigene member of the thyroid–retinoid subgroup of receptors. In the presence of $1,25(OH)_2D$, the VDR and RXR proteins heterodimerize, possibly through one or more of the helical segments located in the distal region of the VDR molecule as described in detail in Chapters 8–10. VDR dimerization with RXR is hypothesized to be essential for VDR transactivation. Hormonal activation of VDR initiates transcription of specific target genes and results in the biological activity induced by the hormone. Other proteins termed corepressors and coactivators have been shown to participate in the transcriptional activity of receptors in the steroid–thyroid–retinoid family. Their potential identity and involvement in

TABLE I *continued*

Family[a]	Patient name/description[b]	Ethnic origin	Consanguinity[c]	Age at onset	Sex	Total affected	Alopecia[d]	$1,25(OH)_2D$ binding	VDR mutation	Ref.
F31	G1	Arab	+		M	2	+	+	Gly33Asp	39,63
	G2		+		M		+	+	Gly33Asp	39,63
F32	Line 11, patient 1	Turkish	+	5 weeks	F	2	+	−	Gln152stop	38,90,113
	Line 11b, patient 2		+		M		+	−	Gln152stop	91,113
F33	Patient 1, patient 1a, patient 4a	Japanese	+	16 months	M	2	+	+	Arg50Gln	40,87
	Patient 2, patient 1b, patient 4b		+	16 months	F		+	+	Arg50Gln	40,87
F34	E1	Arab	+		M	1	+	−	Tyr295stop	64,65
F35	F1	Arab	+		F	1	+	−	Tyr295stop	65
F36	H1	Arab	0		F	1	+	−	Tyr295stop	64,65
F37	J1	Arab	0		M	1	+	−	Tyr295stop	65
F38	K1	Arab	0		M	1	+	−	Tyr295stop	65
F39	L1	Arab	+		F	1	+	−	Tyr295stop	65
F40	Ro-VDR, brother				M	2	0	+		41
	Al-VDR, sister				F		0	+		41
F41	Ab-VDR				M	1	+	−		41
F42	Patient					1		−	Exon 7–9 deletion	110
F43	Child					1			Cys190Trp	110
F44	Patient II, case 2	Tunisian	+		M	1	+	+	Lys45Glu	44,92
F45	Propositus	Japanese-Brazilian	+		M	1	+	+	His35Gln	89
F46	Line 14	Moroccan	+			2	+	−	Arg73stop	91
F47	Patient I	Mauritius	0		F	1	+	+	Phe47Ile	45,92
F48	J.K.	English	0	16 months	F	1	+	+	None	93
F49	N1	Tunisian-Jewish	?	7 months	M	2	+	+	Arg80Gln	85
	N2		?	6 month	F		+	+	Arg80Gln	85
F50	Patient	Greek	0	9 months	F	1	+	−	Exon skipping	111
F51	Patient	Turkish	+	3 months	M	2	0	+	His305Gln	108,109
	Sister		+		F		0	+	His305Gln	108,109
F52	Patient	Saudi	+	20 months	F	3	0	+	Cys44Tyr	Unpublished[e]

[a] Number assigned to family for citation.
[b] Name or description of propositus used in references.
[c] 0, Unrelated; +, related; ?, unknown relationship.
[d] 0, absent; +, present.
[e] Malloy, Al-Ashwal, and Feldman, 1997 (in preparation).

HVDRR are currently under investigation (see also Chapter 10).

IV. CELLULAR BASIS OF HVDRR

A. Use of Cultured Skin Fibroblasts

The syndrome of HVDRR was first recognized as an entity in 1978–1979 [3,4,6,7]. Since then a number of cases of vitamin D resistance have been described [8–45] (Table I). The clinical findings suggested end-organ resistance to vitamin D because the affected children failed to respond to the administration of physiological and even supraphysiological doses of vitamin D. The identification of $1,25(OH)_2D$ as the active form of vitamin D and the identification and characterization of the VDR as the mediator of $1,25(OH)_2D$ action set the stage for understanding the basis of HVDRR. Because the VDR was present in the major target tissues of vitamin D action, a genetically defective receptor was suspected as a likely candidate for causing end-organ resistance to vitamin D [4,6,9,10].

In 1980, Feldman *et al.* [57] demonstrated that receptors for $1,25(OH)_2D_3$ were present in fresh and cultured human foreskin as well as in cultured keratinocytes and dermal fibroblasts grown from adult skin biopsies. In 1982, Feldman *et al.* [15] went on to report the first investigation of the vitamin D system in cultured skin fibroblasts from HVDRR patients (family F11, where the F designation refers to the HVDRR families detailed in Table I). In this study, the authors showed that cultured fibroblasts from the HVDRR patients displayed no detectable ^3H-1,25$(OH)_2D_3$ binding activity. Furthermore, the cells were shown to be resistant to 1,25$(OH)_2D_3$ action by virtue of a lack of induction of the enzyme 25-hydroxyvitamin D 24-hydroxylase (24-hydroxylase) (Fig. 2). Induction of this enzyme has become a well-characterized biomarker for cellular 1,25$(OH)_2D_3$ responsiveness. Several laboratories then analyzed the 1,25$(OH)_2D_3$ binding properties of the VDR and the bioactivity of the hormone in fibroblasts cultured from skin biopsies [13,16–19] or bone [20] of HVDRR patients. In some cases, the lack of ^3H-1,25$(OH)_2D_3$ binding activity suggested the absence of VDR [13,16–20]. These defects were referred to as "receptor negative." In other cases ^3H-1,25$(OH)_2D_3$ binding activity supported the presence of a VDR protein [16,18,20,30], and these were referred to as "receptor positive." Several groups went on to examine the bioactivity or responsiveness of HVDRR cells to 1,25$(OH)_2D_3$. Griffin and Zerwekh [18] and Liberman *et al.* [19,20] similarly used 24-hydroxylase as a marker of 1,25$(OH)_2D_3$ resistance, whereas Clemens *et al.* [17] demonstrated that HVDRR fibroblasts were not growth arrested by hormone treatment. These early observations showed that HVDRR was caused by a cellular resistance or insensitivity to 1,25$(OH)_2D_3$ action and suggested that the resistance was associated with abnormalities in the VDR.

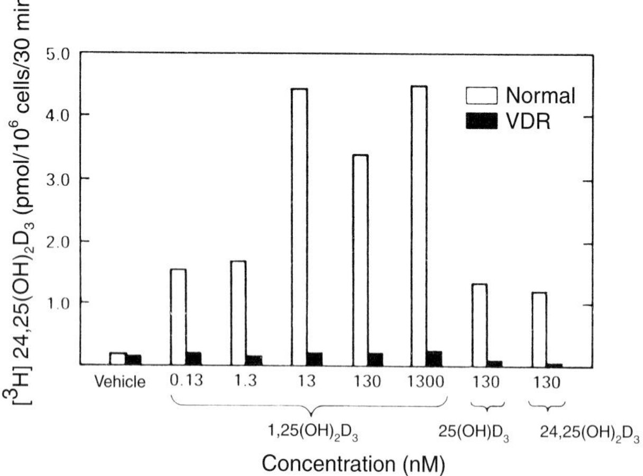

FIGURE 2 Induction of 24-hydroxylase activity in normal fibroblasts and fibroblasts from a patient (F11) with HVDRR. Cells were induced for 16 hr with varying concentrations of 1,25$(OH)_2D_3$, $25OHD_3$, or 24,25$(OH)_2D_3$. 24-Hydroxylase activity was determined by measuring the conversion of ^3H-25$(OH)_2D_3$ to ^3H-24,25$(OH)_2D_3$ by high-performance liquid chromatography (HPLC). Normal control, open bars; HVDRR patient, solid bars. Reprinted with permission from D Feldman, T Chen, C Cone, M Hirst, S Shani, A Benderli, and Z Hochberg. Vitamin D resistant rickets with alopecia: Cultured skin fibroblasts exhibit defective cytoplasmic receptors and unresponsiveness to 1,25$(OH)_2D_3$. *J. Clin. Endocrinol. Metab.* 55:1020–1022;1982. © The Endocrine Society.

B. Receptor-Negative Phenotype

As the number of reports on HVDRR increased, the heterogeneous nature of the defects in the VDR became apparent. Hochberg *et al.* [23,26] reported clinical findings on four patients from two unrelated families (F11, F18) that exhibited 1,25$(OH)_2D_3$ resistance, rickets, and alopecia. Fibroblasts from three of the patients and several of their parents (designated A and C families) as well as an additional unrelated family (F17) (B family) were studied by Chen *et al.* [22] (Table I). In these studies, the authors were unable to detect any ^3H-1,25$(OH)_2D_3$ binding in the HVDRR fibroblasts or to demonstrate that hormonal treatment could induce the 24-hydroxylase enzyme activity (Fig. 3). These individuals were designated as having a hormone binding negative (or receptor negative) phenotype. Interestingly, of the five parents in the study, only the father in the C family (F18) exhibited a phenotype theoretically expected of a heterozygote with one normal allele and one mutant allele. The fibroblasts from the C father contained half of the normal amount of VDR binding activity and exhibited a half-maximal response to 1,25$(OH)_2D_3$ (Fig. 3). In contrast, the fibroblasts of the other four parents showed a normal complement of VDR and a normal response to 1,25$(OH)_2D_3$ treatment (Fig. 3). The latter feature has been the usual pattern found in heterozygotes. The presence of VDR in patients with the hormone binding negative (or receptor negative) phenotype [12,16] was first demonstrated by Pike *et al.* [58]. Using a radioligand immunoassay [59] and a monoclonal antibody to the chick VDR [60–62], an immunoreactive protein was detected in cytosol from fibroblasts of HVDRR patients, which suggested that the defect in the VDR in these patients was due to a structural abnormality in the ligand binding domain (LBD) preventing ^3H-1,25$(OH)_2D_3$ binding and not from defective synthesis of the protein [58].

C. Receptor-Positive Phenotype

Gamblin *et al.* [24] examined 1,25$(OH)_2D_3$ induction of 24-hydroxylase activity in fibroblasts from four unrelated individuals (F1, F3, F5, F20) with HVDRR, all of whom exhibited normal 1,25$(OH)_2D_3$ binding. Complete hormone resistance was seen in two of the cases (F5, F20), whereas two cases (F1, F3) exhibited 1,25$(OH)_2D_3$-inducible 24-hydroxylase activity only when exposed to high concentrations of the hormone. The patients whose fibroblasts showed a response to high concentrations of 1,25$(OH)_2D_3$ *in vitro* also showed a calcemic response to high doses of calciferols *in vivo*. Castells *et al.* [28] described an individual (F22) who

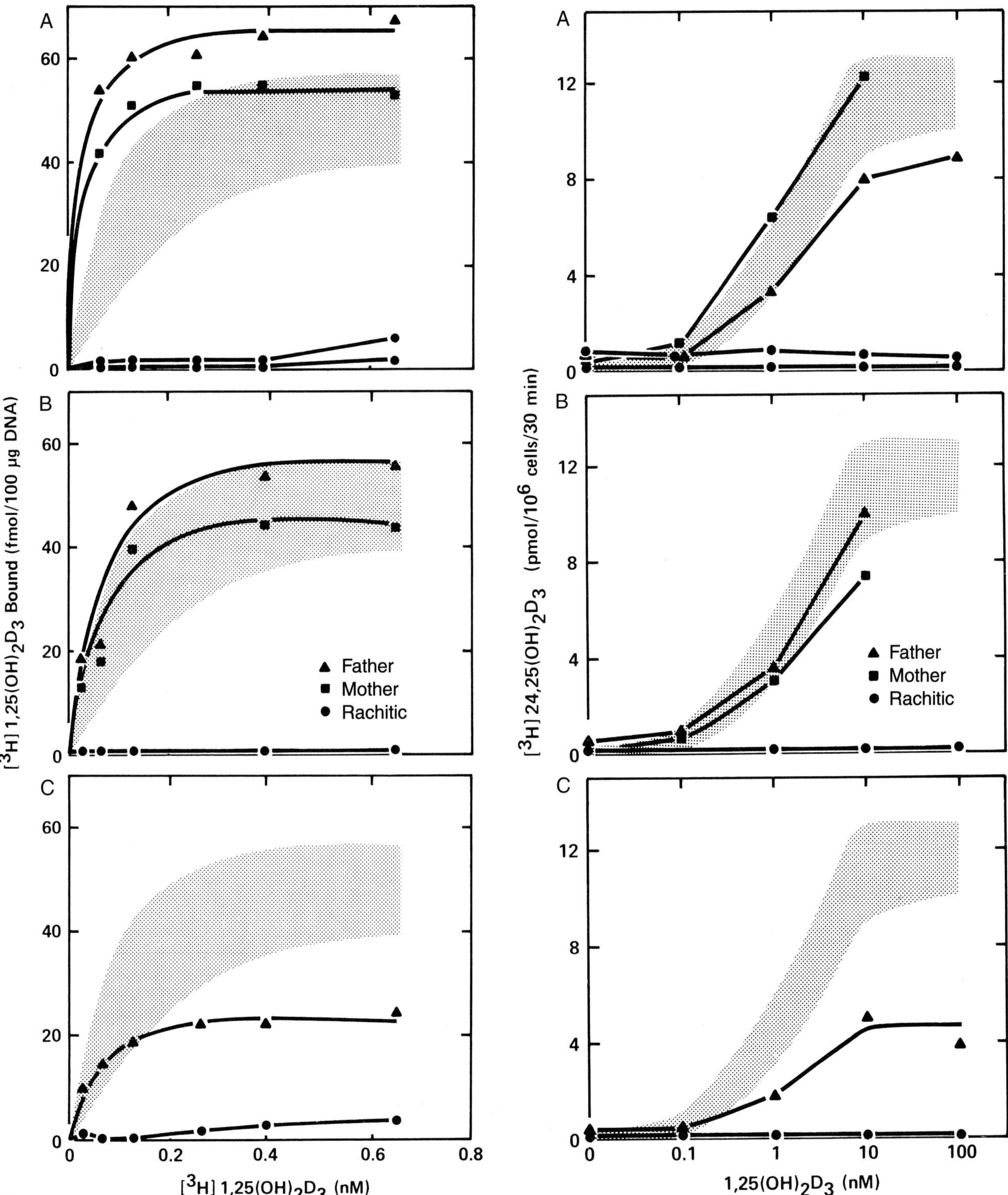

FIGURE 3 ^{3}H-Labeled 1,25(OH)$_2$D$_3$ binding (left) and 24-hydroxylase activity (right) in cultured skin fibroblasts in three families with "receptor-negative" HVDRR. Graphs in A, B, and C represent the A family (F11), B family (F17), and C family (F18), respectively. Shaded areas indicate the normal range. Reprinted with permission from TL Chen, MA Hirst, CM Cone, Z Hochberg, HU Tietze, and D Feldman. 1,25-Dihydroxyvitamin D resistance, rickets, and alopecia: Analysis of receptors and bioresponse in cultured fibroblasts from patients and parents. *J Clin Endocrinol Metab;* 59:383–388;1984. © The Endocrine Society.

had sparse hair, rickets, high circulating 1,25$(OH)_2$D levels, and low affinity 1,25$(OH)_2D_3$ binding. The patient responded to treatment with extremely high doses of 1,25$(OH)_2D_3$, thus overcoming an apparent low hormone affinity abnormality in the VDR.

Hirst *et al.* [25] studied the VDR in fibroblasts from a Haitian family (F19) (D family) with HVDRR. Interestingly, although the fibroblasts from the two sisters with HVDRR were unresponsive to 1,25$(OH)_2D_3$ treatment, each exhibited normal ^{3}H-1,25$(OH)_2D_3$ binding. Sucrose gradient sedimentation revealed that the cells from the affected individuals contained a normal sized VDR; however, the protein showed a decreased ability to form aggregates in low salt conditions as compared to normal receptor. Using DNA–cellulose chromatography, the authors demonstrated that the VDR from the HVDRR fibroblasts exhibited a significant decrease in affinity for heterologous DNA. The normal receptor was shown to elute from the DNA–cellulose at 170–173 m*M* KCl, whereas the mutant receptor eluted at 105–109 m*M* KCl. A second HVDRR family (F31) (the G family) demonstrated a similar defect in the DNA binding properties of the receptor [39]. The VDR from the affected individuals displayed normal ^{3}H-1,25$(OH)_2D_3$ binding and was normal in size as shown by Western blotting. DNA–cellulose chromatography clearly revealed that the VDR had a low affinity for DNA (eluted at 100 m*M* KCl, whereas the normals eluted at 200 m*M* KCl) (Fig. 4). When the VDR-containing extracts from the parent's cells were chromatographed, they were shown to have two forms of the receptor, one form with a high affinity for DNA (eluted at 200 m*M* KCl) and the other with a low affinity for DNA (eluted at 100 m*M* KCl). This was the first clear evidence showing the heterozygous state of the HVDRR parents. The individuals from the D and G families were therefore described as having hormone-binding positive (a receptor-positive) phenotype, but VDR binding to DNA was defective. Liberman *et al.* [30] also described four cases of receptor-positive resistance to 1,25$(OH)_2D_3$. Two of the cases (F5, F20) exhibited VDRs with a low affinity for DNA similar to the D and G families. In the other cases (F1, F3), the VDRs were found to have a reduced ability to localize to the nucleus despite showing normal affinity for DNA.

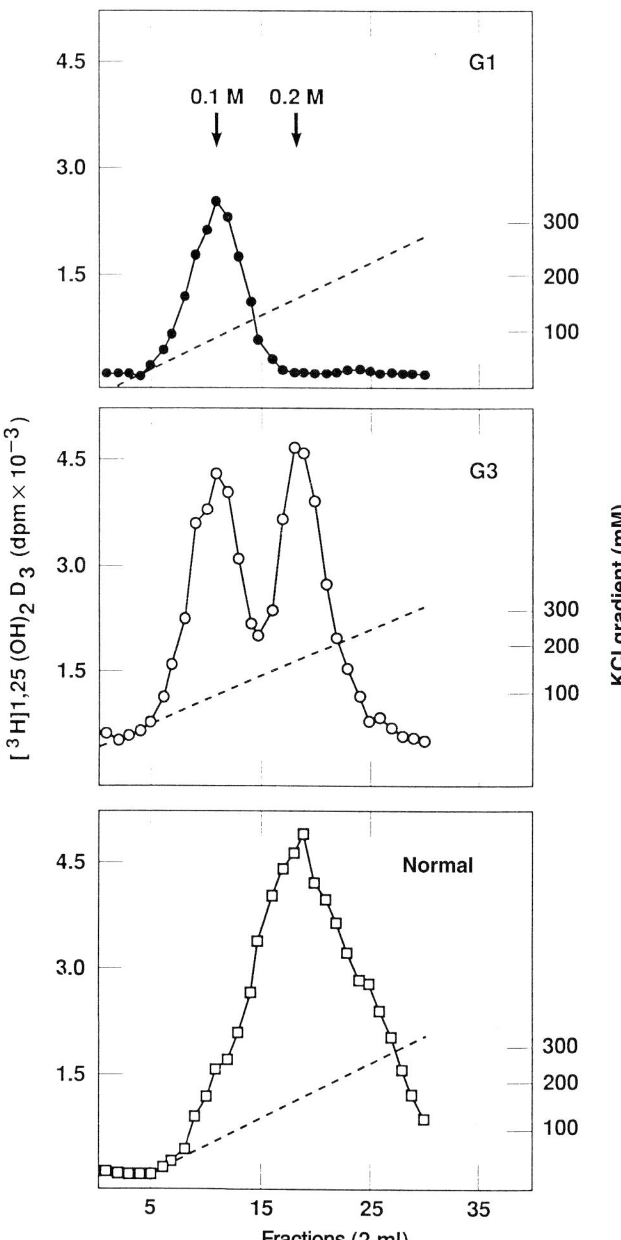

FIGURE 4 DNA–cellulose chromatographies of VDR from a patient (F31) with "receptor-positive" HVDRR and his parent. VDR from fibroblasts prelabeled with ^{3}H-1,25$(OH)_2D_3$ were bound to DNA–cellulose and then eluted with a 0–400 m*M* KCl gradient. The amount of radioactivity in the fractions was determined by liquid scintillation counting. G1, Patient with HVDRR; G3, heterozygotic parent; normal, unaffected control. Reprinted with permission from PJ Malloy, Z Hochberg, JW Pike, and D Feldman. Abnormal binding of vitamin D receptors to deoxyribonucleic acid in a kindred with vitamin D-dependent rickets, type II. J Clin Endocrinol Metab; 68:263–269;1989. © The Endocrine Society.

D. Studies in Other Cell Types

In addition to cultured skin fibroblasts and bone cells, studies on the VDR from patients with HVDRR have been carried out in a number of other cells, including peripheral mononuclear cells [27], phytohemagglutinin (PHA)-stimulated lymphocytes [33,42], myeloid progenitor cells [35], Epstein-Barr virus (EBV)-immortalized B lymphoblasts [39,63–65], and human T-cell leukemia virus type 1 (HTLV-1)-immortalized T lymphoblasts [41]. It is interesting to note that EBV-immortalized B lymphoblasts, which express VDR, do not exhibit any 24-hydroxylase activity or growth inhibition in response to $1,25(OH)_2D_3$ [65]. On the other hand, PHA-stimulated lymphocytes and HTLV-1-immortalized T lymphoblasts are able to respond to $1,25(OH)_2D_3$ [41,66]. In PHA-stimulated lymphocytes, the absence of an inhibitory effect of $1,25(OH)_2D_3$ on DNA synthesis or induction of 24-hydroxylase activity has been used as a marker to rapidly diagnose HVDRR [33,42]. In addition, Takeda *et al.* [42] showed that PHA-stimulated lymphocytes from parents of children with HVDRR expressed intermediate levels of 24-hydroxylase when treated with $1,25(OH)_2D_3$.

V. MOLECULAR BASIS OF HVDRR

A. VDR Domain Structure

The VDR is a member of the thyroid–retinoid group of receptors, a subgroup of the steroid–thyroid–retinoid receptor gene superfamily [67]. Like other members of this group, the VDR acts as a transcriptional regulator of gene expression, which is described in detail in Section II of this volume. Members of this family of proteins are structurally similar in that they are composed of a ligand binding domain (LBD) and a DNA binding domain (DBD). In the VDR, as discussed in Chapter 8, the DBD is located between amino acids 20 and 90 at the amino-terminal portion of the protein [68]. The DBD contains nine cysteine residues that are highly conserved across the superfamily. This region folds into two loops of 12–13 amino acids, each spaced 15 amino acids apart [69]. Four cysteines in each loop bind one molecule of zinc, which allows the formation of a two "zinc finger" structure. It is thought that the first zinc finger module gives specificity to the protein–DNA interaction, whereas the second zinc finger module contributes to its structural stability [69]. The interaction of the VDR with DNA is considered to occur through the zinc fingers at a region of the DBD known as the P box (amino acids 40–46) [70]. The LBD is located in the carboxy-terminal portion of the receptor, and amino acids 387–427 have been shown to play an important role in the ligand binding process [71,72]. Crystallographic studies of thyroid receptor (TR) [73], retinoic acid receptor (RAR) [74], and RXR [75] have shown that the LBD is formed from 11 or 12 α helices and two β sheets. Also common to these receptors, including the VDR, is a 9-heptad repeat region that is located in the LBD [76]. VDR action requires heterodimerization with RXR. How these repeats participate in dimerization of receptors is under intensive investigation. The VDR structure as it relates to dimerization and DNA binding is further discussed in Chapters 9 and 10.

B. VDR Gene Organization

In humans, the VDR gene has been localized to chromosome 12q13-14 [77–79] in the same region as the gene encoding pseudovitamin D deficiency (PDDR, see Chapter 47). The gene spans more than 75 kb of DNA and is composed of eight coding exons and several 5′ noncoding exons [48,63]. As described in detail in Chapter 8, exons 2 and 3 encode the DBD, with each zinc finger encoded by a separate exon. The LBD is encoded by exons 6–9, and a hinge region, the region between the DBD and LBD, is encoded by exons 4–6. The 5′-untranslated region of the VDR transcript is encoded by at least two exons [80]. In this chapter we use the numbering system described by Hughes *et al.* [63], which depicts the VDR gene as containing nine exons. In this numbering system, exon 1 is composed of two or more individual noncoding components (see Chapter 8).

C. Amino Acid Numbering System

The human VDR cDNA was cloned and sequenced by Baker *et al.* [81]. The 4628 nucleotide bases encode a protein of 427 amino acids with a predicted molecular mass of 48,000 daltons. In contrast to Baker *et al.* [81], we [63–65,82–85] and others [86–92] numbered the amino acid residues beginning with a methionine located 3 amino acid residues downstream of the initiating methionine in the Baker *et al.* [81] sequence. This numbering system was implemented because in the first report describing mutations in the VDR gene [63], the HVDRR patients exhibited a polymorphism in the putative initiating methionine codon. This polymorphism altered the putative initiating methionine codon ATG in the sequence of Baker *et al.* [81] to ACG, which in turn reduced the length of the VDR from 427 to 424 amino acids. Since this polymorphism was first described by Saijo *et al.* [87], a number of individuals with HVDRR as well as normal subjects have been shown to have this base change, which results in the loss of the first 3 amino acids at the N terminus of the VDR [90–94]. Nevertheless, because researchers have widely adopted the amino acid numbering system of Baker *et al.* [81], the numbering of all mutations previously described as starting from the second ATG will be increased by 3 amino acids here, to conform to that of the published sequence [81].

D. DNA Binding Domain Mutations

1. INITIAL DESCRIPTION

Two significant events in the late 1980s provided the basis for the subsequent genetic analyses of HVDRR disorders. In 1987, McDonnell *et al.* [95] cloned the avian VDR cDNA using a monoclonal antibody to the VDR originally developed by Pike *et al.* [60]. This discovery eventually lead to the cloning of the human VDR gene [81]. Second, in 1988, Saiki *et al.* [96] showed that it was possible to amplify gene sequences from small amounts of DNA using a thermostable DNA polymerase enzyme and a technique they termed polymerase chain reaction (PCR). The sequence of the VDR gene and the clues obtained from many earlier studies of the HVDRR cases conducted at the biochemical and cellular level were invaluable, as they allowed investigators to focus on defining the molecular basis for this syndrome. From the nucleotide sequence data, a highly conserved zinc finger structure was identified in the VDR gene. Because the zinc finger motif was thought to confer DNA binding properties to the receptor, investigators focused on this region of the VDR gene in HVDRR cases where the hormone binding positive (receptor-positive) phenotype coexisted with a decreased affinity for DNA.

In 1988, Hughes *et al.* [63] used PCR to amplify exons of the VDR gene in DNA samples from the D (F19) and G (F31) families [25,39]. In the D family (F19) [25], a missense mutation was found in exon 3 that encodes the second zinc finger module of the DBD. The mutation substitutes a polar uncharged glutamine (CAA) for a positively charged arginine (CGA) at amino acid 73 (originally designated Arg70Gln) (Fig. 5). In the G family (F31) [39], a missense mutation was identified in exon 2 that encodes the first zinc module of the DBD. This mutation results in the replacement of a glycine (GGC) with a highly charged aspartic acid (GAC) at amino acid 33 (originally designated Gly30Asp) (Fig. 5). Both a normal and mutant allele were found in the parents, which confirmed the recessive nature of the disease. The mutations in the VDR described by Hughes *et al.* [63] represented the first such mutations reported for any member of the steroid–thyroid–retinoid receptor gene superfamily. Using site-directed mutagenesis, each mutation was recreated in the wild-type VDR cDNA and its properties analyzed in COS-1 cells. Both mutant VDRs had normal $1,25(OH)_2D_3$ binding activity but low affinity for heterologous DNA [63,82]. These data confirmed the premise that the identified mutations gave rise to a hormone binding positive (receptor-positive), DNA binding defective phenotype. In cotransfection experiments using CV-1 cells, the mutant VDRs were tested for transcriptional activation from a $1,25(OH)_2D_3$-inducible osteocalcin–CAT reporter plasmid. $1,25(OH)_2D_3$ induction of CAT activity was elicited by the normal VDR but not by the two mutant VDRs, demonstrating that the mutant proteins were transcriptionally inactive [82].

2. Characterization of Additional DBD Mutations

Since this initial report, a number of mutations have been identified in the VDR DBD. These mutations are

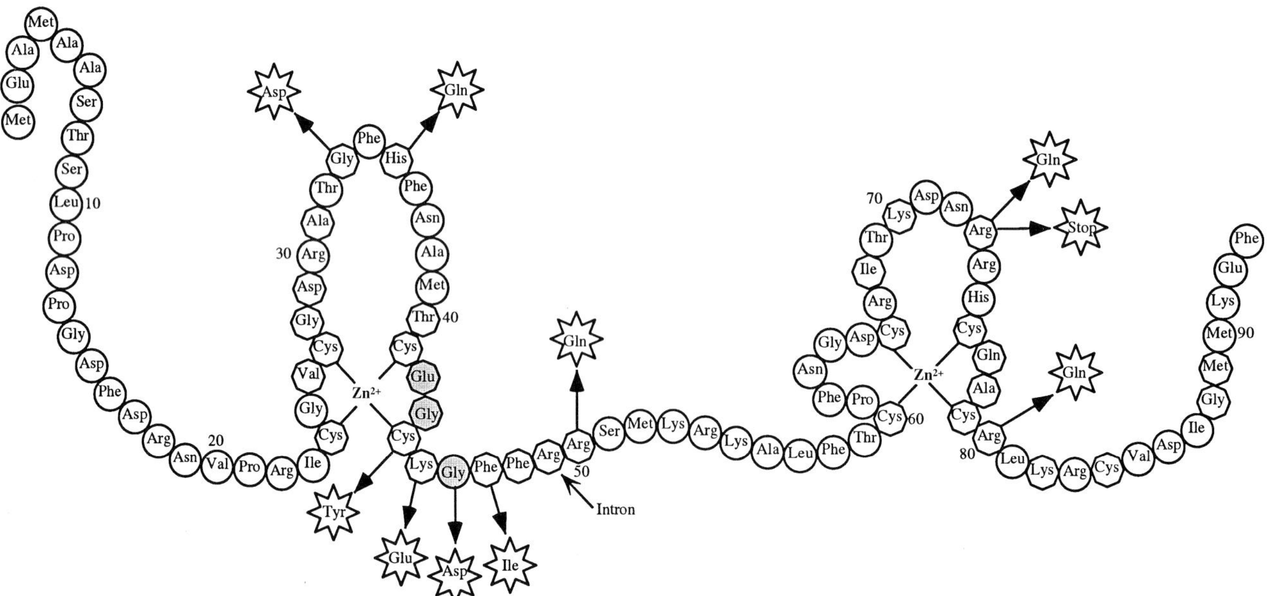

FIGURE 5 Zinc finger structure and location of mutations in the DNA binding domain of the VDR. Amino acids in octagons represent conserved residues. Shaded amino acids indicate DNA contact points. Star-shaped amino acids represent amino acid changes caused by mutation.

depicted in Fig. 5. Sone *et al.* [86] examined the VDR from two unrelated patients (F5 and F20) previously described by Liberman *et al.* [19,30], both of whom exhibited a hormone binding positive (receptor-positive), DNA defective phenotype. In both patients, a G to A missense mutation was identified in exon 3. In this mutation, arginine is replaced by a glutamine at amino acid 80, which is located at the base of the second zinc finger module (originally Arg77Gln) (Fig. 5). The recreated mutant receptor bound hormone but was unable to bind with high affinity to DNA or to activate gene transcription [86].

A fourth mutation was found in the DBD of three HVDRR patients from two unrelated families (F26 and F33) by Saijo *et al.* [87]. As before, prior investigation of fibroblasts from the patients showed normal $1,25(OH)_2D_3$ binding but abnormal nuclear binding of hormone–receptor complex, suggesting that the DBD was defective [33,40,43]. A transition from G to A was identified in exon 3 that converts arginine at position 50 to glutamine (originally Arg47Gln). No mutant VDR cDNA expression data was provided to establish the functional consequences of this defect on the VDR. However, whereas only mutant alleles were found in the homozygous patients, both mutant and normal alleles were detected by single strand conformational polymorphism (SSCP) in the parents, confirming their heterozygosity [87].

A His35Gln (originally His32Gln) mutation in the first zinc finger module of the VDR DBD was identified by Yagi *et al.* [89]. This mutation changed a positively charged amino acid to a neutral one. The VDR from the cells of the patient (F45) exhibited normal $1,25(OH)_2D_3$ binding but, consistent with the location of the mutation in the DBD, was defective for DNA binding. The fibroblasts from the patient were unable to induce gene transcription from a transiently transfected VDRE–CAT reporter plasmid; however, when the cells were transformed with the wild-type VDR, their ability to respond to the hormone was restored.

Two mutations in the VDR DBD were reported by Rut *et al.* [92]. One patient (F47), described in a previous report by Lin and Uttley [45], was shown to have an A to G mutation that resulted in a Lys45Glu change (originally Lys42Glu), whereas the other patient (F44), described in a previous report by Simonin *et al.* [44], had a T to C mutation that resulted in a Phe47Ile change (originally Phe44Ile). In each case, the mutant VDRs exhibited normal $1,25(OH)_2D_3$ binding but were transcriptionally inactive [92].

A missense mutation in the VDR DBD was also identified in two siblings (F49) with HVDRR [85]. This mutation, Arg80Gln (originally Arg77Gln), was the same as the mutation described in an unrelated family reported by Sone *et al.* [86]. In all of the VDR DBD mutations described above, the site of the defect is in a highly conserved amino acid common to all of the steroid receptor superfamily genes.

A patient with HVDRR from a Moroccan family (F46) was examined for mutations in the VDR gene by Wiese *et al.* [91]. At the cellular level, this patient exhibited a hormone binding negative (or receptor-negative) phenotype, suggesting that the mutation would lie in the LDB. The authors discovered an opal mutation (CGA to TGA) in which a C to T substitution introduced a stop at codon 73 (originally Arg70stop). The Arg73stop mutation truncates the receptor in the middle of the second zinc finger module, resulting in the production of a 72-amino acid polypeptide. No evidence for the presence of a truncated VDR in the cells from the patient could be demonstrated because both the LBD and monoclonal antibody binding sites were absent in the mutant protein. Interestingly, the Arg73stop mutation (CGA to TGA) occurs in the same codon that gives rise to the D family (F19) mutation (CGA to CAA) (Arg73Gln) [63] but at a different nucleotide base (Figs. 5 and 6).

Lin *et al.* [97] examined the VDR gene for mutations in a patient (F23) with HVDRR previously described by Sakati *et al.* [32]. The authors found a G to A missense mutation that resulted in a Gly46Asp change. Analysis of an *Mwo*I restriction fragment length polymorphism (RFLP) created by the mutation demonstrated that the patient was homozygous for the mutation and that the father was a carrier of the mutant allele. The Gly46Asp mutant VDR exhibited the characteristics of a DBD mutation in that the mutant receptor bound ^3H-$1,25(OH)_2D_3$ normally but displayed a reduced affinity for DNA. The mutant receptor was also shown to be transcriptionally inactive in reporter gene assays. Unlike the previously reported DBD mutations, this amino acid substitution does not occur at a conserved residue. Gly-46 is located between two highly conserved residues, Lys-45 and Phe-47, and although it is not conserved in all members of the steriod–thyroid–retinoid receptor superfamily, it is conserved among receptors that form heterodimers with RXR proteins and that recognize VDREs that are imperfect direct hexanucleotide repeats.

We have identified a G to A mutation in exon 2 that causes cysteine at residue 44 to be replaced by a tyrosine (P. J. Malloy, A. Al-Ashwal, and D. Feldman, in preparation, 1997). The young girl (F52) with this mutation has extremely severe rickets without alopecia. The Cys44Tyr mutation probably disrupts the zinc finger structure, as Cys-44 contributes to the zinc binding motif of the first zinc finger module. A laboratory-created mutation of this cysteine to serine increased VDR sensi-

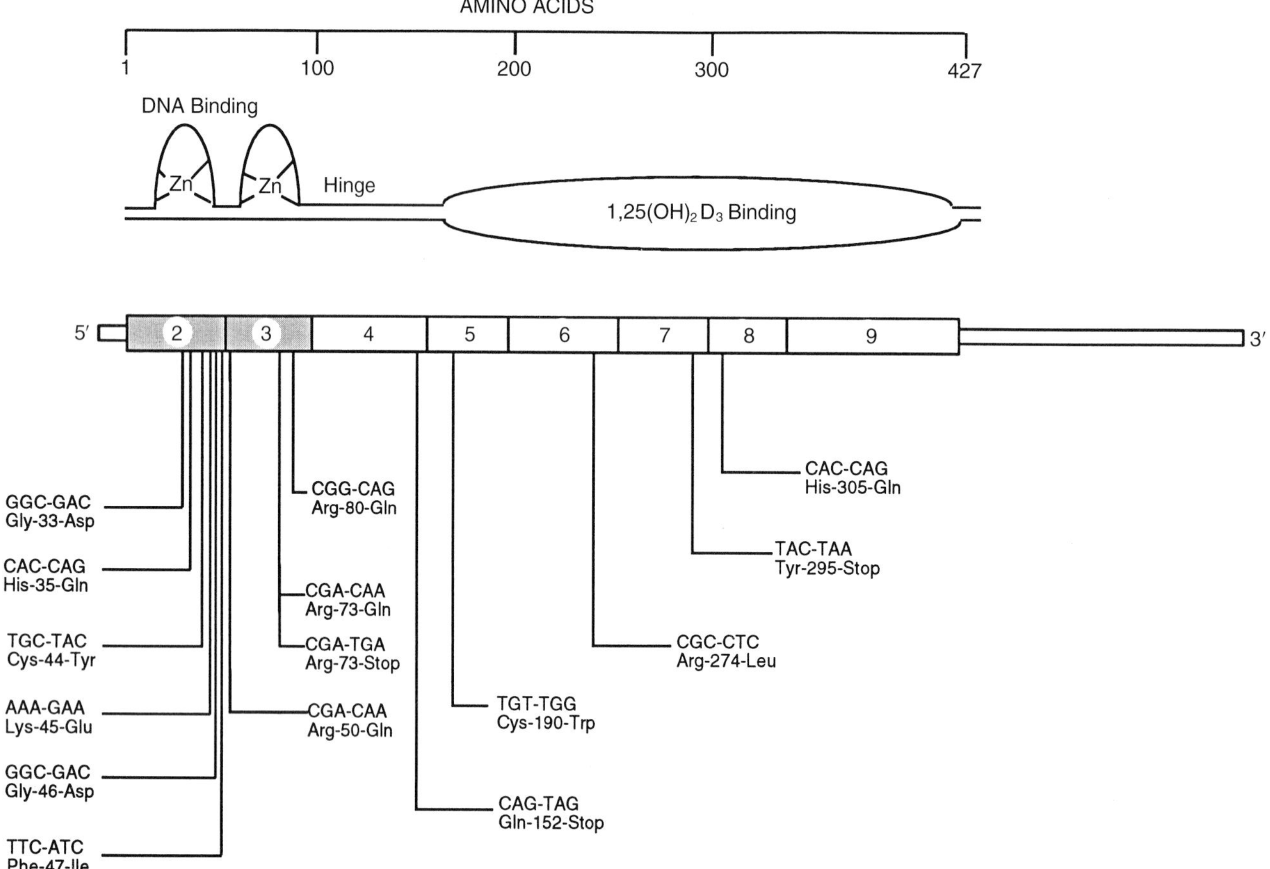

FIGURE 6 Exon organization of the VDR and location of known/published mutations in patients with HVDRR. Adapted from Malloy *et al.* [85] with permission.

tivity to proteolysis and abrogated specific DNA binding and transactivation [98].

3. Structural Analysis

The crystal structure of the VDR has not yet been reported. However, crystallographic studies of the glucocorticoid receptor (GR) [99] and RXR and TR [100] DBD structures have been published, and, on the basis of these studies, the alterations created by the mutations in the VDR DBD can be modeled (also see Chapter 9 and the figures therein). Crystallographic analyses of the GR have demonstrated that amino acids 457–469 (corresponding to residues 38–50 in the VDR) form an α helix which joins the two zinc fingers. This α helix packs perpendicularly with a second α helix at the base of the second zinc finger. Together, the hydrophobic residues of these two α helices comprise the hydrophobic core of the DBD. Lys45Glu and Gly46Asp mutations in VDR are located in the P box, a region of the receptor thought to be important in contacting the DNA bases and determining the specificity of the receptor for specific VDREs (Fig. 5). On the basis of this model, Rut *et al.* [92] proposed that the Lys45Glu mutation would disturb the hydrogen bonding between Lys-45 and a guanine in the VDRE half-site. The conversion of Gly-46 to Asp, a bulky, charged amino acid, might lead to unfavorable electrostatic interactions with the negatively charged phosphate backbone of the DNA helix and prevent contact with specific bases in the VDRE or, alternatively, interfere with the specificity of the interaction [97]. Similarly, because of the negatively charged Asp, the Gly33Asp mutation might also be expected to have a repelling effect on the negatively charged phosphate backbone [92]. The His35Gln mutation might eliminate a hydrogen bond donated from the positively charged histidine to the phosphate of a guanine in the VDRE [92]. The Phe47Ile mutation, although a relatively conservative substitution, might disrupt the hydrophobic core of the DBD and might alter the proposed α-helical structure at the base of the first zinc finger such that the VDR could not bind normally to its VDRE [92].

It is interesting to note that four DBD mutations, Lys45Glu, Gly46Asp, Phe47Ile, and Arg50Gln, occur in a LysXxxPhePhe[Lys/Arg]Arg sequence motif which has been identified as a calreticulin binding site [101–103]. Calreticulin does indeed bind to the VDR [102]. Cotransfection of calrecticulin-expressing plasmids with a VDRE/retinoic acid response element (RARE)–luciferase reporter construct results in decreased activation of the reporter gene in a dose-dependent manner [102]. Thus, it is possible that calreticulin may modulate the ability of VDR to activate transcription of target genes *in vivo*. The effects of these VDR mutations in altering potential calreticulin actions are unknown.

E. Ligand Binding Domain Mutations

1. Initial Description

As discussed in Section IV,B, a number of patients with HVDRR were identified as having a hormone binding negative (or receptor-negative) phenotype. Using a PCR strategy similar to that used by Hughes *et al.* [63] to identify DBD mutations, Ritchie *et al.* [64] examined the VDR gene for mutations in three related families (F18, F34, and F36) (designated C, E, and H families) exhibiting the hormone binding negative (or receptor-negative) phenotype [15,22,23,26]. In all three families, a mutation was identified at nucleotide 970 [64]. This mutation replaces Tyr295 with an ochre (TAA) termination codon (originally Tyr292stop) (Fig. 6), which causes truncation of the VDR by approximately 18,000 daltons. The mutation results in the deletion of a major portion of LBD, thereby producing the hormone binding negative (or receptor-negative) phenotype. As expected, the reconstructed mutant VDR showed no specific ^{3}H-1,25(OH)$_2$D$_3$ binding and failed to activate gene transcription in a VDRE–CAT reporter system. This was the first elucidation of a mutation in the VDR LBD.

These three related families (C, E, and H) were all found to harbor the same Tyr295 stop mutation and together with four additional related families (F35, F37, F38, and F39, F, J, K, and L families) comprise a large kindred in which consanguinous marriages occurred. The entire kindred was fully analyzed by Malloy *et al.* [65]. A total of eight children from this kindred exhibited the same hormone binding negative HVDRR phenotype and the same Tyr295stop genotype. The truncated protein that was predicted to be produced by this mutation was not detected in any members of the kindred by Western blot analysis (Fig. 7A). Interestingly, Northern blots also failed to show that the mutant VDR mRNA was present (Fig. 7B), in all but one case (F35, F family), explaining the absence of the mutant VDR protein. Reduced expression of mutant mRNA transcripts has been reported in other genetic diseases

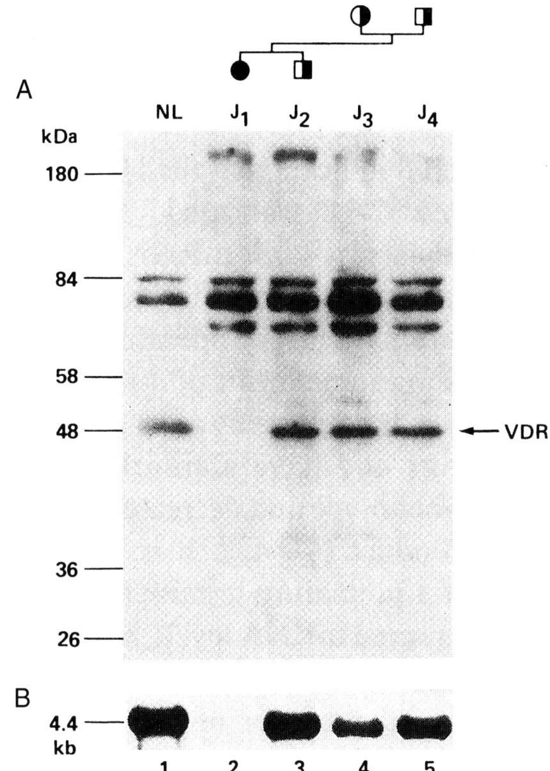

FIGURE 7 Western and Northern analyses of the J family (F37) with Tyr295stop mutation. (A) Western immunoblot of cytosol from normal fibroblasts and fibroblasts from members of the J family. J1 is the patient with HVDRR, J2 is an unaffected sibling, and J3 and J4 are the parents. The VDR was detected with an anti-VDR monoclonal antibody (9A7) by Western immunoblot. (B) Northern blot analysis of RNA isolated from normal fibroblast and fibroblasts from J family members. The VDR message was detected by hybridization with a radiolabeled VDR cDNA probe. Reproduced from *The Journal of Clinical Investigation,* 1990:86:2071–2079, by copyright permission of The American Society for Clinical Investigation.

where a premature stop mutation is the basis for the disease [104–106]. All of the HVDRR individuals were found to be homozygous for the ochre mutation at Tyr-295, and their parents were shown to be heterozygous through an analysis of a *Rsa*I RFLP created by the mutation (Fig. 8) [65].

2. Characterization of Additional LBD Mutations

The A family (F11) described in earlier papers had two children with HVDRR who exhibited the hormone binding negative (or receptor-negative) phenotype [15,22,23,26]. This is a family of Christian Arabs that live in the same town as the extended kindred described above, who are Muslim Arabs. Although there is no known relationship between these two families, the Tyr-295stop mutation that was found in the large kindred was also identified as the cause of HVDRR in the A

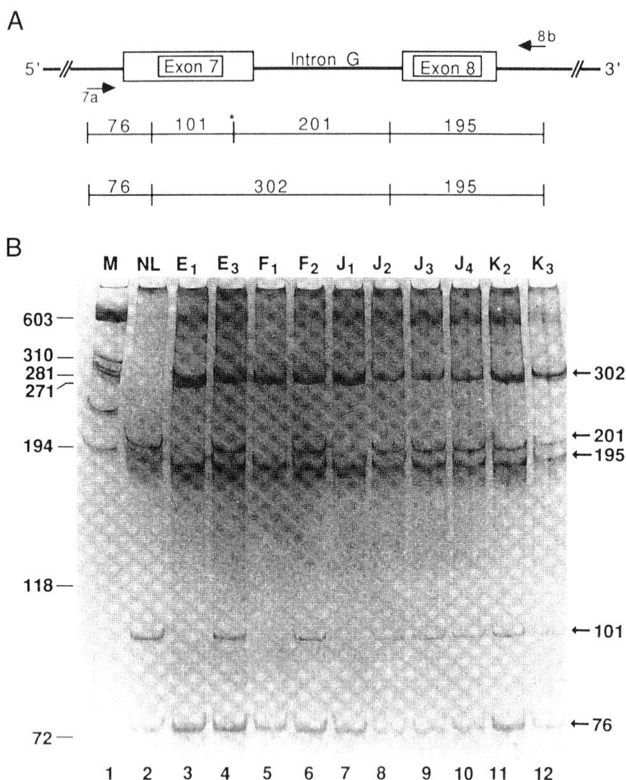

FIGURE 8 Restriction fragment length polymorphism of exons 7 and 8 in HVDRR families (F34, F35, F37, F38) exhibiting the Tyr295 stop mutation. (A) Schematic of PCR of exons 7 and 8 using primers 7a and 8b and locations of *Rsa*I restriction sites in normal and mutant alleles. The asterisk (*) indicates an *Rsa*I site that is present in the normal allele but is deleted in the mutant allele. (B) Gel electrophoresis of PCR products digested with *Rsa*I. Homozygotes (E1, F1, and J1) with HVDRR are identified by having only mutant alleles. Heterozygotes (E3, F2, J2, J3, J4, K2, and K3) are identified by having normal and mutant alleles. Normal control (NL) and molecular weight markers (M) are indicated. Reproduced from *The Journal of Clinical Investigation,* 1990:86:2071–2079, by copyright permission of The American Society for Clinical Investigation.

family [107]. The mutation was shown by the *Rsa*I RFLP and confirmed by DNA sequencing. The Tyr295stop mutation was also identified by Wiese *et al.* [91] in two related patients (F29, F30) from Saudia Arabia that were previously studied by Bliziotes *et al.* [37].

An analysis of VDR abundance from the parents of children with the hormone binding negative (or receptor-negative) phenotype led to an interesting observation. As mentioned previously, fibroblasts from heterozygous parents, with the exception of the father in the C family (F18), have been shown to have a normal complement of VDR (Fig. 3). In contrast, lymphoblasts from the same parents expressed half of the normal level of VDR, which appears consistent with the heterozygous state of these individuals [65]. The reason for the difference in expression of the VDR in the two cell types remains unresolved. It should be noted that the amount of VDR expressed in the lymphoblasts of the C family father is approximately one-fourth of the normal value, which suggests that an additional problem related to VDR expression may have arisen in this individual.

The first missense mutation found in the VDR LBD was in an individual from Kuwait (F21) with HVDRR [29] that was described by Rut *et al.* [88] and Kristjansson *et al.* [90]. Preliminary studies showed that fibroblasts from the patient had normal ^{3}H-1,25(OH)$_2$D$_3$ binding, but the affinity of the receptor for 1,25(OH)$_2$D$_3$ was significantly reduced (K_d 10 × 10^{-10} M) compared to controls (K_d 0.7 × 10^{-10} M) [88]. The fibroblasts from the patient were shown to be resistant to 1,25(OH)$_2$D$_3$, as treatment with hormone failed to induce 24-hydroxylase activity. Using PCR, a G to T missense mutation was identified at nucleotide 821 in exon 7 in these cells [88,90]. This mutation resulted in a positively charged arginine residue being replaced by a neutral leucine at amino acid 274 (originally Arg271Leu) (Fig. 6). Transcriptional activation studies of Arg274Leu VDR were carried out in CV-1 cells [90]. These studies showed that this mutant VDR could activate gene transcription from a VDRE reporter plasmid, but only when the concentration of 1,25(OH)$_2$D$_3$ was raised to levels approximately 1000-fold higher than required to activate the wild-type receptor.

A missense mutation in the VDR LBD has been reported by Malloy *et al.* [108]. The patient in this case exhibited three rare genetic disorders: HVDRR, congenital total lipodystrophy, and persistent Müllerian duct syndrome [109]. The patient, a Turkish boy (F51) who had rickets and high 1,25(OH)$_2$D$_3$ levels, was treated with extremly high doses of calcitriol (12.5 μg/day) that eventually normalized his serum calcium levels and ultimately improved his rickets. Fibroblasts from the patient revealed normal amounts of VDR, but the affinity of the receptor for 1,25(OH)$_2$D$_3$ was decreased by about 2-fold when examined at 0°C. Induction of 24-hydroxylase in the cultured fibroblasts from the patient required approximately a 5-fold increase in 1,25(OH)$_2$D$_3$ compared to control cells. Sequence analysis of the VDR gene uncovered a C to G missense mutation in exon 8. This mutation leads to substitution of a glutamine for a histidine at amino acid 305 (His305Gln) (Fig. 6). Interestingly, ^{3}H-1,25(OH)$_2$D$_3$ binding studies of the reconstructed mutant protein demonstrated an 8-fold lower affinity for 1,25(OH)$_2$D$_3$ compared to the wild-type VDR when the assays were performed at 24°C. In gene transactivation assays, the His305Gln mutant VDR was approximately 5-fold less responsive to 1,25(OH)$_2$D$_3$ compared to the wild-type VDR. RFLP analysis with *AlwN*I demonstrated that the parents were heterozygous but the patient and an affected sibling with HVDRR were homozygous for the mutation.

One other mutation has been described in the VDR LBD in a patient with HVDRR. In this case (F43), a T to G mutation was identified in exon 5 that changed a cysteine to a tryptophan at residue 190 (Cys190Trp) [110].

3. STRUCTURAL ANALYSIS

The crystal structures of the LBDs of the TR, RAR, and RXR have revealed the LBD to consist of a hydrophobic core that is composed of 11 or 12 α helices (H1–H12) and 2 to 4 short β sheets [73–75]. Extrapolating from the TR model to VDR [73], the Cys190Trp, Arg274Leu, and His305Gln mutations would be found within this hydrophobic region (Chapter 9). The Cys190Trp mutation is projected to be located in α helix H1 and the Arg274Leu mutation located in α helix H6 (Fig. 9). The His305Gln mutation would most likely occur in the loop connecting α helices H7 and H8 (Fig. 9). It is possible that the His305Gln mutation would disrupt the contacts between the protein and the ligand or destabilize the binding conformation. The net effect of the His305Gln mutation is to alter the affinity of the VDR for hormone, thus preventing the transformation of the VDR into an active receptor. When higher concentrations of $1,25(OH)_2D_3$ are made available to the VDR, however, this affinity defect can be overcome, leading to full transcriptional activation.

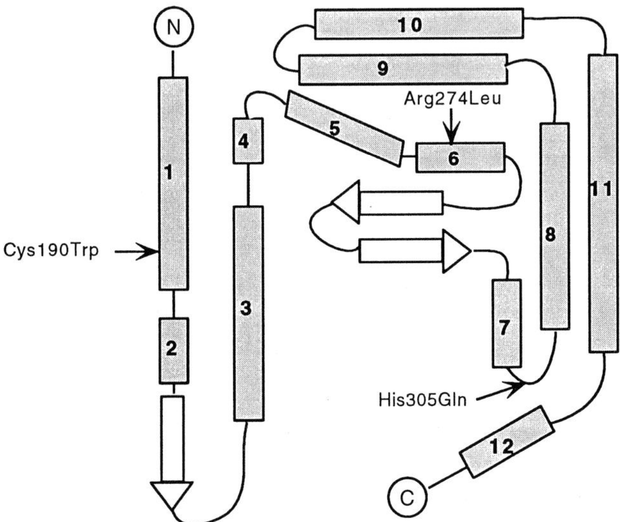

FIGURE 9 Model of the ligand binding domain of the VDR based on the TR model and putative location of mutations. The α helices are shown as shaded rectangles, and β sheets are shown as rectangles with arrowheads. Amino and carboxy termini are marked with N and C, respectively.

F. Other Mutations in the VDR

1. HINGE REGION MUTATION

In a report by Kristjansson *et al.* [90], a Turkish patient (F32) with HVDRR was shown to have a premature stop codon in exon 4 caused by an amber mutation (CAG to TAG). Earlier studies using fibroblasts from this patient had shown that hormone binding and cellular responses to $1,25(OH)_2D_3$ were both absent [38]. This premature stop mutation occurs at amino acid 152 (originally Gln149stop) in the hinge region of the receptor and truncates the VDR by 306 amino acids (Fig. 6), resulting in a hormone binding negative (or receptor-negative) phenotype. As expected, the Gln152stop mutant VDR was unresponsive to $1,25(OH)_2D_3$ in gene activation assays. The Gln152stop mutation was also identified by Wiese *et al.* [91] in a HVDRR patient (F32) previously reported by Barsony *et al.* [38].

2. EXON SKIPPING MUTATION

A single base change in intron 3 in the VDR gene was shown by Hawa *et al.* [111] to be the most probable cause of HVDRR in a young Greek girl (F50). Examination of the VDR message by reverse transcriptase (RT)-PCR showed that the RNA sequence diverged from the normal sequence at nucleotide 147. Where the sequences diverged it was noted that the following sequence was that of exon 4 and that the sequence from exon 3 was deleted. Sequence analysis of the VDR gene uncovered a G to C mutation in the 5′ end of intron 3. This mutation disrupts the consensus sequence for the 5′-donor splice site (normal sequence, GTA/GAGT; mutant sequence, GTA/GACT) and most likely caused exon 3 to be skipped in the processing of the VDR transcript. This exon skipping results in a frameshift that introduces a premature stop codon in the VDR coding sequence. The truncated VDR was unable to bind hormone and failed to induce 24-hydroxylase activity.

3. MAJOR STRUCTURAL MUTATION

There has been one case (F42) reported in which a major structural defect in the VDR gene was found to cause HVDRR [110]. The defect, found by PCR and Southern blotting, was a deletion in the VDR gene that eliminated exons 7, 8, and 9. This is the only case thus far reported in which a partial deletion in the VDR gene has been shown to cause HVDRR.

G. HVDRR without VDR Defects

Since the discovery of HVDRR as a genetic disorder, mutations in the VDR have been suspected as the likely cause of the hormone resistance in the patients. However, it is possible for mutations to occur in proteins involved in the $1,25(OH)_2D_3$ action pathway other than the VDR that could be manifested as HVDRR. Likely candidates for mutations that may cause target organ resistance to $1,25(OH)_2D_3$ include RXR, other transcription factors, and coactivators or corepressors. Mutations in these proteins may block the interaction between the VDR and RXR, block the binding to VDRE, or disrupt the communication between the VDR and other interacting proteins. Interestingly, these type of mutations might be cell or promoter specific.

An HVDRR case (F48) that may be caused by a defect other than in the VDR has been reported. Hewison *et al.* [93] described a case of HVDRR in a young girl who exhibted all the hallmarks of the disease including alopecia. The authors showed that the fibroblasts from the patient expressed a normal sized VDR transcript and that the receptor had a normal affinity for $1,25(OH)_2D_3$; however, the concentration of receptors detected in the cells was approximately half of the amount found in control cells. The hormone-resistant state was demonstrated by failure of the fibroblasts to induce 24-hydroxylase activity when treated with up to 1 μM $1,25(OH)_2D_3$. To determine if this resistance was caused by a mutation in the VDR gene, the VDR coding sequence was amplified by RT-PCR from RNA isolated from fibroblasts. Interestingly, although the cells were clearly resistant to $1,25(OH)_2D_3$, the authors were unable to find a mutation in the coding region of the VDR gene. The VDR cDNA from the patient exhibited a normal transactivation response to $1,25(OH)_2D_3$ in VDRE–CAT reporter assays in CV-1 cells, which indicated that the receptor was normal and that tissue resistance to $1,25(OH)_2D_3$ was not due to a defective VDR. The data suggest that $1,25(OH)_2D_3$ resistance and HVDRR can be caused by mutations in other proteins that participate in the $1,25(OH)_2D_3$ activation pathway or that regulate the level of expression of the VDR.

H. VDR Gene Knock-out Mouse

The HVDRR syndrome can be envisioned as representing a VDR gene knock-out in humans. However, for practical and ethical purposes, only a limited number of tissues can be studied to determine the effects of the gene knock-out in these individuals. A VDR gene knock-out has been created in the mouse by Yoshizawa *et al.* [112]. This knock-out mouse displayed the features of HVDRR including skeletal defects, growth retardation, and some facial alopecia. The animals exhibit low serum calcium and phosphate levels, high $1,25(OH)_2D_3$ levels, and secondary hyperparathyroidism. The mouse model will allow investigators to study the effects of a VDR knock-out in the intestine, kidney, bone, as well as other target tissues not readily available in humans.

VI. THERAPY OF HVDRR

A. Vitamin D

Following the first description of HVDRR, a number of treatment remedies using calcium and active vitamin D metabolities have been tried in attempts to cure the signs and symptoms of the disease. It seems clear that the response to therapy varies widely, no doubt dependent on the nature of the defect in the VDR. In some cases, responsiveness of the patient to high concentrations of $1,25(OH)_2D_3$ can be predicted from analysis of cultured skin fibroblasts. For the most part, it appears that patients with HVDRR that do not exhibit alopecia are generally more responsive to treatment with vitamin D preparations than those with alopecia [46]. In a few of the earlier reports, patients without alopecia responded clinically and radiologically to administration of pharmacological doses of vitamin D ranging from 5000 to 40,000 IU/day [3,4,7]. Patients without alopecia also responded to 20 to 200 μg of $25OHD_3$ per day and 17–20 μg of $1,25(OH)_2D_3$ per day [4]. In two other patients without alopecia, mutations were found in the VDR LBD that decreased the affinity of the receptor for $1,25(OH)_2D_3$ [88,90,108]. The patient (F21) with the Arg274Leu genotype was resistant to treatment with massive doses of $1,25(OH)_2D_3$. Cultured fibroblasts from this patient were also unresponsive to hormone treatment. Interestingly, the re-created Arg274Leu mutant VDR did exhibit functional activity at high doses of hormone. Unfortunately, this patient died as a consequence of his disease [29]. A second patient (F51) with a VDR LBD mutation, genotype His305Gln, was also responsive to high doses of $1,25(OH)_2D_3$ (12.5 μg/day) [109] as were his fibroblasts [108]. This therapy overcame the hormone binding defect and achieved adequate VDR occupancy to mediate normal $1,25(OH)_2D_3$ responses. These studies indicate that some HVDRR patients without alopecia may exhibit LBD defects manifested as diminished $1,25(OH)_2D_3$ affinity for the VDR, and that the defects caused by these mutations can be overcome by treatment with very high doses of vitamin D metabolites.

In general, patients with alopecia appear to be more resistant to treatment with vitamin D metabolites despite the fact that a small number of these patients also have been successfully treated with vitamin D. In one case, although vitamin D and $1,25(OH)_2D_3$ therapies were ineffective, the patient did respond to oral phosphate [6]. Two such patients also showed signs of improvement when given vitamin D or $1\alpha OHD_3$ [11,14], and one patient responded to $25OHD_3$ as well as $1\alpha OHD_3$ [16]. $1\alpha OHD_3$ and $1,25(OH)_2D_3$ also were effective treatments in a number of other cases [8,25,28,36,40]. Two children (F32), each identified as having a Glu152stop mutation in the VDR, showed no calcemic response to high dose vitamin D treatment that raised their circulating $1,25(OH)_2D_3$ levels to more than 100 times the mean normal range. Healing of rickets and suppression of PTH was evident, however, despite the presence of low serum calcium concentrations [113]. When patients fail to respond to $1,25(OH)_2D_3$ therapy, treatment with calcium (as described below) is indicated.

B. Calcium

Calcium administration has been used as a therapy for treating HVDRR patients. The advantage of using high-dose oral calcium therapy in a patient who has failed to respond to calciferols was shown by Sakati *et al.* [32]. The patient was given 3–4 g of elemental calcium orally per day and showed clinical improvement during 4 months of therapy. Balsan *et al.* [114] described the beneficial effects of long-term intravenous calcium infusions in a child with HVDRR and alopecia who had not responded to prior treatment with large doses of vitamin D derivatives or oral calcium supplements. High doses of calcium were infused intravenously during the nocturnal hours over a 9-month period. Clinical improvement was observed within the first 2 weeks of the start of therapy accompanied by relief of bone pain. Within 7 months, the child gained weight and height. Eventually, the hypocalcemia normalized, the secondary hyperparathyroidism was reversed, and the rickets was ultimately cured as assessed by X rays and bone biopsy. The syndrome recurred, however, when the intravenous infusions were discontinued. This alternative therapy apparently bypassed the calcium absorption defect in the intestine caused by the disease. Several other studies have similarly demonstrated beneficial responses in children with HVDRR treated with intravenous calcium infusion [37,115,116]. As shown in Fig. 10, two patients treated with intravenous calcium showed a decrease in serum alkaline phosphatase activity and an increase in their serum calcium and phosphate concentrations over a 1-year period [115]. X-Ray analysis showed resolution of the rickets with the appearance of normal mineralization of bone. After radiological healing of the rickets has been observed, high dose oral calcium therapy has been shown to be effective in maintaining normal serum calcium concentrations [116]. Children that show signs and symptoms of HVDRR are now routinely started on this two-step protocol at 2 years of age [117].

C. Prenatal Diagnosis

A prenatal diagnosis of HVDRR is now possible in pregnant women from high risk families. Cultured cells from chorionic villi samples or amniotic fluid have been used determine if the fetus has HVDRR using ^3H-$1,25(OH)_2D_3$ binding, induction of 24-hydroxylase activity, and RFLP analyses for known mutations [118,119].

D. Spontaneous Healing

It is interesting to note that there have been several cases of HVDRR in which spontaneous improvement in the disease has occurred [22,23,25]. When it occurs, spontaneous healing of rickets usually develops between 7 and 15 years of age, and not necessarily at the time of puberty. Sometimes the recovery occurs after relatively ineffective long-term treatment with vitamin D metabolites and mineral replacement, suggesting that the healing process was spontaneous rather than the result of the treatment. In other cases, the improvement occurred after the treatment was discontinued [25]. In these cases, the patients appear to remain eucalcemic without osteomalacia or rickets and without therapy. Studies of cultured skin fibroblasts from HVDRR patients who apparently have undergone spontaneous healing continued to exhibit resistance to $1,25(OH)_2D_3$ in their cells [25]. Spontaneous improvement has occurred in kindreds exhibiting the hormone binding negative (or receptor-negative) phenotype [22,23] where a Tyr295stop mutation was identified [64,65] as well as in the hormone binding positive (receptor-positive) phenotype [25] where an Arg73Gln mutation was identified [63]. Interestingly, in those patients who showed improvement of their hypocalcemia and rickets, the alopecia persisted [22,23,25].

VII. CONCLUDING REMARKS

HVDRR is a rare disease that results in end-organ resistance to $1,25(OH)_2D$ action. The major effect that this genetic disorder has on the vitamin D endocrine system is to impair intestinal calcium and phosphate absorption, which results in decreased bone mineralization and rickets. Since 1978, more than 50 kindreds exhibiting signs and/or symptoms of HVDRR have

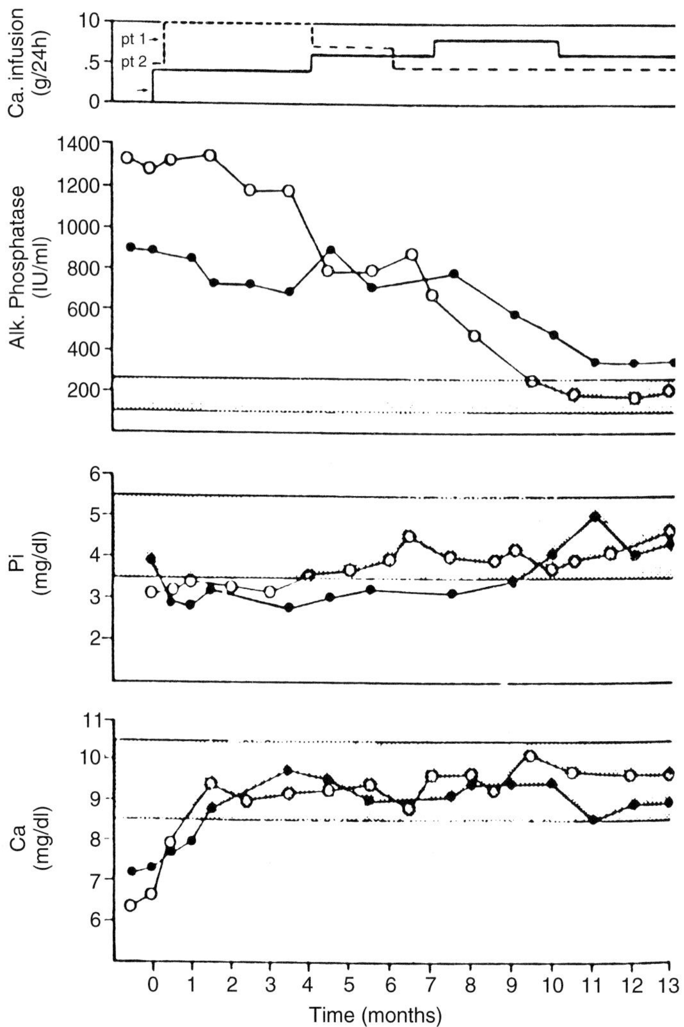

FIGURE 10 Effect of intravenous calcium therapy in two patients with HVDRR. Serum concentrations of alkaline phosphatase, phosphorus, and calcium before and during intravenous calcium infusions administered over a 1-year period are shown. Reprinted by permission of the publisher from Long-term calcium infusion therapy in end-organ resistance to 1,25-dihydroxyvitamin D. Y Weisman, I Bab, D Grazit, Z Spirer, M Jaffe, and Z Hochberg. *Am J Med,* 83:984–990. Copyright 1987 by Exerpta Medica Inc. Exerpta Medica, Inc. P.O. Box 882 Madison Square Station, New York, NY 10159-0882.

been studied. In all cases, the assignment of HVDRR has been based on resistance to vitamin D in combination with high circulating levels of $1,25(OH)_2D$. A number of cases have been analyzed for $1,25(OH)_2D_3$ binding and bioactivity which demonstrated that the disease was caused by heterogeneous defects in the VDR.

In 1988, the human VDR gene was cloned and sequenced, allowing investigators to determine the molecular basis of HVDRR. Thus far, analyses of the VDR gene from 28 families have established that mutations in the receptor cause HVDRR. At the time of this writing, a total of 15 unique single base mutations in the VDR gene have been described (Fig. 6). Ten point mutations have been identified in the DBD, one in the hinge region, and four in the LBD. A single base mutation has also been described in intron 3 that causes exon skipping. In addition to these point mutations, there has been one report of a major structural defect in the VDR gene caused by a partial gene deletion. The studies of the mutations in the VDR gene that cause the HVDRR syndrome provide many interesting insights into structural and functional roles of the VDR in mediating $1,25(OH)_2D$ action. The missense mutations in the VDR DBD appear to totally inactivate the receptor despite the fact that 1,25$(OH)_2D_3$ binding is normal. Conversely, missense mutations in the LBD are less detrimental to the receptor. In some cases, 1,25$(OH)_2D$ responsiveness was restored to individuals by therapy with high doses of hormone. On the other hand, in cases caused by premature stop mutations that produce truncated VDRs, the mutation always appears to lead to complete hormone resistance. Unfortunately, the mutations have not been elucidated in a number of the earlier reported cases of HVDRR. As some of these cases presented late in life, it is possible that an occasional case may be due to nonhereditary resistance to 1,25$(OH)_2D$.

From the genetic analyses of the VDR gene in HVDRR patients, it appears that DBD mutations or premature stop mutations generally result in alopecia or some degree of hair loss. On the other hand, although only a few cases have been reported, patients with LBD mutations do not appear to develop alopecia. The association of alopecia with HVDRR is a curious finding. Alopecia has not been found in other states of vitamin D deficiency, and thus defects in the VDR appear to be the primary cause of this condition. As the VDR has been found in hair follicles [120,121], 1,25$(OH)_2D$ action may be important in the differentiation of this structure during embryogenesis. Furthermore, since hair loss is not restored even after the rickets has healed, either as a result of therapeutic intervention or spontaneous cure, it suggests that a critical time point exists in the early development of the hair follicle that requires the action 1,25$(OH)_2D$. Once this temporal window has passed, the 1,25$(OH)_2D$-dependent stage appears no longer to be hormone sensitive.

The condition of generalized resistance to thyroid hormone (GRTH) is analogous to HVDRR. A large number of mutations have been found in the TR that lead to different classes of thyroid hormone resistance [122]. In contrast to VDR mutations, however, the TR mutations are inherited as autosomal dominant negatives, as affected individuals harbor copies of both a normal and mutant allele. Unlike the case of the VDR, TR mutations are largely confined to the LBD, and none have been found yet in the DBD [122]. A natural dominant negative mutation has not been described in the VDR, and in most cases the heterozygotic parents have been reported to have a normal phenotype. In one particular HVDRR case (F51), however, the parents appear to display a very mild form of hormone resistance, as their serum 1,25$(OH)_2D$ levels are elevated above the normal range [108]. The His305Gln mutation in this family may in fact be acting as a partial dominant negative in the parents. Interestingly, this VDR mutation is similar in location to a glycine residue in TR, which is the site of at least four different dominant negative mutations that lead to GRTH [122]. The fact that VDR and TR both require RXR as a partner for transactivation suggests a common mechanism of action, and it is likely that mutations will be found in the VDR, like TR, that act as dominant negatives.

The correction of hypocalcemia, secondary hyperparathyroidism, and rickets by calcium infusion without calciferol therapy in some cases leads to an interesting consideration of the role of vitamin D. The implication is that 1,25$(OH)_2D$ action is not essential for osteogenesis and its primary function is to act on the intestine to supply adequate amounts of mineral to the bone-forming site. Although there are many well-defined actions of vitamin D on osteoblasts (Chapters 18–20) and osteoclasts (Chaper 21), 1,25$(OH)_2D$ action on the bone may not be necessary in order to form normal bone. In addition, restoration of calcium into the normal range may be adequate to suppress PTH overproduction even though both calcium and 1,25$(OH)_2D_3$ can suppress PTH. Therapy with calcium without phosphate appears sufficient to correct all of the metabolic abnormalities in children with HVDRR, indicating that the secondary hyperparathyroidism is responsible for the hypophosphatemia.

The biochemical and genetic analysis of the VDR in the HVDRR syndrome has yielded important insights into the structure and function of the receptor in mediating 1,25$(OH)_2D$ action. Similarly, studies of affected children with HVDRR continue to provide further insight into the biological role of 1,25$(OH)_2D$ *in vivo*. A concerted investigative approach to HVDRR at the clinical, cellular, and molecular level has proved exceedingly valuable in understanding the mechanism of action of 1,25$(OH)_2D$ and in improving the diagnostic and clinical management of this rare genetic disease.

References

1. Albright F, Butler AM, Bloomberg E 1937 Rickets resistant to vitamin D therapy. Am J Dis Children **54**:531–547.
2. Prader VA, Illig R, Heierli E 1961 Eine besondere form der primaren vitamin-D-resistenten rachitis mit hypocalcamie und autosomal-dominantem erbang: Die hereditare pseudo-mangelrachitis. Helv Paediatr Acta **16**:452–468.
3. Brooks MH, Bell NH, Love L, Stern PH, Orfei E, Queener SF, Hamstra AJ, DeLuca HF 1978 Vitamin-D-dependent rickets type II. Resistance of target organs to 1,25-dihydroxyvitamin D. N Engl J Med **298**:996–999.
4. Marx SJ, Spiegel AM, Brown EM, Gardner DG, Downs RW Jr, Attie M, Hamstra AJ, DeLuca HF 1978 A familial syndrome of decrease in sensitivity to 1,25-dihydroxyvitamin D. J Clin Endocrinol Metab **47**:1303–1310.
5. Balsan S, Garabedian M, Lieberherr M, Gueris J, Ulmann A 1979 Serum 1,25-dihydroxyvitamin D concentrations in two different types of pseudo-deficiency rickets. In: Norman AW, Bouillon R, Thomasset M (eds) Vitamin D: Basic Research and its Clinical

Application. Fourth Workshop on Vitamin D. de Gruyter, New York, pp. 1143–1149.

6. Rosen JF, Fleischman AR, Finberg L, Hamstra A, DeLuca HF 1979 Rickets with alopecia: An inborn error of vitamin D metabolism. J Pediatr **94**:729–735.
7. Zerwekh JE, Glass K, Jowsey J, Pak CY 1979 An unique form of osteomalacia associated with end organ refractoriness to 1,25-dihydroxyvitamin D and apparent defective synthesis of 25-hydroxyvitamin D. J Clin Endocrinol Metab **49**:171–175.
8. Fujita T, Nomura M, Okajima S, Furuya H 1980 Adult-onset vitamin D-resistant osteomalacia with the unresponsiveness to parathyroid hormone. J Clin Endocrinol Metab **50**:927–931.
9. Liberman UA, Samuel R, Halabe A, Kauli R, Edelstein S, Weisman Y, Papapoulos SE, Clemens TL, Fraher LJ, O'Riordan JL 1980 End-organ resistance to 1,25-dihydroxycholecalciferol. Lancet **1**:504–506.
10. Sockalosky JJ, Ulstrom RA, DeLuca HF, Brown DM 1980 Vitamin D-resistant rickets: End-organ unresponsiveness to $1{,}25(OH)_2D_3$. J Pediatr **96**:701–703.
11. Tsuchiya Y, Matsuo N, Cho H, Kumagai M, Yasaka A, Suda T, Orimo H, Shiraki M 1980 An unusual form of vitamin D-dependent rickets in a child: Alopecia and marked end-organ hyposensitivity to biologically active vitamin D. J Clin Endocrinol Metab **51**:685–690.
12. Beer S, Tieder M, Kohelet D, Liberman OA, Vure E, Bar-Joseph G, Gabizon D, Borochowitz ZU, Varon W, Modai D 1981 Vitamin D resistant rickets with alopecia: A form of end organ resistance to 1,25-dihydroxyvitamin D. Clin Endocrinol **14**:395–402.
13. Eil C, Liberman UA, Rosen JF, Marx SJ 1981 A cellular defect in hereditary vitamin-D-dependent rickets type II: Defective nuclear uptake of 1,25-dihydroxyvitamin D in cultured skin fibroblasts. N Engl J Med **304**:1588–1591.
14. Kudoh T, Kumagai T, Uetsuji N, Tsugawa S, Oyanagi K, Chiba Y, Minami R, Nakao T 1981 Vitamin D dependent rickets: Decreased sensitivity to 1,25-dihydroxyvitamin D. Eur J Pediatr **137**:307–311.
15. Feldman D, Chen T, Cone C, Hirst M, Shani S, Benderli A, Hochberg Z, 1982 Vitamin D resistant rickets with alopecia: Cultured skin fibroblasts exhibit defective cytoplasmic receptors and unresponsivenss to $1{,}25(OH)_2D_3$. J Clin Endocrinol Metab **55**:1020–1022.
16. Balsan S, Garabedian M, Liberman UA, Eil C, Bourdeau A, Guillozo H, Grimberg R, Le Deunff MJ, Lieberherr M, Guimbaud P, Broyer M, Marx SJ 1983 Rickets and alopecia with resistance to 1,25-dihydroxyvitamin D: Two different clinical courses with two different cellular defects. J Clin Endocrinol Metab **57**:803–811.
17. Clemens TL, Adams JS, Horiuchi N, Gilchrest BA, Cho H, Tsuchiya Y, Matsuo N, Suda T, Holick MF 1983 Interaction of 1,25-dihydroxyvitamin-D_3 with keratinocytes and fibroblasts from skin of normal subjects and a subject with vitamin-D-dependent rickets, type II: A model for study of the mode of action of 1,25-dihydroxyvitamin D_3. J Clin Endocrinol Metab **56**:824–830.
18. Griffin JE, Zerwekh JE 1983 Impaired stimulation of 25-hydroxyvitamin D-24-hydroxylase in fibroblasts from a patient with vitamin D-dependent rickets, type II. A form of receptor-positive resistance to 1,25-dihydroxyvitamin D_3. J Clin Invest **72**:1190–1199.
19. Liberman UA, Eil C, Marx SJ 1983 Resistance to $1{,}25(OH)_2D_3$: Association with heterogeneous defects in cultured skin fibroblasts. J Clin Invest **71**:192–200.
20. Liberman UA, Eil C, Holst P, Rosen JF, Marx SJ, 1983 Hereditary resistance to 1,25-dihydroxyvitamin D: Defective function of receptors for 1,25-dihydroxyvitamin D in cells cultured from bone. J Clin Endocrinol Metab **57**:958–962.
21. Adams JS, Gacad MA, Singer FR 1984 Specific internalization and action of 1,25-dihydroxyvitamin D_3 in cultured dermal fibroblasts from patients with X-linked hypophosphatemia. J Clin Endocrinol Metab **59**:556–560.
22. Chen TL, Hirst MA, Cone CM, Hochberg Z, Tietze HU, Feldman D 1984 1,25-Dihydroxyvitamin D resistance, rickets, and alopecia: Analysis of receptors and bioresponse in cultured fibroblasts from patients and parents. J Clin Endocrinol Metab **59**:383–388.
23. Hochberg Z, Benderli A, Levy J, Vardi P, Weisman Y, Chen T, Feldman D 1984 1,25-Dihydroxyvitamin D resistance, rickets, and alopecia. Am J Med **77**:805–811.
24. Gamblin GT, Liberman UA, Eil C, Downs RWJ, Degrange DA, Marx SJ 1985 Vitamin D dependent rickets type II: Defective induction of 25-hydroxyvitamin D_3-24-hydroxylase by 1,25-dihydroxyvitamin D_3 in cultured skin fibroblasts. *J Clin Invest* **75**:954–960.
25. Hirst MA, Hochman HI, Feldman D 1985 Vitamin D resistance and alopecia: A kindred with normal 1,25-dihydroxyvitamin D binding, but decreased receptor affinity for deoxyribonucleic acid. J Clin Endocrinol Metab **60**:490–495.
26. Hochberg Z, Gilhar A, Haim S, Friedman-Birnbaum R, Levy J, Benderly A 1985 Calcitriol-resistant rickets with alopecia. Arch Dermatol **121**:646–647.
27. Koren R, Ravid A, Liberman UA, Hochberg Z, Weisman Y, Novogrodsky A, 1985 Defective binding and function of 1,25-dihydroxyvitamin D_3 receptors in peripheral mononuclear cells of patients with end-organ resistance to 1,25-dihydroxyvitamin D. J Clin Invest **76**:2012–2015.
28. Castells S, Greig F, Fusi MA, Finberg L, Yasumura S, Liberman UA, Eil C, Marx SJ 1986 Severely deficient binding of 1,25-dihydroxyvitamin D to its receptors in a patient responsive to high doses of this hormone. J Clin Endocrinol Metab **63**:252–256.
29. Fraher LJ, Karmali R, Hinde FR, Hendy GN, Jani H, Nicholson L, Grant D, O'Riordan JL 1986 Vitamin D-dependent rickets type II: Extreme end organ resistance to 1,25-dihydroxy vitamin D_3 in a patient without alopecia. Eur J Pediatr **145**:389–395.
30. Liberman UA, Eil C, Marx SJ 1986 Receptor-positive hereditary resistance to 1,25-dihydroxyvitamin D: Chromatography of receptor complexes on deoxyribonucleic acid–cellulose shows two classes of mutation. J Clin Endocrinol Metab **62**:122–126.
31. Liberman UA, Eil C, Marx SJ 1986 Clinical features of hereditary resistance to 1,25-dihydroxyvitamin D (hereditary hypocalcemic vitamin D resistant rickets type II). Adv Exp Med Biol **196**:391–406.
32. Sakati N, Woodhouse NJY, Niles N, Harfi H, de Grange DA, Marx S. 1986 Hereditary resistance to 1,25-dihydroxyvitamin D: Clinical and radiological improvement during high-dose oral calcium therapy. Horm Res. **24**:280–287.
33. Takeda E, Kuroda Y, Saijo T, Toshima K, Naito E, Kobashi H, Iwakuni Y, Miyao M 1986 Rapid diagnosis of vitamin D-dependent rickets type II by use of phytohemagglutinin-stimulated lymphocytes. Clin Chim Acta **155**:245–250.
34. Laufer D, Benderly A, Hochberg Z 1987 Dental pathology in calcitriol resistant rickets. J Oral Med **42**:272–275.
35. Nagler A, Merchav S, Fabian I, Tatarsky I, Hochberg Z 1987 Myeloid progenitors from the bone marrow of patients with vitamin D resistant rickets (type II) fail to respond to $1{,}25(OH)_2D_3$. Br J Haematol **67**:267–271.
36. Takeda E, Kuroda Y, Saijo T, Naito E, Kobashi H, Yokota I, Miyao M, 1987 1 alpha-hydroxyvitamin D_3 treatment of three

patients with 1,25-dihydroxyvitamin D-receptor-defect rickets and alopecia. Pediatrics **80**:97–101.

37. Bliziotes M, Yergey AL, Nanes MS, Muenzer J, Begley MG, Viera NE, Kher KK, Brandi ML, Marx SJ 1988 Absent intestinal response to calciferols in hereditary resistance to 1,25-dihydroxyvitamin D: Documentation and effective therapy with high dose intravenous calcium infusions. J Clin Endocrinol Metab **66**:294–300.
38. Barsony J, McKoy W, DeGrange DA, Liberman UA, Marx SJ 1989 Selective expression of a normal action of the 1,25-dihydroxyvitamin D_3 receptor in human skin fibroblasts with hereditary severe defects in multiple actions of that receptor. J Clin Invest **83**:2093–2101.
39. Malloy PJ, Hochberg Z, Pike JW, Feldman D 1989 Abnormal binding of vitamin D receptors to deoxyribonucleic acid in a kindred with vitamin D-dependent rickets, type II. J Clin Endocrinol Metab **68**:263–269.
40. Takeda E, Yokota I, Kawakami I, Hashimoto T, Kuroda Y, Arase S 1989 Two siblings with vitamin-D-dependent rickets type II: No recurrence of rickets for 14 years after cessation of therapy. Eur J Pediatr **149**:54–57.
41. Koeffler HP, Bishop JE, Reichel H, Singer F, Nagler A, Tobler A, Walka M, Norman AW 1990 Lymphocyte cell lines from vitamin D-dependent rickets type II show functional defects in the 1 alpha,25-dihydroxyvitamin D_3 receptor. Mol Cell Endocrinol **70**:1–11.
42. Takeda E, Yokota I, Ito M, Kobashi H, Saijo T, Kuroda Y, 1990 25-Hydroxyvitamin D-24-hydroxylase in phytohemagglutinin-stimulated lymphocytes: Intermediate bioresponse to 1,25-dihydroxyvitamin D_3 of cells from parents of patients with vitamin D-dependent rickets type II. J Clin Endocrinol Metab **70**:1068–1074.
43. Yokota I, Takeda E, Ito M, Kobashi H, Saijo T, Kuroda Y 1991 Clinical and biochemical findings in parents of children with vitamin D-dependent rickets Type II. J Inherit Metab Dis **14**:231–240.
44. Simonin G, Chabrol B, Moulene E, Bollini G, Strouc S, Mattei JF, Giraud F 1992 Vitamin D-resistant rickets type II: Apropos of 2 cases. Pediatrie **47**:817–820.
45. Lin JP, Uttley WS 1993 Intra-atrial calcium infusions, growth, and development in end organ resistance to vitamin D. Arch Dis Children **69**:689–692.
46. Marx SJ, Bliziotes MM, Nanes M 1986 Analysis of the relation between alopecia and resistance to 1,25-dihydroxyvitamin D. Clin Endocrinol **25**:373–381.
47. Reichel H, Koeffler HP, Norman AW 1989 The role of the vitamin D endocrine system in health and disease. N Engl J Med **320**:980–991.
48. Pike JW 1991 Vitamin D_3 receptors: Structure and function in transcription. Annu Rev Nutr **11**:189–216.
49. Walters MR 1992 Newly identified actions of the vitamin D endocrine system. Endocr Rev **13**:719–764.
50. Bikle DD, 1992 Clinical counterpoint: Vitamin D: New actions, new analogs, new therapeutic potential. Endocr Rev **13**:765–784.
51. Darwish H, DeLuca HF 1993 Vitamin D-regulated gene expression. Crit Rev Eukaryotic Gene Expression **3**:89–116.
52. MacDonald PN, Dowd DR, Haussler MR 1994 New insight into the structure and functions of the vitamin D receptor. Semin Nephrol **14**:101–118.
53. Hochberg Z, Borochowitz Z, Benderli A, Vardi P, Oren S, Spirer Z, Heyman I, Weisman Y 1985 Does 1,25-dihydroxyvitamin D participate in the regulation of hormone release from endocrine glands? J Clin Endocrinol Metab **60**:57–61.
54. Even L, Weisman Y, Goldray D, Hochberg Z 1996 Selective modulation by vitamin D of renal response to parathyroid hormone: A study in calcitriol-resistant rickets. J Clin Endocrinol Metab **81**:2836–2840.
55. Manolagas SC, Yu XP, Girasole G, Bellido T 1994 Vitamin D and the hematolymphopoietic tissue: A 1994 update. Semin Nephrol **14**:129–143.
56. Etzioni A, Hochberg Z, Pollak S, Meshulam T, Zakut V, Tzehoval E, Keisari Y, Aviram I, Spirer Z, Benderly A, Weisman Y 1989 Defective leukocyte fungicidal activity in end-organ resistance to 1,25-dihydroxyvitamin D. Pediatr Res **25**:276–279.
57. Feldman D, Chen T, Hirst M, Colston K, Karasek M, Cone C 1980 Demonstration of 1,25-dihydroxyvitamin D_3 receptors in human skin biopsies. J Clin Endocrinol Metab **51**:1463–1465.
58. Pike JW, Dokoh S, Haussler MR, Liberman UA, Marx SJ, Eil C 1984 Vitamin D_3-resistant fibroblasts have immunoassayable 1,25-dihydroxyvitamin D_3 receptors. Science **224**:879–881.
59. Dokoh S, Haussler MR, Pike JW 1984 Development of a radioligand immunoassay for 1,25-dihydroxycholecalciferol receptors utilizing monoclonal antibody. Biochem J **221**:129–136.
60. Pike JW, Donaldson CA, Marion SL, Haussler MR, 1982 Development of hybridomas secreting monoclonal antibodies to the chicken intestinal 1 alpha,25-dihydroxyvitamin D_3 receptor. Proc Natl Acad Sci USA **79**:7719–7723.
61. Pike JW, Marion SL, Donaldson CA, Haussler MR 1983 Serum and monoclonal antibodies against the chick intestinal receptor for 1,25-dihydroxyvitamin D_3. Generation by a preparation enriched in a 64,000-dalton protein. J Biol Chem **258**:1289–1296.
62. Pike JW 1984 Monoclonal antibodies to chick intestinal receptors for 1,25-dihydroxyvitamin D_3. Interaction and effects of binding on receptor function. J Biol Chem **259**:1167–1173.
63. Hughes MR, Malloy PJ, Kieback DG, Kesterson RA, Pike JW, Feldman D, O'Malley BW 1988 Point mutations in the human vitamin D receptor gene associated with hypocalcemic rickets. Science **242**:1702–1705.
64. Ritchie HH, Hughes MR, Thompson ET, Malloy PJ, Hochberg Z, Feldman D, Pike JW, O'Malley BW 1989 An ochre mutation in the vitamin D receptor gene causes hereditary 1,25-dihydroxyvitamin D_3-resistant rickets in three families. Proc Natl Acad Sci USA **86**:9783–9787.
65. Malloy PJ, Hochberg Z, Tiosano D, Pike JW, Hughes MR, Feldman D 1990 The molecular basis of hereditary 1,25-dihydroxyvitamin D_3 resistant rickets in seven related families. J Clin Invest **86**:2071–2079.
66. Takeda E, Yokota I, Saijo T, Kawakami I, Ito M, Kuroda Y 1990 Effect of long-term treatment with massive doses of 1 alpha-hydroxyvitamin D_3 on calcium–phosphate balance in patients with vitamin D-dependent rickets type II. Acta Paediatr Jpn **32**:39–43.
67. Evans RM 1988 The steroid and thyroid hormone receptor superfamily. Science **240**:889–895.
68. McDonnell DP, Scott RA, Kerner SA, O'Malley BW, Pike JW 1989 Functional domains of the human vitamin D_3 receptor regulate osteocalcin gene expression. Mol Endocrinol **3**:635–644.
69. Carson-Jurica MA, Schrader WT, O'Malley BW 1990 Steroid receptor family: Structure and functions. Endocr Rev **11**:201–220.
70. Zilliacus J, Wright AP, Carlstedt-Duke J, Gustafsson JA 1995 Structural determinants of DNA-binding specificity by steroid receptors. Mol Endocrinol **9**:389–400.
71. Nakajima S, Hsieh JC, MacDonald PN, Galligan MA, Haussler CA, Whitfield GK, Haussler MR 1994 The C-terminal region of the vitamin D receptor is essential to form a complex with a receptor auxiliary factor required for high affinity binding to the vitamin D-responsive element. Mol Endocrinol **8**:159–172.

72. Jin CH, Kerner SA, Hong MH, Pike JW 1996 Transcriptional activation and dimerization functions in the human vitamin D receptor. Mol Endocrinol **10**:945–957.
73. Wagner RL, Apriletti JW, McGrath ME, West BL, Baxter JD, Fletterick RJ 1995 A structural role for hormone in the thyroid hormone receptor. Nature **378**:690–697.
74. Renaud JP, Rochel N, Ruff M, Vivat V, Chambon P, Gronemeyer H, Moras D 1995 Crystal structure of the RAR-gamma ligand-binding domain bound to all-*trans*-retinoic acid. Nature **378**:681–689.
75. Bourguet W, Ruff M, Chambon P, Gronemeyer H, Moras D, 1995 Crystal structure of the ligand-binding domain of the human nuclear receptor RXR-alpha. Nature **375**:377–382.
76. Forman BM, Yang CR, Au M, Casanova J, Ghysdael J, Samuels HH 1989 A domain containing leucine-zipper-like motifs mediate novel in vivo interactions between the thyroid hormone and retinoic acid receptors. Mol Endocrinol **3**:1610–1626.
77. Faraco JH, Morrison NA, Baker A, Shine J, Frossard PM 1989 *Apa*I dimorphism at the human vitamin D receptor gene locus. Nucleic Acids Res **17**:2150.
78. Szpirer J, Szpirer C, Riviere M, Levan G, Marynen P, Cassiman JJ, Wiese R, DeLuca HF 1991 The Sp1 transcription factor gene (SP1) and the 1,25-dihydroxyvitamin D_3 receptor gene (VDR) are colocalized on human chromosome arm 12q and rat chromosome 7. Genomics **11**:168–173.
79. Labuda M, Fujiwara TM, Ross MV, Morgan K, Garcia-Heras J, Ledbetter DH, Hughes MR, Glorieux FH 1992 Two hereditary defects related to vitamin D metabolism map to the same region of human chromosome 12q13-14. J Bone Miner Res **7**:1447–1453.
80. Pike JW 1992 Molecular mechanisms of cellular response to the vitamin D_3 hormone. In: Coe FL, Farus MJ (eds) Disorders of Bone and Mineral Metabolism Raven, New York, pp. 163–193.
81. Baker AR, McDonnell DP, Hughes M, Crisp TM, Mangelsdorf DJ, Haussler MR, Pike JW, Shine J, O'Malley BW 1988 Cloning and expression of full-length cDNA encoding human vitamin D receptor. Proc Natl Acad Sci USA **85**:3294–3298.
82. Sone T, Scott RA, Hughes MR, Malloy PJ, Feldman D, O'Malley BW, Pike JW 1989 Mutant vitamin D receptors which confer hereditary resistance to 1,25-dihydroxyvitamin D_3 in humans are transcriptionally inactive in vitro. J Biol Chem **264**:20230–20234.
83. Feldman D, Malloy PJ 1990 Hereditary 1,25-dihydroxyvitamin D resistant rickets: Molecular basis and implications for the role of $1,25(OH)_2D_3$ in normal physiology. Mol Cell Endocrinol **72**:C57–C62.
84. Hughes MR, Malloy PJ, O'Malley BW, Pike JW, Feldman D 1991 Genetic defects of the 1,25-dihydroxyvitamin D_3 receptor. J Receptor Res **11**:699–716.
85. Malloy PJ, Weisman Y, Feldman D 1994 Hereditary 1 alpha-25-dihydroxyvitamin D-resistant rickets resulting from a mutation in the vitamin D receptor deoxyribonucleic acid-binding domain. J Clin Endocrinol Metab **78**:313–316.
86. Sone T, Marx SJ, Liberman UA, Pike JW 1990 A unique point mutation in the human vitamin D receptor chromosomal gene confers hereditary resistance to 1,25-dihydroxyvitamin D_3. Mol Endocrinol **4**:623–631.
87. Saijo T, Ito M, Takeda E, Huq AH, Naito E, Yokota I, Sone T, Pike JW Kuroda Y 1991 A unique mutation in the vitamin D receptor gene in three Japanese patients with vitamin D-dependent rickets type II: Utility of single-strand conformation polymorphism analysis for heterozygous carrier detection. Am J Hum Genet **49**:668–673.
88. Rut AR, Hewison M, Rowe P, Hughes M, Grant D, O'Riordan JLH 1991 A novel mutation in the steroid binding region of the vitamin D receptor (VDR) gene in hereditary vitamin D resistant rickets (HVDRR). In: Norman AW, Bouillon R, Thomasset M (eds) Vitamin D: Gene Regulation, Structure–Function Analysis, and Clinical Application. Eighth Workshop on Vitamin D. de Gruyter, New York, pp. 94–95.
89. Yagi H, Ozono K, Miyake H, Nagashima K, Kuroume T, Pike JW 1993 A new point mutation in the deoxyribonucleic acid-binding domain of the vitamin D receptor in a kindred with hereditary 1,25-dihydroxyvitamin D-resistant rickets. J Clin Endocrinol Metab **76**:509–512.
90. Kristjansson K, Rut AR, Hewison M, O'Riordan JL, Hughes MR 1993 Two mutations in the hormone binding domain of the vitamin D receptor cause tissue resistance to 1,25-dihydroxyvitamin D_3. J Clin Invest **92**:12–16.
91. Wiese RJ, Goto H, Prahl JM, Marx SJ, Thomas M, al-Aqeel A, DeLuca HF 1993 Vitamin D-dependency rickets type II: Truncated vitamin D receptor in three kindreds. Mol Cell Endocrinol **90**:197–201.
92. Rut AR, Hewison M, Kristjansson K, Luisi B, Hughes MR, O'Riordan JL 1994 Two mutations causing vitamin D resistant rickets: Modelling on the basis of steroid hormone receptor DNA-binding domain crystal structures. Clin Endocrinol **41**:581–590.
93. Hewison M, Rut AR, Kristjansson K, Walker RE, Dillon MJ, Hughes MR, O'Riordan JL 1993 Tissue resistance to 1,25-dihydroxyvitamin D without a mutation of the vitamin D receptor gene. Clin Endocrinol **39**:663–670.
94. Gross C, Eccleshall TR, Malloy PJ, Villa ML, Marcus R, Feldman D 1996 The presence of a polymorphism at the translation initiation site of the vitamin D receptor gene is associated with low bone mineral density in postmenopausal Mexican-American women. J Bone Miner Res **11**:1850–1855.
95. McDonnell DP, Mangelsdorf DJ, Pike JW, Haussler MR, O'Malley BW 1987 Molecular cloning of complementary DNA encoding the avian receptor for vitamin D. Science **235**:1214–1217.
96. Saiki RK, Gelfand DH, Stoffel S, Scharf SJ, Higuchi R, Horn GT, Mullis KB, Erlich HA 1988 Primer-directed enzymatic amplification of DNA with a thermostable DNA polymerase. Science **239**:487–491.
97. Lin NU-T, Malloy PJ, Sakati N, Al-Ashwal A, Feldman D 1996 A novel mutation in the deoxyribonucleic acid-binding domain of the vitamin D receptor gene causes hereditary 1,25-dihydroxyvitamin D resistant rickets. J Clin Endocrinol Metab **81**:2564–2569.
98. Sone T, Kerner S, Pike JW 1991 Vitamin D receptor interaction with specific DNA. Association as a 1,25-dihydroxyvitamin D_3-modulated heterodimer. J Biol Chem **266**:23296–23305.
99. Luisi BF, Xu WX, Otinowski Z, Freedman LP, Yamamoto KR, Sigler PB 1991 Crystallographic analysis of the interaction of the glucocorticoid receptor with DNA. Nature **352**:497–505.
100. Rastinejad F, Perlmann T, Evans RM, Sigler PB 1995 Structural determinants of nuclear receptor assembly on DNA direct repeats. Nature **375**:203–211.
101. Burns K, Duggan B, Atkinson EA, Famulski KS, Nemer M, Bleackley RC Michalak M 1994 Modulation of gene expression by calreticulin binding to the glucocorticoid receptor. Nature **367**:476–480.
102. Cao X, Teitelbaum SL, Dedhar S, Zhang L, Ross FP 1994 Retinoic acid and calreticulin, a novel ER-derived transcription factor inhibits $1,25(OH)_2D_3$-induced gene transcription. J Bone Miner Res **9**:S145.
103. Dedhar S, Rennie PS, Shago M, Leung Hagesteijn C-Y, Yang H, Filmus J, Hawley RG, Bruchovski N, Cheng H, Matusik RJ, Giguere V 1994 Inhibition of nuclear hormone receptor activity by calreticulin. Nature **367**:480–483.

104. Orkin SH 1984 The mutation and polymorphism of the human β-globin gene and its surrounding DNA. Annu Rev Genet **18**:131–171.
105. Neufeld EF Natural history and inherited disorders of a lysosomal enzyme, β-hexosaminidase. J Biol Chem **264**:10927–10930.
106. Sicinski P, Geng Y, Ryder-Cook AS, Barnard EA, Darlison MG, Barnard PJ 1989 The molecular basis of muscular dystrophy in the mdx mouse: A point mutation. Science **244**:1578–1580.
107. Malloy PJ, Hughes MR, Pike JW, Feldman D 1991 Vitamin D receptor mutations and hereditary 1,25-dihydroxyvitamin D resistant rickets. In: Normal AW, Bouillon R, Thomasset M (eds) Vitamin D: Gene Regulation, Structure–Function Analysis, and Clinical Application. Eighth workshop on Vitamin D. de Gruyter, New York, pp. 116–124.
108. Malloy PJ, Eccleshall TR, Gross C, Van Maldergem Van Bouillon R, Feldman D 1997 Hereditary vitamin D resistant rickets caused by a novel mutation in the vitamin D receptor that results in decreased affinity for hormone and cellular hyporesponsiveness. J Clin Invest **99**:297–304.
109. Van Maldergem L, Bachy A, Feldman D, Bouillon R, Maassen J, Dreyer M, Rey R, Holm C, Gillerot Y 1996 Syndrome of lipoatrophic diabetes, vitamin D resistant rickets, and persistent müllerian ducts in a Turkish boy born to consanguineous parents. Am J Med Genet **64**:506–513.
110. Thompson E, Kristjansson K, Hughes M 1991 Molecular scanning methods for mutation detection: Application to the 1,25-dihydroxyvitamin D receptor. Eighth workshop on Vitamin D, Paris, France, p. 6.
111. Hawa NS, Cockerill FJ, Vadher S, Hewison M, Rut AR, Pike JW, O'Riordan JL Farrow SM 1996 Identification of a novel mutation in hereditary vitamin D resistant rickets causing exon skipping. Clin Endocrinol **45**:85–92.
112. Yoshizawa T, Handa Y, Uematsu Y, Sekine K, Takeda S, Yoshihara Y, Kawakami T, Sato H, Alioka K, Tanimoto K, Fukamizu A, Masushige S, Matsumoto T, Kato S 1996 Disruption of the vitamin D receptor (VDR) in the mouse. J Bone Miner Res **11**:(Suppl. 1):S124.
113. Kruse K, Feldmann E 1995 Healing of rickets during vitamin D therapy despite defective vitamin D receptors in two siblings with vitamin D-dependent rickets type II. J Pediatr **126**:145–148.
114. Balsan S, Garabedian M, Larchet M, Gorski AM, Cournot G, Tau C, Bourdeau A, Silve C, Ricour C 1986 Long-term nocturnal calcium infusions can cure rickets and promote normal mineralization in hereditary resistance to 1,25-dihydroxyvitamin D. J Clin Invest **77**:1661–1667.
115. Weisman Y, Bab I, Gazit D, Spirer Z, Jaffe M, Hochberg Z 1987 Long-term intracaval calcium infusion therapy in end-organ resistance to 1,25-dihydroxyvitamin D. Am J Med **83**:984–990.
116. Hochberg Z, Tiosano D, Even L 1992 Calcium therapy for calcitriol-resistant rickets. J Pediatr **121**:803–808.
117. Weisman Y, Hochberg Z 1994 Genetic rickets and osteomalacia. Curr Ther Endocrinol Metab **5**:492–495.
118. Weisman Y, Jaccard N, Legum C, Spirer Z, Yedwab G, Even L, Edelstein S, Kaye AM, Hochberg Z 1990 Prenatal diagnosis of vitamin D-dependent rickets, type II: Response to 1,25-dihydroxyvitamin D in amniotic fluid cells and fetal tissues. J Clin Endocrinol Metab **71**:937–943.
119. Weisman Y, Malloy PJ, Krishnan AV, Jaccard N, Feldman D, Hochberg Z 1994 Prenatal diagnosis of calcitriol resistant rickets (CRR) by $1,25(OH)_2D_3$ binding, 24-hydroxylase induction and RFLP analysis. Ninth Workshop on Vitamin D, Orlando, Florida, p. 106.
120. Stumpf WE, Sar M, Reid FA, Tanaka Y, DeLuca HF 1979 Target cells for 1,25-dihydroxyvitamin D_3 in intestinal tract, stomach, kidney, skin, pituitary, and parathyroid. Science **206**:1188–1190.
121. Berger, U, Wilson P, McClelland RA, Colston K, Haussler MR, Pike JW, Coombes RC 1988 Immunocytochemical detection of 1,25-dihydroxyvitamin D receptors in normal human tissues. J Clin Endocrinol Metab **67**:607–613.
122. Refetoff S, Weiss RE, Usala SJ, The syndromes of resistance to thyroid hormone. Endocr Rev **14**:348–399.

CHAPTER 49

Glucocorticoid and Vitamin D Interactions

LAWRENCE G. RAISZ Division of Endocrinology and Metabolism, University of Connecticut Health Center, Farmington, Connecticut

BARBARA P. LUKERT University of Kansas Medical Center, Kansas City, Kansas

I. INTRODUCTION

The importance of interactions between glucocorticoids and vitamin D was first apparent from clinical observations. Vitamin D was used to treat glucocorticoid-induced osteoporosis, and glucocorticoids were used to treat vitamin D intoxication. These interactions were considered to be largely at the level of intestinal absorption. Subsequently, interest turned to the interactions of glucocorticoids and vitamin D in the regulation of osteoblast function. Nevertheless, despite many studies these interactions remain poorly understood. The possible interaction of glucocorticoids and vitamin D in the regulation of osteoclasts is even less well understood (see Chapter 21). In this chapter, we review the available information on the effects of glucocorticoids on vitamin D synthesis, calcium and phosphate transport, osteoblast and osteoclast function, and other aspects of the endocrine and reproductive as well as the hematopoietic and immune systems. Finally, we review the interactions of vitamin D and glucocorticoids in clinical settings (see also Chapter 50). Figure 1 and Table I are designed to summarize some of the major effects of vitamin D and glucocorticoids and the important inhibitory actions of glucocorticoids on vitamin D responses.

II. EFFECTS OF GLUCOCORTICOIDS ON THE SYNTHESIS OF 1,25-DIHYDROXYVITAMIN D

In acute, *in vivo* experiments, high doses of glucocorticoids increase the levels of 1,25-dihydroxyvitamin D [1,25$(OH)_2$D [1–3]. In longer term studies, the results are more variable [4–7]. It seems most likely that the initial increase is mediated by impairment of intestinal calcium absorption and secondary hyperparathyroid-

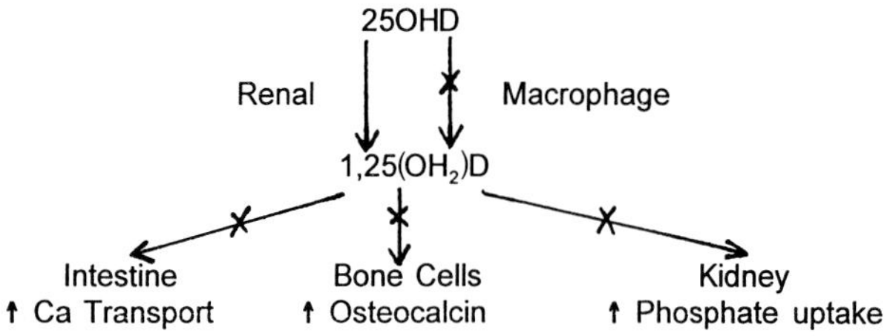

FIGURE 1 Major inhibitory interactions of glucocorticoids on the production and effects of $1{,}25(OH)_2D$. Inhibition of 1-hydroxylase in macrophages, inhibition of the calcium transport effect, and stimulation of osteocalcin production in bone cells by $1{,}25(OH)_2D$ are well-demonstrated effects of glucocorticoids. The interaction with renal phosphate uptake has not been so extensively studied. Glucocorticoids do not directly antagonize the renal synthesis of $1{,}25(OH)_2D$.

ism, but a direct phosphaturic effect of glucocorticoids could also increase $1{,}25(OH)_2D$ levels. The possibility that glucocorticoids also have a separate direct effect on renal 1α-hydroxylase has not been adequately explored. However, the *in vivo* results make it unlikely that there is major direct regulation of this enzyme by glucocorticoids.

In contrast to the lack of direct effect on the renal 1α-hydroxylase, extrarenal synthesis of $1{,}25(OH)_2D$ by cells of the monocyte–macrophage lineage can be markedly affected by glucocorticoids. The 1α-hydroxylase enzyme complex in macrophages may differ from that in renal cells. It can be inhibited directly by addition of dexamethasone, and this may account for the effectiveness of glucocorticoids in the treatment of patients with hypercalcemia due to sarcoidosis and other granulomatous disorders [8–9]. Dexamethasone was also found to inhibit 1α-hydroxylase activity in a chick myelomonocytic cell line [10]. Even here, it is possible that some of the glucocorticoid effect may be indirect. Adams *et al.* [11] found that arachidonic acid metabolites stimulate 1α-hydroxylase in alveolar macrophages from patients with sarcoidosis, whereas leukotriene C_4 increased hydroxylase activity in the presence of dexamethasone. Glucocorticoids could inhibit arachidonic acid release and eicosanoid production. Importantly, glucocorticoids have potent immunosuppressive effects and inhibit the disease process in sarcoidosis. No doubt these actions

TABLE I Some Effects of Vitamin D and Glucocorticoids[a]

	Vitamin D		Glucocorticoid	
Target	Effect	Mechanism	Effect	Mechanism
Pituitary				
GH	Decreased secretion	Inhibits T_3-induced secretion	Decreased secretion	Inhibits GHRH-induced secretion
LH	? Necessary for fertility	?	Decreased secretion	Inhibits LHRH-induced secretion
Gonads				
Estrogen	?	?	Decreased blood levels	Inhibits granulosa cell
Testosterone	?	?	Decreased blood levels	Decreased LH secretion
Gastrointestinal tract, calcium absorption	Increase	Membrane effect, increased CaBP, Ca^{2+}-ATPase	Decrease	Decreased CaBP, Ca receptor (nongenomic), decreased release from mitochondria
Kidney				
Phosphate	Excretion	Response to absorption	Increased excretion	Secondary hyperparathyroidism, Na^+ gradient dependent
Calcium	Excretion	Response to absorption	Increased excretion	Increased bone resorption, decreased tubular reabsorption
Bone				
Osteoclasts	Increased formation and activity	?	Decreased formation and increased activity	?
Osteoblasts	Increased osteocalcin, decreased collagen synthesis	Transcriptional regulation	Decreased osteocalcin, decreased collagen synthesis	Transcriptional regulation, effects on growth factors

[a] GH, Growth hormone; T_3, triiodothyronine; GHRH, growth-hormone-releasing hormone; LH, luteinizing hormone; LHRH, luteinizing hormone-releasing hormone: CaBP, calcium binding protein.

contribute in a major way to their effectiveness in treating the hypercalcemia.

III. GLUCOCORTICOID–VITAMIN D INTERACTIONS ON CALCIUM AND PHOSPHATE TRANSPORT

The effect of the interaction between glucocorticoids and vitamin D on calcium transport was first recognized when it was found that pharmacological doses of glucocorticoids inhibited the gastrointestinal absorption of calcium [4,12–15]. The role of glucocorticoid–vitamin D interactions in glucocorticoid-induced changes in calcium absorption remains unclear. *In vitro* studies show that glucocorticoids decrease intestinal mucosal to serosal flux and increase serosal to mucosal flux [12–13]. The effect is partially independent of vitamin D [16–18], and only 25% of the impaired calcium absorption can be accounted for by slightly reduced $1,25(OH)_2D$ levels [19]. *In vitro*, administration of $1,25(OH)_2D_3$ improves calcium transport but does not return absorption to normal [20].

Intestinal $1,25(OH)_2D$ receptors (VDR) appear to be increased in glucocorticoid-treated rats and decreased in mice [21]. The role of changes in VDR levels as a mechanism for glucocorticoid effects on vitamin D action is discussed in Chapter 11. It has been suggested that glucocorticoids accelerate the breakdown of $1,25(OH)_2D$ at the mucosal receptor site [22], but most studies have shown that neither the localization nor the further mucosal processing of $1,25(OH)_2D$ is altered [12,18,23].

Several observations suggest that glucocorticoid-induced inhibition of calcium transport is caused by alterations of posttranscriptional events. For example, in duodenal organ cultures, dexamethasone in low concentrations increases calcium transport and enhances $1,25(OH)_2D$-dependent calbindin-D_{28K} synthesis by positive cotranscriptional regulation of gene expression [18], whereas high concentrations of dexamethasone mimic the action of pharmacological doses of glucocorticoids administered *in vivo* (inhibition of calcium transport) even though the stimulatory effect on calbindin synthesis is maintained [24].

Several changes in cellular metabolism contribute to the defect in calcium transport caused by glucocorticoids. Calcium release from mitochondria is inhibited by glucocorticoids because of depletion of mitochondrial adenosine triphosphate [17]. Na^+, K^+-ATPase pump is stimulated by glucocorticoids, thus increasing the flow of sodium and water through the cell into the serosal space and then through the pericellular pathway back into the lumen, carrying calcium with it by solvent drag [13,25]. These nongenomic mechanisms may negate the stimulatory effect of vitamin D metabolites on intestinal calcium absorption.

Fasting hypercalciuria and elevated serum concentrations of parathyroid hormone (PTH) are present after only 5 days of glucocorticoid administration. Hypercalciuria is due in part to increased bone resorption, but there is also decreased renal tubular reabsorption of calcium in spite of elevated circulating levels of PTH that should increase tubular reabsorption [26].

Urinary excretion of phosphate is high in patients taking glucocorticoids. This is due in part to secondary hyperparathyroidism. In addition, however, glucocorticoids induce a change in Na^+–H^+ exchange activity that causes a decrease in sodium gradient-dependent phosphate uptake in the proximal tubule and opposes vitamin D-stimulation of phosphate uptake [27,28].

Several findings suggest that a significant, isolated change in the cellular transport of calcium may play a central role in glucocorticoid-induced changes in calcium homeostasis. Secondary hyperparathyroidism and increased levels of $1,25(OH)_2D$ have been observed in patients taking glucocorticoids over long periods [4,27,29–30]. This has been generally attributed to impaired gastrointestinal absorption of calcium and negative calcium balance, which led to increased PTH. PTH then, in turn, induced a rise in the renal synthesis of $1,25(OH)_2D$ with a consequent rise in serum levels of $1,25(OH)_2D$. However, other studies have shown that high doses of glucocorticoids have acute effects on the kidney, causing increased $1,25(OH)_2D$ levels and phosphaturia before an elevation in PTH is observed [1]. Serum levels of PTH and $1,25(OH)_2D$ remain elevated in the presence of persistently high levels of calcium. This suggests that PTH secretion and $1,25(OH)_2D$ synthesis may not be appropriately suppressed by calcium, raising the possibility of a state of calcium resistance caused by glucocorticoids [31]. Thus, changes in $1,25(OH)_2D$ levels may be secondary to a defect in cellular calcium transport rather than being due solely to secondary hyperparathyroidism resulting from negative calcium balance or to a direct effect of steroids on synthesis.

IV. CELLULAR INTERACTIONS BETWEEN GLUCOCORTICOIDS AND 1,25-DIHYDROXYVITAMIN D IN BONE

Glucocorticoids have potent, complex effects on the replication, differentiation, and function of cells of the osteoblastic lineage. Inhibition of bone formation is probably the most important pathogenetic mechanism in glucocorticoid-induced osteoporosis. This may involve direct inhibition of osteoblast replication, differentiation, and collagen synthesis, as well as interactions with local and systemic factors [32]. *In vivo* studies on the interaction of glucocorticoids and $1{,}25(OH)_2D_3$ in regulating osteoblast function are conflicting. At high concentrations, $1{,}25(OH)_2D_3$ inhibits collagen gene transcription in osteoblasts, but this effect is probably not related to the glucocorticoid effect [33]. Glucocorticoids clearly oppose one major effect of $1{,}25(OH)_2D$ on osteoblastic cells, that is, the stimulation of osteocalcin synthesis [34–35]. On this basis, it has been postulated that calcitriol might be useful in the treatment of glucocorticoid osteoporosis, not only for its effects on calcium absorption, but also for its potential effects on osteoblast function [36]. However, in one histomorphometric study in patients with rheumatoid arthritis, addition of calcitriol produced a decrease in the number of osteoblasts and their activity [37]. In this section, we review the many studies on the interaction of glucocorticoids and $1{,}25(OH)_2D_3$ in regulating osteoblast function. These have been carried out in a wide variety of cell models, and it may be for this reason that the results are often conflicting.

In contrast to their inhibitory effect on osteoblasts with prolonged treatment in animals or in organ cultures, glucocorticoids promote the development of the osteoblast phenotype in culture systems consisting largely of osteoblast precursors or marrow stromal cells [38]. In these systems, glucocorticoids can actually increase the stimulatory effect of vitamin D on osteocalcin synthesis. Nevertheless, the effects of dexamethasone and $1{,}25(OH)_2D_3$ on stromal cell differentiation in these culture systems are complex. In fetal rat calvarial derived osteoblasts, $1{,}25(OH)_2D_3$ blocks differentiation of proliferating cells, whereas glucocorticoids enhance this process [39]. Shifts from osteoblastic to adipocytic differentiation have been observed with dexamethasone, which can be either enhanced or opposed by $1{,}25(OH)_2D_3$ [40–42].

Effects on the VDR in bone cells have been reported, but here again the results in different cell models vary. In the rat osteogenic sarcoma cell line ROS 17/2 glucocorticoids were shown to increase the VDR concentration, whereas in the human osteosarcoma cell MG-63 a decrease was reported [43–46]. Changes in mRNA level, however, may not reflect changes in protein levels [47].

The synthesis of osteocalcin is highly sensitive to $1{,}25(OH)_2$ D through a transcriptional effect on a specific vitamin D response element (VDRE). Glucocorticoids reduce osteocalcin levels *in vivo* [48] and inhibit the vitamin D response [49], probably by acting at a separate site on the osteocalcin promoter [50,51]. A negative response element overlapping the TAATA box of the osteocalcin promoter has been invoked to explain the glucocorticoid effect. Additional interactions with factors that bind to the vitamin D response element have also been suggested [52,53]. Another possible mechanism might be enhancement of cyclic AMP responses to prostaglandin E_2 (PGE_2). Endogenous prostaglandins are undoubtedly present in these cell cultures. Glucocorticoids can enhance the level of cyclic AMP in response to PGE_2, and this can in turn inhibit osteocalcin production [54]. The effects of vitamin D and dexamethasone on osteocalcin gene expression can be demonstrated *in vivo* as well as *in vitro* [55]. Glucocorticoids may also accelerate the clearance of osteocalcin from the circulation [56].

Glucocorticoids and $1{,}25(OH)_2D$ may interact in their effects on insulin-like growth factors (IGFs). The initial paradoxical stimulation of collagen synthesis by glucocorticoids in organ culture has been attributed to increased sensitivity to IGF [57]. There is a subsequent decrease in collagen synthesis in these cultures, which has been attributed both to a decrease in the production of other local growth factors and to direct inhibitory effects of glucocorticoids on cell replication and differentiation [32]. $1{,}25(OH)_2D_3$ can oppose the effects of dexamethasone on IGF production in cell culture [58,59]. In addition, dexamethasone can inhibit and $1{,}25(OH)_2D_3$ can enhance the secretion of the IGF binding protein, IGFBP-3, in human osteosarcoma cells [60]. The binding proteins may either enhance or inhibit the response to IGF.

Glucocorticoids can inhibit the production of eicosanoids by blocking phospholipase activity, which causes the release of arachidonic acid from phospholipids, and by preventing stimulation of the inducible prostaglandin G/H synthase (cyclooxygenase-2) [61]. $1{,}25(OH)_2D_3$ can decrease arachidonic acid uptake in human osteoblast-like cells and can inhibit, as can dexamethasone, prostaglandin-induced calcium influx in the rat osteoblastic cell line MC3T3-E1 [62,63]. However, $1{,}25(OH)_2D_3$ can also stimulate prostaglandin production in bone [64]. How these interactions might affect osteoblast regulation is not known.

Glucocorticoids can increase and $1{,}25(OH)_2D_3$ can decrease PTH activation of adenylate cyclase [65–67]. These hormones also have opposing effects on the number of PTH receptors in ROS 17/2.8 cells [68]. The mechanism of these effects is not fully understood, but they appear to be independent of the modulation of PTH-sensitive adenylate cyclase by protein kinase C [67].

Although glucocorticoids are generally considered to impair mineralization and vitamin D to enhance this process, these effects may be largely at the level of the intestine rather than directly on bone. However, toxic doses of 1,25$(OH)_2D_3$ produce a paradoxical inhibition of mineralization, with a marked increase in unmineralized osteoid seams [69]. In contrast, glucocorticoid excess generally is associated with thin osteoid seams, presumably due to impaired matrix synthesis.

Osteoblasts have an ectonucleoside pyrophosphatase activity that may be important in mineralization [70]. 1,25$(OH)_2D_3$ enhances the activity of this enzyme in human osteoblast-like cells, whereas dexamethasone inhibits it. However, the stimulatory effect of 1,25$(OH)_2D_3$ is enhanced in the presence of dexamethasone. 1,25$(OH)_2D_3$ also enhances plasminogen activator activity in clonal rat osteogenic sarcoma cells, and this effect is opposed by glucocorticoids [71]. These effects may influence the role of osteoblasts in activating resorption, as well as their synthetic functions.

Osteoclast formation in cultures that contain both hematopoietic cell precursors and cells of the osteoblast lineage is markedly enhanced by 1,25$(OH)_2D_3$ [72]. There are few studies on the effects of glucocorticoids on these systems. Dexamethasone was found to enhance the effects of 1,25$(OH)_2D_3$ to produce osteoclasts in a mouse bone marrow culture system and, in contrast, to inhibit the effects of granulocyte–macrophage colony-stimulating factor [73]. In a different system using a clonal chondrogenic cell line to support osteoclastogenesis from mouse bone marrow cells, 1,25$(OH)_2D_3$ and dexamethasone had opposing effects [74]. Interpretation of these studies could be complicated by the effects of dexamethasone on differentiation as well as on prostaglandin synthesis which is an important cofactor in osteoclastogenesis in some culture systems [61].

V. REGULATION OF THE ENDOCRINE AND REPRODUCTIVE SYSTEMS

Glucocorticoids and calcitriol both modulate secretion of growth hormone (GH). Calcitriol appears to inhibit triiodothyronine (T_3)-induced GH production due to down-regulation of the nuclear T_3 receptors. Prednisone inhibits secretion of GH in response to GH-releasing hormone (GHRH), but serum concentrations of GH and IGF-1 are normal in patients receiving glucocorticoids [23,24]. The interaction of glucocorticoids and calcitriol on GH secretion and actions has not been studied.

Vitamin D is necessary for normal reproductive function in both male and female rats [75,76]. The effect of calcitriol on reproductive tissue has been most extensively studied in the testis. Luteinizing hormone (LH) and testosterone levels are reduced and spermatogenesis is incomplete in vitamin D-deficient rats. There appears to be a relationship between VDR levels and normal Sertoli cell function and spermatogenic activity. The effect of vitamin D deficiency, vitamin D, or its metabolites on serum testosterone levels and sperm production in humans has not been reported.

Testosterone levels are low in men taking pharmacological doses of glucocorticoids chronically [77–79]. This is due to blunting of pituitary secretion of LH [80,81], a direct effect on testosterone production by the testes [77], and reduced adrenal secretion of testosterone due to suppression of adrenocorticotropin (ACTH). There is no information available about the combined effects of calcitriol and glucocorticoids on male gonadal hormones or reproductive function.

The effect of vitamin D metabolites on serum estrogen levels and reproductive function in female animal models or women has not been so extensively studied. However, it has been shown that calcitriol restores fertility in vitamin D-deficient female rats [75].

Serum estradiol (E_2), estrone (E_1), dehydroepiandrosterone (DHEA), and progesterone concentrations are low in women taking glucocorticoids long term [82,83]. The low serum E_2 and progesterone levels are due to blunting of pituitary secretion of LH and direct inhibition of estrogen production by ovarian granulosa cells [16,80]. E_1 is low because of decreased adrenal secretion of androstenedione, which is aromatized to E_1 in peripheral fat tissues. The reduction in DHEA results from decreased adrenal secretion due to inhibition of ACTH secretion. There are no data available on the interaction of vitamin D metabolites and glucocorticoids on these changes.

VI. VITAMIN D AND GLUCOCORTICOIDS IN CLINICAL SETTINGS

Our understanding of the interaction of vitamin D metabolites and glucocorticoids in clinical situations has been derived largely from observations on the effects of glucocorticoids on vitamin D-dependent hypercalcemia. It has long been known that the hypercalcemia associated with vitamin D intoxication can be controlled by glucocorticoid administration [84]. Hypercalcemia due to granulomatous disease and lymphoma are other classic examples. In patients with granulomas due to sarcoidosis, tuberculosis, or other disease entities associated with granuloma formation, the macrophages within the granuloma produce a 1α-hydroxylase capable of metabolizing calcidiol to calcitriol [11]. The subject is discussed more fully in Chapter 55. Glucocorticoids inhibit the 25-hydroxyvitamin D (25OHD) 1-hydroxylase in pulmonary alveolar macrophages [8], and this probably explains, in part, why glucocorticoids correct hypercalcemia in patients with sarcoidosis.

In some patients with lymphoma, hypercalcemia is due to the production of calcitriol by lymphoma cells [85]. Hypercalcemia is due largely to 1,25$(OH)_2$D-induced increased gastrointestinal absorption of calcium. Treatment with glucocorticoids reduces the levels of serum calcium and 1,25$(OH)_2$D.

VII. CONCLUSIONS

There are a number of important and reasonably well-defined interactions between glucocorticoids and the vitamin D hormone system. However, still more interactions are likely to be discovered with further analysis, particularly at the level of the glucocorticoid and vitamin D receptors. For example, the specific response elements in bone that mediate inhibition of collagen synthesis by both glucocorticoids and 1,25$(OH)_2D_3$ have not been identified. Interaction of trans-acting factors, in addition to the classic receptors for these hormones, may be important in these pathways. Similar interactions may occur in determining cellular responses in the intestine and other target issues. Further studies in these areas may improve our understanding of not only the mechanism of action of these critical hormones, but also the pathogenesis and treatment of disorders of mineral metabolism related to glucocorticoids and vitamin D.

Acknowledgments

We thank Ms. Lisa Godin and Ms. Lynn Limeburner for their careful preparation of the manuscript. This work was supported in part by grants AR18063, AR38933 and M01-RR-06192 from the National Institutes of Health.

References

1. Cosman F, Nieves J, Herbert J, Shen V, Lindsay R 1994 High-dose glucocorticoids in multiple sclerosis patients exert direct effects on the kidney and skeleton. J Bone Miner Res **9**:1097–1105.
2. van der Veen MJ, Bijlsma JW 1992 Effects of different regimes of corticosteroid treatment on calcium and bone metabolism in rheumatoid arthritis. Clin Rheum **11**:388–392.
3. Nielsen HK, Thomsen K, Eriksen EF, Charles P, Storm T, Mosekilde L 1988. The effects of high-dose glucocorticoid administration on serum bone gamma carboxyglutamic acid-containing protein, serum alkaline phosphatase and vitamin D metabolites in normal subjects. Bone Miner **4**:105–113.
4. Bikle DD, Halloran B, Fong L, Steinbach L, Shellito J 1993 Elevated 1,25-dihydroxyvitamin D levels in patients with chronic obstructive pulmonary disease treated with prednisone. J Clin Endocrinol Metab **76**(2):456–461.
5. Lund B, Storm TL, Lund B, Melsen F, Mosekilde L, Andersen RB, Egmose C, Sorensen OH 1985 Bone mineral loss, bone histomorphometry and vitamin D metabolism in patients with rheumatoid arthritis on long-term glucocorticoid treatment. Clin Rheumatol **4**:143–149.
6. Izawa Y, Makita T, Ichiki H 1985 Glucocorticoid-induced bone disorders: Nature and mechanisms of glucocorticoid-induced bone disorders in dogs. Acta Vitaminol Enzymol **7**:77–84.
7. Prummel MF, Wiersinga WM, Lips P, Sanders GT, Sauerwein HP 1991 The course of biochemical parameters of bone turnover during treatment with corticosteroids J Clin Endocrinol Metab **72**:382–386.
8. Adams JS, Gacad MA 1985 Characterization of 1 alpha-hydroxylation of vitamin D_3 sterols by cultured alveolar macrophages from patients with sarcoidosis. J Exp Med **161**:755–765.
9. Adami S, Graziani G, Tartarotti D, Cappelli R, Casati S, Cantaluppi A, Braga V, Lo Cascio V 1987 Extrarenal synthesis of 1,25-dihydroxyvitamin D: Sensitivity to glucocorticoid treatment. Clin Sci **72**:329–334.
10. Adams JS, Ren SY, Arbelle JE, Horiuchi N, Gray RW, Clemens TL, Shany S 1994 Regulated production and intracrine action of 1,25-dihydroxyvitamin D_3 in the chick myelomonocytic cell line HD-11. Endocrinology **134**:2567–2573.
11. Adams JS, Gacad MA, Diz MM, Nadler JL 1990 A role for endogenous arachi-donate metabolites in the regulated expression of the 25-hydroxyvitamin D-1-hydroxylation reaction in cultured alveolar macrophages from patients with sarcoidosis. J Clin Endocrinol Metab **70**:595–600.
12. Favus MJ, Walling MW, Kimberg DV 1973 Effects of 1,25-dihydroxy-cholecalciferol in intestinal calcium transport in cortisone-treated rats. J Clin Invest **52**:1680–1685.
13. Adams JS, Lukert BP 1980 Effects of sodium restriction on ^{45}Ca and ^{22}Na transduodenal flux in corticosteroid-treated rats. Miner Electrolyte Metab **4**:216–226.
14. Adams JS, Wahl TO, Lukert BP 1981 Effects of hydrochlorothiazide and dietary sodium restriction on calcium metabolism in corticosteroid treated patients. Metabolism **30**:217–222.
15. Wajchenberg BL, Pereira VG, Kieffer J, Ursic S 1969 Effect of dexamethasone on calcium and ^{47}Ca kinetics in normal subjects. Acta Endocrinol **61**:173–192.
16. Hsueh AJ, Erickson GF 1978 Glucocorticoid inhibition of FSH-induced estrogen production in cultured rat granulosa cells. Steroids **32**:639–648.
17. Kimaru S, Rasmussen H 1977 Adrenal glucocorticoids, adenine nucleotide translocation, and mitochondrial calcium accumulation. J Biol Chem **252**:1217–1225.
18. Shultz TD, Kumar R 1987 Effect of cortisol on [^{3}H]1,25-dihydroxy vitamin D_3 uptake and 1,25-dihydroxyvitamin D_3-induced DNA-dependent RNA polymerase activity in chick intestinal cells. Calcif Tissue Int **40**(4):224–230.
19. Morris HA, Need AG, O'Loughlin PD, Horowitz M, Bridges A, Nordin BEC 1990 Malabsorption of calcium in corticosteroid-induced osteoporosis. Calcif Tissue Int **46**:305–308.
20. Colette C, Monnier L, Pares Herbute N, Blotman F, Mirouze J 1987 Calcium absorption in corticoid treated subjects effects of a single oral dose of calcitriol. Horm Metab Res **19**:335–338.
21. Hirst M, Feldman D 1982 Glucocorticoid regulation of 1,25$(OH)_2$-vitamin D_3 receptors: Divergent effects on mouse and rat intestine. Endocrinology **111**:1400–1402.
22. Carre M, Ayigbebe O, Miravet L, Rasmussen H 1974 The effect

of prednisolone upon the metabolism and action of 25-hydroxy- and 1,25-dihydroxyvitamin D_3. Proc Natl Acad Sci USA **1**:2996–3000.

23. Kimberg DV, Baerg RD, Gershon E, Graudusius RT 1971 Effect of cortisone treatment on the active transport of calcium by the small intestine. J Clin Invest **50**:1309–1321.
24. Corradino RA, Fullmer CB 1991 Positive cotranscriptional regulation of intestinal calbindin-D28K gene expression by 1,25-dihydroxyvitamin D_3 and glucocorticoids. Endocrinology **128**(2):944–950.
25. Charney AN, Kinsey MD, Myers L, Giannella RA, Gots RE 1975 Na^+–K^+-activated adenosine triphosphatase and intestinal electrolyte transport. Effect of adrenal steroids. J Clin Invest **56**:653–660.
26. Reid IR, Ibbertson HK 1987 Evidence of decreased tubular reabsorption of calcium in glucocorticoid-treated asthmatics. Horm Res **27**:200–204.
27. Lukert BP, Adams JS 1976 Calcium and phosphorus homeostatis in man. Effect of corticosteroids. Arch Intern Med **136**:1249–1253.
28. Frieberg JM, Kinsella J, Sacktor B 1982 Glucocorticoids increase the Na^+–H^+ exchange and decrease the Na^+ gradient-dependent phosphate-uptake systems in renal brush border membrane vesicles. Proc Natl Acad Sci USA **79**:4932–4936.
29. Hahn TJ, Halstead LR, Teitelbaum SL, Hahn BH 1979 Altered mineral metabolism in glucocorticoid-induced osteopenia. Effect of 25-hydroxyvitamin D administration. J Clin Invest **64**:655–665.
30. Suzuki Y, Ichikawa Y, Saito E, Homma M 1983 Importance of increased urinary calcium excretion in the development of secondary hyperparathyroidism of patients under glucocorticoid therapy. Metabolism **32**:151–156.
31. Brown EM, Gamba D, Riccardi M, Lombardi M, Butters R, Kifor O, Sun A, Hediger MA, Lytton J, Hebert SC 1993 Cloning and characterization of an extracellular Ca^{2+}-sensing receptor from bovine parathyroid. Nature **366**:575–580.
32. Lukert BP, Raisz LG 1990 Glucocorticoid-induced osteoporosis: Pathogenesis and management. Ann Intern Med **112**:352–364.
33. Pavlin D, Bedalov A, Kronenberg MS, Kream BE, Rowe DW, Smith CL, Pike JW, Lichtler AC 1994 Analysis of regulatory regions in the COL1A1 gene responsible for 1,25-dihydroxyvitamin D_3-mediated transcriptional repression in osteoblastic cells. J Cell Biochem **56**:490–501.
34. Jowell PS, Epstein S, Fallon MD, Reinhardt TA, Ismail F 1987 1,25-Dihydroxy-vitamin D_3 modulates glucocorticoid-induced alteration in serum bone Gla protein and bone histomorphometry. Endocrinology **120**:531–536.
35. Kasperk C, Schneider U, Sommer U, Niethard F, Ziegler R 1995 Differential effects of glucocorticoids on human osteoblastic cell metabolism in vitro. Calcif Tissue Int **57**:120–126.
36. Sambrook P, Birmingham J, Kelly P, Kempler S, Nguyen T, Pocock N, Eisman J 1993 Prevention of corticosteroid osteoporosis. A comparison of calcium, calcitriol, and calcitonin. N Engl J Med **328**:1747–1752.
37. Dykman TR, Haralson KM, Gluck OS, Murphy WA, Teitelbaum SL, Hahn TJ, Hahn BH 1984 Effect of oral 1,25-dihydroxyvitamin D and calcium on glucocorticoid-induced osteopenia in patients with rheumatic diseases. Arth Rheum **27**:1336–1343.
38. Shalhoub V, Conlon D, Tassinari M, Quinn C, Partridge N, Stein GS, Lian JB 1992 Glucocorticoids promote development of the osteoblast phenotype by selectively modulating expression of cell growth and differentiation associated genes. J Cell Biochem **50**:425–440.
39. Pockwinse SM, Stein JL, Lian JB, Stein GS 1995 Developmental stage-specific cellular responses to vitamin D and glucocorticoids during differentiation of the osteoblast phenotype: Interrelationship of morphology and gene expression by in situ hybridization. Exp Cell Res **216**:244–260.
40. Bellows CG, Wang YH, Heersche JN, Aubin JE 1994 1,25-Dihydroxyvitamin D_3 stimulates adipocyte differentiation in cultures of fetal rat calvaria cells: Comparison with the effects of dexamethasone. Endocrinology **134**:2221–2229.
41. Beresford JN, Bennett JH, Devlin C, Leboy PS, Owen ME 1992 Evidence for an inverse relationship between the differentiation of adipocytic and osteogenic cells in rat marrow stromal cell cultures. J Cell Sci **102**(Part 2):341–351.
42. Beresford JN, Joyner CJ, Devlin C, Triffitt JT 1994 The effects of dexamethasone and 1,25-dihydroxyvitamin D_3 on osteogenic differentiation of human marrow stromal cells in vitro. Arch Oral Biol **39**:941–947.
43. Manolagas SC, Abare J, Deftos LJ 1984 Glucocorticoids increase the $1,25(OH)_2D_3$ receptor concentration in rat osteogenic sarcoma cells. Calcif Tissue Int **36**:153–157.
44. Maenpaa P, Mahonen A, Pirskanen A 1991 Hormonal regulation of vitamin D receptor levels and osteocalcin synthesis in human osteosarcoma cells. Calcif Tissue Int **49**(Suppl.):S85-S86.
45. Mahonen A, Pirskanen A, Maenpaa PH 1991 Homologous and heterologous regulation of 1,25-dihydroxyvitamin D_3 receptor mRNA levels in human osteosarcoma cells. Biochim Biophys Acta **1088**:111–118.
46. Godschalk M, Levy JR, Downs RW Jr 1992 Glucocorticoids decrease vitamin D receptor number and gene expression in human osteosarcoma cells. J Bone Miner Res **7**:21–27.
47. Mahonen A, Maenpaa PH 1994 Steroid hormone modulation of vitamin D receptor levels in human MG-63 osteosarcoma cells. Biochem Biophys Res Commun **205**:1179–1186.
48. Nielsen HK 1994 Circadian and circatrigintan changes in osteoblastic activity assessed by serum osteocalcin. Physiological and methodological aspects. Dan Med Bull **41**:216–227.
49. Pirskanen A, Mahonen A, Maenpaa PH 1991 Modulation of $1,25(OH)_2D_3$-induced osteocalcin synthesis in human osteosarcoma cells by other steroidal hormones. Mol Cell Endocrinol **76**:149–159.
50. Morrison NA, Shine J, Fragonas JC, Verkest V, McMenemy ML, Eisman JA 1989 1,25-Dihydroxyvitamin D-responsive element and glucocorticoid repression in the osteocalcin gene. Science **246**:1158–1161.
51. Morrison N, Eisman J 1993 Role of the negative glucocorticoid regulatory element in glucocorticoid repression of the human osteocalcin promoter. J Bone Miner Res **8**:969–975.
52. Schepmoes G, Breen E, Owen TA, Aronow MA, Stein GS, Lian JB 1991 Influence of dexamethasone on the vitamin D-mediated regulation of osteocalcin gene expression. J Cell Biochem **47**:184–196.
53. Heinrichs AA, Bortell R, Rahman S, Stein JL, Alnemri ES, Litwack G, Lian JB, Stein GS 1993 Identification of multiple glucocorticoid receptor binding sites in the rat osteocalcin gene promoter. Biochemist **32**:11436–11444.
54. Lajeunesse D, Kiebzak GM, Frondoza C, Sacktor B 1991 Regulation of osteo-calcin secretion by human primary bone cells and by the human osteosarcoma cell line MG-63. Bone Miner **14**:237–250.
55. Ikeda T, Kohno H, Yamamuro T, Kasai R, Ohta S, Okumura H, Konishi J, Kikuchi H, Shigeno C 1992 The effect of active vitamin D_3 analogs and dexamethasone on the expression of osteocalcin gene in rat tibiae in vivo. Biochem Biophys Res Commun **189**:1231–1235.
56. Fortune CL, Farrugia W, Tresham J, Scoggins BA, Wark JD 1989 Hormonal regulation of osteocalcin plasma production and clearance in sheep. Endocrinology **124**:2785–2790.
57. Kream BE, Petersen DN, Raisz LG 1990 Cortisol enhances the

anabolic effects of insulin-like growth factor I on collagen synthesis and procollagen messenger ribonucleic acid levels in cultured 21-day fetal rat calvariae. Endocrinology **126**:1576–1583.

58. Chen TL, Chang LY, Bates RL, Perlman AJ 1991 Dexamethasone and 1,25-dihydroxyvitamin D_3 modulation of insulin-like growth factor-binding proteins in rat osteoblast-like cell cultures. Endocrinology **128**:73–80.
59. Chen TL, Mallory JB, Hintz RL 1991 Dexamethasone and 1,25$(OH)_2$ vitamin D_3 modulate the synthesis of insulin-like growth factor-I in osteoblast-like cells. Calcif Tissue Int **48**:278–282.
60. Nakao Y, Hilliker S, Baylink DJ, Mohan S 1994 Studies on the regulation of insulin-like growth factor binding protein 3 secretion in human osteosarcoma cells in vitro. J Bone Miner Res **9**:865–872.
61. Kawaguchi H, Pilbeam CC, Harrison JR, Raisz LG 1995 The role of prostaglandins in the regulation of bone metabolism. Clin Orthop Related Res **313**:36–46.
62. Kozawa O, Tokuda H, Kotoyori J, Suzuki A, Ito Y, Oiso Y 1993 Modulation of prostaglandin E_2-induced Ca^{2+} influx by steroid hormones in osteoblast-like cells. Prostaglandins Leukotrienes Essential Fatty Acids **49**:711–714.
63. Cissel DS, Birkle DL, Whipkey DL, Blaha JD, Graeber GM, Keeting PE 1995 1,25-Dihydroxyvitamin D_3 or dexamethasone modulate arachidonic acid uptake and distribution into glycerophospholipids by normal adult human osteoblast-like cells. J Cell Biochem **57**:599–609.
64. Klein-Nulend J, Pilbeam CC, Raisz LG 1991 Effect of 1,25-dihydroxyvitamin D_3 on prostaglandin E_2 production in cultured mouse parietal bones. J Bone Miner Res **6**:1339–1344.
65. Titus L, Rubin JE, Lorang MT, Catherwood BD 1988 Glucocorticoids and 1,25-dihydroxyvitamin D_3 regulate parathyroid hormone stimulation of adenosine 3′, 5′-monophosphate-dependent protein kinase in rat osteosarcoma cells. Endocrinology **123**:1526–1534.
66. Titus L, Rubin JE, Nanes MS, Catherwood BD 1989 Glucocorticoid and 1,25-dihydroxyvitamin D modulate the degree of adenosine 3′, 5′-monophosphate-dependent protein kinase isoenzyme I and II activation by parathyroid hormone in rat osteosarcoma cells. Endocrinology **125**:2806–2811.
67. Rao LG, Wylie JN 1993 Modulation of parathyroid hormone-sensitive adenylate cyclase in ROS 17/2.8 cells by dexamethasone 1,25-dihydroxyvitamin D_3 and protein kinase C. Bone Miner **23**:35–47.
68. Titus L, Jackson E, Nanes MS, Rubin JE, Catherwood BD 1991 1,25-Dihydroxyvitamin D reduces parathyroid hormone receptor number in ROS 17/2.8 cells and prevents the glucocorticoid-induced increase in these receptors: Relationship to adenylate cyclase activation. J Bone Miner Res **6**:631–637.
69. Hock JM, Gunness-Hey M, Poser J, Olson H, Bell NH, Raisz LG 1986 Stimulation of undermineralized matrix formation by 1,25-dihydroxyvitamin D_3 in long bones of rats. Calcif Tissue Int **38**:79–86.
70. Oyajobi BO, Russell RG, Caswell AM 1994 Modulation of ecto-nucleoside triphosphate pyrophosphatase activity of human osteoblast-like bone cells by 1 alpha, 25-dihydroxyvitamin D_3, 24*R*,25-dihydroxyvitamin D_3, parathyroid hormone, and dexamethasone. J Bone Miner Res **9**:1259–1266.
71. Hamilton JA, Lingelbach S, Partridge NC, Martin TJ 1985 Regulation of plasminogen-activator production by bone-resorbing hormones in normal and malignant osteoblasts. Endocrinology **116**:2186–2191.
72. Suda T, Takahashi N, Martin TJ 1995 Modulation of osteoclast differentiation: Update 1995. Endocr Rev **4**:266–270.
73. Shuto T, Kukita T, Hirata M, Jimi E, Koga T 1994 Dexamethasone stimulates osteoclast-like cell formation by inhibiting granulocyte–macrophage colony-stimulating factor production in mouse bone marrow cultures. Endocrinology **134**:1121–1126.
74. Taylor LM, Turksen K, Aubin JE, Heersche JN 1993 Osteoclast differentiation in cocultures of a clonal chondrogenic cell line and mouse bone marrow cells. Endocrinology **133**:2292–2300.
75. Kwiecinski G, Petrie G, DeLuca H 1989 1,25-Dihydroxyvitamin D_3 restores fertility of vitamin D-deficient female rats. Am J Physiol **256**:E483–487.
76. Kwiecinski G, Petrie G, DeLuca H 1989 Vitamin D is necessary for reproductive functions of the male rat. J Nutr **119**:741–749.
77. Doerr P, Pirke KM 1976 Cortisol-induced suppression of plasma testosterone in normal adult males. J Clin Endocrinol Metab **43**:622–629.
78. Schaison G, Durand F, Mowszowicz I 1978 Effect of glucocorticoids on plasma testosterone in men. Acta Endocrinol **89**:126–131.
79. MacAdams MR, White RH, Chipps BE 1986 Reduction of serum testosterone levels during chronic glucocorticoid therapy. Ann Intern Med **104**:648–651.
80. Sakakura M, Takebe K, Nakagawa S 1975 Inhibition of luteinizing hormone secretion induced by synthetic LRH by long-term treatment with glucocorticoids in human subjects. J Clin Endocrinol Metab **40**:774–779.
81. Luton JP, Thieblot P, Valcke JC, Mahoudeau JA, Bricaire H 1977 Reversible gonadotropin deficiency in male Cushing's disease. J Clin Endocrinol Metab **45**:488–495.
82. Crilly RG, Cawood M, Marshall DH, Nordin BE 1978 Hormonal status in normal, osteoporotic and corticosteroid-treated post-menopausal women. J R Soc Med **71**:733–736.
83. Montecucco C, Caporali R, Caprotti P, Caprotti M, Notario A 1992 Sex hormones and bone metabolism in postmenopausal rheumatoid arthritis treated with two different glucocorticoids. J Rheumatol **19**(12):1895–1899.
84. Streck W, Waterhouse C, Hadded J 1979 Glucocorticol effects in dieting. Archives **139**:974–977.
85. Breslau NA, McGuire JL, Zerwekh JE, Frenkel EP, Pak CY 1984 Hypercalcemia associated with increased serum calcitriol levels in three patients with lymphoma. Ann Intern Med **100**(1):1–7.

CHAPTER 50

Drug and Hormone Effects on Vitamin D Metabolism

ANDREW RAEL BOWMAN Department of Medicine, Albert Einstein Medical Center, Philadelphia, Pennsylvania

SOL EPSTEIN Division of Endocrinology and Metabolism, Albert Einstein Medical Center, and Temple University School of Medicine, Philadelphia, Pennsylvania

I. INTRODUCTION

This chapter deals with drug and hormonal effects on vitamin D metabolism. The level of active vitamin D metabolite and the activity of the renal hydroxylase enzymes that are responsible for stimulating its production are governed by a variety of hormones and cations. Parathyroid hormone (PTH), phosphate, and calcium are the principal controlling factors of the renal hydroxylases (discussed in Chapters 2 and 4–6), yet several other hormones may also play a role in the metabolism of vitamin D. Among those discussed here are parathyroid hormone-related protein (PTHrH), calcitonin, prolactin, growth hormone and insulin-like growth factor (IGF), sex steroids (estrogen, testosterone, and progesterone), insulin, thyroid hormone, prostaglandin, interferon-γ, and tumor necrosis factor. The findings are summarized in Table I. Along with the endogenous regulators of the vitamin D endocrine system are some exogenous ones. Drugs are being increasingly recognized in this regard. Three that have received the most attention are the anticonvulsants, ethanol, and corticosteroids, but many others have also been implicated. The mechanisms by which these drugs affect vitamin D metabolism are discussed in detail and the findings summarized in Table II (see Section III below). It should be realized that the introduction of new drugs and molecular technology may expand this area considerably.

II. HORMONE EFFECTS ON VITAMIN D METABOLISM

A. Parathyroid Hormone and Parathyroid Hormone-Related Protein

This section dealing with parathyroid hormone (PTH) attempts to summarize the relationship between exogenous PTH or stimulated PTH secretion and vitamin D metabolism. For more extensive review the reader is referred to Chapter 23. A large body of evidence has established PTH as one of the main regulating

TABLE I Hormone Effects on Vitamin D Metabolism[a]

Hormone	Study	$25OHD_3$	$1,25(OH)_2D_3$	$24,25(OH)_2D_3$
PTH	*In vitro*		↑	↓
	Animal		↑	↓
	Human		↑	↓
PTHrP	*In vitro*		↑	
	Animal		↑	
	Human		↑	
Calcitonin	*In vitro*		↑, ↓	
	Animal	↔	↑	↓
	Human	↔	↑, ↔	↑, ↔
Growth hormone	*In vitro*		↔	
	Animal		↑	↓
	Human	↔	↑, ↓	
IGF-I	*In vitro*		↑	
	Animal		↑	
Prolactin	*In vitro*		↑	
	Bird		↑	
	Animal		↑, ↔	
	Human	↔	↔, ↑	
Insulin	*In vitro*		↑	↓
	Animal	↔	↑	↓
	Human	↔	↔, ↑	↔, ↓
Estrogen	*In vitro*		↑, ↓, ↔	↓, ↑, ↔
	Animal	↔	↑, ↔	
	Human	↔	↑, ↔	
Testosterone	*In vitro*		↑	↓
	Animal	↔	↔, ↑	
	Human	↔	↑	
Progesterone	*In vitro*		↑	↑
	Animal	↑	↔	
	Human	↔	↔, ↓	
Thyroid hormone	*In vitro*		↓	↑
	Animal	↑	↓	↑
	Human	↔, ↓	↓	↑
Prostaglandins	*In vitro*		↑	↓
	Animal		↑, ↔	
	Human		↑	
Tumor necrosis factor-α	*In vitro*		↑	
Interferon-γ	*In vitro*		↑	
	Animal		↔	

[a] Symbols denote effects on serum levels of vitamin D metabolites: ↑, increased; ↓, decreased; ↔, unchanged. See text for details and references.

hormones of vitamin D metabolism. Historically, in 1972 Garabedian *et al.*, [1] reported that rats fed a low calcium diet were unable to synthesize 1,25-dihydroxyvitamin D_3 [$1,25(OH)_2D_3$] following thyroparathyroidectomy, which could be restored at 48 hr by the administration of small amounts of parathyroid extract. Fraser and Kodicek [2] demonstrated similar *in vivo* regulation of $1,25(OH)_2D_3$ synthesis by PTH. They administered bovine PTH to 1-day-old chicks and observed a 4- to 5-fold increase in enzyme activity. Other investigators showed similar findings [3–7], although some have shown a less dramatic stimulation of 1α-hydroxylase in older chicks and in Japanese quail [8,9].

At about the same time as the *in vivo* experiments, *in vitro* work by Rasmussen *et al.*, [10] demonstrated stimulation of $1,25(OH)_2D_3$ production in isolated renal tubules from vitamin D-deficient chicks by bovine PTH. Larkins *et al.*, [11] and Trechsel *et al.* [12] reported likewise; however, PTH stimulation occurred only if the cells were in a hypocalcemic medium. Others have also showed *in vitro* stimulation of 1α-hydroxylase and inhibition of 24-hydroxylase [13–17]. Not all *in vitro* work is in agreement, however, as Shain [18] reported no stimulatory effect of 1α-hydroxylase following 30 min incubation with high concentrations of PTH.

The mechanism by which PTH enhances renal 1α-hydroxylase and inhibits 24-hydroxylase is most probably via second messengers. PTH increases cyclic AMP (cAMP) production in proximal tubules [19–21] and in renal slices from adult rats treated with PTH [14,15]. Further evidence for a role of cAMP emerged following studies by Horiuchi and co-workers. They found that an infusion of cAMP and dibutyryl cAMP into vitamin D-deficient rats increases the conversion of tritiated 25-hydroxyvitamin D_3 ($25OHD_3$) to $1,25(OH)_2D_3$ [7,22]. Furthermore, cAMP [10,11,16] and forskolin [23,24], a direct activator of adenylate cyclase which increases intracellular cAMP levels [25], have been shown, *in vitro*, to enhance the production of $1,25(OH)_2D_3$ and inhibit 24-hydroxylase activity [7]. PTH has also been shown to decrease 24-hydroxylase activity through suppression of cytochrome P450 24-hydroxylase gene expression [26], thus strengthening the association of PTH, cAMP, and $1,25(OH)_2D_3$.

Along with cAMP, investigators have demonstrated that the phospholipase C/protein kinase C (PKC) second messenger system may also mediate *in vitro* PTH stimulation of $1,25(OH)_2D_3$ secretion by mammalian renal proximal tubules [27,28]. This occurs at concentrations of PTH that are insufficient to raise cAMP content [27]. It is conceivable that PTH activation of $1,25(OH)_2D_3$ may involve both signaling pathways, with the PKC pathway being responsive to lower concentrations of PTH [27,29].

In clinical work, patients with primary hyperparathyroidism have mean plasma concentrations of $1,25(OH)_2D_3$ that are significantly increased in approximately one-third of cases versus controls [30–33]. In patients with Paget's bone disease, $1,25(OH)_2D_3$ elevation in response to mithramycin-induced hypocalcemia also corresponds to an increase in PTH secretion [34].

In addition, treatment with $1,25(OH)_2D_3$ will correct the hypocalcemia found in hypoparathyroid patients [35–37]. Moreover, patients with pseudohypoparathyroidism, a disease characterized by metabolic unresponsiveness to PTH, have low circulating levels of $1,25(OH)_2D_3$ [38].

Finally, with the huge explosion in the field of parathyroid hormone-related protein (PTHrP), it may be informative and timely to look at the role of PTHrP in vitamin D metabolism. PTHrP is a protein that is produced by a number of solid tumors, especially squamous and renal cell carcinomas, and is thought to be responsible for many cases of hypercalcemia of malignancy. However, it is also seen in the normal physiological situation [39]. PTHrP binds to the classic type I PTH/PTHrP receptor, which is expressed in many tissues such as bone and lung, as well as to alternative type II receptors in several nonclassic PTH target tissues such as keratinocytes and squamous carcinoma cell lines [40]. PTHrP has a similar activity to PTH [41,42]; however, unlike in primary hyperparathyroidism where patients have normal or elevated $1,25(OH)_2D_3$ levels (see above), $1,25(OH)_2D_3$ is not elevated and may be decreased in patients with hypercalcemia of malignancy and increased urinary cAMP excretion [43,44]. This is contrast to *in vivo* rodent studies where serum calcium and $1,25(OH)_2D_3$ levels were elevated following tumor transplantation [45] and infusion of synthesized N-terminal fragments of PTHrP [46,47]. Walker *et al.,* also demonstrated direct stimulation of 1α-hydroxylase in rodent kidneys slices *in vitro* by PTHrP(1–36).

Because of the differences observed between hypercalcemia of malignancy and infusion of synthesized PTHrP, Fukomoto *et al.* [48] suggested that additional factors in hypercalcemia of malignancy, other than PTHrP, may be involved which inhibit renal $1,25(OH)_2D_3$ production, whereas Schilling *et al.* [49] suggested that the hypercalcemia itself may be the cause of low $1,25(OH)_2D_3$ levels. However, Nakayama *et al.* [33] reported low $1,25(OH)_2D_3$ levels that were demonstrated at any serum calcium level in patients with hypercalcemia of malignancy compared to patients with primary hyperparathyroidism. This suggests that the reduction in serum $1,25(OH)_2D_3$ observed in these patients cannot be explained solely by an elevation in serum calcium [33]. As in animals, a study by Everhart-Caye *et al.* [50] revealed a dose-related increase in renal production of $1,25(OH)_2D_3$ following a 6-hr infusion of human PTHrP (1–36) into healthy subjects. This confirms a previous report by Fraher and co-workers [51], who observed a similar increase in $1,25(OH)_2D_3$ production following human PTHrP (1–34) infusion in healthy subjects. Although the *in vivo* human studies of Fraher and Everhart-Caye do indeed demonstrate PTHrP stimulation of $1,25(OH)_2D_3$ production, they were conducted in normal healthy subjects and do not mirror the clinical situation as seen in hypercalcemia of malignancy, in which a host of factors may be at play (as discussed below). They were also short-term infusion studies, and it is quite possible that equivalent findings may not be produced with longer term studies.

Possible explanations for the disparate effect of PTHrP in hypercalcemia of malignancy are that other regions or alternatively the full length of the PTHrP (1–141) molecule may act differently from the synthetic molecules administered in previous studies [51]. This may modify or impair the capacity of the N-terminal moiety of PTHrP to stimulate 1α-hydroxylase, as might other factors also produced by tumors, such as the cytokines, interleukin-1 (IL-1), and tumor necrosis factor, and/or growth factors [52].

In summary, available evidence, both *in vitro* and *in vivo* in animal and humans, clearly defines the role of PTH as an activator of $1,25(OH)_2D_3$ synthesis and an inhibitor of $24,25(OH)_2D_3$ production. Amino-terminal fragments of PTHrP also stimulate $1,25(OH)_2D_3$ production experimentally, both *in vitro* and *in vivo* in animals and humans. However, it is as yet unclear as to why patients with hypercalcemia of malignancy and elevated PTHrP have reduced $1,25(OH)_2D_3$.

B. Calcitonin

A vast amount of work both *in vivo* and *in vitro* has been performed with calcitonin in order to ascertain its role in vitamin D metabolism (see Chapter 24). *In vivo* studies by Galante *et al.* [53] first demonstrated a pronounced effect of synthetic salmon calcitonin in vitamin D-deficient rats. They demonstrated that calcitonin infusion could considerably increase $1,25(OH)_2D_3$ and reduce $24,25(OH)_2D_3$ levels both in serum and in intestinal mucosa [53]. Lorenc and co-workers [54] confirmed these findings but showed that following thyroparathyroidectomy (TPTX) the calcitonin effect was eliminated. They concluded that no direct role of calcitonin was evident but rather the response was probably secondary to PTH stimulation [54]. However, a secondary effect of PTH is unlikely as subsequent studies by Horiuchi *et al.* [55] involving vitamin D-deficient TPTX rats demonstrated increased conversion of tritiated $25OHD_3$ to $1,25(OH)_2D_3$ following calcitonin administration. Furthermore the actions of PTH and calcitonin were additive, suggesting independent effects [55]. Further evidence of a direct action of calcitonin that was independent of serum PTH, calcium, phosphate, or vitamin D levels came from a study by Jaeger *et al.* [56]. In vitamin D-replete TPTX rats, calcitonin stimulated

$1,25(OH)_2D_3$ production. This occurred in rats fed regular diets as well as calcium-free diets. There were no significant changes in $25OHD_3$ levels [56]. Thus, this study demonstrates that calcitonin was able to stimulate $1,25(OH)_2D_3$ production *in vivo* without the influence of PTH, calcium, phosphate, or vitamin D levels. Calcitonin may also act as a negative regulator of intestinal 24-hydroxylase. With the inhibition of intestinal 24-hydroxylase activity and expression, $1,25(OH)_2D_3$ may be spared from deactivation, which may therefore result in enhanced $1,25(OH)_2D_3$-mediated activity [57].

Unlike *in vivo* work, *in vitro* results have been varied. Rasmussen *et al.* [10] using isolated renal tubules from vitamin D-deficient chicks found that the *in vitro* conversion of $25OHD_3$ to $1,25(OH)_2D_3$ was actually inhibited by porcine calcitonin [10]. These results were, however, disputed by Larkins *et al.* [11], Armbrecht *et al.* [58], and Kawashima and co-workers [59]. The latter demonstrated that calcitonin stimulated 1α-hydroxylase activity in the proximal straight tubule of vitamin D-deficient rat kidneys. They also noted that post-TPTX calcitonin still stimulated 1α-hydroxylase activity, which was independent of the adenylate cyclase system [59]. This system, as mentioned earlier, is believed to be the mechanism of PTH stimulation of 1α-hydroxylase [22,55]. These findings suggest that calcitonin acts independently of PTH.

In human studies, patients with medullary carcinoma of the thyroid and excess calcitonin had serum $1,25(OH)_2D_3$ levels that were elevated [60]. Therapeutic use of calcitonin has, however, produced mixed results. Injectable [61] and nasal calcitonin [62,63] treatment daily for postmenopausal osteoporosis produces either no changes [61–63] or an increase in $1,25(OH)_2D_3$ levels at 6 months which returned to normal at 1 year, with PTH and $25OHD_3$ levels being unchanged during the study period [64].

No changes in $1,25(OH)_2D_3$ levels were observed in patients with Paget's disease of bone treated for 3 months with calcitonin, although in these patients $24,25(OH)_2D_3$ was elevated, most probably due to decreased consumption by osteoclasts and osteoblasts [65]. Other studies involving the treatment of Paget's disease of bone with nasal calcitonin did show a rise in $1,25(OH)_2D_3$ after 18 months duration, yet longer term therapy had no effect. Levels of other vitamin D metabolites such as $24,25(OH)_2D_3$ and $25OHD_3$ were unchanged [66].

In summary, animal studies and most *in vitro* work provide evidence that calcitonin seems to play an important role in stimulating $1,25(OH)_2D_3$ production. Clinical studies are less convincing, and this may be a dose-related phenomenon or perhaps related to the route of calcitonin administration. The reader is referred to Chapter 24 for further information on calcitonin and vitamin D.

C. Growth Hormone and Insulin-like Growth Factor

The interest in growth hormone (GH) and its action has been spurred by advances in our understanding of its mechanism of action via the intermediaries insulin-like growth factors (IGF) and the IGF binding proteins (IGFBP). One of the many functions of GH is to stimulate intestinal calcium absorption, which has been shown in both rats [67] and humans [68,69]. Although the exact mechanism of this effect is unknown, it may be a $1,25(OH)_2D_3$-dependent phenomenon. Hypophysectomy and subsequent GH deficiency in rats was shown to reduce the level of $1,25(OH)_2D_3$ and to increase the conversion of $25OHD_3$ to $24,25(OH)_2D_3$ [70–73]. It was also discovered that the levels of vitamin D metabolites could be restored to normal by replacing GH [70–73]. No increase in the metabolic clearance or increase in tissue metabolism of $1,25(OH)_2D_3$ occurred, so the most likely explanation is that hypophysectomy decreases the renal 1α-hydroxylase activity [72]. Furthermore, hypophysectomized rats fed a low phosphate diet and replaced with rat GH had increased levels of $1,25(OH)_2D_3$, as a result of stimulation of 1α-hydroxylase activity [74]. In other animal work, exogenous porcine GH also increased $1,25(OH)_2D_3$ in intact pigs [75].

Clinical studies are not so conclusive and demonstrate mixed results. Acromegalic subjects, with endogenous GH excess, have increased levels of $1,25(OH)_2D_3$ [32,76–78] which are reduced by treatment with bromocriptine, which decreases GH [76–78]. Short-term recombinant GH therapy in healthy young [79] or elderly people [80] caused significant increases in $1,25(OH)_2D_3$, and increased vitamin D levels in response to phosphate depletion were found to be dependent on the presence of GH [81]. Other studies, however, have found that chronic GH therapy does not seem to cause a rise in $1,25(OH)_2D_3$ levels [82,83]. Gertner *et al.* [84] found similar results in a study of GH-deficient children aged 9–18 years. GH replacement in this group of patients did not alter $25OHD_3$, $1,25(OH)_2D_3$, or PTH levels [84]. Gertner *et al.* [85] also demonstrated similar results after an overnight GH infusion.

In vitro evidence suggests that the effects of GH *in vivo* on $1,25(OH)_2D_3$ production are indirect, as GH fails to stimulate $1,25(OH)_2D_3$ production in chick renal preparations [86]. Insulin-like growth factor type I (IGF-I) receptors are present on the basolateral membrane of renal proximal tubules [87], and IGF-I is thought to be the stimulator of $1,25(OH)_2D_3$ due to phosphate depletion. In *in vitro* studies, low concentrations of exogenous IGF-I enhance $1,25(OH)_2D_3$ synthesis when phosphate concentrations are low [88], whereas *in vivo* animal studies show that the GH-dependent increases in serum $1,25(OH)_2D_3$ levels induced by dietary phosphate restriction may be mediated by IGF-I, as administration of IGF-I to hypophysectomized rats increases serum $1,25(OH)_2D_3$ to approximately the same degree as GH [89,90]. In addition, Gray [91] showed that serum concentrations of $1,25(OH)_2D_3$ are directly related to the seurm levels of IGF-I in rats fed a low phosphate diet. IGF-I was also shown to stimulate renal 1α-hydroxylase activity in a time- and dose-dependent manner in weanling mice that were phosphate depleted [92], but this was also demonstrated to be independent of changes in serum calcium or phosphate [92]. Currently, no data suggest that IGF-II affects vitamin D metabolism.

Overall, it seems likely that GH, probably via IGF-I and especially during hypophosphatemia, is an important regulator of serum $1,25(OH)_2D_3$ levels in rats, although such adaptation does not require the presence of exogenous IGF-I. Furthermore, it is possible that hormone- and phosphate-dependent enzyme stimulation occur by different mechanisms. However, as stated earlier, the data from human clinical studies are not convincing.

D. Prolactin

Initial work involving prolactin (PRL) as a possible stimulator of $1,25(OH)_2D_3$ synthesis came from *in vitro* studies by Spanos *et al.* [93,94], who demonstrated, in isolated chick renal tubules and primary chicken cell cultures, that ovine PRL could stimulate 1α-hydroxylase. Bickle *et al.* [86] confirmed these results in a subsequent *in vitro* study. Spanos *et al.* [95] in a further study reported that both short-term and long-term administration of physiological doses of ovine PRL stimulated 1α-hydroxylase activity in chick kidney homogenates. Unlike the case in birds, however, in mammals, fish, and amphibians PRL does not seem to play a significant role in vitamin D metabolism. Constant PRL infusion failed to stimulate 1α-hydroxylase activity and elevate $1,25(OH)_2D_3$ levels in TPTX and sham-operated rats [96]. Furthermore, hypophysectomized rats fed a low phosphate diet and given PRL replacement therapy failed to demonstrate a rise of plasma $1,25(OH)_2D_3$ [74], and ovine PRL had no effect on the recovery of ^3H-$1,25(OH)_2D_3$ from tissues of hypophysectomized rats following ^3H-$25(OH)D_3$ administration [70]. Kenney *et al.* [97] also reported that in amphibians and fish PRL showed a lack of stimulation of renal $1,25(OH)_2D_3$ production *in vitro*.

In certain physiological situations such as lactation, however, PRL may play a role in vitamin D metabolism. In rats, PRL stimulates intestinal transport of calcium [98] especially during lactation, when increased serum levels of $1,25(OH)_2D_3$ are found [99,100]. Suppression of PRL by bromocriptine decreases $1,25(OH)_2D_3$ levels in lactating rats, but the drug has no effect in nonlactating controls [101]. Increased vitamin D levels are also found in women during pregnancy and lactation [32,102,103].

In a clinical study a patient with hypoparathyroidism was found to require less exogenous vitamin D supplementation during lactation than at other times [104]. However, other than in lactating females, PRL most probably has no effect on plasma $1,25(OH)_2D_3$ levels or calcium metabolism, as patients with hyperprolactinemia due to functioning pituitary adenomas had $25OHD_3$ and $1,25(OH)_2D_3$ levels in the normal range versus age-matched controls as well as similar serum calcium, phosphate, and PTH levels [32,105–109].

Overall, PRL seems to have species-specific actions and may only play a role in vitamin D metabolism in birds. However, it may also regulate $1,25(OH)_2D_3$ production in normal physiological settings where prolactin is increaesd, such as lactation. However, its contribution to $1,25(OH)_2D_3$ regulation appears minimal.

E. Insulin

Patients with insulin-dependent diabetes mellitus suffer a number of disturbances in bone mineral metabolism, including alterations in vitamin D metabolism [110]. Work in streptozotocin- and alloxan-induced diabetic rats by Schneider and co-workers [111,112] first suggested a role for insulin in the production of $1,25(OH)_2D_3$. This experimental diabetic model is associated with a reduction in duodenal calcium absorption, calcium binding protein, and total and ionized calcium levels [111,112]. Treatment with $1,25(OH)_2D_3$ but not $25OHD_3$ corrects subnormal calcium absorption in these rats [113], as does insulin replacement [114]. These data suggest that a lack of insulin in diabetes impairs 1α-hydroxylation. Further support of this concept emerged after finding that $1,25(OH)_2D_3$ levels in streptozotocin-induced diabetic rats were depressed to one-eighth the level in control rats and were returned to control values with insulin treatment [115]. No changes in serum $25OHD_3$ were found, suggesting either decreased 1α-hydroxylation of $25OHD_3$ or increased catabolism of $1,25(OH)_2D_3$. It is possible that a change in PTH is responsible for decreased conversion of $25OHD_3$. It has been shown that PTH levels increase in diabetic rats, probably secondary to calcium malabsorption resulting from decreased $1,25(OH)_2D_3$ [116–119]; however, some studies have reported low PTH levels in streptozotocin-induced diabetic rats [120]. To further localize the site at which insulin is proposed to act, Spencer *et al.* [121] studied the conversion of ^3H-$25OHD_3$ to ^3H-$1,25(OH)_2D_3$ as well as the metabolic clearance of ^3H-$1,25(OH)_2D_3$ in control, streptozotocin-induced diabetic and insulin-treated streptozotocin-diabetic rats. The results showed that the metabolic clearance was not increased in diabetic rats and that the *in vivo* conversion of ^3H-$25OHD_3$ to ^3H-$1,25(OH)_2D_3$ was reduced by 60% in diabetic rats, normalizing with insulin therapy. No intrinsic intestinal mucosal defect in the incorporation of ^3H-$1,25(OH)_2D_3$ was evident [121]. These results further support the notion that a lack of insulin impairs 1α-hydroxylation. The osteopenia seen in streptozotocin-induced diabetic rats [119,120] is also significantly attenuated by treatment with $1\alpha OHD_3$, suggesting that vitamin D deficiency contributes to this bone loss [122].

Others have shown a similar decrease in 1α-hydroxylase activity together with enhanced 24-hydroxylase activity while 25-hydroxylase activity was similar to controls [123,124]. Sulimovici *et al.* [125], in an *in vitro* study, also suggested that insulin deficiency directly inhibits 25-hydroxylase activity despite finding normal levels of serum $25OHD_3$ in the diabetic animal model. Insulin may also play an important role in the stimulation of renal $1,25(OH)_2D_3$ synthesis in response to phosphate deprivation. In streptozotocin-diabetic rats, $1,25(OH)_2D_3$ increased only slightly following phosphate deprivation as compared with a marked response in rats replaced with insulin [126]. As far as changes in serum vitamin D binding protein (DBP) are concerned, Nyomba *et al.* [127] reported decreased DBP and total $1,25(OH)_2D_3$ with normal levels of free $1,25(OH)_2D_3$ in experimentally induced diabetic rats.

In vitro evidence suggests a permissive role of insulin in vitamin D metabolism. Primary cultures of chick kidney cells in serum-free medium respond to PTH with increased production of $1,25(OH)_2D_3$ only when exposed to insulin [128]. Confirming *in vivo* work, Wongsurawat *et al.* [118], using a renal slice technique, showed reduced $1,25(OH)_2D_3$ and increased $24,25(OH)_2D_3$ levels in streptozotocin-induced diabetic rats which reversed with insulin therapy. PTH levels were found to be higher in diabetic rats relative to controls [118]. Renal resistance to PTH has been suggested as the mechanism by which 1α-hydroxylase is depressed in these rats [129]. However Wongsurawat *et al.* showed a normal cAMP response to PTH in streptozotocin-induced diabetes [118], which makes this theory unlikely. Clinical studies have shown decreased $1,25(OH)_2D_3$ concentrations with increased $24,25(OH)_2D_3$ and normal $25OHD_3$ levels in insulin-dependent diabetic children [130,131] and in poorly controlled African diabetics [132]. However, Heath *et al.* [133] reported contradictory findings in adult insulin-dependent diabetics, on the whole, whether treated or untreated, they showed no abnormalities of calcium or vitamin D metabolism.

In summary, diabetic animals show abnormalities in vitamin D metabolism with depressed levels of $1,25(OH)_2D_3$, increased $24,25(OH)_2D_3$, and normal levels of $25OHD_3$. In human studies findings have been inconsistent, which may be due to variations among the study populations. Alternatively, the animal model of diabetes may differ from diabetes in the clinical setting with regard to changes in vitamin D.

F. Sex Steroids

1. Estradiol

Avian studies by Kenny [134] involving egg-laying Japanese quail first led to the notion that estradiol may play a role in vitamin D metabolism. It was demonstrated that ovulation resulted in enhanced production of $1,25(OH)_2D_3$ [134]. Furthermore, estradiol injections to Japanese quail stimulated *in vitro* renal $1,25(OH)_2D_3$ production in both immature male and female birds [135]. In females $24,25(OH)_2D_3$ was also stimulated, but in males the concentration of this metabolite was reduced. Further studies supported the notion that estradiol might be a regulator of vitamin D. Castillo *et al.* [136] administered 5 mg of estradiol to mature male quail and found that it markedly stimulated 1α-hydroxylase and suppressed 24-hydroxylase activity 24 hr after administration. In castrated male chickens, estradiol injections stimulated $1,25(OH)_2D_3$ production *in vitro* but only in the presence of testosterone or progesterone. If all three hormones were present they acted in a synergistic manner to stimulate $1,25(OH)_2D_3$ synthesis [137]. Others have found that the estrogenic drug stilbestrol injected into immature male chickens stimulated 1α-hydroxylase and suppressed $24,25(OH)_2D_3$ production in chick kidney homogenates [138]. However, there have been some contradictory *in vitro* studies which suggest that estradiol may not exert a stimulatory effect on the hydroxylase enzymes. Trechsel *et al.* [139] reported a dose-dependent inhibition by estradiol of 1α-hydroxylase *in vitro*. Furthermore, Henry [140] showed that estradiol failed to stimulate 1α-hydroxylase or 24-hydroxylase activity in chick kidney cell cultures, which argues against a direct action of estradiol on kidney cells. Thus, the *in vivo* stimulation noted in the above studies may occur via indirect means, possibly via PTH stimulation as discussed below.

In animal studies, female rats treated with estradiol benzoate daily for 8 days increased $1,25(OH)_2D_3$ concentrations in plasma, gut mucosa, and kidneys [141]. Others have not found a rise in $1,25(OH)_2D_3$ levels in rats treated with estradiol but have recorded increased *in vivo* intestinal absorption of calcium [142–144]. In a further study by Ash and Goldin [145], both young and old ovariectomized rats administered ^3H-$25OHD_3$ had reduced ^3H-$1,25(OH)_2D_3$ production, which was increased with estradiol replacement. However, in the same study, parathyroidectomy eliminated the estradiol therapeutic effect on ^3H-$1,25(OH)_2D_3$ recovery, which implicates PTH. Estradiol may either act directly on the parathyroid gland or act by decreasing bone resorption and lowering serum calcium, thus triggering PTH secretion [145]. However, these findings have not been consistently demonstrated in ovariectomized rats by others [142,146].

In the clinical situation, plasma $1,25(OH)_2D_3$ levels are elevated in human pregnancy and remain high postpartum in lactating women [102,103]. Levels of $25OHD_3$ are similar in pregnancy compared to controls [147], and DBP levels are also elevated [148]. Despite increased DBP, free $1,25(OH)_2D_3$ levels are also raised in the pregnant state [148]. Low [149,150] and normal levels [151] of $1,25(OH)_2D_3$ have been reported in early postmenopausal women. The proposed mechanism is that increased bone resorption, leading to a slight elevation of serum calcium, decreases PTH secretion, which subsequently reduces renal 1α-hydroxylase activation and $1,25(OH)_2D_3$ production. This ultimately results in decreased calcium absorption and negative calcium balance [152]. Serum levels of $25OHD_3$ have been reported as unchanged [149,151] or increased [153] in various studies.

In women with postmenopausal osteoporosis, estrogen replacement results in decreased bone resorption that lowers serum calcium, which subsequently stimulates PTH and renal 1α-hydroxylase leading to increased $1,25(OH)_2D_3$ production. The latter increases intestinal calcium absorption [154]. Studies have demonstrated increased $1,25(OH)_2D_3$ levels after both short-term and long-term estrogen therapy [150,155]. However, Stock *et al.* [156], using improved sensitive assays for PTH, failed to show a rise in PTH levels accompanying the elevated $1,25(OH)_2D_3$ in postmenopausal women treated with estradiol. In fact 2 weeks of treatment with estradiol decreased PTH levels. This suggests that the estrogen effect on vitamin D metabolism may not only be secondary to a change in PTH [156].

Estradiol-induced elevation of $1,25(OH)_2D_3$ may occur as a result of increased DBP levels. A higher concentration of DBP would increase total $1,25(OH)_2D_3$ but leave free $1,25(OH)_2D_3$ unchanged [157,158]. However, free levels were shown to be elevated in response to estradiol treatment together with DBP and total $1,25(OH)_2D_3$; PTH levels remained unchanged [159].

At this point it would seem that, unlike the situation in birds, there is no conclusive evidence in mammals for a direct effect of estradiol on vitamin D metabolism. However, estradiol may play an indirect role by altering intestinal calcium absorption, bone resorption, and PTH levels.

2. Testosterone

Evidence of a role for testosterone in vitamin D metabolism first emerged in studies involving chickens. Tanaka *et al.* [137] demonstrated the interdependence of testosterone and estradiol in the *in vitro* stimulation of 1α-hydroxylase in castrated male chickens. Estradiol stimulation of 1α-hydroxylase seems to occur only after testosterone is added, and estradiol, testosterone, and progesterone act synergistically to stimulate 1α-hydroxylase [137]. Castillo *et al.* [136] also demonstrated that testosterone alone administered to the male quail suppressed 24-hydroxylase but produced little change in 1α-hydroxylase. Again the data are not uniform. In mammals, castrated male guinea pigs demonstrated a 50% decline in 1α-hydroxylase activity, which was reversed with testosterone replacement. Ovariectomized female guinea pigs also responded to testosterone therapy with a 50% increase in 1α-hydroxylase activity. Both groups, however, had similar serum levels of $1,25(OH)_2D_3$, DBP, and free $1,25(OH)_2D_3$ levels versus controls [160]. Ohata *et al.* [161] demonstrated elevation of serum $25OHD_3$ levels following androgen treatment in ultraviolet irradiated rats compared to those that did not receive testosterone. Hypoandrogenemia as a result of orchidectomy in rats has produced mixed findings, with either no change in total or free vitamin D metabolites [162] or decreased $1,25(OH)_2D_3$ and DBP with normal free $1,25(OH)_2D_3$ levels [163,164].

Human studies have focused on hypogonadal men. Hagenfeldt *et al.* [165] studied hypogonadal men before and after treatment with testosterone enanthate every 3–4 weeks, for varying lengths of time. They found that basal serum $1,25(OH)_2D_3$, DBP, and free $1,25(OH)_2D_3$ concentrations were similar to those of controls; however, testosterone treatment still increasead total $1,25(OH)_2D_3$ and free $1,25(OH)_2D_3$ significantly. In contrast, Morley *et al.* [166] looked at the effects of testosterone replacement therapy in elderly hypogonadal males (mean age 77.6 ± 2.3 years) and found no effect on PTH or serum vitamin D metabolites pre- or posttreatment. Studies in hypogonadal men with osteoporosis have found both normal [167,168] and low [169] serum $1,25(OH)_2D_3$ levels. In the latter study by Francis *et al.* [169], testosterone replacement therapy increased both total and free $1,25(OH)_2D_3$ levels. However, patients treated with orchidectomy for prostate cancer were found to have no changes in total $1,25(OH)_2D_3$ or DBP concentrations [170]. In studies involving pubertal boys results have been equivocal. Krabbe *et al.* [171] found no changes in the serum level of vitamin D metabolites before and after peak pubertal testosterone surges, whereas Aksnes *et al.* [172] did find an increase in vitamin D.

In summary, it is unlikely that testosterone is a major controlling factor in vitamin D metabolism, but it may play a minor role in overall vitamin D homeostasis. Hypogonadal men, especially those at risk for osteoporosis, should be monitored for $1,25(OH)_2D_3$ vitamin D levels and receive supplementation if necessary.

3. Progesterone

Work by Tanaka *et al.* [137] demonstrated that in castrated male chickens progesterone, like testosterone, supported the stimulation of 1α-hydroxylase by estradiol. A further marked stimulation of 1α-hydroxylase activity occurred with combined progesterone, testosterone, and estradiol treatment [137], demonstrating pronounced synergy among the sex steroids. In Japanese female quail treated with progesterone *in vitro,* renal production of $1,25(OH)_2D_3$ was stimulated; however, this was significantly less than that of estradiol [135]. In the same study immature male quail treated with progesterone had increased $24,25(OH)_2D_3$ production [135]. Similar findings were recorded by Castillo *et al.* [136]. Kalu *et al.* [143] treated ovariectomized rats with progesterone and demonstrated elevated $25OHD_3$ and similar $1,25(OH)_2D_3$ levels versus controls.

Clinical work in postmenopausal women showed that progesterone in combination with estrogen was found to lower estradiol-stimulated increases in total and free vitamin D levels [173] and that norethisterone treatment in postmenopausal women caused a slight decrease in free and total vitamin D [158]. Finally, no effects on vitamin D metabolites following medroxyprogesterone therapy were seen in male patients treated for glucocorticoid-induced osteoporosis [174].

In summary, no conclusive evidence has established progesterone as a major controlling factor in vitamin D metabolism, and the role of both male and female sex hormones in vitamin D homeostasis is clouded with conflicting animal and human data. The reasons for this are unclear, but perhaps the experimental models and varying clinical situations do not allow a definitive conclusion to be reached.

G. Thyroid Hormone

The effect of thyroid hormone on serum concentrations of $1,25(OH)_2D_3$ most probably occurs as a result of changes in serum calcium and phosphate concentrations. Both may affect PTH levels, which in turn alter 1α-hydroxylase activity [175–178]. In an extensive review on bone and mineral metabolism in thyroid disease, Auwerx and Bouillon [178] describe the changes encountered in this disease. In hyperthyroidism, excess thyroid hormone stimulates bone resorption [178–181], which increases serum calcium and phosphate concentration with resultant suppression of PTH [180,182–184] and a decrease in $1,25(OH)_2D_3$ production. This leads to lower intestinal calcium absorption [185,186], which Peerenboom *et al.* [186] had earlier demonstrated to be reversible after treatment of the thyroid abnormality.

Levels of $25OHD_3$ in hyperthyroidism are either unchanged [175–177,186] or decreased [187,188]. Dietary vitamin D intake and exposure to sunlight may contribute to the differences in $25OHD_3$. Circulating levels of $24,25(OH)_2D_3$ are generally increased in hyperthyroid patients [176,177]. Another possible factor contributing to low plasma $1,25(OH)_2D_3$ levels found in hyperthyroidism is that of enhanced metabolic clearance of $1,25(OH)_2D_3$. Karsenty *et al.* [189] studied seven hyperthyroid patients and found increased $1,25(OH)_2D_3$ clearance after administration of tritiated $1,25(OH)_2D_3$. In hypothyroidism the inverse occurs, with decreased bone turnover [190] and low serum calcium levels that activate PTH, which in turn enhances 1α-hydroxylase activity [175,184,191].

In animal studies involving rats, daily injections of L-thyroxine decreased $1,25(OH)_2D_3$ and increased $24,25(OH)_2D_3$ and $25OHD_3$ levels [192]. Hypothyroidism in the rat induced by the drug propylthiouracil, however, produces confusing results in that it is associated with low calcium and phosphate levels and decreased $1,25(OH)_2D_3$ concentration [193].

There is some *in vitro* evidence to suggest that thyroid hormone directly affects renal $25OHD_3$ metabolism. Kano and Jones [194] found that thyroxine, triiodothyronine, and thyrotropin (TSH) decreased $1,25(OH)_2D_3$ synthesis in perfused rat kidneys from vitamin D-deplete rats, whereas $24,25(OH)_2D_3$ synthesis was increased in kidneys from vitamin D-replete rats. Miller and Ghazarian [195] also demonstrated a 50% reduction in 1α-hydroxylase activity in vitamin D-deficient chicks and stimulation of 24-hydroxylase activity in vitamin D-replete chicks following thyroxine administration. Finally, changes in serum binding are unlikely to be involved, as DBP levels were demonstrated to be normal in hyperthyroidism [175] and only slightly increased in hypothyroidism [175].

The majority of evidence suggests that thyroid hormone indirectly affects vitamin D metabolism via alterations in serum calcium, phosphate, and PTH. A direct action of thyroid hormone on renal hydroxylase activity, however, has been suggested by *in vitro* work.

H. Prostaglandins

The role of prostaglandins in the regulation of 1α-hydroxylase activity has been investigated *in vitro* and *in vivo*. Work by Trechsel *et al.* [196] in primary chick kidney cell cultures found that the addition of prostaglandin E_2 (PGE_2) and prostaglandin $F_{2\alpha}$ ($PGF_{2\alpha}$) stimulated 1α-hydroxylase activity in a dose-dependent manner. PGE_2 also significantly decreased 24-hydroxylase activity [196]. Theories involved in the mechanism of action are that prostaglandins may act through an increase in cAMP [196,197], but as discussed below this has not been proved.

Further *in vitro* work by Wark *et al.* [198] in isolated renal tubules, prepared from vitamin D-deficient chicks, showed that the addition of PGE_2 to the tubule incubation medium in the presence of $1,25(OH)_2D_3$ increased $1,25(OH)_2D_3$ production. Acetylsalicylic acid, which inhibits prostaglandin synthesis, decreased the prostaglandin content of the tubule medium and subsequently inhibited $1,25(OH)_2D_3$ production. Frusemide in a dose-dependent manner raised prostaglandin content, which lead to a significant increase in $1,25(OH)_2D_3$ production and decreased $24,25(OH)_2D_3$ synthesis [198]. Kurose *et al.* [199] found similar results *in vitro*.

In vivo work involving vitamin D-deficient TPTX rats by Yamada *et al.* [200] found that, after an intraarterial infusion of PGE_2, $1,25(OH)_2D_3$ production was significantly stimulated. No changes in plasma calcium or phosphate levels or urinary cAMP excretion was observed, suggesting that the effects of prostaglandins are independent of the cAMP system [200].

In a further rat study, the effects of PGE_2 on the actions of PTH and calcitonin and the conversion of ^3H-$25OHD_3$ to ^3H-$1,25(OH)_2D_3$ was investigated. PGE_2 inhibited calcitonin stimulation of $1,25(OH)_2D_3$ but had no effect on PTH-stimulated $1,25(OH)_2D_3$ production. This suggests that PGE_2 may modulate the actions of calcitonin but not those of PTH on 1α-hydroxylase [201]. Reduced plasma levels of $25OHD_3$ and $1,25(OH)_2D_3$ were also found following the administration of indomethacin, a potent inhibitor of prostaglandin synthesis, to rabbits at an early stage of pregnancy [202].

Contrary to the above findings, a lack of effect of prostaglandin *in vivo* has been shown by Katz *et al.* [203], who administered PGE_2 by subcutaneous injection daily for 3 weeks to rats. They found no change in $1,25(OH)_2D_3$ or PTH levels, although a significant increase in bone mass in PGE_2-treated rats was demonstrated [203]. Furthermore, long-term subcutaneous administration of indomethacin failed to alter $1,25(OH)_2D_3$ or affect histomorphometric indices of bone formation and resorption [204].

Very limited clinical studies are available to assess the role of PGE_2 in humans. In one study, elevated levels of PGE_2 were believed to be the cause of hypercalcemia and enhanced 1α-hydroxylase activity in children with Bartter's syndrome [205].

Overall, although it would seem that, *in vitro*, prostaglandins stimulate 1α-hydroxylase, variable results *in vivo* create doubt over the exact contribution that prostaglandins make in vitamin D metabolism. Again, the problem of which particular animal species is used may confound the results.

I. Tumor Necrosis Factor-α and Interferon-γ

Information regarding the effects of particular cytokines on vitamin D metabolism have in the main been derived from *in vitro* cell culture studies. Pryke *et al.* [206] examined the effects of tumor necrosis factor-α (TNFα) on 1α-hydroxylase activity in cultured alveolar macrophages. Incubation for 6 days with 50 IU TNFα resulted in an average 4-fold increase in $1,25(OH)_2D_3$ production. An increase in 1α-hydroxylase activity was maximally reached at 72 hr. This finding may account for the "spontaneous" 1α-hydroxylase activity of sarcoid macrophages encountered in sarcoidosis [206]. A further study by Bickle *et al.* [207] provided evidence that TNFα stimulates $1,25(OH)_2D_3$ production in human keratinocytes. In preconfluent cells, TNFα stimulates $1,25(OH)_2D_3$ production; however, this ceased once the keratinocytes achieved confluence [207]. No effect of TNFα has been shown on bone cell lines, but TNFα was shown to inhibit vitamin D receptor number in osteoblastic cells. This may play a role in certain pathological disease states such as postmenopausal osteoporosis [208].

There are also data concerning interferon-γ (IFN-γ) and its role in vitamin D metabolism. Cultured normal human pulmonary alveolar macrophages in the presence of IFN-γ increased $1,25(OH)_2D_3$ production in a dose-dependent manner [209]. Bickle *et al.* [210] provided confirmatory findings showing that IFN-γ stimulated preconfluent keratinocytes to produce $1,25(OH)_2D_3$. Bone marrow-derived macrophages were also demonstrated to respond to IFN-γ with enhanced $1,25(OH)_2D_3$ production [211]. Because lung T lymphocytes from sarcoidosis patients produce IFN-γ, this may play a role in extrarenal 1,25$(OH)_2D_3$ production *in vivo* in sarcoidosis [212].

Little is known about the *in vivo* effect of IFN-γ on vitamin D metabolism. Mann *et al.* [213] studied the effect of IFN-γ on bone mineral metabolism in rats and reported no changes in serum ionized calcium, PTH, or 1,25$(OH)_2D_3$ levels, whereas IFN-γ had a significant osteopenic effect. In human studies data on the effect of these cytokines are lacking, and the underlying disease for which these factors are therapeutically administered may influence the results (e.g., chronic active hepatitis and malignancy).

In summary, aside from the possibility that TNFα and IFN-γ may play a role in extrarenal 1,25$(OH)_2D_3$ production in certain disease states like sarcoidosis, it is at present uncertain as to whether TNFα and IFN-γ physiologically influence vitamin D metabolism. Please refer to Chapter 55 for further information on sarcoidosis, cytokines, and vitamin D.

III. DRUG EFFECTS ON VITAMIN D METABOLISM

The effects of drugs on vitamin D metabolism are summarized in Table II.

TABLE II Drug Effects on Vitamin D Metabolism[a]

Drug	Study	25OHD$_3$	1,25(OH)$_2$D$_3$	24,25(OH)$_2$D$_3$
Anticonvulsants	Animal	↔	↔	
	Human	↓,↔	↓,↔	↓
Corticosteroids	Animal	↔	↔, ↓	↔
	Human	↓,↔	↑,↓,↔	↔
Ethanol	Animal	↑	↓	
	Human	↓,↔	↓,↔	
Ketoconazole	*In vitro*		↓	↓
	Human	↔	↓	
Hypolipidemic agents	Animal	↓		
	Human	↔, ↓	↔	
Bisphosphonates	*In vitro*		↔	↔
	Human	↔	↑,↓,↔	↓
Thiazide diuretics	Human	↑	↓	↑
Calcium channel blockers	Animal		↓	
	Human	↔	↔	
Heparin	Animal		↓	
	Human	↔	↓	↔
Cimetidine	Animal	↓		
	Human	↔, ↓	↔	↔
Aluminum	Animal	↔	↓,↔	↔
	Human	↔	↓,↑	
Antituberculous agents	Human	↔, ↓	↔, ↓	
Caffeine	*In vitro*		↓	
	Animal		↑	
Theophylline	*In vitro*		↓	↑
	Animal	↓	↔	
Immunosuppressants	Animal		↑	↑
	Human	↔	↔	
Fluoride	Animal	↔	↔	
	Human	↔	↔	
Olestra	Animal	↔		
	Human	↔		
Lithium	Human		↔	

[a] Symbols denote effects on serum levels of vitamin D metabolites: ↑, increased; ↓, decreased; ↔, unchanged. See text for details and references.

A. Anticonvulsants

Historically, the use of anticonvulsants has been associated with a number of alterations of bone mineral metabolism. In 1968 Kruse [214] first reported osteomalacia resulting from the use of anticonvulsants, and since then there have been numerous documentations [215–217]. A number of biochemical changes are found after anticonvulsant use. Serum calcium is reduced [216,218–220], and the overwhelming majority of reports demonstrate low serum levels of 25OHD$_3$ [216,220–230]; however, there are some studies in which 25OHD$_3$ levels are unchanged [217,231–233]. Low calcium concentrations lead to the development of secondary hyperparathyroidism [216,220,226,228] while serum phosphate levels remain unchanged [216,220,224,226,228].

Serum 25OHD$_3$ levels are decreased in patients receiving phenobarbital [221–224,230] and phenytoin [221,223,224,227,230,234,235], whereas there is conflicting evidence regarding the effect of carbamazepine on 25OHD$_3$ levels. Some reports show no change [233,236], and some show decreased levels [221,237] with carbamazepine. No reduction of 25OHD$_3$ concentration has been reported with the use of the anticonvulsant sodium valproate [221,223]. Polytherapy results in the most severe changes [218,221,224,230], and anticonvulsant effects occur both in ambulatory outpatient populations [224,230] as well as in the inpatient hospital setting [223,226]. The effects seem to be time dependent, as patients who receive short-term anticonvulsants have normal 25OHD$_3$ levels [224]. Furthermore, the larger the total daily dose of anticonvulsants received, the more severe is the effect on vitamin D metabolism

[218,224,227,230]. Women also seem to be affected more than men [228].

Phenobarbital, phenytoin, and carbamazepine are well-known inducers of hepatic cytochrome P450 enzymes [238–240]. The reduction in $25OHD_3$ level is thought to arise from anticonvulsants enhancing the hepatic breakdown of vitamin D into inactive polar metabolites other than $25OHD_3$ [222,241,242]. Hahn *et al.* [222] demonstrated, both *in vivo* in humans and *in vitro* in rat liver, that phenobarbital stimulated the conversion of tritiated vitamin D and $25OHD_3$ to more polar metabolites.

Levels of $1,25(OH)_2D_3$ have been reported to be high [229,234,243], normal [227,229,231], or low [228] but $24,25(OH)_2D_3$ levels have been demonstrated as being decreased [243,244] following long-term anticonvulsant use. An increase in $1,25(OH)_2D_3$ concentration may be secondary to increased PTH secretion. Levels of DBP are reported as unchanged [228], and intestinal absorption of vitamin D is not altered by anticonvulsant drugs [216].

Animal studies have also demonstrated enhanced metabolism [245] and biliary excretion [241] of vitamin D. Although Hahn *et al.* [246] showed an initial increase in $25OHD_3$ after phenobarbital treatment in the rat, this was followed by a subsequent decline in levels. Ohta *et al.* [247] reported no effect of low dose phenytoin treatment on serum $25OHD_3$ or $1,25(OH)_2D_3$ under conditions where the rats were fed a vitamin D-supplemented diet. As far as treatment of anticonvulsant-induced alterations of vitamin D is concerned, replacement therapy with 400–4000 units/day of vitamin D_3 has been shown to be effective in normalizing parameters of mineral metabolism and improving bone mass [216,248]. In addition, treatment of anticonvulsant osteomalacia with small doses of oral $25OHD_3$ has also proved effective [225].

In summary, anticonvulsants increase the metabolism of $25OHD_3$, and therefore it is recommended that oral vitamin D plus calcium supplementation should be provided for patients on anticonvulsant therapy for longer than 6 months and that such supplementation be adapted to the requirements of the individual. Those especially at risk include long-term institutionalized patients, those with reduced ultraviolet light exposure, those with poor dietary intake of vitamin D, and those receiving long-term treatment with multiple anticonvulsant drugs [248].

B. Corticosteroids

Pharmacological levels of corticosteroids impair the intestinal absorption of calcium [249–252] and induce hypercalciuria [253,254]. Both lead to a state of secondary hyperparathyroidism [254–257]. In this section we focus on potential actions of corticosteroids to alter vitamin D, although multiple other pathways for these actions are postulated (see Chapter 49). Levels of $25OHD_3$ were initially shown to be low in glucocorticoid-treated men. Avioli and co-workers [258] reported evidence of impaired hepatic conversion of vitamin D to $25OHD_3$. Furthermore, Klein *et al.* [251] demonstrated low $25OHD_3$ levels following high dose prednisone therapy, although in patients receiving low or alternate day doses $25OHD_3$ levels were unchanged. Similar findings were reported by Seeman *et al.* [255] in patients receiving high dose glucocorticoids for the treatment of connective tissue disorders. These findings are plausible in light of the fact that corticosteroids are known to induce hepatic microsomal oxidase enzymes [238], and hence chronic glucocorticoid use could produce effects similar to chronic anticonvulsant use (see Section III,A above).

However, other studies have reported contradictory findings with regard to $25OHD_3$ concentrations. In animals no effect on the conversion of vitamin D to $25OHD_3$ or $25OHD_3$ to $1,25(OH)_2D_3$ have been reported [252,259,260]. In humans no differences in serum $25OHD_3$ levels have been shown in patients with corticosteroid-induced osteopenia [256,261,262] or in healthy [250] or diseased subjects treated with corticosteroids [263–265]. Patients treated with inhaled steroids for asthma also fail to demonstrate any alterations in $25OHD_3$, $1,25(OH)_2D_3$, or PTH levels [266,267]. Furthermore, patients with Cushing's disease were also found to have normal levels of $25OHD_3$ [268,269].

Carré *et al.* [270] suggested that use of prednisolone may enhance the inactivation of $1,25(OH)_2D_3$ to more polar biologically inactive metabolites at the tissue level in rats. The same study showed no alteration in the conversion of $^3H\text{-}25OHD_3$ to $^3H\text{-}1,25(OH)_2D_3$ or $^3H\text{-}24,25(OH)_2D_3$. This is in accordance with the findings of Favus *et al.* [260], who demonstrated normal conversion of tritiated $25OHD_3$ to $1,25(OH)_2D_3$ and also demonstrated normal subcellular localization of $1,25(OH)_2D_3$ in the intestinal mucosa of glucocorticoid-treated rats.

In children treated with glucocorticoids for glomerulonephritis, $1,25(OH)_2D_3$ concentrations have been shown to be reduced [271]. However, these children had proteinuria, which could decrease the levels of DBP. Adolescents with systemic lupus erythematosus also had low $1,25(OH)_2D_3$ levels following treatment with glucocorticoids [272]. However, the effect of corticosteroids on $1,25(OH)_2D_3$ levels remains controversial. Seeman *et al.* [255] reported no changes in $1,25(OH)_2D_3$ levels in 14 patients with either endogenous or exogenous glucocorticoid excess. They also showed no differences in the production or metabolic clearance rate of $1,25(OH)_2D_3$ [255]. Others have confirmed these findings of unchanged $1,25(OH)_2D_3$ levels [249,259,262–264]. However, certain studies have shown increased $1,25(OH)_2D_3$ values following subacute prednisone use [250,273–275], and in patients with Cushing's disease $1,25(OH)_2D_3$ levels are in the normal range but decrease following remission. This suggests that higher, though normal, values of $1,25(OH)_2D_3$ are present in the untreated state [268].

Increased levels of $1,25(OH)_2D_3$ are perceivable in response to the state of secondary hyperparathyroidism produced by glucocorticoids. Cortisol stimulated PTH secretion by rat parathyroid glands *in vitro* [276]. *In vivo* this has been demonstrated both in rats [277] and in humans following acute and chronic administration of glucocorticoids [254–257]. Elevated PTH is not a consistent finding, however, as normal PTH levels have been reported in both endogenous [268] as well as exogenous [250,255,262–265] glucocorticoid excess. Studies have also demonstrated no differences in levels of DBP [249] or concentrations of $24,25(OH)_2D_3$ [250,262–264].

Overall, it seems uncertain as to the exact effect that corticosteroids have on vitamin D metabolism. The large number of discordant findings in the literature may reflect the differences in experimental conditions including variations in dosages, dietary intake, and/or sunlight exposure. Please refer to Chapter 49 for a detailed discussion of glucocorticoids and vitamin D.

C. Ethanol

In alcoholics serum levels of $25OHD_3$ have been reported to be low [278–284] or normal [285–287]. Serum $1,25(OH)_2D_3$ has also been shown to be reduced [283,288,289] or unchanged [285]. This seems to be due to a combination of factors. First, many alcoholics have an inadequate dietary supply of vitamin D. Second, many have insufficient exposure to the ultraviolet light of the sun for adequate synthesis of vitamin D. Malabsorption [290] and increased biliary excretion of $25OHD_3$ [291] are other possible factors involved. However, intestinal absorption of vitamin D has been shown to be normal in patients with alcoholic liver disease [287,292]. Induction by alcohol of the cytochrome P450 system may also occur with subsequent increase in the degradation of vitamin D metabolites in the liver [293]. Ethanol may also inhibit hydroxylase activity in the kidney [294] or liver. However, hepatic hydroxylation was normal in cirrhotic alcoholics [287,292,295]. Decreased DBP may lead to low levels of vitamin D metabolites in patients with cirrhotic liver disease [148,296,297].

Acute ethanol loading in rats has produced both elevated [298] and suppressed [299] serum levels of PTH, whereas alcohol stimulated PTH release from bovine parathyroid cells *in vitro* [300]. In humans, acute alcohol ingestion produces unchanged [301] or increased PTH [300] levels. Serum levels of PTH in chronic alcoholics have been normal [295,302–304] or increased [279,283,285,305], the latter probably secondary to diminished intestinal absorption of calcium [306–309]. Increased PTH secretion leads to inhibition of tubular reabsorption of phosphate and lower serum phosphate levels [310]. However, $1,25(OH)_2D_3$ levels are not increased despite increased PTH and low phosphate, which may be due to inhibition by ethanol of renal 1α-hydroxylase [311].

In contrast to human studies, ethanol in rats seems to increase the production of $25OHD_3$ by the liver and lower $1,25(OH)_2D_3$ levels [312–314]. This is due to the presence of both mitochondrial and microsomal hydroxylases in the rat as opposed to only mitochondrial enzymes in humans [315,316]. Ethanol inhibits mitochondrial enzymes but induces microsomal ones. This might lead to increased $25OHD_3$ in rats by microsomal induction and to decreased $25OHD_3$ in humans by mitochondrial suppression [317]. The rat model thus seems to be inappropriate to assess changes in vitamin D metabolism induced by ethanol.

In summary, in chronic alcoholics, the circulating concentrations of vitamin D metabolites should be interpreted in the light of the overall nutritional and health status of the individual [318]. A well-functioning alcoholic with satisfactory dietary ingestion and adequate sunlight exposure might be expected to have normal levels of circulating $25OHD_3$ or $1,25(OH)_2D_3$. In contrast, nonfunctioning alcoholics with poor nutrition and reduced sunlight exposure are likely to have low $25OHD_3$ levels and $1,25(OH)_2D_3$ [318].

D. Ketoconazole

The antifungal agent ketoconazole has been shown to inhibit cytochrome P450-dependent enzymes [319]. In *in vitro* studies, Loose *et al.* [320] demonstrated a dose-dependent reduction in 24-hydroxylase activity in cultured pig kidney cells. Later, Henry [321] showed that both ketoconazole and a similar antifungal, miconazole, behaved as competitive inhibitors of 1α-hydroxylation of $25OHD_3$ in cultured chick kidney cells [321].

In the clinical situation a reduction in $1,25(OH)_2D_3$ was demonstrated in healthy men treated for 1 week with ketoconazole [322]. No changes in $25OHD_3$, PTH, or serum calcium or phosphate levels were shown, suggesting a direct inhibitory effect of the drug on renal 1α-hydroxylase activity [322]. Subsequently Glass and Eil [323] treated patients with primary hyperparathyroidism and hypercalcemia for 1 week with ketoconazole. The treatment produced a reduction in $1,25(OH)_2D_3$ levels. Serum total calcium but not serum ionized calcium levels fell, with no changes in $25OHD_3$, PTH, or serum phosphate [323]. A slightly longer study of 2 weeks duration, also in hyperparathyroid patients, confirmed low $1,25(OH)_2D_3$ levels but also demonstrated a nonsignificant fall in $25OHD_3$ [324]. This raises the possibility that the 25-hydroxylase enzyme is also inhibited, as it is a cytochrome P450-dependent enzyme. Perhaps studies of a longer duration will demonstrate significant falls in $25OHD_3$ as well. Finally, ketoconazole has been shown to be effective in decreasing serum $1,25(OH)_2D_3$ and calcium concentrations, both in cultured pulmonary alveolar macrophages taken from patients with sarcoidosis [325] and *in vivo* in sarcoid patients [325,326]. This suggests that ketoconazole also inhibits the extrarenal production of $1,25(OH)_2D_3$ known to occur in sarcoidosis.

Current data on the effects of ketoconazole on vitamin D metabolism shows that it decreases $1,25(OH)_2D_3$ levels by direct inhibition of 1α-hydroxylase. Ketoconazole also inhibits adrenal and testicular steroidogenesis [319], which can result in hypogonadism [327] or adrenal insufficiency [319]. These effects may have an impact on vitamin D and bone metabolism. During treatment with ketoconazole for any of its indications, susceptible individuals should be monitored and supplemented with vitamin D if necessary.

E. Hypolipidemics

Drugs that inhibit 3-hydroxy-3-methylglutaryl coenzyme A reductase (HMG-CoA reductase) are commonly used for the treatment of patients with hypercholesterolemia. Because HMG-CoA reductase inhibitors are potent inhibitors of cholesterol synthesis, they may impact vitamin D production because cholesterol is the vitamin D_3 precursor. Ismail *et al.* [328] studied 40 hypercholesterolemic patients treated for 24 weeks with one of the new selective HMG-CoA reductase inhibitors, pravastatin (40–80 mg daily). Results showed that although levels of total and low density lipoprotein (LDL) cholesterol were significantly reduced, no changes in $25OHD_3$, $1,25(OH)_2D_3$, or PTH levels were observed [328]. The capacity of the skin to synthesize vitamin D_3 after ultraviolet light exposure also showed no changes after 3 months of pravastatin therapy [329]. Lack of effect of HMG-CoA reductase inhibitors on vitamin D metabolism has been confirmed by others [330], and indeed combination therapy using both pravastatin (40 mg/day) and the bile acid sequestrant cholestyramine (24 g/day) failed to affect levels of vitamin D metabolites or PTH [328].

Absorption of vitamin D from the gut requires the presence of bile acid. Moreover, vitamin D is excreted in the bile [331], and some degree of enterohepatic circulation of vitamin D occurs [332,333]. It stands to reason that cholestyramine might adversely affect vitamin D absorption. There have been reports in rats of decreased vitamin D absorption following treatment with cholestyramine [334]. Moreover, there are isolated case reports of osteomalacia associated with long-term cholestyramine use, which were attributed to cholestyramine-induced vitamin D deficiency [335,336]. Compston and Thompson [337] also reported decreased levels of $25OHD_3$ and reduced intestinal absorption of vitamin D in patients with primary biliary cirrhosis treated with cholestyramine for greater than 2 years [337]. However, a large double-blind randomized trial using patients treated with either cholestyramine (24 g/day) or placebo for 4 months showed similar levels of vitamin D metabolites, PTH, and serum calcium and phosphate versus placebo [338]. Ismail *et al.* [328] showed equivalent findings after 6 months of therapy [328], as did long-term studies using another bile acid sequestrant, colestipol, for the treatment of children with familial hypercholesterolemia [339,340]. A single report of the effect of fibrates on vitamin D metabolism showed a decline in $25OHD_3$ and a rise in $1,25(OH)_2D_3$ levels [341]. It seems, therefore, that vitamin D is unaffected by the statin drugs and short-term cholestyramine treatment. Long-term cholestyramine therapy may affect $25OHD_3$, and levels need to be routinely monitored in susceptible individuals and vitamin D supplementation provided if necessary.

F. Bisphosphonates

Bisphosphonates are commonly used for the treatment of Paget's bone disease, hypercalcemia of malignancy, and postmenopausal osteoporosis. They inhibit bone resorption [342,343], which leads to an increase in the calcium mineral content of bone and decrease in serum calcium levels [344]. A fall in serum calcium results in the stimulation of PTH secretion. PTH reduces the renal tubular reabsorption of phosphate, which reduces serum phosphate levels. Both increased PTH and low phosphate levels can lead to increased $1,25(OH)_2D_3$ production [345–347].

The first generation bisphosphonate, ethane-1-hydroxy-1,1-diphosphate (EHDP), also known as etidronate, was initially shown to cause a reduction of $1,25(OH)_2D_3$ levels at high doses *in vivo* [348,349]. This effect was probably an indirect one as EHDP failed to stimulate or inhibit 1α- or 24-hydroxylase in primary chick kidney cell cultures [12]. Later, evidence emerged to suggest that bisphosphonates indirectly stimulate 1α-hydroxylase at least in part by some unknown humoral factor. An experimental bisphosphonate compound, YM175, was shown in TPTX and sham-operated rats to increase $1,25(OH)_2D_3$ and 1α-hydroxylase in a dose-dependent manner; however, no increase in 1α-hydroxylase activity was demonstrated *in vitro* [350].

In clinical studies, Paget's disease patients treated with etidronate had higher $1,25(OH)_2D_3$ levels without significant changes in serum PTH or ionized calcium or phosphate [66,351]. Lawson-Matthew *et al.* [352] found increased $1,25(OH)_2D_3$ concentrations following short-term oral etidronate treatment but decreased $1,25(OH)_2D_3$ levels after high dose intravenous etidronate. The effects of etidronate are thus most probably dose related. In rats low dose etidronate inhibits bone resorption and calcium release from bone and stimulates dietary calcium absorption. The lower calcium concentration in turn most probably either directly or indirectly, via PTH, mediates 1α-hydroxylase activation. However, high dose etidronate inhibits bone formation as well as resorption, and decreased calcium absorption from the gut is found [344]. Paget's disease patients treated with another bisphosphonate, aminohydroxypropylidene bisphosphonate (APD), known as pamidronate, also had elevated $1,25(OH)_2D_3$ levels following short-term intravenous [353,354] and oral [355] administration. PTH rose to twice pretreatment levels in a response to a fall in ionized calcium levels [353,355]. $24,25(OH)_2D_3$ concentration declined, but $25OHD_3$ remained unchanged [353]. Concentrations of $1,25(OH)_2D_3$ [356,357] as well as PTH [357–359] were also elevated after treatment with pamidronate in patients with tumor-associated hypercalcemia.

The third generation bisphosphonate 4-aminohydroxybutylidene-1,1-bisphosphonate, or alendronate, is a 100- to 500-fold more potent inhibitor of bone resorption than is etidronate [360]. Postmenopausal women treated with alendronate showed initial rises in $1,25(OH)_2D_3$ and PTH that normalized after chronic administration of the drug [346,361,362], probably because of inhibition of bone resorption and decreased serum calcium concentrations. $1,25(OH)_2D_3$ and PTH levels also rose following intravenous alendronate infusion for the treatment of hypercalcemia of malignancy [357]. It thus seems that bisphosphonates do indeed influence vitamin D metabolism. $1,25(OH)_2D_3$ levels are elevated, but this is most probably due to secondary alterations in serum ionized calcium, phosphate, and PTH levels.

G. Thiazide Diuretics

Clinically the use of thiazide diuretics has been associated with a favorable effect on bone mineral density [363–365]. This has been reported in both males [363] and postmenopausal females [366]. Thiazide use has also been shown to reduce the rate of hip fracture [367–369]. The benefit derived from thiazides may be related to decreased PTH-stimulated bone resorption and decreased bone remodeling [366].

Thiazide use leads to a number of alterations in bone mineral metabolism. Thiazides are well known to decrease urinary calcium excretion [364,370–372], with resultant increases in serum calcium concentration [371,373]. This leads to reduced PTH levels [364,366], which in turn decrease $1,25(OH)_2D_3$ synthesis [364,366,374,375] and leads to a reduction in intestinal calcium absorption [364].

Sakhaee *et al.* [364] found that the administration of 50 mg/day of hydrochlorothiazide to postmenopausal women significantly decreased $1,25(OH)_2D_3$ and increased $25OHD_3$ and $24,25(OH)_2D_3$ levels. Similar findings of low $1,25(OH)_2D_3$ levels were reported by Sowers and co-workers [374]. Riis and Christiansen [376] conducted a double-blind long-term controlled trial in early postmenopausal women and demonstrated a trend toward lower $1,25(OH)_2D_3$ with a significantly elevated $24,25(OH)_2D_3$ concentration [376].

No evidence of a direct effect on the synthesis or degradation of $1,25(OH)_2D_3$ by thiazides is currently available. It thus seems that changes in vitamin D metabolites found after the use of thiazide diuretics are mostly secondary to alterations in serum calcium concentrations and PTH levels.

H. Calcium Channel Blockers

Calcium channel blockers, of which nifedipine, verapamil, and diltiazem are the most well recognized and frequently prescribed, are chemically dissimilar. As such they have different effects on PTH secretion. *In vitro* experiments show that verapamil both increases and decreases PTH secretion depending on extracellular calcium concentrations [377–379]. Seely *et al.* [380] showed that diltiazem inhibited PTH release in bovine parathyroid cells by 40% and in human parathyroid cells by 20%. Others using a chemical analog of nifedipine, known as nitrendipine, showed a similar inhibition of PTH secretion [381]. It has been suggested that in certain circumstances calcium channel blockers can act as calcium agonists, thus leading to an increase rather than a decrease in intracellular calcium concentration with associated inhibition of PTH secretion [380]. Verapamil has been reported to inhibit PTH secretion from rat parathyroid glands *in vitro* [382] and from goat parathyroid glands perfused *in vivo* [379]. In conflicting *in vivo* animal studies, verapamil stimulated PTH secretion in rats [383,384], whereas $1,25(OH)_2D_3$ levels were decreased [383].

Clinical studies, however, have shown no change in PTH levels using either verapamil [385] or nifedipine [386]. Only a short-term (3-day) study with diltiazem reported decreased PTH levels with normal ionized calcium and phosphate concentrations [380]. Further studies that also assessed vitamin D metabolites showed no effects after 16 weeks of diltiazem administration on either PTH or $1,25(OH)_2D_3$ levels [387]. Long-term use of nifedipine also failed to alter PTH or $25OHD_3$ levels or to affect serum parameters of bone turnover or bone mineral density in a group of males treated with calcium channel blockers for coronary heart disease [388]. Finally, calcium channel blockers are known inhibitors of hepatic microsomal cytochrome P450-dependent enzymes, which can lead to alterations in vitamin D metabolites [389]. Despite possible interactions with PTH release, all clinical data thus far have failed to demonstrate any effect on vitamin D metabolites, and thus calcium channel blockers cannot be considered to have a major influence on vitamin D metabolism.

I. Heparin

Long-term use of heparin has been associated with the development of osteopenia in humans [390–393] and in rats [394–396]. Heparin decreases bone formation in cultured fetal rat calvaria [397,398] and stimulates bone resorption by increasing the number and activity of osteoclasts *in vitro* [399]. Case studies in pregnant women treated with heparin for venous thrombosis have demonstrated pronounced bone loss with low $1,25(OH)_2D_3$ levels while serum $25OHD_3$ and $24,25(OH)_2D_3$ levels and calcium and phosphate concentrations were unchanged [400,401].

Mutoh *et al.* [394] treated 4-week-old vitamin D-deficient rats with heparin (2000 IU/day). Significant bone loss developed after 2 weeks, which peaked at 4 weeks. No change in serum total or ionized calcium was observed, but a significant elevation of serum PTH was seen. Furthermore, $1,25(OH)_2D_3$ levels were decreased by 54% versus control [394]. Although in the Mutoh study no change in serum ionized calcium was demonstrated, this is presumed to be the mechanism by which heparin increases serum PTH. Heparin has been reported to have a high affinity for calcium ions [402]. This may lead to lower calcium levels, which would stimulate PTH release. However, this is a doubtful mechanism for heparin-induced osteopenia, as calcium salts of heparin are as effective in inducing osteoporosis as the corresponding sodium salts [403]. The reasons for the low $1,25(OH)_2D_3$ values with heparin are entirely speculative but may involve direct inhibition of the 1α-hydroxylase system, with the low $1,25(OH)_2D_3$ levels in turn influencing receptors on the parathyroid gland, which induces PTH stimulation.

J. Cimetidine

Cimetidine is a frequently employed histamine H_2 receptor antagonist and is one among a class of these agents used for the treatment of peptic ulcer disease. *In vitro* studies have demonstrated histamine H_2 receptors in both normal and adenomatous parathyroid gland tissue [404]. Stimulation of these receptors by histamine increases PTH release [405]. Cimetidine has been shown to decrease serum PTH in patients with either parathyroid adenoma [406,407] or secondary hyperparathyroidism due to chronic renal insufficiency [408,409]. However, there have been equivocal findings regarding the effects on serum calcium accompanying the changes in PTH, with both low [406] and unchanged levels [407,409,410] being found.

Cimetidine was also shown to significantly decrease net intestinal calcium transport either secondary to its effect on PTH or via changes in vitamin D [411]. One possible mechanism by which cimetidine may affect vitamin D is by alterations in its metabolism, as cimetidine has been demonstrated to be an inhibitor of microsomal drug metabolism [412,413]. Hepatic vitamin D 25-hydroxylase, is a cytochrome P450-dependent enzyme, and thus it is quite possible that it, too, may be inhibited.

In animal studies, Bengoa *et al.* [414] using whole liver homogenates from vitamin D-deficient rats analyzed 25-hydroxylase activity in the presence of increasing concentrations of cimetidine. They demonstrated a dose-dependent inhibition of 25-hydroxylase activity [414]. 25-Hydroxylase assays were also performed after *in vivo* administration of cimetidine (120 mg/kg intraperitoneally) 1 hr before sacrifice. A 22% decrease in 25-hydroxylase was found, which was statistically significant [414]. Furthermore, cimetidine was also shown to reduce $25OHD_3$ levels in egg-laying hens [415].

In humans, short-term use of cimetidine (800 mg/day for 4 weeks) did not decrease the level of $25OHD_3$ but prevented the expected seasonal rise in $25OHD_3$. After cessation of cimetidine therapy, levels rose significantly.

Levels of $1,25(OH)_2D_3$ and $24,25(OH)_2D_3$ were not affected, and serum calcium and phosphate concentrations remained normal [416].

Although the above studies implicate cimetidine, other H_2 receptor antagonists may not interfere with vitamin D metabolism. Ranitidine, for example, affects hepatic drug metabolism to a much lower degree compared to cimetidine [413] and as such may not alter hepatic metabolism of vitamin D in the same manner as cimetidine. No studies employing ranitidine or other H_2 receptor antagonists and their effets on vitamin D metabolism are currently available.

The results of studies thus far show that cimetidine has multiple effects on calciotropic hormones. PTH levels are decreased, and $25OHD_3$ is reduced by inhibition of hepatic 25-hydroxylase activity. Despite low PTH levels, however, no changes in serum $1,25(OH)_2D_3$ or serum calcium or phosphate concentrations are evident. Monitoring of $25OHD_3$ levels therefore might be indicated only in susceptible individuals such as those with hepatic insufficiency, poor nutrition, or the elderly, especially during the winter months.

K. Aluminum

Parenteral administration of aluminum to patients receiving total parenteral nutrition (TPN) has been associated with low levels of $1,25(OH)_2D_3$ [417]. Casein hydrolysate, which is used as a protein source in some TPN solutions, was identified as containing substantial amounts of aluminum [418]. These patients also develop a low-turnover osteomalacia with aluminum accumulation in bone [419]. Patients undergoing hemodialysis with water containing high levels of aluminum also have a high incidence of aluminum bone disease [420,421]. Aluminum content in bone is elevated and correlates positively with the development of osteomalacia [422–426]. Patients with renal failure are also susceptible to aluminum accumulation in bone [425,427]. There is some evidence that aluminum may act indirectly on bone by suppressing PTH release. Hemodialysis patients demonstrated lower mean PTH values versus control with normal $25OHD_3$ levels [424,426], and $1,25(OH)_2D_3$ was low because of kidney loss. In rats aluminum was shown to accumulate in parathyroid tissue [428], and reports show that aluminum impairs PTH release *in vitro* [429]. This may lead to low $1,25(OH)_2D_3$ production.

In rats, high dose aluminum injections caused osteomalacia (425,430) and lowered PTH levels in spite of hypocalcemia [430]. However, in a similar experiment in rats using equivalent doses of aluminum, no skeletal changes or alterations in serum vitamin D metabolites were observed [431]. In dogs, serum calcium increased and serum phosphate and PTH did not change significantly following 5 weeks of parenteral aluminum administration daily [432]. $25OHD_3$ was normal, but a marked decline in $1,25(OH)_2D_3$ was demonstrated. Renal function also declined, and this may account for the changes in $1,25(OH)_2D_3$. However, the reduction in $1,25(OH)_2D_3$ occurred prior to the appearance of renal impairment, suggesting a direct inhibitory effect on the synthesis of $1,25(OH)_2D_3$ [433].

Opposite effects on vitamin D occur following oral ingestion of aluminum hydroxide as an antacid. Aluminum is absorbed from the gut and deposits in bone in patients both with [434,435] and without [436] renal impairment. Daylong intake by a group of postmenopausal women resulted in a fall in serum phosphate, which correlated significantly with a rise in $1,25(OH)_2D_3$ levels. Total and ionized calcium and PTH levels were unchanged [437]. This study was very short, however, and longer studies would be required to assess the effects of oral aluminum on vitamin D metabolism. In summary, parenteral aluminum administration causes osteomalacia. PTH release may be impaired by aluminum, which may lead to decreased levels of $1,25(OH)_2D_3$.

L. Antituberculous Agents

There have been anecdotal case reports of rifampicin (rifampin)-induced osteomalacia [438]. Studies by Brodie and colleagues [439] initially suggested that the frequently used first-line antituberculous drugs rifampicin and isoniazid, alone or in combination, could affect vitamin D metabolism. In the first study, short-term use of rifampicin (600 mg/day for 2 weeks) in healthy subjects reduced plasma $25OHD_3$ levels by as much as 70%, whereas $1,25(OH)_2D_3$ and PTH remained unchanged [439]. Rifampicin is a known hepatic enzyme inducer [440], and the enzyme 25-hydroxylase is a cytochrome P450-dependent enzyme located in the liver [441]. It is most likely, therefore, that the decreased levels of $25OHD_3$ represent increased hepatic metabolism of $25OHD_3$ by 25-hydroxylase [439]. Short-term use of isoniazid (300 mg/day) in healthy subjects also produced changes in serum vitamin D metabolites. Levels of $25OHD_3$ as well as $1,25(OH)_2D_3$ were reduced, which was accompanied by a fall in serum calcium and phosphate and rise in PTH [294]. Isoniazid has been shown to inhibit hepatic enzyme activity [442,443], and hepatic 25-hydroxylase and renal 1α-hydroxylase are both cytochrome P450-dependent enzyme systems [444]. Inhibition by isoniazid could explain decreased levels of both $25OHD_3$ and $1,25(OH)_2D_3$ [294]. A further study using both rifampicin and isoniazid also led to decreased $25OHD_3$ and $1,25(OH)_2D_3$ levels together with raised PTH [445]. Similar short-term effects of rifampicin and isoniazid have been reported by others [446,447].

Despite evidence of short-term derangements in vitamin D metabolism, it seems that long-term studies reveal no significant effects [448,449]. In the most recent study, treatment of tuberculous patients with both rifampicin and isoniazid for 9 months produced no significant alterations in $25OHD_3$ and$1,25(OH)_2D_3$ levels [449]. Thus, overall it would seem that only in the short-term would patients be susceptible to vitamin D deficiency, in particular those in developing countries with insufficient vitamin D intake. Those at risk may require vitamin D supplementation, but it is unlikely that antituberculous drugs contribute to the development of overt metabolic bone disease.

M. Caffeine

Daniell [450], studying patients with osteoporosis, noted a high caffeine intake versus age-matched controls. Heany and Recker [451], in studies of perimenopausal women, showed a negative calcium balance in association with caffeine intake. Caffeine has been reported to enhance hepatic microsomal drug metabolism in rats and mice [452,453]. An inhibitory effect on the conversion of $25OHD_3$ to $1,25(OH)_2D_3$ was reported in isolated renal tubules from vitamin D-deficient chicks [454]. Studies in rats following chronic caffeine administration showed normal serum calcium but increased urinary excretion and intestinal endogenous excretion of calcium as well increased intestinal absorption of calcium [455]. Yeh and Aloia [456] studied serial changes of serum calcium, PTH, $1,25(OH)_2D_3$, and calcium balance in young and old adult rats following daily caffeine administration for 4 weeks. In young rats, urinary calcium excretion increased and serum calcium decreased initially but then returned to control levels. Serum PTH and $1,25(OH)_2D_3$ increased after 2 weeks, and intestinal absorption of calcium remained unchanged. In adult rats similar changes occurred except that $1,25(OH)_2D_3$ levels were similar to those in controls [456]. One drawback to this study is that the caffeine content that these rats received is the equivalent of 16 cups of coffee each day, which would be rather excessive for human consumption. A further study demonstrated no effect on bone histomorphometry after administration of caffeine to rats [457].

It seems, therefore, that caffeine is most probably only a contributory risk factor for the development of osteoporosis, especially in elderly subjects with poor nutritional habits. Some discrepancy exists between *in vitro* and *in vivo* studies in regard to effect of caffeine on vitamin D metabolism, and at present no definite effect can be presumed.

N. Theophylline

Taft *et al.* [454], using a system of isolated renal tubules from vitamin D-deficient chicks, showed an inhibitory effect of theophylline on the conversion of $25OHD_3$ to $1,25(OH)_2D_3$. This occurred despite an increase in renal tubule cAMP levels, which have been shown to enhance $1,25(OH)_2D_3$ formation *in vivo* [22]. Furthermore, there are reports of enhanced 24-hydroxylase activity by aminophylline in normal birds [458] and rats [459]. Theophylline *in vivo* increases phosphate and calcium urinary excretion in healthy male subjects [460]. McPherson *et al.* [461] found that patients hospitalized for theophylline toxicity had hypercalcemia that normalized after theophylline was discontinued. PTH levels were unchanged, and thus theophylline may act by enhancing the action of available PTH.

In rat studies the effect of long-term constant subcutaneous theophylline infusion was assessed. Once again, increased urinary calcium excretion was demonstrated together with a reduction in total body calcium. Serum $25OHD_3$ was decreased, but no changes in $1,25(OH)_2D_3$ or PTH levels were observed [462]. One possible explanation for the effects on $25OHD_3$ is that theophylline has been previously demonstrated to induce hepatic microsomal enzymes [453,463], which may lead to enhanced conversion of $25OHD_3$ to other metabolites.

O. Immunosuppressants

The T-cell-specific immunosuppressant cyclosporin A (CsA) produces a high turnover osteopenia in the rat [464–466]. In both the rat and mouse, CsA has been shown to stimulate $1,25(OH)_2D_3$ production in the absence of any changes in serum ionized calcium, phosphate, or PTH levels [465–467]. Kidney homogenates from rats following 14 days of oral CsA treatment (15 mg/kg) showed a significant increase in 24-hydroxylase activity. Furthermore, a significant increase in 1α-hydroxylase activity was demonstrated in mice treated with 30–50 mg/kg CsA for 3 days, which was shown to be due to renal and not extrarenal stimulation of 1α-hydroxylase [467]. There was no evidence of increased metabolic clearance of $1,25(OH)_2D_3$ in CsA-treated rats [467]. Young rats responded to CsA with greater $1,25(OH)_2D_3$ production than older ones [468], and $1,25(OH)_2D_3$ levels normalized following cessation of CsA treatment [466]. Studies in Rowett-nude T-cell-deficient rats also showed a similar increase in $1,25(OH)_2D_3$ production in response to treatment with CsA. This suggests that CsA affects vitamin D independent of the presence T lymphocytes [469]. Cyclosporin G (CsG), an equipotent immunosuppressive molecular analog of CsA, has been shown to be less nephrotoxic than CsA [470,471]. CsG also increased $1,25(OH)_2D_3$ levels in rats, independent of changes in ionized calcium or PTH levels, but to a lesser degree than CsA [472–474].

Other immunosuppressants have also been studied *in vivo* in the rat. The fungal macrolide tacrolimus (FK506) has a similar immunosuppressive action to CsA [475]. In the rat, FK506 appears as deleterious as CsA, in that it too produces severe high turnover osteopenia; however, unlike CsA, it has not been shown to have an effect on $1,25(OH)_2D_3$ production [472,474]. Azathioprine, a thioguanine derivative of mercaptopurine, acts as a purine antagonist and is an effective antiproliferative agent. It, too, failed to influence $1,25(OH)_2D_3$, PTH, or ionized calcium levels in the rat [476]. Finally, administration to rats of another new immunosuppressant, rapamycin (sirolimus), caused either no change [477] or low $1,25(OH)_2D_3$ levels [478] with normal PTH or phosphate and ionized calcium concentrations. No reports of effects on vitamin D by any of the other immunosuppressants are known.

Clinically, long-term CsA treatment of patients with multiple sclerosis revealed similar $1,25(OH)_2D_3$, $25OHD_3$, ionized calcium, and phosphate levels compared to those treated with azathioprine, although the latter group had lower PTH levels [479]. Patients with primary biliary cirrhosis treated for 1 year with CsA also showed no evidence of any changes in vitamin D metabolites or PTH levels [480]. Data on vitamin D metabolites posttransplantation are difficult to interpret, because the underlying disease process may interfere with vitamin D metabolism and because multiple drugs including corticosteroids are given concurrently. Nevertheless, posttransplantation, the use of CsA [481–485] or azathioprine [483–488] does not appear to result in any changes in $1,25(OH)_2D_3$ levels compared to control subjects. Thus, it would seem that only experimentally, in rats and mice, does CsA affect vitamin D and that immunosuppressants, collectively, do not significantly affect vitamin D metabolism in the clinical setting.

P. Fluoride

Turner *et al.* [489] assessed the effects of fluoride, at concentrations of 2.0 and 4.5 m*M* in drinking water, on bone mineral metabolism in rats. Although serum calcium and phosphate levels were reduced by fluoride at the 2.0 m*M* level, no changes in either serum $25OHD_3$ or $1,25(OH)_2D_3$ concentrations were reported [489]. In clinical studies, no changes in serum PTH, $1,25(OH)_2D_3$, $25OHD_3$, or calcium and phosphate levels were found after 1 to 2 years of fluoride treatment for osteoporotic males and females [490,491]. Thus, these data suggest that fluoride does not interfere with vitamin D metabolism.

Q. Olestra

Olestra, formerly known as sucrose polyester (SPE), is a nonabsorbable mixture of hexa-, hepta- and octacarbon fatty acid esters of sucrose. It is an edible material that can be incorporated into the diet as a fat substitute. Olestra has physical properties similar to those of conventional dietary fats [492,493]; however, it is not absorbed [494] or hydrolyzed by gastric lipases [495]. As dietary vitamin D is absorbed from the intestine in association with dietary fats [496] and has an enterohepatic circulation, these processes may be altered by the presence of nondigestible lipid. Although there have been reports of reduced absorption of the fat-soluble vitamins A and E [492,493,497,498], all human studies have thus far showed no significant effects on serum 25OHD levels [497–500] or dietary vitamin D absorption [499]. A 20-month feeding study in dogs also showed no effects on vitamin D status following Olestra ingestion [501]. Photoinduced cutaneous synthesis of vitamin D is the major factor determining vitamin D status [502–504], and thus Olestra ingestion would not be expected to adversely affect vitamin D nutritional status.

R. Lithium

The monovalent cation lithium is an effective treatment for bipolar affective disorders. Lithium is known to have significant effects on systemic calcium metabolism. In humans, serum phosphate may be lowered [505–507] and lithium reduces urinary calcium excretion [507–512]. This occurs as a result of increased tubular reabsorption of calcium and results in hypercalcemia [508,510,511,513–517]. Lithium has been known to cause renal tubular acidosis [518], and PTH levels and parathyroid volume have also been reported to be elevated following a few weeks or months of lithium therapy [506,508,510,513–517,519], which is also reversible following withdrawal of the drug [506]. Parathyroid adenoma has been found in several patients with lithium-induced hyperparathyroidism [520], although another possible mechanism by which lithium may elevate PTH is secondary to lithium-induced nephropathy [521,522].

However, serum creatinine levels were within normal limits in one patient group with increased PTH [506]. There is also evidence that lithium stimulates the release of PTH from human parathyroid tissue *in vitro* [512].

As for the effects of lithium on vitamin D, data are limited because most metabolic studies following lithium administration in humans and animals do not measure vitamin D metabolites. Mallette *et al.* found that PTH was elevated but $1{,}25(OH)_2D_3$ levels were not altered after either short-term (mean, 1.7 months) or long-term (mean, 103 months) lithium carbonate administration [508]. Rosenblatt *et al.* [523] studied 10 patients after 1 month of lithium carbonate therapy and noted elevated serum PTH and reduced $1{,}25(OH)_2D_3$ levels, although $25OHD_3$ and serum calcium levels remained unchanged. This is puzzling because the elevation of PTH levels would be expected to increase rather than decrease $1{,}25(OH)_2D_3$ levels. The authors suggested that lithium perhaps acts by inhibiting renal 1α-hydroxylase, although this has not been proved. Both studies had similar group numbers and mean serum lithium levels in the patients, and thus differences in the studies may relate to different patient populations.

Overall, although lithium certainly affects serum calcium, phosphate, and PTH, all of which can interfere with vitamin D homeostasis, limited data concerning changes of vitamin D metabolites make it difficult to ascribe any direct effect of lithium on vitamin D metabolism.

References

1. Garabedian M, Holick MF, DeLuca HF, Boyle IT 1972 Control of 25-hydroxycholecalciferol metabolism by parathyroid glands. Proc Natl Acad Sci USA **69**:1673–1676.
2. Fraser DR, Kodicek E 1973 Regulation of 25-hydroxycholecalciferol-1-hydroxylase activity in kidney by parathyroid hormone. Nature (New Biol) **241**:163–166.
3. Henry HL, Midgett RJ, Norman AW 1974 Regulation of 25-hydroxyvitamin D_3-1-hydroxylase in vivo. J Biol Chem **249**:7584–7592.
4. Henry HL 1979 Regulation of the hydroxylation of 25-hydroxyvitamin D_3 in vivo and in primary cultures of chick kidney cells. J Biol Chem **254**:2722–2729.
5. Booth BE, Tsai HC, Morris RC Jr 1977 Parathyroidectomy reduces 25-hydroxyvitamin D_3-1α-hydroxylase activity in the hypocalcemic vitamin D-deficient chick. J Clin Invest **60**:1314–1320.
6. Mitlak BH, Williams DC, Bryant HU, Paul DC, Neer RM 1992 Intermittent administration of bovine PTH(1–34) increases serum 1,25-dihydroxyvitamin D concentrations and spinal bone density in senile (23 month) rats. J Bone Miner Res **7**:479–484.
7. Shigematsu T, Horiuchi N, Ogura Y, Miyahara T, Suda T 1986 Human parathyroid hormone inhibits renal 24-hydroxylase activity of 25-hydroxyvitamin D_3 by a mechanism involving adenosine 3′,5′-monophosphate in rats. Endocrinology **118**:1583–1589.
8. Tanaka Y, Lorenc RS, DeLuca HF 1975 The role of 1,25-dihydroxyvitamin D_3 and parathyroid hormone in the regulation of chick renal 25-hydroxyvitamin D_3-24-hydroxylase. Arch Biochem Biophys **171**:521–526.
9. Tanaka Y, Castillo L, DeLuca HF 1976 Control of renal vitamin D hydroxylase in birds by sex hormones. Proc Natl Acad Sci USA **73**:2701–2705.
10. Rasmussen H, Wong M, Bikle D, Goodman DBP 1972 Hormonal control of the renal conversion of 25-hydroxycholecalciferol to 1,25-dihydroxycholecalciferol. J Clin Invest **51**:2502–2504.
11. Larkins RG, MacAuley SJ, Rapoport A, Mertin TJ, Tulloch BR, Byfield PGH, Matthews EW, MacIntyre I 1974 Effects of nucleotides, hormones, ions and 1,25-dihydroxycholecalciferol on 1,25-dihydroxycholecalciferol production in isolated chick renal tubules. Clin Sci Mol Med **46**:569–582.
12. Trechsel U, Bonjour J-P, Fleisch H 1979 Regulation of the metabolism of 25-hydroxyvitamin D_3 in primary cultures of chick kidney cells. J Clin Invest **64**:206–217.
13. Juan D, DeLuca HF 1977 The regulation of 24,25-dihydroxyvitamin D_3 production in cultures of monkey kidney cells. Endocrinology **101**:1184–1193.
14. Armbrecht HJ, Wongsurawat N, Zenser T, Davis BB 1982 Differential effects of parathyroid hormone on the renal 1,25-dihydroxyvitamin D_3 and 24,25-dihydroxyvitamin D_3 production of young and adult rats. Endocrinology **111**:1339–1344.
15. Armbrecht HJ, Wongsurawat N, Zenser TV, Davis BB 1984 Effect of PTH and $1{,}25(OH)_2D_3$ on renal $25(OH)D_3$ metabolism, adenylate cyclase, and protein kinase. Am J Physiol **246**:E102-E107.
16. Rost CR, Bickle DD, Kaplan RA 1981 In vitro stimulation of 25-hydroxycholecalciferol 1α-hydroxylation by parathyroid hormone in chick kidney slices: Evidence for a role for adenosine 3′,5′-monophosphate. Endocrinology **108**:1002–1006.
17. Fukase M, Birge SJ Jr, Rifas L, Avioli LV, Chase LR 1982 Regulation of 25 hydroxyvitamin D_3 1-hydroxylase in serum-free monolayer culture of mouse kidney. Endocrinology **110**:1073–1075.
18. Shain SA 1972 The in vitro metabolism of 25-hydroxycholecalciferol to 1,25-dihydroxycholecalciferol by chick renal tubules. J Biol Chem **247**:4404–4413.
19. Murer H, Werner A, Reshkin S, Wuarin F, Biber J 1991 Cellular mechanisms in proximal tubular reabsorption of phosphate. Am J Physiol **260**:C885–C899.
20. Ro H-K, Tembe V, Krug T, Yang P-YJ, Bushinsky DA, Favus MJ 1990 Acidosis inhibits $1{,}25(OH)_2D_3$ but not cAMP production in response to parathyroid hormone in the rat. J Bone Miner Res **5**:273–278.
21. Chase LR, Aurbach GD 1967 Parathyroid function and the renal excretion of 3′,5′-adenylic acid. Proc Natl Acad Sci USA **58**:518–525.
22. Horiuchi N, Suda T, Takahashi H, Shimazawa E, Ogata E 1977 In vivo evidence for the intermediary role of 3′,5′-cyclic AMP in parathyroid hormone-induced stimulation of 1α,25-dihydroxyvitamin D_3 synthesis in rats. Endocrinology **101**:969–974.
23. Armbrecht HJ, Forte LR, Wongsurawat N, Zenser TV, Davis BB 1984 Forskolin increases 1,25-dihydroxyvitamin D_3 production by rat renal slices in vitro. Endocrinology **114**:644–649.
24. Henry HL 1985 Parathyroid hormone modulation of 25-hydroxyvitamin D_3 metabolism by cultured chick kidney cells is mimicked and enhanced by forskolin. Endocrinology **116**: 503–510.
25. Seamon KB, Daly JW 1982 Forskolin: A unique diterpene activator of cyclic AMP generating systems. J Cyclic Nucleotide Res **71**:201.
26. Shinki T, Jin CH, Nishimura A, Nagai Y, Ohyama Y, Noshiro M, Okuda K, Suda T 1992 Parathyroid hormone inhibits 25-

hydroxyvitamin D_3-24-hydroxylase mRNA expression stimulated by 1α,25-dihydroxyvitamin D_3 in rat kidney but not in intestine. J Biol Chem **267**:13757–13762.

27. Janulis M, Tembe V, Favus MJ 1992 Role of protein kinase C in parathyroid hormone stimulation of renal 1,25-dihydroxyvitamin D_3 secretion. J Clin Invest **90**:2278–2283.
28. Ro HK, Tembe V, Favus MJ 1992 Evidence that activation of protein kinase-C can stimulate 1,25-dihydroxyvitamin D_3 secretion by rat proximal tubules. Endocrinology **131**:1424–1428.
29. Friedlander J, Janulis M, Tembe V, Ro HK, Wong M-S, Favus MJ 1994 Loss of parathyroid hormone-stimulated 1,25-dihydroxyvitamin D_3 production in aging does not involve protein kinase A or C pathways. J Bone Miner Res **9**:339–345.
30. Kaplan RK, Haussler MR, Deftos LJ, Bone H, Pak CYC 1977 The role of 1α,25-dihydroxyvitamin D in the mediation of intestinal hyperabsorption of calcium in primary hyperparathyroidism and absorptive hypercalciuria. J Clin Invest **59**:756–760.
31. Mawer EB, Backhouse J, Hill LF, Lumb GA, De Silva P, Taylor CM, Stanbury SW 1975 Vitamin D metabolism and parathyroid function in man. Clin Sci Mol Med **48**:349–365.
32. Brown DJ, Spanos E, MacIntyre I 1980 Role of pituitary hormone in regulating renal vitamin D metabolism in man. Br Med J **1**:277–278.
33. Nakayama K, Fukumoto S, Takeda S, Takeuchi Y, Ishikawa T, Miura M, Hata K, Hane M, Tamura Y, Tanaka Y, Kitaoka M, Obara T, Ogata E, Matsumoto T 1996 Differences in bone and vitamin D metabolism between primary hyperparathyroidism and malignancy-associated hypercalcemia. J Clin Endocrinol Metab **81**:607–611.
34. Bilezikian JP, Canfield RE, Jacobs TP, Polay JS, D'Adamo AP, Eisman JA, DeLuca HF 1978 Response of 1α,25-dihydroxyvitamin D_3 to hypocalcemia in human subjects. N Engl J Med **299**:437–441.
35. Rosen JF, Fleischman AR, Finberg L, Eisman J, DeLuca HF 1977 1,25-Dihydroxycholecalciferol: Its use in the long-term management of idiopathic hypoparathyroidism in children. J Clin Endocrinol Metab **45**:457–468.
36. Kooh SW, Fraser D, DeLuca HF, Holick MF, Belsey RE, Clark MB, Murray TM 1975 Treatment of hypoparathyroidism and pseudohypoparathyroidism with metabolites of vitamin D: Evidence for impaired conversion 25-hydroxyvitamin D to 1α,25-dihydroxyvitamin D. N Engl J Med **293**:840–844.
37. Davies M, Taylor CM, Hill LF, Stanbury SW 1977 1,25-Dihydroxycholecalciferol in hypoparathyroidism. Lancet **1**:55–59.
38. Drezner MK, Neelon FA, Haussler M, McPherson HT, Lebovitz HE 1976 1,25-Dihydroxycholecalciferol deficiency: The probable cause of hypocalcemia and metabolic bone disease in pseudohypoparathyroidism. J Clin Endocrinol Metab **42**:621–628.
39. Bilezikian JP 1990 Parathyroid hormone-related peptide in sickness and in health. N Engl J Med **322**:1151–1153.
40. Orloff JJ, Kats Y, Urena P, Schipani E, Vasavada RC, Philbrick WM, Behal A, Abou-Samra A-B, Segre GV, Jüppner H 1995 Further evidence for a novel receptor for amino-terminal parathyroid hormone-related protein on keratinocytes and squamous carcinoma cell lines. Endocrinology **136**:3016–3023.
41. Strewler GJ, Stern PH, Jacobs JW, Eveloff J, Klein RF, Leung SC, Rosenblatt M, Nissenson RA 1987 Parathyroid hormone-like protein from human renal carcinoma cells. Structural and functional homology with parathyroid hormone. J Clin Invest **80**:1803–1807.
42. Yates AJP, Gutierrez GE, Smolens P, Travis PS, Katz MS, Aufdemorte TB, Boyce BF, Hymer TK, Poser JW, Mundy GR 1988 Effects of a synthetic peptide of a parathyroid hormone-related protein on calcium homeostasis, renal tubular calcium reabsorption, and bone metabolism in vivo and in vitro in rodents. J Clin Invest **81**:932–938.
43. Stewart AF, Horst R, Deftos LJ, Cadman EC, Lang R, Broadus AE 1980 Biochemical evaluation of patients with cancer-associated hypercalcemia. N Engl J Med **303**:1377–1383.
44. Ralston SH, Cowan RA, Robertson RA, Gardner MD, Boyle IT 1984 Circulating vitamin D metabolites and hypercalcemia of malignancy. Acta Endocrinol **106**:556–563.
45. Insogna KL, Stewart AF, Vignery AMC, Weir EC, Namnum PA, Baron RE, Kirkwood JM, Deftos LM, Broadus AE 1984 Bichemical and histomorphometric characterization of a rat model for humoral hypercalcemia of malignancy. Endocrinology **114**:888–896.
46. Horiuchi N, Caulfield MP, Fisher JE, Goldman ME, McKee RL, Reagan JE, Levy JJ, Nutt RF, Rodan SB, Schofield TL, Clemens TL, Rosenblatt M 1987 Similarity of synthetic peptide from human tumor to parathyroid hormone in vivo and in vitro. Science **238**:1566–1568.
47. Walker AT, Stewart AF, Korn EA, Shiratori T, Mitnick MA, Carpenter TO 1990 Effect of parathyroid homone-like peptides on 25-hydroxyvitamin D-1α-hydroxylase activity in rodents. Am J Physiol **258**:E297–E303.
48. Fukumoto S, Matsumoto T, Yamoto H, Kawashima H, Ueyama Y, Tamaoki N, Ogata E 1989 Suppression of serum 1,25-dihydroxyvitamin D in humoral hypercalcemia of malignancy is caused by an elaboration of a factor that inhibits renal 1,25-dihydroxyvitamin D_3 production. Endocrinology **124**:2057–2062.
49. Schilling T, Pecherstorfer M, Blind E, Leidig G, Ziegler R, Raue F 1993 Parathyroid hormone-related protein (PTHrP) does not regulate 1,25-dihydroxyvitamin D serum levels in hypercalcemia of malignancy. J Clin Endocrinol Metab **76**:801–803.
50. Everhart-Caye M, Inzucchi SE, Guinness-Henry J, Mitnick MA, Stewart AF 1996 Parathyroid hormone (PTH)-related protein(1–36) is equipotent to PTH(1–34) in humans. J Clin Endocrinol Metab **81**:199–208.
51. Fraher LJ, Hodsman AB, Jonas K, Saunders D, Rose CI, Henderson JE, Hendy GN, Goltzman D 1992 A comparison of the in vivo biochemical responses to exogenous parathyroid hormone-(1–34) [PTH-(1–34)] and PTH-related peptide-(1–34) in man. J Clin Endocrinol Metab **75**:417–423.
52. Mundy GR 1988 Hypercalcemia of malignancy revisited. J Clin Invest **82**:1–6.
53. Galante L, Colston KW, Macauley SJ, MacIntyre I 1972 Effect of calcitonin on vitamin D metabolism. Nature **238**:271–273.
54. Lorenc R, Tanaka Y, DeLuca HF, Jones G 1977 Lack of effect of calcitonin on the regulation of vitamin D metabolism in the rat. Endocrinology **100**:468–472.
55. Horiuchi N, Takahashi H, Matsumoto T, Takahashi N, Shimazawa E, Suda T, Ogata E 1979 Salmon calcitonin-induced stimulation of 1α,25-dihydroxycholecalciferol synthesis in rats involving a mechanism independent of adenosine 3′:5′-cyclic monophosphate. Biochem J **184**:269–275.
56. Jaeger P, Jones W, Clemens TL, Hayslett JP 1986 Evidence that calcitonin stimulates 1,25-dihydroxyvitamin D production and intestinal absorption of calcium in vivo. J Clin Invest **78**:456–461.
57. Beckman MJ, Goff JP, Reinhardt TA, Beitz DC, Horst RL 1994 In vivo regulation of rat intestinal 24-hydroxylase: Potential new role of calcitonin. Endocrinology **135**:1951–1955.
58. Armbrecht HJ, Wongsurawat N, Paschal RE 1987 Effect of age on renal responsiveness to parathyroid hormone and calcitonin in rats. J Endocrinol **114**:173–178.
59. Kawashima H, Torikai S, Kurokawa K 1981 Calcitonin selectively stimulates 25-hydroxyvitamin D_3–1α-hydroxylase in proximal straight tubule of rat kidney. Nature **291**:327–329.

60. Emmertsen K, Melsen F, Mosekilde L, Lund B, Lund B, Sørensen OH, Nielsen HE, Sølling H, Hansen HH 1981 Altered vitamin D metabolism and bone remodelling in patients with medullary thyroid carcinoma and hypercalitonemia. Metab Bone Dis Related Res **4**:17–23.
61. Aloia JF, Vaswani A, Kapoor A, Yeh JK, Cohn SH 1985 Treatment of osteoporosis with calcitonin, with and without growth hormone. Metabolism **34**:124–129.
62. Hartwell D, Hassager C, Overgaard K, Riis BJ, Pødenphant J, Christiansen C 1990 Vitamin D metabolism in osteoporotic women during treatment with estrogen, and anabolic steroid, or calcitonin. Acta Endocrinol **122**:715–721.
63. Thamsborg G, Jensen JEB, Kollerup G, Hauge EM, Melsen F, Sørensen OH 1996 Effect of nasal salmon calcitonin on bone remodeling and bone mass in postmenopausal women. Bone **18**:207–212.
64. Thamsborg G, Storm TL, Daugaard H, Schiffer S, Sørensen OH 1991 Circulating levels of calciotropic hormones during treatment with nasal salmon calcitonin. Acta Endocrinol **125**:127–131.
65. Nunziata V, Giannattasio R, Di Giovanni G, Lettera AM, Nunziata CA 1993 Vitamin D status in Paget's bone disease. Effect of calcitonin therapy. Clin Orthop Related Res **293**:366–371.
66. Devlin RD, Gutteridge DH, Prince RL, Retallack RW, Worth GK 1990 Alterations in vitamin D metabolites during treatment of Paget's disease of bone with calcitonin or etidronate. J Bone Miner Res **5**:1121–1126.
67. Mainoya JR 1975 Effects of bovine growth hormone, human placental lactogen and ovine prolactin on intestinal fluid and ion transport in the rat. Endocrinology **96**:1165.
68. Henneman PH, Forbes AP, Moldawer M, Dimpsey EF, Carrol EL 1962 Effects of human growth hormone in man. J Clin Invest **39**:1223.
69. Finkelstein JD, Schachter D 1962 Active transport of calcium by intestine: Effects of hypophysectomy and growth hormone. Am J Physiol **203**:873.
70. Spanos E, Barret D, MacIntyre I, Pike JW, Safilian EF, Haussler MR 1978 Effect of growth hormone on vitamin D metabolism. Nature **273**:246–247.
71. Fontaine O, Pavlovitch H, Balsan S 1978 25-Hydroxycholecalciferol metabolism in hypophysectomized rats. Endocrinology **102**:1822–1826.
72. Spencer EM, Tobiassen O 1981 The mechanism of the action of growth hormone on vitamin D metabolism in the rat. Endocrinology **108**:1064–1070.
73. Pahuja DN, DeLuca HF 1981 Role of the hypophysis in the regulation of vitamin D metabolism. Mol Cell Endocrinol **23**:345–350.
74. Gray RW, Garthwaite TL 1985 Activation of renal 1,25-dihydroxyvitamin D_3 synthesis by phosphate deprivation: Evidence for a role for growth hormone. Endocrinology **116**:189–193.
75. Denis I, Thomasset M, Pointillart A 1994 Influence of exogenous porcine growth hormone on vitamin D metabolism and calcium and phosphorus absorption in intact pigs. Calcif Tissue Int **54**:489–492.
76. Eskildsen PC, Lund B, Sørensen OH, Lund B, Bishop JE, Norman AW 1979 Acromegaly and vitamin D metabolism: Effect of bromocriptine treatment. Endocrinology **49**:484–486.
77. Lund B, Eskildsen PC, Lund B, Norman AW, Sørensen OH 1981 Calcium and vitamin D metabolism in acromegaly. Acta Endocrinol **96**:444–450.
78. Bijlsma JWJ, Nortier JWR, Duursma SA, Croughs RJM, Bosch R, Thijssen JHH 1983 Changes in bone metabolism during treatment of acromegaly. Acta Endocrinol **104**:153–159.
79. Brixen K, Nielsen HK, Bouillon R, Flyvberg A, Mosekilde L 1992 Effects of short-term growth hormone treatment on PTH, calcitriol, thyroid hormones, insulin and glucagon. Acta Endocrinol **127**:331–336.
80. Marcus R, Butterfield G, Holloway L, Gilliland L, Baylink DJ, Hintz RL, Sherman BM 1990 Effects of short-term administration of recombinant human growth hormone to elderly people. J Clin Endocrinol Metab **70**:519–527.
81. Harbison MD, Gertner JM 1990 Permissive action of growth hormone on the renal response to dietary phosphorus deprivation. J Clin Endocrinol Metab **70**:1035–1040.
82. Chipman JJ, Zerwekh J, Nicar M, Marks J, Pak CYC 1980 Effect of growth hormone administration: Reciprocal changes in serum 1α,25-dihydroxyvitamin D and intestinal calcium absorption. J Clin Endocrinol Metab **51**:321–324.
83. Burstein S, Chen IW, Tsang RC 1983 Effects of growth hormone replacement therapy on 1,25-dihydroxyvitamin D and calcium metabolism. J Clin Endocrinol Metab **56**:1246–1251.
84. Gertner JM, Horst RL, Broadus AE, Rasmussen H, Genel M 1979 Parathyroid function and vitamin D metabolism during human growth hormone replacement. J Clin Endocrinol Metab **49**:185–188.
85. Gertner JM, Tamborlane WV, Hintz RL, Horst RL, Genel M 1981 The effects on mineral metabolism of overnight growth hormone infusion in growth hormone deficiency. J Clin Endocrinol Metab **53**:818–822.
86. Bickle DD, Spencer EM, Burke WH, Rost CR 1980 Prolactin but not growth hormone stimulates 1,25-dihydroxyvitamin D_3 production by chick renal preparations in vitro. Endocrinology **107**:81–84.
87. Hammerman MR, Gavin III JR 1986 Binding of IGF-I and IGF-I-stimulated phosphorylation in canine renal basolateral membrane. Am J Physiol **251**:E32–E41.
88. Condamine L, Vztovsnik F, Friedlander G, Menaa C, Garabédian M 1994 Local action of phosphate depletion and insulin-like growth factor 1 on in vitro production of 1,25-dihydroxyvitamin D by cultured mammalian kidney cells. J Clin Invest **94**:1673–1679.
89. Halloran BP, Spencer EM 1988 Dietary phosphate and 1,25-dihydroxyvitamin D metabolism: Influence of insulin-like growth factor 1. Endocrinology **123**:1225–1229.
90. Caverzasio J, Motessuit C, Bonjour JP 1990 Stimulatory effect of insulin-like growth factor-1 on renal Pi transport and plasma 1,25-dihydroxyvitamin D_3. Endocrinology **127**:453–459.
91. Gray RW 1987 Evidence that somatomedins mediate the effect of hypophosphatemia to increase serum 1,25-dihydroxyvitamin D_3 levels in rats. Endocrinology **121**:504–512.
92. Nesbitt T, Drezner MK 1993 Insulin-like growth factor-1 regulation of renal 25-hydroxyvitamin D-1-hydroxylase activity. Endocrinology **132**:133–138.
93. Spanos E, Pike JW, Haussler MR, Colston KW, Evans IMA, Goldner AM, McCain TA, MacIntyre I 1976 Circulating 1α,25-dihydroxyvitamin D in the chicken: Enhancement by injection of prolactin and during egg laying. Life Sci **19**:1751–1756.
94. Spanos E, Brown DJ, Stevenson JC, MacIntyre I 1981 Stimulation of 1,25-dihydroxycholecalciferol production by prolactin and related peptides in intact renal cell preparations in vitro. Biochim Biophys Acta **672**:7–15.
95. Spanos E, Colston KW, Evans IMS, Galante LS, Macauley SJ, MacIntyre I 1976 Effect of prolactin on vitamin D metabolism. Mol Cell Endocrinol **5**:163–167.
96. Matsumoto T, Horiuchi N, Suda T, Takahashi H, Shimazawa E, Ogata E 1979 Failure to demonstrate stimulatory effect of prolactin on vitamin D metabolism in vitamin-D-deficient rats. Metabolism **28**:925–927.

97. Kenny AD, Baksi SN, Galli-Galardo SM 1977 Vitamin D metabolism in amphibia and fish. Fed Proc **36**:1097.
98. Mainoya JR 1975 Further studies on the action of prolactin on fluid and ion absorption by the rat jejunum. Endocrinology **96**:1158–1164.
99. Reichlin S 1974 Neuroendocrinology: Lactation In: William RH (ed) Textbook of Endocrinology. Saunders, Philadelphia, Pennsylvania, pp. 800–801.
100. Boass A, Toverud SA, McCain TA, Pike JW, Haussler MR 1977 Elevated serum levels of 1α,25-dihydroxycholecalciferol in lactating rats. Nature **267**:630–632.
101. MacIntyre I, Colston KW, Szelke M, Spanos E 1978 A survey of the hormonal factors that control calcium metabolism. Ann NY Acad Sci **307**:345–355.
102. Kumar R, Cohen WR, Silva P, Epstein FH 1979 Elevated 1,25-dihydroxyvitamin D plasma levels in normal human pregnancy and lactation. J Clin Invest **63**:342–344.
103. Lund B, Selnes A 1979 Plasma 1,25-dihydroxyvitamin D levels in pregnancy and lactation. Acta Endocrinol **92**:330–335.
104. Cundy T, Haining SA, Guilland-Cumming DF, Butler J, Kanis JA 1987 Remission of hypoparathyroidism during lactation: Evidence for a physiological role for prolactin in the regulation of vitamin D metabolism. Clin Endocrinol **26**:667–674.
105. Kumar R, Abboud CF, Riggs BL 1980 The effect of elevated prolactin levels on plasma 1,25-dihydroxyvitamin D and intestinal absorption of calcium. Mayo Clin Proc **55**:51–53.
106. Adams ND, Garthwaite TL, Gray RW, Hagen TC, Lemann J Jr 1979 The interrelationships among prolactin, 1,25-dihydroxyvitamin D, and parathyroid hormone in humans. J Clin Endocrinol Metab **49**:628–630.
107. Klibanski A, Neer R, Beitins I, Ridgeway E, McArthur J 1981 Decreased bone density in hyperprolactinemic women. N Engl J Med **303**:1511.
108. Schlechte JA, Sherman B, Martin R 1983 Bone density in amenorrheic women with and without hyperprolactinemia. J Clin Endocrinol Metab **56**:1120–1123.
109. Koppelman MCS, Kurtz DW, Morrish KA, Bou E, Susser JK, Shapiro JR, Loriaux DL 1984 Vertebral body bone mineral content in hyperprolactinemic women. J Clin Endocrinol Metab **59**:1050–1053.
110. Bouillon R 1991 Diabetic bone disease. Calcif Tissue Int **49**:155–160.
111. Schneider LE, Schedl HP 1972 Diabetes and intestinal calcium absorption in the rat. Am J Physiol **223**:1319–1323.
112. Schneider LE, Wilson HD, Schedl HP 1974 Effects of alloxan diabetes on duodenal calcium-binding protein in the rat. Am J Physiol **227**:832–838.
113. Schneider LE, Omdahl EJ, Schedl HP 1976 Effects of vitamin D and its metabolites on calcium transport in the diabetic rat. Endocrinology **99**:793–799.
114. Schneider LE, Nowosielski LM, Schedl HP 1977 Insulin-treatment of diabetic rats: Effects on duodenal calcium absorption. Endocrinology **100**:67–73.
115. Schneider LE, Schedl HP, McCain T, Haussler MR 1977 Experimental diabetes reduces circulating 1,25-dihydroxyvitamin D in the rat. Science **196**:1452–1454.
116. Schneider LE, Hargis EK, Schedl HP, Williams GA 1974 Parathyroid function in the alloxan diabetic rat. Endocrinology **95**:749.
117. Schedl HP, Heath H, Wenger J 1978 Serum calcitonin and parathyroid hormone in experimental diabetes: Effects of insulin treatment. Endocrinology **103**:1368–1373.
118. Wongsurawat N, Armbrecht HJ, Zenser TV, Davis BB, Thomas ML, Forte LR 1983 1,25-Dihydroxyvitamin D_3 and 24,25-dihydroxyvitamin D_3 production by isolated renal slices is modulated by diabetes and insulin in the rat. Diabetes **32**:302–306.
119. Glajchen N, Epstein S, Thomas S, Fallon M, Chakrabarti S 1988 Bone mineral metabolism in experimental diabetes mellitus: Osteocalcin as a measure of bone remodeling. Endocrinology **123**:290–295.
120. Shires R, Teitelbaum SL, Bergfeld MA, Fallon MD, Slatopolsky E, Avioli LV 1981 The effect of streptozotocin-induced chronic diabetes mellitus on bone and mineral homeostasis in the rat. J Clin Lab Med **97**:231–240.
121. Spencer EM, Khalil M, Tobiassen O 1980 Experimental diabetes in the rat causes an insulin-reversible decrease in renal 25-hydroxyvitamin D_3-1α-hydroxylase activity. Endocrinology **107**: 300–305.
122. Takeshita N, Yoshino T, Mutoh S, Yamaguchi I 1994 Possible involvement of vitamin D_3-deficiency and relatively enhanced bone resorption in the development of bone loss in streptozotocin-induced diabetic rats. Life Sci **55**:291–299.
123. Hough S, Fausto A, Sonn Y, Dong Jo OK, Birge SJ, Avioli LV 1983 Vitamin D metabolism in the chronic streptozotocin-induced diabetic rat. Endocrinology **113**:790–796.
124. Epstein S, Takizawa M, Stein B, Katz IA, Joffe II, Romero DF, Liang XG, Ke HZ, Jee WSS, Jacobs TW, Berlin J 1994 The effect of cyclosporin A on bone mineral metabolism in experimental diabetes mellitus in the rat. J Bone Miner Res **9**:557–566.
125. Sulimovici S, Roginsky MS 1980 Hepatic metabolism of vitamin D_3 in streptozotocin-induced diabetic rat. Acta Endocrinol **93**:346–350.
126. Matsumoto T, Kawanobe Y, Ezawa I, Shibuya N, Hata K, Ogata E 1986 Role of insulin in the increase in serum 1,25-dihydroxyvitamin D concentrations in response to phosphorus deprivation in streptozotocin-induced diabetic rats. Endocrinology **118**:1440–1444.
127. Nyomba B, Bouillon R, Lissens W, Van Baelen H, De Moor P 1985 1,25-Dihydroxyvitamin D and vitamin D-binding protein are both decreased in streptozotocin-diabetic rats. Endocrinology **116**:2483–2488.
128. Henry HL 1981 Insulin permits parathyroid hormone stimulation of 1,25-dihydroxyvitamin D_3 production in cultured kidney cells. Endocrinology **108**:733–735.
129. Sulimovici S, Roginsky MS, Susser F 1981 Nephrogenous cyclic AMP in streptozotocin-induced diabetic rat. Biochem Biophys Res Commun **100**:471–477.
130. Gertner J, Horst R, Tamborlane W 1979 Mineral metabolism and vitamin D status in juvenile diabetics: Changes following normalization of plasma glucose with a portable infusion pump. Diabetes **28**:354.
131. Frazer TE, White NH, Hough S, Santiago JV, McGee BR, Bryce G, Mallon J, Avioli LV 1981 Alterations in circulating vitamin D metabolites in the young insulin-dependent diabetic. J Clin Endocrinol Metab **53**:1154–1159.
132. Nyomba BL, Bouillon R, Bidingija M, Kandjingu K, De Moor P 1986 Vitamin D metabolites and their binding protein in adult diabetic patients. Diabetes **35**:911–915.
133. Heath III H, Lambert PW, Service FJ, Arnaud SB 1979 Calcium homeostasis in diabetes mellitus. J Clin Endocrinol Metab **49**:462–466.
134. Kenny AD 1976 Vitamin D metabolism: Physiological regulation in egg-laying Japanese quail. Am J Physiol **230**:1609–1615.
135. Baksi SN, Kenny AD 1977 Vitamin D_3 metabolism in immature Japanese quail: Effect of ovarian hormones. Endocrinology **101**:1216–1220.
136. Castillo L, Tanaka Y, DeLuca HF, Sunde ML 1977 The stimula-

tion of 25-hydroxyvitamin D_3-1-α-hydroxylase by estrogen. Arch Biochem Biophys **179**:211–217.

137. Tanaka Y, Castillo L, Wineland MJ, DeLuca HF 1978 Synergistic effect of progesterone, testosterone, and estradiol in the stimulation of chick renal 25-hydroxyvitamin D_3-1α-hydroxylase. Endocrinology **103**:2035–2039.
138. Pike JW, Spanos E, Colston KW, MacIntyre I, Haussler MR 1978 Influence of estrogen on renal vitamin D hydroxylases and serum 1α,25$(OH)_2D_3$ in chicks. Am J Physiol **235**:E338–E343.
139. Trechsel U, Bonjour J-P, Fleisch H 1979 Regulation of the metabolism of 25-hydroxyvitamin D_3 in primary cultures of chick kidney cells. J Clin Invest **64**:206–217.
140. Henry HL 1981 25$(OH)D_3$ metabolism in kidney cell cultures: Lack of direct effect of estradiol. Am J Physiol **240**:E119–E124.
141. Baksi SN, Kenny AD 1978 Does estradiol stimulate in vivo production of 1,25-dihydroxyvitamin D_3 in the rat. Life Sci **22**:787–792.
142. Kalu DN, Liu CC, Hardin RR, Hollis BW 1989 The aged rat model of ovarian hormone deficiency bone loss. Endocrinology **124**:7–16.
143. Kalu DN, Salerno E, Liu CC, Echon R, Ray M, Garza-Zapata M, Hollis BW 1991 A comparative study of the actions of tamoxifen, estrogen and progesterone in the oophorectomized rat. Bone Miner **15**:109–124.
144. Arjmandi BH, Hollis BW, Kalu DN 1994 In vivo effect of 17β-estradiol on intestinal calcium absorption in rats. Bone Miner **26**:181–189.
145. Ash SL, Goldin BR 1988 Effects of age and estrogen on renal vitamin D metabolism in the female rat. Am J Clin Nutr **47**:694–699.
146. Ismail F, Epstein S, Fallon MD, Thomas SB, Reinhardt TA 1988 Serum bone gla protein and the vitamin D endocrine system in the oophorectomized rat. Endocrinology **122**:624–630.
147. Turton CWG, Stamp TCB, Stanley P, Maxwell JD 1977 Altered vitamin-D metabolism in pregnancy. Lancet **1**:222–224.
148. Bickle DD, Gee E, Halloran B, Haddad JG 1984 Free 1,25-dihydroxyvitamin D levels in serum from normal subjects, pregnant subjects, and subjects with liver disease. J Clin Invest **74**:1966–1971.
149. Gallagher JC, Riggs BL, Eisman J, Hamstra A, Arnaud SB, DeLuca HF 1979 Intestinal calcium absorption and serum vitamin D metabolites in normal subjects and osteoporotic patients. J Clin Invest **64**:729–736.
150. Lund B, Sørensen OH, Lund B, Agner E 1982 Serum 1,25-dihydroxyvitamin D in normal subjects and in patients with postmenopausal osteopenia. Influence of age, renal function and oestrogen therapy. Horm Metab Res **14**:271–274.
151. Falch JA, Oftebro H, Haug E 1987 Early postmenopausal bone loss is not associated with a decrease in circulating levels of 25-hydroxyvitamin D, 1,25-dihydroxyvitamin D, or vitamin D-binding protein. J Clin Endocrinol Metab **64**:836–841.
152. Gallagher JC 1990 The pathogenesis of osteoporosis. Bone Miner **9**:215–227.
153. Lóre F, Di Cairano G, Signorini AM, Caniggia A 1981 Serum levels of 25-hydroxyvitamin D in postmenopausal osteoporosis. Calcif Tissue Int **33**:467–470.
154. Gallagher JC, Riggs BL, DeLuca HF 1980 Effect of estrogen on calcium absorption and serum vitamin D metabolites in postmenopausal osteoporosis. J Clin Endocrinol Metab **51**:1359–1364.
155. van Hoof HJC, van der Mooren MJ, Swinkels LMJW, Rolland R, Benraad THJ 1994 Hormone replacement therapy increases serum 1,25-dihydroxyvitamin D: A 2-year prospective study. Calcif Tissue Int **55**:417–419.
156. Stock JL, Coderre JA, Mallette LE 1985 Effects of short course of estrogen on mineral metabolism in postmenopausal women. J Clin Endocrinol Metab **61**:595–600.
157. Bouillon R, Van Assche FA, Van Baelen H, Heyns W, De Moor P 1981 Influence of the vitamin D-binding protein on the serum concentration of 1,25-dihydroxyvitamin D_3. Significance of the free 1,25-dihydroxyvitamin D_3 concentration. J Clin Invest **67**:589–596.
158. Selby PL, Peacock M, Barkworth SA, Brown WB, Taylor GA 1985 Early effects of ethinyloestradiol and norethisterone treatment in post-menopausal women on bone resorption and calcium hormones. Clin Sci **69**:265–271.
159. Cheesma C, Grant BF, Marcus R 1989 Effects of estrogen on circulating "free" and total 1,25-dihydroxyvitamin D and on the parathyroid–vitamin D axis in postmenopausal women. J Clin Invest **83**:537–542.
160. Hagenfeldt Y, Eriksson H, Bjorkhem I 1989 Stimulatory effect of testosterone on renal 25-hydroxyvitamin D_3 1α-hydroxylase in guinea pig. Biochim Biophys Acta **1002**:84–88.
161. Ohata M, Sakagami Y, Fujita T 1977 Elevation of serum 25-hydroxycalciferol levels in androgen-treated and ultraviolet-irradiated rats. Endocrinol Jpn **24**:519–521.
162. Vanderschueren D, Van Herck E, Suiker AMH, Visser WJ, Schot LPC, Bouillon R 1992 Bone and mineral metabolism in aged male rats: Short and long term effects of androgen deficiency. Endocrinology **130**(No. 5):2906–2916.
163. Bouillon R, Vandoren G, Van Baelen H, De Moor P 1978 Immunochemical measurement of the vitamin D-binding protein in rat serum. Endocrinology **102**:1710–1715.
164. Nyomba BL, Bouillon R, DeMoor P 1987 Evidence for an interaction of insulin and sex steroids in the regulation of vitamin D metabolism in the rat. J Endocrinol **115**:295–301.
165. Hagenfeldt Y, Linde K, Sjoberg HE, Zumkeller W, Arver S 1992 Testsosterone increases serum 1,25-dihydroxyvitamin D and insulin-like growth factor-1 in hypogonadal men. Int J Androl **15**:93–102.
166. Morley JE, Perry III HM, Kaiser FE, Kraenzle D, Jensen J, Houston K, Mattammal M, Perry HM Jr 1993 Effects of testosterone replacement therapy in old hypogonadal males: A preliminary study. J Am Geriatr Soc **41**:149–152.
167. Jackson JA, Kleerekoper M, Parfitt M, Sudhaker Rao D, Villanueva AR, Frame B 1987 Bone histomorphometry in hypogonadal and eugonadal men with spinal osteoporosis. J Clin Endocrinol Metab **65**:53–58.
168. Finkelstein JS, Klibanski A, Neer RM, Doppelt SH, Rosenthal DI, Segre GV, Crowley WFJ 1989 Increase in bone density during treatment of men with idiopathic hypogonadotropic hypogonadism. J Clin Endocrinol Metab **69**(No. 4):776–783.
169. Francis RM, Peacock M, Aaron JE, Selby PL, Taylor GA, Thompson J, Marshall DH, Horsman A 1986 Osteoporosis in hypogonadal men: Role of decreased plasma 1,25-dihydroxyvitamin D, calcium malabsorption, and low bone formation. Bone **7**:261–268.
170. Hagenfeldt Y, Carlström K, Berlin T, Stege R 1991 Effects of orchidectomy and different modes of high dose estrogen treatment on circulating "free" and total 1,25-dihydroxyvitamin D in patients with prostatic cancer. J Steroid Biochem Mol Biol **39**:155–159.
171. Krabbe S, Hummer L, Christiansen C 1986 Serum levels of vitamin D metabolites and testosterone in male puberty. J Clin Endocrinol Metab **62**:503–507.
172. Aksnes L, Aarskog D 1982 Plasma concentrations of vitamin D metabolites in puberty: Effect of sexual maturation and implications for growth. J Clin Endocrinol Metab **55**:94–101.

173. Bickle DD, Halloran BP, Harris ST, Potale AA 1992 Progestin antagonism of estrogen stimulated 1,25-dihydroxyvitamin D levels. J Clin Endocrinol Metab **72**:519–523.
174. Grecu EO, Simmons R, Baylink DJ, Haloran BP, Spencer ME 1991 Effects of medroxyprogesterone acetate on some parameters of calcium metabolism in patients with glucocorticoid-induced osteoporosis. Bone Miner **13**:153–161.
175. Bouillon R, Muls E, De Moor P 1980 Influence of thyroid function on the serum concentration of 1,25-dihydroxyvitamin D_3. J Clin Endocrinol Metab **51**:793–797.
176. Jastrup B, Mosekilde L, Melsen F, Lund B, Lund B, Sørensen OH 1982 Serum levels of vitamin D metabolites and bone remodelling in hyperthyroidism. Metabolism **31**:126–132.
177. Macfarlane IA, Mawer EB, Berry J, Hann J 1982 Vitamin D metabolism in hyperthyroidism. Clin Endocrinol **17**:51–59.
178. Auwerx J, Bouillon R 1986 Mineral and bone metabolism in thyroid disease: A review. Q J Med **60**:737–752.
179. Mundy GR, Shapiro JL, Bandelin JG, Canalis EM, Raisz LG 1976 Direct stimulation of bone resorption by thyroid hormones. J Clin Invest **58**:529–534.
180. Mosekilde L, Melsen F, Bagger JP, Myhre-Jensen O, Sørensen NS 1977 Bone changes in hyperthyroidism: Interrelationships between bone morphometry, thyroid function and calcium-phosphorus metabolism. Acta Endocrinol **85**:515–525.
181. Bayley TA, Harrison JE, McNeill KG, Mernagh JR 1980 Effect of thyrotoxicosis and its treatment on bone mineral and muscle mass. J Clin Endocrinol Metab **50**:916–922.
182. Mosekilde L, Christensen MS 1977 Decreased parathyroid function in hyperthyroidism: Interrelationships between serum parathyroid hormone, calcium–phosphorus metabolism and thyroid function. Acta Endocrinol **84**:566–575.
183. Burman KD, Monchik JM, Earll JM, Wartofsky L 1976 Ionized and total serum calcium and parathyroid hormone in hyperthyroidism. Ann Int Med **84**:668–671.
184. Bouillon R, De Moor P 1973 Parathyroid function in patients with hyper-or hypothyroidism. J Clin Endocrinol Metab **38**:999–1004.
185. Shafer RB, Gregory DH 1972 Calcium malabsorption in hyperthyroidism. Gastroenterology **63**:235–239.
186. Peerenboom H, Keck E, Krüskemper HL, Strohmeyer G 1984 The defect of intestinal calcium transport in hyperthyroidism and its response to therapy. J Clin Endocrinol Metab **59**:936–940.
187. Velentzas C, Oreopoulos DG, From G, Porret B, Rapoport A 1977 Vitamin-D levels in thyrotoxicosis. Lancet **1**:370–371.
188. Mosekilde L, Lund B, Sørensen OH, Christensen MS, Melsen F 1977 Serum-25-hydroxycholecalciferol in hyperthyroidism. Lancet **1**:806–807.
189. Karsenty G, Bouchard P, Ulmann A, Schaison G 1985 Elevated metabolic clearance rate of 1α,25-dihydroxyvitamin D_3 in hyperthyroidism. Acta Endocrinol **110**:70–74.
190. Krane SM, Brownell GL, Stanbury JB, Corrigan H 1956 The effect of thyroid disease on calcium metabolism in man. J Clin Invest **35**:874–897.
191. Bijlsma JWJ, Duursma SA, Roelofs JMM, der Kinderen PJ 1983 Thyroid function and bone turnover. Acta Endocrinol **104**:42–49.
192. Weisman Y, Eisenberg Z, Lubelski R, Spirer Z, Edelstein S, Harell A 1981 Decreased 1,25-dihydroxycholecalciferol and increased 25-hydroxy- and 24,25-dihydroxycholecalciferol in tissues of rats treated with thyroxine. Calcif Tissue Int **33**:445–447.
193. Pahuja DN, De Luca HF 1982 Thyroid hormone and vitamin D metabolism in the rat. Arch Biochem Biophys **213**:293–298.
194. Kano K, Jones G 1984 Direct in vitro effect of thyroid hormones on 25-hydroxyvitamin D_3 metabolism in the perfused rat kidney. Endocrinology **114**:330–336.
195. Miller ML, Ghazarian JG 1981 Differential response of kidney mitochondrial calcium-regulating mixed functional oxidase to thyrotoxicosis. J Biol Chem **256**:5643–5645.
196. Trechsel U, Taylor CM, Bonjour J-P, Fleisch H 1980 Influence of prostaglandins and of cyclic nucleotides on the metabolism of 25-hydroxyvitamin D_3 in primary chick kidney cell culture. Biochem Biophys Res Commun **93**:1210–1216.
197. Biddulph DM, Currie MG, Wrenn RW 1979 Effects of interactions of parathyroid hormone and prostaglandins on adenosine 3′,5′-monophosphate concentrations in isolated renal tubules. Endocrinology **104**:1164–1171.
198. Wark JD, Larkins RG, Eisman JA, Wilson KR 1981 Regulation of 25-hydroxy-vitamin D-1α-hydroxylase in chick isolated renal tubules: Effects of prostaglandin E_2, frusemide and acetylsalicyclic acid. Clin Sci **61**:53–59.
199. Kurose H, Sonn YM, Jafari A, Birge SJ, Avioli LV 1985 Effects of prostaglandin E_2 and indomethacin on 25-hydroxyvitamin D_3-1α-hydroxylase activity in isolated kidney cells of normal and streptozotocin-induced diabetic rats. Calcif Tissue Int **37**:625–629.
200. Yamada M, Matsumoto T, Takahashi N, Suda T, Ogata E 1983 Stimulatory effect of prostaglandin E_2 on 1α,25-dihydroxyvitamin D_3 synthesis in rats. Biochem J **216**:237–240.
201. Yamada M, Matsumoto T, Su K-W, Ogata E 1985 Inhibition of prostaglandin E_2 of renal effects of calcitonin in rats. Endocrinology **116**:693–697.
202. Sedrani SH, El-Banna AA 1987 Effect of indomethacin on plasma levels of vitamin D metabolites, oestradiol and progesterone in rabbits during early pregnancy. Comp Biochem Physiol **87A**:635–639.
203. Katz IA, Jee WSS, Joffe II, Stein B, Takizawa M, Jacobs TW, Setterberg R, Lin BY, Tang LY, Ke HZ, Zeng QQ, Berlin JA, Epstein S 1992 Prostaglandin E_2 alleviates cyclosporin A-induced bone loss in the rat. J Bone Miner Res **7**:1191–1200.
204. Boiskin I, Epstein S, Ismail F, Fallon MD, Levy W 1988 Long term administration of prostaglandin inhibitors in vivo fail to influence cartilage and bone mineral metabolism in the rat. Bone Miner **4**:27–36.
205. de Rovetto CR, Welch TR, Hug G, Clark KE, Bergstrom W 1989 Hypercalciuria with Bartter syndrome: Evidence for an abnormality of vitamin D metabolism. J Pediatr **115**:397–404.
206. Pryke AM, Duggan C, White CP, Posen S, Mason RS 1990 Tumor necrosis factor-alpha induces vitamin D-1-hydroxylase activity in normal human alveolar macrophages. J Cell Physiol **142**:652–656.
207. Bickle DD, Pillai S, Gee E, Hincenbergs M 1991 Tumor necrosis factor-α regulation of 1,25-dihydroxyvitamin D production by human keratinocytes. Endocrinology **129**:33–38.
208. Mayur N, Lewis S, Catherwood BD, Nanes MS 1993 Tumor necrosis factor α decreases 1,25-dihydroxyvitamin D_3 receptors in osteoblastic ROS 17/2.8 cells. J Bone Miner Res **8**:997–1003.
209. Koeffler HP, Reichel H, Bishop JE, Norman AW 1985 γ-Interferon stimulates production of 1,25-dihydroxyvitamin D_3 by normal human macrophages. Biochem Biophys Res Commun **127**:596–603.
210. Bickle DD, Pillai S, Gee E, Hincenbergs M 1989 Regulation of 1,25-dihydroxyvitamin D production in human keratinocytes by interferon-γ. Endocrinology **124**:655–660.
211. Reichel H, Koeffler HP, Norman AW 1987 Synthesis in vitro of 1,25-dihydroxyvitamin D_3 and 24,25-dihydroxyvitamin D_3 by interferon-γ-stimulated normal human bone marrow and alveolar macrophages. J Biol Chem **262**:10931–10937.
212. Robinson BWS, McLemore TL, Crystal RG 1985 Gamma interferon is spontaneously released by alveolar macrophages and

lung T lymphocytes in patients with pulmonary sarcoidosis. J Clin Invest **75**:1488–1495.
213. Mann GN, Jacobs TW, Buchinsky FJ, Armstrong EC, Li M, Ke HZ, Ma YF, Jee WSS, Epstein S 1994 Interferon-gamma causes loss of bone volume in vivo and fails to ameliorate cyclosporin A-induced osteopenia. Endocrinology **135**:1077–1083.
214. Kruse R 1968 Osteopathien bei antiepileptisher langzeittherapie. Monatsschrift fur Kinderheilkd **116**:378–381.
215. Ashworth B, Horn DB 1977 Evidence of osteomalacia in an outpatient group of adult epileptics. Epilepsia **18**:37–43.
216. Hahn TJ, Halstead LR 1979 Anticonvulsant drug-induced osteomalacia: Alterations in mineral metabolism and response to vitamin D_3 administration. Calcif Tissue Int **27**:13–18
217. Lau KHW, Nakade O, Barr B, Taylor AK, Hochin K, Baylink DJ 1995 Phenytoin increases markers of osteogenesis for the human species in vitro and in vivo. J Clin Endocrinol Metab **80**:2347–2353
218. Richens A, Rowe DJF 1970 Disturbance of calcium metabolism by anticonvulsant drugs. Br Med J **4**:73–76.
219. Hunter J, Maxwell JD, Stewart DA, Parsons V, Williams R 1971 Altered calcium metabolism in epileptic children on anticonvulsants. Br Med J **4**:202–204.
220. Weinstein RS, Bryce GF, Sappington LJ, King DW, Gallagher BB 1984 Decreased serum ionized calcium and normal vitamin D metabolite levels with anticonvulsant drug treatment. J Clin Endocrinol Metab **58**:1003–1009.
221. Gough H, Goggin T, Bissessar A, Baker M, Crowley M, Callaghan N 1986 A comparative study of the relative influence of different anticonvulsant drugs, UV exposure and diet on vitamin D and calcium metabolism in out-patients with epilepsy. Q J Med **59**:569–577.
222. Hahn TJ, Birge SJ, Scharp CR, Avioli LV 1972. Phenobarbital-induced alterations in vitamin D metabolism. J Clin Invest **51**:741–748.
223. Davie MWJ, Emberson CE, Lawson DEM, Roberts GE, Barnes JLC, Barnes ND, Heeley AF 1983 Low plasma 25-hydroxyvitamin D and serum calcium levels in institutionalized epileptic subjects: Associated risk factors, consequences and response to treatment with vitamin D. Q J Med **205**:79–91.
224. Hahn TJ, Hendin BA, Scharp CR, Haddad JG 1972 Effect of chronic anticonvulsant therapy on serum 25-hydroxycalciferol levels in adults. N Engl J Med **287**:900–904.
225. Stamp TCB, Round JM, Rowe DJF, Haddad JG 1972 Plasma levels and therapeutic effect of 25-hydroxycholecalciferol in epileptic patients taking anticonvulsant drugs. Br Med J **4**:9–12.
226. Bouillon R, Reynaert J, Claes JH, Lissens W, De Moor P 1975 The effect of anticonvulsant therapy on serum levels of 25-hydroxy-vitamin D, calcium, and parathyroid hormone. J Clin Endocrinol Metab **41**:1130–1135.
227. Gascon-Barré M, Villeneuve J-P, Lebrun L-H 1984 Effect of increasing doses of phenytoin on the plasma 25-hydroxyvitamin D and 25-dihydroxyvitamin D and 1,25-dihydroxyvitamin D concentrations. J Am College Nutr **3**:45–50.
228. Välimäki M, Tiihonen M, Laitinen K, Tähtelä R, Kärkkäinen M, Lamberg-Allardt C, Mäkelä P, Tunninen R 1994 Bone mineral density measured by dual-energy x-ray absorptiometry and novel markers of bone formation and resorption in patients on antiepileptic drugs. J Bone Miner Res **9**:631–637.
229. Jubiz W, Haussler MR, McCain TA, Tolman KG 1977 Plasma 1,25-dihydroxyvitamin D levels in patients receiving anticonvulsant drugs. J Clin Endocrinol Metab **44**:617–621.
230. Hahn TJ, Hendin BA, Scharp CR, Boisseau VC, Haddad JG Jr 1975 Serum 25-hydroxycalciferol levels and bone mass in children on chronic anticonvulsant therapy. N Engl J Med **292**:550–554.
231. Takeshita N, Seino Y, Ishida H, Seino Y, Tanaka H, Tsutsumi C, Ogata K, Kiyohara K, Kato H, Nozawa M, Akiyama Y, Hara K, Imura H 1989 Increasing circulating levels of γ-carboxyglutamic acid-containing protein and decreased bone mass in children on anticonvulsant therapy. Calcif Tissue Int **44**:80–85.
232. Wark JD, Larkins RG, Perry-Keene D, Peter CT, Ross DL, Sloman JG 1979 Chronic diphenylhydantoin therapy does not reduce plasma 25-hydroxy-vitamin D. Clin Endocrinol **11**: 267–274.
233. Tjellesen L, Nilas L, Christiansen C 1983 Does carbamazepine cause disturbances in calcium metabolism in epileptic patients? Acta Neurol Scand **68**:13–19.
234. Bell RD, Pak CYC, Zerwekh J, Barilla DE, Vasko M 1979 Effect of phenytoin on bone and vitamin D metabolism. Ann Neurol **5**:374–378.
235. Mosekilde L, Christensen MS, Lund B, Sørensen OH, Melsen F 1977 The interrelationships between serum 25-hydroxycholecalciferol, serum parathyroid hormone and bone changes in anticonvulsant osteomalacia. Acta Endocrinol **84**:559–565.
236. Nielsen HE, Melsen F, Lund B, Sørensen OH, Mosekilde L 1983 Bone histomorphometry, vitamin D metabolites and calcium phosphate metabolism in anticonvulsant treatment with carbamazepine. Calcif Tissue Int **35**(Suppl.)224.
237. Hoikka V, Alhava EM, Karjalainen P, Keranen T, Savolainen KE, Riekkinen P, Korhonen R 1984 Carbamazepine and bone mineral metabolism. Acta Neurol Scand **69**:77–80.
238. Conney AH 1967 Pharmacological implications of microsomal enzyme induction. Pharmacol Rev **19**:317–366.
239. Eichelbaum M, Ekbom K, Bertilsson L, Ringer-Berger VA, Rane A 1975 Plasma kinetics of carbamazepine and its epoxide metabolite in man after single and multiple doses. Eur J Clin Pharmacol **8**:337–341.
240. Perucca E 1978 Clinical consequences of microsomal enzyme-induction by antiepileptic drugs. Pharmacol Ther **2**:285–314.
241. Silver J, Neale G, Thompson GR 1974 Effect of phenobarbitone treatment on vitamin D metabolism in mammals. Clin Sci Mol Med **46**:433–448.
242. Dent CE, Richens A, Rowe DJF, Stamp TCB 1970 Osteomalacia with long-term anticonvulsant therapy in epilepsy. Br Med J **4**:69–72.
243. Zerwekh JE, Homan R, Tindall R, Pak CYC 1982 Decreased 24,25-dihydroxyvitamin D concentration during long-term anticonvulsant therapy in adult epileptics. Ann Neurol **12**:184–186.
244. Weisman Y, Fattal A, Eisenberg Z, Harel S, Spirer Z, Harell A 1979 Decreased serum 24,25-dihydroxy vitamin D concentrations in children receiving chronic anticonvulsant therapy. Br Med J **2**:521–523.
245. Hahn TJ, Scharp CR, Avioli LV 1974 Effect of phenobarbital administration on the subcellular distribution of vitamin D_3-^3H in rat liver. Endocrinology **94**:1489–1495.
246. Hahn TJ, Halstead LR, Scharp CR, Haddad JG Jr 1975 Enhanced biotransformation and biologic efficacy of vitamin D following phenobarbital administration in the rat. Clin Res **23**:111A.
247. Ohta T, Wergedal JE, Gruber HE, Baylink DJ, Lau KHW 1995 Low dose phenytoin is an osteogenic agent in the rat. Calcif Tissue Int **56**:42–48.
248. Collins N, Maher J, Cole M, Baker M, Callaghan N 1991 A prospective study to evaluate the dose of vitamin D required to correct low 25-hydroxyvitamin D levels, calcium, and alkaline phosphatase in patients at risk of developing antiepileptic drug-induced osteomalacia. Q J Med **286**:113–122.
249. Morris HA, Need AG, O'Loughlin PD, Horowitz M, Bridges

A, Nordin BEC 1990 Malabsorption of calcium in corticosteroid-induced osteoporosis. Calcif Tissue Int **46**:305–308.

250. Hahn TJ, Halstead LR, Baran DT 1981 Effects of short term glucocorticoid administration on intestinal calcium absorption and circulating vitamin D metabolite concentrations in man. J Clin Endocrinol Metab **52**:111–115.
251. Klein RG, Arnaud SB, Gallagher JC, DeLuca HF, Riggs BL 1977 Intestinal calcium absorption in exogenous hypercortisonism. J Clin Invest **60**:253–259.
252. Kimberg DV, Baerg RD, Gershon E, Graudusius RT 1971 Effect of cortisone treatment on the active transport of calcium by the small intestine. J Clin Invest **50**:1309–1321.
253. Adams JS, Wahl TO, Lukert BP 1981 Effect of hydrochlorothiazide and dietary sodium restriction on calcium metabolism in corticosteroid treated patients. Metabolism **30**:217–221.
254. Suzuki Y, Ichikawa Y, Saito E, Homma M 1983 Importance of increased urinary calcium excretion in the development of secondary hyperparathyroidism of patients under glucocorticoid therapy. Metabolism **32**:151–156.
255. Seeman E, Kumar R, Hunder G, Scott M, Heath III H, Riggs BL 1980 Production, degradation, and circulating levels of 1,25-dihydroxyvitamin D in health and in glucocorticoid excess. J Clin Invest **66**:664–669.
256. Hahn TJ, Halstead LR, Teitelbaum SL, Hahn BH 1979 Altered mineral metabolism in glucocorticoid-induced osteopenia. J Clin Invest **64**:655.
257. Lukert BP, Adams JS 1976 Calcium and phosphorus homeostasis in man. Arch Intern Med **136**:1249–1253.
258. Avioli LV, Birge SJ, Lee SW 1968 Effects of prednisone on vitamin D metabolism in man. Clin Endocrinol **28**:1341–1346.
259. Jowell PS, Epstein S, Fallon MD, Reinhardt TA, Ismail F 1987. 1,25-Dihydroxyvitamin D_3 modulates glucocorticoid-induced alteration in serum bone gla protein and bone histomorphometry. Endocrinology **120**:531–536.
260. Favus MJ, Kimberg DV, Millar GN, Gershon E 1973 Effects of cortisone administration on the metabolism and localisation of 25-hydroxycholecalciferol in the rat. J Clin Invest **52**:1328–1335.
261. Hahn TJ, Halstead LR, Haddad JG Jr 1977 Serum 25-hydroxyvitamin D concentrations in patients receiving chronic corticosteroid therapy. J Clin Lab Med **90**:399–404.
262. LoCascio V, Bonucci E, Imbimbo B, Ballanti P, Adami S, Milani S, Tartarotti D, DellaRocca C 1990 Bone loss in response to long-term glucocorticoid therapy. Bone Miner **8**:39–51.
263. Prummel MF, Wiersinga WM, Lips P, Sanders GTP, Sauerwein HP 1991 The course of biochemical parameters of bone turnover during treatment with corticosteroids. J Clin Endocrinol Metab **72**:382–386.
264. Zerwekh JE, Emkey RD, Harris ED Jr 1984 Low-dose prednisone therapy in rheumatoid arthritis: Effect on vitamin D metabolism. Arth Rheum **27**:1050–1052.
265. Slovik DM, Neer RM, Ohman JL, Lowell FC, Clark MB, Segre GV, Potts JT Jr 1980 Parathyroid hormone and 25-hydroxyvitamin D levels in glucocorticoid-treated patients. Clin Endocrinol **12**:243–248.
266. Hodsman AB, Toogood JH, Jennings B, Fraher LJ, Baskerville JC 1991 Differential effects of inhaled budesonide and oral prednisolone on serum osteocalcin. J Clin Endocrinol Metab **72**:530–540.
267. Jennings BH, Andersson KE, Johansson SA 1991 The assessment of the systemic effects of inhaled glucocorticosteroids. Eur J Clin Pharmacol **41**:11–16.
268. Findling JW, Adams ND, Lemann J Jr, Gray RW, Thomas CJ, Tyrrell JB 1982 Vitamin D metabolites and parathyroid hormone in Cushing's syndrome: Relationship to calcium and phosphorus homeostasis. J Clin Endocrinol Metab 54:1039–1044.
269. Aloia JF, Roginsky M, Ellis K, Shukla K, Chon S 1974 Skeletal metabolism and body composition in Cushing's syndrome. J Clin Endocrinol Metab **39**:981–985.
270. Carré M, Ayigbedé O, Miravet L, Rasmussen H 1974 The effect of prednisolone upon the metabolism and action of 25-hydroxy- and 1,25-dihydroxyvitamin D_3. Proc Natl Acad Sci USA **71**:2996–3000.
271. Chesney RW, Hamstra AJ, Mazess RB, DeLuca HF, O'Reagan S 1978 Reduction of serum-1,25-dihydroxyvitamin-D_3 in children receiving glucocorticoids. Lancet **2**:1123–1125.
272. O'Regan S, Chesney RW, Hamstra A, Eisman JA, O'Gorman AM, DeLuca HF 1979 Reduced serum 1,25-$(OH)_2$ vitamin D_3 levels in prednisone-treated adolescents with systemic lupus erythematosus. Acta Paediatr Scand **68**:109–111.
273. Bikle DD, Halloran B, Fong L, Steinbach L, Shellito J 1993 Elevated 1,25-dihydroxyvitamin D levels in patients with chronic obstructive pulmonary disease treated with prednisone. J Clin Endocrinol Metab **76**:456–461.
274. Lukert BP, Stanbury SW, Mawer EB 1973 Vitamin D and intestinal transport of calcium: Effects of prednisolone. Endocrinology **93**:718–722.
275. Braun JJ, Juttmann JR, Visser TJ, Birkenhäger JC 1982 Short-term effect of prednisone on serum 1,25-dihydroxyvitamin D in normal individuals and in hyper- and hypoparathyroidism. Clin Endocrinol **17**:21–28.
276. Au WYW 1976 Cortisol stimulation of parathyroid hormone secretion by rat parathyroid glands in organ culture. Science **193**:1015–1017.
277. Williams GA, Peterson WC, Bowser EN, Henderson WJ, Hargis GK, Martinez NJ 1974 Interrelationship of parathyroid and adrenocortical function in calcium homeostasis in the rat. Endocrinology **95**:707–712.
278. Bjørneboe GEA, Johnsen J, Bjørneboe A, Rousseau B, Pedersen JI, Norum KR, Mørland J, Drevon CA 1986 Effect of alcohol consumption on serum concentration of 25-hydroxyvitamin D_3, retinol, and retinol-binding protein. Am J Clin Nutr **44**:678–682.
279. Feitelberg S, Epstein S, Ismail F, D'Amanda C 1987 Deranged bone mineral metabolism in chronic alcoholism. Metabolism **36**:322–326.
280. Mobarhan SA, Russell RM, Recker RR, Posner DB, Iber FL, Miller P 1984 Metabolic bone disease in alcoholic cirrhosis: A comparison of the effect of vitamin D_2, 25-hydroxyvitamin D, or supportive treatment. Hepatology **4**:266–273.
281. Bjørneboe GA, Johnsen J, Bjørneboe A, Mørland J, Drevon C 1987 Effect of heavy alcohol consumption on serum concentrations of fat-soluble vitamins and selenium. Alcohol and Alcoholism Suppl **1**:533–537.
282. Peris P, Parés A, Guañabens N, Del Río L, Pons F, Jesús M, De Osaba M, Monegal A, Caballería J, Rodés J, Muñoz-Gómez J 1994 Bone mass improves in alcoholics after 2 years of abstinence. J Bone Miner Res **9**:1607–1612.
283. Lalor BC, France MW, Powell D, Adams PH, Counihan TB 1986 Bone and mineral metabolism and chronic alcohol abuse. Q J Med **59**:497–511.
284. Verbanck M, Verbanck J, Brauman J, Mullier JP 1977 Bone histology and 25-OH vitamin D levels in alcoholics without cirrhosis. Calcif Tissue Res **22**(Suppl.)538–541.
285. Laitinen K, Lamberg-Allardt C, Tunninen R, Karonen S-L, Ylikahri R, Välimäki M 1991 Effects of 3 weeks' moderate alcohol intake on bone and mineral metabolism in normal men. Bone Miner **13**:139–151.
286. Pepersack T, Fuss M, Otero J, Bergmann P, Valsamis J, Corvilain

J 1992 Longitudinal study of bone metabolism after ethanol withdrawal in alcoholic patients. J Bone Miner Res **7**:383–387.

287. Barragry JM, Long RG, France MW, Wills MR, Boucher BJ, Sherlock S 1979 Intestinal absorption of cholecalciferol in alcoholic liver disease and primary biliary cirrhosis. Gut **20**:559–564.
288. Lindholm J, Steiniche T, Rasmussen E, Thamsborg G, Nielsen IO, Brockstedt-Rasmussen H, Storm T, Hyldstrup L, Schou C 1991 Bone disorder in men with chronic alcoholism: A reversible disease? J Clin Endocrinol Metab **73**:118–124.
289. Bjørneboe GE, Bjørneboe A, Johnsen J, Skylv N, Oftebro N, Gautvik KM, Hoiseth A, Morland J, Drevon CA 1988 Calcium status and calcium-regulating hormones in alcoholics. Alcohol Clin Exp Res **12**:229–232.
290. Meyer M, Wechsler S, Shibolet S, Jedwab M, Harell A, Edelsteim S 1978 Malabsorption of vitamin D in man and rat with liver cirrhosis. J Mol Med **3**:29–37.
291. Gascon-Barré M, Joly J-G 1981 The biliary excretion of [^{3}H]-25-hydroxyvitamin D_3 following chronic ethanol administration in the rat. Life Sci **28**:279–286.
292. Lund B, Sørensen OH, Hilden M, Lund B 1977 The hepatic conversion of vitamin D in alcoholics with varying degrees of liver affection. Acta Med Scand **202**:221–224.
293. Avioli LV, Lee SW, McDonald JE, Lund J, DeLuca HF 1967 Metabolism of D_3-^{3}H in human subjects. Distribution to blood, bile, feces and urine. J Clin Invest **46**:983–992.
294. Brodie MJ, Boobis AR, Hillyard CJ, Abeyasekera G, MacIntyre I, Park BK 1981 Effect of isoniazid on vitamin D metabolism and hepatic monooxygenase activity. Clin Pharmacol Ther **30**:363–367.
295. Posner DB, Russell RM, Absood S, Connor TB, Davis C, Martin L, Williams JB, Norris AH, Merchant C 1978 Effective 25-hydroxylation of vitamin D_2 in alcoholic cirrhosis. Gastroenterology **74**:866–870.
296. Bickle DD, Halloran BP, Gee E, Ryzen E, Haddad JG 1986 Free 25-hydroxyvitamin D levels are normal in subjects with liver disease and reduced total 25-hydroxyvitamin D levels. J Clin Invest **78**:748–752.
297. Bouillon R, Auwerx J, Dekeyser L, Fevery J, Lissens W, De Moor P 1984 Serum vitamin D metabolites and their binding protein in patients with liver cirrhosis. J Clin Endocrinol Metab **59**:86–89.
298. Shah JH, Bowser EN, Hargis GK, Wongsurawat N, Banerjee P, Henederson WJ, Williams GA 1978 Effect of ethanol on parathyroid hormone sectretion in the rat. Metabolism **27**:257–260.
299. Channard J, Lacour B, Drüeke T, Brunois JP, Ruiz JC 1980 Effect of acute ethanol loading on parathyroid gland secretion in the rat. Adv Exp Med Biol **128**:495–504.
300. Williams GA, Bowser EN, Hargis GK, Kukreja SC, Shah JH, Vora NM, Henderson WJ 1978 Effect of ethanol on parathyroid hormone and calcitonin secretion in man. Proc Soc Exp Biol Med **159**:187–191.
301. Ljunghall S, Lundin L, Wide L 1985 Acute effects of ethanol intake on the serum concentrations of parathyroid hormone, calcium and phosphate. Exp Clin Endocrinol **85**:365–368.
302. Bjørneboe GE, Bjørneboe A, Johnsen J, Skylv N, Oftebro H, Gautvik KM, Høiseth A, Mørland J, Drevon CA 1988 Calcium status and calcium regulating hormones in alcoholics. Alcoholism **12**:229–232.
303. Laitinen K, Välimäki M, Lamberg-Allardt C, Kivisaari L, Lalla M, Kärkkäinen M, Ylikahri R 1990 Deranged vitamin D metabolism but normal bone mineral density in Finnish non-cirrhotic male alcoholics. Alcoholism **14**:551–556.
304. Crilly RG, Anderson C, Hogan D, Delaquerrière-Richardson L 1988 Bone histomorphometry, bone mass, and related parameters in alcoholic males. Calcif Tissue Int **43**:269–276.
305. Bickle DD, Genant HK, Cann C, Recker RR, Halloran BP, Strewler GJ 1985 Bone disease in alcohol abuse. Ann Intern Med **103**:42–48.
306. Krawitt EL 1975 Effect of ethanol ingestion on duodenal calcium transport. J Clin Lab Med **85**:665–671.
307. Vodoz JF, Luisier M, Donath A, Courvoisier B, Garcia B 1977 Diminution de l'absorption intestinale de 47-calcium dans l'alcoolisme chronique. Schweiz Med Wochenschr **107**:1525–1529.
308. Avery DH, Overall JE, Calil HM, Hollister LE 1983 Plasma calcium and phosphate during alcohol intoxication. Alcoholics versus non-alcoholics. J Stud Alcohol **44**:205–214.
309. Krawitt EL 1973 Ethanol inhibits intestinal calcium transport in rats. Nature **243**:88–89.
310. Territo MC, Tanaka KR 1974 Hypophosphatemia in chronic alcoholism. Arch Intern Med **134**:445–447.
311. Kent JC, Devlin RD, Gutteridge DH, Retallack RW 1979 Effect of alcohol on renal vitamin D metabolism in chickens. Biochem Biophys Res Commun **89**:155–161.
312. Gascon-Barré M 1982 Plasma 25-hydroxyvitamin D_3 response to pharmacological dose of vitamin D_3 or 25-hydroxyvitamin D_3 during chronic ethanol administration in the rat. Horm Metab Res **14**:332–333.
313. Turner RT, Aloia RC, Segel LD, Hannon KS, Bell NH 1988 Chronic alcohol treatment results in disturbed vitamin D metabolism and skeletal abnormalities in the rat. Alcohol Clin Exp Res **12**:159–162.
314. Gascon-Barré M 1982 Interrelationships between vitamin D_3 and 25-hydroxyvitamin D_3 during chronic ethanol administration in the rat. Metabolism **31**:67–72.
315. Saarem K, Bergseth S, Oftebro H, Pedersen JI 1984 Subcellular localization of vitamin D_3 25-hydroxylase in human liver. J Biol Chem **259**:10936–10940.
316. Saarem K, Pedersen JI 1985 25-Hydroxylation of 1-hydroxyvitamin D_3 in rat and human liver. Biochim Biophys Acta **840**:117–126.
317. Laitinen K (ed) 1993 Alcohol and Bone. Yliopistopaino, Helsinki.
318. Gascon-Barré M 1985 Influence of chronic ethanol consumption on the metabolism and action of vitamin D. J Am College Nutr **4**:565–574.
319. Feldman D 1986 Ketoconazole and other imidizole derivatives as inhibitors of steroidogenesis. Endocr Rev **7**:409–420.
320. Loose DS, Kan PB, Hirst MA, Marcus RA, Feldman D 1983 Ketoconazole blocks adrenal steroidogenesis by inhibiting cytochrome P450-dependent enzymes. J Clin Invest **71**:1495–1499.
321. Henry HL 1985 Effect of ketoconazole and miconazole on 25-hydroxyvitamin D_3 metabolism by cultured chick kidney cells. J Steroid Biochem Mol Biol **23**:991–994.
322. Glass AR, Eil C 1986 Ketoconazol-induced reduction in serum 1,25-dihydroxyvitamin D. J Clin Endocrinol Metab **63**(No. 3):766–769.
323. Glass AR, Eil C 1988 Ketoconazole-induced reduction in serum 1,25-dihydroxyvitamin D and total serum calcium in hypercalcemic rats. J Clin Endocrinol Metab **66**:934–938.
324. Riancho JA, Amado JA, Freijanes J, Otero M, González Marcias J 1989 Ketoconazole and vitamin D metabolism in hyperparathyroidism. Horm Metab Res **21**:51.
325. Adams JS, Sharma OP, Diz MM, Endres DB 1990 Ketoconazole decreases the serum 1,25-dihydroxyvitamin D and calcium concentration in sarcoidosis-associated hypercalcemia. J Clin Endocrinol Metab **70**:1090–1095.
326. Glass AR, Cerletty JM, Elliott W, Lemann J Jr, Gray RW, Eil

C 1990 Ketoconazole reduces elevated serum levels of 1,25-dihydroxyvitamin D in hypercalcemic sarcoidosis. J Endocrinol Invest **13**:407–413.
327. Pont A, Williams PL, Azhar S, Reitz RE, Bochra C, Smith ER, Stevens DA 1982 Ketoconazole blocks testosterone synthesis. Arch Intern Med **142**:2137–2140.
328. Ismail F, Corder CN, Epstein S, Barbi G, Thomas S 1990 Effects of pravastatin and cholestyramine on circulating levels of parathyroid hormone and vitamin D metabolites. Clin Ther **12**:427–430.
329. Dobs A, Levine MA, Margolis S 1991 Effects of pravastatin, a new HMG-CoA reductase inhibitor, on vitamin D synthesis in man. Metabolism **40**:524–528.
330. Montagnani M, Loré F, Di Cairano G, Gonnelli S, Ciuoli C, Montagnani A, Gennari C 1994 Effects of pravastatin treatment on vitamin D metabolites. Clin Ther **16**:824–829.
331. Gascon-Barré M 1982 Biliary excretion of [^{3}H]-25-hydroxyvitamin D_3 in the vitamin D-depleted rat. Am J Physiol **242**:G522–G532.
332. Clements MR, Chalmers TM, Fraser DR 1984 Enterohepatic circulation of vitamin D: A repraisal of the hypothesis. Lancet **1**:1376–1379.
333. Kumar R, Nagubandi S, Mattox VR, Londowski JM 1980 Enterohepatic physiology of 1,25-dihydroxyvitamin D_3. J Clin Invest **65**:277–284.
334. Thompson WG, Thompson GR 1969 Effect of cholestyramine on the absorption of vitamin D_3 and calcium. Gut **10**:717–722.
335. Heaton KW, Lever JV, Barnard D 1972 Osteomalacia associated with cholestyramine therapy for posileectomy diarrhea. Gastroenterology **62**:642–646.
336. Compston JE, Horton LWL 1978 Oral 25-hydroxyvitamin D_3 in treatment of osteomalacia associated with ileal resection and cholestyramine therapy. Gastroenterology **74**:900–902.
337. Compston JE, Thompson RPH 1977 Intestinal absorption of 25-hydroxyvitamin D and osteomalacia in primary biliary cirrhosis. Lancet **1**:721–724.
338. Hoogwerf BJ, Hibbard DM, Hunninghake DB 1992 Effects of long-term cholestyramine administration on vitamin D and parathormone levels in middle-aged men with hypercholesterolemia. J Clin Lab Med **119**:407–411.
339. Schwarz KB, Goldstein PD, Witztum JL, Schonfeld G 1980 Fat-soluble vitamin concentrations in hypercholesterolemic children treated with colestipol. Pediatrics **65**:243–250.
340. Tsang RC, Roginsky MS, Mellies M, Glueck CJ, Kashyap ML 1977 Plasma 25 OH-vitamin D: Familial hypercholesterolemic children on colestipol resin. Clin Res **35**:567A.
341. Wilczek H, Sobra J, Ceska R, Justova V, Prochazkova R, Kvasilova M, Juzova Z 1993 Lecba fibraty a metabolismus vitaminu D. Casopis Lekaru Ceskych **132**:630–632.
342. Fleisch H, Russell RGG, Francis MD 1969 Diphosphonates inhibit hydroxyapatite dissolution in vitro and bone resorption in tissue culture and in vivo. Science **165**:1262–1264.
343. Russell RG, Mühlbauer RC, Bisaz S, Williams DA, Fleisch H 1970 The influence of pyrophosphate, condensed phosphates, phosphonates and other phosphate compounds on the dissolution of hydroxyapatite in vitro and on bone resorption induced by parathyroid hormone in tissue culture and in thyroparathyroidectomized rats. Calcif Tissue Res **6**:183–196.
344. Gasser AB, Morgan DB, Fleisch H, Richelle LJ 1972 The influence of two diphosphonates on calcium metabolism in the rat. Clin Sci **43**:31–45.
345. Kanis JA, Gertz BJ, Singer F, Ortolani S 1995 Rationale for the use of alendronate in osteoporosis. Osteoporosis Int **5**:1–13.
346. Rossini M, Gatti D, Zamberlan N, Braga V, Dorizzi R, Adami S 1994 Long-term effects of a treatment course with oral alendronate on postmenopausal osteoporosis. J Bone Miner Res **9**(11):1833–1837.
347. O'Doherty DP, Bickerstaff DR, McCloskey EV, Hamdy NAT, Beneton MNC, Harris S, Mian M, Kanis JA 1990 Treatment of Paget's disease of bone with aminohydroxybutylidene bisphosphonate. J Bone Miner Res **5**:483–491.
348. Trechsel U, Taylor CM, Eisman JA, Bonjour JP, Fleisch H 1981 Plasma levels of vitamin D metabolites in diphosphonate-treated rats. Clin Sci **61**:471–476.
349. Hill LF, Lumb GA, Mawer EB, Stanbury SW 1973 Indirect inhibition of biosynthesis of 1,25-dihydroxycholecalciferol in rats treated with diphosphonate. Clin Sci **44**:335–347.
350. Nagao Y, Ishitobi Y, Kinoshita H, Fukushima S, Kawashima H 1991 YM175, a new bisphosphonate, increases serum 1,25-dihydroxyvitamin D in rats via stimulating renal 1-hydroxylase activity. Biochem Biophys Res Commun **180**:1172–1178.
351. Ralston SH, Boyce BF, Cowan RA, Fogelman I, Smith ML, Jenkins A, Boyle IT 1987 The effect of 1α-hydroxyvitamin D_3 on the mineralisation defect in disodium-etidronate-treated Paget's disease—A double-blind randomised clinical study. J Bone Miner Res **2**:5–12.
352. Lawson-Matthew PJ, Guilland-Cummings DF, Yates AJP, Russell RGG, Kanis JA 1988 Contrasting effects of intravenous and oral etidronate on vitamin D metabolism. Clin Sci **74**:101–106.
353. Devlin RD, Retallack RW, Fenton AJ, Grill V, Gutteridge DH, Kent GN, Prince RL, Worth GK 1994 Long-term elevation of 1,25 dihydroxyvitamin D after short-term intravenous administration of pamidronate (aminohydroxypropylidene bisphosphonate, APD) in Paget's disease of bone. J Bone Miner Res **9**:81–85.
354. Papapoulos SE, Frolich M, Mudde AH, Harinck HIJ, Berg HVD, Bijvoet OLM 1987 Serum osteocalcin in Paget's disease of bone: Basal concentrations and response to bisphosphonate treatment. J Clin Endocrinol Metab **65**:89–94.
355. Adami S, Frijlink WB, Bijvoet OLM, O'Riordan JLH, Clemens TL, Papapoulos SE 1982 Regulation of calcium absorption by 1,25-dihydroxy-vitamin D—Studies of the effects of a bisphosphonate treatment. Calcif Tissue Int **34**:317–320.
356. Body JJ, Magritte A, Seraj F, Sculier JP, Borkowski A 1989 Aminohydroxypropylidene bisphosphonate (APD) treatment for tumor-associated hypercalcemia: A randomized comparison between a 3-day treatment and single 24-hour infusions. J Bone Miner Res **4**:923–928.
357. Budayr AA, Zysset E, Jenzer A, Thiébaud D, Ammann P, Rizzoli R, Jaquet-Müller F, Bonjour JP, Gertz B, Burckhardt P, Halloran BP, Nissenson RA, Strewler GJ 1994 Effects of treatment of malignancy-associated hypercalcemia on serum parathyroid hormone-related protein. J Bone Miner Res **9**:521–526.
358. Grill V, Murray RML, Ho PWM, Santamaria JD, Pitt P, Potts C, Jerums G, Martin TJ 1992 Circulating PTH and PTHrP levels before and after treatment of tumor induced hypercalcemia with pamidronate disodium (APD). J Clin Endocrinol Metab **74**:1468–1470.
359. Fraser WD, Logue FC, Gallacher SJ, O'Reilly DSJ, Beastall GH, Ralston SH, Boyle IT 1991 Direct and indirect assessment of the parathyroid hormone response to pamidronate therapy in Paget's disease of bone and hypercalcemia of malignancy. Bone Miner **12**:113–121.
360. Sahni M, Guenther HL, Fleisch H, Collin P, Martin TJ 1993 Bisphosphonates act on rat bone resorption through the mediation of osteblasts. J Clin Invest **91**:2004–2011.
361. Chesnut CH, McClung MR, Ensrud KE, Bell NH, Genant HK, Harris ST, Singer FR, Stock JL, Yood RA, Delmas PD, Kher U, Pryor-Tillotson S, Santora II AC 1995 Alendronate treatment

of the postmenopausal osteoporotic woman: Effect of multiple dosages on bone mass and bone remodeling. Am J Med **99**:144–152.

362. Harris ST, Gertz BJ, Genant HK, Eyre DR, Survill TT, Ventura JN, DeBrock J, Ricerca E, Chesnut III CH 1993 The effect of short term treatment with alendronate on vertebral density and biochemical markers of bone remodeling in early postmenopausal women. J Clin Endocrinol Metab **76**(No. 6):1399–1406.
363. Wasnich RD, Benfante RJ, Yano K, Heilbrun L, Vogel JM 1983 Thiazide effect on the mineral content of bone. N Engl J Med **309**:344–347.
364. Sakhaee K, Nicar MJ, Glass K, Zerwekh JE, Pak CYC 1984 Reduction in intestinal calcium absorption by hydrochorothiazide in postmenopausal osteoporosis. J Clin Endocrinol Metab **59**:1037–1043.
365. Wasnich R, Davis J, Ross P, Vogel J 1990 Effect of thiazide on rates of bone mineral loss: A longitudinal study. Br Med J **301**:1303–1305.
366. Dawson-Hughes B, Harris S 1993 Thiazides and seasonal bone changes in healthy postmenopausal women. Bone Miner **21**:41–51.
367. LaCroix AZ, Wienphal J, White LR, Wallace RB, Scherr PA, George LK, Cornoni-Huntley J, Ostfeld AM 1990 Thiazide diuretic agents and the incidence of hip fractures. N Engl J Med **322**:286–290.
368. Ray WA, Downey W, Griffin MR, Melton III LJ 1989 Long-term use of thiazide diuretics and risk of hip fracture. Lancet **1**:687–690.
369. Felson DT, Sloutskis D, Anderson JJ, Anthony JM, Kiel DP 1991 Thiazide diuretics and the risk of hip fracture. JAMA **265**:370–373.
370. Middler S, Pak CYC, Murad F, Bartter C 1973 Thiazide diuretics and calcium metabolism. Metabolism **22**:139–146.
371. Ljunghall S, Backman U, Danielson BG, Fellström B, Johansson G, Wikström B 1981 Calcium and magnesium metabolism during long-term treatment with thiazides. Scand J Urol Nephrol **15**:257–262.
372. Lamberg B-A, Kuhlback B 1959 Effect of chorothiazide and hydrochlorothiazide on the excretion of calcium in urine. Scand J Clin Lab Invest **11**:351–357.
373. Stote RM, Smith LH, Wilson DM, Dube WJ, Goldsmith RS, Arnaud CD 1972 Hydrochorothiazide effects on serum calcium and immunoreactive parathyroid hormone concentrations. Ann Intern Med **77**:587–591.
374. Sowers MR, Wallace RB, Hollis BW 1990 The relationship of 1,25-dihydroxyvitamin D and radial bone mass. Bone Miner **10**:139–148.
375. Zerwekh JE, Pak CYC 1980 Selective effects of thiazide therapy on serum 1α,25-dihydroxyvitamin D and intestinal calcium absorption in renal and absorptive hypercalciurias. Metabolism **29**:13–17.
376. Riis B, Christiansen C 1985 Actions of thiazides on vitamin D metabolism: A controlled therapeutic trial in normal women early in the postmenopause. Metabolism **34**:421–424.
377. Ramp WK, Cooper CW, Ross III AJ, Wells SA Jr 1979 Effects of calcium and cyclic nucleotides on rat calcitonin and parathyroid hormone secretion. Mol Cell Endocrinol **14**:205–215.
378. Larsson R, Åkerström G, Gylfe E, Johansson H, Ljunghall S, Rastad J, Wallfelt C 1985 Paradoxical effects of K^+ and D-600 on parathyroid hormone secretion and cytoplasmic Ca^{2+} in normal bovine and pathological human parathyroid cells. Biochim Biophys Acta **847**:263–269.
379. Hove K, Sand O 1981 Evidence for a function of calcium influx in the stimulation of hormone release from the parathyroid gland in the goat. Acta Physiol Scand **113**:37–43.
380. Seely EW, LeBoff S, Brown EM, Chen C, Posillico JT, Hollenberg NK, Williams GH 1989 The calcium channel blocker diltiazem lowers serum parathyroid hormone levels in vivo and in vitro. J Clin Endocrinol Metab **68**:1007–1012.
381. Cooper CW, Borosky SA, Farrell PE, Steinsland S 1986 Effects of the calcium channel activator BAY-K-8644 on in vitro secretion of calcitonin and parathyroid hormone. Endocrinology **118**:545–549.
382. Ross III AJ, Cooper CW, Ramp WK, Wells Jr SA 1979 Concurrent secretion of calcitonin and parathyroid hormone in vitro: Effects of drugs that alter cellular calcium transport. Surg Forum **30**:102–104.
383. Fox J, Della-Santina CP 1989 Oral verapamil and calcium and vitamin D metabolism in rats: Effect of dietary calcium. Am J Physiol **257**:E632–E638.
384. Bogin E, Chagnac A, Jüppner H, Levi J 1987 Effect of verapamil on plasma parathyroid hormone. J Clin Chem Clin Biochem **25**:83–85.
385. Frishman WH, Klein NA, Charlap S, Klein P, Norsratian F, Strom JA, Sherwood LM 1984 Comparative effects of verapamil and propanolol on parathyroid hormone and serum calcium concentration. In: Packer M, Frishman WH (eds) Calcium Channel Antagonists in Cardiovascular Disease. Appleton-Century-Crofts, Norwalk, Connecticut, pp. 1445–1449.
386. Gozzelino G, Fubini A, Isaia GG, Soagliotti G, Gamna G 1981 Evaluation of calcium metabolism and pituitary gonadotropic secretion during treatment with nifedipine. G Ital Cardiol **11**:1445–1449.
387. Townsend R, Dipette DJ, Evans RR, Davis WR, Green A, Graham GA, Wallace JM, Holland OB 1990 Effects of calcium channel blockade on calcium homeostasis in mild to moderate essential hypertension. Am J Med Sci **300**:133–137.
388. Albers MM, Johnson W, Vivian V, Jackson RD 1991 Chronic use of the calcium channel blocker nifedipine has no significant effect on bone metabolism in men. Bone **12**:39–42.
389. Renton KW 1985 Inhibition of hepatic microsomal drug metabolism by the caclium channel blockers diltiazem and verapamil. Biochem Pharmacol **34**:2549–2553.
390. Avioli LV 1975 Heparin-induced osteopenia: An appraisal. Adv Exp Med Biol **52**:375–387.
391. Sackler JP, Liu L 1973 Heparin-induced osteoporosis. Br J Radiol **46**:548–550.
392. Wise PH, Hall AJ 1980 Heparin-induced osteopenia in pregnancy. Br Med J **2**:110–111.
393. Griffiths HT, Liu DTY 1984 Severe heparin osteoporosis in pregnancy. Postgrad Med J **60**:424–425.
394. Mutoh S, Takeshita N, Yoshino T, Yamaguchi I 1993 Characterization of heparin-induced osteopenia in rats. Endocrinology **133**:2743–2748.
395. Mätzsch T, Bergqvist D, Hedner U, Nilsson B, Ostergaard P 1986 Heparin-induced osteoporosis in rats. Thromb Haemost **56**:293–294.
396. Monreal M, Viñas L, Monreal L, Lavin S, Lafoz E, Angles AM 1990 Heparin-related osteoporosis in rats. Haemostasis **20**:204–207.
397. Hurley MM, Gronowicz G, Kream BE, Raisz LG 1990 Effect of heparin on bone formation in cultured fetal rat calvaria. Calcif Tissue Int **46**:183–188.
398. Hurley MM, Kream BE, Raisz LG 1990 Structural determinants of the capacity of heparin to inhibit collagen synthesis in 21-day fetal rat calvariae. J Bone Miner Res **5**:1127–1133.

399. Chowdhury MH, Hamada C, Dempster DW 1992 Effects of heparin on osteoclast activity. J Bone Miner Res **7**:771–777.
400. Aarskog D, Aksnes L, Lehmann V 1980 Low 1,25-dihydroxyvitamin D in heparin-induced osteopenia. Lancet **2**:650–651.
401. Haram K, Hervig T, Thordarson H, Aksnes L 1993 Osteopenia caused by heparin treatment in pregnancy. Acta Obstet Gynecol Scand **72**:674–675.
402. Hahnemann S 1965 Heparin and osteoporosis. Lancet **2**:855–856.
403. Ellis HA 1965 Effects of long-term administration to animals of dextran sulfate. J Pathol Bacteriol **89**:437–460.
404. Williams GA, Longley RS, Bowser EN, Hargis GK, Kukreja SC, Vora NM, Johnson PA, Jackson BL, Kawahara WJ, Henderson WJ 1981 Parathyroid hormone secretion in normal man and in primary hyperparathyroidism: Role of histamine H_2 receptors. J Clin Endocrinol Metab **52**:122–127.
405. Abboud HE, Zimmerman D, Edis AJ, Dousa TP 1980 Histamine and parathyroid adenoma: Effect on cyclic AMP accumulation. Clin Res **28**:515A.
406. Sherwood JK, Ackroyd FW, Garcia M 1980 Effect of cimetidine on circulating parathyroid hormone in primary hyperparathyroidism. Lancet **1**:616–620.
407. Wiske PS, Epstein S, Norton JA Jr, Bell NH, Johnston CC Jr 1983 The effects of intravenous and oral cimetidine in primary hyperparathyroidism. Horm Metab Res **15**:245–248.
408. Jacob AI, Lanier D Jr, Canterbury J, Bourgoignie JJ 1980 Reduction by cimetidine of serum parathyroid hormone levels in uremic patients. N Engl J Med **302**:671–674.
409. Beehler CJ, Beckner JR, Rosenquist RC, Shankel SW 1980 Parathyroid hormone suppression by cimetidine in uremic patients. Ann Intern Med **93**:840–841.
410. Palmer FJ, Sawywers TM, Wierzbinski SJ 1980 Cimetidine and hyperparathyroidism. N Engl J Med **302**:692.
411. Ghishan FK, Walker F, Meneely R, Patwardhan R, Speeg KV Jr 1981 Intestinal calcium transport: Effect of cimetidine. J Nutr **111**:2157–2161.
412. Speeg KV Jr, Patwardhan RV, Avant GR, Mitchell MC, Schenker S 1982 Inhibition of microsomal drug metabolism by histamine H_2-receptor antagonists studied in vivo and in vitro in rodents. Gastroenterology **82**:89–96.
413. Henry DA, Macdonald IA, Kitchingman G, Bell GD, Langman MJS 1980 Cimetidine and ranitidine: Comparison of effects on hepatic drug metabolism. Br Med J **281**:775–777.
414. Bengoa JM, Bolt MJ, Rosenberg IH 1984 Hepatic vitamin D 25-hydroxylase inhibition by cimetidine and isoniazid. J Clin Lab Med **104**:546–552.
415. Wyatt C, Jensen LS, Rowland III GN 1990 Effect of cimetidine on eggshell quality and plasma 25-hydroxycholecalciferol in laying hens. Poultry Sci **69**:1892–1899.
416. Odes HS, Fraser GM, Krugliak P, Lamprecht SA, Shany S 1990 Effect of cimetidine on hepatic vitamin D metabolism in humans. Digestion **46**:61–64.
417. Klein GL, Horst RL, Norman AW, Ament ME, Slatopolsky E, Coburn JW 1981 Reduced serum levels of 1α,25-dihydroxyvitamin D during long-term total parenteral nutrition. Ann Intern Med **94**:638–643.
418. Klein GL, Alfrey AC, Miller NL, Sherrard DJ, Hazlet TK, Ament ME, Coburn JW 1982 Aluminum loading during parenteral nutrition. Am J Clin Nutr **35**:1425–1429.
419. Ott SM, Maloney NA, Klein GL, Alfrey AC, Ament ME, Coburn JW, Sherrard DJ 1983 Aluminum is associated with low bone formation in patients receiving chronic parenteral nutrition. Ann Intern Med **98**:910–914.
420. Parkinson IS, Feest TG, Ward MK, Fawcett RWP, Kerr DNS 1979 Fracturing dialysis osteodystrophy and dialysis encephelopathy. Lancet **1**:406–409.
421. Pierides AM, Edwards WG Jr, Cullum UX Jr, McCall JT, Ellis HA 1980 Hemodialysis encephalopathy with osteomalacic fractures and muscle weakness. Kidney Int **18**:115–124.
422. Ott SM, Maloney NA, Coburn JW, Alfrey AC, Sherrard DJ 1982 The prevalence of bone aluminum deposition in renal osteodystrophy and its relation to the response to calcitriol therapy. N Engl J Med **307**:709–713.
423. Maloney NA, Ott SM, Alfrey AC, Miller NL, Coburn JW, Sherrard DJ 1982 Histological quantitation of aluminum in iliac bone from patients with renal failure. J Clin Lab Med **99**:206–216.
424. Cournot-Witmer G, Zingraff J, Plachot JJ, Escaig F, Lefèvre R, Boumati P, Bourdeau A, Garabédian M, Galle P, Bourdon R, Drüeke T, Balsan S 1981 Aluminum localization in bone from hemodialyzed patients: Relationship to matrix mineralization. Kidney Int **20**:375–385.
425. Ellis HA, McCarthy JH, Herrington J 1979 Bone aluminium in haemodialysed patients and in rats injected with aluminium chloride: Relationship to impaired bone mineralisation. J Clin Pathol **32**:832–844.
426. Hodsman AB, Sherrard DJ, Alfrey AC, Ott S, Brickman AS, Miller NS, Maloney NA, Coburn JW 1982 Bone aluminum and histomorphometric features of renal osteodystrophy. J Clin Endocrinol Metab **54**:539–546.
427. Alfrey AC, Hegg A, Craswell P 1980 Metabolism and toxicity of aluminum in renal failure. Am J Clin Nutr **33**:1509–1516.
428. Cann CE, Prussin SG, Gordan GS 1979 Aluminum uptake by the parathyroid glands. J Clin Endocrinol Metab **49**:543–545.
429. Morrissey J, Rothstein M, Mayor G, Slatopolsky E 1983 Suppression of parathyroid hormone secretion by aluminum. Kidney Int **23**:699–704.
430. Robertson JA, Felsenfeld AJ, Haygood CC, Wilson P, Clarke C, Llach F 1983 Animal model of aluminum-induced osteomalacia: Role of chronic renal failure. Kidney Int **23**:327–335.
431. Chan YL, Alfrey AC, Posen S, Lissner D, Hills E, Dunstan CR, Evans RA 1983 Effect of aluminum on normal and uremic rats: Tissue distribution, vitamin D metabolites, and quantitative bone histology. Calcif Tissue Int **35**:344–351.
432. Henry DA, Goodman WG, Nudelman RK, DiDomenico NC, Alfrey AC, Slatopolsky E, Stanley TM, Coburn JW 1984 Parenteral aluminum administration in the dog: I. Plasma kinetics, tissue levels, calcium metabolism, and parathyroid hormone. Kidney Int **25**:362–369.
433. Goodman WG, Henry DA, Horst R, Nudelman RK, Alfrey AC, Coburn JW 1984 Parenteral aluminum administration in the dog: II. Induction of osteomalacia and effect on vitamin D metabolism. Kidney Int **25**:370–375.
434. Berlyne GM, Pest D, Ben-Ari J, Weinberger J, Stern M, Gilmore GR, Levine R 1970 Hyperaluminaemia from aluminum resins in renal failure. Lancet **2**:494–496.
435. Clarkson EM, Luck VA, Hynson WV, Bailey RR, Eastwood JB, Woodhead JS, Clements VR, O'Riordan JLH, De Wardener HE 1972 The effect of aluminum hydroxide on calcium, phosphorus and aluminum balances, the serum parathyroid hormone concentration and the aluminum content of bone in patients with chronic renal failure. Clin Sci **43**:519–531.
436. Recker RR, Blotcky AJ, Leffler JA, Rack EP 1977 Evidence for aluminum absorption from the gastrointestinal tract and bone deposition by aluminum carbonate ingestion with normal renal function. J Clin Lab Med **90**:810–815.
437. Villa ML, Packer E, Cheema M, Holloway L, Marcus R 1991 Effects of aluminum hydroxide on the parathyroid–vitamin D

axis of postmenopausal women J Clin Endocrinol Metab **73**:1256–1261.

438. Shah SC, Sharma RK, Chitle H, Chitle AR 1981 Rifampicin induced osteomalacia. Tubercle **62**:207–209.
439. Brodie MJ, Boobis AR, Dollery CT, Hillyard CJ, Brown DJ, MacIntyre I, Park BK 1980 Rifampicin and vitamin D metabolism. Clin Pharmacol Ther **27**:810–814.
440. Onhaus E, Park BK 1979 Measurement of urinary 6-β-hydroxycortisol excretion as an in vivo parameter in the clinical assessment of the microsomal enzyme-inducing capacity of antipyrine, phenobarbitone and rifampicin. Eur J Clin Pharmacol **15**:139–145.
441. Madhok TC, Schnoes HK, DeLuca HF 1978 Incorporation of oxygen-18 into the 25-position of cholecalciferol by hepatic cholecalciferol 25-hydroxylase. Biochem J **175**:479–482.
442. Kutt H, Louis S 1972 Anticonvulsant drugs. II. Clinical pharmacological and therapeutic aspects. Drugs **4**:256–282.
443. Kutt H, Verebely K, McDowell F 1968 Inhibition of diphenylhydantoin metabolism in rats and in rat liver microsomes by antitubercular drugs. Neurology **18**:706–710.
444. Ghazarian JG, DeLuca HF 1977 Kidney microsomal metabolism of 25-hydroxyvitamin D_3. Biochem Biophys Res Commun **75**:550–555.
445. Brodie MJ, Boobis AR, Hillyard CJ, Abeyasekera G, Stevenson JC, MacIntyre I, Park BK 1982 Effect of rifampicin and isoniazid on vitamin D metabolism. Clin Pharmacol Ther **32**:525–530.
446. Toppet M, Vainsel M, Vertongen F, Fuss M, Cantraine F 1988 Evolution sequentielle des metabolites de la vitamine D sous isoniazide et rifampicine. Arch Francaises Pediatr **45**:145–148.
447. Saggese G, Cesaretti G, Bertelloni S, Morganti E, Bottone E 1985 Isoniazid and vitamin D metabolism. In: Norman AW, Schaefer K, Grigoleit H-G, Herrath DV (eds): Vitamin D: Chemical, Biochemical, and Clinical Update. de Gruyter, Berlin, pp. 1123–1124.
448. Perry W, Brown J, Erooga MA, Stamp TCB 1982 Calcium metabolism during rifampicin and isoniazid therapy for tuberculosis. J R Soc Med **75**:533–536.
449. Williams SE, Wardman AG, Taylor GA, Peacock M, Cooke NJ 1985 Long term study of the effect of rifampicin and isoniazid on vitamin D metabolism. Tubercle **66**:49–54.
450. Daniell HW 1976 Osteoporosis of the slender smoker. Arch Intern Med **136**:298–304.
451. Heany RP, Recker RR 1982 Effects of nitrogen, phosphorus, and caffeine on calcium balance in women. J Clin Lab Med **99**:46–55.
452. Mitoma C, Lombrozo L, LeValley SE, Dehn F 1969 Nature of the effect of caffeine on the drug metabolizing enzymes. Arch Biochem Biophys **134**:434–441.
453. Thithapandha A, Chaturapit S, Limlomwongse L, Sobhon P 1974 The effects of xanthines on mouse liver cell. Arch Biochem Biophys **161**:178–186.
454. Taft JL, French M, Danks JA, Larkins RG 1984 Opposing actions of methylxanthines and dibutyryl cAMP on 1,25 dihydroxyvitamin D_3 production and calcium fluxes in isolated chick renal tubules. Biochem Biophys Res Commun **121**:355–363.
455. Yeh JK, Aloia JF, Semla HM, Chen SY 1986 Influence of injected caffeine on the metabolism of calcium and the retention and excretion of sodium, potassium, phosphorus, magnesium, zinc and copper in rats. J Nutr **116**:273–280.
456. Yeh JK, Aloia JF 1986 Differential effects of caffeine administration on calcium and vitamin D metabolism in young and adult rats. J Bone Miner Res **1**:251–258.
457. Glajchen N, Ismail F, Epstein S, Jowell PS, Fallon M 1988 The effect of chronic caffeine administration on serum markers of bone mineral metabolism and bone histomorphometry in the rat. Calcif Tissue Int **43**:277–280.
458. Kulkowski JA, Chan T, Martinez J, Ghazarian JG 1979 Modulation of 25-hydroxyvitamin D_3-24-hydroxylase by aminophylline: A cytochrome P-450 monooxygenase system. Biochem Biophys Res Commun **90**:50–57.
459. Pedersen JI, Shobaki HH, Holmberg I, Bergseth S, Björkhem I 1983 25-Hydroxyvitamin D_3-24-hydroxylase in rat kidney mitochondria. J Biol Chem **258**:742–746.
460. Colin A, Kraiem Z, Kahana KL, Hochberg Z 1984 Effects of theophylline on urinary excretion of cyclic AMP, calcium, and phosphorus. Miner Electrolyte Metab **10**:359–361.
461. McPherson ML, Prince SR, Atamer ER, Maxwell DB, Ross-Clunis H, Estep HL 1986 Theophylline-induced hypercalcemia. Ann Intern Med **105**:52–54.
462. Fortenbery EJ, McDermott MT, Duncan WE 1990 Effect of theophylline on calcium metabolism and circulating vitamin D metabolites. J Bone Miner Res **5**:321–324.
463. Lohmann SM, Miech RP 1976 Theophylline metabolism by the rat liver microsomal system. J Pharmacol Exp Ther **196**:213–225.
464. Movsowitz C, Epstein S, Fallon M, Ismail F, Thomas S 1988 Cyclosporin A in vivo produces severe osteopenia in the rat: Effect of dose and duration of administration. Endocrinology **123**:2571–2577.
465. Movsowitz C, Epstein S, Ismail F, Fallon M, Thomas S 1989 Cyclosporin A in the oophorectomized rat: Unexpected severe bone resorption. J Bone Miner Res **4**:393–398.
466. Schlosberg M, Movsowitz C, Epstein S, Fallon MD, Thomas S 1989 The effect of cyclosporin A administration and its withdrawal on bone mineral metabolism in the rat. Endocrinology **124**:2179–2184.
467. Stein B, Halloran BP, Reinhardt G, Engstrom GW, Bales CW, Drezner MK, Currie KL, Takizawa M, Adams JS, Epstein S 1991 Cyclosporin-A increases synthesis of 1,25-dihydroxyvitamin D_3 in the rat and mouse. Endocrinology **128**:1369–1373.
468. Katz I, Li M, Joffe I, Stein B, Jacobs T, Liang X, Ke H, Jee W, Epstein S 1994 The influence of age on cyclosporin A-induced alterations in bone mineral metabolism in the rat in vivo. J Bone Miner Res **9**:59–67.
469. Buchinsky FJ, Ma YF, Mann G, Rucinski B, Bryer HP, Romero DF, Jee WSS, Epstein S 1996 T-lymphocytes play a critical role in the development of cyclosporine induced osteopenia. Endocrinology **137**:2278–2285.
470. Rooth P, Dawidson I, Diller K, Clothier N 1988 In vivo fluorescence microscopy reveals cyclosporine G to be less nephrotoxic than cyclosporine A. Trans Proc **20**:707–709.
471. Tejani A, Lancman I, Pomarantz A, Khawar M, Chen C 1988 Nephrotoxicity of cyclosporine A and cyclosporine G in a rat model. Transplantation **45**:184–187.
472. Jacobs TW, Katz IA, Joffe II, Stein B, Takizawa M, Epstein S 1991 The effect of FK 506, cyclosporine A, and cyclosporine G on serum 1,25-dihydroxyvitamin D_3 levels. Trans Proc **23**(No. 6):3188–3189.
473. Stein B, Takizawa M, Schlosberg M, Movsowitz C, Fallon M, Berlin JA, Epstein S 1992 Evidence that cyclosporine G is less deleterious to rat bone in vivo than cyclosporine A. Transplantation **53**:628–632.
474. Cvetkovic M, Mann GN, Romero DF, Liang X, Ma YF, Jee WSS, Epstein S 1994 The deleterious effects of long-term cyclosporin A, cyclosporin G, and FK506 on bone mineral metabolism in vivo. Transplantation **57**:1231–1237.
475. Morris R 1994 Modes of action of FK506, cyclosporin A, and rapamycin. Trans Proc **26**(No. 6):3272–3275.
476. Bryer HP, Isserow JA, Armstrong EC, Mann GN, Rucinski B,

Buchinsky FJ, Romero DF, Epstein S 1995 Azathioprine alone in bone sparing and does not alter cyclosporin A-induced osteopenia in the rat. J Bone Miner Res **10**:132–138.
477. Romero DF, Buchinsky FJ, Rucinski B, Cvetkovic M, Bryer HP, Liang XG, Ma YF, Jee SS, Epstein S 1995 Rapamycin: A bone sparing immunosuppressant? J Bone Miner Res **10**(5):760–768.
478. Joffe I, Katz I, Sehgal S, Bex F, Kharode Y, Tamasi J, Epstein S 1993 Lack of change of cancellous bone volume with short term use of the new immunosuppressant rapamycin in rats. Calcif Tissue Int **53**:45–52.
479. Reichel H, Grübinger A, Knehans A, Kühn K, Schmidt-Gayk H, Ritz E 1992 Long-term therapy with cyclosporin A does not influence serum concentrations of vitamin D metabolites in patients with multiple sclerosis. Clin Invest **70**:595–599.
480. Hanley DA, Ayer LM, Gundberg CM, Minuk GY 1991 Parameters of calcium metabolism during a pilot study of cyclosporin A in patients with symptomatic primary biliary cirrhosis. Clin Invets Med **14**:282–287.
481. Saha HHT, Salmela KT, Ahonen PJ, Pietilä KO, Mörsky PJ, Mustonen JT, Lalla MLT, Pasternak AI 1994 Sequential changes in vitamin D and calcium metabolism after successful renal transplantation. Scand J Urol Nephrol **28**:21–27.
482. Riancho JA, de Francisco ALM, del Arco C, Amado JA, Cotorruelo JG, Arias M, Gonzalez-Marcias J 1988 Serum levels of 1,25-dihydroxyvitamin D after renal transplantation. Miner Electrolyte Metab **14**:332–337.
483. Valero MA, Loinaz C, Larrodera L, Leon M, Moreno E, Hawkins F 1995 Calcitonin and bisphosphonates treatment in bone loss after liver transplantation. Calcif Tissue Int **57**:15–19.
484. McDonald JA, Dunstan CR, Dilworth P, Sherbon K, Ross Sheil AG, Evans RA, McCaughan GW 1991 Bone loss after liver transplantation. Hepatology **14**:613–619.
485. Shane E, Rivas M, Staron RB, Silverberg SJ, Seibel MJ, Kuiper J, Mancini D, Addresso V, Michler RE, Factor-Litvak P 1996 Fracture after cardiac transplantation: A prospective longitudinal study. J Clin Endocrinol Metab **81**:1740–1746.
486. Cundy T, Kanis JA, Heynen G, Morris PJ, Oliver DO 1983 Calcium metabolism and hyperparathyroidism after renal transplantation. Q J Med **205**:67–78.
487. Felsenfeld AJ, Gutman RA, Drezner M, Llach F 1986 Hypophosphatemia in long-term renal transplant recipients: Effects on bone histology and 1,25-dihydroxycholecalciferol. Miner Electrolyte Metab **12**:333–341.
488. Sakhee K, Brinker K, Helderman JH, Bengfort JL, Nicar MJ, Hull AR, Pak CYC 1985 Disturbances in mineral metabolism after successful renal transplantation. Miner Electrolyte Metab **11**:167–172.
489. Turner RT, Francis R, Brown D, Garand J, Hannon KS, Bell NH 1989 The effects of fluoride on bone and implant histomorphometry in growing rats. J Bone Miner Res **4**:477–484.
490. Manzke E, Rawley R, Vose G, Roginsky M, Rader JI, Baylink DJ 1977 Effect of fluoride therapy on nondialyzable urinary hydroxyproline, serum alkaline phosphatase, parathyroid hormone, and 25-hydroxyvitamin D. Metabolism **26**:1005–1010.
491. Dure-Smith BA, Farley SM, Linkhart SG, Farley JR, Baylink DJ 1996 Calcium deficiency in fluoride-treated osteoporotic patients despite calcium supplementation. J Clin Endocrinol Metab **81**:269–275.
492. Fallat RW, Glueck CJ, Lutmer R, Mattson FH 1976 Short term study of sucrose polyester a nonabsorbable fa-like material as a dietary agent for lowering plasma cholesterol. Am J Clin Nutr **29**:1204–1215.
493. Crouse JR, Grundy SM 1979 Effects of sucrose polyester on cholesterol metabolism in man. Metabolism **28**:994–1000.
494. Mattson FH, Volpenhein RA 1972 Rate and extent of absorption of the fatty acids of fully esterified glycerol, erythritol, xylitol, and sucrose as measured in thoracic duct cannulated rats. J Nutr **102**:1177–1180.
495. Mattson FH, Volpenhein RA 1972 Hydrolysis of fully esterified alcohols containing from one to eight hydroxyl groups by the lipolytic enzymes of rat pancreatic juice. J Lipid Res **13**:325–328.
496. Kuksis A 1987 Absorption of fat soluble vitamins In: Kuksis A (ed) Fat Absorption, Vol 2. CRC Press, Boca Raton, Florida, pp. 65–86.
497. Glueck CJ, Hastings MM, Allen C, Hogg E, Baehler E, Gartside PS, Phillips D, Jones M, Hollenboch EJ, Braun B, Anastasia JV 1982 Sucrose polyester and covert caloric dilution. Am J Clin Nutr **35**:1352–1359.
498. Mellies MJ, Vitale C, Jandacek RJ, Lamkin GE, Glueck CJ 1985 The substitution of sucrose polyester for dietary fat in obese, hypercholestrolemic outpatients. Am J Clin Nutr **41**:1–12.
499. Jones DY, Miller KW, Koonsvitsky BP, Ebert ML, Lin PYT, Jones MB, DeLuca HF 1991 Serum 25-hydroxyvitamin D concentrations of free-living subjects consuming olestra. Am J Clin Nutr **53**:1281–1287.
500. Mellies MJ, Jandacek RJ, Taulbee JD, Tweksbury MB, Lamkin G, Baehler L, King P, Boggs D, Goldman S, Gouge A, Tsang R, Glueck CJ 1983 A double-blind, placebo controlled study of sucrose polyester in hypercholesterolemic outpatients. Am J Clin Nutr **37**:339–346.
501. Miller KW, Wood FE, Stuard SB, Alden CL 1991 A 20-month olestra feeding study in dogs. Food Chem Toxic **29**:427–435.
502. Lawson DEM, Paul AA, Black AE, Cole TJ, Mandal AR, Davie M 1979 Relative contributions of diet and sunlight to vitamin D state in the elderly. Br Med J **2**:303–305.
503. Poskitt EME, Cole TJ, Lawson DEM 1979 Diet, sunlight and 25-hydroxyvitamin D in healthy children and adults. Br Med J **1**:221–223.
504. Haddad JG, Hahn TJ 1973 Natural and synthetic sources of circulating 25-hydroxyvitamin D in man. Nature **244**:515–516.
505. Mellerup ET, Lauritsen B, Dam H, Rafaelson OJ 1976 Lithium effects on diurnal rhythms of calcium, magnesium and phosphate metabolism in manic–melancholic disorder. Acta Psychiat Scand **53**:360–370.
506. Davis BM, Pfefferbaum A, Krutzik S, Davis KL 1981 Lithium's effect on parathyroid hormone. Am J Psychiatry **138**:489–492.
507. Plenge P, Rafaelsen OJ 1982 Lithium effects on calcium, magnesium and phosphate in man: Effects on balance, bone mineral content, faecal and urinary excretion. Acta Psychiat Scand **66**:361–373.
508. Mallette LE, Khouri K, Zengotita H, Hollis BW, Malini S 1989 Lithium treatment increases intact and midregion parathyroid hormone and parathyroid volume. J Clin Endocrinol Metab **68**:654–660.
509. Nielsen JL, Christiansen MS, Pedersen EB, Darling S, Amdisen A 1977 Parathyroid hormone in serum during lithium therapy. Scand J Clin Lab Invest **37**:369–372.
510. Miller PD, Dubovsky SL, McDonald KM, Arnaud C, Schrier RW 1978 Hypocalciuric effect of lithium on man. Miner Electrolyte Metab **1**:3–11.
511. Bjørum N, Hornum I, Mellerup ET, Plenge PK, Rafaelsen OJ 1975 Lithium, calcium, and phosphate. Lancet **1**:1243.
512. Birnbaum J, Klandorf H, Giuliano A, Van Herle A 1988 Lithium stimulates the release of human parathyroid hormone in vitro. J Clin Endocrinol Metab **66**:1187–1191.
513. Christiansen C, Baastrup PC, Transbøl I 1976 Lithium, hypercalcemia, hypermagnesaemia, and hyperparathyroidism. Lancet **1**:969.

514. Shen F-H, Sherrard DJ 1982 Lithium-induced hyperparathyroidism: An alteration of the "set point." Ann Intern Med **96**: 63–65.
515. Christiansen C, Baastrup PC, Lindgreen P, Transøl I 1978 Endocrine effects of lithium: II. Primary hyperparathyroidism. Acta Endocrinol **88**:528–534.
516. Christiansen C, Baastrup PC, Transbol I 1980 Development of primary hyperparathyroidism during lithium therapy. Neuropsychobiology **6**:280–283.
517. Nordenström J, Elvius M, Bågedahl-Strindlund M, Zhao B, Törring O 1994 Biochemical hyperparathyroidism and bone mineral status in patients treated long-term with lithium. Metabolism **43**:1563–1567.
518. Perez GO, Oster JR, Vaavioucle CA 1975 Incomplete syndrome of renal tubular acidosis induced by lithium carbonate. J Clin Lab Med **86**:386–394.
519. Nordenström J, Strigård K, Perbeck L, Willems J, Bågedahl-Strindlund M, Linder J 1992 Hyperparathyroidism associated with treatment of manic–depressive disorders by lithium. Eur J Surg **158**:207–211.
520. Mallette LE, Eichhorn E 1986 Effects of lithium carbonate on human calcium metabolism. Arch Intern Med **146**:770–776.
521. Hestbech J, Hansen HE, Amdisen A, Olsen S 1977 Chronic renal lesions following long-term treatment with lithium. Kidney Int **12**:205–213
522. Hansen HE, Hestbech J, Sørensen JL, Nørgaard K, Heiskov J, Amidisen A 1979 Chronic interstitial nephropathy in patients on long-term lithium treatment. Q J Med **192**:577–591.
523. Rosenblatt S, Chanley JD, Segal RL 1989 The effect of lithium on vitamin D metabolism. Biol Psychiatry **26**:206–208.

CHAPTER 51

Bone Disorders Associated with Gastrointestinal and Hepatobiliary Disease

E. BARBARA MAWER AND MICHAEL DAVIES

University of Manchester Bone Disease Research Center, Department of Medicine, Manchester Royal Infirmary, Manchester, United Kingdom

I. INTRODUCTION

The bone diseases associated with gastrointestinal and liver disease include both osteomalacia, caused by severe and prolonged deficiency of vitamin D which may more properly be described as vitamin D depletion, and, more commonly, osteoporosis, in which decreased bone density may be partially attributable to low calcium intake or absorption and excessive bone resorption that may be mediated by raised parathyroid hormone (PTH). The latter state may arise as a result of an insufficiency (as opposed to depletion) of vitamin D or of calcium.

The gastrointestinal tract is of primary importance in the biological activity of vitamin D, so it is not surprising that diseases of the liver and intestine have consequences for vitamin D metabolism and function that may result in bone disease. The liver is the site of the first stage of the metabolic activation of vitamin D, namely, the introduction of a hydroxyl group at C-25, and it is also important in the elimination of vitamin D metabolites in the bile (Chapter 4). In addition the liver is the site of synthesis of the plasma vitamin D binding protein (DBP, Chapter 7). Most critically, the mucosa of the small intestine is a major target organ for vitamin D activity, in which the active metabolite, 1,25-dihydroxyvitamin D [$1,25(OH)_2D$], promotes the absorption of calcium from the diet. This is effected by genomic reactions in which $1,25(OH)_2D$, binding to its specific intranuclear receptor (VDR), mediates the transcription of genes for proteins facilitating calcium transport; more recent evidence also supports the concept of a rapid nongenomic action on calcium flux that may involve a membrane receptor (Chapters 13, 14, 16).

II. METABOLIC DISTURBANCES IN GASTROINTESTINAL DISEASE

A. Development of Vitamin D Deficiency and Vitamin D Depletion

1. Supply of Vitamin D: Controlling Factors

The physiological source of vitamin D is solar irradiation of the precursor 7-dehydrocholesterol in skin to give vitamin D_3, but both vitamin D_2 and vitamin D_3 can be supplied in the diet (see Chapters 41 and 42). There is at present no agreed definition of a sufficient level of vitamin D, in terms of serum 25-hydroxyvitamin D (25OHD) concentration (see Chapter 31). Previously any level above that commonly associated with osteomalacia (see Section II,A,1,b) was considered as sufficiency, but more recently it has become recognized that there is an intermediate state in which vitamin D levels are insufficient but not frankly depleted. Such a state is associated with increased levels of PTH and is characterized by increased synthesis of $1,25(OH)_2D$, presumably driven by PTH, when a dose of vitamin D is given [1,2]. The concentration of 25OHD associated with vitamin D sufficiency probably lies in the range 25–50 nmol/liter (~10–20 ng/ml). The recommended daily requirement of vitamin D for people who are not exposed to sunlight and therefore have no cutaneous source of the vitamin is 400 IU or 10 μg per day, and this intake has been shown to achieve plasma 25OHD levels of about 25 nmol/liter (~10 ng-ml) [3–5].

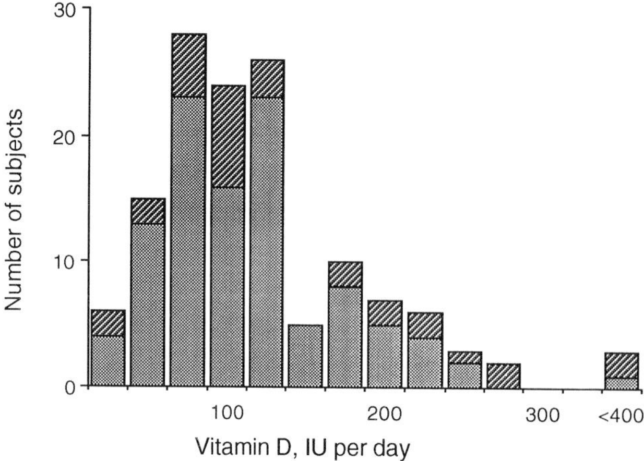

FIGURE 1 Daily dietary intake of vitamin D in a sample of Caucasians (stippled bars) and Asian immigrants (hatched bars) living in northwest England.

a. Role of the Intestine The general problems with absorption of lipids that characterize much gastrointestinal or hepatobiliary disease, whether because of poorly functioning enterocytes, a reduction in their number, or lack of biliary secretion, may include vitamin D as a particular example of a lipid-soluble molecule. Malabsorption of dietary vitamin D may then contribute to vitamin D deficiency [6,7]. However, for people in those many parts of the world where there is little fortification of foodstuffs with the vitamin, endogenous synthesis of vitamin D in the skin, as a result of solar exposure, is the major source of the vitamin [3,8] (Chapters 3 and 33). In this case the diet supplies only a fraction, often less than a quarter or a third, of the daily requirement (see Fig. 1), and so in normal circumstances partial impairment of dietary absorption would not be predicted to have a major effect on vitamin D status, as only a small proportion enters the body by this route. Ill health, such as may be encountered in patients with serious gastrointestinal or liver disease, may coincide with a reluctance or inability to spend time out-of-doors, thus decreasing the endogenous synthesis of vitamin D. When the opportunity to make vitamin D in the skin is minimal, provision of the vitamin in the diet, however little, becomes crucial [9], and any impairment of absorption may then have a significant effect. Factors affecting the elimination of vitamin D by the liver may also play a critical role in the face of gastrointestinal and hepatobiliary disease and may contribute to a state of "acquired vitamin D deficiency" (see Section II,A,2,c).

b. Significance of Changes in 1,25-Dihydroxyvitamin D Concentration Vitamin D deficiency has been defined in various ways, the most extreme of which is the development of osteomalacia or rickets. An earlier, if crude, indication is the measurement of serum 25OHD levels; concentrations associated with the clinical signs of deficiency are usually associated with levels below 12–20 nmol/liter (4.8–8 ng/ml) [10,11]. Biochemical indices such as the plasma concentrations of calcium, inorganic phosphate, alkaline phosphatase, or osteocalcin are sometimes used. Ultimately, vitamin D deficiency must be characterized by the inability to synthesize sufficient $1,25(OH)_2D$ to permit adequate absorption of calcium from the intestine or to stimulate osteoblastic activity, and this state can be defined as vitamin D depletion. During the development of vitamin D deficiency, whether arising from an inadequate dietary supply, intestinal malabsorption, or lack of solar exposure, the concept of a normal range for the serum 1,25$(OH)_2D$ concentration becomes invalid. The process can be envisaged (see Fig. 2) as a progressive descending spiral in which low 25OHD results in a reduction of 1,25$(OH)_2D$ synthesis, thus impairing calcium absorption. The tendency for calcium to fall stimulates parathyroid activity, and the trophic action of PTH on the renal 1α-hydroxylase increases 1,25$(OH)_2D$ synthesis, which will temporarily increase the efficiency of intestinal calcium absorption. Increased 1,25$(OH)_2D$ synthesis depletes still further a diminishing supply of 25OHD, and the process will continue (in the absence of a new supply of vitamin D) until there is insufficient substrate to provide the level of 1,25$(OH)_2D$ required for adequate absorption of calcium, despite an increasing degree of secondary hyperparathyroidism. There is evidence (Section II,A,2,c) that a raised concentration of 1,25$(OH)_2D$ may itself influence the process by shortening the serum half-life of 25OHD, thus accelerating the onset of vitamin D deficiency.

2. ROLE OF THE LIVER

The role of the liver in vitamin D metabolism is 2-fold; 25-hydroxylation of vitamin D occurs in the endoplasmic reticulum of hepatocytes, and the liver is a major excretory organ, both of vitamin D and 25OHD (when these are present in excessive amounts) but mainly of water-soluble compounds, principally glucuronide conjugates, all of which can be demonstrated to be present in bile [12–15]. The 25-hydroxylation function of the liver is affected little if hepatocellular function is maintained (see Section V below), and impairment of 25-hydroxylation is not believed to make a major contribution to the loss of vitamin D activity until a late stage of disease.

a. Biliary Secretion of Vitamin D Metabolites In normal subjects the biliary/fecal route appears to be the major excretory pathway for vitamin D, but precise chemical identification of many of the excretory products is still awaited. The recognition of the importance of the renal 24-hydroxylase reaction in produc-

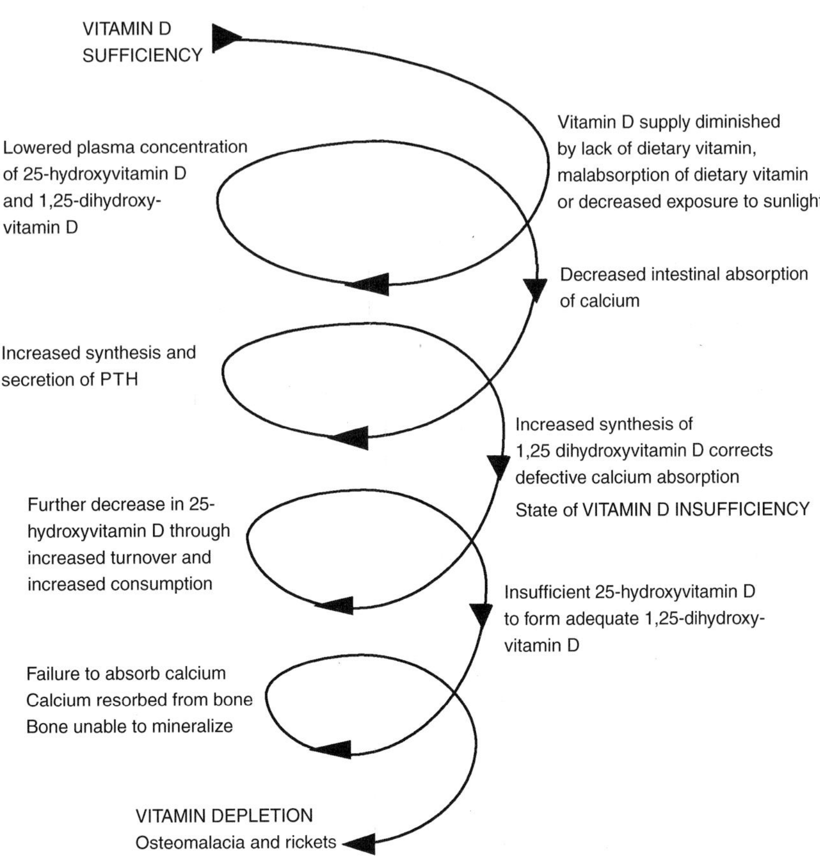

FIGURE 2 Spiral of developing vitamin D deficiency, showing progression through a state of vitamin D insufficiency, characterized by secondary hyperparathyroidism and increased synthesis of 1,25-dihydroxyvitamin D.

ing metabolites such as 24,25-dihydroxyvitamin D [24,25$(OH)_2$D], which can then undergo side chain cleavage and form carboxylic acids similar to bile acids [16,17], raises the question as to whether the urine may be a major excretory route for water-soluble vitamin D metabolites. The results of tracer experiments in humans do not support this argument [12–15], showing, in general, preferential excretion via the intestinal tract. Some experiments in rats seem to indicate considerable urinary excretion in this species [18], but in humans the balance of evidence is that fecal excretion is much more important, except in the presence of biliary obstruction in which case metabolites normally excreted by the biliary–fecal route are instead passed through the kidneys (see Section V,B below).

b. Enterohepatic Circulation of Vitamin D and Its Metabolites The polar vitamin D derivatives that have been reported in bile probably have no significant biological activity, and only insignificant amounts of unchanged vitamin D or 25OHD are usually present [12–15]. Nevertheless, the idea that biologically active metabolites of vitamin D might enter a conservative enterohepatic circulation has proved an attractive one to try to explain the acquired vitamin D deficiency associated with various gastrointestinal diseases in which lipid absorption is impaired. Interruption of such a circulation, it has been argued, would explain the wastage of vitamin D which occurs in conditions such as celiac disease and gastric or intestinal resection. Most studies that appear to support such a hypothesis have used an intravenous bolus dose of labeled vitamin D or 25OHD [19]. In these circumstances, especially if a nonphysiological vehicle has been used, the labeled compound may be seen as a foreign compound and be cleared rapidly by the liver, mainly as hydrophilic metabolites, but also, immediately after injection, in the form of unchanged sterol (see Fig. 3). It has been argued that vitamin D, absorbed from the diet and entering the circulation rapidly, in the form of chylomicrons, constitutes a nonphysiological presentation to the liver. The relative inefficiency of oral vitamin D as opposed to that synthesised in the skin and transported to the liver on DBP has been attributed to such a mechanism [20].

In a study often quoted to support the enterohepatic circulation hypothesis, a high proportion of injected radioactivity given as $25OHD_3$ was recovered in the bile, and was assumed still to be present in that form [19]. In fact, by comparison with other studies, it is most likely that, after the initial equilibration period when the 25OHD would have become bound to DBP, the label would be largely in the form of polar derivatives. Experiments in rats have demonstrated that if biliary excretion products of isotopically labeled vitamin D are collected and given to other animals they can be absorbed and reexcreted in the bile [21]. There is little evidence, however, that these metabolites have biological activity; any conjugates would need to be cleaved enzymatically to release the vitamin D moiety, which is itself likely to be a catabolic product with low activity. It is unlikely, then, that any enterohepatic circulation of vitamin D metabolites which does occur is of a conservative nature, or of physiological significance. Indeed, biliary excretion of vitamin D metabolites appears to be dose dependent, increasing in both humans and rats when vitamin D is plentiful, and being suppressed in vitamin D deficiency [14,21]. An alternative hypothesis for the significance of the biliary secretion of vitamin D compounds is that this mechanism enables the body to dispose of a highly potent biological compound if present in excess amounts and may be seen as a form of detoxication rather than an attempt to conserve a scarce resource [20–22].

c. Development of Acquired Vitamin D Deficiency The concept of interruption of an enterohepatic circulation of vitamin D metabolites is inadequate to explain the development of vitamin D deficiency in patients with gastrointestinal and hepatobiliary disease (see above) [15,20]. In addition to patients with fat malabsorption, various other types of patients also show wasting of vitamin D. These conditions include primary hyperparathyroidism, anticonvulsant therapy, and high cereal and high fiber diets. A theory has been proposed by Clements *et al.* [23] that attempts to explain this phenomenon. In the rat, these authors showed that withdrawal of calcium from the diet increased the plasma clearance of 25OHD; a similar effect was observed when phenobarbitone was administered to animals on a normal calcium diet, and also when fiber was added to the diet. In addition, the plasma half-life of 25OHD was decreased by secondary hyperparathyroidism. Parallel studies on human subjects have shown that both primary and secondary hyperparathyroidism can increase the wastage of $25OHD_3$ [24,25]. In all these studies a strong inverse relationship was demonstrated between the plasma half-life of 25OHD and the prevailing concentration of 1,25$(OH)_2$D (see Fig. 3). In those conditions in which 1,25$(OH)_2$D was raised, ^{3}H-25OHD was removed from the circulation more rapidly, and there was a corresponding increase in fecal radioactivity [24]. The reason for the phenomenon is not fully understood; it has been postulated that 1,25$(OH)_2$D may stimulate, perhaps by a receptor-mediated mechanism, the hepatic conjugation reactions which render vitamin D and 25OHD water soluble [25]. In addition, 1,25$(OH)_2$D is known to stimulate the renal 24-hydroxylase that initiates the side chain cleavage cascade, though the subsequent metabolic fate

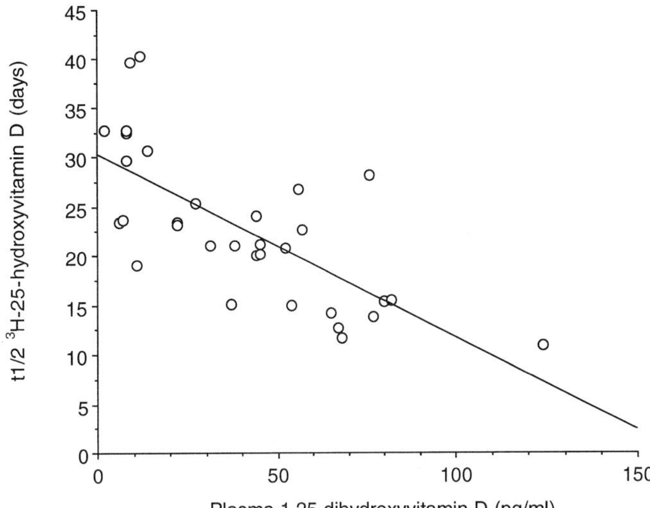

FIGURE 3 Strong inverse relationship between the plasma half-life of 25-hydroxyvitamin D and the prevailing concentration of 1,25-dihydroxyvitamin D; $r_s = -0.706$, $p < 0.001$. Data from studies reported by Clements *et al.* [24,25].

of these compounds in humans is not clear. The relationship between 1,25(OH)$_2$D level and the turnover of 25OHD obtains in a variety of gastrointestinal conditions in which serum 1,25(OH)$_2$D is raised as a result of problems with calcium absorption, leading to secondary hyperparathyroidism (see Sections II,B and III). The phenomenon of increased loss of plasma 25OHD in patients in whom either dietary intake of the vitamin or solar exposure is inadequate could help to explain the prevalence of vitamin D deficiency in gastrointestinal disease.

B. Malabsorption of Calcium

The development of osteomalacia in patients with steatorrhea is usually attributed to malabsorption of dietary vitamin D, but the degree of malabsorption is rarely severe enough to account for the development of vitamin D deficiency in populations with a high oral intake of the vitamin. Because the small intestine is a target organ for vitamin D it is possible that the functional activity of 1,25(OH)$_2$D may be compromised (Chapter 16). Colston *et al.* [26] have concluded that although VDR were abundant in crypt cells from patients with celiac disease, malabsorption of calcium resulted from the loss of vitamin D-regulated proteins and enzymes located in the most mature enterocytes of the mid and tip villous regions. Absorption of dietary calcium has been found to be deficient in various types of gastrointestinal disease including celiac disease and Crohn's disease, in which fat malabsorption may lead to the formation of calcium soaps. The phenomenon may be observed as well in some postgastrectomy and intestinal bypass surgical states in which fat malabsorption is not a problem (see Section III below). The resulting tendency to hypocalcemia with the induction of secondary hyperparathyroidism has been shown in many of these conditions to lead to raised levels of 1,25(OH)$_2$D, the condition that is known to lead to wastage of 25OHD and the development of vitamin D deficiency as discussed above in Section II,A,2,c. Even where the supply of vitamin D is adequate, malabsorption of calcium may lead to secondary hyperparathyroidism. Vitamin D deficiency will not occur, but the skeleton may be subject to the effects of high levels of PTH for many years. It is recognized that PTH excess is detrimental to the skeleton and especially at cortical sites. Any increase in bone turnover will exacerbate the remodeling imbalances that accompany the bone loss associated with aging. Raised levels of 1,25(OH)$_2$D also stimulate resorption and may be deleterious to the skeleton. These phenomena may explain why the principal bone disease in patients with gastrointestinal disease is osteoporosis.

III. ACQUIRED BONE DISEASE IN GASTROINTESTINAL DISORDERS

Osteomalacia has long been known to be associated with diseases of the gastrointestinal tract, but the pathophysiological mechanisms responsible for this bone disease in gut disorders are only now being recognized. Furthermore the mechanisms that lead to osteomalacia may also lead to loss of skeletal tissue, osteoporosis, and easy fracture. Varying degrees of perturbation of calcium and vitamin D metabolism occur in patients with gastrointestinal disease, and the extent, duration, and severity of these derangements are important in determining the nature of any underlying bone disease. Because the major source of vitamin D is cutaneous synthesis rather than the diet (Section II,A,1), adequate solar exposure should protect against vitamin D deficiency.

A. Role of Secondary Hyperparathyroidism

It has only relatively recently been appreciated that certain diseases of the gastrointestinal tract are accompanied by secondary hyperparathyroidism with an increase in the circulating concentration of 1,25(OH)$_2$D. It is assumed that these changes are an adaptation to the calcium malabsorption which has been documented in many diseases affecting the gut. This secondary hyperparathyroidism may be successful in correcting calcium malabsorption provided there is a sufficiency of vitamin D. However, if there is an increase in the catabolism of vitamin D (Section II,A,2,c) the body may become depleted of vitamin D, the amount of $1,25(OH)_2D$ will be insufficient for normal calcium absorption, and so secondary hyperparathyroidism will be intensified (Fig. 2). This leads to an increase in bone resorption and an increase in remodeling imbalance. In addition, the effects of excess PTH on the kidney will produce phosphate wastage and a reduction in the serum phosphate concentration. These changes, if present for a prolonged time, result in the development of osteomalacia because of insufficient $1,25(OH)_2D$ and mineral for normal bone formation. If the supply of vitamin D from the diet or solar exposure is adequate for the increased demands resulting from calcium malabsorption and secondary hyperparathyroidism, then osteomalacia will not occur, but the skeleton will be subjected to the effects of excess PTH perhaps over many years. This can lead to loss of bony tissue, particularly at cortical sites, with the development of osteoporosis and tendency to fracture.

B. Nature of the Bone Disease

Although osteomalacia may occur in patients with gastrointestinal disease, osteoporosis is far more common. This may be explained in part by the state of vitamin D nutrition in patients with gut disease. Where the amount of solar exposure is limited by geographic latitude, vitamin D deficiency and osteomalacia are more likely to be seen.

Following intestinal bypass surgery for obesity, it has been estimated that 12% of patients in Europe and 4% in the United States develop osteomalacia [27]. This reinforces the importance in Northern Europe, including the United Kingdom, of extrinsic (privational) vitamin D deficiency in determining the development of osteomalacia in intestinal [28] and hepatic [29] disorders.

The frequency of osteomalacia in gastrointestinal and hepatobiliary disease has been greatly overestimated. This has resulted from misinterpretation of both biochemical and histological findings. Although a raised serum alkaline phosphatase concentration, often accompanied by serum calcium values in the lower part of the normal range, may be an early sign of underlying osteomalacia, the same biochemical changes are to be found in other clinical situations and may not be due to disturbances in calcium and vitamin D metabolism. Both hepatic and intestinal isoforms of alkaline phosphatase may be increased in certain diseases, and to be certain that a raised level of alkaline phosphatase arises from bone it is necessary to use a bone-specific assay. Hyperosteoidosis due to secondary hyperparathyroidism has been misinterpreted as osteomalacia, as also have been measurements of increased osteoid volume. Clinically, skeletal pain in a patient with primary biliary cirrhosis is more likely to have arisen from multiple osteoporotic related fractures than from osteomalacia.

Osteoporosis is therefore by far the most common metabolic bone disease complicating disorders of the liver and gastrointestinal tract. There is a high incidence of fractures (especially vertebral) in patients with chronic liver disease or a past history of partial gastrectomy. Bone densitometry has demonstrated a significant reduction in bone mineral content in both the axial and appendicular skeleton, even in asymptomatic patients.

C. Role of Factors Other than Vitamin D and Calcium

Gastrointestinal diseases may result in bone disease by mechanisms other than disturbances in calcium and vitamin D metabolism Protein deficiency, which is common in some diseases of the liver and small bowel, may adversely affect bone [30]. Hypoalbuminemia can depress osteoblast function, and other nutrients not immediately considered relevant to the skeleton may be important in bone cell function [31]. Deficiencies of other vitamins, including vitamins C and K, may affect bone although their significance in intestinal disease is not established. Vitamin C is a cofactor for prolyline hydroxylase, which is necessary for the formation of stable collagen polymer, and two proteins, bone Gla protein (osteocalcin) and matrix Gla protein, are generated by vitamin K-dependent enzymes.

D. Development of Metabolic Bone Disease in Gastrointestinal Disease

In many situations, however, an absolute or relative deficiency of calcium or vitamin D will be pivotal to the development of metabolic bone disease. Rao *et al.* [32] studied the histological evolution of vitamin D deficiency in patients with intestinal malabsorption and a plasma 25OHD of less than 25 nmol/liter (~10 ng/ml). In early vitamin D depletion there is a lack of clinical symptoms, although osteopenia may be present with histological evidence of hyperparathyroidism; there may also be an increased risk of fracture. As the severity increases mineralization becomes more defective, bone formation rate declines, osteoid tissue increases in both thickness as well as surface extent, and changes of hyperparathyroidism become more severe [32].

Bone disease resulting from intestinal disease will display features of osteopenia, hyperosteoidosis, osteitis fibrosa, and osteomalacia (see Chapter 37). The extent and degree of these changes will be determined by the duration and severity of vitamin D deficiency, calcium malabsorption, and accompanying secondary hyperparathyroidism. Additional factors such as the use of corticosteroids or parenteral nutrition (see Section IV,F) may also affect bone. Because the mucosal barrier is often compromised in small bowel disease, it may also be possible for unspecified nutritional or perhaps toxic substances to affect bone.

Histomorphometric analysis of transiliac bone biopsy material has revealed several differing abnormalities, some of which represent the evolution of osteomalacia [27]. The first stage in the development of osteomalacia is secondary hyperparathyroidism, where the osteoid surface and volume are increased but osteoid thickness and mineralization lag times are normal. Some patients with vitamin D deficiency show reduced adjusted apposition rate and prolongation of the mineralization lag time. The latter condition is described as hypovitaminosis D osteopathy stage i (HVOi) by Parfitt [27], who divides osteomalacia into HVOii and HVOiii when osteoid thickness exceeds 15 μm and mineralization lag time exceeds 100 days (see Chapter 41). In HVOiii there is no mineralization occurring (using double tetracycline labels), whereas in HVOii mineralization still occurs. Atypical and focal osteomalacia are also seen occasionally. Florid hyperparathyroidism can result in high turnover osteoporosis, but low turnover osteoporosis is perhaps the most common lesion seen. Osteoid thickness is normal or reduced, bone formation and apposition rates are reduced, and there is little or no evidence of hyperparathyroidism. Low turnover osteoporosis is often associated with general undernutrition and protein malnutrition.

IV. GASTROINTESTINAL CONDITIONS ASSOCIATED WITH BONE DISORDERS

The various pathophysiological mechanisms resulting in metabolic bone disease are discussed elsewhere in this volume (Chapters 31, 32, 37, 41, and 42). Whether any gastroenterological disturbance produces bone disease will depend on the extent to which calcium and vitamin D metabolism are disturbed. Thus, disturbances of colonic function are unlikely to affect calcium metabolism, but because inflammatory bowel disease of the colon is commonly treated by systemic steroids, treatment of colonic disease may impact on the skeleton (see Chapters 49, 50).

A. Postgastrectomy Bone Disease

1. Pattern of Bone Disease

a. Incidence of Osteomalacia The changing patterns in both incidence and management of peptic ulcer disease has meant that the problem of postgastrectomy bone disease is in decline, and this trend is likely to continue as the need for stomach resection is reduced. Bone disease related to gastrectomy does not develop for several years after surgery, and because many of those affected are middle aged or elderly it may not be easy to uncouple the effects of menopause and the aging process on the skeleton from those of gastric surgery. Osteopenia or osteoporosis are far commoner than frank osteomalacia. The overall reported incidence of osteomalacia varies in different centers, and this is in part explained by the differing criteria used for diagnosing osteomalacia. In a survey of 1228 patients following partial gastrectomy, Paterson *et al.* [33] found only 6 cases of osteomalacia using clinical, biochemical, and histological criteria. Tovey *et al.* [34] found osteomalacia in 10 of 227 postgastrectomy subjects, but only 15 of the patients had had a bone biopsy performed. When more extensive use is made of bone histology, the incidence of diagnosed osteomalacia increases and in some series is over 20% [35,36]. Despite the numerous explanations for postgastrectomy bone disease, one unanswered question is why the problem of osteomalacia should be greater in women when 80% of gastrectomy patients are men.

b. Vitamin D Metabolism Osteomalacia responds to small doses of vitamin D [37], and vitamin D absorption is normal in the absence of steatorrhea and only reduced by 40% of intake in the most severe malabsorptive states [38]. The etiology of vitamin D deficiency has therefore never been clear. However, it has been shown that some subjects with gastrectomy show a reduced half-life of 25OHD. These patients had evidence of secondary hyperparathyroidism with increased serum $1,25(OH)_2D$ levels. Lowering of the hormone levels by large calcium supplements was accompanied by a prolongation of the half-life of 25OHD [39]. Other workers have also shown evidence of hyperparathyroidism in gastrectomized patients [40,41]. Nilas *et al.* [42] showed reduced serum 25OHD levels, increased $1,25(OH)_2D$ levels, reduced calcium absorption, and osteopenia in subjects several years after gastrectomy. Other potential factors in producing abnormalities in calcium metabolism in this group are the reduced food intake resulting from loss of stomach area, reduced acid secretion which is important for calcium absorption, poor admixture of food with digestive juices, and intestinal hurry.

c. Bone Turnover Vertebral deformity and fracture are commoner in patients with a past history of gastrectomy [43–45] when compared with controls. Rao *et al.* [40] found a past history of gastrectomy in 5% of patients seen with vertebral fracture compared with 1% of controls. Histologically the bone shows thin but extensive osteoid seams, a low apposition rate resulting from decreased collagen synthesis, and a low bone formation rate. Evidence of secondary hyperparathyroidism may be present, and Parfitt argues that these changes are the residua of accelerated bone turnover with net loss of bone. Impaired recruitment and activity of osteoblasts then result in defective bone repair, predisposing to fracture [31,46].

2. Clinical Features of Postgastrectomy Bone Disease

Osteoporosis is in the main without symptoms until a fracture occurs. There may be symptoms from disturbed bowel function, for example, loose motions and steatorrhea, and these should be indications for assessing the patient for bone disease. Multiple vertebral fractures lead to loss of height, thoracic kyphosis, and chronic back pain that is difficult to control. Not surprisingly, accompanying depression is common and may exacerbate the pain.

Osteomalacia presents insidiously with vague bone pain and muscle weakness, particularly affecting the proximal muscles. As the bone disease progresses the patient becomes increasingly incapacitated by bone pains and muscle weakness. Walking becomes labored, with the patient developing a waddling gait and often having to climb stairs "crablike" holding the stair rail with both hands. In the most extreme cases pain around joints from an enthesopathy, owing to the effects of hyperparathyroidism, may simulate active arthritis. Fractures may occur, particularly in chronic cases and especially where pseudofractures have been present; a frequent site is the femoral neck. Insufficiency fractures from osteoporosis can mimic Looser's zones radiologically [47], but the clinical picture is one of acute pain in relation to the fracture rather than the history of chronic vague bone pain that characterizes osteomalacia.

3. Biochemistry

Serum biochemistry in cases of osteoporosis is normal, although the alkaline phosphatase may be slightly increased especially if there has been a recent fracture. Evidence for secondary hyperparathyroidism may be present with raised PTH and $1,25(OH)_2D$ levels. Bone densitometry will show a reduced bone mass, which may be more marked at cortical sites in the forearm and hip than in the spine.

In osteomalacia there is usually a reduction in serum phosphate level with a low normal or frankly reduced serum calcium and elevated alkaline phosphatase concentration. If measured, PTH will be found to be increased, 25OHD reduced, and $1,25(OH)_2D$ low, normal, or increased depending on the amount of vitamin D recently provided from the diet or by solar exposure.

4. Management

The management aim for patients at risk should be to ensure normal plasma biochemistry by providing sufficient calcium and vitamin D as a dietary supplement to suppress any increase in PTH and ensure normal 25OHD levels. When osteomalacia is present, oral vitamin D supplements can be given as a large single bolus using doses of several hundred thousand units or much smaller daily doses of 1000–4000 units (25–100 μg) daily. Patients with steatorrhea will require larger oral doses or parenteral vitamin D, and they should be monitored by measuring serum 25OHD regularly until it can be shown that normal 25OHD values are being achieved. Because there are usually difficulties with calcium absorption, a large calcium supplement (1–2 g of elemental calcium) should also be given long-term. Patients without overt bone disease should have blood assayed for PTH and 25OHD. If there is any evidence of secondary hyperparathyroidism, sufficient oral calcium should be given to suppress the elevated PTH. Vitamin D should also be given if the serum 25OHD is low. Because patients who have had a gastrectomy are at risk of developing osteoporosis, bone densitometry should be performed, and if evidence of osteoporosis is found then appropriate antiresorptive therapy should be considered.

B. Celiac Disease

1. Clinical Features

Symptomatic bone disease is uncommon in celiac disease, which has a prevalence of 1 in 2000–4000 of the population and is particularly common in those of Irish descent. Reduced bone mass is found in a significant proportion of patients with celiac disease compared with age- and sex-matched controls [48–51]. Using antiendomysial antibodies in asymptomatic osteoporotic subjects, Lindh *et al.* [52] showed that the incidence of celiac disease was 10-fold higher than in the normal population. In adults on treatment for celiac disease, osteopenia is twice that expected for a normal population, with a reduction in bone density of between 7 and 13% [53,54]. There are, however, no data showing increased incidence of fracture in celiac disease. Selby *et al.* [55] have shown that cortical bone is more affected than trabecular bone and attributed this to the effects of excess parathyroid hormone on the skeleton. In children, however, institution of treatment with a gluten-free diet results in a normal bone mass when they become teenagers [56]. These data suggest that adults cannot fully restore lost bone, whereas the developing child can fully develop the skeleton if the gut is restored to normality.

Occult celiac disease with no overt signs or symptoms of intestinal malabsorption may occasionally present as an osteomalacic syndrome [57,58], and small bowel biopsy should be considered in the assessment of all patients with osteomalacia in whom the cause is unclear. Alternatively, gastrointestinal symptoms may be so florid as to obscure the symptoms of an underlying osteomalacia. Because celiac disease affects the duodenum and jejunum more severely than the ileum, malabsorption of calcium and vitamin D are common [6,59]. However, not all patients with celiac disease malabsorb vitamin D; in a small study, the celiac disease patient with the most severe malabsorption of vitamin D had the highest serum level of 25OHD, whereas absorption of vitamin D was normal in two patients who showed some degree of vitamin D deficiency [38]. These findings reemphasize the importance of cutaneous synthesis of vitamin D.

2. Development of Vitamin D Deficiency

Evidence of poor vitamin D nutrition can be found in untreated celiac disease. Dibble *et al.* [60] found 2 untreated patients with low serum 25OHD levels (<12 nmol/liter or ~5 ng/ml) but normal values in 12 subjects successfully treated with a gluten free diet. Arnaud *et al.* [61] found low 25OHD values in 3 of 7 celiac patients. In the small series reported by Melvin *et al.* [59], 5 of 9 patients had histological osteomalacia (this study was performed before measurements of vitamin D metabolites were possible).

The mechanisms leading to vitamin D deficiency have been attributed to interruption of a conservative enterohepatic circulation of 25OHD [19], but this concept has been largely discarded (see Section II,A,2,b). A patient with celiac disease, secondary hyperparathyroidism, and elevated serum $1,25(OH)_2D$ level was described in a study by Clements *et al.* [25]. The half-life of radiolabeled $25OHD_3$ was shortened in the hyperparathyroid state compared with the euparathyroid state, which was achieved by using large supplements of calcium. In a study by Selby *et al.* [55] a high proportion of patients with treated celiac disease showed evidence of secondary hyperparathyroidism, suggesting that there may be a continuing problem with calcium absorption despite a gluten-free diet. Support for these last two observations is forthcoming from the paper of Corazza *et al.* [62]. These workers found evidence of low bone mass in celiac patients that was improved but not corrected by a gluten-free diet. Many untreated patients have secondary hyperparathyroidism with significantly raised serum $1,25(OH)_2D$ levels and significantly lower 25OHD levels in the untreated state compared with the treated condition.

3. Management

The treatment of choice for celiac disease is a gluten-free diet, but not all patients will be able to adhere to such a strict regime. Some individuals either overtly or covertly eat a diet containing gluten, which may explain the findings of Selby *et al.* [55]. Additional vitamin D will cure any osteomalacia, and there is probably a good case for giving long-term calcium supplements of 1–2 g per day to guard against the effects of occult secondary hyperparathyroidism on the skeleton.

C. Pancreatic Disease

Pancreatic insufficiency is not normally associated with metabolic bone disease unless other conditions exist such as cystic fibrosis or alcoholism. This paradox implies that the disturbances to mineral and bone metabolism seen in other forms of bowel and liver disease have little to do with steatorrhea. Low 25OHD levels have been reported in cystic fibrosis [63], and both osteoporosis and osteomalacia have been described in this condition [64]. Meredith and Rosenberg [7] have questioned the rarity of bone disease in pancreatic steatorrhea and consider that osteomalacia may be commoner than hitherto believed. Patients should receive pancreatic supplements and also vitamin D if the serum 25OHD concentration is shown to be low.

D. Inflammatory Bowel Disease

1. Clinical and Biochemical Features

Of the two main inflammatory bowel diseases, bone disease is more often seen in Crohn's disease than in ulcerative colitis. Because the colon is principally an organ for conservation of salt and water, disturbance of its function does not normally impact on the skeleton.

However, the use of corticosteroids to control disease activity may cause loss of bone and osteoporosis (see Chapter 49).

Malabsorption of vitamin D and 25OHD have been documented in Crohn's disease [7,38], and calcium is also malabsorbed but perhaps less often than previously considered. In a study by Krawitt *et al.* [65] only 4 of 31 patients with active Crohn's disease were found to have reduced net absorption of calcium and negative calcium balance, and 2 of the 4 had undergone ileal resection. Driscoll *et al.* [66] studied 82 patients with Crohn's disease, from 9 of whom transiliac crest bone biopsies were taken. Six of these patients were found to have osteomalacia, of whom 3 had a repeat biopsy which showed improvement following treatment with vitamin D. There was a high incidence of vitamin D deficiency as assessed by serum 25OHD, with 65% having a low 25OHD level; in 25% of cases the serum 25OHD level was below 25 nmol/liter (~10 ng/ml). It was in this lowest group where the cases of osteomalacia occurred. The lowest 25OHD values were found in those patients who had had a previous ileal resection. Nine of 25 patients with Crohn's disease, assessed by Compston *et al.* [67], had increased osteoid on bone biopsy.

Osteoporosis is more common than osteomalacia. Compston *et al.* [68] found a reduced bone mass in 30% of an unselected group of patients with Crohn's disease. The pathogenesis of osteomalacia may be similar to that in celiac disease, but no reports have shown good evidence of secondary hyperparathyroidism with high serum concentrations of $1,25(OH)_2D$ as has been seen in celiac disease. The fact that calcium absorption is often normal [65] does not exclude excess parathyroid activity because normal calcium absorption may only be maintained at the expense of increased parathyroid drive. However, $1,25(OH)_2D$-induced hepatic wastage of 25OHD seems less common in patients with Crohn's disease.

Glucocorticoids are used frequently to control inflammatory bowel disease and can both impair calcium absorption and reduce osteoblast function, thus exacerbating those processes known to produce bone loss. Protein and calorie malnutrition that are seen in severe cases of inflammatory bowel disease will also adversely affect the skeleton [69].

2. Management

Patients with inflammatory bowel disease should be managed in a similar way to patients with partial gastrectomy. Osteoporosis may pose a problem for management using oral bisphosphonates because of their poor absorption and tendency to cause bowel upset. Intravenous bisphosphonates or nasal calcitonin can be used as an alternative to oral antiresorptive therapy.

E. Jejunoileal Bypass

1. Clinical Features

Intestinal bypass surgery for obesity became popular during the 1970s but was followed by a considerable mortality and morbidity, including a high incidence of skeletal complications. Metabolic bone disease developed after a variable number of years in a significant proportion of patients following bypass surgery for obesity; however, obesity per se is known to be sometimes associated with an increase in PTH [70]. Dano and Christiansen [71] found the bone mineral content to be reduced compared with controls in a group of 37 obese patients before undergoing bypass surgery. Surgery, resulting in severe intestinal malabsorption, may exacerbate preexisting, and perhaps subtle, abnormalities of calcium metabolism. In about half of patients there is a fall in serum calcium and magnesium levels after surgery, and although serum 25OHD levels fall $1,25(OH)_2D$ levels are maintained [72–74]. Osteomalacia is seen on bone biopsy in up to 60% of patients, and in some $1,25(OH)_2D$ levels are low and PTH values are raised [31,72,75]. Profound hypocalcemia may occur secondary to the hypoparathyroidism of magnesium deficiency when osteomalacia is believed to be less severe [31]. Osteomalacia resulting from bypass surgery is seen more frequently in Europe than in America.

The etiology of osteomalacia in bypass patients has been attributed to malabsorption of vitamin D, but the functional abnormalities produced by intestinal bypass surgery should favor hepatic wastage of 25OHD, although this hypothesis has not been tested. Osteopenia is also found following surgery with loss of bone in both the appendicular and axial skeleton [73].

2. Management

Vitamin D and 1α-hydroxyvitamin D have been used with success to treat osteomalacia, but the native vitamin should be adequate to correct any deficiency. The parenteral route may be necessary in view of loss of functional bowel. Liver disease may follow this type of operation, but impaired hepatic hydroxylation of vitamin D is rarely a problem (see Section V below). Treatment failures have responded to antibiotics used to eradicate bacterial overgrowth in the bypass segment [76].

F. Total Parenteral Nutrition

1. Clinical Features

Total parenteral nutrition (TPN) is used for those individuals in whom disease of the bowel is so severe that oral feeding cannot achieve adequate nutrition.

Most patients commencing TPN do so after years of chronic bowel disease and therefore are likely to have pre-existing bone disease. This section is primarily concerned with the effects of TPN per se on the bone.

Total parenteral nutrition may not provide all the nutrients or micronutrients required, and early in the use of this treatment calcium and phosphate were omitted from solutions, which caused rickets in some children [77]. The major iatrogenic bone lesion resulted from the contamination of casein hydrolysate by aluminum at concentrations of up to 1 mg/liter [78]. Serum samples from these patients contained large amounts, and bone contained more than 30 times the normal level of aluminum [77]. It is not surprising, therefore, that most of these patients showed signs of osteomalacia on bone biopsy.

Clinically the problem presented with periarticular bone pains, especially in the legs, back, and ribs, and was only helped by cessation of TPN [79]. The disease has largely disappeared following the substitution of amino acids for casein [80]. Several biochemical abnormalities were found, but these may have been related to the amount of protein present in the feed. Hypercalciuria, low PTH, and low $1,25(OH)_2D$ values returned to normal when purified amino acids were introduced. However, hypercalciuria persists if the amino acid content of the feed is kept too high [81]. The bone disease seen in these patients is mainly osteopenia [82,83], probably related in part to bone changes before TPN was commenced.

2. MANAGEMENT

Bone density, biochemistry, and bone histology should be evaluated at the onset of TPN and any hyperparathyroidism, vitamin D deficiency, or osteomalacia corrected by the appropriate use of calcium and vitamin D. If only osteoporosis is present, parenteral bisphosphonate treatment should be considered, and monitored by bone densitometry. Whenever possible a return to enteral feeding should be encouraged. In cases of severe aluminum poisoning chelation with desferoxamine, as used in dialysis patients, may be effective.

V. LIVER DISEASE

A. Hepatic Osteodystrophy

Both osteomalacia and osteopenia contribute to the syndrome of "hepatic osteodystrophy," a condition characterized by reduced bone mass leading to severe osteoporosis and fractures. Osteomalacia occurs more commonly in the presence of severe long-term cholestasis as in primary biliary cirrhosis (PBC), but even in PBC it is rare and its incidence appears largely restricted to Northern Europe [84–86]. Earlier studies in which a high incidence of osteomalacia in bone biopsies was reported (e.g., 22 of 32 patients, mainly with cholestatic disease) used diagnostic methods that have now been superseded [87,88]. Compston *et al.* [89], using more stringent criteria including the presence of calcification fronts, reported a much lower incidence; only 4 of 32 patients had evidence of osteomalacia. A rather higher incidence was reported by Dibble *et al.* [29], who found osteomalacia in 9 of 29 patients; of the 9, 5 had PBC, 2 had chronic active hepatitis, and one each had sclerosing cholangitis and alcoholic cirrhosis. Indeed, using his strict criteria for diagnosing osteomalacia, Parfitt [27] has argued that only 8 histologically proven cases of hepatic osteomalacia have been published [29,89,90–93]. Many other reported cases have used invalid methods involving decalcified bone [88,94] or not excluding HVOi (see Section III above). By such strict criteria osteomalacia is most uncommon, but when present it is associated with severe vitamin D deficiency and secondary hyperparathyroidism [29,90].

Conflicting reports have appeared on the levels of PTH in severe liver disease; an earlier study [95] claiming a defect in hepatic cleavage of PTH was challenged by Klein *et al.* [96], who failed to find hyperparathyroidism using a range of different assays for the hormone. The latter authors concluded that raised PTH was unlikely to play a part in hepatic osteodystrophy. Dibble *et al.* [29] found evidence of slightly increased PTH in advanced cirrhosis, but osteomalacia was only found together with chronic cholestasis and vitamin D deficiency.

The central role of the liver in maintaining a balance between synthesizing of 25OHD, eliminating excess vitamin and its metabolites, and thus regulating plasma levels of 25OHD is discussed elsewhere (Chapter 4). It had been supposed that osteomalacia in liver disease could be explained by impairment of hepatic 25-hydroxylation, and low plasma levels of 25OHD have been reported in many types of liver disease, including alcoholic hepatitis and cirrhosis, lupoid and cryptogenic cirrhosis, and primary biliary cirrhosis [94]. It is now recognized that where hepatocellular function is maintained, 25-hydroxylation of vitamin D remains normal until a very late stage of disease and that malabsorption of vitamin D as a result of steatorrhea is a more likely cause of deficiency. Where steatorrhea is present, absorption of calcium may well be impaired, leading to osteoporosis. Malabsorption of both vitamin D and calcium may be compounded by the use of bile sequestration agents [97].

Osteoporosis is the most common bone problem seen in chronic liver disease [98] and may be exacerbated by drugs used to control the disease, such as steroids in chronic active hepatitis. The most common histomorphometric parameter is not that of increased bone turnover, as is often associated with intestinal disease, but a low turnover state with low values for osteoid volume and surface extent. Apposition rate is also low when tetracycline labeling is used [86,98]. Although eroded surfaces without osteoclasts may be increased, this may reflect delayed formation rather than increased resorption [27]. This low turnover state reflects depressed osteoblastic function that may be part of the chronic debilitating nature of liver disease or may be due to the persistence of toxins normally removed by the liver. The histological features are, however, similar to those seen in patients after jejunoileal bypass.

B. Primary Biliary Cirrhosis

1. Clinical Features

Vitamin D deficiency has been frequently reported in PBC, at times associated with osteomalacia, and characterized by low plasma levels of 25OHD [94,99] . The explanation is probably poor intestinal absorption [100] from a diet containing little vitamin D; in addition, there may be problems with endogenous synthesis of the vitamin. Sunshine may be deliberately avoided for cosmetic reasons or because of itching and discomfort and it is possible that UV irradiation is less effective because of the jaundice-induced skin pigmentation. Claims that the osteomalacia of PBC was refractory to treatment with the vitamin were based on early studies, the results of which may have been misinterpreted. Reports claiming that parenteral vitamin D therapy was ineffective in PBC may be explained by the use of intramuscular oily preparations of the vitamin, which have been shown to have low bioavailability [101]. Similarly, the availability of vitamin D from oral preparations may depend on the vehicle. Absorption from tablets tends to be poor, especially in the presence of fat malabsorption, whereas absorption from an ethanol solution given in milk is relatively efficient [90]. The deficiency can be corrected by a modest increase in the oral supply of vitamin D or by giving UV irradiation [84,90]. Successful treatment has also been achieved using 25OHD, which, being less lipophilic, is absorbed more efficiently than vitamin D [38,102]. Osteomalacia, as noted above (Section V,A), is rare and may be easily missed because of the severity of symptoms of the liver disease. Changes in alkaline phosphatase, phosphate, and calcium levels may reflect more the liver disease than changes in the skeleton. The clinician needs to be alert to the possibility of vitamin D deficiency and to identify patients at risk by measuring serum 25OHD. Bone pain from rib and vertebral fractures is more likely to result from osteoporosis, and the prevention and treatment of this disease is far more difficult than the prevention or treatment of osteomalacia.

2. Vitamin D Metabolism

Given an adequate supply of vitamin D, the patient with PBC can metabolize the vitamin normally [103]. Reported low levels of 25OHD and $1,25(OH)_2D$ in PBC [93,96,103] seem to be related to malabsorption of the vitamin rather than to any intrinsic defect in metabolism. Subnormal concentrations of vitamin D metabolites are associated with established osteomalacia and provide an adequate explanation for the development of the disease. There is little evidence for raised levels of either $1,25(OH)_2D$ or PTH in PBC, and so the conditions favoring the $1,25(OH)_2D$-induced increased turnover of 25OHD are not present.

Where there is considerable biliary obstruction, water-soluble vitamin D metabolites appear in unusually large amounts in the urine [84,100,103] (see Fig. 4). Jung *et al.* [84] found urinary excretion of metabolites after administering vitamin D but not after 25OHD, and they concluded that the water-soluble metabolites had arisen from vitamin D but not via the 25-hydroxylation pathway. It is doubtful whether this urinary excretion contributes quantitatively to vitamin D deficiency in PBC, since these metabolites would otherwise be excreted via the fecal route and do not appear to constitute an increased net loss.

3. Management

Oral doses of vitamin D have been found to be effective in treating the osteomalacia in PBC patients. Biochemistry and calcium balance were normalized, and bone histology improved after treatment [90]. The precise dose of vitamin D was selected in this study by first measuring the absorption of an ^3H-labeled vitamin D test dose and calculating the administered dose to fit the needs of the patient. The greatest degree of malabsorption was 26% of the administered dose. Doses of 400–4000 IU in solution per day were used, but, in the absence of information on absorption, a dose of 2000 IU per day should be adequate for most patients. Treatment with UV irradiation is also effective [84,90]. Tablets of vitamin D are less well absorbed than solutions, and treatment with high dose tablets needs to be monitored. If, after a trial of 10,000 IU per day, serum 25OHD does not rise toward 50 nmol/liter (~20 ng/ml), then regular monthly or 3-monthly injections of 300,000 IU (7.5 mg) should be tried and the response assessed by measuring serum 25OHD. If a response

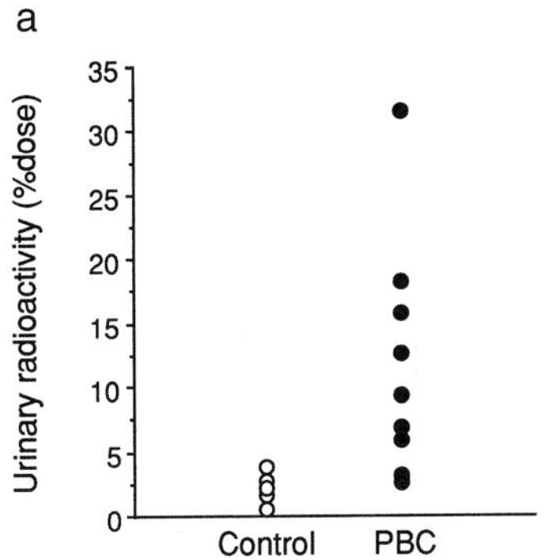

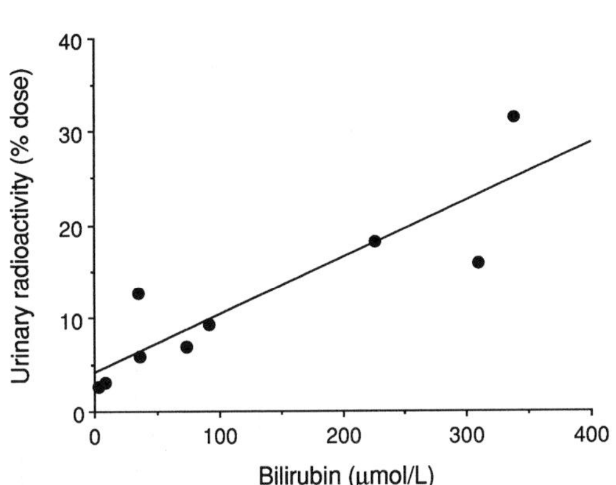

FIGURE 4 Urinary excretion of vitamin D metabolites in patients with primary biliary cirrhosis (PBC) showing (a) percentage dose excreted compared to control subjects and (b) the strong relationship to the degree of biliary obstruction as indicated by plasma bilirubin; Kendall's coefficient of concordance $\tau = 0.788$, $p = 0.001$. Data from Mawer *et al.* [103].

greater than 125 nmol/liter (~50 ng/ml) is produced, then the dose should be reduced. If the extent of steatorrhea or cholestasis increases, the oral dose may need to be increased. At no time should a patient receive more than 10,000 IU per day by mouth without checking 25OHD levels, as eventually intoxication may result. There is no need to use 25OHD or 1α-hydroxylated derivatives, although these are effective in curing bone disease.

Osteoporosis is a difficult problem to address, and any treatment is experimental. Currently treatment with bisphosphonates or hormone replacement therapy may be considered for those with a low or falling bone mineral density, but given that the state is one of low turnover it is difficult to envisage much benefit accruing from antiresorptive treatment. It is possible that fluoride, which stimulates bone formation, might be beneficial, but this awaits a detailed clinical trial.

Despite beneficial effects on the underlying liver disease, treatment with urodeoxycholic acid conferred no benefits on the skeleton in a study of 88 patients with PBC [104]. Supplementation with calcium was superior to treatment with calcitonin in maintaining bone mass, albeit transiently, in a study of 25 women with PBC-associated osteoporosis [105].

C. Alcoholic Liver Disease

1. Clinical Features

Osteoporosis is the more frequent bony abnormality in patients with alcoholic liver disease, as in PBC, but osteomalacia has been reported occasionally [106,107]. The osteoporosis is characterized by a high incidence of fractures [106–108], but these do not follow a classic pattern. In one series, 27 of 76 chronic alcoholic male patients had vertebral fractures, but in only 5 of these was bone mineral density below the fracture threshold [109]. In evaluating fractures it is important to distinguish between bone disease that arises directly from the cirrhosis, ethanol-associated osteopenia [110,111], and increased trauma (including fits) resulting from the lifestyle of some alcoholic patients [107,109]. A study of the histomorphometry of alcohol-induced bone disease established major changes, even though chronic liver damage was not severe [112]. A significant decrease in bone volume was recorded together with increased resorption surfaces and increased osteoclast number.

Low assayed levels of plasma 25OHD have been reported frequently in patients with alcoholic cirrhosis. The development of vitamin D deficiency in alcoholic liver disease probably has causes in common with the same problem in PBC, namely, malabsorption of vitamin D (in patients with cholestasis) and lack of exposure to sunlight. Poor nutrition in alcoholic patients may also reduce the intake of vitamin D.

2. Vitamin D Metabolism

In contrast to the case of PBC, abnormal hepatocellular function in alcoholic liver disease may be associated with disturbances of vitamin D metabolism. Plasma levels of 25OHD were shown to correlate with the degree of hepatic dysfunction as measured by the antipyrine breath test [113], and defective synthesis of 25OHD was demonstrated by Jung *et al.* [114]. In a subsequent study the impairment was shown to be correlated to plasma levels of bilirubin and prothrombin [103] (see Fig. 5). In this study synthesis of 1,25$(OH)_2$D was also shown to be defective in three of the patients studied and was attributed to the low level of precursor 25OHD and to poor renal function. Assayed concentrations of 1,25$(OH)_2$D were low in a minority of alcoholic patients reported by Lalor *et al.* [106], but despite this evidence for defective vitamin D metabolism, osteomalacia is a comparatively rare component of alcoholic liver disease.

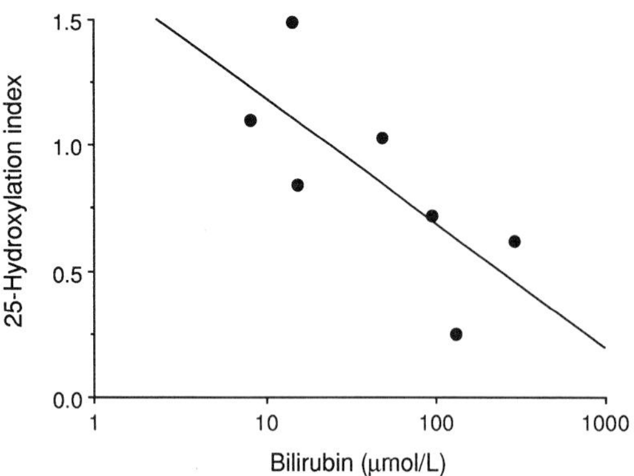

FIGURE 5 Impaired 25-hydroxylation of vitamin D in patients with alcoholic liver disease related to the plasma level of bilirubin; Kendall's coefficient of concordance $\tau = -0.788, p < 0.02$. Data from Mawer *et al.* [103].

3. Management

Alcohol intake should cease, though this is a difficult goal to achieve. Vitamin D and calcium supplements should be given to those with evidence of low serum 25OHD and a poor (<1 g) intake of calcium. This simple maneuver will protect the skeleton from osteomalacia and reduce the tendency to increased bone resorption. If alcohol continues to be consumed to the extent where amnesia is a problem, then compliance with the rigors of bisphosphonate therapy is unlikely. However, reformed alcoholics with osteoporosis seem as likely to respond favorably to antiresorptive therapy as any other group. Monitoring of bone mineral density at the start of therapy with follow-up after 12–18 months will give an indication as to whether treatment should continue long-term.

VI. SUMMARY

It should not be surprising that metabolic bone disease arises when there is chronic disease affecting the gastrointestinal tract. The mineral phase of the skeleton is maintained by the absorption of calcium and other elements from the diet. It has sometimes been difficult to disentangle the effects of the disease itself on the skeleton from any effects which might be exerted by the treatment given for that disease, but understanding of these factors is improving. Interpretation of the pathophysiological mechanisms involved has sometimes been clouded by misinterpretation of data, and there are still many problem areas. The mechanisms that result in low turnover osteoporosis are not fully understood, nor why states of secondary hyperparathyroidism should result in vitamin D wastage. It seems clear, however, that osteomalacia results only from vitamin D deficiency, the predisposition to which may be present in some individuals before development of the gastrointestinal disease, and which can be reversed by appropriate, sometimes parenteral, treatment with the native vitamin. Likewise, HVOi may be managed by large calcium supplements, and perhaps vitamin D, to suppress secondary hyperparathyroidism. It seems that individuals with chronic debilitating bowel or liver disease are programmed to develop osteoporosis through poorly understood disturbances of osteoblast function. Research into this aspect of bone disease needs to be actively pursued, as the lives of many patients with advanced disease can now be prolonged by artificial feeding or by organ transplantation; these procedures themselves may introduce compounding factors into the genesis of metabolic bone disease.

References

1. Peacock M, Selby PL, Francis RM, Brown WB, Hordon L 1985 Vitamin D deficiency, insufficiency and intoxication. What do they mean? In: Norman AW, Schaefer K, Grigoleit H-G, Herrath DV (eds) Vitamin D: Chemical, Biochemical and Clinical Update. de Gruyter, Berlin, pp. 569–570.
2. Krall EA, Sahyoun N, Tannenbaum S, Dallal GE, Dawson-Hughes B 1989 Effect of vitamin D intake on seasonal variations in parathyroid hormone secretion in postmenopausal women. N Engl J Med **321**:1777–1783.
3. Poskitt EME, Cole TJ, Lawson DEM 1979 Diet, sunlight and 25(OH)D in healthy children and adults. Br Med J **1**:221–223.
4. Lawson DEM 1991 Is there a recommended dietary allowance for vitamin D? In: Norman AW, Bouillon R, Thomasset M (eds) Vitamin D: Gene Regulation, Structure–Function Analysis and Clinical Application. de Gruyter, Berlin, pp. 701–707.
5. Stanbury SW, Taylor CM, Lumb GA, Mawer EB, Berry J, Hann J, Wallace J 1981 Formation of vitamin D metabolites following correction of human vitamin D deficiency. Miner Electrolyte Metab **5**:212–227.
6. Thompson GR, Lewis B, Booth CC 1966 Absorption of vitamin D_3-^3H in control subjects and patients with intestinal malabsorption. J Clin Invest **45**:94–102.
7. Meredith SC, Rosenberg IH 1980 Gastro-intestinal–hepatic disorders and osteomalacia. In: Avioli LV, Raisz LG (eds) Clinics in Endocrinology and Metabolism, Vol. 9. Saunders, London, pp. 131–150.
8. Stamp TCB, Haddad JG, Twigg CA 1977 Comparison of oral 25-hydroxycholecalciferol, vitamin D and ultraviolet light as determinants of circulating 25-hydroxyvitamin D. Lancet **1**:1341–1343.
9. Davies M, Mawer EB, Hann JT, Taylor JL 1986 Seasonal changes in the biochemical indices of vitamin D-deficiency in the elderly. A comparison of people in residential homes, long-stay wards and attending a day hospital. Age Aging **15**:77–83.
10. Mawer EB 1980 Clinical implications of measurements of circulating vitamin D metabolites. In: Avioli LV, Raisz LG (eds) Clinics in Endocrinology and Metabolism, Vol. 9. Saunders, London, pp. 63–79.
11. Doppelt SH 1984 Vitamin D, rickets and osteomalacia. Orthop Clin North Am **15**:671–684.

12. Avioli LV, Lee SW, McDonald JE, Lund J, De Luca HF 1967 Metabolism of vitamin D_3-^3H in human subjects: Distribution in blood, bile, feces and urine. J Clin Invest **46**:983–992.
13. De Luca HF, Schnoes HK 1983 Vitamin D: Recent advances. Annu Rev Biochem **52**:411–439.
14. Mawer EB, Backhouse J, Holman CA, Lumb GA, Stanbury SW 1972 The distribution and storage of vitamin D and its metabolites in human tissues. Clin Sci **43**:413–431.
15. Clements MR, Chalmers TM, Fraser DR 1984 Enterohepatic circulation of vitamin D: A reappraisal of the hypothesis. Lancet **1**:1376–1379.
16. Makin G, Lotues D, Buford V, Ray R, Jones G 1989 Target cell metabolism of 1,25-dihydroxyvitamin D_3 to calcitroic acid. Biochem J **262**:173–180.
17. Jones G, Kung M, Kano K 1983 The isolation and identification of the new metabolites of 25-hydroxyvitamin D_3 produced in the kidney. J Biol Chem **258**:12920–12928.
18. Bolt MJG, Jensen WE, Sitrin MD 1992 Metabolism of 25-hydroxyvitamin D_3 in rats: Low calcium diet vs. calcitriol infusion. Am J Physiol (Endocrinol Metab) **262**(25):E359-E367.
19. Arnaud SB, Goldsmith RS, Lambert PW, Go VLW 1975 25-hydroxyvitamin D_3: Evidence of an enterohepatic circulation in man. Proc Soc Exp Biol **149**:570–572.
20. Fraser DR 1983 The physiological economy of vitamin D. Lancet **1**:969–972.
21. Gascon-Barre M 1986 Is there any physiological significance to the enterohepatic circulation of vitamin D sterols? J Am College Nutr **5**:317–324.
22. Mawer EB 1979 The role of the liver in the control of vitamin D metabolism. In: Norman AW (ed) Vitamin D: Basic Research and Its Clinical Application. de Gruyter, Berlin, pp. 533–561.
23. Clements MR, Johnson L, Fraser DR 1987 A new mechanism for induced vitamin D deficiency in calcium deprivation. Nature **325**:62–65.
24. Clements MR, Davies M, Fraser DR, Lumb GA, Mawer EB, Adams PH 1987. Metabolic inactivation of vitamin D is enhanced in primary hyperparathyroidism. Clin Sci **73**:659–664.
25. Clements MR, Davies M, Hayes ME, Hickey CD, Lumb GA, Mawer EB, Adams PH 1992 The role of 1,25-dihydroxyvitamin D in the mechanism of acquired vitamin D deficiency. Clin Endocrinol **37**:17–27.
26. Colston KW, Mackay AG, Finlayson C, Wu JCY, Maxwell JD 1994 Localisation of vitamin D receptor in normal human duodenum and in patients with celiac disease. Gut **35**:1219–1225.
27. Parfitt AM 1990 Osteomalacia and related disorders. In: Avioli LV, Krane SM (eds) Metabolic Bone Disease and Clinically Related Disorders. Saunders, Philadelphia, London, Toronto, Montreal, Sydney, and Tokyo, pp. 329–396.
28. Pittet PG, Davie M, Lawson DEM 1979 Role of nutrition in the development of osteomalacia in the elderly. Nutr Metab **23**:109–116.
29. Dibble JB, Sheridan P, Hampshire R, Hardy GJ, Losowsky MS 1982 Osteomalacia, vitamin D deficiency and cholestasis in chronic liver disease. Q J Med **51**:89–103.
30. Parfitt AM 1983 Dietary risk factors for age related bone loss and fractures. Lancet **2**:1181–1184.
31. Parfitt AM, Podenphant J, Villanueva AR, Frame B 1985 Metabolic bone disease with and without osteomalacia after intestinal bypass surgery. A bone histomorphometric study. Bone **6**:211–220.
32. Rao DS, Villanueva AR, Mathews M, Pumo B, Kleerekoper M, Parfitt AM 1983 Histologic evolution of vitamin D depletion in patients with intestinal malabsorption or dietary deficiency. In: Frame B, Potts JT Jr, (eds) Clinical Disorders of Bone and Mineral Metabolism, International Congress Series 617. Excerpta Medica, Amsterdam, pp. 224–226.
33. Paterson CR, Woods CG, Pulvertaft CN, Fourman P 1965 Search for osteomalacia in 1228 patients after gastrectomy and other operations on the stomach. Lancet **2**:1085–1088.
34. Tovey FI, Karamanolis DG, Godfrey J, Clark CG 1985 Post-gastrectomy nutrition: Methods of outpatient screening for early osteomalacia. Hum Nutr Clin Nutr **39C**:439–446.
35. Eddy RL 1971 Metabolic bone disease after gastrectomy. Am J Med **50**:442–449.
36. Deller DJ, Begley MD, Edwards RG, Addison M 1964 Metabolic effects of partial gastrectomy with special reference to calcium and folic acid. Gut **5**:218–225.
37. Morgan DB, Hunt G, Paterson CR 1970 The osteomalacia syndrome after stomach operations. Q J Med **39**:395–410.
38. Davies M, Mawer EB, Krawitt EL 1980 Comparative absorption of vitamin D_3 and 25-hydroxyvitamin D_3 in intestinal disease. Gut **21**:287–292.
39. Davies M, Heys SE, Selby PS, Berry JL, Mawer EB 1997 Increased catabolism of 25-hydroxyvitamin D in patients with partial gastrectomy and elevated 1,25-dihydroxyvitamin D levels. Implications for metabolic bone disease. J Clin Endocrinol Metab **82**:209–212.
40. Rao RS, Kleerekoper M, Rogers M, Frame B, Parfitt AM 1984 Is gastrectomy a risk factor for osteoporosis? In: Christiansen C, Arnaud CD, Nordin BEC, Parfitt AM, Peck WA, Riggs BL (eds) Osteoporosis: Proceedings of the Copenhagen International Symposium on Osteoporosis. Aalborg Stiftsborgtrykkeri, Copenhagen, pp. 775–777.
41. Bisballe S, Eriksen FF, Melsen F, Mosekilde L, Sorensen O, Hessov I 1991 Osteopenia and osteomalacia after gastrectomy: Interrelations between biochemical markers of bone remodelling, vitamin D metabolites and bone histomorphometry. Gut **32**:1303–1307.
42. Nilas L, Christiansen C, Christiansen J 1985 Regulation of vitamin D and calcium metabolism after gastrectomy. Gut **26**:252–257.
43. Deller DJ, Begley MD 1983 Calcium metabolism and the bones after partial gastrectomy. Clinical features and radiology of bones. Aust Ann Med **12**:282–294.
44. Nilsson BE, Westlin NE 1971 The fracture incidence after gastrectomy. Acta Chir Scand **137**:533–534.
45. Mellstrom D, Johansson C, Johnell O, Lindstedt G, Lundberg PA, Obrant K, Schoon I, Toss G, Ytterberg B 1993 Osteoporosis, metabolic aberrations and increased risk of vertebral fractures after partial gastrectomy. Calcif Tissue Int **53**:370–377.
46. Parfitt AM, Mathews CHE, Villanueva AR, Rao DS, Rogers M, Kleerekoper M, Frame B 1983 Microstructural and cellular basis of age related bone loss and osteoporosis. In: Frame B, Potts JT Jr (eds) Clinical Disorders of Bone and Mineral Metabolism, International Congress Series 617. Excerpta Medica, Amsterdam, pp. 328–332.
47. McKenna MJ, Kleerekoper M, Ellis BI, Rao DS, Parfitt AM, Frame B 1987 Atypical insufficiency fractures confused with Looser zones of osteomalacia. Bone **8**:71–78.
48. Bode S, Hassager C, Gudmand-Hoyer E, Christiansen C 1991 Body composition and calcium metabolism in adult treated celiac disease. Gut **32**:1342–1345.
49. Caraceni MP, Molteni N, Bardella MT, Ortolani S, Gandolini GG, Bianchi P 1988 Bone and mineral metabolism in adult celiac disease. Am J Gastroenterol **83**:274–277.
50. Pistorius LR, Sweidan WH, Purdie DW, Stee SA, Howey S, Bennett JR, Sutton DR 1995 Celiac disease and bone mineral density in adult female patients. Gut **37**:639–642.

51. Walters JRF 1994 Bone mineral density in celiac disease. Gut **35**:150–151.
52. Lindh E, Ljunghall S, Larsson K, Lavo B 1992 Screening for antibodies against gliadin in patients with osteoporosis. J Intern Med **231**:403–406.
53. McFarlane J, Bhalla A, Morgan L, Reeves D, Robertson DAF 1992 Osteoporosis: A frequent finding in treated adult celiac disease. Gut **33**:S48.
54. Butcher GP, Banks LM, Walters JRF 1992 Reduced bone mineral density in celiac disease—the need for bone densitometry estimations. Gut **33**:S54.
55. Selby PL, Davies M, Warnes TW, Adams JE, Mawer EB 1995 Bone metabolism in coeliac disease. J Bone Miner Res **10**(Suppl. 1):S507
56. Molteni N, Caraceni MP, Bardella MT, Ortolani S, Gandolini GG, Bianchi P 1990 Bone mineral density in adult celiac patients and the effect of gluten-free diet from childhood. Am J Gastroenterol **85**:51–53.
57. Moss AJ, Waterhouse C, Terry R 1965 Gluten-sensitive enteropathy with osteomalacia but without steatorrhoea. N Engl J Med **272**:825–830.
58. Hajjar ET, Vincenti F, Salti IS 1974 Gluten-induced enteropathy. Arch Intern Med **134**:565–566.
59. Melvin KEW, Hepner GW, Bordier P, Neale G, Joplin GF 1970 Calcium metabolism and bone pathology in adult celiac disease. Q J Med **39**:83–113.
60. Dibble JB, Sheridan P, Losowsky MS 1984 A survey of vitamin D deficiency in gastrointestinal and liver disorders. Q J Med **209**:119–134.
61. Arnaud SB, Newcomer AD, Dickson ER, Arnaud CD, Go VLW 1976 Altered mineral metabolism in patients with gastrointestinal (GI) diseases. Gastroenterology **70**:860.
62. Corazza GR, Di Sario A, Cecchetti L, Tarozzi C, Corrao G, Bernardi M, Gasbarrini G 1995 Bone mass and metabolism in patients with celiac disease. Gastroenterology **109**:122–128.
63. Hubbard VS, Farrell PM, de Sant'Agnese PA 1979 25-Hydroxycholecalciferol levels in patients with cystic fibrosis. J Pediatr **94**:84–86.
64. Hahn TJ, Squires AE, Halstead LR, Strominger DB 1979 Reduced serum 25-hydroxyvitamin D concentration and disordered mineral metabolism in patients with cystic fibrosis. J Pediatr **94**:38–42.
65. Krawitt EL, Beeken WL, Janney CD 1976 Calcium absorption in Crohn's disease. Gastroenterology **71**:251–254.
66. Driscoll RH, Meredith SC, Sitrin M, Rosenberg IH 1982 Vitamin D deficiency and bone disease in patients with Crohn's. Gastroenterology **83**:1252–1258.
67. Compston JE, Ayers AB, Horton LW, Tighe JR, Creamer B 1978 Osteomalacia after small intestinal resection. Lancet **1**:9–12.
68. Compston JE, Judd D, Crawley EO, Evans, WD, Evans C, Church HA, Reid EM, Rhodes J 1987 Osteoporosis in patients with inflammatory bowel disease. Gut **28**:410–415.
69. Geinoz G, Rapin CH, Rizzoli R, Kraemer R, Buchs B, Slosman D, Michel JP, Bonjour JP 1993 Relationship between bone mineral density and dietary intakes in the elderly. Osteoporosis Int **3**:242–248.
70. Atkinson RL, Dahms WT, Bray GA, Schwartz AA 1978 Parathyroid hormone in obesity: Effects of intestinal bypass. Miner Electrolyte Metab **1**:315–320.
71. Dano P, Christiansen C 1978 Calcium malabsorption and absence of bone decalcification following intestinal shunt operation for obesity. Scand J Gastroenterol **13**:81–85.
72. Teitelbaum SL, Halverson JD, Bates M, Wise L, Haddad JG 1977 Abnormalities of circulating 25-OH vitamin D after jejunoileal bypass for obesity and evidence of an adaptive response. Ann Intern Med **856**:288–293.
73. Rickers H, Christiansen C, Balsev I, Rodbro P 1984 Impairment of vitamin D metabolism and bone mineral content after intestinal bypass for obesity. Scand J Gastroenterol **19**:184–189.
74. Hey H, Stokholm KH, Lund BJ, Lund BI, Sorensen OH 1982 Vitamin D deficiency in obese patients and changes in circulating vitamin D metabolites following jejunoileal bypass. Int J Obesity **6**:473–479.
75. Mosekilde L, Melsen F, Hessov I, Christensen MS, Lund BJ, Lund BI, Sorensen OH 1980 Low serum levels of 1,25 dihydroxyvitamin D and histomorphometric evidence of osteomalacia after intestinal bypass for obesity. Gut **21**:624–631.
76. Compston JE, Horton LWL, Laker MF, Merrett AL, Woodhead JS, Gazet J-C, Pilkington TRE 1980 Treatment of bone disease after jejunoileal by-pass for obesity with oral 1-hydroxyvitamin D_3. Gut **21**:669–674.
77. Klein GL, Chesney RW 1986 Metabolic bone disease associated with total parenteral nutrition. In: Lebenthal E (ed) Total Parenteral Nutrition: Indication, Utilization, Complications and Pathophysiological Considerations. Raven, New York, pp. 431–442.
78. Klein GL, Alfrey AC, Miller ML, Sherrard DJ, Hazlet TK, Ament ME, Coburn JW 1982 Aluminum loading during total parenteral nutrition. Am J Clin Nutr **35**:1425–1429.
79. Klein GL, Targoff CM, Ament ME, Sherrard DJ, Bluestone R, Young JH, Norman AW, Coburn JW 1980 Bone disease associated with total parenteral nutrition. Lancet **2**:1041–1044.
80. Vargas JH, Klein GL, Ament ME, Ott SM, Sherrard DJ, Horst RL, Berquist WE, Alfrey AC, Slatopolsky E, Coburn JW 1988 Metabolic bone disease of total parenteral nutrition: Course after changing from casein to amino acids in parenteral solutions with reduced aluminum content. Am J Clin Nutr **48**:1070–1078.
81. Bengoa JM, Sitrin MD, Wood RJ, Rosenberg IH 1983 Amino acid induced hypercalciuria in patients on total parenteral nutrition. Am J Clin Nutr **38**:264–269.
82. Lipkin EW, Ott SM, Klein GL 1987 Heterogeneity of bone histology in parenteral nutrition patients. Am J Clin Nutr **46**:673–680.
83. Shike M, Shils ME, Heller A, Alcock N, Vigorita V, Brockman R, Holick MF, Lane J, Flombaum C 1986 Bone disease in prolonged parenteral nutrition: Osteopenia without mineralization defect. Am J Clin Nutr **44**:89–98.
84. Jung RT, Davie M, Siklos P, Chalmers TM, Lawson DEM 1979 Vitamin D metabolism in acute and chronic cholestasis. Gut **20**:840–847.
85. Recker RR, Maddrey W, Herlong F, Sorrell M, Russell R 1983 Primary biliary cirrhosis and alcoholic cirrhosis as examples of chronic liver disease associated with bone disease. In: Frame B, Potts JT Jr, (eds) Clinical Disorders of Bone and Mineral Metabolism, International Congress Series 617. Excerpta Medica, Amsterdam, pp. 227–230.
86. Stellon AJ, Webb A, Compston J, Williams R 1987 Low bone turnover state in primary biliary cirrhosis. Hepatology **7**:137–142.
87. Atkinson M, Nordin BEC, Sherlock S 1956 Malabsorption and bone disease in prolonged obstructive jaundice. Q J Med **25**:299–312.
88. Long RG, Varghese Z, Meinhard EA, Skinner RK, Wills MR, Sherlock S 1978 Parenteral 1,25-dihydroxycholecalciferol in hepatic osteomalacia. Br Med J **1**:75–77.
89. Compston JE, Crowe JP, Wells IP, Horton LWL, Hirst D, Merrett AL, Woodhead JS, Williams R 1980 Vitamin D prophylaxis and osteomalacia in chronic cholestatic liver disease. Dig Dis Sci **25**:28–32
90. Davies M, Mawer EB, Klass HJ, Lumb GA, Berry JL, Warnes

TW 1983 Vitamin D deficiency osteomalacia and primary biliary cirrhosis: The response to orally administered vitamin D_3. Dig Dis Sci **28**:145–153.

91. Compston JE, Horton LWL, Thompson RPH 1979 Treatment of osteomalacia associated with PBC with parenteral vitamin D_2 or oral 25-hydroxyvitamin D_3. Gut **20**:133–136.
92. Compston JE, Crowe JP, Horton LWL 1979 Treatment of osteomalacia associated with primary biliary cirrhosis with oral 1 alpha-hydroxyvitamin D_3. Br Med J **2**:309.
93. Danielson A, Lorentzon R, Larsson S-E 1982 Normal hepatic vitamin D metabolism in icteric primary biliary cirrhosis associated with pronounced vitamin D deficiency symptoms. Hepatogastroenterology **29**:6–8.
94. Long RG, Skinner RK, Wills MR, Sherlock S 1976 Serum 25-hydroxy-vitamin-D in untreated parenchymal and cholestatic liver disease. Lancet **2**:650–652.
95. Atkinson MJ, Vido I, Keck E, Hesch RD 1983 Hepatic osteodystrophy in primary biliary cirrhosis: A possible defect in Kupffer cell mediated cleavage of parathyroid hormone. Clin Endocrinol **118**:21–28.
96. Klein GL, Endres DB, Colonna II JD, Berquist WE, Goldstein LI, Busuttil RW, Deftos LJ 1989 Absence of hyperparathyroidism in severe liver disease. Calcif Tissue Int **44**:330–334.
97. Thompson WG, Thompson GR 1969 Effect of cholestyramine on the absorption of vitamin D_3 and calcium. Gut **10**:717–722.
98. Hodgson SF, Dickson ER, Eastell R, Eriksen EF, Bryant SC, Riggs BL 1993 Rates of cancellous bone remodelling and turnover in osteopenia associated with primary biliary cirrhosis. Bone **14**:819–827.
99. Wagonfield JB, Nemchausky BA, Bolt M, Horst JV, Boyer JH, Rosenberg IH 1976 Comparison of vitamin D and 25-hydroxyvitamin D in the therapy of primary biliary cirrhosis. Lancet **2**:391–394.
100. Krawitt EL, Grundman MJ, Mawer EB 1977 Absorption, hydroxylation and excretion of vitamin D_3 in primary biliary cirrhosis. Lancet **2**:1246–1249.
101. Davies M, Mawer EB 1979 The absorption and metabolism of vitamin D_3 from subcutaneous and intramuscular injection sites. In: Norman AW (ed) Vitamin D: Basic Research and Its Clinical Application. de Gruyter, Berlin, pp. 609–612.
102. Reed JS, Meredith SC, Nemchausky BA, Rosenberg IH, Boyer JL 1980 Bone disease in primary biliary cirrhosis: Reversal of osteomalacia by 25-hydroxyvitamin D. Gastroenterology **78**:512–517.
103. Mawer EB, Klass HJ, Warnes TW, Berry JL 1985 Metabolism of vitamin D in patients with primary biliary cirrhosis and alcoholic liver disease. Clin Sci **69**:561–570.
104. Lindor KD, James CH, Crippin JS, Jorgensen RA, Dickson ER 1995 Bone disease in primary biliary cirrhosis: Does urodeoxycholic acid make a difference? Hepatology **21**:389–392.
105. Camisasca M, Crosignani A, Battezzati PM, Albisetti W, Grandinetti G, Pietrogrande L, Biffi A, Zuin M, Podda M 1994 Parenteral calcitonin for metabolic bone disease associated with primary biliary cirrhosis. Hepatology **20**:633–637.
106. Lalor BC, France MW, Powell D, Adams PH, Counihan TB 1986 Bone and mineral metabolism and chronic alcohol abuse. Q J Med **229**:497–511.
107. Israel Y, Orrego H, Holt S, Macdonald DW, Meema HE 1980 Identification of alcohol abuse: Thoracic fractures on routine x-rays as indicators of alcoholism. Alcohol Clin Exp Res **4**:420–422.
108. Nilsson BE 1970 Conditions contributing to fracture of the femoral neck. Acta Chir Scand **136**:383–384.
109. Peris P, Guanabens N, Pares A, Pons F, de Rio L, Monegal A, Suris X, Caballeria J, Rodes J, Munoz-Gomez J 1995 Vertebral fractures and osteopenia in chronic alcoholic patients. Calcif Tissue Int **57**:111–114.
110. Diamond T, Stiel D, Lunzer M, Wilkinson M, Posen S 1989 Ethanol reduces bone formation and may cause osteoporosis. Am J Med **86**:282–288.
111. Crilly RG, Anderson C, Hogan D, Delaquerriere-Richardson L 1988 Bone histomorphometry, bone mass, and related parameters in alcoholic males. Calcif Tissue Int **43**:269–276.
112. Diez A, Puig J, Serrano S, Marinoso M-L, Bosch J, Marrugat J, Mellibovsky L, Nogues X, Knobel H, Aubia J 1994 Alcohol-induced bone disease in the absence of severe chronic liver damage. J Bone Miner Res **9**:825.
113. Hepner GW, Roginsky M, Moo HF 1986 Abnormal vitamin D metabolism in patients with cirrhosis. Dig Dis **21**:527–532.
114. Jung RT, Davie M, Hunter JO, Chalmers TM, Lawson DEM 1978 Abnormal vitamin D metabolism in cirrhosis. Gut **19**:290–293.

CHAPTER 52

Vitamin D and Renal Failure

EDUARDO S. SLATOPOLSKY Renal Division, Department of Medicine, Washington University School of Medicine; Chromalloy American Kidney Center; and Barnes-Jewish Hospital, St. Louis, Missouri

ALEX J. BROWN Renal Division, Department of Medicine, Washington University School of Medicine, St. Louis, Missouri

I. INTRODUCTION

Abnormal vitamin D metabolism and action in chronic renal failure play key roles in the development of secondary hyperparathyroidism and renal osteodystrophy. Research performed since the late 1970s has shown that the kidney is primarily responsible for the production of 1,25-dihydroxyvitamin D [1,25$(OH)_2$D, calcitriol]. Thus, the kidney tubules represent an endocrine tissue that normally carries out the conversion of 25-hydroxyvitamin D (25-OHD), a steroid prohormone, to its active form. As such, the kidney is responsible for a critical metabolic step in calcium and phosphorus homeostasis.

In renal failure a series of alterations characterized by abnormal production, altered metabolism, decreased number of receptors, and resistance to the action of vitamin D are of utmost importance in the pathophysiology of metabolic bone disease and abnormal mineral metabolism. In addition, alterations in vitamin D metabolism in patients with chronic renal failure have been implicated in the impaired growth and development, abnormal immune function, and altered skeletal muscle function frequently seen in these patients.

II. VITAMIN D METABOLISM

A. Reduced 1,25$(OH)_2$D Synthetic Capacity

1. DECREASED RENAL MASS

Because the kidney is the primary organ responsible for the production of 1,25$(OH)_2$D, it is easy to understand that patients with a reduced renal mass will have decreased levels of 1,25$(OH)_2$D production. Although it is well known that patients with advanced renal insufficiency, especially those maintained on chronic hemodialysis, have very low levels or even undetectable levels of $1,25(OH)_2D$, it is still controversial as to the degree of renal insufficiency at which the levels of $1,25(OH)_2D$ begin to fall. Several investigators have shown a progressive decrease in the serum levels of $1,25(OH)_2D$ as renal function deteriorates, although it is still uncertain if the decrease appears at the same degree of renal insufficiency in all patients.

Slatopolsky *et al.* [1] measured $1,25(OH)_2D$ levels in a group of adults with normal renal function [glomerular filtration rate (GFR) of 120 ml/min], patients with mild decreases in GFR (GFR between 40 and 60 ml/min), patients with advanced renal failure (GFR between 5 and 20 ml/min), and patients on dialysis. The plasma levels of $1,25(OH)_2D$ were normal in patients with creatinine clearance rates of 40–60 ml/min and low in patients with more advanced renal disease. Cheung *et al.* [2] found similar results. Other investigators, however, found low circulating levels of $1,25(OH)_2D$ in patients with moderate renal insufficiency [3–5].

If one assumes that a patient with a GFR of 50 ml/min has lost roughly one-half of the renal mass and yet maintains normal levels of $1,25(OH)_2D$, the production of $1,25(OH)_2D$ per unit of renal mass must be increased. Martinez *et al.* [6] studied a group of 165 patients with various degrees of chronic renal failure. These studies revealed that lower levels of $1,25(OH)_2D$ become apparent when the GFR is 65 ml/min or less (Fig. 1). Moreover, patients with mild renal insufficiency do not have calcium malabsorption. Coburn *et al.* [7] measured intestinal calcium absorption using ^{47}Ca isotope techniques in patients with different degrees of renal insufficiency. They found that patients with mild renal failure (serum creatinine <2.5 mg/dl) had the same calcium absorption

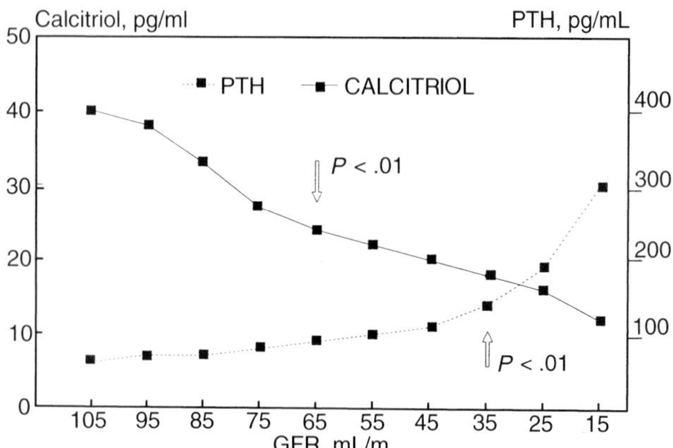

FIGURE 1 Correlation between serum $1,25(OH)_2D$ (calcitriol) and parathyroid hormone (PTH) in a group of 165 patients with different degrees of chronic renal failure. Arrows indicate points where the decrease in $1,25(OH)_2D$ and increase in PTH reach statistical significance. From Martinez *et al.* [6].

as normal volunteers (serum creatinine <1.0 mg/dl). Malluche *et al.* [8] also studied calcium absorption in a large group of patients with renal disease. They found that as the GFR decreased there was a concomitant decline in the absorption of calcium. However, this abnormality became evident only at GFR values below 50 ml/min. Although the levels of $1,25(OH)_2D$ in most patients with advanced renal insufficiency (GFR <20 ml/min) are low, many patients lack bone histological features of vitamin D deficiency.

The discrepant results for the degree of renal insufficiency necessary to produce low levels of $1,25(OH)_2D$ may have several explanations. When a GFR is measured in patients with renal insufficiency, the pathological process responsible for the decrease in renal function may be quite variable. Diseases that affect the glomerulus may not have the same consequence as those that affect primarily the proximal tubule where $1,25(OH)_2D$ is produced. Moreover, parathyroid hormone (PTH) is a key regulator of $1,25(OH)_2D$ synthesis, and the degree of secondary hyperparathyroidism may be very different in patients who have renal failure for short periods of time (e.g., vasculitis) or for 10–20 years as with membranous glomerulonephritis. Other contributing factors such as serum phosphorus and calcium levels and acidosis also may play an important role. For example, Prince *et al.* [9] found that $1,25(OH)_2D$ levels could be augmented in renal failure by calcium restriction, indicating a reserve of synthetic capacity in the remnant kidney, but the increase was much less than in normal subjects, suggesting an active suppression of $1,25(OH)_2D$ production in renal failure. Thus, as most studies to date have not fully characterized all the factors that may influence the production of $1,25(OH)_2D$ it is difficult to compare the results found by different laboratories.

2. SUPPRESSION BY HYPERPHOSPHATEMIA

The body stores, blood levels, or trancellular flux of inorganic phosphate affect the activity of renal 25OHD-1α-hydroxylase. Interest in a possible role of phosphorus on vitamin D metabolism arose because of the similarities between the skeletal findings in phosphate depletion and those seen in osteomalacia due to vitamin D deficiency. Studies in rats [10] and humans [11] indicated that intestinal absorption of calcium is augmented by dietary phosphorus restriction, suggesting an effect on vitamin D metabolism. The total generation of $1,25(OH)_2D$ from 25OHD and activity of the 1α-hydroxylase varied with serum phosphorus concentration [10]. Moreover, dietary phosphorus restriction augmented the activity of the 1α-hydroxylase in intact chicks [12] and increased the plasma concentration of $1,25(OH)_2D$ and its renal synthesis in intact rats [13].

In normal men phosphorus restriction has been shown to increase either the serum concentration or the rate of production of 1,25(OH)$_2$D despite a decrease in the serum levels of PTH. In patients with hyperparathyroidism, those with idiopathic hypercalciuria, and children with moderate renal insufficiency, oral phosphate supplements caused a decrease in the serum concentration of 1,25(OH)$_2$D [14]. It has been proposed that either the intracellular concentration of phosphate or its transcellular flux may be important in regulating the activity of the 1α-hydroxylase. An interesting aspect of the regulation of 1,25(OH)$_2$D by hypophosphatemia is the role of somatomedins (i.e., insulin-like growth factors). Hypophysectomy blocks the stimulating action of dietary phosphate restriction on 1α-hydroxylase activity [15], but injection of growth hormone to hypophysectomized rats restores this response to low dietary phosphorus.

3. Suppression by Acidosis

Metabolic acidosis has also been shown to alter vitamin D metabolism. Patients with renal tubular acidosis may have a normal GFR but defective tubular transport activities, and they may display skeletal abnormalities that are diagnosed as rickets in children and osteomalacia in adults. The bone disease associated with chronic metabolic acidosis can be attributed to a disturbance in mineral metabolism. Although acidosis produces calciuria by reducing renal reabsorption of calcium, there is usually no increase in intestinal absorption of calcium [16a], owing to a decrease in renal synthesis of 1,25(OH)$_2$D [17,18]. Acidosis does not appear to affect the calcitonin-responsive 1α-hydroxylase, but it does blunt the action of PTH on the 1α-hydroxylase of the proximal convoluted tubule [19] as well as other renal responses to PTH such as phosphate reabsorption [20]. However, cyclic AMP restores the 1α-hydroxylase activity to normal in acidotic rats [19]. Because of PTH responsiveness in acidosis is not due to a loss of the renal PTH receptor, the blunted cyclic AMP production is attributed to altered coupling of the receptor with adenylate cyclase [21]. Acidosis has also been shown to increase the activity of the renal 24-hydroxylase [17], which will enhance the rate of degradation of 1,25-(OH)$_2$D. Another consequence of acidosis is an increase in ionized calcium due to the lower blood pH. It is possible that this increase in ionized calcium may be totally responsible for the altered vitamin D metabolism, as EGTA, a calcium chelator, can block the effects of acidosis on 1,25(OH)$_2$D levels without changing blood pH [22]. Thus, acidosis appears to decrease 1,25(OH)$_2$D levels by raising serum ionized calcium and by decreasing the responsiveness of the kidney to PTH.

4. Role of PTH

The conversion of 25OHD to 1,25(OH)$_2$D has been studied in various laboratories and clinical models in the presence of an excess or deficit of PTH. The *in vivo* studies by Garabedian *et al.* [23] in the vitamin D-deficient rat showed a decrease in conversion of radiolabeled 25OHD$_3$ to 1,25(OH)$_2$D$_3$ after parathyroidectomy. Hughes *et al.* [10] found decreased plasma levels of 1,25(OH)$_2$D in parathyroidectomized vitamin D-replete rats compared with levels in rats with intact parathyroid glands. Studies by Henry *et al.* [24] with vitamin D-replete chicks and by Booth *et al.* [25] with vitamin D-deficient chicks showed that the 1α-hydroxylase activity decreased after parathyroidectomy [25]. Booth and associates prevented the decrease in serum calcium that occurs after parathyroidectomy by infusing calcium, thus eliminating the possibility that reduction in serum calcium levels were responsible for altering the enzyme activity. Rasmussen *et al.* [26], studying renal tubules from vitamin D-deficient chicks, found that PTH increased the conversion of labeled 25OHD$_3$ to 1,25(OH)$_2$D$_3$.

Clinical observations have provided the most substantial support for a role of PTH in stimulating the generation of 1,25(OH)$_2$D. Increased plasma levels of 1,25(OH)$_2$D have been found in patients with primary hyperparathyroidism [27], and blood levels of 1,25(OH)$_2$D rose when normal subjects were given PTH [28]. Furthermore, blood levels of 1,25(OH)$_2$D were low in patients with hypoparathyroidism [27]. Gray *et al.* [29] found a correlation between intestinal calcium absorption and increased plasma levels of 1,25(OH)$_2$D in patients with primary hyperparathyroidism, providing evidence that increased intestinal absorption of calcium arises in primary hyperparathyroidism due to increased conversion of 25OHD to 1,25(OH)$_2$D. PTH also decreases 24-hydroxylase activity and slows the degradation of 1,25(OH)$_2$D [30,31]. It is still unclear how cyclic AMP modulates vitamin D hydroxylation, but protein synthesis is required [32]. Cyclic AMP can also stimulate phosphorylation of specific proteins by activation of cyclic AMP-dependent protein kinases, but the role of these phosphoproteins in vitamin D metabolism has not been established.

5. Role of Uremic Toxins

Hsu and Patel [33] have demonstrated a role for uremic toxins in the suppression of 1,25(OH)$_2$D synthesis. They showed that infusion of uremic plasma ultrafiltrate into normal rats reduced the 1,25(OH)$_2$D production rate. Characterization of plasma ultrafiltrates from normal and uremic patients by high-performance liquid chromatography demonstrated low molecular weight substances that suppress 1,25$(OH)_2$D production in normal rats. In addition, the same authors demonstrated, in partially nephrectomized rats, that the admininstration of high protein diets further decreased 1,25$(OH)_2$D production. The authors proposed that the high protein diet increases the production of uremic toxins that suppress 1,25$(OH)_2$D synthesis.

6. Role of Calcitriol

Likely the most important determinant of serum 1,25$(OH)_2$D content is the actual level of 1,25$(OH)_2$D itself. The vitamin D hormone can act through a feedback mechanism to control its serum levels by inhibiting the 1α-hydroxylase and stimulating enzymatic oxidation of its side chain initiated by 24-hydroxylase [34,35]. The action of 1,25$(OH)_2$D on the renal 1α-hydroxylase in cell culture is evident after *only* several hours, requires protein synthesis, and is reversed by removal of the hormone from the medium [31]. In renal failure, loss of renal mass and drop in 1,25$(OH)_2$D production should stimulate the expression or activity of the 1α-hydroxylase in the remnant kidney. However, once GFR drops below 60 ml/min, the stimulating signal can no longer normalize 1,25$(OH)_2$D levels, presumably because of the counterregulatory actions of other factors mentioned above.

7. Extrarenal Production of 1,25$(OH)_2$D

Dusso and collaborators found extrarenal production of 1,25$(OH)_2$D in dogs with experimental renal failure [36] and anephric patients [37]. After pharmacological doses of 25OHD_3 were given to the uremic dogs, the levels of 1,25$(OH)_2D_3$ increased from 16.4 ± 0.9 to 28.0 ± 1.9 pg/ml. Linear regression analysis of the relationship between serum concentration of 1,25$(OH)_2D_3$ versus 25OHD_3 for each dog during this period showed highly significant correlation. To further evaluate the possibility that extrarenal sites contribute to the enhanced production of 1,25$(OH)_2D_3$ synthesis after the administration of pharmacological doses of 25OHD_3, similar studies were performed in four anephric patients undergoing hemodialysis. Serum levels of 1,25$(OH)_2D_3$ were 5.5 ± 2.14 pg/ml and increased to 19.6 ± 5 pg/ml after 25OHD_3 administration (Fig. 2). Again a significant correlation was found for the relationship between the serum levels of 1,25$(OH)_2D_3$ and 25OHD_3 in anephric patients. Of interest is the fact that the administration of 25OHD_3 to four normal volunteers produced no significant changes in the levels of 1,25$(OH)_2D_3$. These studies suggest that supraphysiological levels of 25OHD_3 can augment circulating levels of 1,25$(OH)_2D_3$ in the absence of renal mass, indicating extrarenal production of 1,25$(OH)_2D_3$. In contrast to the renal regulation of the 1α-hydroxylase, the extrarenal enzyme is not

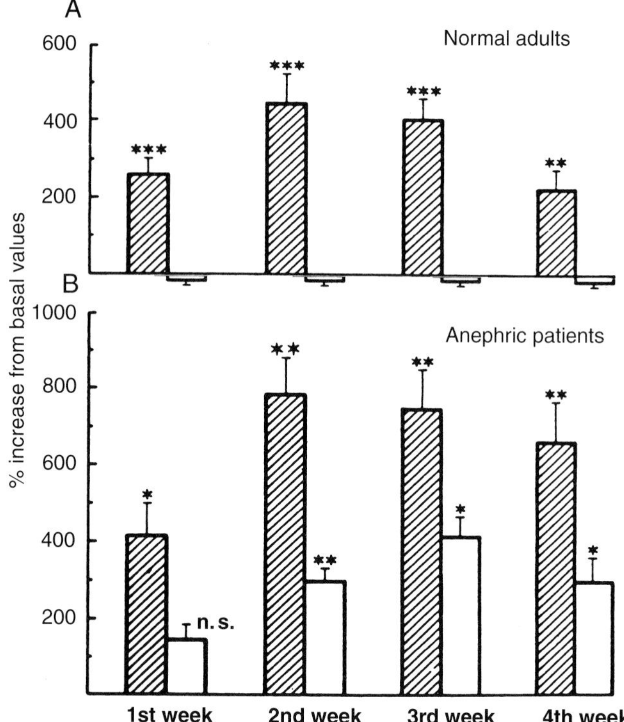

FIGURE 2 Percent increase from basal values of serum 25OHD_3 (hatched bars) and 1,25$(OH)_2D_3$ (open bars) during oral 25OHD_3 administration in (A) four normal adults and (B) four anephric patients. 25OHD_3 was administered during the first 2 weeks. Asterisks denote statistically significant differences compared to basal values: *, $p < 0.05$; **, $p < 0.01$; ***, $p < 0.001$; n.s., not significant. Used with permission from *Kidney International,* vol. 34, p. 368, 1988.

regulated either by PTH or by phosphorus or calcium. It would seem that 1,25$(OH)_2$D per se plays a key role in the regulation of the extrarenal enzyme [37].

III. ALTERATIONS IN 1,25$(OH)_2$D CLEARANCE, METABOLISM, AND ACTION

A. Metabolic Clearance Rate

The levels of serum 1,25$(OH)_2$D depend on a balance between the production rate and the metabolic clearance rate. As discussed above, numerous investigators have demonstrated, in both humans and experimental animals, that the production rate of 1,25$(OH)_2$D is deceased in chronic renal failure (CRF), although disagreement may exist in regard to the level of GFR that usually is accompanied by lower levels of serum 1,25$(OH)_2$D. However, controversial results have been published in regard to the catabolism of 1,25$(OH)_2$D. Dusso and collaborators [38] have performed extensive studies in dogs with experimental renal failure and patients with advanced renal insufficiency and found that the metabolic clearance rate for $1,25(OH)_2D$ is normal. On the other hand, Hsu and collaborators [39] demonstrated a decrease in metabolic clearance rate in rats with renal insufficiency, suggesting a potential compensatory mechanism to maintain the levels of $1,25(OH)_2D$ close to normal. Regardless of these two different points of view (normal versus decreased metabolic clearance rate), it is obvious that even if there is a decrease in the metabolic clearance rate, the predominant factor for the low levels of $1,25(OH)_2D$ is the reduced production rate. In other words, the decrease in metabolic clearance rate cannot compensate for the decrease in production rate of $1,25(OH)_2D$.

B. Vitamin D Responsiveness

The low levels of serum $1,25(OH)_2D$ in CRF may be further compounded by resistance to this vitamin D hormone. Nearly all of the actions of $1,25(OH)_2D$ are mediated by the intracellular vitamin D receptor (VDR). There is now evidence that the VDR content may be decreased in selected target tissues in renal failure. Furthermore, Patel and colleagues have demonstrated decreased DNA binding to vitamin D response elements and reduced transcriptional activity of the VDR in uremia as well [40]. These alterations are detailed below, and the biological implications are discussed in Section IV below.

1. Decreased VDR Expression in Chronic Renal Failure (CRF)

The first report of decreased VDR expression in CRF was by Korkor [41], who found lower $1,25(OH)_2D_3$ binding activity in parathyroid glands from CRF patients than in parathyroid adenomas or parathyroid glands from renal transplant recipients. The decrease was attributed to a decrement in maximal binding with no change in the affinity for $1,25(OH)_2D_3$. The VDR levels correlated positively with renal function and serum calcium and negatively with serum phosphorus and PTH levels. More recently, Fukuda *et al.* [42] performed detailed immunohistochemistry on parathyroid tissue from patients with CRF and found a greater decrement in VDR staining in the nodular portions than in the diffusely hyperplastic regions of the glands. The nature of the histological distinction is as yet unclear, but it has been suggested that the nodules may represent monoclonal growths within the proliferating parathyroid glands.

Decreased parathyroid gland VDR content has been confirmed in several subsequent studies using animal models of experimental uremia, which allowed comparison of VDR levels in CRF versus normal controls. Merke *et al.* [43] found that VDR binding was reduced by approximately 40% in rats 6 days after subtotal nephrectomy. Brown *et al.* [44] reported a greater decrease (75%) in dogs that had been uremic for a longer time (2 to 18 months) (Fig. 3). Not all reports have concurred with these findings. Szabo *et al.* [45] suggested that the decreased VDR content in uremic rat parathyroid glands was due to *ex vivo* degradation. When protease inhibitors were included during tissue preparation, VDR levels were found to be higher in the uremic glands. Denda *et al.* [46], however, reported lower VDR content in uremic rat parathyroid glands despite the use of protease inhibitors (Fig. 4).

The reason for the disparate results in the uremic rat model is unclear. Examination of the serum chemistries in the two studies revealed differences in serum phosphorus. In the study by Szabo *et al.* [45], the uremic rats had lower serum P than the normal controls, whereas the uremic rats in the study by Denda *et al.* [46] had higher serum P. Whether serum P plays a role in the

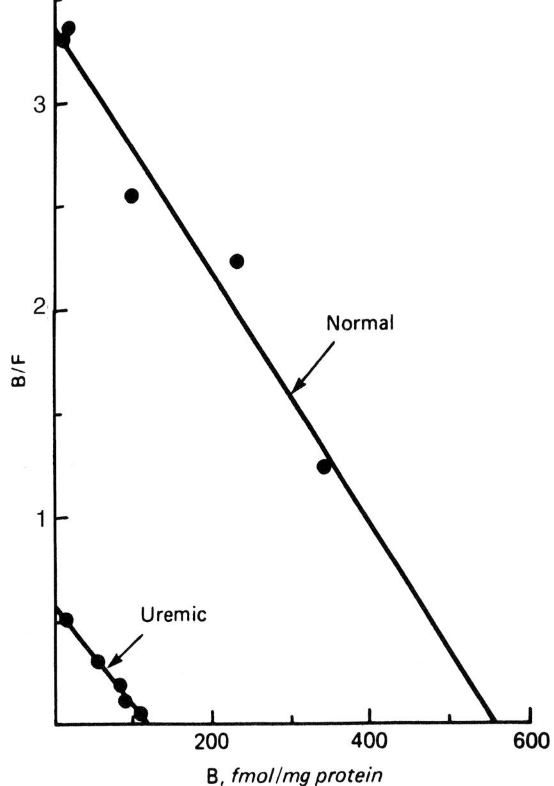

FIGURE 3 Scatchard analysis of specific binding of $1,25(OH)_2D_3$ by cytosols prepared from the parathyroid glands of normal and uremic dogs. Cytosols were incubated with ^3H-$1,25(OH)_2D_3$ (0.1 to 2.0 n*M* for 4 hr at 0°C), and bound and free ligand were separated using the hydroxyapatite method. From Brown *et al.* [44].

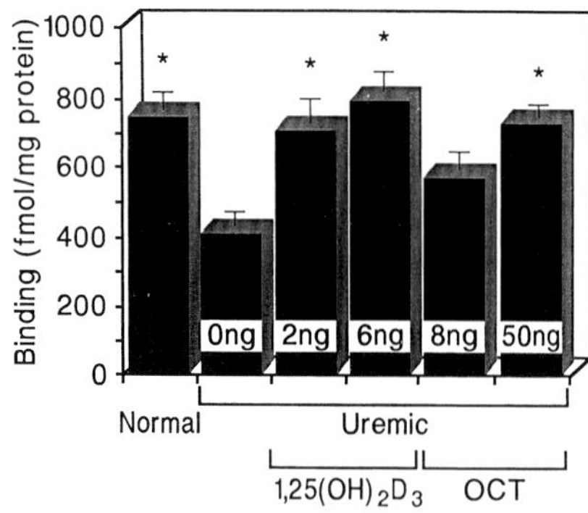

FIGURE 4 Effects of $1,25(OH)_2D_3$ and 22-oxacalcitriol (OCT) on VDR content in parathyroid glands. VDR content was decreased in parathyroid glands of uremic rats compared to normal rats and was corrected by treatment with various doses of $1,25(OH)_2D_3$ or OCT. Used with permission of *Kidney International,* vol. 50, p. 34, 1996.

regulation of parathyroid gland VDR is under investigation, but preliminary studies suggest little difference in VDR expression in uremic rat parathyroid glands with diets containing 0.2 or 0.8% P, dietary conditions that drastically alter PTH secretion rate (E. S. Slatopolsky and A. J. Brown, unpublished data, Sept. 1996).

Down-regulation of the VDR in the parathyroid glands in uremic rats appears to be posttranscriptional. Szabo *et al.* [47] and Shvil *et al.* [48] found no differences in the VDR mRNA in normal and uremic rats. Denda *et al.* [46] observed a strong correlation between serum $1,25(OH)_2D$ levels and the VDR binding activity in the parathyroid glands of uremic rats, suggesting that the decreased VDR levels may be related to the low $1,25(OH)_2D$ levels (Fig. 5). Low doses of $1,25(OH)_2D_3$ (2 or 6 ng, 3 times per week) increased the VDR concentration to normal levels. Because there was no regulation of VDR mRNA by these low doses of $1,25(OH)_2D_3$ (E. S. Slatopolsky and A. J. Brown, unpublished data,

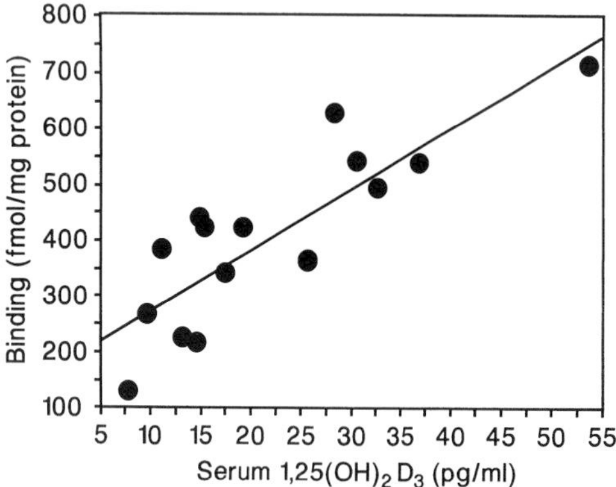

FIGURE 5 Correlation between parathyroid VDR content and serum $1,25(OH)_2D$ levels in uremic and control animals; $r = 0.849$, $p = 0.001$, $n = 15$. Used with permission of *Kidney International,* vol. 50, p. 34, 1996.

Sept. 1996), in agreement with Shvil *et al.* [48], the regulation VDR binding activity by $1,25(OH)_2D_3$ appears to be due to ligand-dependent stabilization of the VDR protein.

VDR levels in the intestine of uremic rats have also been measured by several investigators with disparate results. Szabo *et al.* [45] reported that VDR binding in uremic rat intestine was higher than in normal rat intestine. In contrast, Hsu and colleagues observed decreased VDR binding activity in uremic rat intestine [49,50], despite higher levels of VDR mRNA. Koyama *et al.* [51] found no change in intestinal VDR content by binding assay, or by immunoblot or immunoprecipitation analysis. The reason for these conflicting findings is not obvious.

Examination of the regulation of the VDR in uremic rat intestine by $1,25(OH)_2D_3$ has also produced mixed results. Koyama *et al.* [51] observed increases in VDR mRNA in uremic rat intestine within 5 hr, with a slight increase in VDR protein between 12 and 24 hr. This differed from normal rat intestine, in which VDR mRNA was not significantly affected but VDR protein rose by 12 hr and remained elevated at 24 hr after $1,25(OH)_2D_3$ treatment. The blunted up-regulation of VDR protein in uremic rat intestine by $1,25(OH)_2D_3$ may be due to the inhibitory effects of the high PTH levels in these animals. Reinhardt and Horst [52] reported that induction of hyperparathyroidism by calcium restriction prevents up-regulation of intestinal VDR by $1,25(OH)_2D_3$. However, Patel *et al.* [50] found that $1,25(OH)_2D_3$ treatment of both uremic and normal rats led to similar percent increases in VDR binding after 18 hr that were preceded by transiently increased VDR mRNA. Takahashi *et al.* [53] reported that low doses of $1,25(OH)_2D_3$ increased VDR binding slightly and that decreasing $1,25(OH)_2D_3$ levels by administering the analog 19-nor-$1,25(OH)_2D_2$ led to a drop in intestinal VDR content. The VDR levels are also reduced in peripheral monocytes from renal failure patients [54]. Other tissues have not been examined.

2. DEFECTS IN VDR ACTIVITY

Resistance to $1,25(OH)_2D$ in CRF also appears to occur at the level of VDR activity. Patel and colleagues [55] have presented evidence that components in the blood of renal failure patients diminishes the ability of the VDR to bind to DNA. Intestinal VDR from uremic rats was found to elute from DNA–cellulose at a lower salt concentration, indicating reduced affinity for DNA [55]. They also observed that increasing concentrations of plasma ultrafiltrates from renal failure patients produced a graded decrease in VDR binding affinity for DNA in normal rats. Nuclear uptake studies performed *in vitro* showed that intestinal VDR from uremic rats was taken up to a lesser extent than VDR from normal rat intestine [56]. Incubation of normal VDR with uremic plasma ultrafiltrate causes inhibition of nuclear uptake *in vitro*. Intestinal VDR isolated from normal rats infused with the uremic plasma ultrafiltrate also showed decreased nuclear uptake. Patel *et al.* [40] also determined the effect of the uremic plasma ultrafiltrate on binding of VDR to a known vitamin D response element (VDRE) using an electrophoretic mobility shift assay (EMSA). Analysis of the data indicated that the ultrafiltrate did not affect the affinity of the binding but rather reduced the fraction of VDR capable of binding to the VDRE. Thus, the uremic plasma ultrafiltrate appeared to alter the VDR in a way that prevented DNA binding. It was suggested that components in the ultrafiltrate may chemically derivatize the VDR to produce a nonbinding form; Western blot analysis revealed no change in VDR size, excluding proteolysis as an explanation. The question of whether this *in vitro* effect of the uremic plasma ultrafiltrate could alter VDR activity *in vivo* was answered in part by the demonstration that these plasma fractions could inhibit the $1,25(OH)_2D_3$-mediated transcriptional activation of a reporter gene in a whole cell model. The components present in the blood of renal failure patients that inhibit VDR action have not been identified.

IV. TISSUE-SPECIFIC EFFECTS OF LOW $1,25(OH)_2D$ AND ABNORMAL VDR

The low circulating levels of $1,25(OH)_2D$ in CRF, along with decreased VDR content in some tissues and possible inhibition of the DNA binding activity of the VDR, would be predicted to produce a state of vitamin D deficiency and resistance that would reduce the $1,25(OH)_2D_3$-mediated responses in target tissues. Evidence for decreases in vitamin D responsive processes and the contribution to the pathology of chronic renal failure are discussed below.

A. Intestine

The most important action of $1,25(OH)_2D$ *in vivo* is its regulation of intestinal calcium transport. There is considerable evidence for reduced calcium transport activity in the intestines of patients with end-stage renal disease [57,58] and in animal models of experimental uremia [59,60]. This reduction appears to be due in part to the low serum levels of $1,25(OH)_2D$ as evidenced by the increase in calcium transport following $1,25(OH)_2D_3$ therapy [61,62]. Parallel studies done on normal and uremic subjects or normal and uremic animals showed a blunted stimulation of calcium transport by $1,25(OH)_2D_3$ in uremia [63], suggesting a general resistance to the actions of the vitamin D hormone in the intestine. As discussed above, this resistance may be at the level of VDR expression, although comparative measurements of intestinal VDR levels in normal subjects and CRF patients have not been done. It is also possible that the action of the VDR may be blunted by an inhibition of the DNA binding activity of the VDR by components in the plasma of renal failure patients, as proposed by Hsu and Patel [64]. Consistent with the latter possibility, it has been shown that calcium absorption increases following dialysis [65,66]. However, whether this is due to the removal of uremic toxins is not clear. Other changes such as a decrease in serum P [67] or volume depletion [66] have been suggested to explain the higher fraction of calcium absorption after dialysis.

The other major action of $1,25(OH)_2D$ in the intestine is to increase the rate of phosphate absorption. As with calcium, phosphate absorption is decreased in CRF patients [67,68], and treatment with $1,25(OH)_2D_3$ or with $1\alpha OHD_3$ increased intestinal phosphate transport. Studies comparing the stimulation of phosphate absorption by $1,25(OH)_2D_3$ in normal and uremic subjects have not been reported. Thus, it is unclear whether there is a resistance to $1,25(OH)_2D_3$ in the stimulation of phosphate absorption.

In contrast to the blunted response of CRF patients to the stimulatory effect of $1,25(OH)_2D_3$ on calcium absorption, it has been reported that $1,25(OH)_2D_3$ produces a greater induction of the intestinal 24-hydroxylase in uremic rats than in normal rats [69]. The reason for the disparate responses of calcium transport and 24-hydroxylase in uremia is unclear, but certainly hyperinduction of this catabolic enzyme could reduce the effectiveness of $1,25(OH)_2D_3$ in elevating calcium and phosphate transport.

In summary, there is good evidence for decreased intestinal calcium and phosphate absorption in CRF, and the decrease is due, at least in part, to the vitamin D deficiency and resistance in these patients.

B. Parathyroid Glands

Nearly all patients with end-stage renal disease develop secondary hyperparathyroidism. The abnormal vitamin D metabolism and responsiveness described above play a key role in the pathogenesis of this condition as diagrammed in Fig. 6. Figure 6 also illustrates the complex regulation of parathyroid gland function and the interrelationships between the major factors

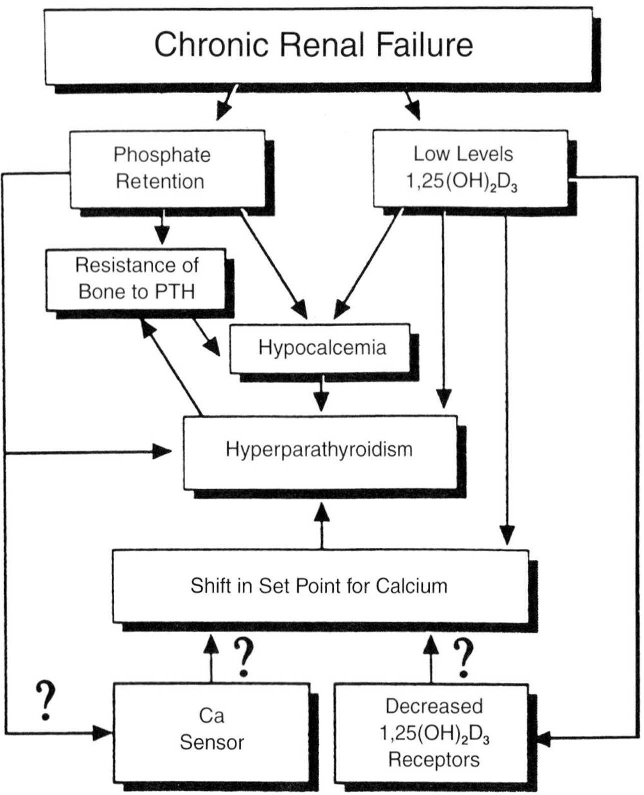

FIGURE 6 Factors involved in the pathogenesis of secondary hyperparathyroidism. From E. Slatopolsky and J. Delmez, *Min. Electrolyte Metab.* **21**:91–96 (1995).

involved: calcium, 1,25(OH)$_2$D, and phosphate. These factors have direct effects on PTH synthesis and/or secretion, and also on parathyroid cell proliferation. The complexity is further increased by the heterologous and homologous regulation of the receptors involved. The role of 1,25(OH)$_2$D in parathyroid physiology is further discussed in Chapter 23.

The primary role of 1,25(OH)$_2$D in controlling PTH synthesis is at the transcriptional level [70]. Studies have identified vitamin D response elements in the human and chicken PTH genes [71,72]. The elements from the two species appear to differ markedly in their structure and binding properties. These studies have been hampered by the lack of a parathyroid cell line, and many questions remain as to the mechanism for the transcriptional suppression of the PTH gene. It is clear, though, that the effect is mediated by the nuclear VDR. Thus, the low serum 1,25(OH)$_2$D levels and decreased VDR content in the parathyroid glands in CRF would lead to overexpression of the PTH gene.

Other actions of 1,25(OH)$_2$D in the parathyroid glands contribute to its suppressive effects on PTH. 1,25(OH)$_2$D$_3$ has been shown to up-regulate VDR binding and VDR mRNA [73] in the parathyroid glands of experimental animals. As discussed above, the low levels of serum 1,25(OH)$_2$D in CRF may be the critical factor in the down-regulated VDR expression observed in the parathyroid glands in this condition.

1,25(OH)$_2$D may also be critical for the full responsiveness of the parathyroid glands to calcium. Heterologous regulation of the calcium-sensing receptor (CaR) of the rat parathyroid glands by 1,25(OH)$_2$D has been reported [74]. Vitamin D deficiency decreased and 1,25(OH)$_2$D$_3$ treatment increased CaR mRNA. However, this finding remains controversial [75]. Immunocytochemical studies [76] have demonstrated a 60% decrease in CaR in parathyroid glands from patients with CRF consistent with previous observations that parathyroid cells isolated from CRF patients display a decreased responsiveness to suppression of PTH secretion by calcium [77]. It should be noted, however, that a study in rats with CRF indicated no change in CaR expression [78]. Clearly, more studies are required to establish whether decreased CaR expression plays a role in the pathogenesis of secondary hyperparathyroidism. However, these early studies on the CaR suggest a potential additional role of vitamin D abnormalities in CRF in the responsiveness of the parathyroid glands to extracellular calcium.

Clinical observations suggest that replacement therapy with 1,25(OH)$_2$D$_3$ to control PTH levels is difficult when CRF patients are hyperphosphatemic [79]. The nature of this resistance to 1,25(OH)$_2$D$_3$ is unclear but it may involve alteration in parathyroid gland VDR expression or activation in these patients.

In CRF patients, the parathyroid glands become hyperplastic. This is due at least in part to chronic hypocalcemia, which may involve vitamin D deficiency and resistance in the intestine and bone as discussed above. It is also possible that low 1,25(OH)$_2$D levels and perhaps decreased VDR expression and activity in the parathyroid glands play a role as well. There is evidence that 1,25(OH)$_2$D can regulate parathyroid cell proliferation. *In vitro* studies have shown that 1,25(OH)$_2$D$_3$ can inhibit the serum-stimulated proliferation of parathyroid cells [80,81]. Vitamin D deficiency leads to parathyroid gland hyperplasia, but this may be due in large part to the concomitant hypocalcemia because calcium supplementation can prevent the development of secondary hyperparathyroidism [82]. In CRF rats, Szabo *et al.* [83] found that secondary hyperparathyroidism could be blocked by prophylactic treatment with 1,25(OH)$_2$D$_3$, but this was associated with increased serum calcium levels. However, suppression of parathyroid gland enlargement by nonhypercalcemic doses of vitamin D analogs has been reported (see Chapter 73).

C. Bone

Renal failure leads to complex bone disorders referred to collectively as renal osteodystrophy [84]. Histologically, these abnormalities can be divided into several general categories that include (1) osteitis fibrosa cystica, (2) osteomalacia and other forms of low bone turnover (e.g., adynamic bone disease), and (3) skeletal defects due to β_2-microglobulin deposition. These abnormalities can occur singly or in combination.

Osteitis fibrosa has been shown to result from hyperparathyroidism. As discussed in Section IV,B, vitamin D deficiency and resistance play a key role in the development of secondary hyperparathyroidism in renal failure patients. A number of other factors, including phosphate retention, hypocalcemia, and skeletal resistance to PTH, all contribute to the overproduction of PTH. A direct role of vitamin D in this disorder has not been demonstrated.

Low $1,25(OH)_2D$ levels in CRF may contribute to the defective bone formation and mineralization as well as the decreased bone turnover observed in some patients. However, although low $1,25(OH)_2D$ levels are found in nearly all patients with advanced renal failure, only a fraction of these patients show evidence of defective bone mineralization [85]. Clearly, other factors are involved. Bone abnormalities in patients on long-term dialysis involving β_2-microglobulin deposition have been described [86]. At present, there is no indication of an involvement of the vitamin D system in this disorder.

Treatment with $1,25(OH)_2D_3$ has been shown to be effective in the management of renal osteodystrophy due to its ability to suppress PTH. The benefits and limitations of this therapy are discussed later in Section V of this chapter.

D. Immune System

The immune system is known to function abnormally in CRF, with increased cases of infections, decreased response to vaccines, and prolonged skin and homograft survival being noted [86,87]. Furthermore, serum from CRF patients has been shown to inhibit cell-mediated immune responses *in vitro*, including lymphocyte blastogenesis, interferon production by normal lymphocytes, and monocyte phagocytosis [86]. These patients also have low lymphocyte counts [88–89]. Alterations in the immune system also have been noted in vitamin D-deficient rickets, with decreased neutrophil phagocytosis [90] and decreased polymorphonuclear leukocyte (PMN) and macrophage activity [91]. These changes can be corrected by treatment of cells with $1,25(OH)_2D_3$ *in vitro* [90,91]. The chemotactic response of PMNs in CRF patients is greatly improved by $1,25(OH)_2D_3$ therapy [92].

Defects in T-cell-mediated immune responses have also been demonstrated in CRF patients that are corrected by $1,25(OH)_2D_3$ treatment. Mitogen activation of lymphocytes from hemodialysis patients is attenuated but can be restored by prior treatment of the patients with $1\alpha OHD_3$, a precursor of $1,25(OH)_2D_3$ [93]. $1,25(OH)_2D_3$ administration to dialysis patients also increases the ratio of T helper to T suppressor cells [94]. In addition, physiological concentrations of $1,25(OH)_2D_3$ have been shown to block proliferation of activated T cells *in vitro* [95]. Natural killer cell activity is depressed in CRF and can be greatly enhanced by treatment with $1,25(OH)_2D_3$ [97,97]. Production of interleukin-2, a T-cell growth factor, by lymphocytes from dialysis patients is decreased but can be normalized by treatment with $1,25(OH)_2D_3$ [98]. This effect was independent of changes in serum calcium, phosphate, and PTH levels.

Other functions of $1,25(OH)_2D_3$ on the immune system have been described, as reviewed in Chapters 29, 55, 69, and 70 of this volume. It is likely that the decreased $1,25(OH)_2D_3$ levels, combined with possible defective VDR expression in immune cells, could lead to other abnormalities in the immune system of CRF patients.

E. Pancreas

Vitamin D deficiency is known to inhibit insulin secretion by the pancreatic β cells [99,100], and this effect is reversed by repletion with the vitamin [101,102]. Whether this effect is direct or is mediated by other factors, notably PTH, remains uncertain. The discovery of VDR [103] and calbindin D [104,105] in β cells suggested a direct effect of $1,25(OH)_2D$ on the regulation of insulin secretion.

Insulin secretion is abnormal in CRF [106,107]; the response of the β cells to a glucose challenge is blunted. Furthermore, there is a systemic resistance to insulin in CRF. The abnormal regulation of insulin in uremia is not fully understood, but it can be normalized by treatment with $1,25(OH)_2D_3$ [108–110]. Allegra *et al.* [111] found that correction of insulin secretion by $1,25(OH)_2D_3$ was not influenced by dietary calcium supplementation, suggesting that this effect of $1,25(OH)_2D_3$ does not involve changes in serum calcium. Furthermore, the doses of $1,25(OH)_2D_3$ used (0.5 μg/day) did not significantly decrease PTH levels, excluding a role for PTH in the improvement of insulin regulation. Thus, vitamin D deficiency in renal failure alters the regulation of insulin secretion. Alterations in β-cell VDR content and activity have not been demonstrated.

V. VITAMIN D THERAPY IN CHRONIC RENAL FAILURE

The ultimate goal of calcitriol therapy is the treatment of osteitis fibrosa, the most common form of renal osteodystrophy, resulting from sustained secondary hyperparathyroidism. Although most patients maintained on chronic dialysis have very low levels of 1,25(OH)$_2$D, the majority of them do not show on bone biopsy the presence of osteomalacia secondary to vitamin D deficiency. Thus, as discussed in this section, not all patients with advanced renal insufficiency or even those maintained on chronic hemodialysis or peritoneal dialysis require 1,25(OH)$_2$D$_3$ treatment. The therapeutic approaches differ for patients before and after they enter a dialysis program and thus are discussed separately.

A. Prevention and Treatment of Secondary Hyperparathyroidism in Patients with Chronic Renal Failure before Treatment with Dialysis

As described above, the serum levels of 1,25(OH)$_2$D decrease with the progression of renal disease, and the majority of patients have low levels when the GFR is less than 50 ml/min. Therefore, it would seem appropriate to replace this hormone. However, the use of calcitriol in patients with moderate to advanced CRF is not completely free of side effects, and many physicians do not prescribe 1,25(OH)$_2$D$_3$ until there are overt manifestations of secondary hyperparathyroidism and bone disease. Although the concern for potential aggravation of renal insufficiency is understandable, careful administration of calcitriol has been beneficial for the majority of patients.

Baker and collaborators [112] studied 16 patients with CRF (creatinine clearance 20–59 ml/min). The patients received either 1,25(OH)$_2$D$_3$ at a dose 0.25 to 0.5 μg/daily or placebo. Bone biopsies were performed before entrance into the study and after 12 months of treatment. Bone histology was abnormal in all patients. Calcitriol treatment was associated with a significant fall in serum phosphorus and alkaline phosphatase concentrations as well as with histological evidence of improvement of hyperparathyroid effects in bone. Over the 12 months of study there was no significant deterioriation of renal function attributable to the treatment. It was recommended that long-term 1,25(OH)$_2$D$_3$ administration in patients with moderate renal failure should be given only to those patients that have high levels of PTH and that patients should be followed closely by their physicians. Hypercalcemia arising from calcitriol treatment could further aggravate the abnormality in renal function. Thus, control of serum calcium and phosphorus and frequent monitoring of PTH levels are imperative in order to prevent potential side effects induced by administration of 1,25(OH)$_2$D$_3$.

Nordal and collaborators [113] studied 13 patients with moderate to terminal renal failure with low doses (average 0.36 μg/day) of calcitriol up to the time of renal transplantation. All patients who started 1,25(OH)$_2$D$_3$ treatment and had a creatinine clearance about 30 ml/min had normal bone histology at the time of renal transplantation. This was not observed, however, when 1,25(OH)$_2$D$_3$ treatment was started with a creatinine clearance below 30 ml/min. The study suggests that the full benefit of 1,25(OH)$_2$D$_3$ at the bone level is obtained only if prophylactic administration is started early in the course of renal failure. It is important to emphasize that these investigators used a small dose of calcitriol. It is possible that such a low dose may not have been sufficient to control secondary hyperparathyroidism and osteitis fibrosa in patients with advanced renal insufficiency.

Although there is extensive literature on the effects of 1,25(OH)$_2$D$_3$ in dialysis patients, there is less experience in patients with moderately advanced renal failure. The results of a large open-ended controlled study on early renal failure with a total of approximately 220 patients has been reviewed [114]. Because of a high incidence of hypercalcemia, many physicians have serious reservations about the use of 1,25(OH)$_2$D$_3$ in early renal insufficiency. However, dose adjustments and frequent determinations of calcium and phosphorus levels may prevent hypercalcemic complications. To obtain further information, Hamdy and collaborators [115] treated a large number of patients with 1αOHD$_3$ (alfacalcidol) in a double-blind, prospective, randomized placebo-controlled study. They studied 176 patients with mild to moderate CRF (creatinine clearance 15–50 ml/min) and with no clinical, biochemical, or radiographic evidence of bone disease. Two years of 1αOHD$_3$ therapy was initiated at a dose of 0.25 μg per day and was titrated according to serum calcium concentration. A total of 132 patients had histological evidence of bone disease at the start of the study; 89 patients received 1αOHD$_3$, and 87 control patients received placebo. After treatment, PTH concentrations had increased by 126% in controls and had not changed in patients given 1αOHD$_3$ ($p < 0.001$). Although hypercalcemic episodes occurred in 10 patients, they resolved rapidly after the dose of the vitamin D derivative was decreased. Histological indices of bone turnover significantly improved in patients given $1\alpha OHD_3$ and significantly deteriorated in controls. There was no difference in the rate of progression of renal failure between the two groups. The investigators concluded that early administration of $1\alpha OHD_3$ can safely and beneficially alter the natural course of renal bone disease in patients with mild to moderate renal failure.

It is recommended that patients with predialysis renal failure be monitored periodically for calcium, phosphorus, and PTH levels. If the levels of PTH are less than 200 pg/ml, the patients likely do not require $1,25(OH)_2D_3$ therapy. Correction of hyperphosphatemia and hypocalcemia are sufficient in these patients. If serum PTH rises above 200 pg/ml, despite controlling calcium and phosphorus levels, the patients should be treated with calcitriol. If the patients are treated on a daily basis the usual dose ranges between 0.25 and 1.0 μg/day. Investigators have shown that oral or intravenous pulse therapy [116–120] could be very effective in controlling secondary hyperparathyroidism. If oral pulse therapy is implemented in patients before dialysis, the dose should range between 1 and 2 μg twice a week. However, it is important to emphasize that there is presently no consensus as to which of the two methods, pulse or daily, is more advantageous. Regardless of the way the patient is treated, serum calcium and phosphorus levels should be measured frequently to prevent metastatic calcification, nephrocalcinosis, and acceleration of renal disease. The levels of PTH should be kept between 200 and 250 pg/ml. If PTH is suppressed to levels below 100 pg/ml, the patients likely will develop adynamic bone disease. On the other hand, if the levels of PTH are allowed to increase to higher levels, the patients may develop severe osteitis fibrosa. Patients with adynamic bone disease usually have very mild symptomatology with a lower prevalence of bone pain, fracture, muscle pain, and rupture of Achilles tendon than in patients with osteitis fibrosa.

B. Treatment of Secondary Hyperparathyroidism in Patients with Chronic Renal Failure Maintained on Hemodialysis

Despite dietary phosphate restriction, the use of phosphate binders, the choice of appropriate levels of calcium in the dialysate, and adequate intake of dietary calcium, a significant number of uremic patients still develop features of osteitis fibrosa. Knowledge of the role of the kidneys in producing $1,25(OH)_2D$ has created interest in the use of active vitamin D metabolites in such patients. In reports of the use of vitamin D analogs for the treatment of uremic bone disease, patients have usually served as their own control, and studies are limited to one form of vitamin D; few observations are available to allow for the comparison of the effectiveness of different vitamin D sterols.

When uremic patients show evidence of adverse secondary hyperparathyroidism (e.g., bone erosions, high PTH levels, and increased alkaline phosphatase) adequate treatment with vitamin D often leads to improvement. Results of numerous studies indicate the efficacy of $1,25(OH)_2D_3$ treatment in patients with symptomatic renal osteodystrophy. These clinical evaluations have shown an improvement in muscle strength and bone pain. In addition, biochemical markers such as plasma alkaline phosphatase have decreased along with a fall in the levels of serum PTH. Bone histology has shown a decrease in marrow fibrosis and other features of secondary hyperparathyroidism such as increased bone resorption and number of osteoclasts. The dose of oral $1,25(OH)_2D_3$ utilized in these trials has varied from 0.25 to 1.0 μg/day, and the major side effect of such treatment is the appearance of hypercalcemia. Hypercalcemia may occur only after many weeks or months of treatment, or it may appear sooner in patients with aluminum-induced osteomalacia or in adynamic bone disease. There is substantial degradation of $1,25(OH)_2D_3$ in the intestine, and it is possible that the oral administration of the vitamin D sterols augments calcium absorption very effectively; however, the delivery of $1,25(OH)_2D_3$ to other target organs may be substantially less.

Slatopolsky and collaborators [119] studied the effects of intravenous administration of $1,25(OH)_2D_3$ in patients maintained on chronic hemodialysis. Twenty patients were given $1,25(OH)_2D_3$ intravenously at the end of each dialysis. The dose was initially 0.5 μg and was gradually increased to a maximum of 4.0 μg per dialysis. Calcitriol was discontinued after 8 weeks of treatment, and blood samples were obtained for an additional 3 weeks. In all patients there was a substantial decrease in the levels of PTH during the period of intravenous $1,25(OH)_2D_3$ treatment, with a mean decrement of 71 $\pm$ 3.2%. After $1,25(OH)_2D_3$ was discontinued, PTH increased rapidly in all patients. There was a significant correlation between the increase in ionized calcium and the decrease in PTH level, showing a crucial role for calcium in the suppression of PTH. However, in addition to the calcemic effect, $1,25(OH)_2D_3$ directly modified the secretion of PTH. The effects of oral and intravenous administration of $1,25(OH)_2D_3$ were compared, and it was shown that the intravenous calcitriol had a greater suppressive effect on the release of PTH than when calcitriol was administered by the oral route.

The intravenous administration of $1,25(OH)_2D_3$ may allow a greater delivery to peripheral tissues such as the parathyroid glands and thereby generate greater expression of biological effects at these sites. Similar results were found by Andress and collaborators [120]; they studied 12 patients on hemodialysis who were not responding to oral $1,25(OH)_2D_3$ and were being considered for parathyroidectomy at the time. All patients exhibited baseline bone formation rates that were above normal, and the rates fell by a mean of 59% during treatment. The results indicate that intravenous administration of $1,25(OH)_2D_3$ is also effective in ameliorating osteitis fibrosa in patients who have moderate to severe secondary hyperparathyroidism. As mentioned before, oral $1,25(OH)_2D_3$ failed to suppress the secretion of PTH adequately in these patients. Parathyroidectomy, originally considered for these patients, became unnecessary due to the implementation of intravenous $1,25(OH)_2D_3$.

Since the original study, more than 50 reports on the effect of intravenous $1,25(OH)_2D_3$ therapy in hemodialyzed CRF patients have been published, including information on more than 1000 patients. In several studies $1\alpha OHD_3$ was administered. Interestingly, Delmez and collaborators [121], in a study of the effect of $1,25(OH)_2D_3$ in a group of patients maintained on hemodialysis, found significant decreases in the levels of serum PTH and an improvement in the calcium set point for PTH secretion after 2 weeks of intravenous calcitriol therapy (Fig. 7). Similar results were found by Dunlay *et al.* [122] in a group of 9 patients maintained on hemodialysis. The patients received 2 μg intravenously after each dialysis. Intravenous calcitriol resulted in a significant decrease within 2 weeks and continued decrease of the high serum PTH levels by the end of 10 weeks. Although these investigators did not find a change in the calcium set point, they found a shift in the calcium–PTH sigmoidal curve toward normal. Thus, both studies [123,124] show an increase in sensitivity of the parathyroid glands to serum calcium after the administration of intravenous $1,25(OH)_2D_3$.

In other countries, owing to an inability to obtain the intravenous preparation of $1,25(OH)_2D_3$, several clinical trials investigated the use of high doses of $1,25(OH)_2D_3$ given orally in an intermittent manner (oral pulse therapy). Several studies in adults and children with renal failure have been published. Most of the results show significant improvement in the suppression of PTH [116,117]. However, in many cases this approach had to be discontinued because of the development of hypercalcemia and hyperphosphatemia Quarles *et al.* [123] published results of a prospective trial of oral pulse versus intravenous calcitriol in the treatment of secondary hyperparathyroidism in hemodialysis patients. These investigators found that episodes of hypercalcemia and hyperphosphatemia were similar in both treatment groups and limited the dose of $1,25(OH)_2D_3$ that could be administered. In this study they found that intermittent $1,25(OH)_2D_3$ therapy, regardless of the route of administration, was poorly tolerated, failed to correct parathyroid gland size and functional abnormalities, and had a limited ability to sustained serum PTH reductions in end-stage renal failure in patients with severe hyperparathyroidism.

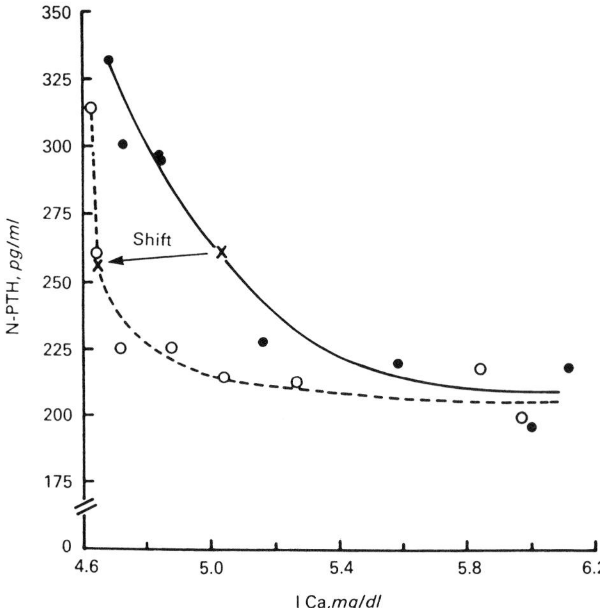

FIGURE 7 Effects of intravenous calcitriol on PTH secretion during calcium infusion in a representative patient. During the control infusion (●) the set point (x) of serum ionized calcium (ICa) was 5.04 mg/dl. After 2 weeks of intravenous calcitriol (○), the PTH levels decreased despite the lower ICa value, and the set point decreased to 4.64 mg/dl. From Delmez *et al.* [121].

Unfortunately, the development of severe hyperphosphatemia in many patients interferes with the beneficial effects of $1,25(OH)_2D_3$. On the other hand, Cannella *et al.* [124] found significant healing of secondary hyperparathyroidism in chronic hemodialysis patients treated with long-term intravenous $1,25(OH)_2D_3$. These investigators followed a group of patients for approximately 35 weeks; the initial dose of $1,25(OH)_2D_3$ was 30 ng/kg of body weight intravenously three times a week after each dialysis. The mean pretreatment serum PTH concentration was 966 ± 160 pg/ml, and the values decreased significantly by the first week and fell by an average of 80% by week 35 (Fig. 8). The ionized calcium concentration was 4.76 ± 0.4 mg/dl and rose slightly to 5.36 mg/dl by the fourth week. There were significant decreases in all bone morphometric indices of secondary hyperparathyroidism. These investigators clearly dem-

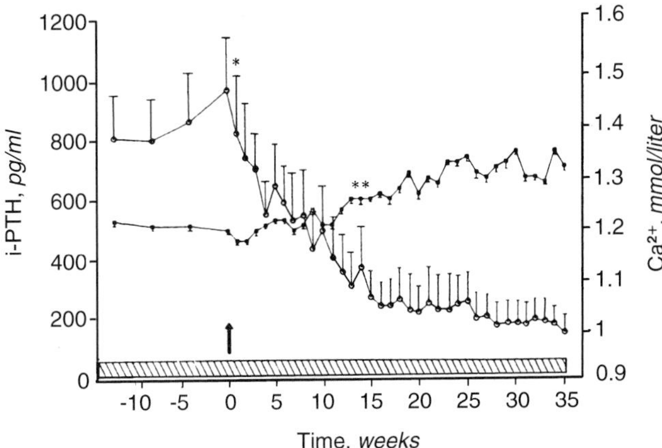

FIGURE 8 Weekly values (means ± SE) for plasma concentrations of intact PTH (○) and ionized calcium (■) in eight hemodialyzed patients, before and after the start (arrow) of intravenous therapy with calcitriol (30 ng/kg body weight thrice weekly for 8 months). Asterisks indicate significant results versus time 0: *, $p < 0.05$; **, $p < 0.01$. Dashed area is the normal PTH range. Used with permission from *Kidney International,* vol. 46, p. 1124, 1994.

onstrated, contrary to the studies of Quarles and collaborators [123], that intravenous $1,25(OH)_2D_3$ is very effective in suppressing secondary hyperparathyroidism. It is of utmost importance to emphasize that Cannella and collaborators [124] were able to control the levels of serum phosphorus, allowing them to provide $1,25(OH)_2D_3$ on a more sustained basis. They also indicated that in some resistant patients higher doses of $1,25(OH)_2D_3$ (4 to 6 μg/treatment) were able to reduce the levels of PTH.

Llach and collaborators [125] also demonstrated the importance in dosing intravenous $1,25(OH)_2D_3$ in dialysis patients with severe hyperparathyroidism. They studied 10 patients with severe hyperparathyroidism (PTH >1200 pg/ml and serum phosphorus <6.5 mg/dl). Ten patients with a mean PTH of 1826 ± 146 pg/ml, were treated for minimum of 48 weeks with an intravenous dose of $1,25(OH)_2D_3$ commensurate with the levels of PTH. The initial $1,25(OH)_2D_3$ dose had to be increased in 7 patients. The mean maximum dose of calcitriol was 3.8 μg three times a week. The authors concluded that patients with severe hyperparathyroidism respond well to intravenous $1,25(OH)_2D_3$; however, the dose of this vitamin D metabolite should be adjusted according to the PTH levels, and hyperphosphatemia should be kept under control. Thus, it seems that although there is no agreement with regard to the route and frequency of administration, the oral pulse and intravenous administration are the most accepted therapies. Further studies are necessary to assess if indeed there is any difference between these two approaches.

Beneficial results also have been observed in patients maintained on continuous ambulatory peritoneal dialysis. Salusky *et al.* [126] studied the pharmokinetics of $1,25(OH)_2D_3$ in continuous ambulatory and cycling peritoneal dialysis patients. The kinetics of $1,25(OH)_2D_3$ was evaluated after a single dose of 60 ng/kg body weight (equivalent to 4.2 μg for a 70-kg man) given orally, intravenously, or intraperitoneally in six patients. The area under the curve for the increment of serum calcitriol concentration above baseline levels for the 24 hr after a single dose of calcitriol was 62% greater following intravenous injection (2340 ± 115 pg/ml) than after either oral (1442 ± 191 pg/ml) or intraperitoneal (1562 ± 195 pg/ml administration. These investigators, using a radioisotope tracer of $1,25(OH)_2D_3$, found that 30 to 40% of the hormone adheres to plastic components of the peritoneal dialysate delivery system. By modifying the technique of intraperitoneal $1,25(OH)_2D_3$ administration, the authors found that they could obtain a dosage effect comparable to intravenous administration. Thus, it would seem that intraperitoneal administration of $1,25(OH)_2D_3$ also is very effective in control of secondary hyperparathyroidism, if precautions to prevent adherence to plastic are taken.

C. Use of New Less Calcemic Analogs of $1,25(OH)_2D_3$

The use of less calcemic analogs of calcitriol is discussed in detail in Chapter 73. Brown and collaborators [127] have shown that 22-oxacalcitriol (OCT) induces significant suppression of PTH levels in rats and dogs with secondary hyperparathyroidism. Similar results were obtained by Slatopolsky *et al.* [128] in rats with CRF using a new analog of calcitriol, 19-nor-$1,25(OH)_2D_2$. These analogs are effective in reducing prepro-PTH mRNA and serum levels of PTH in uremic rats. In addition, 19-nor-$1,25(OH)_2D_2$ does not increase levels of the intestinal vitamin D receptor. In rats these analogs are less calcemic and hyperphosphatemic than the parent compound. Currently, large clinical trials are being conducted with 22-oxacalcitriol in Japan and 19-nor-$1,25(OH)_2D_2$ in the United States. These analogs of $1,25(OH)_2D_3$ could potentially provide a safer means of treating secondary hyperparathyroidism in CRF patients.

Acknowledgments

This work was supported in part by National Institute of Diabetes and Digestive and Kidney Disease, Bethesda, Maryland (Grants DK-09976, DK-30178, and DK-07126). The authors express appreciation to Sue Viviano for assistance in the preparation of the manuscript.

References

1. Slatopolsky E, Gray R, Adams ND, Lewis J, Hruska K, Martin K, Klahr S, DeLuca H, Lemann J 1979 The pathogenesis of secondary hyperparathyroidism in early renal failure. In: Norman A (ed) Fourth International Workshop in Vitamin D. de Gruyter, Berlin, pp. 1209–1213.
2. Cheung AK, Manolagas SC, Catherwood BD, Mosely CA, Mitas JA, Blantz RC, Deftos LJ 1983 Determination of serum $1,25(OH)_2D_3$ levels in renal disease. Kidney Int **24**:104–109.
3. Chesney RW, Hamstra AJ, Mazess RB, Rose P, DeLuca HF 1982 Circulating vitamin D metabolite concentrations in childhood renal disease. Kidney Int **21**:65–72.
4. Juttmann JR, Burman JC, Dekam E, Wisser TJ, Birkenhager JC 1981 Serum concentrations of metabolites of vitamin D in patients with renal failure. Clin Endocrinol **14**:225–232.
5. Wilson L, Felsenfeld A, Drezner MD, Llach F 1985 Altered divalent ion metabolism in early renal failure: Role of $1,25(OH)_2D_3$. Kidney Int **27**:565–570.
6. Martinez I, Saracho R, Montenegro J, Llach F, 1996 A deficit of calcitriol synthesis may not be the initial factor in the pathogenesis of secondary hyperparathyroidism. Nephrol Dial Transplant **11**:22–28.
7. Coburn JW, Koppel MH, Brickman AS, Massry SG 1973 Study of intestinal absorption of calcium in patients with renal failure. Kidney Int **3**:264–272.
8. Malluche H, Werner E, Ritz E 1978 Intestinal absorption of calcium and whole-body calcium retention in incipient and advanced renal failure. Miner Electrolyte Metab **1**:263–270.
9. Prince RL, Hutchison BG, Kent JC, Kent GN, Retallack RW 1988 Calcitriol deficiency with retained synthetic reserve in chronic renal failure. Kidney Int **33**:722–728.
10. Hughes MR, Brumbaugh PF, Haussler MR, Wergedal JE, Baylink DJ 1975 Regulation of serum 1α,25-dihydroxyvitamin D_3 by calcium and phosphorus in the rat. Science **190**:578–580.
11. Portale AA, Halloran BP, Murphy MM, Morris RC Jr 1986 Oral intake of phosphorus can determine the serum concentration of 1,25-dihydroxyvitamin D by determining its production rate in humans. J Clin Invest **77**:7–12.
12. Baxter LA, DeLuca H 1976 Stimulation of 25-hydroxyvitamin D_3-1α-hydroxylase by phosphate depletion. J Biol Chem **251**: 3158–3163.
13. Tanaka Y, DeLuca HF 1973 The control of 25-dihydroxyvitamin D metabolism by inorganic phosphorus. Arch Biochem Biophys **154**:566–574.
14. Portale AA, Booth BE, Haloran BP, Morris RC 1984 Effects of dietary phosphorus on circulating concentrations of $1,25(OH)_2D_3$ and parathyroid hormone in children with moderate renal failure J Clin Invest **73**:1580–1589.
15. Gray RW 1981 Control of plasma 1,25-$(OH)_2$-vitamin D concentrations by calcium and phosphorus in the rat: Effect of hypophysectomy. Calcif Tissue Int **33**:485–488.
16. Gafter U, Kraut JA, Lee DBN, Silis V, Walling NM, Kurokawa K, Haussler MR, Coburn JW 1980 Effect of metabolic acidosis on intestinal absorption of calcium and phosphorus. Am J Physiol **239**:G480–G484.
17. Lee Sw, Russell J, Avioli LV 1977 25-Hydroxycholecalciferol conversion to 1,25-dihydroxycholecalciferol: Conversion to systemic metabolic acidosis. Science **195**:994–996.
18. Sauveur B, Garabedian M, Fellot C, Mongin P, Balsan S 1977 The effect of induced metabolic acidosis on vitamin D_3 metabolism in rachitic chicks. Calcif Tissue Res **23**:121–124.
19. Kawashima KA, Kraut JA, Kurokawa K 1982 Metabolic acidosis suppresses 25-hydroxyvitamin D_2-1α-hydroxylase in the rat kidney. J Clin Invest **70**:135–140.
20. Beck M, Kim HP, Kim KS 1975 Effect of metabolic acidosis on renal action of parathyroid hormone. Am J Physiol **228**:1483–1488.
21. Bellorin-Font E, Humpierres J, Weissinger JR, Milanes Cl, Sylva V, Paz-Martinez V 1985 Effect of metabolic acidosis on the PTH receptor–adenylate cyclase system of canine kidney. Am J Physiol **249**:F566–F572.
22. Bushinsky DA, Rivera GS, Favus MJ, Coe FL 1985 Response of serum 1,25-$(OH)_2D_3$ to variation of ionized calcium during chroninc acidosis. Am J Physiol **249**:F361–F365.
23. Garabedian M, Holick MF, DeLuca HF, Boyle IT 1972 Control of 25-hydroxycholecalciferol metabolism by the parathyroid glands. Proc Natl Acad Sci USA **69**:1673–1676.
24. Henry HL, Midgett RJ, Norman AW 1974 Regulation of 25-hydroxyvitamin D-1-hydroxylase in vivo. J Biol Chem **249**:7584–7590.
25. Booth BE, Tsa HC, Morris RC 1977 Parathyroidectomy reduces 25 hydroxyvitamin D-1-hydroxylase activity in the hypocalcemic vitamin D-deficient chick. J Clin Invest **60**:1314–1320.
26. Rasmussen, H, Wong M, Bikle D, Goodman DBP 1972 Hormonal control of the renal conversion of 25-hydroxycalciferol to $1,25(OH)_2D_3$. J Clin Invest **51**:2502–2510.
27. Haussler MR, McCain TA 1977 Basic and clinical concepts related to vitamin D metabolism and action. N Engl J Med **297**: 974–979.
28. Eisman JA, Wark JD, Prince RL, Moseley JM 1979 Modulation of plasma 1,25-hydroxy vitamin D in man by stimulation or suppression tests. Lancet **2**:931–935.
29. Gray RW, Wilz Dr, Caldos AE, Lemann J 1977 The importance of phosphate in regulating plasma in $1,25(OH)_2D_3$ levels in humans: Studies in healthy subjects in calcium-stone formers and in patients with primary hyperparathyroidism. J Clin Endocrinol Metab **45**:299–306.
30. Henry HL 1981 Insulin permits parathyroid hormone stimulation of 1,25-dihydroxyvitamin D production in cultured kidney cells Endocrinology **108**:733–735.
31. Henry HL 1984 Regulation of the synthesis of 1,25-dihydroxyvitamin D and 24,25-dihydroxyvitamin D in kidney cell culture. In: Kumar R (ed) Vitamin D: Basic and Clinical Aspects. Niijhoff, Boston, pp. 151–174.
32. Korker AB, Gary RW, Henry HL, Kleinman JG, Blumenthal SS, Garancis JC 1987 Evidence that stimulation of $1,25(OH)_2D_3$ production in primary cultures of mouse kidney cells by cyclic AMP requires new protein synthesis. J Bone Miner Res **2**:517–524.
33. Hsu CH, Patel S 1992 Uremic plasma contains factor inhibiting 1α-hydroxylase activity. J Am Soc Nephrol **3**:947–952.
34. Henry HL, Midgett RJ, Norman AW 1974 Studies on calciferol metabolism. Regulation of 25-hydroxyvitamin D_2-1-hydroxylase in vivo. J Biol Chem **249**:7584–7592,
35. Tanaka Y, Lorec RS, DeLuca HF 1975 The role of 1,25-dihydroxyvitamin D and parathyroid hormone in the regulation of chick renal 25-dihydroxyvitamin D-24-hydroxylase. Arch Biochem Biophys **171**:521–526.
36. Dusso A, Lopez-Hilker S, Rapp N, Slatopolsky E 1988 Extrarenal production of calcitriol in chronic renal failure. Kidney Int **34**:368–375.
37. Dusso AS, Finch J, Brown A, Delmez J Schreiner G, Slatopolsky E 1991 Regulation of extrarenal production of calcitriol in normal and uremic humans. J Clin Endocrinol Metab **128**:1687–1692.
38. Dusso A, Lopez-Hilker S, Finch J, Slatopolsky E 1989 Metabolic

clearance rate and production rate of calcitriol in uremia. Kidney Int **35**:860–864.
39. Hsu Ch, Patel S, Young EW, Simpson RV 1987 Production and degradation of calcitriol in renal failure rats. Am J Physiol **22**:F1015–F1019.
40. Patel SR, KE HW, VanHolder R, Koening R, Hsu C 1995 Inhibition of calcitriol receptor binding to vitamin D response elements by uremic toxins. J Clin Invest **96**:50–59.
41. Korkor AB 1987 Reduced binding of [^{3}H]1,25-dihydroxyvitamin D_3 in the parathyroid glands of patients with renal failure. N Engl J Med **316**:1573–1577.
42. Fukuda N, Tanaka H, Tominaga Y, Fukagawa M, Kurokawa K, Seino Y 1993 Decreased 1,25-dihydroxyvitamin D_3 receptor density is associated with a more severe form of parathyroid hyperplasia in chronic uremic patients. J Clin Invest **92**:1436–1443.
43. Merke J, Hugel U, Zlotkowski A, Szabo A, Bommer J, Mall G, Ritz E 1987 Diminished parathyroid $1,25(OH)_2D_3$ receptors in experimental uremia. Kidney Int **32**:3350–3353.
44. Brown AJ, Dusso A, Lopez-Hilker S, Lewis-Finch J, Grooms P, Slatopolsky E 1989 $1,25(OH)_2D_3$ receptors are decreased in parathyroid glands from chronically uremic dogs. Kidney Int **35**:19–23.
45. Szabo A, Merke J, Thomasset M, Ritz E 1991 No decrease of $1,25(OH)_2D_3$ receptors and duodenal calbindin-D_{9k} in uraemic rats. Eur J Clin Invest **21**:521–526.
46. Denda M, Finch J, Brown AJ, Nishii Y, Kubodera N, Slatopolsky E 1996 1,25-Dihydroxyvitamin D_3 and 22-oxacalcitriol prevent the decrease in vitamin D receptor content in the parathyroid glands of uremic rats. Kidney Int **50**:34–39.
47. Szabo A, Ritz E, Schmidt-Gayk H, Reichel H 1996 Abnormal expression and regulation of vitamin D receptor in experimental uremia. Nephron **73**:619–628.
48. Shvil Y, Neveh-Many T, Barach P, Silver J 1990 Regulation of parathyroid cell gene expression in experimental uremia. J Am Soc Nephrol **1**:99–104.
49. Hsu CH, Patel SR, Vanholder R 1993 Mechanism of decreased intestinal calcitriol receptor concentration in renal failure. Am J Physiol **264**:F662–F669.
50. Patel SR, Ke HQ, Hsu CH 1994 Regulation of calcitriol receptor and its mRNA in normal and renal failure rats. Kidney Int **45**:1010–1027.
51. Koyama H, Nishizawa Y, Inaba M, Hino M, Prahl J, DeLuca HF, Morii H 1994 Impaired homologous upregulation of the vitamin D receptor in rats with chronic renal failure. Am J Physiol **266**:F706–F712.
52. Reinhardt TA, Horst RL 1990 Parathyroid hormone down-regulates 1,25-dihydroxyvitamin D receptors (VDR) and VDR messenger ribonucleic acid in vitro and blocks homologous up-regulation of VDR in vivo. Endocrinology **127**:942–948.
53. Takahashi F, Finch J, Denda M, Dusso A, Brown AJ, Slatopolsky E 1996 A new analog of $1,25(OH)_2D_3$, 19-nor-$1,25(OH)_2D_2$, suppresses serum PTH and parathyroid gland growth in uremic rats without elevation of intestinal vitamin D receptor content. J Bone Miner Res **11**:S424.
54. Martinez J, Olmos JM, de Francisco AL, Amado JA, Riancho JA, Gonzales-Macias J 1994 1,25-Dihydroxyvitamin D_3 receptors in peripheral blood mononuclear cells from patients with primary and secondary hyperparathyroidism. Bone Miner **27**:25–32.
55. Patel SR, Ke HQ, Vanholder R, Hsu CH 1994 Inhibition of nuclear uptake of calcitriol receptor by uremic ultrafiltrate. Kidney Int **46**:129–133.
56. Patel SR, Ke HQ, Vanholder R, Koenig RJ, Hsu CH 1995 Inhibition of calcitriol receptor binding to vitamin D response elements by uremic toxins. J Clin Invest **96**:50–59.
57. Ogg CS 1968 The intestinal absorption of ^{47}Ca by patients in chronic renal failure. Clin Sci **34**:467–471.
58. Recker RR, Saville PD 1971 Calcium absorption in renal failure: Its relationsihp to blood urea nitrogen, dietary calcium intake, time on dialysis and other variables. J Lab Clin Med **78**:380–388.
59. Wong RG, Norman AW, Reddy CR, Coburn JW 1972 Biologic effects of 1,25-dihydroxycholecalciferol (a highly active vitamin D metabolite) in acutely uremic rats. J Clin Invest **51**:1287–1291.
60. Hartenbower DL, Coburn JW, Reddy CR, Norman AW 1974 Calciferol metabolism and intestinal calcium transport in the chick with reduced renal function. J Lab Clin Med **83**:38–45.
61. Brickman AS, Coburn JW, Norman AW 1972 Action of 1,25-dihydroxycholecalciferol, a potent, kidney-produced metabolite of vitamin D_3, in uremic man. N Engl J Med **287**:891–895.
62. Brickman S, Coburn JW, Massry SG 1974 1,25-Dihydroxyvitamin D_3 in normal man and patients with renal failure. Ann Intern Med **80**:161–168.
63. Walling MW, Kimberg DV, Wasserman RH, Feinberg RR 1976 Duodenal active transport of calcium and phosphate in vitamin D-deficient rats: Effects of nephrectomy, *Cestrum diunum*, and 1 alpha,25-dihydroxyvitamin D_3. Endocrinology **98**:1130–1134.
64. Hsu CH, Patel SR 1995 Altered vitamin D metabolism and receptor interaction with the target genes in renal failure: Calcitriol receptor interaction with its target gene in renal failure. Curr Opin Nephrol Hyperten **4**:302–306.
65. Juttman JR, Hagenouw-Taal JC, Lameyer LD, Ruis AM, Birkenhager JC 1978 Intestinal calcium absorption, serum phosphate, and parathyroid hormone in patients with chronic renal failure and osteodystrophy before and during hemodialysis. Calcif Tissue Int **26**:119–126.
66. Chanard JM, Drueke T, Zingraff J, Man NK, Russo-Marie F, Funck-Brentano JL 1976 Effects of hemodialysis on fractional intestinal absorption of calcium in uremia. Eur J Clin Invest **6**:261–264.
67. Stanbury SW, Lumb GA 1962 Metabolic studies of renal osteodystrophy. I. Calcium, phosphorus and nitrogen metabolism in rickets, osteomalacia and hyperparathyroidism complicated by chronic uremia and in osteomalacia of Fanconi Syndrome. Medicine **41**:1–10.
68. Brickman AS, Hartenbower DL, Norman AW, Coburn JW 1977 Actions of 1α-hydroxyvitamin D_3 and 1,25-dihydroxyvitamin D_3 on mineral metabolism in man. I. Effects on net absorption of phosphorus. Am J Clin Nutr **30**:1064–1069.
69. Koyama H, Inaba M, Nishizawa Y, Ishimura E, Imanishi Y, Hini M, Furuyama T, Morrii H 1994 Potentiated $1,25(OH)_2D_3$-induced 24-hydroxylase gene expression in uremic rat intestine. Am J Physiol **267**:F926–F930.
70. Silver J, Naveh-Many T 1994 Regulation of parathyroid hormone synthesis and secretion. Semin Nephrol **14**:175–194.
71. Demay MB, Kiernan MS, DeLuca HF, Kronenberg HM 1992 Sequences in the human parathyroid hormone gene that bind the 1,25-dihydroxyvitamin D_3 receptor and mediate transcriptional repression in response to 1,25-dihydroxyvitamin D_3. Proc Natl Acad Sci USA **89**:8097–8101.
72. Liu SM, Koszewski N, Lupex M, Malluche HH, Olivera A, Russell J 1996 Characterization of a response element in the 5′-flanking region of the avian (chicken) PTH gene that mediates negative regulation of gene transcription by 1,25-dihydroxyvitamin D_3 and binds the vitamin D receptor. Mol Endocrinol **10**:206–215.
73. Brown AJ, Zhong M, Finch J, Ritter C, Slatopolsky E 1995 The roles of calcium and 1,25-dihydroxyvitamin D_3 in the regulation

of vitamin D receptor expression by rat parathyroid glands. Endocrinology **136**:1419–1425.
74. Brown AJ, Zhong M, Finch J, Ritter C, McCracken R, Morrissey J, Slatopolsky E 1996 Rat calcium-sensing receptor is regulated by vitamin D but not by calcium. Am J Phyiol **270**:F454–F460.
75. Rogers KV, Dunn CV, Conklin RL, Hadfield S, Petty BA, Brown EM, Hebert SC, Fox J 1995 Calcium receptor expression in the parathyroid glands of vitamin D-deficient rats in not regulated by plasma calcium and 1,25$(OH)_2D_3$. Endocrinology **136**:499–504.
76. Kifor O, Moore FD, Wang P, Goldstein M, Vasselev P, Kifor I, Hebert SC, Brown EM 1996 Reduced immunostaining for the extracellular Ca sensing receptor in primary and uremic secondary hyperparathyroidism. J Clin Endocrinol Metab **81**:1598–1606.
77. Brown EM, Wilson E, Eastman RC, Pallotta J, Marynick SP 1982 Abnormal regulation of parathyroid hormone release by calcium in secondary hyperparathyroidism due to chronic renal failure. J Clin Endocrinol Metab **54**:172–179.
78. Roger KV, Fox J, Dunn CK, Conklin RL, Lowe SH, Petty BA 1995 Parathyroid gland calcium receptor mRNA levels are unaffected by chronic renal insufficiency or low dietary calcium in rats. Endocrine **3**:769–774.
79. Shoji S, Nishizawa Y, Tabata T, Emoto M, Morita A, Goto H, Ishimaura E, Inoue T, Inaba M, Miki T, Morii H 1995 Influence of serum phosphate on the efficacy of oral 1,25-dihydroxyvitamin D_3 pulse therapy. Min Electrolyte Metab **21**:223–228.
80. Nygren P, Larsson R, Johansson H, Ljunghall S, Rastad J, Akerstrom G 1988 1,25$(OH)_2D_3$ inhibits hormone secretion and proliferation but not functional dedifferentiation of cultured bovine parathyroid cells. Calcif Tissue Int **43**:213–218.
81. Kremer R, Bolivar I, Goltzman D, Hendy GN 1989 Influence of calcium and 1,25-dihydroxycholecalciferol on proliferation and proto-oncogene expression in primary cultures of bovine parathyroid cells. Endocrinology **125**:935–941.
82. Kollendirchen U, Fox J, Walters MR 1991 Normocalcemia without hyperparathyroidism in vitamin D-deficient rats. J Bone Miner Res **6**:273–278.
83. Szabo A, Merke J, Beier R, Mall G, Ritz E 1989 1,25$(OH)_2$-vitamin D_3 inhibits parathyroid cell proliferation in experimental uremia. Kidney Int **35**:1049–1056.
84. Gonzalez EA, Martin KJ 1995 Renal osteodystrophy: Pathogenesis and management. Nephrol Dial Transplant **10**(Suppl. 3): 13–21.
85. Malluche HH, Ritz E, Lange HP 1976 Bone histology in incipient and advanced renal failure. Kidney Int **9**:355–362.
86. Goldblum SE, Reed WP 1980 Host defenses and immunologic alterations associated with chronic hemodialysis. Ann Intern Med **93**:597–613.
87. Rice JC, Haverty TP 1990 Vitamin D and immune function in uremia. Semin Dial **3**:237–239.
88. Park K, Ha S, Han DS 1988 Studies of lymphocyte subpopulations and cell-mediated immunity in patients with chronic renal failure. Yonsei Med J **29**:109–116.
89. Lin CY, Huang TP 1988 Serial cell-mediated immunological changes in terminal uremic patients on continuous ambulatory peritoneal dialysis therapy. Am J Nephrol **8**:355–362.
90. Reichel H, Koeffler HP, Norman AW 1989 The role of the vitamin D endocrine system in health and disease. N Engl J Med **320**:980–991.
91. Bar-Shavit Z, Noff D, Edelstein S, Meyer M, Shibolets, Goldman R 1981 1,25-Dihydroxyvitamin D_3 and the regulation of macrophage function. Calcif Tissue Int **33**:673–676.
92. Venezio FR, Kozeny GA, DiVencenzo CA, Hano JE 1988 Effects of 1,25-dihydroxyvitamin D_3 on leukocyte function in patients receiving chronic hemodialysis. J Infect Dis **158**:1102–1105.
93. Tabata T, Suzuki R, Kikunami K, Matsushita Y, Inoue T, Okamoto T, Miki T, Nishizawa Y, Morii H 1986 The effect of 1-hydroxyvitamin D_3 on cell-mediated immunity in hemodialysis patients J Clin Endocrinol Metab **63**:1218–1221.
94. Bargman JM, Kuzniak S, Klein MH 1987 Changes in immune function induced by 1,25-dihydroxyvitamin D_3. Kidney Int **31**:342.
95. Lemire JM, Adams JS, Karmani-Arab V, Bakke AC, Sukai R, Jordan SC 1985 1,25-Dihydroxyvitamin D_3 suppresses human T helper/inducer lymphocyte activity I vitro. J Immunol **134**:3032–3035.
96. Quesada JM, Solano R, Serrano I, Barrio V, Martinez ME, Santemaria M, Martin-Malo A 1989 Immunologic effects of vitamin D N Engl J Med **321**:883.
97. Quesada JM, Serrano I, Borrego F, Martin A, Pena J, Solana R 1995 Calcitriol effect on natural killer cells from hemodialyzed and normal subjects. Calcif Tissue Int **56**:113–117.
98. Tabata T, Shoji T, Kikinami K, Matsushita Y, Inoue T, Tanaka S, Hino M, Miki T, Nishizawa Y, Morii T 1988 In vivo effect of 1α-hydroxyvitamin D_3 on interleukin-2 production in hemodialysis patients. Nephron **50**:295–298.
99. Norman AW, Frankel BJ, Heldt AM Grodsky GM 1980 Vitamin D-deficiency inhibits pancreatic secretion of insulin. Science **209**:823–825.
100. Kadowaki S, Norman AW 1984 Dietary vitamin D is essential for normal insulin secretion from the perfused rat pancreas. J Clin Invest **73**:759–766.
101. Cade C, Norman AW 1986 Vitamin D improves impaired glucose tolerance and insulin secretion in the vitamin D deficient rat in vivo. Endocrinology **119**:84–90.
102. Cade C, Norman AW 1987 Rapid normalization and stimulation by 1,25-dihydroxyvitamin D_3 of insulin secretion and glucose tolerance in the vitamin D deficient rat. Endocrinology **121**:1490–1497.
103. Stumpf WE, Sar M, DeLuca HF 1981 Sites of action of 1,25$(OH)_2$-vitamin D_3 identified by thaw mount autoradiography. In Cohn DV, Talmadge RV, Matthews JL (eds) Hormonal Control of Calcium Metabolism. Excerpta Medica, Amsterdam, pp. 22–229.
104. Morrissey RL, Bucci TJ, Empson RN, Lufkin EG 1975 Calcium-binding protein: Its cellular localization in jejunum, kidney and pancreas. Proc Soc Exp Biol Med **148**:56–60.
105. Roth J, Bonner J, Norman AW, Orci L 1982 Immunohistochemistry of vitamin D-dependent calcium binding protein in chick pancreas: Exclusive location in β cells. Endocrinology **116**: 2216–2218.
106. Lowrie EG, Soeldner JS, Hampers CL, Merrill JP 1970 Glucose metabolism and insulin secretion in uremic, prediabetic and normal subjects. J Lab Clin Med **76**:603–615.
107. Allegra V, Mengozzi G, Martimbianco L, Vasile A 1990 Glucose-induced insulin secretion in uremia: Effects of aminophylline infusion and glucose loads. Kidney Int **38**:1146–1150.
108. Alverstrand A, Mujagic M, Wajngot A, Efendic S 1989 Glucose intolerance in uremic patients: The relative contributions of impaired β-cell function and insulin resistance. Clin Nephrol **31**:175–183.
109. Quesada JM, Martin-Malo A, Santiago J, Hervas F, Martinez ME, Castillo D, Barrio V, Aljama P 1990 Effect of calcitriol on insulin secretion in uremia. Nephrol Dial Transplant **5**:1013–1017.
110. Mak RHK 1989 Insulin secretion in uremia: Effect of parathyroid hormone and vitamin D metabolites. Kidney Int **36**:S227–S230.

111. Allegra V, Luisetto G, Mengozzi G, Martimbianco L, Vasile A 1994 Glucose-induced insulin secretion in uremia: Role of 1,25$(OH)_2$-vitamin D_3. Nephron **68**:41–47.
112. Baker RIL, Abrams SML, Roe CJ, Faugere MC, Fanti P, Subayti Y, Malluche HH 1989 1,25$(OH)_2D_3$ administration in moderate renal failure. A prospective double-blind trial Kidney Int **35**:661–669.
113. Nordal KP, Dahl E, Halse J, Attramadal A, Flatmark A 1995 Long-term low-dose calcitriol treatment in predialysis chronic renal failure: Can it prevent hyperparathyroid bone disease? Nephrol Dial Transplant **10**:203–206.
114. Goodman WG, Coburn JW 1992 The use of 1,25-dihydroxy-vitamin D_3 in early renal failure. Annu Rev Med **43**:226–237.
115. Hamdy NAT, Kannis JA, Beneton MNC, Brown CB, Juttman R, Jordans JGM, Josee S, Meyrier A, Lins RL, Fairey IT 1995 Effect of alfacalcidol on natural course of renal bone disease in mild to moderate renal failure. Br. Med J **310**:358–363.
116. Tsukamoto Y, Nomura M, Takahashi Y, Takagi Y, Yoshida A, Nagaoka T, Togashi K, Kikawada R, Marumo F 1991 Hyperparathyroidism in patients on hemodialysis by intermittent oral administration of 1,25$(OH)_2D_3$. Nephron **57**:23–28.
117. Tsukamoto Y, Mariyo R, Normura Y, Sata N, Faugere MC, Malluche HH 1993 Long-term effect of oral calcitriol pulse therapy on bone in hemodialysis patients. Bone **14**:421–425.
118. Muramoto H, Haruki K, Yoshimura A, Mimo N, Oda K, Tofuku Y 1991 Treatment of refractory hyperparathyroidism in patients on hemodialysis by intermittent oral administration of 1,25$(OH)_2D_3$. Nephron **58**:288–294.
119. Slatopolsky E, Weerts C, Theilan J, Horst R, Harter H, Martin KJ 1984 Marked suppression of secondary hyperparathyroidism by intravenous administration of 1,25-dihydroxycholecalciferol in uremic patients. J Clin Invest **74**:2136–2143.
120. Andress DL, Norris KC, Coburn JW, Slatopolsky EA, Sherrard DJ 1989 Intravenous calcitriol in the treatment of refractory osteitis fibrosa of chronic renal failure. N Engl J Med **321**: 274–279.
121. Delmez JA, Tindira C, Grooms P, Dusso A, Windus DW, Slatopolsky E 1989 Parathyroid hormone suppression by intravenous 1,25-dihydroxyvitamin D. J Clin Invest **83**:1349–1355.
122. Dunlay R, Rodrigues M, Felsenfeld AJ, Llach F 1989 Direct inhibitory effect of calcitriol on parathyroid function (sigmoidal curve) in dialysis patients. Kidney Int **36**:1093–1098.
123. Quarles LD, Yohay DA, carroll BA, Spritzer CE, Minda SA, Bartholomay D, Lobaugh BA 1994 Prospective trial of pulse oral versus intravenous calcitriol treatment of hyperparathyroidism in ESRD Kidney Int **45**:1710–1721.
124. Cannella G, Bonucci E, Roalla D, Ballanti P, Moriero E, De-Grandi R, Augeri C, Claudiani F, DiMaio G 1994 Evidence of healing of secondary hyperparathyroidism in chronically hemodialyzed uremic patients treated with long term intravenous calcitriol. Kidney Int **46**:1124–1132,
125. Llach F, Hervas J, Cerzo S 1995 The importance of dosing intravenous calcitriol in dialysis patients with severe hyperparathyroidism. Am J Kidney Dis **26**:845–851.
126. Salusky IB, Goodman WG, Horst R, Segre GV, Kim L, Norris KC, Adams JS, Holloway M, Fine RN, Coburn JW 1990 Pharmacokinetics of calcitriol in continuous ambulatory and cycling peritoneal dialysis patients. Am J Kidney Dis **16**:126–132.
127. Brown AJ, Dusso A, Slatopolsky E 1994 Selective vitamin D analogs and their therapeutic applications. Semin Nephrol **14**: 156–174.
128. Slatopolsky E, Finch J, Ritter C, Denda M, Morrissey J, Brown A, DeLuca H 1995 A new analog of calcitriol, 19-nor-1,25$(OH)_2D_3$, suppresses parathyroid hormone secretion in uremic rats in the absence of hypercalcemia. Am J Kidney Dis **26**:852–860.

CHAPTER 53

Idiopathic Hypercalciuria and Nephrolithiasis

MURRAY J. FAVUS Nephrology Section and Section of Endocrinology, The University of Chicago Pritzker School of Medicine, Chicago, Illinois

FREDRIC L. COE Nephrology Section, The University of Chicago Pritzker School of Medicine, Chicago, Illinois

I. Introduction

This chapter focuses on idiopathic hypercalciuria (IH) as the cause of hypercalciuria and nephrolithiasis. Other causes of hypercalcemia and hypercalciuria are also likely to increase renal stone formation and are discussed in Chapters 54 and 55. IH is the most common state of 1,25-dihydroxyvitamin D [$1,25(OH)_2D$] excess. IH is found in 5% of the adult population in the United States and is the cause in 50% of calcium oxalate kidney stone formers.

Idiopathic hypercalciuria is characterized by normocalcemia in the absence of known systemic causes of hypercalciuria. Most IH patients have increased intestinal calcium (Ca) absorption, and serum $1,25(OH)_2D$ levels are elevated in one-third to one-half of patients. Elevated levels of serum parathyroid hormone (PTH) are present in less than 5%. However, IH appears to be a heterogeneous disorder, and possible causes include the following: a primary increase in intestinal Ca absorption; a primary overproduction of $1,25(OH)_2D$; and a primary renal tubular Ca transport defect or "renal leak" of Ca. During dietary Ca restriction, in one-half of the patients urine Ca excretion remains too high and negative Ca balance develops. Several lines of evidence indicate that at least some patients with IH have $1,25(OH)_2D$ excess or increased $1,25(OH)_2D$ action. These findings include an observed increase in intestinal Ca absorption, normocalcemia, and negative Ca balance during low Ca intake in healthy subjects given small doses of $1,25(OH)_2D_3$.

An animal model of genetic hypercalciuria has been developed in Sprague-Dawley rats. The mechanism is due to an increase in intestinal, renal, and bone cell vitamin D receptor (VDR) content. A posttranscriptional dysregulation of VDR is suggested in these animals by decreased VDR mRNA and an increased amount of normal VDR protein that has normal binding affinity for $1,25(OH)_2D_3$. The nature of the genetic defect in humans that permits hypercalciuria remains unknown.

II. IDIOPATHIC HYPERCALCIURIA

A. Introduction

Hypercalciuria is common among patients with calcium oxalate nephrolithiasis and is thought to contribute to stone formation by increasing the state of urine supersaturation with respect to Ca and oxalate. Flocks [1] first commented on the frequency of hypercalciuria among patients with Ca stones; however, it was not until the mid-1950s that Albright and Henneman [2,3] defined the condition of IH as hypercalciuria with normal serum Ca, no systemic illness, and no clinical skeletal disease. The definition of hypercalciuria is arbitrary and based on the distribution of urine Ca excretion values among unselected populations of healthy men and women in Western countries [4,5]. The measurements form a continuum clustered about a mean with a long tail of higher values. IH patients are those whose urine Ca exceeds the arbitrary upper limit of normal. Hypercalciuria is most commonly defined as greater than 300 mg/24 hr for men, greater than 250 mg/24 hr for women, or greater than 4 mg/kg body weight for either sex [6]. Using this definition, hypercalciuria is found in about 50% of calcium oxalate stone formers [6,7] and is the most common cause of normocalcemic hypercalciuric stone formation [7]. The diagnosis of IH requires the exclusion of the known causes of normocalcemic hypercalciuria (Table I).

Surveys of stone formers attending kidney stone clinics report a high proportion with kidney stone formation among first degree relatives [8,9]. A genetic basis of IH was further suggested by subsequent surveys [10–12], which revealed a strong familial occurrence of IH with high rates of vertical and horizontal penetrance (Fig. 1) consistent with an autosomal dominant mode of inheritance. IH also occurs in children with the same frequency of occurrence as adults [13]. That hypercalciuria can be genetically based has been clearly demonstrated by breeding experiments in which the offspring of spontaneously hypercalciuric male and female Sprague-Dawley rats are intensely hypercalciuric [14–16]. Other hypercalciuric genetic disorders have been described, but they differ from IH in having either a renal phosphate leak [17] that may lead to rickets (Chapter 46), renal tubular acidosis [18], or X-linked recessive stone formation with early renal failure [19].

TABLE I Causes of Normocalcemic Hypercalciuria

Paget's disease
Sarcoidosis
Hyperthyroidism
Renal tubular acidosis
Cushing's syndrome
Immobilization
Malignant tumor
Furosemide administration

Idiopathic hypercalciuria is a common disorder that affects 2 to 4% of otherwise healthy men and women [4]. If 50% of stone formers have IH [6,7], and the frequency of stone disease among adults is 0.5%, then 80 to 90% of IH is asymptomatic and never associated with stone formation. The increased frequency of osteopenia (see Section B.4) suggests that hypercalciuria may be an important pathogenetic factor for development of bone disease even among those who do not form stones.

B. Pathogenesis of Human Idiopathic Hypercalciuria

1. INCREASED INTESTINAL CALCIUM ABSORPTION

Normally, the quantity of Ca absorbed is determined by dietary Ca intake and the efficiency of intestinal Ca absorption [20]. Absorption of Ca across the intestine is the sum of two transepithelial transport processes: a nonsaturable paracellular pathway and a saturable, cellular active transport system [21,22] (also see Chapter 16). Absorption via the paracellular path is diffusional and driven by the lumen-to-blood Ca gradient [20]. The cellular pathway is vitamin D dependent and is regulated by the ambient concentration of $1,25(OH)_2D$. Thus, Ca translocation via the paracellular and cellular pathways is determined by intraluminal Ca concentration and tissue $1,25(OH)_2D$ levels, respectively.

Increased intestinal Ca absorption has been found in most patients with IH [23–31]. Using either a single oral dose of Ca isotope to measure fecal isotope excretion or double Ca isotope administration in which the intravenous dose adjusts for isotope distribution, IH patients were shown to have an increase in the Ca absorptive flux (Fig. 2). External Ca balance studies conducted while IH patients and normal non-stone formers ingested diets containing comparable amounts of Ca show net intestinal Ca absorption rates to be greater in IH patients [32]. Biopsies of proximal intestine obtained following oral Ca isotopic administration reveal in-

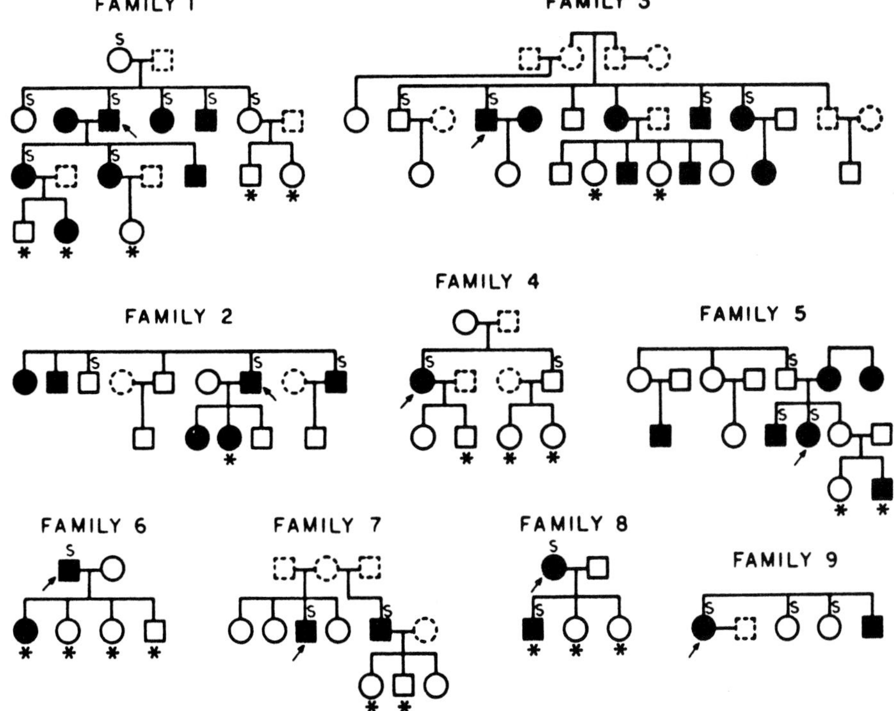

FIGURE 1 Family pedigrees of nine probands with idiopathic hypercalciuria. Solid circles and solid squares are females and males with hypercalciuria; S is stone formation; * indicates children (younger than age 20). Arrows indicate probands from each family. Dashed symbols are relatives who were not studied. Hypercalciuria occurred in 11 of 24 siblings, 7 of 16 offspring, and 1 of 3 parents of probands. Reprinted by permission of *The New England Journal of Medicine* (F. L. Coe, J. H. Parks, and E. S. Moore, Familial idiopathic hypercalciuria, *N Engl. J. Med.* **300**:337–340). Copyright 1979, Massachusetts Medical Society.

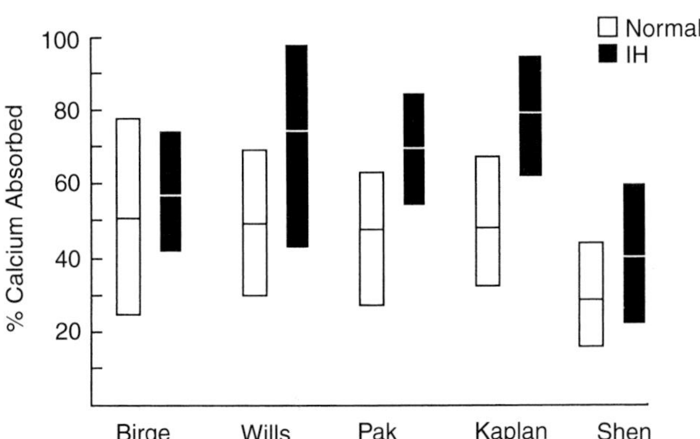

FIGURE 2 Intestinal Ca absorption in healthy volunteers and patients with IH. Absorption rates are expressed as percentages of dietary Ca absorbed as calculated from the appearance of Ca isotopes in blood or fecal collections. Values are means (horizontal bar) ± 2 standard deviations. Names indicate references: Birge [24]; Wills [25]; Pak [28]; Kaplan [29]; and Shen [30].

creased mucosal accumulation of isotope compared to normocalciuric non-stone formers [33]. Thus, by all means of measurement, IH is characterized by increased intestinal Ca absorption.

2. ELEVATED $1,25(OH)_2D$

Kaplan and colleagues [29] first reported elevated serum levels of $1,25(OH)_2D$ in a group of patients with IH. Subsequently, others have confirmed that, on average, serum $1,25(OH)_2D$ levels are greater in IH (Fig. 3). Increased *in vivo* conversion of tritiated 25-hydroxyvitamin D_3 (^3H-$25OHD_3$) to ^3H-$1,25(OH)_2D_3$ and normal metabolic clearance [34] in a group of IH patients with elevated serum $1,25(OH)_2D$ levels indicate that the increase in serum $1,25(OH)_2D$ levels in some IH patients is the result of increased production. Of note is the considerable overlap of serum $1,25(OH)_2D$ levels between IH patients and non-stone formers (Fig. 3). Thus, for many IH patients, increased intestinal Ca transport may be caused by increased circulating $1,25(OH)_2D$, but for others this does not appear to be the case.

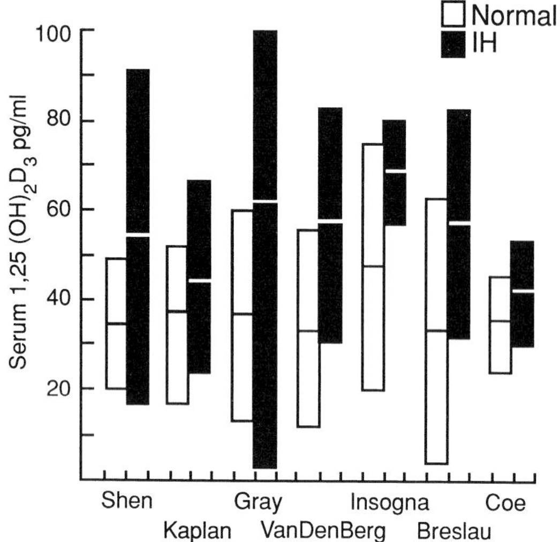

FIGURE 3 Plot of means ± 2 SD of serum $1{,}25(OH)_2D$ in IH patients and non-stone formers. Horizontal bar is mean of group. Names indicate references: Shen [30]; Kaplan [29]; Gray [38]; Van Den Berg [55]; Insogna [34]; Breslau [56]; and Coe [36].

The mechanism whereby $1{,}25(OH)_2D$ production is increased in IH is unknown. The major regulators of renal proximal tubule mitochondrial 25-hydroxyvitamin D 1-hydroxylase (1-hydroxylase) activity include PTH, phosphate depletion, and insulin-like growth factor-I (IGF-I) (see Chapter 5). However, only 5% of IH patients have elevated circulating PTH levels [28,35], and urinary cAMP levels, a surrogate measure of PTH, are also normal in most patients [28,36,37]. Mild hypophosphatemia with reduced renal tubular phosphate reabsorption has been described in as many as one-third of IH patients [29,30,35,38], and a strong inverse association between serum $1{,}25(OH)_2D$ levels and renal tubular phosphate reabsorption has been reported [30,37]. As elevated PTH or hypophosphatemia accompany elevated serum $1{,}25(OH)_2D$ in only a minority of patients, the cause of increased serum $1{,}25(OH)_2D$ in most patients with IH remains unknown. Serum IGF-I levels have not been measured in IH patients.

An extensive overlap of serum $1{,}25(OH)_2D$ values between normals and IH patients is present in each series reported (Fig. 3). Kaplan *et al.* [29] found that in patients with absorptive IH [defined as normal fasting urine Ca and normal or elevated serum $1{,}25(OH)_2D$], intestinal Ca absorption measured by fecal excretion of orally administered ^{47}Ca was increased out of proportion to the simultaneously measured serum $1{,}25(OH)_2D$ concentration (Fig. 4B). In contrast, a strong positive correlation between intestinal Ca absorption and serum $1{,}25(OH)_2D$ is found in normal volunteers, normocalciuric stone formers, patients with primary hyperparathyroidism, and IH patients with fasting hypercalciuria (Fig. 4A). The high intestinal Ca absorption rates with normal or elevated serum $1{,}25(OH)_2D$ levels suggest that the pathogenesis of IH is heterogeneous, with at least one phenotype resulting from $1{,}25(OH)_2D$ overproduction.

3. Decreased Renal Calcium Reabsorption

A defect in the tubular reabsorption of Ca, a so-called renal leak of Ca, has been postulated as a cause of hypercalciuria in IH. Two reports [39,40] found a greater fraction of filtered Ca excreted in the urine of

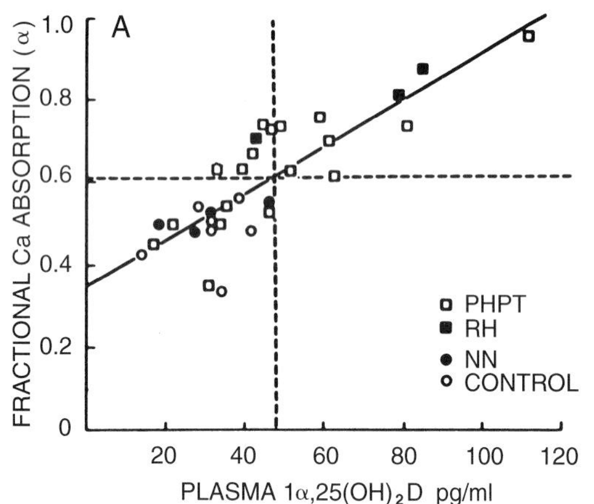

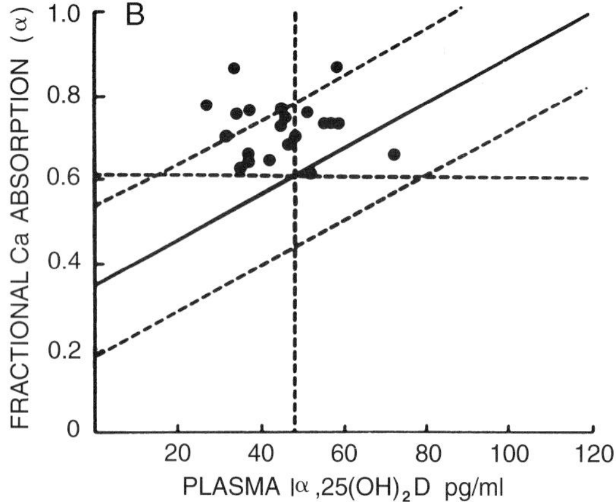

FIGURE 4 Relationship of calcium absorption to $1{,}25(OH)_2D$ levels. (A) Fractional intestinal absorption of oral ^{47}Ca versus serum $1{,}25(OH)_2D$ level in normal controls (open circles), normocalciuric stone formers (NN, filled circles), IH stone formers who have fasting hypercalciuria and elevated PTH (RH, filled squares), and patients with primary hyperparathyroidism (PHPT, open squares). (B) Fractional calcium absorption of IH patients with absorptive hypercalciuria (normal fasting urine Ca) superimposed on the 95% confidence limits for the relationship in controls. Reproduced from *The Journal of Clinical Investigation,* 1977, Vol. 59, pp. 756–760 [29], by copyright permission of The American Society for Clinical Investigation.

IH patients compared to non-stone formers. The values were calculated from inulin clearance or creatinine clearance and used blood ionized Ca as an estimate of ultrafilterable Ca. Although urinary sodium (Na) excretion is a major determinant of Ca excretion in normal and IH patients, there is no evidence that patients overingest or overexcrete Na. Hydrochlorothiazide and acetazolamide increase urine Ca, Na, and magnesium (Mg) excretion compared to normals [41], suggesting a generalized defect in proximal tubule electrolyte and water transport in IH patients. The basis for the abnormal renal transport is not known, but increased activity of the erythrocyte plasma membrane Ca^{2+},Mg^{2+}-ATPase in IH patients and correlation of enzyme activity with urine Ca excretion in families with IH [42] suggest a more widespread genetic defect in mono- and divalent ion transport.

4. Increased Bone Resorption

Abnormal skeletal metabolism in IH has been demonstrated by low bone mineral density of the distal radius [43,44] and lumbar spine [45–47] and by lower Ca content by neutron activation analysis [48]. Reports differ as to possible pathogenesis, as low bone density was found only in those with renal leak hypercalciuria in one study [44], whereas absorptive hypercalciuria was associated with low bone density in another study [46]. Information on bone dynamics is limited to one early study in which ^{47}Ca measurements suggested that bone turnover was increased, with bone resorption and formation both increased [49]. Two studies of bone histology show reduced bone apposition rate and delayed mineralization of osteoid seams [50] and also prolonged mineralization lag time and formation period [51]. These observations suggest a defective mineralization, which may be caused by hypophosphatemia. Measurements of biochemical markers of bone turnover reveal increased urine hydroxyproline excretion in unselected IH patients [52] and increased serum osteocalcin in IH patients with renal but not absorptive hypercalciuria [53].

5. Proposed Pathogenetic Models of Idiopathic Hypercalciuria

On the basis of the constant presence of increased intestinal Ca absorption, normal or elevated serum $1,25(OH)_2D$ levels, and normal or elevated fasting urinary Ca, Pak and colleagues [54] separated IH into three groups: absorptive, renal, and resorptive. In the first, primary intestinal Ca hyperabsorption (Fig. 5A) would transiently raise postprandial serum Ca above normal and increase ultrafilterable Ca. Postprandial hypercalcemia would transiently suppress PTH secretion, resulting in reduced tubular Ca reabsorption and hypercalciuria. In the second, a primary renal tubular leak of Ca (Fig. 5B) would cause hypercalciuria and a transient reduction in serum Ca. Secondary hyperparathyroidism would normalize serum Ca and increase proximal tubule $1,25(OH)_2D$ synthesis, which would stimulate intestinal Ca absorption. PTH secretion would then decline to the extent that serum Ca is normalized. Serum $1,25(OH)_2D$ would be elevated in renal IH and may be normal or elevated in absorptive IH [55,56]. A third possibility is based on a primary overproduction of $1,25(OH)_2D_3$ that will increase intestinal Ca absorption and bone resorption (Fig. 5C) while PTH remains normal and fasting urine Ca excretion may be normal or elevated.

6. Tests of the Models

Knowledge of the pathophysiology of IH is fundamental to developing rational therapy for the prevention of recurrent kidney stones. If the model of primary intestinal overabsorption is correct, then dietary Ca restriction would reduce the amount of Ca absorbed and thus excreted in the urine. If a renal leak of Ca is the primary event, or if urinary Ca originates from bone rather than diet, then restricting dietary Ca will have little effect on urinary Ca excretion, while worsening Ca balance and promoting bone loss. The accuracy of the models of absorptive and renal hypercalciuria have been assessed by testing certain predictions of each model.

a. Fasting Serum PTH, Urine Ca If repeated episodes of postprandial hypercalcemia suppress PTH secretion sufficiently to cause chronic hypoparathyroidism, fasting serum PTH and urine cAMP would be low and fasting urine Ca elevated. Transient suppression of PTH would permit normal serum PTH, urine cAMP, and fasting urine Ca. Renal IH requires increased PTH and urine cAMP and increased fasting urine Ca [57]. As existing PTH radioimmunoassays poorly differentiate normal from low values, most IH patients have been found to have PTH levels in the normal range. Although normal fasting urine Ca is not unusual among IH patients, less than 10% have elevated PTH and, therefore, fail to meet the criteria for renal IH [28,54]. Thus, a primary renal leak of Ca with a secondary increase in PTH cannot account for fasting hypercalciuria in a majority of patients. About 24% of patients meet the criteria of absorptive hypercalciuria by normalizing urine Ca during fasting [57].

b. External Ca Balance The relationship between net intestinal Ca absorption and 24-hr urine Ca excretion calculated from 6-day balance studies is different in IH patients compared to normals (Fig. 6) [58–64]. In non-stone formers, urinary Ca excretion increases with increasing net absorption, and overall Ca balance is

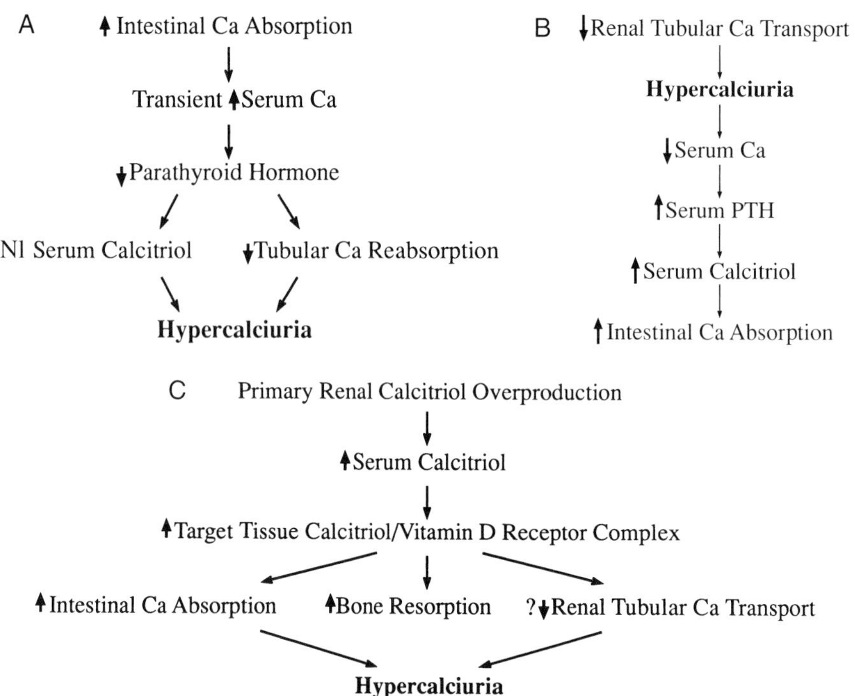

FIGURE 5 Three proposed models of IH. (A) Absorptive IH with primary intestinal overabsorption, postprandial hypercalcemia, suppressed PTH, and normal fasting urine Ca. Serum $1{,}25(OH)_2D$ is normal. (B) Primary renal tubular leak of Ca leads to a transient decrease in serum Ca and elevated PTH with secondary increases in serum $1{,}25(OH)_2D$ and intestinal Ca absorption. Fasting urine Ca is elevated. (C) Primary overproduction of $1{,}25(OH)_2D$ increases serum $1{,}25(OH)_2D$ and stimulates intestinal Ca absorption and bone resorption. Serum PTH is normal or decreased, and fasting urine Ca is normal or increased.

positive when net absorption is greater than 200 mg per 24 hr (see 95% confidence limits calculated from balance studies on normal subjects, Fig. 6). Net Ca absorption tends to be greater in IH patients, and for every level of net absorption, 24-hr urine Ca excretion is higher in the patients compared to healthy subjects. In IH patients, a greater portion of absorbed Ca is excreted in the urine. In normal subjects, net Ca absorption exceeds urine Ca excretion, and balance is positive when net absorption exceeds 200 mg/24 hr. In contrast, almost 50% of the IH patients have urine Ca excretion in excess of net absorption and are in negative Ca balance, even when allowance is made for some variability in the balance data (±50 mg). Thus, at all levels of net Ca absorption negative Ca balance (above the zero balance or above the line of identity) is common in IH patients but not in healthy subjects. Negative Ca balance in the presence of adequate Ca intake is incompatible with a primary hyperabsorption of dietary Ca and cannot, by itself, account for the hypercalciuria.

c. Urine Ca and Ca Balance during Low Ca Diet The hypothesis of primary intestinal Ca overabsorption would lead to the prediction that dietary Ca restriction would reduce the amount of Ca absorbed and would therefore reduce urinary Ca excretion. Like normal subjects, IH patients would be in positive or neutral Ca balance when net absorption is above 200 mg/24 hr (Fig. 6). A low Ca diet would reduce urine Ca excretion through an increase in PTH secretion, which would promote distal tubular Ca reabsorption. In contrast, patients with a primary renal Ca leak would be unable to conserve urine Ca at any level of Ca intake and would maintain an excessive or inappropriately high urine Ca excretion during low Ca diet. As a result, during low Ca diet Ca balance would shift from positive or neutral to negative or become more negative. Serum PTH would be expected to increase to high levels during a low Ca diet.

To test whether the responses of IH patients fit these predictions, Coe *et al.* [36] fed a low Ca diet (2 mg/kg/day) to nine normal volunteers and 26 unselected IH stone formers. After 10 days on the diet, urine Ca excretion decreased to 2.0 mg/kg body weight or less in both patients and controls, but 17 of the 26 IH patients (Fig. 7) showed values greater than the highest value in normal controls. In patients, urine Ca excretion (CaE) ranged from normal to persistently high levels. It exceeded Ca

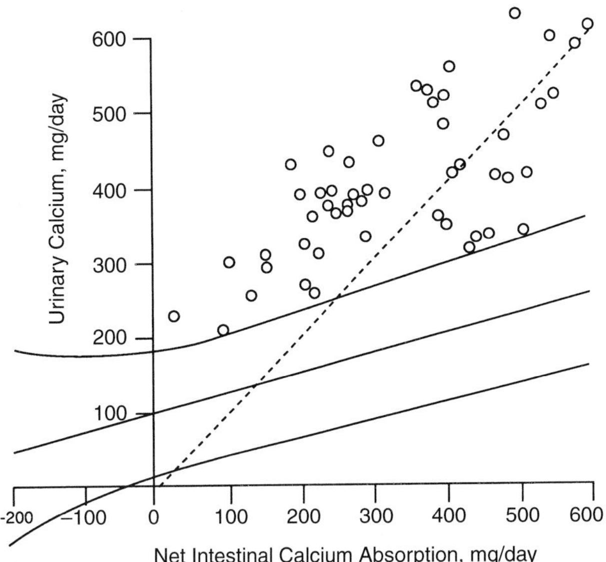

FIGURE 6 Urinary Ca excretion as a function of net intestinal Ca absorption. Data are derived from 6-day external mineral balance studies. Solid lines indicate the 95% confidence limits about the mean regression line derived from the data on 195 adult non-stone formers. Individual balance studies performed on 51 patients with IH are shown as open circles. The dashed line represents equivalent rates of urinary Ca excretion and net intestinal Ca absorption (the line of identity). Normal values are from Refs. 39, 49, 58, and 59. Values from patients are from Refs. 39, 49, 61–64, and J. Lemann (personal communication, 1992). Adapted from Asplin *et al.* [84].

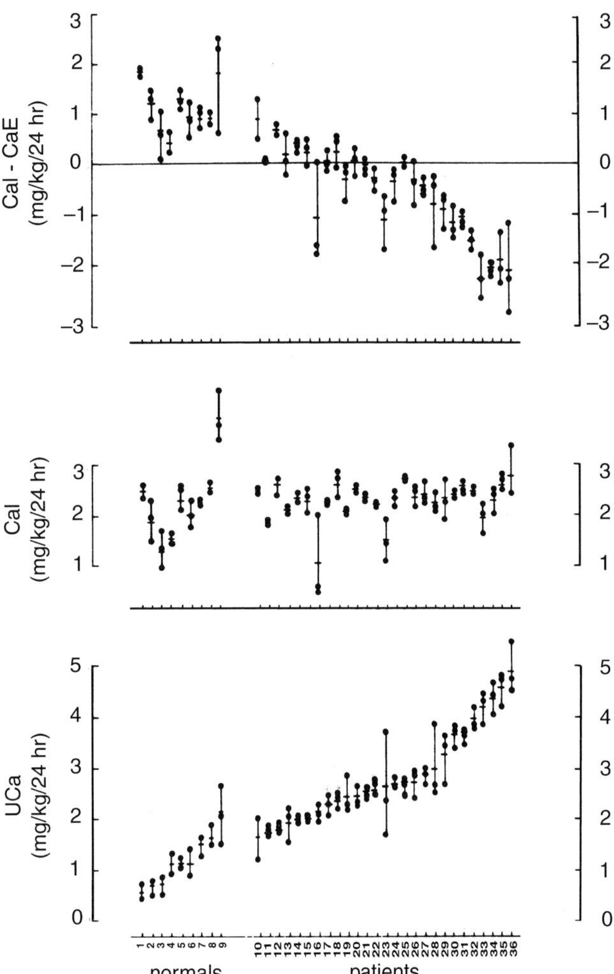

FIGURE 7 Calcium intake and urinary excretion in patients with IH and normal subjects. Urine Ca excretion (UCa) and Ca intake (CaI) are during a low Ca diet. Mean Ca intakes for patients and controls (2.29 ± 0.15 versus 2.31 ± 0.05 mg/kg/day) were not different. Mean urine excretion rates during low Ca intake and values of CaI − CaE (an index of Ca balance) differed significantly between normals and IH patients. Subjects and patients are arranged in ascending order of urinary Ca excretion. Reprinted by permission of the publisher from Coe *et al.* [36]. "Effects of low-calcium diet on urine calcium excretion, parathyroid function and serum $1,25(OH)_2D_3$ levels in patients with idiopathic hypercalciuria and in normal subjects," *American Journal of Medicine,* Vol. 72, pp. 25–32. Copyright 1982 by Excerpta Medica Inc.

intake (CaI) (Fig. 7, CaI − CaE) in 11 of the 26 patients and none of the non-stone-forming controls. Thus, almost 50% of the patients had more Ca in the urine than what was provided by the diet and were clearly in negative Ca balance.

The results indicate that a diet chronically low in Ca may be detrimental for some patients, as the inability to conserve urine Ca during low Ca intake would eventually lead to clinically detectable bone loss. The data also suggest that some patients with IH may have diet-dependent hypercalciuria, whereas others have diet-independent hypercalciuria. The two proposed mechanisms cannot be readily distinguished by any clear break in the continuum of urine Ca values, and serum PTH and $1,25(OH)_2D$ levels do not predict the urine Ca responses during a low Ca diet. Patients with the highest urine Ca and negative Ca balance had serum PTH and $1,25(OH)_2D$ levels that were not different from patients who conserved Ca to the levels found in normal subjects.

d. Role of $1,25(OH)_2D$ Excess The majority of patients are classified as having absorptive hypercalciuria [54,57], yet negative Ca balance during low Ca diet [36] without PTH or $1,25(OH)_2D$ elevation is not predicted by the absorptive model (Fig. 5A). Patients who meet the criteria of renal hypercalciuria tend to have higher serum $1,25(OH)_2D$ levels, but only a small portion have elevated PTH levels. Further, serum $1,25(OH)_2D$ levels do not predict whether patients will be classified as absorptive or renal, and at least one-third of patients have normal serum $1,25(OH)_2D$ levels despite intestinal Ca hyperabsorption. For them, the mechanism of intestinal Ca hyperabsorption remains unexplained.

The model of primary vitamin D excess (Fig. 5C) is supported by elevated $1,25(OH)_2D$ production rates and enhanced biological actions of $1,25(OH)_2D$, including increased intestinal Ca absorption and bone resorption. Creation of a mild form of $1,25(OH)_2D$ excess was achieved by the administration of pharmacological doses of $1,25(OH)_2D_3$ (3.0 μg/day) to healthy men for 10 days while Ca intake varied from low (160 mg) to normal (372 mg) or high (880 mg) [65–67]. Hypercalcemia did not occur despite an increase in urine Ca excretion and net intestinal Ca absorption calculated from 6-day metabolic balance studies (Fig. 8). Dietary Ca strongly influenced the response to $1,25(OH)_2D_3$, as Ca balance was more negative during low Ca intake, and the increase in urine Ca primarily resulted from accelerated bone resorption. At low normal or normal Ca intake, $1,25(OH)_2D_3$ caused greater increases in urine Ca and net intestinal Ca absorption, although Ca balance was not altered. Ca balance became positive during normal Ca diet, and balance was neutral or positive in the presence and absence of $1,25(OH)_2D_3$. Thus, 3 μg/day of $1,25(OH)_2D_3$, which is insufficient to cause hypercalcemia in this 6-day study, has profound effects on intestinal Ca absorption, urine Ca excretion, and Ca balance. Further, $1,25(OH)_2D_3$ administration caused negative Ca balance only during dietary Ca deprivation. These $1,25(OH)_2D_3$-induced changes in normal subjects are similar to those observed in IH patients on comparable levels of Ca intake.

In other experiments, ketoconazole administration to IH patients inhibited renal $1,25(OH)_2D$ biosynthesis [56] and decreased serum $1,25(OH)_2D$ levels, intestinal Ca absorption, and urine Ca excretion. The results of the effects of $1,25(OH)_2D_3$ treatment and the response to ketoconazole provide further support for a primary $1,25(OH)_2D$ excess in at least some patients with IH. The nature of the disordered regulation of renal $1,25(OH)_2D$ production remains to be determined, as neither elevated PTH nor hypophosphatemia were present in responders or were absent in nonresponders to ketoconazole.

III. GENETIC HYPERCALCIURIC RATS

Tests of the absorptive, renal, and vitamin D excess models of IH have been complicated by difficulty in controlling for potential variables such as inheritance and environmental factors that may influence dietary patterns. The availability of an animal model of IH would permit the testing of the three hypotheses under conditions that exclude genetic and dietary influences. The strong familial occurrence of IH in humans and the high frequency of elevated urine Ca in adult men and women suggested that spontaneous hypercalciuria might also be found in animals.

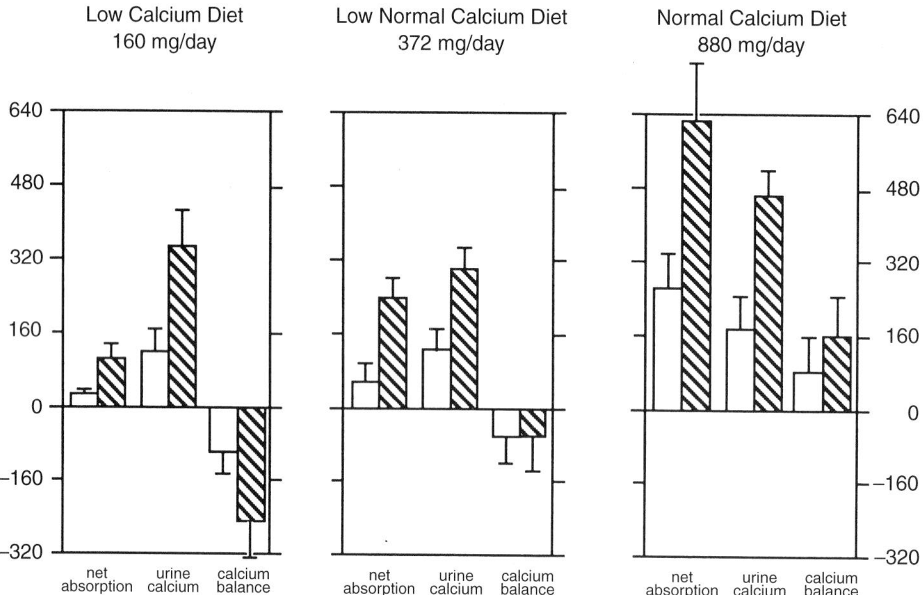

FIGURE 8 Intestinal Ca absorption, urine Ca excretion, and Ca balance in normal men receiving $1,25(OH)_2D_3$ (hatched bars) or controls (open bars) at varying levels of dietary Ca. Values (mg/24 hr) are means ± SEM for 6 men per group. For Ca balance, values above the horizontal line indicate positive balance and those below the line, negative balance. Data from Maierhofer *et al.* [66] and Adams *et al.* [66a]. Reprinted with permission from F. L. Coe and J. H. Parks, Nephrolithiasis: Pathogenesis and Treatment, Second Edition, Year Book Publishers, Chicago, 1988 [65].

1. Establishment of a Colony of Genetically Hypercalciuric Rats

Urine Ca excretion in a population of male Sprague-Dawley rats fed a normal Ca diet (0.8% Ca) was similar to that found in healthy humans in that it followed a non-Gaussian distribution, with values clustering about the mean and a long tail of higher values [14]. Using an arbitrary definition of hypercalciuria as urine Ca greater than 2 standard deviations above the mean value, about 5 to 10% of male and female rats were hypercalciuric. The most hypercalciuric offspring were used for repeated matings, leading to a colony with hypercalciuria that has increased in intensity and frequency with each successive generation [15]. By the twentieth generation, over 95% of males and females were hypercalciuric. By the fortieth generation, mean urine Ca excretion was 7.0 ± 0.3 mg/24 hr compared to the stable mean excretion of less than 0.75 mg/24 hr by wild-type rats [68]. Hypercalciuria may be lifelong, as it can be detected as soon as the animals are weaned (about 50 g body weight). Weight and growth have been comparable to wild-type Sprague-Dawley rats obtained from the same supplier that provided the original spontaneously hypercalciuric animals. No anatomical or structural abnormalities have been identified; however, by 18 weeks of age 100% of the animals have grossly evident Ca-containing kidney stones in the upper and lower urinary tracts [69]. No stones are found in the kidney or urinary tract of wild-type rats.

2. Serum and Urine Chemistries

Serum Ca and Mg are within the normal range in the genetic hypercalciuric (GH) male and female rats [15]. Serum phosphate is lower in female rats, and there is no difference between GH and wild-type males and females. Serum PTH levels in GH rats are not different from controls. Urine volumes are greater in the GH rats.

3. Mineral Balance

Six-day external balance studies performed while the animals were fed a normal Ca diet showed the animals to be in positive balance for Ca, Mg, and phosphorus [15] with greater net Ca absorption in GH rats. The GH rats maintained positive Ca balance because the increased urine Ca excretion was matched by a greater net intestinal Ca absorption.

4. Intestinal Calcium Transport

To investigate the mechanism of the increased Ca absorption, segments of duodenum were mounted *in vitro* in modified Ussing chambers, and transepithelial bidirectional fluxes of Ca were measured in the absence of electrochemical gradients [22]. Under these conditions, [15], duodenal segments from GH rats had a 5-fold increase in the mucosal-to-serosal (absorptive) transepithelial flux of Ca (J_{ms}), whereas the secretory flux of Ca from serosa to mucosa (J_{sm}) was only mildly increased compared to wild type (Table II). As Ca J_{ms} was 10 to 12 times higher than Ca J_{sm}, changes in J_{sm} had a nonsignificant effect on net Ca absorption.

5. Serum $1,25(OH)_2D$

Circulating $1,25(OH)_2D$ levels were lower in the fourth generation GH rats; however, the differences disappeared by the tenth generation [at 190 g, mean ± SD serum $1,25(OH)_2D$ was 135 ± 12 versus 174 ± 19 pg/ml, nonsignificant], and no subsequent differences in serum $1,25(OH)_2D$ levels have been observed [16]. *In vitro* duodenal net flux (J_{net}, equal to $J_{ms} - J_{sm}$) for Ca was positively correlated with serum $1,25(OH)_2D$ in normocalciuric and GH male and female rats (Fig. 9). However, the regression coefficients were different for the wild-type and GH rats, with the latter having a steeper slope. The greater Ca J_{net} in GH rats with serum $1,25(OH)_2D$ levels comparable to the wild-type rats strongly suggests that duodenal Ca-transporting cells in GH rats are more sensitive to $1,25(OH)_2D$.

6. Role of the Vitamin D Receptor

The increased intestinal Ca transport and normal serum $1,25(OH)_2D$ levels in GH rats suggested either that Ca transport was being stimulated by an unidentified, vitamin D-independent process or that $1,25(OH)_2D$ action was being amplified at the level of the target tissues. As $1,25(OH)_2D$ stimulates Ca transport by binding to the vitamin D receptor (VDR) to up-regulate vitamin D-dependent genes that encode for proteins involved in

TABLE II *In Vitro* Bidirectional Duodenal Calcium Active Transport[a]

Flux	NM	GHM	NF	GHF
J_{ms}	51 ± 12	264 ± 27	29 ± 9	258 ± 40
J_{sm}	11 ± 2	19 ± 2	14 ± 2	23 ± 2
J_{net}	40 ± 11	245 ± 28	14 ± 8	235 ± 40

[a] Values are means ± SE for 5 to 11 rats per group. NM and NF are normocalciuric (wild-type) male and female rats, respectively. GHM and GHF are genetic hypercalciuric male and female rats, respectively. J_{ms} and J_{sm} are mucosal-to-serosal and serosal-to-mucosal fluxes of Ca, respectively. J_{net} is net Ca absorption, where $J_{net} = J_{ms} - J_{sm}$. Adapted from Li *et al.* [16] and reproduced with permission from *The Journal of Clinical Investigation,* 1993, Vol. 91, pp. 661–667, by copyright permission of The American Society for Clinical Investigation.

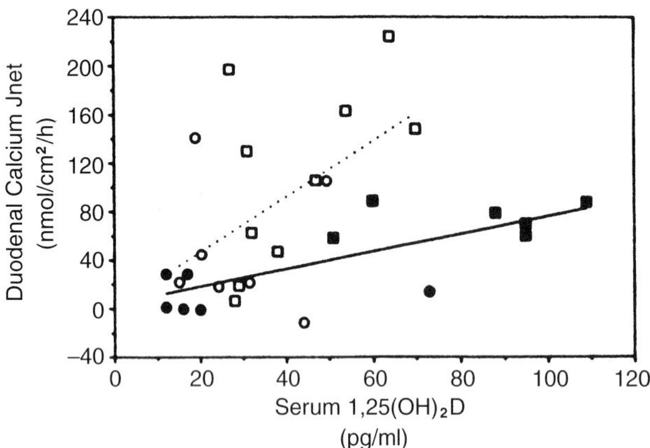

FIGURE 9 Duodenal Ca net flux (J_{net}) as a function of serum $1,25(OH)_2D$ for hypercalciuric and normocalciuric male (open and filled squares, respectively) and female (open and filled circles, respectively) rats. J_{net} and serum $1,25(OH)_2D$ were correlated for male and female normocalciuric rats ($r = 0.789$, $n = 12$, $p < 0.001$, solid line) and for male and female GH rats ($r = 0.500$, $n = 17$, $p < 0.03$, dotted line). The regressions were different (F ratio $= 5.469$, $p < 0.015$). Reproduced from *The Journal of Clinical Investigation,* 1988, Vol. 82, pp. 1585–1591 [15], by copyright permission of The American Society for Clinical Investigation.

transepithelial Ca transport, and because the biological actions of $1,25(OH)_2D$ are directly related to the tissue VDR content [70–72], VDR binding in intestinal epithelial cells was measured. Duodenal cytosolic fractions prepared in high potassium buffer from male GH rats bound more ^3H-$1,25(OH)_2D_3$ than comparable fractions from wild-type control rats [16] (Fig. 10). Cytosolic fractions from kidney cortex and from splenic monocytes also exhibited greater specific binding of $1,25(OH)_2D_3$. Scatchard analysis of the specific binding curves revealed a single class of VDR binding sites in tissues from both wild-type and GH rats. The number of VDR binding sites in GH rat duodenal cells was double that found in cells from wild-type rats (536 ± 73 versus 243 ± 42 fmol/mg protein; $n = 8$ and $n = 14$; $p < 0.001$), with comparable affinity of the receptor for its ligand (0.33 ± 0.01 versus 0.49 ± 0.01 nM; nonsignificant). A 2-fold increase in VDR binding sites was also found in GH rat renal cortical homogenates [16].

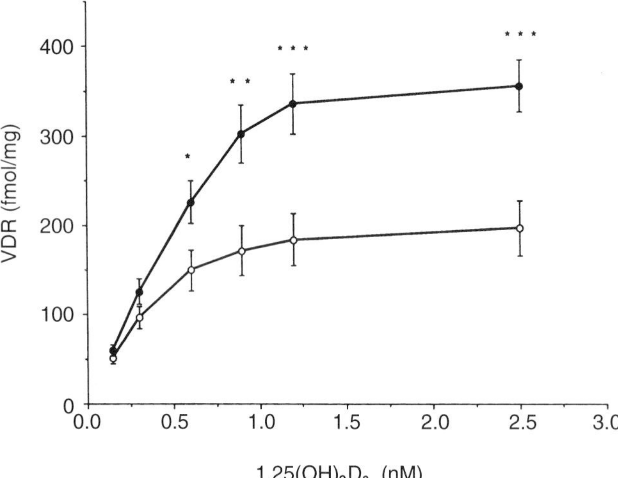

FIGURE 10 Specific binding of ^3H-$1,25(OH)_2D_3$ to duodenal cytosolic fractions (VDR) prepared from GH rats (filled circles) and wild-type controls (open circles) while fed a normal Ca diet. Values are means ± SEM for four observations per concentration point. *, $p < 0.05$; **, $p < 0.01$; ***$p < 0.005$ vs controls. Reproduced from *The Journal of Clinical Investigation,* 1993, Vol. 91, pp. 661–667 [16], by copyright permission of The American Society for Clinical Investigation.

Using Western blotting, homogenates of duodenal mucosa from GH rats contained a band at 50 kDa that comigrated with duodenal extracts from wild-type rats and with human recombinant VDR. The bands from the GH rat tissues were more intense compared to controls, confirming that the increase in specific ^3H-$1,25(OH)_2D_3$ binding was due to an increase in VDR protein. Northern analysis of RNA extracts from GH and wild-type rat tissues revealed a single species of VDR mRNA at 4.4 kb with no difference in migration between the two groups [16]. Duodenal extracts from GH rats contained less VDR mRNA than controls. Estimates of duodenal cell transcription rates using standard nuclear run-on assays found no clear difference between GH rats and controls [16]. The *in vivo* half-life of the VDR mRNA in GH rat duodenum was comparable to that of controls (6 hr) and administration of a small dose of $1,25(OH)_2D_3$ (30 ng as a single dose) resulted in a significant elevation of VDR message and prolongation of message half-life in GH rats but not controls [73]. Thus, in GH rat intestine, the increased VDR level is not due to an increase in VDR gene transcription. The data are consistent with either an increase in VDR mRNA translation efficiency or changes that result in a prolongation of the VDR half-life. The increased accumulation of the vitamin D-dependent calbindin-D_{9K} found in GH rat duodenum [16] is evidence that the increased level of VDR is functional and that the increased Ca transport is likely a vitamin D-mediated process.

Major questions remain as to the genetic basis of the increased VDR activity. A subtle mutation in the VDR gene could have several possible effects on the VDR, including promotion of the synthesis of an abnormal protein leading to an altered rate of protein and degradation that results in a prolongation of the turnover of the VDR.

7. Increased Bone Resorption

In vitro studies of bone resorption using neonatal calvariae from normal and GH rats show that Ca efflux, a measure of bone resorption, increases in a dose-dependent manner in the presence of $1,25(OH)_2D_3$ or PTH [74]. The dose–response curve is much steeper for $1,25(OH)_2D_3$ in calvariae from GH rats, whereas the dose–response curves for PTH-stimulated Ca efflux are not different between control and GH calvariae. Western blotting showed a 4-fold increase in VDR protein from GH neonatal rat calvariae [74]. Thus, the increase in target tissue VDR exerts biological actions that increase $1,25(OH)_2D_3$-dependent bone resorption, which may contribute to the hypercalciuria.

8. Response to Low Calcium Diet

To test whether the hypercalciuria in GH rats is the result of a primary overabsorption of dietary Ca, GH and wild-type control rats were fed diets either normal (0.6% Ca) or low (0.02% Ca) with respect to Ca. During the low Ca diet, urine Ca excretion decreased in both groups (Fig. 11); however, urine Ca remained higher in GH rats and resulted in negative Ca balance [75]. The inability of GH rats to conserve Ca during low Ca intake excludes overabsorption of dietary Ca as the sole cause of hypercalciuria in GH rats.

9. Summary of Pathogenesis in the Genetic Hypercalciuric Rat

Figure 12 summarizes current knowledge of the pathogenesis of hypercalciuria in the GH rats. Breeding by selection for hypercalciuria has emphasized a trait in the offspring that likely involves the expression of several genes for full phenotypic expression. To date, none of the genes has been identified. Studies implicate the increased VDR concentration as part of the primary event(s) and a cause of the hypercalciuria; however, a secondary adaptive increase in VDR to compensate for urinary Ca losses has not been completely excluded. Further information is required regarding the renal handling of Ca in GH rats and whether the GH genotype results in a primary defect in renal Ca transport.

IV. CURRENT VIEW OF HUMAN GENETIC HYPERCALCIURIA

Striking similarities in Ca metabolism between GH rats, IH patients, and human volunteers treated with $1,25(OH)_2D_3$ (Table III) strongly support a primary role of excess $1,25(OH)_2D$ biological action in the pathogenesis of human IH. When deprived of dietary Ca, few patients conserve Ca to the extent that normals do (Fig. 7). The renal IH model predicts ongoing urinary losses of Ca independent of Ca intake, and negative Ca balance during a low calcium diet. However, most patients have normal, not elevated PTH, as renal IH would require. Therefore, the absorptive and renal models of hypercalciuria cannot explain the response of most patients to a low Ca diet. In normal subjects, $1,25(OH)_2D_3$ administration creates the changes in urine Ca and Ca balance observed in a majority of IH patients who have either normal or elevated serum $1,25(OH)_2D$ levels. In some patients, elevated serum $1,25(OH)_2D$ stimulates

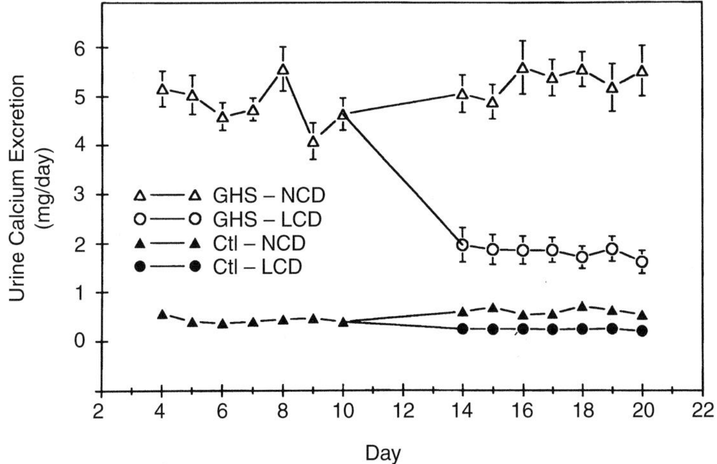

FIGURE 11 Daily urine Ca excretion in nineteenth-generation GH rats (open symbols) or wild-type control rats (filled symbols) fed a normal Ca diet (NCD, 0.6% Ca, triangles) during days 1–10 followed by either continuation of the NCD (triangles) or feeding of a low Ca diet (LCD, 0.02% Ca, circles). Rats were pair-fed to 13 g of diet per day. Reprinted with permission from Kim *et al.* [75].

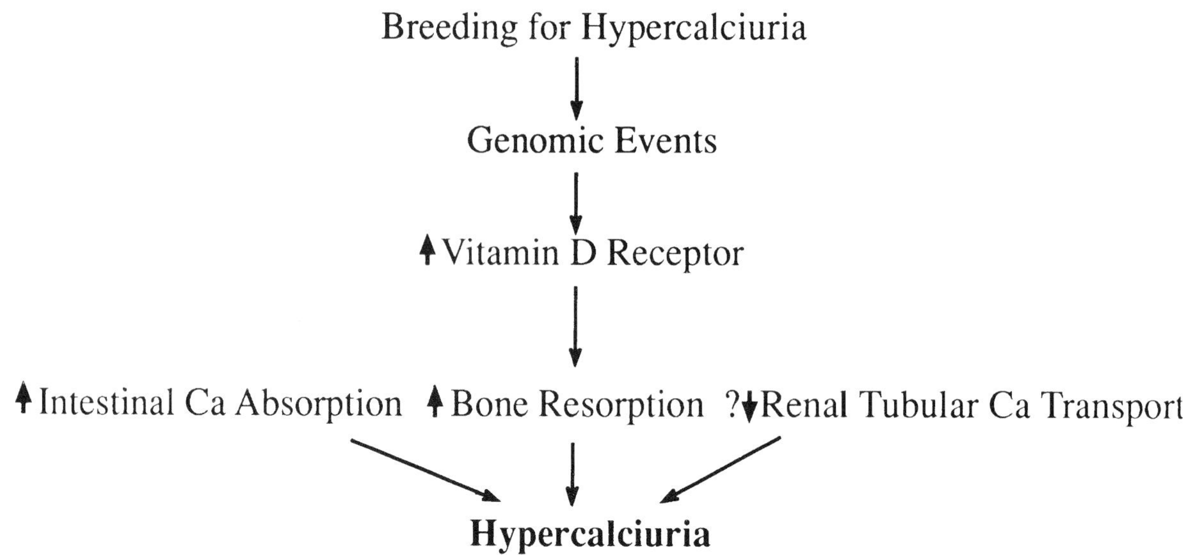

FIGURE 12 Proposed series of events that result from breeding selection for hypercalciuric rats. The renal handling of Ca by GH rats and the role of increased VDR content, if any, in the transport process remain unknown.

intestinal Ca hyperabsorption, hypercalciuria, and negative Ca balance during low Ca intake. For those patients with normal serum $1,25(OH)_2D$ levels, the source of $1,25(OH)_2D$ excess is more elusive. These patients and the GH rats are similar in that both have normal serum $1,25(OH)_2D$, increased intestinal absorption, and enhanced bone resorption during a low Ca diet. Whether these changes in human IH are due to increased intestinal, renal, and bone cell VDR content that can amplify the biological actions of normal circulating $1,25(OH)_2D$ levels remains to be determined.

TABLE III Pathophysiology of Genetic Hypercalciuria[a]

Parameter	Human controls	Human IH	GH rats
Serum Ca	N	N	N
Serum phosphate	N	N–D	N
Serum $1,25(OH)_2D$	I	N–I	N
Urinary Ca on NCD	I	I	I
Urinary Ca on LCD	I	N–I	I
Intestinal Ca absorption	I	I	I
Ca balance on NCD	Pos–N	N–Neg	Pos
Ca balance on LCD	N–Neg	N–Neg	Neg

[a] Values for human controls are responses to treatment with 3 μg $1,25(OH)_2D_3$ daily for 7 days compared to pretreatment. GH, Genetic hypercalciuric; NCD, normal Ca diet; LCD, low Ca diet; N, normal; I, increased; D, decreased; Pos, positive; Neg, negative.

V. THERAPEUTICS OF IDIOPATHIC HYPERCALCIURIA AND EFFECTS ON CALCIUM METABOLISM

A. Dietary Calcium Restriction

Hypercalciuria promotes urine calcium oxalate supersaturation and increases spontaneous crystal formation [76]. The goals of preventive therapy are to reduce supersaturation by increasing urine volume and decreasing urine Ca excretion. If the pathophysiological role of $1,25(OH)_2D$ excess or VDR excess is borne out, then ideal therapy may eventually include either a specific $1,25(OH)_2D$ antagonist or an inhibitor of VDR function. In the absence of such agents, therapies will continue to concentrate on lowering urine Ca through indirect means.

Since the description of IH and its emergence as a major cause of Ca stones, physicians have recommended dietary Ca restriction to lower urine Ca. Dietary Ca restriction or the use of Ca-binding resin to prevent absorption [77] could be efficacious for patients with primary intestinal Ca hyperabsorption (absorptive hypercalciuria). However, it appears that many patients develop negative Ca balance during low Ca intake. For them, chronic dietary Ca restriction and negative Ca balance would eventually cause bone loss and osteoporosis. Reports of lower bone density in IH patients suggest that Ca restriction may only worsen the existing reduction in bone mass. Therefore, treatment with Ca restriction requires knowledge that the patient will normally conserve urine Ca and not develop negative Ca balance during such diet therapy.

B. Thiazides

Thiazide and the related chlorthalidone diuretics reduce urine Ca excretion by inducing a NaCl diuresis, which causes volume contraction and decreases Ca delivery to the distal tubule segments [78]. These agents also stimulate distal tubule Ca reabsorption through a direct interaction with the tubule cells [77,78]. Thiazides may decrease or have no effect [27,31,80] on intestinal Ca transport in IH patients, and serum $1,25(OH)_2D$ and PTH levels are not changed by thiazide. In one study, IH patients treated with chlorthalidone for 6 months improved Ca balance to or toward positive by decreasing both urine Ca and intestinal Ca absorption [81], with urine Ca declining to a greater extent than intestinal absorption. Thus, the epidemiological studies suggesting that chronic thiazide therapy reduces fracture risk [82,83] may result from drug-induced improvement in Ca balance [80] and reduced bone turnover and improved mineralization [51].

The effects of thiazide on urine Ca and bone metabolism are accompanied by a decrease in recurrent Ca nephrolithiasis as compared to placebo controls [84]. The beneficial effect of thiazide is evident during the second and third year of therapy, when stone recurrence is reduced by about 50%. The reduction in new stone formation is due to a decrease in urine calcium oxalate supersaturation, as urine Ca declines while oxalate is unchanged. As thiazides can reduce urine Ca excretion and stone formation rates in all forms of IH [85], knowledge of the pathogenesis of IH in each patient may not be required when selecting thiazide therapy.

VI. RISK OF STONE FORMATION USING VITAMIN D ANALOGS

A growing interest in the cell differentiation and immune modulation effects of vitamin D and analogs may result in their use in a variety of disorders [86–88] (also see Sections VII and VIII of this volume). However, the development of hypercalciuria and hypercalcemia may limit the use of the naturally occurring vitamin D metabolites as well as their synthetic analogs [89,90]. Some vitamin D analogs are reported to have little or no hypercalcemic action; however, the calcemic and calciuric actions may occur at higher doses. Analogs cause hypercalcemia and hypercalciuria through the classic vitamin D actions on intestine, kidney, and bone [88,89]. Low Ca diets have had only modest beneficial effects to limit hypercalciuria and hypercalcemia, and they could promote bone loss. It remains to be determined whether the newer vitamin D analogs with less calcemic activity will, in practice, cause less calciuria with a lower risk of kidney stone formation.

Until such actions of the vitamin D analogs are known, standard approaches to minimize stone formation should be followed. These include (1) assuring sufficient fluid intake to maintain at least 1.5 liters urine output per day; (2) if necessary increasing urine citrate excretion to normal in those with low citrate [76]; and (3) discontinuing or reducing treatment if significant hypercalciuria develops. The addition of a thiazide may avoid or minimize hypercalciuria, but hypercalcemia may occur because of thiazide-induced Ca retention.

References

1. Flocks RH 1939 Calcium and phosphorus excretion in the urine of patients with renal or ureteral calculi. JAMA **13**:1466–1471.
2. Albright F, Henneman P, Benedict PH, Forbes HR 1953 Idiopathic hypercalciuria. A preliminary report. Proc R Soc Med Lond (Biol) **46**:1077.
3. Henneman PH, Benedict PH, Forbes AP Dudley HR 1958 Idiopathic hypercalciuria. N Engl J Med **259**:802–807.
4. Hodgkinson A, Pyrah LN 1958 The urinary excretion of calcium and inorganic phosphate in 344 patients with calcium stones of renal origin. Br J Surg **46**:10–18.
5. Robertson WG, Morgan DB 1972 The distribution of urinary calcium excretions in normal persons and stone-formers. Clin Chim Acta **37**:503–508.
6. Coe FL, Parks JH, Asplin JR 1992 The pathogenesis and treatment of kidney stones. N Engl J Med **327**:1141–1152.
7. Coe FL 1977 Treated and untreated recurrent calcium nephrolithiasis in patients with idiopathic hypercalciuria, hyperuricosuria, or no metabolic disorder. Ann Intern Med **87**:404–410.
8. McGeown MG 1960 Heredity in renal stone disease. Clin Sci **19**:465–471.
9. Resnick M, Pridgen DB, Goodman HO 1968 Genetic predisposition of calcium oxalate renal calculi. N Engl J Med **278**:1313–1318.
10. Coe FL, Parks JH, Moore ES 1979 Familial idiopathic hypercalciuria. N Engl J Med **300**:337–340.
11. Pak CYC, McGuire J, Peterson R, Britton F, Harrod MJ 1981 Familial absorptive hypercalciuria in a large kindred. J Urol **126**:717–719.
12. Mehes K, Szelid Z 1980 Autosomal dominant inheritance of hypercalciuria. Eur J Pediatr **133**:239–242.
13. Moore ES, Coe FL, McMann BJ, Favus MJ 1978 Idiopathic hypercalciuria in children: Prevalence and metabolic characteristics. J Pediatr **92**:906–910.
14. Coe FL, Favus MJ 1981 Hypercalciuric states. Miner Electrolyte Metab **5**:183–200.
15. Bushinsky DA, Favus MJ 1988 Mechanism of hypercalciuria in genetic hypercalciuric rats. Inherited defect in intestinal calcium transport. J Clin Invest **82**:1585–1591.
16. Li X-Q, Tembe V, Horwitz GM, Bushinsky DA, Favus MJ 1993 Increased intestinal vitamin D receptor in genetic hypercalciuric

rats: A cause of intestinal calcium hyperabsorption. J Clin Invest **91**:661–667.
17. Broadus AE, Insogna KL, Lang R, Mallette LE, Oren DA, Gertner JM, Kliger AS, Ellison AF 1984 A consideration of the hormonal basis and phosphate leak hypothesis of absorptive hypercalciuria. J Clin Endocrinol Metab **58**:161–169.
18. Buckalew VM, Purvis ML, Shulman MG, Herndon CN, Rudman D 1974 Hereditary renal tubular acidosis. Medicine **53**:229–254.
19. Frymoyer PA, Scheinman SJ, Dunham PB, Jones DB, Hueber P, Schroeder ET 1991 X-linked recessive nephrolithiasis with renal failure. N Engl J Med **325**:681–686.
20. Klugman VA, Favus MJ 1996 Intestinal absorption of calcium, magnesium and phosphorus. In: Coe FL, Favus MJ, Pak CYC, Parks JH, Preminger GM (eds) Kidney Stones: Medical and Surgical Management, 1st Ed. Lippincott-Raven, Philadelphia, Pennsylvania, pp. 210–221.
21. Bronner F, Pansu D, Stein WD 1986 An analysis of intestinal calcium transport across the rat intestine. Am J Physiol **250**:G561-G569.
22. Favus MJ 1985 Factors that influence absorption and secretion of calcium in the small intestine and colon. Am J Physiol **248**:G147–G157.
23. Caniggia A, Gennari C, Cesari L 1965 Intestinal absorption of ^{45}Ca in stone-forming patients. Br Med J **1**:427–429.
24. Birge SJ, Peck WA, Berman M, Wheadon GD 1969 Study of calcium absorption in man: A kinetic analysis and physiologic model. J Clin Invest **48**:1705–1713.
25. Wills MR, Zisman E, Wortsman J, Evens RG, Pak CYC, Bartter FC 1970 The measurement of intestinal calcium absorption in nephrolithiasis. Clin Sci **39**:95–106.
26. Pak CYC, East DA, Sanzenbacher LJ, Delea CS, Bartter FC 1972 Gastrointestinal calcium absorption in nephrolithiasis. J Clin Endocrinol Metab **35**:261–270.
27. Ehrig U, Harrison JE, Wilson DR 1974 Effect of long-term thiazide therapy on intestinal calcium absorption in patients with recurrent renal calculi. Metabolism **23**:139–149.
28. Pak CYC, Ohata M, Lawrence EC, Snyder W 1974 The hypercalciurias: Causes, parathyroid functions, and diagnostic criteria. J Clin Invest **54**:387–400.
29. Kaplan RA, Haussler MR, Deftos LJ, Bone H, Pak CYC 1977 The role of 1,25-dihydroxyvitamin D in the mediation of intestinal hyperabsorption of calcium in primary hyperparathyroidism and absorptive hypercalciuria. J Clin Invest **59**:756–760.
30. Shen FH, Baylink DJ, Nielsen RL, Sherrard DJ, Ivey JL, Haussler MR 1977 Increased serum 1,25-dihydroxyvitamin D in idiopathic hypercalciuria. J Lab Clin Med **90**:955–962.
31. Barilla DE, Tolentino R, Kaplan RA, Pak CYC 1978 Selective effects of thiazide on intestinal absorption of calcium in absorptive and renal hypercalciurias. Metabolism **27**:125–131.
32. Lemann J Jr 1992 Pathogenesis of idiopathic hypercalciuria and nephrolithiasis. In: Coe FL, Favus MJ (eds) Disorders of Bone and Mineral Metabolism, 1st Ed. Raven, New York, pp. 685–706.
33. Duncombe VM, Watts RWE, Peters TJ 1984 Studies on intestinal calcium absorption in patients with idiopathic hypercalciuria. Q J Med **209**:69–79.
34. Insogna KL, Broadus AE, Dreyer BE, Ellison AF, Gertner JM 1985 Elevated production rate of 1,25-dihydroxyvitamin D in patients with absorptive hypercalciuria. J Clin Endocrinol Metab **61**:490–495.
35. Bataille P, Bouillon R, Fournier A, Renaud H, Gueris J, Idrissi A 1987 Increased plasma concentrations of total and free $1,25(OH)_2D_3$ in calcium stone formers with idiopathic hypercalciuria. Contrib Nephrol **58**:137–142.
36. Coe FL, Favus MJ, Crockett T, Strauss LM, Parks JH, Porat A, Gantt CL, Sherwood LM 1982 Effects of low-calcium diet on urine calcium excretion, parathyroid function and serum $1,25(OH)_2D_3$ levels in patients with idiopathic hypercalciuria and in normal subjects. Am J Med **72**:25–32.
37. Broadus AE, Insogna KL, Lang R, Ellison AF, Dreyer BE 1984 Evidence for disordered control of 1,25-dihydroxyvitamin D production in absorptive hypercalciuria. N Engl J Med **311**:73–80.
38. Gray RW, Wilz DR, Caldas AE, Lemann J Jr 1977 The importance of phosphate in regulating plasma $1,25(OH)_2$-vitamin D levels in humans: Studies in healthy subjects, in calcium-stone formers and in patients with primary hyperparathyroidism. J Clin Endocrinol Metab **45**:299–306.
39. Edwards NA, Hodgkinson A 1965 Metabolic studies in patients with idiopathic hypercalciuria. Clin Sci **29**:143–157.
40. Peacock M, Nordin BEC 1968 Tubular reabsorption of calcium in normal and hypercalciuric subjects. J Clin Pathol **21**:353–358.
41. Sutton RAL, Walker VR 1980 Responses to hydrochlorothiazide and acetazolamide in patients with calcium stones. N Engl J Med **302**:709–713.
42. Bianchi G, Vezzoli G, Cusi D, Cova T, Elli A, Soldati L, Tripodi G, Surian M, Ottaviano E, Rigatti P 1988 Abnormal red-cell calcium pump in patients with idiopathic hypercalciuria. N Engl J Med **319**:897–901.
43. Alhava EM, Juuti M, Karjalainen P 1976 Bone mineral density in patients with urolithiasis. Scand J Urol Nephrol **10**:154–156.
44. Lawoyin S, Sismilich S, Browne R, Pak CYC 1979 Bone mineral content in patients with calcium urolithiasis. Metabolism **28**:1250–1254.
45. Borgi L, Meschi T, Guerra A, Maninetti L, Pedrazzoni M, Macato A, Vescovi P, Novarini A 1991 Vertebral mineral content in diet-dependent and diet-independent hypercalciuria. J Urol **146**:1334–1338.
46. Bataille P, Achard JM, Fournier A, Boudailliez B, Westeel PF, Laval Jeantet MAL, Bouillon R, Sebert JL 1991 Diet, vitamin D and vertebral mineral density in hypercalciuric calcium stone formers. Kidney Int **39**:1193–1205.
47. Pietschmann F, Breslau NA, Pak CYC 1992 Reduced vertebral bone density in hypercalciuric nephrolithiasis. J Bone Miner Res **7**:1383–1388.
48. Barkin J, Wilson DR, Manuel MA, Arnold B, Murray T, Harrison J 1985 Bone mineral content in idiopathic calcium nephrolithiasis. Miner Electrolyte Metab **11**:19–24.
49. Liberman UA, Sperling O, Atsmon A, Frank M, Modan M, deVries A 1968 Metabolic and calcium kinetic studies in idiopathic hypercalciuria. J Clin Invest **47**:2580–2590.
50. Malluche HH, Tschoepe W, Ritz E, Meyer-Sabelle W, Massry SG 1980 Abnormal bone histology in idiopathic hypercalciuria. J Clin Endocrinol Metab **50**:654–658.
51. Steiniche T, Mosekilde L, Christensen MS, Melsen F 1989 A histomorphometric determination of iliac bone remodeling in patients with recurrent renal stone formation and idiopathic hypercalciuria. APMIS **97**:309–316.
52. Sutton RAL, Walker VR 1986 Bone resorption and hypercalciuria in calcium stone formers. Metabolism **35**:465–488.
53. Urivetzky M, Anna PS, Smith AD 1988 Plasma osteocalcin levels in stone disease. A potential aid in the differential diagnosis of calcium nephrolithiasis. J Urol **139**:12–14.
54. Pak CYC, Kaplan R, Bone H, Townsend J, Waters O 1975 A simple test for the diagnosis of absorptive, resorptive and renal hypercalciurias. N Engl J Med **292**:497–500.
55. Van Den Berg CJ, Kumar R, Wilson DM, Heath III H, Smith LH 1980 Orthophosphate therapy decreases urinary calcium excretion and serum 1,25-dihydroxyvitamin D concentrations in idiopathic hypercalciuria. J Clin Endocrinol Metab **51**:998–1001.

56. Breslau NA, Preminger GM, Adams BV, Otey J, Pak CYC 1992 Use of ketoconazole to probe the pathogenetic importance of 1,25-dihydroxyvitamin D in absorptive hypercalciuria. J Clin Endocrinol Metab **75**:1446–1452.
57. Pak CYC, Britton F, Peterson R, Ward D, Northcutt C, Breslau NA, McGuire J, Sakahee K, Bush S, Nicar M, Norman DA, Peters P 1980 Ambulatory evaluation of nephrolithiasis: Classification, clinical presentation and diagnostic criteria. Am J Med **69**:19–30.
58. Knapp EL 1943 Studies on the urinary excretion of calcium. Ph.D. Thesis, Department of Chemistry, State University of Iowa, Ames.
59. Lafferty FW, Pearson OH 1963 Skeletal, intestinal, and renal calcium dynamics in hyperparathyroidism. J Clin Endocrinol Metab **23**:891–902.
60. Nassim JR, Higgins BA 1965 Control of idiopathic hypercalciuria. Br Med J **1**:675–681.
61. Jackson WPU, Dancaster C 1959 A consideration of the hypercalciuria in sarcoidosis, idiopathic hypercalciuria, and that produced by vitamin D. A new suggestion regarding calcium metabolism. J Clin Endocrinol Metab **19**:658–680.
62. Harrison AR 1959 Some results of metabolic investigations in cases of renal stone. Br J Urol **31**:398.
63. Dent CE, Harper CM, Parfitt AM 1964 The effect of cellulose phosphate on calcium metabolism in patients with hypercalciuria. Clin Sci **27**:417–425.
64. Parfitt AM, Higgins BA, Nassim JR, Collins JA, Hilb A 1964 Metabolic studies in patients with hypercalciuria. Clin Sci **27**:463–482.
65. Coe FL, Parks JH 1988 Nephrolithiasis: Pathogenesis and Treatment, 2nd Ed. Year Book Publishers, Chicago, p. 113.
66. Maierhofer WJ, Lemann J Jr, Gray RW, Cheung HS 1984 Dietary calcium and serum 1,25-$(OH)_2$-vitamin D concentrations as determinants of calcium balance in healthy men. Kidney Int **26**:752–759.
66a. Adams ND, Gray RW, Lemann J Jr 1979. The effects of oral $CaCO_3$ loading and dietary calcium deprivation on plasma 1,25-dihydroxyvitamin D concentration in healthy adults. J Clin Endocrinol Metab **48**:1008–1016.
67. Adams ND, Gray RW, Lemann J Jr, Cheung HS 1982 Effects of calcitriol administration on calcium metabolism in healthy men. Kidney Int **21**:90–97.
68. Bashir MA, Nakagawa Y, Riordon D, Coe FL, Bushinsky DA 1995 Increased dietary oxalate does not increase urinary calcium oxalate oversaturation in hypercalciuric rats. J Am Soc Nephrol **6**:943 (abstract).
69. Bushinsky DA, Nilsson EL, Nakagawa Y, Coe FL 1995 Stone formation in genetic hypercalciuric rats. Kidney Int **48**:1705–1713.
70. Costa EM, Hirst MA, Feldman D 1985 Regulation of 1,25-dihydroxyvitamin D_3 receptor by vitamin D analogs in cultured mammalian cells. Endocrinology **117**:2203–2210.
71. Pols HAP, Birkenhager JC, Schlite JP, Visser TJ 1988 Evidence that self-induced metabolism of 1,25-dihydroxyvitamin D_3 limits the homologous up-regulation of its receptor in rat osteosarcoma cells. Biochim Biophys Acta **970**:122–129.
72. Reinhardt TA, Horst RL 1989 Self-induction of 1,25-dihydroxyvitamin D_3 metabolism limits receptor occupancy and target tissue responsiveness. J Biol Chem **264**:15917–15921.
73. Yao J, Wong MS, Tembe V, Bushinsky DA, Favus MJ 1996 Enhanced response of VDR gene expression to $1,25(OH)_2D_3$ in genetic hypercalciuric rats. J Bone Miner Res **11**:S373 (abstract).
74. Krieger NS, Stathopoulos VM, Bushinsky DA 1996 Increased sensitivity to $1,25(OH)_2D_3$ in bone from genetic hypercalciuric rats. Am J Physiol **271**:C130–C135.
75. Kim M, Sessler NE, Tembe V, Favus MJ, Bushinsky DA 1993 Response of genetic hypercalciuric rats to a low calcium diet. Kidney Int **43**:189–196.
76. Parks JH, Coe FL 1996 Pathogenesis and treatment of calcium stones. Semin Nephrol **16**:398–411.
77. Wilson DR, Strauss AL, Manuel MA 1984 Comparison of medical treatments for the prevention of recurrent calcium nephrolithiasis. Urol Res **12**:39–40.
78. Edwards BR, Baer PG, Sutton RA, Dirks JH 1973 Micropuncture study of diuretic effects on sodium and calcium reabsorption in the dog nephron. J Clin Invest **52**:2418–2427.
79. Costanzo LS, Windhager EE 1978 Calcium and sodium transport by the distal convoluted tubule of the rat. Am J Physiol **235**:F492–F506.
80. Zerwekh JE, Pak CYC 1980 Selective effects of thiazide therapy on serum 1-alpha,25-dihydroxyvitamin D and intestinal calcium absorption in renal and absorptive hypercalciurias. Metabolism **29**:13–17.
81. Coe FL, Parks JP, Bushinsky DA, Langman CB, Favus MJ 1988 Chlorthalidone promotes mineral retention in patients with idiopathic hypercalciuria. Kidney Int **33**:1140–1146.
82. Wasnich RD, Benfante RJ, Yano K, Heilbrun L, Vogel JM 1983 Thiazide effect on the mineral content of bone. N Engl J Med **309**:344–347.
82. LaCroix AZ, Wienpahl J, White LR, Wallace RB, Scherr PA, George LK 1990 Thiazide diuretic agents and the incidence of hip fracture. N Engl J Med **322**:286–290.
83. Asplin JR, Favus MJ, Coe FL 1996 Nephrolithiasis. In: Brenner BR (ed) The Kidney, 5th Ed. Saunders, Philadelphia, Pennsylvania, pp. 1893–1935.
85. Ohkawa M, Tokunga S, Nakashima T, Orito M, Hisazumi H 1992 Thiazide treatment for calcium nephrolithiasis in patients with idiopathic hypercalciuria. Br J Urol **69**:571–576.
86. Cheskis B, Lemon BD, Uskokovic M, Lomedico PT, Freedman LP 1995 Vitamin D_3-retinoid X receptor dimerization, DNA binding, and transactivation are differentially affected by analogs of 1,25-dihydroxyvitamin D_3. Mol Endocrinol **9**:1814–1824.
87. Skowronski RJ, Peehl DM, Feldman D 1995 Actions of vitamin D_3, analogs on human prostate cancer cell lines: Comparison with 1,25-dihydroxyvitamin D_3. Endocrinology **136**:20–26.
88. Fleet JC, Bradley J, Reddy GS, Ray R, Wood RJ 1996 1 alpha,25-$(OH)_2$-vitamin D analogs with minimal in vivo calcemic activity can stimulate significant transepithelial calcium transport and mRNA expression in vitro. Arch Biochem Biophys **329**:228–234.
89. Naveh-Many T, Silver J 1993 Effects of calcitriol, 22-oxacalcitriol, and calcipotriol on serum calcium and parathyroid hormone gene expression. Endocrinology **133**:2724–2728.
90. Brown AJ, Finch J, Grieff M, Ritter C, Kubodera N, Nishii Y, Slatopolsky E 1993 The mechanism for the disparate actions of calcitriol and 22-oxacaltriol in the intestine. Endocrinology **133**:1158–1164.

CHAPTER 54

Hypercalcemia Due to Vitamin D Toxicity

SUSAN THYS-JACOBS Department of Medicine, College of Physicians and Surgeons, Columbia University, New York, New York

FREDRIECH K. W. CHAN Department of Medicine, Queen Elizabeth Hospital, Hong Kong

LILIA M. C. KOBERLE Health Sciences Department, Federal University, Sao Carlos, Brazil

JOHN P. BILEZIKIAN Department of Medicine and Pharmacology, College of Physicians and Surgeons, Columbia University, New York, New York

I. INTRODUCTION

Vitamin D toxicity is not a common cause of hypercalcemia. In the differential diagnosis of hypercalcemia, it is often buried amid a long list of other more and less common causes (Table I). Among the more common causes, primary hyperparathyroidism and hypercalcemia of malignancy are the principal etiologies. Many studies have documented that primary hyperparathyroidism and hypercalcemia of malignancy constitute together the overwhelming majority of causes of hypercalcemia. They are in fact so common that the practical issue in the diagnosis of a hypercalcemic individual is to distinguish between these two etiologies.

Patients with primary hyperparathyroidism tend to be asymptomatic, whereas patients with hypercalcemia of malignancy tend to be ill. The diagnosis of primary hyperparathyroidism is established by measuring an elevated concentration of parathyroid hormone (PTH), an association that is made in over 90% of patients with primary hyperparathyroidism. In contrast, patients with hypercalcemia of malignancy, including those whose hypercalcemia is due to the elaboration of parathyroid hormone-related protein (PTHrP), show levels of PTH that are typically suppressed. If the PTH level is suppressed, the diagnosis of primary hyperparathyroidism is ruled out. The diagnosis of malignancy, however, is not necessarily ruled in. Certainly, if a malignancy is detected that is classically associated with hypercalce-

TABLE I Differential Diagnosis of Hypercalcemia

Primary hyperparathyroidism
  Sporadic (adenoma, hyperplasia, or carcinoma)
  Familial
    Isolated
    Cystic
    Multiple endocrine neoplasia type I or II
Malignancy
  Parathyroid hormone-related protein
  Excess production of 1,25-dihydroxyvitamin D
  Other factors (cytokines, growth factors)
Disorders of vitamin D
  Exogenous vitamin D toxicity—parent D compound, 25OHD, $1,25(OH)_2D$
  Endogenous production of 25-hydroxyvitamin D (Williams syndrome)
  Endogenous production of 1,25-dihydroxyvitamin D
    Granulomatous diseases
      a. Sarcoidosis
      b. Tuberculosis
      c. Histoplasmosis
      d. Coccidioidomycosis
      e. Leprosy
      f. Others
    Lymphoma
Nonparathyroid endocrine disorders
  Thyrotoxicosis
  Pheochromocytoma
  Acute adrenal insufficiency
  Vasoactive intestinal polypeptide hormone-producing tumor (VIPoma)
Medications
  Thiazide diuretics
  Lithium
  Estrogens/antiestrogens, testosterone in breast cancer
  Milk–alkali syndrome
  Vitamin A toxicity
Familial hypocalciuric hypercalcemia
Immobilization
Parenteral nutrition
Aluminum excess
Acute and chronic renal disease

mia, such as squamous cell carcinoma of the lung, the etiology becomes clear. However, the longer list of other causes of hypercalcemia is also associated, with rare exceptions, with reduced levels of PTH. This situation, namely, elevated serum calcium concentration with reduced or undetectable levels of PTH, is seen in the various forms of vitamin D toxicity. If primary hyperparathyroidism is ruled out and malignancy is not apparent, the likelihood of vitamin D toxicity looms as an important possible etiology of hypercalcemia. In that long list of other causes, vitamin D toxicity now becomes a major diagnostic consideration (Table I). This chapter reviews the various forms of vitamin D toxicity, mechanisms of hypercalcemia due to vitamin D toxicity, clinical manifestations, diagnosis, and management.

II. FORMS OF EXOGENOUS VITAMIN D TOXICITY

Vitamin D toxicity can be life threatening and associated with high morbidity, if not identified quickly. Hypervitaminosis D with hypercalcemia may be secondary to excessive intake of parent vitamin D, its metabolites 25-hydroxyvitamin D (25OHD), 1,25-dihydroxyvitamin D [$1,25(OH)_2D$], or vitamin D analogs; to increased production of 25OHD or $1,25(OH)_2D$ from exogenous substrate; and even to topical applications of potent vitamin D analogs.

A. Vitamin D and 25-Hydroxyvitamin D Toxicity

The most common etiology of vitamin D toxicity is inadvertent or improper oral use of pharmaceutical preparations. The usual presentation leading to toxicity is in the setting of vitamin D therapy for the hypophosphatemic disorders, hypoparathyroidism, pseudohypoparathyroidism, osteomalacia, renal failure, or osteoporosis. Other cases are due to ingestion of large doses of megavitamins in health conscious adults. Excessive sunlight exposure can raise serum concentrations of 25OHD to as high as 79 ng/ml (normal range 9–52 ng/ml), but there is no evidence that sunlight exposure alone can result in vitamin D toxicity and hypercalcemia in normal individuals [1]. Hypercalcemia associated with granulomatous diseases such as sarcoidosis can be worsened by excessive sunlight exposure. Natural foods, in general, other than fatty fish, eggs, milk, and liver do not contain much vitamin D. Hypervitaminosis D has been associated with drinking milk when erroneously fortified with massive concentrations of vitamin D. One investigation of eight patients manifesting symptoms of nausea, vomiting, weight loss, hyperirritability, or failure-to-thrive revealed markedly elevated mean concentrations of 25OHD of 293 ± 174 ng/ml (nl: 9–52 ng/ml) [2]. Analysis of the milk production facility at the local dairy revealed excessive vitamin D fortification of milk with up to 245,840 IU per liter (232,565 IU of vitamin D_3 per quart). Usual fortification of milk in the United States is 400 IU per quart. In addition to milk, vitamin D fortification of natural foods has included certain breakfast cereals, pasta, baked goods, fats, and oils. There is no documentation that excessive ingestion of any of these other fortified foods has ever resulted in vitamin D toxicity.

Vitamin D_2 and vitamin D_3, although used interchangeably in the treatment of metabolic bone diseases, may differ in toxic potential at higher doses. In general, vitamin D_3 appears to be somewhat more toxic than D_2. Investigations in rats, sheep, pigs, horses, and primates support differences in metabolic clearance rates and in toxicity between the two vitamin D compounds [3]. In horses, vitamin D_2 has a lower toxicity compared to vitamin D_3 [4]. Massive doses of vitamin D_3 administered to Old World primates can cause toxicity and death, whereas equivalent doses of vitamin D_2 are better tolerated [5].

The current recommended dietary allowance (RDA) for vitamin D is 400 IU per day. The smallest dose of parent vitamin D in adults that can produce toxicity and hypercalcemia is not known but is likely to be much higher than the RDA [6]. On the other hand, in infants, daily dosages of 2000 IU or less have been associated with hypercalcemia and nephrocalcinosis [7]. Intermittent oral dosages of 15 mg or 600,000 IU to infants to prevent vitamin deficiency has been shown to be excessive during the first year of life, resulting in transient hypercalcemia and vitamin D overload [8,9]. Lower amounts of 5 mg (200,000 IU) every 6 months or 2.5 mg (100,000 IU) every 3 months appear to be safer and to provide better protection in high risk infants. In adults, doses of greater than 40,000–60,000 IU per day, as commonly used in the treatment of hypoparathyroidism, can be associated with significant toxicity.

Individuals manifest wide variations both in their response to hypercalcemic doses of vitamin D and in the duration of the effect. This variation in individual responsiveness might reflect differences in intestinal absorption and vitamin D metabolism, in the concentration of free vitamin D metabolites, in the rate of degradation of the metabolites and conversion to inactive metabolites, and in the capacity of storage sites for 25OHD [6]. Factors that enhance susceptibility to vitamin D toxicity and hypercalcemia include increased dietary calcium intake, reduced renal function, coadministration of vitamin A, and granulomatous disorders such as sarcoidosis that render subjects more sensitive to vitamin D [10]. The hypercalciuria in hypervitaminosis D usually presents much earlier than the hypercalcemia but is easily missed because it is not routinely measured.

B. 1,25-Dihydroxyvitamin D Toxicity

The greater potency of $1,25(OH)_2D_3$ and its direct actions have resulted in its increased use for a variety of metabolic bone diseases [11]. Its ability to inhibit PTH synthesis and secretion has also made $1,25(OH)_2D_3$ a useful agent in patients with renal osteodystrophy and secondary hyperparathyroidism. Considerable attention, thus, has focused on possible toxic effects of $1,25(OH)_2D_3$ not usually associated with the parent vitamin D compound. The incidence of hypercalcemia and hypercalciuria with $1,25(OH)_2D_3$ use has been reported as very high, with one review citing complications in two-thirds of treated patients [12]. The mechanism of the hypercalcemia is increased intestinal absorption and potentiation of osteoclastic activity [13]. Dosages of $1,25(OH)_2D_3$ above 0.75 μg/day have been associated with toxicity, whereas dosages at or below 0.5 μg/day rarely result in toxicity. One investigation showed that over 90% of patients on doses of $1,25(OH)_2D_3$ between 1.0 and 2.0 μg/day became hypercalcemic, and all had hypercalciuria when calcium intake was set at 1000 mg per day [14]. Accelerated deterioration of renal function was recorded in a number of reports in patients with renal insufficiency receiving $1,25(OH)_2D_3$ therapy [15,16]. Compared to oral therapy, intravenous administration of $1,25(OH)_2D_3$ to renal dialysis patients induces hypercalcemia less frequently, with a smaller increment in the serum calcium concentration and a more effective reduction of PTH levels [17]. More recent studies, however, suggest that intermittent oral pulse administration of $1,25(OH)_2D_3$ may be effective, though not as effective as intravenous $1,25(OH)_2D_3$, in suppressing PTH in uremic patients with secondary hyperparathyroidism [18–20].

C. Toxicity Due to Synthetic Analogs

In one investigation, oral pulse therapy with the 1α-hydroxyvitamin D_3 ($1\alpha OHD_3$) resulted in a rapid control of secondary hyperparathyroidism without causing hypercalcemia or hyperphosphatemia [21]. However, $1\alpha OHD_3$ may harbor similar potential calcemic effects to $1,25(OH)_2D_3$ in the treatment of renal osteodystrophy. Crocker *et al.* [22] investigated the comparative toxicity of vitamin D, $1\alpha OHD_3$, and $1,25(OH)_2D_3$ in weanling male mice at three different doses over a 4-week period. $1\alpha OHD_3$ appeared to be more toxic in the high dose group only, with significantly higher serum calcium levels, higher urinary calcium excretion, and severe nephrocalcinosis [22]. $1\alpha OHD_3$ has been described as less potent than $1,25(OH)_2D_3$ at low doses but equipotent at doses greater than 2.0 μg/day. At the higher doses, there is a delayed onset of action and a prolonged half-life, suggesting a potential for cumulative toxicity in renal insufficiency [23,24].

The potential for hypercalcemia, hypercalciuria, and soft tissue calcifications limits the clinical usefulness of

$1\alpha OHD_3$. Mortensen and colleagues compared the toxicity of both $1\alpha OHD_3$ and $1,25(OH)_2D_3$ in rats fed standard or low calcium diets. High doses of either compound resulted in severe hypercalcemia, with retarded growth, nephrosis, and structural bone changes in the rats fed the standard diet. On the low calcium diet, however, slight hypercalcemia occurred but without growth retardation or bone changes. There was minimal effect on the kidney. Calcium restriction again proved effective in protecting the animals against the toxic effects of the vitamin D analogs. Animals fed the low calcium diet tolerated $1\alpha OHD_3$ at dose levels up to 10 times higher than rats on the standard diets [25]. In human subjects, $1\alpha OHD_3$ causes toxic effects at doses above 1.0 μg/day, but doses of 0.5 to 1.0 μg/day appear to be safe.

Because of the relatively narrow therapeutic window of vitamin D_3 compounds, a synthetic analog of vitamin D_2, $1\alpha OHD_2$, was developed with a concept that the window of therapeutic efficacy to toxicity would be wider. In postmenopausal osteopenic women, doses of $1\alpha OHD_2$ ranging from 1.0 to 5.0 μg/day were administered in 15 subjects. There was no evidence of vitamin D toxicity manifesting as either hypercalciuria or hypercalcemia, whereas significant therapeutic osteoblastic activity was demonstrated [26]. Similar to the vitamin D_3 analog, $1\alpha OHD_2$ requires obligatory hepatic 25-hydroxylation for biological activation. However, $1\alpha OHD_2$ is able to activate its catabolic pathway via hepatic 24-hydroxylation with a lower potential for toxicity [27]. These investigations on synthetic analogs seem to confirm earlier reports that vitamin D_2 compounds, in general, are as efficacious and somewhat better tolerated than D_3 compounds.

Because of our understanding of the nonclassic target tissue effects of vitamin D in the modulation of hormones and cytokines, and in the regulation of cellular differentiation and proliferation, newer clinical uses have been developed (see Section VIII of this book). The clinical applications of these newer properties of vitamin D, however, have also been tempered by the potential for complications such as hypercalcemia and hypercalciuria, prompting the development of other analogs to separate even better the calcemic effects from antiproliferative activity [28] (see also Section VII of this volume). Depending on the chemical modification of the vitamin D compound, some analogs do demonstrate reduced calcemic activity, but others have been developed with increased calcemic activity owing to enhanced intestinal calcium absorption and bone mineral mobilization. Fluorination of C-24, C-26, or C-27 apparently results in markedly increased calcemic activity resulting from altered degradation of the side chain cleavage site. Calcemic potency of $1,25(OH)_2D_3$ and its analogs can be also enhanced at least 2- to 5-fold by epimerization at the C-20 site [29].

Other vitamin D analogs such as topical calcipotriol (MC903) have proved very effective in the treatment of psoriasis (see Chapter 72). Because of its low absorption rate and rapid degradation, calcipotriol is believed to have negligible effects on systemic calcium homeostasis when administered topically. However, isolated cases of hypercalcemia and hypercalciuria have been reported, even in patients taking recommended doses [30]. In one investigation, Bourke and colleagues noted suppression of serum PTH concentrations in all patients within 2 weeks of treatment with calcipotriol. Mean serum and urine calcium levels increased during treatment and fell following withdrawal [31]. The authors concluded that although this particular synthetic analog alters serum and urinary calcium with a dose-dependent effect on systemic calcium homeostasis, it is well tolerated and effective for mild to moderate chronic plaque psoriasis. However, it is potentially hazardous in extensive, unstable, exfoliative disease [32].

III. FORMS OF ENDOGENOUS VITAMIN D TOXICITY

A. Endogenous Production of 25-Hydroxyvitamin D

Rarely, hypervitaminosis D with hypercalcemia is due to endogenous dysregulation of vitamin D metabolites as seen in Williams syndrome [33]. Williams syndrome, an idiopathic infantile form of hypercalcemia, is associated with late psychomotor development, selective mental deficiency, and supravalvular aortic stenosis [34]. The hypercalcemia has been reported to range widely from 12 to 19 mg/dl but usually subsides by 4 years of age. One report suggests an exaggerated production of 25OHD with small doses of vitamin D as a possible etiology of the hypervitaminosis D [35].

B. Production of 1,25-Dihydroxyvitamin D

1. Granulomatous Diseases

In contrast to the megadosage of vitamin D that is usually required to produce vitamin D toxicity, patients with granulomatous diseases can develop hypercalcemia rather easily without excessive intake of exogenous vitamin D. They are said to be hypersensitive to vitamin D. The etiology of the vitamin D toxicity in this syndrome is due to poorly regulated extrarenal synthesis of $1,25(OH)_2D$ by the granulomatous tissue itself (as described in detail in Chapter 55). In contrast to the various presentations of vitamin D toxicity described earlier, the responsible metabolite in granulomatous disease is quite different. In the case of vitamin D toxicity due to overdosage of vitamin D or 25OHD, 25OHD is the active metabolite; renal production of $1,25(OH)_2D$ in this setting is highly regulated. In granulomatous tissue, however, $1,25(OH)_2D$ formation is not subject to control by any recognized regulators, such as PTH, phosphorus, or calcium. Thus, this syndrome is due to ectopic production of $1,25(OH)_2D$ toxicity by the granulomatous tissue itself. The mechanisms by which hypercalcemia occurs, however, are similar to all other vitamin D toxic states, namely, increased intestinal calcium absorption and enhanced osteoclastic bone resorption [36,37]. Many studies have led to greater understanding of the pathophysiology and immunological features associated with this syndrome.

a. Sarcoidosis Abnormalities in calcium metabolism have long been noted in patients with sarcoidosis [38]. Sarcoidosis is also the most common granulomatous disease associated with hypercalcemia. As many as 10 to 20% of patients with sarcoidosis will develop hypercalcemia, and as many as 50% will experience hypercalciuria at some time during the course of the disease [39]. Hypercalciuria is invariably present when patients develop hypercalcemia. In the 1950s, studies had already revealed similarities between hypercalcemia of sarcoidosis and the hypercalcemia of vitamin D toxicity, namely, increased intestinal absorption of calcium, hypercalciuria, and therapeutic efficacy of glucocorticoids [40,41]. The major distinguishing feature was in the amount of vitamin D associated with the hypercalcemia and/or hypercalciuria. Seasonal variation of the serum calcium level in sarcoidosis was correlated with availability of sunlight as a source of vitamin D [42]. In the late 1970s, two independent groups showed that the vitamin D-like principle that appeared to be responsible in sarcoidosis was in fact the active metabolite of vitamin D, $1,25(OH)_2D_3$ [43,44].

Ectopic production of $1,25(OH)_2D_3$ was confirmed by demonstrating high circulating concentrations of $1,25(OH)_2D_3$ in anephric patients with sarcoidosis on hemodialysis during hypercalcemic episodes [45–47]. This observation showed unequivocably that the kidney, usually the sole source of $1,25(OH)_2D_3$ in nonpregnant individuals, could not be the source of $1,25(OH)_2D_3$ in these patients. The serum calcium and $1,25(OH)_2D_3$ levels were positively correlated with indices of disease activity [48–50], namely, the extent of granuloma formation and the angiotensin-converting enzyme level. It was subsequently shown that the granulomatous tissue was, in fact, the site of $1,25(OH)_2D_3$ production. The 1α-hydroxylase enzyme responsible for formation of $1,25(OH)_2D_3$ was present in lymph node homogenates [51]. Moreover, pulmonary alveolar macrophages [52] could be shown to catalyze the formation of an ^3H-labeled $25OHD_3$ metabolite. This metabolite was definitively identified as $1,25(OH)_2D_3$ by high-performance liquid chromatography (HPLC), by the chick intestinal receptor assay for $1,25(OH)_2D_3$, by UV spectroscopy, and by mass spectrometry [53]. The enzyme activity in the sarcoid macrophage is insensitive to feedback inhibition by $1,25(OH)_2D_3$ [54] and is not otherwise regulated, accounting for the uncontrolled synthesis of $1,25(OH)_2D_3$ and the characteristic elevation of this metabolite when hypercalcemia is present. Another property of this enzyme is that it is inhibited in a dose-dependent fashion by dexamethasone and chloroquine [54]. These *in vitro* observations have direct clinical relevance.

There are several mechanisms by which increased circulating concentrations of $1,25(OH)_2D_3$ are able to disturb calcium metabolism in sarcoidosis [55]. First, $1,25(OH)_2D_3$ causes hypercalcemia, in part, by stimulating intestinal calcium absorption. A low calcium diet [56,57], alone or in association with cellulose phosphate [58], was found to normalize the calcium level in some patients with sarcoidosis. Second, $1,25(OH)_2D_3$ directly stimulates osteoclastic-mediated bone resorption; skeletal granulomas are not required for this effect [59,60]. The increased flux of calcium into the extracellular space by these gastrointestinal and skeletal mechanisms, aided by suppression of PTH [44–46], leads to hypercalciuria. Chronic hypercalciuria favors nephrocalcinosis and renal stone formation [61]. When the kidneys are unable to excrete the calcium presented to them, because of either declining renal function, enhanced bone resorption, a sudden influx of dietary calcium, or any combination of these events, hypercalcemia ensues [62].

b. Tuberculosis Longitudinal studies from the United States [63] and India [64] suggested that 16 to 28% of patients with tuberculosis develop hypercalcemia. However, in these early studies vitamin D supplements were employed, increasing the risk and severity of hypercalcemia. A similar study from Greece [65] reported a figure as high as 48% when serum calcium was corrected to a normal albumin level. More recent studies from the United Kingdom [66], Belgium [67], Hong Kong [68], and Malaysia [69] have shown a much lower prevalence of hypercalcemia, in the range of 0 to 2.3%. This difference, over time, may be attributed to changes in vitamin D and calcium intake as well as to reduced sun exposure.

Reports of high circulating levels of $1,25(OH)_2D_3$ in three anephric patients with tuberculosis support an extrarenal source of the active vitamin D metabolite [70,71]. Positive correlation of the albumin-adjusted calcium level with the radiographic extent of the disease has been shown [68]. Hypercalcemia in tuberculosis may occur weeks to months after starting antituberculosis chemotherapy [63,64]. Thus, the hypercalcemia is not related to the presence of viable acid-fast bacilli but rather to the granulomatous process and associated reactions. As with sarcoidosis, hypercalcemia in tuberculosis can be controlled by administration of glucocorticoids [72].

In patients with tuberculous pleuritis, the mean free $1,25(OH)_2D_3$ concentration in pleural fluid was selectively concentrated by 5.3-fold over that in serum [73]. Positive correlation between the concentrations of substrate ($25OHD_3$) and product [$1,25(OH)_2D_3$] in pleural fluid supported the idea that $1,25(OH)_2D_3$ was produced locally by activated inflammatory cells in or adjacent to the pleural space. The pleural fluid was found to have high concentrations of interferon-γ, a cytokine shown subsequently to stimulate activated macrophages *in vitro* to synthesize $1,25(OH)_2D_3$ [74]. Cells obtained from bronchoalveolar lavage in patients with tuberculosis were also found to synthesize $1,25(OH)_2D_3$ *in vitro*. An important source of the active vitamin D metabolite appears to be the CD8+ T lymphocytes at the granulomatous sites [75]. If one wonders teleologically about the production of $1,25(OH)_2D_3$ under these circumstances, the immunomodulatory functions of $1,25(OH)_2D_3$ acting as a beneficial local paracrine factor could be pertinent (see Chapter 55). Viewed in this context, hypercalcemia occurs when $1,25(OH)_2D_3$ is produced in such quantities that it gains entry into the circulation.

Hypercalcemia in tuberculosis is usually mild and asymptomatic. Besides glucocorticoids, ketoconazole administration has been associated with a rapid decline in $1,25(OH)_2D_3$ and normalization of serum calcium levels [76]. Long term antituberculosis therapy with isoniazid and rifampicin can also be effective in treating the hypercalcemia by controlling the disease.

c. Other Granulomatous Diseases Besides the more detailed studies of hypercalcemia in sarcoidosis and tuberculosis, hypercalcemia has also been reported in leprosy [77], coccidioidomycosis [78], histoplasmosis [79], candidiasis [80], eosinophilic granuloma [81], berylliosis [82], and silicone-induced granuloma [83]. The mechanism of increased production of the active vitamin D metabolite is believed to be shared by all of these granulomatous disorders. A possible role for $1,25(OH)_2D_3$ in these granulomatous disorders as noted above includes immunomodulatory features, which are covered in Chapter 55.

2. LYMPHOMA

Hypercalcemia has been reported to occur in 5% [84] and 15% [85] of patients with Hodgkin's disease and non-Hodgkin's lymphoma (NHL), respectively. Up to 80% of patients with human T-cell leukemia virus type 1 (HTLV-1)-associated adult T-cell lymphoma/leukemia (ATLL) will develop hypercalcemia [86]. Not different from other malignancies, hypercalcemia is a poor prognostic feature in lymphoma [87,88], adding substantially to morbidity and mortality. The humoral mediators of hypercalcemia in lymphoma are multiple and heterogeneous. However, evidence has shown $1,25(OH)_2D_3$ to be an important factor in many cases.

Hodgkin's disease is most consistently associated with $1,25(OH)_2D_3$ when hypercalcemia develops. Since the first report of hypercalcemia complicating Hodgkin's disease in 1956 [89], more than 60 cases have been described. In a retrospective review of the literature [90], 84% of patients had a peak serum calcium above 12 mg/dl, 74% of the patients had Ann Arbor stage III or IV disease, and 68% were symptomatic with night sweats, fever, and weight loss. Only 3 of 23 patients had radiological evidence of lytic bone lesions. In 17 hypercalcemic patients, all but one patient had an elevated $1,25(OH)_2D_3$ level. There is no evidence to implicate parathyroid hormone-related peptide (PTHrP) as a mediator of hypercalcemia in Hodgkin's disease. Two patients with Hodgkin's disease [91,92] were reported to have intermittent hypercalcemia during two consecutive summers or on vitamin D challenge. There was a close association between hypercalcemia and the abnormally raised $1,25(OH)_2D_3$ level, but serum $25OHD_3$ was within the normal range. These observations support the idea that the mechanism of the hypercalcemia in Hodgkin's disease is similar to that of the granulomatous diseases, namely, production by the lymphomatous tissue of $1,25(OH)_2D_3$.

There are about 20 case reports of $1,25(OH)_2D_3$-induced hypercalcemia in non-Hodgkin's lymphoma. Most patients had bulky or advanced stage disease but no clinically or radiographically evident bone lesions. Data supporting extrarenal synthesis of $1,25(OH)_2D_3$ are the presence of severe renal failure in a number of instances [93,94]; the demonstration of *in vitro* conversion of $25OHD_3$ to $1,25(OH)_2D_3$ by excised lymph node homogenates [95]; the prompt decline of $1,25(OH)_2D_3$ levels to normal within 24 hr of excision of an isolated splenic lymphoma [96]; and sensitivity to glucocorticoid suppression [94].

Five of ten patients with either AIDS or non-AIDS associated non-Hodgkin's lymphoma and hypercalcemia had frankly elevated serum $1,25(OH)_2D_3$ concentrations [97]. In a prospective study by Seymour *et al.*

[98], a control group was composed of 16 patients with hypercalcemia and multiple myeloma. Using the mean serum $1,25(OH)_2D_3$ level of the control patients plus 3 standard deviations, the investigators defined the upper limit of expected serum $1,25(OH)_2D_3$ during hypercalcemia as 42 pg/ml, well below the upper limit of 76 pg/ml for the normocalcemic reference range. Thus, the typical hypercalcemic patient, if represented by this cohort of patients with multiple myeloma, shows a lower range of normal for $1,25(OH)_2D_3$ concentration. Of the 22 hypercalcemic patients with non-Hodgkin's lymphoma, 12 (55%) had elevated serum $1,25(OH)_2D_3$ levels. Moreover, the serum levels of corrected calcium and $1,25(OH)_2D_3$ were strongly correlated with one another. Even in the normocalcemic group with non-Hodgkin's lymphoma, 71% were hypercalciuric and 18% had elevated serum $1,25(OH)_2D_3$ levels. The precise cell type responsible for the extrarenal synthesis of $1,25(OH)_2D_3$ remains to be established. There are two possibilities. One is the tumor-infiltrating reactive macrophage, recognized by a "starry-sky" appearance [99] in intermediate and high-grade lymphomas, in which hypercalcemia is also most common. Alternatively, it may be that a particular clone of the malignant lymphoma cell synthesizes $1,25(OH)_2D_3$ [100].

1,25-Dihydroxyvitamin D is only one cause of hypercalcemia in lymphoma. About half of the patients with non-Hodgkin's lymphoma and hypercalcemia have suppressed $1,25(OH)_2D_3$ levels. Additional circulating or local osteolytic factors are likely to be involved. Two of 22 patients in the study by Seymour *et al.* had elevated PTHrP levels. A few other cases of hypercalcemia in non-Hodgkin's lymphoma associated with elevated levels of PTHrP have been reported [101,102]. Cytokines such as interleukin-1, tumor necrosis factor-α (TNFα), and transforming growth factor (TGFα) may also play a role in the pathogenesis of lymphoma-associated hypercalcemia.

Although HTLV-1-transformed lymphocytes were shown *in vitro* to possess the capacity to convert $25OHD_3$ to $1,25(OH)_2D_3$ [103], most studies have shown reduced $1,25(OH)_2D_3$ levels in hypercalcemia associated with HTLV-1-related adult T-cell leukemia/lymphoma [104,105]. PTHrP is most strongly implicated as the major mediator in this syndrome [106]. PTHrP messenger RNA has been demonstrated in HTLV-1-infected T cells [107] and tumor cells from adult T-cell lymphoma/leukemia (ATLL) patients with hypercalcemia [108]. Nevertheless, there are two well-documented instances of elevated $1,25(OH)_2D_3$ levels in ATLL [86,94]. In the first case, PTHrP level was not available. In the second case, concomitant elevation of $1,25(OH)_2D_3$ and PTHrP was shown, suggesting the possibility of increased renal 1α-hydroxylase activity secondary to PTHrP. Alternatively, the tissue could be the site of both PTHrP and $1,25(OH)_2D_3$ formation. Most patients with hypercalcemia due to classic squamous cell carcinoma have elevated PTHrP levels and either suppressed or normal $1,25(OH)_2D_3$ levels.

IV. MECHANISMS OF VITAMIN D TOXICITY

A. General Mechanisms

Vitamin D toxicity may occur in patients due to any one of the three forms of vitamin D, namely, the vitamin D parent compound, 25OHD, or $1,25(OH)_2D$. Multiple factors may influence susceptibility to vitamin D toxicity and include the concentration of the vitamin D metabolite itself, vitamin D receptor (VDR) number, activity of 1α-hydroxylase, the metabolic degradation pathway, and the capacity of the vitamin D binding protein (DBP). Vitamin D_2 or D_3 toxicity is more difficult to manage than toxicity due to its metabolites 25OHD or $1,25(OH)_2D$. In part, this is due to the extensive lipid solubility of the parent compund in liver, muscle, and fat tissues and corresponding large storage capacity. As a result, the half-life of vitamin D ranges from 20 days to months. In contrast, the biological half-life of the less lipophilic compound 25OHD is shorter, approximately 15 days [109]. The biological half-life of the least lipophilic compound $1,25(OH)_2D$, is much shorter, approximately 15 hr [110]. In general, duration of toxicity is related to the half-life of the vitamin D compound. Thus, the hypercalcemia of parent vitamin D overdose can last for as long as 18 months, long after dosing is discontinued, because of its slow release from fat depots. Overdosage of 25OHD can persist for weeks also, but excessive $1,25(OH)_2D$ toxicity is more rapidly reversed because $1,25(OH)_2D$ is not stored in appreciable amounts in the body [111].

The toxicity of either parent vitamin D or 25OHD is due to 25OHD. In an investigation examining the concentrations of vitamin D_3 and its metabolites in the rat as influenced by various intakes of vitamin D_3 or 25OHD, Shepard and DeLuca found that large intakes of vitamin D_3, ranging from 0.65 to 6500 nmol/day, resulted in excessive concentrations of vitamin D_3 and $25OHD_3$ but not in $1,25(OH)_2D_3$ (Table II) [112]. Similarly, increased dosages of $25OHD_3$ ranging from 0.46 to 4600 nmol/day resulted in excessive amounts of $25OHD_3$ but not of vitamin D_3 or $1,25(OH)_2D_3$ (Table III). In the setting of toxicity due to overadministration of $1,25(OH)_2D_3$, the active metabolite itself is responsible for the hypercalcemia [113,114]. Unlike $1,25(OH)_2D$ whose production is tightly regulated in the kidney, the

TABLE II Plasma Concentrations of Vitamin D_3 and Metabolites in Rats Given Various Amounts of Vitamin D_3[a]

Amount (nmol/day)	Vitamin D_3 (ng/ml)	$25OHD_3$ (ng/ml)	Lactone (ng/ml)	$24,25(OH)_2D_3$ (ng/ml)	$25,26(OH)_2D_3$ (ng/ml)	$1,25(OH)_2D_3$ (pg/ml)	Plasma calcium (mg/100 ml)
0.65	11.3 ± 6.1	2.3 ± 1.9	<0.06	0.56 ± 0.13	<0.2	80 ± 60	9.0 ± 0.1
6.5	110 ± 43	14.7 ± 8.6	0.35 ± 0.12	3.98 ± 1.90	0.20 ± 0.36	77 ± 64	9.4 ± 0.4
65	368 ± 121	74.2 ± 14.5	10.3 ± 3.9	25.5 ± 5.2	7.60 ± 2.78	88 ± 9	9.7 ± 0.3
650	1339 ± 329[b]	643 ± 93[b]	64.5 ± 19.1[c]	73.5 ± 29.6[d]	16.4 ± 4.7[c]	51 ± 11[c]	12.4 ± 1.0[b]
6500	3108	1111	43.6	86.5	8.4	37	13.8

[a] Rats were orally dosed daily for 14 days with indicated amounts of vitamin D_3. Data are means of 5 rats ± SD. Reprinted with permission from Shepard RM, DeLuca HF. Arch Biochem Biophys 1980;202:43–50.
[b] Differs from control group (0.65 nmol/day) and from group receiving 65 nmol/day at $p < 0.001$.
[c] Differs from group receiving 65 nmol/day at $p < 0.001$.
[d] Differs from control group (0.65 nmol/day) at $p < 0.001$ and from group receiving 65 nmol/day at $p < 0.010$.

production of 25OHD is not tightly controlled by the liver. The high capacity for 25-hydroxylation of vitamin D in the liver as well as poor regulation at this site allows for massive amounts of 25OHD to be generated from large amounts of vitamin D. Thus, excessive concentrations of 25OHD are typically measured in vitamin D toxicity. As would be expected, PTH levels are suppressed in this form of hypercalcemia.

B. Role of Vitamin D Receptor (VDR) in Vitamin D Toxicity

Various investigations have helped to shed light on the interrelationship among vitamin D metabolites, the VDR, and PTH in vitamin D toxicity. The biologically active form of vitamin D, $1,25(OH)_2D$, as is typical of other steroid hormones, binds to a specific intracellular receptor protein (VDR) within its target tissues. The hormone–VDR complex then triggers subsequent transcriptional events by binding to DNA elements.

Regulation of cellular VDR numbers is believed to be an important mechanism by which cellular responsiveness to $1,25(OH)_2D$ is modulated, because the biological activity of $1,25(OH)_2D$ is proportional both to tissue VDR number and concentration of $1,25(OH)_2D$ (see Chapter 11 for a detailed discussion). Increased VDR concentrations imply enhanced tissue responsiveness to 1,25-dihydroxyvitamin D, whereas decreased receptor numbers indicate reduced tissue respon-

TABLE III Plasma Concentrations of Vitamin D_3 and Metabolites in Rats Given Various Amounts of $25OHD_3$[a]

Amount (nmol/day)	Vitamin D_3 (ng/ml)	$25OHD_3$ (ng/ml)	Lactone (ng/ml)	$24,25(OH)_2D_3$ (ng/ml)	$25,26(OH)_2D_3$ (ng/ml)	$1,25(OH)_2D_3$ (pg/ml)	Plasma calcium (mg/100 ml)
0.46	<0.5	6.2 ± 2.3	0.31 ± 0.05	2.29 ± 0.54	<0.2	187 ± 72	9.8 ± 0.5
4.6	<0.5	56.3 ± 11.4	3.02 ± 0.63	11.7 ± 3.3	<0.2	192 ± 65	9.3 ± 0.5
46	<0.5	199 ± 24	32.5 ± 8.5	57.3 ± 19.5	1.19 ± 0.49	82 ± 29	10.2 ± 0.4
460	<0.5	436 ± 58	118 ± 26	170 ± 22	4.02 ± 1.02	33 ± 8	9.7 ± 0.4
4600	<0.5	688 ± 145[b]	110 ± 38[c]	214 ± 117[d]	6.31 ± 1.79[e]	22 ± 7[f]	14.0 ± 0.5[g]

[a] Rats were orally dosed daily for 14 days with indicated amounts of $25OHD_3$. Data are means of 5 rats ± SD. Reprinted with permission from Shepard RM, DeLuca HF. Arch Biochem Biophys 1980;202:43–50.
[b] Differs from control group (0.46 nmol/day) at $p < 0.001$ and from group receiving 460 nmol/day at $p < 0.010$.
[c] Differs from control group (0.46 nmol/day) at $p < 0.001$.
[d] Differs from control group (0.46 nmol/day) at $p < 0.005$.
[e] Differs from group receiving 460 nmol/day at $p < 0.005$.
[f] Differs from control group (0.46 nmol/day) at $p < 0.001$ and from group receiving 460 nmol/day at $p < 0.050$.
[g] Differs from control group (0.46 nmol/day) and from group receiving 460 nmol/day at $p < 0.001$.

siveness. Several investigations have suggested that exogenous $1,25(OH)_2D_3$ can lead to homologous up-regulation of VDR *in vitro* and *in vivo*, in contrast to endogenous production of $1,25(OH)_2D_3$. *In vitro* and *in vivo* administration of $1,25(OH)_2D_3$ to rats has been shown to increase VDR content. *In vitro* exposure of human skin fibroblasts and osteosarcoma cells to $1,25(OH)_2D_3$ has been shown to result in a 3- to 5-fold increase in VDR number [115,116]. Similarly, *in vivo* studies have shown increased VDR with exogenous administration of $1,25(OH)_2D_3$. Costa and Feldman administered 1500 pmol/kg of $1,25(OH)_2D_3$ daily to rats and found a 30% increase in intestinal VDR and a 3-fold increase in renal VDR concentration [117]. Reinhart *et al.* infused rats with 250 pmol/kg of $1,25(OH)_2D_3$ daily for 6 days and noted a 22% increase in VDR levels in the intestine and a 37% increase in bone [118]. Goff and colleagues infused 36 ng of $1,25(OH)_2D_3$ to rats over 7 days and found a 1.5-fold increase in duodenal VDR conent and a 3-fold increase in renal VDR content [119].

Goff *et al.* [119] also demonstrated that endogenously produced $1,25(OH)_2D_3$ has a different effect than exogenous administration of $1,25(OH)_2D_3$ on tissue VDR content. Rats fed a calcium-restricted diet resulting in "nutritional" hyperparathyroidism achieved a similar increase in endogenous $1,25(OH)_2D_3$ concentration as rats administered exogenous $1,25(OH)_2D_3$. However, calcium-restricted rats failed to up-regulate VDR content in the duodenum or kidney, presumably a consequence of the negative control of VDR by PTH [120]. This point has at least conceptual relevance in the case of vitamin D toxicity. Rather than down-regulation occurring during hypervitaminosis D, which is a more typical regulatory and protective event to limit tissue responsiveness, exposure of cells to exogenous $1,25(OH)_2D$ results in enhanced responsiveness by virtue of up-regulation. Such a mechanism would be of particular clinical relevance if the toxicity were due to overexposure of $1,25(OH)_2D$. Moreover, in this setting, the associated suppression of PTH would prevent the regulatory mechanism from being operative.

Evidence suggests that in parent vitamin D toxicity target tissues are responding to high concentrations of 25OHD, not $1,25(OH)_2D$. Concentrations of $1,25(OH)_2D$ are typically only slightly increased, if at all. The hypercalcemia is due to the effects of pharmacologically high levels of 25OHD even though in physiological settings 25OHD has little potency. At high concentrations, 25OHD can compete for binding at VDR sites and thereby produce biological effects similar to those of $1,25(OH)_2D$ on intestine and bone [121]. Beckman and colleagues [122] suggested, furthermore, that hypervitaminosis D, like excessive $1,25(OH)_2D$, is associated with homologous up-regulation of intestinal VDR. Their investigation demonstrated that supraphysiological amounts of vitamin D_2 or vitamin D_3 administered to rats at doses of 25,000 IU daily for 6 days resulted in increasing plasma 25OHD concentrations with significant up-regulation of intestinal VDR concentration and hypercalcemia. Plasma $1,25(OH)_2D$ levels were not altered substantially (see Table IV and Fig. 1).

A comparison between hypervitaminosis D_3 and D_2 was also made [122]. No differences in 25OHD and plasma calcium concentrations were noted between either preparations. Concentrations of 25OHD in each case were markedly higher than the control group. The concentration of $1,25(OH)_2D$ was observed to be only slightly greater in the vitamin D_3-treated group than the vitamin D_2-treated group. Because the 25OHD concentrations were elevated 20- to 25-fold, whereas $1,25(OH)_2D$ showed only minimal increases, the biochemical and clinical changes associated with parent vitamin D toxicity were attributed to 25OHD. The data

TABLE IV Changes in Body Weight, Plasma Calcium, and Plasma Vitamin D Metabolites in Rats Treated for Six Days with Either 25,000 IU/day of Vitamin D_2 or Vitamin D_3[a]

Group	Body weight (g)	Plasma calcium (mg/dl)	25OHD (ng/ml)	$1,25(OH)_2D$ (pg/ml)
Control	251 ± 5	9.5 ± 0.7	20 ± 2	112 ± 11
Vitamin D_2-treated	230 ± 17	11.8 ± 0.6[b]	466 ± 36[b]	123 ± 12
Vitamin D_3-treated	201 ± 18[b]	12.0 ± 0.9[b]	506 ± 67[b]	150 ± 8[c,d]

[a] Data represent means ± SE. Reprinted with permission from Beckman MJ, *et al.* Biochem Biophys Res Commun 1990;169:910–915.
[b] Significant difference at $p < 0.01$ of the treated groups relative to the control group.
[c] Significant difference at $p < 0.05$ of the treated group relative to the control group.
[d] Statistical significance between the D_2- and D_3-treated groups ($n = 6$).

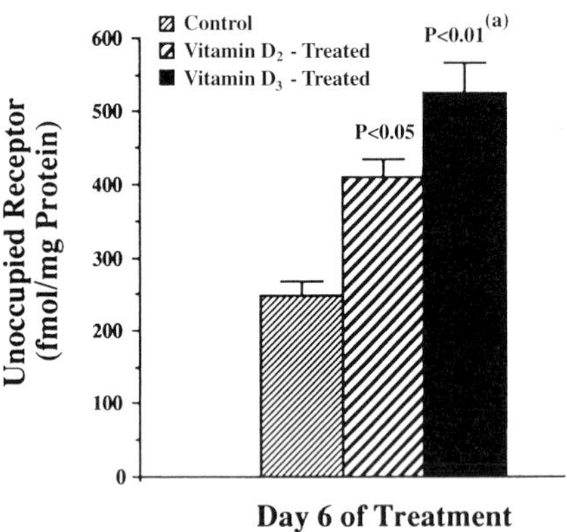

FIGURE 1 Intestinal VDR in rats treated 6 days with 25,000 IU/day of either vitamin D_2 or vitamin D_3 relative to the response in age-matched controls. [a]Response in vitamin D_3-treated rats significantly different from that in vitamin D_2-treated rats ($p < 0.05$). Reprinted with permission from Beckman MJ, *et al.* Biochem Biophys Res Commun 1990;169:910–915.

provided further support for the importance of 25OHD as the major toxic metabolite in vitamin D-associated hypercalcemia as well as for the importance of increased intestinal VDR in the pathophysiological process that leads to enhanced effects of this metabolite.

C. Control of Renal 1α-Hydroxylase in Vitamin D Toxicity

Some investigators have suggested that toxic effects of excessive concentrations of 25OHD may result from PTH suppression and down-regulation of 1α-hydroxylase with increased concentrations of 25OHD. PTH and $1,25(OH)_2D$ have known reciprocal actions on 1-hydroxylase and 24-hydroxylase activities. PTH stimulates 1-hydroxylase activity and down-regulates 24-hydroxylase activity; $1,25(OH)_2D$, on the other hand, down-regulates 1-hydroxylase activity and stimulates 24-hydroxylase activity.

Beckman and colleagues [123] studied the effects of an excess of vitamin D_3 and dietary calcium restriction on tissue 1-hydroxylase and 24-hydroxylase activity in rats. Four groups of rats with different dietary calcium and vitamin D_3 concentrations were studied (normal calcium, NC; low calcium, LC; and the excess vitamin D groups with normal or low calcium, NCT and LCT). The data showed that in the setting of a calcium-restricted diet, a nutritional hyperparathyroidism ensued (Table V). Under conditions of excess vitamin D_3 at doses of 75,000 IU per week and on a calcium-restricted diet, elevations in PTH facilitated the elimination of $25OHD_3$ through its metabolism to $1,25(OH)_2D_3$ and degradation to $24,25(OH)_2D_3$. The elevation in PTH was accompanied by increased activation of renal 1-hydroxylase activity, lower concentrations of $25OHD_3$, increased activation of intestinal 24-hydroxylase activity, and lower renal VDR content compared to the normal calcium group (Table VI). In contrast, the normal calcium diet in the vitamin D_3, excess group contributed to the toxicity by virtue of suppressed PTH concentrations resulting in down-regulation of renal 1-hydroxylase and reduced 1-hydroxylase activity, decreased 24-hydroxylase activity, and, thus, higher $25OHD_3$ concentrations. On the other hand, dietary calcium restriction in the setting of vitamin D_3 excess seemed to be protective in providing less biological stimulation due to higher PTH concentrations with reduced VDR, increased activation of both 1-hydroxylase and 24-hydroxylase activities, greater reductions in $25OHD_3$ concentrations, and lower concentrations of total calcium resulting in a less toxic state. So the low calcium

TABLE V Changes in Plasma Calcium, Phosphorus, PTH, and $1,25(OH)_2D_3$ Concentrations in Response to Dietary Calcium Restriction and Vitamin D_3 Excess[a]

Treatment[b]	Calcium (mg/dl)	Phosphorus (mg/dl)	$25OHD_3$ (ng/ml)	PTH (pg/ml)	$1,25(OH)_2D_3$ (pg/ml)
NC	11.2 ± 0.1	9.3 ± 0.7	15.2 ± 1.7	48.0 ± 2.0	116.0 ± 7.0
NCT	14.6 ± 0.3[c]	9.5 ± 0.5	443 ± 43[c]	44.0 ± 3.0	48.0 ± 8.0
LC	9.1 ± 0.1[c]	8.9 ± 0.5	<1.0[c]	162.0 ± 10.0[c]	615.0 ± 110.0[c]
LCT	9.7 ± 0.4[d,e]	9.0 ± 0.7	244 ± 17[c,e]	154.0 ± 19.0[c,e]	99.0 ± 12.0

[a] Values are means ± SE ($n = 6$). Significant differences were measured by Tukey's multiple range test. Reprinted from Beckman MJ, *et al.* Arch Biochem Biophys 1995;319:535–539 with permission.
[b] NC, 1.0–1.2% Ca, normal D_3; NCT, 1.0–1.2% Ca, excess D_3; LC, 0.02% Ca, normal D_3; LCT, 0.02% Ca, excess D_3.
[c] Significant difference at $p < 0.001$ versus NC group.
[d] Significant difference at $p < 0.01$ versus NC group.
[e] Significant difference at $p < 0.001$ versus NCT group.

Table VI Changes in Renal VDR and Intestinal $25OHD_3$-Hydroxylases in Response to Dietary Calcium Restriction and Vitamin D_3 Excess[a]

Treatment	Renal VDR (fmol/mg protein)	Renal 1α-hydroxylase (pg/hr · mg tissue)	Renal 24-hydroxylase (pg/min · mg tissue)	Intestinal 24-hydroxylase (pg/min · mg tissue)
NC	200 ± 14	1.5 ± 0.2	6 ± 1	7 ± 1
NCT	541 ± 21[c]	1.1 ± 0.1	276 ± 15[b,c]	19 ± 7
LC	115 ± 8[c]	28.7 ± 5.6[b,c]	ND[b]	80 ± 16[d]
LCT	97 ± 6[c,e]	3.6 ± 0.4[d,e]	ND	174 ± 33[c,e]

[a] Values are means ± SE (n = 6–10). Groups NC, NCT, LC, and LCT are defined in Table V. Significant differences were measured by Tukey's multiple range test. Reprinted with permission from Beckman MJ, *et al.* Arch Biochem Biophys 1995;319:535–539.
[b] ND, not detected.
[c] Significant difference at $p < 0.01$ versus NC group.
[d] Significant difference at $p < 0.05$ versus NC group.
[e] Significant difference at $p < 0.01$ versus NCT group.

diet protects, not only by contributing to less hypercalcemia, but also by facilitating metabolic pathways of vitamin D inactivation.

D. Inhibition of the Catabolic Pathway of 24-Hydroxylase

Others have proposed that inhibition of the enzymes that degrade the vitamin D metabolites may have a role in the pathogenesis of hypervitaminosis D. $1,25(OH)_2D$ is a known regulator of its own catabolism and an inhibitor of its synthesis. In the kidney and intestine, $1,25(OH)_2D$ induces the enzyme 24-hydroxylase. This enzyme initiates a catabolic cascade for the side chain oxidation, cleavage, and metabolic elimination of both $1,25(OH)_2D$ and 25OHD, and it accounts for 35–40% of the catabolism of $1,25(OH)_2D$ [124]. The remainder of the metabolic degradation is due to other side chain oxidations and biliary clearance. Reinhart and Horst [125] initially proposed that blunting of the catabolic pathway of $1,25(OH)_2D_3$ with high concentrations of $24,25(OH)_2D_3$ in rat cells would competitively inhibit further inactivation of $1,25(OH)_2D_3$, resulting in an accumulation of $1,25(OH)_2D_3$ and toxicity.

Clinical investigations of the down-regulation of rat intestinal 24-hydroxylase and its inhibition by calcitonin may help to elucidate a role of this hormone in potentiating the toxicity of vitamin D. 24-Hydroxylation is important in the inactivation of both $1,25(OH)_2D_3$ and $25OHD_3$, and in the kidney is largely regulated inversely by 1-hydroxylation [126]. In a study examining the effects of dietary calcium and vitamin D status on the regulation of intestinal 24-hydroxylase enzyme and mRNA expression, rats were fed normal or low calcium diets with variable amounts of vitamin D [127]. Half of the rats on the normal and low calcium diets were administered pharmacological doses of vitamin D_3 (25,000 IU three times weekly). Excess vitamin D_3 resulted in significant elevations in plasma $25OHD_3$ in both calcium groups (LCT and NCT), with a much larger increase noted in the normal calcium group (NCT). Hypercalcemia was most severe in the NCT group, whereas rats in the low calcium and vitamin D_3 excess group (LCT) had plasma calcium levels similar to the NC group (see Table VII). Because the NCT was accompanied by an increased calcitonin concentration compared to the LCT, the authors suggested that the increased calcitonin in the NCT group may have suppressed 24-hydroxylase activity, with resulting higher $25OHD_3$ and calcium concentrations [127]. This concept was further

Table VII Changes in Plasma Calcium, 25OHD, $1,25(OH)_2D$, and Calcitonin Concentrations in Response to Dietary Calcium Restriction and Vitamin D_3 Excess[a]

Treatment	Calcium (mg/dl)	25OHD (ng/ml)	$1,25(OH)_2D$ (pg/ml)	Calcitonin (pg/ml)
NC	10.7 ± 0.2	108 ± 16	106 ± 17	21.2 ± 1.1
NCT	12.6 ± 0.3[b]	1812 ± 165[b]	96 ± 12	36.1 ± 2.5[b]
LC	9.7 ± 0.2[b]	47 ± 10[c]	459 ± 70[c]	15.9 ± 0.9
LCT	10.5 ± 0.2[c]	1130 ± 62[b,c]	188 ± 41	19.4 ± 2.3[c]

[a] Values are means ± SEM (n = 6). Groups NC, NCT, LC, and LCT are defined in Table V. Significant differences were measured by Tukey's multiple range test after analysis of variance. Reprinted from Beckman *et al.* [127 with permission].
[b] $p < 0.05$ versus NC group.
[c] $p < 0.05$ versus NCT group.

TABLE VIII Intestinal 24-Hydroxylase, Plasma Calcium, and $1,25(OH)_2D$ in Response to Vitamin D Excess and Thyroparathyroidectomy[a]

Treatment[b]	Calcium (mg/dl)	$1,25(OH)_2D$ (pg/ml)	24-Hydroxylase (pg/min · mg protein)
NCT	12.9 ± 0.4	86 ± 16	6.7 ± 0.5
NCT/TPTX	13.8 ± 0.3	78 ± 9	14.1 ± 0.2[c]
NCT/CT	9.7 ± 0.6[c]	80 ± 13	5.3 ± 0.2
NCT/TPTX/CT	10.3 ± 0.4[c]	82 ± 11	3.8 ± 0.2[c,d]

[a] Reprinted from Beckman *et al.* [127 with permission].
[b] NCT, Controls given excess vitamin D_3; NCT/TPTX, NCT animals that underwent TPTX; NCT/CT, NCT animals treated with 100 IU calcitonin 4 hr before death; NCT/TPTX/CT, NCT/TPTX animals treated with 100 IU calcitonin 4 hr before death.
[c] Different from NCT ($p < 0.001$).
[d] Different from NCT/TPTX ($p < 0.001$).

supported when rats, subjected to thyroparathyroidectomy (TPTX) which eliminated endogenous calcitonin (Table VIII), were found to have higher concentrations of 24-hydroxylase activity than the NCT group. Through inhibition of intestinal 24-hydroxylase activity, calcitonin could be associated with reduced turnover and catabolism of $25OHD_3$, thereby potentiating its toxicity. Thus, increased expression of 24-hydroxylase activity in cases of pharmacological amounts of $25OHD_3$ may be an important mechanism to counteract vitamin D toxicity.

E. Vitamin D Binding Protein and the Level of Free Metabolite in Vitamin D Toxicity

The vitamin D binding protein (DBP) is a specific transport protein that binds large quantities of the circulating vitamin D metabolites (see Chapter 7). Similar to the situation for other steroid hormones, fat-soluble compounds, and thyroid hormones, only a small fraction of the metabolites circulate free in plasma. The binding affinities of the protein for the vitamin D metabolites is moderate, and the capacity is great. In addition, the various metabolites have different binding affinities for the protein, in the following sequence: 250HD > $24,25(OH)_2D$ > $1,25(OH)_2D$ [128]. Of note is the fact that the potent metabolite $1,25(OH)_2D$ has the least affinity for DBP but the highest affinity for the intracellular VDR that triggers subsequent transcriptional events. Therefore, freeing bound $1,25(OH)_2D$ metabolite from DBP could promote its entry into various tissues and promote biological activity [13,129].

Evidence indicates that the biologically active form of the vitamin D steroid hormone is the free hormone that is accessible to cells [130]. Because of technical difficulties in measuring the free hormone, the determination of vitamin D status involves a measurement combining free vitamin D and DBP concentrations. In normal individuals, 85% of the total $1,25(OH)_2D$ is bound to DBP, 15% is bound to albumin, and 0.4% is free [131]. However, under conditions of altered or reduced albumin and DBP concentrations, as in liver or kidney disease, the free hormone may provide different information compared to the total measured concentration of vitamin D. Theoretically, total hormone concentration in such settings may erroneously suggest deficiency of vitamin D with needless institution of replacement therapy. Bikle and colleagues noted that subjects with liver disease have reduced DBP concentrations with low total $1,25(OH)_2D$ and 25OHD levels, whereas free forms were normal [312,133].

Similarly, in certain forms of renal disease, the concentrations of DBP and vitamin D metabolites are reduced, resulting in measurements of total hormone, providing inaccurate reflections of vitamin D status. Koenig *et al.* [134] investigated free and total $1,25(OH)_2D$ concentrations in subjects with renal disease. Patients with nephrotic syndrome, with varying degrees of renal failure, and on chronic hemodialysis and peritoneal dialysis were studied. The serum concentrations of total and free $1,25(OH)_2D$ correlated well with one another in the patients with renal failure and those undergoing hemodialysis. The concentrations of DBP and 25OHD, thus, were unaffected by renal function. The concentrations of total $1,25(OH)_2D$ accurately reflected free $1,25(OH)_2D$ in patients with varying degrees of renal failure when DBP levels remained normal. However, this did not hold true for the subjects with nephrotic syndrome or those on chronic peritoneal dialysis, who lost DBP and bound vitamin D metabolites into the urine or peritoneal fluid, respectively, with a rise in the percentage free $1,25(OH)_2D$. Measurement of free metabolite in these particular patients may be important to avoid vitamin D toxicity when supplementation is instituted. Thus, in this context, the binding proteins of the vitamin D metabolites not only serve a transport function, but also may provide a buffering mechanism to protect against toxicity [135].

V. CLINICAL MANIFESTATIONS

The clinical manifestations of vitamin D toxicity resulting from hypercalcemia reflect the essential role of calcium in many tissues and targets, including bone, the cardiovascular system, nerves, and cellular enzymes. Initial signs and symptoms of hypervitaminosis D may be similar to other hypercalcemic states and include generalized weakness and fatigue. Central nervous system features may include confusion, difficulty in concentration, drowsiness, apathy, and coma [136]. Neuropsychiatric symptoms include depression and psychosis, which resolve following improvement of the hypercalcemia.

Hypercalcemia can affect the gastrointestinal tract and cause anorexia, nausea, vomiting, and constipation. It can induce hypergastrinemia, but only in men does it appear to be associated with peptic ulcer disease. There is no evidence that peptic ulcers are more common in any other form of hypercalcémia. Rarely, pancreatitis may be a presentation of either acute or chronic hypercalcemia.

In the heart, hypercalcemia may shorten the repolarization phase of conduction reducing the Q-T interval on the electrocardiogram (EKG). A more accurate EKG indication of the level of hypercalcemia is the Q-T interval corrected for rate. Bradyarrhythmias and first degree heart block have been described but are rare. Hypercalcemia may potentiate the action of digitalis on the heart [137].

Kidney function is affected because high concentrations of calcium alter the action of vasopressin on the renal tubules. The net result is reduced urinary concentrating ability and a form of nephrogenic diabetes insipidus. This usually presents as polyuria, but rarely is the volume as high as that associated with central diabetes insipidus. Symptoms may include polydipsia, which is an expected consequence of polyuria [138]. Hypercalciuria is one of the earliest signs of vitamin D toxicity and precedes the occurrence of hypercalcemia. The initial hypercalciuria may be ameliorated as renal failure progresses because of reduced calcium clearance. The pathophysiology of hypercalcemia can be rapidly worsened when dehydration develops. When reduced renal blood flow occurs, less calcium is presented to the renal glomerulus, and hypercalcemia can rapidly progress. Renal impairment from the hypercalcemia is reversible if of short duration. Chronic, uncontrolled hypercalcemia can lead to deposition of calcium phosphate salts in the kidney and permanent damage with eventual nephrocalcinosis. In an investigation of vitamin D-induced nephrocalcinosis, Scarpelli and colleagues [139] noted that cell damage, specifically in mitochondria, preceded intracellular calcium deposition. The hypercalcemia induced in rats by excessive vitamin D administration caused mitochondrial swelling, cell injury, and subsequent calcification.

Ectopic, soft tissue calcification can be a particular problem in hypervitaminosis D. The disposition of soft tissue calcification is compounded by the combination of hypercalcemia and hyperphosphatemia, often exceeding the solubility product of the two ions [140–142]. In rats exposed to excessive vitamin D, Hass and colleagues demonstrated that the pathological processes of vitamin D toxicity were related to dosage, length of time between doses, and duration of exposure [143]. For rats subjected to sublethal doses, generalized calcinosis was seen after only 8 days, when a total of 300,000 units of ergosterol was administered. Pathologically, bones appeared more brittle than normal, with increased cortical bone resorption, increased numbers of osteoclasts, and reduced numbers of osteoblasts. Abnormal calcium deposits were noted in the aorta and its major branches, heart, kidney, muscle, and respiratory tract. The earliest evidence of hypervitaminosis D was in the proximal aorta. Muscle tissue was the least resistant to calcification, with the order of decreasing susceptibility being smooth muscle > cardiac muscle > skeletal muscle [144]. The liver, brain, and pituitary were not affected by high doses of vitamin D.

VI. DIAGNOSIS OF VITAMIN D TOXICITY

With modern assays for calciotropic hormones, PTH, 25OHD, and $1,25(OH)_2D$ (see Chapter 38), one can readily differentiate vitamin D metabolite-mediated hypercalcemia from other causes of hypercalcemia (Table IX). The circulating intact PTH level, preferably measured by the two-site immunoradiometric assay (IRMA) or immunochemiluminometric assay (ICMA), should be suppressed in virtually all hypercalcemic disorders with the exception of primary hyperparathyroidism, familial hypocalciuric hypercalcemia, administration of lithium or thiazides, and renal failure. Although patients with malignancy-associated hypercalcemia tend to have a higher serum calcium concentration than those other causes of hypercalcemia, diminished glomerular filtration rate and subsequent reduction in renal calcium excretion can dramatically increase the serum calcium level in any hypercalcemic patient. In contrast to the low serum phosphorus level in patients with hypercalcemia due to PTH or PTHrP, the serum phosphorus level is at the upper limit of normal or frankly elevated in patients with vitamin D metabolite-mediated hypercalcemia. This is due to increased intestinal absorption and reduced renal clearance of phosphate. An elevated 25OHD concentration with normal $1,25(OH)_2D$ level is indicative of toxicity with exogenously administered vitamin D or 25OHD. The serum $1,25(OH)_2D$ level may be normally increased in patients with primary hyperparathyroidism due to

TABLE IX Calciotropic Hormone Levels in the Differential Diagnosis of Hypercalcemia

Disease	Ca	PO_4	iPTH	25OHD	$1,25(OH)_2D$
Primary hyperparathyroidism	↑	N to ↓	↑	N to ↓	N to ↑
Malignancy					
Solid tumors	↑↑	N to ↓	↓	N	↓
Myeloma	↑↑	N	↓	N	↓
Lymphoma	↑↑	↓ to N to ↑	↓	N	↓ to N to ↑
Granulomatous diseases	↑	N to ↑	↓	N	↑
Toxicity with					
Vitamin D or 25OHD	↑	N to ↑	↓	↑	N
1αOHD or $1,25(OH)_2D$	↑↑	N to ↑	↓	N	↑
DHT or calcipotriol	↑↑	N to ↑	↓	N	N

N, normal.

the induction of renal 1α-hydroxylase by PTH. Abnormally high $1,25(OH)_2D$ levels, in the setting of suppressed PTH and hypercalcemia, indicates dysregulated production of $1,25(OH)_2D$ due to either granulomatous diseases, lymphoma, or toxicity with exogenous $1,25(OH)_2D$ or 1αOHD. In cases of hypercalcemia due to PTHrP or local osteolytic factors, the serum $1,25(OH)_2D$ concentration is usually suppressed. In patients with hypercalcemia due to toxicity with other vitamin D analogs such as dihydrotachysterol (DHT) [145] and calcipotriol, the active metabolites may not be recognized by the conventional competitive protein binding assays for $1,25(OH)_2D$.

The diagnosis of vitamin D toxicity can be made on clinical grounds. Detailed clinical and drug history are of paramount importance in order to make an early diagnosis. Most patients who are suffering from vitamin D toxicity are taking vitamin D for osteoporosis, hypoparathyroidism, pseudohypoparathyroidism, hypophosphatemia, osteomalacia, or renal osteodystrophy in excessive dosages or at too frequent dosing intervals. Therefore, one should have a high index of suspicion in patients who are being treated with pharmacological dosages of vitamin D or its metabolites.

Patients with granulomatous diseases or lymphoma usually have widespread active disease when hypercalcemia develops. In such cases, the diagnosis is obvious at the time of presentation. However, exceptions do exist. In patients with unexplained hypercalcemia, if the $1,25(OH)_2D$ level is elevated and other more easily identifiable causes for this elevation such as primary hyperparathyroidism, pregnancy, and exogenous toxicity (by history) are excluded, measurement of angiotensin-converting enzyme level and a systemic search for lymph node enlargement, pulmonary, renal, hepatosplenic, ocular, central nervous system, and bone marrow granulomas or lymphoma should be made.

VII. TREATMENT OF VITAMIN D TOXICITY

Dietary calcium and vitamin D restriction and avoidance of exposure to sunlight and other ultraviolet light sources should be advised to patients at high risk to develop vitamin D metabolite-mediated hypercalcemia. Those at risk include patients with granulomatous diseases and lymphoma whose disease is widespread and active and patients who are already hypercalciuric. Daily dietary calcium intake should be minimized to approximately 400 mg or less in these patients. Any use of vitamin D supplements should be discontinued. The patient should be encouraged to use sunscreen [sun protection factor (SPF) >15] as much as possible. The calcium level should be monitored closely in patients who have a previous history of hypercalcemia or hypercalciuria, or who have recently taken diets enriched in vitamin D and calcium, or who have a recent history of excessive sunlight exposure.

When hypercalcemia develops, the aforementioned preventive measures will help to ameliorate the severity of hypercalcemia. General measures in those who are symptomatic include hydration with normal saline and the judicious use of a loop diuretic, like furosemide. Specific inhibitors of bone resorption, such as bisphosphonates and calcitonin, can be helpful. Glucocorticoids have proved to be particularly effective in vitamin D intoxication, granulomatous diseases, and lymphoma (see also Chapters 49 and 55). The precise mechanism of action of glucocorticoids in calcium homeostasis is not known. Nonetheless, they are useful because they (1) directly inhibit gastrointestinal absorption of calcium by decreasing the synthesis of calcium binding protein (calbindin-D) and decreasing active transcellular transport [146,147], (2) increase urinary excretion of calcium [148], and (3) may alter hepatic vitamin D metabolism to favor the production of nonactive vitamin D metabolites, resulting in lower concentrations of 25OHD [149]. Evidence also suggests that they may increase the degradation of $1,25(OH)_2D$ at the receptor sites [150]. Glucocorticoids may also limit osteoclastic bone resorption [151]. Institution of glucocorticoid therapy results in prompt decline of the circulating $1,25(OH)_2D$ concentrations within 3 to 4 days [111]. Patients with nonhematological malignancies and those with primary hyperparathyroidism do not usually respond to glucocorticoids.

Chloroquine and hydroxychloroquine are also capable of reducing the $1,25(OH)_2D$ and calcium concentrations in patients with sarcoidosis [152]. Ketoconazole, an antifungal agent, in high dosages can inhibit the mitochondrial cytochrome P450-linked 25OHD 1α-hydroxylase irrespective of whether it is renal [153] or extrarenal as in sarcoidosis [154] and tuberculosis [76]. Owing to the limited experience with these drugs as antihypercalcemic agents and their potential side effects, they should be reserved for patients in whom steroid therapy is unsuccessful or specifically contraindicated.

VIII. SUMMARY AND CONCLUSIONS

Vitamin D toxicity is not a common cause of hypercalcemia, but it can be life threatening if not identified quickly. The major causes of hypercalcemia are primary hyperparathyroidism and malignancy. If these two etiologies are excluded, vitamin D toxicity becomes an important diagnostic consideration. There are many forms of exogenous and endogenous vitamin D toxicity. Inadvertent excessive use of pharmaceutical preparations is the most common etiology of exogenous toxicity. Excessive amounts of the parent compound, vitamin D, can be most difficult to manage as compared to toxicity due to the metabolites 25OHD or $1,25(OH)_2D$. Extensive lipid solubility of vitamin D accounts for its extraordinary half-life and tendency for prolonged hypercalcemia. New clinical applications of $1,25(OH)_2D$ and its synthetic analogs have been accompanied by the increased potential for toxicity. Endogenous etiologies may result from ectopic production of $1,25(OH)_2D$ in granulomatous diseases, such as sarcoidosis and tuberculosis, or in lymphoma. Many different mechanisms have been proposed to account for vitamin D toxicity including the vitamin D metabolite itself, VDR number, activity of 1α-hydroxylase, inhibition of vitamin D metabolism, and the capacity of DBP.

Acknowledgment

This review was facilitated, in part, by support from a Grant from the National Institutes of Health (DK 32333).

References

1. Haddad JG, Chuy KJ 1971 Competitive protein binding radioassay for 25-hydroxycholecalciferol. J Clin Endocrinol Metab **33**:992–995.
2. Jacobus CH, Holick MF, Shao Q, Chen TC, Holm IA, Kolodny JM, Fuleihan GE, Seeley E 1992 Hypervitaminosis D associated with drinking milk. N Engl J Med **326**:1173–1177.
3. Horst RL, Littledike ET, Riley JL, Naploi JL 1981 Quantification of vitamin D and its metabolites and their plasma concentrations in five species of animals. Anal Biochem **116**:189–203.
4. Harrington DD, Page EH 1983 Acute vitamin D_3 toxicosis in horses: Case reports and experimental studies on the comparative toxicity of vitamin D_2 and D_3. J Am Vet Med Assoc **182**:1358–1369.
5. Hunt RD, Garcia FG, Walsh RJ 1972 A comparison of the toxicity of ergocalciferol and cholecalciferol in rhesus monkeys (*Macaca mulatta*). J Nutr **102**:975–986.
6. Parfitt MA, Gallagher JC, Heaney RP, Johnston CC, Neer R, Whedon GD 1982 Vitamin D and bone health in the elderly. Am J Clin Nutr **36**:1014–1031.
7. Food and Drug Administration 1979 Over the Counter Drug Report. Federal Register, March 17, 16164–16169.
8. Markestad TM, Hesse V, Siebenhuner M, Jahreis G, Aksnes L, Plenert W, Aarskog D 1987 Intermittent high dose vitamin D prophylaxis during infancy: Effect on vitamin D metabolites, calcium and phosphorus. Am J Clin Nutr **46**:652–658.
9. Zeghoud F, Ben-Mekbi H, Djeghri N, Garabedian M 1994 Vitamin D prophylaxis during infancy: Comparison of the long term effects of three intermittent doses (15, 5, or 2.5 mg) on 25-hydroxyvitamin D concentrations. Am J Clin Nutr **60**:393–396.
10. Vieth R 1990 The mechanisms of vitamin D toxicity. Bone Miner **11**:267–272.
11. Tilyard MW, Spears GFS, Thompson J, Dovey S 1992 Treatment of postmenopausal osteoporosis with calcitriol or calcium. N Engl J Med **326**:357–361.
12. Massry SG, Goldstein DA 1979 Is calcitriol [$1,25(OH)_2D_3$] harmful to renal function? Am Med Assoc **242**:1875–1876.
13. Key LL, Anast CS 1984 Calcitriol for congenital osteopetrosis. N Engl J Med **311**:55 (letter).
14. Aloia JF, Vaswani A, Yeh JK, Ellis K, Yasmura S, Cohn SH 1988 Calcitriol in the treatment of postmenopausal osteoporosis. Am J Med **84**:401–408.
15. Christiansen C, Rodbro P, Christensen M, Hartnak B, Transbol I 1978 Deterioration of renal function during treatment of chronic renal failure with 1,25-dihydroxycholecalciferol. Lancet **2**:700–703.
16. Christiansen C, Rodbro P, Christensen M, Hartnak B 1981 Is 1,25-dihydroxycholecalciferol harmful to renal function in patients with chronic renal failure? Clin Endocrinol **15**:229–236.
17. Slatopolsky E, Weerts C, Thielan J, Horst R, Harter H, Martin KJ 1984 Marked suppression of secondary hyperparathyroidism

by intravenous administration of 1,25-dihydroxycholecalciferol in uremic patients. J Clin Invest **74**:2136–2143.
18. Caravaca F, Cubero JJ, Jimenez F, Lopez JM, Aparico A, Cid MC, Pizzaro JL, Liso J, Santos I. 1995 Effect of the mode of calcitriol administration on PTH–ionized calcium relationship in uremic patients with secondary hyperparathyroidism. Nephrology **10**:665–670.
19. Monier-Faugere MC 1994 Calcitriol pulse therapy in patients with end stage renal failure. Curr Opin Nephrol Hyperten **3**:615–619.
20. Juergensen P, Santacroce S, Mooraki A, Cooper K, Finkelstein FO, Kliger AS 1994 Pulse oral calcitriol to treat hyperparathyroidism in 43 CAPD patients. Adv Periton Dial **10**:259–260.
21. Feinstein S, Algur N, Drukker A 1994 Oral pulse therapy with vitamin D_3 for control of secondary hyperparathyroidism. Pediatr Nephrol **8**:724–726.
22. Crocker JS, Muhtadie SF, Hamilton DC, Cole DE 1985 The comparative toxicity of vitamin D metabolites in the weanling mouse. Biol Appl Pharmacol **80**:119–126.
23. Brickman AS, Coburn JW, Friedman GR, Okamura WH, Massry SG, Norman AW 1976 Comparison of effect of 1α-hydroxyvitamin D_3 in men. J Lab Clin Invest **57**:1540–1547.
24. Kanis JA, Russell RGG, Smith R 1977 Physiological and therapeutic difference between vitamin D, its metabolities and analogues. Clin Endocrinol **7**:191s–201s.
25. Mortensen JT, Brinck P, Binderup L 1993 Toxicity of vitamin D analogues in rats fed diets with standard or low calcium contents. Pharmacol Toxicol **72**:124–127.
26. Gallagher JC, Bishop CW, Knutson JC, Mazess RB, DeLuca HF 1994 Effects of increasing doses of 1α-hydroxyvitamin D_2 on calcium homeostasis in postmenopausal osteopenic women. J Bone Miner Res **9**:607–614.
27. Sjorden G, Smith C, Lindgren U, DeLuca HF 1985 1α-Hydroxyvitamin D_2 is less toxic than 1α-hydroxyvitamin D_3 in the rat. Proc Soc Exp Biol Med **178**:432–436.
28. Pols HAP, Birkenhager JC, van Leeuwen J 1994 Vitamin D analogues: From molecule to clinical application. Clin Endocrinol **40**:285–291.
29. Bouillon R, Okamura WH, Norman AW 1995 Structure–function relationships in the vitamin D endocrine system. Endocr Rev **16**:200–256.
30. Hardman KA, Heath DA 1993 Hypercalcemia associated with calcipotriol (Dovonex) treatment. Br Med J **306**:896.
31. Bourke JF, Berth-Jones J, Mumford R, Iqbal SJ, Hutchinson PE 1994 High dose topical calcipotriol consistently reduces serum parathyroid hormone levels. Clin Endocrinol **41**:295–297.
32. Dwyer C, Chapman RS 1991 Calcipotriol and hypercalcemia. Lancet **338**:764.
33. Taylor AB, Stern PH, Bell NH 1982 Abnormal regulation of circulating 25-hydroxyvitamin D in the Williams syndrome. N Engl J Med **306**:972–975.
34. Ewart AK, Morris CA, Atkinson D, Jin W, Sternes K, Spallone P, Stock AD, Leppert M, Keating MT 1993 Hemizygosity at the elastin locus in a developmental disorder, Williams syndrome. Nature Genet **5**:11.
35. Taylor AB, Stern PH, Bell NH 1982 Abnormal regulation of circulating 25-hydroxyvitamin D in the Williams syndrome. N Engl J Med **306**:972.
36. Omdahl J, Holick M, Suda T, et al. 1971 Biological activity of 1,25-dihydroxycholecalciferol. Biochemistry **10**:2935.
37. Papapoulos SE, Clemens TL, Fraher LJ, Lewin IG, Sandler LDA, O'Riordan JL 1979 1,25-Dihydroxycholecalciferol in pathogenesis of the hypercalcemia of Sarcoidosis. Lancet **1**:627–630.
38. Harrel GT, Fisher S 1939 Blood chemical changes in Boeck's sarcoid with particular reference to protein, calcium and phosphatase values. J Clin Invest **18**:687–693.
39. Studdy PR, Bird R, Neville E, James, DG 1980 Biochemical findings in sarcoidosis. J Clin Pathol **33**:528–533.
40. Anderson J, Harper C, Dent CE, Philpot GR 1954 Effect of cortisone on calcium metabolism in sarcoidosis with hypercalcemia. Possibly antagonistic action of cortisone and vitamin D. Lancet **2**:720–724.
41. Henneman PH, Dempsey EF, Carroll EL, Albright F 1956 The cause of hypercalciuria in sarcoid and its treatment with cortisone and sodium phytate. J Clin Invest **35**:1229–1242.
42. Taylor RL, Lynch HJ 1963 Seasonal influence of sunlight on the hypercalcemia of sarcoidosis. Am J Med **34**:221–227.
43. Papapoulos SE, Fraher LJ, Sandler LM, Clemens TL, Lewin IG, O'Riordan JLH 1979 1,25-Dihydroxycholecalciferol in the pathogenesis of the hypercalcemia of sarcoidosis. Lancet **2**:627–630.
44. Bell NH, Stern PH, Pantzer E, Sinha TK, DeLuca HF 1979 Evidence that increased circulating 1 alpha,25-dihydroxyvitamin D is the probable cause for abnormal calcium metabolism in sarcoidosis. Lancet **64**:218–225.
45. Barbour GL, Coburn JW, Slatopolsky E, Norman AW, Horst RL 1981 Hypercalcemia in an anephric patient with sarcoidosis: Evidence for extrarenal generation of 1,25-dihydroxyvitamin D. N Engl J Med **305**:440–443.
46. Maesaka JK, Batuman V, Pablo MC, Shakamuri S 1982 Elevated 1,25-dihydroxyvitamin D levels. Occurrence with sarcoidosis with end-stage renal disease. Arch Intern Med **142**:1206–1212.
47. Korkor A, Gray R, Lemann J 1985 Further evidence for extra-renal synthesis of 1,25$(OH)_2$ vitamin D in sarcoidosis. Kidney Int **27**:119.
48. Adams JS, Gacad MA, Anders A, Endres DB, Sharma OP 1986 Biochemical indicators of disordered vitamin D and calcium homeostasis in Sarcoidosis. Sarcoidosis **3**:1.
49. Adams JS 1989 Vitamin D metabolite-mediated hypercalcemia. Endocrinol Metab Clin North Am **18**(No. 3):765–778.
50. Meyrier A, Valeyre D, Bouillon R, Paillard F, Battesti JP, Georges R 1985 Resorptive versus absorptive hypercalcium in sarcoidosis: Correlations with 25-hydroxyvitamin D_3 and 1,25-dihydroxyvitamin D parameters of disease activity. Q J Med **54**:269–281.
51. Mason RS, Frankel T, Chan YL, Lissner D, Posen S 1984 Vitamin D conversion by sarcoid lymph node homogenate. Ann Intern Med **100**:59–61.
52. Adams JS, Sharma OP, Gacad MA, Singh FR 1983 Metabolism of 25-hydroxyvitamin D_3 by cultured pulmonary alveolar macrophages in sarcoidosis. J Clin Invest **72**:1856–1860.
53. Adams JS, Singer FR, Gacad MA, Sharma OP, Hayes MJ, Vountus P, Holeck MF 1985 Isolation and structural identification of 1,25-dihydroxyvitamin D_3 produced by cultured alveolar macrophages in sarcoidosis. J Clin Endocrinol Metab **60**:960–966.
54. Reichel H, Koeffler HP, Barbers R, Norman AW 1987 Regulation of 1,25-dihydroxyvitamin D_3 production by cultured alveolar macrophages from normal human donors and from patients with pulmonary sarcoidosis. J Clin Endocrinol Metab **65**:1201–1209.
55. Fuss M, Gillet PC, Karmali R, Corvilain J 1992 Calcium and vitamin D metabolism in granulomatous diseases. Clin Rheumatol **11**:28–36.
56. Jackson WPV, Doncaster CD 1959 A consideration of the hypercalcemia in sarcoidosis, idiopathic hypercalciuria, and that produced by vitamin D. J Clin endocrinol Metab **19**:658–680.
57. Bell NH, Bartter FC 1967 Studies of ^{47}Ca metabolism in sarcoido-

sis: Evidence for increased sensitivity of bone to vitamin D. Acta Endocrinol **54**:173–180.

58. Dwarakanathan A, Ryan WG 1987 Hypercalcemia of sarcoidosis treated with cellulose sodium phosphate. Bone Miner **2**:333–336.
59. Fallon MD, Perry HM, Teitelbaum SL 1981 Skeletal sarcoidosis with osteopenia. Metab Bone Dis Relat Res **3**:171–174.
60. Alexandre C, Chappard D, Vergnon JM, Emonot A, Riffat J 1987 L'os dans la sarcoidose. Rev Rhum Mal Osteoartic **54**:159–162.
61. Scholz DA, Keating RF 1956 Renal insufficiency, renal calculi and nephrocalcinosis in sarcoidosis. Am J Med **21**:75–84.
62. Sandler LM, Winearls CG, Fraher LJ, Clemens TL, Smith R, O'Riordan JL 1984 Studies of the hypercalcemia of sarcoidosis: Effect of steroids and exogenous vitamin D_3 on the circulating concentrations of 1,25-dihydroxyvitamin D_3. Q J Med **210**: 165–180.
63. Abbasi AA, Chemplavil JK, Farah S, Muller BF, Arnstein AR 1979 Hypercalcemia in active pulmonary tuberculosis. Ann Intern Med **90**:324–328.
64. Sharma SC 1981 Serum calcium in pulmonary tuberculosis. Postgrad Med J **57**:694–696.
65. Kitrou MP, Phytou-Pallikari A, Tzannes SE, Virvidakis K, Mountokalakis TD 1982 Hypercalcemia in active pulmonary tuberculosis. Ann Intern Med **96**:255.
66. Davies PDO, Brown RC, Woodhead JS 1985 Serum concentrations of vitamin D in untreated tuberculosis. Thorax **40**:187–190.
67. Fuss M, Karmali R, Pepersack T, Bergans A, Durcks P, Pregogine T, Bergmann P, Carvilein J 1988 Are tuberculous patients at a greater risk from hypercalcemia? Q J Med **69**: 869–878.
68. Chan TYK, Chan CHS, Shek CC, Davies PDO 1992 Hypercalcemia in active pulmonary tuberculosis and its occurrence in relation to the radiographic extent of disease. Southeast Asian J Trop Med Public Health **23**:702–704.
69. Tan TT, Lee BC, Khalid BAK 1993 Low incidence of hypercalcemia in tuberculosis in Malaysia. J Trop Med Hyg **96**:349–351.
70. Felsenfeld AJ, Drezner MK, Liach F 1986 Hypercalcemia and elevated calcitriol in a maintenance dialysis patient with tuberculosis. Arch Intern Med **146**:1941–1945.
71. Peces R, Alvarez 1987 Hypercalcemia and elevated $1,25(OH)_2D_3$ levels in a dialysis patient with disseminated tuberculosis. Nephron **46**:377–379.
72. Shai F, Baker RK, Addrizzo JR, Wallach S 1972 Hypercalcemia in mycobacterial infection. J Clin Endocrinol Metab **34**:251–256.
73. Adama JS, Barnes PF, Modlin RL, Bikle DD 1989 Transpleural gradient of 1,25-dihydroxyvitamin D in tuberculosis pleuritis. J Clin Invest **83**:1527–1532.
74. Adams JS, Modlin RL, Diz MM, Barne PF 1989 Potentiation of the macrophage 25-hydroxyvitamin D-1-hydroxylation reaction by human tuberculous pleural effusion fluid. J Clin Endocrinol Metab **69**:457–460.
75. Cadranel J, Garabedian M, Milleron B, Guillozo H, Akoun G, Hance AJ 1990 $1,25(OH)_2D_3$ production by T lymphocytes and alveolar macrophages recovered by lavage from normocalcaemic patients with tuberculosis. J Clin Invest **85**:1588–1593.
76. Saggese G, Bertelloni S, Baroncelli GI, Nero GD 1993 Ketoconazole decreases the serum ionized calcium and 1,25-dihydroxyvitamin D levels in tuberculosis-associated hypercalcemia. Am J Dis Children **147**:270–273.
77. Ryzen E, Singer FR 1985 Hypercalcemia in leprosy. Arch Intern Med **145**:1305–1306.
78. Lee JC, Catanzaro A, Parthemore JG, Roach B, Deftos LJ 1977 Hypercalcemia in disseminated coccidioidomycosis. N Engl J Med **297**:431–433.
79. Murray JJ, Heim CR 1985 Hypercalcemia in disseminated histoplasmosis. Aggravation by vitamin D. Am J Med **78**:881–884.
80. Kantarjian HM, Saad MF, Estey EM, Sellin RV, Samaan NA 1983 Hypercalcemia and disseminated candidiasis. Am J Med **74**:721–724.
81. Jurney TH 1984 Hypercalcemia in a patient with eosinophilic granuloma. Am J Med **76**:527–528.
82. Stoeckle JD, Hardy HL, Weber AL 1969 Chronic beryllium disease. Am J Med **46**:545–561.
83. Kozeny GA, Barbato AL, Bansai VK, Vertuno LL, Hano JE 1984 Hypercalcemia associated with silicone-induced granulomas. N Engl J Med **311**:1103–1105.
84. Burt ME, Brennan MF 1980 Incidence of hypercalcemia and malignant neoplasm. Arch Surg **115**:705.
85. Vassilopoulou-Sellin R, Newman BM, Taylor SH, Ensor JE, Guinee VF 1991 Hypercalcemia of malignancy: Incidence in a comprehensive cancer centre [abstract]. Proc Am Soc Clin Oncol **10**:131.
86. Johnston SRD, Hammond PJ 1992 Elevated serum parathyroid hormone related protein and 1,25-dihydroxycholecalciferol in hypercalcemia associated with adult T-cell leukaemia–lymphoma. Postgrad Med J **68**:753–755.
87. Lymphoma Study Group (1984–1987) 1991 Major prognostic factors of patients with adult T-cell leukaemia–lymphoma: A cooperative study. Leuk Res **15**:81.
88. Seymour JF, Estey EH, Kantarjian HM, Pierce SA, Keating MJ 1993 The incidence and prognostic significance of mild hypercalcemia in patients with leukaemia. Proc Am Assoc Cancer Res **34**:194.
89. Plimpton CH, Gellhorn A 1956 Hypercalcemia in malignant disease without evidence of bone destruction. Am J Med **21**:750.
90. Seymour JF, Gagel RF 1993 Calcitriol: The major humoral mediator of hypercalcemia in Hodgkin's disease and non-Hodgkin's lymphomas. Blood **82**:1383–1394.
91. Davies M, Hayes ME, Yin JAL, Berry JL, Mawer EB 1994 Abnormal synthesis of 1,25-dihydroxyvitamin D in patients with malignant lymphoma. J Clin Endocrinol Metab **78**:1202–1207.
92. Karmali R, Barker S, Hewison M, Fraher L, Katz DR, O'Riordan JLH 1990 Intermittent hypercalcemia and vitamin D sensitivity in Hodgkin's disease. Postgrad Med J **66**:757–760.
93. Walker IR 1974 Lymphoma with hypercalcemia. Can Med Assoc J **111**:928.
94. Breslau NA, McGuire JL, Zerwekh JE, Frenkel EP, Pak CYC 1984 Hypercalcemia associated with increased serum calcitriol levels in three patients with lymphoma. Ann Intern Med **100**:1.
95. Mudde AH, Van den Berg H, Boshuis PG, Breedveld FC, Markusse HM, Kluin PM, Bijvoet OLM, Papapoulos SE 1987 Ectopic production of 1,25-dihydroxyvitamin D by B-cell lymphoma as a cause of hypercalcemia. Cancer **59**:1543.
96. Rosenthal N, Insogna KL, Godsall JW, Smaldone L, Waldron JA, Stewart AF 1985 Elevations in circulating 1,25-dihydroxyvitamin D in three patients with lymphoma-associated hypercalcemia. J Clin Endocrinol Metab **60**:29.
97. Adams JS, Fernandez M, Gacad MA, Gill PS, Endres DB, Rasheed S, Singer FR 1989 Vitamin D metabolite-mediated hypercalcemia and hypercalciuria patients with AIDS and non-AIDS associated lymphoma. Blood **73**:235–239.
98. Seymour JF, Gagel RF, Hagemeister FB, Dimopoulos MA, Cabanillae F 1994 Calcitriol production in hypercalcaemic and normocalcaemic patients with non-Hodgkin lymphoma. Ann Intern Med **121**:633–640.
99. The Non-Hodgkin's Lymphoma Pathological Classification Project 1982 National Cancer Institute-sponsored study of classifica-

tions of non-Hodgkin's lymphomas: Summary and description of a working formulation for clinical usage. Cancer **49**:2112–2135.

100. Cox M, Haddad JG 1994 Editoral: Lymphoma, hypercalcemia and the sunshine vitamin. Ann Intern Med **21**:709–712.
101. Dodwell DJ, Abbas SK, Morton AR, Howell A 1991 Parathyroid hormone-related protein and response to pamidronate therapy for tumour-induced hypercalcemia. Eur J Cancer **12**:1629.
102. Wada S, Kitamura H, Matsuura Y, Katayama Y, Ohkawa H, Kugai N, Motoyoshi K, Fuse Y, Nagata N 1992 Parathyroid hormone-related protein as a cause of hypercalcemia in a B-cell type malignant lymphoma. Intern Med **31**:968.
103. Fetchick DA, Bertolini DR, Sarin PS, Weintraub ST, Mundy GR, Dunn JF 1986 Production of 1,25-dihydroxyvitamin D_3 by human T cell lymphotrophic virus-1-transformed lymphocytes. J Clin Invest **78**:592–596.
104. Dodd CD, Winkler CF, Williams ME, Bunn PA, Gray TK 1986 Calcitriol levels in hypercalcaemic patients with adult T-cell lymphoma. Arch Intern Med **146**:1971.
105. Fukumoto S, Matsumoto T, Ikeda K, Yamashita T, Watanabe T, Yamaguchi K, Kiyokawa T, Takatsuki K, Shibuya N, Ogata E 1988 Clinical evaluation of calcium metabolism in adult T-cell leukaemia/lymphoma. Arch Intern Med **148**:921.
106. Senba M, Kawai K 1992 Hypercalcemia and production of parathyroid hormone-like protein in adult T-cell leukaemia-lymphoma. Eur J Haematol **48**:278.
107. Motokura T, Fukumoto S, Takahashi S, Matanabe T, Matsumoto T, Igarashi T, Ogata E 1988 Expression of parathyroid hormone-related protein in a human T-cell lymphotrophic virus type-1 infected T-cell line. Biochem Biophys Res Commun **154**:1182–1188.
108. Moseley JM, Danks JA, Grill V, Lister TA, Horton MA 1991 Immunocytochemical demonstration of PTHrP protein in neoplastic tissue of HTLV-1 positive human adult T cell leukaemia-lymphoma—Implication for the mechanism of hypercalcemia. Br J Cancer **64**:745–748.
109. Haddad JG Jr, Rojanasathit S 1976 Acute administration of 25-hydroxycholecalciferol in man. J Clin Endocrinol Metab **33**:992.
110. Kawakami M, Imawari M, Goodwin DS 1979 Quantitative studies of the interaction of cholecalciferol (vitamin D_3) and its metabolites with different genetic variants of the serum binding protein for the sterols. Biochem J **179**:413–423.
111. Adams JS 1989 Vitamin D metabolite mediated hypercalcemia. Endocrinol Metab **18**:765–778.
112. Shepard RM, DeLuca H 1980 Plasma concentrations of vitamin D_3 and its metabolites in the rat as influenced by vitamin D_3 or 25-hydroxyvitamin D_3 intakes. Arch Biochem Biophys **202**:43–53.
113. Vieth R, McCarten K, Norwich KH 1990 Role of 25-hydroxyvitamin D_3 dose in determining rat 1,25-dihydroxyvitamin D_3 production. Am J Physiol **258**:E780–E789.
114. Huches MR, Baylink DJ, Jones PG, Haussler MR 1976 Radioligand receptor assay for 25-hydroxyvitamin D_2/D_3 and 1α,25-dihydroxyvitamin D. J Clin Invest **58**:61–70.
115. McDonnell DP, Mangelsdorf DJ, Pike JW, Haussler MR, O'Malley BW 1987 Molecular cloning of complementary DNA encoding the avian receptor for vitamin D. Science **235**:1214.
116. Reinhart TA, Horst RL, Engstrom GW, Atkins KS 1989 Ketoconazole inhibits self-induced metabolism of 1,25-dihydroxyvitamin D_3 and amplifies 1,25-dihydroxyvitamin D_3 receptor up-regulation in rat osteosarcoma cells. Arch Biochem Biophys **272**:459.
117. Costa EM, Feldman D 1986 Homologous up regulation of $1,25(OH)_2$ vitamin D_3 receptor in rats. Biochem Biophys Res Commun **137**:742.
118. Reinhart TA, Horst RL, Engstrom GW, Atkins KS 1988 Ketoconazole potentiates $1,25(OH)_2D$-directed upregulation of $1,25(OH)_2D$ receptors in rat intestine and bone. Seventh Workshop on Vitamin D, Rancho Mirage, California. In: Norman AW, Schaefer K, Crigoleit HG, Herrath, DV (eds) Vitamin D: Molecular, Cellular and Clinical Endocrinology. de Gruyter, Berlin, p. 233.
119. Goff JP, Reinhar TA, Beckman MJ, Horst RL 1990 Contrasting effects of exogenous 1,25-dihydroxyvitamin D [$1,25(OH)_2D$] versus endogenous $1,25(OH)_2D$ induced by dietary calcium restriction on vitamin receptors. Endocrinology **126**:1031–1035.
120. Reinhart TA, Horst RL 1990 Parathyroid hormone down-regulates 1,25-dihydroxyvitamin D receptors (VDR) and VDR messenger ribonucleic acid in vitro and blocks homologous up-regulation of VDR in vitro. Endocrinology **127**:942–948.
121. Haussler MR, Cordy PE 1982 Metabolites and analogues of vitamin D; which for what? JAMA **247**:841–844.
122. Beckman MJ, Horst RL, Reinhart TA, Beitz DC 1990 Up-regulation of the intestinal 1,25-dihydroxyvitamin D receptor during hypervitaminosis D: A comparison between vitamin D_2 and vitamin D_3. Biochem Biophys Res Commun **169**:910–915.
123. Beckman MJ, Johnson JA, Goff JP, Reinhart TA, Beitz DC, Horst RL 1995 The role of dietary calcium in the physiology of vitamin D toxicity: Excess dietary vitamin D_3 blunts parathyroid hormone induction of kidney 1-hydroxylase. Arch Biochem Biophys **319**:535–539.
124. Haussler MR 1986 Vitamin D receptor: Nature and function. Annu Rev Nutr **6**:527–562.
125. Reinhart TA, Horst RL 1989 Self induction of 1,25-dihydroxyvitamin D_3 metabolism limits receptor occupancy and target tissue responsiveness. J Biol Chem **264**:15917.
126. Shigematsu T, Horiuchi N, Ogura Y, Miyahara T, Suda T 1986 Human parathyroid hormone inhibits renal 24-hydroxylase activity of 25-hydroxyvitamin D_3 by a mechanism involving adenosine 3′,5′-monophosphate in rats. Endocrinology **118**:1583–1589.
127. Beckman MJ, Goff JP, Reinhart TA, Beitz DC, Horst RL 1994 In vivo regulation of rat intestinal 24-hydroxylase: Potential new role of calcitonin. Endocrinology **135**:1951–1955.
128. Mallon JP, Matuszewski D, Sheppard H 1980 Binding specificity of the rat serum vitamin D transport protein. J Steroid Biochem **13**:409.
129. Adams JS 1984 Specific internalization of 1,25-dihydroxyvitamin D_3 by cultured intestinal epithelial cells. J Steroid Biochem **20**:857–862.
130. Bouillon R, VanAssche FA, van Baelen H, Heyns W, DeMoor P 1981 Influence of vitamin D binding protein on the serum concentration of 1,25-dihydroxyvitamin D_3: Significance of the free 1,25-dihydroxyvitamin D_3 concentration. J Clin Invest **67**:589–596.
131. Bikle DD, Siiteri PK, Ryzen E, Haddad JG, Gee E 1985 Serum protein binding of 1,25-dihydroxyvitamin D: A reevaluation by direct measurement of free metabolite levels. J Clin Endocrinol Metab **61**:969.
132. Bikle DD, Gee E, Halloran B, Haddad JG 1984 Free 1,25-dihydroxyvitamin D levels in serum from normal subjects, pregnant subjects and subjects with liver disease. J Clin Invest **74**:1966.
133. Bikle DD, Halloran BP, Gee E, Ryzen E, Haddad JG 1986 Free 25-hydroxyvitamin D levels are normal in subjects with liver disease and reduced total 25-hydroxyvitamin D levels. J Clin Invest **78**:748.
134. Koenig KG, Lindberg JS, Zerwekh JE, Padalino PK, Cushner HM, Copley JB 1992 Free and total 1,25-dihydroxyvitamin D levels in subjects with renal disease. Kidney Int **41**:161–165.
135. Mendel CM 1989 The free hormone hypothesis: A physiologically based mathematical model. Endoc Rev **10**:232–274.

136. Edelson GW, Kleerekoper M 1995 Hypercalcemic crisis. Med Clin North Am **79**:79–92.
137. Nussbaum S 1993 Pathophysiology and management of severe hypercalcemia. Endocrinol Metab Clin North Am **22**:343–361.
138. Bilezikian JP 1988 Hypercalcemia. In: Bone R (ed) Disease-a-Month. Year Book Medical Publ., Chicago, Illinois.
139. Scarpelli DG, Tremblay G, Pearse AG 1960 A comparative cytochemical and cytologic study of vitamin D induced nephrocalcinosis. Am J Pathol **36**:331–353.
140. Shetty KR, Ajiouni K, Rosenfeld PS, Hagen TC 1975 Protracted vitamin D toxicity. Arch Intern Med **135**:986–988.
141. Rizzoli R, Stoermann C, Ammann P, Bonjour JP 1994 Hypercalcemia and hyperosteolysis in vitamin D toxicity: Effects of clodronate therapy. Bone **15**:193–198.
142. Hefti E, Trechsel U, Fleisch H, Bonjour JP 1983 Nature of calcemic effect of 1,25-dihydroxyvitamin D_3 in experimental hypoparathyroidism. Am J Physiol **244**:E313–E316.
143. Hass G, Truehart R, Taylor B, Stumpe M 1958 An experimental histologic study of hypervitaminosis D. Am J Pathol **3**:395–422.
144. Swierczynski J, Nagel G, Zydowo M 1987 Calcium content in some organs of rats treated with a toxic calciol dosis. Pharmacology **34**:57–60.
145. Taylor A, NormaGlass AR, Eil C 1988 Ketoconazole-induced reduction in serum 1,25-dihydroxyvitamin D and total serum calcium in hypercalcaemic patients. J Clin Endocrinol Metab **66**:934–938.
146. Feher JJ, Wasserman RH 1979 Intestinal calcium binding calcium protein and calcium absorption in cortisol treated chicks: Effects of vitamin D and 1,25-dihydroxyvitamin D_3. Endocrinology **104**:547–551.
147. Finding JW, Adams ND, Lemann J, et al. 1981 Vitamin D metabolites and parathyroid hormone in Cushing's syndrome: Relationship to calcium and phosphorus homeostasis. J Clin Endocrinol Metab **54**:1039.
148. Suzuki Y, Ichikawa Y, Saito E, Homm M 1983 Importance of increased urinary calcium excretion in the development of secondary hyperparathyroidism of patients under glucocorticoid therapy. Metabolism **32**:151–156.
149. Lukert BP, Raisz LG 1990 Glucocorticoid induced osteoporosis: Pathogenesis and management. Ann Intern Med **112**:352–364.
150. Carre M, Ayigbede O, Miravet L, Rasmussen H 1974 The effect of prednisolone upon the metabolism and action of 25-hydroxy- and 1,25-dihydroxyvitamin D_3. Proc Natl Acad Sci USA **1**:2996–3000.
151. Bilezikain JP 1989 Etiologies and therapies of hypercalcemia. Endocrinol Metab Clin North Am **18**(No. 2):389–413.
152. O'Leary TJ, Jones G, Lohnes D, Cohanim M, Yendt ER 1986 The effects of chloroquine on serum 1,25-dihydroxyvitamin D and calcium metabolism in sarcoidosis. N Engl J Med **315**:727–730.
153. Glass AR, Eil C 1988 Ketoconazole induced reduction in serum 1,25-dihydroxyvitamin D and total serum calcium in hypercalcemic patients. J Clin Endocrinol Metab **66**:934–938.
154. Adams JS, Sharma OP, Diz M, Endres D 1989 Ketoconazole decreases the serum 1,25-dihydroxyvitamin D and calcium concentration in sarcoidosis associated hypercalcemia. J Clin Endocrinol Metab **70**:1090–1095.

CHAPTER 55

Extrarenal Production and Action of Active Vitamin D Metabolites in Human Lymphoproliferative Diseases

JOHN S. ADAMS University of California, Los Angeles, School of Medicine, Burns and Allen Research Institute, Cedars-Sinai Medical Center, Los Angeles, California

I. INTRODUCTION

In this chapter we address the pathophysiology of dysregulated vitamin D metabolism that occurs in some patients with granuloma-forming and malignant lymphoproliferative disorders. After this brief introduction, the second section of the chapter reviews the historical aspects of this clinical problem. The third section of the chapter describes what we know about the mechanics and regulation of the vitamin D metabolizing enzymes present in inflammatory cells. The fourth section recapitulates the potential immunological targets for active vitamin D metabolites and proposes a model in which macrophage-derived vitamin D metabolites play a role in the modulation of the local immune response in these diseases. The fifth section provides a comprehensive review of the various human diseases proposed to be associated with the overproduction of active vitamin D metabolites from an extrarenal source. The sixth and final section of this chapter addresses the clinical aspects of disordered calcium homeostasis; this includes a discussion of the diagnosis, treatment, and prevention of hypercalcemia and hypercalciuria in the patient with endogenous vitamin D intoxication.

II. VITAMIN D AND GRANULOMA-FORMING DISEASE: HISTORICAL PERSPECTIVE

A. Evidence of Endogenous Vitamin D Intoxication Associated with Sarcoidosis

A pathophysiological relationship between vitamin D and sarcoidosis was first recognized by Harrell and Fisher in 1939 [1]. Among the six hypercalcemic patients in their initial report, one was observed to experience a steep rise in the serum calcium concentration following ingestion of cod liver oil, known to be enriched in vitamin D. Almost two decades passed before Henneman *et al.* [2] demonstrated in 1956 that the hypercalcemic syndrome of sarcoidosis, characterized by increased intestinal calcium absorption and bone resorption, was remarkably similar to that of exogenous vitamin D intoxication and was treatable by the administration of glucocorticoid. In 1963, Taylor and co-workers [3] performed the first, large-scale seasonal evaluation of serum calcium levels in patients with sarcoidosis. They found that there was a significant increase in the mean serum calcium concentration in 345 patients with sarcoidosis from winter to summer but no such change in over 12,000 control subjects. This was the first evidence that there was an association between enhanced cutaneous vitamin D synthesis, known to occur principally during the summer months, and the blood level of calcium in patients with sarcoidosis. This observation was prospectively confirmed by Dent *et al.* [4], who was able to increase the serum calcium concentration in patients with active sarcoidosis on exposure to whole body ultraviolet radiation. The Dent study also helped validate the earlier work of Hendrix [5], who achieved resolution of hypercalcemia and hypercalciuria in two patients with sarcoidosis by institution of vitamin D-deficient diets and environmental sunlight deprivation.

B. Evidence for the Extrarenal Overproduction of an Active Vitamin D Metabolite

The above mentioned studies led Bell *et al.* [6] to propose in 1964 that development of a clinical abnormality in calcium balance in patients with active sarcoidosis resulted from an increase in target organ responsiveness to vitamin D. This view persisted for more than a decade. However, after the discovery of 1,25-dihydroxyvitamin D [$1,25(OH)_2D$] as the active vitamin D hormone [7,8] and the development of sensitive and specific assays for the hormone in blood [9–13], investigators were quick to determine that the hypercalcemia of sarcoidosis was the result of an increase in the circulating concentrations of a vitamin D metabolite that interacted with the vitamin D receptor (VDR) [14–16]. The fact that a vitamin D hormone was made outside of the kidney in hypercalcemic patients with sarcoidosis was first discovered by Barbour and colleagues in 1981 [17]. These investigators reported high concentrations of a vitamin D metabolite detected as $1,25(OH)_2D$ in the circulation of a hypercalcemic, anephric patient with active sarcoidosis. Two years later, Adams *et al.* [18] determined the macrophage to be the extrarenal source of this active vitamin D metabolite. Unequivocal structural characterization of the metabolite as $1,25(OH)_2D$ was obtained by the same investigators in 1985 [19].

III. PATHOPHYSIOLOGY OF DISORDERED CALCIUM BALANCE IN SARCOIDOSIS: A MODEL FOR THE EXTRARENAL PRODUCTION OF AN ACTIVE VITAMIN D METABOLITE IN HUMAN DISEASE

A. Clinical Evidence for Dysregulated Overproduction of the Vitamin D Hormone

As has been described in detail in earlier chapters, the synthesis of $1,25(OH)_2D$ by the renal vitamin D 1-hydroxylase (1-hydroxylase) is normally strictly regulated, with levels of the hormone product $1,25(OH)_2D$ being some 1000-fold less plentiful in the circulation than that of the principal substrate for the enzyme, 25-hydroxyvitamin D (25OHD). Hormone synthesis in the kidney is stimulated by an increase in the serum parathyroid hormone (PTH) concentration, a decrease in the serum phosphate concentration, and a decrease in the activity of the competing vitamin D 24-hydroxylase. $1,25(OH)_2D$ synthesis is inhibited by a decrease in the level of circulating PTH, an increase in the serum phosphate concentration, and an increase in the activity of the vitamin D 24-hydroxylase. There are now at least four clear lines of clinical evidence to indicate that endogenous $1,25(OH)_2D$ production in hypercalcemic/hypercalciuric patients with sarcoidosis is dysregulated and not bound by the same set of endocrine factors known to regulate $1,25(OH)_2D$ synthesis in the kidney.

First, hypercalcemic patients with sarcoidosis possess a frankly high or inappropriately elevated serum $1,25(OH)_2D$ concentration, although their serum PTH level is suppressed and their serum calcium and phosphate concentration is relatively elevated [20,21]. If $1,25(OH)_2D$ synthesis were under the regulation of PTH, phosphate, and $1,25(OH)_2D$ itself, then $1,25(OH)_2D$ concentrations in such patients should be low.

Second, the serum $1,25(OH)_2D$ concentration in patients with active sarcoidosis is very sensitive to an increase in available substrate [22], whereas the serum $1,25(OH)_2D$ level in normal individuals is not influenced by small or even moderate increments in the circulating 25OHD concentration. Clinically, this aspect of dysregulation is manifest by the long-recognized association of the appearance of hypercalciuria and/or hypercalcemia in sarcoidosis patients in the summer months [23] or following holidays to geographic locations at lower latitudes than those at which the patient normally resides [24]. This link between an increase in cutaneous vitamin D synthesis and the development of a clinical abnormality in calcium balance can be replicated by the oral administration of vitamin D_3 [16] to such patients. It can also be substantiated on a biochemical basis by demonstration of a positive correlation between the serum 25OHD and $1,25(OH)_2D$ concentrations in patients with active sarcoidosis but not in normal human subjects [21].

Third, the rate of endogenous $1,25(OH)_2D$ production, which is significantly increased in patients with sarcoidosis [25], is unusually sensitive to inhibition by factors (i.e., drugs) that do not influence the renal vitamin D 1-hydroxylase at the same doses. Anti-inflammatory concentrations of glucocorticoids have long been recognized as effective combatants of sarcoidosis-associated hypercalcemia [26,27] and more recently have been shown to dramatically lower elevated $1,25(OH)_2D$ levels [21]. On the other hand, administration of the same glucocorticoid doses to patients without sarcoidosis is not associated with a clinically relevant reduction in the serum $1,25(OH)_2D$ or calcium concentration. Chloroquine and its hydroxylated analog hydroxychloroquine are other examples of pharmaceutical agents that appear to act preferentially on the extrarenal vitamin D 1-hydroxylation reaction which is active in patients with sarcoidosis [28–30].

Fourth, the serum calcium and $1,25(OH)_2D$ concentrations are positively correlated to indices of disease activity in patients with sarcoidosis [31–33]. Patients with widespread disease and high angiotensin-converting enzyme (ACE) activity are more likely to be hypercalciuric or frankly hypercalcemic.

B. Correlates *in Vitro* for Dysregulated Vitamin D Hormone *in Vivo*

Investigators have now generated a substantial body of experimental data from cells, including inflammatory cells harvested directly from patients with sarcoidosis, to indicate that the dysregulated vitamin D hormone synthesis in sarcoidosis is due to expression of a 1-hydroxylase that is not different from the renal 1-hydroxylase; rather, expression of the authentic 1-hydroxylase occurs in a macrophage, not a kidney cell. In fact, each of the above-mentioned pieces of clinical evidence for dysregulated vitamin D hormone production in this disease can be borne out *in vitro* in cells from patients with this disease [34].

1. Subcellular Localization, Substrate Selectivity, and Kinetics of the Macrophage Vitamin D 1-Hydroxylase

As is the case with the bona fide 1-hydroxylase of renal origin (see Chapter 5), the macrophage enzyme is a mixed function oxidase confined to mitochondria and associated with detectable cytochrome P450 activity [35] (Fig. 1). Like the renal 1-hydroxylase reconstituted from mitochondrial extracts, the presence of a flavoprotein, ferredoxin reductase, an electron source, and molecular oxygen (O_2) is required for electron transfer to the cytochrome P450 and for the insertion of an oxygen atom in the substrate [35]. Also like the renal 1-hydroxylase, we now know the macrophage 1-hydroxylase is inhibited by the naphthoquinones, molecules which compete with reductase for donated electrons, and by the imidazoles, compounds which compete with the enzyme for receipt of O_2 [36]. Similar to the 1-hydroxylase isolated from the mitochondria of proximal renal tubular epithelial cells, the macrophage 1-hydroxylase requires a secosterol (vitamin D sterol molecule with an open B ring) as substrate [34]. Also similar to the renal 1-hydroxylase, the macrophage enzyme has a particular affinity for secosterols bearing a carbon-25 hydroxyl group as is encountered in the two preferred substrates for this enzyme, 25OHD and 24,25-dihydroxyvitamin D [$24,25(OH)_2D$] [34,37]; the calculated K_m (affinity) of

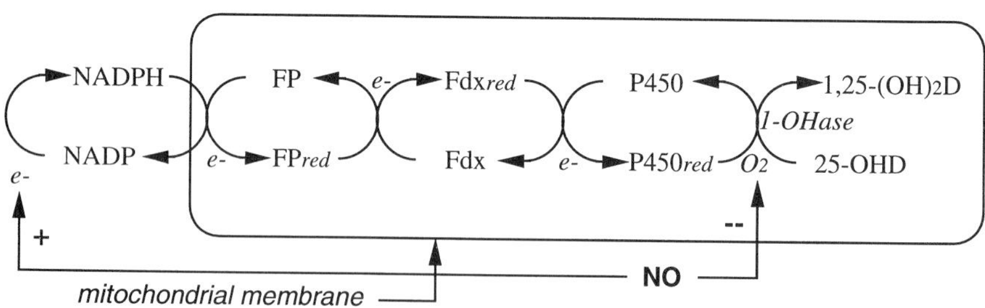

FIGURE 1 Model for the biphasic effect of nitric oxide (NO) on the mitochondrial vitamin D1-hydroxylase in macrophages. A stimulatory effect on the enzyme is mediated by relatively low intracellular NO levels. An electron (*e*-) generated from NO is donated to oxidized NADP+, thus forming NADPH. NADPH, in turn, supplies the electron transport chain of accessory proteins, consisting of a flavoprotein reductase (FP), a ferredoxin (Fdx), and a cytochrome P450, linked to the 1-hydroxylase. On the other hand, the inhibitory effect on the enzyme which occurs at relatively high NO levels in the cell results from competition with O_2 for binding to the P450 heme group, inhibiting the enzyme.

the 1-hydroxylase in pulmonary alveolar macrophages derived directly from patients with active sarcoidosis is in the range of 50–100 n*M* for these two substrates [34,37]. Although neither the cDNA of the 1-hydroxylase of renal origin nor that of macrophage origin has been cloned and expressed, every indication is that the two enzymes are identical. If they are identical, or at the very least closely related, then the dramatic differences in expression of the renal and macrophage 1-hydroxylase *in vivo* must result from the cell in which the enzyme is expressed and the distinct regulatory factors and pathways which impinge on the mitochondrial 1-hydroxylase in these two cell types. As is depicted in Fig. 2 and described below, this is likely to be the case.

2. Macrophages Lack Responsiveness to PTH, Calcium, and Phosphate

In vivo there appear to be three major regulators of the renal 1-hydroxylase, namely, the serum concentrations of calcium, PTH, and phosphate [38] (Fig. 2, left). Hypocalcemia enhances the activity of the renal mitochondrial 1-hydroxylase; however, much of this stimulatory effect may be indirectly mediated through PTH [39]. Any decrease in the serum calcium concentration

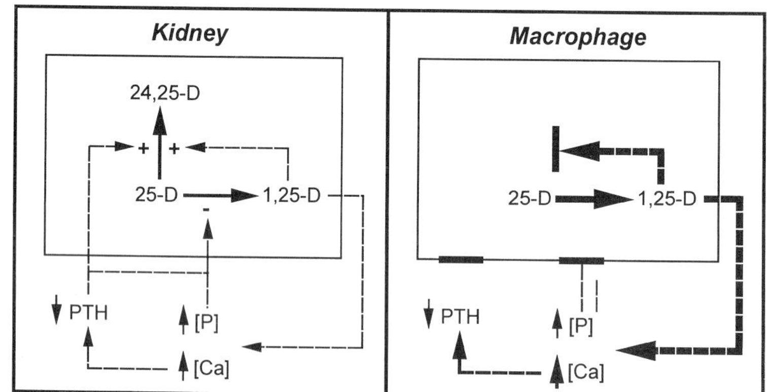

FIGURE 2 Model distinguishing the regulation of the vitamin D 1-hydroxylase in the proximal renal tubular epithelial cell of the kidney (left) and in the granuloma-forming disease-activated macrophage (right). In the kidney the enzymatic conversion of substrate 25-hydroxyvitamin D (25-D) to product 1,25-dihydroxyvitamin D (1,25-D) is subject to negative feedback control with down-regulation of enzyme activity under the influence of (1) a calcium-mediated decrease in the level of circulating parathyroid hormone (PTH), (2) a $1,25(OH)_2D$-mediated increase in the serum phosphate level [P]; and (3) a $1,25(OH)_2D$-mediated increase in vitamin D 24-hydroxylase activity. The macrophage lacks responsiveness to changes in the extracellular [PTH] and [P] and harbors little or no detectable vitamin D 24-hydroxylase.

below normal is a stimulus for increased secretion of PTH [40] which, in turn, is a direct stimulator of the renal 1-hydroxylase [41]. It has been suggested that interaction of PTH with its receptor on cells in the proximal convoluted tubule transduces its stimulatory signal through both the protein kinase A and protein kinase C pathways [42,43], effecting transient changes in the phosphorylation status of the ferredoxin which contributes electrons to the 1-hydroxylase enzyme [44]. A change in the serum phosphate concentration is the other major regulator of renal $1,25(OH)_2D$ production; in adult humans dietary phosphorus restriction causes an increase in circulating concentrations of $1,25(OH)_2D$ to 80% above control values, an increase not due to accelerated metabolic clearance of this hormone [45]. Dietary phosphorus supplementation will have the opposite effect. Although the mechanism by which a drop in the serum phosphate level will increase renal $1,25(OH)_2D$ production remains uncertain, there is no doubt that there exists a concerted, cooperative attempt of the calcium–phosphorus–PTH axis in humans to regulate the conversion of 25OHD to $1,25(OH)_2D$ in the kidney. For example, a drop in the serum calcium concentration will be immediately registered by the parathyroid cell calcium receptor, which will release its inhibition on PTH production and secretion. An increase in circulating PTH will directly stimulate the renal 1-hydroxylase, and a PTH-mediated phosphaturic response and a subsequent decrement in the serum phosphate level will indirectly promote $1,25(OH)_2D$ production.

The macrophage 1-hydroxylase, on the other hand, is immune to the stimulatory effects of PTH and phosphate [36,46] (Fig. 2, right). The macrophage plasma membrane is not enriched with PTH receptors, and there is no evidence that any PTH receptors which are resident in the macrophage membrane are responsive to PTH or PTH-related peptide (PTHrP) in terms of stimulating the protein kinase signaling pathways that are associated with stimulation of the renal 1-hydroxylase [36]. Similarly, the macrophage enzyme appears to be uninfluenced by changes in the extracellular phosphate concentration [36]. Moreover, exposure of activated macrophages expressing the 1-hydroxylase to a calcium ionophore stimulates the hydroxylation reaction [34], whereas increasing the extracellular calcium concentration will have the opposite, inhibitory effect on the renal 1-hydroxylase [38]. These observations appear to confirm the fact that the three most important extracellular signaling systems for the renal 1-hydroxylase, calcium, phosphate, and PTH, are not heeded by the macrophage enzyme and provide an explanation for why $1,25(OH)_2D$ production by the macrophage in diseases like sarcoidosis is not subject to negative feedback control by a drop in the serum PTH concentration and an increase in the circulating calcium and phosphate level. Furthermore, with the possible exception of insulin-like growth factor-I (IGF-I) [47], there is no evidence that the macrophage 1-hydroxylation reaction is influenced by any of the other endocrine factors, including estrogen, prolactin, and growth hormone, purported to increase the renal production of $1,25(OH)_2D$ [48–51].

3. Macrophages Lack $1,25(OH)_2D$-Directed Vitamin D 24-Hydroxylase Activity

The other major contributor to the circulating $1,25(OH)_2D$ level is the activity of the vitamin D 24-hydroxylase. Like the 1-hydroxylase, 24-hydroxylase is a heme-binding mitochondrial enzyme requiring NADPH, molecular oxygen, and magnesium ions [44]. The cDNA and part of the gene for the human, rat, and chicken enzyme, now referred to as P450cc24, have been cloned [52–55] (Chapter 6). As depicted in Fig. 2 (left), expression of P450cc24 is stimulated in kidney cells by $1,25(OH)_2D$, especially if the protein kinase C (PKC) pathway is also up-regulated [56,57]. PTH appears to exert an opposite, inhibitory effect on P450cc24 gene transcription [55] and $24,25(OH)_2D$ synthesis [58,59].

There is dual impact of this mitochondrial, cytochrome P450-linked enzyme system on vitamin D and calcium balance in adult animals including humans. Because P450cc24 is coexpressed in the kidney along with the vitamin D 1-hydroxylase, the first point of impact is on regulation of substrate 25OHD available to the 1-hydroxylase. Like the 1-hydroxylase, the 24-hydroxylase exhibits a preference for 25-hydroxylated secosterol substrates [60]. Although its affinity for 25OHD is reported to be somewhat less than that of renal 1-hydroxylase, its capacity for substrate is substantially greater [44]. Hence, when up-regulated under the influence of circulating or locally produced $1,25(OH)_2D$ or diminished serum PTH levels, the 24-hydroxylase has the capacity to compete with the 1-hydroxylase for substrate 25OHD. Under physiological circumstances, this state of competitive substrate deprivation for the 1-hydroxylase will persist until the serum calcium and PTH concentration are normalized. The second point of impact of the 24-hydroxylase on the circulating $1,25(OH)_2D$ concentration is at the level of catabolism of the $1,25(OH)_2D$ hormone. The affinity of the 24-hydroxylase for $1,25(OH)_2D$ is as great as it is for 25OHD. Considering the fact that the 24-hydroxylase acts as the initial step in the conversion of $1,25(OH)_2D$ to nonbiologically active, water-soluble, excretable metabolites of the hormone, up-regulation of this enzyme will contribute to the lowering of $1,25(OH)_2D$ hormone levels. The macrophage lacks detectable 24-hydroxylase activity (Fig. 2, right) [34]; therefore, this cell, in contrast to the renal tubular epithelial cell, does not possess the capability of shunting substrate 25OHD and 1-hydroxylase product 1,25$(OH)_2D$ down the catabolic 24-hydroxylase pathway. The net result is dysregulated overproduction of 1,25$(OH)_2D$ by the macrophage, escape of the hormone into the general circulation, and the eventual development of hypercalcemia.

4. Macrophage Vitamin D 1-Hydroxylase Exhibits Responsiveness to Immune Cell Regulators

The lack of negative feedback control on the 1-hydroxylase expressed in the macrophage as just described can account for the failure to appropriately inhibit 1,25$(OH)_2D$ synthesis in inflammatory diseases like sarcoidosis. However, this does not adequately explain the fact that 1,25$(OH)_2D$ production rates are increased well above normal in patients with these diseases at a time when the renal 1-hydroxylase is inhibited. This observation suggests that there must be an alternative, "nontraditional" set of factors which stimulate the synthesis of 1,25$(OH)_2D$ by the macrophage but not by the kidney. This is, in fact, a reality.

a. Cytokines Clinical observations from a number of investigative groups around the world indicate that sarcoidosis patients with diffuse, infiltrative pulmonary disease are at greater risk to develop dysregulated vitamin D metabolism. Cultured pulmonary alveolar macrophages (PAM) from such patients were more likely to synthesize more 1,25$(OH)_2D$ *in vitro* on a per cell basis than PAM from a host with less intense or no alveolitis [61,62]. These results led to the conclusion that the specific activity of the 1-hydroxylase reaction in macrophages from patients with active pulmonary sarcoidosis was regulated by endogenously synthesized factors which also modulated the intensity of the host immune response. Of the various bioactive cytokines concentrated in the alveolar space of patients with active sarcoidosis [63], interferon-γ (IFN-γ) was found to be the only potent cytokine stimulator of the sarcoid macrophage 1-hydroxylation reaction [64]; by itself at maximally effective concentrations *in vitro* IFN-γ increased basal hydroxylase activity over 4-fold. Although not a cytokine per se, calcium ionophore was determined to be another stimulator of the macrophage 1-hydroxylase in these same studies [64]. These data suggested that a calcium-dependent, IFN-γ-receptor-linked signal transduction pathway was responsible for stimulation of the reaction in the macrophage.

At the time these initial observations were made, the IFN receptor was known to be linked to two such calcium-dependent pathways in the macrophage, the protein kinase C (PKC) [65] and phospholipase A_2 (PLA_2) pathway [66,67]. Because the macrophage 1-hydroxylase was not influenced by attempts to directly stimulate or inhibit PKC, attention was focused on the PLA_2 pathway and the endogenous arachidonic acid metabolic cascade as the signal transduction pathway of most influence over the macrophage enzyme. Further dissection of the intracellular arachidonate metabolic pathway in this cell demonstrated that signal transduction through the 5-lipoxygenase pathway, specifically with the generation of leukotriene C_4 (LTC_4), was most critical to an increase in 1,25$(OH)_2D$ synthesis [64]. These studies were extended to investigate another compound with potential actions in the PLA_2–arachidonic acid pathway, the 4-aminoquinoline derivative chloroquine. 1,25$(OH)_2D$ synthesis by macrophages was completely inhibited by exposure to 10^{-6} *M* chloroquine *in vitro* [30]. Furthermore, this effect is independent of the apparent ability of chloroquine to alter the pH of intracellular organelles. When given orally to a hypercalcemic patient with sarcoidosis, chloroquine [28,30] or its analog hydroxychloroquine [29] can effectively reduce the serum 1,25$(OH)_2D$ and calcium concentration within a matter of 36 hr.

b. Lipopolysaccharide Lipopolysaccharide (LPS) is a bioactive lipid extractable from the cell wall of infectious microorganisms including mycobacteria. LPS interacts with the cell surface CD14 receptor on the macrophage [68], stimulating the cell. Like the IFN-γ–IFN receptor interaction, the LPS–CD14 interaction is a potent inducer of the 1-hydroxylase in monocyte–macrophage-like cells [36,69] (Fig. 3). In fact, the most reproducibly effective stimulation of the macrophage 1-hydroxylase *in vitro* is achieved by coexposure of macrophages to IFN-γ and LPS [46]. This fact and the knowledge that IFN-γ and LPS are also the two most effective stimulators of nitric oxide (NO) synthesis in macrophages [69,70] led Adams and co-workers to hypothesize that production of NO and 1,25$(OH)_2D$ in macrophage-like cells may be functionally linked [71,72] (see Fig. 3).

c. Nitric Oxide The generation of nitric oxide (NO) in the macrophage is under the control of the enzyme inducible nitric oxide synthase (iNOS) [73]. In contrast to the more stringently regulated, constitutively expressed isoforms of the enzyme (cNOS), which are localized to endothelial cells and neurons, are regulated by calcium, and are capable of producing only modest amounts of NO, the calcium-independent iNOS remains tonically active when "induced" and is capable of generating large quantities of NO in and around the cell. The preferred substrate for all isoforms of NOS, including iNOS, is the semiessential amino acid L-arginine, and

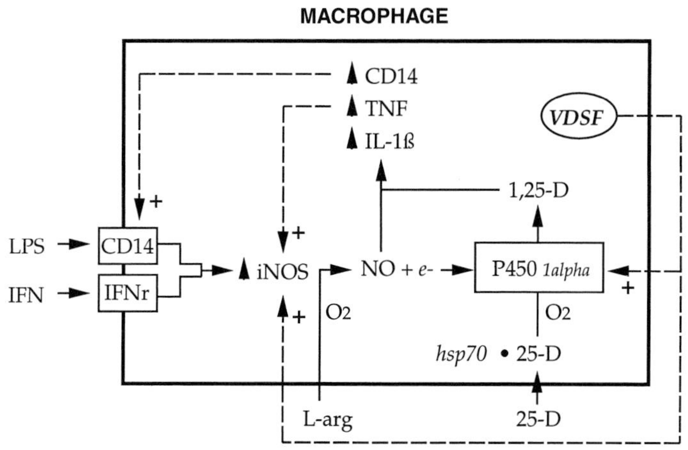

FIGURE 3 Proposed mechanism for the coregulated, autoamplified expression of the macrophage inducible nitric oxide synthase (iNOS) and vitamin D 1-hydroxylase. Interaction of the macrophage stimulatory agents lipopolysacchride (LPS) and interferon-γ (IFN) with their specific cell-surface receptor molecules up-regulates expression of the iNOS gene. Nitric oxide (NO) is synthesized from an extracellular source of molecular oxygen (O_2) and L-arginine (L-arg). NO can serve as an electron-donating source for enzymatic conversion of 25-hydroxyvitamin D (25-D) to 1,25-dihydroxyvitamin D (1,25-D). NO and 1,25$(OH)_2$D act in an intracrine fashion to up-regulate expression of the cytokines interleukin-1β (IL-1β), tumor necrosis factor-α (TNF), and the LPS receptor molecule (CD14); TNF and CD14 promote intracrine stimulation, whereas a macrophage-specific autostimulator (VDSF) promotes autocrine stimulation of NO and 1,25-D synthesis. hsp70, Heat-shock protein 70 or intracellular vitamin D binding protein; e^-, electron; *VDSF,* vitamin D stimulatory factor; P450 *1alpha,* vitamin D 1-hydroxylase.

the principal products of the reaction are citrulline and NO. NO is unique among signaling molecules in that most of its regulatory function is mediated by direct interaction (binding) of this small, highly diffusable molecule with metal centers of larger, more complex molecules (i.e., enzymes) [74]. It is of extreme interest that two of the major stimulators of the human macrophage 1-hydroxylase, IFN-γ and LPS [34,69], are also major regulators of the macrophage iNOS [73,75], itself a heme-containing, cytochrome P450-linked oxidase [74,76]. Interaction of IFN-γ and LPS with their cognate cell surface receptors induces the production of two, cytokine-specific transactivating proteins, IFN-γ regulatory factor-1 (IRF-1) and an LPS-stimulated protein, which bind to specific cis-acting response elements in the iNOS promoter and promote transcription of the iNOS gene [77].

Full expression of the macrophage 1-hydroxylase, like the renal 1-hydroxylase [35,36], is dependent on a source of electrons and molecular oxygen (O_2) (Fig. 1). Although O_2 is freely available to cells in the aerobic environment of oxygenated tissues, a source of electrons must be furnished by a biological intermediate like NADPH. One such potential electron source in the macrophage is NO [76,78]; the NO molecule contains an unpaired electron in its outer shell. NO is freely diffusible across intracellular membranes, including the mitochondrial membrane which houses the macrophage 1-hydroxylase, so it becomes a soluble source of electrons that can be donated to an electron-requiring, mixed function oxidase like the mitochondrial-based 1-hydroxylase. As an initial test of the hypothesis that NO and 1,25$(OH)_2$D synthesis are functionally linked in macrophage-like cells, Adams and colleagues [71] examined the effects of excluding L-arginine, the substrate for endogenous NO production, on expression of the macrophage 1-hydroxylation reaction *in vitro.* L-Arginine deprivation led to a time-dependent decrease in basal as well as IFN-γ-stimulated NO and 1,25$(OH)_2$D production, whereas reintroduction of L-arginine or an iNOS-independent source of intracellular NO into the extracellular medium restored fully ("rescued") 1,25$(OH)_2$D synthetic capacity. Inhibition of iNOS with known competitive inhibitors of the enzyme also significantly inhibited 1,25$(OH)_2$D synthesis. These results confirm that NO generating capacity is indeed functionally linked to the enzymatic synthesis of 1,25$(OH)_2$D.

However, as the amount of NO generated inside the macrophage continues to increase there is a reflex turndown in 1,25$(OH)_2$D production [79] (see Fig. 1), suggesting that there is some kind of a built-in limit on the ability of the cell to produce 1,25$(OH)_2$D. This inhibitory effect of NO on the macrophage 1-hydroxylase is almost certainly due to competition of NO with O_2 for binding to the heme center of the enzyme. A similar effect has been demonstrated for a number of heme-containing enzymes [80–84], including those involved in steroid hormone metabolism [85]. In summary, it appears that NO exerts a biphasic effect on the macrophage 1-hydroxylase. The stimulatory effect on the 1-hydroxylase is mediated by relatively low intracellular NO levels and is initiated by the donation of an electron (*e*-) from NO to oxidized NADP+. This results in the formation of NADPH which, in turn, supplies the electron transport chain linked to the 1-hydroxylase. On the other hand, the inhibitory effect of relatively high NO levels in the cell results from competition of NO with O_2 for binding to the cytochrome P450-linked hydroxylase. The interaction of NO with the heme center is analogous to how carbon monoxide inhibits P450-linked oxidative enzymes.

d. Autocrine Stimulator(s) of the Macrophage 1-Hydroxylase A curious and unexplained observation is the fact that cultured tissue macrophages harvested from patients with active granuloma-forming disease will continue to produce 1,25$(OH)_2$D *in vitro* for days and even weeks [18,19] after removal from the host and long after physical separation from stimulatory cytokines like IFN-γ or inflammatory mediators like LPS that may be circulating in the blood or be present in the local inflammatory microenvironment of the lung. This suggests that these macrophages may be producing an autocrine factor which sustains expression of the 1-hydroxylase *in vitro*. There is now evidence [72] that such a vitamin D stimulatory factor (VDSF) does indeed exist and that its ability to promote 1,25$(OH)_2$D production may be mediated, at least in part, through the generation of NO (see Fig. 3). Incubation of macrophage-like cells with autologous, conditioned medium results in an increase in NO and 1,25$(OH)_2$D synthesis [72]. The increase in NO production under the influence of this factor is mediated by an increase in transcription of the iNOS gene. Although not yet characterized, this autoregulator appears to be a non-endotoxin-related, low molecular mass protein (3000–10,000 kDa) that is exported from the macrophage before interacting with the macrophage cell membrane and stimulating an increase in intracellular NO synthesis [72]. In turn, NO serves as a soluble source of electrons for the NADPH-dependent, mitochondrial, cytochrome P450-associated vitamin D 1-hydroxylase.

e. Heat-shock 70 Proteins Another potential, but not yet proven, autoregulator of macrophage 1,25-$(OH)_2$D synthesis is the stress-induced heat-shock protein 70 (hsp70) family [86]. These proteins are ubiquitously distributed in the nuclear, cytoplasmic, mitochondrial, and endoplasmic reticular compartments of eukaryotic cells. The hsp70 proteins were first recognized as heat-shock-responsive ATP-binding proteins with ATPase activity [87]. They are characterized functionally by their ability to bind and release hydrophobic segments of an unfolded polypeptide chain in an ATP-hydrolytic reaction cycle [86]. This so-called chaperone function of hsp70 is critical for a number of intracellular protein–protein interactions [88]. The hsp70 proteins also are important in the targeting and translocation of proteins across the endoplasmic and mitochondrial membranes [89].

In macrophage-like cells hsp70 expression is known to be induced by both physical (i.e., heat) and cytokine (i.e., IFN-γ) stimuli, and its expression is dramatically enhanced by the vitamin D hormone 1,25$(OH)_2$D [90]. Proteins in the hsp70 family have been shown to be high capacity intracellular binding proteins for 25-hydroxylated vitamin D metabolites [183]. As depicted schematically in Fig. 3, it is possible that up-regulated hsp70 expression in disease-activated macrophages is another key factor in the increased synthesis of 1,25$(OH)_2$D by these cells. Because of its capacity to bind 25OHD, hsp70 may serve to concentrate substrate for the 1-hydroxylase from the general circulation on a relatively low affinity, high capacity binding protein. In fact, by virtue of their organelle-targeting sequences [89], hsp70 proteins or related molecules may be critical in the directed translocation of 25OHD to the inner mitochondrial membrane where the 1-hydroxylase actually resides.

Because of the apparent absence of an accompanying 24-hydroxylase and the possible presence of a facilitating hsp70 25OHD binding protein, the enzymatic product of the macrophage 1-hydroxylase is available at high intracellular concentrations. As such, it is likely that 1,25$(OH)_2$D is present at high enough concentration in the cell to modify, in an intracrine mode, the induction or enhanced transcription of genes characteristic of the activated macrophage phenotype [i.e., CD14, interleukin-1β (IL-1β), and tumor necrosis factor (TNF)] (Fig. 3); considering that none of these genes contains a consensus vitamin D response element (VDRE) in their promoter, it is likely that an intermediary, like the immediate early gene Egr-1, orchestrates the 1,25$(OH)_2$D-directed [91] and perhaps NO-mediated [92] control of

CD14, TNF, and IL-1β expression. Finally, this model suggests that there exists at least two intracrine, positive feedback loops to amplify macrophage reactivity to specific antigens and inflammatory stimulators under the influence of $1,25(OH)_2D$: (1) TNF stimulation of iNOS [93] and (2) induction of CD14 gene expression [94].

IV. LOCAL IMMUNOREGULATORY EFFECTS OF ACTIVE VITAMIN D METABOLITES

A. Intracrine/Autocrine Action on the Monocyte–Macrophage

What is the purpose of $1,25(OH)_2D$ production in patients with granuloma-forming diseases like sarcoidosis? Is local synthesis of the hormone by macrophages beneficial or detrimental to the host? These are important questions that investigators in a number of centers around the world have been addressing since the 1980s when it became known that the activated, circulating monocyte and tissue macrophage expressed the VDR [95,96]. Expression of the receptor for the active vitamin D hormone indicated that the macrophage could actually be a target for the $1,25(OH)_2D$ that the cell itself was making. Indeed, investigators have suggested that $1,25(OH)_2D$ has the potential to interact with the monocyte–macrophage in either an intracrine or autocrine mode [97–99] (Fig. 4, left). For example, incubation with a VDR-saturating concentration of $1,25(OH)_2D$ increases IL-1β expression by 8-fold and decreases by 1000-fold the concentration of stimulator lipopolysaccharide (LPS) required to achieve maximal IL-1β gene expression [99]. This extraordinary priming effect of the sterol for LPS stimulation of the IL-1β gene can also be observed for another monokine gene product, TNF [94], and is due to induction by $1,25(OH)_2D$ of the CD14 gene, the high-affinity receptor for LPS.

The multiplicity and complexity of actions of $1,25(OH)_2D$ on the macrophage are detailed in Chapter 29. Suffice it to say that it is now widely accepted that actions of the hormone are directed toward stimulation of macrophage function. For example, $1,25(OH)_2D$ is known to enhance giant cell formation [100], monokine production [101–103], antigen processing [104], and cytotoxic function [105,106]. These functional consequences of hormone action indicate that elaboration of $1,25(OH)_2D$ by activated macrophages in human diseases like sarcoidosis is important in modulation of the local cellular immune response to the granuloma causing antigen, promoting antigen processing, containment, and destruction [107]. The system is designed for maximal efficiency in that the hormone interacts with the VDR in the same cell in which the hormone is made. Thus, relatively high intracellular and local concentrations of hormone can be achieved to modulate macrophage action without having a generalized, endocrine effect on the host.

B. Paracrine Action on the Lymphocyte

There is also now very strong evidence pointing toward a role for $1,25(OH)_2D$ in the modulation of lymphocyte responsiveness to antigen challenge. If lympho-

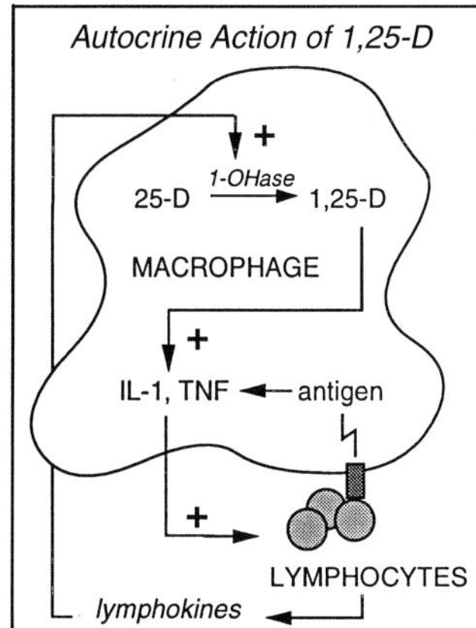

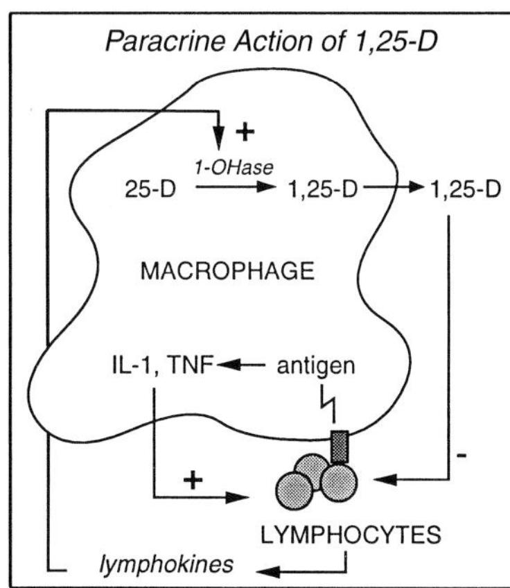

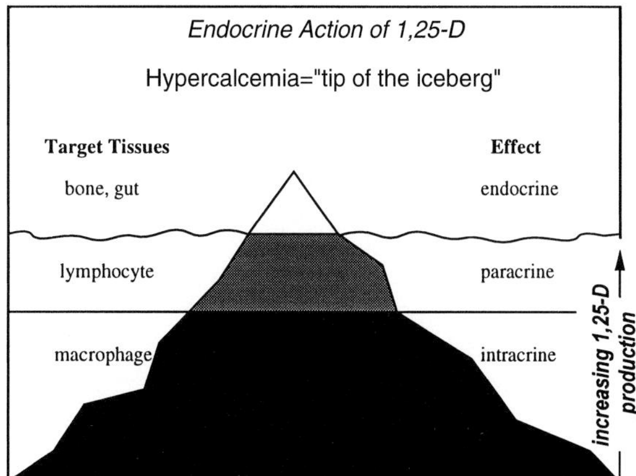

FIGURE 4 Schemes for local production and action of macrophage-derived $1,25(OH)_2D$ (1,25-D). In an autocrine or intracrine mode (left) $1,25(OH)_2D$ promotes antigen handling and monokine production. In a paracrine mode (middle) $1,25(OH)_2D$ acts in a negative feedback fashion to "brake" what may turn out to be an overexuberant lymphocyte response to presented antigen and local monokines. If the immune response and $1,25(OH)_2D$ production are persistent, then the hormone can escape the local inflammatory microenvironment and act in an endocrine mode (right) to alter host calcium balance.

kine stimulation of macrophage 1,25$(OH)_2$D synthesis were persistent, because of difficulty in macrophage-mediated elimination of the offending antigen, then one might conceive of a situation in which the lipid-soluble vitamin D hormone escapes the confines of the macrophage (Fig. 4, middle). Once outside of the macrophage, 1,25$(OH)_2$D would be free to interact in a paracrine fashion with antigen-activated lymphocytes in the local inflammatory microenvironment; as is the case with the monocyte–macrophage, on activation these cells express the VDR [108]. The many reported actions of active vitamin D metabolites on cells of the lymphocyte lineage are also described in Chapter 29.

In general and in contrast to stimulatory effects of the hormone on monocyte–macrophage cells, 1,25$(OH)_2$D and most of the nonhypercalcemic analogs of vitamin D will inhibit lymphocyte responsiveness to mitogen or antigen challenge. Interaction of the VDR with its cognate ligand in activated lymphocytes and natural killer (NK) cells inhibits cellular proliferation [109–111], generally decreases lymphokine production [97,112], and inhibits T-lymphocyte-directed B-cell immunoglobulin synthesis [113,114] and delayed-type hypersensitivity reactions [115,116]. It is postulated [31,107,117] that this apparent paradox in immunoactions of the hormone, to dampen lymphocyte activity while stimulating monocyte–macrophage function, is designed to maximize the ability of the host to combat and contain the granuloma-causing antigen, while controlling the potentially self-destructive lymphocytic response to that offending antigen. In other words, to prevent "overstimulation" of lymphocytes by monokines elaborated at the site of inflammation, some hormone produced by the macrophage will escape the confines of the cell in which it was made, will interact with neighboring, VDR-expressing, activated lymphocytes, and will tend to brake what might be an otherwise overzealous, self-destructive T-cell and B-cell response to the offending antigen. As depicted in Fig. 4 (right), only at times of heightened immunoreactivity (i.e., extraordinary disease activity) does monokine production escape the confines of the site of inflammation and spill over into the general circulation, causing elevated 1,25$(OH)_2$D concentrations. This model would indicate that the endocrine actions of a locally produced vitamin D metabolite that escapes the inflammatory microenvironment is the exception rather than the rule. It also suggests that 1,25$(OH)_2$D is by design an immunomodulatory cytokine in these lymphoproliferative diseases and not a hormone meant to modulate calcium homeostasis in the host.

In summary, currently available information indicates that the extrarenal production of 1,25$(OH)_2$D is a common cause of clinical disturbances of calcium homeostasis in malignant (i.e., lymphoma) and nonmalignant (i.e., sarcoidosis) human lymphoproliferative disorders. Compared to renal 1,25$(OH)_2$D synthesis, production of the hormone by the disease-activated macrophage is dysregulated, because the signal transduction pathways which converge on the 1-hydroxylase in the macrophage differ dramatically from those in kidney cell (see Fig. 2). Endogenous leukotriene production, NO synthesis, hsp70, as well as a paracrine-acting autoregulatory peptide all appear to be key elements in the regulated expression of the macrophage hydroxylase.

C. Accumulation of 1,25$(OH)_2$D at Sites of Inflammation

1. Granuloma-Forming Diseases

Although much effort has been expended to ascertain the immunomodulatory potential of active vitamin D metabolites in human disease, considerably less is known about the immunoactions of these endogenously synthesized molecules *in vivo* in humans and animals. If 1,25$(OH)_2$D is truly a naturally occurring "cytokine," then one should be able to document the accumulation of the metabolite at sites of inflammation and show that the inflammatory cells at this site are under the influence of the locally produced vitamin D metabolite. This was first accomplished by Barnes *et al.* [118], not in sarcoidosis, but in the infectious granuloma-forming disease tuberculosis. They determined that the pleural space in nonhypercalcemic/calciuric patients infected with *Mycobacterium tuberculosis* was one such site of 1,25$(OH)_2$D accumulation. They detected a steep gradient for free, biologically active 1,25$(OH)_2$D across the visceral pleura in patients with tuberculous effusions (but not in patients with nontuberculous effusions), showed that PPD-reacting T-lymphocyte clones from these patients expressed the VDR, and determined that the stimulated proliferation of these T-cell clones was susceptible to 1,25$(OH)_2$D-mediated inhibition. They also showed that the pleural fluid of these patients contained an IFN-γ-like peptide that stimulated the synthesis of 1,25$(OH)_2$D by heterologous sarcoid macrophages [119]. Collectively, these data supported the idea put forward by Rook and colleagues [120] that there exists in the pleural microenvironment of patients with active pulmonary tuberculosis a system whereby (1) mycobacterium-activated macrophages are stimulated to make 1,25$(OH)_2$D; (2) this synthetic reaction is supported by proliferating and lymphokine (i.e., IFN-γ)-producing lymphocytes at the local site of inflammation; and (3) the local accumulation of lymphokines, in turn, acts to further augment the local production of 1,25$(OH)_2$D by the macrophage (see Fig. 4). Investigators [62,121] have viewed this sort of positive feedback effect of IFN on macrophage 1,25(OH)$_2$D production *in vivo* as an efficient mechanism for dealing with antigens, like mycobacteria, the "sarcoid antigen," or certain viruses, that are difficult for the host to irradicate.

2. Other Inflammatory Disease States

Mawer and colleagues [122,123] have demonstrated substrate-dependent accumulation of 1,25(OH)$_2$D in the synovial fluid of patients with "inflammatory arthritis," including subjects with rheumatoid arthritis. These investigators speculate that the local increase in 1,25(OH)$_2$D synthesis may contribute to periarticular bone loss in such individuals. A positive tissue-to-serum gradient for 1,25(OH)$_2$D has also been suggested in the peritoneal space of dialysis patients, particularly when afflicted with peritonitis [124,125]; peritoneal macrophages from such patients have been shown to metabolize 25OHD to 1,25(OH)$_2$D *in vitro* [125].

V. HUMAN DISEASES ASSOCIATED WITH THE EXTRARENAL OVERPRODUCTION OF ACTIVE VITAMIN D METABOLITES

Table I lists diseases associated with the extrarenal overproduction of active vitamin D metabolites.

TABLE I Human Hypercalcemia-Causing Diseases Associated with the Extrarenal Overproduction of an Active Vitamin D Metabolite

Granuloma-forming disease
Sarcoidosis
Tuberculosis
Leprosy
Disseminated candidiasis
Histoplasmosis
Coccidioidomycosis
Crytococcosis
Berylliosis
Silicone-induced granulomatosis
Eosinophilic granuloma
Wegener's granulomatosis
Massive infantile fat necrosis
Malignant lymphoproliferative disease
Non-Hodgkin's lymphoma
Hodgkin's lymphoma
Lymphomatoid granulomatosis

A. Granuloma-Forming Diseases

1. Sarcoidosis

Sarcoidosis is the human disease most commonly complicated by endogenous vitamin D intoxication [126]. In their retrospective, worldwide review of serum calcium concentrations in 3676 patients with sarcoidosis, James *et al.* [127] recorded an 11% incidence of hypercalcemia (serum calcium $\geq$10.5 mg/dl). Studdy *et al.* [128] studied 547 patients with biopsy-proven sarcoidosis in Great Britain and found hypercalcemia to be 38% more frequent in men than women and more common among Caucasian than individuals of West Indian descent. Although not systematically studied, the frequency of hypercalcemia among patients with sarcoidosis tends to be consistently higher in North America than in Northern Europe [126]. This is perhaps due to the lower latitude and more direct sunlight exposure in the United States.

Although fractional intestinal calcium absorption may be increased under the influence of 1,25(OH)$_2$D and fractional urinary calcium excretion may be decreased in patients with renal insufficiency [32], the principal source of calcium which accumulates in the circulation in this disease is the skeleton. This fact is perhaps most strongly confirmed by the observations of Rizzato *et al.* [129] who documented, in serial fashion, a significant decrease in bone mineral density in a group of patients with chronic active sarcoidosis in whom anti-inflammatory agents, including glucocorticoids, were not used in management compared to age- and sex-matched control subjects. This finding is supported by the long-standing observations that hypercalcemia persists in patients with active sarcoidosis in the absence of ingested calcium and may be contributed to by increased bone resorption [27]. The proximal cause of bone loss is increased osteoclast-mediated bone resorption [130] and does not require the presence of extensive granulomata in the bone [131]. These observations suggest that there exists a circulating stimulator of bone resorption in this disease. One such stimulator of bone resorption is of course already known. It is 1,25(OH)$_2$D.

2. Tuberculosis

Of the other human granuloma-forming diseases reported to be associated with vitamin D metabolite-mediated hypercalcemia, tuberculosis is the most common and therefore, aside from sarcoidosis, is the most commonly reported disease to be complicated by an alteration in calcium balance. Hypercalcemia has been recognized as a complication of infection with *Mycobacterium tuberculosis* since the early 1900s [132]. That this disturbance in calcium balance is caused by the extrarenal overproduction of an active vitamin D metabolite was confirmed by investigators in the mid-1980s [133,134]. As is the case with sarcoidosis, the circulating vitamin D metabolite causing hypercalcemia (1) appears to be $1,25(OH)_2D$ [135,136]; (2) is synthesized by disease-activated macrophages [137,138]; (3) is abnormally responsive to small changes in the serum concentration of substrate (25OHD) [139]; and (4) is reducible under the influence of glucocorticoid *in vivo* [140,141]. The prevalence of hypercalcemia in tuberculosis patients has been reported to be as high as 26% [132], and it may be even higher in the era of AIDS owing to the more frequent occurrence of disseminated disease in immunocompromised patients.

3. Other Infectious Diseases

Hypercalciuria or overt hypercalcemia have also been observed in a number of infectious diseases, most characterized by widespread granuloma formation and macrophage proliferation in infected tissue. Included among these diseases are leprosy [142,143], disseminated candidiasis [144], histoplasmosis [145], and coccidioidomycosis [146]. Hypercalcemia in most of these conditions has been documented to be associated with inappropriately elevated serum concentrations of $1,25(OH)_2D$. The true prevalence and incidence of hypercalcemia and hypercalciuria in patients with these diseases is unknown. However, it is likely that this complication of dysregulated vitamin D metabolism and action associated with these diseases will increase in frequency as the number of immunocompromised patients, especially those with AIDS, increases worldwide. For example, hypercalcemia in association with elevated circulating levels of $1,25(OH)_2D$ has been recently reported in an AIDS patient with pneumocystis pneumonia [147]; both serum calcium and $1,25(OH)_2D$ concentrations dropped in this patient with successful treatment of the opportunistic infection.

4. Noninfectious Granuloma-Forming Diseases

The syndrome of extrarenal overproduction of $1,25(OH)_2D$ has also been documented in adult patients with widespread silicone-induced granulomata [148], eosinophilic granuloma [149], and Wegener's granulomatosis [150]. Although the active vitamin D metabolite was not measured, dysregulated calcium balance in the granuloma-forming pulmonary disease berylliosis is also attributed to the extrarenal production of $1,25(OH)_2D$ [151]. In addition, $1,25(OH)_2D$-mediated hypercalcemia has been observed in newborn infants suffering from massive subcutaneous fat necrosis [152]; this is a transient disorder associated with birth trauma and characterized histopathologically by the proliferation of "foreign body-type" giant cells around cholesterol-shaped crystals in necrotizing, subcutaneous adipose tissue. Finally, there are also reports [153] of hypercalcemic patients with elevated serum $1,25(OH)_2D$ levels in whom no underlying disease is encountered; they do, however, regain normal calcium balance in response to glucocorticoid treatment.

B. Malignant Lymphoproliferative Disorders: Lymphoma

By the 1980s data accumulated to suggest that a vitamin D-mediated disturbance in calcium metabolism was not confined to patients with granuloma-forming diseases and could also be observed in patients with lymphoproliferative neoplasms [154–157]. More recent reports [158,159] indicate that the extrarenal overproduction of $1,25(OH)_2D$ is the most common cause of hypercalciuria and hypercalcemia in patients with non-Hodgkin's and Hodgkin's lymphoma, especially in patients with B-cell neoplasms and whether or not the tumor is associated with AIDS [157]. In fact, in the Seymour *et al.* study [159] 71% of normocalcemic patients with non-Hodgkin's lymphoma had hypercalciuria (fasting fractional urinary calcium excretion >0.15 mg/dl glomerular filtrate) and most of these had serum $1,25(OH)_2D$ levels that were above the mid range of normal or frankly elevated. As is the situation with hypercalciuric/calcemic patients with sarcoidosis or other granuloma-forming disease and elevated circulating $1,25(OH)_2D$ levels, the serum concentrations of PTH are suppressed and PTHrP is normal (i.e., not elevated) in lymphoma patients, indicative of the state of dysregulated overproduction of the active vitamin D hormone. Results of clinical studies of hypercalcemic patients with lymphoma before and after successful antitumor therapy [157–160] are most compatible with the tumor being either an immediate source of an active vitamin D metabolite [161] or the source of a soluble factor (i.e., peptide) that stimulates the production of $1,25(OH)_2D$ in the kidney or in other inflammatory cells. This issue has not yet been resolved. An elevated $1,25(OH)_2D$ level has been reported in patients with lymphomatoid granulomatosis [162], a myeloproliferative disorder of unknown etiology.

C. Non-Granuloma-Forming Conditions

There is mounting evidence to indicate that active vitamin D metabolite production at extrarenal sites may occur under normal circumstances in healthy human subjects [163]. Bikle *et al.* have demonstrated cultured human basal epidermal cells (keratinocytes) [164] as well as intact sections of porcine skin [165] to be capable of producing $1,25(OH)_2D$ *in vitro* and *ex vivo*, respectively. In addition to the $1,25(OH)_2D$ produced by circulating monocytes in humans, that produced by the skin may explain why functionally anephric human subjects (i.e., those on chronic dialysis) possess detectable serum levels of $1,25(OH)_2D$ in the absence of the renal 1-hydroxylase. Whether the extrarenal production of an active vitamin D metabolite by inflammatory cells, or skin, or other tissues for that matter, contributes to the inappropriately elevated concentrations of $1,25(OH)_2D$ as it occurs in some patients with idiopathic hypercalciuria [166] remains to be determined. Also unclear is the role of the placenta in the extrarenal production of $1,25(OH)_2D$. Serum $1,25(OH)_2D$ levels go up during the early stages of human pregnancy [167]. *In vitro*, placental tissue can be shown to metabolically convert 25OHD to $1,25(OH)_2D$ [168,169]. Synthesis of $1,25(OH)_2D$ has also been documented in a nephrectomized pregnant rat [170]; however, a similar situation in an anephric human subject has not yet been reported.

Although less well documented, there is also evidence that extrarenal $1,25(OH)_2D$ production can occur in some individuals with solid neoplasms and hypercalcemia. The presumption is that either the tumor itself [171] or tumor-associated macrophages [172] are responsible for hormone synthesis in these malignant diseases.

VI. DIAGNOSIS, PREVENTION, AND TREATMENT OF THE PATIENT WITH ENDOGENOUS VITAMIN D INTOXICATION

A. Diagnosis

The diagnosis of so-called endogenous vitamin D intoxication is made when the following three criteria are met. First is the presence of hypercalciuria and/or hypercalcemia in a patient with an inappropriately elevated serum $1,25(OH)_2D$ level [i.e., the serum $1,25(OH)_2D$ concentration is not suppressed below 20 pg/ml]. Second is the presence in the serum of an appropriately suppressed PTH level if the patient's free (ionized) serum calcium concentration is high; this is evidence that the calcium-sensing receptor in the plasma membrane of the host parathyroid cell is normally operative. This distinguishes the patient with primary hyperparathyroidism and elevated $1,25(OH)_2D$ levels, in whom the calcium-sensing receptor signal transduction pathway to control PTH synthesis and release is disrupted, from the individual with endogenous vitamin D intoxication and elevated $1,25(OH)_2D$ levels. The other major exception here is the patient with absorptive hypercalciuria who possesses, as a primary or secondary abnormality, an inappropriately elevated circulating $1,25(OH)_2D$ concentration [173]. Third is the exclusion of exogenous vitamin D intoxication arising from the oral or parental administration of an active vitamin D metabolite or the substrate for endogenous synthesis of an active vitamin D metabolite.

The most common cause of exogenous vitamin D intoxication occurs with the ingestion or injection of large doses of vitamin D_2 (ergocalciferol) or vitamin D_3 (cholecalciferol) and can usually be detected by measuring a frankly elevated serum 25OHD level; most, if not all, currently available serum assays for 25OHD do not distinguish $25OHD_2$ from $25OHD_3$ [174]. Exogenous vitamin D intoxication may occur in patients taking too much 1α-hydroxyvitamin D, dihydrotachysterol (DHT), or $1,25(OH)_2D$ itself; the two former compounds undergo 25-hydroxylation in the host hepatocyte. In these instances the $1,25(OH)_2D$, concentration will be elevated, not the 25OHD, making distinction from endogenous vitamin D intoxication caused by the extrarenal overproduction of $1,25(OH)_2D$ impossible on strictly biochemical grounds. In these situations a complete knowledge of the medications to which the patient has access is critical for making the correct diagnosis. Examples of such patients would be those receiving relatively large amounts of vitamin D, a vitamin D metabolite, or a vitamin D analog (i.e., patients with hypoparathyroidism, renal failure, and psoriasis, respectively). Most of the newer vitamin D analogs currently in clinical use [175,176] will not be efficiently measured in the serum $1,25(OH)_2D$ assay, so awareness of the use of these kinds of topically and orally administered agents is of particular importance to the diagnosing clinician.

B. Early Detection and Prevention of Hypercalciuria/Hypercalcemia

1. IDENTIFYING PATIENTS AT RISK

Considering the fact that the means of specifically inhibiting the production of active metabolites of vitamin D or of blocking the response of cells to active vitamin D derivatives is not yet available, the best way to treat vitamin D-mediated abnormalities in calcium balance is to prevent their occurrence. The first step is to identify patients at risk. This encompasses primarily patients with granuloma-forming disease as well as patients with malignant lymphoproliferative disorders, especially B-cell and Hodgkin's lymphoma. Disordered calcium balance in these groups of patients results from the endogenous and dysregulated overproduction of 1,25(OH)$_2$D by inflammatory cells. Production of the offending vitamin D metabolite is, in turn, directly related to amount of substrate 25OHD available to the macrophage 1-hydroxylase (see Fig. 2) as well as to the severity and activity of the underlying disease. In terms of sarcoidosis, for example, patients at risk would be those with (1) widespread, active disease; (2) a previous history of hypercalciuria or hypercalcemia; (3) a diet enriched in vitamin D and/or calcium; (4) a recent history of sunlight exposure or treatment with vitamin D; and (5) an intercurrent condition, or medicinal treatment of an intercurrent condition, that increases bone resorption or decreases the glomerular filtration rate.

2. Screening Patients at Risk

Because hypercalciuria almost always precedes the development of overt hypercalcemia in this set of disorders, patients at risk should be checked for the presence of occult hypercalciuria. This is best accomplished by analyzing a fasting 2-hr urine collection for calcium and creatinine. If the calcium:creatinine ratio (grams to grams) is not abnormally high (i.e., is >0.16), then analysis of a 24-hr urine collection for the fractional calcium excretion rate is necessary to establish hypercalciuria. Hypercalciuria is presumably due to increased bone resorption, as many patients with vitamin D-mediated hypercalciuria will be hypercalciuric even in the absence of recent food (calcium) ingestion. If screening is to be done only on an annual basis, then the late summer or early autumn, when 25OHD levels are usually at their peak, is the best time [62].

3. Prevention of Hypercalcemia

Prevention of an overt disorder in calcium balance is obviously preferable to dealing with hypercalcemia. This is especially true for patients (1) in whom one episode of hypercalcuria or hypercalcemia has already been documented and (2) who are taking supplemental vitamin D and calcium preparations for another medical indication (i.e., osteoporosis). Prevention is best achieved by monitoring the serum calcium and urinary calcium excretion rate on a regular basis. If either is high, then the serum concentration of 25OHD and 1,25(OH)$_2$D level should be obtained; the former should be evaluated to rule out the existence or coexistence of exogenous vitamin D intoxication. For patients determined from the appropriate monitoring analyses to be at risk, measures to prevent worsening hypercalciuria and frank hypercalcemia should be instituted. These measures should include (1) the use of UVB-absorbing sunscreens on exposed body parts when patients anticipate being out-of-doors for periods in excess of 20–30 min; (2) caution against ingestion of vitamin and food supplements containing ≥400 IU vitamin D; (3) education on the vitamin D and calcium content of foods, vitamin supplements, and medicinal agents like antacids; (4) caution against the regular ingestion of elemental calcium in excess of 1000 mg daily; and (5) education regarding the earliest signs of hypercalciuria (i.e., nocturia).

C. Treatment of Hypercalciuria and Hypercalcemia

Because hypercalciuria and hypercalcemia reside on the same pathophysiological spectrum, the aim of treatment is the same, namely, normalization of the urinary calcium excretion rate whether the patient is hypercalciuric or frankly hypercalcemic. There are three general therapeutic goals. First is reduction in the serum concentration of the offending vitamin D metabolite or derivative. In patients with exogenous vitamin D intoxication this is usually accomplished by cessation or a reduction in the dose of the vitamin D preparation being used by that patient. Remembering that 25OHD has a serum half-life of months, vitamin D-intoxicated patients may require as much as a year off of therapy. In patients suffering from endogenous intoxication with 1,25(OH)$_2$D made by inflammatory cells, reduction in the serum 1,25(OH)$_2$D level can be most reliably achieved by treatment with anti-inflammatory doses of glucocorticoid (adult dose of 40 mg prednisone or equivalent per day). Glucocorticoids inhibit two key enzymes in macrophage-like cells, phospholipase C [64] and the inducible nitric oxide synthase [71], whose distal products stimulate the 1-hydroxylase. At these doses these drugs have little effect on the renal 1-hydroxylase, so there is little concern for inducing hypocalcemia with glucocorticoid administration. In patients with extrarenal production of the hormone, steroid therapy should result in a drop in the serum 1,25(OH)$_2$D concentration within a matter of 3–4 days, followed shortly thereafter by a decrease in the filtered load of calcium and in the urinary calcium excretion rate provided that the glomerular filtration rate is maintained.

In patients who fail glucocorticoids or in whom glucocorticoids are contraindicated, treatment with the 4-aminoquinoline class of drugs like chloroquine (250 mg twice daily) or hydroxychloroquine (up to 400 mg daily) may be effective [28–30]. A less desirable therapeutic alternative is the cytochrome P450 inhibitor ketoconazole [177]. Ketoconazole will effectively reduce the serum 1,25(OH)$_2$D concentration [178–180], but the therapeutic margin of safety is narrow: doses of the drug that inhibit the macrophage P450 system are very close to those that will also inhibit endogenous glucocorticoid and sex steroid production [181]. The second goal of therapy is to limit the actions of the vitamin D derivative at its target tissues, the gut and bone. A reduction in intestinal calcium absorption is best accomplished by elimination of as much calcium as possible from the diet. Such conservative measures are rarely effective in patients with active, widespread disease, so glucocorticoid administration may also be required to block vitamin D-mediated calcium absorption and bone resorption. Because of the relative effectiveness of glucocorticoid management of this problem, the utility of other skeletal antiresorptive agents, like calcitonin and the bisphosphonates, has not been prospectively evaluated in a large cohort of patients with vitamin D-mediated hypercalcemia. The third goal of therapy is to enhance urinary calcium excretion to a point where the filtered load of calcium is insufficient to cause either hypercalcemia or hypercalciuria. This can be achieved by maintenance of the vascular volume (glomerular filtration rate) and urinary flow rate, and, if needed, by the use of a "loop" diuretic such as furosemide to inhibit calcium reabsorption from the urine. The effects of successfully reducing the serum $1,25(OH)_2D$ concentration and managing hypercalcemia/hypercalciuria on the skeleton long-term are not known. There is preliminary evidence that successful treatment of exogenous vitamin D intoxication may result in a transient increase in bone mineral density [182].

References

1. Harrell GT, Fisher S 1939 Blood chemical changes in Boeck's sarcoid. J Clin Invest **18**:687–693.
2. Henneman PH, Dempsey EF, Carrol EJ, Albright F 1956 The causes of hypercalcemia in sarcoid and its treatment with cortisone. J Clin Invest **35**:1229–1242.
3. Taylor RL, Lynch HJ Jr, Wysor WG 1963 Seasonal influence of sunlight on the hypercalcemia of sarcoidosis. Am J Med **35**:67–89.
4. Dent CE, Flynn FV, Nabarro JDN 1953 Hypercalcemia and impairment of renal function in generalized sarcoidosis. Br Med J **2**:808–810.
5. Hendrix JZ 1963 The remission of hypercalcemia and hypercalciuria in systemic sarcoidosis by vitamin D depletion. Clin Res **11**:220–225.
6. Bell NH, Gill JR Jr, Barter FC 1964 Abnormal calcium absorption in sarcoidosis: Evidence for increased sensitivity to vitamin D. Am J Med **36**:500–513.
7. Holick MF, Schnoes THK, DeLuca HF, Suda T, Cousins RJ 1992 Isolation and identification of 1,25-dihydroxycholecalciferol. A metabolite of vitamin D active in intestine. J NIH Res **4**:88–96.
8. Bell NH 1985 Vitamin D–endocrine system. J Clin Invest **76**:1–6.
9. Hughes MF, Baylink DJ, Jones PG, Haussler MR 1976 Radioligand receptor assay for 25-hydroxyvitamin D_2/D_3 and 1α,25-dihydroxyvitamin D_2/D_3. J Clin Invest **58**:61–70.
10. Clemens TL, Hendy GN, Graham RF 1978 A radioimmunoassay for 1,25-dihydroxycholecalciferol. Clin Sci Mol Med **54**:329–332.
11. Gray TK, McAdoo T 1979 Radioimmunoassay for 1,25-dihydroxyvitamin D_3. In: Norman AW, Schaefer K, Herrath DV (eds) Vitamin D: Basic Research and Its Clinical Application. de Gruyter, Berlin; pp. 763–767.
12. Bouillon R, DeMoor P, Baggiolini EG, Uskokovic MR 1980 A radioimmunoassay for 1,25-dihydroxycholecalciferol. Clin Chem **26**:562–567.
13. Holick MF 1990 The use and interpretation of assays for vitamin D and its metabolites. J Nutr **120**:1464–1469.
14. Bell NH, Stern PH, Pantzer E, Sinha TK, DeLuca HF 1979 Evidence that increased circulating 1,25-dihydroxyvitamin D is the probable cause for abnormal calcium metabolism in sarcoidosis. J Clin Invest **64**:218–225.
15. Papapoulos SE, Clemens TL, Fraher LJ, Lewin IG, Sandler LM, O'Riordan JL 1979 1,25-Dihydroxycholecalciferol in the pathogenesis of the hypercalcemia of sarcoid. Lancet **1**:627–630.
16. Stern PH, Olazabal J, Bell NH 1980 Evidence for abnormal regulation of circulating 1,25-dihydroxyvitamin D in patients with sarcoidosis. J Clin Invest **66**:852–855.
17. Barbour GL, Coburn JW, Slatopolsky E, Norman AW, Horst RL 1981 Hypercalcemia in an anephric patient with sarcoidosis. N Engl J Med **305**:440–443.
18. Adams JS, Sharma OP, Gacad MA, Singer FR 1983 Metabolism of 25-hydroxyvitamin D_3 by cultured alveolar macrophages in sarcoidosis. J Clin Invest **72**:1856–1860.
19. Adams JS, Singer FR, Gacad MA, Sharma OP, Hayes MJ, Vouros P, Holick MF 1985 Isolation and structural identification of 1,25-dihydroxyvitamin D_3 produced by cultured alveolar macrophages in sarcoidosis. J Clin Endocrinol Metab **60**:960–966.
20. Bell NH 1991 Endocrine complications of sarcoidosis. Endocrinol Metab Clin North Am **20**:645–654.
21. Basile JN, Leil Y, Shary J, Bell NH 1993 Increased calcium intake does not suppress circulating 1,25-dihydroxyvitamin D in normocalcemic patients with sarcoidosis. J Clin Invest **91**:1396–1398.
22. Sandler LM, Winearls CG, Fraher LJ, Clemens TL, Smith R, O'Riordan JLH 1984 Studies of the hypercalcemia of sarcoidosis. Q J Med **53**:165–180.
23. Papapoulos SE, Clemens TL, Fraher LJ 1979 Dihydroxycholecalciferol in the pathogenesis of the hypercalcemia of sarcoid. Lancet **1**:627–630.
24. Cronin CC, Dinneen SF, O'Mahony MS, Bredin CP, O'Sullivan DJ 1990 Precipitation of hypercalcaemia in sarcoidosis by foreign sun holidays: Report of four cases. Postgrad Med J **66**:307–309.
25. Insogna KL, Dreyer BE, Mitnich M, Ellison AF, Broadus A 1988 Enhanced production of 1,25-dihydroxyvitamin D in sarcoidosis. J Clin Endocrinol Metab **66**:72–75.
26. Shulman LE, Schoenrich E, Harvey A 1952 The effects of adrenocorticotropic hormone (ACTH) and cortisone on sarcoidosis. Bull John Hopkins Hospital **91**:371–415.
27. Anderson J, Dent CE, Harper C, Philpot GR 1954 Effect of cortisone on calcium metabolism in sarcoidosis with hypercalcemia. Lancet **2**:720–724.
28. O'Leary TJ, Jones G, Yip A, Lohnes D, Cohanim M, Yendt ER 1986 The effects of chloroquine on serum 1,25-dihydroxyvitamin D and calcium metabolism in sarcoidosis. N Engl J Med **315**:727–30.
29. Barre PE, Gascon-Barre M, Meakins JL, Goltzman D 1987 Hydroxychloroquine treatment of hypercalcemia in a patient with sarcoidosis. Am J Med **82**:1259–1262.
30. Adams JS, Diz MM, Sharma OP 1989 Effective reduction in the serum 1,25-dihydroxyvitamin D and calcium concentration in

sarcoidosis-associated hypercalcemia with short-course chloroquine therapy. Ann Intern Med **111**:437–438.
31. Singer FR, Adams JS 1986 Abnormal calcium homeostasis in sarcoidosis. N Engl J Med **315**:755–756.
32. Meyrier A, Valeyre D, Bouillon R, Paillard F, Battesti JP, Georges R 1985 Resorptive versus absorptive hypercalciuria in sarcoidosis. Q J Med **54**:269–281.
33. Adams JS, Gacad MA, Anders A, Endres DB, Sharma DP 1986 Biochemical indicators of disordered vitamin D and calcium homeostasis in sarcoidosis. Sarcoidosis **3**:1–6.
34. Adams JS, Gacad MA 1985 Characterization of 1α-hydroxylation of vitamin D_3 sterols by cultured macrophages from patients with sarcoidosis. J Exp Med **161**:755–765.
35. Shany S, Adams JS 1993 Subcellular localization of the 25-hydroxyvitamin D_3-1-hydroxylase and partial purification from the chick myelomonocytic cell line HD-11. J Bone Miner Res **8**:269–276.
36. Adams JS, Ren S-Y, Arbelle JE, Horiuchi N, Gray RW, Clemens TL, Shany S 1994 Regulated production and intracrine action of 1,25-dihydroxyvitamin D_3 in chick myelomonocytic cell line HD-11. Endocrinology **134**:2567–2573.
37. Reichel H, Koeffler HP, Norman AW 1987 Synthesis in vitro of 1,25-dihydroxyvitamin D_3 and 24,25-dihydroxyvitamin D_3 by interferon-gamma-stimulated normal human bone marrow and alveolar macrophages. J Biol Chem **262**:10931–10937.
38. Fraser DR 1980 Regulation of the metabolism of vitamin D. Physiol Rev **60**:551–613.
39. Reichel H, Koeffler HP, Norman AW 1989 The role of the vitamin D endocrine system in health and disease. N Engl J Med **320**:980–991.
40. Taylor R 1994 A new receptor for calcium ions. J NIH Res **6**:25–27.
41. Garabedian M, Holick MF, DeLuca HF, Boyle IT 1972 Control of 25-hydroxy-cholecalciferol metabolism by parathyroid glands. Proc Natl Acad Sci USA **69**:1973–1976.
42. Janulis M, Wong M-S, Favus MJ 1993 Structure-function requirements of parathyroid hormone for stimulation of 1,25-dihydroxyvitamin D_3 production by rat renal proximal tubules. Endocrinology **133**:713–719.
43. Tang C, Kain SR, Henry HL 1993 The phorbol ester 12-O-tetradecanoyl-phorbol-13-acetate stimulates the dephosphorylation of mitochondrial ferredoxin in cultured chick kidney cells. Endocrinology **133**:1823–1829.
44. Henry HL 1992 Vitamin D hydroxylases. J Cell Biochem **49**:4–9.
45. Portale AA, Halloran BP, Murphy MM, Morris RC Jr 1986 Oral intake of phosphorus can determine the serum concentration of 1,25-dihydroxyvitamin D by determining its production rate in humans. J Clin Invest **77**:7–12.
46. Reichel H, Koeffler HP, Barbers R, Norman AW 1987 Regulation of 1,25-dihydroxyvitamin D_3 production by cultured alveolar macrophages from normal human donors and patients with pulmonary sarcoidosis. J Clin Endocrinol Metab **65**:1201–1209.
47. Nesbitt T, Drezner MK 1993 Insulin-like growth factor-1 regulation of renal 25-hydroxyvitamin D-1-hydroxylase activity. Endocrinology **132**:133–138.
48. Henry HH 1981 25(OH)D_3 metabolism in kidney cell cultures: Lack of a direct effect of estradiol. Am J Physiol **240**:E119–E124.
49. Adams ND, Garthwaite TL, Gray RW 1979 The interrelationship among prolactin, 1,25-dihydroxyvitamin D and parathyroid hormone in humans. J Clin Endocrinol Metab **49**:628–630.
50. Kumar R, Merimee TJ, Silva P, Epstein FH 1979 The effect of chronic excess or deficiency of growth hormone on plasma 1,25-dihydroxyvitamin D levels in man. In: Norman AW, Schaefer K, Herrath DV, (eds) Vitamin D: Basic Research and Its Clinical Application. de Gruyter, Berlin, pp.1005–1009.
51. Brixen K, Nielsen HK, Bouillon R, Flyvbjerg A, Mosekilde L 1992 Effects of short-term growth hormone treatment on PTH, calcitriol, thyroid hormones, insulin and glucagon. Acta Endocrinol **127**:331–336.
52. Ohyama Y, Noshiro M, Okuda K 1991 Cloning and expression of cDNA encoding 25-hydroxyvitamin D_3-24-hydroxylase. FEBS Lett **278**:195–198.
53. Chen K, Goto H, DeLuca HF 1992 Isolation and expression of human 1,25-dihydroxyvitamin D_3 24-hydroxylase cDNA. J Bone Miner Res **7**:S148.
54. Chen KS, Prahl JM, DeLuca HF 1993 Isolation and expression of human 1,25-dihydroxyvitamin D_3-24-hydroxylase. Proc Natl Acad Sci USA **90**:4543–4547.
55. Ismail R, Elaroussi MA, DeLuca HF 1993 Regulation of chicken kidney vitamin D_3 24-hydroxylase mRNA by 1,25-dihydoxyvitamin D_3 and parathyroid hormone. J Bone Miner Res **8**:S208.
56. Uchida M, Shinki T, Ohyama Y, Noshiro M, Okda K, Suda T 1993 Protein kinase C upregulates 1α,25-dihydroxyvitamin D_3 induced expression of the 24-hydroxylase gene. J Bone Miner Res **8**:S171.
57. Chen ML, Boltz MA, Armbrecht HJ 1993 Effects of 1,25-dihydroxyvitamin D_3 and phorbol ester on 25-hydroxyvitamin D_3-24-hydroxylase cytochrome P-450 messenger ribonucleic acid levels in primary cultures of rat renal cells. Endocrinology **132**:1782–1788.
58. Tanaka Y, Lorenc RS, DeLuca HF 1975 The role of 1,25-dihydroxyvitamin D_3 and parathyroid hormone in the regulation of chick renal 25-hydroxyvitamin D_3-24-hydroxylase. Arch Biochem Biophys **171**:521–526.
59. Henry H, Luntao EM 1985 Further studies on the regulation of 25-OH-D metabolism in kidney cell cultures. In: Norman AW, Schaefer K, Grigoleit HG, Herrath DV, (eds) Vitamin D: Chemical, Biochemical and Clinical Update. de Gruyter, Berlin, pp. 505–514.
60. Henry HL 1979 Regulation of the hydroxylation of 25-hydroxyvitamin D_3 *in vivo* and in primary cultures of chick kidney cells. J Biol Chem **254**:2722–2729.
61. Adams JS, Gacad MA, Singer FR, Sharma OP 1986 Production of 1,25-dihydroxyvitamin D_3 by pulmonary alveolar macrophages from patients with sarcoidosis. Ann NY Acad Sci **465**:587–594.
62. Adams JS 1992 Hypercalcemia and hypercalciuria. Semin Respir Med **13**:402–410.
63. Hunninghake GW 1986 Role of alveolar macrophage- and lung T cell-derived mediators in pulmonary sarcoidosis. Ann NY Acad Sci **465**:82–90.
64. Adams JS, Gacad MA, Diz MM, Nadler JL 1990 A role for endogenous arachidonate metabolites in the regulated expression of the 25-hydroxyvitamin D-1-hydroxylation reaction in cultured alveolar macrophages from patients with sarcoidosis. J Clin Endocrinol Metab **70**:595–600.
65. Celada A, Schreiber RD 1996 Role of protein kinase C and intracellular calcium mobilization in the induction of macrophage tumoricidal activity by interferon-gamma. J Immunol **137**:2373.
66. Wightman PD, Humes JL, Davies P, Bonney RJ 1981 Identification of two phospholipase A2 activities in resident mouse peritoneal macrophages. Biochem J **195**:427.
67. Wightman PD, Dahlgren M, Bonney RS 1982 Protein kinase activation of phospholipase A2 in sonicates of mouse peritoneal macrophages. J Biol Chem **257**:6650.
68. Schumann RR, Leong SR, Flaggs GW, Wright SD, Mathison

JC, Tobias PS, Ulevitch RJ 1990 Structure and function of LPS binding protein. Science **249**:1429–1431.
69. Marletta, MA 1994 Nitric oxide synthase: Aspects concerning structure and catalysis. Cell **78**:927–930.
70. Nathan C, Xie Q-W 1994 Nitric oxide synthases: Roles, tolls, and controls. Cell **78**:915–918.
71. Adams, JS, Ren SY, Arbelle JE, Clemens TL, Shany S 1994 A role for nitric oxide in the regulated expression of the 25-hydroxyvitamin D-1-hydroxylation reaction in the chick myelomonocytic cell line HD-11. Endocrinology **134**:499–502.
72. Adams JS, Ren S-Y, Arbelle J, Shany S, Gacad MA 1995 Coordinate regulation of nitric oxide and 1,25-dihydroxyvitamin D production in the avian myelomonocytic cell line HD-11. Endocrinology **136**:2262–2269.
73. Lowenstein CJ, Glatt CS, Bredt DS, Snyder SH 1992 Cloned and expressed macrophage nitric oxide synthase contrasts with the brain enzyme. Proc Natl Acad Sci USA **89**:6711–6715.
74. Valance P, Collier J 1994 Biology and clinical relevance of nitric oxide. Br Med J **309**:453–457.
75. Leone AM, Palmer RMJ, Knowles RG, Francis PL, Ashton DS, Moncada S 1991 Constitutive and inducible nitric oxide synthases incorporate molecular oxygen into both nitric oxide and citrulline. J Biol Chem **266**:23790–23795.
76. Stamler JS, Singel DJ, Loscalzo J 1992 Biochemistry of nitric oxide and its redox-activated forms. Science **258**:1898–1902.
77. Chartrain NA, Geller DA, Koty PP, Sitrin NF, Nussler AK, Hoffman EP, Billiar TR, Hutchinson NI, Mudgett JS 1994 Molecular cloning, structure, and chromosomal localization of the human inducible nitric oxide synthase gene. J Biol Chem **269**:6765–6772.
78. Stamler JS 1994 Redox signaling: Nitrosylation and related target interactions of nitric oxide. Cell **78**:931–936.
79. Adams JS, Ren S-Y 1996 Autoregulation of 1,25-dihydroxyvitamin D synthesis in macrophage mitochondria by nitric oxide. Endocrinology **137**:4514–4517.
80. Griscavage JM, Rogers NE, Sherman MP, Ignarro LJ 1993 Inducible nitric oxide synthase from a rat alveolar macrophage cell line is inhibited by nitric oxide. J Immunol **151**:6329–6337.
81. Khatsenko OG, Gross SS, Rifkind AB, Vane JR 1993 Nitric oxide is a mediator of the decrease in cytochrome P450-dependent metabolism caused by immunostimulants. Proc Natl Acad Sci USA **90**:11147–11151.
82. Griscavage JM, Fukuto JM, Komori Y, Ignarro LJ 1994 Nitric oxide inhibits neuronal nitric oxide synthase by interacting with the heme prosthetic group. J Biol Chem **269**:21644–21649.
83. Stadler J, Trockfeld J, Schmalix WA, Brill T, Siewert JR, Greim H, Doehmer J 1994 Inhibition of cytochromes P450–1A by nitric oxide. Proc Natl Acad Sci USA **91**:3559–3563.
84. Morris SM Jr, Billiar TR 1994 New insights into the regulation of inducible nitric oxide synthesis. Am J Physiol **266**:E829–E839.
85. Van Voorhis BJ, Dunn MS, Snyder GD, Weiner CP 1994 Nitric oxide: An autocine regulator of human granulosa–luteal cell steroidogenesis. Endocrinology **135**:1799–1806.
86. Hartl FU 1996 Molecular chaperones in cellular protein folding. Nature **381**:571–580.
87. Minowada G, Weich WJ 1995 Clinical implications of the stress response. J Clin Invest **95**:3–12.
88. Rutherford SL, Zuker CS 1994 Protein folding and the regulation of signaling pathways. Cell **79**:1129–1132.
89. Berthold J, Bauer MF, Schneider H-C, Klaus C, Dietmeier K, Neupert W, Brunner M 1995 The MIM complex mediates preprotein translocation across the mitochondrial inner membrane and couples it to the mt-Hsp70/ATP driving system. Cell **81**:1085–1093.
90. Polla BS, Healy AM, Wojno WC, Krane SM 1987 Hormone 1α,25-dihydroxyvitamin D_3 modulates heat shock response in monocytes. Am J Physiol **252**:C640–C649.
91. Arbelle JE, Adams JS 1995 Vitamin D-induced CD14 gene expression in the human myelomonocytic cell line U937. J Invest Med **43**:114A.
92. Henderson SA, Lee PH, Aeberhard EE, Adams JW, Ignarro LJ, Murphy WJ, Sherman MP 1994 Nitric oxide reduces early growth response-1 gene expression in rat lung macrophages treated with interferon-γ and lipopolysaccharide. J Biol Chem **269**:25239–25242.
93. Oswald IP, Wynn TA, Sher A, James SL 1992 Interleukin 10 inhibits macrophage microbicidal activity by blocking the endogenous production of tumor necrosis factor required as a costimulatory factor for interferon-gamma-induced activation. Proc Natl Acad Sci USA **89**:8676–8680.
94. Prehn JL, Fagan DL, Jordan SC, Adams JS 1992 Potentiation of lipopolysaccharide-induced tumor necrosis factor-alpha expression by 1,25-dihydroxyvitamin D_3. Blood **80**:2811–2816.
95. Provvedini DM, Tsoukas CD, Deftos LJ, Manolagas SC 1983 1,25-Dihydroxyvitamin D_3 receptors in human leukocytes. Science **221**:1181–1182.
96. Bhalla AK, Amento EP, Clemens TL, Holick MF, Krane SM 1983 Specific high-affinity receptors for 1,25-dihydroxyvitamin D_3 in human peripheral blood mononuclear cells. J Clin Endocrinol Metab 57:1308–1310.
97. Rigby W 1988 The immunobiology of vitamin D. Immunol Today **9**:54–58.
98. Hewison M 1992 Vitamin D and the immune system. J Endocrinol **132**:173–175.
99. Fagan DL, Prehn JL, Jordan SC, Adams JS 1991 The human myelomonocytic cell line U937 as a model for studying alterations in monokine gene expression by 1,25-dihydroxyvitamin D. Mol Endocrinol **5**:179–186.
100. Bar-Shavit Z, Teitelbaum SL, Reitsma P, Hall A, Pegg LE, Trial J, Kahn AJ 1983 Induction of monocytic differentiation and bone resorption by 1,25-dihydroxyvitamin D_3. Proc Natl Acad Sci USA **80**:5907–5910.
101. Lemire JM, Adams JS, Kermani-Arab V, Bakke AC, Sakai R, Jordan SC 1985 1,25-Dihydroxyvitamin D suppresses human T helper-inducer lymphocyte activity in vitro. J Immunol **134**: 219–224.
102. Bhalla AK, Amento EP, Krane SM 1986 Differential effects of 1,25-dihydroxyvitamin D_3 on human lymphocytes and monocyte/macrophages: Inhibition of interleukin-2 and augmentation of interleukin-1 production. Cell Immunol **98**:311–322.
103. Ucla C, Roux-Lombard P, Dayer J-M, Mach B 1990 IFN-γ drastically modifies the regulation of IL-1 genes by endotoxin in U937 cells. J Clin Invest **85**:185–191.
104. Morel PA, Manolagas SC, Provvedini DM, Wegmann DR, Chiller JM 1986 Interferon-gamma-induced IA expression in WEHI-3 cells is enhanced by the presence of 1,25-dihydroxyvitamin D_3. J Immunol **136**:2181–2186.
105. Rook GAW, Steele J, Fraher L, Barker S, Karmali R, O'Riordan J, Stanford J 1986 Vitamin D_3, gamma-interferon, and control of proliferation of *Mycobacterium tuberculosis* by human monocytes. Immunology **57**:159–163.
106. Cohen MS, Mesler DE, Snipes RG, Gray TK 1986 1,25-Dihydroxyvitamin D_3 activates secretion of hydrogen peroxide by human monocytes. J Immunol 136:1049–1053.
107. Jordan SC, Lemire JM, Koeffler PK, Sakai R, Adams JS 1992 Immunoregulatory and prodifferentiating effect of 1,25-dihydroxyvitamin D. In: Cunningham-Rundles (ed) Nutrient

Modulation of the Immune Response. New York, Dekker, pp.3–29.

108. Tsoukas CD, Provvedini DM, Manolagas SC 1984 1,25-Dihydroxyvitamin D_3, a novel immunoregulatory hormone. Science **224**:1438–1440.
109. Manolagas SC, Provvedini DM, Murray EJ, Tsoukas CD, Deftos LJ 1986 The antiproliferative effect of calcitriol on human peripheral blood mononuclear cells. J Clin Endocrinol Metab **63**:394–400.
110. Rigby WF, Noelle RJ, Krause K, Fanger MW 1985 The effects of 1,25-dihydroxyvitamin D_3 on human T lymphocyte activation and proliferation. J Immunol **135**:2279–2286.
111. Nun JD, Katz DR, Barker S, Fraher LJ, Hewison M, Hendy GN, O'Riordan JL 1986 Regulation of human tonsillar T-cell proliferation by the active metabolite of vitamin D_3. Immunology **59**:479–484.
112. Haq AU 1986 1,25-Dihydroxyvitamin D_3 (calcitriol) suppresses IL-2 induced murine thymocyte proliferation. Thymus **8**:295–306.
113. Lemire JM, Adams JS, Sakai R, Jordan SC 1984 1,25-Dihydroxyvitamin D suppresses proliferation and immunoglobulin production by normal human peripheral blood mononuclear cells. J Clin Invest **74**:857–861.
114. Rigby WFC, Denome S, Fanger MW 1987 Regulation of lymphokine production and human T lymphocyte activation by $1,25(OH)_2D_3$. J Clin Invest **79**:1659–1664.
115. Lemire JM, Adams JS 1992 1,25-Dihydroxyvitamin D inhibits delayed-type hypersensitivity mediated by T-cell clones inducing experimental autoimmune encephalomyelitis. J Bone Miner Res **7**:171–178.
116. Lemire JM, Archer DC, Beck L, Spiegelberg HL 1995 Immunosuppressive actions of 1,25-dihydroxyvitamin D_3: Preferential inhibition of Th_1 functions. J Nutr **125**:1704S–1708S.
117. Lemire JM 1995 Immunomodulatory actions of 1,25-dihydroxyvitamin D_3. J Steroid Biochem Mol Biol **53**:599–602.
118. Barnes PF, Modlin RL, Bikle DD, Adams JS 1989 Transpleural gradient of 1,25-dihydroxyvitamin D in tuberculous pleuritis. J Clin Invest **83**:1527–1532.
119. Adams JS, Modlin RL, Diz MM, Barnes PF 1989 Potentiation of the macrophage 25-hydroxyvitamin D-1-hydroxylation reaction by human tuberculous pleural effusion fluid. J Clin Endocrinol Metab **69**:457–460.
120. Rook GAW, Taverne J, Leveton C, Steele J 1987 The role of gamma-interferon, vitamin D_3 metabolites and tumour necrosis factor in the pathogenesis of tuberculosis. Immunology **62**:229–234.
121. Rook GAW 1988 The role of vitamin D in tuberculosis. Am Rev Respir Dis **138**:768–770.
122. Mawer EB, Hayes ME, Still PE, Davies M, Lumb GA, Palit J, Holt PJL 1991 Evidence for nonrenal synthesis of 1,25-dihydroxyvitamin D in patients with inflammatory arthritis. J Bone Miner Res **6**:733–739.
123. Hayes ME, Bayley D, Still P, Palit J, Denton J, Freemont AJ, Cooper RG, Mawer EB 1992 Differential metabolism of 25-hydroxyvitamin D_3 by cultured synovial fluid macrophages and fibroblast-like cells from patients with arthritis. Ann Rheum Dis **51**:220–226.
124. Hayes ME, O'Donoghue DJ, Ballardie FW, Mawer EB 1987 Peritonitis induces the synthesis of 1α,25-dihydroxyvitamin D_3 in macrophages from CAPD patients. FEBS Lett **220**:307–310.
125. Shany S, Rapoport J, Zuili I, Gavriel A, Lavi N, Chaimovitz C 1991 Metabolism of 25-OH-vitamin D_3 by peritoneal macrophages from CAPD patients. Kidney Int **39**:1005–1011.
126. Sharma OP 1996 Vitamin D, calcium, and sarcoidosis. Chest **109**:535–539.
127. James DG, Neville E, Siltzbach LE 1976 A worldwide review of sarcoidosis. Ann NY Acad Sci **278**:321–334.
128. Studdy PR, Bird R, Neville E, James DG 1980 Biochemical findings in sarcoidosis. J Clin Pathol **33**:528–533.
129. Rizzato G, Montemurro L, Fraioli P 1992 Bone mineral content in sarcoidosis. Semin Respir Med **13**:411–423.
130. Vergnon GM, Chappard D, Mounier D 1988 Phosphocalcic metabolism, bone quantitative histomorphometry and clinical activity in 10 cases of sarcoidosis. In: Grassi C, Rizzato G, Pozzi E (eds) Sarcoidosis and Other Granulomatous Disorders. Elsevier, Amsterdam, pp. 499–502.
131. Fallon MD, Perry III HM, Teitelbaum SL 1981 Skeletal sarcoidosis with osteopenia. Metab Bone Dis Res **3**:171–174
132. Need AG, Phillips PJ 1980 Hypercalcaemia associated with tuberculosis. Br Med J **280**:831.
133. Felsenfeld AJ, Drezner MK, Llach F 1986 Hypercalcemia and elevated calcitriol in a maintenance dialysis patient with tuberculosis. Arch Intern Med **146**:1941–1945.
134. Gkonos PJ, London R, Hendler ED 1984 Hypercalcemia and elevated 1,25-dihydroxyvitamin D levels in a patient with end-stage renal disease and active tuberculosis. N Engl J Med **311**:1683–1685.
135. Epstein S, Stern PH, Bell NH, Dowdeswell I, Turner RT 1984 Evidence for abnormal regulation of circulating 1α,25-dihydroxyvitamin D in patients with pulmonary tuberculosis and normal calcium metabolism. Calcif Tissue Int **36**:541–544.
136. Bell NH, Shary J, Shaw S, Turner RT 1985 Hypercalcemia associated with increased circulating 1,25-dihydroxyvitamin D in a patient with pulmonary tuberculosis. Calcif Tissue Int **37**: 588–591.
137. Cadranel J, Garabedian M, Milleron B, Guillozo H, Akoun G, Hance AJ 1990 $1,25(OH)_2D_3$ production by T lymphocytes and alveolar macrophages recovered by lavage from normocalcemic patients with tuberculosis. J Clin Invest **85**:1588–1593.
138. Cadranel JL, Garabedian M, Milleron B, Guillozzo H, Valeyre D, Paillard F, Akoun G, Hance AJ 1994 Vitamin D metabolism by alveolar immune cells in tuberculosis: Correlation with calcium metabolism and clinical manifestations. Eur Res J **7**:1103–1110.
139. Isaacs RD, Nicholson GI, Holdaway IM 1987 Miliary tuberculosis with hypercalcaemia and raised vitamin D concentrations. Thorax **42**:555–556.
140. Shai F, Baker RK, Addrizzo JR, Wallach S 1972 Hypercalcemia in mycobacterial infection. J Clin Endocrinol Metab **34**:251–256.
141. Braman SS, Goldman AL, Schwarz MI 1973 Steriod-responsive hypercalcemia in disseminated bone tuberculosis. Arch Intern Med **90**:327–328.
142. Hoffman VH, Korzeniowski OM 1986 Leprosy, hypercalcemia, and elevated serum calcitriol levels. Ann Intern Med **105**: 890–891.
143. Ryzen E, Rea TH, Singer FR 1988 Hypercalcemia and abnormal 1,25-dihydroxyvitamin D concentrations in leprosy. Am J Med **84**:325–329.
144. Kantarijian HM, Saad MF, Estey EH, Sellin RV, Samaan NA 1983 Hypercalcemia in disseminated candidiasis. Am J Med **74**:721–724.
145. Walker JV, Baran D, Yakub YN, Freeman RB 1977 Histoplasmosis with hypercalcemia, renal failure, and papillary necrosis. Confusion with sarcoidosis. JAMA **237**:1350–1352.
146. Parker MS, Dokoh S, Woolfenden JM, Buchsbaum HW 1984 Hypercalcemia in coccidioidomycosis. Am J Med **76**:341–343.
147. Ahmed B, Jaspan JB 1993 Case report: Hypercalcemia in a

patient with AIDS and *Pneumocystis carinii* pneumonia. Am J Med Sci **306**:313–316.
148. Kozeny G, Barbato A, Bansal VK, Vertuno LL, Hano JE 1984 Hypercalcemia associated with silicone-induced granulomas. N Engl J Med **311**:1103–1105.
149. Jurney TH 1984 Hypercalcemia in a patient with eosinophilic granuloma. Am J Med **76**:527–528.
150. Edelson GW, Talpos GB, Bone III HG, 1993 Hypercalcemia associated with Wegener's granulomatosis and hyperparathyroidism: etiology and management. Am J Nephrol **13**:275–277.
151. Stoeckle JD, Hardy HL, Weber AL 1969 Chronic beryllium disease. Long-term follow-up of sixty cases and selective review of the literature. Am J Med **46**:545–561.
152. Cook JS, Stone MS, Hansen JR 1992 Hypercalcemia in association with subcutaneous fat necrosis of the newborn: Studies of calcium-regulating hormones. Pediatrics **90**:93–96.
153. Kreisberg RA 1994 Stopping short of certainty. N Engl J Med **331**:42–45.
154. Zaloga GP, Eil C, Medbery CA 1985 Humoral hypercalcemia in Hodgkin's disease. Arch Intern Med **145**:155–157.
155. Rosenthal N, Insogna KL, Godsall JW 1985 Elevations in circulating 1,25-dihydroxyvitamin D_3 in three patients with lymphoma-associated hypercalcemia. J Clin Endocrinol Metab **60**:29–33.
156. Davies M, Mawer EB, Hayes ME, Lumb GA 1985 Abnormal vitamin D metabolism in Hodgkin's lymphoma. Lancet **1**:1186–1188.
157. Adams JS, Fernandez M, Endres DB, Gill PS, Rasheed S, Singer FR 1979 Hypercalcemia, hypercalciuria, and elevated serum 1,25-dihydroxyvitamin D concentrations in patients with AIDS- and non-AIDS-associated lymphoma. Blood **73**:235–239.
158. Seymour JF, Gagel RF 1993 Calcitriol: The major humoral mediator of hypercalcemia in Hodgkin's lymphomas. Blood **82**:1383–1394.
159. Seymour JF, Gagel RF, Hagemeister FB, Dimopoulos MA, Cabanillas F 1994 Calcitriol production in hypercalcemic and normoclacemic patients with non-Hodgkin lymphoma. Ann Intern Med **121**:633–640.
160. Davies M, Hayes ME, Liu Yin JA, Berry JL, Mawer EB 1994 Abnormal synthesis of 1,25-dihydroxyvitamin D in patients with malignant lymphoma. J Clin Endocrinol Metab **78**:1202–1207.
161. Fetchick DA, Bertolini DR, Sarin PS 1986 Production of 1,25-dihydroxyvitamin D_3 by human T cell lymphotrophic virus-I-transformed lymphocytes. J Clin Invest **78**:592–596.
162. Schienman SJ, Kelberman MW, Tatum AH, Zamkoff KW 1991 Hypercalcemia with excess serum 1,25-dihydroxyvitamin D in lymphomatoid granulomatosis/angiocentric lymphoma. Am J Med Sci **301**:178–181.
163. Dusso AS, Finch J, Brown A, Ritter C, Delmez J, Schreiner G, Slatopolsky E 1991 Extrarenal production of calcitriol in normal and uremic humans. J Clin Endocrinol Metab **72**:157–164.
164. Bikle DD, Pillai S, Gee E, Hincenbergs M 1991 Tumor necrosis factor-α regulation of 1,25-dihydroxyvitamin D production by human keratinocytes. Endocrinology **129**:33–38.
165. Bikle DD, Halloran BP, Riviere JE 1994 Production of 1,25-dihydroxyvitamin D_3 by perfused pig skin. J Invest Dermatol **102**:796–798.
166. Giannini S, Nobile M, Castrignano R, Pati T, Tasca A, Villi G, Pellegrini F, D'Angelo A 1993 Possible link between vitamin D and hyperoxaluria in patients with renal stone disease. Clin Sci **84**:51–54.
167. Verhaeghe J, Bouillon R 1992 Calciotropic hormones during reproduction. J Steroid Biochem Mol Biol **41**:469–477.
168. Gray TK, Lester GE, Lorenc RS 1979 Evidence for extrarenal 1-hydroxylation of 25-hydroxyvitamin D_3 in pregnancy. Science **204**:1311–1313.
169. Tanaka Y, Halloran B, Schnoes HK, DeLuca HF 1979 In vitro production of 1,25-dihydroxyvitamin D_3 by rat placental tissue. Proc Natl Acad Sci USA **76**:5033–5035.
170. Weisman Y, Vargas A, Duckett G 1978 Synthesis of 1,25-dihydroxyvitamin D in the nephrectomized pregnant rat. Endocrinology **103**:1992–1996.
171. Mawer EB, Hayes ME, Heys SE, Davies M, White A, Stewart MF, Smith GN 1994 Constitutive synthesis of 1,25-dihydroxyvitamin D_3 by a human small cell lung cancer cell line. J Clin Endocrinol Metab **79**:554–560.
172. Hayes ME, Bayley D, Drayson M, Freemont AJ, Denton J, Davies M, Mawer EB 1991 Metabolism of 25-hydroxyvitamin D_3 to 24,25-dihydroxyvitamin D_3 by blood derived macrophages from a patient with alveolar rhabdomyosarcoma during short-term culture and 1a,25-dihydroxyvitamin D_3 after long-term culture. J Steroid Biochem Mol Biol **38**:301–306.
173. Holick MF, Adams JS 1990 Vitamin D metabolism and biological function. In: Avioli LV, Krane SM (eds) Metabolic Bone Disease and Clinically Related Disorders. Saunders, Philadelphia, Pennsylvania, pp. 155–195.
174. Adams JS 1996 Hypercalcemia due to granuloma-forming disorders. In: Favus MJ, Christakos S (eds) Primer on the Metabolic Bone Diseases and Disorders of Mineral Metabolism. Lippincott-Raven, Philadelphia, Pennsylvania, pp. 206–209.
175. Studzinski GP, Mclane JA, Uskokovic MR 1993 Signaling pathways for vitamin D-induced differentiation: Implications for therapy of proliferative and neoplastic diseases. Crit Rev Eukaryotic Gene Expression **3**:279–312.
176. Bouillon R, Okamura WH, Norman AW 1995 Structure–function relationships in the vitamin D endocrine system. Endocr Rev **16**:200–256.
177. Feldman D 1986 Imidazole derivatives as inhibitors of steroidogensis. Endocr Rev **7**:409–430.
178. Glass AR, Eil C 1986 Ketoconazole-induced reduction in serum 1,25-dihydroxyvitamin D. J Clin Endocrinol Metab **63**:766–769.
179. Glass AR, Eil C 1988 Ketoconazole-induced reduction in serum $1,25(OH)_2D_3$ and total serum calcium in hypercalcemic patients. J Clin Endocrinol Metab **66**:934–938.
180. Saggese G, Bertelloni S, Baroncelli GI, DiNero, G 1993 Ketoconazole decreases the serum ionized calcium and 1,25-dihydroxyvitamin D levels in tuberculosis-associated hypercalcemia. AJDC **147**:270–273.
181. Pont A, Williams PL, Loose DS, Feldman D, Reitz RE, Bochra C, Stevens DA 1982 Ketoconazole blocks adrenal steroid synthesis. Ann Intern Med **97**:370–372.
182. Adams JS Lee G 1997 Recovery of bone mineral density following exogenous vitamin D intoxication. Ann Intern Med in press.
183. Gacad MA, Chen H, Arbelle JE, LeBon T, Adams JS 1997 Functional characterization and purification of an intracellular vitamin D-binding protein in vitamin D-resistant New World primate cells. J Biol Chem **272**:8433–8440.

CHAPTER 56

The Hypocalcemic Disorders: Differential Diagnosis and Therapeutic Use of Vitamin D

THOMAS O. CARPENTER Department of Pediatrics, Yale University School of Medicine, New Haven, Connecticut

KARL L. INSOGNA Department of Medicine, Yale University School of Medicine, New Haven, Connecticut

I. PHYSIOLOGY

A. Hypocalcemia and Its Manifestations

Hypocalcemia refers to an abnormally low concentration of ionized calcium in extracellular fluid, almost invariably sampled from the bloodstream. Manifestations of hypocalcemia are related to increased neuromuscular irritability [1]. Tetany is the classic sign of hypocalcemia, yet it is variable in presentation. Paresthesias often occur first around the mouth or in the fingertips and may progress to overt spasm of the muscles of the face and extremities, the latter typified by carpopedal spasm. More subtle presentations have included complaints of writer's cramp or generalized stiffness. Children with tetanic laryngospasm due to hypocalcemia have been mistakenly diagnosed with croup [2]. Infants are more likely than adults to present with jitteriness or twitching, which can progress to overt tonic–clonic seizure activity. Lethargy and cyanosis have also been described in this age group. The term latent tetany refers to signs elicitable with provocative stimuli such as ischemia (Trousseau test) or percussion (e.g., of the facial nerve to elicit Chvostek's sign). Neither the degree of hypocalcemia nor the rapidity with which it develops necessarily correlate with clinical manifestations.

Hypomagnesemia or hyperkalemia may present with similar findings, which can be exacerbated in the setting of hypocalcemia. Conversely, hypermagnesemia or hypokalemia can mask symptoms in a hypocalcemic individual. Abnormalities of repolarization of cardiac musculature may result in a prolonged Q-T interval on the electrocardiogram (EKG). The Q-T interval corrected for heart rate [Q-T_c, which equals Q-T/(R-R interval)$^{1/2}$], is normally less than 0.40 ± 0.04 sec. This abnormality is not always present during hypocalcemia, and it may also be seen in hypokalemia. Cardiac failure may occur in the setting of hypocalcemia [3]. Papilledema has also been attributed to hypocalcemia [4].

Chronic hypocalcemia caused by deficient calcium intake during periods of significant skeletal growth may result in rickets and osteomalacia (see Chapter 42). Severe osteoporosis and dental abnormalities have also been reported in long-standing untreated hypoparathyroidism [5,6]. A mineralization defect has been described in hypoparathyroidism; however, these skeletal consequences appear to be more prevalent in the setting of endemic calcium deficiency, where secondary hyperparathyroidism is evident. Basal ganglia calcifications are typical findings in long-standing hypoparathyroidism as well [7]. Abnormalities in the integument including dry skin, coarse hair, and a form of psoriasis that responds to normalization of the serum calcium concentration [8] have all been described in states of long-standing hypocalcemia.

Regulatory mechanisms maintain the concentration of ionized calcium within a remarkably narrow range of 4.48 to 5.28 mg/dl in whole blood [9]. The ionized fraction of total serum calcium is generally estimated to be approximately 50%, with the remainder of the total serum calcium being bound to serum proteins, most notably albumin, and to a lesser extent complexed with anions, such as citrate or sulfate. Only the ionized fraction of total serum calcium is physiologically important, and it is this component that is regulated on a minute-to-minute basis.

Although it is now possible to measure ionized calcium in many large clinical laboratories, the specimen must usually be obtained anaerobically and analyzed promptly. Therefore, total serum calcium is often used as an indirect assessment of the ionized calcium fraction. A decrease in serum protein concentrations (particularly albumin) often results in reduced total serum calcium concentrations, with preservation of a normal concentration of ionized calcium. Patients in whom this occurs will be asymptomatic, displaying none of the signs or symptoms of hypocalcemia. These findings are often present in patients with nephrotic syndrome, chronic illness, malnutrition, cirrhosis, and volume overexpansion.

A number of clinical guidelines have been suggested that correct for the effects of decreased serum albumin on total serum calcium concentration. One commonly cited rule of thumb is to add 0.8 mg/dl to the total serum calcium for every 1 g/dl decline in serum albumin below 4.0 g/dl. However, these estimates have been shown to be somewhat inaccurate under many circumstances, and it is preferable to directly determine the ionized calcium concentration [10].

B. Role of Parathyroid Hormone in the Acute Defense of Ionized Serum Calcium Concentration

Parathyroid hormone (PTH) is secreted by the parathyroid glands in response to a fall in ionized serum calcium concentration [9]. The relationship between decrements in ionized calcium within the physiological range and increments in PTH secretion is quite steep, permitting rapid and substantial changes in PTH secretion in response to minor fluctuations in ionized calcium [9]. The details of this response have been elucidated with the cloning of the seven transmembrane domain, G-protein-coupled calcium sensor, which is expressed in the parathyroid glands as well as in a variety of other tissues [11]. A rise in ionized calcium suppresses PTH secretion by activating this receptor [11].

Parathyroid hormone acts to regulate ionized calcium through its effects in three principal target tissues, bone, kidney, and intestine. The cellular mechanisms of action of PTH have been clarified by the cloning of the PTH receptor, also a seven transmembrane-domain, G-protein-coupled receptor [12]. Downstream signaling from the PTH receptor involves activation of both protein kinase A- and protein kinase C-dependent pathways [13–15].

Parathyroid hormone acts to increase bone resorption, liberating calcium from the mineralized matrix of bone and thereby increasing the ionized calcium concentration of the extracellular fluid [9]. Details of the cellular mechanism by which this occurs are incompletely understood. The principal target cell for PTH in bone appears to be the osteoblast or osteoblast-like stromal cell, rather than the resorbing cell itself, the osteoclast [16,17]. In response to PTH, osteoblasts or osteoblast-like stromal cells release locally active cytokines that appear to increase the number and activity of osteoclasts [18]. As the resorptive response to PTH is quite rapid, the early effects of the hormone are likely mediated by activation of existing osteoclasts. A miscible pool of incompletely mineralized calcium at the endosteal surface of bone may be the most readily accessible source of calcium liberated in response to this action. It has been speculated that osteocytes may mediate release of calcium from this pool [19]. These effects are evident in 6–12 hr [20].

The renal effects of PTH to defend serum calcium occur within minutes. PTH increases calcium reabsorption in the distal tubule. This effect is greatest in the distal convoluted tubule, where a sodium/calcium exchanger is regulated by PTH [21]. In the proximal renal tubule, PTH acts via a cAMP-dependent mechanism to decrease phosphate reabsorption, resulting in increased phosphaturia. PTH effects this change by inhibiting sodium/phosphate cotransporter activity [22]. These two effects both serve to acutely increase serum calcium; one by causing less calcium to be excreted by the kidney, the other by lowering circulating concentrations of phosphate which favors an increase in ionized calcium.

The third site of action of PTH in the defense of serum calcium is at the intestine. This is an indirect effect, described in detail below, and is a consequence of the ability of PTH to stimulate the renal production of 1,25-dihydroxyvitamin D [1,25$(OH)_2$D].

C. Vitamin D in the Long-term Maintenance of Eucalcemia

Long-term eucalcemia is maintained, in large part, via the vitamin D endocrine system. This system operates in the classic manner of a steroid hormone, resulting in *de novo* protein synthesis directed by vitamin D responsive genes [23] as discussed in detail in Section II of this volume. As noted above, acute changes in serum ionized calcium levels are sensed by G-protein-coupled calcium-sensing receptors located within the parathyroid cell membrane [11]. PTH acts rapidly to correct a fall in serum calcium, and a sustained increase in PTH also stimulates production of 1,25$(OH)_2$D, which enhances intestinal calcium absorption. PTH mediates this change by increasing the activity of the renal 25-hydroxyvitamin D (25OHD) 1-hydroxylase enzyme complex, located in the inner mitochondrial membrane of renal tubular cells. A number of physiological studies have demonstrated increased enzyme activity in animals that were administered PTH [24,25] and decreased activity following parathyroidectomy [26]. The mechanism is discussed in Chapters 5 and 17. PTH also acutely regulates the 1-hydroxylase enzyme complex by altering the phosphorylation state of the associated ferredoxin molecule [27]. Direct stimulation of 1-hydroxylase activity in the absence of PTH can also occur because a low serum calcium level enhances enzyme activity in parathyroidectomized rats [28,29] and in hypoparathyroid humans [5].

Increased synthesis of 1,25$(OH)_2$D results in greater circulating levels of the metabolite, which gain access to specific vitamin D receptors (VDR). The hormone–receptor complex then binds to vitamin D response elements (VDRE) in the regulatory regions of target genes (Chapters 8–10). Of importance to long-term control of calcium homeostasis is the induction by 1,25$(OH)_2$D of the intestinal 9-kDa calcium binding protein, calbindin-D_{9k} (Chapter 14), which is thought to play a role in vitamin D-mediated increases in calcium absorption in the jejunum and duodenum [30] (see also Chapters 14 and 16). Induction of calbindin takes hours to days and is more sustained than the acute compensatory changes that occur with the initial rise in PTH in response to hypocalcemia. These features define a classic feedback loop suitable for long-term calcium homeostasis.

The importance of this system is emphasized by clinical observations in patients with severe vitamin D deficiency (see below). During vitamin D deprivation, the initial decline in ionized serum calcium results in secondary hyperparathyroidism, which maximizes 1,25$(OH)_2$D production and allows for maintenance of eucalcemia in the early stages. Eventually, this compensatory mechanism fails, and intestinal calcium absorption is sufficiently compromised that frank hypocalcemia develops. This may be compounded by an induced resistance to PTH seen in severe hypocalcemic or vitamin D-deficient states [1].

In children with hereditary resistance to vitamin D (HVDRR; see Chapter 48) caused by mutations in the vitamin D receptor, the compensatory changes described above are interrupted by defective vitamin D receptors and the inability of 1,25$(OH)_2$D to signal to the nucleus [31]. Such patients can have severe hypocalcemia leading to convulsions, coma, and death.

The vitamin D system has further complexities that are currently not well understood. For example, some children with vitamin D deficiency (defined by low 25OHD levels) may manifest hypocalcemia even when circulating 1,25$(OH)_2$D levels are actually elevated. Possible explanations for this phenomenon include decrements in expression of calbindin during hypocalcemia, or a requirement for other circulating vitamin D metabolites which are not present. This situation is discussed further in Chapter 31. Whether changes in the concentration of VDR levels play a role in this "resistant" state is not clear, as receptor levels have been reported to increase, decrease, or not change with manipulation of the ambient calcium concentration [32–34] (see Chapters 11 and 54). Finally, the system must return to basal levels of function after calcium availability is restored. The absence of this self-regulating feature would result in hypercalcemia.

D. Biochemical Changes Induced by Hypocalcemia

As noted above, the immediate consequence of hypocalcemia is secretion of PTH. In addition to increasing serum calcium levels, PTH stimulates renal phosphate (P_i) excretion. The fall in serum phosphate may, however, be compensated by sufficient mobilization of phosphate (as well as calcium) from bone, so that circulating phosphate remains largely unchanged. The principle of mass action is thought to maintain the stability of the $Ca \times P_i$ ion product in the blood. As a consequence, local concentrations of the two major mineral components of hydroxyapatite (Ca and P_i) are able to influence the rate of movement in and out of the mineral phase of bone:

$$[Ca] + [P_i] \rightleftharpoons [HA].$$

Thus, a fall in ionized Ca would favor an increase in serum phosphate concentration. Given all of these influences, sustained hypocalcemia usually results in a biochemical picture of secondary hyperparathyroidism, elevated $1,25(OH)_2D$ levels, and variable changes in serum phosphate. If hypocalcemia develops in the setting of diminished or absent PTH function, serum phosphate is usually elevated, owing to increased renal phosphate reabsorption. In this instance, treatment with $1,25(OH)_2D$ would also increase serum phosphate, since this metabolite enhances intestinal phosphate absorption.

The effect of hypocalcemia on circulating vitamin D metabolites is complex. An increase in the biosynthesis of $1,25(OH)_2D$ occurs, as reviewed above. This is largely secondary to the induced increase in circulating PTH but can be a direct consequence of the fall in calcium ion concentration. It has been determined that calcium deprivation results in a general increase in turnover of the parent vitamin D metabolite, 25OHD, such that vitamin D stores are depleted at a more rapid state than normal [35]. The clinical implication of this finding is that susceptibility to vitamin D deficiency may be greater when concomitant calcium deprivation is present.

II. DIFFERENTIAL DIAGNOSIS OF HYPOCALCEMIA

A. Classification

A rapid increase in PTH serves as the major defense against acute hypocalcemia. We therefore classify these disorders as those which manifest hypocalcemia (1) due to abnormalities of PTH availability, (2) due to PTH resistance as a consequence of PTH/PTH-related peptide (PTHrP) receptor or postreceptor defects, or (3) in the setting of normal or increased PTH activity and normal PTH receptor function. Thus, although many of the etiologies relate to abnormalities in PTH, they are very relevant to this book because vitamin D metabolism is always involved and vitamin D is the cornerstone of therapy.

1. Hypocalcemia Due to Abnormalities of PTH Availability

A variety of congenital or acquired disorders can lead to developmental failure of the parathyroid glands, failure of functional hormone production, or destruction of the glands. These all present as hypocalcemia, usually with attendant hyperphosphatemia and undetectable or inappropriately low levels of circulating PTH.

a. Failure of Organogenesis: DiGeorge Sequence DiGeorge sequence is an uncommon developmental disorder that affects the third and fourth branchial clefts and results in dysgenesis of the thymus and the parathyroid glands [36]. Tetany and seizures are common features of the early course of infants with DiGeorge sequence. However, abnormalities in T-cell function with subsequent increased risk for infection often become a major feature of this disorder later in life [37]. Although the gene(s) responsible for the defect in humans is not known, deletion of the homeobox gene, *hox1.5* in mice, recapitulates many of the findings of DiGeorge sequence, suggesting that a similar defect may ultimately prove to be the pathogenesis of the human disorder [38]. It is now known that microdeletions of chromosome 22 are found in many patients with this disorder.

b. Idiopathic Hypoparathyroidism As discussed in Sections II.A.1.c–e, a variety of causes of inherited idiopathic hypoparathyroidism have now been delineated, including one family where the abnormality maps to the X chromosome. Linkage analysis has identified the Xq26-X27 region as the probable site for the molecular abnormality, which presents with failure of parathyroid gland development [39].

c. Molecular Abnormalities in the PTH Gene Parathyroid hormone is secreted and synthesized by a classic secretory pathway. The initial translation product is a prepropeptide, which requires cleavage of the aminoterminal pre and pro sequences before secretion. A family with autosomal dominant inherence of hypoparathyroidism has been reported in which a missense mutation (Cys-18 → Arg) results in an abnormal signal sequence and diminished uptake of preproPTH into the endoplasmic reticulum [40]. Another family with recessively inherited hypoparathyroidism has been reported in which the prepro sequence is deleted by a splicing mutation [41].

d. Molecular Abnormalities in the Calcium-Sensing Receptor Gene The gene for the calcium-sensing receptor has been mapped to chromosome 3 [42]. As mentioned in Section I.B, ionized calcium is a ligand for this receptor, and receptor occupancy suppresses PTH secretion. Thus, as might be anticipated and as has recently been reported, activating mutations of this receptor lead to inherited disorders of PTH secretion and autosomal dominant familial hypocalcemia [43]. One such kindred with a peculiar predisposition to nephrocalcinosis and renal insufficiency has been described [44].

e. Autoimmune Polyglandular Syndrome Type 1 An autoimmune disorder termed autoimmune polyglandular syndrome type 1 is characterized by early development of hypoparathyroidism in association with Addison's disease and mucocutaneous candidiasis. The majority of affected individuals will manifest hypocalcemia by the age of 10 [45]. In addition to Addison's disease, one-third of the patients will develop other endocrine disorders, diabetes mellitus, pernicious anemia, or premature ovarian failure [45].

f. Postsurgery Reduction in PTH Given the close anatomic relationship of the parathyroid glands to the thyroid, complete or near complete extirpation of the thyroid gland as part of the management of either Graves' disease or thyroid cancer can be complicated by destruction or vascular compromise of parathyroid tissue and varying degrees of hypoparathyroidism. This should be a rare complication of thyroid surgery, and with experienced surgeons occurs with a frequency less than 10%.

Even when destruction of the parathyroid glands does not occur following neck surgery, so-called stunned parathyroids with transient declines of approximately 1 mg/dl in total serum calcium are often observed in the first 24 to 48 hr postoperatively. This is presumably due to transient vascular or mechanical damage to the glands. Considerable variability in the degree of hypoparathyroidism following neck surgery occurs, ranging from asymptomatic reduction in parathyroid reserve to frank tetany, requiring chronic therapy with vitamin D and calcium.

g. Infiltrative Diseases and Deposition of Heavy Metals Although uncommon, malignant metastasis to the parathyroid glands with hypoparathyroidism has been reported, usually with breast cancer [46]. It has been postulated that granulomatous involvement of the parathyroids in sarcoidosis can lead to hypoparathyroidism [47]. Patients with transfusion-dependent thalassemia can develop hypoparathyroidism due to hemochromatosis secondary to deposition of iron in the glands [48]. In Wilson's disease hypoparathyroidism can occur, presumably because of copper deposition [49]. Finally, impaired parathyroid reserve has been reported in diabetic patients with uremia [50].

h. Radiation Although the parathyroid glands are quite resistant to radiation, hypoparathyroidism following radioactive iodine treatment for hyperthyroidism has been described [51].

i. Functional Defects in PTH Secretion Magnesium is an important cofactor for parathyroid hormone secretion, apparently required for release of the stored hormone from secretory granules. In severe cases of hypomagnesemia, usually with circulating levels below 1 mg/dl, suppressed parathyroid secretion can occur [52]. This can be seen in the settings of chronic gastrointestinal disease, nutritional deficiency especially in alcoholics, or therapy with cis-platinum. Resistance to the action of PTH at the level of bone and kidney may also contribute to the hypocalcemia seen in the setting of magnesium deficiency. Replenishment of magnesium stores promptly restores parathyroid function to normal.

Transient hypocalcemia in neonates has been reported to be associated with maternal hyperparathyroidism.

2. Hypocalcemia Due to Resistance to the Actions of PTH

Several disorders of PTH action have hypocalcemia as their principal manifestation.

a. Pseudohypoparathyroidism Peripheral tissue insensitivity or resistance to PTH is classically termed pseudohypoparathyroidism (PHP) [53]. The characteristic biochemical manifestations of PHP are hypocalcemia and hyperphosphatemia, as in hypoparathyroidism; however, circulating levels of PTH are elevated, rather than low or undetectable. The renal tubule is the primary site of PTH resistance, although variable degrees of skeletal resistance, depending on treatment status, have also been reported [54]. However, if the skeletal response is unimpaired, lesions characteristic of hyperparathyroidism, including osteitis fibrosa cystica, can develop. PTH stimulates renal cAMP production, and levels of cAMP increase in the urine following administration of the hormone. A direct correlation has been demonstrated between the degree of PTH resistance (as assessed by the magnitude of the change in cAMP excretion or renal phosphate threshold) and the ambient circulating PTH level [55].

The renal cAMP response is the basis of a diagnostic test that allows partial classification of this heterogeneous group of disorders. Individuals with PHP that demonstrate a blunted urinary cAMP response have PHP type I. Those that generate a normal cAMP response have type II PHP.

PHP type I has been further characterized into types Ia, Ib, and Ic. Type Ia describes those individuals with the Albright's hereditary osteodystrophy (AHO) phenotype, which includes short stature and large frame, broad facies, and shortened fourth metacarpals. Soft tissue calcifications and multiple endocrine abnormalities are often present. These individuals often have a mutation in the α subunit of the stimulatory guanine nucleotide binding regulatory protein, G_s [56]. This regulatory protein couples membrane receptors to adenylate cyclase, thereby regulating receptor-dependent cAMP production. The presence of G_s in various cell types accounts for the generalized hormone resistance that may occur. For example, affected patients often have elevated thyrotropin (TSH) levels with a compensated euthyroid state. Variable degrees of gonadotropin, antidiuretic hormone (ADH), adrenocorticotropin (ACTH), and glucagon resistance have been described. Type Ib PHP is manifest solely by PTH resistance, and the AHO phenotype is not present. It has been speculated that altered PTH/PTHrP receptor structure or function might underlie this condition, but no mutation in the coding region of the receptor gene has been found yet. However, altered regulation of the PTH receptor gene by glucocorticoids has been shown in tissues of a patient with type Ib PHP [57]. Patients without a mutation in $G_{s\alpha}$ but with associated hormonal abnormalities and/or AHO have been classified as having type 1c PHP. The catalytic subunit of adenylate cyclase is a possible site for the defect in this condition.

A diagnosis of type II PHP appears to describe a variety of defects distal to cAMP generation in the cascade of hormone action. There is no distinct phenotype, although various autoimmune findings have been described in some patients. Finally, others have suggested that a circulating PTH inhibitor may play a role in the pathogenesis of PHP and have suggested that this inhibitor may be generated by parathyroid tissue itself [58]. Resistance to PTH has also been described in hypomagnesemia, as described below.

b. Hypomagnesemia Magnesium deficiency can interfere with parathyroid secretion and function [52]. Serum magnesium levels are usually moderately to severely depressed (below the range of 1.0–1.4 mg/dl) before this occurs. Despite hypocalcemia, PTH levels may be inappropriately low or only modestly elevated, and tetany refractory to calcium supplementation can ensue. Hypomagnesemia, per se, may cause tetanic symptoms, although concomitant hypocalcemia is more common. Insufficient PTH secretion is the most widely accepted cause of refractory hypocalcemia in magnesium deficiency [52], although it has been suggested that resistance to the calcemic actions of PTH and vitamin D may play a role as well [59]. Impairment of vitamin D synthesis may also be at work [60]. As PTH stimulates conversion of 25OHD to $1,25(OH)_2D$, hypoparathyroidism may result in low circulating $1,25(OH)_2D$ levels, further compromising the body's defense against hypocalcemia [61]. Whether target tissue resistance to infused PTH occurs in magnesium deficiency remains controversial, and it has been suggested that this apparent resistance may simply reflect differences in the basal levels of circulating PTH [62]. To further complicate matters, generalized malnutrition including vitamin D deficiency is often present in hypomagnesemic patients [63]. Hypomagnesemia has been associated with alcohol abuse and may result from inherited disorders of magnesium excretion and/or absorption [64]. It can also be induced by the renal tubular effects of several drugs, including amphotericin B, aminoglycoside antibiotics, chemotherapeutic agents (particularly cis-platinum), diuretics, and cyclosporin.

3. Hypocalcemia in the Setting of Normal or Increased PTH Activity and Normal PTH Receptor Function

Despite normal PTH function and downstream signaling from its receptor, hypocalcemia can still occur due to disturbances in skeletal homeostasis, vitamin D metabolism, and a variety of medical illnesses.

a. Neonatal Hypocalcemia The newborn infant undergoes an acute transition to independently regulated mineral homeostasis at parturition (see also Chapter 34). The maternal source of calcium is eliminated, and the infant's circulating calcium level transiently decreases, with recovery occurring by the third day postpartum. Infants of diabetic or preeclamptic mothers and infants who suffer perinatal asphyxia or other fetal complications may experience an exaggerated fall in serum calcium with a delayed recovery phase. Management with intravenous calcium supplementation is required in the event of symptoms or severely low serum calcium levels. This condition is referred to as "early neonatal hypocalcemia" and is usually transient. It may be associated with transient hypomagnesemia.

Hypocalcemia presenting at 5–10 days of life is referred to as "late neonatal hypocalcemia." This presentation is more typical of term infants after enteral feeding has been established. Infants of hyperparathyroid mothers may present with symptomatic hypocalcemia within this period but have also been reported to present as late as 1 year of life. Familial forms of hypoparathyroidism may present as either "early" or "late" hypocalcemia. Mild to moderate neonatal hypocalcemia commonly occurs in patients with congenital heart disease (apart from those defects common in the DiGeorge sequence) [65] and in some cases can be attributed to transient impairment of parathyroid function.

b. Hypocalcemia Due to Vitamin D Malnutrition Vitamin D synthesis in the skin requires adequate exposure to ultraviolet light. Thus vitamin D deficiency is uncommon in settings where sunlight exposure is abundant. In extremes of latitude (e.g., northern climates in North America), and where industrial pollution can interfere with transmission of UV light, normal vitamin D status is dependent on adequate dietary vitamin D intake. Supplementation of milk products with vitamin D has significantly reduced the incidence of vitamin D deficiency in North America. Despite these measures, certain populations are at risk for development of vitamin D deficiency, and severe hypocalcemia may be a presenting manifestation of the disorder (see also Chapters 33, 41, and 42).

A convergence of various risk factors for development of vitamin D deficiency occurs in breast-fed infants in the first 18 months of life. Presentation appears to be most common during the winter or early spring in northern U.S. cities. The limited direct sunlight exposure during the winter season is a major factor. Black children are at greater risk due to the greater quanta of UV light required to penetrate melanin in the pigmented dermis and induce previtamin D formation [66]. Breast milk contains only small amounts of vitamin D, even when the mother is receiving pharmacological doses of the vitamin. Others have pointed out that dietary practices, including vegetarianism and high grain intake, may place infants at greater risk for the development of this condition [67]. Another group at high risk for development of vitamin D deficiency is found at the other extreme of life, the elderly, because of general nutritional compromise and limited sunlight exposure (see Chapter 43).

Biochemical findings in these conditions vary with the severity or duration of deficiency. Most agree that serum calcium levels in moderate to severe vitamin D deficiency are often normal, compensated by secondary elevations in PTH [68]. In severe vitamin D deficiency, however, overt hypocalcemia is usually manifest, despite elevated circulating PTH. Serum phosphate levels tend to be slightly low or normal. Circulating alkaline phosphatase activity of bone origin is usually markedly elevated in children and can be elevated in adults. The best available test to assess total body vitamin D status is the level of circulating 25OHD. Levels of this metabolite are low in vitamin D deficiency but may rise to normal with recent ingestion of vitamin D or significant sunlight exposure, whereas bone symptoms such as pain, and signs such as leg bowing, persist. In children, radiographs of rachitic extremities at the time of sampling may reveal hyperdense lines of remineralization at the physes, consistent with recent exposure to vitamin D, despite the presence of overt physical findings. Circulating $1,25(OH)_2D$ levels may be low, normal, or elevated during vitamin D deficiency. This may appear paradoxical, but it should be recognized that $1,25(OH)_2D$ circulates in 1000-fold lower concentrations than 25OHD. Furthermore, in the setting of vitamin D deficiency, production of $1,25(OH)_2D$ is maximized. Thus, efficient conversion of small amounts of newly ingested or synthesized 25OHD may markedly increase the circulating $1,25(OH)_2D$ concentration. Perhaps a more intriguing paradox in this setting is the continued malabsorption of calcium despite normal concentrations of $1,25(OH)_2D$.

The skeletal consequence of isolated severe vitamin D deficiency in children is rickets, a disorder of the epiphyseal growth plate. The defective mineralization processes ultimately result in malalignment deformities of the long bones. In adult bone, vitamin D deficiency causes osteomalacia, which is characterized histomorphometrically by excess undermineralized osteoid and a markedly delayed mineralization rate. Adults with osteomalacia may suffer painful pseudofractures, particularly in weight-bearing long bones.

c. Hypocalcemia Due to Vitamin D Malabsorption Because vitamin D is a fat-soluble vitamin, generalized fat malabsorption may result in vitamin D deficiency. Gastrointestinal diseases such as Crohn's disease, celiac sprue, and pancreatic insufficiency can be accompanied by hypocalcemia due to vitamin D malabsorption [69] (see also Chapter 51). We have also encountered children presenting with vitamin D deficiency rickets who have ultimately been diagnosed with cystic fibrosis and fat malabsorption. In addition, interruption of the enterohepatic circulation of both 25OHD and $1,25(OH)_2D$ may lower body vitamin D stores. It is also possible that the diseased bowel may not be able to respond to $1,25(OH)_2D$. Mild hypocalcemia and secondary hyperparathyroidism is also seen in cholestatic liver diseases such as primary biliary cirrhosis [69]. Circulating levels of 25OHD are reduced in this setting owing to impaired hydroxylation of vitamin D in the liver and also because of intestinal malabsorption of vitamin D.

d. Hypocalcemia Due to 1-Hydroxylase Deficiency Impaired metabolism of 25OHD to $1,25(OH)_2D$ is an autosomal recessive condition, in which hypocalcemia and severe rachitic abnormalities occur [70] (see also Chapter 47). The disorder (also termed pseudo-vitamin D-deficiency rickets or vitamin D-dependent rickets, type 1) is inherited in an autosomal recessive manner and is characterized by biochemical features similar to those of vitamin D deficiency rickets, with the exceptions that circulating 25OHD levels are normal and circulating $1,25(OH)_2D$ levels are low. Restoration of eucalcemia and correction of rickets is attainable with physiological doses of $1,25(OH)_2D_3$.

e. Hypocalcemia Due to Hereditary Resistance to $1,25(OH)_2D$ A defect in target tissue responsivity to $1,25(OH)_2D$ was clinically described shortly after the capacity to measure circulating $1,25(OH)_2D$ became available [31]. Patients with hypocalcemia caused by hereditary resistance to $1,25(OH)_2D$ have severe manifestations of vitamin D-deficiency rickets; however, serum 25OHD concentrations are normal, and $1,25(OH)_2D$ levels are usually elevated. This disorder is inherited in an autosomal recessive manner. Additional features in many patients include alopecia totalis and oligodontia. The disease is variably responsive to large doses of $1,25(OH)_2D_3$ and oral calcium therapy. In the most resistant cases, however, long-term parenteral calcium infusions can normalize the serum chemistries and cure the skeletal lesions [71]. The positive therapeutic response to parenteral calcium suggests that mediation of calcium absorption at the intestine is the critical systemic action for $1,25(OH)_2D$.

Several defects in the coding region of the vitamin D receptor (VDR) that impair or prevent either hormone or DNA binding have been described in these patients. Reduced expression of the VDR has also been described. This condition is quite rare but serves as an interesting experiment of nature in which the receptor-mediated function of $1,25(OH)_2D_3$ is specifically ablated (see also Chapter 48).

f. Hypocalcemia Due to Dietary Calcium Deficiency Although uncommon, extremely low calcium intakes have been reported to be associated with mild hypocalcemia. Nigerian and South African children with calcium intakes of 150 mg/day or less were found to have decreased serum calcium values, secondary hyperparathyroidism, and rickets [72–74]. The children were not vitamin D deficient, and their biochemical abnormalities and bone disease responded to treatment with calcium alone.

g. Hypocalcemia Induced by Hyperphosphatemia Since the 1930s, it has been appreciated that oral or parenteral phosphate can induce a decline in serum calcium concentrations. Herbert *et al.* have demonstrated that phosphate infusions lower serum calcium in both the presence and absence of parathyroid glands [75]. Moreover, they reported that the changes in peak urinary calcium excretion during phosphate administration are not sufficient to account for the fall in the serum calcium. The theory they advanced remains the best explanation available for this phenomenon and centers on the hypothesis that the calcium × phosphate molar product, when exceeded, leads to spontaneous precipitation of calcium salts in soft tissues. The Ca × P product, when estimated from total serum ion concentrations (as mg/dl), is normally taken to be ≤ 60 in adults or ≤ 80 in small children.

Hyperphosphatemia sufficient to cause hypocalcemia is usually abrupt in onset and severe in magnitude. Typical clinical settings include (1) excessive enteral or parental phosphate administration, (2) the tumor lysis syndrome, and (3) rhabdomyolysis-induced acute renal failure. Hypocalcemia induced by either oral or parental phosphate administration is often associated with soft tissue calcification. Such ectopic calcification has been observed during the treatment of hypophosphatemia due to either diabetic ketoacidosis or acute alcoholism. Adults receiving phosphate-containing enemas and infants fed "humanized" cow milk rich in phosphate may also become hypocalcemic [76,77]. Under most circumstances discontinuation of exogenous phosphate intake leads to prompt return of the serum calcium level to normal.

Hypocalcemia in the setting of massive tumor lysis results from the release of intracellular phosphate as a consequence of chemotherapy-induced cell death, usually during the treatment of rapidly proliferating neoplasms [78]. The hypocalcemia may continue beyond the period of hyperphosphatemia and appears to be aggravated by suppressed $1,25(OH)_2D$ levels [79]. The use of phosphate binding antacids, oral calcium, and, in severe cases, $1,25(OH)_2D_3$ may help to correct the serum calcium level.

Rhabdomyolysis-induced acute renal failure occurs with trauma and drug or alcohol abuse. Marked hypocalcemia can occur in the early oliguric phase, and moderate to severe hypercalcemia in the subsequent polyuric phase. Llach *et al.* have described hyperphosphatemia and suppressed serum $1,25(OH)_2D$ levels during the initial hypocalcemia, suggesting a mechanism similar to that seen in the tumor lysis syndrome [80]. The appearance of hypercalcemia and high serum $1,25(OH)_2D$ levels during the diuretic phase may result from rapid development of secondary hyperparathyroidism during the initial hypocalcemic period. Treatment includes restriction of phosphate intake and efforts to prevent hypocalcemia during the early stages of the disease.

h. Hypocalcemia Due to Accelerated Skeletal Mineralization Bone remodeling is a controlled process of tissue renewal which, in healthy individuals, results in closely matched rates of bone resorption and formation. If skeletal mineralization exceeds the rate of bone resorption, hypocalcemia can occur. One setting in which this can be observed is following surgical correction of primary or tertiary hyperparathyroidism. The abrupt cessation of PTH-mediated osteoclastic bone resorption with concomitant rapid remineralization of an undermineralized skeleton can lead to "hungry bone syndrome" with severe, even life-threatening hypocalcemia [81]. Postoperative treatment should be instituted when the serum calcium level falls below 8.0 mg/dl, using oral or parenteral calcium supplements and if necessary $1,25(OH)_2D_3$. In general, this condition resolves over the course of several days, although distinguishing it from permanent postoperative hypoparathyroidism can be difficult and requires gradual discontinuation of supportive therapy with careful monitoring. Hypocalcemia also may occur in patients with bony metastases that induce bone formation, as with prostatic and breast cancer [82].

Finally, institution of therapy for vitamin D deficiency osteomalacia or rickets can sometimes lead to a fall in serum calcium associated with rapid mineralization of previously unmineralized osteoid [83]. This is self-limited and can usually be prevented with supplemental calcium.

i. Medical Illness Hypocalcemia in the setting of renal failure results from hyperphosphatemia due to reduced renal phosphate clearance by the failing kidney and is complicated by impaired biosynthesis of $1,25(OH)_2D$ [84].

Hypocalcemia and tetany were first reported in patients with pancreatitis in the early 1940s [85]. Pancreatic lipase released from the damaged gland is believed to liberate free fatty acids that chelate calcium, thereby removing it from the extracellular fluid [86]. Hypomagnesemia resulting from poor oral intake, alcohol use, or vomiting may contribute to the hypocalcemia. Hypocalcemia in the setting of pancreatitis often suggests a poor clinical course. Treatment consists of parental calcium and magnesium when indicated.

Hypocalcemia may also occur in patients with acute sepsis. In one series, 20% of such patients evidenced reductions in ionized serum calcium [87]. Hypocalcemia in this series was associated with a poor prognosis (50% mortality, compared to 30% in eucalcemic patients). This phenomenon is most often reported with gram-negative sepsis but has occurred in toxic shock syndrome caused by staphylococcal infection [88]. The pathophysiology of hypocalcemia in these two settings is unknown. Finally, it has been suggested that parathyroid gland reserve is subnormal in patients with AIDS although hypocalcemia is not a prominent feature of that disorder [89].

j. Medications A variety of medications have been reported to decrease serum ionized calcium concentration. Many of these drugs are used to treat hypercalcemia and/or excessive bone resorption, and hypocalcemia results from their overzealous use. Thus, mithramycin, calcitonin, and the bisphosphonates can all cause hypocalcemia. In susceptible individuals, prolonged therapy with diphenylhydantoin or phenobarbital can lead to hypocalcemia, owing in part to enhanced catabolism of vitamin D metabolites [90] (Chapter 50). Citrated blood products, particularly when used for large volume transfusions or plasma plasmapheresis, can cause hypocalcemia [91]. Radiocontrast agents that contain EDTA (ethylenediaminetetraacetic acid) can also induce falls in serum ionized calcium levels [92]. Finally, foscarnet (trisodium phosphonoformate), used in the treatment of patients with AIDS, has been reported to cause a decline in ionized serum calcium, perhaps through complexing extracellular calcium [93].

III. THERAPY FOR HYPOCALCEMIA

A. Acute Management

1. Newborns

It may be necessary to treat early neonatal hypocalcemia when the circulating concentration of total serum is less than 5–6 mg/dl in premature infants, and less than 6–7 mg/dl in term infants. Appropriate emergency therapy of acute symptomatic hypocalcemia consists of a slow intravenous infusion (<1 ml/min) of calcium gluconate in a 10% (w/v) solution. The calcium gluconate salt consists of 9% elemental calcium. A well-functioning indwelling intravascular catheter should be used, to avoid extravasation. Calcium should never be administered intramuscularly because of local tissue toxicity. It is important to perform cardiac monitoring and careful observation during acute infusions. A total infusion of 1–3 ml will usually arrest convulsions, and no more than 2 mg of elemental calcium per kilogram body weight should be given as a single dose. Such bolus infusions may be repeated up to 4 times in a 24-hr period. If severe hypocalcemia persists, however, it is generally more effective to use a long-term calcium gluconate infusion, such that 20–50 mg of elemental calcium per kilogram body weight is infused over an entire 24-hr period. Calcium chloride is more irritating than calcium gluconate and is not the preferred salt for infusion. Neither bicarbonate nor phosphate should be coinfused with calcium in order to prevent precipitation of their respective calcium salts, either in the infusion line or in the vein.

2. Adults

In adults, emergency management consists of 10–20 ml of 10% calcium gluconate infused over a 10- to 15-min period. In the longer term one can dilute 10 ampules of calcium gluconate in 1 liter of 5% dextrose and, beginning at a rate of 50 ml/hr, titrate the rate to maintain the serum calcium in the low normal range. Finally, in the setting of acute exacerbations of calcium malabsorption, as may typically occur in patients with autoimmune hypoparathyroidism with associated gastrointestinal disorders, nocturnal nasogastric supplementation with calcium carbonate or calcium gluconate has been employed, providing up to 20 mg of elemental calcium per kilogram body weight per 8 hr, as necessary, until the underlying intestinal disturbance has resolved.

3. Role of Magnesium Supplementation

In the setting of hypomagnesemia, magnesium therapy may be required to restore PTH secretion and peripheral activity. Prior to administration of magnesium salts, assessment of renal function and urinary output should be performed.

Magnesium treatment in infancy consists of 5–10 mg of elemental magnesium per kilogram body weight. Although magnesium may be given intramuscularly, the intravenous route is preferred. Magnesium sulfate septahydrate ($MgSO_4 \cdot 7H_2O$) is available as a 50% solution, containing 48 mg/ml of elemental magnesium. These small volumes may be further diluted, but they should be infused slowly; the dose may be repeated every 12–24 hr. In older individuals, up to 2.4 mg of elemental Mg per kilogram body weight can be given over a 10-min period (to a maximum of 180 mg). Others prefer a continuous infusion of 576 mg of elemental magnesium over 24 hr. The length of therapy must be individualized, and maintenance with oral magnesium salts should be implemented in cases where ongoing hypomagnesemia is anticipated.

Magnesium levels should be monitored to avoid toxicity. Deep tendon reflexes can be examined, and therapy should be halted if they diminish. As with calcium therapy, cardiac monitoring should be performed and therapy stopped if EKG changes occur. Intravenous calcium gluconate is a useful antidote for magnesium intoxication, and should be available at the bedside.

B. Long-term Treatment

Many of the causes of hypocalcemia discussed above are corrected by treating the underlying disorder (e.g., vitamin D deficiency, tumor lysis syndrome). Relatively few of these disorders require maintenance therapy for hypocalcemia, and of those the most important are hypoparathyroidism and pseudohypoparathyroidism. Vitamin D-resistant states, although rare, comprise a third group of patients that require long-term treatment.

Hypoparathyroidism, whether primary or secondary to trauma or surgery, is the most frequently encountered condition that requires chronic therapy to maintain eucalcemia. In these individuals, the goal is to maintain serum calcium in the low normal range (8.5 to 9.2 mg/dl as measured by atomic absorption spectrophotometry). This will reduce the likelihood of symptoms such as circumoral tingling, signs such as carpopedal spasm, as well as more long-term complications such as cataracts. Long-term treatment with PTH is not yet practical although short-term therapy has been successful [94].

There is no single best way to achieve stable eucalcemia, although the combination of a vitamin D metabolite with calcium supplements is generally preferred. A wide variety of preparations of both are available (see Tables I and II). Because of the prolonged toxicity that occurs with excessive ingestion of either ergocalciferol or 25OHD, we generally prefer to use rapid acting preparations of vitamin D for the treatment of this disorder. Toxicity, when it occurs with these preparations, corrects more rapidly with discontinuation of the drug. Calcitriol [$1{,}25(OH)_2D_3$] and dihydrotachysterol are two preparations suited to this purpose. Both are fully active *in vivo*. In general, it is best to aim for a stable dose of one of these agents and to further regulate the serum calcium by adjusting the intake of supplemental calcium, rather than by making repeated changes in vitamin D metabolite therapy. We use calcitriol in the majority of our patients. The dose of calcitriol can range from as little as 0.25 up to 2.0 μg/day [95,96]. We have estimated the biological half-life of the drug as 12–14 hr. Hypercalcemia, when it develops during therapy with

TABLE I Oral Calcium Preparations[a]

Name	Calcium content	Cost (1996 U.S.$)/ 100 g Ca
Calcium glubionate		
Neocalglucon	115 mg/5 ml	210
Calcium gluconate		
500 mg tablet	45 mg/tablet	164
Calcium carbonate		
TUMS/Equilet	200 mg/tablet	22
TUMS-EX	300 mg/tablet	15
Calci-Chew/OsCal 500	500 mg/tablet	18
Calci-Mix	500 mg capsule for sprinkling on food	20
Nephro-Calci/Caltrate	600 mg/tablet	15
Titralac tablet	168 mg/tablet	22
Rolaids-Ca Rich	220 mg/tablet	26
Alka-Mint	340 mg/tablet	10
Calcium lactate		
325 mg tablet	42 mg/tablet	62
650 mg tablet	84 mg/tablet	38

[a] Adapted with permission from Carpenter T 1996 Rickets. In: Berg F, Inglefinger J, Wald E (eds) Gellis and Kagan's Current Pediatric Therapy, 15th ed. Saunders, Philadelphia, pp. 363–367.

calcitriol, usually resolves within 3–4 days after discontinuing the drug, although we have had patients in whom it has taken over 1 week for serum calcium to normalize.

In addition to a high calcium diet, calcium supplements are important for the treatment of hypoparathyroidism. Doses of 1000–2000 mg/day of calcium may be necessary. Preparations including the carbonate, citrate, lactate, gluconate, and glucobionate salts are suitable for this purpose. We prefer calcium carbonate because it is inexpensive, well-tolerated, and easily acquired. In some cases of hypoparathyroidism, a thiazide diuretic may be useful in augmenting serum calcium levels and reducing the hypercalciuria that can occur with the institution of treatment.

Magnesium deficiency can occur in patients with hypoparathyroidism, most often secondary to steatorrhea which is seen in the autoimmune forms of this disorder [97,98]. This may render a patient relatively resistant to therapy, and therefore magnesium deficiency should be considered in individuals whose therapeutic requirements unexpectedly increase.

In pseudohypoparathyroidism, the therapeutic approach is similar to that in primary hypoparathyroidism, the principal difference being that hypercalciuria is less of an issue and it is generally easier to maintain eucalcemia in these individuals. Untreated patients with pseudohypoparathyroidism may have varying degrees of osteomalacia and initially may require high dose therapy to achieve eucalcemia as their bones remineralize. Requirements will drop as the bone lesion heals, often heralded by a fall in serum alkaline phosphatase and a rise in serum calcium levels.

Individuals with 1-hydroxylase deficiency have a defect in the ability to generate $1,25(OH)_2D$ from the precursor metabolite 25OHD. In this disorder, eucal-

TABLE II Vitamin D and Related Therapeutic Agents[a,b]

Name	Formulation		Typical daily dose[c]	Approximate cost (1996 U.S.$) for 1 month of treatment[d]
Vitamin D (calciferol), Drisdol	Solution	8,000 IU/ml	2,000 IU	11
	Tablet	25,000 IU	25,000 IU	8
		50,000 IU	50,000 IU	12
25-Hydroxyvitamin D (calcidiol), Calderol	Tablet	20 μg	20 μg	33
		50 μg	50 μg	48
Dihydrotachysterol, Hytakerol, DHT	Solution	0.2 mg/5 ml	0.2 mg	14
	Tablet	0.125 mg	0.125 mg	29
		0.2 mg	0.2 mg	28
		0.4 mg	0.4 mg	50
1,25-Dihydroxyvitamin D (calcitriol) Rocaltrol	Tablet	0.25 μg	0.5 μg	67
		0.50 μg	0.5 μg	52
Calcijex	Intravenous solution	1 μg/ml	0.01 μg/kg/dose 3 times/week	152[e]
		2 μg/ml		266[e]

[a] Adapted with permission from Carpenter T 1996 Rickets. In: Berg F, Ingelfinger J, Wald E (eds) Gellis and Kagan's Current Pediatric Therapy, 15th ed. Saunders, Philadelphia, pp. 363–367.

[b] 1 μg vitamin D_2 = 40 IU.

[c] Daily dose may vary significantly, depending on condition.

[d] Average wholesale prices; retail prices may vary considerably.

[e] Prices assume the use of one ampule per dose.

cemia can be achieved by supplying $1,25(OH)_2D_3$ in physiological dosages [70]. In contrast, hereditary resistance to $1,25(OH)_2D$ represents a spectrum of resistance to therapy, with some individuals responding to doses of calcitriol in the usual therapeutic range and others resistant to even massive doses of the drug [31]. As noted above, chronic therapy with parenteral infusions of calcium have resulted in improvement of the rickets and normalization of all serum biochemical parameters [71].

A variety of stresses such as trauma, infection, and pregnancy can increase the therapeutic requirements of patients with chronic hypocalcemia, and the clinician should be alert to this possibility.

Acknowledgments

Dr. Carpenter is supported by grants from the U.S. Food and Drug Administration (FD-R-000671) and the National Institutes of Health (R55-DK 47337). Dr. Insogna is supported by grants from the NIH (AR39571), U.S. Department of Agriculture (CONR-9400568), and the Patrick and Catherine Weldon Donaghue Medical Research Foundation.

References

1. Harrison H, Harrison H 1979 Hypocalcemia states. In: Disorders of Calcium and Phosphate Metabolism in Childhood and Adolescence. Saunders, Philadelphia, Pennsylvania, pp. 47–99.
2. Sharief N, Matthew DJ, Dillon MJ 1991 Hypocalcaemic stridor in children. How often is it missed? Clin Pediatr **30**:51–52.
3. Wong C, Lau C, Cheng C, Leung W, Freedman B 1990 Hypocalcemic myocardial dysfunction: Short- and long-term improvement with calcium replacement. Am Heart J **120**:381–386.
4. Alpan G, Glick B, Peleg O, Eyal F 1991 Pseudotumor cerebri and coma in vitamin D dependent rickets. Clin Pediatr **20**:254–256.
5. Carpenter TO, Insogna KL, Boulware SD, Mitnick MA 1990 Vitamin D metabolism in chronic childhood hypoparathyroidism: Evidence for a direct regulatory effect of calcium. J Pediatr **116**:252–257.
6. Nikiforuk G, Fraser D 1979 Etiology of enamel hypoplasia and interglobular dentin: The roles of hypocalcemia and hypophosphatemia. Metab Bone Dis Related Res **2**:17–23.
7. Illum F, Dupont E 1985 Prevalence of CT-detected calcification in the basal ganglia in idiopathic hypoparathyroidism and pseudohypoparathyroidism. Neuroradiology **27**:32–37.
8. Stewart AF, Battaglini-Sabetta J, Milstone LM 1984 Hypocalcemia-induced psoriasis of von Zumbusch: New experience with an old syndrome. Ann Intern Med **100**:677–680.
9. Brown EM 1991 Extracellular Ca^{2+} sensing, regulation of parathyroid cell function, and role of Ca^{2+} and other ions as extracellular (first) messengers. Physiol Rev **71**:371–411.
10. Ladenson J, Lewis J, Boyd J 1978 Failure of total calcium corrected for protein, albumin, and pH to correctly assess free calcium status. J Clin Endocrinol Metab **46**:986–993.
11. Chattopadhyay N, Mithal A, Brown EM 1996 The calcium-sensing receptor: A window into the physiology and pathophysiology of mineral ion metabolism. Endocr Rev **17**:289–307.
12. Juppner H, Abou-Samra AB, Freeman MV, Kong X, Schipani E, Richards J, Kolakawski L, Hock J, Kronenberg H, Serge G, 1991 G protein-linked receptor for parathyroid hormone and parathyroid hormone-related peptide. Science **254**:1024–1026.
13. Livesey SA, Kemp BE, Re C, Partridge NC, Martin T 1982 Selective hormonal activation of cyclic AMP-dependent protein-kinase isoenzymes in normal and malignant osteoblasts. J Biol Chem **257**:14983–14987.
14. Hruska KA, Moskowitz D, Esbrit P, Civitelli R, Westbrook S, Huskey M 1987 Stimulation of inositol triphosphate and diacylglycerol production in renal tubular cells by parathyroid hormone. J Clin Invest **79**:230–239.
15. Abou-Samra A, Juppner H, Force T, Freeman M, Knog X, Schipani E, Urena P, Richards J, Bonventre J, Potts J, Kronenberg H, Segre G 1992 Expression cloning of a common receptor for parathyroid hormone and parathyroid hormone-related peptide from rat osteoblast-like cells: A single receptor stimulates intracellular accumulation of both cAMP and inositol triphosphates and increases intracellular free calcium. Proc Natl Acad Sci USA **89**:2732–2736.
16. Lee K, Deeds D, Bond A, Juppner H, Abou-Samra A-B, Segre G 1993 In situ localization of PTH/PTHrP receptor mRNA in the bone of fetal and young rats. Bone **14**:341–345.
17. Rouleau M, Mitchell J, Goltzman D 1990 Characterization of the major parathyroid hormone target cell in the endosteal metaphysis of rat long bones. J Bone Miner Res **5**:1043–1053.
18. McSheehy P, Chambers T 1986 Osteoblast-like cells in the presence of parathyroid hormone release soluble factor that stimulates osteoclastic bone resorption. Endocrinology **119**:1654–1659.
19. Talmage R, Doppelt S, Fondren F 1976 An interpretation of acute changes in plasma ^{45}Ca following parathyroid hormone administration to thyroparathyroidectomized rats. Calcif Tissue Res **22**:117–128.
20. King G, Holtrop M, Raisz L 1978 The relation of ultrastructural changes in osteoclasts to resorption in bone cultures stimulated with parathyroid hormone. Metab Bone Dis Related Res **1**:67–74.
21. Friedman PA, Gesek FA 1993 Calcium transport in renal epithelial cells. Am J Physiol (Renal, Fluid Electrolyte Physiol) **33**:F181–F198.
22. Lötscher M, Kaissling B, Biber J, Murer H, Kempson ST, Levi M 1996 Regulation of rat renal Na/P_i cotransporter by parathyroid hormone: Immunohistochemistry. Kidney Int **49**:1010–1018.
23. Minghetti PP, Norman AW 1988 $1,25(OH)_2$-Vitamin D_3 receptors: Gene regulation and genetic circuitry. FASEB J **2**:3043–3053.
24. Armbrecht J, Wongsurawat N, Zenser TV, Davis BB 1982 Differential effects of parathyroid hormone on the renal 25-dihydroxyvitamin D_3 and 24,25-dihydroxyvitamin D_3 production of young and adult rats. Endocrinology **111**:1339–1344.
25. Walker AT, Stewart AF, Korn EA, Shiratori T, Mitnick MA, Carpenter TO 1989 Effect of parathyroid hormone-like peptides on 25-hydroxyvitamin D-1α-hydroxylase activity in rodents. Am J Physiol **258**:E297–E303.
26. Booth BE, Tsai HC, Morris C 1977 Parathyroidectomy reduces 25-hydroxyvitamin D_3 1α-hydroxylase activity in the hypocalcemic vitamin D-deficient chick. J Clin Invest **60**:1314–1320.
27. Siegel N, Wongsurawat N, Armbrecht J 1986 Parathyroid hormone stimulates dephosphorylation of the renoredoxin component of the 25-hydroxyvitamin D_3-1α-hydroxylase from rat renal cortex. J Biol Chem **261**:16998–17003.
28. Matsumoto T, Ikeda K, Morita K, Fukumoto S, Takahashi H, Ogata E 1987 Blood Ca^{2+} modulates responsiveness of renal $25(OH)D_3$-1α-hydroxylase to PTH in rats. Am J Physiol (Endocrinol Metab **16**) **253**:E503–E507.
29. Bushinsky DA, Riera GS, Favus MJ, Coe FL 1985 Evidence that

blood ionized calcium can regulate serum $1,25(OH)_2D_3$ independently of parathyroid hormone and phosphorus in the rat. J Clin Invest **76**:1599–1604.

30. Kumar R 1991 Vitamin D and calcium transport. Kidney Int **40**:1177–1189.
31. Liberman UA, Marx SJ 1996 Vitamin D dependent rickets. In: Favus M (ed) Primer on the Metabolic Bone Diseases and Disorders of Mineral Metabolism, 3rd Ed. Raven, New York, pp. 311–316.
32. Favus MJ, Mangelsdorf DJ, Tembe V, Coe BJ, Haussler MR 1988 Evidence for *in vivo* upregulation of the intestinal vitamin D receptor during dietary calcium restriction in the rat. J Clin Invest **82**:218–224.
33. van Leeuwen JPTM, Birkenhäger JC, Buurman CJ, Schilte JP, Pols HAP 1990 Functional involvement of calcium in the homologous up-regulation of the 1,25-dihydroxyvitamin D_3 receptor in osteoblast-like cells. FEBS Lett **270**:165–167.
34. Sandgren ME, DeLuca HF 1990 Serum calcium and vitamin D regulate 1,25-dihydroxyvitamin D_3 receptor concentration in rat kidney *in vivo*. Proc Natl Acad Sci USA **87**:4312–4314.
35. Clements MR, Johnson L, Fraser DR 1987 A new mechanism for induced vitamin D deficiency in calcium deprivation. Nature **325**:62–65.
36. Goltzman D, Cole D 1996 Hypoparathyroidism In: Favus M (ed) Primer on the Metabolic Bone Diseases and Disorders of Mineral Metabolism, 3rd Ed. Raven, New York, pp. 220–223.
37. Conley M, Beckwith J, Mancer J, Tenckhoff L 1979 The spectrum of DiGeorge syndrome. J Pediatr **94**:883–890.
38. Chisaka O, Capecchi M 1991 Regionally restricted developmental defects resulting from targeted disruption of the mouse homeobox gene hox-1.5. Nature **350**:473–479.
39. Thakker R, Davies K, Whyte M, Wooding C, O'Riordan J 1990 Mapping the gene causing X-linked recessive idiopathic hypoparathyroidism to Xq26-Xq27 by linkage studies. J Clin Invest **86**:40–45.
40. Arnold A, Horst S, Gardella T, Baba H, Levine M, Kronenberg H 1990 Mutation of the signal peptide-encoding region of the preproparathyroid hormone gene in familial hypoparathyroidism. J Clin Invest **86**:1084–1087.
41. Parkinson D, Thakker R 1992 A donor splice site mutation in the parathyroid gene is associated with autosomal recessive hypoparathyroidism. Nature Genet **1**:149–152.
42. Chou Y, Brown E, Levi T, Crowe G, Atkinson A, Arnqvist H, Toss G, Fuleihan G, Seidman J, Seidman C 1992 The gene responsible for familial hypocalciuric hypercalcemia maps to chromosome 3 in four unrelated families. Nature Genet **1**:295–300.
43. Pollak MR, Brown EM, Step L, McLaine PN, Kifor O, Park J, Hebert SC, Seidman CE, Seidman JG 1994 Autosomal dominant hypocalcaemia caused by a Ca^{2+}-sensing receptor gene mutation. Nature Genet **8**:303–307.
44. Pearce S, Williamson C, Kifor O, Bai M, Coulthard M, Davies M, Lewis-Barned N, McCredie D, Powell H, Kendall-Taylor P, Brown E, Thakker R 1996 A familial syndrome of hypocalcemia with hypercalciuria due to mutations in the calcium-sensing receptor. N Engl J Med **335**:1115–1122.
45. Neufeld M, MacLaren N, Blizzard R 1981 Two types of autoimmune Addison's disease associated with different polyglandular autoimmune (PGA) syndromes. Medicine **60**:355–362.
46. Horwitz C, Myers W, Foote F 1972 Secondary malignant tumors of the parathyroid glands. Report of two cases with associated hypoparathyroidism. Am J Med **52**:797–808.
47. Dill J 1983 Hypoparathyroidism in sarcoidosis. South Med J **76**:414.
48. Brezis M, Shalev O, Leibel B, Bernheim J, Ben-Ishay D 1980 Phosphorus retention and hypoparathyroidism associated with transfusional iron overload in thalassaemia. Miner Electrolyte Metab **4**:57.
49. Carpenter TO, Carnes DL Jr, Anast CS 1983 Hypoparathyroidism in Wilson's disease. N Engl J Med **309**:873–877.
50. Heidbreder E, Gotz R, Schafferhans K, Heidland A 1986 Diminished parathyroid gland responsiveness to hypocalcemia in diabetic patients with uremia. Nephron **42**:285–289.
51. Burch WM, Posillico JT 1983 Hypoparathyroidism after ^{131}I therapy with subsequent return of parathyroid function. J Clin Endocrinol Metab **57**:398–401.
52. Anast C, Mohs J, Kalpan S, Burns P 1972 Evidence for parathyroid failure in magnesium deficiency. Science **177**:606–608.
53. Albright F, Burnett CH, Smith PH, Parson W 1942 Pseudohypoparathyroidism–An example of "Seabright-Bantam syndrome." Endocrinology **30**:922–932.
54. Kruse K, Kracht U, Wohlfart K, Kruse U 1989 Biochemical markers of bone turnover, intact serum parathyroid hormone and renal calcium excretion in patients with pseudohypoparathyroidism and hypoparathyroidism before and during vitamin D treatment. Eur J Pediatr **148**:535–539.
55. Stone M, Hosking D, Garcia-Himmelstine C, White D, Rosenblum D, Worth H 1993 The renal response to exogenous parathyroid hormone in treated pseudohypoparathyroidism Bone **14**:727–735.
56. Patten JL, Johns DR, Valle D, Eil C, Gruppuso P, Steele G, Smallwood PM, Levine MA 1990 Mutation in the gene encoding the stimulatory G protein of adenylate cyclase in Albright's hereditary osteodystrophy. N Engl J Med **322**:1412–1419.
57. Suarez F, Lebrun JJ, Lecossier D, Escoubet B, Coureau C, Silve C 1995 Expression and modulation of the parathyroid hormone (PTH)/PTH-related peptide receptor messenger ribonucleic acid in skin fibroblasts from patients with type Ib pseudohypoparathyroidism. J Clin Endocrinol Metab **80**:965–970.
58. Loveridge N, Tschopp FT, Born W, Devogelaer JP, de Deuxchaisnes CN, Fischer JA 1986 Separation of inhibitory activity from biologically active parathyroid hormone in patients with pseudohypoparathyroidism type I. Biochim Biophys Acta **889**:117–122.
59. Rude RK, Oldham SB, Singer FR 1976 Functional hypoparathyroidism and parathyroid hormone end-organ resistance in human magnesium deficiency. Clin Endocrinol **5**:209–224.
60. Carpenter T 1988 Disturbances of vitamin D metabolism and action during clinical and experimental magnesium deficiency. Magnesium Res **1**:131–139.
61. Rude RK, Adams JS, Ryzen E, Endres DB, Miimi H, Horst RI, Haddad JG, Singer FR 1985 Low serum concentrations of 1,25-dihydroxyvitamin D in human magnesium deficiency. J Clin Endocrinol Metab **61**:933–940.
62. Allgrove J, Adami S, Fraher L, Reuben A, O'Riordan JLH 1984 Hypomagnesaemia: Studies of parathyroid hormone secretion and function. Clin Endocrinol **21**:435–449.
63. Fuss M, Bergmann P, Bergans A, Bagon J, Cogan E, Pepersack T, van Gossum M, Corvilain J 1989 Correction of low circulating levels of 1,25-dihydroxyvitamin D by 25-hydroxyvitamin D during reversal of hypomagnesaemia. Clin Endocrinol **31**:31–38.
64. Anast CS, Gardner DW 1981 Magnesium metabolism. In Bronner F, Coburn J (eds) Disorders of Mineral Metabolism, Vol. 3, Academic Press, New York, pp. 423–506.
65. Robertie PG, Butterworth JF, Prielipp RC, Tucker WY, Zaloga GP 1992 Parathyroid hormone responses to marked hypocalcemia in infants and young children undergoing repair of congenital heart disease. J Am Collect Cardiol **20**:672–677.
66. Clemens TL, Henderson SL, Adams JS, Holick MF 1982 Increased

skin pigment reduces the capacity of skin to synthesize vitamin D_3. Lancet **1**:74–76.
67. Dagnelie PC, Vergote FJVRA, van Staveren WA, van den Berg H, Dingjan PG, Hautvast JGAJ 1990 High prevalence of rickets in infants on macrobiotic diets. Am J Clin Nutr **51**:202–208.
68. Kruse K 1995 Pathophysiology of calcium metabolism in children with vitamin D-deficiency rickets. J Pediatr **126**:736–741.
69. Kumar R 1983 Hepatic and intestinal osteodystrophy and the hepatobiliary metabolism of vitamin D. Ann Intern Med **98**: 662–663.
70. Balsan S 1991 Hereditary pseudo-deficiency rickets or vitamin D-dependency type I. In Glorieux FH (ed) Rickets (Nestle Nutrition Workshop Series, Vol. 21). Raven, New York, pp. 55–165.
71. Balsan S, Garabedian M, Larchet M, Gorski A, Cournot G, Tau C, Bourdeau A, Silve C, Ricour C 1986 Long-term nocturnal calcium infusions can cure rickets and promote normal mineralization in hereditary resistance to 1,25-dihydroxyvitamin D. J Clin Invest **77**:1661–1667.
72. Pettifor JM, Ross FP, Travers R, Glorieux FH, DeLuca HF 1981 Dietary calcium deficiency: A syndrome associated with bone deformities and elevated serum 1,25-dihydroxyvitamin D concentration. Metab Bone Dis Related Res **2**:301–305.
73. Pettifor JM, Ross P, Wang J, Moodley GP, Couper-Smith J 1978 Rickets in children of rural origin in South Africa: Is low dietary calcium a factor? J Pediatr **92**:320–324.
74. Marie P, Pettifor J, Ross F, Glorieux F 1982 Histological osteomalacia due to dietary calcium deficiency in children. N Engl J Med **307**:584–588.
75. Herbert L, Lemann J, Petersen J, Lennon E 1966 Studies of the mechanism by which phosphate infusion lowers serum calcium concentration. J Clin Invest **45**:1886–1894.
76. Chernow B, Rainey T, Georges L, O'Brian J 1981 Iatrogenic hyperphosphatemia: A metabolic consideration in critical care medicine. Crit Care Med **9**:772–774.
77. Venkaraman P, Tsang R, Greer F, Noguchi A, Laskarzewski P, Steichen J 1985 Late infantile tetany and secondary hyperparathyroidism in infants fed humanized cow milk formula. Am J Dis Children **139**:664–668.
78. Arrambide K, Toto R 1993 Tumor lysis syndrome. Semin Nephrol **13**:273–280.
79. Dunlay R, Camp M, Allon M, Fanti P, Malluche H, Llach F 1989 Calcitriol in prolonged hypocalcemia due to the tumor lysis syndrome. Ann Intern Med **110**:162–164.
80. Llach F, Felsenfeld A, Haussler M 1981 The pathophysiology of altered calcium metabolism in rhabdomyolysis-induced acute renal failure. N Engl J Med **305**:117–123.
81. Brasier A, Nussbaum S 1988 Hungry bone syndrome: Clinical and biochemical predictors of its occurrence after parathyroid surgery. Am J Med **84**:654–660.
82. Abramson EC, Gajardo H, Kukreja SC 1990 Hypocalcemia in cancer. Bone Miner **10**:161–169.
83. Anast C, Carpenter T, Key L 1990 Metabolic bone disorders in children. In Avioli L, Krane S (eds) Metabolic Bone Disease and Related Research. Saunders, Philadelphia, Pennsylvania, pp. 850–887.
84. Llach F, Bover J 1996 Renal Osteodystrophy. In Brenner B (ed) The Kidney, 5th Ed. Saunders, Philadelphia, Pennsylvania, pp. 2187–2273.
85. Edmondson H, Berne C 1944 Calcium changes in acute pancreatic necrosis. Surg Gynecol Obstet **79**:240–244.
86. Stewart AF, Longo W, Kreutter D, Jacob R, Burtis WJ 1986 Hypocalcemia due to calcium soap formation in a patient with a pancreatic fistula. N Engl J Med **315**:496–498.
87. Zaloga GP, Chernow B 1987 The multifactorial basis for hypocalcemia during sepsis. Studies of the parathyroid hormone–vitamin D axis. Ann Intern Med **107**:36–41.
88. Chesney RW, McCarron DM, Haddad JG, Hawker CD, DiBella FP, Chesney PJ, Davis JP 1983 Pathogenic mechanisms of the hypocalcemia of the staphylococcal toxic-shock syndrome. J Lab Clin Med **101**:576–585.
89. Jaeger P, Otto S, Speck RF, Villiger L, Horber FF, Casez J-P, Takkinen R 1994 Altered parathyroid gland function in severely immunocompromised patients infected with human immunodeficiency virus. J Clin Endocrinol Metab **79**:1701–1705.
90. Weinstein R, Bryce G, Sappington L, King K, Gallagher B 1984 Decreased serum ionized calcium and normal vitamin D metabolite levels with anticonvulsant drug treatment. J Clin Endocrinol Metab **58**:1003–1009.
91. Tofalletti J, Nissenson RA, Endres D, McGarry E, Mogollon G 1985 Influence of continuous infusion of citrate on responses of immunoreactive PTH, calcium, magnesium components, and other electrolytes in normal adults during plasmapheresis. J Clin Endocrinol Metab **60**:874–879.
92. Mallette LE, Gomez LS 1982 Systemic hypocalcemia after clinical injection of radiographic contrast media: Amelioration by omission of calcium chelating agents. Radiology **147**:677–679.
93. Jacobson MA, Gambertoglio JG, Aweeka FT, Causey DM, Portale AA 1991 Foscarnet-induced hypocalcemia and effects of foscarnet on calcium metabolism. J Clin Endocrinol Metab **72**:1130–1135.
94. Winer K, Yanovski J, Cutler G 1996 Synthetic human parathyroid hormone 1–34 vs calcitriol and calcium in the treatment of hypoparathyroidism: Results of a short-term randomized crossover trial. JAMA **276**:631–636.
95. Russell R, Smith R, Walton R, Preston C, Basson R, Henderson R 1976 1,25-Dihydroxycholecalciferol and 1α-hydroxycholecalciferol in hypoparathyroidism. Lancet **2**:14–17 (July 6).
96. Neer R, Holick M, DeLuca H, Potts J 1975 Effects of 1 α-hydroxyvitamin D_3 on calcium and phosphorus metabolism in hypoparathyroidism. Metabolism **24**:1403–1413.
97. Rosler A, Rabinowitz D 1973 Magnesium-induced reversal of vitamin D-resistance in hypoparathyroidism. Lancet **1**:803–804 (April 4).
98. Ahonen P, Myllarniemi S, Sipila I, Perheentupa J 1990 Clinical variation of autoimmune polyendocrinopathy–candidiasis–ectodermal dystrophy (APECED) in a series of 68 patients. N Engl J Med **322**:1829–1836.

SECTION VII

New Analogs

CHAPTER 57

Chemistry and Design: Structural Biology of Vitamin D Action

WILLIAM H. OKAMURA AND GUI-DONG ZHU
Department of Chemistry, University of California, Riverside, Riverside, California

I. OVERVIEW

The diversity of recent vitamin D research efforts can be easily gleaned from the contents of the present treatise and other review articles [1,2]. The vitamin D molecule, more specifically the hormonally active form 1α,25-dihydroxyvitamin D_3 (**1**, 1,25$(OH)_2D_3$; Fig. 1), seems to have emerged as a universal regulator in a variety of cells of higher animals and is involved in a host of biological processes, including calcium homeostasis, cell differentiation, immunology, and regulation of gene transcription. Some of these biological responses of vitamin D have been applied clinically or have high potential in the treatment of a diverse range of human diseases such as rickets, renal osteodystrophy, osteoporosis, psoriasis, leukemia, breast cancer, prostate cancer, AIDS, and Alzheimer's disease. These biomedical applications continue to stimulate the growing interest in research on the chemistry, biology, and pharmacological applications of vitamin D.

The areas identified in Fig. 1 are among the frontiers of international research efforts in the vitamin D field. Major efforts are under way to develop these areas, at both the basic and applied levels, including the development of new drugs for clinical applications. This review focuses principally on one of these research areas, namely, on the structural biology of vitamin D action. More specifically, this article attempts to provide a detailed review of the three-dimensional structure of the vitamin D ligand and how the topology of the vitamin D molecule might relate to the ligand–protein complex (arguably a necessary intermediate in initiating biological actions) and to suggest some structural bases for the design of vitamin D drugs. A major goal of this chapter is to emphasize the structural flexibility of the ligand, a result of facile interconversion between different conformations (i.e., a result of rotation of single bonds) of vitamin D. Through evaluation of known structure–binding affinity relationships, an analysis is given as to whether a probable binding conformation of vitamin D

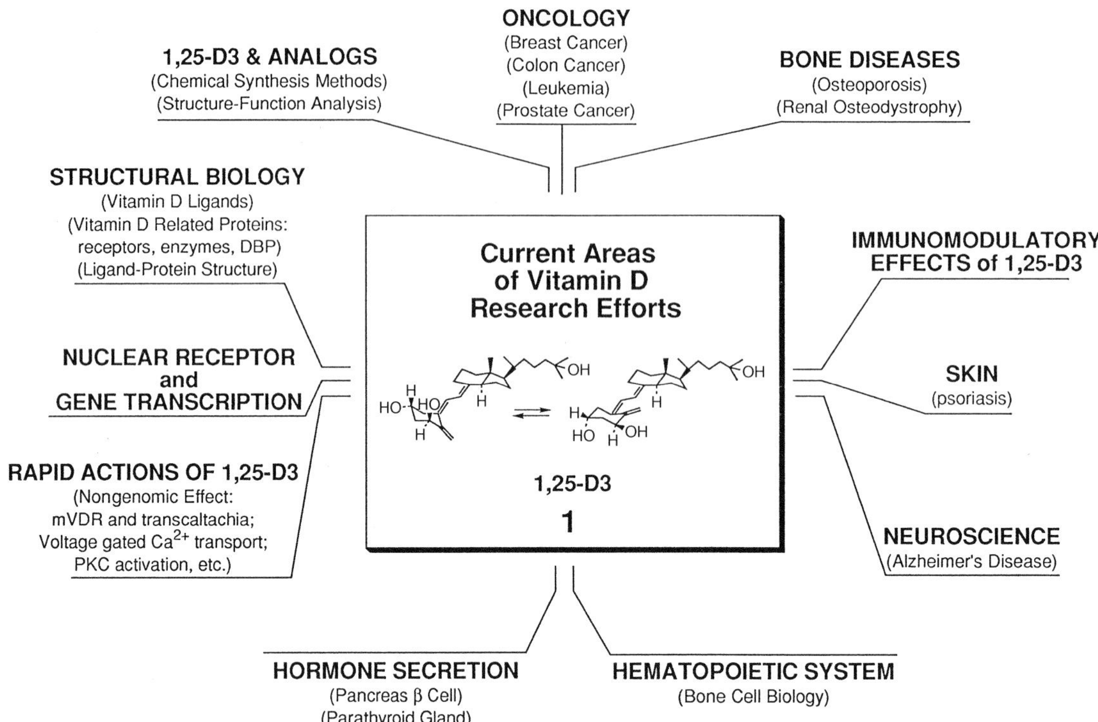

FIGURE 1 Current areas of vitamin D research efforts. Current international efforts on vitamin D research include basic research and clinical applications. The principal focus of this chapter concerns advances in what is known about the structure of D vitamins (the guests) and the various proteins and other systems (the hosts) to which they bind to form host–guest complexes. Knowledge in this area is anticipated to lead to more effective drug design for some of the disease states shown. An understanding of the mechanism of vitamin D action should also emerge.

in the ligand–protein complex can be identified. Structure–function information is needed to develop a basic molecular understanding of the mechanism of action of vitamin D and to provide a more intelligent way of designing analogs with selective physiological action.

A. Vitamin D Metabolism

The principal metabolic pathway for production of vitamin D_3 (**2**, D_3 or 6-s-*trans*-D_3) and its metabolic activation is shown in Fig. 2 by dashed arrows. The natural D_3 is mainly produced in skin on ultraviolet irradiation of 7-dehydrocholesterol (**5**, 7-DHC) or is directly absorbed from the diet. The reaction in skin is believed to be a purely photochemical process [3], first generating previtamin D_3 (**6**, $PreD_3$) which spontaneously isomerizes to D_3. It is to be noted here that the 6-s-cis form **2′** (6-s-*cis*-D_3) is formed first and then rapidly relaxes to the sterically more stable 6-s-trans conformer **2**. At 37°C in hexane, approximately 36 hr is required for the formation of an equilibrium mixture of $PreD_3$ and D_3, with the latter predominating. This isomerization is markedly accelerated in other media such as human skin [4–6], the sea urchin *Psammechinus miliaris* [7,8], phospholipid bilayers [9], cholesteric liquid crystals [10], and aqueous solutions of β-cyclodextrin [11,12]. Entrapment or population enhancement of $PreD_3$ in its s-cis,s-cis conformation might be responsible for the increased rate of conversion of $PreD_3$ to D_3 [4]. Except for serving as a biosynthetic precursor to other metabolites, D_3 itself is believed to be a biologically inert compound to which no single biological function has so far been reported. A plasma protein, vitamin D binding protein or DBP, subsequently transports D_3 to the liver where it interacts with a cytochrome P450 enzyme (D_3–25-hydroxylase, 25-hydroxylase), generating 25-hydroxyvitamin D_3 (**3**, $25OHD_3$), the major circulating form of vitamin D in serum. The $25OHD_3$ is further transported by DBP to the kidney where two other cytochrome P450 enzymes ($25OHD_3$-1α-hydroxylase, 1-hydroxylase; $25OHD_3$-24*R*-hydroxylase, 24-hydroxylase) affords $1,25(OH)_2D_3$, the physiologically active hormonal form of this steroid system, and 24*R*,25-dihydroxyvitamin D_3 (**4**, $24,25(OH)_2D_3$), respectively. The hormone $1,25(OH)_2D_3$ is then transported via the plasma DBP to target cells located, for example, in the intestine, kidney, and bone. In these target cells, $1,25(OH)_2D_3$ interacts with a nuclear vitamin D receptor protein referred to as VDR. The latter, a transcriptional activator which modulates a variety of genomic biological responses, is now generally recognized as a member of a superfamily of transcriptional activators that includes receptors for the classic steroid hormones as well as those for retinoic acid and thyroid hormone among others. As discussed in Chapter 15, the hormone $1,25(OH)_2D_3$ also exerts rapid actions at the membrane level via binding in one case to what has been loosely referred to as m-VDR (a membrane vitamin D receptor), a protein involved in transcaltachia [1,2].

The metabolism of vitamin D is discussed in greater detail in other chapters of this treatise (Chapters 2–6). Here we emphasize the conformational and/or chemical dynamics of vitamin D in the context of its metabolism. In other words, each of the metabolites exists as an equilibrating mixture of 6-s-*trans*-vitamin D, 6-s-*cis*-vitamin D, and previtamin D forms. The 6-s-*cis*-vitamins D **1**′, **2**′, **3**′ and **4**′ (Fig. 2) have never been detected spectroscopically, but they must exist as kinetically competent intermediates because they are required for 6-s-*trans*-vitamin D to back-isomerize to the corresponding previtamins D (**6**, **7**, **8**, and **9**) via [1,7]-sigmatropic hydrogen shifts. Because of the binding force of the various host molecules (usually proteins) in the plasma as well as in cellular compartments, the actual topologies of vitamin D metabolites are still uncertain. Examples can be seen from the perturbation of the isomerization of $PreD_3$ to D_3 such as in skin [4–6], in β-cyclodextrin [11,12], and in other media [7–10] as mentioned earlier. All of the vitamin D species shown in Fig. 2 must be considered as possible candidates for vitamin D actions mediated by the host (receptor proteins, enzymes, membrane components, etc.). A key to fully understanding the mode of vitamin D action at the molecular level resides in developing an understanding of their topologies when bound to protein hosts, etc. The remainder of this section first introduces the major proteins involved in vitamin D action and then provides a general description of other members of the steroid receptor superfamily. It closes with a statement of the ultimate goal of understanding of the structure of ligand–protein complexes and the attendant problems. After a detailed analysis in Section II of the three-dimensional structure of vitamin D ligands, an interpretation and discussion of known structure–binding affinity and structure–function relationships for vitamin D in relation to ligand–protein complexes are given in Section III. Finally, the chapter concludes with suggested approaches to drug design in Section IV followed by a summary (Section V).

B. Proteins Involved in Vitamin D Action

Proteins involved in vitamin D action include plasma carrier proteins (DBP and possibly serum albumin), P450 enzymes (1-hydroxylase, 24-hydroxylase, 25-hydroxylase, and others), nuclear vitamin D receptors (VDR), and, more recently, putative membrane vitamin D receptors (m-VDR). There will be no discussion of other proteins involved with catabolic processes, however. Because of the large excess and high binding affinity of carrier proteins, as well as the inherent insolubility of steroidal ligands in aqueous media, most of the vitamin D species are believed to exist in a protein-bound state. Table I (see Section I,B,3 for references to VDR as well as Refs. [12–26]) summarizes host molecules (proteins and other natural systems as well as artificial host molecules) that associate with various vitamin D metabolites as well as some structural information. However, only the major proteins involved in vitamin D metabolism are discussed in this section.

1. Serum Carrier Proteins

a. Vitamin D Binding Protein (DBP) Vitamin D binding protein (DBP), also known as group specific component or Gc-globulin, belongs to a member of a gene family including albumin and α-fetoprotein [17,18,27]. DBP, synthesized in the liver, is a monomeric 58,000-Da polypeptide possessing one sterol binding site. Human DBP is made up of 458 amino acid residues [28,29], and the predicted molecular weight for the nonmodified protein is 51,355. DBP is extensively discussed in Chapter 7.

The amino acid sequence of DBP [27–29] is homologous to that of serum albumin, α-fetoprotein, and afamin. Although its tertiary structure is still unknown, it is expected to be similar to that of serum albumin which was reported in 1992 [24]. Bouillon's group has most recently targeted this effort and has obtained a DBP single crystal of sufficient quality for an X-ray investigation [30]. By photoaffinity labeling and chemical cleavage of the labeled protein, Ray *et al.* have identified an 11.5-kDa fragment containing the putative $25OHD_3$ binding site [31]. Through chemical modification of Trp and His residues in human DBP, Ray *et al.* have also emphasized the importance of Trp (single residue at position 145) and one His residue (of a total of six) in vitamin D sterol binding [32]. In another study, the sterol binding domain of the human DBP was shown to reside in the amino-terminal portion of the 458-residue protein [33–35]. This was achieved by affinity labeling, limited enzymatic proteolysis, fragment isolation, and amino acid sequencing. By *in vitro* transcription and translation of a rat DBP cDNA [34], Haddad, Cooke, and co-workers [33,35] prepared a 31.5-kDa pro-

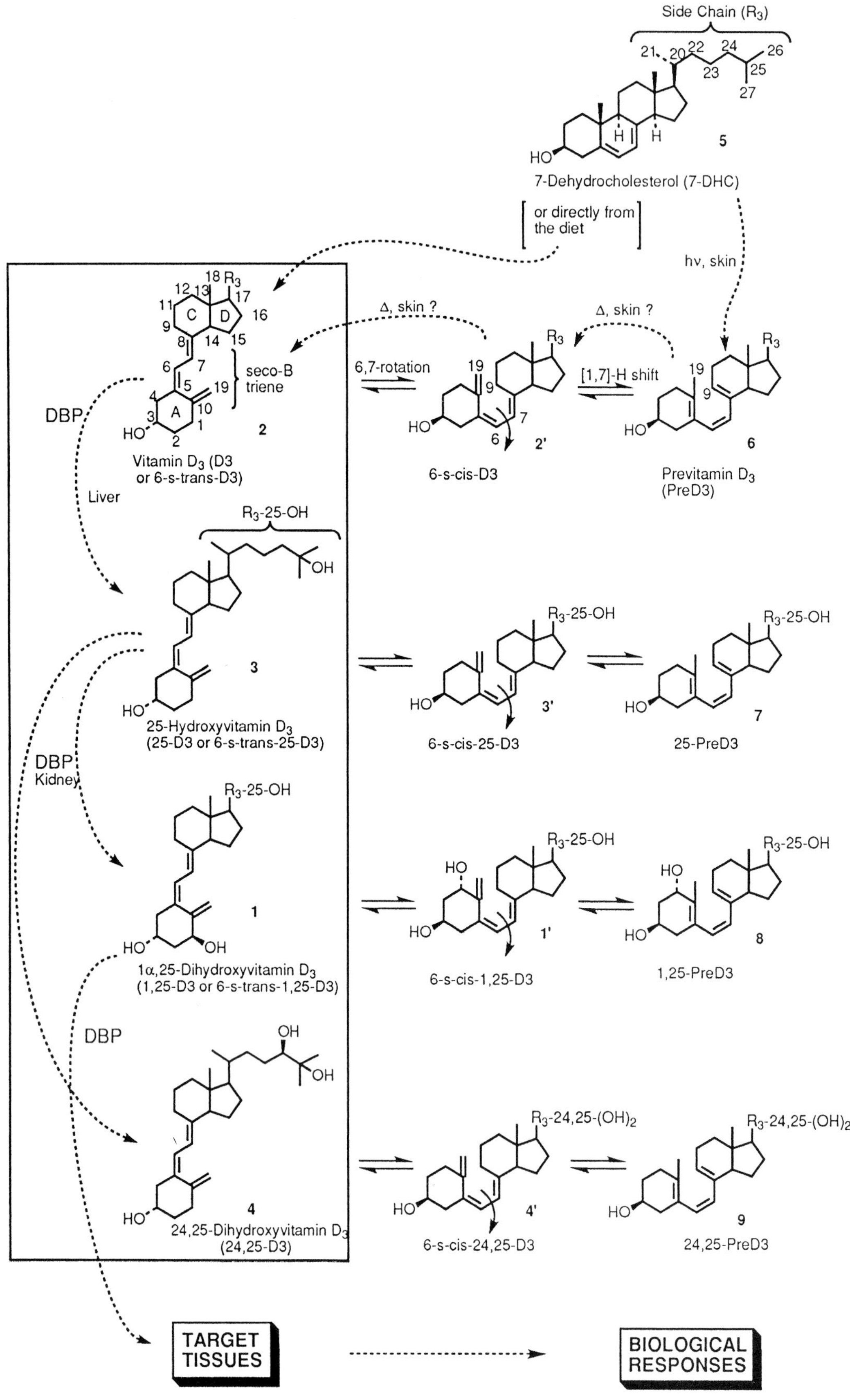
Side Chain (R_3)
7-Dehydrocholesterol (7-DHC)
5
or directly from the diet
hν, skin
Δ, skin ?
Δ, skin ?
6,7-rotation
[1,7]-H shift
seco-B triene
DBP
Liver
Vitamin D_3 (D3 or 6-s-trans-D3)
2
6-s-cis-D3
2'
Previtamin D_3 (PreD3)
6
R_3-25-OH
25-Hydroxyvitamin D_3 (25-D3 or 6-s-trans-25-D3)
3
6-s-cis-25-D3
3'
25-PreD3
7
DBP
Kidney
1α,25-Dihydroxyvitamin D_3 (1,25-D3 or 6-s-trans-1,25-D3)
1
6-s-cis-1,25-D3
1'
1,25-PreD3
8
DBP
24,25-Dihydroxyvitamin D_3 (24,25-D3)
4
R_3-24,25-(OH)$_2$
6-s-cis-24,25-D3
4'
24,25-PreD3
9
TARGET TISSUES
BIOLOGICAL RESPONSES

TABLE I Selected Host Molecules That Bind to Vitamin D Metabolites

Host system	Principal ligand	Host structure	Binding affinity (K_d)	Ref.
VDR (human)	$1,25(OH)_2D_3$	427 amino acids, MW ~48,000	0.5×10^{-9} *M*	See text
m-VDR	$1,25(OH)_2D_3$	1–3 proteins, MW ~60,000	0.72×10^{-9} *M*	13
25-Hydroxylase (pig)	D_3	MW ~50,500	?	14
1-Hydroxylase (chick)	$25OHD_3$	MW ~57,000–59,000	?	15,16
24-Hydroxylase (chick)	$25OHD_3$	MW ~55,000–59,000	?	15,16
DBP (human)	$25OHD_3$	458 amino acids, MW 51,355	10^{-9} *M*	17,18
PKC	$1,25(OH)_2D_3$	—	EC_{50} ~16 n*M*	19,22
B-700	$1,25(OH)_2D_3$	MW ~67,000	1.96×10^{-5} *M*	23
Serum albumin	$25OHD_3$/ (5*E*)-$25OHD_3$	HSA: 585 amino acids, MW ~65,000	~10^{-5} *M*	24–26
		BSA: 582 amino acids, MW ~ 64,000	~10^{-5} *M*	
Membrane bilayer	Various metabolites	Various	?	9
Liposome/ micelles	Various metabolites	Various	?	Unpublished (Zhu & Clemens, 1996)
Cyclodextrin	D_3	MW ~1135	Low	11
Sea urchin shell	$25OHD_3$	Various	Low	7,8

tein that had a binding affinity to $25OHD_3$ similar to that of the native DBP.

Vitamin D binding protein is the major serum transporter for the family of vitamin D sterols in several mammalian and avian species. It selectively binds vitamin D ligands with affinities (K_a) ranging from 10^7 to 10^9 M^{-1} and participates widely in the vitamin D endocrine system. The concentration of DBP in blood (5×10^{-6} *M*) is considerably higher than that of vitamin D metabolites ($25OHD_3$, 5×10^{-8} *M*) [36,37]. DBP plays a major role in the translocation of endogenously synthesized D_3 from skin and appears to modulate metabolites from too rapid or excessive cell entry. DBP does not appear to mediate cellular entry of D sterols, but rather provides an equilibrium with its ligands that permits the availability of ligand to cells in a gradual manner. The relative binding affinity of DBP for vitamin D metabolites [the order of binding affinity is $25OHD_3 = 24,25(OH)_2D_3 = 25,26(OH)_2D_3 > 1,25(OH)_2D_3 > D_3$) provides for differential availability of free ligand. The hormone $1,25(OH)_2D_3$ has 0.4% of its total concentration as "free" sterol, but only 0.04% of $25OHD_3$ is available to tissues.

b. Serum Albumin Serum albumin, a member of the same gene family as DBP, is the most abundant protein in blood plasma and serves as a carrier and storage compartment for numerous endogenous and exogenous compounds [25,26]. Although information on its ability to bind vitamin D is limited, its affinity to vitamin D is believed to be much lower than that of DBP. Human serum albumin (HSA) consists of three homologous domains (I–III) that assemble to form a heart-shaped molecule [24]. Each domain contains a

FIGURE 2 Vitamin D metabolism and the dynamic equilibration of vitamin D_3 and metabolites with 6-s-cis conformers and previtamins. Rotamers and previtamins are formed via rapid 6,7-single bond rotations and slower [1,7]-sigmatropic shifts, respectively. It is emphasized that all of the vitamin D metabolites, either rotamer or previtamin, are equally likely to be involved in guest–host binding. All numbering refers back to the original steroid nucleus from which they are derived.

pair of subdomains that possess common structural motifs. The principal regions of ligand binding to HSA are located in hydrophobic cavities in subdomains IIA and IIIA, which exhibit similar chemistry. A different subdomain, however, shows different binding affinities to different ligands. Subdomains IIA and IIIA share a common interface. Ligand bound to domain III affects conformational changes as well as binding affinity in domain II. Because DBP circulates in remarkably higher titer (30-fold) compared to vitamin D metabolites, the transport of vitamin D by serum albumin is most probably very limited.

2. Enzymes

The metabolic activation of vitamin D is mainly regulated by three cytochrome P450-containing enzymes, the hepatic 25-hydroxylase, the renal 1-hydroxylase, and the renal and intestinal 24-hydroxylase (Fig. 2 and Section I,A), which have been described in detail in several review articles [38–41]. These hydroxylases have been purified from different tissues and the cDNAs for the 24-hydroxylase and the 25-hydroxylase have been cloned and sequenced [14,42–46]. However, their three-dimensional structures and even information concerning their binding to vitamin D ligands are largely unknown. Most recently, the mechanism of action of the chick 1-hydroxylase was investigated at Riverside by the synthesis and biological evaluation of potent inhibitors of this enzyme [47–50]. Earlier investigations in several laboratories have led to the development of inhibitors of the hepatic 25-hydroxylase [51–61].

3. Nuclear and Membrane Receptors

a. Nuclear Vitamin D Receptor (VDR) In the genomic pathway, which has been reviewed in detail in other chapters in this treatise, the nuclear receptor for $1,25(OH)_2D_3$ (VDR) [62–64] has been shown to belong to the same superfamily of transactivating regulators of gene transcription that includes the receptors for estradiol, progesterone, testosterone, dihydrotestosterone, cortisol, aldosterone, thyroid hormone, and retinoic acid. The VDR is a trace protein, and even in its most abundant source, the intestine, the receptor represents only about 0.001% or less of the total soluble protein [62]. The intracellular location of the receptor still remains controversial. An *in vivo* approach using tissue components [65] suggested that the VDR is localized in the cell nucleus, whereas an *in vitro* approach using fluorescence immunocytochemical techniques on fixed tissues from treated and nontreated cells [66] indicated that, although some VDR is located in the nucleus, a majority of the VDR may exist in the cytosol. The latter study remains to be confirmed.

The total size of the deduced amino acid sequence of the receptor protein is 423 residues for rat VDR [67] and 427 residues for human VDR [68]. Because of its low abundance and instability, purification of VDR is very tedious. A number of approaches for overproduction of recombinant VDR have been developed in order to prepare milligram quantities of pure protein for X-ray crystallographic or nuclear magnetic resonance (NMR) studies. An early report of VDR overproduction described the cloning of human VDR cDNA in a high copy yeast expression plasmid, with transcription under the control of the copper-inducible metallothionein promoter [69]. Transformed *Saccharomyces cerevisiae* cells, grown in the presence of copper and $1,25(OH)_2D_3$, were shown to produce VDR at a level of 0.5% of total soluble protein. Reports soon thereafter described the production of recombinant rat [70] and human [71] VDR in insect cells using a baculovirus system, and recombinant human VDR in mammalian cells using an adenovirus expression system [72]. Kumar *et al.* then introduced an expression system for human VDR in a bacterial system [73]. Although the level of rat VDR produced in insect cells was reported to be 5% of the total soluble protein [70], the bacterial source reportedly produces considerably increased levels of the total cellular protein [73]. The ligand binding domain of the human VDR has been obtained in milligram quantities using a glutathione-*S*-transferase fusion protein expression system [74]. Amino acids 105–427 of the truncated human VDR was separated on glutathione-Sepharose and was further purified by Mono Q ion-exchange chromatography. The ligand binding domain thus obtained was characterized as a single band on sodium dodecyl sulfate (SDS)–polyacrylamide gel electrophoresis, and this fragment exhibited high affinity binding to $1,25(OH)_2D_3$ (K_d $\sim 10^{-9}$ *M*), with a binding capacity of 47 pmol/nmol protein.

b. Membrane Vitamin D Receptor (m-VDR) A number of rapid action (nongenomic) effects of $1,25(OH)_2D_3$, including transcaltachia, membrane voltage-gated calcium channel opening, prostaglandin production, and phospholipase C and protein kinase C (PKC) activation [1,2,19–22], have been found at both the cellular and subcellular level in *in vitro* studies. The rapid action of $1,25(OH)_2D_3$ known as transcaltachia is believed to be mediated by a membrane $1,25(OH)_2D_3$ receptor (m-VDR) [13], which is, however, not yet fully characterized. The Norman laboratory at the University of California, Riverside, reported a 4500-fold purification of the m-VDR from the basal lateral membrane (BLM) of vitamin D-replete chick intestinal epithelium [13]. The BLM-VDR thus obtained was believed to be composed of one to three proteins with molecular weight(s) ≥60,000 and exhibited saturable binding for ^3H-1,25$(OH)_2D_3$ with a K_d of 0.72 × 10^{-9} M. The validity of these rapid actions *in vivo* remain to be established.

C. Other Members of the Steroid Receptor Superfamily

Besides the VDR, the steroid receptor superfamily consists of receptors for classic steroid hormones (corticosteroids, estrogens, progestins, and androgens), receptors for thyroid hormone and retinoids, and "orphan receptors" for which ligands have for the most part not yet been identified. The characterization, function, and mechanism of action of these receptors have been extensively reviewed [75–81] and are summarized in Table II. Members of the nuclear hormone receptor superfamily are ligand-dependent transcription factors that modulate a large number of essential cellular processes such as carbohydrate, protein, and lipid metabolism (corticosteroids), female reproduction (estrogens), and calcium homeostasis (vitamin D). All of the nuclear hormone receptors have high affinity for their ligands, with K_d values ranging from 10^{-9} to 10^{-11} M. Ligand binding of these receptors is generally considered to induce a protein conformational change, a critical component in the transcriptional activation of these receptors.

A structural comparison of the classic steroid hormones (with their fully intact ABCD fragments) with vitamin D (with a distinct nonsteroidal seco-B-ring shape) intuitively implies that vitamin D could possess a steroidal shape when bound to VDR [1,13]. This intuition emerges from the notion that there exists a significant level of homology among the ligand binding domains of the various nuclear receptors (Table II). In a situation which is reversed, DeLuca and co-workers reported a seco-B-ring pregnacalciferol analog (with the same triene chromophore as vitamin D) that bound significantly to the progesterone receptor [82].

D. Toward an Understanding of the Binding Conformation of Vitamin D in Ligand–Protein Complexes: Rusting of the Lock and Key Hypothesis

A continuing goal of vitamin D research is to develop a detailed understanding at the molecular level of the biochemical mode of vitamin D action, particularly in gleaning information on how small ligands bind to and activate proteins involved in biochemical transformations. Such a detailed investigation is expected to lead to a more systematic design of a new generation of vitamin D analogs with selective clinical applications. As discussed previously, 1,25$(OH)_2D_3$ exhibits a diverse spectrum of biological activities including its traditional role as a hormone in calcium homeostasis as well as new roles in inducing cell differentiation and inhibiting cellular proliferation. A reasonable hypothesis is that these biological activities of vitamin D may be induced by any one or more of a number of different shapes or forms of the hormone (Fig. 2). This hypothesis is based on the facts that vitamin D is a flexible ligand and that it is not clear at present which topology this ligand assumes when bound to receptor or other hosts.

Current efforts toward the development of an analog with a more useful therapeutic index, specifically directed toward analogs with high cell differentiating ability and low calcitropic action, mainly focus on the structural modification of lead vitamin D analogs. Some lead structures are described later in this chapter (Figs. 15 and 16), but such analogs have often been developed because of synthetic expediency with modest forethought. Future studies will undoubtedly provide more sophisticated approaches. For example, with the rapid development of molecular biology techniques and their application in the vitamin D field (e.g., overexpression of VDR), the molecular mechanism of vitamin D action will be accessible to increasing scrutiny. The various proteins including site-directed mutated forms involved in vitamin D action could be expressed in sufficient quantities for NMR studies (using site specifically labeled protein and/or ligand) as well as for direct X-ray crystallographic applications. The strategy for assessing structure–function relationships will change from systematic studies of conformationally biased analogs in terms of their ability to bind to various proteins to a multifaceted approach including direct structural investigations of ligand–protein complexes. A reasonable hypothesis is that most if not all forms of vitamin D interact selectively with various proteins (receptors, enzymes, and/or transport proteins) and other biological matrices (e.g., lipids), leading to a variety of biological responses. A picture of these vitamin D ligands complexed to proteins and other hosts would provide a more direct basis for the design of pharmacologically selective analogs.

As summarized by Jorgensen [83], it is reasonable to expect that flexible molecules distort to form optimal interactions with binding partners in the lock (protein) and key (ligand) model for designing drugs. A very important practical consequence of this expectation is that any attempt to design drugs by analogy to the structures of flexible, unbound active substances could lead to extreme frustrations. Unbound ligands more generally possess shapes that are quite different from their topology in the bound state. Likewise, bound and un-

TABLE II Sequence Comparison of Members of the Nuclear Steroid Hormone Receptor Superfamily[a]

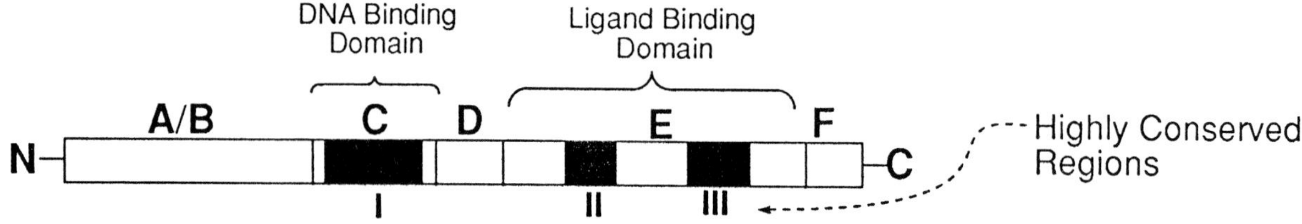

Receptor	Peptide chain and conserved I, II, and III region	Total amino acid residues	Hormone
PR	45 / 38 / 26	933	Progesterone
ER	55 / 40 / 30	595	Estradiol, estrone
AR	47 / 36 / 30	918	Testosterone
GR	50 / 38 / 39	777	Cortisone, cortisol
MR	52 / 31 / 22	984	Aldosterone
TR-α	61 / 26 / 35	410	Triiodothyronine, thyroxine
TR-β	59 / 26 / 35	456	Triiodothyronine, thyroxine
RAR-α	61 / 33 / 43	462	Retinoids, retinoic acid
RAR-β	61 / 33 / 43	448	Retinoids, retinoic acid
RAR-γ	61 / 36 / 39	454	Retinoids, retinoic acid
RXR-α	67 / 48 / 43	467	Retinoids, 9-*cis*-retinoic acid
RXR-β	62 / 45 / 39	410	Retinoids, 9-*cis*-retinoic acid
RXR-γ		463	Retinoids, 9-*cis*-retinoic acid
VDR	50 / 31 / 35	427	$1,25(OH)_2D_3$

[a] Six subregions (regions A through F) are classified based on comparison of sequences of nuclear hormone receptors. The N-terminal portion of the receptor is divided into the A/B/C region, and region C includes the substantially conserved DNA binding region. The C-terminal portion is subdivided as follows: a short region D is referred to as the "hinge" region, region E corresponds to the ligand binding domain, and region F constitutes the remainder. The three highly conserved regions I, II, and III, shown as black bars, are labeled with the number of amino acid residues in each of these regions if known. Region I is part of the DNA binding domain, and regions II and III are found in the hormone binding region. Other chapters in this treatise provide a more detailed description of this topic.

bound protein receptors may also assume quite different shapes in the more general case. Although it may be possible to identify examples where the shape of the ligand and the shape of the receptor protein assume the same shapes in the bound state, one cannot generally expect this to be the case. With regard to the vitamin D molecule, an important matter as indicated above is that it is a conformationally flexible molecule, and, as such, all accessible conformations irrespective of energetic considerations should be considered in the initial analysis. Finally, it should be clearly noted that, even in the event that the ligand–protein structure has been fully established through physicochemical techniques such as X-ray crystallography or through multidimensional NMR analyses, there remains the uncertainty that the specific three-dimensional structure so obtained may have little bearing on the actual structure of the ligand–receptor complex in the *in vivo* situation. Nevertheless, it is still crucial that a detailed structure of the free ligand and the free receptor protein or other hosts be clearly defined. It is for this reason that we delineate in detail what is known at present about the structure of the free ligand.

II. STRUCTURE OF THE VITAMIN D LIGAND

A knowledge of the topology and dynamic behavior of the free steroidal guest molecule (or analog) would be very helpful for the further establishment of binding behavior and shape in appropriate host systems including receptors such as those given in Tables I and II. This section reviews the structure of vitamin D ligands in the solid state and in solution as well as those generated by computational methods.

A. Solid-State Structure of Vitamin D Metabolites and Analogs

The solid-state structure of vitamin D and related compounds was reviewed some time ago by Okamura and Wing [84], but new data have since appeared. The various conformations of vitamin D in the solid state have been firmly established by X-ray crystallographic studies of vitamin D_3 [85], $25OHD_3$ [86], $1,25(OH)_2D_3$ [87], vitamin D_2 [88], a 3,20-bis(ethylenedioxy) analog [89], calciferyl 4-iodo-5-nitrobenzoate [90], 1-(1′-hydroxymethyl)-25-hydroxyvitamin D_3 [91], calcipotriol [92], and 1α-hydroxyvitamin D_2 ($1OHD_2$) (R. Moriarty, R. Gilardi, and co-workers, unpublished, personal communication) [93]. This section focuses on the general features of the vitamin D skeleton in the solid state with emphasis on major conformational differences. Table III summarizes what appears to be all of the crystallographic results reported to date.

TABLE III Selected Torsion Angles of Vitamin D and Analogs in the Solid State

Compound	α[a]	β[a]	C1,2,3,4[b]	C2,3,4,5[b]	C5,6,7,8[b]	C6,5,10,19[b]	Ref.
D_3[c]	+		57.0	−58.2	170.3	−53.6	85
		+	−51.2	48.0	−174.3	55.2	
D_2[c]	+		58	−52	175	−49	88
		+	−58	54	−164	46	
$1OHD_2$[d] (#1)		+	−56.1	50.3	−179.3	50.8	93
$1OHD_2$[d] (#2)	+		57.5	−47.7	168.6	−53.5	93
		+	−51.8	48.8	−176.0	52.9	
$25OHD_3$	+		53.8	−51.8	177.0	−56.7	86
$1,25(OH)_2D_3$		+	−52.8	48.8	−177.2	47.6	87
ECF[e]		+	−53.3	50.8	−168.9	56.3	89
INC[f]			—[i]	—[i]	167	—[i]	90
MC903[g]		+	−57.0	52.5	−179.7	—[i]	92
$1(CH_2OH)25OHD_3$[h]	+		56	−51	158	−54	91

[a] α and β refer to the A-ring conformer as defined in Figs. 3 and 5. A "+" indicates whether the A-ring chair was found to exist as the α or β conformers. In three cases, D_3, D_2, and 1-OH-D_2 (#2), both chair forms were found in the unit cell.

[b] Selected torsion angles of the A ring and triene (steroid numbering).

[c] Both α and β conformers were detected in their crystalline state, and the data are given individually for each conformer.

[d] R. Gilardi, R. M. Moriarty, and co-workers (unpublished observations, 1991) kindly provided the data for two isomorphs (#1 and #2) of 1α-hydroxyvitamin D_2 ($1OHD_2$). Two separate crystals, isomorph 1 and isomorph 2, were isolated from a single crystallization wherein isomorph 1 consisted of only the β conformer [like $1,25(OH)_2D_3$] and isomorph 2 consisted of a 1:1 mixure of α and β conformers like D_3 and D_2. A. Kutner and co-workers have also kindly informed us that $1OHD_2$ and $1OHD_3$ crystallize only as the β conformers according to their more recent x-ray crystallographic studies (unpublished observations, 1997).

[e] 3,20-Bis(ethylenedioxy)-9,10-seco-pregna-5,7,10(19)-triene.

[f] Calciferyl 4-iodo-5-nitrobenzoate.

[g] (24*R*)-1α,24-Dihydroxy-26,27-dehydro-22-ene-vitamin D_3. It is apparent that MC903 (see compound **40** or BT in Fig. 15 later in this chapter) actually crystallizes as the β conformer, not the α form as mentioned in the citation [92].

[h] 1α-Hydroxymethyl-25-hydroxyvitamin D_3. We thank G. Posner for providing the additional details from the recent publication [91].

[i] These torsion angles are not included in the references indicated [90,92].

Because of the existence of two adjacent exocyclic double bonds at C-5 and C-10 (Fig. 3), a pronounced deformation of the A ring from an idealized cyclohexane conformation is expected (the numbering system is shown in Fig. 2). The A ring of vitamin D is somewhat flattened as compared to cyclohexane itself [94]. X-Ray crystallographic studies have shown that in all of the vitamin D compounds studied (Table III), the $\Delta^{10(19)}$ double bond is twisted out of the $\Delta^{5,7}$-diene plane by an average of $\pm 50°$. This large dihedral angle about the $\Delta^{5(10)}$ single bond (C6,5,10,19 torsion angle) imparts a cyclohexane-like chair shape to the A ring. The A ring of all of vitamin D structures studied are frozen into either one or both chair conformations (α and β) in the solid state as depicted in Fig. 3. It is apparent that in the α form, the hydroxyl group at C-3 occupies the equatorial position and the exocyclic CH_2 group lies below the mean ring plane [a $(-)$ torsion angle about the $\Delta^{5(10)}$ bond]. In the β form, the hydroxyl group at C-3′ occupies an axial position, and the exocyclic CH_2 group lies above the mean ring plane [a $(+)$ torsion angle about the $\Delta^{5(10)}$ bond]. Another consequence is that the $\Delta^{5,7}$-diene and the exocyclic $\Delta^{10(19)}$ double bond are largely isolated from the conjugated $\Delta^{5,7}$-*trans*-diene fragment.

This bis-exocyclic diene (the 5,7-diene) of the seco-B-ring triene system is almost planar, with a range of torsion angles between $\pm 158°$ and $180°$, an s-trans orientation. Adoption of the vitamin D A ring to either the α or β chair forms does not markedly influence the near planarity of the seco-B ring in the solid state. The small but distinct twist of the $\Delta^{5,7}$-diene unit occurs in concert with the A-ring conformation, a slight $(+)$ and $(-)$ twist of the torsion angle from 180° in the α and β conformers, respectively. Ring C adopts a somewhat irregular chair conformation, the geometrical distortion of ring C being due to its fusion with the five-membered ring D at C-13 and C-14 coupled with the steric constraints imposed by the planarity induced by the exocyclic double-bond fragment at C-8. The ring D exists in either an envelope or half-chair form. Because of the *trans*-hydrindane fusion of the CD ring, which is relatively rigid, it is useful to think of this bicycle as a semirigid anchor to which a flexible A/seco-B-triene and side chain are attached at C-8 and C-17, respectively. The prime structural characteristic of the 25-hydroxycholesterol side chain is its staggered, zigzag conformation. In contrast, the nonhydroxylated side chains in vitamins D_3 and D_2 are not that regular because of some crystal disordering.

Both α and β forms were found in the vitamin D_3 crystal in an equimolar ratio [85], and the two chair conformers are hydrogen bonded to one another in the form of an infinite spirallike array (Fig. 4a). Like that in the vitamin D_3 crystal structure, both α and β forms were found in the vitamin D_2 X-ray structure unit cell [88]. All other vitamin D crystal structures contain only one chair form (α or β) in the unit cell (Table III) except for the still incomplete structure of an isomorph of $1OHD_2$ [93].

The metabolite $25OHD_3$ crystallizes exclusively into the α form, with the exocyclic CH_2 group situated below the mean cyclohexane-like ring plane [a $(-)$ C6,5,10,19 torsion angle] and the 3-OH occupying the equatorial orientation (Fig. 4b) [86]. In the $25OHD_3$ crystal, infinite chains are formed via identically oriented molecules that are interconnected via a tail-to-head hydrogen bond linkage from a 25-OH group of one molecule to the 3-OH group of an adjacent molecule. The $\Delta^{5,7}$-diene torsional angle of 177.0° in $25OHD_3$ is somewhat larger than that in the α form of D_3 (170.3°) or its β form (174.3°) [85] (Table III).

The conformation of the A ring in the $1,25(OH)_2D_3$ crystal is frozen exclusively in the β form in which the hydroxyl groups at C-1 and C-3 are in the equatorial

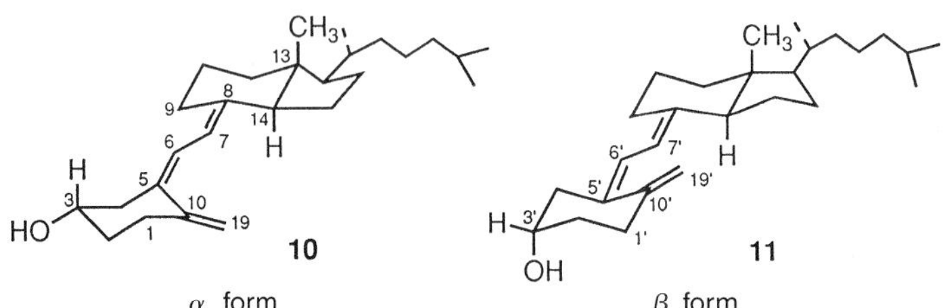

FIGURE 3 Two conformers of vitamin D in the crystalline state. The α form is the vitamin conformer with the 3-hydroxyl group of the A ring oriented in an equatorial position and the $\Delta^{10(19)}$-double bond oriented below the plane defined by the $\Delta^{5,7}$-diene. The β form refers to the conformer in which the 3-hydroxyl group of the A ring is oriented in an axial position and the $\Delta^{10(19)}$-double bond oriented above the plane defined by the $\Delta^{5,7}$-diene. The D ring of vitamin D in both the α and β conformers adopts either half-chair or envelope forms.

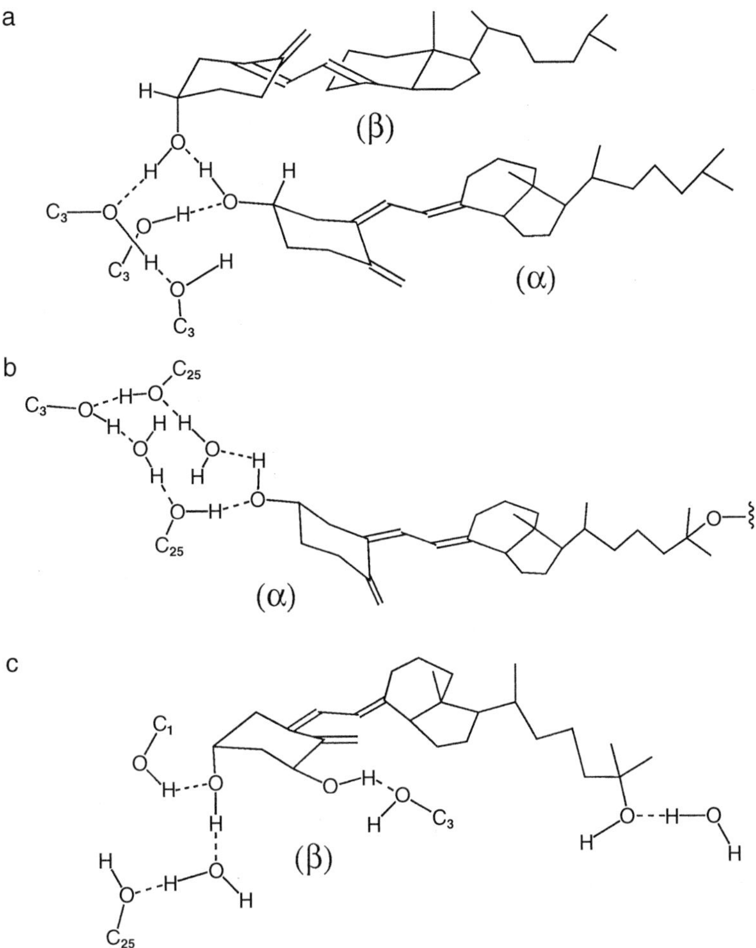

FIGURE 4 X-Ray crystallographic results for vitamin D metabolites. This schematic representation gives the general shape and hydrogen bonding scheme within each unit cell of D_3 (a, showing both α and β conformers hydrogen bonded to one another and the C-3 OH of three other D_3 molecules), 25OHD_3 (b, showing only the α conformer hydrogen bonded at C-3 to two water molecules and three other vitamin molecules, one at C-3 and two at C-25; its C-25 OH group is hydrogen bonded to one water molecule), and 1,25$(OH)_2D_3$ (c, showing only the β conformer hydrogen bonded at the C-3 OH to one water molecule and two other molecules of the steroid, one at C-1 and the other at C-25; in addition, a water molecule is hydrogen bonded at the C-25 OH). The hydrogen bonding scheme was not clear in all cases, and the reader is referred to the original articles cited in the text.

and axial orientation, respectively [87]. Noting again that the α or β label here refers to the equatorial or axial orientation, respectively, of the 3-hydroxyl, it is interesting that among the natural vitamin D metabolites, D_3 crystallizes in both chair forms (α and β), 25OHD_3 only in the α form, and 1,25$(OH)_2D_3$ only in the β form. All three hydroxyl groups of 1,25$(OH)_2D_3$ are involved in a network of hydrogen bonds to a water molecule and are symmetry related to a second 1,25$(OH)_2D_3$ molecule. Of the three water molecules present in the crystal of 1,25$(OH)_2D_3$, one is involved in a network of hydrogen bonds formed with the hydroxyl groups of the compound; the remaining two are disorderly arranged in channels formed along a crystallographic axis. Both cyclohexane-like rings, the A and C rings, adopt slightly distorted chair conformations [95] (for a general review, see Eliel and Wilen [96]), whereas the cyclopentane-like ring D adopts a half-chair conformation with the 2-fold axis of symmetry passing through C-16 and bisecting the C-13 to C-14 bond (Fig. 4c).

Some of the highlights of the other crystallographic results summarized in Table III are as follows. The crystal structures of calciferyl 4-iodo-5-nitrobenzoate (INC [90]) and the 3,20-bis(ethylenedioxy) analog (ECF [89]) were the first vitamin D structures ever reported and should be considered the pioneering studies. The former in particular established unequivocally the absolute stereochemistry of vitamin D as well as significant conformational features. The 1-hydroxymethyl analog {1-(CH_2OH)-25OHD$_3$, [91]} may be noted as the first C-1 substituted derivative of vitamin D subjected successfully to crystallography. It is interesting that this C-1 analog exhibits a solid-state structure whose A ring is in the α form (C-3 hydroxyl substituent equatorial) as opposed to the β form found more recently for the parent hormone 1,25$(OH)_2D_3$ [87].

Most interestingly, Moriarty, Gilardi, and co-workers have obtained two crystalline forms of 1α-hydroxyvitamin D_2 [93]. One crystalline form shows only the β-structure (C-3 and C-1 hydroxyls axial and equatorial, respectively), like 1,25$(OH)_2D_3$. The other crystalline form exhibits a 1:1 mixture of the α and β forms like D_3 [85] and D_2 [88], but the structure has not yet been refined. Although the detailed structural parameters have not been published [93], the Moriarty–Gilardi studies together with the observation that vitamin D seems to crystallize randomly in its α form or β form, or both (Table III), serve to emphasize the importance of crystal packing forces. Very recent studies by Kutner and co-workers (personal communication, 1997) reveal only the β-conformer for 1OHD$_2$ and the corresponding analog with the cholesterol side chain. In our opinion, the crystal packing forces, or particularly binding forces of various host molecules which bind to vitamin D, may lead to higher energy conformers of vitamin D not even apparent in the free ligand when studied by X-ray crystallography (or by NMR methods to be discussed below).

To summarize, X-ray crystallographic studies represent the most effective method for obtaining precise structural parameters including conformational and absolute configurational (stereochemical) information. This method, despite its precision, does not address the important problem of dynamics particularly well, however.

B. Solution Structure of Vitamin D

Vitamin D species are fluxional in solution and exist as an equilibrating mixture of conformations, most of which have not been detected in the solid state. The solution conformations of vitamin D metabolites and their analogs were established mainly by high resolution NMR spectroscopy [97–110] including complete ^1H and ^{13}C chemical shift assignments [110].

In solution, the A ring exists in dynamic equilibrium between mainly the two A-ring chair conformations that are found as α and β forms in the solid state (Fig. 5 shows the A ring possessing the 1α and 3β-hydroxyl groups: in conformer α, the hydroxyl at C-1 is axially oriented whereas that at C-3 is equatorial; in conformer β, the hydroxyl at C-1 is equatorial whereas that at C-3 is axial). Because of the facile conformational isomerism about single bonds, the somewhat flattened twist–boat form **13** may be present in significant amounts. The twist–boat form of the A ring, although likely lower in concentration, is fully accessible and could equally as well be involved in receptor binding as the more stable chair conformers. Wing *et al.* [97,98] (see also the work of La Mar and Budd [99]) examined the ratio between the two chair conformations of D_3, 1,25$(OH)_2D_3$, and dihydrotachysterol (DHT$_3$) mainly through correlation of the observed coupling constants between H-3α and H-4β with the Karplus equation. The NMR studies were assisted by computer simulation of the coupling pattern with or without tris(dipivalomethanato)europium(III) [Eu(dpm)$_3$] shift reagent. In D_3 and 1,25$(OH)_2D_3$, the ratio between the two chair conformations was determined to be approximately ~1:1. In DHT$_3$, however, the equilibrium mixture favors the equatorial (α form) over axial 3β-hydroxyl group by a ratio of 9:1. If the cyclohexanol coupling constants (3 and 11 Hz) accurately apply here [111], the observed coupling constant $J_{3\alpha,4\beta}$ provides an estimate of the 3β-hydroxyl equatorial conformer (α form) population as 88% for DHT$_3$, 57% for D_3, and 44% for 1,25$(OH)_2D_3$ and 1α-hydroxyvitamin D_3.

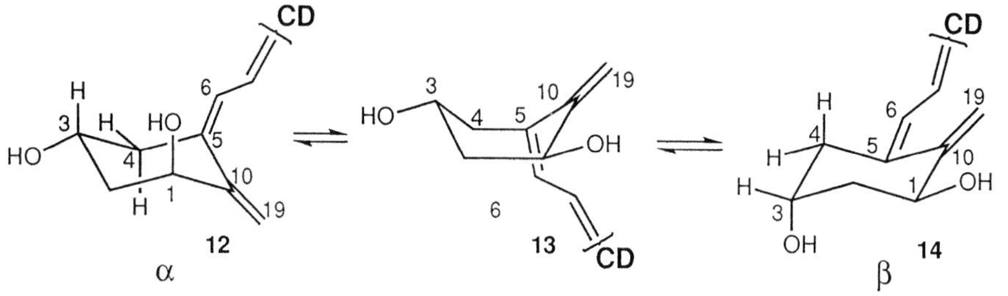

FIGURE 5 Equilibration of 1α,3β-dihydroxylated A rings of vitamin D. The α and β conformers shown (**12** and **14**, respectively) correspond to the chair forms **10** and **11** shown in Fig. 3. In the α form, there is an additional axial hydroxyl at C-1, and in the β form this same hydroxyl is equatorially oriented. In addition, a number of twist–boat conformers of the type **13** are also accessible to this A ring. Half-chair conformers (not shown) are also accessible to the A ring.

The two equilibrating chair conformations of $1,25(OH)_2D_3$ were detected by Eguchi and Ikekawa in 1990 by use of variable temperature ^{1}H-NMR spectroscopy [102]. At a temperature below $-100°C$, splitting into pairs of signals for H-3α, H-4α, and H-6 clearly shows the two conformations in a ratio of 55:45, in good agreement with previous estimates [97]. The two conformers were further examined by the same laboratory by ^{13}C-NMR studies of 1α-[19-^{13}C]hydroxyvitamin D_3 as the A-ring equivalent of the active hormone $1,25(OH)_2D_3$ [100]. The presence of two chair conformers in a ratio of 52:48 was detected at $-100°C$ by ^{13}C-NMR analysis. The rate constant for the equilibrium was estimated by signal shape analysis according to the Gutowsky–Holm equation. The thermodynamic parameters $\Delta G^{\ddagger}_{188K}$, $\Delta H^{\ddagger}$, and $\Delta S^{\ddagger}$ were deduced from the exchange rate to be 8.0 kcal/mol, 6.3 kcal/mol, and -9.0 cal/kmol, respectively.

The A-ring conformations of D_3, vitamin D_2, and $25OHD_3$ were reported to be solvent dependent, with a clear preference for a C-3 equatorial hydroxyl group (α form) in polar solvents such as methanol and dimethyl sulfoxide [101]. The ratios between α and β chair conformers are summarized in Table IV ([101]; W. H. Okamura and G.-D. Zhu, unpublished data, 1996). Acetylation of the 3β-hydroxyl groups of these vitamin D compounds did not significantly affect the equilibrium, but a strong preference for the 3β-equatorial conformer was reported for the *tert*-butyldimethylsilyl ether derivative of D_3 [101]. In SDS micelles, the A rings of D_3 and $25OHD_3$ are completely biased into the α chair conformation in which the 3β-hydroxyl group orients exclusively in an equatorial direction (W. H. Okamura and G.-D. Zhu, unpublished data).

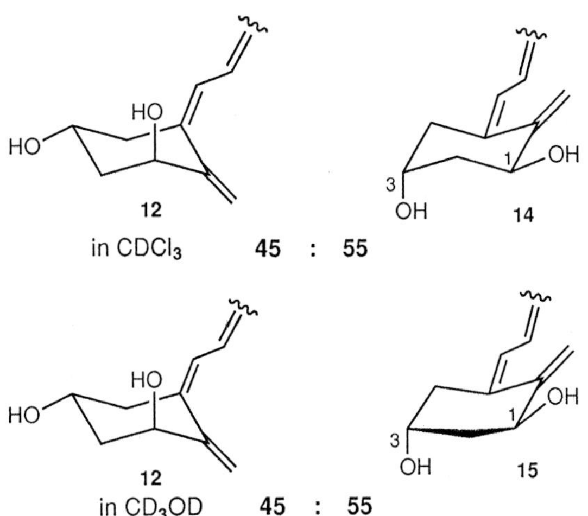

FIGURE 6 Proposed A-ring conformations of $1,25(OH)_2D_3$ under dynamic equilibrium in $CDCl_3$ and CD_3OD. As discussed in the text, in nonpolar solvents such as chloroform there is a near 1:1 mixture of α and β chair conformers (**12** and **14**, respectively). In methanol, however, it has been suggested that the half-chair conformation **15** rather than the β chair conformer **14** is a more dominant conformer and that **12** and **15** are in equilibrium with one another in a 1:1 mixture.

Through analysis of the A-ring coupling constants of $1,25(OH)_2D_3$ in various solvents, Eguchi and Ikekawa proposed that the A ring of $1,25(OH)_2D_3$ undergoes chair–chair (**12**/**14**) interconversion in hydrophobic media and chair–half-chair (**12**/**15**) equilibrium in hydrophilic solvents (Fig. 6) [102]. In deuterated chloroform ($CDCl_3$), for example, the observed vicinal couplings between all the A-ring protons agree well with the averaged values of the model compound, but in deuterated methanol (CD_3OD), the calculated values deviate from

TABLE IV Solvent-Dependent Perturbation of the Chair–Chair Equilibrium Position of the A Rings of Vitamin D_3

Conditions	$J_{3\alpha,4\beta}$	α (eq. 3β-OH)
$CDCl_3$	7.6	58%
Cyclohexane-d_{12}	7.1	50%
Methanol-d_4	9.1	72%
SDS-d_{25} micelle	11.5	100%

observed experimental coupling constants, suggesting that deformation of at least one chair is involved in the equilibration. Karplus–Altona correlation for the proposed chair–half-chair interconversion agrees well with the observed vicinal couplings [112]. The deformed cyclohexane conformation may be attributed to a coplanarity of the conjugated double bond and/or an unfavorable interaction between the exo-methylene group and the 1α-hydroxyl group. MM2 calculations together with a π-system treatment of the $1,25(OH)_2D_3$ A ring afforded two minima of comparable energy, one a chair and the other a half chair, further supporting this conclusion.

DeLaroff *et al.* [106] first established that the $\Delta^{5,7}$-diene unit of D_3 in solution is transoid based on ^{1}H-NMR coupling constants. As in the solid state, the $\Delta^{5(6)}$ and the $\Delta^{7(8)}$ double bonds are nearly coplanar and 6-s-trans (or transoid) in orientation as in **1** in Fig. 7. The exocyclic $\Delta^{10(19)}$ double bond is oriented above or below the plane defined by the $\Delta^{5,7}$-diene as implied in the structures shown in Fig. 5. The ability of $1,25(OH)_2D_3$ to rotate about the $\Delta^{6,7}$ single bond is extremely facile, possessing barriers similar to that observed for 1,3-butadiene. Although direct spectroscopic evidence for the presence of a 6-s-cis conformation is lacking, all of the D vitamins possessing the *Z*-hexatriene unit characteristic of $1,25(OH)_2D_3$ must possess kinetically significant concentrations of the 6-s-cis conformation **1′** (Fig. 7). It is this conformation which is required to rationalize the observation that $1,25(OH)_2D_3$ can equilibrate with the corresponding $1,25(OH)_2PreD_3$ (**8**), formed via a [1,7]-sigmatropic shift by hydrogen migration from C-9 to C-19. Likewise, the back rearrangement of a hydrogen from the C-19 to C-9 of $1,25(OH)_2PreD_3$ to afford $1,25(OH)_2D_3$ is also known to occur. If $1,25(OH)_2D_3$ is allowed to thermally equilibrate, about 95% of $1,25(OH)_2D_3$ exists in the extended transoid form and about 5% in the double bond shifted $1,25(OH)_2PreD_3$ form. To increase the population of the cisoid conformation through fluorine–hydrogen bonding, Zhu *et al.* synthesized a series of 11-fluorinated vitamin D derivatives [103]. However, no conclusive evidence was drawn for the presence of the folded or 6-s-cis form of vitamin D in solution.

The C ring is chairlike, flattened in the vicinity of C-8 due to the presence of the exocyclic double bond, and possesses an unusually large torsion angle at C-13,C-14 due to the CD trans ring junction. The conformation of the D ring may be depicted as a dynamic mixture of envelope and half-chair forms (**16**, **17**, and **18**). Overall, the CD hydrindane system (Fig. 8) is a relatively rigid portion of the $1,25(OH)_2D_3$ molecule with restricted dynamic behavior in solution. To simplify matters, the CD ring of $1,25(OH)_2D_3$ can be depicted as a rigid anchor to which is attached, at C-8 and C-17, the dynamic seco-B-triene/A ring and side chain, respectively.

The 25-hydroxycholesterol side chain is obviously the most flexible structural unit of $1,25(OH)_2D_3$. Six rotatable bonds (represented by the curved arrows in Fig. 9) can lead to a large number of unique staggered conformations ($3^6 = 729$). Dot maps (see Section II,C) have been developed at the University of California at Riverside in order to usefully depict the large number of these energetically similar side chain conformers, to estimate the volume in space occupied by the side chain, and to locate possible occupation sites for the important 25-hydroxyl group.

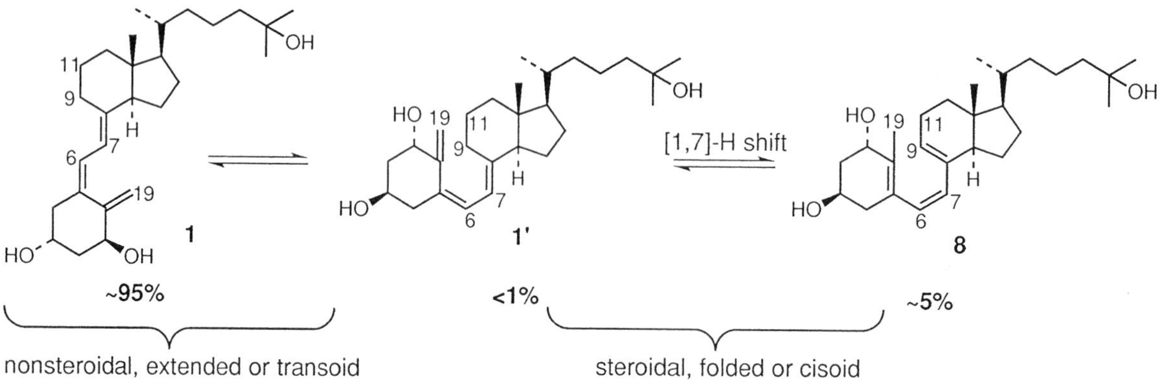

FIGURE 7 Equilibration of 6-s-*trans*-$1,25(OH)_2D_3$ with its previtamin. This is a more detailed picture of the seco-B-ring conformations depicted in Fig. 2. Experimentally, at equilibrium there is a ~95:5 mixture of **1** and **8** when $1,25(OH)_2D_3$ is allowed to stand at ambient temperatures for sufficiently long times. Computational results suggest that there is less than 1% of **1′**, below the normal detection limits of NMR spectroscopy. The native $1,25(OH)_2D_3$ exists in the nonsteroidal, 6-s-trans conformation (also referred to as the extended or transoid form). Species **1′** and **8**, however, can be viewed as steroidal in shape (or as folded or cisoid in conformation).

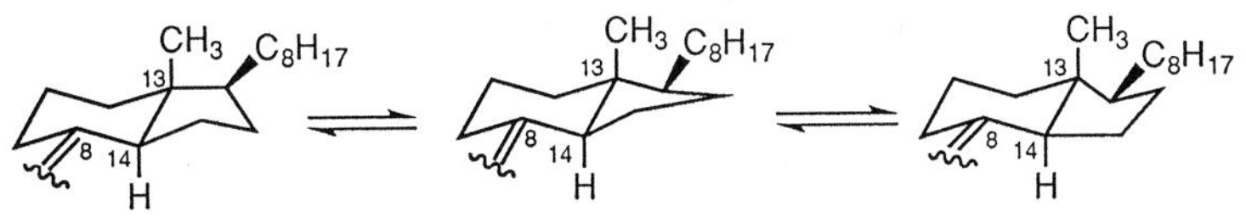

FIGURE 8 Conformational equilibration of the vitamin D CD ring. The chair form of the C-ring portion of the CD ring is relatively rigid. Twist–boat conformations (not shown) are certainly going to be accessible but at much lower concentration. The D-ring part of the CD fragment is rather flexible, equilibrating between half-chair (**17**) and envelope (**16** and **18**) conformations.

It should be summarized here that NMR spectroscopy has been informative regarding the topology and dynamics of the A-ring and seco-B-triene components of vitamin D. The NMR method has thus far been less useful, however, for the CD-ring and side chain dynamic, structural details.

C. Computational Studies of Vitamin D Molecules

The potential utility of computational studies of biologically active molecules including molecular mechanics calculations, *ab initio* computations, and semiempirical treatments in understanding structure–function relationships has been long recognized. Semiempirical or, more increasingly, *ab initio* methods may have predictive value, but molecular mechanics is preferable for larger molecules because computational efforts are less time consuming and therefore less expensive. It is fully realized that D vitamins are highly flexible molecules, able to adopt a wide range of topologies, far more than the conformations observed in the solid state or in solution. Computations using different approaches have been pursued to describe the distribution of possible conformations in order to assess the one responsible for biological function, the so-called active conformer.

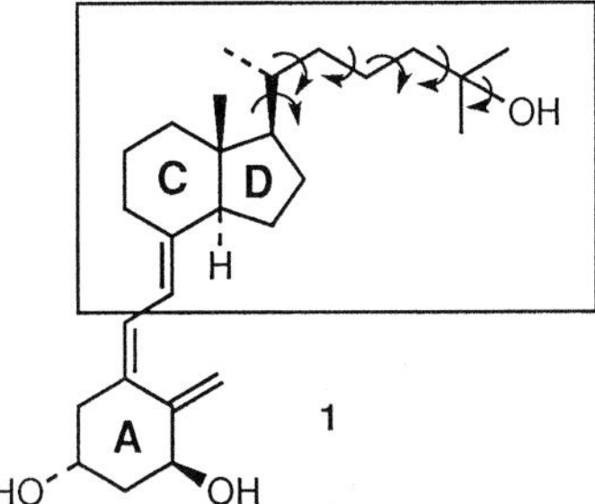

FIGURE 9 Possible side chain orientations of $1,25(OH)_2D_3$. Essentially free rotation is possible about the six σ bonds indicated by the curved arrows superimposed in the box of the structure of $1,25(OH)_2D_3$. If one considers only staggered orientations one can expect 3^6 or 729 forms, but see Fig. 12.

1. THE A RING

A combination of force field calculations, lathacide induced shift (LIS) measurements, and LIS simulations has identified a significant concentration of a twist form of the A ring of D_3 [113] in addition to chair forms identified by X-ray crystallographic and NMR studies. In the computational studies, the two lowest energy conformations are the pair of ring A chair conformations, but the A-ring twist form was reported to lie only ~0.35 kcal above the global minimum, corresponding to about 20% of the total population of conformers. A half-chair A ring conformation was also predicted by Allinger's molecular mechanics (MMPI) calculation for D_3 and $1,25(OH)_2D_3$, which were calculated to exist at quite low energy, less than 0.5 kcal/mol over a global minimum chair conformer [114,115].

2. THE SECO-B RING OR TRIENE

There is no question that the most stable conformation is the 6-s-trans orientation, which is evidenced by X-ray, NMR, and circular dichroism studies [116]. It is also well accepted that rotation about a carbon–carbon single bond is fast, and the energy barrier associated with the torsion around the C-6 to C-7 bond should not be very high. Thus, the cisoid conformation can be accessible to at least some degree for all of the D vitamins. This triene oscillation around the C-6 to C-7 single bond was also suggested by polarization spectroscopy of vitamin D_3 [117]. However, the published theoretical calculations on the population of the cisoid conformer do not agree well with the experimental observations. The molecular mechanics results for D_3 reported by Hofer *et al.* [113] show quite low relative energies of the s-cis conformations, attributing as much as 16% of the total population to the cisoid conformer. The lowest energy conformer, in which the A ring is an α chair and the exo-methylene group lies above the plane of the CD ring, is only 0.85 kcal/mol higher than the global minimum (8.6% population). More surprisingly, calculations on $1,25(OH)_2D_3$ based on a simulated annealing process have predicted the cisoid conformation to be the global energy minimum [118]. Not only is this unrealistic, no NMR evidence for the presence of a cisoid conformation in any vitamin D has been obtained. Even in a seemingly optimal case, no cisoid form was obtained for 11β-fluoro-1α-hydroxyvitamin D_3 (which was predicted to exist mainly in the cisoid conformation because of an accessible fluorine–hydrogen bond) [103].

By 6-s-cis or cisoid, we actually mean the family of conformers that reside within ±90° of the planar s-cis-conformer, which, as far as is apparent, does not exist as an energy minimum. Likewise, by 6-s-trans, we mean the family of conformers that reside within ±90° of the planar s-trans or transoid conformer. Experimental evidence that the 6-s-cis-$1,25(OH)_2D_3$ must exist is based on the view that $1,25(OH)_2D_3$ undergoes thermal equilibrium with its previtamin form; this equilibrium interconversion involves the well-established [1,7]-sigmatropic hydrogen shift, which requires a cyclic transition state, possible for the cisoid conformer but impossible for the transoid conformer [119].

The rotation of the $\Delta^{6,7}$ single bond of $1,25(OH)_2D_3$ through a 360° cycle is depicted three-dimensionally in Fig. 10 [120]. While arbitrarily holding the CD/side chain (SC) in place during the cycle starting from **19**, a 90° rotation about the $\Delta^{6,7}$ bond (step 1) in a downward direction (curved arrow) places the A ring toward the underside of the CD-ring fragment with this ring (A ring) approximately in the plane of the page, labeled as the α-conformer (**20**). A second 90° rotation (step 2) continuing in the same direction produces the planar s-cis conformer (**21**); a third 90° rotation (step 3) again orients the A ring in the plane of the page, but now this ring is above the CD plane in **22** (labeled as the β conformer). A final 90° rotation (step 4) regenerates the starting planar s-trans conformer (**19**). Although there is a continuum of 6,7-single bond rotamers of various torsional energies, only "snapshots" of the selected conformers at 90° increments are presented, starting with the planar s-trans conformer. Within this 360° cycle are numerous energy minima, and an attempt has been made to locate the distribution of these energy minimized conformers by dot maps (Fig. 11) [120]. In Fig. 11, the lowest energy (global minimum) conformer of $1,25(OH)_2D_3$ (an isopropyl side chain model) is given as line drawings in three different orientations. Portions of the D ring and side chain of the higher energy con-

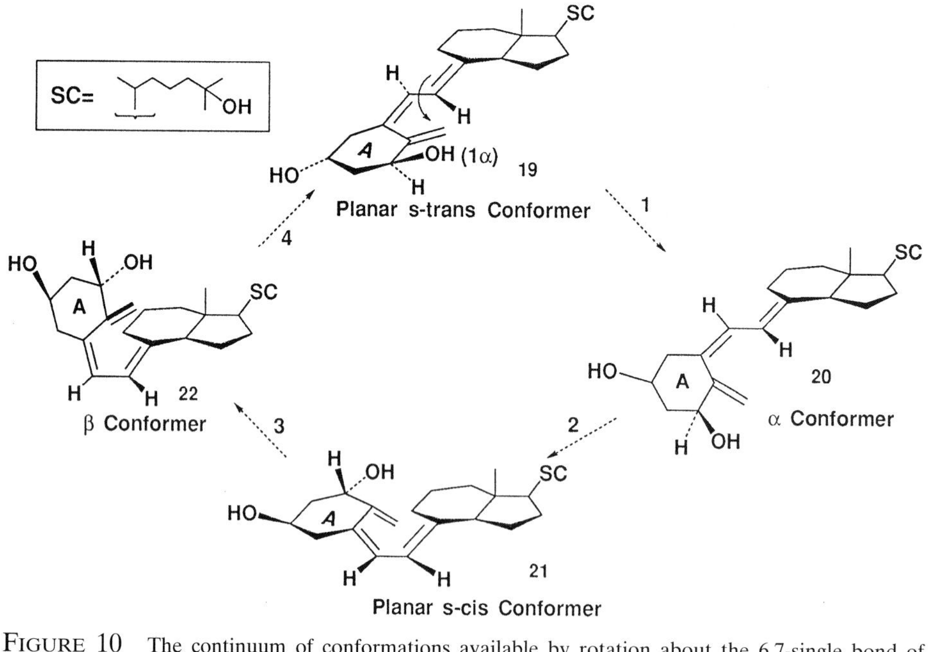

FIGURE 10 The continuum of conformations available by rotation about the 6,7-single bond of $1,25(OH)_2D_3$. This scheme shows snapshots of various limiting conformations of $1,25(OH)_2D_3$ by 6,7-single bond rotations. With respect to the approximate CD plane, snapshots **19** and **21** represent near planar $\Delta^{5,7}$-diene moieties, whereas **20** and **22** represent 6,7-perpendicular topologies of this same diene. The $\Delta^{10,19}$ double bond is likely to always be out of the plane defined by the $\Delta^{5,7}$-diene in this schematic. Further discussion is provided in the text, but the important point is that at the time of protein binding there is sufficient interaction energy between guest and host to force the diene, at least in principle, into any of the instantaneous conformations available through the 360° cycle shown. The global or local minimum conformations of the free $1,25(OH)_2D_3$ ligand may not represent the actual conformation of the steroid in protein active sites.

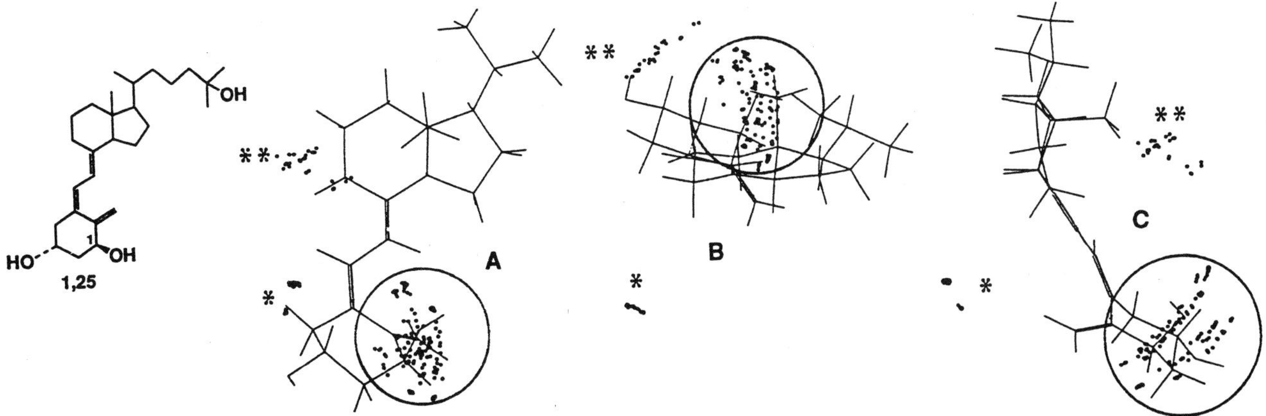

FIGURE 11 Dot maps identifying the location of the C-1 hydroxyl oxygen by 6,7-single bond rotation of the seco-B-triene. Three different views of dot maps for $1,25(OH)_2D_3$ derived from molecular mechanics computations are shown. The line drawings in A, B, and C depict the global energy minimum conformation, and the dots represent the excursions of the C-1 hydroxyl of $1,25(OH)_2D_3$ for the various local minimum conformers. Reprinted from Okamura *et al.* [120] with kind permission from Elsevier Sciences, Ltd, The Boulevard, Langford Lane, Kidlington OX5 1GB, UK.

formers were overlaid on the global minimum to generate the dot maps. The location in space of the various conformers is indicated by the position of the C-1 oxygen atom, depicted as a dot. Three clusters of dots emerge in these maps; those circled represent the excursions of the C-1 oxygen of 6-s-*trans*-$1,25(OH)_2D_3$ for various local minimum energy conformers. The circled set, consisting of both chair forms of the A ring, represents ~99% of the conformations available to $1,25(OH)_2D_3$ within a ~4 kcal window of the global minimum. Two other clusters (* and ** in Fig. 11) represent the set of twisted 6-s-cis-$1,25(OH)_2D_3$ conformational minima, which constitute ~1% of the conformers available to $1,25(OH)_2D_3$ within the same energy window. The 6-s-cis conformers among those in Fig. 11 (*), having the $\Delta^{10(19)}$ double bond positioned on the underside (α face) of the CD ring, are somewhat more stable than those 6-s-cis conformers in Fig. 11 (**), having this same double bond on the β face of the CD ring. These results from this laboratory differ from those obtained by Hofer *et al.* [113].

3. THE SIDE CHAIN

To accommodate the flexibility of the cholesterol side chain of vitamin D, a result rendered by the relatively free rotation about the six single bonds indicated by the curved arrows in Fig. 9, the dot map protocol has been developed by this laboratory as an indication of the volume in space which the 25-OH side chain of vitamin D metabolites or analogs is permitted to occupy (Fig. 12) [121,122] (Wilson and Guarnieri have used an alternative approach to locate the population of different side chain conformations [123]). Using the model system in Fig. 12A, wherein the top view depicted represents the CD/side chain fragment of $25OHD_3$ or $1,25(OH)_2D_3$, a molecular mechanics method was selected for evaluation of side chain conformations. All accessible low energy conformations were generated by a Monte Carlo conformational search using the MM2 force field. Dot maps thus generated indicate the location of the C-25 oxygen relative to the root mean square, best fit overlay of the CD fragment of each energy minimum conformer. Of the many possible mainly staggered rotamers, 394 side chain conformers were determined to reside within 4 kcal of the global minimum. Figure 12B depicts an overlay of the 394 conformations in which the dots represent the C-25 oxygen atom for each conformational minimum. To reflect that volume in space which the side chain can occupy with respect to the rigid CD ring, two nearly perpendicular views of these dot maps have been shown side by side in Fig. 12B,C. In Fig. 12B the CD ring is viewed from the top face, orthogonal to the steroidal plane as in Fig. 12A. Figure 12C views the same dot map from the bottom edge of the approximate plane defined by the CD ring. Stereoviews of the same dot maps are presented in Fig. 12D,E, which can be best viewed separately by the reader with a stereoscope. It must be strongly emphasized that these dot maps for the $1,25(OH)_2D_3$ side chain seek to define an occupancy volume in space or a "parking zone" which the side chain is "permitted" to sweep within a certain energy range (a consequence of van der Waals repulsion, torsional constraints, and restrictions of the covalent bonds of the side chain).

Similar dot maps for 20-epi-$1,25(OH)_2D_3$ (Fig. 12F) were generated by the same treatment, and the results are shown in Fig. 12G,H. In panel Fig. 12G, a top face view similar to that in Fig. 12D is pictured; in Fig. 12H,

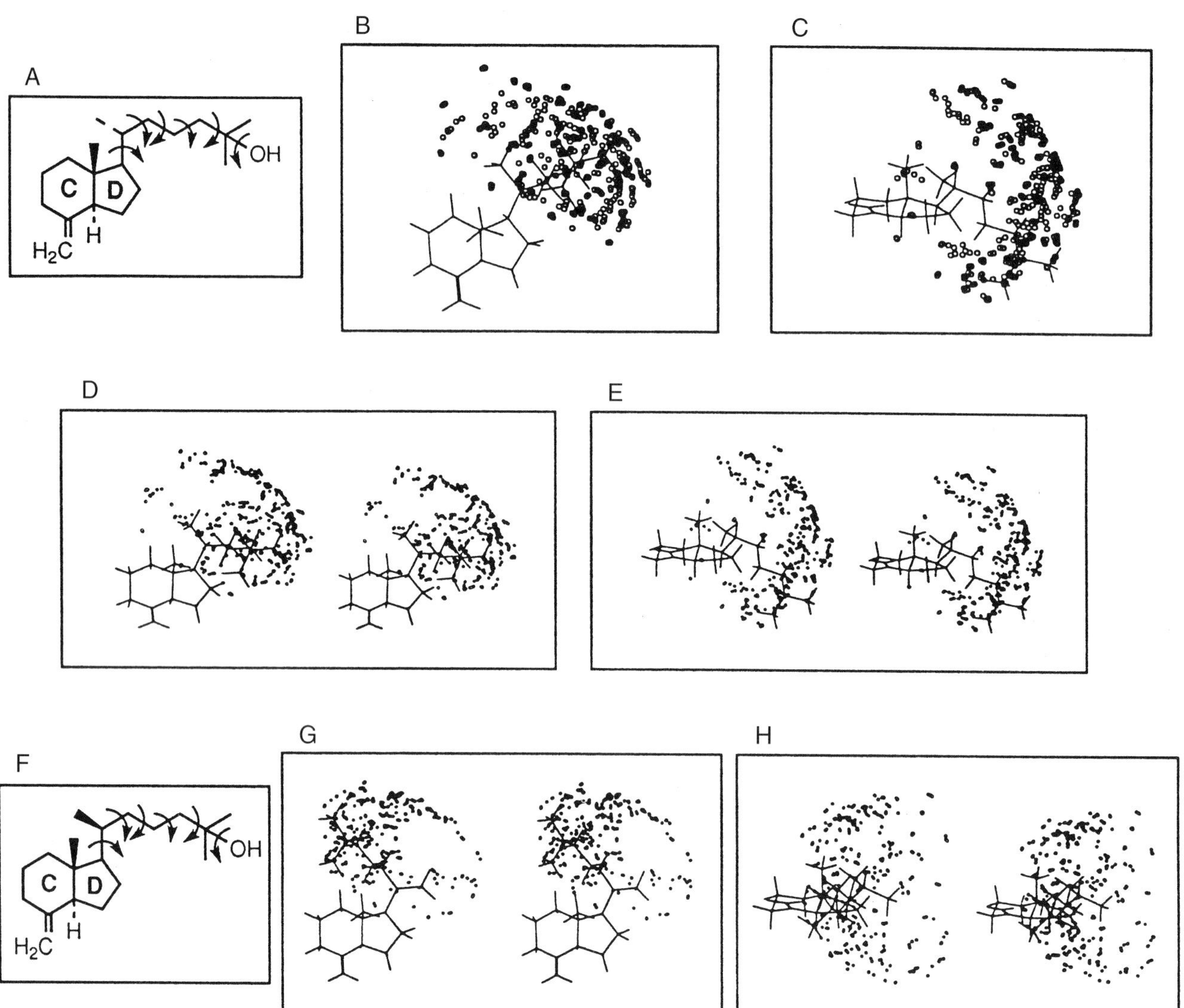

FIGURE 12 Dot maps of side chains of $1,25(OH)_2D_3$ and 20-epi-$1,25(OH)_2D_3$. The dot map system we developed is used to depict the volume in space which the side chain can occupy. The dots locate possible occupation sites of the side chain hydroxyl group considered to be an important factor in hydrogen bonding of the ligand to various host proteins. Using the dot map representation, only a limited number of drawings is needed to display a large number of energy-minimized side chain conformers. The text gives details. Parts A–E pertain to the CD side chain fragment model system shown in A [representative of the side chains of $25OHD_3$ or $1,25(OH)_2D_3$]. In F–H is a similar representation of the 20-epi vitamin (F). As indicated in A and F, in both systems various energy minima are generated via rotation about the six single bonds indicated by the curved arrows. In B and C are the dot maps for a 4 kcal/mol window viewing the natural CD fragment (A) either face on (B) or edge on (C), respectively. D and E are stereoscopic views of B and C, respectively. G and H represent similar stereoscopic dot maps for the 20-epi vitamin model system (F).

an edge-on view similar to Fig. 12E is depicted. It is noteworthy that for the C-20 epimer the side chain with its C-25 hydroxyl group is largely directed toward the northwest (Fig. 12G), whereas with the natural C-20 epimeric vitamin the side chain and 25-OH group are largely located toward the east (Fig. 12D).

The Yamada laboratory in Japan, using a molecular mechanics conjugate gradient method with a consistent valence force field and then a Tripos force field (MAXMIN2 routine in SYBYL) [124,125], has also used a dot map technique to evaluate side chains. Key side chain bonds, namely, the 17,20-, 20,22-, 22,23-, 23,24-, and 24,25-bonds, in crystal structures of molecules possessing the side chain of $25OHD_3$ and its 20-epimer were rotated 360° with 30° steps to yield a total of $(360/30)^5$ = 248,832 conformers for individual compounds. Unacceptable conformers with intramolecular van der Waals repulsions were eliminated by applying a van der Waals bump coefficient of 0.95. Particularly interesting applications were the utilization of computational and biological results for $1,25(OH)_2D_3$, its 20-epi diastereomer, and four diastereomeric 22-methyl derivatives. It was concluded that the near global minimum orientation of $1,25(OH)_2D_3$ (indicated approximately by the side chain line structure in Fig. 12D) and that of its 20-epi isomer (indicated approximately by the side chain line structure in Fig. 12G) are their biologically active conformers toward VDR binding. The Yamada approach actually provided greater definition in preferred side chain orientation than apparent from dot maps in Fig. 12, but it remains to be seen whether these computational results reflect reality. The dichotomy of activity as predicted by Yamada toward the same receptor is perplexing but certainly conceivable.

Although these analyses provide a simple method for evaluating the orientation of the important side chain, including analogs with conformationally perturbed side chains, the application to an understanding of the conformation of the ligand when bound to protein is limited. Studies of conformationally locked analogs would be desirable. The bond rotations (or conformational interconversions) discussed above leading to energy-minimized conformers are extremely rapid. Whether these computations represent reasonable descriptions of the three-dimensional structure of the various energy minimum conformers available to $1,25(OH)_2D_3$ remains uncertain so far. Even if these computational methods lead to precise structural data for free ligand, the more daunting task of accurately determining protein-bound ligand structures remains. The actual topology of $1,25(OH)_2D_3$ in the active site is likely different from any one of the energy minimum structures. In general, the conformational flexibilities of the side chain, seco-B-triene system, A ring, and even the semirigid CD unit of the hormone and analogs, and the equally flexible host protein, are such that computational approaches should be viewed as being only at their infancy. With respect to the notion that steroid hormones function by inducing a conformational change in protein on binding (from an inactive conformer to an active conformer induced by ligand binding), the situation for $1,25(OH)_2D_3$ and its metabolites is unique [126]. Because of the flexibility of vitamin D molecules, there exists the possibility of a mutually induced fit between ligand and protein. As stated earlier, there is considerable truth in Jorgensen's statement that the classic lock and key hypothesis has rusted [83]. Any attempt to design drugs on the basis of structures of free ligand (guest) or free protein (or other hosts) is subject to failure or is apt to lead to disappointment. The host–guest complex must be examined directly.

Nevertheless, experience has revealed that random synthetic and design approaches, and educated guesses based on lead structures, can be successful. For this reason, accurate structures of free hosts and guests will remain a viable area of investigation.

III. APPROACHES TOWARD UNDERSTANDING THE STRUCTURE OF THE VITAMIN D LIGAND IN THE LIGAND–PROTEIN COMPLEX

Basic research on defining the molecular mode of action of vitamin D [1,2] and other members of the steroid hormone superfamily is rapidly developing, particularly with the application of molecular biological techniques to the vitamin D field. Overexpression of hormone binding proteins (DBP and VDRs in particular) in sufficient quantities should allow direct examination of these proteins and hormone–protein complexes by modern spectroscopic methods (X-ray, NMR) to obtain three-dimensional structural information. An example of the rapid progress is shown by reports on the crystal structure of the ligand binding domain of retinoic acid receptors [127–129] and a ligand–protein complex [130] (see also the thyroid hormone–analog complex discussed below). Vitamin D derivatives with various structural modifications have been designed, synthesized, and biologically evaluated both *in vitro* and *in vivo*. Conformationally biased systems including A-ring chair locked, seco-B-ring cisoid and transoid locked, and side chain orientation locked analogs could provide structural insight into the mechanism of vitamin D host binding. It is most informative that certain 20-epi side chain analogs in comparison to the natural hormone $1,25(OH)_2D_3$ induce different conformational changes in VDR [131], suggesting that different shapes of vitamin D induce different conformations of the ligand–protein complex, features probably significant in modulating biological actions. A knowledge of the conformation of vitamin D in ligand–protein complexes is thus likely to be highly significant in the design of vitamin D analogs.

A. Practical Strategies for Probing the Conformation of Vitamin D in the Ligand–Protein Complex

1. Conformationally Locked Analogs

As discussed in Section II,C, the dynamic topological structure of vitamin D is such that no specific conformation corresponding to the biologically active topology of the hormone $1,25(OH)_2D_3$ at the time of binding to host can presently be identified. The only restriction at present is that the topology of ligand in the ligand–protein complex must be within its occupancy volume simply due to the restrictions of strong covalent bonds. The idea here is to incorporate conformationally rigidifying units such as additional rings or double bonds at key locations to identify more precisely the active topology of the vitamin D molecule. Biological evaluation of these conformationally restricted or locked analogs may provide information on the required topology of the vitamin D molecule for active agonist behavior. The danger of course is the possibility that agonist activity may be strongly attenuated due to steric or other effects.

The most practical locations in $1{,}25(OH)_2D_3$ for conformational locking are summarized in the boxes of Fig. 13. Synthesis and biological evaluation of A-ring analogs locked into a single chair form (either α or β chair conformers as given in Fig. 5) would provide structural information on which chair conformer is involved in protein binding at the receptor level. Indeed, opposite chair forms or even a twist–boat conformer (Fig. 5) may be involved at different receptor target sites. For the seco-B ring, any instantaneous rotamer of the continuum shown in Fig. 10 could be the active conformer responsible for a selected biological function. To obtain more complete information, various twisted but locked 6-s-*trans*- and 6-s-*cis*-$1{,}25(OH)_2D_3$ analogs could be designed and biologically evaluated. The CD ring is relatively rigid and could be considered an anchor to which a very flexible side chain and triene moiety are connected. No specific suggestions are made here, and accordingly Fig. 13 indicates this in the form of a dashed box. There is some flexibility here as indicated in Fig. 8, but rotation is limited. Introduction of double bonds at $\Delta^{9(11)}$, Δ^{14}, and Δ^{16} could induce some conformational rigidification, however. Even from a cursory examination of Fig. 12, one can easily discern that the side chain is the most flexible portion of $1{,}25(OH)_2D_3$ and its metabolites. It is apparent from the earlier discussion that the rotation about the $\Delta^{17,20}$ single bond, which biases the side chain to the "northwest" (Fig. 12F–H) in the 20-epi series and to the "east" (Fig. 12A–E) in the natural series, is of exceptional interest. Even so, restriction of rotations about the other single bonds along the remainder of the side chain (C-20, C-22, C-23, C-24, and C-25 in Fig. 9) is of continuing interest as reflected by studies directed toward the introduction of rings and π systems in the side chain (for a review, see [1,2]). On interpreting structure–function information from the biological evaluation of conformationally locked analogs, some caution is necessary. (1) Conformationally locked vitamin D analogs may mimic the parent molecule, but they are structurally perturbed by substituents. A negative biological result is not useful because simple steric effects may attenuate activity. Thus, detection of a significant biological response (agonist behavior) is necessary for assessing topological effects. (2) Biological evaluation should include both *in vitro* (direct binding studies of the locked analogs to selected receptors) and *in vivo* data to define the net biological action. (3) As the overall hypothesis assumes that the conformational perturbations of one part of the

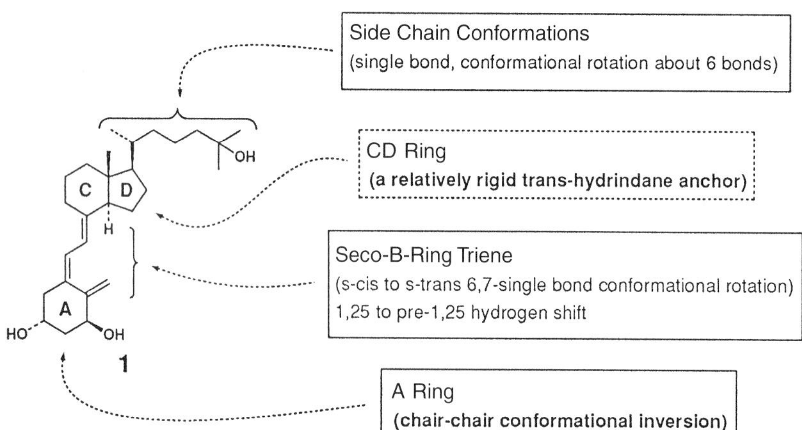

FIGURE 13 Possible structural modifications of $1{,}25(OH)_2D_3$. The vitamin D molecule can be conveniently divided into four geographic regions: the A ring, the seco-B-triene, the CD ring, and the side chain. As the primary focus of this chapter concerns conformational locking of the vitamin, we consider the A ring, the seco-B-triene, and the side chain as possible regions for introducing functional groups or chains that might lock the topology of the molecule into well-defined conformational shapes. The CD ring is considered to be a less attractive region for locking because it is already rather rigid.

molecule would have relatively small effects on other parts of the molecule, it is prudent to make single structural changes during the iterative design and synthesis components of the study. For example, it is fairly obvious that one should compare side chain analogs only where the A, seco-B-triene, and CD structures are retained.

2. Direct Spectroscopic Studies of Ligand–Protein Complexes

a. X-Ray Crystallography Since the first high resolution protein structure [132] was determined by X-ray crystallography, numerous macromolecular structures have been solved, including proteins, ligand–protein complexes, nucleic acids, protein–nucleic acid complexes, viruses, and other larger molecular assemblies [133]. Our knowledge of the three-dimensional structure of these large molecules and/or assemblies has come almost entirely through this solid-state technique [134]. A high resolution X-ray structure ($\geq$2.8 Å) not only gives a well-detailed structure of the entire complex, but can also provide the conformation or shape of the ligand and the ligand–protein interaction at the catalytic or binding site(s). The X-ray technique is still the most accurate and most reliable method to get information on the structure of macromolecules as well as their complexes. One of the recent examples in this regard can be seen by the crystal structure of the lactose operon repressor under three different conditions: alone, bound to two operator sequences on the DNA that it targets, and complexed with a β-galactoside sugar leading to a conformational change [135]. This is the kind of sequential detail that is still lacking in the steroid hormone area.

Regarding the structure of the ligand–protein complexes of the steroid receptor superfamily, the most rapid progress has been for the retinoic acid (RA) receptors (RXR and RAR) [127–130]. The structure of the ligand binding domain (LBD) of the ligand-free RXR and the ligand-bound RAR were both achieved only in 1995. The 2.0-Å crystal structure of RA–RARγ reveals the ligand binding interactions and suggests a dynamic electrostatic guidance mechanism in which the retinoid is attracted to the ligand binding cavity by electrostatic forces and is locked in this position mainly through hydrophobic interactions. The interaction of ligand with surrounding amino acid residues triggers a series of conformational changes which generate the surface that allows transcriptional mediators to bind, leading to the transcription of the target gene. Another major success on elucidation of the structure of the ligand–protein complexes of other members of the steroid receptor superfamily was the crystal structure of the rat α_1 thyroid hormone receptor ligand binding domain bound with a thyroid hormone agonist [136]. Because the LBD of the thyroid hormone receptor participates in several activities, including hormone binding, homo- and/or heterodimerization, and transcriptional activation and repression, the X-ray structure of the ligand–LBD complex reveals an absolute requirement for hormone in inducing the active conformation of the receptor. The hormone completes the hydrophobic core of the active conformation and directs the alignment of secondary structural elements of the receptor needed for functioning. A similar structural role for hormone in the action of the entire class of steroid hormones and retinoids was also speculated on [136].

b. Nuclear Magnetic Resonance Spectroscopy A second method, which emerged in the early 1980s for gleaning information on the three-dimensional structure of macromolecules, employs NMR spectroscopy [137–144]. The structural information obtained by NMR in many ways complements that from X-ray crystallography, expanding our view of macromolecular structure. This is because solution studies may be carried out; moreover, in more recent studies, solid-state NMR investigations of amorphous materials have been shown to be feasible. Thus, the most obvious advantage of NMR is that it can be applied to biopolymers for which no crystallographic grade solids are available. The structural information obtainable by NMR spectroscopy of macromolecules in solution can be extended to dynamic processes ranging from the picosecond to second time scales. For proteins up to ~100 residues, the quality of the structural information obtainable by multidimensional NMR spectroscopy is comparable to 2- to 2.5-Å resolution X-ray structures [138]. Some limitations of the NMR solution method must be mentioned, however. (1) The protein under investigation must be soluble and should not aggregate up to a concentration of at least 1 m*M*. (2) Because of the line broadening for larger molecules, the conventional two-dimensional (2D) NMR technique is limited to structural studies of proteins of up to ~100 residues [138,139]. Introduction of three- and four-dimensional (3D and 4D) NMR, in combination with ^{13}C and ^{15}N isotope labeling [140–143], significantly extends the molecular weight limit of proteins up to 30 kDa (~300 residues). For proteins with more than 300 residues, line widths are often over 45 Hz, comparable to many C–H couplings and larger than many C–C couplings. (3) Considerable time (sometimes months) is required for data acquisition to characterize a single protein in the 25- 30-kDa range. Tens of milligrams of isotopically enriched protein, which must remain stable over long experimental periods, are required for this process.

For ligands or guest molecules that form specific but rapidly associating and dissociating complexes with proteins or other large host molecules in solution, their host-bound conformation can be obtained by binding-induced NMR relaxation enhancements including transferred nuclear Overhauser effects (NOEs) [145,146]. The major advantage of this strategy is that the experiment focuses on the easily detected NMR signals of the free ligand. This technology is particularly useful for those ligands bound to protein targets that are too large (>40 kDa) to be studied directly by NMR. However, the conformation of the host protein or other macromolecules as well as structural information on the ligand–protein interaction cannot be obtained by this technique. Moreover, in the case of steroid hormone-like receptors, ligands are normally bound too tightly and ligand exchange is probably too slow to make the transfer NOE technique useful.

A combination of the use of isotope labeling, heteronuclear NMR correlation spectroscopy, and heteronuclear editing using $^{13}C(\omega_2)$-half-filter techniques [147] yields subspectra containing exclusively intramolecular cross-peaks between protons attached at ^{13}C labels of multi-^{13}C-labeled ligand bound to relatively large proteins. An earlier experiment using a $^{13}C(\omega_1,\omega_2)$-double-half-filter technique provides information on the interaction of protons attached on labels with other protons attached or not attached on labeled carbons. The latter is even more informative for determining the molecular structure of ligand–receptor interactions, but it may be limited due to spectral overlap. An application of this NMR technique was reported by Norris *et al.* [148]. By use of ^{13}C-labeled ligand and a $^{13}C(\omega_2)$-filtered ^1H-NOESY experiment, the conformation of all-*trans*-retinoic acid (**23**, Fig. 14) bound to cellular retinoic acid binding protein I (CRABP-I) and cellular retinoic acid binding protein II (CRABP-II), both 15-kDa proteins, was determined. In CRABP-II, the bound retinoic acid was determined to exist with a 6-s torsion angle of $-60°$ skewed from a cis conformation, whereas the ligand freely rotates about the C-6 to C-7 bond in the RA/CRABP-I complex. Both $\Delta^{7,9}$- and $\Delta^{9,11}$-diene units of retinoic acid are nearly planar and trans in orientation in both proteins (Fig. 14).

Distances ($\leq$5Å) between any pair of ^{13}C labels of ligand–protein complexes in the solid state can be measured with an accuracy of better than 0.5 Å by rotationally resonant magnetization exchange [149]. The 6-s-trans conformation of all-*trans*-retinal (**24**, Fig. 14) in bacteriorhodopsin, a 35-kDa membrane protein, was determined by internuclear distance measurements [149,150] based on ^{13}C–^{13}C dipolar coupling correlations. An even more recent technique to determine internuclear distances, using rotational resonance for restoring homonuclear dipolar couplings with magic angle spinning, gives more accurate measurements of the order of 0.2 Å [151].

B. Status Report on the Structure of the Ligand–Protein Complexes of Other Members of the Steroid Receptor Superfamily

As indicated above in Section II,A,2,a, only the crystal structures of the RA/RARγ LBD and the thyroid hormone agonist analog–thyroid hormone receptor LBD, both published in late 1995 [130,136], provide information on the structure of the ligand–protein complex in the steroid receptor superfamily. Some additional information may be discerned from the crystal structure of the ligand-free LBD of RXR, which was published in mid-1995 [128]. Thus, information is only now beginning to emerge, and this is obviously a fruitful area of future research.

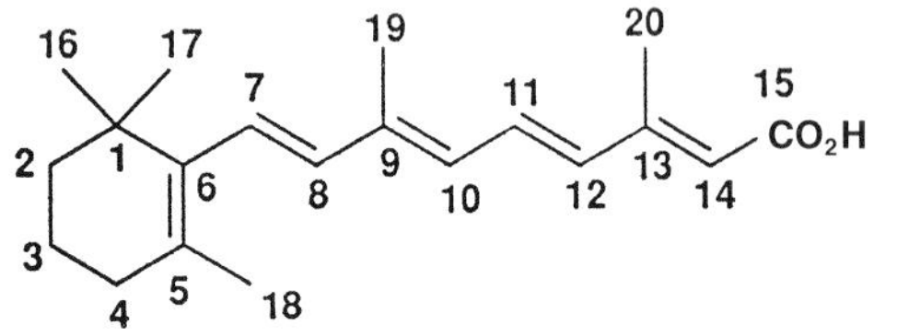

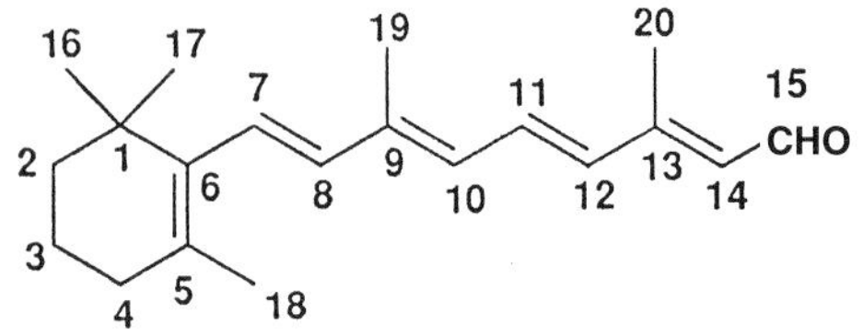

FIGURE 14 Structures and numbering of all-*trans*-retinoic acid and all-*trans*-retinal. All-*trans*-retinoic acid (**23**) and several geometric isomers are considered to be ligands of the RAR or RXR superfamily of steroid hormone-like receptors. All-*trans*-retinal (**24**) is the chromophoric group of bacteriorhodopsin, and a geometric isomer of **24** is a ligand for the visual chromophore rhodopsin. Some of the most extensive NMR studies of ligand-bound proteins pertain to studies of these molecules (see text).

C. Structure–Binding Affinity Relationship of Vitamin D Ligand to Protein

The relative binding affinity of a given analog of vitamin D to DBP, VDR, and m-VDR in comparison to the parent hormone $1,25(OH)_2D_3$ can be obtained via competitive binding assays of the vitamin D derivative versus radiolabeled $1,25(OH)_2D_3$ [152]. Although other *in vitro* binding assays are available, the discussion focuses on competitive binding assays because more extensive data are available (and easily accessible) to the authors. Much of the literature has been reviewed [1,13,153]. Careful evaluation of the binding affinity of vitamin D analogs in relation to the structural modifications can give insight into the nature of the protein binding sites and the optimal topology of the ligand. In this review, the binding affinity of vitamin D analogs is presented as a relative affinity or relative competitive index (RCI) value referenced to that of $1,25(OH)_2D_3$, which is defined as 100% for all proteins discussed. To obtain a more informative comparison, the RCI values of selected vitamin D analogs are given in Fig. 15, most of which are derived from the results tested by the Norman laboratory at Riverside, and some taken from published material [1,13,153]. To make the emphasis on topology clear, all of the compounds shown in Fig. 15 possess hydroxyl groups at C-1 and C-3 except for $25OHD_3$. Because of the flexible nature of the vitamin D molecule, and to focus on the main theme of this review, we have mainly selected those conformationally locked or perturbed analogs that might provide topological information concerning vitamin D in the ligand–protein complexes.

1. Vitamin D Binding Protein

As described in Section II,B, the A ring of vitamin D may exist in an α chair, a β chair, or twist–boat form (Fig. 5), or even half-chair forms (not shown). Unfortunately, there is no analog of $1,25(OH)_2D_3$ locked into any one of these A-ring conformations for which there are adequate biological data. It is very clear, however, that the presence or absence, the configuration, or the orientation of the 1α-hydroxyl group in $1,25(OH)_2D_3$ is very critical. Deletion of the C-1 hydroxyl as in **25** (BO) leads to a ~670-fold increase in its binding affinity to DBP (Fig. 15). Configurational inversion of the C-1 hydroxyl group as shown by **26** (HL) also increases the binding affinity by 4-fold. The orientation of the 3β-hydroxyl group (see **41** and **42**) also seems important for the stereospecific binding of $1,25(OH)_2D_3$ to DBP [153–155]. The seco-B-ring conformation of vitamin D in the vitamin–DBP complex was assessed by evaluation of a series of conformationally locked analogs: **27** (HF), **28** (GF) (cisoid locked); **29** (JM), **30** (JN), **31** (JO), **32** (JP) (planar cisoid locked); **33** (JW) (α twisted cisoid locked); **34** (JV) (β twisted cisoid locked); and **35** (JB), **36** (JD) (transoid locked). These are mimics of the various C-6–C-7 rotamers shown in Fig. 10. The previtamin D **27** is believed to be a good mimic of the cisoid conformer of the vitamin. With the five deuterium atoms present at its C-9 and C-14 positions, the resulting $1,25(OH)_2preD_3$ **27** (HF) is stabilized by a kinetic isotope effect [119]. The binding affinity of the deuterium-labeled analog **27** (HF) reveals that DBP has a lower affinity (~8%) toward this s-cis analog compared to $1,25(OH)_2D_3$ **1** (C). The significant binding affinity of DBP to **33** (JW), an α twisted cisoid conformer mimic of $1,25(OH)_2D_3$, suggests that $1,25(OH)_2D_3$ may actually exist as an α twisted cisoid conformer in the DBP complex. Change of the side chain orientation closer to the plane defined by CD-ring plane, as evidenced by the synthesis and biological evaluation of its 16-ene analog **37** (HM), significantly decreases its binding affinity to DBP [~2% of that of the parent $1,25(OH)_2D_3$], suggesting that the 25-hydroxycholesterol side chain orientation is quite stereospecific in the DBP complex. Over a 20-fold increase in binding affinity was observed by introduction of an aromatic ring in the side chain as in **38** (DF), but other rigidifying moieties in the side chain as in **39** (V) and **40** (BT) lead to a decreased affinity in binding (~20 and 2-fold decrease, respectively). These represent only a limited selection of examples of how topological effects can affect DBP binding (see Bouillon *et al.* [1] and Bishop *et al.* [153] for other effects on DBP binding).

2. Vitamin D Receptor (VDR and m-VDR)

In contrast to the case for DBP, deletion of the 1α-hydroxyl group leads to a >500-fold decrease in binding affinity of $1,25(OH)_2D_3$ to VDR [1]. A study of the affinity of four A-ring diastereomeric vitamin D analogs toward VDR has shown that **26** (HL), **41** (HJ), and **42** (HH) have 0.8, 24, and 0.2% of the binding affinity of $1,25(OH)_2D_3$, suggesting that the configuration of the 1α-hydroxyl group is more important than that of the 3-hydroxyl group for VDR interaction [154]. All of the transoid and cisoid conformationally locked analogs exhibit very weak binding to VDR. No positive structural information can thus be gleaned. However, we speculate now that a twisted cisoid conformation could be the true topology of the triene moiety in the ligand–protein complex (see below). A slight change in conformation of the C and/or D ring does not significantly influence its binding affinity to VDR as evidenced by the data for analogs **43** (BU) and **37** (HM). A comparison of the rigid side chain analogs **38** (DF), **39** (V), and **40** (BT) is most intriguing because their affinities toward VDR

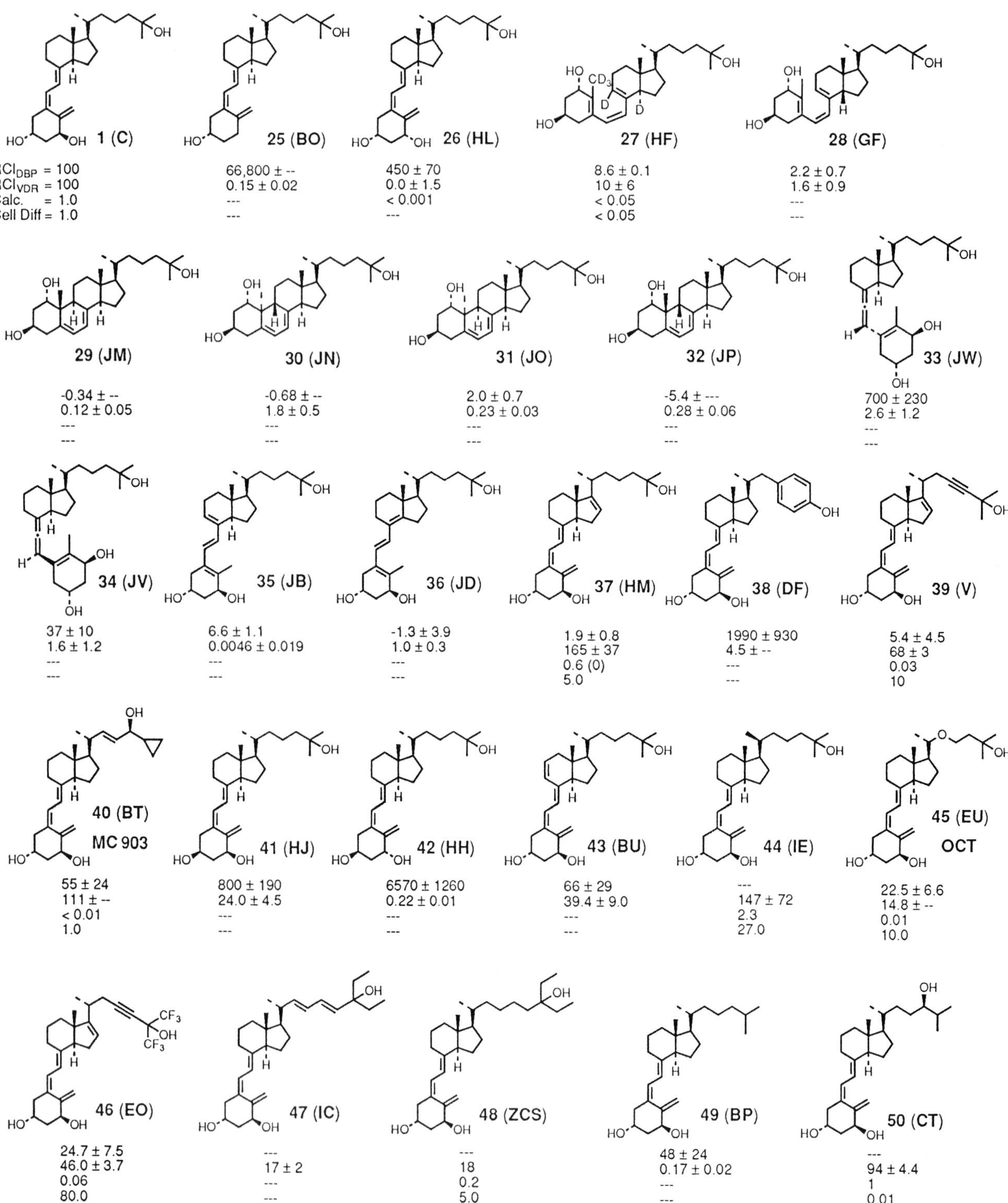

FIGURE 15 Conformationally locked and other selected analogs. These analogs and the data presented were selected mainly from a review by Bouillon *et al.* [1]. In a few instances, data are unpublished results from the laboratory of A. W. Norman at Riverside. The letter codes in parentheses after the compound numbers are those used by the Riverside laboratories for convenient cataloging (see also Bouillon *et al.* [1]). In the data below each structure, the first two rows are for binding of DBP (human) and VDR (mainly from chick intestine) as defined in the text, where all RCI values are normalized to the binding affinity value of 1,25(OH)$_2$D$_3$ (**1** or C), arbitrarily given a score of 100%. The third and fourth row of data are biological activity indices that reflect calcemic (Calc.) and cell differentiation (Cell Diff) effects, where all values are normalized to a standard score of 1.0 for both effects for 1,25(OH)$_2$D$_3$. The latter functional assays are complex in origin, and the reader is referred elsewhere [1] for a more detailed description. Alternative abbreviated names for the above compounds are: **1**, 1α,25-OH)$_2$-D$_3$; **25**, 25-(OH)D$_3$; **26**, 1β,25-(OH)$_2$-D$_3$; **27**, 1α,25-(OH)$_2$-pre-D$_3$-9,14,19,19,19-d$_5$; **28**, 14-Epi-1α,25-(OH)$_2$-pre-D$_3$; **29**, 1α,25-(OH)$_2$-7-DHC; **30**, 1α,25-(OH)$_2$-Lumisterol$_3$; **31**, 1α,25-(OH)$_2$-Pyrocalciferol$_3$; **32**, 1α,25-(OH)$_2$-Isopyrocalciferol$_3$; **33**, (1S,3R,6R)-7,19-Retro-1,25-(OH)$_2$D$_3$; **34**, (1S,3R,6S)-7,19-Retro-1,25-(OH)$_2$-D$_3$; **35**, 1α,25-Dihydroxytachysterol$_3$; **36**, 1α,25-Dihydroxy-trans-Isotachysterol; **37**, 1α,25-(OH)$_2$-16-Ene-D$_3$; **38**, 22-(Para-hydroxyphenyl)-1α,25(OH)$_2$-D$_3$; **39**, 1α,25-(OH)$_2$-16-Ene-23-yne-D$_3$; **40**, 1α,24S-(OH)$_2$-22-Ene-

are reversed in comparison to their affinities towards DBP. Both **39** and **40** bind more effectively to VDR (68 and >100%, respectively) in comparison to **38** (4.5%); in contrast, **39** and **40** bind less effectively to DBP (5 and 55%, respectively) than **38** (>2000%). Note again that $1,25(OH)_2D_3$ is the reference system for DBP and VDR binding, with both RCI values normalized to 100%. Interestingly, the 20-epi analog **44** (IE) (see discussion regarding the dot maps shown in Fig. 12) binds to VDR most effectively among the analogs shown in Fig. 15.

We also take note of the fact that the s-trans locked analogs **35** (JB) and **36** (JD) exhibit extremely weak binding to VDR (0.5 and 1%, respectively), even worse than the s-cis locked analog **27** (HF) and the twisted s-cis locked analogs **33** (JW) and **34** (JV) (10, 2.6, and 1.6%, respectively) (A. W. Norman, unpublished observations). These results have led us to suggest that the $1,25(OH)_2D_3$ may bind to the VDR with a twisted s-cis orientation, somewhat resembling **20** or **22** in Fig. 10 [120, 156].

Finally, we take note of the fact that 6-s-cis locked analogs such as **27** (HF) and **30** (JN) bind equally as well as $1,25(OH)_2D_3$ to the m-VDR (data not given in Fig. 15) associated with transcaltachia [13,120,156–160]. These results provide the clearest evidence yet of a ligand that assumes a higher energy conformer on protein binding. This also provides the clearest implication yet that the topology of vitamin D should be a critical fact in any future structure–function analysis of vitamin D.

Although we cannot propose the specific topology of $1,25(OH)_2D_3$ or metabolites in all of their respective host protein complexes (DBP, VDR, and m-VDR), it is evident that the ligand binding domains of different proteins require different shapes of $1,25(OH)_2D_3$ (for reviews, see Okamura *et al.* [120,156] and Norman *et al.* [157]). This is documented by the consequences of the presence or absence of the three hydroxyls (1α-, 3β-, and 25-OH) in $1,25(OH)_2D_3$; for example, DBP binds to $25OHD_3$ ~670-fold better than to $1,25(OH)_2D_3$, whereas VDR binds $1,25(OH)_2D_3$ ~500-fold better than $25OHD_3$. The presence of an aromatic ring in the side chain enhances binding affinity 20-fold to DBP, but diminishes ligand interaction with VDR by 100-fold. The m-VDR responds effectively to cisoid locked analogs of $1,25(OH)_2D_3$, but these analogs have only 1–2% of the binding affinity to VDR. The 6-s-trans conformation, as evidenced by the evaluation of the transoid locked analogs **35** (JB) and **36** (JD), are not favored by the major proteins studied, but there is some indication that twisted-s-cis analog conformers [as evidenced by the data for **33** (JW) and **34** (JV)] may be important to DBP and VDR binding. It remains for future investigations to assess fully the implications of these (Fig. 15) and other conformationally perturbed analogs in structure–function analyses.

D. Relationship of Ligand–Protein Binding and Biological Activity

It was mentioned in Section I and is discussed in detail in other chapters in this treatise that $1,25(OH)_2D_3$ generates biological responses via both genomic and rapid (nongenomic) pathways. In the well-established genomic pathway, $1,25(OH)_2D_3$ first binds to an intracellular protein (VDR), which subsequently forms a heterodimer, and then binds to a vitamin D response element (VDRE) of a target gene promoter and up- or down-regulates gene transcription, thus generating a series of biological responses. Most of the biological responses of $1,25(OH)_2D_3$ are generated via a transcriptional regulation mechanism, including calcium translocation and induction of cell differentiation, the major topics of the discussion in this section. However, as indicated earlier, there is some evidence for so-called rapid actions of $1,25(OH)_2D_3$ (nongenomic pathways) at the membrane level, although these actions remain to be established *in vivo*. The calcemic and cell differention indices [1] given in Fig. 15 represent biological activities occurring only via the genomic pathway. Because "binding data" to putative membrane receptors are limited [13], the emphasis here concerns whether the RCI_{VDR} data shown in Fig. 15 correlate with genomic calcemic or cell differentiation data.

For calcemic activity, we mainly examine here the relationship between the binding affinity of the conformationally locked vitamin D analogs to VDR and the corresponding calcemic index (see legend to Fig. 15). Besides the reference compound $1,25(OH)_2D_3$, both RCI_{VDR} values and calcemic indices (Fig. 15) are available only for **26** (HL), **27** (HF), **37** (HM), **39** (V), **40** (BT), **44** (IE), **45** (EU), **46** (EO), **48** (ZCS), and **50** (CT). The RCI_{VDR} values should be expected to correlate with the calcemic index, which among other factors reflects intestinal calcium absorption or bone calcium mobilization. The correlation is poor, however, as previously indicated by Norman *et al.* [161]. One aspect here that should be noted is the use of avian VDR to assess

26,27-dehydro-D_3; **41**, 1α,25-$(OH)_2$-3-epi-D_3; **42**, 1β,25-$(OH)_2$-3-epi-D_3; **43**, 9(11)-Dehydro-1α,25-$(OH)_2$-D_3; **44**, 20-Epi-1α,25-$(OH)_2$-D_3; **45**, 22-Oxa-1α,25-$(OH)_2$-D_3; **46**, 1α,25-$(OH)_2$-16-Ene-23-yne-26,27-F_6D_3; **47**, 24a,26a,27a-Trihomo-22,24-diene-1α,25-$(OH)_2$-D_3; **48**, 1α,25-$(OH)_2$-24,26,27-trihomo-D_3; **49**, 1α-(OH)-D_3; **50**, 1α,24R-$(OH)_2$-D_3.

RCI_{VDR}. Species differences between VDRs and their affinities for different vitamin D analogs are likely.

Likewise, besides $1,25(OH)_2D_3$, cell differentiation indices are given for **27** (HF), **37** (HM), **39** (V), **40** (BT), **44** (IE), **45** (EU), **46** (EO), **48** (ZCS), and **50** (CT). RCI values for human HL-60 cells (not shown) have certainly been reported (for review, see Bouillon *et al.* [1]), but the data base is far less extensive as compared to RCI_{VDR} values. As there is moderate correlation between RCI values for HL-60 cells and RCI_{VDR} values [161], it seems reasonable to attempt to correlate the cell differentiation indices and the RCI_{VDR} values of Fig. 15. Again, the correlation is poor.

Perhaps one might have expected the biological response to be directly proportional to the receptor affinity of the ligand. However, the mechanism of action of analogs of $1,25(OH)_2D_3$ at the transcriptional and overall biological response level, whether calcemic or cellular differention effects are being considered, is emerging as being rather complex. A case in point is the observation by the Norman and Peleg laboratories that 20-epi analogs of $1,25(OH)_2D_3$ bind to VDR with a similar or even lower affinity than the natural hormone, but are 100 to 10,000 times more potent than $1,25(OH)_2D_3$ in inducing cell differentiation [131,160]. On the basis of a variety of experiments, it was concluded that different conformations of vitamin D–receptor complexes, induced by $1,25(OH)_2D_3$ or 20-epi analogs, are responsible for the differences in transcriptional activation. Differences in contact sites of the 20-epi analogs and $1,25(OH)_2D_3$ with the VDR generates differences in the ligand-induced conformational change [160, 161]. These coupled equilibrium processes (wherein, after ligand binds to receptor, a series of additional activation processes are involved) can be hypothesized to account for the superactivity of 20-epi ligands. Such multiple events have also been discussed by the O'Malley laboratory [162] in discussing steroid hormone and antihormone activity. Regarding the latter, the possibility of multisite binding can further complicate the correlation [126]. Thus, the expectation, particularly *in vivo*, that biological response is proportional to the affinity of the ligand for VDR may be in error.

In summary, general pathways of calcemic and cell differentiation action of vitamin D should not be expected to correlate with ligand–host protein binding in a linear fashion. This review, however, places emphasis mainly on the structure of the guest–host complex. Nevertheless, biological assays are critical toward developing insight into whether an analog behaves as an agonist or antagonist or whether it is inactive. Hence, the determination of the biological activity (as reflected, e.g., by calcemic or cell differentiation indices) is a critical component toward understanding the mechanism of action of the steroid and interpreting protein binding affinity data. At present, we assume that binding of vitamin D to host protein (and possibly followed by a distinct conformational change in protein) is an initial step in all the biological activation processes associated with vitamin D action. A small difference in the shape of the ligand may lead to a large difference in biological activity, but in the context of the theme of this chapter, it is essential to first determine the ligand structure in the ligand–protein complex for drug design consideration.

IV. VITAMIN D DRUG DESIGN

The traditional medicinal chemistry approach for screening for active analogs has been to structurally modify lead compounds, wherein lead compounds are frequently identified by serendipity. For receptor or other host protein mediated action, a more modern, alternative approach is to gather information about the binding site and then to design analogs with this information. The advent of protein overproduction biotechnology has made this increasingly feasible. In the vitamin D field, the traditional approach has been the primary focus. The lead compounds include $1,25(OH)_2D_3$ and its metabolites, and enormous success has been achieved most notably via side chain modifications (Fig. 16), resulting in a variety of second generation lead compounds exhibiting powerful cell differentiation actions with minimal toxic effects in the form of hypercalcemia. To apply the more modern approach, a central theme of this chapter, the vitamin D field has been limited by the lack of structural information of vitamin D–protein complexes, particularly the shape of vitamin D in the protein active site.

As discussed earlier for Fig. 1, major international efforts have been directed toward the development of clinically useful vitamin D drugs for a diverse set of diseases such as cancers (leukemia as well as breast, colon, prostate, skin, and intestinal cancer), bone disorders (osteoporosis, osteopetrosis, renal osteodystrophy, and hypophosphatemic rickets), autoimmune impairments, and skin abnormalities (see Chapters 64 to 73). We briefly discuss here the design of vitamin D drugs for the selected diseases cancer, psoriasis, and calcemic disruption. Figure 16 lists some of the potential clinical applications of vitamin D analogs in conjunction with useful structural motifs based on the $1,25(OH)_2D_3$ skeleton which have been recognized to perhaps be beneficial in the design of better vitamin D drugs. As is now familiar in the vitamin D field, major attention has focused on dissociation of a desired biological activity (e.g., cell differentiation) from the major toxic side effect attributed to natural D vitamins, namely, hypercalcemia.

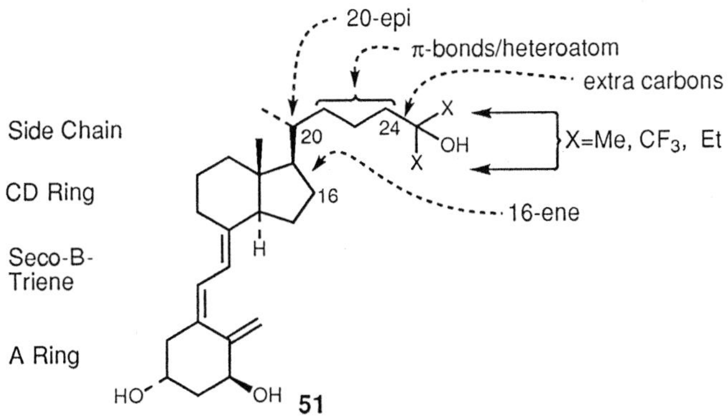

Disease		Code	Structural change relative to **1**	
Cancer	Preleukemia /Leukemia	**39** (V)	16-ene; 23-yne	[163]
	Breast Cancer	**40** (BT)	24-OH; 22-ene; 26,27-Dehydro	[164]
		39 (V)	16-ene; 23-yne	[1]
		45 (EU)	22-oxa	[165]
		46 (EO)	16-ene; 23-yne; 26,27-hexafluoro	[166]
		47 (IC)	24,26,27-trihomo; 22,24-diene	[164]
		48(ZCS)	24,26,27-trihomo	[1]
Bone Diseases	Osteoporosis	**1** (C)		[1]
		49 (BP)	De-25-OH	[1]
	Renal Osteodystrophy	**1** (C)		[1]
		49 (BP)	De-25-OH	[1]
Skin Diseases	Psoriasis	**40** (BT)	24-OH; 22-ene; 26,27-Dehydro	[1]
		50 (CT)	24R-OH	[1]

FIGURE 16 Useful structural motifs for vitamin D drug design. As discussed in the text, certain cancers and bone and skin diseases have been scrutinized in terms of the effectiveness of side chain vitamin D analogs as possible drugs for their treatment. See Fig. 15 and legend for structures and further details. These and other compounds (or lead structures) can be found in a review by Bouillon *et al.* [1], in other chapters of this treatise, and in other references [163–166].

The structure–function relationships for vitamin D have been discussed in detail on the basis of biological data for some 278 analogs which include 820 structural modifications [1]. A number of structural motifs were identified to be useful for modification of vitamin D for dissociation of the beneficial cell differentiation-inducing activity from the toxicity component, calcemic activity. They include homologation of the side chain at C-24, C-26, and/or C-27, $\Delta^{16,17}$-dehydrogenation with or without a 23-yne, 22-oxa modification (either alone or in combination with other side chain modifications), C-20 epimerization, and a combination of a 24*R*-OH, 22-ene, and 26,27-dehydrogenation (**40**, BT). Epimerization at C-20, however, sometimes retains or enhances calcemic potency.

Figure 16, a limited tabulation of drug leads for several disease states [1,163–166], is illustrative of the status of the drug design efforts in the vitamin D field. A more comprehensive description of analogs can be found in other chapters of this treatise as well as in the review article by Bouillon *et al.* [1]. Some key comments are as follows:

1. Synthetic expediency has been a major driving force in drug design efforts. All of the analogs shown in Fig. 16, which include many of the best second gen-

eration potential drug leads for the disease states indicated, are side chain analogs; analysis of some 278 analogs of $1,25(OH)_2D_3$ revealed that 82% are modified in the side chain [1]. Chemical modifications of the side chain of $1,25(OH)_2D_3$ has simply been easier for the synthetic organic chemist.

2. The emphasis of this chapter has been to relate the shape of the ligand or drug to selected disease states. The selection of lead drugs shown in Fig. 16 hardly addresses this point except that some of the analogs have conformationally locked side chains due to the presence of π bonds as in **39** (V), **40** (BT), **46** (EO), and **47** (IC), but this is too limited a list for useful structure–function analysis. More significant is that Fig. 16 represents data for a traditional "functional group" modification approach as depicted in structure **51** (20-epi, π bonds, heteroatoms, etc.). It goes without saying that although this chapter has emphasized the shape of the molecule resulting from conformational changes, the presence of new functional groups has not been neglected. Undoubtedly, however, the conformation of vitamin D in the active site of a protein together with the unnatural functional groups present in the analog (which might interact with amino acid moieties at the active site of the protein host) will both be significant.

3. The discovery of lead structural motifs with potential use in various disease states certainly must not be ignored in future drug design efforts: **39** (V) for leukemia [163]; **39** (V), **40** (BT) [164], **45** (EU) [165], **46** (EO) [166], **47** (IC) [164], and **48** (ZCS) for breast cancer; **1** (C) and **49** (BP) for osteoporosis; **1** (C) and **49** (BP) for renal osteodystrophy; and **40** (BT) and **50** (CT) for psoriasis. This is certainly not meant to be a comprehensive list, and the reader is directed to other chapters in this treatise. It might be wise to incorporate the latter structural motifs together with conformational locking to optimize the effectiveness of biological activity.

4. As the structures in Fig. 16 hardly provide useful information about a unique ligand topology in relation to the activity of a drug, it is interesting to note that Bhatia *et al.* [167] described the ability of the 6-s-cis locked analog **27** (HF) to prime monocytic differentiation of the human cancer cell line NB4. This result is one of the clearest examples of a ligand conformational effect related to cell differentiation. As discussed earlier in Section I,B,3, 6-s-cis topologically biased analogs have thus far been associated mainly with the activation of m-VDR and possibly other membrane protein systems.

V. CONCLUSION: THE THREE-DIMENSIONAL STRUCTURE OF VITAMIN D DETERMINES ITS COMPLEX BIOLOGICAL FUNCTION

Vitamin D drug discovery has concentrated on the traditional medicinal chemistry approach to identifying potential drug candidates by structurally modifying the lead $1,25(OH)_2D_3$ and its metabolites followed by biological evaluation of resulting second generation candidates. This iterative approach has certainly been successful, as evidenced by the properties of some of the drug leads shown in Fig. 16 and as reviewed in other chapters in this treatise and elsewhere [1]. It is suggested here that structural information regarding the ligand–protein complex (particularly the various VDRs, but also DBP and the other host systems shown in Table I) should lead to an even more sophisticated approach to drug design. The structure of the ligand in the active site of various hosts would be of particular importance. This chapter provides a progress report on the structural features of the proteins that bind to vitamin D and the structure and thermodynamic characteristics of the vitamin D ligand. There has obviously not been much progress in this field; indeed, in the steroid hormone field, the first two structures of an occupied ligand binding domain of this superfamily of receptors were published only in December, 1995 [130,136]. There is evidence that the shape of the ligand can affect the overall conformation of the ligand–protein complex, leading to significant amplification of the biological effect (e.g., see Peleg *et al.* [131] and Liu *et al.* [160]). There is evidence that spectroscopically invisible, high energy conformers of vitamin D are involved in vitamin D action (e.g., see Nemere *et al.* [13], Norman *et al.* [158,159], and Bhatia *et al.* [167]). There are also a number of lead vitamin D compounds with potential clinical applications in selected disease states (Fig. 16).

Structure determines biological function, and it is not unreasonable to suggest that different conformations of $1,25(OH)_2D_3$ and metabolites (Figs. 5, 8, 9, and 10) or different analogs are capable of inducing different shapes of protein on binding. Although the binding affinity of vitamin D to receptor does not correlate particularly well with biological activity parameters, and since ligand–protein binding is almost certainly the initial step in inducing a biological response, a combination of structural information concerning the ligand–protein complex and the resulting biological response is essential in understanding the mechanism of vitamin D action. Only in this way can we couple the structural information to drug design. We would certainly like to emphasize the need for increasing efforts in the "modern approach" in drug development, namely, to access information about the binding site and then design analogs around this information.

Acknowledgments

We are most grateful to Robert Moriarty (University of Illinois, Chicago), Richard Gilardi (Naval Research Laboratory), Andrzej Kutner (Pharmaceutical Research Institute, Warsaw, Poland), and Gary Posner (Johns Hopkins University) for providing unpublished X-ray crystallographic information utilized in the preparation of Table III. The National Institutes of Health (DK-16595 and CA-43277) provided generous grant support. Our colleagues at Riverside are also acknowledged for providing input in the preparation of this review.

References

1. Bouillon R, Okamura WH, Norman AW 1995 Structure–function relationships in the vitamin D endocrine system. Endocr Rev **16**:200–257.
2. Zhu G-D, Okamura WH 1995 Synthesis of Vitamin D (calciferol). Chem Rev **95**:1877–1952.
3. Holick MF 1994 McCollum Award Lecture, 1994: Vitamin D—New horizons for the 21st century. Am J Clin Nutr **60**:619–630.
4. Holick MF, Tian XQ, Allen M 1995 Evolutionary importance for the membrane enhancement of the production of vitamin D_3 in the skin of poikilothermic animals. Proc Natl Acad Sci USA **92**:3124–3126.
5. Tian XQ, Chen TC, Matsuoka LY, Wortsman J, Holick MF 1993 Kinetic and thermodynamic studies of the conversion of previtamin D_3 to vitamin D_3 in human skin. J Biol Chem **268**:14888–14892.
6. Tian XQ, Chen TC, Lu Z, Shao Q, Holick MF 1994 Characterization of the translocation process of vitamin D_3 from the skin into the circulation. Endocrinology **135**:655–661.
7. Yates PJ, Hobbs RN, Pennock JF 1988 Precholecalciferol formation by an invertebrate, *Psammechinus miliaris*. In: Norman AW, Schaefer K, Grigoleit H-G, Herrath Dv (eds) Vitamin D: Molecular, Cellular and Clinical Endocrinology. de Gruyter, Berlin, New York, pp. 83–92.
8. Hobbs RN, Hazel CM, Smith SC, Carney DA, Howells AC, Littlewood AJ, Pennock JF 1987 Metabolism of vitamin D_3 by a marine echinoderm, *Psammechinus miliaris*. Chem Scr **27**:199–205.
9. Yamamoto JK, Borch RF 1985 Photoconversion of 7-dehydrocholesterol to vitamin D_3 in synthetic phospholipid bilayers. Biochemistry **24**:3338–3344.
10. Cassis EG, Jr., Weiss RG 1982 Liquid–crystalline solvents as mechanistic probes—V. An investigation of the effect of cholesteric order on the formation rates of vitamin D_3 from the previtamin D_3 and of pre-vitamin D_3 from vitamin D_3. Photochem Photobiol **35**:439–444.
11. Tian XQ, Holick MF 1995 Catalyzed thermal isomerization between previtamin D_3 and vitamin D_3 via β-cyclodextrin complexation. J Biol Chem **270**:8706–8711.
12. Bogoslovsky NA, Kurganov BI, Samochvalova NG, Isaeva TA, Sugrobova NP, Gurevich VM, Valashek IE, Samochvalov GI 1988 Synthesis of 24*R*,25-dihydroxyvitamin D_3 from vitamin D_2 and study on inclusion complexes of vitamin D derivatives with β-cyclodextrin. In: Norman AW, Schaefer K, Grigoleit HG, Herrath Dv (eds) Vitamin D: Molecular, Cellular and Clinical Endocrinology. de Gruyter, Berlin and New York, pp. 1021–1023.
13. Nemere I, Dormanen MC, Hammond MW, Okamura WH, Norman AW 1994 Identification of a specific binding protein for 1α,25-dihydroxyvitamin D_3 in basal–lateral membranes of chick intestinal epithelium and relationship to transcaltachia. J Biol Chem **269**:23750–23756.
14. Axén E, Bergman T, Wikvall K 1992 Purification and characterization of a vitamin D_3 25-hydroxylase from pig liver microsomes. Biochem J **287**:725–731.
15. Mandel ML, Swartz SJ, Ghazarian JG 1990. Avian kidney mitochondrial hemeprotein P-$450_{1\alpha}$: Isolation, characterization and NADPH-ferredoxin reductase-dependent activity. Biochim Biophys Acta **1034**:239–246.
16. Burgos-Trinidad M, Ismail R, Ettinger EA, Prahl JM, DeLuca HF 1992. Immunopurified 25-hydroxyvitamin D 1α-hydroxylase and 1,25-dihydroxyvitamin D 24-hydroxylase are closely related but distinct enzymes. J Biol Chem **267**:3498–3505.
17. Cooke NE, Haddad JG 1989 Vitamin D binding protein (Gc-globulin). Endocr Rev **10**:294–307.
18. Haddad JG 1995 Plasma vitamin D-binding protein (Gc-globulin): Multiple tasks. J Steroid Biochem Mol Biol **53**:579–582.
19. Slater SJ, Kelly MB, Taddeo FJ, Larkin JD, Yeager MD, McLane JA, Ho C, Stubbs CD 1995 Direct activation of protein kinase C by 1α,25-dihydroxyvitamin D_3. J Biol Chem **270**:6639–6643.
20. Bissonnette M, Wali RK, Hartmann SC, Niedziela SM, Roy HK, Tien X-Y, Sitrin MD, Brasitus TA 1995 1,25-dihydroxyvitamin D_3 and 12-*O*-tetradecanoyl phorbol 13-acetate cause differential activation of Ca^{2+}-dependent and Ca^{2+}-independent isoforms of protein kinase C in rat colonocytes. J Clin Invest **95**:2215–2221.
21. Beno DWA, Brady LM, Bissonnette M, Davis BH 1995 Protein kinase C and mitogen-activated protein kinase are required for 1,25-dihydroxyvitamin D_3-stimulated Egr induction. J Biol Chem **270**:3642–3647.
22. Bissonnette M, Tien X-Y, Niedziela SM, Hartmann SC, Frawley BP, Jr., Roy HK, Sitrin MD, Perlman RL, Brasitus TA 1994 $1,25(OH)_2$ vitamin D_3 activates PKC-α in caco-2 cells: A mechanism to limit secosteroid-induced rise in $[Ca^{2+}]$. Am J Physiol **267**:G465–G475.
23. Farzaneh NK, Walden TL, Hearing VJ, Gersten DM 1991 B700, an albumin-like melanoma-specific antigen, is a vitamin D binding protein. Eur J Cancer **27**:1158–1162.
24. He XM, Carter DC 1992 Atomic structure and chemistry of human serum albumin. Nature **358**:209–215.
25. Kragh-Hansen U 1981 Molecular aspects of ligand binding to serum albumin. Pharm Rev **33**:17–53.
26. Peters T, Jr. 1985 Serum albumin. Adv Protein Chem **37**:161–245.
27. Cooke NE, David EV 1985 Serum vitamin D-binding protein is a third member of the albumin and alpha fetoprotein gene family. J Clin Invest **76**:2420–2424.
28. Schoentgen F, Metz-Boutigue M-H, Jollés J, Constans J, Jollés P 1986 Complete amino acid sequence of human vitamin D-binding protein (group-specific component): Evidence of a threefold internal homology as in serum albumin and α-fetoprotein. Biochim Biophys Acta **871**:189–198.
29. Braun A, Kofler A, Morawietz S, Cleve H 1993 Sequence and organization of the human vitamin D-binding protein gene. Biochim Biophys Acta **1216**:385–394.
30. Verboven CC, De Bondt HL, De Ranter C, Bouillon R, Van Baelen H 1995 Crystallization and X-ray investigation of vitamin D-binding protein from human serum. Identification of the crystal content. J Steroid Biochem Mol Biol **54**:11–14.
31. Ray R, Bouillon R, Van Baelen H, Holick MF 1991 Photoaffinity labeling of human serum vitamin D binding protein and chemical

cleavages of the labeled protein: Identification of an 11.5-kDa peptide containing the putative 25-hydroxyvitamin D_3 binding site. Biochemistry **30**:7638–7642.
32. Swamy N, Brisson M, Ray R 1995 Trp-145 is essential for the binding of 25-hydroxyvitamin D_3 to human serum vitamin D-binding protein. J Biol Chem **270**:2636–2639.
33. Haddad JG, Hu YZ, Kowalski MA, Laramore C, Ray K, Robzyk P, Cooke NE 1992 Identification of the sterol- and actin-binding domains of plasma vitamin D binding protein (Gc-globulin). Biochemistry **31**:7174–7181.
34. Cooke NE 1986 Rat vitamin D binding protein. Determination of the full-length primary structure from cloned cDNA. J Biol Chem **261**:3441–3450.
35. McLeod JF, Cooke NE 1989 The vitamin D-binding protein, α-fetoprotein, albumin multigene family: Detection of transcripts in multiple tissues. J Biol Chem **264**:21760–21769.
36. Bikle DD, Gee E, Halloran B, Kowalski MA, Ryzen E, Haddad JG 1986 Assessment of the free fraction of 25-hydroxyvitamin D in serum and its regulation by albumin and the vitamin D-binding protein. J Clin Endocrinol Metab **63**:954–959.
37. Woloszczuk W 1985 Determination of vitamin D binding protein by Scatchard analysis and estimation of a free 25-hydroxy-vitamin D index. Clin Chim Acta **145**:27–35.
38. Okuda K-I, Usui E, Ohyama Y 1995 Recent progress in enzymology and molecular biology of enzymes involved in vitamin D metabolism. J Lipid Res **36**:1641–1652.
39. Okuda K-I 1994 Liver mitochondrial P450 involved in cholesterol catabolism and vitamin D activation. J Lipid Res **35**:361–372.
40. Nelson DR, Kamataki T, Waxman DJ, Guengerich FP, Estabrook RW, Feyereisen R, Gonzalez FJ, Coon MJ, Gunsalus IC, Gotoh O, Okuda K, Nebert DW 1993 The P450 superfamily: update on new sequences, gene mapping, accession numbers, early trivial names of enzymes, and nomenclature. DNA Cell Biol **12**:1–51.
41. Henry HL 1992 Vitamin D hydroxylases. J Cell Biochem **49**:4–9.
42. Armbrecht HJ, Okuda K, Wongsurawat N, Nemani RK, Chen ML, Boltz MA 1992 Characterization and regulation of the vitamin D hydroxylases. J Steroid Biochem Mol Biol **43**:1073–1081.
43. Kawauchi H, Sasaki J, Adachi T, Hanada K, Beppu T, Horinouchi S 1994 Cloning and nucleotide sequence of a bacterial cytochrome $P\text{-}450_{VD25}$ gene encoding vitamin D-3 25-hydroxylase. Biochim Biophys Acta **1219**:179–183.
44. Chen K-S, DeLuca HF 1995 Cloning of the human 1α,25-dihydroxyvitamin D-3 24-hydroxylase gene promoter and identification of two vitamin D-responsive elements. Biochim Biophys Acta **1263**:1–9.
45. Itoh S, Yoshimura T, Iemura O, Yamada E, Tsujikawa K, Kohama Y, Mimura T 1995 Molecular cloning of 25-hydroxy-vitamin D-3 24-hydroxylase (Cyp-24) from mouse kidney: Its inducibility by vitamin D-3. Biochim Biophys Acta **1264**:26–28.
46. Ettinger RA, Ismail R, DeLuca HF 1994 cDNA cloning and characterization of a vitamin D_3 hydroxylase-associated protein. J Biol Chem **269**:176–182.
47. Daniel D, Okamura WH 1994 Syntheses of potential inhibitors of 25-hydroxyvitamin D_3-1α-hydroxylase: A-ring analogs. In: Norman AW, Bouillon R, Thomasset M (eds) Vitamin D, A Pluripotent Steroid Hormone: Structural Studies, Molecular Endocrinology and Clinical Applications. de Gruyter, Berlin and New York, pp. 35–36.
48. Henry HL, Fried S, Shen G-Y, Barrack SA, Okamura WH 1991 Effect of three A-ring analogs of 1α,25-dihydroxyvitamin D_3 on 25-OH-D_3-1α-hydroxylase in isolated mitochondria and on 25-hydroxyvitamin D_3 metabolism in cultured kidney cells. J Steroid Biochem Mol Biol **38**:775–779.
49. Barrack SA, Gibbs RA, Okamura WH 1988 Potential inhibitors of vitamin D metabolism: An oxa analogue of vitamin D. J Org Chem **53**:1790–1796.
50. Daniel D, Middleton R, Henry HL, Okamura WH 1996 Inhibitors of 25-hydroxyvitamin D_3-1α-hydroxylase: A-ring oxa analogs of 25-hydroxyvitamin D_3. J Org Chem **61**:5617–5625.
51. Johnson RL, Okamura WH, Norman AW 1975 Studies on the mode of action of calciferol. X. 24-Nor-25-hydroxyvitamin D_3, an analog of 25-hydroxyvitamin D_3 having "anti-vitamin" activity. Biochem Biophys Res Commun **67**:797–802.
52. Norman AW, Johnson RL, Okamura WH 1979 24-Nor-25-hydroxyvitamin D_3. A specific antagonist of vitamin D_3 action in the chick. J Biol Chem **254**:11450–11456.
53. Norman AW, Hammond ML, Okamura WH 1977 19-Hydroxy-10*S*(19)-dihydrovitamin D_3-II: An analog of vitamin D_3 which possesses "anti-vitamin" activity. Fed Proc, Fed Am Soc Exp Biol **36**:914.
54. Paaren HE, Moriarty RM, Schnoes HK, DeLuca HF 1980 In vivo and in vitro inhibition of rat liver vitamin D_3-25-hydroxylase activity by 19-hydroxy-10(*S*),19-dihydrovitamin D_3. Biochemistry **19**:5335–5339.
55. Onisko BL, Schnoes HK, DeLuca HF 1977. Synthesis of potential vitamin D antagonists. Tetrahedron Lett **13**:1107–1108.
56. Onisko BL, Schnoes HK, DeLuca HF 1979 Inhibitors of vitamin D metabolism and action. In: Norman AW, Schaefer K, Herrath Dv, Grigoleit H-G, Coburn JW, DeLuca HF, Mawer EB, Suda T (eds) Vitamin D, Basic Research and Its Clinical Application. de Gruyter, Berlin, New York, pp. 77–79.
57. Onisko BL, Schnoes HK, DeLuca HF 1979 25-Azavitamin D_3, an inhibitor of vitamin D metabolism and action. J Biol Chem **254**:3493–3496.
58. Onisko BL, Schnoes HK, DeLuca HF 1980 Inhibitors of the 25-hydroxylation of vitamin D_3 in the rat. Bioorg Chem **9**:187–198.
59. Bolt MJG, Holick SA, Holick MF, MacLaughlin J, Rosenberg IH 1988 24-dehydrovitamin D is a potent inhibitor of rat liver microsomal vitamin D-25-hydroxylase. Arch Biochem Biophys **266**:532–538.
60. Kabakoff B, Schnoes HK, DeLuca HF 1983 *In Vitro* inhibitor studies of vitamin D 25-hydroxylase in rat liver microsomes. Arch Biochem Biophys **221**:38–45.
61. Castedo L, Sarandeses L, Granja J, Mascareñas JL, Maestro MA, Mouriño A 1991 Synthesis of potential inhibitors of vitamin D hydroxylases. In: Pandit UK, Alderweireldt FC (eds) Bioorganic Chemistry in Healthcare and Technology. Plenum Press, New York, pp. 251–254.
62. Ross TK, Darwish HM, DeLuca HF 1994 Molecular biology of vitamin D action. Vitam Horm **49**:281–326.
63. Carlberg C 1996 The vitamin D_3 receptor in the context of the nuclear receptor superfamily: The central role of the retinoid X receptor. Endocrine **4**:91–105.
64. Lowe KE, Maiyar AC, Norman AW 1992 Vitamin D-mediated gene expression. Crit Rev Euk Gene Expression **2**:65–109.
65. Nakada M, Simpson RU, DeLuca HF 1984 Subcellular distribution of DNA-binding and non-DNA-binding 1,25-dihydroxy-vitamin D receptors in chicken intestine. Proc Natl Acad Sci USA **81**:6711–6713.
66. Barsony J, Pike JW, DeLuca HF, Marx SJ 1990 Immunocytology with microwave-fixed fibroblasts shows 1α,25-dihydroxyvitamin D_3-dependent rapid and estrogen-dependent slow reorganization of vitamin D receptors. J Cell Biol **111**:2385–2395.
67. Burmester JK, Wiese RJ, Maeda N, DeLuca HF 1988 Structure and regulation of the rat 1,25-dihydroxyvitamin D_3 receptor. Proc Natl Acad Sci USA **85**:9499–9502.
68. Baker AR, McDonnell DP, Hughes M, Crisp TM, Mangelsdorf

DJ, Haussler MR, Pike JW, Shine J, O'Malley BW 1988 Cloning and expression of full-length cDNA encoding human vitamin D receptor. Proc Natl Acad Sci USA **85**:3294–3298.

69. Sone T, McDonnell DP, O'Malley BW, Pike JW 1990 Expression of human vitamin D receptor in *Saccharomyces cerevisiae*. J Biol Chem **265**:21997–22003.
70. Ross TK, Prahl JM, DeLuca HF 1991 Overproduction of rat 1,25-dihydroxyvitamin D_3 receptor in insect cells using the baculovirus expression system. Proc Natl Acad Sci USA **88**:6555–6559.
71. MacDonald PN, Haussler CA, Terpening CM, Galligan MA, Reeder MC, Whitfield GK, Haussler MR 1991 Baculovirus-mediated expression of the human vitamin D receptor. J Biol Chem **266**:18808–18813.
72. Smith CL, Hager GL, Pike JW, Marx SJ 1991 Overexpression of the human vitamin D_3 receptor in mammalian cells using recombinant adenovirus vectors. Mol Endocrinol **5**:867–878.
73. Kumar R, Schaefer J, Wieben E 1992 The expression of milligram amounts of functional human 1,25-dihydroxyvitamin D_3 receptor in a bacterial expression system. Biochem Biophys Res Commun **189**:1417–1423.
74. Craig TA, Kumar R 1996 Synthesis and purification of soluble ligand binding domain of the human vitamin D_3 receptor. Biochem Biophys Res Commun **218**:902–907.
75. O'Malley BW, Tsai M-J 1993 Overview of the steroid receptor superfamily of gene regulatory proteins. In: Parker MG (ed) Steroid Hormone Action. Oxford Univ. Press, Oxford, New York, pp. 45–63.
76. Baniahmad A, Tsai M-J, Burris TP 1994 The nuclear hormone receptor superfamily. In: Tsai M-J, O'Malley BW (eds) Mechanism of Steroid Hormone Regulation of Gene Transcription. CRC Press, Boca Raton, Florida, pp. 1–24.
77. Luisi BF, Schwabe JWR, Freedman LP 1994 The steroid/nuclear receptors: From three-dimensional structure to complex function. Vitam Horm **49**:1–45.
78. O'Malley B 1990 The steroid receptor superfamily: More excitement predicted for the future. Mol Endocrinol **4**:363–369.
79. Evans RM 1988 The steroid and thyroid hormone receptor superfamily. Science **240**:889–895.
80. Parker MG 1995 Structure and function of estrogen receptors. Vitam Horm **51**:267–287.
81. Lindzey J, Kumar MV, Grossman M, Young C, Tindall DJ 1994 Molecular mechanisms of androgen action. Vitam Horm **49**: 383–432.
82. Perlman KL, Darwish HM, DeLuca HF 1994 20-Oxopregnacalciferols: Vitamin D compounds that bind the progesterone receptor. Tetrahedron Lett **35**:2295–2298.
83. Jorgensen WL 1991 Rusting of the lock and key model for protein–ligand binding. Science **254**:954–955.
84. Okamura WH, Wing RM 1980 Conformational analysis of vitamin D and related compounds. In: Norman AW (ed) Vitamin D: Molecular Biology and Clinical Nutrition. Dekker, New York, pp. 59–92 and 685–691.
85. Trinh-Toan, DeLuca HF, Dahl LF 1976 Solid-state conformations of vitamin D_3. J Org Chem **41**:3476–3478.
86. Trinh-Toan, Ryan RC, Simon GL, Calabrese JC, Dahl LF 1977 Crystal structure of 25-hydroxy-vitamin D_3 monohydrate: A stereochemical analysis of vitamin D molecules. J Chem Soc Perkin II 393–401.
87. Suwinska K, Kutner A 1996 Crystal and molecular structure of 1,25-dihydroxycholecalciferol. Acta Cryst **B52**:550–554.
88. Hull SE, Leban I, Main P, White PS, Woolfson MM 1976 The crystal and molecular structure of ergocalciferol (vitamin D_2). Acta Cryst **B32**:2374–2381.
89. Knobler C, Romers C, Braun PB, Hornstra J 1972 The conformation of non-aromatic ring compounds. LXXV. The crystal and molecular structure of the 3,20-bis(ethylenedioxy) analogue of vitamin D. Acta Cryst **B28**:2097–2103.
90. Hodgkin DC, Rimmer BM, Dunitz JD, Trueblood KN 1963 The crystal structure of a calciferol derivative. J Chem Soc 4945–4955.
91. Posner GH, Dai H, Afarinkia K, Murthy NN, Guyton KZ, Kensler TW 1993 Asymmetric total synthesis, X-ray crystallography, and preliminary biological evaluation of 1-(1′-hydroxyethyl)-25-hydroxyvitamin D_3 analogs of natural calcitriol. J Org Chem **58**:7209–7215.
92. Larsen S, Hansen ET, Hoffmeyer L, Rastrup-Andersen N 1993 Structure and absolute configuration of a monohydrate of calcipotriol, ($1\alpha,3\beta,5Z,7E,22E,24S$)-24-cyclopropyl-9,10-secochola-5,7,10(19),22-tetraene-1,3,24-triol. Acta Cryst **C49**:618–621.
93. Moriarty RM, Penmasta R, Awashti A, Kim J, Gilardi R, Bishop CW, Knutson JC 1991 Synthesis and stereochemical studies on 1α and 1β-hydroxy vitamin D_2. Abstracts: Eighth Workshop on Vitamin D, Paris, p. 95.
94. Geise HJ, Buys HR, Mijlhoff FC 1971 Conformation of non-aromatic ring compounds. Part 72. An electron diffraction study of gaseous cyclohexane and methylcyclohexane. J Mol Struct **9**: 447–454.
95. Cremer D, Pople JA 1975 A general definition of ring puckering coordinates. J Am Chem Soc **97**:1354–1358.
96. Eliel EL, Wilen SH 1994 Stereochemistry of Organic Compounds. Wiley, New York.
97. Wing RM, Okamura WH, Pirio MR, Sine SM, Norman AW 1974 Vitamin D in solution: Conformations of vitamin D_3, 1α,25-dihydroxyvitamin D_3, and dihydrotachysterol$_3$. Science **186**: 939–941.
98. Wing RM, Okamura WH, Rego A, Pirio MR, Norman AW 1975 Studies on vitamin D and its analogs. VII. Solution conformations of vitamin D_3 and 1α,25-dihydroxyvitamin D_3 by high-resolution proton magnetic resonance spectroscopy. J Am Chem Soc **97**: 4980–4985.
99. La Mar GN, Budd DL 1974 Elucidation of the solution conformation of the A ring in vitamin D using proton coupling constants and a shift reagent. J Am Chem Soc **96**:7317–7324.
100. Eguchi T, Kakinuma K, Ikekawa N 1991 Synthesis of 1α-[19-^{13}C]hydroxyvitamin D_3 and ^{13}C NMR analysis of the conformational equilibrium of the A-ring. Bioorg Chem **19**:327–332.
101. Helmer B, Schnoes HK, DeLuca HF 1985 ^{1}H nuclear magnetic resonance studies of the conformations of vitamin D compounds in various solvents. Arch Biochem Biophys **241**:608–615.
102. Eguchi T, Ikekawa N 1990 Conformational analysis of 1α,25-dihydroxyvitamin D_3 by nuclear magnetic resonance. Bioorg Chem **18**:19–29.
103. Zhu G-D, Van Haver D, Jurriaans H, De Clercq PJ 1994 11-Fluoro-1α-hydroxyvitamin D_3: The quest for experimental evidence of the folded vitamin D conformation. Tetrahedron **50**: 7049–7060.
104. Berman E, Luz Z, Mazur Y, Sheves M 1977 Conformational analysis of vitamin D and analogues. ^{13}C and ^{1}H nuclear magnetic resonance study. J Org Chem **42**:3325–3330.
105. Berman E, Friedman N, Mazur Y, Sheves M 1978 Conformational equilibria in vitamin D. Synthesis and ^{1}H and ^{13}C dynamic nuclear magnetic resonance study of 4,4-dimethylvitamin D_3, 4,4-dimethyl-1α-hydroxyvitamin D_3, and 4,4-dimethyl-1α-hydroxyepivitamin D_3. J Am Chem Soc **100**:5626–5634.
106. Delaroff V, Rathle P, Legrand M 1963 Étude de La résonance magnétique nucléaire du précalciférol, du tachystérol et du calciférol (NMR study of precalciferol, tachysterol, and calciferol). Bull Soc Chim Fr 1739–1741.
107. Tsukida K, Akutsu K, Saiki K 1975 Carbon-13 nuclear magnetic

resonance spectra of vitamins D and related compounds. J Nutr Sci Vitaminol (*Tokyo*) **21**:411–420.

108. Kotowycz G, Nakashima TT, Green MK, Aarts GHM 1980 A proton magnetic resonance relaxation time study of several vitamin D derivatives. Can J Chem **58**:45–50.
109. Kotovych G, Aarts GHM, Bock K 1980 A proton magnetic resonance nuclear Overhauser enhancement study. Application to vitamin D derivatives D_2 and D_3. Can J Chem **58**:1206–1210.
110. Mizhiritskii MD, Konstantinovskii LE, Vishkautsan R 1996 2D NMR study of solution conformations and complete ^1H and ^{13}C chemical shifts assignments of vitamin D metabolites and analogs. Tetrahedron **52**:1239–1252.
111. Anet FAL 1962 The use of remote deuteration for the determination of coupling constants and conformational equilibria in cyclohexane derivatives. J Am Chem Soc **84**:1053–1054.
112. Haasnoot CAG, de Leeuw FAAM, Altona C 1980 The relationship between proton-proton NMR coupling constants and substituent electronegativities—I. Tetrahedron **36**:2783–2792.
113. Hofer O, Kählig H, Reischl W 1993 On the conformational flexibility of vitamin D. Monatshefte Chem **124**:185–198.
114. Mosquera RA, Rios MA, Tovar CA 1988 Conformational analysis of 5*Z*- and 5*E*-vitamin D_3 dihydroderivatives by molecular mechanics. J Mol Struct (Theochem) **168**:125–133.
115. Mosquera RA, Rios MA, Tovar CA, Maestro M 1989 Conformational analysis of vitamin D_3 derivatives by molecular mechanics. Part II. 1α,25-Dihydroxyvitamin D_3 and analogues. J Mol Struct **213**:297–307.
116. Duraisamy M, Walborsky HM 1983 Circular dichroism of isomeric 10,19-dihydrovitamin D. J Am Chem Soc **105**:3270–3273.
117. Sheves M, Friedman N, Levendis D, Margulies L, Mazur Y 1979 Conformational analysis of flexible *trans*-dienes by polarization spectroscopy. Isr J Chem **18**:359–363.
118. Wilson SR, Unwalla R, Cui W 1988 Computer calculations of the active conformation of 1,25-dihydroxy vitamin D_3. In: Norman AW, Schaefer K, Grigoleit H-G, Herrath Dv (eds) Vitamin D: Molecular, Cellular and Clinical Endocrinology. de Gruyter, Berlin and New York, pp. 78–79.
119. Curtin ML, Okamura WH 1991 1α,25-Dihydroxyprevitamin D_3: Synthesis of the 9,14,19,19,19-pentadeuterio derivative and a kinetic study of its [1,7]-sigmatropic shift to 1α,25-dihydroxyvitamin D_3. J Am Chem Soc **113**:6958–6966.
120. Okamura WH, Midland MM, Hammond MW, Abd.Rahman N, Dormanen MC, Nemere I, Norman AW 1995 Chemistry and conformation of vitamin D molecules. J Steroid Biochem Mol Biol **53**:603–613.
121. Okamura WH, Palenzuela JA, Plumet J, Midland MM 1992 Vitamin D: Structure–function analyses and the design of analogs. J Cell Biochem **49**:10–18.
122. Midland MM, Plumet J, Okamura WH 1993 Effect of C20 stereochemistry on the conformational profile of the side chains of vitamin D analogs. Bioorg Med Chem Lett **3**:1799–1804.
123. Wilson SR, Guarnieri F 1991 Calculation of the side chain flexibility for vitamin D analogs. In: Norman AW, Bouillon R, Thomasset M (eds) Vitamin D: Gene Regulation, Structure–Function Analysis and Clinical Application. de Gruyter, Berlin and New York, pp. 186–187.
124. Yamamoto K, Ohta M, DeLuca HF, Yamada S 1995 On the side chain conformation of 1α,25-dihydroxyvitamin D_3 responsible for binding to the receptor. Bioorg Med Chem Lett **5**: 979–984.
125. Yamamoto K, Sun WY, Ohta M, Hamada K, DeLuca HF, Yamada S 1996 Conformationally restricted analogs of 1,25-dihydroxyvitamin D_3 and its 20-epimer: Compounds for study of the three-dimensional structure of vitamin D responsible for binding to the receptor. J Med Chem **39**:2727–2737.
126. Teutsch G, Nique F, Lemoine G, Bouchoux F, Cérède E, Gofflo D, Philibert D 1995 General structure–activity correlations of antihormones. Ann NY Acad Sci **761**:5–28.
127. Lee MS, Kliewer SA, Provencal J, Wright PE, Evans RM 1993 Structure of the retinoid X receptor α DNA binding domain: A helix required for homodimeric DNA binding. Science **260**:1117–1121.
128. (a) Luisi B, Freedman L 1995 Dymer, dymer binding tight. Nature **375**:359–360. (b) Bourguet W, Ruff M, Chambon P, Gronemeyer H, Moras D 1995 Crystal structure of the ligand-binding domain of the human nuclear receptor RXR-α. Nature **375**:377–382.
129. Lefebvre B, Rachez C, Formstecher P, Lefebvre P 1995 Structural determinants of the ligand-binding site of the human retinoic acid receptor α. Biochemistry **34**:5477–5485.
130. Renaud J-P, Rochel N, Ruff M, Vivat V, Chambon P, Gronemeyer H, Moras D 1995 Crystal structure of the RAR-γ ligand-binding domain bound to all-*trans* retinoic acid. Nature **378**:681–689.
131. Peleg S, Sastry M, Collins ED, Bishop JE, Norman AW 1995 Distinct conformational changes induced by 20-epi analogues of 1α,25-Dihydroxyvitamin D_3 are associated with enhanced activation of the vitamin D receptor. J Biol Chem **270**:10551–10558.
132. Kendrew JC, Dickerson RE, Strandberg BE, Hart RG, Davies DR, Phillips DC, Shore VC 1960 Structure of myoglobin: A three-dimensional Fourier synthesis at 2 Å. resolution. Nature **185**:422–427.
133. Waller DA, Dodson GG 1993 Biological structures obtained by X-ray diffraction methods. In: Diamond R, Koetzle TF, Prout K, Richardson JS (eds) Molecular Structures in Biology. Oxford Univ. Press, Oxford, New York, Tokyo, pp. 1–19.
134. McRee DE 1993 Practical Protein Crystallography. Academic Press, San Diego and London, pp. 1–386.
135. Lewis M, Chang G, Horton NC, Kercher MA, Pace HC, Schumacher MA, Brennan RG, Lu P 1996 Crystal structure of the lactose operon repressor and its complexes with DNA and inducer. Science **271**:1247–1254.
136. Wagner RL, Apriletti JW, McGrath ME, West BL, Baxter JD, Fletterick RJ 1995 A structural role for hormone in the thyroid hormone receptor. Nature **378**:690–697.
137. Wüthrich K 1993 Biopolymers: An NMR survey. In: Diamond R, Koetzle TF, Prout K, Richardson JS (eds) Molecular Structures in Biology. Oxford Univ. Press, Oxford, New York, Tokyo, pp. 20–26.
138. Clore GM, Gronenborn AM 1991 Structures of larger proteins in solution: Three- and four-dimensional heteronuclear NMR spectroscopy. Science **252**:1390–1399.
139. Clore GM, Gronenborn AM 1994 Structures of larger proteins and protein–ligand and protein–DNA complexes by heteronuclear multidimensional NMR. J Protein Chem **13**:441–442.
140. Bax A, Grzesiek S 1993 Methodological advances in protein NMR. Acc Chem Res **26**:131–138.
141. Otting G, Wüthrich K 1990 Heteronuclear filters in two-dimensional [^1H,^1H]-NMR spectroscopy: Combined use with isotope labelling for studies of macromolecular conformation and intermolecular interactions. Q Rev Biophys **23**:39–96.
142. LeMaster DM 1994 Isotope labeling in solution protein assignment and structural analysis. Prog NMR Spectroscopy **26**: 371–419.
143. Clore GM, Gronenborn AM 1993 NMR of Proteins. CRC Press, Boca Raton, Florida and Tokyo, pp. 1–304.
144. Oschkinat H, Müller T, Dieckmann T 1994 Protein structure

determination with three- and four-dimensional NMR spectroscopy. Angew Chem Int Ed Engl **33**:277–293.

145. Ni F, Scheraga HA 1994 Use of the transferred nuclear Overhauser effect to determine the conformations of ligands bound to proteins. Acc Chem Res **27**:257–264.
146. Ni F 1994 Recent developments in transferred NOE methods. Prog NMR Spectroscopy **26**:517–606.
147. Wider G, Weber C, Wüthrich K 1991 Proton–proton Overhauser effects of receptor-bound cyclosporin A observed with the use of a heteronuclear-resolved half-filter experiment. J Am Chem Soc **113**:4676–4678.
148. Norris AW, Rong D, d'Avignon DA, Rosenberger M, Tasaki K, Li E 1995 Nuclear magnetic resonance studies demonstrate differences in the interaction of retinoic acid with two highly homologous cellular retinoic acid binding proteins. Biochemistry **34**: 15564–15573.
149. (a) Creuzet F, McDermott A, Gebhard R, van der Hoef K, Spijker-Assink MB, Herzfeld J, Lugtenburg J, Levitt MH, Griffin RG 1991 Determination of membrane protein structure by rotational resonance NMR: Bacteriorhodopsin. Science **251**:783–786. (b) Raleigh DP, Creuzet F, Das Gupta SK, Levitt MH, Griffin RG 1989 Measurement of internuclear distances in polycrystalline solids: Rotationally enhanced transfer of nuclear spin magnetization. J Am Chem Soc **111**:4502–4503.
150. (a) Jansen FJHM, Kwestro M, Schmitt D, Lugtenburg J 1994 Synthesis and characterization of all-*E* (12,12′-$^{13}C_2$)-, (13,13′-$^{13}C_2$)-, (14,14′-$^{13}C_2$)-, (15,15′-$^{13}C_2$)- and (20,20′-$^{13}C_2$)astaxanthin. Recl Trav Chim Pays-Bas **113**:552–562. (b) Weesie RJ, Askin D, Jansen FJHM, de Groot HJM, Lugtenburg J, Britton G 1995 Protein–chromophore interactions in α-crustacyanin, the major blue carotenoprotein from the carapace of the lobster, *Homarus gammarus*. A study by ^{13}C magic angle spinning NMR. FEBS Lett **362**:34–38. (c) de Groot HJM, Gebhard R, van der Hoef I, Hoff AJ, Lugtenburg J 1992 ^{13}C magic angle spinning NMR evidence for a 15,15′-cis configuration of the spheroidene in the *Rhodobacter sphaeroides* photosynthetic reaction center. Biochemistry **31**:12446–12450.
151. Peersen OB, Groesbeek M, Aimoto S, Smith SO 1995 Analysis of rotational resonance magnetization exchange curves from crystalline peptides. J Am Chem Soc **117**:7228–7237.
152. (a) Wecksler WR, Norman AW 1980 Structural aspects of the binding of 1α,25-dihydroxyvitamin D_3 to its receptor system in chick intestine. Methods Enzymol **67**:494–500. (b) Wecksler WR, Norman AW 1980 Measurement of kinetic rate constants for the binding of 1α,25-dihydroxyvitamin D_3 to its chick intestinal mucosa receptor using a hydroxyapatite batch assay. Methods Enzymol **67**:488–494.
153. Bishop JE, Collins ED, Okamura WH, Norman AW 1994 Profile of ligand specificity of the vitamin D binding protein for 1α,25-dihydroxyvitamin D_3 and its analogs. J Bone Miner Res **9**:1277–1288.
154. Muralidharan KR, de Lera AR, Isaeff SD, Norman AW, Okamura WH 1993 Studies on the A-ring diastereomers of 1α,25-dihydroxyvitamin D_3. J Org Chem **58**:1895–1899.
155. Norman AW, Bouillon R, Farach-Carson MC, Bishop JE, Zhou L-X, Nemere I, Zhao J, Muralidharan KR, Okamura WH 1993 Demonstration that 1β,25-dihydroxyvitamin D_3 is an antagonist of the nongenomic but not genomic biological responses and biological profile of the three A-ring diastereomers of 1α,25-dihydroxyvitamin D_3. J Biol Chem **268**:20022–20030.
156. Okamura WH, Midland MM, Norman AW, Hammond MW, Dormanen MC, Nemere I 1995 Biochemical significance of the 6-s-cis conformation of the steroid hormone 1α,25-dihydroxyvitamin D_3 on the provitamin D skeleton. Ann NY Acad Sci **761**:344–348.
157. Norman AW, Bishop JE, Collins ED, Seo E-G, Satchell DP, Dormanen MC, Zanello SB, Farach-Carson MC, Bouillon R, Okamura WH 1996 Differing shapes of 1α,25-dihydroxyvitamin D_3 function as ligands for the D-binding protein, nuclear receptor and membrane receptor: A status report. J Steroid Biochem Mol Biol **56**:13–22.
158. Norman AW, Okamura WH, Farach-Carson MC, Allewaert K, Branisteanu D, Nemere I, Muralidharan KR, Bouillon R 1993 Structure–function studies of 1,25-dihydroxyvitamin D_3 and the Vitamin D endocrine system. 1,25-Dihydroxy-pentadeuterio-previtamin D_3 (as a 6-s-cis analog) stimulates nongenomic but not genomic biological responses. J Biol Chem **268**:13811–13819.
159. Norman AW, Okamura WH, Hammond MW, Bishop JE, Dormanen MC, Bouillon R, van Baelen H, Ridall AL, Daane E, Khoury R, Farach-Carson MC 1997 Comparison of 6-s-cis and 6-s-trans locked analogs of 1α,25-dihydroxyvitamin D_3 indicates that the 6-s-cis conformation is preferred for rapid nongenomic biological responses and that neither 6-s-*cis* nor 6-s-*trans* locked analogs are preferred for genomic biological responses. Molecular Endocrin accepted for publication pending revision.
160. Liu Y-Y, Collins ED, Norman AW, Peleg S 1996 Differential interaction of 1α,25-dihydroxyvitamin D_3 analogues and their 20-epi homologues with the vitamin D receptor. J Biol Chem **272**:3336–3345.
161. Norman AW, Koeffler HP, Bishop JE, Collins ED, Sergeev I, Zhou L-X, Nemere I, Zhou J, Henry HL, Okamura WH 1991 Structure–function relationships in the vitamin D endocrine system for 1,25$(OH)_2D_3$ analogs. In: Norman AW, Bouillon R, Thomasset M (eds) Vitamin D: Gene Regulation, Structure–Function Analysis and Clinical Application (Proceedings of the Eighth Workshop on Vitamin D, Paris, France), de Gruyter, Berlin, New York, pp. 146–154.
162. Allan GF, Leng X, Tsai SY, Weigel NL, Edwards DP, Tsai M-J, O'Malley BW 1992 Hormone and antihormone induce distinct conformational changes which are central to steroid receptor activation. J Biol Chem **267**:19513–19520.
163. Zhou JY, Norman AW, Chen D, Sun G, Uskokovic MR, Koeffler HP 1990 1,25-Dihydroxy-16-ene-23-yne-vitamin D_3 prolongs time of leukemic mice. Proc Natl Acad Sci USA **87**:3929–3932.
164. Colston KW, Chander SK, Mackay AG, Coombes RC 1992 Effects of synthetic vitamin D analogues on breast cancer cell proliferation *in vivo* and *in vitro*. Biochem Pharmacol **44**: 693–702.
165. Abe-Hashimoto J, Kikuchi T, Matsumoto T, Nishii Y, Ogata E, Ikeda K 1993 Antitumor effect of 22-oxa-calcitriol, a noncalcemic analogue of calcitriol, in athymic mice implanted with human breast carcinoma and its synergism with tamoxifen. Can Res **53**:2534–2537.
166. Anzano MA, Smith JM, Uskokovic MR, Peer CW, Mullen LT, Letterio JJ, Welsh MC, Shrader MW, Logsdon DL, Driver CL, Brown CC, Roberts AB, Sporn MB 1994 1α,25-dihydroxy-16-ene-23-yne-26,27-hexafluorocholecalciferol (Ro24-5531), a new deltanoid (vitamin D analogue) for prevention of breast cancer in the rat. Cancer Res **54**:1653–1656.
167. Bhatia M, Kirkland JB, Meckling-Gill KA 1995 Monocytic differentiation of acute promyelocytic leukemia cells in response to 1,25-dihydroxyvitamin D_3 is independent of nuclear receptor binding. J Biol Chem **270**:15962–15965.

Analog Metabolism

GLENVILLE JONES Departments of Biochemistry and Medicine, Queen's University, Kingston, Ontario, Canada

I. GENERAL CONSIDERATIONS

A. Vitamin D Metabolism

No mention of the metabolism of vitamin D analogs can ignore the rich and varied history of the metabolism of vitamin D itself [1] (see Chapter 1 of this treatise). Metabolic investigations over the last few decades have revealed not only the nature of the hydroxylation/oxidation steps important in the activation of vitamin D but also much about the cytochrome P450-based enzymes involved. In fact, it was the elucidation of the metabolism of vitamin D that sparked the synthesis of the first vitamin D analogs back in the early 1970s. The reader is referred to Chapter 2 for details of natural metabolites and to Chapters 4, 5, and 6 for details of the 25-, 1α-, and 24-hydroxylase enzymes, as this knowledge is important background to a fuller appreciation of the metabolism of vitamin D analogs.

Any review of the metabolism of vitamin D analogs should make an important distinction between (1) that metabolism providing activation of the analog which thereby leads to a more biologically active molecule and (2) that metabolism, otherwise known as catabolism, which deactivates the molecule, leading to its destruction and excretion. In fact, this distinction allows for a basis for a classification of vitamin D analogs into two distinct groups: prodrugs (analogs) requiring one or more step(s) of activation (e.g., one step: 1αOHD$_2$, 1αOHD$_3$, and 25OHD$_3$; multiple steps: vitamin D$_2$ and dihydrotachysterol) and analogs of 1α,25-dihydroxyvitamin D$_3$ (calcitriol) requiring no activation (e.g., calcipotriol, OCT, EB1089, and KH1060).

It is clear that many of the early generations of vitamin D analogs were prodrugs which were designed to take advantage of the enzymes already in place for the metabolic activation of vitamin D itself. Most of these analogs showed a close resemblance to vitamin D since they required sufficient similarity to the natural substance to ensure activation. More recently, chemists in the pharmaceutical industry and university domain have designed more exotic vitamin D analogs which deviate significantly from the basic vitamin D/calcitriol structure. In the majority of cases, it is the vitamin D side chain which has been changed in these analogs, but there exists a significant number of examples stemming from excursions into A-ring modification. Two reviews [2,3] and Chapter 57 offer comprehensive lists of the analogs of vitamin D synthesized to date. In Tables I and II, we have selected, respectively, some of the more interesting prodrugs and calcitriol analogs for which metabolic data are available.

Prodrug activation (e.g., for vitamin D$_2$ or 1αOHD$_3$) may involve the same cytochrome P450-based enzymes (i.e., the 25- and 1α-hydroxylases) as for vitamin D, and consequently this allows us to explore the substrate

TABLE I Vitamin D Prodrugs

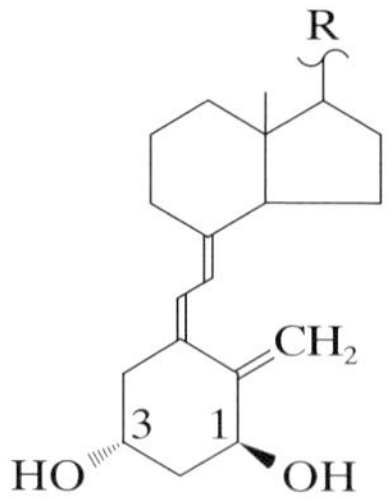

Vitamin D prodrugs [ring structure]	Side chain structure (R)	Company	Possible target diseases	Mode of delivery	Ref.
1αOHD$_3$ [3]	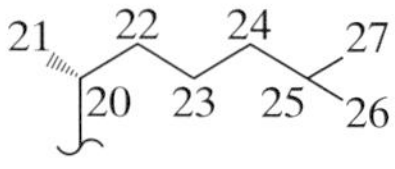	Leo	Osteoporosis	Systemic	111
1αOHD$_2$ [3]	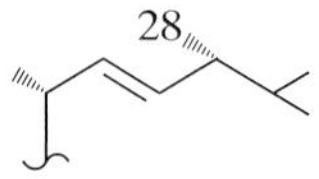	Lunar	Osteoporosis	Systemic	112
Dihydrotachysterol [2]	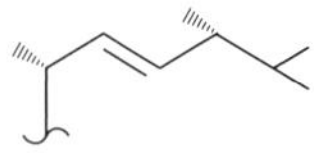	Duphar	Renal failure	Systemic	26
Vitamin D$_2$ [1]	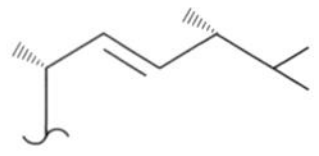	Various	Rickets Osteomalacia	Systemic Systemic	113

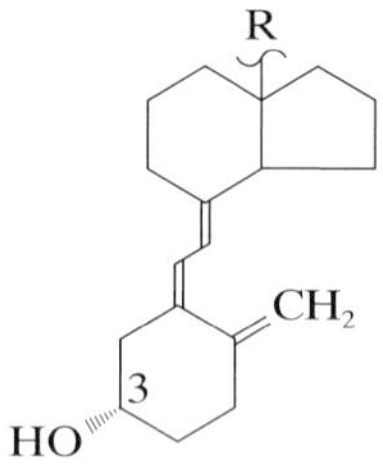

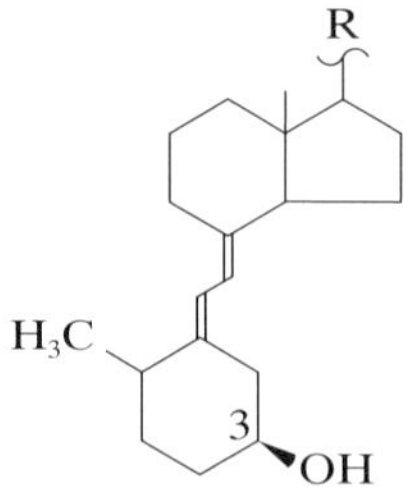

specificity of these vitamin D-related enzymes. Current dogma suggests that 25-hydroxylation of endogenous vitamin D$_3$ is carried out by a bifunctional enzyme also capable of the 25- and 27-hydroxylation steps within pathways of bile acid metabolism [4]. Indeed, this liver mitochondrial cytochrome P450, known as CYP27 (see Chapter 4), when transfected into green monkey kidney COS-1 cells has also been shown to be capable of the 25-hydroxylation (and other hydroxylations at C-24 and C-27) of a number of vitamin D analogs (including vitamin D$_3$), when provided at high substrate concentrations [5]. However, there remains some doubt that CYP27 carries out the 25-hydroxylation of vitamin D$_3$ at the concentrations observed *in vivo*. There is the theoretical possibility that another enzyme exists which is responsible for 25-hydroxylation of low concentrations of endogenous vitamin D$_3$ and which is different from the cytochrome P450 cloned thus far and shown to be capable of the metabolism of vitamin D analogs. Similarly, 1α-hydroxylation of vitamin D analogs lacking a 1α-hydroxyl function may not be the exclusive domain of the renal 1α-hydroxylase described in Chapter 5. It is theoretically possible that the extrarenal 1α-hydroxylase [6] (see also Chapter 55) may activate vitamin D analogs (e.g., dihydrotachysterol), though this author knows of no example where this is the underlying strategy of a clinical treatment protocol.

B. Calcitriol Catabolism

The other side of metabolism is catabolism, and this is where the majority of the emphasis of this chapter is placed. Calcitriol is subject to catabolism by two different pathways.

1. C-24 Oxidation Pathway to Calcitroic Acid

The C-24 oxidation pathway seems to predominate in most target cells because of the inducible nature of the mitochondrial cytochrome P450 involved, known as CYP24, which is the substrate-binding component of the 24-hydroxylase complex [7]. The C-24 oxidation pathway comprises five enzymatic steps (Fig. 1) involving successive hydroxylation/oxidation reactions at C-24 and C-23 followed by cleavage of the molecule between C-23 and C-24 and oxidation of the resultant truncated product to calcitroic acid [8,9]. Reconstitution assays

TABLE II Analogs of $1,25(OH)_2D_3$

[1] [2] [3] [4] [5] [6]

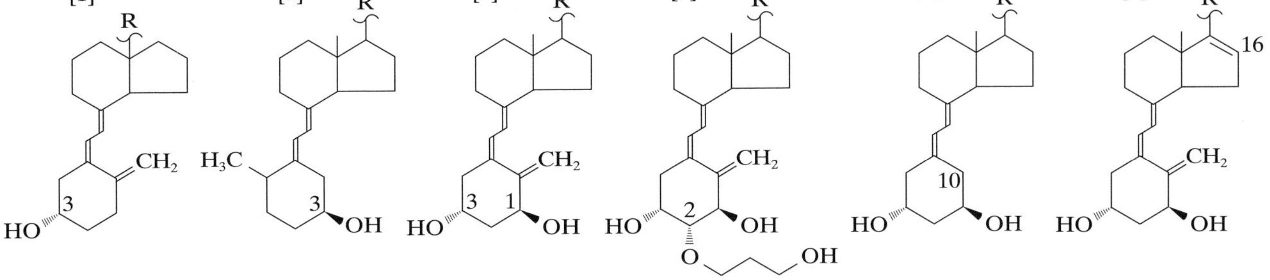

Vitamin D analog [ring structure]	Side chain structure (R)	Company	Possible target diseases	Mode of delivery	Ref.
$1\alpha,25(OH)_2D_3$ [3]	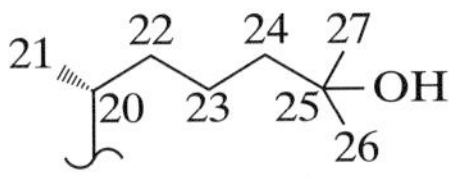 21 22 24 27 20 23 25 OH 26	Roche, Duphar	Hypocalcemia Psoriasis	Systemic Topical	114
$26,27\text{-}F_6\text{-}1\alpha,25(OH)_2D_3$ [3]	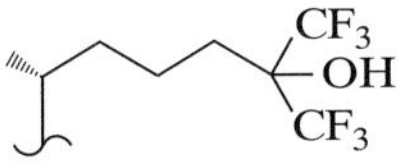 CF_3 OH CF_3	Sumitomo-Taisho	Osteoporosis Hypoparathyroidism	Systemic Systemic	91
$19\text{-Nor-}1\alpha,25(OH)_2D_2$ [5]	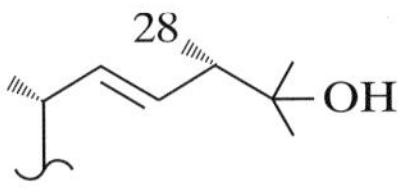 28 OH	Abbott	Hyperparathyroidism	Systemic	57
22-Oxacalcitriol (OCT) [3]	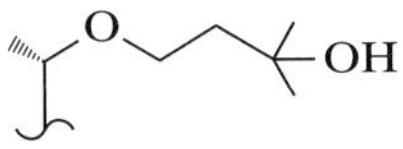 O OH	Chugai	Hyperparathyroidism Psoriasis	Systemic Topical	73
Calcipotriol (MC903) [3]	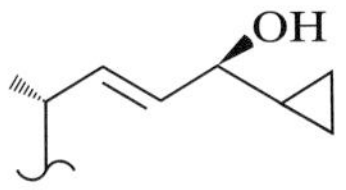 OH	Leo	Psoriasis Cancer	Topical Topical	48
$1\alpha,25(OH)_2\text{-16-ene-23-yne-}D_3$ (Ro 23-7553) [6]	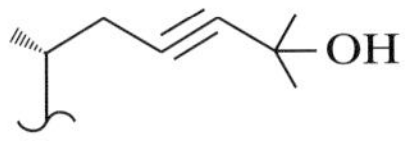 OH	Roche	Leukemia	Systemic	115
EB1089 [3]	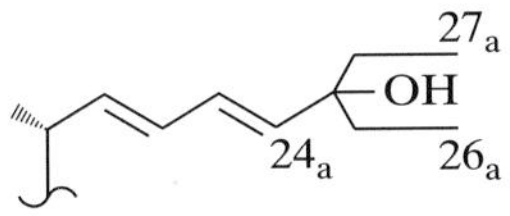 27_a OH 24_a 26_a	Leo	Breast cancer	Systemic	52
$20\text{-Epi-}1\alpha,25(OH)_2D_3$ [3]	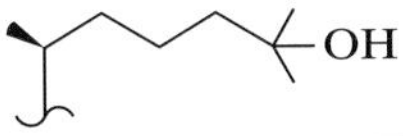 OH	Leo	Immune diseases	Systemic	51
KH1060 [3]	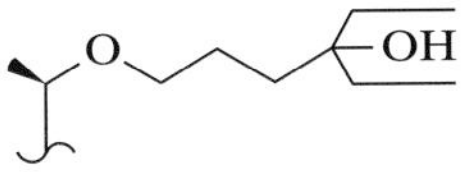 O OH	Leo	Immune diseases	Systemic	74
ED-71 [4]	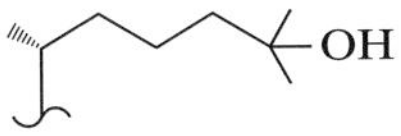 OH	Chugai	Osteoporosis	Systemic	89
$1\alpha,24(S)(OH)_2D_2$ [3]	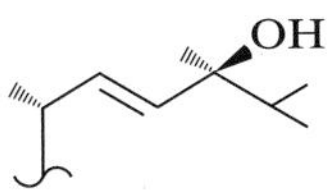 OH	Lunar	Psoriasis	Topical	42
$1\alpha,24(R)(OH)_2D_3$ (TV-02) [3]	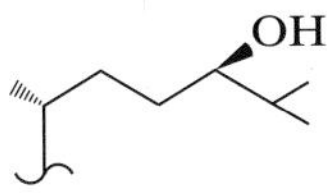 OH	Teijin	Psoriasis	Topical	116

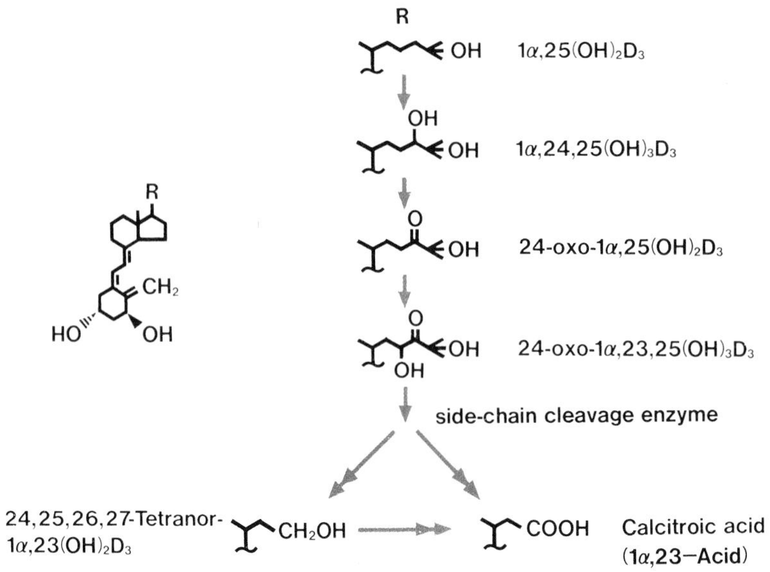

FIGURE 1 C-24-Oxidation pathway. Reprinted from Makin *et al.* [8] with permission.

using recombinant CYP24 produced in *Escherichia coli* and baculovirus systems have shown that the first three enzyme activities of the pathway, and possibly more, reside in a single cytochrome P450 chain [10,11]. The fact that this C-24 oxidation pathway has been connected to a known excretory product of calcitriol, in the form of calcitroic acid, adds credence to the view that the pathway is catabolic in nature. Consistent with this viewpoint is the finding that the mRNA for CYP24 and 24-hydroxylation activity have been found in classic vitamin D target tissues including intestine, kidney, and bone as well as a variety of primary cells and cultured cell lines such as CaCo-2 (colon), UMR-106 (bone), LLC-PK1 (kidney), and HPK1A-ras (keratinocyte) [12–15]. Depending on the structure of the side chain, vitamin D analogs can also be metabolized by these same C-24 oxidation pathway enzymes, usually to molecules with reduced biological activity. Therefore, studies concerned with metabolism of vitamin D analogs have the dual role of defining metabolic products of the analog and allowing for exploration of the substrate preferences of CYP24 or another enzyme(s) involved.

2. C-26 HYDROXYLATION/ 26,23-LACTONE FORMATION

The importance of 26-hydroxylation or the pathway to $1\alpha,25(OH)_2D_3$-26,23-lactone is much less clear. The evidence for the formation of these metabolites is irrefutable, with the pathway to the lactone being well defined [16] (Fig. 2); yet, the enzymes involved remain obscure, both in characterization and in location. Furthermore, it is not clear why the metabolite $1\alpha,25,26(OH)_3D_3$ is formed since this 26-hydroxylated derivative of $1\alpha,25(OH)_2D_3$ does not seem to be the initial precursor to the lactone [17]. Pathway considerations aside, it seems clear that both 26-hydroxylation and 26,23-lactone formation result in molecules with reduced biological activity, suggesting but not conclusively proving that they represent catabolites.

C. Non-Vitamin D-Related Metabolism and Methodology

As the vitamin D molecule is increasingly modified by the organic chemist, it becomes more and more susceptible to metabolism by enzymes that are distinct from those specifically involved in vitamin D metabolism. It is the analogs of calcitriol which are the most likely to be subject to metabolism by non-vitamin D-related enzyme systems. This is because analogs of calcitriol incorporate the greatest number of structural changes and are modified currently in that part of the molecule which is subject to metabolic alteration *in vivo*, specifically the side chain. In the specific examples of vitamin D analogs which follow, it is to be noted that there is strong evidence for involvement of other (thus far undefined) cytochrome P450-based enzyme systems. Given the uncertainty about the identity of the enzymes involved in the metabolism of vitamin D analogs or even their tissue source, it is not surprising that most

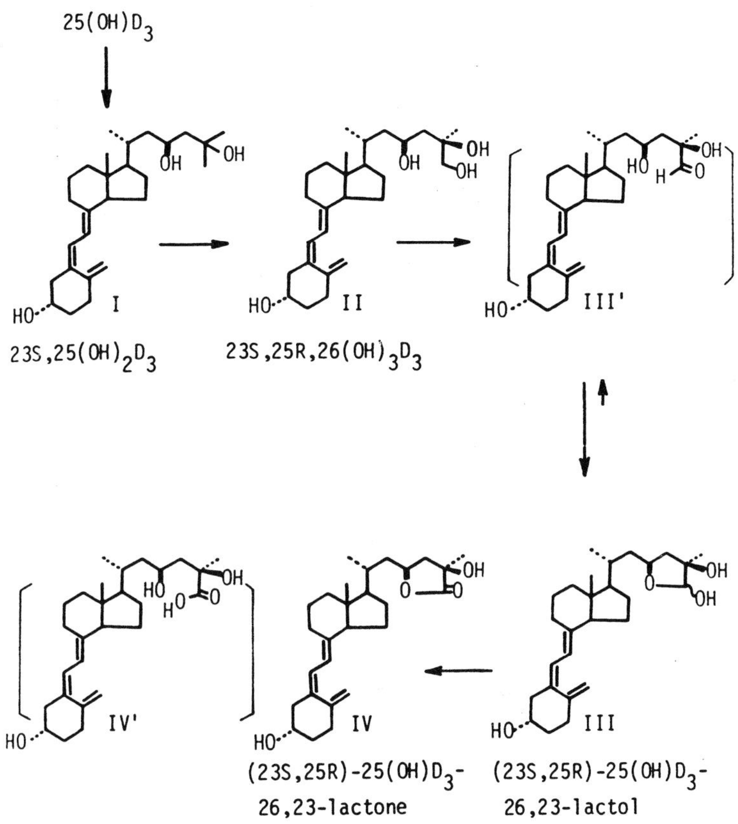

FIGURE 2 26,23-Lactone pathway. The pathway shown is for 25OHD_3. An analogous pathway exists for 1α,25$(OH)_2D_3$. Reprinted from Yamada *et al.* [16] with permission.

investigations involve a cross section of metabolic systems in order to identify the metabolic products and to focus on the enzymes responsible. Metabolic studies utilize a variety of species and *in vivo* and *in vitro* models from broken cell to intact cell systems (Table III). The predominant models used for metabolic studies are liver-based systems and have revealed that this organ is a primary site of vitamin D analog metabolism. As with calcitriol, many vitamin D analogs can also be metabolized by vitamin D target cells. This knowledge is valuable in that it allows for consideration of the possibility of blocking target cell enzymes by specific inhibitors or, alternatively, of studying the substrate preferences of the target cell enzymes in order to design better catabolism-sensitive or catabolism-resistant vitamin D analogs for use as drugs.

Methodology for studying vitamin D analog metabolism not only involves a wide variety of biological systems (Table III) but also usually depends on the availability of a suitably radioactive vitamin D analog. The radioactive analog must possess a sufficiently high specific activity to allow detection of nanomolar concentrations of analog; in addition, the label location must be in a metabolically resistant region of the molecule. This usually means that ^{3}H labeling is the preferred strategy, and a nuclear location of the radioactive tag is best if the label is not to be lost. Occasionally, the susceptibility of the ^{3}H label to metabolic attack is judiciously used as an indicator of the type of metabolism occurring. An excellent example of this is the ingenuity of Chugai chemists working with the Slatopolsky group to compare the metabolism of both [26-^{3}H]OCT and [2β-^{3}H]OCT in their biological systems. Using such a combination of labels, these researchers were able to confirm the loss of a portion of the side chain of OCT (see Chapter 63) during metabolism, thereby suggesting the formation of side-chain-truncated metabolites of OCT [18,19]. Although the availability of radioactive analogs has not been a significant problem when studying most of the well-established compounds developed to date, it remains a significant barrier to the widespread screening of the hundreds of vitamin D analogs currently available.

If radioactive analogs are not readily available, another technique involving detection of metabolites by diode-array spectrophotometry can be used, but as the disadvantage of being forced to employ high substrate concentrations [20]. Although this technique is re-

TABLE III Biological Systems Used in the Study of the Metabolism of Vitamin D and Its Analogs

Biological system	Enzyme activity	Ref.
Isolated perfused organ		
Rat liver	25-Hydroxylation	117
Rat kidney	1α- and 24-hydroxylation	118,119
Primary cells		
Chick kidney	1α- and 24-hydroxylation	120
Bovine parathyroid	24-Oxidation pathway	18
Cultured cell lines[a]		
CaCo-2, human colon ("intestinal")	24-Oxidation pathway	13
UMR-106, rat osteosarcoma	24-Oxidation pathway	14
LLC-PK1, pig kidney	24-Oxidation pathway	121
HPK1A-ras, human keratinocyte	24-Oxidation pathway	15
HD-11, chick myelomonocyte	Extrarenal 1-hydroxylation	30,31
Broken cell systems		28
Rat liver mitochondria	25-Hydroxylation	
Rat kidney mitochondria	1α- and 24-hydroxylation	122
Rat, human, and minipig postmitochondrial supernatant	General metabolism	21
Reconstituted system with cytochrome P450, ferridoxin, reductase	1α- and 25-hydroxylation	87
Transfected cell systems	25-Hydroxylation	5
CYP27 transfected into COS-1 cells		
CYP24 transfected into *E. coli*	24-Oxidation pathway	10
CYP24 transfected into insect cells	24-Oxidation pathway	11

[a] Only a few examples are given.

stricted to high concentrations of analogs, the analysis is quicker and less expensive than using radioactive analogs. The high concentration of substrate results in the generation of large amounts of metabolites, and this, in turn, permits a more rigorous identification of the metabolic products. Our laboratory has used such a procedure very effectively to identify the products of both natural and synthetic vitamin D compounds incubated with a range of *in vitro* biological systems. The quantitative and qualtitative answers that diode-array spectrophotometry provide are, in large part, consistent with those answers resulting from studies employing radioactive vitamin D analogs (cf. Sorensen *et al.* [21] and Masuda *et al.* [15]).

Another metabolic technique worthy of mention is the assay of the ability of a vitamin D analog to compete with $1\alpha,25(OH)_2[1\beta\text{-}^3H]D_3$ for the enzymes of the C-24 oxidation pathway. In this assay, the ability of the nonradioactive vitamin D analog [compared with the control, $1\alpha,25(OH)_2D_3$] to block the metabolism of lipid-soluble $1\alpha,25(OH)_2[1\beta\text{-}^3H]D_3$ to water-soluble $[1\beta\text{-}^3H]$calcitroic acid is measured [22]. As the concentration of the test vitamin D analog is increased, it is metabolized by the enzymes of the C-24 oxidation pathway, resulting in less conversion of $1\alpha,25(OH)_2[1\beta\text{-}^3H]D_3$ to $[1\beta\text{-}^3H]$ calcitroic acid, and thus less radioactivity is found in the water-soluble fraction during a simple Bligh and Dyer [23] lipid extraction. The assay is thus an indirect but sensitive indicator of the susceptibility of an analog to be metabolized by the C-24 oxidation pathway enzymes.

The specific metabolic studies described below provide examples of the use of all three of these approaches to the study of the metabolism of vitamin D analogs.

II. EXAMPLES OF THE METABOLISM OF ANALOGS OF VITAMIN D

A. Dihydrotachysterol

The vitamin D prodrug dihydrotachysterol represents the oldest vitamin D analog and was developed in the 1930s to stabilize the triene structure of one of the photoisomers of vitamin D. The structure of dihydrotachysterol$_2$ shown in Table I contains an A ring rotated through 180°, a reduced C-10 to C-19 double bond, and the side chain structure of ergosterol/vitamin D_2 (see Chapter 57 for numbering system). This side chain is depicted because the clinically approved drug form of dihydrotachysterol is dihydrotachysterol$_2$ (DHT_2). However, it should be noted that dihydrotachysterol$_3$ (DHT_3) can also be synthesized with the side chain of vitamin D_3. The metabolism of both DHT_2 and DHT_3 have been extensively studied [24–27]. Initial studies performed in the early 1970s showed that DHT is efficiently converted to its 25-hydroxylated metabolite [28].

The effectiveness of DHT to relieve the hypocalcemia of chronic renal failure in the absence of a functional renal 1α-hydroxylase led to the hypothesis [29] that 25-OH-DHT might represent the biologically active form of DHT, by virtue of its 3β-hydroxyl group being rotated 180° into a "pseudo 1α-hydroxyl position." It was thus believed that 1α-hydroxylation of 25-OH-DHT was unnecessary. This viewpoint prevailed for at least a decade, but debate was renewed when Bosch *et al.* [25] were able to provide evidence for the existence of a mixture of 1α- and 1β-hydroxylated products of 25-OH-DHT_2 in the blood of rats dosed with DHT_2. Studies involving the perfusion of kidneys from vitamin

D-deficient rats with an incubation medium containing 25-OH-DHT_3, and use of diode-array spectrophotometry to analyze the extracts, showed this molecule to be subject to extensive metabolism by renal enzymes but failed to detect the expected 1-hydroxylated metabolites, opening the possibility that the 1α- and 1β-hydroxylated metabolites observed by Bosch *et al.* [25] might be formed by an extrarenal 1-hydroxylase [26] (Fig. 3). Following the synthesis of appropriate authentic standards, subsequent research [27] has rigorously confirmed the *in vivo* formation and identity of 1α,25$(OH)_2$DHT and 1β,25$(OH)_2$DHT in both rat and human. The ability of these 1α-hydroxylated forms of both DHT_2 and DHT_3 to stimulate a vitamin D response element (VDRE)-inducible growth hormone reporter system exceeded that of 25-OH-DHT and in the process established 1α,25$(OH)_2$DHT and 1β,25$(OH)_2$DHT as the most potent derivatives of DHT identified to date.

The formation of these metabolites also brings into question the importance of the "pseudo 1α-hydroxyl group" hypothesis, though current findings do not rule out the possibility that the biological activity of DHT might be due to the collective action of a group of metabolites including 25-OH-DHT, 1α,25$(OH)_2$DHT, and 1β,25$(OH)_2$DHT. The latest information on the site of biosynthesis of 1-hydroxylated DHT species comes from studies using the cultured chicken myelomonocytic cell line HD-11 [30]. This cell line, which has previously been documented as a rich source of the extrarenal 1α-hydroxylase [31], has also been shown to be capable of the 1-hydroxylation of 25-OH-DHT [30]. These results are consistent with DHT being 25-hydroxylated in the liver and then subject to 1-hydroxylation by a extrarenal hydroxylase of bone marrow origin *in vivo.* The results also suggest that 25-OH-DHT will prove to be a useful tool to differentiate between renal and extrarenal 1-hydroxylases, once these two enzymes are cloned.

Although the enzymes involved in the activation of DHT, especially the 1-hydroxylation step, have an altered specificity toward this molecule, the enzymes involved in the catabolism of DHT_3 appear to treat the molecule as they would $25OHD_3$ or 1α,25$(OH)_2D_3$. Side chain-hydroxylated derivatives of both 25-OH-DHT_3 and 1,25$(OH)_2DHT_3$ have been identified and appear to be analogous to intermediates of the C-24 oxidation and 26,23-lactone pathways of vitamin D_3 metabolism [32,33]. The major difference between the catabolism of DHT_3 and vitamin D_3 is that the 26,23-lactone formation from DHT_3 appears exaggerated, suggesting that either DHT_3 is a better substrate for the enzymes involved in 26,23-lactone formation or else it is discriminated against by CYP24 or other enzymes of the alternative C-24 oxidation pathway.

B. Vitamin D_2 Derivatives

Though vitamin D_2 can be synthesized naturally by irradiation of ergosterol, little finds its way into the human diet unless it is provided as a dietary supplement. Thus one could make a case for considering vitamin D_2 as a prodrug. The complex metabolism of vitamin D_2 has been included as part of Chapter 2 and is not re-

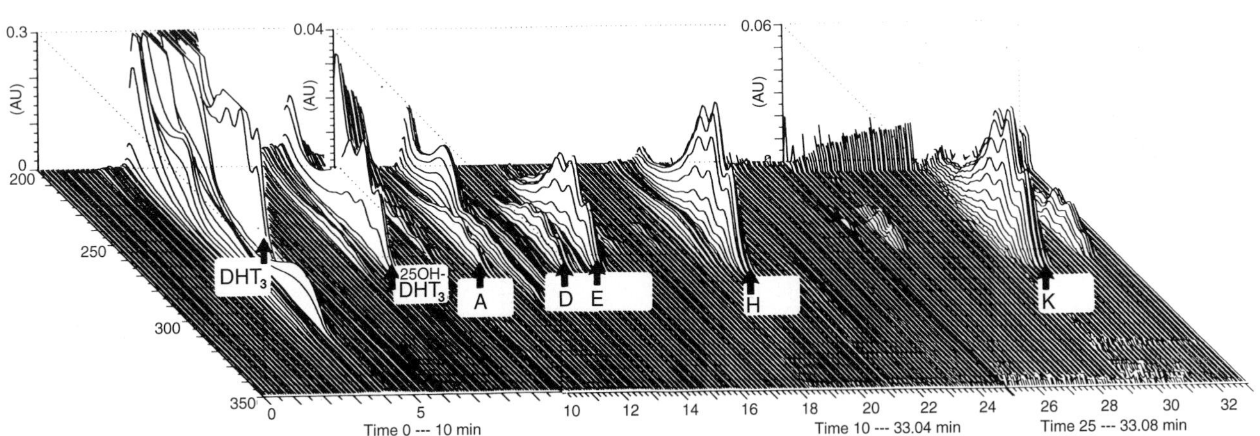

FIGURE 3 *In vivo* metabolism of dihydrotachysterol$_3$ in the rat. High-performance liquid chromatography (HPLC), with diode-array spectrophotometric detection, was used to analyze the plasma extract of a rat administered 1 mg DHT_3 18 hr prior to sacrifice. Metabolites: 25-OH-DHT_3 and peaks A–L were observed, although only some of these are labeled here. All possess the distinctive tricuspid UV spectrum (λ_{max} 242.5, 251, and 260.5 nm). Metabolites A–L were subsequently identified as side chain-modified compounds analogous to vitamin D metabolites of the C-24 oxidation and 26,23-lactone pathways depicted in Figs. 1 and 2. Reprinted from Jones *et al.* [26] with permission.

peated here. However, it is worth noting that vitamin D_2 gives rise to several analogous metabolites to those of vitamin D_3 in the form of 25OHD$_2$ [34], 1α,25(OH)$_2$D$_2$ [35], and 24,25(OH)$_2$D$_2$ [36], as well as several unique metabolites including 24OHD$_2$ [37], 1α,24*S*(OH)$_2$D$_2$ [38], 24,26(OH)$_2$D$_2$ [39], and 1α,25,28(OH)$_3$D$_2$ [40]. These differences in the metabolism of vitamin D_2 have been exploited by synthesizing and using the metabolites unaltered or else creating slightly modified versions [e.g., Roche compound 1α,25,28(OH)$_3$D$_2$; Bone Care compound 1α,24*S*(OH)$_2$D$_2$]. Furthermore, features of the vitamin D_2 side chain, namely, the C-22 to C-23 double bond or the C-24 methyl group (see Table I), have been successfully incorporated into the structure of other analogs (e.g., calcipotriol).

Even a prodrug based on vitamin D_2 has been designed in the form of 1αOHD$_2$ [41]. This molecule is a valuable tool in studying hydroxylation reactions in the liver. At low substrate concentrations, 1αOHD$_2$, like 1αOHD$_3$, is 25-hydroxylated by liver hepatomas Hep3B and HepG2, producing the well-established, biologically active compound 1α,25(OH)$_2$D$_2$. However, when the substrate concentration is increased to micromolar values, the principal site of hydroxylation of 1αOHD$_2$ becomes the C-24 position, the product being 1α,24*S*(OH)$_2$D$_2$, another compound with significant biological activity [42,43]. This metabolite has previously been reported in cows receiving massive doses of vitamin D_2 [38]. Transfection studies using the liver cytochrome P450 (CYP27) expressed in COS-1 cells suggest that 1α,24*S*(OH)$_2$D$_2$ is a product of this cytochrome [5]. Whether the formation of this unique metabolic product of 1αOHD$_2$ is the reason for the relative lower toxicity of 1αOHD$_2$ as compared to 1αOHD$_3$ [44] has not been established definitively.

Active vitamin D_2 compounds, such as 1α,25(OH)$_2$D$_2$ and 1α,24*S*(OH)$_2$D$_2$, are also subject to further metabolism, although the pathway differs from that of calcitriol, essentially because the modifications in the vitamin D_2 side chain prevent the C-23–C-24 cleavage observed during calcitroic acid production. Instead, the principal products are more polar tri- and tetrahydroxylated metabolites such as 1α,24,25(OH)$_3$D$_2$, 1α,25,28(OH)$_3$D$_2$, and 1α,25,26(OH)$_3$D$_2$ from 1α,25(OH)$_2$D$_2$ [40,45,46] and 1α,24,26(OH)$_3$D$_2$ from 1α,24*S*(OH)$_2$D$_2$ [43], produced by as yet undefined enzymes. In the latter case, the rate of 1α,24*S*(OH)$_2$D$_2$ metabolism appears slower than that of 1α,25(OH)$_2$D$_3$ [43]. Some catabolites retain considerable biological activity, and at least one, 1α,25,28(OH)$_3$D$_2$, is patented for use as a drug. The reader is referred to Chapter 2 for further discussion of vitamin D_2 metabolism.

C. Cyclopropane Ring-Containing Analogs of Vitamin D

Cyclopropane ring analogs of vitamin D are modified in their side chains such that C-26 is joined to C-27 to give a cyclopropane ring consisting of C-25, C-26, and C-27. The simplest member of this series is MC969, which possesses the vitamin D_3 chain except for the presence of the cyclopropane ring together with a 1αOHD nucleus [47]. The best-known member of this group of compounds is MC903 or calcipotriol, the structure of which is shown in Table II. In addition to the cyclopropane ring, calcipotriol features a C-22 to C-23 double bond and a 24*S*-hydroxyl group [48]. As is presented in Chapter 61, calcipotriol was the first vitamin D analog to be approved for topical use in psoriasis and is currently used worldwide for the successful control of this skin lesion [49] (see also Chapter 72).

When MC969 is incubated with the hepatoma Hep3B, it is hydroxylated, not at the C-25 position as is 1αOHD$_3$, but at the C-24 position, as is 1α,OHD$_2$, and then further oxidized to a 24-ketone [47]. 25-Hydroxylation of vitamin D analogs containing cyclopropane ring structures is feasible; indeed, the molecule has been synthesized chemically, but it is not produced enzymatically from MC969. It thus appears that the cyclopropane ring directs the hydroxylation site to the C-24 position.

Of course, the other cyclopropane ring-containing analog, calcipotriol, contains a preexisting 24-hydroxyl that has been proposed to act as a surrogate 25-hydroxyl in interactions of the molecule with the vitamin D receptor (VDR). Pharmacokinetic data acquired for calcipotriol showed that it had a very short half-life ($t_{1/2}$) *in vivo,* on the order of minutes; such results are consistent with the lack of a hypercalciuric/hypercalcemic effect when administered *in vivo* [50]. The first metabolism studies [21] revealed that calcipotriol was rapidly metabolized by a variety of different liver preparations from rat, minipig, and human to two novel products. Sorensen *et al.* [21] were able to isolate and identify the two principal products as a C-22=C-23 unsaturated, 24-ketone (MC1046) and a C-22−C-23 reduced, 24-ketone (MC1080). Coincidentally, this is the same product as is formed from MC969. These results were confirmed and extended by Masuda *et al.* [15], who showed that capcipotriol metabolism was not confined to liver tissue but could be carried out by a variety of cells including those cells exposed to topically administered calcipotriol *in vivo*, namely, keratinocytes. Furthermore, Masuda *et al.* [15] proposed further metabolism of the 24-ketone in these vitamin D target cells to side chain-cleaved molecules including calcitroic acid (Fig. 4). The main

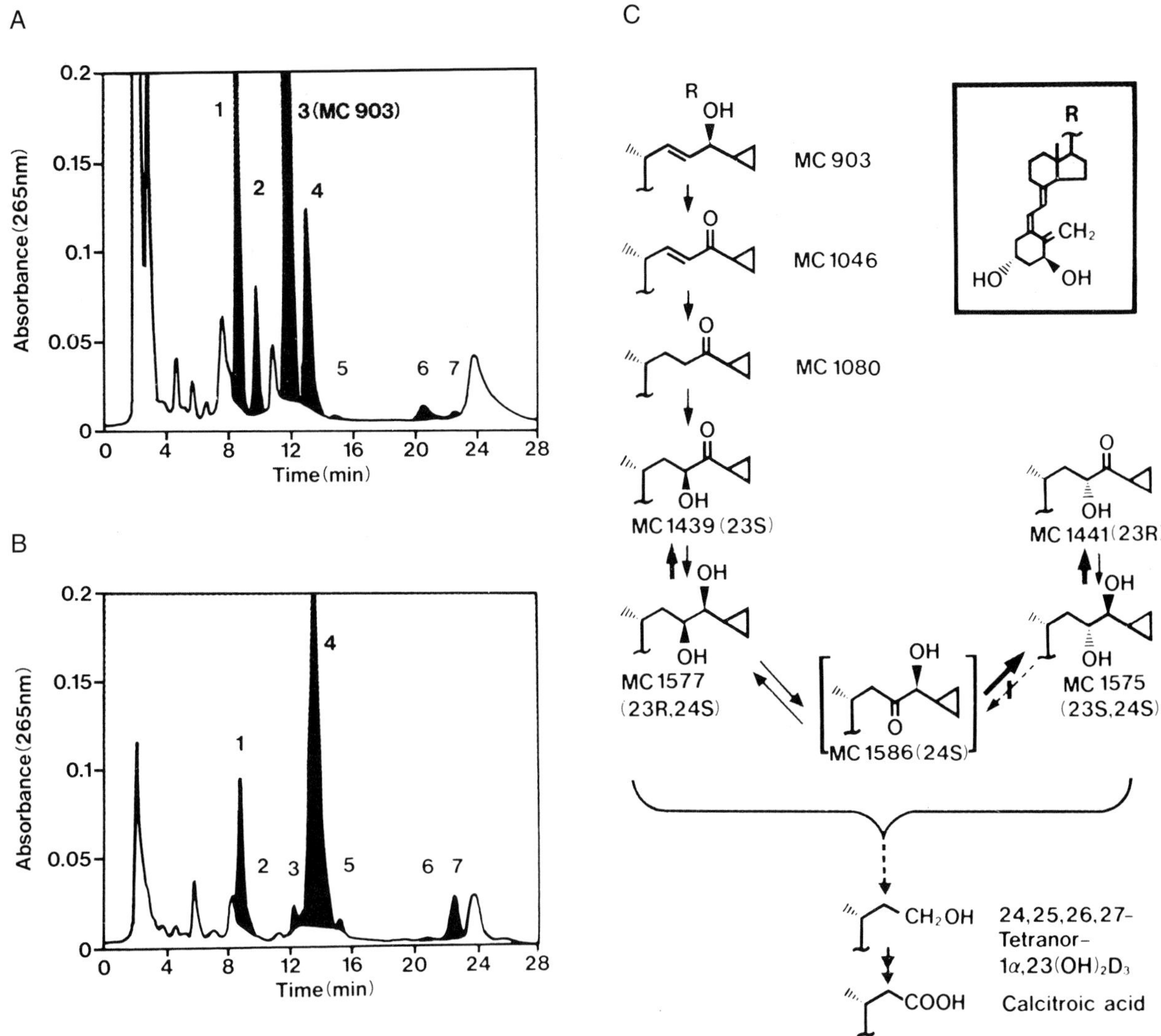

FIGURE 4 (A, B) *In vitro* metabolism of calcipotriol (MC903) by HPK1A-ras cells. Analysis by HPLC of lipid extracts was performed following incubation of MC903 with (A) HPK1A human keratinocytes and (B) HPK1A-ras human keratinocytes. Peak 1, MC1080; peak 2, MC1046; peak 3, MC903 (calcipotriol); peak 4, mixture of MC1439 and MC1441; peak 5, tetranor-1α,23$(OH)_2D_3$; peak 6, MC1577; peak 7, MC1575. (C) Proposed pathway of calcipotriol metabolism in cultured keratinocytes. Reprinted from Masuda *et al.* [15] with permission.

implication of this work is that calcipotriol is subject to rapid metabolism initially by non-vitamin D-related enzymes, then by vitamin D-related pathways to a side chain-cleaved molecule. The catabolites are produced in a variety of tissues and appear to have lower biological activity than the parent molecule.

The reduction of the C-22=C-23 double bond during the earliest phase of calcipotriol catabolism was an unexpected event given that the C-22=C-23 double bond in vitamin D_2 compounds is extraordinarily stable to metabolism. It thus appears that metabolism of calcipotriol provides evidence that the C-24 methyl group in the vitamin D_2 side chain must play a stabilizing role, preventing the formation of the 24-ketone which facilitates the reduction of the C-22=C-23 double bond. However, it is still unknown which enzyme is responsible for this reduction in the side chain of calcipotriol.

D. 20-Epi Analogs

In the early 1990s, Leo organic chemists were first to change the stereochemistry of the side chain at the C-20 position [51,52]. As a result, they were in a position to synthesize a novel class of compounds which were 20-epimers of existing analogs, the simplest being 20-epi-1α,25(OH)$_2$D$_3$ and the most complex being KH1060 (see Table II for structures). Some of these 20-epi analogs are extremely potent in cell differentiation and antiproliferation assays and are thus under development for use in hyperproliferative conditions.

These molecules have been particularly well studied, not only for their susceptibility to metabolism, but also for their ability to transactivate model genes (see Chapter 60). Dilworth *et al.* [53] showed that 20-epi-1,25(OH)$_2$D$_3$ (MC1288) has virtually no binding to vitamin D binding protein (DBP), has slightly improved affinity for the bovine thymus VDR, and competes about 1/36th as effectively as 1,25(OH)$_2$D$_3$ for the C-24 oxidation pathway. Dilworth *et al.* [53] concluded that all these factors (i.e., DBP affinity, VDR affinity, and rate of catabolism in target cells) contribute to the biological activity advantages that 20-epi-1α,25(OH)$_2$D$_3$ appears to possess over 1α,25(OH)$_2$D$_3$ *in vitro* (Fig. 5). The influence of DBP [in fetal calf serum (FCS)] on 20-epi-1,25(OH)$_2$D$_3$- and 1,25(OH)$_2$D$_3$-induced human growth hormone (hGH) reporter gene expression is illustrated in Fig. 5C,D. Although the curve for 20-epi-1,25(OH)$_2$D$_3$ remains relatively unchanged by the presence of DBP, the curve for 1,25(OH)$_2$D$_3$ is reduced by an order of magnitude, exaggerating the potency difference between the two molecules. Dilworth *et al.* [53] postulated that the remaining difference in transactivation activity between MC1288 and 1,25(OH)$_2$D$_3$ is due in part to metabolic differences and in part to

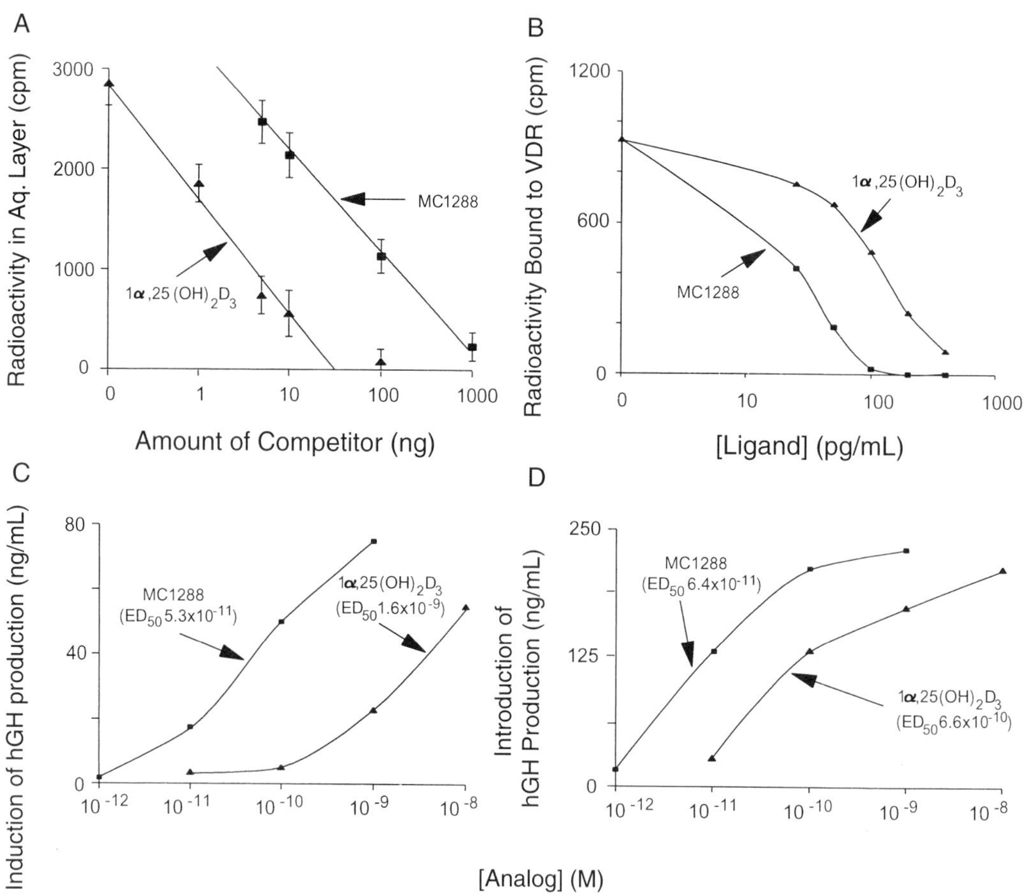

FIGURE 5 Biological parameters for 20-epi-1α,25(OH)$_2$D$_3$. (A) Ability of 20-epi-1α,25(OH)$_2$D$_3$ to compete for C-24 oxidation pathway enzymes. (B) VDR affinity of 20-epi-1α,25(OH)$_2$D$_3$ compared to 1α,25(OH)$_2$D$_3$. (C, with FCS; D, without FCS) Effect of fetal calf serum (containing DBP) on reporter gene induction by 20-epi-1α,25(OH)$_2$D$_3$ in the COS-1 cell line. Reprinted from Dilworth *et al.* [53] with permission.

affinity differences at the transcriptional level (i.e., VDR; VDR–RXR; VDR–RXR–VDRE).

It is interesting to note that Peleg *et al.* [54] (see Chapter 60) have extended this work to study the precise details of the gene transactivation events involved in 20-epi-1α,25$(OH)_2D_3$ action. They found that two 20-epi compounds, 20-epi-1α,25$(OH)_2D_3$ and KH1060, are unusual in that they are able to form highly stable, protease-resistant RXR–VDR–VDRE heterodimeric complexes *in vitro*. Whether the stability of these transactivating complexes is related to the increased transactivation ability in model reporter genes, although promising, is still to be established. Also interesting is the apparent correlation between the stability of 20-epi-1α,25$(OH)_2D_3$-containing transactivation complexes (observed by Peleg *et al.* [54]) and the metabolic stability of 20-epi-1α,25$(OH)_2D_3$ to target cell cytochrome P450 hydroxylation (observed by Dilworth *et al.* [53]). Whether this correlation holds for other potent, metabolically stable vitamin D analogs has yet to be investigated.

The 20-epi compound KH1060 is discussed later in Section II,G.

E. Homologated Analogs

Homologation involves the insertion of carbon atoms into the vitamin D side chain. It can take two different forms: (1) insertion of the carbon atoms in the main side chain between carbons 22 and 25 such that additional carbons are numbered 24a, 24b, 24c, etc., and (2) insertion of extra carbon atoms on the terminal methyl groups, making them into dimethyl (ethyl) groups (likened to lengthening the claws on a crab). These carbons are 26a, 27a, etc.

Homologated compounds were first developed in the early 1980s by Ikekawa and colleagues and initially tested for biological activity by the DeLuca group [55–57]. The Stern group showed that these analogs possess increased biological activity compared to 1α,25$(OH)_2D_3$ when assayed in a cultured bone model *in vitro* [58]. Whether metabolism of these homologated analogs is different from that of 1α,25$(OH)_2D_3$ or might play a role in the increased biological activity was initially unknown. The effect of of lengthening the side chain poses interesting problems for the enzymes involved in hydroxylation of the side chain. The active site of the cytochrome P450 is forced to accommodate a longer side chain, which might alter the efficiency of hydroxylation and, depending on the way the side chain is anchored, might change the site of hydroxylation (termed regioselectivity).

Dilworth *et al.* [59] examined this systematically by studying the effect on metabolism of adding one, two, or three carbons to the main vitamin D side chain. Homologs synthesized by the Leo chemist Martin Calverley and used by Dilworth *et al.* [59] with the 25-hydroxyl group already in place included MC1127 [24a-homo-1α,25$(OH)_2D_3$], MC1147 [24a,24b-dihomo-1α,25$(OH)_2D_3$], and MC1179 [24a,24b,24c-trihomo-1α,25$(OH)_2D_3$]. Only one compound was synthesized without the 25-hydroxyl group, and this was the 1αOHD_3 homolog with two extra carbon atoms: MC1281 (24a,24b-dihomo-1αOHD_3). The results obtained from metabolic studies using HPK1A-ras keratinocytes as a source of target cell 23- and 24-hydroxylase enzymes (presumably CYP24) and using hepatoma cells as a source of 25-hydroxylase (presumably CYP27) suggested that both cytochrome P450 isoforms continued to efficiently hydroxylate all homologs provided. Somewhat surprisingly, CYP24 maintained its hydroxylation sites at C-23 and C-24 despite the extension of the side chain by up to three carbons, seemingly, it preferred not to move down the side chain to C-24a, C-24b, or C-24c to be adjacent to the tertiary carbon, C-25. On the other hand, CYP27 hydroxylated MC1281 terminally at C-25 and C-27, appearing to ignore the longer internal side chain. These homologs therefore offer interesting insights into hydroxylation site selection by the vitamin D-specific cytochromes. Dilworth *et al.* [59] postulated that CYP24 must be directed to its hydroxylation site by the distance from the vitamin D ring structure, whereas CYP27 is directed by the distance of the hydroxylation site from the end of the side chain.

The same authors [60] also examined the effect of the introducing terminal 26,27-dimenthyl groups into the side chain of 1α,25$(OH)_2D_3$ to make the analog 26,27dimethyl-1α,25$(OH)_2D_3$ (MC1548). They found that MC1548 was metabolized at the same rate as 1α,25$(OH)_2D_3$ by the keratinocyte cell line HPK1A-ras. Products included metabolites of the C-24 oxidation pathway, and this was confirmed by the observation that MC1548 and 1α,25$(OH)_2D_3$ were equally effective in blocking the metabolism of [1β-^{3}H]1α,25$(OH)_2D_3$ to [1β-^{3}H]calcitroic acid. However, the products of MC1548 also included 26a-OH-MC1548 suggesting that the introduction of the dimethyl group into the side chain makes it susceptible to attack at a new terminal location by keratinocyte enzymes.

Although MC1548 is the simplest molecule in the dimethyl homologated series, it represents a valuable tool in understanding the relative importance of various modifications within complex homologated molecules such as EB1089 and KH1060. The latter molecules have greatly increased biological activity over 1α,25$(OH)_2D_3$ in cell differentiating systems *in vitro*, and elucidating the importance of metabolism to this increased potency *in vitro* should provide useful insights into vitamin D analog action [61]. The metabolism of EB1089 and KH1060 is discussed in Sections II,F and II,G, respectively.

F. Unsaturated Analogs

The idea of introducing a double bond(s) into the side chain of vitamin D analogs arose from experience with vitamin D_2. Vitamin D_2 metabolites have biological activities similar to those of vitamin D_3, so it appears that the introduction of the double bond is not deleterious. As mentioned already in Sections II,B and II,C, the metabolism of the side chain is significantly altered by this relatively minor change.

The modification has not been confined to the introduction of a C-22 to C-23 double bond. The Roche Company has developed molecules with two novel modifications, namely, introduction of a C-16 to C-17 double bond and introduction of a C-23 to C-24 triple bond. When these modifications are combined, the result is the highly successful 16-ene, 23-yne analog of $1\alpha,25(OH)_2D_3$ [62] (see Table II for structure). As alluded to earlier, Leo Pharmaceuticals has introduced the promising unsaturated analog EB1089, which contains a conjugated double bond system at C-22 to C-23 and C-24 to C-24a, in addition to both main side chain and terminal dimethyl types of homologation [63] (see Table II for structure). These two series of Roche and Leo compounds have shown strong antiproliferative activity both *in vitro* and *in vivo* [62,64,65].

Metabolism of the 16-ene compound by the perfused rat kidney has been studied by Reddy *et al.* [66] and is described in detail in Chapter 62. Reddy *et al.* [66] found that the introduction of the C-16 to C-17 double bond reduces 23-hydroxylation of the molecule, and the implication is that the D-ring modification must alter the conformation of the side chain sufficiently to subtly change the site of hydroxylation by CYP24, the cytochrome P450 thought to be responsible for 23- and 24-hydroxylation. It is worth noting that Dilworth *et al.* [53] also observed the absence of measurable 23-hydroxylation of the analog 20-epi-$1\alpha,25(OH)_2D_3$ in their studies, reinforcing the view that modifications around the C-17 to C-20 bond profoundly influence the rate of 23-hydroxylation.

The metabolism of the 16-ene,23-yne analog of $1\alpha,25(OH)_2D_3$ by WEHI-3 myeloid leukemic cells has been reported [67]. Though one might predict that because this molecule is blocked in the C-23 and C-24 positions it must be stable to C-24 oxidation pathway enzyme(s), it was found experimentally that the 16-ene,23-yne analog has the same $t_{1/2}$ as $1\alpha,25(OH)_2D_3$ when incubated with this cell line (~6.8 hr). The main product of $[25\text{-}^{14}C]1\alpha,25\text{-}(OH)_2\text{-}16\text{-ene-}23\text{-yne-}D_3$ was not identified by these workers but appeared to be more polar than the starting material. It will be interesting to see in any possible follow-up work if the metabolite which they have isolated has lost its C-23 to C-24 triple bond or is simply further hydroxylated at some alternative site in its side chain (e.g., at C-26). On the basis of knowledge emerging from other analogs, this author favors the latter possibility.

Another unsaturated analog that one might predict would be relatively metabolically stable is EB1089 with its conjugated double bond system (discussed more fully in Chapter 61). However, as pointed out earlier, EB1089 contains three structural modifications: the conjugated double bond system is accompanied by two types of side chain homologation. Nevertheless, as expected the conjugated double bond system dominates the metabolic fate of EB1089, there being no C-24 oxidation activity owing to the blocking action of the conjugated diene system. When metabolism is studied with either *in vitro* liver cell systems or the cultured keratinocyte cell line HPK1A-ras, disappearance of EB1089 is much slower than that of $1\alpha,25(OH)_2D_3$ [68,69]. Such data are consistent with the fairly long $t_{1/2}$ in pharmacokinetic studies *in vivo* [70]. Because the conjugated system of EB1089 blocks C-24 oxidation reactions, it is not surprising that a different site in the molecule becomes the target for hydroxylation. Diode-array spectrophotometry has allowed for the identification of the principal metabolic products of EB1089 as 26- and 26a-hydroxylated metabolites (Fig. 6) [69,71]. Note also that these metabolites of EB1089 have been chemically synthesized and shown to retain significant biological activity in cell differentiation and antiproliferation assays [71].

Again, it is interesting to note that with EB1089 and other molecules blocked in the C-23 and C-24 positions such as $1\alpha,24S(OH)_2D_2$ [43], the terminal carbons C-26 and C-26a become the alternative sites of further hydroxylation. However, it should also be noted that even in molecules not blocked in the C-23 and C-24 positions but containing the terminal 26-and 27-dimethyl homologation, such as 26,27-dimethyl-$1\alpha,25(OH)_2D_3$ (MC1548) [60], significant terminal 26a-hydroxylation seems to occur. Therefore, the hydroxylation of EB1089 at C-26 and C-26a may be in part a consequence of the introduction of the conjugated double bond system and in part a consequence of the introduction of the terminal homologation.

When the C-22 to C-23 double bond is present in the side chain in the absence of a C-24 methyl group, as in calcipotriol, the double bond appears vulnerable to reduction. As pointed out earlier, the principal metabo-

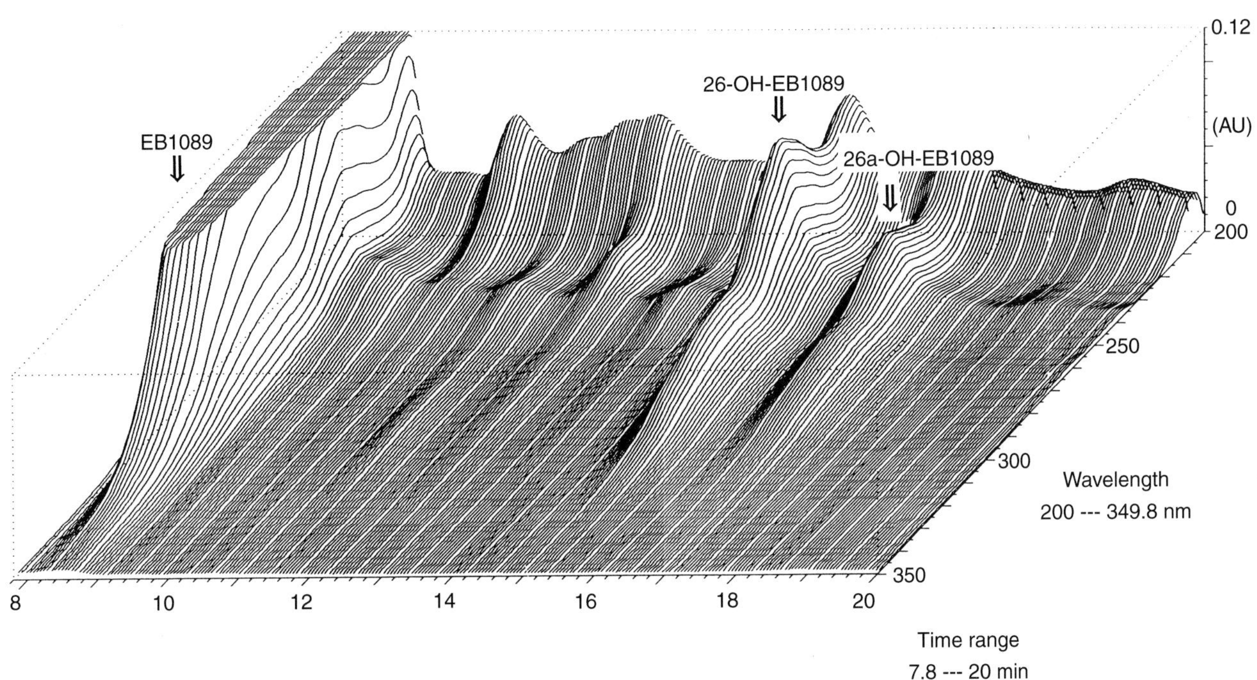

FIGURE 6 *In vitro* metabolism of EB1089 by HPK1A-ras cells, as shown by diode-array HPLC of lipid extracts following incubation of EB1089 (10 μM) with the human keratinocyte line HPK1A-ras for 72 hr. In addition to the substrate at 8.5 min, two metabolites showing the characteristic UV chromophore of EB1089 (λ_{max}235 nm, shoulder 265 nm) are visible in the part of the HPLC profile reproduced here (8–20 min). Metabolite peaks at 15.03 min and 16.55 min were isolated by extensive HPLC and identified [68,69,71] by comparison to synthetic standards on HPLC, gas chromatography–mass spectrometry (GC–MS), and nuclear magnetic resonance (NMR). Peak A at 15.03 min, 26-OH-EB1089; peak B at 16.55 min, 26a-OH-EB1089. From V. N. Shankar, A.-M. Kissmeyer, and G. Jones (unpublished results, 1994).

lites of calcipotriol are reduced in the C-22 to C-23 bond except for one, the C-22=C-23 unsaturated 24-ketone (MC1046) [21,15]. This suggests a C-24 ketone must be present to allow for this reduction to occur. Work of Wankadiya *et al.* [72] using the Roche compound Δ^{22}-1α,25$(OH)_2D_3$, an analog which contains the C-22 to C-23 double bond but lacks a C-24 substituent, tends to indirectly support this theory. When incubated with the chronic myelogenous leukemic cell line RWLeu-4, this molecule, like 1α,25$(OH)_2D_3$, is converted, presumably via Δ^{22}-1α,24,25$(OH)_3D_3$ and Δ^{22}-24-oxo-1α,25$(OH)_2D_3$, metabolites analogous to intermediates in the C-24 oxidation pathway, to the side chain-truncated product 24,25,26,27-tetranor-1,23$(OH)_2D_3$, a molecule which lacks the C-22 to C-23 double bond. However, Wankadiya *et al.* [72] did not identify intermediates in this process, and thus do not know at which stage C-22=C-23 reduction occurred.

G. Oxa Group-Containing Analogs

Oxa analogs involve the replacement of a carbon atom (usually in the side chain) with an oxygen atom. The best known of these are the 22-oxa analogs including 22-oxacalcitriol (OCT) [73] and KH1060 [74] discussed in Chapters 63 and 61, respectively. Lesser known analogs include the 23-oxa series [75–77] and 24-oxa-1α-hydroxyvitamin D_3 [78,79]. All of these molecules are metabolically fascinating to study because the oxa atom makes the molecule inherently unstable should it be hydroxylated at the adjacent carbon atom. Hydroxylation at an adjacent carbon generates an unstable hemiacetal that spontaneously breaks down to eliminate the carbons distal to the oxa group. In the case of the 22-oxa compounds the expected product(s) would be C-20 alcohol/ketone; for the 23-oxa compounds the expected product(s) would be C-22 alcohol/ketone; and in the 24-oxa compounds the expected product(s) would be C-23 alcohol/ketone/acid.

The metabolism of OCT has been extensively studied in a number of different biological systems, including primary parathyroid [18] and keratinocyte cells [80] as well as cultured osteosarcoma, hepatoma, and keratinocyte cell lines [19]. In all these systems, OCT is rapidly broken down. As outlined earlier, the use of two different radioactive labels in [26-^{3}H]OCT and [2β-^{3}H]OCT enabled Brown *et al.* [18] to suggest that the side chain was truncated, though definitive proof of the identity of the products was not immediately forthcoming. It was not until the work of Masuda *et al.* [19] that the principal metabolites were unequivocably identified by gas chromatography–mass spectrometry (GC–MS) as 24-OH-OCT, 26-OH-OCT, and hexanor-1α,20-dihydroxyvitamin D_3. In the case of the keratinocyte HPK1A-ras, an additional product, hexanor-20-oxo-1α-hydroxyvitamin D_3, is formed. The latter two truncated products are suggestive of hydroxylation of OCT at the C-23 position to give the theoretical unstable intermediate postulated at the beginning of the studies. Although all of these products were isolated from *in vitro* systems, there is evidence that the processes also occur *in vivo*; Kobayashi *et al.* [81] have generated data which suggests that the biliary excretory form of OCT in the rat is a glucuronide ester of the truncated 20-alcohol.

The metabolism of 24-oxa-1α-hydroxyvitamin D_3 (MC1090) has been studied at micromolar concentrations using the hepatoma Hep3B [78]. As expected, 24-oxa-1αOHD_3 was found to be converted in high yield to two truncated products; tetranor-1,23$(OH)_2D_3$ and calcitroic acid, again suggesting hydroxylation at the C-25 position adjacent to the 24-oxa group which results in an unstable intermediate. The products were again identified by GC–MS.

Both the above examples of simple oxa analogs provide useful knowledge that can help in predicting the metabolic fate of a complex oxa analog such as KH1060. This highly potent compound, which possesses *in vitro* cell differentiating activity exceeding that of any other analog synthesized to date, has four different modifications to the side chain of 1α,25$(OH)_2D_3$, namely, (1) a 22-oxa group, (2) the 20-epi side chain stereochemistry, (3) 24a-homologation, and (4) 26- and 27-dimethyl homologation (see Table II for structure). Because all these changes are known to affect biological activity *in vitro* and *in vivo* as well as side chain metabolism, it comes as no surprise that the metabolism of KH1060 is extremely complex. KH1060 has a very short $t_{1/2}$ in pharmacokinetic studies *in vivo*, giving a metabolic profile with at least 16 unknown metabolites [70]. Dilworth *et al.* [82] reported the first *in vitro* study using micromolar concentrations of KH1060 incubated with the keratinocyte cell line HPK1A-ras. Dilworth *et al.* [82] were able to discern 22 different metabolites after multiple high-performance liquid chromatography (HPLC) steps and assigned structures to 12 of them (see Fig. 7). As would be expected from consideration of the studies of other oxa compounds, two of these were truncated products identical to the molecules formed from another 22-oxa compound, OCT. As would be expected from consideration of the studies of other homologated compounds (see Section II,E), other products are hydroxylated at specific carbons of the side chain, including C-26 and C-26a. As with EB1089 and MC1548, the presence of dimethyl groups in the terminus of the side chain appears to attract hydroxylation to this site in KH1060. One novel metabolite found only for KH1060 is 24a-OH-KH1060, observed both in broken cell and intact cell models [82,83].

One important facet of this complex metabolic profile is that rather than simplifying our understanding of the mechanism of action of KH1060 the data complicate it. This is because biological assays of each of the metabolic products have shown that several of the principal and long-lived metabolic products of KH1060 retain significant vitamin D-dependent gene-inducing activity in reporter gene expression systems [84]. This point is discussed further in Section III,B) on the implications of metabolic studies to the mechanism of action of analogs.

H. Other Analogs

Several other analogs synthesized to date do not fit readily into the groups discussed thus far and yet have been studied metabolically. These are listed briefly in this section.

1. 1α,24$R(OH)_2D_3$

The compound 1α,24$R(OH)_2D_3$ is marketed as an antipsoriatic drug by Teijin, and its structure is shown in Table II. The metabolism of [1β-^{3}H]1α,24$R(OH)_2D_3$ by isolated perfused rat kidney was described by Reddy *et al.* [85]. Interestingly, Reddy *et al.* [85] found two distinct pathways of metabolism, a major one involving direct C-24 oxidation (24-oxidation, 23-hydroxylation, side chain cleavage) and resulting in calcitroic acid formation and a second minor pathway involving first 25-hydroxylation to 1α,24R,25$(OH)_3D_3$ followed by the same C-24 oxidation steps occurring in the catabolism of 1α,25$(OH)_2D_3$ (see Fig. 1). Because the authors used both pharmacological as well as "physiological" concentrations of the analog, they were able to generate sufficient quantities to identify their products by mass spectrometry. Their finding of 25-hydroxylase activity in the kidney is not a complete surprise, as others [86] have detected the enzyme activity there and CYP27 mRNA is detected in rat kidney tissue [87]. However, a more recent follow-up study by the Reddy group [88] suggests that the renal 25-hydroxylase which is able to hydroxylate 1α,24$R(OH)_2D_3$ is unable to 25-hydroxylate the compounds vitamin D_3 and 1αOHD_3. The implication of this work is that 1α,24$R(OH)_2D_3$ is metabolized by a nonspecific enzyme to intermediates of the C-24 oxida-

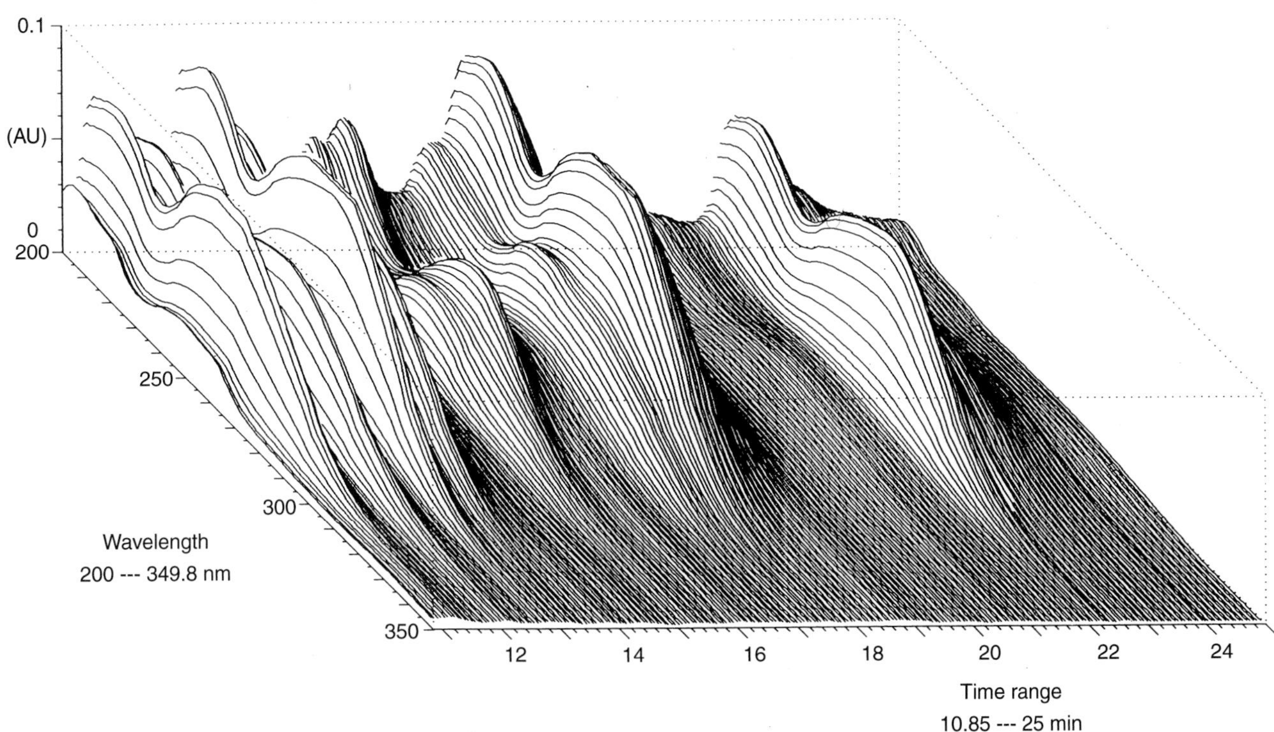

FIGURE 7 *In vitro* metabolism of KH1060 by HPK1A-ras cells, as shown by diode-array HPLC of lipid extracts following incubation of KH1060 (10 μM) with the human keratinocyte line HPK1A-ras for 72 hr. Nine peaks showing the characteristic UV chromophore of vitamin D (λ_{max}265 nm, λ_{min}228 nm) are visible in the part of the HPLC profile reproduced here (11–25 min). Rechromatography on a second HPLC system resulted in the further resolution of the 9 peaks into 22 separate metabolites. Many of the metabolites were identified [82] by comparison to synthetic standards on HPLC and GC-MS. Peak at 13.39 min, 24a-OH-KH1060; peak at 22.14 min, 26-OH-KH1060. From F. J. Dilworth, A.-M. Kissmeyer, and G. Jones (unpublished results, 1995).

tion pathway of $1\alpha,25(OH)_2D_3$. If this is the case *in vivo* one might anticipate the potential for the drug $1\alpha,24R(OH)_2D_3$ to interfere with normal $1\alpha,25(OH)_2D_3$ catabolism.

2. ED-71

The structure of ED-71 is also shown in Table II. It is unusual in that it represents one of the few A-ring-modified analogs whose metabolism has been studied to date. It possesses a unique 2β-hydroxypropoxy group, in addition to the usual 1α-,3β-, and 25-hydroxy groups of $1\alpha,25(OH)_2D_3$. The extra bulky group at the 2β position has the effects of improving the affinity of ED-71 for the plasma binding protein DBP, but as a consequence may make it more difficult for the analog to enter target cells. There is some optimism that these unique properties may make ED-71, a "long-lived vitamin D" and therefore show the necessary properties to be suitable as an antiosteoporosis drug [89]. Because ED-71 has the normal side chain of $1\alpha,25(OH)_2D_3$, it should be susceptible to the same C-24 oxidation sequence and other pathways as the natural hormone. Indeed, initial studies have provided evidence for the formation of several of the same 24- and 26-hydroxylated and 24-oxidized products as $1\alpha,25(OH)_2D_3$ but at a much reduced rate [90]. There is some possibility for metabolism of the 2β group by nonspecific enzymes, although Masuda *et al.* [90] failed to observe such catabolites.

3. 26,26,26,27,27,27-HEXAFLUORO-$1\alpha,25(OH)_2D_3$

The analog 26,26,26,27,27,27-hexafluoro-1α,25-dihydroxy vitamin D_3 [26,27-F_6-$1\alpha,25(OH)_2D_3$] was first synthesized in the early 1980s [91], along with a number of other side chain fluorinated analogs, to test the importance of certain key hydroxylation sites [e.g., C-23, C-24, C-25, C-26(27), C-1] to biological activity. It was noted immediately that 26,27-F_6-$1\alpha,25(OH)_2D_3$ was extremely potent [10-fold higher than $1\alpha,25(OH)_2D_3$] in calcemia assays both *in vitro* and *in vivo* [92–94]. Lohnes and Jones [95] presented evidence using a bone cell line, UMR-106, that 26,27-F_6-$1\alpha,25(OH)_2D_3$ had a longer $t_{1/2}$ inside target cells due to the apparent lack of 24-hydroxylation of 26,27-F_6-$1\alpha,25(OH)_2D_3$. At around the same time, Morii and co-workers noted the appearance of a metabolite of 26,27-F_6-$1\alpha,25(OH)_2D_3$, which they have identified as 26,27-F_6-1α,23,25$(OH)_3D_3$ [96]. This compound has excellent calcemic activity in its own right, but whether this derivative is in part responsible for the biological activity of 26,27-F_6-1α,25$(OH)_2D_3$ is not conclusively proven. Nonetheless, 26,27-F_6-1α,25$(OH)_2D_3$ has undergone clinical trials for hypocalcemia associated with hypoparathyroidism and uremia [97,98].

4. OTHER ANALOGS

As further generations of analogs emerge they will likely be studied metabolically. Each major new modification must be tested for its impact on the metabolic machinery as well as on biological activity. There are some compounds listed in Table II that are currently untested, or for which the test results have not been published. These include A-ring-modified analogs such as 19-nor-1α,25$(OH)_2D_2$ [99] and 1-(hydroxyalkyl)-25OHD_3 [100], as well as the side chain-modified analog 20-methyl-1α,25$(OH)_2D_3$ [101].

III. IMPORTANT IMPLICATIONS DERIVED FROM ANALOG METABOLISM STUDIES

A. Correlations with Pharmacokinetic Information

The susceptibility of a vitamin D analog to metabolism and excretion undoubtedly plays a significant role in determining the biological activity of that analog *in vivo*. The quickest and easiest way to acquire such knowledge is by pharmacokinetic analysis. From a classic vitamin D outlook preferred by this author, pharmacokinetic data reflect a few important parameters regarding each analog, including the following:

1. The affinity of the vitamin D analog for DBP in the bloodstream
2. The rate of target cell uptake and metabolism by target cell enzymes
3. The rate of liver cell uptake, hepatic metabolism, and biliary clearance
4. The rate of storage depot uptake and release

Metabolism, whether target cell or liver, is reflected in only two components of this list of factors measured by pharmacokinetics. Therefore, it cannot be expected that *in vitro* metabolic parameters would exactly correlate with *in vivo* pharmacokinetic parameters. Nevertheless, a comparison of the two might be worthwhile. In the case of some of the analogs shown in Tables I and II, pharmacokinetic data [70,102,103] are available and can be compared to the data provided by *in vitro* metabolic studies.

In Table IV an attempt is made to compare pharmacokinetic data from these sources with DBP binding data and target cell metabolic data. As pointed out by Kissmeyer *et al.* [70], on the basis of their pharmacokinetic parameters the compounds segregate into at least two and perhaps more groups: calcemic and noncalcemic analogs. Calcemic analogs (strong or weak) are those analogs with a long $t_{1/2}$, which is a function of either a strong DBP binding or a reduced rate of metabolism (or both). The analog ED-71 has a strong DBP binding affinity. There appear to be a group of analogs in which a long $t_{1/2}$ is correlated with a slower rate of metabolism [e.g., EB1089, ED-71, 1α,24*S*$(OH)_2D_2$]. With the exception of ED-71, most of these active analogs bind DBP poorly. Noncalcemic analogs are those analogs with a short $T_{1/2}$, which is a function of either poor DBP binding or a rapid rate of metabolism (or both). Examples include calcipotriol, KH1060, and OCT.

It should be noted that although these classifications are used in the vitamin D literature they are somewhat artificial since pure "noncalcemic" analogs do not yet exist. All "noncalcemic" analogs will cause hypercalcemia if their concentration is raised sufficiently. The crucial issue is whether systemically administered, "weakly calcemic" or "noncalcemic" analogs can produce their anti-cell proliferation/pro-cell differentiation effects *in vivo* at concentrations lower than that required to produce calcemia. Various *in vivo* clinical trials currently in progress will be the acid test for this question.

B. Implications for Mechanism of Action of Vitamin D Analogs

There is currently tremendous interest in defining the mechanism of action of vitamin D analogs, particularly for clarifying the difference between calcemic and noncalcemic analogs. Chapters 59 and 60 discuss other aspects of analog action in detail. As is obvious from the amount of space committed to this mechanism, the problem is not simple but multifactorial. Therefore, in this chapter this section focuses on the importance of metabolism to the complex picture.

Metabolism can have an impact on the mode of action in a few different ways:

1. Lack of high affinity binding to DBP in the blood can make the analog vulnerable to liver enzymes,

which may lead to deactivation and excretion of the analog.

2. Target cell enzymes may activate or deactivate the administered analog (a) to metabolites that possess increased, equivalent, or slightly decreased biological activity at target genes, and which may have an extended $t_{1/2}$ inside the cells, or (b) to metabolites that possess much reduced biological activity.
3. The rate of metabolism by target cell enzymes may be influenced by the rate of entry of "free" analog from the cell exterior and the association and dissociation rates of VDR–RXR–DNA complexes [104].

Chapter 59 stresses the importance of DBP binding and pharmacokinetics. It is clear that hepatic metabolism (point 1 above) relates mainly to this chapter. Chapter 60 focuses on the mechanism by which analogs interact with VDR–RXR heterodimeric complexes and ultimately with the vitamin D-dependent gene. Points 2 and 3 on target cell metabolism relate mainly to Chapter 60.

It should be noted with regard to molecular mechanisms of action at the target cell level that metabolism is often disregarded or given too little emphasis. Furthermore, certain metabolic assumptions are made when testing biological activity that are not always valid. These include the following: (1) the analog is biologically active as administered; and (2) the analog is stable in the *in vitro* target cell model used, whether *in vitro* organ culture, cultured target cell, or host cell/reporter gene construct. The validity of this approach is made even more tenuous when data acquired with different

TABLE IV Pharmacokinetic Data on Vitamin D Analogs

Compound[a]	Serum concentration at t = 5 min (ng/ml)	$t_{1/2}$ (hr)	AUC∞ (ng/ml × hr)	Serum clearance (ml/hr/kg)	Binding affinity for DBP (M)	Relative binding affinity for DBP[b]	Rate of metabolism[c]
$25OHD_3$	2040	>2.8	9596	21	9×10^{-9}	33	Very slow
$1\alpha,25(OH)_2D_3$	2429	2.2	7355	27	$1.5–6.0 \times 10^{-7}$	1	Fast
$1\beta,25(OH)_2D_3$	2912	>4	13,228	15	1.7×10^{-8}	17	—
Calcipotriol	121	0.2	27	7407	1.5×10^{-6}	0.1	Very fast
MC 1127	545	1.6	1216	167	5.2×10^{-6}	0.1	Fast
EB 1089	152	2.1	255	784	7.9×10^{-6}	0.03	Slow
CB 966	176	1.8	267	693	3.2×10^{-5}	0.02	—
KH 1139	154	0.7	142	1408	6.5×10^{-5}	0.007	—
KH 1060	103	0.4	46	4348	n.b.	0	Fast
KH 1049	104	0.5	40	5000	n.b.	0	—

Compound[d]	$t_{1/2}$ (hr)	Metabolic clearance rate (ml/min)	Relative binding affinity for DBP[e]	Rate of metabolism[f]
$1\alpha,25(OH)_2D_3$	7.0	5.0	1	Fast
OCT	2.5	48.2	266	Very fast

Compound[a]	Baseline (pg/ml)	$t_{1/2}$ (hr)	AUC∞ (pg/ml × hr)	Relative binding affinity for DBP[h]	Rate of metabolism[i]
$1\alpha,25(OH)_2D_3$	67.0	5.8	3690	1	Fast
$1\alpha,25(OH)_2D_2$	<10	5.1	2676	—	—
$1\alpha,24(OH)_2D_2$	<10	4.9	659	14	Slow

[a] Activity was measured following a single intravenous dose of 200 μg/kg to rats. From Kissmeyer *et al.* [70].
[b] Using human DBP, the relative numbers are compared to the $1\alpha,25(OH)_2D_3$ value. n.b., No binding. From Kissmeyer *et al.* [70].
[c] From Masuda *et al.* [15], Dilworth *et al.* [53,82], and Shankar *et al.* [69].
[d] Following a single intravenous dose of 100 ng/animal to dogs. From Dusso *et al.* [103].
[e] Using rat DBP, the relative numbers are compared to the $1\alpha,25(OH)_2D_3$ value. From Dusso *et al.* [103].
[f] From Masuda *et al.* [19].
[g] Following a single oral dose of 0.39 μg/kg to rats. From Knutson *et al.* [123].
[h] Using rat DBP. From Strugnell *et al.* [42].
[i] From Jones *et al.* [43].

in vitro models where metabolic considerations may or may not apply are compared to data acquired *in vivo* where metabolic considerations definitely apply. The reader is cautioned that invalid comparisons of *in vivo* and *in vitro* data abound in this field.

In the opinion of this author, metabolism will turn out to be a key parameter but not the only important parameter in vitamin D analog action. It is our view that only when we consider all of the parameters which can influence analog action within the overall equation will we then be in a position to fully understand the molecular mechanisms underlying their "noncalcemic" or "calcemic" actions. From this it seems unlikely that there will be two sets of such parameters providing perfect noncalcemic and calcemic analogs, as is presented in Section III,A, but rather several different permutations of the same parameters giving rise to analogs with slightly different applications.

C. Implications for Future Drug Design

A case for the importance of metabolism within vitamin D analog action has been presented throughout this chapter. Much has been learned about modifications to the vitamin D molecule that change metabolism but in the process also improve biological activity. For some applications of vitamin D, this involves the concept of making "metabolism-resistant analogs" (e.g., those blocked in the C-23, C-24, C-26, or C-27 positions) that possess enhanced calcemic activity. For other applications of vitamin D, this involves the concept of making "metabolism-sensitive analogs" (e.g., those with oxa groups at key side chain positions or a C-22=C-23 juxtaposed with a 24-hydroxyl) to localize the biological effect to the site of analog administration. Over the immediate future we can anticipate the following:

1. A search for additional novel synthetic modifications to the vitamin D side chain and ring structure
2. Continuation of the trend to combine "useful" modifications in order to fine-tune the best analogs currently available
3. Synthesis of "smart" molecules where metabolic and structure–activity information gained from earlier generations of molecules is used to improve existing analogs
4. A search for novel chemical entities that mimic certain selective actions of the vitamin D molecule

One can envision that the VDR binding pocket studies and cytochrome P450-substrate binding pocket studies will provide particularly valuable information to the design of further generations of vitamin D analogs or new mimics. The reader is referred to Chapters 8, 9, and 10 for the lastest information on VDR structure and modeling. The final section of this chapter outlines progress in the area of modeling vitamin D related-cytochrome P450 isoforms.

D. Cytochrome P450 Isoform Modeling Studies

The cytochrome P450 superfamily constitutes a group of over 100 proteins, subdivided into microsomal and mitochondrial isoforms, that are responsible for the metabolism (e.g., hydroxylation) of endogenous and exogenous (xenobiotic) compounds [105]. Their structure is well conserved across the superfamily, with domains for heme binding, ferredoxin binding, O_2 binding, and substrate binding. These proteins are membrane associated and are thus not easily studied by X-ray crystallographic means. For several mammalian steroidal cytochrome P450 isoforms (e.g., aromatase, cholesterol side-chain cleavage enzyme, 17-hydroxylase, rat 2B1, human 2D6), modeling studies (e.g., [106,107]) have begun based on information derived from crystal structures of soluble prokaryotic cytochromes P450 (CAM, BM-3, TERP, EryF). Thus, for the mammalian cytochromes P450 this work is in its infancy. Nevertheless, the approach appears highly promising. Such models, despite being crude first approximations, allow for identification of putative active site residues suitable for site-directed mutagenesis studies. Refinements of the model derived from mutant proteins then follow.

In the case of CYP27, such an approach [108] can now be contemplated not only using the information derived from the primary amino acid sequence [109] and other modeled cytochrome P450s, but also using the information derived from natural mutations resulting in the human disease cerebrotendinous xanthomatosis [110]. (Note: No disease has currently been recognized involving mutations of CYP24.) Modeling of the CYP27 protein (Fig. 8; see color insert), together with quantitative structure–activity relationship (QSAR) studies of the vitamin D analogs/steroids docking into the substrate binding pocket of CYP27, and actual studies of the substrate preferences (e.g., [5]) should permit a much clearer picture of the analog–cytochrome P450 interactions occurring *in vivo*. It is hoped that knowledge generated using this approach will eventually result in more rational vitamin D analog design in the future.

Acknowledgments

The author thanks F. Jeffrey Dilworth and David Prosser in compilation of the references, tables, and figures contained in this chapter.

Some of the work cited here is supported through grants to the author from the Medical Research Council of Canada. The collaborative research reviewed here requires the input of the excellent research trainees David Lohnes, Fuad Qaw, Stephen Strugnell, Sonoko Masuda, and V. N. Shankar, as well as the interdisciplinary involvement of talented scientists from around the world. I gratefully acknowledge the essential contributions to our analog work of Hugh L. J. Makin, Martin Calverley, Joyce Knutson, Charles Bishop, Noboru Kubodera, Anne-Marie Kissmeyer, Richard Kremer, Mark R. Haussler, and Hector F. DeLuca.

References

1. DeLuca HF 1988 The vitamin D story: A collaborative effort of basic science and clinical medicine. FASEB J **2**:224–236.
2. Bouillon R, Okamura WH, Norman AW 1995 Structure–function relationships in the vitamin D endocrine system. Endocr Rev **16**:200–257.
3. Calverley MJ, Jones G 1992 Vitamin D. In: Blickenstaff RT (ed) Antitumor Steroids. Academic Press, San Diego, pp. 193–270.
4. Okuda KI, Usui E, Ohyama Y 1995 Recent progress in enzymology and molecular biology of enzymes involved in vitamin D metabolism. J Lipid Res **36**:1641–1652.
5. Guo Y-D, Strugnell S, Back DW, Jones G 1993 Transfected human liver cytochrome P-450 hydroxylates vitamin D analogs at different side-chain positions. Proc Natl Acad Sci USA **90**:8668–8672.
6. Adams JS, Gacad MA 1985 Characterization of 1α-hydroxylation of vitamin D_3 sterols by cultured alveolar macrophages from patients with sarcoidosis. J Exp Med **161**:755–765.
7. Ohyama Y, Noshiro M, Okuda K 1991 Cloning and expression of cDNA encoding 25-hydroxyvitamin D_3 24-hydroxylase. FEBS Lett **278**:195–198.
8. Makin G, Lohnes, D, Byford V, Ray R, Jones G 1989 Target cell metabolism of 1,25-dihydroxyvitamin D_3 to calcitroic acid. Evidence for a pathway in kidney and bone involving 24-oxidation. Biochem J **262**:173–180.
9. Reddy GS, Tserng K-Y 1989 Calcitroic acid, end product of renal metabolism of 1,25-dihydroxyvitamin D_3 through C-24 oxidation pathway. Biochemistry **28**:1763–1769.
10. Akiyoshi-Shibata M, Sakaki T, Ohyama Y, Noshiro M, Okuda K, Yabusaki Y 1994 Further oxidation of hydroxycalcidiol by calcidiol 24-hydroxylase—A study with the mature enzyme expressed in *Escherichia coli.* Eur J Biochem **224**:335–343.
11. Beckman M, Prahl JM, DeLuca HF 1995 Side chain metabolism of [26,27-^{3}H]-25-hydroxycholecalciferol (25-OH-D3) by recombinant human 24-hydroxylase (p450cc24) expressed in baculovirus infected *Spodoptera frugiperda* (SF21) insect cells. J Bone Miner Res **10**:S395 (Abstract M573).
12. Shinki T, Jin CH, Nishimuar A, Nagai Y, Ohyama Y, Noshiro M, Okuda K, Suda T 1992 Parathyroid hormone inhibits 25-hydroxyvitamin D_3-24-hydroxylase mRNA expression stimulated by 1α,25-dihydroxyvitamin D_3 in rat kidney but not in intestine. J Biol Chem **267**:13757–13762.
13. Tomon M, Tenenhouse HS, Jones G 1990 Expression of 25-hydroxyvitamin D_3-24-hydroxylase activity in CaCo-2 cells. An in vitro model of intestinal vitamin D catabolism. Endocrinology **126**:2868–2875.
14. Lohnes D, Jones G 1987 Side-chain metabolism of vitamin D in osteosarcoma cell line UMR-106: Characterization of products. J Biol Chem **262**:14394–14401.
15. Masuda S, Strugnell S, Calverley MJ, Makin HLJ, Kremer R, Jones, G 1994 *In vitro* metabolism of the anti-psoriatic vitamin D analog, calcipotriol, in two cultured human keratinocyte models. J Biol Chem **269**:4794–4803.
16. Yamada S, Nakayama K, Takayama H, Shinki T, Takasaki Y, Suda T 1984 Isolation, identification and metabolism of (23*S*,25*R*)-25-hydroxyvitamin D_3-26,23-lactol: A biosynthetic precursor of (23*S*,25*R*)-25-hydroxyvitamin D_3-26,23-lactone. J Biol Chem **259**:884–889.
17. Horst RL, Wovkulich PM, Baggiolini EG, Uskokovic MR, Engstrom GW, Napoli JL 1984 (23*S*)-1,23,35-Trihydroxyvitamin D_3: Its biological activity and role in 1α,25-dihydroxyvitamin D_3-26,23-lactone biosynthesis. Biochemistry **23**:3973–3976.
18. Brown AJ, Berkoben M, Ritter C, Kubodera N, Nishii Y, Slatopolsky E 1992 Metabolism of 22-oxacalcitriol by a vitamin D-inducible pathway in cultured parathyroid cells. Biochem Biophys Res Commun **189**:759–764.
19. Masuda S, Byford V, Kremer R, Makin HLJ, Kubodera N, Nishii Y, Okazaki A, Okano T, Kobayashi T, Jones G 1996 *In vitro* metabolism of the vitamin D analog, 22-oxacalcitriol, using cultured osteosarcoma, hepatoma and keratinocyte cell lines. J Biol Chem **271**:8700–8708.
20. Jones G 1986 Elucidation of a new pathway of 25-hydroxyvitamin D_3 metabolism. Meth Enzymol **123**:141–154.
21. Sorensen H, Binderup L, Calverley MJ, Hoffmeyer L, Rastrup Anderson N 1990. *In vitro* metabolism of calcipotriol (MC903), a vitamin D analogue. Biochem Pharmacol **39**:391–393.
22. Jones G, Lohnes D, Strugnell S, Guo Y-D, Masuda S, Byford V, Makin HLJ, Calverley MJ 1994 Target Cell Metabolism of Vitamin D and Its Analogs. In: Norman AW, Bouillon R, Thomasset M (eds) Vitamin D. A Pluripotent Steroid Hormone: Structural Studies, Molecular Endocrinology and Clinical Applications. de Gruyter, Berlin, pp. 161–169.
23. Bligh EG, Dyer WJ 1957 A rapid method of total lipid extraction and purification. Can J Biochem **37**:911–917.
24. Suda T, Hallick RB, DeLuca HF, Schnoes HK 1970 25-Hydroxydihydrotachysterol$_3$ synthesis and biological activity. Biochemistry **9**:1651–1657.
25. Bosch R, Versluis C, Terlouw JK, Thijssen JHH, Duursma SA 1985 Isolation and identification of 25-hydroxydihydrotachysterol$_2$, 1α,25-dihydroxydihydrotachysterol$_2$ and 1β,25-dihydroxydihydrotachysterol$_2$. J Steroid Biochem **23**:223–229.
26. Jones G, Edwards N, Vriezen D, Porteous C, Tafford DJH, Cunningham J, Makin HLJ 1988 Isolation and identification of seven metabolites of 25-hydroxydihydrotachysterol$_3$ formed in the isolated perfused rat kidney: A model for the study of side-chain metabolism of vitamin D. Biochemistry **27**:7070–7079.
27. Qaw F, Calverley MJ, Schroeder NJ, Trafford DJH, Makin HLJ, Jones G 1993 *In vivo* metabolism of the vitamin D analog, dihydrotachysterol. Evidence for formation of 1α,25- and 1β,25-dihydroxydihydrotachysterol metabolites and studies of their biological activity. J Biol Chem **268**:282–292.
28. Bhattacharyya MH, DeLuca HF 1973 Comparative studies on the 25-hydroxylation of vitamin D_3 and dihydrotachysterol$_3$. J Biol Chem **248**:2974–2977.
29. Wing RM, Okamura WH, Pirio MP, Sine SM, Norman AW 1974 Vitamin D in solution: Conformations of vitamin D_3, 1,25-dihydroxyvitamin D_3 and dihydrotachysterol$_3$. Science **186**:939–941.
30. Qaw F, Schroeder NJ, Calverley MJ, Maestro M, Mourino A, Trafford DJH, Makin HLJ, Jones G 1992 *In vitro* synthesis of 1,25-dihydroxydihydrotachysterol in the myelomonocytic cell line, HD-11. J Bone Miner Res **7**:S161 (Abstract 274).
31. Shany S, Ren S-Y, Arbelle JE, Clemens TL, Adams JS 1993 Subcellular localization and partial purification of the 25-hy-

droxyvitamin D-1-hydroxylation reaction in the avian myelomonocytic cell line HD-11. J Bone Miner **8**:269–276.
32. Qaw FS, Makin HLJ, Jones G 1992 Metabolism of 25-hydroxydihydrotachysterol$_3$ in bone cells *in vitro*. Steroids **57**:236–243.
33. Schroeder NJ, Qaw F, Calverley MJ, Trafford DJH, Jones G, Makin HLJ 1992 Polar metabolites of dihydrotachysterol$_3$ in the rat: Comparison with *in vitro* metabolites of 1α,25-dihydroxydihydrotachysterol$_3$. Biochem Pharmacol **43**:1893–1905.
34. Suda T, DeLuca HF, Schnoes HK, Blunt JW 1969 Isolation and identification of 25-hydroxyergocalciferol. Biochemistry **8**:3515–3520.
35. Jones G, Schnoes HK, DeLuca HF 1975 Isolation and identification of 1,25-dihydroxyvitamin D$_2$. Biochemistry **14**:1250–1256.
36. Jones G, Rosenthal A, Segev D, Mazur Y, Frolow F, Halfon Y, Rabinovich D, Shakked Z 1979 Isolation and identification of 24,25-dihydroxyvitamin D$_2$ using the perfused rat kidney. Biochemistry **18**:1094–1101.
37. Jones G, Schnoes HK, Levan L, DeLuca HF 1980 Isolation and identification of 24-hydroxyvitamin D$_2$ and 24,25-dihydroxyvitamin D$_2$. Arch Biochem Biophys **202**:450–457.
38. Horst RL, Koszewski NJ, Reinhardt TA 1990 1α-Hydroxylation of 24-hydroxyvitamin D$_2$ represents a minor physiological pathway for the activation of vitamin D$_2$ in mammals. Biochemistry **29**:578–582.
39. Koszewski NJ, Reinhardt TA, Napoli JL, Beitz DC, Horst RL 1988 24,26-Dihydroxyvitamin D$_2$: A unique physiological metabolite of vitamin D$_2$. Biochemistry **27**:5785–5790.
40. Reddy GS, Tserng K-Y 1986 Isolation and identification of 1,24,25-trihydroxyvitamin D$_2$, 1,24,25,28-tetrahydroxyvitamin D$_2$, and 1,24,25,26-tetrahydroxyvitamin D$_2$: New metabolites of 1,25-dihydroxyvitamin D$_2$ produced in the rat kidney. Biochemistry **25**:5328–5336.
41. Lam HY, Schnoes HK, DeLuca HF 1974 1α-Hydroxyvitamin D$_2$: A potent synthetic analog of vitamin D$_2$. Science **186**:1038–1040.
42. Strugnell S, Byford V, Makin HLJ, Moriarty RM, Gilardi R, LeVan LW, Knutson JC, Bishop CW, Jones G 1995 1α,24(*S*)-dihydroxyvitamin D$_2$: A biologically active product of 1α-hydroxyvitamin D$_2$ made in the human hepatoma, Hep3B. Biochem J **310**:233–241.
43. Jones G, Byford V, Kremer R, Makin HLJ, Rice RH, deGraffenreid LA, Knutson JC, Bishop CA 1996 Anti-proliferative activity and target cell catabolism of the vitamin D analog, 1α,24(*S*)-dihydroxyvitamin D$_2$ in normal and immortalized human epidermal cells. Biochem Pharmacol **52**:133–140.
44. Sjoden G, Smith C, Lindgren V, DeLuca HF 1985 1-Alpha-hydroxyvitamin D$_2$ is less toxic than 1-alpha-hydroxyvitamin D$_3$ in the rat. Proc Soc Exp Biol Med **178**:432–436.
45. Horst RL, Koszewski NJ, Reinhardt TA 1988 Species variation of vitamin D metabolism and action: Lessons to be learned from farm animals. In: Norman AW, Schaefer K, Grigoleit H-G, von Herrath D (eds) Vitamin D: Molecular, Cellular and Clinical Endocrinology. de Gruyter, Berlin, pp. 93–101.
46. Clark JW, Reddy GS, Santos-Moore A, Wankadiya KF, Reddy GP, Lasky S, Tserng K-Y, Uskokovic MR 1993 Metabolism and biological activity of 1,25-dihydroxyvitamin D$_2$ and its metabolites in a chronic myelogenous leukemia cell line, RWLEU-4. Bioorg Med Lett **3**:1873–1878.
47. Strugnell S, Calverley MJ, Jones G 1990 Metabolism of a cyclopropane-ring-containing analog of 1α-hydroxy vitamin D$_3$ in a hepatocyte cell model: Identification of 24-oxidised metabolites. Biochem Pharmacol **40**:333–341.
48. Calverley MJ 1987 Synthesis of MC903, a biologically active vitamin D metabolite analog. Tetrahedron **43**:4609–4619.
49. Jones G, Calverley MJ 1993 A dialogue on analogues: Newer vitamin D-drugs for use in bone disease, psoriasis, and cancer. Trends Endocrinol Metab **4**:297–303.
50. Binderup L 1988 MC903—A novel vitamin D analogue with potent effects on cell proliferation and cell differentiation. In: Norman AW, Schaefer K, Grigoleit H-G, von Herrath D (eds) Vitamin D: Molecular: Cellular and Clinical Endocrinology. de Gruyter, Berlin, pp. 300–309.
51. Calverley MJ, Binderup E, Binderup L 1991 The 20-epi modification in the vitamin D series: Selective enhancement of "non-classical" receptor-mediated effects. In: Norman AW, Bouillon R, Thomasset M (eds) Vitamin D: Gene Regulation, Structure–Function Analysis and Clinical Application. de Gruyter, Berlin, pp. 163–164.
52. Binderup L, Latini S, Binderup E, Bretting C, Calverley M, Hansen K 1991 20-Epi-vitamin D$_3$ analogues: A novel class of potent regulators of cell growth and immune responses. Biochem Pharmacol **42**:1569–1575.
53. Dilworth FJ, Calverley MJ, Makin HLJ, Jones G 1994. Increased biological activity of 20-epi-1,25-dihydroxyvitamin D$_3$ is due to reduced catabolism and altered protein binding. Biochem Pharmacol **47**:987–993.
54. Peleg S, Sastry M, Collins ED, Bishop JE, Norman AW 1995 Distinct conformational changes induced by 20-epi analogues of 1α,25-dihydroxyvitamin D$_3$ are associated with enhanced activation of the vitamin D receptor. J Biol Chem **270**:10551–10558.
55. Ostrem VK, Lau WF, Lee SH, Perlman K, Prahl J, Schnoes HK, DeLuca HF, Ikekawa N 1987 Induction of monocytic differentiation of HL-60 cells by 1,25-dihydroxyvitamin D analogs. J Biol Chem **262**:14164–14171.
56. Ostrem VK, Tanaka V, Prahl J, DeLuca HF, Ikekawa N 1987 24- and 26-homo-1,25-dihydroxyvitamin D$_3$: Preferential activity in inducing differentiation of human leukemic cells HL-60 *in vitro*. Proc Natl Acad Sci USA **84**:2610–2614.
57. Perlman K, Kutner A, Prahl J, Smith C, Inaba M, Schnoes HK, DeLuca HF 1990 24-Homologated 1,25-dihydroxyvitamin D$_3$ compounds: Separation of calcium and cell differentiation activities. Biochemistry **29**:190–196.
58. Paulson SK, Perlman K, DeLuca H, Stern PH 1990 24- and 26-homo-1,25-dihydroxyvitamin D$_3$ analogs: Potencies on in vitro bone resorption differ from those reported for cell differentiation. J Bone Miner Res **5**:201–206.
59. Dilworth FJ, Scott I, Green A, Strugnell S, Guo Y-D, Roberts EA, Kremer R, Calverley MJ, Makin HLJ, Jones G 1995a Different mechanisms of hydroxylation site selection by liver and kidney cytochrome P450 species (CYP27 and CYP24) involved in vitamin D metabolism. J Biol Chem **270**:16766–16774.
60. Dilworth FJ, Scott I, Calverley MJ, Makin HLJ, Jones G 1995b Enzymes of side chain oxidation pathway not affected by addition of methyl groups to end of the vitamin D$_3$ side chain. J Bone Miner Res **10**:S388 (Abstract M546).
61. Binderup L, Carlberg C, Kissmeyer AM, Latini S, Mathiasen IS, Mork-Hansen C 1993 The need for new vitamin D analogues: Mechanisms of action and clinical applications. In: Norman AW, Bouillon R, Thomasset M (eds) Vitamin D. A Pluripotent Steroid Hormone: Structural Studies, Molecular Endocrinology and Clinical Applications. de Gruyter, Berlin, pp. 55–63.
62. Zhou J-Y, Norman AW, Chen D-L, Sun G, Uskokovic M, Koeffler HP 1990. 1,25-Dihydroxy-16-ene-23-yne-itamin D$_3$ prolongs survival time of leukemic mice. Proc Natl Acad Sci USA **87**:3929–3932.
63. Binderup E, Calverley MJ, Binderup L 1991 Synthesis and biological activity of 1α-hydroxylated vitamin D analogues with poly-unsaturated side chains. In: Norman AW, Bouillon R, Thomasset M (eds) Vitamin D: Gene Regulation, Structure–

Function Analysis and Clinical Application. de Gruyter, Berlin, pp. 192–193.
64. Colston KW, Mackay AG, James SY, Binderup L, Chandler S, Coombes RC 1992 EB1089: A new vitamin D analogue that inhibits the growth of breast cancer cells *in vivo* and *in vitro*. Biochem Pharmacol **44**:2273–2280.
65. James SY, Mackay AG, Binderup L, Colston KW 1994 Effects of a new synthetic analogue, EB1089, on the oestrogen-responsive growth of human breast cancer cells. J Endocrinol **141**:555–563.
66. Reddy GS, Clark JW, Tserng K-Y, Uskokovic MR, McLane JA 1993 Metabolism of 1,25-$(OH)_2$-16-ene D_3 in kidney: Influence of structural modification of D-ring on side chain metabolism. Bioorg Med Lett **3**:1879–1884.
67. Satchell DP, Norman AW 1996 Metabolism of the cell differentiating agent 1,25-$(OH)_2$-16-ene-23-yne vitamin D_3 by leukemic cells. J Steroid Biochem Mol Biol **57**:117–124.
68. Shankar VN, Makin HLJ, Schroeder NJ, Trafford DJH, Kissmeyer A-M, Calverley MJ, Binderup E, Jones G 1995 Metabolism of the antiproliferative vitamin D analogue, EB1089, in a cultured human keratinocyte model. Bone **17**:326 (abstract).
69. Shankar VN, Dilworth FJ, Makin HLJ, Schroeder NJ, Trafford DAJ, Kissmeyer A-M, Calverley MJ, Binderup E, Jones G 1997 Metabolism of the vitamin D analog EB1089 by cultured human cells: Redirection of hydroxylation site to distal carbons of the side chain. Biochem Pharmacol **53**:783–793.
70. Kissmeyer A-M, Mathiasen IS, Latini S, Binderup L 1995 Pharmacokinetic studies of vitamin D analogues: Relationship to vitamin D binding protein (DBP). Endocrine **3**:263–266.
71. Kissmeyer A-M, Binderup E, Binderup L, Hansen CM, Andersen NR, Schroeder NJ, Makin HLJ, Shankar VN, Jones G 1997 The metabolism of the vitamin D analog EB1089: Identification of *in vivo* and *in vitro* metabolites and their biological activities. Biochem Pharmacol **53**:1087–1097.
72. Wankadiya KF, Uskokovic MR, Clark J, Tserng K-Y, Reddy GS 1992 Novel evidence for the reduction of the double bond in Δ^{22}-1,25-dihydroxyvitamin D_3. J Bone Miner Res **7**:S171 (Abstract 315).
73. Murayama E, Miyamoto K, Kubodera N, Mori T, Matsunaga I 1986 Synthetic studies of vitamin D analogues. VIII. Synthesis of 22-oxavitamin D_3 analogues. Chem Pharm Bull (Tokyo) **34**:4410–4413.
74. Hansen K, Calverley MJ, Binderup L 1991 Synthesis and biological activity of 22-oxa vitamin D analogues. In: Norman AW, Bouillon R, Thomasset M (eds) Vitamin D: Gene Regulation, Structure–Function Analysis and Clinical Application. de Gruyter, Berlin, pp. 161–162.
75. Kubodera N, Miyamoto K, Akiyama M, Matsumoto M, Mori T 1991 Synthetic studies of vitamin D analogs. 9. Synthesis and differentiation-inducing activity of 1α,25-dihydroxy-23-oxa-vitamin D_3, thia-vitamin D_3 and aza-vitamin D_3. Chem Pharm Bull (Tokyo) **39**:3221–3224.
76. Baggiolini EG, Zacobelli JA, Hennessy BM, Batcho AD, Sereno JF, Uskokovic MR 1986 Stereocontrolled total synthesis of 1α,25-dihydroxycholecalciferol and 1α,25-dihydroxyergocalciferol. J Org Chem **51**:3098–3108.
77. Allewaert K, Van Baelen H, Bouillon R, Zhao X-Y, De Clercq P, Vanderwalle M 1993 Synthesis and biological evaluation of 23-oxa-, 23-thia- and 23-oxa-24-oxo-1α,25-dihydroxyvitamin D_3. Bioorg Med Lett **3**:1859–1862.
78. Calverley MJ, Strugnell S, Jones G 1993 The seleno-acetal route to 1α-hydroxy-vitamin D analogues: Synthesis of 24-oxa-1α-hydroxy-vitamin D_3, a useful vitamin D metabolism probe. Tetrahedron **49**:739–746.
79. Sarandeses LA, Valles MJ, Castedo L, Mourino A 1993 Synthesis of 24-oxa-vitamin D_3 and 1α-hydroxy-24-oxa-vitamin D_3. Tetrahedron **49**:731–738.
80. Bikle DD, Abe-Hashimoto J, Su MJ, Felt S, Gibson DFC, Pillai S 1995 22-Oxa calcitriol is a less potent regulator of keratinocyte proliferation and differentiation due to decreased cellular uptake and enhanced catabolism. J Invest Dermatol **105**:693–698.
81. Kobayashi T, Tsugawa N, Okano T, Masuda S, Takeuchi A, Kubodera N, Nishii Y 1994 The binding properties with blood proteins and tissue distribution of 22-oxa-1α,25-dihydroxyvitamin D_3, a noncalcemic analogue of 1α,25-dihydroxyvitamin D_3 in rats. J Biochem (*Tokyo*) **115**:373–380.
82. Dilworth FJ, Calverley MJ, Kissmeyer A-M, Binderup E, Makin HLJ, Jones G 1996 KH1060, a potent vitamin D analog is degraded in cultured keratinocytes via several different pathways. Bone Miner Res **11**:S424 (Abstract T500).
83. Rastrup Anderson N, Buchwald FA, Grue-Sorensen G 1992 Identification and synthesis of a metabolite of KH1060, a new potent 1α,25-dihydroxyvitamin D_3 analogue. Bioorg Med Chem Lett **2**:1713–1716.
84. Dilworth FJ, Williams GR, Jones G 1996 Studies of the biological activity of metabolites of KH1060. Unpublished results.
85. Reddy GS, Ishizuka S, Wandkadiya KF, Tserng K-Y, Yeung B, Vouros P 1992 Metabolism of 1α,24(*R*)-dihydroxyvitamin D_3 into calcitroic acid in rat kidney through two different metabolic pathways. J Bone Miner Res **7**:S170 (Abstract 312).
86. Tucker G, Gagnon RE, Haussler MR 1973 Vitamin D_3-25-hydroxylase: Tissue occurrence and lack of regulation. Arch Biochem Biophys **155**:47–57.
87. Axen E, Postlind H, Wikvall K 1995 Effects of CYP27 mRNA expression in rat kidney and liver by 1α,25-dihydroxyvitamin D_3, a suppressor of renal 25-hydroxyvitamin D_3-1α-hydroxylase activity. Biochem Biophys Res Commun **215**:136–141.
88. Weinstein EA, Siu-Caldera M-L, Ishizuka S, Reddy GS 1995 Evidence of 25-hydroxylation in the rat kidney for 1α,24(*R*)-dihydroxyvitamin D_3 only but not for vitamin D_3 or 1α-hydroxyvitamin D_3. J Bone Miner Res **10**:S497 (Abstract T570).
89. Nishii Y, Sato K, Kobayashi T 1993 The development of vitamin D analogues for the treatment of osteoporosis. Osteoporosis Int **1**:S190–193 (Suppl).
90. Masuda S, Makin HLJ, Kremer R, Okano T, Kobayashi T, Sato K, Nishii Y, Jones G 1994 Metabolism of 2β-(3-hydroxypropoxy)-1α,25-dihydroxyvitamin D_3 (ED-71) in cultured cell lines. J Bone Miner Res **9**:S289 (Abstract B238).
91. Kobayashi Y, Taguchi T, Mitsuhashi S, Eguchi T, Ohshima E, Ikekawa N 1982 Studies on organic fluorine compounds. XXXIX. Studies on steroids. LXXIX. Synthesis of 1α,25-dihydroxy-26,26,26,27,27,27-hexaflurovitamin D_3. Chem Pharm Bull (Tokyo) **30**:4297–4303.
92. Koeffler HP, Armatruda T, Ikekawa N, Kobayashi Y, DeLuca HF 1984 Induction of macrophage differentiation of human normal and leukemic myeloid stem cells by 1α,-25-dihydroxyvitamin D_3 and its fluorinated analogs. Cancer Res **44**:6524–6528.
93. Inaba M, Okuno S, Nishizawa Y, Yukioka K, Otani S, Matsui-Yuasa I, Morisawa S, DeLuca HF, Morii H 1987 Biological activity of fluorinated vitamin D analogs at C-26 and C-27 on human promyelocytic leukemia cells, HL-60. Arch Biochem Biophys **258**:421–425.
94. Kistler A, Galli B, Horst R, Truitt GA, Uskokovic MR 1989 Effects of vitamin D derivatives on soft tissue calcification in neonatal and calcium mobilization in adult rats. Arch Toxicol **63**:394–400.
95. Lohnes D, Jones G 1992 Further metabolism of 1α,25-dihydroxyvitamin D_3 in target cells. J Nutr Sci Vitaminol (Tokyo) Special Issue: 75–78.

96. Inaba M, Okuno S, Nishizawa Y, Imanishi Y, Katsumata T, Sugata I, Morii H 1993 Effect of substituting fluorine for hydrogen at C-26 and C-27 on the side chain of 1α,25-dihydroxyvitamin D_3. Biochem Pharmacol **45**:2331–2336.
97. Nakatsuka K, Imanishi Y, Morishima Y, Sekiya K, Sasao K, Miki T, Nishizawa Y, Katsumata T, Nagata A, Murakawa S 1992 Biological potency of a fluorinated vitamin D analogue in hypoparathyroidism. Bone Miner **16**:73–81.
98. Nishizawa Y, Morii H, Ogura Y, DeLuca HF 1991 Clinical trial of 26,26,26,27,27,27-hexafluro-1α,25-dihydroxyvitamin D_3 in uremic patients on hemodialysis: Preliminary report. Contrib Nephrol **90**:196–203.
99. Perlman KL, Sicinski RR, Schnoes HK, DeLuca HF 1990 1α,25-Dihydroxy-19-nor-vitamin D_3, a novel vitamin D-related compound with potential therapeutic activity. Tetrahedron Lett **31**:1823–1824.
100. Posner GH, Dai H 1993 1-(Hydroxyalkyl)-25-hydroxyvitamin D_3 analogs of calcitriol-I. Synthesis. Bioorg Med Chem Lett **2**:1829–1834.
101. Neef G, Kirsch G, Schwarz K, Wiesinger H, Menrad A, Fahnrich M, Thieroff-Ekerdt, Steinmeyer A 1994 20-Methyl vitamin D analogues. In: Norman AW, Bouillon R, Thomasset M (eds). Vitamin D. A Pluripotent Steroid Hormone: Structural Studies, Molecular Endocrinology and Clinical Applications. de Gruyter, Berlin, pp. 97–98.
102. Bouillon R, Allewaert K, Xiang DZ, Tan BK, Van Baelen H 1991 Vitamin D analogs with low affinity for the vitamin D binding protein: Enhanced *in vitro* and decreased *in vivo* activity. J Bone Miner Res **6**:1051–1057.
103. Dusso AS, Negrea L, Gunawardhana S, Lopez-Hilker S, Finch J, Mori T, Nishii Y, Slatopolsky E, Brown AJ 1991 On the mechanisms for the selective action of vitamin D analogs. Endocrinology **128**:1687–1692.
104. Cheskis B, Lemon BD, Uskokovic MR, Lomedico PT, Freedman LP 1995 Vitamin D_3-retinoid X receptor dimerization. DNA binding, and transactivation are differentially affected by analogs of 1,25-dihydroxyvitamin D_3. Mol Endocrinol **9**:1814–1824.
105. Guengerich FP 1991 Reactions and significance of cytochrome P-450 enzymes. J Biol Chem **266**:10019–10022.
106. Graham-Lorence S, Amarneh B, White RE, Peterson JA, Simpson ER 1995 A three-dimensional model of the aromatase cytochrome P450. Protein Sci **4**:1065–1080.
107. Vijayakumar S, Salerno JC 1992 Molecular modelling of the 3-D structure of cytochrome P-450_{scc}. Biochim Biophys Acta **1160**:281–286.
108. Prosser DE, Dakin KA, Donini OAT, Weaver DF, Jia Z, Jones G 1996 A three-dimensional model of the cytochrome P450, CYP27 and its vitamin D binding site. J Bone Miner Res **11**:S313 (Abstract M525).
109. Cali JJ, Russell DW 1991 Characterization of human sterol 27-hydroxylase. A mitochondrial cytochrome P450 that catalyzes multiple oxidation reactions in bile acid biosynthesis. J Biol Chem **266**:7774–7778.
110. Cali JJ, Hsieh C-L, Francke V, Russell DW 1991 Mutations in the bile acid biosynthetic enzyme sterol 27-hydroxylase underlie cerebrotendinous xanthomatosis. J Biol Chem **266**:7779–7783.
111. Barton DH, Hesse RH, Pechet MM, Rizzardo E 1973 A convenient synthesis of 1α-hydroxy-vitamin D_3. J Am Chem Soc **95**:2748–2749.
112. Paaren HE, Hamer DE, Schnoes HK, DeLuca HF 1978 Direct C-1 hydroxylation of vitamin D compounds: Convenient preparation of 1α-hydroxyvitamin D_3, 1α,25-dihydroxyvitamin D_3 and 1α-hydroxyvitamin D_2. Proc Natl Acad Sci USA **75**:2080–2081.
113. Fraser D, Kooh SW, Kind P, Holick MF, Tanaka Y, DeLuca HF 1973 Pathogenesis of hereditary vitamin D dependancy rickets. N Engl J Med **289**:817–822.
114. Baggiolini EG, Wovkulich PM, Iacobelli JA, Hennessy BM, Uskokovic MR 1982 Preparation of 1-alpha hydroxylated vitamin D metabolites by total synthesis. In: Norman AW, Schaefer K, von Herrath D, Grigoleit H-G (eds) Vitamin D: Chemical, Biochemical and Clinical Endocrinology of Calcium Metabolism. de Gruyter, Berlin, pp. 1089–1100.
115. Baggiolini EG, Partridge JJ, Shiuey S-J, Truitt GA, Uskokovic MR 1989 Cholecalciferol 23-yne derivatives, their pharmaceutical compositions, their use in the treatment of calcium-related diseases, and their antitumor activity, US 4,804,502 [Abstract]. Chem Abstr 111:58160d.
116. Morisaki M, Koizumi N, Ikekawa N, Takeshita T, Ishimoto S 1975 Synthesis of active forms of vitamin D. Part IX. Synthesis of 1α,24-dihydroxycholecalciferol. J Chem Soc Perkin Trans **1**(1)1421–1424.
117. Fukushima M, Suzuki Y, Tohira Y, Nishii Y, Suzuki M, Sasaki S, Suda T 1976 25-Hydroxylation of 1α-hydroxyvitamin D_3 in vivo and in the perfused rat liver. FEBS Lett **65**:211–214.
118. Rosenthal AM, Jones G, Kooh SW, Fraser D 1980 25-Hydroxyvitamin D_3 metabolism by the isolated perfused rat kidney. Am J Physiol (Endocrinol Metab) **239**:E12–E20.
119. Reddy GS, Jones G, Kooh SW, Fraser D, DeLuca HF 1983 Effects of metabolites and analogs of vitamin D_3 on 24(*R*),25-dihydroxyvitamin D_3 synthesis. Am J Physiol **235**:E359–E364.
120. Henry HL 1979 Regulation of the hydroxylation of 25-hydroxyvitamin D_3 in vivo and in primary cultures of chick kidney cells. J Biol Chem **254**:2722–2729.
121. Chandler JS, Chandler SK, Pike JW, Haussler MR 1984 1,25-Dihydroxyvitamin D_3 induces 25-hydroxyvitamin D_3-24-hydroxylase in a cultured monkey kidney cell line (LLC-MK2) apparently deficient in the high affinity receptor for the hormone. J Biol Chem **259**:2214–2222.
122. Vieth R, Fraser D 1979 Kinetic behaviour of 25-hydroxyvitamin D-1-hydroxylase and -24-hydroxylase in rat kidney mitochondria. J Biol Chem **254**:12455–12460.
123. Knutson JC, LeVan LW, Valliere CR, Bishop CW 1997 Pharmacokinetics and systemic effect on calcium homeostasis of 1α,24-dihydroxyvitamin D_2 in rats: Comparison with 1α,25-dihydroxyvitamin D_2, calcitriol and calcipotriol. Biochem Pharmacol **53**:829–837.
124. Jones G 1996 Pharmacological mechanisms of therapeutics: Vitamin D and its analogs. In: Bilezikian J, Raisz L, Rodan G (eds). Principles of Bone Biology. Academic Press, San Diego, pp 1069–1081.

Mechanisms for the Selective Actions of Vitamin D Analogs

ALEX J. BROWN Renal Division, Department of Medicine, Washington University School of Medicine, St. Louis, Missouri

EDUARDO S. SLATOPOLSKY Renal Division, Department of Medicine, Washington University School of Medicine; Chromalloy American Kidney Center; and Barnes-Jewish Hospital, St. Louis, Missouri

Vitamin D was first identified as an essential factor for normal mineral metabolism and skeletal development. These actions are now known to be mediated by metabolites of vitamin D, primarily 1,25-dihydroxyvitamin D_3 [1,25$(OH)_2D_3$] but perhaps 24,25-dihydroxyvitamin D_3 as well. 1,25$(OH)_2D_3$ acts as a steroid hormone and binds to a well-characterized intracellular receptor that is localized in the nucleus and regulates gene transcription by binding to specific motifs in the target genes [1,2] (see Chapters 8–14). There is also *in vitro* evidence that 1,25$(OH)_2D_3$ and 24,25$(OH)_2D_3$ can interact with cell surface receptors and rapidly stimulate signaling from the cell membrane [3] (see Chapter 15). At present, the potential roles for these nongenomic actions *in vivo* are less well understood.

Research from many laboratories has demonstrated that the actions of the vitamin D system extend beyond a role restricted to bone and mineral metabolism [4]. Many of these newly discovered activities have suggested potential therapeutic applications for 1,25$(OH)_2D_3$. The native vitamin D hormone is currently in use for the treament of secondary hyperparathyroidism in renal failure patients [5], psoriasis [6], and X-linked hypophosphatemic rickets [7]. In addition, the ability of 1,25$(OH)_2D_3$ to block proliferation of many cell types, including neoplastic cells, *in vitro* has indicated the potential of this compound for treating various types of cancer [8]. However, a major limitation to 1,25$(OH)_2D_3$ therapy is its potent calcemic and phosphatemic activities. Studies to date suggest that the doses of 1,25$(OH)_2D_3$ necessary to block cell proliferation *in vivo* may also produce profound hypercalcemia and hyperphosphatemia.

I. IDENTIFICATION OF SELECTIVE VITAMIN D ANALOGS

A major breakthrough in vitamin D therapeutics has been the development of vitamin D analogs that retain many of the potential clinically useful activities of

$1,25(OH)_2D_3$ for the newly discovered indications, but have much lower calcemic and phosphatemic activities *in vivo* [9–11]. The promising analogs identified initially displayed high differentiating and antiproliferative activities *in vitro* but were found to have low calcemic activity *in vivo* [12–16]. One explanation for these results could have been that the analogs were completely inactive *in vivo*. Although this may be the case for some analogs, several "noncalcemic" analogs [actually, the analogs are all calcemic but to a lesser degree than $1,25(OH)_2D_3$] have been found to retain some of the activities of $1,25(OH)_2D_3$ in animal models.

Selective actions of several analogs have now been demonstrated *in vivo*. These analogs are discussed in detail in Chapters 61–63, and their uses for specific indications are discussed in Chapters 64–73. Two analogs with low calcemic activity, 22-oxa-$1,25(OH)_2D_3$ (OCT; Chugai Pharmaceuticals) and $1,25(OH)_2$-19-nor-D_2 (Abbott Laboratories), effectively suppressed parathyroid hormone (PTH) levels in hyperparathyroid uremic rats at doses that had minimal effects on serum calcium and phosphate levels [17–19]. This selectivity is illustrated in Fig. 1. 22-Oxa-$1,25(OH)_2D_3$ and the analog EB1089 (Leo Pharmaceuticals) were shown to suppress PTH-related peptide in cancer cells *in vivo* [20,21], suggesting their use in the treatment of hypercalcemia of malignancy. 22-Oxa-$1,25(OH)_2D_3$ and other analogs, calcipotriol and EB1089 (Leo Pharmaceuticals) and 1,25-$(OH)_2$-16-ene-23-yne-26,27-F_6-vitamin D_3 (Hoffmann-LaRoche), have been shown to inhibit the *in vivo* proliferation of breast cancer cells [22–26]. The analog 1,25-$(OH)_2$-16-ene-23-yne-26,27-F_6-vitamin D_3 may be effective in control of androgen-induced carcinoma of the prostate and seminal vesicles [27,28]. 22-Oxa-$1,25(OH)_2D_3$ and 1,25-$(OH)_2$-16-ene-23-yne-26,27-F_6-vitamin D_3 can also inhibit growth of experimentally induced tumors of the small and large intestine [29,30]. Another derivative, 1,25-$(OH)_2$-16-ene-23-yne-vitamin D_3 (Hoffmann-LaRoche), can prolong the survival of mice injected with leukemia cells [31]. Modulation of the immune system has been demonstrated *in vivo* for 22-oxa-$1,25(OH)_2D_3$ [32], for KH1060 (Leo Pharmaceuticals) [33,34], and for $1,25(OH)_2$-16-ene-D_3 and its 24-oxo metabolite [35]. Calcipotriol, which has been approved for the topical treatment of psoriasis, does not alter mineral metabolism at its effective doses [36].

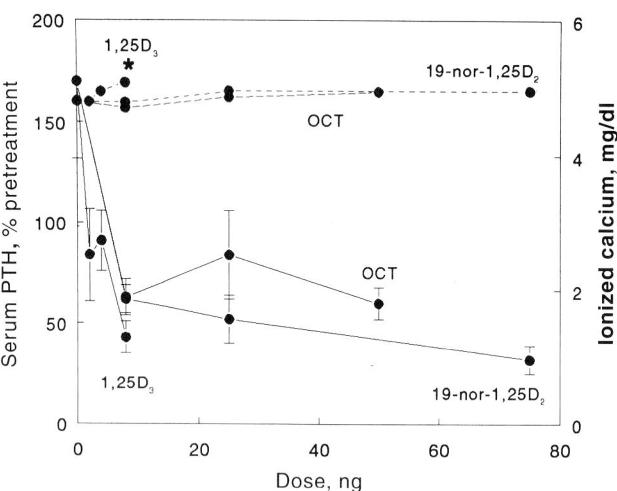

FIGURE 1 PTH suppression and calcemic activity of $1,25(OH)_2D_3$ and its analogs 22-oxa-$1,25(OH)_2D_3$ (OCT) and 19-nor-$1,25(OH)_2D_2$ (19-nor-$1,25D_2$) in uremic rats. Rats were made uremic by 5/6 nephrectomy and maintained for 2 months. The rats were then treated with the specified dose of vitamin D compound every other day, and serum calcium and PTH levels were determined after 8 days of treatment. Data for the 19-nor-$1,25D_2$ rats were adapted from Slatopolsky *et al.* [19] and reprinted with permission.

In general, the low calcemic activity of these analogs permitted the use of doses sufficiently high to obtain the desired effect. Quantification of the degree of selectivity (i.e., potency of the desired effect versus calcemic activity) is difficult to determine in most cases because of lack of data for the relative dose responses for $1,25(OH)_2D_3$ and the analog. In many cases, such data cannot be gathered owing to the hypercalcemic toxicity of $1,25(OH)_2D_3$ at the doses required for desired activity.

Another analog with clinical potential is $1,25(OH)_2$-2β-(3-hydroxypropoxy)-D_3 (ED-71) from Chugai Pharmaceuticals. This analog appears to be much more effective than $1,25(OH)_2D_3$ in stimulating bone mineralization in ovariectomized rats [37] and in corticosteroid-treated rats [38], and it may be useful for treatment of osteoporosis [37]. The selective analogs and their potential applications are summarized in Table I. Further details on the properties and potential applications of these and other analogs are presented elsewhere in this treatise.

The novel aspect of these analogs is their differential actions, compared to $1,25(OH)_2D_3$, *in vivo*. In fact, as these analogs have relatively high affinity for the vitamin D receptor (VDR), usually within one order of magnitude, it is not unexpected that they are able to mimic many of the actions of $1,25(OH)_2D_3$ *in vivo*. Their unique feature is the ability to efficiently support some but not all $1,25(OH)_2D_3$-associated activities. Most commonly, the analogs display decreased potency in enhancing intestinal calcium absorption and/or bone mobilization. In some cases, the analogs have relatively high calcemic effects but tend to produce even higher activities in other specific cells or tissues. The selectivity is not always cell- or tissue-specific, but can be gene- or process-specific within the same tissue. For example, several analogs [$1,25(OH)_2$-22-ene-24-dihomo-D_3,

TABLE I Properties of Vitamin D Analogs That Display *in Vivo* Selectivity[a]

Analog	Application	Calcemia	ICA[b]	BCM[c]	VDR binding	DBP binding
$1,25(OH)_2D_3$	—	100	100	100	100	100
Calcipotriol	Psoriasis, breast cancer	1–5 [26,50,68[d]]	—	—	50–150 [50,57,68]	3–50 [43,50,57]
22-oxa-$1,25(OH)_2D_3$	Hyperparathyroidism, immune regulation	1 [17,32]	1 [48,110]	4 [104,111]	12 [32,44,56]	0.2–0.4 [42–44]
19-nor-$1,25(OH)_2D_2$	Hyperparathyroidism	<10 [9]	—	—	33 [118]	33 [119]
16-ene-$1,25(OH)_2D_3$	Immunosuppression	0 [113]	0 [113]	—	150 [57]	2 [57]
16-ene-23-yne-$1,25(OH)_2D_3$	Leukemia	4 [31]	3–50 [114–116]	2 [116]	50–100 [113]	0.2–5 [44,50,57]
16-ene-23-yne-F_6-$1,25(OH)_2D_3$	Breast cancer	—	6 [117]	6 [117]	31 [57]	25 [57]
EB1089	Breast cancer, hypercalcemia of malignancy	50 [26]	—	—	80 [26]	<1 [26]
KH1060	Immunosuppression	130 [76[d]]	—	—	120 [75]	—
ED-71	Osteoporosis	—	100 [45]	100 [45]	12 [45]	270 [45]

[a] All values are expressed as a percentage of $1,25(OH)_2D_3$ activity.
[b] Intestinal calcium absorption was measured by everted gut sac or *in situ* duodenal loop methods.
[c] Bone calcium mobilization was estimated by increases in serum calcium in animals fed a calcium-deficient diet.
[d] Calcemic activity was assessed by urinary calcium excretion.

$1,25(OH)_2$22-ene-24-trihomo-D_3, and $1,25,28(OH)_3D_2$] have been shown to induce the vitamin D-dependent calcium-binding protein in the intestine without increasing intestinal calcium transport [39,40]. Similarly, 20-epi-$1,25(OH)_2D_3$ was found to be 1000 times more potent than $1,25(OH)_2D_3$ in inducing cell differentiation in a human leukemia cell line (HL-60), but it was equipotent in activating transcription of a vitamin D-responsive reporter construct in the same cells [41].

II. MECHANISMS FOR THE *IN VIVO* SELECTIVITY OF VITAMIN D ANALOGS

Investigation of the potential mechanisms involved in the *in vivo* selectivity of vitamin D analogs has been the focus of research in several laboratories during the early 1990s. A number of potential mechanisms have been considered. The apparent tissue specificity of the analogs led to initial speculation of the existence of tissue-specific forms of the VDR that interact differentially with analogs, but this hypothesis has not been supported with experimental proof. On the other hand, evidence for several other potential mechanisms is accumulating. Figure 2 diagrams the possible steps in the vitamin D action pathways at which differential actions of $1,25(OH)_2D_3$ and its analogs could lead to selectivity at the tissue, cell, or gene level. Studies from a number of laboratories have compared the properties of various vitamin D analogs to those of $1,25(OH)_2D_3$ and have documented alterations at nearly all of the steps highlighted. The sites of differential action include (1) DBP affinity, (2) association with other serum carriers including lipoproteins, (3) cellular uptake, (4) accumulation

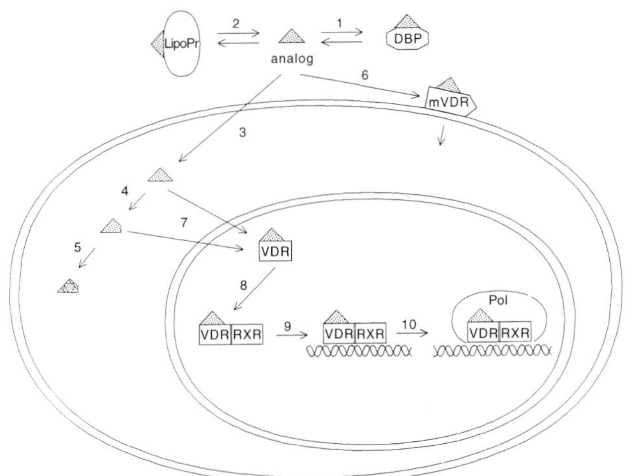

FIGURE 2 Potential sites of differential action of $1,25(OH)_2D_3$ and analogs. The possible steps in the vitamin D activation pathways at which differences in vitamin D analog action could lead to selective activities *in vivo* are shown. The steps diagrammed include (1) DBP affinity, (2) interaction with other serum proteins including lipoproteins, (3) cellular uptake, (4) conversion to active metabolites, (5) catabolic inactivation, (6) activation of the nongenomic pathway through a putative membrane vitamin D receptor (mVDR), (7) interaction with the nuclear vitamin D receptor (VDR), (8) formation of the VDR–RXR complex, (9) binding of the activated complex to DNA, and (10) formation of the preinitiation complex with RNA polymerase II.

of active metabolites, (5) different rates of catabolic inactivation, (6) differential activation of the nongenomic pathway via a distinct membrane-bound receptor (mVDR), and (7) altered binding to the nuclear VDR leading to changes (increases or decreases) in (8) VDR–RXR heterodimerization, (9) DNA binding of the VDR–RXR complex, and (10) formation of the preinitiation complex. Thus, differences in pharmacokinetics, intracellular metabolism, nongenomic activities, and interaction with the VDR may provide potential explanations for the *in vivo* selectivity of any analog. In addition, it is likely that more than one of these mechanisms may contribute to the unique *in vivo* activity profile of each analog. This chapter reviews the evidence for these potential mechanisms and their role in the selectivity of vitamin D analogs.

A. Vitamin D Binding Protein Affinity

The first biochemical property of several analogs shown to differ markedly from that of $1,25(OH)_2D_3$ was a low affinity for the serum vitamin D binding protein (DBP). The subject of DBP is discussed in detail in Chapter 7. Low affinity for DBP was reported initially for 22-oxa-$1,25(OH)_2D_3$ by Okano *et al.* [42]. A competitive binding assay using ^{3}H-$25OHD_3$ and dilute rat plasma as a source of DBP showed that 22-oxa-$1,25(OH)_2D_3$ had approximately 600 times lower affinity than $1,25(OH)_2D_3$ for rat DBP. Subsequent studies by Dusso *et al.* [43] confirmed that 22-oxa-$1,25(OH)_2D_3$ had poor affinity for rat DBP, although the difference between 22-oxa-$1,25(OH)_2D_3$ and $1,25(OH)_2D_3$ was found to be less (250-fold), likely due to methodological differences. Dusso *et al.* [43] and Kobayashi *et al.* [44] reported that 22-oxa-$1,25(OH)_2D_3$ also had low affinity for human and dog DBP. Although the studies agreed that human and rat plasma DBP discriminated against 22-oxa-$1,25(OH)_2D_3$ to the same degree, the study by Dusso *et al.* [43], but not that by Kobayashi *et al.* [43], suggested less discrimination by dog plasma DBP.

The importance of the role of low DBP affinity was underscored by the study of Dusso *et al.* [42] which measured the DBP affinity of all of the analogs available at the time (1991) that displayed high cell differentiating activity *in vitro* but low calcemic activity *in vivo*. They found that all of the "noncalcemic" analogs tested [calcipotriol, $1,25(OH)_2$-16-ene-23-yne-D_3, $1,25(OH)_2$-26,27-dihomo-22-ene-D_3, and $1,25(OH)_2$-24-trihomo-22-ene-D_3] had much lower DBP affinity for rat plasma DBP. Thus, it appeared that low DBP affinity was a common property of these analogs. It was not clear how low DBP affinity led to the selective actions *in vivo*. A clear exception to this is 19-nor-$1,25(OH)_2D_2$. This compound, which effectively suppresses PTH with little hypercalcemia, has a DBP affinity that is only 3 times less than that of $1,25(OH)_2D_3$ (A. J. Brown, unpublished results, 1995). ED-71, which promotes bone mineralization in ovariectomized rats, has been found to have a 2-fold higher affinity than $1,25(OH)_2D_3$ for DBP [45]. The DBP affinities of the selective analogs are listed in Table I.

The role of DBP in vitamin D pharmacokinetics is not fully understood, but DBP has been shown to inhibit the uptake, and therefore the activity, of $1,25(OH)_2D_3$ in cultured cells [46,47]. Therefore, compounds with low DBP affinity should have greater access to target cells. In addition, binding of vitamin D compounds to DBP is thought to reduce the rate of clearance and metabolism *in vivo*. It would be expected, then, that the low DBP affinity of these analogs would cause them to be more rapidly cleared than $1,25(OH)_2D_3$. Finally, the lack of binding to DBP may allow the analogs to associate with other serum proteins, which could alter their delivery to and uptake by various target tissues.

B. Clearance Rates

Rapid clearance *in vivo* has been confirmed for 22-oxa-$1,25(OH)_2D_3$ [43,44,48] and calcipotriol [49]. Clearance of $1,25(OH)_2D_3$ has been shown to be multifactorial, with two or three distinct components. Using the slowest of the three exponential disappearance curves, Kobayashi *et al.* [44] observed half-lives of 30 min for 22-oxa-$1,25(OH)_2D_3$ and 4 hr for $1,25(OH)_2D_3$ in normal rats. Dusso *et al.* [43] found half-lives of 2.5 and 7 hr for 22-oxa-$1,25(OH)_2D_3$ and $1,25(OH)_2D_3$ in dogs. This rapid clearance of 22-oxa-$1,25(OH)_2D_3$ could account for its low calcemic activity but not for the relatively high activity on the parathyroid glands, immune system, and cell proliferation *in vivo*. Similarly, Kissmeyer and Binderup [49] found that intravenously injected calcipotriol was cleared very rapidly. The relative insensitivity of the analyses for intact calcipotriol [high-performance liquid chromatography (HPLC) or bioactivity *in vitro*] allowed estimation of the first exponential [4 min for calcipotriol versus 15 min for $1,25(OH)_2D_3$] but not the slower decay exponential. On the other hand, 19-nor-$1,25(OH)_2D_2$ has been found to have a relatively long half-life of 6 hr (A. J. Brown, unpublished data, 1997), consistent with its higher DBP affinity. Similarly, ED-71 which has a higher DBP affinity than $1,25(OH)_2D_3$ also has a longer half-life of 12 hr [45]. Clearances rates for other vitamin D analogs have not been reported.

C. Cellular Uptake

The low DBP affinity also would be predicted to lead to a higher proportion of the analog in the free, unbound form, and to allow greater access to target cells. This was found to be true for 22-oxa-1,25$(OH)_2D_3$ *in vitro*. Dusso *et al.* [43] measured the percent free 22-oxa-1,25$(OH)_2D_3$ or 1,25$(OH)_2D_3$ after equilibration in media containing increasing amounts of serum. They found that, at all serum concentrations, the percent free 22-oxa-1,25$(OH)_2D_3$ was considerably higher than the percent free 1,25$(OH)_2D_3$ (Fig. 3). The biological relevance of this was demonstrated initially in cell culture. Dusso et al. [43] showed that increasing the serum content in the medium of cultured monocytes had a greater inhibitory effect on induction of the vitamin D catabolic pathway by 1,25$(OH)_2D_3$ than by 22-oxa-1,25$(OH)_2D_3$, presumably owing to greater accessibility of the analog to the cells at all serum concentrations. Bouillon *et al.* [50] demonstrated a similar resistance of calcipotriol, 1,25$(OH)_2$-16-ene-23-yne-D_3, and 1,25$(OH)_2$-26,27-dihomo-D_3 to the inhibitory effects of serum on antiproliferative activity in cultures of lymphocytes. Subsequent studies with cultured cells have shown that other analogs with high VDR affinity but lower DBP affinity, such as 1,25$(OH)_2$-26,27-F_6-D_3 [51], 1,25$(OH)_2$-16-ene-D_3 [52,53], and 20-epi-1,25$(OH)_2D_3$ [54], display higher activity than expected from their VDR affinity only when the treatments are carried out in the presence of serum.

The influence of DBP interaction on tissue uptake *in vivo* has been examined using 22-oxa-1,25$(OH)_2D_3$. Brown *et al.* [48] found that when radiolabeled 22-oxa-1,25$(OH)_2D_3$ or 1,25$(OH)_2D_3$ was injected into vitamin D-deficient rats, the peak levels of 22-oxa-1,25$(OH)_2D_3$ associated with intestinal VDR were 2-fold higher than those for 1,25$(OH)_2D_3$ [48], despite the fact that peak circulating levels of 22-oxa-1,25$(OH)_2D_3$ were 4 times lower in these animals. The dose dependence of intestinal uptake and VDR binding in vitamin D-deficient rats is shown in Fig. 4. Greater accessibility of 22-oxa-1,25$(OH)_2D_3$ to target tissues has been confirmed by Kobayashi *et al.* [44], who also found greater peak levels of 22-oxa-1,25$(OH)_2D_3$ in many other tissues, including parathyroid, liver, kidney, adrenal, and thymus (Fig. 5). Interestingly, less 22-oxa-1,25$(OH)_2D_3$ accumulated in calvaria. This could explain, at least in part, the observation that 22-oxa-1,25$(OH)_2D_3$ has much lower bone mobilizing activity than 1,25$(OH)_2D_3$ *in vivo* [55] but is equally active with 1,25$(OH)_2D_3$ in mobilizing calcium in bone organ cultures [56]. The reason for the lower uptake of 22-oxa-1,25$(OH)_2D_3$ by bone is not known, but it may indicate a unique delivery mechanism for this tissue or rapid intracellular degradation.

D. Association with Lipoproteins

Another consequence of low DBP binding is the freedom of the analogs to associate with other serum proteins. This could potentially alter the delivery of vitamin

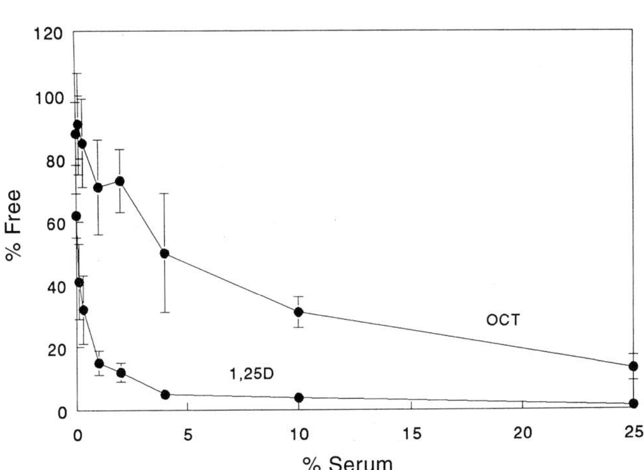

FIGURE 3 Percent free 22-oxa-1,25$(OH)_2D_3$ (OCT) and 1,25$(OH)_2D_3$ (1,25D) in the presence of serum. 1,25$(OH)_2D_3$ and 22-oxa-1,25$(OH)_2D_3$ were incubated for 30 min at 37°C in phosphate-buffered saline (PBS) containing 0.1% bovine serum albumin (BSA) and the specified concentration of human serum, and then bound and free compounds were separated by centrifugal ultrafiltration. Reprinted by permission from Dusso *et al.* [43].

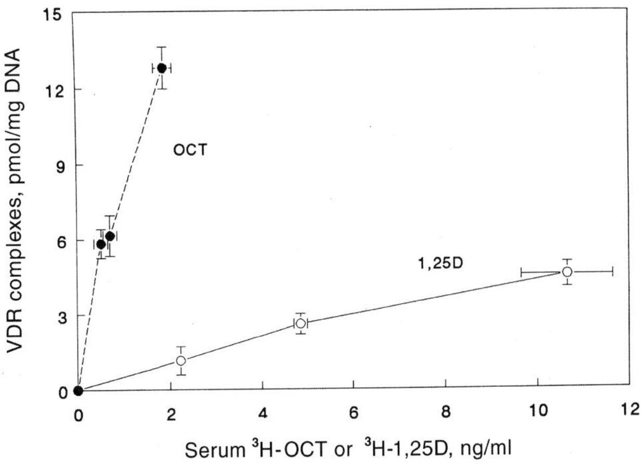

FIGURE 4 Relationship between serum levels *in vivo* and intestinal VDR binding following injection of increasing doses of ^{3}H-1,25$(OH)_2D_3$ (1,25D) or ^{3}H-22-oxa-1,25$(OH)_2D_3$ (OCT). Vitamin D-deficient rats were injected with ^{3}H-labeled 1,25$(OH)_2D_3$ or 22-oxa-1,25$(OH)_2D_3$ at doses of 0.5, 1.0, or 2.0 μg/kg. At the time of peak binding [30 min for 22-oxa-1,25$(OH)_2D_3$, 120 min for 1,25$(OH)_2D_3$] the amount of intact vitamin D compound in the serum and associated with intestinal VDR was determined. Adapted from Brown *et al.* [48] and reprinted with permission.

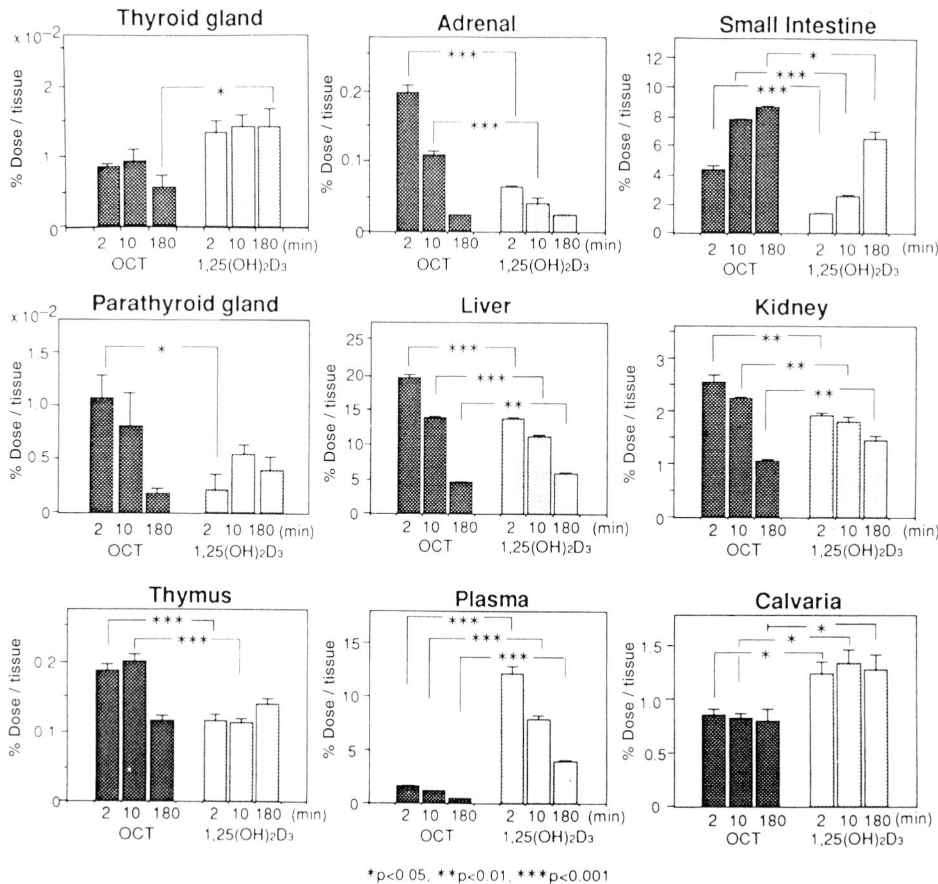

FIGURE 5 Time course of tissue distribution of ^{3}H-22-oxa-1,25$(OH)_2D_3$ and ^{3}H-1,25$(OH)_2D_3$ following intravenous injection into normal rats. Reprinted by permission from Kobayashi *et al.* [44].

D analogs to various target tissues. Currently, there is considerable disagreement as to the degree of binding of vitamin D metabolites and analogs to the various potential carriers. Okano *et al.* [42] and Kobayashi *et al.* [44] reported differences in the association of 22-oxa-1,25$(OH)_2D_3$ and 1,25-$(OH)_2D_3$ with serum lipoproteins. When 22-oxa-1,25$(OH)_2D_3$ was mixed *in vitro* with human plasma, it bound almost exclusively (99%) with the lipoprotein fraction, mainly with chylomicrons and low density lipoproteins (LDL), whereas less 1,25$(OH)_2D_3$ was bound to lipoprotein (60%). A similar study by Teramoto *et al.* [58] reported much less binding of these two compounds to lipoprotein fractions. The significance of the differences in lipoprotein binding in the selectivity of 22-oxa-1,25$(OH)_2D_3$ is unclear, as the role of carrier proteins in the delivery of vitamin D compounds has received only slight attention [59–62]. At present, a comparison of binding of 22-oxa-1,25$(OH)_2D_3$ and 1,25$(OH)_2D_3$ to lipoproteins has been done only in human plasma and only *in vitro*. Therefore, the relevance to the selectivity observed in experimental animals is not clear. Better evidence for a mechanism for selectivity involving carrier protein distribution would be the demonstration of differences in the binding of vitamin D analogs and 1,25$(OH)_2D_3$ to rat lipoproteins *in vivo*.

E. Influence of Altered Pharmacokinetics on Calcemic Activity

The impact of low DBP affinity and altered pharmacokinetics on the calcemic activities has been investigated in greatest detail with 22-oxa-1,25$(OH)_2D_3$. Brown *et al.* [48] followed the changes in intestinal calcium transport following a single injection of 22-oxa-1,25$(OH)_2D_3$ or 1,25$(OH)_2D_3$. Both compounds enhanced calcium absorption by 4 hr; the increase was slightly more rapid with 22-oxa-1,25$(OH)_2D_3$ (Fig. 6). Later time points, however, revealed major differences in the actions of 22-oxa-1,25$(OH)_2D_3$ and 1,25$(OH)_2D_3$. While 1,25$(OH)_2D_3$ elicited a prolonged increase in cal-

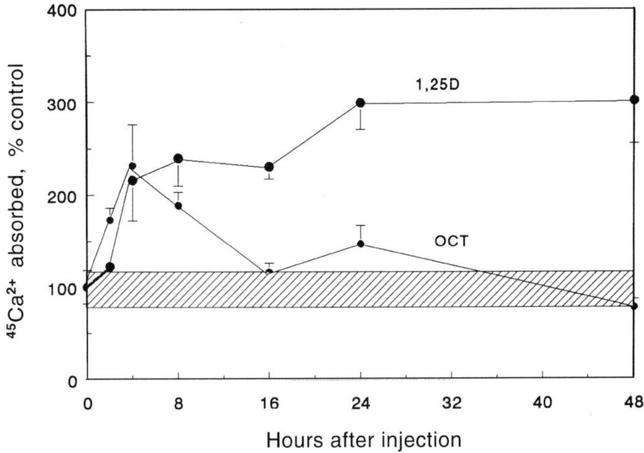

FIGURE 6 Time course for enhancement of intestinal calcium transport by $1,25(OH)_2D_3$ (1,25D) or 22-oxa-$1,25(OH)_2D_3$ (OCT). Vitamin D-deficient rats were injected with 1μg/kg $1,25(OH)_2D_3$ or 22-oxa-$1,25(OH)_2D_3$, and calcium transport was measured by the *in situ* duodenal loop method at the specified time. Reprinted by permission from Brown *et al.* [48].

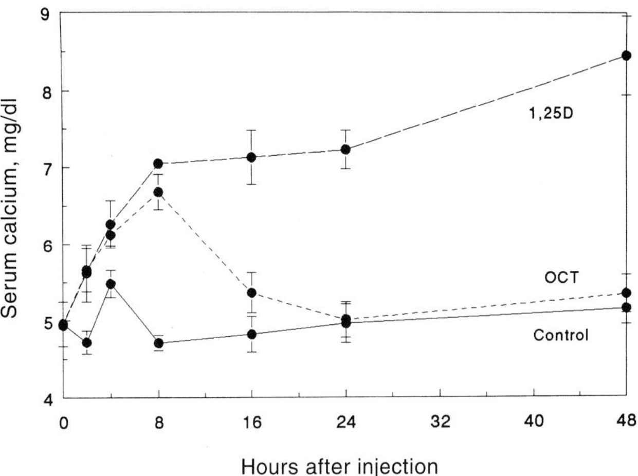

FIGURE 7 Bone calcium mobilization following administration of $1,25(OH)_2D_3$ (1,25D) or 22-oxa-$1,25(OH)_2D_3$ (OCT). Vitamin D-deficient rats were maintained on a calcium-deficient diet for 5 days and then injected intraperitoneally with 1 μg/kg $1,25(OH)_2D_3$ or 22-oxa-$1,25(OH)_2D_3$. Serum calcium, a measure of bone resorption, was measured at the specified time.

cium transport that persisted for at least 48 hr, 22-oxa-$1,25(OH)_2D_3$ was unable to sustain the elevated rate of calcium uptake, which returned to pretreatment levels within 16 hr. On the other hand, when 22-oxa-$1,25(OH)_2D_3$ was infused into the rats for 3 days, calcium transport rates remained high; they returned to basal levels within 24 hr of cessation of the infusion (data not shown). These observations demonstrated that 22-oxa-$1,25(OH)_2D_3$ is active in the intestine and can stimulate calcium transport, but that constant exposure is required for enhanced calcium absorption. The results also suggested the existence of inducible gene products with short half-lives that are critical for maintaining high rates of calcium uptake. This appeared not to be calbindin D_{9k} since transcripts for this vitamin D-inducible calcium-binding protein remained elevated after calcium transport returned to basal levels [48]. Thus, it appears that the rapid clearance of 22-oxa-$1,25(OH)_2D_3$, owing to its low DBP affinity, prevents a sustained stimulation of intestinal calcium absorption.

Rapid clearance may also explain the results of Krisinger *et al.* [39] with the 24-homologs of $1,25(OH)_2D_3$ and those of Wang *et al.* [40] with $1,25,28(OH)_3D_2$, who found induction of calbindin D_{9k} without changes in calcium transport. These analogs all have reduced affinities for DBP [42,56]. ED-71, on the other hand, has been shown to have lower but more long-term effects than $1,25(OH)_2D_3$ on intestinal calcium transport and bone mobilization [45,63], consistent with a longer half-life and lower tissue accessibility resulting from its high DBP affinity.

Bone calcium mobilization is also increased transiently following 22-oxa-$1,25(OH)_2D_3$ (Fig. 7). A single injection of 22-oxa-$1,25(OH)_2D_3$ into vitamin D-deficient rats on a calcium-deficient diet produced an increase in serum calcium that returned to pretreatment levels within 24 hr, whereas the same treatment with $1,25(OH)_2D_3$ increased serum calcium for more than 48 hr. The changes in serum calcium under these conditions are due to release from bone, as there is no dietary contribution. This short-lived effect of 22-oxa-$1,25(OH)_2D_3$ on bone mobilization also is likely the result of its rapid clearance.

Further evidence for the role of pharmacokinetics is supported by several *in vitro* studies in bone cell cultures. It is well established that 22-oxa-$1,25(OH)_2D_3$ has low bone mobilizing activity *in vivo* [55,64], but in organ cultures of rat calvaria 22-oxa-$1,25(OH)_2D_3$ is equipotent to $1,25(OH)_2D_3$ in stimulating calcium release [56]. 22-Oxa-$1,25(OH)_2D_3$ has also been shown to be as active as $1,25(OH)_2D_3$ in stimulating osteocalcin and alkaline phosphatase activity in rat [65,66] and human [67] osteosarcoma cells. Similarly, calcipotriol, which is roughly 100 times less calcemic and calciuric than $1,25(OH)_2D_3$ [68], is equipotent to $1,25(OH)_2D_3$ in inducing alkaline phosphatase and osteocalcin in human osteoblast-like cells [67,69,70] and in osteoclast formation [71] *in vitro*. Thus, the low activity of these compounds in the bone appears to be an *in vivo* phenomenon, and it is probably related to their altered pharmacokinetics.

It is possible that other analogs with low DBP affinity

would display pharmacokinetic properties that are qualitatively similar to those of 22-oxa-1,25$(OH)_2D_3$, and therefore their ability to regulate intestinal calcium transport and bone mobilization may be limited by their rapid clearance. However, additional mechanisms likely play an important role and could account for unique differences among similar analogs. This is clearly true for 19-nor-1,25$(OH)_2D_2$, whose low calcemic activity cannot be attributed to rapid clearance or to low DBP affinity.

F. Differential Metabolism

Another potential mechanism for the selectivity of vitamin D analogs is tissue-specific metabolism (see Chapter 58 for a more detailed discussion). Precedent for this has been demonstrated for other steroid hormone systems [72]. In mineralocorticoid-responsive tissues, the receptor that mediates the actions of aldosterone can bind glucocorticoids with similar affinity. Glucocorticoids display little mineralocorticoid activity because these tissues express the 11β-hydroxylase that selectivity and efficiently metabolizes and inactivates glucocorticoids. Alternatively, hormones can be enzymatically converted to more active metabolites, for example, thyroxine (T_4) to T_3 or estrone (E_1) to estradiol (E_2) in breast tissue [72]. In the case of the vitamin D analogs, selectivity could be achieved by more rapid inactivation than for 1,25$(OH)_2D_3$ in nonresponsive tissues or by conversion to active metabolite in responsive tissues. There is now evidence that 1,25$(OH)_2D_3$ and its analogs are metabolized at different rates, possibly in a cell-specific manner. In addition, metabolism of some analogs terminates at intermediates that have significant biological activities. These differences could play a role in the selectivity of the analogs.

G. Altered Rates of Catabolic Inactivation

Evidence for a role of differential metabolism of vitamin D analogs is beginning to accumulate, although at present it is not possible to assign a clear role for metabolism in the *in vivo* selectivity. For example, the rate of catabolism of 22-oxa-1,25$(OH)_2D_3$ relative to 1,25$(OH)_2D_3$ appears to differ with cell type. Brown *et al.* [73] found that, in cultured bovine parathyroid cells, 22-oxa-1,25$(OH)_2D_3$ and 1,25$(OH)_2D_3$ are catabolized at the same rate and can compete with one another for the same rate-limiting step in the catabolic pathway (presumably 24-hydroxylation). On the other hand, Kamimura *et al.* [74] compared the catabolism of these two compounds in cultured human monocytes and found that 22-oxa-1,25$(OH)_2D_3$ is relatively ineffective in competing with 1,25$(OH)_2D_3$ for the catabolic pathway. Kinetic analysis indicated a higher apparent K_m for 22-oxa-1,25$(OH)_2D_3$ in these cells. The slower rate of metabolism could result in greater biological activity of 22-oxa-1,25$(OH)_2D_3$ in monocytic cells. In contrast, Bikle *et al.* [75] observed that 22-oxa-1,25$(OH)_2D_3$ was catabolized more rapidly than 1,25$(OH)_2D_3$ in cultured human keratinocytes, which could explain, at least in part, its lower potency in regulating keratinocyte proliferation and differentiation. A decreased rate of metabolism of the 20-epi analog may also account for its high differentiating activity [76] *in vitro*. Dilworth *et al.* [54] found that 20-epi-1,25$(OH)_2D_3$ was metabolized 36 times more slowly than 1,25$(OH)_2D_3$ by rat osteosarcoma (UMR-106) cells. ED-71 also appears to be a poorer substrate than 1,25$(OH)_2D_3$ in osteoblast-like (UMR-106) cells based on competition studies; this may provide a partial explanation for the high bone forming activity of this analog [77]. Taken together, these *in vitro* observations provide evidence for a potential role for the catabolic pathway in modulating the activity of vitamin D analogs in a cell-specific manner that could play a role in the *in vivo* selectivity.

H. Accumulation of Active Metabolites of the Analogs

In addition to tissue-specific differences in the rates of metabolism, there may be differences in the end products of metabolism of the analogs. Most of the selective vitamin D analogs are modified in the side chain, the site of oxidation by the enzymes of the catabolic pathway. These modifications could alter enzymatic oxidation and inactivation of the analogs. An example is 1,25$(OH)_2$-16-ene-D_3. This analog can be converted efficiently to the 24-hydroxy and 24-oxo intermediates, but further oxidation at carbon 23 occurs very slowly [78,79]. Accumulation of the 24-oxo metabolite in target cells could have significant biological effects. Lemire *et al.* [35] tested both 1,25$(OH)_2$-16-ene-D_3 and 1,25$(OH)_2$-16-ene-24-oxo-D_3 for immunosuppressive activities in mice with experimentally induced autoimmune encephalomyelitis. The 24-oxo-16-ene metabolite was as active as the parent 16-ene analog, and both 16-ene compounds were more active than 1,25$(OH)_2D_3$. Interestingly, 1,25$(OH)_2$-16-ene-24-oxo-D_3 did not produce hypercalcemia at its effective doses, whereas the parent 1,25$(OH)_2$-16-ene-D_3 appeared to be as calcemic as 1,25$(OH)_2D_3$ in this study. These findings indicate that the 24-oxo metabolite displays even more selectivity than the parent 1,25$(OH)_2$-16-ene-D_3. The reason for the low calcemic activity of 1,25$(OH)_2$-24-oxo-16-

ene-D_3 is still unclear. More recent *in vitro* studies have shown that 1,25$(OH)_2$-16-ene-D_3 and its 24-oxo metabolite are more active in differentiating human leukemia cells and in activating a reporter gene driven by a promoter containing the osteocalcin gene vitamin D response element [80].

Aborted side chain oxidation of 20-epi-1,25$(OH)_2D_3$(MC1288, Leo Pharmaceuticals) may also contribute to the high activity of this analog in cell differentiation [54,76]. Dilworth *et al.* [54] examined the metabolism of 20-epi-1,25$(OH)_2D_3$ in UMR-106 cells and found that this analog, like 1,25$(OH)_2$-16-ene-D_3, was converted to the 24-oxo metabolite, but further oxidation at carbon 23 occurred very slowly. The activity of 20-epi-1,25$(OH)_2$-24-oxo-D_3 was not tested; however, by analogy to other vitamin D compounds, it is likely to retain significant biological activity.

Studies by Reddy and collaborators [81] have identified a new metabolite of 1,25$(OH)_2D_3$: 3α-1,25$(OH)_2D_3$. This metabolite, first identified in keratinocytes [81], is also produced by parathyroid cells [82] and bone cells but not by the kidney or by HL-60 cells (G. S. Reddy, personal communication, 1996). Of particular interest is the observation that 3α-1,25$(OH)_2D_3$ is nearly as active as the native 3β-1,25$(OH)_2D_3$ in suppressing PTH secretion from cultured bovine parathyroid cells but is metabolized more slowly and therefore can accumulate in these cultures. This slower inactivation, if present in the intact animal, may lead to more prolonged effects *in vivo*. At present there is little information about the 3-epimerization of vitamin D compounds in the intestine or the potential role of 3-epimerization on intestinal calcium transport. Also, the ability of vitamin D analogs to undergo 3-epimerization has not been investigated. Potentially, tissue-specific 3-epimerization (or lack of it) could play a role in the selectivity of vitamin D analogs. Further details on the metabolism of vitamin D analogs are presented in Chapter 58.

I. Nongenomic Activity

A number of vitamin D analogs have been shown to possess altered abilities to mimic the nongenomic activities of 1,25$(OH)_2D_3$ [83–86]. The nongenomic actions include rapid changes in lipid metabolism, activation of calcium channels, and increases in intracellular calcium levels *in vitro* [3]. These effects occur within seconds of exposure of cultured cells to 1,25$(OH)_2D_3$ and appear to be mediated by an uncharacterized cell surface receptor that is distinct from the nuclear receptor for 1,25$(OH)_2D_3$ [87,88] (see Chapter 15). Ligand specificity for this putative membrane receptor is also different [83,87], and a number of analogs have been demonstrated to have very low nongenomic activity.

The exact function of the nongenomic actions of 1,25$(OH)_2D_3$ in most cell types is still unclear. In the intestine, it is well established that exposure of the basolateral membrane to 1,25$(OH)_2D_3$ *in vitro* stimulates calcium movement from the lumen to the bloodstream, a process called transcaltachia [83]. In isolated chondrocytes, nongenomic actions of 1,25$(OH)_2D_3$ and 24,25$(OH)_2D_3$ alter membrane lipid turnover, prostaglandin production, and protease activity, which leads to modification of bone matrix and calcification [89]. In other cultured cells, nongenomic events have been proposed to modulate the genomic actions of 1,25$(OH)_2D_3$ [90,91], but this remains highly controversial. Numerous studies have presented evidence that the nongenomic actions are not critical for 1,25$(OH)_2D_3$-mediated gene activation [84,85,92–95] or inhibition of cell proliferation [93,96]. These uncertainties make it difficult to assign a definitive role for nongenomic activity in the *in vivo* selectivity of vitamin D analogs. Thus, we can only speculate regarding a potential scenario involving both pharacokinetics and the nongenomic pathway.

22-Oxa-1,25$(OH)_2D_3$ has been shown to stimulate transcaltachia in the chick intestine [85]. Transcaltachia requires a vitamin D-replete state, indicating that 1,25$(OH)_2D_3$-dependent gene products are necessary. When vitamin D-deficient rats were treated with 22-oxa-1,25$(OH)_2D_3$, only a transient increase in calcium transport was observed (Fig. 6) [48]. The increase indicates that the gene products required for calcium transport have been induced, but the reason for the rapid decline after 22-oxa-1,25$(OH)_2D_3$ is cleared is unknown. Although the return to basal transport rates may be due to turnover of a short-lived gene product, it is also possible that the rapid disappearance of 22-oxa-1,25$(OH)_2D_3$ prevents continued nongenomic activation in the intestine, leading to the decrease in calcium flux. At present, there is no experimental evidence to support (or disprove) this hypothesis.

Other selective analogs such as 1,25$(OH)_2$-16-ene-23-yne-D_3 and calcipotriol have been shown to have very low transcaltachia activity [83], which could explain their low calcemic activity in the intestine. Several other analogs, including 1,25,28-trihydroxyvitamin D_2, 1,25$(OH)_2$-24-dihomo-22-ene-D_3, and 1,25$(OH)_2$-24-trihomo-22-ene-D_3, that stimulate genomic responses (induction of calbindin D_{9K}) in the intestine but not calcium transport [39,40] may lack nongenomic activity, but these analogs have not been tested. Similarly, the low calcemic activity of 19-nor-1,25$(OH)_2D_2$ may be due to an inability to stimulate transcaltachia.

As in other tissues and cells, the role of the nongenomic pathway in bone is not so clear. As described above, the nongenomic effects of $1,25(OH)_2D_3$ and $24,25(OH)_2D_3$ may play a role in bone formation [89]. The role of nongenomic effects of $1,25(OH)_2D_3$ in bone resorption is also not known. It is possible that analogs which are incapable of activating or sustaining a nongenomic response may be unable to efficiently induce or maintain bone resorption. Clearly, much additional information on the nongenomic mechanism of action of vitamin D will be required to assess its contribution to the tissue-selective actions of vitamin D analogs.

J. Interaction with the Nuclear Vitamin D Receptor and Gene Regulation

The selective vitamin D analogs have relatively high affinity for the vitamin D receptor (VDR), usually within one order of magnitude of the affinity of $1,25(OH)_2D_3$ based on equilibrium dissociation constants (K_D). This may not fully predict the functional activity of these compounds, as the K_D is a ratio of the dissociation rate constant (k_d) to the association rate constant (k_a). Thus, compounds with similar K_D values may have different k_d and k_a values. This could lead to differences in overall activity of the vitamin D analogs. In the androgen system, it has been shown that ligands with high k_d (e.g., testosterone) are less active transcriptionally and less able to stabilize the androgen receptor than are ligands with low k_d (e.g., dihydrotestosterone), despite the similar K_D values of the two ligands [97]. We have found (F. Takahashi, unpublished data, 1996) that 19-nor-$1,25(OH)_2D_2$ is unable to increase VDR content in the intestine, a process attributed to ligand stabilization [99–100]. This may indicate a rapid dissociation rate for 19-nor-$1,25(OH)_2D_2$, although other possibilities (e.g., lack of uptake) exist.

Studies have demonstrated that vitamin D analogs can bind to the VDR in a manner that alters VDR complexing with retinoid X receptor (RXR) and vitamin D response elements (VDREs) (see Chapter 60 for an expanded discussion of this topic). Peleg *et al.* [101] demonstrated that two 20-epi analogs of $1,25(OH)_2D_3$ can induce distinct conformational changes in the VDR that increase heterodimerization with the RXR and enhance transcriptional activation, despite displaying K_Ds for VDR similar to that of $1,25(OH)_2D_3$. Sasaki *et al.* found a similar effect of 26,27-hexafluoro-$1,25(OH)_2D_3$ on VDR–RXR interactions with a VDRE [102]. These data [101,102] may also reflect slower rates of dissociation for the analogs, as it is possible for the vitamin D ligands to dissociate during this analysis.

The dissociation kinetics of vitamin D analogs and their influence on analog activity have only relatively recently begun to be examined. Using state-of-the-art surface plasmon resonance, Cheskis *et al.* [103] found that the VDR liganded with any of three analogs [$1,25(OH)_2$-16-ene-23-yne-D_3, $1,25(OH)_2$-16-ene-23-yne-26,27-dihomo-D_3, and $1,25(OH)_2$-16-ene-23-yne-26,27-F_6D_3] had lower affinity (higher K_D) for RXR than VDR liganded with $1,25(OH)_2D_3$. This was due primarily to increased k_d values for the VDR liganded with the analogs. Furthermore, the VDR–RXR complex liganded with $1,25(OH)_2$-16-ene-23-yne-26,27-hexafluoro-D_3 had a higher affinity for the VDRE (lower K_D) than VDR–RXR liganded with $1,25(OH)_2D_3$; this was due primarily to a decrease in the k_d. These findings may shed light on the discrepancy between the VDR affinity of vitamin D analogs and their transcriptional activity. However, other factors must be involved to confer the tissue- or gene-specific selectivity observed *in vivo*.

The control of transcription of target genes involves a complex association of a number of proteins in addition to the VDR and RXR. It is likely that some of the factors involved in transcriptional regulation by vitamin D compounds will be gene-specific. Therefore, it is possible that altered conformations of the VDR induced by the analogs could potentially affect gene transcription in a gene-specific manner, providing an explanation for the selectivity of the vitamin D analogs. For example, in the intestine, where several vitamin D analogs have been shown to induce calbindin D_{9k} but not calcium transport [39,40], it is possible that the ligand may induce a conformation of the receptor that cannot induce transcription of a gene or genes required for increased calcium absorption.

K. Suppression of Calcitropic Hormones

An important contributing factor to the low calcemic activity of vitamin D analogs is their ability to suppress the circulating levels of the endogenous calcitropic hormones PTH and $1,25(OH)_2D_3$. Finch *et al.* [104] demonstrated that at least part of the low calcemic activity of 22-oxa-$1,25(OH)_2D_3$ can be attributed to the decrease in serum PTH levels. 22-Oxa-$1,25(OH)_2D_3$ was significantly more calcemic in rats that had been parathyroidectomized and PTH-repleted by osmotic minipumps. It is likely that similar results would be observed with other vitamin D analogs that suppress PTH.

The natural derivative $1,25(OH)_2D_3$ self-regulates its levels in the blood by inhibiting its own synthesis and inducing its catabolism. Vitamin D analogs have been shown to mimic this action and decrease the levels of endogenous $1,25(OH)_2D_3$. Dusso *et al.* [105] documented that 22-oxa-$1,25(OH)_2D_3$ can produce a profound decrement in serum $1,25(OH)_2D_3$ levels in rats and dogs. The decrease in dogs was shown to be due to both decreased synthesis and increased clearance of $1,25(OH)_2D_3$. Grieff *et al.* [106] demonstrated that inhibition of renal 1α-hydroxylase activity persists for up to 48 hr following injection of 22-oxa-$1,25(OH)_2D_3$. Decreased $1,25(OH)_2D_3$ levels have also been reported following chronic treatment with ED-71 [107]. Furthermore, the report by Akeno *et al.* [108] showed that 22-oxa-$1,25(OH)_2D_3$ was more potent than $1,25(OH)_2D_3$ in inducing intestinal 24-hydroxylase activity, perhaps the result of the greater accessibility of 22-oxa-$1,25(OH)_2D_3$ to the intestinal VDR [48]. Rapid intracellular degradation of endogenous $1,25(OH)_2D_3$ as well as exogenously administered 22-oxa-$1,25(OH)_2D_3$ could play a key role in the low calcemic activity observed with 22-oxa-$1,25(OH)_2D_3$ treatment.

III. CONCLUDING REMARKS

The mechanisms responsible for the *in vivo* selectivity of vitamin D analogs are beginning to emerge. From the discussion above, it is known that certain vitamin D analogs differ from the parent $1,25(OH)_2D_3$ in their pharmacokinetics, intracellular metabolism, nongenomic actions, or interaction with the vitamin D receptor. No single mechanism can be implicated in the selectivity of all analogs, and it is possible that more than one of these differences plays a role in the overall activity observed *in vivo*. However, a few general conclusions can be drawn from the information gathered from numerous studies.

The predominant form of selectivity observed is that the analogs have much lower calcemic activity than expected on the basis of their nuclear VDR affinity and apparent high activity *in vitro*. The majority of these analogs have been found to have low affinity for DBP and therefore are more rapidly cleared from the circulation, but, at the same time, they may be more accessible to the target tissue. There is good *in vivo* evidence that altered pharmacokinetics due to low DBP affinity plays a role in the selectivity of 22-oxa-$1,25(OH)_2D_3$. It has been proposed [48] that the greater uptake and rapid clearance of this analog provides a "pulse" of activity that leads to biological effects with a duration dependent on the half-life of the induced genes involved. Thus, the long-term effects on PTH, 1α-hydroxylase, and calbindin D_{9K} are due to the relatively long half-lives of the mRNAs and/or proteins, whereas the transient effects on intestinal calcium and transport and bone mobilization may reflect the short half-lives of gene products critical for these processes. It is clear that selectivity is lost when 22-oxa-$1,25(OH)_2D_3$ is administered by constant infusion [48]. Furthermore, ED-71, which is more slowly cleared than $1,25(OH)_2D_3$, has more prolonged effects on intestinal calcium transport [45,63].

Clearly, pharmacokinetics can play an important role in the selectivity of vitamin D analogs. Are the other mechanisms described above completely independent, or do their pathways converge? It is possible to conceive scenarios in which the other altered activities of the analogs can produce the same differential effects as pharmacokinetics. For example, analogs that are rapidly degraded intracellularly may produce the same "pulse" of activity produced by 22-oxa-$1,25(OH)_2D_3$, whereas those that are degraded slowly may produce *in vivo* responses similar to those of slowly cleared compounds. Altered VDR–analog conformations may produce heterogeneous induction of target genes that could lead to activation of arrays of genes which differ for various analogs. Variable abilities of the analogs to stimulate (or sustain) nongenomic signaling may influence the genes and/or processes that are activated. Given the complexity of factors involved in determining the overall *in vivo* activity of vitamin D compounds, it is clear that considerable additional research will be required to determine the mechanisms(s) responsible for the unique effects of each analog. These analogs, with their altered properties, may in turn serve as tools for defining previously unappreciated factors that play a critical role in the actions of $1,25(OH)_2D_3$ and other vitamin D compounds.

References

1. Haussler MR, Jurutka PW, Hsieh JC, Thompson PD, Selznick SH, Haussler CA, Whitfield GK 1995 New understanding of the molecular mechanism of receptor-mediated actions of the vitamin D hormone. Bone **17**:33S–38S.
2. Ross TK, Darwish HM, DeLuca HF 1994 Molecular biology of vitamin D action. Vitam Horm **49**:281–326.
3. Baran DT 1994 Nongenomic actions of the steroid hormone 1α,25-dihydroxyvitamin D_3. J Cell Biochem **56**:303–306.
4. Walter MR 1992 Newly identified actions of the vitamin D endocrine system. Endocr Rev **13**:719–764.
5. Slatopolsky E, Weerts C, Thielan J, Horst R, Harter H, Martin KJ 1984 Marked suppression secondary hyperparathyroidism by intravenous administration of 1,25-dihydroxycholecalciferol in uremic patients. J Clin Invest **74**:2136–2143.
6. Smith EL, Pincus SH, Donovan L, Holick MF 1988 A novel approach for the evaluation and treatment of psoriasis, or topical use of 1,25-dihydroxyvitamin D_3 can be safe and effective therapy for psoriasis. J Am Acad Dermatol **19**:516.
7. Drezner MK, Lyles KW, Haussler MR, Harrelson JM 1980 Evaluation of a role for 1,25-dihydroxyvitamin D_3 in the pathogenesis

and treatment of X-linked hypophosphatemic rickets and osteomalacia. J Clin Invest **66**:1020–1032.

8. Studzinski GP, McLane JA, Uskokovic MR 1993 Signalling pathways for vitamin D-induced differentiation: Implications for therapy of proliferative and neoplastic diseases. Crit Rev Euk Gene Expression **3**:279–312.
9. Bikle DD 1992 Clinical counterpoint: Vitamin D: New actions, new analogs, new therapeutic potential. Endocr Rev **13**:765–788.
10. Brown AJ, Dusso A, Slatopolsky E 1994 Selective vitamin D analogs and their therapeutic application. Semin Nephrol **14**:156–174.
11. Norman AW 1995 The vitamin D endocrine system: Manipulation of structure–function relationships to provide opportunities for development of new cancer chemopreventive and immunosuppressive agents. J Cell Biochem **22**:S218–S225.
12. Murayama E, Miyamoto K, Kubordera N, Mori T, Matsunaga I 1987 Synthetic studies of vitamin D_3 analogues. VIII. Synthesis of 22-oxavitamin D_3 analogues. Chem Pharm Bull **34**:4410–4413.
13. Ostrem VK, Tanaka Y, Prahl J, DeLuca HF, Ikekawa N 1987 24- and 26-homo-1,25-dihydroxyvitamin D_3: Preferential activity in inducing differentiation of human leukemia cells HL-60 in vitro. Proc Natl Acad Sci USA **84**:2610–2614.
14. Binderup L, Bramm E 1988 Effects of a novel vitamin D analogue MC903 on cell proliferation and differentiation in vitro and on calcium metabolism in vivo. Biochem Pharmacol **37**:889–895.
15. Zhou JY, Norman AW, Lubbert M, Collins ED, Uskokovic MR, Koeffler HP 1989 Novel vitamin D analogs that modulate leukemic cell growth and differentiation with little effect on either intestinal calcium absorption or bone calcium mobilization. Blood **74**:82–93.
16. Chen TC, Persons K, Uskokovic MR 1992 An evaluation of 1,25-dihydroxyvitamin D_3 analogues on the proliferation and differentiation of cultured human keratinocytes, calcium metabolism and differentiation of HL-60 cells. J Nutr Biochem **4**:49–57.
17. Brown AJ, Ritter CS, Finch JL, Morrissey J, Martin KJ, Murayama E, Nishii Y, Slatopolsky E 1989 The noncalcemic analogue of vitamin D, 22-oxacalcitriol, suppresses parathyroid hormone synthesis and secretion. J Clin Invest **84**:728–732.
18. Naveh-Many T, Silver J 1993 Effects of calcitriol, 22-oxacalcitriol, and calcipotriol on serum calcium and parathyroid hormone gene expression. Endocrinology **133**:2724–2728.
19. Slatopolsky E, Finch J, Ritter C, Denda M, Morrissey J, Brown AJ, DeLuca HF 1995 A new analog of calcitriol, 19-nor-D_2, suppresses parathyroid hormone secretion in uremic rats in the absence of hypercalcemia. Am J Kidney Dis **26**:852–860.
20. Endo K, Ichikawa F, Uchiyama Y, Katsumata K, Ohkawa H, Kumaki K, Ogata E, Ikeda K 1994 Evidence for the uptake of a vitamin D analogue (OCT) by a human carcinoma and its effect of suppressing the transcription of parathyroid hormone-related peptide gene in vivo. J Biol Chem **269**:32693–32699.
21. Haq M, Kremer R, Goltzman D, Rabbani SA 1993 A vitamin D analogue (EB1089) inhibits parathyroid hormone-related peptide production and prevents the development of malignancy-associated hypercalcemia in vivo. J Clin Invest **91**:2416–2422.
22. Abe J, Nakano T, Nishii Y, Matsumoto T, Ogata E, Ikeda K 1991 A novel vitamin D_3 analog, 22-oxa-1,25-dihydroxyvitamin D_3, inhibits the growth of human breast cancer in vitro and in vivo without causing hypercalcemia. Endocrinology **129**: 832–837.
23. Colston KW, Chander SK, MacKay AG, Coombes RC 1992 Effects of synthetic vitamin D analogues on breast cancer cell proliferation in vivo and in vitro. Biochem Pharmacol **44**:693–702.
24. Oikawa T, Yoshida Y, Shimamura M, Ashino-Fuss H, Iwaguchi T 1991 Antitumor effect of 22-oxa-1α,25-dihydroxyvitamin D_3, a potent angiogenesis inhibitor, on rat mammary tumors induced by 7,12-dimethylbenz[*a*]anthracene. Anti-Cancer Drugs **2**:475–480.
25. Anzano MA, Smith JM, Uskokovic MR, Peer CW, Mullen LT, Letterio JJ, Welsh MC, Shrader MW, Logsdon DL, Driver CL, Brown CC, Roberts AB, Sporn MB 1994 1α,25-Dihydroxy-16-ene,23-yne,26,27-hexafluorocholecalciferol (Ro24-5531), a new deltanoid (vitamin D analogue) for prevention of breast cancer in the rat. Cancer Res **34**:1653–1656.
26. Colston KW, MacKay AG, James SY, Binderup L, Chander S, Coombes RC 1992 EB1089: A new vitamin D analogue that inhibits the growth of breast cancer cells in vivo and in vitro. Biochem Pharmacol **44**:2273–2280.
27. Lucia MS, Anzano MA, Slayter MV, Anver MR, Green DM, Shrader MW, Logsdon DL, Driver CL, Brown CC, Peer CW, Roberts AB, Sporn MB 1995 Chemopreventive activity of tamoxifen, *N*-(4-hydroxyphenyl)retinamide, and the vitamin D analogue RO24-5531 for androgen-promoted carcinomas of the rat seminal vesicle and prostate. Cancer Res **55**:5621–5627.
28. Schwartz GG, Hill CC, Oeler TA, Becich MJ, Bahnson RR 1995 1,25-Dihydroxy-16-ene-23-yne-vitamin D_3 and prostate cancer in vivo. Urology **46**:365–369.
29. Otoshi T, Iwata H, Kitano M, Nishizawa Y, Morii H, Yano Y, Otani S, Fukushima S 1995 Inhibition of intestinal tumor development in rat multi-organ carcinogenesis and aberrant crypt foci in rat colon carcinogenesis by 22-oxa-calcitriol, a synthetic analogue of 1α,25-dihydroxyvitamin D_3. Carcinogenesis **16**:2091–2097.
30. Wali RK, Bissonnette M, Khare S, Hart J, Sitrin MD, Brasitus TA 1995 1α,25-Dihydroxy-16-ene-23-yne-26,27-hexafluorocholecalciferol, a noncalcemic analogue of 1α,25-dihydroxyvitamin D_3, inhibits azoxymethane-induced colonic tumorigenesis. Cancer Res **55**:3050–3054.
31. Zhou JY, Norman AW, Chen DL, Sun GW, Uskokovic M, Koeffler HP 1990 1,25-Dihydroxy-16-ene-23-yne-vitamin D_3 prolongs survival time of leukemic mice. Proc Natl Acad Sci USA **87**:3929–3932.
32. Abe J, Takita Y, Nakano T, Miyaura C, Suda T, Nishii Y 1989 A synthetic analogue of vitamin D_3, 22-oxa-1α,25-dihydroxyvitamin D_3, is a potent modulator of in vivo immunoregulating activity without inducing hypercalcemia in mice. Endocrinology **124**:2645–2647.
33. Mathieu C, Waer M, Casteels K, Laureys J, Bouillon R 1995 Prevention of type I diabetes in NOD mice by nonhypercalcemic doses of a new structural analog of 1,25-dihydroxyvitamin D_3, KH1060. Endocrinology **136**:866–872.
34. Mathieu C, Laureys J, Waer M, Bouillon R 1994 Prevention of autoimmune destruction of transplanted islets in spontaneously diabetic NOD mice by KH1060, a 20-epi analog of vitamin D: Synergy with cyclosporin. Transplant Proc **26**:3128–3129.
35. Lemire JM, Archer DC, Reddy GS 1994 1,25-Dihydroxy-24-oxo-16-ene-vitamin D_3, a renal metabolite of the vitamin D analog 1,25-dihydroxy-16-ene-vitamin D_3, exerts immunosuppressive activity equal to its parent without causing hypercalcemia in vivo. Endocrinology **135**:2818–1821.
36. Kragballe K 1995 Calcipotriol: A new drug for topical psoriasis treatment. Pharmacol Toxicol **77**:241–246.
37. Nishii Y, Sato K, Kobayashi T 1993 The development of vitamin D_3 analogues for the treatment of osteoporosis. Osteoporosis Int **1**:S190–S193.
38. Tanaka Y, Nakamura T, Nishida S, Suzuki K, Takeda S, Sato K, Nishii Y 1996 Effects of a synthetic vitamin D analog, ED-

71, on bone dynamics and strength in cancellous and cortical bone in prednisolone-treated rats. J Bone Miner Res **11**:325–336.

39. Krisinger J, Strom M, Darwish H, Perlman K, Smith C, DeLuca HF 1991 Induction of calbindin-D9k mRNA but not calcium transport in rat intestine by 1,25-dihydroxyvitamin D_3 24-homologs. J Biol Chem **266**:1910–1913.
40. Wang YZ, Li H, Bruns ME, Uskokovic M, Truitt GA, Horst R, Reinhardt T, Christakos S 1993 Effect of 1,25-28-trihydroxyvitamin D_2 and 1,24,25-trihydroxyvitamin D_3 on intestinal calbindin-D9k mRNA and protein: Is there a correlation with intestinal calcium transport? J Bone Miner Res **8**:1483–1490.
41. Elstner E, Lee YY, Hashiya M, Pakkala S, Binderup L, Norman AW, Okamura WH, Koeffler HP 1994 1α,25-Dihydroxy-20-epi-vitamin D_3: An extraordinarily potent inhibitor of leukemic cell growth in vitro. Blood **84**:1960–1967.
42. Okano T, Tsugawa N, Masuda S, Takeuchi A, Kobayashi T, Nishii Y 1989 Protein-binding properties of 22-oxa-1α,25-dihydroxyvitamin D_3, a synthetic analogue of 1α,25-dihydroxyvitamin D_3. J Nutr Sci Vitaminol (Tokyo) **35**:529–533.
43. Dusso A, Gunawardhana S, Negrea L, Finch JL, Lopez-Hilker S, Mori T, Nishii Y, Slatopolsky E, Brown AJ 1991 On the mechanisms for the selective action of vitamin D analogs. Endocrinology **128**:1687–1692.
44. Kobayashi T, Tsugawa N, Okano T, Masuda S, Takeuchi A, Kubodera N, Nishii Y 1994 The binding properties, with blood proteins, and tissue distribution of 22-oxa-1,25-dihydroxyvitamin D_3, a noncalcemic analogue of 1,25-dihydroxyvitamin D_3, in rats. J Biochem (Tokyo) **115**:373–380.
45. Okano T, Tsugawa N, Masuda S, Takeuchi A, Kobayashi T, Takita Y, Nishii Y 1989 Regulatory activities of 2β-(3-hydroxypropoxy)-1α,25-dihydroxyvitamin D_3, a novel synthetic vitamin D derivative, on calcium metabolism. Biochem Biophys Res Commun **163**:1444–1449.
46. Vanham G, van Baelen H, Tan BK, Bouillon R 1988 The effect of vitamin D analogs and of vitamin D binding protein on lymphocyte proliferation. Steroid Biochem **29**:381–386.
47. Bikle DD, Gee E 1989 Free, and not total, 1,25-dihydroxyvitamin D regulates 25-hydroxyvitamin D metabolism by keratinocytes. Endocrinology **124**:649–654.
48. Brown AJ, Finch J, Grieff M, Ritter C, Kubodera N, Nishii Y, Slatopolsky E 1993 The mechanism for the disparate actions of calcitriol and 22-oxacalcitriol in the intestine. Endocrinology **133**:2719–2724.
49. Kissmeyer AM, Binderup L 1991 Calcipotriol (MC903): Pharmacokinetics in rats and biological activities of metabolites. Biochem Pharmacol **41**:1601–1606.
50. Bouillon R, Allewaert K, Xiang DZ, Tan BK, van Baelen H 1991 Vitamin D analogs with low affinity for the vitamin D binding protein: Enhanced in vitro and decreased in vivo activity. J Bone Miner Res **6**:1051–1057.
51. Okuno S, Inaba M, Nishizawa Y, Morii H 1995 Biological activities of 26,26,26,27,27,27-hexafluoro-1,25-dihydroxyvitamin D_3 on human promyelocytic leukemic HL-60 cells: Effect of fetal bovine serum and of incubation time. Miner Electrolyte Metab **21**:211–216.
52. Jung SJ, Lee YY, Pakkala S, de Vos S, Elstner E, Norman AW, Green J, Uskokovic M, Koeffler HP 1994 1,25-$(OH)_2$-16-ene-vitamin D_3 is a potent antileukemic agent with low potential to cause hypercalcemia. Leuk Res **18**:453–463.
53. Skowronski RJ, Peehl DM, Feldman D 1995 Actions of vitamin D_3 analogs on human prostate cancer cell lines: Comparison with 1,25-dihydroxyvitamin D_3. Endocrinology **136**:20–26.
54. Dilworth FJ, Calverley MJ, Makin HL, Jones G 1994 Increased biological activity of 20-epi-1,25-dihydroxyvitamin D_3 is due to reduced catabolism and altered protein binding. Biochem Pharmacol **47**:987–993.
55. Takizawa M, Falton M, Stein B, Epstein S 1992 The effect of new vitamin D analog, 22-oxa-1α,25$(OH)_2D_3$, on bone mineral metabolism in normal male rats. Calcif Tissue Int **50**:521–523.
56. Sato K, Nishii Y, Woodiel FN, Raisz LG 1993 Effects of two new vitamin D_3 derivatives, 22-oxa-1α,25-dihydroxyvitamin D_3 (OCT) and 2β-(3-hydroxypropoxy)-1α,25-dihydroxyvitamin D_3 (ED-71), on bone metabolism in organ culture. Bone **14**:47–51.
57. Bishop JE, Collins ED, Okamura WH, Norman AW 1994 Profile of ligand specificity of the vitamin D binding protein for 1α,25-dihydroxyvitamin D_3 and its analogs. J Bone Miner Res **9**:1277–1288.
58. Teramoto T, Endo K, Ikeda K, Kubodera N, Kinoshita M, Yamanaka M, Ogata E 1995 Binding of vitamin D to low-density-lipoprotein (LDL) and LDL receptor-mediated pathway into cells. Biochem Biophys Res Commun **215**:199–204.
59. Haddad JG, Jennings AS, Aw TC 1988 Vitamin D uptake and metabolism by perfused rat liver: Influences of carrier proteins. Endocrinology **123**:498–504.
60. Haddad JG, Aden DP, Aw TC 1989 Plasma carrier proteins influence the uptake of cholecalciferol by human hepatoma-derived cells. J Bone Miner Res **4**:243–247.
61. Keenan MJ, Holmes RP 1991 The uptake and metabolism of 25-hydroxyvitamin D_3 and vitamin D binding protein by cultured porcine kidney cells (LLC-PK1). Int J Biochem **23**:1225–1230.
62. Esteban C, Geuskens M, Ena JM, Mishal Z, Macho A, Torres JM, Uriel J 1992 Receptor mediated uptake and processing of vitamin D-binding protein in human B-lymphoid cells. J Biol Chem **267**:10117–10183.
63. Brown AJ, Finch J, Ritter C, Grieff M, Dusso A, Kubodera N, Nishii Y, Slatopolsky E 1993 The importance of the serum vitamin D binding protein (DBP) in the calcemic activity of vitamin D analogs. J Am Soc Nephrol **4**:S718.
64. Abe J, Morikawa M, Miyamoto K, Kaiho S, Fukushima M, Miyaura C, Abe E, Suda T, Nishii Y 1987 Synthetic analogues of vitamin D_3 with an oxygen atom in the side chain. A trial of the development of vitamin D compounds which exhibit potent differentiation-inducing activity without inducing hypercalcemia. FEBS Lett **226**:58–62.
65. Pernalete N, Mori T, Nishii Y, Slatopolsky E, Brown AJ 1991 The activity of 22-oxacalcitriol in osteoblast-like (ROS 17/2.8) cells. Endocrinology **129**:778–784.
66. Morrison NA, Eisman JA 1991 Nonhypercalcemic 1,25$(OH)_2D_3$ analogs potently induce the human osteocalcin gene promoter stably transfected into rat osteosarcoma cells (ROSCO-2). J Bone Miner Res **6**:893–899.
67. Valaja T, Mahonen A, Pirskanen A, Maenpaa PH 1990 Affinity of 22-oxa-1,25$(OH)_2D_3$ for 1,25-dihydroxyvitamin D receptor and its effects on the synthesis of osteocalcin in human osteosarcoma cells. Biochem Biophys Res Commun **169**:629–635.
68. Binderup L, Bramm E 1988 Effects of a novel vitamin D analogue MC903 on cell proliferation and differentiation in vitro and on calcium metabolism in vivo. Biochem Pharmacol **37**:889–895.
69. Evans DB, Thavarajah M, Binderup L, Kanis JA 1989 Comparison of the actions of MC903 and 1,25$(OH)_2D_3$ on human osteoblast-like cells in vitro. J Bone Miner Res **4**:S334.
70. Marie PJ, Connes D, Hott M, Miravet L 1990 Comparative effects of a novel vitamin D analogue MC-903 and 1,25-dihydroxyvitamin D_3 on alkaline phosphatase activity, osteocalcin and DNA synthesis by human osteoblastic cells in culture. Bone **11**:171–179.
71. Thavarajah M, Evans DB, Binderup L, Kanis JA 1990 1,25$(OH)_2D_3$ and calcipotriol (MC903) have similar effects on

the induction of osteoclast-like cell formation in human bone marrow cultures. Biochem Biophys Res Commun **171**:1056–1063.

72. Stewart PM, Shappard MC 1992 Novel aspects of hormone action: Intracellular ligand supply and its control by a series of tissue-specific enzymes. Mol Cell Endocrinol **83**:C13–C18.
73. Brown AJ, Berkoben M, Ritter C, Mori T, Nishii Y, and Slatopolsky E 1992 In vitro metabolism of 22-oxacalcitriol: Side chain cleavage in parathyroid cells by a vitamin D-inducible pathway. Biochem Biophys Res Commun **189**:759–764.
74. Kamimura S, Gallieni M, Kubodera N, Nishii Y, Brown AJ, Slatopolsky E, Dusso A 1993 Differential catabolism of 22-oxacalcitriol and 1,25-dihydroxyvitamin D_3 by normal human peripheral monocytes. Endocrinology **133**:1158–1164.
75. Bikle DD, Abe-Hashimoto J, Su MJ, Felt S, Gibson DFC, Pillai S 1995 22-Oxacalcitriol is a less potent regulator of keratinocyte proliferation and differentiation due to decreased cellular uptake and enhanced catabolism. J Invest Dermatol **105**:693–698.
76. Binderup L, Latini S, Binderup E, Bretting C, Calverley M 1991 20-Epi-vitamin D_3 analogues: A novel class of potent regulators of cell growth and immune responses. Biochem Pharmacol **42**:1569–1575.
77. Masuda S, Makin HLJ, Kremer R, Okano T, Kobayashi T, Sato K, Nishii Y, Jones G 1994 Metabolism of 2β(3-hydroxypropoxy)-1α,25-dihydroxyvitamin D_3 (ED-71) in cultured cell lines. J Bone Miner Res **9**:S289.
78. Reddy GS, Clark JW, Tserng KY, Uskokovic MR, McLane JA 1993 Metabolism of 1,25-$(OH)_2$-16-ene D_3 in kidney: Influence of structural modification of D-ring on side chain metabolism. Bioorg Med Chem Lett **3**:1879-1884.
79. Yeung B, Vouros P, Siu-Caldera M, Reddy GS 1995 Characterization of the metabolic pathway of 1,25-dihydroxy-16-ene vitamin D_3 in rat kidney by on-line high performance liquid chromatography–electrospray tandem mass spectrometry. Biochem Pharmacol **49**:1099–1110.
80. Siu-Caldera ML, Clark JW, Santos-Moore A, Peleg S, Liu Y, Uskokovic MR, Sharma S, Reddy GS 1996 1α,25-Dihydroxy-24-oxo-16-ene vitamin D_3, a metabolite of a synthetic vitamin D analog, 1α,25-dihydroxy-16-ene vitamin D_3, is equipotent to its parent in modulating growth and differentiation of human leukemic cells. J Steroid Biochem Mol Biol **59**:405–412.
81. Reddy GS, Muralidharan KR, Okamura WH, Tserng KY, McLane JA 1994 Metabolism of 1α,25-dihydroxyvitamin D_3 and one of its A-ring diastereomers 1α,25-dihydroxy-3-epivitamin D_3 in neonatal human keratinocytes. Ninth Workshop on Vitamin D, Orlando, Florida.
82. Brown AJ, Ritter CS, Slatopolsky E, Okamura WH, Muralidharan KR, Reddy GS 1994 Natural metabolites of 1,25-dihydroxyvitamin D_3 produced by parathyroid cells suppress parathyroid hormone secretion. Ninth Workshop on Vitamin D, Orlando, Florida.
83. Zhou LX, Nemere I, Norman AW 1992 1,25-Dihydroxyvitamin D_3 analog structure–function assessment of the rapid stimulation of intestinal calcium absorption (transcaltachia). J Bone Miner Res **7**:457–463.
84. Khoury R, Ridall AL, Norman AW, Farach-Carson MC 1994 Target gene activation by 1,25-dihydroxyvitamin D_3 in osteosarcoma cells is independent of calcium influx. Endocrinology **135**:2446–2453.
85. Farach-Carson MC, Abe J, Nishii Y, Khoury R, Wright C, Norman AW 1993 22-Oxacalcitriol: Dissection of 1,25-$(OH)_2D_3$ receptor and Ca^{2+} entry-stimulating pathways. Am J Physiol **34**:F705–F711.
86. Tien XY, Brasitus TA, Norman AW, Sitrin MD 1994 Effects of 1,25-dihydroxyvitamin D_3 and its analogs on cytosolic calcium and inositol trisphosphate in CaCo-2 cells: Correlation with nuclear receptor binding affinity. Gastroenterology **104**:649.
87. Nemere I 1995 Nongenomic effects of 1,25-dihydroxyvitamin D_3: Potential relation of a plasmalemmal receptor to the acute enhancement of intestinal calcium transport in chick. J Nutr **125**:1695S–1698S.
88. Baran DT, Ray R, Sorensen AM, Honeyman T, Holick MF 1994 Binding characteristics of a membrane receptor that recognizes 1α,25-dihydroxyvitamin D_3 and its epimer, 1β,25-dihydroxyvitamin D_3. J Cell Biochem **56**:510–517.
89. Boyan BD, Dean DD, Sylvia VL, Schwartz Z 1994 Nongenomic regulation of extracellular matrix events by vitamin D metabolites. J Cell Biochem **56**:331–339.
90. Baran DT, Sorensen AM 1994 Rapid actions of 1α-25-dihydroxyvitamin D_3: Physiologic role. Proc Soc Exp Biol Med **207**:175–179.
91. Baran DT, Sorensen AM, Shalhoub V, Owen T, Stein G, Lian J 1992 The rapid nongenomic actions of 1α,25-dihydroxyvitamin D_3 modulate the hormone-induced increments in osteocalcin gene transcription in osteoblast-like cells. J Cell Biochem **50**:124–129.
92. Khoury RS, Weber J, Farach-Carson MC 1995 Vitamin D metabolites modulate osteoblast activity by Ca^{2+} influx-independent genomic and Ca^2-influx-dependent nongenomic pathways. J Nutr **125**:1699S-1703S.
93. Norman AW, Bouillon R, Farach-Carson MC, Bishop JE, Zhou LX, Nemere I, Zhao J, Muralidharan KR, Okamura WH 1993 Demonstration that 1β,25-dihydroxyvitamin D_3 is an antagonist of the nongenomic but not the genomic biological responses and biological profile of the three A-ring diastereomers of 1α,25-dihydroxyvitamin D_3. J Biol Chem **268**:20022–20030.
94. Jurutka PW, Terpening CM, Haussler MR 1993 The 1,25-dihydroxyvitamin D_3 receptor is phosphorylated in response to 1,25-dihydroxyvitamin D_3 and 22-oxacalcitriol in rat osteoblasts, and by casein kinase II in vitro. Biochemistry **32**:8184–8192.
95. Zhou LX, Norman AW 1995 1,25-Dihydroxyvitamin D_3 analog structure–function assessment of intestinal nuclear receptor occupancy with induction of calbindin D28k. Endocrinology **136**:1145–1152.
96. Hedlund TE, Moffatt KA, Miller GJ 1996 Stable expression of the nuclear vitamin D receptor in the human prostatic carcinoma cell line JCA-1: Evidence that the antiproliferative effects of 1α,25-dihydroxyvitamin D_3 are mediated exclusively through the genomic signaling pathway. Endocrinology **137**:1554–1561.
97. Zhou ZX, Lane MV, Kemppainen JA, French FS, Wilson EM 1995 Specificity of ligand-dependent androgen receptor stabilization: Receptor domain interactions influence ligand dissociation and receptor stability. Mol Endocrinol **9**:208–218.
98. Davoodi F, Brenner RV, Evans SR, Schumaker LM, Shabahang M, Nauta RJ, Buras RR 1995 Modulation of vitamin D receptor and estrogen receptor by 1,25-$(OH)_2$-vitamin D_3 in T-47D human breast cancer cells. J Steroid Biochem Mol Biol **54**:147–53.
99. Wiese RJ, Uhland-Smith A, Ross TK, Prahl JM, DeLuca HF 1992 Up-regulation of the vitamin D receptor in response to 1,25-dihydroxyvitamin D_3 results from ligand-induced stabilization. J Biol Chem **267**:20082–20086.
100. Arbour NC, Prahl JM, DeLuca HF 1993 Stabilization of the vitamin D receptor in rat osteosarcoma cells through the action of 1,25-dihydroxyvitamin D_3. Mol Endocrinol **7**:1307–1312.
101. Peleg S, Sastry M, Collins ED, Bishop JE, Norman AW 1995 Distinct conformational changes induced by 20-epi analogues of 1α,25-dihydroxyvitamin D_3 are associated with enhanced activation of the vitamin D receptor. J Biol Chem **270**:10551–10558.

102. Sasaki H, Harada H, Handa Y, Morino H, Suzawa M, Shimpo E, Katsumata T, Masuhiro Y, Matsuda K, Ebihara K 1995 Transcriptional activity of a fluorinated vitamin D analog on VDR–RXR-mediated gene expression. Biochemistry **34**:370–377.
103. Cheskis B, Lemon BD, Uskokovic M, Lomedico PT, Freedman LP 1995 Vitamin D_3–retinoid X receptor dimerization, DNA binding, and transactivation are differentially affected by analogs of 1,25-dihydroxyvitamin D_3. Mol Endocrinol **9**:1814–1824.
104. Finch JL, Brown AJ, Mori T, Nishii Y, Slatopolsky E 1992 Suppression of PTH and decreased action on bone are partially responsible for the low calcemic activity of 22-oxacalcitriol relative to $1,25(OH)_2D_3$. J Bone Miner Res **7**:835–839.
105. Dusso AS, Negrea L, Finch J, Kamimura S, Lopez-Hilker S, Mori T, Nishii Y, Brown AJ, Slatopolsky E 1992 The effect of 22-oxacalcitriol on serum calcitriol. Endocrinology **130**:3129–3134.
106. Grieff M, Dusso A, Mori T, Nishii Y, Slatopolsky E, Brown AJ 1992 22-Oxacalcitriol suppresses 25-hydroxycholecalciferol-1-hydroxylase in rat kidney. Biochem Biophys Res Commun **185**:191–196.
107. Okano T, Tsugawa N, Murano M, Masuda S, Takeuchi A, Kobayashi T, Sato K, Nishii Y 1992 ED-71 increases bone mineral density, mineral content and mechanical strength in ovariectomized rats. J Bone Miner Res **7**:S156.
108. Akeno N, Saikatsu S, Kimura S, Horiuchi N 1994 Induction of vitamin D 24-hydroxylase messenger RNA and activity by 22-oxacalcitriol in mouse kidney and duodenum. Possible role in decrease of plasma 1α,25-dihydroxyvitamin D_3. Biochem Pharmacol **48**:2081–2090.
109. Masuda S, Strugnell S, Calverley MJ, Makin HL, Kremer R, Jones G 1994 In vitro metabolism of the anti-psoriatic vitamin D analog, calcipotriol, in two cultured human keratinocyte models. J Biol Chem **269**:4794–4803.
110. Abe J, Morikawa M, Takita Y, Miyamoto K, Kaiho S, Fukushima M, Miyaura C, Abe E, Suda T, Nishii Y 1988 1α,25-Dihydroxy-22-oxavitamin D_3: A new synthetic analogue of vitamin D having potent differentiation-inducing activity without inducing hypercalcemia in vivo and in vitro. In: Norman AW, Schaefer K, Grigoleit HG, von Herrath D (eds) Vitamin D: Molecular, Cellular and Clinical Endocrinology. de Gruyter, Berlin, pp. 310–319.
111. Finch J, Brown AJ, Kubodera N, Nishii Y, Slatopolsky E 1993 Differential effects of $1,25(OH)_2D_3$ and 22-oxacalcitriol on phosphate and calcium metabolism. Kidney Int **43**:561–566.
112. Kobayashi T, Okano T, Tsugawa N, Masuda S, Takeuchi A, Nishii Y 1991 Metabolism and transporting system of 22-oxacalcitriol. Contrib Nephrol **91**:129–133.
113. Bouillon R, Okamura WH, Norman AW 1995 Structure–function relationships in the vitamin D endocrine system. Endocr Rev **16**:200–257.
114. Uskokovic MR, Baggiolini E, Shiuey SJ, Iacobelli J, Hennessy B, Kiegel J, Daniewski AR, Pizzolato G, Courtney LF, Horst RL 1991 The 16-ene analogs of 1,25-dihydroxycholecalciferol: Synthesis and biological activity. In: Norman AW, Bouillon R, Thomasset M (eds) Vitamin D: Gene Regulation Structure–Function Analysis and Clinical Application. de Gruyter, Berlin, pp. 139–145.
115. Norman AW, Zhou J, Henry HL, Uskokovic MR, Koeffler HP 1990 Structure–function studies on analogues of 1α,25-dihydroxyvitamin D_3: Differential effects on leukemia cell growth, differentiation and intestinal calcium absorption. Cancer Res **50**:6857–6864.
116. Zhou JY, Norman AW, Lubbert M, Collins ED, Uskokovic MR, Koeffler HP 1989 Novel vitamin D analogs that modulate leukemic cell growth and differentiation with little effect on either intestinal calcium absorption or bone calcium mobilization. Blood **74**:82–93.
117. Zhou JY, Norman AW, Akashi M, Chen DL, Uskokovic MR, Aurrecoechea JM, Dauben WG, Okamura WH, Koeffler HP 1991 Development of a novel $1,25(OH)_2$-vitamin D_3 analog with potent ability to induce HL-60 cell differentiation without modulating calcium metabolism. Blood **78**:75–82.
118. DeLuca HF personal communication, 1995.
119. Brown AJ unpublished data, 1995.

CHAPTER 60

Molecular Basis for Differential Action of Vitamin D Analogs

SARA PELEG Department of Medical Specialties, The University of Texas, M. D. Anderson Cancer Center, Houston, Texas

I. INTRODUCTION

The most prominent physiological role of the hormonally active metabolite of vitamin D_3, 1α,25-dihydroxyvitamin D_3 [$1,25(OH)_2D_3$], is the regulation of calcium and phosphorous homeostasis as well as bone remodeling through its actions in the intestine, kidney, and bone [1–4]. The hormone also contributes to growth and differentiation of epidermal cells [5] and the bone marrow precursors of osteoclasts [6–9]. As discussed in detail in subsequent chapters, at pharmacological concentrations $1,25(OH)_2D_3$ also has immunoregulatory effects [10] and induces differentiation and inhibits growth of psoriatic skin [11] and a variety of malignant cell types [12–17].

Genomic and nongenomic signal transduction pathways are believed to mediate the diverse effects of this hormone. The genomic pathway has been well established and is mediated by the nuclear vitamin D receptor (VDR) [18]. The VDR belongs to a large family of transcription factors that contain two highly conserved zinc-binding finger structures in their DNA binding domain [19]. Transcriptional activity of the VDR depends primarily on its binding with $1,25(OH)_2D_3$ [20]. This interaction induces conformational changes in the receptor [20,21] and facilitates self-dimerization [22] or heterodimerization with the retinoid X receptor (RXR) [23,24], promotes binding to specific DNA responsive elements [23,25,26], and eventually leads to regulation of gene expression [27].

Several lines of evidence suggest that the broad range of physiological and pharmacological activities induced by $1,25(OH)_2D_3$, both those related to regulation of calcium metabolism and those associated with the regulation of growth and differentiation, are mediated by transcriptional activities of VDR. For example, in vitamin D-dependent rickets type II, defects in VDR cause a severe bone disease and deficient calcium absorption [20]. In the same disease, the growth inhibitory effect of $1,25(OH)_2D_3$ on lectin-induced T cells is diminished [28]. Finally, a link between normal development of skin and functional VDR is suggested by the frequent presence of hairlessness (alopecia) in rickets patients with defective VDR [20]. In cell culture systems, the growth regulatory responses to $1,25(OH)_2D_3$ are correlated with the levels of VDR [29]. In the heterogeneous human myeloid leukemia cell line HL-60, $1,25(OH)_2D_3$ regulates the growth of clones with high VDR levels, but clones with few or no VDRs do not respond to $1,25(OH)_2D_3$ [30]. In animal models, growth inhibition is induced by $1,25(OH)_2D_3$ only in solid tumors containing VDR [17]. In conclusion, VDR appears to be essential for both physiological and pharmacological responses to $1,25(OH)_2D_3$.

Since the early 1980s, several studies have suggested mechanisms for vitamin D action that appear to be independent of VDR (the nongenomic pathway). One of these mechanisms (discussed in detail in Chapter 15) is the induction of rapid calcium mobilization associated with an increase in intracellular calcium [31]. This process occurs in a variety of cells, including osteoblasts, parathyroid cells, enterocytes, keratinocytes, muscle cells, and colonic epithelium [32,33]. It is presumed that these actions of the hormone are mediated through a plasma membrane receptor coupled to voltage-gated calcium channel, and evidence is accumulating for the presence of such a protein in chicken intestinal cell membranes [34]. The physiological role of this putative receptor in the intestine may be to regulate certain facets of the transport of calcium across the intestinal border and into the bloodstream, a process called transcaltachia [35]. The contribution of calcium mobilization in other cell types to the physiological actions of $1,25(OH)_2D_3$ is not known.

This chapter focuses on the molecular mechanisms by which chemically and stereochemically modified ligands produce different effects with respect to activation of the nuclear receptor.

II. LIGAND STRUCTURE AND SIGNAL PREFERENCE

The idea that vitamin D may act through more than one signaling pathway evolved from studies of the natural hormone but has been substantiated by studies with synthetic analogs of vitamin D [36]. An extension of this idea is that the structural features of the ligands which are important for interaction with the VDR or with a putative plasma membrane receptor may be different. Support for this hypothesis comes from extensive studies on ligand structure–function relationships [36] that will be described in detail elsewhere in this book. However, the basic principles for these relationships are briefly described here to provide the chemical background for the differential action of vitamin D analogs.

Analogs of vitamin D can be expected to act through each effector pathway system as agonists (compounds with activity similar to that of the parent hormone), superagonists (compounds with an activity significantly greater than that of the parent hormone), or antagonists (compounds lacking activity on their own but having the ability to block the biological activity of the natural hormone). If ligand structural requirements for each protein effector pathway are different, it should also be possible to design analogs with segregated activities. For example, a compound that is a potent agonist in one effector pathway might be a poor agonist in the other or vice versa. Alternatively, given the complexity of the genomic pathway and the multiple gene targets that are affected, it is also possible to achieve tissue-selective activities via this single pathway. Ligand modifications that lead to segregated activities through differential activation of individual pathways or through selective modulation of the genomic pathway are described below.

A. Structural Requirements for VDR Activation

Ligand-mediated transcriptional activity of VDR depends primarily on the following three structural features: the A ring, the side chain, and the D ring (Fig. 1) [36]. Because transcriptional activities of VDR are mediated initially by binding to ligand, the minimal requirement for agonist activity is substantial affinity for the VDR. The structural features of the ligand that are important for binding to VDR were elucidated through studies of naturally occurring metabolites of vitamin D_3 and synthetic analogs of $1,25(OH)_2D_3$ [36,37].

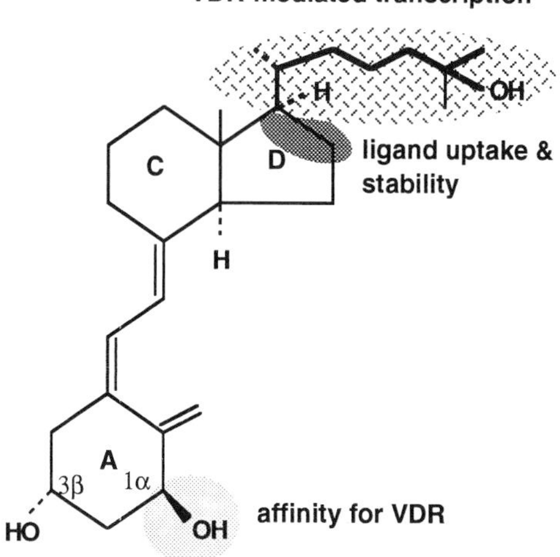

FIGURE 1 Structural formula of $1,25(OH)_2D_3$. Shaded areas indicate structural features that regulate the actions of VDR.

1. The A Ring

The backbone of 1,25$(OH)_2D_3$ is similar to that of cholesterol and the steroid hormones except that the four-ring structure has been disrupted by ultraviolet irradiation of the second (B) ring between carbons 9 and 10 (Fig. 1) [2–4] This change makes the already flexible organic compound even more flexible, giving the A ring the freedom to rotate from the steroidlike ("folded") conformation to a vitamin ("extended") conformation [38]. It has been shown that the steroidlike conformer does not interact with VDR, whereas the vitamin conformer does [38]. Therefore, chemical modifications that reduce the mobility of the A ring so that it prefers the extended conformation may increase the affinity of a ligand for VDR and thus increase the stability of ligand–VDR complexes.

High affinity of the vitamin conformer for VDR also depends on the hydroxyl groups in the A ring, particularly the 1α-hydroxyl [36,37]. For example, the precursor of 1,25$(OH)_2D_3$, 25-hydroxyvitamin D_3, binds poorly to VDR, but hydroxylation of this metabolite at the 1α position induces high affinity for the receptor [36]. Likewise, another natural metabolite of vitamin D_3, 24*R*,25-dihydroxyvitamin D_3, also binds poorly to the VDR, but its 1α hydroxyl metabolite has a significantly greater affinity [39]. Similarly, 1,25$(OH)_2D_3$ analogs that have modified side chains and lack the 1α-hydroxyl group bind poorly to VDR, whereas addition of hydroxyl or fluoride groups restores binding to VDR [40].

2. The Side Chain

Although it is expected that the VDR-mediated transcriptional activities of natural and synthetic ligands will be proportional to their affinity for VDR, it is possible to increase the transcriptional potency of a ligand up to 100- to 10,000-fold without changing its affinity through chemical or stereochemical modifications of the side chain (Table I). The side chain is also flexible, and in 1,25$(OH)_2D_3$ it can assume 1083 conformations in a general "northeast" orientation [36]. Transcriptional superagonists can be generated by introducing a 180° change in the general orientation of the side chain to "northwest," by epimerization of carbon 20 [41]. In the 20-epi analogs this stereochemical change slightly reduces affinity for VDR, but it increases the potency of VDR-mediated transcriptional activity more than a hundredfold [41]. This puzzling discrepancy between affinity for VDR and transcriptional potency is common in analogs of 1,25$(OH)_2D_3$ with longer side chains [41] and with fluorides added at positions 26 and 27 [42]. Furthermore, several hybrid analogs with both 20-epi

Table I Structure–Function Relationships of 1α,25-Dihydroxyvitamin D_3 and Its Analogs

Ligand	Receptor binding[a]	Transcription[b]
Parent compound		
1α,25$(OH)_2D_3$	1×10^{-9}	2×10^{-9}
A ring-modified analogs		
1β,25$(OH)_2D_3$	2×10^{-7}	2×10^{-7}
1β-(Hydroxymethyl)-3α-25OHD_3	8×10^{-8}	$>10^{-6}$
Side chain-modified analogs[c]		
20-Epi-1α,25$(OH)_2D_3$ (IE)	3×10^{-9}	5×10^{-12}
24a,26a,27a-Trihomo-1α,25$(OH)_2D_3$ (CB)	5×10^{-9}	5×10^{-11}
20-Epi-24a,26a,27a-trihomo-1α,25$(OH)_2D_3$ (MC)	1×10^{-9}	1×10^{-12}
22-Oxa-24a,26a,27a-trihomo-1α,25$(OH)_2D_3$ (KH)	3×10^{-9}	1×10^{-10}
20-Epi-22-oxa-24a,26a,27a-trihomo-1α,25$(OH)_2D_3$ (ID)	3×10^{-9}	5×10^{-11}
Hybrid analogs		
1β-20-Epi-24a,26a,27a-trihomo-25$(OH)_2D_3$	1×10^{-7}	2×10^{-9}
1β-(Hydroxymethyl)-3α-20-epi-22-oxa-24a,26a,27a-trihomo-25OHD_3	$>10^{-6}$	1×10^{-8}
Analogs with double bond between C-16 and C-17		
1α,25$(OH)_2$-16-ene-D_3	1×10^{-9}	1×10^{-10}
1α,25$(OH)_2$-16-ene-23-yne-D_3	0.8×10^{-9}	6×10^{-10}

[a] Concentration of ligand (M) required to reach 50% displacement of ^3H-1,25$(OH)_2D_3$ binding to recombinant human VDR from transfected COS-1 cells (S. Peleg, unpublished data).

[b] Concentration of ligand (M) required to produce 50% of maximal transcription activation of a reporter gene containing the osteocalcin VDRE in ROS 17/2.8 cells in serum-free culture medium. The transcriptional activities of 1,25$(OH)_2D_3$ and the last two analogs (i.e., the 16-ene analogs) were also examined in cells grown in 10% serum. Under these conditions, the ED_{50} of the three compounds were 5×10^{-10}, 6×10^{-12}, and 2×10^{-11} *M*, respectively (S. Peleg, unpublished data).

[c] Abbreviations in parentheses are short names for the side chain-modified analogs.

side chains and an A-ring modification that diminishes ligand affinity for VDR still have VDR-mediated transcriptional activities ([43] and Table I).

3. The D Ring

Another modification that enhances VDR-mediated transcriptional activity of the hormone is insertion of a double bond between carbons 16 and 17 in the D ring [44]. Compounds containing this modification, with or without additional chemical modifications in the side chain, have transcriptional activity significantly greater than that of $1{,}25(OH)_2D_3$ without a significant increase in their affinity for VDR. In addition, this group of analogs exhibits diminished calcium regulating activity. The unique properties of these compounds include lower affinity for vitamin D binding protein and slower catabolism [45,46], both of which may increase potency without changing the actual mode of interaction with the receptor (Table I).

In conclusion, the A ring regulates high affinity for VDR, the D ring controls ligand uptake and metabolism, and the side chain regulates VDR transactivation (Fig. 1). No ligand structures that antagonize VDR-mediated transcriptional activities of $1{,}25(OH)_2D_3$ have yet been identified.

B. Structural Requirements for Use of Alternative Signaling Pathways

Because the biochemical and molecular nature of the vitamin D-binding component in the plasma membrane is not known, it is not possible to assess ligand affinity for this putative receptor. Furthermore, because the hormone induces several plasma membrane-mediated signals, it is not clear whether they all are mediated through a single binding component or through several distinct, cell-specific ones. Ligand structural requirements for plasma membrane-mediated signals therefore must be assessed by means of a biochemical assay that measures the induction of a single, well-defined plasma membrane-mediated event that is believed to reflect a physiological response *in vivo*. One of these assays is the induction of calcium transport through the intestinal epithelial barrier into the blood (transcaltachia) [47]. For this activity the A ring is the most important structural feature [36]. For example, when the A ring is in the folded (steroidlike) configuration, the ligand induces a significant transcaltachia (in the absence of VDR-mediated transcription). On the other hand, in the extended configuration, the ligand strongly induces both transcaltachia and VDR actions [38]. The chemistry and stereochemistry at position 1 of the A ring are also very important. For example, having a hydroxyl group in the 1α position is essential for transcaltachia, and for nuclear receptor binding as well [6]. Changing the stereochemistry of this hydroxyl group to the 1β position generates a compound that binds poorly to VDR but antagonizes the plasma membrane-mediated action of the parent hormone *in vitro* [48].

In conclusion, both side chain modifications and A-ring modifications appear to differentially regulate nuclear receptor-mediated and plasma membrane-mediated responses to $1{,}25(OH)_2D_3$. Analogs with A-ring modifications that diminish affinity for VDR might still exhibit significant agonistic or antagonistic activity through the plasma membrane. Conversely, there are side chain modifications that preferentially increase VDR-mediated transcriptional activity but do not increase rapid calcium transport.

III. DIFFERENTIAL ACTIVATION OF THE VDR

The finding that VDR-mediated transcriptional activity of certain analogs with modified side chains does not correlate with their affinity for VDR suggests that another level of diversity in their action can be achieved by fine-tuning of nuclear receptor activation. For optimal functioning, VDR must interact with dimerization partners [49]. Furthermore, studies of other nuclear receptors suggest that for optimal transcriptional activation VDR must also interact with transcription coactivators and with factors in the basal transcriptional machinery [49]. Therefore, diversity in transcriptional responses is theoretically possible through selective recruitment of different factors by VDR–hormone and VDR–analog complexes.

A. The Ligand Binding Domain and Potential Contact Sites in the Ligand Binding Pocket

The ligand binding domain of VDR (like that of other steroid receptors) is multifunctional: it regulates dimerization [50], interaction with transcription factors, and general transcriptional activity [50–53]. Therefore, analogs that bind differently and induce distinct conformational changes in VDR may induce transcriptional activities different from those of the natural hormone.

The exact structure of the ligand binding pocket of VDR and the location of ligand contact points with the VDR are not known. However, tentative structural predictions can be made by comparing the ligand binding domain of VDR with the ligand binding domains of nuclear receptors for which the crystal structures are already known. In retinoid X receptor (RXR), thryoid hormone receptor (TR), and retinoic acid receptor (RAR), this domain contains a cluster of 11 or 12 α helices and two to four β strands that together form a sandwich structure into which the ligand enters along an electrostatic gradient [54–56]. A secondary structure prediction for the ligand binding domain of VDR reveals clear similarities. First, there is a cluster of 11 helices and eight β strands between amino acids 204 and 427 (Fig. 2). Second, a comparison of this region with analogous regions of RAR (Fig. 2) and TR (not shown, [55]) shows where potential common contact sites for ligands of nuclear receptors are localized. One of these sites is between amino acid residues 396 and 410 in VDR. Deletion analysis has shown that this region is essential for binding of $1,25(OH)_2D_3$ to VDR (Figs. 2 and 3) [57]. Another site is the conserved transcription activation function 2 (AF-2) domain at the C-terminal region. In TR and RAR, this domain seals the ligand binding pocket after ligand entry, and also forms contact with the ligand [55,56]. In VDR this region is important for binding of $1,25(OH)_2D_3$ (Figs. 2 and 3).

In addition to these similarities, there are also structural differences between the ligand binding domains of VDR and other ligand-binding nuclear receptors. For example, the C-terminal cluster of 11–12 α helices (~28 kDa) in RAR, TR and RXR, which is structurally analogous to the region between amino acid residues 204 and 427 in VDR, is sufficient for ligand binding [54–56]. However, it appears that additional N-terminal residues may be essential for ligand binding to VDR [57], although this additional requirement for VDR may be structural as opposed to essential for ligand contact. An

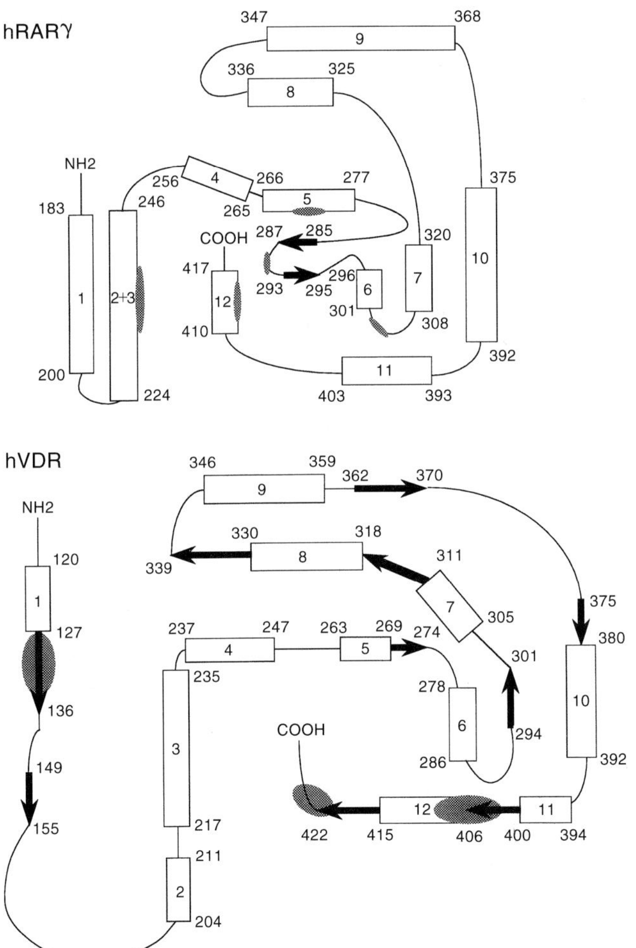

FIGURE 2 Topology of the ligand binding domains of human RARγ and human VDR. Secondary structures were predicted with the Genetics Computer Group (GCG) peptide prediction program using Chou–Fasman parameters. The bars indicate α helices, and the arrows indicate β strands. The shaded areas in the RARγ map indicate ligand contact sites identified by X-ray crystallography of RARγ–ligand complexes [56]. The shaded areas in the VDR map indicate amino acid residues essential for $1,25(OH)_2D_3$ binding and for the ligand-induced transcriptionally active conformation. These amino acid residues were identified by deletions and site-directed mutagenesis [50,51,57,58].

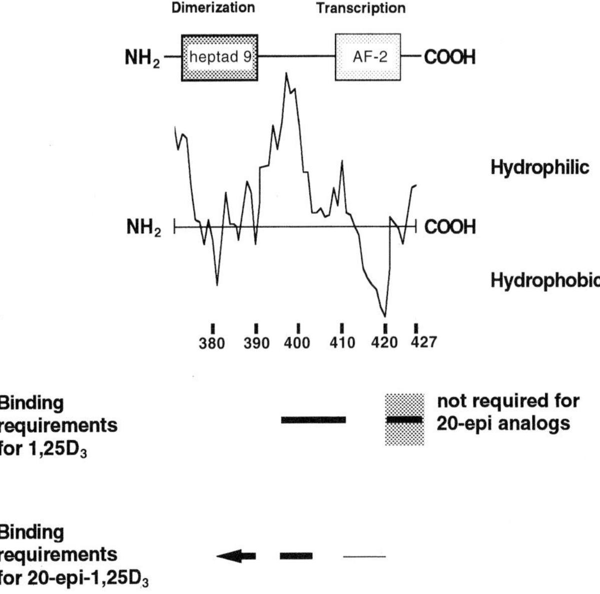

FIGURE 3 Differential binding requirements of $1,25(OH)_2D_3$ and 20-epi analogs in the C-terminal region of VDR. The binding requirements for $1,25(OH)_2D_3$ and its analogs were analyzed by deletions and site-directed mutagenesis [58]. At top is shown a schematic illustration of the functional domains that are within the mapped region [heptad 9 and transcription activation function 2 (AF-2)]. The hydropathy profile of the C-terminal residues is shown in the middle, and the bars at bottom indicate the locations of residues important for binding of the natural hormone and for binding of the 20-epi analogs. The shaded area indicates the seven C-terminal residues important for high-affinity binding of $1,25(OH)_2D_3$ and analogs with naturally oriented side chains. These residues are not required for binding of 20-epi analogs.

extended secondary structure analysis predicts that this N-terminal region (amino acid residues 114–204) contains one α helix and two β strands (Fig. 2). Removing the N-terminal β strand by internal deletion of residues 127–138 abolishes ligand binding (S. Peleg and Y.-Y. Liu, unpublished). It is possible that this β strand forms hydrogen bonds with a β strand in the C-terminal region and that this interaction is essential for stabilization of the ligand binding pocket. These additional features in the ligand binding domain of VDR may be necessary to accommodate a structurally complex ligand with two highly mobile groups (the A ring and the C–D side chain). The N-terminal and the C-terminal regions may each have a low affinity site for a different chemical group of the ligand (e.g., hydroxyl groups in the A ring may form contact with one region, whereas hydroxyl groups in the side chain may form contacts with another), and high affinity may require that both groups contact the receptor. Superagonists may use completely different or only partially overlapping contact sites. Likewise, putative antagonists of $1,25(OH)_2D_3$ may use alternative contact sites within the ligand binding pocket.

B. Implications of Analog Binding at Alternative Contact Sites on Dissociation Rate and VDR Conformation

The mapping of C-terminal residues required for binding of $1,25(OH)_2D_3$ and several analogs with modified side chains is summarized in Fig. 3. Preliminary mutational analysis demonstrated that the binding of analogs with natural side chain stereochemistry requires the C-terminal residues of VDR, whereas analogs with side chains with 20-epi stereochemistry do not require these residues [58]. This observation provides evidence that alternative structures within the ligand binding domain are used and suggests that the 20-epi analogs may be more deeply buried in the ligand binding pocket than $1,25(OH)_2D_3$. Support for this interpretation comes from the finding that the dissociation rates of 20-epi analogs from VDR are significantly slower than those for the natural hormone and for analogs with natural side chain orientation (Table II) [58]. This difference in binding properties may have an impact on the stability of the transcriptionally active VDR–ligand complex and partly explains the puzzling discrepancy between the equilibrium dissociation constants and the potencies of these compounds.

To further evaluate the consequence of a shift in binding requirements at the C-terminal region of VDR, it is important to elaborate on the role of these residues

TABLE II Dissociation of Analogs with Modified Side Chains from VDR

Ligand	Net increase in % unoccupied VDR sites[a]
$1\alpha,25(OH)_2D_3$	60
24a,26a,27a-Trihomo-$1\alpha,25(OH)_2D_3$	40
22-Oxa-24a,26a,27a-trihomo-$1\alpha,25(OH)_2D_3$	45
20-Epi-$1\alpha,25(OH)_2D_3$	20
20-Epi-24a,26a,27a-trihomo-$1\alpha,25(OH)_2D_3$	5
20-Epi-22-oxa-24a,26a,27a-trihomo-$1\alpha,25(OH)_2D_3$	20

[a] COS-1 cells expressing recombinant human VDR were incubated with receptor-saturating ligand, in serum-free medium for 1 hr. The ligands were then removed, and the number of unoccupied binding sites was determined immediately and 3 hr later. The difference between the number of unoccupied binding sites at the two time points reflects the dissociation of the ligand in intact cells.

in the actions of nuclear receptors. The C-terminal region contains a cluster of hydrophobic amino acid residues called AF-2 [59]. This region is conserved in ligand-binding nuclear receptors, and it functions to modulate transcriptional activity. AF-2 is not part of the ligand binding pocket itself [54], but its organization is clearly different between unoccupied and ligand-occupied receptors; in the latter it seems to "seal" the ligand binding pocket and provides contact sites for the ligand [55,56]. It has been speculated that when the ligand enters the binding pocket, it induces conformational changes that modify the steric position of the hydrophobic residues in AF-2 so that they become available for interaction with transcription coactivators [56]. If AF-2 plays a similar role in VDR, it is possible that ligands with natural side chain orientation may actually make contact with these hydrophobic residues, thus preventing their optimal folding and exposure for interaction with coactivators. On the other hand, in 20-epi analog–VDR complexes (which do not use these residues for binding), the conformation of the hydrophobic residues may allow for a more efficient interaction with coactivators and thereby increase transcriptional activity.

Protease clipping assays (Y.-Y. Liu and S. Peleg, unpublished) have provided physical proof for the role of the C-terminal residues in stabilizing the conformation of ligand–VDR complexes (Fig. 4). These assays demonstrate that $1,25(OH)_2D_3$ and its analogs each induce a distinct protease-resistant conformation in wild-type VDR. Site-directed mutagenesis of AF-2 (residues 421 and 422) abolishes transcriptional activity of VDR, diminishes the $1,25(OH)_2D_3$-induced endoproteinase Glu-C-resistant conformation (Fig. 4B), and drastically changes the $1,25(OH)_2D_3$- and analog-induced trypsin-resistant conformation (Fig. 4A). Therefore, it seems

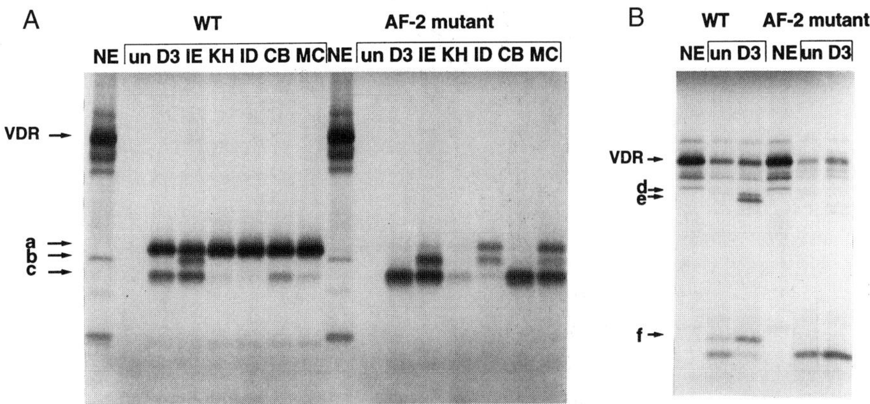

FIGURE 4 Regulation of receptor conformation by ligand and by AF-2 residues. ^{35}S-Labeled synthetic wild-type (WT) and transcriptionally inactive VDR (AF-2 mutant) were incubated with or without ligands and then subjected to proteolytic digestion with trypsin (A) or with endoproteinase Glu-C (B). The proteolytic products were separated by sodium dodecyl sulfate (SDS) gel electrophoresis and visualized by autoradiography. NE, No enzyme; un, no ligand; D3, 1,25$(OH)_2D_3$; the chemical names of the 20-epi analogs IE, ID, and MC and their natural side chain homologs KH and CB are shown in Table 1. Fragments a, b, c, d, e, and f are 34, 30, 28, 43, 42, and 17 kDa, respectively. Note that AF-2 mutation diminishes the intensity of the 34-kDa trypsin-resistant fragment and abolishes the 43- and 42-kDa endoproteinase Glu-C-resistant fragments.

that AF-2 residues are essential for stabilization of ligand–VDR complexes in the appropriate, transcriptionally active conformation.

C. Analog Regulation of VDR Dimerization

Dimerization of many nuclear receptors is necessary for stable binding to DNA responsive elements containing either indirect repeats (as in steroid hormone receptors) or direct repeats (as in TR, RAR, RXR, and VDR [19,49]). Nuclear receptors of the latter group bind to their cognate response elements either as homodimers or as heterodimers with RXR. Dimerization of TR, RAR, and VDR is regulated through three regions (Fig. 5). The first dimerization interface (DI1) is provided through amino acid residues in the first zinc finger (F1) and through residues in the carboxyl-terminal extension (T box) downstream from the second zinc finger [60]. The ligand-binding domain provides two conserved regions that are important for dimerization. One is encoded by nine hydrophobic heptad repeats (DI3) [50,51] that in RAR and TR contribute most of the contact points for the ligand [55,56], and the other (DI2), which is located immediately upstream from DI3, does not contribute directly to ligand binding [61,62].

VDR forms several kinds of dimers *in vitro:* homodimers [22] and heteodimers with RXRα [63], RXRβ [24], and RXRγ [25]. Several studies also suggest that it heterodimerizes with RAR and TR [64,65]. Because VDR and its dimerization partner are coexpressed in large quantities in recombinant studies, it is not clear

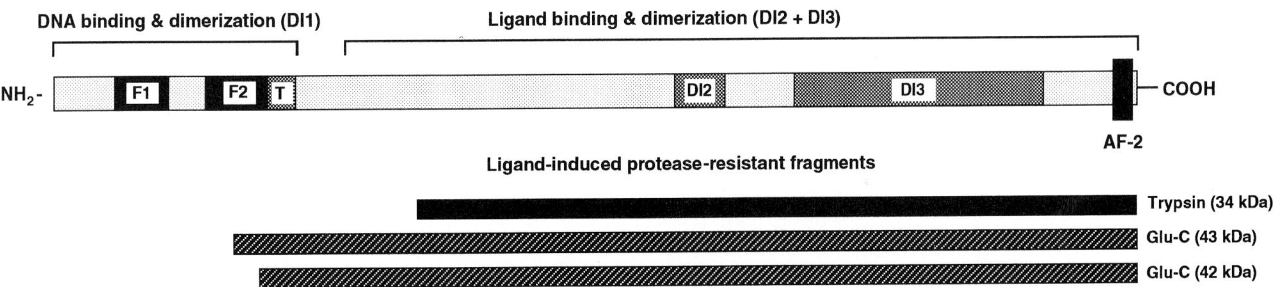

FIGURE 5 Alignment of ligand-induced protease-resistant fragments with functional domains of VDR. A schematic illustration of the functional domains of VDR is shown at top: F1, F2, and T are the first and second zinc fingers and the C-terminal extension of the DNA binding domain, respectively. DI1, DI2, and DI3 are the three dimerization interfaces of VDR. N-terminal deletion mutants were used to map the ligand-induced trypsin-resistant and endoproteinase Glu-C-resistant fragments.

which of these dimers are transcriptionally active and which are artifacts of cell-free assay systems. Low levels of transactivation in yeast cells lacking RXR but transformed with a VDR expression vector suggest that in high concentration VDR homodimers can assemble on direct repeats and form transcriptionally active complexes [26]. However, heterodimerization of VDR and RXR has been confirmed in many studies [24,25,63]. RXR is the clear and preferential dimerization partner of VDR in crude nuclear extracts used in all DNA binding assays [21]. Therefore, the most probable dimerization events for VDR are those with one or more of the three isoforms of RXR.

What is the role of the natural hormone and analogs in the dimerization of VDR? Because the dimerization interface of DI3 overlaps with the ligand binding pocket [50,55–57], it is reasonable to predict that binding of the natural hormone will directly affect the conformation of this region and that analogs which use alternative contact sites may have different effects on DI3. Mutations in DI2 do not significantly affect ligand binding [62]. However, because DI2 is in a region that undergoes conformational changes in response to ligand binding (i.e., the 34-kDa polypeptide that is protected against trypsin digestion, Fig. 5), it is likely that the function of DI2 is also regulated by various ligands.

Finally, the ligand may also modulate the functions of DI1, although this dimerization interface is definitely outside the ligand binding domain. Protease clipping assays strongly suggest that ligand-induced conformational changes occur within the T box and the second zinc finger (Figs. 4 and 5). These amino acid residues are important for base recognition, definition of spacing between DNA half-sites, and dimerization [23]. Therefore, conformational changes in these regions may affect both the dimerization interface of DI1 and the site of physical contact of VDR with DNA responsive elements.

Changing ligand structure may affect various aspects of the dimerization process including dimer stability and preferences for dimerization partners. Additional aspects are downstream actions of the dimer such as the DNA-binding preferences of VDR, the stability of VDR–DNA complexes, and their transcriptional activity. Because only a few *in vitro* assays for VDR dimerization conserve ligand responsiveness, this aspect of the action of $1,25(OH)_2D_3$ and the actions of its analogs has not been investigated to a great extent. Using recombinant human VDR expressed in COS-1 cells, Peleg *et al.* [21] were able to show that heterodimerization of RXR with VDR–20-epi analog complexes was favored in DNA binding assays over heterodimerization with VDR-$1,25(OH)_2D_3$ complexes. These experiments suggest that the 20-epi analogs either increase the preference of VDR for RXR or stabilize VDR–RXR complexes more effectively than does the natural hormone. On the other hand, analogs with the 16-ene-23-yne modification reduced the affinity of bacteria-expressed VDR for RXR [66], even though these analogs exhibited a 1000-fold increase in transcriptional activity compared to $1,25(OH)_2D_3$ (Table I). The 16-ene-23-yne analogs are exceptionally poor substrates for the hydroxylases and have poor affinity for vitamin D binding protein [45,46]. Therefore, it is possible that the transcriptional potency of these compounds, unlike that of the 20-epi analogs, is due more likely to their uptake and stability in cells than to their mode of interaction with VDR (Table I). Another possible reason for the apparent lack of effect of 16-ene-23-yne modification on the dimerization properties of VDR is the assay system used (receptor expressed in bacteria instead of in mammalian cells). It is possible that some aspects of the dimerization process are regulated by cellular mechanisms such as post-translational modification of VDR and therefore cannot be reproduced adequately *in vitro* or in bacterial cells [67].

D. Analog Modulation of DNA Binding Properties and Transcriptional Activity of VDR

1. Diversity of DNA Binding Sites for VDR Complexes

The range of transcriptional activities of nuclear receptors depends on their DNA binding specificity and on their mode of interaction with the basal transcriptional apparatus. Nuclear receptors that form several types of dimers [68,69] have some flexibility in their DNA binding preferences. For example, the DNA motifs selected by the RAR homodimer are clearly different from the motifs selected by the RAR–RXR heterodimers. The TR homodimers and TR–RXR heterodimers have similar flexibility.

As described above, several kinds of VDR dimers may exist, and each has somewhat different DNA binding preferences. Table III summarizes the binding preferences of five forms of VDR for various synthetic DNA motifs. The data suggest that VDR dimers select DNA binding sites in two ways. The first is through sequence recognition; for example, the VDR homodimer preferentially recognizes motifs containing the half-site GGTTCA, whereas VDR–RXR heterodimer prefers motifs containing the half-site AGGTCA [70]. A second means of selecting a binding site is through spacing preference. For example, VDR–RXRβ preferentially recognizes DR3 motifs (direct repeats with spacing of 3

TABLE III DNA Binding Preferences of VDR Complexes[a]

VDR complex	DNA binding site
VDR monomer[b]	DR3 (GGTTCA) > half-site (GGTTGCA), not DR3 (AGGTCA) [70]
VDR homodimer	DR3 (AGTTCA) but not DR3 (AGGTCA) [70]; DR5, DR6 [64]
VDR–RXRα	DR5 > DR3 > DR6 [25], DR3 > DR6 [26], DR3 only (AGGTCA) [70]
VDR–RXRβ	DR3 > DR5, not DR6 [25], DR3 > natural DR4 (MHC)[c], natural DR5 (RARβ-RE)[c] (AGGTCA) [24]
VDR–RXRγ	DR5 > DR3, not DR6 (AGGTCA) [25]

[a] Binding of VDR complexes to synthetic DNA motifs was studied by the mobility shift assay. DR4, DR5, and DR6 are direct repeats of the half-site AGGTCA with spacing of four, five, and six random nucleotides, respectively. The experiments were performed with two types of DR3 motifs: one containing the half-site A/GTTCA and the other AGGTCA. According to these studies, the latter half-site has a preference for the VDR–RXR heterodimer whereas the former has a preference for the VDR homodimer. All data were collected from studies performed in the absence of ligand, and so they may not represent the DNA binding preferences of VDR–ligand complexes. These binding preferences do not necessarily reflect *in vivo* interactions.

[b] Binding of VDR monomer was examined by site-selection experiments in which immobilized glutathione-*S*-transferase–VDR was used to select DNA elements from a random pool of short DNA sequences. DNA binding of VDR was also examined by the mobility shift assay in which the VDR monomer and VDR homodimer complexes with DNA were distinguished by their electrophoretic mobilities.

[c] Binding of VDR–RXRβ complexes to a natural DR4 (from the myosin heavy chain promoter, MHC) and natural DR5 (the RARβ response element, RARβ-RE) was demonstrated by mobility shift assays using synthetic VDR and RXRβ.

base pairs), and VDR–RXRγ preferentially recognizes DR5 (direct repeats with spacing of 5 base pairs) [24,25].

To evaluate the physiological relevance of DNA binding flexibility of VDR complexes on the synthetic elements shown in Table III, it is important to review the natural responsive elements for $1,25(OH)_2D_3$. Table IV shows that most of the natural vitamin D response elements (VDREs) are DR3 motifs [71–78]. However, there are examples of diversity in sequence of natural VDREs. One is the negative response element of the human parathyroid hormone (hPTH) promoter: this element may actually be a half-site for VDR binding and transactivation [77]. The negative response elements in the interleukin 2 promoter also stand out because they do not resemble any known motifs for VDR action, despite the fact that they function as weak binding sites for VDR monomers and homodimers [78]. Finally, VDR–RXRβ heterodimers bind to a natural DR4 (thyroid hormone responsive element of the myosin heavy chain) and to a natural DR5 (RARβ response element) [23], and may function through these or similar motifs [79].

In conclusion, diversity in VDR binding sites can be demonstrated by using artificial response elements and natural ones. The nature of a response element may determine the type of VDR complex assembled on it. For example, VDR–RXR will bind effectively to all natural DR3 motifs shown, whereas VDR monomers

TABLE IV Natural Response Elements for VDR[a]

Pronater	DNA Sequence		
r24-OH (distal)	GGTTCA	GCG	GGTGCG
r24-OH (proximal)	AGGTGA	GTG	AGGGCG
rOC	GGGTGA	ATG	AGGACA
hOC	GGGTGA	ACG	GGGGCA
mOP	GGTTCA	CGA	GGTTCA
rCA9K	GGGTGT	CGG	AAGCCC
mCA28K	GGGGGA	TGT	GAGGAG
CDKI21	AGGGAG	ATT	GGTTCA
hPTH (nVDRE)	GGTTCA		
IL-2 (nVDRE)	AGAAAG		
IL-2 (nVDRE)	GTTTCA	TAC	AGAAGG

[a] The response elements shown were isolated from the promoters of the following genes: r24-OH, rat 24-hydroxylase [24]; rOC, rat osteocalcin [71]; hOC, human osteocalcin [72]; mOP, mouse osteopontin [73]; rCA9K, rat calbindin-D_{9k} [74]; mCA28K, mouse calbindin-D_{28K} [75]; CDKI21, cyclin-dependent kinase inhibitor p21 [76]; hPTH, human parathyroid hormone [77]; IL-2, interleukin-2 [78], nVDRE, negative vitamin D response element. The first eight promoter elements up-regulate transcription in response to $1,25(OH)_2D_3$. The hPTH and IL-2 sequences down-regulate transcription in response to vitamin D. Direct binding of VDR or VDR–RXR has been demonstrated to all of these sequences.

or homodimers may bind preferentially to response elements containing PuGTTCA half-sites [the osteopontin VDRE, the hPTH negative VDRE, the cyclin-dependent kinase inhibitor, p21 (CDKI21) VDRE, and perhaps the interleukin 2 negative VDRE] (Table IV). How selective homodimerization on these elements can occur in the presence of RXR is unknown.

2. Effect of Ligand Structure on DNA Binding and Transcriptional Activity of VDR

Binding of VDR to natural and synthetic motifs is ligand modulated. Therefore, a question is raised as to whether analogs modulate the DNA binding properties of VDR differently from the natural hormone *in vitro* and, by extension, whether they can change the spectrum of transcriptional activities of VDR *in vivo*.

The 20-epi analogs change several aspects of VDR binding to DNA. For example, maximal binding of VDR to the osteopontin VDRE occurs at a 10- to 100-fold lower concentration of analog than of the natural hormone [21]. Furthermore, analog-induced DNA binding does not correlate with the apparent affinity of each analog for VDR in cell-free systems and in intact cells (Fig. 6) [21]. This implies that the affinity of the analogs for VDR complexes assembling as dimers on DNA response elements is higher than their affinity for VDR in solution. Alternatively, VDR–20-epi complexes may selectively recruit unoccupied VDR to bind to DNA (Fig. 6). Another change induced by 20-epi analogs is an increase in the proportion of RXR in the DNA-binding VDR complexes [21]. Overall, these changes in DNA binding and dimerization are associated with a potency increase in DR3-mediated transcription of up to 200-fold relative to $1,25(OH)_2D_3$. Because these ligands seem to alter dimer composition, it is possible that they modulate the preference of VDR complexes for DNA motifs with different base compositions (Tables III and IV).

Studies with the hexafluorinated analog $1\alpha,25(OH)_2$-$26,27$-F_6-D_3 have shown that this side chain modification does not have a significant effect on the half-life of VDR–RXR–VDRE complexes [42]. However, an analog concentration one-tenth that of $1,25(OH)_2D_3$ is required to produce maximal binding of VDR–RXR complexes to DNA and to induce maximal transcription activation. Another effect on this analog is that, at receptor-saturating concentrations, maximal DNA binding and transcriptional activity induced by it are somewhat higher than that for the $1,25(OH)_2D_3$-induced activities [42]. This response pattern suggests either that VDR–analog complexes recruit unoccupied VDR to bind DNA or that the maximal number of VDR binding sites available for the analog is greater than the maximal number of binding sites available for $1,25(OH)_2D_3$.

Although the dissociation rate of the hexafluorinated analog–VDR complexes from VDRE is not different from that of $1,25(OH)_2D_3$–VDR complexes, the combination of hexafluorides with a triple bond between carbons 23 and 24 and a double bond at the D ring generates a new compound [$1\alpha,25(OH)_2$-16-ene-23-yne-26,27-F_6-D_3] that slightly reduces the dissociation rate of VDR–RXR complexes from VDRE [66]. This slight change, however, cannot explain why this analog is 3-fold more efficacious in transcriptional activity than $1,25(OH)_2D_3$

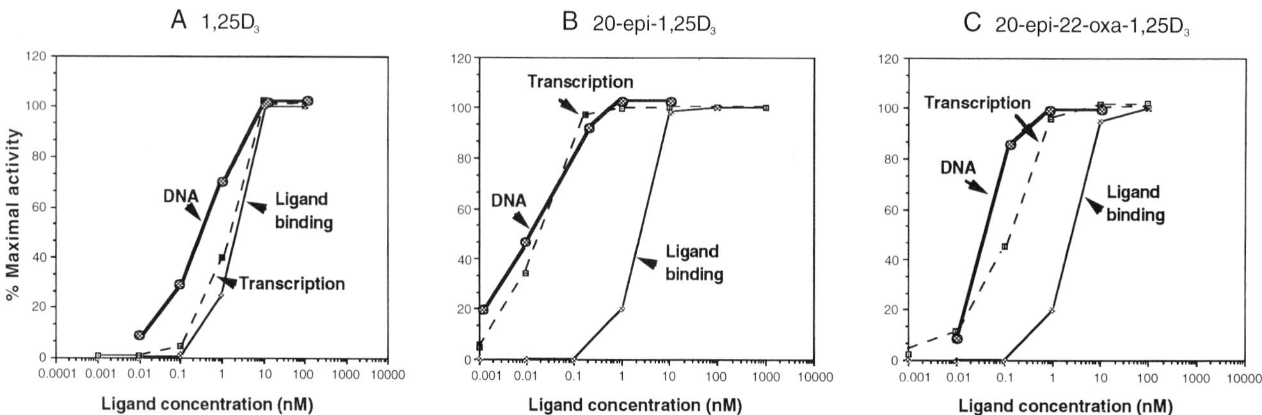

FIGURE 6 Relationship between analog binding, DNA binding, and transcriptional activity of VDR. COS-1 cells expressing recombinant human VDR were incubated in serum-free medium with increasing concentrations of the indicated ligands for 1 hr. Excess ligand was then removed, and the unoccupied ligand binding sites were quantified by incubating cell homogenates with ^3H-$1,25(OH)_2D_3$. Binding of VDR to DNA containing a DR3 motif was assessed by a mobility shift assay using extracts of ligand-treated VDR-transfected COS-1 cells and the osteopontin VDRE. Transcriptional activity was assessed by measuring reporter gene expression in ligand-treated cells transfected with a VDRE–thymidine kinase–growth hormone reporter gene. $1,25D_3$, $1\alpha,25(OH)_2D_3$; 20-epi-$1,25D_3$, 20-epi-$1\alpha,25(OH)_2D_3$; 20-epi-22-oxa-$1,25D_3$, 20-epi-22-oxa-24a,26a,27a-trihomo-$1\alpha,25(OH_2D_3$.

and 1000-fold more potent than $1,25(OH)_2D_3$. It is possible that the modified activity of this analog is due to a combination of direct physical effect on VDR and unique pharmacological properties, such as poor binding to serum vitamin D binding protein (DBP) and slow catabolism. This possibility is supported by the results of other studies that show that the enhanced transcriptional activity of 16-ene-23-yne analogs is completely dependent on presence of DBP (Table I) [80].

Evidence for preferential selection of synthetic response elements by $1,25(OH)_2D_3$ and its analogs has been reported by Nayeri *et al.* [81]. The side chain-modified analog EB1089 transactivates VDR through an IP9 (an inverted repeat motif with a spacing of 9 bp) more effectively than through the DR3 motif, whereas the transcriptional activity of $1,25(OH)_2D_3$ from these two motifs is similar. These results have not yet been corroborated with data on dimerization and DNA binding properties of VDR–analog complexes. Importantly, an IP9 sequence has not been identified in any vitamin D responsive gene promoter.

Evidence for cell-specific gene regulation by $1,25(OH)_2D_3$ and the analog EB1089 in an animal model has been reported by Roy *et al.* [82]. In that study, EB1089 induced transcription of the 24-hydroxylase gene in the kidney but not in the intestine. In contrast, the natural hormone induced transcription of this gene with similar efficacy and potency in both kidney and intestine. These intriguing experiments suggest that the VDR–analog complexes are significantly different in their transcriptional properties from VDR–$1,25(OH)_2D_3$ complexes, and that this difference may be mediated by the cellular environment.

IV. FUTURE PROSPECTS

A. Proposed Mechanisms for Differential Regulation of VDR Action by Analogs

Figure 7 summarizes several possible mechanisms for differential transcriptional activation by structurally

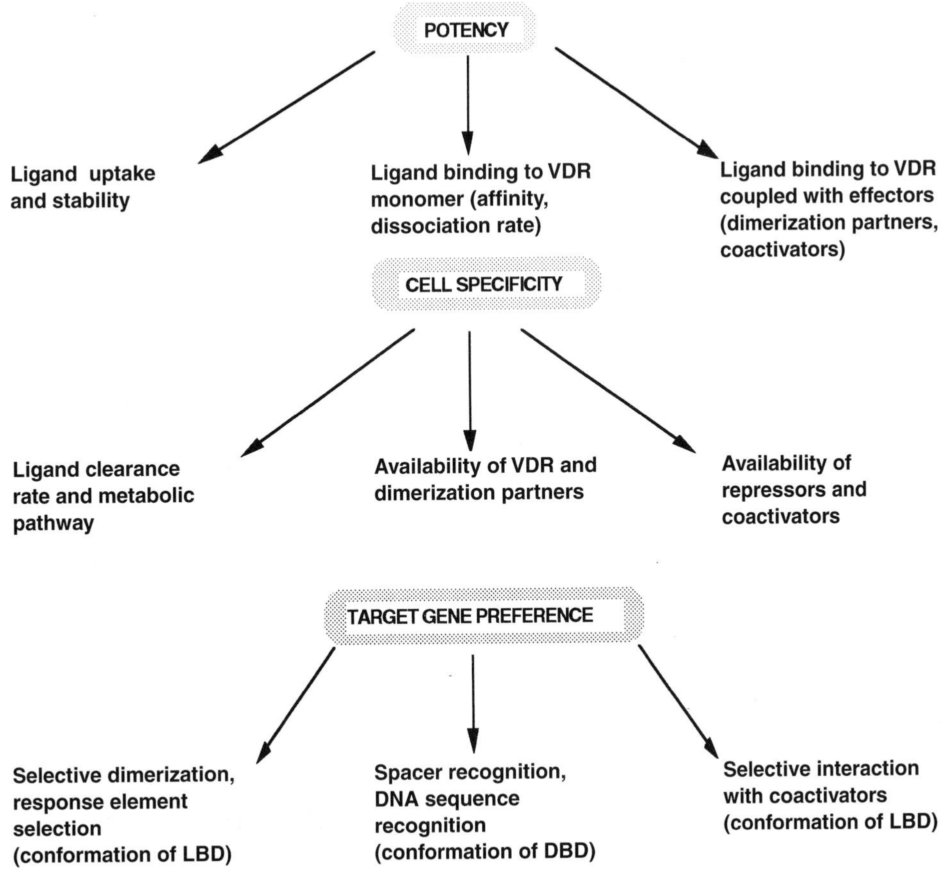

FIGURE 7 Three proposed mechanisms for differential activation of VDR by the natural hormone and synthetic analogs. DBD, DNA binding domain; LBD, ligand binding domain.

modified ligands of VDR. There is experimental evidence (described earlier in this chapter) for a subset of these mechanisms, although others will require further investigation. The differential activation of VDR by the natural hormone and its analogs may occur at the following three levels.

1. POTENCY

The potency of analogs is defined by ligand properties that determine the cellular level of receptor–ligand complexes. These include pharmacological properties of the ligand (e.g., uptake and clearance rate), its affinity for VDR, and its rate of dissociation from it. Although the affinity and dissociation rate of ligands are traditionally measured for VDR monomer, possible changes in the ligand binding properties of dimerized VDR or changes in ligand-binding properties of VDR–DNA complexes and VDR complexes with transcription coactivators should be considered. All of these parameters modulate the overall levels of receptor–ligand complexes available for transcriptional activity but not the physical properties and composition of these complexes or their target gene preference.

2. CELL SPECIFICITY

Target cell preference may be differently regulated by default: limiting levels of VDR, dimerization partners, and coactivators in specific tissues may be sufficient to elicit a response from potent analogs but not from the natural hormone. Likewise, analogs that are poor substrates for enzymes that catabolize $1{,}25(OH)_2D_3$ may have greater potency than the natural hormone in target cells that have low levels of these enzymes; this advantage will be lost in cells that have high levels of catabolic enzymes.

3. TARGET GENE PREFERENCE

Structural changes in the ligands may affect target gene preference by induction of distinct conformational changes in dimerization interfaces that regulate the selection of dimerization partners (DI2 and DI3, Fig. 5) and the selection of DNA response elements for VDR (DI1). Target gene preference may also be regulated by unique ligand-induced conformational changes in the ligand binding domain that selectively modulate interactions with coactivators and other transcriptional effectors. This level of fine tuning depends more on the composition of promoter-specific factors than on the DNA response element to which VDR binds. Differential modulation of interaction with coactivators by agonists and antagonists of other nuclear receptors has already been described, and several accessory factors required for optimal transcriptional activity of nuclear receptors have been identified [49]. However, it is still necessary to determine which if any of these factors are required for transcriptional activity of VDR and to what extent their functions are dependent on the structure of ligands for VDR.

B. Clinical Significance of Differential Gene Regulation by Vitamin D Analogs

The therapeutic usefulness of vitamin D analogs has been considered for hyperproliferative skin disorders, secondary hyperparathyroidism, posttransplantation immunosuppression, and various malignancies. At the present time $1{,}25(OH)_2D_3$ and an analog are in clinical use for treatment of the skin disease psoriasis, whereas the natural hormone is used for treatment of bone disease and renal failure (reviewed by Bouillon *et al.* [36]). However, a wider application of $1{,}25(OH)_2D_3$ to treat other clinical conditions is deterred by the toxicity associated with the pharmacological doses required to achieve a significant response. The general approach for screening of new compounds has been to examine their effect on ionized calcium levels in plasma *in vivo* and their growth inhibitory action in cell culture and animal models. This approach, however, is somewhat simplistic, because it overlooks the specific gene regulatory events associated with each of these biological responses. For that reason, the mechanism(s) of action of the few compounds that exhibit a segregated activity (i.e., low calcemic effect and significant growth inhibitory action) is unclear. As a consequence, it is difficult to develop a strategy to improve analog function by predicting favorable structural modifications.

To achieve functional segregation, potency, and target cell specificity, it may be advantageous to screen analogs by target cell and target gene preferences. For example, calcium homeostasis depends on VDR transcriptional activities in intestine and kidney (regulation of calbindins and 24-hydroxylase gene transcription and possibly transcriptional up-regulation of the plasma membrane calcium pump) [25,75,76]. Therefore, analogs that fail to induce VDR-mediated transcription in intestine and kidney may exhibit low calcemic activity. Likewise, growth inhibition and differentiation of leukemia cells require transcriptional up-regulation of cyclin-dependent kinase inhibitors [76]. Analogs that preferentially increase the transcriptional activity through half-sites with the sequence GGTTCA (which is part of the cyclin-dependent kinase inhibitor p21 VDRE) may be particularly useful for treatment of myeloid leukemia. The same type of analogs may also be useful for treatment of secondary hyperparathyroidism, by repression of the PTH gene through the half-site GGTTCA [77]. Finally, effective immunosuppression requires transcriptional down-regulation of several cytokines (e.g., IL-2). Because this repression appears to be mediated by interaction of VDR with Jun–Fos, analogs that preferentially enhance interaction of VDR with the Jun–Fos complex may be particularly useful for treating clinical conditions that require down-regulation of interleukin-2 [78].

Long-term end-organ effects should also be considered when using highly potent analogs. For example, the differentiating action of $1,25(OH)_2D_3$ is reversible, but using analogs that stabilize the transcriptionally active VDR complex (either because they dissociate very slowly from VDR or because they slow down the dissociation of VDR from RXR or from DNA response elements) may lead to irreversible terminal differentiation and apoptosis. This shift from cytostatic to cytotoxic action has already been demonstrated in breast cancer cells treated with 20-epi analogs *in vitro* [83].

In conclusion, it is possible that the therapeutic efficacy of current vitamin D analogs can be improved by carefully matching analog structure with target cells and target genes and by identifying analogs that induce irreversible changes in the fate of the affected cells.

References

1. Norman AW 1979 Vitamin D: The Calcium Homeostatic Steroid Hormone. Academic Press, New York, pp. 1–490.
2. Lawson DEM, Fraser DR, Kodicek E, Morris HR, Williams DH 1971 Identification of 1,25-dihydroxycholecalciferol, a new kidney hormone controlling calcium metabolism. Nature **230**:228–230.
3. Holick MF, Schnoes HK, DeLuca HF 1971 Identification of 1,25-dihydroxycholecalciferol, a form of vitamin D_3 metabolically active in the intestine. Proc Natl Acad Sci USA **68**:803–804.
4. Norman AW, Myrtle JF, Midgett RJ, Nowicki HG, William V, Popjak G 1971 1,25-Dihydroxycholecalciferol: Identification of the proposed active form of vitamin D_3 in the intestine. Science **173**:51–54.
5. Bikle DD, Pillai S 1993 Vitamin D, calcium and epidermal differentiation. Endocr Rev **14**:3–19.
6. Takahashi N, Akatsu T, Sasaki T, Nicholson GC, Moseley JM, Martin TJ, Suda T 1988 Induction of calcitonin receptor by 1α,25-dihydroxy vitamin D_3 in osteoclast-like multinucleated cells from mouse bone marrow cells. Endocrinology **123**:1504–1510.
7. Udagawa N, Takahashi N, Akatsu T, Tanaka H, Sasaki T, Nishihara T, Koga T, Martin TJ, Suda T 1990 Origin of osteoclasts: Mature monocytes and macrophages are capable of differentiating into osteoclasts under suitable microenvironment prepared by bone marrow-derived stromal cells. Proc Natl Acad Sci USA **87**:7260–7264.
8. Roodman GD, Ibbotson KJ, MacDonald BR, Kuehl TJ, Mundy GR 1985 1,25-Dihydroxy vitamin D_3 causes formation of multinucleated cells with several osteoclast characteristics in cultures of primate marrow. Proc Natl Acad Sci USA **82**:8213–8217.
9. Takahashi N, Yamana H, Yoshiki S, Roodman GD, Mundy GR, Jones SJ, Boyde A, Suda T 1988 Osteoclast-like cell formation and its regulation by osteotropic hormones in mouse bone marrow cultures. Endocrinology **122**:1373–1382.
10. Amento EP 1987 Vitamin D and the immune system. Steroids **49**:55–72.
11. Kragballe K 1993 Vitamin D_3 Analogues in psoriasis—Clinical use and mode of action. In: Bernard BA, Schroot B (eds) Molecular Biology to Therapeutics. Birkhaeuser, Basel, pp. 174–181.
12. Abe E, Miyaura C, Sakagami H, Takeda M, Konno K, Yamazaki T, Yoshiki S, Suda T 1981 Differentiation of mouse myeloid leukemia cells induced by 1,25-dihydroxyvitamin D_3. Proc Natl Acad Sci USA **78**:4990–4994.
13. Bar-Shavit Z, Teitelbaum SL, Reitsma P, Hall A, Pegg LE, Trial J, Kahn AJ 1983 Induction of monocytic differentiation and bone resorption by 1,25-dihydroxyvitamin D_3. Proc Natl Acad Sci USA **80**:5907–5911.
14. Walters MR 1992 Newly identified actions of the vitamin D endocrine system. Endocr Rev **13**:719–764.
15. Frampton RJ, Omond SA, Eisman JA 1983 Inhibition of human cancer cell growth by 1,25-dihydroxyvitamin D_3 metabolites. Cancer Res **43**:4443–4447.
16. Saez S, Falette N, Gullot C, Meggouh F, Lefebvre MF, Crepin M 1993 $1,25(OH)_2D_3$ modulation of mammary tumor cell growth in vitro and in vivo. Breast Cancer Res Treat **27**:69–81.
17. Eisman JA, Barkla DH, Tutton PJM 1987 Suppression of in vivo growth of human cancer solid tumor xenografts by 1,25-dihydroxy-vitamin D_3. Cancer Res **46**:21–25.
18. Haussler MR, Mangelsdorf DJ, Komm BS, Terpenning CM, Yamaoka K, Allegretto EA, Baker AR, Shine J, McDonnell DR, Hughes M, et al. 1988 Molecular biology of the vitamin D hormone. Recent Prog Horm Res **44**:263–305.
19. Mangelsdorf DJ, Thummel C, Beato M, Herrlick P. Schutz G, Umensono K, Blumberg B, Kastner P, Mark M, Chambon P, Evans RM 1995 The nuclear receptor superfamily: The second decade. Cell **83**:835–839.
20. Pike JW 1992 Molecular mechanism of cellular response to the vitamin D_3 hormone. In: Coe FL, Favus MJ (eds) Disorders of Bone and Mineral Metabolism. Raven, New York, pp. 163–193.
21. Peleg S, Sastry M, Collins ED, Bishop JE, Norman AW 1995 Distinct conformational changes induced by 20-epi analogues of 1α,25-dihydroxyvitamin D_3 are associated with enhanced activation of the vitamin D receptor. J Biol Chem **270**:10551–10558.
22. Towers TL, Luisi BF, Asianov A, Freedman LP 1993 DNA target selectivity by the vitamin D_3 receptor: Mechanism of dimer binding to an asymmetric repeat element. Proc Natl Acad Sci USA **90**:6310–6314.
23. Yu V, Delsert C, Anderson B, Holloway JM, Davary OV, Naar AM, Kim SY, Bouton J-M, Glass CK, Rosenfeld MG 1991 RXRβ: A coregulator that enhances binding of retinoic acid, thyroid hormone and vitamin D receptors to their cognate response elements. Cell **67**:1251–1266.
24. Cheskis B, Freedman LP 1994 Ligand modulates the conversion of DNA-bound vitamin D_3 receptor (VDR) homodimers into VDR–retinoid X receptor heterodimers. Mol Cell Biol **14**:3329–3338.
25. Kephart DD, Walfish PG, DeLuca H, Butt TR 1996 Retinoid X receptor isotype identity directs human vitamin D receptor heterodimer transactivation from the 24-hydroxylase vitamin D response elements in yeast. Mol Endocrinol **10**:407–419.
26. Cheng HJ, Pike JW 1996 Human vitamin D receptor-dependent transactivation in *Saccharomyces cerevisiae* requires retinoid X receptor. Mol Endocrinol **10**:196–205.
27. Hannah SS, Norman AW 1994 $1\alpha,25(OH)_2$-vitamin D_3 regulated expression of the eukaryotic genome. Nutr Rev **52**:376–381.
28. Koren R, Ravid A, Liberman UA, Hochberg Z, Weisman Y 1985 Defective binding and function of 1,25-dihydroxyvitamin D_3 receptors in peripheral mononuclear cells from patients with end-

organ resistance to 1,25-dihydroxyvitamin D_3. J Clin Invest **76**:2012–2015.

29. Haussler MR, Donaldson CA, Marion SL, Allgretto EA, Kelly MA, Mangelsdorf DJ, Pike JW 1986 Receptors for the vitamin D hormone: Characterization and functional involvement in 1,25-dihydroxyvitamin D_3 regulated events. In: Gotto M, O'Malley BW (eds) The Role of Receptors in Biology and Medicine. Raven, New York, pp. 91–104.
30. Mangelsdorf DJ, Koeffler HP, Donaldson CA, Pike JW, Haussler MR 1984 1,25-Dihydroxyvitamin D_3-induced differentiation in a human promyelocytic leukemic cell line (HL-60): Receptor-mediated maturation to macrophage-like cells. Endocrinology **98**:391–398.
31. Nemere J, Yoshimoto Y, Norman AW 1984 Studies on the mode of action of calciferol. LIV. Calcium transport in perfused duodena from normal chicks: Enhancement with 14 minutes of exposure to 1,25-dihydroxyvitamin D_3. Endocrinology **115**:1476–1483.
32. Caffrey JM, Farach-Carson MC 1989 Vitamin D_3 metabolites modulate dihydropyridine-sensitive calcium currents in clonal rat osteosarcoma cells. J Biol Chem **264**:20265–20274.
33. Studzinski GP, McLane JA, Uskokovic MR 1993 Signaling pathways for vitamin D-induced differentiation: Implications for therapy of proliferative and neoplastic disease. Crit Rev Euk Gene Expression **3**:279–312.
34. Nemere I, Dormanen MC, Hammond MW, Okamura WH, Norman AW 1994 Identification of a specific binding protein for 1,25-dihydroxyvitamin D_3 in basal–lateral membranes of chick intestinal epithelium and relationship to transcaltachia. J Biol Chem **38**:23750–23756.
35. Nemere I, Norman AW 1987 The rapid hormonally stimulated transport of calcium (transcaltachia). J Bone Miner Res **2**:167–169.
36. Bouillon R, Okamura WH, Norman AW 1995 Structure–function relationship in the vitamin D endocrine system. Endocr Rev **16**:200–257.
37. Fraser DR 1980 Regulation of the metabolism of vitamin D. Physiol Rev **60**:551–613.
38. Norman AW, Okamura WH, Farach-Carson MC, Allewaret K, Branisteanu D, Nemere I, Raman-Muralidharan K, Boullion R 1993 Structure–function studies of 1,25-dihydroxyvitamin D_3 and the vitamin D endocrine system. 1,25-Dihydroxy-pentadeuterio-previtamin D_3 (as a 6-s cis analog) stimulates nongenomic but not genomic biological responses. J Biol Chem **268**:13811–13819.
39. Napoli JL, Pramanik BC, Royal PM, Reinhardt TA, Horst RL 1983 Intestinal synthesis of 24-keto 1,25 dihydroxyvitamin D_3. J Biol Chem **158**:9100–9107.
40. Zhou JY, Norman AW, Akashi M, Chen D-L, Uskokovic MR, Aurrecoechea JM, Dauben WG, Okamura WH, Koeffler HP 1991 Development of a novel 1,25$(OH)_2$-vitamin D_3 analog with potent ability to induce HL-60 cell differentiation without modulating calcium metabolism. Blood **12**:75–82.
41. Binderup L, Latini S, Binderup E, Bretting C, Calverley M, Hansen K 1991 20-EPI-Vitamin D_3 analogues: A novel class of potent regulators of cell growth and immune responses. Biochem Pharmacol **42**:1569–1575.
42. Sasaki H, Harada H, Hand Y, Morino H, Suzawa M, Shimpo E, Katsumata T, Masuhiso Y, Matsuda K, Ebihara K, Ono T, Masushige S, Kato S 1995 Transcriptional activity of a fluorinated vitamin D analog on VDR–RXR-mediated a gene expression. Biochemistry **34**:370–377.
43. Peleg S, Lui Y-Y, Reddy S, Horst RL, White C, Posner GH 1996 A 20-epi side chain restores growth-regulatory and transcriptional activities of an A ring-modified hybrid analog of 1α,25-dihydroxyvitamin D_3 without increasing its affinity to the vitamin D receptor. J Cell Biochem **63**:149–161.
44. Ferrara J, McCuaing K, Hendy GN, Uskokovic M, White JH 1994 Highly potent transcriptional activation by 16-ene derivatives of 1,25-dihydroxyvitamin D_3. J Biol Chem **269**:2971–2981.
45. Uskokovic MR, Baggiolini E, Shuiey S-J, Lacobelli J, Hennessy B, Kriegiel J, Daniewski AR, Pizzolato G, Courtney LF, Horst RL 1991 The 16-ene analogs of 1,25-dihydroxycholecalciferol. Synthesis and Biological Activity. In: Norman AW, Bouillon R, Thomasset M (eds) Vitamin D: Gene Regulation, Structure–Function Analysis and Clinical Applications. de Gruyter, New York, pp. 139–156.
46. Reddy GS, Clark JW, Tseng K-Y, Uskokovic MR, McLane JA 1993 Metabolism of 1,25$(OH)_2$-16-ene D_3 in kidney: Influence of structural modification of D-ring on side chain metabolism. Bioorg Med Chem Lett **3**:1879–1884.
47. De Boland AR, Norman AW 1990 Influx of extracellular calcium mediates 1,25-dihydroxyvitamin D_3-dependent transcaltachia (the rapid stimulation of duodenal Ca^{++} transport). Endocrinology **127**:2475–2480.
48. Norman AW, Bouillon R, Farach-Carson MC, Bishop JE, Zhou L-X, Nemere I, Zhao J, Muralidharan RK, Okamura WH 1993 Demonstration that 1β,25-dihydroxyvitamin D_3 is an antagonist of the nongenomic but not genomic biological responses and biological profile of the three A-ring diastereomers of 1α,25-dihydroxyvitamin D_3. J Biol Chem **268**:20022–20030.
49. Mangelsdorf DJ, Evans RM 1995 The RXR heterodimers and orphan receptors. Cell **83**:841–850.
50. Nakajima S, Hsieh JC, MacDonald PN, Galligan MA, Haussler CA, Whitfield GK, Haussler MR 1994 The C-terminal region of the vitamin D receptor is essential to form a complex with a receptor auxiliary factor required for high affinity binding to the vitamin D-responsive element. Mol Endocrinol **8**:159–172.
51. Nishikawa J-I, Kitaura M, Imagawa M, Nishikara T 1995 Vitamin D receptor contains multiple dimerization interfaces that are functionally different. Nucleic Acids Res **23**:606–611.
52. Blanco JC, Wang I-M, Tsai SY, Tsai M-J, O'Malley BW, Jurutka PW, Haussler MR, Ozato K 1995 Transcription factor TFIIB and the vitamin D receptor cooperatively activate ligand-dependent transcription. Proc Natl Acad Sci USA **92**:1535–1539.
53. MacDonald PN, Sherman DR, Dowd DR, Jefcoat SC, DeLisle RK 1995 The vitamin D receptor interacts with general transcription factor IIB. J Biol Chem **270**:4748–4752.
54. Bourguet W, Ruff D, Chambon P, Gronemeyer H, Moras D 1995 Crystal structure of the ligand binding domain of the human nuclear receptor RXRα. Nature **375**:377–382.
55. Wagner RL, Apriletti JW, McGrath ME, West BL, Baster JD, Fletterick RJ 1995 A structural role for hormone in the thyroid hormone receptor **378**:690–697.
56. Renaud J-P, Natacha R, Ruff M, Vivat V, Chambon P, Gronemeyer H, Moras D 1995 Crystal structure of the RAR-gamma ligand-binding domain bound to all-trans retinoic acid. Nature **378**:681–689.
57. McDonnell DP, Scott RA, Kerner SA, O'Malley BW, Pike JW 1989 Functional domains of the human vitamin D_3 receptor regulate osteocalcin gene expression. Mol Endocrinol **3**:635–644.
58. Liu Y-Y, Collins E, Reddy S, Norman AW, Peleg S 1995 The choice of contact points with the vitamin D receptor modulates the half-life of the vitamin D receptor complex with 1,25-dihydroxyvitamin D_3 and its analogs. J Bone Miner Res **10**(Suppl 1):S166.
59. Denielian PS, White R, Lees JA, Parker MG 1992 Identification of a conserved region required for hormone dependent transcrip-

tional activation by steroid hormone receptors. EMBO J **11**:1025–1033.

60. Rastinejad F, Perlmann T, Evans RM, Sigler PB 1995 Structural determinants of nuclear receptor assembly on DNA direct repeats. Nature **375**:203–211.
61. Rosen ED, Beinghof EG, Koenig RJ 1993 Dimerization interfaces of thyroid hormone, retinoic acid, vitamin D and retinoid X receptors. J Biol Chem **268**:11534–11541.
62. Whitfield GK, Hsieh J-C, Nakajima S, MacDonald PN, Thompson PD, Jurutka PW, Haussler CA, Haussler MR 1995 A highly conserved region in the hormone binding domain of the human vitamin D receptor contains residues vital for heterodimerization with retinoid X receptor and for transcriptional activation. Mol Endocrinol **9**:1166–1179.
63. Umensono K, Murakami KK, Thompson CC, Evans RM 1991 Direct repeats as selective response elements for the thyroid hormone, retinoic acid, and vitamin D_3 receptors. Cell **65**:1255–1266.
64. Schrader M, Bendik I, Becker-Andre M, Carlberg C 1993 Interaction between retinoic acid and vitamin D signaling pathway. J Biol Chem **268**:17830–17836.
65. Schrader M, Muller KM, Carlberg C 1994 Specificity and flexibility of vitamin D signaling. J Biol Chem **269**:5501–5504.
66. Cheskis B, Lemon BD, Uskokovic M, Lemedico PT, Freedman LP 1995 Vitamin D_3–retinoid X receptor dimerization, DNA binding and transactivation are differentially affected by analogs of 1,25-dihydroxyvitamin D_3. Mol Endocrinol **9**:1814–1824.
67. Liu Y-Y, Peleg S 1996 The importance of cellular environment and conformational changes of vitamin D receptor (VDR) in ligand-induced dimerization with retinoid X receptors (RXR). Proceedings of the 10th International Congress of Endocrinology. ORG-3, p. 59.
68. Zechel C, Shen X-Q, Chen J-Y, Chen Z-P, Chambon P, Gronemeyer H 1994 The dimerization interfaces formed between DNA binding domains of RXR, RAR and TR determine the binding specificity and polarity of the full length receptors to direct repeats. EMBO J **13**:1425–1433.
69. Kurokawa R, Yu VC, Naar A, Kyakumoto S, Han Z, Silverman S, Rosenfeld MG, Glass CK 1993 Differential orientations of the DNA-binding domain and carboxy-terminal dimerization interface regulate binding site selection by nuclear receptors heterodimers. Genes Dev **7**:1423–1435.
70. Freedman JP, Perez-Fernandez R, Arce V 1994 DNA sequences that act as high affinity targets for the vitamin D_3 receptor in the absence of retinoid X receptor. Mol Endocrinol **8**:265–273.
71. Demay MB, Gerardi JM, DeLuca HF, Kronenberg HM 1990 DNA sequences in the rat osteocalcin gene that bind the 1,25-dihydroxyvitamin D_3 receptor and confer responsiveness to 1,25-dihydroxyvitamin D_3. Proc Natl Acad Sci USA **87**:369–373.
72. Ozono K, Liao J, Kerner SA, Scott RA, Pike JW 1990 The vitamin D-responsive element in the human osteocalcin gene. J Biol Chem **265**:21881–21888.
73. Noda M, Vogel RL, Craig AM, Prahl J, DeLuca HF, Denhardt DT 1990 Identification of a DNA sequence responsible for binding of the 1,25-dihydroxyvitamin D_3 enhancement of mouse secreted phosphoprotein 1 (Spp-1 or osteopontin) gene expression. Proc Natl Acad Sci USA **87**:9995–9999.
74. Darwish HM, DeLuca HF 1992 Identification of a 1,25-dihydroxyvitamin D_3-response element in the 5′-flanking region of the rat calbindin D-9k gene. Proc Natl Acad Sci USA **89**:603–607.
75. Gill RK, Christakos S 1993 Identification of sequence elements in mouse calbindin-D28k that confer 1,25-dihydroxyvitamin D_3- and butyrate-inducible responses. Proc Natl Acad Sci USA **90**:2984–2988.
76. Liu M, Lee M-H, Cohen M, Bommakanti M, Freedman LP 1996 Transcriptional activation of the Cdk inhibitor p21 by vitamin D_3 leads to the induced differentiation of the myelomonocytic cell line U937. Genes Dev **10**:142–153.
77. Demay MB, Kiernan MS, DeLuca HF, Kronenberg HM 1992 Sequences in the human parathyroid hormone gene that bind the 1,25-dihydroxyvitamin D_3 receptor and mediate transcriptional repression in response to 1,25-dihydroxyvitamin D_3. Proc Natl Acad Sci USA **89**:8097–8101.
78. Alroy I, Towers TL, Freedman LP 1995 Transcriptional repression of the interleukin-2 gene by vitamin D_3: Direct inhibition of NFATp/AP1 complex formation by a nuclear hormone receptor. Mol Cell Biol **15**:5789–5799.
79. Garcia-Villalba P, Jimenez-Lara AM, Aranda A 1996 Vitamin D interferes with transactivation of the growth hormone gene by thyroid hormone and retinoic acid. Mol Cell Biol **16**:318–327.
80. Imai Y, Pike JW, Koeffler HP 1995 Potent vitamin D_3 analogs: Their abilities to enhance transactivation and to bind to the vitamin D_3 response element. Leuk Res **19**:147–158.
81. Nayeri S, Danielsson C, Kahlen J-P, Schrader M, Mathiasen IS, Binderup L, Carlberg C 1995 The anti-proliferative effect of vitamin D_3 analogues is not mediated by inhibition of the AP-1 pathway, but may be related to promoter selectivity. Oncogene **11**:1853–1858.
82. Roy S, Martel J, Tenenhouse HS 1995 Comparative effects of 1,25-dihydroxyvitamin D_3 and EB1089 on mouse renal and intestinal 25-hydroxyvitamin D_3-24-hydroxylase. J Bone Miner Res **10**:1951–1959.
83. Elstner E, Linker-Israeli M, Said M, Umiel T, deVos S, Shintaku P, Heber D, Binderup L, Uskokovic M, Koeffler HP 1995 20-Epi vitamin D_3 analogues: A novel class of potent inhibitors of proliferation and inducers of differentiation of human breast cancer cell lines. Cancer Res **55**:2822–2830.

CHAPTER 61

Development of New Vitamin D Analogs

LISE BINDERUP, ERNST BINDERUP, AND WAGN OLE GODTFREDSEN
Departments of Biochemistry and Chemical Research, Leo Pharmaceutical Products, Ballerup, Denmark

I. INTRODUCTION

The involvement of Leo Pharmaceutical Products in the synthesis and evaluation of new vitamin D analogs and metabolites dates back to the early 1970s, with the development of 1α-hydroxycholecalciferol (1αOHD$_3$) for the treatment of renal osteodystrophy and hyperparathyroidism. Our interest was further stimulated in the early 1980s with the appearance of reports describing receptors for 1α,25-dihydroxyvitamin D$_3$ [1,25(OH)$_2$D$_3$] in nonclassical target tissues and the demonstration of the role of 1,25(OH)$_2$D$_3$ in regulating growth and differentiation of various cancer cell lines, *in vitro* and *in vivo* [1,2]. At the same time, a number of reports suggested that 1,25(OH)$_2$D$_3$ might also influence various functions of activated lymphocytes and thereby play a role as a physiological regulator of the immune system [3,4].

These findings suggested new therapeutic possibilities for 1αOHD$_3$ and 1,25(OH)$_2$D$_3$, especially in neoplastic and immune-mediated diseases. In 1983, Leo took steps to initiate clinical trials with 1αOHD$_3$ in leukemia and non-Hodgkin's lymphomas [5,6]. The therapeutic usefulness of 1αOHD$_3$ and 1,25(OH)$_2$D$_3$ was, however, likely to be limited by their potent effects on calcium metabolism, leading to side effects such as hypercalcemia and soft tissue calcifications. It was therefore decided to try to develop new analogs with a more favorable therapeutic profile.

In 1985, the preliminary testing of a small series of new synthetic analogs led to the discovery of a promising candidate, later named calcipotriol. At the same time, clinical observations suggested that 1αOHD$_3$ and 1,25(OH)$_2$D$_3$ might exert antipsoriatic effects [7,8]. It was therefore decided to test calcipotriol in patients with psoriasis and to further expand our engagement in vitamin D chemistry. This chapter reviews our main efforts and achievements in this field since the mid-1980s.

II. STRATEGY FOR DEVELOPMENT OF NEW VITAMIN D ANALOGS

A. Basic Screening Strategy

At the start of the program for synthesis and evaluation of new vitamin D analogs, the primary aim was to identify compounds that were potent regulators of cell proliferation and differentiation but which had a reduced ability to exert the classic effects of $1,25(OH)_2D_3$ on calcium homeostasis. To study structure–activity relationships of a relatively large number of analogs, a fairly simple screening system was needed.

For studies on cell proliferation and differentiation, the human histiocytic lymphoma cell line U937 was chosen. This cell line possesses well-characterized, high affinity receptors for $1,25(OH)_2D_3$ [9]. *In vitro*, $1,25(OH)_2D_3$, at a concentration of approximately 10^{-8} *M*, was found to inhibit the proliferation of U937 cells and to induce cell differentiation along the monocyte–macrophage pathway [10]. New analogs were tested in the U937 cell culture system for 4 days, at concentrations ranging from 10^{-12} to 10^{-7} *M*. At the end of the incubation the cells were counted and cell survival was assayed by eosin Y exclusion. Induction of cell differentiation was assessed by scoring of cells positive for membrane-associated nonspecific esterase activity [11].

Another parameter that was considered to be of interest for the evaluation of new vitamin D analogs was the binding affinity of the compounds for the vitamin D receptor (VDR). The binding of vitamin D analogs was assessed by displacement of bound ^3H-$1,25(OH)_2D_3$ from receptor protein obtained from the intestinal epithelium of rachitic chickens, using an adaptation of the method originally described by Mellon *et al.* [12].

To assess the effects of new vitamin D analogs on calcium metabolism, an *in vivo* model was chosen [10]. The analogs were administered orally to rats, daily for 7 days. Urine was collected daily, and blood was collected by cardiac puncture at the end of the experiment. Metaphyseal bone was prepared from tibiae. Calcium levels were determined in urine and serum samples, and the calcium content in bone was assessed after ashing. To assure detection of even small differences in potency between various analogs, the rats were given a vitamin D-replete diet with a high calcium content (1%), in contrast to many of the older studies in rats and chickens given low calcium and/or low vitamin D diets [13,14]. Analogs with low calcemic activities were also tested with intraperitoneal (i.p.) administration, in order to avoid errors in evaluation due to differences in absorption.

As a follow-up to the screening systems described above, a test for the assessment of the half-life of new analogs *in vivo* was introduced a few years later [15]. Each analog was administered as a high single intravenous (i.v.) dose (200 μg/kg) to rats, and blood samples were collected at 0, 5, 15, and 30 min and at 1, 2, and 4 hr after administration. The serum level of the analog was determined by high-performance liquid chromatography (HPLC).

B. Synthesis Strategy

From the beginning of the project it was decided to concentrate our efforts on the synthesis of analogs of $1,25(OH)_2D_3$ (**I**), in which the C-17 side chain is modified while the seco-steroid ring system is kept intact. This decision was partly dictated by the fact that this part of the molecule is more easily accessible to chemical manipulation than the ring system, but it was also decisive that the side chain is known to play a crucial role in the binding of $1,25(OH)_2D_3$ to its receptor [16]. It is beyond the scope of this chapter to detail the synthesis of the more than 800 analogs that have been made in the Leo laboratories, but some general pathways are outlined in the following sections.

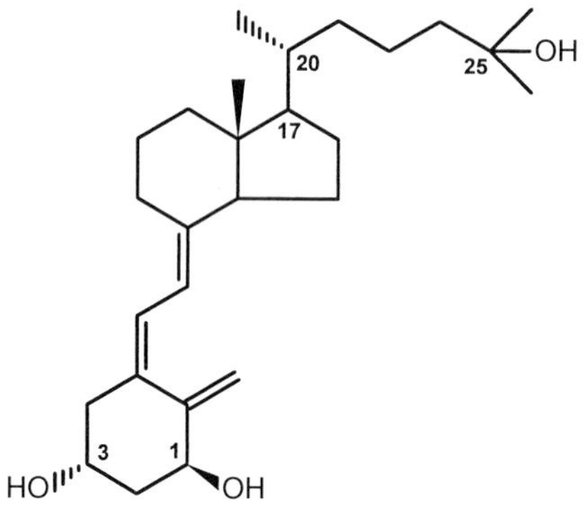

(I)

Our starting material has been the readily available ergocalciferol (**II**) (Scheme 1), which according to a method originally devised by Hesse [17,18] can be converted to a 1α-hydroxylated *trans* vitamin derivative conveniently isolated as the pure, crystalline, bis-silyl ether (**III**) [19]. After protection of the conjugated triene system as the sulfur dioxide adduct (**IV**), the side chain can be cleaved selectively by ozonolysis with formation of the aldehyde (**V**). By heating this in the presence of $NaHCO_3$, SO_2 is expelled, and the key intermediate **VI** can be isolated as a crystalline compound [19]. Compound **VI** has been a cornerstone in our synthetic work.

By means of the Wittig reaction new side chains could be introduced. Scheme 2 illustrates the synthesis of the analog EB1089 (**IX**), which contains two conjugated double bonds in the side chain [20]. The product of the Wittig reaction (**VII**) is reacted with ethyl lithium to give **VIII**, which is then isomerized to the *cis* form by ultraviolet irradiation in the presence of the photosensitizer anthracene [21]. Finally, the hydroxyl groups are deprotected with tetrabutylammonium fluoride in tetrahydrofuran (THF) to give **IX**. A Wittig reaction with

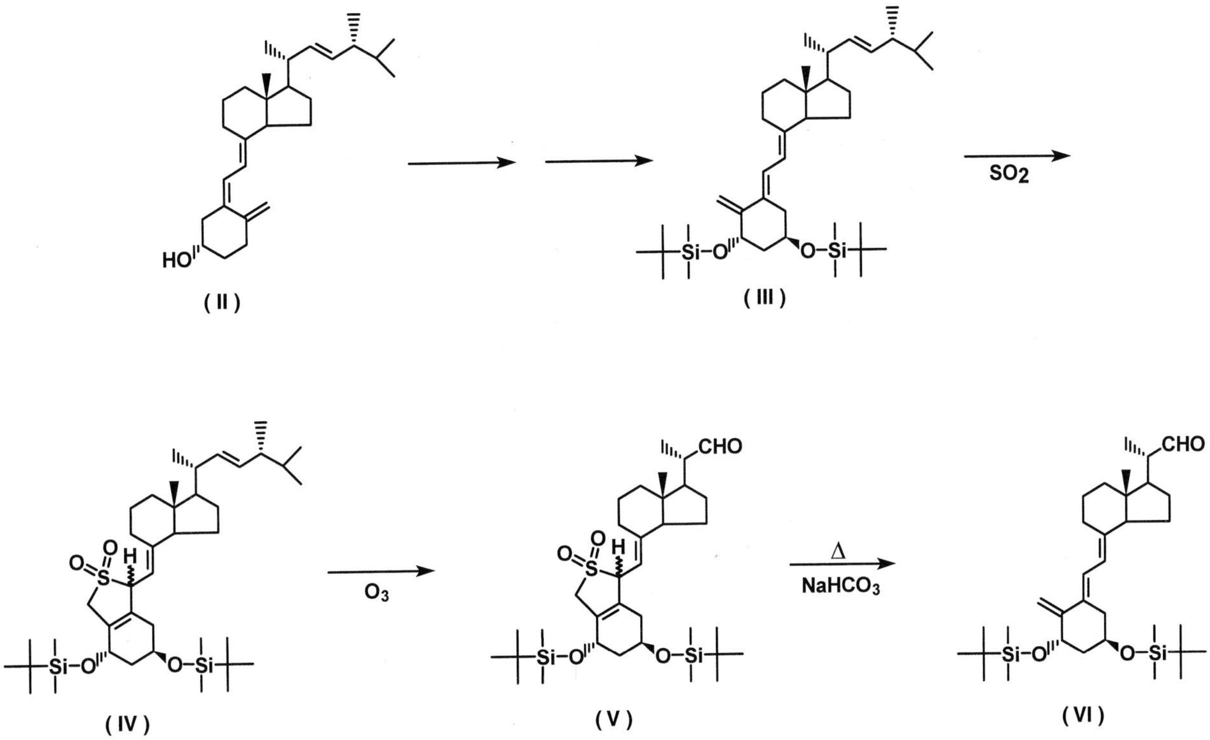

SCHEME 1

VI is also a step in the synthesis of the important antipsoriatic drug calcipotriol (MC903, see Table 3) [19].

Another route to new analogs involves $NaBH_4$ reduction of the aldehyde (**VI**), followed by tosylation of the resulting alcohol to give **X**, which is subsequently reacted with a Grignard reagent to form **XI**. Finally, **XI** is isomerized and deprotected to provide the analog CB966 (**XII**) [22].

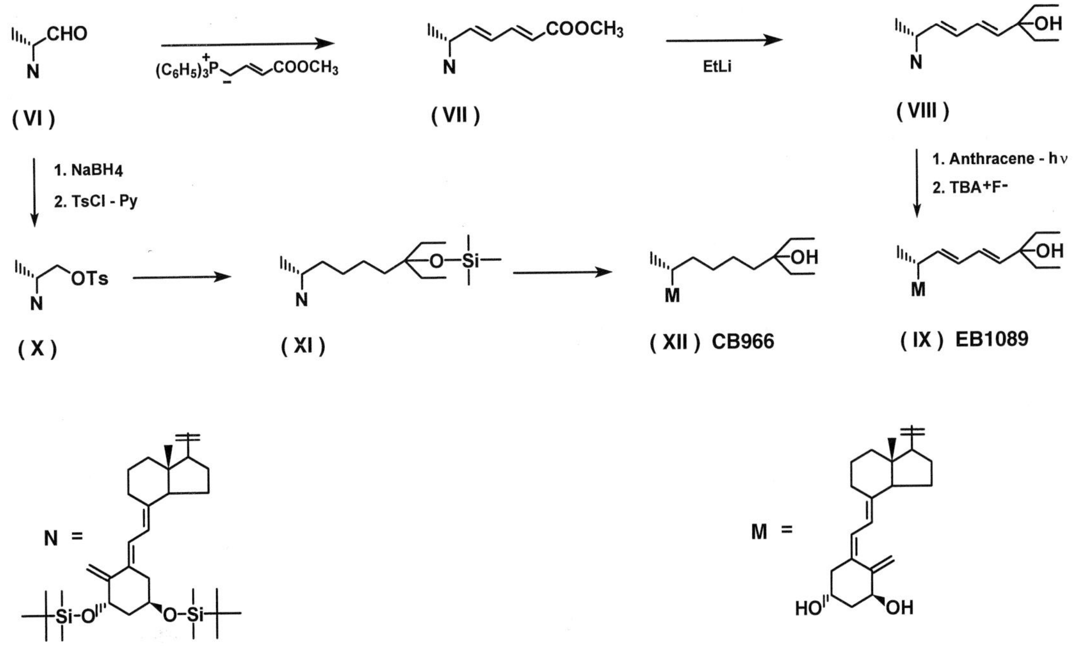

SCHEME 2

Analogs containing an oxygen atom in the 22 position (22-oxa analogs) can be synthesized as shown in Scheme 3, which depicts the synthesis of the C-20 epimeric compounds KH1139 and KH1060 [23]. Oxidation of the aldehyde (**VI**) with air in the presence of a copper catalyst yields the methylketone (**XIII**). Reduction of this with $NaBH_4$ leads to a mixture of the two epimeric alcohols (**XIVa** and **XIVb**), where the 20-epi isomer (**XIVb**) dominates. Alkylation of the two alcohols, followed by isomerization and deprotection as described above, yields the cis analogs **XVa** (KH1139) and **XVb** (KH1060), respectively. As described in Section III,E, epimerization at C-20 has a profound influence on the biological properties of the analogs.

Another route to 20-epi compounds (Scheme 3) starts with an epimerization of the aldehyde (**VI**) to the 20-epi aldehyde (**XVI**), which then, by the same sequence of reactions used in the 20-normal series (Scheme 2), can be converted to 20-epi analogs, such as 20-epi-1,25$(OH)_2D_3$ (**XVII**) (MC1288) [24]. The tosylate **XVIII**, used in the synthesis of MC1288, has also been used to synthesize the 20-epi-23-thia analog (**XIX**) (GS1500) [25].

The reactions depicted in Schemes 1–3 are typical of the pathways used in the synthesis of a wide variety of side chain analogs, but it is obvious that many variations have been necessary.

III. STRUCTURE–ACTIVITY RELATIONSHIPS

In this section the effect of systematic chemical modifications of the 1,25$(OH)_2D_3$ side chain on various biological parameters is presented. All the analogs discussed here have been tested for calcemic activity, antiproliferative activity, and ability to induce cell differentiation, as described in Section II,A. However, because the antiproliferative and differentiation-inducing properties run parallel, only the antiproliferative potencies are shown in the tables. All values are given in relation to 1,25$(OH)_2D_3$.

A. Variation of Chain Length

In Table I the biological activities of a number of analogs which differ from 1,25$(OH)_2D_3$ with respect to the length of the C-17 side chain are listed [22]. It is seen that if the chain length is increased with one methylene group as in MC1127, the ability to inhibit proliferation is increased, whereas the calcemic activity is reduced to about one-third of that of 1,25$(OH)_2D_3$. This compound has also been described by Ostrem *et al.* [26]. If the two terminal methyl groups in MC1127 are replaced by ethyl

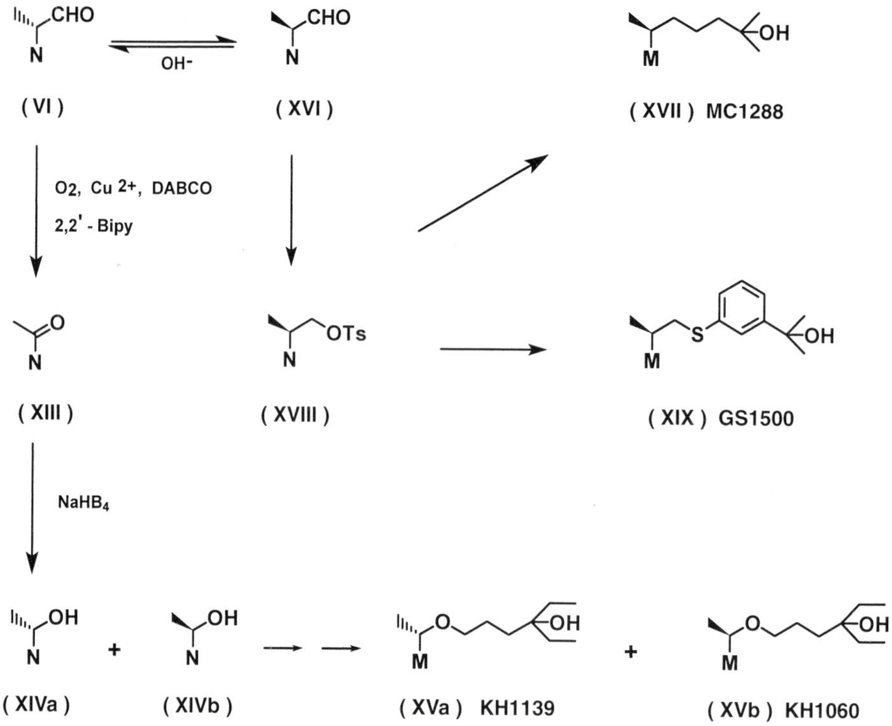

SCHEME 3. N and M have the same meaning as in Scheme 2.

TABLE I Variation of Chain Length

Compound	Side chain structure	Inhibition of U937 cell proliferation, IC_{50} (M)	Calcemic activity relative to $1,25(OH)_2D_3$ (%)
$1,25(OH)_2D_3$	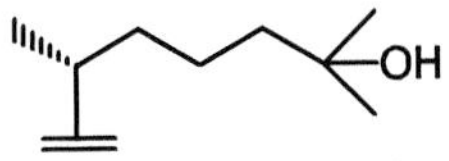 OH	3×10^{-8} **[1×]**[a]	100
MC1127	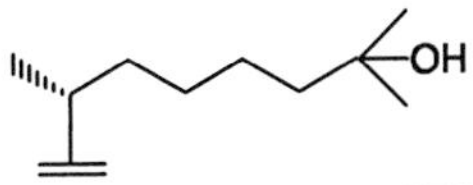 OH	5×10^{-9} **[6×]**	38
CB966	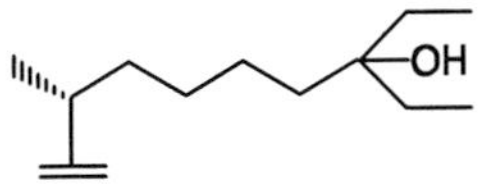 OH	1×10^{-9} **[30×]**	17
CB973	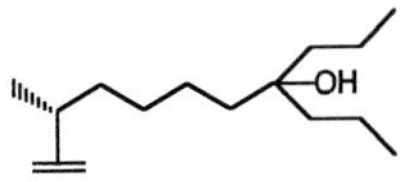 OH	5×10^{-8} **[0.6×]**	n.d.[b]
MC1147	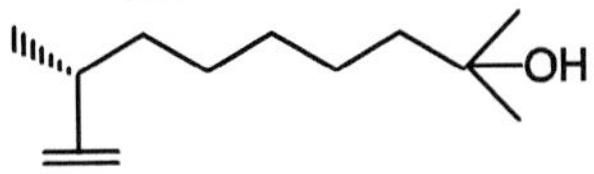 OH	5×10^{-9} **[6×]**	4
CB953	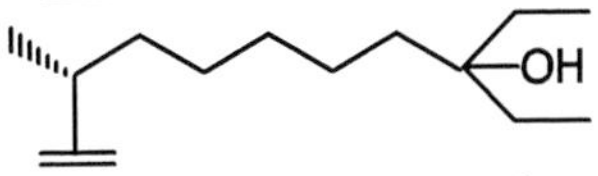 OH	2×10^{-8} **[1×]**	2
MC1179	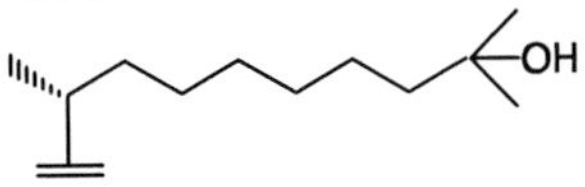 OH	4×10^{-8} **[0.7×]**	2

[a] Boldface figures indicate activity relative to $1,25(OH)_2D_3$.
[b] n.d., Not determined.

groups as in CB966, the antiproliferative potency is increased and the calcemic activity further reduced. On the other hand, if propyl groups are substituted for the methyl groups, the antiproliferative potency is reduced to about the same level as that of $1,25(OH)_2D_3$. An increase of the $1,25(OH)_2D_3$ side chain with two methylene groups (MC1147), also described by Kutner *et al.* [27], causes a further reduction of the calcemic activity, whereas the antiproliferative potency remains the same as in MC1127. However, with the introduction of one more methylene group (MC1179), the antiproliferative activity is reduced. In other words, the optimal number of methylene groups between C-20 and the tertiary hydroxyl group seems to be four or five.

B. Introduction of Double and Triple Bonds

The effect of introducing one or more double bonds in the C-17 side chain is illustrated in Table II. Whereas introduction of Δ^{22} double or triple bonds in CB966 has little effect on the biological activities [24,28], introduction of a further (Δ^{24}) double bond gives a compound (EB1089) [20] which, with respect to cell proliferation, is the most active in the series, being 100 times more potent than $1,25(OH)_2D_3$. Because the calcemic activity of EB1089 is three times lower, a substantial separation of the effects has been achieved. This compound is under clinical development as an anticancer drug (see Section V,B).

C. Calcipotriol and Related Analogs

One of the first $1,25(OH)_2D_3$ analogs synthesized at Leo for which a clear separation between the calcemic activity and the effects on cell regulation was achieved was MC903 [19] (Table III), which later received the United States Adopted Name (USAN) name calcipotriene and the International Nonproprietary Name (INN) name calcipotriol. In this compound a Δ^{22} double bond is introduced in the side chain, the 25-hydroxyl is moved to the 24 position (with the indicated stereochemistry), and a cyclopropane ring is substituted for the isopropyl group in $1,25(OH)_2D_3$. Table 3 shows that MC903 has retained the cell regulating potency of $1,25(OH)_2D_3$, whereas its calcemic activity has been reduced by a factor 200.

As described in Section V,A of this chapter and in Chapter 72, calcipotriol has become an important antipsoriatic drug. Its 24-epimer MC900 [19] has a considerably lower antiproliferative potency, and the same holds

TABLE II Double and Triple Bonds

Compound	Side chain structure	Inhibition of U937 cell proliferation, IC_{50} (M)	Calcemic activity relative to $1,25(OH)_2D_3$ (%)
$1,25(OH)_2D_3$	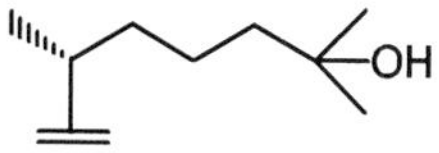 OH	3×10^{-8} **[1×]**[a]	100
CB966	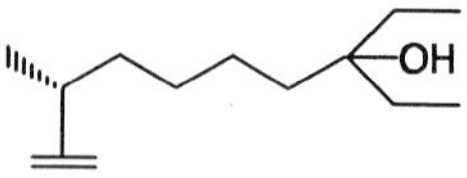 OH	1×10^{-9} **[30×]**	17
MC1473	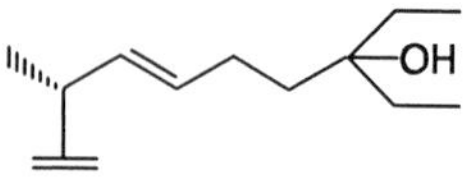 OH	2×10^{-8} **[1.5×]**	24
CB1309	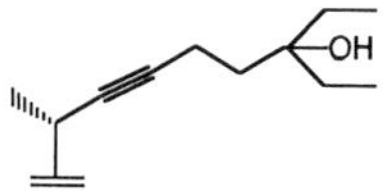 OH	3×10^{-8} **[1×]**	n.d.[b]
EB1089	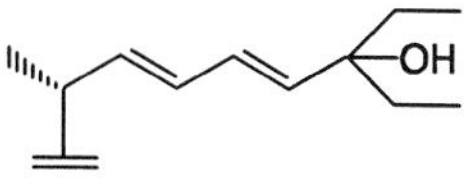 OH	3×10^{-10} **[100×]**	31
MC1147	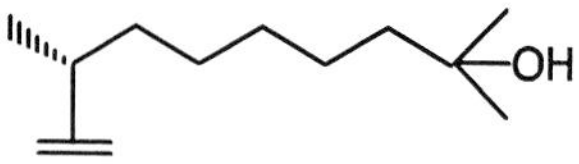 OH	5×10^{-9} **[6×]**	4
MC1226	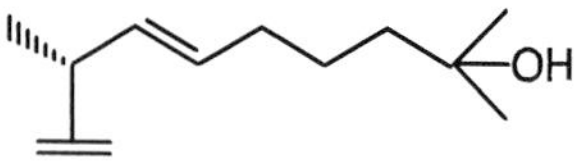 OH	6×10^{-9} **[5×]**	12
EB1180	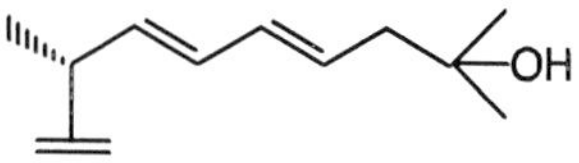 OH	2×10^{-9} **[15×]**	30

[a] Boldface figures indicate activity relative to $1,25(OH)_2D_3$.
[b] n.d., Not determined.

TABLE III Calcipotriol and Analogs

Compound	Side chain structure	Inhibition of U937 cell proliferation, IC_{50} (M)	Calcemic activity relative to $1,25(OH)_2D_3$ (%)
$1,25(OH)_2D_3$	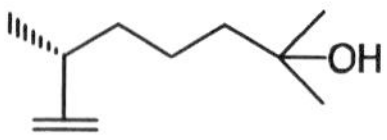 OH	3×10^{-8} **[1×]**[a]	100
MC903 (calcipotriol)	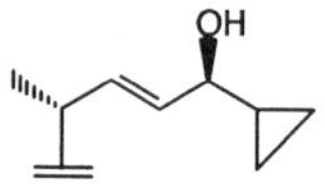 OH	2×10^{-8} **[1.5×]**	0.5
MC900	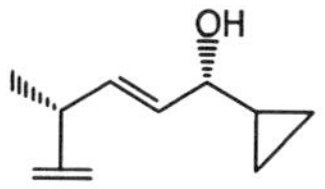 OH	$>1 \times 10^{-7}$ **[<0.3×]**	<1
MC1046	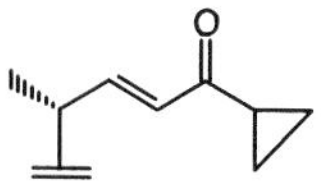 O	$>1 \times 10^{-7}$ **[<0.3×]**	<1
MC1080	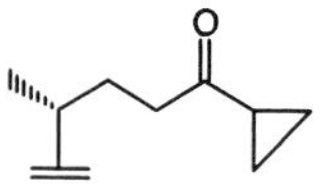 O	$>1 \times 10^{-7}$ **[<0.3×]**	<1

[a] Boldface figures indicate activity relative to $1,25(OH)_2D_3$.

true for MC1046 and MC1080, the two main metabolites of MC903 [29].

The effect of the size of the terminal ring was studied by Calverley [30], who synthesized three pairs of 24-epimeric analogs in which the cyclopropyl group in calcipotriol is replaced by cyclobutyl, cyclopentyl, and cyclohexyl groups, respectively. Although the stereochemistry of the 24-hydroxyl group in these compounds has not been rigorously established, their polarities suggest that in MC1070, MC1052, and MC1048 the 24-hydroxyl group has the same stereochemistry as in MC903, whereas in MC1069, MC1050, and MC1033 the stereochemistry of the 24-hydroxyl group is as in MC900. It was found that the antiproliferative potencies of the cyclobutyl and cyclopentyl analogs MC1070 and MC1050 were similar to that of calcipotriol (MC903), whereas the cyclohexyl analog MC1048 was less potent. All these compounds were, however, significantly more calcemic than MC903.

D. Introduction of Heteroatoms

The first indication that introduction of a heteroatom in the side chain of $1,25(OH)_2D_3$ could increase the cell-regulating potency without a corresponding increase of calcemic activity came in 1987, when Abe *et al.* [31] reported that 22-oxa-$1,25(OH)_2D_3$, a compound developed by Chugai Pharmaceutical Co. in Japan (see Chapter 63), was one order of magnitude more effective than $1,25(OH)_2D_3$ in suppressing cell growth and inducing phagocytic activity in a mouse leukemia cell line, whereas its calcemic activity appeared to be 50–100 times less than that of $1,25(OH)_2D_3$.

As shown in Table IV, we have synthesized a small series of 22-oxa analogs [23], including 22-oxa-$1,25(OH)_2D_3$ (MC1275) for reference purposes under our experimental conditions. The most interesting of these analogs is KH1139, which is 150 times more potent than $1,25(OH)_2D_3$ in the antiproliferative test, whereas its calcemic activity is only 30% of that of $1,25(OH)_2D_3$.

E. Epimerization at C-20

A seemingly minor modification of the side chain in $1,25(OH)_2D_3$ that has a dramatic effect on its biological activities is epimerization at C-20. As Table V shows, the 20-epimer of $1,25(OH)_2D_3$ (MC1288) [24] is about 100 times more potent than the natural hormone as an inhibitor of cell proliferation, whereas its calcemic activity has increased by a factor of only 2. Even more pronounced is the effect of 20-epimerization on immunosuppressive properties. In the mouse thymocyte activation test, where T lymphocytes are activated by costimulation with interleukin-1 (IL-1) and phytohemagglutinin (PHA), MC1288 is 7300 times more potent than $1,25(OH)_2D_3$. In the mixed lymphocyte reaction (MLR) test, where T-lymphocyte activation is induced by allogeneic cells, MC1288 is 270 times more potent than $1,25(OH)_2D_3$ [32].

In view of these results we found it mandatory to investigate the effect of 20-epimerization more broadly [23,24]. Table V shows the activities of pairs of 20-epimers. Both the antiproliferative potency and the calcemic activity are generally higher in the 20-epi than in the 20-normal series. A particularly noteworthy compound is the 22-oxa analog KH1060, which is 3000 times

TABLE IV 22-Oxa Analogs

Compound	Side chain structure	Inhibition of U937 cell proliferation, IC_{50} (M)	Calcemic activity relative to $1,25(OH)_2D_3$ (%)
$1,25(OH)_2D_3$	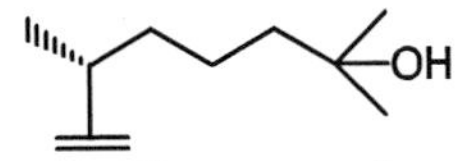 OH	3×10^{-8} **[1×]**[a]	100
MC1275	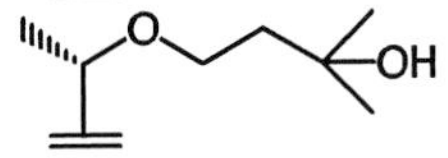 O, OH	5×10^{-8} **[0.6×]**	31
MC1273	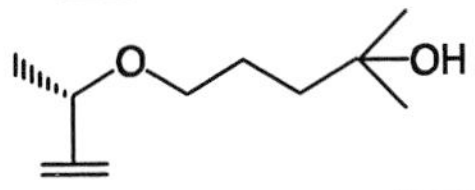 O, OH	1×10^{-8} **[3×]**	71
KH1139	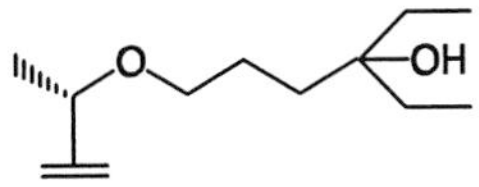 O, OH	2×10^{-10} **[150×]**	30
MC1270	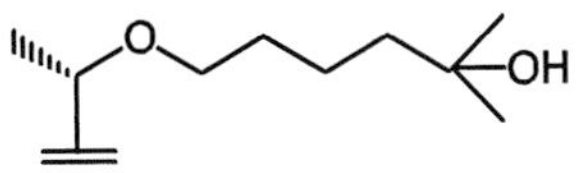 O, OH	2×10^{-8} **[1.5×]**	7

[a] Boldface figures indicate activity relative to $1,25(OH)_2D_3$.

TABLE V Effect of 20-Epimerization

Compound	Side chain structure	Inhibition of U937 cell proliferation, IC_{50} (M)	Calcemic activity relative to $1,25(OH)_2D_3$ (%)
$1,25(OH)_2D_3$	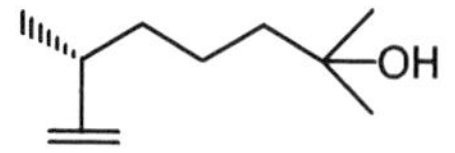 OH	3×10^{-8} **[1×]**[a]	100
MC1288	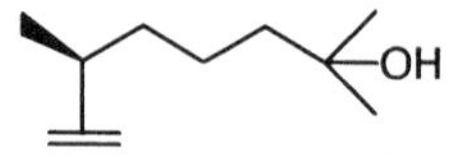 OH	3×10^{-10} **[100×]**	200
MC1127	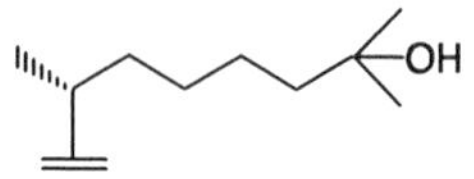 OH	5×10^{-9} **[6×]**	38
EB1231	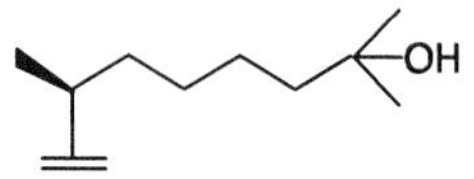 OH	6×10^{-11} **[500×]**	500
CB966	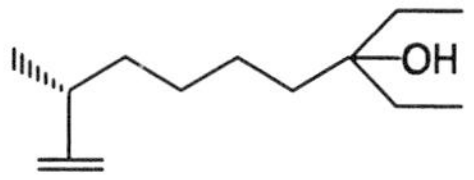 OH	1×10^{-9} **[30×]**	17
MC1301	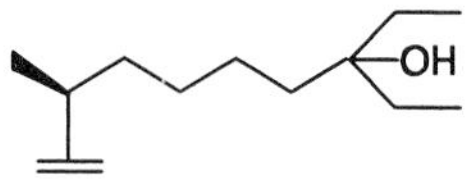 OH	3×10^{-10} **[100×]**	200
MC1275	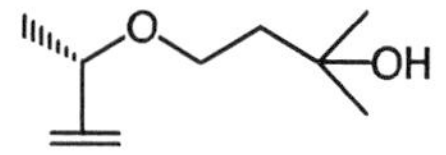 O, OH	5×10^{-8} **[0.6×]**	31
MC1292	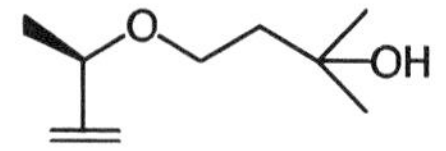 O, OH	4×10^{-9} **[8×]**	10
MC1273	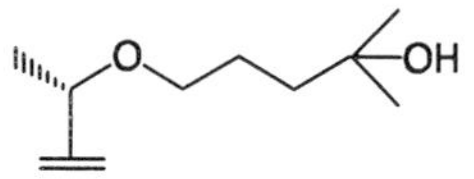 O, OH	1×10^{-8} **[3×]**	71
KH1059	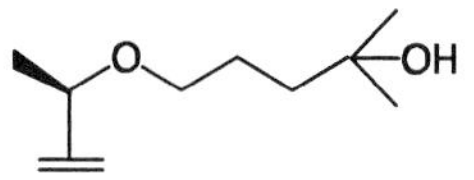 O, OH	2×10^{-10} **[150×]**	270
KH1139	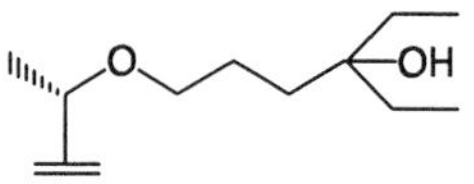 O, OH	2×10^{-10} **[150×]**	30
KH1060	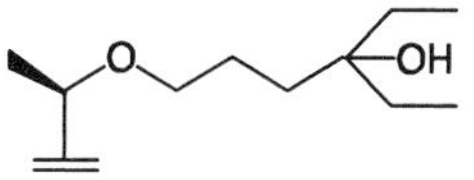 O, OH	1×10^{-11} **[3000×]**	130

[a] Boldface figures indicate activity relative to $1,25(OH)_2D_3$.

more potent than $1,25(OH)_2D_3$ as a proliferation inhibitor but only slightly more calcemic.

F. Introduction of an Aromatic Ring

The effect of introducing a benzene ring in the side chain is illustrated in Table VI. The analogs listed in Table VI are all 20-epi compounds, where the benzene ring is attached to an oxygen atom or a sulfur atom in the 23 position and further substituted in the ortho, meta, or para position with a 2-hydroxypropyl or a 3-hydroxypentyl group [25]. Whereas the ortho- and para-substituted analogs have antiproliferative potencies of the same order of magnitude as $1,25(OH)_2D_3$, the meta-substituted analogs (EB1213, EB1219, GS1500, and GS1730) are considerably more potent. All the compounds tested are at least one order of magnitude less calcemic than $1,25(OH)_2D_3$.

IV. BIOLOGICAL ACTIVITIES

Since the establishment of the initial screening strategy for new vitamin D analogs (described in Section II,A), research on the effects of $1,25(OH)_2D_3$ at the cellular and molecular level has greatly expanded. The availability of new synthetic analogs has played an im-

TABLE VI Analogs with an Aromatic Ring

Compound	Side chain structure	Inhibition of U937 cell proliferation, IC_{50} (M)	Calcemic activity relative to $1,25(OH)_2D_3$ (%)
$1,25(OH)_2D_3$	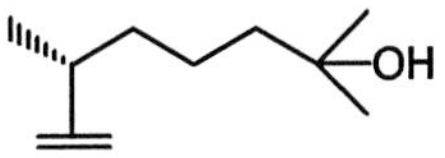 OH	3×10^{-8} **[1×]**[a]	100
EB1224	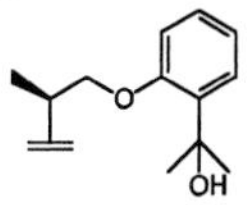 O, OH	2×10^{-8} **[1.5×]**	3
EB1213	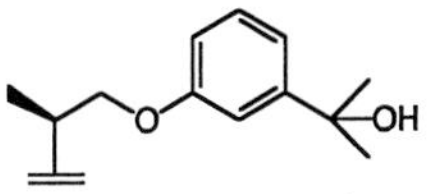 O, OH	1×10^{-10} **[300×]**	7
EB1220	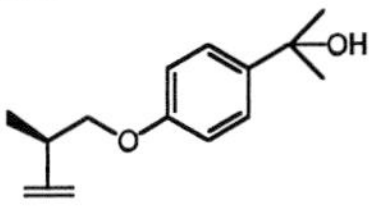 O, OH	8×10^{-8} **[0.3×]**	n.d.[b]
EB1219	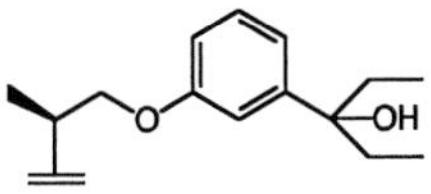 O, OH	8×10^{-11} **[375×]**	6
GS1780	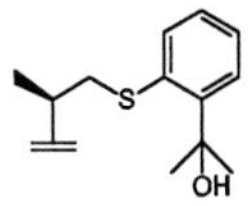 S, OH	6×10^{-8} **[0.5×]**	n.d.
GS1500	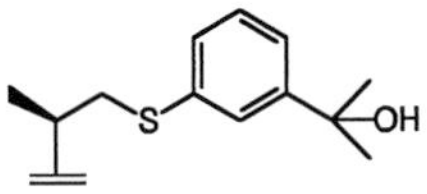 S, OH	1×10^{-10} **[300×]**	<1
GS1790	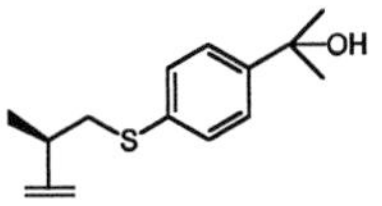 S, OH	5×10^{-8} **[0.6×]**	n.d.
GS1730	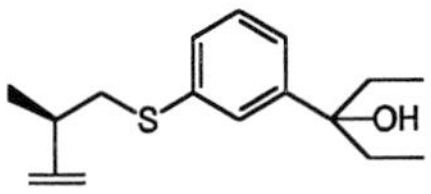 S, OH	1×10^{-9} **[30×]**	<1

[a] Boldface figures indicate activity relative to $1,25(OH)_2D_3$.
[b] n.d., Not determined.

portant role in this development. At the same time, new information has permitted a more in-depth evaluation of the therapeutic potential of the analogs. This section briefly summarizes some of the main findings involving $1,25(OH)_2D_3$ and new Leo analogs in the fields of cell growth and regulation, immune regulation, receptor binding and gene expression, and calcium and bone metabolism.

A. Effects on Cell Proliferation and Differentiation

Skin cells (keratinocytes and fibroblasts) and many cancer cells, both from primary tumors and established cell lines, constitutively express the VDR and respond to $1,25(OH)_2D_3$ *in vitro* by inhibition of cell proliferation and induction of differentiation (see Chapter 25). The antiproliferative effects in keratinocytes are accompanied by a decrease in c-*myc* mRNA and an increase in transforming growth factor-β (TGFβ) levels [33,34], whereas induction of differentiation is characterized by increased transglutaminase activity and formation of cornified envelopes [35]. These findings have provided a rationale for the use of new vitamin D analogs such as calcipotriol for the treatment of psoriasis. Hyperproliferative fibroblasts may also be a target for growth regulation by $1,25(OH)_2D_3$ and its analogs, as suggested by the ability of these compounds to inhibit proliferation of fibroblasts isolated from patients with scleroderma [36].

A large number of cancer cells respond to the growth inhibitory effects of $1,25(OH)_2D_3$ *in vitro* (see Chapter 64). For many years the human leukemia cell lines U937 (histiocytic lymphoma) and HL-60 (myelocytic leukemia) have been the model systems of choice in identifying the basic pathways involved in growth regulation by $1{,}25(OH)_2D_3$ and its analogs. Cell cycle analysis in the HL-60 cells has shown that growth arrest is associated with a block in G_0/G_1 and G_2/M [37]. Additional studies have shown that cells derived from solid tumors are also responsive to vitamin D analogs such as EB1089, both *in vitro* and in tumor-bearing animals (see Section V,B). More recent studies suggest that vitamin D analogs may exert part of their antitumor effects via inhibition of angiogenesis [38] and/or by induction of apoptosis [39]. A study with the 20-epi vitamin D_3 analog KH1060 has shown that induction of differentiation in a number of breast cancer cell lines is accompanied by cell cycle arrest in G_0/G_1 and a concomitant decrease in the expression of the apoptosis-protective gene *bcl-2* [40].

B. Effects on Cells of the Immune System

Substantial evidence indicates that vitamin D analogs modulate a number of immunological functions (see Chapter 29). Monocytes and macrophages constitutively express VDR, whereas lymphocytes require prior activation with antigen or mitogen before expressing the receptor. Activation of T lymphocytes *in vitro* is inhibited by $1{,}25(OH)_2D_3$, presumably via inhibition of interleukin-2 (IL-2) production [41]. $1{,}25(OH)_2D_3$ reportedly exerts its main effects on the T-helper 1 (TH1) subset of lymphocytes that preferentially produce IL-2 and interferon-γ (IFN-γ) [42]. A number of vitamin D analogs with the 20-epi vitamin D_3 structure (Table V, Section III) have been shown to exert potent inhibitory effects on lymphocyte activation and IL-2 production *in vitro* [32].

In animal models of autoimmunity, $1{,}25(OH)_2D_3$ inhibits the development of murine experimental autoimmune encephalomyelitis (EAE) [43] and attenuates the clinical manifestations of the lupus syndrome in MRL/1 mice [44]. Prevention of EAE, type I diabetes in NOD mice, and $HgCl_2$-induced autoimmune nephritis in rats is also reported with several vitamin D analogs, including the 20-epi vitamin D_3 analog KH1060 [45–48]. The vitamin D analogs CB966 and KH1060 are effective in prolonging the survival of skin allografts in mice [49], whereas KH1060 has been found to be inactive in preventing acute renal allograft rejection in rats [50]. Further details can be found in Chapters 69 and 70.

C. VDR Binding and Gene Expression

Most of the biological effects of $1{,}25(OH)_2D_3$ *in vitro* and *in vivo* are believed to be mediated via binding to the VDR, but the activity of many vitamin D analogs is not directly correlated with their binding affinity for the VDR, as assessed by displacement of ^{3}H-$1{,}25(OH)_2D_3$. These observations suggest that different analogs may confer different conformational changes to the VDR–ligand complex. This may in turn lead to an altered stability of the complex and/or an altered ability to interact with DNA binding sites or accessory nuclear factors. An additional level of complexity results from the ability of the VDR to form heterodimers with other nuclear hormone receptors, such as the retinoid receptor (RXR), which binds 9-*cis*-retinoic acid as ligand [51]. The VDR–RXR heterodimer interacts with classic vitamin D response elements (VDREs) of the DR3 type, characterized by direct repeats of hexameric core binding motifs, spaced by three nucleotides (see Chapter 9). Work with the 20-epi vitamin D_3 analogs KH1060 and MC1288 has shown that these analogs increase the transcriptional activity of the VDR, presumably by producing conformational changes that enhance dimerization of VDR with RXR [52]. Besides the DR3 type, a number of other natural and synthetic VDREs have been identified, including DR4, DR6, and inverted palindrome (IP) elements [53]. Vitamin D analogs such as EB1089 have been shown to stimulate the transcriptional activity from IP9-type elements at lower concentrations than from DR3-type elements [54]. Further structure–activity studies with a selected series of 20-epi vitamin D analogs have recently been published [55].

In addition to its genomic effects mediated via binding to the VDR, $1{,}25(OH)_2D_3$ also exerts rapid nongenomic effects in a variety of cells (see Chapter 15). *In vitro*, $1{,}25(OH)_2D_3$ increases intracellular calcium in osteoblasts, via opening of voltage-gated calcium channels and activation of phospholipase C [56]. *In vivo* in the chicken, $1{,}25(OH)_2D_3$ stimulates the rapid intestinal transport of calcium, known as transcaltachia. A number of vitamin D analogs with potent effects on VDR-mediated pathways have been shown to be less effective than $1{,}25(OH)_2D_3$ in mediating the nongenomic effects, whereas other analogs with a low affinity for the VDR are highly effective [57]. This specificity suggests that the increase in cellular calcium levels initiates a distinct signal transduction pathway, which is regulated by a different mechanism than that of the VDR-mediated pathway.

D. Effects on Calcium Metabolism

The classic physiological role of $1{,}25(OH)_2D_3$ is to regulate calcium homeostasis and promote bone mineralization. Over the years, the existence of a closely regulated network that controls the endogenous formation and degradation of $1{,}25(OH)_2D_3$ has been established [58]. The role of $1{,}25(OH)_2D_3$ in the treatment of bone diseases such as rickets and renal osteodystrophy is well established, and newer reports indicate that $1\alpha OHD_3$ and $1,25(OH)_2D_3$ may also be beneficial for patients with osteoporosis [59]. Surprisingly, efforts to develop new synthetic analogs for bone diseases have been modest. To date, only $1,25(OH)_2D_3$ and $1\alpha OHD_3$ are in clinical use. In contrast to the field of bone diseases, considerable efforts have been directed at finding new analogs with low calcemic activity but with strong effects on cell growth and immune reactivity. This search has mainly focused on two main pathways: the design of analogs with modified pharmacokinetic and metabolic profiles and the design of tissue-specific analogs, along the lines that are currently used in the development of estrogen receptor agonists and antagonists.

The first part of this strategy has met with considerable success in the synthesis of numerous analogs with good *in vitro* activity but with low calcemic effects *in vivo* (see Section III). Many of these analogs have a short half-life in serum and a low affinity for the vitamin D binding protein (DBP) that functions as transport protein for $1,25(OH)_2D_3$ and $25OHD_3$ in blood [60] (see also Chapter 59). These findings have led to a closer investigation of the role of DBP in the kinetics and metabolism of a series of new analogs [61]. It was shown that the affinity for DBP correlated well with the initial serum concentration of the analogs after intravenous administration, and hence with the rate of clearance from the circulation. In contrast, no correlation was found between the affinity for DBP and the rate of metabolic clearance, as assessed by the serum half-life. Theoretically, it should thus be possible to design analogs with a short half-life and rapid metabolism, useful for topical application, and analogs with a longer half-life but with lower calcemic activity, due to rapid uptake in the tissues, for systemic administration. Further details on specific analogs are found in Section V.

The realization of the last part of the strategy, the design of tissue-specific analogs, has seemed distant for many years. However, with the discovery of a growing number of different vitamin D response elements in the promoter region of genes specifically involved in the regulation of calcium metabolism or cell growth and differentiation, this task should become more approachable in the near future.

V. PROFILE OF LEO ANALOGS UNDER DEVELOPMENT

A. Psoriasis

1. PRECLINICAL DEVELOPMENT OF CALCIPOTRIOL

Calcipotriol, synthesized in 1985, is a 1α,24-dihydroxyvitamin D analog containing a cyclopropane ring and a double bond in the side chain [19] (see Table III and Section III). In the basic screening system described in Section II,A, calcipotriol was found to be a potent inducer of cell differentiation and inhibitor of cell proliferation in U937 cells. Its activity was comparable to that of $1,25(OH)_2D_3$, and it bound to the VDR with the same affinity as $1,25(OH)_2D_3$ [10]. In addition, calcipotriol was found to induce terminal differentiation in cultures of human keratinocytes and to inhibit cellular DNA synthesis, at concentrations comparable to those of $1,25(OH)_2D_3$ [62].

Several studies indicate that calcipotriol exerts immunological effects *in vitro*, similar to those reported for $1,25(OH)_2D_3$. These effects include inhibition of T-lymphocyte activation induced by IL-1 [63], reduction of immunoglobulin production through interference with T-helper cell functions [64], and modulation of release of immune mediators from activated peripheral blood mononuclear cells and keratinocytes [65–68]. These studies indicate that calcipotriol is able to down-regulate immune responsiveness and thereby control some of the immunological reactions that are involved in the pathogenesis of psoriasis.

In vivo, however, calcipotriol was shown to be 100–200 times less active than $1,25(OH)_2D_3$ in causing calcemic effects when administered orally or intraperitoneally to rats [10]. More recent studies have compared the calcemic effects of calcipotriol with those of $1,25(OH)_2D_3$ and $1,24(OH)_2D_3$, after systemic or topical administration to rats [69,70]. Regardless of the route of administration, calcipotriol had a much lower calcemic potency than the reference compounds.

The low calcemic activity of calcipotriol was shown to be associated with a rapid rate of clearance. After intravenous administration to rats, calcipotriol had a half-life of less than 10 min, whereas $1,25(OH)_2D_3$ and $1,24(OH)_2D_3$ had half-lives of 2.3 and 1.4 hr, respectively [15,71]. Two major metabolites of calcipotriol, MC1046 and MC1080, were identified, using liver homogenates from rats, minipigs, and humans [28]. The metabolic pathway involved oxidation at carbon 24 in the side chain and reduction of the Δ^{22} double bond. Formation of the 24-oxidized metabolites was found to constitute a deactivation pathway for calcipotriol, as the biological effects of MC1046 and MC1080 on cell proliferation and differentiation were much weaker than those exerted by calcipotriol [15] (see also Table III and Section III). Initial metabolism of calcipotriol to MC1046 and MC1080 was also observed in cultured human keratinocytes, resulting in the final formation of calcitroic acid, the end product of the catabolic pathway of $1,25(OH)_2D_3$ [72]. The rapid rate of metabolic inactivation of calcipotriol has been associated with its low affinity for DBP [73]. However, as discussed in Section IV,D, the affinity of various vitamin D analogs for DBP has been found to be more closely correlated with the rate of clearance from the serum than with the rate of metabolism.

2. Clinical Development of Calcipotriol

The efficacy and safety of topically applied calcipotriol to patients with mild to moderate psoriasis was established in double-blind, placebo-controlled studies initiated in 1987 [74,75]. Later studies showed that treatment with calcipotriol compared favorably with betamethasone [76,77]. The first approval of calcipotriol was obtained in 1991, and numerous studies have since established the usefulness of this treatment as monotherapy, in combination with other topical or systemic treatments, and for long-term management of chronic plaque psoriasis [78] (see Chapter 72).

3. New Analogs for the Treatment of Psoriasis

A new generation of vitamin D analogs for the treatment of psoriasis is presently under development. The aim has been to improve the therapeutic efficiency of the analogs and to eliminate side effects such as skin stinging, itching, and irritation, while retaining a low calcemic activity. Another aim has been to select a candidate useful for systemic treatment of psoriasis.

Two analogs, EB1213 and GS1500, have been selected for clinical trials in chronic plaque psoriasis, using topical application. These analogs belong to the 20-epi series and are characterized by the presence of an aromatic ring in the side chain (see Table VI and Section III). The two analogs are very similar in structure: the only difference is that GS1500 contains a sulfur atom in the 23 position, whereas EB1213 has an oxygen atom in this position. The compounds were selected on the basis of structure–activity studies and were further characterized with regard to their effects on keratinocyte cell proliferation, T-cell activation, and serum half-life [79].

The most recent series of analogs selected for topical treatment of psoriasis includes the 20-epi analogs KH1218, KH1230, and KH1266 [80]. These analogs are similar to calcipotriol with regard to inhibition of keratinocyte proliferation and calcemic activity, but they have a lower potential for inducing skin irritation in animal models. In addition, one of the compounds, KH1230, is a potent inhibitor of IL-2 and INF-γ production from activated human T lymphocytes (Table VII). Another vitamin D analog, EB1089 (see also Table II and Sections III and V,B), has been selected as a candidate for oral treatment of psoriasis.

B. Cancer

1. Preclinical Development of EB1089

The rationale for the use of vitamin D analogs in cancer is briefly described in Section IV,A and more fully discussed in Chapters 64–68. The search for new analogs with potential clinical usefulness has been directed at finding a candidate with good systemic bioavailability, potent effects on cell proliferation, and low calcemic activity. From our screening program, we selected the analog EB1089, which fulfilled these criteria (see Table II and Section III). EB1089 is characterized by having two double bonds in the side chain, which makes it less susceptible to metabolic degradation. Two reports describe the metabolism of EB1089 *in vivo* [81,82]. Compared to $1,25(OH)_2D_3$ and the antitumor antibiotic daunomycin, EB1089 was shown to be a potent inhibitor of the growth of various cancer cells *in vitro* (see Table VIII). The effects of EB1089 have been studied in further detail in MCF-7 cells [83–85].

In vivo, EB1089 has been tested for antitumor effects in rats bearing tumors induced by nitrosomethylurea [83]. Oral (p.o.) treatment with EB1089 at 0.5 μg/kg/day for 4 weeks resulted in a significant inhibition of tumor growth, without changes in the serum calcium

TABLE VII Effects on Human Keratinocytes and on IL-2 and INF-γ Production in Human Lymphocytes *in Vitro*

Test Compound	Skin cell proliferation	Calcemic effects (%)	Skin irritation	IL-2 production, IC_{50} (M)	INF-γ production, IC_{50} (M)
$1,25(OH)_2D_3$	5×10^{-8}	100	+++	2×10^{-8}	3×10^{-10}
Calcipotriol	3×10^{-8}	0.5	+++	2×10^{-8}	9×10^{-11}
KH1218	3×10^{-8}	0.1	+	1×10^{-8}	3×10^{-10}
KH1230	3×10^{-9}	14	++	1×10^{-11}	2×10^{-12}
KH1266	3×10^{-9}	0.5	++	2×10^{-10}	3×10^{-11}

TABLE VIII Effects of EB1089 on Tumor Cell Proliferation *in Vitro*

Test compound	Cell line	Inhibition of proliferation, IC_{50} (*M*)
EB1089	MCF-7	2×10^{-10}
$1,25(OH)_2D_3$	(human breast cancer)	1×10^{-8}
Daunomycin		2×10^{-8}
EB1089	HT-29	8×10^{-10}
$1,25(OH)_2D_3$	(human colon cancer)	4×10^{-8}
Daunomycin		8×10^{-9}
EB1089	B16	6×10^{-11}
$1,25(OH)_2D_3$	(mouse melanoma)	6×10^{-9}
Daunomycin		5×10^{-8}

TABLE IX Effects of MC1288 on Cardiac Graft Rejection in Rats

Test compound	Dosage (route of administration)	Graft survival (range, days)
Vehicle	—	8 (7–9)
MC1288	0.05 μg/kg/day (i.p.)	12 (8–22)
MC1288	0.1 μg/kg/day (i.p.)	22 (16–27)
Cyclosporin A	10 mg/kg/day (p.o.)	19 (15–27)

levels. The same dose of $1,25(OH)_2D_3$ did not affect tumor growth but caused hypercalcemia. Further analysis of the effects of EB1089 in tumor-bearing animals can be found in Chapter 65.

2. CLINICAL DEVELOPMENT OF EB1089

A clinical program for the study of EB1089 in the treatment of cancer has been initiated in 1995. Phase I/II studies are in progress in various cancer forms including leukemia, breast cancer, and pancreatic cancer. So far, a small phase I study in patients with advanced cancer has shown that dosages up to 12 μg/m²/day are well tolerated [86].

C. Immune-Mediated Diseases

As described in Sections III,E and IV,B, modification of the stereochemistry of the methyl group at C-20 in the side chain of $1,25(OH)_2D_3$ has led to a series of analogs (the 20-epi vitamin D_3 analogs) with potent inhibitory effects on T-lymphocyte activation *in vitro*. One of these analogs, 20-epi-$1,25(OH)_2D_3$ (MC1288), has been selected as our candidate for studying the potential of this class of compounds as therapeutic agents in graft rejection and autoimmune diseases.

In animal studies, MC1288 at 0.1 μg/kg/day i.p. was shown to prolong the survival of cardiac allografts in rats [87]. The effects were comparable to those obtained with the well-known immunosuppressive agent cyclosporin A (Table IX) [88]. In the same model, combined treatment with MC1288 (0.1 μg/kg/day i.p.) and cyclosporin A (5 mg/kg/day p.o.) produced a significantly prolonged survival time of the cardiac grafts, compared with therapy with either agent alone [89].

MC1288 was also tested in a model of small bowel transplantation in rats, in which high doses of immunosuppressants are generally required to control rejection. In this model, the intestinal intraluminal secretion of hyaluronic acid in the graft is used as a measure of graft rejection. In animals treated with MC1288, the hyaluronic acid secretion was significantly decreased, compared to untreated animals undergoing rejection [88].

MC1288 has also been shown to prolong the survival of skin allografts in mice in doses from 0.02 to 0.2 μg/kg/day i.p. [90]. MC1288 is currently being evaluated for its effects in a number of animal models of autoimmune diseases, such as experimental autoimmune encephalomyelitis (EAE) in rats and experimental autoimmune diabetes in NOD mice, as described in Chapters 69 and 70.

The mechanism of action by which MC1288 inhibits graft rejection is not fully known. It is likely that inhibition of the production of IL-2 and of other cytokines involved in immune regulation plays a major role, in analogy with what is known for cyclosporin A [41]. The effects of MC1288 in animal models are, however, not identical to those observed with cyclosporin A. For example, MC1288 was more potent than cyclosporin A in reducing interstitial inflammation and vascular intimal proliferation during chronic renal allograft rejection in rats [91]. These findings suggest that MC1288, in addition to its effects on T-cell activation and IL-2 production, also affects antigen-presenting cells such as monocytes and/or B lymphocytes.

In all the animal models studied so far, monotherapy with MC1288 was optimally effective only at doses that caused a small, but significant, increase in serum calcium levels. It is therefore to be envisaged that if this compound proves successful, its therapeutic place will be in combination with other immunosuppressive agents, thus reducing the dosage and risk of side effects of all the administered drugs.

VI. Conclusions and Future Aspects

Synthesis of new vitamin D analogs and increased understanding of their mechanism of action have led to the selection of specific analogs as candidates for investigation in a variety of diseases, such as psoriasis, cancer, graft rejection, and autoimmune diseases. Up to the present, this selection has mainly been based on the ability of the analogs to regulate differentiation and proliferation in various cell types, to interact with the vitamin D receptor, to exert low calcemic effects *in vivo*, and to display a suitable pharmacokinetic profile.

In the future, it is hoped that the identification of new genes which are directly or indirectly regulated by $1,25(OH)_2D_3$ will allow for the identification of more selective analogs. Studies should be directed at identifying new VDREs and the accessory factors that facilitate or interfere with VDR binding to DNA. In addition, studies of VDR–ligand binding will be greatly facilitated when the crystal structure of the VDR becomes available.

With regard to the genes involved in the regulation of cell growth by $1,25(OH)_2D_3$, recent interest has been directed toward the p21 gene [92] and the genes involved in the apoptotic pathway. Similarly, the regulation of the IL-2 gene by $1,25(OH)_2D_3$ [93] points to new directions in our understanding of the immunoregulatory role of vitamin D. However, it is important that our enthusiasm for these new possibilities should not make us forget the classic role of $1,25(OH)_2D_3$ in mediating bone cell differentiation and bone mineralization. The development of new vitamin D analogs for the treatment of bone diseases such as osteodystrophy and osteoporosis should be one of the important goals of future research in the vitamin D field.

References

1. DeLuca HF 1988 The vitamin D story: A collaborative effort of basic science and clinical medicine. FASEB J **2**:224–236.
2. Minghetti PP, Norman AW 1989 $1,25(OH)_2$-vitamin D_3 receptors: Gene regulation and genetic circuitry. FASEB J **2**:3043–3053.
3. Rigby WFC 1988 The immunobiology of vitamin D. Immunol Today **9**:54–58.
4. Manolagas SC, Hustmyer FG,Yu XP 1990 Immunomodulating properties of 1,25-dihydroxyvitamin D_3. Kidney Int **38**:9–16.
5. Wieslander SB, Mortensen BT, Binderup L, Nissen NI 1987 $1\alpha(OH)D_3$ (Etalpha)® treatment and receptor studies in 16 patients with chronic and myeloproliferative disorders. Eur J Haematol **39**:35–38.
6. Cunningham D, Gilchrist NL, Cowan RA, Forrest GJ, McArdle C, Soukop M 1985 Alfacalcidol as a modulator of growth of low grade non-Hodgkin's lymphomas. Br Med J **291**:1153–1155.
7. Morimoto S, Kumahara Y 1985 A patient with psoriasis cured by 1α-hydroxy-vitamin D_3. Med J Osaka Univ **35**:51–54.
8. Dikstein S, Hartzshtark A 1984 Cosmetic and dermatological compositions containing 1-alpha-hydroxycholecalciferol. European Patent No. 0129003B2.
9. Mezzetti G, Bagnara G, Monti MG, Bonsi L, Brunelli MA, Barbiroli B 1984 1α,25-Dihydroxycholecalciferol and human histiocytic lymphoma cell line (U-937): The presence of a receptor and inhibition of proliferation. Life Sci **34**:2185–2191.
10. Binderup L, Bramm E 1988 Effects of a novel vitamin D analogue MC 903 on cell proliferation and differentiation in vitro and on calcium metabolism in vivo. Biochem Pharmacol **37**:889–895.
11. Yam LT, Li CY, Crosby WH 1971 Cytochemical identification of monocytes and granulocytes. Am J Clin Pathol **55**:283–290.
12. Mellon WS, Franceschi RT, DeLuca HF 1980 An in vitro study of the stability of the chicken intestinal cytosol 1,25-dihydroxyvitamin D_3-specific receptor. Arch Biochem Biophys **202**:83–92.
13. Gallagher JA, Beneton M, Harvey L, Lawson DEM 1986 Response of rachitic rat bones to 1,25-dihydroxyvitamin D_3: Biphasic effects on mineralization and lack of effect on bone resorption. Endocrinology **119**:1603–1609.
14. Norman AW, Wong RG 1972 Biological activity of the vitamin D metabolite 1,25-dihydroxycholecalciferol in chickens and rats. J Nutr **102**:1709–1718.
15. Kissmeyer AM, Binderup L 1991 Calcipotriol (MC 903): Pharmacokinetics in rats and biological activities of metabolites. A comparative study with $1,25(OH)_2D_3$. Biochem Pharmacol **41**:1601–1606.
16. Kream BE, Jose JL, DeLuca HF 1977 The chick intestinal cytosol binding protein for 1,25-dihydroxyvitamin D_3: A study of analog binding. Arch Biochem Biophys **179**:462–468.
17. Hesse RH 1983 Intermediates in the synthesis of vitamin D derivatives. European Patent Application No. 0078704.
18. Andrews DR, Barton DHR, Hesse RH, Pechet MM 1986 Synthesis of 25-hydroxy- and 1α,25-dihydroxyvitamin D_3 from vitamin D_2 (calciferol). J Org Chem **51**:4819–4828.
19. Calverley MJ 1987 Synthesis of MC 903, a biologically active vitamin D metabolite. Tetrahedron **43**:4609–4619.
20. Binderup E, Calverley MJ, Binderup L 1991 Synthesis and biological activity of 1α-hydroxylated vitamin D analogues with polyunsaturated side chains. In: Norman AW, Bouillon R, Thomasset M (eds) Vitamin D: Gene Regulation, Structure–Function Analysis and Clinical Application. de Gruyter, Berlin, New York, pp. 192–193.
21. Gielen JWJ, Koolstra RB, Jacobs HJC, Havinga E 1977 Triplet-sensitized interconversion and photooxygenation of vitamin D and *trans*-vitamin D. Recl Trav Chim Pays-Bas **113**:306–311.
22. Bretting C, Calverley MJ, Binderup L 1991 Synthesis and biological activity of 1α-hydroxylated vitamin D_3 analogues with hydroxylated side chains, multi-homologated in the 24 or 24,26,27 positions. In: Norman AW, Bouillon R, Thomasset M (eds) Vitamin D: Gene Regulation, Structure–Function Analysis and Clinical Application. de Gruyter, Berlin, New York, pp. 159–160.
23. Hansen K, Calverley MJ, Binderup L 1991 Synthesis and biological activity of 22-oxa vitamin D analogues. In: Norman AW, Bouillon R, Thomasset M (eds) Vitamin D: Gene Regulation, Structure–Function Analysis and Clinical Application. de Gruyter, Berlin, New York, pp. 161–162.
24. Calverley MJ, Binderup E, Binderup L 1991 The 20-epi modification in the vitamin D series: Selective enhancement of "non-classical" receptor-mediated effects. In: Norman AW, Bouillon R, Thomasset M (eds) Vitamin D: Gene Regulation, Structure–Function Analysis and Clinical Application. de Gruyter, Berlin, New York, pp. 163–164.
25. Grue-Sørensen G, Binderup E, Binderup L 1994 Chemistry and biology of 23-oxa-aro- and 23-thia-aro-vitamin D analogues with

high antiproliferative and low calcemic activity. In: Norman AW, Bouillon R, Thomasset M (eds) Vitamin D—A Pluripotent Steroid Hormone: Structural Studies, Molecular Endocrinology and Clinical Applications. de Gruyter, Berlin, New York, pp. 75–76.

26. Ostrem VK, Tanaka Y, Prahl J, DeLuca HF, Ikekawa N 1987 24- and 26-homo-1,25-dihydroxyvitamin D_3: Preferential activity in inducing differentiation of human leukemia cells HL-60 in vitro. Proc Natl Acad Sci USA **84**:2610–2614.
27. Kutner A, Perlman KL, Lago A, Sicinski RR, Schnoes HK, DeLuca HF 1988 Novel convergent synthesis of side-chain modified analogues of 1α,25-dihydroxycholecalciferol and 1α,25-dihydroxyergocalciferol. J Org Chem **53**:3450–3457.
28. Bretting C, Mørk Hansen C, Rastrup Andersen N 1994 Chemistry and biology of 22,23-yne analogs of calcitriol. In: Norman AW, Bouillon R, Thomasset M (eds) Vitamin D—A Pluripotent Steroid Hormone: Structural Studies, Molecular Endocrinology and Clinical Applications. de Gruyter, Berlin, New York, pp. 73–74.
29. Sørensen H, Binderup L, Calverley MJ, Hoffmeyer L, Andersen NR 1990 In vitro metabolism of calcipotriol (MC 903), a vitamin D analogue. Biochem Pharmacol **39**:391–393.
30. Calverley MJ 1992 Novel vitamin D analogues in the calcipotriol (MC 903) series: Synthesis and effect of side chain structure on the selective biological actions. In: Sarel S, Mechoulam R, Agranat I (eds) Trends in Medicinal Chemistry '90. Blackwell, Oxford, pp. 299–306.
31. Abe J, Morikawa M, Miyamoto K, Kaiho S, Fukushima M, Miyaura C, Abe E, Suda T, Nishii Y 1987 Synthetic analogues of vitamin D_3 with an oxygen atom in the side chain skeleton. A trial of the development of vitamin D compounds which exhibit potent differentiation-inducing activity without inducing hypercalcemia. FEBS Lett **226**:58–62.
32. Binderup L, Latini S, Binderup E, Bretting C, Calverley MJ, Hansen K 1991 20-epi-vitamin D_3 analogues: A novel class of potent regulators of cell growth and differentiation. Biochem Pharmacol **42**:1569–1575.
33. Matsumoto K, Hashimoto K, Nishida Y, Hashiro M, Yoshikawa K 1990 Growth-inhibitory effects of 1,25-dihydroxyvitamin D_3 on normal human keratinocytes cultured in serum-free medium. Biochem Biophys Res Commun **166**:916–923.
34. Kim HJ, Abdelkader N, Katz M, McLane JA 1992 1,25-Dihydroxy-vitamin-D_3 enhances antiproliferative effect and transcription of TGF-β1 on human keratinocytes in culture. J Cell Physiol **151**:579–587.
35. Smith EL, Walworth NC, Holick MF 1986 Effect of 1α,25-dihydroxyvitamin D_3 on the morphologic and biochemical differentiation of cultured human epidermal keratinocytes grown in serum-free conditions. J Invest Dermatol **86**:709–714.
36. Bottomley WW, Jutley J, Wood EJ, Goodfield MDJ 1995 The effect of calcipotriol on lesional fibroblasts from patients with active morphoea. Acta Derm Venereol (Stockh) **75**:364–366.
37. Zhang F, Godyn JJ, Uskokovic M, Binderup L, Studzinski GP 1994 Monocytic differentiation of HL60 cells induced by potent analogs of vitamin D_3 precedes the G_1/G_0 phase cell cycle block. Cell Proliferation **27**:643–654.
38. Oikawa T, Hirotani K, Ogasawara H, Katayama T, Nakamura O, Iwaguchi T, Hiragun A 1990 Inhibition of angiogenesis by vitamin D_3 analogues. Eur J Pharmacol **178**:247–250.
39. Welsh JE 1994 Induction of apoptosis in breast cancer cells in response to vitamin D and antiestrogens. Biochem Cell Biol **72**:537–545.
40. Elstner E, Linker-Israeli M, Said J, Umiel T, de Vos S, Shintaku IP, Heber D, Binderup L, Uskokovic M, Koeffler HP 1995 20-Epi-vitamin D_3 analogues: A novel class of potent inhibitors of proliferation and inducers of differentiation of human breast cancer cell lines. Cancer Res **55**:2822–2830.
41. Bouillon R, Garmyn M, Verstuyf A, Segaert S, Casteels K, Mathieu C 1995 Paracrine role for calcitriol in the immune system and skin creates new therapeutic possibilities for vitamin D analogs. Eur J Endocrinol **133**:7–16.
42. Lemire JM, Archer DC, Beck L, Spiegelberg HL 1995 Immunosuppressive actions of 1,25-dihydroxyvitamin D_3: Preferential inhibition of Th_1 functions. J Nutr **125**:1704–1708.
43. Lemire JM, Archer DC 1991 1,25-Dihydroxyvitamin D_3 prevents the in vivo induction of murine experimental autoimmune encephalomyelitis. J Clin Invest **87**:1103–1107.
44. Lemire JM, Ince A, Takashima M 1992 1,25-Dihydroxyvitamin D_3 attenuates the expression of experimental murine lupus of MRL/1 mice. Autoimmunity **12**:143–148.
45. Lemire JM, Archer DC, Reddy GS 1994 1,25-Dihydroxy-24-oxo-16-ene-vitamin D_3, a renal metabolite of vitamin D analog 1,25-dihydroxy-16-ene-vitamin D_3, exerts immunosuppressive activity equal to its parent without causing hypercalcemia in vivo. Endocrinology **135**:2818–2821.
46. Mathieu C, Waer M, Casteels K, Laureys J, Bouillon R 1995 Prevention of type I diabetes in NOD mice by nonhypercalcemic doses of a new structural analog of 1,25-dihydroxyvitamin D_3, KH 1060. Endocrinology **136**:866–872.
47. Lillevang ST, Rosenkvist J, Andersen CB, Larsen S, Kemp E, Kristensen T 1992 Single and combined effects of the vitamin D analogue KH 1060 and cyclosporin A on mercuric-chloride-induced autoimmune disease in the BN rat. Clin Exp Immunol **88**:301–306.
48. Vendeville B, Baran D, Gascon-Barré M 1995 Effects of vitamin D_3 and cyclosporin A on $HgCl_2$-induced autoimmunity in Brown Norway rats. Nephrol Dial Transplant **10**:2020–2026.
49. Veyron P, Pamphile R, Binderup L, Touraine JL 1993 Two novel vitamin D analogues, KH 1060 and CB 966, prolong skin allograft survival in mice. Transplant Immunol **1**:72–76.
50. Lewin E, Olgaard K 1994 The in vivo effect of a new, in vitro, extremely potent vitamin D_3 analog KH 1060 on the suppression of renal allograft rejection in the rat. Calcif Tissue Int **54**:150–154.
51. Freedman LP, Luisi BF 1993 On the mechanism of DNA binding by nuclear hormone receptors: A structural and functional perspective. J Cell Biochem **51**:140–150.
52. Peleg S, Sastry M, Collins ED, Bishop JE, Norman AW 1995 Distinct conformational changes induced by 20-epi analogues of 1α,25-dihydroxyvitamin D_3 are associated with enhanced activation of the vitamin D receptor. J Biol Chem **270**:10551–10558.
53. Schräder M, Müller KM, Becker-André M, Carlberg C 1994 Response element selectivity for heterodimerization of vitamin D receptors with retinoic acid and retinoid X receptors. J Mol Endocrinol **12**:327–339.
54. Nayeri S, Danielsson C, Kahlen JP, Schräder M, Mathiasen IS, Binderup L, Carlberg C 1995 The anti-proliferative effect of vitamin D_3 analogues is not mediated by inhibition of the AP-1 pathway, but may be related to promoter selectivity. Oncogene **11**:1853–1858.
55. Nayeri S, Mathiasen IS, Binderup L, Carlberg, C 1996 High affinity nuclear receptor binding of 20-epi analogues of 1,25 dihydroxyvitamin D_3 correlates well with gene activation. J Cell Biochem **62**:325–333.
56. Baran DT 1994 Nongenomic actions of the steroid hormone 1α,25-dihydroxy-vitamin D_3. J Cell Biochem **56**:303–306.
57. Zhou LX, Norman AW 1995 $1\alpha,25(OH)_2$-vitamin D_3 analog struc-

ture–function assessment of intestinal nuclear receptor occupancy with induction of calbindin-D_{28K}. Endocrinology **136**:1145–1152.
58. Kawashima H, Kurokawa K 1986 Metabolism and sites of action of vitamin D in the kidney. Kidney Int **29**:98–107.
59. Civitelli R 1995 The role of vitamin D metabolites in the treatment of osteoporosis. Calcif Tissue Int **57**:409–414.
60. Bouillon R, Allewaert K, Xiang DZ, Tan BK, van Baelen H 1991 Vitamin D analogs with low affinity for the vitamin D binding protein: Enhanced in vitro and decreased in vivo activity. J Bone Miner Res **6**:1051–1057.
61. Kissmeyer AM, Mathiasen IS, Latini S, Binderup L 1995 Pharmacokinetic studies of vitamin D analogues: Relationship to vitamin D binding protein (DBP). Endocrine **3**:263–266.
62. Kragballe K, Wildfang IL 1990 Calcipotriol (MC 903), a novel vitamin D analogue, stimulates terminal differentiation and inhibits proliferation of cultured human keratinocytes. Arch Dermatol Res **282**:164–167.
63. Müller K, Svenson M, Bendtzen K 1988 1α,25-dihydroxyvitamin D_3 and a novel vitamin D analogue MC 903 are potent inhibitors of human interleukin 1 in vitro. Immunol Lett **17**:361–366.
64. Müller K, Heilmann C, Poulsen LK, Barington T, Bendtzen K 1991 The role of monocytes and T cells in 1,25-dihydroxyvitamin D_3 mediated inhibition of B cell function in vitro. Immunopharmacology **21**:121–128.
65. Hustmyer FG, Benninger L, Manolagas SC 1991 Comparison of the effects of 22-oxa-1,25$(OH)_2D_3$ and MC 903 on the production of IL-6, gamma-IFN and lymphocyte proliferation in peripheral blood mononuclear cells. J Bone Miner Res **6**:292.
66. Kemény L, Kenderessy AS, Olasz E, Michel G, Ruzicka T, Farkas B, Dobozy A 1994 The interleukin-8 receptor: A potential target for antipsoriatic therapy? Eur J Pharmacol **258**:269–272.
67. Zhang JZ, Maruyama K, Ono I, Iwatsuki K, Kaneko F 1994 Regulatory effects of 1,25-dihydroxyvitamin D_3 and a novel vitamin D_3 analogue MC 903 on secretion of interleukin-1 alpha (IL-1α) and IL-8 by normal human keratinocytes and a human squamous cell carcinoma cell line (HSC-1). J Dermatol Sci **7**:24–31.
68. Maruyama K, Zhang JZ, Nihei Y, Ono I, Kaneko F 1995 Regulatory effects of antipsoriatic agents on interleukin-1α production of human keratinocytes stimulated with gamma interferon in vitro. Skin Pharmacol **8**:41–48.
69. Binderup L 1993 Comparison of calcipotriol with selected metabolites and analogues of vitamin D_3: Effects on cell growth regulation in vitro and calcium metabolism in vivo. Pharmacol Toxicol **72**:240–244.
70. Mortensen JT, Lichtenberg J, Binderup L 1996 Toxicity of 1,25-dihydroxyvitamin D_3, tacalcitol and calcipotriol after topical treatment in rats. J Invest Dermatol Symp Proc **1**:60–63.
71. Kissmeyer AM, Binderup L 1994 The in vitro metabolism of calcipotriol, 1,24$(OH)_2D_3$ and 1,25$(OH)_2D_3$. In: Norman AW, Bouillon R, Thomasset M (eds) Vitamin D—A Pluripotent Steroid Hormone: Structural Studies, Molecular Endocrinology and Clinical Applications. de Gruyter, Berlin, pp. 178–179.
72. Masuda S, Strugnell S, Calverley MJ, Makin HLJ, Kremer R, Jones G 1994 In vitro metabolism of the anti-psoriatic vitamin D analog, calcipotriol, in two cultured human keratinocyte models. J Biol Chem **269**:4794–4803.
73. Binderup L 1988 MC 903—A novel vitamin D analogue with potent effects on cell proliferation and cell differentiation. In: Norman AW, Bouillon R, Thomasset M (eds) Vitamin D: Molecular, Cellular and Clinical Endocrinology. de Gruyter, Berlin, New York, pp. 300–309.
74. Kragballe K, Beck HI, Søgaard H 1988 Improvement of psoriasis by a topical vitamin D_3 analogue (MC 903) in a double-blind study. Br J Dermatol **119**:223–230.
75. Staberg B, Roed-Petersen J, Menné T 1989 Efficacy of topical treatment in psoriasis with MC 903, a new vitamin D analogue. Acta Derm Venereol (Stockh) **69**:147–150.
76. Kragballe K, Karlsmark T, Nieboer C, Gjertsen BT, van de Kerkhof PCM, Roed-Petersen J, Tikjøb G, de Hoop D, Larkö O, Strand A 1991 Double-blind, right/left comparison of calcipotriol and betamethasone valerate in treatment of psoriasis vulgaris. Lancet **337**:193–196.
77. Cunliffe WJ, Berth-Jones J, Claudy A, Fairiss G, Goldin D, Gratton D, Henderson CA, Holden CA, Maddin WS, Ortonne JP, Young M 1992 Comparative study of calcipotriol (MC 903) ointment and betamethasone 17-valerate ointment in patients with psoriasis vulgaris. J Am Acad Dermatol **26**:736–743.
78. Kragballe K 1995 The use of vitamin D_3 analogues in dermatology. Dermatology **2**:198–203.
79. Binderup L, Carlberg C, Kissmeyer AM, Latini S, Mathiasen IS, Hansen CM 1994 The need for new vitamin D analogues: Mechanisms of action and clinical applications. In: Norman AW, Bouillon R, Thomasset M (eds) Vitamin D—A Pluripotent Steroid Hormone: Structural Studies, Molecular Endocrinology and Clinical Applications. de Gruyter, Berlin, New York, pp. 55–63.
80. Binderup L 1996 Development of new vitamin D analogues for treatment of psoriasis: Effects on hyperproliferative cells and on immune cells. Abstract, 3rd International Calcipotriol Symposium: Vitamin D in Dermatology.
81. Kissmeyer AM, Binderup E, Binderup L, Hansen CM, Andersen NR, Makin HLJ, Shankar VN, Jones G 1997 The metabolism of the vitamin D analog EB 1089: Identification of in vivo and in vitro liver metabolites and their biological activities. Biochem Pharmacol **53**:1087–1097.
82. Shankar VN, Dilworth FJ, Makin HLJ, Schroeder NJ, Trafford DJH, Kissmeyer AM, Calverley MJ, Binderup E, Jones G 1996 Metabolism of the vitamin D analog EB 1089 by cultured human cells: Conjugated system in side chain directs hydroxylation site to distal carbons. Biochem Pharmacol **53**:783–793.
83. Colston KW, Mackay AG, James SY, Binderup L, Chander S, Coombes RC 1992 EB 1089: A new vitamin D analogue that inhibits the growth of breast cancer cells in vivo and in vitro. Biochem Pharmacol **44**:2273–2280.
84. Mathiasen IS, Colston KW, Binderup L 1993 EB 1089, a novel vitamin D analogue, has strong antiproliferative and differentiation inducing effects on cancer cells. J Steroid Biochem Mol Biol **46**:365–371.
85. Narvaez CJ, Vanweelden K, Byrne I, Welsh JE 1996 Characterization of a vitamin D_3-resistant MCF-7 cell line. Endocrinology **137**:400–409.
86. Gulliford T, English J, Colston K, Sprøgel P, Coombes RC 1996 A phase I study of EB 1089, a vitamin D analogue. Proc Am Assoc Cancer Res **37**:164.
87. Johnsson C, Tufveson G 1994 MC 1288—A vitamin D analogue with immuno- suppressive effects on heart and small bowel grafts. Transplant Int **7**:392–397.
88. Johnsson C, Binderup L, Tufveson G 1996 Immunosuppression with the vitamin D analogue MC 1288 in experimental transplantation. Transplant Proc **28**:888–891.
89. Johnsson C, Binderup L, Tufveson G 1995 The effects of combined treatment with the novel vitamin D analogue MC 1288 and cyclosporine A on cardiac allograft survival. Transplant Immunol **3**:245–250.
90. Veyron P, Pamphile R, Binderup L, Touraine JL 1995 New 20-epi-

vitamin D_3 analogs: Immunosuppressive effects on skin allograft survival. Transplant Proc **27**:450.

91. Kallio E, Häyry P, Pakkala S 1996 The effect of MC 1288, a vitamin D analogue, on chronic renal allograft rejection in the rat. Abstract, 2nd International Conference on New Trends in Clinical and Experimental Immunosuppression.
92. Liu M, Lee MH, Cohen M, Bommakanti M, Freedman LP 1996 Transcriptional activity of the Cdk inhibitor p21 by vitamin D_3 leads to the induced differentiation of the myelomonocytic cell U 937. Genes Dev **10**:142–153.
93. Alroy I, Towers TL, Freedman LP 1995 Transcriptional repression of the interleukin-2 gene by vitamin D_3: Direct inhibition of $NFAT_p$/AP-1 complex formation by a nuclear hormone receptor. Mol Cell Biol **15**:5789–5799.

CHAPTER 62

The 16-Ene Vitamin D Analogs

MILAN R. USKOKOVIĆ Hoffmann-La Roche, Nutley, New Jersey

GEORGE P. STUDZINSKI University of Medicine and Dentistry of New Jersey—New Jersey Medical School, Newark, New Jersey

SATYANARAYANA G. REDDY Brown University School of Medicine, Providence, Rhode Island

I. INTRODUCTION

Widespread research efforts are directed at identifying therapeutic areas for the use of 1,25-dihydroxyvitamin D_3 [$1,25(OH)_2D_3$, calcitriol] and its analogs. These efforts are based on evidence implicating an essential role of $1,25(OH)_2D_3$ as a hormone in a host of cellular processes involved in calcium homeostasis, inhibition of cell growth, and induction of cell differentiation. $1,25(OH)_2D_3$ induces differentiation of myeloid leukemia cells along the monocyte–macrophage lineage, and inhibits the proliferation of breast, prostate, colon, and T-leukemia cell lines. $1,25(OH)_2D_3$ also acts as an immunosuppressive agent by inhibiting the release of interleukin-12 (IL-12), interferon-γ, and IL-2 in activated T lymphocytes. $1,25(OH)_2D_3$ and some of its analogs are used in treatment of renal osteodystrophy, secondary hyperparathyroidism, osteoporosis, psoriasis, and scleroderma. Also, it has potential for use in treatment of leukemia, solid tumors, and various autoimmune diseases. However, the clinical use of $1,25(OH)_2D_3$ is limited to doses that do not produce hypercalcemia.

In a search of vitamin D analogs that will exhibit high potency in cell growth inhibitory and cell differentiation activities, but with decreased calcemic properties, alterations of the ring A, ring D, and side chain of $1,25(OH)_2D_3$ have been explored. To produce various spatial orientations of the side chain, replacement of specific single bonds with double and triple bonds was examined. It was also expected that double and triple bonds can prevent metabolic hydroxylations known to occur at relevant carbons in the saturated side chain of $1,25(OH)_2D_3$ (**1**). (Bold numerals indicate structures shown in illustrations.) Prevention of these metabolic modifications was also expected from the substitution of hydrogens with fluorines at the same carbons.

Derivatives of vitamin D_3 with the 16-ene modification of the seco-steroid skeleton have produced highly encouraging results in a number of systems that serve as models of human neoplastic diseases. In addition to

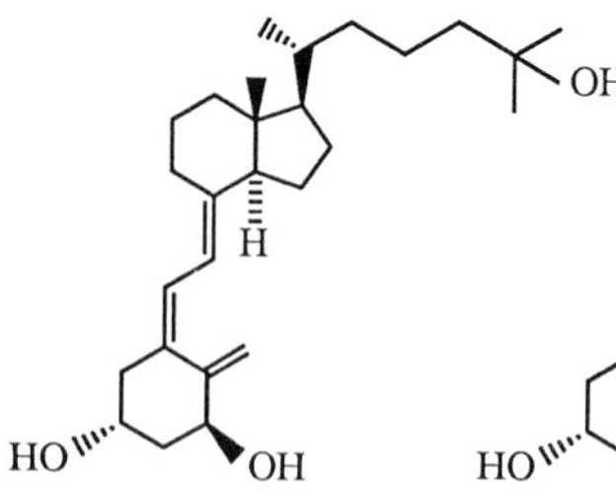

$1,25(OH)_2D_3$ (**1**) $1\alpha,25(OH)_2$-16,23E-diene D_3 (**7**) $1\alpha,25(OH)_2$-16-ene-23-yne D_3 (**9**)

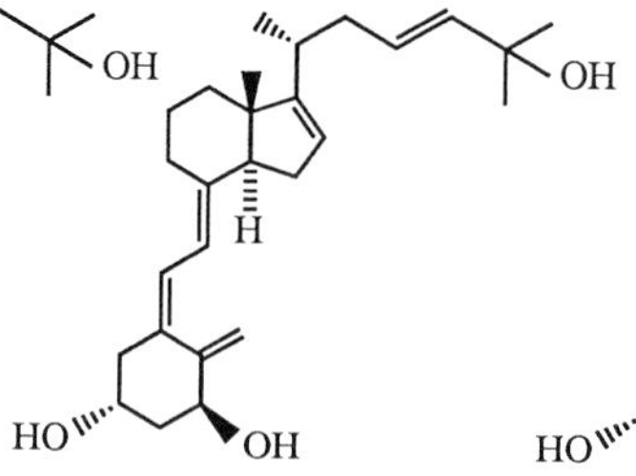

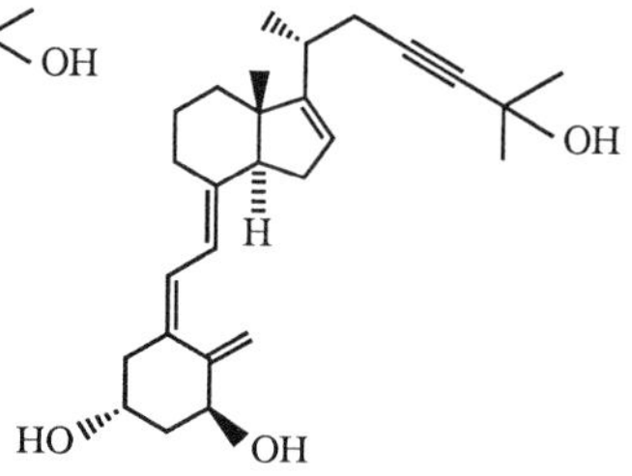

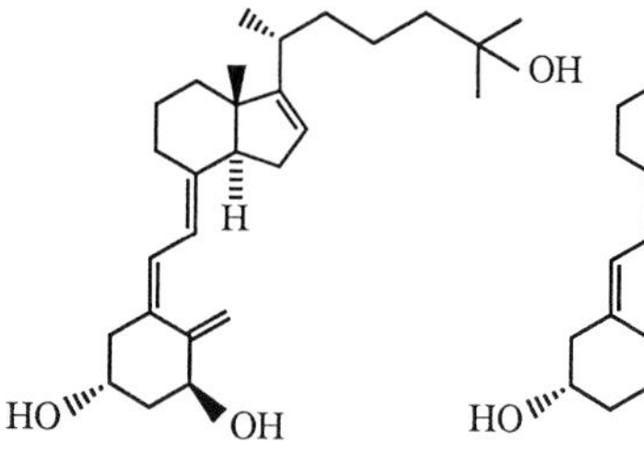

$1\alpha,25(OH)_2$-16-ene D_3 (**2**) $1\alpha,25(OH)_2$-16-ene-23-yne-26,27-F_6 D_3 (**33**) 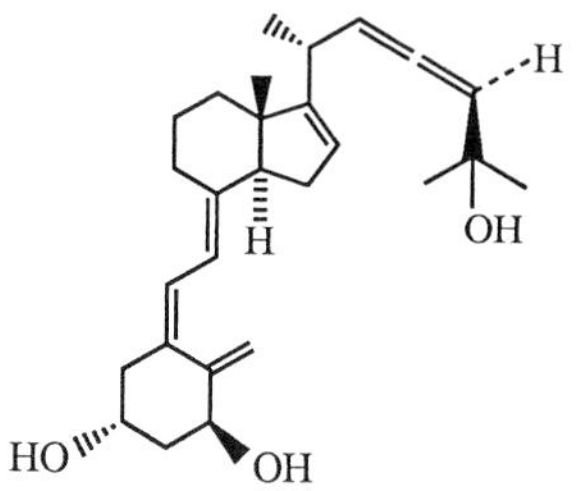 22(R)-$1\alpha,25(OH)_2$-16,22,23-triene D_3 (**12**)

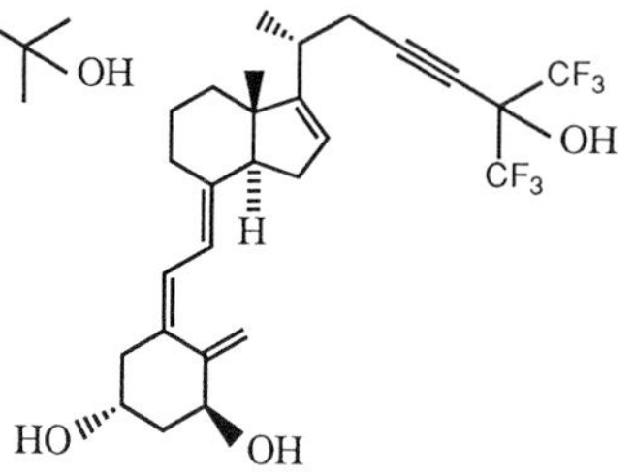

a marked antiproliferative and differentiation-inducing activity on human myeloid leukemia cells, this group of compounds offers the promise of therapeutic success against common human malignancies such as cancers of the prostate, breast, and colon. Further, 16-ene analogs have shown strong antiproliferative effects on human keratinocytes and activated T lymphocytes. However, in contrast to a relatively large number of studies from different laboratories describing these effects, there is a paucity of reports addressing the underlying mechanisms that would explain the effectiveness of these analogs. This may be due to the current lack of precise understanding of the detailed molecular events that result from exposure of target cells to steroid and steroidlike hormones in general. However, recent studies of altered controls of the cell cycle and of target tissue metabolism of 16-ene analogs offer hope that the mystery surrounding the basis of the increased potency of these compounds will soon be unraveled.

II. ANTIPROLIFERATIVE AND CELL DIFFERENTIATION ACTIVITIES

A. Leukemia Cells

There are many derivatives of vitamin D, including $1,25(OH)_2D_3$, that are potent inducers of differentiation of neoplastic hematopoietic stem and progenitor cells of nonlymphoid origin, but their use is limited by major disturbances of calcium homeostasis which accompany their action. Interestingly, several 16-ene analogs have high antiproliferative activity on myeloid leukemia cells but relatively low hypercalcemia-inducing effects. This places these compounds among the most promising derivatives of vitamin D for potential inclusion in the armamentarium of the hematological oncologist. Of all the 16-ene analogs, 1α-$25(OH)_2$-16-ene-23-yne D_3 (**9**) and $1\alpha,25(OH)_2$-16-ene D_3 (**2**) are the two analogs that were studied in detail.

1. 1α,25-DIHYDROXY-16-ENE-23-YNE-VITAMIN D_3

Several laboratories reported the high potency of the antiproliferative, differentiation-inducing activity of the analog **9** on human and murine myeloid leukemia cells. It was first reported [1] that among a large group of analogs of $1,25(OH)_2D_3$ tested, analog **9** was the most promising as an antileukemia agent. It was 4 to 12 times more potent than $1,25(OH)_2D_3$ in inducing differentiation and inhibiting colony formation in soft agar of a number of human myeloid leukemia cell lines but, surprisingly, slightly increased the cloning efficiency of normal human myeloid stem cells. More importantly, compound **9** was 30 times less potent than $1,25(OH)_2D_3$ in tests of intestinal calcium absorption, and 50 times less potent as bone calcium mobilizing agent [1]. This implied that **9** was an analog of $1,25(OH)_2D_3$ worthy of study in animal models.

This suggestion was tested using 8- to 10-week-old inbred BALB/c male mice implanted with syngeneic cells of a murine myeloid leukemia line WEHI 3BD+ [2]. The mice were maintained on calcium-free food beginning 2 days before the experiments, and 1 day after implantation of the leukemic cells, and every other day thereafter, the mice were injected intraperitoneally (i.p.) with varying doses of $1,25(OH)_2D_3$ or analog **9**. Mice receiving 1.6 μg of $1,25(OH)_2D_3$ all died within 2 weeks, and even the mice administered 0.1 μg of $1,25(OH)_2D_3$ had marked (>12 mg/dl) hypercalcemia. In contrast, 1.6 μg of analog **9** was well tolerated by the animals; compound **9** only moderately elevated the levels of serum calcium, and it significantly prolonged survival of animals inoculated with varying numbers of tumor cells. When the tumor cell inoculum was low, approximately 40% of the inoculated mice survived tumor-free when they were maintained on low calcium diets and received analog **9**.

Other laboratories have confirmed that analog **9** is a potent differentiation agent for human leukemia cells in culture. For instance, Clark *et al.* [3] evaluated the effects of a series of analogs on the proliferation and differentiation of the chronic myelogenous leukemia cell line RWLeu-4 and found that two analogs, each with a double bond at position 16 in the D ring and either a double or a triple bond in position 23, had enhanced antiproliferative activity against these cells. One of these compounds was $1,25(OH)_2$-16,23E-diene D_3 (**7**),* the other $1,25(OH)_2$-16-ene-23-yne D_3 (**9**). In view of the reported low hypercalcemia-inducing activity of **9**, its potential for clinical use as an antineoplastic agent was reinforced.

In another study [4] analog **9** was found to induce differentiation of HL-60G cells, a well-differentiated subclone of human promyelocytic leukemia HL-60 cells [5], with essentially similar potency to that in RWLeu-4 cells (i.e., ED_{50} of 0.7 nM at 72 hr). However, its antiproliferative activity on HL-60 cells in suspension culture, measured by arrest of cell cycle progression in the G_1 phase, was only slightly greater than the activity of $1,25(OH)_2D_3$ and was not as marked as the antiproliferative activity of 22-oxo or 20-epi analogs of $1,25(OH)_2D_3$ [4]. Thus, some aspects of the action of this compound are puzzling, and clarification of the mechanisms that disassociate induction of differentiation from the ability to arrest cells in the G_1 phase of the cell cycle, and from elevation of serum calcium levels, is still awaited.

* E and Z stand for *trans* and *cis* configurations of double bonds.

2. 1α,25-Dihydroxy-16-ene-vitamin D_3

Although exhibiting fewer modifications than molecule **9**, since it lacks the 23-triple bond, $1,25(OH)_2$-16-ene D_3 (**2**) has been reported to be even more potent than **9** in assays of clonogenic growth of HL-60 cells, without inhibiting the clonogenic potential of normal myeloid colony-forming cells [6]. Again, there was some dissociation of antiproliferative and differentiation-inducing activities on HL-60 cells: when differentiation markers were assayed, analog **2** appeared 10 times less potent than **9** in the studies of Jung *et al.* [6]. Further, **2** showed only slight and generally nonsignificant activity increases over the activity of $1,25(OH)_2D_3$ (**1**) when differentiation markers and cell cycle arrest were measured in two clones of HL-60 cells [7]. This suggests that the 16-double bond is particularly critical in clonogenic assays, whereas the addition of the 23-double or -triple bond increases the potency to induce the differentiated phenotype.

The studies referred to above also showed that unsaturation at the 23 position, in the absence of the 16-double bond, had little or no effect on increasing the potency of $1,25(OH)_2D_3$. The 23-ene modification of $1,25(OH)_2D_3$ studied by Jung *et al.* [6] resulted in only a 10-fold increase in potency over $1,25(OH)_2D_3$ in the clonogenic growth assay, and a 6-fold increase in the differentiation assay, whereas the 23-yne grouping appeared to reduce the activity of $1,25(OH)_2D_3$ in the studies of Zhang *et al.* [4]. Thus, it is clear that the 16-ene group confers a great increase in antiproliferative potency of $1,25(OH)_2D_3$ and that only in its presence does the introduction of a double or triple bond at position 23 further increase the differentiation-inducing activity.

3. 16-Ene Vitamin D Analogs with Other Modifications of the Side Chain

Several modifications of the side chain other than unsaturation at position 23 have been shown to result in compounds with high antiproliferative activity on breast or prostate cancer cells, but these modifications have so far not been found to produce antileukemic agents more effective than analog **9** or **2**. For instance, replacement of the six hydrogen atoms at the end of the side chain with fluorines in **9** produces 1α,$25(OH)_2$-16-ene-23-yne-26,27-F_6 D_3 (**33**), which in the studies of Zhang *et al.* [4] showed no significant differences from **9** in differentiation-inducing or cell cycle traverse blocking activity in HL-60 cells. Also, on the basis of inhibition of colony formation of HL-60 cells, analog **9** actually appears to have a higher potency (ED_{50} of 2×10^{-11} M) [6] than its hexafluorinated derivative **33** (ED_{50} of 20×10^{-11} M) [8], although in an earlier study

33 was stated to be more potent than **9** [4]. Another highly modified 16-ene derivative, 22(*R*)-1α,25(OH)$_2$-16,22,23-triene D$_3$ (**12**), has an ED$_{50}$ for clonogenicity inhibition of HL-60 cells similar to the ED$_{50}$ values of **2** and **9**. This compound also has been tested using stem cells from three patients with acute myeloid leukemia, and the mean ED$_{50}$ of the inhibition of clone formation by these cells was 8×10^{-11} *M*. If **12** does not have similarly high hypercalcemia-inducing activity, it should be a valuable adjunct to antileukemia therapy.

B. Prostate Cancer Cells

Prostate cancer has become the most frequently occurring non-skin cancer among men, with over 300,000 estimated new cases in the United States in 1996 [9]. Approximately 41,000 Americans die annually from prostate cancer, making it the second leading cause of cancer mortality among males. No successful long-term therapy exists once the cancer progresses beyond the prostate capsule. The absence of a cure for such a widespread disease has provided an impetus to develop new therapeutic approaches. One area that has been intensively researched since the early 1980s is the use of natural hormones or their analogs to retard the clonal growth or induce differentiation of prostate cancer cells to a mature, nondividing stage. These studies included the natural metabolites and analogs of vitamin D [10,11].

It has been demonstrated that 1,25(OH)$_2$D$_3$ and its 16-ene-23-yne analog **9** inhibit proliferation of prostate primary malignant tissue [12,13] and prostate cell lines [14–16]. Koeffler and colleagues have focused their investigation of the 16-ene vitamin D analogs on three prostate cancer cell lines: LNCap obtained from a lymph node metastasis of a patient with hormonally refractory prostate cancer, PC-3 derived from a primary adenocarcinoma of the prostate, and DU-145 from prostate cancer metastatic to the brain [17]. LNCap cells have wild-type p53, whereas the other two cell types have mutant p53 genes. LNCap cells are androgen responsive, but PC-3 and DU-145 are not. These three prostate cancer cell lines have been demonstrated to have functional 1,25(OH)$_2$D$_3$ nuclear vitamin D receptor (VDR) [18,19]. The cell lines were incubated with vitamin D analogs in a dose-dependent manner for 14 days before colonies were counted. Results of this study are presented in Table I.

The natural hormone 1,25(OH)$_2$D$_3$ (**1**) exhibited weak clonal growth inhibitory activity in the PC-3 cell line, but no ED$_{50}$ was achieved in LNCap or DU-145 cells. It should be noted that in other studies 1,25(OH)$_2$D$_3$ was very active in inhibiting LNCap cell growth [19]. The 16-ene-23-yne analog **9** was active in LNCap cells but inactive in the other two cell lines. Replacement of two C-25 methyl groups in **9** with ethyl groups produced the analog **39**, which retained LNCap activity but was also highly active in PC-3 cell line. The two corresponding 23-double bond epimers **37** and **38** showed very similar activity profiles. Of the two 16-ene-22-allenes only the 22(*R*)-epimer **12** shows weak activity in the PC-3 cell line. Activity in all three cell lines was achieved when the side chain end methyl hydrogens of **9** were substituted by fluorine to give the analog **33**. In this case the corresponding 23-double bond epimers **31** and **32** were also active, **31** being the most active in

Table I Inhibition of Clonal Growth of Prostate Cancer Cell Lines

Analog	ED$_{50}$ (n*M*)		
	LNCap	PC-3	DU-145
1,25(OH)$_2$D$_3$ (**1**)	na[a]	100	na
1,25(OH)$_2$-16-ene-23-yne D$_3$ (**9**)	2.5	na	na
1,25(OH)$_2$-16-ene-23-yne-26,27-bishomo D$_3$ (**39**)	8.0	0.5	na
1,25(OH)$_2$-16,23*E*-diene-26,27-bishomo D$_3$ (**37**)	15	5.0	na
1,25(OH)$_2$-16,23*Z*-diene-26,27-bishomo D$_3$ (**38**)	17	1.5	na
1,25(OH)$_2$-16-ene-22(*R*)-allene D$_3$ (**12**)	na	100	na
1,25(OH)$_2$-16-ene-22(*S*)-allene D$_3$ (**13**)	na	na	na
1,25(OH)$_2$-16-ene-23-yne-26,27-F$_6$ D$_3$ (**33**)	15	35	100
1,25(OH)$_2$-16,23*E*-diene-26,27-F$_6$ D$_3$ (**31**)	—	—	2.5
1,25(OH)$_2$-16,23*Z*-diene-26,27-F$_6$ D$_3$ (**32**)	1.5	20	100
1,25(OH)$_2$-16-ene-23-yne-19-nor-26,27-F$_6$ D$_3$ (**66**)	6.0	4.5	20
1,25(OH)$_2$-16,23*E*-diene-19-nor-26,27-F$_6$ D$_3$ (**64**)	0.025	0.75	1.0
1,25(OH)$_2$-16,23*Z*-diene-19-nor-26,27-F$_6$ D$_3$ (**65**)	0.075	0.5	1.0

[a] na, Not active (compound does not reach ED$_{50}$).

the notoriously resistant DU-145 cell line. The highest specificity for inhibition of the clonal growth of prostate cancer cells was observed in the 19-nor series, when the 19-nor modification of ring A was combined with a 16-double bond, 23*E*- and 23*Z*-double bonds, or 23-triple bond and 26,27-hexafluoro substitution. Compounds **64**, **65**, and **66** were fully active in LNCap, PC-3, and DU-145 cell lines at low nanomolar and subnanomolar levels and merit further investigation as potential clinical candidates.

C. Other Cancer Cell Targets

Studies also suggest therapeutic uses of 16-ene analogs in the management of several other common human cancers. For instance, a panel of 16-ene-containing vitamin D analogs was tested on thymidine incorporation by cells derived from human colon adenocarcinoma, CaCo-2 [20]. Two analogs with low hypercalcemia-inducing activity, the 16-ene-23-yne **9** and its 26,27-hexafluoro derivative **33**, were found to be most potent in this regard. The latter compound, $1\alpha,25(OH)_2$-16-ene-23-yne-26,27-hexafluoro-vitamin D_3, was found to inhibit azoxymethane-induced colon carcinogenesis in rats [21] and, also in rats, *N*-nitroso-*N*-methylurea-induced mammary carcinogenesis [22]. Among other examples of the potential clinical importance of these compounds McElwain *et al.* [23] have shown that the 16-ene-23-yne analog **9** has strong antiproliferative effects in squamous cell carcinoma model systems. Thus, these two compounds appear to have widespread activity on human cancer cells, although the experience with prostate cancer cells discussed above shows that there also appears to be selectivity of the analogs on cell-specific targets.

III. SUMMARY OF STRUCTURE–ACTIVITY RELATIONSHIPS IN DIVERSE SYSTEMS

The structure–activity correlation of the 16-ene vitamin D_3 analogs is based on a large number of these compounds being available by synthesis, on their ability to induce HL-60 cell differentiation and to decrease the tendency to induce hypercalcemia in comparison with the natural hormone $1,25(OH)_2D_3$ (**1**). Calcium liability is a consequence of changes in calcium–phosphate homeostasis, and it can be estimated by retardation in weight gain or by weight loss of mice used in a primary test for tolerance. The discussion that follows is organized according to the structural modifications in the side chain and in the ring A of the 16-ene-$1,25(OH)_2D_3$ (**2**) as a basic model.

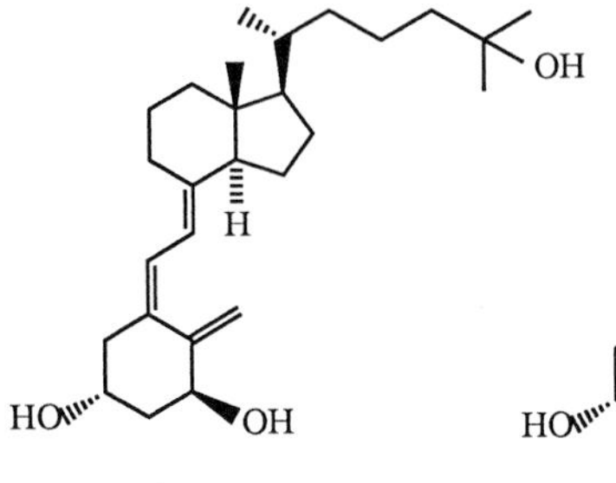

HO OH OH H **1**

HO OH OH H **2**

HO OH OH H **3**

HO OH OH H **4**

HO OH OH H **5**

HO OH OH H **6**

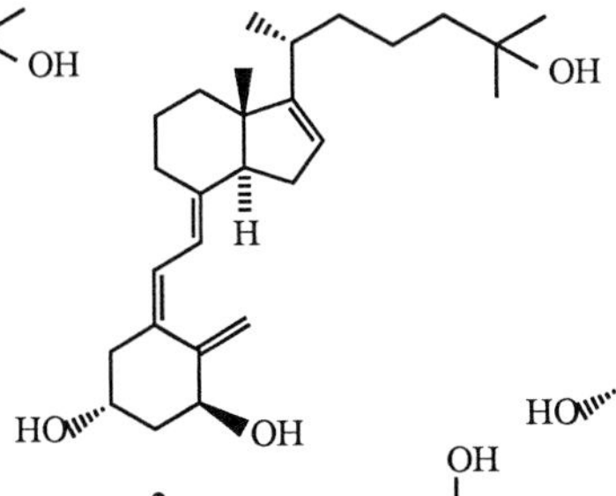

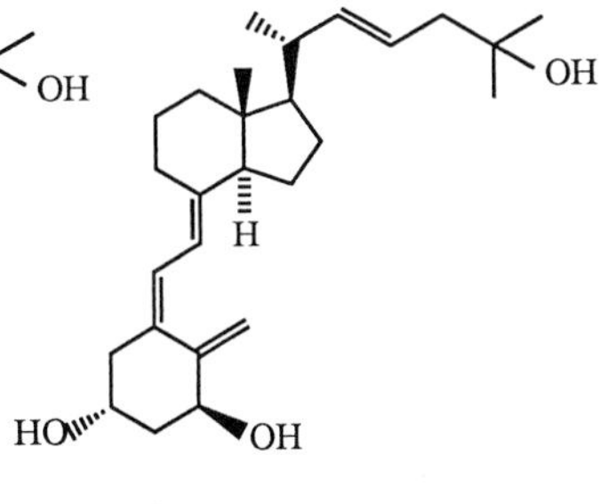

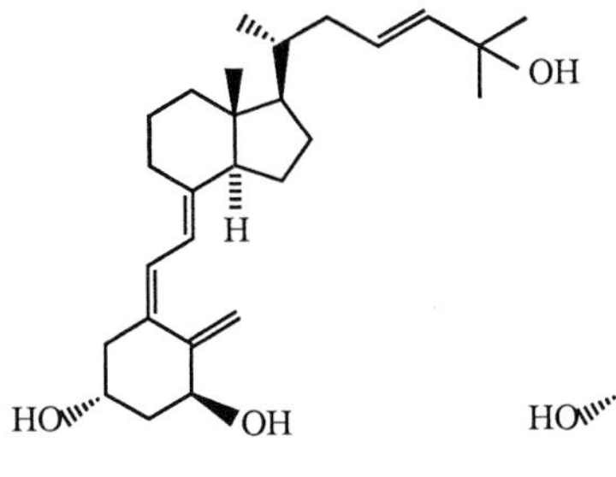

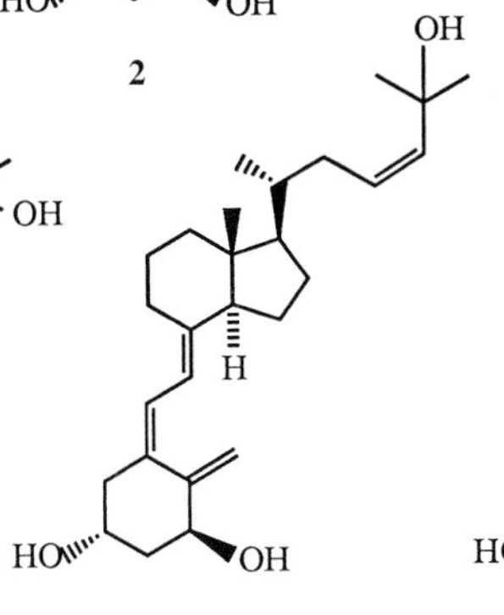

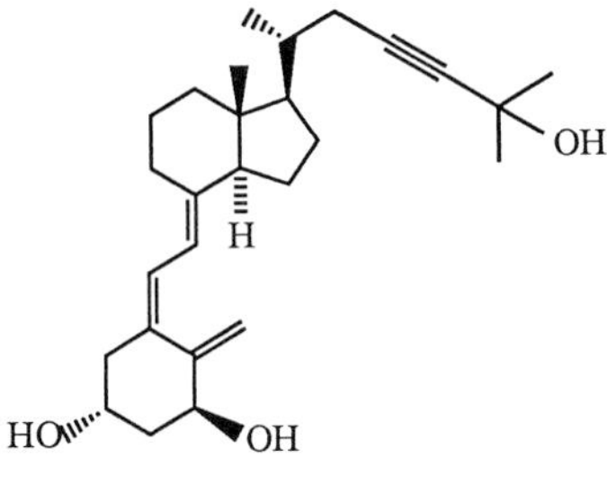

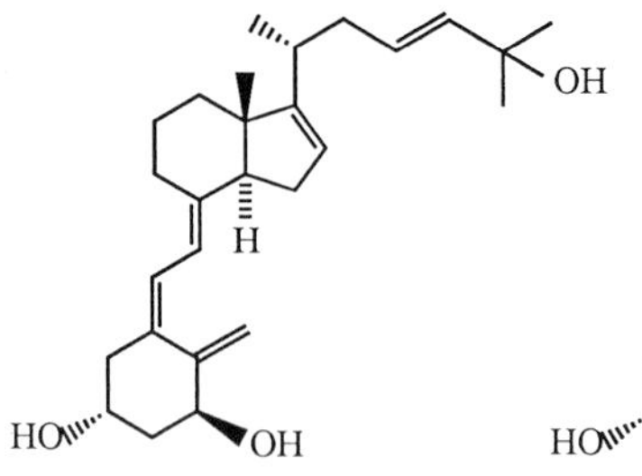

OH
OH
OH
HO OH HO OH HO OH
7 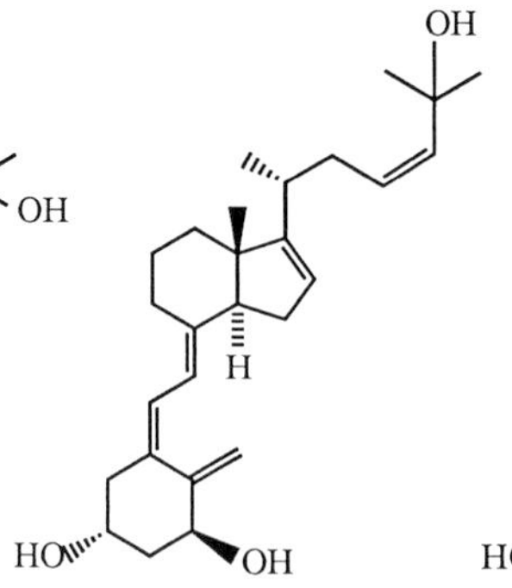 8 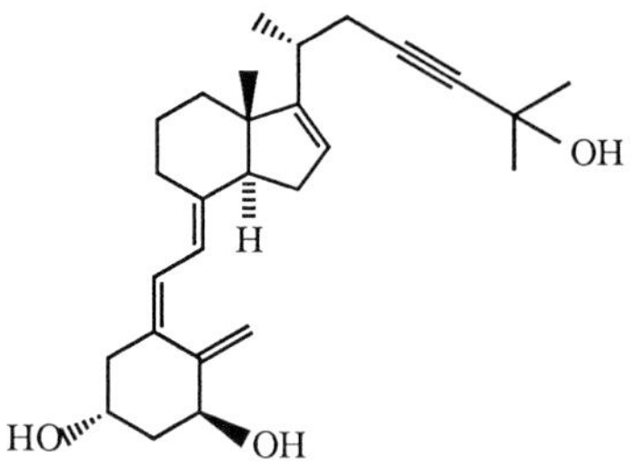 9

A. Introduction of Double and Triple Bonds

The spatial orientation of the $1,25(OH)_2D_3$ side chain has a decisive influence on binding affinity for the VDR, and on the conformation of the VDR–retinoid X receptor (RXR)–$1,25(OH)_2D_3$ complex bound to vitamin D response elements (VDREs) of the specific genes. Introduction of the double and triple bonds in the molecular frame of $1,25(OH)_2D_3$ was expected to effect changes in VDR binding and protein conformation, and thus to produce changes in cell differentiation and calcemic activities of the various analogs (**2–6**).

Introduction of the 16-double bond dramatically alters the profile of these activities. Compound **2** does not produce any change of the $1,25(OH)_2D_3$ (**1**) calcemic effects, but it is 100 times more active than **1** in inhibition of HL-60 cell clonal growth and 7 times more active in induction of cellular differentiation [6]. The 22*E*-double bond analog **3** is equipotent to $1,25(OH)_2D_3$, with respect to both cell differentiation and calcemic properties [24]. The 23*E*-double bond compound **4** is 11 times more potent in inhibition of clonal growth and 6 times so in differentiation assays, and it is 10 times less hypercalcemic than $1,25(OH)_2D_3$. Although no HL-60 cell data are available for the 23*Z*-double bond isomer **5**, this analog is 320 times less hypercalcemic than

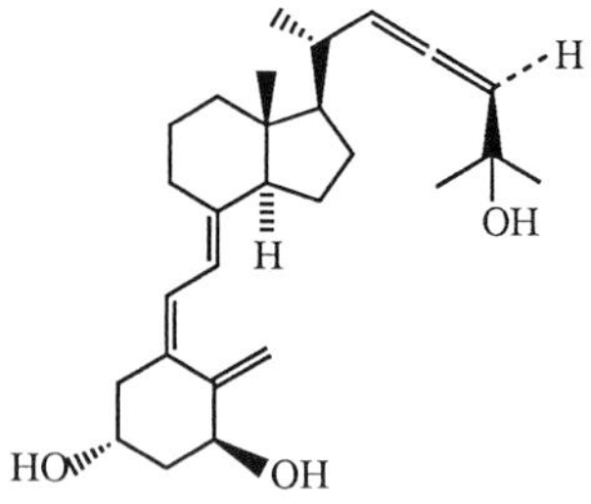

H OH
H
HO OH HO OH
10 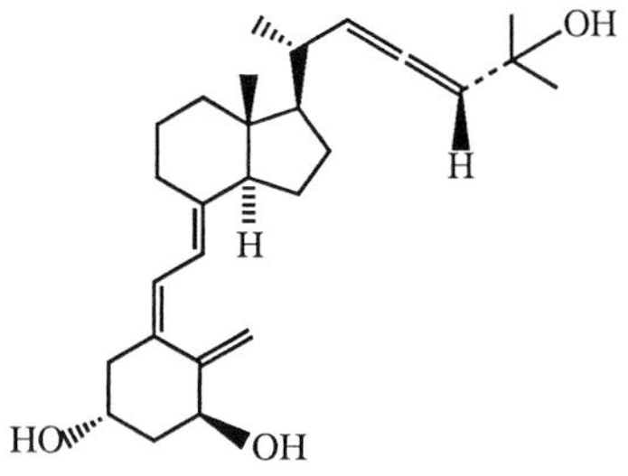 11

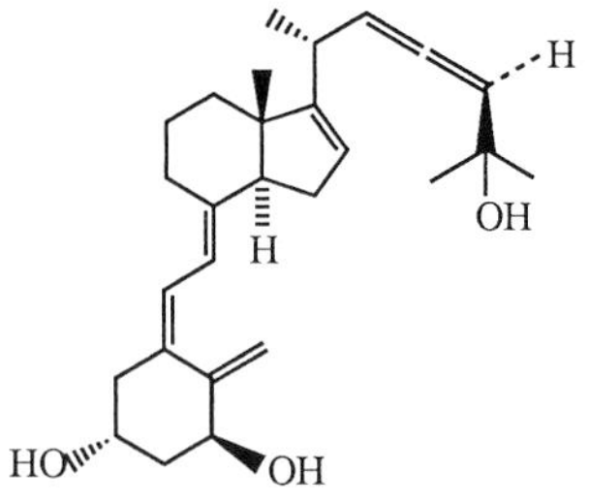

H OH
H
HO OH HO OH
12 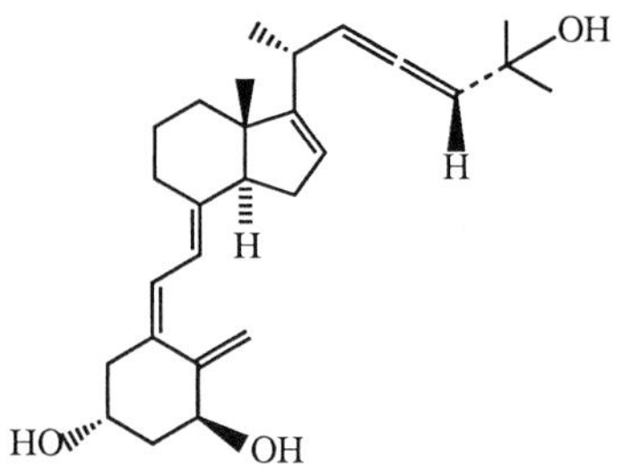 13

$1,25(OH)_2D_3$. Finally, the 23-triple bond compound **6** is 3 times more active in cell differentiation tests and 10 times less hypercalcemic than the parent compound **1** [25].

Subsequently, combinations of the 16-double bond with various unsaturation patterns of the side chain have been investigated. The analog **7**, bearing 16- and 23*E*-double bonds, was 5 times more active in the HL-60 cell differentiation assay and 70 times less hypercalcemic than $1,25(OH)_2D_3$. The corresponding 23*Z*-analog **8** was also 5 times more effective in cell differentiation and 25 times less calcemic. Combination of the 16-double bond and the 23-triple bond gives compound **9**, which is 20 times more active in cell differentiation and 15 times less hypercalcemic than $1,25(OH)_2D_3$ [2,6,26, 27].

B. Allene Side Chains

Powerful combinations of unsaturations in the ring D and side chain came from investigations of 22-allenes. The 22(*R*)-allene analog **10** was equally potent to $1,25(OH)_2D_3$ in the HL-60 assay, unlike the 22(*S*)-epimer **11**, which was 10 times less active [28]. Addition of the 16-double bond had a profound effect on this activity profile. The 22(*R*)-16-ene-allene **12** appears to be one of the most active vitamin D analogs presently known [8]. It was reported to be 40 times more active in HL-60 cell differentiation, 400 times more active in inhibition of HL-60 clonal growth, and 5 times less hypercalcemic than $1,25(OH)_2D_3$. The stereochemistry at C-22 is very important in this case. The 22(*S*)-16-ene-allene **13** was equal to $1,25(OH)_2D_3$ in the HL-60 differentiation assay but 100 times less hypercalcemic *in vivo*.

C. Epimerization at C-20

The major reorientation of the $1,25(OH)_2D_3$ side chain is affected by epimerization at C-20. The two Leo Pharmaceuticals analogs **14** and **15** exhibit dramatically increased growth inhibitory and cell differentiation-inducing effects on various cancer cell lines, accompanied also by significantly increased hypercalcemic properties [29]. Peleg and co-workers, using the limited proteolytic digestion assay, showed that two 20-epi analogs exhibited distinct protease digestion patterns and significantly greater effect on a VDR–RXR regulated gene transcription than $1,25(OH)_2D_3$ [30].

Interestingly, 16-ene-20-epi-$1,25(OH)_2D_3$ (**16**) retained the high differentiation-inducing property characteristic of the 20-epi parent compound **14**, but with dramatically decreased hypercalcemic effects, 80 times

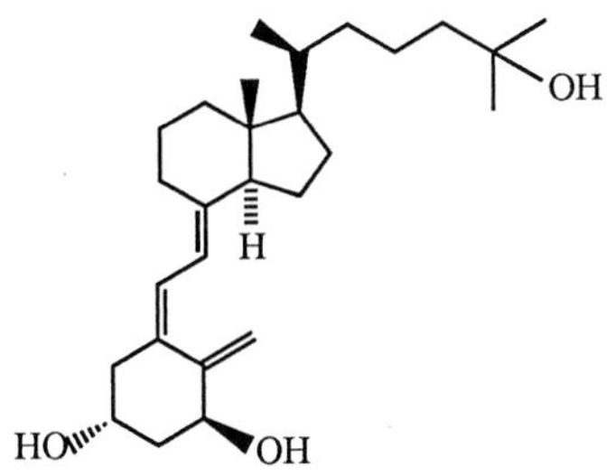

14 MC1288

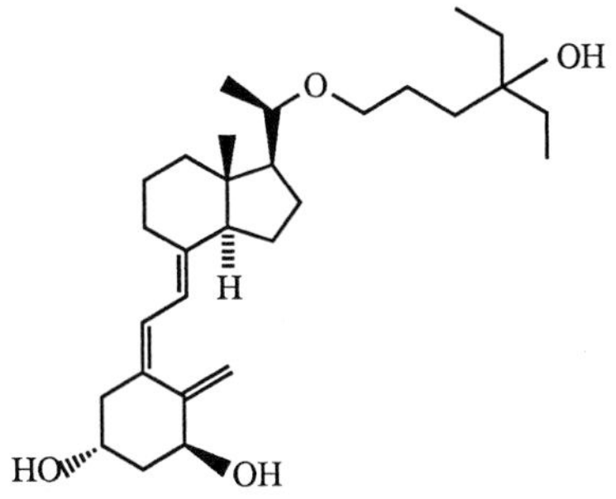

15 KH1060

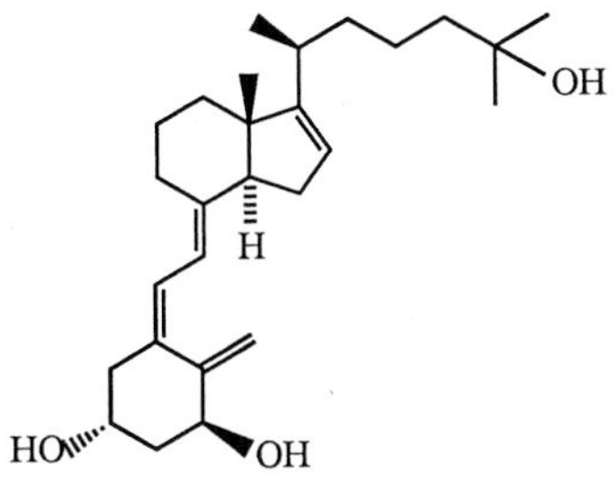

16

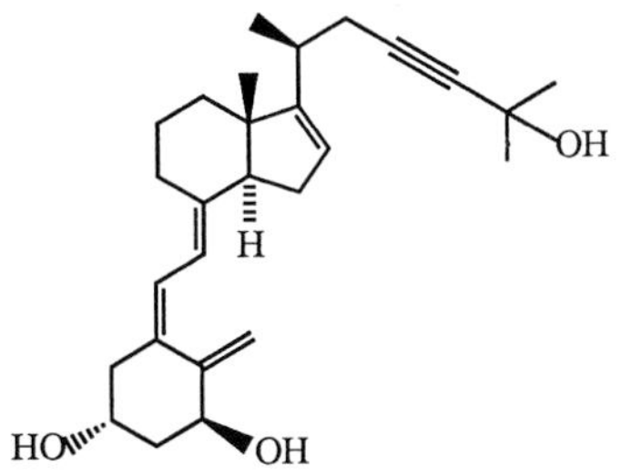

17

lower than that of 1,25$(OH)_2D_3$. The analog with an additional 23-triple bond, in this case compound **17**, was 200 times less calcemic but only half as effective as 1,25$(OH)_2D_3$ in HL-60 cell differentiation tests.

D. Substitution of the Side Chain Hydrogens with Fluorine

Awareness of the catabolic inactivation of 1,25$(OH)_2D_3$ by hydroxylation at various positions of the side chain led to the imitation of the known 24(*R*)-hydroxy,24-keto, and 26-hydroxy metabolites of 1,25$(OH)_2D_3$ with corresponding fluorine surrogates. Substitution of one of the C-24 hydrogens with fluorine as in compound **18** resulted in one-third lesser activity both in HL-60 cell differentiation and in calcemic properties [24]. Installment of two fluorines at C-24, however, increased 10 times the HL-60 cell differentiation but without any increase in hypercalcemia. The two epimeric 26-trifluoro derivatives **20** and **21** are 3 times more active than 1,25$(OH)_2D_3$ in the HL-60 cell assay; the 25(*R*)-epimer **20** is 2 times more calcemic than 1,25$(OH)_2D_3$. The 26,27-hexafluoro analog **22** is 20 times more active as a differentiation agent than 1,25$(OH)_2D_3$, without a significant increase in hypercalcemia. These studies suggest that the substitution of side chain hydrogens with fluorine at carbons susceptible to metabolic hydroxylation stabilize the framework of 1,25$(OH)_2D_3$ against catabolic inactivation. The polar effects of fluorines on their own may contribute to an independent increase in intrinsic activity on cell differentiation. To supplement the increased growth inhibitory and cell differentiation activity of the fluoro analogs with a reduction in calcemic properties, various unsaturations in the side chains of 26-trifluoro and 26,27-hexafluoro analogs were investigated.

It was shown in the saturated series with compounds **20** and **21** that the absolute stereochemistry at C-25 does not influence biological activities. However, in the case of 22*E*-double bond epimers **23** and **24** significant selectivity was observed. As its saturated congener **20**, the 25(*R*)-epimer **23** was still only 3 times more active than 1,25$(OH)_2D_3$ in HL-60 cell differentiation assays, but the 25(*S*)-epimer **24** was 40 times more potent in the same test. Calcemic data for **23** and **24** are not available. Another pair of 25-epimeric 26-trifluoro analogs was acquired in the 16,23*E*-diene series. In this case the 25(*R*)-epimer **25** was 27 times more active in HL-60 cell differentiation and 10 times more calcemic than 1,25$(OH)_2D_3$. However, its 25(*S*)-epimer **26** was 40 times more active but equally calcemic as 1,25$(OH)_2D_3$.

The 26-trifluoro compound **27** bearing 16-double and 23-triple bonds resisted separation into pure epimers. This epimeric mixture was 50 times more active in HL-

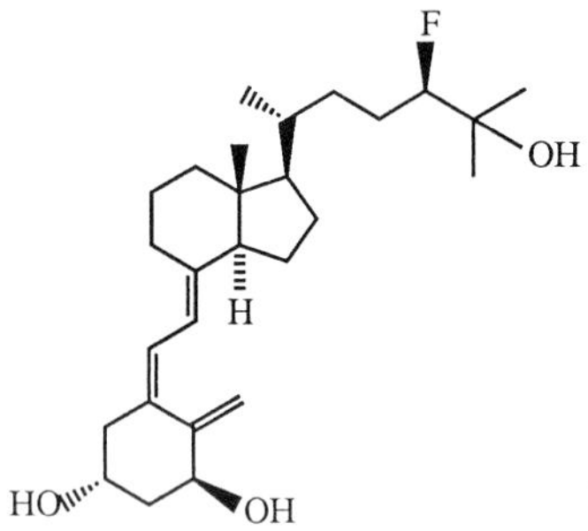

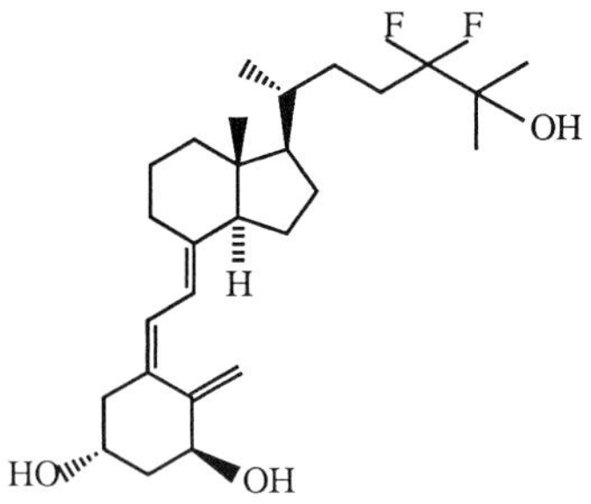

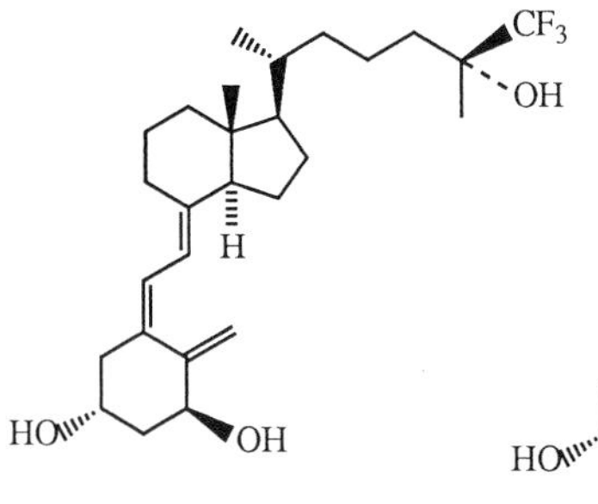

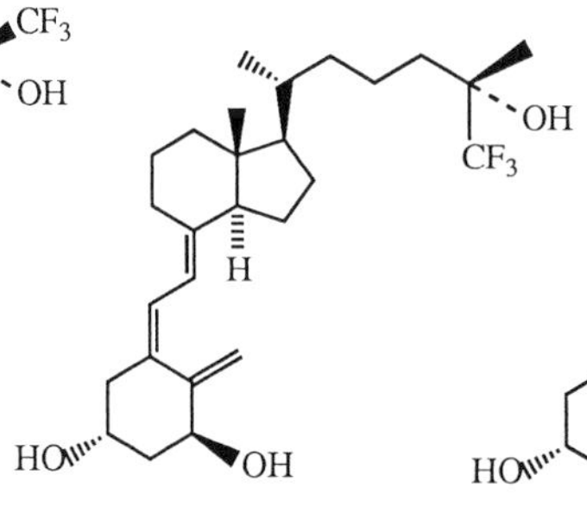

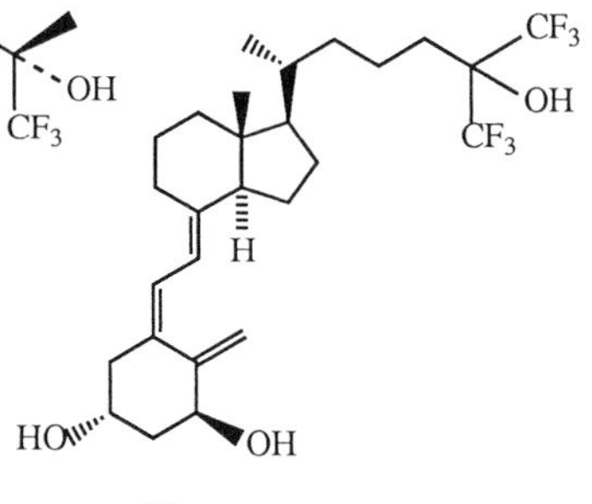

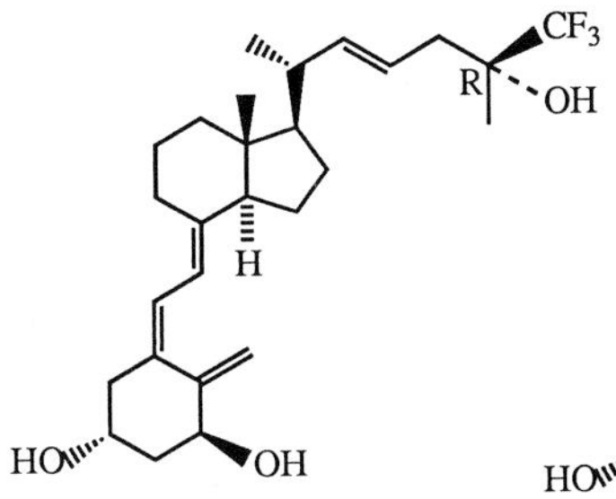

60 cell differentiation but 4 times less hypercalcemic than $1{,}25(OH)_2D_3$.

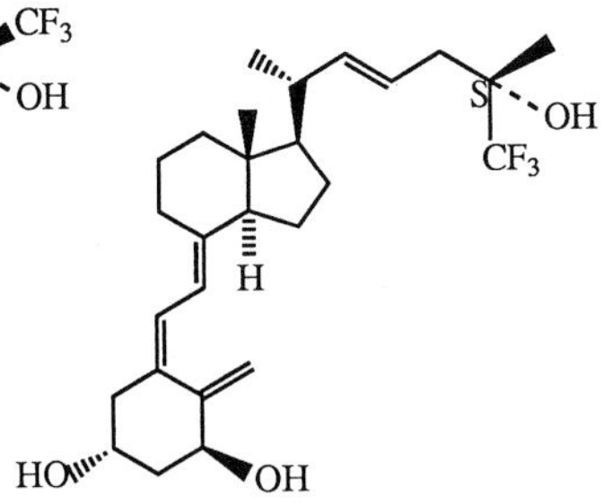

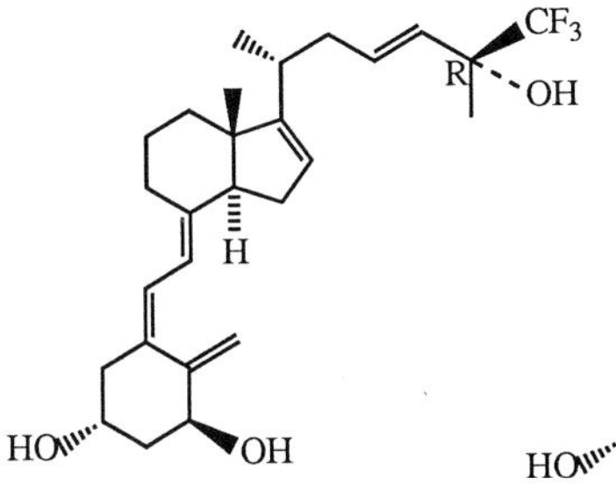

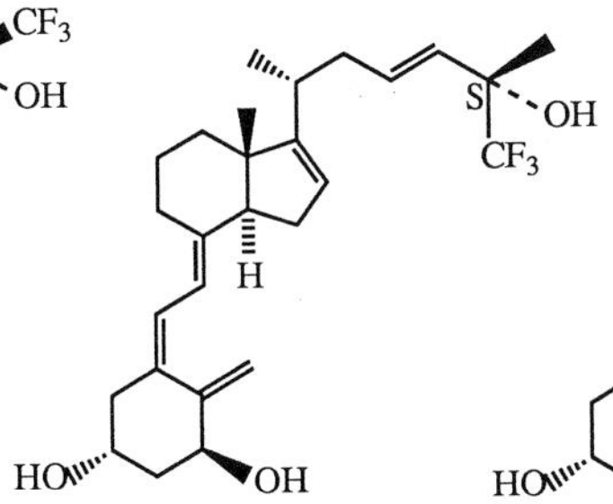

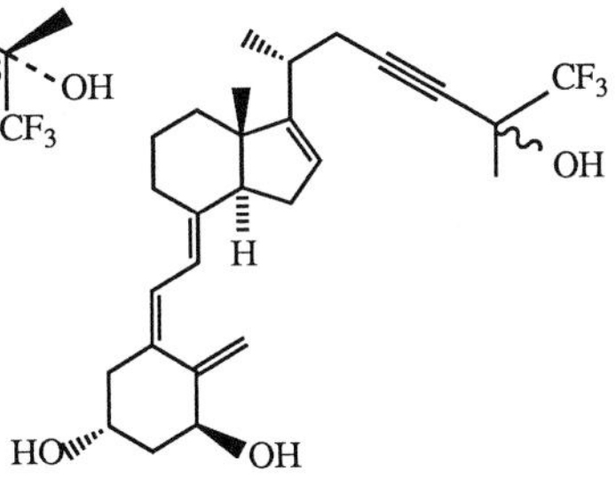

As previously discussed, 26,27-hexafluoro-1,25-$(OH)_2D_3$ (**22**) was 20 times more active in HL-60 cell differentiation but not more hypercalcemic than the parent $1{,}25(OH)_2D_3$. Insertion of the 22*E*-double bond gives compound **28**, which was 30 times more active in cell differentiation but 3 times less hypercalcemic than $1{,}25(OH)_2D_3$. The corresponding 23*E*-double bond analog **29** was only 10 times more active than $1{,}25(OH)_2D_3$ in cell differentiation; no calcemic data for this compound are available. The hexafluoro 23-triple bond analog **30** was also 10 times more active in cell differentiation assays, but it was 2 times more hypercalcemic than $1{,}25(OH)_2D_3$.

No improvement in separation of cell differention and calcemic activities was observed in hexafluoro analogs bearing both 16- and 23-double bonds. The 23*E*-compound **31**, although 40 times more active than 1,25-$(OH)_2D_3$ in cell differentiation, was also 5 times more hypercalcemic. The 23*Z*-congener **32** was 30 times more cell differentiating and at least 2 times more hypercalcemic. However, the analog **33** bearing 16-double and 23-triple bonds in hexafluoro series was 80 times more potent in cell differentiation assays and 17 times less hypercalcemic than $1{,}25(OH)_2D_3$ [31]. The corresponding 20-epi congeners **34**, **35**, and **36** have not offered any advantage over the compounds with natural configuration in this case.

E. The 26,27-Bishomo Analogs

It was originally observed by Ikekawa *et al.* [32] that the substitution of C-25 methyl groups of 1,25-$(OH)_2D_3$ with ethyl groups increased cell differentiation activity without substantially changing calcemic properties [33]. Later this substitution was exemplified by researchers at Leo Pharmaceuticals in the preparation of highly active analogs such as the presently clinically studied EB1089 and KH1060 (**15**). We have applied the same substitution in combination with 16-double and 23-double and -triple bonds.

Presently no in-depth comparison of the ethyl congeners **37**, **38**, and **39** to the corresponding methyl compounds **7**, **8**, and **9** is available. In the HL-60 cell clonal growth test the ethyl analogs were 90, 40, and 100 times more inhibitory than $1{,}25(OH)_2D_3$ [8].

F. Ring A Modified Analogs

In the continued pursuit of improvement in the separation of cell differentiating and calcemic activities of vitamin D analogs, the above-described ring D side chain modifications were combined with several alterations in ring A. The 1α-hydroxy group is believed to exercise a crucial role in activities mediated by binding to VDR and involving gene transcriptional mechanisms. These include cell differentiation, inhibition of cell

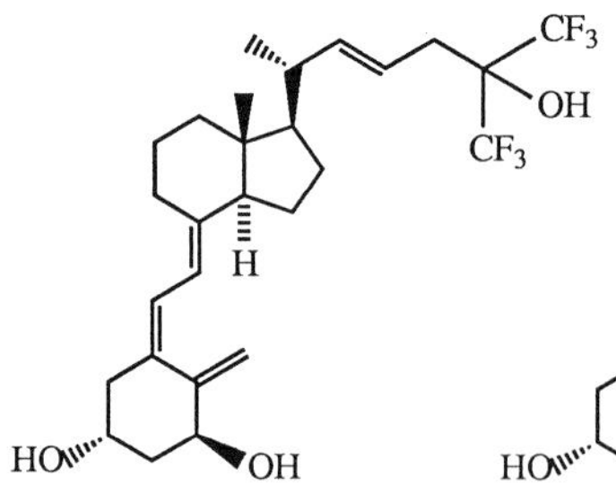

28 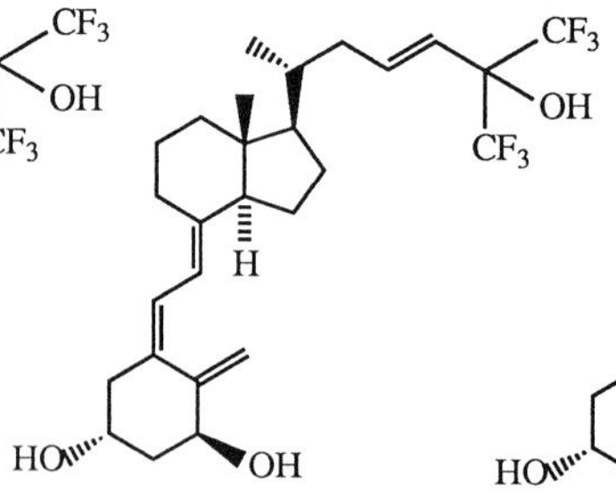 29 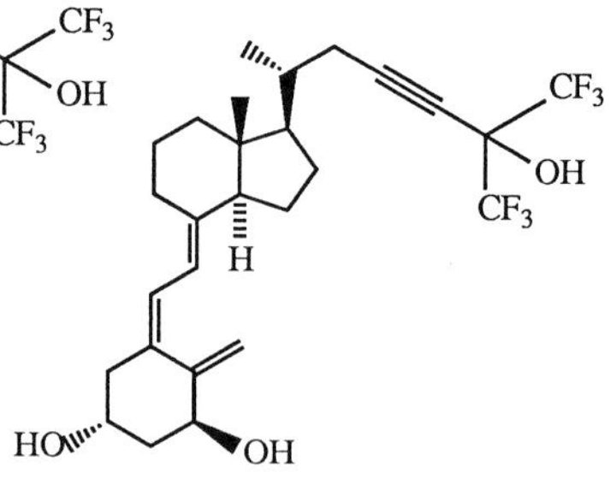 30

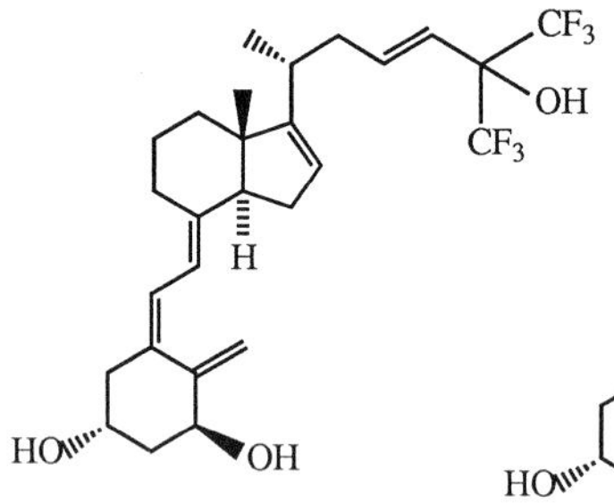

31 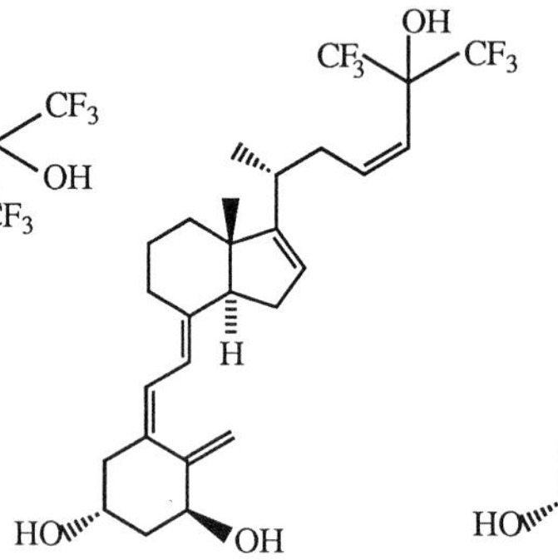 32 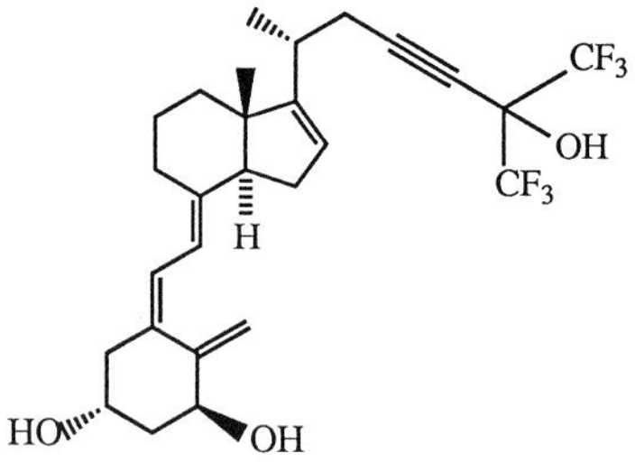 33

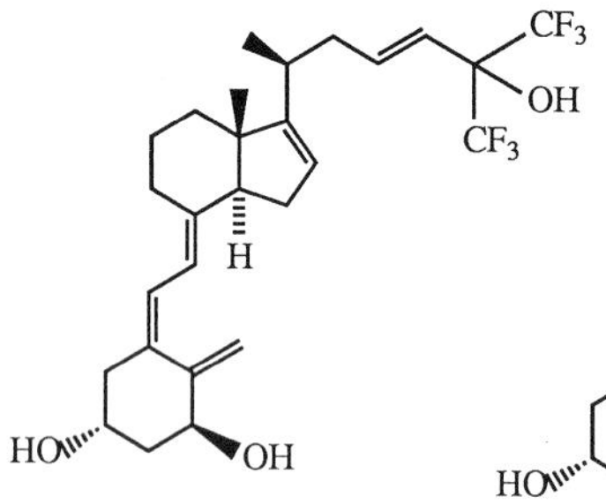

34 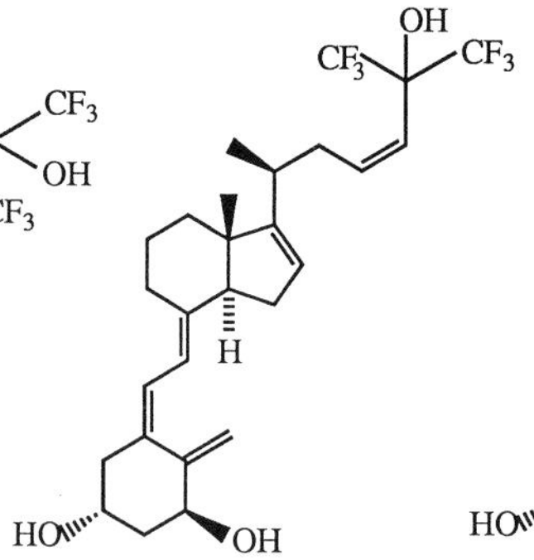 35 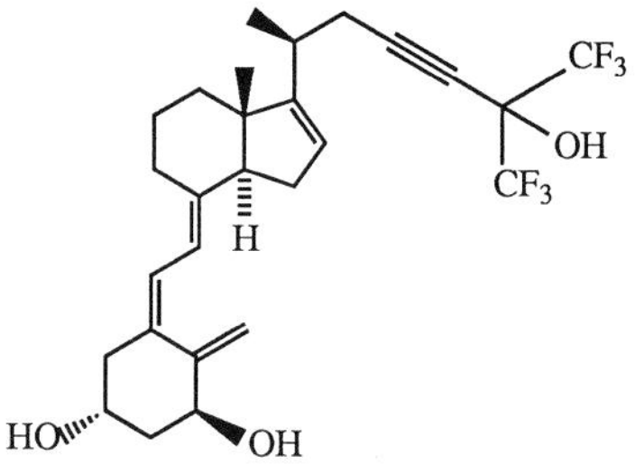 36

clonal growth, and various aspects of intracellular calcium transport.

Inversion of the configuration at C-1 of the natural $1,25(OH)_2D_3$ hormone produces $1\beta,25(OH)_2D_3$ (**40**), which is 50 times less active in induction of HL-60 cell differentiation. It is also 180 times less hypercalcemic than the natural hormone, and it was found to be a specific inhibitor of the nongenomic responses of both transcaltachia and the uptake of calcium in ROS 17/2.8 cells [34,35]. Inversion of the configuration of C-1 of the most active 16-ene-23-yne-26,27-hexafluoro analog **33** produces 1β-epimer **41**, which is 1600 times less active than the analog **33** in induction of cell differentiation.

Replacement of the 1α-hydroxyl group with fluorine in the case of highly active ring D side chain analogs **25**, **26**, **27**, **31**, **32**, and **33** furnished compounds **42**–**47**. Interest in these 1α-fluoro analogs grew out of an earlier finding that $1\alpha F,25(OH)$-16-ene-23-yne D_3 (**48**) was only slightly less active in cell differentiation assays but 100 times less calcemic than the parent $1,25(OH)_2$-16-ene-23-yne D_3 (**9**) [26]. Results of the HL-60 cell differentiation assay indicate that the 1α-fluoro analogs **42**–**47** (Fig. 1) are active at the one-digit nanomolar level, but still one order of magnitude less than the corresponding 1α-hydroxyl congeners. However, they were 4 (**46**) to 80 times (**42** and **47**) better tolerated *in vivo* with respect to their calcemic properties [26,31]. Thus, the 1α-fluoro

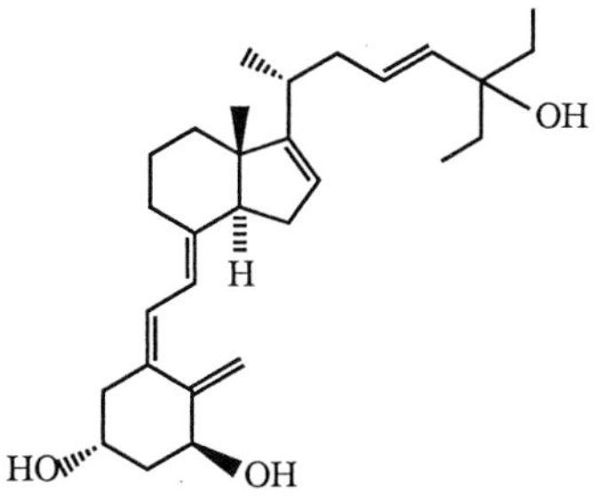

37

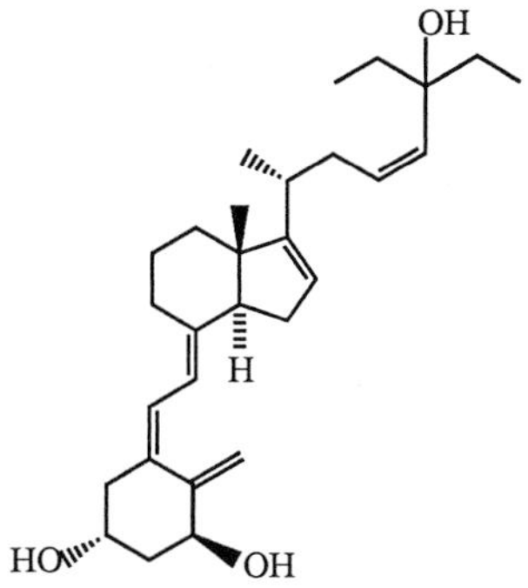

38

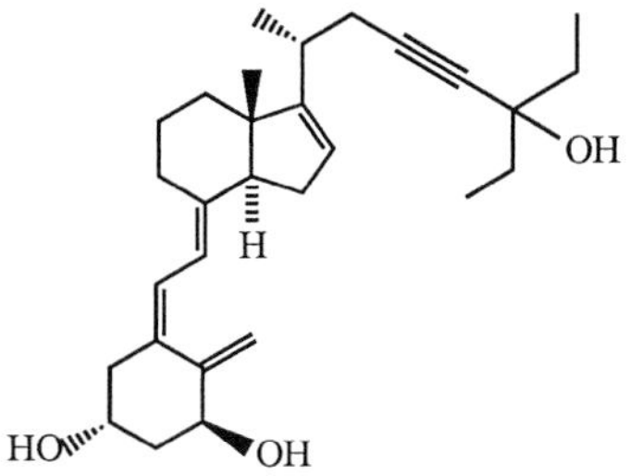

39

group may be a useful substitute for the 1α-hydroxyl function.

Removal of the 1α-hydroxyl group by replacement with hydrogen has been found to cause a dramatic reduction in gene transcriptional activities as measured by HL-60 cell differentiation [24,36]. Illustrative examples of these effects are presented in Fig. 2. See Section VI for more recent results.

Replacement of the 3β-hydroxyl group with hydrogen produces 3-deoxy analogs shown in Fig. 3. These compounds are approximately one order of magnitude less active than their parent 1α,3β-dihydroxy analogs but 5 to 100 times better tolerated *in vivo*.

The most interesting modification of the ring A moiety is removal of the 10-methylene group to furnish 19-nor analogs. In the case of the natural hormone $1,25(OH)_2D_3$ the 19-nor analog is equally active in cell differentiation but at least 10 times better tolerated *in vivo* [37,38]. A similar comparison was achieved with more active 16-ene derivatives as shown in Fig. 4 [17]. The 19-nor vitamin D series presently offer the most promising clinical candidates in several therapeutic applications of vitamin D compounds.

IV. MECHANISMS OF ANTIPROLIFERATIVE AND CELL DIFFERENTIATION ACTIVITIES

In addition to the differential metabolism and the rates of inactivation (discussed below), the mechanisms

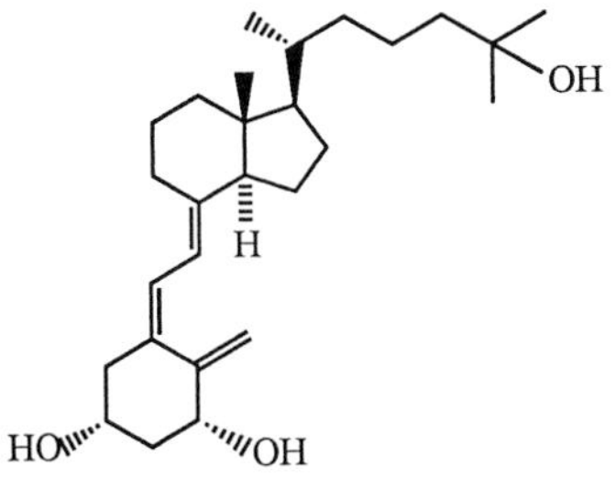

40

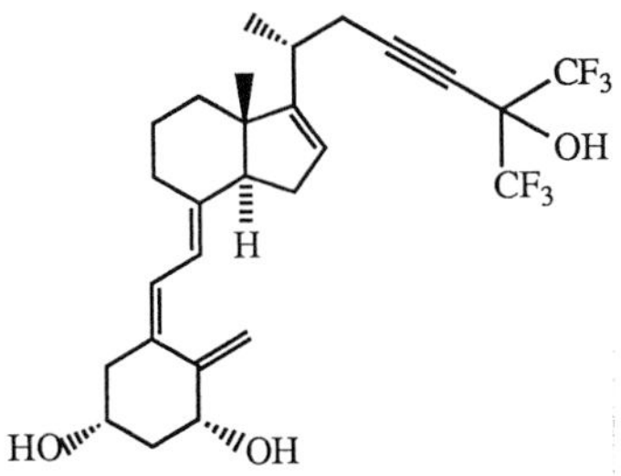

41

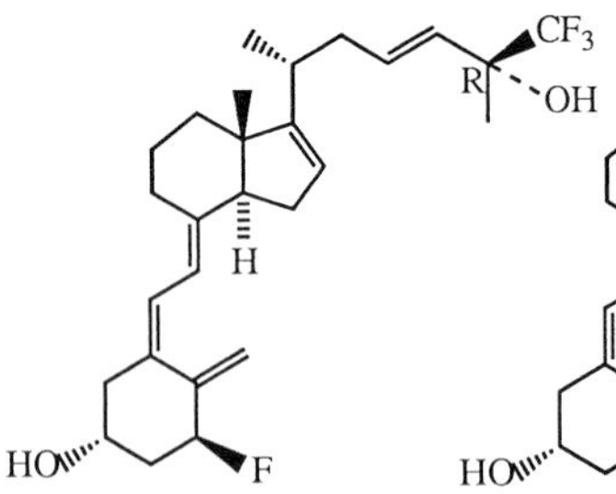

42 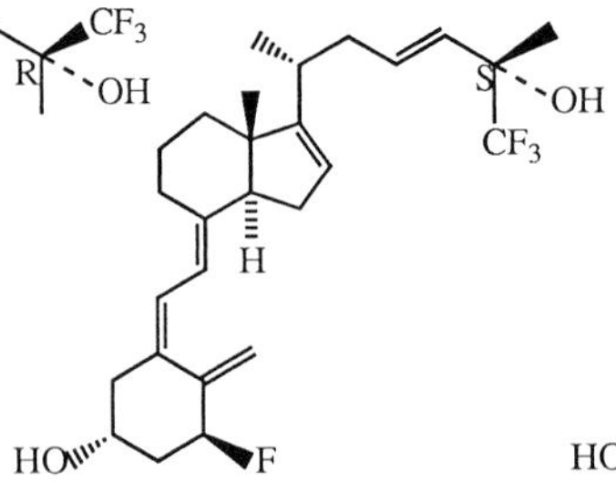 43 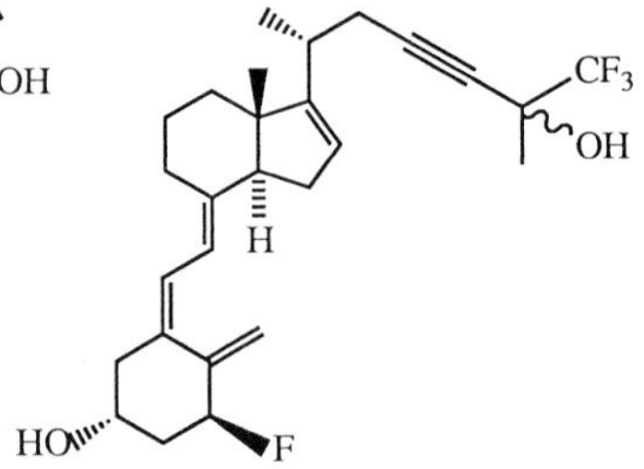 44

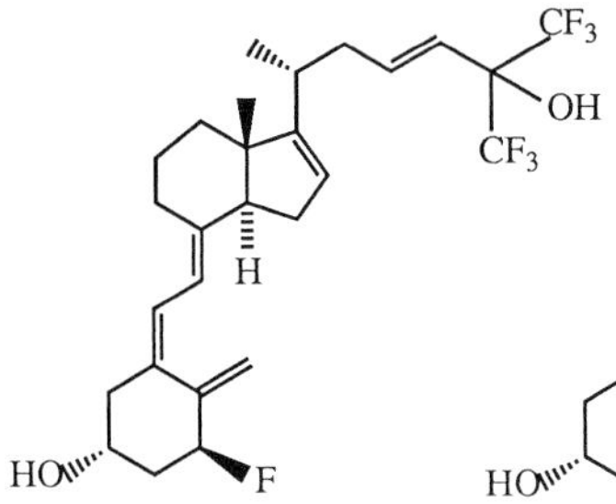

45 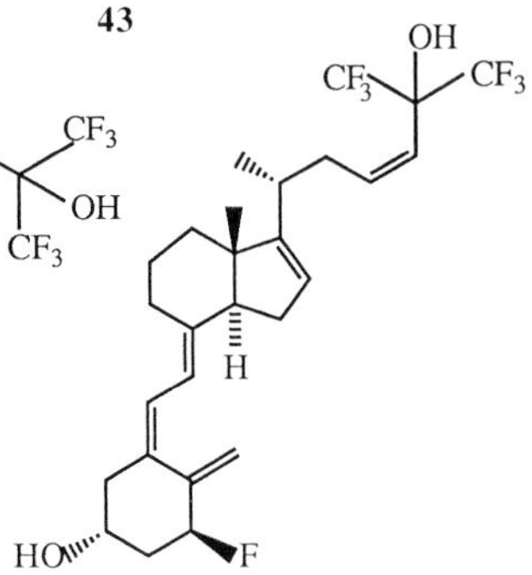 46 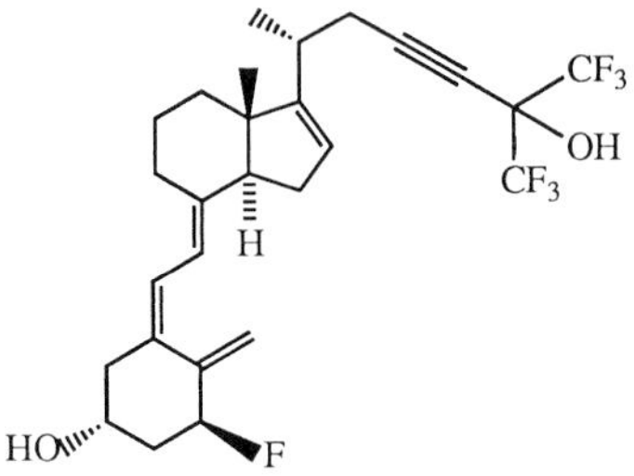 47

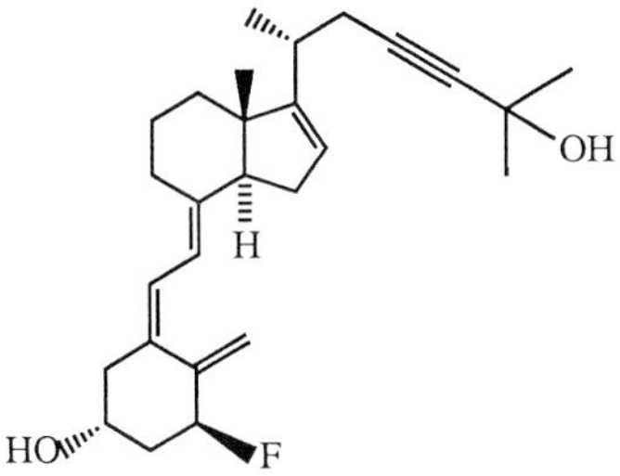

48 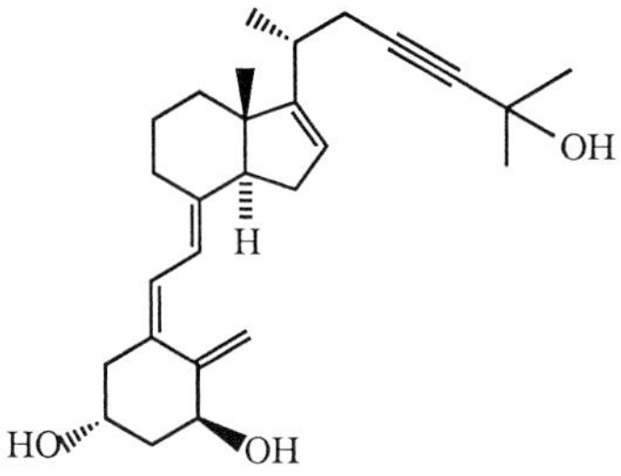 9

that are potentially responsible for the enhanced differentiation-inducing to hypercalcemia-inducing ratio of $1,25(OH)_2D_3$ analogs include (1) avidity of binding of the analogs to the serum vitamin D binding protein (DBP), (2) avidity of binding to the $1,25(OH)_2D_3$ specific nuclear receptor (VDR) in cells of different types, (3) ability to alter dimerization of the VDR with another VDR or with one of three RXR proteins, and to alter the interaction of the liganded VDR–VDR and VDR–RXR complexes with the DNA elements binding VDR (VDREs) and with the accessory proteins, and (4) cellular events downstream from the altered regulation of the primary target genes. These subjects are discussed in detail in Chapters 59 and 60.

A. Avidity of the 16-Ene Analogs for DBP

Some of the 16-ene analogs have been shown to bind less strongly to the plasma transport protein DBP than $1,25(OH)_2D_3$. For instance, $1,25(OH)_2$-16-ene D_3 (**2**) appears to bind to DBP with 1.8% of the avidity of $1,25(OH)_2D_3$ for this protein [6]. Therefore, more free ligand is able to enter the cells when **2** is administered. Under *in vitro* conditions the absence of serum will affect comparisons of the potency of an analog in a way that may have misleading implications for the *in vivo* situation.

B. Binding Affinity of the 16-Ene Analogs to VDR

Binding affinities of analogs as ligands for the isolated VDR protein from chick intestinal epithelium or from HL-60 cells have been measured and extensively tabulated by Bouillon *et al.* [25], who also list a number of binding affinities to DBP. Surprisingly, the strength of binding to VDR by an analog does not in general correlate with its potency. For instance, Munker *et al.* [8]

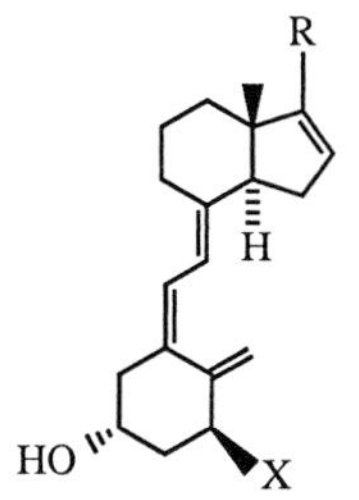

R	HL-60 cell differentiation IC_{50} 10^{-9}M	
	X = OH	X = F
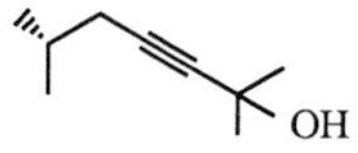	0.5 (**9**)	2.0 (**48**)
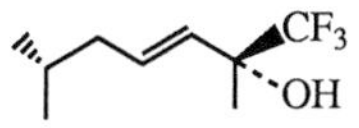	0.3 (**25**)	12.0 (**42**)
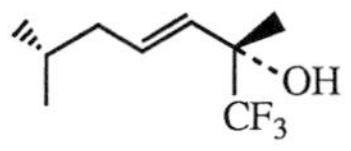	0.2 (**26**)	3.4 (43)
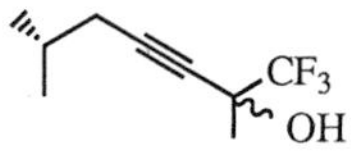	0.15 (**27**)	3.7 (**44**)
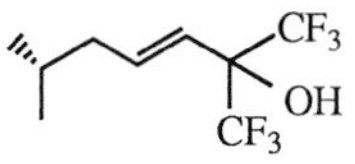	0.19 (**31**)	3.0 (**45**)
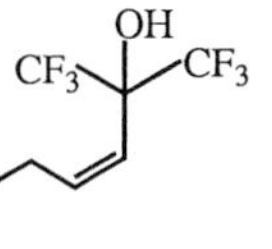	0.25 (**32**)	1.8 (**46**)
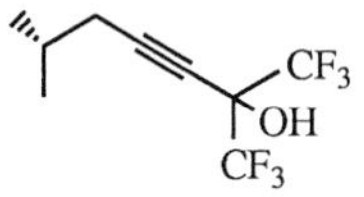	0.1 (**33**)	2.0 (**47**)

FIGURE 1 Effect of 1α-hydroxyl and 1α-fluoro derivatives on differentiation of HL-60 cells.

found that 25(OH)-16,23*E*-diene-26,27-F_6 D_3 (**53**) binds very poorly to VDR, approximately 400 times less well than 1,25$(OH)_2D_3$ (**1**), but was 2–3 times more potent than **1** in inhibition of clonal growth of HL-60 cells. Thus, the strength of VDR binding does not appear to be important for growth inhibition; however, in view of the low effect of **53** on calcium homeostasis in HL-60 cells found by Gardner *et al.* [39], it may be necessary for the activation of genes that up-regulate calcium influx pathways.

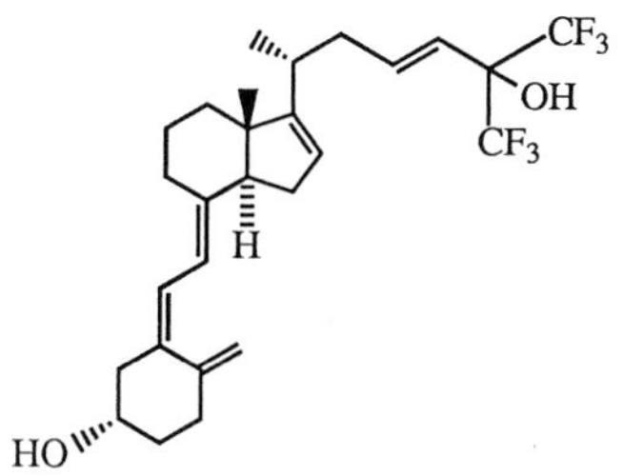

25(OH)-16,23E-diene-26,27-F_6 D_3 (**53**)

C. Effects of 16-Ene Analogs on Receptor Dimerization and DNA Binding

The mechanism of 1,25$(OH)_2D_3$ cellular activity involves binding to a unique nuclear vitamin D receptor (VDR). In addition to intestine, parathyroid gland, and kidney, the classic organs participating in mineral homeostasis, VDR receptor has been found in more than 30 tissues including prostate, skin, testes, breast, muscle, pancreas, endocrine glands, thymus, and bone marrow. The 1,25$(OH)_2D_3$–VDR complex binds to the responsive DNA elements (VDREs) of specific genes usually

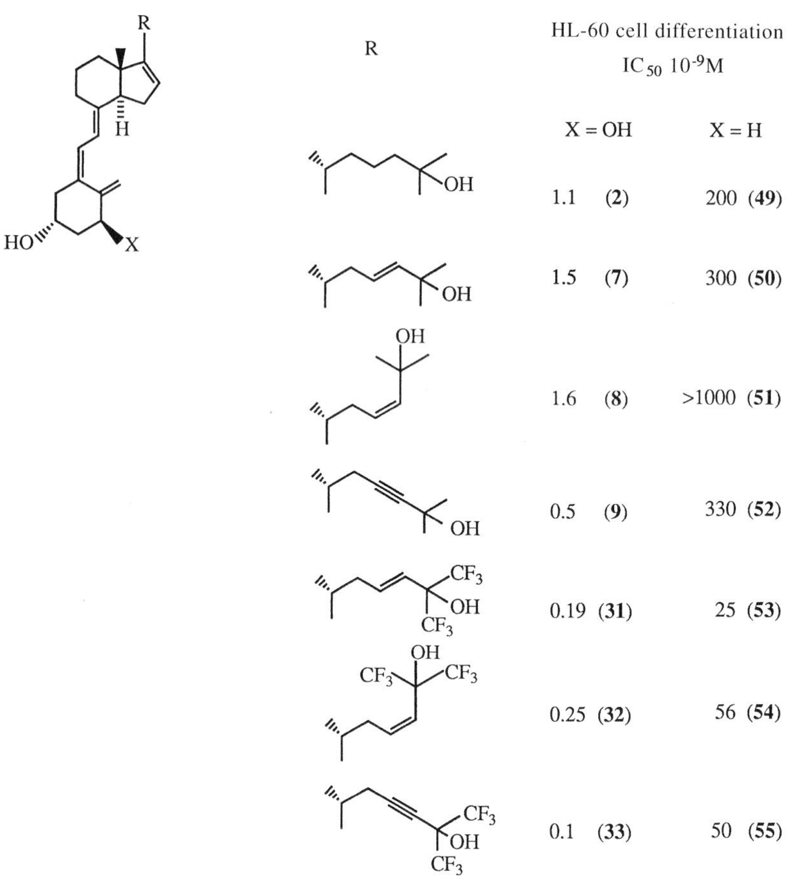

R
HO
X
H
HL-60 cell differentiation
IC_{50} 10^{-9}M
X = OH
X = H
OH
CF_3
1.1 (**2**)
200 (**49**)
1.5 (**7**)
300 (**50**)
1.6 (**8**)
>1000 (**51**)
0.5 (**9**)
330 (**52**)
0.19 (**31**)
25 (**53**)
0.25 (**32**)
56 (**54**)
0.1 (**33**)
50 (**55**)

FIGURE 2 Effect of 1α-hydroxyl and 1α-deoxy derivatives on differentiation of HL-60 cells.

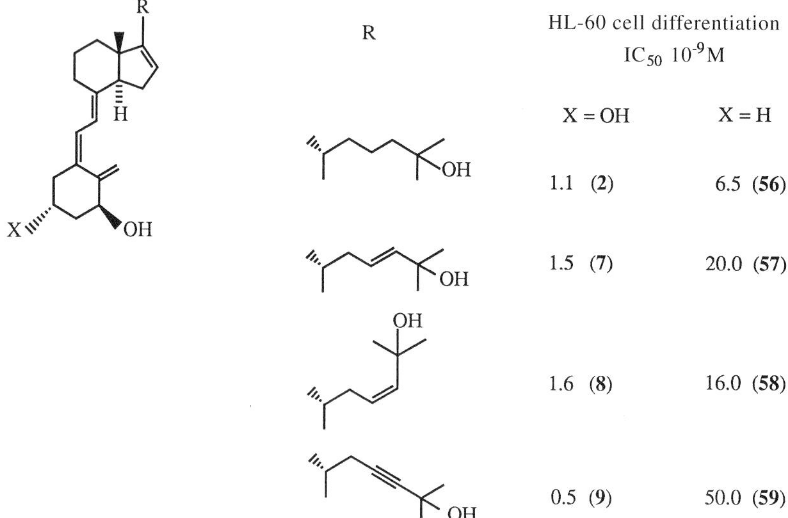

R
X
OH
H
HL-60 cell differentiation
IC_{50} 10^{-9}M
X = OH
X = H
1.1 (**2**)
6.5 (**56**)
1.5 (**7**)
20.0 (**57**)
1.6 (**8**)
16.0 (**58**)
0.5 (**9**)
50.0 (**59**)

FIGURE 3 Effect of 3β-hydroxyl and 3-deoxy analogs on differentiation of HL-60 cells.

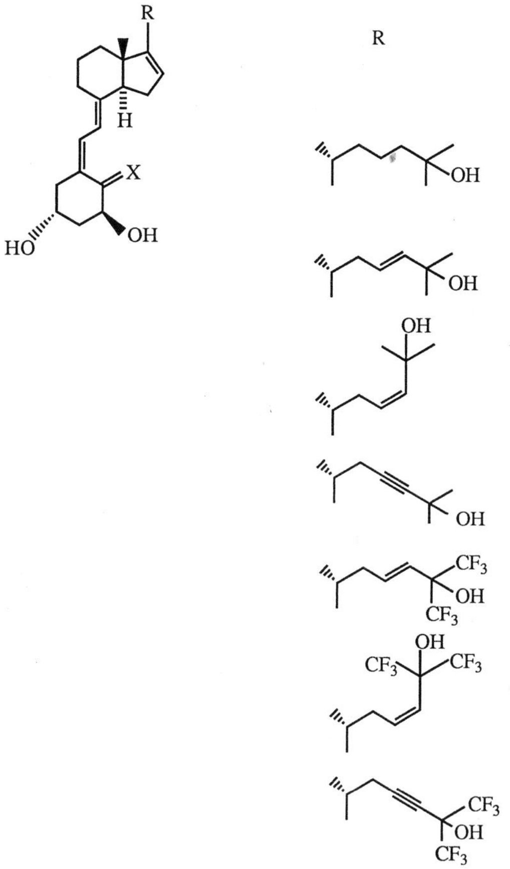

R	HL-60 cell differentiation IC_{50} $10^{-9}M$ X = OH	$X = H_2$
	1.1 (**2**)	0.7 (**60**)
	1.5 (**7**)	2.0 (**61**)
	1.6 (**8**)	1.4 (**62**)
	0.5 (**9**)	1.5 (**63**)
	0.19 (**31**)	0.3 (**64**)
	0.25 (**32**)	0.26 (**65**)
	0.1 (**33**)	0.2 (**66**)

FIGURE 4 Effect of removal of the 10-methylene group on differentiation of HL-60 cells.

as a preformed heterodimer with retinoid X receptor (RXR). These VDREs are typically located upstream of promoters of responsive genes, and their interactions with the ligand result in transcriptional activation or repression. A number of 16-ene vitamin D analogs have been described above that uncouple growth inhibitory and differentiation-inducing activities from hypercalcemic effects. Because of structural modifications, these analogs may undergo or resist the well-recognized metabolic modifications of $1,25(OH)_2D_3$. In some cases metabolites of the analogs can be more suitable than the parent compounds to participate in the events leading to specific gene transcriptional activation or repression [40]. The distinct structural features of analogs, or of their metabolites acting as ligands, induce conformational changes on the VDR–RXR heterodimer bound to DNA of the target genes, which are different than those produced by $1,25(OH)_2D_3$. The results of these conformational effects are often an increased affinity of the VDR–RXR complexes for VDRE and ultimately higher transcriptional activity of the target genes [41].

Increased transactivation by 16-ene analogs of a VDRE reporter gene transfected into murine osteosarcoma cells and monkey kidney cells was noted by Ferrara *et al.* [42]. However, there was no correlation between the affinities of these analogs for VDR and the ability to transactivate gene transcription once bound to the ligand. This lack of parallelism between the affinity of the 16-ene analogs for VDR and the ability to mediate its biological activity was confirmed using HL-60 cells [43]. Thus, other factors appear to be important for the regulation of gene activity by 16-ene analogs.

The Freedman laboratory has made a major contribution to this area of research. Using the technique of surface plasmon resonance they compared the effects of three 16-ene analogs with $1,25(OH)_2D_3$ on VDR–RXR heterodimerization and DNA binding [44]. Interestingly, $1,25(OH)_2D_3$ was the strongest inducer of heterodimerization among the ligands tested, but $1,25(OH)_2D_3$-liganded VDR–RXR was bound to VDRE with the lowest stability. It was suggested that the introduction of a strongly nucleophilic group, such as the one provided by the hexafluro group in analog **33**, into the hydrophobic pocket in the ligand binding domain of VDR results in changes in the receptor that increase its affinity for the VDRE.

D. Events Downstream to the Interaction with Vitamin D Response Elements of Liganded VDR Heterodimers

What happens after interaction of VDREs with liganded VDR–RXR heterodimers is a largely unexplored area. Among the few reports regarding the specific genes regulated by 16-ene analogs in a manner different from 1,25(OH)$_2$D$_3$ (**1**) are the studies of Zhou *et al.* [1,31] showing that analogs **9** and **33** down-regulate c-*myc* expression more markedly than does 1,25(OH)$_2$D$_3$ in HL-60 cells. As the down-regulation of a c-*myc* by 1,25(OH)$_2$D$_3$ has been shown to be an early effect of 1,25(OH)$_2$D$_3$, and is associated with a decline in DNA replication, these results provide a potential mechanism for the more potent effects of the 16-ene analogs to inhibit the clonogenic growth of HL-60 cells.

Munker *et al.* [8] reported that 1,25(OH)$_2$D$_3$ (**1**) and 16-ene analogs induce the expression of a protein inhibitor of the cell cycle progression p21 WAF1/CIP1 in HL-60 cells, and Liu *et al.* [45] found that 1,25(OH)$_2$D$_3$ (**1**) transcriptionally activates the expression of this gene in U937 (human histiocytic lymphoma) cells. However, both the data in the above reports and the report by Wang *et al.* [46] show that increased expression of p21 WAF1/CIP1 peaks early and is not fully maintained when the cell cycle arrest is evident. Wang *et al.* [46] found an upregulation of p27 KIP1, another inhibitor of cyclin-dependent kinases which regulates the G$_1$ to S phase transition, to parallel the 1,25(OH)$_2$D$_3$-induced G$_1$ arrest. These exciting studies of cell cycle control by 1,25(OH)$_2$D$_3$ and its analogs are likely to bring new insights into the mechanisms of action of these compounds.

V. METABOLISM OF 16-ENE VITAMIN D$_3$ ANALOGS

From prior discussions it becomes clear that the introduction of the 16-ene modification into the D ring of the hormone 1,25(OH)$_2$D$_3$ alters significantly the ability of the molecule to interact with the classic vitamin D interacting proteins, namely, vitamin D binding protein (DBP) and vitamin D receptor (VDR). Therefore, it may also be anticipated that there will be some differences in the way 16-ene D$_3$ analogs interact with other vitamin D related proteins, namely, the enzymes that metabolize vitamin D. This altered interaction with vitamin D metabolizing enzymes in turn may lead to some specific differences in the metabolism of 16-ene D$_3$ analogs from the well established metabolic pathways of 1,25(OH)$_2$D$_3$ (also see Chapter 58). Before we begin to understand the differences that exist in the pathways of metabolism between these analogs and the native hormone 1,25(OH)$_2$D$_3$ it is important to review briefly target tissue metabolism of 1,25(OH)$_2$D$_3$ (**1**).

Since the mid-1970s, following the discovery of the steroid hormone 1,25(OH)$_2$D$_3$, there has been great interest in isolating and identifying the various natural metabolites of both the hormone and its precursor, 25(OH)D$_3$. As a result, to date more than 30 natural metabolites of vitamin D$_3$ are known to exist, and they are formed in almost all the target tissues that possess receptors for 1,25(OH)$_2$D$_3$ and respond to the hormone [25]. This subject is discussed in detail by Horst and Reinhardt in Chapter 2 of this book. The major known natural metabolites of 1,25(OH)$_2$D$_3$ are depicted in Scheme 1. All the metabolites are formed as a result of multiple side chain oxidations, which can be divided into three pathways. The C-24 oxidation pathway initiated by C-24 hydroxylation leads to the formation of calcitroic acid, which is the major end product of the hormone [47,48]. This pathway has been extensively studied, and 24-hydroxylase, the key enzyme of the pathway, has been cloned (discussed in Chapter 6). The regulation of 24-hydroxylase at the molecular level has been extensively investigated, and it was also found that the same enzyme catalyzes 24-hydroxylation, 24-oxidation, and subsequent 23-hydroxylation of the C-24 oxo compound [49,50]. The remaining two pathways initiated by C-23 and C-26 hydroxylations lead to the formation of the final product calcitriol lactone [51]. Little is known regarding the enzymology of these latter pathways. At present it is the general belief that the only major function of the target tissue metabolism of 1,25(OH)$_2$D$_3$ (**1**) is to inactive the hormone. However, the possibility of the intermediary metabolites possessing biological activities of their own cannot be ruled out. In this regard, calcitriol lactone was found to have unique biological activities that are different from its parent hormone. This natural metabolite has been shown to decrease serum calcium and promote bone formation [52]. Thus, the importance of understanding the biological activities of the various natural intermediary metabolites of the hormone as well as its analogs should not be underestimated.

A. Target Tissue Metabolism of 1α,25-Dihydroxy-16-ene-vitamin D$_3$

A comparative metabolism study between 1α,25(OH)$_2$-16-ene D$_3$ (**2**) and 1,25(OH)$_2$D$_3$ (**1**) in rat

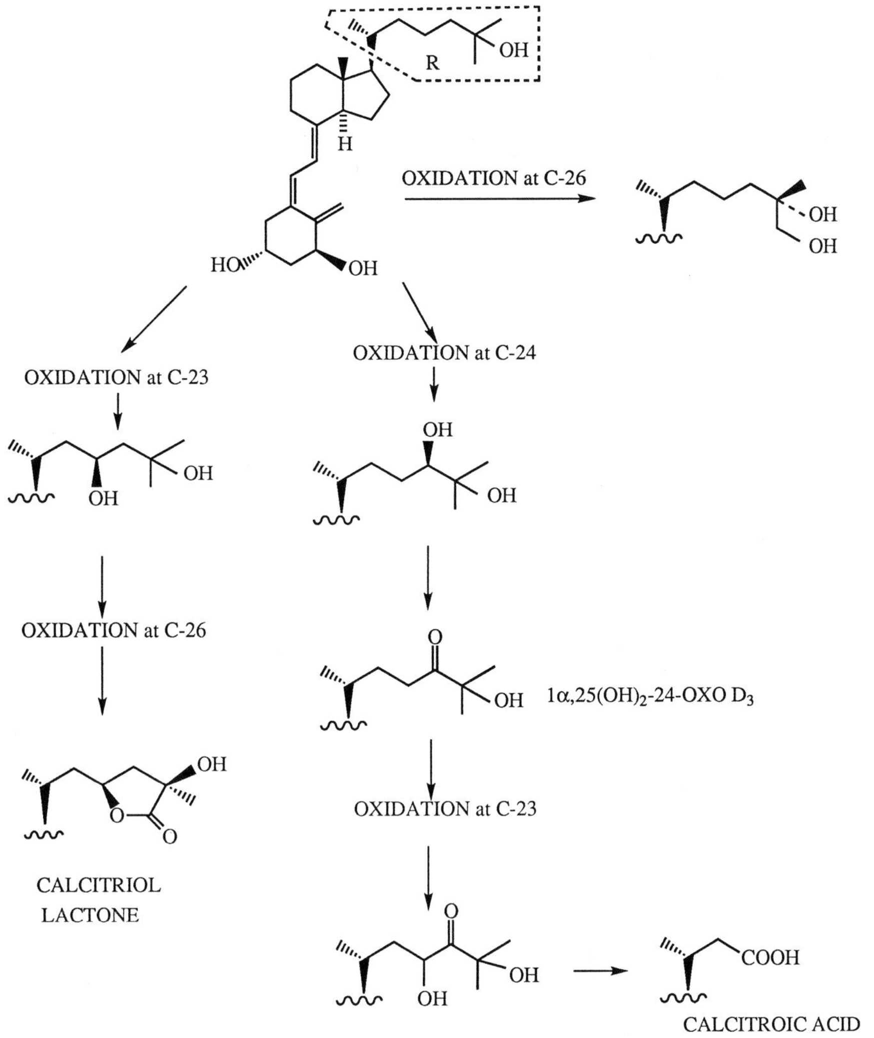

OXIDATION at C-26

OXIDATION at C-23

OXIDATION at C-24

OXIDATION at C-26

$1\alpha,25(OH)_2$-24-OXO D_3

OXIDATION at C-23

CALCITRIOL LACTONE

CALCITROIC ACID

SCHEME 1 Metabolism of $1,25(OH)_2D_3$.

kidneys was performed using the isolated perfused rat kidney system, which has been widely used previously to study target tissue metabolism of $1,25(OH)_2D_3$ (**1**).

1. Isolated Perfused Rat Kidneys

The metabolic fates of both $1,25(OH)_2D_3$ (**1**) and $1\alpha,25(OH)_2$-16-ene D_3 (**2**) were studied in isolated perfused rat kidneys by introducing 400 nmol of each compound into 100 ml of the perfusate, and each kidney perfusion was continued for 8 hr [53]. The lipid extract of the 50 ml of final perfusate was subjected directly to high-performance liquid chromatography (HPLC) using the chromatographic conditions described in the legend to Fig. 5. The various vitamin D metabolites were monitored by a photodiode array detector, and the metabolites were identified based on the elution positions of the previously known standards. The unknown metabolites of $1\alpha,25(OH)_2$-16-ene D_3 (**2**) were identified through mass spectrometry and specific chemical reactions as described previously [53]. From these studies it became obvious that both $1,25(OH)_2D_3$ (**1**) and $1\alpha,25(OH)_2$-16-ene D_3 (**2**) are metabolized through the C-24 oxidation pathway and are converted to their respective 24-hydroxyl and 24-oxo metabolites. However, as expected, whereas $1\alpha,25(OH)_2$-24-oxo D_3, the natural metabolite of the hormone, is further metabolized to $1\alpha,23,25(OH)_3$-24-oxo D_3 as a result of C-23 hydroxylation, $1\alpha,25$-$(OH)_2$-16-ene-24-oxo D_3 (**71**), the metabolite of the analog, appears to resist C-23 hydroxylation (Scheme 2). As a result, $1\alpha,25(OH)_2$-24-oxo-16-ene D_3 (**71**) accumulates in increasing amounts when compared to $1\alpha,25(OH)_2$-24-oxo D_3 (Fig. 6). In a later study [54] of the metabolic pathway of $1\alpha,25(OH)_2$-16-ene D_3 (**2**) in isolated perfused rat kidney the actual formation of $1\alpha,23,25(OH)_3$-16-ene-24-oxo D_3 (**73**) as a minor metabolite of **2** was characterized by on-line

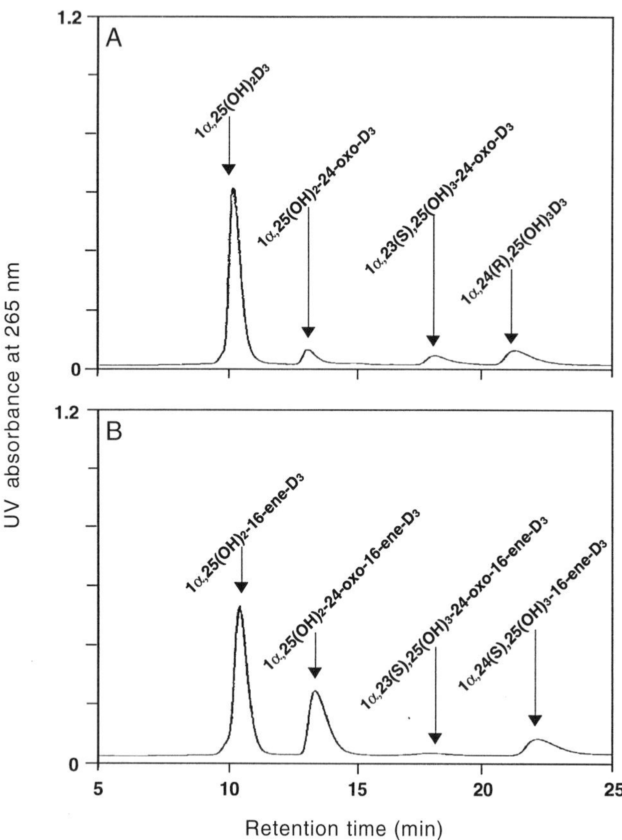

FIGURE 5 HPLC profiles of various vitamin D metabolites produced in isolated rat kidneys perfused for 8 hr with either the hormone 1,25$(OH)_2D_3$ (**1**) (A) or the analog 1α,25$(OH)_2$-16-ene D_3 (**2**) (B). Lipid extracts from 50 ml of perfusate were analyzed using a straight phase HPLC system. Chromatographic conditions: Zorbax Sil column (25 cm × 4.6 mm); 10% 2-propanol in hexane, flow rate 2 ml/min.

HPLC–electrospray tandem mass spectrometry. These observations indicated that 1α,25$(OH)_2$-16-ene-24-oxo D_3 (**71**), unlike 1α,25$(OH)_2$-24-oxo D_3, resists 23-hydroxylation and is converted to 1α,23,25$(OH)_3$16-ene-24-oxo D_3 (**73**) at a very slow rate. As a consequence, the metabolite **71** accumulates in the kidney perfusate. In a following study [40], the metabolism of 1α,25$(OH)_2$-16-ene D_3 (**2**) was compared with that of 1α,25$(OH)_2D_3$ using human myeloid leukemic cells (RWLeu-4 cells) as a model cell line.

2. Human Myeloid Leukemic Cells

Human myeloid leukemia (RWLeu-4) cells were grown in roller bottles in a minimal medium (300 ml and 10% fetal calf serum) in a 5% CO_2 atmosphere at 37°C. Both compounds (150 nmol each) were added to the cultures when the cells reached saturation density. The cells were incubated with each compound in 300 ml of medium for 24 hr. Lipid extracts of the medium and the cells were analyzed by HPLC using the straight phase HPLC system, using a Zorbax-Sil column eluted with 6% 2-propanol in hexane (2 ml/min). Figure 6 shows the HPLC profiles of the substrate and the metabolites produced. All the major metabolites of both 1,25$(OH)_2D_3$ (**1**) and 1α,25$(OH)_2$-16-ene D_3 (**2**) were purified on several HPLC systems and were identified by the techniques of diode array spectrophotometry and mass spectrometry with the aid of specific chemical reactions. The results suggested that analog **2** is metabolized via the C-24 oxidation pathway in a manner analogous to the hormone 1,25$(OH)_2D_3$ (**1**), as illustrated in Fig. 6. However, it is significant to note that whereas the unmetabolized substrate concentrations of 1α,25$(OH)_2$-16-ene D_3 (**2**) and 1,25$(OH)_2D_3$ (**1**) are almost equal, the amounts of intermediary metabolites produced from **2** are different from the amounts of the corresponding intermediary metabolites produced from **1**. The difference is more significant in the case of the intermediary metabolite 1,25$(OH)_2$-16-ene-24-oxo D_3 (**71**), the concentration of which is about 15-fold higher than that of the corresponding natural intermediary metabolite, 1α,25$(OH)_2$-24-oxo D_3. In addition, the identification of 1,24(S),25$(OH)_3$-16-ene D_3 (**72**) in this study is also significant, as it is known that the natural configuration of C-24 oxidation product is 24(R), whereas the 24(S)-hydroxyl stereoisomer has been shown to be derived from the reduction of the 24-oxo group. Thus, as shown in Scheme 2, it appears that as a result of a partial block in the conversion of 16-ene-24-oxo D_3 (**71**) to 1,23,25$(OH)_3$-16-ene-24-oxo D_3 (**73**), the formation of 1α,24(S),25$(OH)_3$-16-ene D_3 (**72**) takes place as an alternate pathway for inactivation of the analog 1α,25$(OH)_2$-16-ene D_3 (**2**). The metabolism studies in RWLeu-4 cells indicated that the intermediary metabolite 1α,25$(OH)_2$-16-ene-24-oxo D_3 (**71**) accumulates in the cells as a predominant metabolite, and the concentration of this metabolite exceeds even the concentration of the starting substrate.

From the target tissue metabolism studies of 1α,25$(OH)_2$-16-ene D_3 (**2**) in both rat kidney and RWLeu-4 cells, it becomes obvious that a modification in the D-ring structure such as insertion of a double bond between C-16 and C-17 possibly produces a conformational change in the side chain, which contributes to efficient C-24 hydroxylation and C-24 oxidation, as in the case of the native hormone, but not to the 23-hydroxylation.

B. Target Tissue Metabolism of 1α,25-Dihydroxy-16-ene-23-yne-vitamin D_3

Of all the 16-ene analogs, as discussed in this chapter, 1α,25$(OH)_2$-16-ene-23-yne D_3 (**9**) was studied most ex-

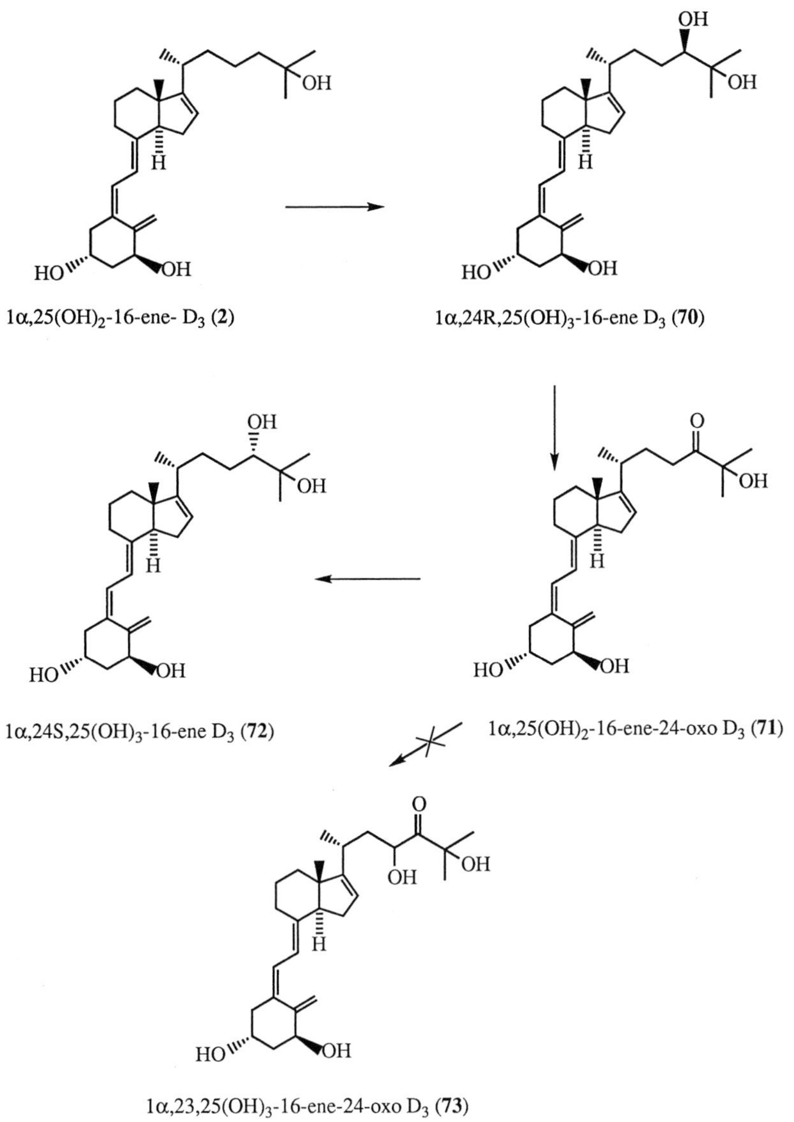

SCHEME 2

tensively in different biological systems because of its great potential to become a drug for the treatment of leukemia. Presently, there is a paucity of information with regard to the target tissue metabolism of this important vitamin D analog.

Indirect evidence for the possibility that the target tissue metabolism pathways for this analog are not the same in different target tissues came from a study [55] which was designed to demonstrate the concept that certain vitamin D analogs can generate selective and different biological actions in different target tissues. In this study, intestinal calcium absorption, bone calcium resorption, induction of intestinal and renal calcium binding proteins (CaBP), and occupancy of the intestinal and renal nuclear $1,25(OH)_2D_3$ receptor (VDR) in vitamin D-deficient chicks after a single dose of either $1,25(OH)_2D_3$ (**1**) or $1\alpha,25(OH)_2$-16-ene-23-yne D_3 (**9**) were evaluated. The $1,25(OH)_2D_3$ doses (0.075–1.2 nmol) generated responses in intestinal calcium absorption, bone calcium resorption, intestinal CaBP, and renal CaBP. When $1,25(OH)_2$-16-ene-23-yne D_3 (**9**) (1.2–300 nmol) was administered, increases in bone calcium resorption and renal CaBP were noted. However, a significant response in intestinal calcium absorption and intestinal CaBP appeared only after a 300-nmol dose. Unoccupied VDR in the intestine and kidney was determined *in vivo* after dosing with $1,25(OH)_2D_3$ (**1**) and $1\alpha,25(OH)_2$-16-ene-23-yne D_3 (**9**). Doses (0.25–6.0 nmol) of $1,25(OH)_2D_3$ (**1**) reduced unoccupied receptor to 24% in intestine and 13% in kidney. $1\alpha,25(OH)_2$-16-ene-23-yne D_3 (**9**) (6.0–600 nmol) decreased unoccupied receptor significantly in kidney; in intestine, however,

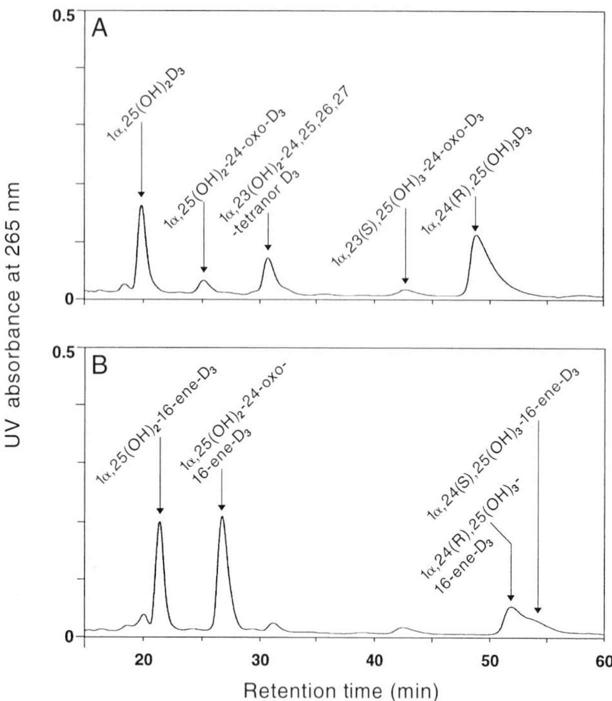

FIGURE 6 HPLC profiles of various vitamin D metabolites produced in RWLeu-4 cells incubated with either the hormone $1,25(OH)_2D_3$ (**1**) (A) or the analog $1\alpha,25(OH)_2$-16-ene D_3 (**2**) (B). Lipid extracts were analyzed using a straight phase HPLC system. Chromatographic conditions: Zorbax-Sil column (25 cm × 4.6 mm); 10% 2-propanol in hexane, flow rate 2 ml/min.

the analog **9** reduced unoccupied receptor only to 75%. These results confirm that this vitamin D analog can generate selective biological responses because of different levels of target organ receptor occupancy. These studies provide indirect evidence for the possibility of rapid inactivation of $1\alpha,25(OH)_2$-16-ene-23-yne D_3 (**9**) in intestine in contrast to kidney and bone in which the analog is inactivated slowly and differently. This can lead to increases in the concentration of analog in kidney and bone when compared to intestine, resulting in blunting of the appearance of a biological response in intestine while a better biological response is obtained in kidney and bone. Thus, from these studies, it may be hypothesized that the analog **9** is metabolized differently in different target tissues, producing different spectra of biological activities. However, this hypothesis has to be viewed with caution until more direct evidence becomes available.

A preliminary study [56] was performed to determine the tissue accumulation and metabolism of $1\alpha,25(OH)_2$-16-ene-23-yne D_3 (**9**) relative to $1,25(OH)_2D_3$ (**1**) in BALB/c leukemic mice tissues and WEHI-3 cells, as this analog has been shown to induce cell differentiation and inhibit cell proliferation in WEHI-3 cells up to 7-fold more efficiently than $1,25(OH)_2D_3$ (**1**) when injected into BALB/c mice. Pharmacologic doses of 1,25-$[27,27\text{-}^3H](OH)_2D_3$ and $1\alpha,25\text{-}[25\text{-}^{14}C](OH)_2$-16-ene-23-yne D_3 were either injected intravenously into BALB/c mice or incubated with WEHI-3 cells in culture. Lipid extracts of the tissues and the cell cultures were analyzed by normal phase HPLC. Results from monitoring the serum levels of the analog and the hormone in leukemic BALB/c mice indicated differential catabolism of these two compounds. The approximate half-lives of $1\alpha,25(OH)_2$-16-ene-23-yne D_3 (**9**) and $1,25(OH)_2D_3$ (**1**) are 33 min and 4 hr, respectively. This is the first study to show the rapid clearance of $1\alpha,25(OH)_2$-16-ene-23-yne D_3 (**9**) from plasma due to its low binding affinity to DBP, and this property appears to be common to most of the analogs that were found to be therapeutically useful. For example, calcipotriol and 22-oxo-calcitriol have both exhibited similar properties. Tissue analysis in BALB/c mice and lipid extracts of WEHI-3 cells indicated further metabolism of the $1\alpha,25(OH)_2$-16-ene-23-yne D_3 (**9**) into polar metabolites, but no attempt has been made to identify the metabolites. The extent of metabolism of these compounds was unclear.

In a different study [57] the metabolism of $1\alpha,25(OH)_2$-16-ene-23-yne-vitamin D_3 (**9**) was investigated in the isolated perfused rat kidney system. It was noted that this analog, unlike $1\alpha,25(OH)_2$-16-ene D_3 (**2**) is resistant to further metabolism and is converted to only a single polar metabolite, which was identified unequivocally as $1\alpha,25,26(OH)_3$-16-ene-23-yne D_3 through mass spectrometry and its sensitivity to sodium periodate ($NaIO_4$). Thus, evidence was provided, at least in rat kidney, that $1\alpha,25(OH)_2$-16-ene-23-yne D_3 (**9**) is very stable. It resists side chain oxidation because of the presence of the triple bond between carbons 23 and 24, which prevented hydroxylations at both C-24 and C-23. It is of interest to note that C-26 hydroxylation is still possible, but at this point the metabolic fate and the biological activity of $1\alpha,25,26(OH)_3$-16-ene-23-yne D_3 is still unknown. However, it has to be realized that the data presented here with respect to rat kidney may not be universal to all other target tissues. It is possible that this compound may be metabolized differently and rapidly into unidentified metabolites in other tissues such as liver.

C. Biological Activity Studies of 1α,25-Dihydroxy-16-ene-24-oxo-vitamin D_3, the Major Intermediary Metabolite of 1α,25-Dihydroxy-16-ene-vitamin D_3

There have been several other studies [58–60] investigating the target tissue metabolism of different vitamin

D analogs such as calcipotriol, 22-oxacalcitriol, and 20-epi-1α,25(OH)$_2$D$_3$ that were shown to have great promise for clinical use. One compound especially, calcipotriol, is already in clinical use for the treatment of psoriasis, as discussed elsewhere in this book (see Chapter 72). The metabolism of these unique compounds have been extensively studied in the laboratory of Glenville Jones, using cultured cell models representing liver, keratinocytes, and osteoblasts (Chapter 58). It was demonstrated that all of these analogs are extensively metabolized in these tissues and are degraded into a variety of hydroxylated and side chain truncated metabolites. The biological activities of all the principal intermediary metabolites of these compounds formed in the target tissues before their final inactivation were assessed using a growth hormone reporter gene, transcription activation system, and a vitamin D receptor assay. It was reported that the biological activities of all the intermediary metabolites were lower when compared to the respective parent analogs. From these extensive target tissue metabolism and biological activity studies, these investigators have developed the viewpoint that all the intermediary metabolites of analogs produced in target tissues are indeed catabolites possessing minimal biological activity. Therefore, it has been proposed that the biological activity expressed by a given analog is due to the parent analog itself rather than its metabolite or metabolites. For more discussion of this subject, we refer the reader to Chapter 58 by Jones. However, this viewpoint does not seem to hold true with respect to the analog 1α,25(OH)$_2$-16-ene D$_3$ (**2**), as one of its intermediary metabolites, 1α,25(OH)$_2$-16-ene-24-oxo D$_3$ (**71**) accumulates significantly in the target tissues during the process of its further metabolism. Therefore, it is important to determine the biological activity of 1α,25(OH)$_2$-16-ene-24-oxo D$_3$ (**71**) in order to determine the extent of contribution of this intermediary metabolite to the overall biological activity of the parent analog 1α,25(OH)$_2$-16-ene D$_3$ (**2**) itself.

1. *In Vitro* Biological Activity of 1α,25-Dihydroxy-16-ene-24-oxo-vitamin D$_3$

1,25(OH)$_2$D$_3$ (**1**), 1α,25(OH)$_2$-16-ene D$_3$ (**2**), and its 24-oxo metabolite (**71**) inhibit the proliferation of RWLeu-4 cells in a dose-dependent manner. However, the effective dose of 1,25(OH)$_2$D$_3$ (**1**) needed in these cells to produce the same effect as the analog **2** and its 24-oxo metabolite **71** is significantly higher. The dose of 1,25(OH)$_2$D$_3$ (**1**) to reduce RWLeu-4 cell growth by 50% (ED$_{50}$) was approximately 1.5 n*M*, whereas the ED$_{50}$ of both 16-ene D$_3$ (**2**) and 16-ene-24-oxo D$_3$ (**71**) was approximately 0.1 n*M*. Later, the ability of the vitamin D compounds to induce differentiation of myeloid leukemic cells into monocytic cells was evaluated at a dose of 0.32 n*M*. At this dose, 1,25(OH)$_2$D$_3$ (**1**) enabled 40% of RWLeu-4 cells to differentiate, whereas a similar dose of 1α,25(OH)$_2$-16-ene D$_3$ (**2**) or its metabolite 1α,25(OH)$_2$-16-ene-24-oxo D$_3$ (**71**) induced differentiation of 80–90% of cells. As the cell growth regulating and differentiating activities of 1,25(OH)$_2$D$_3$ and its analogs are primarily receptor mediated, VDR-mediated transactivation activities of each compound were assessed by measuring this ability to induce the growth hormone production in ROS 17/2.8 cells transfected with an osteocalcin VDRE–growth hormone gene construct. Vitamin D receptor-mediated transcriptional activity through the osteocalcin VDRE was dose dependent for 1,25(OH)$_2$D$_3$ (**1**) as well as for 1α,25(OH)$_2$-16-ene D$_3$ (**2**) and its metabolite 1α,25(OH)$_2$-16-ene-24-oxo D$_3$ (**71**). The ED$_{50}$ for transcriptional activity of 1,25(OH)$_2$D$_3$ was 5×10^{-10} *M*, the ED$_{50}$ for 1α,25(OH)$_2$-16-ene D$_3$ was 6×10^{-12} *M*, and the ED$_{50}$ for 1α,25(OH)$_2$-16-ene-24-oxo D$_3$ was 3×10^{-11} *M*. Thus, 1α,25(OH)$_2$-16-ene D$_3$ (**2**) and its metabolite 1α,25(OH)$_2$-16-ene-24-oxo D$_3$ (**71**) were significantly more potent inducers of VDR-mediated transcription than the natural hormone 1,25(OH)$_2$D$_3$ (**1**).

2. *In Vivo* Biological Activity of 1α,25-Dihydroxy-16-ene-24-oxo-vitamin D$_3$

In a previous study [61] it was shown that the steroid hormone 1,25(OH)$_2$D$_3$, when given *in vivo*, prevents the induction of murine autoimmune diseases such as experimental autoimmune encephalomyelitis (EAE) [62]. The hormone inhibits both the cellular and humoral responses when administered prior to induction or early in the disease process (see Chapter 69). We have taken advantage of the EAE model to evaluate the immunosuppressive properties of 1α,25(OH)$_2$-16-ene-24-oxo D$_3$ (**71**) in comparison to its parent analog **2** and 1,25(OH)$_2$D$_3$ (**1**). Naive SJL/J mice, 4 weeks old, were immunized with neuroantigen in adjuvant to induce experimental autoimmune encephalitis. Treatment with 1α,25(OH)$_2$-16-ene-24-oxo D$_3$ (**71**) was given at 0.05, 0.15, and 0.3 μg i.v. on alternate days starting 3 days prior to and for up to 5 days postimmunization, and compared to a similar treatment with 0.1 μg 1,25(OH)$_2$D$_3$ (**1**) or the parent analog 1α,25(OH)$_2$-16-ene D$_3$ (**2**). Suppression of EAE was obtained with 0.15 μg 1α,25(OH)$_2$-16-ene-24-oxo D$_3$ (**71**), comparable to the suppression with the parent compound **2** and more potent than 1,25(OH)$_2$D$_3$ (**1**). However, no hypercalcemia was seen in mice treated with 0.5 μg of oxo metabolite **71**, in contrast to 1α,25(OH)$_2$D$_3$ (**1**) and 1α,25(OH)$_2$-16-ene D$_3$ (**2**). These results suggested that 1α,25(OH)$_2$-16-ene-24-oxo D$_3$ (**71**) is also active *in vivo*, as is its parent analog, in terms of its ability to exert immunosuppressive activity without causing hypercal-

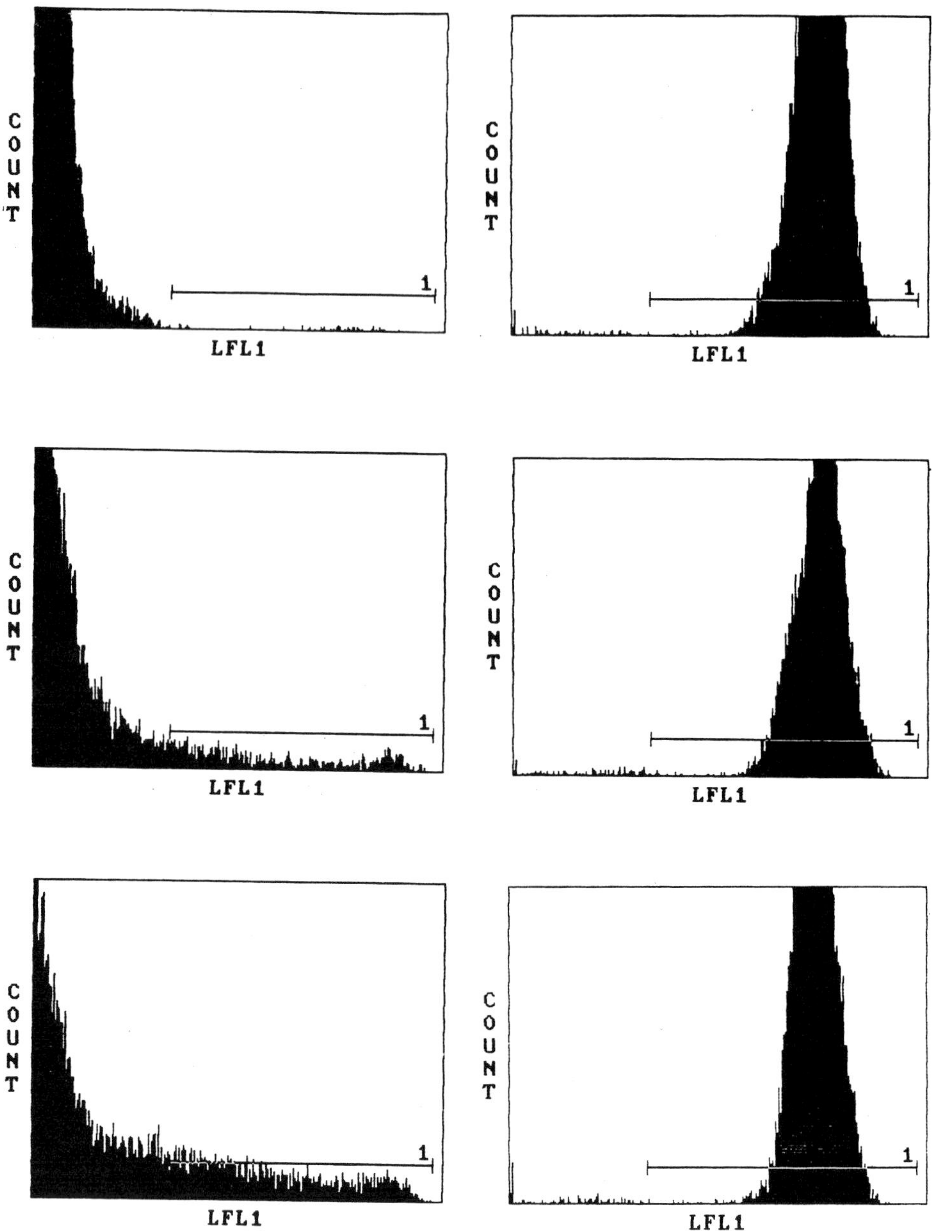

FIGURE 7 Comparison of the effects of removal of the 1α-OH group from three nonfluorinated derivatives of vitamin D_3 on flow cytometric determination of the surface marker of monocytic differentiation, CD14. The left-hand side shows HL-60 cells treated with each compound for 96 hr with 10^{-7} *M* concentration of 25(OH)-vitamin D_3 (top), 25(OH)-16,23*E*-diene-vitamin D_3 (**50**) (middle), or 25(OH)-16-ene-23-yne vitamin D_3 (**52**) (bottom). Note the absence of differentiation following such treatment. The right-hand side shows expression of the CD14 marker following a similar treatment with the corresponding analogs $1,25(OH)_2D_3$ (**1**), $1\alpha,25(OH)_2$-16,23*E*-diene D_3 (**7**), and $1\alpha,25(OH)_2$-16-ene-23-yne D_3 (**9**) which have the 1α-OH group. Strong fluorescence in these cells shows monocytic differentiation.

cemia *in vivo* and this activity is equal to that of its parent **2**.

Through these *in vitro* and *in vivo* studies, it was shown that the target tissue metabolism of $1\alpha,25(OH)_2$-16-ene D_3 (**2**) results in intracellular production and accumulation of $1\alpha,25(OH)_2$-16-ene-24-oxo D_3 (**71**), which in turn appears to play a critical role in the final expression of the full spectrum of biological activities generated by $1\alpha,25(OH)_2$-16-ene D_3 (**2**). Thus, the metabolism and biological activity studies of $1\alpha,25(OH)_2$-

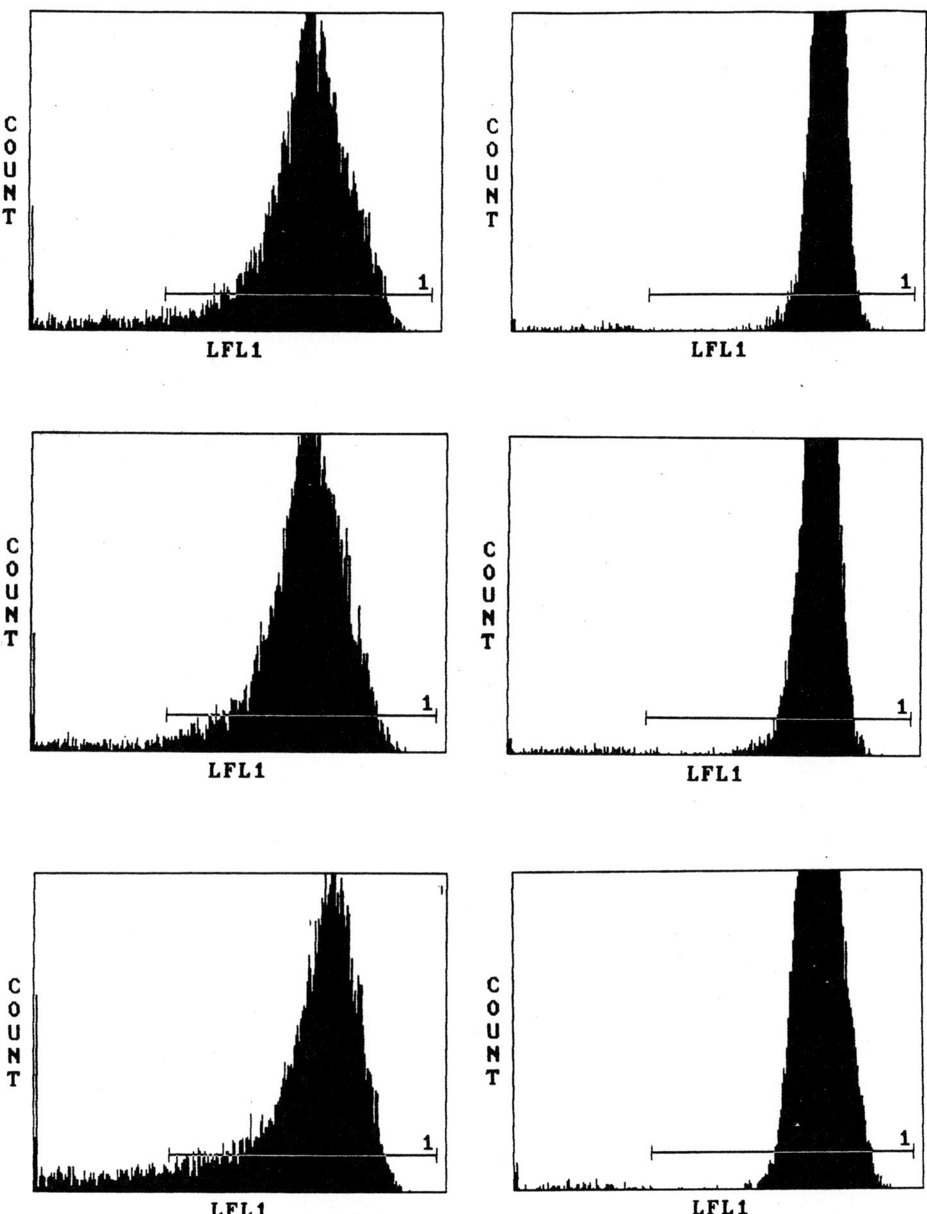

FIGURE 8 Comparison of the effects of removal of the 1α-OH group from three derivatives of vitamin D_3 with hexafluorinated substitutions of the terminal methyl groups (26,27) in the side chain, namely, 25(OH)-16,23*E*-diene-26,27-F_6 D_3 (**53**), 25(OH)-16,23*Z*-diene-26,27-F_6 D_3 (**54**), and 25(OH)-16-ene-23-yne-26,27-F_6 D_3 (**55**). Monocytic differentiation is shown by expression of the CD14 surface marker as determined by flow cytometry. The left-hand side shows effects of treatment for 96 hr with 10^{-7} *M* concentration. Note that CD14 positivity is indicative of monocytic differentiation. The right-hand side shows expression of the CD14 marker following treatment with the corresponding 1α-OH analogs, namely, 1α,25$(OH)_2$-16,23*E*-diene-26,27-F_6 D_3 (**31**), 1α,25$(OH)_2$-16,23*Z*-diene-26,27-F_6 D_3 (**32**), and 1α,25$(OH)_2$-16-ene-23-yne-26,27-F_6 D_3 (**33**).

16-ene D_3 (**2**) and its major intermediary metabolite 1α,25$(OH)_2$-16-ene-24-oxo D_3 (**71**) have given convincing evidence that a further metabolite of an analog in some cases can be better than the analog itself. However, even though it appears that the target tissue metabolism of some analogs plays an important role in the final expression of the biological activity of those analogs, this concept may not hold true for all the analogs. The mechanisms responsible for the unique biological actions of analogs can differ from one analog to another.

VI. DISSOCIATION OF DIFFERENTIATION ACTIVITY FROM EFFECTS ON CELLULAR CALCIUM HOMEOSTASIS

The accumulated experience with analogs of $1,25(OH)_2D_3$ discussed above clearly shows that there are several compounds with extraordinarily high potency against human leukemic cells. However, it is equally clear that although the calcium mobilizing activity of these analogs is reduced, it is not negligible, and a considerable risk of dangerously high levels of serum calcium can prevent their use in clinical trials. One source of difficulty is that the effects of $1,25(OH)_2D_3$ and analogs on intestinal absorption of calcium and its mobilization from bone cannot be assessed directly in humans. Indeed, until recently, it has not been shown whether the differentiation-inducing and antiproliferative effects of $1,25(OH)_2D_3$ analogs can be dissociated from altered calcium homeostasis in the target cell. One study does, however, suggest that they can [39]. Five pairs of 16-ene analogs of $1,25(OH)_2D_3$, one without and one with the 1α-OH group in each pair, were compared with respect to their differentiation and cell cycle arrest inducing activities, and the extent of the accompanying disturbance of intracellular calcium homeostasis in HL-60 cells was measured. It was found that $1,25(OH)_2D_3$ derivatives without the 1α-OH group were essentially inert as differentiation-inducing agents in HL-60 cells (Fig. 7), but if the terminal six hydrogen atoms in the side chain were replaced with fluorines the compounds had strong differentiation-inducing activity even in the absence of the 1α-OH group (Fig. 8). Even more importantly, one of these latter compounds, 25(OH)-16,23*E*-diene-26,27-F_6 D_3 (**53**), had only minimal effects on the cellular calcium homeostasis. These data offer promise that successive generations of $1,25(OH)_2D_3$ analogs will be identified which also have minimal effects on calcium homeostasis of other cells as well.

VII. SUMMARY

The investigations of $1,25(OH)_2D_3$ analogs described in this chapter were conducted in part as a search for novel anticancer agents. Several structural modifications have been uncovered that contribute to improvement in the stimulation of HL-60 cell differentiation, inhibition of HL-60 cell proliferation, and reduction of calcemic properties *in vivo*. They include the introduction of 16-, 22*E*-, 23*E*-, and 23*Z*-double bonds, a 23-triple bond, or a 22(*R*)-allene, and substitution of C-26 and C-27 hydrogens with fluorines or methyl groups. The biggest gains have been achieved by combination of a 16-double bond with a 23-double or -triple bond and 26-trifluoro or 26,27-hexafluoro substitution patterns. Separately, the combination of the 16-double bond with the 22(*R*)-allene has produced a highly active analog. With respect to modifications in ring A, high activities in cell differentiation and inhibition of cell proliferation with significant reduction of calcemic properties were observed in the 1α-fluoro, 3-deoxy, and 19-nor series. It was also shown that the lack of the 1α-hydroxyl group can be overcome by optimized modifications in ring D and the side chain; 25(OH)-16,23*E*-diene-26,27-F_6 D_3 (**53**, Fig. 2) is fully active in HL-60 cell differentiation assays but has only minimal effects on cellular calcium homeostasis. Target tissue metabolism studies of the 16-ene analogs indicated that in some instances, as in the case of $1\alpha,25(OH)_2$-16-ene D_3 (**2**), it is important to evaluate the target tissue metabolism of a given analog. Differences in the target tissue metabolism between the hormone and analog can be one of the important factors in determining the superiority of the analog over the hormone in terms of its ability to produce beneficial therapeutic effects without producing adverse toxic effects.

Acknowledgments

This work was supported by National Institutes of Health Grant R01-CA44722 from the National Cancer Institute to G.P.S. S.G.R. acknowledges the kind assistance of Mei-Lung Siu-Caldera in preparing the figures.

References

1. Zhou JY, Norman AW, Lubbert M, Collins ED, Uskokovic MR, Koeffler HP 1989 Novel vitamin D analogs that modulate leukemic cell growth and differentiation with little effect on either intestinal calcium absorption or bone calcium mobilization. Blood **74**:82–93.
2. Zhou JY, Norman AW, Chen DL, Sun G, Uskokovic M, Koeffler HP 1990 1,25-Dihydroxy-16-ene-23-yne-vitamin D_3 prolongs survival time of leukemic mice. Proc Natl Acad Sci USA **87**:3929–3932.
3. Clark JW, Posner MR, Marsella JM, Santos A, Uskokovic M, Eil C, Lasky SR 1992 Effects of analogs of $1,25(OH)_2$ vitamin D_3 on the proliferation and differentiation of the human chronic myelogenous leukemia cell line, RWLeu-4. J Cancer Res Clin Oncol **118**:190–194.
4. Zhang F, Godin JJ, Uskokovic M, Binderup L, Studzinski GP 1994 Monocytic differentiation of HL60 cells induced by potent analogs of vitamin D_3 precedes the G1/G0 phase cell cycle block. Cell Proliferation **27**:643–654.
5. Studzinski GP, Bhandal AK, Brelvi ZS 1985 A system for mono-

cytic differentiation of leukemic cells HL60 by 1,25-dihydroxycholecalciferol. Proc Soc Exp Biol Med **179**:288.
6. Jung SJ, Lee YY, Pakkala S, deVos S, Elstner E, Norman AW, Green J, Uskokovic M, Koeffler HP 1994 1,25$(OH)_2$-16-ene-vitamin D_3 is a potent antileukemic agent with low potential to cause hypercalcemia. Leuk Res **18**:453–463.
7. Godyn JJ, Xu HM, Zhang F, Kolla SS, Studzinski GP 1994 A dual block to cell cycle progression in HL60 cells exposed to analogs of vitamin D_3. Cell Proliferation **27**:37–46.
8. Munker R, Zhang W, Elstner E, Norman AW, Uskokovic M, Andreeff M, Koeffler HP 1996 A new series of vitamin D analogs are highly active for clonal inhibition, differentiation and induction of WAF1 in myeloid leukemia. Blood in press.
9. Parker SL, Tong T, Bolden S, Wings PA 1996 Cancer statistics, 1996. CA (Cancer J Clin) **65**:5–27.
10. Feldman D, Skowronski RJ, Peehl DM 1995 Vitamin D and prostate cancer. Adv Exp Med Biol **375**:53–63.
11. Niles RM 1995 Use of vitamin A and D in chemoprevention and therapy of cancer: Control of nuclear receptor expression and function. Vitamins, cancer and receptors. Adv Exp Med Biol **375**:1–15.
12. Lucia MS, Anzano MA, Slayter MV, Anver MR, Green DM, Shrader MW, Logsdon DL, Driver CL, Brown CC, Peer CW 1995 Chemopreventive activity of tamoxifen, *N*-(4-hydroxyphenyl) retinamide, and vitamin D analogue Ro 24-5531 for androgen promoted carcinomas of the rat vesicle and prostate. Cancer Res **55**(23):5621–5627.
13. Peehl DM, Skowronski RJ, Leung GK, Wong ST, Stamey TA, Feldman D 1994. Antiproliferative effects of 1,25-dihydroxy-vitamin D_3 on primary cultures of human prostatic cells. Cancer Res **54**(3):805–810.
14. Skowronski RJ, Peehl DM, Feldman D 1995 Actions of vitamin D_3 analogs on human prostate cancer cell lines: Comparison with 1,25-dihydroxy-vitamin D_3. Endocrinology **136**:20–26.
15. Schwartz GG, Oeler TA, Uskokovic MR, Bahnson RR 1994 Human prostate cancer cells: Inhibition of proliferation by vitamin D analogs. Anticancer Res **14**(3A):1077–1081.
16. Schwartz GG, Hill CC, Oeler TA, Becich MJ, Bahnson RR 1995 1,25-Dihydroxy-16-ene-23-yne vitamin D_3 and prostate cancer cell proliferation *in vivo*. Urology **46**(3):365–369.
17. Campbell MJ, Hirama T, Elstner E, Holden S, Norman AW, Uskokovic M, Koeffler Hp 1996 19-Nor-hexafluoride analogs of vitamin D_3 are potent inhibitors of *in vitro* clonal proliferation of prostate cancer cells: PC-3, DU-145 and LNCaP. Unpublished.
18. Miller GJ, Stapleton GE, Ferrara JA, Lucia MS, Pfister S, Hedlund TE, Upadhya P 1992. The human prostatic carcinoma cell line LNCap expresses biologically active, specific receptors for 1alpha,25-dihydroxyvitamin D_3. Cancer Res **52**(3):515–520.
19. Skowronski RJ, Peehl DM, Feldman D 1993. Vitamin D and prostate cancer: 1,25-Dihydroxy-vitamin D_3 receptors and action in human prostate cancer cell lines. Endocrinology **132**(5):1952–1960.
20. Bischof MG, Redlich K, Schiller C, Chirayath MV, Uskokovic M, Peterlik M, Cross HS 1995 Growth inhibitory effects on human colon adenocarcinoma-derived caco-2 cells and calcemic potential of 1α,25-dihydroxyvitamin D_3 analogs: structure–function relationships. *J Pharmacol Exp Ther* **275**:1254–1260.
21. Wali RK, Bissonnette M, Khare S, Hart J, Sitrin MD, Brasitus TA 1995 1α,25-Dihydroxy-16-ene-23-yne-26,27-hexafluoro cholecalciferol, a noncalcemic analogue of 1α,25-dihydroxyvitamin D_3, inhibits azoxymethane-induced colonic tumorigenesis. Cancer Res **55**:3050–3054.
22. Anzano MA, Smith JM, Uskokovic MR, Peer CW, Mullen LT, Letterio JJ, Welsh MC, Shrader MW, Logsdon DL, Driver CL, Brown CC, Roberts AB, Sporn MB 1994 1α,25-Dihydroxy-16-ene-23-yne-26,27-hexafluorocholecalciferol (Ro 24-5531), a new deltanoid (vitamin D analogue) for prevention of breast cancer in the rat. Cancer Res **54**:1653–1656.
23. McElwain MC, Dettelbach MA, Modzelewski RA, Russell DM, Uskokovic MR, Smith DC, Trump DL, Johnson CS 1995 Antiproliferative effects *in vitro* and *in vivo* of 1,25-dihydroxy vitamin D_3 analog in a squamous cell carcinoma model system. Mol Cell Differentiation **3**:31–50.
24. Chen TC, Persons K, Uskokovic MR, Horst RL, Holick MF 1993 An evaluation of 1,25-dihydroxyvitamin D_3 analogs on the proliferation and differentiation of cultured human keratinocytes, calcium metabolism and the differentiation of human HL-60 cells. J Nutr Biochem **4**:49–57.
25. Bouillon R, Okamura WH, Norman AW 1995 Structure–function relationships in the vitamin D endocrine system. Endocr Rev **16**:200–257.
26. Norman AW, Zhou JY, Henry HL, Uskokovic MR, Koeffler HP 1990 Structure–function studies on analogues of 1,25-dihydroxy vitamin D_3: Differential effects on leukemic cell growth, differentiation, and intestinal calcium absorption. Cancer Res **50**:6857–6864.
27. Elstner E, Dawson MI, deVos S, Pakkala S, Binderup L, Okamura W, Uskokovic M, Koeffler HP 1996 Myeloid differentiation mediated through new potent retinoids and vitamin D_3 analogs. Acute leukemias 5:439–450.
28. Craig AS, Norman AW, Okamura WH 1992 Two novel allenic side chain analogues of 1α,25-dihydroxy-vitamin D_3. J Org Chem **57**:4374–4380.
29. Elstner E, Linker-Israeli M, Said J, Umiel T, deVos S, Shintaku IP, Heber D, Binderup L, Uskokovic M, Koeffler HP 1995 20-Epi-vitamin D_3 analogues: A novel class of potent inhibitors of proliferation and inducers of differentiation of human breast cancer cell lines. Cancer Res **55**:2822–2830.
30. Peleg S, Sastry M, Collins ED, Bishop JE, Norman AW 1995 Distinct conformational changes induced by 20-epi analogues of 1,25-dihydroxyvitamin D_3 are associated with enhanced activation of the vitamin D receptor. J Biol Chem **270**:10551–10558.
31. Zhou JY, Norman AW, Akashi M, Chen DL, Uskokovic MR, Aurrecoechea JM, Dauben WG, Okamura WH, Koeffler HP 1991 Development of a novel 1,25-$(OH)_2$-vitamin D_3 analog with potent ability to induce HL-60 cell differentiation without modulating calcium metabolism. Blood **78**:75–82.
32. Ikekawa N, Eguchi T, Hara N, Takatsuto S, Honda A, Mori Y, Otomo S 1987 26,27-Diethyl-1α,25-dihydroxy-vitamin D_3 and 24,24-difluoro-24-homo-1α,25-dihydroxyvitamin D_3. Chem Pharm Bull (Tokyo) **35**:4362–4365.
33. Eguchi T, Ikekawa N, Sumitani K, Kumegawa M, Higuchi S, Otomo S 1990 Effect of carbon lengthening at the side chain terminal of 1α,25-dihydroxy-vitamin D_3 for calcium regulating activity. Chem Pharm Bull (Tokyo) **38**:1246–1249.
34. Norman AW, Nemere I, Muralidharan RK, Okamura WH 1992 1β,25-$(OH)_2$-vitamin D_3 is an antagonist of 1α,25-$(OH)_2$-vitamin D_3 stimulated transcaltachia (the rapid hormonal stimulation of intestinal calcium transport). Biochem Biophys Res Commun **189**:1450–1456.
35. Norman AW, Bouillon R, Farach-Carson MC, Bishop JE, Zhou LX, Nemere I, Zhao J, Muralidharan RK, Okamura WH 1993 Demonstration that 1β,25-dihydroxyvitamin D_3 is an antagonist of the non-genomic but not genomic biological responses and biological profile of the three A-ring diastereomers of 1α,25-dihydroxyvitamin D_3. J Biol Chem **268**:20022–20030.
36. Uskokovic MR, Baggiolini E, Shiuey SJ, Iacobelli J, Hennessy B, Kiegel J, Daniewski AR, Pizzolato G, Courtney LF, Horst RL

1991 The 16-ene analogs of 1,25-dihydroxycholecalciferol, synthesis and biological activity. In: Norman AW, Bouillon R, Thomasset M (eds) Vitamin D: Gene Regulation, Structure–Function Analysis and Clinical Application. de Gruyter, Berlin, pp. 139–145.

37. Perlman KL, Swenson RE, Paaren HE, Schnoes HK, DeLuca HF 1991 Novel synthesis of 19-nor-vitamin D compounds. Tetrahedron Lett **32**:7663–7666.
38. Bouillon R, Sarandeses LA, Allewaert K, Zhao J, Mascarenas JL, Mourino A, Vrielyuck S, DeClercq P, Vandewalle M 1993 Biologic activity of dihydroxylated 19-nor-(pre)vitamin D_3. J Bone Miner Res **8**:1009–1015.
39. Gardner JP, Zhang F, Uskokovic MR, Studzinski GP 1996 Vitamin D analog 25(OH)-16,23*E*-diene-26,27-hexafluoro vitamin D_3 induces differentiation of HL-60 cells with minimal effects on the cellular calcium homeostasis. J Cell Biochem in press.
40. Siu-Caldera M-L, Clark JW, Santos-Moore A, Peleg S, Liu YY, Uskokovic MR, Sharma S, Reddy GS 1996 1α,25-Dihydroxy-24-oxo-16-ene vitamin D_3, a metabolite of a synthetic vitamin D_3 analog, 1α,25-hydroxy-16-ene vitamin D_3, is equipotent to its parent in modulating growth and differentiation of human leukemic cells. J Steroid Biochem Mol Biol in press.
41. Cheskis B, Freedman LP 1994 Ligand modulates the conversion of DNA-bound vitamin D_3 receptor (VDR) homodimer into VDR–RXR heterodimers. Mol Cell Biol **14**:3329–3338.
42. Ferrara J, McCraig K, Hendy GN, Uskokovic M, White JH 1994 Highly potent transcriptional activation by 16-ene derivatives of 1,25-dihydroxyvitamin D_3. J Biol Chem **269**:2971–2981.
43. Imai Y, Pike JW, Koeffler HP 1995 Potent vitamin D_3 analogs: Their abilities to enhance transactivation and to bind the vitamin D response element. Leuk Res **19**:147–158.
44. Cheskis B, Lemon BD, Uskokovic M, Lomedico PT, Freedman LP 1995 Vitamin D_3–retinoid X receptor dimerization, DNA binding, and transactivation are differentially affected by analogs of 1,25-dihydroxyvitamin D_3. Mol Endocrinol **9**:1814–1824.
45. Liu M, Lee MH, Cohen M, Bommakanti, Freedman LP 1996 Transcriptional activation of the Cdk inhibitor p21 by vitamin D_3 leads to the induced differentiation of the myelo-monocytic cell line U937. Gene Dev **10**:142–153.
46. Wang QM, Jones JB, Studzinski GP 1996 The cyclin-dependent kinase inhibitor p27 as a mediator of the G1/S phase block induced by 1,25-dihydroxyvitamin D_3 in HL60 cells. Cancer Res **56**: 264–267.
47. Reddy GS, Tserng K-Y 1989 Calcitroic acid, end product of renal metabolism of 1,25-dihydroxyvitamin D_3 through C-24 oxidation pathway. Biochemistry **28**:1763–1769.
48. Makin G, Lohnes D, Byford V, Ray R, Jones G 1989 Target cell metabolism of 1,25-dihydroxyvitamin D_3 to calcitroic acid, Evidence for a pathway in kidney and bone involving 24-oxidation. Biochem J **262**:173–180.
49. Akiyoshi-Shibata M, Sakaki Y, Ohyama Y, Noshiro M, Okuda K, Yabusaki Y 1994 Further oxidation of hydroxycalcidiol by calcidiol 24-hydroxylase—A study with the mature enzyme expressed in *Escherichia coli*. Eur J Biochem **224**:335–343.
50. Okuda K-I, Usui E, Ohyama Y 1995 Recent progress in enzymology and molecular biology of enzymes involved in vitamin D metabolism. J Lipid Res **36**:1641–1652.
51. Ishizuka S, Norman AW 1987 Metabolic pathways from 1α,25-dihydroxyvitamin D_3 to 1α,25-dihydroxyvitamin D_3-26,23-lactone. J Biol Chem **262**:7165–7170.
52. Seino Y, Ishizuka S 1991 23(*S*),25(*R*)-1,25-Dihydroxyvitamin D_3-26,23-lactone in bone formation. In: Norman AW, Bouillon R, Thomasset M (eds) Vitamin D: Gene Regulation, Structure–Function Analysis and Clinical Application. de Gruyter, New York, pp. 565–571.
53. Reddy GS, Clark JW, Tserng K-Y, Uskokovic MR, McLane JA 1993 Metabolism of $1,25(OH)_2$-16-ene D_3 in kidney: Influence of structural modification of D-ring on side chain metabolism. Bioorg Med Chem Lett **3**:1879–1884.
54. Yeung B, Vouros P, Siu-Caldera ML, Reddy GS 1995 Characterization of the metabolic pathway of 1,25-dihydroxy-16-ene vitamin D_3 in rat kidney by on-line high performance liquid chromatography–electrospray tandem mass spectrometry. Biochem Pharmacol **49**:1099–1110.
55. Norman AW, Sergeev IN, Bishop JE, Okamura WH 1993 Selective biological response by target organs (intestine, kidney, and bone) to 1,25-dihydroxyvitamin D_3 and two analogues. Cancer Res **53**:3935–3942.
56. Satchell DP, Norman AW 1996 Metabolism of the cell differentiating agent 1α-25-$(OH)_2$-16-ene-23-yne vitamin D_3 by leukemic cells. J Steroid Biochem Mol Biol **57**:117–124.
57. Dantuluri PK, Haning C, Uskokovic MR, Tserng K-Y, Reddy GS 1994 Isolation and identification of $1,25,26(OH)_3$-16-ene-23-yne D_3, a metabolite of $1,25(OH)_2$-16-ene-23-yne D_3 produced in the kidney. Ninth Workshop of Vitamin D, Abstract 043.
58. Masuda S, Strugnell S, Calverley MJ, Makin HLJ, Kremer R, Jones G 1994 *In vitro* metabolism of the anti-psoriatic vitamin D analog, calcipotriol, in two cultured human keratinocyte models. J Biol Chem **269**:4794–4803.
59. Masuda S, Byford V, Kremer R, Makin HLJ, Kubodera N, Nishii Y, Okazaki A, Okano T, Kobayashi T, Jones G 1996 *In vitro* metabolism of the vitamin D analog, 22-oxacalcitriol, using cultured osteosarcoma, hepatoma and keratinocyte cell lines. J Biol Chem **271**:8700–8708.
60. Dilworth FJ, Calverley MJ, Makin HLJ, Jones G 1994 Increased biological activity of 20-epi-1,25-dihydroxyvitamin D_3 is due to reduced catabolism and altered protein binding. Biochem Pharmacol **47**:987–993.
61. Lemire JM, Archer DC, Reddy GS 1994 1,25-Dihydroxy-24-oxo-16-ene-vitamin D_3, a renal metabolite of the vitamin D analog 1,25-dihydroxy-16-ene-vitamin D_3, exerts immunosuppressive activity equal to its parent without causing hypercalcemia *in vivo*. Endocrinology **135**:2818–2821.
62. Lemire JM, Clay Archer D 1991 1,25-Dihydroxyvitamin D_3 prevents the *in vivo* induction of murine experimental autoimmune encephalomyelitis. J Clin Invest **87**:1103–1107.

CHAPTER 63

Characteristics of 22-Oxacalcitriol (OCT) and 2β-(3-Hydroxypropoxy)-calcitriol (ED-71)

NOBORU KUBODERA, KATSUHIKO SATO, AND YASUHO NISHII
Chugai Pharmaceutical Co., Ltd., Tokyo, Japan

I. INTRODUCTION

Biological and pharmacological research on newly synthesized vitamin D analogs took a significant turn in 1981. This turning point can be attributed to the fact that Suda and colleagues discovered in that year that 1α,25-dihydroxyvitamin D_3 [1,25$(OH)_2D_3$] has a significant effect on the induction of differentiation of various cell lineages [1]. Subsequently, research on vitamin D analogs surged forward following reports of the immunoregulatory effects of 1,25$(OH)_2D_3$. At present, more than 300 synthetic vitamin D analogs have been reported, and considerable efforts are being exerted to develop them into pharmaceutical products [2]. Chugai has made special efforts to use structural modifications in order to separate these kinds of biological activities from the potent hypercalcemic activity of 1,25$(OH)_2D_3$. In this chapter, we deal with two characteristic analogs, 1α,25-dihydroxy-22-oxavitamin D_3 (22-oxacalcitriol; OCT) [3] and 1α,25-dihydroxy-2β-(3-hydroxypropoxy)vitamin D_3 [2β-(3-hydroxypropoxy)calcitriol; ED-71] [4]. OCT has an oxygen atom at position 22 in the side chain and is characterized by potent differentiation-inducing activity and low calcemic liability. OCT is now undergoing phase III clinical studies as a candidate drug for the treatment of secondary hyperparathyroidism in an injectable form and late phase II studies as an antipsoriatic ointment. ED-71 has a hydroxypropoxy substituent at position 2β and is characterized by high calcemic activity and strong binding affinity to DBP. It is undergoing clinical investigation for the treatment of osteoporosis. The chemical

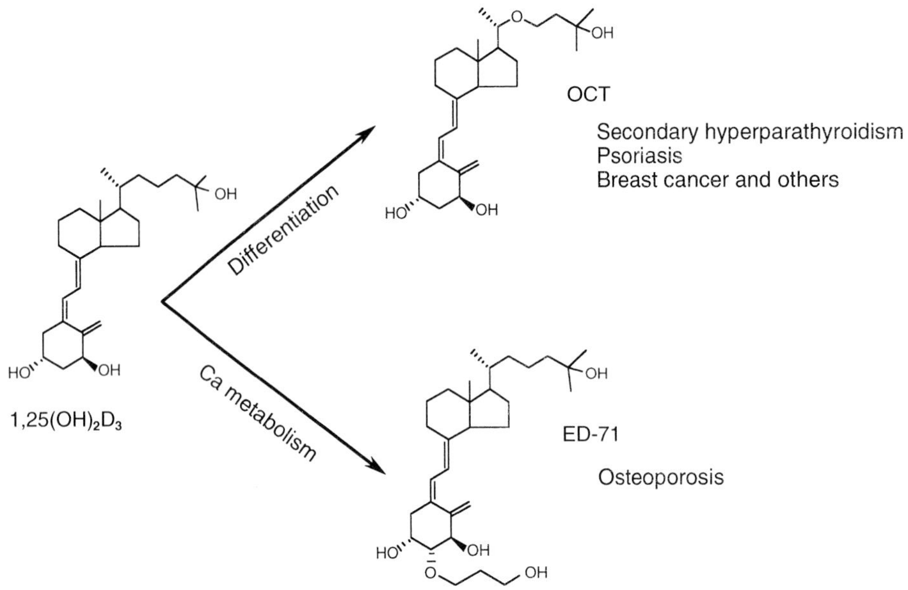

FIGURE 1 Chemical structures and expected indications for OCT and ED-71.

structures and indications for therapy are shown in Fig. 1.

This chapter includes the following subjects. First, we explain our basic strategy for the biological evaluation of new analogs, including screening systems, modification of the side chain to yield OCT, and modification of the A ring to yield ED-71. Second, the general biological properties of OCT and ED-71 are described. Third, we focus our attention on the therapeutic potential of OCT, including both current trials and future prospects. Fourth, the therapeutic potential of ED-71 for the treatment of osteoporosis is also explained in comparison with active vitamin D_3. Finally, we refer to the expectation of new analogs with fully distinct biological characteristics as the next generation of vitamin D analogs.

II. BASIC STRATEGIES FOR BIOLOGICAL EVALUATION OF NEW ANALOGS

A. Screening Systems

When we began our studies to make modifications in $1,25(OH)_2D_3$, our aim was to synthesize new analogs of $1,25(OH)_2D_3$ that would be capable of separating the differentiation-inducing activities on HL-60 (human myeloid leukemia) cells from their *in vivo* calcemic effect. Intense interest was also directed at developing analogs that were more active than $1,25(OH)_2D_3$ in their regulatory effects on calcium metabolism. We used the following assay systems to evaluate the biological properties of the new analogs synthesized [5]:

1. *In vitro* differentiation-inducing activity in HL-60 or WEHI-3 (murine myeloid leukemia) cells
2. *In vitro* binding activity to chick embryonic intestinal vitamin D receptor (VDR)
3. *In vitro* binding activity to rat plasma vitamin D binding protein (DBP)
4. *In vivo* calcemic activity in vitamin D-deficient rats, by oral or intravenous administration

B. Modification of the Side Chain of $1,25(OH)_2D_3$

From the point of view of bioisosterism, we were interested in analogs bearing heteroatoms on the side chain of $1,25(OH)_2D_3$. When we initiated the structural modification of $1,25(OH)_2D_3$ in 1982, it was already known that $1,25(OH)_2D_3$ is hydroxylated at the 23, 24, and 26 positions in the side chain as the first step in its metabolism. Therefore, we anticipated that, in addition to bioisosteric effects, the introduction of heteroatoms into the side chain might have a great influence on the metabolic features of the analogs. The relationships between the structural and biological activities obtained by our modification studies of $1,25(OH)_2D_3$ are summarized in Fig. 2 [6–13]. In 1985, we obtained OCT with an oxygen atom at position 22 [3]. OCT is characterized by increased potency to inhibit cellular proliferation

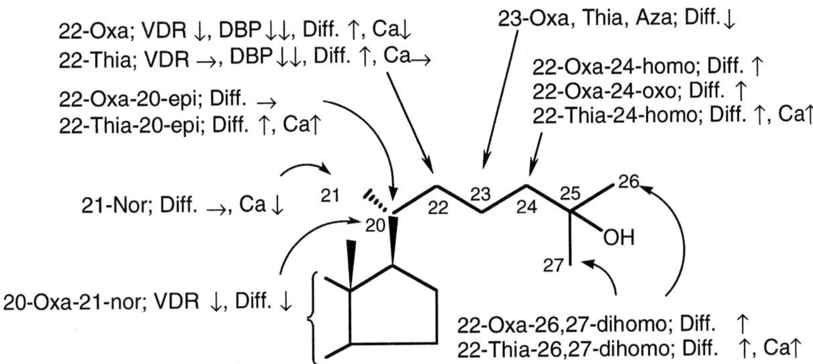

FIGURE 2 Relationship between structural modification of the side chain and biological responses compared to $1,25(OH)_2D_3$. Arrows indicate activities stronger than $1,25(OH)_2D_3$ (↑), weaker than $1,25(OH)_2D_3$ (↓), or similar to $1,25(OH)_2D_3$ (→). VDR, Affinity to vitamin D receptor; DBP, affinity to vitamin D binding protein; Diff., differentiation-inducing activity; Ca, calcemic activity.

and induce cellular differentiation, while having low calcemic activity.

C. Modification of the A Ring of $1,25(OH)_2D_3$

Modification studies on the A ring of $1,25(OH)_2D_3$ were based on the α-epoxide, which is readily prepared by industrial-scale synthesis of the intermediate of 1α-hydroxyvitamin D_3 (alphacalcidol: $1OHD_3$) (R = H in Fig. 3). On the basis of the steroidal α-epoxide, we introduced substituents at the 2β position, as shown in Fig. 3. In 1985, we prepared ED-71, with the hydroxypropoxy substituent at the 2β position. ED-71 is characterized by high calcemic activity and strong binding affinity to DBP [4,14].

D. Properties of OCT and ED-71

The characteristic properties of OCT and ED-71, in comparison with those of $1,25(OH)_2D_3$, which were obtained in our screening systems, are shown in Table I and Figs. 4 and 5. Figure 4 shows results of comparative studies on the differentiation-inducing activity of OCT, ED-71, and $1,25(OH)_2D_3$ as assessed by the inhibition of proliferation and the induction of phagocytosis [15]. OCT inhibited proliferation and induced phagocytic activity in WEHI-3 cells. These effects were stronger than those observed with $1,25(OH)_2D_3$ [3]. In contrast, ED-71 showed weaker effects on both proliferation and phagocytosis relative to OCT and $1,25(OH)_2D_3$. Figure 5 shows a comparison of the effects of OCT, ED-71, and $1,25(OH)_2D_3$ on calcium absorption in the intestines of vitamin D-deficient rats. Biphasic stimulation of calcium absorption by the intestines was observed after $1,25(OH)_2D_3$ administration, with the first phase occurring after 6 hr and the second phase after 24 hr. When ED-71 was administered intravenously to vitamin D-deficient rats, intestinal calcium transport showed a similar pattern as that due to $1,25(OH)_2D_3$, although in the first phase ED-71 was slightly less potent than $1,25(OH)_2D_3$. On the other hand, stimulation of calcium absorption by OCT in the second phase was markedly diminished compared to $1,25(OH)_2D_3$ or ED-71 [16].

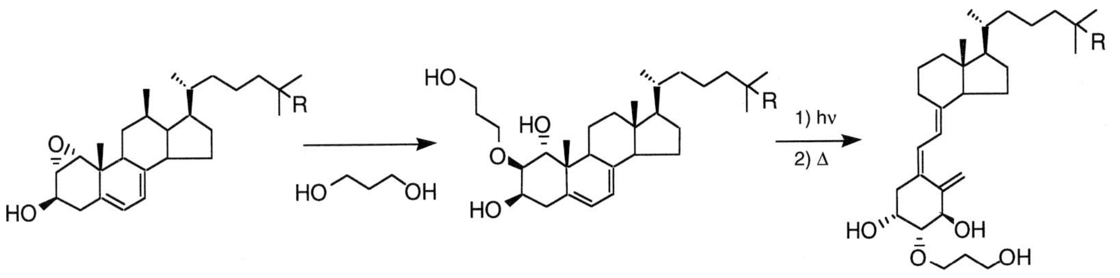

(R = H or OH)

FIGURE 3 Synthetic procedure for the vitamin D_3 derivative modified at the 2β position of the A ring.

TABLE I Properties of OCT and ED-71 Compared with Those of $1,25(OH)_2D_3$

	Binding affinity to VDR	Binding affinity to DBP	Metabolic clearance from serum or $t_{1/2}$
$1,25(OH)_2D_3$	1	2	$t_{1/2}$ = 1.5 days in human
OCT	1/8	1/580	Very rapidly excreted into bile
ED-71	1/8	2	Very long duration in serum

The differences in intestinal calcium absorption between OCT and ED-71 might be explained by the differences in binding affinity for BDP and the different half-life in serum after *in vivo* administration, as shown in Table I [15].

The low *in vivo* hypercalcemic character of OCT is believed to be due to its short residence in the bloodstream, which results from its weak binding affinity for DBP compared to $1,25(OH)_2D_3$. This pattern may give OCT a great advantage over $1,25(OH)_2D_3$ with respect to clinical use for the treatment of diseases such as secondary hyperparathyroidism and psoriasis. In the next section, we focus our attention on the potential therapeutic uses for OCT, including both current trials and future prospects.

III. BIOLOGICAL FEATURES OF OCT

A separation of the differentiation activity from the calcitropic activity of $1,25(OH)_2D_3$ has been realized with OCT. This distinction in biological characteristics of OCT from $1,25(OH)_2D_3$ has been demonstrated in a number of studies, for example, in (1) inhibition of proliferation of normal and psoriatic fibroblasts and keratinocytes [17], (2) prevention of immunological disorders and immunoregulatory activities [18,19], (3) healing effects on hind paw edema in rats with type II collagen-induced arthritis [20], (4) inhibition of growth of human breast cancer [21,22], prostate cancer [23], and pancreatic cancer [24], (5) inhibition of humoral hypercalcemia of malignancy [25], (6) antiangiogenic activity in chorioallantoic membranes [26,27], (7) antitumor effects on rat mammary tumors induced by 7,12-dimethylbenz[*a*]anthracene (DMBA) [28], (8) regulation of parathyroid hormone (PTH) synthesis and secretion [29], (9) suppression of transcription of parathyroid hormone-related peptide (PTHrP) [25],

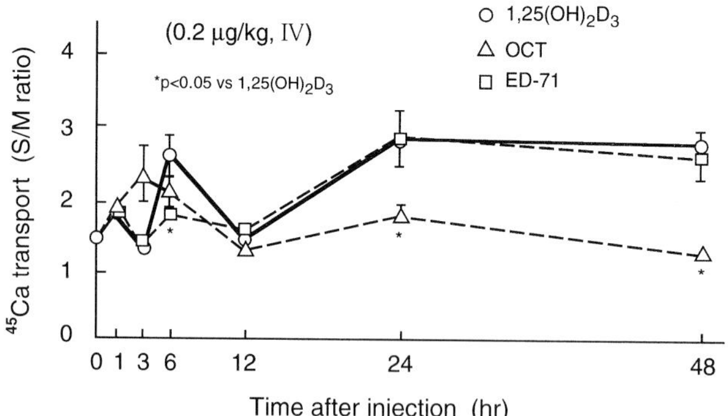

FIGURE 5 Effects of $1,25(OH)_2D_3$, OCT, and ED-71 on intestinal calcium transport in vitamin D-deficient rats.

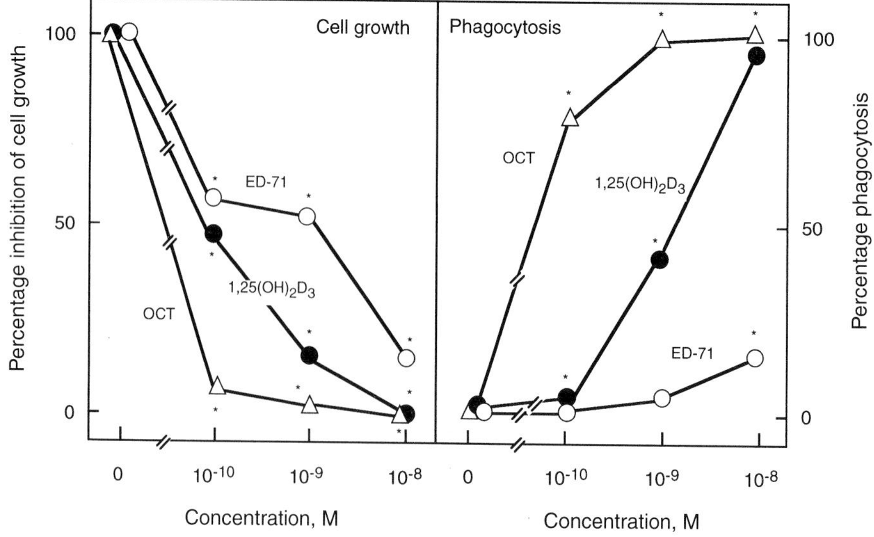

FIGURE 4 Comparison of growth-inhibiting and differentiation-inducing activities of OCT, $1,25(OH)_2D_3$, and ED-71 in WEHI-3 cells. *$p < 0.05$ vs controls.

and (10) stimulation of production of prostaglandin I_2 [30].

Some of these biological activities have already been shown to be different from those of $1,25(OH)_2D_3$. Brown *et al.* have demonstrated that the binding efficacy of OCT to the intestinal vitamin D receptor is stronger than that of $1,25(OH)_2D_3$, and that the effect of OCT on the induction of calbindin 9K mRNA in the intestine is comparable to that due to $1,25(OH)_2D_3$ [31]. However, OCT had a very weak effect on intestinal calcium transport *in vivo* [31]. These observations suggest that the mechanism of the OCT-induced calcemic effect may be different from that due to $1,25(OH)_2D_3$, with regard to the genomic mechanism that follows $1,25(OH)_2D_3$ binding to the VDR [31]. One possibility is a predominant nongenomic action of OCT on calcium transport processes. This remains to be determined. Thus, the pattern of OCT activity may exhibit a nongenomic component.

A. Secondary Hyperparathyroidism

Slatopolsky and colleagues have demonstrated that OCT has similar activities to $1,25(OH)_2D_3$ in suppressing PTH secretion and synthesis both *in vivo* and *in vitro* [29]. Since that demonstration, we have been engaged in the development of OCT given by injection for the treatment of secondary hyperparathyroidism in chronic renal failure. We evaluated the dose-dependent effect of OCT on PTH secretion and serum calcium in uremic rats which developed secondary hyperparathyroidism after 3 months of renal failure (5/6 nephrectomy). Uremic rats were given daily intraperitoneal (i.p.) injections of the vehicle control or 0.001, 0.005, 0.025, or 0.125 μg/kg of OCT for 14 days. PTH decreased in a dose-dependent manner, and total serum calcium did not increase in any group except in the rats receiving 0.125 μg/kg (data not shown). Intravenous (i.v.) administration of OCT suppressed PTH secretion in a dose-dependent manner, without causing hypercalcemia. When rats were given daily i.v. injections of the control or 0.125 or 0.625 μg/kg of OCT for 14 days, PTH levels decreased. However, serum calcium remained normal (Fig. 6). Thus, when OCT administration is i.v. rather than i.p., a much greater dose (0.625 versus 0.125 μg/kg) is tolerated prior to the development of hypercalcemia.

In early phase II clinical studies, it is now confirmed that OCT is highly effective in suppressing PTH in dialysis patients with secondary hyperparathyroidism [32]. Phase III clinical trials in patients with chronic renal failure are now in progress. Details of the OCT clinical studies are discussed in Chapter 73 of this book.

B. Psoriasis

Since $1,25(OH)_2D_3$ analogs have proved to be effective in the treatment of psoriatic skin lesions (see Chapter 72), two characteristic analogs of $1,25(OH)_2D_3$, MC903 and TV-02, have been used clinically as antipsoriatic ointments. Because OCT has more potent inhibitory effects on the proliferation of normal human

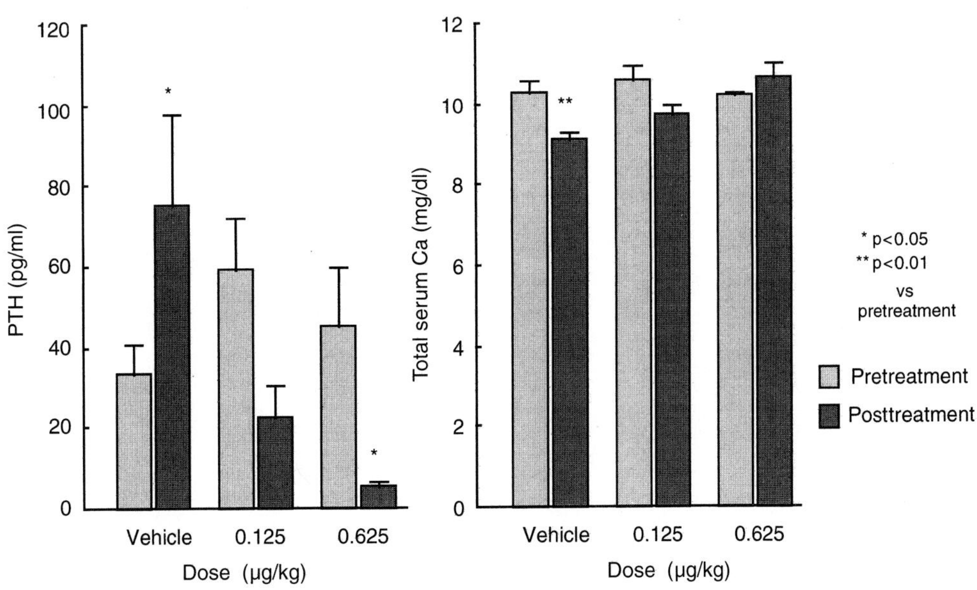

FIGURE 6 Suppression of serum PTH levels by intravenous injection of OCT in 5/6 nephrectomized rats.

keratinocytes and psoriatic human fibroblasts than $1,25(OH)_2D_3$ [17], it was anticipated that OCT may also be a useful therapeutic agent for psoriasis. We evaluated the calcemic response to OCT after 14 days of topical administration in rats, using an equivalent application to skin as for $1,25(OH)_2D_3$. As shown in Fig. 7, OCT showed less calcemic activity than $1,25(OH)_2D_3$, and only at high doses was an elevation in serum calcium levels evident. The noncalcemic character of topical OCT at clinical doses has been confirmed in phase I clinical studies aimed at the treatment of psoriasis. Late phase II clinical trials are now in progress.

C. Cancer-Associated Hypercalcemia

When radioactive OCT was administered intravenously into nude mice bearing FA-6 tumors (human pancreas carcinoma line), OCT was cleared from the circulation more rapidly than $1,25(OH)_2D_3$ [25]. The uptake of OCT into the tumor tissue, relative to the radioactivity in circulation, was greater than that of $1,25(OH)_2D_3$. Intravenous or oral administration of OCT reduced the steady-state levels of PTHrP mRNA in the FA-6 tumors, and nuclear run-off assays demonstrated that the effect of OCT on PTHrP gene expression occurred at a transcriptional level. RNase mapping analysis revealed that both upstream and downstream promoters of the human PTHrP gene were down-regulated by OCT (data now shown). Intravenous administration of OCT (6.25 μg/kg) was started 24 days after tumor implantation, when blood calcium concentrations were markedly elevated (day 0). OCT caused a reduction in blood calcium levels, whereas vehicle-treated animals developed further hypercalcemia. The therapeutic effect of OCT on cancer-associated hypercalcemia was apparent as early as day 3 of treatment. Treatment with OCT was also shown to prolong the survival time of tumor-bearing animals by approximately 1.5-fold, presumably as a result of mitigating cancer-associated hypercalcemia (data not shown). These results suggest that OCT is effectively taken up by a VDR-positive human carcinoma *in vivo* and exhibits a therapeutic potential for treating cancer-associated hypercalcemia through suppression of PTHrP gene transcription.

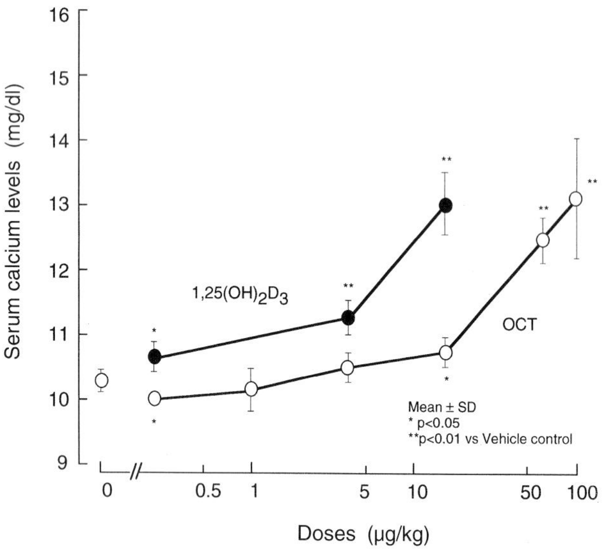

FIGURE 7 Calcemic responses in rats treated topically with OCT and $1,25(OH)_2D_3$ for 14 days.

D. Breast Cancer

We have previously reported that OCT inhibited the proliferation of both estrogen receptor (ER)-positive (MCF-7, T47-D, and ZR-75-1) and ER-negative (MDA-MB-231 and BT-20) cells *in vitro*, and that this activity was more potent than that of $1,25(OH)_2D_3$ [21]. OCT was also examined *in vivo* in female athymic mice implanted with human breat carcinomas that were either positive or negative for ER [22]. In ER-positive MCF-7 tumors, oral administration of OCT, as well as the antiestrogen tamoxifen, five times a week for 4 weeks suppressed tumor growth in a dose-dependent fashion (Fig. 8A). The antitumor effect of 1.0 μg/kg OCT was comparable to that of 2.0 μg/kg tamoxifen. In addition, a synergistic antitumor effect of submaximal doses of OCT and tamoxifen was observed in MCF-7 tumors (Fig. 8B). Oral administration of OCT three times a week for 4 weeks also suppressed the growth of ER-negative MX-1 tumors in a dose-dependent manner without raising serum calcium concentrations (data not shown).

These results indicate that OCT suppresses the growth of ER-negative as well as ER-positive breast carcinomas *in vivo*, without causing hypercalcemia, and that the antitumor effect of OCT can be enhanced by tamoxifen in ER-positive tumors. Thus, OCT may provide a new strategy, either alone or in combination with other anticancer drugs, for systemic adjuvant therapy of breast carcinoma, regardless of ER status.

E. Pancreatic Cancer

Kawa and colleagues demonstrated that OCT may provide a more useful tool than $1,25(OH)_2D_3$ for the chemotherapy of pancreatic cancer [24]. As shown in Fig. 9, daily oral administration of OCT (2 μg/kg) significantly delayed the growth of BxPC-3 tumors after

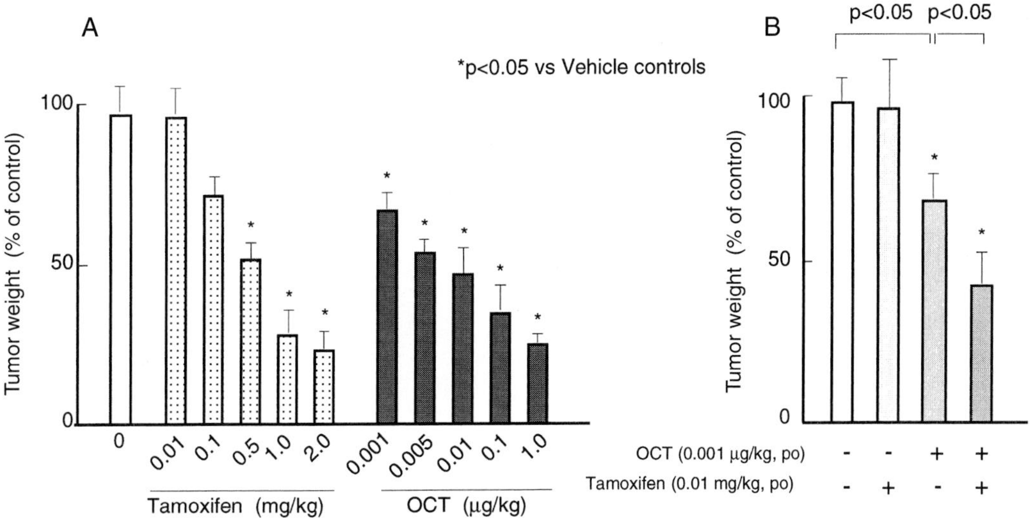

FIGURE 8 Effects of OCT and tamoxifen on the growth of ER-positive breast carcinoma MCF-7 in athymic mice.

10 days of treatment, compared with that of the control. The tumor volume of the treated group was 38% of that in the controls at the end of experiment (on day 31). On the other hand, no significant growth inhibition was observed with $1,25(OH)_2D_3$, (2 μg/kg, daily oral administration) treatment. OCT exerted no hypercalcemic effect, whereas $1,25(OH)_2D_3$ induced a significant elevation in serum calcium levels; serum calcium concentration in the OCT group was 9.8 ± 0.3 mg/dl, that in the vehicle group was 9.6 ± 0.2 mg/dl, and that in the $1,25(OH)_2D_3$ group was 11.5 ± 0.5 mg/dl. In addition, OCT had no inhibitory effect on body weight gain, compared with the other two groups (data not shown).

F. Antitumor Effects on Rat Mammary Tumors Induced by DMBA

The effect of OCT on the growth of autochthonous rat mammary tumors induced by 7,12-dimethylbenz-[*a*]anthracene (DMBA) was examined [28]. Two doses of OCT, 0.1 and 1 μg/kg, were used. Daily administration of OCT at a dose of 1 μg/kg resulted in significant inhibition of the growth of mammary tumors at 1, 2, and 3 weeks after administration, although the agent caused little or no regression of tumors. After daily administration for 3 weeks, significant antitumor effects were also observed in the group treated with 0.1 μg/kg

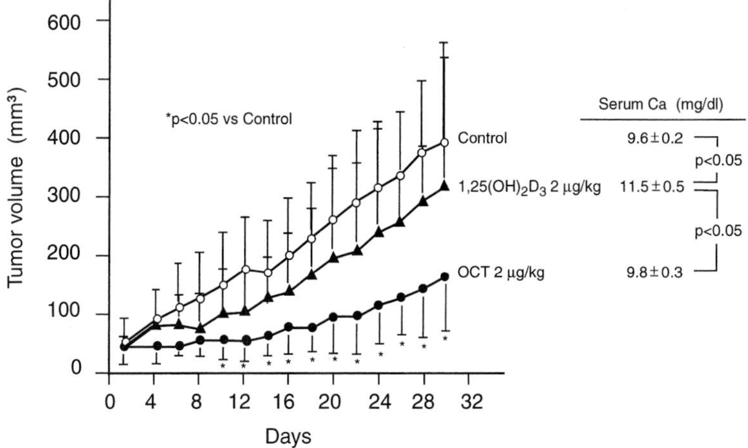

FIGURE 9 *In vivo* effects of OCT and $1,25(OH)_2D_3$ on the growth of BxPC-3 xenografts implanted in athymic mice.

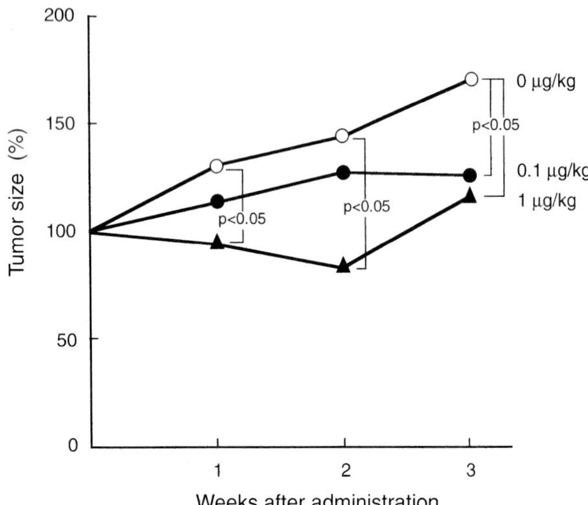

FIGURE 10 Effect of OCT on autochthonous rat mammary tumors induced by DMBA.

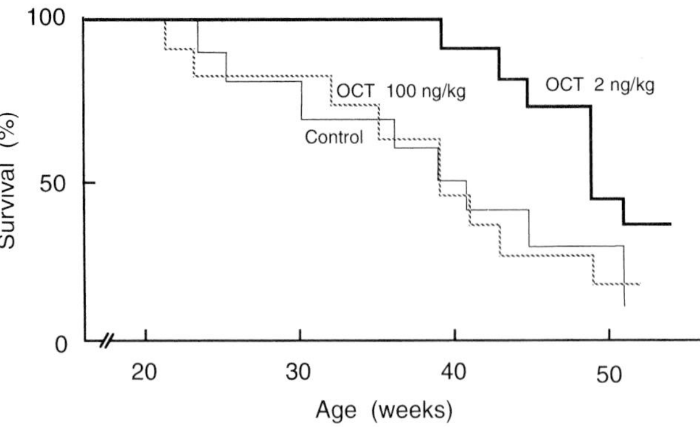

FIGURE 11 Effect of OCT on survival of MRL/l mice.

(Fig. 10) Treatment with OCT did not affect serum calcium levels (data not shown). These results suggest that OCT has a significant growth inhibitory effect on DMBA-induced autochthonous mammary tumors in rats, without severe side effects such as hypercalcemia. Additionally, it is possible that the antitumor effect of OCT is due, in part, to the antiangiogenic activity of OCT, since tumor growth depends on an angiogenic response by the host tissues and OCT exhibits marked antiangiogenic activity [28].

G. Immunological Disorders

We examined the immunoregulatory effects of OCT on spontaneously developing autoimmune disorders in MRL/Mp-lpr/lpr mice (MRL/l mice) [18,19]. MRL/l mice were orally given either the vehicle or 2 or 100 ng/kg OCT, 5 days a week, from 6 weeks of age until death. As shown in Fig. 11, although the median survival time for both the control group and the high dose group (100 ng/kg) was 39 weeks, that for the group given low dose (2 ng/kg) of OCT was 51 weeks, significantly longer than the other groups. The appearance of proteinuria was monitored periodically at 6, 11, 16, 21, 26, and 30 weeks of age, as it is an indicator of renal dysfunction. The mice in the control group clearly showed increased protein excretion in urine. Approximately 50% were positive for proteinuria at 21 weeks of age. The group treated with 2 ng/kg of OCT showed significant inhibition of proteinuria at 12, 26, and 30 weeks of age (data not shown). The 100 ng/kg-treated group also showed considerable inhibition of proteinuria at 26 and 30 weeks of age, although the effect was weaker than that induced by 2 ng/kg of OCT (data not shown). In the control and 100 ng/kg treatment groups, all the mice that died showed high levels of proteinuria prior to death. In the 2 ng/kg treatment group, all mice lived until 30 weeks of age. Histopathological investigations also revealed that pathological conditions such as renal arthritis, granuloma, or arthritis of the knee joints were less marked in the 2 ng/kg group than the other groups (data not shown). These results suggest that OCT inhibits the development of lupus nephritis in MRL/l mice and may be therapeutically effective for autoimmune disorders.

H. Atherosclerosis

We examined the effects of OCT and $1{,}25(OH)_2D_3$ on prostaglandin (PG) I_2 synthesis by analysis of rat aortic rings and serum calcium levels [30]. Normal Wistar rats were treated orally with OCT (1 or 10 μg/kg) or $1{,}25(OH)_2D_3$ (1 μg/kg) for 2 weeks. Blood samples from all rats were taken for measuring serum calcium, and their thoracic aortas were removed. Aortic rings were incubated in Krebs-Ringer solution at 37°C for 30 min, and the amount of PGI_2 released into the medium was measured as 6-keto-$PGF_{1\alpha}$ by radioimmunoassay. As shown in Fig. 12, PGI_2 synthesis by aortic rings of rats treated with either OCT or $1{,}25(OH)_2D_3$ was significantly higher than in the control group. There were no significant increases in plasma calcium levels even in the 10 μg/kg OCT group. In contrast, 1 μg/kg $1{,}25(OH)_2D_3$ induced a significant increase in plasma calcium level. These results suggest that OCT may possibly protect against the development of atherosclerosis by modulating prostaglandin metabolism.

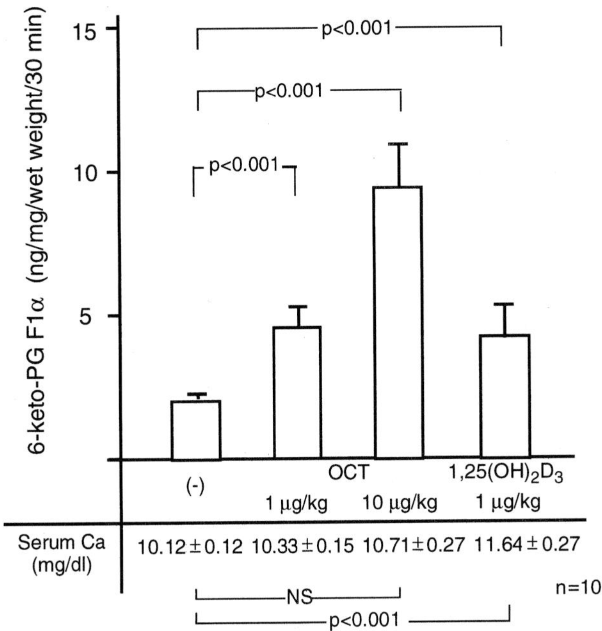

FIGURE 12 Effects of OCT and $1,25(OH)_2D_3$ on PGI_2 synthesis by rat aortic rings and serum calcium levels.

IV. FEATURES OF ED-71

A. Pharmacological Properties

1. Antirachitic Effects

The hormone $1,25(OH)_2D_3$ is well known not only as a calcium homeostasis regulator *in vivo*, but also as an important antirachitic factor. In studies of $1,25(OH)_2D_3$ and ED-71 in rickets, the widths of the epiphyseal growth plates in rachitic rats were significantly greater in comparison to normal controls. Both ED-71 and $1,25(OH)_2D_3$ dose-dependently reduced the extended thickness of the growth plate (Fig. 13C). The efficacy of ED-71 on growth plate thickness was estimated to be about 15- or 20-fold greater than that of $1,25(OH)_2D_3$. Bone mineral density (BMD) measured by DXA in the rachitic controls was significantly lower than in the normal controls. ED-71, but not $1,25(OH)_2D_3$, increased the BMD values of the femurs. Significant increases in femoral BMD were observed at doses of ED-71 over 0.125 μg/kg/day (Fig. 13D). Serum calcium levels tended to be dose-dependently elevated by administration of ED-71 or $1,25(OH)_2D_3$. However, serum calcium levels, even at a dose of 0.5 μg/kg/day ED-71, which had the maximal effect on the thickness of the growth plate, were not significantly different from the rachitic controls (Fig. 13A). Serum phosphate levels were also dose-dependently elevated by ED-71 administration, but a significant elevation was observed only with more than 0.25 μg/kg/day ED-71 (Fig. 13B).

2. Effects on Serum Calcium Elevation

The calcemic effects of ED-71 were assessed using vitamin D-deficient rats on a low calcium diet (0.003%). ED-71 and $1,25(OH)_2D_3$ were given orally by either a single administration or daily administration for 7 days. Serum calcium levels were dose-dependently elevated by both the single and repeated administrations of vitamin D derivatives (Fig. 14). Whereas ED-71 showed a weaker effect on calcium elevation than $1,25(OH)_2D_3$ with single administration (Fig. 14A), the calcium-elevating effect of ED-71 appeared stronger than that of $1,25(OH)_2D_3$ with repeated administration (Fig. 14B). Significant effects of ED-71 on calcium elevation were observed at doses of more than 2.5 and 0.5 μg/kg/day, by both single and repeated administration, respectively.

3. Effects on Bone Resorption and Collagen Synthesis *in Vitro*

To clarify the effects of ED-71 on bone metabolism *in vitro*, we further examined the effects on bone resorption and collagen synthesis in two organ culture systems [33]. Whereas $1,25(OH)_2D_3$ significantly stimulated bone resorption in a dose-dependent manner at concentrations of 10^{-10} to 10^{-8} *M*, ED-71 significantly stimulated bone resorption only at 10^{-9} and 10^{-8} *M*. The maximum response at 10^{-8} *M* ED-71 was slightly, but not significantly, greater than with $1,25(OH)_2D_3$ (Fig. 15A). On the other hand, the collagen synthesis activity assessed as the percentage collagen synthesis was decreased by both $1,25(OH)_2D_3$ and ED-71. $1,25(OH)_2D_3$ showed significant inhibitory effects on collagen synthesis at concentrations of 10^{-10} to 10^{-8} *M* whereas ED-71 showed a significant effect only at 10^{-8} *M* (Fig. 15B). These results indicate that ED-71 has lesser effects than $1,25(OH)_2D_3$ on both stimulation of bone resorption and inhibition of collagen synthesis.

4. Summary of Biological Properties

Some biological characteristics of ED-71 in comparison with $1,25(OH)_2D_3$ determined in our laboratory and by our colleagues [34] are summarized in Table II. Judging from these results, the *in vitro* effects of ED-71 are rather weaker than those of $1,25(OH)_2D_3$, whereas the *in vivo* effects of ED-71 are almost identical to those of $1,25(OH)_2D_3$. Pharmacokinetic differences may explain the marked differences in response *in vivo* and *in vitro* to these two agents. ED-71 was confirmed to have a long half-life in blood; the $t_{1/2}(\beta)$ is approximately 70 hr in rats and 35 hr in dogs (unpublished observation).

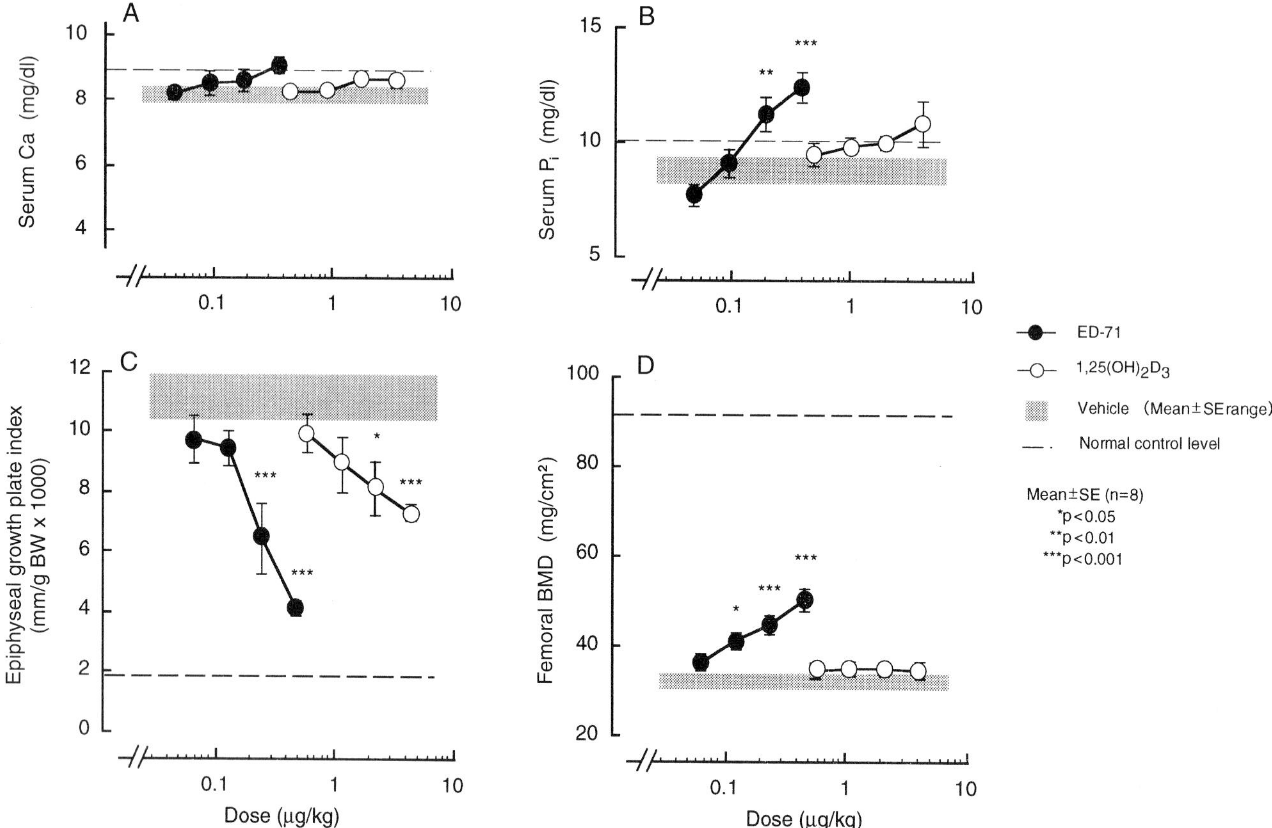

FIGURE 13 Effects of ED-71 and $1,25(OH)_2D_3$ on serum calcium (A), inorganic phosphate (B), epiphyseal growth plate indices (C), and femoral bone mineral densities (D) in rachitic rats.

In addition, the binding affinity of ED-71 to vitamin D binding protein (DBP) is thought to be approximately twice that of $1,25(OH)_2D_3$ [14]. Besides the binding affinity to DBP, More than 90% of the intact form of ED-71 binds to plasma proteins and is thought to circulate in the blood as the intact form (unpublished observation). The strong binding affinity of ED-71 to blood proteins probably leads to the long half-life in blood, resulting in the *in vivo* effects of ED-71 becoming more marked on repeated administration. On the other hand,

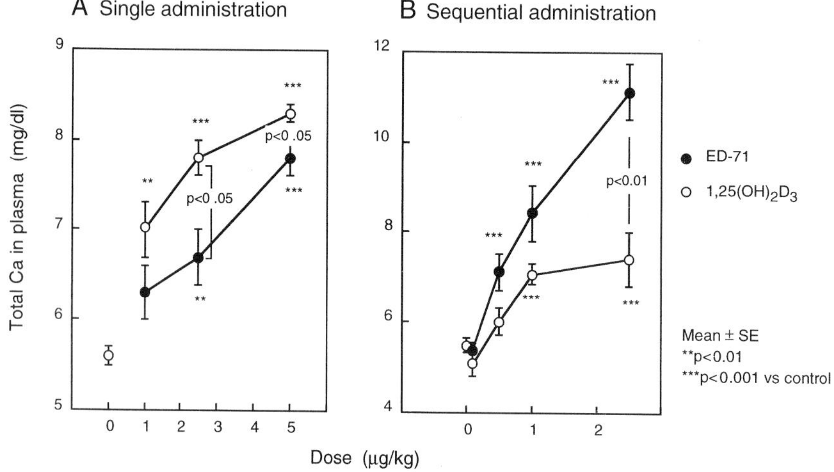

FIGURE 14 Effects of ED-71 and $1,25(OH)_2D_3$ on serum calcium levels in vitamin D-deficient rats.

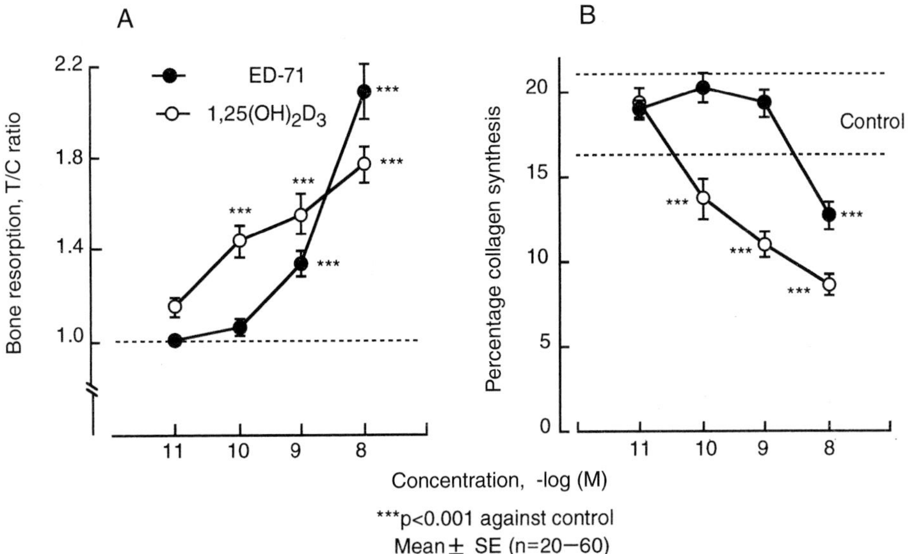

FIGURE 15 Effects of ED-71 and $1,25(OH)_2D_3$ on bone resorption (A) and collagen synthesis (B).

it has been reported that intestinal calcium transport was stimulated by large doses (6.25 μg/kg) of ED-71 and $1,25(OH)_2D_3$, and that the efficacies were almost identical [16].

B. Effects on Bone

1. Effects on Bone Mineral Density in Ovariectomized Rat Model

To assess the preventive effects of ED-71 on bone loss in osteoporosis, we used two osteoporotic rat models, ovariectomy (OVX) and ovariectomy combined with immobilization induced by sciatic neurotomy (OVX–NOX). The BMD values of the femurs in both OVX and OVX–NOX rats significantly decreased 3 months after surgery, and ED-71 increased the BMD levels in a dose-dependent manner. Significant effects of ED-71 on bone losses were observed at doses of more than 0.1 μg/kg/week and doses of 0.2 and 0.4 μg/kg/week in OVX and OVX–NOX rats, respectively (Fig. 16). These results indicate that ED-71 can preserve bone loss induced by ovarian dysfunction or immobilization in rats. Okumura and colleagues also reported that ED-71 was capable of preventing the osteopenic decrease of bone mass in the immobilized rat model induced by OVX and hemicordotomy [35].

We further performed dosing experiments to elucidate whether ED-71 restores bone mass once it has decreased, or improves the mechanical properties of bone in established osteopenia, after ovariectomy in rats (an intervention study). The spinal BMD levels in OVX controls decreased significantly to levels of approximately 80 to 70% compared to sham-operated animals, 16 and 28 weeks after surgery, respectively (Fig. 17). The decreased spinal BMD levels gradually increased back to the levels in sham-operated rats after 6 weeks of treatment with ED-71. At the end of the experiment (12 weeks of treatment with ED-71), the spinal BMD levels in the group treated with 0.4 μg/kg/week of ED-71 were greater than the levels in sham-operated rats (Fig. 17). In a separate study, we determined that the increases in spinal BMD involved structural changes, especially increases in bone volume and trabecular thickness. ED-71 markedly increased the mechanical strength of the vertebral body and increased vertebral mineral density as well (data not shown).

TABLE II Biological Characteristics of ED-71 *in Vivo* and *in Vitro*

Characteristics	$1,25(OH)_2D_3$	ED-71
In vitro		
Binding affinity to DBP	1	2.7
Binding affinity to VDR	1	1/8
Stimulation of bone resorption	Strong	Weak
Inhibition of collagen synthesis	Strong	Weak
Differentiation-inducing activity	Strong	Weak
In vitro		
Intestinal Ca absorption	Strong	Strong
Bone mobilization activity	Strong	Strong
Half-life in plasma	Short	Long

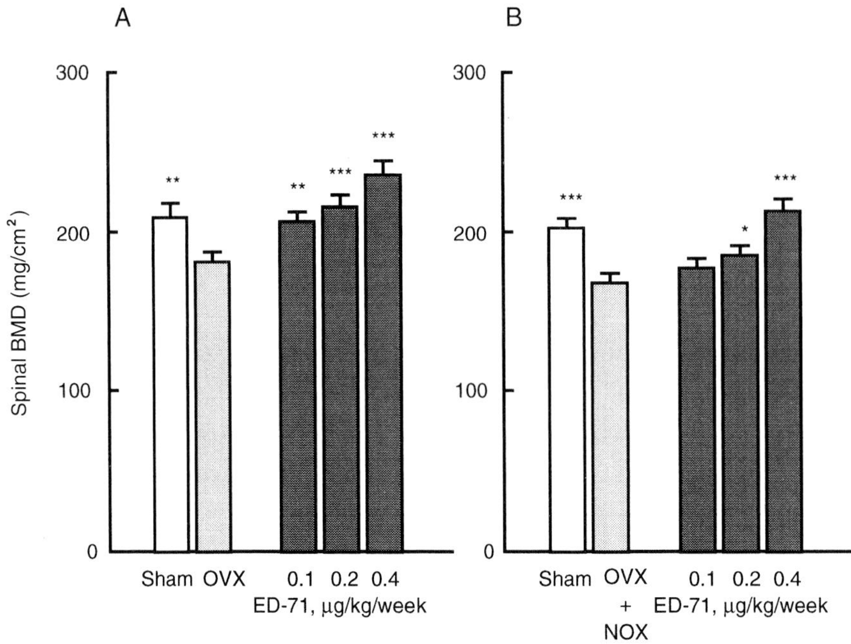

FIGURE 16 Effects of ED-71 and $1,25(OH)_2D_3$ on bone loss in OVX (A) and OVX–NOX rats (B).

Femoral BMD and mechanical strength are shown in Table III. BMD values in femurs treated with ED-71 were significantly greater than those in OVX controls, for proximal, medial, and distal femur sites. The results of a three-point bending test were that either structural strength or stiffness increased to the sham levels with low dose ED-71 administration and apparently exceeded those in sham controls and those with high dose treatment. Cross-sectional cortical bone areas were also increased by ED-71 administration, reflecting increases in the mechanical strength of the cortical bone region.

These observations indicate that ED-71 completely prevents bone loss in OVX rats and also has recovery effects on bone mass in animals with established osteoporosis. It should be noted that these effects of ED-71 were found not only in the trabecular bones, but also in the cortical bones.

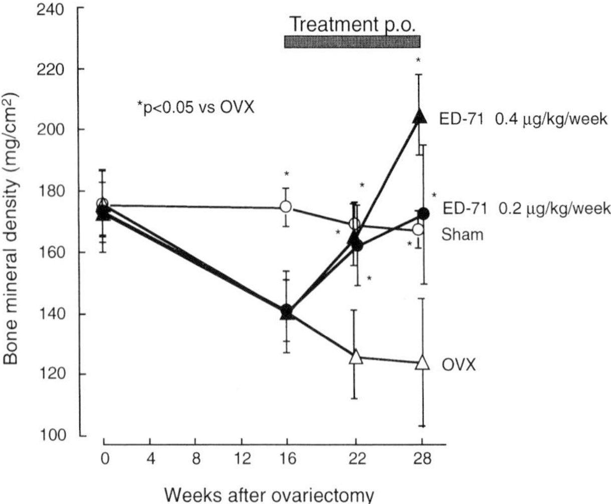

FIGURE 17 Recovery effect of ED-71 on decreased bone mass in rats with established osteoporosis.

2. Effects on Bone Mineral Density in Other Osteopenic Rat Models

Tanaka *et al.* reported the effects of ED-71 on bone dynamics and strength in rats treated with prednisolone (PSL) [36]. Although age-dependent increases of spinal BMD, compressive strength, bone volume, and trabecular thickness were significantly reduced in prednisolone (PSL)-treated rats, ED-71 dose-dependently increased these parameters. Moreover, bone formation rates (BFR/BS) and perimeter ratios of double labels over single labels (dLS/sLS) decreased by PSL treatment were increased by ED-71 in a dose-dependent manner. Finally, it was concluded that PSL affects age-related changes in bone metabolism and that ED-71 counteracts these effects by increasing intestinal calcium absorption, reducing bone resorption, and enhancing mineralization [36].

We examined the effects of ED-71 on BMD increases in an orchidectomized osteopenic model (ORCH). Vertebral BMD values in ORCH controls were significantly

TABLE III Femoral Bone Mineral Densities and Biomechanical Properties in Ovariectomized Rats Administered with ED-71[a]

	Sham, vehicle	Ovariectomy		
		Vehicle	ED-71 (0.2 μg/kg/week)	ED-71 (0.4 μg/kg/week)
Femoral length (cm)	3.75 ± 0.06	3.71 ± 0.07	3.66 ± 0.05	3.69 ± 0.06
Average BMD (mg/cm^2)				
Proximal third	188.4 ± 7.4*	157.4 ± 10.9	190.4 ± 19.7*	202.4 ± 10.0*
Medial third	180.2 ± 7.5*	145.4 ± 7.1	173.1 ± 19.3*	173.5 ± 10.3*
Distal third	181.9 ± 8.8*	135.8 ± 5.6	176.6 ± 23.8*	196.9 ± 14.3*
Whole	183.1 ± 6.8*	146.1 ± 6.7	179.4 ± 20.4*	189.2 ± 10.3*
Mechanical strength (three-point bending test)				
Maximum load (kgf)	29.2 ± 3.9***	20.7 ± 3.4	28.7 ± 4.3*	31.4 ± 4.4*
Stiffness (kgf/mm)	10.7 ± 2.4**	7.8 ± 1.0	10.5 ± 2.0*	11.1 ± 2.7***
Cortical area (mm^2)	8.7 ± 0.6*	7.1 ± 0.7	9.4 ± 1.2***	9.3 ± 1.3**

[a] Values are means ± SD. *, $p < 0.05$; **, $p < 0.01$; ***, $p < 0.001$ different from the OVX controls.

lower than in the sham controls, and ED-71 lowered bone losses to the sham control levels (N. Kubodera, K. Sato, and Y. Nishii, unpublished observation). These results indicate that ED-71 preserves or increases bone mass in several osteopenic animal models, such as steroid-induced and orchidectomized rats, as well as in ovariectomized rats.

3. Differences in the Mechanism of Bone Mass Increase Between 1α-Hydroxyvitamin D_3 and ED-71

a. Increase of Bone Mass in Normal Rats We further examined the effects of ED-71 on bone mass increase in normal rats. Figure 18 shows the time course change of spinal BMD values in normal rats sequentially monitored by DXA. The spinal BMD in the vehicle-administered control group showed no marked changes throughout the experimental period. ED-71 gradually increased the spinal BMD during the administration period, and the values of spinal BMD in the group treated with 0.2 or 0.4 μg/kg/week of ED-71 increased by approximately 20% against those in control group by the twelfth week of administration. 1αOHD_3 increased spinal BMD, but significant effects were observed only in the eighth week in the group administered with 0.6 μg/kg/week of 1αOHD_3.

Bone morphometric analysis indicated that ED-71 is capable of stimulating bone formation in rats and that its activity is higher than that of 1αOHD_3. The relative osteoid surface, osteoid volume, and the values of both the double-labeling surface and the bone formation rate

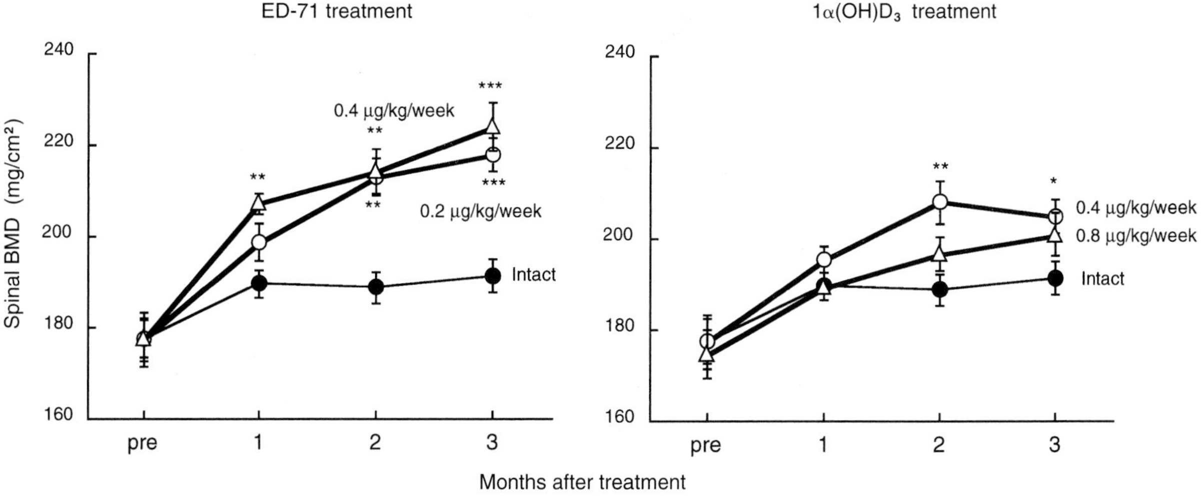

FIGURE 18 Time course of changes in spinal BMD values in rats treated with ED-71 or $1OHD_3$.

were markedly increased by the ED-71 administration in a dose-dependent manner (data not shown).

b. Increase of Bone Formation in Ovariectomized Rats It has been reported that the *in vivo* administration of active vitamin D_3 decreases osteoclast recruitment and reduces bone turnover in rats [37], dogs [38], and humans [39–41]. It has also been reported that vitamin D metabolites, such as $1,25(OH)_2D_3$, $24R,25(OH)_2D_3$, and $23S,25R$-1α,25-dihydroxyvitamin D_3-26,23-lactone, were capable of increasing bone volume by stimulating bone formation in normal vitamin D-replete rats [42–45], rabbits [46], and mice [47], as described in the previous section. However, these metabolites were found to be incapable of increasing bone formation in an estrogen-deficient condition after ovariectomy [48–50].

We therefore performed a dosing study of ED-71 to elucidate whether ED-71 increases bone mass in OVX rats compared with $1\alpha OHD_3$ [51]. The mineralized surfaces and the bone formation ratio were significantly extended in the ED-71 treatment groups when compared with the OVX controls; however, these values in the $1\alpha OHD_3$ treatment group did not differ from the OVX controls. The osteoclastic surface and number of osteoclasts tended to decrease in both the ED-71 and $1\alpha OHD_3$-treatment groups, and marked decreases were observed in the $1\alpha OHD_3$ treatment group.

These observations taken together with the results of the normal rat study indicate that ED-71 administration increases bone formation at the tissue and cellular levels. In the estrogen-deficient condition, $1\alpha OHD_3$ administration did not increase bone formation at either the tissue or cellular levels. ED-71, however, tended to dose-dependently decrease the indices of bone resorption, such as the number of osteoclasts, whereas $1\alpha OHD_3$ gave a marked reduction in the bone resorption indices. The changes observed in animals treated with ED-71 seem to be different from those of animals treated with other vitamin D derivatives. First, the reduction in bone resorption was not as marked as in animals administered with other vitamin D derivatives. Second, augmentation of bone formation seemed to be more potent, and ED-71 increased bone formation surfaces in an estrogen-deficient condition. However, the mechanism by which ED-71 increases bone formation without greatly reducing bone turnover is not clear. A phase I clinical trial of ED-71 with the ultimate aim of treating osteoporosis is now in progress.

V. FUTURE POTENTIAL OF VITAMIN D ANALOGS

Although more than 40 metabolites of vitamin D have been found to occur naturally, $1,25(OH)_2D_3$ has been the primary target of investigation. We considered that, in addition to $1,25(OH)_2D_3$, other naturally occurring metabolites which have been regarded as mostly biologically inactive intermediates in the metabolic pathway of vitamin D may be responsible for certain physiological activities. This possibility might guide the design of new analogs.

Whereas $1,25(OH)_2D_3$ and $1\alpha OHD_3$ are the first generation of drugs from vitamin D research, the com-

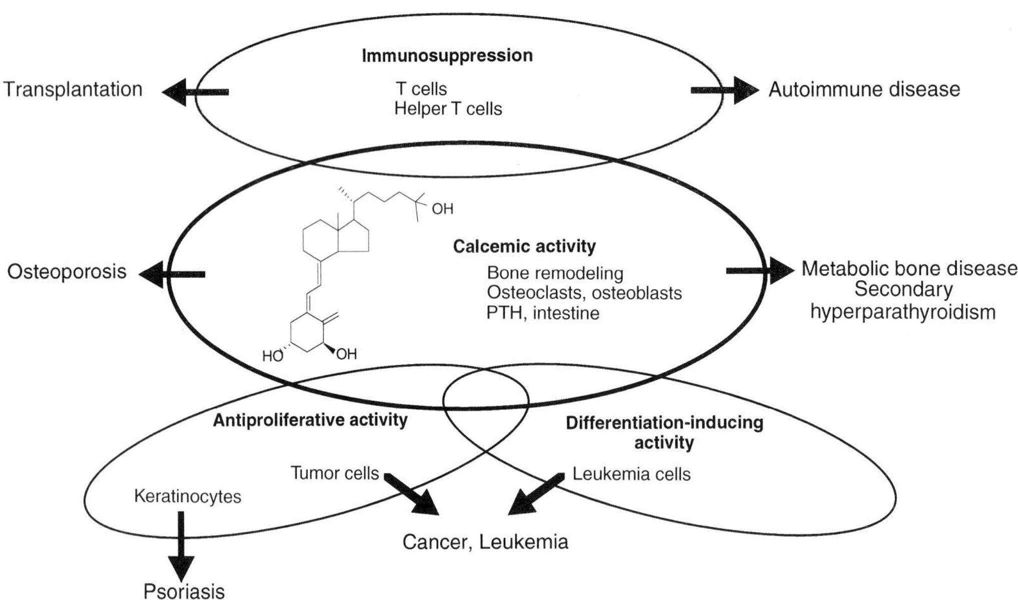

FIGURE 19 Relationship between the multiple functions of vitamin D analogs and their therapeutic applications.

pounds OCT and ED-71 derived from research to separate differentiation-inducing activities from calcemic activities should be considered to be second generation analogs. During the development of these second generation candidates as useful drugs, the detailed mechanisms of the biological actions of these vitamin D analogs should be clarified by progress in molecular biological methodology. We expect that these insights will then lead to the discovery of new analogs with fully distinct biological characteristics, that is, the third generation of vitamin D analogs.

Figure 19 shows the biological activities of vitamin D analogs and their therapeutic applications. It may be possible to identify one specific action from among other biological effects by modifying the structure of $1,25(OH)_2D_3$. It has frequently been reported that hypercalcemia, a side effect which is well known to clinicians, is induced by raising the doses of vitamin D analogs. Consequently, if it is possible to design specific analogs to induce a specific biological activity at a dose below that at which hypercalcemia occurs, the effect of that vitamin D analog would undoubtedly be a great improvement in the treatment of the target disease.

Acknowledgments

We are very grateful to the following people for contributing their effort and ideas: E. Ogata and colleagues (Cancer Institute Hospital, Tokyo, Japan), T. Suda and colleagues (Showa University, Tokyo, Japan), T. Kobayashi and colleagues (Kobe Pharmaceutical University, Kobe, Japan), T. Nakamura and colleagues (University of Occupational and Environmental Health, Fukuoka, Japan), T. Akizawa and colleagues (Showa University, Tokyo, Japan), K. Kurokawa and colleagues (Tokai University, Kanagawa, Japan), L. G. Raisz and colleagues (University of Connecticut Health Center, Farmington, CT), and E. Slatopolsky and colleagues (Washington University, St. Louis, MO). We also thank our many co-workers in the Research Laboratories and Clinical Department of Chugai Pharmaceutical Co., Ltd.

References

1. Abe E, Miyaura C, Sakagami H, Takeda M, Konno K, Yamazaki T, Yoshiki S, Suda T 1981 Differentiation of mouse myeloid leukemia cells induced by 1α,25-dihydroxyvitamin D_3. Proc Natl Acad Sci USA **78**:4990–4994.
2. Bouillon R, Okamura WH, Norman AW 1995 Structure–function relationships in the vitamin D endocrine system. Endocr Rev **16**:200–257.
3. Murayama E, Miyamoto K, Kubodera N, Mori T, Matsunaga I 1986 Synthetic studies of vitamin D_3 analogues. VIII. Synthesis of 22-oxavitamin D_3 analogues. Chem Pharm Bull (Tokyo) **34**:4410–4413.
4. Miyamoto K, Murayama E, Ochi K, Watanabe H, Kubodera N 1993 Synthetic studies of vitamin D analogues. XIV. Synthesis and calcium regulating activity of vitamin D_3 analogues bearing a hydroxyalkoxy group at the 2β-position. Chem Pharm Bull (Tokyo) **41**:1111–1113.
5. Abe J, Morikawa M, Miyamoto K, Kaiho S, Fukushima M, Miyaura C, Abe E, Suda T, Nishii Y 1987 Synthetic analogues of vitamin D_3 with an oxygen atom in the side chain skeleton. FEBS LEtt **226**:58–62.
6. Kawase A, Hirata M, Endo K, Kubodera N 1995 Synthesis of 1α,25-dihydroxy-22-thiavitamin D_3 and related analogs. BioMed Chem Lett **5**:279–282.
7. Kubodera N, Miyamoto K, Akiyama M, Matsumoto M, Mori T 1991 Synthetic studies of vitamin D analogues. IX. Synthesis and differentiation-inducing activity of 1α,25-dihydroxy-23-oxa-, thia-, and azavitamin D_3. Chem Pharm Bull (Tokyo) **39**:3221–3224.
8. Kubodera N, Miyamoto K, Matsumoto M, Kawanishi T, Ohkawa H, Mori T 1992 Synthetic studies of vitamin D analogues. X. Synthesis and biological activities of 1α,25-dihydroxy-21-norvitamin D_3. Chem Pharm Bull (Tokyo) **40**:648–651.
9. Kubodera N, Miyamoto K, Ochi K, Matsunaga I 1986 Synthesis studies of vitamin D analogues. VII. Synthesis of 20-oxa-21-norvitamin D_3 analogues. Chem Pharm Bull (Tokyo) **34**:2286–2289.
10. Kubodera N, Watanabe H, Kawanishi T, Matsumoto M 1992 Synthetic studies of vitamin D analogues. XI. Synthesis and differentiation-inducing activities of 1α,25-dihydroxy-22-oxavitamin D_3 analogues. Chem Pharm Bull (Tokyo) **40**:1494–1499.
11. Kubodera N, Watanabe H, Kawanishi T, Matsumoto M 1992 Synthetic studies of vitamin D analogues. XI. Synthesis and differentiation-inducing activity of 1α,25-dihydroxy-22-oxavitamin D_3 analogues. Chem Pharm Bull (Tokyo) **40**:1494–1499.
12. Kubodera N, Watanabe H, Miyamoto K, Matsumoto M, Matsuoka S, Kawanishi T 1993 Synthetic studies of vitamin D analogues. XVII. Synthesis and differentiation-inducing activity of 1α,24-dihydroxy-22-oxavitamin D_3 analogues and their 20(*R*)-epimers. Chem Pharm Bull (Tokyo) **41**:1659–1663.
13. Kubodera N, Watanabe H, Miyamoto K, Matsumoto M 1993 Synthesis and differentiation-inducing activity of 1α,24-dihydroxy-22-oxavitamin D_3 analogues. BioMed Chem Lett **3**:1869–1872.
14. Okano T, Tsugawa N, Masuda S, Takeuchi A, Kobayashi T, Takita Y, Nishii Y 1989 Regulatory activities of 2β-(3-hydroxypropoxy)-1[α,25-dihydroxyvitamin D_3, a novel synthetic vitamin D_3 derivative, on calcium metabolism. Biochem Biophys Res Commun **163**:1444–1449.
15. Nishii Y, Abe J, Sato K, Kobayashi T, Okano T, Tsugawa N, Slatopolski E, Brown AJ, Dusso AD, Raisz LG 1991 Characteristic of two novel vitamin D_3 analogues; 22-oxa-1α,25-dihydroxyvitamin D_3 [OCT] and 2β-(3-hydroxypropoxy)-1α,25-dihydroxyvitamin D_3 [ED-71]. In: Norman AW, Bouillon R, Thomasset M (eds) Vitamin D: Gene Regulation, Structure–Function Analysis and Clinical Application. de Gruyter, Berlin, New York, pp. 289–297.
16. Tsugawa N, Okano T, Masuda S, Kobayashi T, Kubodera N, Sato K, Nishii Y 1994 Effects of vitamin D_3 analogues on parathyroid hormone secretion and calcium metabolism in vitamin D-deficient rats. J Bone Miner Metab **12**(Suppl. 1):S13–S17.
17. Morimoto S, Imanaka S, Koh E, Shiraishi T, Nabata T, Kitano S, Miyashita Y, Nishii Y, Ogihara T 1989 Comparison of the inhibitions of proliferation of normal and psoriatic fibroblasts by 1α,25-dihydroxyvitamin D_3 and synthetic analogues of vitamin D_3 with an oxygen atom in the side chain. Biochem Int **19**:1143–1149.
18. Abe J, Takita Y, Nakano T, Miyaura C, Suda T, Nishii Y 1989 A synthetic analogue of vitamin D_3, 22-oxa-1α,25(OH)-dihydroxyvitamin D_3, is a potent modulator of in vivo immunoregulat-

ing activity without inducing hypercalcemia in mice. Endocrinology **124**:2645–2647.

19. Abe J, Nakamura K, Takita Y, Nakano T, Irie H, Nishii Y 1990 Prevention of immunological disorders in MRL/l mice by a new synthetic analogue of vitamin D_3: 22-Oxa-1α,25-dihydroxyvitamin D_3. J Nutr Sci Vitaminol (Tokyo) **36**:21–31.
20. Kitamura A 1994 Study of the effects of a new vitamin D_3 derivative on type II collagen-induced arthritis in an experimental rat model of rheumatoid arthritis. Jpn Orthop Assoc **68**:1068–1080.
21. Abe J, Nakano T, Nishii Y, Matsumoto T, Ogata E, Ikeda K 1991 A novel vitamin D_3 analog, 22-oxa-1,25-dihydroxyvitamin D_3, inhibits the growth of human breast cancer in vitro and in vivo without causing hypercalcemia. Endocrinology **129**:832–837.
22. Abe-Hashimoto J, Kikuchi T, Matsumoto T, Nishii Y, Ogata E, Ikeda K 1993 Antitumor effect of 22-oxacalciferol, a noncalcemic analogue of calcitriol, in athymic mice implanted with human breast carcinoma and its synergism with tamoxifen. Cancer Res **53**:2534–2537.
23. Skowronski RJ, Peehl DM, Feldman D 1995 Actions of vitamin D_3 analogs on human prostate cancer cell lines: Comparison with 1,25-dihydroxyvitamin D_3. Endocrinology **136**:20–26.
24. Kawa S, Yoshizawa K, Kotoo M, Imai H, Oguchi H, Kiyosawa K, Homma T, Nikaido T, Furihata K 1996 Inhibitory effect of 22-oxa-1,25-dihydroxyvitamin D_3 on the proliferation of pancreatic cancer cell lines. Gastroenterology **110**:1605–1613.
25. Endo K, Ichikawa F, Uchiyama Y, Katsumata K, Ohkawa H, Kumaki K, Ogata E, Ikeda K 1994 Evidence for uptake of vitamin D analogue (OCT) by human carcinoma and its effect of suppressing the transcription of parathyroid hormone-related peptide gene *in vivo*. J Biol Chem **269**:32693–32699.
26. Oikawa T, Hirotani K, Ogasawara H, Katayama T, Nakamura O, Iwaguchi T, Hiragun A 1990 Inhibition of angiogenesis by vitamin D_3 analogues. Eur J Pharmacol **178**:247–250.
27. Oikawa T, Shimamura M, Ashino H, Morita I, Murata S, Abe J, Nishii Y, Tominaga T 1993 22-Oxa-1,25$(OH)_2D_3$, a potent angiostatic vitamin, exerts an antitumor effect without producing serious side effect such as hypercalcemia. Drug News & Perspectives **6**:157–162.
28. Oikawa T, Yoshida Y, Shimamura M, Ashino-Fuse H, Iwaguchi T, Tominaga T 1991 Antitumor effect of 22-oxa-1α,25-dihydroxyvitamin D_3, a potent angiogenesis inhibitor, on rat mammary tumors induced by 7,12-dimethylbenz[*a*]anthracene. Anti-Cancer Drugs **2**:475–480.
29. Brown AJ, Ritter CR, Finch JL, Morrissey J, Martin KJ, Murayama E, Nishii Y, Slatopolski E 1989 The noncalcemic analogue of vitamin D, 22-oxacalcitriol, suppresses parathyroid hormone synthesis and secretion. J Clin Invest **84**:728–732.
30. Inoue M, Wakasugi M, Wakao R, Gan N, Tawata M, Nishii Y, Onaya T 1992 A synthetic analogue of vitamin D_3, 22-oxa-1,25-dihydroxyvitamin D_3, stimulates the production of prostacyclin by vascular tissues. Life Sci **51**:1105–1112.
31. Brown AJ, Finch J, Griff M, Ritter C, Kubodera N, Nishii Y, Slatopolski E 1993 The mechanism for the disparate actions of calcitriol and 22-oxacalcitriol in the intestine. Endocrinology **133**:1158–1164.
32. Kurokawa K, Akizawa T, Suzuki M, Akiba T, Ogata E, Slatopolski E 1996 Effect of 22-oxacalcitriol on hyperparathyroidism of dialysis patients: Results of a preliminary study. Nephrol Dial Transplant **11** (Suppl. 3):121–124.
33. Sato K, Nishii Y, Woodiel FN, Raisz LG 1993 Effects of two new vitamin D_3 derivatives, 22-oxa-1α,25-dihydroxyvitamin D_3 (OCT) and 2β-(3-hydroxypropoxy)-1α,25-dihydroxyvitamin D_3 (ED-71), on bone metabolism in organ culture. Bone **14**:47–51.
34. Kobayashi T, Okano T, Tsugawa N, Murano M, Masuda S, Takeuchi A, Sato K, Nishii Y 1993 2β-(3-Hydroxypropoxy)-1α,25-dihydroxyvitamin D_3 (ED-71), preventive and therapeutic effects on bone mineral loss in ovariectomized rats. BioMed Chem Lett **13**:1815–1819.
35. Okumura H, Yamamuro T, Kasai R, Iwashita Y, Ikeda T 1990 Immobilization combined with ovariectomy and effect of active vitamin D_3 analogues in the rat. Cells and Materials (Suppl. 1):125–130.
36. Tanaka Y, Nakamura T, Nishida S, Suzuki K, Takeda S, Sato K, Nishii Y 1996 Effects of a synthetic vitamin D analog, ED-71, on bone dynamics and strength in cancellous and cortical bone in prednisolone-treated rats. J Bone Miner Res **11**:325–336.
37. Faugere MC, Okamoto S, DeLuca HF, Malluche HH 1986 Calcitriol corrects bone loss induced by oophorectomy in rats. Am J Physiol **250**:E35–E38.
38. Malluche HH, Faugere MC, Friedlar RM, Fanti P 1988 1,25-Dihydroxyvitamin D_3 corrects bone loss but suppresses bone remodeling in ovariohysterectomized beagle dogs. Endocrinology **122**:1998–2009.
39. Aloia JF, Vaswani A, Yeh KEJK, Yasumura S, Cohn SH 1988 Calcitriol in the treatment of postmenopausal osteoporosis. Am J Med **84**:401–408.
40. Gallagher JC, Jerpbak CM, Jee WSS, Johnson KA, DeLuca HF, Riggs BL 1982 1,25-Dihydroxyvitamin D_3: Short- and long-term effects on bone and calcium metabolism in patients with postmenopausal osteoporosis. Proc Natl Acad Sci USA **79**:3325–3329.
41. Riggs BL, Nelson KI 1985 Effect of long-term treatment with calcitriol on calcium absorption and mineral metabolism in postmenopausal osteoporosis. J Clin Endocrinol Metab **61**:457–461.
42. Boyce RW, Weisbrode SE 1983 Effect of dietary calcium on the response of bone to 1,25$(OH)_2D_3$. Lab Invest **48**:683.
43. Boyce RW, Weisbrode SE 1985 The histogenesis of hyperosteoidosis in 1,25$(OH)_2D_3$-treated rats fed high levels of dietary calcium. Bone **6**:105.
44. Nakamura T, Kurokawa T, Orimo H 1988 Increase of bone volume in vitamin D-repleted rats by massive administration of 24*R*,25$(OH)_2D_3$. Calcif Tissue Int **43**:235–243.
45. Wronski TJ, Halloran BP, Bikle DD, Morey-Holton ER 1986 Chronic administration of 1,25-dihydroxyvitamin D_3 increased bone but impaired mineralization. Endocrinology **119**:2580–2585.
46. Nakamura T, Suzuki K, Hirai T, Kurokawa T, Orimo H 1992 Increased bone volume and reduced bone turnover in vitamin D-replete rabbits by the administration of 24*R*,25-dihydroxyvitamin D_3. Bone **13**:229–236.
47. Shima M, Tanaka H, Norman AW, Yamaoka K, Yoshikawa H, Toakaoka K, Ishizuka S, Seino Y 1990 23(*S*),25(*R*)-1,25-dihydroxyvitamin D_3-26,23-lactone stimulates murine bone formation in vivo. Endocrinology **126**:832–836.
48. Lindgren JU, DeLuca HF 1982 Role of parathyroid hormone and 1,25-dihydroxyvitamin D_3 in the development of osteopenia in oophorectomized rats. Calcif Tissue Int **34**:510–514.
49. Lindgren JU, Lindholm TS 1979 Effect of 1α-hydroxyvitamin D_3 on osteoporosis in rats induced by oophorectomy. Calcif Tissue Int **27**:161–164.
50. Nakamura T, Nagai Y, Yamato H, Suzuki K, Orimo H 1991 Regulation of bone turnover and prevention of bone atrophy in ovariectomized beagle dogs by the administration of 24*R*,25$(OH)_2D_3$. Calcif Tissue Int **50**:221–227.
51. Tsurukami H, Nakamura T, Suzuki K, Sato K, Higuchi Y, Nishii Y 1994 A novel synthetic vitamin D analogue, 2β-(3-hydroxypropoxy)1α,25-dihydroxyvitamin D_3 (ED-71), increases bone mass by stimulating the bone formation in normal and ovariectomized rats. Calcif Tissue Int **54**:142–149.

SECTION VIII

Emerging Therapeutic Uses

Chapter 64

Vitamin D: Anticancer and Differentiation

Johannes P. T. M. van Leeuwen and Huibert A. P. Pols
Department of Internal Medicine III, Erasmus University Medical School, Rotterdam, The Netherlands

I. INTRODUCTION

The seco-steroid hormone 1,25-dihydroxyvitamin D_3 [$1,25(OH)_2D_3$] is the most potent metabolite of vitamin D_3 and is an important regulator of calcium homeostasis and bone metabolism via actions in intestine, bone, kidney, and parathyroid glands. $1,25(OH)_2D_3$ exerts its effects via an intracellular receptor that is a member of the steroid hormone receptor family (see Chapters 8–10 in this book). In recent years it has become evident that a wide variety of cells and tissues not primarily related to calcium and bone metabolism also contain the vitamin D receptor (VDR). As exemplified in Table I, the VDR has also been demonstrated in a broad range of tumors and malignant cell types. For colon and breast cancer cells an inverse relationship between VDR level and degree of differentiation has been described by some investigators [1,2]. However, no associations between VDR and clinical and biochemical parameters of breast cancer were found in several other studies [3–5].

The widespread distribution of the VDR in malignant cells indicates that regulation of cancer cell function might be a new target in the action of $1,25(OH)_2D_3$. In this chapter we describe the present knowledge on vitamin D regulation of tumor cells, its possible mechanism, and potential clinical applications.

II. VITAMIN D AND CANCER

A. Growth and Development

In 1980 an epidemiological study based on indirect evidence suggested a relationship between vitamin D and cancer. This was derived from analyses of death rates from colon cancer which tended to increase with increasing latitude and decreasing sunlight [6]. Later more direct evidence about a relation between vitamin D and colon cancer came from the inverse relationship between levels of serum 25-hydroxyvitamin D_3 [a $1,25(OH)_2D_3$ precursor] and incidence of colonic cancer [7,8]. In addition, a similar relationship between sunlight exposure, vitamin D, and the risk for fatal breast and prostate cancer has been suggested [9–13]. However, a Canadian study noted similar vitamin D intakes in breast cancer patients and control subjects [14]. The relationship between sunlight exposure and cancer, es-

TABLE I VDR in Tumors and Malignant Cell Types

Breast carcinoma	Osteogenic sarcoma
Cervical carcinoma	Ovarian carcinoma
Colonic adenocarcinoma	Pancreatic carcinoma
Colorectal carcinoma	Pituitary adenoma
Gall bladder carcinoma	Prostate carcinoma
Lung carcinoma	Renal cell carcinoma
Lymphocytic leukemia	Squamous cell carcinoma
Malignant melanoma	Transitional cell bladder carcinoma
Medullary thyroid carcinoma	Uterine carcinosarcoma
Myeloid leukemia	

pecially with respect to vitamin D, has been carefully reviewed by Studzinski and Moore [15]. Also, the relationship between cancer, diet, and calcium intake and vitamin D has been addressed in several studies [16–19]. However, in a mouse model no relationship was found between dietary intake of a wide range of doses of calcium or vitamin D on carcinogen-induced skin tumors [20]. From these studies it is clear that sunlight exposure, vitamin D intake, and other dietary components such as calcium and fat should be considered as possibly interacting with one another when the relationship between vitamin D and cancer risk is assessed.

In addition to these epidemiological studies, since the early 1980s an increasing amount of cell biological data have demonstrated that at high concentrations (10^{-9}–10^{-7} *M*) $1,25(OH)_2D_3$ inhibits the growth of tumor cells *in vitro*. It was demonstrated as early as 1981 that $1,25(OH)_2D_3$ inhibits the growth of malignant melanoma cells and stimulates the differentiation of immature mouse myeloid leukemia cells in culture [21,22]. $1,25(OH)_2D_3$ also induces differentiation of normal bone marrow cells. Immature bone marrow cells of the monocyte–macrophage lineage are believed to be the precursors of osteoclasts, and $1,25(OH)_2D_3$ induces differentiation of immature myeloid cells toward monocytes–macrophages and also stimulates the activation and fusion of some macrophages (discussed in Chapters 21, 29, and 68). From these results it has been postulated that $1,25(OH)_2D_3$ stimulates differentiation and fusion of osteoclast progenitors into osteoclasts [23–25]. Also, in the intestine, $1,25(OH)_2D_3$ has important effects on cellular proliferation and differentiation [26]. Thus the differentiation inducing capacity of bone and interstitial cells, $1,25(OH)_2D_3$ may play an important role in the regulation of calcium and bone metabolism. These *in vitro* findings were followed by the *in vivo* observation that $1,25(OH)_2D_3$ prolongs the survival time of mice inoculated with myeloid leukemia cells [27]. As shown in Table II, over the years $1,25(OH)_2D_3$ has been shown to have beneficial effects in several other *in vivo* models of various types of cancers [28–42].

TABLE II *In Vivo* Effects of $1,25(OH)_2D_3$ and $1\alpha OHD_3$ in Animal Models of Cancer[a]

Tumor	Model	Effect	Refs.
Breast	NMU-induced breast cancer in rats	Tumor suppression	28
Breast	DMBA induced breast cancer in rats	Tumor suppression	29
Colon	Human colon cell line implanted into nude mice	Tumor suppression	30,31
Colon	DMH-induced colon cancer in rats	Reduction of the incidence of colon adenocarcinomas	32
Leydig tumor	Leydig cell tumor implanted into rats	Tumor suppression	33
Lung	Implantation of Lewis lung carcinoma into mice	Reduction of the number of metastases (without suppression of primary tumor)	34
Lung	Implantation of Lewis lung carcinoma into mice	Tumor suppression	35
Lung	Implantation of lung carcinoma cells into mice	Reduction of the number of metastases (without suppression of primary tumor)	36
Melanoma	Human melanoma cells implanted into nude mice	Tumor suppression	30
Osteosarcoma	Human osteosarcoma cells implanted into nude mice	Tumor suppression	37
Retinoblastoma	Retinoblastoma cell line implanted into nude mice	Tumor suppression	38
Retinoblastoma	Transgenic mice with retinoblastoma	Tumor suppression	39
Walker carcinoma	Walker carcinoma cells injected in rats	Tumor suppression	40
Skin	DMBA/TPA-induced skin tumors in mice	Inhibition of tumor formation	41
Skin	DMBA/TPA-induced skin tumors in mice	Inhibition of tumor formation	42

[a] The dosage, duration of treatment, diet, and effects on serum/urinary calcium vary among the studies. NMU, Nitrosomethylurea; DMBA, 7,12-dimethylbenz[*a*]anthracene; DMH, 1,2-dimethylhydrazine dihydrochloride; TPA, 12-*O*-tetradecanoylphorbol-13-acetate.

Up to now a few clinical trials of vitamin D in cancer have been performed. Alfacalcidol (1α-hydroxyvitamin D_3; 1αOHD_3), which is converted to 1,25(OH)$_2$D$_3$ *in vivo*, caused a beneficial response in low-grade non-Hodgkin's lymphoma patients [43]. Also, with alfacalcidol, transient improvement in peripheral blood counts was seen in patients with myelodysplasia, however, half of the patients developed hypercalcemia [44]. Another study reported a sustained hematological response in six myelodysplasia patients treated with high doses of alfacalcidol [45]. These patients were restricted in their dietary calcium intake; nevertheless, four patients developed hypercalcemia due to increased bone resorption. With respect to treatment of cutaneous T-cell lymphoma with a combination of 1,25(OH)$_2$D$_3$ and retinoids, contrasting results have been obtained. It has been suggested that the variability was due to differences in phenotype of the various lymphomas [46–50].

An important aspect and limitation of the treatment of cancer with 1,25(OH)$_2$D$_3$ was revealed by this limited set of clinical trials; to achieve growth inhibition, high doses are needed (confirming the *in vitro* data), which can cause the side effect of hypercalcemia. This has prompted the development of analogs of 1,25(OH)$_2$D$_3$ in order to dissociate the antiproliferative effect from the calcemic and bone metabolism effects (see Chapters 57–63) [51,52]. Although the precise mechanism is not completely understood, at the moment several 1,25(OH)$_2$D$_3$ analogs are available that seem to fulfil these criteria. In Table III the *in vivo* animal studies using 1,25(OH)$_2$D$_3$ analogs on various cancer types are summarized [54–66]. In humans the analog calcipotriol (MC903) has been used for topical treatment of advanced breast cancer, however, several of the patients still developed hypercalcemia [53]. In Chapter 65 results of a phase I trial of the 1,25(OH)$_2$D$_3$ analog EB1089 in advanced breast and colon cancer is discussed, and at the moment phase I and II studies on pancreatic and breast cancer are in progress.

B. Angiogenesis and Metastasis

For the tumor suppressive activity of vitamin D_3 compounds *in vivo*, besides growth inhibition, two additional aspects may be involved. First, angiogenesis is an essential requirement for the growth of solid tumors. Compounds that inhibit angiogenesis might therefore contribute to antitumor therapy. Antiangiogenic drugs may lead to inhibition of tumor progression, stabilization of tumor growth, tumor regression, and prevention of metastasis. Antiangiogenic effects may play a role in the tumor suppressive activity of vitamin D_3 compounds. Two studies reported an antiangiogenic effect of 1,25(OH)$_2$D$_3$ and the analog 22-oxacalcitriol using different experimental model systems [56,67]. In addition

TABLE III *In Vivo* Effects of 1,25(OH)$_2$D$_3$ Analogs in Animal Models for Cancer[a]

Vitamin D_3 analog	Model	Antitumor effect	Rise in serum Ca	Refs.
OCT	Breast	Tumor suppression	None	54,55
OCT	Breast	Tumor suppression	None	56
OCT	Breast	Tumor suppression	None	55
OCT	Colon	Decreased tumor incidence	None	57
OCT	Intestine	Decreased tumor incidence	Marked	58
MC903	Breast	Tumor suppression	Mild	59
EB1089	Breast	Tumor suppression	None	60,61
EB1089	Leydig cell tumor	Tumor suppression	None	33
CB966	Breast	Tumor suppression	Mild	61
DD-003	Colon	Tumor suppression	None	62
Ro 23-7553	Prostate	Tumor suppression	None	63
Ro 23-7553	Leukemia	Increased survival	None	64
Ro 24-5531	Breast	Decreased tumor incidence	None	65
Ro 24-5531	Colon	Decreased tumor incidence	None	66

[a] OCT, 22-Oxacalcitriol; MC903, see Table III of Chapter 61; EB1089, see Table II of Chapter 61; CB966, see Table I of Chapter 61; DD-003, 22(*S*)-24-homo-26,26,26,27,27,27-hexafluoro-1α,22,25-trihydroxyvitamin D_3; Ro 23-7553, 1,25-dihydroxy-16-ene-23-yne-vitamin D_3; Ro 24-5531, 1,25-dihydroxy-16-ene-23-yne-26,27-hexafluorovitamin D_3.

more recent data have demonstrated that 1,25$(OH)_2D_3$ inhibits angiogenesis induced by the human papilloma virus type 16 (HPV16)- or HPV18-containing cell lines HeLa, Skv-e2, and Skv-e12 when intradermally injected into immunosuppressed mice [68]. Also, with the non-virus-transformed human cell lines T47-D (breast carcinoma) and A431 (vulva carcinoma), similar results were obtained [69]. In these studies the mice were treated for 5 days with 1,25$(OH)_2D_3$ prior to the injection of tumor cells. The effect of 1,25$(OH)_2D_3$ on angiogenesis can be due to inhibition of tumor cell proliferation, resulting in less angiogenic cells. However, inhibition of angiogenesis could also be observed when the tumor cells were treated *in vitro* with 1,25$(OH)_2D_3$ and, after cell washing, were injected into mice [69]. Under these conditions both control and 1,25$(OH)_2D_3$-treated mice were injected with similar numbers of cells. Therefore, these data indicate that 1,25$(OH)_2D_3$ inhibits the release of angiogenic factors (vascular endothelium growth factor, transforming growth factor-α, basic fibroblast growth factor, epidermal growth factor, etc.) or stimulates antiangiogenic factors. Finally, in another model, 1,25$(OH)_2D_3$ has also been shown to reduce angiogenesis in retinoblastomas in mice [70].

The second mechanism of antitumor activity relates to metastasis, which is the primary cause of the fatal outcome of cancer diseases. A study by Mork Hansen *et al.* indicated that 1,25$(OH)_2D_3$ may be effective in reducing the invasiveness of breast cancer cells [71]. They showed that 1,25$(OH)_2D_3$ inhibited the invasion and migration of a metastatic human breast cancer cell line (MDA-MB-231) using the Boyden chamber invasion assay. In an *in vivo* study it was shown that 1,25$(OH)_2D_3$ reduces the metastasis to the lung of subcutaneously implanted Lewis lung carcinoma cells [35]. A fact to be considered in relation to metastasis is that bone is the most frequent site of metastasis of advanced breast and prostate cancer. There are some indications from clinical studies that bone metastases develop preferentially in areas with high bone turnover [72,73]. In contrast, agents that inhibit bone resorption have been reported to reduce the incidence of skeletal metastasis [74]. As 1,25$(OH)_2D_3$ is an important stimulator of bone resorption and consequently of bone turnover, treatment of cancer with 1,25$(OH)_2D_3$ might theoretically increase the risk of skeletal metastases. So far, in most *in vivo* animal studies, this possibility has not been addressed. Only Krempien has reported, in abstract form, that following intraarterial injection of Walker 256 tumor cells rats treated with 1,25$(OH)_2D_3$ developed significantly more bone metastases than untreated controls [75]. Therefore, this aspect of 1,25$(OH)_2D_3$ therapy certainly needs further study. In this report, the use of vitamin D_3 analogs with reduced calcemic activity or treatment with vitamin D_3 in combination with other compounds to reduce bone turnover (see Section IV) may be helpful. The data obtained so far on angiogenesis and metastasis show that these two processes might be involved in the anticancer activities of vitamin D_3.

C. Parathyroid Hormone-Related Peptide

1,25$(OH)_2D_3$ and parathyroid hormone (PTH) mutually regulate synthesis and secretion of one another. Production and secretion of PTH are inhibited by 1,25$(OH)_2D_3$ via a transcriptional effect, and a vitamin D responsive element (VDRE) in the promoter of the PTH gene has been identified [76,77]. Parathyroid hormone-related peptide (PTHrP) was initially isolated from several carcinomas and is responsible for the humoral hypercalcemia of malignancy [78]. Although originally identified in carcinomas, PTHrP has also been identified in normal cells.

In normal human mammary epithelial cells, 1,25$(OH)_2D_3$ did not affect basal but inhibited growth factor-stimulated PTHrP expression via an effect on transcription [79]. In normal keratinocytes 1,25$(OH)_2D_3$ had no effect on PTHrP secretion in basal culture conditions [80] but did inhibit growth factor-stimulated PTHrP production as well [81]. Likewise, 1,25$(OH)_2D_3$ as well as the analogs 22-oxacalcitriol and MC903 inhibited PTHrP secretion in immortalized human keratinocytes (HPK1A), but this inhibition was less in the more malignant *ras*-transfected clone HPK1A-*ras* [82,83]. In addition to keratinocytes, 1,25$(OH)_2D_3$ and 22-oxacalcitriol inhibit PTHrP gene expression and PTHrP secretion in the human T-cell lymphotrophic virus type I (HTLV-I)-transfected T-cell line MT-2 [84] and in rat H-500 Leydig tumor cells [85]. *In vivo* observations comparable to these *in vitro* observations have also been made. When these H-500 Leydig tumor cells were implanted in Fisher rats, treatment with 1,25$(OH)_2D_3$ and the analog EB1089 resulted in reduced levels of tumor PTHrP mRNA and PTHrP serum levels [33]. In Fisher rats implanted with the Walker carcinoma, 1,25$(OH)_2D_3$ caused a decrease in serum PTHrP but the ratio of PTHrP, levels and tumor weight was similar in rats receiving vehicle or 1,25$(OH)_2D_3$. The data point to an indirect effect on PTHrP via growth inhibition. However, the PTHrP mRNA levels appeared to be decreased by 1,25$(OH)_2D_3$ [40]. In nude mice bearing the FA-6 cell line of a pancreas carcinoma lymph node metastasis, 22-oxacalcitriol inhibits PTHrP gene expression, which is related to inhibition of tumor-induced hypercalcemia [86]. In contrast to these inhibitory effects, a stimulatory effect of 1,25$(OH)_2D_3$ on PTHrP production by a canine oral squamous carcinoma cell line (Scc 2/88) has been observed [87]. Together, the overall picture that emerges from these studies is that an important additional anticancer effect of vitamin D_3 and analogs could be the inhibition of the humoral hypercalcemia of malignancy.

III. VITAMIN D EFFECTS ON TUMOR CELLS

A. Cell Cycle

To gain more insight into the actions of $1{,}25(OH)_2D_3$ on proliferation, several studies on the effect of $1{,}25(OH)_2D_3$ on cell cycle kinetics have been performed. Proliferating cells progress through the cell cycle, which comprises the G_0/G_1 phase (most differentiated, nondividing cells are in the G_1 phase), the S phase in which new DNA is synthesized, and the G_2 phase, which is followed by mitosis (M phase) whereon the cells reenter the G_0/G_1 phase. Within these major cell cycle phases, several other specific checkpoints have been described. However, thus far these have not been addressed in relation to the antiproliferative action of vitamin D. Table IV summarizes the data obtained so far on the cell cycle regulation by vitamin D_3 in various cancer cells [88–108]. In Fig. 1 the effect of $1{,}25(OH)_2D_3$ on the cell cycle of the rat osteosarcoma cell line UMR-

TABLE IV Effect of $1{,}25(OH)_2D_3$ on Cell Cycle Kinetics of Tumor Cells of Various Origins and Skin Cells[a]

Cell line	G_0/G_1 phase	S phase	G_2/M phase	Refs.
Breast				
MCF-7	↑	↓	↓ [89], = [88,91]	88–91
BT-20	↑ [89], = [90]	↑ [89], = [90]	↑ [89], = [90]	89,90
BT-474	↑	↓	n.d	90
MDA-MB-231	↑	↓	n.d.	90
MDA-MD-436	=	=	n.d.	90
SK-BR-3	↑	↓	n.d.	90
T47-D	↑	↓	↑	31,92
Hematopoietic				
HL-60	↑	↓	↑ [93], = [94]	93–96
HIMeg	↑	↓	↓	97
WEHI-3	↑	↓	n.d.	98
RWLeu-4	↑	n.d.	n.d.	99
$JMRD_3$	=	n.d.	n.d.	99
T lymphocytes[b]	↑	↓	n.d.	100,101
Intestine				
HT-29	↑			102
CaCo-2	↑	n.d.	n.d.	103
Miscellaneous				
KU-2 (renal carcinoma)	=[c]	=	=	104
SK-N-SH (neuroblastoma)	=[d]	=	=	105
Skin				
Keratinocytes	↑	↓	↑	106
Psoriatic lesion	↑	↓	n.d.	107
HPK1A-*ras*	↑	↓	n.d.	108

[a] Effects of $1{,}25(OH)_2D_3$ analogs are not specifically identified. All responses were not measured in all references. ↑, Increase; ↓, decrease; =, no effect; n.d., not determined. If different effects were observed, the references are cited in the particular columns.

[b] Normal T lymphocytes.

[c] No effect on specific cell cycle phases but doubling time increased.

[d] Low VDR level.

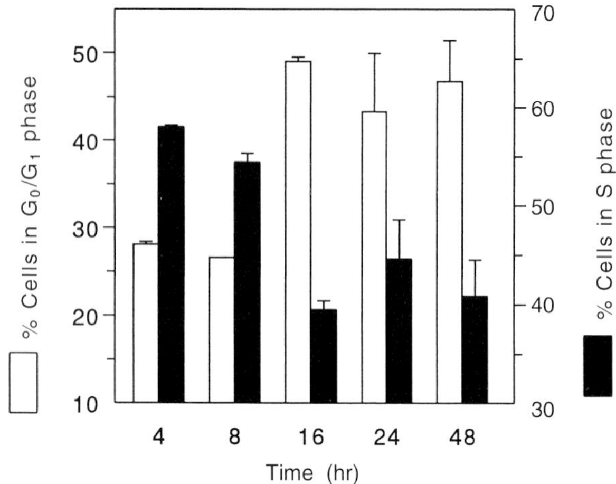

FIGURE 1 Effect of $1,25(OH)_2D_3$ on the cell cycle in UMR-106 osteosarcoma cells. Dual-parameter flow propidium iodide uptake in DNA [109] was used to examine the effects of 10^{-8} *M* $1,25(OH)_2D_3$ on cell cycle kinetics after 48 hr of incubation.

106 is shown [109]. In most of the cells studied treatment with $1,25(OH)_2D_3$ results in blockade of the cells in the G_0/G_1 phase and reduction of the number of cells in S phase. Some studies also have examined the effect on the G_2 phase, but these results are somewhat more diverse. In general it can be concluded that blocking the transition from the G_0/G_1 phase to the S phase plays an important role in the growth inhibitory effect of $1,25(OH)_2D_3$.

B. Apoptosis

A process that is directly related to the cell cycle is apoptosis. Apoptosis (programmed cell death) is an orderly and characteristic sequence of biochemical, molecular, and structural changes resulting in the death of the cell [110]. It is suggested that early G_1 phase may be the point at which switching between cell cycle progression and induction of apoptosis occurs [111,112]. Induction of apoptosis may be a mechanism by which $1,25(OH)_2D_3$ inhibits tumor cell growth. As summarized in Table V, $1,25(OH)_2D_3$ has been shown to induce apoptosis of cancer cells of various origins, including breast cancer. A central role for apoptosis in the action of $1,25(OH)_2D_3$ is unclear because growth inhibition of several other breast cancer cells appeared to be independent of apoptosis [90]. Also, MCF-7 cells that showed growth inhibition by $1,25(OH)_2D_3$ could after removal of the hormone again be stimulated to grow, implying transient growth inhibition and not cell death [113]. Stable transfection of leukemic U937 cells with the wild-type p53 tumor suppressor gene resulted in a reduced growth rate and produced cells that can undergo either apoptosis or maturation. In these cells $1,25(OH)_2D_3$ protects against p53-induced apoptosis and enhances p53-induced maturation [114]. In two independent studies with HL-60 cells, $1,25(OH)_2D_3$ was found either to protect against or to have no effects on apoptosis [94,115]. In the latter study it was shown that in the presence of 9-*cis*-retinoic acid, $1,25(OH)_2D_3$ did induce apoptosis.

In summary, the data obtained so far show that $1,25(OH)_2D_3$-induced growth inhibition can be related to apoptosis in some cases but that growth inhibition also can be observed independent of apoptosis. Possibly in these latter cases induction of differentiation is more prominent. What decides whether cells undergo apoptosis or differentiation is unclear but is probably dependent on cell cycle stage, presence of other factors, and levels of expression of oncogenes and tumor suppressor genes.

C. Oncogenes and Tumor Suppressor Genes

In relation to growth inhibition and apoptosis, the expression and regulation of several oncogenes and tumor suppressor genes have been investigated. At the moment the *bcl-2*, c-*myc*, and retinoblastoma genes and gene products are the most widely studied with respect to $1,25(OH)_2D_3$. The effects of $1,25(OH)_2D_3$ in various cell types are shown in Table V. The *bcl-2* oncogene is unique because its expression does not enhance the growth rate but decreases the rate of programmed cell death [116]. The *bcl-2* gene product has been shown to inhibit the process of apoptosis. However, protection of HL-60 cells against apoptosis (Section III,B) occurred despite down-regulation of *bcl-2* gene expression [94]. In several breast cancer cell lines (MCF-7, BT-474, MDA-MB-231) $1,25(OH)_2D_3$ and the analog KH1060 and EB1089 decreased *bcl-2* expression [90,117]. However, only in MCF-7 cells was this change in *bcl-2* expression accompanied by apoptosis.

In view of the suggested intimate relationship between regulators of the cell cycle, apoptosis, and differentiation, the oncogene c-*myc* is interesting. c-*Myc* has been postulated to play an early role in the following cascade of events in G_1: cyclins activate cyclin-dependent kinases (CDKs) which in turn can phosphorylate the retinoblastoma tumor suppressor gene product ($p110^{RB}$), resulting in transition from G_1 to S phase. In HL-60, breast cancer, and several other cell types $1,25(OH)_2D_3$ has been reported to decrease c-*myc* oncogene expression (Table V) [118–148]. Analysis of HL-60 sublines showed a relation between reduction of c-*myc* expression and inhibition of proliferation [127]. We did not observe a $1,25(OH)_2D_3$-induced change in

TABLE V Effect of $1,25(OH)_2D_3$ on Proliferation, Differentiation, Apoptosis, and Expression or Regulation of *bcl*-2, c-*myc*, and Retinoblastoma[a]

Cell line	Proliferation	Differentiation	Apoptosis	*bcl-2*	c-*myc*	Rb[b]	Refs.
Breast							
MCF-7	↓	↑	↑	↓	↓ [120], = [121]	↓P [91]	90,91,118–123
BT-20	=[c]	↑		=			90
BT-474	↓	↑	=	↓			90
MDA-MB-231	↓	↑	=	↓			90,118
MDA-MD-436	=	↑		=			90
SK-BR-3	↓	↓	=	=			90
CAMA-IEe	↓					↓P	124
T47-D	↓				↓		120
Hematopoietic							
HL-60	↓, = [127]	↑	↑ [115], ↓ [94]	↓ [94]	↓, = [127]	↓P, ↓	94–96,115, 125–135
U-937	↓	↑	↓		↓		114,136
RWLeu-4	↓	↑				↓P	99,137
$JMRD_3$	=	↑				= P	99,137
KH-2	↓				↓		138
M1	↓	↑			↓		139
K562	=	↓			=		140
Intestine							
HT-29	↓	↑	↑				102
CaCo-2	↓				=		103
Miscellaneous							
C6 (rat glioma)	↓		↑				141
LA-N-5 (human neuroblastoma)	↓	↑			↓ (N-*myc*)		143
NIH:OVCAR3 (ovarian)	↓				↓		120
HPK1A-*ras* (keratinocyte)	↓				↓, =		108
C3H/10T1/2 CI 8[d]	↓				↑ [143], = [144]		143,144
C3H/10T1/2 CI 16[d]	=				↑ [143], = [144]		143,144
C3H/10T1/2 TPA 482[d]	=				=		144
MG-63 (osteosarcoma)	↓ [145]	↑			↑ [146]		145,146
TT (human C-cell carcinoma)	↑				↑		147,148

[a] Effects of $1,25(OH)_2D_3$ analogs are not specifically identified. Only cells in which proliferation and at least one of the other parameters was examined are included. All responses were not measured in all references. ↑, Increase; ↓, decrease; =, no effect. If different effects were observed, the references are cited in the particular column.

[b] Rb, Retinoblastoma; ↓P indicates dephosphorylation of the retinoblastoma gene product.

[c] No growth effect was observed; however, Chouvet *et al.* [89] (see Table IV) did demonstrate growth inhibition of BT-20 cells.

[d] C3H/10T1/2 CI 8 cells are nontransformed embryonic fibroblasts, and C3H/10T1/2 CI 16 and C3H/10T1/2 TPA 482 are transformed tumorigenic embryic fibroblasts.

c-*myc* expression in MCF-7 and ZR-75.1 breast cancer cells while they were both growth inhibited [121], and a similar observation has been made for the colon-adenocarcinoma CaCo-2 cell line [103]. Nontransformed embryonic fibroblasts are growth inhibited by $1,25(OH)_2D_3$, whereas c-*myc* is not changed or is even increased [143,144]. In the MG-63 osteosarcoma cell line, $1,25(OH)_2D_3$ has been shown to enhance c-*myc* expression [146], whereas we observed growth inhibition by $1,25(OH)_2D_3$ [145].

In contrast to the *bcl*-2 and c-*myc* genes, the effects of $1,25(OH)_2D_3$ on the retinoblastoma tumor suppressor gene were similar in all cells examined (Table V). The $p110^{RB}$ retinoblastoma gene product can either be phosphorylated or dephosphorylated. In the phosphorylated form it can activate several transcription factors and cause transition to S phase and DNA synthesis. In human chronic myelogenous leukemia cells [99], breast cancer cells [124], and HL-60 cells [95,96], $1,25(OH)_2D_3$ caused a dephosphorylation of $p110^{RB}$, which is related to growth inhibition and cell cycle arrest in G_0/G_1 and in one study also in G_2 [95]. In the leukemic cells $1,25(OH)_2D_3$ also caused a reduction in the cellular level of $p110^{RB}$ [96,99]. In nontransformed keratinocytes $1,25(OH)_2D_3$ induced dephosphorylation of $p110^{RB}$ as well [106].

Data on the effect of $1,25(OH)_2D_3$ on other oncogenes (c-*myb*, c-*fos*, c-*fms*, *junD*, c-Ha-*ras*, N-*ras*) on p53, and also directly on the cyclin p34cdc2 [149] have been reported but are often limited to a single observation in one particular cell line. Nevertheless, it is clear that $1,25(OH)_2D_3$ has effects on the expression of various oncogenes and tumor suppressor genes. The data so far are not conclusive with respect to which genes are crucial in the growth inhibitory action of $1,25(OH)_2D_3$. This can be attributed to the fact that their postulated role is often complex. For example, increased c-*myc* expression can be related to induction of apoptosis but also to stimulation of cell cycle progression.

D. Growth Factors and Growth Factor Receptors

Besides regulation of cell cycle-related oncogenes and tumor suppressor genes, interaction with tumor- or stroma-derived growth factors is important for growth inhibition. Stimulation of breast cancer cell proliferation by coculture with fibroblasts is inhibited by $1,25(OH)_2D_3$ [150]. A good candidate to interact with the $1,25(OH)_2D_3$ action is transforming growth factor-β (TGFβ). TGFβ is involved in cell cycle control and apoptosis [151,152]. TGFβ can interfere with the cascade of events in the G_1 phase described above and inhibit the ability of cells to enter S phase when it is present during the G_1 phase. TGFβ has been shown to suppress c-*myc*, cyclin A, cyclin E, and cdk2 and cdk4 expression [152]. In line with this, TGFβ has been reported to inhibit phosphorylation of $p110^{RB}$ [153]. Vitamin D_3 compounds induce dephosphorylation of the retinoblastoma gene product, and vitamin D_3 growth inhibition of MCF-7 breast cancer cells is inhibited by a TGFβ neutralizing antibody [91]. $1,25(OH)_2D_3$ and several analogs stimulated the expression of TGFβ mRNA and secretion of active and latent TGFβ_1 by the breast cancer cell line BT-20 [154]. $1,25(OH)_2D_3$ enhanced TGFβ_1 gene expression in human keratinocytes [155] and the secretion of TGFβ in murine keratinocytes [156]. In both studies antibodies against TGFβ inhibited the growth inhibitory effect of vitamin D_3. Further evidence for a vitamin D_3–TGFβ interaction is that bone matrix of vitamin D-deficient rats contains substantially less TGFβ than controls [157]. Therefore, on the basis of these consistent findings, TGFβ is a likely candidate to play a role in the $1,25(OH)_2D_3$-induced growth inhibition.

Interactions with the insulin-like growth factor (IGF) system have also been described. IGFs are potent growth stimulators of various cells, and their effect is regulated via a series of IGF binding proteins (IGFBPs). $1,25(OH)_2D_3$ and the analog EB1089 inhibit the IGF-I-stimulated growth of MCF-7 breast cancer cells [122]. In prostate cancer cell lines, $1,25(OH)_2D_3$ induced expression of IGF-BP6 but not IGF-BP4 [158]. In human osteosarcoma cell lines, $1,25(OH)_2D_3$ and the analog 1α-dihydroxy-16-ene-23-yne-26,27-hexafluorocholecalciferol potently stimulated the expression and secretion of IGF-BP3 [149,159,160]. In one study an association has been made between increased IGF-BP3 levels and $1,25(OH)_2D_3$ growth inhibition [149]. Interestingly, in the human osteosarcoma cell line MG-63, $1,25(OH)_2D_3$ and TGFβ synergistically increased IGF-BP-3 secretion [160]. An example of growth factor receptor regulation by $1,25(OH)_2D_3$ concerns the epidermal growth factor receptor. This receptor is down-regulated in T47-D breast cancer cells and up-regulated in BT-20 breast cancer cells. Nevertheless, $1,25(OH)_2D_3$ inhibits the growth of both cell lines [161,162]. These data provide evidence that interactions with growth factors are part of the $1,25(OH)_2D_3$ action on tumor cells.

As described above, it is clear that $1,25(OH)_2D_3$ has effects on the expression of various oncogenes and tumor suppressor genes and that multiple interactions with various growth factors exist. However, the data on these aspects separately as well as in combination are still too limited to define a distinct mechanism of action for the $1,25(OH)_2D_3$ anti-cancer effects. However, with respect to growth inhibition, at this time two models of action can be postulated. In the first one $1,25(OH)_2D_3$ directly interferes with a crucial gene(s) involved in the control of the cell cycle. In this case, in view of the general pattern of the genes involved in cell cycle control, this mechanism of action will be similar in all types of cancer cells. However, the effect on cell cycle genes will be dependent on the presence or absence of additional growth factors. This will determine, depending on which growth factors are present, the differences in $1,25(OH)_2D_3$ action between cancer types of different origin but also within cancer types of similar origin. The second model is based on an indirect effect of $1,25(OH)_2D_3$ on cell cycle progression and tumor growth. In this case $1,25(OH)_2D_3$ may either inhibit or potentiate the effect of growth stimulatory or inhibitory factors, respectively, via, for example, effects on growth factor production, growth factor binding protein levels, or receptor regulation. It is also conceivable that a combination of both models forms the basis of $1,25(OH)_2D_3$ regulation of tumor cell growth.

E. Differentiation

In addition to proliferation and apoptosis, the third major cellular process is differentiation. As described above for the classic actions of $1,25(OH)_2D_3$ related to calcium homeostasis, effects on cell differentiation and proliferation are involved. The coupling between proliferation and differentiation has been most widely studied for cells of the hematopoietic system and keratinocytes. In general, $1,25(OH)_2D_3$ inhibits proliferation and induces differentiation along the monocyte–macrophage lineage. Rapidly proliferating and poorly differentiated keratinocytes can be induced to differentiate by $1,25(OH)_2D_3$. A further relationship between the vitamin D_3 system and differentiation is demonstrated by the fact that in poorly differentiated keratinocytes $1,25(OH)_2D_3$ production and vitamin D receptor levels are high, whereas after induction of differentiation these levels decrease [163]. Effects on differentiation have also been reported for other cell types. Inhibition of prostate cancer cell proliferation is paralleled by an increased production of prostate specific antigen [164,165]. In several breast cancer cell lines the stimulation of differentiation has been established by determining lipid production by the cells [90]. In this study, Elstner *et al.* demonstrated an uncoupling between effects on proliferation and differentiation. In two breast cancer cell lines $1,25(OH)_2D_3$ and various analogs induced differentiation even though the cells were resistant to cell cycle and antiproliferative effects. This together with data obtained with human myelogenous leukemia cells [99] suggest a dissociation between the cellular vitamin D_3 pathways involved in regulation of differentiation and proliferation (see also Section V). For a HL-60 subclone a similar observation was made [127], and in another HL-60 subclone the induction of differentiation was found to precede the G_0/G_1 cell cycle block [93]. In contrast to the above-mentioned observations on stimulation of differentiation, $1,25(OH)_2D_3$ inhibits erythroid differentiation of the erythroleukemia cell line K562 [140]. Although precise relationships among growth inhibition, cell cycle effects, and apoptosis are unclear, it can be concluded that an important effect of vitamin D_3 on both normal and malignant cells is induction of differentiation.

IV. COMBINATION THERAPY

The data obtained with $1,25(OH)_2D_3$ and $1,25(OH)_2D_3$ analogs on growth inhibition and stimulation of differentiation offer promise for their use as an endocrine anticancer treatment. Single agent treatment with low calcemic $1,25(OH)_2D_3$ analogs could be useful; however, combination therapy with other tumor effective drugs may provide an even more beneficial effect. Up to now several *in vitro* and *in vivo* studies have focused on possible future combination therapies with $1,25(OH)_2D_3$ and $1,25(OH)_2D_3$ analogs.

For breast cancer cells the combination of the presently most widely used endocrine therapy, the antiestrogen tamoxifen, with $1,25(OH)_2D_3$ and $1,25(OH)_2D_3$ analogs resulted in a greater growth inhibition of MCF-7 and ZR-75-1 cells than treatment with either compound alone [55,113,121]. In combination with tamoxifen, the cells were more sensitive to the antiproliferative action of $1,25(OH)_2D_3$ and the analogs; that is, the EC_{50} values of the vitamin D_3 compounds in the presence of tamoxifen were lower than those in the absence of tamoxifen. Studies with MCF-7 cells suggested a synergistic effect of $1,25(OH)_2D_3$ and tamoxifen on apoptosis [119]. In addition, in *in vivo* breast cancer models a synergistic effect of the tamoxifen–$1,25(OH)_2D_3$ analogs combination was observed [55,65].

Combination of vitamin D_3 and retinoids has been examined in various systems. A combination of retinoic acid and $1,25(OH)_2D_3$ resulted in a more profound inhibition of both T47-D breast cancer cells [166] and LA-N-5 human neuroblastoma cells [142]. 9-cis-Retinoic acid augmented $1,25(OH)_2D_3$-induced growth inhibition and differentiation of HL-60 cells [167]. Besides growth inhibition and differentiation effects, the combination of $1,25(OH)_2D_3$ and various isomers of retinoic acid were more potent in reducing angiogenesis than either compound alone [67,69,168]. The background of the interaction between retinoids and $1,25(OH)_2D_3$ may be attributed to heterodimer formation of the respective receptors [169].

For several cytokines, interactions with $1,25(OH)_2D_3$ have been described. Interferon-γ and $1,25(OH)_2D_3$ synergistically inhibited the proliferation and stimulated the differentiation of HL-60 and WEHI-3 myelocytic leukemia cells [170,171], and in the mouse myeloid leukemia cell line M1 interleukin-4 enhanced $1,25(OH)_2D_3$-induced differentiation [172]. $1,25(OH)_2D_3$ and tumor necrosis factor synergistically induced growth inhibition and differentiation of HL-60 cells [130]. For MCF-7 cells an interaction between $1,25(OH)_2D_3$ and tumor necrosis factor has also been reported [173], and in the presence of granulocyte–macrophage colony-stimulating factor lower concentrations of $1,25(OH)_2D_3$ could be used to achieve a similar antiproliferative effect in these cells [174].

Furthermore, combinations of vitamin D_3 compounds with cytotoxic drugs have been studied. *In vivo* adriamycin and *in vitro* carboplatin and cisplatin interacted synergistically with $1,25(OH)_2D_3$ to inhibit breast cancer cell growth [54,175]. In a carcinogen-induced rat mammary tumor model, treatment with 1α-hydroxyvitamin D_3 and 5-fluorouracil, however, did not result in enhanced anti-tumor effects [29].

The data on combinations of $1,25(OH)_2D_3$ and $1,25(OH)_2D_3$ analogs with various other anticancer compounds are promising and merit further analyses. The development of effective combination therapies may result in better response rates and lower required dosages, thereby reducing the risk of negative side effects.

V. VITAMIN D RESISTANCE

Classic vitamin D resistance concerns the disease hereditary vitamin D-resistant rickets, which is characterized by the presence of a nonfunctional VDR and consequently aberrations in calcium and bone metabolism (see Chapter 48). For cancer cells the presence of a functional VDR is also a prerequisite for a growth regulatory response, and a relationship between VDR level and growth inhibition has been suggested for osteosarcoma, colon carcinoma, breast cancer, and prostate cancer cells [1,2,124,176–178]. However, the presence of the VDR is not always coupled to a growth inhibitory response of $1,25(OH)_2D_3$. Results from studies with transformed fibroblasts [144], myelogenous leukemia cells [99,127,179], transformed keratinocytes [108], and various breast cancer cell lines [90,180] demonstrated a lack of growth inhibition by $1,25(OH)_2D_3$ even in the presence of VDR. In this situation the designation "resistant" is based on the lack of growth inhibition, even though, as discussed earlier in Section III,E, some of these cells are still capable of being induced to differentiate [90,99]. This points to a specific defect in the growth inhibitory pathway. In the resistant MCF-7 cells this defect is not located at a very common site in the growth inhibitory pathway of the cell, because the growth could still be inhibited with the antiestrogen tamoxifen [180].

For VDR-independent resistance to growth inhibition the underlying mechanism(s) is unknown. For the resistant MCF-7 clone this is not related to up-regulation of the P-glycoprotein [180]. In the resistant leukemia $JMRD_3$ cell line, altered regulation and DNA binding activity of *junD* as part of the AP-1 complex has been reported [137]. An inverse relationship between cellular metabolism of $1,25(OH)_2D_3$ via 24-hydroxylation and growth inhibition of prostate cancer cells has been suggested [177]. The latter observation is intriguing, the more so as an inverse relationship between VDR level and induction of 24-hydroxylase activity was reported. In general, there may exist a direct relationship between VDR level and induction of 24-hydroxylase activity [178,181]. The $1,25(OH)_2D_3$ sensitive and resistant cell clones provide interesting models to examine the molecular mechanisms of $1,25(OH)_2D_3$-induced growth inhibition.

VI. STIMULATION OF PROLIFERATION

Over the years a limited number of studies have demonstrated that, in contrast to growth inhibition, $1,25(OH)_2D_3$ can also stimulate tumor cell growth and tumor development. In several cells $1,25(OH)_2D_3$ has been reported to have a biphasic effect, that is, at lower concentrations ($<10^{-9}$ M) it stimulates proliferation and at higher concentrations (10^{-9} to 10^{-7} M) it inhibits proliferation. However, clear growth stimulation can sometimes be observed not only at low concentrations but also at the concentrations generally found to inhibit tumor cell proliferation and tumor development. $1,25(OH)_2D_3$ has been shown to stimulate the growth of a human medullary thyroid carcinoma cell line [147]. Not only cancer cells but also several normal cells, for example, human monocytes [182], smooth muscle cells [183], and alveolar type II cells [184], are stimulated to grow by $1,25(OH)_2D_3$.

Skin is another organ in which different effects of $1,25(OH)_2D_3$ have been observed. *In vivo* studies demonstrated that $1,25(OH)_2D_3$ and analogs stimulate keratinocyte proliferation in normal mice [185–188] and enhance anchorage-independent growth of preneoplastic epidermal cells [189]. In contrast, other studies showed $1,25(OH)_2D_3$ inhibition of proliferation of mouse and human keratinocytes [107,190], and $1,25(OH)_2D_3$ is also effective in the treatment of the hyperproliferative disorder psoriasis [191]. Moreover, *in vivo* studies demonstrated that, depending on the carcinogen, $1,25(OH)_2D_3$ can either reduce [41] or enhance the induction and development of skin tumors in mice [192]. In addition, $1,25(OH)_2D_3$ enhances the chemically induced transformation of BALB 3T3 cells and hamster embryo cells [193,194].

Another example comes from research on osteosarcoma cells. In 1986 it was shown that $1,25(OH)_2D_3$ stimulated the growth of tumors in athymic mice inoculated with the ROS 17/2.8 osteosarcoma cell line [195]. Earlier the same group reported growth stimulation *in vitro* of these osteosarcoma cells at low concentrations but growth inhibition by 10^{-8} M [176]. They speculated that this discrepancy resulted from limited *in vivo* availability of $1,25(OH)_2D_3$ for the tumor cells, resulting in concentrations shown to be growth stimulatory *in vitro*. However, in other experiments with nude mice the availability of $1,25(OH)_2D_3$ did not seem to be a factor, as growth inhibition was observed (see Table II). In particular, in nude mice implanted with human osteosarcoma cells (MG-63), growth inhibition and tumor suppression by $1,25(OH)_2D_3$ were observed [37]. In two different *in vitro* studies, growth inhibition of MG-63 and growth stimulation of ROS 17/2.8 cells was reported [196,197]. For smooth muscle cells it has been demonstrated, for example, that growth inhibition or stimulation can depend on the presence of additional growth factors in the culture medium [183]. We followed up on this concept by comparing the effects of $1,25(OH)_2D_3$ and analogs on the growth and osteoblastic characteristics of the two osteosarcoma cell lines under identical culture conditions. At concentrations 10^{-10} to 10^{-7} *M* 1,25-$(OH)_2D_3$ caused an increase in cell proliferation by 100% in ROS 17/2.8 cells, whereas the proliferation of MG-63 cells was inhibited (Fig. 2) [145]. In contrast, in both cell lines $1,25(OH)_2D_3$ stimulated osteoblastic differentiation characteristics such as production of osteocalcin and alkaline phosphatase activity [145,196]. Analyses with another steroid hormone demonstrated that glucocorticoids inhibited the growth of both osteosarcoma cell lines [198,199]. These data indicate specific differences between these cell lines, especially with respect to the $1,25(OH)_2D_3$ growth regulatory mechanisms.

Taken together, the data on growth stimulation and tumor development, although detected in only a minority of cancer cells, demonstrate that treatment with $1,25(OH)_2D_3$ or analogs may not always cause growth inhibition and tumor size reduction. It is therefore of utmost importance to identify the mechanism(s) by which $1,25(OH)_2D_3$ exerts its inhibitory and stimulatory effects on cell growth. This may provide tools to assess whether treatment of a particular tumor will be beneficial. Moreover, purely from a mechanistic point of view, the presence of growth-stimulated and growth-inhibited cells, like the $1,25(OH)_2D_3$ sensitive and resistant cells, may provide tools to examine the $1,25(OH)_2D_3$ mechanism of growth regulation.

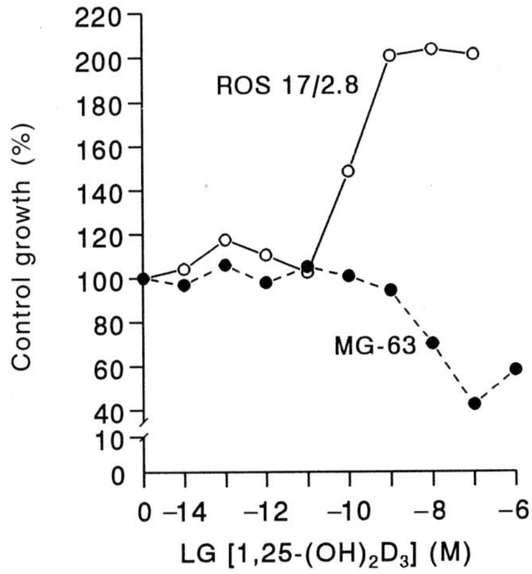

FIGURE 2 Effect of $1,25(OH)_2D_3$ on proliferation of the osteosarcoma cell lines ROS 17/2.8 and MG-63. Effects on proliferation were examined as described by van den Bemd *et al.* [145].

VII. CONCLUSIONS

The data obtained so far, on (1) the distribution of the VDR in a broad range of tumors and (2) the inhibition of cancer cell growth, angiogenesis, metastasis, and PTHrP synthesis by $1,25(OH)_2D_3$, all hold promise for the development of treatment strategies based on vitamin D_3 use in a wide range of cancers. This possibility is enhanced by the development of $1,25(OH)_2D_3$ analogs with potent growth inhibitory actions and reduced hypercalcemic activity. Moreover, combination of vitamin D_3 with other antitumor drugs or growth factors is an additional therapeutic option. At the moment clinical studies are needed in order to establish whether vitamin D_3 and especially vitamin D_3 analogs have therapeutical potential. In the meantime it is crucial to understand the mechanism(s) by which vitamin D_3 exerts its effects on tumor cell growth so that these drugs may be employed more effectively.

References

1. Shabahang M, Buras RR, Davoodi F, Schumaker LM, Nauta RJ, Evans SRT 1993 1,25-Dihydroxyvitamin D_3 receptor as a marker of human colon carcinoma cell line differentiation and growth inhibition. Cancer Res **53**:3712–318.
2. Buras RR, Schumaker LM, Davoodi F, Brenner RV, Shabahang M, Nauta RJ, Evans SRT 1994 Vitamin D receptors in breast cancer cells. Breast Cancer Res Treat **31**:191–202.
3. Freake HC, Abeyasekera G, Iwasaki J, Marcocci C, MacIntyre I, McClelland RA, Skilton RA, Easton DF, Coombes RC 1984 Measurement of 1,25-dihydroxyvitamin D_3 receptors in breast cancer and their relationship to biochemical and clinical indices. Cancer Res **44**:1677–1681.
4. Eisman JA, Suva LJ, Martin TJ 1986 Significance of 1,25-dihydroxyvitamin D_3 receptor in primary breast cancers. Cancer Res **46**:5406–5408.
5. Berger U, McClelland RA, Wilson P, Greene GL, Haussler MR, Pike JW, Colston K, Easton D, Coombes Rc 1991 Immunocytochemical determination of estrogen receptor, progesterone receptor, and 1,25-dihydroxyvitamin D_3 receptor in breast cancer and relationship to prognosis. Cancer Res **51**:239–244.

6. Garland CF, Garland FC 1980 Do sunlight and vitamin D reduce the likelihood of colon cancer? Int J Epidemiol **9**:227–231.
7. Garland CF, Comstock GW, Garland FC, Helsing KJ, Shaw EK, Gorham ED 1989 Serum 25-hydroxyvitamin D and colon cancer: Eight year prospective study. Lancet **2**:1176–1178.
8. Garland CF, Garland FC, Gorham ED 1991 Can colon cancer incidence and death rates be reduced with calcium and vitamin D. Am J Clin Nutr **54**:193S–201S.
9. Garland FC, Garland CF, Gorham ED, Young JF 1990 Geographic variation in breast cancer mortality in the United States: A hypothesis involving exposure to solar radiation. Prev Med **19**:614–622.
10. Schwartz GG, Hulka BS 1990 Is vitamin D deficiency a risk factor for prostatic cancer? (Hypothesis.) Anticancer Res **10**:1307–1311.
11. Gorham ED, Garland FC, Garland CF 1990 Sunlight and breast cancer incidence in the USSR. Int J Epidemiol **19**:820–924.
12. Hanchette Cl, Schwartz GG 1992 Geographic patterns of prostate cancer mortality. Cancer **70**:2861–2869.
13. Ainsleigh HG 1993 Beneficial effects of sun exposure on cancer mortality. Prev Med **22**:132–140.
14. Simard A, Vobecky J, Vobecky JS 1991 Vitamin D deficiency and cancer of the breast: an unprovocative ecological hypothesis. Can J Public Health **82**:300–303.
15. Studzinski GP, Moore DC 1995 Sunlight—Can it prevent as well as cause cancer? Cancer Res **55**:4014–4022.
16. Jacobson EA, James KA, Newmark HL, Carroll KK 1989 Effects of dietary fat, calcium, and vitamin D on growth and mammary tumorigenesis induced by 7,12-dimethylbenz[*a*]anthracene in female Sprague-Dawley rats. Cancer Res **49**:6300–6303.
17. Barger-Lux MJ, Heaney RP 1994 The role of calcium intake in preventing bone fragility, hypertension, and certain cancers. J Nutr **124**:1406S–1411S.
18. Newmark HL 1994 Vitamin D adequacy: A possible relationship to breast cancer. Adv Exp Med Biol **364**:109–114.
19. Kleibeuker JH, van der Meer R, de Vries EGE 1995 Calcium and vitamin D: Possible protective agents against colorectal cancer. Eur J Cancer **31A**:1081–1084.
20. Pence BC, Richard BC, Lawlis RS, Kuratko CN 1991 Effects of dietary calcium and vitamin D_3 on tumor promotion in mouse skin. Nutr Cancer **16**:171–181.
21. Colston KW, Colston JM, Feldman D 1981 1,25-Dihydroxyvitamin D_3 and malignant melanoma: The presence of receptors and inhibition of cell growth in culture. Endocrinology **108**:1083–1086.
22. Abe E, Miyaura C, Sakagami H, Takeda M, Konno K, Yamazaki T, Yoshiki S, Suda T 1981 Differentiation of mouse myeloid leukemia cells induced by 1,25-dihydroxyvitamin D_3. Proc Natl Acad Sci USA **78**:4990–4995.
23. Suda T, Miyaura C, Abe E, Kuroki T 1986 Modulation of cell differentiation, immune responses and tumor promotion by vitamin D compounds. In: Peck WA (ed) Bone and Mineral Research, Volume 4. Elsevier, Amsterdam, pp. 1–48.
24. Suda T, Takahashi N, Abe E 1992 Role of vitamin D in bone resorption. J Cell Biochem **49**:53–58.
25. Suda T, Udagawa N, Nakamura I, Miyaura C, Takahashi N 1995 Modulation of osteoclast differentiation by local factors. Bone **17**(Suppl 2):87S–91S.
26. Wu JCY, Smith MW, Lawson DEM 1992 Time dependency of 1,25-$(OH)_2D_3$ induction of calbindin mRNA and calbindin expression in chick enterocytes during their differentiation along the crypt–villus axis. Differentiation **51**:195–200.
27. Honma Y, Hozumi M, Abe E, Konno K, Fukushima M, Hata S, Nishii Y, DeLuca HF, Suda T 1983 1α,25-Dihydroxyvitamin D_3 and 1α-hydroxyvitamin D_3 prolong the survival time of mice inoculated with myeloid leukemia cells. Proc Natl Acad Sci USA **80**:201–204.
28. Colston KW, Berger U, Coombes RC 1989 Possible role for vitamin D in controlling breast cancer cell proliferation. Lancet **1**:188–191.
29. Iino Y, Yoshida M, Sugamata N, Maemura M, Ohwada S, Yokoe T, Ishikita T, Horiuchi R, Morishita Y 1992 1alpha-Hydroxyvitamin D_3, hypercalcemia, and growth suppression of 7,12-dimethylbenz[*a*]anthracene-induced rat mammary tumors. Breast Cancer Res Treat **22**:133–140.
30. Eisman JA, Barkla DH, Tutton PJM 1987 Suppression of in vivo growth of human cancer solid xenografts by 1,25-dihydroxyvitamin D_3. Cancer Res **47**:21–25.
31. Eisman JA, Koga M, Sutherland RL, Barkla DH, Tutton PJM 1989 1,25-Dihydroxyvitamin D_3 and the regulation of human cancer cell replication. Proc Soc Exp Med **191**:221–226.
32. Belleli A, Shany S, Levy J, Guberman R, Lamprecht SA 1992 A protective role of 1,25-dihydroxyvitamin D_3 in chemically induced rat colon carcinogenesis. Carcinogenesis **13**:2293–2298.
33. Haq M, Kremer R, Goltzman D, Rabbani SA 1993 A vitamin D analogue (EB1089) inhibits parathyroid hormone-related peptide production and prevents the development of malignancy-associated hypercalcemia in vivo. J Clin Invest **91**:2416–2422.
34. Sato T, Takusagawa K, Asoo N, Konno K 1982 Antitumor effect of 1alpha-hydroxyvitamin D_3. Tohuku J Exp Med **138**:445–446.
35. Young MRI, Ihm J, Lozano HY, Wright MA, Prechel MM 1995 Treating tumor-bearing mice with vitamin D_3 dimishes tumor-induced myelopoiesis and associated immunosuppression, and reduces tumor metastasis and recurrence. Cancer Immunol Immunother **41**:37–45.
36. Young MRI, Halpin J, Jussain R, Lozano Y, Djordjevic A, Devata S, Matthews JP, Wright MA 1993 Inhibition of tumour production of granulocyte–macrophage colony-stimulating factor by 1alpha,25-dihydroxyvitamin D_3 reduces tumour motility and metastasis. Invasion Metastasis **13**:169–177.
37. Tsuchiya H, Morishita H, Tomita K, Ueda Y, Tanaka M 1993 Differentiating and antitumour activities of 1,25-dihydroxyvitamin D_3 in vitro and 1alpha-hydroxyvitamin D_3 in vivo on human osteosarcoma. J Orthop Res **11**:122–130.
38. Cohen SM, Saulenas AM, Sullivan CR, Albert DM 1988 Further studies of the effect of vitamin D on retinoblastoma. Arch Ophthalmol **106**:541–543.
39. Albert DM, Marcus DM, Gallo JP, O'Brien JM 1992 The antineoplastic effect of vitamin D in transgenic mice with retinoblastoma. Invest Ophthalmol Vis Sci **33**:2354–2364.
40. Cohen-Solal ME, Bouizar Z, Denne MA, Graulet AM, Gueris J, Bracq S, Jullienne A, de Vernejoul MC 1995 1,25-Dihydroxyvitamin D_3 and dexamethasone decrease in vivo Walker carcinoma growth, but not parathyroid hormone-related protein secretion. Horm Metab Res **27**:403–407.
41. Wood AW, Chang RL, Huang MT, Uskokvic M, Conney AH 1983 1alpha,25-Dihydroxyvitamin D_3 inhibits phorbol ester-dependent chemical carcinogenesis in mouse skin. Biochem Biophys Res Commun **116**:605–611.
42. Chiba K, Hashiba H, Fukushima M, Suda T, Kuroki T 1985 Inhibition of tumour promotion in mouse skin by 1alpha-hydroxyvitamin D_3. Cancer Res **45**:5426–5430.
43. Cunningham D, Gilchrist NL, Cowan RA, Soukup K 1985 Alfacalcidol as a modulator of growth of low-grade non-Hodgkin's lymphomas. Br Med J **291**:1153–1155.
44. Koeffler HP, Hirji K, Itri L, Southern California Leukemia Group 1985 1,25-Dihydroxyvitamin D_3 *in vitro* and *in vivo* effects

on human preleukemic and leukemic cells. Cancer Treat Rep **69**:1399–1407.

45. Kelsey SM, Newland AC, Cunningham J, Makin HLJ, Coldwell RD, Mills MJ, Grant IR 1992 Sustained haematological response to high-dose oral alfacalcidol in patients with myelodysplastic syndromes. Lancet **340**:316–317.
46. French LE, Ramelet AA, Saurat J-H 1994 Remission of cutaneous T-cell lymphoma with combined calcitriol and acitretin. Lancet **344**:686–687.
47. Majewski S, Skopinska M, Bollag W, Jablonska S 1994 Combination of isotretin and calcitriol for precancerous and cancerous skin lesions. Lancet **344**:1510–1511.
48. Thomsen K 1995 Cutaneous T-cell lymphoma and calcitriol and isotretinoin treatment. Lancet **345**:1583.
49. French LE, Saurat J-H 1995 Treatment of cutaneous T-cell lymphoma by retinoids and calcitriol. Lancet **346**:376.
50. Bagot M 1995 Treatment of cutaneous T-cell lymphoma by retinoids and calcitriol. Lancet **346**:376–377.
51. Pols HAP, Birkenhäger JC, van Leeuwen JPTM 1994 Vitamin D analogues: from molecule to clinical application. Clin Endocrinol (Oxf) **40**:285–291.
52. Bouillon R, Okamura WH, Norman AW 1995 Structure–function relationships in the vitamin D endocrine system. Endocrine Rev **16**:200–257.
53. Bower M, Colston KW, Stein RC, Hedley A, Gazet J-C, Ford HT, Coombes RC 1991 Topical calcipotriol treatment in advanced breast cancer. Lancet **337**:701–702.
54. Abe J, Nakano T, Nishii Y, Matsumoto T, Ogata E, Ikeda K 1991 A novel vitamin D_3 analog, 22-oxa-1,25-dihydroxyvitamin D_3, inhibits the growth of human breast cancer in vitro and in vivo without causing hypercalcemia. Endocrinology **129**:832–837.
55. Abe-Hashimoto J, Kikuchi T, Matsumoto T, Nishii Y, Ogata E, Ikeda K 1993 Antitumor effect of 22-oxa-calcitriol, a noncalcemic analogue of calcitriol, in athymic mice implanted with human breast carcinoma and its synergism with tamoxifen. Cancer Res **53**:2534–2537.
56. Oikawa T, Yoshida Y, Shimamura M, Ashino-Fuse H, Iwaguchi T, Tominaga T 1991 Antitumor effect of 22-oxa-1alpha,25-dihydroxyvitamin D_3, a potent angiogenesis inhibitor, on rat mammary tumors induced by 7,12-dimethylbenz[*a*]anthracene. Anti-Cancer Drugs **2**:475–480.
57. Yano Y, Otani S, Fukushima S 1995 Inhibition of intestinal tumor development in rat multi-organ carcinogenesis and aberrant crypt foci in rat colon carcinogenesis by 22-oxa-calcitriol, a synthetic analogue of 1-alpha,25-dihydroxyvitamin D_3. Carcinogenesis **16**:2091–2097.
58. Otoshi T, Iwata H, Kitano M, Nishizawa Y, Morii H, Yano Y, Otani S, Fukushima S 1995 Inhibition of intestinal tumor development in rat multi-organ carcinogenesis and aberrant crypt foci in rat colon carcinogenesis by 22-oxa-calcitriol, a synthetic analogue of 1alpha,25-dihydroxyvitamin D_3. Carcinogenesis **16**:2091–2097.
59. Colston KW, Chander SK, Mackay AG, Coombes RC 1992 Effects of synthetic vitamin D analogues of breast cancer cell proliferation in vivo and in vitro. Biochem Pharmacol **44**:693–702.
60. Colston KW, Mackay AG, James SY, Binderup L 1992 EB1089, a new vitamin D analogue that inhibits the growth of breast cancer cell in vivo and in vitro. Biochem Pharmacol **44**:2273–2280.
61. Colston KW, Mackay AG, Chandler S, Binderup L, Coombes RC 1991 Novel vitamin D analogues suppress tumour growth in vivo. In: Norman AW, Bouillon R, Thomasset M (eds) Vitamin D: Gene Regulation, Structure–Function Analysis and Clinical Application. de Gruyter, Berlin, pp. 465–466.
62. Tanaka Y, Wu AY, Ikekawa N, Iseki K, Kawai M, Kobayashi Y 1994 Inhibition of HT-29 human colon cancer growth under the renal capsule of severe combined immunodeficient mice by an analogue of 1,25-dihydroxyvitamin D_3, DD-003. Cancer Res **54**:5148–5153.
63. Schwartz GG, Hill CC, Oeler TA, Becich MJ, Bahnson RR 1995 1,25-Dihydroxy-16-ene-23-yne-vitamin D_3 and prostate cancer cell proliferation in vivo. Urology **46**:365–369.
64. Zhou J-Y, Norman AW, Chen D-L, Sun G, Uskokovic M, Koeffler HP 1990 1,25-Dihydroxy-16-ene-23-yne-vitamin D_3 prolongs survival time of leukemic mice. Proc Natl Acad Sci USA **87**:3929–3932.
65. Anzano MA, Smith JM, Uskokovic MR, Peer CW, Mullen LT, Letterio JJ, Welsh MC, Shrader MW, Logsdon JL, Driver CL, Browh CC, Roberts AB, Sporn MB 1994 1alpha,25-Dihydroxy-16-ene-23-yne-26,27-hexafluorocholecalciferol (R024–5531), a new deltanoid (vitamin D analogue) for prevention of breast cancer in the rat. Cancer Res **54**:1653–1656.
66. Wali RK, Bissonnette M, Khare S, Hart J, Sitrin MD, Brasitus TA 1995 1alpha,25-Dihydroxy-16-ene-23-yne-26,27-hexafluorocholecalciferol, a noncalcemic analogue of 1alpha,25-dihydroxyvitamin D_3, inhibits azoxymethane-induced colonic tumorigenesis. Cancer Res **55**:3050–3054.
67. Majewski S, Szmurlo A, Marczak M, Jablonska S, Bollag W 1993 Inhibition of tumor cell-induced angiogenesis by retinoids, 1,25-dihydroxyvitamin D_3 and their combination. Cancer Lett **75**:35–39.
68. Bollag W, Majewski S, Jablonska S 1994 Cancer combination chemotherapy with retinoids: Experimental rationale. Leukemia **8**:1453–1457.
69. Majewski S, Marczak M, Szmurlo A, Jablonska S, Bollag W 1995 Retinoids, interferon α, 1,25-dihydroxyvitamin D_3 and their combination inhibit angiogenesis induced by non-HPV-harboring tumor cell lines. RARα mediates the antiangiogenic effect of retinoids. Cancer Lett **89**:117–124.
70. Shokravi MR, Marcus DM, Alroy J, Egan K, Saornil MA, Albert DM 1995 Vitamin D inhibits angiogenesis in transgenic murine retinoblastoma. Invest Ophthalmol Vis Sci **36**:83–87.
71. Mork Hansen CM, Frandsen TL, Brünner N, Binderup L 1994 1α,25-Dihydroxyvitamin D_3 inhibits the invasive potential of human breast cancer cells in vitro. Clin Exp Metastasis **12**:195–202.
72. Agha FP, Norman A, Hirschl S, Klein R 1976 Paget's disease coexistence with metastatic carcinoma. NY State J Med **76**:734–735.
73. Orr FW, Varani J, Gondek MD, Ward PA, Mundy GR 1979 Chemotactic responses of tumor cells to products of resorbing bone. Science **203**:176–179.
74. Sasaki A, Boyce BF, Story B, Wright KR, Chapman M, Boyce R, Mundy GR 1995 Bisphosphonate risedronate reduces metastatic human breast cancer burden in bone in nude mice. Cancer Res **55**:3551–3557.
75. Krempien B 1984 The Walker carcinoma–sarcoma 256 as an experimental model of bone metastases: Influence of local and metabolic factors on incidence and pattern of metastases. Calcif Tissue Int **36**:S26 (abstract).
76. Silver J, Naveh-Many T, Mayer H, Schmeizer HJ, Popovtzer MM 1986 Regulation by vitamin D metabolites of parathyroid hormone gene transcription in vivo in the rat. J Clin Invest **78**:1296–1301.
77. Mackey SL, Heymont JL, Kronenberg HM, Demay MB 1996 Vitamin D receptor binding to the negative human parathyroid

hormone vitamin D response element does not require the retinoid X receptor. Mol Endocrinol **10**:298–305.

78. Moseley JM, Gillespie MT 1995 Parathyroid hormone-related protein. Crit Rev Clin Lab Sci **32**:299–343.
79. Sebag M, Henderson J, Goltzman D, Kremer R 1994 Regulation of parathyroid hormone-related peptide production in normal human mammary epithelial cells in vitro. Am J Physiol **267**:C723-C730.
80. Werkmeister JR, Merryman JI, McCauley LK, Horton JE, Capen CC, Rosol TJ 1993 Parathyroid hormone-related protein production by normal human keratinocytes in vitro. Exp Cell Res **208**:68–74.
81. Kremer R, Karaplis AC, Henderson J, Gulliver W, Banville D, Hendy GN, Goltzman D 1991 Regulation of parathyroid hormone-like peptide in cultured normal human keratinocytes. J Clin Invest **87**:884–893.
82. Henderson J, Sebag M, Rhim J, Goltzman D, Kremer R 1991 Dysregulation of parathyroid hormone-like peptide expression and secretion in a keratinocyte model of tumor progression. Cancer Res **51**:6521–6528.
83. Yu J, Papavasiliou V, Rhim J, Goltzman D, Kremer R 1995 Vitamin D analogs: New therapeutic agents for the treatment of squamous cancer and its associated hypercalcemia. Anti-Cancer Drugs **6**:101–108.
84. Inoue D, Matsumoto T, Ogata E, Ikeda K 1993 22-Oxa-calcitriol, a non-calcemic analogue of calcitriol, suppresses both cell proliferation and parathyroid hormone-related peptide gene expression in human T cell lymphotrophic virus, type I-infected T cells. J Biol Chem **268**:16730–16736.
85. Liu B, Goltzman D, Rabbani SA 1993 Regulation of parathyroid hormone-related peptide production in vitro by the rat hypercalcemic Leydig cell tumor H-500. Endocrinology **132**:1658–1664.
86. Endo K, Ichikawa F, Uchiyama Y, Katsumata K, Ohkawa H, Kumaki K, Ogata E, Ikeda K 1994 Evidence for the uptake of vitamin D analogue (OCT) by a human carcinoma and its effect of suppressing the transcription of parathyroid hormone-related peptide gene in vivo. J Biol Chem **269**:32693–32699.
87. Merryman JI, Capen CC, McCauley LK , Werkmeister JR, Suter MM, Rosol TJ 1993 Regulation of parthyroid hormone-related protein production by a squamous carcinoma cell line in vitro. Lab Invest **69**:347–354.
88. Pols HAP, Birkenhäger JC, Foekens JA, van Leeuwen JPTM 1990 Vitamin D: A modulator of cell proliferation and differentiation. J Steroid Biochem Mol Biol **37**:873–876.
89. Chouvet C, Vicard E, Devonec M, Saez S 1986 1,25-Dihydroxyvitamin D_3 inhibitory effect on the growth of two human breast cancer cell lines (MCF-7, BT-20). J Steroid Biochem **24**:373–376.
90. Elstner E, Linker-Israeli M, Said J, Umiel T, de Vos S, Shintaku PI, Heber D, Binderup L, Uskokovic M, Koeffler HP 1995 20-Epi-vitamin D_3 analogues: A novel class of potent inhibitors of proliferation and inducers of differentiation of human breast cancer cells. Cancer Res **55**:2822–2830.
91. Simboli-Campbell M, Welsh J 1995 1,25-Dihydroxyvitamin D_3: Coordinate regulator of active cell death and proliferation in MCF-7 breast cancer cells. In: Tenniswood M, Michna H (eds) Apoptosis in Hormone Dependent Cancers. Springer-Verlag, Berlin, pp. 181–200.
92. Eisman JA, Sutherland RL, McMenemy ML, Fragonas JC, Musgrove EA, Pang GYN 1989 Effects of 1,25-dihydroxyvitamin D_3 on cell cycle kinetics of T47D human breast cancer cells. J Cell Physiol **138**:611–616.
93. Zhang F, Godyn JJ, Uskokovic M, Binderup L, Studzinski GP 1994 Monocytic differentiation of HL60 cells induced by potent analogs of vitamin D_3 precedes the G_1/G_0 phase cell cycle block. Cell Proliferation **27**:643–654.
94. Xu H-M, Tepper CG, Jones JB, Fernandez CE, Studzinski GP 1993 1,25-Dihydroxyvitamin D_3 protects HL-60 cells against apoptosis but down-regulates the expression of the *bcl-2* gene. Exp Cell Res **209**:367–374.
95. Yen A, Varvayanis S 1994 Late dephosphorylation of the RB protein in G2 during the process of induced cell differentiation. Exp Cell Res **214**:250–257.
96. Yen A, Chandler S, Forbes ME, Fung Y-K, T'Ang A, Pearson R 1992 Coupled down-regulation of the RB retinoblastoma and c-myc genes antecedes cell differentiation: Possible role of RB as a "status quo" gene. Eur J Cell Biol **57**:210–221.
97. Song LN, Cheng T 1992 Effects of 1,25-dihydroxyvitamin D_3 analogue 1,24$(OH)_2$-22-ene-24-cyclopropyl D_3 on proliferation and differentiation of a human megakaryoblastic leukemia cell line. Biochem Pharmacol **43**:2292–2295.
98. Abe J, Moriya Y, Saito M, Sugawara Y, Suda T, Nishii Y 1986 Modulation of cell growth, differentiation, and production of interleukin-3 by 1alpha,25-dihydroxyvitamin D_3 in the murine myelomonocytic leukemia cell line WEHI-3. Cancer Res **46**:6316–3621.
99. Lasky SR, Posner MR, Iwata K, Santos-Moore A, Yen A, Samuel V, Clark J, Maizel AL 1994 Characterization of a vitamin D_3-resistant human chronic myelogenous leukemia cell line. Blood **84**:4283–4294.
100. Rigby WF Yirinec B, Oldershaw RL, Fanger MW 1987 Comparison of the effects of 1,25-dihydroxyvitamin D_3 on T lymphocyte subpopulations. Eur J Immunol **17**:563–566.
101. Rigby WF, Noelle RJ, Krause K, Fanger MW 1985 The effects of 1,25-dihydroxyvitamin D_3 on human T lymphocyte activation and proliferation: A cell cycle analysis. J Immunol **135**:2279–2286.
102. VandeWalle B, Wattez N, Lefebvre J 1995 Effects of vitamin D_3 derivatives on growth, differentiation and apoptosis in tumoral colonic HT 29 cells: Possible implication of intracellular calcium. Cancer Lett **97**:99–106.
103. Hulla W, Kallay E, Krugluger W, Peterlik M, Cross HS 1995 Growth control of human colon-adenocarcinoma-derived Caco-2 cells by vitamin D compounds and extracellular calcium in vitro: Relation to c-*myc* oncogene and vitamin D receptor expression. Int J Cancer **62**:711–716.
104. Nagakura K, Abe E, Suda T Hayakawa M, Nakamura H, Tazaki H 1986 Inhibitory effect of 1alpha,25-hydroxyvitamin D_3 on the growth of the renal carcinoma cell line. Kidney Int **29**:834–840.
105. Goplen DP, Brackman D, Aksnes L 1994 Effects of 1,25-dihydroxyvitamin D_3 and retinoic acid on the proliferation and cell cycle phase distribution of neuroblastoma SK-N-SH cells. Pediatr Hematol Oncol **11**:173–179.
106. Kobayaski T, Hashimoto K, Yoshikawa K 1993 Growth inhibition of human keratinocytes by 1,25-dihydroxyvitamin D_3 is linked to dephosphorylation of retinoblastoma gene product. Biochem Biophys Res Commun **196**:487–493.
107. Kitano Y, Ikedat N, Okano M 1991 Suppression of proliferation of human epidermal keratinocytes by 1,25-dihydroxyvitamin D_3. Analysis of its effect on psoriatic lesion and of its mechanism using human keratinocytes in culture. Eur J Clin Invest **21**:53–58.
108. Sebag M, Henderson J, Rhim J, Kremer R 1992 Relative resitance to 1,25-dihydroxyvitamin D_3 in a keratinocyte model of tumor progression. J Biol Chem **267**:12162–12167.
109. Bontebal M, Sieuwert AM, Kleij JGM, Peters HA, Krijnen HLJM, Sonneveld P, Foekens JA 1989 Effect of hormonal manipulation and doxorubicin administration on cell cycle kinetics of human breast cancer cells. Br J Cancer **60**:688–692.

110. Kerr JFR, Wyllie AG, Currie AR 1972 Apoptosis: A basic biological phenomenon with wide ranging implications in tissue kinetics. Br J Cancer **26**:239–257.
111. Walker PR, Kwast-Welfeld J, Gourdeau H, Leblanc J, Neugebauer W, Sikorska M 1993 Relationship between apoptosis and the cell cycle in lymphocytes: Roles of protein kinase C, tyrosine phosphorylation and AP1. Exp Cell Res **207**:142–151.
112. Denmeade SR, Lin XS, Isaacs JT 1996 Role of programmed (apoptotic) cell death during the progression and therapy for prostate cancer. Prostate **28**:251–265.
113. Vink-van Wijngaarden T, Pols HAP, Buurman CJ, Birkenhäger JC, van Leeuwen JPTM 1993 Combined effects of 1,25-dihydroxyvitamin D_3 and tamoxifen on the growth of MCF-7 and ZR-75–1 human breast cancer cells. Breast Cancer Res Treat **29**:161–168.
114. Ehinger M, Bergh G, Olofsson T, Baldetorp B, Olsson I, Gullberg U 1996 Expression of the p53 tumor suppressor gene induces differentiation and promotes induction of differentiation by 1,25-dihydroxycholecalciferol in leukemic U-937 cells. Blood **87**:1064–1074.
115. Wallington LA, Bunce CM, Durham J, Brown G 1995 Particular combinations of signals, by retinoid acid and 1alpha,25-dihydroxyvitamin D_3, promote apoptosis of HL60 cells. Leukemia **9**:1185–1190.
116. Hockenberry D, Nunez G, Milliman C, Schreiber RD, Korsmeyer SJ 1990 Bcl-2 is an inner mitochrondrial membrane protein that blocks programmed cell death. Nature **348**:334–336.
117. James SY, Mackay AG, Colston KW 1995 Vitamin D derivatives in combination with 9-*cis*-retinoic acid promote active cell death in breast cancer cells. J Mol Endocrinol **14**:391–394.
118. VandeWalle B, Hornez L, Wattez N, Revillion F, Lefebvre J 1995 Vitamin-D_3 derivatives and breast-tumor cell growth: effect on intracellular calcium and apoptosis. Int J Cancer **61**:806–811.
119. Welsh J 1994 Induction of apoptosis in breast cancer cells in response to vitamin D and antiestrogens. Biochem Cell Biol **72**:537–545.
120. Saunders DE, Christensen C, Wappler NL, Schultz JF, Lawrence WD, Malviya VK, Malone JM, Deppe G 1993 Inhibition of c-*myc* in breast and ovarian carcinoma cells by 1,25-dihydroxyvitamin D_3, retinoic acid and dexamethasone. Anti-Cancer Drugs **4**:201–208.
121. Vink-van Wijngaarden T, Pols HAP, Buurman CJ, van den Bemd GJCM, Dorssers LCJ, Birkenhäger JC, van Leeuwen JPTM 1994 Inhibition of breast cancer cell growth by combined treatment with vitamin D_3 analogues and tamoxifen. Cancer Res **54**:5711–5717.
122. Vink-van Wijngaarden T, Pols HAP, Buurman CJ, Birkenhäger JC, van Leeuwen JPTM 1996 Inhibition of insulin- and insulin-like growth factor-I-stimulated growth of human breast cancer cells by 1,25-dihydroxyvitamin D_3 and the vitamin D_3 analogue EB1089. Eur J Cancer **32**:842–848.
123. Mathiasen IS, Colston KW, Binderup L 1993 EB1089, a novel vitamin D analogue, has strong antiproliferative and differentiation inducing effects on cancer cells. J Steroid Biochem Mol Biol **46**:63–69.
124. Fan FS, Yu WCY 1995 1,25-Dihydroxyvitamin D_3 suppresses cell growth, DNA synthesis, and phosphorylation of retinoblastoma protein in a breast cancer cell line. Cancer Invest **13**:280–286.
125. Biskobing DM, Rubin J 1993 1,25-Dihydroxyvitamin D_3 and phorbol myristate acetate produce divergent phenotypes in a monomyelocytic cell line. Endocrinology **132**:862–866.
126. Sariban E, Mitchell T, Kufe D 1985 Expression of the c-*fms* protooncogene during human monocytic differentiation. Nature **316**:64–66.
127. Taoka T, Collins ED, Irino S, Norman AW 1993 1,25-$(OH)_2$-vitamin D_3 mediated changes in mRNA for c-*myc* and 1,25-$(OH)_2D_3$ receptor in HL-60 cells and related subclones. Mol Cell Endocrinol **95**:51–57.
128. Mangasarian K, Mellon WS 1993 1,25-Dihydroxyvitamin D_3 destabilizes c-*myc* mRNA in HL-60 leukemic cells. Biochim Biophys Acta **1172**:55–63.
129. Simpson RU, Hsu T, Wendt MD, Taylor JM 1989 1,25-Dihydroxyvitamin D_3 regulation of c-*myc* protooncogene transcription. Possible involvement of protein kinase C. J Biol Chem **264**:19710–19715.
130. Katakami Y, Nakao Y, Katakami N, Koizumi T, Ogawa R, Yamada H, Takai Y, Fujita T 1988 Cooperative effects of tumor necrosis factor-alpha and 1,25-dihydroxyvitamin D_3 on growth inhibition, differentiation, and c-*myc* reduction in human promyelocytic leukemia cell line HL-60. Biochem Biophys Res Commun **152**:1151–1157.
131. Simpson RU, Hsu T, Begley DA, Mitchell BS, Alizadeh BN 1987 Transcriptional regulation of the c-*myc* protooncogene by 1,25-dihydroxyvitamin D_3 in HL-60 promyelocytic leukemia cells. J Biol Chem **262**:4104–4108.
132. Brelvi ZS, Studzinski GP 1986 Changes in the expression of oncogenes encoding nuclear phosphoproteins but not c-Ha-*ras* have a relationship to monocytic differentiation of HL-60 cells. J Cell Biol **102**:2234–2243.
133. Watanabe T, Sariban E, Mitchell T, Kufe D 1985 Human c-*myc* and N-*ras* expression during induction of HL-60 cellular differentiation. Biochem Biophys Res Commun **126**:999–1005.
134. Reitsma PH, Rothberg PG, Astrin SM, Trial J, Bar-Shavit Z, Hall A, Teitelbaum SL, Kahn AJ 1983 Regulation of myc gene expression in HL-60 leukemia cells by a vitamin D metabolite. Nature **306**:492–494.
135. Brelvi ZS Christakos S, Studzinski GP 1986 Expression of monocyte-specific oncogenes c-*fos* and c-*fms* in HL60 cells treated with vitamin D_3 analogs correlates with inhibition of DNA syntehsis and reduced calmodulin concentration. Lab Invest **55**:269–275.
136. Karmali R, Bhalla AK, Farrow SM, Williams MM, Lal S, Lydyard PM, O'Riordan JL 1989 Early regulation of c-*myc* mRNA by 1,25-dihydroxyvitamin D_3 in human myelomonocytic U937 cells. J Mol Endocrinol **3**:43–48.
137. Lasky SR, Iwata K, Rosmarin AG, Caprio DG, Maizel AL 1995 Differential regulation of JunD by dihydroxycholecalciferol in human chronic myelogenous leukemia cells. J Biol Chem **270**:19676–19679.
138. Koizumi T, Nakao Y, Kawanishi M, Maeda S, Sugiyama T, Fujita T 1989 Suppression of c-*myc* mRNA expression by steroid hormones in HTLV-I-infected T-cell line, KH-2. Int J Cancer **44**:701–706.
139. Kasukabe T, Okabe-Kado J, Hozumi M, Honma Y 1994 Inhibition by interleukin 4 of leukemia inhibitory factor-, interleukin 6-, and dexamethasone-induced differentiation of mouse myeloid leukemia cells: role of c-*myc* and *junB* proto-oncogenes. Cancer Res **54**:592–597.
140. Moore DC, Carter DL, Studzinski GP 1992 Inhibition by 1,25-dihydroxyvitamin D_3 of c-*myc* down regulation and DNA fragmentation in cytosine arabinoside-induced erythroid differentiation of K562 cells. J Cell Physiol **151**:539–548.
141. Naveilhan P, Berger F, Haddad K, Bardot N, Benabib AL, Brachet P, Wion P 1994 Induction of glioma cell death by 1,25-$(OH)_2$-vitamin D_3: Towards an endocrine therapy of brain tumors. J Neurosci Res **37**:271–277.
142. Moore TB, Sidell N, Chow VJ, Medzoyan RH, Huang JI, Yamashiro JM, Wada RK 1995 Differentiating effects of 1,25-dihy-

droxycholecalciferol (D_3) on LA-N-5 human neuroblastoma cells and its synergy with retinoic acid. J Pediatr Hematol Oncol **17**:311–317.

143. Paatero GI, Trydal T, Karlstedt KA, Aarskog D, Lillehaug JR 1994 Time-course study of 1,25-$(OH)_2$-vitamin D_3 induction of homologous receptor and c-*myc* in non-transformed and transformed C3H/10T1/2 cell clones. Int J Biochem **26**:367–374.
144. Trydal T, Lillehaug JR, Aksnes L, Aarskog D 1992 Regulation of cell growth, c-*myc* mRNA, and 1,25-$(OH)_2$-vitamin D_3 receptor in C3H/10T1/2 mouse embryo fibroblasts by calcipotriol and 1,25-$(OH)_2$-vitamin D_3. Acta Endocrinol **126**:75–79.
145. van den Bemd GJCM, Pols HAP, Birkenhäger JC, Kleinekoort WMC, van Leeuwen JPTM 1995 Differential effects of 1,25-dihydroxyvitamin D_3-analogs on osteoblast-like cells and on in vitro bone resorption. J Steroid Biochem Mol Biol **55**:337–346.
146. Mahonen A, Pirskanen A, Mäenpää PH 1991 Homologous and heterologous regulation of 1,25-dihydroxyvitamin D_3 receptor mRNA levels in human osteosarcoma cells. Biochim Biophys Acta **1080**:111–118.
147. Baier R, Grauer A, Lazaretti-Castro M, Ziegler R, Raue F 1994 Differential effects of 1,25-dihydroxyvitamin D_3 on cell proliferation and calcitonin gene expression. Endocrinology **135**:2006–2011.
148. Grauer A, Baier R, Ziegler R, Raue F 1995 Crucial role of c-*myc* in 1,25-$(OH)_2D_3$ control of C-cell-carcinoma proliferation. Biochem Biophys Res Commun **213**:922–927.
149. Velez-Yanguas MC, Kalebic T, Maggi M, Chavez Kapel C, Letterio J, Uskokovic M, Helman LJ 1996 1alpha,25-Dihydroxy-16-ene-23-yne-26,27-hexafluorocholecalciferol (Ro24–5531) modulation of insulin-like growth factor-binding protein-3 and induction of differentiation and growth arrest in a human osteosarcoma cell line. J Clin Endocrinol Metab **81**:93–99.
150. Lefebvre MF, Guillot C, Crepin M, Saez S 1995 Influence of tumor derived fibroblasts and 1,25-dihydroxyvitamin D_3 on growth of breast cancer cell lines. Breast Cancer Res Treat **33**:189–197.
151. Rotello RJ, Lieberman RC, Purchio AF, Gerschenson LE 1991 Coordinated regulation of apoptosis and cell proliferation by transforming growth factor $\beta 1$ in cultured uterine epithelial cells. Proc Natl Acad Sci USA **88**:3412–3415.
152. Alexandrow MG, Moses HL 1995 Transforming growth factor β and cell cycle regulation. Cancer Res **55**:1452–1457.
153. Laiho M, DeCaprio JA, Ludlow JW, Livingston DM, Massague J 1990 Growth inhibition by TGF-β linked to suppression of retinoblastoma protein phosphorylation. Cell **62**:175–185.
154. Koli K, Keski-Oja J 1995 1,25-Dihydroxyvitamin D_3 enhances the expression of transforming growth factor $\beta 1$ and its latent form binding protein in cultured breast carcinoma cells. Cancer Res **55**:1540–1546.
155. Kim HJ, Abdelkader N, Katz M, McLane JA 1992 1,25-Dihydroxyvitamin D_3 enhances antiproliferative effect and transcription of TGF-$\beta 1$ on human keratinocytes in culture. J Cell Physiol **151**:579–587.
156. Koli K, Keski-Oja J 1993 Vitamin D_3 and calcipotriol enhance the secretion of transforming growth factor-$\beta 1$ and -$\beta 2$ in cultured murine keratinocytes. Growth Factors **8**:153–163.
157. Finkelman RD, Linkhart TA, Mohan S, Lau K-HW, Baylink DJ, Bell NH 1991 Vitamin D deficiency causes a selective reduction in deposition of transforming growth factor β in rat bone: Possible mechanism for impaired osteoinduction. Proc Natl Acad Sci USA **88**:3657–3660.
158. Drivdahl RH, Loop SM, Andress DL, Ostenson RC 1995 IGF-binding proteins in human prostate tumor cells: Expression and regulation by 1,25-dihydroxyvitamin D_3. Prostate **26**:72–79.
159. Moriwake T, Tanaka H, Kanzaki S, Higuchi J, Seino Y 1992 1,25-Dihydroxyvitamin D_3 stimulates the secretion of insulin-like growth factor binding protein 3 (IGFBP-3) by cultured human osteosarcoma cells. Endocrinology **130**:1071–1073.
160. Nakao Y, Hilliker S, Baylink DJ, Mohan S 1994 Studies on the regulation of insulin-like growth factor binding protein 3 secretion in human osteosarcoma cells in vitro. J Bone Miner Res **9**:865–872.
161. Koga M, Eisman JA, Sutherland RL 1988 Regulation of epidermal growth factor receptor levels by 1,25-dihydroxyvitamin D_3 in human breast cancer cells. Cancer Res **48**:2734–2739.
162. Falette N, Frappart L, Lefebvre MF, Saez S 1989 Increased epidermal growth factor receptor level in breast cancer cells treated by 1,25-dihydroxyvitamin D_3. Mol Cell Endocrinol **63**:189–198.
163. Bikle DD, Pillai S 1993 Vitamin D, calcium, and epidermal differentiation. Endocrine Rev **14**:3–19.
164. van Leeuwen JPTM, van Steenbrugge GJ, Veldscholte J, van den Bemd GJCM, Oomen MHA, Pols HAP, Birkenhäger JC 1994 Growth regulation of three sublines of the human prostatic carcinoma line LNCaP by an interrelated action of 1,25-dihydroxyvitamin D_3 and androgens. In: Norman AW, Bouillon R, Thomasset M (eds) Vitamin D, a Pluripotent Steroid Hormone: Structural Studies, Molecular Endocrinology and Clinical Application. de Gruyter, Berlin, pp. 510–511.
165. Esquenet M, Swinnen JV, Heyns W, Verhoeven G 1996 Control of LNCaP proliferation and differentiation: Actions and interactions of androgens, 1α,25-dihydroxycholecalciferol, all-*trans*-retinoic acid, 9-*cis*-retinoic acid and phenylacetate. Prostate **28**:182–194.
166. Koga M, Sutherland RL 1991 Retinoic acid acts synergistically with 1,25-dihydroxyvitamin D_3 or antiestrogen to inhibit T47-D human breast cancer cell proliferation. J Steroid Biochem Mol Biol **39**:455–460.
167. Bunce CM, Wallington LA, Harrixon P, WIlliams GR, Brown G 1995 Treatment of HL-60 cells with various combinations of retinoids and 1α,25-dihydroxyvitamin D_3 results in differentiation towards neutrophils or monocytes or a failure to differentiate and apoptosis. Leukemia **9**:410–418.
168. Bollag W 1994 Experimental basis of cancer combination chemotherapy with retinoids, cytokines, 1,25-dihydroxyvitamin D_3, and analogs. J Cell Biochem **56**:427–435.
169. Kliewer SA, Umesono K, Mangelsdorf DJ, Evans RM 1992 Retinoid X receptor interacts with nuclear receptors in retinoic acid, thyroid hormone, and vitamin D_3 signaling. Nature **335**:446–449.
170. Matsui T, Takahashi R, Mihara K, Nakagawa T, Koizumi T, Nakao Y, Sugiyama R, Fujita T 1985 Cooperative regulation of c-*myc* expression in differentiation of human promyelocytic leukemia induced recombinant gamma-interferon and 1,25-dihydroxyvitamin D_3. Cancer Res **45**:4366–4371.
171. Morel PA, Manolagas SC, Provedini DM, Wegmann DR, Chiller JM 1986 Interferon-gamma-induced IA expression in WEHI-3 cells is enhanced by the presence of 1,25-dihydroxyvitamin D_3. J Immunol **136**:2181–2186.
172. Kasukabe T, Okabe-Kado J, Honma Y, Hozumi M 1991 Interleukin-4 inhibits the differentiation of mouse myeloid leukemia M1 cells induced by dexamethasone, D-factor/leukemia inhibitory factor and interleukin-6, but not by 1α,25-dihydroxyvitamin D_3. FEBS Lett **291**:181–184.
173. Rocker D, Ravid A, Liberman UA, Garach-Jehoshua, Koren R 1994 1,25-Dihydroxyvitamin D_3 potentiates the cytotoxic effects of TNF on human breast cancer cells. Mol Cell Endocrinol **106**:157–162.

174. Hassan HT, Eliopoulos A, Maurer HR, Spandidos DA 1992 Recombinant human GM-CSF enhances the anti-proliferative activity of vitamin D in MCF-7 human breast cancer clonogenic cells. Eur J Cancer **28A**:1588–1589.
175. Cho YL, Christensen C, Saunders DE, Lawrence WD, Deppe G, Malviya VK, Malon JM 1991 Combined effects of 1,25-dihydroxyvitamin D_3 and platinum drugs on the growth of MCF-7 cells. Cancer Res **51**:2848–2853.
176. Dokoh S, Donaldson CA, Haussler MR 1984 Influence of 1,25-dihydroxyvitamin D_3 on cultured osteogenic sarcoma cells: Correlation with the 1,25-dihydroxyvitamin D_3 receptor. Cancer Res **44**:2103–2109.
177. Miller GJ, Stapleton GE, Hedlund TE, Moffatt KA 1995 Vitamin D receptor expression, 24-hydroxylase activity, and inhibition of growth by 1α,25-dihydroxyvitamin D_3 in seven prostatic carcinoma cell lines. Clin Cancer Res **1**:997–1003.
178. Hedlund TE, Moffatt KA, Miller GJ 1996 Stable expression of the nuclear vitamin D receptor in the human prostatic carcinoma cell line JCA-1: Evidence that the antiproliferative effects of 1α,25-dihydroxyvitamin D_3 are mediated exclusively through the genomic signaling pathway. Endocrinology **137**:1554–1561.
179. Xu H-M, Kolla SS, Goldenberg NA, Studzinski GP 1992 Resistance to 1,25-dihydroxyvitamin D_3 of a deoxycytidine kinase deficient variant of human leukemia HL-60 cells. Exp Cell Res **203**:244–250.
180. Narvaez CJ, Vanweelden K, Byrne I, Welsh J 1996 Characterization of a vitamin D-resistant MCF-7 cell line. Endocrinology **137**:400–409.
181. Chen TL, Hauschka PV, Feldman D 1986 Dexamethasone increases 1,25-dihydroxyvitamin D_3 receptor level and augments bioresponses in rat osteoblast-like cells. Endocrinology **118**: 1119–1126.
182. Ohta M, Okabe T, Ozawa K, Urabe A, Takaku F 1985 1α,25-Dihydroxyvitamin D_3 (calcitriol) stimulates proliferation of human circulating monocytes. FEBS Lett **185**:9–13.
183. Mitsuhashi T, Morris RC, Ives HE 1991 1,25-Dihydroxyvitamin D modulates growth of vascular smooth muscle cells. J Clin Invest **87**:1889.
184. Edelson JD, Chan S, Jassal D, Post M, Tanswell AK 1994 Vitamin D stimulates DNA synthesis in alveolar type-II cells. Biochim Biophys Acta **1221**:159–166.
185. Lutzow-Holm C, De Angelis P, Grosvik H, Clausen OP 1993 1,25-Dihydroxyvitamin D_3 and the vitamin D analogue KH1060 induce hyperproliferation in normal mouse epidermis. A BrdUrd/DNA flow cytometric study. Exp Dermatol **2**:113–120.
186. Gniadecki R 1994 A vitamin D analogue KH1060 activates the protein kinase C-c-*fos* signalling pathway to stimulate epidermal proliferation in murine skin. J Endocrinol **143**:521–525.
187. Gniadecki R, Serup J 1995 Stimulation of epidermal proliferation in mice with 1,25-dihydroxyvitamin D_3 and receptor-active 20-epi analogues of 1,25-dihydroxyvitamin D_3. Biochem Pharmacol **49**:621–624.
188. Gniadecki R, Gniadecka M, Serup J 1995 The effects of KH1060, a potent 20-epi analogue of the vitamin D_3 hormone, on hairless mouse skin *in vivo*. Br J Dermatol **132**:841–852.
189. Hosoi J, Abe E, Suda T, Colburn NH, Kuroki T 1986 Induction of anchorage-independent growth of JB6 mouse epidermal cells by 1,25-dihydroxyvitamin D_3. Cancer Res **46**:5582–5586.
190. Hosomi J, Hosoi J, Abe E, Suda T, Kuroki T 1983 Regulation of terminal differentiation of cultured mouse epidermal cells by 1,25-dihydroxyvitamin D_3. Endocrinology **113**:1950–1957.
191. Kragballe K 1992 Vitamin D_3 and skin diseases. Arch Dermatol Res **284**:S30–S36.
192. Kuroki T, Sasaki K, Chida K, Abe E, Suda T 1983 1,25-Dihydroxyvitamin D_3 markedly enhances chemically induced transformation in Balb 3T3 cells. Gann **74**:611–614.
193. Wood AW, Chang RL, Huang M-T, Baggiolini E, Partridge JJ, Uskokovic M, Conney AH 1985 Stimulatory effect of 1α,25-dihydroxyvitamin D_3 on the formation of skin tumours in mice treated chronically with 7,12-dimethylbenz[*a*]anthracene. Biochem Biophys Res Commun **130**:924–931.
194. Jones CA, Callaham MF, Huberman E 1984 Enhancement of chemical carcinogen-induced cell transformation in hamster embryo cells by 1,25-dihydroxycholecalciferol, the biologically active metabolite of vitamin D_3. Carcinogenesis **5**:1155–1159.
195. Yamaoka K, Marion SL, Gallegos A, Haussler MR 1986 1,25-dihydroxyvitamin D_3 enhances the growth of tumours in athymic mice inoculated with receptor rich osteosarcoma cells. Biochem Biophys Res Commun **139**:1292–1298.
196. Franceschi RT, James WM, Zerlauth G 1985 1α,25-Dihydroxyvitamin D_3 specific regulation of growth, morphology, and fibronectin in a human osteosarcoma cell line. J Cell Physiol **123**:401–409.
197. Gronowicz G, Egan JJ, Rodan GA 1986 The effect of 1,25-dihydroxyvitamin D_3 on the cytoskeleton of rat calvaria and rat osteosarcoma (ROS 17/2.8) osteoblastic cells. J Bone Miner Res **1**:441–455.
198. Hodge BO, Kream BE 1988 Variable effects of dexamethasone on protein synthesis in clonal rat osteosarcoma cells. Endocrinology **122**:2127–2133.
199. Abbadia Z, Amiral J, Trzeciak M-C, Delmas PD, Clezardin P 1993 The growth-supportive effect of thrombospondin (TSP1) and the expression of TSP1 by human MG-63 osteoblastic cells are both inhibited by dexamethasone. FEBS Lett **335**:161–166.

CHAPTER 65

Vitamin D and Breast Cancer: Therapeutic Potential of New Vitamin D Analogs

KAY COLSTON Steroid Biochemistry Group, Clinical Biochemistry, St. George's Hospital Medical School, London, England

I. INTRODUCTION

A. Breast Cancer—Prevalence and Current Therapeutic Approaches

Breast cancer is the most common cause of death from cancer in women, affecting 1 in 10 women in the United Kingdom. Every year approximately 26,000 women are diagnosed as having breast cancer and 15,000 die of their disease. Although significant improvements in the management of the disease have been made since the 1970s, with surgery, radiotherapy, endocrine therapy, and chemotherapy providing useful treatment modalities, breast cancer remains an essentially incurable disease. Adjuvant endocrine and chemotherapy induces a significant survival advantage in only a proportion of women. In patients with advanced disease, although good responses can be achieved with sequential hormone and other chemotherapies, the response duration remains low. A major problem in trying to treat cancer cells is that they are often poorly differentiated and hence resistant to normal endocrine and immunological controls. In 1896 Beaston demonstrated striking tumor regression following oophorectomy in young women with breast cancer and pioneered endocrine therapy as a major therapy for this disease.

B. Approaches to Endocrine Therapy

The final common pathway in different forms of endocrine treatment for breast cancer is either lowering of plasma estrogen or blocking of the effects of estrogen in the target tissues. It is not entirely clear how estrogens modulate cell growth. The expression of mitogenic growth factors by tumor cells (autocrine) or by adjacent stromal cells (paracrine) are of great importance, and characterization of these factors has helped to identify some of the regulatory mechanisms involved in breast cancer cell growth. Tamoxifen, a triphenylethylene derivative, is a potent antiestrogen with additional estrogenic activity at some target tissues [1], and it is a widely used drug in breast cancer. Its mode of action is considered to be the blocking of estrogen receptor (ER) activity. However, a paradox arises with some clinical observations. First, approximately 10% of estrogen receptor negative tumors respond to tamoxifen, and, second tamoxifen has been shown in two major trials to prolong the relapse-free interval and survival in patients with ER negative tumors [2,3].

The main source of estrogen in premenopausal women is the ovary. In patients with advanced disease, an approach to reducing estrogen levels is the use of leutenizing hormone-releasing hormone (LHRH) analogs. These synthetic compounds generally have amino acid substitutions or deletions at the 6 and 10 positions. Effects are achieved via decreased production of gonadotropins, leading to reduced ovarian estrogen synthesis and resulting in a reversible "medical oophorectomy." LHRH analogs do not have a place in the treatment of postmenopausal women in whom circulating estrogen is low and is predominately synthesized by peripheral conversion of adrenal androgens by the aromatase enzyme, principally in adipose tissue. A number of aromatase inhibitors such as aminoglutethimide and 4-hydroxyandrostenedione have been developed. Although aromatase activity is reduced by at least 95% with this type of treatment, circulating estrogens are reduced by only 50–60%. The source of the residual plasma estrogen is largely unknown but may be related to dietary factors.

Research has been aimed at producing "pure" antiestrogens without additional estrogenic activity [4]. These include the compounds ICI 164,384 and ICI 182,780. These compounds possess pure estrogen antagonist activity with potent antiproliferative effects on human breast cancer models *in vitro* and *in vivo* [5]. However, it is possible that the partial agonist activity displayed by tamoxifen may not be entirely disadvantageous, as studies on patients receiving long-term treatment have so far failed to show detrimental effects on bone mineral density or the cardiovascular effects known to be associated with estrogen deprivation. Trials of tamoxifen in breast cancer prevention are currently underway. Current endocrine therapies such as tamoxifen only bring about objective remissions in approximately 60% of ER positive breast tumors [6]. Furthermore, all tumors eventually become resistant to the therapy after a median duration of remission of about 18 months [7].

II. VITAMIN D AND BREAST CANCER

A. Vitamin D Receptors in Breast Cancer Cells

1. Detection of Receptors in Normal and Malignant Breast Tissue

Evidence that the hormonal form of vitamin D_3, 1,25α-dihydroxyvitamin D_3 [1,25$(OH)_2D_3$], possesses the activity to inhibit the proliferation and promote the differentiation of transformed cells has led to the suggestion of its use as a therapeutic agent in malignancy. The obvious risk of development of hypercalcemia and hypercalciuria associated with the use of conventional vitamin D derivatives has prompted efforts to develop synthetic analogs that retain the ability to inhibit cell replication but are reduced in their calcemic activity and thus display a more favorable profile of activity. The first experimental evidence linking vitamin D with breast cancer was the demonstration that the established human breast cancer cell line MCF-7 possesses specific receptors for 1,25$(OH)_2D_3$ [8]. These cells are estrogen responsive but 1,25$(OH)_2D_3$ receptors are also present in estrogen independent breast-cancer cells such as MDA-MB-231 and Hs 587T [9].

Receptors for 1,25$(OH)_2D_3$ have also been demonstrated in the mammary gland of pregnant and lactating rats [10] and in carcinogen induced rat mammary tumors [11,12]. Vitamin D receptors (VDR) are also present in normal human breast and many other epithelial tissues [13]. In addition, a large proportion of breast tumor biopsy specimens express VDR protein, as assessed by specific 1,25$(OH)_2D_3$ binding in tumor extracts [14,15] or positive immunostaining using antibodies to the vitamin D receptor [16–18]. A radioimmunometric assay has also been developed by Sandgren and associates [19]. These authors demonstrated that a high proportion of lung tumors and colorectal carcinomas had significant receptor levels. There is as yet no evidence that vitamin D receptors in neoplastic tissues differ from the wild type. Vitamin D receptors studied in colorectal cancer cells exhibited no difference in binding characteristics as compared to the high affinity receptor [20].

2. Relationship of Receptor Status and Disease Outcome

In breast cancer the relationship between VDR levels and clinical indices has been determined. Tumor VDR status does not appear to be related to overall survival

[14,15,18] or to survival after relapse [18]. However, an association was found between tumor VDR positivity and disease-free survival time. In a study of 136 patients with primary breast cancer, it was found that those patients with VDR negative tumors relapsed significantly earlier than those with VDR positive tumors [17,18]. However the study by Eisman and associates [15] reported that the presence of the receptor is correlated with late lymph node metastases.

3. Relationship to Other Hormone Receptors

Depending on the method of analysis and the cutoff level for receptor positivity, approximately 60% of breast tumors contain estrogen receptors. Breast tumors with ER positive cells are strongly related to a higher response rate to hormonal treatment and to a longer survival time after relapse [18,21]. Estrogen stimulation of breast cancer cells leads to the induction of a number of proteins, among which is the progesterone receptor (PR). The presence of progesterone receptor indicates functionality of the ER signaling pathway, and detection of this receptor improves the predictive accuracy for hormonal responsiveness to approximately 75% [22]. Although the presence of the ER and PR in breast tumors are correlated, several studies have noted that no apparent association exists between the presence of VDR and ER or PR [14,16,23].

B. Epidemiological Studies

Epidemiological studies have pointed to a relationship between vitamin D and cancer. It was suggested in 1980 that calcium and vitamin D may reduce the risk of colon cancer, the assertion being derived from the geographic epidemiology of death rates for colon cancer which tend to increase with increasing latitude and decreasing sunlight intensity [24] (see Chapter 67). Furthermore, an inverse relationship between serum 25-hydroxyvitamin D and incidence of colonic cancer has been demonstrated [25]. It has been noted that mortality and incidence of breast cancer in North America and other areas of the world increases with increasing latitude. Garland and associates [26] have reported that the risk of fatal breast cancer in major urban areas of the United States appears to be inversely proportional to intensity of sunlight and that synthesis of vitamin D from sunlight exposure may be associated with low risk for fatal breast cancer. These authors further suggest that differences in ultraviolet irradiation across the United States may account for the observed regional differences in breast cancer mortality. In addition, mean serum levels of 1,25$(OH)_2D_3$ were found to be lower in postmenopausal women with early breast cancer than in a group of matched controls [27]. Schwartz and Hulka [28] have suggested that mortality from prostate cancer might also be linked to vitamin D because of a geographic relationship between mortality from this malignancy and UV light intensity similar to that for breast and colon cancer (Chapter 66).

III. VITAMIN D AS AN ANTIPROLIFERATIVE AGENT IN BREAST CANCER

A. Actions of 1,25-Dihydroxyvitamin D_3 and Its Analogs in Cultured Breast Cancer Cells

The antiproliferative effect of 1,25$(OH)_2D_3$ on cultured human cancer cells was first demonstrated in 1981 [29]. These first experiments were performed in amelanotic melanoma cells and showed that 1,25$(OH)_2D_3$ at concentrations in the nanomolar range significantly inhibited cell proliferation. Later the same year the observation was made that 1,25$(OH)_2D_3$ can promote the differentiation of leukemic cells [30]. Subsequently it has been demonstrated that the growth of a variety of cultured human cancer cell lines can be inhibited by 1,25$(OH)_2D_3$, including cell lines derived from breast cancers [31–35].

A variety of structural modifications to the vitamin D molecule have been adopted in an attempt to separate hypercalcemic activity from antiproliferative and differentiation promoting actions displayed by 1,25$(OH)_2D_3$. Many of the analogs developed to date have modifications in the C-17 side chain of the vitamin D molecule, while keeping the 1α-hydroxylated A ring and *cis*-triene the same as in the 1,25$(OH)_2D_3$ molecule [36]. Modifications leading to compounds with reduced hypercalcemic activity have primarily focused on carbons 22, 23, and 24, as receptor binding is not substantially altered by this change but metabolic degradation as well as interaction with vitamin D binding protein (DBP) are reduced.

Initially, the effectiveness of the various analogs has been evaluated with *in vitro* methods. The human leukemic cell lines HL-60 and U937 have been extensively utilized to assess effects of conventional vitamin D metabolites and synthetic analogs on cell growth and differentiation [37], whereas actions of these compounds on calcium homeostasis *in vivo* have generally been studied in chickens [38] or rodents [39,40]. The compound 1,25-$(OH)_2$-22-oxa-vitamin D_3 (OCT), in which an oxygen atom is substituted for the methylene group at C-22, has been reported to be equipotent with 1,25$(OH)_2D_3$

in promoting cell differentiation coupled with reduced effects on calcium mobilization [41]. A series of 26- and 24-homologated 1,25-dihydroxyvitamin D_3 compounds and their delta derivatives have been synthesized and appear to have enhanced activity in HL-60 cells [42]. Norman and associates [43] reported a comparison of effects of eight synthetic analogs on HL-60 cells. One analog, 1,25-$(OH)_2$-16-ene 23-yne-vitamin D_3, and its hexadeutero form were reported to be severalfold more potent than the parent compound.

Certain of these compounds have been tested for their effectiveness in inhibiting the growth of breast cancer cells. The compound MC903 (calcipotriol from Leo Pharmaceutical Products) contains a cyclopropyl substitution in the side chain [39] and appears to be equipotent with 1,25$(OH)_2D_3$ in inhibiting the proliferation of MCF-7 breast cancer cells *in vitro* [44]. In contrast, a 100 to 200-fold greater dose of MC903 than 1,25$(OH)_2D_3$ is required *in vivo* to give an equivalent effect on serum and urinary calcium in experimental animals [32,44]. MCF-7 human breast cancer cells have also been shown to be responsive to growth inhibition by EB1089, a second generation analog from Leo Pharmaceutical Products that contains a conjugated double bond system (Fig. 1) and is approximately 50 times more effective than 1,25$(OH)_2D_3$ *in vitro* [45,46]. A further series of analogs produced by Leo, the 20-epi-vitamin D_3 analogs such as KH1060, are characterized by altered stereochemistry at the C-20 position [47]. These are also potent inhibitors of human breast cancer cells *in vitro* (Table I), but certain of these compounds display potent immunosuppressive activity *in vivo* [48]. Studies with a series of analogs from Hoffmann-La Roche have confirmed increased potency relative to 1,25$(OH)_2D_3$ in human breast cancer as well as leukemic cells [49]. In addition, Abe and associates [50] have documented potent antiproliferative effects of OCT in ER positive and negative breast cancer cells.

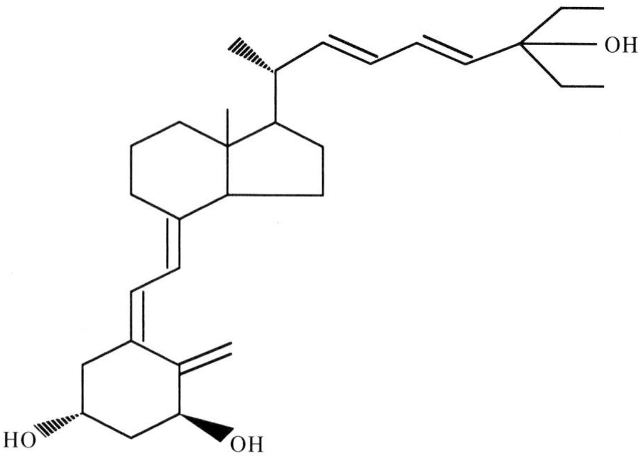

OH
HO OH

FIGURE 1 Structure of the vitamin D analog EB1089.

B. Preclinical Trials of Vitamin D Analogs

In animal studies, administration of 1,25$(OH)_2D_3$ and its analog alfacalcidol [which is converted to 1,25$(OH)_2D_3$ *in vivo*] prolonged the survival time of mice inoculated with myeloid leukemia cells [51]. In another study, immunodeficient mice bearing tumor xenografts of colon carcinomas and melanomas were treated with 1,25$(OH)_2D_3$, and inhibition of tumor growth was reported [52]. However, these animals were maintained on a low calcium diet to limit intestinal absorption of calcium. In the nitrosomethylurea (NMU)-induced rat mammary tumor model of hormone-dependent breast cancer, alfacalcidol given intraperitoneally (i.p.) at a dose of 0.5 μg per kg body weight significantly inhibited the tumor progression, but with increasing doses marked hypercalcaemia developed, even when the animals were maintained on a low calcium diet [44]. These laboratory studies highlight the drawback to considering conventional vitamin D metabolites as therapeutic agents because of their potent calcemic activity. Consequently, interest has more recently focused on the potency of new vitamin D analogs with reduced calcemic activity in animal models of cancer.

In vivo antitumor effects of calcipotriol have been assessed in rats bearing mammary tumors induced by the carcinogen NMU, and a modest inhibition of tumor progression was observed [44]. Further Leo analogs have also been tested for antitumor activity using this animal model of breast cancer, with analog dose being chosen on the basis of the known hypercalcemic activity. No significant inhibition of tumor progression was seen with 1,25$(OH)_2D_3$ or the 20-epi-compounds KH1060, KH1049, or MC1301 (Fig. 2), but all produced a significant rise in serum calcium concentration [53]. Three analogs caused tumor regression, namely, CB966, EB1089, and CB1093. In particular, EB1089 at 0.5 μg/kg/day over a 4-week period resulted in significant inhibition of tumor growth in the absence of changes in serum calcium levels [44,54]. Effects of this analog were found to be dose dependent and treatment with 2.5 μg/kg/day caused a reduction in mean tumor volume of more than 75%, although at the expense of a significant increase in serum calcium (Fig. 3). Furthermore, treatment with this dose of the analog prevented development of new tumors in the treated group while a substantial number of new tumors developed in control animals (Fig. 4). The analog OCT retards the growth of human breast cancer xenografts developed in athymic mice from the ER negative MX-1 tumor [50] and from ER

Table I Structure–Activity Relationships of Vitamin D Analogs[a]

Analog	MCF-7 cell growth inhibition, IC_{50}	Activity relative to $1,25(OH)_2D_3$	Side chain
$1,25(OH)_2D_3$	1.6×10^{-8} *M*	1	OH
KH1230	3.5×10^{-8} *M*	0.5	O, OH
EB1089	4.0×10^{-10} *M*	40	OH
CB1093	3.7×10^{-10} *M*	43	O, C≡C, OH
KH1060	3.6×10^{-11} *M*	444	O, OH
ICI 182,780	4.5×10^{-10} *M*	36	—

[a] MCF-7 cells (2×10^4 cells/ml) were seeded into 24-well plates and cultured for 7 days in the presence of $1,25(OH)_2D_3$ or vitamin D analogs. All compounds were tested in three separate experiments. Cell proliferation was determined by the crystal violet assay. Effects with the antiestrogen ICI 182,780 are shown for comparison.

positive MCF-7 cells [55]. Also found was a synergistic antitumor effect of a submaximal dose of OCT and tamoxifen. This synthetic vitamin D derivative has also been demonstrated to exert growth inhibitory effects in rat mammary tumors induced by 7,12-dimethylbenz-[*a*]anthracene (DMBA) [56]. In addition to inhibiting the growth or promoting regression of established tumors, vitamin D derivatives may also prevent mammary tumor development in response to carcinogen treatment. Anzano and associates [57] have examined the effectiveness of 1α,25-dihydroxy-16,27-hexafluorocholecalciferol alone and in combination with tamoxifen in the prevention of tumor induction by NMU and have demonstrated an extended tumor latency and lessened tumor burden in treated animals. At the doses tested, the treatment regimens did not induce severe hypercalcemia.

C. Receptor Binding, Serum Transport, and Metabolic Stability of Vitamin D Analogs

From the current preclinical trials of vitamin D analogs in animal models of breast cancer, it appears that these compounds can be given in doses which achieve significant effects on tumor growth while displaying reduced hypercalcemic activity relative to the natural hormonal form of vitamin D. There are a number of reasons why vitamin D analogs might display altered profiles of activity. (This subject is discussed in detail in Chapters 58–63.) Availability of ligand for the VDR due to differences in serum transport and delivery to the target cell or altered intracellular metabolism could play a part. Several analogs, for example, MC903, OCT, and EB1089, have decreased affinity for the vitamn D serum transport protein (DBP). This property has implications for metabolic clearance rates. OCT binds mainly to lipoproteins and has a plasma half-life approximately one-third that of $1,25(OH)_2D_3$. Preliminary data indicate that, in the rat, EB1089 has a half-life approximately equal to $1,25(OH)_2D_3$, although it binds poorly to transport proteins in rat serum [45]. Pharmacokinetic studies of a series of new analogs from Leo Pharmaceuticals have been undertaken; results show that the initial serum analog level correlated with the affinity for DBP, but no correlation with metabolic rate (as assessed by measurement of serum half-life) was demonstrated [58]. The low calcemic activity of calcipotriol can be attrib-

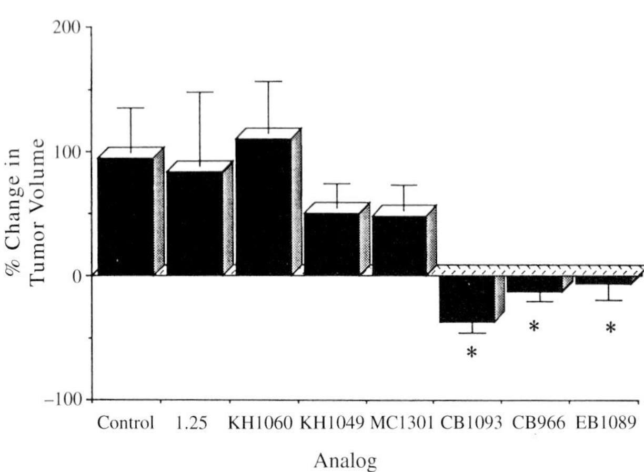

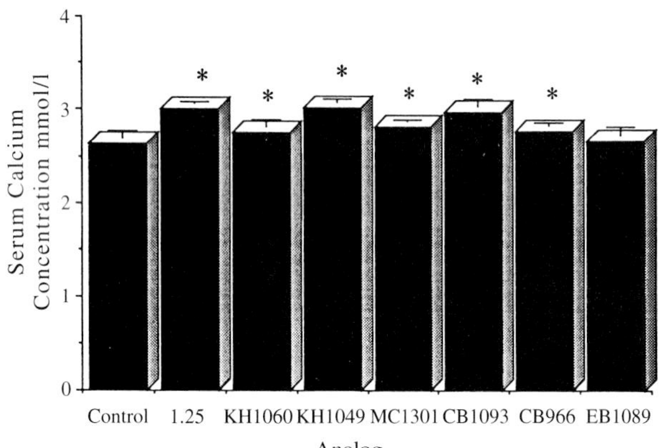

FIGURE 2 Comparison of the effects of vitamin D analogs on the growth of nitrosomethylurea (NMU)-induced rat mammary tumors and serum calcium concentration. Tumor-bearing rats were given the following doses of the vitamin D compounds orally daily for 28 days (per kilogram body weight): 0.5 μg 1,25(OH)$_2$D$_3$, 0.1 μg KH1060, 0.5 μg KH1049, 0.1 μg MC1301, 1.0 μg CB1093, 1.0 μg CB 966, 0.5 μg EB1089. Positive values indicate tumor progression; negative values indicate tumor regression. Mean serum calcium concentrations at the end of the treatment period for each group are shown at bottom. An asterisk (*) denotes results significantly different from the control group at $p < 0.05$. Adapted from Colston *et al.* [53].

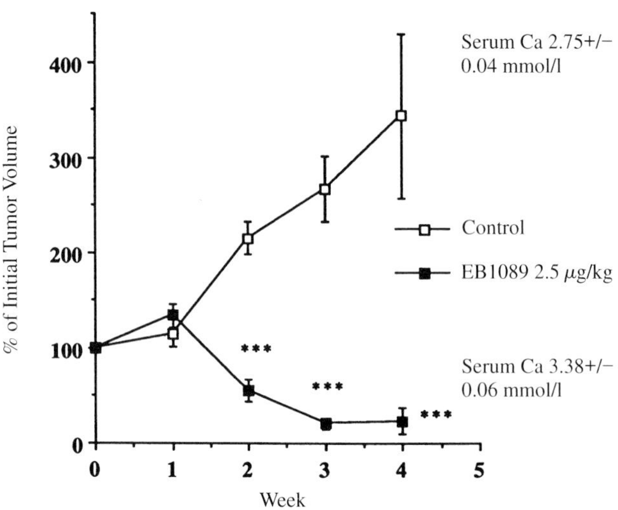

FIGURE 3 Mammary tumor regression in rats treated with EB1089 (2.5 μg/kg body weight daily. Tumors in the control group progressed during the 4-week experimental period. Results are expressed as percentages of initial tumor volume. It should be noted that the EB1089 treatment caused some hypercalcemia. ***$p < 0.001$.

uted to its rapid metabolic clearance via formation of 24-oxidized metabolites in the liver [59]. As a consequence of these findings, calcipotriol was considered more suitable for topical than for systemic use and has been developed as a safe and effective topical therapy for the treatment of psoriasis [60]. Its value in topical therapy in cutaneous nodules in patients with advanced breast cancer has also been assessed [61].

Although differences in binding to serum transport proteins may explain in part the apparent increased potency of vitamin D analogs on cell differentiation and proliferation *in vitro* (due to increases in the "free" fraction), these properties do not explain the apparent selective actions of analogs seen *in vivo*. It has also been reported that the binding affinity of vitamin D derivatives for VDR is not necessarily correlated with the potency of their biological activity *in vitro*. Thus EB1089 binds less well to VDR in MCF-7 cells than does 1,25(OH)$_2$D$_3$ while being at least one order of magnitude more potent in inhibiting cell proliferation [46]. The receptor binding affinity of OCT for VDR in HL-60 cells is weaker than that of 1,25(OH)$_2$D$_3$, whereas the analog is an order of magnitude greater in inducing differentiation of these leukemic cells [41].

D. Mechanism Implicated in Antiproliferative Effects of Vitamin D Analogs in Breast Cancer Cells

1. Interaction with Estrogen Response Pathways

In estrogen responsive breast cancer cells, a possible mechanism for growth inhibitory effects of vitamin D derivatives is through a modulation of the growth stimulatory effects of estrogens. Effects of vitamin D analogs on estrogen response pathways have been assessed by a number of groups. EB1089 down-regulates the expression of the estrogen receptor in MCF-7 cells and limits their responsiveness both to the mitogenic effects of 17β-estradiol and to the induction by this steroid of the progesterone receptor protein and pS$_2$ mRNA

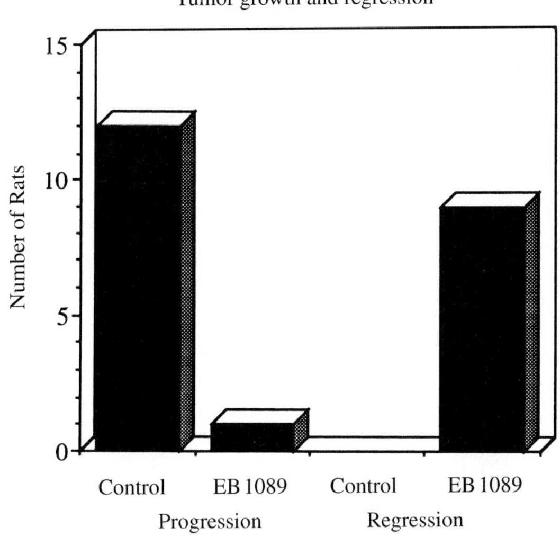

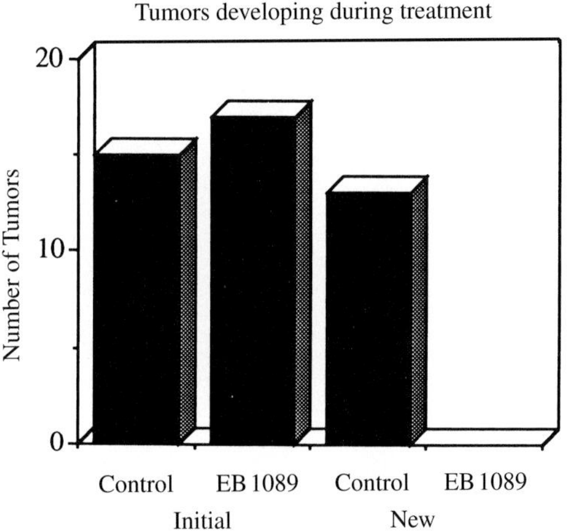

FIGURE 4 Effects of EB1089 (2.5 μg/kg) on mammary tumor regression and development. Of the 12 animals in the EB1089 treatment group (Fig. 3), 11 showed >50% regression of total tumor volume. No control animals showed tumor regression. Furthermore, although the number of initial tumors was similar in control and treated groups, 13 new tumors developed in the control group during the 4-week experimental period but no new tumors developed in EB1089-treated animals.

[53,62]. Demirpence and associates [64] have reported that both $1,25(OH)_2D_3$ and all-*trans*-retinoic acid inhibit estrogen-induced growth of MCF-7 cells and that $1,25(OH)_2D_3$ inhibits the estrogen-induced transcription of the pS_2 gene. This group did not observe any decrease in estrogen receptor expression by $1,25(OH)_2D_3$ but have suggested that the molecular mechanism by which the hormone exerts its antiestrogenic effects appears to involve the activity of the ligand binding, dimerization, and ligand-dependent transactivating domains of ER. Thus vitamin D derivatives may act at several points along the estrogen response pathway, including effects on both the abundance of ER protein and its ability to function as a transcriptional activator [63].

The interaction of EB1089 and the pure antiestrogen ICI 182,780 [5] on the estradiol-stimulated growth of MCF-7 cells has been investigated. Treatment of cell cultures with EB1089 in combination with ICI 182,780 and in the presence of 17β-estradiol produced an augmented inhibition of proliferation compared to the actions of either compound alone [62]. Cooperative effects of combined treatment with vitamin D analogs and tamoxifen have also been demonstrated in both MCF-7 and ZR-75-1 cells [55,64]. However, it was suggested that vitamin D derivatives exert their effects primarily in an estrogen-independent manner because no changes in ER expression were demonstrated in the studies. Further evidence that effects on cell growth may be mediated independently of estrogen response pathways has been obtained by studies in estrogen-independent cells. Both OCT and EB1089 have been shown to inhibit the growth of ER negative breast cancer cells in culture, indicating that effects of these compounds are not solely mediated by limiting cellular responsiveness to estrogen [50,62].

2. MODULATION OF GROWTH FACTOR RESPONSES

An additional way in which vitamin D derivatives may inhibit breast cancer cell growth is via interactions with growth factor-mediated mechanisms, as breast cancer cells both elaborate and respond to a variety of paracrine/autocrine growth factors. The analog EB1089 at low concentrations can reverse the growth stimulatory effects of epidermal growth factor (EGF) [65] and insulin-like growth factor type 1 (IGF-1) (Fig. 5a), indicating that this vitamin D derivative can modulate cellular responsiveness to such mitogens. Similar effects have also been shown with other vitamin D analogs [66], including CB1093 and KH1060 (Fig. 5b). Regulation of epidermal growth factor receptor levels by $1,25(OH)_2D_3$ in human breast cancer cells has been demonstrated [67]. Our own studies have shown that vitamin D derivatives decrease IGF-1 receptor levels in MCF-7 cells [66].

Unlike EGF, the IGFs, and transforming growth factor-α (TGFα), the effect of transforming growth factor-β (TGFβ) is to inhibit the growth of breast cancer cells. It is thus of interest that $1,25(OH)_2D_3$ has been shown to enhance the expression of TGFβ1 and its latent form binding protein in cultured breast carcinoma cells [68]. Studies utilizing a neutralizing antibody to TGFβ have also suggested that the antiproliferative effect of the analog EB1089 could in part be mediated by increased expression of this growth inhibitory peptide [69]. Effects of mitogenic growth factors on cell proliferation are associated with changes in expression of early response

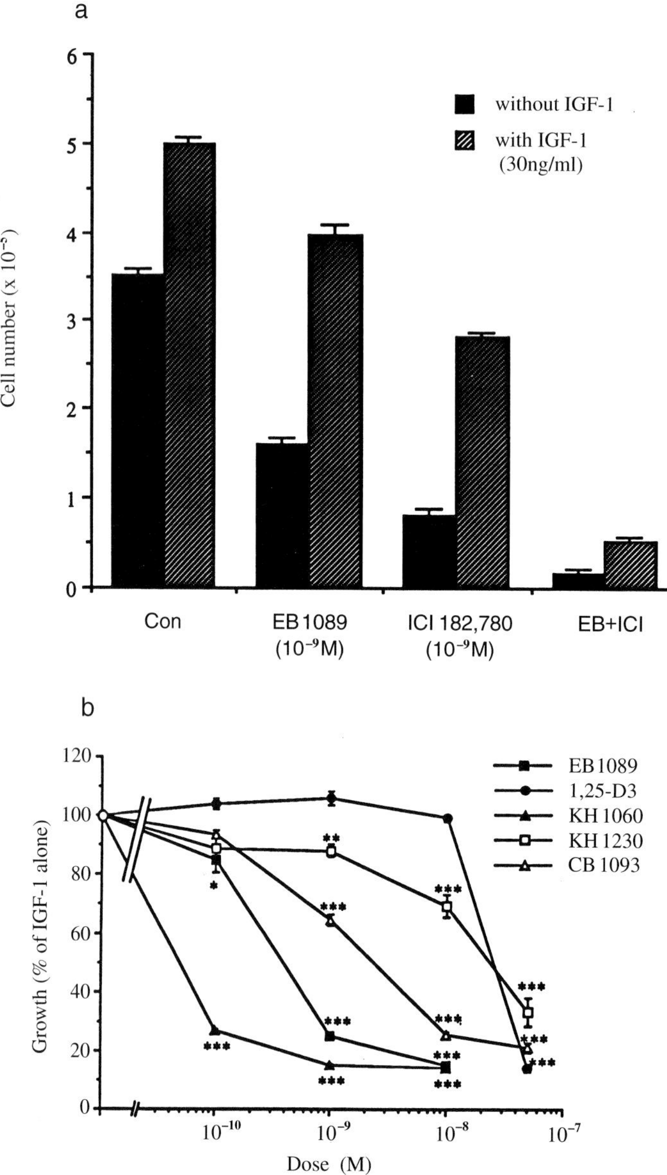

FIGURE 5 (a) Effects of EB1089 and ICI 182,780 (each at a concentration of 1 n*M*) alone and in combination on the autonomous and IGF-1-stimulated growth of MCF-7 breast cancer cells. Cell proliferation was assessed by crystal violet assay after 7 days of treatment with EB1089, ICI 182,780, or both agents in the presence or absence of IGF-1 (30 ng/ml). This dose of IGF-1 was previously determined to maximally stimulate cell growth in the absence of vitamin D compounds or antiestrogen. (b) Effects of vitamin D analogs on the growth of MCF-7 cells in the presence of IGF-1. Cell proliferation was assessed after 7 days of treatment with IGF-1 alone (30 ng/ml) or together with increasing concentrations of the indicated vitamin D analogs. Results are expressed as percentages of mean cell numbers in the presence of IGF-1 alone. *$p < 0.05$; **$p < 0.01$, ***$p < 0.005$ compared with IGF-1 alone. Adapted from Ref. [66].

genes, which include the protooncogenes c-*fos* and c-*myc*. A number of reports have documented changes in oncogene expression [70–72]. The effect of EB1089 on regulation of growth and differentiation of MCF-7 cells have been further studied at the level of expression of the c-*myc* and c-*fos* protooncogenes. EB1089 decreased the expression of c-*myc* mRNA and transiently increased c-*fos* expression, being approximately 50 times more potent than $1,25(OH)_2D_3$ [46]. In contrast, Vink-van Wijngaarden and associates [64] reported no modulation of either basal or 17β estradiol-induced c-*myc* expression by $1,25(OH)_2D_3$ in MCF-7 cells.

3. Induction of Apoptosis

a. Characteristics of Apoptosis (Programmed or Active Cell Death) Apoptosis is an active, energy-dependent process in which a distinct series of biochemical and molecular events leads to cell death by specific signals. It is characterized by altered gene expression, cytoplasmic and nuclear condensation, chromatin cleavage, and finally fragmentation of the cell into membrane-bound apoptotic bodies [73]. Studies have indicated that induction of apoptosis may be a feature of the antitumor effects of certain vitamin D analogs. Tumor regression occurs when the rate of cell death is greater than the rate of cell proliferation. Many chemotherapeutic agents induce tumor regression through their ability to activate the apoptotic pathway. Apoptosis is induced in estrogen-dependent human breast tumors during regression in response to antiestrogens and is associated with enhanced expression of a series of genes [74].

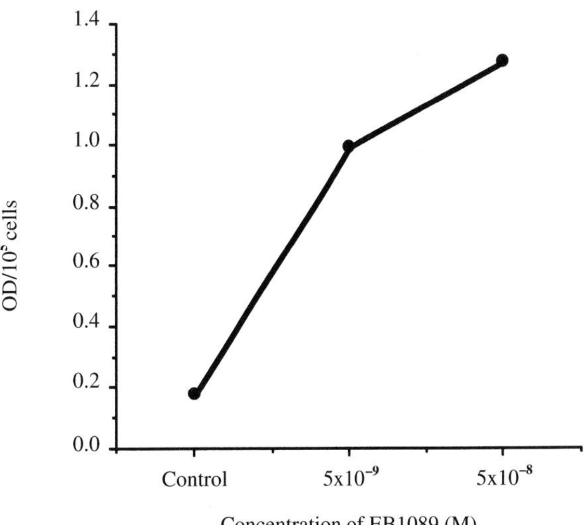

FIGURE 6 Detection of cytoplasmic oligonucleosomes in MCF-7 cells exposed to EB1089. MCF-7 cultures were treated for 6 days with ethanol vehicle or EB1089 (5–50 n*M*). After correction for cell number, cells were lysed and assessed for the presence of cytoplasmic histone–oligonucleosomes using the Boehringer-Mannheim cell death enzyme-linked immunosorbent assay (ELISA) method according to the manufacturer's instructions.

b. Morphological Changes and DNA Fragmentation Effects of $1,25(OH)_2D_3$ on MCF-7 cell growth and morphology have been characterized (75). Treated cells exhibit cytoplasmic condensation and hyperchromatic, pycnotic nuclei resulting from chromatin condensation typical of apoptotic cells. Cell growth indices assessed in relation to the onset of morphological evidence of apoptosis following $1,25(OH)_2D_3$ treatment demonstrated that the major effects of the hormone after 48 hr of treatment is to induce accumulation of cells in G_0/G_1 phase, similar to reported effects of the hormone on cell cycle kinetics in T-47D breast cancer cells [33]. $1,25(OH)_2D_3$ also increases the proportion of MCF-7 cells detected as a discrete population in the sub-G_0 peak, which represents apoptotic cells [76]. Evidence indicates that early G_1 may be the point at which the switch occurs between cell cycle progression or induction of apoptosis [77].

Apoptosis in response to antiestrogens is also associated with arrest of cells in G_0/G_1. Morphological assessment of MCF-7 cells treated with $1,25(OH)_2D_3$ and tamoxifen in combination showed enhanced induction of apoptosis [76]. The ability of new synthetic analogs of vitamin D to induce apoptosis in breast cancer cells has also been examined, and several groups have presented evidence for the presence of DNA fragmentation, a key feature of apoptosis, in breast cancer cells treated with vitamin D derivatives [78–81] (Fig. 6).

c. Changes in Expression of Apoptosis-Related Genes The molecular mechanisms by which vitamin D derivatives may induce active cell death are not understood, but there is some evidence that these compounds may modulate the expression of apoptosis-related genes. The induction of apoptosis is associated with a specific set of genes. Both the *p53* and *bcl-2* genes have been implicated in apoptosis. Wild-type p53 plays a crucial role in regulating growth arrest pathways and apoptosis, whereas mutant p53 inhibits apoptosis [82]. Wild-type p53 functions as a cell cycle checkpoint regulator, and its effects are mediated through transcriptional activation of genes such as *WAF-1* which codes for a potent G_1/G_2 cyclin-dependent kinase inhibitor ($p21^{CIP1/WAF-1}$). However, apoptosis may be induced by either p53-dependent or -independent pathways [83]. The *bcl-2* gene, originally identified as the breakpoint of a chromosomal translocation characteristic of follicular lymphoma [84], has been reported to confer resistance to apoptosis in certain systems [85]. Moreover, the protein products of other members of the *bcl-2* gene family, including *bax*, *bad*, and *bcl*-X_L can interact to form homo- and heterodimers that can either accelerate or inhibit apoptosis. The effects the analog EB1089 on the expression of *bcl-2* and p53 proteins have been examined in cultured MCF-7 cells, and decreases in *bcl-2* expression and increases in p53 expression have been demonstrated [78]. The analog KH1060 at high concentrations also induces p53 expression in MCF-7 cells [81]. Increases in p21 protein levels, the product of the *WAF-1* gene, have also been reported in response to vitamin D derivatives [79]. This protein plays an integral role in cell cycle regulation as a cyclin-dependent kinase inhibitor and is a potent and reversible inhibitor of cell cycle progression at both the G1 and G2 checkpoints.

In addition to the observation that vitamin D compounds may modulate expression of these cell cycle checkpoint regulators, there is some evidence for regulation of proteins associated with the later stages of the apoptotic process. Increased clusterin (*TRPM-2*) gene expression is associated with apoptotic regression in a variety of cell types [86]. Its expression is dramatically induced during mammary gland involution following weaning [87]. Clusterin mRNA and/or protein has been shown to be increased in a variety of experimental animal tumors regressing as a consequence of active cell death [74,88]. Although the function of clusterin is not completely understood it is thought to play a role in remodeling of membranes during apoptotic body formation. Studies to examine the effects of vitamin D derivatives on clusterin expression have shown enhanced expression of both unprocessed (60 kDa) and mature (40 kDa) forms of the protein in vitamin D-sensitive MCF-7 cells treated with $1,25(OH)_2D_3$, a response which was absent in a vitamin D-resistant subline [80]. Similar studies of the analog EB1089 on *TRPM-2* expression in (wild-type) MCF-7 cells have shown increases in the levels of protein and mRNA in treated cells compared to control cultures [89,90]. Interestingly, vitamin D derivatives also increase the level of *TRPM-2* transcripts in estrogen receptor negative MDA-MB-231 cells (Fig. 7). The expression of cathepsin B, a lysosomal protease, is dramatically induced in regressing mammary gland [91] and during antiestrogen-induced regression of mammary tumors [92]. Increases in mature cathepsin B protein and steady state mRNA levels have been demonstrated in cultured breast cancer cells treated with $1,25(OH)_2D_3$, suggesting that induction of apoptosis by the hormone is associated with enhanced lysosomal activity as is observed during mammary gland involution [75].

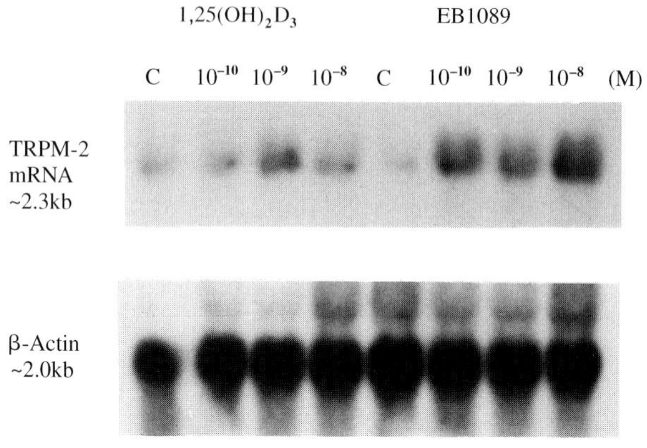

FIGURE 7 Effect of vitamin D derivatives on clusterin (*TRPM-2*) mRNA levels in MDA-MB-231 cells. Cultures were treated with 0.1% ethanol vehicle (lanes C), $1,25(OH)_2D_3$, or EB1089 (10^{-10} to 10^{-8} *M*) for 2 days prior to extraction of total RNA and Northern blot analysis.

d. IGF-1 Signaling Pathways The susceptibility of a tumor cell to undergo apoptosis is likely to be dependent on a wide variety of factors. The molecular basis for development of resistance to cytotoxic drug treatment may originate from genetic lesions that alter the apoptotic set point. In addition, autocrine/paracrine secretion of a variety of growth factors, some of which act as potent cell survival signals, may alter the balance between the cell death and cell survival pathways. Insulin-like growth factors are potent mitogens and inhibitors of apoptosis for many normal and neoplastic cells including breast cancer cells. Furthermore, IGF-1 is frequently produced by stromal cells in breast tumors. Vitamin D analogs, as well as antiestrogens, limit responsiveness of MCF-7 cells to the mitogenic effects of IGF-1 (Fig. 15). Thus, a possible mechanism by which these compounds promote active cell death in breast cancer cells may be via abrogation of the survival signals generated through the IGF-1 signaling pathway. The underlying mechanisms by which vitamin D derivatives are able to limit responsiveness to IGF-1 are not clear, but they may be related to altered expression of its membrane receptor or IGF binding proteins (IGF-BPs). Effects of IGF-1 are mediated via its receptor, IGF-1R, to which it binds with high affinity. IGF-1 can also interact with insulin and IGF-II receptors, but with lower affinity. Breast cancer cells also produce a variety of IGF-BPs which, by binding to IGF-1, modulate its ability to interact with its receptor. Thus, the effects of IGF-1 on cell growth are dependent on the level of expression and affinity of target cell receptors as well as the circulating and extracellular levels of IGF-BPs [93].

In preliminary studies we have found that binding of ^{125}I-IGF-1 to MCF-7 cell membranes is decreased in cells incubated with increasing doses of $1,25(OH)_2D_3$ and its analog EB1089 compared to control cultures (K. Colston, unpublished observations), indicating that these compounds may decrease the abundance of IGF-1R and/or alter the profile of expression of IGF-BPs. Similar effects were seen in cultures treated with the antiestrogen ICI 182,780. Previous studies demonstrated that both antiestrogens and retinoids decrease expression of IGF-BP3 in MCF-7 cells [94,95]. IGF-BP3 binds IGF-1 with high affinity, decreasing interaction with its membrane receptor and inhibiting IGF-1 signaling pathways. The gene encoding IGF-BP3 is induced by wild-type p53 [96]. It remains to be determined whether vitamin D derivatives may alter expression of this IGF binding protein either directly or via changes in p53 levels.

Clearly, much additional research will be needed in order to define the pathways by which vitamin D derivatives may activate cell death. Preliminary studies have principally been carried out with MCF-7 cells which express wild-type *p53*, although the protein may be dysfunctional as it is excluded from the nucleus [97]. *p53* is the most frequently affected gene in human solid tumors, being involved in most types of solid neoplasms. Mutant *p53* is present in 30–50% of breast cancers, whereas dysfunctional wild-type *p53* is found in a further 30% of cases [97]. It remains to be established whether similar effects of vitamin D derivatives on induction of apoptosis may be seen in cells expressing the mutant *p53*. Identification of genes transcriptionally regulated by the VDR that are involved in apoptosis will aid in the development of novel therapeutic stategies which target these pathways in breast cancer cells.

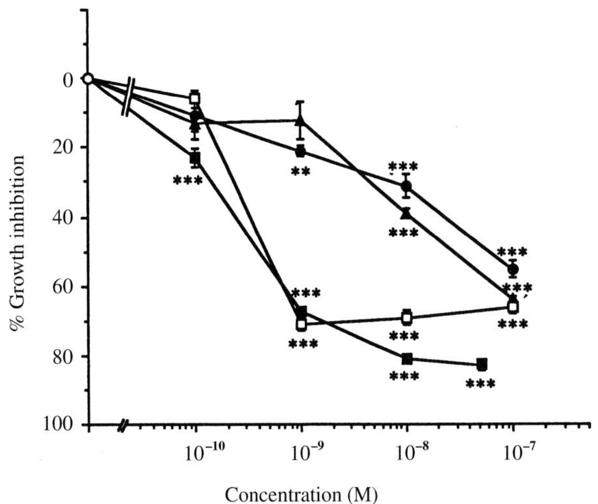

FIGURE 8 Effects of EB1089 (■), all-*trans*-retinoic acid (●), and 9-*cis*-retinoic acid (▲) on proliferation of MCF-7 cells. Cultures, grown in 48-well plates, were treated with the indicated concentrations of EB1089, retinoids, and ICI 182,780 (□). Cell numbers were assessed after 7 days of treatment. $^{**}p < 0.01$; $^{***}p < 0.005$.

E. Interactions with Retinoids

Retinoids are also reported to inhibit the growth of experimental rat mammary tumors *in vivo* and breast cancer cell *in vitro* [98,99]. The retinoid X receptor (RXR) binds and is activated by the retinoic acid isomer 9-*cis*-retinoic acid (9-*cis*-RA) [100] in addition to all-*trans*-retinoic acid (ATRA). Evidence suggests that the VDR can bind to DNA in the form of a heterodimeric partnership with RXR [101], resulting in an increase in transcriptional activation and enhanced induction of vitamin D responsive genes. The apparent role of RXR as an auxiliary factor required for efficient binding of VDR to its specific response elements suggests a molecular mechanism for cooperative effects between vitamin D derivatives and retinoids. Studies have shown that both 9-*cis*-RA and ATRA inhibit the *in vitro* growth of ER positive human breast cancer cells and that effects induced by 9-*cis*-RA are equal to or greater than those induced by equimolar concentrations of ATRA (Fig. 8).

Previously reported findings have demonstrated that ATRA acts synergistically with $1,25(OH)_2D_3$ and antiestrogens to inhibit the growth of ER positive T-47D breast cancer cells [35]. Cooperative effects between retinoids and $1,25(OH)_2D_3$ on inhibition of tumor angiogenesis have been reported, indicating the possible therapeutic efficacy of this combination of agents [102]. There is evidence that 9-*cis*-RA may augment the effects of EB1089 on the induction of apoptosis in cultured MCF-7 cells [78]. Cooperative effects of EB1089 and 9-*cis*-RA on the expression of p53 protein and *TRPM-2* mRNA have been noted, whereas combined treatment of MCF-7 cells also led to enhanced down-regulation of *bcl-2* protein levels coupled with increased DNA fragmentation, a key feature of apoptosis. These findings may have therapeutic implications in breast cancer.

IV. CLINICAL TRIALS IN CANCER

There have been few clinical trials of vitamin D in cancer. A trial of alfacalcidol in patients with low grade non-Hodgkin's lymphoma showed response in a proportion of patients [103]. This compound was also assessed in a trial of patients with myelodysplasia. A transient improvement in peripheral blood counts was seen, but half of the patients developed hypercalcemia [104]. A second report from Kelsey and colleagues [105] showed sustained hematological response in 6 patients treated with high dose alfacalcidol. In this study patients were given dietary advice to limit calcium intake. Four patients developed hypercalcemia and received treatment with pamidronate to reduce bone resorption.

With regard to breast cancer, the efficacy of topical treatment with a vitamin D analog of cutaneous nodules in women with locally advanced or cutaneous metastatic disease has been evaluated [61]. In this study, 19 women with locally advanced or cutaneous metastatic breast cancer were treated topically with calcipotriol ointment (MC903, 100 μg daily). Treated and, where possible, control nodules were identified. Surface area was measured by computer-assisted planimetry. Of the 14 patients who completed 6 weeks of treatment, 3 showed partial response (50% reduction in the bidimensional diameter of the treated lesion) and 1 a minimal response. The presence of VDR in tumor cells (from fine needle aspirates or, where insufficient material was available, in sections of primary tumor) was evaluated by immunocytochemistry. Receptor protein was detected in tumor cells from 7 of 13 patients including all 4 of the responders (Table II, Fig. 9).

A phase I clinical trial of oral EB1089 in 36 patients with advanced breast and colorectal cancer has been completed. This trial was an open, noncontrolled single-center study with sequentially assigned dose levels [106]. Twenty-five females had breast cancer, and four females and seven males had colorectal cancer. All patients had received standard chemotherapy or hormone therapy in the past. Patients received the analog twice daily for 5 days with a 3-week postdosing follow-up. Compassionate treatment was allowed after this postdosing interval. During compassionate treatment patients were seen at 2- to 4-week intervals. Effects of the vitamin D analog treatment on calcium metabolism were measured by total serum calcium concentration (corrected for albumin levels), urinary calcium excretion, and circulating parathyroid hormone (PTH) concentration. The study showed that the analog EB1089 can be given to patients with advanced cancer with predictable effects on calcium metabolism. In assessing changes in biochemical parameters over the 5-day treatment period, dose-dependent increases in serum calcium and urinary calcium excretion and decreases in serum PTH levels were observed. The maximum tolerable dose is in the range 9–12.5 $\mu g/m^2$ (total daily dose 10–20 μg), whereas the maximum tolerated dose for $1,25(OH)_2D_3$ is on the order of 1 to 2 μg daily. Twenty patients received compassionate treatment for between 10 and 234 days (mean 90 ± 62 days). Doses ranged from 0.2 to 31 μg daily. No clear antitumor effects were seen in this study. Five patients showed stabilization of their disease for longer than 3 months.

TABLE II Characteristics and Treatment Outcome of Assessable Patients with Cutaneous Metastatic Breast Cancer[a]

				VDR Status	
Patient	Age (years)	Objective response	Planimetry ratio[b]	Skin nodule	Primary tumor
1	72	PR	0.36	+	
2	65	PR	0.36	+	
3	51	NC	0.9	−	
4	69	NC	0.68	−	
5	55	NC	N/A[c]	−	
6	70	NC	N/A[c]	−	
7	64	PR	0.42	+	
8	64	PD	N/A[c]	−	
9	56	MR	0.73	N/A	+
10	38	PD	0.98	N/A	+
11	75	PD	N/A[c]	N/A	N/A
12	88	PD	1.13	N/A	+
13	73	NC	1.5	+	
14	64	PD	3.9	−	

[a] Cutaneous breast cancer nodules were treated with calcipotriol ointment. MR, Minimal response; PR, partial response; PD, progressive disease; N/A, not available/inadequate sample; NC, no change; +, positive; −, negative. Adapted from Bower *et al.* [61].

[b] Surface area control/treated nodule at start divided by same ratio at end of trial. Values <1.0 indicate differential change in favor of treated nodule.

[c] In four patients only a single nodule was present and no control was used.

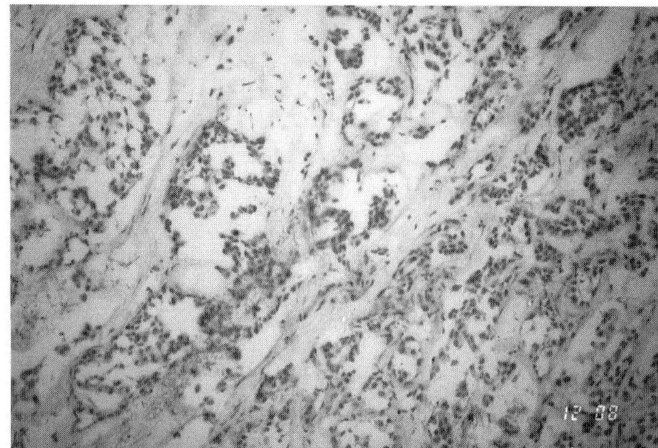

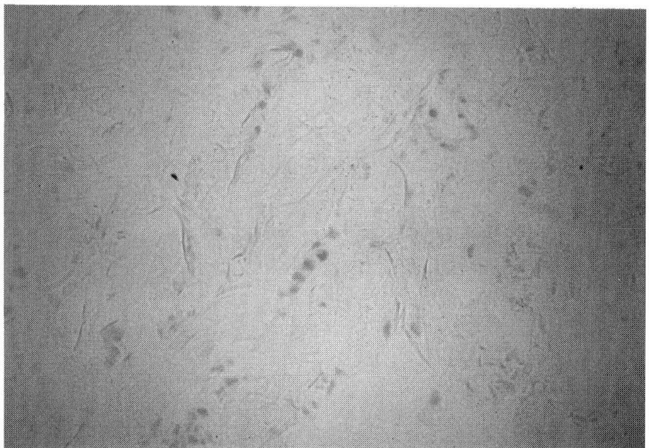

FIGURE 9 Presence of immunoreactive VDR in nuclei of tumor cells in a section of breast carcinoma (top, original magnification ×200) and in a section of a dermal metastatic deposit from a patient with advanced breast cancer.

V. CONCLUSIONS

A. Molecular Basis for Selective Actions of Vitamin D Analogs

An important aim in developing new vitamin D compounds as therapeutic agents in malignancy is separation of hypercalcemic from antiproliferative actions. In the analogs currently available for study this aim has, at least in part, been achieved. The analog EB1089, when tested in rats to assess its effects on calcium handling, is found to be less potent than $1,25(OH)_2D_3$ despite displaying strong direct effects on cell growth and differentiation. This compound causes striking regression of experimental mammary tumors, exerting effects that are comparable to tamoxifen.

The mechanisms underlying the selective actions of these structurally modified vitamin D derivatives are not clear. In contrast to the case for the closely related thyroid and retinoic acid receptors, there is little evidence to suggest the existence of multiple forms of the vitamin D receptor. Current data indicate that there is a single gene for VDR [107] and a single translated receptor molecule, although differences in phosphorylation have been reported [108]. However, differences in postbinding events could be important in the selective actions of vitamin D analogs. Binding of structurally modified vitamin D derivatives to VDR could result in alternative conformational changes, allowing binding of the receptor ligand complex to a different spectrum of gene regulatory sequences. Interactions of VDR with transcriptional cofactors may be improved by binding of vitamin D analogs as compared with $1,25(OH)_2D_3$, and the expression of such cofactors may be different in intestine, bone, or tumor tissue. Genomic actions require binding of $1,25(OH)_2D_3$ to the VDR, heterodimerization with RXR, and interaction with specific vitamin D response elements (VDREs) in the promoter region of target genes. A consensus VDRE consisting of two hexanucleotide half-sites arranged as direct repeats and interspaced by three nucleotides (DR3 motif) has been identified in several genes involved in the classic actions of vitamin D.

Alternative VDRE conformations may exist in genes involved in growth regulation. In this regard, Nayeri and associates [109] have identified a putative VDRE, comprising an inverted palindromic arrangement of half-sites with a nine-nucleotide spacing (IP9 motif), and have shown that EB1089 stimulates transcriptional activation from this type of element more effectively than from DR3-type elements. Although it remains to be established whether altered VDR phosphorylation and heterodimerization play a role in the selective actions of vitamin D analogs, studies assessing receptor binding to the osteocalcin VDRE have shown that EB1089 induces VDR to form selective receptor complexes depending on cell type [110].

B. Future Prospects

Research with new synthetic analogs clearly shows the possibility of developing vitamin D analogs with enhanced modulatory effects on the nonclassic actions of $1,25(OH)_2D_3$. These compounds are potent regulators of cell growth and differentiation and are now considered to have a wide variety of clinical applications including treatment of osteoporosis, secondary hyperparathyroidism due to chronic renal failure, autoimmune disorders, and alopecia associated with chemotherapy [111] as well as in psoriasis.

The discovery that the VDR is detectable in cancer cells of both epithelial and hematopoietic origin and that $1,25(OH)_2D_3$ and its active analogs display the ability to modulate cellular function has established these compounds as potential therapeutic agents in malignancy. Phase I/II trials of one analog (EB1089) are currently underway in patients with advanced pancreatic carcinoma as well as breast cancer and myelodysplasia. It is likely that similar trials in other types of malignancy will follow. In addition, consideration is likely to be given to the value of combination therapy, as laboratory findings have indicated the potential efficacy of this approach. An increasing body of evidence indicates that an important aspect of the antitumor effects of certain vitamin D analogs is activation of the cell death pathway. Further characterization of the apoptosis-related genes that are directly regulated by vitamin D derivatives may provide a basis for the design of new compounds which can target these pathways in breast cancer cells.

Acknowledgments

The colleagues whose collaboration made this chapter possible include Sharon James, Alan Mackay and Shaoping Xie. We are grateful to Lise Binderup, Leo Pharmaceuticals, Denmark, for providing vitamin D analogs for these studies. Clinical trials were carried out under the direction of R. C. Coombes, Medical Oncology, Charing Cross Hospital London.

References

1. Jordan VC, Murphy CS 1990 Endocrine pharmacology of antiestrogens as antitumour agents. Endocr Rev **11**:578–610.
2. Scottish Breast Cancer Trials Committee 1987 Adjuvant tamoxi-

fen in the management of operable breast cancer. Lancet **2**:171–175.
3. NATO 1988 Controlled trial of tamoxifen as a single adjuvant agent in management of early breast cancer. Br J Cancer **57**:608–611.
4. Wakeling AE, Bowler J 1988 Novel anti-estrogens without partial agonist activity. J Steroid Biochem **31**:645–653.
5. Wakeling AE, Dukes M, Bowler J 1991 A potent specific pure antiestrogen with clinical potential. Cancer Res **51**:3867–3873.
6. Early Breast Cancer Trialists' Collaborative Group 1992 Systemic treatment of early breast cancer by hormonal, cytotoxic or immune therapy. Lancet **339**:1–15.
7. Patterson JS 1981 Clinical aspects and development of antioestrogen therapy: A review of the endocrine effects of tamoxifen in animals and man. J Endocrinol **89**:67P–75P.
8. Eisman JA, Martin TJ, MacIntyre I, Frampton RJ, Moseley JH, Whitehead R 1980 1,25-Dihydroxyvitamin D receptor in a cultured human breast cancer cell line (MCF-7). Biochem Biophys Res Commun **93**:9–15.
9. Colston K, Colston MJ, Fieldsteel AH, Feldman D 1982 1,25-Dihydroxyvitamin D_3 receptors in human epithelial cancer cell lines. Cancer Res **42**:856–861.
10. Colston K, Berger U, Wilson P, Hadcocks L, Naeem I, Earl HM, Coombes RC 1988 Mammary gland 1,25-dihydroxyvitamin D receptor content during pregnancy and lactation. Mol Cell Endocrinol **60**:15–22.
11. Colston K, Wilkinson JR, Coombes RC 1986 1,25-Dihydroxyvitamin D_3 receptor in estrogen responsive rat breast tumor. Endocrinology **119**:397–403.
12. Sahota SS Edgar AJ, Colston K, Coombes RC 1991 Analysis of 1,25-dihydroxyvitamin D_3 receptor mRNA in human breast cancer tissues by polymerase chain reaction and in rat mammary tumours by Northern blots. In: Norman AW, Bouillon R, Thomasset M (eds) Vitamin D: Gene Regulation, Structure–Function Analysis and Clinical Application. de Gruyter, Berlin, pp. 459–460.
13. Berger U, Wilson P, McClelland RA, Colston K, Haussler MR, Pike JW, Coombes RC 1988 Immunocytochemical detection of 1,25-dihydroxyvitamin D receptor in normal human tissues. J Clin Endocrinol Metab **67**:607–613.
14. Freake HC, Abeyasekera G, Iwasaki J, Marocci C, MacIntyre I, McClelland RA, Skilton RA, Easton DF, Coombes RC 1984 Measurement of 1,25-dihydroxyvitamin D $_3$ receptors in breast cancer and the relationship to biochemical and clinical indices. Cancer Res **44**:1677–1681.
15. Eisman JA, Suva LJ, Martin TJ 1986 Significance of 1,25-dihydroxyvitamin D_3 receptor in primary breast cancers. Cancer Res **46**:5406–5408.
16. Berger U, Wilson P, McClelland R, Colston K, Haussler MR, Pike JW, Coombes RC 1987 Immunocytochemical detection of 1,25-dihydroxyvitamin D_3 receptor in breast cancer. Cancer Res **47**:6793–6795.
17. Colston KW, Berger U, Coombes RC 1989 Possible role for vitamin D in controlling breast cancer cell proliferation. Lancet **1**:185–191.
18. Berger U, McClelland RA, Wilson P, Greene GL, Haussler MR, Pike JW, Colston K, Easton D, Coombes RC 1991 Immunocytochemical determination of estrogen receptor, progesterone receptor and 1,25-dihydroxyvitamin D_3 receptor in breast cancer and relationship to prognosis. Cancer Res **51**:239–244.
19. Sandgren M, Danforth L, Plasse TF, Deluca HF 1991 1,25-dihydroxyvitamin D_3 receptors in human carcinomas, a pilot study. Cancer Res **51**:2021–2024.
20. Meggouh F, Lointier P, Saez S 1991 Sex steroid and 1,25-dihydroxyvitamin D_3 receptors in human colorectal adenocarcinoma and normal mucosa. Cancer Res **51**:1227–1233.
21. Howat JMT, Harris M, Swindell R, Barnes DM 1985 The effect of estrogen and progesterone receptors on recurrence and survival in patients with carcinoma of the breast. Br J Cancer **81**:263–270.
22. Horwitz K, Wei LL, Sedlacek SM, D'Arville CN 1985 Progestin action and progesterone receptor structure in breast cancer: A review. Recent Prog Horm Res **41**:249–316.
23. Ulmann A, Brami M, Corcos D, Bader C, Delarue J, Conteso C 1984 Systematic search for 1,25-dihydroxyvitamin D3 receptors in human breast carcinomas. Biomed Pharmacother **38**:204–208.
24. Garland CF, Garland FC 1980 Do sunlight and vitamin D reduce the likelihood of colon cancer? Inter J Epidemiol **9**:227–231.
25. Garland CF, Comstock GW, Garland FC, Helsing KJ, Shaw EK, Gorham ED 1989 Serum 25-hydroxyvitamin D and colon cancer: An eight year prospective study. Lancet **2**:1176–1178.
26. Garland FC, Garland CF, Gorham MPH, Young BA 1990 Geographic variation in breast cancer mortality in the United States: a hypothesis involving exposure to solar radiation. Prev Med **19**:616–622.
27. Janowsky EC, Hulka BS, Lester GE 1994 Vitamin D levels as a risk for female breast cancer. In: Norman AW, Bouillon R, Thomasset M (eds) Vitamin D. A Pluripotent Steroid Hormone: Structural Studies, Molecular Endocrinology and Clinical Applications. de Gruyter, Berlin, pp. 496–497.
28. Schwartz GG, Hulka BS 1990 Is vitamin D deficiency a risk factor for prostatic cancer? (Hypothesis). Anticancer Res **10**:1307–1311.
29. Colston KW, Colston MJ, Feldman D 1981 1,25-Dihydroxyvitamin D_3 and malignant melanoma: The presence of receptors and inhibition of cell growth in culture Endocrinology **108**:1083–1086.
30. Abe E, Miyaura C, Sakagami H, Takeda M, Konno K, Yamazaki T, Yoshiki S, Suda T 1981 Differentiation of mouse myeloid leukemia cells induced by 1,25-dihydroxyvitamin D_3. Proc Nat Acad Sci USA **78**:4990–4995.
31. Frampton RJ, Omond SA, Eisman JA 1983 Inhibition of human cancer cell growth by 1,25-dihydroxyvitamin D_3 metabolites. Cancer Res **43**:4443–4447.
32. Simpson RU, Arnold AJ 1986 Calcium antagonizes 1,25-dihydroxyvitamin D_3 inhibition of breast cancer cell proliferation. Endocrinology **119**:2284–2289.
33. Eisman JA, Sutherland RI, McMenemy ML, Fragonas JC, Musgrove EA, Pang GYN 1989 Effects of 1,25-dihydroxyvitamin D_3 on cell cycle kinetics of T47D human breast cancer cells. J Cell Physiol **138**:611–616.
34. Chouvet C, Vicard E, Devonee M, Saez S 1986 1,25-Dihydroxyvitamin D_3 inhibitory effect on the growth of two human breast cancer cell lines (MCF-7, BT-20). J Steroid Biochem **24**:373–376.
35. Koga M, Sutherland RL 1991 Retinoic acid acts synergistically with 1,25-dihydroxyvitamin D_3 or antioestrogen to inhibit T-47D human cancer cell proliferation. J Steroid Biochem Mol Biol **39**:455–460.
36. Jones G, Calverly MJ 1993 A dialogue on analogs, newer vitamin-D drugs for use in bone disease, psoriasis and cancer. Trends in Endocrin and Metab **4**:297–292.
37. Deluca HF, Ostrem VK 1988 Analogs of the hormonal form of vitamin D and their possible use in leukemia. Prog Clin Biol Res **259**:41–55.
38. Zhou JY, Norman AW, Lubbert M, Collins ED, Uskokovic MR, and Koeffler HP 1989 Novel vitamin D analogs that modulate leukemic cell growth and differentiation with little effect on

either intestinal calcium absorption or bone mobilization. Blood **74**:82–93.

39. Binderup L, Bramm E 1988 Effects of a novel vitamin D analogue MC903 on cell proliferation and differentiation *in vitro* and on calcium metabolism *in vivo*. Biochem Pharmacol **37**:887–895.
40. Ostrem VK, Lau WF, Lee SH, Perlman K, Prahl J, Schnoes HK, DeLuca HF 1987 Induction of monocytic differentiation of HL-60 cells by 1,25-dihydroxyvitamin D analogs. J Biol Chem **262**:14164–14171.
41. Abe J, Morikawa M, Miyamoto K, Kaiho S, Fukushima M, Miyaura C, Abe E, Suda T, Nishii Y 1987 Synthetic analogs of vitamin D_3 with an oxygen atom in the side chain skeleton. A trial of the development of vitamin D compounds which exhibit potent differentiation-inducing activity without inducing hypercalcemia. FEBS Lett **226**:58–62.
42. Perlman K, Kutner A, Prahl J, Smith C, Inaba M, Schnoes HK, DeLuca HF 1990 24-Homologated 1,25-dihydroxyvitamin D_3 compounds: separation of calcium and cell proliferation activities. Biochemistry **29**:190–196.
43. Norman AW, Zhou JY, Henry HL, Uskokovic MR, Koeffler HP 1990 Structure–function studies on analogues of 1alpha,25-dihydroxyvitamin D_3: Differential effects on leukemic cell growth, differentiation and intestinal absorption. Cancer Res **50**:6857–6864.
44. Colston KW, Chander SK, Mackay AG, Coombes RC 1992 Effects of synthetic vitamin D analogs in breast cancer cell proliferation *in vivo* and *in vitro*. Biochem Pharmacol **44**:673–702.
45. Colston KW, Mackay AG, James SY, Binderup L, Chander S, Coombes RC 1992 EB1089: A new vitamin D analogue that inhibits the growth of breast cancer cells *in vivo* and *in vitro*. Biochem Pharmacol **44**:2273–2280.
46. Mathiasen IS, Colston KW, Binderup L 1993 EB1089, a novel vitamin D analogue, has strong antiproliferative and differentiation inducing effects on cancer cells. J Steroid Biochem Mol Biol **46**:365–371.
47. Binderup L, Latini S, Binderup E, Bretting C, Calverley M, Hansen K 1991 20-Epi-vitamin D_3 analogs: A novel class of potent regulators of cell growth and immune responses. Biochem Pharmacol **42**:1569–1575.
48. Lillevang ST, Rosenkvist J, Andersen CB, Larsen S, Kemp E, Kristensen T 1992 Single and combined effects of the vitamin D analogue KH1060 and cyclosporin A on mercuric chloride-induced autoimmune disease in the BN rat. Clin Exp Immunol **88**:301–306.
49. Brenner RV, Shabahang M, Schumaker LM, Nauta RJ, Uskokovic MR, Evans SRT, Buras RR 1996 The anti-proliferative effect of vitamin D analogs on MCF-7 human breast cancer cells Cancer Lett **92**:77–82.
50. Abe J, Nakano T, Nishii Y, Matsumoto T, Ogata E, Ikeda K 1991 A novel vitamin D_3 analog, 22-oxa-1,25-dihydroxyvitamin D_3, inhibits the growth of human breast cancer *in vitro* and *in vivo* without causing hypercalcemia. Endocrinology **129**: 832–837.
51. Honma Y, Hozumi M, Abe E, Konno K, Fukushima M, Hata S, Nishil Y, DeLuca HF, Suda T 1983 1,25-Dihydroxyvitamin D_3 and 1a hydroxyvitamin D_3 prolong survival time of mice inoculated with myeloid leukemia cells. Proc Nat Acad Sci USA **80**:201–206.
52. Eisman JA, Barkla DH, Tutton PJ 1987 Suppression of *in vivo* growth of human cancer solid tumor xenografts by 1,25-dihydroxyvitamin D_3. Cancer Res **47**:21–25.
53. Colston KW, Mackay AG, James SY 1995 Vitamin D_3 derivatives and breast cancer. In: Tenniswood M, Michna H (eds) Schering Foundation Workshop Volume 14. Springer-Verlag, Heidelberg, pp. 201–224.
54. Colston KW, Mackay AG, Chander S, Binderup L, Coombes RC 1991 Novel vitamin D analogues suppress tumor growth *in vivo* In: Norman AW, Bouillon R, Thomasset M (eds) Vitamin D: Gene Regulation, Structure–Function Analysis and Clinical Applications. de Gruyter, Berlin, pp. 465–466.
55. Abe-Hashimoto J, Kikuchi T, Matsumoto T, Nishii Y, Ogata E, Ikeda K 1993 Anti-tumor effect of 22-oxa-calcitriol, a noncalcaemic analog of calcitriol in athymic mice implanted with human breast carcinoma and its synergism with tamoxifen. Cancer Res **53**:2534–2537.
56. Oikawa T, Yoshida Y, Shimamra M, Ashino Fuse H, Iwaguchi T, Tominaga T 1991 Antitumour effect of 22-oxa-1α-25-dihydroxyvitamin D_3, a potent angiogenesis inhibitor of rat mammary tumours induced by 7,12-dimethylbenz[*a*]anthracene. Anti-Cancer Drugs **2**:475–481.
57. Anzano MA, Byers SM, Smith JM, Peer CW, Mullen LT, Brown CC, Roberts AB, Sporn MB 1994 Prevention of breast cancer in the rat with 9-*cis*-retinoic acid as a single agent and in combination with tamoxifen. Cancer Res **52**:4614–4617.
58. Kissmeyer AM, Mathiasen IS, Latini S, Binderup L 1995 Pharmacokinetic study of vitamin D analogues; relationship to vitamin D binding protein (DBP). Endocrine **3**:263–266.
59. Sorenson H, Binderup L, Calverley MJ, Hoffmeyer L, Andersen NR 1990 *In vivo* metabolism of calcipotriol (MC903), a vitamin D analogue. Biochem Pharmacol **39**:391–393.
60. Kragballe K 1989 Treatment of psoriasis by the topical application of the novel analogue calcipotriol (MC903). Arch Dermatol **125**:1647–1652.
61. Bower M, Colston KW, Stein RC, Hedley A, Gazet JC, Ford HT, Coombes RC 1991 Topical calcipotriol treatment in advanced breast cancer. Lancet **337**:701–702.
62. James SY, Mackay AG, Binderup L, Colston KW 1994 Effects of a new synthetic analogue, EB1089, on the oestrogen responsive growth of human breast cancer cells. J Endocrinol **141**:555–563.
63. Demirpence E, Balaguer P, Trousse F, Nicolsas JC, Pons M, Gagne D 1994 Antiestrogenic effects of all-*trans*-retinoic acid and 1,25-dihydroxyvitamin D_3 in breast cancer cells occur at the estrogen response element level but through different molecular mechanisms. Cancer Res **54**:1458–1464.
64. Vink-van Wijngaarden T, Pols HAP, Bourman CJ, van den Bend GJCM, Dorssers CJ, Birfenhager JC, van Leeuwen JPTM 1994 Inhibition of breast cancer cell growth by combined treatment with vitamin D analogues and tamoxifen. Cancer Res **54**:5711–5717.
65. Saez S, Meggough F, Lefebvre M-F, Descotes F, Pampile R, Adam L, Crepin M 1994 Potential direct and indirect influence of $1,25(OH)_2D_3$ on the growth of human colonic and breast carcinoma. In: Norman AW, Bouillon R, Thomasset M (eds) Vitamin D. A Pluripotent Steroid Hormone: Structural Studies, Molecular Endocrinology and Clinical Applications. de Gruyter, Berlin, pp. 469–476.
66. Xie SP, James SY, Colston KW 1997 Vitamin D derivatives inhibit the mitogenic effects of IGF-1 on MCF-7 human breast cancer cells. J Endocrinol in press.
67. Koga M, Eisman JA, Sutherland RL 1988 Regulation of epidermal growth factor receptor levels by $1,25(OH)_2D_3$ in human breast cancer cells. Cancer Res **48**:2734–2739.
68. Koli K, Keski-Oja J 1994 1,25-Dihydroxyvitamin D_3 has been shown to enhance the expression of transforming growth factor β_1 and its latent form binding protein in cultured breast carcinoma cells. Cancer Res **55**:1540–1457.
69. Simboli-Campbell M, Welsh J E 1994 Comparative effects of

$1,25(OH)_2D_3$ and EB 1089 on cell cycle kinetics in MCF-7 cells. In: Norman AW, Bouillon R, Thomasset M (eds) Vitamin D. A Pluripotent Steroid Hormone: Structural Studies, Molecular Endocrinology and Clinical Applications de Gruyter, Berlin, pp. 528–529.

70. Reitsma PH, Rothberg PG, Astrin SM, Trial J, Barshavit Z, Hall A, Teitelbaum SL, Kahn AJ 1983 Regulation of *myc* gene expression in HL-60 leukemia cells by a vitamin D metabolite. Nature **306**:492–494.

71. Simpson RU, Hseu T, Begley DA, Mitchell BS, Alizadeh BN 1983 Transcriptional regulation of the c-*myc* protooncogene by 1,25-dihydroxyvitamin D_3 in promyelocytic leukemia cells. J Biol Chem **262**:4104–4108.

72. Brelvi ZS, Christakos S, Studzinski GP 1986 Expression of monocyte-specific oncogenes c-*fos* and c-*fms* in HL-60 cells treated with vitamin D analogs correlates with inhibition of DNA synthesis and reduced calmodulin concentration. Lab Invest **55**:269–275.

73. Wyllie AH, Morris RG, Smith AL, Dunlop D 1984 Chromatin cleavage in apoptosis, association with condensed chromatin morphology and dependence on macromolecular synthesis. J Pathol **142**:67–77.

74. Kyprianou N, English HF, Davidson NE, Isaacs JT 1991 Programmed cell death during regression of the MCF-7 human breast cancer following estrogen ablation. Cancer Res **51**: 162–166.

75. Simboli-Campbell M, Welsh JE 1995 1,25-Dihydroxyvitamin D_3: Coordinate regulator of active cell death and proliferation in MCF-7 breast cancer cells. In: Tenniswood M, Michna H (eds) Schering Foundation Workshop Volume 14. Springer Verlag, Heidelberg, pp. 181–200.

76. Welsh JE 1995 Induction of apoptosis in breast cancer cells in response to vitamin D and antiestrogens. Biochem Cell Biol **72**:537–545.

77. Walker PR, Kwast-Welfeld J, Gourdeau H, Leblanc J, Neugehauer, W, Sikorsska M 1993 Relationship between apoptosis and the cell cycle in lymphocytes: role of protein kinase C, tyrosine phosphorylation and AP1. Exp Cell Res **207**:142–151.

78. James SY, Mackay AG, Colston KW 1995 Vitamin D derivatives in combination with 9-*cis*-retinoic acid promote active cell death in breast cancer cells. J Mol Endocrinol **14**:391–394.

79. James SJ, Mackay AG, Colston KW 1996 Effects of 1,25-dihydroxyvitamin D_3 and its analogues on induction of apoptosis in breast cancer cells. J Steroid Biochem Mol Biol **58**:395–401.

80. Narvez CJ, VanWeelden K, Byrne I, Welsh J 1996 Characterization of a vitamin D_3-resistant MCF-7 cell line. Endocrinology **137**:400–409.

81. Elstner E, Linkner-Israeli M, Said J, Umiel T, de Vos S, Shintaku IP, Heber D, Binderup L, Uskokovic M, Koeffler HP 1995 20-Epi-vitamin D analogues: A novel class of potent inhibitors of proliferation and inducers of differentiation of human breast cancer cell lines. Cancer Res **55**:2822–2830.

82. Shaw P, Bovey R, Tardy S, Sahil R, Sordat B, Costa J 1992 Induction of apoptosis by wild-type p53 in a human colon tumour derived cell line. Proc Nat Acad Sci USA **89**:4492–4499.

83. Clarke AR, Purdie CA, Harrison DJ, Morris RG, Bird CC, Hooper ML, Wyllie AH 1993 Thymocyte apoptosis induced by p53-dependent and independent pathways. Nature **362**:849–851.

84. Tsujimoto Y, Finger L, Yunis J, Nowell PC, Croce CM 1984 Cloning of the chromosome breakpoint of neoplastic B cells with the t(14;18) chromosome translocation. Science **226**:1097–1099.

85. Reed J 1994 Bcl-2 and the regulation of programmed cell death. J Cell Biol **124**:1–6.

86. Monpetit ML, Lawless KR, Tenniswood M 1986 Androgen repressed messages in the rat ventral prostate. Prostate **8**:25–36.

87. Guenette RS, Corbeil H, Leger J, Wong K, Mezl V, Mooibroek M. Tenniswood M 1993 Induction of gene expression during involution of the lactating mammary gland of the rat. J Mol Endocrinol **12**:47–60.

88. Rennie PS, Bruchovsky N, Buttyan R, Benson M, Cheng H 1988 Gene expression during the early phases of regression of the androgen-dependent Shionogi mouse mammary carcinoma. Cancer Res **48**:6309–6312.

89. Mackay AG, James SY, Ofori-Kuragu EA, Binderup L, Colston KW 1993 Down-regulation of the oestrogen receptor and induction of apoptosis in breast cancer cells by EB 1089, a novel vitamin D analogue. J Endocrinol **139**(Suppl.):14.

90. Welsh J, Simboli-Campbell M, Tenniswood M 1994 Induction of apoptotic cell death by $1,25(OH)_2D_3$ in MCF-7 breast cancer cells. In: Norman AW, Bouillon R, Thomasset M (eds) Vitamin D. A Pluripotent Steroid Hormone: Structural Studies, Molecular Endocrinology and Clinical Applications. de Gruyter, Berlin, pp. 526–527.

91. Guenette RS, Mooibroek M, Wong K, Wong P, Tenniswood M 1994 Cathepsin B, a cysteine protease implicated in metastatic progression, is also expressed during regression of the rat prostate and mammary glands. Eur J Biochem **226**:311–321.

92. Warri AM, Huovinen RL, Laine AM, Martikainen PM, Harkonen PL 1993 Apoptosis in toremifene-induced growth inhibition of human breast cancer cells *in vivo* and *in vitro*. J Natl Cancer Inst **85**:1412–1418.

93. Winston R, Kao PC, Kiang DT 1994 Regulation of insulin-like growth factors by antiestrogen. Breast Cancer Res Treat **31**:107–115.

94. Huynh H, Yang X, Pollak M 1996 Estradiol and antiestrogens regulate a growth inhibitory insulin-like growth factor binding protein 3 autocrine loop in human breast cancer cells. J Biol Chem **271**:1016–1021.

95. Gucev ZS, Oh Y, Kelley KM, Rosenfeld RG 1996 Insulin-like growth factor binding protein 3 mediates retinoic acid and transforming growth factor β_2-induced growth inhibition in human breast cancer cells. Cancer Res **56**:1545–1550.

96. Buckbinder L, Talbott R, Velasco-Miguel S, Takenaka I, Faha B, Seizinger BR, Kley N 1995 Induction of the growth inhibitor IGF-binding protein 3 by p53. Nature **377**:646–649.

97. Moll UM, Rion G, Levine AJ 1992 Two distinct mechanisms alter p53 in breast cancer: Mutation and nuclear exclusion. Proc Natl Acad Sci USA **89**:7262–7266.

98. Moon RC, Grubbs CJ, Sporn MB 1976 Inhibition of 7,12-dimethylbenz[*a*]anthracene-induced mammary carcinogenesis by retinyl acetate. Cancer Res **36**:2527–2360.

99. Anzano MA, Byers SM, Smith JM, Peer CW, Mullen LT, Brown CC, Roberts AB, Sporn MB 1994 Prevention of breast cancer in the rat with 9-*cis*-retinoic acid as a single agent and in combination with tamoxifen. Cancer Res **52**:4614–4617.

100. Levine AA, Sturzenbecker LJ, Kazmer S, Bosakowski T, Huselton C, Allenby G, Speck J, Kratzeisen CL, Rosenberger M, Lovey A, Grippo JF 1992 Cis retinoic acid stereoisomer binds and activates the nuclear receptor RXRα Nature **355**:359–361.

101. Yu VC, Delsert C, Andersen B, Holloway JM, Devary OV, Naar AM, Kim SY, Boutin J-M, Glass CK, Rosenfeld MG 1991 RXRβ: A coregulator that enhances binding of retinoic acid, thyroid hormone and vitamin D receptors to their cognate response elements. Cell **67**:1251–1266.

102. Majewski S, Szmurlo A, Marczak M, Jablonska S, Bollag W 1993 Inhibition of tumor cell-induced angiogenesis by retinoids, 1,25-

dihydroxyvitamin D_3 and their combination. Cancer Lett **75**:35–39.

103. Cunningham D, Gilchrist NL, Cowan RA, Soukup K 1985 Alfacalcidol as a modulator of growth of low grade non-Hodgkin's lymphomas. Br Med J **291**:1153–1155.
104. Koeffler HP, Hirji K, Itri L, Southern California Leukemia Group 1985 1,25-Dihydroxyvitamin D_3 *in vitro* and *in vivo* effects on human preleukemic and leukemic cells. Cancer Treat Rep **69**:1399–1407.
105. Kelsey SM, Newland AC, Cunningham J, Makin HLJ, Coldwell RD, Mills MJ, Grant IR 1992 Sustained haematological response to high-dose oral alfacalcidol in patients with myelodysplastic syndromes. Lancet **340**:316–317.
106. Gulliford T, English J, Colston K, Sprogel P, Coombes RC 1996 A phase I study of EB 1089, a vitamin D analogue. Proceedings of the 87th Annual Meeting of the American Association for Cancer Research, Washington D.C. **37**:(Abstract 1130).
107. Baker AR, McDonnell D, Hughes MR, Crisp TM, Manglesdorf DJ, Haussler MR, Pike JW, Shine J, O'Malley BW 1988 Cloning and expression of full-length cDNA encoding human vitamin D receptor. Proc Natl Acad Sci USA **85**:3294–3298.
108. Brown TA, DeLuca HF 1990 Phosphorylation of the 1,25-dihydroxyvitamin D_3 receptor. J Biol Chem **264**:20265–20274.
109. Nayeri S, Danielsson C, Kahlen J-P, Schrader M, Mathiasen IS, Binderup L, Carlberg C 1995 The anti-proliferative effects of vitamin D_3 analogues is not mediated by inhibition of the AP-1 pathway but may be related to promoter selectivity Oncogene **11**:1853–1858.
110. Mathiasen IS, Mackay AG, Colston KW, Binderup L 1994 EB 1089, a vitamin D analogue, induces the vitamin D receptor to form selective receptor complexes depending on cell type. In: Norman AW, Bouillon R, Thomasset M (eds) Vitamin D. A Pluripotent Steroid Hormone: Structural Studies, Molecular Endocrinology and Clinical Applications. de Gruyter, Berlin, pp. 255–256.
111. Pols HAP Birkenhager JC, van Leewan JPTM 1994 Vitamin D molecules; from molecule to clinical application. Clin Endocrinol **40**:285–291.

CHAPTER 66

Vitamin D and Prostate Cancer

COLEMAN GROSS Department of Medicine, Stanford University School of Medicine, Stanford, California

DONNA M. PEEHL Department of Urology, Stanford University School of Medicine, Stanford, California

DAVID FELDMAN Department of Medicine, Stanford University School of Medicine, Stanford, California

I. INTRODUCTION

A. Scope of the Problem

Adenocarcinoma of the prostate gland is the most commonly diagnosed malignancy in American males, excluding skin cancer. It is estimated that there will be 317,000 new cases of prostate cancer diagnosed in the United States in 1996 [1], which will account for 41% of newly diagnosed malignancies, excluding basal cell and squamous cell carcinomas of the skin. These numbers are indeed staggering and may actually be substantial underestimates of the problem because clinically silent prostate cancer is very common. In men over the age of 50 years, subclinical prostate cancer is found in as many as 40% of individuals [2]. It has been estimated that there are 11 million men in the United States who have foci of adenocarcinoma in their prostate glands [3]. Although prostate carcinoma is generally a slow growing malignancy, mortality from the disease is nonetheless considerable. In contrast to many other malignancies, mortality from prostate cancer has continued to rise each year; over 41,000 men in the United States will die of prostate cancer in 1996. This represents 14% of all cancer deaths projected for 1996 and makes prostate cancer the second leading cause of cancer death among U.S. males [1]. Since the 1970s the age-adjusted mortality rate from prostate cancer increased 7% among U.S. Caucasian men. As prostate cancer rates increase with advancing age, one can expect that prostate cancer will become an even greater problem as life expectancy continues to increase. As a result prostate cancer has rapidly become a major public health concern not only in the United States but worldwide.

B. Epidemiology

On the basis of epidemiological studies, several risk factors for prostate cancer have been identified including age, race, and genetic factors [4]. However, environmental factors may have a strong influence on the expression of the disease. For example, when compared to Nigerian men, African American men have a 6-fold increased risk of developing clinically detectable prostate cancer [4]. It has long been appreciated that solar radiation can decrease the mortality rates of noncutaneous malignancies [5]. Of particular interest is the hypothesis put forward by Schwartz and colleagues [6,7]. Based on the observation that prostate cancer mortality rates in the United States are inversely proportional to the geographically determined incident UV radiation exposure from the sun, and that UV light is essential for vitamin D synthesis, these authors suggested a role for vitamin D in decreasing the risk of developing prostate cancer. This hypothesis is not without precedent, as vitamin D may have a role in the prevention of colon cancer as well (see Chapter 67). It may also offer a potential explanation of why African American men have a higher incidence of prostate cancer than Caucasian men [8] as African American individuals have lower serum vitamin D levels as a result of their darker skin pigmentation [9].

Corder *et al.* [10] have measured levels of vitamin D and its metabolites in the stored serum of 181 men who later went on to develop prostate cancer. The samples were obtained several years prior to the study, and controls were selected from blood samples taken the same day from men who did not go on to develop prostate cancer. Mean levels of 1,25-dihydroxyvitamin D [$1,25(OH)_2D$], the active metabolite of vitamin D were slightly but significantly lower in the cancer patients compared to controls. The risk of prostate cancer decreased with higher $1,25(OH)_2D$ levels, especially in men with low 25OHD levels. In men over the age of 57, $1,25(OH)_2D$ was an important risk factor for palpable and anaplastic tumors but not for incidentally discovered tumors or well-differentiated cancers. The case–control study of Corder *et al.* [10] looked at a nested group of sera from men who ultimately developed prostate cancer out of a larger population of 250,000 individual samples. On the other hand, Braun *et al.* [11] were unable to confirm the results of Corder *et al.* in a prediagnostic study in a group of 20,305 men in Maryland. However, in a further analysis of their data, Corder *et al.* [12] subsequently found that $1,25(OH)_2D$ levels showed a seasonal variation in prostate cancer cases but not controls, with a nadir in summer months. This may explain the lack of an effect in the Braun study, in which serum collections occurred mainly in the fall and as a result may have missed a nadir in levels among individuals that went on to develop prostate cancer.

The vitamin D binding protein (DBP) may modulate vitamin D action by controlling free levels of the hormone (see Chapter 7). As such, it may play a role in the etiology of prostate cancer. However, studies in this area have produced divergent results. Corder *et al.* [12] failed to find an association with DBP levels and prostate cancer in their cohort of subjects. In a group of 68 men with prostate cancer, Schwartz *et al.* [13] found that DBP levels were significantly higher in 68 individuals with prostate cancer compared to 42 controls. Individuals with DBP levels above 350 mg/liter had a greater than 5-fold increased risk of prostate cancer.

Although the mechanism of action of $1,25(OH)_2D$ on the prostate is poorly understood, it is likely that the hormone acts through the classic ligand-dependent activation of genes via the vitamin D receptor (VDR) (see Chapters 8–10). Although still controversial, it has been demonstrated that polymorphisms in the VDR gene may contribute to the risk of osteoporosis [14–17] (also see Chapter 45). This may be the result of differential activity of the VDR which contributes to more or less efficient transactivation of $1,25(OH)_2D$ target genes. Investigators have now begun to address the relevance of this observation in prostate cancer as well as in osteoporosis. Two preliminary studies have shown that polymorphisms in exon 9 and the 3′ untranslated region of the VDR gene may be associated with an increased risk for the development of prostate cancer [18,19]. Moreover, Ingles *et al.* [19] suggest that the VDR polymorphisms may correlate with disease stage. Interestingly, it appears that the alleles which may be protective against prostate cancer in men are predictive of low bone mass in some groups of women. The clinical utility of these VDR polymorphisms in predicting prostate cancer risk is unclear at present, and the subject requires more investigation [102].

C. Etiology—Hormonal Factors

It is generally accepted that androgens are integrally involved in the regulation of prostate growth and cell proliferation [3,4,20]. Although several studies have demonstrated a correlation between serum testosterone levels and increased risk of prostate cancer [4], it is unclear if androgen plays a dominant role in neoplastic transformation. Androgen receptors are detectable in nuclear fractions of approximately 80% of prostate tumors [21]. Moreover, evidence is now accumulating that mutations in the androgen receptor gene may play an etiologic role in the pathogenesis of prostate cancer [22].

Circulating androgens may serve as factors that promote tumor growth and perhaps neoplastic transformation. Indeed, the withdrawal of androgens causes involution and apoptosis of normal as well as malignant prostatic epithelial cells [3,4]. This provides the basis for the use of androgen ablation in the treatment of clinically advanced prostate cancer [23].

Evidence that progestins and estrogens play a role in prostate cancer is less clear. Receptors for both progesterone and estrogen have been observed in prostatic tumors; however, no correlation has been found between tumor estrogen receptor levels and response to hormonal therapy [24]. More recently, estrogens have actually been shown to stimulate the growth of cultured human prostate cancer cells (LNCaP) [25], the clinical implications of which are unknown at the present time.

II. PROSTATE AS A TARGET FOR VITAMIN D

A. Vitamin D is an Antiproliferative and Differentiation Agent

Although the role of vitamin D in maintaining calcium homeostasis has been understood for a long while (see Chapter 1), it is only relatively recently that investigators have begun to understand the broader scope of vitamin D action. Not only have studies documented potent immunomodulatory activities of this hormone, but other studies have shown $1,25(OH)_2D$ to have important antiproliferative and differentiating properties [26]. An up-to-date review is provided by van Leeuwen and Pols (see Chapter 64).

The idea that the prostate represents a vitamin D target organ is suggested by the epidemiological evidence described above. The discovery of VDR within the epithelial cells of the prostate [27], however, was an important confirmation that $1,25(OH)_2D$ might play a direct role in prostate biology.

B. Vitamin D in Normal Prostate

Although the initial description of VDR in prostate and most of the subsequent investigation has centered around prostate cancer cell lines, $1,25(OH)_2D$ appears to play an important role in normal prostate tissue. Peehl *et al.* [28] reported the presence of VDR in freshly obtained surgical prostate specimens as well as primary cultures of epithelial and stromal elements of the prostate grown in serum-free medium (Table I). Moreover primary cultures from surgical specimens of benign prostatic hyperplasia (BPH) also demonstrated VDR. Although VDR were present in both epithelial and stromal elements cultured separately, there appeared to be lower levels of VDR in the fibroblastic stromal cells (N_{max} = 0–27 fmol/mg protein) compared to the glandular epithelium (N_{max} = 10–79 fmol/mg protein). The region of origin within the prostate tissue did not influence the abundance of VDR, as both the peripheral zone and central zone cultures had similar amounts of VDR. The affinity of the prostate VDR was the same as that reported for other target tissues ($K_d = 10^{-10}$ M). In clonal growth assays, epithelial cells were inhibited by $1,25(OH)_2D_3$ at concentrations of 0.1 nM and were completely growth-arrested at a $1,25(OH)_2D_3$ concentration of 25 nM (ED_{50} = 0.25–1 nM, Fig. 1). Moreover, the inhibitory effect on growth appeared to be irreversible, as removal of $1,25(OH)_2D_3$ from the media failed to reverse the growth inhibitory effect. Because apoptosis was not observed, it was felt that $1,25(OH)_2D_3$ induced cell cycle arrest. These effects were less prominent in the prostatic fibroblast cells, consistent with their lower VDR content [28]. Interestingly, Esquenet *et al.* [29] failed to demonstrate an irreversible effect of $1,25(OH)_2D_3$ on cell growth in the LNCaP human prostate cancer cell line, suggesting that this effect is dependent on cell type and culture conditions.

III. VITAMIN D IN PROSTATE CANCER

A. Cancer Cell Lines

Miller and co-workers [27] described VDR in the LNCaP cell line. They found these human prostate cancer epithelial cells to contain a single class of high affinity VDR ($K_d = 1.4 \times 10^{-9}$ M) present in moderate amounts (2500 sites/cell). They documented the presence of the 4.6-kb mRNA for VDR in these cells by Northern analysis, and furthermore they demonstrated VDR immunologically with a monoclonal antibody to the chick VDR [30]. Interestingly, they documented a mitogenic effect of $1,25(OH)_2D_3$ on LNCaP cells at concentrations from 10^{-11} to 10^{-9} M when these cells were cultured in a medium supplemented with charcoal-stripped serum rather than unstripped fetal bovine serum. However, the effect was biphasic, as higher doses of $1,25(OH)_2D_3$ failed to stimulate growth. Under these culture conditions prostate cells grow poorly, and this may not reflect *in vivo* conditions. In comparison to the findings of Miller and colleagues, Skowronski *et al.* [31] demonstrated VDR not only in LNCaP cells, but also in two other human prostate cancer cell lines DU-145 and PC-3. In this study, all three cell lines were found to have

TABLE I VDR and 1,25(OH)$_2$D Effects in Various Prostate Cells[a]

Cell type	VDR K_d (M)	VDR Abundance (fmol/mg protein)	Growth inhibition[b]	24OHase induction	PSA induction	Ref.
Fresh Tissue						
Normal/CA		10–22				28
Primary cell cultures						
Normal—E	1×10^{-10}	13–55	+++			28
Normal—F		9–27	++			28
BPH—E		24–51	+++			28
BPH—F		0–11	++			28
CA—E	1×10^{-10}	21–79	+++			28
CA—F		19	++			28
Virally transformed cells						
SV40-T	5×10^{-10}	8–15	+	Yes		51,52
HPV-T		17–20	++	Yes		51,52
Cancer cell lines						
LNCaP	8×10^{-11}	26	+++	No	Yes	31
PC-2	6×10^{-11}	78	++	Yes	No	31
Du-145	5×10^{-11}	31	±	Yes	No	31
Cancer cell lines—whole cell binding[c]						
PPC-1	3×10^{-9}	600	+	Yes		32
PC-3	1×10^{-9}	800	±	Yes		32
DU-145	0.6×10^{-9}	1100	+	Yes		32
TSU-Prl	1×10^{-9}	1200	±	Yes		32
LNCaP	1×10^{-9}	2900	+++	Yes	Yes	27,32
ALVA-31	0.5×10^{-9}	9700	++	Yes		32
JCA-1	—	N/D	±	No		32,75

[a] CA, Cancer; BPH, benign prostatic hyperplasia; K_d, dissociation constant; 24OHase, 24-hydroxylase; PSA, prostatic specific antigen; E, epithelial; F, fibroblastic; SV40-T, simian virus 40-transformed; HPV-T, human papillomavirus-transformed; N/D, not detected.

[b] Growth inhibition: ±, insignificant; +, minimal; ++, moderate; +++, extensive.

[c] VDR analysis by whole cell binding assay, abundance in receptor number/cell.

high affinity VDR ($K_d = 5\text{–}8 \times 10^{-11}$ M) in moderate amounts (26–78 fmol/mg protein). As is found for other vitamin D target tissues, the VDR appeared to be distributed mainly in the nucleus as was shown by immunohistochemistry. Interestingly, Skowronski and co-workers found 1,25(OH)$_2$D$_3$ to have a striking growth-inhibitory effect on LNCaP cells when the cells were cultured in unstripped fetal bovine serum (Fig. 2). These cells were inhibited by over 60% of control values when cultured in the presence of increasing concentrations of 1,25(OH)$_2$D$_3$ up to 100 nM. Although PC-3 cells were also growth inhibited they were not as sensitive as LNCaP; proliferation was decreased to approximately 50% of control in the presence of 1,25(OH)$_2$D$_3$. DU-145 cells were only minimally growth inhibited by 1,25(OH)$_2$D$_3$.

The discrepancy in the growth profiles of LNCaP cells in response to 1,25(OH)$_2$D$_3$ may be due to the culture conditions. When Skowronski *et al.* [31] replicated the charcoal-stripped serum conditions used by Miller *et al.* [27], a mild stimulatory effect was seen. However, the growth inhibitory effect was the most predominant finding in unstripped serum. Clearly, the conditions of the growth assays affect the results. In charcoal-stripped serum, prostate cells grow poorly and 1,25(OH)$_2$D$_3$ improves growth slightly. However, when prostate cells grow well, 1,25(OH)$_2$D$_3$ produces a substantial growth inhibitory effect. Indeed, Peehl *et al.* [28] used a serum-free medium that allows robust cell growth and demonstrated growth inhibition in primary cultures of prostate cells. Although it is at present not possible to replicate exactly the local *in vivo* conditions in the prostate in an *in vitro* experiment, one would expect that the use of unstripped serum would be most representative. As such, the mild stimulatory effect reported by Miller *et al.* [27] may not be predictive of an *in vivo* response. Indeed the published data, including subsequent work by Miller *et al.* [32], shows that the effect of 1,25(OH)$_2$D$_3$ on prostatic cells is generally growth inhibitory (Table I).

Skowronski *et al.* also investigated other 1,25-(OH)$_2$D$_3$-induced functions [31]. They found that

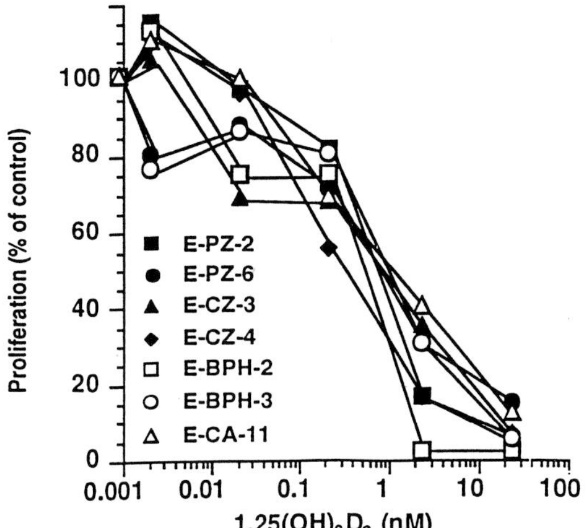

FIGURE 1 $1,25(OH)_2D_3$ causes growth inhibition in normal and malignant primary prostatic epithelial cells. Proliferation was determined by clonogenic assays in serum-free medium after 10 days of treatment with increasing doses of $1,25(OH)_2D_3$ as indicated. E-PZ and E-CZ are normal prostatic epithelial cell lines. E-BPH are epithelial cell cultures derived from benign prostatic hyperplasia specimens. E-CA refers to epithelial cell cultures generated from prostatic carcinoma. Adapted with permission from DM Peehl, RJ Skowronski, GK Leung, ST Wong, TA Stamey, and D Feldman. Antiproliferative effects of 1,25-dihydroxyvitamin D_3 on primary cultures of human prostatic cells. *Cancer Res* 54;805–810;1994.

mRNA for 24-hydroxylase, a vitamin D responsive gene [33], was induced by $1,25(OH)_2D_3$ in both DU-145 and PC-3 cells (Table I). Interestingly, induction of 24-hydroxylase mRNA was not detected in LNCaP cells.

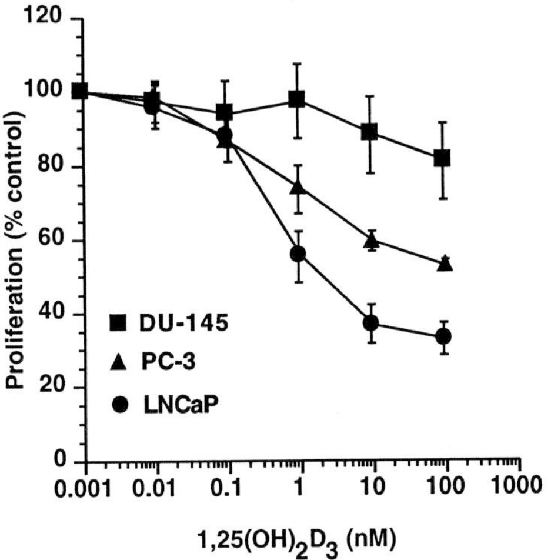

FIGURE 2 $1,25(OH)_2D_3$ causes growth inhibition in human prostate cancer cell lines. Proliferation was determined by attained mass of DNA after 6 days of treatment with increasing doses of $1,25(OH)_2D_3$ as indicated. Adapted with permission from RJ Skowronski, DM Peehl, and D Feldman. Vitamin D and prostate cancer: 1,25-Dihydroxyvitamin D_3 receptors and actions in human prostate cancer cell lines. *Endocrinology* 132;1952–1960;1993. © The Endocrine Society.

This pattern may perhaps explain the remarkable sensitivity of LNCaP cells to $1,25(OH)_2D_3$ in terms of growth inhibition and the resistance to such an effect in DU-145 cells. 24-Hydroxylase catalyzes the hydroxylation of $1,25(OH)_2D_3$ to $1,24,25(OH)_3D_3$ and is therefore an important step in the metabolic inactivation of the active hormone (see Chapter 6). As such, the induction of 24-hydroxylase would be expected to speed the clearance of $1,25(OH)_2D_3$ from the culture medium and as a result mitigate the growth inhibitory effect of the hormone. Miller *et al.* [32] were able to detect induction of 24-hydroxylase by $1,25(OH)_2D_3$ in LNCaP cells using an enzymatically based assay that is substantially more sensitive but less specific than assessing mRNA induction by Northern blot analysis as Skowronski *et al.* performed [31]. However Miller and coworkers [32] found the least induction of 24-hydroxylase induction in response to $1,25(OH)_2D_3$ in LNCaP cells, of the seven prostate cancer cell lines studied. VDR abundance, as determined by Scatchard analysis in whole cell binding assays, predicted growth inhibitory response in these different cell lines. Cell lines with the greatest amount of VDR (LNCaP and ALVA-31) were the most sensitive to the growth inhibitory action of $1,25(OH)_2D_3$. Interestingly, these investigators found that the pattern of growth inhibitory responses to $1,25(OH)_2D_3$ of the cell lines was inversely correlated with the relative amount of 24-hydroxylase induction by $1,25(OH)_2D_3$. It is not clear why the pattern of 24-hydroxylase was inversely correlated with the abundance of VDR in the various cell lines, however.

Prostate specific antigen (PSA) is also induced by physiolological concentrations of $1,25(OH)_2D_3$ in LNCaP cells [27,29,31]. Total PSA measured in media from cultures treated with $1,25(OH)_2D_3$ is less than that from control cultures; however, when expressed on a per cell basis (DNA content) the PSA production increased 2- to 2.5-fold after 72 hr of treatment [31]. Moreover, PSA mRNA is induced approximately 2-fold after 48 hr of $1,25(OH)_2D_3$ treatment [34]. This relatively long delay in the appearance of PSA mRNA suggests that the induction may not be a direct effect of $1,25(OH)_2D_3$. PSA production can be considered a phenotypic marker of a differentiated prostatic cell [35], and $1,25(OH)_2D_3$ may induce differentiation and thereby induce PSA secretion indirectly. Although $1,25(OH)_2D_3$ increases PSA production in the absence of androgen, $1,25(OH)_2D_3$-mediated PSA production is augmented in the presence of androgens to levels higher than those achieved with androgen alone [34]. Interestingly, $1,25(OH)_2D_3$ up-regulates androgen receptor levels in LNCaP cells [29,34], and this may be responsible for the apparent synergistic effect of $1,25(OH)_2D_3$ and androgen on PSA secretion.

B. Primary Cancer Cell Cultures

Cancer cell lines provide an abundant source of material to investigate the biology of vitamin D in the prostate. However, results from studies using cancer cell lines may not represent true *in vivo* phenomena due to culture conditions, mutated receptors, prolonged passaging, and uncertain origin of cell lines, among other problems. The use of primary cultures of cancer cells provides an important tool to study cancer biology as it may be more closely related to the clinical setting than cell culture using established cancer cell lines. Because of the difficulty in generating primary cultures of prostate cancer cells [36], little has been published with regard to the action of vitamin D in primary prostate cancer cultures. Peehl *et al.* [28] reported on their investigation of a series nine primary prostate cancer cell cultures. Epithelial cultures were generated from surgical specimens obtained during prostatectomy. $1,25\text{-}(OH)_2D_3$ ligand binding experiments revealed variable levels of VDR abundance among the different cultures (21–79 fmol/mg protein, see Table I). The VDR abundance did not correlate with histologic tumor grade. The affinity (K_d) of the VDR for $1,25(OH)_2D_3$, as determined by Scatchard analysis, was similar to that reported for established prostate cancer cell lines and other vitamin D target organs. These primary cultured prostate cells are profoundly growth inhibited by $1,25(OH)_2D_3$ (ED_{50} = 0.25–1 n*M*, Fig. 1). Interestingly, in clonogenic assays, these cells were irreversibly growth inhibited. When prostate cancer cells were exposed to $1,25(OH)_2D_3$ for as little as 2 hr, growth did not resume even after a 10-day period in media lacking vitamin D. Moreover, apoptosis was not observed. This observation suggests that $1,25(OH)_2D_3$ induces an irreversible cell cycle block and prevents further cell division (see Section VI,C below). The finding of a growth inhibitory action of $1,25(OH)_2D_3$ in primary cultures of malignant prostatic epithelial cells suggests a potential role for $1,25(OH)_2D_3$ and its analogs in the therapy of prostate cancer patients (see Section III,E).

Peehl *et al.* [28] also studied cultures of stromal fibroblasts generated from the prostate surgical specimens and found that these cells also contained VDR and were growth inhibited by $1,25(OH)_2D_3$ (Table I). The stromal fibroblasts contained less VDR (19 fmol/mg protein) and were less growth inhibited by $1,25(OH)_2D_3$ when compared to the epithelial cultures. The significance of the presence of VDR in the fibroblastic component of prostate cancer is uncertain. However, there is evidence that fibroblasts may affect prostate cancer epithelial cell growth through the production of growth factors that act in a paracrine manner [37,38]. $1,25(OH)_2D_3$ may potentially regulate stromal production of some of these growth factors and thereby influence the growth of prostatic epithelium.

C. Virally Transformed Cells

Viruses may play a role in the pathogenesis of prostate cancer. Human papillomavirus (HPV) has been detected in naturally occurring human prostate adenocarcinoma tissue [39,40]. A postulated mechanism of how viral infection may lead to the development of carcinoma involves the interactions of viral oncogene products with tumor supressor genes such as *p53* and the retinoblastoma gene (*Rb*). Interestingly, mutations in *p53* and *Rb* have been observed in clinical prostate cancer specimens [41–45]. The large T antigen of simian virus 40 (SV40) and the E6 and E7 proteins of HPV cause disruptions in *p53* and *Rb* [46,47] and have been used to transform prostate cells into immortalized cell lines [48–50]. Studies of such virally transformed prostate cell lines are useful models to help gain an understanding of the role of *p53* and *Rb* in the pathogenesis of prostate cancer and to potentially elucidate the mechanism of the growth inhibitory action of $1,25(OH)_2D_3$.

Investigators have evaluated VDR and $1,25(OH)_2D_3$ action in both SV40- and HPV-transformed cell lines [49,51,52]. High affinity VDR were present in the SV40 and HPV cell lines (N_{max} = 8–20 fmol/mg protein) [52]. 24-Hydroxylase mRNA was similarly induced by $1,25(OH)_2D_3$ treatment in the virally transformed cell lines; however, the growth inhibitory pattern was different in the SV40- and HPV-transformed cell lines. Whereas the HPV-transformed cell lines were growth inhibited by $1,25(OH)_2D_3$, the SV40-transformed cell lines were resistant to $1,25(OH)_2D_3$ action in terms of an antiproliferative response. Skowronski *et al.* have described a similar discrepancy in growth inhibition, as well as in 24-hydroxylase mRNA induction, in LNCaP and DU-145 cells [31]. The observation that prostate cancer cell lines respond differently to $1,25(OH)_2D_3$ suggests that there is differential activation of vitamin D responsive genes and that there are cell-specific profiles of this activation. This may provide a clue to candidate genes that mediate the antiproliferative action of $1,25(OH)_2D_3$ (see Section VI). $1,25(OH)_2D_3$-mediated growth inhibition may be the result of the activation of a number of vitamin D responsive genes unrelated to those thus far identified, such as 24-hydroxylase.

D. *In Vivo* Animal Studies

Unfortunately little has been published with regard to the *in vivo* effect of vitamin D in prostate cancer.

R. J. Skowronski, D. M. Peehl, and D. Feldman (unpublished data) inoculated nude mice with either LNCaP cells or PC-3 cells. Animals received intraperitoneal injections of 1,25$(OH)_2D_3$ (100 ng) or vehicle on alternate days starting 2 days prior to cell inoculation and continuing for 6 weeks. In the LNCaP xenografts, tumors weights in control, vehicle-treated mice (n = 9) were 364 ± 33 mg, whereas tumors in 1,25$(OH)_2D_3$-treated mice (n = 9) were 260 ± 24 mg. In the PC-3 xenografts, tumors in control mice (n = 8) were 284 ± 38 mg, whereas tumors in 1,25$(OH)_2D_3$-treated mice (n = 8) were 169 ± 30 mg. Although preliminary, the observed reduction in tumor weight supports findings in cell culture experiments, demonstrating an antiproliferative action of 1,25$(OH)_2D_3$ on prostate cancers *in vivo*.

Schwartz *et al.* [53] treated 12 immunodeficient (nu:nu BALB/c) mice harboring PC-3 xenografts with an analog of 1,25$(OH)_2D_3$, 1,25-dihydroxy-16-ene-23-yne-vitamin D_3 (see Chapter 62). The mice were inoculated with PC-3 cells 18 days prior to the initiation of treatment to ensure tumor take, which occurred in 100% of the animals. Twelve animals received alternate day intraperitoneal injections of 1.6 μg of the vitamin D analog and 12 animals were given injections of vehicle only. As discussed below (see Section IV), many analogs of vitamin D are more potent and less toxic (hypercalcemic) than 1,25$(OH)_2D_3$ and therefore may have a greater therapeutic effect in inhibiting prostate cancer cell growth *in vivo*. Schwartz and colleagues found that although tumors continued to grow in control as well as treated animals, tumor volumes were approximately 15% less in the treated animals. Although there was a significantly smaller mean increase in tumor volume in the treated animals compared to control animals at 4 days after initiation of treatment, the rate of increase in tumor volume subsequently was similar in both groups. Serum calcium levels were similar in treated and control animals. These early results are encouraging; however, much more work is required to establish a role for vitamin D and its analogs in inhibiting the growth of prostate cancer in patients.

E. Clinical Studies

Because of the importance of prostate cancer morbidity and mortality and because of the relative safety of vitamin D therapy, investigators have begun clinical trials evaluating the role of 1,25$(OH)_2D_3$ in the treatment of prostate cancer patients despite the paucity of animal data. Osborn *et al.* [54] reported a small phase II trial in 13 patients with hormone refractory metastatic prostate carcinoma. Of the 13 patients enrolled, 11 completed the dose escalation portion of the study and achieved a final dose of 1.5 μg per day. No objective responses (>50% reduction in PSA or >30% reduction in measurable tumor mass) were observed, and median time to progression was 10.6 weeks. Two subjects, however, had 25 and 45% reductions in PSA that lasted for 30 and 45 days, respectively. As discussed above, 1,25$(OH)_2D_3$ induces PSA production on a per cell basis in LNCaP cells, although PSA levels are decreased overall in cell cultures [31]. This suggests that the interpretation of PSA levels in subjects treated with 1,25$(OH)_2D_3$ may not be straightforward. As Osborn *et al.* [54] point out, PSA levels may not be as clinically useful in monitoring disease progression in individuals treated with 1,25$(OH)_2D_3$ as it is in the general population of individuals with prostate cancer treated with standard modalities. Clearly, more experience is needed to understand the effect of 1,25$(OH)_2D_3$ on serum PSA levels *in vivo*. The subjects in this study had advanced disease, and the lack of benefit in this study does not exclude potential therapeutic utility in early stages of the disease. Similarly, the maximal dose of 1,25$(OH)_2D_3$ (1.5 μg) used in this study may be insufficient to produce an antiproliferative effect on prostate cancer *in vivo* in humans. Therefore, one cannot rule out a potential beneficial clinical effect of this hormone at higher doses.

Gross *et al.* [103] treated 7 individuals with minimally recurrent prostate cancer. These men had early stage disease (A or B), were treated initially with surgery or radiation, and demonstrated recurrence as indicated by rising serum PSA levels. All subjects had small tumor burdens and none had evidence of metastasis. The men were treated with 1,25$(OH)_2D_3$ (1.5–2.25 μg/day) for up to 1 year. All of the subjects demonstrated a slowing of the rate of rise of PSA levels during treatment when compared to the pretherapy period; this was statistically significant in four of the men.

Much controversy exists regarding the timing of androgen ablation therapy after surgical prostatectomy and/or radiation therapy [55]. Hormonal therapy has a limited period of efficacy that averages 12 to 18 months for initial responses [56]. Androgen ablation therapy also decreases quality of life as it causes hot flashes and impotence in many men, as well as increasing the risk of osteoporosis. As a result, many physicians will observe asymptomatic patients with slowly rising PSA levels without therapy. An effective, relatively nontoxic agent such as 1,25$(OH)_2D_3$ would be an ideal therapy to use at this point in the treatment course of a prostate cancer patient.

Another important finding in the study by Osborn *et al.* [54] was the rather high incidence of hypercalcemia (11 of 13 subjects). This finding underscores the fact that hypercalcemia constrains the maximal dose of $1,25(OH)_2D_3$ that may be safely given, which potentially limits its therapeutic antiproliferative and differentiating benefit (which may only be realized at higher doses). Numerous analogs of vitamin D have been developed that are less calcemic and are more potent antiproliferative agents than $1,25(OH)_2D_3$. As discussed below, these agents are attractive candidates for the potential treatment of prostate cancer.

IV. VITAMIN D ANALOGS

A. Potency Versus Toxicity

Several less hypercalcemic analogs of $1,25(OH)_2D_3$ have been synthesized by various groups, and they are fully discussed in Section VII of this book. As alluded to above, these compounds hold great promise for the treatment of a variety of neoplastic and hyperproliferative conditions including prostate cancer. There has been great interest in developing agents that have less hypercalcemic potential yet maintain the antiproliferative and differentiating activities of $1,25(OH)_2D_3$. Boullion *et al.* [57] and Okamura and Zhu (see Chapter 57) have extensively reviewed the structure–function relationships that determine relative calcemic and differentiating potencies of the vitamin D analogs. Although it is generally believed that $1,25(OH)_2D_3$ and its analogs act through the same receptor to transactivate genes, there may be subtle conformational changes induced in the VDR–ligand complex by these different ligands that in turn determine the particular profile of responsive genes that are activated (see Chapter 60). Additionally, metabolic clearance rates may show tissue specific differences with regard to different analogs and may allow for the dissociation of the hypercalcemic and antiproliferative action of the analogs (see Chapter 58). Similarly, tissue and cell specific uptake and vitamin D binding protein (DBP) affinity may determine the relative potencies of the hypercalcemic and differentiating properties of the various analogs (see Chapter 59). Clearly these processes are complex, and they likely act in concert to determine the profile of relative hypercalcemic and antiproliferative activities of each analog.

B. Vitamin D Analogs and Prostate Cancer

Schwartz *et al.* [58] investigated the effect of three vitamin D analogs on the growth of the human prostate cancer cell lines LNCaP, PC-3, and DU-145. Subconfluent cultures were pulse-treated for 2 hr with $1,25(OH)_2D_3$, 1α,25-dihydroxy-16-ene-vitamin D_3 (Ro 24-2637), 1α,25-dihydroxy-16,23*E*-diene-vitamin D_3 (Ro 24-2201), 25-hydroxy-16,23*E*-diene-vitamin D_3 (Ro 24-2287), or ethanol vehicle as a control. As with previous studies [31], LNCaP cells were strongly growth inhibited (60%) by $1,25(OH)_2D_3$, PC-3 cells were growth inhibited to a lesser degree (25%), and DU-145 cells were not growth inhibited. The various analogs produced similar growth inhibitory responses in the three cell lines; however, 1α,25-dihydroxy-16,23*E*-diene-vitamin D_3 inhibited the growth of DU-145 cells. This analog did not appear to have greater potency in the other cell lines.

Skowronski *et al.* [59] studied the effects of five analogs on LNCaP and PC-3 cells. They also studied Ro 24-2287 and Ro 24-2637 in addition to the following analogs: 1α,24*S*-dihydroxy-22-25,26,27-cyclopropylvitamin D_3 (calcipotriol, MC903), 1α,25-dihydroxy-22,24-diene-24,26,27-trihomovitamin D_3 (EB1089), 22-oxa-1α,25-dihydroxyvitamin D_3 (22-oxacalcitriol). Similar to the results of Schwartz *et al.* [58], Ro 24-2637 inhibited LNCaP proliferation by 60% when cells were treated for 6 days. Skowronski and co-workers [59] found that the other analogs significantly inhibited LNCaP growth, with ED_{50} values ranging from 0.5 to 0.9 n*M* compared to an ED_{50} for $1,25(OH)_2D_3$ of 2 n*M*. Ro 24-2287, which does not have a 1α-hydroxyl group, did not bind to the VDR in competitive binding experiments with $1,25(OH)_2D_3$, and it was used as a negative control in proliferation assays. EB1089 appeared to be the most potent antiproliferative agent in this study and also demonstrated the greatest affinity for VDR, even higher than $1,25(OH)_2D_3$. Relative VDR binding affinity did not correlate with the antiproliferative activity for the following analogs: Ro 24-2637, 22-oxacalcitriol, and MC903. These analogs were more potent than $1,25(OH)_2D_3$ in inhibiting LNCaP growth yet showed less affinity for VDR than $1,25(OH)_2D_3$. Interestingly, these investigators found that the analogs were more potent inducers of PSA production than $1,25(OH)_2D_3$. In terms of induction of 24-hydroxylase mRNA, the analogs were more potent than $1,25(OH)_2D_3$, with the exception of MC903, which was the most potent antiproliferative agent, and Ro 24-2287 which did not bind VDR. It appears that the level of antiproliferative potency of $1,25(OH)_2D_3$ or an analog does not correlate well with affinity. Although VDR affinity for ligand is an important and essential factor in the transactivation of vitamin D responsive genes (see Chapters 8 and 9), affinity is not sufficient to determine potency. Thus, there must be other critical determinants of the antiproliferative potency of a given analog [104].

Two studies investigating the *in vivo* effect of vitamin D analogs in prostate cancer have been published [53,60]. The study of Schwartz and co-workers [53] showed a modest antiproliferative effect of 1,25-dihydroxy-16-ene-23-yne-vitamin D_3 on immunodeficient mice xenografted with PC-3 cells. These results are more fully described above in the section on *in vivo* animal studies (see Section III,D). In the second study, Lucia *et al.* [60] evaluated the chemopreventive activity of a related compound, 1α,25-dihydroxy-16-ene-23-yne-26,27-hexafluorovitamin D_3 (Ro 24-5531), in an *N*-nitroso-*N*-methylurea (NMU), androgen-promoted prostate carcinoma rat model. In two of three experiments, the incidence of prostate carcinomas was reduced in Ro 24-5531-treated animals compared to control animals 45 versus 78% and 39 versus 72%, respectively. In contrast to the study by Schwartz *et al.* [53], treatment was initiated prior to tumor development (1 week after NMU treatment and prior to androgen treatment) and therefore suggests a potential chemopreventive role for vitamin D. Interestingly, tumor incidence was decreased in the anterior rather than the dorsal prostate of these rats. The dorsal prostate of the rat more closely resembles the human prostate [61] than does the anterior prostate. Such a divergence in antitumor activity in different prostate zones suggests that there are intraprostatic geographic differences as well as potential species differences in prostate biology that may affect the antiproliferative and differentiating activity of vitamin D analogs.

V. VITAMIN D IN COMBINATION WITH OTHER AGENTS

The efficacy of an active drug is directly related to the dose administered. However, toxicity becomes more frequent at higher doses and therefore limits the maximum dose that can be given safely. This window or therapeutic index is defined by the ratio of the largest dosage that produces no toxicity to the smallest dosage that is efficacious. Ideally, one would prefer to use an agent with a large therapeutic index. The development of vitamin D analogs (see above) is an example of a class of agents that has a higher therapeutic index than the parent compound, $1,25(OH)_2D_3$. Another avenue to increase efficacy and decrease toxicity is to use a combination of agents that act by different mechanisms, at doses that are less than usual when the agents are administered individually. This drug combination strategy has the advantage of limiting the toxicity associated with the individual drugs while obtaining additive and potentially synergistic therapeutic effects. With this latter concept in mind, investigators have begun studying the antiproliferative effect of vitamin D on prostate cancer in combination with other agents. Although there is a paucity of available data, interesting results have been published and are discussed below.

A. Vitamin D and Retinoids

The retinoids comprise vitamin A and its derivatives, including all-*trans*-retinoic acid, 9-*cis*-retinoic acid, and 13-*cis*-retinoic acid. These compounds are ligands for the retinoid receptors (RARs and RXRs) that act as nuclear transcription factors. RXRs are transcription factors that heterodimerize with RAR, VDR, and a variety of other nuclear transcription factors. In turn these ligand activated complexes control a variety of cellular processes including growth and differentiation. This subject has been reviewed in Chapters 8–10 [62–64].

Because of their role in controlling growth and differentiation, retinoids have been examined for potential anticancer activity. Retinoids have been studied in epithelial and nonepithelial malignancies, and they may be particularly active in certain hematologic malignancies [65]. Prostate tissue has been shown to contain RARs and RXRs [66,67], indicating a potential role for retinoids in prostate physiology. Interestingly, prostate cancer tissue appears to have lower concentrations of endogenous all-*trans*-retinoic acid than normal prostate tissue, suggesting a possible rationale for a chemopreventive role for retinoids in prostate cancer [68]. Moreover, retinoids have demonstrated efficacy in controlling prostate cancer growth in animal models [69–71].

In an effort to decrease toxicity associated with retinoid therapy and to take advantage of possible synergy, retinoids in combination with other agents have been evaluated for their effects on various malignant cells *in vitro* as well as *in vivo* [65]. Peehl *et al.* [72] demonstrated a synergistic growth inhibitory effect of all-*trans*-retinoic acid and $1,25(OH)_2D_3$ on primary cultures of normal and malignant prostatic epithelial cells. At low concentrations of $1,25(OH)_2D_3$ (0.25 n*M*) proliferation of the malignant cells was inhibited by only 20%; however, with the addition to the culture medium of a small and ineffective concentration (0.3 n*M*) of all-*trans*-retinoic acid, the growth inhibitory effect was augmented to 84% of control. On the other hand, Esquenet *et al.* [29] failed to demonstrate an interactive effect of $1,25(OH)_2D_3$ and all-*trans*-retinoic acid or 9-*cis*-retinoic acid on the growth of cultured LNCaP cells in the presence of androgen stimulation. In the absence of androgen, however, the retinoids stimulated LNCaP growth by approximately 70%, and this stimulation was attenuated by 50% with the addition of 1 n*M* $1,25(OH)_2D_3$. Although these two studies differed in many respects, including cell type, culture conditions, and proliferation assays used, it appears that there may be $1,25(OH)_2D_3$–retinoid interaction which may augment the independent effectiveness of these agents in controlling prostate cell growth *in vitro*. Clearly, more studies are required to examine the vitamin D–retinoid interaction in the prostate.

B. Vitamin D and Other Agents

The antiparasitic drug suramin has shown potential efficacy in the treatment of hormone refractory prostate cancer [73]. *In vitro*, 50 μg/ml of suramin has been demonstrated to completely inhibit prostate cell growth [74]. Peehl *et al.* [72] recently investigated the potential synergism of suramin and $1,25(OH)_2D_3$ in inhibiting prostate cell growth. They demonstrated a small *in vitro* growth inhibitory effect (<20%) of a suboptimal dose of suramin (5 μg/ml) on primary cultures of normal and malignant epithelial prostatic epithelial cells. Similarly, the growth inhibitory effect of a low dose of $1,25(OH)_2D_3$ (0.25 n*M*) was limited to 20–40% of control. However, when cells were cultured in the presence of both agents, at these low concentrations, growth inhibition was synergistically augmented to 62% of control.

VI. MECHANISM OF VITAMIN D ACTION IN PROSTATE CANCER

Although there is much observational data regarding the antiproliferative action of vitamin D in a variety of tissues including the prostate, significantly less is known about the mechanism of this effect. In Chapter 64, van Leeuwen and Pols review current knowledge of the mechanisms of the antiproliferative and differentiating effect of vitamin D. In the next sections, we will comment on the few published studies that address these issues in the prostate.

A. VDR Abundance

Miller and co-workers [32] found that VDR abundance, as determined by Scatchard analysis in whole cell binding assays, predicted growth inhibitory responses in seven different human prostate carcinoma cell lines (Table I). Cells with the highest level of VDR were the most sensitive to the growth inhibitory effects of $1,25(OH)_2D_3$ (LNCaP and ALVA-1). Conversely, the human prostatic cell line JCA-1 does not normally contain VDR and are not growth inhibited by $1,25(OH)_2D_3$. Hedlund *et al.* [75] transfected JCA-1 cells with VDR and obtained two stable clones that expressed different levels of VDR. These investigators found that the clone with the higher VDR level was more sensitive to $1,25(OH)_2D_3$ in terms of growth inhibition and 24-hydroxylase induction, whereas the clone with the lower levels of VDR was less sensitive. Those cells transfected with vector alone expressed no VDR and demonstrated no growth inhibition or 24-hydroxylase induction in response to $1,25(OH)_2D_3$. Hedlund *et al.* [76] have also shown similar results in the ALVA-31 human prostatic cancer cell line. ALVA-31 cells, which express endogenous VDR, were transfected with a VDR-antisense plasmid. The magnitude of the growth-inhibititory effect of $1,25(OH)_2D_3$ correlated inversely with the level of VDR expressed in the transfected cells. 24-Hydroxylase induction by $1,25(OH)_2D_3$ was also attenuated in the transfected cells. These studies suggest that VDR levels determine the growth-inhibitory response of prostate cancer cells, as has been described in other systems [77]. This phenomenon may be cell specific as certain cell lines with ample levels of VDR, such as DU-145, fail to be growth inhibited by $1,25(OH)_2D$ (see discussion above in Section III,A).

B. Growth Factors

1. INSULIN-LIKE GROWTH FACTORS

The role of the insulin-like growth factors (IGFs) and their receptors in the prostate has been reviewed [78]. It appears that a complete IGF axis exists within the prostate. Both IGF-I and IGF-II [79–81] are produced by prostate epithelial cell lines, and IGF-II production has been demonstrated in stromal cells [82]. Receptors for IGF-I have been detected in normal and malignant prostatic epithelial cells [79,80,83,84] as well as in stromal cells [82]. Several IGF binding proteins (IGFBP-2, -3, -4, and -6) are produced by prostate epithelial cells [78,85,86]. Moreover, the IGFs appear to be mitogenic for prostate epithelium, and the dysregulation of the normal prostate–IGF axis may play a role in the pathogenesis of prostate cancer [78,87]. Interestingly, Drivdahl *et al.* [85] showed that $1,25(OH)_2D_3$ caused a dose-dependent increase in the level of IGFBP-6 mRNA in the human prostate carcinoma cell lines ALVA-31 and ALVA-101. Although the importance of this observation in prostate physiology is unknown, the authors suggest that the $1,25(OH)_2D_3$-mediated upregulation of IGFBP-6 might suppress uncontrolled prostatic epithelial cell growth by inhibition of IGF-mediated mitogenesis.

Another layer of complexity is added to the IGF story by virtue of prostate tissue being a source of a variety of proteases including cathepsin D and PSA, both of which appear to act on IGFBP-3 [88,89]. The resulting cleavage of IGFBP-3 decreases its affinity for IGF-I and thereby allows the growth factor access to the prostate cell to act as a mitogen [89,90]. It is interesting to note that $1,25(OH)_2D_3$ mediates PSA production in LNCaP cells [27,29,31,34]. Clearly, $1,25(OH)_2D_3$ can influence the complex prostate–IGF axis and may influence prostate cell growth in part through regulation of the IGFBPs.

2. TRANSFORMING GROWTH FACTOR-β

Transforming growth factor-β (TGF-β) is composed of a family of dimeric proteins that control cellular differentiation and growth. They tend to be inhibitory toward epithelial cell growth in a variety of tissues including prostate [91]. In the nontumorigenic rat prostate epithelial cell line NRP-152, Danielpour [92] demonstrated $1,25(OH)_2D_3$-mediated induction of TGF-β2 and TGF-β3 mRNA, but not TGF-β1 mRNA, after 25 hr of treatment with hormone. Moreover, TGF-β2 and TGF-β3 proteins were also increased by $1,25(OH)_2D_3$. Interestingly, the all-*trans*-retinoic acid-mediated increase in TGF-β3 protein was augmented by the addition of $1,25(OH)_2D_3$ to the culture medium, whereas there was no such additive effect on TGF-β2 protein levels. These findings correlate with the synergistic effect of $1,25(OH)_2D_3$ and all-*trans*-retinoic acid on inhibiting cell growth and with the growth inhibitory action of the TGFs on these cells noted in an earlier paper [93]. Danielpour [92] demonstrated that $1,25(OH)_2D_3$-mediated induction of thrombospondin-1 and fibronectin mRNA could be blocked by a pan-TGF-β antibody; however, the evidence that TGF-β mediates prostate epithelial cell differentiation is still controversial, and the effect may be cell specific [94]. Nonetheless, TGF-β appears to play a role in prostate cell growth and may be an important factor in the growth-inhibitory action of $1,25(OH)_2D_3$.

3. PARATHYROID HORMONE-RELATED PROTEIN

Parathyroid hormone-related protein (PTHrP) shares a high degree of N-terminal homology with parathyroid hormone (PTH) and acts via the same PTH/PTHrP receptor. In addition to its role in humoral hypercalcemia of malignancy, PTHrP is an important growth factor that mediates processes of growth and differentiation in a variety of tissues [95]. PTHrP has been detected in clinical specimens of prostatic carcinoma and prostate cancer cell lines [96–99] and may play a role in prostate growth and differentiation. Although $1,25(OH)_2D_3$ regulates PTHrP expression in a number of cell types [100], Cramer *et al.* [99] found no effect of $1,25(OH)_2D_3$ on PTHrP mRNA levels in primary cultures of human prostatic epithelial cells derived from BPH or carcinoma despite the presence of VDR in these cells. These investigators showed that the epidermal growth factor (EGF), which acts as a prostate mitogen, was responsible for up-regulation of PTHrP mRNA and protein but that other growth stimulatory factors did not increase PTHrP levels. Because the other mitogens studied did not up-regulate PTHrP, the EGF-mediated increase in PTHrP in these primary prostatic cultures was not responsible for the growth stimulatory effect of EGF. However, PTHrP has been shown to stimulate growth in several other prostatic epithelial cell lines, suggesting a role for PTHrP in prostatic cell growth [98].

C. Oncogenes and Tumor Supressor Genes

$1,25(OH)_2D_3$ has been shown to regulate a number of oncogenes and tumor supressor genes in epithelial and nonepithelial cells (see Chapter 64). Indeed, $1,25(OH)_2D_3$ has been observed to directly transactivate the cyclin-dependent kinase inhibitor p21 [101]. Although it is likely that $1,25(OH)_2D_3$ interacts with similar genes in prostate as well, there are few published data to support this hypothesis. Clues to identifying the involved genes can be obtained by studying cells that demonstrate divergent responses to $1,25(OH)_2D_3$. DU-145 cells provide such a model system because they are resistant to the antiproliferative effects of $1,25(OH)_2D_3$, yet retain their responsiveness to the hormone in regard to 24-hydroxylase induction [31] (see above). Interestingly, DU-145 cells have a disrupted retinoblastoma gene (*Rb*) [41], which normally plays an important role in cell cycle control. $1,25(OH)_2D_3$-mediated differentiation is thought to involve *Rb* in other cell lines as discussed by van Leeuwen and Pols (see Chapter 64). Bookstein *et al.* [41] transfected a normal *Rb* into DU-145 cells (DU-145/Rb) and demonstrated that the restoration of normal *Rb* function reduced the tumorgenicity of the cells when xenografted into nude mice. However, when the DU-145/*Rb* cells were evaluated for their response to $1,25(OH)_2D_3$, Gross *et al.* found that they were not growth inhibited, similar to the parental DU-145 cell line [52]. Although this suggests that $1,25(OH)_2D_3$ acts to inhibit the growth of prostate cells through an *Rb*-independent mechanism, *Rb* may still be an important element in controlling prostate cell growth. It is clear from studies in other cell lines that numerous gene products are influenced by $1,25(OH)_2D_3$ and that much work remains to be done to identify $1,25(OH)_2D_3$-regulated genes which control prostate cell growth and differentiation.

VII. FUTURE DIRECTIONS AND CONCLUSIONS

The role of $1,25(OH)_2D_3$ as an antiproliferative and differentiating agent has been established in a variety of cell types including prostate cancer. Among the challenges remaining for investigators is elucidation of the mechanism of $1,25(OH)_2D_3$-mediated growth inhibition and in particular, the identification of the $1,25(OH)_2D_3$-regulated genes that are important in controlling cell growth and differentiation. Another related challenge is to understand the mechanism of how various vitamin D analogs maintain potent antiproliferative effects while being less hypercalcemic. It is hoped that this will pave the way to rational drug design which might lead to potent agents able to control many malignant tumors including prostate cancer. Although clinical studies have begun to address the role of $1,25(OH)_2D_3$ in treating patients with prostate cancer, hypercalcemia is a major limiting factor to efficacy. Therefore, the less hypercalcemic analogs provide an appealing alternative to $1,25(OH)_2D_3$.

In conclusion, epidemiological studies have suggested a role for vitamin D in the pathogenesis of prostate cancer. Given the high prevalence of prostate cancer in the general population, the importance of such a relationship cannot be overestimated. Studies subsequent to the epidemiological findings have clearly established the prostate as a vitamin D target tissue. As with classic vitamin D target tissues, such as intestine, bone, and kidney, $1,25(OH)_2D_3$ acts via the VDR to transactivate vitamin D responsive genes in the prostate. $1,25(OH)_2D_3$ demonstrates potent antiproliferative activity in a variety of normal and malignant prostate epithelial cells. Although the actions of $1,25(OH)_2D_3$ in the prostate are generally growth-inhibitory, this effect is cell specific and is not entirely explained by VDR abundance. It appears that certain prostate cells may retain sensitivity to the growth inhibitory effects of the hormone yet have decreased responses to $1,25(OH)_2D_3$ in other ways, such as 24-hydroxylase induction (e.g., LNCaP). The reverse pattern is also possible, and this divergence in $1,25(OH)_2D_3$ action may provide clues to elucidating the antiproliferative mechanisms of vitamin D in the prostate. The vitamin D analogs may take advantage of these diverging pathways and produce a greater antiproliferative response while causing less of a calcemic response. These compounds have considerable promise for becoming an important addition to the therapeutic arsenal available for the treatment of prostate cancer.

References

1. Parker SL, Tong T, Bolden S, Wingo PA 1996 Cancer statistics, 1996. CA (Cancer J Clin) **46**:5–27.
2. Stamey TA, Freiha FS, McNeal JE, Redwine EA, Whittemore AS, Schmid H-P 1993 Localized prostate cancer. Cancer **71**(Suppl.):933–938.
3. Carter H, Coffey DS 1988 Prostate cancer: The magnitude of the problem in the United States. In: Coffey DS, Resnick MI, Dorr FA, Karr JP (eds) A Multidisciplinary Analysis of Controversies in the Management of Prostate Cancer. Plenum, New York, pp. 1–9.
4. Kozlowski JM, Grayhack JT 1995 Carcinoma of the prostate. In: Gillenwater JY, Grayhack JT, Howards SS, Duckett JW (ed) Adult and Pediatric Urology, 2nd Ed. Mosby Year Book, St. Louis, Missouri, pp. 1575–1713.
5. Apperly FL 1941 The relation of solar radiation to cancer mortality in North America. Cancer Res **1**:191–195.
6. Schwartz GG, Hulka BS 1990 Is vitamin D deficiency a risk factor for prostate cancer? (Hypothesis). Anticancer Res **10**:1307–1311.
7. Hanchette CL, Schwartz GG 1992 Geographic patterns of prostate cancer mortality. Evidence for a protective effect of ultraviolet radiation. Cancer **70**:2861–2869.
8. Studzinski GP, Moore DC 1995 Sunlight—Can it prevent as well as cause cancer? Cancer Res **55**:4014–4022.
9. Bell NH, Greene A, Epstein S, Oexmann MJ, Shaw S, Shary J 1985 Evidence of alteration of the vitamin D–endocrine system in blacks. J Clin Invest **76**:470–473.
10. Corder EH, Guess HA, Hulka BS, Friedman GD, Sadler M, Vollmer RT, Lobaugh B, Drezner MK, Vogelman JH, Orentreich N 1993 Vitamin D and prostate cancer: A prediagnostic study with stored sera. Cancer Epidemiol Biomarkers Prev **2**:467–472.
11. Braun MM, Helzlsouer KJ, Hollis BW, Comstock GW 1995 Prostate cancer and prediagnostic levels of serum vitamin D metabolites (Maryland, United States). Cancer Causes Control **6**:235–239.
12. Corder EH, Friedman GD, Vogelman JH, Orentreich N 1995 Seasonal variation in vitamin D, vitamin D-binding protein, and dehydroepiandrosterone: Risk of prostate cancer in black and white men. Cancer Epidermol Biomarkers Prev **4**:655–659.
13. Schwartz GG, Hulka BS, Morris D, Mohler JL 1992 Prostate cancer and vitamin (hormone) D: A case control study. J Urol **147**(Suppl):294a.
14. Morrison NA, Qi JC, Tokita A, Kelly PJ, Crofts L, Nguyen TV, Sambrook PN, Eisman JA 1994 Prediction of bone density from vitamin D receptor alleles. Nature **367**:284–287.
15. Peacock M 1995 Vitamin D receptor gene alleles and osteoporosis: A contrasting view. J Bone Miner Res **10**:1294–1297.
16. Eisman JA 1995 Vitamin D receptor gene alleles and osteoporosis: An affirmative view. J Bone Miner Res **10**:1289–1293.
17. Gross C, Eccleshall TR, Feldman D 1996 Vitamin D receptor gene alleles and osteoporosis. In: Bilezikian JP, Raisz LG, Rodan GA (eds) Principles of Bone Biology. Academic Press, San Diego, pp. 917–933.
18. Taylor JA, Hirvonen A, Watson M, Pittman G, Mohler JL, Bell DA 1996 Association of prostate cancer with vitamin D receptor gene polymorphism. Cancer Res **56**:4108–4110.
19. Ingles SA, Ross RK, Yu MC, Irvine RA, La Pera G, Haile RW, Coetzee GA 1997 Association of prostate cancer with genetic polymorphisms in vitamin D receptor and androgen receptor. J Natl Cancer Inst **89**:166–170.
20. McConnell JD 1991 Physiologic basis of endocrine therapy for prostatic cancer. Urol Clin North Am **18**:1–13.
21. Ekman P, Brolin J 1991 Steroid receptor profile in human prostate cancer metastases as compared with primary prostatic carcinoma. Prostate **18**:147–153.
22. Taplin M-E, Bubley G, Shuster T, Frantz M, Spooner A, Ogata G, Keer H, Balk S 1995 Mutation of the androgen-receptor gene in metastatic androgen-independent prostate cancer. N Engl J Med **332**:1393–1398.

23. Huggins C, Stevens RE, Hodges CV 1941 Studies on prostatic cancer II. The effects of castration on advanced carcinoma of the prostate gland. Arch Surg **43**:209–223.
24. Wolf RM, Schneider SL, Pontes JE, Englander L, Karr JP, Murphy GP, Sandberg AA 1985 Estrogen and progestin receptors in human prostatic carcinoma. Cancer **455**:2477–2481.
25. Castagnetta LA, Miceli MD, Sorci CMG, Pfeffer U, Farruggio R, Oliveri G, Calabro M, Carruba G 1995 Growth of LNCaP human prostate cancer cells is stimulated by estradiol via its own receptor. Endocrinology **136**:2309–2319.
26. Bikle DD 1992 Clinical counterpoint: Vitamin D: New actions, new analogs, new therapeutic potential. Endocr Rev **13**:765–784.
27. Miller GJ, Stapleton GE, Ferrara JA, Lucia MS, Pfister S, Hedlund TE, Upadhya P 1992 The human prostatic carcinoma cell line LNCaP expresses biologically active, specific receptors for 1α,25-dihydroxyvitamin D_3. Cancer Res **52**:515–520.
28. Peehl DM, Skowronski RJ, Leung GK, Wong ST, Stamey TA, Feldman D 1994 Antiproliferative effects of 1,25-dihydroxyvitamin D_3 on primary cultures of human prostatic cells. Cancer Res **54**:805–810.
29. Esquenet M, Swinnen JV, Heyns W, Verhoeven G 1996 Control of LNCaP proliferation and differentiation: Actions and interactions of androgens, 1α,25-dihydroxycholecalciferol, all-*trans*-retinoic acid, 9-*cis*-retinoic acid, andphenylacetate. Prostate **28**:182–194.
30. Pike JW, Marion SL, Donaldson CA, Haussler MR 1983 Serum and monoclonal antibodies against the chick intestinal receptor for 1,25-dihydroxyvitamin D_3. Generation by a preparation enriched in a 64,000-dalton protein. J Biol Chem **258**:1289–1296.
31. Skowronski RJ, Peehl DM, Feldman D 1993 Vitamin D and prostate cancer: 1,25-Dihydroxyvitamin D_3 receptors and actions in human prostate cancer cell lines. Endocrinology **132**:1952–1960.
32. Miller GJ, Stapelton GE, Hedlund TE, Moffatt KA 1995 Vitamin D receptor expression, 24-hydroxylase activity, and inhibition of growth by 1α,25-dihydroxyvitamin D_3 in seven human prostatic carcinoma cell lines. Clinical Cancer Research **1**:997–1003.
33. Ohyama Y, Ozono K, Uchida M, Shinki T, Kato S, Suda T, Yamamoto O, Noshiro M, Kato Y 1994 Identification of a vitamin D-responsive element in the 5′-flanking region of the rat 25-hydroxyvitamin D_3 24-hydroxylase gene. J Biol Chem **269**:10545–10550.
34. Zhao X-Y, Ly LH, Peehl DM, Feldman DF 1997 1α,25-Dihydroxyvitamin D_3 actions in LNCaP human prostrate cancer cells are androgen-dependent. Endocrinology **138** in press.
35. Sinha AA, Wilson MJ, Gleason DF 1987 Immunoelectron microscopic localization of prostatic-specific antigen in human prostate by the protein A–gold complex. Cancer **60**:1288–1293.
36. Peehl DM 1992 Culture of human prostatic epithelial cells. In: Freshney IA (ed) Culture of Specialized Cells: Epithelial Cells. pp. 159–180 Wiley, New York.
37. Chung LWK 1991 Fibroblasts are critical determinants in prostatic cancer growth and dissemination. Cancer Metastasis Rev **10**:263–274.
38. Chung LWK 1995 The role of stromal–epithelial interaction in normal and malignant growth. Cancer Surv **23**:33–42.
39. Tu H, Jacobs SC, Mergner WJ, Kyprianou N 1994 Rare incidence of human papillomavirus types 16 and 18 in primary and metastatic human prostate cancer. Urology **44**:726–731.
40. Sarkar FH, Sakr WA, Li YW, Sreepathi P, Crissman JD 1993 Detection of human papillomavirus (HPV) DNA in human prostatic tissues by polymerase chain reaction (PCR). Prostate **22**:171–180.
41. Bookstein R, Shew JY, Chen PL, Scully P, Lee WH 1990 Suppression of tumorigenicity of human prostate carcinoma cells by replacing a mutated RB gene. Science **47**:712–715.
42. Bookstein R, Rio P, Madreperla SA, Hong F, Allred C, Grizzle WE, Lee WH 1990 Promoter deletion and loss of retinoblastoma gene expression in human prostate carcinoma. Proc Natl Acad Sci USA **87**:7762–7766.
43. Bookstein R, MacGrogan D, Hilsenbeck SG, Sharkey F, Allred DC 1993 p53 is mutated in a subset of advanced-stage prostate cancers. Cancer Res **53**:3369–3373.
44. Effert PJ, Neubauer A, Walther PJ, Liu ET 1992 Alterations of the p53 gene are associated with the progression of a human prostate carcinoma. J Urol **147**:789–793.
45. Effert PJ, McCoy RH, Walther PJ, Liu ET 1993 p53 gene alterations in human prostate carcinoma. J Urol **150**:257–261.
46. Ludlow J 1993 Interactions between SV40 large-tumor antigen and the growth suppressor proteins pRB and p53. FASEB J **7**:866–871.
47. Vousden K 1993 Interactions of human papillomavirus transforming proteins with the products of tumor suppressor genes. FASEB J **7**:872–879.
48. Cussenot O, Berthon P, Berger R, Mowszowicz I, Faille A, Hojman F, Teillac P, Le Duc A, Calvo F 1991 Immortalization of human adult normal prostatic epithelial cells by liposomes containing large T-SV40 gene. J Urol **146**:881–886.
49. Lee M-S, Garkovenko E, Yun JS, Weijerman PC, Peehl DM, Chen L-S, Rhim JS 1994 Characterization of adult human prostatic epithelial cells immortalized by polybrene-induced DNA transfection with a plasmid containing an origin-defective SV40 genome. Int J Oncol **4**:821–830.
50. Weijerman PC, Konig JJ, Wong ST, Niesters HG, Peehl DM 1994 Lipofection-mediated immortalization of human prostatic epithelial cells of normal and malignant origin using human papillomavirus type 18 DNA. Cancer Res **54**:5579–5583.
51. Peehl DM, Wong ST, Rhim JS 1995 Altered growth regulation of prostatic epithelial cells by human papillomavirus-induced transformation. Int J Oncol **6**:1177–1184.
52. Gross C, Skowronski RJ, Plymate SR, Rhim JS, Peehl DM, Feldman D 1996 Simian virus 40-, but not human papillomavirus-, transformation of prostatic epithelial cells results in loss of growth-inhibition by 1,25-dihydroxyvitamin D_3. Int J Oncol **8**:41–47.
53. Schwartz GG, Hill CC, Oeler TA, Becich MJ, Bahnson RR 1995 1,25-Dihydroxy-16-ene-23-yne-vitamin D_3 and prostate cancer cell proliferation *in vivo*. Urology **46**:365–369.
54. Osborn JL, Schwartz GG, Smith DC, Bahnson R, Day R, Trump DL 1995 Phase II trial of oral 1,25-dihydroxyvitamin D (calcitriol) in hormone refractory prostate cancer. Urologic Oncology **1**:195–198.
55. Crawford ED, DeAntonio EP, Labrie F, Schroder FH, Geller J 1995 Endocrine therapy of prostate cancer: optimal form and appropriate timing. J Clin Endocrinol Metab **80**:1062–1078.
56. Santen RJ 1992 Endocrine treatment of prostate cancer. J Clin Endocrinol Metab **75**:685–689.
57. Bouillon R, Okamura WH, Norman AW 1995 Structure–function relationships in the vitamin D endocrine system. Endocr Rev **16**:200–257.
58. Schwartz GG, Oeler TA, Uskokovic MR, Bahnson RR 1994 Human prostate cancer cells: Inhibition of proliferation by vitamin D analogs. Anticancer Res **14**:1077–1081.
59. Skorwonski RJ, Peehl DM, Feldman D 1995 Actions of vitamin D_3, analogs on human prostate cancer cell lines: Comparison with 1,25-dihydroxyvitamin D_3. Endocrinology **136**:20–26.
60. Lucia MS, Anzano MA, Slayter MV, Anver MR, Green DM, Shrader MW, Logsdon DL, Driver CL, Brown CC, Peer CW,

Roberts AB, Sporn MB 1995 Chemopreventive activity of tamoxifen, *N*-(4-hydroxyphenyl)retinamide, and the vitamin D analogue Ro24-5531 for androgen-promoted carcinomas of the rat seminal vesicle and prostate. Cancer Res **55**:5621–5627.
61. Cunha GR, Donjacour AA, Cooke PS, Mee S, Bigsby RM, Higgins SJ, Sugimura Y 1987 The endocrinology and developmental biology of the prostate. Endocr Rev **8**:338–362.
62. Chambon P 1996 A decade of molecular biology of retinoic acid receptors. FASEB J **10**:940–954.
63. Manglesdorf DJ, Evans RM 1995 The RXR heterodimers and orphan receptors. Cell **83**:841–850.
64. Giguere V 1994 Retinoic acid receptors and cellular retinoid binding proteins: Complex interplay in retinoid signaling. Endocr Rev **15**:61–79.
65. Tallman MS, Wiernik PH 1992 Retinoids in cancer treatment. J Clin Pharmacol **32**:868–888.
66. Petkovich M, Brand NJ, Krust A, Chambon P 1987 A human retinoic acid receptor which belongs to the family of nuclear receptors. Nature **330**:444–450.
67. de The H, Marchio A, Tiollais P, Dejean A 1989 Differential expression and ligand regulation of the retinoic acid receptor alpha and beta genes. EMBO J **8**:429–433.
68. Pasquali D, Thaller C, Eichele G 1996 Abnormal level of retinoic acid in prostate cancer tissues. J Clin Endocrinol Metab **81**:2186–2191.
69. Pienta KJ, Nguyen NM, Lehr JE 1993 Treatment of prostate cancer in the rat with the synthetic retinoid fenretinide. Cancer Res **53**:224–226.
70. Slawin K, Kadmon D, Park SH, Scardino PT, Anzano M, Sporn MB, Thompson TC 1993 Dietary fenretinide, a synthetic retinoid, decreases the tumor incidence and the tumor mass of *ras* + *myc*-induced carcinomas in the mouse prostate reconstitution model system. Cancer Res **53**:4461–4465.
71. Stearns ME, Wang M, Fudge K 1993 Liarazole and 13-*cis*-retinoic acid anti-prostatic tumor activity. Cancer Res **53**:3073–3077.
72. Peelh DM, Wong ST, Cramer SD, Gross C, Feldman D 1995 Suramin, hydrocortisone, and retinoic acid modify inhibitory effects of 1,25-dihydroxyvitamin D_3 on prostatic epithelial cells. Urologic Oncology **1**:188–194.
73. Eisenberger MA, Reyno L, Sinibaldi V, Sridhara R, Carducci M, Egorin M 1995 The experience with suramin in advanced prostate cancer. Cancer **75**(Suppl.):1927–1934.
74. Peehl DM, Wong ST, Stamey TA 1991 Cytostatic effects of suramin on prostate cencer cells cultured form primary tumors. J Urol **145**:624–630.
75. Hedlund TE, Moffatt KA, Miller GJ 1996 Stable expression of the nuclear vitamin D receptor in the human prostatic carcinoma cell line JCA-1: Evidence that the antiproliferative effects of 1α,25-dihydroxyvitamin D_3 are mediated exclusively through the genomic signaling pathway. Endocrinology **137**:1554–1561.
76. Hedlund TE, Moffatt KA, Miller GJ 1996 Vitamin D receptor expression is required for growth modulation by 1α,25-dihydroxyvitamin D_3 in the human prostatic carcinoma cell line ALVA-31. J Steroid Biochem Mol Biol **58**:277–288.
77. Krishnan AV, Cramer SD, Bringhurst FR, Feldman D 1995 Regulation of 1,25-dihydroxyvitamin D_3 receptors by parathyroid hormone in osteoblastic cells: Role of second messenger pathways. Endocrinology **136**:705–712.
78. Peehl DM, Cohen P, Rosenfeld RG 1995 The insulin-like growth factor system in the prostate. World J Urol **13**:306–311.
79. Kaicer EK, Blat C, Harel L 1991 IGF-I and IGF-binding proteins: Stimulatory and inhibitory factors secreted by human prostatic adenocarcinoma cells. Growth Factors **4**:231–237.
80. Pietrzkowski Z, Mullholland G, Gomella L, Jameson BA, Wernike D, Baserga R 1993 Inhibition of growth of prostatic cancer cell lines by peptide analogues of insulin-like growth factor I. Cancer Res **53**:1102–1106.
81. Figueroa JA, Lee AV, Jackson JG, Yee D 1995 Proliferation of cultured human prostate cancer cells is inhibited by insulin-like growth factor (IGF) binding protein-1: Evidence for an IGF-II autocrine growth loop. J CLin Endocrinol Metab **80**:3476–3482.
82. Cohen P, Peehl P, Baker B, Liu F, Hintz RL, Rosenfeld RG 1994 Insulin-like growth factor axis abnormalities in prostatic stromal cells from patients with benign prostatic hyperplasia. J Clin Endocrinol Metab **79**:1410–1415.
83. Cohen P, Peehl DM, Lamson G, Rosenfeld RG 1991 Insulin-like growth factors (IGFs), IGF receptors and IGF binding proteins in primary cultures of prostate epithelial cells. J Clin Endocrinol Metab **73**:401–407.
84. Iwamura M, Sluss PM, Casamento JB, Cockett ATK 1993 Insulin-like growth factor I: Action and receptor characterization in human prostate cancer cell lines. Prostate **22**:243–252.
85. Drivdahl RH, Loop SM, Andress DL, Ostenson RC 1995 IGF-binding proteins in human prostate tumor cells: Expression and regulation by 1,25-dihydroxyvitamin D_3. Prostate **26**:72–79.
86. Tennant MK, Thrasher JB, Twomey PA, Birnbaum RS, Plymate SR 1996 Insulin-like growth factor-binding proteins (IGFBP)-4, -5, and -6 in the benign and malignant human prostate: IGFBP-5 messenger ribonucleic acid localization differs from IGFBP-5 protein localization. J Clin Endocrinol Metab **8**:3783–3792.
87. Tennant MK, Thrasher JB, Twomey PA, Drivdahl RH, Birnbaum RS, Plymate SR 1996 Protein and messenger ribonucleic acid (mRNA) for the type 1 insulin-like growth factor (IGF) receptor is decreased and IGF-II mRNA is increased in human prostate carcinoma compared to benign prostate epithelium. J Clin Endocrinol Metab **81**:3774–3782.
88. Cohen P, Graves HCB, Peehl DM, Kamarei M, Giudice LC, Rosenfeld RG 1992 Prostate-specific antigen (PSA) is an insulin-like growth factor binding protein-3 protease found in seminal plasma. J Clin Endocrinol Metab **75**:1046–1053.
89. Conover CA, Perry JE, Tindall DJ 1995 Endogenous cathepsin D-mediated hydrolysis of insulin-growth factor-binding proteins in cultured human prostatic carcinoma cells. J Clin Endocrinol Metab **80**:987–993.
90. Cohen P, Peehl DM, Graves HCB, Rosenfeld RG 1994 Biological effects of prostate specific antigen as an insulin-like growth factor binding protein-3 protease. J Endocrinol **142**:407–415.
91. Peehl DM, Wong ST, Bazinet M, Stamey TA 1989 In vitro studies of human prostatic epithelial cells: Attempts to identify distinguishing features of malignant cells. Growth Factors **1**:237–250.
92. Danielpour D 1996 Induction of transforming growth factor-beta autocrine activity by all-*trans*-retinoic acid and 1α,25-dihydroxyvitamin D_3 in NRP-152 rat prostatic epithelial cells. J Cell Physiol **166**:231–239.
93. Danielpour D, Kadomatsu K, Anzano MA, Smith JM, Sporn MB 1994 Development and characterization of nontumorigenic and tumorigenic epithelial cell lines from rat dorsal–lateral prostate. Cancer Res **54**:3413–3421.
94. Peehl DM, Leung GK, Wong ST 1994 Keratin expression: A measure of phenotypic modulation of prostatic epithelial cells by growth inhibitory factors. Cell Tissue Res **277**:11–18.
95. Orloff JJ, Reddy D, De Papp AE, Yang KH, Soifer NE 1994 Parathyroid hormone-related protein as a prohormone: Post-translational processing and receptor interactions. Endocr Rev **15**:40–60.
96. Rodan SB, Insogna KL, Vignery AMC, Stewart AF, Broadus AE, D'Souza SM, Bertolini DR, Mundy GR, Rodan GA 1983

Factors associated with humoral hypercalcemia of malignancy stimulate adenylate cyclase in osteoblastic cells. J Clin Invest **72**:1511–1515.

97. Kao PC, Klee GG, Taylor RL, Heath H 1990 Parathyroid hormone-related peptide in plasma of patients with hypercalcemia and malignant lesions. Mayo Clin Proc **65**:1399–1406.
98. Iwamura M, Abrahamsson P, Foss KA, Wu G, Cockett AT, Deftos LF 1994 Parathyroid hormone-related protein: A potential autocrine growth regulator in human prostate cancer cell lines. Urology **43**:675–679.
99. Cramer CD, Peehl DM, Edgar MG, Wong ST, Deftos LJ, Feldman D 1996 Parathyroid hormone-related protein (PTHrP) is an epidermal growth factor-regulated secretory product of human prostatic epithelial cells. Prostate **29**:20–29.
100. Feldman D, Malloy PJ, Gross C 1996 Vitamin D: Metabolism and action. In: Marcus R, Feldman D, Kelsey J (eds) Osteoporosis. Academic Press, New York, pp. 205–235.
101. Liu M, Lee M-H, Cohen M, Bommakanti M, Freedman LP 1996 Transcriptional activation of the cdk inhibitor p21 by vitamin D_3 leads to the induced differentiation of the myelomonocytic cell line U937. Genes Dev **10**:142–153.
102. Feldman D 1997 Androgen and vitamin D receptor gene polymorphisms: The long and short of prostate cancer risk. J Natl Cancer Inst **89**:109–111.
103. Gross C, Feldman D 1997 Calcitriol treatment of early recurrent prostate cancer: A pilot trial. Proceedings of the 79th Annual Meeting of the Endocrine Society. 261(Abstract PI-508).
104. Zhao X-Y, Eccleshall TR, Krishnan AV, Gross C, Feldman D 1997 Analysis of vitamin D analog-induced heterodimerization of vitamin D receptor with retinoid X receptor using the yeast two-hybrid system. Mol Endocrinol **11**:366–378.

CHAPTER 67

Chemoprevention of Colon Cancer by Vitamin D_3 and Its Metabolites/Analogs

THOMAS A. BRASITUS AND MICHAEL D. SITRIN
Department of Medicine, The University of Chicago Hospitals & Clinics, Chicago, Illinois

I. INTRODUCTION

Colorectal cancer is the second leading cause of cancer mortality in the United States, and in spite of advances in surgery, chemotherapy, and radiotherapy since the 1950s there has been no significant reduction in death from this disorder. Although much progress has been made in identifying genetic factors predisposing to colorectal cancer, it is estimated that only 5–10% of colorectal cancers are due to inherited genetic abnormalities, with the vast majority of these neoplasms caused by environmental factors, particularly the diet. Much epidemiologic and experimental evidence has accumulated indicating that dietary factors such as high fat, red meat, and lack of fruit and vegetable consumption may increase the risk of colon cancer. More recently, certain dietary components, including vitamin D and calcium, have been suggested as possibly protective against colon cancer. Interest in vitamin D as a chemopreventive agent has been further spurred by the development of vitamin D analogs that produce less hypercalcemia than the parent compound, therefore reducing the major cause of toxicity. In this chapter, we critically review the available data regarding the potential role of vitamin D metabolites and analogs as chemopreventive agents against colorectal cancer. Information from epidemiologic investigations, experimental animal cancer models, and studies of human colonic cancer cell lines in culture is considered. In addition, the possible cellular and molecular mechanisms whereby vitamin D secosteroids might influence colonic carcinogenesis are discussed.

II. ACTIONS OF 1,25-DIHYDROXYVITAMIN D_3 AND OTHER METABOLITES/ANALOGS OF VITAMIN D_3 IN NORMAL RAT COLONOCYTES AND IN CULTURED COLONIC CANCER CELLS

A. Genomic Effects

Until relatively recently, the cellular actions of 1,25-dihydroxyvitamin D_3 [1α,25$(OH)_2D_3$] and other active metabolites of vitamin D_3 were classically all thought to occur via genomic mechanisms initiated by their binding to the nuclear vitamin D_3 receptor (VDR). Through association with this high affinity receptor and subsequent binding of the VDR–secosteroid complex to unique sequences in the genome, 1,25$(OH)_2D_3$ and other metabolites of vitamin D_3 have been shown to either up- or down-regulate the transcription of a large number of genes involved in not only mineral homeostasis, but also DNA replication and cell differentiation [1,2]. In support of this paradigm, the VDR has now been localized to a large number of normal and malignant cells/tissue, including normal mammalian colon [3,4] and sporadic human [3,4] and chemically induced rat colonic tumors [5–20], in addition to the classic vitamin D-responsive organs, namely, bone, kidney, and intestine [1,2].

The regulation of specific genes in colonocytes mediated by the VDR is just now being elucidated. Several conflicting studies have addressed whether the expression of the VDR in human colon cancer is altered compared to the normal colonic mucosa [11,21]. Similarly, conflicting data exist regarding the level of the VDR in human cultured colonic cell lines as a function of their state of differentiation [5,9,14,15,22].

B. Rapid/Nongenomic Effects

It has become clear, however, that several metabolites and synthetic analogs of vitamin D_3, in addition to their well-accepted actions involving the VDR, can also elicit very rapid (seconds to minutes) biochemical effects, which may not be mediated by the VDR, in both nonmalignant and malignant cells [23,24], including epithelial cells of the colon. Evidence has now accumulated from several sources to suggest that the rapid, "nongenomic" actions of these secosteroids may, in fact, occur via binding to a putative plasma membrane receptor, distinct from the classic nuclear VDR (see below and Chapter 15).

On the basis of these considerations, the actions of these secosteroids have been considered to be either genomic or nongenomic in nature. It bears emphasis, however, that this terminology, which now is commonly used in the literature, may at times be too simplistic. Thus, for example, just because a metabolite/analog of vitamin D_3 has a very low binding affinity for the VDR, does not necessarily mean that its actions are not mediated via the VDR (i.e., are nongenomic). Studies have demonstrated, in fact, that control of genes by these secosteroids is very complex and is regulated not only by binding of secosteroids to the VDR, but also by heterodimerization of the VDR with the retinoid X receptor (RXR) [25]. In addition, the compounds have the ability to induce phosphorylation of serine residues on the VDR, independent of their binding to this receptor, via the activation of multiple serine kinases, including protein kinase C (PKC), cAMP-dependent protein kinase, and casein kinase II [26]. Many of the apparent nongenomic actions of 1,25$(OH)_2D_3$ may, therefore, ultimately be found to be mediated by the VDR via these latter mechanisms.

In an attempt to characterize the biochemical and physiological effects of these secosteroids, as well as to elucidate their actions in the prevention of colon cancer (see below), our laboratory has studied the effects of 1,25$(OH)_2D_3$ in rat colonocytes and in CaCo-2 cells, a cell line derived from a human colonic adenocarcinoma. These latter cultured cells, despite being transformed, have proved to be extremely useful for these purposes, as they possess very similar signal transduction elements and respond almost identically with respect to the actions of 1,25$(OH)_2D_3$ as their normal vitamin D-sufficient rat colonocyte counterparts.

Our laboratory has demonstrated that 1,25$(OH)_2D_3$, in a concentration-dependent manner (10^{-10} to 10^{-8} M), but not 25$(OH)D_3$ or 24,25$(OH)_2D_3$ (at final concentrations up to 10^{-6} M), rapidly (seconds to minutes) induced the hydrolysis of membrane polyphosphoinositides, thereby generating inositol 1,4,5-trisphosphate (IP_3) and 1,2-diacylglycerol (DAG) and increasing intracellular calcium ($[CA^{2+}]_i$) in a biphasic manner in isolated rat colonocytes [27] and in CaCo-2 cells [28]. As a consequence of the rise in DAG and $[Ca^{2+}]_i$, 1,25$(OH)_2D_3$ also activated PKC, as assessed by the translocation of this kinase from the cytosolic to membrane fraction of these cells [27,29]. Moreover, as an additional consequence of the increased $[Ca^{2+}]_i$, this secosteroid increased the internal pH (pH_i) of both of these cell types, by inhibiting Na^+–H^+ exchange, via a Ca^{2+}–calmodulin-dependent process [27,29].

In subsequent studies our laboratory has extended these observations by demonstrating that (1) the *in vitro* addition of $1,25(OH)_2D_3$ to isolated colonocytes harvested from vitamin D-deficient rats failed to induce any of the aforementioned biochemical changes noted in cells from D-sufficient animals [29]; (2) these *in vitro* rapid effects of $1,25(OH)_2D_3$, however, were all partially, but significantly restored after *in vivo* repletion of D-deficient animals with this secosteroid for 7–10 days [29]; (3) the *in vitro* rapid responses to $1,25(OH)_2D_3$ could also not be elicited in "premalignant" colonocytes harvested from animals treated with the colonic procarcinogen 1,2-dimethylhydrazine (DMH) [30]; (4) the initial (transient) rise in $[Ca^{2+}]_i$ was secondary to the release of intracellular stores of this divalent cation by the increase in IP_3 induced by $1,25(OH)_2D_3$, whereas the second (sustained) rise in $[Ca^{2+}]_i$ was found to be due to the opening by this secosteroid of a relatively nonselective cation channel ($Ca^{2+} > Na^+$) in the plasma membranes of CaCo-2 cells [31]; (5) the hydrolysis of polyphosphoinositides induced by $1,25(OH)_2D_3$ in both rat colonocytes and CaCo-2 cells was restricted to the basolateral region of the plasma membranes of these cells [32]; (6) several isoforms of PKC were present in these cells, but only PKC-α (alpha) and -β_{II} (beta$_{II}$) were activated by $1,25(OH)_2D_3$ [33–35]; (7) the activation of PKC-α by $1,25(OH)_2D_3$, in turn, was found to limit the rise in $[Ca^{2+}]_i$ induced by increases in IP_3 at a step distal to its formation, possibly by phosphorylating the IP_3 receptor, in CaCo-2 cells [33]; and (8) $1,25(OH)_2D_3$, in a dose-dependent manner (10^{-9} to 10^{-7} *M*) inhibited the proliferation and induced the differentiation of CaCo-2 cells [5]. Taken together, these observations indicated that $1,25(OH)_2D_3$ could influence many important biochemical and physiological processes in these cells. Moreover, these studies also demonstrated that the vitamin D status, as well as the malignant transformation process, at least in the DMH model, appeared to modulate many of the biochemical effects of this secosteroid in rat colonocytes.

More recent studies in our laboratory have indicated that the biochemical actions of $1,25(OH)_2D_3$ in both rat colonocytes and in CaCo-2 cells may, in fact, be even more complex than originally thought. In this regard, in preliminary studies [36] we found that $1,25(OH)_2D_3$, in a time- and concentration-dependent manner, activated c-Src, a nonreceptor tyrosine kinase, in rat colonocytes. This activation of c-Src, in turn, appeared to cause a significant increase in the biochemical activity, protein abundance, and tyrosine phosphorylation of phosphoinositide–phospholipase C-γ (PI-PLCγ), specifically in the basolateral membranes (BLM) of these cells [36]. These findings indicate that PI-PLCγ activation by $1,25(OH)_2D_3$ via c-Src stimulation, at least in part, was responsible for the hydrolysis of polyphosphoinositides induced by this secosteroid previously noted in the BLM of these cells (see above). Moreover, as in the case of other rapid actions of these secosteroids, the activation of c-Src by $1,25(OH)_2D_3$ was lost in D deficiency and reversibly restored in the *in vivo* repletion of the D-deficient animals with $1,25(OH)_2D_3$ (T. A. Brasitus and M. D. Sitrin, unpublished observations).

Because, as noted earlier, $1,25(OH)_2D_3$ stimulated PKC-α in these cells [33], which had been shown to activate Raf-1, a mitogen-activated protein kinase (MAPK) kinase, we also investigated the potential effect of this secosteroid on members of the MAPK family. In preliminary studies [36], we recently found that $1,25(OH)_2D_3$ increased the activities and tyrosine phosphorylation of two members of this family, ERK-1 and ERK-2, which have been implicated in the regulation of cellular growth in many cell types [36].

In addition, our laboratory demonstrated that several of the aforementioned rapid biochemical effects of $1,25(OH)_2D_3$ could be mimicked in CaCo-2 cells by synthetic analogs of this secosteroid that had much lower affinities (~1000-fold less) for the VDR [37]. These findings suggested that structural requirements for the secosteroid-induced rapid actions on signal transduction in these cells were different from those for VDR binding and VDR-induced transcription regulation. Taken together with our previous observations, which showed that the hydrolysis of polyphosphoinositides induced by $1,25(OH)_2D_3$ was restricted to the BLM of CaCo-2 cells and rat colonocytes, these findings suggested that these cells may possess on their BLM a VDR which is distinct from the classic nuclear VDR. In preliminary studies [38,39], we have also demonstrated that specific binding between rat BLM and ^3H-$1,25(OH)_2D_3$ was present at concentrations as low as 10^{-10} *M* and increased in a concentration-dependent manner. This binding was also found to be dependent on time, temperature, and pH, as well as to be reversible and sensitive to trypsin digestion of the membranes. In addition, the binding to the BLM was 2.5-fold greater than that seen with purified rat brush-border membranes. Moreover, Western blots of the membranes revealed minimal contamination with the nuclear VDR, insufficient to explain the observed binding [38,39].

In keeping with the contention that a unique receptor for $1,25(OH)_2D_3$ may exist in BLM of rat colonocytes and CaCo-2 cells, Nemere *et al.* [40] demonstrated binding of ^3H-$1,25(OH)_2D_3$ to a protein extracted with detergents from purified chick intestinal BLM *in vitro*. Baran *et al.* [41] also reported specific binding of radiolabeled $1,25(OH)_2D_3$ to membrane preparations from

ROS 24/1 cells, an osteoblast-like cell line lacking the nuclear VDR. In addition, Slater *et al.* [42] reported direct activation of several PKC isoforms by $1,25(OH)_2D_3$, raising the possibility that one or more of these membrane-associated PKCs could serve as a receptor for this secosteroid. In view of studies that have implicated alterations in the abundance of the classic nuclear VDR in colonic malignant transformation [3,4,21,43,44], it will be of interest to further purify, characterize, and quantity this putative BLM receptor for $1,25(OH)_2D_3$, as well as to determine its potential role in mediating the anticarcinogenic actions of these secosteroids in this organ.

III. EVIDENCE THAT VITAMIN D_3 AND ITS METABOLITES/ANALOGS MAY PREVENT THE DEVELOPMENT OF COLORECTAL CANCER

A. Epidemiologic Studies

Studies examining the geographical patterns of colon cancer throughout the world have suggested to some investigators that vitamin D synthesized in the skin from sunlight exposure may reduce cancer risk [45–49]. Colon cancer risk and mortality have been demonstrated to increase with increasing latitude [47,48]. Countries with low prevalence of colon cancer are generally located within 20 degrees latitude of the equator, whereas countries at high latitudes generally have high colon cancer rates. A notable exception is Japan, which has relatively little colon cancer but is located at a relatively high latitude. A potential explanation for this discrepancy offered by some investigators is that the Japanese diet is very rich in vitamin D because of the popularity of fish containing large amounts of the vitamin, providing an estimated intake of 800–1000 IU per day [47].

Emerson and Weiss used data from nine population-based cancer registries in the United States to examine incidence rates for colon and rectal cancer as a function of variations in solar irradiation [46]. This analysis was limited to Caucasians aged 25–84 years, and only individuals born and diagnosed in the same state were considered in order to minimize biases due to migration. Solar radiation data consisted of long-term annual averages for daily global radiation provided by the National Weather Service. Age-adjusted colon and rectal cancer incidence rates for men showed an inverse relationship with solar radiation, so that the rates in areas with low solar radiation such as Michigan and western Washington state were 50–80% higher than New Mexico and Utah. For women, a similar trend which was not statistically significant was noted for colon cancer, but no effect of solar irradiation on rectal cancer was found. The authors speculated that differences in occupation and recreational activities might explain why a greater effect of solar radiation was observed in men than in women. Spitz *et al.*, however, reported significantly higher colon cancer rates in white and black men and women residing in areas with low solar radiation [49]. Garland and Garland found that age-adjusted death rates from colon cancer in Caucasian males were nearly three times higher in the northeastern United States than in sunnier regions [45].

Other types of ecological studies have also suggested that cutaneous vitamin D synthesis was protective against colon cancer. Atmospheric conditions, such as the concentration of ozone and the severity and type of air pollution, can affect the transmission of ultraviolet B light through the atmosphere. As discussed in Chapter 3, ultraviolet light in the range of 290–310 nm is optimal for the production of vitamin D in the skin. In a study of 18 Canadian cities, Gorham *et al.* reported a positive association between the level of acid haze, composed mainly of sulfur dioxide that absorbs ultraviolet light and ammonium dioxide particles that scatter it, and colon cancer mortality rates [50]. A similar effect of variation in acid-haze levels have been offered as an explanation for differences in colon cancer incidence and mortality rates in different regions of Italy and the former Soviet Union [51,52]. Within the United States and other western countries, colonic cancer is more prevalent in urban metropolitan regions than in rural areas [48]. Differences in occupations, lifestyle patterns, housing conditions, and atmospheric pollution affecting exposure to sunlight have been suggested as possible explanations for this urban–rural gradient. Physical activity has been reported to have a protective effect against colon cancer, which could possibly relate to more time spent outdoors and increased sunlight exposure [47,53].

Several epidemiologic studies have examined the relationship between dietary vitamin D intake and colorectal cancer [48,54–57]. This type of research is complicated by the close correlation between dietary intakes of vitamin D and calcium as fortified dairy foods; calcium and milk products have also been suggested in many studies to be protective against colorectal cancer [48,54,55]. Garland *et al.* studied almost 2000 male workers at the Western Electric Company in Chicago [54]. Nutritionists obtained two 28-day diet histories in the late 1950s, and the health status of the workers was monitored over the subsequent 19 years. Men who developed colorectal cancer were found to have significantly lower dietary intakes of both vitamin D and calcium. This association persisted after adjustment for other possible risk factors, such as age, smoking, body mass index, alcohol consumption, and dietary fat intake. Men in the lowest two quartiles of vitamin D intake had a risk of colorectal cancer that was approximately twice that of men in the highest two quartiles.

Bostick *et al.* investigated more than 41,000 women aged 55–69 years in the Iowa Women's Health Study [55]. In 1986, these women completed a mailed questionnaire that included a semiquantitative food frequency questionnaire and questions concerning the use of vitamin and mineral supplements. Colon cancer incidence was determined through the State Health Registry of Iowa. During a 5-year follow-up period, 212 cases of colon cancer were documented. These cases had lower intakes of total (dietary plus supplemental) vitamin D and calcium than controls who did not develop colon cancer. Dietary consumption of vitamin D, calcium, and milk products tended to be lower in the colon cancer cases, but these data were not statistically significant. No significant differences were noted with respect to the use of vitamin/mineral supplements in general between cases and controls, but significantly more women who developed colon cancer did not use any supplement containing vitamin D or calcium. Age-adjusted relative risk of colon cancer was inversely related to total and supplemental vitamin D and calcium intakes, with those in the highest categories of intake of these nutrients having about half the risk of those in the lowest categories. Using multivariate models controlling for risk factors, the trends for inverse relationships between vitamin D and calcium intakes and colon cancer risk were still present, but they were no longer statistically significant.

Other studies of diet as a risk factor for colorectal neoplasms have not confirmed an important influence of vitamin D consumption. Kampman *et al.* examined data from two large United States cohort studies [57]. In 1986, the Health Professionals Follow-Up Studies enrolled more than 50,000 male health care workers aged 40–75 years. In 1976, the Nurses' Health Study enrolled more than 100,000 female registered nurses. Dietary evaluations were performed in 1980 and 1984 in the Nurses' Health Study and in 1986 in the Health Professionals Follow-Up study using a mailed semiquantitative food frequency questionnaire. Men and women who had undergone sigmoidoscopy or colonoscopy during the follow-up periods were used for this study, and since sigmoidoscopy was the more common examination the analyses were restricted to left-sided colonic or rectal adenomas. Thus, 9490 men were examined during 1986–1990, and 8945 women during 1980–1988. Slight, nonsignificant inverse relationships were noted between total vitamin D intake and colorectal adenoma risk, which disappeared in multivariate analysis. A significant inverse association with total vitamin D intake was observed in nurses in the 1980–1988 follow-up for rectal adenomas. This finding, however, could not be confirmed using the 1984–1988 follow-up period, and the apparent protective effect in the 1980–1988 follow-up appeared to be mainly attributable to multivitamin use. The authors speculated that intake of other nutrients, such as folate, in the multivitamins may account for this finding. No relationships were found between calcium or dairy food intakes and risk for colorectal adenomas.

Kearney *et al.* used data from the Health Professionals Follow-Up study to examine relationships between vitamin D, calcium, dairy products and colonic carcinoma [56]. Inverse relationships were found between total, dietary, and supplemental vitamin D intakes and age-adjusted colon cancer-risk, but these were significantly attenuated in multivariate analyses using other potential risk factors. In particular, red meat and alcohol use largely accounted for the differences between the simple age-adjusted and full multivariate models. Similar, but even less impressive, associations were found between calcium or dairy food consumption and colon cancer risk.

To more directly examine the combined effect of vitamin D from both the diet and cutaneous synthesis on colon cancer risk, Garland *et al.* studied the relationship between serum 25-hydroxyvitamin D (25OHD) levels and colon cancer [52]. They used a bank of blood samples collected in 1974 from volunteers in Maryland. Thirty-four subjects with colon cancer were identified over the subsequent 8 years and matched to 67 noncancer controls. Serum 25OHD concentrations were significantly lower in cancer cases than controls (30.5 ng/ml versus 33.3 ng/ml, $p = 0.05$). Colon cancer risk was determined according to quintiles of serum 25OHD. Individuals with a serum 25OHD ≥ 20 ng/ml had about one-third the risk of colon cancer compared to those with a level of 0–19 ng/ml. Braun *et al.* used the same bank of serum samples to examine the association between serum vitamin D metabolite levels and colon cancer [58]. They identified 57 cases of colon cancer during the period 1984–1991, 10–17 years after collection of the blood samples, and matched them to 114 controls. Serum 25OHD and $1,25(OH)_2D_3$ levels were not significantly different in cases versus controls. Analysis by quintiles of serum vitamin D metabolite levels found no significant differences in odds ratios for colon cancer and no dose–response effect. It was noted, however, that those in the lowest quintile of serum 25OHD (<17.2 ng/ml) tended to have a higher risk of colon cancer than individuals in the other quintiles, raising the possibility of a threshhold effect. Lashner *et al.* reported that patients with chronic ulcerative colitis who had dysplasia or colon cancer identified during surveillance colonoscopy had lower serum 25OHD levels than matched controls who were free of neoplastic changes [59].

B. Cultured Colon Cancer Cell Studies

As noted earlier, many cell lines derived from human colonic adenocarcinomas have been shown to possess the VDR, as well as to undergo growth inhibition and/or differentiation by *in vitro* treatment with $1,25(OH)_2D_3$, and/or other metabolites/analogs of vitamin D_3 [5–20]. In addition to decreasing proliferation in cell culture systems, a $1,25(OH)_2D_3$ analog, 22(*S*)-24-homo-26,26,26,27,27,27-hexafluoro-1α,22,25-trihydroxyvitamin D_3, inhibited the growth and invasiveness of a human colon cancer cell line implanted under the renal capsule of immunodeficient mice in a dose-dependent manner. Withdrawal of the analog, moreover, lead to a resumption of tumor growth [60]. It bears emphasis that although studies in many malignant cell lines, particularly HL-60 leukemic cells, have indicated that the VDR is intimately involved in the mediation of the aforementioned effects of $1,25(OH)_2D_3$ [22,61,62], studies by Bhatia *et al.* [63], in NBH acute promyelocytic leukemia cells, have indicated that their $1,25(OH)_2D_3$-induced differentiation was independent of the VDR. These findings suggest that, at least in some malignant cell lines, the actions of $1,25(OH)_2D_3$, and/or its synthetic analogs, may, in fact, not require the VDR. Further studies to address this important issue are therefore warranted, particularly since, as noted above, interactions of the VDR with the RXR and the phosphorylation status of the VDR, in addition to VDR binding, may influence gene regulation by these secosteroids. It is also important to note that, in general, pharmacological doses of $1,25(OH)_2D_3$ are frequently necessary to induce these cellular effects [5]. In certain instances, for as yet unexplained reasons, *in vitro* treatment of these cells with physiological concentrations of this secosteroid may, in fact, induce proliferation rather than cause growth arrest [18].

As described in Section III,D, a number of analogs of vitamin D_3 have been synthesized, and several have been found to have a more pronounced effect on cellular proliferation and/or differentiation than their parent compounds [24,64–67]. Interestingly, many of these synthetic analogs also have markedly reduced effects on bone mineral resorption and on intestinal calcium absorption compared to $1,25(OH)_2D_3$ [24,64–67], thereby, at least in part, accounting for their reduced ability to induce hypercalcemia in animals chronically (months) fed these secosteroids [68,69]. The mechanisms responsible for the accentuated effects on cellular proliferation and differentiation, as well as for their reduced hypercalcemic activity compared to $1,25(OH)_2D_3$, are unclear at this time but may involve alterations in the free (active) concentrations, owing to a decrease in their affinity for the serum vitamin D binding protein [70], and/or alterations in metabolism, thereby leading to differences in the duration of action compared to $1,25(OH)_2D_3$ [25,67]. (This subject is extensively discussed in Chapters 58–60.) In addition, as shown by Cheskis *et al.* [25], these analogs, as assessed by surface plasmon resonance techniques, may differ from $1,25(OH)_2D_3$ in their ability to induce VDR–RXR heterodimerization and DNA binding to specific vitamin D response elements. Regardless of their mechanism(s) of action, however, several of these synthetic analogs, whose structure–function relationships have been reviewed [24,64,66,67,71], have shown promise as chemotherapeutic and/or chemopreventive agents in *in vitro* and *in vivo* studies of colonic malignant transformation (see below).

In addition to the effects of metabolites and analogs of vitamin D_3 on proliferation and differentiation, studies in breast cancer [72] and HL-60 cells [73] have indicated that the secosteroids may also induce these cells to undergo apoptosis, one form of programmed cell death (see also Chapter 65). In agreement with these studies, our laboratory has found that $1,25(OH)_2D_3$, as well as a noncalcemic, fluorinated analog of this secosteroid, 1,25-dihydroxy-16-ene-23-yne-26,27-hexafluorocholecalciferol, could also induce the apoptosis of CaCo-2 cells (T. A. Brasitus and M. D. Sitrin, unpublished observations). The proapoptotic effects of these secosteroids may, therefore, contribute to their potential chemotherapeutic and chemopreventive actions with respect to colon cancer.

C. Effects of Vitamin D Metabolites on Experimental Colon Cancer

Several groups have examined the effects of vitamin D metabolites or analogs on experimental animal models of colonic cancer. The most commonly used model has been 1,2-dimethylhydrazine (DMH)-induced colon cancer in the rat. DMH is a procarcinogen that is excreted in bile and is extensively metabolized to several intermediates, including azoxymethane which can also been administered as a procarcinogen, ultimately generating methyl carbonium ions that cause DNA methylation and mutations, including guanine to adenine transitions in the K-*ras* oncogene. DMH mainly causes adenomas and adenocarcinomas of the colon, with a lesser number of small intestinal cancers. DMH-induced colon cancer has generally been accepted as a good model for the human disease because it causes cancers with a histologic appearance, distribution in the colon, expression of tumor markers such as carcinoembryonic antigen, and prevalence of K-*ras* mutations that are similar to those in human colonic cancer. In addition, DMH-induced colonic cancer can be modulated by factors such as dietary fat, bile salts, selenium, and nonsteroidal anti-inflammatory agents that also have been suggested to influence colonic carcinogenesis in humans. It is important to recognize, however, that various investigators have used different doses and schedules of DMH administration, and have employed different experimental diets and methods of vitamin D supplementation.

Sitrin *et al.* studied the effect of dietary calcium supplementation with or without concomitant vitamin D deficiency on DMH-induced colonic carcinogenesis in the rat [74]. Rats were placed on one of three diets: (1) normal calcium (0.87%) and vitamin D_3 (2.2 IU/g); (2) high calcium (1.8%), vitamin D_3 sufficient (2.2 IU/g); or (3) high calcium, vitamin D deficient. All animals were raised in a dark room to prevent cutaneous vitamin D synthesis. After 6 weeks on the experimental diets, the rats received weekly subcutaneous injections of DMH for 26 weeks, and then were sacrificed for histologic examination. An important feature of this experimental design was that the three experimental groups had comparable weight gains, and that even animals in group 3 maintained normal serum calcium and phosphorus concentrations. The high calcium, vitamin D-sufficient dietary group has the same incidence of colonic cancer as the control diet group; however, the number of rats with multiple tumors was significantly reduced, and a smaller mean tumor size was observed. Concomitant vitamin D deficiency, however, abolished the protective effects of calcium supplementation, and very large, metastatic cancers occurred only in the vitamin D-deficient group.

Pence and Buddingh used a $2 \times 2 \times 2$ factorial design to investigate the effects of supplemental calcium or vitamin D_3 on DMH-induced colonic carcinogenesis promoted with a 20% corn oil diet in mole rats [75]. Animals on the high fat diet had an increased tumor incidence compared to rats on a low fat diet (86 versus 53%), and supplemental calcium or vitamin D separately given reduced the tumor incidence in animals fed high fat to equal to or less than that observed in rats on the low fat chow. For example, supplementation with 2000 IU/kg diet of vitamin D_3 versus 1000 IU/kg diet reduced the tumor incidence to 47%. For unclear reasons, supplementation with both vitamin D and calcium given together did not significantly reduce tumor incidence in rats fed the 20% corn oil diet. Vitamin D and calcium supplementation had no inhibitory effect on tumorigenesis in rats on a low fat diet. Belleli *et al.* studied the effect of subcutaneous administration of $1,25(OH)_2D_3$ before, during, and after procarcinogen administration on DMH-induced colon cancers [76]. The number of cancers was reduced by 50% in rats that received five weekly 400-ng doses of $1,25(OH)_2D_3$ prior to administration of DMH. There was no protective effect of $1,25(OH)_2D_3$ treatment during or after carcinogen administration. These investigators also reported that DMH treatment reduced by 50% the number of $1,25(OH)_2D_3$ binding sites in the colonic mucosa.

In contrast to these studies suggesting a protective effect of vitamin D, Comer *et al.* found no effect of vitamin D supplementation of DMH-induced colon cancer [77]. They gave a single dose of DMH, and 2 weeks later placed groups of rats on diets containing 250–10,000 IU of vitamin D_3/kg diet. Tumor incidence and number in the proximal or distal colon did not differ in the various dietary groups. The rats on the diet containing 10,000 IU/kg of vitamin D_3, in fact, had an expansion of the proliferative compartment of the colonic crypt, a change that has been related in some studies to an increased cancer risk. In a later study, this group examined the effect of three levels of dietary calcium and three levels of vitamin D_3 in a 3×3 factorial design [78]. The rats were all fed 20% fat and were started on the experimental diets after receiving a single dose of DMH. Animals fed the highest dietary calcium (15 g/kg diet) and vitamin D_3 (0.1 mg/kg diet) had a 45% lower incidence of both total tumors and carcinoma *in situ* in the distal colon ($p = 0.12$). No differences were noted in the other dietary groups. The rats receiving the diet highest in calcium and vitamin D had the lowest weight gain of the nine experimental groups, and low weight gain per se has been correlated with low tumor yield. In this study, high levels of vitamin D and calcium were correlated with a smaller crypt proliferative zone.

D. Effects of Vitamin D Analogs on Experimental Colon Cancer

More recent studies have focused on the effects of synthetic analogs of vitamin D on experimental colonic cancer. These compounds are of interest because they produce less hypercalcemia than natural vitamin D metabolites, thereby decreasing the major toxicity of vitamin D secosteroids. Wali *et al.* studied the effects of 1,25-dihydroxy-16-ene-23-yne-26,27-hexafluorocholecalciferol (Ro 24-5531, described in Chapter 62) on azoxymethane-induced colonic tumorigenesis in the rat [69]. The analog was given in the rat chow (2.5 nmol/kg feed) before and during azoxymethane administration (initiation phase) or after the procarcinogen treatment (promotion phase). In the control dietary group, 50%

of the animals developed colonic neoplasms, and more than half were invasive carcinomas. Treatment with Ro 24-5531 during the initiation phase significantly reduced the tumor incidence to 15%, and all tumors were benign adenomas. Administration of the vitamin D analog during the promotional phase tended to decrease tumor incidence (30%); however, this was not statistically significant, and again all the tumors occurring in this treatment group were benign adenomas. Rats treated with Ro 24-5531 during the initiation or promotion phases had normal weight gain and serum calcium, phosphorus, and 25OHD levels.

Otoshi *et al.* examined the effect of another nonhypercalcemic vitamin D analog, 22-oxacalcitriol (OCT, described in Chapter 63), on colonic cancer [79]. They used a model of rats treated with a combination of five kinds of carcinogens and demonstrated that administration of OCT intraperitoneally after carcinogen exposure significantly decreased the incidence of small bowel cancer and tended to reduce the incidence of colonic cancer. The number of small and large intestinal tumors per rat was significantly reduced with OCT treatment. In additional experiments, these investigators demonstrated that treatment with OCT during the promotional phase of DMH-induced carcinogenesis significantly reduced the formation of aberrant colonic crypt foci, changes that are believed to be preneoplastic lesions. In addition, expression of the proliferating cell nuclear antigen, a marker for colonocyte proliferation, was reduced in OCT-treated rats.

Kawaura *et al.* investigated the effect of supplementation with 1α-hydroxyvitamin D_3 on colonic carcinogenesis induced by intrarectal injection of *N*-methyl-*N*-nitrosourea (MNU) in rats [80,81]. This compound, unlike DMH or azoxymethane, is a direct acting carcinogen, and, thus, any observed influence on cancer formation is not due to effects of the vitamin D secosteroid on procarcinogen activation or metabolism. It was found that administration of 1α-hydroxyvitamin D_3, via the intragastric route during carcinogen treatment, significantly reduced cancer incidence in experiments where lithocholic acid was given as a promoter of colon cancer, or when an intense schedule of MNU was employed, but not when a more modest MNU dosing regimen was used.

Taken together, these various studies of the influence of vitamin D metabolites and analogs on experimental colon cancer generally indicated a protective effect of vitamin D secosteroids against colonic carcinogenesis. However, the magnitude of the response was markedly influenced by other dietary components (e.g., calcium, fat), the particular dose and administration schedule for the carcinogen and the vitamin D secosteroid, and other factors influencing tumor promotion.

E. Human Studies

Limited information is available concerning the effect of vitamin D metabolites or analogs on the human colonic mucosa. Thomas *et al.* examined the effect of vitamin D secosteroids on the growth characteristics of sigmoidoscopic rectal biopsies maintained in short-term organ culture [82,83]. Cellular proliferation was assessed using metaphase arrest to determine the crypt cell production rate and by immunohistochemical staining using a Ki-67 monoclonal antibody against an antigen present in proliferating cells. $1,25(OH)_2D_3$ (10^{-10}–10^{-6} *M*) reduced colonocyte proliferation, although a clear dose–response effect was not documented [82]. Ergocalciferol (10^{-8} *M*) and the less hypercalcemic vitamin D analog calcipotriol (10^{-7} *M*) (see Chapter 61) also had antiproliferative effects in this model system [82]. Rectal biopsies from patients with familial adenomatous polyposis, who are at extremely high risk for colorectal cancer, showed accelerated colonocyte proliferation compared to normal mucosa, and 10^{-10} *M* $1,25(OH)_2D_3$ decreased the crypt cell production rate by approximately 50% [82]. These investigators also reported that colonocyte proliferation was increased in rectal biopsies from patients with chronic ulcerative colitis, which also predisposes to colonic cancer. Calcipotriol (10^{-6} *M*) reduced the crypt cell production rate in the rectal mucosa of ulcerative colitis patients by 62% [83].

IV. POTENTIAL MECHANISMS INVOLVED IN THE COLONIC CHEMOTHERAPEUTIC AND CHEMOPREVENTIVE ACTIONS OF VITAMIN D_3 METABOLITES/ANALOGS

As reviewed in Section III of this chapter, several lines of evidence strongly suggest that $1,25(OH)_2D_3$, and/or one or more of the synthetic analogs of vitamin D_3, may ultimately prove useful in the treatment and prevention of colon cancer in humans. Moreover, it would appear that the colonic anticarcinogenic effects of these secosteroids may involve their actions on proliferation, differentiation, and/or apoptosis.

As noted earlier, $1,25(OH)_2D_3$ and many synthetic analogs of vitamin D_3 have been shown to inhibit the growth of a number of human colon cancer cell lines [5–15,18–20]. Moreover, in basic agreement with the results of studies performed in other malignant cell types [65,84,85], studies in CaCo-2 cells [13] have demonstrated that the antiproliferative effects of $1,25(OH)_2D_3$, as well as a number of synthetic analogs of this secosteroid, at least in part appeared to involve the arrest of cells in the G_1 phase of the cell cycle. Moreover, the antiproliferative effects of these secosteroids appeared to be modulated by extracellular Ca^{2+} [13,16] but, unlike the case in HL-60 cells, were independent of changes in the expression of the protooncogene c-*myc* [5,13].

Cell cycle progression in mammalian cells is regulated by sequential formation, activation, and subsequent inactivation of a series of cyclin–cyclin kinase-dependent (Cdk) complexes [86,87]. The major cyclin–Cdk complexes involved in the progression of the G_0/G_1 phase to the S phase of the cycle was cyclin D–Cdk 4 and cyclin E–Cdk 2, acting in G_1 and at G_1/S, respectively [86,87]. The formation of these latter complexes depends on the cell cycle-regulated expression of cyclins, which assemble with preexisting Cdks. They are also regulated positively and negatively by phosphorylation [86,87]. In most cells, the progression from G_0/G_1 to S phase is also negatively regulated by the retinoblastoma susceptibility gene product pRb, by the tumor suppression gene product p53, as well as by members of the Cdk inhibitor family, including $p21^{waf1}$ and $p27^{kip1}$ [85].

Although studies on the effects of $1,25(OH)_2D_3$ on the aforementioned cell cycle regulatory proteins have not been performed, to date, in any human colonic malignant cell line to our knowledge, $1,25(OH)_2D_3$ has been shown to increase the expression of both $p27^{kip1}$ and, more transiently, $p21^{waf1}$ in HL-60 cells [85]. In addition, in human keratinocytes [88], $1,25(OH)_2D_3$ has been shown to inhibit phosphorylation of the pRb. Each of these latter secosteroid-induced biochemical events, if also found to be present in human colon cancer cell lines, would at least in part explain why these secosteroids cause the G_1 arrest of these cells. In MC7 breast cancer cells, $p21^{waf1}$ expression is regulated by Raf-1 activation [89]. As noted earlier, prior studies have shown that $1,25(OH)_2D_3$ activated Raf-1 in rat hepatic Ito cells [90], as well as two MAPK, downstream effectors of Raf-1, in rat colonocytes [36]. It will therefore also be of interest to determine whether the secosteroid-induced block of the cell cycle of malignant colonic cells in the G_1 phase involves Raf-1 stimulation and, if so, by what mechanisms. In this regard, Raf-1 has been shown to be activated by stimulation of PKC-α [91] and by increases in ceramide [92]. These findings are of interest, as $1,25(OH)_2D_3$ has been shown to activate PKC-α in many cell types, including CaCo-2 cells [33], as well as stimulate a neutral sphingomyelinase in HL-60 cells, thereby increasing their intracellular levels of ceramide [93]. Moreover, in preliminary studies [94,95] our laboratory has demonstrated that alterations in the expression of PKC-α in CaCo-2 cells was accompanied by alterations in the growth, differentiation, contact inhibition, and transformation status of these cells.

As noted earlier, $1,25(OH)_2D_3$ has also been shown to induce the differentiation of a number of human colonic malignant cell lines. In CaCo-2 cells, for example, our laboratory and others have shown that $1,25(OH)_2D_3$ significantly enhanced the temporal rise in alkaline phosphatase activity [5,15] and in mRNA abundance of this enzyme [5], which normally occurs in their postconfluent state; these findings are consistent with the induction of differentiation of these cells by this secosteroid [5]. Moreover, in keeping with its effects on differentiation, $1,25(OH)_2D_3$ also increased the mean dome diameter and microvillus length and density of these cells [5]. In contrast to these findings, however, our laboratory has reported that the activity and mRNA abundance of sucrase–isomaltase, another brush-border hydrolase commonly used as a marker of differentiation, was actually decreased, not increased, in postconfluent CaCo-2 cells [5]. Giuliano *et al.* [15] have also reported that $1,25(OH)_2D_3$ treatment decreased sucrase activity in undifferentiated CaCo-2 cells and had no effect on its activity in differentiated cells. These findings would therefore indicate that the induction of differentiation of these colonic malignant cells by $1,25(OH)_2D_3$ may be very complex.

In support of this contention, although $1,25(OH)_2D_3$ has been shown to inhibit the growth of HT-29 colon cancer cells [10,20], when administered alone this secosteroid has not been found to induce their differentiation [10,20]. $1,25(OH)_2D_3$, however, has been shown to enhance sodium butyrate-induced differentiation in these cells [10,20]. Moreover, three analogs of this secosteroid, namely, 26,26,26,27,27,27-hexafluoro-$1,25(OH)_2D_3$, 24,24-difluoro-24-homo-$1,25(OH)_2D_3$, and 26,27-dimethyl-$1,25(OH)_2D_3$, were found to be 100-, 10-, and 5-fold, respectively, more effective than $1,25(OH)_2D_3$ in enhancing the differentiation of HT-29 cells [10]. The mechanism(s) involved in this latter phenomenon, however, remains elusive [10].

Although the biochemical events involved in the induction of differentiation of these colonic malignant cell types are unclear, observations in other cell types may be relevant to this important issue. $1,25(OH)_2D_3$ and/or its analogs have been shown to cause a number of biochemical alterations in HL-60 cells, which have been implicated in the induction of their differentiation into monocytes–macrophages. These include increases in intracellular Ca^{2+} and pH, increases in ceramide, as well as alterations in the calcium-dependent PKC isoforms, PKC-α and -β [96–101]. These findings are of interest because $1,25(OH)_2D_3$ has been shown by our laboratory to increase $[Ca^{2+}]_i$ and pH_i, as well as to activate PKC-α [28,33], in CaCo-2 cells, indicating that these biochemical events may also play a role in the differentiation of these colonic maligant cells by $1,25(OH)_2D_3$, by as yet unexplained mechanisms.

In MCF-7 breast cancer cells, $1,25(OH)_2D_3$ and $1,25(OH)_2$-16-ene-23-yne-vitamin D_3 induced these cells to undergo apoptosis, at least in part, by modifying their $[Ca^{2+}]_i$ [72]. KH1060, a 20 epi-vitamin D_3 analog (see Chapter 61), when concomitantly administered *in vitro* with 9-*cis*-retinoic acid, but not alone, also markedly stimulated the apoptotic process in HL-60 cells [73], presumably via VDR–RXR heterodimerization-mediated events [25,73]. In the latter cells, moreover, this combination markedly decreased the expression of *bcl-2* and increased the Bax : Bcl-2 ratio, thus, at least in part, accounting for their proapoptotic actions [73]. In agreement with these prior findings, in preliminary experiments our laboratory has found that $1,25(OH)_2D_3$ and 1,25-dihydroxy-16-ene-23-yne-26,27-hexafluoro vitamin D_3 caused CaCo-2 cells to undergo apoptosis (T. A. Brasitus and M. D. Sitrin, unpublished observations). On the basis of these findings, it would appear that the proapoptotic effects of these secosteroids may also play a potential role in their anticarcinogenic actions in the colon. Whether the latter effects are widespread or are confined to a few such cell types, as well as whether they involve alterations in $[Ca^{2+}]_i$, in *bcl-2* expression, or other mechanisms, however, are unclear at this time.

In this regard, however, as noted earlier, $1,25(OH)_2D_3$ has been shown to activate RAF-1 in other cells [90], and this MAPK kinase, in turn, has been shown to regulate the expression of *bcl-2* in MCF-7 cells [102]. In addition, Raf-1 is activated by PKC-α [91] and ceramide [92], both of which have been shown to be regulated by $1,25(OH)_2D_3$ in CaCo-2 and HL-60 cells, respectively [33,93]. Further studies along these lines will therefore be of interest in defining the exact mechanisms involved in the induction of apoptosis by these secosteroids.

Finally, it bears emphasis that, as described in Section III,C, our laboratory has described at least two potential anticarcinogenic actions of these secosteroids which have broad implications with respect to colonic tumor proliferation, differentation, and/or apoptosis. Indeed, we have demonstrated that the vitamin D status of animals modulated not only the frequency of tumors and their size in tumor-bearing animals in the DMH model [74], but also the frequency of k-*ras* mutations detectable in tumors induced by this carcinogen [103]. Although the mechanisms responsible for these phenomena remain unclear, these findings are potentially significant given the importance of these genetic mutations in the development of human colon cancer [103], presumably by leading to the dysregulation of one or more of the aforementioned fundamental cellular processes.

In addition, as noted in Section III,D, we have also shown that supplemental dietary 1α,25-dihydroxy-16-ene-23-yne-26,27-hexafluoro vitamin D_3 not only decreased the number of azoxymethane (AOM)-induced colonic tumors and totally abolished colonic carcinomas [69], but it also concomitantly preserved the expression of PKC-ζ in adenomas from animals fed this analog compared to tumors (adenomas and carcinomas) from carcinogen-treated, control fed rats [104]. These findings suggest that differences in the levels of expression of PKC-ζ between these adenomas might underlie the propensity of the adenomas from dietary unsupplemented animals to progress to carcinomas, at least in this model. In addition, we have demonstrated that two other structurally unrelated agents that are chemoprotective in the AOM model, namely, piroxicam [105,106], a nonsteroidal anti-inflammatory drug, and ursodeoxycholic acid [69,105], a secondary bile acid, also preserved PKC-ζ expression in colonic tumors induced by this carcinogen. The preservation of this PKC isoform by these structurally distinct chemopreventive agents, therefore, provides the most compelling, albeit circumstantial, evidence implicating this isoform in chemoprevention. Because dietary supplementation with each of the other two agents, like supplementation with 1α,25-dihydroxy-16-ene-23-yne-26,27-hexafluoro vitamin D_3, also abolished the development of carcinomas, this lends support to our contention that PKC-α down-regulation may be critical to the progression of adenomas to carcinomas. These findings are also of considerable interest in view of the results of a number of studies that have implicated PKC-α in the regulation of cell growth, differentiation, and possibly malignant transformation [107–112]. Additional studies to elucidate the potential role of PKC-α in the chemoprevention of vitamin D-related compounds in experimental models of colon cancer, as well as perhaps in human colon cancer, will therefore be of high interest.

References

1. Minghetti PP, Norman AW 1988 $1,25(OH)_2$-vitamin D_3 receptors: Gene regulation and genetic circuitry. FASEB J **2**:3043–3053.
2. Walters MR 1992 Newly identified actions of the vitamin D endocrine system. Endocr Rev **13**:719–763.
3. Lointier P, Meggouh F, Dechelotte P, Pezet D, Ferrier C, Chipponi J, Saez S 1991 1,25-Dihydroxyvitamin D_3 receptors and human colon adenocarcinoma. Br J Surg **78**:435–439.
4. Meggouh F, Lointier P, Pezet D, Saez S 1990 Evidence of 1,25-dihydroxyvitamin D_3-receptors in human digestive mucosa and carcinoma tissue biopsies taken at different levels of the digestive tract in 152 patients. J Steroid Biochem **36**:143–147.

5. Halline AG, Davidson NO, Skarosi SF, Sitrin MD, Tietze C, Alpers DH, Brasitus TA 1994 Effects of 1,25-dihydroxyvitamin D_3 on proliferation and differentiation of CaCo-2 cells. Endocrinology **134**:1710–1717.
6. Cross HS, Hulla W, Tong W-M, Peterlik M 1995 Growth regulation of human colon adenocarcinoma-derived cells by calcium, vitamin D and epidermal growth factor. J Nutr **125**:2004S–2008S.
7. Lointier P, Wargovich MJ, Saez S, Levin B, Wildrick DM, Boman BM 1987 The role of vitamin D_3 in the proliferation of a human colon cancer cell line in vitro. Anticancer Res **7**:817–821.
8. Wargovich MJ, Lointier PH 1987 Calcium and vitamin D modulate mouse colon epithelial proliferation and growth characteristics of a human colon tumor cell line. Can J Physiol Pharmacol **65**:472–477.
9. Zhao X, Feldman D 1993 Regulation of vitamin D receptor abundance and responsiveness during differentiation of HT-29 human colon cancer cells. Endocrinology **132**:1808–1814.
10. Tanaka Y, Bush K, Eguchi T, Ikekawa N, Taguchi T. Kobayashi Y, Higgins P 1990 Effects of 1,25-dihydroxyvitamin D_3 and its analogs on butyrate-induced differentiation of HT-29 human colonic carcinoma cells and on the reversal of the differentiated phenotype. Arch Biochem Biophys **276**:415–423.
11. Shabahang M, Buras RR, Davoodi F, Schumaker LM, Nauta RJ, Evans SR 1993 1,25-Dihydroxyvitamin D_3 receptor as a marker of human colon carcinoma cell line differentiation and growth inhibition. Cancer Res **53**:3712–3718.
12. Shabahang M, Buras RR, Davoodi F, Schumaker LM, Nauta RJ, Uskokovic MR, Brenner RV, Evans SR 1994 Growth inhibition of HT-29 human colon cancer cells by analogues of 1,25-dihydroxyvitamin D_3. Cancer Res **54**:4057–4064.
13. Hulla W, Kallay E, Krugluger W, Peterlik M, Cross HS 1995 Growth control of human colon-adenocarcinoma-derived Caco-2 cells by vitamin-D compounds and extracellular calcium in vitro: Relation to c-myc-oncogene and vitamin-D-receptor expression. Int J Cancer **62**:711–716.
14. Harper KD, Iozzo RV, Haddad JG 1989 Receptors for and bioresponses to 1,25-dihydroxyvitamin D in a human colon carcinoma cell line (HT-29). Metabolism **38**:1062–1069.
15. Giuliano AR, Franceschi RT, Wood RJ 1991 Characterization of the vitamin D receptor from the Caco-2 human colon carcinoma cell line: Effect of cellular differentiation. Arch Biochem Biophys **285**:261–269.
16. Cross HS, Huber C, Peterlik M 1991 Antiproliferative effect of 1,25-dihydroxyvitamin D_3 and its analogs on human colon adenocarcinoma cells (CaCo-2): Influence of extracellular calcium. Biochem Biophys Res Commun **179**:57–62.
17. Cross HS, Farsoudi KH, Peterlik M 1993 Growth inhibition of human colon adenocarcinoma-derived Caco-2 cells by 1,25-dihydroxyvitamin D_3 and two synthetic analogs: Relation to in vitro hypercalcemic potential. Naunyn Schmiedebergs Arch Pharmacol **347**:105–110.
18. Frampton RJ, Osmond SA, Eisman JA 1983 Inhibition of human cancer cell growth by 1,25-dihydroxyvitamin D_3 metabolites. Cancer Res **43**:4443–4447.
19. Brehier A, Thomasset M 1988 Human colon cell line HT-29: Characterisation of 1,25-dihydroxyvitamin D_3 receptor and induction of differentiation by the hormone. J Steroid Biochem **29**:265–270.
20. Tanaka Y, Bush KK, Klauck TM, Higgins PJ 1989 Enhancement of butyrate-induced differentiation of HT-29 human colon carcinoma cells by 1,25-dihydroxyvitamin D_3. Biochem Pharmacol **38**:3859–3865.
21. Sandgren M, Danforth L, Plasse TF, DeLuca HF 1991 1,25-Dihydroxyvitamin D_3 receptors in human carcinomas: A pilot study. Cancer Res **51**:2021–2024.
22. Feldman J, Federico MHH, Sonohara S, Katayama MLH, Koike MAA, Roela RA, Da Silva MRP, Brentani MM 1993 Vitamin D_3 binding activity during leukemic cell differentiation. Leuk Res **17**:97–101.
23. Berry DM, Antochi R, Bhatia M, Meckling-Gill KA 1996 1,25-Dihydroxyvitamin D_3 stimulates expression and translocation of protein kinase Cα and Cδ via a nongenomic mechanism and rapidly induces phosphorylation of a 33-kDa protein in acute promyelocytic NB4 cells. J Biol Chem **271**:16090–16096.
24. Norman AW, Okamura WH, Farach-Carson MC, Allewaer K, Branisteanu D, Nemere I, Muralidharan KR, Bouillon R 1993 Structure–function studies of 1,25-dihydroxyvitamin D_3 and the vitamin D endocrine system. J Biol Chem **268**:13811–13819.
25. Cheskis B, Lemon BD, Uskokovic M, Lomedico PT, Freedman LP 1995 Vitamin D_3–retinoid X receptor dimerization, DNA binding, and transactivation are differentially affected by analogs of 1,25-dihydroxyvitamin D_3. Mol Endocrinol **9**:1814–1824.
26. Haussler MR, Jurutka PW, Hsieh J-C, Thompson PD, Selznick SH, Haussler CK, Whitfield GK 1994 Receptor mediated genomic actions of $1,25(OH)_2D_3$: Modulation by phosphorylation. In: Norman AW, Bouillon R, Thomasset M (eds) Vitamin D. de Gruyter, New York, pp. 209–216.
27. Wali RK, Baum CL, Sitrin MD, Brasitus TA 1990 $1,25(OH)_2$ vitamin D_3 stimulates membrane phosphoinositide turnover, activates protein kinase C, and increases cytosolic calcium in rat colonic epithelium. J Clin Invest **85**:1296–1303.
28. Wali RK, Baum CL, Bolt MJG, Brasitus TA, Sitrin MD 1992 1,25-Dihydroxyvitamin D_3 inhibits Na^+–H^+ exchange by stimulating membrane phosphoinositide turnover and increasing cytosolic calcium in CaCo-2 cells. Endocrinology **131**:1125–1133.
29. Wali RK, Baum CL, Sitrin MD, Bolt MJ, Dudeja PK, Brasitus TA 1992 Effect of vitamin D status on the rapid actions of 1,25-dihydroxycholecalciferol in rat colonic membranes. Am J Physiol **262**:G945–G953.
30. Baum CL, Wali RK, Sitrin MD, Bolt MJG, Brasitus TA 1990 1,2-Dimethylhydrazine-induced alterations in protein kinase C activity in the rat preneoplastic colon. Cancer Res **50**:3915–3920.
31. Tien XY, Katnik C, Qasawa BM, Sitrin MD, Nelson DJ, Brasitus TA 1993 Characterization of the 1,25-dihydroxycholecalciferol-stimulated calcium influx pathway in CaCo-2 cells. J Membr Biol **136**:159–168.
32. Wali RK, Bolt MJ, Tien XY, Brasitus TA, Sitrin MD 1992 Differential effect of 1,25-dihydroxycholecalciferol on phosphoinositide turnover in the antipodal plasma membranes of colonic epithelial cells. Biochem Biophys Res Commun **187**:1128–1134.
33. Bissonnette M, Tien XY, Niedziela SM, Hartmann SC, Frawley BP Jr, Roy HK, Sitrin MD, Perlman RL, Brasitus TA 1994 $1,25(OH)_2$vitamin D_3 activates PKC-α in Caco-2 cells: A mechanism to limit secosteroid-induced rise in $[Ca^{2+}]_i$. Am J Physiol **267**:G465–G475.
34. Bissonnette M, Wali RK, Hartmann SC, Niedziela SM, Roy HK, Tien X-Y, Sitrin MD, Brasitus TA 1995 1,25-Dihydroxyvitamin D_3 and 12-*O*-tetradecanol phorbol 13-acetate cause differential activation of Ca^{2+}-dependent and Ca^{2+}-independent isoforms of protein kinase C in rat colonocytes. J Clin Invest **95**:2215–2221.
35. Wali RK, Bissonnette M, Starvarkos J, Sitrin MD, Brasitus TA 1996 $1,25(OH)_2D_3$ causes persistent activation of PKC-β_{II} in rat colonocytes and increases the baso-lateral membrane association of this isoform. Gastroenterology **110**:A1131.
36. Khare S, Wali RK, Roy H, Bolt M, Niedziela S, Scaglione-Sewell B, Sitrin MD, Brasitus TA, Bissonnette M 1996 $1,25(OH)_2D_3$

activates c-SRC and PLC-γ in rat colonocytes. Gastroenterology **110**:A1088.

37. Tien XY, Brasitus TA, Qasawa BM, Norman AW, Sitrin MD 1993 Effect of $1,25(OH)_2D_3$ and its analogues on membrane phosphoinositide turnover and $[Ca^{2+}]_i$ in Caco-2 cells. Am J Physiol **265**:G143–G148.
38. Mailloux RJ, Wali R, Brasitus TA, Sitrin MD 1994 Specific binding of 1,25-dihydroxyvitamin D_3 to rat colonocyte basolateral membranes. Gastroenterology **106**:A619.
39. Mailloux RJ, Bolt MJG, Wali RK, Brasitus TA, Sitrin MD 1994 Specific binding of 1,25-dihydroxyvitamin D_3 to rat colonocyte basolateral membranes. In: Norman AW, Bouillon R, Thomasset M (eds) Proceedings of the Ninth Workshop on Vitamin D, Volume 9. de Gruyter, New York, pp. 361–362.
40. Nemere I, Dormanen MC, Hammond MW, Okamura WH, Norman AW 1994 Identification of a specific binding protein for 1α,25-dihydroxyvitamin D_3 in basal–lateral membranes of chick intestinal epithelium and relationship to transcaltachia. J Biol Chem **269**:23750–23756.
41. Baran DT, Ray R, Sorensen AM, Honeyman T, Holick MG 1994 Binding characteristics of a membrane receptor that recognizes 1,25-dihydroxyvitamin D_3. J Cell Biochem **56**:510–517.
42. Slater SJ, Kelly MB, Taddeo FJ, Larkin JD, Yeager MD, McLane JA, Ho C, Stubbs CD 1995 Direct activation of protein kinase C by 1α,25-dihydroxyvitamin D_3. J Biol chem **270**:6639–6643.
43. Kane KF, Langman MJ, Williams GR 1995 1,25-Dihydroxyvitamin D_3 and retinoid X receptor expression in human colorectal neoplasms. Gut **36**:255–258.
44. Meggouh F, Lointier P, Saez S 1991 Sex steroid and 1,25-dihydroxyvitamin D_3 receptors in human colorectal adenocarcinoma and normal mucosa. Cancer Res **51**:1227–1233.
45. Garland CF, Garland FC 1980 Do sunlight and vitamin D reduce the likelihood of colon cancer? Int J Epidemiol **9**:227–231.
46. Emerson JC, Weiss NS 1992 Colorectal cancer and solar radiation. Cancer Causes and Control **3**:95–99.
47. Gorham ED, Garland CF, Garland FC 1989 Physical activity and colon cancer risk. Int J Epidemiol **18**:728–729.
48. Garland CF, Garland FC, Gorham Ed 1991 Can colon cancer incidence and death rates be reduced with calcium and vitamin D? Am J Clin Nutr **54**:193S–201S.
49. Spitz MR, Paolucci MJ, Newell GR 1988 Occurrence of colorectal cancer: Focus on Texas. Cancer Bull **40**:187–191.
50. Gorham ED, Garland CF, Garland FC 1989 Acid haze air pollution and breast and colon cancer mortality in 20 Canadian cities. Can J Public Health **80**:96–100.
51. Gorham ED, Garland CF, Garland FC 1990 Sunlight and breast cancer incidence in the USSR. Int J Epidemiol **19**:820–824.
52. Garland CF, Comstock GW, Garland FC, Helsing KJ, Shaw EK, Gorham Ed 1989 Serum 25-hydroxyvitamin D and colon cancer: Eight-year prospective study. Lancet **2**:1176–1178.
53. Gerhurdsson H, Huderus B, Norel SE 1988 Physical activity and colon cancer risk. Int J Epidemiol **17**:743–746.
54. Garland C, Shekelle RB, Barrett-Connor E, Criqui MH, Rossof AH, Paul O 1985 Dietary vitamin D and calcium and risk of colorectal cancer: A 19 year prospective study in men. Lancet **1**:307–309.
55. Bostick RM, Potter JD, Sellers TA, McKenzie DR, Kushi LH, Folsom AR 1993 Relation of calcium, vitamin D, and dairy food intake to incidence of colon cancer among older women. Am J Epidemiol **137**:1302–1317.
56. Kearney J, Giovannucci E, Rimm EB, Ascherio A, Stampfer MJ, Colditz GA, Wing A, Kampman E, Willett WC 1996 Calcium, vitamin D, and dairy foods and the occurrence of colon cancer in men. Am J Epidemiol **143**:907–917.
57. Kampman E, Giovannucci E, van't Veer P, Rimm E, Stampfer MJ, Dolditz GA, Kok FJ, Willett WC 1994 Calcium, vitamin D, dairy foods, and the occurrence of colorectal adenomas among men and women in two prospective studies. Am J Epidemiol **139**:16–29.
58. Braun MM, Helzlsouer KJ, Hollis BW, Comstock GW 1995 Colon cancer and serum vitamin D metabolite levels 10–17 years prior to diagnosis. Am J Epidemiol **142**:608–611.
59. Lashner BA 1993 Red blood cell folate is associated with the development of dysplasia and cancer in ulcerative colitis. J Cancer Res Clin Oncol **119**:549–554.
60. Tanaka Y, Wu A-YS, Ikekawa N, Iseki K, Kawai M, Kobayashi Y 1994 Inhibition of HT-29 human colon cancer growth under the renal capsule of severe combined immunodeficient mice by an analogue of 1,25-dihydroxyvitamin D_3 DD-003. Cancer Res **54**:5148–5153.
61. Kuribayashi T, Tanaka H, Abe E, Suda T 1983 Functional defect of variant clones of a human myeloid leukemia cell line (HL-60) resistant to 1α-25-dihydroxyvitamin D_3. Endocrinology **113**: 1992–1998.
62. Ostrem VK, Lau WF, Lee SH, Perlman K, Prahl J, Schnoes HK, DeLuca HF, Ikekawa N 1987 Induction of monocytic differentiation of HL-60 cells by 1,25-dihydroxyvitamin D analogs. J Biol Chem **262**:14164–14171.
63. Bhatia M, Kirkland JB, Meckling-Gill KA 1995 Monocytic differentiation of acute promyelocytic leukemia cells in response to 1,25-dihydroxyvitamin D_3 is independent of nuclear receptor binding. J Biol Chem **270**:15962–15965.
64. Bikle DD 1992 Vitamin D: New actions, new analogs, new therapeutic potential. Endocr Rev **13**:765–784.
65. Kawa S, Yoshizawa K, Tokoo M, Imai H, Oguchi H, Kiyosawa K, Homma T, Nikaido T, Furihata K 1996 Inhibitory effect of 22-oxa-1,25-dihydroxyvitamin D_3 on the proliferation of pancreatic cancer cell lines. Gastroenterology **110**:1605–1613.
66. Pols HA, Birkenhager JC, van Leeuwen JPTM 1994 Vitamin D analogues: From molecule to clinical application. Clin Endocrinol **40**:285–292.
67. Brown AJ, Dusso A, Slatopolsky E 1994 Selective vitamin D analogs and their therapeutic applications. Semin Nephrol **14**:156–174.
68. Anzano MA, Smith JM, Uskokovic MR, Peer CW, Mullen LT, Letterio JJ, Welsh MC, Shrader MW, Logsdon DL, Driver CL, Brown CC, Roberts AB, Sporn MB 1994 1α,25-Dihydroxy-16-ene-23-yne-26,27-hexafluorocholecalciferol (RO24-5531), a new deltanoid (vitamin D analogue) for prevention of breast cancer in the rat. Cancer Res **54**:1653–1656.
69. Wali RK, Bissonnette M, Khare S, Hart J, Sitrin MD, Brasitus TA 1995 1α,25-Dihydroxy-16-ene-23-yne-26,27-hexafluorocholecalciferol, a noncalcemic analogue of 1α,25-dihydroxyvitamin D_3, inhibits azoxymethane-induced colonic tumorigenesis. Cancer Res **55**:3050–3054.
70. Dilworth FJ, Calverley MJ, Makin HL, Jones G 1994 Increased biological activity of 20-epi-1,25-dihydroxyvitamin D_3 is due to reduced catabolism and altered protein binding. Biochem Pharmacol **47**:987–993.
71. Bouillon R, Allewaert K, van Leeuwen JPTM, Tan BK, Xiang DZ, De Clereq P, Vandewalle M, Pols HA, Bos MP, Van Baelen H, Birkenhager JC 1992 Structure function analysis of vitamin D analogs with C-ring modifications. J Biol Chem **267**:3044–3051.
72. Vandewalle B, Hornez L, Wattez N, Revillion F, Lefebvre J 1995 Vitamin-D_3 derivatives and breast-tumor cell growth: Effect on intracellular calcium and apoptosis. Int J Cancer **61**:806–811.
73. Elstner E, Linker-Israeli M, Umiel T, Le J, Grillier I, Said J, Shintaku IP, Krajewski S, Reed JC, Binderup L, Koeffler HP

1996 Combination of a potent 20-epi-vitamin D_3 analogue (KH 1060) with 9-*cis*-retinoic acid irreversibly inhibits clonal growth, decreases bcl-2 expression, and induces apoptosis in HL-60 leukemic cells. Cancer Res **56**:3570–3576.

74. Sintrin MD, Halline AG, Abrahams C, Brasitus TA 1991 Dietary calcium and vitamin D modulate 1,2-dimethylhydrazine-induced colonic carcinogenesis in the rat. Cancer Res **51**:5608–5613.
75. Pence BC, Buddingh F 1988 Inhibition of dietary fat-promoted colon carcinogenesis in rats by supplemental calcium or vitamin D_3. Carcinogenesis **9**:187–190.
76. Belleli A, Shany S, Levy J, Guberman R, Lamprecht SA 1992 A protective role of 1,25-dihydroxyvitamin D_3 in chemically induced rat colon carcinogenesis. Carcinogenesis **13**:2293–2298.
77. Comer PF, Clark TD, Glauert HP 1993 Effect of dietary vitamin D_3 (cholecalciferol) on colon carcinogenesis induced by 1,2-dimethylhydrazine in male Fischer 344 rats. Nutr Cancer **19**: 113–124.
78. Beaty MM, Lee EY, Glauert HP 1993 Influence of dietary calcium and vitamin D on colon epithelial cell proliferation and 1,2-dimethylhydrazine-induced colon carcinogenesis in rats fed high fat diets. J Nutr **123**:144–152.
79. Otoshi T, Iwata H, Kitano M, Nishizawa Y, Morii H, Yano Y, Otani S, Fukushima S 1995 Inhibition of intestinal tumor development in rat multi-organ carcinogenesis and aberrant crypt foci in rat colon carcinogenesis by 22-oxa-calcitriol, a synthetic analogue of 1α,25-dihydroxyvitamin D_3. Carcinogenesis **16**:2091–2097.
80. Kawaura A, Takahashi A, Tanida N, Oda M, Sawada K, Sawada Y, Maekawa S, Shimoyama T 1990 1α-Hydroxyvitamin D_3 suppresses colonic tumorigenesis induced by repetitive intrarectal injection of *N*-methyl-*N*-nitrosourea in rats. Cancer Lett **55**: 149–152.
81. Kawaura A, Tanida N, Sawada K, Oda M, Shimoyama T 1989 Supplemental administration of 1α-hydroxyvitamin D_3 inhibits promotion by intrarectal instillation of lithocholic acid in *N*-methyl-*N*-nitrosourea-induced colonic tumorigenesis in rats. Carcinogenesis **10**:647–649.
82. Thomas MG, Tebbutt S, Williamson RCN 1992 Vitamin D and its metabolites inhibit cell proliferation in human rectal mucosa and a colon cancer cell line. Gut **33**:1660–1663.
83. Thomas MG, Nugent KP, Forbes A, Williamson RC 1994 Calcipotriol inhibits rectal epithelial cell proliferation in ulcerative proctocolitis. Gut **35**:1718–1720.
84. Godyn JJ, Xu H, Zhang F, Kolla S, Studzinski GP 1994 A dual block to cell cycle progression in HL60 cells exposed to analogues of vitamin D_3. Cell Proliferation **27**:37–46.
85. Wang QM, Jones JB, Studzinski GP 1996 Cyclin-dependent kinase inhibitor p27 as a mediator of the G1–S phase block induced by 1,25-dihydroxyvitamin D_3 in HL60 cells. Cancer Res **56**: 264–267.
86. Toyoshima H, Hunter T 1994 p27, a novel inhibitor of G1 cyclin–Cdk protein kinase activity, is related to p21. Cell **78**:67–74.
87. Ohtsubo M, Roberts JM 1993 Cyclin-dependent regulation of G1 in mammalian fibroblasts. Science **259**:1908–1912.
88. Kobayashi T, Hashimoto K, Yoshikawa K 1993 Growth inhibition of human keratinocytes by 1,25-dihydroxyvitamin D_3 is linked to dephosphorylation of retinoblastoma gene product. Biochem Biophys Res Commun **196**:487–493.
89. Blagosklonny MV, Schulte TW, Nguyen P, Mimnaugh EG, Trepel J, Neckers L 1995 Taxol induction of p21WAF1 and p53 requires c-raf-1. Cancer Res **55**:4623–4626.
90. Beno DWA, Brady LM, Bissonnette M, Davis BH 1995 Protein kinase C and mitogen-activated protein kinase are required for 1,25-dihydroxyvitamin D_3-stimulated *egr* induction. J Biol Chem **270**:3642–3647.
91. Kolch W, Heldecker G, Kochs G, Hummel R, Vahidi H, Mischak H, Finkenzeller G, Marmé D, Rapp UR 1993 Protein kinase C α activates RAF-1 by direct phosphorylation. Nature **364**:249–252.
92. Belka C, Wiegmann K, Adam D, Holland R, Neuloh M, Herrmann F, Kronke M, Brach MA 1995 Tumor necrosis factor (TNF)-α activates c-raf-1 kinase via the p55 TNF receptor engaging neutral sphingomyelinase. EMBO J **14**:1156–1165.
93. Okazaki T, Bell RM, Hannun YA 1989 Sphingomyelin turnover induced by vitamin D_3 in HL-60 cells. Role in cell differentiation. J Biol Chem **264**:19076–19080.
94. Scaglione-Sewell BA, Abraham C, Davidson NO, Davis B, Bissonnette M, Brasitus TA 1995 Alterations in expression of protein kinase C-α in CaCo-2 cells result in changes in cellular proliferation. Gastroenterology **108**:A534.
95. Scaglione-Sewell BA, Abraham C, Davidson NO, Davis B, Bissonnette M, Brasitus TA 1996 Alterations in expression of protein kinase Cα in CaCo-2 cells results in changes in both cell growth and contact inhibition. Gastroenterology **110**:A588.
96. Levy R, Nathan I, Chaimovitz C, Shany S 1987 The involvement of calcium ions in the effect of 1,25-dihydroxyvitamin D_3 on HL-60 cells. In: Hurvits S, Sela J (eds) Current Advances in Skeletogenesis. Heiliger Publishing House, Jerusalem, pp. 240–249.
97. Levy R, Nathan I, Barnea E, Chaimovitz C, Shany S 1988 The involvement of calcium ions in the effect of 1,25-dihydroxyvitamin D_3 on HL-60 cells. Exp Hematol **16**:290–294.
98. Hazav P, Shany S, Moran A, Levy R 1989 Involvement of intracellular pH elevation in the effect of 1,25-dihydroxyvitamin D_3 on HL-60 cells. Cancer Res **49**:72–75.
99. Obeid LM, Okazaki T, Karolak LA, Hannun YA 1990 Transcriptional regulation of protein kinase C by 1,25-dihydroxyvitamin D_3 in HL-60 cells. J Biol Chem **265**:2370–2374.
100. Simpson RU, Hsu T, Begley DA, Mitchell BS, Alizadeh BN 1987 Transcriptional regulation of the c-myc protooncogene by 1,25-dihydroxyvitamin D_3 in HL-60 promyelocytic leukemia cells. J Biol Chem **262**:4104–4108.
101. Solomon DH, O'Driscoll K, Sosne G, Weinstein IB, Cayre YE 1991 1α,25-Dihydroxyvitamin D_3-induced regulation of protein kinase C gene expression during HL-60 cell differentiation. Cell Growth Differ **2**:187–194.
102. Blagosklonny MV, Schulte T, Nguyen P, Trepel J, Neckers LM 1996 Taxol-induced apoptosis and phosphorylation of Bcl-2 protein involves c-Raf-1 and represents a novel c-Raf-1 signal transduction pathway. Cancer Res **56**:1851–1854.
103. Llor X, Jacoby RF, Teng BB, Davidson NO, Sitrin MD, Brasitus TA 1991 K-ras mutations in 1,2-dimethylhydrazine-induced colonic tumors: Effects of supplemental dietary calcium and vitamin D deficiency. Cancer Res **51**:4305–4309.
104. Wali R, Bissonnette M, Khare S, Aquino B, Niedziela S, Sitrin M, Brasitus T 1996 Protein kinase C isoforms in the chemopreventive effects of a novel vitamin D_3 analogue in rat colonic tumorigenesis. Gastroenterology **111**:118–126.
105. Earnest DL, Holubec H, Wali RK, Jolley CS, Bissonette M, Bhattacharyya AK, Roy H, Khare S, Brasitus TA 1994 Chemoprevention of azoxymethane-induced colonic carcinogenesis by supplemental dietary ursodeoxycholic acid. Cancer Res **54**:5071–5074.
106. Roy HK, Bissonnette M, Frawley BP, Jr, Wali RK, Niedziela SM, Earnest D, Brasitus TA 1995 Selective preservation of protein kinase C-ζ in the chemoprevention of azoxymethane-induced colonic tumors by piroxicam. FEBS Lett **366**:143–145.
107. Diaz-Meco MT, Berra E, Municio MM, Sanz L, Lozano J, Dom-

inguez I, Diaz-Golpe V, Lain de Lera MT, Alcami J, Paya CV, Arenzana-Seisdedos F, Virelizier J-L, Moscat J 1993 A dominant negative protein kinase C ζ subspecies blocks NF-κB activation. Mol Cell Biol **13**:4770–4775.

108. Dominguez I, Diaz-Meco MT, Municio MM, Berra E, De Herreros AG, Cornet ME, Sanz L, Moscat J 1992 Evidence for a role of protein kinase C ζ subspecies in maturation of *Xenopus laevis* oocytes. Mol Cell Biol **12**:3776–3783.
109. Nakanishi H, Brewer KA, Exton JH 1993 Activation of the ζ isoform of protein kinase C by phosphatidylinositol 3,4,5-trisphosphate. J Biol Chem **268**:13–16.
110. Diaz-Meco MT, Lozano J, Municio MM, Berra E, Frutos S, Sanz L, Moscat J 1994 Evidence for the *in vitro* and *in vivo* interaction of Ras with protein kinase C ζ. J Biol Chem **269**:31706–13710.
111. Ways DK, Posekany K, deVente J, Garris T, Chen J, Hooker J, Qin W, Cook P, Fletcher D, Parker P 1994 Overexpression of protein kinase C-ζ stimulates leukemic cell differentiation. Cell Growth Differ **5**:1195–1203.
112. Wooten MW, Zhou G, Seibenhener ML, Coleman ES 1994 A role for ζ protein kinase C in nerve growth factor-induced differentiation of PC12 cells. Cell Growth Differ **5**:395–403.

CHAPTER 68

Vitamin D Compounds and Analogs: Effects on Normal and Abnormal Hematopoiesis

ROBERTA MOROSETTI Instituto di Semeiotica Medica, Universita Cattolica del S. Cuore, Policlinico Agostino Gemelli, Rome, Italy

H. PHILLIP KOEFFLER Division of Hematology/Oncology, Department of Medicine, Cedars–Sinai Medical Center, UCLA School of Medicine, Los Angeles, California

I. OVERVIEW OF HEMATOPOIESIS

Hematopoiesis is the process that leads to the formation of the highly specialized circulating blood cells from bone marrow pluripotent progenitor stem cells. These stem cells are the most primitive blood cells, and they have the ability to self-replicate or differentiate. They are regulated by a feedback system and are affected by various stimuli such as bone marrow depletion, hemorrhage, infection, and stress. They produce more mature "committed" cells that are able to proliferate and differentiate into cells of different lineages, acquiring specific functional properties (Fig. 1).

The pluripotent stem cell common to granulocytes, erythrocytes, monocytes, and megakaryocytes is called the colony-forming unit-GEMM (CFU-GEMM), and the committed cells giving rise to the lineage specific cells are assayed *in vitro* as erythroid burst-forming units (BFU-E), megakaryocyte colony-forming units (CFU-MK), and granulocyte–monocyte colony-forming units (CFU-GM). Each of these stem cells has cell surface receptors for specific cytokines. Binding of cytokine to these receptors stimulates secondary intracellular signals that deliver a message to the nucleus to stimulate proliferation, differentiation, and/or activation.

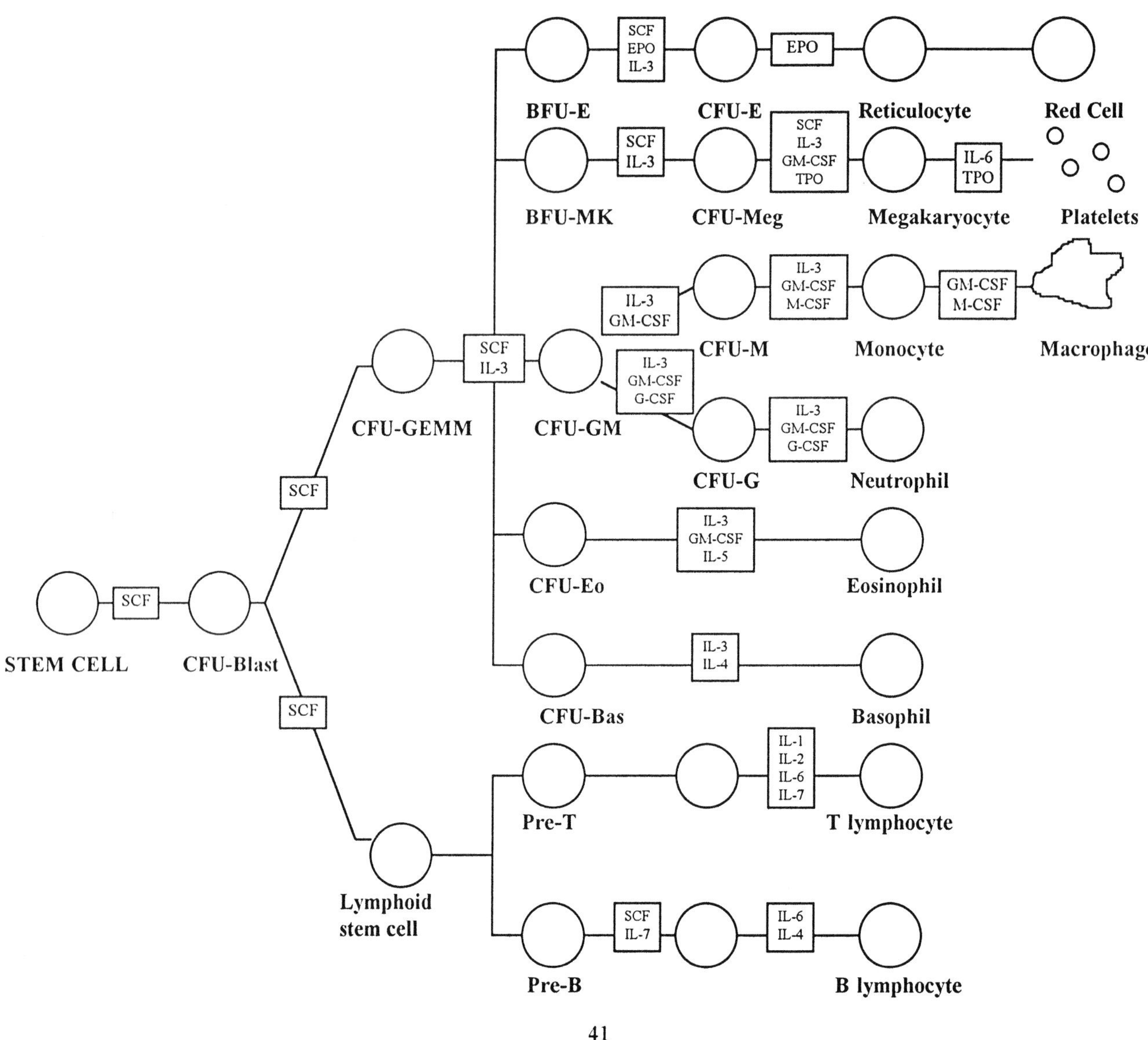

FIGURE 1 Scheme of hematopoiesis. The key progenitor cells and their growth factors are shown. CFU-Blast, colony-forming unit—blast; CFU-GEMM, CFU—granulocyte, erythrocyte, megakaryocyte, macrophage; BFU-E, burst-forming unit—erythroid; CFU-E, CFU—erythroid; BFU-MK, BFU—megakaryocyte; CFU-Meg, CFU—megakaryocyte; CFU-GM, CFU—granulocyte–monocyte; CFU-Eo, CFU—eosinophil; CFU-Bas, CFU—basophil; SCF, stem cell factor; IL-3, interleukin-3; GM-CSF, granulocyte—monocyte colony-stimulating factor; EPO, erythropoietin; TPO, thrombopoietin.

The CFU-GMs, in the presence of cytokines, undergo a differentiation program progressing to granulocytes and monocytes. The growth factors acting primarily on the granulocyte–macrophage pathway are granulocyte–macrophage colony-stimulating factor (GM-CSF), granulocyte colony-stimulating factor (G-CSF), and macrophage colony-stimulating factor (M-CSF). The GM-CSF also stimulates eosinophils, enhances megakaryocytic colony formation, and increases erythroid colony formation in the presence of erythropoietin (Epo). *In vivo*, it causes an increase in granulocytes, monocytes, and eosinophils. It can activate these cells to efficiently fight microbes.

The G-CSF stimulates the formation of granulocyte colonies *in vitro*. It is able to act synergistically with interleukin-3 (IL-3), GM-CSF, and M-CSF. It is active *in vivo*, stimulating an increase of peripheral blood granulocytes count.

The M-CSF stimulates the formation of macrophage colonies *in vitro*. It maintains the survival of differentiated macrophages *in vitro* and increases their antitumor activities and secretion of oxygen reduction products and plasminogen activators. This cytokine binds to a receptor that is the product of the protooncogene c-*fms*.

Interleukin-3 has multilineage stimulating activity and acts directly on the granulocyte–macrophage pathway, but with an important enhancement of development of erythroid, megakaryocytic, and mast cells, and possibly T lymphocytes. In synergy with Epo, IL-3 stimulates the formation of early erythroid stem cells, promoting the formation of colonies of red cells in soft gel culture known as BFU-E. In addition, it supports the formation of early multilineage blast cells *in vitro*. IL-3 also induces leukemic blasts to enter the cell cycle and induces, either alone or in combination with other growth factors, the production of all myeloid cells *in vivo*.

The growth factor Epo stimulates the formation of erythroid colonies (CFU-E) *in vitro* and is the primary hormone of erythropoiesis in animals and humans *in vivo*. It binds to a specific receptor (Epo-R). Production of erythroblasts is hormonally regulated by a feedback mechanism mediated by the linear correlation between tissue oxygenation of Epo-producing cells in the kidney mediated by oxygen-carrying hemoglobin in red blood cells. Anemia causes tissue hypoxia, resulting in an increase of serum Epo levels.

II. VITAMIN D RECEPTORS IN BLOOD CELLS

Since the mid-1980s, many target tissues of 1,25-dihydroxyvitamin D [$1,25(OH)_2D$] have been found that contain functioning vitamin D receptors (VDRs). The VDR belongs to the superfamily of steroid–thyroid receptors; it mediates the action of its ligand $1,25(OH)_2D$ and regulates gene transcription by binding to specific responsive elements (VDREs) as a heterodimer with retinoid X receptor (VDR–RXR) [1–5]. The human VDR was cloned and sequenced by Baker *et al.* [6]. The VDR is highly homologous to the thyroid receptor and is characterized by a highly conserved DNA binding region rich in cysteine, lysine, and arginine and a carboxyl-terminal hydrophobic ligand binding region. After binding to its ligand, this receptor activates the transcription of specific genes. The mechanism of action of $1,25(OH)_2D_3$ via the VDR is discussed in detail in Chapters 8–10.

Cells of the hematolymphopoietic tissue contain VDR. It is expressed constitutively in monocytes, in certain subsets of thymocytes, and after *in vitro* activation of B and T lymphocytes [7–10]. Expression of VDR is induced in the lymphocytes of patients with rheumatoid arthritis, in human tonsillar lymphocytes, and in pulmonary lymphocytes of patients with tuberculosis and sarcoidosis [11–13]. In addition, lymphocytes of patients with hereditary vitamin D-resistant rickets type II (HVDRR) have various alterations of the VDR [14,15]. Also, fewer receptors have been detected in the peripheral blood mononuclear cells of patients with X-linked hypophosphatemic rickets [16]. Examination of a large array of myeloid leukemia cell lines blocked at various stages of maturation showed that they all expressed VDR, albeit at different levels [9]. Bone marrow-derived stromal cells express VDR and show a reduction of their proliferation that occurs after their exposure to $1,25(OH)_2D_3$. Both T-helper and T-suppressor lymphocytes express similar concentrations of VDR. In particular, T lymphocytes express high levels of VDR mRNA, whereas resting B lymphocytes express very low or nondetectable levels of VDR transcripts [9,17]. Nevertheless, $1,25(OH)_2D_3$ inhibits the synthesis of immunoglobulins (Ig) by B lymphocytes *in vitro* [18,19] This suppression, however, could be the result of the inhibition of T-helper activity [19]. Production of lymphokines, including IL-2, is markedly decreased by $1,25(OH)_2D_3$ in activated T lymphocytes, and this could cause the suppression of Ig synthesis [20–23]. The effects of vitamin D on the immune system are discussed in Chapters 29, 69, and 70.

The HL-60 myeloblastic cell line cultured in the presence of $1,25(OH)_2D_3$ (10^{-7} *M*) has a 50% decrease of VDR protein levels at about 24 hr which returned to normal levels after 72 hr; no change of VDR mRNA expression occurred [9]. These data suggested that a major site of regulation of VDR expression occurs at the posttranscriptional level. The same cell line exposed to a lower dose of $1,25(OH)_2D_3$ for 12 hr appeared to have an increased number of VDRs, as determined by immunoprecipitation, which returned to normal levels after 72 hr [24].

The HL-60 myeloblasts cultured with retinoic acid (RA) and dimethyl sulfoxide (DMSO) or 12-*O*-tetradecanoylphorbol-13-acetate (TPA) terminally differentiate into granulocytes or macrophages, respectively. The differentiation is associated with induction of high expression of VDR mRNA. Also, normal human nondividing macrophages express VDR mRNA, and these levels do not change after exposure to activating factors such as tumor necrosis factor-α (TNFα).

The expression of VDR mRNA was not detectable in nonproliferating lymphocytes harvested from normal human peripheral blood, but VDR mRNA expression increased in proliferating lymphocytes after a 24 hr exposure to the lectin phytohemagglutinin-A (PHA), suggesting that in lymphocytes a major site of regulation of VDR expression is at the transcriptional level [9,25]. Moreover, low levels of VDR expression were detected in low-grade non-Hodgkin's lymphoma (NHL) tumor samples and in the follicular lymphoma B-cell lines SU-DHL4 and SU-DHL5 [26].

The VDR can bind to the osteocalcin response element along with the activator protein-1 (AP1) complexes [27]. In addition, Jun and Fos protooncogenes are up-regulated by $1,25(OH)_2D_3$ [28]. Jun-D DNA binding activity is increased during cell cycle arrest in the human chronic myelogenous leukemia RWLeu-4 cultured with $1,25(OH)_2D_3$, suggesting that Jun D binding activity may play a role in the regulation of cell proliferation by $1,25(OH)_2D_3$ [27].

III. MACROPHAGES CAN MAKE $1,25(OH)_2D$; PATHOLOGY OF THESE CELLS CAN RESULT IN HYPERCALCEMIA

The most active form of vitamin D is $1,25(OH)_2D$ and it plays a fundamental role in the regulation of calcium metabolism in humans [29]. The major source of production of $1,25(OH)_2D$ is the kidney which converts 25OHD to $1,25(OH)_2D$ with the enzyme 25-hydroxy-D-1α-hydroxylase (see Chapter 5). The 1α-hydroxylase activity is tightly regulated by serum calcium levels and parathyroid hormone (PTH) [30]. Extrarenal synthesis of $1,25(OH)_2D$ can occur under both physiological and pathological conditions. Ectopic production of $1,25(OH)_2D$ has been demonstrated by Gray [31] during pregnancy by the placenta and decidua, probably contributing to the increase of plasma levels of $1,25(OH)_2D$ detectable in this condition.

In 10 to 20% of patients with sarcoidosis, hypercalcemia is a typical complication, and these patients can have elevated serum levels of $1,25(OH)_2D$. This subject is discussed in detail in Chapter 55. Reports of hypoparathyroid and anephric patients with sarcoidosis having high serum levels of $1,25(OH)_2D$ suggested that this could have an extrarenal origin [32]. Pulmonary alveolar macrophages (PAM) from patients with sarcoidosis can synthesize $1,25(OH)_2D$ in culture [33,34]. Pulmonary T lymphocytes in sarcoidosis patients often synthesize increased levels of IL-2. Therefore, we examined the capacity of normal human macrophages (BMM) to metabolize 25-hydroxyvitamin D_3 ($25OHD_3$) and showed that these cells and PAM after exposure to interferon-γ (IFN-γ) were able to produce a metabolite identified as $1,25(OH)_2D_3$ by several techniques including mass spectroscopy. Furthermore, the purified material had the ability to induce the differentiation of the HL-60 myeloblastic cell line down the macrophage pathway and inhibit IFN-γ production by normal human lymphocytes [35]. These results together with the results obtained from expression studies of VDR in macrophages and T lymphocytes, described above, would suggest a possible interaction between macrophages and T lymphocytes in a paracrine fashion using $1,25(OH)_2D_3$ and IFN-γ as signals [30,35,36]. In another study, we tested the effects of lipopolysaccharide (LPS), a microbial product, on $25OHD_3$ metabolism of hematopoietic cells, including macrophages, and showed that LPS stimulated normal human monocytes and PAM to synthesize a metabolite identical to $1,25(OH)_2D_3$ [22].

The 1α-hydroxylase in activated macrophages is comparable to its renal counterpart in substrate specificity; on the other hand, however, it is inducible by interferon-γ (IFN-γ) and bacterial LPS, and it is not sensitive to PTH [30]. Also, we showed that in the presence of $1,25(OH)_2D_3$, activated macrophages selectively synthesized $24,25(OH)_2D_3$, suggesting that macrophages were sensitive to the feedback control by $1,25(OH)_2D_3$.

Other granulomatous disorders can be accompanied by an alteration of calcium metabolism. These include extrarenal production of $1,25(OH)_2D_3$ in tuberculosis, end-stage renal disease, leprosy, disseminated candidiasis, silicone-induced granuloma, and lymphomas [37,38].

IV. EFFECTS OF $1,25(OH)_2D_3$ ON NORMAL AND LEUKEMIC HEMATOPOIESIS

A. $1,25(OH)_2D_3$ and Normal Stem Cells

The role of $1,25(OH)_2D_3$ in cell differentiation was first described by Abe *et al.* [39] in the murine leukemia cell line M1, which was induced to differentiate into more mature cells by $1,25(OH)_2D_3$. Normal human bone marrow committed stem cells cultured in soft agar with $1,25(OH)_2D_3$ (10^{-7} *M*) or in liquid culture with $1,25(OH)_2D_3$ (5×10^{-9} *M* for 5 days) and monocytes cultured in serum-free medium with $1,25(OH)_2D_3$ (10^{-8} *M* for 7 days) differentiate into macrophages [40,41]. In further studies, these macrophages were functionally competent and able to release large amounts of TNFα and IL-6 and showed reduction of IL-1 [42]. Furthermore, the terminal differentiation of monocytes into mature macrophages can be obtained *in vitro* by culturing these cells in the presence of serum or in a serum-free medium with the addition of vitamin D_3 compounds [42–44].

As mentioned earlier, $1,25(OH)_2D_3$ is able to inhibit IL-2 synthesis and the proliferation of peripheral blood lymphocytes [19,21,22,45]. Indeed, $1,25(OH)_2D_3$ appears to be able to regulate many lymphokines. For example, Tobler *et al.* [46] showed that expression of the lymphokine GM-CSF is regulated by $1,25(OH)_2D_3$ through cellular VDR by a process independent of IL-2 production. In particular, it was demonstrated that $1,25(OH)_2D_3$ is able to inhibit both GM-CSF mRNA and protein expression in PHA-activated normal human peripheral blood lymphocytes (PBL). This down-regulation was obtained at concentrations similar to those reached *in vivo*. A 50% reduction of GM-CSF activity occurred at 10^{-10} *M* $1,25(OH)_2D_3$. In addition, IL-2 did not affect the modulation of GM-CSF production by $1,25(OH)_2D_3$ in the PBL cocultured with $1,25(OH)_2D_3$ (10^{-10}–10^{-7} *M*) and high concentrations of IL-2. Interestingly, the $1,25(OH)_2D_3$ decreased accumulation of GM-CSF mRNA; this occurred at least in part by destabilizing and shortening the half-life of the GM-CSF mRNA [46].

B. $1,25(OH)_2D_3$ and Leukemia Cell Lines

The derivative $1,25(OH)_2D_3$ and related vitamin D_3 compounds have a similar, potent effect on inducing differentiation and inhibiting proliferation of several acute myeloid leukemia cell lines such as HL-60, U937, THP-1, HEL, and NB4. In contrast, more immature myeloid leukemia cell lines such as HL-60 blasts, KG1, KG1a, and K562 do not respond to the metabolite. HL-60 cells treated with $1,25(OH)_2D_3$ acquire the morphology and functional characteristics of macrophages (Table I). They become adherent to charged surfaces, develop pseudopodia, stain positively for nonspecific esterase (NSE) with a reduction of nitroblue tetrazolium (NBT), and acquire the ability to phagocytose yeast during incubation with $1,25(OH)_2D_3$ (10^{-10}–10^{-7} *M* for 7 days) [41,47]. In addition, the treated cells acquired the ability to degrade bone marrow matrix *in vitro*, raising the possibility that the cells may have acquired some osteoclast-like characteristics. The proliferation of HL-60 cells is also inhibited by $1,25(OH)_2D_3$; in fact, colony formation in soft agar is reduced by 50% (ED_{50}) in the presence of about 10^{-9} *M* $1,25(OH)_2D_3$ [48]. Cells of other leukemia cell lines, such as U937, HEL, THP-1, and M1, are also inhibited in their clonal growth after their exposure to $1,25(OH)_2D_3$, with ED_{50} values ranging from 4×10^{-9} to 3×10^{-8} *M* [48]. Scatchard analysis data suggest to some investigators that the diverse responsiveness of the cell lines is due to the higher number of receptors present on the more differentiated cell lines [47]. However, further confirmation of this hypothesis is required.

Leukemic cells from acute myelogenous leukemia (AML) patients respond to vitamin D_3 compounds when cultured *in vitro*; however, they are often less sensitive than the cell lines. They are still able to undergo monocytic differentiation assessed by NBT reduction, morphology, and phagocytic ability. Furthermore, their clonal growth is often inhibited [41,49].

Because of the potential toxicity of $1,25(OH)_2D_3$ at the concentrations required *in vivo*, various attempts have been made to use $1,25(OH)_2D_3$ with another compound that might act synergistically for an antileukemic effect capable of promoting cell differentiation yet with an acceptable toxicity (Table II). Vitamin D_3 compounds may cooperate with other differentiating agents

TABLE I Relative Competitive Binding to Vitamin D-Binding Protein (DBP) and VDR by Vitamin D Compounds and Analogs, and Their Effect on Proliferation and Differentiation of the HL-60 Leukemic Cell Line[a]

Compound	RCI		ED_{50} ($\times 10^{-9}$ mol/liter)	
	DBP	VDR	Clonal growth inhibition	NBT
$1,25(OH)_2D_3$	100	100	16	37
$1,25(OH)_2$-16-ene-D_3	1.8	147	0.015	9
$1,25(OH)_2$-16-ene-23-yne-D_3	33	79	0.02	1
KH1060		120	0.005	40
MC903	3	50–100		

[a] DBP RCI, Relative competitive index for the analog binding to the vitamin D-binding protein (the RCI for $1,25(OH)_2D_3$ is set at 100%); VDR RCI, relative competitive index for the analog binding VDR (the RCI for $1,25(OH)_2D_3$ is set at 100%). Clonal growth inhibition and nitroblue tetrazolium (NBT) reduction (measurement of differentiation) used HL-60 myeloblasts as target cells. From Elstner [51,101], Pakkala [95], Jung [99], and Brown [91].

TABLE II Vitamin D Compounds and Analogs That Enhance Myeloid Leukemia Cell Lines to Differentiate to Monocytes

Cell line	Agent	Enhancer[a]
HL-60	$1,25(OH)_2D_3$	IFN-γ, ATRA, 9-*cis*-RA
	$1,25(OH)_2$-16-ene-23-yne D_3	ATRA
	$1,25(OH)_2$-20-epi D_3	
	KH1060	9-*cis*-RA
U937	$1,25(OH)_2D_3$	ATRA
	KH1060	
	Calcipotriol	
THP-1	$1,25(OH)_2D_3$	
NB4	$1,25(OH)_2D_3$	TPA

[a] Compounds that enhance the activity of the agent (i.e., clonal growth inhibition and induction of differentiation) are listed. IFN, Interferon; 9-*cis*-RA, 9-*cis*-retinoic acid; ATRA, all-*trans*-retinoic acid; TPA, 12-*O*-tetradecanoylphorbolacetate.

such as retinoids, tissue plasminogen activator (TPA), and interferons. For example, $1,25(OH)_2D_3$ can potentiate IFN-γ action to induce the expression of CD11b and CD14. Also, the combination of $1,25(OH)_2D_3$ and either all-*trans*-retinoic acid (ATRA) or 9-*cis*-retinoic acid (9-*cis*-RA) can potentiate the terminal differentiation process of HL-60 cells down the monocyte–macrophage pathway (~100% cells $CD14^+$) [50,51]. These findings have also been demonstrated by other investigators [52,53]. Cells cultured in the presence of the combination of $1,25(OH)_2D_3$ and ATRA developed atypically, having a neutrophilic morphology, but in other properties were typical of monocytes (e.g., CD14 expression, ability to bind to bacterial LPS, and ability to develop sodium fluoride-inhibited NSE) [50,51].

A synergistic effect of ATRA and $1,25(OH)_2D_3$ to induce monocytic differentiation was also observed in the promonocytic cell line U937 [54]. The same group observed that U937 cells exposed to a moderate thermal stress responded with increased differentiation after the addition of $1,25(OH)_2D_3$ and ATRA [55].

In the promyelocytic cell line NB4, carrying the translocation t(15;17) typical of acute promyelocytic leukemia (APL), $1,25(OH)_2D_3$ as single agent was weakly able to induce monocytic differentiation [56]. Bathia *et al.* [57] showed that the combination of $1,25(OH)_2D_3$ and TPA resulted in a synergistic response in NB4 cells, which exhibited complete differentiation to fully functional macrophages with a rapid arrest of cell growth in the first 24 hr and an increasing proportion of adherent cells. Remarkable inhibition of proliferation and induction of differentiation occurred when NB4 cells were cultured with both 9-*cis*-RA and KH1060 (a 20-epivitamin D_3 analog) [58]. An antiproliferative effect of $1,25(OH)_2D_3$ was reported by Hickish *et al.* [26] in the NHL cell lines SU-DHL4 and SU-DHL5; however, this effect was observed only using high concentrations of the compound.

V. ASSESSMENT OF EFFECTS OF VITAMIN D_3 ON CELL CYCLE PARAMETERS IN HEMATOPOIETIC CELLS

Identifying the genes that are either activated or repressed directly by $1,25(OH)_2D_3$ and its analogs and that are, therefore, responsible for ligand-induced cell differentiation has been a long-term goal of investigation. The c-*myc*, c-*jun*, c-*fms*, *p53*, N-*ras*, and protein kinase C genes are modulated by $1,25(OH)_2D_3$, but this modulation may not be a direct effect, and may simply reflect the entire process of differentiation [59–63].

Activation of the protooncogene c-*myc* by retroviral insertion or chromosomal rearrangement is a typical feature of human leukemias. The HL-60 leukemia cell line is characterized by high levels of expression of c-*myc* due to gene amplification [64,65]. Treatment of this cell line with $1,25(OH)_2D_3$ results in a down-regulation of expression of this oncogene related to cell differentiation.

The derivative $1,25(OH)_2D_3$ has a protective effect against apoptosis in HL-60 cells [66]. This effect lends support to the observation that monocytic differentiation interferes with programs leading to apoptotic death. In other cell types, inhibition of apoptosis correlates with elevated levels of Bcl-2, but this does not appear to be the case with myeloid cells. In fact, after culture with $1,25(OH)_2D_3$ a down-regulation of Bcl-2 was observed both at the mRNA and protein levels [66]. Exposure of HL-60 cells to $1,25(OH)_2D_3$ induces the expression of the protooncogene c-*fms*, which occurs in parallel with the induction of CD14 expression (marker of monocytes) and a block of their cell cycle in G_0/G_1 phase [67]. In the chronic myelogenous leukemia (CML) cell line RWLeu-4, an inhibition of proliferation was observed after $1,25(OH)_2D_3$ treatment. Moreover, the binding activity of the protooncogene *junD* is enhanced by $1,25(OH)_2D_3$ in these cells during their cell cycle arrest, whereas it was not decreased in a $1,25(OH)_2D_3$-resistant variant cell line [27].

The cell lines HL-60 and U937 have been used to attempt to identify early response genes directly regulated by VDR. Bories *et al.* [68] identified a serine protease, myeloblastin, that was down-regulated by phorbol esters in promyelocytic cells, causing growth arrest and cell differentiation. They also isolated cDNAs coding for fructose 1,6-biphosphatase, whose expression is up-regulated by $1,25(OH)_2D_3$ in HL-60 cells and peripheral blood monocytes.

The *p53* gene is one of the most frequently altered genes in cancer. Its inactivation can occur through mutations, formation of complexes with viral oncoproteins, or binding to a cellular protein encoded by MDM-2 oncogene [69]. After exposure to DNA damaging agents, an up-regulation of expression of p53 protein is observed, and it is associated with a block of the cell cycle in the G_1 phase. Furthermore, wild-type p53 (wt-p53), by interacting with TATA binding protein, is able to repress the activity of various gene promoters such as c-*jun*, c-*fos*, and IL-6 [70]. The p53 also binds to DNA consensus sites, transactivating the expression of GADD45, MDM-2, and $p21^{WAF1}$ (p21) [71–73]. The p21 gene, also known as WAF1, CIP1, and SDI1, belongs to a family of proteins called the cyclin-dependent kinase inhibitors (CDKIs) that are important regulators of the cell cycle. The p53-induced growth arrest in response to DNA damage is mediated, at least in part, by induction of p21. Also, p53-independent pathways of p21 induction have also been widely described as alternative mechanisms of control of cell proliferation [74]. Differentiating cells are characterized by an elongation of the G_1 phase of the cell cycle with an inhibition of G_1/S transition. An arrest of the cell cycle in the G_1 or G_0/G_1 phase has been observed in hematopoietic cell lines including U937, HL-60, and M1 after their exposure to differentiating agents [75–79]. The induction of p21 expression has been observed in HL-60 cells, which lack endogenous p53, when induced to differentiate down either the granulocytic or monocytic pathway, including after their exposure to $1,25(OH)_2D_3$ [80,81]. A strong correlation appears to exist between early induction of p21 and the beginning of the differentiation program. The marked increase of p21 protein expression may be due to enhanced posttranscriptional stabilization of p21 mRNA [81]. The up-regulation of p21 mRNA occurred independently of *de novo* protein synthesis, further supporting the hypothesis that p21 is an early gene. Indeed, the p21 promoter has a vitamin D response element, and induction requires the presence of VDR. Also, experiments using a variety of cell lines showed that $1,25(OH)_2D_3$ and other differentiating agents could mediate their induction of p21-independent of an intact p53 gene.

Using differential hybridization, Liu *et al.* [82] showed that p21 is differentially expressed in response to $1,25(OH)_2D_3$ in the myelomonocytic cell line U937. A transient overexpression of p21 and p27 in U937 cells promoted the expression of the cell surface molecules CD14 and CD11b, which are specific for monocytes–macrophages. One series of experiments showed that the p21, p27, p15, p16, and p18 CDKIs were all up-regulated in a time-dependent manner after the addition of $1,25(OH)_2D_3$ [82]. This induction occurred within 4 hr of the addition of $1,25(OH)_2D_3$ in the presence of cycloheximide (CHX), suggesting a direct transcriptional activation by VDR.

In another study, the protein expression of different G_1-phase regulators has been examined in HL-60 cells exposed to different concentrations of $1,25(OH)_2D_3$. A strong up-regulation of p27 protein expression was evident after 72 hr of exposure to the compound, and it was dependent on $1,25(OH)_2D_3$ concentration. This up-regulation was also associated with increased levels of D- and E-cyclins, coinciding with the G_1 arrest. These results suggested a prominent role of p27 in mediating the antiproliferative activity of $1,25(OH)_2D_3$ in this cell line [83].

VI. EFFECTS OF NEW ANALOGS ON HEMATOPOIETIC CELLS

All of the studies conducted so far with $1,25(OH)_2D_3$ emphasize the need for new vitamin D_3 analogs with greater antileukemic effects and less toxicity. In spite of the promising data obtained from *in vitro* studies, results of clinical trials with $1,25(OH)_2D_3$ are limited in scope and thus far have exhibited only mediocre results. For example, the myelodysplastic syndrome (MDS) is associated with anemia, thrombocytopenia, and leukopenia and an increased number of myeloid progenitor cells in the bone marrow. Some patients with MDS go on to develop acute myeloid leukemia. We treated 18 MDS patients with increasing doses of $1,25(OH)_2D_3$ up to a maximum of 2 μg/day for 12 weeks. Although an improvement of at least one hematologic parameter occurred in 8 patients after more than 4 weeks, the response was not durable and not detectable at the end of the study at 12 weeks [49]. Nine patients developed hypercalcemia, which was the dose-limiting toxicity. In another study, seven MDS patients were treated with $1,25(OH)_2D_3$ (2.5 μg/day, for at least 8 weeks), with no beneficial effects [84].

To date, more than 50 vitamin D_3 analogs have been identified that have higher differentiating and antiproliferative activities than $1,25(OH)_2D_3$ but lower ability to cause hypercalcemia as compared to $1,25(OH)_2D_3$ (see Chapters 61–63). These analogs could provide a larger therapeutic window for the treatment of hematologic malignancies, retaining the useful properties of $1,25(OH)_2D_3$ [85].

The first attempts using analogs focused on 1α-hydroxyvitamin D_3 (1αOHD3), a vitamin D_3 analog that is efficiently converted to $1,25(OH)_2D_3$ *in vivo* by D_3-25-hydroxylase. This compound was administered to mice previously inoculated with the M1 leukemia cell line, and results revealed that the compound was therapeutically more active than $1,25(OH)_2D_3$ [86]. Its conversion to the active form resulted in a more prolonged elevation of plasma levels of $1,25(OH)_2D_3$, and the dose (25 pmol, every other day) produced only a slight and not significant elevation of serum calcium. In addition, survival of the leukemic mice was increased by 50–60%; however, the more effective doses produced hypercalcemia. Also, the administration of $1\alpha OHD_3$ has been shown to produce tumor regression in follicular NHLs in rats, but hypercalcemia was the dose-limiting factor [26].

In one study, six patients with MDS were treated with $1\alpha OHD_3$ at 1 μg/day for a minimum of 3 months, but neither a good clinical response nor toxicity was observed in these cases [87]. In another clinical study thirty MDS patients were included in two different groups: one group received 1α-OHD_3 at 4–6 μg/day and the other group received placebo; the patients were treated for a median of 17 weeks [88]. An improvement of hematologic parameters was detected in only one patient, but the authors felt the treated group had a greater proportion of patients who did not progress to leukemia as compared to the control group. Hypercalcemia and increased serum creatinine were observed in two patients, and these values regressed with reduction of the dose [88]. A case has been reported of an individual with chronic myelomonocytic leukemia (subtype of MDS) who achieved complete remission with 25-hydroxyvitamin D_3 therapy for 15 months; this remission was sustained 15 months after the end of the treatment [89]. These results are surprising because $25OHD_3$ has low activity by itself and *in vitro* has little antileukemic activity.

Calcipotriol (MC903) has a cyclopropyl group at the end of the side chain formed by the fusion of C-26 and C-27, a hydroxyl group at C-24, and a double bond at C-22 (see Chapter 61 for details). This compound was equipotent to $1,25(OH)_2D_3$ in inhibiting the proliferation and inducing the differentiation of the monoblastic cell line U937 [90,91]. In bone marrow cultures, this compound was able to promote the formation of multinucleated osteoclast-like cells, a vitamin D function. The effects of this compound on the immune system were very similar to those induced by $1,25(OH)_2D_3$. By interfering with T-helper cell activity, calcipotriol reduced immunoglobulin production and blocked the proliferation of thymocytes induced by IL-1 [92,93]. Exposure of the follicular NHL B-cell lines SU-DUL4 and SU-DUL5, carrying the t(14;18) translocation characteristic of the disease, to MC903 resulted in an inhibition of proliferation only at high concentrations of the compound (10^{-7} *M*) [26]. At the same time, calcipotriol was 100-fold less active than $1,25(OH)_2D_3$ in inducing hypercalcemia and mobilizing bone calcium in rats [94]. However, as discussed in Chapter 61, it is rapidly inactivated in the intact animal and therefore has been developed as a topical agent (see Chapter 72 for discussion of its use as a topical agent in psoriasis).

The analog $1,25(OH)_2$-16-ene-23-yne D_3 has potent antiproliferative and differentiating effects on leukemic cells *in vitro* [95] (see Chapter 62 for discussion of 16-ene analogs). In blocking HL-60 clonal growth, $1,25(OH)_2$-16-ene-23-yne D_3 has a potency about four times higher than $1,25(OH)_2D_3$. This compound administered to vitamin D-deficient chicks is about 30 times less effective than $1,25(OH)_2D_3$ in stimulating intestinal calcium absorption and about 50 times less effective in inducing bone calcium mobilization [91]. Further experiments have demonstrated the therapeutic potential of $1,25(OH)_2$-16-ene-23-yne D_3 by its ability to markedly prolong the survival of mice inoculated with the myeloid leukemic cell line WEHI $3BD^+$ and treated with a high dose (1.6 μg every other day) of the compound [96]. A synergistic antineoplastic effect of this compound and ATRA has been shown in HL-60 cells [97]. A HL-60 clone resistant to ATRA was more than 20-fold more sensitive to inhibition of proliferation by $1,25(OH)_2$-16-ene-23-yne D_3 than was $1,25(OH)_2D_3$. In addition, the induction of differentiation of these cells by $1,25(OH)_2$-16-ene-23-yne D_3 was much stronger in these cells than in the wild-type HL-60 cells [98].

Another promising analog is $1,25(OH)_2$-16-ene D_3, which is over 1000-fold more active than $1,25(OH)_2D_3$ in inhibiting clonal growth of HL-60 cells and 5-fold more potent in inducing their differentiation [99]. This compound has high binding affinity for VDR and has about 50-fold lower affinity for the vitamin D binding proteins (DBP) present in the serum, therefore increasing the availability of this compound for target tissues. However, the ability of this compound to induce hypercalcemia in mice was comparable to that of $1,25(OH)_2D_3$. This analog, as well as other vitamin D compounds, did not inhibit the clonal growth of normal human granulocyte–macrophage committed stem cells [99]. The 16-ene analogs are discussed more fully in Chapter 62.

The compound $1,25(OH)_2$-20-epi D_3 is characterized by an inverted stoichiometry at C-20 of the side chain (see Chapter 61 for discussion of 20-epi analogs). The monoblastic cell line U937 cultured with this compound showed a strong induction of differentiation [100]. It was also a potent modulator of cytokine-mediated T-lymphocyte activation and exerted calcemic effects comparable to $1,25(OH)_2D_3$ in rats. A study by ourselves suggested that $1,25(OH)_2$-20-epi D_3 is the most potent vitamin D_3 compound at inhibiting the clonal growth of HL-60 cells and at inducing cell differentiation. In fact, it was about 2600-fold more potent than $1,25(OH)_2D_3$ in inhibiting the clonal growth of HL-60 cells and about 5000-fold more effective in preventing clonal growth of fresh human leukemic myeloid cells [101]. $1,25(OH)_2$-20-epi D_3 exerts its effects by binding directly to VDR as shown by a T-lymphocytic cell line established from a patient with vitamin D-dependent rickets type II (HVDRR) lacking a functional VDR. Clonal growth was not inhibited after their treatment of these cells with high doses of either $1,25(OH)_2$-20-epi D_3 or $1,25(OH)_2D_3$ (10^{-7} *M*). In contrast, control experiments showed that these compounds [$1,25(OH)_2$-20-epi $D_3 > 1,25(OH)_2D_3$] were powerful inhibitors of a human T-cell leukemia virus type I (HTLV-I) transformed T-cell line that possessed VDR.

KH1060 is a potent vitamin D_3 20-epi analog with an oxygen in place of C-22 and three additional carbons in the side chain. It is about 14,000-fold more potent than $1,25(OH)_2D_3$ in inhibiting the clonal growth of the monoblastic cell line U937 [95]. It also has a powerful effect on other leukemic cells [51,100,101]. However, it has the same hypercalcemic activity and the same receptor binding affinity as $1,25(OH)_2D_3$ [95]. Further discussion of 20-epi analogs can be found in Chapter 61.

Most of the more potent vitamin D_3 analogs identified so far have a lower affinity for the serum DBP than $1,25(OH)_2D_3$. In theory this would facilitate the availability of the unbound compounds for the VDR of the target cells, thereby explaining their potent activity on specific tissues *in vivo*. In addition, the free compounds might also be cleared faster, explaining in part the lower calcemic activity. The role of DBP in determining analog potency is discussed in Chapter 59. In addition, these analogs might have a different binding affinity for critical VDRE-containing genes such as the VDRE for osteocalcin and the VDREs specific for hematopoiesis [99].

Combining two effective drugs is a mechanism for increasing potency while decreasing toxicity (Table II). The combination of ATRA (10^{-9} *M*) and the vitamin D_3 analogs $1,25(OH)_2$-16-ene-23-yne D_3 or $1,25(OH)_2$-23-yne D_3 (10^{-9} to 10^{-10} *M*) showed a synergistic effect on the induction of differentiation and inhibition of proliferation of HL-60 cells [97]. In addition, the inhibition of DNA synthesis of HL-60 cells correlated well with their reduction in soft-gel clonogenicity assay. A decrease of c-*myc* expression was also observed in the presence of ATRA and $1,25(OH)_2$-16-ene-23-yne D_3. This down-regulation of c-*myc* was stronger than that observed using single agents and correlated with the initiation of differentiation [97].

In conclusion, new vitamin D analogs have been shown to be characterized by potent antileukemic activity and lower hypercalcemic effect and should be considered for the treatment of hematologic malignancies either alone or in combination with other differentiating agents. However, more phase I trials are still necessary to assess the safety and effectiveness of these treatments.

VII. SUMMARY AND CONCLUSIONS

The hormone $1,25(OH)_2D$ plays a role in normal hematopoiesis, enhancing the activity of monocytes–macrophages and inhibiting cytokine production by T lymphocytes. It can also inhibit proliferation and induce differentiation of various myeloid leukemia cell lines. Its activity is mediated by vitamin D receptors that belong to the superfamily of steroid–thyroid receptors. However, the antileukemic activity of vitamin D *in vivo* is associated with high toxicity and the onset of hypercalcemia as the dose-limiting effect. Limited clinical trials have been performed for the treatment of preleukemia with differentiating agents including $1,25(OH)_2D_3$, but the *in vitro* effective dose caused hypercalcemia *in vivo*. Since the mid-1980s, over 50 vitamin D_3 analogs have been identified with reduced hypercalcemic activity and high potential to induce cell differentiation and to inhibit proliferation of leukemic cells. Further studies have been performed *in vitro* and *in vivo* using these analogs with other differentiating agents such as retinoids and phorbol esters, in the hopes that the combination of agents working through different pathways could lead to synergistic activity. Proof of principle that $1,25(OH)_2D_3$ and analogs are beneficial in cancer has occurred in experiments conducted *in vitro* and in laboratory animals; however, clinical trials in patients using vitamin D_3 analogs are only in their infancy.

Acknowledgments

Support is gratefully acknowledged to National Institutes of Health Grants CA-42710, CA-43277, CA-70675-01, CA-26038, and DK42792, as well as grants from the U.S. Army MRMC, the Parker Hughes Trust, and the Concern Foundation. H. P. Koeffler is a member of the Jonsson Comprehensive Cancer Center and holds the Mark Goodson Chair of Oncology Research.

References

1. Pike JW 1985 Intracellular receptors mediate the biologic action of 1,25-dihydroxyvitamin D_3, Nutr Rev **43**:161–168.
2. Carlberg C, Bendik I, Wyss A, Meier E, Sturzenbecker LJ, Grippo JF, Hunziker W 1993 Two nuclear signaling pathways for vitamin D. Nature **361**:657–660.

3. Schrader M, Bendik I, Becker-André M, Carlberg C 1993. Interaction between retinoic acid and vitamin D signaling pathways. J Biol Chem **268**: 17830–17836.
4. Schrader M, Muller KM, Carlberg C 1994 Specificity and flexibility of vitamin D signaling. J Biol Chem **269**:5501–5505.
5. Imai Y, Pike JW, Koeffler HP 1995 Potent vitamin D_3 analogs: Their abilities to enhance transactivation and to bind the vitamin D3 response element. Leuk Res **19**:147–158.
6. Baker AR, McDonnell DP, Hughes M, Crisp TM, Mangelsdorf DJ, Haussler MR, Pike JW, Shine J, O'-Malley BW 1988 Cloning and expression of full-length cDNA encoding human vitamin D receptor. Proc Natl Acad Sci USA **85**:3294–3298.
7. Provvedini DM, Tsoukas CD, Deftos LJ, Manolagas SC 1986 1,25-Dihydroxyvitamin D_3-binding macromolecules in human B lymphocytes: Effects on immunoglobulin production. J Immunol **136**:2734–2740.
8. Provvedini DM, Manolagas SC 1989 1,25-Dihydroxyvitamin D_3 receptor distribution and effects in subpopulation of normal human T lymphocytes. J Clin Endocrinol Metab **124**:1532–1538.
9. Kizaki M, Norman AW, Bishop JE, Karmakar A, Koeffler HP 1991 1,25-Dihydroxyvitamin D_3 receptor RNA: Expression in hematopoietic cells. Blood **77**:1238–1247.
10. Zerwekh JE, YU X-P, Breslau NA, Manolagas S, Pak CYC 1993 Vitamin D receptor quantitation in human blood mononuclear cells in health and disease. Mol Cell Endocrinol **96**:1–6.
11. Manolagas SC, Werntz DA, Tsoukas CD, Provvedini DM, Vaughan JH 1986 Dihydroxyvitamin D_3 receptors in lymphocytes from patients with rheumatoid arthritis. J Lab Clin Med **108**:596–600.
12. Provvedini DM, Ruiot CM, Sobol RE, Tsoukas CD, Manolagas SC 1987 1,25-Dihydroxyvitamin D_3 receptors in human thymic and tonsillar lymphocytes. J Bone Miner Res **2**:239–247.
13. Biyoudi-Vouenze R, Cadranel J, Valeyre D, Milleron B, Hance AJ, Soler P 1991 Expression of $1,25(OH)_2D_3$ receptors in alveolar lymphocytes from patients with pulmonary granulomatous diseases. Am Rev Respir Dis **143**:1376–1380.
14. Koren R, Ravid A, Liberman UA, Hochberg Z, Weisman Y, Novogrodsy A 1985 Defective binding and function of 1,25-dihydroxyvitamin D_3 receptors in peripheral mononuclear cells of patients with end-organ resistance to 1,25-dihydroxyvitamin D_3. J Clin Invest **76**:2012–2015.
15. Reichel H, Koeffler HP, Norman AW 1990 Production of 1,25-dihydroxyvitamin D_3 by hematopoietic cells. In Paterlik M, Bronner F (eds) Progress in Clinical and Biological Research. Volume 332. Molecular and Cellular Regulation of Calcium and Phosphate Metabolism. Alan R Liss, New York, pp. 81–97.
16. Nakajima S, Yamaoka K, Yamamoto T, Okada S, Tanaka H, Seino Y 1990 Decreased concentration of 1,25-dihydroxyvitamin D_3 receptors in peripheral mononuclear cells of patients with X-linked hypophosphatemic rickets: Effect of phosphate supplementation. Bone Miner **10**:201–209.
17. Kizaki M, Koeffler HP, Lin CW, Miller CW 1990 Expression of retinoic acid receptor mRNA in hematopoietic cells. Leuk Res **14**:645–655.
18. Iho S, Takahashi T, Kura F, Sugiyama H, Hoshino T 1986 The effect of 1,25-dihydroxyvitamin D_3 on in vitro immunoglobulin production in human B cells. J Immunol **236**:4427–4431.
19. Lemire JM, Adams JS, Kermani-Arab V, Bakke AC, Sakai R, Jordan SC 1985 1,25-Dihydroxyvitamin D_3 suppresses human T helper/inducer lymphocyte activity in vitro. J Immunol **134**: 3032–3035.
20. Rigby WFC, Denome S. Fanger MW 1984 Regulation of lymphokine production and human T lymphocyte activation by 1,25-dihydroxyvitamin D_3: Specific inhibition at the level of messenger RNA. J Clin Invest **79**:1659.
21. Tsoukas CD, Provvedini DM, Manolagas SC 1984 1,25-Dihydroxyvitamin D_3: A novel immunoregulatory hormone. Science **224**:1438–1440.
22. Reichel H, Koeffler HP, Tobler A, Norman AW 1987 1,25-Dihydroxyvitamin D_3 inhibits gamma-interferon synthesis by normal human peripheral blood lymphocytes. Proc Natl Acad Sci USA **84**:3385–3389.
23. Tobler A, Miller CW, Norman AW, Koeffler HP 1988 1,25-Dihydroxyvitamin D_3 modulates the expression of a lymphokine (granulocyte–macrophage colony-stimulating factor) posttranscriptionally. J Clin Invest **81**:1819.
24. Lee Y, Inaba M, De Luca H, Mellon WS 1989 Immunological identification of 1,25-dihydroxyvitamin D_3 receptors in human promyelocytic leukemia cells (HL-60) during homologous regulation. J Biol Chem **264**:13701–13705.
25. Provvedini DM, Tsoukas CD, Deftos LJ, Manolagas SC 1983 1,25-Dihydroxyvitamin D_3 receptors in human leukocytes. Science **221**:1181–1183.
26. Hickish T, Cunningham D, Colston K, Millar BC, Sandle J, Mackay AG, Soukop M, Sloane J 1993 The effect of 1,25-dihydroxyvitamin D_3 on lymphoma cell lines and expression of vitamin D receptor in lymphoma Br J Cancer **68**(4):668–672.
27. Lasky SR, Iwata K, Rosmarin AG, Caprio DG, Maizel AL 1995 Differential regulation of JunD by dihydroxycholecalciferol in human chronic myelogenous leukemia cells. J Biol Chem **270**:19676–19679.
28. Gaynor R, Simon K, Koeffler HP 1991 Expression of c-jun during macrophage differentiation of HL-60 cells. Blood **77**:2618–2623.
29. Norman AW, Roth J, Orci L 1982 The vitamin D endocrine system: Steroid metabolism, hormone receptors, and biological response (calcium binding proteins). Endocr Rev **3**:331–366.
30. Reichel H, Koeffler HP, Norman AW 1989 The role of the vitamin D endocrine system in health and disease. N Engl J Med **320**:980–991.
31. Gray TK, Lester GE, Lorenc RS 1979 Evidence for extrarenal 1α-hydroxylation in pregnancy. Science **204**:1311–1313.
32. Barbour GL, Coburn JW, Slatopolsky E, Norman AW, Horst RW 1981 Hypercalcemia in an anephric patient with sarcoidosis. Evidence of extrarenal generation of 1,25-dihydroxyvitamin D_3. N Engl J Med **305**:440–443.
33. Adams JS, Sharma OP, Gacad MA, Singer FR 1984 Metabolism of 25-hydroxyvitamin D_3 by cultured pulmonary alveolar macrophages in sarcoidosis. J Clin Invest **72**:1856–1860.
34. Adams JS, Gacad MA 1985 Characterization of 1-hydroxylation of vitamin D_3 sterols by cultured alveolar macrophages from patients with sarcoidosis. J Exp Med **161**:755–765.
35. Reichel H, Koeffler HP, Bishop JE, Norman AW 1987 25-Hydroxyvitamin D_3 metabolism by lipopolysaccharide-stimulated normal human macrophages. J Clin Endocrinol Metab **64**:1–9.
36. Koeffler HP, Reichel H, Bishop JE, Norman AW 1985 Gamma-interferon stimulates production of 1,25-dihydroxyvitamin D_3 by normal human macrophages. Biochem Biophys Res Commun **127**:596–603.
37. Felsenfeld AJ, Drezner MK, Llach F 1986 Hypercalcemia and elevated calcitriol in a maintenance dialysis patient with tuberculosis. Arch Intern Med **146**:1941–1945.
38. Kozeny GA, Barbato AL, Barbato VK, Vertuno LL, Hano JE 1984 Hypercalcemia associated with silicone-induced granulomas. N Engl J Med **311**:1103–1105.
39. Abe E, Miyaurak C, Sakagami H, Takeda M, Konno K, Yamazaki T, Yoshiki S, Suda T 1981 Differentiation of mouse myeloid

leukemia cells induced by 1,25-dihydroxyvitamin D_3. Proc Natl Acad Sci USA **78**:4990–4994.

40. Koeffler HP, Amatruda T, Ikekawa N, Kobayashi Y, De Luca HF 1984 Induction of macrophage differentiation of human normal and leukemic myeloid stem cells by 1,25-dihydroxyvitamin D_3 and its fluorinated analogs. Cancer Res **44**:5624–5628.
41. Paquette R, Koeffler HP 1992 Differentiation therapy. In: Myelodysplastic Syndromes. Hematol/Oncol Clin North Am **6**:687–706.
42. Kreutz M, Andreesen R 1990 Induction of human monocyte into macrophage maturation by 1,25-Dihydroxyvitamin D_3. Blood **76**:2457–2461.
43. Provvedini DM, Deftos LJ, Manolagas SC 1986 1,25-Dihydroxyvitamin D_3 promotes in vitro morphologic and enzymatic changes in normal human monocytes consistent with their differentiation into macrophages. Bone **7**:23–28.
44. Choudhuri U, Adams JA, Byrom N, McCarthy DM, Barett J 1990 1,25-Dihydroxyvitamin D_3 induces normal mononuclear blood cells to differentiate in the direction of monocyte–macrophage Haematology **23**:9–19.
45. Rigby WFC, Stacy T, Fanger MW 1984 Inhibition of T-lymphocyte mitogenesis by 1,25-dihydroxyvitamin D_3 (calcitriol). J Clin Invest **74**:1451–1455.
46. Tobler A, Gasson J, Reichel H, Norman AW, Koeffler HP 1987 Granulocyte–macrophage colony-stimulating factor: Sensitive and receptor-mediated regulation by 1,25-dihydroxyvitamin D_3 in normal human peripheral blood lymphocytes. J Clin Invest **79**:1700–1705.
47. Mangelsdorf DJ, Koeffler HP, Donaldson CA, Pike JW, Haussler MR 1984 1,25-Dihydroxyvitamin D_3-induced differentiation in a human promyelocytic cell line (HL-60): Receptor-mediated maturation to macrophage-like cells. J Cell Biol **98**:391–398.
48. Munker R, Norman AW, Koeffler HP 1986 Vitamin D compounds. Effect on clonal proliferation and differentiation of human myeloid cells. J Clin Invest **78**:424–430.
49. Koeffler HP, Hirji K, Itri L 1985 1,25-Dihydroxyvitamin D_3: In vivo and in vitro effects on human preleukemic and leukemic cells. Cancer Treat Rep **69**:1399–1407.
50. Masciulli R, Testa U, Barberi T, Samoggia P, Tritarelli E, Pustorino R, Mastroberardino G, Camagna A, Peschle C 1995 Combined vitamin D_3/retinoic acid induction of human promyelocytic cell lines: Enhanced phagocytic cell maturation and hybrid granulomonocytic phenotype. Cell Growth Differ **6**: 493–503.
51. Elstner E, Linker-Israeli M, Le J, Grillier I, Said J, Shintaku P, Krajewski S, Reed JC, Binderup L, Koeffler HP 1996 Combination of potent 20-epi-vitamin D_3 analog (KH1060) and 9-*cis*-retinoic acid inhibits irreversibly clonal growth, decreases bcl-2 expression and induces apoptosis in HL-60 leukemic cells. Cancer Res **56**:3570–3576.
52. Brown G, Bunce CM, Rowlands DC, Williams GR 1994 All-*trans*-retinoic acid and 1,25-dihydroxyvitamin D_3 co-operate to promote differentiation of the human promyeloid leukemia cell line HL60 to monocytes. Leukemia **8**:806–815.
53. Bunce CM, Wallington LA, Harrison P, Williams GR, Brown G 1995 Treatment of HL60 cells with various combinations of retinoids and 1,25-dihydroxyvitamin D_3 results in differentiation towards neutrophils or monocytes or a failure to differentiate and apoptosis. Leukemia **9**:410–418.
54. Taimi M, Chateau MT, Cabane S, Marti J 1991 Synergistic effect of retinoic acid and 1,25-dihydroxyvitamin D_3 on the differentiation of the human monocytic cell line U937. Leuk Res **15**:1145–1152.
55. Mathieu FM, Cellier M, Taimi M, Chateau M-T, Cannat A, Marti J 1993 Thermal stress as an inducer of differentiation of U937 cells. Leuk Res **17**:649–656.
56. Hu ZB, Ma W, Uphoff CC, Lanotte M, Drexler HG 1993 Modulation of gene expression in the acute promyelocytic cell line NB4. Leukemia **7**:1817–1823.
57. Bathia M, Kirkland JB, Meckling-Gill KA 1994 M-CSF and 1,25-dihydroxyvitamin D_3 synergize with 12-*O*-tetradecanoylphorbol-13-acetate to induce macrophage differentiation in acute promyelocytic leukemia NB4 cells. **8**:1744–1749.
58. Elstner E, Linker-Israeli M, Le J, Umiel T, Michl P, Said JW, Binderup L, Reed JC, Koeffler HP 1997 Synergistic decrease of clonal proliferation, induction of differentiation and apoptosis in APL cells after combined treatment with novel 20-epi vitamin D_3 analogs and 9-*cis*-retinoic acid. J Clin Invest **99**:349–360.
59. Reitsma PH, Rothberg PG, Astrin SM, Trial J, Bar-Shavit Z 1983 Regulation of myc gene expression in HL-60 leukemia cells by a vitamin D metabolite. Nature **306**:492–494.
60. Grosso LE, Pitot HC 1985 Transcriptional regulation of c-myc during chemically induced differentiation. Cancer Res **45**:847–850.
61. Sariban E, Mitchell T, Kufe D 1985 Expression of the c-fms proto-oncogene during human monocytic differentiation. Nature **316**:64–66.
62. Obeid L, Okazaki T, Karoloak LA, Hannun YA 1990 Transcriptional regulation of protein kinase C by 1,25-dihydroxyvitamin D_3 in HL-60 cells. J Biol Chem **265**:2370–2374.
63. Datta R, Sherman ML, Stone RM, Kufe D 1991 Expression of the jun-B gene during induction of monocytic differentiation. Cell Growth Differ **2**:43–49.
64. Collins S, Groudine M 1982 Amplification of endogenous myc-related DNA sequences in a human leukaemia cell line. Nature **298**:679–681.
65. Dalla Favera P, Wong F, Gallo RC 1982 Oncogene amplification in promyelocytic leukaemia cell line HL-60 and primary leukaemic cells of the same patient. Nature **299**:61.
66. Xu HM, Tepper CG, Jones JB, Fernandez CE, Studzinski GP 1993 1,25-Dihydroxyvitamin D_3 protects HL60 cells against apoptosis but down-regulates the expression of the bcl-2 gene. Exp Cell Res **209**:367–374.
67. Rowley PT, Farley B, Giuliano R, LaBella S, Leary JF 1992 Induction of the fms proto-oncogene product in HL-60 cells by vitamin D: A flow cytometric analysis. Leuk Res **16**:403–410.
68. Bories D, Raynal M-C, Solomon DH, Darzynkiewicz Z, Cayre YE 1989 Downregulation of a serine protease, myeloblastin, causes growth arrest and differentiation of promyelocytic leukemia cells. Cell **59**:959–968.
69. Vogelstein B, Kinzler KW 1992 p53 function and dysfunction. Cell **70**:523–526.
70. Ginsberg DF, Mechtor M, Yaniv M, Oren M 1991 Wild-type p53 can down-modulate the activity of various promoters. Proc Natl Acad Sci USA **88**:9979–9983.
71. Kastan MBQ, Zhan WS, El-Deiry F, Carrier T, Jacks WV, Walsh BS, Plunkett B, Vogelstein B, Fornace AJ Jr 1992 A mammalian cell cycle checkpoint pathway utilizing p53 and GADD45 is defective in ataxia–teleangectasia. Cell **71**:587–597.
72. Barak YT, Juven R, Hafner R, Oren M 1993 MDM2 expression is induced by wild-type p53 activity. EMBO J **12**:461–468.
73. El-Deiry WS, Tokino T, Velculescu VE, Levy DB, Parsons R, Trent JM, Lin D, Mercer WE, Kinzler KW, Vogelstein B 1993 WAF1, a potential mediator of p53 tumor suppression. Cell **75**:817–825.
74. Zhang W, Grasso L, McLain CD, Gambel AM, Cha Y, Travali S, Deisseroth AB, Mercer WE 1995 p53-independent induction of WAF1/CIP1 in human leukemia cells is correlated with growth

arrest accompanying monocyte/macrophage differentiation. Cancer Res **55**:668–674.

75. Furukawa Y, Uenoyama S, Ohta M, Tsunoda A, Griffin JD, Saito M 1992 Transforming growth factor beta inhibits phosphorylation of the retinoblastoma susceptibility gene product in human monocytic cell line JOSK-1. J Biol Chem **267**:17121–17127.
76. Kiyokawa H, Richon VM, Venta-Perez G, Rifkind RA, Marks PA 1993 Hexamethylenbisacetamide-induced erythroleukemia cell differentiation involves modulation of events required for cell cycle progression through G1. Proc Natl Acad Sci USA **90**: 6746–6750.
77. Yen A, Varvayanis S, Platko JD 1993 12-*O*-Tetradecanoylphorbol-acetate and staurosporine induce increased retinoblastoma tumor suppressor gene expression with megakaryocytic differentiation of leukemic cells. Cancer Res **53**:3085–3091.
78. Oberg F, Larsson LG, Anton R, Nilsson K 1991 Interferon gamma abrogates the differentiation block in v-myc-expressing U-937 monoblasts. Proc Natl Acad Sci USA **88**:5567–5571.
79. Antoun GR, Re GG, Terry NH, Zipf TF 1991 Molecular genetic evidence for a differentiation–proliferation coupling during DMSO-induced myeloid maturation of HL-60 cells: Role of the transcription elongation block in the c-myc gene. Leuk Res **15**:1029–1036.
80. Jiang H, Lin J, Su Z-Z, Collart FR, Huberman E, Fisher PB 1994 Induction of differentiation in human promyelocytic HL-60 leukemia cells activates p21, WAF1/CIP1, expression in the absence of p53. Oncogene **9**:3397–3406.
81. Schwaller J, Koeffler HP, Niklaus G, Loetscher P, Nagel S, Fey MF, Tobler A 1995 Posttranscriptional stabilization underlies p53-independent induction of $p21^{WAF1/CIP1/SDI1}$ in differentiating human leukemic cells. J Clin Invest **95**:973–979.
82. Liu M, Lee MH, Cohen M, Bommakanti M, Freedman LP 1996 Transcriptional activation of the Cdk inhibitor p21 by vitamin D_3 leads to the induced differentiation of the myelomonocytic cell line U937. Genes Dev **10**:142–153.
83. Wang QM, Jones JB, Studzinski GP 1996 Cyclin-dependent kinase inhibitor p27 as a mediator of the G1–S phase block induced by 1,25-dihydroxyvitamin D_3 in HL-60 cells. Cancer Res **56**: 264–267.
84. Richard C, Mazo E, Cuadrado MA, Iriondo A, Bello C, Gandarillas MA, Zubizarreta A 1986 Treatment of myelodysplastic syndrome with 1,25-dihydroxyvitamin D_3. Am J Hematol **73**: 175–178.
85. Zhou JY, Norman AW, Lubbert M, Collins ED, Uskokovic MR, Koeffler HP 1989 Novel vitamin D analogs that modulate leukemic cell growth and differentiation with little effect on either intestinal calcium absorption or bone calcium mobilization. Blood **74**:82–93.
86. Honma Y, Hozumi M, Abe E, Konno K, Fuku S, Hima M, Hata S, Nishii Y, DeLuca HF, Suta T 1983 1,25-Dihydroxyvitamin D_3 prolong survival time mice inoculated with myeloid leukemia cells. Proc Natl Acad Sci USA **80**:201–204.
87. Metha AB, Kumaran TO, Marsh GW 1984 Treatment of myelodysplastic syndrome with alfacalcidol [letter]. Lancet **2**:761.
88. Motomura S, Kanamori H, Maruta A, Kodama F, Ohkubo T 1991 The effect of 1-hydroxyvitamin D_3 for prolongation of leukemic transformation-free survival in myelodysplastic syndromes. Am J Hematol **38**:67–68.
89. Mellibovsky L, Diez A, Aubia J, Nogues X, Perez-Vila E, Serrano S, Recker RR 1993 Long-standing remission after 25-OH-D_3 treatment in a case of chronic myelomonocytic leukaemia. Br J Haematol **85**:811–812.
90. Binderup L, Bramm E 1988 Effects of a novel vitamin D analogue MC903 on cell proliferation and differentiation in vitro and on calcium metabolism in vivo. Biochem Pharmacol **37**:889–895.
91. Brown AJ, Dusso A, Slatopolsky E 1994 Selective vitamin D analogs and their therapeutic applications. Semin Nephrol **14**:156–174.
92. Muller K, Svenson M, Bendtzen K 1988 1,25-Dihydroxyvitamin D_3 and a novel vitamin D analogue MC903 are potent inhibitors of human interleukin 1 in vitro. Immunol Lett **17**:361–366.
93. Muller K, Heilmann C, Poulsen LK, Barington T, Bendtzenk K 1991 The role of monocytes and T-cells in 1,25-dihydroxyvitamin D_3 mediated inhibition of B cell function in vitro. Immunopharmacology **21**:121–128.
94. Rebel VI, Ossenkoppele GJ, van de Loosdrecht AA, Wijermans PW, Beelen RH, Langenhuijsen MM 1992 Monocytic differentiation induction of HL-60 cells by MC 903, a novel vitamin D analogue. Leuk Res **16**:443–451.
95. Pakkala S, de Vos S, Elstner E, Ruder K, Uskokovic M, Binderup L, Koeffler HP 1995 Vitamin D_3 analogs: Effect on leukemic clonal growth and differentiation, and on serum calcium levels. Leuk Res **19**:65–72.
96. Zhou JY, Norman AW, Chen DL, Sun GW, Uskokovic MR, Koeffler HP 1990 1,25$(OH)_2$-16ene-23yne vitamin D_3 prolongs survival time of leukemic mice. Proc Natl Acad Sci USA **87**:3929–3932.
97. Doré BT, Uskokovic MR, Monparler RL 1993 Interaction of retinoic acid and vitamin D_3 analogs on HL-60 myeloid leukemic cells. Leuk Res **17**:749–757.
98. Doré BT, Uskokovic MR, Monparler RL 1994 Increased sensitivity to a vitamin D_3 analog in HL-60 myeloid leukemic cells resistant to all-*trans*-retinoic acid. Leukemia **8**:2179–2182.
99. Jung SJ, Lee YY, Pakkala S, de Vos S, Elstner E, Norman AW, Green J, Uskokovic M, Koeffler HP 1994 1,25$(OH)_2$-16ene-vitamin D_3 is a potent antileukemic agent with low potential to cause hypercalcemia. Leuk Res **18**:453–463.
100. Binderup L, Latini S, Binderup E, Bretting C, Calverley M, Hansen K 1991 20-Epi-vitamin D_3 analogues: A novel class of potent regulators of cell growth and immune response. Biochem Pharmacol **42**:1569–1575.
101. Elstner E, Lee YY, Hashiya M, Pakkala S, Binderup L, Norman AW, Okamura WH, Koeffler HP 1994 1,25-Dihydroxy-20-epi-vitamin D_3: An extraordinarily potent inhibitor of leukemic cell growth in vitro. Blood **84**:1960–1967.

CHAPTER 69

The Role of Vitamin D_3 in Immunosuppression: Lessons from Autoimmunity and Transplantation

JACQUES LEMIRE Division of Pediatric Nephrology, Department of Pediatrics, University of California, San Diego, La Jolla, California

I. INTRODUCTION

Since the mid-1980s, a variety of new properties and applications have been discovered for the hormone 1,25-dihydroxyvitamin D_3 [$1,25(OH)_2D_3$]. Almost simultaneously the antiproliferative, prodifferentiating, and immunosuppressive activities of this metabolite of vitamin D were defined. It became obvious from the early investigations that in order to achieve maximal immunosuppressive activity *in vitro*, $1,25(OH)_2D_3$ was required at a concentration higher than that needed to obtain antiproliferative activity. This observation explains in part the early success of the hormone when used for the treatment of psoriasis, whereas, currently, the inherent hypercalcemic properties of $1,25(OH)_2D_3$ still prevent its clinical use for immunosuppression in humans. However, the development of analogs of the active metabolite has now broadened the potential clinical applications of the hormone. The optimal compound would be one that exerts maximal immunosuppressive activity while sparing the recipient of hypercalcemic complications. To achieve that goal, a variety of animal models of autoimmunity and transplantation have been studied using $1,25(OH)_2D_3$ and related analogs. A comprehensive analysis of the effects of vitamin D_3 in cellular systems involved in autoimmunity and transplantation will then lead to its application in animal models *in vivo*.

II. AUTOIMMUNITY

A. Immune Mechanisms Operational in Autoimmunity

Further understanding of the characteristics of cells of the immune lineage has drastically improved our knowledge of the mechanisms leading to the immune response in normal or pathological conditions (Fig. 1; see also Chapter 29 for further discussion). Three different T helper cell subsets have been well described: Th0, Th1, and Th2 cells. The first subset, Th0, represents what appears to be an early precursor or transitory cell that subsequently differentiates into Th1 or Th2 cells on the appropriate stimulus, which is most likely provided by antigen-presenting cells (monocytes, macrophages, B cells, or dendritic cells). One of these stimuli is interleukin-12 (IL-12), which can be produced by both monocytes–macrophages and B cells [1] and promotes Th0 cells to differentiate along the Th1 pathway [2]. The helper T-cell subsets Th1 and Th2 are defined by their cytokine secretion patterns. The Th0 cells produce an unrestricted pattern of cytokines. Th1 cells produce IL-2 and interferon-γ (IFN-γ), whereas Th2 cells produce IL-4, -5, -6, -10, and -13 [3]. Furthermore, Th1 cells can transfer delayed-type hypersensitivity (DTH) [4] and provide help to B cells to produce the antibody isotype immunoglobulin G_{2a} (IgG_{2a}), whereas Th2 cells help B cells for IgG_1 and IgE secretion [5]. Because of their IFN-γ production, Th1 cells can also interact with macrophages to increase bactericidal properties [6]. These Th subsets also cross-regulate one another: IFN-γ produced by Th1 cells can down-modulate Th2 cells, and IL-4 and IL-10 produced by Th2 cells inhibits Th1 cells [7]. Interestingly, IL-10 produced by Th2 cells can indirectly inhibit Th1 cell responses by acting on monocytes that are required by Th1 cells for antigen-specific proliferation and lymphokine secretion [8], most likely by inhibiting IL-12 secretion [9]. The lymphokine IFN-γ produced by Th1 cells also enhances class II antigen expression [10].

This dichotomy between Th1 and Th2 cells has been confirmed in humans. A similar pattern of cytokine secretion for both Th subsets is present [11]. Th1 cells express cytolytic activity against antigen-presenting cells and provide helper function for IgM, IgG, and IgA synthesis at low T cell/B cell ratios. At T/B ratios higher than 1 : 1, a decline in B-cell help is observed, related to the lytic activity of Th1 cells against autologous antigen-presenting B cells [12]. This down-regulation of antibody responses could be operational *in vivo*. In contrast,

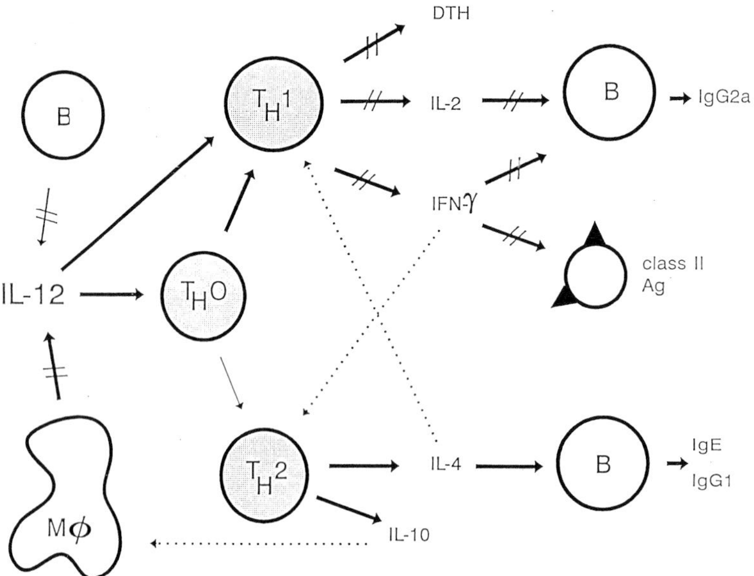

FIGURE 1 Immunoregulation of helper T-cell subsets, showing potential immunosuppressive mechanisms of $1,25(OH)_2D_3$. Solid arrows denote activation/differentiation, dotted arrows represent inhibition, and blocked arrows denote inhibitory effects of $1,25(OH)_2D_3$. B, B cell; T_H0, Th0 subset; T_H1, Th1 subset; T_H2, Th2 subset; Mϕ, macrophage; IL-2, -4, -10, -12, interleukin -2, -4, -10, -12; DTH, delayed-type hypersensitivity; IFN-γ, interferon-γ; Ig, immunoglobulin; class II Ag, class II major histocompatibility antigen. From Lemire *et al.* [98], © J Nutrit (vol. 125, p. 17075), American Institute of Nutrition.

Th2 cells develop in response to allergens or parasites, provide help for all immunoglobulin classes including IgE, and lack cytolytic potential [13]. The absence of lytic activity of Th2 cells may account for the long-term IgE responses of patients with atopy or parasitic infections [13]. High efficiency cloning of peripheral blood $CD4^+$ T cells from healthy individuals generates the Th1, Th0, and Th2 cytokine profiles roughly distributed according to a 2:4:1 ratio [13]. The heterogeneity of the cytokine profile in humans is not restricted to $CD4^+$ cells; $CD8^+$ cells, which have the phenotype of cytotoxic and suppressor cells, can also be further defined by analysis of their lymphokine profile [14].

What triggers the development of a particular Th subset? The answer is far from definitive. At this time, it is clear that IFN-γ promotes Th1 development while IL-4 and IL-10 contribute to Th2 expansion. However, the source of the polarization signal remains to be elucidated. The type of antigen-presenting cell (APC) interacting with the Th cell does not seem to be as critical as previously thought [15]. However, because IL-12 produced by APC promotes Th1 differentiation, a similar pro-Th2 cytokine yet to be discovered could be operational. The initial triggering factor for those existing (IL-12) or potential (pro-Th2) cytokines remains to be discovered.

1. ANIMAL MODELS OF AUTOIMMUNITY

A long-standing useful model for the study of autoimmune diseases has been the animal model of multiple sclerosis, experimental autoimmune encephalomyelitis (EAE). In this model, immunization of susceptible mice or rats with central nervous system proteins will induce a progressive paralysis in the recipients within 2 weeks. Developments in peptide technology have lead to a higher rate of disease induction in the susceptible recipients [16]. There is strong evidence that EAE is a Th1-mediated disease. Antigen-specific Th1 cells can transfer disease [17]. Moreover, at the peak of the disease, there is a predominance of Th1 cytokines (IL-2 and IFN-γ) in the central nervous system of the mice; during remission, IL-10 prevails, suggesting a Th2 predominance [18].

Another animal model is autoimmune diabetes in NOD mice (discussed in Chapter 70). There is no definite evidence for Th1 cells being pathogenic in this disease process, but T-cell clones generated in the presence of islets can produce Th1 cytokines and transfer disease [19]. Additional models such as experimental autoimmune thyroiditis and uveitis might be Th1 mediated, but the data are not conclusive at this point.

A particularly interesting model is the experimental lupus of MRL/lpr mice. A potential role of Th1-mediated IgG_{2a} in the pathogenesis of the disease was suggested by treatment of MRL/l mice with anti-IgM antisera from birth. This resulted in a depletion of IgG_{2a} antibodies and prevented the development of skin but not glomerular lesions [20]. A significant increase (8-fold) in IgG_{2a}-producing cells is observed in MRL/l mice between 2 to 5 months of age [21]. In response to thymus-dependent antigens, the IgG subclass profiles in all systemic lupus erythematosus (SLE) mice differ from those of normals, with a predominance of IgG_{2b} and IgG_{2a} rather than IgG_1. In addition, sera of the majority of MRL/l mice contain rheumatoid factors that react most strongly with IgG_{2a} [22]. The dependence of IgG_{2a} secretion on Th1 cells [5] as well as class II expression secondary to INF-γ secretion by Th1 cells would suggest an important role for the Th1 cell subset in the pathogenesis of experimental lupus.

2. HELPER T-CELL SUBSETS IN HUMAN AUTOIMMUNE DISEASES

Although the role of helper T-cell subsets in the pathogenesis of human autoimmune disorders remains to be more extensively studied, there is evidence to suggest a potential participatory role of Th1 cells in some of these diseases. For example, T-cell clones derived from patients with multiple sclerosis produce a Th1 cytokine profile [23], and the administration of IFN-γ to these patients exacerbates the disease [24].

Similarly, T-cell lines derived from retroorbital infiltrates of patients with Graves' thyroiditis can secrete a Th1 cytokine pattern [25]. In patients with rheumatoid arthritis, cloning of T cells obtained from synovial fluid and synovial membrane revealed a strong predominance of Th1 clones [26]. Finally, strong expression of IFN-γ has been observed in islet cells of patients with type I diabetes [27]. However, a heterogeneity of cytokine expression has been found in patients with SLE, and the emergence of a predominant helper subset has not yet been observed [28].

B. Effects of $1,25(OH)_2D_3$ and Analogs on Cells Derived from Autoimmune Diseases

1. ANIMAL CELLS

The availability of cloned T cells with known antigen specificity has allowed for a more reproducible and consistent analysis of the effects of novel vitamin D analogs on the proliferation of these cells. This system has also allowed for better screening of newer analogs than the antiproliferative assay using neoplastic cells such as HL-60 since the clones selected could exert immunologic activity. One needs to remain cautious, however, before assuming that antiproliferative activity equates to im-

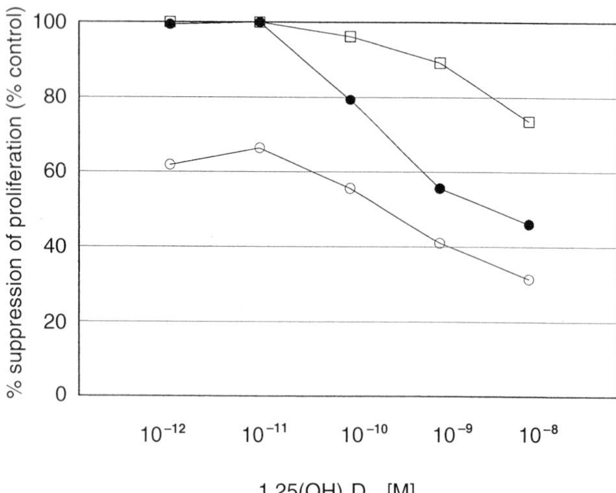

FIGURE 2 Inhibitory effect of vitamin D_3 compounds on proliferation of an antigen-specific (PLP peptide) T-cell clone. Each point represents the mean of three experiments of proliferation assessed by [^{3}H]thymidine incorporation and is expressed as the percentage of control cells incubated without vitamin D_3 compound. □, $1{,}25(OH)_2D_3$; ○, $1{,}25(OH)_2$-16-ene D_3; (●), $1{,}25(OH)_2$-24-oxo-16-ene D_3.

munosuppressive activity, as the latter may be observed in the absence of the former [29].

For *in vitro* studies of the antiproliferative effects of $1{,}25(OH)_2D_3$ and its analogs, T cells were cloned from animals immunized with neuroantigen. Initially myelin basic protein (MBP) was used as antigen [30], but, with the availability of neuropeptides, a synthetic preparation of proteolipid protein (PLP) was used, peptide 139–151 [16], to generate the T-cell clones. Susceptible SJL mice were immunized with 100 μg of PLP peptide in complete Freund's adjuvant followed by an intravenous injection of pertussigen. Ten days later, mice were sacrificed and lymph nodes harvested, and then cell lines were generated by weekly propagation with antigen (MBP or neuropeptide), irradiated syngeneic antigen-presenting cells, and growth factor (IL-2). Next, T-cell clones were established by dilution techniques (single-cell suspension) and propagated once again with antigen-presenting cells and growth factor. After a few weeks, growing clones were tested for antigen specificity and repropagated. The cytokine profile of the clones was then determined. Interestingly, all clones generated were of the Th1 type.

The antiproliferative effects of $1{,}25(OH)_2D_3$ and two well-studied analogs on a PLP-specific T-cell clone are depicted in Fig. 2. A dose-dependent inhibition of proliferation is observed for each compound, but the degree of inhibition varies according to the vitamin D preparation. The compound $1{,}25(OH)_2$–24-oxo-16-ene D_3 is a natural metabolite of the analog $1{,}25(OH)_2$-16-ene D_3. This metabolite was obtained by enzymatic synthesis in isolated perfused rat kidneys [31]. Contrary to $1{,}25(OH)_2D_3$, which undergoes a series of conversions in the kidney before final conversion into calcitroic acid, the analog $1{,}25(OH)_2$-16-ene D_3 is metabolized at a slower rate than the equivalent metabolite of $1{,}25(OH)_2D_3$ (Fig. 3), and its presence in the circulation for a longer time may account for some of the immuno-

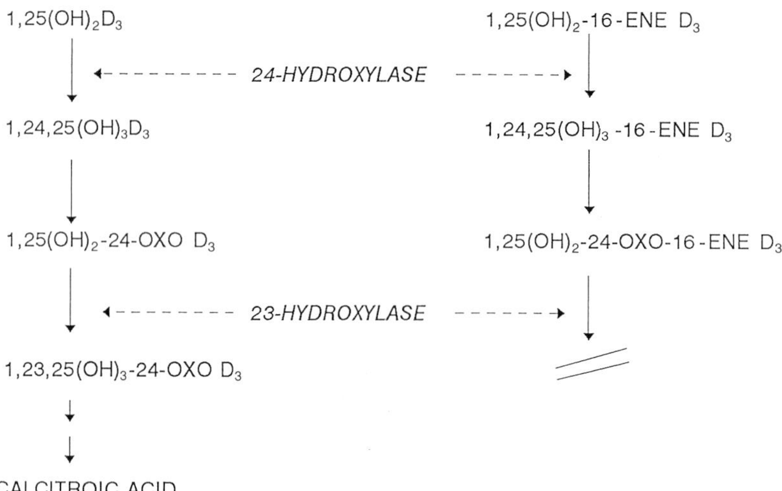

FIGURE 3 Renal metabolism of $1{,}25(OH)_2D_3$ and $1{,}25(OH)_2$-16-ene-D_3. From J. M. Lemire, D. C. Archer, and G. S. Reddy, 1,25-Dihydroxy-16ene-vitamin D_3, a renal metabolite of the vitamin D analog 1,25-dihydroxy-16ene-vitamin D_3, exerts immunosuppressive activity equal to its parent without causing hypercalcemina *in vivo*. Endocrinology **135**(6):2818–2821(1994). © The Endocrine Society.

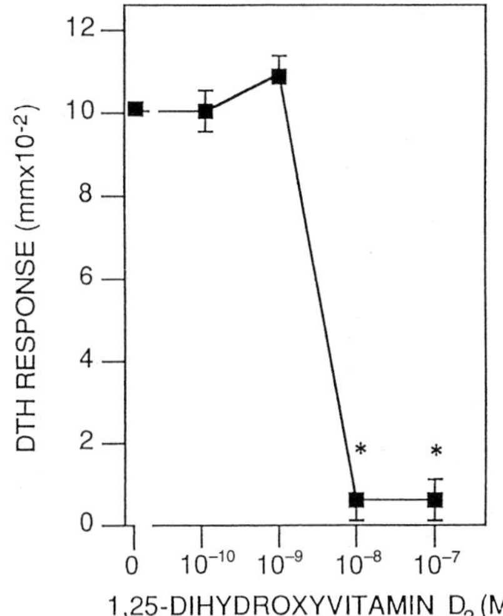

FIGURE 4 Delayed-type hypersensitivity (DTH) response in histocompatible recipients of cocultures of MBP-activated, cloned T cells and syngeneic spleen cells exposed to 20 μg/ml MBP. Cocultures were incubated for 5 days in the presence or absence of varying concentrations of $1,25(OH)_2D_3$ before transfer. Each value is the mean $\pm$ SD of the DTH response value in five individual mice. An asterisk signifies a significant ($p < 0.005$) decrease in the DTH response compared to cells not exposed to hormone. From Lemire and Adams [29] with permission.

suppressive activities of the metabolite (see also Chapter 62 on the 16-ene analogs).

Further evidence for an immunosuppressive effect of $1,25(OH)_2D_3$ on a biological activity of a T-cell clone was provided by the passive transfer of MBP-specific Th1 clones. A characteristic of Th1 cells is their ability to transfer delayed-type hypersensitivity (DTH) [4]. MBP-reactive T-cell clones were activated with syngeneic spleen cells and antigen (MBP) in the presence or absence of $1,25(OH)_2D_3$ before being washed and transferred to the footpads of naive mice. Swelling, as an index of DTH, was measured before and 18 hr after cell transfer using a pressure-sensitive caliper. A complete inhibition of the passive transfer of DTH was observed with 10 nM $1,25(OH)_2D_3$ (Fig. 4) [29]. These results suggested that the hormone could directly interfere with functional activity of Th1 cells.

2. HUMAN CELLS

The heterogeneity of helper T-cell clones generated in humans allowed for an ideal situation to analyze the effects of $1,25(OH)_2D_3$ and analogs on such subsets. Helper T-cell clones were isolated from atopic patients sensitive to the rye grass antigen Lol pI and characterized as Th0, Th1, and Th2 based on their lymphokine secretion pattern [32]. The Th subsets were activated with Lol pI and antigen-presenting cells in the presence or absence of $1,25(OH)_2D_3$ or the analog $1,25(OH)_2$-16-ene D_3, and the effect of the vitamin D compounds on the lymphokine production was analyzed (Fig. 5). Both $1,25(OH)_2D_3$ and $1,25(OH)_2$-16-ene D_3 suppressed the production of IFN-γ by Th1 cells in a dose-dependent manner, but these compounds had minimal effect on IL-4 production by Th2 cells and only at the highest concentrations tested. Interestingly, Th0 cells, producer of both cytokines, showed a profound reduction in IFN-γ in the presence of the vitamin D compounds, whereas IL-4 secretion was less inhibited, suggesting once again a pro-Th1 effect of the hormone and its analog.

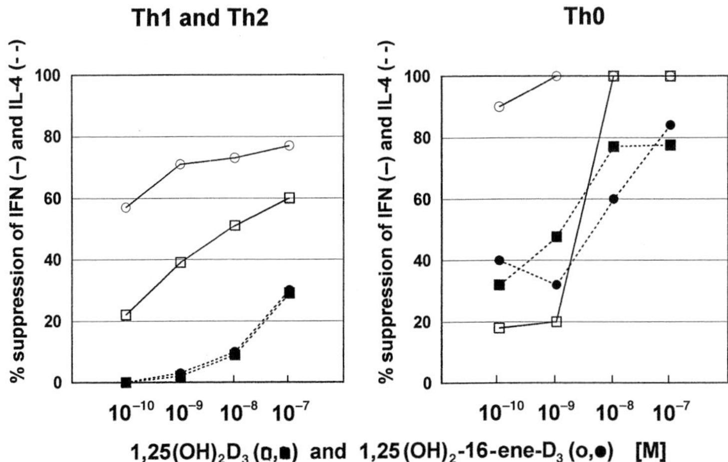

FIGURE 5 Effect of $1,25(OH)_2D_3$ (○, ●) and analog $1,25(OH)_2$-16-ene D_3 (□, ■) on lymphokine production by Th1 (left, open symbols), Th2 (left, filled symbols), and Th0 (right) cell clones. Th clones were activated with antigen, APC, and the lymphokines IL-4 and IFN-γ and assayed by radioimmunoassay. The data are expressed as percentage suppression.

In patients with rheumatoid arthritis, CD4$^+$ cells isolated from peripheral blood showed a more pronounced inhibition of proliferation and IL-2 production than cells from normal volunteers in the presence of $1,25(OH)_2D_3$, and the action of the hormone was synergistic with cyclosporin [33]. A subset of CD4$^+$ cells, CD45R0$^+$, implicated in rheumatoid arthritis and multiple sclerosis [34,35], is particularly sensitive to the action of the metabolite; proliferation and cytokine production (lymphotoxin, IL-2, IFN-γ) was reduced in the presence of 10 n*M* $1,25(OH)_2D_3$. The data suggest that Th cells implicated in certain human autoimmune diseases might potentially be suppressed by the action of $1,25(OH)_2D_3$, and, therefore, the compound could be of therapeutic value in these patients.

C. Effects of $1,25(OH)_2D_3$ and Analogs on Autoimmune Diseases *in Vivo*

1. Animal Models

At the beginning of the 1990s, it became evident that the immunosuppressive properties of the hormone

TABLE 1 $1,25(OH)_2D_3$ and Analogs in Autoimmunity

Organ	Model/Species	Vitamin D_3 or analog	Dose (μg/kg)[a]	Outcome, treated/controls (measure)	Serum calcium (mg/dl)	Ref.
Nervous system	EAE/mouse	$1,25(OH)_2D_3$	5/2d	80%/5% (survival)	9.7	Lemire and Archer [36]
	EAE/mouse	$1,25(OH)_2$-16-ene D_3	5/2d	1/4 (disease activity)	11.2	Lemire *et al.* [37]
	EAE/mouse	$1,25(OH)_2$-24-oxo-16-ene D_3	7.5/2d	1/4 (disease activity)	9.7	
	EAE/mouse	$1,25(OH)_2$-16-ene-23-ene-26,27-hexafluroro D_3	2.5/2d	3.3/5 (disease activity)	8.7	Lemire *et al.* [38]
	EAE/mouse	MC1288	0.2/2d	25%/92.8% (disease incidence)	10.2	Lemire *et al.* [39]
Thyroid	Experimental auto immune thyroiditis/mouse	$1,25(OH)_2D_3$	0.2/d	50%/85.7% (histologic incidence)	N/A[b]	Fournier *et al.* [40]
Joint	Adjuvant arthritis/rat	$1,25(OH)_2D_3$	0.2/d	11.9/16.9 (arthritic score)	12.0	Boissier *et al.* [41]
Kidney	Heymann nephritis/rat	$1,25(OH)_2D_3$	0.5/d	80/210 (mg urinary protein/day)	11.8	Branisteau *et al.* [42]
		KH1060	0.5/d	210/210 (mg urinary protein/day)	11	
	Mercuric chloride-induced nephritis/rat	$1,25(OH)_2D_3$	0.1/d	180/780 (mg urinary protein/day)	>14	Lillevang *et al.* [43]
		KH1060	0.3/d	<20/780 (mg urinary protein/day)	>14	
	Nephrotoxic serum nephritis/rat	$1,25(OH)_2D_3$	0.5/d	<50/300 (mg protein/day)	11	Hattori [44]
	Lupus nephritis/mouse	$1,24R(OH)_2D_3$	0.1/d	1.0/3.3 (dipstick)	N/A	Koizumi *et al.* [45]
	Lupus nephritis/mouse	OCT	0.002–0.1/d	10–40%/80% (proteinuria incidence)	10.4	Abe *et al.* [46]
	Lupus nephritis/mouse	$1,25(OH)_2D_3$	5/2d	<4/≥6 (urinary protein/creatinine ratio)	8	Lemire *et al.* [47]
Skin	Lupus/MRL mouse	$1,25(OH)_2D_3$	5/2d	None/present (skin lesions)	8	
Pancreas	Diabetes/NOD mouse	$1,25(OH)_2D_3$	5/2d	58%/80% (insulitis incidence)	10	Mathieu *et al.* [48]
	Diabetes/NOD mouse	KH1060	0.4/2d	11.1%/54.8% (disease incidence)	10.8	Mathieu *et al.* [49]

[a] d, Daily; 2d, every second day.
[b] N/A, Not available.

described *in vitro* could be applied *in vivo*. The availability of animal models of autoimmunity has rapidly allowed investigators to determine whether the hormone could be effective in preventing or improving existing diseases. Table I provides a summary of the effects of $1,25(OH)_2D_3$ and related analogs on various animal models classified by the autoimmune target organ [36–49]. Furthermore, many of these models were used to study the potential synergistic or additive effect of $1,25(OH)_2D_3$ and analogs with cyclosporin. Their combined effects are summarized in Table II [40,41,43,50].

A few lessons can be learned from these studies:

1. The immunosuppressive activity of $1,25(OH)_2D_3$ is limited by its hypercalcemic effect.
2. The effectiveness of $1,25(OH)_2D_3$ is enhanced in Th1-mediated diseases such as EAE.
3. Optimal activity of the compound is achieved when administered prior to antigen-induced autoimmune disease (EAE, thyroiditis, nephritis).
4. Prevention of a spontaneous murine model of autoimmunity (lupus, diabetes) is more effective when the compounds are administered from an early age (1 month of age).
5. $1,25(OH)_2D_3$ is not effective in inducing remission of already established disease.
6. Some target organs might be more susceptible to the action of $1,25(OH)_2D_3$ than others (skin > kidneys in lupus).
7. The action of analogs is similar to that of the natural hormone but with enhanced potency due to reduced hypercalcemic effects.
8. The activity of $1,25(OH)_2D_3$ can be potentiated by cyclosporin.

In many respects, the immunoregulatory properties of $1,25(OH)_2D_3$ appear to be similar to those of cyclosporin. Cyclosporin acts preferentially on T lymphocytes during initial activation by antigen [51,52], possibly by interfering with antigen recognition processes [53]. Cyclosporin suppresses the release of IL-2 by T cells [54,55], potentially by inhibition of gene expression at the level of mRNA transcription [56]. Cyclosporin prevents the differentiation of T-cytotoxic precursors into mature suppressor T cells [57,58] but does not reduce the cytotoxic/suppressor activity of suppressor T cells already generated [58,59]. Both $1,25(OH)_2D_3$ and cyclosporin appear to modulate early events of antigen presentation by preferentially inhibiting helper T-cell activity and interfering with the production or release of IL-2 and by sparing suppressor T-cell activity. Therefore, it is not unreasonable to speculate that $1,25(OH)_2D_3$ and related analogs might be useful in conditions where cyclosporin effectiveness has been demonstrated.

2. HUMANS

The natural metabolite $1,25(OH)_2D_3$ is part of the therapeutic arsenal used in patients with end-stage renal disease as described in Chapter 59. In this condition, it is used for its effects on calcium, parathyroid hormone (PTH), and bone. The usual dosage recommendation is between 0.25 and 1 μg/day; many patients receive vitamin D dosage in this range but only three times a week following hemodialysis treatments. Thus far, the

TABLE II Combined Effects of $1,25(OH)_2D_3$ and Analogs with Cyclosporin in Autoimmunity[a]

Model	Vitamin D_3 or analog, dose and outcome	Cyclosporin dose and outcome	Combination therapy outcome	Serum calcium (mg/dl) for D_3/CYA/both
Adjuvant arthritis [41]	$1,25(OH)_2D_3$ 0.2 μg/kg/day 0.2 days[1]	5 mg/kg/day 7.4 days	9.4 days	12/10.7/11.3
Mercuric chloride nephritis [43]	KH1060 0.1 μg/kg/day <20 mg[2]	3 mg/kg/day 180 mg	<10 mg	13.5/—/14
Experimental autoimmune encephalomyelitis [50]	$1,25(OH)_2D_3$ 2 μg/kg/every second day 78%[3]	10 mg/kg/every second day 54%	0%	13/11/12.5
Experimental autoimmune thyroiditis [40]	$1,25(OH)_2D_3$ 0.2 μg/kg/day 100%[3]	10 mg/kg/day 100%	66.6%	N/A

[a] Outcomes: [1] delay of onset of first clinical signs of arthritis; [2] 24-hour urinary protein excretion; [3] disease incidence. CYA, Cyclosporin; N/A, not available.

use of $1,25(OH)_2D_3$ for immunosuppressive activity has been very limited. However, there is already some evidence that $1,25(OH)_2D_3$ can indeed be used for this purpose in humans. In scleroderma, a connective tissue disease, progressive fibrotic lesions develop in the skin and various internal organs. A localized skin variant of the disorder is referred to as morphea. Evidence for increased helper T-cell activity has been shown in these patients [60–62]. Administration of up to 1.75 μg/day from 6 months to 3 years in such patients has led to clinical improvement (skin flexibility and extensibility) [63] or complete clearing of physical lesions [64]. The serum calcium level was reported not to have exceeded 10.5 mg/dl [63].

No other immunosuppressive application of vitamin D compounds, with the exception of psoriasis (see chapter 72), has yet been confirmed in humans. However, there are autoimmune disorders in which vitamin D metabolism has been studied. For example, levels of $1,25(OH)_2D_3$ have been found to be reduced or suppressed in systemic lupus erythematosus [65] and in Graves' disease [66] and increased in rheumatoid arthritis [67]. It remains to be seen if the reduced levels of the hormone could account for the "immune hyperactivity" observed.

III. TRANSPLANTATION

A. Immune Mechanisms Operational in Transplantation

The various immunologic processes involved in the rejection of allografts are somewhat different than the ones implicated in autoimmune reactions. Even though the T cells play a major role in the acute rejection process, other cells of immune origin such as monocytes–macrophages, natural killer (NK) cells, dendritic cells, and B cells participate in the process. It is beyond the scope of this summary to review all the pathogenic mechanisms involved in the rejection process. We concentrate on cellular mechanisms of acute rejection, the most common cause of allograft loss. This form of rejection is best prevented by the optimal use of cyclosporin, a drug sharing many immunosuppressive properties with $1,25(OH)_2D_3$.

The role of T cells is evident at various levels of the immune response: antigen recognition, immunoregulation, and cytolytic effector function. The problem of transplant rejection is complex: a progressive donor-directed T-cell activity may consist, on one hand, of direct cytotoxicity mediated by antigen-specific cytolytic T cells and, on the other hand, of inflammatory T-cell activity primarily related to the secretion of a variety of lymphokines by noncytotoxic T cells. The lymphokines can regulate (amplify or suppress) a specific or nonspecific antidonor response as well as recruit and activate other cells, such as macrophages, to execute antidonor cytotoxicity [68].

The participation of T cells in the rejection process is suggested by their ability to induce a delayed-type hypersensitivity (DTH) reaction [69]. By producing IL-2, T cells can promote the differentiation of cytotoxic T-cell precursors into functionally active cytotoxic T cells [70] as well as enhance the functional activity of NK cells and killer cells mediating antibody-dependent cytotoxicity (ADCC) [71]. In addition, the production of IFN-γ by T cells activates macrophages to be cytotoxic [72] and enhances the expression of class II major histocompatibility complex (MHC) antigen on resident cells [73]. Cytotoxic T cells respond mainly to class I MHC alloantigens [74]. The role of suppressor cell activity in the maintenance of allograft function remains complex, with an intricate network of contrasuppression and veto cells [75,76].

The mixed lymphocyte reaction (MLR) constitutes an *in vitro* model of allograft response. In this system, alloantigen-specific helper–inducer and cytotoxic–suppressor cells interact, and lymphocyte subsets, or their precursors, differentiate into effectors of cytotoxicity or suppression [77].

Although one is tempted to speculate that rejection or tolerance of organ transplant is under the regulation of Th1 and Th2, respectively, the evidence to imply a causal role for either population of Th cells at this point is inconclusive [78].

B. Effects of $1,25(OH)_2D_3$ on the Mixed Lymphocyte Reaction

A primary MLR was generated in the presence or absence of $1,25(OH)_2D_3$. After 7 days, the effector cells recovered were added to an autologous MLR (same donors). As shown in Fig. 6, when $1,25(OH)_2D_3$ was added at the initiation of the effector MLR, the day 7 effector cells exerted enhanced suppressor activity when added to a primary MLR, with maximal suppressor effect of 67.7 $\pm$ 2.7% MLR inhibition observed at 10 nM $1,25(OH)_2D_3$. Furthermore, this suppressor activity was not antigen specific, as effector cells generated in the presence of the hormone were still able to effectively reduce the MLR response where the responders and stimulators were heterologous to the primary MLR [79].

The same MLR results in the generation of a cytotoxic response against ^{51}Cr-labeled stimulator targets at

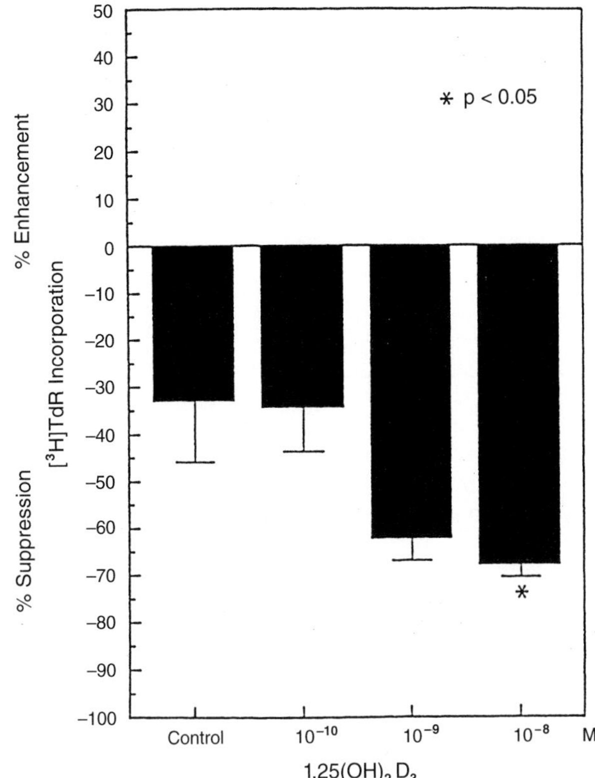

FIGURE 6 Antigen-specific suppression by $1,25(OH)_2D_3$-treated effector cells of autologous primary MLR. Primed lymphocytes from control and $1,25(OH)_2D_3$ (10^{-10} to 10^{-8} M)-treated MLR cultures were used as effectors (2×10^6 cells/ml, 2500 R) in a primary MLR (2×10^6 responder cells/ml; 2×10^6 stimulator cells/ml, 3000 R). Cells were cultured for 4 days and pulsed with [^{3}H]thymidine for an additional 18 hr. Data represent the effect (% enhancement or % suppression) of effectors on the primary autologous MLR compared to control MLR in the absence of effector cells. From Meehan *et al.* [79] with permission.

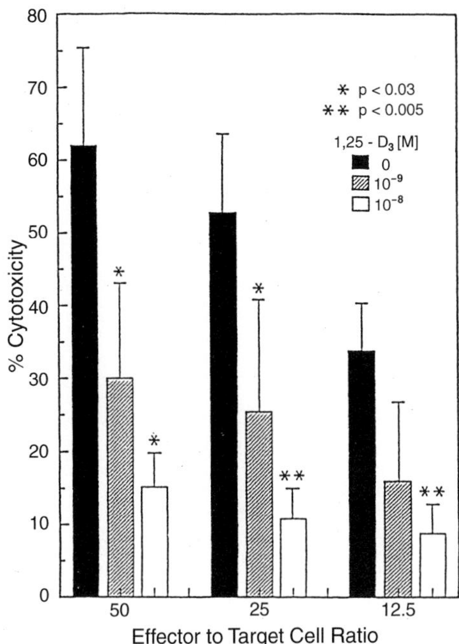

FIGURE 7 Inhibition of generation of cytotoxic cells by $1,25(OH)_2D_3$. After a 7-day MLR culture in the presence or absence of $1,25(OH)_2D_3$ (10^{-9} to 10^{-7} M), the lymphocytes were used as effector cells against ^{51}Cr-labeled phytohemagglutinin (PHA)-stimulated lymphocytes. Data represent the percentage cytotoxicity at effector to target cell ratios ranging from 50 : 1 to 12.5 : 1. From Meehan *et al.* [79] with permission.

7 days from the initiation of culture. As shown in Fig. 7, cells incubated with $1,25(OH)_2D_3$ exerted a significant reduction in the degree of cytotoxicity. The suppressor activity resulting from the addition of the hormone to the cultures was not secondary to the selection of cells of cytotoxic phenotypes; however, a reduction of class II antigen (HLA-DR) was seen after treatment with $1,25(OH)_2D_3$ when cells were analyzed with cell sorting techniques [79]. Therefore, in the human MLR, $1,25(OH)_2D_3$, like cyclosporin, enhances suppressor cell activity and reduces cytotoxic cell generation and the expression of class II antigen.

Using the human MLR, there is also some evidence to suggest that $1,25(OH)_2D_3$, when added to cyclosporin or FK 506, allows for a 10- and 5-fold reduction of dosage, respectively, for those immunosuppressants [80].

C. Effects of $1,25(OH)_2D_3$ on the Human Cytotoxic Response

1. NATURAL KILLER CELL ACTIVITY AND ANTIBODY-DEPENDENT CYTOTOXICITY

Human peripheral blood mononuclear cells were isolated from five normal volunteers, fractionated into adherent and nonadherent cells, and incubated in the presence or absence of $1,25(OH)_2D_3$ or $24,25(OH)_2D_3$ for 48 hr. The various cell populations were then analyzed for cytotoxicity against herpes simplex virus (HSV)-infected targets in the presence [antibody-dependent cytotoxicity (ADCC)] or absence [NK cytotoxicity (NKC)] of antibody [81]. As shown in Fig. 8, $1,25(OH)_2D_3$ had no effect on ADCC function but significantly inhibited NK activity of the nonadherent cells, with maximal activity at 10 nM of the hormone [82]. Phenotypic analysis by surface markers of the hormone-sensitive effector cell population was identical for treated or untreated groups, suggesting a functional inhibition by the metabolite. Additional studies have also revealed that $1,25(OH)_2D_3$ can prevent the development of new cytotoxic NK cells, but the inhibitory effect was not observed on cells already generated [83,84].

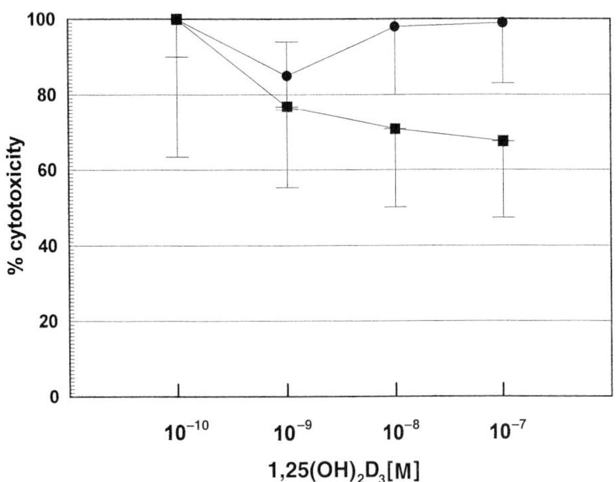

FIGURE 8 Effect of $1,25(OH)_2D_3$ on antibody-dependent cytotoxicity (●) and NK cytotoxicity (■) activity after 48 hr of incubation. Cytotoxicity was tested in a 60:1 effector to target cell ratio, and the results are expressed as percent cytotoxicity ± SD. From Lemire *et al.* [82] with permission.

2. Human Cytotoxic $CD4^+$ and $CD8^+$ T-Cell Lines

Acute rejection of human allografts depends on the generation of cytotoxic T cells. Both $CD4^+$ and $CD8^+$ cells with cytotoxic capabilities have been identified in rejected allografts [85]. The contribution of each subset to the rejection process is controversial. The evidence so far has suggested that $1,25(OH)_2D_3$ could inhibit $CD4^+$ helper function, but could it interfere with $CD4^+$ cytotoxic function?

In an attempt to provide an answer to this question, cells were isolated from the renal parenchyma of a rejected live-related renal allograft, and a $CD8^+$ cell line was generated by weekly stimulation with an irradiated Epstein-Barr virus (EBV)-transformed B-lymphoblastoid cell line from the donor and human recombinant IL-2 twice a week. Limiting dilution of this primary cell line at 1 cell per well with the same weekly stimulation resulted in a secondary $CD4^+$ cell line. Cells were then incubated in the presence or absence of $1,25(OH)_2D_3$ for 4 days prior to the cytotoxicity assay. Cell-mediated cytotoxicity (CMC) was tested using the standard ^{51}Cr release assay at effector to target cell ratios of 10:1 to 0.1:1 (Table III).

A significant suppression of $CD4^+$ but not $CD8^+$ cytotoxicity was observed in cells incubated with $1,25(OH)_2D_3$ during activation; the hormone did not affect cytotoxicity of cells when added to effectors 2 hr before CMC or at CMC, or to targets 24 hr before CMC. These results suggest that the hormone can modulate cells with the $CD4^+$ phenotype. These data may have importance not only in the allograft rejection phenomenon but in other situations such as tumor-specific $CD8^+$ cells whose activation is dependant on $CD4^+$ cells [86].

TABLE III Effect of $1,25(OH)_2D_3$ on Cell-Mediated Cytotoxicity

Effector: target cell ratio	% Cytotoxicity			
	CD8 line		CD4 line	
	Control	$1,25(OH)_2D_3$	Control	$1,25(OH)_2D_3$
10:1	67.9	77.4	51.4	27.3
5:1	68.8	64.0	49.2	23.6
1:1	45.6	51.9	33.2	15.3
0.1:1	22.1	23.6	14.6	6.1

D. Effects of $1,25(OH)_2D_3$ and Analogs on Animal Models of Transplantation

A variety of organs (skin, heart, kidney) have been transplanted into various species (mouse, rat, dog) under immunosuppression with $1,25(OH)_2D_3$ or analogs. The vitamin D compounds have been administered intraperitoneally; however, oral administration of analogs for transplant immunosuppression was also found to be effective [87]. In most instances, the natural hormone $1,25(OH)_2D_3$ cannot exert immunosuppression to the degree required to prevent allograft rejection without significant toxicity, mainly hypercalcemia. Therefore, vitamin D analogs have opened a new pathway to transplant immunosuppression.

An example of the mode of action of both $1,25(OH)_2D_3$ and the analog $1,25(OH)_2$-16-ene D_3 in the context of heart allograft in the mouse is illustrated in Fig. 9. The dosage of $1,25(OH)_2D_3$ used was ineffective in significantly prolonging graft survival but still caused hypercalcemia [88]. The analog $1,25(OH)_2$-16-ene D_3 could prolong heart graft survival up to 27 ± 4 days, compared to 11.57 ± 0.5 days for control mice not receiving any immunosuppression. The efficacy of the analog could not be observed without some degree of hypercalcemia [88]. Table IV summarizes the studies of vitamin D compounds as immunosuppressants of allografts in animal models [89–95].

In humans, cyclosporin remains the pivotal immunosuppressive agent for optimal allograft tolerance despite its toxicity (nephrotoxicity, hypertension, and hirsutism among others). Because $1,25(OH)_2D_3$ and analogs exert immunosuppression by similar and potentially synergistic mechanisms with cyclosporin, the possibility of adding a vitamin D analog to a transplant regimen is appealing since it might reduce the dose of cyclosporin and its toxicity, on one hand, or replace prednisone as an adjuvant to cyclosporin, on the other. In an attempt to provide some insight into that possibility, combined regimens of vitamin D analogs and cyclosporin were

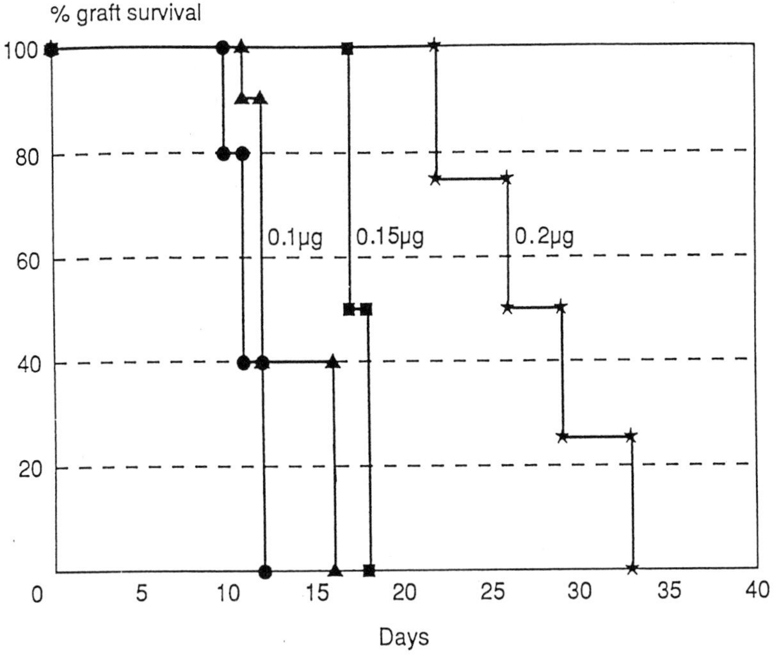

Treatment	Dose (μg)	Graft Survival Time (Days)	Mean Survival Time (days ± S.D.)	P
Controls	0.00	11,11,11,12,12,12,12	11.57 ± 0.53	
1,25-D_3	0.10	11,11,11,12,12,12,16,16	12.62 ± 2.13	NS
1,25-Δ^{16}-D_3	0.15 0.20	17,18,18,19,19 22,26,26,29,33	18.2 ± 0.83 27.2 ± 4.0	0.001 0.001

FIGURE 9 Time course of allograft heart survival. Recipient BALB/c mice received vehicle only (●), $1,25(OH)_2D_3$ at 0.1 μg (▲), or $1,25(OH)_2$-16-ene D_3 (1,25-Δ^{16}-D_3) at 0.15 μg (■) or 0.2 μg (★), given intraperitoneally every other day starting 3 days before donor C_3H heart transplantation and until graft rejection. Rejection was defined by the disappearance of the clearly visible heart graft pulsation. Statistical significance of differences between the drug-treated and vehicle control groups was calculated using Student's *t* test. From Lemire *et al.* [88] with permission.

studied in animal models. Table V illustrates the results obtained in skin allograft transplantation in mice and in cardiac allograft transplantation in rats [90,96]. In the two models of transplantation studied, the results suggest an additive rather than synergistic effect of the vitamin D analogs with cyclosporin, with superior efficacy in heart rather than skin transplantation. Higher dosage of the analog did not improve the immunosuppressive effect of cyclosporin and was associated with hypercalcemia.

IV. CONCLUSIONS

The initial *in vitro* observations that $1,25(OH)_2D_3$ inhibited helper T-cell function [97] were confirmed *in vivo* in animal models of autoimmunity and transplantation where a predominant helper T-cell activity had been found to be present. Moreover, studies of helper T-cell subset differentiation, lymphokine secretion, and influence of cytokines (mainly IL-12) on these T-cell responses have further characterized and defined the mode of action of $1,25(OH)_2D_3$ and predicted a role for the steroid in conditions when an overactivity of helper T-cell type 1 was observed. Because this T-cell subset was found to be sensitive to the action of cyclosporin, it prompted studies of $1,25(OH)_2D_3$ in disease states that were known to respond to cyclosporin in addition to studies analyzing the additive or synergistic effect of both drugs.

The evidence suggests that $1,25(OH)_2D_3$ and related analogs exert similar actions on the immune system but differ in the degree of immunosuppression induced and their abilities to mobilize calcium resulting in hypercalcemia. Unfortunately, in many instances, optimal immunosuppression could not be achieved without some de-

TABLE IV $1,25(OH)_2D_3$ and Analogs in Transplantation

Animal	Organ	Vitamin D_3 or analog	Dose (μg/kg/day)	Outcome, treated/control	Serum calcium (mg/dl)[a]	Ref.
Mouse	Skin	$1,25(OH)_2D_3$	0.7	9.7–14.5 days/9.7 days	N/A	Jordan *et al.* [89]
		$1,25(OH)_2D_3$	0.2–0.4	13.9–15.5 days/10.3 days	N/A	Veyron *et al.* [90]
		KH1060	0.2–0.4	19.6–24.5 days/10 days	9.9–11.4	
		MC1288	0.2	22.2 days/10 days	10	Veyron *et al.* [91]
		MC1301	0.2	19.6 days/10 days	10	
		MC1357	0.2	19.6 days/10 days	8	
		CB966	0.2–0.4	14.3–18.7 days/10.3 days	N/A	Veyron *et al.* [90]
	Heart	$1,25(OH)_2D_3$	5 every second day	12.6 days/11.6 days	10.5	Lemire *et al.* [88]
		$1,25(OH)_2$-16-ene D_3	7.5–10	18.2–27.2 days/11.6 days	10.2–11.8	
Rat	Heart	$1,25(OH)_2D_3$	1.0	10 days/6 days	Toxicity (death)	Jordan *et al.* [92]
		KH1060	0.2–6.0	7–9 days/6 days	≥12.2	Lewin and Olgaard [93]
		MC1288	0.1	22 days/	—	Johnsson and Tufveson [94]
	Small bowel	MC1288	0.1	9.5 days/14 days (death)	N/A	
Dog	Kidney	$1,25(OH)_2$-16-ene D_3	0.025	5 days/5.5 days	10.6	Lemire [95]

[a] Serum calcium of treated animals. N/A, Not available.

gree of hypercalcemia. It seems very unlikely that the natural hormone $1,25(OH)_2D_3$ could ever be used as a sole agent for profound immunosuppression such as is required in transplantation, owing to its hypercalcemic properties. However, the hormone could have a role in conditions where the immune activity is less intense or in conditions involving a more localized site such as the skin (scleroderma, psoriasis). The search for the ideal vitamin D_3 analog with superior immunosuppression without hypercalcemic activity is in progress. Historically, steroid therapy has been used effectively but with significant toxicities; the replacement of prednisone with a vitamin D_3 analog without toxicity would be a major contribution to the field of transplantation and autoimmunity.

TABLE V Combined Effects of Vitamin D_3 Analog with Cyclosporin on Transplantation

Analog (μg/kg/day)	Cyclosporin (μg/kg/day)	Graft survival (days)	Serum calcium (mg/dl)
KH1060 (skin/mouse) [90]			
0	0	10.9	9.2
0.2	0	18.2	11.4
0	20	15.1	9.2
0.1	20	23.4	10.4
MC1288 (heart/rat) [96]			
0.05	0	12.0	N/A[a]
0.10	0	22.0	N/A
0	5	17.0	N/A
0	10	19.0	N/A
0.05	5	21.5	N/A
0.05	10	25.0	12.4
0.1	5	27.0	N/A

[a] N/A, Not available.

References

1. D'Andrea A, Rengaraju M, Valiante NM, Chehimi J, Kubin M, Aste M, Chan M, Kobayashi M, Young D, Nickbarg E, Chizzonite R, Wolf SF, Trinchieri G 1992 Production of natural killer cell stimulatory factor (interleukin-12) by peripheral blood mononuclear cells. J Exp Med **176**:1387–1398.
2. Manetti R, Parronchi P, Guidizi MG, Piccinni M-P, Maggi E, Trinchieri G, Romagnani S 1993 Natural killer cell stimulatory factor [interleukin 12(IL-12)] induces T helper type 1 (Th_1)-specific immune responses and inhibits the development of IL-4 producing Th cells. J Exp Med **177**:1199–1204.
3. Mosmann TR, Coffman RL 1989 Th_1 and Th_2 cells: Different patterns of lymphokine secretion lead to different functional properties. Annu Rev Immunol **7**:145–173.
4. Cher DJ, Mosmann TR 1987 Two types of murine helper T cell clone. II. Delayed-type hypersensitivity is mediated by Th_1 clones. J Immunol **138**:3688–3694.
5. Koizumi T, Nakao Y, Matsui T, Nakagawa I, Matsuda S, Komoriya K 1985 Effects of corticosteroid and 1,24R-dihydroxy-vitamin D_3 administration on lymphoproliferative and autoimmune disorders in MRL/MP-Lpr/lpr mice. Int Arch Allergy Appl Immunol **77**(4):396–404.
6. Gazzinelli R, Oswald I, Hieny S, James S, Sher A 1992 The microbicidal activity of interferon-γ treated macrophages against *Trypanosoma cruzi* involves an L-arginine-dependent, nitrogen oxide-mediated mechanism inhibitable by interleukin-10 and transforming growth factor β. Eur J Immunol **22**:2501–2506.
7. Mosmann TR 1991 Cytokine secretion patterns and crossregulation of T cell subsets. Immunol Res **10**:183–188.

8. Mosmann TR, Moore KW 1991 The role of IL-10 in crossregulation of TH1 and TH2 responses. Immunol Today **12A**:49–53.
9. D'Andrea A, Aste-Amerzaga M, Valiante NM, Ma X, Kubin M, Trinchieri G 1993 Interleukin 10 (IL-10) inhibits human lymphocyte interferon γ-production by suppressing natural killer cell stimulatory factor/IL-12 synthesis in accessory cells. J Exp Med **178**:1041–1048.
10. Basham TY, Merigan TC 1984 Recombinant interferon-γ increases HLA-DR synthesis and expression. J Immunol **130**:1492–1494.
11. Salgame P, Abrams JS, Clayberger C, Goldstein H, Convitt J, Modlin RL, Bloom BR 1991 Differing lymphokine profiles and functional subsets of human CD4 and CD8 T cell clones. Science **254**:279–281.
12. Del Prete G, De Carli M, Ricci M, Romagnani S 1991 Helper activity for immunoglobulin synthesis of T helper type 1 (Th1) and Th2 human T cell clones is limited by their cytolytic capacity. J Exp Med **174**:809–813.
13. De Carli M, D'Elios MM, Zancuoghi G, Romagnani S, Del Prete G 1994 Human Th1 and Th2 cells: Functional properties, regulation of development and role in autoimmunity. Autoimmunity **18**:301–308.
14. Cox FEG, Liew FY 1992 T-cell subsets and cytokines in parasitic infections. Immunol Today **13**:445–448.
15. Seder RA, Paul WE, Davis MM, Fazekas de St Groth B 1992 The presence of interleukin 4 during in vitro priming determines the lymphocyte-producing potential of CD4+ T cells from T cell receptor transgenic mice. J Exp Med **176**:1091–1098.
16. Tuohy VK, Lu Z, Sobel RA, Laursen RA, Lees MB 1989 Identification of an encephalitogenic determinant of myelin proteolipid protein for SJL mice. J Immunol **142**:1523–1527.
17. Baron JL, Madri JA, Ruddle NH, Hashim G, Janeway CA Jr 1993 Surface expression of α4 integrin by CD4 T cells is required for their entry into brain parenchyma. J Exp Med **177**:57–68.
18. Kennedy MK, Torrance DS, Picha KS, Mohler KM 1992 Analysis of cytokine mRNA expression in the central nervous system of mice with experimental allergic encephalomyelitis reveals that IL-10 mRNA expression correlates with recovery. J Immunol **149**:2496–2505.
19. Bach JF 1994 Insulin-dependent diabetes mellitus as an autoimmune disease. Endocr Rev **15**:516–541.
20. Berney T, Fulpius T, Shibata T, Reininger L, Van Snick J, Shan H, Weigert M, Marshak-Rothstein A, Izuu S 1992 Selective pathogenicity of murine rheumatoid factors of the cryoprecipitable IgG3 subclass. Int Immunol **4**:93–99.
21. Slack JH, Hang LM, Barkley J, Fulton RJ, O'Hoostelaere L, Robinson A, Dixon FJ 1984 Isotypes of spontaneous and mitogen-induced autoantibodies in SLE-prone mice. J Immunol **132**:1271–1275.
22. Theofilopoulos AN, Balderas RS, Hang LM, Dixon FJ 1983 Monoclonal IgM rheumatoid factors derived from arthritic MRL/Mp-lpr/lpr mice. J Exp Med **158**:901–906.
23. Zhang J, Markovic-Plese S, Lacet B, Raus J, Weiner H, Hafler DA 1994 Increased frequency of interleukin 2-responsible T cells specific for myelin basic protein and proteolipid protein in peripheral blood and cerebrospinal fluid of patients with multiple sclerosis. J Exp Med **179**:973–984.
24. Panitch HS, Hirsch RL, Schindler J, Johson KP 1987 Treatment of multiple sclerosis with γ interferon. Exacerbations associated with activation of the immune system. Neurology **37**:1097–1102.
25. De Carli M, D'Elios MM, Mariotti S, Marcocci C, Pinchera A, Ricci M, Romagnani S, Del Prete GF 1993 Cytolytic T cells with Th1-like cytokine profile predominate in retroorbital lymphocytic infiltrates of Graves' ophthalmopathy. J Clin Endocrinol Metab **77**:1120–1124.
26. Quayle AJ, Chomarat P, Miossec P, Kjeldsen-Kragh J, Forre O, Natvig JB 1993 Rheumatoid inflammatory T-cell clones express mostly Th1 but also Th2 and mixed (ThO) cytokine patterns. Scand J Immunol **38**:75–82.
27. Foulis AK, McGill M, Farquharson MA 1991 Insulitis in type I (insulin-dependent) diabetes mellitus in man. Macrophage lymphocytes, and interferon-γ-containing cells. J Pathol **165**:97–103.
28. Al-Janadi M, Al-Balla S, Al-Dalaan A, Raziuddin S 1993 Cytokine profile in systemic lupus erythematosus, rheumatoid arthritis, and other rheumatic diseases. J Clin Immunol **13**:58–67.
29. Lemire JM, Adams JS 1992 1,25-Dihydroxyvitamin D_3 inhibits the passive transfer of cellular immunity by a myelin basic protein specific T cell clone. J Bone Miner Res **7**:171–177.
30. Lemire JM, Weigle WO 1986 Passive transfer of experimental allergic encephalomyelitis by myelin basic protein specific L3T4+ T cell clones possessing several functions. J Immunol **37**:3169–3174.
31. Reddy GS, Clark JW, Tserng KY, Uskokovic MR, McLane JA 1993 Metabolism of 1,25$(OH)_2$16ene D_3 in kidney: Influence of structural modification of D-ring on side chain metabolism. Bioorg Med Chem Lett **3**:1879–1884.
32. Spiegelberg HL, Beck L, Stevenson DD, Ishioka GY 1994 Recognition of T cell epitopes and lymphokine secretion by rye grass allergen *lolium perenne* I-Specific human T cell clones. J Immunol **152**:4706–4711.
33. Gepner P, Bernard A, Fournier C 1989 1,25-Dihydroxyvitamin D_3 potentiates the in vitro inhibitory effects of cyclosporin A on T cells from rheumatoid arthritis patients. Arthritis Rheum **32**:31–36.
34. Lasky HP, Bauer K, Pope RM 1988 Increased helper inducer and decreased suppressor inducer phenotypes in the rheumatoid joint. Arthritis Rheum **31**:52–59.
35. Chofflon M, Gonzalez V, Weiner HL, Hafler DA 1989 Inflammatory cerebrospinal fluid T cells have activation requirements characteristic of CD4+ CD45RA-T cells. Eur J Immunol **19**:1791–1795.
36. Lemire JM, Archer DC 1991 1,25-Dihydroxyvitamin D_3 prevents the *in vivo* induction of murine experimental autoimmune encephalomyelitis. J Clin Invest **87**:1103–1107.
37. Lemire JM, Archer DC, Reddy GS 1994 1,25-Dihydroxy-24-oxo-16ene-vitamin D_3, a renal metabolite of the vitamin D analog 1,25-dihydroxy-16ene-vitamin D_3, exerts immunosuppressive activity equal to its parent without causing hypercalcemia *in vivo*. Endocrinology **135**:2818–2821.
38. Lemire JM, Archer DC, Beck L, Reddy GS, Uskokovic MR, Spiegelberg HL 1994 The role and mechanism of vitamin D analogs in immunosuppression. In: Norman AW, Bouillon R, Thomasset M (eds) Vitamin D. A Pluripotent Steroid Hormone: Structural Studies, Molecular Endocrinology and Clinical Applications. de Gruyter, Berlin, New York, pp. 531–539.
39. Lemire JM, Archer DC, Binderup L 1997 The vitamin D analog, 20-epi-1,25$(OH)_2D_3$ (MC1288) prevents the onset of autoimmunity with reduced calcemic properties. Submitted for publication.
40. Fournier C, Gepner P, Sadouk M, Charreire J 1990 In vivo beneficial effects of cyclosporin A and 1,25-dihydroxyvitamin D_3 on the induction of experimental autoimmune thyroiditis. Clin Immunol Immunopathol **54**:53–63.
41. Boissier MC, Chiocchia G, Fournier C 1992 Combination of cyclosporin A and calcitriol in the treatment of adjuvant arthritis. J Rheumatol **19**:754–757.
42. Branisteau DD, Leenaerts P, van Damme B, Bouillon R 1993

Partial prevention of active Heymann nephritis by 1α,25-dihydroxyvitamin D_3. Clin Exp Immunol **94**:412–417.
43. Lillevang ST, Rosenkvist J, Andersen CB, Larsen S, Kemp E, Kristensen T 1992 Single and combined effects of the vitamin D analogue KH1060 and cyclosporin A on mercuric chloride-induced autoimmune disease in the BN rat. Clin Exp Immunol **88**:301–306.
44. Hattori M 1990 Effect of 1α,25$(OH)_2D_3$ on experimental rat nephrotoxic serum nephritis. Nippon Jinzo Gakkai Shi **32**:147–159.
45. Koizumi T, Nakao Y, Matsui T, et al. 1985 Effects of corticosteroid and 1,24*R*-dihydroxy-vitamin D_3 administration on lymphoproliferative and autoimmune disorders in MRL/MP-Lpr/lpr mice. Int Arch Allergy Appl Immunol **77**:396–404.
46. Abe J, Takita Y, Nakano T, Miyaura C, Suda T, Nishii Y 1990 22-Oxa-1α-25-dihydroxyvitamin D_3: A new synthetic analogue of vitamin D having a potent immunoregulating activity without inducing hypercalcaemia in mice. In: Cohn DV, Glorieux FH, Martin TJ (eds) Calcium regulation and bone metabolism. Amsterdam, Elsevier, 146–151.
47. Lemire JM, Ince A, Takasima M 1992 1,25-Dihydroxyvitamin D_3 attenuates the expression of experimental murine lupus of MRL/l mice. Autoimmunity **12**:143–148.
48. Mathieu C, Laureys J, Sobis H, Vandeputte M, Waer M, Bouillon R 1992 1,25-Dihydroxyvitamin D_3 prevents insulitis in NOD mice. Diabetes **41**(11):1491–1495.
49. Mathieu C, Waer M, Casteels K, Laureys J, Bouillon R 1995 Prevention of type I diabetes in NOD mice by nonhypercalcemic doses of a new structural analog of 1,25-dihydroxyvitamin D_3, KH1060. Endocrinology **136**(3):866–872.
50. Branisteanu DD, Waer M, Sobis M, Marcelis S, Vandeputte M, Bouillon R 1995 Prevention of murine experimental allergic encephalomyelitis: Cooperative effects of cyclosporine and 1α1,25-$(OH)_2$-D_3. J Neuroimmunol **61**:151–160.
51. White DJG, Plumb AM, Pawelec G, Brons G 1979 Cyclosporine A: An immunosuppressive agent preferentially active against proliferating T cells. Transplantation **27**:55–58.
52. Wiesinger D, Borel JF 1980 Studies on the mechanism of action of cyclosporine A. Immunobiology **156**:454–463.
53. Kaufmann Y, Chang AE, Robb RJ, Rosenberg SA 1984 Mechanism of action of cyclosporin A: Inhibition of lymphokine secretion studied with antigen-stimulated T cell hybridomas. J Immunol **133**:3107–3111.
54. Bunjes D, Hardt C, Rollinghoff M, Wagner H 1981 Cyclosporine A mediates immunosuppression of primary cytotoxic T cell responses by impairing the release of interleukin 1 and interleukin 2. Eur J Immunol **11**:657–661.
55. Andrus L, Lafferty KJ 1982 Inhibition of T-cell activity by cyclosporin A. Scand J Immunol **15**:449–458.
56. Kronke M, Leonard WJ, Depper JM, Araya SK, Wongstaal F, Gallo RC, Waldmann TA, Greene WC 1984 Cyclosporin A inhibits T-cell growth factor gene expression at the level of mRNA transcription. Proc Natl Acad Sci USA **81**:5214–5218.
57. Morris RJ, Mason DW, Hutchinson IV 1983 The effect of cyclosporine A on lymphocytes in animal models of tissue transplantation. Transplant Proc **15**:2287–2289.
58. Hess AD, Tutschka PJ 1980 Effect of cyclosporin A on human lymphocyte responses *in vitro:* I. CsA allows for the expression of alloantigen-activated suppressor cells while preferentially inhibiting the induction of cytolytic effector lymphocytes in MLR. J Immunol **124**:2601–2608.
59. Burckhardt JJ, Guggenheim B 1979 Cyclosporin A: *In vivo* and *in vitro* suppression of rat T-lymphocyte function. Immunology **36**:753–757.
60. Umehara H, Kumagai S, Ishida H, Suginoshita T, Maeda M, Imura H 1988 Enhanced production of interleukin-2 in patients with progressive systemic sclerosis: Hyperactivity of CD4-positive T cells? Arthritis Rheum **31**:401–407.
61. Whiteside T, Kumagai Y, Roumm A, Almendinger R, Rodnan G 1983 Suppressor cell function and T lymphocyte subpopulations in peripheral blood of patients with progressive systemic sclerosis. Arthritis Rheu **26**:841–847.
62. Inoshita T, Whiteside T, Rodnan G, Taylor F 1981 Abnormalities of T lymphocyte subsets in patients with progressive systemic sclerosis (PSS, scleroderma). J Lab Clin Med **97**:264–277.
63. Humbert P, Dupond JL, Agache P, Laurent R, Rochefort A, Drobacheff C, de Wazieres B, Aubin F 1993 Treatment of scleroderma with oral 1,25-dihydroxyvitamin D_3: Evaluation of skin involvement using non-invasive techniques. Acta Derm Venereol (Stockh) **73**:449–451.
64. Humbert P, Delaporte E, Dupond JL, Rochefort A, Laurent R, Drobacheff C, De Wazieres B, Bergoend H, Agache P 1994 Treatment of localized scleroderma with oral 1,25-dihydroxyvitamin D_3. Eur J Dermatol **4**:21–3.
65. Muller K, Kriegbaum NJ, Baslund B, Sorensen OH, Thymann M, Bentzen K 1995 Vitamin D_3 metabolism in patients with rheumatic diseases: Low serum levels of 25-hydroxyvitamin D_3 in patients with systemic lupus erythematosus. Clin Rheumatol **14**:397–400.
66. Czernobilsky H, Scharla S, Schmidt-Gayk H, Ziegler R 1988 Enhanced suppression of 1,25$(OH)_2D_3$ and intact parathyroid hormone in Graves' disease as compared to toxic nodular goiter. Calcif Tissue Int **42**:5–12.
67. Gates S, Shary J, Turner RT, Wallach S, Bell NH 1986 Abnormal calcium metabolism caused by increased circulating 1,25-dihydroxyvitamin D in a patient with rheumatoid arthritis. J Bone Miner Res **1**:221–226.
68. Thomas FT, Thomas JM, Ganghoff O, Gross U 1986 Mechanisms of cell-mediated rejection. In: Kidney Transplant Rejection. Dekker, New York, pp. 29–53.
69. Brent L, Brown I, Medawar P 1962 Quantitative studies on tissue transplantation immunity. IV. Hypersensitivity reactions associated with the rejection of homografts. Proc R Soc Lond [Biol] **156**:187–209.
70. Farrar J, Benjamin W, Hilfiker M, Howard M, Farrar W, Fuller-Farrar J 1967 The biochemistry, biology and role of interleukin 2 in the induction of cytotoxic T cell and antibody forming B cell responses. Immunol Rev **63**:129–166.
71. Kewase I, Brooks C, Kuribayashi K, Glabuenaga S, Newman W, Gillis S, Henney CS 1983 Interleukin 2 induces interferon production: Participation of macrophages and NK-like cells. J Immunol **131**:228–292.
72. Fischer D, Golightly G, Koren H 1983 Potentiation of the cytolytic activity of peripheral blood monocytes by lymphokines and interferon. J Immunol **130**:1220–1225.
73. Tokuda N, Mano T, Levy RB 1990 1,25-Dihydroxyvitamin-D_3 antagonizes interferon-gamma-induced expression of class II major histocompatibility antigens on thyroid follicular and testicular Leydig cells. Endrocrinology **127**:1419–1427.
74. Schwartz R 1984 The role of gene products of the major histocompatibility complex in T cell activation and cellular interactions. In: Paul WE (ed) Fundamental Immunology. Raven Press, New York, pp. 379–437.
75. Fink PJ, Rammensee H-G, Benedetto JD, Staerz UD, Lefrancois L, Bevan MJ 1984 Cloned cytolytic T cells can suppress primary cytotoxic responses directed against them. J Immunol **133**:1775–1781.
76. Fink PJ, Rammensee H-G, Bevan MJ 1984 Studies on the mechanism of suppression of primary cytotoxic responses by cloned cytotoxic T lymphocytes. J Immunol **133**:1769–1774.

77. Thomas FT, Lee HM, Lower RR, Thomas JM 1979 Immunological monitoring as a guide to the management of transplant recipients. Surg Clin North Am **59**:253.
78. Dallman, MJ 1995 Cytokines and transplantation: Th1/Th2 regulation of the immune response to solid organ transplants in the adult. Curr Opin Immunol **7**:632–638.
79. Meehan MA, Kerman RH, Lemire JM 1992 1,25-Dihydroxyvitamin D_3 enhances generation of non-specific suppressor cells while inhibiting the induction of cytotoxic cells in a human MLR. Cell Immunol **140**:400–409.
80. Mathieu C, Bouillon R, Rutgeerts O, Vandeputte M, Waer M 1994 Potential role of 1,25-$(OH)_2$ vitamin D_3 as a dose-reducing agent for cyclosporine and FK 506. Transplant Proc **26**(6):3130.
81. Kohl S, Starr SE, Oleske JM, Shore SL, Ashman RB, Nahmias AJ 1977 Human monocyte–macrophage-mediated antibody-dependent cytotoxicity to herpes simplex virus-infected cells. J Immunol **118**:729–734.
82. Lemire JM, Ince A, Cox P, Kohl S 1990 1,25-Dihydroxyvitamin D_3 selectively interferes with cellular mechanisms of cytotoxicity. Kidney Int **37**:420.
83. Merino F, Alvarez-Mon M, De la Hera A, Ales JR, Bonilla F, Durantez A 1989 Regulation of natural killer cytotoxicity by 1,25-dihydroxyvitamin D_3. Cell Immunol **118**:328–336.
84. Leung KH 1989 Inhibition of human natural killer cell and lymphokine-activated killer cell cytotoxicity and differentiation by vitamin D_3. Scand J Immunol **30**:199–208.
85. Trentin L, Zambello R, Faggian G, Livi E, Thiene G, Gasparotto G, Agostini C 1992 Phenotypic and functional characterization of cytotoxic cells derived from endomyocardial biopsies in human cardiac allografts. Cell Immunol **141**:332–341.
86. Selvan RS, Nagarkatti PS, Nagarkatti M 1990 Role of IL-2, IL-4 and IL-6 in the growth and differentiation of tumor-specific CD4+ T helper and CD8+ T cytotoxic cells. Int J Cancer **45**:1096–1104.
87. Johnsson C, Binderup L, Tufveson G 1995 Vitamin D analogues for immunosuppression: A comparison between various routes of administration. Transplant Proc **27**:3538–3539.
88. Lemire JM, Archer DC, Khulkarni A, Ince A, Uskokovic MR, Stephkowsky S 1992 The vitamin D_3 analogue, 1,25-dihydroxy-Δ^{16}-cholecalciferol prolongs the survival of murine cardiac allografts. Transplantation **54**:762–763.
89. Jordan SC, Shibuka R, Mullen Y 1988 1,25-Dihydroxyvitamin D_3 prolongs skin graft survival in mice. In: Norman AW, Schaefer K, Grigoleit H-G, von Herrath D (eds) Vitamin D: Molecular, Cellular and Clinical Endocrinology. de Gruyter, Berlin, pp. 346–347.
90. Veyron P, Pamphile R, Binderup L, Touraine J-L 1993 Two novel vitamin D analogues, KH 1060 and CB 966, prolong skin allograft survival in mice. Transplant Immunol **1**:72–76.
91. Veyron P, Pamphile R, Binderup L, Touraine J-L 1995 New 20-epi-vitamin D3 analogs: Immunosuppressive effects on skin allograft survival. Transplant Proc **27**:450.
92. Jordan SC, Nigata M, Mullen Y 1988 1,25-Dihydroxyvitamin D_3 prolongs rat cardiac allograft survival. In: Norman AW, Schaefer K, Grigoleit H-G, von Herrath D (eds) Vitamin D: Molecular, Cellular and Clinical Endocrinology. de Gruyter, Berlin, pp. 334–335.
93. Lewin E, Olgaard K 1994 The *in vivo* effect of a new, *in vitro*, extremely potent vitamin D_3 analog KH1060 on the suppression of renal allograft rejection in the rat. Calcif Tissue Int **54**:150–154.
94. Johnsson C, Tufveson G 1994 MC 1288. A vitamin D analogue with immunosuppressive effects on heart and small bowel grafts. Transplant Int **7**:392–397.
95. Lemire JM 1996 Unpublished observations.
96. Johnsson C, Binderup L, Tufveson G 1995 The effects of combined treatment with the novel vitamin D analogue MC1288 and cyclosporine A on cardiac allograft survival. Transplant Immunol **3**:245–250.
97. Lemire JM, Adams J, Kermani-Arab V, Bakke A, Sakai R, Jordan C 1985 1,25-Dihydroxyvitamin D_3 suppresses human helper/inductor lymphocyte activity in vitro. J Immunol **34**:3032–3035.
98. Lemire JM, Archer DC, Beck L, Spiegelberg HL 1995 Immunosuppressive actions of 1,25-Dihydroxyvitamin D_3. Preferential inhibition of Th_1 functions. J Nutr **125**:1704S–1708S.

CHAPTER 70

Vitamin D and Diabetes

CHANTAL MATHIEU, KRISTINA CASTEELS, AND ROGER BOUILLON
Laboratory for Experimental Medicine and Endocrinology (LEGENDO),
Catholic University of Leuven, Belgium

I. INTRODUCTION

The influence of insulin on vitamin D and bone homeostasis is complex. Insulin enhances the 1α-hydroxylation of 25-hydroxyvitamin D *in vitro* [1], whereas insulin-like growth factor (IGF) also mediates the effects of hypophosphatemia on renal 1,25-dihydroxyvitamin D_3 [$1,25(OH)_2D_3$] production [2]. Diabetic rats with either spontaneous diabetes [3,4] or streptozotocin-induced diabetes [5–7] have decreased serum concentrations of $1,25(OH)_2D_3$.

Insulin-dependent diabetic children, and also poorly controlled diabetic adults, have decreased serum concentrations of $1,25(OH)_2D_3$ [8–10], whereas good diabetes control normalizes circulating $1,25(OH)_2D_3$ [9,11]. Most of the effects of insulin deficiency on serum $1,25(OH)_2D_3$ can, however, be explained by decreased concentrations of the vitamin D binding protein (DBP) [5,12]. Indeed, the DBP-corrected or calculated free concentrations of $1,25(OH)_2D_3$ are either normal or slightly increased in both diabetic animals and humans [12]. Kinetic studies in spontaneously diabetic BB rats clearly showed increased metabolic clearance (+50%) of $1,25(OH)_2D_3$ but normal production rates. No such studies have been performed in human diabetics.

The classic vitamin D target tissues are, however, differently affected by insulin deficiency: the basal intestinal active calcium absorption and duodenal calbindin D concentration are decreased [3,13–15], but calbindin and intestinal calcium absorption are still responsive to low calcium intake or $1,25(OH)_2D_3$ therapy [12]. The renal calbindin D_{28K} is not affected by insulin deficiency. At the bone level, however, a marked decrease in osteoblast number and function has repeatedly been observed in both diabetic animals and humans [16,17]. Therapy with $1,25(OH)_2D_3$, even with high doses, clearly shows marked resistance of the osteoblasts to $1,25(OH)_2D_3$ in contrast to the near normal responsiveness of the intestinal mucosa [12]. The osteoblast response, however, can be quickly normalized by insulin infusion [18], whereas IGF-I therapy is partly effective both on $1,25(OH)_2D_3$ concentration [19] and osteoblast function [17]. Insulin and IGF-1 thus have marked effects on the kinetics of $1,25(OH)_2D_3$ (production and catabolism) as well as on the sensitivity of classic target tissues (bone, intestine, kidney) to this secosteroid hormone.

This chapter mainly focuses on the effects of vitamin D and its activated form, $1,25(OH)_2D_3$, on the pathogenesis of the disease diabetes itself. Diabetes mellitus is a common disease in the Western world, with an estimated prevalence of 4 to 5%. The majority (90%) of diabetic patients suffer from type II diabetes or non-insulin-dependent diabetes mellitus (NIDDM), a metabolic syndrome characterized by insulin resistance and relatively inadequate insulin production by the β cell in the islets of Langerhans of the pancreas [20]. In this metabolic syndrome it is still unclear whether the primary dysfunction is localized to the peripheral target organs of insulin (being mainly muscle) or to the β cell [21]. Insulin resistance is induced by obesity and sedentary life style and is also involved in the pathogenesis of cardiovascular disease. This insulin resistance is probably the major determinant of the NIDDM, but β cell dysfunction is always present and will determine the severity of the clinical presentation. Type I diabetes, also known as juvenile or insulin-dependent diabetes mellitus (IDDM), is etiologically a totally different disease. It has become clear that IDDM is an autoimmune disorder, related, for example, to Graves' disease and characterized by a destruction of the insulin-producing β cells in the pancreas by the body's own immune system [22]. Whereas NIDDM or type II diabetes is a typical disease of the obese and aging patient, IDDM or type I diabetes mainly occurs in children and adolescents.

Since vitamin D receptors (VDR) were demonstrated in the main cells involved in the pathogenesis of both types of diabetes [23,24], scientific and clinical interest has focused on these molecules with respect to their potential role in the pathogenesis of the diseases, but even more with respect to a therapeutic potential in their prevention [25,26]. The actions of $1,25(OH)_2D_3$ on the β cell that may have direct implications for the pathogenesis and prevention of mainly type II but also type I diabetes are described and discussed. In addition, the role of $1,25(OH)_2D_3$ in the prevention of type I diabetes and its effects on the immune system of the animal models for the disease are covered.

II. VITAMIN D AND THE β CELL

A. Metabolic Effects

Early observations in 1980 by Norman *et al.* [27] indicated that pancreatic insulin secretion is selectively inhibited by vitamin D deficiency. Since then several reports have demonstrated an active role for the vitamin D activated form, $1,25(OH)_2D_3$, in the regulation of the function of the endocrine pancreas, especially the β cell. Receptors for $1,25(OH)_2D_3$ have been described in β cells [28]. Not only is the classic VDR present, but other observations, based on the transcaltachia phenomenon (see Chapter 15), also raise the possibility of the existence of a receptor localized in the plasma membrane [28–32]. In addition to VDR, the effector part of the vitamin D machinery is present in β cells, including the demonstration of immunoreactive vitamin D-dependent calbindin D_{28K} (CaBP) [33].

1. Effects *in Vivo* in Animal Models

The early observations in this field mainly focused on the effects of vitamin D deficiency on insulin secretion and glucose tolerance *in vivo* in several animal models and humans [27,34–40]. Vitamin D deficiency clearly impairs secretion of insulin (but not the other pancreatic hormones) and induces glucose intolerance, whereas addition of vitamin D to the diet restored the abnormalities to normal [29,34–39]. The effects of vitamin D deficiency and repletion on glucose homeostasis *in vivo* are, however, diverse and not only involve the β cell. Rachitic animals do not have normal food intake, and they lose weight and cannot maintain a normal plasma calcium level. Moreover, these metabolic changes by themselves cause hypocalcemia which directly impairs Ca^{2+} handling in the β cell and provokes β cell dysfunction and glucose intolerance [40,41].

2. Effects *in Vitro*

The results of *in vivo* experiments failed to conclusively resolve whether the effects of vitamin D deficiency on glucose tolerance were direct effects on the β cell or were due to hypocalcemia. However, more conclusive data were obtained by *in vitro* experiments. Islets isolated from rachitic animals show impaired insulin release when cultured *in vitro* and challenged with glucose [42,43]. These abnormalities in insulin secretion can be abolished by coculturing the islets in the presence of high concentrations of $1,25(OH)_2D_3$ [29,42–44]. The data on islets isolated from normal animals are even more convincing. Most papers show an enhancement of insulin release on glucose challenge in the presence of high doses of $1,25(OH)_2D_3$ [32,44,45]. Interpretation of the findings is sometimes made difficult by the many different methods used, including variations in incubation time with $1,25(OH)_2D_3$, animal source of islets, and type of glucose challenge, but overall an improvement in β cell function is realized in the presence of $1,25(OH)_2D_3$. Interestingly, some authors also studied the effects of $1,25(OH)_2D_3$ on insulin synthesis and even this compound was normalized or increased by $1,25(OH)_2D_3$ further contributing to normal β cell function [34, C. Mathieu, unpublished data, 1996].

Progressively, insight has been gained into the mechanism by which $1,25(OH)_2D_3$ might act on insulin secretion: a significant rise in cytosolic calcium levels is observed after incubation with vitamin D and precedes insulin secretion. Controversy still exists as to whether only an influx of extracellular calcium is responsible for this rise or whether mobilization of intracellular calcium reserves is also involved [28,32,42,43]. In the older studies, mainly the genomic pathway was examined, and effects on calcium mobilization and insulin secretion were only observed after long treatment periods [42–

45]. However, the nongenomic pathway, involving a putative plasmalemmal receptor inducing rapid calcium fluxes (transcaltachia), has also been implicated in the effects of vitamin D on insulin secretion [28–30,32] (see Chapter 15).

3. EFFECTS *in Vivo*—CLINCAL IMPLICATIONS

On the basis of observations *in vitro* and in animal models of vitamin D deficiency *in vivo*, clinical trials with vitamin D or $1,25(OH)_2D_3$ have been performed. The most obvious situation where vitamin D repletion would affect glucose tolerance was the rachitic state. As expected, a correction of the rachitic state with vitamin D (and calcium) restores glucose tolerance to normal [46]. More interesting are studies of the effects of $1,25(OH)_2D_3$ repletion in the relatively $1,25(OH)_2D_3$-deficient state of uremia [47]. In the study by Allegra *et al.* [47], uremic patients with lowish $1,25(OH)_2D_3$ serum levels were treated with 0.5 μg of $1,25(OH)_2D_3$ ($\pm$500 mg calcium) per day for 21 days. Intravenous glucose tolerance tests (IVGTTs) were performed prior to and after treatment. Interestingly, $1,25(OH)_2D_3$ ($\pm$500 mg calcium) caused an increment in glucose-induced insulin response only in the first few minutes of the stimulation test. In uremia this rapid phase of insulin release is especially disturbed. In this study repletion of $1,25(OH)_2D_3$ could not completely reverse the glucose intolerance. Orwoll *et al.* perfomed an interesting pilot study on possible clinical applications of $1,25(OH)_2D_3$ in a situation of impaired insulin secretion without vitamin D deficiency [48]. Type II diabetic patients received $1,25(OH)_2D_3$ (1 μg/day for 4 days) in a double blinded, placebo controlled, crossover trial. Clear effects of $1,25(OH)_2D_3$ treatment were noted on parameters of calcium metabolism, but no alteration of glucose tolerance or insulin secretion was seen. Since the lack of effect on insulin secretion could be due to the short duration of $1,25(OH)_2D_3$ treatment, this approach is not conclusive. An additional interesting intervention would be to test the newer structural analogs of $1,25(OH)_2D_3$, after screening them *in vitro* for their β cell effects, in the various models of defective insulin secretion.

B. Effects of Vitamin D on β Cell Characteristics

In the pathogenesis of type I diabetes, β cell damage by cytokines and other inflammatory agents might play an important role. Sandler *et al.* demonstrated that $1,25(OH)_2D_3$ and some of its new structural analogs counteract the suppressive effects of interleukin-1β (IL-1β) on β cell function, such as insulin synthesis and insulin secretion [49]. Similar effects were observed in the case of impairment of β cell function induced by interferon-γ (IFN-γ). Hahn *et al.* demonstrated that $1,25(OH)_2D_3$ and some of its analogs can prevent the IFN-γ-induced decrease in insulin synthesis and secretion (H. J. Hahn, personal communication, 1995). These effects were observed at high concentrations of $1,25(OH)_2D_3$ (5–10 n*M*). On the other hand, Mauricio *et al.* could not demonstrate any effects of $1,25(OH)_2D_3$ on IL-1β-induced β cell dysfunction [50]. The main difference between the paper describing a protective effect and this latter study is the incubation time; in the former study incubation periods ranging between 48 and 72 hr were used, whereas in the latter study islets were incubated with $1,25(OH)_2D_3$ for only 24 hr. Considering other observations made under similar conditions [42–45], the timing of the incubation might be the crucial point in determining whether there is an effect of $1,25(OH)_2D_3$.

Preliminary data from our laboratory (H. J. Hahn, personal communication, 1995) indicate that not only are the effects of cytokines on β cell function altered by $1,25(OH)_2D_3$, but also the induction of surface markers by these cytokines appears to be blocked. When neonatal rat islets were incubated with IFN-γ, several surface markers such as major histocompatibility complex (MHC) class II molecules and adhesion molecules (ICAM1) were up-regulated as expected. Coincubating the islets with $1,25(OH)_2D_3$ or one of its analogs decreased markedly the up-regulation of MHC class II molecules after IFN-γ stimulation (15% MHC class II positive β cells versus 30% in the presence of IFN-γ only, $p < 0.001$).

These data showing β cell protection against inflammatory agents involved in the pathogenesis of type I diabetes may have direct implications for the observed *in vivo* effects of $1,25(OH)_2D_3$ and its analogs in the prevention of type I diabetes in animal models.

III. VITAMIN D AND THE IMMUNE SYSTEM IN DIABETES MELLITUS

Prevention of type I juvenile diabetes (IDDM) is one of the major goals of health care for the future. When strategies for prevention of a disease are to be developed, a first requisite is knowledge of the pathogenesis of the disease. From studies in two animal models for type I diabetes, the NOD mouse and the BB rat, but also in humans, it has become clear that type I diabetes can be considered an autoimmune disease. As such, a central role in the destruction of the β cell is played by the body's own immune system. In this scheme of destruction of β cells almost all immune cells [monocytes–macrophages, T lymphocytes, B lymphocytes, natural killer (NK) cells] play a role.

Most prevention studies up to now have been carried out in the NOD mouse and can be divided into four major categories: pure immunosuppression, immunomodulation, antigen (specific) tolerance induction, and β cell protection. Results in NOD mice are promising for many of these treatments, but many obstacles to human applications still exist. Therapeutic plans that involve long-term immunosuppression are not a conceivable strategy for the prevention of a chronic disease that strikes mainly children. Moreover, the preliminary results of immunosuppressive drug trials following the recent onset of diabetes have been disappointing. At the moment, a major approach is focused on immunomodulation and β cell protection, two characteristics exhibited by $1,25(OH)_2D_3$ and many of its newer analogs.

Because $1,25(OH)_2D_3$ can be produced by monocytes–macrophages, and because VDR are present in several immune cells, a physiological role for this locally produced hormone as a messenger or cytokine-like molecule between cells of the immune system is probable (see Chapter 55). This raises the therapeutic possibility that $1,25(OH)_2D_3$ or its analogs might be useful in the prevention of this autoimmune disease. Moreover, as described above, clear β cell protective effects of $1,25(OH)_2D_3$ against several inflammatory agents involved in β cell destruction have been observed.

A. Primary Prevention of Type I Diabetes in Animal Models by $1,25(OH)_2D_3$ and its Analogs—The NOD Mouse

1. Spontaneous Disease

a. Early Intervention—Effects of $1,25(OH)_2D_3$ on Diabetes Prevention Two valid animal models for type I diabetes in humans have been described up to now: the BB rat (BioBreeding Laboratories, Ottawa) and the NOD (nonobese diabetic) mouse. Both strains develop a disease quite similar to human diabetes in a spontaneous manner; however, the BB rat is characterized by severe immune abnormalities, and therefore most studies on pathogenesis and the development of therapeutic strategies have been performed in the NOD mouse. The NOD mouse was derived from a cataract-developing substrain of the outbred JcI-ICR mouse by selective inbreeding between 1974 and 1980 [51]. Diabetes develops spontaneously between the twelfth and thirtieth weeks of age, preceded by the histological lesion of the disease, called insulitis. Insulitis is the reflection of the infiltration of the islets of Langerhans by a mixture of immune cells.

Our group has demonstrated that chronic administration of pharmacological doses of $1,25(OH)_2D_3$ can reduce the incidence of both insulitis and diabetes in NOD mice [52,53]. Briefly, $1,25(OH)_2D_3$ was administered intraperitoneally at a dose of 5 μg per kg body weight every 2 days from the age of weaning (21 days) until the end of the life of the NOD mice (200 days). By 200 days of age, control mice developed insulitis, the histopathological lesion of type I diabetes, in 80% and overt diabetes in 56%. These rates are comparable to the incidences in the stock colony at this time. $1,25(OH)_2D_3$ reduced insulitis to 58% ($p < 0.05$) and diabetes to 8% of treated mice ($p < 0.001$). Although treatment was generally well tolerated, effects on calcium and bone metabolism were dramatic. The mice developed increases in serum calcium (2.5 ± 0.2 versus 2.2 ± 0.2 mmol/liter* in controls, $p < 0.005$) and osteocalcin levels (17.2 ± 5.7 versus 9.3 ± 1.4 nmol/liter in controls, $p < 0.001$) and decreases in bone calcium (2.9 ± 0.6 versus 6.2 ± 1.1 mg calcium/tibia in controls, $p < 0.01$). Extensive immunological screening tests, including fluorescence-activated cell sorting (FACS) analysis of major lymphocyte subsets, evaluation of T-cell proliferation (after CD3 or concanavalin A stimulation), and analysis of NK cell function, were unable to uncover major changes in the treated versus control mice. A major finding, however, was the restoration to normal of a well-known defect in the NOD mouse: the absence of suppressor cell function. This effect could be demonstrated both *in vitro* and *in vivo*. *In vitro*, cells generated in an autologous mixed lymphocyte reaction (MLR) in control NOD mice were unable to suppress a read out allogeneic MLR. Cells taken from $1,25(OH)_2D_3$-treated NOD mice displayed a suppressive capacity after an autologous MLR that was comparable to that found in a reference, non-autoimmune mouse strain (i.e., C3H mice). The presence of these suppressor cells in the $1,25(OH)_2D_3$-treated mice was even more clear *in vivo*. Splenocytes taken from $1,25(OH)_2D_3$-treated NOD mice could almost completely block the transfer of diabetes into young irradiated NOD mice by splenocytes taken from diabetic NOD mice, whereas splenocytes taken from control NOD mice had no effect on diabetes transfer.

b. Late Intervention—Effects of $1,25(OH)_2D_3$ on Diabetes Prevention The NOD mouse is considered a good model for human diabetes and not only allows

* Calcium in mmol/liter can be converted to mg/dl by dividing by 0.2495.

for testing new therapeutic drugs, but also offers the opportunity to elaborate on optimal and achievable treatment modalities. A relevant question is, for instance, whether long-term $1,25(OH)_2D_3$ treatment is necessary or whether a short-term intervention would suffice for disease prevention, and, if so, when should this short-term treatment be administered. We therefore designed an experiment in which NOD mice received $1,25(OH)_2D_3$ in different time windows: one group received $1,25(OH)_2D_3$ for their whole lifetime (from weaning until 200 days of age), another group was treated only during early life (from weaning until 100 days of age), and a third group was treated only from 100 until 200 days of age. All mice were followed up until 200 days, and the dose of $1,25(OH)_2D_3$ was 5 μg per kg body weight in all groups. A control group received vehicle (arachis oil).

As expected, insulitis incidence at 200 days was lowered from 93% (13 of 14) to 54% (8 of 15) ($p < 0.025$) in the long-term treatment group, whereas diabetes incidence was also reduced from 86% (12 of 14) in the control group to 13% (2 of 15) in the long-term $1,25(OH)_2D_3$-treated group ($p < 0.001$) (Fig. 1). In mice

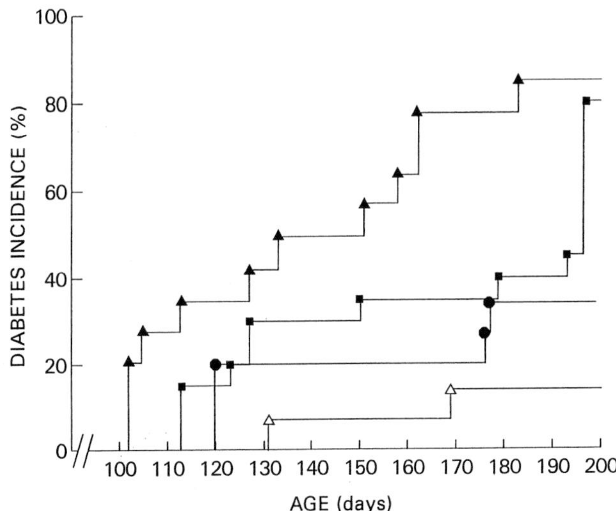

FIGURE 1 Diabetes incidence in NOD mice treated with $1,25(OH)_2D_3$ during different time windows of treatment. Mice were tested three times weekly for glycosuria and were considered diabetic when having positive glycosuria and blood glucose exceeding 12 mmol/liter (equivalent to 216 mg/dl; glucose in mmol/liter can be converted to mg/dl by dividing by 0.0555) on two consecutive days. Open triangles (△) represent the $1,25(OH)_2D_3$ group treated from 21 to 200 days of age ($n = 15$), whereas filled triangles (▲) represent the vehicle (arachis oil)-treated group ($n = 14$). Circles (●) represent the $1,25(OH)_2D_3$ group treated from 21 to 100 days of age ($n = 15$), and squares (■) represent the $1,25(OH)_2D_3$ group treated from 100 to 200 days of age ($n = 20$). Statistical analysis was performed using the chi-square test: $1,25(OH)_2D_3$ for 21–200 days or 21–100 days versus control ($p < 0.001$) and $1,25(OH)_2D_3$ for 100–200 days versus control (not significant).

treated before 100 days of age there was also a significantly lower insulitis incidence [46% (7 of 15), $p < 0.025$ versus the control group], and diabetes occurred in 33% (5 of 15) of these mice ($p < 0.001$). When therapy was initiated at day 100 (when ~75% of control mice already have insulitis), insulitis was present in 90% (18 of 20), and 80% (16 of 20) of these mice developed diabetes by day 200. These values are not significantly different from the control values. When, however, not the end points but the timing of diabetes onset was compared, the treated animals showed a delayed onset of diabetes compared to the control group ($p < 0.01$) (log rank statistics). In the group treated from 100 until 200 days of age, effects on serum calcium levels were comparable to the long-term treated group; in the group treated from 21 until 100 days of age, serum calcium levels were normal as expected (Table I). Bone turnover (reflected by serum osteocalcin levels) in the animals treated with $1,25(OH)_2D_3$ from 21 until 200 days of age was increased as previously demonstrated, whereas the bone turnover in the two groups with the shorter treatment duration was closer to the normal range. In the group receiving the treatment in early life, however, these values were still elevated. Determination of bone calcium content reflected the severe impact of $1,25(OH)_2D_3$ treatment in young animals, as their bone calcium content, even 100 days after stopping the treatment, was still significantly decreased compared to the control group ($p < 0.001$). This period seems to be crucial in bone remodeling, and interference with bone modeling in this period leaves traces for the rest of the animal's life. On the

TABLE I Effects of Different Treatment Schedules of $1,25(OH)_2D_3$ on Calcium Metabolism[a]

	Serum calcium (m*M*)	Serum osteocalcin (n*M*)	Bone calcium (mg/tibia)
Control ($n = 14$)	2.2 ± 0.2	8.8 ± 1.5	7.6 ± 0.3 ($n = 3$)
$1,25(OH)_2D_3$ days 21–200 ($n = 15$)	2.6 ± 0.2 $p < 0.005$	20.2 ± 3.3 $p < 0.0001$	4.0 ± 0.4 ($n = 3$) $p < 0.0005$
$1,25(OH)_2D_3$ days 21–100 ($n = 15$)	2.3 ± 0.1 NS	14.9 ± 1.6 $p < 0.005$	5.3 ± 0.2 ($n = 3$) $p < 0.001$
$1,25(OH)_2D_3$ days 100–200 ($n = 20$)	2.6 ± 0.2 $p < 0.0005$	13.1 ± 7 NS	4.6 ± 0.3 ($n = 2$) $p < 0.01$

[a] On the day of diabetes diagnosis or at 200 days of age, animals were bled by heart puncture. Statistical analysis was made by Student's *t* test for unpaired data and Mann Whitney *U* test. The *p* values shown are in comparison to the control values. NS, Not significant.

other hand, when treatment was initiated at 100 days of age, important bone loss with $1,25(OH)_2D_3$ treatment was also seen, although less than with the long-term treatment ($p < 0.01$).

2. Cyclophosphamide-Induced Diabetes—Insight into Mechanisms of Action of $1,25(OH)_2D_3$

Meehan *et al.* [53] observed an enhancement of suppressor cells by $1,25(OH)_2D_3$ *in vitro*, and Mathieu *et al.* [54] described a restoration of the naturally defective suppressor cell system in NOD mice protected against diabetes by $1,25(OH)_2D_3$. The question remains, however, whether this restoration of suppressor cells is the main mechanism involved in protection against diabetes by $1,25(OH)_2D_3$, because protection not only against diabetes, but also against insulitis was seen, pointing toward interference with the induction of autoimmunity itself.

Diabetes occurring in NOD mice after cyclophosphamide administration is believed to be due to elimination of regulator cells [55,56]. The time needed for recovery of the immune cells after injection of a high dose of cyclophosphamide is different among various T-cell populations. Long-lived effector T cells recover more promptly than the short-lived suppressor or regulator cells [57]. Several potential preventive therapies for diabetes have already been tested in this cyclophosphamide model. Protection against diabetes by therapeutic interventions that are believed to induce suppressor T cells but that have no effect on autoimmune effector cells can be broken by cyclophosphamide. For example, this was shown for the protection against diabetes achieved by complete Freund's adjuvant (CFA) or for the prevention of recurrence of diabetes in syngeneic islet grafts achieved by bacillus Calmette-Guérin (BCG) [58,59]. On the other hand, cyclophosphamide could not reverse the protection against diabetes in animals treated with the streptococcal preparation OK-432, as this substance prevents diabetes by inhibiting the generation of effector cells for β cell destruction [60]. In the protection against diabetes by $1,25(OH)_2D_3$ and its analogs suppressor cells also were observed, but it remained unclear whether these cells were essential in the protection.

Casteels *et al.* demonstrated that $1,25(OH)_2D_3$ can also prevent diabetes induced by cyclophosphamide in the NOD mouse [61]. In this study male NOD mice, age- and litter-matched, were treated with $1,25(OH)_2D_3$ (5 μg/kg body weight every 2 days) during different time intervals. In some groups $1,25(OH)_2D_3$ treatment was initiated at weaning, whereas in other groups treatment was started just before or after cyclophosphamide was administered. All animals received a single cyclophosphamide injection of 200 mg/kg body weight at 70 days of age.

Cyclophosphamide induced diabetes (78%) and insulitis (100%) in control animals. When $1,25(OH)_2D_3$ was administered from day 21 until day 69, protection against diabetes (17%, $p < 0.01$) and insulitis (42%, $p < 0.005$) was observed (Fig. 2). In the other groups, where treatment was initiated at weaning, protection against diabetes was also seen. This protection was achieved despite the total elimination of suppressor cells in the $1,25(OH)_2D_3$-treated group by cyclophosphamide, as shown by cotransfer experiments. In addition, $1,25(OH)_2D_3$ treatment did not interfere with the quantitative and qualitative recovery of the major lymphohematopoietic cells after cyclophosphamide injection. The most striking finding was the absence of insulitis in most animals treated with $1,25(OH)_2D_3$. The combination of these data (resistance against cyclophosphamide and reduction of insulitis), together with the absence of protection in cotransfer experiments, suggested that indeed cyclophosphamide had eliminated all suppressor cells but that these suppressor cells were not responsible for the main protective mechanism in the $1,25(OH)_2D_3$-treated NOD mice. The basis of protection by these substances seems to be more a reshaping of the immune repertoire with elimination of effector cells; however, the direct β cell protective effects of $1,25(OH)_2D_3$ might play a major role in disease prevention.

3. Streptozotocin-Induced Diabetes

Studies on diabetes prevention in the NOD mouse model, which is characterized by spontaneous develop-

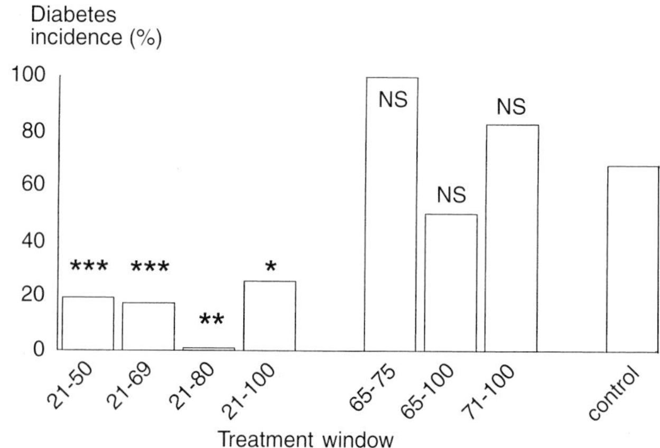

FIGURE 2 Incidence of cyclophosphamide-induced diabetes. A significant protection against diabetes was present in all groups receiving $1,25(OH)_2D_3$ from the time of weaning (left-hand side). No significant protection was seen in animals where $1,25(OH)_2D_3$ treatment was started just before or after cyclophosphamide (right-hand side). Asterisks signify statistically significant results versus the control group: ***, $p < 0.005$; **, $p < 0.01$; *, $p < 0.025$; NS, not significant.

ment of type I diabetes, are probably the most relevant for direct application of the findings in the human situation. However, the fluctuating incidences of diabetes in the stock colonies and the long duration of most intervention studies make this model less than optimal for the screening of large groups of new potentially therapeutic agents. Some researchers have therefore looked for a more convenient and especially more rapid model for screening new drugs. Such a rapid model is the multiple low dose streptozotocin model [62]. Streptozotocin is an antibiotic produced by *Streptomyces achromogenes* that has a specific β cell toxic effect. A single high dose of streptozotocin (70–250 mg/kg body weight) causes a rapid and complete destruction of β cells in most species. The administration of multiple subdiabetic doses of streptozotocin causes more subtle β cell damage, thus triggering a nonspecific inflammatory reaction in the islets that is then followed by insulin deficiency. A major criticism of this model is the fact that the complete scenario of β cell destruction in this model is unclear and is most probably due to nonspecific inflammatory damage of the β cell together with other islet cells. Therefore, this diabetes model is not a true autoimmune model. However, the question remains whether, in some humans, type I diabetes is a true autoimmune disease and not a consequence of nonspecific β cell destruction by one or another cause of inflammation (e.g., viral infection).

Inaba *et al.* used the high and low dose streptozotocin model to test the effect of 1α-hydroxyvitamin D_3, a precursor of 1,25$(OH)_2D_3$, on diabetes prevention [63]. When diabetes was chemically induced by a single iv injection (200 mg/kg body weight) of streptozotocin, no protection against diabetes was seen. However, when multiple low doses of streptozotocin were administered, 1α-hydroxyvitamin D_3 reduced the diabetes incidence dramatically in a dose-dependent manner; control mice developed diabetes in 100% of cases (13/13), whereas administration of 0.4 μg of 1α-hydroxyvitamin D_3 per kg body weight reduced the diabetes incidence to 46% (6/13, $p < 0.01$). Administration of 0.3 μg also provided protection [diabetes incidence 61% (8/13), $p < 0.025$], but administration of 0.2 μg/kg of 1α-hydroxyvitamin D_3 did not prevent diabetes (13/13). Data on toxicity in this study were unfortunately limited to evaluation of body weight (unchanged); no indication of calcemic or bone effects were given. Histologic examination of the pancreas of the experimental mice demonstrated that 1α-hydroxyvitamin D_3 also reduced insulitis in this model.

4. ANALOGS OF 1,25$(OH)_2D_3$

A major obstacle to human application of 1,25$(OH)_2D_3$ are its important effects on calcium and bone metabolism. As described in Chapters 57–63 new structural analogs of 1,25$(OH)_2D_3$ with less effects on calcium metabolism but more pronounced immunological effects have been developed, especially through side chain modifications [64–66]. The actions of this family of molecules on the immune system are exerted via VDR that are present in several immune cells, such as monocytes–macrophages, activated T lymphocytes, and B lymphocytes. *In vivo* these immune effects result in protection against autoimmunity and prolongation of allograft survival [52,53,67–71]. Several of the most promising of these analogs coming from different chemical laboratories have been tested in the model of spontaneous type I diabetes in the NOD mouse (Tables II and III). The most promising data up to now were obtained with an analog synthesized in the Leo laboratories in Denmark; namely, KH1060 [1α,25$(OH)_2$-20-epi-22-oxa-24,26,27-tris-homo-vitamin D]. This analog, which has superagonistic activity on the immune system *in vitro* [72] and effects on calcium and bone metabolism that are comparable to 1,25$(OH)_2D_3$ (see Chapter 61), can prevent type I diabetes in NOD mice without side effects on calcium and bone homeostasis [73]. Indeed, KH1060 given in two different doses (0.4 μg and 0.2 μg per kg body weight) every 2 days from 21 until 200 days of age reduced the incidence of insulitis and diabetes drasti-

TABLE II Prevention of Insulitis by Analogs of 1,25$(OH)_2D_3$[a]

	Insulitis, *n* (%)	Serum calcium (m*M*)	Serum osteocalcin (n*M*)
Controls ($n = 36$)	38 (86)	2.3 ± 0.1	12.9 ± 2.9
1,25$(OH)_2D_3$ ($n = 43$)	18 (42) $p < 0.005$	2.5 ± 0.3 $p < 0.01$	14.4 ± 3.6 $p < 0.01$
Ro I ($n = 39$)	15 (38) $p < 0.005$	2.7 ± 0.1 $p < 0.0001$	22.1 ± 2.5 $p < 0.00001$
MC I ($n = 19$)	9 (47) $p < 0.005$	2.3 ± 0.1 NS	13.2 ± 2.6 NS
MC II ($n = 18$)	10 (55) $p < 0.025$	2.1 ± 0.1 NS	11.8 ± 2.2 NS

[a] For the sake of presentation, data from different experiments have been pooled. Pancreatic biopsies were performed at 100 days of age in NOD mice treated with the products from 21 days of age onward. Statistical analysis for insulitis incidence was performed using the chi-square test, and that for calcemia and osteocalcin was performed with Student's *t* test for unpaired data. The *p* values shown are in comparison to the control group. NS, Not significant. The analogs used in the experiments [Ro I, 1,25$(OH)_2$-16-ene-vitamin D_3 (Hoffmann-LaRoche) at 5 μg/kg every 2 days; MC I, 1,25(*S*)$(OH)_2$-22-ene-26,27-cyclo-vitamin D_3 (Leo Pharmaceutical) at 50 μg/kg/day; MC II, calcipotriol (Leo Pharmaceutical) at 50 μg/kg every 2 days] are sometimes more effective than 1,25$(OH)_2D_3$ (5 μg/kg every 2 days) in insulitis prevention.

TABLE III Prevention of Insulitis by Analogs of $1,25(OH)_2D_3$[a]

	Diabetes, n (%)	Serum calcium (mM)	Serum osteocalcin (nM)
Controls ($n = 137$)	80 (58)	2.4 ± 0.2	10.9 ± 2.4
$1,25(OH)_2D_3$ (5 μg/kg) ($n = 146$)	33 (23) $p < 0.000001$	2.4 ± 0.1 NS	17.7 ± 2.1 $p < 0.00001$
KH1060 (0.2 μg/kg) ($n = 44$)	11 (25) $p < 0.0001$	2.4 ± 0.1 NS	10.4 ± 2.3 NS
KH1060 (0.4 μg/kg) ($n = 71$)	13 (18) $p < 0.00001$	2.7 ± 0.2 $p < 0.0001$	9.9 ± 1.7 NS
Analog 1 (10 μg/kg) ($n = 33$)	14 (42) NS	2.3 ± 0.1 NS	14.4 ± 4.0 $p < 0.05$
Analog 1 (25 μg/kg) ($n = 28$)	9 (32) $p < 0.025$	2.1 ± 0.2 $p < 0.01$	12.4 ± 4.0 NS
Analog 2 (30 μg/kg) ($n = 33$)	16 (48) NS	2.4 ± 0.3 NS	11.9 ± 4.5 NS
Analog 2 (60 μg/kg) ($n = 34$)	14 (41) NS	2.6 ± 0.4 $p < 0.01$	15.2 ± 2.9 $p < 0.0001$

[a] For the sake of analysis, data from different experiments have been pooled. In every experiment an appropriate control group was included. Mice were treated with arachis oil or $1,25(OH)_2D_3$ or analogs every 2 days from 21 to 200 days of age. Statistical analysis for insulitis incidence was performed using the chi-square test. For calcemia and osteocalcin, Student's t test for unpaired data was used. The p values shown are in comparison to the control group. NS, Not significant. Note that some of the tested analogs are as effective as $1,25(OH)_2D_3$ in diabetes prevention, without, however, the detrimental effects on calcium and bone metabolism. Analog 1: 14-epi-19-nor-20-epi-23-yne-$1,25(OH)_2D_3$. Analog 2: 26,27-bishomo-$1,25(OH)_2D_3$, as described in Bio Org Medicin Chem Lett 6:1697–1702, 1996.

cally; insulitis dropped to 48% (13/27) and 44% (12/27) respectively, compared to 84% (26/31) in controls ($p < 0.01$ in all groups versus controls). A dose-dependent prevention of diabetes was achieved by the analog KH1060, as diabetes occurred in only 25 and 18% of mice treated with the low and high dose of KH1060, respectively ($p < 0.0001$ and $p < 0.00001$ versus the control group).

Serum calcium levels were raised only in mice treated with the high dose of KH1060, whereas osteocalcin levels were perfectly normal in both groups. Calcium content in the tibiae of mice treated with either dose of KH1060 was completely maintained (6.5 ± 0.8 and 6.8 ± 0.7 mg/tibia, not significant versus controls). Even careful evaluation of other parameters of calcium and bone metabolism, such as urinary excretion of collagen cross-links, did not reveal significant effects of the analog on bone turnover (Table IV). A search for effects on the immune system induced by the analog again demonstrated the presence of immune regulator cells in treated mice in contrast to control mice.

5. Combination of Vitamin D Analogs with Other Immunomodulators

In animal models of type I diabetes, such as the NOD mouse, disease prevention can be achieved by chronic use of immunosuppressants such as cyclosporin A (CyA) [74]. Such an approach cannot be applied to humans because of chronic side effects. In NOD mice, diabetes can also be prevented by $1,25(OH)_2D_3$ and its nonhypercalcemic analogs when treatment is started before insulitis, the histological lesion of the disease, is present [53,73]. A critical question for the applicability of these analogs to the human situation, however, is whether the analogs of $1,25(OH)_2D_3$ can arrest progression to clinically overt diabetes if administered when active β cell destruction is already present, which is the situation in prediabetic subjects in whom immune intervention is considered [75].

Casteels *et al.* demonstrated that some analogs, when combined with a short induction course of a classic immunosuppressant such as cyclosporin A, can arrest progression of the disease even when administered after autoimmunity has already started [76]. In an early study, the protective potential of MC1288 [20-epi-1,25-dihydroxyvitamin D_3 (see Chapter 61)] in combination with a short induction course of CyA was demonstrated in NOD mice that already showed active β cell destruction. To prove that this process of autoimmune β cell destruction had already started, pancreatic biopsies were taken before treatment was started. Biopsies were taken in female NOD mice at an age of 70 days (when 60–70% of control mice have insulitis). Subsequently, mice were randomized into three treatment groups receiving, respectively, CyA at 7.5 mg/kg/day from days 85 to 105 ($n = 19$), MC1288 at 0.1 μg/kg every other day from days 85 to 200 ($n = 20$), or the combination of the two regimens ($n = 20$). Diabetes incidence by 200 days of age was 74% (14/19) in the CyA-treated group, which is comparable to the diabetes incidence in the stock colony. Treatment with MC1288 could not reduce diabetes incidence (14/20, 70%), but the combination therapy reduced diabetes incidence to 35% (7/20, $p < 0.01$ versus CyA group, $p < 0.025$ versus MC1288). The occurrence of diabetes in animals presenting with insulitis before treatment was 86% in the group treated with CyA, 67% in the group treated with MC1288 ($p < 0.05$), and 53% in the group receiving MC1288 and CyA ($p < 0.01$). Insulitis-free animals at the time of biopsy

TABLE IV Effects of $1,25(OH)_2D_3$ and KH1060 on Calcium Metabolism[a]

	Serum calcium (mM)	Serum osteocalcin (nM)	Bone calcium (mg/tibia)	PYD (nM/mM creatinine)	DEOXYPYD (nM/mM creatinine)	Calcium binding protein (μg/mg protein)
Controls ($n = 31$)	2.3 ± 0.1	16 ± 4	6.4 ± 0.5	338 ± 30	24.25 ± 1.1	1.7 ± 1.1
$1,25(OH)_2D_3$ ($n = 38$)	2.3 ± 0.1 NS	27.3 ± 3 $p < 0.00001$	4.1 ± 0.7 $p < 0.0001$	500 ± 56 $p < 0.05$	33.25 ± 2.09 $p = 0.05$	5.2 ± 2.3 $p < 0.002$
KH1060 0.2 μg/kg ($n = 27$)	2.3 ± 0.1 NS	16 ± 3 NS	6.8 ± 0.7 NS	382 ± 36 NS	25.6 ± 1.2 NS	3.3 ± 1.9 NS
KH1060 0.4 μg/kg ($n = 27$)	2.6 ± 0.2 $p < 0.0001$	15 ± 3 NS	6.5 ± 0.8 NS	439 ± 31 $p < 0.05$	28.2 ± 3 NS	6.7 ± 3.3 $p < 0.001$

[a] Extensive screening of effects of treatment with $1,25(OH)_2D_3$ (5 μg/kg) or its analogs on calcium and bone metabolism was performed. Student's t test for unpaired data was used. The p values shown are in comparison to the control group. Note that the lower dose of KH1060 (which is still effective in diabetes prevention) hardly has any effects on calcium and bone metabolism. PYD, urinary pyridinolines; DEOXYPYD, urinary deoxypyridinolines.

(day 70) developed insulitis in all cases when treated with monotherapy only, in contrast to none of the animals treated with the combination therapy ($p < 0.01$). All treatment regimens were well tolerated, without any side effects on calcium and bone metabolism as demonstrated by analysis of bone turnover parameters and bone calcium content. Immunological screening (concanavalin A, MLR) at 200 days could not detect immunosuppression in either treatment group. Other analogs have also been tested, and some share the potential of disease prevention with MC1288.

The approach of combining a short induction course of a classic immunosuppressant with nonhypercalcemic analogs of $1,25(OH)_2D_3$ is very promising and might open new perspectives in the prevention of autoimmune diabetes in humans.

B. Primary Prevention of Type I Diabetes in Animal Models by $1,25(OH)_2D_3$ and Its Analogs—The BB Rat

The BB rat is the second animal model for type I diabetes. Data on the immunopathogenesis of type I diabetes in the BB rat are less abundant than in the NOD mouse, and observations are hampered by the severe lymphopenia and T-cell dysfunction occurring in these animals [77–79]. The existence of different pathogenic mechanisms for autoimmune diabetes in animal models suggests the possible existence of different pathogenic pathways in the human situation as well. This is also suggested by the sometimes contradictory findings in different patient populations, for example, different genetic characteristics between Japanese and Caucasian diabetic patients, differences in immune characteristics between the patients with very early onset versus later onset of the disease, and finally the differences observed between patients with isolated type I diabetes and patients with multiorgan involvement.

The effects of $1,25(OH)_2D_3$ on diabetes prevention were also evaluated in the BB rat model (C. Mathieu, K. Casteels, J. Lawreys, and R. Bouillon, unpublished data, 1996). BB/Pfd rats, inbred in the animal facility of the Katholieke Universiteit Leuven, were treated following a protocol similar to the one used in NOD mice [53]. Rats ($n = 17$), male and females, were treated from weaning until 120 days of age with the highest tolerable dose of $1,25(OH)_2D_3$, 0.8 μg/kg intraperitoneally every 2 days. Control BB rats ($n = 18$) received the treatment vehicle, arachis oil. No difference in diabetes incidence between control (6/18, 33%) and treated rats (4/17, 24%, NS) was observed. Similar results have been obtained in another BB rat substrain, the BB/OK, residing in the animal facility of the Institut für Diabetes in Karlsburg, Germany (B. Kuttler, personal communication, 1996). Extensive screening of the severely impaired immune system of the BB rats did not reveal any differences between controls and treated rats (evaluated by MLR, concanavalin A and phytohemagglutinin proliferation, NK cell function). On the other hand, the effects of $1,25(OH)_2D_3$ on bone turnover and calcium metabolism were clear, with hypercalcemia (2.9 ± 0.3 mM versus 2.4 ± 0.1 mM in controls, $p < 0.00001$) and increased serum levels of osteocalcin (12.8 ± 1.3 nM versus 7.1 ± 0.5 nM in controls, $p < 0.001$). Bone calcium content, however, was not significantly altered.

These findings in the BB rat again confirm the basic differences in disease pathogenesis that can be found between the two available animal models for type I diabetes. They also indicate that caution is warranted when transferring findings from either of these models to the human situation.

C. Secondary Prevention of Type I Diabetes in the NOD Mouse by Analogs of $1,25(OH)_2D_3$—Prevention of Recurrence of Autoimmune Diabetes after Islet Transplantation

Type I diabetes is characterized not only by an autoimmune destruction of the body's own β cells, but also by the formation of an autoimmune memory. The latter phenomenon is responsible for the destruction of MHC matched or syngeneic β cells, transplanted in the form of isolated β cells, islets, or whole pancreas [80–86]. This disease recurrence explains why, in clinical pancreas and islet transplantation in type I diabetic patients, extensive immunosuppression is needed. Relatively high doses of several immunosuppressants are required not only to prevent allograft rejection, but also to break the autoimmune memory [87–89].

Some analogs of $1,25(OH)_2D_3$ have been tested for their capacity to prevent disease recurrence after islet transplantation in spontaneously diabetic NOD mice. The most dramatic results were obtained with a combination of KH1060 [1α,25(OH)$_2$-20-epi-22-oxa-24,26,27-trishomo-vitamin D] and subtherapeutic doses of cyclosporin A [90] (Table V). Administration of high doses of the analog (1 μg/kg every 2 days) alone was as effective in delaying islet loss as the highest tolerable dose of cyclosporin A (15 mg/kg/day), but eventually the disease recurred in all mice. In the group receiving KH1060 (0.5 μg/kg every 2 days) together with cyclosporin A (7.5 mg/kg/day), a synergistic effect between both drugs was seen. Four of seven mice maintained a functioning graft for 60 days, and, more importantly, these animals did not show recurrence for at least 30 days after stopping the treatment. All treatment was administered from the day before transplantation until diabetes recurrence or in case of normoglycemia until 60 days after transplantation. Insulin content determinations of the graft and native pancreas of the recipient clearly demonstrated that normoglycemia was the result of graft survival and not of recovery of the β cells of the recipient's own pancreas: insulin content in pancreases of recurring and nonrecurring mice were comparable and showed no regeneration of the original β cells (0.0125 $\pm$ 0.012 pmol/mg in recurring versus 0.008 $\pm$ 0.004 pmol/mg in nonrecurring mice, difference not significant), whereas the insulin content in the grafts showed a clear difference between recurring and nonrecurring mice (45 $\pm$ 27 pmol/graft in recurring versus 1285 $\pm$ 106 pmol/graft in nonrecurring mice, $p < 0.00001$).

The highest dose of KH1060 as well as the highest dose of cyclosporin A had clear toxic effects on the general condition of the animals, as reflected by changes in body weight. However, by giving KH1060 in a fractionated way (1 μg/kg every 2 days instead of 0.5 μg/kg/day), these cumulative high doses were well tolerated for long periods. Calcemic values of all treatment groups, measured 24 hr after the last injection, were

TABLE V Islet Graft Survival in Syngeneic Islet Transplantations[a]

	Number of animals	Survival of islets (days)	Mean survival time (days)	p
Control	8	5, 7, 7, 8, 8, 9, 10, 13	8	
KH1060 0.5 μg/kg/2 days	5	4, 8, 11, 19, 19	12	NS
KH1060 1 μg/kg/2 days	8	20, 39, 41, 56, 60, 63, 70, 70	55	$p < 0.0001$
Cyclosporin A 7.5 μg/kg/day	8	4, 5, 7, 8, 13, 14, 37, 42	16	NS
Cyclosporin A 15 mg/kg/day	6	22, 45, 55, 67, 69, >90	>58	$p < 0.0001$
KH1060 0.5 μg/kg/2 days + cyclosporin A 7.5 mg/kg/day	7	7, 33, 35, >90, >90, >90, >90	>60	$p < 0.001$

[a] All mice were spontaneously diabetic NOD mice transplanted with 500 NOD islets under the left kidney capsule. After transplantation, mice were tested three times weekly for glycosuria and blood glucose and were considered diabetic when having positive glycosuria and blood glucose >12 mmol/liter on two consecutive days. Survival of the islets was evaluated by their function: the day of recurrence was the first day of hyperglycemia. All treatment regimens were initiated the day before transplantation and were discontinued on the day of disease recurrence or 60 days after transplantation. Statistical analysis was performed using Student's t test for unpaired data. The p values shown are in comparison to the control group. NS, Not significant.

TABLE VI Effects of KH1060 and Cyclosporin A on Calcium Metabolism[a]

	Serum calcium (mM)	Serum osteocalcin (nM)	Bone calcium (mg/tibia)
Control (n = 8)	2.6 ± 0.1	5.4 ± 2.2	7.2 ± 0.3 (n = 7)
KH1060 0.5 μg/kg/2 days (n = 5)	2.6 ± 0.2	8.7 ± 3.6	6.5 ± 0.2 (n = 4) $p < 0.05$
KH1060 1 μg/kg/2 days (n = 8)	2.7 ± 0.5	10.9 ± 3.9	6.3 ± 0.7 (n = 2) NS
Cyclosporin A 7.5 mg/kg/day (n = 8)	2.7 ± 0.3	6.8 ± 1.7	6.6 ± 0.5 (n = 4) NS
Cyclosporin A 15 mg/kg/day (n = 6)	2.7 ± 0.1	6.8 ± 2.9	6.8 ± 0.2 (n = 4) NS
KH1060 0.5 μg/kg/2 days + cyclosporin A 7.5 mg/kg/day (n = 7)	2.7 ± 0.3	10.7 ± 4.2	5.9 ± 0.8 (n = 2) $p < 0.025$

[a] On the day of diabetes recurrence or 90 days after transplantation, serum calcium and osteocalcin were determined as well as bone calcium content per tibia. Mean values ± SD are shown. Statistical analysis for calcium and osteocalcin levels was made by Student's t test for unpaired data. No value was statistically different from the control value. Note the low osteocalcin levels in control animals. For bone calcium, statistics were performed using the Mann Whitney U test. The p values shown are in comparison to the control group. NS, Not significant.

comparable in all groups (Table VI). Note that cyclosporin A also had clear effects on calcium metabolism. This effect of cyclosporin A on calcium and bone metabolism has already been described by others [91] (see also Chapter 50). The subtherapeutic doses of KH1060 and cyclosporin A were nontoxic and had minor effects on serum calcium and osteocalcin levels. The combination of both drugs was also well tolerated and led to similar effects on serum calcium, but it resulted in clear effects on osteocalcin levels, indicating a synergistic effect on bone remodeling as well. Bone calcium content in all treatment groups was decreased. In the KH1060 groups this effect was more important and dose dependent. Unfortunately, as already reflected by the osteocalcin levels, the combined subtherapeutic doses of KH1060 and cyclosporin A also act synergistically on bone remodeling. Caution in interpreting these results is warranted, however, since the treatment duration in tested animals in the combination group lasted for 60 days while the animals tested in the other groups received treatment for shorter periods (~20 days).

IV. CLINICAL PERSPECTIVES

Clear effects of $1,25(OH)_2D_3$ and its newer analogs on the different major players in the pathogenesis of diabetes mellitus, both type I and type II diabetes, have been described. *In vitro* as well as *in vivo*, a modest stimulation of insulin synthesis and insulin secretion by $1,25(OH)_2D_3$ is observed [29,32,42–45]. This positive effect is observed not only on repletion of $1,25(OH)_2D_3$ in the vitamin D-deficient state [29,42–44], but also in the vitamin D-replete state [32,44,45]. Moreover, a direct β cell protection by $1,25(OH)_2D_3$ and its analogs against metabolic and inflammatory stress has been demonstrated [49]. On the other hand, major effects on the immune system, involved in the pathogenesis of type I diabetes, have been described *in vitro* as well as *in vivo* [23,25,26], and prevention of type I diabetes and its recurrence after islet transplantation by $1,25(OH)_2D_3$ and analogs can be achieved (alone or in combination with other immunomodulators) [52,54,63,73,90].

A major problem with using $1,25(OH)_2D_3$ or the currently available analogs in prevention or cure of diabetes are the hypercalcemic and bone remodeling effects when the compounds are administered in the doses needed for immune or β cell protective effects. Future applications of this therapy in human diabetes are conceivable; through chemical alterations of the $1,25(OH)_2D_3$ molecule even better analogs, with an optimal dissociation between calcemic and immunomodulatory effects, can be synthesized [66].

A place for these analogs in the treatment (prevention or cure) of diabetes can be conceived first of all in terms of β cell protective and stimulating agents added to the current treatment modalities of type II diabetes. Further, these substances could play a major role in prevention strategies for type I diabetes in humans, because of their ideal profile as β cell protective and especially immune active drugs. Before these drugs are applied in humans, however, more information should be gathered not only on their mechanism of action but especially on the safety of the products in long-term use.

References

1. Henry HL 1981 25-OH-D_3 metabolism in kidney cell cultures: Lack of a direct effect of estradiol. Am J Physiol **240**:E119–E124.
2. Condamine L, Vztovsnik F, Friedlander G, Menaa C, Garabedian M 1994 Local action of phosphate depletion and insulin-like growth factor 1 on in vitro production of 1,25-dihydroxyvitamin D by cultured mammalian kidney cells. J Clin Invest **94**:1673–1679.

3. Nyomba BL, Verhaeghe J, Thomasset M, Lissens W, Bouillon R 1989 Bone mineral homeostasis in spontaneously diabetic BB rats. I. Abnormal vitamin D metabolism and impaired active intestinal calcium absorption. Endocrinology **124**:565–572.
4. Verhaeghe J, Van Herck E, Visser WJ, Suiker AMH, Thomasset M, Einhorn TA, Faierman E, Bouillon R 1990 Bone and mineral metabolism in BB rats with long-term diabetes. Diabetes **39**: 477–482.
5. Nyomba BL, Bouillon R, Lissens W, Van Baelen H, De Moor P 1985 1,25-Dihydroxyvitamin-D and vitamin-D-binding protein are both decreased in streptozotocin-diabetic rats. Endocrinology **116**:2483–2488.
6. Hough S, Russell JE, Teitelbaum SL, Avioli LV 1982 Calcium homeostasis in chronic streptozotocin-induced diabetes mellitus in the rat. Am J Physiol **242**:E451–E456.
7. Schneider LE, Schedl HP, McCain T, Haussler MR 1977 Experimental diabetes reduces circulating 1,25-dihydroxyvitamin D in the rat. Science **196**:1452–1454.
8. Storm TL, Sorensen OH, Lund B, Christiansen JS, Andersen AR, Lumholtz IB, Parving HH 1983 Vitamin D metabolism in insulin-dependent diabetes mellitus. Metab Bone Dis Relat Res **5**:107–110.
9. Nyomba BL, Bouillon R, Bidingij M, Kandjing K, De Moor P 1986 Vitamin-D metabolites and their binding-protein in adult diabetic-patients. Diabetes **35**:911–915.
10. Rodland O, Markesta T, Aksnes L, Aarskog D 1985 Plasma concentrations of vitamin-D metabolites during puberty of diabetic children. Diabetology **28**:663–666.
11. Auwerx J, Dequeker J, Bouillon R, Geusens P, Nijs J 1988 Mineral metabolism and bone mass at peripheral and axial skeleton in diabetes mellitus. Diabetes **37**:8–12.
12. Verhaeghe J, Suiker AMH, Van Bree R, Van Herck E, Jans I, Visser WJ, Thomasset M, Allewaert K, Bouillon R 1993 Increased clearance of $1,25(OH)_2D_3$ and tissue-specific responsiveness to $1,25(OH)_2D_3$ in diabetic rats. Am J Physiol **265**:E215–E223.
13. Schneider LE, Schedl HP 1972 Diabetes and intestinal calcium absorption in the rat. Am J Physiol **223**:1319–1323.
14. Schneider LE, Wilson HD, Schedl HP 1974 Effects of alloxan diabetes on duodenal calcium-binding protein in the rat. Am J Physiol **227**:832–838.
15. Stone LA, Weaver VM, Bruns ME, Christakos S, Welsh J 1991 Vitamin-D receptors and compensatory tissue growth in spontaneously diabetic BB rats. Ann Nutr Metab **35**:196–202.
16. Bouillon R, Verhaeghe J, Bex M, Suiker A, Visser W 1991 Diabetic bone disease. In: Rifkin H, Colwell JA, Taylor SI (eds) Diabetes. Elsevier, Amsterdam, pp. 1271–1274.
17. Verhaeghe J, Suiker AMH, Visser WJ, Van Herck E, Van Bree R, Bouillon R 1992 The effects of systemic insulin, insulin-like growth factor-I and growth hormone on bone growth and turnover in spontaneously diabetic BB rats. J Endocrinol **134**:485–492.
18. Verhaeghe J, Bouillon R 1994 Actions of insulin and the IGFs on bone. NIPS **9**:20–22.
19. Binz K, Schmid C, Bouillon R, Froesch ER, Jürgensen K, Hunziker EB 1994 Interactions of insulin-like growth factor I with dexamethasone on trabecular bone density and mineral metabolism in rats. Eur J Endocrinol **130**:387–393.
20. De Fronzo RA, Bonadonna RC, Ferrannini E 1992 Pathogenesis of NIDDM. Diabetes Care **15**:318–368.
21. Taylor SI, Accili D, Imai Y 1994 Insulin resistance or insulin deficiency—Which is the primary cause of NIDDM? Diabetes **43**:735–740.
22. Castano L, Eisenbarth GS 1990 Type-I diabetes: A chronic autoimmune disease of human, mouse, and rat. Annu Rev Immunol **8**:647–679.
23. Hewison M 1992 Vitamin D and the immune system. J Endocrinol **132**:173–175.
24. Christakos S, Norman AW 1979 Studies on the mode of action of calciferol XVII. Evidence for a specific high affinity binding protein for 1,25-dihydroxyvitamin D_3 in chick kidney and pancreas. Biochem Biophys Res Commun **89**:56–63.
25. Bouillon R, Garmyn M, Verstuyf AM, Segaert S, Casteels K, Mathieu C 1995 A paracrine role for $1,25(OH)_2D_3$ in the immune system and skin creates new therapeutic possibilities for vitamin D analogues. Eur J Endocrinol **133**:7–16.
26. Casteels K, Bouillon R, Waer M, Mathieu C 1995 Immunomodulatory effects of $1,25(OH)_2D_3$. Current Opinion in Nephrology and Hypertension **4**:313–318.
27. Norman AW, Frankel BJ, Heldt AM, Grodsky GM 1980 Vitamin D deficiency inhibits pancreatic secretion of insulin. Science **209**:823–825.
28. Lee S, Clark SA, Gill RK, Christakos S 1994 1,25-Dihydroxyvitamin D_3 and pancreatic β-cell function: Vitamin-D receptors, gene expression, and insulin secretion. Endocrinology **134**:1602–1610.
29. Cade C, Norman AW 1987 Rapid normalization/stimulation by $1,25(OH)_2$-vitamin D_3 of insulin secretion and glucose tolerance in the vitamin D-deficient rat. Endocrinology **120**:1490–1497.
30. de Boland AR, Norman AW 1990 Influx of extracellular calcium mediates 1,25-dihydroxyvitamin D_3-dependent transcaltachia (the rapid stimulation of duodenal Ca^{2+} transport). Endocrinology **127**:2475–2480.
31. Norman AW 1994 Editorial: The vitamin-D endocrine system: Identification of another piece of the puzzle. Endocrinology **134**:A1601–C1601.
32. Sergeev IN, Rhoten WB 1995 1,25-Dihydroxyvitamin D_3 evokes oscillations of intracellular calcium in a pancreatic β-cell line. Endocrinology **136**:2852–2861.
33. Roth J, Bonner-Weir S, Norman AW, Orci L 1982 Immunocytochemistry of vitamin D-dependent calcium binding protein in chick pancreas: Exclusive localization in B-cells. Endocrinology **110**:2216–2218.
34. Chertow BS, Sivitz WI, Baranetsky NG, Clark SA, Waite A, DeLuca HF 1983 Cellular mechanisms of insulin release: The effects of vitamin D deficiency and repletion on rat insulin secretion. Endocrinology **113**:1511–1518.
35. Nyomba BL, Bouillon R, De Moor P 1984 Influence of vitamin D status on insulin secretion and glucose tolerance in the rabbit. Endocrinology **115**:191–197.
36. Kadowaki S, Norman AW 1984 Dietary vitamin D is essential for normal insulin secretion from the perfused rat pancreas. J Clin Invest **73**:759–766.
37. Cade C, Norman AW 1986 Vitamin D_3 improves impaired glucose tolerance and insulin secretion in the vitamin D-deficient rat *in vivo*. Endocrinology **119**:84–90.
38. Gedik O, Akalin S 1986 Effects of vitamin D deficiency and repletion on insulin and glucagon secretion in man. Diabetologia **29**:142–145.
39. Tanaka Y, Seino Y, Ishida M, Yamaoka K, Yabuuchi H, Ishida H, Seino S, Imura H 1984 Effect of vitamin D_3 on the pancreatic secretion of insulin and somatostatin. Acta Endocrinol **105**: 528–533.
40. Tanaka Y, Seino Y, Ishida M, Yamaoka K, Satomura K, Yabuuchi H, Imura H 1986 Effect of 1,25-dihydroxyvitamin D_3 on insulin secretion: Direct or mediated? Endocrinology **118**:1971–1976.
41. Beaulieu C, Kestekian R, Havrankova J, Gascon-Barré M 1993 Calcium is essential in normalizing intolerance to glucose that accompanies vitamin D depletion in vivo. Diabetes **42**:35–43.
42. Billaudel BJL, Faure AG, Sutter BCJ 1990 Effect of 1,25-dihy-

droxyvitamin D_3 on isolated islets from vitamin D_3-deprived rats. Am J Physiol **258**:E643–E648.

43. Billaudel BJL, Delbancut APA, Sutter BCJ, Faure AG 1993 Stimulatory effect of 1,25-dihydroxyvitamin D_3 on calcium handling and insulin secretion by islets from vitamin D_3-deficient rats. Steroids **58**:335–341.
44. Kadowaki S, Norman AW 1985 Time course study of insulin secretion after 1,25-dihydroxyvitamin D_3 administration. Endocrinology **117**:1765–1771.
45. d'Emden MC, Dunlop M, Larkins RG, Wark JD 1989 The in vitro effect of 1α,25-dihydroxyvitamin D_3 on insulin production by neonatal rat islets. Biochem Biophys Res Commun **164**:413–418.
46. Kumar S, Davies M, Zakaria Y, Mawer EB, Gordon C, Olukoga AO, Boulton AJM 1994 Improvement in glucose tolerance and beta-cell function in a patient with vitamin D deficiency during treatment with vitamin D. Postgrad Med J **70**:440–443.
47. Allegra V, Luisetto G, Mengozzi G, Martimbianco L, Vasile A 1994 Glucose-induced insulin secretion in uremia: Role of $1\alpha,25(OH)_2$-vitamin D_3. Nephron **68**:41–47.
48. Orwoll E, Riddle M, Prince M 1994 Effects of vitamin D on insulin and glucagon secretion in non-insulin-dependent diabetes mellitus. Am J Clin Nutr **59**:1083–1087.
49. Sandler S, Buschard K, Bendtzen K 1994 Effects of 1,25-dihydroxyvitamin D_3 and the analogues MC903 and KH1060 on interleukin-1 beta induced inhibition of rat pancreatic islet beta-cell function in vitro. Immunol Lett **41**:73–77.
50. Mauricio D, Andersen HU, Karlsen AE, Mandrup-Poulsen T, Nerup J 1995 Effect of steroidal hormones on interleukin-1β action on rat islets Diabetologia **38S**:A80.
51. Makino S, Kunimoto K, Muraoka Y, Mizushima Y, Katagiri K, Tochino Y 1980 Breeding of a non-obese, diabetic strain of mice. Exp Anim **29**:1–13.
52. Mathieu C, Laureys J, Sobis H, Vandeputte M, Waer M, Bouillon R 1992 1,25-Dihydroxyvitamin D_3 prevents insulitis in NOD mice. Diabetes **41**:1491–1495.
53. Meehan MA, Kerman RH, Lemire JM 1992 1,25-Dihydroxyvitamin D_3 enhances the generation of non specific suppressor cells while inhibiting the induction of cytotoxic cells in a human MLR. Cell Immunol **140**:400–409.
54. Mathieu C, Waer M, Laureys J, Rutgeerts O, Bouillon R 1994 Prevention of type I diabetes in NOD mice by 1,25-dihydroxyvitamin D_3. Diabetologia **37**:552–558.
55. Harada M, Makino S 1984 Promotion of spontaneous diabetes in non-obese diabetes-prone mice by cyclophosphamide. Diabetologia **27**:604–606.
56. Yasunami R, Bach JF 1988 Anti-suppressor effect of cyclophosphamide on the development of spontaneous diabetes in NOD mice. Eur J Immunol **18**:481–484.
57. Mitsuoka A, Baba M, Morikawa S 1976 Enhancement of delayed hypersensitivity by depletion of suppressor T cells with cyclophosphamide in mice. Nature **262**:77–78.
58. Qin HY, Sadelain M, Hitchon C, Lauzon J, Singh B 1993 Complete Freund's adjuvant-induced T cells prevent the development and adoptive transfer of diabetes in nonobese diabetic mice. J Immunol **150**:2072–2080.
59. Lakey J, Singh B, Warnock G, Elliott J, Rajotte R 1995 Long-term survival of syngeneic islet grafts in BCG-treated diabetic NOD mice can be reversed by cyclophosphamide. Transplantation **59**:1751–1753.
60. Shintani S, Satoh J, Seino H, Goto Y, Toyota T 1990 Mechanism of action of a streptococcal preparation (OK-432) in prevention of autoimmune diabetes in NOD mice. J Immunol **144**:136–141.
61. Casteels KM, Bouillon R, Waer M, Laureys J, Mathieu C 1997 Protection against cyclophosphamide-induced diabetes in the NOD mouse by 1,25-dihydroxyvitamin D_3. Submitted for publication.
62. Kolb H, Kröncke KD 1993 IDDM—Lessons from the low-dose streptozocin model in mice. Diabetes Review **1**:116–126.
63. Inaba M, Nishizawa Y, Song K, Tanishita H, Okuno S, Miki T, Morii H 1992 Partial protection of 1α-hydroxyvitamin D_3 against the development of diabetes induced by multiple low-dose streptozotocin injection in CD-1 mice. Metabolism **41**:631–635.
64. Bikle DD 1992 Clinical counterpoint: Vitamin D: New actions, new analogs, new therapeutic potential. Endocr Rev. **13**:765–784.
65. Bouillon RK, Allewaert K, Xiang DZ, Tan BK, Van Baelen H 1991 Vitamin D analogs with low affinity for the vitamin D binding protein: Enhanced in vitro and decreased in vivo activity. J Bone Miner Res **6**:1051–1057.
66. Bouillon R, Okamura WH, Norman AW 1995 Structure–function relationships in the vitamin D endocrine system. Endocr Rev **16**:200–257.
67. Branisteanu DD, Leenaerts P, Van Damme B, Bouillon R 1993 Partial prevention of active Heymann nephritis by 1α,25-dihydroxyvitamin D_3. Clin Exp Immunol **94**:412–417.
68. Lemire JM, Archer C 1991 1,25-Dihydroxyvitamin D_3 prevents the in vivo induction of murine experimental autoimmune encephalomyelitis. J Clin Invest **87**:1103–1107.
69. Lemire JM, Adams JS 1992 1,25-Dihydroxyvitamin D_3 inhibits the passive transfer of cellular immunity by a myelin basic protein-specific T cell clone. J Bone Miner Res **7**:171–177.
70. Fournier C, Gepner P, Sadouk M, Charriere J 1990 In vivo beneficial effects of cyclosporine A and 1,25-dihydroxyvitamin D_3 on the induction of experimental autoimmune thyroiditis. Clin Immunol Immunopathol **54**:53–63.
71. Jordan SC, Nigata M, Mullen Y 1988 1,25-Dihydroxyvitamin D_3 prolongs rat cardiac allograft survival. In: Schaeffer K, Norman AW, Grigoleit HG, Herrath DV (eds) Vitamin D: Molecular, Cellular and Clinical Endocrinology. de Gruyter, Berlin, New York, pp. 334–335.
72. Latini S, Binderup L 1991 KH1060 A new vitamin D analogue with potent effects on cell-mediated immune responses in vitro. In: Norman AW, Bouillon R, Thomasset M (eds) Vitamin D: Gene Regulation, Structure–Function Analysis and Clinical Application. de Gruyter, Berlin, pp. 516–519.
73. Mathieu C, Waer M, Casteels K, Laureys J, Bouillon R 1995 Prevention of type I diabetes in NOD mice by non-hypercalcemic doses of a new structural analogue of $1,25(OH)_2D_3$, KH1060. Endocrinology **136**:866–872.
74. Mori Y, Suko M, Okudaira H, Matsuba I, Tsuruoka A, Sasaki A, Yokoyama H, Tanase T, Shida T, Nishumira M, Terada E, Ikeda Y 1986 Preventive effects of cyclosporine on diabetes in NOD mice. Diabetologia **29**:244–247.
75. Bach JF 1994 Predictive medicine in autoimmune diseases: From the identification of genetic predisposition and environmental influence to precocious immunotherapy. Clin Immunol Immunopathol **72**:156–161.
76. Casteels K, Waer M, Bouillon R, Allewaert K, Laureys J, Mathieu C 1996 Prevention of type I diabetes by late intervention with non-hypercalcemic analogues of vitamin D_3 in combination with cyclosporin A. Transplant Proc **28**:3095.
77. Guttmann RD, Colle E, Michel F, Seemayer T 1983 Spontaneous diabetes mellitus syndrome in the rat—T lymphopenia and its association with clinical disease and pancreatic lymphocytic infiltration. J Immunol **130**:1732–1735.
78. Georgiou HM, Lagarde AC, Bellgrau D 1988 T cell dysfunction in the diabetes-prone BB rat. J Exp Med **167**:132–147.
79. Elder ME, Maclaren NK 1983 Identification of profound periph-

eral T lymphocyte immunodeficiencies in the spontaneously diabetic BB rat. J Immunol **130**:1723–1731.
80. Weringer EJ, Like AA 1986 Diabetes mellitus in the BB/W rat—Insulitis in pancreatic islet grafts after transplantation in diabetic recipients. Am J Pathol **125**:107–112.
81. Mandel TE, Koulmanda M, Loudovaris T, Bacelj A 1989 Islet grafts in NOD mice: A comparison of iso-, allo-, and pig xenografts. Transplant Proc **21**:3813–3814.
82. Lafferty KJ 1989 Islet transplantation as a therapy for type I diabetes mellitus. Diabetes Nutr Metab **2**:323–332.
83. Bartlett ST, Killewich LA 1991 Recurrent diabetes in pancreas transplants in the BB rat: Evidence for a defect extrinsic to the beta cell. Transplant Proc **23**:689–690.
84. Pipeleers D, Pipeleers-Marchal M, Markholst H, Hoorens A, Klöppel G 1991 Transplantation of purified islet cells in diabetic BB rats. Diabetologia **34**:390–396.
85. Woehrle M, Pullmann J, Bretzel RG, Federlin K 1992 Recurrence of diabetes after islet transplantation in the BB rat. Transplant Proc **24**:999.
86. Woehrle M, Pullmann J, Bretzel RG, Federlin K 1992 Prevention of recurrent autoimmune diabetes in the BB rat by islet transplantation under the renal kidney capsule. Transplantation **53**:1099–1102.
87. Lacy PE 1993 Status of islet cell transplantation. Diabetes Review **1**:76–92.
88. Ricordi C, Tzakis AG, Carroll PB, Zeng Y, Rodriguez Rilo HL, Alejandro R, Shapiro R, Fung JJ, Demetris AJ, Mintz DH, Startzl TE 1992 Human islet isolation and allotransplantation in 22 consecutive cases. Transplantation **53**:407–414.
89. Sharp DW, Lacy PE, Santiago JV, McCullough CS, Weide LG, Boyle PJ, Falqui L, Marchetti P, Ricordi C, Gingerich RL, Jaffe AS, Cryer PE, Hanto DW, Anderson CB, Wayne Flye M 1991 Results of our first nine intraportal islet allografts in type I, insulin-dependent diabetic patients. Transplantation **51**:76–85.
90. Mathieu C, Bouillon R, Laureys J, Waer M 1994 Prevention of autoimmune destruction of transplanted islets in spontaneously diabetic NOD mice by KH1060, a 20-epi analogue of vitamin D—Synergy with cyclosporin A. Transplant Proc **26**:3128–3129.
91. Katz I, Epstein S 1992 Posttransplantation bone disease. J Bone Miner Res **7**:123–126.

CHAPTER 71

Vitamin D in the Nervous System: Actions and Therapeutic Potential

SUSAN CARSWELL Cephalon, Inc., West Chester, Pennsylvania

I. INTRODUCTION

The role of vitamin D in the nervous system has received limited attention until relatively recently. This is despite the fact that it has been known, in many cases for decades, that there is an association between effective levels of vitamin D and neurological manifestations, suggesting possible actions of the hormone on the nervous system. Although these clinical findings point to a potential link between levels of vitamin D or its receptor and neurological effects, they cannot distinguish whether the manifest neurologic dysfunction is the result of indirect effects, such as hypercalcemia, or direct, local actions of the altered levels of circulating vitamin D (and/or its metabolites or receptors). Many actions on the nervous system appear to be indirect. Rickets, for example, a disease of vitamin D deficiency, was reported in 1929 to be associated with seizures [1], which subside after administration of vitamin D [2]. In addition, reversible myopathies are associated with vitamin D deficiency [3–6]. It has also been documented that chronic hypocalcemia, which can be caused by vitamin D deficiency, is often accompanied by irritability, depression, psychosis, tetany, convulsions, and seizures [7]. These patients also manifest decreased sensory and motor neuron fiber conduction velocity [8]. Defects in mental development have been reported in cases of both abnormally low and high intake of vitamin D [9]. On the other hand, other evidence suggests possible direct actions of vitamin D. Vitamin D deficiency reversibly increases glucose metabolism and acetylcholinesterase activity in the cerebral cortex [10]. It was reported that vitamin D receptor (VDR) mRNA levels are reduced in the hippocampus of Alzheimer's disease patients [11].

Together, these findings suggest that vitamin D may play a role in neuronal systems, both indirectly and directly. However, proof for this must come from detailed biochemical and histopathological analyses, as well as functional studies.

A few early studies, to be described in Section II, addressed directly whether vitamin D plays a role in the central nervous system (CNS). Yet, little further work exploring its actions in either the central or peripheral nervous system (PNS) was reported at the time. However, a renewed interest in the role of vitamin D in the nervous system has emerged, as evidenced by a notable increase in the number of active laboratories working in the field. This chapter reveals, however, that our understanding of how vitamin D acts in the nervous system is still quite limited.

This chapter is divided into two major parts. In Section II, data pointing to the nervous system as a site of action of vitamin D are reviewed. Evidence is presented for blood–brain barrier permeability of 1,25-dihydroxyvitamin D_3 [1,25$(OH)_2D_3$] and VDR localization in the adult nervous system and in primary neuronal and glial cell culture systems, along with evidence for activity of vitamin D metabolic enzymes and metabolites in the CNS. *In vitro* and *in vivo* experiments revealing vitamin D-elicited gene inductions and overall functional effects in neuronal systems are then described. The focus of the second portion of this chapter (Section III) is on data suggesting the potential role of vitamin D compounds to prevent neurodegeneration, with emphasis on their potential utility for the treatment of Alzheimer's disease.

II. THE NERVOUS SYSTEM AS A SITE OF ACTION FOR VITAMIN D

It was originally thought that the brain was not a target tissue for vitamin D because early studies failed to observe radiolabeled 1,25$(OH)_2D_3$ binding to extracts of total brain [12,13]. This failure to detect specific binding most likely reflects the low levels of VDR in the adult rat brain. It was later shown that VDR localizes to the nucleus in both the occupied and unoccupied states [14–16]. Yet, this author could find no published attempts to observe 1,25$(OH)_2D_3$ receptor binding to extracts containing nuclear fractions of brain tissue, which would be enriched for VDR. Instead, subsequent studies, reviewed in Sections II,A–C below, used other experimental approaches to generate data suggesting that 1,25$(OH)_2D_3$ can enter and has specific binding sites in the brain and that the nervous system expresses VDR.

A. Blood–Brain Barrier Penetrance of 1,25$(OH)_2D_3$

To determine whether 1,25$(OH)_2D_3$ can enter the brain, experiments were performed that followed the fate of radiolabel in the brain and spinal cord of adult rodents after peripheral administration of tritiated 1,25$(OH)_2D_3$. Although tritium was detected in the brain, the levels were quite low [17,18], suggesting detectable but limited blood–brain barrier penetrance. In one of the experiments, for instance, less than 0.2% of total label and tissue concentrations of only approximately 5% of plasma levels could be measured in the brain 30 min after carotid injection of labeled hormone into adult rats [17]. Preexisting levels of vitamin D had no effect on these findings because similar levels of uptake were observed in vitamin D-deficient and replete animals [17].

Alone, these results are suggestive but do not provide convincing proof that 1,25$(OH)_2D_3$ has limited access to the brain. On the one hand, each study analyzed levels of radioactivity at only a single time point, leaving open the possibility that much higher levels might have been observed at a different time. In addition, these were assessments of whole brain extracts; it is conceivable that much higher concentrations of label might be detectable in selected regions of the brain that contain relatively high levels of VDR. On the other hand, the levels of label in the brain that were reported in these studies were so low that they raise the question of their biological relevance. Furthermore, these studies did not demonstrate whether the tritium measured in brain corresponded to intact 1,25$(OH)_2D_3$ or was merely an inactive metabolic breakdown product. The most convincing evidence that peripherally administered 1,25$(OH)_2D_3$ is active in the brain comes, instead, from measurements of changes in several parameters in the brain, as described in detail in Section II,H, that occur after systemic administration of 1,25$(OH)_2D_3$. These include the demonstration in the brain of specific binding of radiolabeled 1,25$(OH)_2D_3$ (Section II,B) and induction of selected target genes (Section II,H,2).

B. Vitamin D Binding Sites in the Adult Nervous System

A number of labeling experiments attempted to localize the regions of the brain and spinal cord to which 1,25$(OH)_2D_3$ distributes. In these similarly designed studies [19–22], autoradiography was performed on brains and spinal cords of vitamin D-deficient rodents treated after two intravenous injections of tritiated

TABLE I Summary of Data Localizing Binding Sites for either $1,25(OH)_2D_3$ or VDR in Adult Rodent Brain

	Assessment of VDR[a]			
Brain region	^3H-$1,25(OH)_2D_3$ autoradiography (rodent)[b]	VDR mRNA (RT-PCR) (rat)[c]	VDR Immunohistochemistry (rat)[d]	VDR Immunohistochemistry (human)[e]
Cortex	n.d.	+	n.d.	+
Hippocampus	+	+	n.d.	+
Basal forebrain	+	+	n.d.	n.d.
Striatum	+	+	n.d.	n.d.
Hypothalamus	+	+	n.d.	n.d.
Mesencephalon	n.d.	+	n.d.	n.d.
Cerebellum	+	+	+	+
Thalamus	+	n.d.	n.d.	n.d.
Pons/medulla	+	+/−	n.d.	n.d.
Spinal cord	+	n.d.	n.d.	n.d.

[a] +, Readily detectable; +/−, negligible levels detected; n.d., not done.
[b] Results from Stumpf *et al.* [19–22], described in Section II,B.
[c] Preliminary unpublished results from our laboratory, discussed in Section II,C. RNA was prepared from dissected brain regions (as shown) of an adult male Sprague Dawley rat and then subjected to semiquantitative reverse transcriptase polymerase chain reaction (RT-PCR) analysis, using primers specific for rat VDR.
[d] Results from Clemens *et al.* [15,16], described in Section II,C.
[e] Results from Sutherland *et al.* [11], described in Section II,C. Data are from analysis of the brains of Alzheimer's and Parkinson's disease patients.

$1,25(OH)_2D_3$, spaced 1 hr apart, and then sacrificed 2–4 hr later. The results revealed widespread distribution of tritium in the central and peripheral nervous system of adult rodents. Labeling was shown to reflect specific binding by its ability to be competed with unlabeled $1,25(OH)_2D_3$. Virtually every region of the brain retained label (summarized in Table I). The most intensely labeled structures included many limbic system nuclei and their extensions in the brain, particularly the interstitial nucleus of the stria terminalis, the central nucleus of the amygdala, and the periventricular nucleus of the hypothalamus. A portion of cells in virtually all other sections of brain and spinal cord regions retained at least some radioactivity. These include the arcuate nucleus, supramammillary nucleus, the ventral portion of the hippocampus, the central gray area of the pons, parabrachial nuclei, the substantia gelatinosa of the trigeminal nucleus, the ventromedial and reticular nuclei of the thalamus, the caudate nucleus, Golgi type II cells of the cerebellum, and sensory and motor nuclei of the peripheral nervous system. The regions of the brain involved in Alzheimer's disease, diagrammed in Fig. 1 and discussed in Section III, that were examined in these studies were labeled. However, these studies did not examine the cortex and some portions of the basal forebrain, two regions that are important in Alzheimer's disease.

Interpretation of the autoradiographic binding studies is limited, however. Without the radioactive species that shows binding being identified as intact

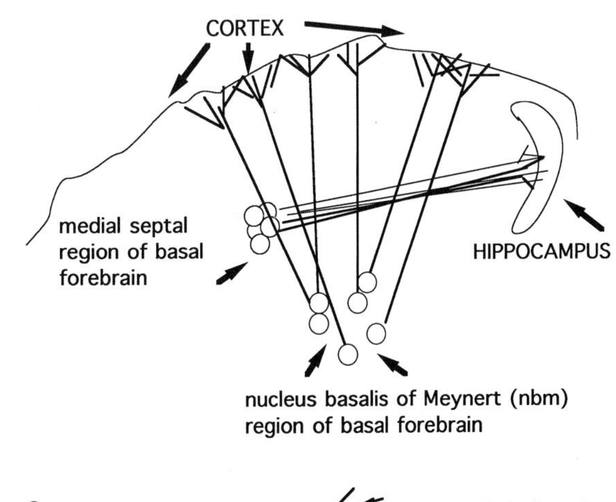

FIGURE 1 Basal forebrain cholinergic system depicting the cell bodies of neurons that project from the septal and nucleus basalis regions of the basal forebrain to the hippocampus and the cortex.

$1,25(OH)_2D_3$, the possibility remains that the observed labeling might not represent actual ligand bound to VDR and, thus, might not be biologically relevant. (This is the same caveat as in the interpretations of the blood–brain barrier entry studies, cited in Section II,A). Another drawback of autoradiography is its lack of sensitivity, which may prevent the detection of cells that express low levels of VDR. Indeed, later immunohistochemical experiments mapping VDR to cells of discrete brain regions (discussed next in Section II,C) detected VDR in cell populations of the cerebellum and hippocampus that did not retain tritium in the radiolabeled ligand binding studies. In sum, then, these early works using labeled $1,25(OH)_2D_3$ suggest potential blood–brain barrier entry and a regionally ubiquitous distribution of VDR in the brain and spinal cord. However, because of limitations in the methodology, these interpretations are not definitive and require further supporting evidence.

C. VDR Expression in the Adult Nervous System

After the vitamin D receptor protein had been purified and cloned, VDR-specific antibodies and probes became available [23,24]. With these reagents more direct characterizations of the distribution of VDR in the brain were possible. Using immunohistochemistry, Clemens and co-workers [15,16] demonstrated a widespread, but not ubiquitous, distribution of VDR in several brain regions of adult rats fed a normal diet (vitamin D replete). As in the radiolabeled $1,25(OH)_2D_3$ studies (see Section II,B above), staining was exclusively nuclear. Particularly strong staining was observed in cerebellar and hippocampal granule cells, and in pyramidal neurons of the CA1 and CA2 regions of the hippocampus. Negligible VDR immunoreactivity was observed in the CA3 section of the hippocampus, in cerebellar Purkinje cells, and in glia. These negative results may, however, be a function of the limited sensitivity of the assay, rather than reflect an absence of VDR in those cells. The purpose of these studies was not to map VDR throughout the adult brain, but rather to compare the localization of VDR expression in the brain to that of the vitamin D-dependent calcium-binding protein, calbindin-D_{28K}. Consequently, the results described were focused on VDR expression in only the above-mentioned brain regions. Thus, although VDR was observed in at least some cells in each region examined, much of the brain was not characterized. Another study [11] used *in situ* hybridization to examine VDR mRNA expression in the temporal cortex, cerebellum, and CA1 and CA2 pyramidal cells of human Alzheimer's and Parkinson's diseased brains. Some cells were positive in all regions examined, but, again, only these selected brain regions were analyzed.

Inasmuch as these are the only existing studies that directly measure VDR expression in the nervous system, a systematic analysis with detailed localization of VDR in the developing and adult nervous system is much needed in order to define regions of the brain and the selected neuronal and glial cell types that might be vitamin D responsive. Our laboratory is undertaking this task presently. Thus far, relative levels of VDR mRNA have been quantitated in various regions of the adult rat brain, using semiquantitative reverse transcriptase polymerase chain reaction (RT-PCR) [24a]. Interestingly, readily detectable—but not equivalent—levels of VDR mRNA were observed in all regions examined except the brain stem, where it was expressed at very low levels (see Table I). Although the specific cell types expressing VDR mRNA cannot be identified by this technique, these results confirm the previous findings suggesting that VDR is expressed throughout the brain.

An important finding for the therapeutic implications of vitamin D compounds in Alzheimer's disease, discussed in Section III of this chapter, is that the areas of the central nervous system that degenerate in this disease—the basal forebrain (both the septal and nucleus basalis regions), the hippocampus, and the cortex—all contain either radiolabeled $1,25(OH)_2D_3$ or VDR-positive cells (summarized in Table I). These brain regions are illustrated in Fig. 1, and their relevance to Alzheimer's disease is discussed in more detail in Section III.

D. VDR Expression in Primary, Embryonic Neuronal Cell Culture Systems

The developmental program for VDR expression in the nervous system has not been characterized. Preliminary work from our laboratory [24a] suggests that VDR expression may commence at differing times in development, depending on the cell type and/or brain region.

Similarly, a systematic examination of VDR expression in primary cultures of embryonic neuronal and glial cells has not been performed. However, VDR expression in a few selected culture systems of neuronal origin has been investigated. A carefully performed study [25] showed that VDR mRNA, analyzed on Northern blots, is expressed [and up-regulated by $1,25(OH)_2D_3$] in cultures enriched for glia prepared from the brains of newborn rats. (Glia are dividing cells in the CNS that are thought to play a supporting role for neurons by, for example, producing and secreting growth factors.) Cell depletion and enrichment experiments in this study suggest that the $1,25(OH)_2D_3$-responsive glial cell type is most likely astrocytes.

In another study, cell cultures were prepared from dorsal root ganglia, which contain peripheral sensory neurons and glia, between embryonic days 13 and 21. Immunohistochemical colocalization data showed that neurons, but not glia, express both calbindin-D_{28K} and VDR [26]. However, neurons were distinguished from glia only by observation of morphologic differences, which may be misleading.

Two other types of cultured cells of neuronal origin have been documented to respond to treatment with $1,25(OH)_2D_3$, strongly suggesting that they express VDR, although the receptor has not been formally identified in them. These are cultured adult human retinal glial cells [27] and the neuroblastoma cell line LA-N-5 [28]. The vitamin D-stimulated effects observed in these cells are described later in Section II,F.

E. Vitamin D Metabolites and Metabolic Activity in the Nervous System and in Neuronal Cell Cultures

The observation of the presence of vitamin D metabolites and the demonstration of activity of its metabolic enzymes in the nervous system provide further support that $1,25(OH)_2D_3$ plays a role there. Human cerebrospinal fluid was shown to contain $1,25(OH)_2D$ as well as its precursor, 25-hydroxyvitamin D, and the vitamin D metabolite of unknown function, 24,25-dihydroxyvitamin D [29].

Another study demonstrated the presence and induction of a vitamin D metabolic enzyme in primary glial cultures prepared from newborn rats. Naveilhan *et al.* [30] showed by Northern blot analysis that the mRNA of 25-hydroxyvitamin D_3 24-hydroxylase (24-hydroxylase), the enzyme that produces 24,25-hydroxyvitamin D_3, is expressed in these embryonic cultures, but not in C6 glioma, a glial cell line. Furthermore, the enzyme is upregulated in these primary cultures after administration of $1,25(OH)_2D_3$.

Results using cultured brain microglia derived from newborn rats suggest that the brain also may be a site of synthesis of $1,25(OH)_2D_3$. Microglia are phagocytic cells that reside in the CNS but are of a macrophage–monocyte lineage. Activation of these cells with either lipopolysaccharide or interferon-γ and subsequent treatment with the $1,25(OH)_2D_3$ precursor 25-hydroxyvitamin D_3 results in the synthesis of $1,25(OH)_2D_3$, as identified by high-performance liquid chromatography and binding affinity to chick intestinal vitamin D receptor [31]. Thus, in response to immune stimulation, microglia from newborn rats synthesize $1,25(OH)_2D_3$, suggesting a possible immunomodulatory role of $1,25(OH)_2D_3$ in the CNS.

F. $1,25(OH)_2D_3$-Elicited Effects in Neuronal Cell Culture

Further evidence confirming the nervous system as a target for vitamin D comes from the demonstration, both *in vitro* and *in vivo*, that $1,25(OH)_2D_3$ mediates functional responses. Dose-dependent inductions of several genes by $1,25(OH)_2D_3$ have been reported in primary cultures of glia prepared from the brains of newborn rats. These include the up-regulation of mRNA of VDR [31], 24-hydroxylase [30], and three members of the neurotrophin family, namely, nerve growth factor (NGF) [31], neurotrophin-3 (NT-3), and NT-4 [32]. The maximal induction of NGF mRNA was observed 24 and 48 hr after treatment with 100 n*M* $1,25(OH)_2D_3$ in serum-free medium. Secreted NGF protein was measured 24 hr after treatment, and maximal increases were approximately 2-fold. Interestingly, however, when the cells were pretreated with submaximally inducing concentrations of $1,25(OH)_2D_3$ (either 1 or 10 n*M*) for 3 days and then treated with various doses of $1,25(OH)_2D_3$ for 24 hr, NGF protein was maximally induced to a much greater extent (5-fold) at 100 n*M* than had been observed in cells without the pretreatment. (The dose response for NGF induction was the same under both conditions.) The authors postulate that this enhancement in maximal efficacy may occur because, as they demonstrated, VDR levels had been induced in the cells during the pretreatment, which allowed for an increase in induced levels of NGF [31].

Induction of NT-3 and NT-4/5 in primary astrocytes was observed at 10 n*M* $1,25(OH)_2D_3$ in serum-free medium, but no other doses were tested [32]. Another neurotrophin, brain-derived neurotrophic factor (BDNF) was not induced by $1,25(OH)_2D_3$ at 10 n*M*, but a complete dose–response experiment was not performed [32], leaving open the possibility that induction of BDNF might occur at higher doses.

There are two reports of vitamin D eliciting functional effects in cultured cells of neuronal origin. In the first, $1,25(OH)_2D_3$ (10 n*M*) was shown to enhance the phagocytic activity of human retinal glia obtained postmortem from adult eyes, as measured by flow cytometric analysis of the uptake of fluorescein-labeled beads [27]. This enhanced phagocytic activity may be due to an increase in calcium influx into the cell, as reduction of extracellular calcium or treatment of the cells with a calcium channel blocker inhibited the phagocytic activity. However, this was not conclusively established, because it was not determined whether treatment with calcium channel blockers reduced the $1,25(OH)_2D_3$-induced phagocytic activity.

In the second study, $1,25(OH)_2D_3$ induced differentiation of the human neuroblastoma cell line LA-N-5 into a neuronal lineage [28]. Differentiation was measured by induction of neurite outgrowth, increased activity of the differentiation-specific marker gene, induction of acetylcholinesterase, inhibition of cell growth, and decreased N-*myc* expression. Maximal effects were observed at 2.4 n*M* of $1,25(OH)_2D_3$ or at approximately 10-fold lower doses with two $1,25(OH)_2D_3$ analogs, KH1060 [33] and EB1089 [34], which were reported to be potent differentiating compounds in cancer cell lines (see Chapter 61). These results suggest that $1,25(OH)_2D_3$ may be able to act directly on neurons to induce their differentiation. To support this hypothesis, however, experiments are needed that demonstrate that $1,25(OH)_2D_3$ can induce differentiation in cultures of primary neurons.

G. Activation of VDR by Dopamine

Matkovits and Christokos [35] demonstrated that treatment of CV-1 cells (monkey kidney cells) with dopamine can activate VDR, as measured by vitamin D response element-driven reporter gene transcription. Thus, VDR-mediated transcription was evoked in the absence of $1,25(OH)_2D_3$ by a neurotransmitter. Although the physiological relevance of this phenomenon is not known, the authors point out that VDR and the dopamine D1 receptor colocalize in many cells in the CNS. This opens the possibility that activation of dopaminergic pathways might coordinately activate VDR signaling. These data thus suggest another way in which VDR may act in the brain.

H. Effects of $1,25(OH)_2D_3$ in the Adult Nervous System

A limited number of investigations have shown that treatment with $1,25(OH)_2D_3$ can elicit biological changes in the adult PNS and CNS. These are the most compelling data yet indicating that vitamin D plays a role in the nervous system and may have therapeutic utility for neurological indications. Two types of data have been generated: the first demonstrates functional effects of $1,25(OH)_2D_3$, whereas the second shows that treatment with $1,25(OH)_2D_3$ induces specific genes in the CNS.

1. Functional Effects of $1,25(OH)_2D_3$

In an *ex vivo* study in which avian intestinal nerves from vitamin D-replete and vitamin D-deficient chickens were compared, vitamin D deprivation was shown to decrease conduction velocity [36]. This finding suggests that, at least in the chick, $1,25(OH)_2D_3$ may be required for normal peripheral nerve function. This effect appears to be a function of the activities of vitamin D in modulating calcium homeostasis, as administration of excess calcium to vitamin D-deficient animals restored conductance velocity to control levels. Thus, the effects of vitamin D on conductance velocity may have been exclusively an effect that was secondary to calcium regulation. Even if this is so, vitamin D, nonetheless, altered peripheral nerve function in this study, albeit indirectly.

An important set of studies demonstrating that vitamin D has activity in the brain examined the effects of $1,25(OH)_2D_3$ in two models of multiple sclerosis, both acute [37–39] and chronic, relapsing [39,40] experimental allergic encephalomyelitis (EAE), effects which appear to involve the immune system (discussed in more detail in Chapter 69). The EAE is initiated by immunization of animals with myelin basic protein from guinea pig. This elicits an autoimmune inflammatory response in the nervous system, the ultimate effect of which is the demyelination of axons in both the peripheral and central nervous systems. In the acute model of EAE, treatment of mice with $1,25(OH)_2D_3$, using various dosing paradigms, prior to and also after immunization, prevents development of disease symptoms [37,39], a rise in titer of antibodies to myelin basic protein [37], and histopathological changes [37]. Two other reports, using rat [40] and mouse [39] models of recurring EAE, extend these findings by showing that administration of $1,25(OH)_2D_3$ after disease onset can halt or retard its progression. DeLuca and co-workers [39] treated mice with $1,25(OH)_2D_3$ [300 ng intraperitoneal (i.p.)] after they began to show symptoms and then fed them a $1,25(OH)_2D_3$-containing diet (20 ng/day). Treatment with $1,25(OH)_2D_3$ halted progression of symptoms for the duration of the experiment (40 days), although symptoms resumed within 10 days of discontinuing $1,25(OH)_2D_3$. Also shown in this work is that making the animals vitamin D deficient prior to induction of EAE results in earlier onset of symptoms, suggesting that $1,25(OH)_2D_3$ may play a role in maintaining normal immune function.

In another study using a rat model of EAE, Brachet and co-workers [40] showed that clinical scores are improved with administration of $1,25(OH)_2D_3$ (5 μg/kg, i.p.) on days 11 and 13 after administration of antigen (during the first attack, as assessed clinically) and again at a lower dose (1 μg/kg) on days 19, 21, and 23 (after remission and at the onset of the second attack). Treatment with 1,25(OH)$_2$D$_3$ also reduces the number of cells in the brain that are immunostained for OX42, a marker of invading macrophages and activated microglia. A possible mechanism(s) by which 1,25(OH)$_2$D$_3$ efficacy is achieved in this model has not been defined, although the investigators postulate that it is through immunosuppressive actions. If 1,25(OH)$_2$D$_3$ exerts its effects in this model by suppressing activation of peripheral T cells that invade the CNS when activated, then it could be argued that the effects of 1,25(OH)$_2$D$_3$ on the nervous system are only indirect. However, it is also possible that 1,25(OH)$_2$D$_3$ directly suppresses activation of resident microglia in the brain, which would indicate it has direct effects in the CNS. Additionally, the possibility exists that 1,25(OH)$_2$D$_3$ might also act directly on compromised cells in the nervous system, a concept that has not been addressed experimentally. Regardless, these results in EAE models demonstrate that peripheral administration of 1,25(OH)$_2$D$_3$ has functional effects in the adult nervous system.

2. *In Vivo* Induction of Genes by Treatment with 1,25(OH)$_2$D$_3$

One of the earliest and perhaps the most convincing studies providing evidence that 1,25(OH)$_2$D$_3$ plays a role in the CNS was conducted by Christakos and coworkers [41]. Vitamin D-deficient adult rats were treated with 1,25(OH)$_2$D$_3$ either by peripheral daily dosing (100 and 200 ng/day i.p. for 8 days) or by continuous infusion directly into the brain [25 ng/day, intracerebral ventricular (I.C.V.) via osmotic minipumps for 7 days]. Homogenates were prepared from isolated brain nuclei from various regions, including the basal forebrain, hippocampus, as well as several other regions previously shown to contain vitamin D binding by autoradiography (described in Section II,B). These extracts were assessed for calbindin D$_{28K}$ by radioimmunoassay (RIA) and for enzymatic activity of choline acetyltransferase (ChAT) and monoamine oxidase. Significant increases in ChAT activity were observed in the arcuate-median eminence and in the stria terminalis under all dosing conditions. Nonstatistically significant increases in ChAT activity were observed in the CA1 region of the hippocampus and the amygdala following the i.p. 200 ng dose and the i.c.v. minipump infusion. Monoamine oxidase was increased significantly only in the ventral medial nucleus of the basal forebrain. In all other samples, and including the assessments of calbindin-D$_{28K}$, treatment with 1,25(OH)$_2$D$_3$ had no effect. Because these were not full dose–response studies, the possibility remains that induction of these genes might be observable with higher doses in additional brain regions. The therapeutic significance of these findings is discussed in Section III.

Another investigation demonstrated that 1,25-(OH)$_2$D$_3$ can elicit the induction of a specific gene in the brain [42,43]. NGF was shown to be inducible in adult rat hippocampus and cortex after either acute [42] or chronic [43] i.c.v. administration of 1,25(OH)$_2$D$_3$. These findings correlate with the results showing 1,25(OH)$_2$D$_3$ induction of NGF in primary glial cultures [31]. It was later shown that NGF could be synthesized and induced by 1,25(OH)$_2$D$_3$ in the basal forebrain [44]. Figure 2 shows the NGF induction observed in the hippocampus, cortex, and septal regions of the basal forebrain after a single i.c.v. dose of 1,25(OH)$_2$D$_3$. It will be recalled that these three brain regions—the cortex, hippocampus, and basal forebrain—are the most affected in Alzheimer's disease. (The therapeutic implications of these results are discussed in Sections III,A and III,C below.) With acute dosing of 1,25(OH)$_2$D$_3$, maximal (~1.5- to 2-fold) inductions of NGF mRNA, measured by RNase protection, were observed at doses between 1 and 100 nmol. A time course of this effect revealed that highest levels of NGF mRNA were achieved 4 hr after administration, decreased by 8 hr, and returned to control levels at 24 hr. Although induction of NGF mRNA was detectable after a single treatment, no increase in NGF protein could be measured by an NGF enzyme-linked immunosorbent assay (ELISA). In the chronic dosing paradigm [43], however, elevations in both NGF mRNA and protein were observed. In these studies, 1,25(OH)$_2$D$_3$ was administered via continuous infusion into the brain at various doses for 6 days. As in the acute study, an approximately 2-fold induction in mRNA was observed, with the maximal effect measured between 5 and 10 ng/hr. Small (~20%) but significant increases in NGF protein were

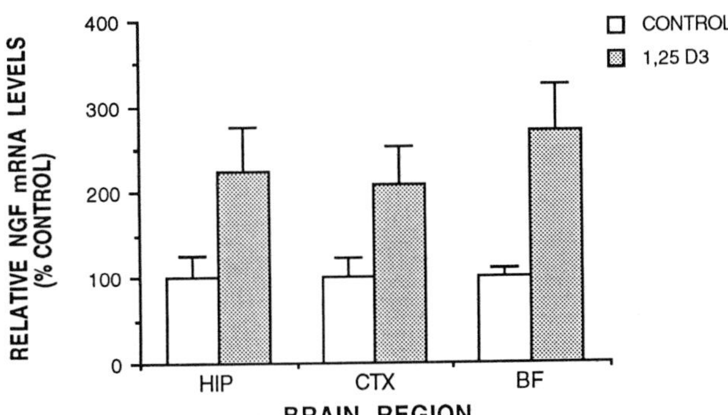

FIGURE 2 Comparative induction of NGF mRNA by 1,25(OH)$_2$D$_3$ (1,25 D3) in the septal region of the hippocampus (HIP), cortex (CTX), and basal forebrain (BF). Rats were administered 10 nmol 1,25(OH)$_2$D$_3$ or vehicle (n = 6) i.c.v., as described previously [42]. Eight hours after treatment RNA was purified from the brain regions shown and analyzed by RNase protection assays. Error bars represent standard deviations. All results obtained from treated animals were statistically different from untreated controls. Adapted from Saporito and Carswell [44], *The Journal of Neuroscience,* with permission.

also seen at these doses. The doses that induce NGF are also those that elicited weight loss, which correlates with elevations in serum calcium (S. Carswell, unpublished results).

I. Summary

A substantial body of data suggest that the nervous system is a target for vitamin D. Several regions of the CNS and PNS have been shown to either contain specific binding sites for $1,25(OH)_2D_3$ or express VDR, or both. In addition, two types of primary cells of neuronal cultures, derived from dorsal root ganglia and brain glia, express VDR. Adult human retinal glia and the neuroblastoma cell line LA-N-5 are vitamin D responsive. Further indication that vitamin D acts in the brain comes from evidence of the presence of vitamin D metabolites and its metabolic enzymes in both human cerebrospinal fluid and cultured rat glial cells.

The most convincing data supporting the notion that $1,25(OH)_2D_3$ plays a role in the nervous system comes from studies showing that treatment of either neuronally derived cultured cells or adult animals with $1,25(OH)_2D_3$ results in a biological effect in the nervous system. These results include $1,25(OH)_2D_3$ induction in primary cultures of brain glia of VDR, NGF, NT-3, NT-4/5, and 24-hydroxylase. $1,25(OH)_2D_3$ also enhances phagocytic activity of adult human retinal glia and induces the differentiation of a neuronal phenotype in LA-N-5 neuroblastoma cells.

Several studies also have shown functional effects in the adult nervous system. Conductance velocity of avian intestine nerves was enhanced by treatment with $1,25(OH)_2D_3$. Administration of $1,25(OH)_2D_3$ to rats or mice prevents the onset or halts the progression of EAE, depending on the dosing schedule.

Gene inductions in the brain have also been observed after treatment with $1,25(OH)_2D_3$. Peripheral administration of $1,25(OH)_2D_3$ results in an induction in ChAT enzyme activity in certain brain nuclei, and i.c.v. administration of $1,25(OH)_2D_3$ induces NGF in three regions of the CNS.

The combined data make a strong case that $1,25(OH)_2D_3$ acts on multiple, disparate cell populations in the nervous system, probably in selective ways depending on the cell type and condition. One role that vitamin D has been postulated to play in the nervous system is to prevent the loss of neuronal function in neurodegenerative diseases. The potential for vitamin D compounds to be used in the treatment of neurodegenerative disorders is the topic of the remainder of this chapter.

III. THERAPEUTIC POTENTIAL OF VITAMIN D IN NEURODEGENERATIVE DISEASES

Neurons have an absolute requirement for neurotrophic factors for their development, maintenance, and survival. Accordingly, the term "neurotrophic" may be broadly defined as generating or maintaining a neuronal phenotype or enhancing the survival of neurons. Two ways in which neurons are classified are by the neurotransmitter that they produce (e.g., cholinergic neurons synthesize acetylcholine, and dopaminergic neurons make dopamine) and the region in the nervous system in which they are found (e.g., basal forebrain cholinergic neurons). Specific classes of neurons express receptors for and are dependent on one or more selected neurotrophic factor(s). For example, the cholinergic neurons of the basal forebrain and striatal regions of the brain are dependent on NGF and BDNF, whereas cholinergic spinal cord motor neurons respond selectively to another group of factors [45], including BDNF but not NGF.

The best characterized neurotrophic factors are in the neurotrophin family (reviewed by Barbacid [46]). The neurotrophins include NGF, BDNF, NT-3, and NT-4/5. These are small, secreted, homodimeric peptides, each of which is a specific ligand for a given member of the Trk family of receptors (Trk A, B, C, and D, respectively). Trk is a classic plasma membrane-bound tyrosine kinase receptor (reviewed by Green and Kaplan [47]). On binding of the receptor to its cognate neurotrophic factor via its extracellular domain, the intracellular portion of the Trk molecule becomes an activated tyrosine kinase, which phosphorylates several intracellular substrates. This begins a cascade of poorly understood signaling events that culminate in neurotrophic effects. Examples of such neurotrophic effects include the production of a differentiated phenotype during development, maintenance of a mature neuronal phenotype, and protection of the neuron against damaging events. The activities of neurotrophins that mediate their ability to prevent the loss of injured neurons have engendered great interest in their therapeutic potential for the treatment of various neurodegenerative diseases. In the case of Alzheimer's disease, NGF has been shown to protect degenerating basal forebrain cholinergic neurons. As elaborated below, interest in $1,25(OH)_2D_3$ as a therapeutic agent in neurodegenerative disease is based on its ability to induce NGF in the CNS.

It is important to note that induction of NGF is one of several potential mechanisms by which vitamin D compounds might prevent neurodegeneration. Other mechanisms might include induction of other growth factors, induction of differentiation-specific genes, and activation of the sphingomyelin cycle. Although, as yet, there are no direct data to support the concept that $1,25(OH)_2D_3$ might be neurotrophic through these activities, several clues exist in the literature in nonneuronal systems that suggest that $1,25(OH)_2D_3$ might have activities in addition to NGF induction that could enhance the survival and function of neurons. These are discussed in Section III,E.

A. Alzheimer's Disease

Alzheimer's disease (reviewed by Braak and Braak [48]) is a slowly developing dementia that is characterized by multiple cognitive deficits. Histopathologically, Alzheimer's disease is defined by the presence of amyloid plaques and neurofibrillary tangles in the cortex and subcortical regions. These lesions are thought to elicit the atrophy and ultimate death of neighboring neurons. The most vulnerable neurons appear to be the cholinergic cells of the basal forebrain and the glutamatergic neurons of the entorhinal cortex. Furthermore, the loss of these populations of neurons is closely associated with the cognitive impairments of the disease [49]. The focus of this discussion is on the potential of $1,25(OH)_2D_3$ to spare degenerating cholinergic cells in the basal forebrain, although it remains of interest to examine whether vitamin D compounds can also protect entorhinal cortical neurons.

Basal forebrain cholinergic neurons are cells that produce the neurotransmitter acetylcholine. Their axons extend principally from their cell bodies in the basal forebrain to the cortex and hippocampus (illustrated in Fig. 1). The basal forebrain comprises two subregions: the septum (also known as the medial septum) and the nucleus basalis of Meynert (nbm). The cholinergic cells in these subregions differ only in the location of their cell bodies and in the regions to which their axons project. Cholinergic neurons in the medial septum extend to the hippocampus, whereas the nbm cells principally innervate the cortex (see Fig. 1). In Alzheimer's disease, there is a reduction in the activity of the acetylcholine-producing enzyme, choline acetyltransferase (ChAT), in the basal forebrain, cortex, and hippocampus. These biochemical decreases reflect the functional loss of basal forebrain cholinergic neurons.

B. NGF and Its Potential in Alzheimer's Disease

Nerve growth factor (reviewed by Barbacid [46] and Greene and Kaplan [47]) is a small (13 kDa), secreted protein synthesized in many tissues throughout the body, including the brain. It is required for the maintenance and survival of sympathetic and sensory neurons in the peripheral nervous system, as well as of basal forebrain cholinergic neurons in the CNS. Although NGF was discovered in the 1950s, its function in nonneuronal tissues such as the prostate, submandibular gland, and skin remains unknown.

Nerve growth factor exists primarily as a dimer. Its high affinity receptor is Trk A, a tyrosine kinase receptor found on the plasma membrane of basal forebrain cholinergic cells in the CNS as well as other NGF-responsive cells. Another membrane-bound protein, known as p75 or the low affinity NGF receptor, binds NGF with low affinity and also colocalizes to basal forebrain cholinergic neurons. The respective roles of Trk and p75 in eliciting NGF-mediated responses are incompletely understood but are under intense investigation.

The literature indicates that there is a strong rationale for the use of NGF in Alzheimer's disease (reviewed elsewhere [50–52]). This is based on the ability of NGF to prevent the loss of basal forebrain cholinergic cells. It is thought that retarding or preventing the loss of these cells should impact disease progression. NGF is essential for the survival and maintenance of basal forebrain cholinergic neurons in culture and during development [53,54]. In the adult, NGF is known to be synthesized in the regions of the adult CNS that contain the cell bodies and projections of these cholinergic cells (reviewed by Saporito and Carswell [44]). Furthermore, NGF supplied exogenously to the adult attenuates basal forebrain cholinergic neuronal loss and associated cognitive deficits in animal models of cholinergic damage. The best characterized of these models is the fimbria–fornix lesion of the adult rat or monkey. In this paradigm, some of the axons of cholinergic neurons in the septal region of the basal forebrain that project to the hippocampus are severed, resulting in a gradual loss of markers of the cell bodies and axons of cholinergic neurons, as well as an impairment of performance on learning and memory tasks. When NGF is delivered to the brain via a catheter implanted in the skull—an i.c.v. route of administration—to animals with fimbria–fornix lesions, the number of septal cells expressing cholinergic markers and cholinergic activity in the hippocampus are restored to near control levels [52]. It is thought that NGF acts to prevent the atrophy and death of axotomized cell bodies and also to up-regulate cholinergic phenotypic expression in uninjured cholinergic neurons.

The nbm lesion of the rat and primate is another NGF-responsive model of Alzheimer's disease (reviewed by Saporito and Carswell [55]). In this model, the nbm region of the basal forebrain (see Fig. 1) is partially damaged by direct administration of a neurotoxin. The effect is to destroy a significant portion of nbm neurons and thereby reduce cholinergic markers in both the nbm and the cortex. In addition, the lesion causes behavioral deficits in such cognitive skills as spatial learning and attention, as measured by water maze performance. Treatment with NGF does not prevent the loss of cholinergic cells, but it does appear to up-regulate cholinergic markers in cells not destroyed by the insult. This up-regulation of activity in uninjured (and possibly nonlethally injured) cells leads to an improvement in behavioral performance.

In addition, administration of NGF (2 weeks, i.c.v.) has been shown to reverse age-associated cognitive deficits in aged rats in a water maze task of spatial learning (reviewed by Saporito and Carswell [55]). Together, these preclinical data suggest that relatively small increases in NGF may significantly attenuate the neuronal degeneration associated with Alzheimer's disease, thereby slowing or halting progression of both the neuronal and cognitive losses. In fact, on the strength of these findings, NGF is presently being tested in pilot human trials for Alzheimer's disease in Sweden and is also being considered for this application in the United States. Preliminary reports from the Swedish study suggest that increasing NGF levels in the CNS does not pose major safety hazards [52].

The use of NGF as a therapeutic agent, however, has serious practical limitations because it is unable to cross the blood–brain barrier and thus must be administered directly to the brain by an i.c.v. route of administration. This is an invasive and cumbersome surgical procedure, which requires drilling a hole in the skull and implanting a pump in the body. Alternative methods of delivery would be highly desirable.

There are several research programs with alternative approaches designed to exploit the neuroprotective potential of NGF for Alzheimer's disease, but all have limitations (reviewed by Olson [52]). Examples include transplantation into the CNS of genetically engineered cells or viral vectors that express NGF and peripheral administration of NGF conjugated to a transferrin antibody, a small portion of which directly enters the brain.

An alternative approach is to up-regulate constitutively expressed NGF in the brain using a systemically administered small organic molecule capable of crossing the blood–brain barrier. Since it is known that NGF synthesis in the CNS is unaffected in Alzheimer's disease (reviewed by Olson [52]), it is reasonable to expect that its up-regulation can be achieved. Also in support of this approach are studies, discussed below and reviewed by Carswell [56], demonstrating that NGF is pharmacologically inducible by multiple agents in both *in vitro* and *in vivo* neuronal systems, and that NGF induction in the brain elicits NGF-mediated effects.

Nerve growth factor has been shown to be inducible by a number of structurally and functionally unrelated compounds in several *in vitro* and *in vivo* systems (reviewed by Carswell [56]). In cell culture, for instance, NGF is inducible by a variety of growth factors, cytokines, catechol derivatives, ligands of nuclear receptors [including dexamethasone, aldosterone, retinoic acid, and $1,25(OH)_2D_3$], and miscellaneous molecules. Most of these compounds induce NGF in mouse fibroblast L cells, which has been shown to be reasonably predictive of NGF induction in the brain [56]. The maximal induction in L cells observed with any individual compound is approximately 2-fold, although combinations of compounds can produce greater effects [57].

C. *In Vitro* and *In Vivo* Induction of NGF by $1,25(OH)_2D_3$

Brachet and co-workers [58] first demonstrated that $1,25(OH)_2D_3$ induces NGF mRNA and secreted NGF protein in L cells. In comparative studies performed in our laboratory ([57] and S. Carswell, unpublished data), $1,25(OH)_2D_3$ was found to be among the most potent and efficacious NGF inducers in this system, having an EC_{50} of approximately 1 n*M*. MC903, an analog of $1,25(OH)_2D_3$ with reduced calcemic activity *in vivo* [59], induces NGF *in vitro* with efficacy and potency equivalent to $1,25(OH)_2D_3$ [60]. These data suggest that the induction of NGF by a vitamin D analog can be uncoupled from its calcemic effects, thus providing the opportunity for clinical utility in a neurodegenerative disorder such as Alzheimer's disease.

Intracerebroventricular administration of $1,25(OH)_2D_3$ to adult rats induces NGF mRNA approximately 2-fold in the basal forebrain [44], hippocampus, and cortex [42,43] after both acute and chronic treatment, as described earlier in this chapter (Section II,H,2) and shown in Fig. 2. The importance of these regions lies in the anticipation that increases in NGF in each of them may be needed to achieve maximal neuroprotective effects on basal forebrain cholinergic neurons. This is because the cell bodies and axons of these cells, both of which express NGF receptors, are located in these areas (refer to Fig. 1). Small but significant increases in NGF protein are also observed in these regions after chronic administration of $1,25(OH)_2D_3$ (Section II,H,2). This is an important result because it provides assurance that when induction of NGF mRNA by $1,25(OH)_2D_3$ is observed, it is indicative that translation of NGF protein is also occurring.

That NGF induction is observed after 6 days of continuous i.c.v. infusion [43] is an important indicator that the induction of NGF by $1,25(OH)_2D_3$ can be sustained, an essential criterion for its consideration as a drug candidate. Also critical for its therapeutic potential is that $1,25(OH)_2D_3$ is quite potent, even after chronic dosing, with maximal induction of NGF mRNA being observed between doses of 120 and 240 ng/day. However, the data also reveal that the doses of $1,25(OH)_2D_3$ which result in detectable induction of NGF mRNA also reduce weight gain in the animals [43]. Furthermore, at doses of 240 ng/day and above, the animals appeared to be severely compromised. They were lethargic and appeared wasted and dehydrated. In subsequent dosing studies, when serum calcium levels were also measured (S. Carswell, unpublished data), serum calcium was found to be elevated even at 60 ng/day, a dose which shows little to no NGF induction [43]. These results suggest, then, that there is no separation between hypercalcemic toxicity and NGF induction with administration of $1,25(OH)_2D_3$, making this compound an unlikely drug candidate. An analog of $1,25(OH)_2D_3$ that provides a further separation between NGF induction and hypercalcemia is required.

Although the ability of $1,25(OH)_2D_3$ to induce NGF within the brain after systemic administration has not yet been demonstrated, positive results would be predicted on the basis of previous work showing induction of ChAT activity in the brain after peripheral dosing [41].

A major question is whether NGF-inducing compounds have NGF-like efficacy in a model of Alzheimer's disease. There is a report that an NGF-inducing compound, idebenone [61], increases cholinergic markers and attenuates behavioral deficits in rats with nbm lesions. These findings strongly support the use of an NGF-inducing compound for the treatment of Alzheimer's disease, but they have yet to be replicated. A theoretical concern is that the small inductions of NGF that are observed after administration of $1,25(OH)_2D_3$ are not sufficient to elicit the NGF-mediated effects in the brain required to protect injured cholinergic neurons. However, two lines of evidence argue that small increases in NGF levels in the CNS would be therapeutically relevant. First, it is known that continuous i.c.v. administration of amounts of NGF that are so small that they cannot be detected biochemically (<0.4 pmol/day) can elicit a maximal biochemical response, as assessed by increased hippocampal ChAT activity, in the fimbria–fornix model [62]. Second, it was demonstrated that treatment with a single dose of dexamethasone, which also induces NGF mRNA in the brain approximately 2-fold, elicits activation of Trk A, the NGF receptor, as determined in a Trk autophosphorylation assay [63]. Furthermore, the magnitude of Trk autophosphorylation stimulated by the NGF-inducing compound is similar to that elicited by maximally activating doses of i.c.v. administered NGF [63]. Finally, treatment with a compound reported to induce NGF *in vivo* has been shown to elicit NGF-like activities in primary cell culture and animal models of ischemia [64]. Clenbuterol was reported to induce NGF in the brain, to protect rat hippocampal neurons in culture from excitotoxic damage, and also to decrease infarct size in the cerebral cortex of mice subjected to an ischemic insult.

D. Summary

NGF has promise for preventing cholinergic neuronal degeneration in Alzheimer's disease, but its therapeutic potential is limited because it cannot freely enter the brain and, therefore, must be administered i.c.v. Induction of NGF with a vitamin D analog is a viable strategy to exploit the neurotrophic effects of NGF. $1,25(OH)_2D_3$ [58], as well as an analog with reduced calcemic effects [60], can potently induce NGF. $1,25(OH)_2D_3$ also induces NGF mRNA and protein in the brain after several days of i.c.v. administration, demonstrating that it can elicit sustained NGF induction. However, doses that induce NGF *in vivo* are hypercalcemic. Together, these data support the therapeutic potential for the treatment of Alzheimer's disease with analogs of $1,25(OH)_2D_3$ with reduced calcemic activity that induce NGF in the CNS at nontoxic doses.

E. Other Potential Neurotropic Mechanisms of Vitamin D Compounds

The combined data suggest that analogs of $1,25(OH)_2D_3$ with sufficiently reduced calcemic activity have promise as NGF-inducing agents for the treatment for Alzheimer's disease. In addition, $1,25(OH)_2D_3$-related compounds could have additional activities that may also rescue various classes of degenerating neurons. Several of these potential activities are related to actions of $1,25(OH)_2D_3$ that have been reported in nonneuronal systems. It should be stressed, however, that none of the putative activities discussed below has yet been demonstrated in neuronal systems.

Induction of calbindin-D_{28K} in rat brain is one such potential vitamin D-mediated neuroprotective mechanism. Calbindin-D_{28K} (discussed in Chapter 13) is induced by vitamin D in the kidney [14, 65]. Thus, it is logical to anticipate that it can also be induced in the CNS, where it is expressed in several cell populations, including septal cholinergic neurons [68]. There has been great interest in demonstrating its induction in the CNS because there is a wealth of literature supporting the role of calbindin-D_{28K} in neuroprotection (reviewed by Miller [69]). It is believed that one function of calbindin-D_{28K} might be to buffer the toxic effects of elevated concentrations of intracellular calcium, and elevations in intracellular calcium mediate some forms of neuronal damage. In the brains of patients with Parkinson's, Alzheimer's, and Huntington's diseases, there is a correlation between loss of calbindin-D_{28K} expression and pathology [70]. Also, neurons that express the highest levels of calbindin have been shown to be the least susceptible to neurodegeneration. It is reasoned, then, that induction of calbindin-D_{28K} could protect vulnerable neurons, slowing or altering the course of disease. However, induction of calbindin-D_{28K} in the CNS of rodents has not been demonstrated, although there have been numerous attempts to do so. Although the results produced to date suggest that this mechanism is not operative in the CNS, the topic may deserve further attention because the studies performed thus far may not have used techniques capable of detecting subtle changes in calbindin-D_{28K} expression, such as increases within small populations of neurons.

Vitamin D deficiency does not decrease nor does administration of excess vitamin D or $1,25(OH)_2D_3$ increase calbindin-D_{28K} expression in (1) the cerebellum, as measured by RIA [71] and densitometry [68]; (2) whole brain extracts, assessed by ELISA and Northern blots [67]; and (3) multiple brain regions, assessed by Northern blots [41,66] and RIA [72]. In addition, there is a dissociation between cells in the nervous system that express calbindin-D_{28K} and VDR [15,16,21,73], strongly suggesting that calbindin-D_{28K} is not regulated by vitamin D, at least in many of the cells in the CNS in which it is expressed. Despite these negative findings, if $1,25(OH)_2D_3$ induction of calbindin-D_{28K} occurred in small numbers or specific populations of cells of the CNS, these changes might not have been observed in the above studies. Furthermore, parvalbumin, another calcium-buffering protein, was induced in the caudate putamen region of the brain after long-term administration of high doses of vitamin D [72]. This result raises the possibility that vitamin D might play a role in protecting neurons from calcium-mediated damage.

An additional possible mechanism by which $1,25(OH)_2D_3$ might prevent cell death in neuronal populations other than those that are NGF-responsive is through its ability to induce other growth factors. Vitamin D is known to induce multiple growth factors in various systems (reviewed by Lowe *et al.* [74]). As detailed in Section II of this chapter, $1,25(OH)_2D_3$ has been shown to induce NT-3 and NT-4/5 in cultures of primary glia [32]. These neurotrophic factors selectively support the maintenance and survival of specific classes of neurons. Their induction *in vivo* could broaden the spectrum of neurodegenerative diseases for which vitamin D compounds could be applicable. There is, in fact, substantial evidence that $1,25(OH)_2D_3$ can act on several types of neuronal cells (see Section II). These data include the demonstration that numerous cell types of neuronal origin express $1,25(OH)_2D_3$ receptors and that *in vitro* and *in vivo* treatment with $1,25(OH)_2D_3$ has functional effects on glia and multiple classes of neurons.

Another potentially neurotrophic mechanism of $1,25(OH)_2D_3$, suggested by its activity in other systems, could be through activation of the sphingomyelin cycle. $1,25(OH)_2D_3$ activates this cycle in myeloid cells [75]. Another study demonstrates that NGF induces differentiation of T9 glioma cells through activation of the sphingomyelin cycle [76]. Together, these results may suggest that an additional neurotrophic mechanism of vitamin D compounds may be through activation of the sphingomyelin cycle.

The hormone $1,25(OH)_2D_3$ may also be found to be neurotrophic through induction of genes that cause differentiation. Vitamin D compounds are known to cause multiple cell types, including a number of progenitor and cancer cells, to differentiate (reviewed by Reichel *et al.* [77]). Recent data suggest that at least part of the basis for this differentiating activity may be through blocking entry of the cell into mitosis via induction of specific cell cycle inhibitors, including the cyclin-dependent kinase inhibitors p21, p27, and INK4 family members [78]. These proteins inhibit cyclins and thereby block cell cycle progression from G_1 to S. Studies in primary neuronal cultures suggest that neurons undergoing apoptosis make an abortive attempt to enter the cell cycle [79]. Treatment of these cells with compounds that block the G_1 to S progression can prevent them from dying [80]. Therefore, there is a potential link between this activity and the ability to protect cells from dying. It is possible, then, that vitamin D might exert neurotrophic effects through the same mechanism(s) by which it differentiates nonneuronal cells.

IV. CONCLUSION

There are potentially a number of ways, in addition to NGF induction, through which vitamin D compounds might be able to rescue degenerating neurons. These mechanisms might differ for individual classes of neurons. If any of these potential modes of action prove to be operative in models of neurodegeneration, the types of neurodegenerative diseases for which $1,25(OH)_2D_3$ and its analogs might have therapeutic utility would be greatly expanded. An argument could also be made that $1,25(OH)_2D_3$ might act on a single population of injured neurons through a combination of several independent and potentially synergistic mechanisms. However, NGF induction is the only $1,25(OH)_2D_3$-mediated neurotrophic mechanism that has been directly examined to date. Much work is needed to determine the biological and therapeutic relevance of both NGF induction and these potentially alternative mechanisms.

Acknowledgments

We thank F. Haun, N. Neff, M. Miller, and D. Feldman for helpful discussions and comments.

References

1. Hess AF 1929 Pathogenesis of rickets. In: Rickets Including Osteomalacia and Tetany. Lea & Febiger, Philadelphia, pp. 130–142.
2. Christiansen C, Rodbro P, Sjo O 1974 "Anticonvulsant" action of vitamin D in epileptic patients? A controlled pilot study. Br Med J **2**:258–259.
3. Mallette LE, Patten BM, Engel WK 1975 Neuromuscular disease in secondary hyperparathyroidism. Ann Intern Med **82**:474–483.
4. Heyburn PJ, Peacock M, Casson IF, Crilly RG, Taylor GA 1983 Vitamin D metabolism in post-menopausal women and their relationship to the myopathic electromyogram. Eur J Clin Invest **13**:41–44.
5. Peacock M 1977 The treatment of proximal myopathy with 1,25-dihydroxyvitamin D3 or its synthetic analogue 1-alpha hydroxyvitamin D in various endocrinological diseases. Q J Med **56**:556.
6. Schott GD, Wills MR 1976 Muscle weakness in osteomalacia. Lancet **1**:626–629.
7. Krane, SM, Holick, MF 1994 Metabolic bone diseases. In: Isselbacher KJ, Braunwald E, Wilson JD, Martin JB, Fauci AS, Kasper D, Harrison L (eds) Harrison's Principles of Internal Medicine. McGraw Hill, New York, pp. 2172–2183.
8. Ivic MA, Kostic S, Strahinjic S, Stefanovic V 1989 The effect of parathyroid hormone and $1,25(OH)_2D_3$ on nerve conductance in patients on maintenance hemodialysis. Acta Med Iugosl **43**: 349–355.
9. Seelig MS 1969 Vitamin D and cardiovascular, renal, and brain damage in infancy and childhood. Ann NY Acad Sci **147**:537–582.
10. Stio M, Lunghi B, Iantomasi T, Vincenini MT, Treves C 1993 Effect of vitamin D deficiency and 1,25-dihydroxyvitamin D_3 on the metabolism and D-glucose transport in rat cerebral cortex. J Neurosci Res **35**:559–566.
11. Sutherland MK, Somerville MJ, Yoong LKK, Bergeron C, Haussler MR, Crapper DR, McLachlan C 1992 Reduction of vitamin-D hormone receptor mRNA levels in Alzheimer as compared to Huntington hippocampus: Correlation with calbindin-28K mRNA levels. Mol Brain Res **13**:239–250.
12. Pike JW, Gooze LL, Haussler MR 1980 Biochemical evidence for 1,25-dihydroxyvitamin D receptor macromolecules in parathyroid, pancreatic, pituitary and placental tissues. Life Sci **26**:407–414.
13. Colston K, Hirst M, Feldman D 1980 Organ distribution of the cytoplasmic 1,25-dihydroxycholecalciferol receptor in various mouse tissues. Endocrinology **107**:1916–1922.
14. Walters MR, Hunziker W, Norman AW 1980 Unoccupied 1,25-dihydroxyvitamin D3 receptors: Nuclear/cytosol ratio depends on ionic strength. J Biol Chem **255**:6799–6805.
15. Clemens TL, Zhou XY, Pike JW, Haussler MR, Sloviter RS 1985 1,25-Dihydroxyvitamin D receptor and vitamin D-dependent calcium-binding protein in rat brain: Comparative immunocytochemical localization. Sixth Workshop on Vitamin D. de Gruyter, Berlin, pp. 95–96.
16. Clemens T, Garrett KP, Zhou XY, Pike JW, Hassler MR, Dempster DW 1988 Immunocytochemical localization of the 1,25-dihydroxyvitamin D3 receptor in target cells. Endocrinology **122**:1224–1230.
17. Gascon-Barre M, Huet P-M 1983 Apparent $[^3H]$1,25-dihydroxyvitamin D_3 uptake by canine and rodent brain. Am J Physiol **244**:E266–E271.
18. Partridge WM, Sakiyama R, Coty WA 1985 Restricted transport of vitamin D and A derivatives through the rat blood–brain barrier. J Neurochem **44**:1138–1141.
19. Stumpf WE, Sar M, Clark SA, DeLuca HF 1982 Brain target sites for 1,25-dihydroxyvitamin D_3. Science **215**:1403–1405.
20. Stumpf WE, O'Brien LP 1987 $1,25(OH)_2$ vitamin D_3 sites of action in the brain. Histochemistry **87**:393–406.
21. Stumpf WE, Clark SA, O'Brien LP, Reid FA 1988 $1,25(OH)_2$ vitamin D_3 sites of action in spinal cord and sensory ganglion. Anat Embryol **177**:307–310.
22. Musiol IM, Stumpf WE, Bidmon HJ, Heiss C, Mayerhofer A, Bartke A 1992 Vitamin D nuclear binding to neurons of the septal, substriatal and amygdaloid area in the Siberian hamster (*Phodopus sungorus*) brain. Neuroscience **48**:841–848.
23. Pike JW, Donaldson CA, Marion SL, Haussler MR 1982 Development of hybridomas secreting monoclonal antibodies to the chicken intestinal 1alpha,25-dihydroxyvitamin D_3 receptor. Proc Natl Acad Sci USA **79**:7719–7723.
24. Baker AR, McDonnell DP, Hughes M, Crisp TM, Mangelsforf DJ, Haussler MR, Pike JW, Shine J, O'Malley BW 1988 Cloning and expression of full-length cDNA encoding human vitamin D receptor. Proc Natl Acad Sci USA **85**:3294–3298.

24a. Dobrzanski P, Haun F, Carswell S 1997 Localization of the vitamin D receptor in the developing and adult CNS. Manuscript in preparation.

25. Neveu I, Naveilhan P, Jehan F, Baudet C, Wion D, De Luca HF, Brachet P 1994 1,25-dihydroxyvitamin D_3 regulates the synthesis of nerve growth factor in primary cultures of glial cells. Brain Res Mol Brain Res **24**:70–76.
26. Johnson JA, Grande JP, Windebank AJ, Kumar R 1996 1,25-Dihydroxyvitamin D_3 receptors in developing dorsal root ganglia of fetal rats. Dev Brain Res **92**:120–124.
27. Mano T, Puro DG 1990 Phagocytosis by human retinal glial cells in culture. Invest Ophthalmol Vis Sci **31**:1047–1055.
28. Moore TB, Koeffler HP, Yamashiro JM, Wada RK 1996 Vitamin D_3 analogs inhibit growth and induce differentiation in LA-N-5 human neuroblastoma cells. Clin Exp Metastasis **14**:239–245.
29. Balabanova S, Richter HP, Antoniadis G, Homoki J, Kremmer N, Hanle J, Teller WM 1984 25-Hydroxyvitamin D, 24,25-dihydroxyvitamin D and 1,25-dihydroxyvitamin D in human cerebrospinal fluid. Klin Wochenschr **62**:1086–1090.
30. Naveilhan P, Neveu I, Baudet C, Ohyama KY, Brachet P, Wion D 1993 Expression of 25(OH) vitamin D_3 24-hydroxylase gene in glial cells. Neuroreport **5**:255–257.
31. Neveu I, Naveilhan C, Menaa D, Wion D, Brachet P, Garbedian M 1994 Synthesis of 1,25-dihydroxyvitamin D_3 by rat brain macrophages in vitro. J Neurosci Res **38**:214–220.
32. Neveu I, Naveilhan P, Baudet C, Brachet P, Metsis M 1994 1,25-Dihydroxyvitamin D_3 regulates NT-3, NT-4 but not BDNF mRNA in astrocytes. Neuroreport **6**:124–126.

33. Binderup L, Latini S, Binderup E 1991 20-Epi-vitamin D_3 analogues: A novel class of potent regulators of cell growth and immune response. Biochem Pharmacol **12**:1569–1575.
34. Colston KW, Mackay AG, James SY, Binderup L, Chander S, Coombes RC 1992 EB1089: A new vitamin D analogue that inhibits the growth of breast cancer cells *in vivo* and *in vitro*. Biochem Pharmacol **44**:2273–2280.
35. Matkovits T, Christakos S 1995 Ligand occupancy is not required for vitamin D receptor and retinoid-mediated transcriptional activation. Mol Endocrinol **9**:232–242.
36. Cai Q, Tapper DN, Gilmour RF, de Talamoni N, Aloia RC, Wasserman RH 1994 Modulation of the excitability of avian peripheral nerves by vitamin D: Relation to calbindin-D_{28K}, calcium status and lipid composition. Cell Calcium **15**:401–410.
37. Lemire JM, Archer DC 1991 1,25-Dihydroxyvitamin D_3 prevents the *in vivo* induction of murine experimental autoimmune encephalomyelitis. J Clin Invest **87**:1103–1107.
38. Branisteanu DD, Waer M, Sobis H, Marcelis S, Vandeputte M, Boullion R 1995 Prevention of murine experimental allergic encephalomyelitis: Cooperative effects of cyclosporine and 1alpha,25-$(OH)_2D_3$. J Neuroimmunol **61**:151–160.
39. Cantorna MT, Hayes CE, DeLuca HF 1996 1,25-Dihydroxyvitamin D_3 reversibly blocks the progression of relapsing encephalomyelitis, a model of multiple sclerosis. Proc Natl Acad Sci USA **93**:7861–7864.
40. Nataf S, Garcion E, Darcy F, Chabannes D, Muller JY, Brachet P 1996 1,25-Dihyroxyvitamin D_3 exerts regional effects in the central nervous system during experimental allergic encephalomyelitis. J Neuropathol Exp Neurol **55**:904–914.
41. Sonnenberg J, Luine VN, Krey LC, Christakos S 1986 1,25-Dihydroxyvitamin D3 treatment results in increased choline acetyltransferase activity in specific brain nuclei. Endocrinology **118**:1433–1439.
42. Saporito MS, Wilcox HM, Hartpence KC, Lewis ME, Vaught JL, Carswell SC 1993 Pharmacological induction of nerve growth factor mRNA in adult rat brain. Exp Neurol **123**:295–302.
43. Saporito MS, Brown ER, Hartpence KC, Wilcox HM, Vaught JL, Carswell S 1994 Chronic 1,25-dihydroxyvitamin D3-mediated induction of nerve growth factor mRNA and protein in L929 fibroblasts and in adult rat brain. Brain Res **633**:189–196.
44. Saporito MS, Carswell S 1995 High levels of synthesis and local effects of nerve growth factor in the septal region of the adult rat brain. J Neurosci **15**:2280–2286.
45. Oppenheim RW 1996 Neurotrophic survival molecules for motoneurons: An embarrassment of riches. Neuron **17**:195–197.
46. Barbacid M 1995 Neurotrophic factors and their receptors. Curr Opin Cell Biol **7**:148–155.
47. Greene LA, Kaplan DR 1995 Early events in neurotrophin signalling via Trk and p75 receptors. Curr Opin Neurobiol **5**:579–587.
48. Braak H, Braak E 1991 Neuropathological staging of AD-related changes. Acta Neuropathol **82**:239–259.
49. Whitehouse PJ, Price DL, Struble RG, Clark AW, Delong MR 1982 Alzheimers disease and senile dementia: Loss of neurons in the basal forebrain. Science **215**:237–1239.
50. Hefti F, Weiner J 1986 NGF and Alzheimer's disease. Ann Neurol **20**:275–281.
51. Saffran BN 1992 Should intracerebroventricular NGF be used to treat Alzheimer's disease? Perspect Biol Med **35**:471–486.
52. Olson L 1993 NGF and the treatment of Alzheimer's disease. Exp Neurol **124**:5–15.
53. Hefti F 1986 Nerve growth factor promotes survival of septal cholinergic neurons after fimbrial transections. J Neurosci **6**:2155–2162.
54. Mobley WC, Rutkowski JL, Tennekoon GI, Gemski J, Buchanan K, Johnston MV 1986 Nerve growth factor increases choline acetyltransferase activity in developing basal forebrain neurons. Mol Brain Res **1**:53–62.
55. Saporito MS, Carswell S (1997). Therapeutic potential of nerve growth factor in Alzheimer's disease. In: LeRioth D, Bondy C (eds) Growth Factors and Cytokines in Health and Disease. JAI Press, Greenwich, Connecticut, 439–458.
56. Carswell S 1993 The potential for treating neurodegenerative disorders with NGF-inducing compounds. Exp Neurol **124**:36–42.
57. Carswell S, Hoffman EK, Clopton-Hartpence K, Wilcox HM, Lewis ME 1992 Induction of NGF by isoproterenol, 4-methylcatechol and serum occurs by three distinct mechanisms. Mol Brain Res **15**:145–150.
58. Wion D, MacGrogan D, Neveu I, Jehan F, Houglatte R, Brachet P 1991 1,25-Dihydroxyvitamin D_3 is a potent inducer of NGF synthesis. J Neurosci Res **28**:110–114.
59. Thavarajah M, Evans DB, Binderup L, Kanis JA 1990 $1{,}25(OH)_2D_3$ and calcipotriol (MC 903) have similar effects on the induction of osteoclast-like cell formation in human bone marrow cultures. Biochem Biophys Res Commun **171**:1056.
60. Jehan F, Neveu I, Barbot N, Binderup L, Brachet P, Wion D 1991 MC903, an analogue of 1,25-dihydroxyvitamin D_3, increases the synthesis of nerve growth factor. Eur J Pharmacol **208**:189–191.
61. Nitta A, Murakami Y, Furukawa Y, Kawatsura W, Hayashi K, Yamada K, Hasegawa T, Nabeshima T 1994 Oral administration of idebenone induces nerve growth factor in the brain and improves learning and memory in basal forebrain-lesioned rats. Naunyn Schmiedebergs Arch Pharmacol **349**: 401–407.
62. Robbins E, Saporito MS, Hartpence KC, Battle J, Kromer LF, Vaught JL, Carswell S 1994 The effect of dose and time on the efficacy of NGF in the fimbria fornix lesion model. Abstracts, Soc Neurosci **20**:1097.
63. Saporito MS, Brown EM, Clopton-Hartpence K, Wilcox HM, Robbins E, Vaught JL, Carswell S 1994 Systemic dexamethasone administration increases septal Trk autophosphorylation in adult rats via an induction of NGF. Mol Pharmacol **45**:395–401.
64. Semkova I, Schilling M, Henrich-Noack P, Rami A, Krieglstein J 1996 Clenbuterol protects mouse cerebral cortex and rat hippocampus from ischemic damage and attenuates glutamate neurotoxicity in cultured hippocampal neurons by induction of NGF. Brain Res **717**:44–54.
65. Norman AW, Roth J, Orci L 1982 The vitamin D endocrine system: Steroid metabolism, hormone receptors, and biological response (calcium binding proteins). Endocr Rev **3**:331–366.
66. Varghese S, Lee S, Huang Y, Christakos S 1988 Analysis of rat vitamin D-dependent calbindin-D_{28K} gene expression. J Biol Chem **263**:9776–9784.
67. Hall AK, Norman AW 1991 Vitamin D-independent expression of chick brain calbindin-D_{28K}. Mol Brain Res **9**:9–14.
68. Pasteels JL, Pochet R, Surardt L, Hubeau C, Chirnoaga M, Parmentier M, Lawson DEM 1986 Ultrastructural localization of brain vitamin D-dependent calcium binding protein. Brain Res **384**:294–303.
69. Miller RJ 1995 Regulation of calcium homeostasis in neurons: The role of calcium-binding proteins. Biochem Soc Trans **23**:629–632.
70. Iacopino AM, Christakos S 1990 Specific reduction of calcium-binding protein (28-kilodalton calbindin-D) gene expression in aging and neurodegenerative diseases. Proc Natl Acad Sci USA **87**:4078–4082.
71. Thomasset M, Parkes CO, Cuisinier-Gleizes P 1982 Rat calcium-binding proteins: Distribution, development, and vitamin D dependence. Am J Physiol **243**:E483–E488.
72. Viragh PA, Haglid KG, Celio MR 1989 Parvalbumin increases

in the caudate putamen of rats with vitamin D hypervitaminosis. Proc Natl Acad Sci USA **86**:3887–3890.
73. Stumpf WE, O'Brien LP 1987 1,25$(OH)_2$ vitamin D_3 sites of action in the brain. An autoradiographic study. Histochemistry **87**:393–406.
74. Lowe KE, Maiyar AC, Norman AW 1992 Vitamin D-mediated gene expression. Crit Rev Eukaryotic Gene Expression **2**: 65–109.
75. Okazaki T, Bell RM, Hannun YA 1989 Sphingomyelin turnover induced by vitamin D_3 in HL-60 cells. J Biol Chem **264**:19076–19080.
76. Dobrowsky RT, Werner MH, Castellino AM, Chao MV, Hannun YA 1994 Activation of the sphingomyelin cycle through the low-affinity neurotrophin receptor. Science **265**:1596–1599.
77. Reichel H, Koeffler HP, Norman AW 1989 The role of vitamin D in health and disease. N Engl J Med **320**:980–991.
78. Liu M, Lee M, Cohen M, Bommakanti M, Freedman LP 1996 Transcriptional activation of the Cdk inhibitor p21 by vitamin D_3 leads to the induced differentiation of the myelomonocytic cell line U937. Genes Dev **10**:1–12.
79. Freeman RS, Estus S, Johnson EM 1994 Analysis of cell cycle-related gene expression in postmitotic neurons: Selective induction of cyclin D1 during programmed cell death. Neuron **12**:343–355.
80. Farinelli SE, Greene LA 1996 Cell cycle blockers mimosine, ciclopirox, and deferoxamine prevent the death of PC12 cells and postmitotic sympathetic neurons after removal of trophic support. J Neurosci **16**:1150–1162.

CHAPTER 72

Psoriasis and Other Skin Diseases

KNUD KRAGBALLE Department of Dermatology, Marselisborg Hospital, University of Aarhus, Aarhus, Denmark

I. INTRODUCTION

When the first purified preparations of vitamin D became available in the 1930s, they were used therapeutically in a variety of skin diseases, including psoriasis. The rationale for this treatment was the hypothesis that sunlight improves psoriasis by stimulating vitamin D_3 synthesis in the skin. Although oral preparations of vitamin D were reported to improve psoriasis, the treatment was abandoned because of the associated toxicity [1].

The potential role of vitamin D in skin diseases was then almost forgotten, until the mid-1980s when three independent lines of evidence suggested the possible use of vitamin D for the treatment of psoriasis. The first evidence was a chance observation that a patient with osteoporosis, who received oral 1α-hydroxyvitamin D_3 ($1\alpha OHD_3$, 0.75 μg/day), had a dramatic improvement of her severe psoriasis [2]. Second, the skin was discovered as a target organ for vitamin D_3. Thus, the bioactive form of vitamin D_3 was shown to inhibit keratinocyte proliferation and to promote keratinocyte differentiation [3]. Finally, various groups achieved the synthesis of vitamin D_3 analogs with potent effects on the cellular level, but at the same time with lesser hypercalcemic activity than naturally occurring vitamin D_3 [4].

II. ACTIONS OF VITAMIN D_3 IN THE SKIN

As determined by ligand binding and immunoblotting, the vitamin D receptor (VDR) has been detected in most cell types in the skin, including cultured keratinocytes [5], Langerhans cells [6], melanocytes [7], fibroblasts [8], and endothelial cells [9]. Furthermore, monocytes and activated T cells participating in inflammatory skin reactions express the VDR [10]. These findings strongly support the idea that the skin is a target for 1,25-dihydroxyvitamin D_3 [$1,25(OH)_2D_3$] (see Chapter 25).

The effects of $1,25(OH)_2D_3$ on epidermal keratinocytes have been extensively investigated. In cultured keratinocytes committed to terminal differentiation, $1,25(OH)_2D_3$ stimulates proliferation, whereas proliferation of undifferentiated keratinocytes is inhibited by

$1,25(OH)_2D_3$ [11,12]. A similar paradoxical response to $1,25(OH)_2D_3$ is observed *in vivo*. Topical application of $1,25(OH)_2D_3$ and its analogs to normal mouse skin increases keratinocyte proliferation and causes epidermal hyperplasia [13,14]. In contrast, petrolatum-induced epidermal hyperplasia in guinea pigs is inhibited by the analog $1,24(OH)_2D_3$ [15]. This inhibitory activity corresponds to the antiproliferative effect observed in psoriatic hyperplasia. It is believed that the level of the VDR determines the responsiveness to vitamin D_3 (see Chapter 11). It is therefore of interest that $1,25(OH)_2D_3$ and calcipotriol up-regulate VDR levels in human epidermal keratinocytes *in vitro* [16] as well as *in vivo* [17].

The metabolite $1,25(OH)_2D_3$ is recognized to have immunoregulatory properties *in vitro* (see Chapter 29). Defined actions of the hormone on human peripheral blood mononuclear cells include inhibition of mitogen/antigen-stimulated proliferation [18]. $1,25(OH)_2D_3$ also modulates the ability of monocytes to provide signals important in T-lymphocyte activation. Thus, $1,25(OH)_2D_3$ decreases monocyte HLA-DR expression, monocyte antigen presentation, and monocyte promotion of T-cell proliferation [19]. Accordingly, topical application of the vitamin D_3 analog calcipotriol twice daily for 4 days suppresses the number of Langerhans cells, the antigen-presenting cells (APCs) of the epidermis, as well as their accessory cell function [6].

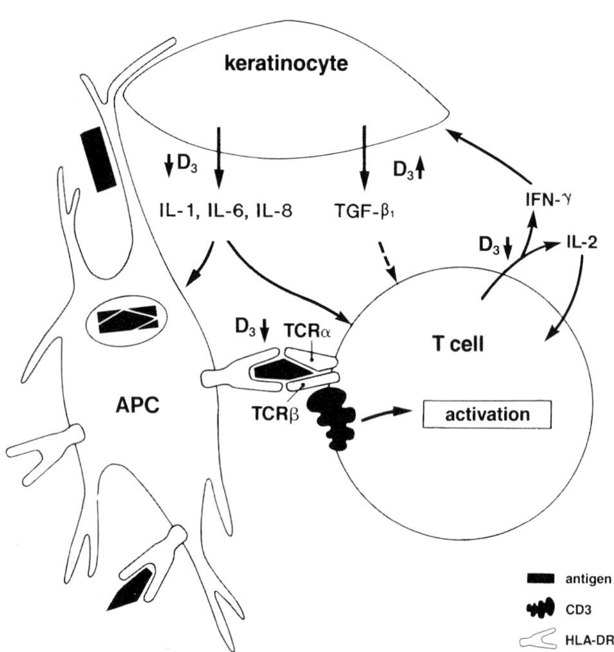

FIGURE 1 Targets of vitamin D_3 in psoriatic skin lesions. The direction of the arrows indicates whether D_3 stimulates or inhibits the activity.

III. PATHOGENESIS OF PSORIASIS

Psoriasis is a chronic or chronically relapsing skin disease characterized by erythematous patches covered by thick scales. A substantial amount of epidemiologic data supports the concept that psoriasis susceptibility is heritable, although the mode of inheritance is unclear. Psoriasis is a result of epidermal stem cell growth, initiated by lymphokines released from activated memory T cells within the skin [20] (Fig. 1). Release of lymphokines such as interleukin-2 (IL-2) and interferon-γ (IFN-γ) from activated T cells is believed to be caused by increased activity of APCs in the skin lesions. The APCs present an as yet unidentified autoantigen to memory T cells, causing their activation and subsequent lymphokine release. The current aim of therapeutic intervention is, therefore, to disable this ternary complex (APC–autoantigen–T cell) or to inhibit epidermal stem cell growth.

IV. VITAMIN D_3 METABOLISM AND ACTION IN PSORIASIS

Because the available anti-VDR antibodies cross-react with non-VDR proteins, it has been difficult to assess the VDR levels in psoriatic skin. In an immunohistochemical study, lesional psoriatic epidermis showed increased staining with the monoclonal 9A7γ anti-VDR antibody compared to nonlesional epidermis and normal epidermis [21]. Using the same antibody, Western blot analysis showed normal levels of VDR in lesional as well as in nonlesional psoriatic skin [22]. These results were confirmed using a polyclonal rabbit anti-VDR antibody. Furthermore, the VDR mRNA levels were normal in both acute and chronic psoriatic lesions [22].

The levels of retinoid X receptor α (RXRα), the heterodimeric partner for VDR, are also normal in psoriatic skin [23]. Furthermore, the normal binding of psoriatic skin extracts to a vitamin D response element of the DR-3 type suggests that the genomic signaling pathway is normal is this skin disease [23].

There exist conflicting results regarding vitamin D metabolism in psoriatic patients. Morimoto *et al.* [24] observed no difference in the mean levels of circulating $1,25(OH)_2D_3$ between psoriatic and normal subjects, whereas Staberg *et al.* [25] reported reduced $1,25(OH)_2D_3$ concentrations in psoriatics with disseminated disease. Smith *et al.* [26] and Guilhou *et al.* [27] found normal serum $1,25(OH)_2D_3$ levels in patients with moderate to extensive psoriasis. An inverse relationship between the severity of psoriasis and the serum $1,25(OH)_2D_3$ level may explain these conflicting results [24].

Severe psoriasis has been observed in association with hypoparathyroidism [28] and hypocalcemia [29]. In these patients psoriasis improved when serum calcium levels were restored to normal. Although these case reports demonstrate that fluctuation in serum calcium can precipitate psoriasis, all parameters of calcium and bone metabolism are normal in larger groups of psoriatic patients [26,30]. Taken together the available data do not support the idea that psoriasis is a manifestation of abnormal vitamin D or calcium metabolism.

A related question is whether psoriatic skin is less sensitive to $1,25(OH)_2D_3$ than normal skin. It has been reported that cultures of dermal fibroblasts and epidermal keratinocytes from psoriatics are relatively resistant to the antiproliferative effect of $1,25(OH)_2D_3$, despite normal binding to the VDR [31,32]. However, a subsequent study was unable to confirm these results [26], indicating that there is apparently no intrinsic insensitivity of psoriatic fibroblasts or keratinocytes to $1,25(OH)_2D_3$. It is also of interest that the cutaneous formation of vitamin D_3 after exposure to a single dose of UVB is normal in psoriasis [33].

V. PHARMACODYNAMIC ACTIONS OF VITAMIN D_3 IN PSORIASIS

Together with epidermal hyperproliferation and incomplete terminal differentiation, activated immunocytes are key features of the psoriatic lesion. Treatment with vitamin D_3 analogs has been shown to modulate each of these processes (Fig. 2). It is, however, unclear which of these actions are most important for the antipsoriatic effect of vitamin D_3.

In skin biopses stained with hematoxylin–eosin, treatment with topical $1,25(OH)_2D_3$ results in reappearance of the granular layer, regression of acanthosis and parakeratosis, and resumption of the orthokeratotic horny layer. Also, the infiltrating leukocytes disappear from the epidermis, but they may still be detected around superficial blood vessels and in the papillary dermis [34]. Immunohistopathological evaluation shows normalization of the epidermal staining with the monoclonal antibodies CD15, CD16, CD36, CD1a, and Ki67 during $1,25(OH)_2D_3$ therapy [34]. These results suggest that treatment with $1,25(OH)_2D_3$ has an essential effect on epidermal proliferation and differentiation in psoriasis.

During treatment with topical calcipotriol, keratin levels have been assessed by electrophoretic and immunohistochemical analysis of skin biopsies. Improvement of the psoriatic lesions is accompanied by a reduction in the amount of keratins 5, 16, and 18 (markers of basal

A

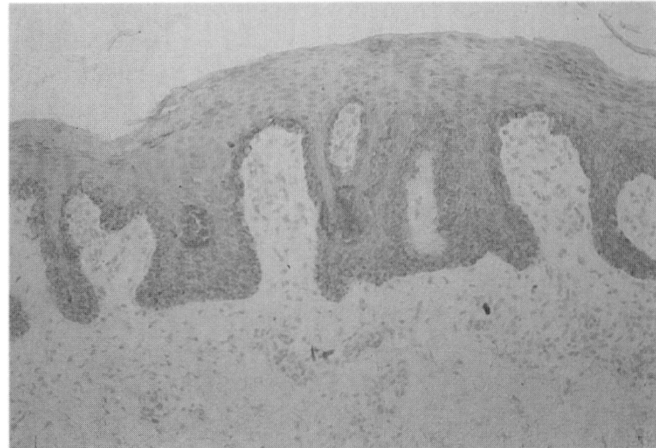

B

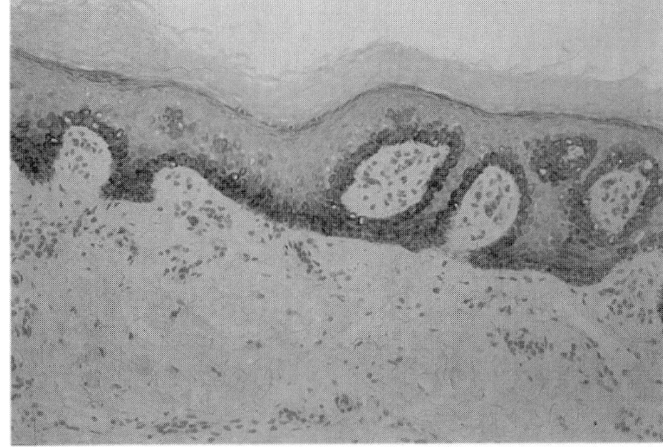

FIGURE 2 Micrographs of a psoriatic lesion stained with antibody BG3C8 against basal keratinocytes before (A) and after (B) treatment with calcipotriol ointment (50 μg/g for 8 weeks).

and hyperproliferating keratinocytes) and an increase in keratins 1, 2, and 10 (markers of differentiating keratinocytes) [35–40]. The changes in these markers of epidermal growth correlate with the degree of clinical improvement. The changes in levels of markers of differentiation during calcipotriol therapy are similar to those observed during photochemotherapy, namely, small changes in keratin 1 levels and large increases in keratin 2 levels. The keratin changes contrast with those observed during retinoid therapy, namely, large increase in keratin 1 levels with no increase in keratin 2 levels [41]. Calcipotriol treatment also produces a profound effect on immunocytes in psoriatic epidermis and dermis. The number of T lymphocytes is reduced, relatively more for $T4^+$ cells than for $T8^+$ cells [38,39]. In contrast, the epidermal $T6^+$ cells (Langerhans cells) may increase during treatment, although clinical resolution is accompanied by a decrease in $T6^+$ cells [38]. The biphasic effect on $T6^+$ cells has also been reported with psoralen–

ultraviolet A (PUVA) and cyclosporin A treatment. Using elastase activity as a marker of neutrophil infiltration, there is a significant decrease in neutrophils during calcipotriol treatment. These changes are already detectable after treatment for 1 week and may precede the decrease of T lymphocytes [37].

Topical treatment of normal human skin with calcipotriol is accompanied by an increase in VDR expression in the epidermis as revealed by Western blot analysis [17]. A similar increase of VDR protein and mRNA has been found in cultured human keratinocytes after $1{,}25(OH)_2D_3$ treatment [16]. Because the *in vitro* responsiveness to $1{,}25(OH)_2D_3$ correlates with the VDR levels, it is of interest that the induction of VDR mRNA expression in psoriatic plaques correlates with the clinical response to $1{,}25(OH)_2D_3$ [42].

Taken together these results indicate that treatment with vitamin D_3 analogs has effects on the proliferation and differentiation of keratinocytes as well as on the infiltration and activation of neutrophils and immunocytes in psoriatic skin lesions. From the existing data it is controversial whether the effects on epidermal keratinocytes precede or are more marked than the effect on the cellular infiltrate. A primary target cell has not been identified, and it is likely that topical treatment with vitamin D_3 analogs has multiple target points in psoriatic skin.

VI. USE OF $1{,}25(OH)_2D_3$ (CALCITRIOL) IN PSORIASIS

Treatment of psoriasis with topical $1{,}25(OH)_2D_3$ (calcitriol) (Table I) has produced mixed results. The main reason for the variability is that different drug concentrations and vehicles were used in the various studies. In an early, open study, improvement was reported in 16 of 19 patients who applied $1{,}25(OH)_2D_3$ ointment at a concentration of 0.5 $\mu g/g$ [43]. However, in double-blind studies a solution at 2 $\mu g/g$ produced no improvement compared to placebo [44,45]. The concentration of $1{,}25(OH)_2D_3$ was raised in the subsequent studies. Applied twice daily in petrolatum, 3 $\mu g/g$ [33] as well as 15 $\mu g/g$ of $1{,}25(OH)_2D_3$ [46] is superior to vehicle. Furthermore, 15 $\mu g/g$ of $1{,}25(OH)_2D_3$ is more efficacious than 3 $\mu g/g$ [47].

TABLE I Vitamin D_3 Analogs Used for Treatment of Psoriasis

Code name	Generic name	Trade name
MC903	Calcipotriol, calcipotriene (United States)	Daivonex, Dovonex, Psorcutan
$1{,}25(OH)_2D_3$	Calcitriol	
$1{,}24(OH)_2D_3$	Tacalcitol	

Treatment with $1{,}25(OH)_2D_3$ ointment at 3 $\mu g/g$ in amounts ranging from 35 to 142 g/week appears to be safe [48]. However, $1{,}25(OH)_2D_3$ ointment at 15 $\mu g/g$ can produce hypercalcemia even when applied to limited skin areas [46]. In patients receiving $1{,}25(OH)_2D_3$ at 3 or 15 $\mu g/g$, skin irritation appears to be uncommon.

Although long-term results are still lacking, it can be concluded from these results that treatment of psoriasis with $1{,}25(OH)_3D_3$ ointment at 3 $\mu g/g$ is well tolerated and safe. This form of treatment was, however, less effective than calcipotriol at 50 $\mu g/g$ in a comparative trial [49].

Although the natural form of vitamin D_3, $1{,}25(OH)_2D_3$, is safe to use topically, there remains concern that oral $1{,}25(OH)_2D_3$ will be of limited value for treating psoriasis, because of its potent hypercalcemic effect. In the first studies with orally administered $1{,}25(OH)_2D_3$, too few patients were treated for too short a time to assess short-term and long-term effects on calcium metabolism [50,51]. In a more recent study, 85 patients were enrolled in an open trial to evaluate the efficacy and safety of oral $1{,}25(OH)_2D_3$ used for 6 to 36 months [52]. Patients were instructed to ingest no more than 800 mg per day of dietary calcium and were started on 0.5 μg of calcitriol at bedtime. The calcitriol dose was increased in increments of 0.5 μg every 2 weeks, as long as serum and 24-hr urinary calcium concentrations remained in the normal range. Eighty-eight percent of the patients had some improvement in their disease. Serum calcium concentration increased slightly, whereas 24-hr urinary calcium excretion increased by almost 150%. Despite the increased urinary calcium excretion, no significant change in bone mineral density was detected after 24 months. To assess kidney function, renal clearance was determined for creatinine, inulin, and *para*-aminohypurate. Only creatinine clearance decreased (23%) during oral $1{,}25(OH)_2D_3$ therapy. This led the investigators to conclude that oral $1{,}25(OH)_2D_3$ alters creatinine metabolism or secretion, but not renal function [52].

Despite the apparent safety of oral calcitriol, it is recommended that blood and urine calcium levels be monitored frequently and that yearly renal ultrasound be performed in psoriatic patients on oral calcitriol. It should also be borne in mind that psoriasis is not an approved indication for oral $1{,}25(OH)_2D_3$ (calcitriol, Rocaltrol) according to the U.S. Food and Drug Administration. Calcitriol is not yet marketed for topical use (Table I).

VII. USE OF CALCIPOTRIOL (CALCIPOTRIENE) IN PSORIASIS

Calcipotriol (calcipotriene, MC903) is a synthetic vitamin D analog with potent cell-regulating properties but with lower risk of inducing calcium-related side effects [4]. It is a $1,25(OH)_2D_3$ analog containing a double bond and a ring structure in the side chain (see Chapter 61). As a consequence of this modification of the side chain, calcipotriol is rapidly transformed into inactive metabolites [4]. Therefore, calcipotriol is about 200 times less potent than $1,25(OH)_2D_3$ in producing hypercalcemia and hypercalciuria after oral and intraperitoneal administration in rats. In contrast, calcipotriol and $1,25(OH)_2D_3$ are equipotent in their affinity for the VDR and in their *in vitro* effects [4].

The antipsoriatic effect of calcipotriol ointment at 50 μg/g has been documented in a number of clinical trials involving several thousand patients. Calcipotriol ointment at 50 μg/g is marketed for the treatment of plaque-type psoriasis vulgaris under the trade names Daivonex, Dovonex, and Psorcutan (Table I). A calcipotriol cream [53–55] and solution [56,57] have also been found to be effective.

A

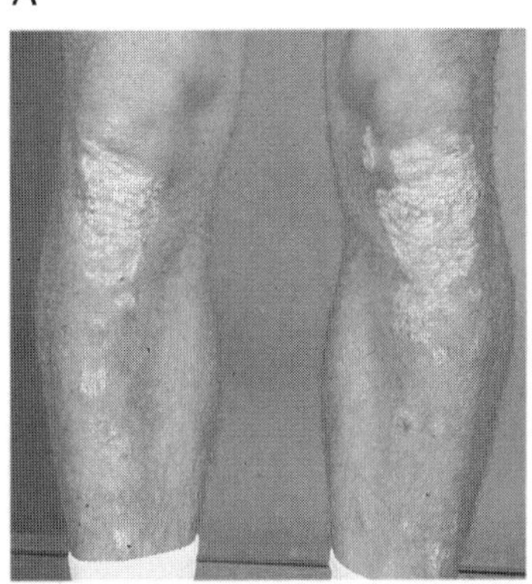

B

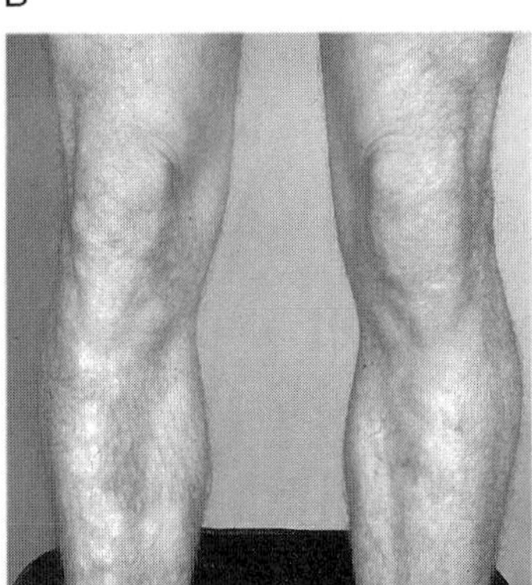

FIGURE 3 Psoriatic skin lesion before (A) and after (B) treatment with calcipotriol ointment (50 μg/g for 8 weeks).

A. Comparison with Placebo

In a dose–response study calcipotriol ointment (25, 50, and 100 μg/g) and vehicle were compared in 50 patients with psoriasis vulgaris [58]. The study was designed as a randomized, double-blind right–left comparison. After treatment for 8 weeks calcipotriol at 50 μg/g had a significantly greater antipsoriatic effect than vehicle. Calcipotriol at 50 μg/g was more effective than calcipotriol ointment at 25 μg/g, whereas no difference was found between 50 and 100 μg/g calcipotriol. It was concluded that a calcipotriol concentration of 50 μg/g in an ointment is optimal for the treatment of psoriasis. Treatment with this calcipotriol concentration induces significant improvement as soon as 1 week, and a marked improvement is observed in about two-thirds of patients after 8 weeks (Fig. 3). The efficacy and safety of calcipotriol ointment at 50 μg/g were later confirmed in a multicenter, double-blind, placebo-controlled, right–left study including 66 psoriatic patients [59].

B. Comparison with Corticosteroid, Dithranol, and Occlusion

Treatment with calcipotriol ointment at 50 μg/g has been compared with 0.1% betamethasone 17-valerate ointment in a multicenter, randomized double-blind, right–left comparison [60]. Three hundred forty-five patients with symmetrical, stable plaque-type psoriasis were randomized to either treatment for 6 weeks. Both treatments produced a time-dependent improvement, and after 6 weeks the psoriasis area and severity index (PASI) was reduced by 68.8% with calcipotriol and 61.4% with betamethasone. At all visits the differences between the two treatments was statistically highly significant in favor of calcipotriol. Furthermore, calcipotriol has been compared with 0.1% betamethasone 17-valerate ointment in a parallel-group comparison involving 409 patients [61]. After treatment for 6 weeks the PASI reduction was similar in the calcipotriol group (54.5%) and in the betamethasone group (48.8%). However, according to patients' overall assessment of improvement, a significantly higher percentage of calcipotriol-treated patients showed a marked or better improvement (62.2%) than did betamethasone 17-valerate-treated patients (50.1%). In addition, calcipotriol ointment was shown be more effective than the potent corticosteroids betamethasone dipropionate [62] and fluocinonide [63]. Taken together these results document that treatment with calcipotriol ointment at 50 μg/g is efficacious for psoriasis.

Calcipotriol ointment at 50 μg/g has been compared with dithranol (anthralin) used either as short-contact therapy in outpatients [64] or as conventional inpatient therapy [65]. In a multicenter, open, randomized parallel-group comparison patients were treated for 8 weeks with either calcipotriol ointment at 50 μg/g applied twice daily or Dithrocream applied once daily for 30 min according to a short-contact regimen [64]. The dose of dithranol (0.1, 0.25, 0.5, 1, and 2%) was increased stepwise every 7 days. A significant fall in PASI was observed in both groups after treatment for 2, 4, and 8 weeks. Both the percentage reduction of PASI and the absolute reduction of PASI was significantly greater in the calcipotriol group (58.1%) than in the dithranol group (41.6%) at the end of treatment [64]. Using a left–right comparative study design, 10 patients hospitalized with refractory psoriasis showed a better therapeutic response to calcipotriol ointment at 50 μg/g than to dithranol after 2 weeks [65]. Despite the low number of patients and the lack of follow-up of these patients, these results challenge the superiority of one of the gold standards of antipsoriatic therapy.

The therapeutic response to calcipotriol ointment can be increased by occlusion under a polythene film [66] or a hydrocolloid dressing [67]. The beneficial effect of occlusion is mainly due to greater penetration and delivery of calcipotriol. Because of the risk of increased systemic absorption, calcipotriol ointment should only be occluded in small areas. In particular, very thick plaques benefit from occlusive therapy. Some dermatologists advocate the use of descaling agents such as 40% propylene glycol in aqueous cream or 20% urea in white soft paraffin to remove the scale and thereby produce better results. However, there are no scientific data available yet to support the use of descaling agents in combination with calcipotriol.

Although there are no published posttreatment recurrence data after calcipotriol therapy, the clinical experience has been that the average remission time is rather short after stopping calcipotriol therapy. This means that a large proportion of patients may require repeated courses of treatment. It is important to note that calcipotriol treatment can be stopped without resulting in an exacerbation of psoriasis.

C. Long-term Therapy

Because psoriasis is a chronic and relapsing disease, it is important to determine whether the beneficial effect of calcipotriol seen in the short-term studies can be maintained when patients are treated on a long-term basis. This question has been assessed in two prospective, noncomparative, open studies. Included were patients who had a good clinical response to calcipotriol previously. In the first study, 15 patients from a single center were treated with calcipotriol ointment at 50 μg/g twice daily (maximally 100 g ointment per week) for at least 6 months [68]. Assessment of efficacy at the end of therapy showed at least a moderate improvement in 80% of treated patients. These results have been confirmed and extended in a multicenter study that included 167 patients [69]. In most patients the beneficial effect seen in short-term trials was maintained over the course of 1 year. Although approximately 10% of the patients had to be withdrawn because of an insufficient beneficial effect of calcipotriol, the dose required to maintain efficacy did not have to be increased with time. Thus, the mean quantity of ointment used fell from 35 to 23 g/week during the last 6 months of therapy [69]. Therefore, there was no suggestion of development of a pharmacological tolerance to calcipotriol. The major advantage of calcipotriol over corticosteroids may be that the former does not induce skin atrophy during long-term treatment [68].

D. Combination Therapy

There has been a considerable interest in combining topical calcipotriol with UVB phototherapy. In trials that examined this combination, the two treatments were started at the same time. Calcipotriol was then applied twice daily and UVB given 3 times weekly. To avoid the risk of inactivation of calcipotriol by the UV light, patients applied calcipotriol ointment either at least 2 hr before UVB exposure or after the UVB exposure. In the first studies the combination therapy was compared to calcipotriol monotherapy. After treatment for 8 weeks the combination therapy was more effective than calcipotriol alone [70,71]. Broadband UVB light sources were used in these studies, but the superiority of the combination therapy was also observed with narrowband UVB light [72].

In other trials the combination therapy was compared with UVB phototherapy alone. In a single-center study that included 19 patients, the combination therapy apparently had a stronger antipsoriatic effect [73]. However, in a larger, multicenter study, the advantage of the combination therapy was not statistically significant [74]. These results, therefore, seriously question the advocation of calcipotriol/UVB combination therapy. One reason why UVB monotherapy was found to be almost as effective as the combination with calcipotriol may be that an aggressive UVB protocol was used [74]. This UVB treatment may, by itself, be so effective that an additional therapeutic effect of calcipotriol may be difficult to detect. However, in the case of suberythematogenic UVB doses, the addition of calcipotriol does improve the therapeutic response to UVB, in particular during the first treatment weeks. There are no follow-up data from UVB studies. It is therefore unknown whether the combination of UVB and calcipotriol improves remission times compared with monotherapy.

It has consistently been found that the skin irritation (see below) induced by calcipotriol is not enhanced when UVB therapy is added. Furthermore, increased photosensitivity has not been reported in these prospective trials. This is apparently in contrast to a case report of photosensitivity when combining UVB and calcipotriol [75]. In these cases calcipotriol was added to ongoing UVB therapy. It is thus conceivable that the "photosensitivity" was a calcipotriol-induced irritation of skin that had been subclinically irritated by the aggressive UVB therapy. In general, calcipotriol and UVB should be started simultaneously. If calcipotriol is added to ongoing UVB therapy, patients should be aware of the potential to develop skin irritation. The risk of skin irritation can be reduced if the UVB dose is temporarily reduced when calcipotriol therapy is started.

Calcipotriol improves the response to photochemotherapy with psoralen–ultraviolet A (PUVA). In a bilateral comparison of calcipotriol and ointment vehicle ($n = 13$), topical treatment was commenced on the first day of PUVA therapy [76]. Lesions treated with calcipotriol cleared earlier or were consistently judged to respond better. In a larger, multicenter study, calcipotriol or vehicle treatment was started 2 weeks prior to PUVA and then maintained during the therapy [77]. Using this protocol, the combination of calcipotriol and PUVA reduced the cumulative UVA dose and the number of PUVA treatments required for clearance of psoriasis [77]. Because the combination of PUVA and calcipotriol can reduce UVA exposure and thereby the potential long-term hazards of PUVA (skin carcinomas), this combination therapy may be considered to be real progress in the management of psoriasis.

Treatment with cyclosporin (3–5 mg/kg/day) is very effective in patients with severe psoriasis. Because of the associated risk of nephrotoxicity, there has been an interest in combining low dose (2 mg/kg/day) cyclosporin with calcipotriol. In a double-blind, placebo-controlled, multicenter study that included 69 patients, calcipotriol/cyclosporin was more effective than placebo/cyclosporin [78]. It is unknown whether the superiority of the combination therapy is due to an additive effect or to a synergistic effect of calcipotriol on the cyclosporin-induced suppression of interleukin-2 secretion [79].

E. Tolerability

Skin irritation is the only important local side effect seen with calcipotriol therapy. It is reported in about 15% of treated patients, but only 1–2% stop treatment for this reason. The skin irritation usually develops within the first few weeks of treatment, and no additional skin irritation is observed during long-term treatment. The lesional or perilesional irritation consists of a burning or stinging sensation. It is, in general, transient and has no clinical consequence. In more severe cases, erythema and scaling are present. The face is particularly sensitive to calcipotriol [58]. Facial irritation may be seen not only after local application, but also after transfer of calcipotriol ointment applied elsewhere. Facial irritation can nearly be avoided if the face is not treated, and if the patients are instructed to wash hands after applying the ointment [60,61]. In a patch test study, calcipotriol was confirmed to be a weak irritant [80]. Unfortunately, it is not possible to predict which patients will become irritated by calcipotriol.

The reaction to the irritant patch test with calcipotriol may be difficult to distinguish from delayed-type contact allergy, but allergy to calcipotriol is apparently extremely rare. There are four reported cases of suspected allergic contact dermatitis to calcipotriol [81–84]. Because patch testing with calcipotriol carries a high risk of false-positive reactions [80], in general it is not recommended to patch test patients who develop a dermatitis during calcipotriol therapy. The explanation of why calcipotriol more often than other vitamin D_3 analogs causes skin irritation is, in all likelihood, that calcipotriol is used in the highest concentration [80].

F. Safety

The potential effect on calcium and bone metabolism is the dose-limiting factor in the use of calcipotriol. In the clinical trials patients were provided with 100 or 120 g ointment per week. Serum calcium did not change in either short-term or long-term trials, except in one patient [61]. This patient applied approximately 400 g of calcipotriol ointment at 50 μg/g during 10 days, that is, about 3 times the amount permitted according to the protocol. Another case of hypercalcemia during excessive use of calcipotriol ointment has been reported [85]. This psoriatic patient had a "moderate" degree of renal impairment. After applying approximately 200 g calcipotriol ointment to her extensive disease during the first week of treatment, she developed nausea, muscle weakness, and abdominal pain, and the serum calcium rose from 2.44 to 3.51 mmol/liter. After stopping calcipotriol therapy serum calcium normalized within 1 week. No sequelae were reported. Two additional cases of hypercalcemia developed after application of 150 and 200 g/week of calcipotriol ointment (data on file with Leo Pharmaceutical Products, Denmark).

Although serum calcium levels are unchanged when calcipotriol ointment is used according to the guidelines, the four cases of hypercalcemia that developed after excessive use of calcipotriol raise the question of whether changes in bone and calcium metabolism can take place, even when the serum calcium concentration remains unchanged. This question has been assessed in a randomized, double-blind, placebo-controlled, parallel group comparison [86]. Thirty-four psoriasis patients were randomized to receive treatment with either calcipotriol ointment at 50 μg/g or placebo ointment for 3 weeks. Calcipotriol-treated patients (n = 17) used on average 40.3 g ointment per week. During treatment there was no difference between the calcipotriol group and the placebo group in urinary calcium excretion or in the other biochemical indices of calcium and bone metabolism assessed. The absence of an effect of calcipotriol treatment on bone and calcium metabolism has been confirmed in an open noncomparative study involving 12 psoriatics, who were treated for 4 weeks with an average of 100 g of ointment per week [87]. In contrast to these results, it was reported that the application of calcipotriol ointment at 100 g/week for 4 weeks induced a slight, but statistically significant increase in urinary calcium excretion [88] and decrease in serum parathyroid hormone levels [89]. These results seem to indicate that the weekly dose of calcipotriol ointment should be kept below 100 g.

It has been a concern that long-term treatment with calcipotriol might have a cumulative effect on calcium and bone metabolism. Fortunately, the available data indicate that long-term calcipotriol treatment is apparently as safe as short-term therapy. Thus, serum calcium levels and urinary calcium excretion did not increase in patients treated for 1 year with a mean of 30 g of ointment per week [88]. Measurements of bone density have not yet been reported.

Patients with renal impairment may be at risk of developing hypercalcemia even when applying less than 100 g calcipotriol ointment per week. Thus, at least two of the three reported cases of hypercalcemia occurring after applying 70–80 g of calcipotriol ointment per week had some degree of renal impairment [90,91]. These results may indicate that the small amounts of calcipotriol absorbed systemically (<5%) may be sufficient to cause hypercalcemia in patients with a decreased renal capacity to regulate urinary calcium excretion.

Taken together the available evidence suggests that treatment with calcipotriol ointment at 50 μg/g is safe when used in amounts up to 100 g per week. In patients requiring greater amounts of ointment, it is recommended that serum parathyroid hormone (PTH) levels be monitored. A drop in serum PTH indicates a change in calcium homeostasis. In contrast, serum calcium is not sensitive enough to pick up minor changes of calcium metabolism.

Calcipotriol is not teratogenic (data on file with Leo Pharmaceutical Products, Denmark), but there is no clinical experience from the treatment of pregnant patients. It is not known whether calcipotriol or its metabolites enter breast milk. In case a patient becomes pregnant while being treated with calcipotriol, treatment should be stopped. Elective abortion is not indicated. Studies in children with psoriasis have been conducted, but results have not yet been reported.

VIII. USE OF TACALCITOL IN PSORIASIS

The compound $1,24(OH)_2D_3$ (tacalcitol) is another synthetic vitamin D analog that has been developed for topical use in psoriasis (Table I). $1,24(OH)_2D_3$ is equipotent with $1,25(OH)_2D_3$ in its affinity for the VDR [92] and in its capacity to inhibit keratinocyte proliferation and stimulate keratinocyte differentiation *in vitro* [93]. Also, tacalcitol is as effective as $1,25(OH)_2D_3$ in increasing cytosolic calcium levels in cultured keratinocytes [93]. The advantage over $1,25(OH)_2D_3$ is that talcalcitol induces a smaller increase of serum calcium levels after a single intravenous dose to rats [93]. However, the concentrations of tacalcitol and $1,25(OH)_2D_3$ that induce hypercalcemia are similar. Although tacalcitol may be advantageous over $1,25(OH)_2D_3$ for clinical use, it is much less selective than calcipotriol in its effects on calcium metabolism.

As early as 1986 it was reported that tacalcitol ointment at 2 and 4 μg/g improved psoriasis in an open label study [94]. In a subsequent controlled study, in which patients applied tacalcitol ointment twice daily to a single lesion for 4 weeks, a concentration of 2 μg/g was superior to 1 μg/g and as effective as 4 μg/g [95]. In another dose–response study tacalcitol ointment was applied once daily to 2-cm^2 areas of lesional psoriasis for 4 weeks [96]. Using this study design, 4 μg/g was more effective than 2 μg/g and similar to 8 and 16 μg/g. The differences in application frequency may be the reason why the optimal tacalcitol concentration was different in these two studies. When different body regions were treated with tacalcitol ointment at 2 μg/g, the treatment was more effective on the face than in other skin areas [95].

There is limited knowledge about the efficacy of tacalcitol compared with other topical antipsoriatic agents. Apparently treatment with tacalcitol ointment at 2 μg/g twice daily is as effective as hydrocortisone butyrate, but slightly less effective than betamethasone 17-valerate [95]. In the dose–response study [96] calcipotriol ointment at 50 μg/g was included for comparison, but the results were not reported.

Used at a low concentration (2 μg/g), tacalcitol ointment is well tolerated, and skin irritation is uncommon [95,97]. Regarding safety, hypercalcemia was not observed in two studies that included 210 patients [95,97]. There is, however, no information on the amount of tacalcitol ointment used by these patients. In a smaller study that included 12 patients, an average cumulative dose of 340 μg/g (approximately 42 g of tacalcitol ointment at 2 μg/g per week) did not increase serum calcium levels [98]. There is no information available on the more sensitive markers of calcium metabolism (24-hr calcium excretion and serum PTH) during tacalcitol therapy.

IX. VITAMIN D_3 THERAPY IN OTHER SKIN DISEASES

The presence of VDR in most cell types in the skin including keratinocytes [5], Langerhans cells [6], melanocytes [7], fibroblasts [8], endothelial cells [9], T lymphocytes [10], and hair follicle cells [99] suggests that vitamin D_3 analogs may have a role in skin diseases other than psoriasis. In this context it is of interest that some synthetic vitamin D_3 analogs selectively affect the activity of immune-competent cells [100]. A number of case reports indicate positive effects of calcipotriol in a variety of skin diseases characterized by disordered epidermal keratinization such as pityriasis rubra pilaris [101], disseminated superficial actinic porokeratosis [102], inflammatory linear verrucous epidermal naevus [103], and Grover's disease [104] (Table II). All these indications have to be further evaluated in proper clinical trials.

TABLE II Skin Diseases Other than Psoriasis Improved by Treatment with Vitamin D_3 Analogs

Ichthyosis (hyperproliferating variants)
Pityriasis rubra pilaris
Porokeratosis (superficial actinic type)
Inflammatory linear verrucous epidermal naevus
Grover's disease
Seborrheic keratosis
Localized scleroderma
Cutaneous T-cell lymphoma

The efficacy and tolerability of topical calcipotriol in disorders of keratinization (ichthyoses, Darier's disease, palmoplantar keratoderma, follicular keratosis) have been assessed in a randomized, double-blind, vehicle-controlled, right–left comparative study [105]. Skin lesions were treated twice daily with calcipotriol ointment at 50 μg/g and placebo (vehicle of calcipotriol) for up to 12 weeks. At the end of treatment certain ichthyoses, particularly the hyperproliferative variants, had improved. No therapeutic effect was observed in palmoplantar keratoderma or keratosis pilaris, whereas some patients with Darier's disease had a worsening of their disease [105]. Flow cytometric investigations showed that clinical improvement to some extent correlated with enhanced expression of differentiation markers during calcipotriol treatment [106]. In the amounts applied (up to 120 g/week) there was no change in serum calcium levels [105]. Patients should, however, be aware of the theoretical risk of hypercalcemia from absorption of calcipotriol after application to extensive skin areas. This is underscored by a case report of symptomatic hypercalcemia in an ichthyotic patient applying excessive amounts of calcipotriol ointment at 50 μg/g [107].

Seborrheic keratosis is the most common benign neoplasm of the epidermis. The effect of oral $1,25(OH)_2D_3$ on widespread seborrheic keratosis was examined in 51 patients [108]. Patients were divided into two groups. One group received a high dose (0.5 μg/day) and the other group a low dose (0.25 μg/day). Patients showing improvement after 12 weeks continued treatment with decreasing doses for 1 year. Both doses were effective in clearing or reducing the size of the tumors, particularly the small ones [108]. There is also an interest in assessing the therapeutic effect of vitamin D_3 in malignant and premalignant skin conditions [109]. A variety of tumor cell lines, including melanoma and squamous carcinoma cells, possess VDR. It is, however, important to emphasize that the presence of VDR in these cell lines does not necessarily imply that their proliferation is affected by $1,25(OH)_2D_3$ [110].

There are theoretical reasons why vitamin D_3 analogs may prove to be effective in the treatment of some forms of hair loss. There is a high density of VDR in the hair follicle [99], and patients affected with heritable rickets, due to alterations in the structure of VDR, often present with complete alopecia (see Chapter 48). It was also reported that $1,25(OH)_2D_3$ has a biphasic effect on human hair follicle growth and hair fiber production in whole-organ cultures [111]. Furthermore, chemotherapy-induced alopecia in neonatal rats is prevented by pretreatment with topical $1,25(OH)_2D_3$ [112]. Using an alternative mouse model of chemotherapy-induced alopecia, there was no suppression of alopecia, but acceleration of normal hair regrowth was noted [113]. So far there is no evidence that hair loss in humans responds to vitamin D_3. Thus, topical calcipotriol did not improve alopecia areata [114].

Scleroderma is an autoimmune disease that shows increased amounts of dermal collagen. Topical application of calcipotriol to normal human skin stimulates the synthesis of collagens type I and type III [115]. Although the proliferation of cultured normal fibroblasts is not affected by calcipotriol, cultured scleroderma fibroblasts are inhibited by calcipotriol [116]. These *in vitro* results give grounds for optimism that vitamin D_3 analogs might prove therapeutically useful in scleroderma. Thus far, vitamin D_3 therapy in scleroderma has been confined to three reports. A woman with a 2-year history of scleroderma localized to the skin improved after receiving oral $1{,}25(OH)_2D_3$ at 0.5 μg/day for 4 months [117]. In a another report three patients treated with oral $1{,}25(OH)_2D_3$ at 0.50–0.75 μg/day showed improvement of the skin lesions after treatment for 3 to 7 months [118]. Finally, Humbert *et al.* [119] found that 11 patients with systemic scleroderma had improvement in skin changes after treatment with high dose $1{,}25(OH)_2D_3$ (1.75 μg/day) for 3 to 36 months. There was no increase of serum calcium level with this high $1{,}25(OH)_2D_3$ dose. More extensive, clinically controlled studies are needed to verify these promising results.

There are several mechanisms by which vitamin D_3 analogs may exert an antitumor effect in lymphoma [18,19] (see Chapter 68). In the clinical setting, cutaneous T-cell lymphoma has been successfully treated in one case with topical calcipotriol [120]. Although the beneficial effect of calcipotriol was less clear in three similar patients [121], these results justify further investigation.

References

1. Wright C 1941 Vitamin D therapy in dermatology. Arch Dermatol Syph **43**:145–154.
2. Morimoto S, Kumahara Y 1985 A patient with psoriasis cured by 1-alpha-hydroxyvitamin D_3. Med J Osaka Univ **35**:51.
3. Hosomi J, Hosoi J, Abe E, Suda T, Kuroki T 1983 Regulation of terminal differentiation of mouse cultured epidermal cells by 1,25-dihydroxyvitamin D_3. Endocrinology **113**:1950–1957.
4. Binderup L, Kragballe K 1992 Origin of the use of calcipotriol in psoriasis treatment. Rev Contemp Pharmacother **3**:401–409.
5. Feldman D, Chen T, Hirst M, Colston K, Karasek M, Cone C 1980 Demonstration of 1,25-dihydroxyvitamin D_3 receptor in human skin biopsies. J Clin Endocrinol Metab **51**:1463–1465.
6. Dam TN, Møller B, Hindkjaer J, Kragballe K 1996 The vitamin D analog calcipotriol suppresses the number and antigen-presenting function of Langerhans cells in normal human skin. J Invest Dermatol Symp Proc **1**:72–77.
7. Ranson M, Posen S, Manson R 1988 Human melanocytes as a target tissue for hormones. In vitro studies with 1,25-dihydroxyvitamin D_3, alpha-melanocyte stimulating hormone, and beta-estradiol. J Invest Dermatol **91**:593–598.
8. Eil C, Marx S 1981 Nuclear uptake of 1,25-dihydroxycholecalciferol in dispersed fibroblasts cultured from normal human skin. Proc Natl Acad Sci USA **78**:2562–2566.
9. Merke J, Milde P, Lewicka S, Hügel U, Klaus G, Mangelsdorf DJ, Haussler MR, Rauterberg EW, Ritz E 1989 Identification and regulation of 1,25-dihydroxyvitamin D_3 receptor activity and biosynthesis of 1,25-dihydroxyvitamin D_3. J Clin Invest **83**:1903–1915.
10. Bhalla AK, Amento EP, Clemens TL 1983 Specific high-affinity receptors for 1,25-dihydroxy-vitamin D_3 in human peripheral blood mononuclear cells: Presence in monocytes and induction in T-lymphocytes following activation. J Clin Endocrinol Metab **57**:1308–1310.
11. Gniadecki R 1996 Stimulation versus inhibition of keratinocyte growth by 1,25-dihydroxyvitamin D_3: Dependence on cell culture conditions. J Invest Dermatol **106**:510–516.
12. Svendsen ML, Geysen J, Binderup L, Kragballe K 1997 Proliferation and differentiation of cultured human kerationocytes is modulated by $1{,}25(OH)_2D_3$ and synthetic vitamin D_3 analogs in a cell-density-, calcium- and serum-dependent manner. Pharmacol Toxicol **80**:49–56.
13. Lützow-Holm C, De Angelis P, Grosvik H, Clausen OP 1993 1,25-Dihydroxyvitamin D_3 and the vitamin D_3 analogue KH 1060 induced hyperproliferation in normal mouse epidermis. Exp Dermatol **2**:113–120.
14. Gniadecki R, Serup J 1995 Stimulation of epidermal proliferation in mice with 1α,25-dihydroxyvitamin D_3 and receptor-active 20-epi analogues of 1α,25-dihydroxyvitamin D_3. Biochem Pharmacol **49**:621–624.
15. Kato T, Terui T, Tagami H 1987 Topically active vitamin D_3 analogue, 1α,24-dihydroxy cholecalciferol, has an anti-proliferative effect on the epidermis of guinea pig skin. Br J Dermatol **117**:528–530.
16. Sølvsten H, Svendsen ML, Fogh K, Kragballe K 1997 Upregulation of vitamin D receptor levels by 1,25-dihydroxyvitamin D_3 in cultured human keratinocytes. Arch Dermatol Res, in press.
17. Sølvsten H, Jensen TJ, Sørensen S, Kragballe K 1997 Application of calcipotriol to normal human skin increases the level of the vitamin D receptor (VDR) and the binding of the VDR–retinoid X receptor heterodimer to a vitamin D response element. Submitted for publication.
18. Tsoukas DD, Provvedini DM, Manolagas SC 1984 1,25-Dihydroxy vitamin D_3: A novel immunoregulatory hormone. Science **14**:38–40.
19. Rigby WFC, Waugh MG 1992 Decreased accessory cell function and costimulatory activity by 1,25-dihydroxyvitamin D_3 treated monocytes. Arthritis Rheum **35**:110–119.
20. Griffiths CEM 1994 Cutaneous leukocyte trafficking and psoriasis. Arch Dermatol **130**:494–498.
21. Milde P, Hauser U, Simon T, Mall G, Ernst V, Haussler MR, Frosch P, Rauterberg EW 1991 Expression of 1,25-dihydroxyvitamin D_3 receptors in normal and psoriatic skin. J Invest Dermatol **97**:230–236.
22. Sølvsten H, Fogh K, Svendsen M, Kristensen P, Åstrøm, Kumar R, Kragballe K 1996 Normal levels of the vitamin D receptor and its message in psoriatic skin. J Invest Dermatol Symp Proc **1**:28–32.
23. Jensen TJ, Sølvsten H, Kragballe K Correlation between vitamin D receptor and retinoid X receptor levels and binding to DNA in psoriatic skin. Br J Dermatol, in press.
24. Morimoto S, Yoshikawa K, Fukuo K, et al. 1990 Inverse relation-

ship between severity of psoriasis and serum 1,25-dihydroxyvitamin D level. J Dermatol Sci **1**:277–282.

25. Staberg B, Oxholm A, Klemp P, Christiansen C 1987 Abnormal vitamin D metabolism in patients with psoriasis. Acta Derm Venereol (Stockh) **67**:65–68.
26. Smith EL, Pincus SH, Donovan L, Holick MF 1988 A novel approach for the evaluation and treatment of psoriasis. J Am Acad Dermatol **10**:360–364.
27. Guilhou JJ, Colette C, Monpoint S, Lancrenon E, Guillot B, Monnier L 1990 Vitamin D metabolism in psoriasis before and after phototherapy. Acta Derm Venereol (Stockh) **70**:351–354.
28. Risum G 1973 Psoriasis excerbated by hypoparathyroidism with hypocalcemia. Br J Dermatol **89**:309–312.
29. Stewart AF, Battaglini-Sabetta J, Millstone L 1984 Hypocalcemia-induced pustular psoriasis of von Zumbusch. Ann Intern Med **100**:677–680.
30. Mortensen L, Kragballe K, Wegman E, Schifter S, Risteli J, Charles P 1993 Treatment of psoriasis vulgaris with topical calcipotriol has no short-term effect on calcium and bone metabolism. Acta Derm Venereol (Stockh) **73**:300–304.
31. Abe J, Kondo S, Nishii Y, Kuroki T 1989 Resistance to 1,25-dihydroxyvitamin D_3 of cultured psoriatic epidermal keratinocytes from involved and uninvolved skin. J Clin Endocrinol Metab **68**:851–854.
32. McLaughlin JA, Gange W, Taylor D 1985 Cultured psoriatic fibroblasts from involved and uninvolved sites have partial, but not absolute resistance to the proliferation-inhibition activity of 1,25 dihydroxyvitamin D_3. Proc Natl Acad Sci USA **82**:5409–5412.
33. Matsouka LY, Wortsman J, Haddad J, Hollis B 1990 Cutaneous formation of vitamin D_3 in psoriasis. Arch Dermatol **126**:1107–1108.
34. Langner A, Verjans H, Stapor V, Mol M, Fraczykowska M 1992 1alpha,25-Dihydroxyvitamin D_3 ointment in psoriasis. J Dermatol Treat **3**:177–180.
35. Holland DB, Roberts SG, Russell A, Wood EJ, Cunliffe WJ 1990 Changes of epidermal keratin levels during treatment of psoriasis with topical vitamin D_3 analogue MC 903. Br J Dermatol **122**:284.
36. De Mare S, De Jong EGJM, van de Kerkhof PCM 1990 DNA content and Ks 8.12 binding of the psoriatic lesions during treatment with the vitamin D_3 analogue MC 903 and betamethasone. Br J Dermatol **123**:291–295.
37. Mallett RB, Coulson IH, Purkis PE 1990 An immunohistochemical analysis of the changes in the immune infiltrate and keratin expression in psoriasis treated with calcipotriol compared with betamethasone ointment. Br J Dermatol **123**:837.
38. De Jong EMGH, van de Kerkhof PCM 1991 Simultaneous assessment of inflammation and epidermal proliferation plaques during long-term treatment with the vitamin D_3 analogue MC 903: Modulations and interrelations. Br J Dermatol **124**:221–229.
39. Berth-Jones J, Fletcher A, Hutchinson PE 1992 Epidermal cytokine and immunocyte responses during treatment of psoriasis with calcipotriol and betamethasone valerate. Br J Dermatol **126**:356–361.
40. Nieboer C, Verburgh CA 1992 Psoriasis treatment with vitamin D_3 analogue MC 903. Br J Dermatol **126**:302–303.
41. Holland DB, Wood EJ, Cunliffe WJ, Turner DM 1989 Keratin gene expression during the resolution of psoriatic plaques: Effect of dithranol, PUVA, etretinate and hydroxyurea regimes. Br J Dermatol **120**:9–19.
42. Chen ML, Perez A, Sanan DK, Heinrich G, Chen TC, Holick M 1996 Induction of vitamin D receptor mRNA expression in psoriatic plaques correlates with clinical response to 1,25-dihydroxyvitamin D_3. J Invest Dermatol **106**:637–641.
43. Morimoto S, Yoshikawa K, Kozuka T, Kitano Y, Imanaka S, Fukuo K, Koh E, Kumahara Y 1986 An open study of vitamin D_3 treatment in psoriasis vulgaris. Br J Dermatol **115**:421–429.
44. van de Kerkhof PCM, van Bokhoven M, Zultak M, Czarnetzki MB 1989 A double-blind study of topical 1,25-dihydroxyvitamin D_3 in psoriasis. Br J Dermatol **120**:661–664.
45. Henderson CA, Papworth-Smith J, Cunliffe WJ, Highet AS, Shamy HK, Czarnetzki BM 1989 A double-blind placebo-controlled trial of topical 1,25-dihydroxycalciferol in psoriasis. Br J Dermatol **121**:493–496.
46. Langner A, Verjans H, Stapar V, Mol M, Fraczy-Kowska A 1993 Topical calcitriol in the treatment of chronic plaque psoriasis: A double-blind study. Br J Dermatol **128**:566–571.
47. Langner A, Verjans H, Stapor V, Mol M, Elzerman J 1991 Treatment of chronic plaque psoriasis by 1alpha,25-dihydroxyvitamin D_3 ointment. In: Norman AW, Bouillon R, Thomasset M (eds) Vitamin D: Gene Regulation Structure–Function Analysis and Clinical Application. de Gruyter, Berlin, pp. 430–431.
48. Wishart JM 1994 Calcitriol (1,25-dihydroxy-vitamin D_3) ointment in psoriasis: A safety, tolerance and efficacy multicentre study. Dermatology **188**:135–139.
49. Bourke JF, Featherstone S, Iqbal SJ, Hutchinson PE 1997 A double-blind comparison of topical calcitriol (3 μg/g) and calcipotriol (50 μg/g) in the treatment of chronic plaque psoriasis vulgaris. Acta Derm-Venereol (Stockh) **77**:228–230.
50. Morimoto S, Yoshikawa K 1989 Psoriasis and vitamin D_3. A review of our experience. Arch Dermatol **125**:231–234.
51. Smith EL, Pincus SH, Donovan L, Holick MF 1988 A novel approach for the evaluation of treatment of psoriasis. J Am Acad Dermatol **19**:516–528.
52. Perez A, Raab R, Chen TC, Turner A, Holick HF 1996 Safety and efficacy of oral calcitriol (1,25-dihydroxy vitamin D_3) for the treatment of psoriasis. Br J Dermatol **134**:1070–1078.
53. Kragballe K, Beck HI, Sogaard H 1988 Improvement of psoriasis by a topical vitamin D_3 analogue (MC903) in a double-blind study. Br J Dermatol **199**:223–230.
54. Staberg B, Roed-Petersen J, Menne T 1989 Efficacy of topical treatment in psoriasis with MC 903, a new vitamin D analogue. Acta Derm Venereol (Stockh) **69**:147–150.
55. Harrington CI, Goldin D, Lovell CR, van de Kerkhof P, Nieboer C, Austad J, Molin L, Clareus BW, Rask-Petersen E 1996 Comparative effects of two different calcipotriol (MC903) cream formulations versus placebo in psoriasis vulgaris. A randomized, double-blind, placebo-controlled, parallel group multi-centre study. J Eur Acad Dermatol Venereol **6**:152–158.
56. Green C, Ganpule M, Harris D, Kavanagh G, Kennedy C, Mallett R, Rustin M, Downes N 1994 Comparative effects of calcipotriol (MC 903) solution and placebo (vehicle or MC 903) in the treatment of psoriasis of the scalp. Br J Dermatol **130**:483–487.
57. Klaber MR, Hutchinson PE, Pedvis-Leftick, Kragballe K, Reunala TL, van de Kerkhof PCM, Johnsson MK, Molin K, Corbett MS, Downes N 1994 Comparative effects of calcipotriol solution (50 μg/ml) and betamethasone 17-valerate solution (1 mg/ml) in the treatment of scalp psoriasis. Br J Dermatol **131**:678–683.
58. Kragballe K 1989 Treatment of psoriasis by the topical application of the novel cholecaliferol analogue calcipotriol (MC903). Arch Dermatol **125**:1647–1652.
59. Dubertret L, Wallach D, Souteyrand P, Perussel M, Kalis B, Meyaadier J, Chevrant-Breton J, Beylot E, Bazex JA, Jürgensen HJ 1992 Efficacy and safety of calcipotriol (MC903) ointment in psoriasis vulgaris. J Am Acad Dermatol **27**:983–988.
60. Kragballe K, Gjertsen BT, De Hoop D, Karlsmark T, van de

Kerkhof PCM, Larkö O, Nieboer C, Roed-Petersen J, Strand A, Tikjøb G 1991 Double-blind, right–left comparison of calcipotriol and betamethasone valerate in treatment of psoriasis vulgaris. Lancet **337**:193–196.

61. Cunliffe WJ, Claudy A, Fairiss G, Goldin D, Gratton D, Henderson CA, Holden CA, Maddin WS, Ortonne JP, Young M 1992 A multicenter comparative study of calcipotriol and betamethasone 17-valerate in patients with psoriaisis vulgaris. J Am Acad Dermatol **26**:736–743.
62. Ortonne JP 1994. Psoriasis: New therapeutic modality by calcipotriol and betamethasone dipropionate. Nouv Dermatol **13**: 746–751.
63. Bruce S, Epinette WW, Funicella 1994 Comparative study of calcipotriene (MC 903) ointment and fluocinonide ointment in the treatment of psoriasis. J Am Acad Dermatol **31**:755–759.
64. Berth-Jones J, Chu AC, Dodd WAH, Ganpule M, Griffths WAD, Haydey RO, Klaber MR, Murray SJ, Rogers S, Jürgensen HJ 1992 A multicenter parallel-group comparison of calcipotriol ointment and short-contact dithranol therapy in chronic plaque psoriasis. Br J Dermatol **127**:266–271.
65. Van der Vleuten CJM, de Jong EMGJ, Ruto EHFC, Gertsen M-JP, van der Kerkhof PCM 1995 In-patient treatment with calcipotriol versus dithranol in refractory psoriasis. Eur J Dermatol **5**:676–679.
66. Bourke JF, Berth-Jones J, Hutchinson PE 1993 Occlusion enhances the efficacy of topical calcipotriol in the treatment of psoriasis vulgaris. Clin Exp Dermatol **18**:504–506.
67. Nielsen PG 1993 Calcipotriol or clobetasol propionate occluded with a hydrocolloid dressing for treatment of nummular psoriasis. Acta Derma Venereol (Stockh) **73**:394.
68. Kragballe K, Fogh K 1991 Long-term efficacy and tolerability of topical calcipotriol in psoriasis. Acta Derma Venereol (Stockh) **71**:475–478.
69. Ramsay CA, Berth-Jones J, Brundin G, Cunliff WJ, Dubertret L, van de Kerkhof PCM, Menné T, Wegmann E 1994 Long-term use of topical calcipotriol in chronic plaque-type psoriasis. Dermatology **189**:260–264.
70. Kragballe K 1990 Combination of topical calcipotriol (MC 903) and UVB radiation for psoriasis vulgaris. Dermatologica **181**:211–214.
71. Molin L 1995 Calcipotriol combined with phototherapy (UVB and PUVA) in the treatment of psoriasis. J Eur Acad Dermatol **5**(Suppl):S 184 (abstract).
72. Kerscher M, Volkenandt M, Plewig G 1993 Combination phototherapy of psoriasis with calcipotriol and narrowband UVB. Lancet **343**:923 (letter).
73. Kokelj F, Lavaroni G, Guadagnini A 1995 UVB versus UVB plus calcipotriol (MC 903) therapy for psoriasis vulgaris. Acta Derm Venereol (Stockh) **75**:386–387.
74. Koo J, 1996 Phototherapy in combination with vitamin D therapy. Proc Clinical Dermatology 2000, 28–31 May 1996, Vancouver, p. 88 (abstract).
75. McKenna KE, Stern RS 1995 Photosensitivity associated with combined UV-B and calcipotriene therapy. Arch Dermatol **131**:1305–1307.
76. Speight EL, Farr PM 1994 Calcipotriol improves the response of psoriasis to PUVA. Br J Dermatol **130**:79–82.
77. Frappaz A, Thivolet J 1993 Calcipotriol in combination with PUVA: A randomized double-blind placebo study in severe psoriasis. Eur J Dermatol **3**:351–354.
78. Grossman RM, Thivolet J, Claudy A, Souteyrand P, Guilhou JJ, Thomas P, Amblard P, Belaich S, De Belilovsky C, de la Brassine M, Martinet C, Bazex JA, Beylot C, Combemale P, Lambert D, Ostojic A, Denooeux JP, Lauret P, Vaillant L, Weber M, Pamphile R, Dubertret L 1994 A novel therapeutic approach to psoriasis with combination calcipotriol ointment and very low-dose cyclosporine: Results of a multicenter placebo-controlled study. J Am Acad Dermatol **31**:68–74.
79. Gupta S, Fass D, Shimizu M 1989 Potentiation of immunosuppressive effects of cyclosporine A by 1alpha,25-dihydroxyvitamin D_3. Cell Immunol **121**:290–297.
80. Fullerton A, Avnstorp C, Agner T, Dahl JC, Olsen LO, Serup J 1996 Patch test study with calcipotriol ointment in different patient groups, including psoratic patients with and without adverse dermatitis. Acta Derm Venereol (Stockh) **76**:194–202.
81. Bruynzel DP, Hol CW, Nieboer C 1992 Allergic contact dermatitis to calcipotriol. Br J Dermatol **126**:66.
82. De Groot AC 1994 Contact allergy to calcipotriol. Contact Dermatitis **30**:242–243.
83. Yip J, Goodfield M 1991 Contact dermatitis from MC 903, a topical vitamin D_3 analogue. Contact Dermatitis **25**:139–140.
84. Steinkjer B 1994 Contact dermatitis from calcipotriol. Contact Dermatitis **31**:122.
85. Dwyer C, Chapman RS 1991 Calcipotriol and hypercalcemia. Lancet **338**:764–765.
86. Mortensen L, Kragballe K, Wegmann E, Schifter S, Risteli J, Charles P 1993 Treatment of psoriasis vulgaris with topical alcipotriol has no short-term effect on calcium or bone metabolism. Acta Derm Venereol (Stockh) **73**:300–304.
87. Gumowski-Sunek D, Rizzoli R, Saurat JH 1991 Effects of topical calcipotriol on calcium metabolism in psoriatic patients: Comparison with oral calcitriol. Dermatologica **183**:275–79.
88. Berth-Jones J, Bourke JF, Iqbal SJ, Hutchinson PE 1993 Urine calcium excretion during treatment of psoriasis with topical calcipotriol. Br J Dermatol **129**:411–414.
89. Bourke JF, Berth-Jones J, Mumford R, Iqbal SJ, Hutchinson PE 1994 High dose topical consistently reduces serum parathyroid hormone levels. Clin Endocrinol **41**:295–297.
90. Hardman KH, Heath DA 1993 Hypercalcaemia associated with calcipotriol (Dovonex) treatment. Br Med J **306**:896 (letter).
91. Russell S, Young MJ 1994 Hypercalcemia during treatment of psoriasis with calcipotriol. Br J Dermatol **130**:795–796.
92. Matsumoto K, Hashimoto K, Kiyoki M, Yamamoto M, Yoshikawa K 1990 Effect of 1,24-dihydroxyvitamin D_3 on the growth of human keratinocytes. J Dermatol **17**:97–103.
93. Matsunaga T, Yamamoto M, Mimura H, Ohta T, Kiyoki M, Ohba T, Naruchi T, Hosoi J, Kuroki T 1990 1,24*R*-Dihydroxyvitamin D_3, a novel active form of vitamin D_3 with high activity for inducing epidermal differentiation but decreased hypercalcemia activity. J Dermatol **17**:135–142.
94. Kato T, Rokogu M, Teuri T, Tagami H 1986 Successful treatment of psoriasis with topical application of the active vitamin D_3 analogue, 1,24-dihydroxy cholecalciferol. Br J Dermatol **115**:431–433.
95. Nishimura M, Hori Y, Nishiyama S, Nakimizo Y 1993 Topical 1,24-dihydroxyvitamin D_3 for the treatment of psoriasis. A review of the literature. Eur J Dermatol **3**:255–261.
96. Baadsgaard O, Traulsen J, Roed-Petersen J, Jakobsen HB 1995 Optimal concentration of tacalcitol in once-daily treatment of psoriasis. J Dermatol Treat **6**:145–150.
97. TV-02 Ointment Research Group 1991 A placebo controlled double-blind, right–left comparison study on the efficacy of TV-02 ointment for the treatment of psoriasis. Nishinihon J Dermatol **53**:1252–1256.
98. Nishimura M, Makino Y, Matugi H 1994 Tacalcitol ointment for psoriasis. Acta Derm Venereol (Stockh) **186**:(Suppl.) 166–168.
99. Stumpf WE, Clark SA, Sar M, DeLuca HF 1984 Topographical

and developmental studies on target sites of 1,25(OH)$_2$-vitamin D_3 in skin. Cell Tissue Res **238**:489–496.
100. Binderup L 1992 Immunological properties of vitamin D analogues and metabolites. Biochem Pharmacol **43**:1885–1892.
101. van de Kerkhof PCM, Jong EMGJ 1992 Topical treatment with the vitamin D_3 analogue MC 903 improves pityriasis rubra pilaris. Clinical and immunohistochemical observations. Br J Dermatol **125**:293–294.
102. Harrison PV, Stollery N 1994 Disseminated superficial actinic porokeratosis responding to calcipotriol. Clin Exp Dermatol **19**:95.
103. Gatti S, Carrozzo AM, Orlandi A, Nini G 1995 Treatment of inflammatory linear verucous epidermal naevus with calcipotriol. Br J Dermatol **32**:837–839.
104. Keohane SG, Cork MJ 1995 Treatment of Grover's disease with calcipotriol. Br J Dermatol **132**:832–833.
105. Kragballe K, Steijlen PM, Ibsen HH, van de Kerkhof PCM, Esmann J, Sørensen LH, Axelsen MB 1995 Efficacy, tolerability, and safety of calcipotriol ointment in disorders of keratinization. Arch Dermatol **131**:556–560.
106. Lucker GPH, Steijlen PM, Suykerbuyk JA, Kragballe K, Brandrup F, van de Kerkhof PCM 1996 Flow-cytometric investigation of epidermal cell characteristics in monogenic disorders keratinization and their modulation by topical calcipotriol. Acta Derm Venereol (Stockh) **76**:97–101.
107. Hoeck HC, Laurberg G, Laurberg P 1994 Hypercalcaemic crisis alter excessive topical use of a vitamin D_3 derivative. J Intern Med **235**:281–282.
108. Asagami C, Muto M, Hirota T, Shimizo T, Hamamoto Y 1996. Anti-tumor effects of 1,25-dihydroxyvitamin D_3 in seborrheic keratosis. J Invest Dermatol Symp Proc **1**:94–96.
109. Majewski S, Skopinska M, Marczak M, Samuto A, Bollag W, Jablonska S 1996 Vitamin D_3 is a potent inhibitor of tumor cell-induced angiogenesis. J Invest Dermatol Symp Proc **1**:97–102.
110. Ratnam AV, Buhle DD, Su MJ, Pillai S 1996 Squamous carcinoma cell lines fail to respond to 1,25-dihydroxyvitamin D despite normal levels of the vitamin D receptor. J Invest Dermatol **106**:522–525.
111. Harmon CS, Nevins TD 1994 Biphasic effects of 1,25-dihydroxyvitamin D_3 on human hair follicle growth and hair fiber production in whole-organ cultures. J Invest Dermatol **103**:318–322.
112. Jimenez JJ, Yunis AA 1992 Protection from chemotherapy induced alopecia by 1,25-dihydroxyvitamin D_3. Cancer Res **52**: 5123–5125.
113. Paus R, Schilli MB, Handjiski B, Menrad A, Reichrath J, Czarnetzki B, Plonka P 1996 Calcitriols as modulators of chemotherapy induced alopecia in mice: No suppression of alopecia, but acceleration of normal hair re-growth. J Invest Dermatol Symp Proc **1**:105 (abstract).
114. Berth-Jones J, Hutchinson PE 1991 Alopecia totalis does not respond to the vitamin D analogue calcipotriol. J Dermatol Treat **1**:293–294.
115. Kim SE, Heickendorff L, Bjerring P, Kragballe K 1996 Metabolism of collagens type I and type III in normal skin after application of calcipotriol: Stimulation of collagen synthesis but potentiation of the corticosteroid-induced inhibition of collagen synthesis. J Invest Dermatol Symp Proc **1**:106 (abstract).
116. Bottomley WW, Jutley J, Wood EJ, Goodfield MDJ 1995 The effect of calcipotriol on lesional fibroblasts from patients with active morphea. Acta Derm Venereol (Stockh) **75**:364–366.
117. Humbert PG, Dupond JL, Rochefort A, Vasselet R, Lucas A, Laurent R 1990 Localised scleroderma responds to 1,25-dihydroxyvitamin D_3. Clin Exp Dermatol **15**:396–398.
118. Hulshof MM, Pavel S, Breedveld FC, Dijkmans JAN, Vermeer BJ 1994 Oral calcitriol as a new therapeutic modality for generalized morphea. Arch Dermatol **130**:1290–1293.
119. Humbert PG, Dupond JL, Agache P, Laurent R, Rochefort A, Drobacheff J 1993 Treatment of scleroderma with oral 1,25-dihydroxyvitamin D_3: Evaluation of skin involvement using non-invasive techniques. Acta Derm Venereol (Stockh) **73**:449–451.
120. Scott-Mackie P, Hickish T, Mortimer P, Sloane J, Cunningham D 1993 Calcipotriol and regression in T-cell lymphoma of skin. Lancet **342**:172 (letter).
121. Harwood CA, Sloane J, Cook MG, Mortimer PS 1995 Calcipotriol in mycosis fungoides. Br J Dermatol **133**(Suppl. 45):43 (abstract).

CHAPTER 73

Renal Failure and Hyperparathyroidism

MASAFUMI FUKAGAWA First Department of Internal Medicine, University of Tokyo School of Medicine, Bunkyo-ku, Tokyo, Japan

MASAFUMI KITAOKA Division of Endocrinology and Metabolism, Showa General Hospital, Kodaira-shi, Tokyo, Japan

KIYOSHI KUROKAWA Tokai University Faculty of Medicine, Boseidai, Isehara-shi, Kanagawa, Japan

I. INTRODUCTION

Hyperparathyroidism is one of the most common causes of bone diseases in chronic dialysis patients [1]. In patients with hyperparathyroidism, excess parathyroid hormone (PTH) accelerates bone turnover, which, in severe cases, leads to a typical abnormality known as osteitis fibrosa [2].

As therapeutic agents, vitamin D metabolites have been used primarily to suppress the secretion of PTH by correcting the hypocalcemia as well as, in part, by acting directly on parathyroid cells. However, despite several therapeutic modalities available, it is still often difficult to suppress parathyroid hyperfunction, especially in patients with marked parathyroid hyperplasia [3]. Another problem is that successful suppression of parathyroid function does not necessarily mean normalization of bone disease in terms of both quality and quantity of bone [4]. In addition, relative hypoparathyroidism partially due to excessive treatment, has been recognized as a major problem that may lead to adynamic bone [5].

In this chapter, we focus on these problems and summarize the new therapeutic uses of vitamin D metabolites. Conventional uses of vitamin D in mild and advanced renal failure, including $1,25(OH)_2D_3$ pulse therapy, are discussed in Chapter 52; direct actions of 1,25-dihydroxyvitamin D_3 [$1,25(OH)_2D_3$] on parathyroid cells are discussed in Chapter 23. Here, we discuss the mechanisms of parathyroid gland resistance to

$1,25(OH)_2D_3$ and go on to discuss three treatment strategies: (1) selective percutaneous ethanol injection, (2) direct $1,25(OH)_2D_3$ injection into parathyroid hyperplasia, and (3) the use of vitamin D analogs to treat hyperparathyroidism in renal failure.

II. RESISTANCE TO $1,25(OH)_2D_3$ AS A CAUSE OF HYPERPARATHYROIDISM IN RENAL FAILURE

As decreased ionized calcium and/or decreased production of $1,25(OH)_2D$ are the major stimuli for PTH secretion in chronic renal failure, treatment of hyperparathyroidism has been aimed mainly at correcting these abnormalities. Thus, calcium carbonate as a phosphate binder is used to ameliorate hypocalcemia as well as to correct the hyperphosphatemia, and oral active vitamin D sterols are routinely used to maintain the physiological plasma concentration of $1,25(OH)_2D$ [1,3]. Despite normal plasma concentrations of calcium and $1,25(OH)_2D$ achieved by these treatment modalities, there are still many patients with high plasma PTH levels and with bone disease. Some of these patients may respond to supraphysiological concentrations of $1,25(OH)_2D_3$ achieved either by intermittent high doses of $1,25(OH)_2D_3$ or by so-called $1,25(OH)_2D_3$ pulse therapy [6–8]. These observations support the notion that the resistance of parathyroid cells to $1,25(OH)_2D$ may play a major role in the pathogenesis of secondary hyperparathyroidism in chronic renal failure [9].

Resistance to physiological concentrations of $1,25(OH)_2D$ may exist even from the early phase of chronic renal failure as shown in an animal model. In a rat model of mild renal failure, PTH secretion, PTH synthesis, and parathyroid cell proliferation were all enhanced even in the presence of normal plasma concentrations of calcium and $1,25(OH)_2D$ [10]. Hyperparathyroidism returned to normal with the administration of pharmacological doses of $1,25(OH)_2D_3$ without hypercalcemia. In these rats, $1,25(OH)_2D$ receptor (VDR) density in parathyroid glands, detected by Western blot, was reduced compared to normal rats.

The reduction of VDR density in parathyroid glands is currently considered the central abnormality responsible for the resistance of parathyroid glands to $1,25(OH)_2D$ treatment in chronic renal failure [11–13]. As shown in the rat model, VDR density may decrease even in the early phase of chronic renal failure. The mechanism for such a decrease in VDR density has been considered the result of disturbed up-regulation of this receptor, for which several steps of up-regulation have been implicated [14–20]. The subject of VDR regulation is discussed in Chapter 11.

III. TREATMENT WITH SUPRAPHYSIOLOGICAL DOSES OF $1,25(OH)_2D_3$ IN RENAL FAILURE

A. Advantages and Limitations of $1,25(OH)_2D_3$ Pulse Therapy

On the basis of a pathophysiological model of $1,25(OH)_2D_3$ resistance, a therapeutic strategy known as pulse therapy has been developed utilizing supraphysiological concentrations of $1,25(OH)_2D_3$, achieved either by intravenous or oral bolus administration of the metabolite. This therapy has been shown to be effective in suppressing PTH secretion in chronic dialysis patients resistant to conventional oral active vitamin D therapy, that is, those resistant to physiological concentrations of $1,25(OH)_2D_3$ [6–8]. However, the potential advantages of intravenous bolus administration over oral bolus administration still remain controversial [21]. Although 1α-hydroxyvitamin D_3 ($1\alpha OHD_3$) had also been tried in the same manner, the advantages of $1\alpha OHD_3$ pulse therapy over $1,25(OH)_2D_3$ pulse therapy have not been fully determined yet [22,23].

Because the peak concentration of $1,25(OH)_2D_3$ is more important in achieving suppression of PTH secretion than the total dose of $1,25(OH)_2D_3$, as shown in dialysis patients [6] and in experimental animals [24], higher doses of $1,25(OH)_2D_3$ might theoretically be more effective. However, high doses of $1,25(OH)_2D_3$ often cause hypercalcemia, which may cause ectopic calcification; pulse therapy should be discontinued if hypercalcemia develops. Thus, supraphysiological doses of $1,25(OH)_2D_3$ may not be without significant side effects for the treatment of severe hyperparathyroidism in chronic dialysis patients.

B. Parathyroid Size as a Marker for the Resistance to $1,25(OH)_2D_3$

Marked parathyroid hyperplasia is a unique feature of hyperparathyroidism in chronic dialysis patients [25]. The four glands differ in size even in the same patient.

However, a consensus has been recognized that the size of the largest gland roughly correlates with the length and severity of uremia and with the degree of prevailing plasma and stimulated peak PTH levels [26,27]. The size of the largest gland also correlates with the degree of abnormal control of PTH secretion [28,29], which may be normalized by $1,25(OH)_2D_3$ pulse therapy [30].

As we have reported earlier, these hyperplastic glands may regress to some extent with $1,25(OH)_2D_3$ pulse therapy [31]. Such regression in dialysis patients was not only confirmed by ultrasonography [32] but also suggested by observations made by scintigraphy [33]. Similar regression has also been demonstrated in experimental animals treated with intermittent $1,25(OH)_2D_3$ administration [24]. However, not all studies have found similar results [21,34]. To confirm our initial observation, we examined the changes in parathyroid size after $1,25(OH)_2D_3$ pulse therapy in another patient group with a longer history of dialysis and thus with more severe parathyroid hyperfunction. We confirmed that enlarged glands did regress as long as PTH secretion was suppressed by $1,25(OH)_2D_3$ pulse therapy [35].

Longer observation of changes in parathyroid size by ultrasonography revealed that the initial size of the largest gland is the critical marker for the long-term prognosis of vitamin D therapy [36]. If the largest gland is larger than 1 cm in diameter or about 0.5 cm^3 in volume, it is usually difficult to suppress PTH secretion by $1,25(OH)_2D_3$ pulse therapy. In such patients, enlarged glands never regress to normal size, and parathyroid hyperfunction always persists or relapses, even if it initially responded to $1,25(OH)_2D_3$ pulse therapy. In contrast, patients with only smaller glands usually respond to $1,25(OH)_2D_3$ pulse therapy well, and parathyroid function can be controlled with active vitamin D sterols. Thus, the size of the parathyroid glands may have more relevance than only the plasma PTH levels in assessing the efficacy of $1,25(OH)_2D_3$ pulse therapy [37].

Data from chronic dialysis patients who underwent parathyroidectomy also suggested that gland size is correlated with the degree of resistance to medical therapy [38]. The larger parathyroid glands tended to show higher proliferative potentials [39,40]. Tominaga and Takagi demonstrated that autoimplantation of tissue fragments from glands heavier than 0.5 g resulted in frequent relapse of parathyroid hyperfunction [41]. This weight exactly corresponds to a gland volume of 0.5 cm^3 or diameter of 1 cm as detected by ultrasonography. Thus, the size of the largest parathyroid gland reflects the resistance to active vitamin D therapy. The critical size to determine the management strategy for hyperparathyroidism in chronic dialysis patients seems to be 0.5 cm^3 in volume.

C. Role of Resistance to $1,25(OH)_2D$ in the Pathogenesis of Parathyroid Hyperplasia

The correlation between parathyroid gland size and resistance to $1,25(OH)_2D$ can also be explained by the degree of decrease in VDR density. In patients with severe parathyroid hyperfunction, large parathyroid glands are usually composed of nodular hyperplasia, a more advanced type than diffuse hyperplasia seen in smaller glands [42]. It has been reported that cells from nodular hyperplasia have more proliferative potential [39,40] and more abnormal regulation of PTH secretion [28] than those from diffuse hyperplasia.

We have shown that VDR content is more decreased in nodular hyperplasia than in diffuse hyperplasia [43]. Furthermore, VDR density is inversely correlated with the weight of the enlarged glands [43]. We have also demonstrated more direct evidence of increased parathyroid cell proliferation with a decrease in VDR density [44]. Because nodular hyperplasia is almost always the case in glands heavier than 0.5 g [41], the difference in the response to $1,25(OH)_2D_3$ that is dependent on gland size may be explained by this difference in histology. In other words, the difference in the response to vitamin D therapy based on gland size is due to the difference in VDR content, which is inversely correlated with the gland size. The large glands have nodular hyperplasia, decreased VDR content and resistance to $1,25(OH)_2D_3$ action.

Parathyroid cell growth is stimulated in chronic renal failure by several factors such as decreases in plasma concentrations of ionized calcium and $1,25(OH)_2D$ and increases in serum phosphate [45]. Along with these stimuli, VDR density in parathyroid cells decreases, and parathyroid cells become resistant to $1,25(OH)_2D$ and continue to grow, leading to diffuse hyperplasia. Cells in small nodules, formed within areas of diffuse hyperplasia, tend to have reduced VDR content compared with cells in the surrounding tissue [43,44]. Thus, cells with a more severe decrease inVDR proliferate more vigorously to eventually form nodular hyperplastic areas. Each nodular hyperplastic gland is initially composed of several monoclonal nodules of different origin [46,47], one of which may grow more vigorously than others and may finally form most of the enlarged gland. The biochemical and molecular mechanisms responsible for the differences of proliferative potency among these nodules remain to be elucidated. In a small number of patients, gene rearrangement may occur in some cells, probably within nodular hyperplasia, leading to autonomous proliferation [48]. Such gene rearrangements in parathyroid glands in dialysis patients have not been analyzed in sufficient depth compared with those of primary hyperparathyroidism [41,49].

IV. MANAGEMENT OF HYPERPARATHYROIDISM IN PATIENTS WITH MARKED PARATHYROID HYPERPLASIA

A. Principles for the Management of Hyperparathyroidism

Based on experimental and clinical observations, three principles should be applied to the general management of hyperparathyroidism in chronic dialysis patients. The first principle is to prevent parathyroid hyperplasia, starting from the early phase of renal failure [37,45]. This can be achieved by dietary phosphate restriction, early use of phosphate binders, and cautious use of active vitamin D sterols [50].

The second principle is to routinely evaluate parathyroid size in order to decide on the therapeutic options [36,37]. It is of no use or sometimes even dangerous to continue $1,25(OH)_2D_3$ pulse therapy in patients with glands that are larger than 0.5 cm^3. Thus, gland size should be checked before or during $1,25(OH)_2D_3$ pulse therapy if suppression of hyperparathyrodism, judged by monitoring plasma PTH levels, is not satisfactory.

The third principle is not to suppress PTH too much [4,51]. This is very important in order to avoid the risk of developing adynamic bone disease, which is recognized as a serious problem [5,52]. Relative hypoparathyroidism due to $1,25(OH)_2D_3$ pulse therapy has been considered responsible, at least in part, for this type of bone abnormality. In addition, direct effects of $1,25(OH)_2D_3$ on osteoblasts may play a role in the development of adynamic bone disease as suggested by *in vitro* studies [53,54]. Thus, intact PTH levels should be maintained at not less than 200–250 pg/ml [51], and serum markers for bone turnover, such as serum alkaline phosphatase activity, should be monitored.

B. Selective Percutaneous Ethanol Injection as an Adjunct to $1,25(OH)_2D_3$ Pulse Therapy

As patients with very enlarged parathyroid glands are resistant to $1,25(OH)_2D_3$, can these patients be managed medically without surgical parathyroidectomy? Even in such patients, the small glands should still be responsive to $1,25(OH)_2D_3$. By improving the technique of percutaneous ethanol injection (PEIT) into parathyroid glands under ultrasonographic guidance, we have selectively destroyed large glands that are resistant to $1,25(OH)_2D_3$ pulse therapy, thus leaving only small glands which are responsive to $1,25(OH)_2D_3$ [55]. PEIT was originally introduced as a practical procedure by Gianglande and co-workers and has shown a 60% success rate over 10 years [56–59]. In our protocol, flow mapping by Doppler ultrasonography, specifically tuned for detection of the blood supply to parathyroid glands, is routinely used to confirm the destruction of parathyroid tissues and to detect the remaining tissues and the relapse of cell proliferation. After successful destruction of large glands by several ethanol injections, PTH secretion became controllable within desirable ranges (intact PTH 150–250 pg/ml) by the following $1,25(OH)_2D_3$ pulse therapy and conventional oral active vitamin D therapy [55]. Thus, this therapy can be an alternative for total parathyroidectomy, which is usually done with autoimplantation of the fragments from the smallest gland, thus composed of the least proliferative cells [40].

There are several possible side effects of PEIT therapy. Recurrent nerve palsy caused by excess ethanol is a critical problem reportedly seen in 10% of patients [57]. We have successfully avoided this complication by injecting smaller amounts of ethanol in repeated injections. Another concern is the possibility that subsequent surgical removal of parathyroid glands which had undergone ethanol injection, should it become necessary, may be more difficult due the fibrosis of the glands and other surrounding tissues. By reducing the amount of ethanol to be injected and by optimizing the site of injection, we believe the risk of such complications should be minimal. In addition, excessive suppression of parathyroid function was not seen in the long term because we have destroyed only minimal amounts of parathyroid tissue. The only problem that we noted was the transient but severe "hungry bone syndrome" after PEIT in patients with a single very large glands. This complication should be managed by large doses of active vitamin D sterols.

In spite of such favorable results, we still note several areas that can be improved from the initial PEIT protocol. First, in our experience, all glands larger than 0.5 cm^3 need to be destroyed for the long-term control of this condition by medical therapy. Second, residual cells that have survived the first ethanol injection may proliferate within a few weeks before the next injection. Third, medical therapy after PEIT was also very important. Considering these problems, we have adopted a new protocol, which we refer to as repeated intensive

PEIT of parathyroid (RIPP) [60]. In this protocol all glands larger than 0.5 cm^3 are destroyed within a 1-week interval by no more than three injections. Usually, the first step is to inject a volume of ethanol that corresponds to 70% of the calculated gland volume. Next, a small amount of ethanol is injected into the optimum site determined by color Doppler flow mapping. Third, complete destruction of the tissue is also confirmed by this technique. In our preliminary trial of this protocol, all patients became responsive to $1,25(OH)_2D_3$ pulse therapy without any complications. Thus, selective PEIT should be tried in patients with parathyroid hyperplasia larger than 0.5 cm^3, who are refractory to $1,25(OH)_2D_3$ pulse therapy. The algorithm for the medical management of hyperparathyroidism in dialysis patients is shown in Fig. 1.

C. Direct $1,25(OH)_2D_3$ Injection into Parathyroid Hyperplasia

Theoretically, a higher concentration of $1,25(OH)_2D_3$ should be needed to suppress the function of parathyroid glands with lower numbers of $1,25(OH)_2D$ receptors; however, higher doses of $1,25(OH)_2D_3$ usually lead to marked hypercalcemia [1,3]. One solution for this problem is to develop new vitamin D analogs that are more specific for parathyroid gland suppression with less effect on serum calcium [61]. Another solution is to develop a method to deliver $1,25(OH)_2D_3$ exclusively to the parathyroid glands, thus avoiding the systemic effects of $1,25(OH)_2D_3$.

To achieve high concentrations of $1,25(OH)_2D_3$ only locally within parathyroid glands, we repeatedly injected $1,25(OH)_2D_3$ solution (1 μg/ml) directly into glands larger than 0.5 cm^3 [62]. Six injections within 2 weeks suppressed PTH hypersecretion in all patients, as shown in Fig. 2. In most patients, responsiveness to $1,25(OH)_2D_3$ therapy was restored after direct $1,25(OH)_2D_3$ injections. Furthermore, during the oral $1,25(OH)_2D_3$ pulse therapy which followed the direct $1,25(OH)_2D_3$ injection, significant regression of enlarged glands was also observed. On the other hand, the uninjected glands, which were smaller and thus should have been more responsive to the pulse therapy, regressed only minimally during the same period.

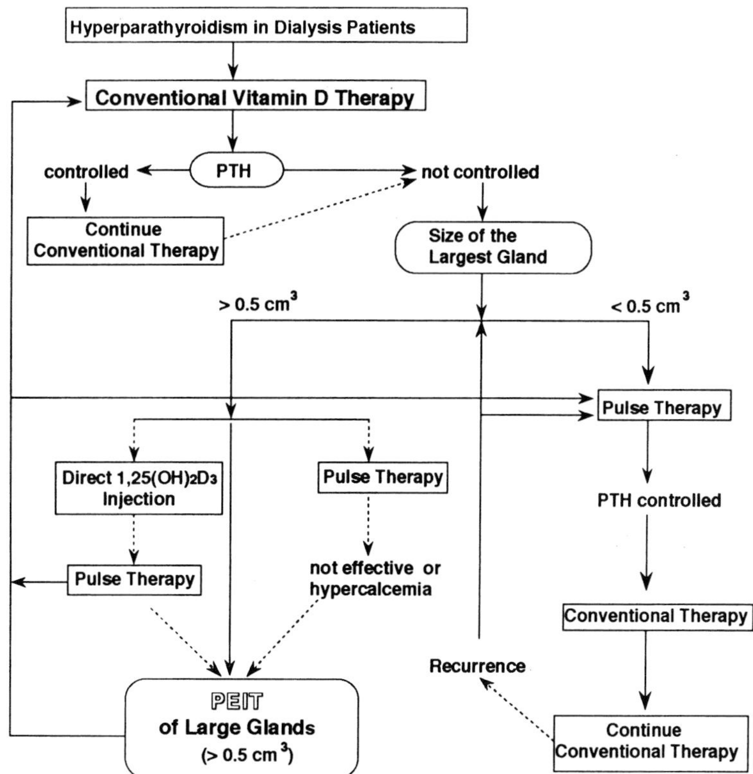

FIGURE 1 Medical management of hyperparathyroidism in chronic dialysis patients. The goal of PTH suppression should be circulating PTH levels between 150 to 200 pg/ml as determined by an intact PTH assay. Total parathyroidectomy with autoimplantation of the fragment from the smallest gland can also be used instead of PEIT of glands larger than 0.5 cm^3.

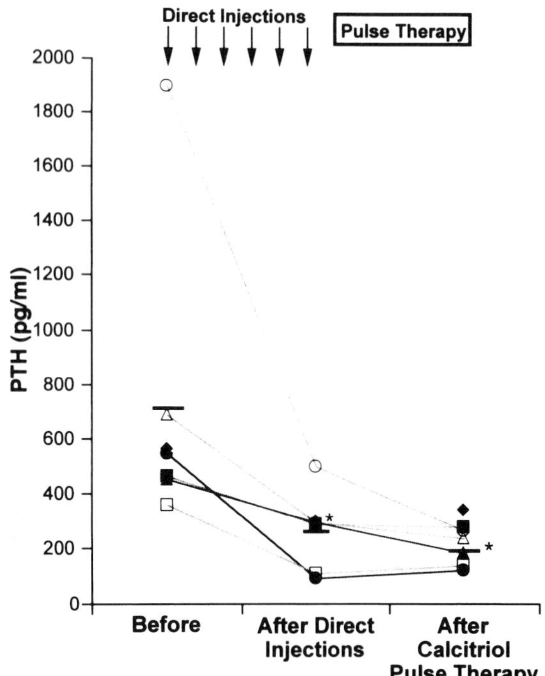

FIGURE 2 Suppression of PTH by direct $1,25(OH)_2D_3$ injections. PTH decreased after six direct $1,25(OH)_2D_3$ injections and was further decreased by subsequent pulse therapy. Data from eight patients [62] are shown.

Thus, in this protocol of direct injection into large parathyroid glands, suppression of hyperparathyroidism as judged by a fall in serum PTH and regression of parathyroid hyperplasia were considered mainly to be the result of the direct injection of $1,25(OH)_2D_3$ rather than systemic effects of the injected hormone.

In our direct injection protocol, parathyroid cells were exposed to 1 μg/ml of $1,25(OH)_2D_3$ at least transiently. This is a much higher concentration than the peak concentration achievable by the usual intravenous pulse therapy, which is 500 pg/ml peak induced by 2 μg $1,25(OH)_2D_3$ injected intravenously [6]. We chose the concentration of 1 μg/ml because it was the only preparation available to us for human use and because lower doses used in a pilot study, by injecting diluted preparations, were not effective. Although we did not monitor the serum concentration of $1,25(OH)_2D_3$ serially after direct injection in every patient, we do not consider that the increase in $1,25(OH)_2D_3$ in the circulation diffusing out from injected glands should impose significant clinical problems. This is becuse the dose of injected $1,25(OH)_2D_3$ per day was comparable to the dose used in intravenous $1,25(OH)_2D_3$ pulse therapy [6,7]. Also, in our experience thus far no evidence of hypercalcemia was observed, in sharp contrast to oral or intravenous $1,25(OH)_2D_3$ pulse therapy.

The underlying mechanisms for the effectiveness of this therapy can be divided into two parts. First, parathyroid cells with lower VDR density within enlarged glands responded to the high local concentration of $1,25(OH)_2D_3$, although we have no information on how long such a high local concentration could be maintained. A more important finding is that hyperparathyroidism, which was not controllable by $1,25(OH)_2D_3$ pulse therapy previous to this direct $1,25(OH)_2D_3$ injection, became responsive to the pulse therapy. Thus, as the second mechanism, the high concentration of $1,25(OH)_2D_3$ may have up-regulated VDR density [14,15] in parathyroid cells, a mechanism which has been shown to be impaired in chronic renal failure [15–20]. This leads these cells to become responsive to $1,25(OH)_2D_3$ pulse therapy.

In addition, two other theoretical mechanisms may play a small role in this therapy. One mechanism is direct parathyroid cell toxicity of the small amounts of organic solvent and detergent contained in the vehicle used to dilute the $1,25(OH)_2D_3$. Ethical considerations prevented us from performing vehicle injections. Thus, the possibility could not be completely ruled out. However, this vehicle was different from that used in Calcijex and has been developed to comply with Japanese regulations on drug safety. In addition, because repeated injections of gradually increasing doses of $1,25(OH)_2D_3$ with saline-diluted preparations or a single injection of undiluted preparation during the pilot study had no effect on parathyroid function, the contribution of this potential mechanism seems to be minor. The second theoretical mechanism is the possible elevation of intracapsular pressure caused by the injected fluid. However, this also seems unlikely because injections of diluted $1,25(OH)_2D_3$ preparations with the same calculated volume during the pilot study did not suppress hyperparathyroidism.

As for the side effects, there were no episodes of severe pain or recurrent nerve palsy after the direct $1,25(OH)_2D_3$ injections so far. Such side effects have been encountered, often transiently, in a few patients after ethanol injection [55,57]. Thus, direct injection of $1,25(OH)_2D_3$ is theoretically and practically an effective and safe method to control severe parathyroid hyperfunction in chronic dialysis patients.

Nevertheless, there are several aspects of the protocol which need further improvement. The first point is that multiple injections were necessary to control hyperparathyroidism. This observation may suggest the possibility that only a portion of the parathyroid cells are responsive to $1,25(OH)_2D_3$ at any given time, probably depending on their cell cycle. A second point is that the effect of the direct injection itself appeared transient, and $1,25(OH)_2D_3$ pulse therapy should follow after a series of direct injections. Third, this therapy may not be effective in those patients with the most severe secondary hyperparathyroidism accompanied by multiple very large glands. Such patients should be controlled either by ethanol injection or parathyroidectomy. However, these three problems might be solved by new preparations with higher $1,25(OH)_2D_3$ concentration; alternatively, future new analogs of vitamin D may be developed that are more selective for the parathyroid cell or may stay within parathyroid glands longer.

Despite these limitations, direct injection of $1,25(OH)_2D_3$ into parathyroid glands should remain an important means to control hyperparathyroidism. Genetic mutation leading to marked parathyroid hyperplasia has been clearly elucidated in primary hyperparathyroidism, in which cyclin D overexpression by gene rearrangement leads to uncontrolled proliferation of parathyroid cells. In such cases, the cyclin D gene is under control of the 5′ promoter sequence of the PTH gene which contains the vitamin D responsive elements [49]. Thus, it may be theoretically possible to suppress cyclin D gene transcription through this translocated promoter sequence of the PTH gene with analogs of $1,25(OH)_2D_3$ that have higher affinity for the VDR, thus avoiding the systemic effects of vitamin D. Furthermore, this direct route of administration of therapeutic agents to parathyroid glands may also be suitable for selective gene therapy of parathyroid glands in the future, as we have reported in preliminary form [63].

V. CLINICAL USE OF NEW VITAMIN D ANALOGS IN RENAL FAILURE

A. Ideal Vitamin D Analogs for the Treatment of Hyperparathyroidism in Renal Failure

Several endogenous or synthetic vitamin D analogs have become available, and some of them appear to be promising for the treatment of hyperparathyroidism in renal failure. Disadvantages of the available active vitamin D sterols, namely, $1,25(OH)_2D_3$ and $1\alpha OHD_3$, are mainly due to their systemic effects, especially on the intestine, which often lead to hypercalcemia and hyperphosphatemia [1,3]. Hyperphosphatemia or phosphate load itself may directly stimulate PTH secretion and may increase the risk of ectopic calcification in combination with hypercalcemia [64–67]. It has also been suggested, in addition to calcium and $1,25(OH)_2D$ [68], that phosphate may modulate VDR density [69], thereby modulating target cell sensitivity to active vitamin D therapy. Thus, the development of new vitamin D analogs that exhibit a more specific effect to inhibit parathyroid function would provide an improvement in the management of hyperparathyroidism in chronic renal failure.

On the other hand, some data suggest that the suppression of PTH does not necessarily lead to normalization of bone histology [2,21]. Rather, relative hypoparathyroidism caused by such therapy may lead to adynamic bone disease [5,51,52]. Thus, vitamin D analogs that have the capacity to normalize the abnormality of chronic renal failure and preferably exhibit less suppressive effects on parathyroid function are also needed. In this regard, 22-oxa-$1,25(OH)_2D_3$ and $24,25(OH)_2D_3$ are discussed. Because detailed data on the chemical and biological properties of these and other analogs are discussed in the chapters in Section VII of this book, we focus here on the uses of the drugs in uremic patients.

B. 22-Oxa-$1,25(OH)_2D_3$

The compound 22-oxa-$1,25(OH)_2D_3$ (OCT) was discovered initially as a synthetic analog of vitamin D that may mimic certain actions of $1,25(OH)_2D_3$ on cell differentiation [70,71] (see Chapter 63). This analog was soon recognized as a nonhypercalcemic agent for the treatment of hyperparathyroidism. In addition to the initial reports by Brown and associates in normal rats [61] and in uremic dogs [72], the usefulness of this derivative in the management of hyperparathyroidism in chronic renal failure was confirmed by other groups [10,73]. Thus, enhanced PTH secretion and elevated PTH mRNA levels in uremic rats were suppressed to normal by 22-oxa-$1,25(OH)_2D_3$ without causing hypercalcemia [10]. In contrast to the original report by Brown and associates, some data suggest that the effect of 22-oxa-$1,25(OH)_2D_3$ to suppress PTH secretion is slightly weaker than $1,25(OH)_2D_3$, partially because of the absence of hypercalcemia with 22-oxa-$1,25(OH)_2D_3$ at the dose used [73].

Such selective action of 22-oxa-$1,25(OH)_2D_3$ on the parathyroids has been partly explained by its rapid metabolic clearance due to the very low affinity for vitamin D-binding protein (DBP) [74]. In addition, it is possible that the suppression of PTH may contribute to the reduced calcemic effect of 22-oxa-$1,25(OH)_2D_3$, has been suggested [75]. It has also been shown that 22-oxa-$1,25(OH)_2D_3$ does not affect phosphorus metabolism [76]. Thus, 22-oxa-$1,25(OH)_2D_3$ is likely to be quite useful for the management of hyperparathyroidism in chronic renal failure.

On the basis of these experimental data, clinical trials using 22-oxa-$1,25(OH)_2D_3$ in chronic dialysis patients are now under way. In a preliminary report from Japan

[77], the compound in doses as low as 5.5 μg per hemodialysis, was effective in suppressing PTH, alkaline phosphatase, and tartate-resistant acid phosphatase in three patients (Fig. 3). However, hypercalcemia was observed with higher doses of 22-oxa-1,25$(OH)_2D_3$ in two patients. In cases with severe hyperparathyroidism, a very high dose was needed to suppress PTH secretion. Thus, the risk/benefit ratio and the dose must be further evaluated and optimized by additional clinical studies.

C. 24*R*,25$(OH)_2D_3$

24*R*,25-Dihydroxyvitamin D [24,25$(OH)_2$D] is a major endogenous vitamin D metabolite, and its plasma concentration is 1000 times greater than that of 1,25$(OH)_2$D. The physiological role of 24,25$(OH)_2$D has still not been fully elucidated, though this vitamin D metabolite may be essential for normal bone formation [78,79] (see also Chapters 18 and 19). It has less effect on parathyroid function than 1,25$(OH)_2$D [80,81]. Although somewhat controversial, reports suggest that 24,25$(OH)_2$D may have unique direct effects on bone, causing bone mass gain in vitamin D-repleted rats [82] and in the Hyp mouse [83]. Such effects were also demonstrated to a lesser degree in patients with X-linked hypophosphatemic rickets [84]. As discussed in Chapter 18, data on abnormal bone development in 24-hydroxylase knock-out mice also supports the unique effects of this analog on bone [85].

In advanced chronic renal failure, the production of 24,25$(OH)_2$D, as well as that of 1,25$(OH)_2$D, is extremely reduced because high levels of PTH and decreased levels of 1,25$(OH)_2$D suppress 24-hydroxylase activity within the remaining renal tissue [86]. As suggested by Lambrey and associates, treatment with 1αOHD$_3$ may be effective in normalizing bone resorption only in the presence of normal serum levels of 24,25$(OH)_2$D [87]. Thus, the reduction of circulating 24,25$(OH)_2$D in chronic renal failure may play some role in the pathogenesis of uremic bone disease. However, probably due to the heterogeneous nature of renal osteodystrophy, the usefulness of 24*R*,25$(OH)_2D_3$ alone or in combination with 1,25$(OH)_2D_3$ or 1αOHD$_3$ has been controversial in terms of both the suppression of parathyroid function and the amelioration of bone abnormalities [81,88–91].

Reports by Popovtzer and associates showed in both

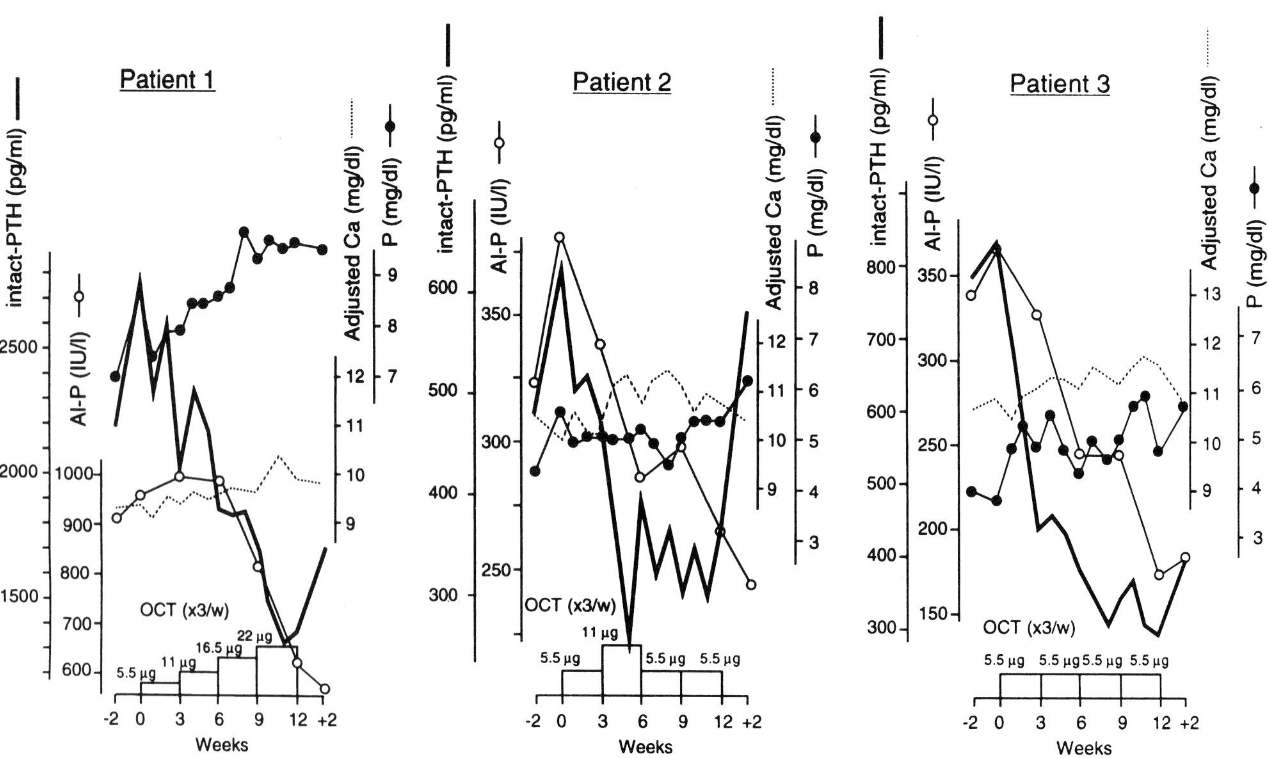

FIGURE 3 Treatment of hyperparathyroidism with 22-oxa-1,25$(OH)_2D_3$ (OCT) in three dialysis patients. Therapy with 22-oxa-1,25$(OH)_2D_3$ was initiated at a dose of 5.5 μg after each dialysis. The dose of 22-oxa-1,25$(OH)_2D_3$ was able to be increased to 22.0 μg in patient 1, but it could not be increased in patients 2 and 3 due to the development of hypercalcemia. PTH levels and serum alkaline phophatase (Al-P) activity were significantly suppressed in all three patients while phosphate levels rose. Adapted from Kurokawa *et al.* [77] with permission.

dialysis patients [92,93] and uremic rats [94] that the combined use of 24*R*,$25(OH)_2D_3$ with an active vitamin D compound effectively reduced bone resorption without affecting bone formation. Renormalization of bone formation and mineralization have been also reported by Birkenhager-Frenkel and associates [95]. The cellular mechanism of 24*R*,$25(OH)_2D_3$ action remains to be elucidated. The metabolite has been shown to exhibit inhibitory effects on the differentiation and function of osteoclastic cells *in vitro* [96]. Also, this compound might possibly regulate bone metabolism by changing the sensitivity of bone cells to other calcium-regulating factors, including PTH [97].

Whatever the mechanisms involved, it seems quite clear that 24*R*,$25(OH)_2D_3$, at high doses, is able to positively influence the net turnover of bone formation and resorption in chronic renal failure. Microfocus X-ray tomography of the femoral bone clearly revealed that 24*R*,$25(OH)_2D_3$ dramatically corrected the cancellous bone loss in uremic rats [98,99] (Fig. 4), an effect that could not be detected by conventional measurement of bone mineral density. Another possible advantage of 24*R*,$25(OH)_2D_3$ is that it may ameliorate hypercalcemia and help to increase the doses of $1,25(OH)_2D_3$ or $1\alpha OHD_3$ that can be given [88]. In addition, weaker effects on parathyroid function [80] may also be an advantage with less risk of inducing adynamic bone disease. Thus, 24*R*,$25(OH)_2D_3$ may be an effective and safe adjunct to the use of active vitamin D therapy.

The compound 24*R*,$25(OH)_2D_3$ has already been used therapeutically in Israel, and clinical trials are in progress in the United States and in Japan. However, to prove the clinical advantages of this analog, careful analysis of these clinical trials is essential. Also, a more sensitive method for the measurement of bone mass than conventional DEXA would be helpful.

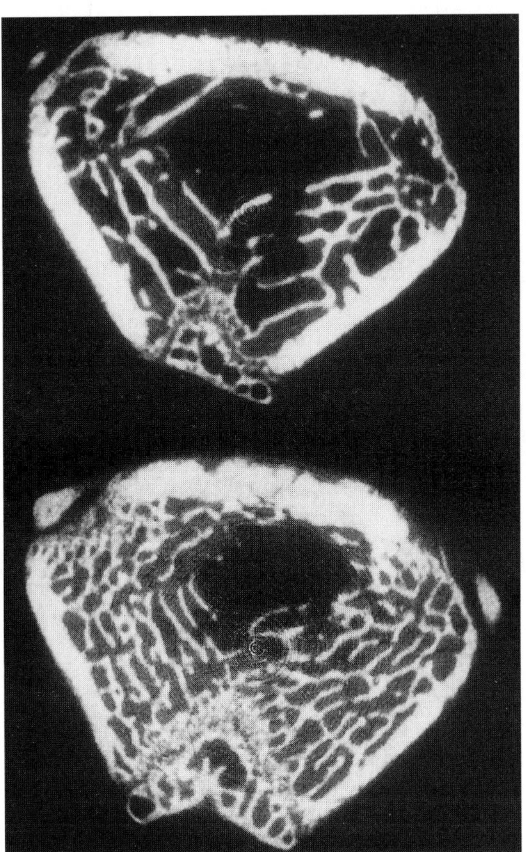

FIGURE 4 Prevention of bone loss in uremic rats with 24*R*,$25(OH)_2D_3$ therapy using microfocus X-ray tomography. Significant trabecular bone loss was observed by microfocus X-ray tomography in the femur of uremic rats (top) at 24 weeks after 7/8 nephrectomy. Therapy with 500 μg/kg of 24*R*,$25(OH)_2D_3$ was given for 16 weeks beginning at 8 weeks after partial nephrectomy. Trabecular bone volume was significantly increased (bottom), even when compared to normal rats (not shown). Courtesy of Kureha Chemical Industry Co. Ltd., Tokyo, and NS Elex Co. Ltd., Sakai, Japan.

D. Other Metabolites

Several new vitamin D analogs, such as 26,27-hexafluoro-$1\alpha,25(OH)_2D_3$, $1,24(OH)_2D_3$, 23(*S*),25(*R*)-$1\alpha,25(OH)_2D_3$-26,23-lactone. 1α,25-dihydroxy-16-ene-23-yne-D_3, 2β-(3-hydroxypropoxy)-$1\alpha,25(OH)_2D_3$, and $1,24(OH)_2$-Δ^{22}-25,26,27-cyclopropyl-D_3, have been proposed for the use in the management of hyperparathyroidism in chronic renal failure. Biological and chemical features of these vitamin D analogs are discussed elsewhere in this book. Their clinical advantages remain to be critically determined by well-designed laboratory and clinical studies.

The analog 26,27-hexafluoro-$1\alpha,25(OH)_2D_3$ is a new vitamin D metabolite whose biological activity has been shown to be 10 times stronger and longer than $1,25(OH)_2D_3$ [100–102]. It has been suggested that 26,27-hexafluoro-$1\alpha,25(OH)_2D_3$ is metabolized more slowly within the target cells, and thus that it acts longer with lower doses [103]. It was also shown in nonclassic target cells (HeLa) as well as classic target cells (UMR-106) that the transcriptional activity of 26,27-hexafluoro-$1\alpha,25(OH)_2D_3$ was two times stronger than $1,25(OH)_2D_3$ [104] despite its lower binding affinity for VDR [105]. Thus, 26,27-hexafluoro-$1\alpha,25(OH)_2D_3$ may be a potent agent to suppress parathyroid function at lower doses than $1,25(OH)_2D_3$ and might be less likely to cause hypercalcemia. In preliminary experiments with uremic rats, elevated PTH secretion and synthesis returned to normal after a 6-day treatment period with either 12.5 or 25 pg/kg body weight of 26,27-hexafluoro-$1\alpha,25(OH)_2D_3$. There were no changes noted in serum concentrations of ionized calcium and phosphate. However, the risk of hypercalcemia with higher doses or by longer administration of this analog could not be excluded. Although preliminary clinical data have already been reported [106], the potential advantages of this vitamin D metabolite over $1,25(OH)_2D_3$ remain to be elucidated by further study.

Acknowledgments

This work was in part supported by grants from the Ministries of Education, Science and Culture and of Health and Welfare of Japan. We are greatful to Hung Yi, Naoko Fukuda, and Junichiro Kazama for their contribution to our works cited in this review and to Yoshiki Seino, Yoshihiro Tominaga, and Hidehiro Ozawa for collaboration.

References

1. Coburn JW, and Llach F 1994 Renal osteodystrophy. In: Narins NG (ed) Clinical Disorders of Fluid and Electrolyte Metabolism, 5th Ed, McGraw-Hill, New York, pp. 1299–1377.
2. Malluche H, Faugere MC 1990 Renal bone disease 1990: An unmet challenge for the nephrologist. Kidney Int **38**:193–211.
3. Akizawa T, Fukagawa M, Koshikawa S, Kurokawa K 1993 Recent progress in management of secondary hyperparathyroidism of chronic renal failure. Curr Opin Nephrol Hypertens **2**:558–565.
4. Quarles LD, Lobaugh B, Murphy G 1992 Intact parathyroid hormone overestimates the presence and severity of parathyroid-mediated osseous abnormalities in uremia. J Clin Endocrinol Metab **75**:145–150.
5. Hercz G, Pei Y, Greenwood C, Manuel A, Saiphoo C, Goodman WG, Segre GV, Fenton S, Sherrard DJ 1993 Aplastic osteodystrophy without aluminum: The role of "suppressed" parathyroid function. Kidney Int **44**:860–866.
6. Slatopolsky EA, Weerts C, Thielan J, Horst R, Harter H, Martin KJ 1984 Marked suppression of secondary hyperparathyroidism by intravenous administration of 1,25-dihydroxycholecalciferol in uremic patients. J Clin Invest **74**:2136–2143.
7. Andress DL, Norris KC, Coburn JW, Slatopolsky EA, Sherrard DJ 1989 Intravenous calcitriol in the treatment of refractory osteitis fibrosa of chronic renal failure. N Engl J Med **321**:274–279.
8. Tsukamoto Y, Nomura M, Takahashi Y, Takagi Y, Yoshida A, Nagaoka T, Togashi K, Kitawada R, Marumo F 1990 The 'oral 1,25-dihydroxyvitamin D_3 pulse therapy' in hemodialysis patients with severe secondary hyperparathyroidism. Nephron **56**: 368–373.
9. Fukagawa M, Fukuda N, Yi H, Kurokawa K 1995 Resistance to calcitriol as a cause of parathyroid hyperfunction in chronic renal failure. Nephrol Dial Transplant **10**:316–319.
10. Fukagawa M, Kaname S, Igarashi T, Ogata E, Kurokawa, K 1991 Regulation of parathyroid hormone synthesis in chronic renal failure in rats. Kidney Int **39**:874–881.
11. Korkor AB 1987 Reduced binding of $[^3H]$1,25-dihydroxyvitamin D_3 in the parathyroid glands of patients with renal failure. N Engl J Med **316**:1573–1577.
12. Merke J, Hugel U, Zlotowski A, Szabo A, Bommer J, Mall G, Ritz E 1987 Diminished parathyroid $1,25(OH)_2D_3$ receptors in experimental uremia. Kidney Int **32**:350–353.
13. Brown A, Dusso A, Lopez-Hilker S, Lewis-Finch J, Grooms P, Slatopolsky EA 1989 $1,25(OH)_2D$ receptors are decreased in parathyroid glands from chronically uremic dogs. Kidney Int **35**:19–23.
14. Strom M, Sandgrin ME, Brown TA, DeLuca HF 1989 1,25-Dihydroxyvitamin D_3 up-regulates the 1,25-dihydroxyvitamin D_3 receptor in vivo. Proc Natl Acad Sci USA **86**:9770–9773.
15. Naveh-Many T, Marx R, Keshet E, Pike JW, Silver J 1990 Regulation of 1,25-dihydroxyvitamin D_3 receptor gene expression by 1,25-dihydroxyvitamin D_3 in the parathyroid in vivo. J Clin Invest **86**:1968–1975.
16. Shvil Y, Naveh-Many T, Brach P, SIlver J 1990 Regulation of parathyroid cell gene expression in experimental uremia. J Am Soc Nephrol **1**:99–104.
17. Koyama H, Nishizawa Y, Inaba M, Hino M, Prahl JM, DeLuca HF, Morii H 1994 Impaired homologous upregulation of vitamin D receptor in rats with chronic renal failure. Am J Physiol **266**:F706–712.
18. Patel SR, Ke HQ, Hsu CH 1994 Regulation of calcitriol receptor and its mRNA in normal and renal failure rats. Kidney Int **45**:1020–1027.
19. Hsu CH, Patel SR, Young EW, Vanholder R 1994 The biological action of calcitriol in renal failure. Kidney Int **46**:605–612.
20. Patel SR, Ke H-Q, Vanholder R, Koenig RJ, Hsu CH 1995 Inhibition of calcitriol receptor binding to vitamin D responsive elements by uremic toxins. J Clin Invest **96**:50–59.
21. Quarles LD, Yohay DA, Carroll BA, Spritzer CE, Minda SA, Bartholomay D, Lobaugh BA 1994 Prospective trial of pulse oral versus intravenous calcitriol treatment of hyperparathyroidism in ESRD. Kidney Int **45**:1710–1721.
22. Lgunghall S, Alttoff P, Fellstrom B, Wide L 1990 Effect on serum PTH of intravenous treatment with alfacalcidol in patients on chronic hemodialysis. Nephron **35**:380–385.
23. Brandi L, Daugaard H, Tuedegaard E, Nielsen PK, Olgaard K 1992 Long term suppression of secondary hyperparathyroidism by intravenous 1α-OH vitamin D_3 in patients on chronic hemodialysis. Am J Nephrol **12**:311–318.
24. Reichel H, Szabo A, Uhl J, Pesian S, Schmultz A, Schmidt-Gayk H, Ritz E 1993 Intermittent versus continuous administration of 1,25-dihydroxyvitamin D_3 in experimental renal hyperparathyroidism. Kidney Int **44**:1259–1265.
25. Parfitt AM 1994 Parathyroid growth: Normal and abnormal. In: Bilezikian JP, Marcus R, Levine MA (eds) The Parathyroids: Basic and Clinical Concepts. Raven, New York, pp. 373–406.
26. Malmaeus J, Glimelius L, Johansson G, Akerstrom G, Ljunghall S 1984 Parathyroid pathology in hyperparathyroidism secondary to chronic renal failure. Scand J Urol Nephrol **18**:157–166.
27. McCarron DA, Muther RS, Lenfesty B, Bennett WM 1982 Parathyroid function in persistent hyperparathyroidism: Relationship to gland size. Kidney Int **22**:662–670.
28. Wallfelt CH, Larsson R, Gylfe E, Ljunghall S, Rasted J, Akerstrom 1988 Secretory disturbance in hyperplastic parathyroid nodules of uremic hyperparathyroidism: Implication for parathyroid autotransplantation. World J Surg **12**:431–438.
29. Felsenfeld AJ, Llach F 1993 Parathyroid function in chronic renal failure. Kidney Int **43**:771–789.
30. Delmez JA, Tindra C, Grooms P, Dusso A, Windus DW, Slatopolsky E 1989 Parathyroid hormone suppression by intravenous $1,25(OH)_2$ vitamin D. A role for increased sensitivity to calcium. J Clin Invest **83**:1349–1355.
31. Fukagawa M, Okazaki R, Takano K, Kaname S, Ogata E, Kitaoka M, Harada S, Sekine N, Matsumoto T, Kurokawa K 1990 Regresson of parathyroid hyperplasia by calcitriol-pulse

therapy in patients on long-term dialysis. N Engl J Med **323**:421–422,
32. Hyodo T, Koumi T, Ueda M, Miyagawa I, Kodani K, Doi S, Ishibashi M, Takemoto M 1991 Can oral 1,25$(OH)_2D_3$ therapy reduce parathyroid hyperplasia? Nephron **59**:171–172.
33. Cannella G, Bonucci E, Rolla D, Ballanti P, Moriero E, De Grandi R, Augeri C, Claudiani F, Di Maio G 1994 Evidence of healing of secondary hyperparathyroidism in chronically hemodialyzed uremic patients treated with long-term intravenous calcitriol. Kidney Int **46**:1124–1132.
34. Szabo A, Merke J, Beier E, Mall G, Ritz E 1989 1,25$(OH)_2$ vitamin D_3 inhibits parathyroid cell proliferation in experimental uremia. Kidney Int **35**:1049–1056.
35. Fukagawa M, Fukuda N, Yi H, Kitaoka M, Kurokawa K 1996 Derangement of parathyroid function in renal failure: Biological and clinical aspects. J Nephrol **9**:219–224.
36. Fukagawa M, Kitaoka M, Yi H, Fukuda N, Matsumoto T, Ogata E, Kurokawa K 1994 Serial evaluation of parathyroid size by ultrasonography is another useful marker for the long-term prognosis of calcitriol pulse therapy in chronic dialysis patients. Nephron **68**:221–228.
37. Fukagawa M, Kitaoka M, Fukuda N, Yi H, Kurokawa K 1995 Pathogenesis and management of parathyroid hyperplasia in chronic renal failure: Role of calcitriol. Miner Electrolyte Metab **21**:97–100.
38. Akerstrom G, Malmaeus J, Grimelius L, Ljunghall S, Bergstrom R 1984 Histological changes in parathyroid glands in subclinical and clinical renal disease. Scand J Urol Nephrol **18**:75–84.
39. Tominaga Y, Grimelius L, Falkmer UG, Johansson H, Falkmer S 1991 DNA ploidy pattern of parathyroid parenchymal cells in renal secondary hyperparathyroidism with relapse. Anal Cell Pathol **3**:325–333.
40. Tominaga Y, Tanaka Y, Sato K, Numano M, Uchida K, Falkmer U, Grimelius L, Johansson H, Takagi H 1992 Recurrent renal hyperparathyroidism and DNA analysis of autografted parathyroid tissue. World J Surg **16**:595–603.
41. Tominaga Y, Takagi H 1996 Molecular genetics of hyperparathyroid disease. Curr Opin Nephrol Hypertens **5**:336–341.
42. Tominaga Y, Sato K, Tanaka Y, Numano M, Uchida K, Takagi H 1995 Histopathology and pathophysiology of secondary hyperparathyroidism due to chronic renal failure. Clin Nephrol **44**:S42–S47.
43. Fukuda N, Tanaka H, Tominaga Y, Fukagawa M, Kurokawa K, Seino Y 1993 Decreased 1,25-dihydroxyvitamin D_3 receptor density is associated with a more severe form of parathyroid hyperplasia in chronic uremic patients. J Clin Invest **92**:1436–1443.
44. Koike T, Fukuda N, Fukagawa M, Ota K, Kurokawa K 1996 Decreased density of vitamin D receptor directly correlates with the proliferative potency of parathyroid cells in hyperplasia. J Am Soc Nephrol **7**:1815 (Abstract).
45. Druecke TB 1995 The pathogenesis of parathyroid gland hyperplasia in chronic renal failure. Kidney Int **48**:259–272.
46. Tominaga Y, Kohara S, Namii Y, Nagasaka T, Haba T, Uchida K, Numano M, Tanaka Y, Takagi H 1996 Clonal analysis of nodular hyperplasia in renal hyperparathyroidism. World J Surg **20**:744–750.
47. Arnold A, Brown MF, Urena P, Gaz RD, Sarfati E, Drueke TB 1995 Monoclonality of parathyroid tumors in chronic renal failure and in primary parathyroid hyperplasia. J Clin Invest **95**:2047–2053.
48. Falchetti A, Bale AE, Amorosi A, Bordi C, Cicchi P, Bandini S, Marx SJ, Brandi ML 1993 Progression of uremic hyperparathyroidism involves alleic loss on chromosome 11. J Clin Endocrinol Metab **76**:139–144.
49. Arnold A 1993 Genetic basis of endocrine disease 5: Molecular genetics of parathyroid gland neoplasia. J Clin Endocrinol Metab **7**:1108–1112.
50. Goodman WG, Coburn JW 1992 The use of 1,25-dihydroxyvitamin D_3 in early renal failure. Annu Rev Med **43**:227–237.
51. Sherrard DJ, Herez G, Pei Y, Chan W, Maloney N 1994 Parathyroid hormone (PTH): Goal value to prevent symptomatic renal osteodystrophy (ROD). J Am Soc Nephrol **5**:888 (abstract).
52. Goodman WG, Ramirez JA, Belin TR, Chon Y, Gales B, Segre GV, Salusky IB 1994 Development of adynamic bone in patients with secondary hyperparathyroidism after intermittent calcitriol therapy. Kidney Int **46**:1160–1166.
53. Kubota M, Ng KW, Martin TJ 1985 Effect of 1,25-dihydroxyvitamin D_3 on cyclic AMP responses to hormone in clonal osteogenic sarcoma cells. Biochem J **231**:11–17.
54. Ikeda K, Imai Y, Fukase M, Fujita T 1990 The effect of 1,25-dihydroxyvitamin D_3 on osteoblast-like osteosarcoma cell: Modification of response to PTH. Biochem Biophys Res Commun **168**:889–897.
55. Kitaoka M, Fukagawa M, Ogata E, Kurokawa K 1994 Reduction of functioning parathyroid cell mass by ethanol injection in chronic dialysis patients. Kidney Int **46**:1110–1117.
56. Solbiati L, Giangrande A, Pra LD, Belloti E, Cantu P, Raavetto C 1985 Ultrasound-guided percutaneous fine-needle ethanol injection into parathyroid glands in secondary hyperparathyroidism. Radiology **155**:607–610.
57. Giangrande A, Castiglioni A, Sorbiati L, Allaria P 1992 Ultrasound guided percutaneous fine needle ethanol injection into parathyroid glands in secondary hyperparathyroidism. Nephrol Dial Transplant **7**:412–421.
58. Page B, Zingraff J, Souberbielle JC, Courtris G, Sarfati E, Drueke T, Moreau JF 1992 Correction of severe secondary hyperparathyroidism in two dialysis patients: Surgical removal versus percutaneous ethanol injection. Am J Kidney Dis **19**:378–381.
59. Takeda S, Michigishi T, Takakura E 1992 Successful ultrasonically guided percutaneous ethanol injection for secondary hyperparathyroidism. Nephron **62**:100–103.
60. Fukagawa M, Kitaoka M, Kurokawa K 1996 A new intensive and safer protocol for percutaneous ethanol injection therapy (PEIT) of severe parathyroid hyperplasia. J Am Soc Nephrol **9**:1813 (abstract).
61. Brown AJ, Ritter CR, Finch JL, Morrissey J, Martin KJ, Murayama E, Nishii Y, Slatopolsky E 1991 The noncalcemic analogue of vitamin D, 22-oxacalcitriol, suppresses parathryoid hormone synthesis and secretion. J Clin Invest **84**:728–732.
62. Kitaoka M, Fukagawa M, Kurokawa K 1995 Direct injection of calcitriol into parathyroid hyperplasia in chronic dialysis patients with severe parathyroid hyperfunction. Nephrology **1**:563–568.
63. Fukuda N, Katoh T, Yi H, Fukagawa M, Saito I, Kanegae Y, Chang H, Kurokawa K 1995 Successful in vivo gene transfer into parathyroid glands with adenoviral vector delivered by selective injection in rats. J Am Soc Nephrol **6**:961 (abstract).
64. Lopez-Hilker S, Dusso AS, Rapp NS, Martin KJ, Slatopolsky E 1990 Phosphorus restriction reverses secondary hyperparathyroidism independent of changes in calcium and calcitriol. Am J Physiol **259**:F432–F437.
65. Yi H, Fukagawa M, Yamato H, Kumagai M, Watanabe T, Kurokawa K 1995 Prevention of enhanced parathyroid hormone secretion, synthesis and hyperplasia by mild dietary phosphorus restriction in early chronic renal failure in rats: Possible direct role of phosphorus. Nephron **70**:242–248.
66. Slatopolsky E, Finch J, Denda M, Ritter C, Zhong M, Dusso A, McDonald PN, Brown AJ 1996 Phosphorus restriction prevents

parathyroid gland growth—High phosphorus directly stimulates PTH secretion in vitro. J Clin Invest **96**:2534–2540.
67. Almaden Y, Canalejo A, Hernandez A, Ballesteros E, Garcia-Navarro S, Torres A, Rodriguez M 1996 Direct effect of phosphorus on PTH secretion from whole rat parathyroid glands in vitro. J Bone Miner Res **7**:970–976.
68. Russell J, Bar A, Sherood LM, Hurwitz 1993 Interaction between calcium and 1,25-dihydroxyvitamin D_3 in the regulation of preproparathyroid hormone and vitamin D receptor messenger ribonucleic acid in avian parathyroids. Endocrinology **132**:2639–2644.
69. Sriussadaporn S, Wong MS, Pike JW, Favus MJ 1995 Tissue specificity and mechanism of vitamin D receptor up-regulation during dietary phosphorus restriction in the rat. J Bone Miner Res **10**:271–280.
70. Abe J, Morikawa M, Miyamoto K, Kaiho S, Fukushima M, Miyaura C, Abe E, Suda T, Nishii Y 1987 Synthetic analogues of vitamin D_3 with an oxygen atom in the side chain skeleton. FEBS Lett **226**:58–62.
71. Abe J, Takita Y, Nakano T, Miyaura C, Suda T, Nishii Y 1989 A synthetic analogue of vitamin D_3, 22-oxa-1α,25-dihydroxyvitamin D_3, is a potent modulator of in vivo immunoregulating activity without inducing hypercalcemia in mice. Endocrinology **124**:2645–2647.
72. Slatopolsky E, Beckoben M, Kelber J, Brown A, Delmez J 1992 Effect of calcitriol and non-calcemic vitamin D analogs on secondary hyperparathyroidism. Kidney Int **42**:S43–S49.
73. Kubrusly M, Gagne ER, Urena P, Hanrotel C, Chabanis S, Lacour B, Druecke TB 1993 Effect of 22-oxacalcitriol on calcium metabolism in rats with severe secondary hyperparathyroidism. Kidney Int **44**:551–556.
74. Dusso AS, Negrea L, Gunawardhana S, Lopez-Hilker S, Finch J, Mori T, Nishii Y, Slatopolsky E 1991 On the mechanisms for the selective action of vitamin D analogs. Endocrinology **128**: 1687–1692.
75. Finch JL, Brown AJ, Mori T, Nishii Y, Slatopolsky E 1992 Suppression of PTH and decreased action on bone are partially responsible for the low calcemic activity of 22-oxacalcitriol relative to $1,25(OH)_2D_3$. J Bone Miner Res **7**:835–839.
76. Finch JL, Brown AJ, Kubodera N, Nishi Y, Slatopolsky E 1993 Differential effect of $1,25(OH)_2D_3$ and 22-oxacalcitriol on phosphate and calcium metabolism. Kidney Int **43**:561–566.
77. Kurokawa K, Akizawa T, Suzuki M, Akiba T, Ogata E, Slatopolsky E 1996 Effect of 22-oxacalcitriol on hyperparathyroidism of dialysis patients: Results of a preliminary study. Nephrol Dial Transplant **11**:121–129.
78. Ornoy A, Goodwin D, Noff D, Edelstein S 1978 24,25-Dihydroxyvitamin D is a metabolite of vitamin D essential for bone formation. Nature **276**:517–519.
79. Endo H, Kiyoki M, Kawashima K, Naruuchi T, Hashimoto Y 1980 Vitamin D_3 metabolites and PTH synergistically stimulate bone formation of chick embryo femur *in vitro*. Nature **286**:262–264.
80. Silver J, Naveh-Many T, Mayer H, Schmelzer HJ, Popovtzer MM 1986 Regulation by vitamin D metabolites of parathyroid hormone gene transcription in vivo in the rat. J Clin Invest **78**:1296–1301.
81. Olgaard K, Finco D, Schwartz J 1984 Effect of 24,25-dihydroxyvitamin D_3 on PTH levels and bone histology in dogs with chronic uremia. Kidney Int **25**:791–797.
82. Nakamura T, Kurokawa T, Orimo H 1988 Increase of bone volume in vitamin D-repleted rats by massive administration of $24R,25(OH)_2D_3$. Calcif Tissue Int **43**:235–243.
83. Yamate T, Tanaka H, Nagai Y, Yamato H, Taniguchi N, Nakamura T, Seino Y 1994 Bone forming ability of 24*R*,25-dihydroxyvitamin D_3 in the hypophosphatemic mouse. J Bone Miner Res **9**:1967–1974.
84. Carpenter TO, Keller M, Schwartz D, Mitnick M, Smith C, Ellison A, Carey D, Comite F, Horst R, Travers R, Glorieux FH, Gundberg CM, Poole R, Insogna KL 1996 24,25-Dihydroxyvitamin D supplementation corrects hyperparathyroidism and improves skeletal abnormalities in X-linked hypophosphatemic rickets—A clinical research center study. J Clin Endocrinol Metab **81**:2381–2388.
85. St-Arnaud R, Arabian AA, Glorieux FH 1996 Abnormal bone development in mice deficient for the vitamin D 24-hydroxylase gene. J Bone Miner Res **11**:S126 (abstract).
86. Henry HL 1992 Vitamin D hydroxylases. J Cell Biol **49**:4–9.
87. Lambrey G, N'Guyen TM, Garabedian M, Sebert JL, de Fremont JF, Marie P, Caillens C, Gueris, J, Meunier P, Balsan S, Fournier A 1982 Possible link between changes in plasma 24,25-dihydroxyvitamin D and healing of bone resorption in dialysis osteodystrophy. Metab Bone Dis Relat Res **4**:25–30.
88. Hodsman AB, Sherrard DJ, Brickman AS, Lee DBN, Norman AW, Coburn JW 1983 Preliminary trials with 24,25-dihydroxyvitamin D_3 in dialysis osteomalacia. Am J Med **74**:407–414.
89. Dustan CR, Hills E, Norman AW, Bishop JE, Mayer E, Wong SYP, Eade Y, Johnson JR, George CRP, Collett P, Kalowski S, Wyndham RN, Lawrence JR, Alfrey AC, Evans RA 1985 Treatment of hemodialysis bone disease with $24,25(OH)_2D_3$ and $1,25(OH)_2D_3$ alone or in combination. Miner Electrolyte Metab **11**:358–368.
90. Ben-Ezer D, Shany S, Conforty A, Rapoport J, Edelstein S, Bdolah-Abram T, Kafka DR, Chaimovitz C 1991 Administration of $24,25(OH)_2D_3$ suppresses the serum parathyroid hormone levels of dialysis patients. Nephron **58**:283–287.
91. Varghese Z, Moorhead JF, Farrington K 1992 Effect of 24,25-dihydroxycholecalciferol on intestinal absorption of calcium and phosphate and on parathyroid hormone secretion in chronic renal failure. Nephron **60**:286–291.
92. Popovtzer MM, Levi J, Bar-Khayim Y, Shasha SM, Benheim J, Chaimovitz C, Bab I 1992 Assessment of combined $24,25(OH)_2D_3$ and $1\alpha(OH)D_3$ therapy for bone disease in dialysis patients. Bone **13**:369–377.
93. Rubinger D, Moscovitz A, Popovtzer MM, Bernheim J, Bab I, Gazit D 1990 $24,25(OH)_2D_3$ in combination with calcitriol reverses osteoclastic hyperactivity in chronic renal failure Kidney Int **37**:451 (abstract).
94. Rubinger D, Moscovitz A, Popovtzer MM, Bab I, Gazit D 1989 $24,25(OH)_2D_3$ in combination with $1,25(OH)_2D_3$ ameliorates renal osteodystrophy in rats with chronic renal failure. Kidney Int **35**:379 (abstract).
95. Birkenhager-Frenkel DH, Pols HA, Zeelenberg J, Eijgelsheim JJ, Birkenheimer JC 1995 Effects of 24*R*,25-dihydroxyvitamin D_3 in combination with 1alpha-hydroxyvitamin D_3 in predialysis renal insufficiency: Biochemistry and histomorphometry of cancellous bone. J Bone Miner Res **10**:197–204.
96. Yamato H, Okazaki R, Ishii T, Ogata E, Sato T, Kumegawa M, Akaogi K, Taniguchi N, Matsumoto T, 1993 Effect of 24*R*,25-dihydroxyvitamin D_3 on the formation and function of osteoclastic cells. Calcif Tissue Int **52**:225–260.
97. Mortensen BM, Aareth HP, Ganss R, Haug E, Gautvik KM, Gordeladze JO 1993 24,25-Dihydroxy vitamin D_3 treatment inhibits parathyroid-stimulated adenylate cyclase in iliac crest biopsies from uremic patients. Bone **14**:125–131.
98. Kazama J, Fukagawa M, Yi H, Kumagai M, Yamato H, Taniguchi N, Geijyo F, Arakawa M, Ozawa H, Kurokawa K 1996 24*R*,25-Dihydroxyvitamin D_3 ameliorates the high-turnover

bone diseases without suppressing parathyroid function in chronic renal failure in rats. Nephrology **2**:361–366.

99. Kazama JJ, Fukagawa M, Yi H, Kumagai M, Yamato H, Taniguchi N, Koshikawa A, Arakawa M, Ozawa H, Kurokawa K 1996 Fine structure of trabecular bone in uremic rats analyzed by microfocus X-ray imaging. J Bone Miner Res **11**:S490 (abstract).
100. Tanaka Y, DeLuca HF, Kobayashi Y, Ikekawa N 1984 26,26,26,27,27,27-Hexafluoro-1,25-dihydroxyvitamin D_3: A highly potent, long-lasting analog of 1,25-dihydroxyvitamin D_3. Arch Biochem Biophys **229**:348–354.
101. Inaba M, Okuno S, Nishizawa Y, Yukioka K, Otani S, Matsui-Yuase I, Morisawa S, DeLuca HF, Morii H 1987 Biological activity of fluorinated vitamin D analogs at C-26 and C-27 on human promyelocytic leukemia cells. Arch Biochem Biophys **258**:421–425.
102. Yukioka K, Otani S, Matsui-Yuasa I, Goto H, Morisawa S, Okuno S, Inaba M, Nishizawa Y, Morii H 1988 Biological activity of 26,26,26,27,27,27-hexafluorinated analogs of vitamin D_3 in inhibiting interleukin-2 production by peripheral blood mononuclear cells stimulated by phytohemagglutinin. Arch Biochem Biophys **260**:45–50.
103. Honda A, Nakashima N, Shida Y, Mori Y, Nagata A, Ishizuka S 1993 Modification of 1alpha,25-dihydroxyvitamin D_3 metabolism by introduction of 26,26,26,27,27,27-hexafluoro atoms in human promyelocytic leukaemia (HL-60) cells: Isolation and identification of a novel bioactive metabolite, 26,26,26,27,27,27-hexafluoro-1α,23(*S*),25-trihydroxyvitamin D_3. Biochem J **295**:509–516.
104. Sasaki H, Harada H, Handa Y, Morino H, Suzawa M, Shimpo E, Katsumata T, Masuhiro Y, Matsuda K, Ebihara K, Ono T, Masushige S, Kato S 1995 Transcriptional activity of a fluorinated vitamin D analog on VDR–RXR-mediated gene expression. Biochemistry **34**:370–377.
105. Inaba N, Okuno S, Inoue A, Nishizawa Y, Morii H, DeLuca HIF 1989 DNA binding property of vitamin D_3 receptors associated with 26,26,26,27,27,27-hexafluoro-1,25-dihydroxyvitamin D_3. Arch Biochem Biophys **121**:520–524.
106. Nishizawa Y, Morii H, Ogura Y, DeLuca HF 1991 Clinical trial of 26,26,26,27,27,27-hexafluoro-1,25-dihydroxyvitamin D_3 in uremic patients on hemodialysis: Preliminary report. Contrib Nephrol **90**:196–203.

Index

A

B

C

E

H

J

K

L

M

N

O

P

Q

R

S

T

U

V

X

Z

ISBN 0-12-252685-6